CHARGED PARTICLE AND PHOTON INTERACTIONS WITH MATTER

RECENT ADVANCES, APPLICATIONS, AND INTERFACES

Charged Particle and Photon Interactions with Matter

Recent Advances, Applications, and Interfaces

Edited by

Yoshihiko Hatano
Yosuke Katsumura
A. Mozumder

CRC Press
Taylor & Francis Group
Boca Raton London New York

CRC Press is an imprint of the
Taylor & Francis Group, an **informa** business

CRC Press
Taylor & Francis Group
6000 Broken Sound Parkway NW, Suite 300
Boca Raton, FL 33487-2742

First issued in paperback 2020

© 2011 by Taylor and Francis Group, LLC
CRC Press is an imprint of Taylor & Francis Group, an Informa business

No claim to original U.S. Government works

ISBN-13: 978-0-367-57707-0 (pbk)
ISBN-13: 978-1-4398-1177-1 (hbk)

Visit the Taylor & Francis Web site at
http://www.taylorandfrancis.com

and the CRC Press Web site at
http://www.crcpress.com

This book is dedicated to
the late Dr. Mitio Inokuti
for his sterling contributions
to radiation research and
atomic collision research.

Contents

Preface

The editors of *Charged Particle and Photon Interactions with Matter: Chemical, Physicochemical, and Biological Consequences with Applications,* eds., A. Mozumder and Y. Hatano, Marcel Dekker, New York (2004), received, soon after its publication, highly supportive comments from international communities of radiation research distributed widely into physics, chemistry, biology, medicine, and technology, and also from other broad areas of science and technology, concerned, in part, with the common phenomena of ionization and excitation of matter. These comments strongly motivated the editors to bring forth a new book that includes more detailed scientific contents, such as recent advances, future perspectives, and information on applications, than those only briefly summarized in the first book. There have been recently two kinds of applications: one is the relatively direct application used mainly by those in radiation research fields, and the other is the interface between radiation research and other fields. Thus, the subtitle of this book is *Recent Advances, Applications, and Interfaces.*

The first book was published in 2004, and each chapter had referred to papers published before 2000. For this book, a further analysis of recent advances in these respective research fields has been strongly requested. A detailed survey of the applications in these fields, which were introduced only briefly in the first book (in supplementary Chapters 20 through 26, as discussed in Chapter 1, Introduction, in this book) has also been strongly suggested, since the applications have progressed much in recent years. Active interfaces with different research fields have been in evidence. Here, there are two important points of caution in further activating the applications and interfaces in radiation research fields. One is that the applications and interfaces are produced not only in technology but also in basic science. The other is that these applications and interfaces certainly activate the traditionally important core part of radiation research fields, so that the real activation of the core part is essential for the newer applications and interface production in future.

With this scope and focus, we began our editorial work for this book. To select the chapters and their authorship, we planned an international symposium on the topics we had in mind. Thus, the symposium on Charged Particle and Photon Interactions with Matter was held as the *7th International Symposium on Advanced Science Research* (ASR2007) at the Advanced Science Research Center of the Japan Atomic Energy Agency, Tokai, Japan, November 6–9, 2007, hosted by Y. Katsumura. We wish to record our appreciation to the Japan Atomic Energy Agency and to the Japanese Society for the Promotion of Science for their generous financial and organizational support (see the special issue for this symposium published in *Radiat. Phys. Chem.*, 77, 1119–1339, 2008).

Our plans, together with a list of the candidates for the chapter titles and their authors, were reviewed by scientists and by the editors at Taylor & Francis. These were then modified in accordance with their comments and suggestions.

In the preparatory stage of our editorial work, we decided that the chapters should be written with the following necessary conditions. We asked the contributors not to overlap with any scientific contents addressed in the previous book. We also requested them to be careful for referring to chapters in the earlier book and to other chapters in that book by other authors.

We also asked them to avoid using jargon or special technical terms in their own research fields, since, as clearly shown in the table of contents in this book, the topics and research fields are distributed across a wide variety of scientific and technological fields. If any specialized terminology is used at all, it should be briefly explained in plain terms.

In Chapter 1, we have incorporated the introduction from the 2004 book and the scientific background that brought it to publication. We also discuss the relation between the scientific contents of the first book and the present one, as well as the internal relation among the chapters in this book.

We, the editors, acknowledge the cooperative work of senior chemistry editor, Barbara Glunn, editorial assistant, Jennifer Derima, and project coordinator, David Fausel, during the initial stages of production. We would also like to thank project editor, Richard Tressider, and project manager, Dr. Sedumadhavan Vinithan, for reviewing the proofs comprehensively just before printing.

We are greatly indebted to all the contributors for their scientifically excellent contributions to this book and for their cooperation during the editorial work.

Finally, we hope that this book, as well as the first one, will contribute much to the advancement of the research fields that address charged particle and photon interactions with matter, and also, more generally, to great progress in science and technology.

Yoshihiko Hatano
Japan Atomic Energy Agency
Tokai, Japan

Yosuke Katsumura
The University of Tokyo
Tokyo, Japan

Asokendu Mozumder
University of Notre Dame
Notre Dame, Indiana

Editors

Yoshihiko Hatano, director general, Advanced Science Research Center, Japan Atomic Energy Agency (JAEA), Tokai, and professor emeritus, Tokyo Institute of Technology, Japan, received his PhD in chemistry from Tokyo Institute of Technology in 1968. Since then he has worked in different capacities in the Department of Chemistry at Tokyo Institute of Technology, as follows: associate professor, 1970–1984; professor, 1984–2000; and dean of the Faculty of Science, 1997–1999. After retiring from Tokyo Institute of Technology in 2000, he worked as a professor in the Department of Molecular and Material Sciences at Kyushu University, Kasuga, Japan, from 2000 to 2003 and as distinguished professor at the Synchrotron Radiation Research Center at Saga University, Saga, Japan, from 2004 to 2009. Since 2005, he has served as the director general of the Advanced Science Research Center, JAEA until his retirement from it at the end of March 2010.

Dr. Hatano has been a visiting scientist/professor at many institutions, including the University of Notre Dame, Indiana; the University of Kaiserslautern, Germany; the University of California, Berkeley; the University of Science and Technology of China, Hefei; International Atomic Energy Agency (IAEA); and High Energy Accelerator Research Organization (KEK) at Photon Factory. He has also served in other scientific capacities: chairperson of the Japanese Society of Radiation Chemistry and the Society for Atomic Collision Research; advisory editor of *Chemical Physics Letters*; coeditor of *Charged Particle and Photon Interactions with Matter: Chemical, Physicochemical, and Biological Consequences with Applications*, Marcel Dekker, New York (2004); councilor of the International Association for Radiation Research, the Chemical Society of Japan, and the Japanese Society of Synchrotron Radiation Research; and chair/cochair of the *International Symposium on Chemical Applications of Synchrotron Radiation*, the *International Symposium on Electron-Molecule Collisions and Swarms*, the *International Conference on Photonic, Electronic, and Atomic Collisions*; and the *International Symposium on Advanced Science Research*.

His research interests include (1) primary and fundamental processes in charged particle and photon interactions with matter, (2) spectroscopy and dynamics of molecular superexcited states in the dissociative excitation of molecules in photonic or electronic collisions with molecules, (3) electron attachment and recombination, and (4) collisional de-excitation of excited rare gas atoms. He is the author or coauthor of more than 280 refereed journal articles, scientific papers, and books.

Yosuke Katsumura, professor, received his PhD from the Department of Nuclear Engineering and Management, School of Engineering, the University of Tokyo, Tokyo, in 1981. He is also a group leader of Basic Radiation Research, Advanced Science Research Center, Japan Atomic Energy Agency. He has previously worked as a research associate, Nuclear Engineering Research Laboratory, 1972–1984; as an associate professor, Department of Nuclear Engineering, 1984–1994; as a professor, Department of Quantum Engineering and Systems Sciences, 1994–1996; Nuclear Engineering Research Laboratory, 1996–2004; Department of Nuclear Engineering and Management, the University of Tokyo, 2005–present.

Dr. Katsumura has been a visiting professor/fellow at many universities and institutions worldwide, including the Swiss Federal Institute of Technology (ETH, Zürich), Switzerland; the University of Science and Technology of China; the University of Sherbrooke, Canada; and the University of Paris-Sud, Orsay, France. He was the president of the Japan Society of Radiation Chemistry in 2007 and 2008, and is the division head of the Water Chemistry Division, Atomic Energy Society of Japan, since 2009. He received an award from the Japan Society of Radiation Chemistry in 2005. He has published more than 220 articles in peer-reviewed journals and books. He has also organized several international symposia of radiation chemistry and edited special issues of *Radiation Physics and Chemistry*.

Dr. Katsumura has been working on subjects related to nuclear engineering such as radioly-sis of high-temperature water, radiation effects in spent fuel reprocessing, and radiation effects in high-level waste repository. His recent interests include radiolysis of supercritical water, ultrafast pulse radiolysis, and heavy ion beam radiolysis of water.

Asokendu Mozumder, research professor emeritus, Radiation Laboratory and Department of Chemistry, University of Notre Dame, received his PhD in theoretical physics from the Indian Institute of Technology (IIT), Kharagpur, India, in 1961. Since then he has worked as a lecturer in physics at IIT, 1961–1962; as a postdoctoral research associate at the Radiation Laboratory, University of Notre Dame, 1962–1965; as a scientist, 1965–1969; as an associate faculty fellow and research associate professor, Radiation Laboratory and Department of Chemistry, University of Notre Dame, 1969–1986; and as a research professor, 1986–1996, when he retired from active employment.

He has been a visiting professor/fellow at many universities and institutions worldwide, including the Bhabha Atomic Research Centre, Trombay, India; Kyoto University, Japan; Waseda University, Tokyo, Japan; Japan Atomic Energy Agency; University of Paris-Sud; and Oxford University, the United Kingdom (Harwell Fellow at Wolfson College and the Department of Physical Chemistry).

He has been chairperson/principal speaker at several international conferences, including Gordon conferences in radiation chemistry; Tihany symposium on radiation chemistry; and the recently held ASR 2007 conference on charged particle and photon interactions with matter, Tokai, Japan. He is a coeditor of *Charged Particle and Photon Interactions with Matter: Chemical, Physicochemical, and Biological Consequences with Applications*, Marcel Dekker, New York (2004).

His research interests include (1) theoretical aspects of radiation chemistry, (2) early stages of radiolysis, (3) theories of electron localization and trapping, and (4) free-ion yield and mobility in liquid hydrocarbons. He is the author or coauthor of nearly 125 articles in refereed journals; a comprehensive book, *Fundamentals of Radiation Chemistry* (Academic Press, 1999); an article in *Encyclopedia Britannica*; and several chapters in other books.

Contributors

Amitava Adhikary
Department of Chemistry
Oakland University
Rochester, Michigan

Joseph M. Ajello
Jet Propulsion Laboratory
California Institute of Technology
Pasadena, California

Gérard Baldacchino
Institut Rayonnement Matière Saclay
Commissariat à l' Énergie Atomique et aux
    Énergies Alternatives
Saclay, France

and

Laboratoire Claude Fréjacques
Centre National de la Recherche Scientifique
Gif-sur-Yvette, France

David Becker
Department of Chemistry
Oakland University
Rochester, Michigan

Ortwin Brede
Faculty of Chemistry and Mineralogy
University of Leipzig
Leipzig, Germany

Vincent De Waele
Laboratoire de Chimie Physique
Centre National de la Recherche Scientifique
Université Paris-Sud
Orsay, France

Kentaro Fujii
Advanced Science Research Center
Japan Atomic Energy Agency
Tokai, Japan

Khashayar Ghandi
Department of Chemistry
Mount Allison University
Sackville, New Brunswick, Canada

Gregory A. Grieves
School of Chemistry and Biochemistry
Georgia Institute of Technology
Atlanta, Georgia

Akira Harata
Faculty of Engineering Sciences
Kyushu University
Kasuga, Japan

Yoshihiro Hase
Quantum Beam Science Directorate
Japan Atomic Energy Agency
Takasaki, Japan

Yoshihiko Hatano
Advanced Science Research Center
Japan Atomic Energy Agency
Tokai, Japan

Hisashi Hayashi
Department of Chemical and Biological Sciences
Japan Woman's University
Tokyo, Japan

Tetsuya Hirade
Nuclear Science and Engineering Directorate
Japan Atomic Energy Agency
Tokai, Japan

and

Institute of Applied Beam Science
Ibaraki University
Mito, Japan

Kenzo Hiraoka
Clean Energy Research Center
University of Yamanashi
Kofu, Japan

Koichi Hirota
Quantum Beam Science Directorate
Japan Atomic Energy Agency
Takasaki, Japan

Akira Hitachi
Molecular Biophysics
Kochi Medical School
Nankoku, Japan

Gordon L. Hug
Radiation Laboratory
University of Notre Dame
Notre Dame, Indiana

and

Faculty of Chemistry
Adam Mickiewicz University
Poznan, Poland

Toshio Ishioka
Faculty of Engineering Sciences
Kyushu University
Kasuga, Japan

Yukikazu Itikawa
Department of Basic Space Science
Institute of Space and Astronautical Science
Sagamihara, Japan

Jun-Ichi Iwata
Center for Computational Sciences
    and Institute of Physics
University of Tsukuba
Tsukuba, Japan

Jean-Paul Jay-Gerin
Département de Médecine Nucléaire
    et de Radiobiologie
Université de Sherbrooke
Sherbrooke, Québec, Canada

Tadashi Kamada
Research Center for Charged Particle Therapy
National Institute of Radiological Sciences
Chiba, Japan

Yosuke Katsumura
Department of Nuclear Engineering
    and Management
The University of Tokyo
Tokyo, Japan

and

Advanced Science Research Center
Nuclear Science Research Institute
Japan Atomic Energy Agency
Tokai, Japan

Yosuke Kawashita
Graduate School of Pure and
    Applied Sciences
University of Tsukuba
Tsukuba, Japan

Yugo Kimoto
Aerospace Research and Development
    Directorate
Japan Aerospace Exploration Agency
Tsukuba, Japan

Kazuo Kobayashi
The Institute of Scientific and Industrial
    Research
Osaka University
Osaka, Japan

Noriyuki Kouchi
Department of Chemistry
Tokyo Institute of Technology
Tokyo, Japan

Takahiro Kozawa
The Institute of Scientific and Industrial
    Research
Osaka University
Osaka, Japan

Isabelle Lampre
Laboratoire de Chimie Physique
Centre National de la Recherche Scientifique
Université Paris-Sud
Orsay, France

Jay A. LaVerne
Radiation Laboratory
and
Department of Physics
University of Notre Dame
Notre Dame, Indiana

Mingzhang Lin
Advanced Science Research Center
Japan Atomic Energy Agency
Tokai, Japan

Yasunari Maekawa
Quantum Beam Science Directorate
Japan Atomic Energy Agency
Takasaki, Japan

Rao S. Mangina
Jet Propulsion Laboratory
California Institute of Technology
Pasadena, California

Jason L. McLain
School of Chemistry and Biochemistry
Georgia Institute of Technology
Atlanta, Georgia

Jintana Meesungnoen
Département de Médecine Nucléaire
    et de Radiobiologie
Université de Sherbrooke
Sherbrooke, Québec, Canada

Robert R. Meier
Department of Physics and Astronomy
George Mason University
Fairfax, Virginia

Bruce J. Mincher
Aqueous Separations and Radiochemistry
    Department
Idaho National Laboratory
Idaho Falls, Idaho

Yasuhiro Miyake
Japan Proton Accelerator Complex
High Energy Accelerator Research Organization
Tokai, Japan

Mehran Mostafavi
Laboratoire de Chimie Physique
Centre National de la Recherche Scientifique
Université Paris-Sud
Orsay, France

Asokendu Mozumder
Radiation Laboratory
University of Notre Dame
Notre Dame, Indiana

Takashi Nakatsukasa
RIKEN Nishina Center
Wako, Japan

Sergej Naumov
Department of Chemistry
Leibniz Institute of Surface Modification
Leipzig, Germany

Koji Noda
Research Center for Charged
    Particle Therapy
National Institute of Radiological Sciences
Chiba, Japan

Takeshi Odagiri
Department of Chemistry
Tokyo Institute of Technology
Tokyo, Japan

Takeshi Ohshima
Quantum Beam Science Directorate
Japan Atomic Energy Agency
Takasaki, Japan

Shinobu Onoda
Quantum Beam Science Directorate
Japan Atomic Energy Agency
Takasaki, Japan

Thomas M. Orlando
School of Chemistry and Biochemistry
and
School of Physics
Georgia Institute of Technology
Atlanta, Georgia

Kevin M. Prise
Centre for Cancer Research and
    Cell Biology
Queen's University Belfast
Belfast, United Kingdom

K. Indira Priyadarsini
Radiation and Photochemistry Division
Bhabha Atomic Research Centre
Mumbai, India

Akinori Saeki
The Institute of Scientific and Industrial
    Research
Osaka University
Osaka, Japan

Kimiaki Saito
Quantum Beam Science Directorate
Japan Atomic Energy Agency
Tokai, Japan

Miki Sato
Faculty of Engineering Sciences
Kyushu University
Kasuga, Japan

Giuseppe Schettino
Centre for Cancer Research and
    Cell Biology
Queen's University Belfast
Belfast, United Kingdom

Shu Seki
Division of Applied Chemistry
Osaka University
Osaka, Japan

Michael D. Sevilla
Department of Chemistry
Oakland University
Rochester, Michigan

Naoya Shikazono
Advanced Science Research Center
Japan Atomic Energy Agency
Tokai, Japan

Isao H. Suzuki
Photon Factory
High Energy Accelerator Research Organization
Tsukuba, Japan

and

National Metrology Institute of Japan
National Institute of Advanced Industrial
    Science and Technology
Tsukuba, Japan

Satoshi Suzuki
Advanced Research Institute for Science
    and Engineering
Waseda University
Tokyo, Japan

Seiichi Tagawa
The Institute of Scientific and Industrial
    Research
Osaka University
Osaka, Japan

Mitsumasa Taguchi
Quantum Beam Science Directorate
Japan Atomic Energy Agency
Takasaki, Japan

Junichi Takagi
Chemical System Design and Engineering
    Department
Toshiba Corporation
Yokohama, Japan

Kenji Takahashi
Department of Chemistry and Chemical
    Engineering
Kanazawa University
Kanazawa, Japan

Masao Tamada
Quantum Beam Science Directorate
Japan Atomic Energy Agency
Takasaki, Japan

Atsushi Tanaka
Quantum Beam Science Directorate
Japan Atomic Energy Agency
Takasaki, Japan

Hiroshi Tanaka
Department of Physics
Sophia University
Tokyo, Japan

Yasuo Udagawa
Institute of Multidisciplinary Research
    for Advanced Materials
Tohoku University
Sendai, Japan

Masatoshi Ukai
Department of Applied Physics
Tokyo University of Agriculture
    and Technology
Tokyo, Japan

James F. Wishart
Chemistry Department
Brookhaven National Laboratory
Upton, New York

Kazuhiro Yabana
Center for Computational Sciences
    and Institute of Physics
University of Tsukuba
Tsukuba, Japan

Makoto Yamaguchi
Geological Isolation Research
    and Development Directorate
Japan Atomic Energy Agency
Tokai, Japan

Shinichi Yamashita
Advanced Science Research Center
Japan Atomic Energy Agency
Tokai, Japan

Akinari Yokoya
Advanced Science Research Center
Japan Atomic Energy Agency
Tokai, Japan

1 Introduction

Yoshihiko Hatano
Japan Atomic Energy Agency
Tokai, Japan

Yosuke Katsumura
The University of Tokyo
Tokyo, Japan
and
Japan Atomic Energy Agency
Tokai, Japan

Asokendu Mozumder
University of Notre Dame
Notre Dame, Indiana

CONTENTS

In Chapter 1 of *Charged Particle and Photon Interactions with Matter: Chemical, Physicochemical, and Biological Consequences with Applications* (Mozumder and Hatano, 2004), early investigations were briefly described with respect to photochemistry and radiation chemistry. The Bethe theory was also discussed briefly in terms of the quantitative similarity of the excitation and ionization processes of a molecule by photon and charged particle impacts, with emphasis on the importance of optical oscillator strength. It was further pointed out that the applications in the field of charged particle and photon interactions with matter have a relatively short history, except for some medical applications. However, their importance has gradually increased in science and technology since the 1960s.

The chapter briefly discussed the timescale of charged particle and photon interactions with matter, for example, in liquid water. Fundamental processes in the physical, physicochemical, and chemical stages of these interactions were delineated.

The 2004 book was organized into 26 chapters, and a list of chapters along with their contributors follows:

1. Introduction (A. Mozumder and Y. Hatano)
2. Interaction of Fast Charged Particles with Matter (A. Mozumder)
3. Ionization and Secondary Electron Production by Fast Charged Particles (L. H. Toburen)

4. Modeling of Physicochemical and Chemical Processes in the Interactions of Fast Charged Particles with Matter (S. M. Pimblott and A. Mozumder)
5. Interaction of Photons with Molecules: Photoabsorption, Photoionization, and Photodissociation Cross Sections (N. Kouchi and Y. Hatano)
6. Reactions of Low-Energy Electrons, Ions, Excited Atoms and Molecules, and Free Radicals in the Gas Phase as Studied by Pulse Radiolysis Methods (M. Ukai and Y. Hatano)
7. Studies of Solvation Using Electrons and Anions in Alcohol Solutions (C. D. Jonah)
8. Electrons in Nonpolar Liquids (R. A. Holroyd)
9. Interactions of Low-Energy Electrons with Atomic and Molecular Solids (A. D. Bass and L. Sanche)
10. Electron–Ion Recombination in Condensed Matter: Geminate and Bulk Recombination Processes (M. Wojcik, M. Tachiya, S. Tagawa, and Y. Hatano)
11. Radical Ions in Liquids (I. A. Shkrob and M. C. Sauer, Jr.)
12. The Radiation Chemistry of Liquid Water: Principles and Applications (G. V. Buxton)
13. Photochemistry and Radiation Chemistry of Liquid Alkanes: Formation and Decay of Low-Energy Excited States (L. Wojnarovits)
14. Radiation Chemical Effects of Heavy Ions (J. A. LaVerne)
15. DNA Damage Dictates the Biological Consequences of Ionizing Irradiation: The Chemical Pathways (W. A. Bernhard and D. M. Close)
16. Photon-Induced Biological Consequences (K. Kobayashi)
17. Track Structure Studies of Biological Systems (H. Nikjoo and S. Uehara)
18. Microdosimetry and Its Medical Applications (M. Zaider and J. F. Dicello)
19. Charged Particle and Photon-Induced Reactions in Polymers (S. Tagawa, S. Seki, and T. Kozawa)
20. Charged Particle and Photon Interactions in Metal Clusters and Photographic System Studies (J. Belloni and M. Mostafavi)
21. Applications of Radiation Chemical Reactions to the Molecular Design of Functional Organic Materials (T. Ichikawa)
22. Applications to Reaction Mechanism Studies of Organic Systems (T. Majima)
23. Applications of Radiation Chemistry to Nuclear Technology (Y. Katsumura)
24. Electron Beam Applications to Flue Gas Treatment (H. Namba)
25. Ion-Beam Therapy: Rationale, Achievements, and Expectations (A. Wambersie, J. Gueulette, D. T. L. Jones, and R. Gahbauer)
26. Food Irradiation (J. Farkas)
27. New Applications of Ion Beams to Material, Space, and Biological Science and Engineering (M. Fukuda, H. Itoh, T. Ohshima, M. Saidoh, and A. Tanaka)

The 2004 book was motivated by two projects. One was a long-term IAEA international project, from 1985 to 1995, that surveyed the accomplishments in basic radiation research over the past 100 years following the discovery of ionizing radiation by Curie and Roentgen in the late nineteenth century (Inokuti, 1995). The other was a textbook of radiation chemistry published in 1999 (Mozumder, 1999). Since the activities of the former project, summarized in an IAEA report, were unfortunately not well known among the international science and technology communities, the participants of the IAEA project agreed that the scientific results of the activities should be published elsewhere, in a book with wider circulation. Furthermore, the IAEA project focused mainly on primary interactions, that is, the physical stage of the fundamental processes of radiation chemistry. Therefore, Mozumder and Hatano collaborated to edit a new book that would include the physicochemical and chemical stages, in addition to the physical stage, of the fundamental processes of radiation chemistry, and, consequently, those of radiation biology.

TABLE 1.1

Fundamental Processes of Radiation Chemistry

$AB \longrightarrow\!\!\!\bigwedge\!\!\bigwedge\!\!\bigwedge\!\!\longrightarrow AB^+ + e^-$	Direct ionization
$\longrightarrow\!\!\!\bigwedge\!\!\bigwedge\!\!\bigwedge\!\!\longrightarrow AB^{**}$	Superexcitation (direct excitation)
$\longrightarrow\!\!\!\bigwedge\!\!\bigwedge\!\!\bigwedge\!\!\longrightarrow AB^*$	Excitation (direct excitation)
$AB^{**} \rightarrow AB^+ + e^-$	Autoionization
$\rightarrow A + B$	Dissociation
$AB^+ \rightarrow A^+ + B$	Ion dissociation
$AB^+ + AB$ or $S \rightarrow$ Products	Ion–molecule reaction
$AB^+ + e^- \rightarrow AB^*$	Electron–ion recombination
$AB^+ + S^- \rightarrow$ Products	Ion–ion recombination
$e^- + S \rightarrow S^-$	Electron attachment
$e^- + nAB \rightarrow e^-_s$	Solvation
$AB^* \rightarrow A + B$	Dissociation
$\rightarrow AB$	Internal conversion and intersystem crossing
$\rightarrow BA$	Isomerization
$\rightarrow AB + h\nu$	Fluorescence
$AB^* + S \rightarrow AB + S^*$	Energy transfer
$AB^* + AB \rightarrow (AB)_2^*$	Excimer formation
$2A \rightarrow A_2$	Radical recombination
$\rightarrow C + D$	Disproportionation
$A + AB \rightarrow A_2B$	Addition
$\rightarrow A_2 + B$	Abstraction

Source: Mozumder, A. and Hatano, Y. (eds.), *Charged Particle and Photon Interactions with Matter: Chemical, Physicochemical, and Biological Consequences with Applications*, Marcel Dekker, New York, 2004.

The fundamental processes of radiation chemistry are shown in Table 1.1 (Mozumder and Hatano, 2004), in which the first three constitute the physical stage or the primary process of the charged particle and photon interactions with matter (Hatano, 2003). Here "primary" means the earliest stage that is conceivable either theoretically or experimentally. Sometimes "initial" is used for measured yields at the shortest time that is possible in a given experimental setup. The primary stage is followed by the physicochemical and chemical stages. These are followed by the biological stages. See also Chapter 1 of the 2004 book.

The 2004 book succeeded in surveying critically and in detail the comprehensive features of the physical, physicochemical, chemical, and biological stages. Further, the applications of charged particle and photon interactions with matter were treated briefly. Most of the papers referred in the 2004 book were published before 2000.

1.1 THEORETICAL STUDIES AS NEW APPROACHES TO PRIMARY PROCESSES THAT MOTIVATE NEW EXPERIMENTS

From the late nineteenth century to the first half of the twentieth century, studies of the interaction of ionizing radiation with matter were mainly phenomenological in character. A new theoretical approach, particularly for the primary process, that is, the physical stage, was initiated during 1955–1965 by Platzman, Fano, and Inokuti. They considered the interaction to be the collision of high-energy particles with matter, that is, basically molecules. The important findings of their studies are summarized as follows (Platzman, 1962a,b; Hatano, 1999, 2003; Mozumder and Hatano, 2004, Chapters 1 and 5; and Chapter 2):

1. Ionizing radiation is generally classified according to high-energy (a) photons; (b) electrons; (c) heavy charged particles; and (d) other particles such as neutrons, positrons, muons, etc. Although the initial interaction of each of these particles with a molecule depends largely on the kind and energy of the particle, common features among the initial interactions and further the following electron–molecule collisions should be the formation of electrons in a wide energy range. These are called "secondary electrons." The secondary electrons are classified according to their energy in the middle- and high-energy ranges, and those in the subexcitation energy region. It was concluded that the essential features of the interaction of ionizing radiation with molecules in the primary process is electron–molecule collisions in a wide range of collision energies, which are followed by cascades of multiple electron–molecule collisions in matter (Spencer and Fano, 1954; Mozumder and Hatano, 2004, Chapters 3, 6, and 9; and Chapter 3).

2. Secondary electron collisions with molecules in the middle- and high-energy ranges may be treated approximately by the Bethe theory (Inokuti, 1971), resulting in the important conclusion that the generalized oscillator strength and, further, the optical oscillator strength are of great importance in interpreting the primary result of the interaction of ionizing radiation with matter. Thus, G-values have been estimated theoretically from the optical oscillator strength by "the optical approximation" (Platzman, 1962a,b). That is, the energy deposition spectra in the interaction of ionizing radiation with molecules can be estimated from the optical oscillator strength (Hatano, 2003, 2009; Mozumder and Hatano, 2004, Chapter 5; and Chapter 2).

3. Since the optical oscillator strength, which is of the great importance in basic sciences, had not yet been calculated either theoretically or experimentally, Platzman and Fano realized and pointed out for the first time in the late 1950s that synchrotron radiation should be a powerful photon source in a wide span of photon energies from UV-visible to hard x-rays (Hatano, 1995).

4. After a scientifically careful and intuitive analysis of the primary interaction of ionizing radiation with molecules, as obtained from the Bethe theory and optical approximation, Platzman realized that for most molecules there is a big difference between the ionization threshold energy and the energy region where the major part of the oscillator strength distribution is located, as deduced from optical data and sum rules. He presented his idea of "superexcited states," which are neutral excited states located in the energy region above the ionization threshold (Platzman, 1962a,b; Hatano, 2003; Mozumder and Hatano, 2004, Chapter 5; and Chapter 2).

The theoretical studies, summarized above in (1) through (4), have motivated much new experimental research since the late 1960s as described below (Hatano, 1999; Mozumder and Hatano, 2004, Chapter 5; and Chapter 2):

1. Experimental evidence was first obtained, independent of these theoretical studies, for the reaction of hot hydrogen atoms formed from the dissociation of highly excited states, produced by direct excitation during the radiolysis of liquid olefins (Hatano and Shida, 1967). The experimental results were analyzed in terms of the theoretical studies to compare the experimental G-values with the theoretical ones for the superexcited states estimated by the optical approximation, giving the first experimental evidence for the important role of superexcited states in radiolysis (Hatano et al., 1968).

2. To obtain experimentally the electronic states of superexcited molecules and their dissociation dynamics to form hot hydrogen atoms, which could be electronically and/or translationally excited, Doppler spectroscopy combined with an electron–molecule collision apparatus was developed (Ito et al., 1976, 1977; Hatano, 1983; Kouchi et al., 1997). In this experiment using molecular hydrogen, the doubly excited and singly excited (with vibrational/rotational excitation) high Rydberg states converging individually to each of the ionized states were observed as superexcited states for the first time. For other molecules such as HF, H_2O, NH_3, and CH_4, the doubly and inner-core excited states were also observed.

3. To obtain more detailed information with state selectivity and higher-energy resolution, synchrotron radiation (SR) has been used as an excitation source for this kind of investigation (Hatano, 1999). The measurements in these SR experiments are classified into two types: One is the absolute measurements of photoabsorption cross sections (optical oscillator strengths), photoionization cross sections, photodissociation cross sections, and photoionization quantum yields. The other is the measurement, with high-energy resolution, of state-specified dissociation fragments formed from state-specified superexcited states. It was concluded that the electronic states and the dissociation dynamics of molecular superexcited states were experimentally evidenced for the first time in these investigations (Hatano, 1999). Accordingly, an important role of the superexcited states in the primary process of the interaction of ionizing radiation with matter has been well substantiated (Hatano, 2003). The results obtained are summarized as follows. Further, these investigations have made great progress recently, which are described in Chapter 2.

    a. Superexcited states are (i) vibrationally/rotationally excited high Rydberg states, (ii) doubly excited states, or (iii) inner-core excited states, giving conclusive experimental evidence for Platzman's idea.

    b. They dissociate into neutral fragments, with excess electronic or translational energies, in competition with autoionization.

    c. Their dissociation dynamics, as well as the dissociation products, are quite different from those for the lower excited states below ionization thresholds.

    d. Molecules are not easily ionized, which is an unexpected phenomenon.

    e. New information obtained has motivated fresh investigations of quantum theories applied to the spectroscopy and dynamics of such highly excited molecules (see Chapter 2), as well as explaining the oscillator strengths (see Chapter 4).

    f. The new information obtained has also substantiated, to a great extent, the superexcited states considered as a collision complex in some important processes such as Penning ionization, electron–ion recombination, and electron attachment to molecules.

    g. The new information has greatly motivated the reanalysis of various other kinds of phenomena, besides radiolysis, for the ionization and excitation of molecules, such as reactive plasmas, plasmas in the upper atmosphere and space, and so on.

    h. The new information was previously almost limited to molecules in the gas phase. The oscillator strengths in the condensed phase were discussed briefly (Hatano and Inokuti, 1995) and have recently been measured using a newly developed method (see Chapter 5).

4. Measurements of optical oscillator strengths (photoabsorption cross sections and photoionization cross sections) as deduced from electron–molecule collision experiments have been made in which the optimum conditions were selected using the Bethe theory. This method has been called the "poor man's synchrotron experiments" or "imaginary-photon experiments" as opposed to "real-photon experiments" using synchrotron radiation, which requires the construction of big-scale facilities. These two types of experiments were compared in detail in Chapter 5 of the 2004 book. The data obtained by these methods have been critically evaluated and compiled elsewhere as recommended ones (Kameta et al., 2003).

An important part of the new theories of the primary processes was obtained from W-value studies (Platzman, 1961). New directions in these studies have been made possible by using synchrotron radiation and are reviewed in Chapter 6.

Remarkable progress has recently been made of the interaction of positrons and muons with matter, which are surveyed in Chapters 7 and 8, respectively.

With regard to the information summarized above, future perspectives and future research programs that need more work on the theoretical and experimental aspects of the primary processes have recently been discussed elsewhere (Hatano, 2009).

1.2 ADVANCES IN THE THEORETICAL AND EXPERIMENTAL STUDIES OF THE PHYSICOCHEMICAL, CHEMICAL, AND BIOLOGICAL STAGES

Virtually all important studies published in or before 2000 of the physical, physicochemical, chemical, and biological stages of the charged particle and photon interactions with matter were surveyed critically and in detail, both theoretically and experimentally, in the 2004 book. Some of these are detailed in the next paragraph.

The theoretical studies were surveyed in Chapters 2, 4, 10, and 17. Reactions of electrons, ions, excited atoms and molecules, and also of free radicals in the gas phase, as studied by pulse radiolysis methods, were surveyed in Chapter 6, while those in the condensed phase in Chapters 7, 8, 10, and 11. The radiation chemistry of liquid water, liquid alkanes, polymers, and metal clusters/photographic systems was surveyed in Chapters 12, 13, 19, and 20, respectively. Radiation chemistry at high-LET was reviewed in Chapter 14. Biological consequences were followed up in Chapters 15 and 16. Applications in medical microdosimetry, molecular designing, organic chemistry, nuclear technology, flue gas treatment, ion-beam therapy, food irradiation, and other new material, space, and biological science and engineering were surveyed in Chapters 18, 21 through 27, respectively.

New advances in the studies of these stages, which were not covered in the 2004 book, have been remarkable since 2000. This has been pointed out in the preface. Furthermore, great progress has recently been made in the applications and interface formation.

The outline of recent advances in the studies of primary processes (the physical stage) is described briefly in Section 1.2 (also refer Chapters 2 through 8). Those of the physicochemical, chemical, and biological stages, as well as of the applications and the interface formation, are briefly described below.

New theoretical studies of the physicochemical and chemical stages are introduced in Chapters 9 and 14, respectively; these studies describe the behavior of electrons in liquid hydrocarbons and for the high-LET radiolysis of liquid water. In Chapter 9, the authors make the first application of the Anderson localization concept for electron mobility in liquid hydrocarbons. New experimental research in the physicochemical and chemical stages are described in Chapters 10 through 13, 15 through 18 for each of the specific characteristics of matter to be studied or under their specific experimental conditions. New experimental studies of the biological stage are introduced in Chapters 19 through 22. The applications in health physics and cancer therapy are found in Chapters 23 and 24, respectively. Applications to polymers are discussed in Chapters 25 through 27. The applications and the interface formation in space science and technology are introduced in Chapters 28 through 30. Applications for the research and development of radiation detectors, environmental conservation, plant breeding, and nuclear engineering are further available in Chapters 31 through 34, respectively.

With regard to the information summarized above, future perspectives and research programs that need more theoretical and experimental work on the physicochemical and chemical stages of the fundamental processes have recently been discussed elsewhere (Hatano, 2009).

REFERENCES

Hatano, Y. 1983. Electron impact dissociation of simple molecules. *Comments Atom. Mol. Phys.* 13: 259–273.

Hatano, Y. 1995. Applications of synchrotron radiation to radiation research. In *Radiation Research* (Congress Lecture, the *10th International Congress of Radiation Research*, Wurzburg, Germany), U. Hagen, D. Harder, H. Jung, and C. Streffer (eds.), Vol. II, pp. 86–92. Wurzburg, Germany: Universitatsdrukerei, H. Strutz AG.

Hatano, Y. 1999. Interaction of vacuum ultraviolet photons with molecules. Formation and dissociation dynamics of molecular superexcited states. *Phys. Rep.* 313: 109–169.

Hatano, Y. 2003. Spectroscopy and dynamics of molecular superexcited states. Aspects of primary processes of radiation chemistry. *Radiat. Phys. Chem.* 67: 187–198.

Hatano, Y. 2009. Future perspectives of radiation chemistry. *Radiat. Phys. Chem.* 78: 1021–1025.

Hatano, Y. and Inokuti, M. 1995. Photoabsorption, photoionization, and photodissociation cross sections. In *Atomic and Molecular Data for Radiotherapy and Radiation Research*, IAEA-TECDOC-799, M. Inokuti (ed.), Chapter 5. Vienna, Austria: IAEA.

Hatano, Y. and Shida, S. 1967. Hydrogen formation in the radiolyses of liquid butene-1 and trans-butene-2. *J. Chem. Phys.* 46: 4784–4788.

Hatano, Y., Shida, S., and Inokuti, M. 1968. Hydrogen formation and superexcited states in the radiolysis of liquid olefins. *J. Chem. Phys.* 48: 940–941.

Inokuti, M. 1971. Inelastic collisions of fast charged particles with atoms and molecules. The Bethe theory revisited. *Rev. Mod. Phys.* 43: 297–347.

Inokuti, M. (ed.). 1995. *Atomic and Molecular Data for Radiotherapy and Radiation Research*, IAEA-TECDOC-799. Vienna, Austria: IAEA.

Ito, K., Oda, N., Hatano, Y., and Tsuboi, T. 1976. Doppler profile measurements of Balmer-α radiation by electron impact on H_2. *Chem. Phys.* 17: 35–43.

Ito, K., Oda, N., Hatano, Y., and Tsuboi, T. 1977. The electron energy dependence of the Doppler profiles of Balmer-α emission from H_2, D_2, CH_4 and other simple hydrocarbons by electron impact. *Chem. Phys.* 21: 203–210.

Kameta, K., Kouchi, N., and Hatano, Y. 2003. Cross sections for photoabsorption, photoionization, and photodissociation of molecules. In *Landolt-Boernstein*, Y. Itikawa (ed.), New Series, Vol. I/17C. Berlin, Germany: Springer.

Kouchi, N., Ukai, M., and Hatano, Y. 1997. Dissociation dynamics of superexcited molecular hydrogen. *J. Phys. B: Atom. Mol. Opt. Phys.* 30: 2319–2344.

Mozumder, A. 1999. *Fundamentals of Radiation Chemistry*. San Diego, CA: Academic Press.

Mozumder, A. and Hatano, Y. (eds.). 2004. *Charged Particle and Photon Interactions with Matter: Chemical, Physicochemical, and Biological Consequences with Applications*. New York: Marcel Dekker.

Platzman, R. L. 1961. Total ionization in gases by high energy particles: An appraisal of our understanding. *Int. J. Appl. Radiat. Isot.* 10: 116–127.

Platzman, R. L. 1962a. Superexcited states of molecules, and the primary action of ionizing radiation. *Vortex* 23: 372–385.

Platzman, R. L. 1962b. Superexcited states of molecules. *Radiat. Res.* 17: 419–425.

Spencer, L. V. and Fano, U. 1954. Energy spectrum resulting from electron slowing down. *Phys. Rev.* 93: 1172–1181.

2 Oscillator Strength Distribution of Molecules in the Gas Phase in the Vacuum Ultraviolet Range and Dynamics of Singly Inner-Valence Excited and Multiply Excited States as Superexcited States

Takeshi Odagiri
Tokyo Institute of Technology
Tokyo, Japan

Noriyuki Kouchi
Tokyo Institute of Technology
Tokyo, Japan

CONTENTS

2.1 INTRODUCTION

The interaction of light with an atom or molecule is classified into absorption, scattering, and pair production (Hatano, 1999: 109). In this chapter, the trivial interaction of light of moderate photon energy, in particular, the interaction of vacuum ultraviolet light, is discussed, and therefore, only the absorption process is considered. We follow the definition of the vacuum ultraviolet light by Samson (Samson, 1980: 1–3), that is, the light of wavelength ranging from 200 to 0.2 nm. The photon energy of the vacuum ultraviolet light thus ranges from approximately 6 eV to 6 keV. The reason why we focus on vacuum ultraviolet light is that the maximum value of the photoabsorption cross section is usually given in this range.

We proceed by using a semiclassical model in which light is treated classically while the atom or molecule is described by quantum mechanics. However, we often use the term "photon" for convenience. The semiclassical model is valid when the light intensity is so high that the number of photons in a single mode can be treated as a continuous variable. Consequently, the absorption and stimulated emission of light are usually described well by the model, while the spontaneous emission of light is not, because only one photon is concerned. The electric dipole approximation (Bransden and Joachain, 2003: 183–236) is valid in the present range of incident light wavelength. The rate of transition from state |i> with energy E_i to state |k> with energy E_k by the photoabsorption of light of angular frequency ω_{ki} in the electric dipole approximation, W_{ki}, is expressed by (Bransden and Joachain, 2003: 183–236)

$$W_{ki} = \frac{4\pi^2}{c\hbar^2}\left(\frac{e^2}{4\pi\varepsilon_0}\right)I(\omega_{ki})\left|\hat{\varepsilon}\cdot\mathbf{r}_{ki}\right|^2, \tag{2.1}$$

$$\mathbf{r}_{ki} = \left\langle k\left|\sum_s \mathbf{r}_s\right|i\right\rangle, \tag{2.2}$$

$$\omega_{ki} = \frac{(E_k - E_i)}{\hbar}, \tag{2.3}$$

where

     c is the velocity of light in vacuum
     $\hbar$ is Planck's constant divided by 2π
     e is the elementary charge
     ε_0 is the permittivity of free space
     $I(\omega_{ki})$ is the energy flux density per unit range of the angular frequency of light
     $\hat{\varepsilon}$ is the unit polarization vector
     $\mathbf{r}_s$ is the position vector of the electron in an atom or molecule

A ket-vector |j> is normalized as

$$\left\langle j''\,|\,j'\right\rangle = \delta_{j''j'}, \tag{2.4}$$

where $\delta_{j''j'}$ is Kronecker's delta. Equation 2.1 is derived for the case in which the absorption line is so narrow that only the light of the angular frequency ω_{ki} is absorbed. It is convenient to introduce a dimensionless quantity, f_{ki}, called the oscillator strength for the transition from state |i> to state |k>. It is defined by

$$f_{ki} = \frac{2m(E_k - E_i)}{3\hbar^2}\left|\left\langle k\left|\sum_s \mathbf{r}_s\right|i\right\rangle\right|^2, \tag{2.5}$$

where m is the mass of the electron. Equation 2.5 is rewritten in a more useful form as

$$f_{ki} = \frac{(E_k - E_i)}{R}\frac{\left|\left\langle k\left|\sum_s \mathbf{r}_s\right|i\right\rangle\right|^2}{3a_0^2}, \tag{2.6}$$

where

     R is the Rydberg energy
     a_0 is the Bohr radius

We can readily understand from Equation 2.6 that f_{ki} is dimensionless. In fact the initial energy level, E_i, and the final energy level, E_k, are often degenerate, and we define the oscillator strength for the transition from energy level E_i to energy level E_k by averaging the right-hand side of Equation 2.6 for initial states $|i, a>$ and summing for the final states $|k, b>$ as

$$f_{ki} = \frac{(E_k - E_i)}{R} \frac{1}{g_i} \sum_a \sum_b \frac{\left| \left\langle k,b \left| \sum_s \mathbf{r}_s \right| i,a \right\rangle \right|^2}{3a_0^2}, \qquad (2.7)$$

where

g_i is the degree of degeneracy of energy level E_i

a and b specify each member of the degenerate states

We assume that the population of each initial state $|i, a>$ is uniform. A ket-vector $|j, c>$ is normalized as

$$\left\langle j'',c'' \middle| j',c' \right\rangle = \delta_{j''j'} \delta_{c''c'}. \qquad (2.8)$$

The initial energy level, E_i, is usually the ground energy level, E_0, and so, let us use f_k instead of f_{k0}. A set of $E_k - E_0$ and f_k characterizes a discrete spectrum of photoabsorption.

To discuss a transition from the ground energy level, E_0, to continuous energy, E, which is due to photoionization, we define the density of the oscillator strength per unit range of energy E, simply called oscillator strength distribution, by

$$\left(\frac{df}{dE} \right)_c = \frac{E}{R} \frac{1}{g_i} \sum_a \sum_\varsigma \int d\xi \frac{\left| \left\langle E,\xi,\varsigma \left| \sum_s \mathbf{r}_s \right| 0,a \right\rangle \right|^2}{3a_0^2}. \qquad (2.9)$$

The subscript c in Equation 2.9 is different from the symbol c in Equation 2.8. The final state $|E, \xi, \zeta>$ is specified by the continuous energy, E, and sets of other quantum numbers, ξ and ζ. Note that ξ is the set of continuous quantum numbers, for example, the direction of the ionized electron, while ζ is the set of discrete quantum numbers. The origin of energy is taken at the ground energy level, E_0, in Equation 2.9, which we follow from now on. The initial state $|0, a>$ is normalized following Equation 2.8, while the final state $|E, \xi, \zeta>$ is normalized as

$$\left\langle E'',\xi'',\varsigma'' \middle| E',\xi',\varsigma' \right\rangle = \delta(E'' - E')\delta(\xi'' - \xi')\delta_{\varsigma''\varsigma'}, \qquad (2.10)$$

where δ is the Dirac delta function. The quantity $(df/dE)_c$ has the dimension of the reciprocal of energy.

Let us then define the oscillator strength distribution also for the discrete transitions from the ground energy level, E_0, to $\{E_k\}$, a set of discrete energy levels other than the ground energy level, as

$$\left(\frac{df}{dE} \right)_d = \sum_k f_k \delta(E_k - E). \qquad (2.11)$$

The quantity $(df/dE)_d$ has also the dimension of the reciprocal of energy. The oscillator strength distribution for all the transitions from the ground energy level, E_0, is defined as

$$\frac{df}{dE} = \left(\frac{df}{dE} \right)_c + \left(\frac{df}{dE} \right)_d. \qquad (2.12)$$

The energy of an atom or molecule, E, ranges from zero to infinity. The quantity df/dE has the dimension of the reciprocal of energy. The oscillator strength distribution for all the transitions, df/dE, which is a function of E, is often called the oscillator strength distribution for photoabsorption. It represents the magnitude of the interaction of a photon of energy E with an atom or molecule. The most remarkable character of the oscillator strength distribution for photoabsorption is

$$\int \left(\frac{df}{dE} \right) dE = Z, \tag{2.13}$$

where Z is the number of electrons in the atom or molecule of interest. Equation 2.13 is well known as the Thomas–Kuhn–Reiche sum rule.

As supposed from Equation 2.1, the cross section, σ, for the absorption of a photon of energy E, that is, the photoabsorption cross section, is proportional to the oscillator strength distribution for photoabsorption, df/dE, defined as

$$\sigma(E) = 4\pi^2 \alpha a_0^2 \frac{df}{d(E/R)}, \tag{2.14}$$

where α is the fine structure constant. Note that the photon energy is equivalent to the energy of an atom or molecule of interest. Equation 2.14 is more conveniently written as

$$\sigma(E) = 1.098 \times 10^{-16} \left(\frac{df}{dE} \right), \tag{2.15}$$

where $\sigma(E)$ is expressed in cm^2 and df/dE in eV^{-1}.

The partial oscillator strength distribution of a channel q, $(df/dE)_q$, is also defined as

$$\left(\frac{df}{dE} \right)_q = \eta_q(E) \left(\frac{df}{dE} \right), \tag{2.16}$$

where $\eta_q(E)$ is the branching ratio of the channel q following the excitation of an atom or molecule by the absorption of a photon of energy E within the electric dipole approximation. Accordingly, the partial cross section of channel q originating from the absorption of a photon of energy E, $\sigma_q(E)$, is defined by

$$\sigma_q(E) = \eta_q(E)\sigma(E). \tag{2.17}$$

Equation 2.14 holds for $\sigma_q(E)$ and $(df/dE)_q$, and thus the oscillator strength distribution, df/dE or $(df/dE)_q$, has a corresponding cross section, $\sigma(E)$ or $\sigma_q(E)$, respectively. The summations of $(df/dE)_q$ and $\sigma_q(E)$ over all channels give (df/dE) and $\sigma(E)$, respectively:

$$\frac{df}{dE} = \sum_q \left(\frac{df}{dE} \right)_q, \tag{2.18}$$

$$\sigma(E) = \sum_q \sigma_q(E), \tag{2.19}$$

since $\displaystyle\sum_q \eta_q(E) = 1$.

In this chapter, we discuss the formation and decay of singly inner-valence excited and multiply excited states as superexcited states produced in the interaction of light in the vacuum ultraviolet range with an isolated molecule, that is, a molecule in the gas phase. Their dynamics is probed by the partial oscillator strength distribution defined by Equation 2.16 or the partial cross section defined by Equation 2.17, as described in detail in Section 2.3. The oscillator strength distribution for photoabsorption, that is, total oscillator strength distribution, has previously been discussed by Kouchi and Hatano (Kouchi and Hatano, 2004: 105–120), and thus this chapter is a continuation of it. In Section 2.2, an overview of the superexcited states of molecules is given.

2.2 SUPEREXCITED STATES OF MOLECULES

Superexcited states of atoms and molecules are electronically excited states of energies higher than the first ionization potentials (Hatano, 1999: 109). They are hence degenerate with ionization continua or, in other words, embedded in ionization continua. The concept of superexcited states was first introduced by Platzman to understand the primary interaction of ionizing radiation with matter (Platzman, 1962a: 372; Platzman, 1962b: 419). There are two distinct kinds of superexcited states depending on how they have energy higher than the first ionization potential (Nakamura, 1991: 123). In Figure 2.1, we take a diatomic molecule AB. The solid curves are potential energy curves of neutral electronic states of AB, and the curves with shadows represent the continua of potential energy curves of $AB^+ + e^-$, where e^- is a free electron and, thus, it has continuous energy. As seen in Figure 2.1a, the potential energy curve, that is, the solid line, is embedded in the

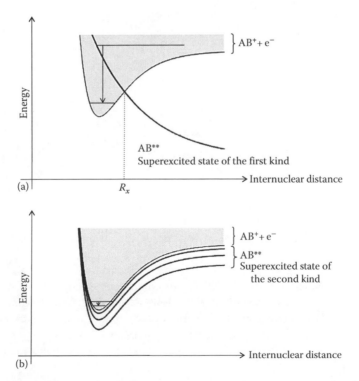

FIGURE 2.1 Potential energy curves of the diatomic molecule AB for the superexcited states of the first (a) and second (b) kinds (solid curves) and those of $AB^+ + e^-$ (solid curves with shadows). Horizontal lines indicate the vibrational–rotational levels, and the downward arrows connecting them indicate the autoionization. The length of the arrow shows the kinetic energy of the emitted electron with respect to the nuclei-fixed frame. (Based on Figure 19 of Nakamura, H., *Int. Rev. Phys. Chem.*, 10, 123, 1991. With permission.)

electronic continuum. Such states are called superexcited states of the first kind. Their electronic energy exceeds that of AB$^+$ + e$^-$ with zero kinetic energy of e$^-$ at least in a certain range of the internuclear distance. On the other hand, the potential energy curves, that is, the solid lines, in Figure 2.1b are below the electronic continuum. Hence, some readers may think that such states are not superexcited states. However, the vibrational–rotational level can be higher than the zero point energy associated with the potential energy of AB$^+$ + e$^-$ with zero kinetic energy of e$^-$, that is, the lower edge of the continuum. Such states are called superexcited states of the second kind. Their energies exceed the lowest energy of AB$^+$ + e$^-$ by vibrational or rotational excitation, and thus they are embedded in the ionization continuum. The superexcited state of the first kind is a superexcited state when we take only the degree of freedom of electronic motion into account. The superexcited state of the second kind, on the other hand, is a superexcited state when we take account of the degree of freedom of nuclear motion as well as that of electronic motion.

As shown in Figure 2.1, the superexcited molecule can spontaneously ionize since it degenerates with the ionization continuum. This process is called autoionization, which is indicated by the downward arrow connecting upper and lower vibrational–rotational levels. The length of the arrow shows the kinetic energy of the emitted electron with respect to the nuclei-fixed frame. The autoionization of superexcited states of the first kind is attributed to the breakdown of the independent electron model, that is, the electron correlation. In other words, a mixing of discrete and continuous energy levels of the electronic system causes the autoionization. Such autoionization is called electronic autoionization. The superexcited state of the first kind is described by local complex potential energy, $U(R)$ (Nakamura, 1991: 123):

$$U(R) = W_d(R) - \frac{i}{2} \Gamma_d(R), \tag{2.20}$$

$$W_d(R) = \langle d | H_{el} | d \rangle, \tag{2.21}$$

$$\Gamma_d(R) = 2\pi \sum_{cont} \left| \langle cont | H_{el} | d \rangle \right|^2, \tag{2.22}$$

where
   $|d\rangle$ is the electronic state of the superexcited AB molecule
   $|cont\rangle$ is the continuous electronic state of AB$^+$ + e$^-$
   H_{el} is the electronic Hamiltonian defined at internuclear distance R in the framework of the
      Born–Oppenheimer approximation

The summation in Equation 2.22 is carried out for continuous electronic states, $|cont\rangle$, that are degenerate with the superexcited state, $|d\rangle$, at R. Both $W_d(R)$ and $\Gamma_d(R)$ have the dimension of energy. The real part of $U(R)$, $W_d(R)$, is the potential energy of the superexcited state $|d\rangle$ and shown by the solid curve in Figure 2.1a. The imaginary part, $-(1/2)\Gamma_d(R)$, expresses decay of the superexcited molecule in the $|d\rangle$ state through autoionization. The lifetime of the autoionization from the superexcited state $|d\rangle$, $\tau_d(R)$, is given by

$$\tau_d(R) = \frac{\hbar}{\Gamma_d(R)} . \tag{2.23}$$

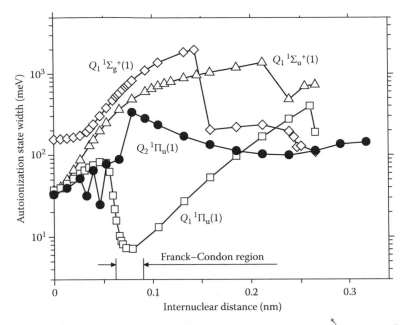

FIGURE 2.2 The autoionization state widths of the doubly excited states of H_2 as a function of internuclear distance calculated by Martín's group. (From Sánchez, I. and Martín, F., *J. Chem. Phys.*, 106, 7720, 1997; Sánchez, I. and Martín, F., *J. Chem. Phys.*, 110, 6702, 1999; Fernández, J. and Martín, F., *J. Phys. B*, 34, 4141, 2001.)

The quantity $\Gamma_d(R)$ is hence referred to as the autoionization state width or resonance width of the superexcited state $|d\rangle$. It is difficult to calculate the autoionization state width as a function of internuclear distance except for the lower-lying doubly excited states of hydrogen molecules. Figure 2.2 shows their results. The state width of 100 meV gives a lifetime of 6.6×10^{-15} s according to Equation 2.23.

The autoionization mechanism of the superexcited state of the second kind is much different from that of the superexcited state of the first kind. It originates from the breakdown of the Born–Oppenheimer approximation. In other words, energy transfer from the degree of freedom of nuclear motion, that is, vibration and rotation, to that of electronic motion brings about this autoionization. This kind of autoionization is thus called vibrational or rotational autoionization. The state widths for vibrational autoionization in H_2 are shown as a function of a vibrational quantum number, v_i, of the superexcited $(1s\sigma_g)(np\sigma_u)$ $^1\Sigma_u^+$ states in Figure 2.3. The state width of 10 cm^{-1} gives the lifetime of 5.3×10^{-13} s according to Equation 2.23. Vibrational autoionization is, in general, much slower than electronic autoionization, as seen in Figures 2.2 and 2.3, which is reasonable since vibrational autoionization requires energy transfer from the degree of freedom of the vibration to that of the electronic motion, while electronic autoionization only requires energy transfer within the electronic system. The typical value of the state width for rotational autoionization in H_2 is 2.3 cm^{-1}, which gives a lifetime of 2.3×10^{-12} s (Lefebvre-Brion and Field, 2004: 578). Rotational autoionization is, in general, slower than vibrational autoionization, which is reasonable since the period of rotation is much longer than the period of vibration.

Superexcited molecules decay through not only autoionization but also neutral dissociation. In other words, neutral dissociation competes with autoionization. The lifetime of neutral dissociation is considered to be in the same order of magnitude of the vibrational period, which is typically $\approx 10^{-14}$ s. Hence, we can easily understand that the competition takes place based on the estimation of the autoionization lifetimes mentioned above. The most remarkable feature of the dynamics of superexcited molecules is this competition. In general, the lifetime of the emission of fluorescence is longer than nanoseconds, and thus it competes with neither

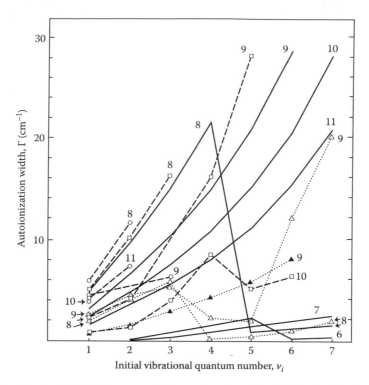

FIGURE 2.3 Vibrational autoionization state widths as a function of the initial vibrational quantum number, v_i, of H_2 $(1s\sigma_g)(np\sigma_u)$ $^1\Sigma_u^+$ states, the superexcited states of the second kind. The numbers attached to the lines indicate the principal quantum number, n. (From Nakamura, H., *Int. Rev. Phys. Chem.*, 10, 123, 1991. With permission.)

autoionization nor neutral dissociation. The formation and decay of a superexcited molecule is schematically shown below (Kouchi and Hatano, 2004: 105):

$$AB + hv \rightarrow AB^+ + e^- : \text{direct ionization} \tag{2.24}$$

$$\rightarrow AB^{**} : \text{superexcitation} \tag{2.25}$$

$$\rightarrow AB^+ + e^- : \text{autoionization} \tag{2.26}$$

$$\rightarrow A + B : \text{neutral dissociation} \tag{2.27}$$

$$\rightarrow \text{Other minor processes, for example, ion-pair formation} \tag{2.28}$$

The interference effect between direct ionization (Equation 2.24) and autoionization (Equation 2.26) following superexcitation (Equation 2.25) is expected. In fact the effect results in the appearance of the asymmetric peak shape, called the Fano profile or the Beutler–Fano profile, in the ionization cross-section curves against incident photon energy (Fano, 1961: 1866). We stress that superexcited molecules play an important role as reaction intermediates in radiation fields, processing plasma, upper atmosphere, and so forth, where electrons, ions, and excited atoms and molecules exist (Nakamura, 1991: 123; Hatano, 1999: 109; Ukai and Hatano, 2004: 121–157).

In this chapter, we take singly inner-valence excited and multiply excited states as superexcited states of the first kind, since they are not amenable to the independent electron model and cannot be eigenstates of the electronic Hamiltonian, H_{el}, defined at a fixed internuclear distance because of the degeneracy with the electronic continuum, as shown in Figure 2.1a (Nakamura, 1991: 123).

It is a subject of current interest to reveal the formation and decay dynamics of the singly inner-valence excited and multiply excited states of molecules. The coupling between the superexcited state of the first kind, $|d\rangle$, and the continuous electronic state, $|cont\rangle$, through H_{el} or the electron correlation is a remarkable feature, and is expressed as

$$V_{cont,d}(R) = \langle cont | H_{el} | d \rangle. \tag{2.29}$$

$V_{cont,d}(R)$ is called electronic coupling and is related to the autoionization state width, $\Gamma_d(R)$, as shown in Equation 2.22.

2.3 OBSERVING THE SINGLY INNER-VALENCE EXCITED AND MULTIPLY EXCITED STATES

The usual way to observe the excited states of a molecule is measuring the photoabsorption cross sections, that is, the oscillator strength distribution for photoabsorption, df/dE, in Equation 2.12, as a function of incident photon energy, where they appear as peaks. This is certainly the case in the energy range below the first ionization potential of the molecule, for example, as shown in Figure 2.4 for methane (Kameta et al., 2002: 225). The vertical ionization potentials of CH_4 are indicated in this figure by short vertical bars, which were measured by means of photoelectron spectroscopy (Bieri and Åsbrink, 1980: 149). Photoabsorption cross sections well above the first ionization potential are dominated by the transition to continuous electronic states of CH_4, that is, the direct ionization in Equation 2.24, and thus direct ionization prevents the superexcited states from being observed, as seen in Figure 2.4. The cross sections just decrease with the increasing incident photon energy except for the small undulation in the range of 20–23 eV. The key to observing superexcited states of molecules, in particular, the singly inner-valence excited and multiply excited states, is measuring cross sections free from ionization as a function of incident photon energy. As shown in Equations 2.24 through 2.28, neutral dissociation is an ionization-free process. It is more difficult to detect neutral fragments directly than charged species. However, they are often electronically excited, and thus emit fluorescence. Hence, we can observe singly inner-valence excited and

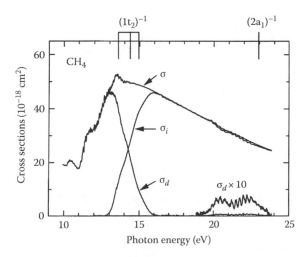

FIGURE 2.4 Photoabsorption (σ), photoionization (σ_i), and neutral-dissociation (σ_d) cross sections of CH_4 as a function of incident photon energy measured with a wavelength resolution of 0.1 nm, which corresponds to the energy resolution of 32 meV at an incident photon energy of 20 eV. The relation among σ, σ_i, and σ_d is $\sigma_d = \sigma - \sigma_i$. The vertical ionization potentials are indicated by the vertical bars (Bieri and Åsbrink, 1980: 149). (From Kameta, K. et al., *J. Electron Spectrosc. Relat. Phenom.*, 123, 225, 2002. With permission.)

multiply excited states in the cross sections for the emission of fluorescence from neutral fragments, that is, the oscillator strength distribution for the emission of fluorescence, as a function of incident photon energy. This sort of oscillator strength distribution is of course partial (see Equation 2.16 as well). There is another way to measure ionization-free cross sections as a function of incident photon energy: the photoabsorption cross sections and photoionization cross sections are measured as a function of incident photon energy by means of the double ionization-chamber method or the fast electron impact dipole-simulation method, and the difference of both cross sections gives the neutral-dissociation cross sections, as shown in Figure 2.4 (Kouchi and Hatano, 2004: 105–120). However, this method is not suitable for observing the singly inner-valence excited and multiply excited states since the difference between photoabsorption cross sections and photoionization cross sections is smaller in the energy range of superexcited states of the first kind than in the energy range of those of the second kind (see Figure 2.4).

2.4 OSCILLATOR STRENGTH DISTRIBUTION FOR THE EMISSION OF FLUORESCENCE FROM NEUTRAL FRAGMENTS

In this section, the cross sections or oscillator strength distributions for the emission of fluorescence in the visible and ultraviolet range from neutral fragments in the photoexcitation of CH_4 (Kato et al., 2002: 4383), NH_3 (Kato et al., 2003: 3541), and H_2O (Kato et al., 2004: 3127) as a function of incident photon energy are discussed. All of them have 10 electrons and are different from each other in terms of symmetry: CH_4 has T_d symmetry in the ground electronic state, NH_3 has C_{3v} symmetry, and H_2O has C_{2v} symmetry. The ground electronic state of CH_4 is

$$CH_4 : \tilde{X}^1 A_1 \quad (1a_1)^2 (2a_1)^2 (1t_2)^6, \tag{2.30}$$

where

the $1a_1$ orbital is an inner-shell orbital
the $2a_1$ orbital is an inner-valence orbital
the $1t_2$ orbital is an outer-valence orbital (Herzberg, 1991: 348)

The ground electronic state of NH_3 is

$$NH_3 : \tilde{X}^1 A_1 \quad (1a_1)^2 (2a_1)^2 (1e)^4 (3a_1)^2, \tag{2.31}$$

where

the $1a_1$ orbital is an inner-shell orbital
the $2a_1$ orbital is an inner-valence orbital
the $1e$ and $3a_1$ orbitals are outer-valence orbitals (Herzberg, 1991: 609)

The ground electronic state of H_2O is then

$$H_2O : \tilde{X}^1 A_1 \quad (1a_1)^2 (2a_1)^2 (1b_2)^2 (3a_1)^2 (1b_1)^2, \tag{2.32}$$

where

the $1a_1$ orbital is an inner-shell orbital
the $2a_1$ orbital is an inner-valence orbital
the $1b_2$, $3a_1$, and $1b_1$ orbitals are outer-valence orbitals (Herzberg, 1991: 585)

An outline of the apparatus is shown in Figure 2.5. The synchrotron radiation monochromatized by a 3 m normal incidence monochromator (BL-20A, Photon Factory, KEK, Tsukuba, Japan)

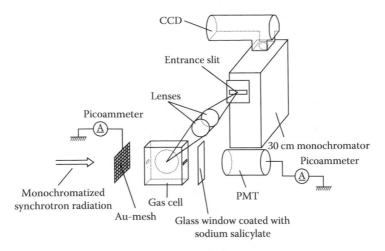

FIGURE 2.5 The outline of the apparatus used to measure the oscillator strength distributions for the emission of the fluorescence in the visible and ultraviolet range from neutral fragments in the photoexcitation of molecules. (From Kato, M. et al., *J. Phys. B*, 35, 4383, 2002. With permission.)

(Ito et al., 1995: 2119) was introduced into the gas cell filled with sample gas. Fluorescence due to the photoexcitation of the sample molecule was collected with a pair of lenses and focused onto the entrance slit of the 30 cm monochromator for ultraviolet and visible light. The 30 cm monochromator was set such that the longitudinal direction of the entrance slit was parallel to the incident photon beam to collect more fluorescence photons, and thus fluorescence was dispersed vertically. The dispersed fluorescence was detected by a liquid-nitrogen-cooled and back-illuminated charge-coupled device (CCD), which consists of 1340×400 pixels. The counts read out from the pixels aligning perpendicular to the dispersion direction were summed, and a wavelength-linear one-dimensional image covering the range of 280 nm width was obtained, which is a fluorescence spectrum. An example of fluorescence spectra is shown in Figure 2.6, which were measured for the photoexcitation of H_2O (Kato et al., 2004: 3127). Such fluorescence spectra were recorded at a wide range of the incident photon energies.

The pressure in the gas cell was carefully chosen such that the fluorescence count is proportional to the sample gas pressure even in the range of the incident photon energy where photoelectrons have enough energy to produce fluorescence through colliding with sample molecules. Such a secondary effect was seen for CO (Ehresmann et al., 1997: 1907). The intensities of the fluorescence spectra obtained were normalized for the flux of incident photons, the sample gas pressure, and the accumulation time, and then the normalized intensities were converted into absolute values of the fluorescence cross sections. Details of this procedure were described by Kato et al. (2002: 4383). In brief, the spectra of N_2^+ ($B^2\Sigma_u^+ v' \rightarrow X^2\Sigma_g^+ v''$) fluorescence in the photoionization of N_2 were measured in the range of the fluorescence wavelength of 357–522 nm as a reference to obtain conversion factors from normalized intensities to cross sections as a function of the fluorescence wavelength, since the cross sections for this fluorescence have been known (Samson et al., 1977: 1749; Woodruff and Marr, 1977: 87). We note that the conversion factors were measured online. These were measured in the range of the fluorescence wavelength of 357–522 nm, as mentioned above, so that the cross sections for the emission of fluorescence out of this range were not placed on an absolute scale, but on a relative scale.

The clearly observed fluorescences are as follows:

1. CH_4
   a. Balmer-α, -β, -γ, -δ, and -ϵ
   b. CH($A^2\Delta \rightarrow X^2\Pi$) and CH($B^2\Sigma^- \rightarrow X^2\Pi$)

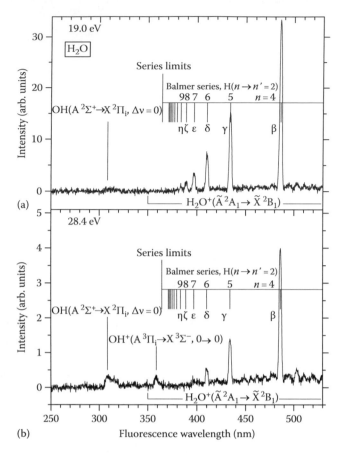

FIGURE 2.6 Fluorescence spectra in the photoexcitation of H_2O measured with a spectral resolution of 2.9 nm. The incident photon energies were (a) 19.0 eV and (b) 28.4 eV. The fluorescence intensities were normalized for the incident photon flux, the target gas pressure (≈ 1.6 Pa), and the accumulation time, and then the normalized intensities were plotted against the fluorescence wavelength without correcting for the sensitivity of a whole system. This correction was, of course, carried out to obtain the absolute values of the cross section for the emission of each fluorescence, as was described in detail by Kato et al. (2002: 4383). (From Kato, M. et al., *J. Phys. B*, 37, 3127, 2004. With permission.)

2. NH_3
    a. Balmer-α, -β, -γ, -δ, and -ϵ
    b. NH(A $^3\Pi_i \rightarrow$ X $^3\Sigma^-$), NH(c $^1\Pi \rightarrow$ a $^1\Delta$), and NH(c $^1\Pi \rightarrow$ b $^1\Sigma^+$)
3. H_2O
    a. Balmer-α, -β, -γ, -δ, -ϵ, -ζ, and -η
    b. OH(A $^2\Sigma^+ \rightarrow$ X $^2\Pi_i$)
    c. OH^+ (A $^3\Pi_i \rightarrow$ X $^3\Sigma^-$)
    d. H_2O^+ ($\tilde{A}\,^2A_1(3a_1)^{-1} \rightarrow \tilde{X}\,^2B_1(1b_1)^{-1}$)

As an example of the fluorescence cross-section curves that we measured, that is, the absolute values of the cross sections for the emission of the Balmer-β fluorescence, H($n = 4 \rightarrow n' = 2$), in the photoexcitation of CH_4 as a function of incident photon energy (Kato et al., 2002: 4383) are shown in Figure 2.7 together with experimental (Göthe et al., 1991: 2536) and theoretical (Cederbaum et al., 1978: 1591) photoelectron spectra. In Figure 2.7a, the left and right vertical axes represent the cross section and oscillator strength distribution for the emission of the Balmer-β fluorescence, respectively. The peaks around 22 and 29 eV in Figure 2.7a are clearly attributed to

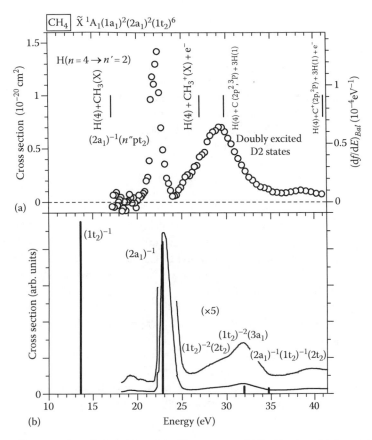

FIGURE 2.7 (a) Cross sections (left axis) and oscillator strength distributions (right axis) for the emission of the Balmer-β fluorescence, $H(n = 4 \rightarrow n' = 2)$, in the photoexcitation of CH_4 as a function of incident photon energy (Kato et al., 2002: 4383). The dissociation limits for the formation of $H(n = 4)$ are indicated by the short vertical bars. (b) Experimental and theoretical photoelectron spectra: The photoelectron spectrum of CH_4 taken at a photon energy of 65 eV and an angle of 90° with respect to the electric vector of the incident polarized light (solid curve) (Göthe et al., 1991: 2536) and the corresponding spectrum calculated by Cederbaum et al. (vertical bars) (Cederbaum et al., 1978: 1591). The horizontal axes indicate the incident photon energy in (a) and the binding energy in (b). The ground electronic state of CH_4 in T_d symmetry is shown in the upper part. (From Kouchi, N., Measurements of the fluorescence cross sections in the photo-excitation of CH_4, NH_3 and H_2O in the vacuum ultraviolet range, in *Atomic and Molecular Data and Their Applications: Joint Meeting of 14th International Toki Conference on Plasma Physics and Controlled Nuclear Fusion (ITC14); and 4th International Conference on Atomic and Molecular Data and Their Applications (ICAMDATA2004), AIP Conference Proceedings*, T. Kato, H. Funaba, and D. Kato (eds.), Vol. 771, 199–208, 2005. With permission.)

the superexcited states of CH_4 since they are above the $(1t_2)^{-1}$ ionic state of $CH_4{}^+$. It is still difficult to substantiate these superexcited states because there are no theoretical investigations as far as we know. Thus, let us rely on the knowledge of the ionic states of $CH_4{}^+$ that were investigated much better than the superexcited states of CH_4 by means of photoelectron spectroscopy, as shown in Figure 2.7b. There exist four double-hole one-electron states above the single-hole $(2a_1)^{-1}$ state around 23.0 eV, that is, the $(1t_2)^{-2}(2t_2)$ state around 29.2 eV, the $(1t_2)^{-2}(3a_1)$ state around 32.1 eV, the $(2a_1)^{-1}(1t_2)^{-1}(2t_2)$ state around 38.5 eV, and the $(2a_1)^{-2}(3a_1)$ state around 43.3 eV, the highest of which is out of range in Figure 2.7a. The neutral superexcited molecule is considered a molecule in which one electron is bound on the ion as an ion-core. Let us substantiate the superexcited states around 22 and 29 eV along this line; it hence follows that these superexcited states seem to be built on the single-hole $(2a_1)^{-1}$ state around 23.0 eV and $(1t_2)^{-2}(3a_1)$ state around 32.1 eV, respectively. Superexcited states around 22 eV are thus singly inner-valence excited states with the electron configuration of $(2a_1)^{-1}(mo)$, and those around 29 eV are doubly excited states with the electron configuration of $(1t_2)^{-2}(3a_1)(mo')$, where mo and mo' refer to molecular orbitals. The singly inner-valence excited $(2a_1)^{-1}(mo)$ states around 22 eV were also observed by measuring the excitation spectra for the formation of H^- (Mitsuke et al., 1991: 6003; Mitsuke et al., 1993: 6642) and threshold photoelectrons (Dutuit et al., 1990: 223; Furuya et al., 1994: 2720), and the excitation spectra for the emission of undispersed fluorescence (Sorensen et al., 1995: 554). Considering their results, the singly inner-valence excited states around 22 eV were assigned the $(2a_1)^{-1}(n''pt_2)$ $^1T_2$ Rydberg states ($n'' = 3$ and 4). On the other hand, the doubly excited $(1t_2)^{-2}(3a_1)(mo')$ states around 29 eV have not been understood well and were labeled D2. The doubly excited states labeled D1 and D3 were observed in the cross-section curves for the emission of CH(A, B → X) fluorescences in the photoexcitation of CH_4. Interestingly, the doubly excited D2 states are not responsible for the emission of CH(A, B → X) fluorescences. More recently, the doubly excited D4 states with the electron configuration of $(2a_1)^{-2}(3a_1)(mo'')$ were found in the cross-section curve for the emission of the Lyman-α fluorescence in the photoexcitation of CH_4 (Fukuzawa et al., 2005: 565). The $2a_1$ orbital in the D4 states is fully unoccupied, and hence they are hollow molecular states. This was the first report of finding hollow states of molecules.

The most remarkable feature of Figure 2.7a is that the cross sections originating from doubly excited D2 states seem to be much larger than those expected within the independent electron model, where double photoexcitation is much weaker than single photoexcitation since the electric dipole moment, $\sum_s \mathbf{r}_s$, in Equation 2.2, is a sum of single-electron coordinates. It is more convenient to discuss this interesting feature in terms of oscillator strength. The integration of each peak in the cross-section curve in Figure 2.7a over the range of incident photon energy gives the oscillator strength for the emission of the Balmer-β fluorescence originating from each superexcited state, as in the following equation:

$$\sigma_q(E) = 4\pi^2 \alpha a_0^2 R \left(\frac{df}{dE} \right)_q . \tag{2.33}$$

Equation 2.33 is easily derived from Equations 2.14, 2.16, and 2.17. Channel q in Figure 2.7a is the emission of the Balmer-β fluorescence, and thus $(df/dE)_q$ is written as $(df/dE)_{Bal}$, the scale of which is shown on the right vertical axis of Figure 2.7a following Equation 2.33. The oscillator strength, f_{Bal}, for the emission of the Balmer-β fluorescence in CH_4 originating from the singly inner-valence excited $(2a_1)^{-1}(n''pt_2)$ $^1T_2$ states ($n'' = 3$ and 4) and doubly excited D2 states were obtained as follows:

$$f_{Bal}((2a_1)^{-1}(n''pt_2)) = 1.7 \times 10^{-4}, \tag{2.34}$$

$$f_{Bal}(D2) = 3.4 \times 10^{-4}. \tag{2.35}$$

Equations 2.34 and 2.35 are indicative of the breakdown of the independent electron model, which is probably due to a mixing between the doubly excited D2 states and nearby singly inner-valence excited $(2a_1)^{-1}(n''pt_2)$ $^1T_2$ states, that is, due to the electron correlation. We note that the oscillator strength for the emission of the Balmer-β fluorescence is given by the product of the oscillator strength for excitation to the precursor superexcited state and the branching ratio of the formation of H($n = 4$) fragment atoms. This point was discussed in detail by Kato et al. (2002: 4383), and the above conclusion, that is, the breakdown of the independent electron model, was drawn. The experimental photoelectron spectrum in Figure 2.7b also shows the breakdown of the independent electron model: the single-hole $(2a_1)^{-1}$ main peak due to one-electron ionization is accompanied by the satellite peaks due to one-electron ionization and one-electron excitation. The degree of the breakdown in the photoexcitation process (Figure 2.7a) is much larger than in the photoionization process (Figure 2.7b) since the "satellite" peak around 29 eV is stronger than the "main" peak around 22 eV in Figure 2.7a.

The cross sections for the emission of the Balmer-β fluorescence, $H(n = 4 \rightarrow n' = 2)$, in the photoexcitation of NH_3 and H_2O as a function of incident photon energy are shown in Figure 2.8

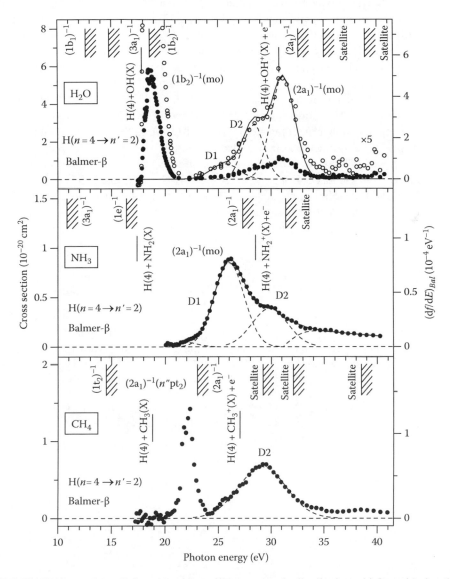

FIGURE 2.8 Cross sections (left axis) and oscillator strength distributions (right axis) for the emission of the Balmer-β fluorescence, $H(n = 4 \rightarrow n' = 2)$, in the photoexcitation of CH_4 (Kato et al., 2002: 4383), NH_3 (Kato et al., 2003: 3541), and H_2O (Kato et al., 2004: 3127) as a function of incident photon energy. The vertical ionization potentials for the single-hole states and double-hole one-electron states (satellites) are indicated together with the dissociation limits for the formation of $H(n = 4)$. D1 and D2 refer to doubly excited states and "mo" refers to a molecular orbital. $CH_4 : \tilde{X} \, ^1A_1 (1a_1)^2 (2a_1)^2 (1t_2)^6$ in T_d symmetry (Herzberg, 1991: 348), $NH_3 : \tilde{X} \, ^1A_1 (1a_1)^2 (2a_1)^2 (1e_2)^4 (3a_1)^2$ in C_{3v} symmetry (Herzberg, 1991: 609), and $H_2O : \tilde{X} \, ^1A_1 (1a_1)^2 (2a_1)^2 (1b_2)^2 (3a_1)^2 (1b_1)^2$ in C_{2v} symmetry (Herzberg, 1991: 585). (From Kato, M. et al., *J. Phys. B*, 37, 3127, 2004. With permission.)

together with those in CH_4. The right vertical axis represents the oscillator strength distributions for the emission of the Balmer-β fluorescence. The cross-section curves of NH_3 and H_2O were analyzed in the same way as CH_4, and the doubly excited states labeled D1 and D2 as well as the singly inner-valence excited $(2a_1)^{-1}(n''pt_2)$ or $(2a_1)^{-1}(mo)$ states were found. The contribution from a superexcited state is approximated by a Gaussian function according to the multidimensional reflection approximation (Schinke, 1993: 109–120), and the result of the fitting is shown by the dashed lines in Figure 2.8. Such a procedure was needed for NH_3 and H_2O since peaks are not separated well. The fitting was also done for CH_4. The peak due to the singly outer-valence excited $(1b_2)^{-1}(mo)$ states is seen only in the cross-section curve of H_2O because of energetic reasons (see the vertical ionization potentials indicated by vertical bars with hatches, and dissociation limits for the formation of $H(n = 4)$ indicated by vertical bars). We note that the superexcited states in Figure 2.8 were observed by measuring cross-section curves or oscillator strength distributions free from ionization, and that doubly excited states were first found by Kato et al. (2002: 4383; 2003: 3541; 2004: 3127). Interestingly, as shown in this figure, energies of the singly inner-valence excited $(2a_1)^{-1}(n''pt_2)$ or $(2a_1)^{-1}(mo)$ states become higher on going from CH_4 to H_2O, while those of the doubly excited D2 states do not change quite so much. Figure 2.8 also seems indicative of the breakdown of the independent electron model. This is well illustrated in Figure 2.9, where oscillator strengths for the emission of Balmer fluorescences, $H(n \rightarrow n' = 2)$, in CH_4 (Kato et al., 2002: 4383), NH_3 (Kato et al., 2003: 3541), and H_2O (Kato et al., 2004: 3127) are plotted against n on logarithmic scales for both axes. The independent electron model seems violated in the energy range near the ionization potential of the inner-valence $2a_1$ electron in CH_4, NH_3, and H_2O. In particular, the result of CH_4 in Figure 2.9, that is,

$$f_{Bal}(D2) = 2 f_{Bal}((2a_1)^{-1}(n''pt_2)),\qquad(2.36)$$

is remarkable. We again note that $f_{Bal}(D2) \ll f_{Bal}((2a_1)^{-1}(n''pt_2))$ is expected within the independent electron model. There seems to be a tendency that the degree of breakdown of the independent electron model becomes lower on going from CH_4 to H_2O.

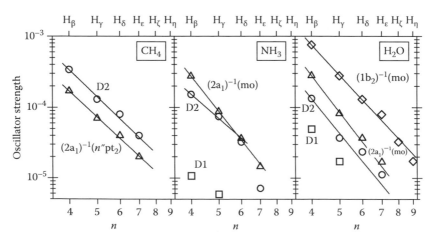

FIGURE 2.9 Oscillator strengths for the emission of the Balmer fluorescences, $H(n \rightarrow n' = 2)$, in CH_4 (Kato et al., 2002: 4383), NH_3 (Kato et al., 2003: 3541), and H_2O (Kato et al., 2004: 3127) plotted against the principal quantum number of the upper level of the hydrogen atom, n, on logarithmic scales for both axes. Open diamonds: $(1b_2)^{-1}(mo)$, open triangles: $(2a_1)^{-1}(mo)$ or $(2a_1)^{-1}(n''pt_2)$, open squares: D1, and open circles: D2. D1 and D2 refer to doubly excited states and "mo" refers to a molecular orbital. (From Kato, M. et al., *J. Phys. B*, 37, 3127, 2004. With permission.)

2.5 CONCLUSION

The oscillator strength distribution for photoabsorption (df/dE) or the photoabsorption cross section (σ) as a function of incident photon energy gives the basic feature of the interaction of photons, as discussed in Section 2.1 and by Kouchi and Hatano (2004: 105–120). However, the peaks due to superexcited states, in particular, singly inner-valence excited and multiply excited states, are not noticeable in the df/dE or the σ curve because of a large contribution of ionization. The key to observing superexcited states is thus measuring the "partial" oscillator strength distribution or cross section free from ionization as a function of incident photon energy. The typical ionization-free process is a neutral dissociation following the superexcitation, and thus we can observe the superexcited states by measuring the oscillator strength distribution or cross section for the emission of fluorescence from neutral fragments as a function of incident photon energy. The results of such measurements carried out for 10-electron molecules, that is, CH_4, NH_3, and H_2O, are presented as an example. We stress that it is significant to measure absolute values of the oscillator strength distribution or cross section for a detailed and quantitative discussion. The singly inner-valence excited and doubly excited states were clearly observed, which shows that the above-mentioned method is a powerful tool. A strong indication of the breakdown of the independent electron model was found. The observed peaks are not classified into main peaks due to singly excited states and into satellite peaks due to doubly excited states. The effect of electron correlation has to be considered in a proper way to describe the dynamics of superexcited states as well as energy eigenvalues.

ACKNOWLEDGMENTS

The authors wish to thank Drs. Masahiro Kato, Kosei Kameta, and Makoto Murata, and Prof. Yoshihiko Hatano for their excellent contribution to the investigations described in Section 2.4. The successful results presented in Section 2.4 are due to the high stability of the synchrotron radiation at the Photon Factory, Institute of Materials Structure Science, KEK, Tsukuba, Japan, and the negligibly low intensity of the higher-order component mixed in incident light at the BL-20A of the Photon Factory.

REFERENCES

Bieri, G. and Åsbrink, L. 1980. 30.4 nm He(II) photoelectron spectra of organic molecules: Part I. Hydrocarbons. *J. Electron Spectrosc. Relat. Phenom.* 20: 149–167.

Bransden, B. H. and Joachain, C. J. 2003. *Physics of Atoms and Molecules.* Edinburgh, U.K.: Pearson Education Limited.

Cederbaum, L. S., Domcke, W., Schirmer, J., von Niessen, W., Diercksen, G. H. F., and Kraemer, W. P. 1978. Correlation effects in the ionization of hydrocarbons. *J. Chem. Phys.* 69: 1591–1603.

Dutuit, O., Aït-Kaci, M., Lemaire, J., and Richard-Viard, M. 1990. Dissociative photoionization of methane and its deuterated compounds in the A state region. *Phys. Scr.* T31: 223–226.

Ehresmann, A., Machida, S., Ukai, M., Kameta, K., Kitajima, M., Kouchi, N., Hatano, Y., Ito, K., and Hayaishi, T. 1997. CO Rydberg series converging to the CO $^+$ D and C states observed by VUV-fluorescence spectroscopy. *J. Phys. B* 30: 1907–1926.

Fano, U. 1961. Effects of configuration interaction on intensities and phase shifts. *Phys. Rev.* 124: 1866–1878.

Fernández, J. and Martín, F. 2001. Autoionizing $^1\Sigma_u^+$ and $^1\Pi_u$ states of H_2 above the third and fourth ionization threshold. *J. Phys. B* 34: 4141–4153.

Fukuzawa, H., Odagiri, T., Nakazato, T., Murata, M., Miyagi, H., and Kouchi, N. 2005. Doubly excited states of methane produced by photon and electron interactions. *J. Phys. B* 38: 565–578.

Furuya, K., Kimura, K., Sakai, Y., Takayanagi, T., and Yonekura, N. 1994. Dissociation dynamics of CH_4^+ core ion in the $^2A_1$ state. *J. Chem. Phys.* 101: 2720–2728.

Göthe, M. C, Wannberg, B., Karlsson, L., Svensson, S., Baltzer, P., Chau, F. T., and Adam, M.-Y. 1991. X-ray, ultraviolet, and synchrotron radiation excited inner-valence photoelectron spectra of CH_4. *J. Chem. Phys.* 94: 2536–2542.

Hatano, Y. 1999. Interaction of vacuum ultraviolet photons with molecules. Formation and dissociation dynamics of molecular superexcited states. *Phys. Rep.* 313: 109–169.

Herzberg, G. 1991. *Molecular Spectra and Molecular Structure III. Electronic Spectra and Electronic Structure of Polyatomic Molecules*. Malabar, FL: Krieger Publishing Company.

Ito, K., Morioka, Y., Ukai, M., Kouchi, N., Hatano, Y., and Hayaishi, T. 1995. A high-flux 3-M normal incidence monochromator at beamline 20A of the photon factory. *Rev. Sci. Instrum.* 66: 2119–2121.

Kameta, K., Kouchi, N., Ukai, M., and Hatano, Y. 2002. Photoabsorption, photoionization, and neutral-dissociation cross sections of simple hydrocarbons in the vacuum ultraviolet range. *J. Electron Spectrosc. Relat. Phenom.* 123: 225–238.

Kato, M., Kameta, K., Odagiri, T., Kouchi, N., and Hatano, Y. 2002. Single-hole one-electron superexcited states and doubly excited states of methane in the vacuum ultraviolet range as studied by dispersed fluorescence spectroscopy. *J. Phys. B* 35: 4383–4400.

Kato, M., Odagiri, T., Kameta, K., Kouchi, N., and Hatano, Y. 2003. Doubly excited states of ammonia in the vacuum ultraviolet range. *J. Phys. B* 36: 3541–3554.

Kato, M., Odagiri, T., Kodama, K., Murata, M., Kameta, K., and Kouchi, N. 2004. Doubly excited states of water in the inner valence range. *J. Phys. B* 37: 3127–3148.

Kouchi, N. 2005. Measurements of the fluorescence cross sections in the photoexcitation of CH_4, NH_3 and H_2O in the vacuum ultraviolet range. In *Atomic and Molecular Data and Their Applications: Joint Meeting of 14th International Toki Conference on Plasma Physics and Controlled Nuclear Fusion (ITC14); and 4th International Conference on Atomic and Molecular Data and Their Applications (ICAMDATA2004)*, 5–8 October 2004, Toki, Japan. *AIP Conference Proceedings*, T. Kato, H. Funaba, and D. Kato (eds.), Vol. 771, pp. 199–208.

Kouchi, N. and Hatano, Y. 2004. Interaction of photons with molecules: Photoabsorption, photoionization, and photodissociation cross sections. In *Charged Particle and Photon Interactions with Matter*, A. Mozumder and Y. Hatano (eds.), pp. 105–120. New York: Marcel Dekker.

Lefebvre-Brion, H. and Field, R. W. 2004. *The Spectra and Dynamics of Diatomic Molecules*. Amsterdam, the Netherlands: Elsevier.

Mitsuke, K., Suzuki, S., Imamura, T., and Koyano, I. 1991. Negative-ion mass spectrometric study of ion-pair formation in the vacuum ultraviolet. IV. $CH_4 \rightarrow H^- + CH_3^+$ and $CD_4 \rightarrow D^- + CD_3^+$. *J. Chem. Phys.* 94: 6003–6006.

Mitsuke, K., Hattori, H., and Yoshida, H. 1993. Ion-pair formation from saturated hydrocarbons through photoexcitation of an inner-valence electron. *J. Chem. Phys.* 99: 6642–6652.

Nakamura, H. 1991. What are the basic mechanisms of electronic transitions in molecular dynamic processes? *Int. Rev. Phys. Chem.* 10: 123–188.

Platzman, R. L. 1962a. Superexcited states of molecules, and the primary action of ionizing radiation. *Vortex* 23: 372–389.

Platzman, R. L. 1962b. Superexcited states of molecules. *Radiat. Res.* 17: 419–425.

Samson, J. A. R. 1980. *Techniques of Vacuum Ultraviolet Spectroscopy*. Lincoln, NE: Pied Publications.

Samson, J. A. R., Haddad, G. N., and Gardner, J. L. 1977. Total and partial photoionization cross sections of N_2 from threshold to 100 Å. *J. Phys. B* 10: 1749–1759.

Sánchez, I. and Martín, F. 1997. The doubly excited states of the H_2 molecule. *J. Chem. Phys.* 106: 7720–7730.

Sánchez, I. and Martín, F. 1999. Doubly excited autoionizing states of H_2 above the second ionization threshold: The Q_2 resonance series. *J. Chem. Phys.* 110: 6702–6713.

Schinke, R. 1993. *Cambridge Monographs on Atomic, Molecular and Chemical Physics 1. Photodissociation Dynamics*. Cambridge, U.K.: Cambridge University Press.

Sorensen, S. L., Karawajczk, A., Strömholm, C., and Kirm, M. 1995. Dissociative photoexcitation of CH_4 and CD_4. *Chem. Phys. Lett.* 232: 554–560.

Ukai, M. and Hatano, Y. 2004. Reactions of low-energy electrons, ions, excited atoms and molecules, and free radicals in the gas phase as studied by pulse radiolysis methods. In *Charged Particle and Photon Interactions with Matter*, A. Mozumder and Y. Hatano (eds.), pp. 121–157. New York: Marcel Dekker.

Woodruff, P. R. and Marr, G. V. 1977. The photoelectron spectrum of N_2, and partial cross sections as a function of photon energy from 16 to 40 eV. *Proc. R. Soc. Lond. A* 358: 87–103.

3 Electron Collisions with Molecules in the Gas Phase

Hiroshi Tanaka
Sophia University
Tokyo, Japan

Yukikazu Itikawa
Institute of Space and Astronautical Science
Sagamihara, Japan

CONTENTS

3.1 INTRODUCTION

Radiation interaction with matter is a subject of radiation science (i.e., radiation physics, radiation chemistry, and radiation biology). This subject is roughly divided into two parts. The first is the fate of incident radiation, or, more specifically, the amount of energy deposited in matter (the linear energy transfer (LET) or stopping power). The second is the change of matter due to the absorption of radiation energy. Both aspects of radiation–matter interaction are understood microscopically in terms of collisions among the radiation particle, the medium molecules, and the products of the interaction itself. Thus, knowledge of the relevant atomic collision processes is an essential ingredient of radiation science. As a part of this knowledge, this chapter provides a general picture of the collisions between electrons and molecules. The roles that these electron–molecule collisions play in radiation–matter interaction are described in other chapters.

Now, we consider ionizing radiation (i.e., fast charged particles and high-energy photons). In the interaction of ionizing radiation with matter, the most dominant microscopic process that occurs is the ionization of medium molecules (production of ions and free electrons). If the ejected electron has a large enough energy, it can ionize the medium molecules further. The free electrons thus produced are called "secondary electrons." Many of the chemical or biological effects of radiation are triggered by the action of these secondary electrons. For example, a part of the radiation damage to DNA (called the direct damage) is due to the direct interaction of secondary electrons with the DNA molecule. As is shown in Section 3.2.1, most of the secondary electrons have energies less than 100 eV. These electrons are often called low-energy electrons, compared with the fast electrons in the primary radiation. Section 3.3 summarizes the recent advances in the study of the collisions of these low-energy electrons with molecules.

The behavior of the secondary (and the primary, if any) electron is studied in two different ways: calculation of the degradation spectrum and the Monte Carlo simulation. The distribution function of the electron energy satisfies a kinetic equation. On the basis of the kinetic theory, Spencer and Fano derived an equation governing the degradation process of the electron energy spectrum (see the review by Kimura et al., 1993). Once the spectrum is obtained, one can calculate the yield of a particular species produced during radiation interaction with matter. Another approach is the Monte Carlo simulation (see Section 3.4). It numerically traces all the trajectories of a number of representative electrons in matter. It reveals the three-dimensional structure of the radiation track, leading to the understanding of biochemical and biological effects of radiation (Nikjoo et al., 2006). Both these methods require the information of collision processes between electrons and matter molecules (i.e., cross-section data). Particularly, in the Monte Carlo method, this information is necessary on the outgoing direction of the electron after collision (i.e., the differential cross section [DCS]).

To advance on with radiation science, it is essential to have a reliable set of cross-section data for the collision processes involved. According to Inokuti, for these cross-section data to be applicable to any practical problem, they must fulfill the trinity requirement, that is, the data should be *correct*, *absolute*, and *comprehensive*. Here, the term "comprehensive" means that the data must cover a wide enough range of variables (e.g., incident energy and scattering angle). On the contrary, data are often available only over a limited range of variables. There have been several attempts to compile a reliable database of the cross section for electron–molecule collisions. Some examples are given in Section 3.4.

Some specific processes of electron–molecule collisions are of special importance in the interaction of radiation with matter. Recently, collisions of low-energy electrons with biological molecules are being extensively studied. The dissociation of molecules via collision with electrons of very low energies ($\lesssim 10\,eV$) has been highlighted in relation to radiation damage of DNA. This topic is briefly discussed in Section 3.3.

Electron–molecule collisions play fundamental roles in various application fields. For example, they are important in the study of the upper atmosphere (ionosphere) of the Earth and other planets. The understanding of low-temperature plasmas (e.g., those used for plasma processing) is based on the knowledge of electron–molecule collisions. This chapter serves also as a guide to electron–molecule collisions in these fields. More details of the electron–molecule collision data and their applications are given in a book by Itikawa (2007). Finally, it should be pointed out that the discussion in this chapter is restricted to the collision processes in the gas phase. Of course, collision processes in the liquid phase (or in other condensed phases) are also important in the study of radiation interaction with matter. These topics are treated in other chapters.

3.2 FUNDAMENTAL ASPECTS OF COLLISION PROCESSES

3.2.1 SOURCE OF SECONDARY ELECTRONS

The interaction of a charged particle with a molecule can be broadly divided into two categories: distant (or soft) collision and close (or hard) collision. When the charged particle is located far from the target molecule, it exerts an electromagnetic force over the entire molecule, and the interaction resembles the photoabsorption of the molecule. On the other hand, when the incident particle comes close to the target, a knock-on collision occurs between the particle and the bound electron in the molecule. In this case, the interaction can be described by a binary-encounter model. Ionization cross sections are approximately constructed by combining the models in these two categories (Kim and Rudd, 1994). More details of the ionization process are given by Toburen (2004).

A secondary electron is characterized by two parameters: its kinetic energy and its moving direction with respect to the beam of incident particles. The distribution of energy and angles of the electron ejected after an ionizing collision are specified by the so-called doubly differential cross section (DDCS) of ionization. When we integrate the DDCS over the angles, we obtain the energy distribution of the ejected electron (called the singly differential cross section, SDCS). One can easily note that the integration of the SDCS over the energy of the ejected electron leads to the ionization cross section.

As an example, Figure 3.1 (Ohsawa et al., 2005) shows the DDCS for the ionization of water (H_2O) vapor by the impact of alpha particles. Figure 3.2 (Bolorizadeh and Rudd, 1986) gives the SDCS for the ionization of H_2O by electron collisions. Both figures show that the energy distribution of the secondary electron has a peak at around zero energy. In other words, the energy of most of the secondary electrons is less than about 100 eV. These figures show the spectra of the first generation of secondary electrons. Electrons with impact energy above the ionization potential of H_2O (12.6 eV) can ionize H_2O molecules further. The resulting free electrons are the second generation of secondary electrons. Thus, the mean energy of the secondary electron beam is lower than that of the spectra shown in the figures.

Collisions of fast photons with molecules also generate free electrons. When the photon energy is lower than about 10 keV, these electrons are ejected through the photoelectric effect. Above this energy, Compton scattering is the dominant mechanism used for electron emission. As an example, Uehara and Nikjoo (1996) studied the energy spectrum of ejected electrons in the process of photon interaction with water vapor. In this case, the kinetic energy of the ejected electron is not necessarily small. But the first-generation secondary electrons quickly lose their energy through collision with matter molecules.

3.2.2 COLLISION CROSS SECTION AND RELATED QUANTITIES

All physical phenomena discussed in radiation sciences are analyzed in terms of relevant collision cross sections. Here, the cross section for an inelastic collision process is defined as follows: Suppose that I_0 electrons of energy E_0, per unit area per unit time, are incident on a molecular gas with number density N. The number of electrons scattered per unit time into the solid angular element

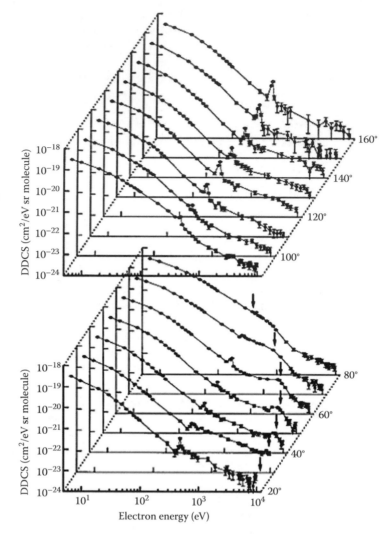

FIGURE 3.1 DDCS for 6.0 MeV/u He^{2+} impact on water vapor. The secondary electrons are measured at 7–10,000 eV and 20°–160°. Arrows indicate the binary-encounter peaks. (Reprinted from Ohsawa, D. et al., *Nucl. Instrum. Methods B*, 227, 431, 2005. With permission.)

$d\Omega$ in the direction $\Omega(\theta, \phi)$, measured from the polar axis taken along the direction of the incident electron, can be written as

$$I_{0n}(\Omega) = \frac{NI_0 d\sigma_{0n}(E_0, \Omega)}{d\Omega} \tag{3.1}$$

The subscript $0n$ indicates a transition from the ground state 0 to an excited or ionized state n. The quantity $d\sigma_{0n}(E_0, \Omega)/d\Omega$ is called the differential cross section for the excitation $0 \rightarrow n$.

Theoretically, the DCS is expressed in terms of scattering amplitude $f_{0n}(E_0, \Omega)$, which is derived from the asymptotic behavior of the electron wave function, that is,

$$\frac{d\sigma_{0n}(E_0, \Omega)}{d\Omega} = \left(\frac{k_n}{k_0}\right)\left|f_{0n}(E_0, \Omega)\right|^2 \tag{3.2}$$

where k_0 and k_n are magnitudes of the electron momentum before and after collision, respectively.

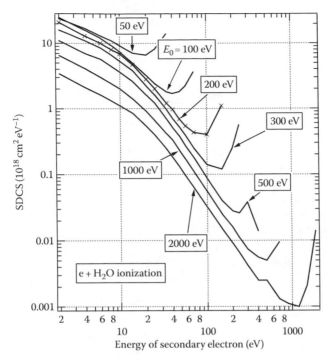

FIGURE 3.2 SDCS for electron collision with H_2O measured by Bolorizadeh and Rudd (Bolorizadeh and Rudd 1986). Incident electron energies (E_0) are indicated.

The integral of the DCS over all scattering angles, namely,

$$q_{0n}(E_0) = \iint \frac{d\sigma_{0n}(E_0,\Omega)}{d\Omega} \sin\theta\, d\theta\, d\phi \qquad (3.3)$$

is called the (integral) cross section for the excitation $0 \to n$.

The elastic-scattering cross section, $q_0(E_0)$, is defined similarly, by replacing the final state n with the ground state 0 in Equations 3.1 through 3.3. The effect of elastic scattering on electron transport phenomena is normally explained by the momentum-transfer cross section defined by

$$q_0^M(E_0) = \iint \frac{d\sigma_0(E_0,\Omega)}{d\Omega}(1-\cos\theta)\sin\theta\, d\theta\, d\phi \qquad (3.4)$$

The sum of the cross section (3.3) over all possible excitation processes (including the elastic one), namely,

$$Q(E_0) = q_0(E_0) + \sum q_{0n}(E_0) \qquad (3.5)$$

is called the total scattering cross section (TCS). From this definition, it is clear that the TCS gives an upper bound of the cross section for any individual process.

If the incident particle has a velocity distribution given by $F(v)$, then the reaction rate constant for a process with a cross section q_{0n} is calculated as

$$\kappa_{0n} = \int q_{0n}F(v)v\,dv \qquad (3.6)$$

Finally, an average energy loss per unit path length of the incident electron is given as

$$\frac{-dE}{dx(E)} = \sum S(E;i)N(i) \tag{3.7}$$

$$S(E;i) = \sum (dE)_{if} q_{if}(E) \tag{3.8}$$

where

$N(i)$ is the number density of the target molecule in its ith state
$S(E;i)$ is called the stopping cross section

The summation on the right-hand side of (3.8) is taken over all possible transitions $i \to f$ with an energy loss $(dE)_{if}$. Strictly speaking, elastic scattering also has a contribution to the stopping cross section. The contribution is given by

$$S_{\text{elas}}(E) = \left(\frac{2m_e}{M}\right) E q_0^M (E) \tag{3.9}$$

Here, m_e and M are masses of the electron and the molecule, respectively, while q_0^M is the momentum-transfer cross section defined by Equation 3.4. When the energy of the incident electron is above the threshold of the first electronic excited state, the elastic stopping cross section can be ignored in comparison with inelastic ones. But for subexcitation electrons, the elastic contribution should be considered in the evaluation of the energy-loss rate in Equation 3.7.

3.2.3 Theory and Its Role

Electron collisions with molecules have been studied theoretically for many years. Several elaborate methods have been developed to calculate collision cross sections (Huo and Gianturco, 1995). This section does not address the details of these theoretical methods. Instead, general aspects of the theory as a tool to complement experimental results are presented in the following text.

One of the main objectives of theory is to prepare a framework to interpret the results obtained experimentally (Schneider, 1994). In other words, theory provides the means with which we can gain insight into the physics underlying collision processes. Furthermore, it is only with theory that we can predict the results of future experiments. One of the typical examples is the interpretation of the resonant structure in the energy dependence of the cross section (see Section 3.3). This structure is thought to be caused by a temporary capture of the incident electron to the molecule. This interpretation is confirmed only through theory. Whenever any structure is found in the experimental cross section, it tends to be assigned to a resonance. But the assignment should be tested against theory. Another example is the asymptotic form of the cross section in the limit of high impact energy. The Born–Bethe theory shows that the cross section for any dipole-allowed transition has the asymptote $(\ln E)/E$ as a function of collision energy, E (Inokuti, 1971). This is a universal law and often used as a check of the validity of experimental data.

Another important role of collision theory is to supplement the cross-section data obtained experimentally. In many cases, experimental data do not satisfy the trinity requirement mentioned in Section 3.1. Normally, relative measurements are much easier than absolute ones. In some cases, the experimental cross section on the relative scale is normalized to an absolute value theoretically obtained at some point of energy. Most of the measurements of the cross section are performed at a limited number of independent parameters (e.g., incident energies and scattering angles). In this sense, most experimental data are not comprehensive but fragmentary. Theoretical models have been proposed to make these data more comprehensive.

As is mentioned in Section 3.2.1, a close collision between an electron and a molecule can be approximately described by a binary-encounter theory for collision of the incoming electron and the molecular one. Kim and his colleagues (Kim and Rudd, 1994; Hwang et al., 1996; Kim et al., 1997) derived an ionization cross section from the classical two-body collision theory. To make the resulting formula more physically reasonable, they combined this ionization cross section with an asymptotic form of the quantum mechanical cross section in the high-energy limit (i.e., the Bethe formula). To take into account the characteristics of the molecule, they incorporated into the resulting formula the binding and kinetic energies of the molecular electrons. They called this the binary-encounter-dipole (BED) model. The practical application of this model is shown in Section 3.3.4.1. Kim (2007) also proposed a simple model to produce cross sections for the excitation of the electronic state. This (called the BEf-scaled Born cross section) is given in Section 3.3.3.3 in relation to the excitation of H_2O.

For their Monte Carlo study, Garcia and his group (Muñoz et al., 2007, 2008) obtained elastic cross sections (particularly DCS) with a theoretical model. First, they prepared a model optical potential for an electron–atom collision. With this potential, they approximately took into account the effects of inelastic processes on elastic scattering. Then they adopted the independent atom model (IAM), that is, the elastic cross section for an electron–molecule collision was obtained by an incoherent sum of the elastic cross sections for electron scattering from the constituent atoms. In so doing, they approximately included screening effects to consider the molecular nature of the target. They showed that, at least for the integral cross section, their model reproduces the experimental data available.

In principle, theory can produce cross sections on an absolute scale. When no experimental data are available, theoretical cross sections are used for application. However, it is very difficult to evaluate the accuracy of these cross sections. The reliability of the theory used can be sometimes judged, but it is generally impossible to numerically estimate the possible error of the cross sections calculated. Great care should be taken when any theoretical values are adopted in the database for application.

3.3 RECENT ADVANCES IN LOW-ENERGY ELECTRON COLLISIONS WITH MOLECULES

3.3.1 OVERVIEW OF THE CROSS SECTIONS FOR ELECTRON COLLISIONS WITH MOLECULES

Two typical examples of the cross-section set are presented here: one for H_2O (Figure 3.3) (Itikawa and Mason, 2005) and the other for CH_4 (Figure 3.7) (Kurachi and Nakamura, 1990). The electron-impact energy is covered from 0.1 to 1000 eV for H_2O, and from 0.01 to 100 eV for CH_4. As is described in Section 3.2.2, collision is classified into two kinds, namely, elastic ($0 \to 0$) and inelastic ($0 \to n$) processes. An elastic collision, in which internal energy of the molecule is not changed during the collision, takes place at any energy. Strictly speaking, a small part, ΔE, of the kinetic energy of the electron is transferred to the target molecule. The relative amount of energy transfer is given by $\Delta E/E \sim m_e/M \sim 10^{-4}$, where m_e is the electron mass and M is the mass of the molecule. In an inelastic collision, internal energy is changed according to the processes, such as rotational, vibrational, and electronic excitations; dissociation; ionization; and electron attachment. The relative amounts of energy transfer to rotational, vibrational, and electronic degrees of freedom are roughly of the order of $(m_e/M)^{1/2}$: $(m_e/M)^{1/4}$: 1. From this, it is clear that the adiabatic approximation holds for rotational and vibrational motions. In the following text, some details of the cross-section set for H_2O (Figure 3.3) are given. Cross sections for CH_4 are discussed in Section 3.3.2.2 in relation to the swarm experiment.

Figure 3.3 shows the recommended values of the cross sections for total scattering, elastic scattering, momentum transfer, rotational transition, vibrational excitations of bending and stretching modes, and total ionization (Itikawa and Mason, 2005). The figure also shows cross sections for several different processes of molecular dissociation. They are dissociative attachment to produce H^-, dissociative emissions to produce Ly alpha and Balmer alpha lines of H, and A-X emission of OH, and finally the production of OH in its ground (X) state and O in ^1S state. Each inelastic cross section has a threshold and a specific shape of energy dependence.

To determine a cross section of a certain process over a wide range of energies, several different experimental methods described below (i.e., an attenuation method, a swarm technique, and a

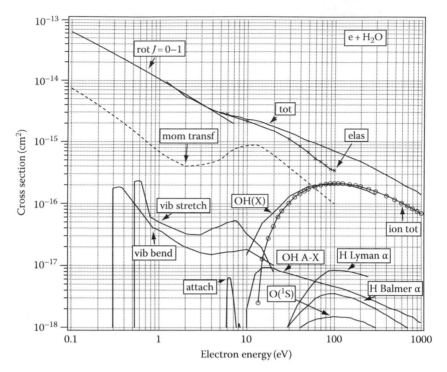

FIGURE 3.3 Summary of the recommended data on electron collision cross sections for H_2O. (Reprinted from Itikawa, Y. and Mason, N.J., *J. Phys. Chem. Ref. Data*, 34, 1, 2005. With permission.)

crossed-beam method) must be used in combination. To obtain a reliable (though not necessarily complete) set of the cross sections, there is no standard way to employ. As one can see in an extensive data compilation, such as is shown in Figure 3.3, it takes efforts of many workers in many different institutions to produce cross sections for one molecule. In spite of the efforts, some of the data sets may be fragmentary. Furthermore, results from different laboratories are often discordant. It is therefore necessary to collect as many sets of data as possible from the literature, to assess their reliability and to determine the most trustworthy set of data for use in applications. Efforts toward such data compilations and analysis are being made by various groups, as described in Section 3.4.

As for H_2O, the cross-section set recommended by Itikawa and Mason (Figure 3.3) does not necessarily satisfy every user. The most serious problem is the excitation of the electronic state. At the time of compilation, no reliable experimental data were available for the process so that nothing was recommended for the excitation of the electronic state, as is seen in Figure 3.3. However, the situation is now improved (see Section 3.3.3.3). Total and elastic-scattering cross sections and the ionization process have been well studied. No experimental data are available for rotational excitation, and values plotted in Figure 3.3 are the result of a theoretical calculation. Some of the other cross sections are now being revised, which are discussed in the relevant sections below. Particularly for dissociation processes, a recent review by McConkey et al. (2008) is very informative and useful.

Once the cross-section data set is determined, useful physical quantities like Equations 3.6 and 3.8 are evaluated. For instance, the stopping cross sections thus calculated are shown for H_2O in Figure 3.4 (Itikawa, 2007). In this figure, the stopping cross sections are presented only over the energy region below 10 eV. In this region, rotational transition is the most dominant process for stopping the electron. The two curves for vibrational excitation correspond to the bending (vib2) and stretching (vib13) modes, respectively. The stopping cross section for the excitation of the electronic state (designated as exc.) is based on a theoretical cross section so that its quantitative accuracy is uncertain. For electron energies above 10 eV, ionization and electronic excitation dominate. We need more detailed information about these processes (particularly about the excitation of electronic states) to evaluate the stopping cross section.

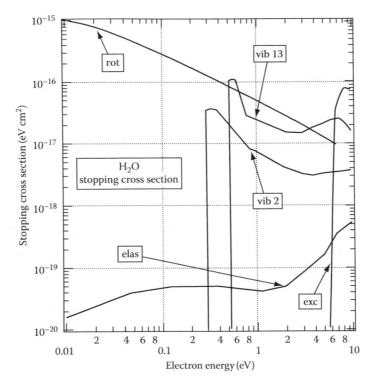

FIGURE 3.4 Stopping cross sections for electron collisions with H_2O. Only those for low-energy electrons are shown. (Reprinted from Itikawa, Y., *Molecular Processes in Plasmas: Collisions of Charged Particles with Molecules*, Springer, Berlin, Germany, 2007. With permission.)

3.3.2 Measurements by Electron-Beam Attenuation and Transport Methods

3.3.2.1 Total Scattering Cross Section—An Upper Bound of Cross Section

The TCS, Q, defined by Equation 3.5 is obviously an upper bound of any of the individual cross sections. It is normally determined with a beam attenuation method. With this method, however, no information is available for each $q_{0n}(E_0)$ component on the right-hand side of Equation 3.5. Each individual cross section has to be determined by the measurement of the crossed-beam type described in Section 3.3.3.

To determine Q, one may use the *Lambert–Beer* law, commonly used in photoabsorption measurements. Suppose that one sends an electron beam (having uniform velocity) of intensity I_0 per unit area into a gas consisting of N molecules (of a single species) per unit volume. If one determines the number I of these electrons passing through unit area at a distance L in the gas, then one may write $I/I_0 = \exp(-NQL)$. This relation is valid under a condition of single collision during the passage of the electron in the gas cell. Such a measurement is feasible for electrons of kinetic energies between 0.1 and 1000 eV or higher.

The apparatus for the attenuation experiment appears to be simple in principle, but a great deal of ingenuity and care is required to achieve a precision of a few percent or better in the results. In this experiment, the electrons are assumed to be lost from the beam once they collide with molecules. Some of the electrons, however, move in the forward direction even after collision. The intensity, I, should not include these forward-scattering electrons. The reliability of the measured value of Q critically depends on the care taken about this point. Sometimes, a uniform magnetic field is applied in parallel to the electron beam so as to limit the spatial divergence of the electron beam. When a magnetic field is applied, its influence on a transmitted beam must be compensated by a calculation.

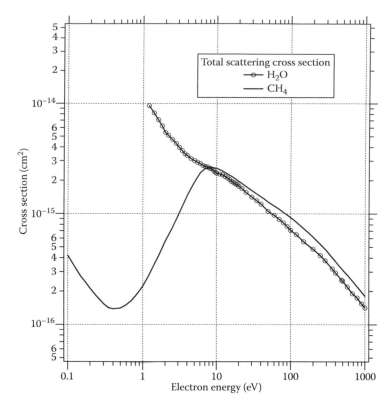

FIGURE 3.5 TCSs for electron collisions with H_2O and CH_4. The values for H_2O are the same as those shown in Figure 3.3.

Care is needed also in the correction for the effective cell length, because some gas leaks from both the entrance and the exit orifice of the cell.

Assessing the total scattering cross-section data available in the literature, Itikawa and Mason determined the recommended values for H_2O, as shown in Figure 3.3, which is reproduced in Figure 3.5. Due to the permanent dipole moment of H_2O, the TCS increases steeply with decreasing electron energy. For a comparison, the TCS for CH_4 (collected from a data compilation by Karwasz et al., 2003) is also plotted in Figure 3.5. In the energy region below about 10 eV, the two sets of cross sections differ very much from each other. The minimum in the Q of CH_4 at around 0.5 eV corresponds to the Ramsauer–Townsend minimum, discussed in Section 3.3.2.2.

The beam attenuation method has been recently extended to extremely-low-energy electrons (Field et al., 2001). By using photoelectrons, instead of electrons from hot filaments, electron beams of ~10 meV of energy are successfully produced with a very high resolution of a few meV. A beam of monochromatized light of 786.5 Å from a synchrotron radiation source is used to excite Ar to an autoionizing state, $Ar^+(^2P_{3/2})$, and emit electrons of 5 meV to 4.0 eV, with a resolution of about 5 meV. However, the beam intensity is as weak as about 10^{-10} A. Great care is necessary to account for the contact potential of the analyzer material, and to prevent the leakage of the electric field outside the electrode region. Until now, only one laboratory has produced the TCSs in this way, and it has been suggested to confirm their result with other experiments.

Figure 3.6 shows the TCS measured by Field and his group using their very-low-energy electron beams (Čurík et al., 2006). It should be noted that the electron energy goes down to about 0.02 eV, where the cross section has a value as large as 10^{-13} cm². One also should note that the measured cross section is not the real value of the TCS. Actually, the measured value corresponds to the cross section excluding a certain part of the forward-scattering electrons. In this sense, the values shown

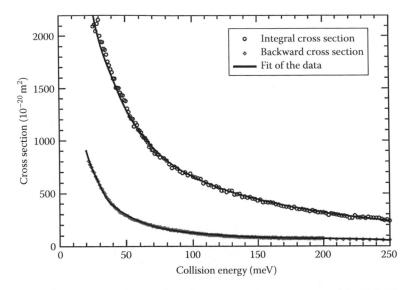

FIGURE 3.6 Results of attenuation experiment with very-low-energy electrons in H_2O. The upper curve is the TCS, but with a part of forward scattering excluded (see Section 3.3.2.1). The lower curve is a cross section for total backward scattering. (Reprinted from Čurík, R. et al., *Phys. Rev. Lett.*, 97, 123202-1, 2006. With permission.)

in Figure 3.6 do not exactly correspond to the ones in Figure 3.5. Further experimentation is needed for the TCS of H_2O at very low energies.

3.3.2.2 The Swarm Method to Determine a Cross-Section Set

When electrons are emitted from a source and flow through a gas under an applied uniform electric field, they undergo many collisions with gaseous molecules. Macroscopic properties of this electron swarm can be measured. These properties are the drift velocity, the diffusion coefficient, and other transport coefficients, as well as the rate constants for excitations, ionization, and electron attachment. These are functions of the ratio of the electric-field strength to the gas density, and determined by the electron energy distribution function (EEDF). Thus, the measured values of the transport properties can be related to electron–molecule collision cross sections through the EEDF, which is obtained by solving the Boltzmann equation. The swarm method provides a set of cross sections, especially at low electron energies (down to 0.01 eV), where the electron-beam method is difficult to apply.

As is easily understood, electron transport involves various kinds of collision processes. It is difficult to derive cross sections without any ambiguity. Consequently, an analysis of swarm data usually takes into account some information from other measurements and determines a full set of cross sections in such a way that the set is consistent with all the available data. Figure 3.7 shows a set of cross sections thus determined for CH_4 (Kurachi and Nakamura, 1990). Methane is a prototype of the single-bond hydrocarbon molecules and is used in many applications, such as radiation counters. The momentum-transfer cross section, q_0^M, defined by Equation 3.4 dominates over all the electron energies shown. In the case of CH_4, q_0^M has a minimum at around 0.34 eV. This means that at the energy around the minimum, electrons pass through the gas almost freely. This minimum arises from the Ramsauer–Townsend effects, observed also for heavier rare gases (Ar, Kr, and Xe). To be specific, the phase shift of the s wave ($l = 0$) of the scattered electron is an integral multiple of π at this energy, whereas higher partial waves have negligible contributions at such a low energy. Cross sections for inelastic processes are also obtained, as is shown in Figure 3.7. They are excitations of vibrational modes, dissociative electron attachment (DEA), dissociation to produce neutral fragments, and ionization. Details of vibrational excitation are discussed in Section 3.3.3.3.

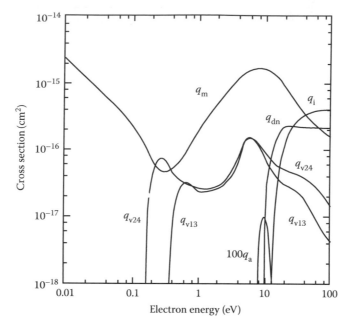

FIGURE 3.7   A set of electron collision cross sections for CH_4, determined with a swarm method. (Reprinted from Kurachi, M. and Nakamura, Y., Electron collision cross sections for SiH_4, CH_4, and CF_4 molecules, in *Proceedings of the 13th Symposium on Ion Sources and Ion-Assisted Technology* (*ISAT'90*), Tokyo, Japan, pp. 205–208, 1990. With permission.)

3.3.3   Crossed-Beam Methods for the Measurement of Cross Section

3.3.3.1   Crossed-Beam Methods and the Measurement of DCS

In a crossed-beam method, one sends a well-collimated beam of electrons of a fixed kinetic energy, E_0, into a molecular target (normally in a beam), and analyses the kinetic energy (i.e., the so-called residual energy, E_r) of the scattered electrons. This method provides much more detailed information than that provided by the attenuation method and certainly reflects the dynamics of the collision process. The electron energy analyzers commonly used are the 127° electrostatic cylinder and the 180° electrostatic hemisphere (Celotta and Huebner, 1979). These are also used as monochromators of the energy of the incident electron beams. A hot filament is commonly used as a source of electrons, with an energy spread of 0.3–0.5 eV. After energy selection with a monochromator, a resolution of about 30 meV and a beam of intensity of about 10^{-9} A (~17 meV and ~10^{-10} A at best) are usually obtained. In order to maintain the single-collision condition, the pressure of a gas target must be kept at about 10^{-3} torr. This limits the scattering intensity available. This situation is in sharp contrast with the electron spectroscopy of a solid surface, which has a much higher atomic density, and hence readily accomplishes an energy resolution of several meV.

The electron spectrometer can be operated in several different ways. One of the standard ways of measurement (called the energy-loss mode, EL) is to derive an energy-loss spectrum. In this spectrum, one plots the intensity of electrons scattered into a fixed angle as a function of the energy loss, ΔE (i.e., $\Delta E = E_0 - E_r$), for a fixed incident electron energy. The location of the energy-loss peak indicates the excited state of the target molecule, and the height of the peak is proportional to the corresponding cross section for the excitation of the state. Figure 3.8 (Dillon et al., 1989) shows energy-loss spectra of one of the DNA bases, adenine. Other ways to operate the spectrometer are as follows: (1) Detecting only the electrons with a specified energy loss with the analyzer, and then sweeping the incident energy, to observe the energy dependence of a specific excitation process (the excitation

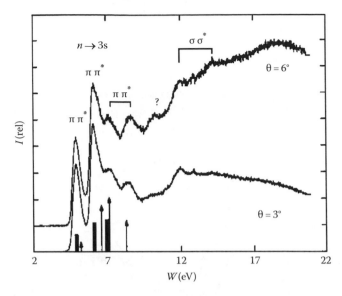

FIGURE 3.8 Electron energy-loss spectra for adenine. The incident electron energy is 200 eV. Data for two scattering angles (3° and 6°) are shown. See Dillon et al. (1989) for details. (Reprinted from Dillon, M.A. et al., *Radiat. Res.*, 117, 1, 1989. With permission.)

function mode, EF, see Figure 3.15). The resulting intensity distribution can reveal the structure due to resonance. (2) Keeping the residual energy of the scattered electron constant and changing both the incident energy and the energy loss, to detect the excitation function (the constant residual energy mode, CRE). With this method, each energy-loss feature is recorded at the same energy above threshold. Thus, it can avoid the effect of different efficiencies, if possible, of the detector for different-energy electrons. (3) Optimizing the spectrometer to detect scattered electrons with near-zero kinetic energy and sweeping the incident energy (the threshold electron spectroscopy mode, TES [see Figure 3.17]).

Electron scattering from molecules of biological significance, or biomolecules, is an area that has been actively studied in line with the recent trends in science and technology. Crossed-beam methods have been employed for these studies, because these biomolecules are now available commercially and have sublimation characteristics that make a molecular beam easily producible. These molecules are, however, often in a condition different from the circumstances of real biomolecules. For example, the target molecules in the experiment are not hydrated and are at a high temperature. Therefore, there have been serious discussions among atomic physicists, radiation scientists, and biologists as to how such elementary collision processes can be effective in understanding the radiation effect. As frequently repeated in this book, it is obvious that the OH radical dissociated from H_2O is an essential reactant for the radiation damages. But, in this chapter, an electron scattering from biomolecules is descried as an example of the general study of electron–molecule collisions, although it may contribute little to the actual radiation damage.

Typical examples of the DCS measurement are presented in Figures 3.8 (Dillon et al., 1989) and 3.9 (Vizcaino et al., 2008). The first measurement of an electron collision with a gas-phase DNA-base molecule was performed for adenine by Dillon et al. (1989). The electron energy-loss spectrum was recorded over a range of excitation energy of 3–22 eV for scattering angles of 3° and 6°. In addition to observing accurate positions of the energy-loss peaks, the data solved the long-held contention that the group of peaks in the 6–10 eV range belong to transitions originating from valence π orbitals. In the vapor-phase spectrum, a Rydberg transition corresponding to an $n \rightarrow 3s$ excitation is readily observed with a term value of 3.45 eV relative to the first lone-pair ionization potential.

Since early 1990, the bond-breaking process in the molecular constituents of DNA has been extensively studied by the dissociative attachment method, as described in Section 3.3.4.3.

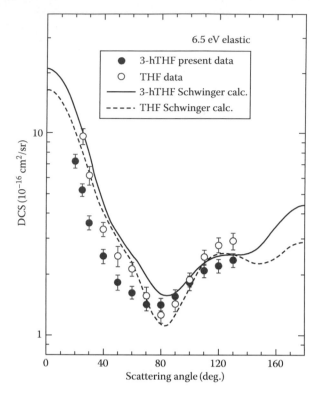

FIGURE 3.9 DCS for elastic electron scattering at 6.5 eV. The experimental data are shown for 3-hTHF (filled circles) compared with those for THF (open circles). The lines are the corresponding theoretical values. (Reprinted from Vizcaino, V. et al., *New J. Phys.*, 10, 053002-1, 2008. With permission.)

In Figure 3.9, the DCSs for elastic scattering of electrons are presented for 3-hydroxytetrahydro-furan (3-hTHF) (Vizcaino et al., 2008) compared to the DCS for tetrahydrofuran (THF) (Colyer et al., 2007). In this figure, DCS measured with a crossed-beam method are compared with a theoretical calculation. These two molecules have a similar structure and are common analogues to the deoxyribose sugar component in DNA. The experimental results show no essential difference between the elastic-scattering cross sections for 3-hTHF and THF. This is in contrast with the dissociative attachment experiment, where a large difference is found in the two molecules.

3.3.3.2 Measurements of DCS over the Full Range of Scattering Angles

The electron-beam method has a limitation of scattering angles covered for the following three reasons. First, it is difficult to extend measurements to backward scattering, usually beyond a maximum scattering angle of about 160°, because of the geometrical restriction of the electron analyzer. Second, it is difficult to distinguish the electrons elastically scattered at 0° from the unscattered incident electrons. Finally, good angular and energy resolutions require a minimization of magnetic fields present, such as the Earth's magnetic field and the fields due to the residual magnetization of electron-analyzer materials. The magnetic field, if any, influences electrons of low kinetic energies.

To overcome these difficulties, an innovation has been proposed and implemented (Read and Channing, 1996). The basic idea is to introduce into a collision region a suitably controllable magnetic field generated by several electromagnetic coils. The field guides those electrons scattered at inaccessible forward- and backward-scattering angles to a direction acceptable to the analyzer. A combination of the coils is designed so that the magnetic field does not leak outside the collision region. This technique is called a magnetic angle changer (MAC).

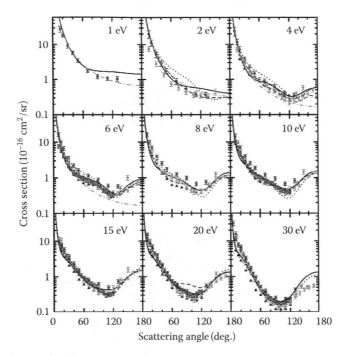

FIGURE 3.10 Elastic electron scattering DCSs for H_2O at various impact energies. The filled circles are those measured by Khakoo et al. Other details are given in Khakoo et al. (2008). (Reprinted from Khakoo, M.A. et al., *Phys. Rev. A*, 78, 052710-1, 2008. With permission.)

Cho et al. (2004) measured an absolute value of DCS for elastic scattering by H_2O at incident electron energies between 5 and 40 eV. They obtained the DCS over the scattering angles from 10° to 180°, using an MAC. Extrapolating their DCS in the forward directions, Cho et al. obtained an integral elastic cross section (ICS). The resulting values, however, are very small at electron energies below about 20 eV, compared with theoretical results. (See the discussion by Itikawa and Mason, 2005.) In 2008, Khakoo et al. (2008) reported their measurement. They obtained the elastic DCS for H_2O at incident energies of 1–100 eV for scattering angles from 5° to 130° (Figure 3.10). To obtain integral cross sections, they extrapolated their DCS in the following manner. Due to the strong dipole moment of water molecules, the electrons are dominantly scattered in the forward direction. Therefore, the DCS, particularly at low incident energies, should have a very sharp peak in the vicinity of 0°. Khakoo et al. carefully considered this fact in their extrapolation in the small-angle region. In the backward direction, they referred to the measurement by Cho et al. The resulting integral cross section of Khakoo et al. becomes larger than that of Cho et al. and consistent with the theoretical value (on which the recommended value in Figure 3.3 is based) (Figure 3.11).

Cho et al. (2008) measured the elastic DCS in the backward directions also for CH_4. In Figure 3.12, the DCS at 5 eV for CH_4 are presented for the full angular range. The experimental data are compared with theoretical values. It is noted that in the backward direction, theory cannot reproduce the experimental result at all. From the definition (3.4), momentum-transfer cross sections are much affected by large-angle scattering. The DCS measurement in the backward direction, such as that shown in Figure 3.12, is critical to evaluate accurate values of momentum-transfer cross sections.

3.3.3.3 Electron Spectroscopy at High Resolution

Once properly normalized, the electron energy-loss spectra easily enable the evaluation of DCSs for any collision process. To assign each energy-loss peak to a specific elastic or inelastic process, the energy resolution of the electron beam should be sufficiently high. A polyatomic molecule normally has complicated energy levels of molecular rotation and vibration. In the analysis of the energy-loss

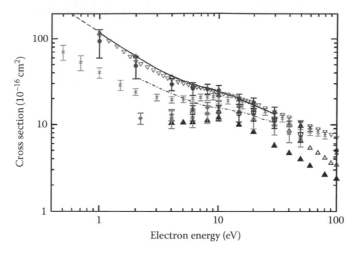

FIGURE 3.11 ICS and TCSs for H_2O. Filled circles with error bars are those derived by Khakoo et al. Inverted triangles are the TCSs recommended by Itikawa and Mason. For further details, see Khakoo et al. (2008). (Reprinted from Khakoo, M.A. et al., *Phys. Rev. A*, 78, 052710-1, 2008. With permission.)

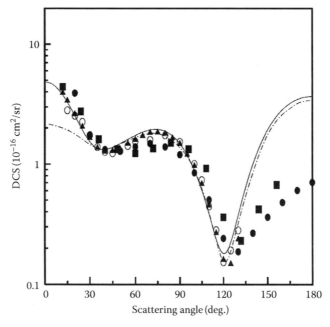

FIGURE 3.12 DCS for elastic electron scattering from CH_4 at 5 eV. Filled circles are the data measured by Cho et al. Lines are the corresponding theoretical results. See Cho et al. (2008). (Reprinted from Cho, H. et al., *J. Phys. B: At. Mol. Opt. Phys.*, 41, 45203-1, 2008. With permission.)

spectra of molecules, a high-resolution experiment is intrinsically needed. In the following text, excitations of rotational, vibrational, and electronic states are discussed separately.

3.3.3.3.1 Deduction of Rotational Excitation Cross Section

When an electron approaches a molecule, it feels an electric force induced by a nonspherical charge distribution of the molecule. When the electron is located far from the molecule, for instance, the electron–molecule interaction is represented by the interaction between the electron and the electric multipole (i.e., dipole, quadrupole,...) moments of the molecule. These

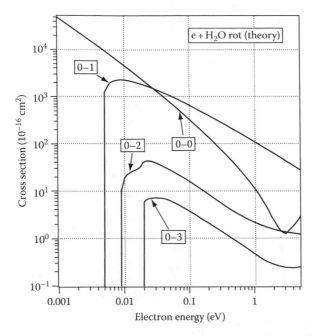

FIGURE 3.13 Rotational excitation cross sections for H_2O, calculated by Tennyson's group (Faure et al. 2004). (Reprinted from Itikawa, Y. and Mason, N.J., *J. Phys. Chem. Ref. Data*, 34, 1, 2005. With permission.)

moments include an electrostatic moment (due to initial molecular charge distribution) and an induced moment (due to polarization by the approaching electron). Through nonspherical interaction, the electron exerts a torque on the molecule, leading to rotational excitation (or de-excitation). Rotational excitation occurs over a wide range of collision energy. Unfortunately, the electron-beam method at present is hardly capable of resolving individual energy levels of rotation (except for hydrogen and hydride molecules). Measurements so far have dealt with an envelope of a rotational structure in an energy-loss spectrum, which is analyzed numerically for gross information with the deconvolution procedure (Jung et al., 1982). Thus, in almost all cases, only theoretical calculations produce rotational cross sections. The theoretical cross sections for H_2O obtained by Tennyson's group are shown in Figure 3.13 (Faure et al., 2004). A comparison of rotational excitations for H_2, N_2, HCl, H_2O, and CH_4 are presented in Figure 3.14 (Itikawa, 2007).

3.3.3.3.2 Resonant Vibrational Excitation

An electron approaching a molecule exerts also a force causing changes in internuclear distances due to the deformation of the electron charge density in the molecule. This leads to vibrational excitation of the molecule. In the CH_4 molecule, which is tetrahedral, there are four normal modes of vibration with frequencies of v_1, v_2, v_3, and v_4, whose excitation energies are 0.362, 0.190, 0.374, and 0.162 eV, respectively. Normally, it is difficult to separate v_1 and v_3 modes and v_2 and v_4 modes in the energy-loss measurement (however, see below). Figure 3.15 (Čurík et al., 2008) shows the combined cross sections for the excitations of the v_1 and v_3 modes (designated as $v_1 + v_3$) and v_2 and v_4 modes (as $v_2 + v_4$). A resonance feature emerges in the vibrational cross section at around 7 eV, as shown in Figure 3.15 (Čurík et al., 2008). This is the so-called shape resonance, which is thought of as a temporary bound state of an electron in the molecule. The bound state is formed by an effective potential well. This well might be, for instance, due to the combination of the (repulsive) centrifugal force for a certain orbital angular momentum and an (attractive) molecular potential. The temporary bound state may be viewed as an excited state of a negative ion, that is, the system of the neutral molecule plus the incoming electron. If the

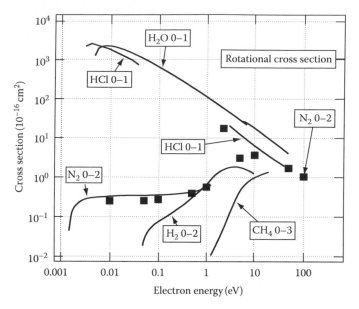

FIGURE 3.14 Comparison of rotational cross sections for H_2, N_2, HCl, H_2O, and CH_4. Theoretical values for the lowest transitions from the ground state are shown. (Reprinted from Itikawa, Y., *Molecular Processes in Plasmas: Collisions of Charged Particles with Molecules*, Springer, Berlin, Germany, 2007. With permission.)

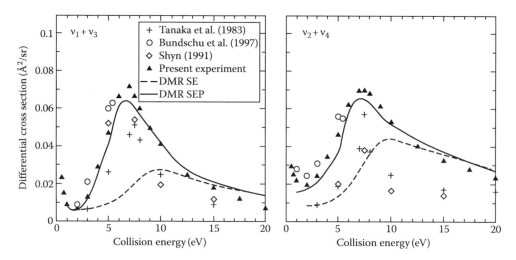

FIGURE 3.15 DCS (at $\theta = 90°$) for the vibrational excitation of CH_4. Combined cross sections for the excitations of the first and third modes and for the second and fourth modes are shown. Both are the cross sections for the excitation from the ground to the first excited states of the respective modes. For details, see Čurík et al. (2008). (Reprinted from Čurík, R. et al., *J. Phys. B: At. Mol. Opt. Phys.*, 41, 115203-1, 2008. With permission.)

temporary bound state is sufficiently stable against autodetachment, and has a lifetime much longer than the period of a molecular vibration, then the nuclei experience forces different from those in the original molecule. This causes a conversion of a part of the electron's energy to the nuclear vibrational motion. This resonant mechanism of vibrational excitation can sometimes enhance the corresponding cross section by orders of magnitude over the nonresonant case. Finally, Figure 3.16 shows the energy-loss spectra in which individual vibrational modes are partially resolved. It is noted that a theory confirms the relative magnitudes of the vibrational cross sections.

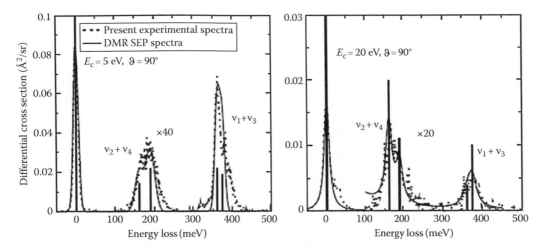

FIGURE 3.16   Energy-loss spectra of CH_4 at a scattering angle $\theta = 90°$ for incident electrons of 5 and 20 eV. Vertical bars represent theoretical values of the corresponding DCSs. (Reprinted from Čurík, R. et al., *J. Phys. B: At. Mol. Opt. Phys.*, 41, 115203-1, 2008. With permission.)

3.3.3.3.3   Energy-Loss Spectra for Electronic Excitation of H_2O

Figure 3.17 shows the threshold electron spectrum for H_2O (Jureta, 2005). This spectrum was obtained by a high-resolution electron spectrometer combined with the penetrating field method for scattered electrons with energies close to zero electron volts. This technique is sensitive to observing a resonance feature. In the H_2O molecule, one clearly sees resonant structures emerging near the thresholds of the first and the second band. These structures are not observed in the ordinary energy-loss spectra (Figure 3.18) (Kato et al., 2005). Note that resonances are important in the DEA producing negative-ion fragments (see Section 3.3.4.3).

As is shown in Section 3.3.1, a summary of the available cross-section data for electron scattering from H_2O was given by Itikawa and Mason (2005). These authors noted that "to date no

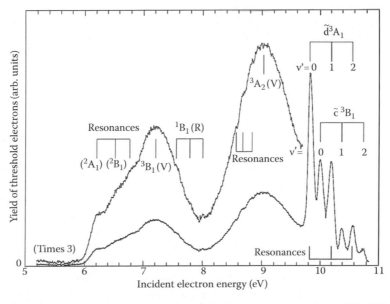

FIGURE 3.17   Threshold electron-impact spectrum of H_2O in the energy region of 5.2–10.8 eV. (Reprinted from Jureta, J.J., *Eur. Phys. J. D*, 32, 319, 2005. With permission.)

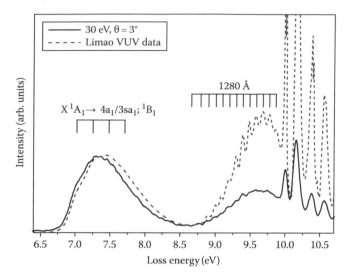

FIGURE 3.18 Electron energy-loss spectrum of H_2O at an impact energy of 30 eV for a scattering angle of 3°. The result from Kato et al., 2005 is compared with the VUV photoabsorption data obtained by Limao (cited in Mota et al., 2005).

electron-beam measurements have reported absolute values of the (electronic state) excitation cross section of H_2O." This fact prevented Itikawa and Mason from providing a recommended set of values, which they considered to be a serious problem since "electronic excitation is important in planetary atmospheres, plasmas, and radiation chemistry." As a consequence, they noted that "experiments and refined theory are urgently needed." The results of more recent experiments described here serve as one step in trying to overcome this deficiency.

With the use of a high-resolution technique, Tanaka and his collaborators obtained the energy-loss spectra of H_2O. One example is shown in Figure 3.18 in comparison with the optical absorption spectrum. The first broadband with no structure is located from 6.8 to 8.5 eV, leading to the dissociation channel. The second with a small vibration structure on the left side and with a sharp spike due to the Rydberg states on the right is seen from 8.5 to 11 eV. From this and other spectra, the authors successfully obtained cross sections for the excitations of low-lying six electronic states of H_2O (Thorn et al., 2007a,b; Brunger et al., 2008). As is shown by Itikawa and Mason, there are a number of theoretical calculations reporting the excitation cross sections. In the above papers, the newly obtained experimental cross sections are compared with these calculations to find a large disagreement between theory and experiment.

3.3.3.3.4 Application of the Scaled Born Model in Optically Allowed Transition

An excitation of the $\tilde{A}\,^1B_1$ electronic state of water is an important channel for the production of the OH $(X^2\Pi)$ radical, as well as for the emission of the atmospheric Meinel bands. As a part of their series of the work on electron-impact electronic excitations (see Section 3.3.3.3.3), Thorn et al. measured the cross section for the $\tilde{A}\,^1B_1$ state (Thorn et al., 2007a). Their measurement was performed over incident energies of 20–200 eV and scattering angles of 3.5°–90°. This is the first time for such data to be reported in the literature. To test the accuracy of the experimental data, Thorn et al. deduced the optical oscillator strength (OOS) for the $\tilde{A}\,^1B_1 - \tilde{X}\,^1A_1$ transition. First they calculated the generalized oscillator strength (GOS) from the DCS measured at 100 and 200 eV. In the limit of zero momentum transfer, they obtained the OOS from the GOS. The resulting OOS was found to agree well with the value derived from the photoabsorption experiment. Then they applied the scaled Born method to make the data on the excitation cross section more comprehensive.

It is well known that the Born approximation gives fairly well an excitation cross section for optically allowed states at high collision energies. Modifying the Born result, Kim developed a

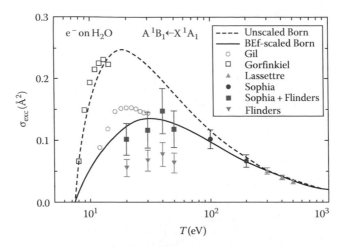

FIGURE 3.19 Comparison of experimental and theoretical integral cross sections for excitation of the $\tilde{A}\,^1B_1$ electronic state in H_2O. The solid line is the BEf-scaled Born result. For details, see Thorn et al. (2007a). (Reprinted from Thorn, P.A. et al., *J. Chem. Phys.*, 126, 064306-1, 2007a. With permission.)

new, simple method of calculation of the excitation cross section of atoms over the entire range of collision energy (Kim, 2001). He extended this to the case of a molecular target. It is called the BEf-scaled Born method (Kim, 2007). The BEf-scaled Born cross section, σ_{BEf}, is given by

$$\sigma_{BEf}(T) = \frac{f_{accur}T}{f_{Born}(T+B+E)}\sigma_{Born}(T) \qquad (3.10)$$

where
 T is the energy of the incident electron
 B is the binding energy of the electron being excited
 E is the excitation energy

The OOS is denoted by f. The quantity f_{accur} is an accurate value obtained from experiments or calculations with accurate wave functions, and f_{Born} is the value obtained from calculation with the wave function employed in the calculation of the unscaled Born cross section, σ_{Born}. The f-scaling process has the effect of replacing the wave function used for σ_{Born} with an accurate one. *BE* scaling corrects the deficiency of the Born approximation at low T, without losing its well-known validity at high T. Thorn et al. obtained σ_{Born} and f_{Born} from their GOS, and then substituted them into Equation 3.10. The resulting σ_{BEf} is shown in Figure 3.19 in comparison with the measured values by Thorn et al. It is clear that the BEf-scaled Born method gives fairly good values of the cross section over all collision energies. This method is now applied to H_2 (Kim, 2007; Kato et al., 2008a), CO (Kato et al., 2007), and CO_2 (Kawahara et al., 2008). Thus, once the Born cross section is obtained theoretically or experimentally, a fairly reasonable cross section can be produced at any incident energy with this approach.

3.3.4 Cross Sections for the Production of the Secondary Species

3.3.4.1 Ionization Cross Section
The ionization cross section is extremely important in the study of radiation sciences. Therefore, experimental data on the ionization cross section have been accumulated for a very long time. In total ion measurement, a collection of all product ions should be guaranteed. Particular care must be taken to avoid discrimination against energetic (fragment) ions. Furthermore, a preference is

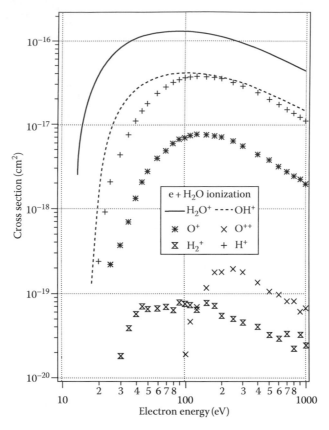

FIGURE 3.20   Recommended values of partial ionization cross sections of H_2O for the production of H_2O^+, OH^+, O^+, O^{++}, H_2^+, and H^+. (Reprinted from Itikawa, Y. and Mason, N.J., *J. Phys. Chem. Ref. Data*, 34, 1, 2005. With permission.)

placed on the experiment adopting no normalization to other data. To measure partial ionization cross sections, one can use a parallel-plate apparatus with a time-of-flight (TOF) mass spectrometer and a position-sensitive detector. The partial ionization cross sections can be made absolute independently for each mass number, but, in some cases, are normalized to the total ion measurement. Recommended data of partial cross sections of H_2O are presented for the products of H_2O^+, OH^+, O^+, O^{++}, H_2^+, and H^+ in Figure 3.20 (Itikawa and Mason, 2005).

Along the increase of research activity in application fields, the need of ionization cross section is extended to new molecular species, some of which are not accessible to experiment at present. To produce ionization cross sections, several theoretical methods have been proposed. One of these is the binary-encounter-Bethe (BEB) model of Kim and his colleagues (Hwang et al., 1996). This is a simplified version of the BED model (see Section 3.2.3), particularly applicable to molecular targets. The ionization cross section in the model is given by

$$\sigma_{BEB} = \frac{S}{t+u+1}\left[\frac{\ln t}{2}\left(1-\frac{1}{t^2}\right)+\left(1-\frac{1}{t}-\frac{\ln t}{t+1}\right)\right] \qquad (3.11)$$

where
  $t = T/B$
  $u = U/B$
  $S = 4\pi a_0^2 N (R/B)^2$

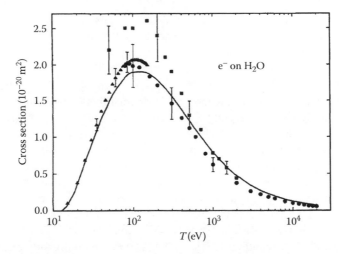

FIGURE 3.21 Total ionization cross section of H_2O. The BEB cross section (solid line) is compared with experimental data. (Reprinted from Kim, Y.-K. and Rudd, M.E., *Phys. Rev. A*, 50, 3954, 1994. With permission.)

The quantities U, B, and N are the kinetic and binding energies and the occupation number of molecular electrons, respectively, and T is the energy of the incident electron. The numerical constants are $a_0 = 0.52918$ Å and $R = 13.6057$ eV. (In some cases, a more sophisticated formula with dipole oscillator strengths is also called a BEB model. See the paper by Hwang et al., 1996.) Note that formula (3.11) gives the total ionization cross section. The resulting cross sections for small atoms, a variety of large and small molecules, and radicals are accurate within an error of 5%–20% from threshold to $T \sim 1$ keV (NIST). Examples are shown in Figures 3.21 (Kim and Rudd, 1994) and 3.22 (Kim et al., 1997) for H_2O and CH_4, respectively.

3.3.4.2 Dissociation to Produce Neutral Fragments

Many of the neutral fragments of molecular dissociation are active species (e.g., radicals). Furthermore, most of the fragments have a considerable kinetic energy. Hence, these dissociation products may play an active role in radiation interaction with matter. When the fragments are

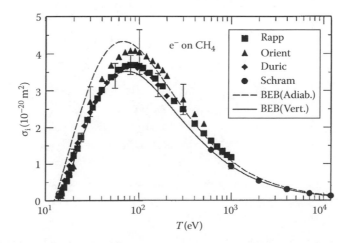

FIGURE 3.22 Total ionization cross section of CH_4. The BEB cross section (solid line) is compared with experimental data. (Reprinted from Kim, Y.-K. et al., *J. Chem. Phys.*, 106, 1026, 1997. With permission.)

produced in their excited states and emit radiation, they are easily detected. But, when the fragments are in the ground or metastable states, they cannot be easily detected. Two examples of the detection of neutral fragments are given as follows:

3.3.4.2.1 Neutral Radical Measurement for CH_4

All electronically excited states of CH_4 are repulsive states, leading to dissociation. This implies that in any plasma containing CH_4, chemically active neutral molecules (or radicals) are formed with high efficiency. For example, we expect to have CH, CH_2, and CH_3 in large quantities. However, it is very difficult to measure these neutral fragments. Tanaka and his group have developed a new technique to measure an absolute cross section for the production of CH_3 through electron-impact dissociation of CH_4. This method is based on the combination of the crossed-beam method and the threshold ionization technique (for details, see the paper Makochekanwa et al., 2006). Figure 3.23 (Makochekanwa et al., 2006) presents the cross section measured. In this figure, the threshold for the neutral CH_3 formation is observed at 7.5 ± 0.3 eV. Comparing with photoabsorption data, the authors conclude that all the excitations of CH_4 below 12.5 eV predominantly result in dissociation via the production of CH_3.

3.3.4.2.2 Production of OH Radicals in H_2O

In radiation interaction with matter, OH radicals are of particular importance. It is a fundamental question, therefore, that how many OH molecules are produced when electrons collide with water molecules. Harb et al. (2001) measured the cross section for the production of OH in its ground state through electron-impact dissociation of H_2O. They applied the laser-induced fluorescence technique to detect neutral OH. To obtain an absolute cross section, they compared the channel of production H + OH to that of H^- + OH. The absolute cross section of the latter process is available in the literature. The resulting cross section is shown in Figure 3.3. It is noted that, compared with other processes, the OH(X) production has a large cross section.

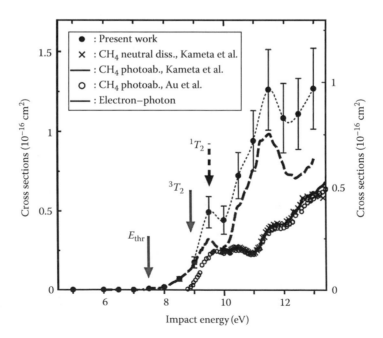

FIGURE 3.23 Absolute cross sections for the production of CH_3 radicals upon electron collisions with CH_4. Also included are the photon impact results from the literature. For more details, see Makochekanwa et al. (2006).

3.3.4.3 Negative-Ion Production

Electron interactions with biomolecules have been the subject of considerable interest, since it was discovered that low-energy electrons can cause significant DNA strand break (Boudaiffa et al., 2000). Electrons with a specific energy readily attach to the molecular constituents of DNA to form transient negative ions, which then dissociate to produce fragment negative ions and neutrals. This bond-breaking process, called dissociative electron attachment, has been demonstrated to occur in various molecules, such as water, DNA bases, amino acids, and fluorocarbons.

Dissociative attachment is a kind of resonance process. It proceeds through a (doubly) excited state of the negative molecular ion in a way as follows:

$$e + AB \rightarrow (AB^-)^{**} \rightarrow A + B^-$$

The intermediate negative-ion state is unstable and called the "resonance state."

Figure 3.24 shows one representative scheme of the DEA. In this case, the negative-ion state is repulsive and crosses to the ground state of the neutral molecule at the internuclear distance $R = R_c$. When R is smaller than R_c, the negative ion is unstable against the autodetachment of the electron (i.e., $AB^- \rightarrow AB + e$). Once R exceeds R_c, dissociation to $A + B^-$ takes place automatically. When the neutral molecule AB is initially in its vibrationally ground state, the attachment occurs only in the Franck–Condon region. Depending on the steepness of the repulsive curve, dissociative attachment has a finite value of the cross section only in a very narrow range of electron energy. When the negative-ion state has an attractive potential, dissociative attachment can occur through the excitation to the vibrational continuum of the negative ion.

Electron attachment to a molecule is studied by crossing an electron beam with a molecular beam. The negative ions formed are recorded mass spectrometrically as a function of the electron energy. Figure 3.25 (Itikawa and Mason, 2005) shows the cross sections of the dissociative attachment to H_2O to produce three different ions (H^-, O^-, and OH^-). This is cited from the review article by Itikawa and Mason (2005). After the publication of this article, a new measurement has been reported by Rawat et al. (2007). Their results are a little, but not much, different from the values shown in Figure 3.25.

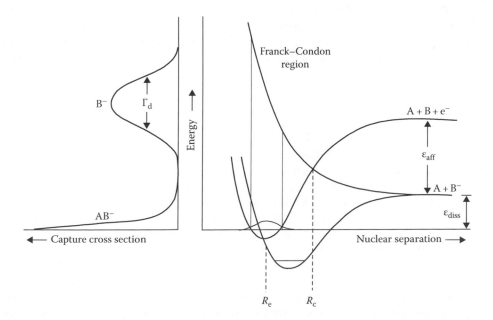

FIGURE 3.24 Potential diagram of a diatomic molecule for the mechanism of dissociative attachment.

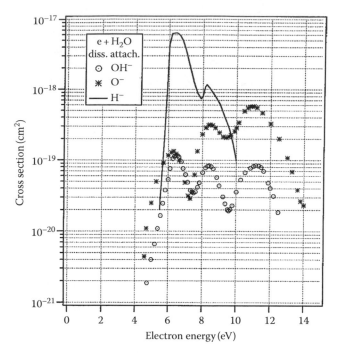

FIGURE 3.25 Dissociative attachment cross sections for H_2O. Partial cross sections for the production of H^-, O^-, and OH^- are shown. (Reprinted from Itikawa, Y. and Mason, N.J., *J. Phys. Chem. Ref. Data*, 34, 1, 2005. With permission.)

One example of the DEA study of biomolecules is given here. Figure 3.26 shows the electron energy dependence of the yield of a variety of negative-ion fragments, induced by a resonant attachment of subionization electrons to thymine (Huels et al., 1998). To produce an effusive molecular beam, high-purity (>99.5%) thymine is sublimated at 120°C–180°C, i.e., well below its decomposition temperature of about 320°C. The molecular beam is crossed at right angles with an electron beam generated by a trochoidal electron monochromator, and the negative ions produced are mass-analyzed by a quadrupole mass spectrometer. Table 3.1 (Huels et al., 1998) summarizes the peak positions observed, and Figure 3.27 (Huels et al., 1998) shows the channels for the production of each ion from gaseous thymine. This subject has been extended to the studies of fragmentation of elementary compounds from condensed H_2O, hydrated DNA, sugar analogues, oligonucleotides, and so on. For further information about these DEA fragmentations, see Sanche (2005).

3.3.5 Collision Processes Involving Vibrationally Excited Molecules

In most of the experiments of electron–molecule collision, target molecules are in their (vibrationally and electronically) ground states. In the nature, there are many molecules in their (particularly, vibrationally) excited states. Sometimes, the interaction between electrons and excited molecules is significantly different from that of the ground-state molecules, leading to a new subject of study in atomic and molecular physics (Christophorou and Olthoff, 2001). One typical example is the DEA. The DEA process is very sensitive to the internal degrees of freedom of the initial molecule (Christophorou and Olthoff, 2001). The distribution of the rotational–vibrational states of molecules depends on the temperature of the molecular gas. Hence, the cross section of the DEA depends sensitively on the temperature of the target molecular gas (Christophorou et al., 1984; Ruf et al., 2007). For other collision processes, little has been known about the target-temperature dependence of the cross section. It is usually difficult to perform any experiment of electron collisions with vibrationally excited molecules. This is mainly due to the difficulty in producing excited

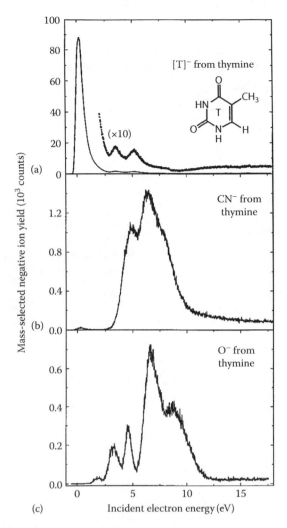

FIGURE 3.26 Yields of representative negative ions produced in the electron irradiation of gas-phase thymine, as a function of incident electron energy. Panel (a) shows the yield of the negative ion of the parent molecule (thymine), panel (b) is the CN-yield, and panel (c) is the O-yield. (Reprinted from Huels, M.A. et al., *J. Chem. Phys.*, 108, 1309, 1998. With permission.)

species in a sufficient number density. Vibrationally excited molecules can be produced by heating, laser photoabsorption, or electron bombardment of the molecule.

The first quantitative experiment to study low-energy electron scattering from vibrationally excited CO_2 was performed by Buckman et al. (1987). They measured the TCSs as a function of temperature using a linear attenuation TOF spectrometer. They observed a substantial increase in the TCS at electron energies below 2 eV, which they attributed to the enhanced scattering by a bent CO_2. No significant change in the TCS was observed at higher energies. Ferch et al. (1989) repeated the experiment of Buckman et al., also using a TOF spectrometer, and confirmed the earlier result. They also found a significant change in the resonance structure of the TCS at around 3–5 eV, due to vibrational excitation of the target molecule.

A DCS measurement for an electron collision with hot CO_2 was performed by Johnstone et al. (1993), but only at one energy point (4 eV). Johnstone et al. (1999) repeated the experiment at 3.8 eV. Recently, a similar but more extensive experiment was performed by Kato et al. (2008b). They covered the collision energies in the range of 1–9 eV. This includes the resonance region around 3–5 eV.

TABLE 3.1

Peak Positions (in eV, with Uncertainty of ±0.15 eV) of the Yield of Negative Fragment Ions from Electron Collisions with Thymine and Cytosine

Thymine (T)—$C_5H_6N_2O_2$–126 amu

T⁻	OCN⁻	CN⁻	OCNH⁻	O⁻	H⁻	OCNH₂⁻	CH₂⁻
0.18	0.20*	0.28*		1.8			
3.4	2.8*		2.8*	3.2			
	4.6	4.8	4.6	4.5	4.0	4.6	
5.2	5.8*		6.0*		5.2		
		6.4		6.6	6.6		6.6
	7.7	7.8	8.0	8.7	8.0*		

Cytosine (C)—$C_4H_5N_3O$–111 amu

C⁻	OCN⁻	H⁻	C₄H₃N₃O⁻	CN⁻	O⁻ or NH₂⁻	C₄H₅N₃⁻ or C₄H₃N₂O⁻	C₃H₃N₂⁻
(0.1*)1.4	1.0	(0.1)1.4	1.5	1.2	2.3*		
					3.8		4.5
5.3	5.0	5.6	5.3*	5.1	5.2	5.2	
6.9*					7.4	7.3	7.3*
	7.7			7.8		8.2*	
		9.5			9.2		

Source: Reprinted from Huels, M.A. et al., *J. Chem. Phys.*, 108, 1309, 1998. With permission.

Note: The asterisks indicate a faint peak or a shoulder.

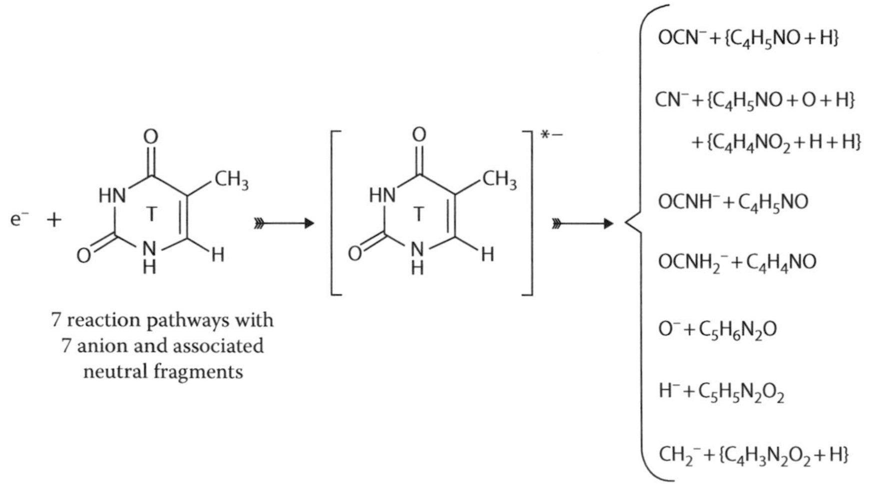

FIGURE 3.27 Pathways of producing negative fragment ions from electron collisions with thymine. (Reprinted from Huels, M.A. et al., *J. Chem. Phys.*, 108, 1309, 1998. With permission.)

They obtained absolute cross sections for the inelastic (i.e., vibrational excitation) and the super-elastic (i.e., vibrational de-excitation) transitions of CO_2 in its first excited state of the bending mode of vibration (as shown in Figure 3.28). The resonance structure of these cross sections has been revealed for the first time. This may be helpful in theoretically understanding the resonance in the electron–molecule collision. Further extension of this kind of study is desirable also from the point of view of application. It is well known that vibrationally excited molecules are abundant in the upper atmosphere of the Earth and other planets and discharge plasmas.

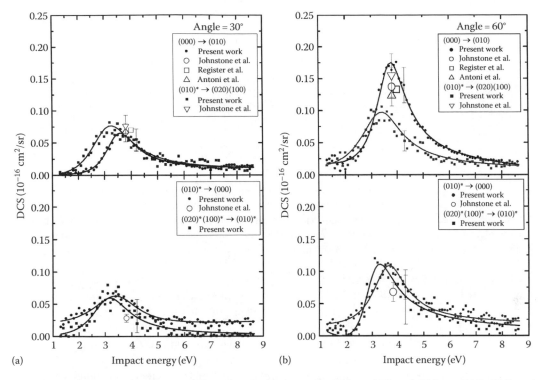

FIGURE 3.28 Vibrational excitation functions for the inelastic (000)→ (010) and (010) → (020) (100) and the superelastic (010) → (000) and (020) (100) → (010) transitions of CO_2 at impact energies of 1–9 eV and at scattering angles of (a) 30° and (b) 60°. See Kato et al. (2008b) for further details. (Reprinted from Kato, H. et al., *Chem. Phys. Lett.*, 465, 31, 2008b. With permission.)

3.3.6 COMPLETE COLLISION DYNAMICS IN (e, 2e) EXPERIMENTS FOR H_2O

As described in Section 3.3.4.1, partial and total ionization cross sections are of great importance for practical applications, but these provide only limited insight into the ionization process itself. Conversely, a great deal of information may be obtained from the study of DCSs (Märk, 1984), that is,

$$\text{SDCS}: \quad \frac{d\sigma(T,W)}{dW}$$

$$\text{DDCS}: \quad \frac{d^2\sigma(T,W,\theta)}{dW\,d\Omega}$$

$$\text{Triple DCS}: \quad \frac{d^3\sigma(T,W,\theta_1,\theta_2,\varphi_2)}{dW\,d\Omega_1\,d\Omega_2}$$

where
W is the energy
θ and φ are the polar angles of detection of the electrons after collision (see Figure 3.29; Märk, 1984)

In addition to this electron–electron spectroscopy, an ion kinetic energy spectroscopy and the measurement of angular distributions of fragment ions are necessary for a full picture of the ionization event. It is outside the scope of this chapter to give a detailed description on the subject, but a brief comment about the recent advances is given here.

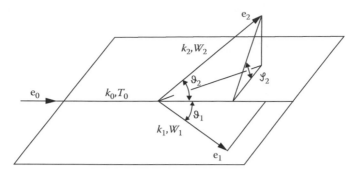

FIGURE 3.29   Schematic view of the kinematics of an electron-impact ionization: incident electron e_0 with kinetic energy T_0 and momentum k_0; and scattered and ejected electrons e_1 and e_2 with kinetic energies W_1 and W_2 and momenta k_1 and k_2, respectively. (Reprinted from Märk, T.D., Ionization of molecules by electron impact, in Christophorou, L.G. (ed.), *Electron-Molecule Interactions and Their Applications*, Vol. 1, Academic Press, New York, 1984, 251–334. With permission.)

Electron momentum spectroscopy (EMS), also known as (e, 2e) spectroscopy, has made a substantial contribution to the study of the electronic structure of matter. The basic principle of EMS is a complete measurement of the momenta of the two electrons emerging after collision. It is a kinematically complete study of electron-impact ionization events. Figure 3.30 shows typical examples of the results of the (e, 2e) experiment for gaseous water (Hafied et al., 2007). This experiment was performed in a symmetric noncoplanar geometry. A high-energy incident electron ($E_0 = 1200\,eV$) knocks out one electron from H_2O, and the two outgoing electrons are detected in coincidence at the

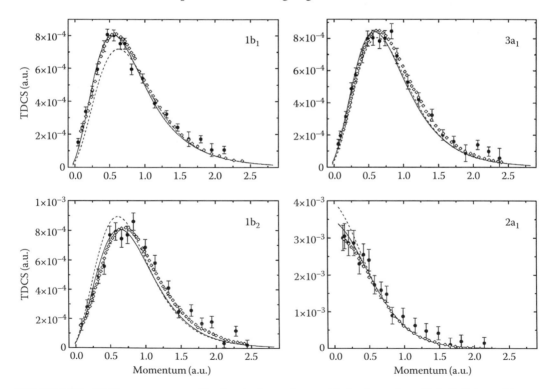

FIGURE 3.30   EMS momentum profiles for the four valence orbitals of gas-phase H_2O. Full dots with error bars are the experimental data taken from Bawagan et al. (1987). Lines are the results of calculations using different levels of wavefunctions. (Reprinted from Hafied, H. et al., *Chem. Phys. Lett.*, 439, 55, 2007. With permission.)

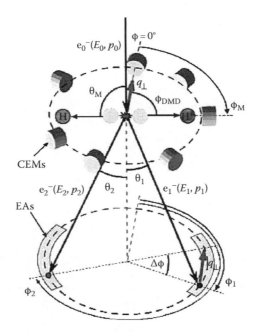

FIGURE 3.31 Schematic diagram of an (e, 2e + M) experiment for a symmetric noncoplanar geometry. This shows seven channel electron multipliers (CEMs), and a pair of entrance apertures (EAs) of a spherical analyzer with a pair of position-sensitive detectors. (Reprinted from Takahashi, M. et al., *Phys. Rev. Lett.*, 94, 213202-1, 2005. With permission.)

same high energies ($E_1 = E_2 \simeq 600\,\text{eV}$) and the same polar angles ($\theta_1 = \theta_2 = 45°$). The experimental result (Bawagan et al., 1987) is compared with several quantum chemical calculations. All the theoretical profiles in Figure 3.30 have been folded with the instrumental momentum resolution. The results of these studies have been applied practically to the charged-particle track structure analysis in both vapor and liquid waters.

The (e, 2e) experiments, however, only give the result averaged out over molecular directions. Of late, Takahashi and his colleagues (Takahashi et al., 2005) have started a new experiment that provides the information of a molecule fixed in space. They measure the momentum of a fragment ion in a dissociative ionization, in coincidence with the two outgoing electrons (i.e., a triple coincidence experiment). In Figure 3.31, the scheme of the measurement is given. For further details, see Takahashi et al. (2005).

3.4 APPLICATIONS OF CROSS-SECTION DATA AND DATABASES

3.4.1 ENERGY DEPOSITION MODEL BASED ON COLLISION CROSS SECTIONS

The aim of the study of electron–molecule collisions is not only to learn the physics of collision dynamics but also to provide a set of recommended data of the cross section for various practical applications. As an example of applications of the cross-section data, a model of energy deposition of radiation in water vapor is shown here. As is mentioned earlier, it is understood now that even sub-ionizing electrons can produce damages (in terms of strand breaks and molecular dissociation). We need to consider collision processes over a very wide range (from meV to MeV) of electron energies.

Garcia and his colleagues have developed a new simulation program based on the Monte Carlo method (Muñoz et al., 2008). For this program, they prepared an extensive set of cross-section data for elastic-scattering and inelastic (ionization and excitation) processes. They also

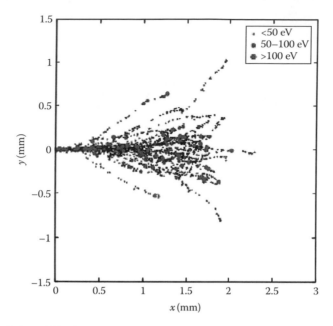

FIGURE 3.32 Simulation of 5 keV electron tracks in 1 atm of water vapor. Three different amounts of energy deposition are separately indicated. (Reprinted from Muñoz, A. et al., *J. Phys.: Conf. Ser.*, 133, 012002-1, 2008. With permission.)

used the TCS to estimate the mean free path of incoming electrons. When no comprehensive data were available experimentally, they resorted to theoretical calculations. For instance, the angular dependence (i.e., DCS) of the elastic cross section was derived from a model potential calculation. Figure 3.32 shows one sample result of their simulation of 5 keV electron tracks in 1 atm water vapor. This shows one aspect of the track structure: local distribution of energy deposition events. They can identify each event with a specific collision process. They plan to extend this model to water in the liquid phase.

3.4.2 Electron Transport in High-Pressure Xe Doped with CH_4

High-pressure Xe has been studied extensively in relation to the development of a new-generation gamma-ray camera for applications in planetary science (Pushkin et al., 2006), as well as of compact and faster versions of radionuclide imaging in nuclear medicine (Barr et al., 2002). As a more familiar example, the Xe gas has been commonly used in the plasma display panel (PDP) (Sobel, 1998) for producing, efficiently and steadily, near-UV light (at a wavelength of 147 nm or photon energy of 8.43 eV). The light is emitted as a result of a transition from the resonant state to the ground state ($5P^5$ $[^2P_{3/2}]$ $6s \rightarrow 5P^6$ $[^1S_0]$). For lowering the discharge voltage, a mixture of Xe and a few percent of He or Ne is used to produce metastable states of He* and Ne*.

In developing the next-generation gamma-ray imager (Pushkin et al., 2006), it was found that the electron diffusion and scintillation properties in a high-pressure Xe gas (Hunter and Christophorou, 1984) are greatly influenced by adding molecules such as CH_4 and H_2. Scintillation in xenon, initiated by a Compton electron, is used to generate a start signal, which enables the point of interaction of the γ-ray quantum with the detector material to be reconstructed by the drift time of the ejected secondary electrons. Xenon is characterized by a low drift velocity of "hot" electrons, which limits the diffusion of the electron cloud drifting toward the electrode (the anode). An addition of methane (~0.2%) to xenon significantly increases the drift velocities of electrons, as shown in Figure 3.33.

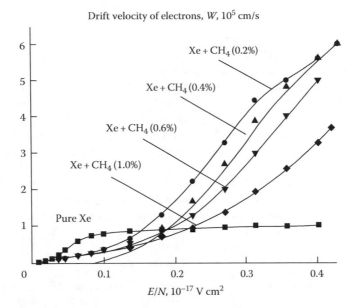

FIGURE 3.33 Drift velocities of electrons in pure gaseous xenon and xenon–methane mixtures with different methane concentrations (0.2%–1%), as a function of reduced electric field E/N. (Reprinted from Pushkin, K.N. et al., *Instrum. Exp. Tech.*, 49, 489, 2006. With permission.)

Thus, a doping of methane helps in improving the time and coordinate characteristics of the Xe time projection chamber counter.

3.4.3 DATABASES

Electron–molecule collisions have been extensively studied for a long time. Cross sections have been reported in a wide range of publications. Sometimes it takes time to survey the literature to find suitable data. It is therefore useful to compile cross-section data and publish them. A list of data compilations of electron–molecule collision cross sections is given in the book by Itikawa (2007). This book also summarizes the methods to find the cross-section data.

One of the most comprehensive compilations of cross-section data for electron–molecule collisions has been published as a volume of the Landolt–Börnstein series of data books (Itikawa, 2003). This includes the cross sections for ionization, electron attachment, total scattering, elastic scattering, momentum transfer, and excitations of rotational, vibrational, and electronic states of more than 70 molecular species. For individual species of a molecule, Itikawa and his colleagues published a series of data compilations in the *Journal of Physical and Chemical Reference Data* (Itikawa, 2002, 2006, 2009; Itikawa and Mason, 2005; Yoon et al., 2008).

Almost all the compilations of cross-section data published so far are the collection of an integral cross section (i.e., the cross section integrated over scattering angles). As is emphasized earlier, the Monte Carlo simulation of radiation tracks needs the information of the scattering angle distribution of electrons (i.e., the DCS). Normally, the numerical data on DCS are too detailed to be presented in a compact form. Tanaka and his group recently published a collection of DCS for elastic scattering of electrons from polyatomic molecules, which they experimentally obtained (Hoshino et al., 2008). This paper provides at least a part of the information needed in the Monte Carlo simulation.

Those who use the databases should notice one thing: any recommended values of cross-section data can be changed due to the development of experimental techniques and theoretical methods. One should be careful to use the most recent version of the databases.

3.5 CONCLUSIONS

To understand the details of radiation interaction with matter, we need information about the elementary collision processes between electrons and molecules. The study of electron–molecule collisions has a long history, but recently it deals with (large) biomolecules. In this chapter, the recent advances in experimental and theoretical studies of electron collisions with molecules are reviewed in conjunction with a view on radiation science. The collision processes considered are elastic scattering; momentum transfer; excitations of rotational, vibrational, and electronic states; ionization; dissociation; and electron attachment to the molecule. Total scattering and stopping cross sections are also presented in relation to their importance in the study of radiation effects. Sample cross-section data are graphically shown, mainly for two representative molecules, H_2O and CH_4. Cross-section databases currently available are briefly mentioned. The present status of the study of electron–molecule collisions is not completely satisfactory. Available cross-section data are not comprehensive. Further efforts are needed to obtain more complete and more accurate cross sections.

ACKNOWLEDGMENTS

One of the authors, Hiroshi Tanaka, is grateful to Profs. S. J. Buckman, M. J. Brunger, and H. Cho for their kind international collaboration among Japan, Australia, and Korea with the experimental studies. Also, we would like to thank Drs. M. Hoshino and H. Kato for their kind assistance in preparing the figures.

REFERENCES

Antoni, Th., Jung, K., Ehrhardt, H., and Chang, E. S. 1986. Rotational branch analysis of the excitation of the fundamental vibrational modes of CO_2 by slow electron collisions, *J. Phys. B: At. Mol. Phys.* 19: 1377–1396.
Barr, A., Bonaldi, L., Carugno, G., Charpak, G., Iannuzzi, D., Nicoletto, M., Pepato, A., and Ventura, S. 2002. A high-speed, pressurised multi-wire gamma camera for dynamic imaging in nuclear medicine. *Nucl. Instrum. Methods A* 477: 499–504.
Bawagan, A. O., Brion, C. E., Davidson, E. R., and Feller, D. 1987. Electron momentum spectroscopy of the valence orbitals of H_2O and D_2O: Quantitative comparisons using Hartree-Fock limit and correlated wavefunctions. *Chem. Phys.* 113: 19–42.
Bolorizadeh, M. A. and Rudd, M. E. 1986. Angular and energy dependence of cross sections for ejection of electrons from water vapor. I. 50–2000-eV electron impact. *Phys. Rev. A* 33: 882–887.
Boudaiffa, B., Cloutier, P., Hunting, D., Huels, M. A., and Sanche, L. 2000. Resonant formation of DNA strand breaks by low-energy (3 to 20 eV) electrons. *Science* 287: 1658–1660.
Brunger, M. J., Thorn, P. A., Campbell, L., Diakomichalis, N., Kato, H., Kawahara, H., Hoshino, M., Tanaka, H., and Kim, Y.-K. 2008. Excitation of the lowest lying $^3B_1$, $^1B_1$, $^3A_2$, $^1A_2$, $^3A_1$ and $^1A_1$ electronic states in water by 15 eV electrons. *Int. J. Mass Spectrom.* 271: 80–84.
Buckman, S. J., Elford, M. T., and Newman, D. S. 1987. Electron scattering from vibrationally excited CO_2. *J. Phys. B: At. Mol. Phys.* 20: 5175–5182.
Bundschu, C. T., Gibson, J. C., Gulley, R. J., Brunger, M. J., Buckman, S. J., Sanna, N., and Gianturco, F. A. 1997. Low-energy electron scattering from methane. *J. Phys. B: At. Mol. Opt. Phys.* 30: 2239–2260.
Celotta, R. J. and Huebner, R. H. 1979. Electron impact spectroscopy: An overview of the low-energy aspects. In *Electron Spectroscopy: Theory, Techniques and Applications*, C. R. Brundle and A. D. Baker (eds.), pp. 41–125. New York: Academic Press.
Cho, H., Park, Y. S., Tanaka, H., and Buckman, S. J. 2004. Measurements of elastic electron scattering by water vapour extended to backward angles. *J. Phys. B: At. Mol. Opt. Phys.* 37: 625–634.
Cho, H., Park, Y. S., Castro, E. A.-Y., de Souza, G. L. C., Iga, I., Machado, L. E., Brescansin, L. M., and Lee, M.-T. 2008. A comparative experimental–theoretical study on elastic electron scattering by methane. *J. Phys. B: At. Mol. Opt. Phys.* 41: 45203-1–45203-7.
Christophorou, L. G. and Olthoff, J. K. 2001. Electron interactions with excited atoms and molecules. *Adv. At. Mol. Opt. Phys.* 44: 155–293.

Christophorou, L. G., McCorkle, D. L., and Christodoulides, A. A. 1984. Electron attachment processes. In *Electron-Molecule Interactions and Their Applications*, L. G. Christophorou (ed.), Vol. 1, pp. 477–617. New York: Academic Press.

Colyer, C. J., Vizcaino, V., Sullivan, J. P., Brunger, M. J., and Buckman, S. J. 2007. Absolute elastic cross-sections for low-energy electron scattering from tetrahydrofuran. *New J. Phys.* 9: 41–51.

Čurík, R., Ziesel, J. P., Jones, N. C., Field, T. A., and Field, D. 2006. Rotational excitation of H_2O by cold electrons. *Phys. Rev. Lett.* 97: 123202-1–123202-4.

Čurík, R., Čársky, P., and Allan, M. 2008. Vibrational excitation of methane by slow electrons revisited: Theoretical and experimental study. *J. Phys. B: At. Mol. Opt. Phys.* 41: 115203-1–115203-7.

Dillon, M. A., Tanaka, H., and Spence, D. 1989. The electronic spectrum of adenine by electron impact methods. *Radiat. Res.* 117: 1–7.

Faure, A., Gorfinkiel, J. D., and Tennyson, J. 2004. Electron-impact rotational excitation of water. *Mon. Not. R. Astron. Soc.* 347: 323–333.

Ferch, J., Masche, C., Raith, W., and Wiemann, L. 1989. Electron scattering from vibrationally excited CO_2 in the energy range of the $^2\Pi_u$ shape resonance. *Phys. Rev. A* 40: 5407–5410.

Field, D., Lunt, S. L., and Ziesel, J.-P. 2001. The quantum world of cold electron collisions. *Acc. Chem. Res.* 34: 291–298.

Hafied, H., Eschenbrenner, A., Champion, C., Ruiz-López, M. F., Dal Cappello, C., Charpentier, I., and Hervieux, P.-A. 2007. Electron momentum spectroscopy of the valence orbitals of the water molecule in gas and liquid phase: A comparative study. *Chem. Phys. Lett.* 439: 55–59.

Harb, T., Kedzierski, W., and McConkey, J. W. 2001. Production of ground state OH following electron impact on H_2O. *J. Chem. Phys.* 115: 5507–5512.

Hoshino, M., Kato, H., Makochekanwa, C., Buckman, S. J., Brunger, M. J., Cho, H., Kimura, M., Kato, D., Murakami, I., Kato, T., and Tanaka, H. 2008. Elastic differential cross sections for electron collisions with polyatomic molecules, Research Report NIFS-DATA Series. NIFS-DATA-101 (National Institute for Fusion Science, Toki, Japan), pp. 1–60.

Huels, M. A., Hahndorf, I., Illenberger, E., and Sanche, L. 1998. Resonant dissociation of DNA bases by subionization electrons. *J. Chem. Phys.* 108: 1309–1312.

Hunter, S. R. and Christophorou, L. G. 1984. Electron motion in low- and high-pressure gases. In *Electron-Molecule Interactions and Their Applications*, L. G. Christophorou (ed.), Vol. 2, pp. 89–219. New York: Academic Press.

Huo, W. M. and Gianturco, F. A. (eds.). 1995. *Computational Methods for Electron-Molecule Collisions*. New York: Plenum Press.

Hwang, W., Kim, Y.-K., and Rudd, M. E. 1996. New model for electron-impact ionization cross sections of molecules. *J. Chem. Phys.* 104: 2956–2966.

Inokuti, M. 1971. Inelastic collisions of fast charged particles with atoms and molecules—The Bethe theory revisited. *Rev. Mod. Phys.* 43: 297–347.

Itikawa, Y. 2002. Cross sections for electron collisions with carbon dioxide. *J. Phys. Chem. Ref. Data* 31: 749–767.

Itikawa, Y. (ed.). 2003. *Photon and Electron Interactions with Atoms, Molecules and Ions*, Landolt-Börnstein, Vol. I/17, Subvolume C. Berlin, Germany: Springer.

Itikawa, Y. 2006. Cross sections for electron collisions with nitrogen molecules. *J. Phys. Chem. Ref. Data* 35: 31–53.

Itikawa, Y. 2007. *Molecular Processes in Plasmas: Collisions of Charged Particles with Molecules*. Berlin, Germany: Springer.

Itikawa, Y. 2009. Cross sections for electron collisions with oxygen molecules. *J. Phys. Chem. Ref. Data* 38: 1–20.

Itikawa, Y. and Mason, N. J. 2005. Cross sections for electron collisions with water molecules. *J. Phys. Chem. Ref. Data* 34: 1–22.

Johnstone, W. M., Mason, N. J., and Newell, W. R. 1993. Electron scattering from vibrationally excited carbon dioxide. *J. Phys. B: At. Mol. Opt. Phys.* 26: L147–L152.

Johnstone, W. M., Brunger, M. J., and Newell, W. R. 1999. Differential electron scattering from the (010) excited vibrational mode of CO_2. *J. Phys. B: At. Mol. Opt. Phys.* 32: 5779–5788.

Jung, K., Antoni, T., Muller, R., Kochem, K. H., and Ehrhardt, H. 1982. Rotational excitation of N_2, CO and H_2O by low-energy electron collisions. *J. Phys. B: At. Mol. Phys.* 15: 3535–3555.

Jureta, J. J. 2005. The threshold electron impact spectrum of H_2O. *Eur. Phys. J. D* 32: 319–328.

Karwasz, G. P., Brusa, R. S., and Zecca, A. 2003. *Photon and Electron Interactions with Atoms, Molecules and Ions*, Landolt-Börnstein, Y. Itikawa (ed.), Vol. I/17, Subvolume C, Chapter 6.1. Berlin, Germany: Springer.

Kato, H., Makochekanwa, C., Hoshino, M., Kitajima, M., Tanaka, H., Thorn, P. A., Campbell, L., Brunger, M. J., Teubner, P. J. O., and Cho, H. 2005. Generalized oscillator strengths for electron scattering from H_2O. In *24th International Conference on Photonic, Electronic, and Atomic Collisions (ICPEAC*, July 20, 2005, Rosario, Argentina) *Book of Abstracts*, Vol. 1, p. 281, and P. Limão-Vieira: private communication.

Kato, H., Kawahara, H., Hoshino, M., Tanaka, H., Brunger, M. J., and Kim, Y.-K. 2007. Cross sections for electron impact excitation of the vibrationally resolved A $^1\Pi$ electronic state of carbon monoxide. *J. Chem. Phys.* 126: 064307-1–064307-13.

Kato, H., Kawahara, H., Hoshino, M., Tanaka, H., Campbell, L., and Brunger, M. J. 2008a. Electron-impact excitation of the B $^1\Sigma_u^+$ and C $^1\Pi_u$ electronic states of H_2. *Phys. Rev. A* 77: 062708-1–062708-7.

Kato, H., Kawahara, H., Hoshino, M., Tanaka, H., Campbell, L., and Brunger, M. J. 2008b. Vibrational excitation functions for inelastic and superelastic electron scattering from the ground-electronic state in hot CO_2. *Chem. Phys. Lett.* 465: 31–35.

Kawahara, H., Kato, H., Hoshino, M., Tanaka, H., Campbell, L., and Brunger, M. J. 2008. Integral cross sections for electron impact excitation of the $^1\Sigma_u^+$ and $^1\Pi_u$ electronic states in CO_2. *J. Phys. B: At. Mol. Opt. Phys.* 41: 085203-1–085203-6.

Khakoo, M. A., Silva, H., Muse, J., Lopes, M. C. A., Winstead, C., and McKoy, V. 2008. Electron scattering from H_2O: Elastic scattering. *Phys. Rev. A* 78: 052710-1–052710-10.

Kim, Y.-K. 2001. Scaling of plane-wave Born cross sections for electron-impact excitation of neutral atoms. *Phys. Rev. A* 64: 032713-1–032713-10.

Kim, Y.-K. 2007. Scaled Born cross sections for excitations of H_2 by electron impact. *J. Chem. Phys.* 126: 064305-1–064305-8.

Kim, Y.-K. and Rudd, M. E. 1994. Binary-encounter-dipole model for electron-impact ionization. *Phys. Rev. A* 50: 3954–3967.

Kim, Y.-K., Hwang, W., Weinberger, N. M., Ali, M. A., and Rudd, M. E. 1997. Electron-impact ionization cross sections of atmospheric molecules. *J. Chem. Phys.* 106: 1026–1033.

Kimura, M., Inokuti, M., and Dillon, M. A. 1993. Electron degradation in molecular substances. *Adv. Chem. Phys.* 84: 192–291.

Kurachi, M. and Nakamura, Y. 1990. Electron collision cross sections for SiH_4, CH_4, and CF_4 molecules. In *Proceedings of the 13th Symposium on Ion Sources and Ion-Assisted Technology (ISAT'90)*, Tokyo, Japan, pp. 205–208.

Makochekanwa, C., Oguri, K., Suzuki, R., Ishihara, T., Hoshino, M., Kimura, M., and Tanaka, H. 2006. Experimental observation of neutral radical formation from CH_4 by electron impact in the threshold region. *Phys. Rev. A* 74: 042704-1–042704-4.

Märk, T. D. 1984. Ionization of molecules by electron impact. In *Electron-Molecule Interactions and Their Applications*, L.G. Christophorou (ed.), Vol. 1, pp. 251–334. New York: Academic Press.

McConkey, J. W., Malone, C. P., Johnson, P. V., Winstead, C., McKoy, V., and Kanik, I. 2008. Electron impact dissociation of oxygen-containing molecules—A critical review. *Phys. Rep.* 466: 1–103.

Mota, R., Parafita, R., Giuliani, A., Hubin–Franskin, M. J., Lourenço, J. M. C., Garcia, G., Hoffmann, S. V., Mason, N. J., Ribeiro, P. A., Raposo, M., and Limão-Vieira, P. 2005. Water VUV electronic state spectroscopy by synchrotron radiation. *Chem. Phys. Lett.* 416: 152–159.

Muñoz, A., Oller, J. C., Blanco, F., Gorfinkiel, J. D., Limão-Vieira, P., and Garcia, G. 2007. Electron-scattering cross sections and stopping powers in H_2O. *Phys. Rev. A* 76: 052707-1–052707-7.

Muñoz, A., Oller, J. C., Blanco, F., Gorfinkiel, J. D., Limão-Vieira, P., Maria-Vidal, A., Borge, M. J. G., Tengblad, O., Huerga, C., Tĕllez, M., and Garcia, G. 2008. Energy deposition model based on electron scattering cross section data from water molecules. *J. Phys.: Conf. Ser.* 133: 012002-1–012002-13.

Nikjoo, H., Uehara, S., Emfietzoglou, D., and Cucinotta, F. A. 2006. Track-structure codes in radiation research. *Rad. Meas.* 41: 1052–1074.

NIST: More information is available from http://physic.nist.gov/PhysRefData/Ionization/intro.html

Ohsawa, D., Kawauchi, H., Hirabayashi, M., Okada, Y., Honma, T., Higashi, A., Amano, S., Hashimoto, Y., Soga, F., and Sato, Y. 2005. An apparatus for measuring the energy and angular distribution of secondary electrons emitted from water vapor by fast heavy-ion impact. *Nucl. Instrum. Methods B* 227: 431–449.

Pushkin, K. N., Hasebe, N., Tezuka, C., Kobayashi, S., Mimura, M., Hosojima, T., Doke, T., Miyajima, M., Miyachi, T., Shibamura, E., Dmitrenko, V. V., and Ulin, S. E. 2006. A scintillation response and an ionization yield in pure xenon and mixtures of it with methane. *Instrum. Exp. Tech.* 49: 489–493.

Rawat, P., Prabhudesai, V. S., Aravind, A., Rahman, M. A., and Krishnakumar, E. 2007. Absolute cross sections for dissociative electron attachment to H_2O and D_2O. *J. Phys. B: At. Mol. Opt. Phys.* 40: 4625–4636.

Read, F. H. and Channing, J. M. 1996. Production and optical properties of an unscreened but localized magnetic field. *Rev. Sci. Instrum.* 67: 2372–2377.

Register, D. F., Nishimura, H., and Trajmar, S. 1980. Elastic scattering and vibrational excitation of CO_2 by 4, 10, 20, and 50 eV electrons, *J. Phys. B: At. Mol. Phys.* 13: 1651–1662.

Ruf, M.-W., Braun, M., Marienfeld, S., Fabrikant, I. I., and Hotop, H. 2007. High resolution studies of dissociative electron attachment to molecules: Dependence on electron and vibrational energy. *J. Phys.: Conf. Ser.* 88: 012013-1–012013-10.

Sanche, L. 2005. Low energy electron-driven damage in biomolecules. *Eur. Phys. J. D* 35: 367–390.

Schneider, B. I. 1994. The role of theory in the evaluation and interpretation of cross-section data. *Adv. At. Mol. Opt. Phys.* 33: 183–214.

Shyn, T. W. 1991. Vibrational excitation cross sections of methane by electron impact. *J. Phys. B: At. Mol. Opt. Phys.* 24: 5169–5174.

Sobel, A. 1998. Television's bright new technology. *Sci. Am.* (May): 48–55.

Takahashi, M., Watanabe, N., Khajuria, Y., Udagawa, Y., and Eland, J. H. D. 2005. Observation of a molecular frame (e, 2e) cross section: An (e, 2e+M) triple coincidence study on H_2. *Phys. Rev. Lett.* 94: 213202-1–213202-4.

Tanaka, H., Kubo, M., Onodera, N., and Suzuki, A. 1983. Vibrational excitation of CH_4 by electron impact: 3–20 eV. *J. Phys. B: At. Mol. Phys.* 16: 2861–2870.

Thorn, P. A., Brunger, M. J., Teubner, P. J. O., Diakomichalis, N., Maddern, T., Bolorizadeh, M. A., Newell, W. R., Kato, H., Hoshino, M., Tanaka, H., Cho, H., and Kim, Y.-K. 2007a. Cross sections and oscillator strengths for electron-impact excitation of the $\tilde{A}$ 1B1 electronic state of water. *J. Chem. Phys.* 126: 064306-1–064306-10.

Thorn, P. A., Brunger, M. J., Kato, H., Hoshino, M., and Tanaka, H. 2007b. Cross sections for the electron impact excitation of the $\tilde{a}^3$ B_1, $\tilde{b}^3$ A_1 and $\tilde{B}^1$ A_1 dissociative electronic states of water. *J. Phys. B: At. Mol. Opt. Phys.* 40: 697–708.

Toburen, L. H. 2004. Ionization and secondary electron production by fast charged particles. In *Charged Particle and Photon Interactions with Matter-Chemical, Physicochemical, and Biological Consequences with Applications*, A. Mozumder and Y. Hatano (eds.), pp. 31–74. New York: Marcel Dekker.

Uehara, S. and Nikjoo, H. 1996. Energy spectra of secondary electrons in water vapour. *Radiat. Environ. Biophys.* 35: 153–157.

Vizcaino, V., Roberts, J., Sullivan, J. P., Brunger, M. J., Buckman, S. J., Winstead, C., and McKoy, V. 2008. Elastic electron scattering from 3-hydroxytetrahydrofuran: Experimental and theoretical studies. *New J. Phys.* 10: 053002-1–053002-10.

Yoon, J.-S., Song, M.-Y., Han, J.-M., Hwang, S. H., Chang, W.-S., Lee, B.-J., and Itikawa, Y. 2008. Cross sections for electron collisions with hydrogen molecules. *J. Phys. Chem. Ref. Data* 37: 913–931.

4 Time-Dependent Density-Functional Theory for Oscillator Strength Distribution

Kazuhiro Yabana
University of Tsukuba
Tsukuba, Japan

Yosuke Kawashita
University of Tsukuba
Tsukuba, Japan

Takashi Nakatsukasa
RIKEN Nishina Center
Wako, Japan

Jun-Ichi Iwata
University of Tsukuba
Tsukuba, Japan

CONTENTS

4.1 INTRODUCTION

The oscillator strength distribution is a basic physical quantity that characterizes the interaction of photons with atoms and molecules. Most of the oscillator strength of atoms and molecules distributes in the excitation energy region above the ionization threshold, in which the oscillator strength is a continuous function of energy. Below the ionization threshold, the oscillator strength distribution is composed of discrete transitions characterized by the values of the oscillator strength and the excitation energy.

In the last decade, the theoretical description of the oscillator strength distribution has shown remarkable progress by virtue of the development of the time-dependent density-functional theory (TDDFT). Nowadays, the density-functional theory (DFT) is recognized as a standard computational method for the various properties of matters including atoms, molecules, nano-materials, solids, and so on. The DFT allows us to describe and predict the various properties of these matters from the first principles. However, the application of the ordinary DFT is limited to the electronic ground state. The TDDFT is an extension of the DFT so as to describe the electron dynamics induced by a time-dependent external potential. Since the oscillator strength distribution is related to a density change induced by a spatially uniform electric field, the TDDFT is applicable for it.

The first application of the TDDFT for the oscillator strength distribution appeared in 1980 for rare gas atoms (Zangwill and Soven, 1980). It was found that the TDDFT provides an accurate description of the oscillator strength distribution of rare gas atoms for a wide energy region. The application was then extended to the oscillator strength distributions of small molecules (Levine and Soven, 1984), Mie plasmon (a surface plasma oscillation of metallic clusters) (Ekardt, 1984), nonlinear polarizabilities (Senatore and Subbaswamy, 1987), dielectric functions (Bertsch et al., 2000), and so on. The TDDFT is nowadays implemented in many computational chemistry programs as a standard tool to calculate electronic excitations (Dreuw and Head-Gordon, 2005). However, they are limited to discrete excitations.

We have been developing a real-space computational method to solve the time-dependent Kohn–Sham (TDKS) equation, which is the basic equation of the TDDFT (Yabana and Bertsch, 1996, 1999; Nakatsukasa and Yabana, 2001; Yabana et al., 2006). In this method, we express single-electron orbitals on uniform grid points in the three-dimensional Cartesian coordinate. One of the advantages of the real-space method is the flexibility to impose the boundary condition, both for isolated and infinitely periodic systems. To describe photoionization processes, one must impose an outgoing boundary condition for photoemitted electrons. We have developed the real-space method imposing the outgoing boundary condition for electrons, and have applied the method for the oscillator strength distribution of small molecules, metallic clusters (Nakatsukasa and Yabana, 2001), and fullerenes (Kawashita et al., 2009). We have also developed a similar method for infinitely periodic systems and applied the method to describe the dielectric functions of crystalline solids as well (Bertsch et al., 2000).

In this chapter, we would like to present theoretical results for the oscillator strength distribution of molecules based on the TDDFT, taking several molecules as example. We concentrate on the theoretical descriptions of electronic excitations, freezing the positions of the atomic nuclei. We will show the calculations of small molecules, N_2, O_2, H_2O, and CO_2; several hydrocarbon molecules of small and medium size, acetylene, ethylene, benzene, and naphthalene; and a fullerene C_{60} as an example of a large molecule. These results will show that the TDDFT works well to reproduce the measured oscillator strength distributions accurately, in both aspects of the absolute value and the individual structures that reflect bound-to-bound and bound-to-continuum excitations.

The organization of this chapter is as follows: In Section 4.2, we present a basic formalism of the TDDFT for the oscillator strength distribution. In Section 4.3, we present the computational details of our approach. In Section 4.4, we present the results of the calculations of oscillator strength distributions. In Section 4.5, we discuss the extensions of our framework for other optical observables including nonlinear polarizabilities, optical activities of chiral molecules, and dielectric functions of infinitely periodic systems. In Section 4.6, a summary will be presented.

4.2   FORMALISM

4.2.1   DEFINITION OF THE OSCILLATOR STRENGTH DISTRIBUTION

We take $\hbar = 1$ throughout this chapter. We consider a molecule in which the positions and the charge numbers of atomic nuclei are denoted as $\mathbf{R}_a$ ($a = 1 - N_I$) and Z_a, respectively, and the positions of electrons are denoted as $\mathbf{r}_i$ ($i = 1 - N_e$). We freeze the positions of the atomic nuclei $\mathbf{R}_a$ throughout the present analyses, ignoring the coupling of the electrons with molecular vibrations. The electronic Hamiltonian of a molecule is given as follows:

$$H = \sum_{i=1}^{N_e} \frac{\mathbf{p}_i^2}{2m} - \sum_{i=1}^{N_e} \sum_{a=1}^{N_I} \frac{Z_a e^2}{|\mathbf{r}_i - \mathbf{R}_a|} + \sum_{i<j}^{N_e} \frac{e^2}{|\mathbf{r}_i - \mathbf{r}_j|}. \tag{4.1}$$

For bound states, we denote the eigenvalues and eigenfunctions of the Hamiltonian as $E_n(E_n < 0)$ and Φ_n, respectively, which satisfy $H\Phi_n = E_n\Phi_n$. We denote the ground state as Φ_0. Above the ionization threshold, we denote the eigenvalues and eigenfunctions as $E(E > 0)$ and Φ_E, respectively, which satisfy $H\Phi_E = E\Phi_E$. The wave function Φ_E satisfies the appropriate scattering boundary condition. The eigenfunctions Φ_n and Φ_E satisfy the orthonormal relationship, $\langle \Phi_n | \Phi_{n'} \rangle = \delta_{nn'}$ for bound states and $\langle \Phi_E | \Phi_{E'} \rangle = \delta(E - E')$ for scattering states, respectively.

The linear optical properties of molecules are characterized by the frequency-dependent dipole polarizability. Consider that an external electric field of a fixed frequency ω is applied to a molecule. The polarizability $\alpha_{\mu\nu}(\omega)$ relates the polarization in μ-direction, $p_\mu(t)$, with the electric field in ν-direction, $E_\nu(t) = E_0 e^{-i\omega t}$:

$$p_\mu(t) = \alpha_{\mu\nu}(\omega) E_\nu(t). \tag{4.2}$$

The explicit expression of the polarizability $\alpha_{\mu\nu}(\omega)$ is given by

$$\alpha_{\mu\nu}(\omega) = e^2 \sum_n \left(\frac{1}{E_{n0} - \omega - i\delta} + \frac{1}{E_{n0} + \omega + i\delta} \right) \left\langle \Phi_0 \left| \sum_i r_{i\mu} \right| \Phi_n \right\rangle \left\langle \Phi_n \left| \sum_i r_{i\nu} \right| \Phi_0 \right\rangle, \tag{4.3}$$

where we define $E_{n0} = E_n - E_0$, and the sum over excited states n extends both bound and scattering states.

We define the oscillator strength for the bound states n as

$$f_n = \frac{2mE_{n0}}{3} \left| \left\langle \Phi_n \left| \sum_i \mathbf{r}_i \right| \Phi_0 \right\rangle \right|^2 . \tag{4.4}$$

Similarly, for the scattering states, we define the oscillator strength distribution at energy $E(E > 0)$ as

$$\frac{df(\omega)}{d\omega} = \frac{2m\omega}{3} \left| \left\langle \Phi_E \left| \sum_i \mathbf{r}_i \right| \Phi_0 \right\rangle \right|^2 , \tag{4.5}$$

where $\omega = E - E_0$ holds. Combining them, we introduce the strength function $S(\omega)$ that is defined for the whole energy region,

$$S(\omega) = \sum_n \delta(\omega - E_{n0}) f_n + \frac{df(\omega)}{d\omega} \theta(\omega + E_0) . \tag{4.6}$$

The Thomas–Reiche–Kuhn (TRK) sum rule for the strength function may be given as follows:

$$\int d\omega \, S(\omega) = \sum_n f_n + \int_{|E_0|}^{\infty} d\omega \, \frac{df(\omega)}{d\omega} = N_e . \tag{4.7}$$

The photoabsorption cross section $\sigma(\omega)$ is related to the strength function $S(\omega)$ by

$$\sigma(\omega) = \frac{2\pi^2 e^2}{mc} S(\omega) . \tag{4.8}$$

The imaginary part of the polarizability is related to the strength function by

$$S(\omega) = \frac{2m\omega}{e^2 \pi} \frac{1}{3} \sum_{\mu=1}^{3} \mathrm{Im}\, \alpha_{\mu\mu}(\omega) . \tag{4.9}$$

4.2.2 Density–Density Response Function and the Polarizability

We introduce a density operator of electrons defined by

$$\hat{n}(\mathbf{r}) = \sum_{i=1}^{N_e} \delta(\mathbf{r}_i - \mathbf{r}) . \tag{4.10}$$

The ground state density $n_0(\mathbf{r})$ is then expressed as $n_0(\mathbf{r}) = \langle \Phi_0 | \hat{n}(\mathbf{r}) | \Phi_0 \rangle$. When a weak and time-dependent external potential, $V_{\mathrm{ext}}(\mathbf{r}, t)$, is applied to a molecule, it induces a change of the electron density, which is a linear functional of the external potential. We denote the electron density at time t as $n(\mathbf{r}, t)$. The density–density response function, $\chi(\mathbf{r}, \mathbf{r}', t - t')$, relates the density change $\delta n(\mathbf{r}, t) = n(\mathbf{r}, t) - n_0(\mathbf{r})$ with the external potential,

$$\delta n(\mathbf{r},t) = \int\limits_{-\infty}^{t} dt' \int d\mathbf{r}' \chi(\mathbf{r},\mathbf{r}',t-t')V_{ext}(\mathbf{r}',t'), \tag{4.11}$$

where we assume that the system is independent of time in the absence of the external potential.

One may derive an explicit expression of the density–density response function in quantum mechanics employing an elementary time-dependent perturbation theory. The result is as follows:

$$\chi(\mathbf{r},\mathbf{r}',t-t') = \frac{1}{i}\theta(t-t')\langle\Phi_0|\,[\hat{n}(\mathbf{r},t),\,\hat{n}(\mathbf{r}',t')]\,|\Phi_0\rangle, \tag{4.12}$$

where $\theta(t)$ is the step function. $\hat{n}(\mathbf{r},t)$ is defined by

$$\hat{n}(\mathbf{r},t) = e^{iHt}\hat{n}(\mathbf{r})e^{-iHt}. \tag{4.13}$$

For a perturbation of a fixed frequency ω, the relevant response function is given by the Fourier transform of Equation 4.12. We write the explicit expression of the response function as follows:

$$\chi(\mathbf{r},\mathbf{r}',\omega) = \int\limits_{-\infty}^{\infty} dt\, e^{i\omega t}\chi(\mathbf{r},\mathbf{r}',t)$$

$$= \sum_n \frac{\langle\Phi_0|\hat{n}(\mathbf{r})|\Phi_n\rangle\langle\Phi_n|\hat{n}(\mathbf{r}')|\Phi_0\rangle}{\omega+i\delta-E_{n0}} - \frac{\langle\Phi_0|\hat{n}(\mathbf{r}')|\Phi_n\rangle\langle\Phi_n|\hat{n}(\mathbf{r})|\Phi_0\rangle}{\omega+i\delta+E_{n0}}, \tag{4.14}$$

where the sum over n extends all the electronic excited states of the system including scattering states. δ is a small positive number. The density–density response function in the frequency representation thus includes information on the electronic excited states: excitation energies as poles and the transition densities as residues.

We consider a dipole field in ν-direction with frequency ω. The potential is given by $V_{ext}(\mathbf{r},t) = eE_0 r_\nu e^{-i\omega t}$, where r_ν is one of the Cartesian coordinates, x, y, z. Since the polarizability $\alpha_{\mu\nu}(\omega)$ relates the polarization in μ-direction, $p_\mu(t)$, with the electric field in ν-direction, $E_\nu(t) = E_0 e^{-i\omega t}$,

$$p_\mu(t) = \int d\mathbf{r}(-e)r_\mu \delta n(\mathbf{r},t) = \alpha_{\mu\nu}(\omega)E_\nu(t). \tag{4.15}$$

The polarizability may be expressed in terms of the density–density response function as

$$\alpha_{\mu\nu}(\omega) = -e^2 \int d\mathbf{r}\, d\mathbf{r}'\, r_\mu\, r_\nu'\, \chi(\mathbf{r},\mathbf{r}',\omega). \tag{4.16}$$

4.2.3 TIME-DEPENDENT DENSITY-FUNCTIONAL THEORY FOR POLARIZABILITY

In the TDDFT, one expects that the density change $\delta n(\mathbf{r},t)$ induced by an external potential $V_{ext}(\mathbf{r},t)$ is accurately described by the TDKS equation,

$$i\frac{\partial}{\partial t}\psi_i(\mathbf{r},t) = \{h_{KS}[n(t)] + V_{ext}(\mathbf{r},t)\}\psi_i(\mathbf{r},t), \tag{4.17}$$

where the Kohn–Sham Hamiltonian $h_{KS}[n(t)]$ is given by

$$h_{KS}[n(t)] = -\frac{1}{2m}\nabla^2 + V_{ion}(\mathbf{r}) + \int d\mathbf{r}' \frac{e^2}{|\mathbf{r} - \mathbf{r}'|} n(\mathbf{r}',t) + \mu_{xc}[n(\mathbf{r},t)]. \qquad (4.18)$$

The time-dependent density is given by

$$n(\mathbf{r},t) = \sum_i |\psi_i(\mathbf{r},t)|^2. \qquad (4.19)$$

The exchange-correlation potential $\mu_{xc}[n(t)]$ is a key quantity in the TDDFT. It can be a nonlocal function of the time-dependent density $n(\mathbf{r},t)$ in both space and time. If an accurate exchange-correlation potential is adopted, we expect that the time-dependent density $n(\mathbf{r},t)$ calculated by the TDKS equation coincides with that in the exact many-body quantum theory. In practice, however, we can at most employ an approximate one. Indeed, in the calculations below, we will employ a simple adiabatic approximation in which the exchange-correlation potential employed to describe the ground state properties is employed for the time-dependent dynamics as well. We will later mention the spatial dependence of the exchange-correlation potential.

To obtain an expression for the polarizability in the TDDFT, we apply a perturbation theory to the TDKS equation, expanding the Kohn–Sham Hamiltonian up to a linear order in the density change around the density in the ground state. In the adiabatic approximation, we have

$$h_{KS}[n(\mathbf{r},t)] = h_{KS}[n_0(\mathbf{r})] + \int d\mathbf{r}' \frac{\delta h_{ks}[n(\mathbf{r})]}{\delta n(\mathbf{r}')} \delta n(\mathbf{r}',t), \qquad (4.20)$$

where the functional derivative of the Kohn–Sham Hamiltonian, $\delta h_{ks}/\delta n$ has the following form in the local-density approximation (LDA),

$$\frac{\delta h_{ks}[n(\mathbf{r})]}{\delta n(\mathbf{r}')} = \frac{e^2}{|\mathbf{r} - \mathbf{r}'|} + \frac{\delta \mu_{XC}[n(\mathbf{r})]}{\delta n} \delta(\mathbf{r} - \mathbf{r}'). \qquad (4.21)$$

The potential, $(\delta h_{ks}/\delta n)\delta n(t)$, includes effects caused by a change of the electron density and is called the dynamical screening potential. If we regard a sum of the external potential and the induced potential as a perturbation, one may describe the response of the system in terms of the density–density response function without two-body interaction, which we call the independent-particle response function and we denote it as $\chi_0(\mathbf{r},\mathbf{r}',t - t')$. The density change is expressed as

$$\delta n(\mathbf{r},t) = \int_{-\infty}^{t} dt' \int d\mathbf{r}' \chi_0(\mathbf{r},\mathbf{r}',t - t') \left\{ V_{ext}(\mathbf{r}',t') + \int d\mathbf{r}'' \frac{\delta h_{ks}[n(\mathbf{r}')]}{\delta n(\mathbf{r}'')} \delta n(\mathbf{r}'',t') \right\}. \qquad (4.22)$$

For a perturbation with a fixed frequency, ω, we can again separate the time variable. Expressing $\delta n(\mathbf{r},t)$ as $\delta n(\mathbf{r},\omega)e^{-i\omega t}$ and $V_{ext}(\mathbf{r},t)$ as $V_{ext}(\mathbf{r},\omega)e^{-i\omega t}$, we have

$$\delta n(\mathbf{r},\omega) = \int d\mathbf{r}' \chi_0(\mathbf{r},\mathbf{r}',\omega) V_{scf}(\mathbf{r}',\omega), \qquad (4.23)$$

$$V_{scf}(\mathbf{r}, \omega) = V_{ext}(\mathbf{r}, \omega) + \int d\mathbf{r}' \frac{\delta h[n(\mathbf{r})]}{\delta n(\mathbf{r}')} \delta n(\mathbf{r}', \omega). \tag{4.24}$$

This equation should be solved for $\delta n(\mathbf{r}, \omega)$ at each frequency, ω.

An explicit form of the independent-particle response function is given by

$$\chi_0(\mathbf{r}, \mathbf{r}', \omega) = \sum_{i \in occ.} \sum_{m \in unocc.} \left\{ \frac{\phi_i(\mathbf{r})\phi_m(\mathbf{r})\phi_m(\mathbf{r}')\phi_i(\mathbf{r}')}{\varepsilon_i - \varepsilon_m - \omega - i\delta} + \frac{\phi_i(\mathbf{r})\phi_m(\mathbf{r})\phi_m(\mathbf{r}')\phi_i(\mathbf{r}')}{\varepsilon_i - \varepsilon_m + \omega + i\delta} \right\}, \tag{4.25}$$

where ϕ_i and ϕ_m are occupied and unoccupied Kohn–Sham orbitals with eigenenergies ε_i and ε_m, respectively. They satisfy $h_{KS}[n_0] \phi_k = \varepsilon_k \phi_k$.

The dipole polarizability in the TDDFT may be defined in the same way as Equation 4.16. We first set the external potential as $V_{ext}(\mathbf{r}, \omega) = r_\nu$. Then $\delta n(\mathbf{r}, \omega)$ is calculated by solving Equation 4.22 at each frequency ω. We then obtain the polarizability by

$$\alpha_{\mu\nu}(\omega) = -e^2 \int dr r_\mu \delta n(\mathbf{r}, \omega). \tag{4.26}$$

The oscillator strength function $S(\omega)$ can be defined in the same manner as Equation 4.9. We note that the TRK sum rule is satisfied exactly in the TDDFT if one employs the adiabatic approximation for the exchange-correlation potential and if one solves the TDKS equation without any approximation.

We can express the oscillator strength distribution as a sum of contributions of occupied orbitals (Zangwill and Soven, 1980; Nakatsukasa and Yabana, 2001). We first express the oscillator strength distribution as follows:

$$\frac{df(\omega)}{d\omega} = -\frac{2m\omega}{\pi} \frac{1}{3} \sum_{\mu=1}^{3} \mathrm{Im} \int d\mathbf{r} V_{scf}^*(\mathbf{r}, \omega) \chi_0(\mathbf{r}, \mathbf{r}', \omega) V_{scf}(\mathbf{r}, \omega). \tag{4.27}$$

Then, noting that the independent-particle response function $\chi_0(\mathbf{r}, \mathbf{r}', \omega)$ defined in Equation 4.25 is written as a sum over occupied orbitals, one may decompose the oscillator distributions into orbitals.

4.2.4 Exchange-Correlation Potential

In practical calculations, the choice of the exchange-correlation potential significantly affects the calculated oscillator strength distribution. In the TDDFT, we need to specify the frequency (time) and the spatial dependence of the exchange-correlation potential. Regarding the frequency dependence, we employ the adiabatic approximation, in which the exchange-correlation potential does not depend on the frequency as mentioned above, because reliable frequency-dependent energy functionals are not yet available. We employ the same exchange-correlation potential for both ground state and response calculations. We also need to specify the spatial dependence of the exchange-correlation potential. The simplest choice is the LDA, where it is assumed that the electrons at each position behave as if they were in a uniform electron gas with the electron density at each point. It has been known that even the LDA, the simplest functional, works reasonably for the oscillator strength distribution. However, it is not accurate enough, especially around the ionization threshold. It is well known that the spurious self interaction does not vanish in the LDA and, consequently, the long-range $-e^2/r$ tail lacks in the LDA. Because of this failure, the magnitude of the orbital energy of the highest occupied molecular orbital (HOMO) is much smaller than the ionization potential. Thus, the oscillator strength distribution around the ionization threshold cannot be described accurately in the LDA. To remedy the failure of the LDA, a gradient corrected potential, the LB94, was proposed (Van Leeuwen and Baerends, 1994). In the LB94, a term depending on the gradient of the density, $\nabla n(\mathbf{r})$, produces the potential that shows the correct asymptotic behavior. The ionization potentials are also reproduced rather accurately. In our calculations presented below, we will employ the LB94 potential unless otherwise stated.

For an accurate description of low-lying excitations in molecules, assessment regarding the accuracy of exchange-correlation potentials has been extensively achieved in quantum chemistry calculations. We quote a few references for such researches (Casida et al., 1998; Furche and Ahlrichs, 2002; Dreuw and Head-Gordon, 2005).

4.3 COMPUTATIONAL DETAILS

The calculations of the oscillator strength distribution with the linear response TDDFT have been first achieved in the early 1980s for spherical systems, such as atoms and metallic clusters. For spherical systems, $\chi_0(\mathbf{r}, \mathbf{r}', \omega)$ defined by Equation 4.25 is a rotationally invariant kernel, and Equations 4.23 and 4.24 for $\delta n(\mathbf{r}, \omega)$ may be solved for each multipole component separately. The equation for each multipole component is an integral equation of radial variable only, and can easily be solved numerically. Also, the outgoing boundary condition for emitted electrons may easily be incorporated in the partial wave expansion for the wave functions.

For systems without spherical symmetry, on the other hand, an explicit construction of the independent-particle response function given by Equation 4.25 is numerically demanding, since it is a function of two coordinate variables, $\mathbf{r}$ and $\mathbf{r}'$. This is especially true in the Cartesian-coordinate grid representation that we adopted here. For this reason, several efficient computational approaches that avoid an explicit construction of the response function χ_0 have been developed. In this section, we outline briefly the computational methods that we adopted in the calculations presented later.

4.3.1 REAL-SPACE GRID REPRESENTATION

To express orbital wave functions $\psi_i(\mathbf{r}, t)$, we employ a three-dimensional Cartesian-coordinate grid representation (Chelikowsky et al., 1994). We treat valence electrons explicitly, employing the so-called, norm-conserving pseudopotential for electron–ion interaction (Troullier and Martins, 1991). We treat H^+, C^{4+}, N^{5+}, and O^{6+} as cores. The pseudopotential is further approximated to be a separable form, known as the Kleinman–Bylander form (Kleinman and Bylander, 1982). This is a standard prescription in the first-principles DFT calculations with grid representation in either momentum or real-space grid representation. We thus freeze the core electrons ignoring their polarization effects. The typical grid spacing is 0.5 a.u.: this is fine enough to describe valence electron dynamics. We take grid points inside a cubic box area. The typical size of the box is 60 a.u. in one side, which includes 120^3 grid points.

4.3.2 REAL-TIME METHOD

The real-time method solving the TDKS equation in time domain is one of the straightforward ways to calculate the polarizability of the whole spectral region (Yabana and Bertsch, 1996). As explained in Equation 4.2, the dipole polarizability $\alpha_{\mu\nu}(\omega)$ is defined as a proportion coefficient between the induced polarization, $p_\mu(t)$, and the applied external electric field, $E_\nu(t)$, of a fixed frequency, ω, applied to ν-direction, $E_\nu(t) = E_0 e^{-i\omega t}$. Now let us consider an induced dipole moment $p_\mu(t)$ for an electric field of arbitrary time profile in ν-direction. In the linear response regime, each ω component of $p_\mu(t)$ and $E_\nu(t)$ should be related by the dipole polarizability, $\alpha_{\mu\nu}(\omega)$, since there holds a principle of superposition,

$$\int dt\, e^{i\omega t} p_\mu(t) = \alpha_{\mu\nu}(\omega) \int dt\, e^{i\omega t} E_\nu(t). \tag{4.28}$$

This relation tells us that for an applied electric field of any time profile, we can calculate the dipole polarizability by taking the ratio of the Fourier transforms,

$$\alpha_{\mu\nu}(\omega) = \frac{1}{\tilde{E}_\nu(\omega)} \int dt\, e^{i\omega t} p_\mu(t), \qquad (4.29)$$

where $\tilde{E}_\nu(\omega)$ is the Fourier transform of the applied electric field, $\tilde{E}_\nu(\omega) = \int dt\, e^{i\omega t} E_\nu(t)$.

In the practical calculation, we employ an impulsive dipole field,

$$V_{\text{ext}}(\mathbf{r},t) = I\delta(t)r_\nu, \qquad (4.30)$$

where I is the magnitude of the impulse. We assume that a molecule is in the ground state described by the static solution of the DFT before applying the external potential at $t = 0$. The impulsive force is then applied to all the electrons in the molecule at $t = 0$. In classical mechanics, each electron gets an initial velocity $v = I/m$ by this impulse. In quantum mechanics, the same situation is described as follows: every orbital is multiplied by the plane wave with the momentum I.

$$\psi_i(\mathbf{r}, t = 0_+) = e^{i I r_\nu/\hbar} \phi_i(\mathbf{r}). \qquad (4.31)$$

We evolve the orbitals starting from this initial condition. During the time evolution, we do not apply any external potential. The dipole moment is calculated by $d_\mu(t) = \int d\mathbf{r}\, r_\mu \sum_i |\psi_i(\mathbf{r}, t)|^2$. The Fourier transform of the impulse field is just the constant I. Then the polarizability is calculated by

$$\alpha_{\mu\nu}(\omega) = \frac{e^2}{I} \int_0^\infty dt\, e^{i\omega t} d_\mu(t). \qquad (4.32)$$

Since the impulsive field includes the frequency components of the whole spectral region uniformly, one may obtain the polarizability for the whole spectral region from a single time evolution calculation.

In the polarization function in time, $p_\mu(t)$, transitions to bound excited states appear as steady oscillations that persist without any damping. However, the real-time evolution is achieved for a certain finite period and, consequently, the Fourier spectrum of such oscillations shows wiggles around each excitation energy because of the abrupt cut of the steady oscillations within a finite time period. The wiggles may be diminished to some extent by introducing a damping function in the Fourier transformation of Equation 4.32. A simple choice for the damping function is the exponential function $e^{-\gamma t}$. Since this is equivalent to introducing a small imaginary part, $i\gamma$, in the frequency, this gives the bound transitions a Lorentzian line shape if the time evolution is achieved for a sufficiently long period. Another choice is the damping function of a third order polynomial in time,

$$f(t) = 1 - 3\left(\frac{t}{T}\right)^2 + 2\left(\frac{t}{T}\right)^3. \qquad (4.33)$$

Since this function does not change the slope of $p_\mu(t)$ at $t = 0$, it does not change the TRK sum.

The typical time step is $\Delta t = 0.027$ a.u. We perform the time evolution for 5×10^4 time steps. Thus, the total period of the time evolution is given by $T = N\Delta t = 1350$ a.u. The energy resolution in the real-time calculation is determined by the inverse of the time period. In the present case, it is $\Delta E = 2\pi/T = 0.13$ eV. We employ the width for smoothing, $\gamma = 0.25$ eV.

4.3.3 Treatment of Outgoing Boundary Condition for Photoionization Process

To describe the photoionization process, we need to impose the scattering boundary condition for the emitted electrons. For electrons emitted from a molecule, it is not at all obvious how to impose the scattering boundary condition computationally. A lot of methods have been developed to treat the problem. (See references in Cacelli et al., 1991; Nakatsukasa and Yabana, 2001.)

In the real-time calculation, the Kohn–Sham orbitals are distorted, by the impulsive external potential. They include components corresponding to the unbound orbitals of the static Kohn–Sham Hamiltonian. During the time evolution, electrons excited to the unbound orbitals are emitted outside the molecule. Once the electrons are emitted outside the molecule, they never return to the region around the molecule. They will not contribute to the transition density and to the polarizability since the density change $\delta n(\mathbf{r}, t)$ only exists in the spatial region of the ground state density distribution.

In the three-dimensional grid representation that we adopt in our calculation, we solve the TDKS equation inside a certain box area. If we simply solve the equation, the electrons emitted outside the molecule are reflected at the box boundary and return to the region of the molecule. Then the reflected wave makes a standing wave in the region of the molecule and makes a spurious contribution to the polarizability. To describe the ionization process adequately, one must remove electrons that are emitted outside the molecule during the time evolution.

An approximate removal of emitted electrons is feasible by placing a negative imaginary potential outside the molecule. In the presence of the negative imaginary potential, the flux is absorbed during the propagation. This is the so-called absorbing boundary condition. The absorbing potential should be smooth enough so that the flux getting into the region of the absorbing potential is not reflected. On the other hand, the absorbing potential should be strong enough so that all the flux is absorbed efficiently. These two conditions may be satisfied if one employs a sufficiently long distance of absorbing potential. This, however, increases the number of grid points to be employed in the numerical calculation. We must find a compromise of these contrary conditions.

In the practical calculations presented below, we employ the absorbing potential with a linear coordinate dependence. The potential is placed at a spatial region where the radial coordinate r is greater than a certain radius R, beyond which the electron density in the ground state can be negligible. Denoting the thickness of the absorbing region as ΔR, the absorbing potential is placed in the region $R < r < R + \Delta R$,

$$-iW(\mathbf{r}) = \begin{cases} 0 & (0 < r < R) \\ -iW_0 \dfrac{r - R}{\Delta R} & (R < r < R + \Delta R) \end{cases}. \tag{4.34}$$

Here the height W_0 and the width ΔR must be carefully chosen. We adopt $W_0 = 0.147$ a.u. and $\Delta R = 18.9$ a.u. in the calculations presented below.

In the real-time method, the absorbing potential is simply added to the Kohn–Sham Hamiltonian in the time evolution,

$$i\frac{\partial}{\partial t}\psi_i(\mathbf{r}, t) = \left\{ h[n(t)] - iW(\mathbf{r}) + V_{\text{ext}}(\mathbf{r}, t) \right\} \psi_i(\mathbf{r}, t). \tag{4.35}$$

In the presence of the absorbing potential, the time evolution is no more unitary. However, since we apply sufficiently weak perturbation $V_{\text{ext}}(\mathbf{r}, t)$, the loss of the flux by the absorbing potential is small. We have found that the time evolution is stable for a sufficiently long period to obtain the polarizability with a desired accuracy.

4.4 OSCILLATOR STRENGTH DISTRIBUTIONS OF MOLECULES

We present the TDDFT calculations of the oscillator strength distributions for a variety of molecules. We will show the calculations of small molecules, N_2, O_2, H_2O, and CO_2; several hydrocarbon molecules of small and medium size, acetylene, ethylene, benzene, and naphthalene; and a fullerene C_{60} as an example of a large molecule. We will show that the TDDFT is capable of describing the overall features of the oscillator strength distribution for all molecules studied. We will also show the decompositions of the oscillator strength distributions into occupied orbitals and clarify the character of the discrete transitions. As the molecular size increases, the measured oscillator strength distributions show fewer structures than those in the calculation. This should be due to the coupling of electrons with molecular vibration that we ignored in the present calculation.

We make a comparison of our calculation with only a few measurements, usually the most recently published values. We also quote only a few recently published calculations among many others in the past.

4.4.1 DIATOMIC AND TRIATOMIC MOLECULES

4.4.1.1 N_2 Molecule

We show the TDDFT calculation for the oscillator strength distribution of N_2 molecule below 55 eV in Figure 4.1, in comparison with measurement (Chan et al., 1993a). The decomposition of the strength into occupied orbitals is shown in Figure 4.2 for 10–20 eV energy region. The calculated energy of HOMO orbital is 15.63 eV, which is in good agreement with the measured ionization potential, 15.58 eV. As seen from Figure 4.1, the overall features of the oscillator strength distribution are nicely reproduced by the calculation. There are two transitions below the ionization threshold. The transition at 13 eV is composed of two transitions, from two orbitals, $3\sigma_g$ and $2\sigma_u$. The transition at 14 eV originates mainly from the $1\pi_u$ orbital. At around 16 eV, there are two transitions seen in the TDDFT calculation. Measurement shows a bump at around the energy. This mainly comes from the $1\pi_u$ orbital. At around 22 eV, the measurement shows a small bump. However, it is not visible in the TDDFT calculation.

A similar TDDFT calculation has been presented by Stener and Decleva (2000), with a different numerical method. The results presented here is consistent with those by Stener's. Levine and Soven (1984) also made a calculation of N_2. Montuoro and Moccia (2003) made a calculation with the K-matrix method.

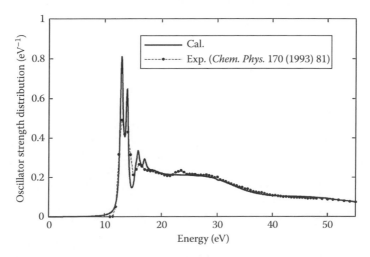

FIGURE 4.1 Oscillator strength distribution of N_2 molecule. Solid line, TDDFT calculation; circles connected with dotted line, measurement. (Data from Chan, W.F. et al., *Chem. Phys.*, 170, 81, 1993a.)

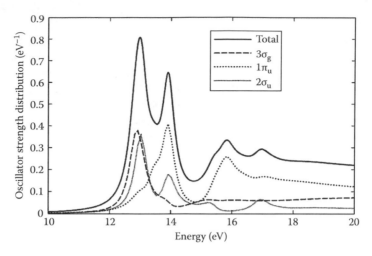

FIGURE 4.2   Decomposition of the calculated oscillator strength distribution of N_2 molecule into orbitals.

4.4.1.2 O₂ Molecule

We show the TDDFT calculation for the oscillator strength distribution of O_2 molecule below 70 eV in Figure 4.3, in comparison with measurement (Brion et al., 1979). The decomposition of the strength into occupied orbitals is shown in Figure 4.4 for 5–30 eV energy region. The ground state of the O_2 molecule is spin triplet, the last two electrons occupy the two degenerate orbitals of $1\pi_g$ aligning their spin directions. However, in the present calculation, we ignore the spin structure, assuming that the two degenerate $1\pi_g$ orbitals of spin-up and spin-down are occupied equally by two electrons, 0.5 electron for each. We thus employ the energy functional with saturated spin structure. The energy eigenvalue of the HOMO is at 12.01 eV, in good agreement with the measured ionization potential, 12.07 eV.

Two transitions appear below the ionization threshold. They originate from the occupied orbitals of $1\pi_g$ and $1\pi_u$. Above the threshold, three sharp structures appear at around 15 eV in the calculation that is mainly from $1\sigma_u$, $1\pi_u$, and $3\sigma_g$ orbitals. However, in the measured spectrum, they are not well separated.

At around 35 and 40 eV, the calculated oscillator strength shows a peak. In the same energy region (40 eV), a slight bump is seen in the measured spectrum. In the calculation, these peaks come

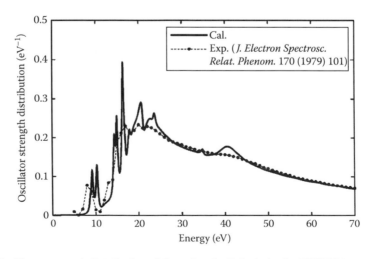

FIGURE 4.3   Oscillator strength distribution of O_2 molecule. Calculation by TDDFT is compared with measurement. (Data from Brion, C.E. et al., *J. Electron Spectrosc. Relat. Phenom.*, 17, 101, 1979.)

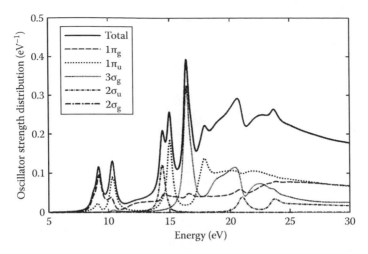

FIGURE 4.4   Decomposition of the calculated oscillator strength distribution of O_2 molecule into orbitals.

mainly from $2\sigma_g$ orbital. The orbital energy of the $2\sigma_g$ is 37.55 eV. Therefore, the energy of 40 eV is slightly higher than the ionization threshold of this orbital.

For the O_2 molecule, Lin and Lucchese (2002) made a calculation employing multichannel Schwinger method with configuration interaction.

4.4.1.3 H_2O Molecule

We show the TDDFT calculation for the oscillator strength distribution of H_2O molecule below 55 eV in Figure 4.5, in comparison with measurement (Chan et al., 1993b). The decomposition of the strength into occupied orbitals is shown in Figure 4.6 for the same energy region. The calculated energy of HOMO orbital is 13.06 eV, which is slightly larger than the measured ionization potential, 12.61 eV. As seen from Figure 4.5, the overall features of the oscillator strength distribution are well reproduced by the calculation. The transition at around 8 eV comes from the $1b_1$ orbital. The next one around 10 eV is $3a_1$. These two transitions are visible in the measurement. In the calculation, next transition appears around 11 eV, which is from $1b_1$. The large transition at around 12 eV, which is visible in the measurement, comes from two orbitals, $3a_1$ and $1b_2$. In the calculation, a Fano-like structure is seen at around

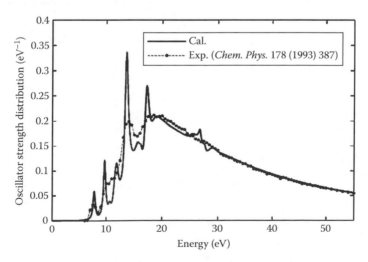

FIGURE 4.5   Oscillator strength distribution of H_2O molecule. Calculation by TDDFT is compared with measurement. (Data from Chan, W.F. et al., *Chem. Phys.*, 178, 38, 1993b.)

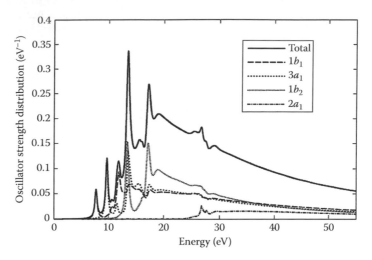

FIGURE 4.6 Decomposition of the calculated oscillator strength distribution of H_2O molecule into orbitals.

28 eV, which comes from $2a_1$ orbital and is close to the threshold of $2a_1$ orbital whose orbital energy is 30.08 eV. In the measurement, such feature is not visible.

For the H_2O molecule, Dupin et al. (2002) made a calculation employing the Stieltjes moment method. Cacelli et al. (1991) made a calculation with the K-matrix method.

4.4.1.4 CO_2 Molecule

We show the TDDFT calculation for the oscillator strength distribution of CO_2 molecule below 55 eV in Figure 4.7, in comparison with measurement (Chan et al., 1993c). The decomposition of the strength into occupied orbitals is shown in Figure 4.8 for the same energy region. The calculated energy of HOMO orbital is 14.96 eV, which is slightly larger than the measured ionization potential 13.78 eV. As seen from Figure 4.7, the overall features of the oscillator strength distribution are again well reproduced by the calculation. The transition at around 12 eV is composed of several transitions. At least four transitions are seen in the calculation that originate from $1\pi_g$, $1\pi_u$, $3\sigma_u$, and $4\sigma_g$.

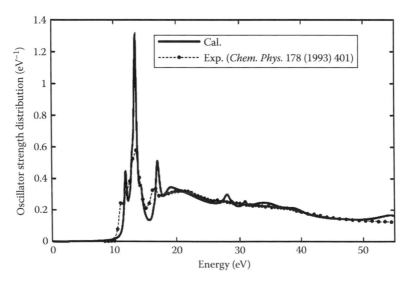

FIGURE 4.7 Oscillator strength distribution of CO_2 molecule. Calculation by TDDFT is compared with measurement. (Data from Chan, W.F. et al., *Chem. Phys.*, 178, 401, 1993c.)

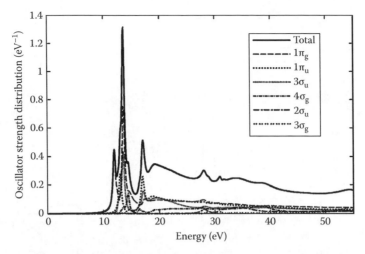

FIGURE 4.8   Decomposition of the calculated oscillator strength distribution of CO_2 molecule into orbitals.

The next one around 17 eV comes from $1\pi_u$ and $3\sigma_u$ orbitals. In the calculation, there appear a few transitions around 30 eV, which is originated from the $2\sigma_u$ orbital whose orbital energy is 33 eV. However, they are not visible in the measured spectrum.

For the CO_2 molecule, Olalla and Martin (2004) made a calculation with the molecular quantum defect orbital method.

4.4.2   Hydrocarbon Molecules

We report the oscillator strength distributions of several hydrocarbon molecules of small and medium sizes. In addition to molecules shown below, we reported the oscillator strength distributions of C_3H_6 isomers in (Nakatsukasa and Yabana, 2004).

4.4.2.1   Acetylene C_2H_2

We show the TDDFT calculation for the oscillator strength distribution of the acetylene, C_2H_2, molecule below 40 eV in Figure 4.9, in comparison with measurement (Ukai et al. 1991, Cooper et al.

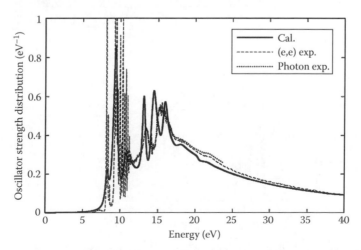

FIGURE 4.9   Oscillator strength distribution of acetylene, C_2H_2, molecule. Calculation by TDDFT is compared with measurement. (Data from Ukai, M. et al. *J. Chem. Phys.*, 95, 4142, 1991; Cooper, G. et al., *J. Electron Spectrosc. Relat. Phenom.*, 73, 139, 1995b.)

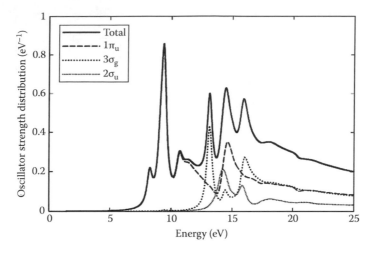

FIGURE 4.10 Decomposition of the calculated oscillator strength distribution of C_2H_2 molecule into orbitals.

1995b). The decomposition of the strength into occupied orbitals is shown in Figure 4.10 below 25 eV. The calculated energy of HOMO orbital is 12.0 eV, which is slightly larger than the measured ionization potential, 11.4 eV. For this molecule, a detailed analysis has previously been reported in Nakatsukasa and Yabana (2001).

4.4.2.2 Ethylene C_2H_4

We show the TDDFT calculation for the oscillator strength distribution of the ethylen; C_2H_4, molecule below 40 eV in Figure 4.11, in comparison with measurement (Holland et al., 1997; Cooper et al., 1995a). The decomposition of the strength into occupied orbitals is shown in Figure 4.12 below 25 eV. The calculated energy of HOMO orbital is 11.67 eV, which is slightly larger than the measured ionization potential, 11.0 eV. For this molecule, a detailed analysis has previously been reported in Nakatsukasa and Yabana (2001).

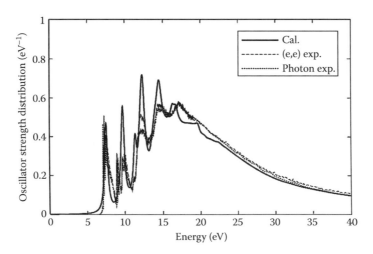

FIGURE 4.11 Oscillator strength distribution of ethylene, C_2H_4, molecule. Calculation by TDDFT is compared with measurement. (Data from Holland, D.M.P. et al. *Chem. Phys.*, 219, 91, 1997; Cooper, G. et al., *Chem. Phys.*, 194, 175, 1995a.)

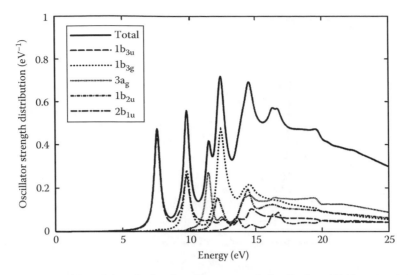

FIGURE 4.12 Decomposition of the calculated oscillator strength distribution of C_2H_4 molecule into orbitals.

4.4.2.3 Benzene C_6H_6

We show the TDDFT calculation for the oscillator strength distribution of the benzene, C_6H_6, molecule below 60 eV in Figure 4.13, in comparison with measurement (Feng et al. 2002). The decomposition of the strength into occupied orbitals is shown in Figure 4.14 for the energy region of 5–13 eV. The calculated energy of HOMO orbital is 11.18 eV, which is larger than the measured ionization potential, 9.24 eV. As seen from Figure 4.13, the overall shape of the oscillator strength distribution is very nicely reproduced by the calculation. Although several structures appear in the TDDFT calculation, only a few shoulders appear in the measurement above the ionization threshold. In the spectrum, the intense transition of the $1e_{1g}$ orbital appears at 6.9 eV. There appear several transitions around 10–12 eV region, which are seen as a single shoulder in the measurement. In the TDDFT calculation, they come from several occupied orbitals. Among them, the orbitals of $3e_{2g}$, $3e_{1u}$, and $1a_{2u}$ contribute substantially in this energy region.

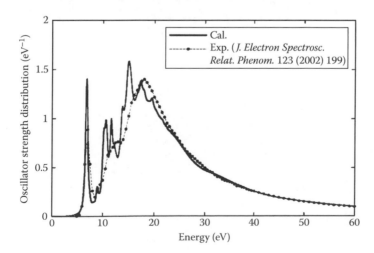

FIGURE 4.13 Oscillator strength distribution of benzene, C_6H_6, molecule. Calculation by TDDFT is compared with measurement. (Data from Feng, R. et al. *J. Electron Spectrosc. Relat. Phenom.*, 123, 199, 2002.)

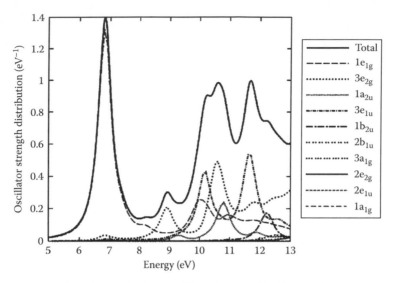

FIGURE 4.14 Decomposition of the calculated oscillator strength distribution of C_6H_6 molecule into orbitals.

Recently, Gokhberg et al. (2009) reported the oscillator strength distribution of benzene molecule by the Stieltjes–Chebyshev moment theory.

4.4.2.4 Naphthalene $C_{10}H_8$

We show the TDDFT calculation for the oscillator strength distribution of the naphthalene, $C_{10}H_8$, molecule below 50 eV in Figure 4.15, in comparison with measurement (de Souza et al., 2002). The spectrum of naphthalene resembles that of benzene. The calculated spectrum of naphthalene is somewhat smoother than that of benzene, probably because of the higher density of states in the excited state. In the naphthalene spectrum, a bump is seen around 23 eV, which is not seen in benzene. However, it is not clear in the measured spectrum. The absolute value of the measured spectrum is somewhat lower than the calculation.

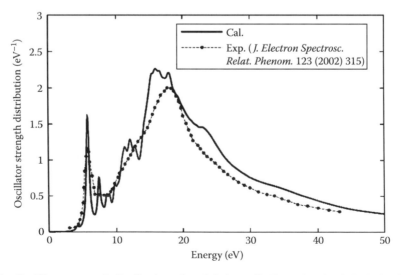

FIGURE 4.15 Oscillator strength distribution of naphthalene, $C_{10}H_8$, molecule. Calculation by TDDFT is compared with measurement. (Data from de Souza, G.G.B. et al., *J. Electron Spectrosc. Relat. Phenom.*, 123, 315, 2002.)

4.4.3 FULLERENE C_{60}

As a final example of our calculations, we show the TDDFT calculation for the fullerene molecule, C_{60} (Kawashita et al., 2009). Here, we employ the LDA for exchange-correlation potential. The C_{60} molecule has a quite high spatial symmetry of icosahedron. Reflecting the symmetry, the density of states of dipole-allowed transitions is rather low in comparison with the other molecules of similar size. We show in Figure 4.16 the calculated oscillator strength distribution of C_{60}. We take a sufficiently large spatial area in the calculation so that the spectrum converges with respect to computational parameters such as the box size and mesh spacing. Below the ionization threshold, four transitions are seen in the calculation, which show a reasonable agreement with measurements where three transitions are observed. One of the interesting features of the spectrum of C_{60} is the presence of a number of sharp transitions in the spectrum, even up to 35 eV. In the measurements, however, such fine structures are smeared out, probably by the electron-vibration coupling in the molecule. In Figure 4.17, we show a

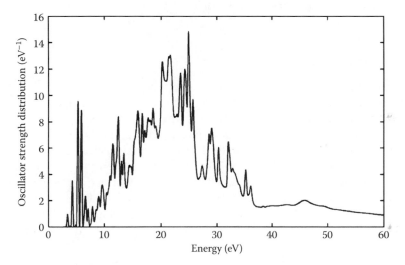

FIGURE 4.16   Oscillator strength distribution of fullerene, C_{60}, molecule. (Taken from Kawashita, Y. et al., *J. Mol. Struct. THEOCHEM*, 914, 130, 2009.)

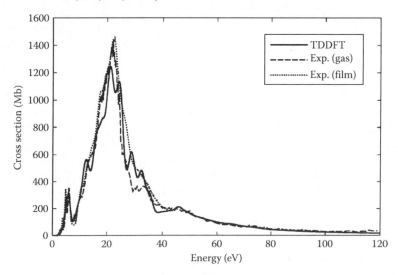

FIGURE 4.17   Oscillator strength distribution of C_{60} molecule. Calculated result by TDDFT is shown, in comparison with measurements in gas phase and in film. (Data from Kafle, B.P. et al., *J. Phys. Soc. Jpn.*, 77, 014302, 2008; Yagi, H. et al., *Carbon* 47, 1152, 2009; and Kawashita, Y. et al., *J. Mol. Struct. THEOCHEM*, 914, 130, 2009.)

comparison of the TDDFT oscillator strength distribution with recent measurement (Kafle et al., 2008; Yagi et al., 2009). For the TDDFT calculation, we made a smoothing using a Gaussian function with a width of $0.27 \, eV^2$. As is seen in Figure 4.17, the overall features of the oscillator strength distribution is well reproduced by the calculation. There are several shoulders and plateaus in the measurements, at 10, 13, 18, 29–34, and 40–50 eV. At corresponding positions, structures are also seen in the TDDFT calculation. However, the detailed assignment of these structures has not been made yet.

4.5   OTHER OPTICAL QUANTITIES STUDIED WITH TDDFT

The applications of the TDDFT are not limited to the oscillator strength distributions, which we have shown in the previous section, but are extended to a lot of optical observables. In this section, we outline several other applications of the TDDFT briefly.

4.5.1   NONLINEAR POLARIZABILITIES

As the intensity of the optical field increases, materials come to show the nonlinear optical responses, which are related to technologically important nonlinear optical phenomena such as second-harmonic generation, degenerate four-wave mixing, two-photon absorption, and so on (Brédas et al., 1994). The nonlinear responses of molecules are characterized by the nonlinear polarizabilities, which are obtained by a straightforward extension of the linear response theory to the nonlinear response theory (higher-order perturbation theory). The first nonlinear response calculation based on the TDDFT was performed for rare gas atoms in 1987 (Senatore and Subbaswamy, 1987). The TDDFT calculation for molecular hyperpolarizabilities has been performed in 1998 based on a quantum chemical approach (Van Gisbergen et al., 1998a,b, 1999), and in 2001 the same calculation has been performed based on the real-space approach (Iwata et al., 2001).

4.5.2   DIELECTRIC FUNCTIONS OF BULK MATERIALS

The TDDFT has also been applied for infinitely periodic systems as well as the isolated systems such as molecules. The linear optical response of bulk materials is characterized by the dielectric function $\varepsilon_{\mu\nu}(\omega)$. The real-space and real-time TDDFT calculation on the dielectric functions was performed by us, as detailed in Bertsch et al. (2000).

The TDDFT describes well the overall features of the dielectric function for a wide energy range. However the TDDFT gives poor results for the dielectric response of semiconductors and insulators in the energy range near the bandgap. It is well known that the LDA systematically underestimates the bandgap energies. The failure of the LDA cannot be recovered by the TDDFT with the adiabatic LDA. Thus, several attempts have been achieved to incorporate correlation effects that are lacked in the TDDFT. The Bethe–Salpeter theory based on the GW approximation (Onida et al., 2002) is one of the successful approaches for the dielectric function, but the computational cost is much higher than that of the TDDFT.

4.5.3   OPTICAL ACTIVITIES OF CHIRAL MOLECULES

For chiral molecules, linear optical properties are characterized by the optical activity as well as the polarizability. The TDDFT is straightforwardly applicable to them, and the calculations of circular dichroism have been extensively discussed in the literature (Yabana and Bertsch, 1999).

4.6   SUMMARY

We have presented a theoretical description for the oscillator strength distribution of molecules for a full spectral region of valence electrons below 100 eV. We rely upon the TDDFT, which is an extension of the DFT, so as to describe electron dynamics in the first-principles calculation. We first describe a theoretical basis of the TDDFT to calculate the oscillator strength distribution. We then show the calculated results of the oscillator strength distributions for molecules, taking diatomic and triatomic molecules—N_2, O_2, H_2O, and CO_2; several hydrocarbon molecules of small and medium size—acetylene, ethylene, benzene, and naphthalene; and fullerene C_{60} as an example of a large molecule. It has been shown that the TDDFT provides accurate and quantitative descriptions for the oscillator strength distributions of these molecules. Finally, we mention briefly the applications of the TDDFT for other optical quantities, including hyperpolarizabilities, the optical activities of chiral molecules, and the dielectric function for bulk materials.

REFERENCES

Bertsch, G. F., Iwata, J.-I., Rubio, A., and Yabana, K. 2000. Real-space, real-time method for the dielectric function. *Phys. Rev. B* 62: 7998–8002.

Brédas, J. L., Adant, C., Tackx, P., Persoons, A., and Pierce, B. M. 1994. 3rd-order nonlinear-optical response in organic materials—Theoretical and experimental aspects. *Chem. Rev.* 94: 243–278.

Brion, C. E., Tan, K. H., van der Wiel, M. J., and van der Leeuw, Ph. E. 1979. Dipole oscillator strengths for the photoabsorption, photoionization and fragmentation of molecular oxygen. *J. Electron Spectrosc. Relat. Phenom.* 17: 101–119.

Cacelli, I., Carravetta, V., Rizzo, A., and Moccia, R. 1991. The calculation of photoionisation cross section of simple polyatomic molecules by L^2 methods. *Phys. Rep.* 205: 283–351.

Casida, M. E., Jamorski, C., Casida, K. C., and Salahub, D. R. 1998. Molecular excitation energies to high-lying bound states from time-dependent density-functional response theory: Characterization and correction of the time-dependent local density approximation ionization threshold. *J. Chem. Phys.* 108: 4439–4449.

Chan, W. F., Cooper, G., Sodhi, R. N. S., and Brion, C. E. 1993a. Absolute optical oscillator strengths for discrete and continuum photoabsorption of molecular nitrogen (11–200 eV). *Chem. Phys.* 170: 81–97.

Chan, W. F., Cooper, G., and Brion, C. E. 1993b. The electronic spectrum of water in the discrete and continuum regions. Absolute optical oscillator strengths for photoabsorption (6–200 eV). *Chem. Phys.* 178: 387–400.

Chan, W. F., Cooper, G., and Brion, C. E. 1993c. The electronic spectrum of carbon dioxide. Discrete and continuum photoabsorption oscillator strengths (6–203 eV). *Chem. Phys.* 178: 401–413.

Chelikowsky, J. R., Troullier, N., Wu, K., and Saad, Y. 1994. Higher-order finite-difference pseudopotential method: An application to diatomic molecules. *Phys. Rev. B* 50: 11355–11364.

Cooper, G., Olney, T. N., and Brion, C. E. 1995a. Absolute UV and soft x-ray photoabsorption of ethylene by high resolution dipole (e,e) spectroscopy. *Chem. Phys.* 194: 175–184.

Cooper, G., Grodon, R., Burton, R., and Brion, C. E. 1995b. Absolute UV and soft x-ray photoabsorption of acetylene by high resolution dipole (e,e) spectroscopy. *J. Electron Spectrosc. Relat. Phenom.* 73: 139–148.

de Souza, G. G. B., Boechat-Roberty, H. M., Rocco, M. L. M., and Lucas, C. A. 2002. Generalized oscillator strength for the 1B2u←1Ag transition and the observation of forbidden processes at the C 1s spectrum of the naphthalene molecule. *J. Electron Spectrosc. Relat. Phenom.* 123: 315–321.

Dreuw, A. and Head-Gordon, M. 2005. Single-reference ab initio methods for the calculation of excited states of large molecules. *Chem. Rev.* 105: 4009–4037.

Dupin, H., Baraille, I., Larrieu, C., Rerat, M., and Dargelos, A. 2002. Theoretical determination of the ionization cross-section of water. *J. Mol. Struct. (Theochem)* 577: 17–33.

Ekardt, W. 1984. Dynamical polarizability of small metal particles: Self consistent spherical jellium background model. *Phys. Rev. Lett.* 52: 1925–1928.

Feng, R., Cooper, G., and Brion, C. E. 2002. Dipole (e,e) spectroscopic studies of benzene: Quantitative photoabsorption in the UV, VUV and soft x-ray regions. *J. Electron Spectrosc. Relat. Phenom.* 123: 199–209.

Furche, F. and Ahlrichs, R. 2002. Adiabatic time-dependent density functional methods for excited state properties. *J. Chem. Phys.* 117: 7433–7447.

Gokhberg, K., Vysotskiy, V., Cederbaum, L. S., Storchi, L., Tarantelli, F., and Averbukh, V. 2009. Molecular photoionization cross section by Stieltjes-Chebyshev moment theory applied to *Lanczos pseudospectra*. *J. Chem. Phys.* 130: 064104–064108.

Holland, D. M. P., Shaw, D. A., Hayes, M. A., Shpinkova, L. G., Rennie, E. E., Karisson, L., Baltzer, P., and Wannberg, B. 1997. A photoabsorption, photodissociation and photoelectron spectroscopy study of C_2H_4 and C_2D_4. *Chem. Phys.* 219: 91–116.

Iwata, J.-I., Yabana, K., and Bertsch, G. F. 2001. Real-space computation of dynamic hyperpolarizabilities. *J. Chem. Phys.* 115: 8773–8783.

Kafle, B. P., Katayanagi, H., Prodhan, Md. S. I., Yagi, H., Huang, C. Q., and Mitsuke, K. 2008. Absolute total photoionization cross section of C_{60} in the range of 25–120 eV: Revisited. *J. Phys. Soc. Jpn.* 77: 014302.

Kawashita, Y., Yabana, K., Noda, M., Nobusada, K., and Nakatsukasa, T. 2009. Oscillator strength distribution of C_{60} in the time-dependent density functional theory. *J. Mol. Struct.: THOECHEM* 914 (1–3): 130–135.

Kleinman, L. and Bylander, D. 1982. Efficacious form for model pseudopotentials. *Phys. Rev. Lett.* 48: 1425–1428.

Levine, Z. H. and Soven, P. 1984. Time-dependent local-density theory of dielectric effects in small molecules. *Phys. Rev. A* 29: 625–635.

Lin, P. and Lucchese, R. R. 2002. Theoretical studies of cross section and photoelectron angular distribution in the valence photoionization of molecular oxygen. *J. Chem. Phys.* 116: 8863–8875.

Montuoro, R. and Moccia, R. 2003. Photoionization cross sections calculation with mixed L^2 basis set: STOs plus B-Splines. Results for N_2 and C_2H_2 by KM-RPA method. *Chem. Phys.* 293: 281–308.

Nakatsukasa, T. and Yabana, K. 2001. Photoabsorption spectra in the continuum of molecules and atomic clusters. *J. Chem. Phys.* 114: 2550–2561.

Nakatsukasa, T. and Yabana, K. 2004. Oscillator strength distribution in C_3H_6 isomers studied with the time-dependent density functional method in the continuum. *Chem. Phys. Lett.*, 374: 613–619.

Olalla, E. and Martin, I. 2004. Theoretical study of the valence and K-shell spectra of atmospherically relevant CO_2. *Int. J. Quantum Chem.* 99: 502–510.

Onida, G., Reining, L., and Rubio, A. 2002. Electronic excitations: Density-functional versus many-body Green's-function approaches. *Rev. Mod. Phys.* 74: 601–659.

Senatore, G. and Subbaswamy, K. R. 1987. Nonlinear response of closed-shell atoms in the density-functional formalism. *Phys. Rev. A* 35: 2440–2447.

Stener, M. and Decleva, P. 2000. Time-dependent density functional calculations of molecular photoionization cross sections: N_2 and PH_3. *J. Chem. Phys.* 112: 10871–10879.

Troullier, N. and Martins, J. L. 1991. Efficient pseudopotentials for plane-wave calculations *Phys. Rev. B* 43: 1993–2006.

Ukai, M., Kameta, K., Chiba, R., Nagao, K., Kouchi, N., Shinsaka, K., Hatano, Y., Umemoto, H., and Ito, Y. 1991. Ionizing and nonionizing decays of superexcited acetylene molecules in the extreme-ultraviolet region. *J. Chem. Phys.* 95: 4142–4153.

Van Gisbergen, S. J. A., Snijders, J. G., and Baerends, E. J. 1998a. Calculating frequency-dependent hyperpolarizabilities using time-dependent density functional theory. *J. Chem. Phys.* 109: 10644–10656.

Van Gisbergen, S. J. A., Snijders, J. G., and Baerends, E. J. 1998b. Accurate density functional calculations on frequency-dependent hyperpolarizabilities of small molecules. *J. Chem. Phys.* 109: 10657–10668.

Van Gisbergen, S. J. A., Snijders, J. G., and Baerends, E. J. 1999. Erratum: "Calculating frequency-dependent hyperpolarizabilities using time-dependent density functional theory" [*J. Chem. Phys.* 109, 10644 (1998)]. *J. Chem. Phys.* 111: 6652.

Van Leeuwen, R. and Baerends, E. J. 1994. Exchange-correlation potential with correct asymptotic behavior. *Phys. Rev. A* 49: 2421–2431.

Yabana, K. and Bertsch, G. F. 1996. Time-dependent local-density approximation in real time. *Phys. Rev. B* 54: 4484–4487.

Yabana, K. and Bertsch, G. F. 1999. Time-dependent local-density approximation in real time: Application to conjugated molecules. *Int. J. Quantum Chem.* 75: 55–66.

Yabana, K., Nakatsukasa, T., Iwata, J.-I., and Bertsch, G. F. 2006. Real-time, real-space implementation of the linear response time-dependent density-functional theory. *Phys. Stat. Sol. (b)* 243: 1121–1138.

Yagi, H., Nakajima, K., Koswattage, K. R., Nakagawa, K., Huang, C., Prodhan, Md. S. I., Kafle, B. P., Katayanagi, H., and Mitsuke, K. 2009. Photoabsorption cross section of C_{60} thin films from the visible to vacuum ultraviolet. *Carbon* 47: 1152–1157.

Zangwill, A. and Soven, P. 1980. Density-functional approach to local-field effects in finite systems: Photoabsorption in the rare gases. *Phys. Rev. A* 21: 1561–1572.

5 Generalized Oscillator Strength Distribution of Liquid Water

Hisashi Hayashi
Japan Women's University
Tokyo, Japan

Yasuo Udagawa
Tohoku University
Sendai, Japan

CONTENTS

5.1 INTRODUCTION

Water is the most common liquid substance on the earth's surface. Many fundamental processes, in particular those related to the development of life, take place in aqueous solutions. Accordingly, water is the most extensively studied liquid from theoretical as well as experimental points of view; several chapters of this book have the word "liquid water" in their title. Nevertheless, our knowledge about liquid water is far from complete; even now there is a debate on such a fundamental issue as to whether a water molecule is hydrogen-bonded with two other molecules or four in the liquid state (Smith et al., 2004; Wernet et al., 2004).

The optical properties of liquid water in the vacuum ultraviolet (VUV) region, like extinction coefficient (κ), refractive index (n), and reflectivity (R), are some of those which are not fully explored in spite of their importance. They all depend on photon energy, $\hbar\omega$, and are connected with each other and also to the complex dielectric response function ($\varepsilon(\omega) = \varepsilon_1(\omega) + i\varepsilon_2(\omega)$) by the following equations: $\varepsilon(\omega)^{1/2} = n + i\kappa$ and $R = [(n - 1)^2 + \kappa^2]/[(n + 1)^2 + \kappa^2]$. It is the dielectric response function that governs interactions between matter and photons or charged particles, which is the theme of this chapter. If one of those optical or dielectric properties is known for a wide energy range, not only can other optical properties be calculated, but also various other physical properties can be evaluated

(Inokuti, 1983; Williams et al., 1991). For example, the mean excitation energy of a material, the most important quantity required to describe the interaction with ionizing radiations, can be calculated on absolute scale. Unfortunately, such studies have been hampered by the lack of optical data in the VUV. To the best of our knowledge, optical data available in this energy region are a series of reflectance measurements at Oak Ridge (Painter et al., 1969; Kerr et al., 1971, 1972; Heller et al., 1974), but they are not wide enough, only up to 26 eV, and it has been estimated that the measurements are accurate to perhaps 30% (Kutcher and Green, 1976).

The dielectric response function ε is in fact a function of momentum transfer q also, and should be expressed as $\varepsilon(q, \omega)$. Not only are the available optical data not wide enough, but also what reflectance measurements provide are merely $\varepsilon(0, \omega)$, because the momentum of a VUV photon is very small. If $\varepsilon(q, \omega)$ of a medium is known over a wide energy and momentum range, the interaction of a charged particle with the medium can be fully understood. Such quantities as the mean free path of charged particles, cross sections for various interactions, and energy deposition, all of which are crucial in radiation chemistry/biology in vivo but are difficult to determine experimentally, can be evaluated. They ultimately lead to a track-structure analysis, which is one of the main subjects of radiation research as well as that of the previous version of this book (Nikjoo and Uehara, 2003; Nikjoo et al., 2006). Hence, many efforts have been made to estimate ω and q-dependence of $\varepsilon(q, \omega)$ of liquid water. In most of such studies, the Tennessee group's $\varepsilon(0, \omega)$ are extrapolated to a higher energy region and combined with some assumptions to extend them to finite momentum transfers (Dingfelder et al., 1998). Lacking experimental results to compare with, it has not been possible to evaluate how accurate those calculations are.

Experimentally, q-dependence of dielectric functions $\varepsilon(q, \omega)$ can be determined on gases and solids with inelastic electron scattering or electron energy loss spectroscopy (EELS) (Bonham and Fink, 1974; Egerton, 1996). For example, Lassettre et al. and Takahashi et al. have made extensive studies about dielectric functions of gaseous water (Lassettre and Skerbele, 1974; Lassettre and White, 1974; Takahashi et al., 2000) and Daniels of solid water and ice (Daniels, 1971). However, because of experimental difficulties, EELS measurements have never been carried out on volatile liquids like water.

We have pointed out that inelastic x-ray scattering (IXS) is free from difficulties inherent to optical as well as EELS measurements, the most serious of which is the need of vacuum, and can provide $\varepsilon(q, \omega)$ of volatile liquids for a wide energy and momentum-transfer range (Watanabe et al., 1997, 2000; Hayashi et al., 1998; Hayashi et al., 2000). IXS studies do not require target materials to be placed in vacuum and hence volatile liquids can be studied. Since momentum of a hard x-ray photon is large, q-dependence can easily be determined by changing the scattering angle. In the following, the manner in which photoabsorption is related to the interaction between charged particles and materials is described first, then the principles of IXS and EELS are reviewed, and finally IXS studies on liquid water are detailed.

5.2 PHOTOABSORPTION AND OPTICAL OSCILLATOR STRENGTH

It is well known that there is a close connection between the interaction of fast moving charged particles with a material and photoabsorption (Ritchie, 1982; Inokuti, 1986). In order to understand the relation heuristically, let us consider the following model depicted in Figure 5.1, where a particle with charge ze and velocity v is traveling along near an atom or a molecule with an impact parameter b. The particle exerts an electric field upon the target. Although the electric field has two components, as illustrated in the figure, only $E_\perp$ is important because $E_\parallel$ changes its sign at the distance of closest approach ($t = 0$), and as a result, the effect of $E_\parallel$ almost vanishes if integrated over the whole period. The electric field $E_\perp$ has a bell-shaped distribution in time and a simple calculation shows that its FWHM is about $2.6 b/v$. Suppose the particle is an electron with a kinetic energy of 10 keV and $b = 1$ nm; the velocity v is calculated to be 6×10^7 m/s and hence the FWHM of the bell is 4×10^{-17} s, short enough to approximate the electric field as a δ-function in time. From the Fourier principle, a δ-function-like electric field in the time domain is equivalent to white light in the frequency domain. That is, the effect of the fast charged particle traveling nearby is almost the same as that of white light illuminated on the target.

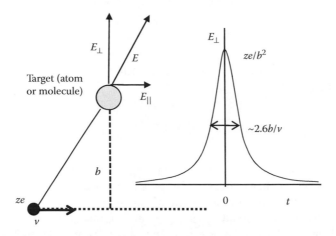

FIGURE 5.1 A particle with charge ze and velocity v passes by a target at an impact parameter b, and exerts the electric field E that depends on time t. The two components of E are given by $E_\perp = zeb/(v^2t^2 + b^2)^{3/2}$ and $E_\parallel = zevt/(v^2t^2 + b^2)^{3/2}$.

The interaction between white light and a target results in an absorption of a part of the white light, accompanied by excitation of the target. From the time-dependent perturbation theory, the absorption cross section, σ_{abs}, for transition from the ground state $|0\rangle$ with energy E_0 to an excited state $|n\rangle$ with energy E_n, defined as energy absorbed per unit time by the atom, divided by energy flux of the radiation field, is given by (Sakurai, 1994)

$$\sigma_{abs} = \frac{4\pi^2\hbar}{m^2\omega}\left(\frac{e^2}{\hbar c}\right)\left|\langle n|e^{i\mathbf{k}\cdot\mathbf{r}}\boldsymbol{\varepsilon}\cdot\mathbf{p}|0\rangle\right|^2 \delta(E_n - E_0 - \hbar\omega), \tag{5.1}$$

where
  $\mathbf{k} = (\omega/c)\mathbf{n}$ is the wave number vector of the monochromatic light propagating toward $\mathbf{n}$ with polarization vector $\boldsymbol{\varepsilon}$
  $\mathbf{p}$ the electron momentum operator
  $\mathbf{r}$ the coordinate of the electron involved in the transition

By the use of dipole approximation combined with the commutation relation $[x, H_0] = i\hbar p_x/m$, Equation 5.1 is transformed to

$$\sigma_{abs} = 4\pi^2\alpha\omega_{n0}\left|\langle n|x|0\rangle\right|^2 \delta(\omega - \omega_{n0}). \tag{5.2}$$

Here, the radiation field is assumed to propagate in the z direction with the polarization vector along the x-axis, $\alpha = e^2/\hbar c$ is the dimensionless fine-structure constant, and $\hbar\omega_{n0} = E_n - E_0$. Atoms and molecules have many excited states, and hence their response to white light is represented by simply integrating the above over all the excited states:

$$\int \sigma_{abs}d\omega = \sum_n 4\pi^2\alpha\omega_{n0}\left|\langle n|x|0\rangle\right|^2. \tag{5.3}$$

In atomic and molecular physics it is common to use a dimensionless quantity called optical oscillator strength, or simply oscillator strength, defined by

$$f_{n0} \equiv \frac{2m\omega_{n0}}{\hbar} \left| \langle n | x | 0 \rangle \right|^2. \tag{5.4}$$

In terms of the oscillator strength, the Thomas–Reiche–Kuhn sum rule can be expressed in the following simple manner:

$$\sum_n f_{n0} = 1. \tag{5.5}$$

Since Equation 5.5 applies for each electron, the right-hand side becomes N if the system has N electrons. Hence, if oscillator strength distribution (the relative values of f_{n0} for wide energy range) is measured, it can be brought into absolute scale by the use of the sum rule. The energy range is, in principle, from zero to infinity. In practice, measurements can be made only for a limited energy range, and hence we have to know how wide it should be in order to justify the use of the sum rule, as seen in Equation 5.5.

Figure 5.2 shows an IXS or x-ray Raman scattering spectrum of liquid water that, as is described later, is theoretically expected to be similar to the optical absorption spectrum. The signal due to valence electron excitation starts at around energy loss (E) of 7 eV, has a maximum at around 22 eV, and monotonically decays to higher energy. At around 540 eV, a scattering corresponding to oxygen K absorption starts, but the peak intensity is about two orders of magnitude less than that of valence electron excitation. Global profiles of VUV absorption spectra of organic compounds are somewhat similar, except that K absorption of carbon starts at 284 eV; most of the oscillator strength by valence electron excitations distributes between 0 and about 200 eV (Williams et al., 1991).

Photoabsorption spectra on liquids, gases, and solids are routinely measured in the infrared, visible, and UV regions. However, photoabsorption experiments in the VUV (above about 7 eV)

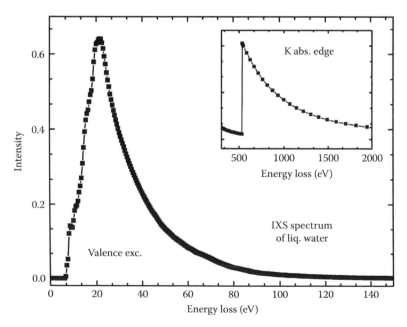

FIGURE 5.2 A hard x-ray inelastic-scattering spectrum of liquid water at small momentum transfer, which is essentially identical with VUV absorption spectrum (Hayashi et al., 2000). See text following Equation 5.7. The inset, which corresponds to K absorption of oxygen, is based on the NIST photoelectric cross section database for the water molecule (http://physics.nist.gov/ffast).

impose serious difficulties because air strongly absorbs VUV photons and hence measurements have to be carried out in a vacuum. In addition, absorbance of most substances is so high in the VUV that almost no window material is available above around 10 eV. Consequently, the direct absorption method is applicable only for low-pressure gases or very thin films, combined with differential pumping technique. Therefore, instead of direct absorption, reflectance measurements are conventionally employed for VUV studies on condensed phase substances (Seki et al., 1981; Kobayashi, 1983; Ikehata et al., 2008). Even with the reflectance method, however, measurements of the optical spectra of volatile liquids present a further difficulty, namely, how to keep them in vacuum. In an effort to obtain optical functions in the VUV, Heller et al. measured the reflectance spectrum from free-water surface kept in an open dish cooled to 1°C in near-vacuum conditions made with two stages of differential pumping; each stage included a cryopump capable of pumping 80,000 L of water vapor per second (Heller et al., 1974). Still, the spectral range measured was limited to below 25.6 eV. Hence, in order to evaluate optical functions, they had to resort to extrapolation, assuming either exponential or power functions. They estimated errors in optical constants due to the extrapolation to be as large as 20% above 20 eV. To the best of our knowledge, no VUV absorption study on liquid water for wide energy range has been reported since the 1970s, in spite of recent advancements in VUV technology.

In this respect, however, it may be worth mentioning here that very recently an absorption spectrum of liquid water in a very narrow range (530–545 eV, corresponding to the onset of K absorption in Figure 5.2) was observed by monitoring Kα fluorescence from oxygen at 525 eV (Myoneni et al., 2002). Liquid water was kept in a He atmosphere at a pressure of 760 Torr and was separated from the high vacuum of a beam line by a Si_3N_4 window. This method can be employed neither for valence electron excitation nor for observation of wide energy range in question here, but it suggests that some day improvements of light sources, detectors, or window materials may make direct observation of VUV absorption possible.

5.3 INELASTIC X-RAY SCATTERING AND GENERALIZED OSCILLATOR STRENGTH

Inelastic x-ray as well as electron scatterings can provide a wealth of information, a part of which is equivalent to the optical oscillator strength distribution. The basic principle of a typical IXS process is sketched in Figure 5.3. A photon of energy $\hbar\omega_0$, momentum $\hbar k_0$, and polarization vector ε_0 impinges upon a target and is inelastically scattered by an angle θ into a photon of energy $\hbar\omega_1$, momentum $\hbar k_1$, and polarization vector ε_0. Concomitantly, the target undergoes a transition from the initial state $|0\rangle$ with energy E_0 to an excited state $|n\rangle$ with energy E_n. The energy $E = \hbar(\omega_0 - \omega_1)$ and the momentum $\hbar\mathbf{q} = \hbar(\mathbf{k}_0 - \mathbf{k}_1)$ are transferred to the target.

The double differential scattering cross section for IXS of isotropic materials such as gases and liquids is expressed as follows (Bonham, 2000):

$$\frac{d^2\sigma}{d\Omega dE} = r_0^2 \left(\frac{\omega_1}{\omega_0}\right)(\varepsilon_0 \cdot \varepsilon_1)^2 \left(\frac{q^2}{E}\right)\frac{df(q,E)}{dE}, \tag{5.6}$$

where

$$\frac{df(q,E)}{dE} = \frac{E}{q^2}\sum_n \left\langle \left| \left\langle n \left| \sum_j^N \exp(i\mathbf{q}\cdot\mathbf{r}_j) \right| 0 \right\rangle \right|^2 \right\rangle_\Omega \delta(E - (E_n - E_0)) \tag{5.7}$$

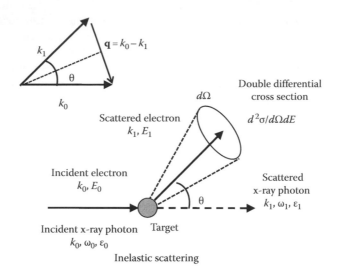

FIGURE 5.3   A schematic diagram of inelastic-scattering processes of x-ray photons and electrons.

defines the generalized oscillator strength (GOS), which is an extension of the optical oscillator strength and is often used in atomic and molecular physics. Here r_0 is the classical electron radius ($=e^2/mc^2$), $\langle - - -\rangle_\Omega$ means orientation average, and $d\Omega$ corresponds to the solid angle. The summation about n is over all the excited states, discrete as well as continuum. N is the number of electrons in the target system and $\mathbf{r}_j$ is the instantaneous position of the jth electron.

Equation 5.6 is quite general. Depending on the magnitude of $qr_j = |\mathbf{q} \cdot \mathbf{r}_j|$, however, IXS spectra show entirely different features. In the case $qr_j \gg 1$, that is, the momentum transfer $q = |\mathbf{q}|$ is large, and/or the electron involved in the process is loosely bound and consequently $r_j = |\mathbf{r}_j|$ is large, the corresponding IXS is the well-known Compton scattering; binding energies of the electrons can be neglected and the spectra are approximated by symmetric parabolas. On the contrary, if the electron is tightly bound and/or q is small and accordingly $qr_j \ll 1$, the binding energy cannot be neglected but energy transfer can. Expanding the exponential into the power series and making use of the orthogonality of the states 0 and n, one can easily see the first non-zero term in the bracket of Equation 5.7 reduces to the same form as that of Equation 5.4. Hence at the limit $qr_j \to 0$, the scattering spectrum becomes identical with the corresponding x-ray absorption spectrum. IXS under these conditions are sometimes called x-ray Raman scattering and can be a substitute of soft x-ray and VUV absorption (Tohji and Udagawa, 1989; Bowron et al., 2000). In between, that is, $qr_j \sim 1$, both the binding energy and momentum transfer should be taken into consideration, providing GOS that depends both on momentum transfer as well as energy.

The GOS at any momentum transfer can be made absolute by using the following Bethe sum rule (Bethe and Jackiw, 1968; Inokuti, 1971):

$$\int \frac{df(q, E)}{dE} dE = N, \tag{5.8}$$

where N is the total number of electrons in the target.

In solid state physics, it is more common to use a quantity slightly different from GOS called the dynamic structure factor $S(\mathbf{q}, E)$, defined by the following equation:

$$S(\mathbf{q}, E) = \frac{q^2}{E} \frac{df(\mathbf{q}, E)}{dE} = \sum_n \left| \left\langle n \left| \sum_j \exp(i\mathbf{q} \cdot \mathbf{r}_j) \right| 0 \right\rangle \right|^2 \delta(E - (E_n - E_0)). \tag{5.9}$$

In isotropic substances, the dielectric function depends only on the magnitude of the momentum transfer q. Dynamic structure factor can be written in terms of the macroscopic dielectric response function $\varepsilon(q, E)$ through the fluctuation-dissipation theorem (Pines, 1964) by

$$S(q, E) = \frac{\hbar q^2}{4\pi^2 e^2 n_e} \, \text{Im} \left[\frac{-1}{\varepsilon(q, E)} \right]_{\hbar\omega = E}, \tag{5.10}$$

where n_e is the average electron density in the material. The function at the right-hand side of Equation 5.10, namely,

$$\text{Im} \left[\frac{-1}{\varepsilon(q, E)} \right] = \frac{\varepsilon_2(q, E)}{\varepsilon_1^2(q, E) + \varepsilon_2^2(q, E)}, \tag{5.11}$$

plays a central role in the theory of the interaction between charged particles and the media, and is often called energy loss function (ELF). While the numerator in Equation 5.11 corresponds to single-particle transitions of an isolated atom or molecule, the denominator accounts for the influence of the condensed phase, that is, shielding or screening effect. The real part of the dielectric function can be derived from ELF by making use of the well-known Kramers–Kronig transformation as follows:

$$Re \left[\frac{1}{\varepsilon(q, E)} \right] = 1 + \frac{2}{\pi} P \int_0^\infty \frac{\text{Im} \left[\varepsilon^{-1}(q, E') \right]}{E'^2 - E^2} E' dE'. \tag{5.12}$$

In short, IXS can provide q- and E-dependent GOS or ELF, ultimately leading to real and imaginary parts of complex dielectric function. In a special case, $qr \rightarrow 0$, the spectrum is essentially identical to the optical spectrum.

5.4 INELASTIC ELECTRON SCATTERING AND GENERALIZED OSCILLATOR STRENGTH

As was described already, photoabsorption of a matter is closely related to interactions between moving charged particles and the matter. Figure 5.1 is, however, an oversimplified picture and in fact energy as well as momentum is transferred to the target. Such a phenomenon is called collision, a schematic diagram of which is also included in Figure 5.3. Theoretically it is well described by the first Born approximation, and the double differential scattering cross section for isotropic substances is given by the following equation (Bethe and Jackiw, 1968; Bonham and Fink, 1974):

$$\frac{d^2\sigma}{d\Omega dE} = \frac{k_1}{k_0 q^2} \frac{4}{E} \frac{df(q, E)}{dE}, \tag{5.13}$$

where the notation is the same as Equation 5.6. That is, GOS can also be obtained from EELS studies. It should be noted that Equation 5.13 is inversely proportional to q^2, and hence the cross section becomes smaller with increasing momentum transfer. EELS spectroscopy is now an established technique (Bonham and Fink, 1974; Egerton, 1996) and has been widely employed for studies on gaseous atoms, molecules, and solid surfaces, and provides data to complement those obtained by IXS as is demonstrated later.

5.5 GOS MEASUREMENTS BY IXS

As already stated, IXS spectroscopy has unique experimental advantages: a vacuum is not required, various kinds of window materials are available, it is free from charge-up phenomenon, contamination of higher order reflection is insignificant, and bulk properties can be obtained. In the past, however, very low scattering intensities hindered one from obtaining accurate IXS spectra.

Recent advancements of synchrotron radiation (SR) facilities have made it possible to carry out a number of experiments that were impossible with conventional x-ray sources, and IXS measurement is one of them. At beam lines of SR facilities, intense, brilliant, polarized, and monochromatic hard x-rays are available. Still, inelastically scattered x-rays are so weak that they must be collected and monochromatized as efficiently as possible. Since the GOS is a normalized quantity, no absolute measurements of the scattering intensities, or solid angle, or sample concentration are required to obtain absolute values. Instead, what is required is the accuracy of relative intensities within a scattering spectrum and is the observation over a wide-enough energy range.

To obtain quality IXS spectra, two approaches have been employed: the use of a spherically or cylindrically bent dispersing crystal to collect as large a solid angle as possible, and the use of a multidimensional detector to improve sensitivity. The two can be combined too. Two examples, a schematic of the beamline X21 of National Synchrotron Light Source in the United States and of BL16X at Photon Factory at KEK in Japan are shown in Figure 5.4.

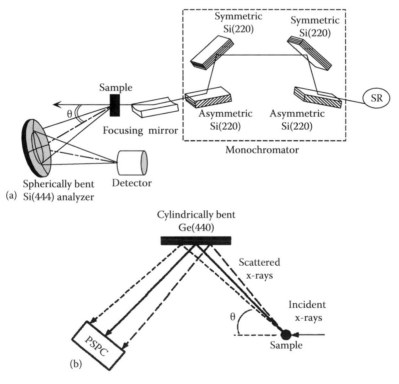

FIGURE 5.4 (a) Schematic diagram of the IXS system at National Synchrotron Light Source beam line X21 of NSLS, Brookhaven National Laboratory, United States (Hayashi et al., 2000). White x-rays from storage ring are monochromatized with a monochromator consisting of four Si(220) crystals and are focused on the sample. A spherically bent Si(444) analyzer is employed as an analyzer. The pass energy of the analyzer is fixed and incident x-ray energy is scanned. (b) Schematic diagram of the analyzer employed at BL16A of Photon Factory, KEK, Japan (Watanabe et al., 1997). Incident x-ray energy is fixed and scattered x-rays are vertically focused and horizontally dispersed with respect to the scattering plane with a cylindrically bent Ge(440) crystal. They are detected with a position-sensitive proportional counter (PSPC) combined with a multichannel analyzer, thus eliminating any effects due to variation in intensity of the incident x-rays.

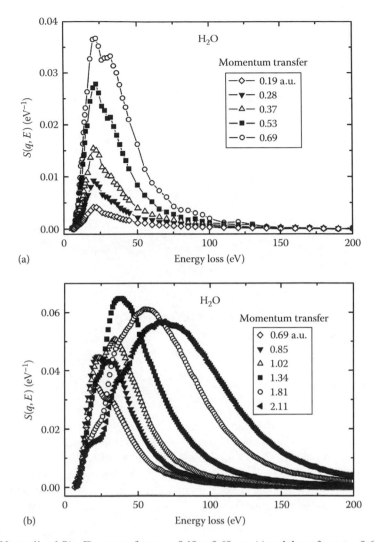

FIGURE 5.5   Normalized $S(q, E)$ spectra from $q = 0.19$ to 0.69 a.u. (a) and those from $q = 0.69$ to 2.79 a.u. (b).

Figure 5.5a and b shows normalized $S(q, E)$ spectra from $q = 0.19$ to 0.69 atomic unit (1 a.u. = $\hbar/a_0$, a_0: Bohr radius = 0.529×10^{-11} m) collected at X21 of NSLS and those from $q = 0.69$ to 2.79 a.u. at BL16A of Photon Factory. For those shown in Figure 5.5a, normalization was made by using the theoretically calculated static structure factor $S(q) = \int S(q,E)dE$ (Wang et al., 1994; Hayashi et al., 1998). In order to normalize the spectra shown in Figure 5.5b where the tails extend to higher energy loss, the data were least-squares fitted to the function A/E^b over the energy region 100–250 eV for q less than 2.9 a.u. and 300–420 eV for q larger than 2.9 a.u. (not shown in the figure), and extrapolated to infinity. The total area was then normalized to a value of 8.34, which corresponds to the total number of valence electrons (8) plus a small correction (0.34) for the Pauli-excluded transitions from the oxygen K shell electrons to the already occupied valence shell orbitals (Chan et al., 1993). The spectral shapes of $S(q,E)$ at $q = 0.69$ a.u. in Figure 5.5a and b agree quite well with each other, indicating that the measurements as well as normalizations were properly made at both SR facilities.

In atomic and molecular physics, it is more common to use GOS or ELF than dynamic structure factor. The three-dimensional representation of ELF or GOS vs. E and q, like the one shown in

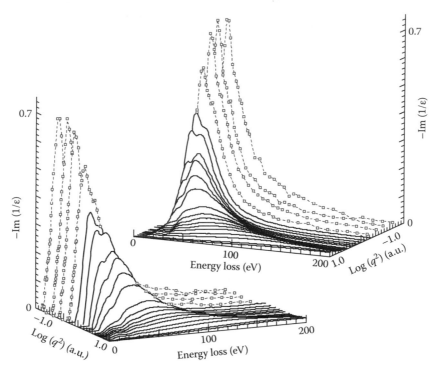

FIGURE 5.6 Experimental Bethe surface of liquid water viewed from two different directions. Squares, data obtained at X21 of NSLS; solid lines, data obtained at BL16A of PF.

Figure 5.6, gives a surface, named Bethe surface by Inokuti, which contains all the information about the inelastic scattering of fast charged particles by the atom or molecule within the first Born approximation (Inokuti, 1971). For small values of momentum transfer, the form of the surface resembles a photoabsorption spectrum, having a peak at around 22 eV. With increasing momentum transfer, the Bethe ridge, that is, the range of the peaks at around $E = \hbar q^2/2m$, shifts to the high energy and the shape gets broader and closer to symmetric, that is, it becomes Compton scattering like.

5.6 OPTICAL LIMIT

Now we have experimentally obtained energy and momentum dependence of dynamic structure factor $S(q, E)$, GOS, or ELF of liquid water in absolute scale for a wide range of energy and momentum transfers. They can easily be converted to dielectric function through Equations 5.10 and 5.12. Unfortunately, however, no other experimental data exist that can be utilized to make comparisons in order to examine the accuracy of the results presented in Figure 5.6. However, comparisons can be made in a special case for $q = 0$, because Equation 5.7 predicts that GOS, ELF, and dielectric function all converge as q approaches zero and become identical with those deduced from optical measurements. In the following, whether or not the optical limit is achieved under the experimental conditions employed is first examined, and subsequently optical constants and functions derived from IXS data under the optical limit are compared with corresponding ones at visible and near-UV regions.

The ELFs of liquid water are plotted for several q's in Figure 5.7, which clearly demonstrates that they converge with decrease in q, approaching the optical limit. Those for $q = 0.28$ a.u. and for $q = 0.19$ a.u. overlap within the experimental error; thus, they both can be regarded as the ELF at $q = 0$, that is, Im $[-1/\varepsilon(0, E)]$. Figure 5.8 shows the ELF at $q = 0.28$ a.u. for a much wider range together with the one derived from a reflectance measurement (Heller et al., 1974). The shape of the two resemble each other, but the maximum values differ considerably, 0.65 vs. 1.1. The difference has a significant effect on the absolute values of dielectric response functions.

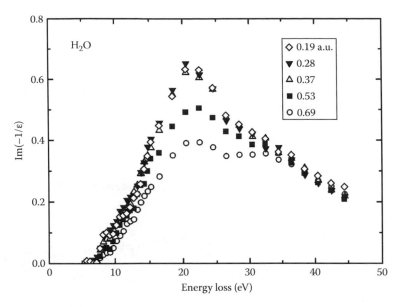

FIGURE 5.7   The ELF, $Im[-1/\varepsilon(q, E)]$, of liquid water plotted for 0.19 a.u. $\leq q \leq$ 0.69 a.u.

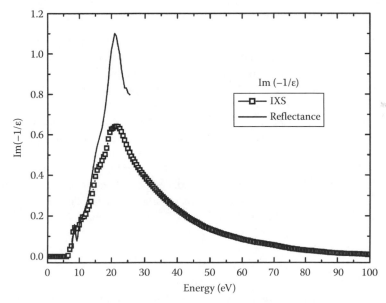

FIGURE 5.8   The ELF, $Im[-1/\varepsilon(0, E)]$, derived from IXS. That from the reflectance measurement (Heller et al., 1974) is also shown.

Real (ε_1) as well as imaginary (ε_2) parts of the dielectric response function $\varepsilon(0, E)$ of liquid water derived by the use of Kramers–Kronig relation are shown in Figure 5.9. Several values of ε_1 calculated from the well-documented refractive index of water in the visible and near-UV region are indicated in the figure by the solid squares. All the squares fall almost exactly on the observed curve, which endorses the accuracy of the present results.

Also shown in Figure 5.9 are the real and imaginary parts of the dielectric response functions derived from reflectance (Heller et al., 1974). As far as the global shapes are concerned, they show a qualitative agreement with those of ours, but again quantitatively there are substantial differences. The most significant difference lies in the behavior of ε_1 where ELF shows a maximum.

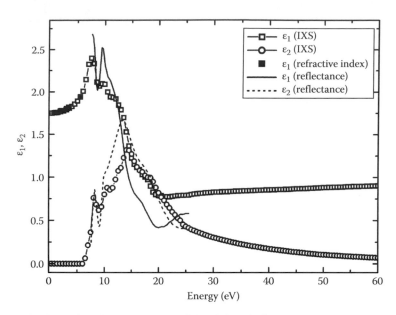

FIGURE 5.9   Real (ε_1) and imaginary (ε_2) parts of the dielectric function of liquid water determined with IXS (Watanabe et al., 1997) and those from reflection (Heller et al., 1974). Closed squares are ε_1 calculated from index of refraction.

While in Heller's results ε_1 displays a distinct valley with a minimum value of 0.42, ours has barely observable minimum with a value of 0.75. It has been a matter of debate whether or not the peak of ELF at 21–22 eV is plasmon-like collective excitation (Heller et al., 1974; Kaplan, 1995). The observation here supports the hypothesis that the peak, though of some collective character, is not due to a plasmon excitation (La Verne and Mozumder, 1993), because ε_1 does not make a deep valley and cannot be regarded to be much smaller than 1 at the peak energy of ELF. As the energy increases, ε_1 gradually approaches 1 from below and ε_2 monotonically decreases toward zero until K absorption starts.

The optical oscillator strength distribution of liquid water from IXS is shown in Figure 5.10 together with estimated uncertainties. The ordinate scale is absolute. Except for a small shoulder at around 8 eV and another less distinct one at around 11 eV, the oscillator strength distribution of liquid water increases sharply with increasing energy, reaches a peak at around 22 eV, and then decreases monotonically.

Figure 5.10 also shows the optical oscillator strength distribution of gaseous water obtained by EELS (Chan et al., 1993). The gas phase spectrum is characterized by sharp absorption bands followed by continuum above ionization threshold (12.6 eV), but the data shown here were obtained at 1 eV resolution and hence the structures are a convolution of many sharp lines. In the liquid state, sharp absorption features are lost and the entire oscillator strength distribution shifts toward the higher energy side. This reflects the fact that in condensed phases, excited electrons are somewhat limited in space (Inokuti, 1991). As the energy increases, the oscillator strength distributions of gaseous water and liquid water get closer as expected. This is because the shielding factor $\varepsilon_1^2 + \varepsilon_2^2$ reaches almost 1 at high energies, reflecting the fact that ε_1 and ε_2 monotonically approach 1 and 0 at higher energies until K absorption starts at 540 eV.

The absolute oscillator strength distributions of two forms of ice can be calculated from the reflection spectra on hexagonal ice at 80 K (Seki et al., 1981; Kobayashi, 1983) and that from the EELS spectrum on amorphous ice at 78 K (Daniels, 1971), both up to about 28 eV. They are shown in the inset of Figure 5.8 together with those of liquid as well as gaseous water. For hexagonal ice, a distinct peak is observed at 8.7 eV, followed by several other structures at higher energy. In contrast, these structures are broadened and less well-defined in amorphous ice. In fact, the oscillator strength distribution of amorphous ice almost overlaps with that of liquid water; both the peak

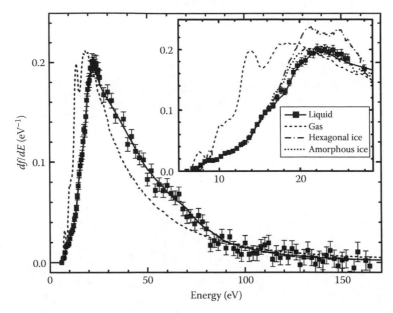

FIGURE 5.10 Oscillator strength distribution of liquid water, gas phase water (Chan et al., 1993), hexagonal ice, and amorphous ice.

energy of the oscillator strength distribution and its magnitude are almost identical for the two systems. This similarity has been confirmed elsewhere (Emfietzoglou et al., 2006a). Amorphous ice bears a similar molecular layout to that of liquid water, and thus the optical spectra of these two media are expected to be rather similar. In other words, taking into consideration that one is from EELS on ice and the other is on liquid water from IXS, the agreement cannot be fortuitous and it presents one of the evidences to prove an accuracy of the present IXS measurements.

5.7 COMPARISON OF GOS FROM EELS AND IXS

GOS can be obtained from EELS experiments on gaseous molecules. GOS studies on gas phase water have been first reported by Lassettre et al. and later by Takahashi et al. (Lassettre and Skerbele, 1974; Lassettre and White, 1974; Takahashi et al., 2000). In the former study, the incident kinetic energy was about 500 eV, and hence the momentum transfer and energy loss ranges were limited to below 1.5 a.u. and 75 eV, respectively. The elastic scattering intensities were employed for normalization. In the latter, the incident electron energy was 3 keV, and hence measurements were carried out for much wider momentum and energy transfer range, that is, up to 3.5 a.u. and 400 eV. The data were normalized by the use of the sum rule. Figure 5.11a and b compares GOS of the two studies at energy losses of 25 and 45 eV. In spite of the difference in incident energy and normalization method, the results agree quite well wherever the comparison can be made.

Figure 5.11a–e also compares GOSs of the gas phase water by EELS and liquid water by IXS for a wider momentum and energy transfer range. It is clear from the figure that they almost coincide in absolute scale at every energy loss examined. Considering that EELS and IXS are completely different experimental techniques and made on water in different phases, the agreement is rather striking and endorses that both measurements were properly carried out. Furthermore, it can be concluded that single-particle excitation prevails over collective excitation in the momentum transfer range studied here (0.69 a.u. $< q <$ 3.59 a.u.). It is rational, because 0.69 a.u. in momentum space corresponds to about 0.08 nm in position space, which is much smaller than the van der Waals radius of water molecule. The results here suggest that "gas phase approximation," where a simple extrapolation of gas phase data to unit density is employed (Paretzke, 1987), is fairly good except for very small q.

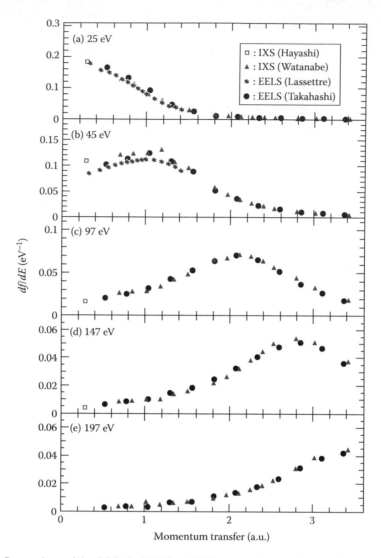

FIGURE 5.11   Comparison of the GOSs by EELS and IXS at several energy loss values.

5.8   BETHE SURFACE OF LIQUID WATER

Dielectric response function $\varepsilon(\omega, q)$ is, as has been stressed already, the central quantity governing various properties of a material and is best presented as a Bethe surface. Bethe surfaces of molecules can in principle be obtained by ab initio calculations, but in practice, the calculation from the first-principles is a formidable task for realistic materials. Hence, apart from the atomic hydrogen and the free electron gas, theoretical construction of Bethe surface of liquid water has so far been made semi-empirically (Dingfelder and Inokuti, 1999; Emfietzoglou et al., 2005, 2006b, 2008a,b). Most studies start with an analytic representation of the experimental optical data $\text{Im}[-1/\varepsilon(0, E)]$, which fulfill the sum rule. Subsequently, momentum dependence is introduced by assuming various dispersion models. Though physically plausible models are contrived and elaborated, calculations based on the processes described above had generally been viewed as tentative because there had been no way to test the accuracy of the q-dependence until the data presented in Figure 5.6 were reported.

Now experimental Bethe surface of liquid water is known between 0.19 a.u. $< q < 2.79$ a.u and $0 < E < 150\,\text{eV}$. Figure 5.12a shows the experimental Bethe surface with much finer meshes

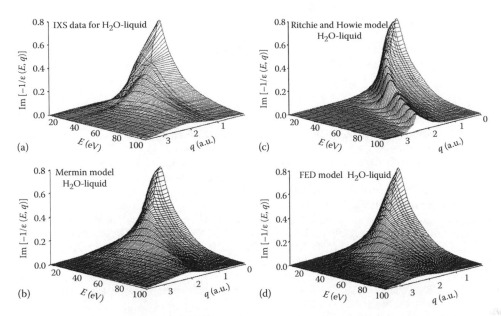

FIGURE 5.12 Observed Bethe surface of liquid water by IXS (a), and calculated ones by Mermin model (b), Ritchie and Howie model (c), and fully extended Drude model (d). (Reprinted from Emfietzoglou, D. et al., *Nucl. Instrum. Methods B*, 266, 1154, 2008b. With permission).

calculated by extrapolation of the IXS data. Figure 5.12b–d presents three examples of calculated Bethe surface of liquid water (Emfietzoglou et al., 2008b). In all the calculations, the ELF shown in Figure 5.8 is fitted by a superposition of five Drude functions (corresponding to excited states) and five derivative Drude functions (ionized states), and subsequently different q-dependence models were assumed. Although all the models reproduced the global characteristics of the experimentally observed Bethe surface, there are obvious differences in the way the Bethe ridge develops.

Figure 5.12c employs the very popular Ritchie and Howie model that assumes a simple quadratic dispersion (Ritchie and Howie, 1977). It is evident that the Bethe ridge is emphasized too much compared with that from IXS observation, Figure 5.12a. That the Bethe ridge is not so steep is supported also from EELS data in Figure 5.11, and hence other models must be looked for. Two model calculations that almost reproduce the experimental Bethe surface are presented and are shown in Figure 5.12b and d. Dependences of some physical quantities on varied optical data and on momentum dispersion models have been studied (Emfietzoglou and Nikijoo, 2005).

5.9 CONCLUDING REMARKS

The dielectric response function $\varepsilon(q, \omega)$ fully describes the interaction of matter and photons or charged particles. For volatile liquids, only IXS experiment can provide $\varepsilon(q, \omega)$ data for finite values of momentum transfer, but because of weak intensity of IXS, not many studies have been reported so far. In condensed phases, intermolecular interactions tend to broaden and smear characteristic low-energy peaks often observed for gas phase molecules. Hence, what is required to extend IXS studies to various other condensed systems is not high resolution but high sensitivity over a wide energy region. In this decade, impressive progress has been made in various aspects of x-ray techniques, and every year, new, more improved monochromators have been commissioned (Bergmann et al., 2002; Hayashi et al., 2004, 2008; Welter et al., 2005; Fister et al., 2006; Huotari et al., 2006; Hudson et al., 2007). It will now be possible to apply IXS to probe dielectric properties of various liquids of interest such as supercritical fluids, ionic fluids, electrolytes, as well as high-pressure gases.

ACKNOWLEDGMENTS

The authors are grateful to Dr. N. Watanabe of IMRAM, Tohoku University and Dr. C.-C. Kao of NSLS for their collaboration.

REFERENCES

Bergmann, U., Glatzel, P., and Cramer, S. P. 2002. Bulk-sensitive XAS characterization of light elements: From x-ray Raman scattering to x-ray Raman spectroscopy. *Microchem. J.* 71: 221–230.

Bethe, H. A. and Jackiw, R. 1968. *Intermediate Quantum Mechanics*, 2nd edn. New York: Benjamin.

Bonham, R. A. 2000. Corrections to the x-ray incoherent scattering factor to obtain the total inelastic x-ray scattering. *J. Mol. Struct. (Theochem)* 527: 103–111.

Bonham, R. A. and Fink, M. 1974. *High Energy Electron Scattering*. New York: van Nostrand Reinhold.

Bowron, D. T., Krisch, M. H., Barnes, A. C., Finney, J. L., Kaprolat, A., and Lorenzen, M. 2000. X-ray-Raman scattering from the oxygen K edge in liquid and solid H_2O. *Phys. Rev. B* 62: R9223–R9227.

Chan, W. F., Cooper, G., and Brion, C. E. 1993. The electronic spectrum of water in the discrete and continuum regions. Absolute optical oscillator strengths for photoabsorption (6–200 eV). *Chem. Phys.* 178: 387–400.

Daniels, J. 1971. Bestimmung der optischen konstanten von eis aus energie-verlustmenssungen von schnellen elektronen. *Opt. Commun.* 3: 240–243.

Dingfelder, M. and Inokuti, M. 1999. The Bethe surface of liquid water. *Radiat. Environ. Biophys.* 38: 93–96.

Dingfelder, M., Hantke, D., Inokuti, M., and Paretzke, H. G. 1998. Electron inelastic-scattering cross sections in liquid water. *Rad. Phys. Chem.* 53: 1–18 and references therein.

Egerton, R. F. 1996. *Electron Energy-Loss Spectroscopy in the Electron Microscope*, 2nd edn. New York: Plenum Press.

Emfietzoglou, D. and Nikijoo, H. 2005. The effect of model approximations on single-collision distributions of low-energy electrons in liquid water. *Radiat. Res.* 163: 98–111.

Emfietzoglou, D., Cucinotta, F. A., and Nikjoo, H. 2005. A complete dielectric response model for liquid water: A solution of the Bethe Ridge problem. *Radiat. Res.* 164: 202–211.

Emfietzoglou, D., Nikjoo, H., Petsalakis, I. D., and Pathak, A. 2006a. A consistent dielectric response model for water ice over the whole energy-momentum plane. *Nucl. Instrum. Methods B* 256: 141–147.

Emfietzoglou, D., Nikjoo, H., and Pathak, A. 2006b. A comparative study of dielectric response function models for liquid water. *Radiat. Prot. Dosim.* 122: 61–65.

Emfietzoglou, D., Pathak, A., and Moscovitch, M. 2008a. Modeling the energy and momentum dependent loss function of the valence shells of liquid water. *Nucl. Instrum. Methods B* 230: 77–84.

Emfietzoglou, D., Abril, I., Garcia-Molina, R., Petsalakis, I. D., Nikjoo, H., Kyriakou, I., and Pathak, A. 2008b. Semi-empirical dielectric descriptions of the Bethe surface of the valence bands of condensed water. *Nucl. Instrum. Methods B* 266: 1154–1161.

Fister, T. T., Seidler, G. T., Wharton, L., Battle, A. R., Ellis, T. B., Cross, J. O., Macrander, A. T., Elam, W. T., Tyson, T. A., and Qian, Q. 2006. Multielement spectrometer for efficient measurement of the momentum transfer dependence of inelastic x-ray scattering. *Rev. Sci. Instrum.* 77: 063901-1–063901-7.

Hayashi, H., Watanabe, N., Udagawa, Y., and Kao, C.-C. 1998. Optical spectra of liquid water in vacuum UV region by means of inelastic x-ray scattering spectroscopy. *J. Chem. Phys.* 108: 823–825.

Hayashi, H., Watanabe, N., Udagawa, Y., and Kao, C.-C. 2000. The complete optical spectrum of liquid water measured by inelastic x-ray scattering. *Proc. Natl. Acad. Sci.* 97: 6264–6266.

Hayashi, H., Kawata, M., Takeda, R., Udagawa, Y., Watanabe, Y., Takano, T., Nanao, S., and Kawamura, N. 2004. A multi-crystal spectrometer with a two-dimensional position-sensitive detector and contour maps of resonant Kβ emission in Mn compounds. *J. Electron Spectrosc. Relat. Phenom.* 136: 191–197.

Hayashi, H., Azumi, T., Sato, A., and Udagawa, Y. 2008. A cartography of Kβ resonant inelastic x-ray scattering for lifetime-broadening-suppressed spin-selected XANES of α-Fe_2O_3. *J. Electron Spectrosc. Relat. Phenom.* 168: 34–39.

Heller, Jr., J. M., Hamm, R. N., Birkhoff, R. D., and Painter, L. R. 1974. Collective oscillation in liquid water. *J. Chem. Phys.* 60: 3483–3486.

Hudson, A. C., Stolte, W. C., Lindle, D. W., and Guiemin, R. 2007. Design and performance of a curved-crystal x-ray emission spectrometer. *Rev. Sci. Instrum.* 78: 053101-1–053101-5.

Huotari, S., Albergamo, F., Vanko, G., Verbeni, R., and Monaco, G. 2006. Resonant inelastic hard x-ray scattering with diced analyzer crystals and position-sensitive detectors. *Rev. Sci. Instrum.* 77: 053102-1–053102-6.

Ikehata, A., Higashi, N., and Ozaki, Y. 2008. Direct observation of the absorption bands of the first electronic transition in liquid H_2O and D_2O by attenuated total reflectance far-UV spectroscopy. *J. Chem. Phys.* 129: 234510 and references therein.

Inokuti, M. 1971. Inelastic collisions of fast charged particles with atoms and molecules—The Bethe theory revisited. *Rev. Mod. Phys.* 43: 297–347.

Inokuti, M. 1983. Radiation physics as a basis of radiation chemistry and biology. In *Applied Atomic Collision Physics*, Vol. 4, *Condensed Matter*, S. Datz (ed.), pp. 179–236. New York: Academic Press, Inc.

Inokuti, M. 1986. VUV absorption and its relation to the effects of ionizing corpuscular radiation. *Photochem. Photobiol.* 44: 279–285.

Inokuti, M. 1991. How is radiation energy absorption different between condensed phase and the gas phase. *Radiat. Eff. Defects Solids* 117: 163–167.

Kaplan, I. G. 1995. The track structure in condensed matter. *Nucl. Instrum. Methods B* 105: 8–13.

Kerr, G. D., Cox, Jr., J. T., Painter, L. R., and Birkhoff, R. D. 1971. A reflectometer for studying liquids in the vacuum ultraviolet. *Rev. Sci. Instrum.* 42: 1418–1422.

Kerr, G. D., Hamm, R. N., Williams, M. W., Birkhoff, R. D., and Painter, L. R. 1972. Optical and dielectric properties of water in the vacuum ultraviolet. *Phys. Rev. A* 5: 2523–2527.

Kobayashi, K. 1983. Optical spectra and electronic structure of ice. *J. Phys. Chem.* 87: 4317–4321.

Kutcher, G. L. and Green, A. E. S. 1976. Model for energy deposition in liquid water. *Radiat. Res.* 67: 408–425.

La Verne, J. A. and Mozumder, A. 1993. Concerning plasmon excitation in liquid water. *Radiat. Res.* 133: 282–288.

Lassettre, E. N. and Skerbele, A. 1974. Generalized oscillator strengths for 7.4 eV excitation of H_2O at 300, 400, and 500 eV kinetic energy. Singlet-triplet energy differences. *J. Chem. Phys.* 60: 2464–2469.

Lassettre, E. N. and White, E. R. 1974. Generalized oscillator strengths through the water vapor spectrum to 75 eV excitation energy; electron kinetic energy 500 eV. *J. Chem. Phys.* 60: 2460–2463.

Myneni, S., Luo, Y., Naslund, L. A., Cavalleri, M., Ojamae, L., Ogasawara, H., Pelmenschikov, A., Wernet, P., Vaterlein, P., Heske, C., Hussain, Z., Pettersson, L. G. M., and Nilsson, A. 2002. Spectroscopic probing of local hydrogen-bonding structures in liquid water. *J. Phys. Condens. Matter* 14: L213–L319.

Nikjoo, H. and Uehara, S. 2003. Track structure studies of biological systems. In *Charged Particle and Photon Interactions with Matter*, A. Mozunderand and Y. Hatano (eds.), pp. 491–531. New York: Marcel Dekker.

Nikjoo, H., Uehara, S., Emfietzoglou, D., and Cucinotta, F. A. 2006. Track-structure codes in radiation research. *Radiat. Meas.* 41: 1052–1074 and references therein.

Painter, L. R., Birkhoff, R. D., and Arakawa, E. T. 1969. Optical measurements of liquid water in the vacuum ultraviolet. *J. Chem. Phys.* 51: 243–251.

Paretzke, H. G. 1987. Radiation track structure theory. In *Kinetics of Nonhomogeneous Processes*, G. R. Freeman (ed.), pp. 89–170. New York: John Wiley & Sons.

Pines, D. 1964. *Elementary Excitations in Solids*. New York: Benjamin.

Ritchie, R. H. 1982. Energy losses by swift charged particles in the bulk and at the surface of condensed matter. *Nucl. Instrum. Methods* 198: 81–91.

Ritchie, R. H. and Howie, A. 1977. Electron-excitation and optical-potential in electron-microscopy. *Philos. Mag.* 36: 463–481.

Sakurai, J. J. 1994. *Modern Quantum Mechanics, Rev. Edn.* Reading, MA: Addison-Wesley.

Seki, M., Kobayashi, K., and Nakahara, J. 1981. Optical spectra of hexagonal ice. *J. Phys. Soc. Jpn.* 50: 2643–2648.

Smith, J. D., Cappa, C. D., Wilson, K. R., Messer, B. M., Cohen, R. C., and Saykally, R. J. 2004. Energetics of hydrogen bond network rearrangements in liquid water. *Science* 306: 851–853.

Takahashi, M., Watanabe, N., Wada, Y., Tsuchizawa, S., Hirosue, T., Hayashi, H., and Udagawa, Y. 2000. Bethe surfaces and x-ray incoherent scattering factor for H_2O studied by electron energy loss spectroscopy. *J. Electron Spectrosc. Relat. Phenom.* 112: 107–114.

Tohji, K. and Udagawa, Y. 1989. X-ray Raman scattering as a substitute for soft-x-ray extended x-ray-absorption fine structure. *Phys. Rev. B* 39: 7590–7594.

Wang, J., Tripathi, A. N., and Smith, Jr., V. H. 1994. Chemical binding and electron correlation effects in x-ray and high energy electron scattering. *J. Chem. Phys.* 101: 4842–4854.

Watanabe, N., Hayashi, H., and Udagawa, Y. 1997. Bethe surface of liquid water determined by inelastic x-ray scattering spectroscopy and electron correlation effects. *Bull. Chem. Soc. Jpn.* 70: 719–726.

Watanabe, N., Hayashi, H., and Udagawa, Y. 2000. Inelastic x-ray scattering study on molecular liquid. *J. Phys. Chem. Solid* 61: 407–409.

Welter, E., Machek, P., Drager, G., Bruggmann, U., and Froba, M. 2005. A new x-ray spectrometer with large focusing crystal analyzer. *J. Synch. Radiat.* 12: 448–454.

Wernet, P., Nordlund, D., Bergmann, U., Cavalleri, M., Odelius, M., Ogasawara, H., Naslund, L. A., Hirsch, T. K., Ojamae, L., Glatzel, P., Pettersson, L.G.M., and Nilsson, A. 2004. The structure of the first coordination shell in liquid water. *Science* 304: 995–999.

Williams, M. W., Arakawa, E. T., and Inagaki, T. 1991. Optical and dielectric properties of materials relevant to biological research. In *Handbook on Synchrotron Radiation*, Vol. 4, S. Ebashi, S. M. Kochand, and E. Rubenstein (eds.), pp. 95–145. Amsterdam, the Netherlands: Elsevier Science Publisher.

6 New Directions in *W*-Value Studies

Isao H. Suzuki

High Energy Accelerator Research Organization
Tsukuba, Japan
and
National Institute of Advanced Industrial Science and Technology
Tsukuba, Japan

CONTENTS

6.1 INTRODUCTION

The interaction of energetic particles with matter has different kinds of effects on the biological system, materials of industrial importance, environments, and so forth. Thus, researches using various detection techniques have been performed for many years in order to obtain a clear understanding of these effects on several ionizing radiations. The number of electron–ion pairs formed in matter upon the complete slowing down of ionizing radiation is an important quantity characterizing the ability of initiating a chemical reaction chain within a gas and/or of inducing certain radiation effects in biological systems. Therefore, the study of expected values of ion pairs has drawn the attention of many researchers toward fields such as radiation chemistry, charged-particle spectroscopy, and dosimetry (Inokuti, 1975; Samson and Haddad, 1976; ICRU, 1979; Combecher, 1980;

Inokuti et al., 1980; Waibel and Grosswendt, 1983; Grosswendt, 1984; Krajcar-Bronic et al. 1988; Kowari et al. 1989; Kimura et al., 1991; Pansky et al., 1996; Krajcar-Bronic, 1998). A fundamental quantity, W-value, is defined as the average energy expended by radiation to produce an electron–ion pair during complete slowing-down in the matter. A variety of measurements have been performed on W-values of atoms and molecules for different radiations (Samson and Haddad, 1976; ICRU, 1979; Combecher, 1980; Waibel and Grosswendt, 1983; Krajcar-Bronic et al., 1988; Pansky et al., 1996). Theoretical approaches have pursued clarification of physical insight into W-values using the Monte Carlo technique together with an analytical method using the Spencer–Fano equation (Inokuti et al., 1980; Grosswendt, 1984; Kowari et al., 1989; Kimura et al., 1991). These studies have shown that the W-value is insensitive to quality and energy for radiations above several keV (Inokuti, 1975; ICRU, 1979; Inokuti et al., 1980; Grosswendt, 1984; Kowari et al., 1989; Kimura et al., 1991; Krajcar-Bronic, 1998). A limited number of measurements were reported for W-values of some gases in energy regions below 1 keV (Samson and Haddad, 1976; Combecher, 1980; Waibel and Grosswendt, 1983; Krajcar-Bronic et al., 1988; Pansky et al., 1996). The W-value of electrons increases with decreasing electron energy. This behavior is explained as a phenomenon originating from the finding that the ratio of the probability of excitation to that of ionization in a single collision of an electron with a gas molecule increases as the electron energy decreases. Further, W-values of photons exhibit a variation in energy dependence in sub-keV regions where electrons ejected upon photoabsorption have low energies. This variation arises from the contributions of different electron orbitals responsible for the initial photoionization process, which critically depends on the photon energy.

Samson and Haddad (1976) measured the W-value of Xe for photons in the vacuum ultraviolet radiation region and found a distinct increase near the 4d ionization threshold. A Japanese group reported nonlinearity effects in the pulse height distribution from the Xe-filled proportional counter near the Xe K-shell photoabsorption edge (Tsunemi et al., 1993). These nonlinearities had also been found by other groups (Dias et al., 1997). A Portuguese group ascribed these phenomena as atomic shell effects using a simulation calculation, which takes into account photoabsorption from different electron orbitals. A variation in photon W-values in the C K-edge region was discovered for hydrocarbon molecules using monochromatized synchrotron radiation (Suzuki and Saito, 1985a,b, 1987; Saito and Suzuki, 1986). The observed oscillatory structure was well reproduced with a model based on the 1s electron transition, which yields Auger electrons with energies lower than photoelectrons emitted from valence orbitals. These results emphasized a significant effect of the initial photoionization step on the W-value in the soft x-ray region. However, a thin-window proportional counter was used in the measurements, and thus the W-values obtained in these studies were not on an absolute scale. In order to measure W-values on an absolute scale in the region of the soft x-ray, studies on photon W-values of rare gas atoms were carried out using a multielectrode ion chamber, together with monochromatized synchrotron radiation (Saito and Suzuki, 2001a,b; Suzuki and Saito, 2001). The multielectrode ion chamber technique allowed measurements of absolute photon intensity for soft x-rays. The W-value obtained was compared with the data for low-energy electrons (Combecher, 1980) and with the values calculated by a model based on multiple photoionization effects of atoms.

In this chapter, new directions in researches on W-values of several gases are discussed. Section 6.2 briefly outlines the historical background of these studies, and the W-values in the low-energy region are described to some extent in Section 6.3. Section 6.4 presents the results of hydrocarbon molecules near the C K-edge in detail together with a simple model of oscillatory behavior in the energy dependence. Then, absolute W-values for rare gas atoms are exhibited in the sub-keV region, combined with a precise model of oscillatory variation in the energy dependence.

6.2　HISTORICAL BACKGROUND

Some decades ago, an international commission had surveyed a number of studies on W-values of matter from various points of view and published a report on this survey as ICRU Report No. 31 (1979). Table 6.1 provides the data of several atoms and molecules for some ionizing radiations,

TABLE 6.1
W-Values of Electron, Proton, and
α-Particle for Several Gases Quoted
from ICRU Report No. 31 (1979) (in eV)

Gas	Electron	Proton	α-Particle
H_2	36.5	—	36.4
CH_4	27.3	30.5	29.1
C_2H_4	25.8	28.8	27.9
Ar	26.4	26.7	26.3
Kr	24.4	23.0	24.1
Air	33.9	35.2	35.1

Note: See Report No. 31 by ICRU (1979) regarding
uncertainties of listed values.

which were extracted from this report on account of relevance to this chapter. It is found that the *W*-value is almost insensitive to radiation quality and is close to or slightly higher than twice of the first ionization energy. This finding suggests that energy transfer from radiation to gases takes place in excitation into neutral states at about half the probability, as well as in ionization also at about half the probability.

In order to explain the characteristics of *W*-values theoretically, Inokuti and coworkers (1975, 1980) derived the formula for degradation spectra of electrons using the Spencer–Fano equation. They obtained deep insight into the *W*-value behavior, because the degradation spectra could be utilized for the evaluation of various important physical quantities for radiation effects. Further, they pointed out that an analytical expression of the electron *W*-value (W_e), given in the following, was very useful for understanding energy dependences of *W*-values:

$$W_e(E) = \frac{E \cdot W}{(E - U)},$$ (6.1)

where
    E denotes incident electron energy
    W indicates the *W*-value at a sufficiently high energy
    U means a sub-ionization electron energy, which depends on the character of electron collisions
        with target gases

The character of the term *U* was discussed in consideration of the first ionization energy and some electronically excited states. Moreover, an oscillatory variation as a function of the number of carbon atoms in normal alkanes was pursued, resulting in the finding that super-excited states of these molecules play an important role in branching into neutral dissociation or ionization (Kimura et al., 1991). The Monte Carlo simulation technique was applied for the interpretation of dependence of *W*-values on incident electron energy by Waibel and Grosswendt (1978, 1983, 1991) and Grosswendt (1984). This technique was also used for the analysis of nonlinearity in a proportional scintillation counter filled with rare gases in the keV energy region, by Santos et al. (1991) and Dias et al. (1997).

Krajcar-Bronic, Srdoc, and their coworkers (Srdoc, 1973; Srdoc and Obelic, 1976; Krajcar-Bronic and Srdoc, 1994; Krajcar-Bronic, 1998) observed *W*-values of photons for several atoms and molecules using some proportional counters, together with electron *W*-values. They mainly utilized characteristic x-rays from metal targets and examined the *W*-values of gas mixtures as well as pure gases in low-energy regions. Then it was found that obtained *W*-values indicate higher values than those for high-energy photons like γ-rays (Srdoc, 1973; Srdoc and Obelic, 1976). Further, they evaluated several models for *W*-values of gas mixtures with regard to the applicability of an additivity rule (Krajcar-Bronic and Srdoc, 1994).

Energy dependences of *W*-values were precisely measured for several gases using an electron beam in the low-energy region, in particular below 1 keV (see Section 6.3). Combecher (1980) found that *W*-values increase with decreasing incident energy for all gases studied. This behavior was clarified elaborately using a Monte Carlo calculation by Waibel and Grosswendt (1978, 1983, 1991).

On the photon *W*-values in keV energy regions, several research groups observed nonlinearity effects (discontinuities) in the pulse height distribution from rare-gas-filled proportional counters near the inner-shell photoabsorption edges (Kowalski et al., 1985; Jahoda and McCammon, 1988; Santos et al., 1991; Tsunemi et al., 1993; Dias et al., 1997). Santos and colleagues (1991) carried out Monte Carlo simulation calculations on the measured average electron yields, which were obtained using a gas proportional scintillation counter filled with Xe. In their calculation, ionization and excitation collisions with Xe atoms for all paths of produced electrons were taken into consideration, together with the photoionization of Xe, vacancy cascade Auger electron emission, electron shake-off, and K-shell fluorescence. They ascribed these discontinuities as atomic shell effects through photoabsorption from different electron shells and clarified variations in *W*-values, Fano factors, and widths of pulse height distributions at inner-shell photoabsorption edges. Pansky and coworkers (1996, 1997) obtained accurate *W*-values for energies around 1 keV using an electron-counting technique combined with a low-pressure proportional counter, in which Fano factors and *W*-values were derived through a Monte Carlo simulation on the detection process. Their results of hydrocarbon molecules and their mixtures with rare gases indicated a variation near inner-shell absorption edges of constituent gases. The *W*-value of Xe for photons was measured in the vacuum ultraviolet radiation region using a double ion chamber technique, in which a distinct increase was found at the 4d ionization threshold (Samson and Haddad, 1976).

W-values for ions were studied for several gases as targets using different ion beams, including heavy ions in keV to MeV energy regions (Miller and Boring, 1974; Huber et al., 1985). Huber et al. (1985) obtained characteristic behavior in the ion *W*-values for some gases in the use of various ion beams at low energies down to 0.5 keV. Further, they compared measured data with the *W*-value behavior calculated using the Thomas–Fermi potential.

6.3   CHARACTERISTICS OF *W*-VALUES IN THE ENERGY REGION BELOW 1 keV

Low-energy electrons play an important role in a variety of fields of radiation. They are created in large numbers during the passage of energetic photons and all sorts of charged particles through matter. They have an energy transfer capability comparable to that of protons and α-particles, and are responsible for the greater part of radiation damage in all kinds of ionizing radiation. Therefore, understanding the *W*-value for low-energy electrons is particularly important for the following reasons. First, they are necessary in calculating the absorbed energy from the detected number of ions produced in micro-dosimetric experiments. Second, they are a useful tool to evaluate theoretical radiation-transport calculations for low-energy electrons, because sufficient and accurate data sets of excitation and differential ionization cross sections of electron–molecule collisions are not available. Third, using the stopping power and the differential ionization cross sections for high-energy ions, they enable us to calculate high-energy ion *W*-values, which are required for dosimetric measurements. Several measurements were carried out for the determination of the *W*-value in low-energy regions (Cole, 1969; Srdoc, 1973; Smith and Booz, 1978). Owing to difficulty of handling electron beams in low energies, experimental data were scattered considerably and seemed to have large uncertainties.

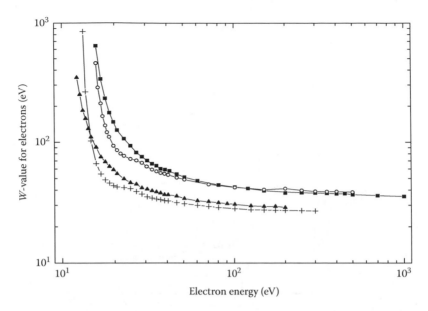

FIGURE 6.1 Electron *W*-values for several gases in the sub-keV region. +: Kr, ○: H₂, ▲: ethylene, ■: air. Curves connecting data points are drawn for better presentation. This figure is made from the tables given by Combecher (1980).

Combecher (1980) measured energy dependences of *W*-values for a number of atoms and molecules including mixed gases using an electron beam in the energy region below 0.5 keV. He examined and tuned very carefully several experimental factors, like operation of the electron gun and pulsed ion collection, which resulted in the measurement of accurate *W*-values with uncertainties smaller than 2%. It was found that *W*-values increase with decreasing incident energy for all gases that were utilized as target samples. Typical results are shown in Figure 6.1, which exhibits energy dependences of electron *W*-values for Kr, H₂, C₂H₄, and a mixed gas (air) in the low-energy region (Combecher, 1980). These energy dependences indicate very similar behavior, suggesting that the analytical equation, Equation 6.1, holds for these gases.

Waibel and Grosswendt observed electron *W*-values for some gases as well as spatial energy dissipation profiles and electron ranges using an ion chamber technique. Figure 6.2 exhibits a schematic view of the parallel-plate chamber utilized by them (Waibel and Grosswendt, 1978). Electron beams emerge from the electron gun, pass through some diaphragms, and enter the chamber, which is equipped with a set of collecting plates and guard rings. On examination of several values for gas

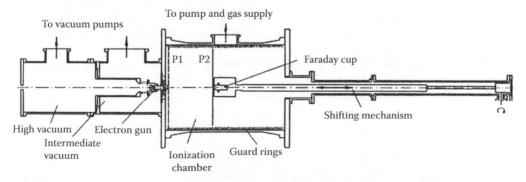

FIGURE 6.2 Experimental setup for measurements of electron *W*-values of gases. P1 and P2: charge-collecting plates. (Reproduced from Waibel, E. and Grosswendt, B., *Radiat. Res.*, 76, 241, 1978.)

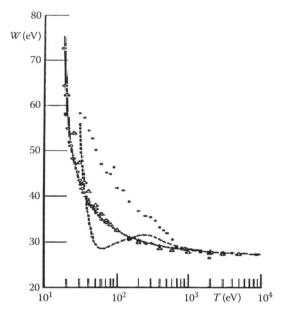

FIGURE 6.3 *W*-values (in eV) for methane as a function of electron energy. +: Experiment (Waibel and Grosswendt, 1983), ○: experiment (Combecher, 1980), ●: experiment (Smith and Booz, 1978), △ and solid curve: calculation (Waibel and Grosswendt, 1983), broken curve: calculation (Dayashankar, 1977), horizontal arrow (27.3 eV): *W*-value at high energies (ICRU, 1979). (Reproduced from Waibel, E. and Grosswendt, B., *Nucl. Instrum. Method*, 211, 487, 1983.)

density, plate-to-plate distance, and so forth, *W*-values were elaborately estimated from the measured currents at the plates. In particular, they precisely examined the backscattering effects of incident electron beams, which resulted in the determination of accurate *W*-values with uncertainties smaller than 1%. The obtained characteristics of energy dependence for some atoms and molecules (Waibel and Grosswendt, 1978, 1983, 1991) have similar features to those found by Combecher (1980). Further, they carried out a Monte Carlo simulation for measured *W*-values on the basis of data for ionization cross sections and excitation cross sections for electron collisions with molecules. The characteristic behavior of increasing *W*-value with decreasing energy was well clarified by their simulation, in particular, for the *W*-value of methane. Figure 6.3 exhibits a comparison of *W*-values for methane between experimental and calculated data (Waibel and Grosswendt, 1983). The experimental data obtained by Waibel and Grosswendt, and by Combecher (1980) are in good agreement with the simulation, but earlier measured data by Smith and Booz (1978) and previous results calculated by Dayashankar (1977) are different from this calculation. According to these studies, this behavior in the low-energy region is clearly understood by the fact that the neutral excitation probability increases with decreasing electron energy in comparison with that for ionization in collision of electrons with molecules.

By using the ion chamber technique, Samson and Haddad (1976) measured the *W*-value of Xe for electrons that were ejected from Xe by the irradiation of photons of 24–90 eV from a spark-discharge source. They reduced the acceleration of ejected electrons by the applied field for the sake of the ion collection technique, which utilized the cylindrical chamber having an off-axis collector electrode. The result obtained was well reproduced with the analytic equation by Inokuti (1975), Equation 6.1. Further, Samson and Haddad obtained the *W*-value for photons using the same apparatus. Figure 6.4 exhibits the photon *W*-value for Xe as a function of energy. It has been found that the *W*-value remains constant between 22 and 65 eV, with a distinct increase at the 4d ionization threshold of 67.5 eV. The measured result was discussed in consideration of energy levels of Xe ionic states and of partial photoionization cross sections for different shells.

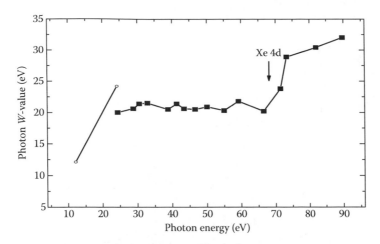

FIGURE 6.4   Photon *W*-value of Xe in the vacuum ultraviolet radiation region. ■: Data points. Curves connecting data are drawn for better presentation. The solid line in the low-energy region denotes expected values from physical consideration. (Modified from Samson, J.A.R. and Haddad, G.N., *Radiat. Res.*, 66, 1, 1976.)

6.4   VARIATION IN ENERGY DEPENDENCE OF PHOTON *W*-VALUES FOR HYDROCARBON MOLECULES

6.4.1   INTRODUCTION

On the basis of progress in electron beam techniques, studies on electron energy loss spectra provided excitation spectra of several molecules with a high resolution in the energy region of inner-shell transitions, as well as those for rare gas atoms (Backx and van der Wiel, 1975; Hitchcock and Brion, 1977; Key et al., 1977; King et al., 1977; Ungier and Thomas, 1984, 1985). According to these spectra, the first loss peak at the lowest energy is usually observed for a vacant molecular orbital, which is the lowest unoccupied valence orbital, for example, π^* orbital in a linear molecule. The second peak comes from the transition into a Rydberg orbital in most cases, which is followed by those into higher Rydberg orbitals and ionization continuum (Backx and van der Wiel, 1975; Hitchcock and Brion, 1977; Key et al., 1977; King et al., 1977). Similar spectra were also measured using monochromatized soft x-rays from synchrotron radiation sources (Eberhardt et al., 1976; Brown et al., 1978; Hayaishi et al., 1984). Several years ago, highly resolved photoabsorption spectra were obtained, which were compared with the electron energy loss spectra. Synchrotron radiation studies have often shown energy resolving powers higher than those obtained using the electron beam technique. An example of the synchrotron radiation spectra is displayed in Figure 6.5, where a photoabsorption spectrum of N_2 is exhibited in the N K-shell threshold region (Chen et al., 1989). The spectrum shows a variety of peak structures corresponding to transitions of the π^* orbital, Rydberg orbitals, and ionization continua. Some peak structures have components of vibrational excitation; in particular, π^* excitation shows six or more vibrational components. Above 409.94 eV (N 1s ionization energy), some structures induced from double excitation and shape resonance are found. Further electron emission processes were studied for many gases (Dixon et al., 1978; Rye et al., 1978, 1980; Tamenori et al., 2004); in particular, photoelectron spectra using monochromatic synchrotron radiation were observed for a number of atoms and molecules. Figure 6.6 shows a photoemission spectrum of Kr with a moderate resolution at a photon energy of 410 eV, where many peaks are seen for some inner-shell photoelectrons (3s, 3p doublet, 3d, and satellites) and for Auger electrons from the M_{23}-shell hole states (Tamenori et al., 2004). These Auger electrons, that is, electrons ejected through Coster–Kronig transitions, correspond to the decays of $3p^{-1}$ states into $3d^{-1}4p^{-1}$ and $3d^{-1}4s^{-1}$ states.

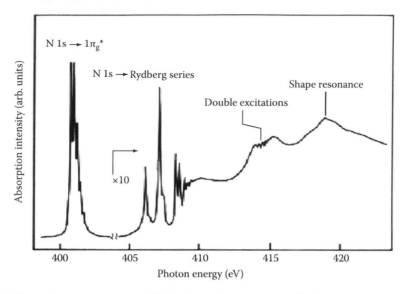

FIGURE 6.5 Photoabsorption spectra of N_2 in the soft x-ray region. Excitation of 1s electrons into the π^* orbital, and 3s, 3p, and 3d Rydberg orbitals, as well as structures by double excitation and shape resonance are seen. Note that the ordinate scale is expanded above 404 eV. (Based on Chen, C.T. et al., *Phys. Rev. A*, 40, 6737, 1989.)

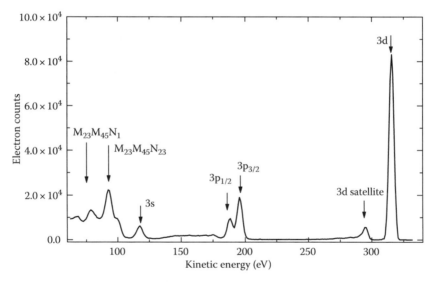

FIGURE 6.6 Photoelectron spectra of Kr observed in the soft x-ray region. The photon energy is 410 eV. Arrows denote 3s, $3p_{1/2}$, $3p_{3/2}$, and 3d photoelectron peaks and structures by Auger electrons. (Based on Tamenori, Y. et al., *J. Phys. B*, 37, 117, 2004.)

As described in Section 6.3, Samson and Haddad (1976) identified a distinct variation of the photon *W*-value for Xe around 70 eV. This variation was ascribed to a change in the ejected electron energy, which was caused by photoionization of the inner-shell, Xe 4d orbital. A similar variation is expected for hydrocarbon molecules in the C K-edge region, because photoexcitation is supposed to create several types of excited states in this energy region, yielding electrons emitted with various energies. Recent advance in synchrotron radiation researches provides us with monochromatic soft x-rays having continuous wavelength ranges. Therefore, we can perform studies on photon *W*-values using monochromatic soft x-rays combined with precise steps of energy scan.

Here, we present results of photon *W*-values for ethylene, methane, and propane obtained using a proportional counter, because hydrocarbon molecules used in a gas counter indicate a clear pulse height distribution in the detection of ionizing radiation (Suzuki and Saito, 1985a,b, 1987; Saito and Suzuki, 1986). First, the outcome for ethylene is presented and explained in detail, and then those for the other molecules are given.

6.4.2 MEASUREMENTS AND ANALYSIS

Synchrotron radiation from the storage ring at the Advanced National Institute of Industrial Science and Technology (to which the Electrotechnical Laboratory was reorganized in 2001) was mono-chromatized by a plane-grating monochromator (resolving power: about 150) (Tomimasu et al., 1983; Saito and Suzuki, 1986). The storage ring was operated in a lower-energy mode than the normal mode for suppressing higher-order radiation. Figure 6.7 shows the experimental setup for measurements of photon *W*-values for molecules (Suzuki and Saito, 1985b, 1987; Saito and Suzuki, 1986). The monochromatized photons entered a gas-flow proportional counter (Manson-4) with a thin window (thickness: about 1500 A). Sample gases, ethylene, methane, and propane, were supplied to the counter at about 5×10^3 Pa. Signals from the counter were amplified and accumulated in a multichannel analyzer. The resultant pulse height distributions of the signals were transferred to a personal computer. Using this computer, the pulse height distributions were analyzed to obtain the average pulse height. An example of pulse height distribution is displayed in Figure 6.8, where the distribution was observed under a methane pressure of 5×10^3 Pa at a photon energy of 388 eV.

The pulse height distribution can be approximately expressed by the following equation (Breyer, 1973):

$$f(z) = \frac{z^{x-1} \cdot \exp(-z/z_a)}{z_a^x} \cdot \Gamma(x). \tag{6.2}$$

where
    z indicates the number of electrons amplified by the electric field within the counter
    z_a denotes the average gas amplification factor
    x is the average number of the electrons initially produced by the radiation effect of a photon
    Γ is the gamma function

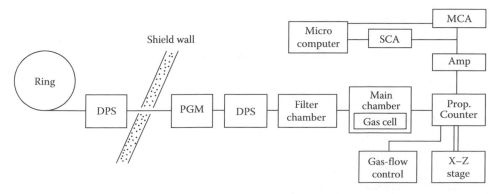

FIGURE 6.7 Schematic experimental setup of measurements of photon *W*-values using a proportional counter. DPS, differential pumping system; PGM, plane-grating monochromator; SCA, single-channel analyzer; Prop.Counter, proportional counter; Amp, amplifier; MCA, multichannel analyzer.

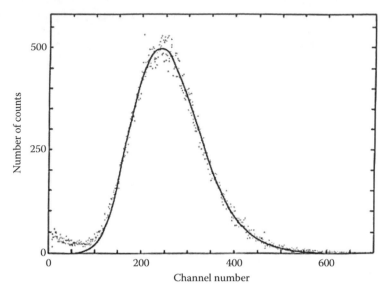

FIGURE 6.8 Pulse height distribution of a proportional counter for a monochromatic soft x-ray (388 eV). Dots denote measured data, and the solid curve indicates the profile calculated with a fitting technique. (Reproduced from Saito, N. and Suzuki, I.H., *Chem. Phys.*, 108, 327, 1986.)

The distributions calculated using Equation 6.2 were matched with those obtained by observation. The average pulse height, P, is given by

$$P(E_p) = \int z \cdot f(z)dz = z_a \cdot x = \frac{z_a \cdot E_p}{W_p(E_p)}, \tag{6.3}$$

where
E_p indicates the energy of a photon
W_p denotes the W-value of photons (Suzuki and Saito, 1985a,b, 1987; Saito and Suzuki, 1986)

Using Equation 6.3, relative W-values were obtained for photons over the energy range of C K-edges. Any change in P owing to a drift in the gas pressure was canceled by measuring the P several times at the same photon energy in the course of the experiment. There may be photoabsorption by hydrocarbon molecules in which only neutral fragments are produced, but in which no ion pair is formed. This process should alter the W-value, although the present technique using a proportional counter cannot count these types of processes. At present, these processes are assumed to be negligible, in consideration of the results of the ionization yield of methane and other molecules in the vacuum ultraviolet radiation region (Backx and van der Wiel, 1975).

6.4.3 RESULT OF ETHYLENE

The energy dependence of the W-value is shown with solid squares for photon energies from 269 to 324 eV in Figure 6.9 (Suzuki and Saito, 1987). The solid curve indicates a calculated profile based on a model described below. The measured W-value gradually decreases from 269 to 283 eV as the photon energy increases. The W-value rises to a peak at 285 eV, drops to a sharp valley at 287 eV, and again goes up a broad peak at 290 eV. Above a minimum at 293 eV, the data show a near-linear dependence on the photon energy between 293 and 301 eV. Above the latter energy, the W-value seems to have a slightly decreasing trend.

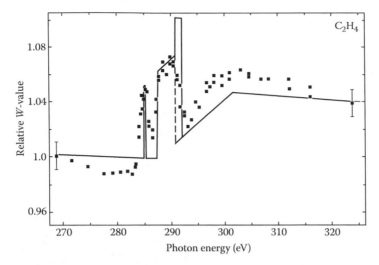

FIGURE 6.9 Photon *W*-value for C_2H_4 as a function of soft x-ray energy. Solid squares indicate measured data. The solid curve and a broken line denote the calculated profile based on the model explained in Section 6.4.4. (Reproduced from Suzuki, I.H. and Saito, N., *Bull. Chem. Soc. Jpn.*, 60, 2989, 1987.)

It is important to consider the oscillatory variation of the *W*-value of ethylene in connection with the photoexcitation of the C 1s electron. Hitchcock and Brion (1977) observed electron energy loss spectra of ethylene for the 1s electron excitation in the energy resolution of 0.2–0.5 eV. The observed spectra show strong resonant transitions of the 1s electron to the $1b_{2g}$ orbital with vibrational excitations at 284.68, 285.04, and 285.50 eV. Ethylene is a pseudo-linear molecule, which has a π-type orbital. This b_{2g} is often called a π* orbital. Transitions into the 3s orbital and the 3p orbital were found to occur at 287.4 and 287.8 eV, respectively. The transition into the 3s orbital is forbidden in a spherical field. However, this transition is allowed in a molecule although the transition strength is weak. Transitions to higher Rydberg orbitals and to the ionized state appeared above 288.3 and 290.6 eV, respectively. They further found shake-up states at 292.6 and 295.2 eV. (A shake-up state is an ionized state having an excited valence electron.) Eberhardt and coworkers (1976) observed an electron yield spectrum of ethylene in the same energy region by the use of monochromatized synchrotron radiation. The same transitions were identified although the resolution was lower and the observed values deviated slightly (0.3–0.7 eV).

6.4.4 Simple Model for Explanation of Variation in *W*-Value

A comparison between the inner-shell excited states and the energy dependence of the *W*-value provides us with a model for interpretation as follows. For simplicity of explanation, energy regions are divided into seven parts. In each energy region, only one photoionization process is considered, and others are neglected owing to very small probability. Figure 6.10 illustrates a change in electron configuration during subsequent electron emission following initial photoabsorption, together with the final charge state of the present molecule. Table 6.2 lists energy values necessary for the calculation of the *W*-value on the basis of the present model. These energies have been estimated from available data in the literature, for example, Auger electron spectra and photoelectron spectra (Rye et al., 1978; Liegener, 1985; Kimura et al., 1981; Suzuki and Saito, 1987).

(i) *Below 284.7 eV*: Only a valence electron is ejected by photoabsorption, because the photon cannot excite an inner-shell electron. The average energy of this ejected electron is $E_p - E_v$, where E_v is the average binding energy of the valence electrons. Ethylene has six orbitals for valence electrons, which are $1b_{1u}$, $1b_{1g}$, $3a_g$, $1b_{2u}$, $2b_{3u}$, and $2a_g$ (Dixon et al., 1978;

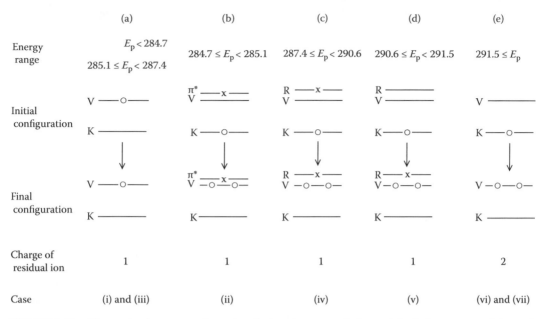

FIGURE 6.10 Change in electron configuration during electron emission and final charge state of the molecule. K, V, R, and π^* indicate the K-shell, the valence orbitals, the Rydberg orbitals, and the π^* orbital, respectively. x: occupation by an electron, ○: hole in the valence orbital or the K-shell.

TABLE 6.2
List of Energy Values Necessary for Calculation of the Photon *W*-Value of Ethylene in the Model at Section 6.4.4 (in eV)

W-value for sufficiently high energy radiation	W	25.8
Average energy of sub-ionization electron for an electron with an energy E		
U_1	for $E \geq 200$	22
U_2	for $E \leq 33$	11
First ionization energy	E_f	10.5
Average of ionization energy for valence electron	E_v	16.1
Ionization energy of the K-shell	E_K	290.6
Average energy of Auger electron	E_A	245.8
Average energy of de-excitation electron from the $(1s)^{-1}(\pi^*)^1$ state	$E_{A\pi}$	255
Average energy of de-excitation electron from the $(1s)^{-1}(R)^1$ state	E_{AR}	255
Average energy gain of Auger electron in a post-collision interaction	α	$E_p - 288$

Note: π^* and R indicate $1b_{2g}$ and Rydberg orbitals, respectively.
E_p denotes photon energy.

Kimura et al., 1981). The value of E_v is assumed to be 16.1 eV. The ejected photoelectrons ionize ambient molecules and produce a number of ion pairs. On the other hand, the W-value of low-energy electrons (W_e) can be approximately given by the following equation, as indicated in Section 6.2 (Inokuti, 1975; Samson and Haddad, 1976; Combecher, 1980):

$$W_e(E) = \frac{E \cdot W}{(E - U)}, \tag{6.4}$$

where
   W denotes the W-value for radiation with a sufficiently high energy for ethylene (25.8 eV) (ICRU, 1979)
   U indicates the average energy of sub-ionization electrons
   E is the energy of an emitted electron

The sub-ionization electrons are the electrons that do not contribute to ionization in a system receiving irradiation. In the present instance, the average number of ion pairs, N_e, is given by

$$N_e = \frac{E}{W_e} = \frac{(E - U)}{W}. \tag{6.5}$$

Using Equations 6.4 and 6.5, the photon W-value is expressed by the following equation:

$$W_p(E_p) = \frac{E_p}{(1 + N_e)} = \frac{E_p \cdot W}{(E_p + W - E_v - U_1)}, \tag{6.6}$$

where U_1 is assumed to be 22 eV from the data for the W-value of low-energy electrons (Combecher, 1980). Multiple photoionization of the valence electrons is assumed to occur only at negligibly low probability at present.

(ii) *Between 284.7 and 285.1 eV*: The 1s electrons can be excited to the π^* orbital ($1b_{2g}$), and a hole is formed in the C K-shell. After this excitation, an Auger transition is supposed to occur, because fluorescence yield is extremely low in a light element. A variety of Auger transitions have been observed in ethylene (Rye et al., 1978; Liegener, 1985). The average energy of Auger electrons, E_A, is estimated here to be 245.8 eV, according to the spectrum by Rye et al. (1978). However, the average Auger electron energy in the present instance is supposed to be somewhat higher than 245.8 eV owing to the existence of that electron in the π^* orbital. This average energy, $E_{A\pi}$, is assumed to be 255 eV, by analogy from a resonance-type Auger electron spectrum of CO (Ungier and Thomas, 1984, 1985). The W-value is expressed by

$$W_p(E_p) = \frac{E_p}{(1 + N_e)} = \frac{E_p \cdot W}{(E_{A\pi} + W - U_1)}. \tag{6.7}$$

When Equation 6.7 is compared with Equation 6.6, $E_{A\pi}$ (255 eV) is lower than $E_p - E_v$ (e.g., 270.9 eV for the photon with an energy of 287 eV). This fact implies that the W-value at 284.7 eV becomes higher than that at 287 eV. In the present model, it is assumed that the photoabsorption probability involving valence electrons can be negligibly low in this photon energy region on account of the spectra of the inner-shell excitation (Eberhardt et al., 1976; Hitchcock and Brion, 1977; Brown et al., 1978; Henke et al., 1993). This assumption is also applied to the energy region above 287.4 eV (case iv to case vii).

(iii) *Between 285.1 and 287.4 eV*: Photons can excite only valence electrons, because there is no inner-shell excited state in this region. Ejected photoelectrons ionize ambient molecules and produce a number of ion pairs. W_p is expressed by Equation 6.6.

(iv) *Between 287.4 and 290.6 eV*: Photons can excite inner-shell electrons to the Rydberg orbitals, and then the formed inner-hole is filled with a valence electron through the subsequent Auger transition. The average energy of the Auger electrons, E_{AR}, is assumed to be 255 eV.

(v) *Between 290.6 and 291.5 eV*: Inner-shell electrons can be ejected from the molecule by photons, but these ejected photoelectrons leave slowly. In the case of CO, the K-shell photoelectron with low energy is retrapped by the molecular ion when a second electron with higher energy is ejected by an Auger transition (post-collision interaction [PCI]) (Key et al., 1977; Hayaishi et al., 1984). Since there has been no detailed study on these molecules, the slow photoelectron in the present case is assumed to be retrapped at the instant of the Auger transition. The Auger electron obtains a slightly higher energy owing to the existence of the slow photoelectron than the normal Auger electron. This energy gain of the Auger electron is equal to the energy loss of the trapped electron. The average energy of the Auger electron is assumed here to be $245.8 + \alpha$ eV ($\alpha = E_p - 288$ eV), because of no available data in molecules. This PCI is tentatively assumed to occur between 290.6 and 291.5 eV, in consideration of the energy giving a minimum W-value (293 eV) and of the photon energy resolution.

(vi) *Between 291.5 and 301.1 eV*: The inner-shell electrons can be ionized and can promptly escape from the molecular field. This photoelectron ejection is followed by a normal Auger transition. The photoelectron has an energy of $E_p - 290.6$ eV. However, this electron cannot ionize other ambient molecules because the electron energy is lower than the first ionization energy of the valence electron ($E_f = 10.5$ eV). The Auger electron (E_A) has an average energy of 245.8 eV.

(vii) *Above 301.1 eV*: The photoelectron released from the inner shell has an energy higher than the first ionization energy of ethylene. This electron and the Auger electron produced subsequently can ionize ambient molecules. The W-value is expressed by the following equation:

$$W_p(E_p) = \frac{E_p}{(2 + N_{e1} + N_{e2})} = \frac{E_p \cdot W}{(E_p + E_A + 2W - E_K - U_1 - U_2)}, \tag{6.8}$$

where U_2 denotes the average energy of the sub-ionization electrons for the electron with kinetic energy between 11 and 33 eV. U_2 is assumed to be 11 eV on the basis of the study by Combecher (1980).

By using the present model, the relative W-value has been calculated and shown with the solid curve in Figure 6.9. The resolving power of the monochromator was not included in this calculation. Energy dependence of the calculated W-value is in agreement with the experimental results. Therefore, the present model is essentially correct.

There is a slight discrepancy between the experimental data and the calculated result at a few energies. The discrepancy at 286 eV is ascribed to the low resolving power of the present monochromator, while that between 291 and 293 eV is presumed to originate from two reasons, other than low resolving power. First, although the energy of the electron may be lowered through a post-collision interaction below 291.5 eV, there is a possibility that some fraction of slow photoelectrons are not retrapped by the molecular ion. If this is the case, the final configuration is that having two holes in the valence orbital. The calculated W-value becomes lower than that shown in Figure 6.9 between 290.6 and 291.5 eV. Second, slightly distorted pulse height distributions in the proportional counter around 291 eV have been observed. The width of the distribution was broader than those at other energies. This phenomenon is supposed to be brought about by a contamination of the used grating in the soft x-ray monochromator, which resulted in a decrease of output intensity near the C K-edge. This should cause the effective resolving power to be lower. From the minimum at 293 eV (not at 290.6 eV), it can be said that the post-collision interaction takes place just above the C K-edge. However, it is impossible at present to estimate the size of the energy region where it occurs. The broken line between 290.6 and 291.5 eV in Figure 6.9 shows the calculated result in case of no occurrence of the post-collision interaction.

The reason why the experimental data are slightly lower than the calculated curve near 280 eV is not clear at present. The photoabsorption cross section near 280 eV is very low owing to no excitation of inner-shell electrons. The electron cloud induced by photoabsorption is distributed more homogeneously in the proportional counter near 280 eV than other energies. This fact may have a small effect on the average pulse height distribution through a possible change in the gas amplification factor.

The shake-up states were reported to exist at 292.6 and 295.2 eV in the literature (Hitchcock and Brion, 1977). In contrast to the transition to the π^* orbital, the transition probability to these states was considerably lower than to the normal continuum state. Thus, this transition was not expected to have an appreciable effect on the *W*-value. The experimental *W*-value does not show a change at the energies in Figure 6.9 within experimental uncertainty.

6.4.5 METHANE AND PROPANE

Methane is usually used as a counter gas for the measurement of ionizing radiation. Since the photoabsorption spectrum of methane shows a fine structure near the C K-edge due to transitions of C 1s electrons (Eberhardt et al., 1976; Brown et al., 1978), the *W*-value of methane may also have a fine structure related to these transitions.

Figure 6.11 shows the relative *W*-value of methane for photon energies between 240 and 400 eV (Saito and Suzuki, 1986). Open and solid circles denote the data obtained by using Al and the copolymer of vinyl chloride and vinyl acetate (VYNS) windows, respectively. The *W*-value gradually

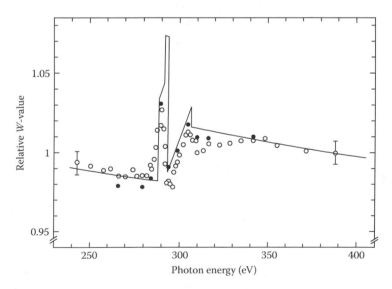

FIGURE 6.11 Photon *W*-value for CH_4 as a function of soft x-ray energy. Solid and open circles indicate measured data using the VYNS window and the Al window, respectively. The solid curve denotes the calculated profile based on the model explained in Section 6.4.5. (Reproduced from Saito, N. and Suzuki, I.H., *Chem. Phys.*, 108, 327, 1986.)

decreases from 240 to 280 eV as the photon energy increases. Two maxima can be seen at about 290 and 305 eV, accompanied by a minimum at about 295 eV. From 310 to 400 eV, the W-value remains essentially constant.

An electron energy loss spectrum and a photoabsorption spectrum of methane show the four peaks at 287, 288, 289.4, and 289.8 eV (Brown et al., 1978; Wight and Brion, 1974). A strong resonance is located at 288 eV, and the three other peaks are weak in intensity. The transitions of an electron from the inner C K-shell to Rydberg orbitals give rise to these peaks. The K-shell ionization energy is 290.8 eV (Pireaux et al., 1976). The oscillatory structure shown by the circles in Figure 6.11 is presumed to connect with the structure caused by transitions of the K-shell electron.

A model, which is the same as that for ethylene, seems applicable for interpreting the structure in the energy dependence of the W-value (Saito and Suzuki, 1986; Suzuki and Saito, 1987). Since methane has no transition into the π^* orbital, the oscillatory structure is simpler than that for ethylene. Then, the energy region of methane classified by the photoabsorption process is 5, which is smaller than that for ethylene. The W-value calculated is illustrated with a solid curve in Figure 6.11. The calculation has reproduced the measured data essentially well, although it overestimates the effect on PCI just above the C K-edge.

Propane has K-shell ionization energy of 290.6 eV (Pireaux et al., 1976). Transitions of the 1s electrons to unoccupied orbitals are supposed to take place slightly below the ionization threshold. These types of excitation and ionization probably cause molecules to change into different types of states of the molecular ion. The W-value is expected to be affected by these ion states near the K-edge. The relative W-values derived from the measurements are shown with open triangles for photon energies from 260 to 360 eV in Figure 6.12 (Suzuki and Saito, 1985b). The solid curve indicates a calculated profile based on the model mentioned below. The measured W-value slightly decreases with increasing energy from 260 to 280 eV. The W-value shows a sharp rise at 286 eV, goes to a maximum at 291 eV, and then decreases to a minimum at 294 eV. A linear dependence of the W-value data on the photon energy is seen from 294 to 304 eV. The W-value is essentially constant above 304 eV.

In order to clarify the origin of the variation in the W-value shown in Figure 6.12, Suzuki and Saito (1985b) observed the photoabsorption spectrum of propane around the C K-edge. Although the inner-shell excited states are not separated from one another owing to the low resolving power of the monochromator used, the following items are possibly discernible. The absorption efficiency is

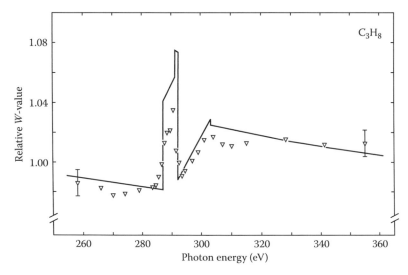

FIGURE 6.12 Photon W-value for C_3H_8 as a function of soft x-ray energy. Open triangles indicate measured data. The solid curve denotes the calculated profile based on the model explained in Section 6.4.5. (Reproduced from Suzuki, I.H. and Saito, N., *Bull. Chem. Soc. Jpn.*, 58, 3210, 1985b.)

low below 285 eV, becomes high at about 287 eV, and reaches a maximum at around 291 eV. Above 291 eV, this efficiency gradually decreases with increasing photon energy. Judging from the inner-shell excitation spectra of methane and ethane (Eberhardt et al., 1976), the high efficiency at 287 eV can probably be ascribed to a transition of C-atomic orbitals, 1s → 3s. In the cases of methane and ethane, the transition (1s → 3s) is low in intensity. This type of transition is supposed to occur with a higher probability in propane because of the poorer symmetry than those for methane and ethane. This transition is followed by transitions to higher excited states (e.g., 1s → 3p) and to ionized states. Bearing in mind the photoabsorption spectrum, a model similar to that for methane is proposed for the interpretation of the energy dependence of the *W*-value (Suzuki and Saito, 1985b). The calculated curve has well reproduced the measured data and the model is essentially correct.

6.5 OSCILLATORY VARIATION IN ENERGY DEPENDENCE OF PHOTON *W*-VALUES FOR RARE GAS ATOMS AND ATOMIC SHELL EFFECTS

6.5.1 INTRODUCTION

Based on the progress in synchrotron radiation researches, many photoabsorption spectra have been measured in a high resolution of photon energy, and electron emission spectra have also been observed in a high resolution of electron energy (Svensson et al., 1988; Chen et al., 1989; Okada et al., 2005; Kato et al., 2007). Further, several ions with various charges, produced by the photo-ionization of atoms and molecules, have been observed by means of mass spectrometry (Hayaishi et al., 1984; Saito and Suzuki, 1992, 1994, 1997; Tamenori et al., 2004; Suzuki et al., 2006). In some of these studies, average charge states have been obtained for ions produced from rare gas atoms irradiated with soft x-rays (Saito and Suzuki, 1992, 1994; Tamenori et al., 2004). Some measurements were attempted for the determination of the absolute intensity of soft x-rays using a multi-electrode ion chamber and other techniques as per the requirements in photon science and photon engineering (Rabus et al., 1997; Saito and Suzuki, 1998, 1999). Photoionization cross sections of rare gas atoms were obtained on an absolute scale in the soft x-ray region using a sophisticated ion chamber technique (Suzuki and Saito, 2002, 2005). These developments have enabled us to determine absolute photon *W*-values of some gases in the sub-keV energy regions, even though it is very difficult for the proportional counter technique to provide accurate *W*-values on the absolute scale.

In this section, photon *W*-values for rare gas atoms are presented, which were obtained using monochromatic synchrotron radiation combined with the multielectrode ion chamber (Saito and Suzuki, 2001a,b; Suzuki and Saito, 2001). In deriving these quantities from observed photoion currents, average charge states, which are called γ-values here, have been utilized. Branching ratios of produced ion charges in individual photon energies are used in the calculation of the *W*-value on the basis of a precise model, which takes atomic shell effects into consideration. First, a measurement method and an analysis technique are given, and the result for Kr is discussed in detail in comparison with a calculated profile (Saito and Suzuki, 2001b). Then, results for Ar and Xe are presented (Saito and Suzuki, 2001a; Suzuki and Saito, 2001).

6.5.2 MEASUREMENTS

Synchrotron radiation at the AIST (National Institute of Advanced Industrial Science and Technology) was dispersed by a Grasshopper grating monochromator (Saito and Suzuki, 2001a,b; Suzuki and Saito, 2001). The monochromatic photons entered the multielectrode ion chamber, whose shape is cylindrical, 65 mm in diameter and 1300 mm in length. In order to reduce higher-order radiation and scattered lights, the electron energy of the storage ring was often lowered from 750 MeV to 250–600 MeV, and some thin films were inserted into the soft x-ray beam. Figure 6.13 shows the schematic diagram of the measurement system for *W*-values (Saito and Suzubi, 2001b). The window of the ion chamber consists of a thin VYNS foil and a circular aperture of 2 mm diameter.

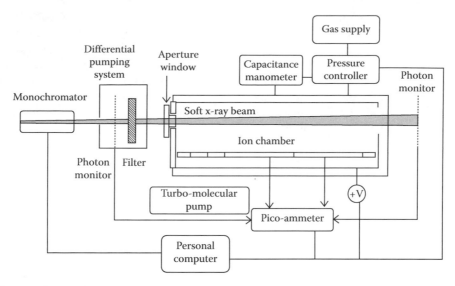

FIGURE 6.13 Schematic of an experimental system for measurements of the photon W-value using a multi-electrode ion chamber. The personal computer controls the monochromator, the voltage supply (+V), and the pressure controller. (Reproduced from Saito, N. and Suzuki, I.H., *Radiat. Res.*, 156, 317, 2001b.)

The chamber contains a set of six electrodes for detection of photoion currents and for suppression of end distortion in electric fields. The sample gases of Kr, Ar, and Xe (research grade purity) were supplied at about 10^{-2}–10^3 Pa, the densities of which were monitored with a capacitance manometer and controlled with an automatic valve system. A positive potential was applied to the outer cylindrical electrode. The photoion current at each inner electrode was transferred to a pico-ammeter and then to a personal computer. The incident intensity of the soft x-rays was continuously monitored with the Au-plated Ni-mesh, illustrated in Figure 6.13.

6.5.3 ANALYSIS

Let us consider that a total of N electrons are produced upon absorption of an incident soft x-ray of energy E_p in a sufficiently high density gas (Saito and Suzubi, 2001b). Then, the photon W-value, W_p, is expressed by

$$W_p = \frac{E_p}{N}. \qquad (6.9)$$

The number of electrons, n, produced in the ion chamber with a certain gas density, p, is the sum of contributions from two factors. The first factor is the initial ionization through atomic multiple photoionization effects, and the second is the secondary ionization effects originating from collisions of emitted electrons with ambient gases in the chamber. Thus,

$$n(p) = \gamma + \delta(p), \qquad (6.10)$$

where
  γ denotes the number of electrons ejected from the atom having absorbed a soft x-ray quantum
  δ is the number of electrons secondarily produced

The value of γ was determined previously for several rare gas atoms in the soft x-ray region using a time-of-flight mass spectrometer technique (Saito and Suzuki, 1992; Suzuki and Saito, 1992).

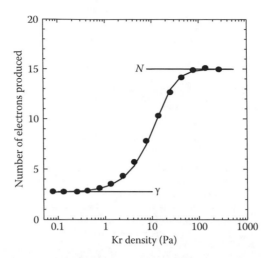

FIGURE 6.14 Number of electrons produced in the ion chamber as a function of Kr gas density. Data are plotted at a photon energy of 400.3 eV. See Section 6.5.3 for *N* and γ. (Reproduced from Saito, N. and Suzuki, I.H., *Radiat. Res.*, 156, 317, 2001b.)

As the gas density decreases, the parameter value δ approaches zero and the *n* value becomes equal to the γ value. On the other hand, the δ value approaches a constant value and the *n* value reaches a plateau value of *N*, as the gas density increases. An example of results measured at a photon energy of 400 eV is shown in Figure 6.14, where the number of electrons produced is represented as a function of the Kr density (Saito and Suzubi, 2001b). From the curve in Figure 6.14, constant values at low- and high-density limits are easily estimated with a high precision. When the purity of the soft x-ray beam was not very high, the plateau value did not show up clearly. This situation has been overcome by the improvement of the photon purity, as indicated above.

As described previously on the absolute measurement of soft x-rays (Saito and Suzuki, 1998, 1999), the photoion current, *i*, in the ion chamber under a certain gas density is given as

$$i = enI \exp(-l\sigma p)\{1 - \exp(-L\sigma p)\}. \tag{6.11}$$

Here, *e*, *I*, σ, *L*, and *l* denote the elementary charge, the photon absolute intensity, the photoabsorption cross section, the length of the electrode, and the length of the insensitive region at the front end, respectively. Then, the photoion current at a sufficiently low gas density is

$$i = e\gamma I \exp(-l\sigma p)\{1 - \exp(-L\sigma p)\}. \tag{6.12}$$

Similarly, the photoion current at a sufficiently high density is

$$i = eNI \exp(-l\sigma p)\{1 - \exp(-L\sigma p)\}. \tag{6.13}$$

The photoabsorption cross section has been obtained with ion currents from the two successive electrodes, i_1 and i_2, under appropriate gas densities as

$$\sigma = \frac{1}{L}\frac{d}{dp}\left\{\ln\left(\frac{i_1}{i_2}\right)\right\}. \tag{6.14}$$

By using Equations 6.12 and 6.14, we have obtained the absolute intensity of the incident soft x-ray. Then, the value of *N*, that is, the total number of electrons produced, has been derived from the

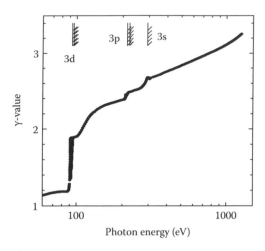

FIGURE 6.15 γ-Value (average charge state) of Kr as a function of photon energy. Bars with hatching denote the ionization thresholds of 3d, 3p, and 3s electrons. (Reproduced from Saito, N. and Suzuki, I.H., *Radiat. Res.*, 156, 317, 2001b.)

obtained photon intensity using Equation 6.13. Finally, the photon W-value has been calculated from the derived N value, according to Equation 6.9.

Figure 6.15 shows the value of γ for Kr in the region of 60–1100 eV (Saito and Suzuki, 2001b). These values have been derived from branching ratios into several charge states in multiple photoionization (Suzuki and Saito, 1992). It is found that the γ-value shows a steep jump at the 3d electron ionization threshold (93.8 eV). This finding can be interpreted in terms of the formation of doubly charged Kr ions through the 3d electron ionization. Even below the 3d threshold, the γ-value is slightly higher than unity, showing that two valence electrons are appreciably ejected through absorption of one photon. The value of γ increases gradually above the 3d threshold, and then exhibits small upward steps at the 3p (214.4 eV) and 3s (292.8 eV) electron ionization thresholds. Above the 3s threshold, the γ-value acquires an additional feature. This feature probably comes from a shake-off process induced by the inner-shell ionization (Saito and Suzuki, 1994, 1997; Suzuki et al., 2006).

6.5.4 RESULT OF Kr

The photon W-value of Kr is shown on an absolute scale as a function of photon energy from 85 to 1000 eV in Figure 6.16 (Saito and Suzuki, 2001b). The present data are denoted with solid circles. The bars with hatching indicate the ionization thresholds of 3d, 3p, and 3s electrons. The open square data points give the W-value for electrons, W_e, which were reported earlier and are shown in Figure 6.1 (Combecher, 1980). The solid curve represents the result calculated from the model described below. The measured W_p exhibit considerably lower values than those for electrons below 91 eV. The sharp peak structure appears just below the 3d threshold, which is shown in Figure 6.17.

The value of W_p steeply increases at the 3d ionization threshold and decreases slightly between 120 and 210 eV. This W_p shows a small oscillation near the 3p and 3s ionization thresholds, and then essentially reaches a constant value above 400 eV. The W_e for electrons is close to the present measured value of W_p in energies between 200 and 300 eV. The variation in W_p near the ionization thresholds is supposed to originate from inner-shell ionization effects, which is similar to those for hydrocarbon molecules near the C 1s ionization thresholds, as described in Section 6.4.4 (Suzuki and Saito, 1985a,b, 1987; Saito and Suzuki, 1986). Then, the mechanism underlying these phenomena is assumed to have a close relation with the model proposed in regard to these molecules.

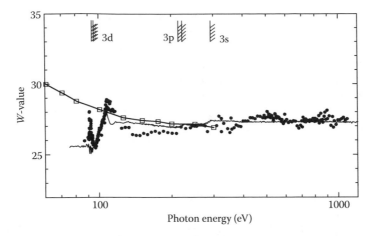

FIGURE 6.16 Photon *W*-value for Kr as a function of photon energy. Solid circles show data by Saito and Suzuki, and open squares are data for electrons by Combecher (1980). The solid curve represents the photon *W*-values calculated using the model explained in Section 6.5.5. The bars with hatching indicate the ionization thresholds of 3d, 3p, and 3s electrons. (Reproduced from Saito, N. and Suzuki, I.H., *Radiat. Res.*, 156, 317, 2001b.)

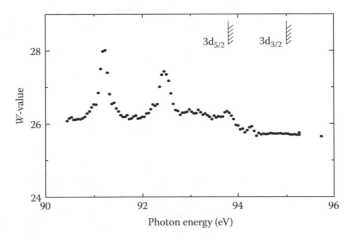

FIGURE 6.17 Photon *W*-value for Kr in the region of 3d thresholds as a function of photon energy. Solid circles show measured data. The bars with hatching denote ionization thresholds for $3d_{3/2}$ and $3d_{5/2}$ electrons. (Reproduced from Saito, N. and Suzuki, I.H., *Radiat. Res.*, 156, 317, 2001b.)

6.5.5 Precise Model of the Atomic Shell Effect on Variation in the *W*-Value of Kr

The model for explaining the oscillatory variation in the measured W_p value is described as follows (Saito and Suzuki, 2001b): The number of secondary electrons produced by the electrons emitted from a Kr atom is calculated using the experimental data of Combecher (1980).

(i) *Below the 3d threshold*: Single, double, and triple valence photoionizations take place. In the case of single ionization, a valence electron of 4p or 4s with an energy of E_{4p} or E_{4s}, respectively, is ejected (Svensson et al., 1988; Schmidt, 1997). The numbers of electrons produced from collisions of the 4p and 4s photoelectrons with ambient atoms can be calculated to be $E_{4p}/W_c(E_{4p})$ and $E_{4s}/W_c(E_{4s})$, respectively. The branching ratio of the 4p and 4s ionizations is simply assumed to be the multiplicity of the two orbitals, that is, 3:1. Then the total number of electrons produced by single valence ionization, N_s, is given by $1 + 0.75E_{4p}/W_e(E_{4p}) + 0.25E_{4s}/W_e(E_{4s})$. In the double photoionization case, two electrons with energies E_1 and E_2 are emitted through photoabsorption. The excess energy $(E_1 + E_2)$ is calculated by assuming that two 4p electrons are ejected in double photoionization. The distribution of excess energy into the two electrons is simply assumed so that one electron has all excess energy and the other has no energy $(E_2 = 0)$. Then the total number of electrons produced via double ionization, N_d, is expressed by $2 + E_1/W_e(E_1)$. In the triple photoionization case, three electrons, E_1, E_2, and E_3, are ejected through photoabsorption, and E_2 and E_3 are assumed to be zero. The total number of electrons, N_t, is expressed as $3 + E_1/W_e(E_1)$. The branching ratios of Kr$^+$, Kr^{2+}, and Kr^{3+} below the 3d threshold, that is, R_1, R_2, and R_3, respectively, which can be cited from the literature (Saito and Suzuki, 1992, 1994), correspond to the branching ratios of single, double, and triple photoionizations, respectively. The photon W-value is derived as

$$W_p = \frac{E_p}{R_1 N_s + R_2 N_d + R_3 N_t}. \tag{6.15}$$

Table 6.3 lists the term symbols used in the present model, along with their explanation.

(ii) *Above the 3d threshold*: Photoionization of a 3d electron mainly takes place (Henke et al., 1993). The formed 3d hole is dominantly filled through the normal Auger transition, and a double Auger decay slightly contributes to multiple ionization. The total number of electrons in the normal Auger decay, N_{nA}, is given by $2 + E_{3d}/W_e(E_{3d}) + E_{nA}/W_e(E_{nA})$, where E_{nA} is the kinetic energy of Auger electrons, which is approximately calculated to be the

---

TABLE 6.3

List of Term Symbols Used in the Model Proposed for Photon W-Value

Symbol	Content
N_s	Total number of electrons through single valence ionization
N_d	Total number of electrons through double valence ionization
N_t	Total number of electrons through triple valence ionization
N_{nA}	Total number of electrons through the normal Auger decay
N_{dA}	Total number of electrons through the double Auger decay
N_{sh}	Total number of electrons through the photoionization shake-off followed by the normal Auger decay
N_{shdA}	Total number of electrons through the photoionization shake-off followed by the double Auger decay
R_1	Branching ratio of single ionization
R_2	Branching ratio of double ionization
R_3	Branching ratio of triple ionization
R_4	Branching ratio of quadruple ionization
R_{iv}	Fraction of valence i-fold ionization among the total valence ionization
R_{nd}	Ratio of the double Auger decay to the normal Auger decay
E_i	Energy of the ith photoelectron
E_{nA}	Energy of the normal Auger electron
E_{dAi}	Energy of the ith electron formed through the double Auger decay
E_{sh}	Energy of the ejected electron through photoionization shake-off
E_{shA}	Energy of the normal Auger electron after the photoionization shake-off
E_{shdA}	Energy of the double Auger electron after the photoionization shake-off

energy difference between the ground state of 3d hole states and that of doubly charged states $(4p^{-2})$ (Radzig and Smirnov, 1985; Schmidt, 1997). The total number of electrons in the double Auger decay, N_{dA}, is given by $3 + E_{3d}/W_e(E_{3d}) + E_{dA1}/W_e(E_{dA1})$, where E_{dA1} denotes the kinetic energy for one of the double Auger electrons, and the energy of another Auger electron, E_{dA2}, is assumed to be zero. The energy E_{dA1} is assumed to be the energy difference between the ground state of 3d hole states and that of triply charged states. Since normal Auger and double Auger processes produce Kr^{2+} and Kr^{3+}, the branching ratios of these ions can be utilized for the ratios of normal to double Auger processes, which was observed using a coincidence technique (Saito and Suzuki, 1997). We cannot neglect the contribution of valence photoionization. The ratio of the single valence ionization corresponds to the branching ratio of Kr^+. We assume that the ratios of single, double, and triple valence ionization above the 3d threshold are equal to those just below the 3d threshold, say R_{1v}, R_{2v}, and R_{3v}, respectively. Then the ratios of the double and triple valence ionizations are calculated to be $R_1 R_{2v}/R_{1v}$ and $R_1 R_{3v}/R_{1v}$, respectively (Saito and Suzuki, 1992, 1994). W_p is expressed as follows:

$$W_p = \frac{E_p}{R_1 N_s + R_1 \dfrac{R_{2v}}{R_{1v}} N_d + R_1 \dfrac{R_{3v}}{R_{1v}} N_t + \left(R_2 - R_1 \dfrac{R_{2v}}{R_{1v}}\right) N_{nA} + \left(R_3 - R_1 \dfrac{R_{3v}}{R_{1v}}\right) N_{dA}}. \tag{6.16}$$

(iii) *At energies considerably higher than the 3d ionization threshold and lower than the 3p threshold*: A photoionization shake-off process occurs, in which a 3d electron and a valence electron are ejected at the initial photoabsorption step. The atom having two holes turns into a triply charged ion in most cases via Auger decay and slightly into a quadruply charged ion via double Auger decay. The kinetic energies for the initially ejected electrons and the Auger electron are assumed to be E_{sh}, and zero and E_{shA}, respectively. The total number of electrons via the photoionization shake-off and Auger decay (N_{sh}) is given by $3 + E_{sh}/W_e(E_{sh}) + E_{shA}/W_e(E_{shA})$. In the case of the double Auger decay, the kinetic energies of the two Auger electrons are assumed to be E_{shdA} and zero. Then the total number of electrons via the photoionization shake-off and double Auger decay (N_{shdA}) is given by $4 + E_{sh}/W_e(E_{sh}) + E_{shdA}/W_e(E_{shdA})$. In these cases, the energies of E_{sh}, E_{shA}, and E_{shdA} are assumed to be the differences between the corresponding ground states among the related hole states (Radzig and Smirnov, 1985; Schmidt, 1997). The branching ratios of normal Auger, double Auger, shake-off + Auger, and shake-off + double Auger can be calculated using the branching ratios of Kr ions and the ratio of the double to normal Auger processes. The ratio of the double to normal Auger transitions (R_{nd}) is cited from the literature (Saito and Suzuki, 1997). Then W_p is represented as

$$W_p = \frac{E_p}{\left[\begin{array}{c} R_1 N_s + R_1 \dfrac{R_{2v}}{R_{1v}} N_d + R_1 \dfrac{R_{3v}}{R_{1v}} N_t + \left(R_2 - R_1 \dfrac{R_{2v}}{R_{1v}}\right) N_{nA} + \left(R_2 - R_1 \dfrac{R_{2v}}{R_{1v}}\right) R_{nd} N_{dA} \\ + \left\{R_3 - R_1 \dfrac{R_{3v}}{R_{1v}} - \left(R_2 - R_1 \dfrac{R_{2v}}{R_{1v}}\right) R_{nd}\right\} N_{sh} + R_4 N_{shdA} \end{array}\right]} \tag{6.17}$$

In this equation, R_4 indicates the branching ratio of the double Auger decay after the photoionization shake-off process.

(iv) *Above the 3p ionization threshold and the 3s threshold*: Similar assumptions have been made for the calculation of W_p. Partial photoabsorption cross sections of 3p and 3s electrons are small, and the details of the assumptions have not considerably affected the calculated values of W_p.

Let us compare the measured data with the result calculated by this model. The curve for W_p calculated in this model has been plotted as the solid line in Figure 6.16. This curve has well reproduced the energy dependence of the experimental photon W-value. This finding suggests that the present model seems quantitatively valid. However, a slight difference between experimental and calculated W-values is seen in the region of 125–160 eV. A possible reason may be the effect of impurity contained in the sample gas. If this is the case, the Penning ionization may create additional ions. This effect will lead to a lower W-value than real. However, this kind of effect is supposed to induce lower W-values at other photon energies. Figure 6.16 essentially indicates the agreement between the experiment and the calculation in other energy regions. Therefore, the impurity effect does not contribute much to the discrepancy near 140 eV. Another reason may be the mixing of higher-order radiation in the incident soft x-ray. The second-order radiation can create ions in the chamber about two times of the first because of similar values of W_p near 140 and 280 eV. The energy of the electron beam in the storage ring was lowered to 350 MeV in the measurement of W_p in this region, and the percentage of the second-order light was supposed to be extremely low. This supposition was confirmed from the saturation curve with an increase in the gas density in the ion chamber, as shown in Figure 6.14. Another possibility is related to the assumption of the ratio of photoionization shake-off in the proposed model. This shake-off process forms many electrons simultaneously, and thus makes secondary ionizations less efficient. A smaller number of the total electrons is given by a larger ratio of this shake-off process, that is, a larger W_p is derived through this assumption.

When multiply charged Kr ions collide with neutral atoms, their charge states change due to the collision:

$$Kr^{n+} + Kr \rightarrow Kr^{(n-1)+} + Kr^+ + KE \tag{6.18}$$

$$\rightarrow Kr^{(n-1)+} + Kr^{2+} + e - KE' \tag{6.19}$$

The terms KE and KE′ denote energies of exothermic and endothermic reactions. In most cases, the total charge of the two species is taken to be the same, as given in Equation 4.18, before and after these collisions, because the ion does not have a high kinetic energy. The collision process expressed by Equation 6.19 takes place only with an extremely low probability. This process contributes to the increase in the total charge. Therefore, this process has a slight possibility of lowering the measured W_p near 140 eV. This possibility is probably supported by the fact that the ground state of a quadruply charged Kr ion is positioned at 130 eV. However, the photon energy for peaking the Kr^{4+} fraction in the initial photoionization step is about 290 eV and that for Kr^{5+} is about 400 eV, according to the data from the literature (Saito and Suzuki, 1992). The process expressed by Equation 6.19 should play a more significant role at energies of about 280 eV than about 140 eV. Finally, the discrepancy near 140 eV cannot be clarified at present, and several factors possibly contribute to this discrepancy.

The W_p obtained at 1 keV in the present study, which is 27.2 ± 1.0 eV, is slightly higher than 24.0 ± 0.7 eV at 5.9 keV that was obtained previously using the proportional counter (Borges and Conde, 1996). Since the curve of W_p in Figure 6.16 shows a tendency of decreasing with increasing photon energy, this trend seems to take the present W_p value close to the previous one at that energy. However, Kr has the L-shell absorption edges around 1.7 keV, and these edges probably yield upward jumps in the W_p. If the jump is not large, energy dependence will explain the difference of W_p at the two photon energies.

Figure 6.17 shows some sharp peaks of the experimental W_p in the energy region for excitation of the 3d electron into Rydberg orbitals. According to the photoabsorption spectrum (Hayaishi et al., 1984; Saito and Suzuki, 1992), each peak corresponds to a transition of the 3d electron into a Rydberg orbital, that is, the peak at 91.2 eV comes from the transition to the $3d_{5/2}^{-1}$ 5p state and the peak at 92.5 eV originates from transitions to the $3d_{3/2}^{-1}$ 5p state and to the $3d_{5/2}^{-1}$ 6p state. These peaks are connected to the energy of resonant Auger electrons. Based on the studies on resonant Auger transitions, the decay processes are approximately classified into two types (Carlson et al., 1989; Okada et al., 2005). The first type is a spectator decay, in which the excited electron remains in the orbital just occupied, and the second is a participator. Usually the spectator decay takes place at a high probability when the relevant orbital is a Rydberg. Then the energy of the electron formed through the spectator is considerably lower than the valence photoelectron at the same photon energy. This lower-energy electron induces a lower ionization yield in the ion chamber than the photoelectron. This consideration assumes that a higher W_p is obtained at the Rydberg excitation than the valence orbital ionization. The peak width for this excitation is very narrow, and the width measured has been governed by the monochromator resolution. Therefore, the maximum value of W_p at the Rydberg excitation is not conclusive, owing to the moderate resolution of the present soft x-ray monochromator. Slightly lower values for W_p are derived just above the ionization thresholds. This finding is presumed to come from the emission of 3d electrons, which have very low energies at these energies. The low energy emission serves to produce efficient yields of ionization in the gas system.

6.5.6 Ar AND Xe

The photon W-value of Ar is shown in the absolute scale as a function of the photon energy from 50 to 1000 eV in Figure 6.18 (Saito and Suzuki, 2001a). The present data are exhibited with solid circles. The arrow denotes the 2p electron ionization threshold. The open squares indicate the W-value for electrons, W_e, which was reported previously (Combecher, 1980). The measured W_p exhibits lower values than those for electrons below 250 eV. The W_p increases sharply near the 2p threshold, which shows slightly higher or almost the same values as W_e. The absolute values for W_p measured here are presumed to be correct in consideration of the agreement with the data for electrons, W_e, in the

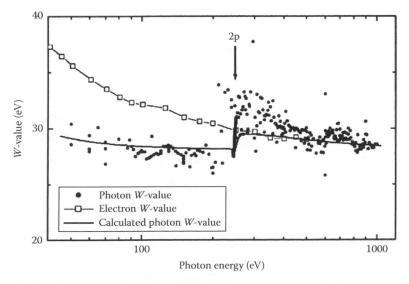

FIGURE 6.18 Photon W-value for Ar as a function of photon energy. Solid circles show data by Saito and Suzuki, and open squares are data for electrons by Combecher (1980). The solid curve represents the photon W-values calculated using the model explained in Section 6.5.6. The arrow indicates the ionization threshold of 2p electrons. (Reproduced from Saito, N. and Suzuki, I.H., *Radiat. Phys. Chem.*, 60, 291, 2001a.)

region above 400 eV. The sharp increase near the 2p threshold for W_p seems to originate from the 2p photoionization, which has a close connection with the model proposed previously in the case of hydrocarbon molecules (Suzuki and Saito, 1985a,b, 1987; Saito and Suzuki, 1986).

The solid curve in Figure 6.18 denotes the result calculated from a model similar to that for Kr (Saito and Suzuki, 2001b). This calculated curve agrees well with the experimental W-values in the region of 50–1000 eV. This result postulates that the present model is semiquantitatively valid.

A slight difference between experimental and calculated W-values is seen near the 2p electron ionization threshold. One possible reason is the assumption that energy is shared between two electrons that are simultaneously ejected. This assumption causes the ionizing collisions to occur a little more frequently than in the real situation. This effect induces lowering of calculated W-values. Another reason probably originates from the difficulty encountered in the measurement of photoion currents. In this energy region, the photoabsorption cross section is large, and ion production takes place inhomogeneously in the ion chamber, which possibly results in the loss of collection of photoions owing to the recombination reaction of ions with electrons. This loss causes the experimental W-value to be higher than the real one. If the recombination reaction makes a considerable effect, the saturation curve on the electron number produced in the chamber should show a decrease at higher gas densities. This decrease was not found to be appreciable in the present experiment.

The W_p obtained at 1 keV in the present study, which is 28.5 ± 1.0 eV, is slightly higher than 25.8 eV ± 0.6 eV at 5.9 keV previously obtained (Borges et al., 1996). Since the curve of W_p in Figure 6.18 shows a tendency of decreasing with increasing photon energy, this trend seems to take the present W_p value close to the previous one at that energy. However, Ar has the K-shell absorption edge at 3.2 keV, and this edge probably yields a jump in the W_p. If the jump is not large, it will explain the difference of W-value at the two photon energies.

The photon W-value of Xe is shown on the absolute scale as a function of photon energy from 120 to 1000 eV in Figure 6.19 (Suzuki and Saito, 2001). The data are denoted with solid circles. Bars with hatching indicate the ionization thresholds of 4p, 4s, 3d, and 3p electrons. The data points of open squares give the W-value for electrons, W_e, which was reported earlier (Combecher, 1980). The measured W_p increases slightly near the 4p ionization thresholds, and then shows a decreasing trend with photon energy below 670 eV although the data scatter appreciably. The value of W_p steeply increases at the 3d thresholds and remains at similar values above 800 eV. Although the data seem to rise near the 3p thresholds, this feature is not decisive owing to the sizes of data scatters.

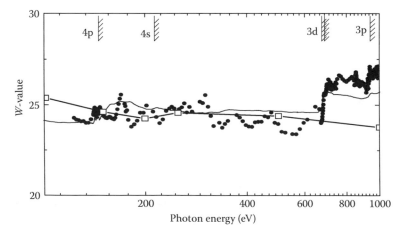

FIGURE 6.19 Photon W-value for Xe as a function of photon energy. Solid circles show data by Suzuki and Saito, and open squares are data for electrons by Combecher (1980). The solid curve represents the photon W-values calculated using the model explained in Section 6.5.6. The bars with hatching indicate the ionization thresholds of 4p, 4s, 3d, and 3p electrons. (Reproduced from Suzuki, I.H. and Saito, N., *J. Electron Spectrosc. Relat. Phenom.* 119, 147, 2001.)

The measured data of W_p agree with those for electrons between 200 and 600 eV, while above 700 eV, W_p is higher than W_e. The agreement suggests that absolute values of the measured W_p are reliable. The variations near the 4p and 3d thresholds in the W_p are probably due to the ionization of these inner-shell electrons. This phenomenon is very similar to those for Ar 2p thresholds and Kr 3d thresholds (Saito and Suzuki, 2001a,b). Inner-shell ionization increases the ratio of multiple photoionization considerably, which results from Auger electron emission filling inner-shell vacancy.

The solid curve in Figure 6.19 shows the result calculated from a model similar to that for Kr (Saito and Suzuki, 2001b). The W_p value calculated has well reproduced the characteristics of the measured W_p, in particular, the steep increase at the 3d ionization thresholds. The model proposed here is essentially valid. However, a slight discrepancy between the calculated and experimental W_p values is seen near 500 and 800 eV. A possible reason originates from the mixing of higher-order radiation in the incident photon beam. The second-order radiation can yield ions in the ion chamber about two times of the first because the W_p does not change largely in the energy region of interest. Since the energy of the electron beam in the storage ring was lowered to 500 MeV in the measurement of W_p, the ratio of the second-order light was supposed to be extremely low. This supposition makes us find another reason for the discrepancy. Scattered light with a longer wavelength produces ions with a quantity lower than the first-order photon, and then induces a higher W_p than real. The mixing of the scattered light can explain that experimental W_p values near 800 eV are higher than the calculated results. It is suggested that a thicker filter be inserted into the incident photon beam.

The discrepancy near 500 eV may be related to the assumption of the ratio of the photoionization shake-off process in the model calculation. This shake-off process forms many electrons simultaneously, and thus makes secondary ionization less efficient. A smaller number of the total electrons is given by a larger ratio of this shake-off process, that is, a higher W_p is obtained through this assumption. Finally, the small discrepancy between the measurement and the calculation cannot be clarified at present, and more precise work is expected to be carried out.

The W_p at 1 keV in the present study, which is 26.8 eV, is higher than the previous value of 24.2 eV at 5.9 keV (Borges et al., 1996). The curve of W_p in Figure 6.19 shows a tendency of decreasing with increasing photon energy, which is similar to those for Ar and Kr (Saito and Suzuki, 2001a,b). This trend seems to take the present W_p value close to the previous one at that energy.

6.6 OUTLOOK

As described above, we can now measure the absolute intensity of soft x-rays and the total charges produced in rare gas atoms using the multielectrode ion chamber technique. This technique seems to be applicable to molecular samples. However, charge states of molecules just after irradiation have not been determined so far in most cases, because it is not certain that kinetic ions are detected completely. Also, branching ratios for ion production and/or γ-values have not been obtained in the soft x-ray region for the case of molecules. Photoionization shake-off processes and multiple Auger decays have not been evaluated so far in molecules, because several fragments with some kinetic energies complicate the situation, making measurements very difficult. At present, it is difficult to establish a precise model for *W*-values of molecules in the sub-keV energy region. Thus, it is desired that a novel technique is devised for the measurement of all photoions and that molecular data on the charge state, that is, multiple photoionization ratio, are accumulated in the course of progress in synchrotron radiation researches.

At the peak energies of resonance transitions into Rydberg orbitals, the absolute value of the photon *W*-value for Kr has some ambiguity because the photoabsorption efficiency is very high, suggesting that a number of charges are formed near the window of the chamber. In order to correct *W*-values in the absolute scale, it is necessary that the charges formed are completely detected and that the monochromator has an energy width narrower than the natural lifetime width for the resonance peak.

In these years, a cryogenic radiometer has been utilized for the absolute measurement of soft x-rays (Rabus et al., 1997; Morishita et al., 2005; Kato et al., 2010). In this instrument, all energy of incident x-rays is received with a sensor unit and the temperature of the sensor rises. The temperature increase is detected and converted to absorbed energy. An accuracy realized with this instrument is usually higher than that expected for the ion chamber technique. The metrology group of the AIST has attempted to obtain the W-value for a mixed gas (dry air) in the keV energy region (Kato et al., 2010). The result is almost in agreement with the electron W-value on the absolute scale and on the energy dependence. However, the energy dependence shows a slight jump at 3.2 keV, where Ar contained in air (about 1%) has the K-shell photoabsorption edge. Using a similar model to that described in Section 6.5.5, this jump is well reproduced. The effect of the Ar K-shell photoabsorption has been observed as nonlinearity by another group who utilized the gas-scintillation proportional counter (Monteiro et al., 2003). Further, electron W-values for air have been redetermined through a precise measurement, and then utilized for the accurate determination of a mass energy transfer coefficient in the medium of free air (Buiermann et al., 2006). This coefficient is a significantly important quantity for dosimetric measurements using the ion chamber with free air, although this was estimated so far only from the calculation using data on radiation–atom interactions. Moreover, W-values for electronegative gases and mixtures are being evaluated for advanced utilization of time projection chambers (Pushkin and Snowden-Ifft, 2009).

6.7 SUMMARY

The developments in W-value studies are described, in particular, the studies in low-energy regions are explained in some detail in this chapter. Although the W-value is insensitive to energy and radiation quality in high-energy regions, it varies considerably with decreasing energy in low-energy regions. About 20 years ago, the photon W-value was found to show distinct variation in keV energy regions in investigations using rare-gas proportional counters, and this finding was ascribed to the atomic shell effect. On the basis of recent advance in synchrotron radiation researches, a variation in photon W-values for hydrocarbon molecules was derived in the C 1s electron transition region by means of the narrow-step energy scan. The observed oscillatory structure was well reproduced with the model based on the valence and 1s electron transitions including Auger decays, which yield various energies for emitted electrons. These results clarified the effect of the initial photoionization step on the variation in the energy dependence of the W-value. Absolute values for the photon W-value of rare gas atoms were determined using the multielectrode ion chamber, which provided results of oscillatory variations near inner-shell excitation regions. The model based on precise atomic shell effects has provided agreement with the measured data.

ACKNOWLEDGMENTS

The author sincerely thanks Dr. Norio Saito of the National Metrology Institute of Japan, National Institute of Advanced Industrial Science and Technology (NMIJ/AIST) for his close cooperation in the study of W-values. He is indebted to the accelerator group of AIST for providing synchrotron radiation. He would also like to thank Drs. Yuichirou Morishita and Masahiro Kato for performing measurements with the cryogenic radiometer. He appreciates the support of Prof. Kenji Ito of Photon Factory during the writing of this chapter.

REFERENCES

Backx, C. and van der Wiel, M. J. 1975. Electron-ion coincidence measurements of CH_4. *J. Phys. B* 8: 3020–3033.
Borges, F. I. G. M. and Conde, C. A. N. 1996. Experimental W-values in gaseous Xe, Kr and Ar for low energy x-rays. *Nucl. Instrum. Method Phys. Res. A* 381: 91–96.

Breyer, B. 1973. Pulse height distribution in low energy proportional counter measurements. *Nucl. Instrum. Method* 112: 91–93.

Brown, F. C., Bachrach, R. Z., and Bianconi, A. I. 1978. Fine structure above the carbon K-edge in methane and in the fluoromethane. *Chem. Phys. Lett.* 54: 425–429.

Buiermann, L., Grosswendt, B., Kramer, H.-M., Selbach, H.-J., Gerlach, M., Hoffmann, M., and Krumrey, M. 2006. Measurement of the x-ray mass energy-absorption coefficient of air using 3 keV to 10 keV synchrotron radiation. *Phys. Med. Biol.* 51: 5125–5150.

Carlson, T. A., Mullins, D. R., Beall, C. E., Yates, B. W., Taylor, J. W., Lindle, D. W., and Grimm, F. A. 1989. Angular distribution of ejected electrons in resonant Auger processes of Ar, Kr, and Xe. *Phys. Rev. A* 39: 1170–1185.

Chen, C. T., Ma, Y., and Sette, F. 1989. K-shell photoabsorption of the N_2 molecule. *Phys. Rev. A* 40: 6737–6740.

Cole, A. 1969. Absorption of 20-eV to 50,000-eV electron beams in air and plastic. *Radiat. Res.* 38: 7–33.

Combecher, D. 1980. Measurement of *W* value of low-energy electrons in several gases. *Radiat. Res.* 84: 189–218.

Dayashankar. 1977. Mean energy expended per ion pair formed in methane by electrons. *J. Mysore Univ. Sect. B* 26: 181–185.

Dias, T. H. V. T., dos Santos, J. M. F., Rachinhas, P. J. B. M., Santos, F. P., Conde, C. A. N., and Stauffer, A. D. 1997. Full-energy absorption of x-ray energies near the Xe L- and K-photoionization thresholds in xenon gas detectors: Simulation and experimental results. *J. Appl. Phys.* 82: 2742–2753.

Dixon, A. J., Hood, S. T., and Weigold, E. 1978. Correlation effects and electron momentum distributions in the valence orbitals of ethylene. *J. Electron Spectrosc. Relat. Phenom.* 14: 267–275.

Eberhardt, W., Haelbach, R. P., Iwan, M., Koch, E. E., and Kunz, G. 1976. Fine structure at the carbon 1s edge in vapors of simple hydrocarbons. *Chem. Phys. Lett.* 40: 180–184.

Grosswendt, B. 1984. Statistical fluctuations of the ionization yield of low-energy electrons in He, Ne and Ar. *J. Phys. B* 17: 1391–1404.

Hayaishi, T., Morioka, Y., Kageyama, K., Watanabe, M., Suzuki, I. H., Mikuni, A., Isoyama, G., Asaoka, S., and Nakamura, M. 1984. Multiple photoionization of the rare gases in the XUV region. *J. Phys. B* 17: 3511–3527.

Henke, B. L., Gullikson, E. M., and Davis, J. C. 1993. x-Ray interactions: Photoabsorption, scattering, transmission, and reflection at E = 50–30,000 eV, Z = 1–92. *At. Data Nucl. Data Tables* 54: 181–342.

Hitchcock, A. P. and Brion, C. E. 1977. Carbon K-shell excitation of C_2H_2, C_2H_4, C_2H_6 and C_6H_6 by 2.5 keV electron impact. *J. Electron Spectrosc. Relat. Phenom.* 10: 317–330.

Huber, R., Combecher, D., and Buger, G. 1985. Measurement of average energy required to produce an ion pair (*W* value) for low-energy ion in several gases. *Radiat. Res.* 101: 237–251.

ICRU. 1979. Average energy required to produce an ion pair. ICRU Report 31. International Commission on Radiation Units and Measurements, Washington, D.C.

Inokuti, M. 1975. Ionization yields in gases under electron irradiation. *Radiat. Res.* 64: 6–22.

Inokuti, M., Douthat, D. A., and Rau, A. R. P. 1980. Statistical fluctuations in the ionization yield and their relation to the electron degradation spectrum. *Phys. Rev. A* 22: 445–453.

Jahoda, K. and McCammon, D. 1988. Proportional counters as low energy photon detectors. *Nucl. Instrum. Method Phys. Res. A* 272: 800–813.

Kato, M., Morishita, Y., Oura, M., Yamaoka, H., Tamenori, Y., Okada, K., Matsudo, T., Gejo, T., Suzuki, I. H., and Saito, N. 2007. Absolute photoionization cross sections with ultra-high energy resolution for Ar, Kr, Xe and N_2 in the inner-shell ionization regions. *J. Electron Spectrosc. Relat. Phenom.* 160: 39–48.

Kato, M., Suzuki, I. H., Nohtomi, A., Morishita, Y., Kurosawa, T., and Saito, N. 2010. Photon *W*-value of dry air determined using a cryogenic radiometer combined with a multi-electrode ion chamber for soft X-rays. *Radiat. Phys. Chem.* 79: 397–404.

Key, R. B., van der Leeuw, Ph. E., and van der Wiel, M. J. 1977. Ionic fragmentation and post-collision interaction in the Auger decay of carbon-K ionized CO. *J. Phys. B* 10: 2521–2529.

Kimura, K., Katsumata, S., Achiba, Y., Yamazaki, T., and Iwata, S. 1981. *Handbook of HeI Photoelectron Spectra of Fundamental Organic Molecules*. Japan Scientific Societies Press, Tokyo, Japan.

Kimura, M., Kowari, K., Inokuti, M., Krajcar-Bronic, I., Srdoc, D., and Obelic, B., 1991. Theoretical study of *W* values in hydrocarbon gases. *Radiat. Res.* 125: 237–242.

King, G. C., Tronc, M., Read, F. H., and Bradford, R. C. 1977. An investigation of the structure near the $L_{2,3}$ edges of argon, the $M_{4,5}$ edges of krypton and the $N_{4,5}$ edges of xenon using electron impact with high resolution. *J. Phys. B* 10: 2479–2495.

Kowalski, T. Z., Smith, A., and Peacock, A. 1985. Fano factor implications from gas scintillation proportional counter measurements. *Nucl. Instrum. Method Phys. Res. A* 279: 567–572.

Kowari, K., Kimura, M., and Inokuti, M. 1989. Electron degradation and yields of initial products: V. Degradation spectra, the ionization yield, and the Fano factor for argon under electron irradiation. *Phys. Rev. A* 39: 5545–5553.

Krajcar-Bronic, I. 1998. *W* values and Fano factors for electrons in rare gases and rare gas mixtures. *Hoshasen (Ionizing Radiation) (Tokyo)* 24: 101–125.

Krajcar-Bronic, I. and Srdoc, D. 1994. A comparison of calculated and measured *W* values in tissue-equivalent gas mixtures. *Radiat. Res.* 137: 18–24.

Krajcar-Bronic, I., Srdoc, D., and Obelic, B. 1988. The mean energy required to form an ion pair for low-energy photons and electrons in polyatomic gases. *Radiat. Res.* 115: 213–222.

Liegener, G. M. 1985. Calculations of the Auger spectra of ethylene and acetylene. *Chem. Phys.* 92: 97–101.

Miller, M. S. and Boring, J. W. 1974. Total inelastic energy loss by heavy ions stopped in a gas. *Phys. Rev. A* 9: 2421–2433.

Monteiro, C. M. B., Simoes, P. C. P. S., Veloso, J. F. C. A., dos Santos, J. M. F., and Conde, C. A. N. 2003. Energy non-linearity effects in argon gaseous detectors in the region of the Ar K-absorption edge: Experimental results. *Nucl. Instrum. Method Phys. Res. A* 505: 233–237.

Morishita, Y., Saito, N., and Suzuki, I. H. 2005. Comparison of the absolute soft x-ray intensity between a cryogenic radiometer and an ion chamber. *J. Electron Spectrosc. Relat. Phenom.* 144/147: 1071–1073.

Okada, K., Kosugi, M., Fujii, A., Nagaoka, S., Ibuki, T., Samori, S., Tamenori, Y., Ohashi, H., Suzuki, I. H., and Ohno, K. 2005. Variation in resonant Auger yields into the ^1G_4nl states of Kr across the L_3 threshold. *J. Phys. B* 38: 421–431.

Pansky, A., Breskin, A., and Chechik, R. 1996. A new technique for studying the Fano factor and the mean energy per ion pair in counting gases. *J. Appl. Phys.* 79: 8892–8898.

Pansky, A., Breskin, A., and Chechik, R. 1997. Fano factor and the mean energy per ion pair in counting gases at low x-ray energies. *J. Appl. Phys.* 82: 871–877.

Pireaux, J. J., Svensson, S., Basillier, E., Malmqvist, P.-A., Gelius, U., Caudano, R., and Siegbahn, K. 1976. Core-electron relaxation energies and valence-band formation of linear alkanes studies in the gas phase by means of electron spectroscopy. *Phys. Rev. A* 14: 2133–2145.

Pushkin, K. and Snowden-Ifft, D. 2009. Measurements of W-value, mobility and gas gain in electronegative gaseous CS_2 and CS_2 gas mixtures. *Nucl. Instrum. Method Phys. Res. A* 606: 569–577.

Rabus, H., Persch, V., and Ulm, G. 1997. Synchrotron-radiation-operated cryogenic electrical-substitution radiometer as the high-accuracy primary detector standard in the ultraviolet, vacuum-ultraviolet, and soft-x-ray spectral ranges. *Appl. Opt.* 36: 5421–5440.

Radzig, A. A. and Smirnov, B. M. 1985. *Reference Data on Atoms, Molecules, and Ions*. Springer-Verlag, Berlin, Germany.

Rye, R. R., Madey, T. E., Houston, J. E., and Holloway, P. H. 1978. Chemical-state effects in Auger electron spectroscopy. *J. Chem. Phys.* 69: 1504–1512.

Rye, R. R., Jennison, D. R., and Houston, J. E. 1980. Auger spectra of alkanes. *J. Chem. Phys.* 73: 4867–4874.

Saito, N. and Suzuki, I. H. 1986. Fine structure near the C K-edge in the photon *W*-value of methane. *Chem. Phys.* 108: 327–333.

Saito, N. and Suzuki, I. H. 1992. Multiple photoionization in Ne, Ar, Kr and Xe from 44 eV to 1,300 eV. *Int. J. Mass Spectrom. Ion Process.* 115: 157–168.

Saito, N. and Suzuki, I. H. 1994. Shake-off processes in photoionization and Auger transition for rare gases irradiated by soft x-rays. *Phys. Scr.* 49: 80–85.

Saito, N. and Suzuki, I. H. 1997. Double Auger probabilities from Xe $4d_j$, Kr $3d_j$, and Ar $2p_j$ hole states. *J. Phys. Soc. Jpn.* 66: 1979–1985.

Saito, N. and Suzuki, I. H. 1998. Absolute soft x-ray measurements using an ion chamber. *J. Synchrotron Radiat.* 5: 869–871.

Saito, N. and Suzuki, I. H. 1999. Absolute fluence rates of soft x-rays using a double ion chamber. *J. Electron Spectrosc. Relat. Phenom.* 101/103: 33–37.

Saito, N. and Suzuki, I. H. 2001a. Photon *W*-value for Ar in the sub-keV x-ray region. *Radiat. Phys. Chem.* 60: 291–296.

Saito, N. and Suzuki, I. H. 2001b. Photon *W* value for krypton in the M-shell transition region. *Radiat. Res.* 156: 317–323.

Samson, J. A. R. and Haddad, G. N. 1976. Average energy loss per ion pair formation by photon and electron impact on xenon between threshold and 90 eV. *Radiat. Res.* 66: 1–10.

Santos, F. P., Dias, T. H. V. T., Stauffer, A. D., and Conde, C. A. N. 1991. Variation of energy linearity and *w* value in gaseous xenon radiation detectors for x-rays in the 0.1 to 25 keV energy range: A Monte Carlo simulation study. *Nucl. Instrum. Method Phys. Res. A* 307: 347–352.

Schmidt, V. 1997. *Electron Spectrometry of Atoms Using Synchrotron Radiation.* Cambridge University Press, Cambridge, U.K.

Smith, B. G. R. and Booz, J. 1978. Experimental results on *W*-values and transition of low energy electrons in gases. In *Proceedings of the 6th Symposium on Microdosimetry,* London, U.K., pp. 759–775.

Srdoc, D. 1973. Dependence of the energy per ion pair on the photon energy below 6 keV in various gases. *Nucl. Instrum. Method* 108: 327–332.

Srdoc, D. and Obelic, B. 1976. Measurement of *W* at very low photon energy. In *Proceedings of the 5th Symposium on Microdosimetry,* Luxembourg, Germany, pp. 1007–1021.

Suzuki, I. H. and Saito, N. 1985a. Energy dependence of *W*-value of methane in the ultra-soft x-ray region. *Radiat. Phys. Chem.* 26: 305–307.

Suzuki, I. H. and Saito, N. 1985b. Effect of inner-shell excitation on the *W*-value of propane. *Bull. Chem. Soc. Jpn.* 58: 3210–3214.

Suzuki, I. H. and Saito, N. 1987. Oscillatory variation near the C K-edge in the photon *W*-value of ethylene. *Bull. Chem. Soc. Jpn.* 60: 2989–2992.

Suzuki, I. H. and Saito, N. 1992. γ-Value of rare gases for soft x-ray absolute measurement (in Japanese). *Bull. Electrotech. Lab.* 56: 688–711.

Suzuki, I. H. and Saito, N. 2001. Photon *W*-value for Xe in the soft x-ray region. *J. Electron Spectrosc. Relat. Phenom.* 119: 147–153.

Suzuki, I. H. and Saito, N. 2002. Photoabsorption cross-section for Kr in the sub-keV energy region. *J. Electron Spectrosc. Relat. Phenom.* 123: 239–245.

Suzuki, I. H. and Saito, N. 2005. Total photoabsorption cross-section of Ar in the sub-keV energy region. *Radiat. Phys. Chem.* 73: 1–6.

Suzuki, I. H., Tamenori, Y., Morishita, Y., Okada, K., Oyama, T., Yamamoto, K., Tabayashi, K., Ibuki, T., and Moribayashi, K. 2006. Formation of multi-charged Kr ions through photoionization of 2p electrons studied with a coincidence technique. *Radiat. Phys. Chem.* 75: 1778–1783.

Svensson, S., Eriksson, B., Martensson, N., Wendin, G., and Gelius, U. 1988. Electron shake-up and correlation satellites and continuum shake-off distributions in x-ray photoelectron spectra of the rare gas atoms. *J. Electron Specrosc. Relat. Phenom.* 47: 327–384.

Tamenori, Y., Okada, K., Tanimoto, S., Ibuki, T., Nagaoka, S., Fujii, A., Haga, Y., and Suzuki, I. H. 2004. Branching ratios of multiply charged ions formed through photoionization of Kr 3d, 3p and 3s sub-shells using a coincidence technique. *J. Phys. B* 37: 117–129.

Tomimasu, T., Noguchi, T., Sugiyama, S., Yamazaki, T., Mikado, T., and Chiwaki, M. 1983. A 600-MeV ETL electron storage ring. *IEEE Trans. Nucl. Sci.* NS-30: 3133–3136.

Tsunemi, H., Hayashida, K., Torii, K., Tamura, K., Miyata, E., Murakami, H., and Ueno, S. 1993. Nonlinearity at the K-absorption-edge in the Xe-filled gas proportional counter. *Nucl. Instrum. Method Phys. Res. A* 336: 301–303.

Ungier, L. and Thomas, T. D. 1984. Resonance-enhanced shakeup in near-threshold core excitation of CO and N_2. *Phys. Rev. Lett.* 53: 435–438.

Ungier, L. and Thomas, T. D. 1985. Near threshold excitation of *KVV* Auger spectra in carbon monoxide using electron–electron coincidence spectroscopy. *J. Chem. Phys.* 82: 3146–3151.

Waibel, E. and Grosswendt, B. 1978. Determination of *W* values and backscatter coefficients for slow electrons in air. *Radiat. Res.* 76: 241–249.

Waibel, E. and Grosswendt, B. 1983. Spatial energy dissipation profiles, *W* values, backscatter coefficients, and ranges for low-energy electrons in methane. *Nucl. Instrum. Method* 211: 487–498.

Waibel, E. and Grosswendt, B. 1991. Degradation of low-energy electrons in carbon dioxide; Energy loss and ionization. *Nucl. Instrum. Method Phys. Res. B* 53: 239–250.

Wight, G. R. and Brion, C. E. 1974. K-shell excitation of CH_4, NH_3, H_2O, CH_3OH, CH_3OCH_3 and CH_3NH_2 by 2.5 keV electron impact. *J. Electron Spectrosc.* 4: 25–42.

7 Positron Annihilation in Radiation Chemistry

Tetsuya Hirade
Japan Atomic Energy Agency
Tokai, Japan
and
Ibaraki University
Mito, Japan

CONTENTS

7.1 INTRODUCTION

A positron is the antiparticle of an electron. It has the same mass but an opposite charge to that of the electron. Dirac introduced an idea of the positive electron in 1930 (Dirac, 1930: 361), and, soon after, Anderson found it, that was the positron, in 1932 during observation of radiations from the universe, as shown in Figure 7.1 (Anderson, 1933: 491). With this discovery came into existence the study of positrons. Many techniques, such as electron paramagnetic resonance (EPR)

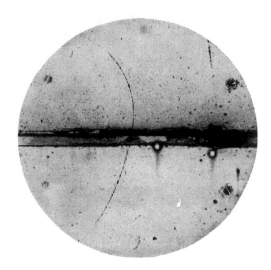

FIGURE 7.1 A 63 million volt positron passing through a 6 mm lead plate and emerging as a 23 million volt positron. (Reprinted from Anderson, C.D., *Phys. Rev.*, 43, 491, 1933. With permission.)

and electron diffraction, and instruments, such as the electron microscope, are found based on the electron. Similarly, the positron microscope (Van House and Rich, 1988: 169; Oshima et al., 2009: 194104) and positron diffraction (Kawasuso and Okada, 1998: 2695; Fukaya et al., 2009: 193310) are found based on the positron. However, the biggest difference between the two is that free positrons do not exist in our world. Therefore, the number of positrons that can be obtained is quite limited. Recently, it has become possible to store a large amount of positrons and inject them into a sample. The formation of positronium (Ps) molecules is now possible (Cassidy et al., 2007: 062511). Ps is a bound state of a positron and an electron. It may be considered that Ps is an isotope of the hydrogen atom.

In 1974, Mogensen proposed a Ps formation model for liquids, solids, and dense gases (Mogensen, 1974: 998). This proposal initiated the deep connection between radiation chemistry and Ps chemistry. The binding energy of Ps is 6.8 eV, and therefore it is impossible to pick off one of the electrons from the molecules, because the ionization potentials of molecules are more than 6.8 eV. If a positron is placed in liquids, solids, or gases, Ps does not form. The energetic positrons interact with matter similarly as the energetic electrons. Then the injected positron in insulating materials forms spurs, which thermalize in the terminal spur at the end of the track. It is called the positron spur. The positron has a chance to capture one of the excess electrons in the positron spur to form Ps. Once the positron and the excess electrons are trapped or solvated, Ps formation is usually very difficult, because the lifetime of positrons in insulating materials is about 400 ps. Hence, Ps formation takes place as fast as the geminate recombination of cations and free electrons, which usually takes about 1 ps. Thus, this process provides information of very fast reactions in the spurs. There are three methods—electron pulse radiolysis, laser flash photolysis, and the positron annihilation method—to study very fast reactions.

There is another process of Ps formation at low temperatures, which is very different from Mogensen's model. This mechanism was clarified in 1998. The details are given in Section 7.6.6. The triplet Ps, that is, ortho-Ps (o-Ps), has the longest lifetime of 1–10 ns in insulating materials, and it can be used as a probe of reactive species, such as cation radicals and hydrated electrons. This is explained in Section 7.6.8.

Here, we introduce the basic knowledge and the methods to help realize the possibilities in the field of radiation chemistry by using positrons. It is strongly recommended to read Mogensen's book (Mogensen, 1995) to understand Ps chemistry in detail. Even now, the arguments in the book have retained their relevance and are of interest.

7.2 POSITRON ANNIHILATION IN INSULATORS

The positron and electron annihilation can take place when these overlap each other. When they annihilate, the energy equivalent to their total rest mass is emitted as γ-rays. These annihilation γ-rays contain information of momenta of the electron and the positron just before the annihilation and, hence, can be a powerful tool. The permitted number of γ-rays is governed by the conservation law of charge parity. The permitted number of annihilation γ-rays is even for singlet positron–electron pairs and odd for triplet pairs. The annihilation probability decreases by increasing the number of annihilation γ-rays, and hence almost all of the free positrons annihilate with electrons that have an anti-spin of the positron spin, with a lifetime of about 400 ps, by emitting two γ-rays. A positron and an electron form a bound state, that is, the positronium (Ps). There are singlet and triplet Ps. The singlet Ps is called para-Ps (p-Ps) that consists of the electron and the positron having anti-parallel spins. The triplet Ps is called ortho-Ps, having parallel spins. The intrinsic annihilation lifetime of p-Ps is 125 ps when two γ-rays are emitted, and that of o-Ps is 142 ns when three γ-rays are emitted. The intrinsic annihilation of o-Ps has lower probability than the annihilation with one of the electrons in the surrounding molecules that has anti-spins of the positron spin. This annihilation process is called pick-off annihilation, and the lifetime of o-Ps pick-off annihilation is usually 1–10 ns. Therefore, almost all of the positron annihilations occur mainly through free-positron annihilation, p-Ps intrinsic annihilation, and o-Ps pick-off annihilation, giving two γ-rays for annihilation in condensed materials.

7.3 POSITRONIUM

Ps is a bound state of a positron and an electron. It is considered to be a very light isotope of the hydrogen atom (H). The reduced mass of Ps is half of the reduced mass of H. Therefore, Ps has a Bohr radius twice as large, 0.105 nm, and half the ionization potential, 6.8 eV, of H.

Ps has negative work functions in most materials and is repelled by most materials. The electron in Ps is repelled by the electrons in the molecules because of the Coulomb forces and the exchange effect. The latter is a result of the Pauli principle. The space in the materials is occupied by the electrons that are in the orbital of molecules. Only two electrons can occupy one orbital when they have opposite spins, because of the Pauli principle. Therefore, they need to be in higher energy states, rather than filled states, to penetrate into electron clouds. This causes repulsion, as the electron in Ps needs to be in a higher energy state for penetration into other electrons. The positron is only repelled by the positive atomic core by penetrating deep into molecules. However, the positron is kept out of the electron clouds of the molecules because of the Coulomb attraction of the Ps electron that is repelled by electrons of molecules, as mentioned above. This is the reason why the annihilation probability of free positrons is larger than the pick-off annihilation probability of o-Ps.

The lifetime of o-Ps in condensed materials is controlled by the overlapping of the positron in Ps and the electrons in the surrounding molecules. Therefore, it is possible to know the information of the size of small holes in materials. As mentioned above, Ps has negative work functions in most materials and prefers to go away from the surface. If there are holes in the material, Ps will be trapped. When the hole size is 0.1–50 nm, the lifetime of o-Ps can give information of the hole size using the Tao–Eldrup model (Tao, 1972: 5499; Eldrup et al., 1981: 51) for smaller holes and using the extended Tao–Eldrup model (Ito et al., 1999: 4555; Dull et al., 2001: 4657) for larger holes, as indicated in Figure 7.2.

7.3.1 Ps IN LIQUIDS

Ps is repelled by molecules, and hence it creates a bubble in liquids. The liquid pressure and surface tension squeeze the bubble and Ps pushes the wall of the bubble. The total energy of the bubble state is

$$E = E_0(U, r) + 4\pi r^2 \gamma + \frac{4}{3\pi r^3 p} \qquad (7.1)$$

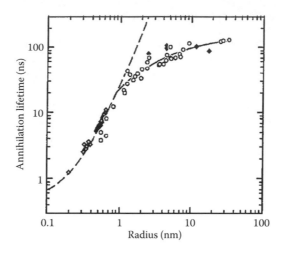

FIGURE 7.2 Annihilation lifetime of the o-Ps measured in various porous materials as a function of average pore radius. The dashed line is a calculated correlation curve. (Reprinted from Ito, K. et al., *J. Phys. Chem. B*, 103, 4555, 1999. With permission.)

where
    U is the depth of the potential for Ps
    r is the radius of the potential
    γ is the surface tension
    p is the pressure

$E_0(U, r)$ is the Ps zero-point energy in the bubble. The second term is the energy of the surface, and the third term is the volume energy. The bubble is stable at the minimum of the total energy, E. Therefore, the bubble size is larger at smaller surface tensions, which means that it will be larger also at higher temperatures. Figure 7.3 shows the temperature dependence of o-Ps on the positron annihilation lifetime (PAL) in neopentane (Jacobsen et al., 1982: 71). Since, at higher temperatures, the bubble size is larger as the surface tension is lower, the lifetime of o-Ps is also longer. Above the critical temperature, T_c, there is no effect of the surface tension, and hence there is no temperature dependence on the o-Ps lifetime, that is, the bubble size.

At lower temperatures, Ps is squeezed out from the bubble due to large surface tension. Very interesting phenomena were observed for one liquid and some solids (Mogensen, 1994, 377; Goworek, 2007, 318). In liquid CS_2, Ps is squeezed out from the bubble because of the large surface tension and large electron and positron affinities of CS_2. The zero-point energy of Ps elevates due to large surface tension, and then the Ps electron and positron prefer to smear out on the molecules, even though the positron and the electron interact with each other due to Coulomb attraction. This is called the fourth positron state. The transition from the Ps bubble state to the fourth positron state is indicated in Figure 7.4 (Hirade, 1996: 2153). This state was confirmed by adding 1 M of methanol to CS_2. The energy level of the fourth positron state is lower in methanol, and therefore the transition to this state appeared at higher temperatures, as shown in Figure 7.4 (Hirade, 1996: 2153). Pressure also affects the bubble size and Ps formation (Kobayashi, 1992: 1869).

7.3.2 Ps in Solids

The repulsion between Ps and molecules exists even in molecular solids. If there are open volumes like vacancies or vacancy clusters, Ps will be trapped. Moreover, positrons and electrons can be trapped by these defects, which will affect Ps formation. Probably, Ps chemistry in solids is more complicated than in liquids.

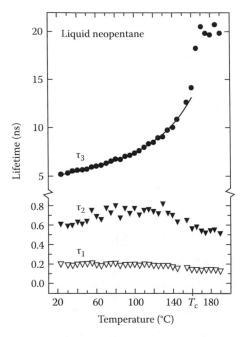

FIGURE 7.3 Temperature dependence of the positron lifetimes in liquid neopentane as obtained from lifetime spectra resolved into three exponential components. T_c is the critical temperature. (Reprint from Jacobsen, F.M. et al., *Chem. Phys.*, 69, 71, 1982. With permission.)

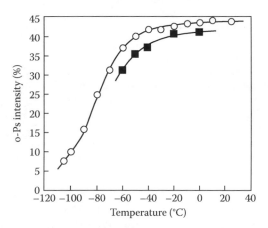

FIGURE 7.4 Intensity of the longest-lifetime component, that is, o-Ps, versus measuring temperature. Open circles indicate the results for pure CS_2 and filled squares are for 1 M methanol/CS_2.

Several very interesting effects can be observed in solids. One of them is the Ps Bloch state in crystals. It was found only in some very brittle solids. The first experimental evidence of the Bloch function state for Ps was found in quartz (Brandt, 1969: 522), MnF_2 (Coussot, 1970), and later for H_2O and D_2O ice (Mogensen et al., 1971: 71), and in several alkali halides at LN_2 and LHe temperatures (Hyodo and Takahashi, 1977: 1065; Kasai et al., 1988: 329). Ice is the only molecular crystal in which the Ps Bloch function was found, as shown in Figure 7.5 (Mogensen and Eldrup, 1978: 85). The Bloch function state can be detected by angular correlation of annihilation radiations (ACAR) (see Section 7.4.4). The curve shown in Figure 7.5 is only the component of the p-Ps

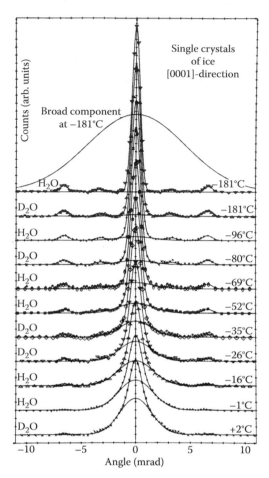

FIGURE 7.5 Temperature dependence of the ACAR spectra for single crystals of H_2O ice oriented along the a-axis. The broad component was subtracted. (Reprinted from Mogensen, O.E. and Eldrup, M., Positronium Bloch function, and trapping of positronium in vacancies, in ice, Risø Report No. 366, 1977. With permission.)

intrinsic annihilation, that is, the broad components of free positron and o-Ps pick-off annihilation are subtracted. Another broad component appears at higher temperatures. It is the component of the intrinsic annihilation of p-Ps localized on vacancies.

Irradiated ice was also studied using positron techniques (Eldrup, 1976: 5283). Figure 7.6 shows the measured temperature dependence on PAL for irradiated ice at −196°C. In this case, the lifetime spectra were analyzed with four lifetime components, because there were two different vacancies that give two different lifetimes of the o-Ps pick-off annihilation. One of these had $\tau_3 = 1.20$ ns that was considered to be the o-Ps pick-off annihilation in mono-vacancies. The other had τ_4, that is, the o-Ps pick-off annihilation in vacancy clusters. The change of τ_4 can provide the information of change of the cluster size. I_1 is the intensity of p-Ps, I_3 is the intensity of the pick-off annihilation of o-Ps in mono-vacancies, and I_4 is the intensity of the pick-off annihilation of o-Ps in vacancy clusters. There was inhibition of Ps formation below −180°C, which was caused by hydroxyl radicals formed by irradiation. τ_4 was initially about 2.3 ns, probably the lifetime of the divacancies, and it increased by elevating the temperature. This indicates the increasing size of the clusters. Then I_3 and I_4 decreased simultaneously because of the decrease in the number density of vacancy clusters. Above 130°C, the concentration of clusters became so low that Ps trapping could not be detected.

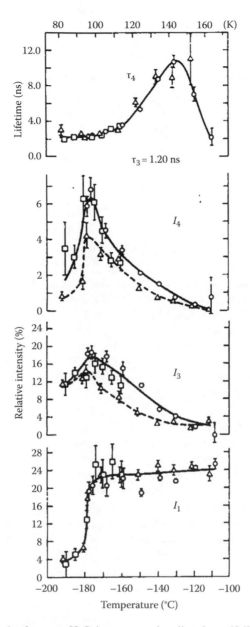

FIGURE 7.6 Lifetime results for pure H_2O ice gamma-irradiated at $-196°C$ and measured as a function of increasing temperature. ($\circ$) Polycrystalline samples (11 Mrad), ($\triangle$) single crystals (4.1 Mrad), and ($\square$) another single crystal (11 Mrad). (Reprinted from Eldrup, M., *J. Chem. Phys.*, 64, 5283, 1976. With permission.)

7.4 EXPERIMENTAL TECHNIQUES

Positron methods can be one of the important tools to study radiation chemistry, as mentioned above, because Ps formation is the result of the fast spur reactions and very short lifetime of positrons in condensed matter. Hence, positron annihilation methods are very useful for radiation chemistry. Therefore, only the methods related to the positron annihilation techniques used for the study of radiation chemistry will be discussed here.

TABLE 7.1
Characteristics of Some Common Positron-Emitting Radioisotopes

Radioisotopes	Fraction of e+	Half Lifetime	Maximum Energy (MeV)	Prompt γ Energy (MeV)
C-11	99%	20 months	0.97	—
Na-22	90%	2.7 years	0.54	1.28
Ti-44 (as Sc-44)	88%	47 years	1.47	1.16
Ni-57	46%	36 h	0.40	1.4
Co-58	15%	71 days	0.48	0.81
Cu-64	19%	12.8 h	0.66	—
Zn-65	1.7%	245 days	0.33	—
Ga-68 (from Ge-68)	88%	275 days	0.98	—

7.4.1 Positron Source

A positron is the antiparticle of an electron. We need to employ some special methods to obtain positrons. The most convenient method is the use of radioisotopes. There are some radioisotopes that emit positrons. A list of positron emitters is given in Table 7.1. ^{22}Na is the most common radioisotope. There are some reasons why it is often used. A γ-ray of 1.28 MeV emitted almost simultaneously with a positron is one of the reasons. A γ-ray of 1.28 MeV is very convenient because it gives information of the time of the positron emission, and hence the time interval between the 1.28 MeV γ-ray and 511 keV annihilation γ-rays gives information of the lifetime of the positron. Another reason why ^{22}Na is often used is its longer half-life of 2.6 years. It is possible to use the same source for several years.

The maximum energy of positrons emitted from ^{22}Na is 540 keV. Radioisotopes are usually shielded by polymer or metal films. An ^{22}Na source of about 400 kBq shielded by Kapton or Mylar films is often used for PAL, Doppler broadening (DB), and age-momentum correlation (AMOC) measurements. Kapton is a polyimide and has no Ps formation. It means that there is just one annihilation-lifetime component of the free positron that is about 380 ps. It is easy to subtract the PAL component from the measured spectra. Kapton films having a thickness of about 7.5 μm are often used. It is necessary to know how many positrons will annihilate in the film to subtract the annihilation component in the source films from the spectra. It is possible to estimate this by calculation or experiment. For example, the lifetime in well-annealed metals has just one short-lifetime component and the fraction of positron annihilation in the film can be estimated. There seems to be the effect of back-scattered positrons, that is, the fraction of positron annihilation in the film will be larger for heavier samples. It is difficult to estimate it experimentally for samples with a free-positron lifetime of about 400 ps. For these samples, it is necessary to assume the fraction of positron annihilation in the films. Usually, a fraction of 10%–12% is used.

Recently, it is possible to buy sources for these methods. A shielded ^{22}Na source of less than 1 MBq can be used in laboratories that are not specially installed for handling radioisotopes.

For angular correlation measurements or mono-energetic positron beams, stronger shielded positron sources (400–4000 MBq) are used. Pair production by high-energy radiation is also used for creating a large amount of positrons for intense positron beams (Suzuki et al., 2000: 603; 1991: L532).

7.4.2 Positron Annihilation Lifetime Measurement

The most important and common positron annihilation method is the PAL measurement. For PAL, ^{22}Na is the most common positron source, because it emits a positron and a prompt γ-ray of 1.27 MeV, almost simultaneously. The dominant decay scheme is indicated in Figure 7.7. The detection of

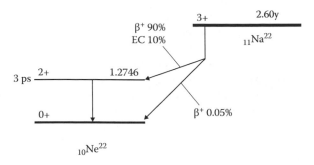

FIGURE 7.7 The dominant decay scheme of ^{22}Na.

a 1.27 MeV γ-ray indicates the emission of a positron. The annihilation of a positron with an electron can be observed by the detection of one or two annihilation γ-rays. In condensed mater, the energy of the annihilation γ-ray is about 511 keV, as almost all of the annihilations give two γ-rays. The detection of a 511 keV γ-ray indicates the annihilation of a positron. Therefore, the time interval of a coincident event of two γ-rays, 1.27 MeV and 511 keV, will be the measured PAL.

This is the gamma–gamma coincidence method. There is one more method that can be applied for measuring the PAL. It is the beta–gamma coincidence method. A positron injecting into a sample is observed by a beta-particle detector. If the energy of the positron is very high, scintillators can be used as the detector (Castellaz et al., 1996: 457). If radioisotopes are used, thin photodiodes can be used as the detector (Chalermkarnnon et al., 2002: 1004). There are some advantages of this method. One is that a prompt γ-ray is not necessary. The second advantage is a lower random coincidence background on the PAL spectra. However, there are some disadvantages too. One of them is the difficulty in obtaining a good time resolution because of the time-of-flight of the positron coming into the sample (Chalermkarnnon et al., 2002: 1004).

The most common PAL setup is shown in Figure 7.8. Fast scintillation detectors should be used to obtain a good time resolution. Therefore, the scintillators used should be BaF_2 or fast plastic scintillators. The BaF_2 scintillators have larger efficiencies but a slower luminescence component. When the counting rate is high, piling up of fast and slow luminescence components will shift the energy spectra; therefore, the energy windows selected by the single-channel analyzers (SCAs) of the

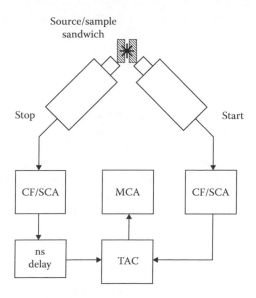

FIGURE 7.8 The most common PAL setup.

constant fraction (CF) cannot detect the objective signals. The photomultiplier (PM) tubes need quartz windows because the fast luminescence component is ultraviolet light. The PM also should have a fast rise time of the signal. The PM tube used most often is H3378-51 (Hamamatsu) and has a rise time of 0.7 ns. The most important module for PAL is a CF discriminator, which is supplied by many companies. The signals from PM tubes come into CF/SCA directly without inserting preamps. These signals should be large enough to be selected by SCA. Then, two sets of detectors and CF/SCA supply the timing signals for positron birth and death. The time interval should be converted into amplitude of the output using a time-to-amplitude converter (TAC). Then the lifetime spectra can be viewed using a multi-channel analyzer. This is the conventional PAL measurement technique.

Some important information is introduced here. It is most important to reject the piling up of the 1.27 MeV γ-ray and one of the annihilation γ-rays. The geometry one can obtain for a good counting rate is to place the sample and the source assembly in the middle of the detectors. One of the detectors is used for the 1.27 MeV γ-ray and the other is used for one of the annihilation γ-rays. One should not forget that there is one more annihilation γ-ray that will go in the opposite direction of the other annihilation γ-ray. It means that it will enter the detector for the 1.27 MeV γ-ray, and then two signals will easily pile up. This will affect the lifetime spectra, and the artificial short-lifetime component will easily appear. It is also important to have enough knowledge of the interaction between γ-rays and the materials. When one prefers to have a large counting rate, one tends to apply the wider energy windows. If wider energy windows are applied for the 1.27 MeV γ-ray including the Compton area, scattered γ-rays can enter the detector for the annihilation γ-rays. It means that artificial counts appear near time zero on the lifetime spectra.

Recently, it has become possible to use digital storage oscilloscopes (DSOs) to store all of the waves from detectors. One of the advantages of applying DSOs is that several analyses with different parameters can be performed. For example, it is possible to select the wave data by adjusting energy windows. Moreover, when one finds something strange on the spectra, it may be possible to find the cause. The biggest advantage is that the triple coincidence measurement is quite easy. This measurement has two stop detectors for both of the annihilation γ-rays. It provides better time resolution (Saito et al., 2002: 612).

7.4.3 DOPPLER BROADENING MEASUREMENT

The peak width of the whole absorption peak of annihilation γ-rays observed by Ge detectors is wider than those of the other γ-rays. It is caused by the Doppler shift by the momenta of an electron and a positron just before the annihilation. It means that it is possible to obtain information of electron and/or positron momenta in the materials. p-Ps annihilation is intrinsic, and the momenta of the electron and the positron cancel out each other. And hence, the energy of the annihilation γ-rays is very close to 511 keV, which is equivalent to the rest mass of an electron (or a positron). On the other hand, o-Ps and free positrons annihilate by picking off one of the electrons from the surrounding molecules. This means that the energy distribution of the annihilation γ-rays is wider because of the Doppler shift. A so-called S parameter is often used to indicate the annihilation γ-ray energy distribution. The value of S is the ratio between the counts appearing in a fixed central area and those in the entire peak area. The narrower the peak, the larger is the S value.

With spin conversion reactions, p-Ps annihilation increases and the distribution of the annihilation γ-ray energy becomes narrower. These reactions are difficult to detect by just the lifetime measurement. There are many researches carried out using the lifetime and Doppler shift measurements (Komuro et al., 2007: 330).

7.4.4 AGE–MOMENTUM CORRELATION MEASUREMENT

AMOC is the combination of PAL and DB, as shown in Figure 7.9. The most important use of AMOC in radiation chemistry is in determining the time-resolved DB. Fortunately, there are two

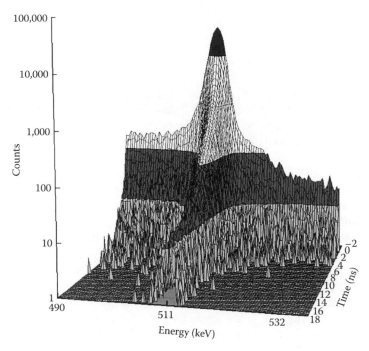

FIGURE 7.9   AMOC spectrum of water.

annihilation γ-rays in many cases. One of them is used for PAL and the other is used for DB, and hence the DB measurement with the annihilation time stamp is possible. Afterward, it is possible to construct a DB spectrum at the specific range of the positron annihilation time stamps. As mentioned above, DB is an effective tool for the investigation of the annihilation from p-Ps. This method can provide the information of the positron annihilation process, for example, the Ps formation process and spin conversion reactions (Castellaz et al., 1996: 457; Komuro et al., 2007: 330). The details are given in Section 7.6.7.

7.4.5   OTHER TECHNIQUES

ACAR is also a very important method. It measures the angular correlation between two annihilation γ-rays. The shift from 180° can indicate the shift on the sum of the momenta of the annihilated electron and the annihilated positron. The obtained information is almost the same as DB, but the resolution is much better. On the other hand, it is necessary to use much stronger sources so that the positron irradiation effect is very large. Therefore, it may be difficult to apply ACAR in studies of polymers.

The Ps Bloch state in crystals can be observed as shown in Figure 7.5. Many liquids are measured, and the ACAR measurement could give very important information on the positron state in liquids. Unfortunately, now there may be no ACAR method that can be applied for liquid measurements.

Positrons annihilate with one of the electrons from molecules or atoms, which means that a positron can be used to create cations. This phenomenon was used in some methods. One of them is the Auger electron measurement. Usually, Auger electrons are obtained by the ionization produced by electron irradiation. However, the background is usually very large, because many secondary electrons also enter the detectors. Nevertheless, there exists no background for the positron Auger method, because low-energy positrons are used instead of high-energy electrons. Sometimes oxygen exists as an impurity on the surface, but it can be detected effectively by the positron Auger method due to its large positron affinity (Ohdaira et al., 1997: 177).

7.5   POSITRONIUM FORMATION

If a positron is placed in a material, Ps does not form. The binding energy of Ps is 6.8 eV in vacuum, which is smaller than the ionization potentials of molecules or atoms. This means that positrons in materials cannot pick off electrons to form Ps. A fast Ps formation model, the so-called Ore model, was proposed by Ore in 1949 (Ore, 1949). As mentioned above, extra energy is necessary to pick off an electron from molecules or atoms; therefore, the idea of the Ore model is to supply the necessary energy from the kinetic energy of the positron. This model can explain the Ps formation in rare gases.

However, this model cannot explain the Ps formation in condensed materials. In 1974, Mogensen proposed the spur model (Mogensen, 1974: 998) that could explain the Ps formation in condensed materials. In the case of the Ore model, extra kinetic energy from the positrons can make Ps formation possible, while in the case of the spur model, the production of free or quasi-free electrons makes Ps formation possible. Injected positrons will make spurs on the track of positrons and will be thermalized at the end of the track. The structure is similar to that indicated in Figure 7.10 (Mogensen, 1974: 998). Therefore, it is quite possible to have free electrons near the thermalized positrons, which allows Ps formation. This model could well explain qualitatively the experimental results.

However, in 1980s, a very large enhancement of Ps formation at low temperatures was reported for many materials (Kindl and Reiter, 1987: 707). This enhancement could not be explained by the spur model, which was the main reason why some people did not accept this model. Until almost the end of the twentieth century, this Ps formation enhancement had not been explained, and so, some people used other models. One of these models was called the free volume model of Ps formation, which was often applied to discuss polymer physics (Jean, 1993: 569). The basic idea of this model was as follows. The measured o-Ps lifetime was always the longest and clearly identified by the PAL measurement. The long o-Ps lifetime indicates that there were empty spaces for o-Ps in the materials (see Figure 7.2). Hence, something like a hole (which does not mean the positively charged species) was needed for Ps formation, and thus, more holes could produce more Ps. Some people believed that the total free volume could be obtained by the free volume hole size observed by the o-Ps lifetime (see Figure 7.2) and the number of holes observed by the Ps formation probability.

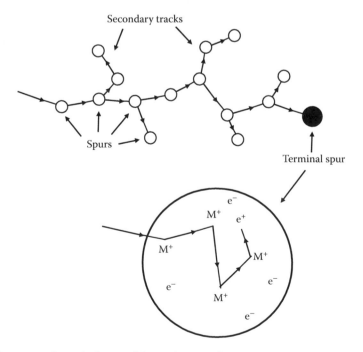

FIGURE 7.10   Structure of terminal spur of the positron track.

The measured slow enhancements of Ps formation at low temperatures were used for the study of polymer physics. It was believed that the slow enhancement observed at low temperatures showed the physical change of polymers that could not be observed by other methods.

Ps can be in the delocalized state in crystals. It means that the empty spaces are not needed for Ps formation. There is no Ps formation in some polymers, such as Kapton, which is a polyimide, even though there must exist some free volume. Indeed, any reasonable explanation has not been obtained by the free volume model until now.

In 1998, a new explanation of the appearance of Ps formation enhancement at low temperatures was given by Hirade (Hirade et al., 1998: 89; 2000: 465; Wang, 1998: 4654). The Ps formation enhancement, observed by the PAL measurement at low temperatures, appeared slowly. He explained it by the accumulation of shallowly localized electrons, such as trapped electrons. However, shallowly localized electrons are stable and live for a long time at low temperatures because the molecular motions are frozen. Positrons in materials cannot pick off electrons from molecules or atoms to form Ps, as mentioned above, because of the larger ionization potentials of molecules or atoms than the Ps binding energy. The binding energy of shallowly localized electrons is usually 0.5–3 eV, which is smaller than the Ps binding energy. It means that positrons can pick off shallowly localized electrons without the need of extra energy to form Ps. Thus, Ps can form just by placing positrons in the materials where many shallowly localized electrons exist.

This Ps formation process proposed by Hirade can explain many phenomena, such as Ps formation quenching by visible light (Hirade et al., 1998: 89; 2000: 465), relationship between the density of the shallowly localized electrons and Ps formation enhancement (Hirade et al., 2000: 465), and delayed Ps formation caused by the diffusion of positrons (Suzuki et al., 2003: 647; Hirade et al., 2007: 3714).

The free volume model often used for Ps formation at low temperatures cannot explain the Ps formation quenching by visible light, while the Ps formation process proposed by Hirade and the spur model proposed by Mogensen could explain Ps formation at any temperatures. It was accepted that radiation chemistry processes are very important for Ps formation (Hirade, 2007: 84).

7.6 RADIATION CHEMISTRY STUDIES BY POSITRON ANNIHILATION

7.6.1 COMPARISON OF THE YIELDS OF HYDRATED ELECTRONS OBSERVED BY PULSE RADIOLYSIS

Positron annihilation methods can provide some information regarding fast reactions in spurs, as pulse radiolysis experiments can. Duplatre and Jonah tried to compare the positron and pulse radiolysis experiments. The electron scavenger effects observed in aqueous solutions are studied for both these methods. The inhibition of hydrated electron formation and Ps formation showed very similar tendencies, as shown in Figure 7.11, and hence it was clarified that the precursor of Ps was mainly free electrons or quasi-free electrons (Duplatre, 1985: 557). As the Ps formation time is about 1 ps in common materials, the yields of Ps formation will give information of free or quasi-free electrons similarly as the yields of hydrated electrons observed by pulse radiolysis experiments. The most important detection method applied for pulse radiolysis is light absorption, which can make very fast experiments possible. This implies that only the species that absorb light can be detected by the fast pulse radiolysis experiments. On the other hand, Ps formation yields can be obtained as the longest-lifetime component by the PAL measurement, which means that the precursor of Ps, that is, free or quasi-free electrons, can be detected by positrons.

7.6.2 COMPARISON OF ELECTRON MOBILITY EXPERIMENTS

There is one more advantage of applying positron methods, which is the short lifetime of positrons. In condensed materials, the longest lifetime is given by o-Ps, which is usually less than 4–5 ns. This means that small amounts of impurities do not affect the yields of Ps very much. Therefore, it is not necessary to have very pure samples. In some cases, very pure samples are needed, for example

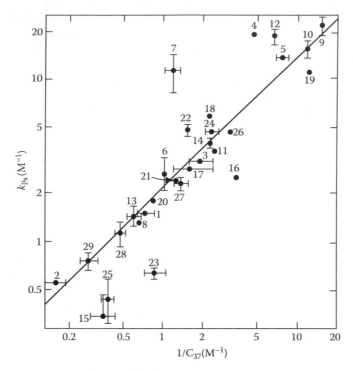

FIGURE 7.11 Plot of $1/C_{37}(M^{-1})$ for the scavengers versus the Ps inhibition constant, $k_{Ps}(M^{-1})$. (Reprinted from Duplatre, G. and Jonah, C.D., *Radiat. Phys. Chem.*, 24(5/6), 557, 1985. With permission.)

for the electron mobility measurements in radiation chemistry studies. Carbon disulfide, CS_2, has a high electron affinity. Therefore, it was expected that the electron mobility in CS_2 should not be high. In the initial experiments, a small mobility of electrons was obtained, as expected (Gee and Freeman, 1989: 5399). However, very high Ps formation yields were obtained in CS_2. The CS_2 concentration dependence on Ps formation was measured in many nonpolar liquids, as shown in Figure 7.12 (Jansen and Mogensen, 1977: 75). Ps formation was inhibited by the addition of a small amount of CS_2, as observed for many other electron scavengers. However, Ps formation enhancement was observed at higher concentration. The only reasonable explanation for this phenomenon is the high electron mobility at higher concentration of CS_2. It was completely opposite of the results obtained by the electron mobility measurement; it was then measured again, and high electron mobility was finally obtained (Gee and Freeman, 1989: 5399).

7.6.3 Electron Thermalization in Water

Jonah tried to investigate the electron thermalization in water by pulse radiolysis experiments (Jonah and Chernovitz, 1990: 935). The decay curves of the absorption of hydrated electrons in water were measured as a function of the concentration of heavy water, as shown in Figure 7.13. The heavy water concentration dependence of decay rates of hydrated electrons shows a linear relation with the concentration of water with two Ds, as indicated in Figure 7.14. Therefore, it was mentioned that H_2O and HOD showed similar effects on the electron thermalization.

Ps formation is caused by free or quasi-free electrons, as mentioned in Section 7.6.1. Ps formation probability will increase with the decreasing distance of electron–positron pairs, because the competitive phenomenon of Ps formation is the hydration in water. Therefore, Ps formation will be capable of providing some information regarding the thermalization distance. Figure 7.15 shows that Ps formation yields showed linear dependence on the concentration of D, and not on the water

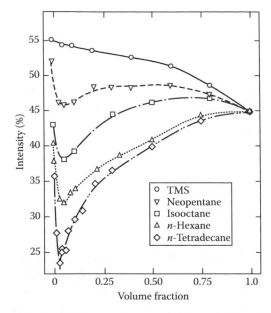

FIGURE 7.12 Intensity, I_3, of the long-lived component versus volume fraction of carbon disulfide. (Reprinted from Jansen, P. and Mogensen, O.E., *Chem. Phys.*, 25, 75, 1977. With permission.)

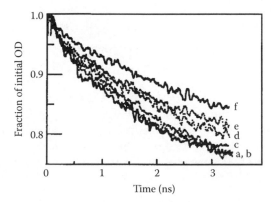

FIGURE 7.13 Decay of the hydrated electron at 600 nm in different mixtures of H_2O and D_2O: (a) 100 mol% H_2O, (b) 60 mol% H_2O, (c) 50 mol% H_2O, (d) 40 mol% H_2O, (e) 20 mol% H_2O, and (f) 0 mol% H_2O. (Reproduced from Jonah, C.D. and Chernovitz, A.C., *Can. J. Phys.*, 68, 935, 1990.)

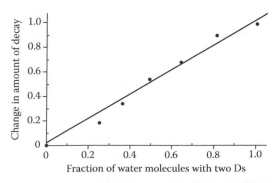

FIGURE 7.14 Change in the amount of decay at 3.4 ns versus fraction of D_2O molecules in the solution. (Reprinted from Jonah, C.D. and Chernovitz, A.C., *Can. J. Phys.*, 68, 935, 1990. With permission.)

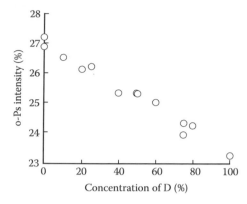

FIGURE 7.15   D concentration dependence on o-Ps intensity.

molecules with two Ds (Hirade, 1995: 675). This shows that the thermalization distance is larger in heavy water. However, it does not show that H_2O and HOD have the same effect on the thermalization distance. Crowell et al. indicated that the diffusion coefficients of reactive species in D_2O influenced the reactions of the hydrated electrons, and that the electron thermalization distance in D_2O did not need to be greater than that in H_2O (Crowell and Bartels, 1996: 17713). Further studies are needed to understand the thermalization of electrons in water.

7.6.4   ELECTRIC FIELD EFFECT (BLOB MODEL OF PS FORMATION)

The application of external electric fields can quench Ps formation, as shown in Figure 7.16 (Stepanov et al., 2005: 054205). Stepanov explained the electric field dependence on Ps formation using the modified spur model, that is, the blob model. He introduced the distribution of excess electrons and positrons in the terminal spur of the positron track, because the number of species in the terminal spur is large. Probably, there are more than 20–30 excess electrons. He employed the method of Debye screening of the positron in the terminal spur. He then divided these positrans into two positron species. One of these is thermalized in the excess electron blob, and the other is out of that blob, as indicated in Figure 7.17 (Stepanov et al., 2005: 054205). In the blob, there are many excess

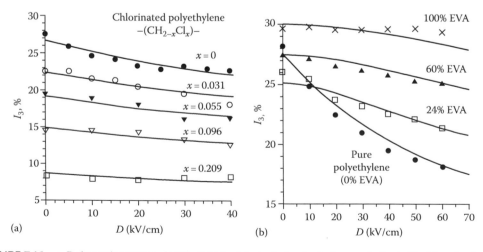

FIGURE 7.16   o-Ps intensity versus electric field in chlorinated polyethylene as indicated in (a) and polyethylene (PE) and ethylene vinyl acetate copolymer (EVA) as indicated in (b). Lines are fits obtained by the model explained in the text. (Reprinted from Stepanov, S.V. et al., *Phys. Rev. B*, 72, 054205, 2005. With permission.)

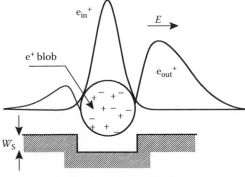

Effective potential well for e^+

FIGURE 7.17 Spatial distribution of the positron density at the Ps formation stage in the presence of an external electric field, E. e_{in}^+ is the density of the positrons bounded within the blob and not perturbed by E. The other part, e_{out}^+, is biased by the field. (Reprinted from Stepanov, S.V. et al., *Phys. Rev. B*, 72, 054205, 2005.)

electrons and the external electric field is not effective. Stepanov succeeded in explaining many experimental results using the blob model, as shown in Figure 7.16 (Stepanov et al., 2005: 054205).

7.6.5 SLOW PS FORMATION IN SPURS (YOUNG-AGE BROADENING EFFECTS)

A very interesting attempt has been made to clarify the Ps formation mechanism. The Stuttgart group produced many experimental results with the use of AMOC. They showed that there exists young-age broadening on $S(t)$ in many materials (Stoll et al., 1995: 17). One example of $S(t)$ without the young-age broadening is shown as filled circles in Figure 7.18 (Komuro et al., 2007: 330). At the positron age of zero, large $S(t)$ appeared. It was caused by the p-Ps intrinsic annihilation. One of the $S(t)$ observed by the Stuttgart group is shown in Figure 7.19 (Stoll et al., 1995: 17). At very young ages, smaller $S(t)$ appeared. Smaller $S(t)$ means that the energy distribution is broader. This is the reason why it is called "young-age broadening." They interpreted that young-age broadening

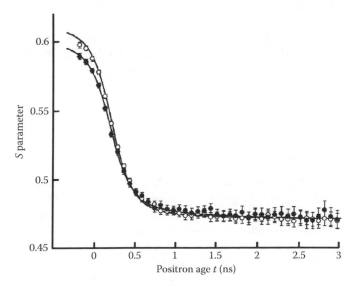

FIGURE 7.18 $S(t)$ curves for a synthetic fused quartz. The solid lines were simulated by assuming that Ps is formed at the time of positron injection. Open circles were measured under no electric fields, and filled circles were measured under an electric field of 48 kV/cm. (Reprint from Komuro, Y. et al., *Radiat. Phys. Chem.*, 76, 330, 2007. With permission.)

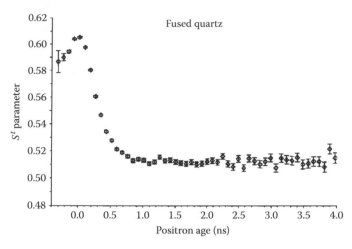

FIGURE 7.19 Annihilation γ-rays line-shape parameter, S^t, measured at room temperature for fused quartz. (Reprinted from Stoll, H. et al., *Appl. Surf. Sci.*, 85, 17, 1995. With permission.)

was caused by the incompletely thermalized p-Ps (Stoll et al., 1995: 17). They also claimed that it is possible to observe the process of Ps thermalization by young-age broadening. On the other hand, Komuro interpreted that young-age broadening is caused by the slow Ps formation in the terminal spur. For the electron–positron pair that acquires a larger initial distance after the thermalization, Ps formation time is slower. If there is slow Ps formation, the fraction of annihilation from the free positron is large near the positron age of zero; therefore, young-age broadening appears. Komuro applied a weak external electric field to test the model. The young-age broadening, if caused by slow Ps formation, will disappear because a weak external electric field will affect the slow Ps formation. She used very pure fused quartz to observe the young-age broadening. The measured $S(t)$ is shown in Figure 7.18 (Komuro et al., 2007: 330). Open circles showed probably the first young-age broadening except for the results form the Stuttgart group. Although the effect was very small, the external electric field quenched the young-age broadening as the filled circles indicated (Komuro et al., 2007: 330).

The slow Ps formation reactions were obtained for ionic liquids. The result for *N,N,N*-trimethyl-*N*-propylammonium bis(trifluoromethanesulfonyl)imide (TMPA-TFSI) is shown in Figure 7.20 (Hirade, 2009b: 232). The solid line indicates the $S(t)$ curve assuming Ps formation at the positron age of zero. The dashed line indicates the $S(t)$ curve with an electron solvation time of about 30 ps. It is a clear indication of the long lifetime of free or quasi-free electrons in ionic liquids, as indicated by Katoh (Katoh et al., 2007: 4770).

7.6.6 Ps Formation with Trapped Electrons

As explained in Section 7.5, the large and slow enhancements of Ps formation at low temperatures are caused by the accumulation of long-lived shallowly localized electrons, such as trapped electrons, by the irradiation of injected positrons. The most famous experimental result was observed for branched polyethylene by Kindle, as shown in Figure 7.21, in 1987 (Kindle 1987: 707). As explained in Section 7.5, some people believed that it was a special phenomenon for polymers. However, the details of the mechanism of the slow enhancement of Ps formation were explained by the accumulation of trapped electrons in 1998, and it was not a special phenomenon for polymers (Hirade et al., 1998: 89; 2000: 465; Wang et al., 1998: 4654). Indeed the slow enhancement of Ps formation in cyclohexane at low temperatures had been measured by Eldrup in 1980 (Eldrup et al., 1980: 175). This was the first experimental result that showed a slow Ps enhancement caused by the accumulation of trapped electrons, as indicated in Figure 7.22 (Hirade, 2003: 375).

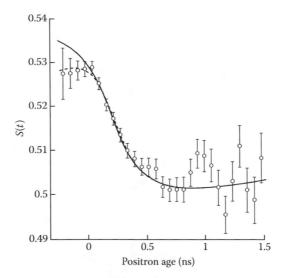

FIGURE 7.20 $S(t)$ curve for N,N,N-trimethyl-N-propylammonium TMPA-TFSI. Solid lines are obtained by an assumption that Ps forms at the positron age = 0. Dashed lines are obtained by slow reactions in ionic liquids. (Reprinted from Hirade, T., *Mater. Sci. Forum*, 607, 232, 2009b. With permission.)

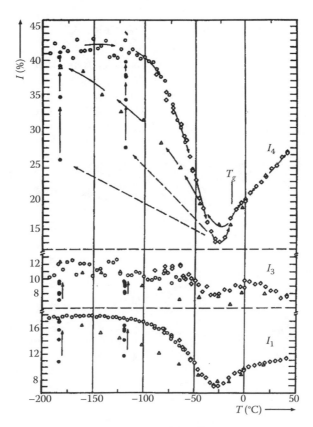

FIGURE 7.21 The relative intensities obtained by fourth-component analysis as a function of temperature. I_4 indicates the relative intensity of o-Ps. Open triangles were measured with faster cooling rates than open circles. Filled circles were measured with fixed temperatures. (Reproduced from Kindl, P. and Reiter, G., *Phys. Status Soldii (a)*, 104, 707, 1987. With permission.)

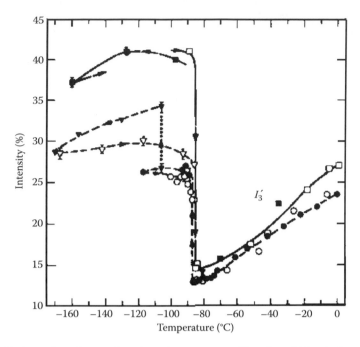

FIGURE 7.22 o-Ps intensity, I_3', as a function of temperature. The dotted line indicates the isothermal measurement. (Reproduced from Eldrup, M. et al., *Faraday Discuss. Chem. Soc.*, 69, 175, 1980.)

Some experiments exhibited Ps formation by trapped electrons. At first, the visible light effect was used. In this case, the Ps formation was inhibited immediately, because the visible light quenched the trapped electrons. This effect was tested by Hirade at first, and subsequently many other investigators obtained the same effect in many materials, as shown in Figure 7.23 (Hirade et al., 2000: 465). It is obtained by the use of ^{60}Co γ-irradiation.

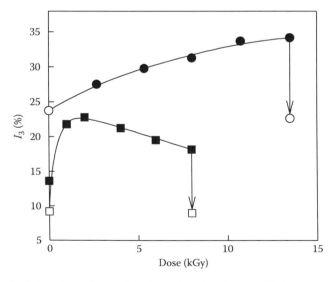

FIGURE 7.23 Absorbed dose dependence of the intensity of the longest-lifetime component, I_3, in PE (■, in darkness; □, under visible light) and PMMA6N (●, in darkness; ○, under visible light) at 77 K. (Reprinted from Hirade, T. et al., *Radiat. Phys. Chem.*, 58, 465, 2000. With permission.)

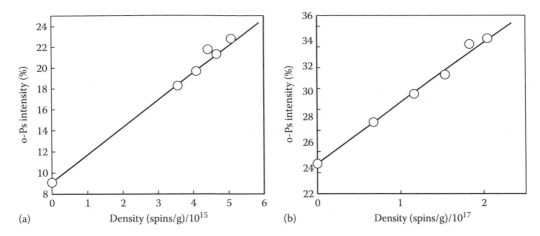

(a)

(b)

FIGURE 7.24 Relation between trapped electron density and intensity of the longest-lifetime component, I_3, at 77 K, (a) in PE and (b) in PMMA6N. (Reprinted from Hirade, T., et al., *Radiat. Phys. Chem.*, 58, 465, 2000. With permission.)

The trapped electron density dependence of Ps formation enhancement was also measured by Hirade, as shown in Figure 7.24 (Hirade et al., 2000: 465). A combination of EPR and PAL measurements gives linear relations in polyethylene (PE) and polymethyl methacrylate (PMMA). This is the only result of measuring the Ps formation enhancements caused by the irradiation of ^{60}Co γ-rays.

Ps formed by the spur process and that formed by trapped electrons could be separated by observing the formation time. The Ps formation by the spur process occurred within about 1 ps, and that by trapped electrons was delayed because positron diffusion was needed. p-Ps annihilation gives a narrower distribution of the annihilation γ-ray, as explained in Section 7.4.2, and a fast annihilation lifetime of 125 ps. The $S(t)$ curve observed for PE at 20 K is indicated in Figure 7.25 (Hirade et al., 2007: 3714). While a simple decrement of $S(t)$ was observed for PE with no trapped electrons, the $S(t)$ curve of the Ps formation by trapped electrons showed increment at first. This means that there exists the

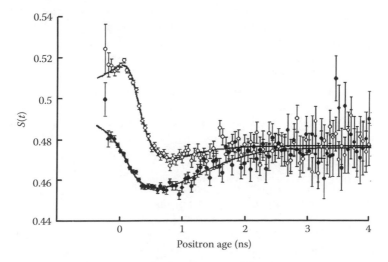

FIGURE 7.25 Positron age dependence of the S parameter. The open circles were measured in darkness, and the filled circles were measured under visible light. Solid lines are obtained by the least square fits by considering the positron localization. (From Hirade, T., *Radiat. Phys. Chem.*, 76, 84, 2007. With permission.)

delayed Ps formation by trapped electrons. The solid line in Figure 7.25 shows the fitted curve taking positron trapping into consideration. It is impossible to fit the $S(t)$ curve without positron trapping. The competitive phenomenon of Ps formation with trapped electrons, that is, positron trapping, was successfully observed and the positron trapping rate could be obtained (Hirade et al., 2007: 3714).

There are several methods for the detection of trapped electrons or anions formed at low temperatures. EPR, light absorption, and glow curve measurements are often applied in radiation chemistry, and now the positron annihilation method can be used. The binding energy of trapped electrons is usually 0.5–3 eV, and the Ps binding energy is 6.8 eV in vacuum and probably 4–5 eV in materials. Therefore, just by placing a positron in the materials, Ps can form, by picking off the trapped electrons. This means that positrons have two chances of Ps formation. The first chance is given by the spur reactions, and the positrons that escape from Ps formation by the spur process have the next chance of Ps formation with trapped electrons.

The observation of trapped electrons by the positron methods yields some advantages. The injected positrons induce irradiation and act as a probe. Continuous irradiation takes place during measurement. Some interesting measurements are possible, for example, the measurement of the apparent activation energies of local motions. The Arrhenius plots can give the apparent activation energies, for which the isothermal decay rates should be measured at several temperatures. This means that several samples should be prepared. However, the apparent activation energies can also be obtained just by measuring PAL at several temperatures, because continuous irradiation exists during measurements and the measured values are under the equilibrium condition of thermal decay and formation that takes place through positron irradiation. The formation rate is constant; therefore, Ps formation yields under the equilibrium condition at several temperatures can give the Arrhenius plots, through which the apparent activation energies can be obtained (Hirade, 2003: 375).

The saturated density of trapped electrons in PE is much lower than that in PMMA, as shown in Figure 7.24. However, Ps formation yields by the accumulation of trapped electrons are almost the same for both of these (Hirade et al., 1998: 89). This means that the positron diffusion length in PE is larger than that in PMMA.

On the other hand, the saturation of trapped electron density is controlled by the electron diffusion length. A larger diffusion length gives larger probability of electron–cation recombination, and so the saturated density becomes small. For example, a smaller diffusion length of electrons at lower temperatures gives higher density, because the positron mobility in PE is lower at lower temperatures (Brusa et al., 1995: 447). If the temperature dependence of diffusion lengths for electrons and positrons are similar, Ps formation yields will have very small temperature dependence.

An interesting experiment was conducted. At a certain low temperature, PAL was measured until the Ps formation yield saturated in a long-chain alkane. After a long time of saturation, the temperature was elevated and measured. The density of trapped electrons was not changed immediately, but the diffusion lengths of positrons and electrons had to be changed, as these are larger at higher temperatures. The density saturated at lower temperature, and then Ps formed at higher temperature. The Ps yields should be larger than the saturated value at that temperature. This phenomenon was successfully revealed by Zgardzinska (Zgardzinska et al., 2007: 309).

7.6.7 REACTIONS OF o-Ps WITH SPUR SPECIES

The fourth-lifetime components in pure benzene (Consolati et al., 1991: 7739), the scintillator NE104 (Consolati et al., 1992: 131), and a 1 M solution of 2,5-diphenyloxazole (PPO) in toluene (Consolati et al., 1992: 131) have been measured by Consolati et al., who showed that the fourth component is quite possibly caused by a Ps state due to the use of three-gamma measurements. The lifetime spectra of many liquids showed the fourth-lifetime components, as shown in Table 7.2 (Mogensen, 1995: 73). These fourth lifetimes are shorter than the longest lifetimes, and are caused by the reactions of o-Ps and the spur species (Mogensen and Jacobsen, 1982: 223). The most important reactions are those of oxidation of o-Ps by cation radicals. Mogensen proposed that

TABLE 7.2
Results of Four-Term Analysis of Lifetime Spectra for Liquids

Liquids	I_1 (%)	τ_1 (ns)	I_2 (%)	τ_2 (ns)	I_3 (%)	τ_3 (ns)	I_4 (%)	τ_4 (ns)
Hexane	15.2	0.127	39.8	0.410	6.0	1.7	39.0	4.18
2,2-Dimethylebutane	15.9	0.122	28.8	0.372	8.1	0.8	47.2	4.47
diethylether	15.6	0.155	49.4	0.410	8.4	2.1	26.6	4.60
3-Methylpentane	16.8	0.131	39.1	0.429	5.7	2.03	38.4	4.31
3-Methyloctane	14.9	0.132	41.7	0.407	4.5	1.70	38.9	3.87
2,2-Dimethylpentane	12.8	0.109	37.3	0.373	7.2	1.08	42.8	4.34
2,2-Dimethylhexane	13.6	0.122	46.4	0.430	4.2	1.96	35.8	4.14
Methanol	14.0	0.184	63.1	0.446	6.8	2.3	16.1	4.02

Source: Based on data from Mogensen, O.E., *Positron Annihilation in Chemistry*, Springer, Berlin, Germany, 1995, 73.

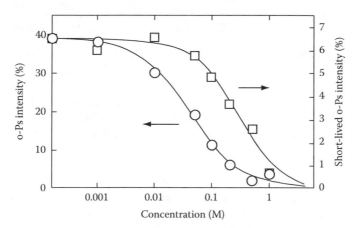

FIGURE 7.26 Intensities of the two o-Ps components in the lifetime spectra versus CCl$_4$ concentration in n-hexane. Open circles indicate the long-lifetime o-Ps, and open squares indicate the short-lifetime o-Ps. The lines are given by Equation 7.1 with α, β equal to 24 M^{-1}, 1.0, and 3.5 M^{-1}, 1.2, respectively. (Reprinted from Hirade, T. and Mogensen, O.E., *Chem. Phys.*, 170, 249, 1993a. With permission.)

electron scavengers should inhibit the fourth-lifetime component as the o-Ps component could be inhibited. This effect was measured directly for CCl$_4$/hexane mixtures, as indicated in Figure 7.26 (Hirade and Mogensen, 1993a: 249). The interesting feature of the inhibition of the fourth component is that the inhibition parameter, β, is larger than 1. The inhibition of Ps formation is normally described by the expression (Levay and Mogensen, 1980: 131; Mogensen and Jacobsen, 1982: 223)

$$I(C) = \frac{I(0)}{1+(\alpha C)^{\beta}} \qquad (7.2)$$

where
  $I(C)$ $(I(0))$ is the intensity at concentration $C(0)$
  α and β are fitting parameters

The Equation 7.2 was proposed by Levay and Mogensen (Levay and Mogensen, 1977) by applying empirical expression using a adjustable parameter β. The parameter β for the inhibition of the fourth component is 1.2. The β values for 59 different combinations of inhibitors and nonpolar liquids are found to be $\beta \leq 1$ in all cases (Mogensen, 1995: 122). The reason of this large β is that there exist

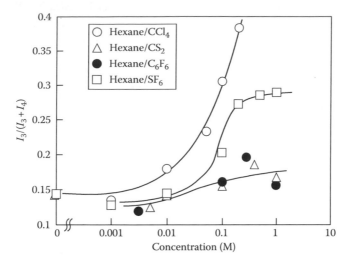

FIGURE 7.27 Effects on the fraction of short-lived o-Ps, I_3, in whole o-Ps, $I_3 + I_4$, by electron scavengers. (Reprinted from Hirade, T. and Mogensen, O.E., *J. Phys. IV* C4, II(3), 127, 1993b. With permission.)

enhancement and inhibition effects. The electron scavengers capture the excess electrons and the Ps formation is inhibited. Simultaneously, the electron scavengers inhibit the cation–excess electron recombination, and the cation concentration in the positron spur is enhanced. This is the reason why the electron scavenging effect is not visible for the fourth component around 0.01 M in Figure 7.26. Figure 7.27 indicates the fraction of the fourth components in all of the o-Ps components, that is, the sum of the fourth and the longest-lifetime components (Hirade and Mogensen, 1993b: 127; 1994: 491). It is smaller for CS_2 and C_6F_6. These are not just electron scavengers but also Ps formation enhancers. These electron scavengers capture electrons and the cation–electron recombination is inhibited, but freely moving positrons can arrive at these electron scavengers to pick off electrons to form Ps. The positron formed by the Ps oxidation by cations can also arrive at these scavengers to form Ps once more. Three-fourths of newly formed Ps will be o-Ps, and hence the oxidation is not visible.

The other important feature is a spin conversion reaction of o-Ps by radical species, such as cations or solvated electrons. p-Ps caused by the spin conversion reaction annihilates intrinsically with a lifetime of about 125 ps. It can be one of the reasons of the appearance of the fourth-lifetime component.

The temperature dependence of the lifetime of the o-Ps pick-off annihilation in water showed a strange tendency. The o-Ps pick-off annihilation lifetime should be larger at higher temperatures, as explained in Section 7.3.1. However, it is shorter at higher temperatures in water (Kotera et al., 2005: 184). This could be explained by the reaction between o-Ps and spur species (Stepanov et al., 2007: 90). There is some possibility of a reaction of geminate pairs of o-Ps and a radical in the positron spur. The geminate pair reactions show quantum beats caused by the spin-dependent reactions, and the details are given in Section 7.6.8.

7.6.8 POSITRONIUM AS A PROBE OF FREE RADICALS

AMOC measurements are a very useful tool to investigate the reactions of positron and Ps. As shown in Figure 7.28, AMOC can detect Ps oxidation, p-Ps/o-Ps spin conversion, and complex formation and positron reactions (Castellaz et al., 1996: 457). An example of p-Ps/o-Ps spin conversion was indicated in Figure 7.29 (Stoll et al., 1995: 17). While it is very difficult to know the existence of the p-Ps/o-Ps spin conversion by PAL, $S(t)$ obtained by AMOC is a very useful tool for this purpose.

Recently, a new method for the use of Ps reactions has been established by Hirade (2009a: 132). It is possible to observe the quantum beats caused by the hyperfine coupling in radicals. The complex formation and the oxidation of Ps should show spin dependence. The complex formation, that is, the radical reaction, can occur while the two unpaired electrons are in a singlet state, because both the

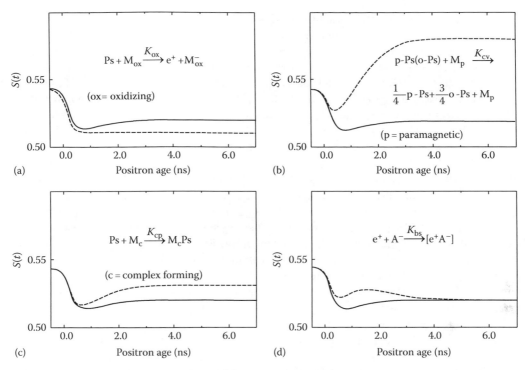

FIGURE 7.28 The age dependence of the line-shape functions, $S(t)$, for different chemical reactions. Solid lines: Ps formation without reactions. Dashed lines: Inclusion of reactions as indicated in the figures: (a) oxidation of Ps, (b) spin conversion of Ps, (c) Ps complex formation, and (d) e + bound-state formation. (From Castellaz, P., *J. Radioanal. Nucl. Chem.*, 210(2), 457, 1996, Figure 3. With permission.)

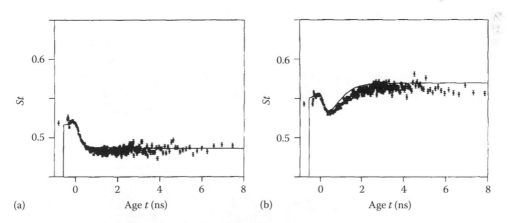

FIGURE 7.29 $St(t)$ for 0.1 M HTEMPO in benzene obtained by AMOC measured at 266 K (a, solid) and 313 K (b, liquid). (Reprinted from Siegle, A. et al., *Appl. Surf. Sci.*, 116, 140, 1997. With permission.)

electrons will be in the same orbital after the reaction. The same can be said for the oxidation. Ps is a bound state of an electron and a positron, that is, it has an unpaired electron, which means that Ps is a free radical. This electron has been one of the excess electrons in the terminal spur of the positron track. Therefore, there is a radical with an unpaired electron that has spin correlation with the electron in Ps, that is, the radical and the Ps are a geminate pair. There is one more possibility of reaction between the Ps and the radical, that is, the spin conversion reaction, as mentioned above. If the Ps is o-Ps, it has the longest lifetime and it is very easy to detect it, as indicated in Figure 7.29.

The geminate radical–Ps reaction is a diffusion-controlled reaction, and the radical reaction, that is, the complex formation or the oxidation, and the spin conversion reaction are the competitive reactions. When the probability of one reaction goes down, that of the other goes up. Now the radical reaction of the geminate radical–Ps pairs depends on the spin state of the two unpaired electrons.

The unpaired electron in the free radical, such as the hydroxyl radical in water, will have Larmor precession because of the hyperfine coupling. In liquids, only isotropic hyperfine coupling, that is, the Fermi contact term, is detectable. The Larmor precession frequency caused by a proton having a spin of 1/2 will be about 50–500 MHz. Ps has a very large hyperfine coupling constant, 203 GHz in vacuum. A positron spin of 1/2 gives the Larmor precession frequency of the electron in Ps, that is, 101.5 GHz. Although it is a value for Ps in vacuum, the frequency is very fast. Now, we applied 500 MHz for the radical and 101.5 GHz for Ps as an example. The geminate radical pair is in the singlet state initially, and these Larmor precessions make a transition between the singlet state and the triplet state. There are two frequencies of the singlet–triplet transition, 101.5 GHz + 500 MHz and 101.5 GHz − 500 MHz. Therefore, the observable singlet-state probability will have a swell, as shown in Figure 7.30. There is no fast oscillation at the positron ages of 0.5, 1.5, 2.5 ns,…, and half of the radical–Ps pairs are in the singlet state. Therefore, more than half of the radical–Ps pairs perform the p-Ps/o-Ps spin conversion reaction that can be detected by the AMOC measurement. On the other hand, the singlet-state probability at the swells, that is, at 0, 1, 2 ns,…, oscillates with about a 10 ps cycle. Although the average singlet-state probability is 0.5, most of the radical–Ps pairs will be in a singlet state every 10 ps at the swells. It means that the possibility of the radical reactions, such as complex formation or oxidation, at the swells will be more than between the swells, and hence the probability of the p-Ps/o-Ps spin conversion reaction will be less at the swells. Hence, there will be beats of enhancement on the $S(t)$ caused by the spin conversion of o-Ps between the swells, that is, at the positron ages of 0.5, 1.5, 2.5 ns,…, in the case of this example, with 500 MHz for the radical and 101.5 GHz for Ps. $S(t)$ were measured in water at 18°C and 25°C, and the quantum beats caused by the spin conversion of p-Ps/o-Ps were successfully observed, as in Figure 7.31 (Hirade, 2009a: 132). The hyperfine coupling constants of the hydroxyl radicals in water were 870 and 758 MHz at 18°C, and 816 and 656 MHz at 25°C (Hirade, 2009a: 132). These large hyperfine coupling constants indicate that it is not a simple hydroxyl radical. The quantity of the isotropic hyperfine coupling constant, that is, the Fermi contact term, of the OH radical is 73.25 MHz, while that of the OH–H_2O complex is 155.3 MHz (Brauer et al., 2004: 420). The large hyperfine coupling constants are probably caused by the radicals in water clusters or water cluster cation radicals. The beats appeared at the expected positron ages, and this indicates that the structures are quite stable

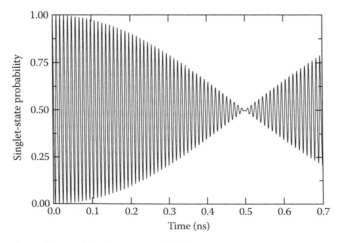

FIGURE 7.30 Time dependence of singlet-state probability of the geminate radical pair with Larmor precession frequencies, 500 MHz for the radical and 101.5 GHz for o-Ps.

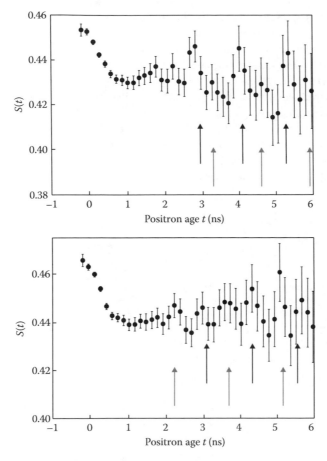

FIGURE 7.31 $S(t)$ curves obtained by AMOC for pure water after nitrogen gas bubbling. The upper curve is measured at 18°C and the lower curve is measured at 25°C. (From Hirade, T., *Chem. Phys. Lett.*, 480, 132, 2009a. With permission.)

at 0–6 ns. If there are many different structures that exist continuously, there will be many different hyperfine couplings, and so the beats cannot be observed. This means that there are only two structures and these show almost no changes up to about 6 ns. Every structure changes continuously with the elevation of temperature. The hyperfine coupling constants are smaller at higher temperatures.

The fourth-lifetime component was explained in Section 7.6.7. It is mainly caused by the reaction of o-Ps and cations. It is found in many liquids. This means that the quantum beats can be measured in many liquids and the direct observation of cation radicals is possible by this method. In the case of water, the hyperfine coupling constants seem to depend on the temperature. However, there will not be large temperature effects on the hyperfine coupling constants for radicals in other liquids. Figure 4 in Stoll et al. (1995: 17) shows $S(t)$ for *n*-hexane, and it is clear that there are reactions between o-Ps and cation radicals (Levay and Mogensen, 1980: 131; 1993a: 249; 1993b: 127; 1994: 491). The quantum beats appeared at around 2.7 and 3.8 ns. The spin state of the geminate pair, o-Ps and *n*-hexane cation radical, is a singlet at the positron age of zero; thus, the ages for having the beats are quite reasonable.

7.7 POSITRON ANNIHILATION ON MOLECULES

The most important process that we can derive from positron studies is annihilation. Positron annihilation can provide information of the interaction between positrons and molecules or atoms in gases. This information is very important in many areas of science and technology. Annihilation γ-rays from the universe are becoming a very interesting object for astrophysics (Milne, 2006: 548). At the incident positron energies below the threshold of Ps formation by the Ore model (Ore, 1949), the annihilation rates with some molecules are far in excess of the expected rates by simple two-body collisions (Paul and Saint-Pierre, 1963: 493). Many experiments and theoretical studies have been conducted since the initial experiments (Paul and Saint-Pierre, 1963: 493). The positron–molecule annihilation, studied by positrons with a well-controlled and tunable energy, was sufficient to resolve structures on the scale of molecular vibrational energies (Gilbert et al., 1997: 1944). Subsequently, large resonances in annihilation rates, Z_{eff}, were discovered, as shown in Figure 7.32 (Gilbert et al., 2002: 043201). Z_{eff} is the measured annihilation rate divided by that calculated for an uncorrelated electron gas at a density equal to the molecular number density. If there is no correlation between the molecular electrons and the positron, Z_{eff}, is the number of electrons in molecule.

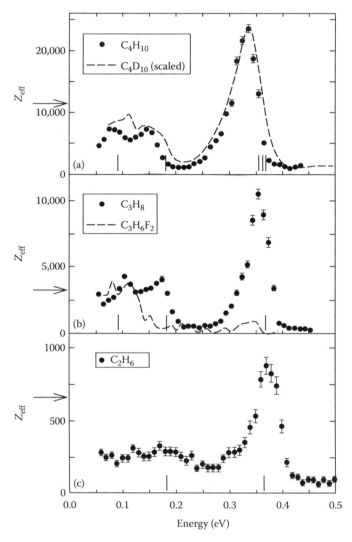

FIGURE 7.32   Positron annihilation rate, Z_{eff}, for (a) butane, (b) propane, and (c) ethane, as a function of positron energy, ε, in the range of $50 < \varepsilon < 450$ meV. Vertical bars along the abscissas indicate the strongest infrared-active vibrational modes. Arrows on the ordinate indicate Z_{eff} for a Maxwellian distribution of positrons at 300 K. Z_{eff} for d-butane is shown by the dashed curve in (a), with the amplitude normalized, and the energy scaled by the appropriate reduced-mass factor. The annihilation rate for 2,2-difluoropropane ($C_3H_6F_2$) is indicated by the dashed curve in (b). (Reprinted from Gilbert, S.J. et al., *Phys. Rev. Lett.*, 88(4), 043201, 2002. With permission.)

Recently, it has been understood that this phenomenon is due to positron attachment to molecules. This attachment enhances the overlap of the positron with molecular electrons, which increases the annihilation rates. The attachment is caused by vibrational Feshback resonances (Young and Surko, 2009: 9).

REFERENCES

Anderson, C. D. 1933. The positive electron. *Phys. Rev.* 43: 491–494. http://link.aps.org/abstract/PR/v4355/p491

Brandt, W., Coussot, G., and Paulin, R. 1969. Positron annihilation and electronic lattice structure in insulator crystals. *Phys. Rev. Lett.* 23: 522–524.

Brauer, C. S., Sedo, G., Grumstrup, E. M., Leopold, K. R., Marshall, M. D., and Leung, H. O. 2004. Effects of partially quenched orbital angular momentum on the microwave spectrum and magnetic hyperfine splitting in the OH–water complex. *Chem. Phys. Lett.* 401(4–6): 420–425.

Brusa, R. S., Duarte Naia, M., Margoni, D., and Zecca, A. 1995. Positron mobility in polyethylene in the 60–400 K temperature range. *Appl. Phys. A* 60: 447–453.

Cassidy, D. B., Deng, S. H. M., and Mills, A. P. 2007. Evidence for positronium molecule formation at a metal surface. *Phys. Rev. A* 76: 062511.

Castellaz, P., Major, J., Mujica, C., Schneider, H., Sebger, A., Siegle, A., Stoll, H., and Billard, I. 1996. Chemical reactions of positronium studied by age-momentum correlation (AMOC) using a relativistic positron beam. *J. Radioanal. Nucl. Chem.*, Articles, 210(2): 457–467.

Chalermkarnnon, P., Shishido, I., Yuga, M., Araki, H., and Shirai, Y. 2002. b+-g Coincidence positron lifetime spectrometer with positron energy selection by electromagnetic lens. *J. Jpn. Inst. Met.* 66(10): 1004–1008.

Consolati, G., Gerola D., and Quasso, F. 1991. An unusual three-quantum yield from positrons annihilated in benzene. *J. Phys.: Condens. Matter* 3: 7739–7742.

Consolati, G., Gerola, D., and Quasso, F. 1992. A positron annihilation study in some amorphous scintillators. *Z. Phys. B: Condens. Matter* 88: 131–133.

Coussot, G. 1970. Contribution a l'etudu de la structure electronique des isolants par annihilation du positon. PhD thesis, Faculté des Sciences d'Orsay, Universite Paris-Sud, Orsay, France.

Crowell R. A. and Bartels D. M. 1996. H_2O/D_2O isotope effect in geminate recombination of the hydrated electron. *J. Phys. Chem.* 100: 17713–17715.

Dirac, P. A. M. 1930. Annihilation of electrons and protons. *Proc. Camb. Philos. Soc.* 26: 361–375.

Dull, T. L., Frieze, W. E., Gidley, D. W., Sun J. N., and Yee, A. F. 2001. Determination of pore size in mesoporous thin films from the annihilation lifetime of positronium. *Phys. Chem. B* 105: 4657–4662.

Duplatre, G. and Jonah, C. D. 1985. Reactions of electrons in high-concentration water solutions—A comparison between pulse-radiolysis and positron annihilation lifetime spectroscopy data. *Radiat. Phys. Chem.* 24(5/6): 557–565.

Eldrup, M. 1976. Vacancy migration and void formation in g-irradiated ice. *J. Chem. Phys.* 64: 5283–5290.

Eldrup, M., Lightbody, D., and Sherwood, J. 1980. Studies of phase transformation in molecular crystals using the positron annihilation technique. *Faraday Discuss. Chem. Soc.* 69: 175.

Eldrup, M., Lightbody, D., and Sherwood, J. N. 1981. The temperature dependence of positron lifetimes in solid pivalic acid. *Chem. Phys.* 63: 51–58.

Fukaya, Y., Kawasuso, A., and Ichimiya, A. 2009. Inelastic scattering processes in reflection high-energy positron diffraction from a Si(111)-7×7 surface. *Phys. Rev. B* 79: 193310.

Gee, N. and Freeman, G. R. 1989. Electron and ion mobility in carbon disulfide liquid from 163 to 501 K. *Chem. Phys.* 90(10): 5399–5405.

Gilbert, S. J., Kurz, C., Greaves, R. G., and Surko, C. M. 1997. Creation of a monoenergetic pulsed positron beam. *Appl. Phys. Lett.* 70: 1944–1946.

Gilbert, S. J., Barnes, L. D., Sullivan, J. P., and Surko, C. M. 2002. Vibrational-resonance enhancement of positron annihilation in molecules. *Phys. Rev. Lett.* 88(4): 043201.

Goworek, T. 2007. Free volumes in solids under high pressure. *Radiat. Phys. Chem.* 76: 318–324.

Hirade, T. 1995. Positronium formation in H_2O, D_2O and HDO mixture. *Mater. Sci. Forum* 175–178: 675–678.

Hirade, T. 1996. The study of the electron-positron pair in CS_2. *Bull. Chem. Soc. Jpn.* 69(8): 2153–2157.

Hirade, T. 2003. Positronium formation in low temperature polymers. *Radiat. Phys. Chem.* 68: 375–379.

Hirade, T. 2007. Connections between radiation and positronium chemistry. *Radiat. Phys. Chem.* 76: 84–89.

Hirade, T. 2009a. Age-momentum correlation measurements of positron annihilation in water: Possibility of quantum beats on ortho-positronium reactions. *Chem. Phys. Lett.* 480: 132–135.

Hirade, T. 2009b. Positronium formation in room temperature ionic liquids. *Mater. Sci. Forum* 607: 232–234.

Hirade, T. and Mogensen, O. E. 1993a. Dependence of the yield of the short-lived ortho-Ps state on the concentration of CCl_4 in hexane. *Chem. Phys.* 170: 249–256.

Hirade, T. and Mogensen, O. E. 1993b. Effects of scavengers on short-lived ortho-Ps. *J. Phys. IV* C4, II, 3: 127–130.

Hirade, T. and Mogensen, O. E. 1994. Secondary reaction of Ps in the positron spur. *Hyperfine Interact.* 84: 491–498.

Hirade, T., Wang, C. L., Maurer, F. H. J., Eldrup, M., and Pedersen, N. J. 1998. Visible light effect on positronium formation in PMMA at low temperature. In *Abstract book for The 35th Annual Meeting on Radioisotopes in the Physical Science and Industries*, June 30–July 1, Tokyo, Japan, p. 89.

Hirade, T., Maurer, F. H. J., and Eldrup, M. 2000. Positronium formation at low temperatures: The role of trapped electrons. *Radiat. Phys. Chem.* 58: 465–471.

Hirade, T., Suzuki, N., Saito, F., and Hyodo, T. 2007. Reactions of free positrons in low temperature polyethylene-delayed Ps formation and positron localization. *Phys. Status Solidi (c)* 4, 10: 3714–3717.

Hyodo, T. and Takahashi, Y. 1977. Evidence for positronium-like states in NaF and NaCl. *J. Phys. Soc. Jpn.* 42: 1065–1066.

Ito, K., Nakanishi, H., and Ujihira, Y. 1999. Extension of the equation for the annihilation lifetime of ortho-positronium at a cavity larger than 1 nm in radius. *J. Phys. Chem. B* 103: 4555–4558.

Jacobsen, F. M., Mogensen, O. E., and Trumpy, G. 1982. Correlation of positronium yield with excess electron mobility in liquid neopentane. *Chem. Phys.* 69: 71–80.

Jansen, P. and Mogensen, O. E. 1977. Further studies of the spur process of positronium formation in mixtures of organic liquids. *Chem. Phys.* 25: 75–86.

Jean, Y. C. 1993. Positron spectroscopy of solids. In *Proceedings of the International School of Physics* "Enrico Fermi" Course CXXV, A. Dupasquier and A. P. Mills Jr. (eds), Varena on Lake Como, pp. 569–571.

Jonah, C. D. and Chernovitz, A. C. 1990. The mechanism of electron thermalization in H_2O, D_2O, and HDO. *Can. J. Phys.* 68: 935–939.

Kasai, J., Hyodo, T., and Fujiwara, K. 1988. Positronium in alkali halides. *J. Phys. Soc. Jpn.* 57: 329–341.

Katoh, R., Yoshida, Y., Katsumura, Y., and Takahashi, K. 2007. Electron photodetachment from iodide in ionic liquids through charge-transfer-to-solvent (CTTS) band excitation. *J. Phys. Chem. B* 111: 4770–4774.

Kawasuso, A. and Okada, S. 1998. Reflection high energy positron diffraction from a Si (111) surface. *Phys. Rev. Lett.* 81: 2695–2698.

Kindl, P. and Reiter, G. 1987. Investigations on the low-temperature transitions and time effects of branched polyethylene by the positron lifetime technique. *Phys. Status Solidi (a)* 104: 707–713.

Kobayashi, Y. 1992. Pressure dependence of ortho-positronium lifetimes in molecular liquids. *Ber. Bunsen-Ges.:-Phys. Chem. Chem. Phys.* 96, 12: 1869–1872.

Komuro, Y., Hirade, T., Suzuki, R., Ohdaira, T., and Muramatsu, M. 2007. Positronium formation in fused quartz: Experimental evidence of delayed formation. *Radiat. Phys. Chem.* 76: 330–332.

Kotera, K., Saito, T., and Yamanaka, T. 2005. Measurement of positron lifetime to probe the mixed molecular states of liquid water. *Phys. Lett. A* 345: 184–190.

Levay, B. and Mogensen, O. E. 1977. Correlation between the inhibition of positronium formation by scavenger molecules, and chemical reaction rate of electrons with these molecules in nonpolar liquids, *J. Phys. Chem.* 81: 373–377.

Levay B. and Mogensen, O. E. 1980. Positron spur reactions with excess electrons and anions in liquid organic mixtures of electron acceptors. *Chem. Phys.* 53: 131–139.

Milne, P. A. 2006. Distribution of positron annihilation radiation. *New Astron. Rev.* 50, 7–8: 548–552.

Mogensen, O. E. 1974. Spur reaction model of positronium formation. *J. Chem. Phys.* 60: 998–1004.

Mogensen, O. E. 1994. Positron states between positronium and "free positron" states in condensed matter. *Hyperfine Interactions*, 84, 1: 377–387.

Mogensen, O. E. 1995. *Positron Annihilation in Chemistry*. Springer, Berlin, Germany.

Mogensen, O. E. and Eldrup, M. 1977. Positronium Bloch function, and trapping of positronium in vacancies, in ice. Risø Report No. 366.

Mogensen, O. E. and Eldrup, M. 1978. Vacancies in pure ice studied by positron annihilation techniques. *J. Glaciol.* 21, 85: 85–99.

Mogensen, O. E. and Jacobsen, F. M. 1982. Positronium yields in liquids determined by lifetime and angular correlation measurements. *Chem. Phys.* 73: 223–234.

Mogensen, O. E., Kvajic, G., Eldrup, M., and Milosevic-Kvajic, M. 1971. Angular correlation of annihilation photons in ice single crystals. *Phys. Rev. B* 4: 71–73.

Ohdaira, T., Suzuki, R., Mikado, T., Ohgaki, H., Chiwaki, M., and Yamazaki, T. 1997. An apparatus for high-resolution positron-annihilation induced Auger-electron spectroscopy using a time-of-flight technique. *Appl. Surf. Sci.* 116: 177–180.

Ore, A. 1949. *Univ. of Bergen Aarb. Naturvit. Rekke* 9.

Oshima, N., Suzuki, R., Ohdaira, T., Kinomura, A., Narumi, T., Uedono, A., and Fujinami, M. 2009. Rapid three-dimensional imaging of defect distributions using a high-intensity positron microbeam. *Appl. Phys. Lett.* 94: 194104.

Paul, D. A. L. and Saint-Pierre, L. 1963. Rapid annihilations of positrons in polyatomic gases. *Phys. Rev. Lett.* 11: 493–496.

Saito, H., Nagashima, Y., Kurihara, T., and Hyodo, T. 2002. A new positron lifetime spectrometer using a fast digital oscilloscope and BaF$_2$ scintillators. *Nucl. Instrum. Methods Phys. Res. A* 487: 612–617.

Siegle, A., Stoll, H., Castellaz, P., Major, J., Schneider, H., and Seeger, A. 1997. Two-dimensional analysis of positron age-momentum correlation (AMOC) data. *Appl. Surf. Sci.* 116: 140–144.

Stepanov, S. V., Byakov, V. M., and Kobayashi, Y. 2005. Positronium formation in molecular media: The effect of the external electric field. *Phys. Rev. B* 72: 054205.

Stepanov, S. V., Byakov, V. M., and Hirade, T. 2007. To the theory of Ps formation. New interpretation of the e lifetime spectrum in water. *Radiat. Phys. Chem.* 76: 90–95.

Stoll, H., Koch, M., Lauff, U., Maier, K., Major, J., Schneider, H., Seeger, A., and Siegle, A. 1995. Annihilation of incompletely thermalized positronium studied by age-momentum correlation. *Appl. Surf. Sci.* 85: 17–21.

Suzuki, R., Kobayashi, Y., Mikado, T., Ohgaki, H., Chiwaki, M., Yamazaki, T., and Tomimasu, T. 1991. Slow positron pulsing system for variable energy positron lifetime spectroscopy. *Jpn. J. Appl. Phys.* 30: L532–L534.

Suzuki, R., Ohdaira, T., and Mikado, T. 2000. A positron lifetime spectroscopy apparatus for surface and near-surface positronium experiments. *Radiat. Phys. Chem.* 58, 5–6: 603–606.

Suzuki, N., Hirade, T., Saito, F., and Hyodo, T. 2003. Positronium formation reaction of trapped electrons and free positrons: Delayed formation studied by AMOC. *Radiat. Phys. Chem.* 68: 647–649.

Tao, S. J. 1972. Positronium annihilation in molecular substances. *J. Chem. Phys.* 56: 5499–5510.

Van House, J. and Rich, A. 1988. Arthur, first results of a positron microscope. *Phys. Rev. Lett.* 60: 169–172.

Wang, C. L., Hirade, T., Maurer, F. H. J., Eldrup, M., and Pedersen, N. J. 1998. Free-volume distribution and positronium formation in amorphous polymers: Temperature and positron-irradiation-time dependence. *J. Chem. Phys.* 108 11: 4654–4661.

Young, J. A. and Surko, C. M. 2009. Resonant positron annihilation on molecules. *Mater. Sci. Forum* 607: 9–16.

Zgardzinska, B., Hirade, T., and Goworek, T. 2007. Positronium formation on trapped electrons in n-heptadecane. *Chem. Phys. Lett.* 446: 309–312.

8 Muon Interactions with Matter

Khashayar Ghandi
Mount Allison University
Sackville, New Brunswick, Canada

Yasuhiro Miyake
High Energy Accelerator Research Organization
Tokai, Japan

CONTENTS

8.1 THE INTERACTION OF MUONS WITH MATTER IN THE GAS PHASE

8.1.1 GENERAL INTRODUCTION

Muons (μ^+, μ^-) occur naturally in cosmic ray showers, but not in great enough intensity for study. Therefore, they are produced artificially with MeV kinetic energies (4.1 MeV for surface muons and larger energies for muons from decay channels) at nuclear accelerators like TRIUMF in Canada, Rutherford Appleton Laboratory (RAL) in the United Kingdom, The Paul Scherrer Institute in Switzerland, and the Japan Proton Accelerator Research Complex.

In the decay sequence $\pi^+ \rightarrow \mu^+ \rightarrow e^+$, the muon is produced 100% spin polarized from pion decay and the (detected) positron is subsequently emitted preferentially along the muon-spin direction.

Both provide, due to parity violation, a sensitive measure of the interactions of the muon spin with its environment (Walker, 1983; Roduner, 1988, 1993; Kempton et al., 1991; Fleming and Senba, 1992; Smigla and Belousov, 1994; Mozumder, 1999; Percival et al., 1999; Johnson et al., 2005).

Studies on muon interactions with matter in the gas phase are relevant to the fields of atomic, nuclear, and particle physics; as well as radiation chemistry, hot-atom chemistry, and studies of linear energy transfer (LET), which is the average energy released per unit path length due to ionization and excitation processes as shown in Mozumder, 1999.

To maximize the lifetime of materials used in the reaction vessels for applications of radiation, radiation chemistry in the gas phase needs to be understood. The main purpose of this section is to review the thermalization effects and radiation chemistry of positive muons in the gas phase.

There are some differences in the thermalization processes of electrons and heavy charged particles in the gas phase, affecting the LET and hence affecting the nature of the radiation effects involved. The extent of carrier gas decomposition due to ionizing radiation is a strong function of the absorbed energy density of the incident radiation, with higher LET radiation causing decomposition of the medium, while low LET radiation leads to little or no decomposition. Differing LETs and stopping distances can give rise to differences in measured yields of products shown in Mozumder (1999).

8.1.2 Muon Polarization and Its Measurements

The initial spin polarization of the μ^+, at observation times, splits into three principal environments, depending on the nature of the stopping medium:

- Paramagnetic muonium (Mu $\equiv \mu^+ e^-$) with polarization P_{Mu} (Walker, 1983; Smigla and Belousov, 1994).
- Diamagnetic molecules, including molecules (like MuH, in water, shown in Percival et al., 1999 or in hydrocarbons, shown in Kempton et al., 1991) and molecular ions, AMu^+, where A is the molecule of the carrier gas (Johnson et al., 2005), with polarization P_D.
- Muoniated free radicals like CH_2MuCH_2 (Walker, 1983), with polarization P_R.

In the realm of radiation chemistry, it is important to be able to distinguish between these different environments (P_{Mu}, P_D, and P_R) and the timescale of their formation, provided by measurement of their differing spin precession frequencies in a transverse magnetic field (TF) (Walker, 1983; Smigla and Belousov, 1994; Kempton et al., 1991; Percival et al., 1999) or by RF-μSR measurements in the work of Johnson et al. (2005).

In contrast to conventional magnetic resonance studies, where bulk spin polarization is a consequence of differing Boltzmann populations in high magnetic fields, in muon techniques (collectively called μSR) the muon polarization is intrinsic to the probe, and is a direct consequence of the nuclear weak interaction. Throughout this work, an external static magnetic field (such as in TF muon methods) is taken along the positive z-direction $\vec{B} = (0, 0, B)$. The initial muon-spin direction at time $t_0 = 0$ is specified by the direction cosines:

$$\hat{S}_0 = (\cos\alpha_0, \cos\beta_0, \cos\gamma_0), \tag{8.1}$$

where α_0, β_0, and γ_0 are the angles between the initial spin and the x-, y-, and z-axes, respectively. Depending on which component of the muon-spin polarization is measured at observation time, one can consider the following two configurations separately. In the first, called the longitudinal detector/field configuration, the expectation value of the z component of the muon spin $\sigma_z^\mu \hbar/2$ is the observable quantity, where $\sigma_\mu = (\sigma_x^\mu, \sigma_y^\mu, \sigma_z^\mu)$ represents the Pauli spin matrices for the positive muon. The second is the transverse detector/field configuration in which the expectation value of

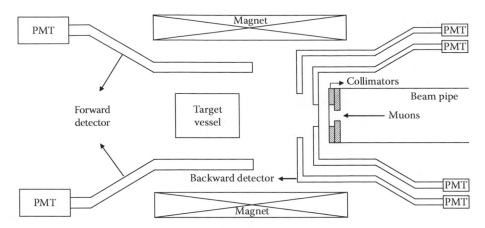

FIGURE 8.1   A schematic of the setup for μSR experiments.

the complex muon-spin polarization in the xy plane, $\sigma_\mu^+ = \sigma_x^\mu + i\sigma_y^\mu$, is relevant at observation time, where σ_μ^+ is the muon-spin raising operator.

Muon beams enter and stop in the sample vessel. Detectors in coincidence detect muons entering the sample and positrons emitted when the stopped muons decay. Positron detectors are arranged around the sample vessel (Figure 8.1).

In a μSR experiment at a continuous beam facility in the time-differential mode, the elapsed time between the stop of each muon registered through a muon coincidence (and with a time to digital converter, TDC) and the detection of its decay positron (stopping the TDC) is measured, and the data are collected and binned in a histogram of counts as a function of time (Figure 8.2). In the transverse detector/static field configuration, the probability of detecting the decay positron in a given direction varies as the expectation value of the complex muon-spin polarization in the xy plane oscillates in the magnetic field. Thus, μSR histograms obtained from each positron detector contain oscillations in the muon decay spectrum, and these oscillations correspond to the time dependence of the muon polarization.

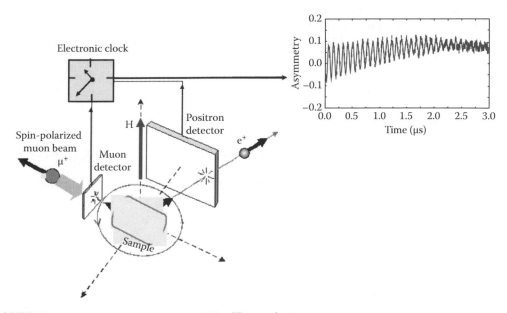

FIGURE 8.2   A schematic of the setup for TF-μSR experiments.

Each histogram has the following form:

$$N(t) = N_0 e^{-t/\tau_\mu} \left[1 + A(t)\right] + b, \tag{8.2}$$

where

N_0 is an overall normalization that depends on a number of factors, such as the solid angle of the positron detectors and the number of stopped muons

b represents random accidental (background) events

$\tau_\mu = 2.197\,\mu s$ (the muon lifetime)

$A(t)$ represents the "asymmetry," which represents the μSR signal of interest, and is similar to free induction decay (FID) in magnetic resonance

The asymmetry parameter includes contributions from *all* muon environments, paramagnetic Mu and free-radical species, as well as diamagnetic species:

$$A(t) = \Sigma_i A_i \exp(-\lambda_i t) \cos(w_i t + \varphi_i), \tag{8.3}$$

where

A_i is the initial amplitude of the muon fraction i in a given environment

λ_i is the relaxation rate of the muon spin in that environment

w_i is the corresponding precession frequency

φ_i is the initial phase for this fraction

The diamagnetic muon relaxation rate is invariably slow enough to be ignored in studies of Mu reactivity.

The parameters of interest (A_i, λ_i, and w_i) are extracted from fits of Equations 8.2 and 8.3 to experimental data and give information, respectively, on Mu formation, kinetics, and hyperfine interactions (in Mu or muoniated radicals), depending on the focus of a given experiment.

The total (nonrelativistic) Hamiltonian describing the isotropic hyperfine and Zeeman interactions of the muon and electron spin (in Mu) with an external magnetic field $\vec{B} = (0,0,B)$ is given by

$$H = -\left(\frac{\hbar\omega_\mu}{2}\right)\sigma_\mu^z + \left(\frac{\hbar\omega_e}{2}\right)\sigma_e^z + \left(\frac{\hbar\omega_0}{4}\right)\vec{\sigma}_\mu \cdot \vec{\sigma}_e, \tag{8.4}$$

where

$\vec{\sigma}_e$ represent the Pauli spin matrices for the electron

$\omega_0/2\pi$ is the Mu hyperfine frequency

the quantity ω_e is defined by $\omega_e = \gamma_e B$ in terms of the gyromagnetic factor γ_e of the electron $\gamma_e/2\pi = 28.02421$ GHz/T

the quantity ω_μ is defined by $\omega_\mu = \gamma_\mu B$ in terms of the gyromagnetic factor γ_μ of the muon

The energy eigenvalues $\hbar\omega_1$, $\hbar\omega_2$, $\hbar\omega_3$, and $\hbar\omega_4$ are labeled in the decreasing order in the low field regime:

$$\omega_1 = \frac{\omega_0}{4} + x\left(\frac{\omega_0}{2}\right)\frac{(\gamma_e - \gamma_\mu)}{(\gamma_e + \gamma_\mu)}, \quad \omega_2 = -\frac{\omega_0}{4} + \left(\frac{\omega_0}{2}\right)\sqrt{x^2 + 1},$$

$$\omega_3 = \frac{\omega_0}{4} - x\left(\frac{\omega_0}{2}\right)\frac{(\gamma_e - \gamma_\mu)}{(\gamma_e + \gamma_\mu)}, \quad \omega_4 = -\frac{\omega_0}{4} - \left(\frac{\omega_0}{2}\right)\sqrt{x^2 + 1}.$$

The energy eigenstates $|1\rangle$, $|2\rangle$, $|3\rangle$, and $|4\rangle$, corresponding to the energy values $\hbar\omega_1$, $\hbar\omega_2$, $\hbar\omega_3$, and $\hbar\omega_4$, labeled in the decreasing order in the low field regime, can be expressed as superposition of the Mu spin basis set $|\alpha_\mu\alpha_e\rangle$, $|\alpha_\mu\beta_e\rangle$, $|\beta_\mu\alpha_e\rangle$, and $|\beta_\mu\beta_e\rangle$ as

$$
\begin{pmatrix} |1\rangle \\ |2\rangle \\ |3\rangle \\ |4\rangle \end{pmatrix} = \begin{pmatrix} 1 & 0 & 0 & 0 \\ 0 & s & c & 0 \\ 0 & 0 & 0 & 1 \\ 0 & c & -s & 0 \end{pmatrix} \begin{pmatrix} |\alpha_\mu\alpha_e\rangle \\ |\alpha_\mu\beta_e\rangle \\ |\beta_\mu\alpha_e\rangle \\ |\beta_\mu\beta_e\rangle \end{pmatrix} \equiv U \begin{pmatrix} |\alpha_\mu\alpha_e\rangle \\ |\alpha_\mu\beta_e\rangle \\ |\beta_\mu\alpha_e\rangle \\ |\beta_\mu\beta_e\rangle \end{pmatrix},
\tag{8.5}
$$

where c and s are positive parameters defined by

$$
c^2 = \left(\frac{1}{2}\right)\left(\frac{1+x}{\sqrt{x^2+1}}\right) \quad \text{and} \quad s^2 = \left(\frac{1}{2}\right)\left(\frac{1-x}{\sqrt{x^2+1}}\right)
$$

in terms of the reduced magnetic field

$$
x = \frac{B}{B_0} \quad \text{with} \quad B_0 = \frac{\omega_0}{\gamma_\mu + \gamma_e}.
$$

By solving the above equation for $|\alpha_\mu\alpha_e\rangle$, $|\alpha_\mu\beta_e\rangle$, $|\beta_\mu\alpha_e\rangle$, and $|\beta_\mu\beta_e\rangle$, one obtains

$$
\begin{pmatrix} |\alpha_\mu\alpha_e\rangle \\ |\alpha_\mu\beta_e\rangle \\ |\beta_\mu\alpha_e\rangle \\ |\beta_\mu\beta_e\rangle \end{pmatrix} = \begin{pmatrix} 1 & 0 & 0 & 0 \\ 0 & s & 0 & c \\ 0 & c & 0 & -s \\ 0 & 0 & 1 & 0 \end{pmatrix} \begin{pmatrix} |1\rangle \\ |2\rangle \\ |3\rangle \\ |4\rangle \end{pmatrix} \equiv U^{-1} \begin{pmatrix} |1\rangle \\ |2\rangle \\ |3\rangle \\ |4\rangle \end{pmatrix}.
\tag{8.6}
$$

Assuming $\phi_{\alpha\alpha}(t)$, $\phi_{\alpha\beta}(t)$, $\phi_{\beta\alpha}(t)$, and $\phi_{\beta\beta}(t)$ are the Mu spin states that coincide, at $t = 0$, with $|\alpha_\mu\alpha_e\rangle$, $|\alpha_\mu\beta_e\rangle$, $|\beta_\mu\alpha_e\rangle$, and $|\beta_\mu\beta_e\rangle$, respectively. By using the time-dependent Schrödinger equation, one can write

$$
\begin{pmatrix} \phi_{\alpha\alpha}(t) \\ \phi_{\alpha\beta}(t) \\ \phi_{\beta\alpha}(t) \\ \phi_{\beta\beta}(t) \end{pmatrix} = U^{-1} \begin{pmatrix} e^{-i\omega_1 t}|1\rangle \\ e^{-i\omega_2 t}|2\rangle \\ e^{-i\omega_3 t}|3\rangle \\ e^{-i\omega_4 t}|4\rangle \end{pmatrix} = \begin{pmatrix} G_{11}(t) & 0 & 0 & 0 \\ 0 & G_{22}(t) & G_{23}(t) & 0 \\ 0 & G_{32}(t) & G_{33}(t) & 0 \\ 0 & 0 & 0 & G_{44}(t) \end{pmatrix} \begin{pmatrix} |\alpha_\mu\alpha_e\rangle \\ |\alpha_\mu\beta_e\rangle \\ |\beta_\mu\alpha_e\rangle \\ |\beta_\mu\beta_e\rangle \end{pmatrix},
\tag{8.7}
$$

where $G_{11}(t) = e^{-i\omega_1 t}$,

$$
G_{22}(t) = s^2 e^{-i\omega_2 t} + c^2 e^{-i\omega_4 t},
$$

$$
G_{23}(t) = cse^{-i\omega_2 t} - cse^{-i\omega_4 t}, \quad G_{32}(t) = cse^{-i\omega_2 t} - cse^{-i\omega_4 t},
$$

$$
G_{33}(t) = c^2 e^{-i\omega_2 t} + s^2 e^{-i\omega_4 t}, \quad G_{44}(t) = e^{-i\omega_3 t}.
$$

It should be noted that the matrix $[G_{jk}(t)]$ is a unit matrix at $t = 0$: $G_{jk}(0) = \delta_{jk}$.

The state in which the spin points in the positive x-direction at time $t = 0$ is given by

$$\psi(0) = \left(\frac{1}{\sqrt{2}}\right)(\alpha_\mu + \beta_\mu). \tag{8.8}$$

In magnetic fields <200 G, both w_{12} and w_{23} ($w_{ij} = w_j - w_i$) are easily observable, but w_{14} and w_{34}, whose parentage is mainly the "singlet" state, are comparable to ω_0, too fast to be seen, and hence are averaged to zero by the experimental time resolution. The difference in frequency between w_{12} and w_{23} vanishes in fields ~10 G, such that only a single frequency, w_{Mu}, is observed. In this limit, Equation 8.7 reduces to the simple result of coherently precessing "triplet" Mu (Equation 8.3) with the "Larmor precession," $v_{Mu} = 1.39$ MHz G^{-1}. At such a low field, the diamagnetic Larmor precession frequency, $v_\mu = 13.55$ kHz G^{-1}, cannot be seen over the short time range but shows up characteristically over longer time ranges and at fields ~20 G. At slightly larger magnetic fields, the spectra show the fast oscillations and slower beat frequency characteristic of two-frequency Mu precession, defined by Equations 8.9 and 8.10, and reflecting allowed transitions between muon-electron energy states in a magnetic field (Roduner, 1993):

$$v_{12} = \frac{1}{2}\left[(v_e - v_\mu) - \left[(v_e + v_\mu)^2 + A_\mu^2\right]^{1/2} + A_\mu\right], \tag{8.9}$$

$$v_{23} = \frac{1}{2}\left[(v_e - v_\mu) + \left[(v_e + v_\mu)^2 + A_\mu^2\right]^{1/2} - A_\mu\right]. \tag{8.10}$$

where

v_e is the electron Larmor frequency
v_μ the muon Larmor frequency of the coupled two-spin system, with A_μ the corresponding muonium hyperfine frequency

The total Mu amplitudes are most easily found in weak fields, <10 G, giving a single amplitude, while the diamagnetic amplitudes are usually found from larger fields (Ghandi et al., 2004).

Since knowing the absolute muon polarization is important, studies are usually first carried out in the same sample vessel with a standard sample, usually N_2 for gas phase studies, since the absolute polarizations for N_2 are well established over a wide range of densities as shown in Kempton et al. (1991). Comparison of the ratio of measured muonium/diamagnetic amplitudes with these known fractions in N_2 provides a determination both of the diamagnetic signal due to muon stops in the cell window and walls of the sample vessel, as well as the total muon asymmetry corresponding to the full polarization. The measured amplitudes in the samples of interest are then corrected according to established procedures (Roduner, 1993):

$$P_D = \frac{(A_D - A_W)}{(A_S - A_W)}, \tag{8.11}$$

and

$$P_{Mu} = \frac{2A_{Mu}}{(A_S - A_W)}, \tag{8.12}$$

where

A_D is the diamagnetic amplitude
A_W is the sum of the wall and window amplitude
A_S is the amplitude of the standard (N_2)
A_{Mu} is the muonium amplitude

The factor 2 in Equation 8.12 accounts for non-observed singlet muonium.

The initial stage of the energy loss process of any ion in matter is known as the Bethe–Bloch ionization (Inokuti, 1971; Walker, 1983; Senba, 1988; Roduner, 1993; Smigla and Belousov, 1994; Pimblott and LaVerne, 1997; Mozumder, 1999; Hatano, 2003; Senba et al., 2006). After passing through the entrance window of the target cell (Figure 8.1), muons entering the sample still have the MeV kinetic energies noted earlier, many orders of magnitude higher than those of chemical interest. Most of this initial energy is dissipated in ionization and excitation processes, described by the Bethe–Bloch stopping-power formula for $-dE/dX$ (Walker, 1983; Senba, 1988; Smigla and Belousov, 1994; Mozumder, 1999; Hatano, 2003; Senba et al., 2006). During this regime, there is no loss of muon polarization.

In a low density gas, this is followed by a regime of cyclic charge exchange with moderator "M" beginning around 100 keV for the positive muon and described by

$$\mu^+ + M \xrightarrow{\sigma_{10}} Mu + M^+$$

$$Mu + M \xrightarrow{\sigma_{01}} \mu^+ + M + e^-$$

(8.13)

with average electron capture (σ_{10}) and loss (σ_{01}) cross sections, together with those for elastic and inelastic energy moderation, determining the outcome as shown in the works of Fleming and Senba (1992), Kempton et al. (2005), and Senba et al. (2006). In this regime, lasting down to an energy $E_{min} \sim 10\,eV$, the muon undergoes many cycles in less than 1 ns (at 1 atm in the gas phase) as described in Fleming and Senba (1992); Senba et al. (2006), emerging as either a bare μ^+ or as the Mu atom, prior to entering its third thermalization stage, from E_{min} to $k_B T$. Near the end of the cyclic charge-exchange regime while the muon slows down, one cycle of charge exchange can involve a series of complex collisions, including (1) the electron capture collision by the muon to form muonium (Mu), (2) repeated spin exchange collisions with paramagnetic species, during the time the projectile is in the neutral Mu state. This is when there are most likely secondary electrons present in the radiation track, and (3) an electron loss collision by Mu to form μ^+ again.

The times taken by positive muons to slow down from initial energies in the range of ~1 MeV, to the energy of the last muonium formation, approximately 10 eV, at the end of cyclic charge exchange, have been measured in the pure gases H_2, N_2, Ar, and in the gas mixtures Ar-He, Ar-Ne, Ar-CF_4, H_2-He, and H_2-SF_6 as shown by Senba et al. (2006). At 1 atm pressure, these slowing-down times, in Ar and N_2, vary from 14 ns at the highest initial energy of 2.8 MeV, to 6.5 ns at 1.6 MeV. Similar variations were seen in the gas mixtures, depending also on the total charge and nature of the mixture consistent with Bragg's additivity rules. The slowing-down times were also used to determine the stopping powers, dE/dX, of the positive muon in N_2, Ar, and H_2, at kinetic energies near 2 MeV. The results demonstrated that the positive muon and proton have the same stopping power at the same projectile velocity, as expected from the Bethe–Bloch formula (Hatano, 2003). The energy of the first neutralization collision that forms muonium (hydrogen), which initiates a series of charge-exchanging collisions, was also calculated for He, Ne, and Ar. The slowing-down times through the first two regimes are controlled by the relevant ionization and charge-exchange cross sections, whereas the final thermalization regime is most sensitive to the forwardness of the elastic scattering cross sections in the low density gas phase. In this regime, the slowing-down times (to kT) at nominal pressures are expected to be less than or similar to 100 ns. In this final stage of energy loss,

Mu may undergo hot-atom reactions and the μ^+ as well may continue to form Mu by charge exchange and/or form muon molecular ions, MMu^+, both in competition with elastic and inelastic scattering.

Although most studies of diamagnetic and muonium (Mu) fractions formed in low pressure gases have used transverse field-μSR (TF-μSR) that provides information on sub-microsecond radiation processes, these studies have recently been extended to the microsecond timescale using the technique of time-delayed radio frequency muon-spin resonance (RF-μSR) in Johnson et al. (2005).

In the RF technique, a static magnetic field, B, is applied collinearly with the muon-spin polarization, with an oscillating B_1 field applied transversely to it. Similar to NMR, adjusting the static field strength and RF frequency to meet the resonance condition for the system causes the probe spin polarization to experience a torque and precesses about the B_1 field for its duration. μSR data are collected with and without the RF field. The experiment alternates between the two conditions, typically every 10 s, to identify the effect of the RF field on the system and reduce the effects due to equipment instability (Johnson et al., 2005).

Results obtained with inert gases (nitrogen and noble gases) at pressures higher than 1 atm established the validity of the TF-μSR results, proving that formation of these species is due only to prompt processes (sub-nanosecond timescale) resulting from charge-exchange and thermalization processes from keV to epithermal energies. The result was also consistent with the view that the diamagnetic environment is due to a muon molecular ion, MMu^+, and not a bare μ^+. The coupling between the μ^+ and the gas suggested that the MMu^+ molecular ion is a stable species with no conversion to a bare μ^+. In addition, this work showed that the RF-μSR technique provides polarization fractions in good accord with those obtained using conventional transverse-field muon-spin resonance measurements. This is a result that will no doubt be exploited in the years to come.

8.1.3 FREE-RADICAL FORMATION IN THE GAS PHASE

The identity (i.e., the molecular structure) of muoniated free radicals is usually unambiguous, provided hyperfine constants can be determined by TF-μSR or RF-μSR, and avoided level crossing (μALCR). The spin polarization of the muon beam is perpendicular to the applied magnetic field for the TF-μSR experiments, perpendicular during the RF radiation and RF-μSR experiments, and parallel to the field for avoided level crossing-μSR (ALC-μSR) experiments (Roduner, 1988, 1993; Smigla and Belousov, 1994). Positron counters are positioned appropriately in each case.

The precession frequency of the spin in the TF-μSR varies as a function of the strength of the magnetic field and the electronic environment, through the hyperfine interaction. At high magnetic fields where Zeeman states are sufficiently separated, only two transitions, v_{12} and v_{34}, are observable (Roduner, 1988, 1993). Fourier transformation of the time spectrum shows three frequency peaks, two of which correspond to the generated radical (v_{12} and v_{34}) and one to muons in a diamagnetic environment (v_d). The spectrum in Figure 8.3 is due to the muoniated ethyl radical in CO_2 (Cormier et al., 2008).

At high magnetic fields all energy transitions degenerate to only two, and the associated two frequencies can be used to determine the muon hyperfine coupling constant (HFC), A_μ according to (Roduner, 1988, 1999):

$$v_{ij} = v_\mu \pm \frac{1}{2}A_\mu, \tag{8.14}$$

where v_μ is the muon Larmor frequency. The HFC is proportional to the unpaired spin density at the nucleus.

It is possible to observe "delayed species" that are formed up to 1 μs following muon implantation using ALC-μSR, and this makes it feasible to study samples with low concentrations of the free-radical precursors. The spin states in large magnetic fields are the products of the unpaired

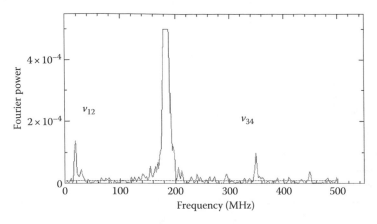

FIGURE 8.3 The TF-μSR spectra of ethyl radical in a low concentration ethene in supercritical CO_2 at 305 K at 85 bar.

electron and the nuclear Zeeman states; consequently, the muon spin is time independent and the asymmetry is independent of the applied magnetic field. At different values of the applied magnetic field, near-degenerate pairs of spin states can be mixed through the isotropic and anisotropic components of the hyperfine interaction. The muon polarization oscillates between the two mixing states of the avoided crossing. This results in a resonant-like change in the asymmetry as the magnetic field is swept. There are three kinds of resonances, classified by the selection rule $|\Delta M| = 0$, 1, and 2, where M is the sum of the quantum numbers for the z components of the muon and nuclear spins.

In the case of isotropic environments ($|\Delta M| = 0$), once the muon HFC is known using the TF-μSR technique, proton HFC (A_p) or the hyperfine coupling constants of other spin-active nuclei can be obtained from ALC-μSR measurements through the following equation (Roduner, 1988, 1993):

$$B_{LCR} = \frac{1}{2}\left[\frac{A_\mu - A_p}{\gamma_\mu - \gamma_p} - \frac{A_\mu + A_p}{\gamma_e}\right], \tag{8.15}$$

where

B_{LCR} is the magnetic field on resonance
γ_p and γ_e are the proton and electron gyromagnetic ratio, respectively

In ALC-μSR experiments, the muons are injected into the sample with their spin parallel to the magnetic field. The integrated number of positrons emitted in the forward and backward directions with regard to the incoming muon beam is measured as a function of the applied magnetic field. The muon polarization is proportional to the experimental asymmetry. In contrast to TF-μSR, there is no limit on the number of muons in the sample at one time; consequently, it is possible to run at a high incident muon rate. The field is scanned in a series of small steps. It is possible to do both TF-μSR and ALC-μSR experiments on the same spectrometer, just by moving the counters and changing the beam polarization. Sometimes a modulation field (~80 G) is applied to compensate for the background. This gives the resonance a differential lineshape; however, this is only useful for studies of isotropic environments where the resonances are not very wide.

Despite the superiority of μSR to characterize free radicals (Dilger et al., 1997) and their electronic structure in the gas phase (polyatomic radicals are generally not amenable to study by electron spin resonance in the gas phase), gas phase studies of free radicals over the last 10 years were not very productive. However, there has been an important study on the mechanism of the muoniated ethyl radical formation in the gas phase by Percival et al. (2003).

FIGURE 8.4 Reaction of the muonium with ethene to produce the muon-substituted ethyl radical.

The study tackled the following questions: Is muonium formation the first step, followed by addition to the alkene (Figure 8.4)? Or is there an ionic process, such as muon attachment to give a cation, followed by electron addition? The work in reference Percival et al. (2003) explored such questions by studying how the spin polarizations of the incident muons were transferred to the radical product. It was possible to fit the TF-μSR signal amplitudes with a model involving a single reaction step—Mu addition to ethene. However, the rate constant deduced from the model fit was significantly higher than the literature value for a thermal reaction (Garner et al., 1990), as determined by direct study of Mu decay in low partial pressures of ethene. The conclusion was that at partial pressures above 1 bar, the reaction (Figure 8.4) occurs before the incoming (beam) muons are fully thermalized, that is, via a hot-Mu reaction. Comparing the experimental rate constant for Mu + ethene with the predictions of canonical variational transition state theory (Villa et al., 1998) suggested that Mu would have an effective temperature of about 2000 K, that is, ~0.25 eV kinetic energy. Using the reactive cross section from the thermal reaction (Garner et al., 1990) and adjusting for epithermal energies, as shown by Senba et al. (2000), led to 4000 K as an estimate of an effective reaction temperature (~0.5 eV). The first-order muonium decay rate is 7–50×10^9 s^{-1}, depending on ethene pressure, and therefore the reaction timescale is of the order 10–100 ps. This meant Mu could still be epithermal; tens of picoseconds after energetic muons entered the gas at pressures below 10 bar. This is consistent with expectations based on measurements of muon thermalization in low pressure gases by Senba et al. (2006), where the slowing-down time to ~10 eV is of the order of 10 ns at 1 bar.

8.2 MUON RADIATION CHEMISTRY IN MOLECULAR LIQUIDS

There is consensus about the mechanism of muon thermalization in the gas phase, due to the simplicity of the low density gas phase interactions. However, structures are more complex in the liquid phase and there is no consensus regarding radiation chemistry in molecular liquids. There have been several theoretical and experimental studies of liquid muon radiation chemistry over the last 10 years, especially in muonium formation in liquefied inert gases.

The spatial distribution of electron–muon pairs in superfluid helium (He-II) was determined theoretically for reconstructing the muonium (Mu) formation probability by Kosarev and Krasnoperov (1999). It was shown that because a gap is present in the excitation spectrum of He-II, the thermalization time of muons and secondary electrons increases with decreasing temperature. As a result, the average distance in the electron–muon pairs increases and, correspondingly, the muonium formation rate decreases (Kosarev and Krasnoperov, 1999). Such theoretical predictions of radiolysis effects in molecular liquids during muon thermalization were later put on firm ground using the technique of muon-spin relaxation in frequently reversed electric fields (Eshchenko et al., 2000, 2001, 2002, 2003). Such experiments proved that the excess electrons liberated in the μ+ ionization track converge upon the positive muons and form Mu atoms. This process was shown to be crucially dependent upon the electron's interaction with its environment (i.e., whether it occupies the conduction band or becomes localized) and upon its mobility in these states (Eshchenko et al., 2000, 2001). The characteristic lengths involved are quite long and in the range of 10^{-6}–10^{-4} cm; the characteristic times range from nanoseconds to tens of microseconds. At the nanosecond timescale, the thermalization timescale is similar to the gas phase thermalization time but the longer timescales are much longer than in the gas phase.

In general, there have been two types of theoretical (computational) modeling of muon thermalization in molecular liquids. One type is based on electron number density calculations (Kosarev and Krasnoperov, 1999; Gorelkin et al., 2000) and is similar to the kinetic theory of plasma, while the other one is based on a stylized initial track structure comprised of many ion pairs, where the trajectory of each charged particle is followed under the competing influences of Coulomb forces and Brownian diffusion (Siebbeles et al., 1999, 2000). Both types of studies are limited to the low permittivity liquids, where Coulomb forces are long range. The later works of Siebbeles et al. (1999, 2000) were on liquid hydrocarbons that had accessible electron mobility and electron thermalization distances. The former works were on liquid rarefied gases by Gorelkin et al. (2000) and Kosarev and Krasnoperov (1999). The studies on liquid hydrocarbons suggested that delayed muonium formation could account entirely for the missing polarization or lost fraction (Siebbeles et al., 1999, 2000), $P_L = 1 - P_{Mu} - P_D$ in the absence of free-radical formation, as opposed to the interaction of Mu with solvated electrons in the spur from the muon radiation track. For liquids with no unsaturated bonds, $P_R = 0$ because free radicals are not formed. For most saturated hydrogenated materials, P_L has been found to be similar in magnitude to P_{Mu} according to Walker et al. (2003a,b).

Such theoretical studies set the stage for both experimental and more advanced theoretical investigation of muon radiolysis effects. The theoretical extension is certainly necessary, in particular a computational study where the track structure and radiation chemical kinetics simulation would be extended to include charge cycling, muon trapping as RMu where R is the alkyl radical, and the effect of the R group on trapping, hot muonium formation, and hot muonium reactions.

There have been preliminary experimental tests carried out by Walker et al. (2003a,b) of the theoretical predictions of the radiolysis processes in hydrocarbons performed by Siebbeles et al. (1999, 2000). In the experiments that were carried out by Walker et al. (2003a,b), instead of using Equations 8.6 and 8.7, which is the most accurate way to study the diamagnetic and Mu fraction, the computer-fitted A_D values were converted to fractional muon yields, P_D, using $P_{D(x)} = (A_{D(x)}/A_{D(water)}) \times 0.62$, based on the diamagnetic yield in water of 0.62. Although such a method does not give a definitive test of the delayed muonium formation predicted by Siebbeles et al. (1999, 2000), the experimental results suggest that muonium forms on a much shorter timescale than the proposed delayed mechanism (microseconds) for a fraction of formed Mu. Certainly a more accurate measurement of both diamagnetic and muonium fractions, a study of magnetic and electric field dependence and the effect of scavengers on both Mu and diamagnetic fractions, and RF-μSR investigation of liquid hydrocarbons are needed as definitive tests of the theoretical studies by Siebbeles et al. (1999, 2000). Such studies are also useful to distinguish between the spur and hot-atom models in liquids. Indeed, if electron–muon (or muoniated molecular ion) recombination has a significant role in muonium formation, the muonium amplitude in a transverse magnetic field could be magnetic field-dependent. Such investigations along with the measurements of muon polarizations as a function of electric field, RF, and laser frequencies and delay times will be useful to shed light on mechanism of muon thermalization in molecular liquids.

Two things that are agreed upon regarding the muon thermalization process in liquids between scientists (radiolysis model proponents such as Roduner, 1988; Kosarev and Krasnoperov, 1999; Percival et al., 1999; Siebbeles et al., 1999; vs. hot-atom model proponents such as Walker, 1983; Walker et al., 2003a,b) are that (1) the initial stage of the thermalization is ionization and excitation, and (2) the charge-exchange process (Equation 8.8) follows the established understanding of radiolysis in molecular liquids, where, when a charged particle (except for electrons of energy >100 MeV) is being thermalized, it loses energy by ionizing and exciting the molecules of the medium (Salmon, 2003).

The question now is, what is the thermalization process after this initial stage? There are two schools of thought. One suggests the final thermalization steps are only due to hot-atom reactions of Mu* that determine the final diamagnetic and muonium fractions proposed by Walker (1983) and Walker et al. (2003a,b). The other school of thought is based on ionic processes that may arise at the end of the radiation track (Eshchenko et al., 2000, 2001, 2002, 2003; Ghandi et al., 2007).

Such ionic processes involve muons, muoniated ions, and secondary electrons (formed along the tracks, with a distribution of energies ranging from a few eV upwards) that are able to bring about further ionizations.

There is a significant gap in our knowledge about stopping power and energy loss events in liquids for muons. Such knowledge is important to understand the thermalization processes of ionizing radiation, including muon radiation. Since muon LET is higher than electron LET, this knowledge is also of interest for medical and technological applications. There is a great need for simulations of the inelastic collisions between bound electrons in liquids and the muons moving through it, especially for energies below 100 keV.

There is a limited number of published works over the last 10 years on radiation effects on muons in molecular liquids except for Kosarev and Krasnoperov (1999) and Walker et al. (2003a). There are previous works, however, on muonium formation in molecular liquids. The mechanism of Mu formation in water, as opposed to other molecular liquids, has been well-studied with many definitive and accurate experiments that are reviewed in Percival (1990).

In the case of muon irradiation of water, the key competitive reactions among transient species in radiolysis are the following:

$$\mu^+ + e^- \rightarrow Mu \tag{8.16}$$

$$\mu^+ + H_2O \rightarrow MuH_2O^+ \tag{8.17}$$

$$MuH_2O^+ + H_2O \rightarrow MuOH + H_3O^+ \tag{8.18}$$

$$Mu + e_{aq}^- \rightarrow MuH + OH^- \tag{8.19}$$

$$Mu + e_{aq}^- \rightarrow \text{spin depolarized Mu} \tag{8.20}$$

$$Mu + OH \rightarrow MuOH \tag{8.21}$$

$$Mu + H \rightarrow MuH \tag{8.22}$$

Reactions (8.16) through (8.18) determine the initial distribution of muons between Mu and the diamagnetic fraction and mostly take place in less than a picosecond. Muon attachment to water, Reaction (8.17), and its solvation does not prevent Mu formation. Reaction (8.16) could involve μ^+ in a variety of solvated molecular ions. The secondary encounter of muonium with radiolysis transients, such as hydrated electrons or H atoms, in Reactions (8.19) through (8.22) extends into the sub-microsecond range and leads to the "missing fraction" of muon-spin polarization in water.

As can be seen from the reaction scheme in this model (known as the spur or radiolysis model), the hydrated electron plays a significant role in the processes that determine the distribution of different muon fractions in water.

In a recent work, a novel technique was introduced. Laser muon-spin spectroscopy can be used to study the excited-state chemistry of muonium and muoniated free radicals and to investigate radiation effects in muon thermalization. Experiments were performed at 298 K and 1 bar pressure in water, in a 20 G transverse magnetic field, using 532 nm laser excitation with the excitation pulse delayed relative to the muon pulse. The diamagnetic muon signal, which precesses at the muon Larmor frequency, $\nu_\mu = 13.55$ kHz G^{-1}, was detected at 20 G. Changes in μSR amplitudes,

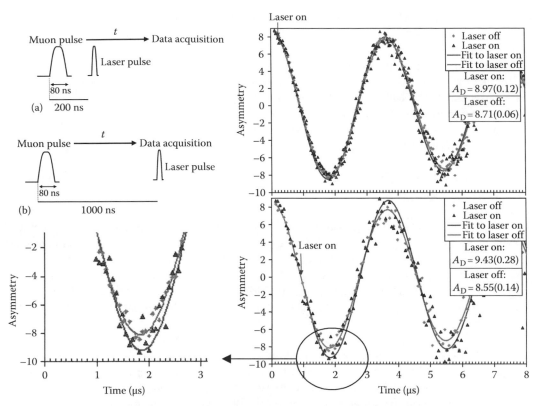

FIGURE 8.5 (a) The asymmetry in water following excitation at 532 nm with a delay of 0.2 μs relative to the muon pulse, (b) the asymmetry in water with a delay of 1 μs relative to the muon pulse. The triangles are the "laser on" data points and the dark curves are fits of Equation 8.1 to the data. The diamonds are the "laser off" data points and the light curves are fits of Equation 8.1 to the data. A range around 2 μs (minimum in the asymmetry) is magnified to demonstrate the increase in the diamagnetic fraction due to laser irradiation.

corresponding to the diamagnetic precession frequency in water at 20 G during laser irradiation, were detected at time delays up to 1.2 μs; selected laser delays of 0.2 and 1 μs are depicted in Figure 8.5.

The percent asymmetry is shifted by 10.5% ± 3.7% after a 1 μs delay and by 2.7% ± 1.5% after a 0.2 μs delay, showing the rise of asymmetry upon laser irradiation.

The result suggests that the laser excitation can be used to manipulate the processes that determine the distribution of muon-spin polarization among the different muon fractions. This confirms the radiolysis model as opposed to hot-atom model for muonium formation mechanism in water. Also, the technique can shine light on the process of muonium and free-radical formation in other molecular liquids.

8.3 THE INTERACTION OF MUONS WITH SUPERCRITICAL FLUIDS

A supercritical fluid (SCF) is any substance above its critical temperature (T_c) and pressure (P_c) where only one phase exists. In SCFs, the solvent properties (density, dielectric constant, etc.) vary with the temperature and pressure continuously. These easily tunable properties make SCFs (particularly supercritical CO_2 and H_2O) an excellent media for chemical processes, as described in Noyori (1999) and Rayner (2007). The change in the solvent properties can significantly affect the radiation chemistry, electronic structures of the reactive intermediates, and the corresponding transition states. In the last decade, muon methods have been used to investigate such tunable solvent effects by Cormier et al. (2008), Ghandi and Percival (2003), Ghandi et al. (2000, 2002, 2003), Ghandi (2002, 2004), McKenzie and Roduner (2009), and Percival et al. (1999, 2000, 2007).

Understanding the radiolysis of water in subcritical and supercritical water (SCW) is essential for the design of the next generation of pressurized water nuclear reactors. Understanding the radiation chemistry of supercritical CO_2 is relevant to radiation-induced polymerization, waste management applications (Clifford et al., 2001), and development of supercritical CO_2 Brayton cycles in nuclear reactors (Moisseytsev and Sienicki, 2008). There are only a handful of studies on electron-induced radiation chemical processes in SCFs, and there is very little report on any higher LET radiation. The radiation chemical processes due to higher LET radiation might be very different, such as with α-particle radiation from the radioactive decay of heavy nuclei in waste management applications.

The studies of muon radiation chemistry can be used to explore the effects of ionizing radiation in SCFs on a particle of intermediate mass. Studying radiolysis processes during muonium formation as a function of density (which can be easily done in SCFs) can shed light on the mechanism of muonium formation. A recent study by Ghandi et al. (2010) has compiled the results of three previous studies on supercritical CO_2 by Ghandi et al. (2004), H_2O by Percival et al. (1999), and C_2H_6 (mostly gas phase) by Kempton et al. (1991). The corrected results for C_2H_6 using the data in Kempton et al. (1991) and new data from Ghandi et al. (2010) are presented in Figures 8.6 through 8.8.

Based on a model in which hot-atom reactions of Mu and ethane determine the diamagnetic fraction (Kempton et al. 1991), the second derivative of density of this fraction should always be negative. Thus, when the hot-atom reactions determine the diamagnetic fraction, the diamagnetic fraction versus density can never curve upward, in contrast to the data shown in Figure 8.6. Founded on this model, the diamagnetic fraction in ethane can be explained on the basis of hot-atom reactions up to ~0.3 g cm^{-3}. This is in contrast to the case in SCW that above 0.2 g cm^{-3}, the radiolysis effects have the major role in the process of muon thermalization (Figure 8.8) (Percival et al. (1999).

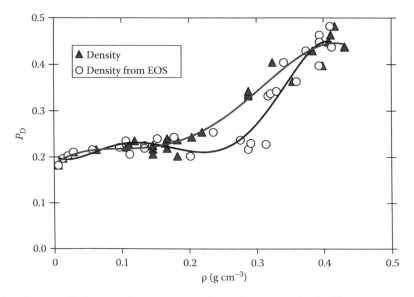

FIGURE 8.6 Diamagnetic (P_D) fractions in pure C_2H_6 as a function of density (for clarity the error bars are not shown). The data set from Kempton et al. (1991) is presented with triangles, while the dataset, presented with open circles, is based on corrected densities from Ghandi et al. (2010).

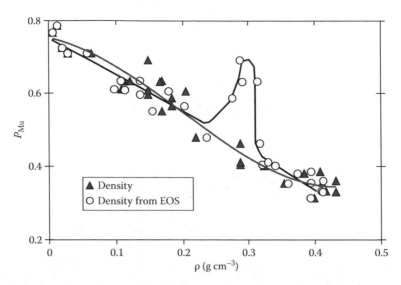

FIGURE 8.7 Muonium (P_{Mu}) fractions in pure C_2H_6 as a function of density (for clarity the error bars are not shown). The data set from Kempton et al. (1991) is presented with triangles while the dataset, presented with open circles, is based on corrected densities from Ghandi et al. (2010).

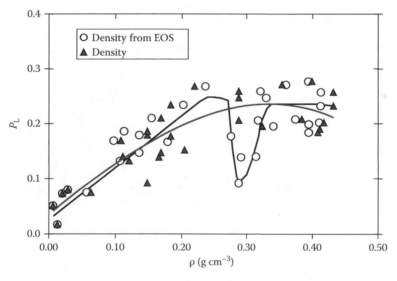

FIGURE 8.8 Missing (P_L) fractions in pure C_2H_6 as a function of density (for clarity the error bars are not shown). The data set from Kempton et al. (1991) is presented with triangles while the dataset, presented with open circles, is based on corrected densities from Ghandi et al. (2010).

The steep increase and then decrease of the muonium fraction as a function of density in near-critical ethane is clear in Figure 8.7. Also, there is a concomitant reverse behavior at the same density range for the missing fraction as a function of density (Figure 8.8). This is the only case that has revealed such drastic effects. There is a reverse density dependence of P_L in supercritical CO_2, although not as pronounced, and there is no such effect in SCW (Figures 8.9 through 8.11).

This result suggests that density inhomogeneity in near-critical conditions affects the radiolysis processes involved in muonium formation, the missing fraction in ethane (and probably in other hydrocarbons), and in supercritical CO_2 in an opposite, yet less pronounced way. There is a

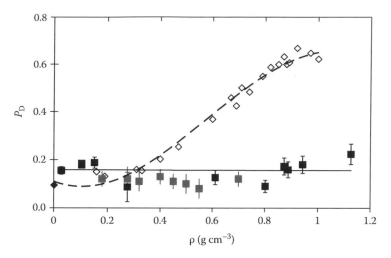

FIGURE 8.9 Diamagnetic (P_D) fractions in pure H_2O and CO_2 as a function of density. Values of P_D for H_2O (empty diamonds) are taken from Percival et al. (1999). The light squares are data on CO_2 from Ghandi et al. (2010) while the dark squares are the data taken from Ghandi et al (2010).

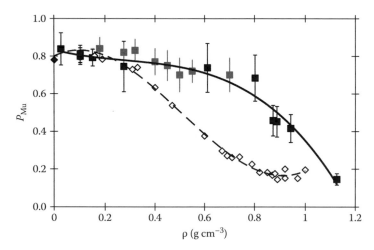

FIGURE 8.10 Muonium (P_{Mu}) fractions in pure H_2O and CO_2 as a function of density. Values of P_D for H_2O (empty diamonds) are taken from Percival et al. (1999). The light squares are data on CO_2 from Ghandi et al. (2010) while the dark squares are the data taken from Ghandi et al (2004).

maximum in the muonium fraction at a density close to the critical density at the expense of a minimum in the missing fraction in ethane. This is in contrast to the near-critical water and CO_2. The data in near-critical CO_2 show a small maximum in the lost fraction. This, along with the density dependence of P_D in ethane, suggests different mechanisms for the missing fractions in CO_2 and ethane. The missing fraction in ethane is due to the interaction of hot Mu in low density regions with alkyl radicals in high density regions in spurs, while in CO_2 it is due to the interaction of Mu and solvated electrons (both in high density regions). In water there is no sign of density inhomogeneity effects on muon polarizations. This could mean that the density inhomogeneity does not affect radiolysis processes in near-critical water, or it could mean that the different effects cancel each other; for example, the effect of diffusion of Mu and a hydrated electron cancels the effect of clustering around Mu and the hydrated electron.

Observing maxima or minima in the missing fraction as a function of thermodynamic parameters gives us a hint about the nature of the missing fraction (and therefore the process of muon

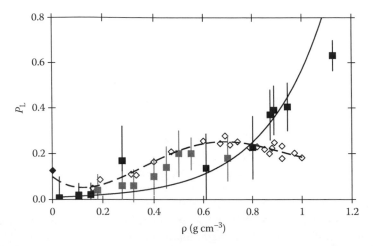

FIGURE 8.11 Missing fractions (P_L) in pure H_2O and CO_2 as a function of density. Values of P_D for H_2O (empty diamonds) are taken from Percival et al. (1999). The light squares are data on CO_2 from Ghandi et al. (2010) while the dark squares are the data taken from Ghandi et al. (2004).

thermalization). As an example (Ghandi et al., 2010), the delayed muonium formation mechanism is unlikely to be responsible for the minimum in P_L in ethane, based on significant clustering around positive ions. Such a clustering would decrease the mobility of these ions and therefore have the opposite effect to the one observed in ethane.

8.4 MUON RADIATION CHEMISTRY IN IONIC LIQUIDS

This is a new field being investigated by Ghandi's group (Lauzon et al., 2008; Taylor et al., 2009). The objective of these studies is to establish an understanding, at the microscopic level, of the solvent effects in ionic liquids (ILs) on the dynamics of free-radical reactions and free-radical formation, especially the mechanism of free-radical formation and the radiolysis processes in ILs in comparison with molecular solvents. ILs are two-component systems, composed of only equimolar amounts of anions and cations. The contrasting nature of cations and anions, and the proximity of close sites of high electron deficiency and high electron richness, suggest that IL solvents could enable chemistry that may not be possible in normal molecular solvents. ILs have a wide range of applications and their properties are tunable by small changes in their structure (see Chapter 11). Understanding many chemical processes in ILs requires probing reactive intermediates, such as free radicals. Kinetics and dynamics of solvated and pre-solvated electrons and few other reactive intermediates (see Chapter 11) in a few ILs have been studied. However, the selectivity of ILs as a media for chemical (particularly, free-radical) reactions remains unknown. Very little is known about the structure and dynamics of non-persistent free radicals in ILs and the mechanism of free-radical formation. This is surprising since these materials offer the most disparate local environments for solvent effects on radicals.

Most μSR studies have examined the cyclohexadienyl radical, most probably formed by the reaction of Mu with benzene (Scheme 8.1), as shown in Taylor et al. (2009) and Lauzon et al. (2008). Benzene was selected because the muon and proton HFCs of C_6H_6Mu have been measured in a variety of reaction media (including the gas phase) and over a wide temperature range. One of the important results is the enhanced free radical formation in trihexyl (tetradecyl) phosphonium chloride as compared to molecular solvents. After performing careful calibration with water in the sample cells, the diamagnetic and free-radical fractions were measured by investigating the transverse-field amplitudes in Lauzon et al. (2008). The diamagnetic and free-radical fractions

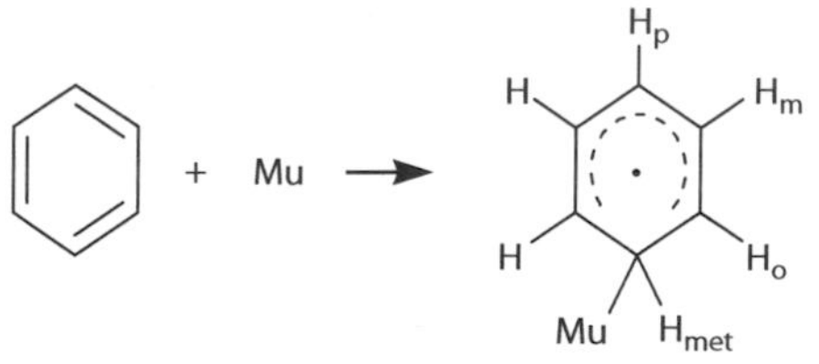

SCHEME 8.1   Reaction of muonium with benzene to produce the cyclohexadienyl radical.

in the three solvents [propanol as a representative protic polar solvent; *n*-heptane, a nonpolar solvent; and trihexyl (tetradecyl) phosphonium chloride (IL 101)] were compared, and the data suggests free-radical formation is more enhanced in ILs compared to the other two molecular solvents. Similar experimental conditions were used for all of the comparison experiments, specifically temperature, magnetic field–strength, and benzene concentration. In these works, the enhanced free-radical formation in phosphonium IL is clear and dramatic. A similar result was obtained in other aromatic compounds, which may suggest that this is a general trend of addition to aromatic rings in tetra-alkyl phosphonium ILs. In order to investigate this more directly, we measured the rate constant by using the Mu decay rate in low concentrations of benzene in IL 101. The rate constant at room temperature and 1 bar was ~1.5×10^{10} M^{-1} s^{-1}, almost an order of magnitude larger than the rate constant in water. Such a large rate constant is suggestive of a diffusion-controlled reaction.

8.5   REVIEW OF THE MODERN ASPECTS OF MUONIUM REACTION DYNAMICS AND FREE-RADICAL STUDIES BY MUONS

8.5.1   Introduction and Techniques

Muonium, from a chemical point of view, is simply a light isotopic analogue of the H atom. Hydrogen is the simplest atom in nature and consequently the study of its interactions and chemical reaction rates has been central to the field of reaction dynamics.

Over the past 10 years, studies of Mu chemical kinetics have been extended to SCW, a medium of interest for a variety of chemical applications (including green chemistry and waste destruction) and of specific importance in some designs of future generation nuclear reactors (Gen IV reactors) (Ghandi and Percival, 2003; Ghandi et al., 2002, 2003), Ghandi (2002). The supercritical-water-cooled reactor (SCWR) would operate at much higher temperatures than existing pressurized water-cooled reactors (PWRs), and a major technology gap for SCWR development has been the lack of knowledge of water radiolysis under supercritical conditions. Accurate modeling of aqueous chemistry in the heat transport systems of PWRs and the SCWR requires data on the rate constants of reactions involved in the radiolysis of water and the action of water treatment additives. Unfortunately, most experimental data do not even extend to the temperatures used in current PWRs (typically around 320°C), which is well short of the supercritical conditions envisaged in generation IV designs. Many types of Mu reactions have been studied in subcritical and SCW, and Mu kinetics have been used to deduce the nature of H-atom chemical dynamics under extreme conditions that are very difficult to study by other techniques.

The conventional methods for detection of radicals, intermediates, and products of the radiolysis of water are optical spectroscopy and electron spin resonance (see Chapters 12, 13, and 15). Some intermediates have been detected by optical spectroscopic methods (see Chapter 15) under supercritical conditions, but the material composition of the optical windows impose significant limitations on all forms of optical spectroscopy in SCW. The limitations are imposed by the necessity to resist corrosion and to transmit a wide spectral range under a variety of extreme conditions.

Electron paramagnetic resonance (EPR), the most conventional option for organic free-radical characterization, is even more limited at high temperatures and pressures. Unlike the optical detection methods usually employed in pulse radiolysis studies, muon-spin spectroscopy is sensitive only to the transient species under study (Mu in this case) and is unaffected by environmental effects (scattering of light, change in extinction coefficient, etc.). Indeed, the only magnetic resonance techniques used to study reactive free radicals in high temperature and pressure SCFs were, until recently, μSR techniques used by Cormier et al. (2009), Ghandi and Percival (2003), Ghandi et al. (2002, 2003, 2004), Ghandi (2002), Kruse and Dinjus (2007), and Percival et al. (2003). These include the TF-μSR technique described above and the ALC-μSR technique. It is possible to observe "delayed species" using ALC-μSR that are formed up to one microsecond after muon implantation and this makes it feasible to study samples with low concentrations of the free-radical precursors.

8.5.2 STUDIES IN SUPERCRITICAL FLUIDS AND LIQUIDS

Part of this section will focus on the work on muonium reaction kinetics in water from standard conditions to SCW conditions as a typical study of reaction kinetics where muonium can be used to probe the chemical kinetics of H atom in a complex system. To put this into context, we will first review the changes in the properties of water as its thermodynamic conditions change towards supercritical conditions. At around room temperature, water behaves abnormally (compared to simple fluids) due to the angular correlation existing between neighboring H_2O molecules, as shown in Kusalik and Svishchev (1994) and Soper et al. (1997). Some examples of abnormal behavior of water around room temperature can be seen in Table 8.1, where some properties of water are compared with the same properties of other solvents.

Water has a high dielectric constant that enables electrolytes to dissociate completely (Table 8.1). The high dielectric constant also makes water a very good solvent for polar molecules. On the other hand, the angular correlation between neighboring H_2O molecules of the hydrogen bond network demands a large amount of work to produce a cavity to accommodate a solute (note that the surface tension of water is much larger than other conventional solvents). Therefore, if the molecule is not polar or ionic, its solubility at room temperature is very small.

By increasing the temperature and/or decreasing the density of water in the liquid or supercritical state, the angular correlation between neighboring H_2O molecules decreases, as shown by Svishchev et al. (1996), Liew et al. (1998), and Matubayasi et al. (1997). This causes a change in

TABLE 8.1
Comparison of Physicochemical Properties of Water with Other Solvents under Room Conditions

	Surface Tension[a] (N m^{-1})	Density[b] (g cm^{-3})	Compressibility[c] (GPa^{-1})	Dielectric Constant[d]	Viscosity[e] (mPa s)
Water	0.072	1.00	0.5	78.4	0.9
Methanol	0.022	0.79	1.2	32.7	0.5
Hexane	0.018	0.65	1.7	1.9	0.3
Pentane	0.015	0.62	2.2	1.8	0.2

[a] Data taken from White et al. (1995), Riddick et al. (1986).
[b] Data taken from Lemmon et al. (1998), Riddick et al. (1986).
[c] Data taken from Harvey et al. (2001), Riddick et al. (1986).
[d] Data taken from Mesmer et al. (1988), Riddick et al. (1986).
[e] Data taken from Sengers and Kamgar-Parsi (1984), Riddick et al. (1986).

the solvent properties of water, such that under supercritical conditions, most nonpolar molecules are completely soluble in water. The dramatic changes in the properties of water at the microscopic level should make its solvent properties very sensitive to temperature and pressure. The study of solvent effects on the chemical kinetics of *transient intermediates* in subcritical and SCW is one of the purposes of the work on muonium chemistry in subcritical and SCW. At the outset of these studies, there was no study on transient species in SCW. Mechanistic studies were mainly based on end product analysis. All the kinetics information needed for modeling complex processes such as radiolysis of water and SCW oxidation were based on extrapolation of the low temperature data. The questions that these studies undertook were as follows:

Would the solvent cage change significantly with the temperature and/or pressure?

If it changes, then is the cage property under extreme conditions more like the cage properties of typical hydrocarbon solvents or is there no significant cage effect as in high-pressure gases?

Is there any significant cage effect at temperatures >300°C?

The dielectric constant of water decreases with temperature and at 200°C–300°C it reaches the values of alcohols under room conditions (Table 8.1). Under supercritical conditions it decreases to the typical values of hydrocarbons under room conditions (Table 8.1 and Figure 8.12). Would such a large decrease of the dielectric constant affect radical chemistry or radiolysis of water?

A purpose of the research of this thesis was to explore the effect of pressure on radical chemistry. A key parameter that affects the pressure tuning of radical chemistry is the isothermal compressibility. Isothermal compressibility is defined as

$$\kappa(T,P) = -\frac{1}{\overline{V}}\left(\frac{\partial \overline{V}}{\partial P}\right)_T, \tag{8.23}$$

where
    κ is the isothermal compressibility
    $\overline{V}$ is the partial molar volume
    P is the pressure
    T is the temperature

The isothermal compressibility along three isobars is presented in Figure 8.13. It increases slowly at low temperatures without any significant pressure dependence at $T < 300°C$. At higher

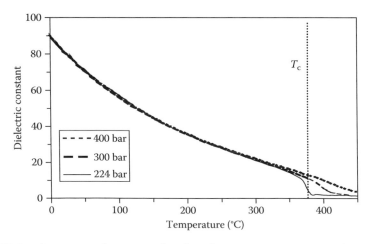

FIGURE 8.12 Dielectric constant of water as a function of temperature at pressures above the critical point. T_c is the critical temperature of water.

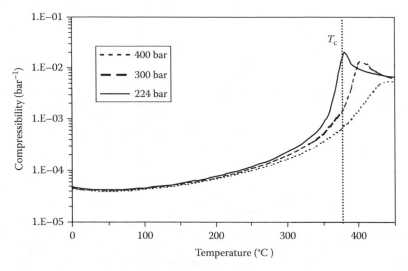

FIGURE 8.13 Isothermal compressibility as a function of temperature at constant pressures.

temperatures; however, the isothermal compressibility increases sharply until it goes through a maximum. The maximum becomes broader and shifts to higher temperatures with pressure. Would that mean a maximum pressure effect at temperatures close to the critical point and at lower pressures? Are equilibrium or dynamic properties affected and to what degree? Is pressure the significant parameter in the radiation or radical chemistry?

Densities under supercritical conditions are typical of high-pressure gas densities while at 360°C the density of water has changed only (at 224 bar) to ~600 kg m^{-3}, the density of n-pentane (Table 8.1, Figure 8.14). Would the solvent cage change significantly with the temperature and/or pressure? If it changes, then is the cage property under extreme conditions more like the cage properties of typical hydrocarbon solvents or is there no significant cage effect as in high-pressure gases?

Due to the fast reactions of radicals and intermediates in the radiation chemistry (radiolysis of water), many reactions of radicals and other intermediates are believed to be diffusion controlled, or nearly diffusion controlled. Would they remain diffusion controlled at high temperatures? There is significant pressure dependence of viscosity only between 360°C and 450°C (Figure 8.15). If a reaction is diffusion controlled, then a negative pressure dependence is expected, which would be significant in the temperature range of 360°C < T < 450°C.

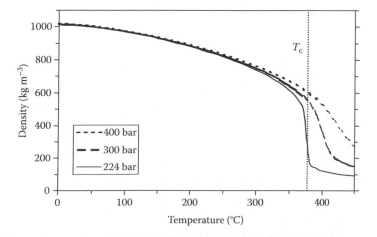

FIGURE 8.14 Temperature dependence of density at pressures above critical point.

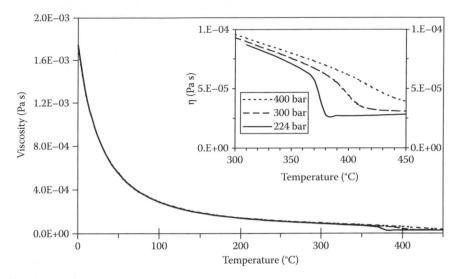

FIGURE 8.15 Viscosity as a function of temperature at pressures above critical point.

The other question that was addressed by these works was as follows: How would the temperature or pressure dependence of the self-dissociation of water (Figure 8.16) affect the radiolysis of water and radical chemistry?

For most of these works, The TF-μSR technique has been employed. It was expected that the hyperfine coupling constant of muonium would give information regarding interactions of Mu with the water molecules comprising the solvent cage around it. This expectation was realized and it was found that the cage around muonium changes significantly under subcritical and supercritical conditions, as shown by Ghandi et al. (2000).

The HFCs were determined from the splitting of the muonium precession frequencies in an intermediate field, typically 200 G. The results for the HFCs were consistent with a break in the cage structure at higher temperatures. This could be evidence for smaller cooperative interactions among water molecules at higher temperatures. Indeed, the hyperfine coupling constant of Mu was found to correlate with the bimolecular collision frequency.

Very close to the critical point, the hyperfine coupling constant of Mu was found to deviate from a smooth model. This was attributed to a local density enhancement. It is important because the

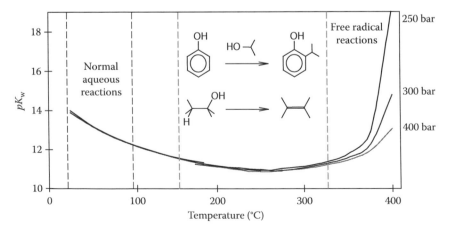

FIGURE 8.16 The dissociation constant of water as a function of temperature at different pressures and the typical chemistry involved in each region.

hydrophobic nature of Mu under standard conditions makes it similar to nonpolar organic compounds and gases used in SCW oxidation reactors.

Theses studies have led to the work on muoniated organic free radicals and muonium kinetics in subcritical and SCW. For example, the µSR technique has been used to study radical formation in aqueous solutions of acetone by Ghandi et al. (2003), Ghandi (2002). These investigations proved that the µSR technique is uniquely able to provide mechanistic information for radical reactions in SCW. The study of muon hyperfine coupling constants in muoniated radicals in SCFs by Cormier et al. (2009), Ghandi et al. (2003), and Percival et al. (2000, 2003) has also demonstrated the unique ability of µSR to probe temperature, pressure, and solvent dependence of hyperfine coupling constants under extreme conditions. For the first time, radical addition to the enol form of acetone has been observed. The studies in supercritical CO_2 by Cormier et al. (2009) also demonstrated the value of µSR to test theories of the temperature and density dependence of hyperfine coupling constants.

The investigations with acetone by Ghandi et al. (2003) and Ghandi (2002) could be considered as simple examples of multi-step synthesis. These examples show that it is possible to tune the properties of water so as to produce the precursors of a radical under one thermodynamic condition and then tune to another thermodynamic condition, which gives the highest concentration of the radical. Full characterization of free radicals in SCFs was carried out by a combination of TF-µSR and ALC-µSR by Cormier et al. (2009) and Percival et al. (2003, 2005). This gave the ability to provide both the magnitude and sign of the coupling constants in dilute solutions (Percival et al., 2003; Cormier et al., 2009) and as well provided a tool to explore the effect of concentration of radical precursor under extreme conditions of subcritical fluids and SCFs.

The studies on kinetics in subcritical and SCW by Ghandi and Percival (2003), Ghandi et al. (2002) and Ghandi (2002), and Percival et al. (2007) could be considered as an extension of the work of Troe and his coworkers (see Schwarzer et al., 1998). Troe and his group have extensively studied kinetics over a wide range of conditions from the liquid phase to the gas phase in simple molecular fluids such as nitrogen and simple hydrocarbons.

Although qualitatively the findings of the works in Ghandi and Percival (2003), Ghandi et al. (2002) and Ghandi (2002), were similar to the findings of Troe and his group, it was found that Troe's model could not describe our results in subcritical and SCW. It was also found that the analysis of the temperature dependence by means of the model widely used by Buxton et al. (1988) and his coworkers for the reactions of H and other radiolysis transient species does not give sensible parameters. The experimental strategy was to study kinetics to separate the effect of transport (viscosity) from other solvent effects. To do this they studied a reaction without a significant activation barrier. This was the spin exchange between Mu and Ni^{2+} (Ghandi et al., 2002). Spin exchange reactions are expected to be diffusion controlled, but at elevated temperatures the rate constants were found to have values far below those predicted by Stokes–Einstein–Smoluchowski theory. For spin exchange, the finding that the rate constants go thorough a maximum under subcritical conditions was attributed to a decrease of the encounter time under those conditions. Such a decrease resulted in a transition from the strong exchange to the weak exchange limit. This should be a general phenomenon for all spin exchange reactions in subcritical and SCW. The study of temperature dependence of near-diffusion-controlled reactions also gave similar results; the rate constants pass through a maximum under near-critical conditions and then fall with an increase in temperature. In the *fall-off* region, the rate constant increases with pressure. It was found that the efficiency of the reactions depends on the number of collisions over the duration of the encounter (Ghandi et al., 2002).

If this effect is general (since the temperature dependence of the encounter time depends mainly on solvent properties), similar observations should be expected for different types of reactions. Therefore, further reactions were studied based on three criteria: experiments to test the stability of reactants under a wide range of temperatures and pressures, theoretical calculations, and importance of the reaction for industry. The reactions between Mu and $HCOO^-$, and Mu and OH^- were studied because they were expected to show strong solvent dependence (Ghandi 2002). Also, the reactions between Mu and benzene in Ghandi (2002) and Mu and methanol in Ghandi et al. (2003) were chosen since they were expected to have small solvent effects, and therefore the effect of multiple collisions was expected to be more important. There are some general trends among all these reactions: it is clear from Figures 8.17 and 8.18 that pressure dependence is more significant at temperatures close to the critical point; there is positive pressure dependence under near-critical conditions. For all reactions, the rate constants pass through a maximum in temperature under near-critical conditions; both temperature dependence and pressure dependence under near-critical conditions were consistent with the cage effect suggested for near-diffusion-controlled

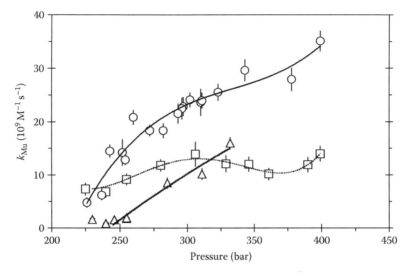

FIGURE 8.17 Rate constants for the reaction of Mu with formate in water at various temperatures: 290°C (□), 378°C (○), and 385°C (△). The lines through the data only serve to guide the eyes. (From Ghandi, K., Muonium kinetics in sub- and supercritical water, PhD thesis, Simon Fraser University, Burnaby, British Columbia, 2002.)

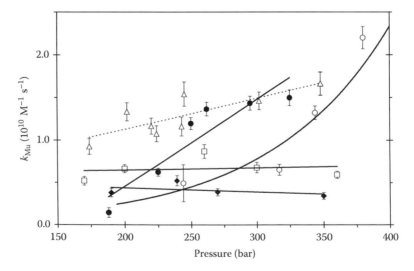

FIGURE 8.18 Rate constants for the reaction of Mu with hydroxide in water at various temperatures: 205°C (◆), 250°C (□), 305°C (△), 362°C (•), and 390°C (○). The lines through the data only serve to guide the eyes. (From Ghandi, K., Muonium kinetics in sub- and supercritical water, PhD thesis, Simon Fraser University, Burnaby, British Columbia, 2002.)

reactions (Ghandi et al., 2002). At lower temperatures the solute–solvent effects were found to be more important. In particular, in studies on reactions between Mu and OH⁻ and HCOO⁻ it was found that the variation of the dielectric constant of water with temperature plays an important role in the subcritical region (Ghandi 2002). Although both reactions between Mu and OH⁻ and HCOO⁻ showed the evidence of the cage effect under near-critical conditions, the reaction of formate showed a positive deviation from the Arrhenius extrapolation while the reaction with OH⁻ showed a negative deviation.

The study of kinetics of the Mu reaction with formate can be considered as an extension of the work of Lossack et al. (2001) to supercritical conditions. Lossack showed that in the reaction between H and formate in water, the dipole moment of the transition state is significantly smaller than the dipole moment of the reactants in Lossack et al. (2001). This causes significant destabilization of the transition state compared to the reactants. Positive deviation from the Arrhenius temperature dependence for the reaction of Mu and formate is consistent with this solvent effect under hydrothermal conditions. At higher temperatures, the dielectric constant of water drops (Figure 8.12), and therefore its effect on the reaction path becomes smaller. This leads to larger rate constants compared to the predictions from Arrhenius temperature dependence with activation energy calculated from the data under room conditions by Ghandi (2002).

The ab initio calculations at the density functional theory level UB3LYP with a Gaussian 6-31+G(d,p) basis set for the reaction of $Mu + OH^- \Leftrightarrow MuOH^- \Leftrightarrow MuOH + (e^-)_{aq}$ in the presence of three water molecules (keeping this distance between O on OH⁻ and Mu fixed and optimizing other variables at each step) (Figure 8.19) suggest the dipole moment increases along the reaction path. The larger dipole moment of the transition state compared to the reactants suggests a negative deviation from predictions based on the Arrhenius model. The result of these calculations suggests that water molecules are involved in the transition state explicitly and they are part of the reaction coordinate. In addition, the effect of the dielectric constant and the cage effect are in the same direction.

The above studies led to development of a new model consistent with the temperature and pressure dependence of rate coefficients over a broad range of conditions in water. The results are useful for modeling radiolysis of water in pressurized water nuclear reactors.

Another new development in the field of muonium kinetics and reaction dynamics in liquids has been the extension of muonium kinetics to the realm of excited-state reaction dynamics by Ghandi et al. (2007).

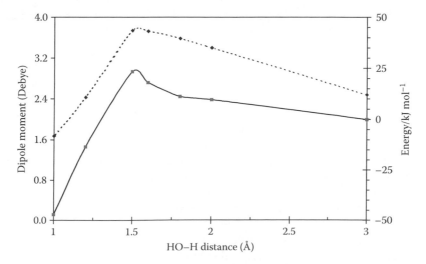

FIGURE 8.19 The solid curve is the energy of the reaction complex relative to the energy of separated reactants (OH⁻ and H), and the dotted curve is the dipole moment of the reaction complex, as a function of HO–H distance.

In this work, laser-muon-spin spectroscopy in liquids has been developed, which is a technique to study the excited-state chemical dynamics of transient species. The work has more applications than reaction dynamics but on the reaction dynamics front, as a proof of concept reaction kinetics of muonium and Rose Bengal in the ground and excited electronic state (triplet state) has been studied. This work also opens the way to study chemistry of excited-state muoniated free radicals. This spectroscopic technique has made new directions possible both for Mu chemists and the reaction dynamics community by studying kinetic isotope effects of reactions of Mu with laser pumped molecules.

8.5.3 Studies in the Gas Phase

The most recent and important developments in the studies of reactions of Mu and muoniated free radicals in gas phase are as follows: (1) The time-delayed RF-μSR studies of Mu reactions with O_2 by Johnson et al. (2005). (2) The preliminary studies of a new class of chemical bonds (rovibrational bonds), where the combination of vibrational and rotational motion of nuclei would lead to transformation of a saddle point on the electronic potential energy surface (PES) to a stable species in the work of Ghandi et al. (2006). This was in the context of a muoniated free radical formed from reaction of Mu with Br_2. (3) Studies of direct abstraction reactions and kinetic isotope effects (KIEs) in comparison with the H-atom analogue and with "heavy muonium," Heμ (Arseneau et al., 2009). Heμ is the muonic atom, $\alpha^{++}\mu^-e^-$, where the electron is in a $1s$ orbital, similar to its position in the ordinary H atom (or Mu). In the "heavy muonium," the negative muon is in the small $1s$ orbital, 400 times closer to α. The atom can be treated as a heavy isotope of H, with a mass 4.116 amu, that is, 37 times Mu mass! In heavy muonium, $(\alpha^{++}\mu^-)^+$ acts as the nucleus. $(\alpha^{++}\mu^-)^+$ is formed by the negative muon capture onto He, with the ejection of two electrons by the Auger process. To form the neutral Heμ, charge exchange with an easily ionized reactant, such as Xe or NH_3, is required; therefore, studies of chemical reactions of "heavy muonium" are usually performed in the presence of one of these species. Studies of this nature demonstrate most clearly the unique sensitivity of the muon-based-atoms to quantum mass effects in reaction dynamics, in particular to zero-point energy (ZPE) shifts at the transition state, exemplified by late-barrier reactions with H_2. (4) The studies of Mu reactions with vibrational excited H_2 in the work of Bakule et al. (2009a,b) that is an extension of the work of Ghandi et al. (2007), that is, chemical dynamics of Mu reaction with excited-state molecules in the liquid phase to the gas phase. Although the laser effect was not significantly above the noise level in this work, it can be an exciting development since along with the work by Ghandi et al. (2007), they generated a new class of experimental techniques for studies of reaction dynamics.

In addition to the above-mentioned novel works in the gas phase, there have been several theoretical and conventional experimental studies over the last 10 years on different systems, mostly to investigate the KIEs. These works demonstrate the unique sensitivity of the Mu atom to quantum mass effects in reaction dynamics, to both ZPE shifts at the transition state, exemplified by late-barrier reactions such as the Heμ + H_2 reaction described above (Arseneau et al., 2009), Mu + CH_4 (Pu and Truhlar, 2002) (KIEs $k_{Mu}/k_H \ll 1$), and quantum tunneling exemplified by "early-barrier" reactions like Mu + Br_2 (Ghandi et al., 2006) (KIEs $k_{Mu}/k_H \gg 1$) at lowest temperatures and concomitantly with much smaller activation energies for Mu reactions.

There have been studies of Mu addition reactions, both in the high-pressure limit exemplified by the Mu + C_2H_4 (Villa et al., 1999) reaction where the high-pressure limit is easily realized at moderator pressures ~1 bar, and in the low pressure (termolecular) regime where much higher pressures (>1000 bar) are required for stabilization, in the case of small molecules with only a few degrees of freedom, such as Mu + NO (Pan et al., 2000), Mu + O_2 (Himmer and Roduner, 2000), and Mu + CO (Pan et al., 2006) reactions, in which the experimental data have been compared with the predictions of unimolecular (Troe) theory. In these indirect reactions, rates for addition and stabilization are in competition with those for unimolecular dissociation, with the overall (effective) rate constant then exhibiting different pressure limits. Such studies are important model systems for theoretical studies (Harding et al., 2000; Marques and Varandas, 2001).

In general, the experimental results of Mu reactivity in these studies continue to present challenges to reaction theory. Therefore, an interface between experimental and theoretical data on Mu is very useful to refine reaction dynamics theories. In the realm of collision dynamics, atomic and molecular beam studies produced under single-collision conditions at epithermal (~1 eV) energies, with their ability to probe ro-vibrational cross sections at the state-to-state level (Althorpe et al., 2002), play an ever increasingly important role in assessing the interface between experiment and reaction theory. However, at these kinds of impact energies, well above threshold, the general features of reactive collision processes are often well described by classical (or quasi-classical) dynamics, so that the study of chemical kinetics and bulk thermal rate constants, which are generally much more sensitive to quantum mass effects in probing the details of the PES at the barrier, continue to be important. This is truer in the case of sensitive isotopic mass probes that now extend from Mu to Heμ, a ratio of 1/37, and hence the above-mentioned studies are unique probes of quantum mass effects in chemical reactivity.

8.6 MODERN SLOW MUON BEAM PRODUCTION TECHNIQUES

The conventional muon beams are obtained from proton accelerators with proton energies higher than 400 MeV. According to the momentum of the generated muon beam, these beams are referred to as "decay muon" or "surface muon" beams. The decay muons are obtained through the in-flight decay of π^+/π^-, confined by a strong longitudinal field of several tesla from a superconducting solenoid magnet. Muons obtained this way have high momentum, between 40 to several hundreds of MeV c^{-1}. Not only positive but also negative muon beams are available (Nagamine, 1981).

The surface muons are obtained from the decay of positive pions (π^+) stopped near the surface of the pion production target in the proton beam line (Pifer et al., 1976). Compared to the decay muons that have high momentum due to the in-flight decay of pions, the surface muons obtained from the decay of π^+ at rest have a low and unique kinetic energy of 4.1 MeV (corresponding to a momentum of 29.8 MeV c^{-1}). Since the negative pions (π^-) are easily captured by the nucleus, only positive surface muon beams are available. The spin-polarized surface muons are widely used as magnetic microprobes in materials (Schenck, 1985). The surface muon beam has a lower energy compared to the decay muon beam, and when injected into a bulk sample, they have, therefore, an implantation depth of just 0.1–1 mm. Despite the name surface muons, it is typically used as a probe to bulk phenomena rather than surface phenomena.

Lately, the study of surface/subsurface nanomaterials and multilayered thin films is becoming increasingly important. However, for that purpose, we must have muon beams that have sufficiently low energy to stop on or near the surface of the sample. To perform such studies, so called slow (low energy) muons are required with energy that is of the order of several eV to a few tens of keV, far lower than the energies available from the conventional muon beams. In this section we introduce the novel techniques to generate such slow muon beams.

8.6.1 THE GENERATION OF LOW ENERGY μ⁺ THROUGH THE CRYOGENIC MODERATION METHOD

The generation of low energy μ⁺ through the cryogenic moderation method using van der Waals solids, such as solid Ar or N_2, has been developed in the following four stages.

[Stage 1, low energy μ⁺] The first observation of the emission of low energy positive μ⁺ was performed by Harshman et al. at TRIUMF. They could extract the so-called epithermal μ⁺ with a kinetic energy less than 10 eV from solid lithium fluoride (LiF), quartz, and copper with an efficiency of $(1.6 \pm 0.25) \times 10^{-7}$ per incident surface muon (Harshmann et al., 1986a,b).

[Stage 2, low energy μ⁺] Soon after this first successful extraction of the epithermal μ⁺, Harshman et al. also succeeded in observation of the epithermal muons below ~10 eV with a tail extending to higher energies, from solid neon (Ne), argon (Ar), krypton (Kr), and xenon (Xe) moderators exposed to the surface muons. Among the moderators, Ar was discovered to produce the highest yield of 10^{-5} low energy μ⁺ per incident surface muon, as shown by Harshmann et al. (1986a,b).

[Stage 3, low energy μ⁺] At Paul Scherrer Institute (PSI), Morenzoni et al. (1994) succeeded in the first measurement of the spin polarization of the low energy muon beam emitted from solid Ar and Kr, utilizing the world's strongest direct current (DC, not pulsed) surface muon beam. They demonstrated that the low energy μ⁺ beam generated through the cryogenic moderation method using the van der Waals solids, such as Ar and N_2, is a very powerful spin-polarized probe and that there is no observable spin depolarization during the slowing-down process in the cryogenic moderator.

The importance of this discovery should be stressed, since not only did they develop the practical experimental technique of the low energy muon generation through the cryogenic moderation of the surface muon beam, but also opened and established, in reality, various kinds of scientific applications, such as surface and interface studies, using the low energy μ⁺.

[Stage 4, low energy μ⁺] Recently, Prokscha et al. (2008) constructed the new μE4 beamline dedicated for the low energy μ⁺ generation at PSI. With a solid-angle acceptance of 135 msr, they obtained the highest flux of 4.2×10^8 surface muons s⁻¹. By constructing the most intense DC surface muon beam channel, they have now prepared an intense and excellent low energy μ⁺ source at PSI and opened up many new applications for the low energy muons.

8.6.1.1 Experimental Technique and Features of the Low Energy μ⁺ Beam through the Cryogenic Moderation Method

At PSI, the low energy μ⁺ beams had been developed at πE1 and πE3 beam lines (Morenzoni et al., 1994) and now have been extended at the μE4 by Prokscha et al. (2008). The principle of the experimental setup to obtain epithermal low energy μ⁺ through the cryogenic moderation method at PSI is beautifully simple and is shown in Figure 8.20 (Morenzoni et al., 1994).

The intense surface muon beam is introduced into the cryogenic moderator consisting of a high-purity (99.999%) aluminum (Al) support foil of 250 mm thickness and a thin layer of deposited solid Ar or N_2. In the case of Ar or N_2 (Ne), about 5×10^{-5} (1.5×10^{-4}) of the surface muons are converted into the epithermal μ⁺. The epithermal μ⁺ emitted from the moderator are then re-accelerated up to 20 kV and transported and focused by electrostatic Einzel lenses and a mirror into the sample. On the way towards the sample, the accelerated μ⁺ pass through a carbon foil of 10 nm thickness, which provides a trigger signal for the TOF measurements (implantation time) at DC muon facility. The use of such a trigger is unavoidable, but has the unwanted consequence of introducing a small amount of energy and temporal spread. Finally, the kinetic energy of the low energy μ⁺ (implantation energy) can be adjusted from 0.5 to 30 kV, by applying an accelerating or decelerating potential up to 12 kV to the sample.

8.6.1.2 Yield of the Low Energy Muons at PSI

According to Prokscha et al. (2008), at the newly built μE4 beam channel, the maximum 16×10^3 low energy μ⁺ s⁻¹ at the Ne production target were obtained and 7.7×10^3 s⁻¹ were extracted at the sample.

8.6.1.3 Aspects of the Charged Particle Interactions on the Low Energy μ⁺ Generation

The energy distribution of the low energy μ⁺ generated through the cryogenic moderation method has a peak at around 15 eV with a tail extending to higher energies. In that sense, the low energy μ⁺ through the cryogenic moderation method can be regarded as an epithermal μ⁺ with the full spin polarization. The mechanism of the emission of the epithermal μ⁺ may include important features of the "Charged Particle Interactions with Matter." In the matter, the slowing down of μ⁺ can be proceeded by (1) Bethe–Bloch ionization, (2) cyclic charge-exchange muonium (designated as Mu, consisting of a bound μ⁺ and an e⁻, effectively a hydrogen-like atom) formation and break-up process, and (3) thermalization (Senba et al., 2006).

However, in the wide-band-gap insulators such as Ar, N_2, or Ne with band-gap energy of between 11 and 22 eV, once μ⁺ reached a kinetic energy of the order of their band-gap level, the

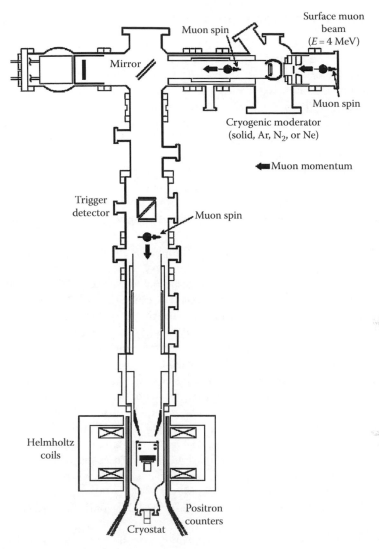

FIGURE 8.20 The principle of the experimental setup for epithermal low energy μ^+ through the cryogenic moderation method developed at PSI (Morenzoni et al., 1994; Prokscha et al., 2008). The intense surface muon beam is introduced into the cryogenic moderator consisting of a high-purity (99.999%) aluminum (Al) support foil of 250 mm thickness and a thin layer of deposited solid Ar or N_2. The epithermal μ^+ emitted from the moderator are then re-accelerated up to 20 kV and transported and focused by electrostatic Einzel lenses and a mirror into the sample.

processes (2) and (3) are significantly suppressed. At the last process of (2), epithermal μ^+ may escape from the ionization track where excess e^- are abundant, resulting in suppressing Mu formation and efficient emission of epithermal μ^+ into vacuum. Prokscha et al. (2008) examined Mu and diamagnetic muon fraction in Ar, N_2, and Ne, etc. as a function of implantation energy utilizing the low energy μ^+. In those cryogenic solid samples, Mu fraction was found to be suppressed near the surface, in contrast to the diamagnetic muon fraction increasing near the surface, where the excess e^- are not abundant. These views occurring in the ionization track can also be reconfirmed by the dependence of the external electric field on the Mu and μ^+ amplitudes in the solid N_2 at 20 K by Storchak et al. (1999). Both experimental results show the importance of charged particle interactions with matter, in particular the behavior of excess e^- in the ionization track.

8.6.2 Generation of Ultraslow μ⁺ through the Resonant Ionization of Thermal Mu

The second method is the ultraslow muon generation through the resonant ionization of thermal Mu atoms, developed at KEK (High Energy Accelerator Research Organization) and RIKEN-RAL (RIKEN Muon Facility of Rutherford Appleton Laboratory) pulsed muon facilities. This method has been developed by the following five stages:

[Stage 1, ultraslow μ⁺] Thermal muonium formation in vacuum.

[Stage 2, ultraslow μ⁺] Experiments for the precise measurement of Mu's 1s-2s state for the purpose of the quantum electro dynamics (QED) test at KEK-MSL (Muon Science Laboratory).

[Stage 3, ultraslow μ⁺] The ultraslow muon generation by 1s-2p-unbound transitions, placing a Mu production target W in a primary proton beam at KEK-MSL.

[Stage 4, ultraslow μ⁺] The ultraslow μ⁺ generation by 1s-2p-unbound transitions, utilizing an intense surface muon beam at RIKEN-RAL.

[Stage 5, ultraslow μ⁺] Japan Proton Accelerator Research Complex (J-PARC) ultraslow μ⁺ project.

[Stage 1, ultraslow μ⁺] When surface muons are stopped at the hot tungsten (W) in the ultra high vaccum (UHV) condition, thermal energy Mu atoms were found to be evaporated into vacuum from a clean and hot W foil. The total Mu yield was found to increase with increasing temperature and about 0.04 Mu yield per stopped surface muons (i.e., efficiency of 4%) was observed at 2300 K, according to the Mills et al. (1986) experiments. Note that it is this extremely high efficiency of converting surface muons to ~0.2 eV thermal energy muons (albeit bound to an electron) that can be exploited for low energy muon generation through the cryogenic moderation.

[Stage 2, ultraslow μ⁺] The second stage consisted of the experiments for the precise measurement of Mu's 1s-2s transition frequency for the purpose of the precise QED test, performed by Chu et al. (1988). Since Mu consists of muon (lepton) and electron (lepton) pair compared with hydrogen that is a pair of hadrons (consisting of three quarks) and electron (lepton), Mu is considered to be the ideal and simplest system to test QED theory. In this experiment, thermal Mu atoms were produced in vacuum by utilizing a SiO_2 powder sample. The 1s-2s transition as well as the 2s-unbound transitions were induced by light from a frequency doubled pulsed dye laser. The $1S(F=1)$–$2S(F=1)$ transition frequency was experimentally determined to be within 300 MHz of the QED theory prediction (Chu et al., 1988; Meyer et al., 2000). In these experiments, for the first time, the ultraslow μ⁺ was generated by the laser resonant ionization of Mu, although it was not intended.

[Stage 3, ultraslow μ⁺] The third stage consisted of the ultraslow μ⁺ generation via 1s-2p-unbound two-photon ionization of thermal Mu at KEK. In this step, boron nitride (BN) followed by W target was installed in the proton line. In BN, π production and πμ decay takes place. In W, a fraction of decayed μ⁺ stop and diffuse into the rear surface of W, and then pick up electrons at surface, following which Mu atoms evaporate into vacuum. The Mu atoms that are evaporated from W into vacuum are then ionized by intense laser pulses to generate the ultraslow μ⁺. This concept, a new method to generate the ultraslow μ⁺, was successfully established by Nagamine et al. (1995).

[Stage 4, ultraslow μ⁺] The fourth stage consisted of the experiments at RIKEN-RAL at ISIS in UK where very intense pulsed surface muon beam is available. Here, the experiment was setup directly at the beamline delivering the intense surface muon beam, taking the advantage of the fact that the intensity of the primary proton beam at ISIS of 200 μA is about 40 times larger than that of KEK-MSL (Miyake et al., 2000). At the RIKEN-RAL muon facility, finally, extraction of 20 ultraslow μ⁺ s⁻¹ out of 1.2×10^6 s⁻¹ surface muons was achieved with an overall efficiency comparable to the cryogenic moderator method and many unique parameters.

[Stage 5, ultraslow μ⁺] The fifth step is the ultraslow muon source at J-PARC. The muon science facility (MUSE), along with the neutron, hadron, and neutrino facilities, is one of the experimental areas of J-PARC, which was approved for construction at the Tokai JAEA site. The MUSE facility is located in the Materials and Life Science Facility (MLF), which is a building integrated to include both neutron and muon science programs. Construction of the MLF building was started in the beginning of 2004, and the first muon beam was delivered in the autumn of 2008. A super omega muon channel with a large acceptance to extract the world's strongest pulsed surface muon beam is planned to be installed. Its goal is to extract 4×10^8 surface muons s^{-1} for the generation of the intense ultraslow μ^+, utilizing laser resonant ionization of Mu by applying an intense pulsed VUV laser system. As a maximum, 1×10^6 ultraslow muons s^{-1} will be expected, which will allow for the extension of μSR into the field of thin film and surface science (Miyake et al., 2009). All the experimental apparatus that have been installed at RIKEN-RAL muon facility are planned to be transferred to J-PARC when the super omega muon beam channel is ready at J-PARC.

8.6.2.1 The Ultraslow μ^+ Generation by Resonant Ionization of Mu

The ultraslow pulsed muons are generated by the resonant ionization of thermal muonium atoms generated from the surface of a hot W foil, placed at the intense surface muon beam line. In order to efficiently ionize the Mu near the W surface, a resonant ionization scheme via the {1S-2P-unbound} transition was adopted. This requires a laser system to generate Lyman-α photons (in the vacuum ultraviolet (VUV) region around 120 nm) and 355 nm photons (for the ionization step from the 2P state). Since there are only very few solids that are partly transparent in VUV region, the generation of the 120 nm photons is the most challenging part of such experiment. For the experiments both at KEK-MSL and at RIKEN-RAL, a very efficient technique of nonlinear up-conversion in Kr gas, the so-called resonant four-wave frequency mixing ($\omega_{VUV} = 2\omega_r - \omega_t$), has been adopted. This wave mixing technique generates the Lyman-α photon through a parametric process of adding energies of two 212.55 nm photons (ω_r) and subtracting energy of one tunable infrared photon (ω_t). This allows the generation of a VUV pulse that is easily tunable and has the bandwidth of ~200 GHz required to cover the Doppler broadening of the Mu atoms moving at thermal velocities corresponding to 2000 K. The 212.55 nm UV beam is generated as a fourth harmonic of the fundamental laser operating at 850 nm. This single-mode 850 nm light with a bandwidth of 0.5–1.0 GHz is generated from an OPO system (Continuum Mirage800) and amplified by series of Ti:sapphire amplifiers. The output energy of the single-mode beam at 850 nm is as high as 300 mJ p^{-1} at 25 Hz repetition rate and pulse duration of 5 ns. The amplified 850 nm light is quadrupled by using two β-Ba$_2$BO$_4$ crystals, generating two beams at 212.55 nm wavelength with pulse energy of 10 mJ each. The difference tunable infrared beam was generated by a broadband OPO laser system (Continuum Mirage3000). Finally, in order to ionize the Mu atom from the excited 2P state, a third harmonic of a Nd:YAG laser (Continuum Powerlite 9025) with a wavelength of 355 nm was used (Miyake et al., 1995).

Figure 8.21 (top) shows a resonant ionization scheme via the 1s-2p-unbound transition for the hydrogen atom isotopes and the scheme for the Lyman-α generation via the four-wave frequency mixing method using two 212.55 nm (ω_r) photons for the two-photon resonant excitation of the 4P^{5}5P[1/2,0] state in krypton, subtracted by a photon with a tunable difference wavelength.

Experiments have been performed using "slow-ion optics," which consists of an immersion lens (SOA), a magnetic bend for mass separation, an electrostatic mirror, and five sets of electrostatic quadrupoles. Charged particles ionized by the laser pulses are extracted by the SOA lens with an acceleration voltage of 9.0 kV (20 kV to be prepared), and transported to a micro-channel plate (MCP) detector located about 3 m from the target. Any ions produced at the W target region can be identified through the mass/charge (Q) ratio by setting the bending magnet in the ion optics, and by observing the time-of-flight (TOF) spectrum, consequently minimizing the background significantly. Additionally, the electrostatic deflector selects the energy of the particles that are transported to the sample, thus significantly suppressing the background. The laser pulses are delayed by typically 400 ns relative to the surface muon pulse to allow time for Mu to move away from the target, since the mean thermal velocity in only 20 mm μs^{-1}. A high-purity (99.9999%) W foil with 50 μm thickness (obtained from Metallwerk Plansee GmbH) is placed inside the SOA target chamber. The W foil is heated up to 2300 K by a pulsed DC current that is turned off for

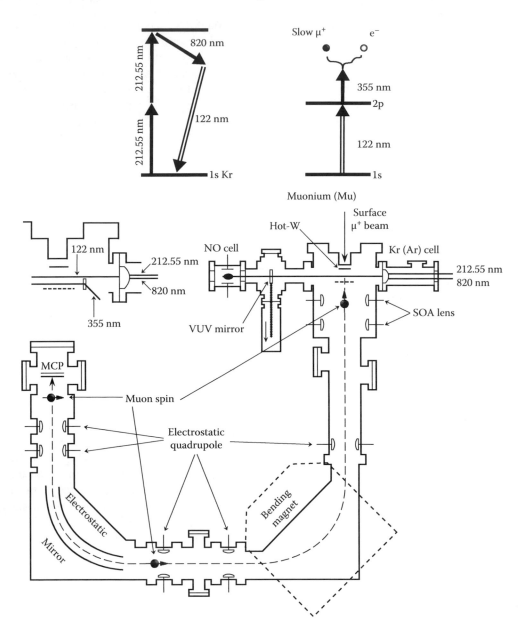

FIGURE 8.21 The top figure is a resonant ionization scheme via the {1S-2P-unbound} transition for the Muonium (Mu) and the scheme for the Lyman-α generation via the four-wave frequency mixing method using two 212.55 nm (ω_r) photons for the two-photon resonant excitation of the 4P55P [1/2,0] state in krypton, subtracted by a photon with a tunable difference wavelength (Nagamine et al., 1995; Bakule et al., 2008). The bottom figure is a layout of "slow-ion optics" consisting of an immersion lens (SOA), a magnetic bend for mass separation, an electrostatic mirror, and five sets of electrostatic quadrupoles.

1 ms in order to avoid any interaction of the magnetic field produced by the DC current with the resonantly ionized ions (muons). The target chamber is evacuated down to ultrahigh vacuum level of 2×10^{-7} Pa at 2300 K. The W target is cleaned via surface treatment in 3.0×10^{-5} Pa of oxygen at 1800 K for about 10 h, and then heated up to 2000 K for the experiments. Figure 8.21 (bottom) shows a layout of "slow-ion optics" consisting of an immersion lens (SOA), a magnetic bend for mass separation, an electrostatic mirror, and five sets of electrostatic quadrupoles.

8.6.2.2 Unique Features of the Pulsed Ultraslow μ⁺ Beam

8.6.2.2.1 Pulse Width

One of the most important features of the ultraslow μ⁺ beam generated by the resonant laser ionization of Mu is its short pulse width. Regardless of the temporal profile of the surface muon pulse—with its double bunch structure consisting of ~100 ns (full width at half maximum [FWHM]) pulses separated by 320 ns at RIKEN-RAL—the pulse duration of the ultraslow muon beam is determined only by the duration of the laser pulse and by the transport properties of the low energy ion transport beamline. Consequently, at RIKEN-RAL the duration of the ultraslow μ⁺ pulse in the time-of-flight spectrum at the end of the transport beamline is just 7.5 ns (FWHM). In principle, the pulse duration can be reduced further to ~1 ns by using a laser with shorter pulse duration and optimizing the beamline transport. There is another advantage of this pulsed method—the capability to externally trigger when the muonium is ionized, that is, to trigger the ultraslow μ⁺ pulse width. This is possible because of the relatively low velocity of the thermal Mu cloud. The Mu atoms that are emitted from the W target foil stay near the surface of the foil, with their concentration not significantly changing over a timescale of ~500 ns. Within this timescale, one can then choose when to trigger the laser pulse to generate the low energy muons. Such a feature is extremely useful for pump-probe type experiments where synchronization of the muon implantation with some other sample excitation (e.g., by another laser, rf pulse, etc.) is required.

8.6.2.2.2 Beam Size and Energy Resolution

Since Mu atoms evaporate from the hot W (2000 K) at thermal energies, the initial kinetic energy of the ultraslow μ⁺ generated by the laser resonant ionization is only 0.2 eV. Therefore, the beam emittance of the ultraslow μ⁺ beam accelerated by SOA lens to 9 keV is so small that the beam can be focused onto a tiny spot on the sample. Compared to the size of the incident surface muon beam of 40 mm (FWHM), the resulting ultraslow μ⁺ beam could be focused to a spot of just 3.3 mm (FWHM, horizontally) by 4.1 mm (FWHM, vertically) presently at RIKEN-RAL. In principle, the beam size is not limited by the beam emittance but by the transport characteristics of the beamline and by better optimization of the beamline, spots as small as 1 mm (FWHM) can be potentially obtained, perhaps at a cost of reducing the transport efficiency. According to the evaluation by Bakule et al. (2008, 2009a,b), energy resolution of the low energy muons is only $\sigma_E = 14$ eV (33 eV at FWHM) at RIKEN-RAL, which is in contrast with $\sigma_E = 400$ eV (950 eV FWHM) at PSI because of energy spread caused by the thin trigger counter at PSI. Having such an essentially monoenergetic beam is very important for near-surface studies on depth scales of 0–10 nm.

8.6.2.2.3 Polarization of the Ultraslow Muons by the Laser Resonant Ionization Method

Although the surface muons are 100% polarized initially, the polarization of the singlet Mu, in which the muon and electron spins are parallel, is depolarized due to the hyperfine interaction of the paired electron. Therefore, the ultraslow μ⁺ produced by the laser resonant ionization method is only 50% spin polarized only by the contribution of the triplet Mu (Miyake et al., 1997). Possibly, re-polarization of the singlet Mu by the external strong longitudinal field might be performed with future developments.

8.6.2.2.4 Yield of Ultraslow Muons by the Laser Resonant Ionization Method

At the RIKEN-RAL muon facility, 20 ultraslow μ⁺ s⁻¹ (Bakule et al., 2008) can be extracted out of 1.2×10^6 s⁻¹ surface muons from the sample. At the J-PARC MUSE super omega muon channel, 4×10^8 surface muons s⁻¹ can be extracted from a large acceptance of 400 msr. From the viewpoint of the intensity of the surface muon, we can gain a factor of 300. Taking into account the repetition rate of the pulsed laser system and the muon beam, we can gain a factor of two, since at J-PARC, both the laser and muon beams can be synchronized to 25 Hz, whereas at the RIKEN-RAL

TABLE 8.2
Comparison of Features of Slow Muon Beam Obtained by the Cryogenic Moderator Using Solid Ne or Ar and Laser Resonant Ionization

	Ultraslow μ^+ Generation by Laser Resonant Ionization of Mu	Low-Energy μ^+ Generation by Cryogenic Moderator of Ar
MUON Facility	KEK, RIKEN-RAL and J-PARC	PSI
Energy	0.1–30 keV	0.5–30 keV
Monochromacity	14 eV (RIKEN-RAL), (started with 0.2 eV)	400 eV
Beam size	ϕ 1–4 mm	ϕ 10–15 mm
Time resolution	sub ns ~ ns	~10 ns
Polarization	50%	92%
Beam intensity	20 s^{-1} (RIKEN-RAL), 10$^{4-6}$ s^{-1} (J-PARC)	10$^{3-4}$ s^{-1}
Synchronization	Possible	Not possible

facility, the muon beam (50 Hz) is synchronized to every second laser pulse. Therefore, we can expect 1.3×10^4 s^{-1} of the ultraslow μ^+ s^{-1} without any additional laser development at J-PARC (Miyake et al., 2009).

Although more than 71 μJ cm^{-2} of the Lyman-α light is needed in order to saturate the transition of Mu from the 1S state to the 2P state, we are, at present, producing <1 μJ cm^{-2} of the Lyman-α light through the resonant four-wave frequency mixing ($\omega_{VUV} = 2\omega_r - \omega_t$; 212.55 nm ($\omega_r$) 820 nm ($\omega_t$)). As a possible promising method to obtain a more intense Lyman-α laser source, we are planning to adopt the tripling of 366 nm photons (non-resonant tripling: $\omega_{VUV} = 3\omega_{366}$) with pico second pulse width to match the Doppler broadening of the Mu at 2000 K, expecting more than 100 μJ cm^{-2} of the Lyman-α light. With sufficient laser development, we expect about 100 times increase in gain. Finally, we expect a maximum rate of 1.3×10^6 s^{-1} ultraslow μ^+ s^{-1} (Miyake et al., 2009).

Scientists at PSI are developing a slow muon source by adopting a cryogenic moderator and have started various kinds of surface/interface science programs, for example, those by Jackson et al. (2000), Khasanov et al. (2004), Morenzoni et al. (2008), Senba et al. (2006), and Suter et al. (2004). Although the J-PARC project is lagging behind that of PSI due to more time required for the construction of the super omega channel, more intense ultraslow μ^+ with a smaller beam size, a shorter temporal width, and lower background is expected at J-PARC. Table 8.2 shows a comparison of the slow muon beam between the PSI cryogenic method and the J-PARC resonant ionization of Mu.

8.6.2.2.5 *Aspects of the Charged Particle Interactions on the Ultraslow μ^+ Generation*

The mechanism of the emission of the thermal Mu may include important features of the charged particle interactions with matter. Laser resonant ionization is one of the most powerful methods to study the mechanism for the emission of thermal neutral atoms originating from high-energy nuclear reactions, owing to the capability of isotope/particle selection as well as a high efficiency. This idea has been used as a chemically selected laser-ionized ion source for unstable heavy nuclei at CERN-ISOLDE (On-line Isotope Mass Separator at CERN) (Mishin et al., 1993). By only changing the difference wavelength (ω_t) in the present four-wave frequency mixing method without changing any other experimental condition, the Lyman-α wavelength can be easily tuned for any hydrogen isotopes. A very important capability of the laser resonant ionization technique is that it enables us to very efficiently detect and distinguish even hydrogen isotopes existing at the same location by adjusting the individual Lyman-α wavelength.

Figure 8.22 (top) summarizes the resonance spectra for all hydrogen isotopes (Mu, H, D, and T). The observed resonance peaks agree with the calculated values. The width of the Mu resonance is consistent with the FWHM expected due to Doppler broadening and a Mu temperature of 2000 K. In the case of H, the resonance curve was taken as a cold run from the residual hydrogen in the UHV chamber. Figure 8.22 (bottom) shows the temperature dependence of the total yield of Mu and T. The Mu evaporation was negligible at the temperature below 1200 K, started to increase above 1200 K, and increased monotonically with temperature and leveled off at ~2000 K. The T evaporation was also negligible below 1500 K and increased with temperature

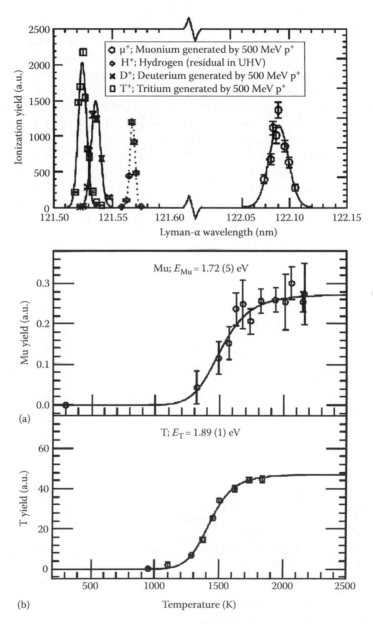

FIGURE 8.22 The top figure is a resonance curve of the laser resonantly ionized Mu, H, D, and T, where Mu, D, and T were generated by nuclear reactions of 500 MeV protons and the H atoms were extracted from residual hydrogen in the UHV chamber. A bottom figure shows the temperature dependence of the total yield of Mu and T. The solid lines are fits to the temperature dependence Equation 8.24.

above 1500 K. The solid lines in Figure 8.22 (bottom) are fits to the temperature dependence Equation 8.24 shown below:

$$y(T) = \frac{c_1 \sqrt{T} e^{-\frac{E_H}{kT}}}{1 + c_2 \sqrt{T} e^{-\frac{E_H}{kT}}}, \tag{8.24}$$

where E_H is the activation energy of hydrogen-like atom in eV. A least square fit of Equation 8.24 to the data gives $E_{Mu} = 1.72$ (5) eV and $E_T = 1.89$ (1) eV. The binding energy of surface H atom relative to the free Hydrogen in vacuum is 2.95 (2), 3.03 (3), and 2.94 (2) eV for W(100), W(110), and W(111) (Tamm and Schmidt, 1971), respectively. Therefore, H atoms are bound to the W surface by ~2.97 (2) eV averaged over W(100), W(110), and W(111). In contrast, since the molecular H_2 binding energy $E(H_2)$ and the solution enthalpy of H in W is given to be 2.24 eV (Herzberg and Monfils, 1960) and -0.68 eV (McClellan and Oats, 1973), respectively, the work function of hydrogen to the solid is estimated to be ~1.56 eV (Mills et al., 1986).

Thermionic emission is a process that competes with trapping at the surface and with a loss due to μ^+-decay for Mu and trapping at impurities. However, our experimental evidences indicate that both the lightest and heaviest hydrogen isotopes, Mu and T, are thermionically emitted from the bulk not likely to be trapped at any surface. The difference of ~0.2 eV between E_{Mu} and E_T can be explained by a contribution of the ZPE to the Mu work function (Miyake et al., 1999).

REFERENCES

Althorpe, S. C., Alonso, F. F., Bean, B. D. et al. 2002. Observation and interpretation of a time-delayed mechanism in the hydrogen exchange reaction. *Nature* 416: 67–70.
Arseneau, D. J., Fleming, D. G., Sukhorukov, O. et al. 2009. The muonic He atom and a preliminary study of the ^{4}Heμ + H_2 reaction. *Physica B* 404: 946–949.
Bakule, P., Matsuda, Y., Miyake, Y. et al. 2008. Pulsed source of ultra low energy positive muons for near-surface μSR studies. *Nucl. Instrum. Meth. B.* 266: 335–346.
Bakule, P., Matsuda, Y., Miyake, Y., and Nagamine, K. 2009a. Prospects for ultra-low-energy muon beam at J-PARC. *Nucl. Instrum. Meth. A* 600: 35–37.
Bakule, P., Sukhorukov, O., Matsuda, Y. et al. 2009b. Toward the first study of chemical reaction dynamics of Mu with vibrational-state-selected reactants in the gas phase. *Physica B* 404: 1013–1016.
Buxton, G. V., Greenstock, C. L., Helman, W. P., and Ross, A. B. 1988. Critical review of rate constants for reactions of hydrated electrons, hydrogen atoms and hydroxyl radicals (OH/O$^-$) in aqueous solution. *J. Phys. Chem. Ref. Data.* 17: 513–886.
Chu, S., Mills, A. Jr., Yodh, A. G. et al. 1988. Laser excitation of the 1S 2S transition in muonium. *Phys. Rev. Lett.* 60: 101–104.
Clifford, A. A., Zhu, S., Smart, N. G. et al. 2001. Modelling of the extraction of uranium with supercritical carbon dioxide. *J. Nucl. Sci. Tech.* 38: 433–438.
Cormier, P., Arseneau, D. J., Brodovitch, J. C. et al. 2008. Free radical chemistry in supercritical carbon dioxide. *J. Phys. Chem. A.* 112: 4593–4600.
Cormier, P., Taylor, B. A., and Ghandi, K. 2009. Hyperfine interactions of a muoniated ethyl radical in supercritical CO_2. *Physica B* 404: 930–932.
Dilger, H., Roduner, E., Stolmar, M. et al. 1997. Why ALC μSR is superior for gas-phase radical spectroscopy. *Hyperfine Interact.* 106: 137–142.
Eshchenko, D. G., Storchak, V. G., Brewer, J. H. et al. 2000. Muonium formation in condensed neon and argon. *Physica B* 289: 418–420.
Eshchenko, D. G., Storchak, V. G., Brewer, J. H. et al. 2001. Muonium diffusion in solid CO_2. *J. Low Temp. Phys.* 27: 854–857.
Eshchenko, D. G., Storchak, V. G., Brewer J. H. et al. 2002. Excess electron transport and delayed muonium formation in condensed rare gases. *Phys. Rev. B* 66: 035105–035121.

Eshchenko, D. G., Storchak, V. G., Brewer, J. H. et al. 2003. Excess electron transport in cryoobjects. *J. Low Temp. Phys.* 29: 185–195.

Fleming, D. G. and Senba, M. 1992. In *Perspectives in Meson Science*, T. Yamazaki, K. Nakai, and K. Nagamine (eds.), pp. 219–264. Amsterdam, the Netherlands: North Holland Press.

Garner, D. M., Fleming, D. G., Arseneau, D. J. et al. 1990. Muonium addition reactions in the gas phase: Quantum tunnelling in Mu + C_2H_4 and Mu + C_2D_4. *J. Chem. Phys.* 93: 1732–1740.

Ghandi, K. and Percival, P. W. 2003. Prediction of rate constants for reactions of the hydroxyl radical in water at high temperatures and pressures. *J. Phys. Chem. A* 107: 3005–3008.

Ghandi, K., Brodovitch, J. C., Addison-Jones, B. et al. 2000. Hyperfine coupling constants of muonium in sub and supercritical water. *Physica B* 289: 476–481.

Ghandi, K., Brodovitch, J. C., Addison-Jones, B. et al. 2002. Near-diffusion-controlled reactions of muonium in sub- and supercritical water. *PCCP* 4: 586–595.

Ghandi, K., 2002. Muonium chemistry in sub- and supercritical water. PhD thesis, Simon Fraser University, Burnaby, British Columbia.

Ghandi, K., Addison-Jones, B., Brodovitch, J. C. et al. 2003. Enolization of acetone in superheated water detected via radical formation. *J. Am. Chem. Soc.* 125: 9594–9595.

Ghandi, K., Arseneau, D. J., Bridges, M. D., and Fleming, D. G. 2004. Muonium formation as a probe of radiation chemistry in sub- and supercritical carbon dioxide. *J. Phys. Chem. A* 52: 11613–11625.

Ghandi, K., Cottrell, S. P., Fleming, D. G., and Johnson, C. 2006. The first report of a muoniated free radical formed from reaction of Mu with Br_2. *Physica B* 374: 303–306.

Ghandi, K., Clark, I. P., Lord, J. S., and Cottrell, S. P. 2007. Laser-muon spin spectroscopy in liquids—A technique to study the excited state chemistry of transients. *PCCP* 9: 353–359.

Ghandi, K., Cormier, P., and Alcorn, C. 2010. Effects of density inhomogeneity in near critical fluids on Muonium formation. *Radiat. Phys. Chem.*, submitted.

Gorelkin, V. N., Soloviev, V. R., Konchakov, A. M., and Baturin, A. S. 2000. Muonium and muonium-like systems formation in spur model. *Physica B* 289: 409–413.

Harding, L. B., Troe, J., and Ushakov, V. G. 2000. Classical trajectory calculations of the high pressure limiting rate constants and of specific rate constants for the reaction H + O_2 → HO_2: Dynamic isotope effects between tritium + O_2 and muonium + O_2. *Phys. Chem. Chem. Phys.* 2: 631–642.

Harshmann, D. R., Mills, A. Jr., Beveridge, J. L. et al. 1986a. Generation of slow positive muons from solid rare-gas moderators. *Phys. Rev. B* 36: 8850–8853.

Harshmann, D. R., Warren, J. B., Beveridge, J. L. et al. 1986b. Observation of low-energy μ^+ emission from solid surfaces. *Phys. Rev. Lett.* 56: 2850–2853.

Harvey, A. H., Peskin, A. P., and Klein, S. A. 2001. *NIST/ASME Steam Properties Database: version 2.2. NIST Standard Reference Database 10.* Gaithersburg, MD: National Institute of Standards and Technology.

Hatano, Y. 2003. Spectroscopy and dynamics of molecular superexcited states. Aspects of primary processes of radiation chemistry. *Radiat. Phys. Chem.* 67: 187–198.

Herzberg, G. and Monfils, A. 1960. The dissociation energies of the H_2, HD, and D_2 molecules. *J. Mol. Spectrosc.* 5: 482–498.

Himmer, U. and Roduner, E. 2000. The addition reaction of X to O_2 (X = Mn, H, D): Isotope effects in intra- and intermolecular energy transfer. *PCCP* 2: 339–347.

Inokuti, M. 1971. Inelastic collisions of fast charged particles with atoms and molecules—The Bethe theory revisited. *Rev. Mod. Phys.* 43: 297–347.

Jackson, T. J., Riseman, T. M., Forgan, E. M. et al. 2000. Depth-resolved profile of the magnetic field beneath the surface of a superconductor with a few nm resolution. *Phys. Rev. Lett.* 84: 4958–4961.

Johnson, C., Cottrell, S. P., Ghandi, K., and Fleming, D. G. 2005. Muon implantation in inert gases studied by radio frequency spectroscopy. *J. Phys. B: At. Mol. Opt. Phys.* 38: 119–134.

Kempton, J. R., Senba, M., Arseneau, D. J. et al. 1991. Hot muonium and muon spur processes in nitrogen and ethane. *J. Chem. Phys.* 94: 1046–1059.

Khasanov, R., Eshchenko, D. G., Luetkens, H. et al. 2004. Direct observation of the oxygen isotope effect on the in-plane magnetic field penetration depth in optimally doped $YBa_2Cu_3O_7$-delta. *Phys. Rev. Lett.* 92: 057602–057604.

Kosarev, E. L. and Krasnoperov, E. P. 1999. Kinetics of muonium formation in liquid helium. *JETP Lett.* 69: 273–280.

Kruse, A. and Dinjus, E. 2007. Hot compressed water as reaction medium and reactant properties and synthesis reactions. *J. Supercrit. Fluids* 39: 362–380.

Kusalik, P. G. and Svishchev, I. M. 1994. The spatial structure in liquid water. *Science* 265: 1219–1221.

Lauzon, J. M., Arseneau, D. J., Brodovitch, J. C. et al. 2008. The formation of non-persistent free radicals in tetra-alkyl phosphonium ionic liquids. *PCCP* 10: 5957–5962.

Lemmon, E. W., McLinden, M. O., and Friend, D. J. 1998. *NIST Chemistry Webbook, NIST Standard Reference Database*. Gaithersburg, MD: National Institute of Standards and Technology.

Liew, C. C., Inomata, H., and Arai, K. 1998. Flexible molecular models for molecular dynamics study of near and supercritical water. *Fluid Phase Equilib.* 144: 287–298.

Lossack, A. M., Roduner, E., and Bartels, D. M. 2001. Solvation and kinetic isotope effects in H and D abstraction reactions from formate ions by D, H and Mu atoms in aqueous solution. *PCCP* 3: 2031–2037.

Marques, J. M. C. and Varandas, J. C. 2001. On the high pressure rate constants for the H/Mu+O_2 addition reactions. *PCCP* 3: 505–507.

Matubayasi, N., Wakai, C., and Nakahara, M. 1997. Structural study of supercritical water.1. Nuclear magnetic resonance spectroscopy. *J. Chem. Phys.* 107: 9133–9140.

McClellan, H. E. and Oats, W. A. 1973. The solubility of hydrogen in rhodium, ruthenium, iridium and nickel. *Acta Metall.* 21: 181–185.

McKenzie, I. and Roduner, E. 2009. Using polarized muons as ultrasensitive spin labels in free radical chemistry. *Naturwissenschaften*, 96: 873–887.

Mesmer, R. E., Marshal, W. L., Palmer, D. A. et al. 1988. Thermodynamics of aqueous association and ionization reactions at high temperatures and pressures. *J. Solution Chem.* 17: 699–718.

Meyer, V., Bagayev, S. N., Baird, P. E. G. et al. 2000. Measurement of the 1s-2s energy interval in muonium. *Phys. Rev. Lett.* 84: 1136–1139.

Mills, A. Jr., Imazato, J., Saitoh, S. et al. 1986. Generation of thermal muonium in vacuum. *Phys. Rev. Lett.* 56: 1463–1466.

Mishin, V. I., Fedoseyev, V. N., Kluge, H. J. et al. 1993. Chemically selective laser ion-source for the Cern-isolde online mass separator facility. *Nucl. Instrum. Meth. B* 73: 550–560.

Miyake, Y., Marangos, J. P., Shimomura, K. et al. 1995. Laser system for the resonant ionization of hydrogen-like atoms produced by nuclear-reactions. *Nucl. Instrum. Meth. B* 95: 265–275.

Miyake, Y., Shimomura, K., Mills, A. Jr., and Nagamine, K. 1997. Ultra slow muons generated by laser resonant ionization of thermal muonium produced by 500-MeV protons with hot tungsten. *Hyperfine Interact.* 106: 237–244.

Miyake, Y., Shimomura, K., Mills, A. Jr. et al. 1999. Thermionic emission of hydrogen isotopes (Mu, H, D and T) from W and interpretation of a role of the hot W as an atomic hydrogen source. *Surf. Sci.* 433–435: 785–789.

Miyake, Y., Shimomura, K., Matsuda, Y. et al. 2000. Construction of the experimental set-up for ultra slow muon generation by thermal Mu ionization method at RIKEN-RAL. *Physica B* 289–290: 666–669.

Miyake, Y., Nakahara, K., Shimomura, K. et al. 2009. Ultra slow muon project at J-PARC, MUSE. *AIP Conf. Proc.* 1104: 47–52.

Moisseytsev, A. and Sienicki, J. J. 2008. Transient accident analysis of a supercritical carbon dioxide Brayton cycle energy converter coupled to an autonomous lead-cooled fast reactor. *Nucl. Eng. Des.* 238: 2094–2105.

Morenzoni, E., Kottmann, F., Maden, D. et al. 1994. Generation of very slow polarized positive muons. *Phys. Rev. Lett.* 72: 2793–2797.

Morenzoni, E., Luetkens, H., Prokscha, T. et al. 2008. Depth-dependent spin dynamics of canonical spin-glass films: A low-energy muon-spin-rotation study. *Phys. Rev. Lett.* 100: 147205–147601.

Mozumder, A. 1999. *Fundamentals of Radiation Chemistry*. London, U.K.: Academic Press.

Nagamine, K. 1981. Pulsed MU-SR facility at the KEK booster. *Hyperfine Interact.* 8: 787–795.

Nagamine, K., Miyake, Y., Shimomura, K. et al. 1995. Ultraslow positive-muon generation by laser ionization of thermal muonium from hot tungsten at primary proton beam. *Phys. Rev. Lett.* 74: 4811–4814.

Noyori, R. (ed.) 1999. Supercritical fluids. *Chem. Rev.* 99: 353–354.

Pan, J. J., Arseneau, D. J., Senba, M., Fleming, D. G., Himmer, U., and Suzuki, Y. 2000. Measurements of Mu + NO termolecular kinetics up to 520 bar: Isotope effects and the Troe theory. *PCCP* 2: 621–629.

Pan, J. J., Arseneau, D. J., Senba, M. et al. 2006. Termolecular kinetics for the Mu+CO+M recombination reaction: A unique test of quantum rate theory. *J. Chem. Phys.* 1: 125.

Percival, P. W. 1990. Current trends in muonium chemistry. *Hyperfine Interact.* 65: 901–911.

Percival, P. W., Brodovitch, J. C., Ghandi, K. et al. 1999. Muonium in sub- and supercritical water. *PCCP* 1: 4999–5004.

Percival, P. W., Ghandi, K., Brodovitch, J. C. et al. 2000. Detection of muoniated organic free radicals in supercritical water. *PCCP* 2: 4717–4720.

Percival, P. W., Brodovitch, J. C., Arseneau, D. J. et al. 2003. Formation of the muoniated ethyl radical in the gas phase. *Physica B* 326: 72–75.

Percival, P. W., Brodovitch, J. C., Ghandi, K. et al. 2005. Organic free radicals in superheated water studied by muon spin spectroscopy. *J. Am. Chem. Soc.* 127: 13714–13719.

Percival, P. W., Brodovitch, J. C., Ghandi, K. et al. 2007. H atom kinetics in superheated water studied by muon spin spectroscopy. *Radiat. Phys. Chem.* 76: 1231–1235.

Pifer, A. E., Bowen, T., and Kendall, K. R. 1976. A high stopping density Mu^+ beam. *Nucl. Instrum. Meth.* 135: 39–46.

Pimblott, S. M. and LaVerne, J. A. 1997. Stochastic simulation of the electron radiolysis of water and aqueous solutions. *J. Phys. Chem. A* 101: 5828–5838.

Prokscha, T., Morenzoni, E., Deiters, K. et al. 2008. The new μE4 beam at PSI: A hybrid-type large acceptance channel for the generation of a high intensity surface-muon beam. *Nucl. Instrum. Meth. A* 595: 317–331.

Pu, J. and Truhlar, D. G. 2002. Validation of variational transition state theory with multidimensional tunneling contributions against accurate quantum mechanical dynamics for $H + CH_4 \rightarrow H_2 + CH_3$ in an extended temperature interval. *J. Chem. Phys.* 117: 1479–1481.

Rayner, C. M. 2007. The potential of carbon dioxide in synthetic organic chemistry. *Org. Process Res. Dev.* 11: 121–132.

Riddick, J. A., Bunger, W. B., and Sakaro, T. K. 1986. *Organic Solvents*, 4th edn. New York: Wiley InterScience.

Roduner, E. 1988. *The Positive Muon as a Probe in Free Radical Chemistry, Lecture Notes in Chemistry.* Vol. 49. Heidelberg, Germany: Springer.

Roduner, E. 1993. Polarized positive muons probing free-radicals—A variant of magnetic-resonance. *Chem. Soc. Rev.* 22: 337–346.

Roduner, E., 1999. *Muon Science.* Berlin, Germany: NATO Advanced Study Institute.

Salmon, G. 2003. The radiation chemistry connection. *Physica B* 326: 46–50.

Schenck, A. 1985. *Muon Spin Rotation Spectroscopy.* London, U.K.: Adam Hilger Press.

Schwarzer, D., Troe, J., and Zerezke, M. 1998. Preferential solvation in the collisional deactivation of vibrationally highly excited azulene in supercritical xenon/ethane mixtures. *J. Phys. Chem. A* 102: 4207–4212.

Senba, M. 1988. Muon spin depolarization in noble gases during slowing down in a longitudinal magnetic field. *J. Phys. B: At. Mol. Opt. Phys.* 21: 5233–5260.

Senba, M., Fleming, D. G., Arseneau, D. J., and Mayne, H. R. 2000. Hot atom reaction yields in $Mu^* + H_2$ and $T^* + H_2$ from quasiclassical trajectory cross sections on the Liu–Siegbahn–Truhlar–Horowitz surface. *J. Chem. Phys.* 112: 9390–9400.

Senba, M., Arseneau, D. J., Pan, J. J., and Fleming, D. G. 2006. Slowing-down times and stopping powers for ~2-MeV μ^+ in low-pressure gases. *Phys. Rev. A* 74: 042708–042725.

Sengers, J. V. and Kamgar-Parsi, B. 1984. Representative equations for the viscosity of water substance. *J. Phys. Chem. Ref. Data* 13: 185–205.

Siebbeles, L. D. A., Pimblott, S. M., and Cox, S. F. J. 1999. Simulation of muonium formation in liquid hydrocarbons. *J. Chem. Phys.* 111: 7493–7499.

Siebbeles, L. D. A., Pimblott, S. M., and Cox, S. F. J. 2000. Muonium formation dynamics in radiolytic tracks. *Physica B* 289: 404–408.

Smigla, V. P. and Belousov, Y. M. 1994. *Muon Method in Science.* New York: Nova Science.

Soper, A. K., Bruni, F., and Ricci, M. A. 1997. Site–site pair correlation functions of water from 25 to 400°C: Revised analysis of new and old diffraction data. *J. Chem. Phys.* 106: 247–254.

Storchak, V. G., Brewer, J. H., Morris, G. D. et al. 1999. Muonium formation via electron transport in solid nitrogen. *Phys. Rev. B* 59: 10559–10572.

Suter, A., Morenzoni, E., Khasanov, R. et al. 2004. Direct observation of nonlocal effects in a superconductor. *Phys. Rev. Lett.* 92: 087001–087004.

Svishchev, I. M., Kusalik, P. G., Wang, G., and Boyd, R. J. 1996. Polarizable point-charge model for water: Results under normal and extreme conditions. *J. Chem. Phys.* 105: 4742–4750.

Tamm, P. W. and Schmidt, L. D. 1971. Binding states of hydrogen as Tungsten. *J. Chem. Phys.* 54: 4775.

Taylor, B. A., Cormier, P., Lauzon, J. M., and Ghandi, K. 2009. Investigating the solvent and temperature effects on the cyclohexadienyl radical in an ionic liquid. *Physica B* 404: 936–939.

Villa, J., Corchado, J. C., Gonzalez-Lafont, A., Lluch, J. M., and Truhlar, D. G. 1998. Explanation of deuterium and muonium kinetic isotope effects for hydrogen atom addition to an olefin. *J. Am. Chem. Soc.* 120: 12141–12142.

Villa, J., Corchado, J. C., Gonzalez-Lafont, A. G. et al. 1999. Variational transition-state theory with optimized orientation of the dividing surface and semiclassical tunneling calculations for deuterium and muonium kinetic isotope effects in the free radical association reaction $H+C_2H_4 \rightarrow C_2H_5$. *J. Phys. Chem. A* 103: 5061–5074.

Walker, D. C. 1983. *Muon and Muonium Chemistry.* Cambridge, U.K.: Cambridge University Press.

Walker, D. C., Karolczak, S., Gillis, H. A., and Porter, G. B. 2003a. Solvent-dependent rate constants of muonium atom reactions. *Can. J. Chem.* 81: 175–178.

Walker, D. C., Karolczak, S., Porter, G. B., and Gillis, H. A. 2003b. No "delayed" muonium-formation in organic liquids. *J. Chem. Phys.* 118: 3233–3236.

White, H. J., Sengers, J. V., Neumann, D. B., and Bellows, J. C. 1995. Physical chemistry of aqueous systems: Meeting the needs of industry. In *International Conference on the Properties of Water and Steam*, IAPWS Release on the Surface Tension of Ordinary Water Substance. New York: Begell House.

9 Electron Localization and Trapping in Hydrocarbon Liquids

Gordon L. Hug
University of Notre Dame
Notre Dame, Indiana
and
Adam Mickiewicz University
Poznan, Poland

Asokendu Mozumder
University of Notre Dame
Notre Dame, Indiana

CONTENTS

9.1 INTRODUCTION

In the interaction of charged particles and photons with matter, electrons occupy a central position, both as an incident radiation and as a chemical reactant. Of particular interest are electrons liberated from geminate cations, which may be free, quasi-free, or trapped. Such electrons are extremely important for consideration, *inter alia*, of radiation-induced conductivity and dosimetry. Since some energy states of the electron must be delocalized to confer upon it a nonzero mobility (long-range transport), electron localization and trapping have emerged as important topics of investigation (Anderson, 1958; Hug and Mozumder, 2008).

In this chapter, our aim is to provide a consistent theoretical framework for electron localization and trapping in hydrocarbon liquids. Saturated hydrocarbon molecules do not have positive electron affinities. Yet, in the liquid phase, electron trapping is ubiquitous, as evidenced from the activated mobility in these liquids (Allen, 1976; Nishikawa, 1991). So far, most of the theoretical treatments have been phenomenological, in that a certain two-state model is assumed and its parameters are adjusted to reproduce the correct order of magnitude of the room temperature mobility and its activation energy. In these models, electron motion is assumed to be divided between those in the quasi-free and trapped states, in direct proportion to the respective mobilities in these states (Davis and Brown, 1975; Allen, 1976; Nishikawa, 1991). The activation energy is similar to, if not identical with, the binding energy in the trap. The magnitude of the mobility is then given through the quasi-free mobility for which often a uniform value of ~100 cm^2 V^{-1} s^{-1} is assigned. Mozumder (1993, 1995) introduced an essential improvement of the model by explicitly recognizing the competition between trapping and momentum relaxation processes in the quasi-free state. Yet, the most important question remained unanswered, that is, what characteristics of the individual molecule and/or of the liquid structure are responsible for initial localization and, subsequently, for trapping. Sometimes, the Anderson model has been invoked, but no calculation has been provided to indicate the physical parameters responsible for localization (Funabashi, 1974; Cohen, 1977; Chandler and Leung, 1994). Recently, however, Hug and Mozumder (2008) have employed the Anderson theory with the anisotropy of the electron–molecule polarizability interaction as the source of diagonal energy fluctuation and with the liquid x-ray structure to provide the connectivity. Thus, they were able to evaluate approximately the fraction of the initial electron states that would be delocalized in the liquid alkane series. Further work is in progress along these lines (Hug and Mozumder, forthcoming).

The relationship between electron mobility and molecular shape, especially for the isomeric pentane series, has often been alluded to, but without any specific conclusion; in the same manner, certain intramolecular properties have been addressed, again without proven specifics (Hug and Mozumder, 2008). Our assertion is that both intra- and intermolecular factors are responsible for electron localization and trapping, for example, the electron–molecule anisotropic polarizability interaction and the liquid structure factor. We will develop these ideas in this chapter for localization, which is considered first. Some models of trapping and their consequences would then be provided.

In Section 9.2, the Anderson model of localization of electrons is discussed, with special reference to liquid alkanes. Section 9.3 is devoted to trapping, following initial localization. In Section 9.4, mobility theories are briefly considered. We summarize our conclusions in Section 9.5.

9.2 LOCALIZATION

9.2.1 LOCALIZATION AND TRAPPING

A distinction should be made between localization and trapping. Localization, at least in the sense of Mott (1967) and Anderson (1958), implies that the overall envelope of the wave functions for an electron is confined within a microscopic volume of the space without rearrangement of the medium molecules. Trapping requires medium rearrangement, with a concomitant abrupt change in the energy and position of the electron (Cohen, 1977). According to Cohen (1973, 1977), the free energy is irreversibly lowered on trapping. The self-trapped configuration is locally stable, and a mixture of quasi-free and trapped states can occur, with little probability of intermediate states. No stable self-trapped state may be found within the deformation potential approximation. Along with self-trapped states, there is always a possibility of preexisting traps, created by random equilibrium fluctuation in the density and/or potential in the liquid. Unfortunately, in the literature, the distinction between localization and trapping is not always carefully maintained, and the terms are used interchangeably.

9.2.2 DISORDER: THE ANDERSON MODEL

Localization in a condensed phase is intimately connected with disorder, which does not mean lack of all order. It implies absence of translational symmetry, which destroys long-range order. There is always a short-range order dictated by intermolecular chemical bonding. Although many of the crystal properties are lost in disordered systems (e.g., the Bloch states), two of them remain (Ziman, 1979). The first is homogeneity. Thus, a translation from a molecule at $\vec{r}_i$ to another at $\vec{r}_j$ may be associated with a phase factor, $\exp[i\vec{k} \cdot (\vec{r}_i - \vec{r}_j)]$, which is the closest analogue to Bloch waves for homogeneously disordered systems. However, it does not hold for k compared to the reciprocal of average intermolecular spacing. The second is that almost all disordered liquids are relatively close-packed, subject to the constraints of chemical bonding and local structure, implying that the Wigner–Seitz cell must be replaced by Voronoi polyhedra, both of which are space-filling. Further, Ioffe and Regel point out that in cases where the coordination number is preserved upon melting, the liquid roughly inherits the band structure of the solid.

Two important properties of the medium are the coordination number, Z, and the connectivity, K. The coordination number is the unique number of nearest neighbors in an atomic solid. The maximum value of Z, that is, $Z = 12$, is found in face-centered cubic (fcc) crystals (many metals) and also in liquefied rare gases (LRGs) and solid methane. In molecular media, Z may be defined by a small set specifying nearest-neighbor environments for each distinct site type. In practice, Z is obtained by integrating the molecular pair distribution function, $g(r)$, which in turn is based on the small-angle x-ray scattering data (Hug and Mozumder, 2008), as given below:

$$Z = \int_{r_{min}}^{r_{max}} 4\pi r^2 g(r)\, dr. \tag{9.1}$$

In Equation 9.1, r_{min} is automatically obtained from x-ray data by the elimination of intramolecular scattering. About r_{max}, there is some choice depending on the method employed.

We have consistently used a criterion by which $g(r)$ returns to 1 for the first time beyond the principal peak (Hug and Mozumder, 2008).

The connectivity, K, determines how well the elements of the lattice (not necessarily crystalline) are connected together. If the lattice has only nearest-neighbor connections, K may be defined by $K = L^{-1} \ln[N(L)]$, where $N(L)$ is the number of non-repeating paths of length L emanating from a site. Classically, for a regular lattice, it is $Z - 1$ or $Z - 2$, depending on the circumstances (Anderson, 1958). For a Bethe–Peierls lattice (see Section 9.2.4, Figure 9.2), $N(L) = (K + 1)K^{L-1}$. To verify this, consider $N(1) = K + 1$ and $N(L + 1)/N(L) = K(L \geq 1)$, both of which are obvious; then iterate. A quantum analogue is also available, but seldom used except in model numerical studies (Root et al., 1988). For liquid hydrocarbons, we have consistently used $K = Z - 1$ (Hug and Mozumder, 2008). It is then clear that the coordination number enters the localizability criterion in two ways. First, it determines the random site energy fluctuation, due to each electron–molecule interaction. Second, it spreads the electron wave function from one molecule to the next. The increase of Z tends to localize the electron by the first effect, and then tends to delocalize it by the second effect. In a given situation, the extent of localization is a result of the competition between these two effects.

Originally, the Anderson model (1958) was devised to explain the lack of spin diffusion in low-temperature pure Si with donor impurities (P, As, etc.). Experiments indicated spin localization in

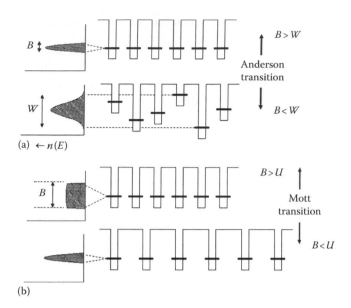

FIGURE 9.1 Schematic contrast of the (a) Anderson and (b) Mott transitions. B is the electronic bandwidth, U is the intersite electron–electron energy, W is the width of the disorder in the site energies in the one-electron tight-binding picture of Anderson.

the timescale of seconds or longer (Feher et al., 1955), by which time the spin should have completely diffused out by ordinary transport theory. Anderson argued that spin energy at a given location (called the *site energy*) is subject to random fluctuation due to the hyperfine interaction. This generates what has since been known as *diagonal disorder*. When the site energy fluctuation has a narrow width of distribution, W, relative to the bandwidth, B, caused by transfer from one location to the next ($W \ll B$), transport is easy, because of the facility of finding energy-matched states. On the other hand, when $W \gg B$, a state picked at random will have energy that has little chance to match at the transferred location. The sites are then effectively decoupled, and the transport stops. Figure 9.1 gives a qualitative description of this phenomenon.

Soon after Anderson's pioneering work, it was realized that, in a random lattice, the transport of other species, for example, electrons and excitations, can be described in the same way by constructing a suitable physical model for diagonal disorder and for transfer energy. It turns out that the statistical distribution of site energy is the most important thing. The transfer energy may have some fluctuation, but it is not too relevant. Actually, in most tight-binding approximations (TBAs), it is treated as a constant.

We now state Anderson's theorem in a qualitative form. For a sufficiently large disorder ($W/B \gg 1$), all states in the band are localized and no transport at all can take place. Later, we derive a quantitative criterion. A-localization, for brevity, is purely a quantum mechanical consequence of disorder in an independent particle picture. It is important to realize that A-localization describes lack of quantum diffusion, not that of thermal diffusion. Stated in this manner, it is exactly valid at the absolute zero of temperature. At a finite temperature, a small transport is always allowed by thermal activation. Mott (1967) adds that, even at smaller disorders, some states in the band tail are localized. Another kind of localization, often invoked in alloys, is due to an electron–electron correlation energy fluctuation (U). This phenomenon, called a Mott transition, occurs when $U > B$ in a necessarily two-electron picture.

9.2.3 RELEVANCE TO HYDROCARBON LIQUIDS

In liquid alkanes, electron trapping is ubiquitous, with the exception of liquid methane. Clear evidence of it comes from the activation energy of mobility, for which values of the order of several times the thermal energy, $k_B T$, have often been found experimentally (Allen, 1976; Nishikawa, 1991). Since localization is a prerequisite for trapping, it becomes an established fact.

The theoretical description of electron localization and trapping in liquid hydrocarbons has been so far largely phenomenological (Dodelet and Freeman, 1972, 1977; Davis et al., 1973; Dodelet et al., 1976; Holroyd, 1977; Mozumder, 1993, 1995). In the often-used two-state models, electron mobility is seen as a combination of mobilities in the trapped and quasi-free states, in direct proportion to the lifetimes in these states. Given a trap density, a binding energy in the trap, and the mobility of the electron in the quasi-free state, one can calculate the effective mobility of the electron and its activation energy, which compare well with experiments for reasonable values of the parameters. Certain other features, such as solute reaction rates with the electron and the field dependence of the mobility, can also be explained within the context of these models.

Despite the apparent success of the two-state models, the basic problem remained unanswered, that is, what properties of the molecule and of the liquid structure give rise to electron localization or delocalization. Sometimes the Anderson localization has been cited as a possible mechanism (Cohen, 1973; Funabashi, 1974; Chandler and Leung, 1994), but no calculations or realistic models have been proposed so far. Only recently, Hug and Mozumder (2008) have employed the anisotropy of the electron–molecule polarizability interaction as the explicit cause of diagonal disorder in these systems. Although some progress has been made in this manner for the localization problem, many details remain to be filled in, especially those concerned with the relationship of transfer energy to V_0, the lowest energy of the conducting state in the liquid. Finally, the problem of trapping ensuing upon localization has not been adequately addressed. We consider these problems in some detail in this chapter.

A well-known, but little understood, problem relates to molecular shape dependence of mobility. Generally speaking, liquids of nearly spherical molecules have electron mobility (μ_e) much larger than in those where the molecules are less spherical (Schmidt and Allen, 1970; Dodelet and Freeman, 1972, 1977; Bakale and Schmidt, 1973; Allen, 1976; Holroyd and Cipollini, 1978). A spectacular contrast among isomeric pentanes has often been alluded to and discussed (Freeman, 1986; Stephens, 1986; and references therein). Electron mobility in neopentane is ~60 times that in isopentane and ~400 times as great as in n-pentane. In many other respects, the molecules are very similar. A corresponding situation exists for tetramethylsilane (TMS) and its variants, tetramethylgermanium (TMGe), and tetramethyltin (TMSn) (Holroyd et al., 1991). Various correlations, but no real explanation, have been proposed between μ_e and some intramolecular properties, for example, polarizability anisotropy (Dodelet and Freeman, 1972, 1977; Funabashi, 1974; Gyorgy and Freeman, 1979), molecular symmetry (Shinsaka et al., 1975), and sphericity (Dodelet and Freeman, 1972, 1977). Stephens (1986), and references therein, argued against the intramolecular correlation by pointing out that the density-normalized mobility curves for isomeric pentanes reverse their order in the gas phase, while crossing over near the critical density. He stressed the relevance of the peak position of the liquid scattering function, $S(k)$, determined by x-ray measurement. Freeman (1986), and references therein, was generally supportive of this idea, but criticized the direct correlation, because the peaks in $S(k)$ for n-, iso-, and neopentane did not follow the sequence of μ_e. He also pointed out that the details relied in part on the differences in the structure functions that were likely to be the same within experimental uncertainty, especially between iso- and neopentane. The central idea of this chapter is that the entire liquid structure needs to be considered for localization, and both intra- and intermolecular properties are involved.

9.2.4 BASIC DERIVATIONS

Anderson's model of localization is of general validity for any inherent species, electrons, spin, and excitation states, even though in the present case we are dealing with electrons (Anderson, 1958). Site energy, ε_n, is a random variable characterized by a distribution function, $p(\varepsilon_n)$. The matrix elements, V_{nm}, negotiating transfer between nearest neighbors only, are taken to be a constant V. This is a tight-binding model for noninteracting electrons. Anderson shows that if the variation in ε_n is large enough compared to V, then an electron on a particular site will not be able to easily hop to its nearest neighbor, because energy cannot be conserved in the absence of a heat bath. This implies localization.

The theory can be developed in several different ways. The question of localization can be reduced to whether there is any chance that an electron starting out at time $t = 0$ on site n will remain there at time $t \to \infty$. To this end, Anderson (1958) starts with the time-dependent Schrödinger equation for the Hamiltonian $H = H_0 + V$, where the unperturbed Hamiltonian, H_0, already contains the statistically distributed energy, ε_n. The Laplace-transformed state amplitude is then developed as a perturbation series in V. While Anderson insists that the entire perturbed series be treated as a probability variable, its convergence is identified with localization without a direct correspondence between perturbed and unperturbed states.

In Anderson's treatment, repeated indices are allowed in the perturbation series, which is computationally inconvenient. To avoid this, he employs Watson's method (Watson, 1957), which replaces site energies by "renormalized energies," to be determined self-consistently. Graphically, this procedure is equivalent to a "self-avoiding random walk" problem. Following a fairly complicated mathematical reasoning, Anderson concludes that the perturbation series almost always converges, that is, localization occurs, if W/V is greater than a critical value, where W is the "width" of the statistical distribution of site energies. Two criteria have been derived, one for an upper limit and the other for the best estimate, both depending on the lattice connectivity.

Abou-Chacra et al. (1973) point out several flaws in Anderson's original treatment, of which two are important: (1) approximation of the renormalized perturbation series with complicated denominators may be inadequate and (2) while the terms of the series may be written as sums of contributions with probability distributions, these contributions are not wholly independent. Instead of correcting these flaws, they provide a somewhat different procedure, still within the context of the Anderson model.

Their method is based on a self-consistent solution for the equation of self-energy in second-order perturbation (vide infra). This solution can be purely real almost everywhere (localized states), or can be complex everywhere (delocalized states). The present method replaces the difficulty of finding the convergence of Anderson's series of high-order perturbation by a self-consistent procedure on the self-energy, which is a distinct advantage. However, it applies essentially to an infinite Cayley tree (Bethe lattice), for which their theory is exact and a useful approximation for a real lattice in the case of the Ising (1925) model. In principle, Anderson's method should apply to any lattice. Localized states are those in which the imaginary part of self-energy $\to 0$, with probability unity as $E \to$ real axis. For delocalized states, the imaginary part of self-energy remains finite and nonzero as $E \to$ real axis.

In the Anderson model, there is only one energy level, ε_i, statistically distributed at site i (diagonal disorder), while the electron is transferred from i to j by the matrix element (overlap integral) $-V_{ij}$ (often taken as a constant V for nearest neighbors). Schrödinger's equation for the eigenstate α is characterized by an amplitude a_i^α and an eigenvalue E^α, which satisfy

$$\varepsilon_i a_i^\alpha - \sum_j V_{ij} a_j^\alpha = E^\alpha a_i^\alpha.$$

Matrix elements of Green's function, defined operationally by $G = (E - H)^{-1}$, are given by $(E - \varepsilon_i) G_{ik} \sum_j V_{ij} G_{jk}(E) = \delta_{ik}$, while, by definition, self-energy is written as follows:

$$S_i(E) = E - \varepsilon_i - [G_{ii}(E)]^{-1}.$$

Note that self-energy is site-specific at this level and connects with only the diagonal matrix elements of G. Self-energy is a renormalized energy including the interactions. In Anderson's original method (Anderson, 1958), the self-energy is expanded as a perturbation series, each term of which contains a modified self-energy up to a certain approximation, corresponding to self-avoiding paths. It is a complicated expression even for the relatively simple model of Anderson, as shown in the following equation:

$$S_n(E) = \sum_{j \neq n} \frac{V_{nj} V_{jn}}{E - \varepsilon_j - S_j^{(n)}(E)} - \sum_{j \neq n} \sum_{k \neq n, j} \frac{V_{nk} V_{kj} V_{jn}}{\left(E - \varepsilon_k - S_k^{(nj)}(E)\right)\left(E - \varepsilon_j - S_j^{(n)}(E)\right)} + \cdots$$

In this equation, the V_{nn} term is dropped. The S's in the denominator are self-energies of the sites specified in the subscript with specific sites named in the superscript not being considered. Then the convergence and the statistical properties of the sum are investigated. In the present method (called AAT), only the first term of the series, corresponding to second-order perturbation, is retained, however, insisting that self-energy distribution must be the same on both sides of the equation. This gives the fundamental equation:

$$S_i(E) = \sum_j V_{ij}[E - \varepsilon_j - S_j(E)]^{-1} V_{ji}. \tag{9.2}$$

Taking $V_{ij} = V$, a constant for nearest neighbors, and noting that ε_i are random and independent, it is argued that $S_i(E)$ would be a random variable independent of ε_i and that the probability distribution of $S_i(E)$ and $S_j(E)$ would be identical.

A Cayley-tree (Bethe lattice, see Figure 9.2) is characterized by a connectivity $K = Z - 1$, where $Z =$ coordination number of nearest neighbors. The number of nonintersecting paths emanating from a given site is $(K + 1)K^{L-1}$, where L is the step size. For nearest-neighbor interactions only, the perturbation series reduces to a single term, thus offering simplification over a regular lattice. Nonintersecting paths starting and ending in i have only two steps, $i \rightarrow j \rightarrow i$. Thus, the above equation for self-energy has K terms on the right-hand side. There are K independent random variables, $S_j(E)$, and K independent random energies, ε_j, giving $2K$ independent random variables. These generate a distribution of $S_i(E)$, identical to that of $S_j(E)$, in a self-consistent fashion.

AAT first separate the real and imaginary parts of $E = R + i\eta$ and $S_i = E_i - i\Delta_i$, and obtain from Equation 9.2,

$$E_i = \sum_j \frac{|V_{ij}|^2 (R - \varepsilon_j - E_j)}{(R - \varepsilon_j - E_j)^2 + (\eta + \Delta_j)^2} \quad \text{and} \quad \Delta_i = \sum_j \frac{|V_{ij}|^2 (\eta + \Delta_j)}{(R - \varepsilon_j - E_j)^2 + (\eta + \Delta_j)^2}. \tag{9.3}$$

Taking $p(\varepsilon_j)$ as the site energy probability and calling $f(E_j, \Delta_j)$ as the joint probability of (E_j, Δ_j), AAT derive a nonlinear integral equation for the (double) Fourier transform of f, which is very difficult to

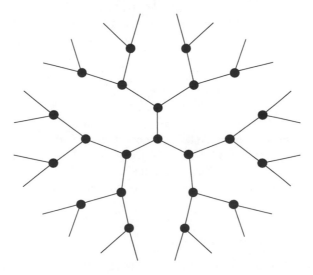

FIGURE 9.2 Bethe lattice with $Z = 3$.

solve. Considerable simplification is achieved when η is very small (the important case), and Δ_js are very small (the localized case). In this approximation, one gets from (9.3)

$$E_i = \sum_j \frac{|V_{ij}|^2}{(R - \varepsilon_j - E_j)} \text{ (which does not involve } \Delta) \quad \text{and} \quad \Delta_i = \sum_j \frac{|V_{ij}|^2 (\Delta_j + \eta)}{(R - \varepsilon_j - E_j)^2}. \tag{9.4}$$

The inhomogeneous equations (9.4) have a solution if the homogeneous equations, $\lambda^2 \Delta_i = \sum_j \left(|V_{ij}|^2 \big/ (R - \varepsilon_j - E_j)^2 \right)$, have the largest eigenvalue $\lambda^{-2} < 1$. $\lambda = 1$ separates the stability of localized states. To determine the upper limit of the width of site energy distribution for localized states, AAT ignore the real part of self-energy, as in Anderson (1958), thus obtaining $\Delta_i = \sum_j |V_{ij}|^2 (\eta + \Delta_j)/(R - \varepsilon_j)^2$ (cf. Equation 9.4). The corresponding Laplace transform of its probability is given by

$$f(s) = \left[\int dx\, p(R - x) f\left(\frac{sV^2}{x^2} \right) \exp\left(\frac{-sV^2 \eta}{x^2} \right) \right]^K. \tag{9.5}$$

The Laplace transform is more convenient for studying long time behavior ($s \to 0$). A critical value of V (or of W/V) can be obtained from $\mathrm{Lim}_{s \to 0} f(s)$. The probability of Δ_i falls off as $|\Delta_i|^{-3/2}$, if $p(R) \neq 0$. The tails of the distribution can fall off no faster than this. Thus, assuming $\mathrm{Lim}_{s \to 0} f(s) \approx 1 - As^\beta$, with $0 < \beta < 1/2$, and substituting in Equation 9.5, one obtains

$$KV^{2\beta} \int \frac{dx\, p(R - x)}{x^{2\beta}} = 1. \tag{9.6}$$

Differentiating Equation 9.6 with respect to β for given energy R and site energy distribution $p(\varepsilon)$, one obtains a critical value of V as follows:

$$\ln V_c = \frac{\int p(R - x) |x|^{-2\beta_c} \ln |x|\, dx}{\int p(R - x) |x|^{-2\beta_c} dx}, \tag{9.7}$$

where V_c and β_c also satisfy Equation 9.6. Taking $R = 0$ at the band center and uniform site energy distribution between $-W/2$ and $+W/2$ (Anderson model), AAT obtain, from Equations 9.6 and 9.7, $\ln V_c = \ln(W/2) - (1 - 2\beta_c)^{-1}$ and $(2V/W)^{2\beta} K/(1 - 2\beta) = 1$, which on eliminating β give the upper limit equation of Anderson (1958):

$$\left(\frac{W}{2V_c} \right) = fK \ln\left(\frac{W}{2V_c} \right), \tag{9.8}$$

with $f = e$, the base of natural logarithms. Anderson's best estimate, with $f = 2$ in Equation 9.8, can also be obtained similarly when $R \ll W/2$. We have consistently used a form of Equation 9.8 (vide infra).

The theory of Abou-Chacra et al. (1973) is regarded exact for an infinite Cayley tree, or a good approximation for a real lattice at the level of mean field. It is supposed to be free of approximations, but not free from assumptions.

9.2.5 Mobility Edge: Scaling Theory Results

The term "mobility edge" was coined by Cohen, Fritzsche, and Ovshinsky (Cohen et al., 1969). From the previous discussion, it is clear that the central idea in electron localization in a condensed medium is the interplay between the diagonal disorder (W), the transfer integral (V), and the connectivity (K) (cf. Equation 9.8). As stated in Section 9.2.4, the numerical factor f depends, inter alia, on the nature of the random lattice. For Anderson's (1958) uniform site energy distribution between $-W/2$ and $+W/2$, the f-values have been substantiated by the more accurate self-consistent solution of the equation of self-energy by Abou-Chacra et al. (1973). According to Phillips (1993), f should equal 4 on a Cayley tree, which is completely unacceptable for liquid hydrocarbons (Hug and Mozumder, 2008). If we take $f = 4$ in Equation 9.8, $(W/2V)_c$ would be so large that almost all states would be delocalized. While the connectivity K is predominantly a medium property, Shante and Kirkpatrick (1971) have determined numerically $K = 4.7$ for percolation on a cubic lattice. For the same lattice, Chang et al. (1990) found $(W/2V)_c = 6.0$ by a scaling calculation. (Simply stated, such a powerful numerical technique relies on scaling the Hamiltonian and the consequent wave functions by factors that can be used repeatedly to determine if or not the position probability distribution remains localized in the long time limit.) Substituting the results of Shante and Kirkpatrick (1971) and of Chang et al. (1990) in Equation 9.8, Hug and Mozumder (2008) obtained $f = 0.7125$, which they have used consistently and satisfactorily for all hydrocarbon liquids. Figure 9.3 shows the solution of $(W/2V)_c$ versus K for $f = 0.7125$, which demonstrates the sensitivity on K.

From the discussion in Section 9.2.4, we have learned that the most important criterion for localization is the imaginary part of self-energy. If it is zero, the state would be localized. If not, the state would be delocalized. Since the imaginary part of the self-energy cannot be zero and nonzero at the same time, a state at a given energy is either localized or delocalized, but cannot be both. At a given energy, more and more states would get localized as the disorder is increased, until at a critical disorder (designated $\sigma_c \equiv (W/2V)_c$) all eigenstates would be so localized that no diffusion can ensue. This is called Anderson's theorem. It applies for static disorder, without contact with a heat bath, in a TBA of the independent particle picture. The associated phenomenon is known as Anderson localization.

Anderson's localization theory has been employed in several important applications using the scaling procedure. Our purpose here is twofold (Hug and Mozumder, 2008): (1) to estimate the fraction of delocalized states in a fluid as a function of the disorder parameter and (2) to give a quantitative measure of the extent, ξ, of localization of the excess electron's spatial envelope,

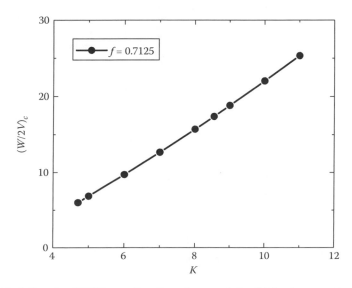

FIGURE 9.3 Critical disorder $(W/2V)_c$ as a function of connectivity (K) for the numerical factor $f = 0.7125$. Agreement of the results of a percolation model (Shante and Kirkpatrick 1971) with the scaling theory of Chang et al. (1990) is obtained for $f = 0.7125$.

exp($-r/\xi$). The latter has been written by Mott (1967) as sin(kr)exp($-r/\xi$), where k is proportional to the momentum of the localized electron.

For disorders less than the critical value ($\sigma \leq \sigma_c$), the eigenstates are either localized or delocalized, in the ascending order of energy, that is, the parameter ξ of the electron's wave function envelope is a monotonically increasing function of energy E. Since long transport requires some delocalization of the electron to acquire significant mobility, it is apparent that, for this purpose, the electron must have energy $E \geq E_c$, a critical value for the disordered medium. Therefore, the critical energy is a function of the disorder $E_c(\sigma)$, or, by inversion, the critical disorder is a function of energy $\sigma_c(E)$, which is called the mobility edge trajectory.

The parameter ξ of the wave function envelope, often called the correlation length of the localized eigenstates, diverges ($\xi \to \infty$) at a given disorder as the energy approaches the critical value from below according to an index law given by the following equation:

$$\xi(E) \sim [E_c(\sigma) - E]^{-\nu} \quad (E \leq E_c). \tag{9.9}$$

At a given energy, it also diverges as the disorder approaches the critical value from above ($\sigma \to \sigma_c$), as given in the following equation:

$$\xi(\sigma) \sim [\sigma - \sigma_c]^{-\nu} \quad (\sigma \geq \sigma_c). \tag{9.10}$$

It is an important fundamental result of the scaling theory that the critical exponent, ν, is the same in Equations 9.9 and 9.10 (Chang et al., 1990 and references therein). However, there is considerable disagreement in the literature about the values of ν obtained by numerical and scaling methods, indicating only a range between ~0.6 and 1.95. It probably depends somewhat upon the nature of the lattice, but it should be the same in a given class of liquids. Chayes et al. (1986) have presented theoretical argument to show that in three dimensions, $\nu \geq 2/3$. Following Chang et al. (1990) and some other authors, we have consistently used $\nu \approx 1$ for liquid hydrocarbons (Hug and Mozumder, 2008).

Using a symmetry argument, Chang et al. (1990) proposed a quadratic expression for the mobility edge trajectory in their numerical study, that is, $\sigma_c(E) = \sigma_c - AE^2$, where A is a positive constant and E is not very large (note that in their case, the mobility edge is positive). For our negative mobility edge trajectory, we have used the following ansatz:

$$\frac{\sigma_c(E)}{\sigma_c} = 1 - \exp\left(-\frac{E}{E_1}\right)^2, \tag{9.11}$$

where E_1 is a characteristic energy that has been interpreted for alkane liquids (Hug and Mozumder, 2008). This ansatz is convenient in the sense that it gives $\sigma_c(0) = 0$ (no state localized at zero disorder) and $\sigma_c(\infty) = \sigma_c$ (all states localized at critical disorder). Additionally, at small values of E, $\sigma_c(E) = \sigma_c[E/E_1]^2$, which satisfies the quadratic symmetry requirement of Chang et al. (1990). This quadratic symmetry has been supported by a numerical study by Bulka et al. (1987) for diagonally disordered systems with non-pathological site energy distribution. Further, the quadratic mobility edge trajectory has been construed by Chang et al. (1990) to imply that the fraction of delocalized states at disorder σ should be given by

$$f(\sigma) = \left(1 - \frac{\sigma}{\sigma_c}\right)^{1/2} \quad (\sigma \leq \sigma_c). \tag{9.12}$$

Note that Equation 9.12 gives just the fraction of initially delocalized states, irrespective of their occupation. These states have been called the *doorway* states, from which their final transition to quasi-free or trapped states may ensue (Hug and Mozumder, 2008).

9.2.6 APPLICATION TO LIQUID HYDROCARBONS

The starting point for application to liquid hydrocarbons is conveniently chosen to be the calculation of connectivity, $K = Z - 1$, where Z is the number of nearest neighbors. The latter is obtainable from the pair correlation function $g(r)$, for which we prefer to rely upon the experimental data of x-ray or neutron scattering at small angles. There are several equivalent and alternative mathematical forms connecting $g(r)$ with the experimental, normalized scattered intensity, $I(k)$, of the incident radiation at momentum transfer $\hbar k$. They all originate from the early work of Zernike and Prins (1927), of which a simplified version given below is useful (Warren, 1933):

$$g(r) = \rho + \left(\frac{1}{2\pi^2 r}\right)\int_0^\infty k[I(k) - 1]\sin(kr)dk, \tag{9.13}$$

where

ρ is the liquid density per unit volume
$g(r)$ is the radial pair density distribution function at separation r
$k = 4\pi \sin \theta/\lambda$
2θ is the scattering angle
λ is the wavelength of the radiation

$I(k)$ is obtainable from experiment, after careful elimination of intramolecular scattering, so that $g(r)$ may correctly refer to the intermolecular distribution (Warren, 1933; Kuchitsu, 1968; Narten, 1979). The upper limit of integration in Equation 9.13 may be taken as $2\pi/\lambda$, or often less, without appreciable error (Pierce, 1935). Some examples of $g(r)$, so obtained for liquid hydrocarbons, have been discussed by Hug and Mozumder (2008). Figure 9.4 gives the radial distribution function in neopentane at three different temperatures obtained by Narten (1979), while Figure 9.5 shows our calculation for isopentane, using the experimental data of Stephens (1986), following the procedure of Narten (1979). In these figures, $g(r)$ is normalized to the liquid density. From knowledge of $g(r)$, the number of nearest neighbors, Z, is computed according to Equation 9.1. There is no difficulty about r_{min}, that is, the lower limit of the integral over r in Equation 9.1, as eliminating the intramolecular scattering automatically

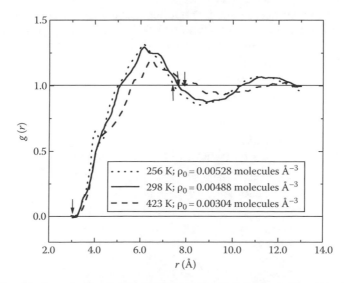

FIGURE 9.4 Intermolecular pair (density) distribution in liquid neopentane from the experimental data of Narten, normalized to the average number density. Number of nearest neighbors is obtained by integration between the vertical lines. The results are 8.6, 8.7, and 7.1 molecules Å^{-3}, for 256, 298, and 423 K, respectively.

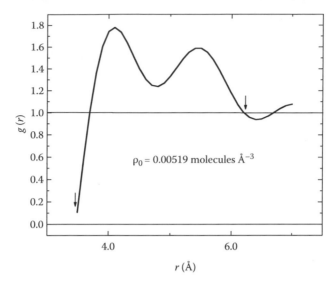

FIGURE 9.5 Intermolecular pair (density) distribution in liquid isopentane, normalized to the average number density. $g(r)$ was calculated from the structure factor, $S(k)$, in Stephens (1986) using the analysis of Narten (1979).

sets $g(r) = 0$ at a small value of r. Thus, in neopentane at 25°C, one gets $r_{min} = 3.0\,\text{Å}$ in Figure 9.4. However, there is some flexibility in determining r_{max}, depending on the method of integration and pre-symmetrization of the data for $g(r)$ (Mikolaj and Pings, 1967). We have consistently used such a value of r_{max} where the normalized value of $g(r)$ returns to one for the first time beyond the principal peak (Hug and Mozumder, 2008). This procedure automatically gives nearest-neighbor separation. For liquid methane, it correctly predicts $Z = 12$, a well-known value for this liquid (Petz, 1965; Karnicky and Pings, 1976). Methane has the highest value of Z that is theoretically possible, while the situation is expected to be similar for TMS and its analogues. The so-determined $K = Z - 1$ and the nearest-neighbor separation (r_1) are given in Table 9.1 for different hydrocarbon liquids. Also given are the critical values of $(W/2V)_c$ for the given connectivity, K, of the liquid, obtained from Equation 9.8.

In the Anderson theory of localization, it is not the diagonal energy per se, but the width of its fluctuation that is the determining factor (Anderson, 1958). Of the various possibilities of the source of this fluctuation, the involvement of the electron-bond (micro) dipole interaction has been discounted by Funabashi (1974), on the ground that it is too small, despite the possibility of internal rotation. It has also been discounted by Kimura et al. (1977) because it would predict less stability with the availability of CH_2 groups, contrary to experimental evidence on spin-echo data. Freeman and his

TABLE 9.1
Connectivity (K) and Nearest-Neighbor Separation (r_1) in Hydrocarbon Liquids

Molecule/T (K)	K	r_1 (Å)	$(W/2V)_c \equiv \sigma_c$
Ethane (105)	8.55	5.5	17.37
Propane (300)	7.1	6.0	12.95
n-Pentane (300)	5.0	5.0	6.86
Isopentane (300)	5.1	6.22	7.14
Neopentane (256)	7.6	7.4	14.56
Neopentane (298)	7.7	7.7	14.83
Neopentane (423)	6.1	7.9	9.93

associates (Dodelet and Freeman, 1977; Gyorgy and Freeman, 1979) have given experimental evidence of the involvement, among many other things, of the anisotropy of molecular polarizability. This has been supported theoretically by Funabashi (1974). The anisotropy of molecular polarizability can be measured by depolarized Rayleigh scattering, or by the Kerr effect (Gelbart, 1974), of which the first gives the difference between the parallel and perpendicular components $(\alpha_\parallel - \alpha_\perp)$, while the latter can, in principle, give all three components. For the diagonal energy fluctuation, one needs only the difference between the maximum and minimum polarizability components (vide infra), as the two perpendicular components are equal for alkanes. For methane, neopentane, tetramethylsilane, etc., all three components are equal giving no anisotropy. Consequently, one does not expect much localization in these liquids. Table 9.2 gives the values of the mean polarizability and the difference between the maximum and minimum components in the *n*-alkane series, along with certain other parameters explained later. Both of these increase with the carbon number, the first because of the size effect and the second indicating that the molecule is increasingly becoming "cigar like." Cases of isopentane and neopentane have been included for intercomparison within the isomeric series. The polarizability components of isopentane are not readily available in the literature. These were obtained from the mean polarizability $\left[\bar{\alpha} = 1/3(\alpha_\parallel + 2\alpha_\perp)\right]$, using the Lorentz–Lorenz equation and a measured value of the Rayleigh depolarization ratio, $\Delta = 0.053$ (Stuart, 1962). The latter is related to $\delta^2 \equiv 10\Delta/(6-7\Delta) = (2/9)(\alpha_\parallel - \alpha_\perp)^2/[(\bar{\alpha})^2]$. For isopentane, this yields $\bar{\alpha} = 10.1\ \text{Å}^3$ and $|\alpha_\parallel - \alpha_\perp| = 6.57\ \text{Å}^3$. Therefore, $\alpha_\parallel = 14.48\ \text{Å}^3$ and $\alpha_\perp = 7.91\ \text{Å}^3$. The polarizability ellipsoid for isopentane is shown in Figure 9.6 and compares it with those of *n*-pentane and neopentane. The progressive sphericity, or loss of anisotropy, is apparent as one goes from *n*- to iso- to neopentane. However, notice that it is due mainly to the contraction of the major axis, rather than by the expansion of the minor axes.

We may now prescribe the width of fluctuation of the electron–molecule polarizability interaction in the liquid as

$$W = Z \frac{e^2}{2\left(\dfrac{r_1}{2}\right)^4} \left|\alpha_\parallel - \alpha_\perp\right|, \tag{9.14}$$

TABLE 9.2
Polarizability, Diagonal Disorder, and Transfer Energy in Hydrocarbon Liquids

Molecule	$\bar{\alpha}$ (Å³)	$\lvert\alpha_\parallel - \alpha_\perp\rvert$ (Å³)	W (eV)	V_0 (eV)	V (eV)	ξ_{max} (Å)	% Initially Delocalized States
Ethane	4.47	1.51	1.20	0.2	0.74	55	97.6
Propane	6.29	1.92	1.38	−0.1	0.18	96	83.9
n-Pentane	9.95	3.13	3.46	0.0		95	0.0
Isopentane	10.68	6.57	1.63	−0.2	0.25	112	73.7
Neopentane	9.8	0.0	0.0	−0.4			100

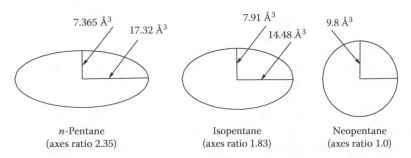

FIGURE 9.6 Relative shapes of polarizability spheroids for pentane isomers.

where e is the electronic charge. Using $Z = K + 1$ from Table 9.1 and $|\alpha_{\parallel} - \alpha_{\perp}|$ from Table 9.2, third column, we calculate W, which is given in the fourth column of Table 9.2. With the increase of the carbon number, molecular anisotropy increases, but Z decreases up to pentane. These two effects partially cancel each other, so that one can expect the variation in r_1 to reflect in W. Beyond pentane, both Z and r_1 remain sensibly constant, causing W to be roughly proportional to $|\alpha_{\parallel} - \alpha_{\perp}|$, that is, to the size of the molecule. For highly symmetric molecules, such as methane and neopentane, we may use $W \sim 0$, due only to the polarizability interaction. A nonzero contribution may be expected due to electron–microdipole interaction, which is currently neglected.

At present, our knowledge of the electronic site energy is not sufficiently accurate to give a reliable estimate of the transfer energy, V. However, Funabashi (1974) hypothesized a value of ~0.1 eV for a typical hydrocarbon liquid. Model calculations by Hug and Mozumder (2008) for the percentage of initially delocalized states as a function of V in the alkane series show progressive delocalization with an increase of V for a given carbon number and progressive localization with the carbon number for a given V. Both are expected effects (cf. Equation 9.12).

Taking a cue from the mobility edge trajectory, we can set $E_1 = V$ in Equation 9.11. Further, for a given liquid with specific disorder $\sigma \equiv (W/2V)$, we must have $E_c = V_0$, since all states would be delocalized above this energy. Since $\sigma(E_c) = \sigma$, an implicit equation for V from Equation 9.11 is obtained, that is,

$$|V_0| = V\left[-\ln\left(1 - \frac{\sigma}{\sigma_c}\right)\right]^{1/2} = V\left[-\ln\left(1 - \frac{W}{2V\sigma_c}\right)\right]^{1/2}. \qquad (9.15)$$

Given σ_c and W from Tables 9.1 and 9.2, respectively, V can be calculated from Equation 9.15 when V_0 is known experimentally (Allen, 1976; Nishikawa, 1991). Some values of V obtained in this manner are shown in Table 9.2, but no such V could be obtained beyond n-butane, since in these liquids already $\sigma > \sigma_c$, implying that all states are initially localized in the Anderson sense. If we formally apply Equation 9.15 when W is very small (e.g., in methane), we would obtain $VW = 2\sigma_c V_0^2 = 4.56$ (eV)^2, by inserting appropriate values for methane from Table 9.1 and Nishikawa (1991). This means $V \gg 4.56$ eV since W is small. Therefore, localized states would not result in such cases.

The fraction of initially delocalized (doorway) states at low energy (near thermal or epithermal energy) may now be computed with values of σ and σ_c from Tables 9.2 and 9.1, respectively, using Equation 9.12. The fraction of initially delocalized states falls precipitously from propane, due to both the increase of σ and the decrease of σ_c. It is zero from n-pentane onward. In the case of isomeric pentanes, the effect of molecular symmetry is apparent, contributed both by the isotropicity of polarizability and the connectivity. For ethane and propane, the initial states are highly delocalized; only trapping and thermal equilibrium produce eventual localization, as is evident from studies of mobility and its activation energy.

Of the three pentane isomers, isopentane occupies an intermediate position in respect of polarizability, V_0, and electron mobility. Its computed connectivity (K) is close to that of n-pentane, nearest-neighbor distance (r_1) is intermediate, while the percentage of initially delocalized states is close to that of neopentane (Mozumder and Hug, 2009).

Disorder-induced Anderson localization gives electron wave functions with random amplitude and phase, having an exponentially decaying envelope, $\sim \exp(-r/\xi)$ (vide supra). Denoting the maximum and minimum energies of the localized electron by E_{max} and E_{min}, respectively, equating E_c with V_0 (vide supra), and employing Equation 9.9, we get

$$\frac{\xi_{max}}{\xi_{min}} = \left[\frac{(V_0 - E_{max})}{(V_0 - E_{min})} \right]^{-\nu}, \tag{9.16}$$

where, according to Chang et al. (1990), $\nu = 1$. This, however, is not well established. Further, Equation 9.16 is strictly not applicable to equilibrated trapped electrons, since trapping involves lattice relaxation, which is not considered in the Anderson localization. Nevertheless, Hug and Mozumder (2008) obtained some idea of the spatial extent of the wave function envelope by setting $V_0 - E_{min} = \Delta E_{act}$, the activation energy of mobility. For $V_0 - E_{max}$, an intermolecular vibrational quantum, $h\nu$, of 0.01 eV is used following Mozumder and Magee (1967) and Rassolov and Mozumder (2001). Any higher value will make the electron delocalized ($\xi \to \infty$). Thus, roughly, $\xi_{max}/\xi_{min} = \Delta E_{act}/h\nu$. Values of $\xi_{min} = r_1$ are taken from Table 9.1 and the corresponding values of ξ_{max} may be computed using the experimental activation energies (Allen, 1976; Nishikawa, 1991). The last two columns of Table 9.2 give ξ_{max} and the percentage of initially delocalized states. Another estimate of the extent of ground-state wave function is obtainable from the field dependence of mobility, using the quasi-ballistic theory of Mozumder (2006). Generally, these are ~1.5–2.5 times larger than the r_1 values listed. Currently, it can only be surmised that ξ_{max}, in most liquid hydrocarbons, will be bracketed between ~50 to a couple of hundred Å. The number of molecules, n_{tr}, interacting with the ground state of the electron may be taken as $Z = K + 1$, from Table 9.1. Then the number n_{qf} interacting with the quasi-free electron will be given by $n_{qf}/n_{tr} = \xi_{max}/\xi_{min} = \Delta E_{act}/h\nu$. Using the data in Table 9.1, one obtains $n_{qf} = 130$, 114, and 120, respectively, for propane, pentane, and hexane. These may be compared with an estimate by Berlin and Schiller (1987) in the range of ~60–120 by a completely different line of reasoning.

9.3 TRAPPING

9.3.1 GENERAL FEATURES

In the literature, there is considerable confusion as to what constitutes localization and trapping. Often these terms are used synonymously, including that used in a recent benchmark paper on electron solvation in ammonia cages (Lee et al., 2008) and in another seminal paper on the stability criterion of an excess electron in nonpolar fluids (Springett et al., 1968). As we have remarked before, Anderson localization refers to no medium rearrangement, whereas trapping always involves some rearrangement of medium molecules, the extent of which may be debated.

Generally, a distinction is made between preexisting and self-dug traps. Electrons in liquid hydrocarbons and in LRGs may be divided into three classes. The first class is where the electron is highly delocalized at all times and where its mobility is large ($\geq$ several hundred cm^2 V^{-1} s^{-1}) without any evidence of positive activation energy. In such cases, as in LAr, LKr, LXe, liquid methane, and TMS, the drift mobility and the Hall mobility are about the same (Munoz, 1991). The second class is where the electron is initially delocalized, but trapping is evidenced by positive activation energy, resulting in a significant difference between drift and Hall mobilities. Such cases occur in neopentane, isooctane, and possibly in ethane and propane (Hug and Mozumder, 2008). Electron mobility in these liquids may be small or large, but seldom exceeding 100 cm^2 V^{-1} s^{-1}. The third class is where the electron is initially localized, the electron motion is activated, and the room temperature mobility is small ($\leq$10 cm^2 V^{-1} s^{-1}). Such cases occur in the liquid alkane series starting from butane, and in many other liquids. The Hall signal is generally unobservable in these liquids because the drift mobility is small. Nevertheless, at least in the two-state theories of transport, a quasi-free state with a high mobility has been postulated (Holroyd and Schmidt, 1989; Munoz, 1991). Liquids in the first class are essentially trap-free. It stands to reason that traps are self-dug in the second class. In the third class, traps may be either preexisting or self-dug, but probably preexisting in the appropriate cases of liquid hydrocarbons.

9.3.2 MODELS OF PREEXISTING TRAPS

There are several models of trapping, preexisting and self-dug. Space does not permit us to go into details. Here, we discuss two cases briefly. The theory of Springett et al. (1968) obtains a criterion for the stability of an electron in a bubble embedded in a nonpolar fluid. Without any positive electron affinity, the existence of the bubble depends on the electron–medium interaction, which is mainly repulsive. The authors distinguish between three classes: (1) a stable quasi-free state, (2) a metastable bubble, and (3) an energetically stable bubble, conditioned by $E_t < V_0$, where E_t is the ground-state energy of the electron bubble and V_0 is the lowest energy of the electron in the quasi-free state. After a complicated set of reasoning, the stability criterion is given as follows. A quasi-free state will be stable when $F \equiv 4\pi\beta\gamma/V_0^2 \geq 0.087$, where γ is the liquid surface tension, $\beta = \hbar^2/2m$, in which, $\hbar$ is the reduced Planck's constant and m is the electron mass. A metastable state exists when $0.047 \leq F \leq 0.087$, and a stable bubble state requires $F \leq 0.047$. Examples occur for delocalized quasi-free states in LAr, LKr, and LXe with negative V_0 values. With positive V_0, stable bubble states are found in liquid He, H_2, and D_2, while metastable states occur in LNe.

Both the theories of Kestner and Jortner (1973) and of Schiller et al. (1973) relate the observed electron mobility in a liquid hydrocarbon with its V_0 value and the allowed fraction C of the liquid volume. In the allowed region, the electron energy $E \geq V$, where V is the local potential. Transport is forbidden when $E < V$. However, there are some fundamental differences between these theories. In the Kestner–Jortner model, the medium is treated as *inhomogeneous* with rotational fluctuation resulting in high- and low-mobility regions, characterized respectively by $\mu_0 \sim 100$ and $\mu_1 \sim 0.1\,cm^2\,V^{-1}\,s^{-1}$. Making some drastic assumptions, and adjusting the parameters of the Gaussian, the authors calculate the allowed fraction, C, of volume available for transport. A critical value, C^*, separates open and closed channels for transport, as in percolation models. For $C > 0.4$, the mobility is simply given by $\mu = \mu_0(3C/2 - 1/2)$. For $C < 0.4$, the mobility drops sharply, ultimately reaching μ_1 when $C \approx 0$. In the application of this model to liquid hydrocarbons, Kestner and Jortner make parametric adjustments to obtain the best agreement with experiments. The resultant mobility, μ, as a function of the quasi-free energy, V_0, shows a precipitous fall around $V_0 = -0.2\,eV$. This model is partially successful in explaining the variation of electron mobility in liquid hydrocarbons with the quasi-free energy, but a serious objection has been raised with respect to the Hall mobility (Munoz, 1991).

In the model of Schiller et al. (1973), the medium is treated as *homogeneous*. Electron energy, E_t, in the bubble is the sum of the quantum mechanical electron energy, E_e, a surface bubble energy, and a pressure–volume work. Together, they are the medium reorganization energy, of which the pressure–volume work is neglected being too small, in the same manner as in the Kestner–Jortner theory. To these, a term $-(e^2/R)(\varepsilon - 1)/2\varepsilon$ is added representing the reversible work needed to bring the electron from vacuum into the bubble. This is estimated to be $\sim -0.80\,eV$. The localization (actually trapping) criterion is taken as $E_t < V_0$. The probability $p(E)dE$ that the electron in the bubble would have an energy between E and $E + dE$ is assumed to be Gaussian, given by

$$p(E) = (2\pi\sigma^2)^{-1/2} \exp\left[-\frac{(E - E_t)^2}{2\sigma^2}\right].$$

The total trapping probability is now expressed as follows:

$$P = \int_{-\infty}^{V_0} p(E)\,dE.$$

The observed mobility is written as $\mu = \mu_f(1 - P)$, where μ_f is the quasi-free mobility. The Gaussian parameter, σ, of the probability distribution is not adjustable, but is given for a canonical ensemble with constant bubble radius by $\sigma^2 = k_B T C_V$, where k_B is the Boltzmann constant, T is the absolute temperature, and C_V is the specific heat at constant volume. At $T = 300\,\text{K}$, the authors estimate $\sigma = 0.12\,\text{eV}$, a constant value in all liquids, which does not appear to be very likely. Comparison with experiment requires the parametric values of μ_f and the total energy E_t of the electron in the bubble. The former is taken to be equal to the measured mobility in TMS, that is, $90\,\text{cm}^2\,\text{V}^{-1}\,\text{s}^{-1}$, and the latter is adjusted to $-0.26\,\text{eV}$ to be in agreement with the measured mobility in n-hexane. Therefore, in this model, the only other non-adjusted quantity is the quasi-free energy. Taking V_0 from experiment gives only a moderate agreement with mobility measurements. Notice that the factor $1 - P$ is not quite comparable to the allowed volume fraction, C, of the Kestner–Jortner model (1973). Even in the high-mobility regime, the electron mobility is not given simply by the product of C and the quasi-free mobility in the Kestner–Jortner model, but in the model of Schiller et al. (1973) the mobility is always given by $\mu_f(1 - P)$ (vide supra). Among other reasons, the use of the percolation model and an effective medium theory are responsible for the difference.

9.3.3 STRUCTURE OF TRAPS IN NONPOLAR MEDIA

Davis et al. (1973) proposed a possible model for the electron trap in nonpolar hydrocarbons. It is assumed that the electron is trapped in a cavity (preexisting) and the interaction potential arises from both the electron-bond dipoles of the molecule and the electron-induced dipole (polarizability) of the entire molecule as follows:

$$V(r) = \frac{-N\bar{\alpha}e^2}{2r_d^4} - \frac{ND_t e}{r_d^2} \quad \text{for } r < R,$$

$$V(r) = V_0 \quad \text{for } r > R,$$

(9.17)

where,

N is the coordination number of the electron

$\bar{\alpha}$ is the mean polarizability of the molecule (spherically averaged)

r_d is the separation of the center of the cavity from the center of the nearest-neighbor molecules

$R = r_d - \tilde{a}$, in which $\tilde{a}$ is the rationalized molecular radius

The quantity D_t represents the total dipole moment of the molecule, arising mainly from CH_2 groups that do not add up to zero because only a part of the molecule is involved. The potential of Equation 9.17 is of the square-well type in three dimensions, for which the binding energy, the difference between the ground-state energy, and V_0 are rationalized with the activation energy of electron mobility in n-hexane ($4.06\,\text{kcal}\,\text{mol}^{-1}$) for $N = 6$ and $\bar{\alpha} = 11.8\,\text{Å}^3$ (the value for n-hexane). This requires D_t to be 0.78 Debye, for which the potential is $V(r) = -2.4\,\text{eV}$ ($r < R$). Considering the bond dipole moment of CH to be 0.4 Debye, the authors considered the model to be plausible.

Davis et al. (1973) criticized Dodelet and Freeman's (1972, 1977) conjecture relating the decrease of electron mobility with the anisotropy of molecular polarizability. According to these authors, the correlation is weak and uncertain. It has recently been shown by Hug and Mozumder (2008) that the anisotropy of molecular polarizability, together with lattice connectivity, is central for determining the initial electron localization. This, however, does not preclude the final trapping being due to the isotropic part of the polarizability with or without the interaction of the electron and bond dipoles. On the other hand, the suggestion of Davis et al. (1973) that only the interaction of the electron with the CH_2 group of the molecule is important should not be taken literally, as it would tend to produce similar activation energy in very different liquid hydrocarbons, while ignoring other correlations. Also, the coordination number of six is not always supported by x-ray determination (Hug and Mozumder, 2008).

The measurement of the Hall mobility in a hydrocarbon liquid can give some useful information about electron trapping. In the presence of mutually perpendicular electric ($\vec{E}$) and magnetic ($\vec{B}$) fields, the average electron velocity $\langle\vec{v}\rangle$ is given by

$$\langle\vec{v}\rangle = \mu\vec{E} + \mu\mu_H\vec{E}\times\vec{B}, \tag{9.18}$$

where
  μ is the ordinary drift mobility
  μ_H is the so-defined Hall mobility

Equation 9.18 derives from the Lorentz force on a (nearly) free electron. It, therefore, only applies to the quasi-free electrons. The Hall mobility should approach the drift mobility in liquids where the transport is mostly in the quasi-free state, although some difference is expected anyhow, because these mobilities represent different moments of the momentum relaxation time, τ (Holroyd and Schmidt, 1989; Munoz, 1991).

It is clear from Equation 9.18 that the Hall signal is proportional to the drift mobility. Therefore, as yet, significant measurements are limited to liquids in which the drift mobility is greater than ~10 cm^2 V^{-1} s^{-1}. In the case of TMS, the two mobilities are virtually the same over the entire temperature range studied (Holroyd and Schmidt, 1989). This liquid is then considered to be trap-free for electrons. In neopentane, the Hall mobility is about the same as the drift mobility up to ~140°C, but systematically higher. Since there is a small activation energy of mobility in this liquid, it could be interpreted in the two-state model as due to a small population of electrons residing in shallow traps. Beyond 140°C, the Hall ratio in neopentane rises significantly above unity, being ~5 at the critical temperature. This has been attributed to localization due to density fluctuation (Holroyd and Schmidt, 1989).

9.3.4 Timescale of Trapping and Trap Size

Since trapping requires some medium rearrangement (vide supra), it is evident that trapping from either a localized or delocalized state would need a finite time. Closely associated with it is the question of the trap size. At present, neither of these is available from fundamental theory, but considerable information is available from experiments and from phenomenological theory. Due to space limitation, we will not elaborate on this aspect.

9.4 MOBILITY

9.4.1 Basic Considerations

It should be remembered that the primary reason for the investigation of electron localization and trapping is their relationships with mobility. Certain other considerations related to electron mobility, such as its variation with density, pressure, and external electric and magnetic fields (Hall effect), are clearly outside the purview of this chapter.

When excess electrons are generated in an isotropic liquid in the presence of an external electric field, the steady-state electron velocity (v) will be parallel to the field (E) and, for not too high an external field, will be proportional to it. The constant of proportionality, $\mu = v/E$, is defined to be drift mobility. The importance of mobility derives from the fact that the electron is a very sensitive probe of interaction with molecules and also with the liquid structure. Therefore, it can vary greatly from one liquid to another within the same class of hydrocarbons having very similar densities. Mobility study can be utilized to extract information about the state of the electron in the liquid (Holroyd and Schmidt, 1989; Munoz, 1991). When the mobility is low and increases with temperature, the activation energy, ΔE_{act}, may be defined by the relation $\mu(T) = A \exp(-\Delta E_{act}/k_B T)$, where $\mu(T)$ is the electron mobility at absolute temperature T, A is a certain constant, and k_B is the Boltzmann constant. Usually, the activation energy is extracted by measuring the mobility over a range of temperature of ~30°C or so. Different methods for measuring electron mobility in hydrocarbon liquids have been discussed by Munoz (1991) with their relative advantages and disadvantages.

Extensive tabulations exist for electron mobilities and their activation energies in many liquid hydrocarbons, as well as in other liquids including LRGs (Allen, 1976; Munoz, 1991; Nishikawa, 1991). From these, it is apparent that electron mobility in liquid hydrocarbons can vary by three orders of magnitude or more under similar conditions of temperature and density (Allen, 1976; Nishikawa, 1991). Activation energy can also vary, but not so much; typical values range from ~5 to 20 kJ mol^{-1} (Nishikawa, 1991). Typical mobility values in straight-chain hydrocarbons are ~0.1–0.2 cm^2 V^{-1} s^{-1} at room temperature, with activation energies ~15–20 kJ mol^{-1}. Highly branched hydrocarbons exhibit much higher mobilities, often with negative activation energies (i.e., mobility decreasing with temperature). For example, electron mobility in neopentane and in TMS at room temperature has been reported to be 70 and 100 cm^2 V^{-1} s^{-1}, respectively (Nishikawa, 1991). Some intermediate cases are also known. Generally speaking, smaller mobility and higher activation energy are indicative of considerable electron trapping, while the reverse is true for electrons existing mainly in the quasi-free state. In either situation, some electrons may be in the trapped state while the remaining electrons are in the quasi-free state, with an established equilibrium between them.

9.4.2 EMPIRICAL RELATIONSHIPS

Although the experimentally measured mobility in a hydrocarbon liquid decreases with V_0, a satisfactory theoretical explanation is hard to come by. The inhomogeneous percolation theory (Kestner and Jortner, 1973) and the homogeneous fluctuation theory (Schiller et al., 1973; Berlin and Schiller, 1987) (vide supra, Section 9.3.2) aim to do just that, but the quantitative agreement is not good, except in the high- or low-mobility regimes (Holroyd and Schmidt, 1989).

Notwithstanding the theoretical situation, an empirical relationship between V_0 and mobility has been advanced by Wada et al. (1977) as follows:

$$\mu = \frac{125}{1 + 360 \exp(15 V_0)}. \tag{9.19}$$

In Equation 9.19, the mobility, μ, is expressed in cm^2 V^{-1} s^{-1} and the quasi-free electron energy, V_0, in eV. This equation encompasses a surprisingly large amount of experimental data, although, understandably, not all the data are equally well fitted by it. As yet, no theoretical derivation has been proposed for this equation.

There is yet another correlation between electron mobility and free-ion yield, G_{fi}, proposed by Jay-Gerin et al. (1993). The free-ion yield is the number of electrons escaping geminate recombination and homogeneously collected by a vanishingly small electric field, normalized to 100 eV of the deposited energy. Experimentally, it has been known for a long time that the free-ion yield usually increases with electron mobility. In the theory of Onsager (1938), the probability of escaping geminate recombination depends on the initial separation, interpreted to be the thermalization distance. In this theory, it is independent of the electron's diffusion coefficient, and therefore of the mobility. The experimental observation then must be rationalized on the dependence of the thermalization distance with the mobility. In any case, the importance of the finding of Jay-Gerin et al. lies in establishing $G_{fi} \propto \mu^{0.31}$ (albeit approximately, and with some deviations), when the mobility ranges from ~0.1 to ~100 cm^2 V^{-1} s^{-1}, but the same remains essentially constant at ~0.1 for very low electron mobility ($<0.1 \, cm^2 \, V^{-1} \, s^{-1}$). This total correlation cannot be explained by the commonly accepted theories of electron transport, namely, the hopping model or the two-state model. The quasi-ballistic model of electron transport, advanced by Mozumder (1993, 1995, 2002), can give a reasonable explanation of the phenomena in terms of the elastic and inelastic mean free paths of epithermal electrons. These determine both the thermalization distance distribution and the thermal quasi-free electron mobility. The former gives the free-ion yield, and the latter can be converted to an effective mobility using the electron-trap concentration and the binding energy. Thus, a relationship can be obtained between free-ion yield and mobility. When the electron mobility is very low, transport is governed by random trapping and detrapping only, being independent of the quasi-free mobility. Then the mobility and the free-ion yield become essentially decoupled.

Over many years, Freeman and his associates have strived to correlate the variation of electron mobility with some intramolecular property of the liquid (Freeman, 1963a,b, 1986; Tewari and Freeman, 1968a,b; Robinson et al., 1971; Dodelet and Freeman, 1972, 1977; Fuochi and Freeman, 1972; Dodelet et al., 1973, 1975, 1976; Freeman and Huang, 1979; Gyorgy and Freeman, 1979; Gee and Freeman, 1983, 1987; Gee et al., 1988). These correlations are suggestive, however, without providing a theoretical basis. The early investigations (Freeman, 1963a,b; Tewari and Freeman, 1968a,b) correlated the conductance (current overshoot) with molecular sphericity. The more spherical a molecule is, the greater would be the free-ion yield. The free-ion yield of LMe was measured to be 0.8, which is much greater than that of other hydrocarbons (Robinson et al., 1971). It was inferred that electrons penetrate further in LMe, and this was attributed to its sphere-like structure. The mobility in LMe was measured to be $300 \, cm^2 \, V^{-1} \, s^{-1}$; it is much less in other hydrocarbons with decreasing sphericity (Fuochi and Freeman, 1972). A similar correlation was found in both mobility and free-ion yield in $C_5 - C_{12}$ alkane liquids (Dodelet and Freeman, 1972, 1977). The presence of a small molecular dipole moment (≤ 0.5 Debye) did not seem to affect electron mobility (Dodelet et al., 1973).

Dodelet et al. (1975) sought and found a correlation between the experimental activation energy of mobility and V_0 in various hydrocarbons and in TMS (Dodelet et al., 1975). Generally, the activation energy increased with V_0, negating a theoretical expectation by Kestner and Jortner (1973), based on percolation theory, that predicted a maximum at $V_0 \sim -0.15 \, eV$ and a minimum at $V_0 \sim -0.27 \, eV$. The correlation between molecular structure on one hand, and mobility and electron range on the other hand, was sought in various liquid hydrocarbons by Dodelet et al. (1976). The penetration range, b_{GP}, was found from the free-ion yield by fitting to a modified Gaussian distribution with a power tail. The results were interpreted by suggesting that less rigid molecules would allow longer penetration range, thus partially modifying the earlier correlation with sphericity. It was further suggested that the epithermal electron interaction with a molecule is essentially confined to two C–C bonds in series. The authors analyze their findings in terms of a model, which depends, in part, on Schiller et al.'s theory (1973), which in turn is based on the equilibrium energy fluctuation in the liquid (vide supra).

A particularly important study is the intercomparison of electron mobilities in liquid pentane isomers: normal pentane, isopentane (2-methyl butane), and neopentane (Gyorgy and Freeman, 1979). The mobilities in the liquid phase increase by orders of magnitude going from less to more spherical molecules. The values (in $cm^2 \, V^{-1} \, s^{-1}$) range from ~0.1 in n-pentane, to ~1.0 in isopentane, to ~50–70 in neopentane. These values clearly indicate the effect of the molecular shape. The authors offer alternative reasons: (1) the Ramsauer–Townsend effect, implying minimum scattering cross section occurring at lower velocity for less spherical molecules, and (2) a low-lying transient negative-ion state. None of these could be verified, especially because a Ramsauer–Townsend minimum has never been found in the liquid state. However, the authors succeeded in demonstrating that electron transport, in all these cases, is activated and that the process is similar in n-pentane and isopentane, but rather different in neopentane.

A similar correlation between electron mobility and molecular shape has also been seen by Bakale and Schmidt (1973), Holroyd and Cipollini (1978), and Schmidt and Allen (1970). However, Stephens (1986) pointed out that the mobility in isomeric pentanes reverses its order in the gas phase and that the density-normalized mobility curves cross over near the critical density. Therefore, intramolecular symmetry, or some other property, cannot be the sole determinant. To our knowledge, he first suggested the involvement of the liquid structure factor, although it is well known for electron mobility in LRGs. Stephens (1986) stressed the peak of the liquid structure factor as determined by low-angle x-ray scattering. Freeman (1986) acknowledged this as a useful procedure, but correctly pointed out that detailed correlation depended, in part, on the differences in the structure factors that are likely to be within experimental uncertainty. In a recent article, we have shown that the entire liquid structure needs to be taken into account, not just the peak position (Hug and Mozumder, 2008). Actually, this only gives the initial localization in the Anderson sense. Further considerations are needed to connect trapping with localization.

Several other correlations have been proposed to connect molecular shape with mobility, such as the anisotropy of molecular polarizability (Dodelet and Freeman, 1972, 1977; Funabashi, 1974; Gyorgy and Freeman, 1979), which is an important factor. Together with connectivity, it determines the extent of Anderson localization probability (Hug and Mozumder, 2008). However, it is the absolute difference between the maximum and minimum values of the polarizability that determines the width of diagonal energy fluctuation, which is important, rather than the mean polarizability, as was earlier thought to provide for the trapping potential.

In the face of many correlations and conjectures regarding free-ion yield, mobility, molecular shape, etc., a few things stand out. These are the anisotropy of molecular polarizability, the liquid structure, and the scattering of low-energy electrons. At a deeper level, these may be interrelated, but, as yet, not much is known about this.

9.4.3 THEORY

Since electron mobility in the condensed phase varies over several orders of magnitude, and since it depends sensitively on the interaction of the electron with both the molecule and the medium structure, it is not reasonable to expect that all experimental results will be explained by the same theoretical model. We shall now focus on two theoretical models specific to liquid hydrocarbons. These are the hopping model and the two-state models.

9.4.3.1 Hopping Models

In this model, the electron is assumed to exist in a preexisting trap. Transport occurs by trap-to-trap hopping, sometimes with phonon assistance (Holstein, 1959; Funabashi and Rao, 1976; Schmidt, 1977). Random hopping generates a Markov process with the resultant mobility given by the Nernst–Einstein equation as follows:

$$\mu = eL^2 \bar{w} / k_B T, \tag{9.20}$$

where

L is the mean distance between nearest-neighbor traps
$\bar{w}$ is the corresponding mean transition frequency

Classically, this transition probability at low electric fields is given by $\bar{w} = v \exp(-\varepsilon_0/k_B T)$, where v is the attempt frequency and ε_0 is the energy barrier to hopping. A quantum mechanical version of the transition probability is also available (Funabashi, 1974), which however is not applicable to hydrocarbon liquids. Schmidt (1977) has reviewed the mobility in several liquid hydrocarbons where electron mobility is low (<1 cm² V⁻¹ s⁻¹). By fitting into Equation 9.20, over the temperature interval of 100–300 K, it was found that for ethane, propane, butane, and cyclopentane, good agreement with experiments could be found for reasonable energy barriers (~0.1–0.16 eV) and also for reasonable attempt frequencies (1.4×10^{13}–2.0×10^{14} s⁻¹). However, the resultant jump lengths were very large (~10–45 Å), and their temperature variation could not be justified by any model. For example, the jump length in ethane increased rapidly from 10 Å at 100 K to 45 Å at 200 K, while that in propane was virtually independent of temperature. The situation is expected to be similar in other low-mobility liquid hydrocarbons, as has been commented on by Davis et al. (1973) in the case of n-hexane. On the other hand, Funabashi and Rao (1976) consider electron hopping over a fluctuating energy barrier, instead of a constant, ε_0. They were able to justify the field-dependent mobility in ethane and propane with a plausible distribution of an energy barrier and a reasonable value of $L = 7.5$ Å. One must remember, however, that this model is one-dimensional. No three-dimensional generalization has yet been made. This point is important because the electron can avoid a high barrier in three dimensions by going around it. Funabashi (1974) points out that for the validity of the hopping model, it is required that $\bar{w} < 10^{13}$ s^{-1}. If we take $L = 10$ Å, we get from Equation 9.20 $\mu \leq 4$ cm^2 V^{-1} s^{-1} at room temperature, which is the upper limit of validity of the hopping model.

9.4.3.2 Two-State Models

A fundamental assumption of two-state models is that an electron can exist in two different states of vastly different mobilities. One important case, sometimes called the trapping model, was first formulated by Frommhold (1968) for electron motion in a gas, interrupted by random attachment–detachment processes. It was then applied to liquid hydrocarbons by Minday et al. (1971, 1972). In this model, an electron can exist either in a quasi-free state of high mobility or in a trapped state of much lower mobility. Thermal equilibrium between these states is brought about by frequent and random trapping and detrapping processes. This results in a probability, P_f, of the electron of being in the quasi-free state and a probability, $P_t \equiv 1 - P_f$, of it being in the trapped state. In terms of the mean lifetimes in the quasi-free and trapped states, denoted respectively by $\tau_f = k_{ft}^{-1}$ and $\tau_t = k_{tf}^{-1}$, P_f is given by $P_f = \tau_f / (\tau_f + \tau_t)$. Here, k_{ft} and k_{tf}, respectively, are the rates of transition from the quasi-free state to the trapped state and vice versa. In the simple trapping model, the overall mobility is given by

$$\mu = P_f \mu_{qf} + (1 - P_f)\mu_t \approx P_f \mu_{qf}, \tag{9.21}$$

where μ_{qf} and μ_t are the mobilities in the quasi-free and trapped states, respectively. The extreme right-hand side of Equation 9.21 is obtained because, in most cases, the drift velocity is determined by the motion in the quasi-free state even though the electron stays in this state only momentarily. Nevertheless, this equation implies that in all situations, the lifetime in the quasi-free state is sufficiently long for a stationary drift velocity to ensue.

The two-state trapping model is intended for low- and intermediate-mobility liquids, in which case $P_f \approx k_{tf}/k_{ft}$, since $\tau_t \gg \tau_f$. This ratio of rates may be obtained from detailed balancing (Ascarelli and Brown, 1960) as $k_{tf}/k_{ft} = (1/n_t h^3)(2\pi m k_B T)^{3/2} \exp(-\varepsilon_0/k_B T)$, where n_t is the trap density, m is the electron mass, and ε_0 is the binding energy in the trap, that is, the amount by which the trapped-state energy lies below the quasi-free energy, V_0. From the magnitude and temperature dependence of the observed mobility in many liquid hydrocarbons, the quasi-free mobility has often been taken as a constant ($\mu_{qf} = 100$ cm^2 V^{-1} s^{-1}) (Minday et al., 1971, 1972; Davis and Brown, 1975). Although there is no fundamental reason for the constancy of the quasi-free mobility, some justification derives from the same value in TMS, which is considered to be trap free. With a known or assumed value of μ_{qf} and the ratio of detrapping to trapping rates as given above, the mobility in the two-state trapping model may be written as follows:

$$\mu = \mu_{qf}\left(\frac{1}{n_t h^3}\right)(2\pi m k_B T)^{3/2} \exp\left(\frac{-\varepsilon_0}{k_B T}\right). \tag{9.22}$$

Equation 9.22 can be fitted for electron mobility in a large class of hydrocarbons in which the electron mobility is <10 cm^2 V^{-1} s^{-1}. So-determined values of activation energy lie between 0.10 and 0.25 eV with a few exceptions. Trap densities are on the order of ~10^{19} cm^{-3} (Allen, 1976; Munoz, 1991;

Nishikawa, 1991; Mozumder, 1999, Chapter 10). Since the dominant temperature effect in Equation 9.22 is in the exponential term, the activation energy in this model equals the trap binding energy, if $\varepsilon_0 \gg k_B T$. A thermodynamic interpretation of electron transport in the two-state model has been given by Baird and Rehfield (1987).

Schmidt (1977) first pointed out that the validity of Equation 9.21 depends on a sufficiently long residence of the electron in the quasi-free state so that a steady drift velocity can be obtained. Mozumder (1993, 1995) introduced the quasi-ballistic model to correct for the competition between trapping and velocity randomization in the quasi-free state. Some conclusions of this model may now be summarized as follows: (1) Typically, the activation energy slightly exceeds the binding energy with $\varepsilon_0/E_{act} \approx 0.89$. (2) For relatively high-mobility liquids, the trap-controlled mobility dominates. For intermediate mobilities, both ballistic and trap-controlled mobilities make comparable contributions. For very-low-mobility liquids ($<0.1 \, cm^2 \, V^{-1} \, s^{-1}$), diffusion results only by random trapping and detrapping, irrespective of motion in the quasi-free state. This is an important concept, and it is required to explain the dependence of the free-ion yield on the mobility (vide infra). (3) With few exceptions, agreement with experiments for normal mobility and activation energy can be obtained (Mozumder, 1999, Chapter 10, and references therein), with the binding energy between ~ 0.1 and $0.28 \, eV$ and with the trap density between ~ 0.1 and $1.0 \times 10^{19} \, cm^{-3}$. (4) A consistent correlation between free-ion yield and mobility has been obtained by Mozumder (2002) using the quasi-ballistic model. The elastic- and inelastic-scattering mean free paths of epithermal electrons ($\leq 0.2 \, eV$) determine the thermalization distance distribution and, therefore, the free-ion yield. The quasi-free mobility (μ_{qf}) is also obtained from the same cross-sections using the Lorentz model. The effective mobility is then calculated from the quasi-free mobility, the trap concentration, and the binding energy. Thus, a relationship is established between the free-ion yield and the mobility in the quasi-ballistic model (vide supra). When the mobility is very low ($<0.1 \, cm^2 \, V^{-1} \, s^{-1}$), diffusion is governed by random trapping and detrapping, being independent of the quasi-free mobility. As found empirically by Jay-Gerin et al. (1993), the free-ion yield is then virtually independent of mobility. This is clearly explained in the quasi-ballistic model, but not in the usual trapping model. By comparing with experiment, it has been found that the elastic mean free path, L, in low-mobility liquids is ~ 1–$5 \, \text{Å}$ and the probability of an inelastic collision per elastic scattering is ~ 0.1–0.3 (Mozumder, 2002). These values increase with mobility, while, in the high-mobility case, L is approximately a few tens of Å and the inelastic events outnumber the elastic ones by a factor of ~ 2–4.

Comparing the hopping and two-state models, it can be said that the hopping model is applicable only to low-mobility liquids (vide supra). The jump lengths are usually large and vary erratically with the nature of the liquid and with temperature. The two-state models have a wider range of applicability, but they require additional parameters. Consistency between free-ion yield and mobility has been achieved only in the quasi-ballistic model. In either type of model, it is hard to reconcile the activation energy of mobility with the peak of the optical absorption spectrum, where observable. Sometimes it has been attributed to two kinds of trapped electrons, shallowly trapped and deeply trapped. However, no consistent workable theory has yet been proposed along this line.

9.5 CONCLUDING REMARKS

In this chapter, we have considered electron localization and trapping in hydrocarbon liquids. We have proposed and developed the application of the Anderson model for initial localization. However, the transformation from the initially localized or delocalized doorway state to eventual trapping, or promotion to the quasi-free state, has not yet been achieved theoretically. Comparing with experiments, it seems that the two-state model has wider applicability with reasonable values for quasi-free mobility, trap concentration, and binding energy in the trap. Consistency between free-ion yield and mobility is possible within the context of the quasi-ballistic version of the two-state model.

In both hopping and two-state models, as applied to liquid hydrocarbons, it is a reasonable assumption that the traps are preexisting. The situation is quite different in polar media, or when the mobility is very high. In polar media and in dense cold gases, such as He, H_2, and Ne, there is evidence of self-trapping.

ACKNOWLEDGMENTS

The authors are thankful to Dr. G. Ferraudi for numerous discussions on molecular and liquid structures. This paper is Document No. NDRL-4808 from the Notre Dame Radiation Laboratory, which is supported by the Office of Basic Energy Sciences of the U.S. Department of Energy.

REFERENCES

Abou-Chacra, R., Anderson, P. W., and Thouless, D. J. 1973. Self-consistent theory of localization. *J. Phys. C* 6: 1734–1752.
Allen, A. O. 1976. *Drift Mobilities and Conduction Band Energies of Excess Electrons in Dielectric Liquids.* Vol. 58, NSRDS-NBS. Gaithersburg, MD: NBS.
Anderson, P. W. 1958. Absence of diffusion in certain random lattices. *Phys. Rev.* 109: 1492–1505.
Ascarelli, G. and Brown, S. C. 1960. Recombination of electrons and donors in *n*-type germanium. *Phys. Rev.* 120: 1615–1626.
Baird, J. K. and Rehfeld, R. H. 1987. Thermodynamics of electron-transport in amorphous insulators. *J. Chem. Phys.* 86: 4090–4095.
Bakale, G. and Schmidt, W. F. 1973. Excess electron-transport in liquid hydrocarbons. *Chem. Phys. Lett.* 22: 164–166.
Berlin, Y. A. and Schiller, R. 1987. Number of atoms interacting with a quasi-free electron in nonpolar liquids. *Radiat. Phys. Chem.* 30: 71–73.
Bulka, B., Schreiber, M., and Kramer, B. 1987. Localization, quantum interference, and the metal-insulator-transition. *Z. für Phys. B* 66: 21–30.
Chandler, D. and Leung, K. 1994. Excess electrons in liquids: Geometrical perspectives. *Annu. Rev. Phys. Chem.* 45: 557–591. [H. L. Strauss, G. T. Babcock, and S. R. Leone (eds.). Palo Alto, CA: Annual Reviews.]
Chang, T.-M., Bauer, J. D., and Skinner, J. L. 1990. Critical exponents for Anderson localization. *J. Chem. Phys.* 93: 8973–8982.
Chayes, J. T., Chayes, L., Fisher, D. S., and Spencer, T. 1986. Finite-size scaling and correlation lengths for disordered-systems. *Phys. Rev. Lett.* 57: 2999–3002.
Cohen, M. H. 1973. The electronic structures of disordered materials. In *Electrons in Fluids. The Nature of Metal-Ammonia Solutions*, J. Jortner and N. R. Kestner (eds.), pp. 257–285. New York: Springer-Verlag.
Cohen, M. H. 1977. Electrons in fluids: Role of disorder. *Can. J. Chem.* 55: 1906–1915.
Cohen, M. H., Fritzsche, H., and Ovshinsky, S. R. 1969. Simple band model for amorphous semiconducting alloys. *Phys. Rev. Lett.* 22: 1065–1068.
Davis, H. T. and Brown, R. G. 1975. Low-energy electrons in nonpolar fluids. In *Advances in Chemical Physics*, I. Prigogine and S. A. Rice (eds.), pp. 329–464. New York: Wiley.
Davis, H. T., Schmidt, L. D., and Brown, R. G. 1973. Mobility studies of excess electrons in nonpolar hydrocarbons. In *Electrons in Fluids. The Nature of Metal-Ammonia Solutions*, J. Jortner and N. R. Kestner (eds.), pp. 393–411. New York: Springer-Verlag.
Dodelet, J.-P. and Freeman, G. R. 1972. Mobilities and ranges of electrons in liquids: Effect of molecular structure in C5-C12 alkanes. *Can. J. Chem.* 50: 2667–2679.
Dodelet, J.-P. and Freeman, G. R. 1977. Electron mobilities in alkanes through liquid and critical regions. *Can. J. Chem.* 55: 2264–2277.
Dodelet, J.-P., Shinsaka, K., and Freeman, G. R. 1973. Electron mobilities in liquid olefins: Structure effects. *J. Chem. Phys.* 59: 1293–1297.
Dodelet, J.-P., Shinsaka, K., and Freeman, G. R. 1975. Electron-mobility models for liquid hydrocarbons. *J. Chem. Phys.* 63: 2765–2766.

Dodelet, J.-P., Shinsaka, K., and Freeman, G. R. 1976. Molecular-structure effects on electron ranges and mobilities in liquid hydrocarbons: Chain-branching and olefin conjugation: Mobility model. *Can. J. Chem.* 54: 744–759.

Feher, G., Fletcher, R. C., and Gere, E. A. 1955. Exchange effects in spin resonance of impurity atoms in silicon. *Phys. Rev.* 100: 1784–1786.

Freeman, G. R. 1963a. Gamma-radiation-induced conductance in liquid hydrocarbons. *J. Chem. Phys.* 38: 1022–1023.

Freeman, G. R. 1963b. Yield of free ions in gamma-irradiated liquid saturated hydrocarbons. *J. Chem. Phys.* 39: 988–996.

Freeman, G. R. 1986. Electron-mobility in the liquid isomeric pentanes: Reply. *J. Chem. Phys.* 84: 4723–4723.

Freeman, G. R. and Huang, S. S. S. 1979. Localized excess-electron states in simple classical fluids: Quasilocalization. *Phys. Rev. A* 20: 2619–2620.

Frommhold, L. 1968. Resonance scattering and the drift motion of electrons through gases. *Phys. Rev.* 172: 118–125.

Funabashi, K. 1974. Excess-electron processes in radiation chemistry of disordered materials. In *Advances in Radiation Chemistry*, M. Burton and J. L. Magee (eds.), pp. 103–180. New York: Wiley.

Funabashi, K. and Rao, B. N. 1976. Field-dependent mobility in liquid hydrocarbons. *J. Chem. Phys.* 64: 1561–1563.

Fuochi, P. G. and Freeman, G. R. 1972. Molecular structure effects of free-ion yields and reaction-kinetics in radiolysis of methyl-substituted propanes and liquid argon: Electron and ion mobilities. *J. Chem. Phys.* 56: 2333–2341.

Gee, N. and Freeman, G. R. 1983. Electron thermalization ranges and free-ion yields in dielectric fluids: Effects of density and molecular shape. *Phys. Rev. A* 28: 3568–3574.

Gee, N. and Freeman, G. R. 1987. Electron mobilities, free ion yields, and electron thermalization distances in liquid long-chain hydrocarbons. *J. Chem. Phys.* 86: 5716–5721.

Gee, N., Senanayake, P. C., and Freeman, G. R. 1988. Electron-mobility, free ion yields, and electron thermalization distances in *n*-alkane liquids: Effects of chain-length. *J. Chem. Phys.* 89: 3710–3717.

Gelbart, W. M. 1974. Depolarized light scattering by simple fluids. In *Advances in Chemical Physics*, I. Prigogine and S. A. Rice (eds.), pp. 1–106. New York: Wiley.

Gyorgy, I. and Freeman, G. R. 1979. Effects of density and temperature on mobilities of electrons in vapors and liquids of pentane isomers: Molecular-structure effects. *J. Chem. Phys.* 70: 4769–4777.

Holroyd, R. A. 1977. Equilibrium reactions of excess electrons with aromatics in non-polar solvents. *Ber. Bunsen-Gesellschaft Phys. Chem.* 81: 298–304.

Holroyd, R. A. and Cipollini, N. E. 1978. Correspondence of conduction-band minima and electron-mobility maxima in dielectric liquids. *J. Chem. Phys.* 69: 501–503.

Holroyd, R. A. and Schmidt, W. F. 1989. Transport of electrons in nonpolar fluids. *Annu. Rev. Phys. Chem.* 40: 439–468.

Holroyd, R. A., Geer, S., and Ptohos, F. 1991. Free-ion yields for several silicon-containing, germanium-containing, and tin-containing liquids and their mixtures. *Phys. Rev. B* 43: 9003–9011.

Holstein, T. 1959. Studies of polaron motion. 2. The small polaron. *Ann. Phys.* 8: 343–389.

Hug, G. L. and Mozumder, A. 2008. Electron localization in liquid hydrocarbons: The Anderson model. *Radiat. Phys. Chem.* 77: 1169–1175.

Hug, G. L. and Mozumder, A. forthcoming.

Ising, E. 1925. Report on the theory of ferromagnetism. *Z. für Phys.* 31: 253–258.

Jay-Gerin, J.-P., Goulet, T., and Billard, I. 1993. On the correlation between electron-mobility, free-ion yield, and electron thermalization distance in nonpolar dielectric liquids. *Can. J. Chem.* 71: 287–293.

Karnicky, J. F. and Pings, C. J. 1976. Recent advances in the study of liquids by x-ray diffraction. In *Advances in Chemical Physics*, I. Prigogine and S. A. Rice (eds.), pp. 157–202. New York: Wiley.

Kestner, N. R. and Jortner, J. 1973. Conjecture on electron mobility in liquid hydrocarbons. *J. Chem. Phys.* 59: 26–30.

Kimura, T., Fueki, K., Narayana, P. A., and Kevan, L. 1977. Semicontinuum model with a structured 1st solvation shell for excess electrons in liquid and glassy alkanes. *Can. J. Chem.* 55: 1940–1951.

Kuchitsu, K. 1968. Comparison of molecular structures determined by electron diffraction and spectroscopy. Ethane and diborane. *J. Chem. Phys.* 49: 4456–4462.

Lee, I.-R., Lee, W., and Zewail, A. H. 2008. Dynamics of electrons in ammonia cages: The discovery system of solvation. *ChemPhysChem* 9: 83–88.

Mikolaj, P. G. and Pings, C. J. 1967. Structure of liquids. III. An x-ray diffraction study of fluid argon. *J. Chem. Phys.* 46: 1401–1411.

Minday, R. M., Schmidt, L. D., and Davis, H. T. 1971. Excess electrons in liquid hydrocarbons. *J. Chem. Phys.* 54: 3112–3125.

Minday, R. M., Schmidt, L. D., and Davis, H. T. 1972. Mobility of excess electrons in liquid hydrocarbon mixtures. *J. Phys. Chem.* 76: 442–446.

Mott, N. F. 1967. Electrons in disordered structures. *Adv. Phys.* 16: 49–144.

Mozumder, A. 1993. Quasi-ballistic model of electron-mobility in liquid hydrocarbons. *Chem. Phys. Lett.* 207: 245–249.

Mozumder, A. 1995. Electron-mobility in liquid hydrocarbons: Application of the quasi-ballistic model. *Chem. Phys. Lett.* 233: 167–172.

Mozumder, A. 1999. *Fundamentals of Radiation Chemistry*. San Diego, CA: Academic Press.

Mozumder, A. 2002. Free-ion yield and electron mobility in liquid hydrocarbons: A consistent correlation. *J. Phys. Chem. A* 106: 7062–7067.

Mozumder, A. 2006. The quasi-ballistic model of electron mobility in liquid hydrocarbons: Effect of an external electric field. *Chem. Phys. Lett.* 420: 277–280.

Mozumder, A. and Hug, G. L. 2009. Electron mobility in liquid isomeric pentanes: Rationalization based on Anderson localization. *Chem. Phys. Lett.* 480: 71–74.

Mozumder, A. and Magee, J. L. 1967. Theory of radiation chemistry. VIII. Ionization of nonpolar liquids by radiation in the absence of external electric field. *J. Chem. Phys.* 47: 939–945.

Munoz, R. C. 1991. Mobility of excess electrons in dielectric liquids: Experiment and theory. In *Excess Electrons in Dielectric Media*, C. Ferradini and J.-P. Jay-Gerin (eds.), pp. 161–210, Chapter 6. Boca Raton, FL: CRC Press.

Narten, A. H. 1979. X-ray-diffraction study of liquid neopentane in the temperature range 17 to 150°C. *J. Chem. Phys.* 70: 299–304.

Nishikawa, M. 1991. Electrons in condensed media. In *CRC Handbook of Radiation Chemistry*, Y. Tabata, Y. Ito, and S. Tagawa (eds.), pp. 395–438, Chapter VII. Boca Raton, FL: CRC Press.

Onsager, L. 1938. Initial recombination of ions. *Phys. Rev.* 54: 554–557.

Petz, J. I. 1965. X-ray determination of the structure of liquid methane. *J. Chem. Phys.* 43: 2238–2243.

Phillips, P. 1993. Anderson localization and the exceptions. *Annu. Rev. Phys. Chem.* 44: 115–144. [H. L. Strauss, G. T. Babcock, and S. R. Leone (eds.). Palo Alto, CA: Annual Reviews.]

Pierce, W. C. 1935. Scattering of x-rays by polyatomic liquids. *n*-heptane. *J. Chem. Phys.* 3: 252–255.

Rassolov, V. A. and Mozumder, A. 2001. Monte Carlo simulation of electron thermalization distribution in liquid hydrocarbons: Effects of inverse collisions and of an external electric field. *J. Phys. Chem. B* 105: 1430–1437.

Robinson, M. G., Fuochi, P. G., and Freeman, G. R. 1971. Yields of free ions in x radiolysis of some simple saturated and unsaturated hydrocarbon liquids: Effects of molecular structure and temperature. *Can. J. Chem.* 49: 3657–3664.

Root, L. J., Bauer, J. D., and Skinner, J. L. 1988. New approach to localization: Quantum connectivity. *Phys. Rev. B* 37: 5518–5521.

Schiller, R., Vass, S., and Mandics, J. 1973. Energy of the quasi-free electrons and the probability of electron localization in liquid hydrocarbons. *Int. J. Radiat. Phys. Chem.* 5: 491–503.

Schmidt, W. F. 1977. Electron-mobility in nonpolar liquids: Effect of molecular-structure, temperature, and electric-field. *Can. J. Chem.* 55: 2197–2210.

Schmidt, W. F. and Allen, A. O. 1970. Mobility of electrons in dielectric liquids. *J. Chem. Phys.* 52: 4788–4794.

Shante, V. and Kirkpatrick, S. 1971. An introduction to percolation theory. *Adv. Phys.* 20: 325–357.

Shinsaka, K., Dodelet, J.-P., and Freeman, G. R. 1975. Electron mobilities and ranges in liquid hydrocarbons: Cyclic and polycyclic, saturated and unsaturated-compounds. *Can. J. Chem.* 53: 2714–2728.

Springett, B. E., Jortner, J., and Cohen, M. H. 1968. Stability criterion for the localization of an excess electron in a nonpolar fluid. *J. Chem. Phys.* 48: 2720–2731.

Stephens, J. A. 1986. Comment on electron-mobility in the liquid isomeric pentanes. *J. Chem. Phys.* 84: 4721–4723.

Stuart, H. A. 1962. Depolarisationsgrade bei der molekularen Lichtstreuung an Flüssigkeiten. In *Landolt-Börnstein*, Vol. II, Part 8, pp. 5-815–5-826. Berlin, Germany: Springer-Verlag.

Tewari, P. H. and Freeman, G. R. 1968a. Radiolysis of liquid hydrocarbons: Dependence of free-ion yield on molecular structure. *J. Chem. Phys.* 49: 954–955.

Tewari, P. H. and Freeman, G. R. 1968b. Dependence of radiation-induced conductance of liquid hydrocarbons on molecular structure. *J. Chem. Phys.* 49: 4394–4399.

Wada, T., Shinsaka, K., Namba, H., and Hatano, Y. 1977. Electron reactivity in liquid-hydrocarbon mixtures. *Can. J. Chem.* 55: 2144–2155.

Warren, B. E. 1933. X-ray diffraction in long chain liquids. *Phys. Rev.* 44: 969–976.

Watson, K. M. 1957. Multiple scattering by quantum-mechanical systems. *Phys. Rev.* 105: 1388–1398.

Zernike, F. and Prins, J. A. 1927. Die Beugung von Röntgenstrahlen in Flüssigkeiten als Effekt der Molekulanordnung. *Z. Phys.* 41: 184–194.

Ziman, J. M. 1979. *Models of Disorder. The Theoretical Physics of Homogeneously Disordered Systems.* Cambridge, U.K.: Cambridge University Press.

10 Reactivity of Radical Cations in Nonpolar Condensed Matter

Ortwin Brede
University of Leipzig
Leipzig, Germany

Sergej Naumov
Leibniz Institute of Surface Modification
Leipzig, Germany

CONTENTS

10.1 INTRODUCTION

As a consequence of the radiation-induced ionization of matter, electrons and parent radical cations are primarily formed. Whereas in polar media the parent ions are extremely unstable, in nonpolar systems (alkanes, alkyl chlorides, freons, etc.) such radical cations may persist for longer times, depending on the temperature and viscosity. The lifetime of radical cations in neat systems is determined by neutralization, deprotonation, and fragmentation. This is known since decades (see, e.g., Mehnert, 1991; Tabata et al., 1991), and will be briefly reviewed in this chapter.

In the presence of solutes having lower ionization potential (IP) than the matrix (liquid solvents, solid-state glasses, etc.), electron transfer (ET) from the solute to the solvent parent radical cations happens. This is often applied for the generation and characterization of radical cations of various solutes. These studies were preferably performed with the solid-state matrix isolation technique (Hamill, 1968; Shida, 1988) as well as electron pulse radiolysis (Baxendale and Busi, 1982; Fartahaziz and Rodgers, 1987). The detailed investigation of the ET from substituted aromatics to solvent parent cations implies mechanistic peculiarities insofar that in the liquid phase, molecule dynamics governs the product formation. Some of the ET phenomena, such as free electron transfer (FET) in liquids (Brede and Naumov, 2006) and special transformations of radical cations in low-temperature glasses (Naumov et al., 2003a,b,c,d, 2004a,b,c, 2005), will be described here.

10.2   IONIZATION OF NONPOLAR SYSTEMS

Because of its high energy excess, ionizing radiation does not act on matter very selectively. Two pathways are generally passed, namely, ionization and excitation of the material. Concerned with nonpolar substances represented here by alkanes (RH) and alkyl halides (RX), it implies that in the course of gamma or electron irradiation, besides the direct ionization products, fragmentation and dissociation products originating from excitation of matter also appear (cf. Equation 10.1):

$$RX \rightarrow \diagup \rightarrow RX^{\bullet+} + e_{solv.}^{-}; \quad R^{\bullet} + H + \text{Neutral fragments}, \quad X = H, Cl, \text{etc.} \quad (10.1)$$

We will focus on the ionization products given in boldface. The radical cations derived from pure σ-bonded saturated molecules represent structures where the spin density is distributed over the whole molecule (Brede and Naumov, 2006). This can be seen in Figure 10.1 where spin densities and charge distributions of some $RX^{\bullet+}$ are demonstrated.

Radical cations derived from n-alkanes are metastable above a particular molecule size ($\geq C_7$). Below this size, the n-alkane radical cations are less stable and fragmentize in a less specific manner in olefin radical cations or carbenium ions (Mehnert et al., 1984; Le Motais and Jonah, 1989) (e.g., see Equation 10.2). This happens also with vibronically excited radical cations existing at a very early stage after radiation action.

$$n\text{-}C_6H_{14}^{\bullet+} \rightarrow n\text{-}C_nH_{2n}^{\bullet+} + RH; \quad n\text{-}C_nH_{2n}^{\bullet} + R^{+} \ldots \quad (10.2)$$

As a characteristic of radiation chemistry, the ionization products are primarily inhomogeneously distributed in local entities called spurs, blobs, etc., which have a size around 5–10 Å. In a competition between neutralization in the spurs and diffusive escape, in nonpolar media, most of the ionic species decay within the spurs (geminate ions) and only a small part (free ions) results in a homogeneous distribution. So, of the primarily formed yield of $G \approx 4\ldots5$, only a fraction of $G_{fi} = 0.05–0.50$ results in a homogeneous distribution (Tabata et al., 1991). Here, the G-value represents the yield of the species per 100 eV absorbed energy. The low homogeneous ion yield is a characteristic of the radiation chemistry of nonpolar media. Table 10.1 gives free ion yields of some nonpolar liquids (Tabata et al., 1991). It should be taken into account that these yields were determined through

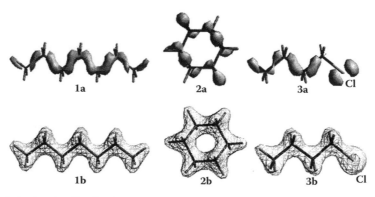

FIGURE 10.1   Spin density distribution (**a**) and charge distribution (**b**) of the radical cations derived from (**1**) n-heptane, (**2**) cyclohexane, and (**3**) n-butyl chloride.

TABLE 10.1

Yields and Mobilities of Free Ions in Nonpolar Liquids, G_{fi} as Ions per 100 eV, μ in cm^2 V^{-1} s^{-1}

Substance	G_{fi} as Ions per 100 eV	μ in cm^2 V^{-1} s^{-1}
Neopentane	0.9–1.0	55
Tetramethylsilane	0.74	90
2,2,4-Trimethylpentane	0.33	7
2,2-Dimethylbutane	0.30	10
n-Butane	0.19	0.4
Isopentane	0.17	
Cyclohexene	0.15–0.20	
Cyclohexane	0.15–0.19	0.35
n-Butyl chloride	≈0.20	
n-Hexane	0.10–0.18	0.09
Cyclopentane	0.16	1.1
n-Pentane	0.15	0.16
Carbon tetrachloride	0.05–0.09	
Benzene	0.05–0.08	

conductivity techniques, which is unspecific, because this sums up parent radical cations as well as fragment radical cations and carbenium ions (Tabata et al., 1991).

The parent radical cations have been observed by matrix insulation experiments in gamma-irradiated solid samples (glasses, polycrystalline matter) at low temperatures. The detection was performed by steady state UV–vis spectroscopy (Hamill, 1968; Shida, 1988) or EPR spectroscopy (Shida et al., 1984; Knight, 1986). From these experiments, structural information originated, which confirmed the mechanistic postulations of steady state irradiations combined with product analytics (Földiak, 1981). Although the parent ions of alkenes did not exhibit as remarkable optical properties as those of the solvated electrons (Baxendale et al., 1973; Allen and Holroyd, 1974), the observation and kinetic characterization of alkane radical cations in liquids at room temperature, with electron pulse radiolysis, succeeded in 1977 (Mehnert et al., 1977, 1979a,b). Subsequently, the Delft, Argonne, and Tokyo pulse radiolysis laboratories also got engaged in this matter. For example, from recent measurements, Figure 10.2 shows optical absorption spectra of an alkane and an alkyl chloride radical cation detected by pulse radiolysis at room temperature (Brede and Naumov, 2006).

The kinetic behavior and reactions of the parent alkane radical cations was intensely studied (Mehnert, 1991). Similar investigations were performed also with carbon tetrachloride (Mehnert et al., 1979a,b; Brede et al., 1980) and n-butyl chloride (Arai et al., 1976; Mehnert et al., 1982). Because of high sensitivity, optical spectroscopy was usually preferred for detection (Baxendale and Busi, 1982), but conductivity methods (Hummel and Schmidt, 1974; Beck, 1979), the microwave absorption technique (Infelta et al., 1977), as well as electron spin resonance spectroscopy (Trifunac et al., 1979) were also used for this purpose. The results are presented in detail in some monographs (Baxendale and Busi, 1982; Fartahaziz and Rodgers, 1987; Lund and Shiotani, 1991).

The fate of alkane radical cations in the liquid phase is determined by neutralization with the counter charge (solvated electrons or subsequent anions), deprotonation with the solvent, and fragmentation to olefin radical cations (analogous to reaction 10.2) (see reactions 10.3a through 10.3c):

$$n\text{-C}_n\text{H}_{2n+2}{}^{\bullet+} + e_{solv.}^{-} \quad \text{or} \quad \text{Cl}^- \rightarrow \text{Neutral products} \tag{10.3a}$$

$$n\text{-C}_n\text{H}_{2n+2}{}^{\bullet+} + \text{RH} \rightarrow n\text{-C}_n\text{H}_{2n+1}{}^{\bullet} + \text{RH}_2{}^{+} \tag{10.3b}$$

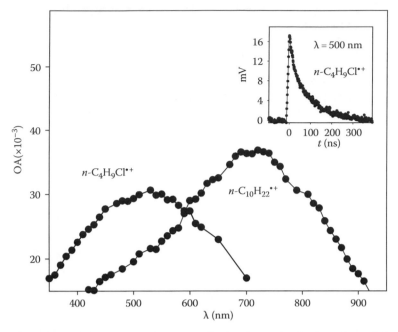

FIGURE 10.2 Optical absorption spectra of parent radical cations taken in the pulse radiolysis of the pure solvents n-decane or n-butyl chloride. The inset shows a time profile of the n-butyl chloride radical cation at a characteristic wavelength.

$$n\text{-}C_nH_{2n}^{\bullet+} \rightarrow n\text{-}C_{n-1}H_{2n-2}^{\bullet+} + CH_4 \tag{10.3c}$$

As a result of these reactions, the lifetime of alkane radical cations at room temperature falls in the range of 1–20 ns. Branched alkanes were found to form less stable parent ions, which is caused by unimolecular fragmentation such as that in (10.3c). The radical cations of alkyl chlorides exhibit lifetimes up to 200 ns. Certainly, this depends, in each case, also on the purity of the substances.

Alkene radical cations formed by reaction type (10.3c) are of minor importance. They are formed in lower yields as the parent ions and exhibit a lower reactivity (Mehnert et al., 1982) (see also Section 10.3). Now we focus on the reactivity of the parent ions of alkanes and alkyl chlorides and present some recently found kinetic phenomena.

10.3 REACTION OF PARENT SOLVENT CATIONS WITH ADDED SCAVENGERS

The ionization of alkanes and alkyl chlorides results mainly in parent radical cations where the σ-bond skeleton is preserved. The spin and the charge are uniformly distributed in such species (see Figure 10.1) (Brede and Naumov, 2006). A small quantity of alkene radical cations is also formed either as direct by-products of the ionization or as fragments of the decay of the parent ions (10.3c). These alkene radical cations have a more localized spin and charge distribution (cf. Figure 10.3), and therefore, a lower reactivity, because, in practical terms, they are derived from an unsaturated system.

In the presence of any additives (scavengers), both types of solvent-derived radical cations tend to undergo ion–molecule reactions, such as proton transfer (PT) (reactions 10.4a and 10.4b) or ET (reaction 10.4c). PT mainly involves polar and protic scavengers, and hence the rate constants depend on polarity parameters, such as nucleophile power, and perhaps dielectric constants (Mahalaxmi et al., 2001). Quenching of transient signals by such reactions was often used for the

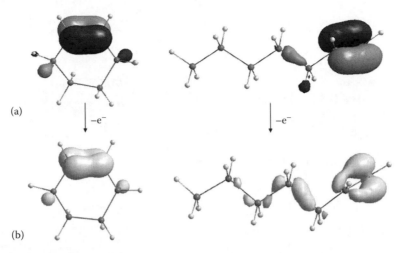

FIGURE 10.3 Electron density and spin distribution of cyclohexene and *n*-heptene and of the parent alkene radical cations derived from them.

identification of radical cations. Apparently, anions like chloride also act as nucleophiles and efficiently quench parent radical cations.

$$RHX^{\bullet+} + RYH \rightarrow RX^{\bullet} + RYH_2^{+}, \quad Y = O, N, \quad RHX = \text{alkanes, alkyl chlorides} \quad (10.4a)$$

$$RHX^{\bullet+} + Cl^{-} \rightarrow RX^{\bullet} + HCl, \quad Cl^{-} \text{ from, for example, } (Bu)_4 N^{+} Cl^{-} \quad (10.4b)$$

$$RX^{\bullet+} + D \rightarrow RX + D^{\bullet+}, \quad k_{4b} \approx 10^{10} \text{ dm}^3 \text{ mol}^{-1} \text{ s}^{-1}, \text{ that is, diffusion controlled,}$$
$$RX = \text{alkanes, alkyl chlorides} \quad (10.4c)$$

ET as the ion–molecule reaction (10.4c) plays a dominant role. Here, the additives act as electron donors (D). The reason for this is the high IP of the RX compounds, which is often 1–3 eV higher than that of the donors. To judge the energetics of the ET reactions, the well-known gas-phase ionization potentials (Wedenejew et al., 1971; Meot-Ner et al., 1981), IP_{gas}, are used, neglecting the polarization energy part in condensed matter (Brede et al., 1987). Hence, the ionization potential difference, ΔIP, between the IPs of RX and D can be used as a measure of the free energy of the ET (10.4c).

For a bimolecular reaction, apart from the energetics, the mobility of the reaction partners should also be considered. At $\Delta IP \geq 0.5$ eV, the ET was found to be diffusion controlled (Brede et al., 1987), that is, under normal conditions (RT, relaxed parent ions), possessing a rate constant of $k \approx 10^{10}$ dm^3 mol^{-1} s^{-1}. So far as to the principle of ET (10.4c).

Traditionally, steady state irradiation experiments at low temperature (77 K) glasses of D dissolved in RX were performed to observe donor radical cations. The nature of these species can be tentatively characterized by bleaching as well as stepwise warming up (heating) experiments (Hamill, 1968; Shida, 1988). This causes changes in the viscosity and, therefore, slightly accelerates the diffusion. Hence, a qualitative identification of the donor radical cations and, sometimes, of their decay products can be performed. Furthermore, optical properties as absorption maxima and extinction coefficients of the donor cations can be obtained. As an example, Figure 10.4 shows the optical absorption spectra, taken by the matrix insulation technique, of a sample containing the sterically hindered 4-methyl-2,6-*tert*-butyl-phenol (Zubarev and Brede, 1997). At room temperature

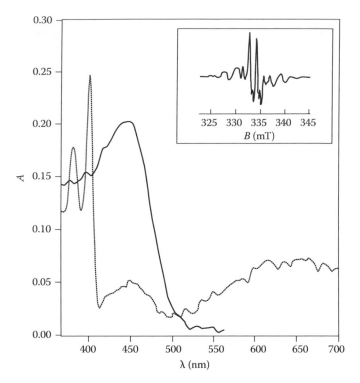

FIGURE 10.4 UV–vis optical absorption spectra of ArOH$^{\bullet+}$ (boldfaced curve) and ArO$^{\bullet}$ (dotted curve) in an irradiated s-BuCl glassy matrix. The inset shows the EPR spectrum at 77 K; the central part of the spectrum shows the characteristic quartet due to ArOH$^{\bullet+}$. ArOH = 4-methyl-2,6-di-*tert*-butyl-phenol.

and in polar solvents, phenol radical cations are extremely unstable (pK_a around −1) (Dixon and Murphy, 1978). In nonpolar glass at 77 K, first the spectrum of ArOH$^{\bullet+}$ is observed; on heating up the sample, the well-known phenoxyl radical appears. Additionally, the EPR spectrum of ArOH$^{\bullet+}$ taken at 77 K is shown in the inset of Figure 10.4.

The formation of donor radical cations by ET in rigid matrices at low temperatures, however, cannot be explained by normal diffusion and, therefore, assumes something like a nonclassical transport process. A similar conclusion has been drawn from pulse radiolysis measurements with liquid solutions of aromatic donors in cyclohexane (Zador et al., 1973; Brede et al., 1974) and *n*-alkanes (Brede et al., 1974), where an unusually rapid formation of donor species has been found. This was interpreted as a reaction of extremely mobile positive holes (+), which are postulated to be precursors of the relaxed parent radical cations (see reaction 10.5):

$$(+)_{RH} + D \rightarrow D^{\bullet+} + RH \quad k_5 \leq 10^{12} \text{ dm}^3 \text{ mol}^{-1} \text{ s}^{-1} \tag{10.5}$$

The mechanism of this rapid migration is not completely clear. It is hypothesized that something like a resonance transfer or a special hole migration takes place (Mehnert, 1991). For most of the alkanes and alkyl chlorides, after about 1 ns, relaxed parent radical cations with normal diffusion-conditioned mobility exist. These species are the subject matter of our further elucidations.

At donor concentrations of up to some 10^{-3} mol dm^{-3}, the reaction of free and homogeneously distributed ions is observed. If the concentration is increased ($c > 10^{-3}$ mol dm^{-3}), then the product cation yield increases dramatically. This is explained by scavenging the inhomogeneously distributed parent ions (geminate ions), which normally decay by inhomogeneous neutralization in the spurs (reaction 10.3a). This behavior can be described by scavenger curves (see Figure 10.5). The shape of the

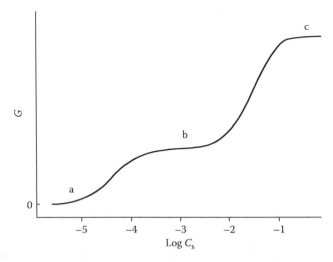

FIGURE 10.5 Common shape of a scavenger curve describing the yield of donor radical cations dependent on the scavenger concentration: a—range of competition between reactions (10.4c) and (10.3a), b—range of reaction (10.4c) with free ions, c—range of additionally scavenging geminate ions.

scavenger curves depends on the efficiency of scavenging, that is, on the rate constant and free energy of the reaction and on the molecule size and other donor-specific parameters (Warman et al., 1969), and is preferably used for the description of steady state experiments. The described concentration-dependent yield, however, does not influence markedly the kinetics of the ET reaction of type (10.4c).

The ET (10.4c) from donors to relaxed parent solvent radical cations has been often used for the generation of donor cations and the subsequent investigation of their properties. Insofar, it is a valuable tool for donor cation formation and identification, also in comparison with photochemical investigations. Among the many others, a few examples from our own research are mentioned. For the most part, unknown radical cations of

- Pyrimidines (Lomoth et al., 1999), of various phenols (Mahalaxmi et al., 2000)
- HALS compounds as cationic intermediates of photostabilization (Brede et al., 1998)
- Styrene type as polymerizing cationic species (Brede et al., 1982)

were found and kinetically characterized.

10.4 FREE ELECTRON TRANSFER

Using substituted aromatics as donors in the ET (10.4c), a surprising anomaly was observed, which is discussed in the following text.

10.4.1 THE PHENOMENON

In pulse radiolysis experiments, it has been found that in the ET from substituted aromatic donors to parent solvent radical cations, two products are simultaneously formed, for example, the expected donor radical cations as well as (unexpected) radicals derived from them (for phenol, cf. Equation 10.6). Also for diverse additionally substituted phenols, the product ratio was found to be 50:50, which means that it is not simply a deprotonation effect (Mahalaxmi et al., 2000):

$$C_4H_9Cl^{\bullet+} + ArOH \rightarrow C_4H_9Cl + ArOH^{\bullet+}(50\%), \quad ArO^{\bullet}(50\%) + H_{solv.}^{+} \tag{10.6}$$

From an experimental point of view, it should be mentioned that to avoid any disturbing radical anion formation, an internal electron scavenger is used (here, *n*-butyl chloride, BuCl) or an efficient

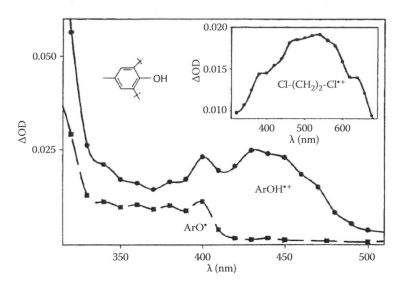

FIGURE 10.6 Transient optical absorption spectra of ArOH$^{\bullet+}$ and ArO$^{\bullet}$ taken 50 ns after the electron pulse on a solution of 4-methyl-2,6-di-*tert*-butylphenol in 1,2-dichlorethane (•). After adding 0.1 mol dm^{-3} ethanol only ArO$^{\bullet}$ (■) remains. The inset shows the spectrum of the 1,2-dichlorethane radical cation taken in the pure solvent.

electron scavenger is added (e.g., CCl$_4$), (cf. reaction 10.7). All the reported effects appear in alkanes as well as alkyl chlorides (Mahalaxmi et al., 2000; Brede et al., 2001a). For practical reasons, in most of the subsequently reported ET experiments, BuCl was preferred as the solvent.

$$RCl + e_{solv.}^{-} \rightarrow R^{\bullet} + Cl^{-} \qquad (10.7)$$

Giving an example from pulse radiolysis at room temperature, Figure 10.6 shows optical absorption spectra representing the products of ET (10.6) (Brede et al., 1996 2002a), using the same sterically hindered phenol as described above for matrix insulation experiments (see Figure 10.4). The spectrum taken immediately after the ET (•) consists of a superposition of both the product species ArOH$^{\bullet+}$ and ArO$^{\bullet}$. As for the identification, a polar additive (ethanol) quenches ArOH$^{\bullet+}$, and ArO$^{\bullet}$ remains. So the phenol transients can be well distinguished. We repeat that both transients appeared from the very beginning.

Using a variety of phenols (Brede et al., 1996, 2001a, 2002a; Mahalaxmi et al., 2000), biphenols (Brede et al., 2002b), naphthols (Mohan et al., 1998; Baidak et al., 2008a), and other chalcogenols (Hermann et al., 2000; Brede et al., 2001b) an analogous behavior has been observed. This was also the case for primary and secondary aromatic amines (Brede et al., 2005; Maroz et al., 2005). As already mentioned, the radical cations of chalcogenols are relatively unstable (very low pK_a values, Dixon and Murphy, 1978) in polar media and, therefore, tend to deprotonate, whereas the radical cations of aromatic amines are persistent (Alkaitis et al., 1975). Nevertheless, the two-product situation exists also in the amine case. This is concluded in Table 10.2 for donors of the type Ar-Y-H, where Y = O, S, Se, N.

Considering the yields of the two kinds of ionization products, it is obvious that the products are synchronously formed in a constant ratio for each class of compounds. This does not depend on steric and electronic effects. But the intramolecular mobility of the substituent seems to be an important factor. In the case of restricted mobility, because of hydrogen bonding (*ortho*-salicylate) and a rigid molecule structure (carbazol), only one product of the ET (10.4c) was found, that is, only donor radical cations were generated.

TABLE 10.2

Product Ratio of the Electron Transfer (10.6), Calculated Times of One Motion of Bending (Wagging) and Valence Vibration (Stretching) Calculated with (B3LYP/6-31G(d))

Electron Donor	$t_{bending}$ [10^{-15} s] Ground	$t_{valence}$ [10^{-15} s] State	E_a, kJ mol^{-1} Singlet (Cation)	Product Ratio $(\cdot+)/(\cdot)$	Remarks
HO—⟨phenyl⟩	97	9.3	12.6 (62)	50:50	Also deuterated: Ar-OD
HO—⟨phenyl⟩—Cl	98	9.3	13.0 (55)	50:50	
HO—⟨phenyl⟩—OMe	110	9.3	10.0 (57)	50:50	
HO—⟨di-tert-butyl-methyl-phenyl⟩	35	9.1	9.8 (46)	50:50	Sterically hindered phenols
HO—⟨phenyl⟩—OH	100	9.3	9.7 (60)	40:60	All biphenols
HO—⟨phenyl⟩—C(=O)OMe	89	9.3	17	50:50	*Para*-salicylate
⟨Me-O···H-O substituted phenyl-R⟩	294	12.8	34 (50)	62:38	Curcumin, weak hydrogen bond
⟨OMe, O=C···H-O substituted phenyl⟩	294	12.8	61	100:0	*Ortho*-salicylate, strong hydrogen bond
HS—⟨phenyl⟩	294	12.8	1.7 (89)	35:65	Various thiophenols
HSe—⟨phenyl⟩	347	14.8	2.9 (82)	10:90	Selenophenol
ArHN—⟨phenyl⟩	608	9.2	13	50:50	Also primary and secondary aromatic amines
⟨carbazole⟩	—	9.1	—	100:0	Carbazol, rigid skeleton

Note: E_a denotes the energy barrier of rotation along the Ar–X axis for the singlet ground state, as well as for the radical cation (in brackets).

10.4.2 Molecule Dynamics

With the above experimental behavior in mind, a reasonable explanation for the two-product phenomenon should be found. In the used nonpolar liquids, solvation (solvent molecule shells) of ions and polarized molecules plays only a minor role. Under normal conditions, the donor molecules are mobile and can move without the influence of a solvent shell. This also holds for the intramolecular mobility. Molecule dynamics takes place in the femtosecond time range and involves different kinds of oscillations: stretching of bonds, bending motions (torsion, wagging, etc.) are simply taken here as rotations.

The rotation of a polarizing group in a molecule of Ar-Ⓞ-Y-H (Y = hetero atom) is accompanied with shifts of the electron density within the highly occupied π- and n-orbitals (Brede and Naumov, 2006). For the donor molecules, this effect can be visualized by quantum chemical calculation and results of the electron density dependent on the angle of deformation. So the critical bond (marked with a rotating arrow) is twisted out of the plane of the stable state (here, 0°) in a stepwise manner. For the phenol molecule, the results of the calculations are given in Figure 10.7. It can be seen that the electron distribution changes in a stepwise manner from an even one (0°) to a localization (90°) on the oxygen hetero atom. Under molecule dynamics conditions, it is implied that with the rotational motion in the femtosecond time range, a continuous change in the electron density in the ground-state molecules takes place. In chemical terms, it means that a dynamic mixture of rotational conformers exists (see feature article Brede and Naumov, 2006).

Focusing the interest on the ET, the elementary reaction (10.4c) can be put into the following steps (reaction sequence 10.8):

$$RX^{\bullet+} + D \underset{\longleftarrow}{\overset{diffusion}{\longrightarrow}} RX^{\bullet+} \cdots D \underset{\longleftarrow}{\overset{e\text{-}jump}{\longrightarrow}} RX \cdots D^{\bullet+} \underset{\longleftarrow}{\overset{diffusion}{\longrightarrow}} RX + D^{\bullet+} \qquad (10.8)$$

Diffusion brings the reactants together. During the encounter of the reactants, the electron jumps to the parent solvent ion, and diffusion separates the products. According to the Born–Oppenheimer approximation, the molecular geometry remains stiff during the electron jump, which is assumed to be an instantaneous and, therefore, extremely rapid process. If the ET is not hindered by a solvent shell and if the jump occurs in the first encounter of the reactants, then the dynamic mixture of donor conformers is chemically noticed. The molecule is ionized in the momentum of a distinct deformation state, and therefore, initially, donor radical cations of different spin distributions are formed (see Born–Oppenheimer approximation). A differing spin (and charge) distribution means that donor cations of different stabilities exist.

Simplifying this complex situation, we distinguish product radical cations into two kinds, namely, metastable and dissociative. The dissociation should occur during the first vibration motion

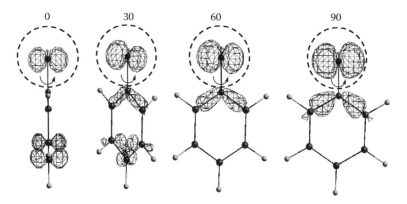

FIGURE 10.7 Transformation of the *n*-orbital of the Ar-OH singlet ground state (isocontour = 0.1) dependent on the angle of the OH-group rotation.

of the critical bond. Otherwise, relaxation immediately forms stable donor cations. Concluding this deduction, as products of the ET (10.4c, 10.8) we distinguish metastable as well as dissociative product radical cations. They are derived from different types of dynamic conformers of the ground-state molecules, distinguished as being near to the plane or twisted. Now, we generalize and rewrite reaction (10.6) and explain it in terms of molecule dynamics (10.9):

$$[\text{Ar-Y-H}]_{\text{plane}} \rightarrow \text{RX} + \textbf{Ar-Y-H}^{\bullet +}_{\text{(metastable)}} \qquad (10.9a)$$

$$\updownarrow$$

$$\textbf{RX}^{\bullet +} + \text{Ar-}\circlearrowleft\text{-Y-H}$$

$$\updownarrow$$

$$[\text{Ar-Y-H}]_{\text{twisted}} \rightarrow [\text{Ar-Y-H}^{\bullet +}]_{\text{(dissociative)}} \rightarrow \text{RX} + \textbf{ArY}^{\bullet} + \text{H}^{+}_{\text{solv.}} \qquad (10.9b)$$

where
  X = H, Cl
  Ar = aromatic ring
  Y = O, S, Se, N
  species in boldface can be directly observed

Because of the elucidated special properties—immediate electron jump and dynamic control—we call this reaction type "free electron transfer." So, FET stands for an electron transfer process where the molecule oscillations of the donor are reflected in a bimolecular reaction. Thus, "free" means *unhindered* and refers to the *transfer* and not to the electron.

10.4.3 ENERGETICS

A critical point of the FET hypotheses is the interpretation of the rapid dissociation of the twisted donor radical cation (reaction 10.9b). For a clear understanding of this point, we made an energetic analysis of the reaction path.

The ET from the neutral ground-state substrate (ArOH) to the solvent cation radical (BuCl$^{\bullet +}$) is energetically favored by about 2.05 eV. For an energetic analysis, the following primary reaction steps should be considered (Brede and Naumov, 2006):

$$\text{ArOH} + \text{BuCl}^{\bullet +} \xrightarrow[\text{e-transfer}]{\Delta H = -1.32\,\text{eV}} \left\{ \text{ArOH}^{\bullet +} + \text{BuCl} \right\}_{\text{unrelaxed}} \xrightarrow[\text{relaxation}]{\Delta H = -0.73\,\text{eV}} \text{ArOH}^{\bullet +} + \text{BuCl} \qquad (10.10)$$

FET occurs between two electronically relaxed reaction partners. After the electron jump, both products, namely, ArOH$^{\bullet +}$ and the solvent molecule (BuCl), are primarily in an unrelaxed state, due to the change in the electronic configuration within the same molecular structure. This first step is exergonic by 1.32 eV. It follows a rapid relaxation via the adjustment of the molecular geometry to the new electron distribution. The relaxation (reorganization) energy of about 0.73 eV is distributed between ArOH$^{\bullet +}$ (0.19 eV) and BuCl (0.53 eV).

Primarily, before reaching the thermal equilibrium, the excess energy of the ET ($\Delta H = -1.32$ eV) is partitioned off the reactants. Considering the proton dissociation energy of about 8.7 eV in the gas phase and 2.6 eV in the liquid state (BuCl as the proton acceptor), the excess energy is far from being enough for achieving deprotonation in the planar state of the phenol radical cation. Because of the strong electron–vibration coupling, the excess electron energy is accepted by its nearest vibrations, first of all by the C-H valence vibrations ($\sim$3200 cm^{-1} = 0.39 eV), and then is dissipated by interaction

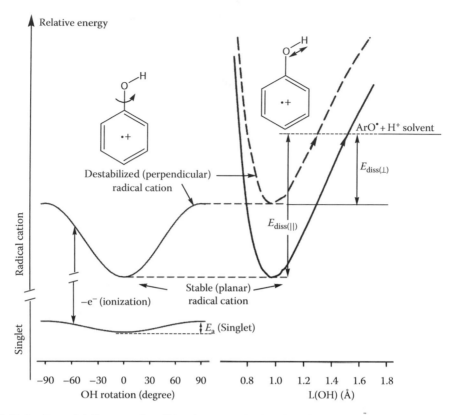

FIGURE 10.8 Potential diagrams describing the energetic situation of phenol as a donor in the ground state and for the different conformer radical cations. The left part shows the potentials dependent on the rotation angle, and the right part illustrates energy changes dependent on the bond length (L) of Ar-OH.

with the surroundings (Brede et al., 1987). However, the additional destabilization of the deformed (twisted) donor radical cations (see reactions 10.9a and 10.9b) helps to overcome easily the dissociation barrier (cf. also Figure 10.8).

Any deprotonation in the sense of self-dissociation is energetically unfavorable. Therefore, for a mechanistic explanation of the prompt deprotonation within FET, we have to take into account the solvent as the proton acceptor and stabilizer (reaction 10.11):

$$\left[\text{Ar-O}^{\bullet+}\text{-H} \right]^{*}_{\text{perpend.}} \xrightarrow{+\text{BuCl}} \text{ArO}^{\bullet} + \text{H}^{+}(\text{BuCl}) \tag{10.11}$$

The fact that the proton affinity of many solvents (Wedenejew et al., 1971), such as butyl chloride, lies just between those of the metastable planar radical cation structure and of the twisted unstable one (see Figure 10.8) fully corroborates our view.

A potential energy diagram concerning the energetics of the solvent-driven deprotonation (10.9b, 10.11) of dissociative cations is displayed in Figure 10.8. The ground-state (singlet) potential is rather flat and shows the energy changes during the rotation of the Ar-Ö-OH group, which amounts to <15 kJ mol^{-1}. After ionization, the potential of the radical cation [Ar-Ö-OH]$^{\bullet+}$ becomes considerably deeper and the rotational motion is hindered by an activation barrier of ≈60 kJ mol^{-1}.

The planar radical cation is metastable and is placed at the bottom of the vibration potential (Figure 10.8, right-hand side). The perpendicular radical cation state, however, has a much higher energy. At distances $L > 1.3$ Å, the potential of the latter crosses the level of the proton affinity of the solvent. This critical length is easily reached during the first vibration motion and, as a consequence, dissociation (deprotonation) takes place.

The energetic data used in the discussion originated from quantum chemical calculations (DFT) using the program B3LYP/6-31G(d) (Becke, 1993, 1996). Although the results of quantum chemical calculations have already been used for rationalizing the experimentally observed unusual product distribution in the FET (of phenols), the calculated parameters should be listed here: The information of interest included, in particular, (a) the dynamic behavior of the donor molecules (frequencies of vibration and bending motions), (b) the electron distribution in the affected orbitals and the consequences of its ionization, and (c) the detailed mechanism of the dissociation of the unstable donor radical cations as far as this is a reflection of the energetics of the reaction and of some structural details (changes in bond lengths, etc.).

10.4.4 GENERALIZATION OF THE FET PHENOMENON

The experimental characteristics of FET consist in the appearance of two different products, which are formed synchronously in the bimolecular ion–molecule reaction (cf. reactions 10.4 and 10.9). These products are derived from two kinds of donor radical cations, one stable and another dissociative. The background is the electron distribution changing in connection with the bending motion of the hetero substituents at the aromatic moiety.

Aromatic molecules with hetero substituents other than those capable of deprotonation should exhibit a comparable dynamic behavior. Hence, FET involving donors with other leaving groups was studied (cf. Table 10.3). The experiments are discussed in the following text.

10.4.4.1 Benzylsilanes

In benzyltrimethylsilane-type compounds (Ar-Ⓞ-CH$_2$SiMe$_3$), the substituent is freely mobile versus the aromatic ring, which causes dynamic changes in the electron distribution similar to those explained above. In this case, however, an inverse electron shift exists insofar that π-electrons from the aromatic ring are shifted toward the substituents. This is different from the phenol situation (Brede et al., 2004; Karakostas et al., 2005) (see Figure 10.9). In particular, changes in the ArCH$_2$– SiMe$_3$ bond have to be taken into account.

For benzylsilanes, FET (10.12) results in the formation of metastable donor radical cations as well as unstable radical cations, which immediately dissociate into benzyl radicals and silyl cations.

$$(10.12)$$

TABLE 10.3
FET with Highly Substituted Aromatic Compounds, Product Ratios According to Reaction (10.9) and Dynamics Data*

Donor	$t_{rot,def}$ [10^{-15} s]	$t_{valence}$ [10^{-15} s]	E_a(S), Singlet, (Cation), kJ mol^{-1}	Product Ratio (•+)/(•)	Remarks
R_3-Ar-CR_1R_2-$SiMe_3$	1330–2380	37	14–24 (57–87)	40:60	$R_{1,2}$ = H, Me, Ph; R_3 = H, PHCO-; 10 compounds
	600	40	—	25:75	Xanthenyl R = Me, Ph; 6 examples
	—	40	—	100:0	Fluorenyl R = Me, Ph; 2 examples
				Ph_3C^+/ ($Ph_3C^•$)*	*Endothermic reaction product
	440	33 O–CPh_3	5.9 (29.3)	90: 10	
	1393	40 S–CPh_3	4.2 (43.9)	65:35	Also Ph-S-CPh_3

* Calculated t-times of one cycle of torsion (wagging) and valence (stretching) ($v_{valence}$) oscillations (B3LYP/6-31G(d), scale factor ($f = 0.96$); E_a(S) denotes the energy barrier of rotation along the Ar–CY axis, and bending motions for the singlet ground state, as well as for the radical cation (in brackets).

In pulse radiolysis experiments, the benzylsilane radical cations ($\lambda \approx 310, 500...650$ nm) and the benzyl radicals ($\lambda \approx 320...350$ nm) are visible, but the trimethylsilyl cations are out of the observation range (Brede et al., 2004). This behavior has been observed for a variety of differently substituted silanes, and thus can be generalized (Brede and Naumov, 2006). Corresponding data are given in Table 10.3. In the case of cyclic substituents (fluorenyl and xanthenyl trimethylsilanes) (Karakostas et al., 2005), a convincing relationship with the flexibility of the silane substituent can be established, which, in turn, may allow a calibration of the ratio between rapidly formed radicals and metastable radical cations. This correlation is in fact one of the main arguments supporting the hypothesis that the products of the FET are a direct reflection of the molecular dynamics. The examples involving benzyltrimethylsilane-type compounds clearly manifest the generality of the FET concept. Concerning the rapid dissociation of the primary radical cation, this means that not only protons, as in phenol, but also any other suitable leaving group may be cleaved instantaneously if the radical cation prevails in the right stereo-electronic state for this process. In the case of the above-mentioned silane compounds, the leaving entity is the trimethylsilyl cation.

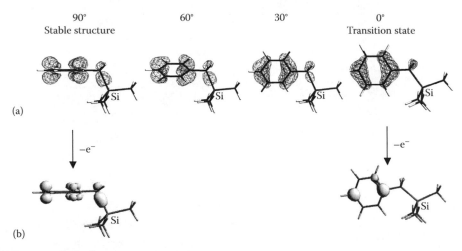

FIGURE 10.9   (a) Transformation of HOMO of the $Ar\text{-}CH_2\text{-}Si(Me)_3$ singlet ground state (isocontour = 0.06) dependent on $C\text{-}SiMe_3$-group rotation and (b) atomic spin density distribution of the $Ar\text{-}CH_2\text{-}SiMe_3$ radical cation (isospin = 0.01) as calculated with B3LYP/6-31G(d).

10.4.4.2 Trityl-Hetero-Substituted Aromatics

So far, the FET examples described were those in which the dissociation pattern of the unstable donor radical cations was uniform and in which only one of the two formed species could be observed, whereas other species, such as protons or silyl cations, were invisible. Using a massive substitution at the hetero atom, a complete identification of the dissociation products could be demonstrated. Hence, we took into account the $Ar\text{-}Y\text{-}CPh_3$ structures with the hetero atoms Y = O, S, N substituted with the trityl rest. The effect of the deformation motion of such molecules is quite similar to that shown for phenol (cf. Figure 10.7) (Baidak et al., 2008b). In the planar resonant structure, the π-electrons are distributed over the whole Ar-Y structure, whereas in the twisted state, the electrons are more localized at the hetero atom. If, in the ground state, the electron density in the twisted conformation is localized on the hetero (sulfur) atom, then the structure is extremely unstable in the ionized form. In contrast, the planar state leads to metastable radical cations.

In the FET experiments with the trityl-substituted compounds, partially unexpected results were obtained. In the case of the example of the highly substituted 2-naphthyltrimethylsulfide, we obtained the complete transient spectrum shown in Figure 10.10.

The metastable (but, nevertheless, relatively short-lived) $2\text{-}Np\text{-}S\text{-}CPh_3^{\cdot+}$ absorbs around $\lambda = 650\,nm$. It decays with $\tau_{1/2} \approx 200\,ns$. Furthermore, naphthylthiyl radicals (500 nm) and the well-known trityl species (Faria and Steenken, 1990) $^+CPh_3$ (410 nm) and $^{\cdot}CPh_3$ (330 nm) are observed. The trityl species can be easily identified because of their high extinction coefficients and the marked peaks. The appearance of the trityl cation is expected because of the exothermicity of the dissociation of the C–S bond, whereas the observation of the trityl radical is astonishing because of the highly endothermic dissociation energy (see reactions 10.13a and 10.13b). Table 10.3 sums up the relevant data.

$$\rightarrow 2\text{-}NpS^{\cdot} + {}^+CPh_3 \quad \Delta H = -10 \text{ kJ mol}^{-1}, 65\% \text{ yield} \tag{10.13a}$$

$$2\text{-}Np - S\text{-}CPh_3^{\cdot+} \rightarrow$$

$$\rightarrow 2\text{-}NpS^+ + {}^{\cdot}CPh_3 \quad \Delta H = 176 \text{ kJ mol}^{-1}, 35\% \text{ yield} \tag{10.13b}$$

An analogous effect has been found for trityl-phenylsulfide (Karakostas et al., 2007) and also for trityl-phenyloxide (Baidak et al., 2008b) (see Table 10.3). For the example of the sulfide $Np\text{-}S\text{-}CPh_3$, the FET reaction scheme should be adapted (reaction 10.14). In this reaction sequence, it is assumed

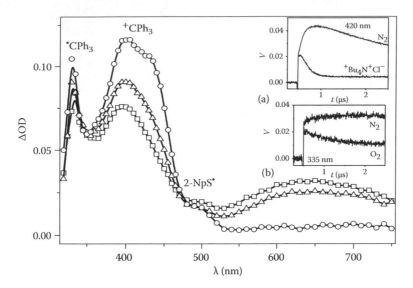

FIGURE 10.10 Transient optical absorption spectra obtained in the pulse radiolysis of a N_2-saturated solution of 2-NpSCPh$_3$ (2×10^{-3} mol dm^{-3}) in BuCl. Spectra correspond to times (□) 45 ns, (△) 140 ns, and (○) 700 ns after the pulse. Experimental time profiles demonstrating the influence of (a) Bu$_4$N$^+$Cl$^-$ and (b) O$_2$ on the kinetic behavior of the transients are given in insets.

that the unusual dissociation products are formed by the prompt dissociation of the twisted radical cation. The observed species are given in boldface:

$$[Np\text{-}S\text{-}CPh_3]_{planar} \rightarrow \underset{metastable}{Np\text{-}S\text{-}CPh_3^{\cdot\,+}} \underset{delayed}{\rightarrow\rightarrow\rightarrow} \mathbf{NpS^{\cdot}} + {}^+\mathbf{CPh_3} \qquad (10.14a)$$

$$\updownarrow$$

$$Np\text{-}S\text{-}\overset{\cdot}{O}\text{-}CPh_3 + RX^{\cdot\,+} RX$$

$$\updownarrow$$

$$[Np\text{-}S\text{-}CPh_3]_{twisted} \rightarrow \underset{dissociative}{[Np\text{-}S\text{-}CPh_3^{\cdot\,+}]} \rightarrow \underset{immediately}{\mathbf{NpS^+} + {}^{\cdot}\mathbf{CPh_3}}, \ (NpS^{\cdot} + {}^+CPh_3) \qquad (10.14b)$$

The assignment of the endergonic products to the dissociation of the twisted radical cation needs an energetic justification, which is given with the help of a potential energy scheme (see Figure 10.11).

The low-lying potential describes the energetics of the ground state—the planar state in the valley, and the twisted state on the wall sides. After ionization, the upper deep potential valley is reached, either in the stable state (black point) or in the twisted situation placed at the walls. The metastable radical cation (block circle) can undergo *delayed fragmentation*, which is an exergonic process. The twisted radical cation on the top of the wall (open circle), however, will undergo an immediate endergonic fragmentation. Some arguments support this hypothesis:

1. Kinetic control instead of the thermodynamic one: the vibration motion of the critical bond is more than one order of magnitude faster than the rotation (Table 10.3). Hence, dissociation happens within the first vibration motion and yields unfavorable products.
2. The cation affinity of the solvent BuCl decreases the dissociation energy compared to gas-phase conditions, which are used for calculations.
3. FET is a very exergonic process, which also causes vibration excitation.

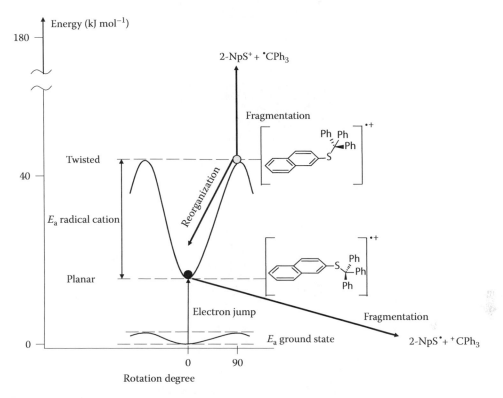

FIGURE 10.11 Potential energy diagram describing energetic profiles of the planar and twisted radical cation conformers and subsequent fragmentation reactions.

Considering all of these factors, an immediate fragmentation can explain the energetically unfavorable products, such as NpS$^+$ and $^\bullet$CPh$_3$. Clearly, rapid relaxation (reorganization) to the metastable radical cation can also happen. Because of the excess energy, this would result in an immediate dissociation in the energetically favorable channel, forming NpS$^\bullet$ and $^+$CPh$_3$, in (10.14b) given in brackets. Hence, the yield of the unusual dissociation products does not give a quantitative measure, but is a crucial information about the FET mechanism.

10.4.5 DRIVING FORCE OF FET

The energetic aspects of the dissociation of the unstable donor radical cations (reaction path 10.9b) have already been discussed. Now, the dependence of the ET on the free energy of the gross reaction (10.4c) will be analyzed. Here, we use the difference between the IPs of the nonpolar solvent and the donor (ΔIP) as a measure of ΔG. In earlier ET studies, we observed the expected dependence of the rate constants on ΔIP (see Figure 10.12) (Mehnert et al., 1985).

The investigations were performed without product control, because only the influence of the added donors on the time profiles of the primary solvent radical cations was evaluated to determine rate constants. The rate constants were normalized on the diffusion-controlled limit for the particular solvent. Although interpreted with another theoretical model (Brede et al., 1987), the plots show the typical shape of the well-known Marcus plots (Eberson, 1982). This macroscopic behavior can be understood in terms of the usual sum kinetics. From the plot, it can be seen that the ET shows activation-controlled rate constants up to ΔIP ≤ 0.3 eV. Then the diffusion-controlled plateau is reached.

Now, we vary the ΔIP values in another way, that is, by *stepwise ET*. The concept is that primary stable aromatic radical cations should be formed, which could then be used as electron acceptors for an aromatic sulfide. This is shown in the reaction sequence (10.15):

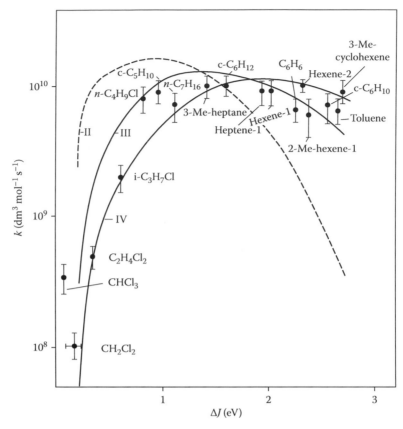

FIGURE 10.12 Dependence of the rate constant, k_{ET}, on ΔIP. The donors are indicated. Electron acceptors were different alkane or alkyl chloride parent radical cations. Solid and dotted lines are calculated for different deformation energies, as described by Brede et al. (1987).

$$BuCl^{\bullet+} + C_6H_6 \rightarrow BuCl + C_6H_6^{\bullet+} \tag{10.15a}$$

$$C_6H_6^{\bullet+} + Ar\text{-}S\text{-}CPh_3 \rightarrow C_6H_6 + \textbf{products} \tag{10.15b}$$

So, the ΔIP value has been varied from about 3 eV down to 0.45 eV. As a criterion for judging the reaction course, we looked for the product pattern derived from the sulfide, that is, whether trityl cations as well as trityl radicals were formed (Karakostas et al., 2007). This concept is illustrated in Figure 10.13.

As a result of the ET from the sulfide to the radical cations of butyl chloride, benzene, or butylbenzene, both trityl species were found in each case. The ET from biphenyl to the sulfide, however, resulted only in trityl cations. This led to a conclusion that in the first case, the FET took place with an electron jump occurring immediately after the first collision of the partners. In the second case, it is hypothesized that in the ion–molecule reaction, an encounter complex has been passed, and, as a consequence, only the way via the metastable intermediate sulfide radical cation was followed (Karakostas et al., 2007; Baidak et al., 2008b). Here, *metastable* refers to a short-lived species, but not a dissociative one, which would be formed in femtoseconds. These findings are also in accord with stepwise ET experiments using phenol as the final donor and monitor (Brede et al., 2003). One can conclude that at ΔIP below 0.5 eV, a process similar to a classical diffusion-controlled ET happens, that is, an ET with a high rate but within an encounter complex situation.

In summary, the following can be drawn: (a) FET with its typical product situation takes place at free energies of the gross reaction of $-\Delta G$ higher than 0.5 eV. (b) Below this value, the gross reaction

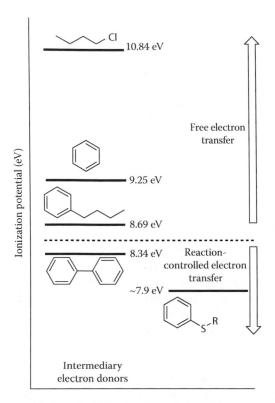

FIGURE 10.13   The scheme symbolizes the ET under the variation of free energy of the reaction by stepwise ET (see reactions 10.15a and 10.15b).

also exhibits diffusion-controlled rate constants ($k \approx 10^{10}\,\mathrm{dm^3\,mol^{-1}\,s^{-1}}$), but seems to proceed via an encounter complex. (c) At energies of $-\Delta G$ below 0.3 eV, activation-controlled rate constants are observed.

10.4.6   RELATION BETWEEN MOLECULE DYNAMICS AND REACTION KINETICS

The dynamic electron shifts in the donor orbitals documented above (Figures 10.7 and 10.9) do not play a role in common bimolecular chemical reactions. In normal ET reactions, the timescale is usually longer than a nanosecond (this is the time range of diffusion). However, in the exotic case, where the reaction-determining step is faster than the bending (rotation) motions, the discrete conformations of the deformed molecules in their momentary states strongly influence the reaction. This can be experimentally observed, because different products are formed for different encounters. It involves the assumption that the deciding step of the reaction happens in the first collision of the reaction partners, which also means that no defined encounter complex is passed.

Generally, an encounter complex is defined as an entity formed after the approach of the reaction partners, in which a more or less rigid and energetically optimized structure is reached (Weller, 1982; Klessinger and Michl, 1995; Rosokha and Kochi, 2008). In this structure, for common ET processes (one-electron oxidation, photoinduced triplet- or singlet-sensitized ET, proton-coupled ET, etc.), a process similar to a time-delayed, but rapid, electron jump from the donor part to the acceptor happens (Klessinger and Michl, 1995). These ET processes follow the usual sum and equilibrium kinetics placed in the time range ≥ 1 ns.

For the analysis of FET, however, a dynamically controlled regime has to be used instead of the mentioned description as an elementary reaction with thermodynamic parameters. Understanding the FET as a reaction of a *diversity of (rotational) conformers* (instead of one uniform molecule type), which by the electron jump forms a *diversity of ionization products*, replaces the commonly used picture of *mesomerism* (Beyer and Walter, 1991) of substituted aromatic molecules in this case. If at all such a mesomerism exists, then it is a property of the encounter complex, for which a rigid and energetically favored structure should exist.

The main feature of FET consists in the fact that the femtosecond dynamic motions of (donor) molecules are reflected in a diffusion-controlled reaction taking place at the nanosecond timescale. Therefore, with the used time-resolved spectroscopy, a steady state observation of the transients, which appear in nanoseconds (see bimolecular reaction), is performed. However, their structure is determined by femtosecond dynamics. Insofar, the FET is a new phenomenon of reaction kinetics.

10.4.7 COMPARISON OF ET THEORIES

Here, the classical ET models are sketched briefly, mainly under the aspect of the distribution of excess energy of ET.

As already mentioned, common interpretations of ET reactions are based on sum kinetics and use equilibria for the energetic description (cf. reaction 10.8). To describe the dependence of the ET transfer rate constant, k_s (rate of the gross reaction), on the free energy change, ΔG°, two empirical attempts formulated by Marcus (Marcus, 1964; Marcus and Sutin, 1964) and alternatively by Rehm and Weller (1969) are used. In both models, the excess energy is distributed either within the reactant molecules or in the solvent surroundings. In polar media, this reorganization energy concerns primarily the solvent rearrangement, taking into account mainly excitation of the low-energy modes of the solvent.

In contrast to the situation discussed above, for nonpolar media, the solvent reorganization energy is expected to play a minor role, and, hence, the reorganization of high-frequency intramolecular vibrational modes will determine the ET rate. Taking into account the quantum nature of these vibrations, the description of k_s was modified. Using the concept of Ulstrup and Jortner (1975), the ET in alkanes and alkyl chlorides (Brede et al., 1987) has been described. Taking C-H modes of the solvent radical cation/solute couple, and treating the solvent part in a classical way as proposed by Levich and Dogonadze (1961), an expression for k_s could be obtained.

In principle, the FET phenomenon can use parts of both theories discussed above. However, the theory developed for nonpolar systems is more appropriate (Brede et al., 1987). It should be noted that in FET, molecular dynamics plays a decisive role (Brede and Naumov, 2006) compared to the classical descriptions of ET, which are based on a kinetic analysis using thermodynamic data. The FET process, however, can only be understood in terms of intramolecular mobility (vibrations and bending motions) of the donor. With the peculiarity of a prompt electron jump taking place in the first approach of the reaction partners, instead of one type of donor molecules, we have to take into account a dynamic diversity of conformers with different electron distributions in the momentum of the electron jump. This leads to the distinction of two types of rotation conformers. These conformers react with solvent radical cations, forming at first donor radical cations of different stability (dissociative and metastable), which finally result in different products (fragments and metastable radical cations). Most of these events occur in the femtosecond time domain (dissociation), or even faster (electron jump). Considering bimolecular FET, a kinetic paradox is observed, that is, instead of the slowest step, the fastest one determines the reaction path.

Overall, FET is an electron transfer process that is governed by molecular dynamics. Apart from the unusual "two-product" situation, it is also noteworthy that the endergonic fragmentation products derived from the dissociative radical cations do not follow the rules of the classical energetic analysis (Baidak et al., 2008a).

To illustrate the discussion on the time range of FET and common ETs, in Figure 10.14, relevant processes are assigned to a logarithmic timescale. This should visualize the crucial distinction

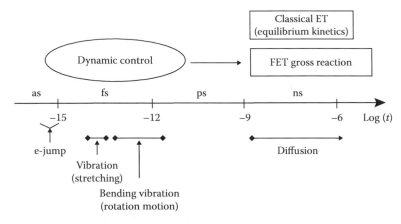

FIGURE 10.14 Time correlation of the ET processes discussed in this chapter.

between ET sum kinetics under equilibrium conditions on one side and the dynamically controlled FET on the other side. Because both ET processes are bimolecular, the observation is placed in the nanosecond real-time spectroscopy, which is an adequate technique for studying such diffusion-controlled reactions. Concerning the dynamics of the femtosecond events mentioned above, the investigations were carried out under quasi steady state aspects.

10.5 TRANSFORMATION OF DONOR RADICAL CATIONS IN FREON MATRICES

10.5.1 REACTION PRINCIPLES

Rapid transformations of donor radical cations formed by ET to freon parent ions (reaction 10.3c) can be observed by matrix insulation studies in electron-irradiated frozen freon matrices at low temperature. Under these conditions, FET cannot take place because of the very limited mobility of the donor molecule. For this purpose, low-temperature EPR measurements in combination with quantum chemical calculations were performed (Knolle et al., 1999; Janovský et al., 2003; Naumov et al., 2003a,b,c,d). Depending on the experimental conditions, the transformation of the primary radical cations proceeds via monomolecular or bimolecular reactions. At high solute concentrations and in the $CFCl_2CF_2Cl$ (freon F-13) matrix, the transformations often happen by intermolecular H-abstraction or proton transfer from the parent molecule, forming a deprotonated radical and a carbocation, or performing an addition reaction, which leads to a dimer radical cation (Naumov et al., 2003b; Janovský et al., 2005). The dimer radical cation, which could be a distonic one, characterized by separation of charge and radical site, can react further through intermolecular H-abstraction from the parent molecule, forming a dimer carbocation and a radical (Naumov et al., 2003c, 2005). Subsequently, the dimer radical cation can perform an addition reaction with the parent molecule, forming a trimer radical cation, etc.

10.5.2 INTRAMOLECULAR TRANSFORMATIONS

At low solute concentrations and using a CF_3CCl_3 matrix (which generally does not form complexes with the solute cations), the transformation can proceed in a monomolecular way, such as intramolecular H-shift, ring closing (opening), conformational changes, and decomposition. Here, some examples of monomolecular rearrangements of donor radical cations by intramolecular H-shift and ring closing are demonstrated.

The reaction pathways of experimentally observed intramolecular transformation of the primary radical cations of 2,5-dihydrofuran (2,5-DHF) (Janovský et al., 1999; Naumov et al., 2003a,d) and

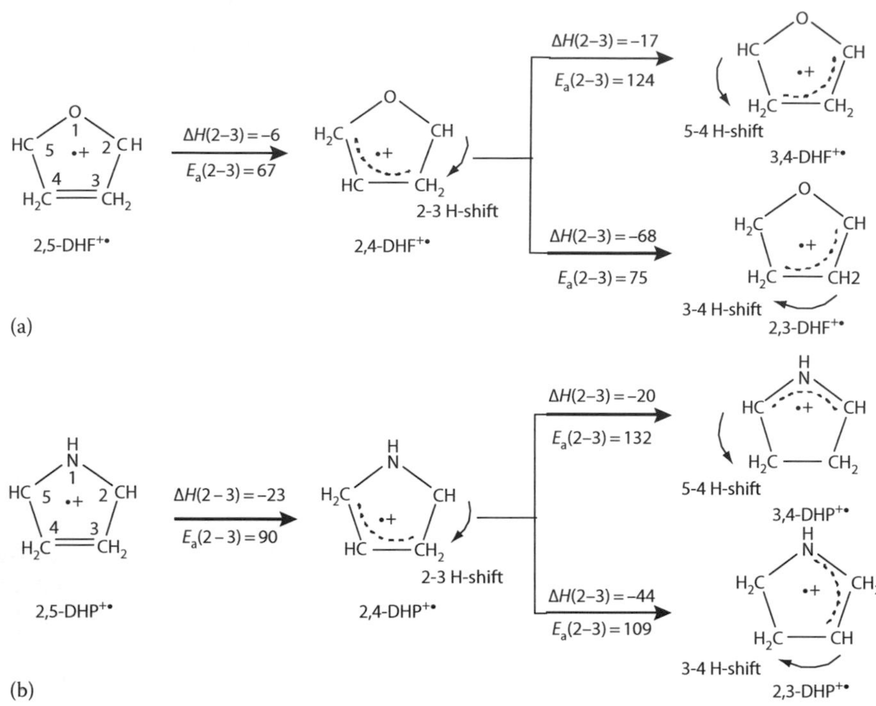

FIGURE 10.15 Experimentally observed intramolecular transformations of primary (a) 2,5-DHF$^{+•}$ and (b) 2,5-DHP$^{+•}$ through the H-shift within the molecular ring. ΔH and E_a are the calculated reaction enthalpies and the activation energies (kJ mol^{-1}), respectively, for the H-shift.

of 2,5-dihydropyrrole (2,5-DHP) (Naumov et al., 2004a) are shown in Figure 10.15. The primary oxygen-centered 2,5-DHF$^{+•}$ transforms into 2,4-DHF$^{+•}$, which is stable at 77 K. A subsequent transformation of 2,4-DHF$^{+•}$ can be induced by illumination with visible light. This process is an intramolecular H-shift, leading to the formation of 2,3-DHF$^{+•}$ and 3,4-DHF$^{+•}$, the latter being a newly identified species, stable in the range of 77–145 K, characterized by hfs (hyperfine splitting constants) a(2H) = 1.59 mT and a(4H) = 2.82 mT.

This transformation pattern was observed also in the case of 2,5-DHP$^{+•}$ (Figure 10.15b). The primary nitrogen-centered radical cation, 2,5-DHP$^{+•}$, in freon matrices is relatively stable (in CF$_3$CCl$_3$ up to 140 K); however, on illumination with visible light of wavelength $\lambda > 600$ nm, it is transformed into 2,4-DHP$^{+•}$ by an intramolecular 5 $\rightarrow$ 4 H-shift. The latter is stable up to 143 K, but its further intramolecular rearrangements through 2 $\rightarrow$ 3 and 4 $\rightarrow$ 3 H-shifts can be induced by illumination with visible light of shorter wavelengths (400 nm < λ < 600 nm). Both transformations proceed simultaneously, with the same yields of the two isomers of the DHP radical cation, namely, 2,3-DHP$^{+•}$ and 3,4-DHP$^{+•}$. These transformations are most feasible also due to the strong localization of the unpaired electron in position 4 in 2,4-DHP$^{+•}$.

The thermal and photochemical transformations of primary amine radical cations generated radiolytically in freon matrices have been comprehensively investigated using low-temperature EPR spectroscopy (Janovský et al., 2004; Knolle et al., 2006). The primary nitrogen-centered amine radical cations are very unstable even at 77 K and undergo a monomolecular transformation through an H-shift into a more stable structure (Figure 10.16). It should be noted that, in agreement with the calculated high activation energy, a rearrangement of the primary nitrogen-centered amine radical cations by a hypothetical H-atom shift from C1 to the ionized NH$_2$ group was not observed. The H-shift goes in the direction of the lower barrier.

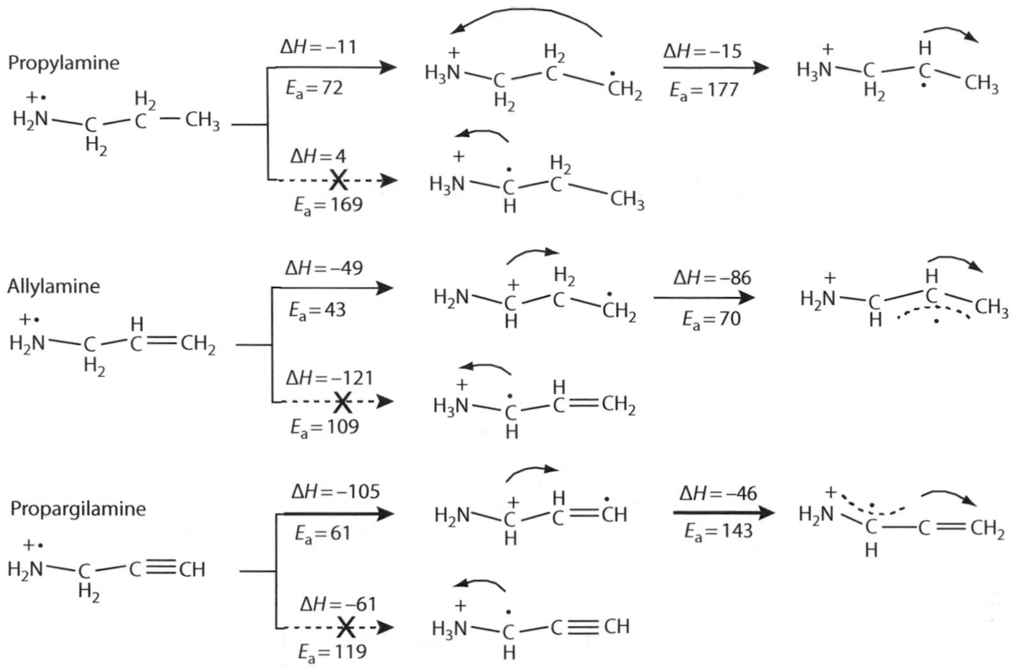

Propylamine

$\overset{+\bullet}{H_2N} - \underset{H_2}{C} - \overset{H_2}{C} - CH_3$

$\Delta H = -11$
$E_a = 72$

$\overset{+}{H_3N} - \underset{H_2}{C} - \overset{H_2}{C} - \overset{\bullet}{C}H_2$

$\Delta H = -15$
$E_a = 177$

$\overset{+}{H_3N} - \underset{H_2}{C} - \overset{H}{\underset{\bullet}{C}} - CH_3$

$\Delta H = 4$
$E_a = 169$ ✗

$\overset{+}{H_3N} - \underset{H}{\overset{\bullet}{C}} - \overset{H_2}{C} - CH_3$

Allylamine

$\overset{+\bullet}{H_2N} - \underset{H_2}{C} - \overset{H}{C} = CH_2$

$\Delta H = -49$
$E_a = 43$

$H_2N - \underset{H}{\overset{+}{C}} - \overset{\bullet}{C} - \overset{H_2}{C}H_2$

$\Delta H = -86$
$E_a = 70$

$H_2N - \underset{H}{\overset{+}{C}} - \overset{H}{C} \cdots CH_3$

$\Delta H = -121$
$E_a = 109$ ✗

$\overset{+}{H_3N} - \underset{H}{\overset{\bullet}{C}} - \overset{H}{C} = CH_2$

Propargilamine

$\overset{+\bullet}{H_2N} - \underset{H_2}{C} - C \equiv CH$

$\Delta H = -105$
$E_a = 61$

$H_2N - \underset{H}{\overset{+}{C}} - \overset{H}{C} = \overset{\bullet}{C}H$

$\Delta H = -46$
$E_a = 143$

$H_2N - \underset{H}{\overset{+}{C}} - C = CH_2$

$\Delta H = -61$
$E_a = 119$ ✗

$\overset{+}{H_3N} - \underset{H}{\overset{\bullet}{C}} - C \equiv CH$

FIGURE 10.16 Experimentally observed intramolecular transformations of amine radical cations through the intramolecular H-shift. ΔH and E_a are the calculated reaction enthalpy and the activation energy (kJ mol^{-1}), respectively, for the H-shift.

A further monomolecular rearrangement of a primary donor radical cation was studied in the case of lactones of different sizes (Naumov et al., 2004b). While the primary radical cation of a four-membered ring β-butyrolactone is unstable at 77 K and undergoes ring opening and fragmentation, ending at the deprotonated neutral (CH$_2$CHCH$_2$)$^\bullet$ radical, the primary radical cations of the five-, six-, and seven-membered ring lactones could be directly observed. Here, the transformation of the primary carbonyl-centered radical cation proceeds by an intramolecular H-shift to the carbonyl oxygen (Figure 10.17). Calculations as well as experiments show that the H-shift occurs always from C1 in the α-position to the carbonyl group. The stability of the primary radical cation toward transformation depends on the ring size, which determines the activation energy of the transformation and the distance, L(H–O), of the carbonyl oxygen from the nearest H-atom on the ring. The larger the ring, the smaller the L(H–O) and also the activation energy of the H-shift, making the transformation of the primary radical cation more feasible.

The transformations of radical cations of ethyl acrylate (EtA) radiolytically generated in freon (CFC$_{12}$CF$_2$Cl, CF$_3$CC$_{13}$, and CFC$_{13}$) and in argon matrices were investigated using low-temperature EPR spectroscopy (Knolle et al., 2002). The primary cation of EtA could be trapped below 40 K in a CFCl$_3$ matrix only. In all the freon matrices studied, the intramolecular rearrangement of this cation occurs via the H-shift from the CH$_3$ group to carbonyl oxygen, leading to the formation of a distonic radical cation (Figure 10.18).

Although the calculated reaction enthalpy for the H-shift from the CH$_2$ group to carbonyl oxygen is larger than that for the H-shift from the CH$_3$ group, the reaction proceeds in the direction of the lower energetic barrier, apparently due to the more favorable six-ring transition structure. In the case of the CF$_3$CCl$_3$ matrix, warming the samples to temperatures above 130 K results in the simultaneous formation of two new species, which were assigned to six- and five-member ring structures (a(H)/mT: 2.36 (Hα), 5.13 (Hβ1), 1.9 (Hβ2), and 2.27 (2Hα) and 2.6 (Hβ), respectively) formed by intramolecular cycloaddition (ring closing) of the terminal radical to the vinyl double bond.

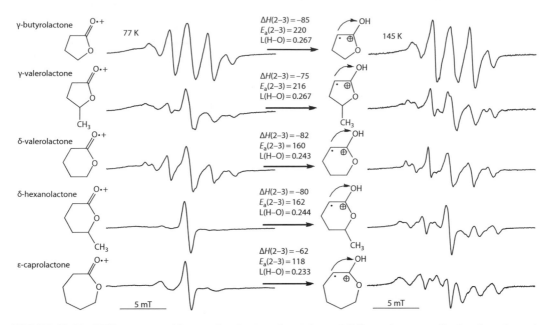

FIGURE 10.17 EPR spectra and intramolecular transformations of different lactone radical cations through the H-shift from C^1 in the α-position on the ring to carbonyl oxygen. ΔH and E_a are the calculated reaction enthalpy and the activation energy (kJ mol^{-1}), respectively, for the intramolecular H-shift; L(H–O) is the distance (nm) between a H-atom on C^1 and carbonyl oxygen in a primary radical cation.

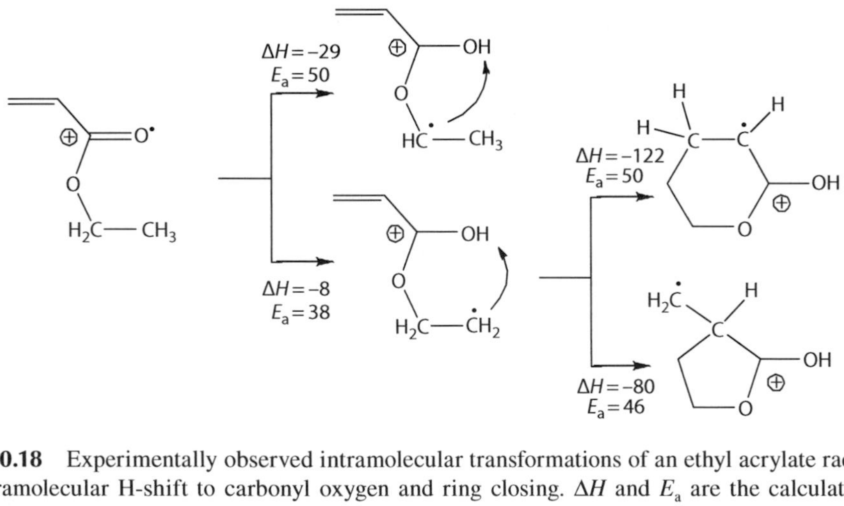

FIGURE 10.18 Experimentally observed intramolecular transformations of an ethyl acrylate radical cation via an intramolecular H-shift to carbonyl oxygen and ring closing. ΔH and E_a are the calculated reaction enthalpy and the activation energy (kJ mol^{-1}), respectively.

These few examples demonstrate that the radical cations produced directly from functional neutral molecules by ionization are highly reactive. Under selected experimental conditions (absence of reaction partners and choice of a matrix preventing bimolecular reaction), they can undergo various intramolecular transformations, leading to more stable radical cation structures, which may not exist as parent ground-state molecules. The rearrangement can occur through an intramolecular H-shift and ring closing. The calculations of the thermochemical parameters of possible reaction pathways and electronic parameters of different structures are often helpful for a better understanding of the studied mechanism.

REFERENCES

Alkaitis, S. A., Beck, G., and Grätzel, M. 1975. Laser photoionization of phenothiazine in alcoholic and aqueous micellar solution. Electron transfer from triplet states to metal ion acceptors. *J. Am. Chem. Soc.* 97: 5723–5729.

Allen, A. O. and Holroyd, R. A. 1974. Chemical reaction rates of quasi free electrons in nonpolar liquids. *J. Phys. Chem.* 78: 796–803.

Arai, S., Kira, A., and Imamura, M. 1976. Mechanism of the formation of cationic species in the radiolysis of butyl chlorides. 1. *J. Phys. Chem.* 80: 1968–1973.

Baidak, A., Naumov, S., Hermann, R., and Brede, O. 2008a. Ionization of amino-, thio- and hydroxyl-naphthalenes via free electron transfer. *J. Phys. Chem. A* 112: 11036–11043.

Baidak, A., Naumov, S., and Brede, O. 2008b. Kinetic and energetic analysis of the free electron transfer. *J. Chem. Phys. A* 112: 10200–10209.

Baxendale, J. H. and Busi, F. (eds.) 1982. *The Study of Fast Processes and Transient Species by Electron Pulse Radiolysis.* Dordrecht, the Netherlands: D. Reidel Publishing Company.

Baxendale, J. H., Bell, C., and Wardman, P. 1973. Observations on solvated electrons in aliphatic hydrocarbons at room temperature by pulse radiolysis. *J. Chem. Soc. Faraday Trans. I* 69: 776–786.

Beck, G. 1979. Transient conductivity measurements with subnanosecond time resolution. *Rev. Sci. Instrum.* 50: 1147–1151.

Becke, A. D. 1993. Density-functional thermochemistry. III. The role of exact exchange. *J. Chem. Phys.* 98: 5648–5652.

Becke, A. D. 1996. Density-functional thermochemistry. IV. A new dynamical correlation functional and implications for exact-exchange mixing. *J. Chem. Phys.* 104: 1040–1046.

Beyer, H. and Walter, W. (eds.) 1991. *Lehrbuch der Organischen Chemie*, pp. 486–491. Stuttgart, Germany: S. Hirzel, or any other textbook of Organic Chemistry.

Brede, O. and Naumov, S. 2006. Femtodynamics reflected in nanoseconds: Bimolecular free electron transfer in non-polar media. *J. Phys. Chem. A* 110: 11906–11918, feature article.

Brede, O., Helmstreit, W., and Mehnert, R. 1974. Fast formation of aromatic cations in cyclohexane observed by pulse radiolysis. *Chem. Phys. Lett.* 28: 43–45.

Brede, O., Bös, J., and Mehnert, R. 1980. Pulse radiolytic investigation of the radiation-induced cationic primary processes in carbon tetrachloride. *Ber. Bunsenges. Phys. Chem.* 84: 63–68.

Brede, O., Bös, J., Helmstreit, W., and Mehnert, R. 1982. Primary processes of the radiation-induced polymerization of aromatic olefins studied by pulse radiolysis. *Radiat. Phys. Chem.* 19: 1–15.

Brede, O., Mehnert, R., and Naumann, W. 1987. Kinetics of excitation and charge transfer reactions in nonpolar media. *Chem. Phys.* 115: 279–296.

Brede, O., Orthner, H., Zubarev, V., and Hermann, R. 1996. Radical cations of sterically hindered phenols as intermediates in radiation-induced electron transfer processes. *J. Phys. Chem.* 100: 7097–7105.

Brede, O., Beckert, D., Windolph, C., and Göttinger, H. A. 1998. One electron oxidation of sterically hindered amines to nitroxyl radicals: Intermediate amine radical cations, aminyl, α-aminoalkyl and aminylperoxyl radicals. *J. Phys. Chem.* 102: 1457–1464.

Brede, O., Mahalaxmi, G. R., Naumov, S., Naumann, W., and Hermann, R. 2001a. Localized electron transfer in non-polar solution: reaction of phenols and thiophenols with free solvent radical cations. *J. Phys. Chem. A* 105: 3757–3764.

Brede, O., Hermann, R., Naumov, S., and Mahal, H. S. 2001b. Discrete ionization of two rotation-conditioned selenophenol conformers by rapid electron transfer to solvent parent radical cations. *Chem. Phys. Lett.* 350: 165–172.

Brede, O., Hermann, R., Naumann, W., and Naumov, S. 2002a. Monitoring of the hetero-group twisting dynamics in phenol-type molecules via different characteristic free-electron-transfer products. *J. Phys. Chem. A* 106: 1398–1405.

Brede, O., Kapoor, S., Mukherjee, T., Hermann, R., and Naumov, S. 2002b. Diphenol radical cations and semiquinone radicals as direct products of the free electron transfer from catechol, resorcinol and hydroquinone to parent solvent radical cations. *Phys. Chem. Chem. Phys.* 4: 5096–5104.

Brede, O., Naumov, S., and Hermann, R. 2003. Monitoring molecule dynamics by free electron transfer. *Radiat. Phys. Chem.* 67: 225–230.

Brede, O., Hermann, R., Naumov, S., Perdikomatis, G. P., Zarkadis, A. K., and Siskos, M. G. 2004. Free electron transfer mirrors rotational conformers of substituted aromatics: Reaction of benzyltrimethylsilanes with n-butyl chloride parent radical cations. *Phys. Chem. Chem. Phys.* 6: 2267–2275.

Brede, O., Maroz, A., Hermann, R., and Naumov, S. 2005. Ionization of cyclic aromatic amines by free electron transfer: Products are governed by molecule flexibility. *J. Phys. Chem. A* 109: 8081–8087.

Dixon, W. T. and Murphy, D. 1978. Determination of acid dissociation constants of some phenol radical cations. *J. Chem. Soc. Faraday Trans. II* 74: 432–439.

Eberson, L. 1982. Electron-transfer reactions in organic chemistry. *Adv. Phys. Org. Chem.* 18: 79–185.

Faria, J. L. and Steenken, S. 1990. Photoionization (.lambda. = 248 or 308 nm) of triphenylmethyl radical in aqueous solution. Formation of triphenylmethyl carbocation. *J. Am. Chem. Soc.* 112: 1277–1279.

Fartahaziz and Rodgers, M. A. J. (eds.) 1987. *Radiation Chemistry, Principles and Applications*. Weinheim, Germany: VCH Publishers.

Földiak, G. (ed.) 1981. *Radiation Chemistry of Hydrocarbons*. Amsterdam, the Netherlands: Elsevier.

Hamill, W. H. 1968. *Radical Ions*, Kaiser, E. T. and Kevan, L. (eds.), p. 321. New York: Interscience Publishers.

Hermann, R., Dey, G. R., Naumov, S., and Brede, O. 2000. Thiol radical cations and thiyl radicals as direct products of the free electron transfer from aromatic thiols to n-butyl chloride radical cations. *Phys. Chem. Chem. Phys.* 2: 1213–1220.

Hummel, A. and Schmidt, W. F. 1974. Ionization of dielectric liquids by high energy radiation studied by means of electrical conductivity measurements. *Radiat. Res. Rev.* 5: 199–300.

Infelta, P. P., de Haas, M. P., and Warman, J. M. 1977. The study of the transient conductivity of pulse irradiated dielectric liquids on a nanosecond timescale using microwaves. *Radiat. Phys. Chem.* 10: 353–365.

Janovský, I., Naumov, S., Knolle, W., and Mehnert, R. 1999. Low-temperature EPR study of radical cations of 2,5- and 2,3-dihydrofuran and their transformations in a freon matrices. *J. Chem. Soc. Perkin Trans.* 2: 2447–2453.

Janovský, I., Naumov, S., Knolle, W., and Mehnert, R. 2003. Investigation of the molecular structure of radical cation of s-trioxane: Quantum chemical calculations and low-temperature EPR results. *Rad. Phys. Chem.* 67: 237–241.

Janovský, I., Knolle, W., Naumov, S., and Williams, F. 2004. EPR studies of amine radical cations: I. Thermal and photo-induced rearrangements of n-alkylamine radical cations to their distonic forms in low-temperature freon matrices. *Chem. Eur. J.* 10: 5524–5534.

Janovský, I., Naumov, S., Knolle, W., and Mehnert, R. 2005. Radiation-induced polymerisation of 2,3-dihydrofuran: Free-radical or cationic mechanism? *Rad. Phys. Chem.* 72(2–3): 125–133.

Karakostas, N., Naumov, S., Siskos, M. G., Zarkadis, A. K., Hermann, R., and Brede, O. 2005. Free electron transfer from xanthenyl- and fluorenylsilanes (Me-3 or Ph-3) to parent solvent radical cations: Effect of molecule dynamics. *J. Phys. Chem. A* 109: 11679–11686.

Karakostas, N., Naumov, S., and Brede, O. 2007. Ionization of aromatic sulfides in non-polar media: Free vs. reaction-controlled electron transfer. *J. Phys. Chem. A* 111: 71–78.

Klessinger, M. and Michl, J. 1995. *Excited States and Photochemistry of Organic Molecules*, p. 278, 285, 291. Weinheim, Germany: VCH Publishers.

Knight, L. B. 1986. ESR investigations of molecular cation radicals in neon matrices at 4 K: Generation, trapping, and ion-neutral reactions. *Acc. Chem. Res.* 19: 313–321.

Knolle, W., Feldman, V. I., Janovský, I., Naumov, S., Mehnert, R., Langguth, H., Sukhov, F. F., and Orlov, A. Yu. 2002. EPR study of methyl and ethyl acrylate radical cations and their transformations in low-temperature matrices. *J. Chem. Soc. Perkin Trans.* 2: 687–699.

Knolle, W., Janovský, I., Naumov, S., and Mehnert, R. 1999. Low-temperature EPR study of radical cations of 2,5- and 2,3-dihydrofuran and their transformations in a freon matrices. *J. Chem. Soc. Perkin Trans.* 2: 2447–2453.

Knolle, W., Janovský, I., Naumov, S., and Williams, F. 2006. EPR studies of amine radical cations. Part 2. Thermal and photo-induced rearrangements of propargylamine and allylamine radical cations in low-temperature freon matrices. *J. Phys. Chem. A* 110: 13816–13826.

Le Motais, B. C. and Jonah, C. D. 1989. Picosecond pulse-radiolysis studies of the primary processes in hydrocarbon radiation-chemistry. *Radiat. Phys. Chem.* 33: 505–517.

Levich, V. G. and Dogonadze, R. R. 1961. Adiabatic theory of the electron processes in solutions. *Coll. Czech. Chem. Commun.* 26: 193–214.

Lomoth, R., Naumov, S., and Brede, O. 1999. Genuine pyrimidine radical cations generated by radiation-induced electron transfer to butyl chloride or acetone parent ions. *J. Phys. Chem. A* 103: 2641–2648.

Lund, A. and Shiotani, M. (eds.) 1991. *Radical Ionic Systems, Properties in Condensed Phases*. Dordrecht, the Netherlands: Kluwer Academic Publishers.

Mahalaxmi, G. R., Hermann, R., Naumov, S., and Brede, O. 2000. Free electron transfer from several phenols to radical cations of non-polar solvents. *Phys. Chem. Chem. Phys.* 2: 4947–4955.

Mahalaxmi, G. R., Naumov, S., Hermann, R., and Brede, O. 2001. Nucleophilic effects on the deprotonation of phenol radical cations. *Chem. Phys. Lett.* 337: 335–340.

Marcus, R. A. 1964. Chemical and electrochemical electron-transfer theory. *Ann. Rev. Phys. Chem.* 15: 155–196.

Marcus, R. A. and Sutin, N. 1964. Electron transfers in chemistry and biology. *Biochim. Biophys. Acta* 811: 265–322.

Maroz, A., Hermann, R., Naumov, S., and Brede, O. 2005. Ionization of aniline and its N-methyl and N-phenyl substituted derivatives by free electron transfer to n-butyl chloride parent radical cation. *J. Phys. Chem. A* 109: 4690–4696.

Mehnert, R. 1991. Radical cations in pulse radiolysis. In *Radical Ionic Systems, Properties in Condensed Phases*, Lund, A. and Shiotani, M. (eds.), pp. 231–284. Dordrecht, the Netherlands: Kluwer Academic Publishers.

Mehnert, R., Brede, O., and Bös, J. 1977. Nanosekunden-Pulsradiolyse von Kohlenwasserstoffen by Raumtemperatur. *Z. Chem. (Leipzig)* 17: 268–270.

Mehnert, R., Bös, J., and Brede, O. 1979a. Nanosecond pulse radiolysis study of radical cations in liquid alkanes. *Radiochem. Radioanal. Lett.* 38: 47–52.

Mehnert, R., Brede, O., Bös, J., and Naumann, W. 1979b. Charge transfer from the carbon tetrachloride radical cation to alkyl chlorides, alkanes and aromatics. *Ber. Bunsenges. Phys. Chem.* 83: 992–996.

Mehnert, R., Brede, O., and Naumann, W. 1982. Charge transfer from solvent radical cations to solutes studied in pulse-irradiated liquid n-butyl chloride. *Ber. Bunsenges. Phys. Chem.* 86: 525–529.

Mehnert, R., Brede, O., and Naumann, W. 1984. Spectral properties and kinetics of cationic transients generated in electron pulse irradiated C7- to C16 alkanes. *Ber. Bunsenges. Phys. Chem.* 88: 71–76.

Mehnert, R., Brede, O., and Naumann, W. 1985. Charge transfer from n-hexadecane radical cation to cycloalkanes and aromatics. *Ber. Bunsenges. Phys. Chem.* 89: 1031–1035.

Meot-Ner, M., Sieck, L. W., and Ausloos, P. 1981. Ionization of normal alkanes: Enthalpy, entropy, structural and isotope effects. *J. Am. Chem. Soc.* 103: 5342–5348.

Mohan, H., Hermann, R., Naumov, S., Mittal, J. P., and Brede, O. 1998. Two channels of electron transfer observed for the reaction of n-butyl chloride parent cations with naphthols and hydroxybiphenyls. *J. Phys. Chem.* 102: 5754–5762.

Naumov, S., Janovský, I., Knolle, W., and Mehnert, R. 2003a. Formation of 3,4-dihydrofuran radical cation through intramolecular H-shift: Quantum chemical calculations and low-temperature EPR study. *Rad. Phys. Chem.* 67: 243–246.

Naumov, S., Janovský, I., Knolle, W., and Mehnert, R. 2003b. Radical cation, dimer radical cation and neutral radical of 2,3-dihydropyran - Possible initiators of its polymerisation? *Macromol. Chem. Phys.* 204: 2099–2104.

Naumov, S., Janovský, I., Knolle, W., and Mehnert, R. 2003c. Distonic dimer radical cation of 2,3-dihydrofuran: Quantum chemical calculations and low temperature EPR results. *Nucl. Instrum. Meth. Phys. Res. B* 208: 385–389.

Naumov, S., Janovský, I., Knolle, W., and Mehnert, R. 2003d. Radical cations of tetrahydropyran and 1,4-dioxane revisited: Quantum chemical calculations and low-temperature EPR results. *Phys. Chem. Chem. Phys.* 5: 3133–3139.

Naumov, S., Janovský, I., Knolle, W., and Mehnert, R. 2004a. Transformations of 5-membered heterocyclic radical cations as studied by low-temperature EPR and quantum chemical methods. *Phys. Chem. Chem. Phys.* 6: 3933–3937.

Naumov, S., Janovský, I., Knolle, W., Mehnert, R., and Turin, D. A. 2004b. Low-temperature EPR and quantum chemical study of lactone radical cations and their transformations. *Rad. Phys. Chem.* 73: 206–212.

Naumov, S., Janovský, I., Knolle, W., and Mehnert, R. 2004c. On the radiation-induced polymerisation of 2,3-dihydropyran. *Macromol. Chem. Phys.* 205: 1530–1535.

Naumov, S., Janovský, I., Knolle, W., and Mehnert, R. 2005. Role of distonic dimer radical cations in the radiation-induced polymerisation of vinyl ethers. *Nucl. Instr. Meth. Phys. Res. B* 236: 461–467.

Rehm, D. and Weller, A. 1969. Kinetik und Mechanismus der Elektronenübertragung bei der Fluoreszenzlöschung in Acetonitril. *Ber. Bunsenges. Phys. Chem.* 73: 834–839.

Rosokha, S. and Kochi, J. K. 2008. Fresh look at electron-transfer mechanisms via the donor/acceptor bindings in the critical encounter complex. *Acc. Chem. Res.* 41: 641–653.

Shida, T. 1988. *Electronic Absorption Spectra of Radical Ions. Physical Science Data 34.* Amsterdam, the Netherlands: Elsevier.

Shida, T., Haselbach, E., and Bally, T. 1984. Organic radical ions in rigid systems. *Acc. Chem. Res.* 17: 180–186.

Tabata, Y., Ito, Y., and Tagawa, S. (eds.) 1991. Radical ions in condensed media. In *CRC Handbook of Radiation Chemistry*, p. 439, 409, 591. Boca Raton, FL: CRC Press.

Trifunac, A. D., Norris, J. R., and Lawler, R. G. 1979. Nanosecond time-resolved EPR in pulse radiolysis via the spin echo method. *J. Chem. Phys.* 71: 4380–4391.

Ulstrup, J. and Jortner, J. 1975. The effect of intramolecular quantum modes on free energy relationships for electron transfer reactions. *J. Chem. Phys.* 63: 4358–4369.

Warman, J. M., Asmus, K.-D., and Schuler, R. H. 1969. Electron scavenging in the radiolysis of cyclohexane solutions of alkyl halides. *J. Phys. Chem.* 73: 931–936.

Wedenejew, W. J., Gurwitsch, L., Kondratjew, W. H., Medwedew, W. A., and Frankiewitsch, E. L. 1971. *Energien chemischer Bindungen, Ionisationspotentiale und Elektronenaffinitäten*, Leipzig, Germany: VEB Deutscher Verlag für die Grundstoffindustrie.

Weller, A. 1982. Exciplex and radical pairs in photochemical electron transfer. *Pure Appl. Chem.* 54: 1885–1888.

Zador, E., Warman, J. M., and Hummel, A. 1973. Anomalously high rate constants for the reaction of solvent positive ions with solutes in irradiated cyclohexane and methylcyclohexane. *Chem. Phys. Lett.* 23: 363–388.

Zubarev, V. E. and Brede, O. 1997. Direct detection of the cation radical of 2,6-di-tert-butyl-4-methylphenol generated by electron transfer oxidation with matrix alkylhalogenated cation radicals. *Acta Chem. Scand.* 51: 224–227.

11 Radiation Chemistry and Photochemistry of Ionic Liquids

Kenji Takahashi
Kanazawa University
Kanazawa, Japan

James F. Wishart
Brookhaven National Laboratory
Upton, New York

CONTENTS

11.1 INTRODUCTION

Ionic liquids are commonly defined as molten salts with melting points below 100°C. Typically, they are combinations of organic and inorganic ions selected to have poor packing or weakened interionic interactions in the solid state, and hence, low melting points. Ionic liquids have been considered as an ideal medium for a wide range of applications in the areas of energy, synthesis, electrochemistry, and biocatalysis, among others, and the field of ionic liquids continues to grow as new types of ionic liquids continue to be developed (Welton, 2004; MacFarlane et al., 2007; van Rantwijk and Sheldon, 2007; Plechkova and Seddon, 2008; Wishart, 2009). The potential safety and environmental benefits of ionic liquids, as compared to conventional solvents, have drawn interest in their use as processing media for the nuclear fuel cycle (Earle and Seddon, 2000; Allen et al., 2002; Jensen et al., 2003; Chaumont and Wipff, 2004; Nikitenko et al., 2005; Cocalia et al., 2006; Dietz, 2006; Stepinski et al., 2006; Luo et al., 2006; Binnemans, 2007; Han and Armstrong, 2007). Therefore, an understanding of the interactions of ionizing radiation and photons with ionic liquids is strongly needed.

In 2001, Neta and coworkers began to publish a series of investigations on ionic liquid radiation chemistry and radiation-induced chemical kinetics in ionic liquids (Behar et al., 2001; 2002; Grodkowski and Neta, 2002a,b,c). Contemporaneously, groups in Poland and the United Kingdom studied the radiolysis of ionic liquids containing the imidazolium cation (Marcinek et al., 2001; Allen et al., 2002). Since then, picosecond pulse radiolysis on ionic liquids and static electron paramagnetic resonance (EPR) measurements on irradiated ionic liquid glasses have been used to characterize the initial radiolysis products of ionic liquids and their reactivity (Grodkowski et al., 2003; Wishart, 2003; Wishart and Neta, 2003; Funston and Wishart, 2005; Wishart et al., 2005, 2006; Shkrob et al., 2007; Shkrob and Wishart, 2009; Wishart and Shkrob, 2009). Electron photodetachment studies have also been performed using UV lasers to study solvated electrons in ionic liquids (Katoh et al., 2007; Takahashi et al., 2008). Through these studies, it has been revealed that the reactivities of pre-solvated and solvated electrons, and the kinetics of radical reactions, are quite different from those found in molecular solvents. In this chapter, we introduce recent advances in radiation chemistry and photochemistry of ionic liquids. We also include studies related to the physicochemical properties, structure, and solvation dynamics of ionic liquids that are important to understand the radiation chemistry of ionic liquids.

11.2 PHYSICOCHEMICAL PROPERTIES OF IONIC LIQUIDS

Because ionic liquids are binary ionic compounds, their physicochemical properties can vary significantly from those of molecular solvents. Ionic liquids are typically characterized by their melting point, viscosity, conductivity, and ionic self-diffusion coefficients. Conventional salts like sodium chloride (NaCl, melting point 801°C) form high-melting solids because of the very strong Coulombic attractions between ions in the close-packed lattice of the solid. Weakening the Coulombic interactions makes it easier for the solid to melt. Replacing sodium with the larger cesium cation reduces the melting point of CsCl to 645°C, and substituting the large organic tetrapropylammonium cation $[(Pr)_4N]^+$ lowers it to 241°C. Going further, the melting point of ethylmethylimidazolium chloride (Emim$^+$ Cl$^-$) is 87°C, allowing it to qualify as an ionic liquid.

Melting points may be influenced not only by the size and shape of the cation but also by the anion. The melting point of Emim$^+$ NO$_3^-$ is 38°C, while those of Emim$^+$ BF$_4^-$ and Emim$^+$ N(SO$_2$CF$_3$)$_2^-$ are 15°C and −18°C, respectively, making them both room-temperature ionic liquids (RTILs). It is important to clarify, however, that ion size is only one of the many factors that determine the melting point of a specific combination of ions. Another important factor is the distribution of charges on cations and anions. If the charges are localized on specific parts of the cation and the anion, the electrostatic interaction between them can be strong, leading to a higher melting point. On the other hand, if the charges are delocalized widely over the cation and the anion, the attraction between the ions would be weaker.

The presence of aliphatic, aromatic, or fluorous side chains, as well as the addition of functional groups, also influences the lattice-packing properties and interionic interactions that control melting points. These features also control the solubility of solutes in ionic liquids and the miscibility of ionic liquids with other solvents. Frequently, the hydrophobicity or hydrophilicity of the ionic liquid is determined by the affinity of the anion for water. For example, Bmim$^+$ Cl and Bmim$^+$ BF$_4$ are water-miscible, as are most ionic liquids containing Cl$^-$, BF$_4^-$, Br$^-$, I$^-$, CF$_3$CO$_2^-$, and CH$_3$OSO$_3^-$, unless the cation contains a lot of organic side chains. On the other hand, Bmim$^+$ N(SO$_2$CF$_3$)$_2^-$ and Bmim$^+$ PF$_6^-$ contain hydrophobic anions and are water-immiscible.

It is well known that viscosities of ionic liquids are much higher than those of normal molecular solvents. For example, the viscosity of Bmim$^+$ N(SO$_2$CF$_3$)$_2^-$ at 303 K is 40 mPa s (Jacquemin et al., 2006), almost a factor of 40 higher than that of water. The viscosities of three ionic liquids as a function of temperature are shown in Figure 11.1 (Jacquemin et al., 2006). Ionic liquids generally show glassy behavior and their viscosities decrease in a "super-Arrhenius" manner with temperature, that is, their viscosities decrease faster with temperature than predicted by a linear Arrhenius plot.

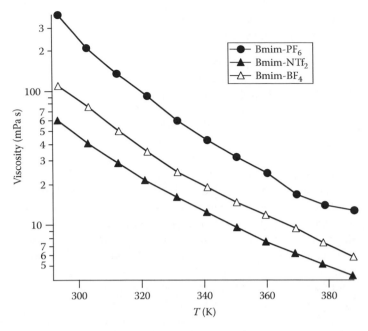

FIGURE 11.1   Viscosities of the ionic liquids $Bmim^+ PF_6^-$, $Bmim^+ N(SO_2CF_3)_2^-$ (NTf_2), and $Bmim^+ BF_4^-$ as a function of temperature. (Data from Jacquemin, J. et al., *Green Chem.*, 8, 172, 2006.)

Thus, at moderately elevated temperatures, many ionic liquids are fluid enough for applications in replacing conventional solvents.

Recently, the heterogeneous nature of ionic liquids has begun to be examined in detail. There are in fact two types of heterogeneities, static and dynamic. The static heterogeneity is manifested by the existence of structurally distinct domains of polarity and nonpolarity occupied by the charged and uncharged parts of the constituent ions and their side chains, respectively. The volumes and interconnectivity of these heterogeneous domains can depend on the lengths of the aliphatic side chains (Del Popolo and Voth, 2004; Wang and Voth, 2005; Lopes and Padua, 2006; Triolo et al., 2007). Various solutes will distribute themselves differently among the domains depending on their characteristics. Consequently, attempts to characterize the polarity of ionic liquids using solvato-chromic probes resulted in disparate answers depending on the probe molecule used (Weingartner, 2008). Hamaguchi et al. showed through Raman spectral analysis that two distinct conformational structures of the $Bmim^+$ ion exist simultaneously in the ionic liquid state (Hayashi et al., 2003; Ozawa et al., 2003). They also have obtained evidence of the mesoscopic local structure in ionic liquids by coherent anti-Stokes Raman scattering measurements (Shigeto and Hamaguchi, 2006).

The second type of heterogeneity exhibited in ionic liquids is the dispersive nature of their dynamics and kinetics (Giraud et al., 2003; Holbrey et al., 2004; Castriota et al., 2005; Shirota and Castner, 2005; Shirota et al., 2005, 2007). For example, the Maroncelli group observed that the time-dependent response to a solute electronic perturbation extends over three or more decades in time in ionic liquids (Arzhantsev et al., 2007; Jin et al., 2007). In the case of conventional super-cooled liquids, it is generally agreed that dispersive kinetics are the results of heterogeneity (Ediger, 2000). In such situations, molecules in different local environments relax at significantly differ-ent rates. The presence of such heterogeneity in ionic liquids has been clearly demonstrated both experimentally and in computational studies (Del Popolo and Voth, 2004; Mandal et al., 2004; Wang and Voth, 2005; Hu and Margulis, 2006; Lopes and Padua, 2006; Triolo et al., 2007). Such inhomogeneous environments may have significant effects on the formation and reactions of solvated electrons in ionic liquids.

11.3 FORMATION OF EXCESS ELECTRONS IN IONIC LIQUIDS: PULSE RADIOLYSIS AND ELECTRON PHOTODETACHMENT

The earliest pulse radiolysis experiments on ionic liquids were performed on imidazolium cation-based ionic liquids (Allen et al., 2002; Behar et al., 2001, 2002; Grodkowski and Neta, 2002a,b,c; Marcinek et al., 2001). Pulse radiolysis transient absorption spectroscopy of neat, deoxygenated $Bmim^+ PF_6^-$ revealed a transient product peak at 325 nm and a shoulder around 375 nm (Behar et al., 2001; Marcinek et al., 2001). The identities of the species responsible for the observed transients were determined by pulse radiolysis on aqueous solutions of the $Bmim^+$ salt. The hydrated electron reacts with $Bmim^+$ very rapidly ($k = 1.9 \times 10^{10} M^{-1} s^{-1}$) (Behar et al., 2001). The product is the neutral $Bmim^\bullet$ radical, which has an absorption peak at 323 nm ($e = 5900 M^{-1} cm^{-1}$) and a reduction potential of $-2 V$ vs. normal hydrogen electrode (NHE) in aqueous solution. Reactions of the $Bmim^+$ cation with OH and H radicals produce the corresponding radical adducts, and oxidation of $Bmim^+$ by $SO_4^{-\bullet}$ produces the ring-centered $Bmim^{2+\bullet}$ radical dication. All three of these products have similar absorption spectra to the $Bmim^\bullet$ radical. Thus, in the neat $Bmim^+ PF_6^-$ liquid, the transient absorption seen at 325 nm contains contributions from both the reduction and oxidation products of the $Bmim^+$ cation (Behar et al., 2001). Because the excess electrons produced by ionizing radiation are rapidly scavenged by imidazolium cations, solvated electrons in such imidazolium-based ionic liquids have not yet been conclusively observed.

On the other hand, if the cations (and anions) of the ionic liquid are sufficiently unreactive toward reduction, solvated electrons can exist for microseconds as stable transient species. Quaternary ammonium and phosphonium cations meet this criterion. The electrochemical window of ionic liquid methyltributylammonium bis(trifluoromethylsulfonyl)imide ($MeBu_3N^+N(SO_2CF_3)_2^-$) is almost 6 V at a Pt electrode, with a cathodic limit of about $-3.2 V$ vs. ferrocene/ferrocenium. The first observation of the solvated electron in a neat ionic liquid was made by pulse radiolysis transient absorption measurements on $MeBu_3N^+N(SO_2CF_3)_2^-$ (Wishart and Neta, 2003). The absorption spectrum of the solvated electron in this ionic liquid is shown in Figure 11.2. The spectrum is very broad and the absorption peak is located around 1400 nm ($e = 22,000 M^{-1} cm^{-1}$) with a long tail toward the visible and UV regions. Spectra of the solvated electron in ionic liquids without functional groups tend to peak in the region between 1000 and 1500 nm (Wishart and Neta, 2003; Funston and Wishart, 2005; Wishart et al., 2005; Takahashi et al., 2008, 2009b).

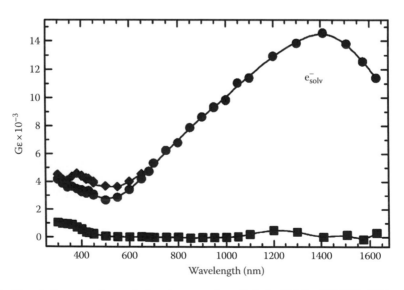

FIGURE 11.2   Transient spectra observed after electron pulse radiolysis of $MeBu_3N^+N(SO_2CF_3)_2^-$. (Reproduced from Wishart, J.F. and Neta, P., *J. Phys. Chem. B*, 107, 7261, 2003. With permission.)

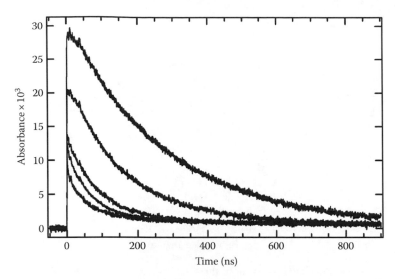

FIGURE 11.3 Transient absorbance at 900 nm for electron capture by (top to bottom) 0, 24, 48, 64, and 90 mM benzophenone in $MeBu_3N^+N(SO_2CF_3)_2^-$. (Reproduced from Wishart, J.F. and Neta, P., *J. Phys. Chem. B*, 107, 7261, 2003. With permission.)

The reactivity of solvated electrons with some aromatics, such as benzophenone, pyrene, and phenanthrene, was examined in the same study (Wishart and Neta, 2003). Figure 11.3 shows examples of the decay of the solvated electron absorption at 900 nm with increasing scavenger concentration. The reaction rate constants are in a range of 1.3–$1.7 \times 10^8 \, M^{-1} \, s^{-1}$. The theoretical diffusion-controlled rate limit is estimated to be $1.5 \times 10^7 \, M^{-1} \, s^{-1}$ from the measured viscosity and the equation $k_{diff} = 8000RT/3\eta$. The calculated rate limit is one order of magnitude slower than the experimental values. A similar difference was observed for diffusion-limited electron-transfer reactions from pyridinyl radical to duroquinone in several ionic liquids (Behar et al., 2002; Skrzypczak and Neta, 2003), and in numerous other cases of diffusion-controlled reactions in ionic liquids. This trend indicates that reactant solutes may experience different microviscosities that are lower than the viscosity of the medium, the latter being determined by the movement of the anions and cations subject to the constraints of their Coulombic interactions. It is also worth noting that the diffusion-controlled reactions of the solvated electron with the aromatic scavengers are approximately two orders of magnitude slower than in common molecular solvents and about ten times slower than the reaction of the hydrogen atom with the same reactants in the same ionic liquid (Grodkowski et al., 2003).

Solvated electrons can also be produced in ionic liquids by electron photodetachment by the excitation of the charge-transfer-to-solvent (CTTS) band of iodide (Katoh et al., 2007). In trimethyl-n-propylammonium bis(trifluoromethanesulfonyl)imide ($TMPA^+N(SO_2CF_3)_2^-$), the CTTS absorption maximum is located around 225 nm. Upon photoexcitation in this wavelength range, electron photodetachment occurs efficiently:

$$I^- \xrightarrow{\ h\nu\ } I + e^-_{hot}$$

where e^-_{hot} is the electron before it has become thermalized by losing its excess kinetic energy through interaction with the ionic liquid medium. (The thermalized electron subsequently undergoes a solvation process that is discussed later in this chapter.) The quantum yield of the photodetachment from iodide to $TMPA^+N(SO_2CF_3)_2^-$ (F_{ion}) is estimated to be 0.34 (Katoh et al., 2007). This high quantum yield of the photodetachment may be due to the high mobility of the pre-solvated electron (see below) and/or effective screening of the geminate electron–hole pair by the ionic liquid.

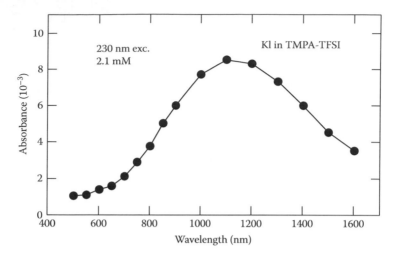

FIGURE 11.4 Transient absorption spectrum of the solvated electron in $TMPA^+N(SO_2CF_3)_2^-$ by electron photodetachment. (Reproduced from Katoh, R. et al., *J. Phys. Chem. B*, 111, 4770, 2007. With permission.)

Figure 11.4 shows the transient absorption spectrum of a $TMPA^+N(SO_2CF_3)_2^-$ solution of KI recorded just after excitation by a 230 nm light pulse. The spectrum is similar to that obtained previously by the pulse radiolysis technique, indicating that an electron can be ejected from iodide into $TMPA^+N(SO_2CF_3)_2^-$ by CTTS band excitation.

As mentioned above, solvated electrons have not been observed in nanosecond pulse radiolysis measurements on ionic liquids based on the imidazolium cation, instead forming radical species by attachment to an imidazolium cation (Behar et al., 2001; Marcinek et al., 2001), which could potentially lead to permanent degradation (Shkrob et al., 2007; Shkrob and Wishart, 2009). For this reason, neat imidazolium cation–based ionic liquids might not be suitable for the nuclear fuel cycle process application, but they could be used in mixtures. Therefore, we investigated the reaction of solvated electrons with $Bmim^+$ cations in $TMPA^+ N(SO_2CF_3)_2^-$ in the hope that the rate constant of the reaction would give us a strategy for designing ionic liquid systems that are stable under irradiation. The rate constant was determined to be $5.3 \times 10^8 M^{-1} s^{-1}$, indicating that the half-life of solvated electrons in neat $Bmim^+ N(SO_2CF_3)_2^-$ would be 350 ps. In reality, the excess electrons may be scavenged even before they become solvated—femtosecond experiments suggest that the electrons may be scavenged in less than 5 ps in imidazolium iodide ionic liquids (Chandrasekhar et al., 2008). More detailed studies are warranted.

11.4 SOLVATION DYNAMICS AND REACTIONS OF PRE-SOLVATED (DRY) ELECTRONS

The sections above have discussed radiation- and photon-induced ionization of ionic liquids and the reactivity of solvated electrons; however, from the radiation chemistry standpoint, one of the major differences between ionic liquids and conventional solvents is the fact that dynamic timescales are hundreds to thousands of times longer in ionic liquids (Giraud et al., 2003; Holbrey et al., 2004; Castriota et al., 2005; Shirota and Castner, 2005; Shirota et al., 2005, 2007; Arzhantsev et al., 2007; Jin et al., 2007). Moreover, the response of an ionic liquid to the introduction of an excess electron is subject to static and dynamic heterogeneities that depend on its composition. Although the initial stages of solvent response involve inertial motions that operate on very short timescales, even in ionic liquids, the slower response components scale with the viscosity of the medium. Consequently, it takes much longer to solvate an excess electron in an ionic liquid as compared to normal liquids, and the mobility and reactivity of pre-solvated electron states become much larger determinants of radiolysis product yields, as a result.

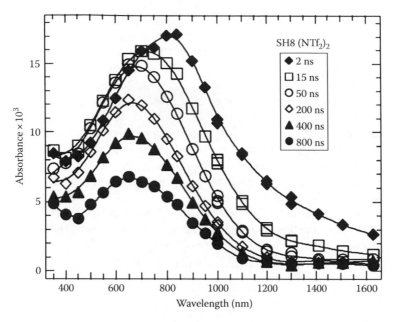

FIGURE 11.5 Transient spectra of solvated electron in an alcohol-functionalized ionic liquid SH8^{2+}(N(SO$_2$CF$_3$)$_2$$^-$)$_2$. (Reproduced from Wishart, J.F. et al., *Radiat. Phys. Chem.*, 72, 99, 2005. With permission.)

An extreme example of the slow response of ionic liquids is depicted in Figure 11.5, where the transient spectra of the excess electron in the highly viscous alcohol-functionalized ionic liquid SH8^{2+}(N(SO$_2$CF$_3$)$_2$$^-$)$_2$ at different times following the electron pulse are shown (Wishart et al., 2005). The molecular structures of the alcohol-functionalized ionic liquids in question are shown in Figure 11.6. The spectrum at 2 ns has a shoulder in the range of 1200 nm characteristic of an electron solvated in an alkane ionic liquid environment. Over the course of tens of nanoseconds, the shoulder disappears and the remaining spectrum peaks at 650 nm, characteristic of an electron solvated by alcohol functionalities. The average electron solvation time constants in methanol, ethanol, and ethylene glycol are about 2–9 ps and that of viscous 1,2,6-trihydroxyhexane (2490 mPa s) is ~1 ns, while in the viscous ionic liquid alcohols SH6^{2+}(N(SO$_2$CF$_3$)$_2$$^-$)$_2$ and SH8^{2+}(N(SO$_2$CF$_3$)$_2$$^-$)$_2$ (viscosities 6860 and 4570 mPa s, respectively), the observed time constants for absorption decay are 25 ns (SH8^{2+}) to 40 ns (SH6^{2+}). In the case of these ionic liquid alcohols, dynamic hindrance of hydroxypropyl side chain reorientation imposed by the electrostatic constraints of the ionic lattice plays the greatest role in slowing the electron solvation process.

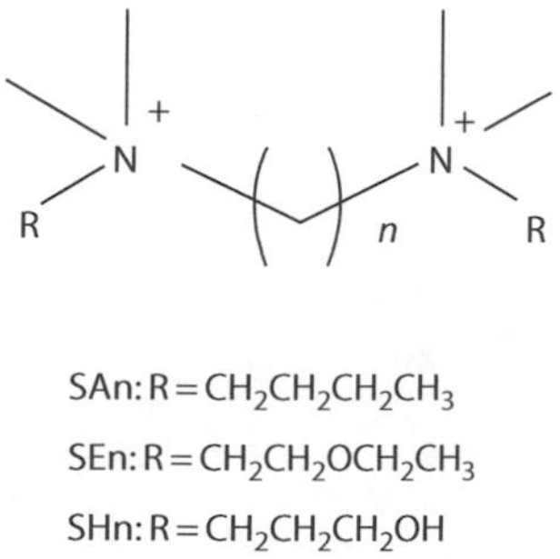

SAn: R = CH$_2$CH$_2$CH$_2$CH$_3$

SEn: R = CH$_2$CH$_2$OCH$_2$CH$_3$

SHn: R = CH$_2$CH$_2$CH$_2$OH

FIGURE 11.6 Molecular structure of alcohol-functionalized ionic liquids. (Reproduced from Wishart, J.F. et al., *Radiat. Phys. Chem.*, 72, 99, 2005. With permission.)

A more typical example involves the popular ionic liquid *N*-butyl-*N*-methylpyrrolidinium bis(trifluoromethanesulfonyl)imide (bmpyrr$^+$ N(SO$_2$CF$_3$)$_2$$^-$), which has a relatively low viscosity (94 mPa s at 20°C) and a wide electrochemical window. Picosecond pulse-probe transient absorption spectroscopy was performed at Brookhaven National Laboratory's (BNL) Laser-Electron Accelerator Facility (Wishart et al., 2004) to observe the temporal blueshift of the excess electron spectrum associated with the solvation process in bmpyrr$^+$ N(SO$_2$CF$_3$)$_2$$^-$. The spectral data, measured from 600 to 1600 nm with a time resolution of 15 ps, extending to 10 ns, showed a strong absorption around 1400 nm at short time intervals that shifted to the equilibrated spectrum of the solvated electron peaking at 1100 nm. Factor analysis of the data showed a multiexponential or distributed solvation response with an average solvation time of 260 ps (Wishart et al., manuscript in preparation). In comparison, the average solvation time for the fluorescent solvation probe counarin-153 is 350 ps in the same ionic liquid (Funston et al., 2007). Electron solvation dynamics in ionic liquids is often difficult to measure because of the need to exchange (flow) the sample to avoid the accumulation of radiolytic damage. Solvatochromic dye solvation measurements, which are much more extensively studied in ionic liquids (Arzhantsev et al., 2007; Funston et al., 2007; Jin et al., 2007), are a useful proxy for estimating electron solvation times.

Since electron solvation processes in normal liquids occur on the time scale of picoseconds, excess electrons in ionic liquids can persist hundreds to thousands of times longer in mobile, pre-solvated states with reactivity profiles that differ from fully solvated states. As already shown in Figure 11.3, the solvated electron yield near time zero decreased with the concentration of added electron scavenger. In water and most other molecular solvents, such decreases in the "initial" yield cannot be observed in nanosecond pulse radiolysis measurements. This decreased yield is due to scavenging of the pre-solvated electron by the solute (Lam and Hunt, 1975; Jonah et al., 1977, 1989; Glezen et al., 1992; Lewis and Jonah, 1986). For each solvent and solute, the efficiency of pre-solvated electron scavenging can be quantified by the constant C_{37} defined as follows:

$$\frac{G_c}{G_0} = \exp\left(\frac{-c}{C_{37}}\right) \tag{11.1}$$

where
  G_c is the yield of solvated electrons at a given scavenger concentration c
  G_0 is the yield in the absence of scavenger
  C_{37} is the concentration where only $1/e$ (37%) of the electrons survive to be solvated

Examples of C_{37} values and reaction rate constants of solvated electrons with scavengers $k(e_s^-)$ are listed in Table 11.1. In $MeBu_3N^+ \, N(SO_2CF_3)_2^-$, the values of C_{37} for benzophenone, pyrene, and phenanthrene are 0.062, 0.063, and 0.084 M, respectively. The C_{37} value for $Bmim^+$ is 0.057 M in $TMPA^+ \, N(SO_2CF_3)_2^-$, suggesting that the pre-solvated electron reacts with $Bmim^+$ at least as efficiently as with the aromatics.

It is reported that C2-alkylation of imidazolium cations is effective in extending the electrochemical redox windows of the imidazolium-based ionic liquids (Hayashi et al., 2005), suggesting that C2-alkylation may reduce the reaction rate with solvated electrons. The solvated electron

TABLE 11.1
Rate Constants and C_{37} Values for the Reactions of Solvated and Pre-Solvated Electrons with Several Scavengers

Ionic Liquid	Scavenger	$k(e_s^-)$ ($M^{-1} s^{-1}$)	C_{37} (M)
$MeBu_3N^+ \, N(SO_2CF_3)_2^-$	Benzophenone	$(1.6 \pm 0.1) \times 10^8$	0.062
$MeBu_3N^+ \, N(SO_2CF_3)_2^-$	Pyrene	$(1.7 \pm 0.1) \times 10^8$	0.063
$MeBu_3N^+ \, N(SO_2CF_3)_2^-$	Phenanthrene	$(1.3 \pm 0.1) \times 10^8$	0.084
$MeBu_3N^+ \, N(SO_2CF_3)_2^-$	Indole	$(4.3 \pm 0.3) \times 10^7$	0.22
$TMPA^+ \, N(SO_2CF_3)_2^-$	$Bmim^+$	5.3×10^8	0.057
$TMPA^+ \, N(SO_2CF_3)_2^-$	$EDmim^+$	3.8×10^8	0.081
$TMPA^+ \, N(SO_2CF_3)_2^-$	$BDmim^+$	3.0×10^8	0.13
$TMPA^+ \, (N(SO_2CF_3)_2^-$	$HDmim^+$	3.0×10^8	0.12

capture rate constants and C_{37} values for *N*-ethyl-2,3-dimethylimidazolium (EDmim$^+$), *N*-butyl-2,3-dimethylimidazolium (BDmim$^+$), and *N*-hexyl-2,3-dimethylimidazolium (HDmim$^+$) are summarized in Table 11.1 (Takahashi et al., 2008). For each of the C2-alkylated imidazolium cations, there is only a slight reduction of the rate constant compared to Bmim$^+$. This is quite reasonable because the rate constants are considered to be diffusion limited. On the other hand, the C_{37} values for BDmim$^+$ and HDmim$^+$ are significantly larger than that for non-alkylated imidazolium Bmim$^+$. These results have interesting implications for the reaction mechanisms of pre-solvated and solvated electron capture and the nature of the electron–cation adduct, which may suggest a strategy for designing ionic liquids that are more stable under irradiation.

11.5 PHOTOEXCITATION OF SOLVATED ELECTRONS

There have been several studies for dielectrons in molten salts and solid salts, such as KCl (Lynch and Robinson, 1968). A recent *ab initio* simulation for electrons in LiF (Zhang et al., 2008) also suggested the presence of two types of dielectrons, namely, the singlet dielectron and the triplet dielectron. The stabilization of two electrons in one trap site is considered to be due to not only electrostatic interactions but also through hole–orbital coupling among solvent molecules (Zhang et al., 2008). Because ionic liquids can be considered to have similar characteristics as molten salts, we can postulate the presence of the dielectron in ionic liquids. To examine this possibility, an experiment was performed in which the solvated electron was excited by light pulses at 532 and 1064 nm. If the dielectron exists, photoexcitation might produce two solvated electrons, which could be detected as an increase in the absorption signal in the near-infrared wavelength region. The solvated electrons were produced by electron photodetachment from iodide in an ionic liquid solution by a 248 nm KrF excimer laser pulse, as described in Section 11.3. After a variable delay, a 532 nm Nd-YAG laser pulse irradiated the sample (Takahashi et al., 2009b).

Figure 11.7 shows a transient absorption signal observed at 1000 nm. When a 532 nm laser pulse was used to irradiate the sample during the lifetime of the solvated electron, the transient absorption signal displayed an instrument-limited decrease, and the decrease of absorption did not recover within the lifetime of the solvated electron. Therefore, a fraction of the solvated electron population was permanently removed by a 532 nm excitation. Transient absorption signals were monitored at several different wavelengths, and a similar permanent bleaching was observed at each wavelength.

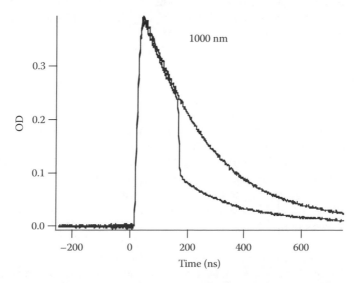

FIGURE 11.7 Transient absorption signals observed in TMPA$^+$N(SO$_2$CF$_3$)$_2^-$ at 1000 nm with and without a 532 nm pulse. (Reproduced from Takahashi, K. et al., *Radiat. Phys. Chem.*, 78, 1129, 2009b. With permission.)

The above results suggest that a portion of the solvated electrons react and disappear during the 532 nm laser pulse. One possible explanation may be the reaction of the dry electron with its parent iodine atom via the following reaction scheme:

$$I^- \xrightarrow{h\nu(248\,nm)} I + e_p^- \tag{11.2}$$

$$e_p^- \longrightarrow e_s^- \tag{11.3}$$

$$e_s^- \xrightarrow{h\nu(532\,nm)} (e_s^-)^* \tag{11.4}$$

$$(e_s^-)^* + I \longrightarrow I^- \tag{11.5}$$

$$I^- + I \longrightarrow I_2^- \tag{11.6}$$

$$I_2^- + I_2^- \longrightarrow I_3^- + I^- \tag{11.7}$$

where

    e_p^- is the pre-solvated electron
    e_s^- is the solvated electron
    $(e_s^-)^*$ indicates the electron after excitation with a 532 nm light pulse

If permanent bleaching of the solvated electron takes place due to the above reaction scheme, it may be possible to observe a decrease in I_2^- concentration, because the iodine atoms would disappear on reacting with $(e_s^-)^*$. Therefore, the transient absorption of I_2^- was measured with a wide time window. The absorption spectrum of I_2^- in TMPA-NTf$_2$ is similar to that in water, with absorption maxima located at 400 and 740 nm. The evolution of the absorption signal from I_2^- was monitored at 830 nm. At this wavelength, both the I_2^- species and also the reactivity of solvated electrons can be observed. The results are shown in Figure 11.8. In the early time regime ($t < 500$ ns), the absorption signal occurs mostly due to the solvated electron. After the decay of the solvated electron, a slow buildup of I_2^- was seen. As shown in Figure 11.8, when the sample was irradiated by a 532 nm laser pulse approximately 200 ns after the production of the solvated electron, a permanent bleaching was observed; however, in the microsecond regime, where the absorption is mainly due to I_2^-, there was no significant change in the absorption. It can be concluded that the iodine atom is not a reaction partner of the excited electrons, since the yield of I_2^- at longer time periods remains unchanged.

Although at the present time we cannot identify the reaction mechanism(s) for the photo-induced bleaching of solvated electrons with a 532 nm light pulse, we can gain some insight from the previous work done on water and alcohols (Silva et al., 1998; Yokoyama et al., 1998; Kambhampati et al., 2001). In these studies, pump-probe spectroscopy of the solvated electron in water, methanol, and ethanol was studied with a 300 fs time resolution. At low pump power, the observed dynamics were assigned to s → p excitation and subsequent relaxation of the localized solvated electron. In contrast, at high pump power, two-photon absorption produces mobile conduction band electrons, which are subsequently trapped and relax at a resonant site far away from the initial equilibrated electron site. Interestingly, in the two-photon excitation, they observed an ultrafast proton-transfer reaction from the alcohol molecules, resulting in a permanent bleach. Although we cannot conclusively say that we are observing the same process, we tentatively conclude that the excitation of the solvated electron results in the generation of a reactive dry electron or pre-solvated electron in the ionic liquid.

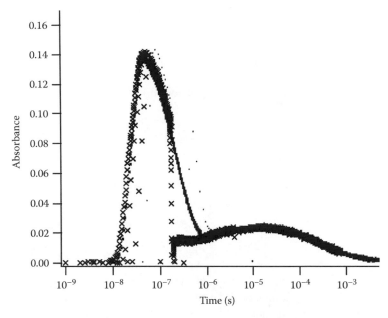

FIGURE 11.8 Time courses of the transient absorbance at 830 nm after 248 nm excitation of KI in $TMPA^+N(SO_2CF_3)_2^-$ with (cross symbols) and without (dots) a subsequent 532 nm pulse. The 532 nm pulse was delayed by 200 ns after the 248 nm pulse. (Reproduced from Takahashi, K. et al., *Radiat. Phys. Chem.*, 78, 1129, 2009b. With permission.)

11.6 RADICAL REACTIONS AND KINETIC SALT EFFECTS

11.6.1 REACTIONS BETWEEN DIIODIDE ANION RADICALS IN IONIC LIQUIDS

Ionic liquids are finding increasing uses in electrochemical devices, such as electric double-layer capacitors, lithium batteries, fuel cells, and dye-sensitized solar cells (DSSCs). In DSSCs, a photo-oxidized dye molecule must be reduced by a redox couple (Papageorgiou et al., 1996; Wang et al., 2004). The I^-/I_3^- redox couple has been used for this purpose and plays an important role in charge (hole) transport in DSSCs. The diffusion of ions in ionic liquids is generally slow because of the high viscosity of ionic liquids, and this fact is a disadvantage when an ionic liquid is used as an electrolyte. Recently, using an ultramicroelectrode technique, it has been proposed that in the I^-/I_3^- redox couple system, a Grotthuss-like mechanism involving the exchange of I_2 between I_3^- and I^- is responsible for efficient hole transport in ionic liquid electrolytes for DSSCs (Kawano and Watanabe, 2003, 2005). The exchange reaction would occur by the following process:

$$I^- + I_3^- \rightarrow I^- \cdots I_2 \cdots I^- \rightarrow I_3^- + I^-$$ (11.8)

To have this type of exchange take place, I^- and I_3^- must be in close proximity. However, the close approach of I^- and I_3^- is hindered in molecular solvents because of Coulombic repulsion between the two ions. On the other hand, the ions of an ionic liquid may weaken the electrostatic repulsion by Coulombic shielding, allowing I^- and I_3^- to approach each other without a large Coulombic energy barrier. The effect of Coulombic shielding was examined through the reaction between two diiodide (I_2^-) anion radicals. The I_2^- species can be formed by the 248 nm laser photolysis of iodide through the following reactions:

$$I^- + h\nu \rightarrow I + e_s^-$$ (11.9)

$$I + I^- \xrightarrow{\ k_2\ } I_2^-$$ (11.10)

$$I_2^- + I_2^- \xrightarrow{\ k_3\ } I_3^- + I^-$$ (11.11)

In molecular solvents, the diffusion-controlled rate for an ionic reaction can be calculated by the Debye–Smoluchowski equation (Debye, 1942):

$$k_{\mathrm{diff}} = 4\pi R_{\mathrm{ab}} D_{\mathrm{ab}} f(\varepsilon)$$ (11.12)

$$f(\varepsilon) = \frac{r_{\mathrm{c}}/R_{\mathrm{ab}}}{\exp\left(r_{\mathrm{c}}/R_{\mathrm{ab}}\right) - 1}$$ (11.13)

$$r_{\mathrm{c}} = \frac{Z_{\mathrm{a}} Z_{\mathrm{b}} e^2}{4\pi\varepsilon_0 \varepsilon k_{\mathrm{B}} T}$$ (11.14)

where
R_{ab} is the reaction distance of the closest approach
D_{ab} is the relative diffusion coefficient of reactants
$f(\varepsilon)$ is the so-called Debye correction factor
e is the charge on the ions
k_{B} is Boltzmann's constant
ε_0 is the permittivity of free space
ε is the dielectric constant
T is the absolute temperature

If the diffusion coefficient can be expressed by a simple hydrodynamic model, that is, by the Stokes–Einstein equation, $D_{\mathrm{a}} = k_{\mathrm{B}} T/(6\pi\eta R_{\mathrm{a}})$, the diffusion-limited rate constant (Equation 11.12) may be written as follows:

$$k_{\mathrm{diff}} = \frac{8000 R T f(\varepsilon)}{3\eta}$$ (11.15)

where
η is the viscosity (in Pa s)
R is the gas constant (8.3144 J K^{-1} mol^{-1})

The Debye correction factors, $f(\varepsilon)$, for H_2O, MeOH, and EtOH were calculated from the dielectric constants of the solvents, and are listed in Table 11.2. The Debye correction factor for an aqueous solution was calculated to be 0.52, whereas the values for methanol and ethanol were calculated to be 0.18 and 0.091, respectively, indicating that the Coulombic repulsion between the I_2^- radical anions in the alcohol solvents is greater than that in water. These calculated Debye correction factors predict that the reaction between I_2^- anion radicals in the alcohol solvents is much slower than that in water. As indicated in Table 11.2, the reaction rate constants in MeOH and EtOH are four to five times lower than in water.

The decay rate constants, k_3, for reaction 11.11 in the molecular solvents and the ionic liquids are plotted in Figure 11.9 as a function of the reciprocal of viscosity. The solid line in Figure 11.9 is the plot for the diffusion-limited rate constants calculated from Equation 11.15 with $f(\varepsilon) = 1$, that is, the calculations for reaction between neutral molecules (denoted as k_0). For the molecular solvents, the k_3/k_0 ratios for H_2O, MeOH, and EtOH are 0.37, 0.064, and 0.086, respectively, reflecting the importance of electrostatic repulsion between the diiodide anion radicals. In contrast, the reaction rate constants for the ionic liquids are very close to the solid line. The k_3/k_0 ratios for the

TABLE 11.2
Solvent Properties and Rate Constants for the Bimolecular
Reaction between I_2^- Anion Radicals

No.	Solvent	η^a	ε	$k_0{}^b$	$k_3{}^c$	$f(\varepsilon)^d$	k_3/k_0
1	H_2O	0.89	78.4	7.4×10^9	$(2.8 \pm 0.2) \times 10^9$	0.52	0.37
2	MeOH	0.55	32.7	1.2×10^{10}	$(7.7 \pm 0.5) \times 10^8$	0.18	0.064
3	EtOH	1.01	24.6	6.1×10^9	$(5.3 \pm 0.7) \times 10^8$	0.091	0.086
4	40% Gly^e	3.1	68	2.1×10^9	$(9.1 \pm 0.8) \times 10^8$	0.47	0.43
5	60% Gly^e	8.6	61	7.7×10^8	$(3.0 \pm 0.2) \times 10^8$	0.43	0.37
6	70% Gly^e	17	56	3.8×10^8	$(1.7 \pm 0.3) \times 10^8$	0.39	0.44
7	80% Gly^e	43	51	1.5×10^8	$(6.8 \pm 0.9) \times 10^7$	0.35	0.45
8	TMPA$^+$ TFSI^{-f}	69		9.6×10^7	$(1.1 \pm 0.2) \times 10^8$		1.15
9	P13$^+$ TFSI^{-f}	55		1.2×10^8	$(1.3 \pm 0.2) \times 10^8$		1.08
10	P14$^+$ TFSI^{-f}	69		9.6×10^7	$(9.1 \pm 0.5) \times 10^7$		0.95
11	PP13$^+$ TFSI^{-f}	135		4.9×10^7	$(6.6 \pm 0.4) \times 10^7$		1.35
12	DEME$^+$ TFSI^{-f}	63		1.0×10^7	$(1.2 \pm 0.5) \times 10^8$		1.14
13	DEME$^+$ BF$_4{}^-$	300		2.2×10^7	$(4.0 \pm 0.7) \times 10^7$		1.82

Source: Data from Takahashi, K. et al., *J. Phys. Chem. B*, 111, 4807, 2007.

^{a} Viscosity, in mPa s.
^{b} Calculated diffusion-limited rate constant with $f(\varepsilon) = 1$, in M^{-1} s^{-1}.
^{c} Experimental rate constant for reaction 3, in M^{-1} s^{-1}.
^{d} Debye correction factor.
^{e} Glycerol–H_2O mixture.
^{f} TFSI$^-$ = N(SO$_2$CF$_3$)$_2{}^-$.

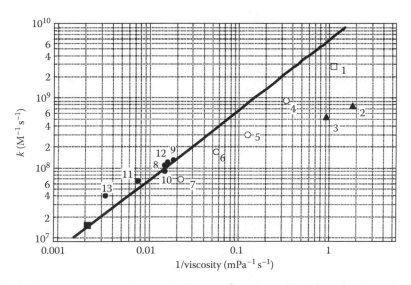

FIGURE 11.9 Plots of rate constants for reaction between I_2^- anion radicals in molecular solvents and ionic liquids as a function of the inverse of viscosity: (□) H_2O, (▲) alcohols, (○) glycerol–H_2O mixtures, (●) ionic liquids. The numbers correspond to the numbers in Table 11.2. The filled square is data for reaction between Br_2^- anion radicals in an ionic liquid. The line plots the calculated diffusion-limited rate constants for reaction between neutral molecules. (Reproduced from Takahashi, K. et al., *J. Phys. Chem. B*, 111, 4807, 2007. With permission.)

ionic liquids (except $PP13^+N(SO_2CF_3)_2^-$ and $DEME^+BF_4^-$) are almost unity (Table 11.2), indicating that the Debye factors, $f(\varepsilon)$, for the ionic liquids are close to unity and that the electrostatic repulsion between the diiodide anion radicals is unimportant.

The coincidence of the experimental rate constants of I_2^- in the ionic liquids with those calculated based on the viscosity was, to a certain extent, unusual. In previous pulse radiolysis experiments on the reaction of butylpyridinyl radical with duroquinone (Skrzypczak and Neta, 2003), the experimental reaction rates deviated significantly from calculated values based on viscosities with the simple hydrodynamic assumption. In that work, the experimental rates were almost an order of magnitude higher than the values estimated from the viscosities of the ionic liquids. Skrzypczak and Neta (2003) suggested that this difference was due to the voids that exist in ionic liquids and the possibility that reacting species diffuse through the movement of small groups of ions, whereas viscosity is related to simultaneous movement of all ions. Furthermore, Maroncelli et al. have pointed out the limitation of the continuum approximation of solvents and the importance of considering the molecularity of solvents in solvation dynamics (Maroncelli and Fleming, 1987; Papazyan and Maroncelli, 1995; Ito et al., 2004). The reason the simple hydrodynamic model can predict the present results for the reaction between I_2^- anion radicals could be due to the relatively simple molecular shape of the diiodide anion radicals. However, for the reaction of I_2^- in $PP13^+N(SO_2CF_3)_2^-$ and $DEME^+BF_4^-$, the agreement between the measured reaction rate and the calculated rate was not good. These results could reflect a limitation of the Stokes–Einstein approximation and indicate the importance of considering the molecularity of ionic liquids.

The diffusion coefficients of I_2^- anion radicals in ionic liquids have been measured recently by Kimura and Nishiyama by the transient grating method (Nishiyama et al., 2009). The diffusion coefficients of I_2^-, calculated rate constants by the Smoluchowski and Debye–Smoluchowski treatments, and the experimental rate constants for comparison are listed in Table 11.3. An important point is that the calculated values by the Smoluchowski equation, $k_{(Smoluchowski)}$, are not consistent with the experimental rate constants in the molecular solvents, while the calculated values by the Smoluchowski equation are nearly consistent with the experimental rate constants in ionic liquids. These results suggest that the Coulombic screening effect efficiently shields the electrostatic interaction between the I_2^- anion radicals in ionic liquids.

Kimura and Nishiyama also compared the measured diffusion coefficients of I_2^- with the calculated values from the Stokes–Einstein relation (Nishiyama et al., 2009). The calculated diffusion coefficients in $TMPA^+N(SO_2CF_3)_2^-$ and $PP13^+N(SO_2CF_3)_2^-$ are 1.2 and $0.62 \times 10^{11}\,m^2\,s^{-1}$, respectively, whereas the measured values are 2.5 and $1.17 \times 10^{11}\,m^2\,s^{-1}$, respectively. Therefore, the hydrodynamics approximation predicts a smaller value by a factor of 1/2 than the measured value in ionic liquids. Hence, we can say again that the viscosity is not a measure for the diffusion rate of molecules in ionic liquids.

TABLE 11.3
Diffusion Coefficients of I_2^- and Calculated Rate Constants in Molecular Solvents and Ionic Liquids

Solvent	$D_{I_2^-}^a$	$k_{(Smoluchowski)}^b$	$k_{(Debye-Smoluchowski)}^b$	$k_{(exp)}^b$
Methanol	120	550	100	77
Ethanol	79	380	35	53
$TMPA^+\ N(SO_2CF_3)_2^-$	2.5	12	0.036	11
$PP13^+\ N(SO_2CF_3)_2^-$	1.17	5.6	0.017	6.6

Source: Data from Nishiyama, Y. et al., *J. Phys. Chem. B*, 113, 5188, 2009.

[a] The unit is $10^{-11}\,m^2\,s^{-1}$.

[b] The unit is $10^7\,M^{-1}\,s^{-1}$.

11.6.2 Kinetic Salt Effects in Ionic Liquid/Methanol Mixtures

It is well known that the addition of salt increases or decreases the rates of ionic reactions, depending on whether the reactant charges are like or unlike each other. These effects have been treated with Debye–Hückel theory, although the applicable ion concentration range is limited to about 0.01 mol L^{-1}. In contrast, the concentrations of ions in ionic liquids may be in the range of 2–5 mol L^{-1}. Therefore, ionic reactions under such high ionic strength conditions are of interest. In Section 11.6.1, the disproportionation of diiodide anion radicals I_2^- in ionic liquids was discussed. In this section, we examine the kinetic salt effect on the reaction between diiodide anion radicals in methanol using a wide concentration of ionic liquid salts. To extract the specific salt effect of ionic liquids in organic solvents, we also used the inorganic salt lithium bis(trifluoromethylsulfonyl) imide ($Li^+N(SO_2CF_3)_2^-$) for comparison. Using ionic liquids, it is possible to examine the salt effect under very high mole fraction conditions (Takahashi et al., 2009a).

Figure 11.10 shows examples of transient absorption signals at 700 nm with different concentrations of $TMPA^+N(SO_2CF_3)_2^-$ in methanol. These decay signals correspond to the disproportion reaction of I_2^-. As the concentration of $TMPA^+N(SO_2CF_3)_2^-$ increases, the decay rate becomes faster, indicating that the addition of the $TMPA^+N(SO_2CF_3)_2^-$ salt accelerates the reaction rate between I_2^- anion radicals. Because the I_2^- anions are expected to be surrounded by $TMPA^+$ cations, the Coulombic repulsion between the I_2^- anion radicals is screened. In Figure 11.11, $\log(k/k_0)$ was plotted against the square root of the ionic strength, I, for three different salts, where k and k_0 are rate constants with and without salt, respectively. At low ionic strength, the values of $\log(k/k_0)$ are proportional to $\sqrt{I}$, indicating that the Debye–Hückel limiting law could be applicable. However, as the ionic strength increases, the plots deviate from a straight line. As can be seen in Figure 11.11, the effect of ionic strength on the rate constants depends on the kind of salts used. Because these salts are composed of the common anion NTf_2^-, the difference in the effect of ionic strength can be attributed to specific effects of the cations. At low ionic strength, the $TMPA^+$ cation increases the reaction rate most effectively, while the Li cation is less effective. There are a few possibilities to explain this result, such as screening effects, reduction of I_2^- anion mobility, and electrolyte ion association. At high salt concentrations, it can be expected that the degree of dissociation of $Li^+N(SO_2CF_3)_2^-$ is lower than for ionic liquids; hence, the kinetic salt effect may be less effective in $Li^+N(SO_2CF_3)_2^-$ than in ionic liquids. The screening effect on the electrostatic repulsion between I_2^- anions may depend on the size of the cations. A larger cation could effectively screen

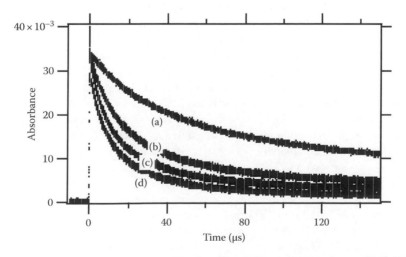

FIGURE 11.10 Transient absorption signals at 700 nm with different concentrations of $TMPA^+N(SO_2CF_3)_2^-$ in methanol. (Reproduced from Takahashi, K. et al., *Chem. Lett.*, 38, 236, 2009a. With permission.)

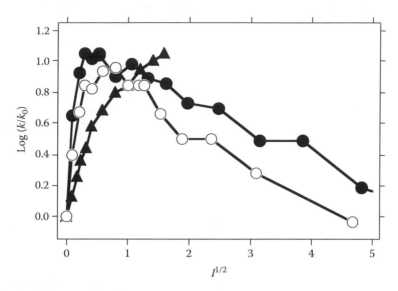

FIGURE 11.11   Relation between $\log(k/k_0)$ and the square root of the ionic strength. ($\bullet$) $TMPA^+N(SO_2CF_3)_2^-$, ($\circ$) $PP13^+N(SO_2CF_3)_2^-$, ($\blacktriangle$) $Li^+N(SO_2CF_3)_2^-$. (Reproduced from Takahashi, K. et al., *Chem. Lett.*, 38, 236, 2009a. With permission.)

the electrostatic repulsion between I_2^- anions. On the other hand, the mobility of I_2^- may be reduced by asymmetric effects of ion solvation around I_2^-. Because the smallest cation among the salts used is Li^+, the Li^+ cation may have the strongest electrostatic interaction with the I_2^- anion. This may lead to a decrease in the mobility of I_2^-. The effect of salt on the diffusion coefficient of the hydrated electron has been studied previously, and the diffusion rate of the hydrated electron decreased with the addition of salt.

A notable contrast in the kinetic salt effects for $TMPA^+N(SO_2CF_3)_2^-$ and $PP13^+N(SO_2CF_3)_2^-$ at higher salt concentrations (0.5–1.5 M) is that the values of $\log(k/k_0)$ decrease with increasing salt concentration, whereas no such decrease was observed for $Li^+N(SO_2CF_3)_2^-$ in the same concentration range. Since the ionic liquids are miscible with methanol, it is possible to explore the entire range of mole fraction. The solution viscosity rises significantly at higher proportions of ionic liquid. Accordingly, the rate constants were plotted against the inverse of the viscosity, η, in Figure 11.12. The solid line in the figure is the expectation according to Equation 11.15 for diffusion-limited rate constants of reactions between neutral molecules. In the case of ionic liquids, there is a significant change in the reaction rate with salt addition under low-viscosity conditions ($0.5 < \eta < 1$ mPa s). The solution viscosities barely increased upon the addition of ionic liquids up to almost 1 M, while greater screening of the electrostatic repulsion between diiodide anions caused the reaction rate constants to increase over one order of magnitude from 3×10^8 to $5 \times 10^9 M^{-1} s^{-1}$. However, at higher viscosities ($\eta > 2$ mPa s), there was no further increase of the reaction rates, which remained close to the behavior of the diffusion-limited rate constant, as predicted by Equation 11.15. This transition represents a change in regimes between a methanol solution of a salt and an ionic liquid containing methanol as a cosolvent. The behavior shown in Figure 11.12 indicates that the diffusion coefficients of diiodide anions in the ionic liquid/methanol mixtures follow those of the constituent ions of the ionic liquid.

For the $Li^+N(SO_2CF_3)_2^-$ salt, in the lowest viscosity region ($\eta < 0.7$ mPa s) there is no significant difference between $Li^+N(SO_2CF_3)_2^-$ and the ionic liquids. However, at a slightly higher viscosity region ($0.7 < \eta < 2.0$ mPa s), the viscosity dependence of the reaction rate is different from that for the ionic liquids. At the same viscosity level, the reaction rate for the $Li^+N(SO_2CF_3)_2^-$ solution is slower than those for ionic liquid salts. As the reaction rate constant is very close to the diffusion limit for neutral molecules calculated from Equation 11.15 at $\eta = 2.0$ mPa s,

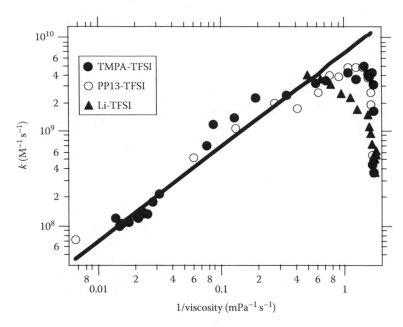

FIGURE 11.12 Plots of rate constants for reaction between anion radicals as a function of the inverse of viscosity: (•) $TMPA^+N(SO_2CF_3)_2^-$, (○) $PP13^+N(SO_2CF_3)_2^-$, (▲) $Li^+N(SO_2CF_3)_2^-$. (Reproduced from Takahashi, K. et al., *Chem. Lett.*, 38, 236, 2009a. With permission.)

it can be said that the electrostatic repulsion between I_2^- anions is almost entirely screened by $Li^+N(SO_2CF_3)_2^-$ at this concentration.

11.6.3 Reactions of Excited-State Radicals in Ionic Liquids

The photochemistry and radical reactions of benzophenone have been widely investigated in conventional solvents over the past 50 years. One can obtain insights into the characteristics of ionic liquids as solvents by comparing benzophenone reaction dynamics in ionic liquids with those in various molecular solvents. Recently, Wakasa has studied the magnetic field effects on photoinduced hydrogen abstraction reactions of triplet benzophenone with thiophenol in ionic liquids (Wakasa, 2007; Hamasaki et al., 2008). The yield of the ketyl radical decreased even with small magnetic field intensities. The high viscosity of ionic liquids could not be the reason of such a large magnetic field effect. He speculated that the cage effect could be important because a confined system is necessary for the spin conversion. This speculation is consistent with recent findings on the local structure or domains in ionic liquids.

In this section, excited-state reactions of benzophenone ketyl radical (BPH) are investigated using a two-color, twin-pulse laser. Because the absorption maximum of BPH is located around 540 nm, it is easy to excite BPH using the second harmonic of an Nd-YAG laser. In molecular solvents, a considerable number of studies of the excited benzophenone ketyl radical ($^2BPH^*$) have been carried out with absorption and emission spectroscopies (Nagarajan and Fessenden, 1984; Cai et al., 2003, 2005; Sakamoto et al., 2004). The dipole moment of $^2BPH^*$ was estimated to be 7 D, indicating that $^2BPH^*$ is highly polarized. It is reported that $^2BPH^*$ decays via radiative and nonradiative relaxations, and unimolecular and bimolecular chemical reaction processes. For the chemical reaction processes, O–H bond cleavage of $^2BPH^*$ and electron transfer from $^2BPH^*$ to benzophenone have been reported (Nagarajan and Fessenden, 1984; Cai et al., 2003, 2005; Sakamoto et al., 2004, 2005). These characteristics of $^2BPH^*$ are interesting enough to be studied in ionic liquids.

In Figure 11.13, transient kinetics of the excited state of BPH were examined in (a) methanol and (b) $Bmim^+N(SO_2CF_3)_2^-$, respectively. The transient kinetics were monitored at 500 nm, and

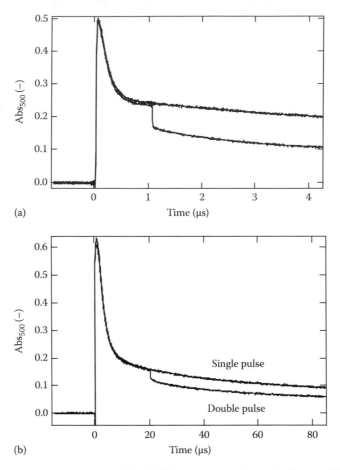

(a)

(b)

FIGURE 11.13 Transient absorption signal for BPH in (a) methanol and (b) Bmim$^+$ N(SO$_2$CF$_3$)$_2^-$ with the second 532 nm pulse.

CCl$_4$ was added as a reaction partner of the excited state of BPH. The BPH formed by irradiation with a 355 nm pulse was excited with a 532 nm pulse at 20 ms after the first 355 nm pulse. When the sample containing CCl$_4$ was irradiated with the 532 nm laser pulse, the transient absorption signal displayed an instrument-limited decrease, and the absorption decrease did not recover within the lifetime of BPH. As the concentration of CCl$_4$ increased, the amount of permanent bleaching was increased. Therefore, the bleaching can be attributed to the reaction of the excited state of BPH (^{2}BPH*) with CCl$_4$. It has been reported that ^{2}BPH* decays via both radiative and nonradiative relaxations, and unimolecular and bimolecular chemical reaction processes. The unimolecular chemical reaction process has been assigned to the O–H bond cleavage. For the bimolecular reaction of ^{2}BPH*, electron transfer from ^{2}BPH* to BP has been suggested. By considering the previously suggested reaction paths, the following reactions may be considered to explain the present permanent bleaching:

$$^2\text{BPH*} \xrightarrow{k_1} \text{BPH} \tag{11.16}$$

$$^2\text{BPH*} \xrightarrow{k_2} \text{BP} + \text{H} \tag{11.17}$$

$$^2\text{BPH*} + \text{BP} \xrightarrow{k_3} \text{BPH}^+ + \text{BP}^- \tag{11.18}$$

$$^2\mathrm{BPH}^* + \mathrm{CCl}_4 \xrightarrow{\;k_4\;} \mathrm{P} \tag{11.19}$$

Because the chemical reactions (11.16) through (11.19) cannot reproduce BPH, the following relation may be established between the observed bleaching and the above reactions:

$$\frac{\mathrm{OD}_0 - \mathrm{OD}_t}{F \times \mathrm{OD}_0} = \frac{k_2[^2\mathrm{BPH}^*] + k_3[^2\mathrm{BPH}^*][\mathrm{BP}] + k_4[^2\mathrm{BPH}^*][\mathrm{CCl}_4]}{k_1[^2\mathrm{BPH}^*] + k_2[^2\mathrm{BPH}^*] + k_3[^2\mathrm{BPH}^*][\mathrm{BP}] + k_4[^2\mathrm{BPH}^*][\mathrm{CCl}_4]} \tag{11.20}$$

where OD_0 and OD_t are the optical densities immediately before and after the 532 nm pulse, respectively. F is a constant that depends on the experimental conditions such as laser intensity and quantum efficiency for the formation of $^2\mathrm{BPH}^*$ by a 532 nm irradiation to BPH. When the concentration of CCl_4 is high enough for the condition $k_4[\mathrm{CCl}_4] \gg (k_2 + k_3[\mathrm{BP}])$ to be satisfied, Equation 11.20 may be written as follows:

$$\frac{\mathrm{OD}_0}{\mathrm{OD}_0 - \mathrm{OD}_t} = \frac{1}{F} + \frac{k_1}{F \times k_4[\mathrm{CCl}_4]} \tag{11.21}$$

Therefore, at high CCl_4 concentrations, the value of F and the ratio of k_1/k_4 can be determined from a plot of $\mathrm{OD}_0/(\mathrm{OD}_0 - \mathrm{OD}_t)$ versus $1/[\mathrm{CCl}_4]$. Consequently, the factor F in both ethanol and $\mathrm{Bmim}^+\,\mathrm{N(SO_2CF_3)_2}^-$ was determined to be 0.7 ± 0.02. Once factor F is determined, the ratio k_1/k_4 can be elucidated by a least squares fitting of the data. The fitting results are $k_1/k_4 = 7.8 \times 10^{-2}$ (in methanol) and 9.5×10^{-2} (in $\mathrm{Bmim}^+\,\mathrm{N(SO_2CF_3)_2}^-$). k_1 was estimated to be 4×10^9 and $2 \times 10^8\,\mathrm{s}^{-1}$ from the lifetime of $\mathrm{BPH(D_1)}$ in methanol and $\mathrm{Bmim}^+\,\mathrm{N(SO_2CF_3)_2}^-$, respectively. Therefore, k_4 in methanol and $\mathrm{Bmim}^+\mathrm{N(SO_2CF_3)_2}^-$ is found to be 5.1×10^{10} and $2.1 \times 10^9\,\mathrm{M}^{-1}\,\mathrm{s}^{-1}$, respectively, suggesting that the reaction of $\mathrm{BPH(D_1)}$ with CCl_4 may be a diffusion-limited reaction. However, despite the fact that the viscosity of $\mathrm{Bmim}^+\,\mathrm{N(SO_2CF_3)_2}^-$ is about 100 times higher than that of methanol, the rate in $\mathrm{Bmim}^+\,\mathrm{N(SO_2CF_3)_2}^-$ is only 20 times slower than that in methanol. Therefore, this result suggests that the diffusion of $^2\mathrm{BPH}^*$ cannot be described by the hydrodynamic approximation using the viscosity of ionic liquid $\mathrm{Bmim}^+\mathrm{N(SO_2CF_3)_2}^-$, as discussed earlier in this chapter.

11.7 SUMMARY

This chapter provides an overview of current topics in the radiation chemistry and photochemistry of ionic liquids. The photochemistry of benzophenone in ionic liquids was also covered, but a complete review of photochemical investigations in ionic liquids is well beyond the scope of this chapter.

The ionic liquid radiation chemistry described here has focused on electron dynamics and reactivity. However, the reactions of hole species and radiation-induced radicals are equally important, although they are often harder to study due to weak or nonexistent optical spectroscopic features. Recently, EPR studies of irradiated and photolyzed ionic liquid glasses revealed important information about the initial distributions of holes in several classes of ionic liquids, and their implications for pathways of radiolytic damage accumulation (Shkrob et al., 2007; Shkrob and Wishart, 2009; Wishart and Shkrob, 2009). The results are consistent with the findings of the elegant and detailed radiolytic product studies using electrospray mass spectroscopy by the Moisy group at CEA Marcoule (Berthon et al., 2006; Bosse et al., 2008; Le Rouzo et al., 2009). Nevertheless, further time-resolved investigations of hole and radical chemistry in ionic liquids are sorely needed. It is expected that transient EPR and vibrational spectroscopic radiolysis studies on ionic liquids will be conducted in the near future. Since ionic liquids show potential as a medium for the processing of spent nuclear fuel, further understanding of the interactions of charged particles and photons with ionic liquids could make a significant contribution to future world energy supplies.

ACKNOWLEDGMENTS

This work was supported in part at the Brookhaven National Laboratory (BNL) by the U.S. Department of Energy Office of Basic Energy Sciences, Division of Chemical Sciences, Geosciences, and Biosciences under contract # DE-AC02-98CH10886. Kenji Takahashi acknowledges the Ministry of Education, Culture, Sports, Science and Technology (MEXT) of Japan for a Grant-in-Aid for Scientific Research (Priority Area 452 "Science of Ionic Liquids").

REFERENCES

Allen, D., Baston, G., Bradley, A., Gorman, T., Haile, A., Hamblett, I., Hatter, J., Healey, M., Hodgson, B., Lewin, R., Lovell, K., Newton, B., Pitner, W., Rooney, D., Sanders, D., Seddon, K., Sims, H., and Thied, R. 2002. An investigation of the radiochemical stability of ionic liquids. *Green Chem.* 4: 152–158.

Arzhantsev, S., Jin, H., Baker, G. A., and Maroncelli, M. 2007. Measurements of the complete solvation response in ionic liquids. *J. Phys. Chem. B* 111: 4978–4989.

Behar, D., Gonzalez, C., and Neta, P. 2001. Reaction kinetics in ionic liquids: Pulse radiolysis studies of 1-butyl-3-methylimidazolium salts. *J. Phys. Chem. A* 105: 7607–7614.

Behar, D., Neta, P., and Schultheisz, C. 2002. Reaction kinetics in ionic liquids as studied by pulse radiolysis: Redox reactions in the solvents methyltributylammonium bis(trifuluoromethylsulfonyl)imide and n-butylpyridinium tetrafluoroborate. *J. Phys. Chem. A* 106: 3139–3147.

Berthon, L., Nikitenko, S. I., Bisel, I., Berthon, C., Faucon, M., Saucerotte, B., Zorz, N., and Moisy, P. 2006. Influence of gamma irradiation on hydrophobic room-temperature ionic liquids [BuMeIm]PF_6 and [BuMeIm]$(CF_3SO_2)_2$N. *Dalton Trans.* 2526–2534.

Binnemans, K. 2007. Lanthanides and actinides in ionic liquids. *Chem. Rev.* 107: 2592–2614.

Bosse, E., Berthon, L., Zorz, N., Monget, J., Berthon, C., Bisel, I., Legand, S., and Moisy, P. 2008. Stability of [MeBu$_3$N][Tf$_2$N] under gamma irradiation. *Dalton Trans.* 924–931.

Cai, X., Hara, M., Kawai, K., Tojyo, S., and Majima, T. 2003. Sensitized reactions by benzophenones in the higher excited state. *Chem. Phys. Lett.* 371: 68–73.

Cai, X., Sakamoto, M., Fujitsuka, M., and Majima, T. 2005. Higher triplet excited states of benzophenones and bimolecular triplet energy transfer measured by using nanosecond-picosecond two-color/two-laser flash photolysis. *Chem. Eur. J.* 11: 6471–6477.

Castriota, M., Caruso, T., Agostino, R. G., Cazzanelli, E., Henderson, W. A., and Passerini, S. 2005. Raman investigation of the ionic liquid *N*-methyl-*N*-propylpyrrolidinium bis(trifluoromethanesulfonyl)imide and its mixture with LiN(SO$_2$CF$_3$)$_2$. *J. Phys. Chem. A.* 109: 92–96.

Chandrasekhar, N., Schalk, O., and Unterreiner, A. N. 2008. Femtosecond UV excitation in imidazolium-based ionic liquids. *J. Phys. Chem. B.* 112: 15718–15724.

Chaumont, A. and Wipff, G. 2004. Solvation of uranyl(II) and europium(III) cations and their chloro complexes in a room-temperature ionic liquid. A theoretical study of the effect of solvent "humidity". *Inorg. Chem.* 43: 5891–5901.

Cocalia, V. A., Gutowski, K. E., and Rogers, R. D. 2006. The coordination chemistry of actinides in ionic liquids: A review of experiment and simulation. *Coord. Chem. Rev.* 250: 755–764.

Debye, P. 1942. Reaction rates in ionic solutions. *Trans. Electrochem. Soc.* 82: 265–272.

Del Popolo, M. G. and Voth, G. A. 2004. On the structure and dynamics of ionic liquids. *J. Phys. Chem. B* 108: 1744–1752.

Dietz, M. L. 2006. Ionic liquids as extraction solvents: Where do we stand? *Sep. Sci. Technol.* 41: 2047–2063.

Earle, M. J. and Seddon, K. S. 2000. Ionic liquids. Green solvents for the future. *Pure Appl. Chem.* 72: 1391–1398.

Ediger, M. D. 2000. Spatially heterogeneous dynamics in supercooled liquids. *Annu. Rev. Phys. Chem.* 51: 99–128.

Funston, A. M. and Wishart, J. F. 2005. Dynamics of fast reactions in ionic liquids. In *Ionic Liquids IIIA: Fundamentals, Progress, Challenges, and Opportunities, Properties and Structure*, R. D. Rogers and K. R. Seddon, (eds.), pp. 102–116. Washington, DC: American Chemical Society.

Funston, A. M., Fadeeva, T. A., Wishart, J. F., and Castner, E. W. Jr. 2007. Fluorescence probing of temperature-dependent dynamics and friction in ionic liquid local environments. *J. Phys. Chem. B* 111: 4963–4977.

Giraud, G., Gordon, C. M., Dunkin, I. R., and Wynne, K. 2003. The effects of anion and cation substitution on the ultrafast solvent dynamics of ionic liquids: A time-resolved optical Kerr-effect spectroscopic study. *J. Chem. Phys.* 119: 464–477.

Glezen, M. M., Chernovitz, A. C., and Jonah, C. D. 1992. Pulse radiolysis of dextran-water solutions. *J. Phys. Chem.* 96: 5180–5183.

Grodkowski, J. and Neta, P. 2002a. Reaction kinetics in the ionic liquid methyltributylammonium bis(trifluoromethylsufonyl)imide. Pulse radiolysis study of CF_3 radical reactions. *J. Phys. Chem. A* 106: 5468–5473.

Grodkowski, J. and Neta, P. 2002b. Reaction kinetics in the ionic liquid methyltributylammonium bis(trifluoromethylsulfonyl)imide. Pulse radiolysis study of 4-mercaptobenzoic acid. *J. Phys. Chem. A* 106: 9030–9035.

Grodkowski, J. and Neta, P. 2002c. Formation and reaction of Br_2^- radicals in the ionic liquid methyltributylammonium bis(trifluoromethylsulfonyl)imide and in other solvents. *J. Phys. Chem. A* 106: 11130–11134.

Grodkowski, J., Neta, P., and Wishart, J. F. 2003. Pulse radiolysis study of the reactions of hydrogen atoms in the ionic liquid methyltributylammonium bis(trifluoromethyl)sulfonyl]imide. *J. Phys. Chem. A* 107: 9794–9799.

Hamasaki, A., Yago, T., Takamasu, T., Kido, G., and Wakasa, M. 2008. Anomalous magnetic field effects on photochemical reactions in ionic liquid under ultrahigh fields of up to 28 T. *J. Phys. Chem. B* 112: 3375–3379.

Han, X. and Armstrong, D. W. 2007. Ionic liquids in separations. *Acc. Chem. Res.* 40: 1079–1086.

Hayashi, S., Ozawa, R., and Hamaguchi, H. 2003. Raman spectra, crystal polymorphism, and structure of a prototype ionic-liquid [bmim]Cl. *Chem. Lett.* 32: 498–499.

Hayashi, K., Nemoto, Y., Akuto, K., and Sakurai, Y. 2005. Alkylated imidazolium salt electrolyte for lithium cells. *J. Power Source* 146: 689–692.

Holbrey, J. D., Reichert, W. M., and Rogers, R. D. 2004. Crystal structures of imidazolium bis(trifluoromethanesulfonyl)imide ionic liquid salts: The first organic salt with a *cis*-TFSI anion conformation. *Dalton Trans.* 15: 2267–2271.

Hu, Z. and Margulis, C. J. 2006. A study of the time-resolved fluorescence spectrum and red edge effect of ANF in a room-temperature ionic liquid. *J. Phys. Chem. B* 110: 11025–11028.

Ito, N., Arzhantsev, S., Heitz, M., and Maroncelli, M. 2004. Solvation dynamics and rotation of coumarin 153 in alkylphosphonium ionic liquids. *J. Phys. Chem. B* 108: 5771–5777.

Jacquemin, J., Husson, P., Padua, A. A. H., and Majer, V. 2006. Density and viscosity of several pure and water-saturated ionic liquids. *Green Chem.* 8: 172–180.

Jensen, M. P., Neuefeind, J., Beitz, J. V., Skanthakumar, S., and Soderholm, L. 2003. Mechanisms of metal ion transfer into room-temperature ionic liquids: The role of anion exchange. *J. Am. Chem. Soc.* 125: 15466–15473.

Jin, H., Li, X., and Maroncelli, M. 2007. Heterogeneous solute dynamics in room temperature ionic liquids. *J. Phys. Chem. B* 111: 13473–13478.

Jonah, C. D., Miller, J. R., and Matheson, M. S. 1977. The reaction of the precursor of the hydrated electron with electron scavengers. *J. Phys. Chem.* 81: 1618–1622.

Jonah, C. D., Bartels, D. M., and Chernovitz, A. C. 1989. Primary processes in the radiation chemistry of water. *Int. J. Radiat. Appl. Inst. Part C. Radiat. Phys. Chem.* 34: 145–156.

Kambhampati, P., Son, D. H., Kee, T. W., and Barbara, P. F. 2001. Delocalizing electrons in water with light. *J. Phys. Chem. A* 105: 8269–8272.

Katoh, R., Yoshida, Y., Katsumura, Y., and Takahashi, K. 2007. Electron photodetachment from iodide in ionic liquids through charge-transfer-to-solvent band excitation. *J. Phys. Chem. B* 111: 4770–4774.

Kawano, R. and Watanabe, M. 2003. Equilibrium potentials and charge transport of and I-/I3- redox couple in ionic liquid. *Chem. Commun.* 330–331.

Kawano, R. and Watanabe, M. 2005. Anomaly of charge transport of an iodide/tri-iodide redox couple in an ionic liquid and its importance in dye-sensitized solar cells. *Chem. Commun.* 2107–2109.

Lam, K. Y. and Hunt, J. W. 1975. Picosecond pulse radiolysis VI. Fast electrons in concentrated solutions of scavengers in water and alcohols. *Int. J. Radiat. Phys. Chem.* 7: 317–338.

Le Rouzo, G., Lamouroux, C., Dauvois, V., Dannoux, A., Legand, S., Durand, D., Moisy P., and Moutiers, G. 2009. Anion effect on radiochemical stability of room-temperature ionic liquids under gamma irradiation. *Dalton Trans.* DOI: 10.1039/b903005k.

Lewis, M. A. and Jonah, C. D. 1986. Evidence for two electron states in solvation and scavenging process in alcohols. *J. Phys. Chem.* 90: 5367–5372.

Lopes, J. N. C. and Padua, A. A. H. 2006. Nanostructural organization in ionic liquids. *J. Phys. Chem. B* 110: 3330–3335.

Luo, H., Dai, S., Bonnesen, P. V., Haverlock, T. J., Moyer, B. A., and Buchanan, A. C. III, 2006. A striking effect of ionic-liquid anions in the extraction of Sr^{2+} and Cs^+ by dicyclohexano-18-crown-6. *Solvent Extr. Ion Exch.* 24: 19–31.

Lynch, D. W. and Robinson, D. A. 1968. Study of the F' center in several alkali halides. *Phys. Rev.* 174: 1050–1059.

MacFarlane, D. R., Forsyth, M., Howlett, P. C., Pringle, J. M., Sun, J., Annat, G., Neil, W., and Izgorodina, E. I. 2007. Ionic liquids in electrochemical devices and processes: Managing interfacial electrochemistry. *Acc. Chem. Res.* 40: 1165–1173.

Mandal, P. K., Sarkar, M., and Samanta, A. 2004. Excitation-wavelength-dependent fluorescence behavior of some dipolar molecules in room-temperature ionic liquids. *J. Phys. Chem. A* 108: 9048–9053.

Marcinek, A., Zielonka, J., Gebicki, J., Gordon, C. M., and Dunkin, I. R. 2001. Ionic liquids: Novel media for characterization of radical ions. *J. Phys. Chem. A* 105: 9305–9309.

Maroncelli, M. and Fleming, G. R. 1987. Picosecond solvation dynamics of coumarin 153: The importance of molecular aspects of solvation. *J. Chem. Phys.* 86: 6221–6239.

Nagarajan, V. and Fessenden, R. W. 1984. Flash photolysis of transient radicals. Benzophenone ketyl radical. *Chem. Phys. Lett.* 112: 207–211.

Nikitenko, S. I., Cannes, C., Le Naour, C., Moisy, P., and Trubert, D. 2005. Spectroscopic and electrochemical studies of U(IV)-hexachloro complex in hydrophobic room-temperature ionic liquids [BuMeIm][Tf_2N] and [$MeBu_3N$][Tf_2N]. *Inorg. Chem.* 44: 9497–9505.

Nishiyama, Y., Terazima, M., and Kimura, Y. 2009. Charge effect on the diffusion coefficient and the bimolecular reaction rate of diiodide anion radical in room temperature ionic liquids. *J. Phys. Chem. B* 113: 5188–5193.

Ozawa, R., Hayashi, S., Saha, S., Kobayashi, A., and Hamaguchi, H. 2003. Rotational isomerism and structure of the 1-butyl-3-methylimidazolium cation in the ionic liquid state. *Chem. Lett.* 32: 948–949.

Papageorgiou, N., Athanassov, Y., Armand, M., Bonhôte, P., Pettersson, H., Azam, A., and Grätzel, M. 1996. The performance and stability of ambient temperature molten salts for solar cell applications. *J. Electrochem. Soc.* 143: 3099–3108.

Papazyan, A. and Maroncelli, M. 1995. Rotational dielectric friction and dipole solvation: Tests of theory based on simulations of simple model solutions. *J. Chem. Phys.* 102: 2888–2919.

Plechkova, N. V. and Seddon, K. R. 2008. Application of ionic liquids in the chemical industry. *Chem. Soc. Rev.* 37: 123–150.

Sakamoto, M., Cai, X., Hara, M., Tojyo, S., Fujitsuka, M., and Majima, T. 2004. Transient absorption spectra and lifetime of benzophenone ketyl radicals in the excited state. *J. Phys. Chem. A* 108: 8147–8150.

Shigeto, S. and Hamaguchi, H. 2006. Evidence for mesoscopic local structures in ionic liquids: CARS signal distribution of $C_n$mim[PF6] (n = 4, 6, 8). *Chem Phys. Lett.* 427: 329–332.

Shirota, H. and Castner, E. W. Jr. 2005. Physical properties and intermolecular dynamics of an ionic liquid compared with its isoelectronic neutral binary solution. *J. Phys. Chem. A* 109: 9388–9392.

Shirota, H., Funston, A. M., Wishart, J. F., and Castner, E. W. Jr. 2005. Ultrafast dynamics of pyrrolidinium cation ionic liquids. *J. Chem. Phys.* 122: 84512(12 pages).

Shirota, H., Wishart, J. F., and Castner, E. W. Jr. 2007. Intermolecular interactions and dynamics of room temperature ionic liquids that have silyl- and siloxy-substituted imidazolium cations. *J. Phys. Chem. B* 111: 4819–4829.

Shkrob, I. A., Chemerisov, S. D., and Wishart, J. F. 2007. The initial stages of radiation damage in ionic liquids and ionic liquid-based extraction systems. *J. Phys. Chem. B* 111: 11786–11793.

Shkrob, I. A. and Wishart, J. F. 2009. Charge trapping in imidazolium ionic liquids. *J. Phys. Chem. B* 113: 5582–5592.

Silva, C., Walhout, P. K., Reid, P. J., and Barbara, P. F. 1998. Detailed investigation of the pump-probe spectroscopy of the equilibrated solvated electron in alcohol. *J. Phys. Chem. A* 102: 5701–5707.

Skrzypczak, A. and Neta, P. 2003. Diffusion-controlled electron-transfer reactions in ionic liquids. *J. Phys. Chem. A* 107: 7800–7803.

Stepinski, D. C., Young, B. A., Jensen, M. P., Rickert, P. G., Dzielawa, J. A., Dilger, A. A., Rausch, D. J., and Dietz, M. L. 2006. Application of ionic liquids in actinide and fission product separations: Progress and prospects. In *Separations for the Nuclear Fuel Cycle in the 21st Century*, G. Lumetta, K. I. Nash, S. B. Clark, and J. I. Friese, (eds.), pp. 233–245. Washington, DC: American Chemical Society.

Takahashi, K., Sakai, S., Tezuka, H., Hiejima, Y., Katsumura, Y., and Watanabe, M. 2007. Reaction between diiodide anion radicals in ionic liquids. *J. Phys. Chem. B* 111: 4807–4811.

Takahashi, K., Sato, T., Katsumura, Y., Yang, J., Kondoh, T., Yoshida, Y., and Katoh, R. 2008. Reactions of solvated electrons with imidazolium cations in ionic liquids. *Rad. Phys. Chem.* 77: 1239–1243.

Takahashi, K., Tezuka, H., Saoth, T., Katsumura, Y., Watanabe, M., Crowell, R. A., and Wishart, J. F. 2009a. Kinetic salt effects on an ionic reaction in ionic liquid/methanol mixtures: Viscosity and coulombic screening effects. *Chem. Lett.* 38: 236–237.

Takahashi, K., Suda, K., Seto, T., Katsumura, Y., Katoh, R., Crowell, R. A., and Wishart, J. F. 2009b. Photo-detrapping of solvated electrons in an ionic liquid. *Radiat. Phys. Chem.* 78: 1129–1132.

Triolo, A., Russina, O., Bleif, H.-J., and Di Cola, E. 2007. Nanoscale segregation in room temperature ionic liquids. *J. Phys. Chem. B* 111: 4641–4644.

van Rantwijk, F. and Sheldon, R. A. 2007. Biocatalysis in ionic liquids. *Chem. Rev.* 107: 2757–2785.

Wang, Y. T. and Voth, G. A. 2005. Unique spatial heterogeneity in ionic liquids. *J. Am. Chem. Soc.* 127: 12192–12193.

Wang, P., Zakeeruddin, S. M., Moser, J. E., Humphry-Baker, R., and Grätzel, M. 2004. A solvent-free, $SeCN^-/(SeCN)_3^-$ based ionic liquid electrolyte for high-efficiency dye-sensitized nanocrystalline solar cells. *J. Am Chem. Soc.* 126: 7164–7165.

Wakasa, M. 2007. The magnetic field effects on photochemical reactions in ionic liquids. *J. Phys. Chem. B* 111: 9434–9436.

Weingartner, H. 2008. Understanding ionic liquids at the molecular level: Facts, problems, and controversies. *Angew. Chem. Int. Ed.* 47: 654–670.

Welton, T. 2004. Ionic liquids in catalysis. *Coord. Chem. Rev.* 248: 2459–2477.

Wishart, J. F. 2003. Radiation chemistry of ionic liquids: Reactivity of primary species. In *Ionic Liquids as Green Solvents: Progress and Prospects*, R. D. Rogers and K. R. Seddon, (eds.), pp. 381–396. Washington, DC: American Chemical Society.

Wishart, J. F. 2009. Energy applications of ionic liquids. *Energy Environ. Sci.* 2: 956–961.

Wishart, J. F. and Neta, P. 2003. Spectrum and reactivity of the solvated electron in the ionic liquid methyltri-butylammonium bis(trifluoromethylsulfonyl)imide. *J. Phys. Chem. B* 107: 7261–7267.

Wishart, J. F. and Shkrob, I. A., 2009. The radiation chemistry of ionic liquids and its implications for their use in nuclear fuel processing. In *Ionic Liquids: From Knowledge to Application*. R. D. Rogers, N. V. Plechkova, and K. R. Seddon, (eds.). ACS Symp. Ser. 1030, American Chemical Society, Washington, DC, pp. 119–134.

Wishart, J. F., Cook, A. R., and Miller, J. R. 2004. The LEAF picosecond pulse radiolysis facility at Brookhaven National Laboratory. *Rev. Sci. Instrum.* 75: 4359–4366.

Wishart, J. F., Lall-Ramnarine, S. I., Raju, R., Scumpia, A., Bellevue, S., Ragbir, R., and Engel, R. 2005. Effects of functional group substitution on electron spectra and solvation dynamics in a family of ionic liquids. *Radiat. Phys. Chem.* 72: 99–104.

Wishart, J. F., Funston, A. M., and Szreder, T., 2006. Radiation chemistry of ionic liquids. In *Molten Salts XIV*, R. A. Mantz, P. C. Trulove, H. C. De Long, G. R. Stafford, R. Hagiwara, and D. A. Costa, (eds.), pp. 802–813. Pennington, NJ: The Electrochemical Society.

Wishart, J. F., Szreder, T., and Funston, A. M., manuscript in preparation.

Yokoyama, K., Silva, C., Son, D. H., Walhout, P. K., and Barbara, P. F. 1998. Detailed investigation of the femtosecond pump-probe spectroscopy of the hydrated electron. *J. Phys. Chem. A* 102: 6957–6966.

Zhang, L., Yan, S., Cukier, R. I., and Bu, Y. 2008. Solvation of excess electrons in LiF ionic pair matrix: Evidence for a solvated dielectron from *ab initio* molecular dynamics simulations and calculations. *J. Phys. Chem. B* 112: 3767–3772.

12 Time-Resolved Study on Nonhomogeneous Chemistry Induced by Ionizing Radiation with Low Linear Energy Transfer in Water and Polar Solvents at Room Temperature

Vincent De Waele
Université Paris-Sud
Orsay, France

Isabelle Lampre
Université Paris-Sud
Orsay, France

Mehran Mostafavi
Université Paris-Sud
Orsay, France

CONTENTS

12.1 INTRODUCTION

Among all the radiation–chemical reactions that have been studied in solution, the most important is the decomposition of pure water, the "life solvent" itself. At the beginning of the last century, fruitful observations were made on the radiation chemistry of water (Giesel, 1900; Curie and Debierne, 1901; Kernbaum, 1909; Duane and Scheuer, 1913). An important breakthrough is due to Fricke's studies on x-ray effects that appeared to be very similar to those of γ-rays (Fricke, 1934). Since that demonstration, a wealth of data has been obtained to understand the mechanisms of water decomposition under irradiation, particularly after the advent of nuclear energy and the extensive research program around the Manhattan Project (Allen, 1989). One of the singular aspects of radiation chemistry is energy deposition in the media, characterized by linear energy transfer (LET). In contrast to UV–visible and IR light absorption that is homogeneous, the energy of ionizing radiation is nonhomogeneously deposited in solutions. The direct visualization of α-particles branch tracks in the Wilson cloud chamber demonstrated the nonhomogeneity of the energy deposition. The macroscopic visualization of such an effect can also be performed with the electron beam of an accelerator focused on a transparent block of polymers (Figure 12.1) (Gross, 1957, 1958). This important aspect in radiation physics and chemistry led to the development of kinetics of nonhomogeneous processes at the beginning of the last century (Freeman, 1987). When an ionizing radiation passes through a solution, it loses energy and generates excited or ionized molecules. The ejected electrons, called secondary electrons, possess enough kinetic energy to further ionize and excite the molecules and to break and build chemical bonds. The initial event, energy deposition, classified as radiation

FIGURE 12.1   This Captured Lightning® sculpture is created by irradiating a block of acrylic (Plexiglas) by a beam of electrons from a 3 MV particle accelerator. The first Lichtenberg figures were actually two-dimensional patterns formed in dust on the surface of charged insulating plates in the laboratory of their discoverer, German physicist Georg Christoph Lichtenberg (1742–1799). Pioneering research on the detailed behavior of charge storage and movement within dielectrics was performed by Bernhard Gross in the early 1950s. Gross confirmed that internal Lichtenberg figures could be created within a variety of polymers and glasses by injecting them with high energy electrons using a linear accelerator (LINAC).

physics, is followed by a sequence of events such as ionization and excitation processes that lead to the final products of solvent decomposition. The yields of the final products depend on the energy distribution in the solvent. In solvents, the primary free radicals formed are assumed to be distributed in small volumes, called spurs, favoring radical–radical reactions according to the mechanism first suggested by Weiss (1944) and Allen (1948). For low LET, it is assumed that the spurs are separated by large distances relative to their diameter that is around a few nanometers, and that in each spur a few tens of eV energy is deposited. If we consider that the average energy in a spur is around 40 eV, the average separation of spurs along the track is 80/LET. For low LET (0.2 eV nm^{-1}) the spurs are hundreds of nanometers apart (Magee and Chatterjee, 1987). To elucidate this special feature of energy deposition, one of the decisive points was the observation that the yield of the final products depends on the LET, that is, on the energy distribution in the solvent. During World War II, the idea of the spur and track structure was successfully supported by finding smaller radical yields (for e_s^-, H$^{\cdot}$, and OH$^{\cdot}$) and larger molecular yields (H$_2$ and H$_2$O$_2$) with high LET radiolysis.

To follow the fast radiation-induced reactions in solvents, researchers have developed tools to deliver ionizing radiations for their scientific studies. The advent of electron accelerators generated the first pulse-radiolysis experiments in the 1960s. Then, the knowledge on water decomposition allowed the use of water radiolysis to trigger other reactions in a controlled manner. So, water and more generally solvent radiolysis provide a very versatile method of generating radicals in a quantitative way. The species issued from the solvent radiolysis can produce, in solution, different radicals, redox states, or new chemical bonds through reactions with solutes or other solvent radicals. The key point in these studies is the knowledge of the yields of the radical and molecular products at a given time after energy deposition. The reactions with solutes at low concentration occur usually in the homogeneous steps, that is, a hundred nanoseconds after energy deposition by ionizing radiation with low LET such as γ-rays or accelerated electrons. The yields at the homogeneous step depend on the initial distribution and chemical reactivity of the different radicals and molecules formed by irradiation, that is revealed by the knowledge of the yield in the time window before the homogeneous step. Today, the yields at the homogeneous step are accurately known for the different radicals and molecules issued from water irradiation, but their values in the subnanosecond time window are still debated.

To elucidate the mechanism of solvent ionization and spur reactions, two approaches are possible. In the first approach, the various radiation–chemical yields at different times can be estimated by the use of solutes at high concentrations that selectively scavenge the radicals induced by irradiation in the solvent, and by the analysis of the products. The second method consists of time-resolved measurements to probe the media at different times after energy deposition and to follow the radicals or the product of the radical scavenging using the spectrophotometric technique. In his book published in 2004, Buxton (2004) presented an overview of the radiation chemistry of water at room temperature and atmospheric pressure with low LET from the start of its systematic study. To demonstrate the time evolution of the radical's yields, he presented mostly the results concerning the scavenging method. Since the publication of Buxton's chapter, new results within the short-time domain after energy deposition in the solvent have been reported thanks to the development of new accelerators dedicated to pulse radiolysis and the amelioration of the detection systems.

In this chapter, we focus on the chemistry induced by irradiation before the homogeneous step when the spurs are present in the solvent. The kinetics in water and also in some polar solvents, ranging from a few hundred femtoseconds to a few nanoseconds, is discussed. After a general discussion of solvent decomposition under ionizing radiation, the recent time-resolved measurements concerning the solvation dynamics of thermalized electron generated in water and other polar solvents are discussed first. New results are reported to discuss this very fast step that is usually studied by femtosecond laser spectroscopy. Then, recent progress in the picoseconds pulse-radiolysis technique is presented and new data concerning the time-dependent yields of hydrated electron, e_{hyd}^-, and hydroxyl radical, OH$^{\cdot}$, are reviewed.

12.2 SOLVENT DECOMPOSITION

A cascade of events is initiated by the absorption of energy by solvent from the incident ionizing radiation. The first step, shorter than 10^{-16} s, corresponds to the formation of an excited-state electron and parent cation. For a given solvent denoted as RH, the ionizing radiation with low LET induces RH^{+}, RH^{*}, and e^{-} within this step (Figure 12.2) in small clusters called spurs that on average contain two or three ion pairs. The ejected electron generally has sufficient energy to ionize or further excite solvent molecules. Mozumder (2002) reported that ionization potential in liquid water defined as the minimum energy required to generate a quasi-free electron in the band gap has a value of 11.7 ± 0.2 eV (see Figure 12.3). Nevertheless, low-energy processes, down to 6.5 eV, also result in a small yield of e_{aq}^{-} via optical charge transfer or photoinduced electron transfer

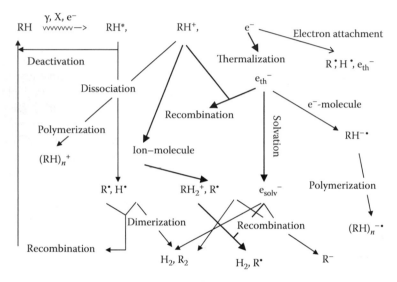

FIGURE 12.2 General scheme for the chemical events occurring after energy deposition in RH solvent. The main reactions in polar solvent are indicated with bold arrows.

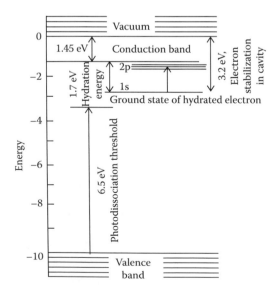

FIGURE 12.3 Energy levels of liquid water and hydrated electron.

(Sander et al., 1993). Laser experiments performed by Bartels and Crowell (2000) with photon energy up to 9.3 eV support a concerted proton-coupled electron transfer mechanism, first proposed by Keszei and Jay-Gerin (1992). When the secondary electron loses the rest of its excess energy, the so-called thermalized electron is formed. The excited states of the solvent molecule undergo dissociation processes and the radical cation reacts quickly with another solvent molecule to form RH_2^+ and $R^{\bullet}$. Solvation processes of electrons and recombination reactions between different species take place in the spurs in parallel with diffusion of radicals out of spurs. This competition controls the yields of the different products in the homogeneous step. The time of the end of spur reactions depends on the solvent and temperature.

In the case of water, the current state of knowledge can be summarized by the reactions reported in Figure 12.4. In less than 1 as (attosecond) after energy deposition, H_2O^*, $H_2O^{\bullet+}$, and e^- are generated in the solvent. As the rate constant of the ion–molecule reaction is known to occur in the gas phase with a rate of 8×10^{12} dm^3 mol^{-1} s^{-1}, the positive radical ion $H_2O^{\bullet+}$ is believed to undergo the ion–molecule reaction in approximately 10^{-14} s. In fact, the reaction between $H_2O^{\bullet+}$ and H_2O occurs without any diffusion in the prethermal step. In the past, the observation of contribution of the water cation, $H_2O^{\bullet+}$, in bulk water at 460 nm was stated (Gauduel et al., 1990; Gauduel, 1995). But we know that the very fast proton transfer reaction leads into H_3O^+ and OH$^{\bullet}$ radical in less than 10 fs. Therefore, perhaps an ultrashort pulse of a few attoseconds or femtoseconds is needed to observe the H_2O^+ cation in bulk water. Up to now, such an experiment has not yet been carried out. The ejected electron remains in interaction with OH$^{\bullet}$ and H_3O^+ before its full solvation. Electron solvation dynamics has been studied by several groups and will be discussed in Section 12.3. The electronically excited states H_2O^* are known to dissociate, as in the vapor phase, mainly into OH$^{\bullet}$ and H$^{\bullet}$ radicals, but the contribution of this channel is less important than the ionization of water molecule. Another way of water dissociation is electron attachment occurring with electrons having an energy of 5–15 eV and giving H$^-$ and OH$^{\bullet}$. The hot H$^-$ ion can produce an electron by detachment, resulting in dissociative electron attachment of a water molecule. But in that case, H$^{\bullet}$ and OH$^{\bullet}$ are generated separately at a greater distance compared to a direct dissociation process. The H$^-$ ion can also react

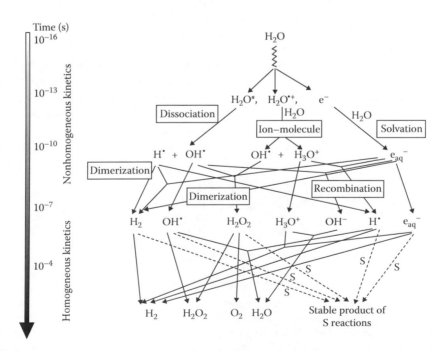

FIGURE 12.4   Mechanism of water decomposition under ionizing radiation.

directly with H_2O, producing unscavengable H_2 (Goulet and Jay-Gerin, 1989; LaVerne and Pimblott, 2000; Mozumder, 2005).

The H_3O^+, $OH^·$, $H^·$, and e_{hyd}^- species formed inside the spurs a few picoseconds after energy deposition begin to diffuse randomly; so, a fraction of them encounter one another and react to form molecular and secondary radical products, while the remainder escape into the bulk liquid and become homogeneously distributed with respect to reaction with solutes acting as radical scavengers. In water, almost 100 ns are needed to complete the nonhomogeneous kinetics and obtain a homogeneous distribution of the different radicals and ions in the bulk solvent.

12.3 SOLVATION PROCESSES OF ELECTRONS IN POLAR SOLVENTS

Two centuries ago, the solvated electron was first observed in liquid ammonia (Thomas et al., 2008), and a century later the idea of electrons trapped in cavities created by solvent molecules and behaving like an anion was introduced (Kraus, 1908). It was only in 1962 that pulse-radiolysis experiments gave evidence of the existence of the solvated electron by the observation of its optical absorption spectrum (Hart and Boag, 1962). This discovery triggered many experimental and theoretical studies, as the solvated electron is a major transient species in condensed phase radiation chemistry. However, the solvated electron continues to raise several fundamental questions, in particular on its precursors, the nature of the "cavity" structure, and dynamics of the electron–solvent interactions. In a simplified picture, the solvated electron structure resembles that of the hydrogen atom with a ground state of s-type character and three nondegenerate p-type excited states.

The solvated electrons are generally formed by photoemission with ionizing radiation (radiolysis or photolysis), and the early steps in electron solvation are depicted in a simplified manner in Figure 12.5. Knowledge of the timescale involved in solvated electron formation is crucial to understand electron solvation and reactivity. Indeed, among the electron scavengers, some such as selenate or molybdate anions are known to scavenge precursors to the solvated electron efficiently and the solvated electron poorly (Figure 12.6) (Pommeret et al., 2001). Others react with both precursors and solvated electrons with similar efficiencies, for instance NO_3^- or Cd^{2+} (Kee et al., 2001), while the remaining electron scavengers react poorly with the precursors but efficiently with the solvated electron (Pastina et al., 1999).

Studies on electron solvation began in the 1970s, first in liquid alcohols at low temperatures by pulse-radiolysis measurements (Baxendale and Wardman, 1971; Chase and Hunt, 1975), because by decreasing the temperature, the molecular movements are slowed down and the electron dynamics last several nanoseconds. Then, the emergence of ultrashort laser pulses made the investigation of electron formation and early reactivity at room temperature possible (Nikogosyan et al., 1983). The pioneering pump-probe laser experiments were performed in 1987 by Migus et al. (1987) in water. They showed that the hydration process lasts less than 1 ps and they gave evidence of precursors

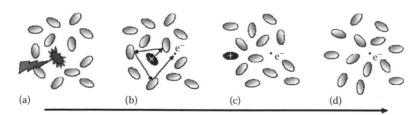

FIGURE 12.5 Formation of solvated electron by photoemission with ionizing radiation: (a) ionization, ejection of an electron from a molecule; (b) thermalization, loss of the excess kinetic energy of the electron by collisions with solvent molecules; (c) localization, trapping of the electron in a solvent cavity; (d) solvation, relaxation, and reorganization of the solvent molecules around the electron to reach the equilibrium configuration.

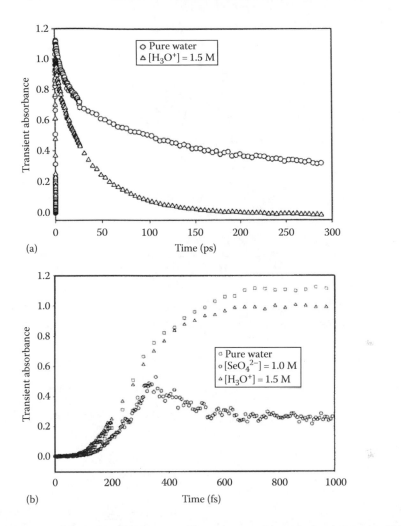

FIGURE 12.6   Reactivity of the hydrated electron and its precursor with the hydronium cation H_3O^+ (a) and the selenate anion SeO_4^{2-} (b); transient kinetics measured at 800 nm after the photoexcitation of aqueous solutions at 266 nm. (Reprinted from Pommeret, S. et al., *J. Phys. Chem. A*, 105, 11400, 2001. With permission.)

of the hydrated electron absorbing in the infrared spectral domain (Figure 12.7). Since then, many light-absorbing species or states were identified using decomposition analysis of transient absorption spectra: "hot" s-like states, excited p-like states, "dry"/ "wet" electrons, "weakly bound" electrons, etc. For the kinetic/spectral analyses, two main solvation models have been usually employed: stepwise mechanisms (Hirata and Mataga, 1990; Long et al., 1990; Sander et al., 1992; Shi et al., 1995; Reuther et al., 1996; Assel et al., 1999; Holpar et al., 1999; Kambhampati et al., 2002; Scheidt and Laenen, 2003; Thaller et al., 2006) and continuous blueshift model (Hertwig et al., 1999; Kloepfer et al., 2000; Madsen et al., 2000). The first mechanisms consider a small number of species/states with well-defined, time-independent spectra (Figure 12.8), while the second model consists of only one species with a spectrum shifting to shorter wavelength but keeping its shape. In some instances, both kinds of models were combined to interpret the data (Pépin et al., 1994, 1997; Turi et al., 1997; Goulet et al., 1999). However, the assumptions made in such models are not obvious.

Recently, the formation of solvated electrons was investigated by photoionization in four polyalcohols, ethane-1,2-diol (Soroushian et al., 2006), propane-1,2-diol, propane-1,3-diol (Bonin et al., 2007), and propane-1,2,3-triol (Bonin et al., 2008). In the diols, just after the pump pulse, the

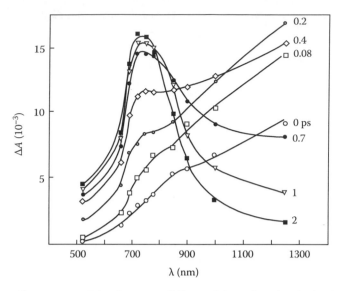

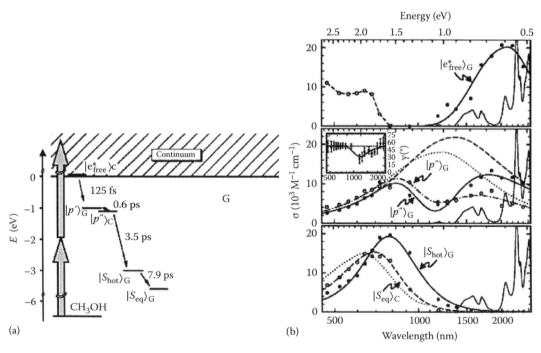

FIGURE 12.7 Absorption spectra of the electron at different delays after photoionization of liquid water at 21°C by two-photon excitation with pulses of 100 fs duration at 310 nm. (Reprinted from Migus, A. et al., *Phys. Rev. Lett.*, 58, 1559, 1987. With permission.)

FIGURE 12.8 (a) Level diagram for the relaxation channel G of the kinetic model relevant for the photoionization experiment of methanol with two-photon energy of 9.1 eV. Terminating states of the probe absorption are omitted. Vertical double arrow marks the excitation transition. (b) Spectral signatures of four intermediates in the photoionization process and of the final solvated electron as deduced from the signal transients (isotropic component) with the help of the kinetic model G. All spectra are normalized to the molar extinction coefficient of solvated electrons in methanol; experimental points, calculated curves. (Reprinted from Thaller, A. et al., *J. Chem. Phys.*, 124, 024515, 2006. With permission.)

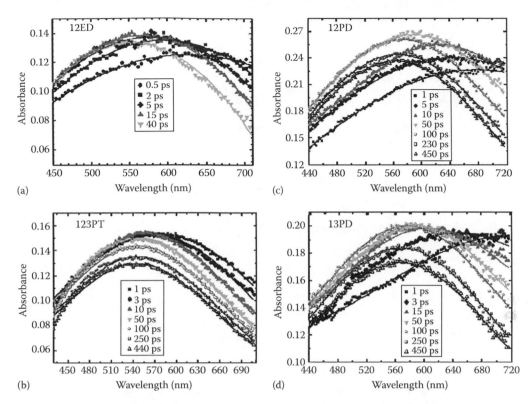

FIGURE 12.9 Time evolution of the absorption spectrum obtained in four polyols upon two-photon ionization of the solvent at 263 nm: (a) 12ED, ethane-1,2-diol; (b) 123PT, propane-1,2,3-triol; (c) 12PD, propane-1,2-diol; (d) 13PD, propane-1,3-diol. (Reprinted from Lampre, I. et al., *J. Mol. Liq.*, 141, 124, 2008a. With permission.)

electron absorbs in the visible and near-infrared (NIR) domain and then, for the first few picoseconds or tens of picoseconds, the red part of the spectrum goes down while the blue part increases slightly, inducing a blueshift of the spectra. In contrast, in the triol, the electron absorption spectrum is situated early on in the visible domain and evolves only on the red side (Figure 12.9). The kinetics signals observed for the four polyols at 700 nm and at the wavelength of the maximum of the fully solvated electron band focus on the change in the electron spectra on the red side and highlight distinct evolutions for the four polyols (Figure 12.10). More closely, the time evolution of the peak position and the width of the electron spectra show that the solvation dynamics is different in the two propanediols and is much faster in the propanetriol despite its higher viscosity (Figure 12.11). Those results clearly show that the solvation dynamics cannot be explained by the viscosity alone but also depends on the molecular structure of the solvent, in particular on the OH groups. To extract the characteristic times of the solvation process, it is essential to analyze the full set of transient absorption spectra. For that, Bayesian data analysis is a powerful tool (Sivia, 1996; D'Agostini 2003). This probabilistic method, mostly used in environmental modeling (Brun et al., 2001) and biokinetics (D'Avignon et al., 1998, 1999; Holtzer et al., 2001), provides a consistent framework for optimization and identification of independent parameters, determination of uncertainties, and model comparison. So, the results of the Bayesian data analyses performed for the four polyols and comparing various models are in favor of a modified continuous "blueshift" model considering a unique species with an absorption band changing continuously not only in position but also in shape. These analyses are corroborated by the analogy between the time evolution of the electron absorption band at a given temperature and the change in the spectrum of the solvated electron by decreasing the temperature. A similar type of dynamics was observed in liquid water by Lian and

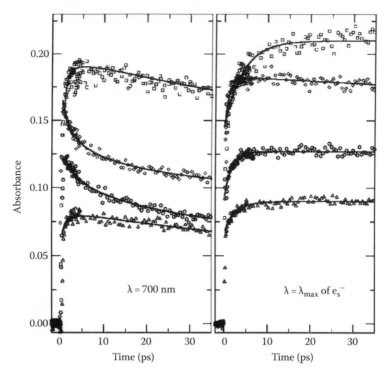

FIGURE 12.10   Transient absorption signals recorded at 700 nm and at the wavelength of the solvated electron absorption band maximum for the four studied polyols (Figure 12.9): ◇ 123PT, ○ 12ED, △ 13PD, and □ 12PD. (Reprinted from Lampre, I. et al., *Radiat. Phys. Chem.*, 77, 1183, 2008b. With permission.)

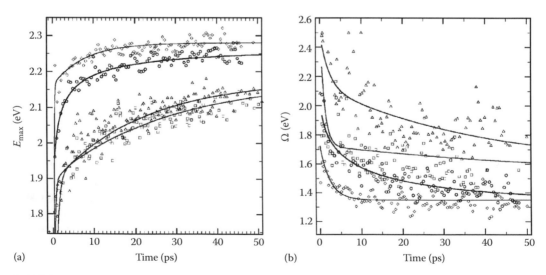

FIGURE 12.11   Time evolution of the peak position E_{max} (a) and the width (b) of the absorption band of the photogenerated excess electron in the four studied polyols (Figure 12.9): ◇ 123PT, ○ 12ED, △ 13PD, and □ 12PD. (Reprinted from Bonin, J. et al., *J. Phys. Chem. A*, 112, 1880, 2008. With permission.)

coworkers (Lian et al., 2005) with two regimes: first, a rapid shift of the spectrum with a change in shape on the subpicosecond timescale, then, a slower narrowing of the spectrum on the red side while the position of the maximum remains fixed. Such behavior looks like vibrational relaxation in photoexcited molecules and the presence of a short lived peak at $1.2\,\mu m$, overtone of the OH stretch mode, indicates a strong vibronic coupling between the electron and the water solvent.

Those results strongly suggest that in femtosecond solvent photoionization (or electron photodetachment), electrons are very quickly trapped and only localized electrons in s-like states are observed. Indeed, while probing the electron delocalization in liquid water and ice at attosecond timescales, Nordlund et al. found that the trapping of the electron at a broken H-bond provides a timescale (20 fs) long enough for librational response of the water molecules and early-time dynamics leading to the hydrated electron (Nordlund et al., 2007). Moreover, time-resolved resonance Raman experiments performed in water gave evidence of enhanced Raman signals for the electron precursor (Mizuno et al. 2005). Since intensity enhancement is due to s-p transition, the precursor is assumed to be a nonequilibrium s-state electron. It is worth noticing that resonance Raman experiments have provided valuable data on the structure of the solvated electron in water (Tauber and Mathies, 2002, 2003; Mizuno and Tahara, 2003) and primary alcohols (Tauber et al., 2004), revealing strong vibronic couplings between the solvated electron and the normal modes of the solvent. In the case of alcohols, couplings to the methyl/methylene deformations indicate that the electronic wave function of the electron extend on the alkyl group of the alcohols. Those experimental results agreed with the theory developed by Abramczyk and coworkers in which the coupling between the excess electron and the solvent vibrational modes plays a significant role in the electron dynamics and absorption spectra (Abramczyk, 1991; Abramczyk and Kroh, 1992, 1994). The other conceptual picture, suggesting that the solvated electron is an unusual kind of solvent multimer anions in which the excess electron density occupies voids and cavities between the molecules and frontier p-orbitals in the heteroatoms in the solvating groups, can also account for these experimental observations (Symons, 1976; Shkrob, 2006). Ultrafast relaxation dynamics of the solvated electron in liquid ammonia solutions were also investigated by femtosecond NIR pump-probe absorption spectroscopy (Lindner et al., 2006). Immediately, after photoexcitation, the absorption spectrum of the electron is substantially redshifted with respect to its stationary spectrum, and a subsequent blue shift occurs on a timescale of 150 fs. The data are understood in terms of a temperature-jump model and ground-state "cooling." Spectral fingerprints of the excited p-state were neither unambiguously observed nor required to explain the observed dynamical response indicating that its lifetime is very short (<100 fs).

If p-like states do not seem detectable upon photoionization or electron photodetachment, such states may be generated by excitation of solvated electrons (Walhout et al., 1995; Assel et al., 1998, 1999; Silva et al., 1998a,b; Kambhampati et al., 2002; Thaller et al., 2006) and large water anions (Bragg et al., 2004; Paik et al., 2004; Lee et al., 2008; Ehrler and Neumark, 2009). In bulk solvents, two main schemes were proposed for the relaxation of p-like states: (1) slow adiabatic internal conversion (IC) and (2) fast nonadiabatic IC, leading to a "hot" s-like state that relaxes subsequently. In water, according to the first scheme, the p-state lifetime is determined to be a few hundreds of femtoseconds (Assel et al., 1998; Silva et al., 1998b; Yokoyama et al., 1998) after a fast excited-state solvation dynamics, while with the second scenario, the IC time is around 50 fs and the hundreds femtosecond component is attributed to the initial relaxation of the "hot" s-like state (Pshenichnikov et al., 2004). The second scenario is supported by photon echo and resonant transient grating measurements (Emde et al., 1998) as well as time-resolved photoelectron experiments on anion clusters. Indeed, cluster experiments indicate p $\rightarrow$ s IC time depending on cluster size n (Figure 12.12). Extrapolation of these values to $n \rightarrow \infty$ yields IC lifetimes of 50 and 157 fs for water and methanol, respectively (Ehrler and Neumark, 2009). Paik and coworkers reported bi-exponential ground-state relaxation, after IC (Paik et al., 2004). They found that the solvation time (300 fs) was similar to that of bulk water, indicating the dominant role of the local water structure in the dynamics of hydration; but in contrast, the relaxation in other nuclear coordinates was on a much longer timescale (2–10 ps)

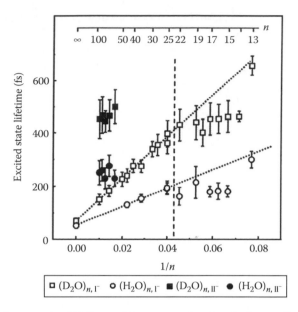

FIGURE 12.12 Internal conversion lifetimes of the p → s relaxation for isomer I and II water clusters as a function of cluster size. (Reprinted from Ehrler, O.T. and Neumark, D.M., *Acc. Chem. Res.*, 42, 769, 2009. With permission.)

and depended on cluster size (Paik et al., 2004). Study of the dynamics of electrons in clusters of ammonia by time-resolved photoelectron spectra also showed ultrafast p → s conversion that results in the formation of a coherent wave packet on a "hot" s-like state, but the observed solvation dynamics differs from that of electrons in water cages (Lee et al., 2008). This difference is explained by the different solvent motions for electron interaction in water (libration), in ammonia (phonon-like), and the extent of cage rigidity.

Over the last decade, technological advancements have generated many results and rapid progress in understanding the properties of the solvated electron. The simplified picture of a "particle in a box" must be revised. Recent results showed that, for electron solvation dynamics, the stepwise mechanism involving different excited states and/or species is not accurate. The electron solvation dynamics appears as a complex relaxation process with nonhomogeneous kinetics and, to get common features and reliable time constants, global spectro-kinetics analysis is necessary. Moreover, the answers of several questions dealing with the nature of the solvated electron and the dynamics of its formation require improvements in quantum chemical and molecular dynamics simulations in order to get a more realistic picture on the electron solvation process in many solvents.

12.4 SPUR REACTIONS AND TIME-DEPENDENT RADICAL YIELDS

The ionization–excitation processes induced by energetic electrons, by nature, exhibit a stochastic character. As a consequence, the radiolytic species are heterogeneously distributed in small clusters, called spurs, formed along the radiation tracks. The description of these initial spurs in terms of nature, numbers, and spatial repartition of primary radicals, as well as the distribution of these spurs along the radiation tracks are the aims of many experimental and theoretical research efforts in order to predict the evolution of a system irradiated by an ionizing radiation. In this section, we present the recent contributions to our understanding of the spurs chemistry that have been obtained by picosecond pulse-radiolysis methods. We first present the significant progress achieved during the last decade and the actual state-of-the-art status of picosecond pulse-radiolysis setups, most of them based on laser-driven photocathode accelerators. The new emerging generation of ultrafast pulse-radiolysis facilities is also presented. The second part of this section gives a survey of the most recent experimental results obtained concerning the initial radiolytic yields, G-values, and the spurs reactivity in aqueous solutions, alcohols, and tetrahydrofuran.

12.4.1 ULTRAFAST PULSE-RADIOLYSIS SETUPS FOR THE STUDY OF SPUR REACTIONS

12.4.1.1 Goals of Ultrafast Pulse-Radiolysis Measurements

Two different approaches can be used to model the spurs reactions chemistry, the deterministic (LaVerne and Pimblott, 1991; Swiatla-Wojcik and Buxton, 1995) and the stochastic (Pimblott and LaVerne, 1997; Cobut et al., 1998; Frongillo et al., 1998; du Penhoat et al., 2000) approach. The deterministic model is based on the concept of an average spur at the end of the physicochemical stage. It contains all the initial radiolytic products in certain yields and spatial distributions and in thermal equilibrium with the liquid. For low LET radiation, the spurs are treated as isolated spheres with a Gaussian distribution of each species. In this approach, the initial G-value and the spatial distribution of each radiolytic species are empirical parameters adjusted to fit the experimental data. A set of coupled differential equations are solved to obtain the nonhomogeneous kinetics of the spur processes and diffusion. So, the deterministic simulations are well adapted to study the radiolytic events at the nonhomogeneous stage and, providing a realistic track description, they are capable of good predictions of the time-dependent G-values. However, as the model is based on the concept of average spurs, it fails to take into account nonnegligible effects like overlapping of spurs that is important for tracks of high LET radiation. The deterministic approach also gives a poor description of micro-heterogeneous media.

On the other hand, the stochastic approaches are based on Monte Carlo calculations to directly generate the initial spatial distribution and yields of the radiolytic species at the pre-chemical stage using the cross section of the interactions between the charged particles and the solvents. Hence, despite a higher computing cost, the stochastic simulations provide much more fundamental insight into charged-particle interactions and track-structures. Recent works have shown the ability to model the complex radiation-track and chemistry of high LET radiation. However, the accuracy of these simulations strongly depends on the value of the interaction cross sections. So far, uncertainties still remain for the inelastic cross sections, particularly, those associated with the low energetic secondary electrons ($E < 100\,\mathrm{eV}$) that play an important role in radiolytic events.

In order to validate and improve the models of the radiation-induced chemistry, there is a need for providing experimental kinetics measured with a high temporal resolution and a very good signal-to-noise (S/N) ratio. Indeed, many reported studies illustrate how the spur models are strongly sensitive to experimental parameters. For example, the first deterministic simulations were based on the initial G-value of hydrated electrons $G(e_{aq}^-) = 5 \times 10^{-7}\,\mathrm{mol\,J^{-1}}$ and, lead to an initial spur radius for the e_{aq}^-, $r = 2.3\,\mathrm{nm}$ (Swiatla-Wojcik and Buxton, 1995). Later, with the revised G-value, $G(e_{aq}^-) = 4.3 \times 10^{-7}\,\mathrm{mol\,J^{-1}}$, this radius was increased to $3.8\,\mathrm{nm}$ (Swiatla-Wojcik and Buxton, 2000).

The time evolution of the radiolytic yield can be measured by scavenging methods or time-resolved measurements. The scavenging methods are particularly well suited for the studies of the radiolysis of diluted solutions (typically up to $10^{-3}\,\mathrm{mol\,L^{-1}}$). Using a high concentration of scavenger, or a high scavenging power, intra-spur scavenging is also possible. Buxton has exhaustively reviewed the considerable contribution of scavenging studies to our understanding of the initial G-values and spurs reactions (Buxton, 2004). The scavenging method remains the only one capable of providing the evolution of yield of non-absorbing species like H•, H_2, or H_2O_2, but it is limited mostly because a given scavenger or a product of the scavenging reaction might react with several primary radicals in the spurs. Alternatively, picosecond pulse-radiolysis experiments allow for the direct measurement of initial G-values at the earliest point in time during the physicochemical stage of the radiolytic events and can provide the full temporal evolution of the radiolytic species for comparison with the simulations.

12.4.1.2 Picosecond Pulse-Radiolysis Systems: A Short Story and the State of the Art

The first part of the ultrafast pulse-radiolysis story started during the early 1970s and extended to the mid-1980s. At that time, the electron bunches were emitted by thermionic guns. The first pulse-radiolysis experiments were performed using a bunch of picosecond electrons pulses that were separated by 350 ps as the pump beam (Bronskil et al., 1970). This type of system was also implemented later during the mid-1980s in Japan (Sumiyoshi and Katayama, 1982; Sumiyoshi et al., 1985). The Cherenkov light pulses generated in the air by the accelerated electrons were used as the optical probe. The time resolution and the temporal window of the measured kinetics were limited by the temporal structure of electron bunches. In 1975, at the Argonne National Laboratory, United States, the temporal window was extended to several nanoseconds by using single electron pulse generation (Jonah, 1975; Jonah et al., 1976). An analogue but more advanced configuration named "twin linac pulse-radiolysis system" (Kobayashi and Tabata, 1985; Tabata et al., 1985) was developed later at the Tokyo University, Japan; a first linac was used for the pulse irradiation, while a second generated the synchronized Cherenkov optical probe.

The second phase of the ultrafast pulse-radiolysis history ran from the 1990s to the end of the 2000s and was associated with remarkable progress in shortening the electron bunch duration and the time resolution of the experiments. Shortening of the electron bunch duration down to 800 fs (Uesaka et al., 1994) was achieved by the implementation of magnetic pulse compressor on the S-band linac at NERL (Tokyo). Then, a similar device was developed at ISIR (Osaka) to compress the width of 30 ps electrons bunch down to 125 fs (Kozawa et al., 1999), and subpicosecond pulse-radiolysis measurements were successfully performed (Okamoto et al., 2003). The setup consisted of a sub-picosecond L-band linac as an irradiation source and a femtosecond Ti:Sa laser, synchronized with the RF field as an analyzing light (Kozawa et al., 2000). The time jitter between the optical laser probe and the electron pulse was however several picoseconds and had to be corrected shot to shot. Nevertheless, this system allowed pulse-radiolysis measurements with 800 fs time resolution. Bunch compression is still in use on the current generation of picosecond electrons pulse accelerators. The performances and the significant achievements obtained by these systems were described by Wojcik et al. (2004).

This current generation is based on the laser-triggered photocathode RF gun. The electron bunches are then no more generated by thermionic effect, but are produced by photoemission from a metal or semi-conductor photocathode hinted by a short UV laser pulse. This laser-triggered RF gun technology was developed in order to improve the beam quality, notably in terms of more reduced emittance levels. The laser-driven RF guns immediately became attractive alternatives for the design of new picosecond pulse-radiolysis systems. First, the lower emittance allows for a better transport and focusing of the accelerated electron beam, and thus a better pulse-probe overlap in pulse-radiolysis experiments. Second, the RF-gun design is compact and permits the construction of smaller facilities at lower costs and easier operation and maintenance conditions. And last, the electron bunches delivered by the photocathode RF gun can be precisely synchronized with tunable femtosecond optical laser pulses to perform pulse-probe measurements at a very accurate time resolution that is only limited by the electron pulse duration. Six ultrafast pulse-radiolysis systems based on a photocathode RF gun and a femtosecond laser source are in operation all over the world: at Brookhaven National Laboratory (the LEAF facility) (Wishart et al., 2004), at Osaka University (Kozawa et al., 1999; Saeki et al., 2005; Yang et al., 2006), at Sumitomo Heavy Industries (SHI, Tokyo) (Aoki et al., 2000), at Waseda University (Tokyo) (Kawaguchi et al., 2005; Nagai et al., 2007), at the University of Tokyo (NERL) (Muroya et al., 2001a,b, 2005a, 2008), and at the University Paris Sud (ELYSE) (Belloni et al., 2004; Marignier et al., 2006).

The performances and usual operating conditions of the laser-triggered RF guns and of their associated time-resolved experiments for studying the spurs reactions are described below. They may be considered as the state-of-the-art pulse-radiolysis systems when this chapter was written (Muroya et al., 2008). However, new facilities that exploit the laser-plasma acceleration process for the generation of ultrashort bunches of electrons and charged particles are under tests and developments and already featured exciting research fields for pulse radiolysis (Brozek-Pluska et al., 2005; Uesaka et al., 2005; Oulianov et al., 2007).

12.4.1.3 Picosecond Pulse-Radiolysis Experiments Using Laser-Driven Photocathode RF Gun

The six above-mentioned picosecond pulse-radiolysis, apart from variations specific to each facility, all work on the same technical scheme, and at comparable levels of performances. A typical picosecond pulse-probe experimental setup is illustrated in Figure 12.13, in the case of ELYSE. ELYSE is composed of a photocathode RF gun and an accelerating section (booster), both powered by a single 20 MW klystron, and three beam-lines to transport and focus the electron bunches to the samples. Details about the RF part of the machine can be found elsewhere (Belloni et al., 2004), as well as the description of the installed pulse-radiolysis setups (Marignier et al., 2006). We focus here on the picosecond pulse-probe setup. In the gun, the photoelectrons are emitted from a Cs_2:Te photocathode illuminated by UV laser pulses at 262 nm. The original laser system of ELYSE has been recently upgraded. The actual configuration is composed of a customized Trident laser system from Amplitude Technologies. Briefly, it consists of a femtosecond Ti:Sa laser oscillator (Femtolaser) synchronized at 78.9 MHz on the phase of the RF field and used to seed a two stage amplifier pumped at 100 Hz synchronously with the accelerator. By driving the Pockels cells of the amplifier, the continuous 100 Hz laser pulses train can be modulated to extract at the repetition rate of the accelerator, either single laser pulses, or bunches of laser pulses separated in time by 10 ms. The energy per pulse is 2.3 mJ, the pulse-width is 70–80 fs and the emission wavelength is centered at 785 nm. Two thirds of this initial energy is converted at 262 nm to generate the UV laser pulse that is then seeded into the photocathode RF gun. If needed, the UV pulse duration can be adapted before the injection into the accelerator vacuum tube by the means of appropriate dispersive optics. Usually, a pulse with a duration in the range 1–5 ps is preferred for the extraction and transport of electron bunches with a high charge per pulse (Wishart et al., 2004; Yang et al., 2006). The electron

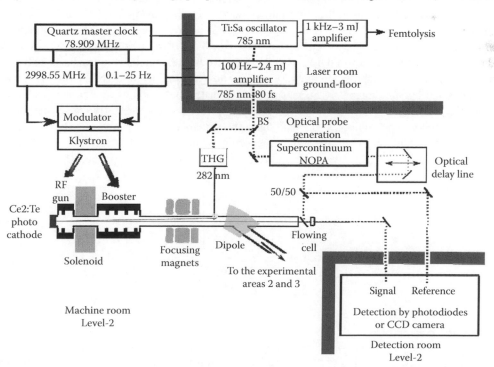

FIGURE 12.13 Ultrafast pulse-radiolysis system based on pulse-probe transient absorption spectroscopy.

bunches generated at ELYSE possess an energy that can be continuously adjusted from 2 to 9 MeV. The charge in the bunch during pulse-radiolysis experiments is usually in the range 2–7 nC for an electron pulse duration in the range 7–15 ps.

The rest of the initial laser pulse is used to create tunable or broadband optical probes synchronized with the electron pulses. Two kinds of femtosecond optical probes are available at ELYSE: (1) a white-light supercontinuum probe, generated in a sapphire plate, with a typical spectral extension from 450 nm up to 850 nm, and (2) an optical pulse generated by parametric amplification using a NOPA stage tunable between 470 and 1600 nm in the visible–NIR and, that can be frequency-doubled to reach the near-UV (250–500 nm) spectral region. The probe beam is propagated along a motorized optical delay line and overlaps with the electron beam into the optical cell for pulse-probe transient absorption measurements.

Transient absorption spectroscopy in the UV–visible–NIR spectral domain is until now the method widely used for picosecond pulse-radiolysis experiments. In a transient absorption experiment, the time-dependent change in the absorbance $A(t)$ of a species is measured and is directly connected with the time-dependent G-value of this species, $G(t)$, by the relation

$$G(t) = \frac{A(t)}{\varepsilon l D} \tag{12.1}$$

with
ε (L mol^{-1} cm^{-1}), the species molar extinction coefficient
l (cm^{-1}), the optical path
D (J kg^{-1}) the dose

Providing that D and ε are known, the absolute $G(t)$ function is directly deduced from measurements. However, the accurate determination of the dose in the pulse-probe overlap volume is very tedious. Therefore, the radiolytic yields are usually obtained indirectly from the knowledge of the G-value at longer time delays or by comparison with the known G-value of another transient species formed in the sample within the temporal window covered by the pulse-probe delay line of the experiment. For this reason, the optical delay lines installed at picosecond pulse-radiolysis facility preferentially extend up to 5–15 ns in order to recover with the data measured at the ns timescale.

Transient absorption is acquired by consecutively recording the intensity of the laser with (I) and without (I^0) electron pulse applied to the sample. The shot-to-shot fluctuation of the probe laser must be compensated by simultaneously measuring the intensity fluctuation of a reference probe beam (I_{ref} and I^0_{ref}, respectively) created by a 50/50 beam splitter placed on the optical probe pathway before the sample. Synchronized contributions to the signals from the electronic devices or from parasite light (Cherenkov …) are rejected by recording the intensity on the detectors in the absence of optical probe (I^{dark}, I^{dark}_{ref}). The probe and reference signals, $(I, I^0, I^{dark})_{mes}$ and $(I, I^0, I^{dark})_{ref}$, are measured by using either two photodiodes or two separated areas of a multichannel CCD camera. The transient absorbance is then given by the formula

$$A(t) = \frac{\left(\dfrac{(I - I^{dark})}{(I^0 - I^{dark})}\right)_{mes}}{\left(\dfrac{(I - I^{dark})}{(I^0 - I^{dark})}\right)_{ref}} \tag{12.2}$$

where t is the pulse-probe time delay.

This acquisition scheme is classical for pump-probe transient absorption spectroscopy setup, and is also implemented at LEAF and NERL. However, it must be mentioned that the environmental conditions in a picosecond pulse-radiolysis facility cannot be so finely controlled than in standard laser labs and so, the experiments are much more affected by vibrations and thermal fluctuations.

To reduce the influence of these factors, the acquisition scheme at ELYSE is based on the generation of a probe laser bunch containing two laser pulses separated by 10 ms, corresponding to I and I^0, respectively. By this way, the perturbation induced by the noise in the hertz frequency range is reduced. The signal $A(t)$ can be acquired at a repetition rate up to 25 Hz using photodiodes or a CCD camera. The sensitivity of the detection setup is typically better than 1 mOD for an average of 10 laser probe pulses, that is, in the order of magnitude of the shot-noise limit. Figures 12.14 and 12.18 illustrate the sensitivity routinely obtained with the multichannel (CCD) or the monochannel (photodiodes) pulse-probe setups installed at ELYSE, respectively.

At Osaka University, Yang et al. (2006) have developed an alternative acquisition scheme. The RF gun is triggered by a Nd:YLF picosecond laser, while a second Ti:Sa oscillator is mode-locked at the repetition rate of 79.3 MHz in phase with the 36th harmonic of the 2856 accelerating RF field and is used to generate the synchronized optical probe beam. A so-called double pulse can be extracted from the laser train to create the I^0, and I probe pulse, respectively. For this configuration, no references are needed, considering that the noise in the 80 MHz bandwidth is very low. The same group also performs pulse-probe measurements using optical probe pulse delivered by OPAs, or by generation of a white-light supercontinuum. In the latter case, they operate a 1 kHz regenerative amplifier, and the same double-pulse acquisition scheme is still applied at 1 kHz (Saeki et al., 2006).

The time resolution in transient absorption spectroscopy depends on three parameters: the electron pulse duration, the jitter between the arrival times of the electron pulse and the laser probe pulse, and the velocity mismatch between the relativistic electron bunch traveling at a speed close to the light

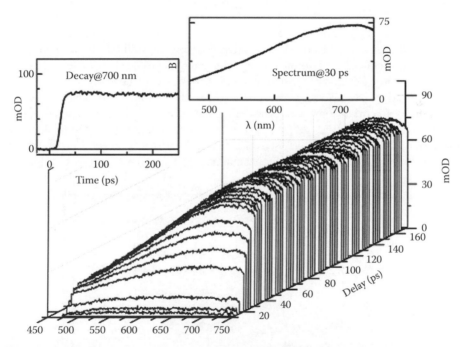

FIGURE 12.14 Transient absorption spectra and decay of the solvated electrons in water measured by pulse-probe spectroscopy. The charge and the energy of electrons pulses were 3 nC and 8.3 MeV, respectively. The electron pulse duration, deduced from the rise time of the absorption signal, was shorter than 10 ps. The optical probe beam consisted of supercontinuum pulses generated in a sapphire plate and was detected on a multichannel CCD camera. At every step along the pulse-probe delay scan, the single-shot transient absorption spectra were measured at the repetition rate of the accelerator (on this case 10 Hz, but the acquisition setup can work up to repetition rate of 50 Hz). In the 3D graph, each spectrum has been obtained by averaging 10 single-shot acquisitions. The full scan extended over a pulse-probe delay between −20 and 250 ps, with a 2 ps step. Left inset: Time-dependent absorbance at λ_{probe} = 700 nm, reconstituted from the same series of transient spectra. Right inset: Spectrum of solvated electrons recorded 30 ps after the electron pulse and averaged over 50 electron pulses.

celerity, c, and the optical probe traveling at the speed c/n with n the refractive index of the sample. The time jitter between the laser and the electron pulses is governed by the phase stability between the mode-locked oscillator and the RF accelerating wave. Shot-to-shot jitter in the range of 1 ps rms, are routinely obtained at picosecond pulse-radiolysis facilities. At ELYSE, the typical electron pulse duration for pulse-probe experiments is 8 ps for a charge of 2–5 nC. Similar bunch properties are available at LEAF (>7 ps for 6–8 nC). Both facilities are operated with beam energy slightly below 9 MeV and without a magnetic bunch compressor inserted in the electron beam transportation. NERL and ISIR facilities are equipped with inserted compressing devices, and deliver electron bunches accelerated up to 30 MeV. A significant advantage of acceleration at higher energy is a better focusing of the electron beam that allows the deposition of dose as high as 40 Gy/shot and, at the same time, keeping the charge per bunch in the range of 2 nC and a duration compressed down to 2 ps. Electron bunches shorter than 500 fs, for a charge of 1.25 nC have also been reported. While short electron bunches are available, for many studies it is often preferred to record signals at a high S/N ratio. Toward this end, the optical path in the sample is kept around 5 mm, and the time resolution is reduced due to the velocity mismatch between the laser and the electron bunch. As an example, the time-of-flight difference between an electron bunch and a laser pulse at 600 nm, traveling collinearly in water, is about 1 ps mm^{-1} of sample. Figure 12.15 illustrates the changes in the time resolution due to the increase in the optical path length. A good estimation of the time resolution, τ_{min}, of a transient absorption measurement in collinear geometry is given by (Kashiwagi et al., 2005; Muroya et al., 2008)

$$\tau_{min} = \tau_{vel} + \left(\tau_{puls}^2 + \tau_{las}^2 + \tau_{jitter}^2\right)^{0.5} \tag{12.3}$$

where τ_{vel}, τ_{puls}, τ_{las} and τ_{jitter} represent the time-of-flight difference, the electron pulse duration, the laser probe pulse duration, and the arrival time jitter, respectively. Methods are available to directly measure the shot-to-shot jitters and the bunch length longitudinal profile (De Waele et al., 2009), from which the exact response function of the detection system can be calculated (Wojcik et al., 2004).

Transient absorption spectroscopy is well adapted to study the time dependence of the G-values, due to its sensitivity and the absolute characters of the measured absorbance. In practice, the record of kinetics is often limited by the stability of the electron accelerator itself due to slow time drift during the acquisition. So, a lot of effort is also put to stabilize the machines, and notably the phase of the klystron and the long-term laser beam pointing. These parameters are strongly coupled

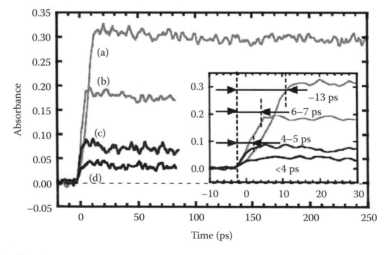

FIGURE 12.15 Time behavior of hydrated electron measured at 700 nm. Inset: enlarged part of the figure within 30 ps. Optical path lengths of the cells are (a) 10 mm, (b) 5 mm, (c) 2 mm, and (d) 1 mm. Pulse-probe measurements at the NERL pulse-radiolysis facility. (Reprinted from Muroya, Y. et al., *Radiat. Phys. Chem.*, 77, 1176, 2008. With permission.)

with the environmental conditions of the facility (air temperature, cooling systems stability, etc.). Alternatively, setups for transient absorption spectroscopy are under development with the aims of decreasing the acquisition time. At Brookhaven, A. Cook and J. Wishart have developed a setup based on a temporally dispersed optical probe beam by using a bundle of 100 optical-fibers of different lengths as delay line to simultaneously record the transient absorbance at 100 different pulse-probe delays, keeping the high temporal resolution permitted by the pulse-probe method. At ELYSE, J.L. Marignier is exploiting a high-dynamic range streak-camera with picosecond time resolution coupled with a powerful home-made flash lamp working at 25 Hz, to simultaneously record the spectral changes and the kinetics induced by the electron beam. One advantage of this system is that the same setup allows measurements on different temporal windows from 0–500 ps to several microseconds (Marignier et al., 2006). Another approach has been proposed by Bartels and coworkers, based on a 30 ps/20 MeV linac, combined with an asynchronous Ti:Sa used for performing time-correlated absorption spectroscopy (Bartels et al., 2000). The resulting time resolution is lower than that of the previously described systems, and the spurs dynamics can be investigated with a high S/N ratio over a large temporal window of tens of nanoseconds.

12.4.2 Recent Advances in Spurs Chemistry Obtained by Picosecond Pulse-Radiolysis

Water decomposes under ionizing radiation into e_{aq}^-, $OH^\bullet$, H_3O^+, $H^\bullet$, H_2, and H_2O_2. These radicals and molecular products are formed in the spurs during the physicochemical stage according to the following primary reactions:

$$H_2O \rightarrow e^- + H_2O^+ \tag{I}$$

$$H_2O \rightarrow H_2O^* \tag{II}$$

Followed by

$$e^- + H_2O \rightarrow OH^\bullet + H^- \rightarrow H_2 + OH^\bullet + OH^- \tag{III}$$

$$e^- + nH_2O \rightarrow e_{aq}^- \tag{IV}$$

$$H_2O^* \rightarrow H + OH^\bullet \tag{Va}$$

$$H_2O^* + H_2O \rightarrow H_2 + 2OH^\bullet \text{ (or } H_2O_2) \tag{Vb}$$

$$H_2O^{\bullet+} + H_2O \rightarrow OH^\bullet + H_3O^+ \tag{VI}$$

The secondary electrons formed during reaction (I) that escape geminate recombination with their $H_2O^{\bullet+}$ parent cation can react with water molecules according to reaction (III) and are the source for the so-called unscavengable H_2 production. The main part of the low-energy electrons is thermalized and becomes hydrated according to (IV), as discussed in Section 12.3. The excited water molecules decompose directly or by reaction with a neighbor molecule to form H, H_2, $OH^\bullet$, and H_2O_2 species. $H_2O^{\bullet+}$ reacts according to (VI) and is the major source of $OH^\bullet$. The reactions (I) through (VI) are responsible for the initial yields, that is, G-values for the primary radiolytic products of water. Then, the species evolve in the spurs, according to the spurs reactions listed in Table 12.1.

12.4.2.1 Time-Dependent G-Value of Solvated Electron in Water

The time-dependent measurement of solvated electron yield at a short-time range was carried out in water by Bronskil et al. (1970) with a time resolution of 30 ps. Then, several groups focused on measuring time-dependent escape probability with different fast techniques. By using pulse techniques

TABLE 12.1
Main Reactions Involved in the Radiolysis
of Cl⁻ Concentrated Solutions

Reactions	Rate Constant (L M⁻¹ s⁻¹)
$e_{hyd}^- + e_{hyd}^- + 2H_2O \longrightarrow H_2 + 2HO^-$	5.5×10^9
$e_{hyd}^- + H_3O^+ \longrightarrow H^\bullet + H_2O$	1.3×10^{10}
$e_{hyd}^- + H^\bullet \longrightarrow H_2 + HO^-$	2.5×10^{10}
$HO^\bullet + e_{hyd}^- \longrightarrow HO^-$	3×10^{10}
$e_{hyd}^- + H_2O_2 \longrightarrow HO^\bullet + HO^-$	1.1×10^{10}
$H_3O^+ + HO^- \longrightarrow 2H_2O$	1.4×10^{11}
$H^\bullet + H^\bullet \longrightarrow H_2$	7.8×10^9
$HO^\bullet + H^\bullet \longrightarrow H_2O$	2×10^{10}
$H^\bullet + H_2O_2 \longrightarrow H_2O + HO^-$	9×10^7
$HO^\bullet + HO^\bullet \longrightarrow H_2O_2$	5.5×10^9
$e_{hyd}^- + H_3O^+ \longrightarrow H^\bullet + H_2O$	1.3×10^{10}
$H_3O^+ + HO^- \longrightarrow 2H_2O$	1.4×10^{11}
$H_3O^+ + O_2^{\bullet -} \longrightarrow HO_2^\bullet + H_2O$	3.8×10^{10}
$HO^\bullet + Cl^- \longrightarrow ClOH^{\bullet -}$	4.3×10^9
$ClOH^{\bullet -} \longrightarrow HO^\bullet + Cl^-$	6.1×10^9
$ClOH^{\bullet -} + H^+ \longrightarrow Cl^\bullet + H_2O$	2.1×10^{10}
$Cl^\bullet + Cl^- \longrightarrow Cl_2^{\bullet -}$	8.9×10^9
$Cl_2^{\bullet -} + H^\bullet + H_2O \longrightarrow H_3O^+ + 2Cl^-$	8.9×10^9

(Hunt et al., 1973; Jonah et al., 1973, 1976; Wolff et al., 1973, 1975; Sumiyoshi and Katayama, 1982; Sumiyoshi et al., 1985), different values were reported for $G(t)$. The initial yield of formation of the hydrated electron was revisited in the last decade with the state-of-the-art pulse-radiolysis systems.

Bartels and coworkers measured the decay of hydrated electrons in H_2O and D_2O on a continuous time window up to 200 ns by combining time-correlated absorption spectroscopy and standard time-resolved absorption measurements with ns time resolution (Bartels et al., 2000). The initial formation yield of hydrated electron was obtained by extrapolation of the fitted decay and found to be $G = 4.2 \times 10^{-7}$ mol J^{-1} in light water. In heavy water, the value of 4.3×10^{-7} mol J^{-1} and a slower decay during the first 10 ns were reported (Bartels et al., 2001). This isotopic effect had already been observed and is associated with a larger escape distance covered by the electron due to lower vibrational modes of heavy water compared to light water. The initial G-value for hydrated electron in H_2O was measured directly with a higher time resolution by Muroya et al. who reported a value of $G = (4.2 \pm 0.2) \times 10^{-7}$ mol J^{-1} (Muroya et al., 2005b). These two new measurements of the G-values are in good agreement but significantly lower (20%) than the previously adopted value of 4.8×10^{-7} mol J^{-1}. Monte Carlo calculations have been parameterized to take into account this new data (Muroya et al., 2002). According to Jay-Gerin and coworkers, an initial yield equal to 4.0×10^{-7} mol J^{-1} cannot be fitted by the stochastic model, and a value of $(4.4–4.5) \times 10^{-7}$ mol J^{-1} is proposed instead, associated with an average initial thermalization distance equal to 13.9 nm (equivalent to a spur radius of 8.7 nm) and a yield for the electron–cation recombination of 18%, in good agreement with experimental data. The time-dependent G-value measured by Muroya et al. and the comparison with the Monte Carlo simulation are reported in Figure 12.16, together with the decay measured by Bartels and coworkers (Bartels et al., 2000).

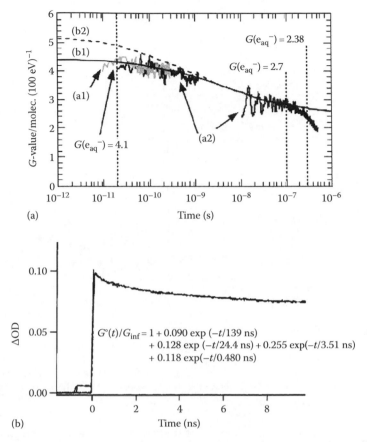

FIGURE 12.16 Recent revisited time-dependent G-values, $G(t)$, of hydrated electrons in water at 25°C, measured by pulse-radiolysis technique. (a) Spur decay kinetics of the hydrated electron following picosecond pulse radiolysis of water measured using a time-correlated transient absorption technique, and the deduced functional expression for $G(t)$. (Adapted from Bartels, D.M. et al., *J. Phys. Chem. A*, 104, 1686, 2000. With permission.) (b) Time-dependent G-value measured by picosecond pulse-probe technique and compared with Monte Carlo calculations. (Reprinted from Muroya, Y. et al., *Radiat. Phys. Chem.*, 72, 169, 2005b. With permission.)

The initial G-value is governed by the ultrafast physical processes, in particular geminate recombination, that could take place within the first few hundreds of femtoseconds. A new type of accelerator based on the laser-plasma acceleration technique could in principle give access to this temporal window. At the moment, the time resolution is not better than that of linac accelerators, but these pioneering works are encouraging despite the difficulties of these types of measurements. Electron pulses of relativistic energies (1.5–20 MeV) generated by a terawatt laser-wakefield accelerator (LWA) have been used for ultrafast pulse-radiolysis experiments. Brozek-Pluska et al. (2005) and Oulianov et al. (2007) reported the first pulse-radiolysis experiments using LWA technique. In both experiments the decay of hydrated electron is probed in the NIR by pulse-probe transient absorption. LWA technique is capable of generating subpicosecond electron bunches with a charge of a few nanocoulombs. However, due to the geometric configuration of the pulse-probe setups (sample thickness, non-collinear geometry) and the dispersion in energy of the electron bunch, the effective time resolution was limited to a few picoseconds. The signals obtained by the two groups are reported in Figure 12.17. The data from Brozek-Pluska et al. (2005) (Figure 12.17, left) show an instantaneous formation of the signal within the time resolution of the setup (3.5 ps), followed by a strong decay of the absorbance that is divided by a factor two during the first 50 ps after the irradiation by the electron bunch pulse. The associated G-value, at 3.5 ps, is 7×10^{-7} mol J^{-1}. This value and the kinetics of the decay are in disagreement

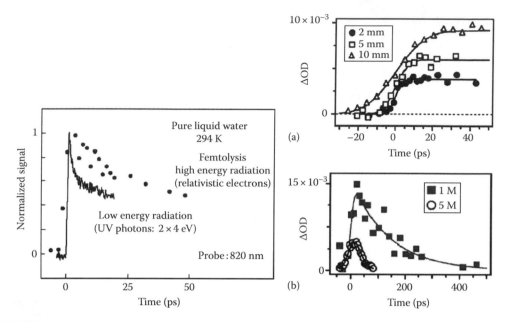

FIGURE 12.17   Pulse-radiolysis experiments performed using relativist electron bunches generated by laser-wakefield acceleration as irradiation beam and a laser pulse around 800 nm as a probe beam. Left: comparison of the decays of solvated electrons formed by two-photon ionization, and pulse radiolysis of water. (Reprinted from Brozek-Pluska, B. et al., *Radiat. Phys. Chem.*, 72, 149, 2005.) Right: decay of solvated electrons formed by pulse radiolysis of water measured for different cell thicknesses (a) and in the presence of 1 and 5 M acid (b). (Adapted from Oulianov, D.A. et al., *J. Appl. Phys.*, 101, 053102, 2007.)

with the published experimental and theoretical studies discussed above. The changes observed in the 0–50 ps timescale are supported according to the authors by ultrafast photolysis experiments that also show an initial decay assigned to a "dispersive dynamics of ultrafast radicals events." The results obtained by Oulianov et al., (2007) are also reported in Figure 12.17 (right): in contrast, as expected for the solvated electron, a quasi-flat absorption signal is observed during the first tens of picoseconds. To obtain these results, the authors conducted the experiments with 10 Hz LWA systems, and combined their pulse-probe setup with the measurement of hydrated electron decay at the nanosecond timescale as a normalization source to correct the dose fluctuations shot to shot. The authors also outline the difficulty to keep this normalization constant during the series of experiments because of laser pointing instabilities.

12.4.2.2   Solvent Effect on the Time-Dependent *G*-Value of Solvated Electron

Tetrahydrofuran (THF) is an aprotic solvent that is largely used in organic chemistry. Considering the spurs reactivity, the interest of THF comes from its dielectric constant that is 10 times lower than that of water. In THF, the Coulombic field is very strong and favors long-range interactions between charged reactive centers. It is also known that ionized THF formed by irradiation is rapidly deprotonated and that solvated electrons can recombine with both the proton adduct and the residual C-centered radical (Salmon et al., 1974). Based on a systematic study of solvated electron scavenging by methyl bromide and ethyl bromide with different concentrations in THF, it was concluded (Kadhum and Salmon, 1984) that the initial formation yield of solvated electron in THF is around $(4.2 \pm 0.4) \times 10^{-7}$ mol J^{-1} and that, after spur reactions, the value is decreased to $(0.36 \pm 0.02) \times 10^{-7}$ mol J^{-1} at microsecond timescale (or $(0.4 \pm 0.02) \times 10^{-7}$ mol J^{-1} depending on data treatment). The transient absorption signals of solvated electron in water and THF recorded at 790 nm after radiolysis by a picosecond electron pulse are reported in Figure 12.18 (De Waele et al., 2006, 2007). The ratio between the initial absorbance (at 30 ps) and the absorbance at

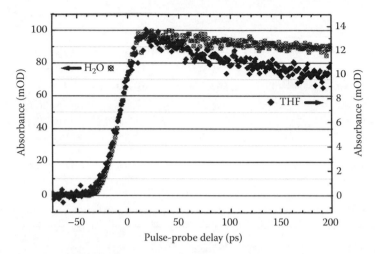

FIGURE 12.18 Decay of solvated electrons measured in water (right scale) and in THF (left scale), see (De Waele et al., 2006) for details. (Reprinted from De Waele, V. et al., *Chem. Phys. Lett.*, 423, 30, 2006.)

200 ps is 0.95 and 0.79 in water and in THF, respectively. Figure 12.19 (inset) presents the decay of solvated electrons in THF at longer times, up to 2.5 ns. The G-value of solvated electrons in THF is much more time dependent than those in polar solvents like water and alcohols. From the fit of solvated electron decay, it was found that the initial separation distance between solvated electrons and the cation is around 4 nm (De Waele et al., 2007). This distance is shorter than the Onsager radius for this solvent (7 nm). The low yield of solvated electron in THF can be

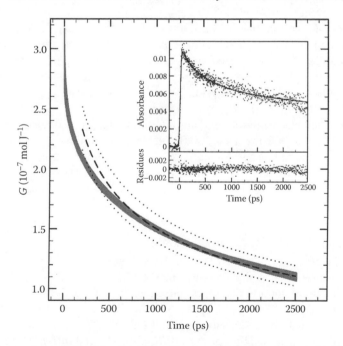

FIGURE 12.19 Comparison of the time-dependent G-value of solvated electrons in THF, determined by scavenging method and direct measurement of the time-dependent decay. The dashed line corresponds to the time-dependent G-value recalculated from the scavenging data from Salmon et al. (1974), the dotted-line gives the data uncertainty on these values. The grey area is the direct time-dependent G-value obtained from the fitted experimental decay given in inset and represents the 99% confidence interval, given data uncertainty. Details are given in De Waele et al. (2006).

TABLE 12.2
G-Values Measured for the Solvated Electron in Alcohols, Water and THF

	$\eta^{25°C}$ (mPa·s)	Dielectric Constant	λ_{max} of e_s^- (nm)	ε (M^{-1}·cm^{-1})	G (μmol·J^{-1})	G at 200 ns (μmol·J^{-1})
12ED	16.1	41.4	570	9,000	0.43 (at 30 ps)	0.17
12PD	40.4	27.5	565	9,700	0.35 (at 100 ps)	0.17
13PD	39.4	35.1	575	10,000	0.38 (at 100 ps)	0.22
123PT	934	42	530	10,800	0.37 (at 250 ps)[a]	0.16
Water	0.89	78.4	720	18,800	0.44 (at 30 ps)	0.27
THF	0.55	7.6	2100	38,000	0.27 (at 30 ps)	0.04

[a] Unpublished results from Mostafavi et al.

understood in terms of the short thermalization distance of the electron compared to the Onsager radius in this solvent.

The time-dependent G-value obtained from the scavenging data of (Kadhum and Salmon, 1984) is compared with the fit of the time-resolved measurement in Figure 12.19. A very good agreement between the kinetics decay and the value obtained by the scavenging method is obtained between 500 ps and 2.5 ns. At short times (less than 500 ps), there is a discrepancy, which is understandable because the scavenging method is not appropriate at a very short time. According to the time-resolved measurements in THF, the G-value at 30 ps is around 2.7×10^{-7} mol J^{-1} and from the deconvolved signal at zero time it is around 3×10^{-7} mol J^{-1} that is noticeably smaller than the value reported by scavenging method (4×10^{-7} mol J^{-1}). This suggests that either the initial yield in THF (at zero time) is lower than that in water or that a very fast decay occurs within the electron pulse. Results reported by Scwartz et al. (Martini et al., 2000) and obtained by femtosecond laser photolysis show that the decay in the first hundred picoseconds is indeed very fast. Recently, the reactivity between Biphenyl molecules and the precursors of the solvated electron in THF was studied by picosecond pulse-radiolysis (Saeki et al., 2007). The results also support an initial G-value of solvated electrons equal to 2.7×10^{-7} mol J^{-1} at 30 ps in THF.

The yield of solvated electron was also measured in different alcohols by picosecond pulse-radiolysis (Lin et al., 2006; Muroya et al., 2008). The data are summarized in Table 12.2, together with the values for water and THF. Compared to the situation in water, the initial G-values in alcohols are also found close to the value of 4.0×10^{-7} mol J^{-1}. However, the decays are more pronounced than that in water, in good agreement with the dielectric constant lower in alcohols than in water.

12.4.2.3 Time-Dependent G-Value of OH˙ Radical

The G-value of OH˙ after 100 ns is well known. Nevertheless, its initial G-value is still controversial. Direct measurements of the G-values of the OH˙ radical were performed (Jonah and Miller, 1977). The authors reported the time dependence of G(OH˙) by following the decay of its weak absorption band in the UV. A value of 6.1×10^{-7} mol J^{-1} was estimated at 200 ps. Other evaluations were obtained from scavenging studies using formate or hexacyanoferrate ions. These data are limited to a scavenging power of 0.7×10^{10} s^{-1}, and different stochastic and deterministic modelings of water radiolysis were performed (Figure 12.20). The often-cited value of G(OH˙) = 6.1×10^{-7} mol J^{-1} for OH˙ radical at 200 ps was based on an estimated $G_{200ps}(e_s^-) = 4.7 \times 10^{-7}$ mol J^{-1} for solvated electrons. The OH˙ yield was reduced to $G_{200ps}(OH˙) = 5.3 \times 10^{-7}$ mol J^{-1} on the basis of the recent G-value of hydrated electron. Jay-Gerin and Ferradini reported that the G-value for OH˙ at 100 ps is 4.6×10^{-7} mol J^{-1} (Jay-Gerin and Ferradini, 2000). This controversy

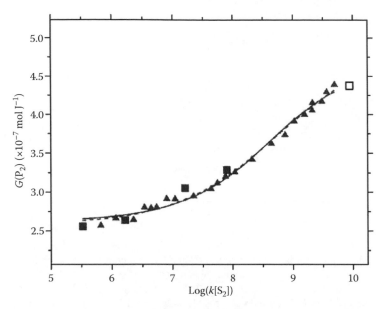

FIGURE 12.20 Experimental and predicted values of yield $G(P_2)$ of the product of HO$^\bullet$ scavenging reaction versus the scavenging power $k_{11}[S_2]$ where S_2 represents (■) HCO_2^- and (▲) $Fe(CN)_6^{4-}$ reacting with HO$^\bullet$. (Adapted from Buxton, G. W., *Charged Particle and Proton Interaction with Matter*, A. Mozumder and Y. Hatano (eds.), Marcel Dekker, New York, 2004.) The hollow square (□) corresponds to the value obtained by scavenging HO$^\bullet$ in 2 M Cl$^-$ solution and measured by picosecond pulse-radiolysis. (Data from Atinault, E. et al., *Chem. Phys. Lett.*, 460, 461, 2008.)

on the radiolytic yield of OH$^\bullet$ at short time was already discussed (Buxton, 2004). The difficulties for a direct determination of the time-dependent G-value of OH$^\bullet$ result from the weak absorption coefficient of OH$^\bullet$ (535 L^{-1} mol cm^{-1} at 250 nm) (Janik et al., 2007). So far, no picosecond pulse-radiolysis study of OH$^\bullet$ has been reported yet. Up to now, the time-dependent G-value of OH$^\bullet$ was only obtained by scavenging methods, as reviewed by Buxton (2004).

It is well known that OH$^\bullet$ can be efficiently scavenged by halide X$^-$ anions, (Br$^-$, Cl$^-$, or I$^-$), leading to the formation of long-lived and even stable radicals (X$_2^{\bullet-}$, X$_3^-$) that can be easily detected by transient or continuous absorption spectroscopy. So, the radiolysis of concentrated halide solutions was intensively studied in the past to understand the direct effects (at salt concentration higher than 2 M), but also to scavenge the OH$^\bullet$ radicals directly in the spurs. However, the reaction schemes of X$^-$ anion are very complex involving several reactions and equilibriums as illustrated for the Cl$^-$ anion in Table 12.1, from Atinault et al. (2008) and references therein, or Br$^-$ in Mirdamadi-Esfahani et al. (2009). Many of these reactions have rate constants close to diffusion controlled values and take place during the heterogeneous phase in the spurs. These complex reaction schemes limit the application of these systems for direct scavenging in the spurs due to the important possibility of back reactions with time.

Lately, we have studied the radiolysis of high concentrated 2 M Cl$^-$ solutions by picosecond pulse-radiolysis experiments. At the picosecond timescale, most of the reactions given in Table 12.2 do not take place and the chemistry of the systems in the first hundreds of picoseconds is simplified, and then allows the determination of the G-value of OH$^\bullet$ by scavenging methods. The study was performed in neutral and acidic conditions at a constant ionic strength, in 2 M NaCl, 1 M HCl + 1 M NaCl, and in 2 M HCl. In Figure 12.21, the decays recorded in these different aqueous solutions and in pure water are reported. It is worth noting that the time-dependent G-values of e$_{aq}^-$ are the same in pure water and in 2 M NaCl solution, showing that neither the high ionic strength, nor the conversion of OH$^\bullet$ into ClOH$^{\bullet-}$ significantly modify the spur reactivity of hydrated electrons during the first

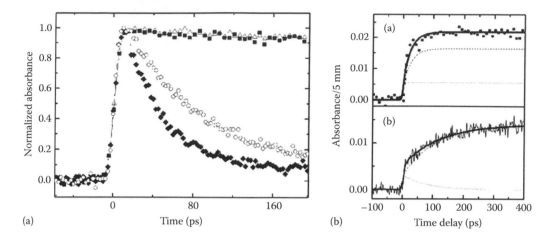

FIGURE 12.21 Pulse radiolysis in concentrated NaCl and HCl aqueous solution. (a) decays of the hydrated electron measured at 650 nm in water ($\triangle$), and in 2 M NaCl ($\blacksquare$), 1 M NaCl + 1 M HCl ($\bigcirc$), 2 M HCl ($\blacklozenge$) solutions. (b) Transient absorption recorded at 345 nm in 2 M NaCl (a), and in 1 M NaCl + 1 M HCl (b), and corresponding to the formation of $ClOH^{\bullet-}$ and $Cl_2^{\bullet-}$, respectively. The dashed lines show the contribution of the solvated electron at this probe wavelength. (Reprinted from Atinault, E. et al., *Chem. Phys. Lett.*, 460, 461, 2008. With permission.)

200 ps. Same conclusions were drawn from ultrafast photolysis measurements (Sauer et al., 2004; Lian et al., 2005). With the initial G-value of 4.3×10^{-7} mol J^{-1} (Muroya et al., 2005b), the $G_{200\,ps}(e_{aq}^-)$ is found to be 4.1×10^{-7} mol J^{-1}. In acidic conditions, the hydrated electron reacts in less than 100 ps with the hydronium cation. From the decays, the rate constant of the reaction in 2 M ionic-strength conditions was determined ($k = 1.3 \times 10^{10}$ L mol^{-1} s^{-1}); this value is lower than half of the value at infinite dilution, as reported earlier, and in good agreement with the Davis model of ionic solutions (Atinault et al., 2008).

The transient absorption signals recorded at 345 nm in 2 M Cl^- solution, in neutral and acidic conditions (1 M HCl) are reported in Figure 12.21 (b). The signals correspond mostly to the formation of $ClOH^{\bullet-}$ in neutral media and $Cl_2^{\bullet-}$ in acidic media, respectively. From the decay analysis (Atinault et al., 2008), the value of $G(OH^{\bullet})_{100\,ps} = 5.0 \times 10^{-7}$ mol J^{-1} is deduced. This value is in good agreement with the theoretical prediction from Jay-Gerin and Ferradini (2000), $G(OH^{\bullet}) = (4.76 \pm 0.26) \times 10^{-7}$ mol J^{-1} at 100 ps.

12.4.2.4 Time-Dependent *G*-Value of H• Atom

The case of H• atoms differs from that of hydrated electron and OH• radical. In fact, the H• atom absorbs only in deep UV with a very low extinction coefficient and its radiolytic yield is also low compared to that of the hydrated electron and the OH• radical. Moreover, the H• atom behaves like a hydrated electron in some cases (reducing species) and like OH• in others (oxidizing species). For these reasons, the H• atom that is produced by the fragmentation of a excited water molecule at sub-picosecond range is followed by the scavenging method with the measurements of H_2 yield. Huerta et al. (2008) used formate solutions up to 1 M to probe the time-dependent yield of H• atom in water radiolysis. The H• atom yields can be obtained by subtracting from the total amount of H_2 that of H_2 produced without scavenger (0.47×10^{-7} mol J^{-1}). The total H_2 yield increases from 0.66×10^{-7} mol J^{-1} to 1.05×10^{-7} mol J^{-1} for 5 μs to 5 ns scavenging time, respectively (Figure 12.22). The yield of H• atoms at a few nanoseconds is around 0.57×10^{-7} mol J^{-1}. These experiments confirm that the production of H• atoms in radiolysis of water is relatively small.

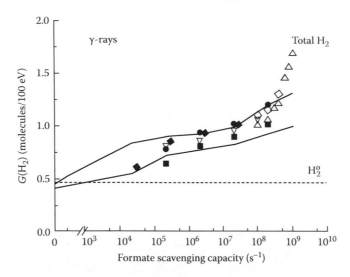

FIGURE 12.22 Production of H_2 in the γ-radiolysis of aqueous formate solutions as a function of the formate scavenging capacity for H atoms: (●) HCO_2^-, 1 mM NO_3^-; (■) HCO_2^-, 24 mM NO_3^-; (◆) DCO_2^-, 1 mM NO_3^-; (◇) HCO_2^-, 0.25 mM NO_3^-; (△) HCO_2^-, 24 mM N_2O; (▽) HCO_2^-, 16 mM N_2O. (The dashed line is the yield determined in water with 1 mM NO_3^- and no formate. Solid lines are the results of the Monte Carlo calculations for formate with 1 or 24 mM NO_3^-.) (Reprinted from Huerta Parajon, M. et al., *Radiat. Phys. Chem.*, 77, 1203, 2008. With permission.)

12.5 EFFICIENT SYSTEM TO SCAVENGE ALL PRIMARY RADICALS IN SPURS

In the framework of researches on the release of uranium in water under irradiation, studies found that an acidic solution of the $U^{IV}Cl_6^{2-}$ complex constitutes an interesting system for determining the radiolytic yield of radicals induced by irradiation. Here, such a solution is shown to scavenge almost all the radicals formed by water radiolysis, resulting in a better understanding of spur reactions and providing a good estimate of the initial radiolytic yield, at the picosecond timescale, of the radicals formed by the radiolytic decomposition of water (Atinault et al., 2009b).

Solutions containing 5×10^{-3} mol L^{-1} U^{IV} in the presence of 2 mol L^{-1} Cl^- at pH = 0 were prepared in saturated O_2 (0.7×10^{-3} mol L^{-1}) and were irradiated with γ-rays using a ^{60}Co source at a dose rate of 33.8 ± 0.4 Gy min^{-1}. The decrease in intensity of the U^{IV} complex absorbance at 648 nm is shown in Figure 12.23 as a function of the irradiation dose. The slope of the dependence of the concentration of the UCl^{3+} complex on the irradiation dose gives a radiolytic yield for the oxidation of U^{IV} of 8.7×10^{-7} mol J^{-1}. A concurrent increase in the formation rate of U^{VI} indicates that the U^{IV} is oxidized to U^{VI}. In fact, U^V is known to be unstable in solution. The high yield for the oxidation of UCl_6^{2-} complex is the result of the scavenging of all the radicals produced under irradiation. Such a finding has never been reported in radiation chemistry. In the case of the deaerated Fricke dosimeter in which there is no amplification step and in which H$^•$, OH$^•$, and H_2O_2 react with Fe^{2+}, the oxidizing yield is reported to be 8.5×10^{-7} mol J^{-1} (Barr and Schuler, 1959). In the U^{IV} case, the predominant oxidizing species are $HO_2^•$ and $Cl_2^{•-}$ (or Cl_3^-):

$$G(Ox)_{Total} \approx G(Cl_2^{-•}) + G(HO_2^•) = G(e_s^-) + G(H^•) + G(OH^•) \tag{12.4}$$

The most important reactions occurring on the submicrosecond timescale under these conditions are listed in Table 12.1. Hydrated electrons are rapidly converted to H$^•$ atoms in a highly acidic medium. All of the other e_{hyd}^- reactions are negligible, especially the important spur reaction between OH$^•$ and e_{hyd}^-, which is normally the main radical termination step. In oxygenated

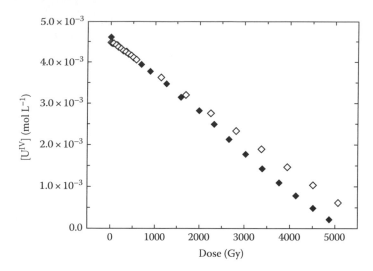

FIGURE 12.23 Concentration of U^{IV} in O_2-saturated aqueous solutions containing initially 5×10^{-3} mol L^{-1} of U^{IV} and 2 mol L^{-1} of Cl^- at pH = 0 as a function of irradiation dose: experimental data (closed symbols) and model simulation (open symbols). (Reprinted from Atinault, E. et al., *J. Phys. Chem. A*, 114, 2080, 2009a; Atinault, E. et al., *J. Phys. Chem. A*, 113, 949, 2009b. With permission.)

solutions, H$^•$ atoms are scavenged by molecular oxygen to form HO$_2^•$. The OH$^•$ radicals are efficiently scavenged by Cl$^-$ and converted to Cl$_2^{•-}$. A small amount of disproportionation reaction of Cl$_2^{•-}$ occurs within the microsecond timescale. The net result of these scavenging reactions is the production of the three species Cl$_2^{•-}$, Cl$_3^-$, and HO$_2^•$ that readily oxidize U^{IV}. Similar results are found for the yields of the radiolytic oxidation of U^{IV} and of the U^{VI} formation in N$_2$O-aqueous solutions in the presence of 2 mol L^{-1} Cl$^-$ at pH = 0 (HCl) (Atinault et al., 2009a).

The chemistry in the isolated spur was examined using a nonhomogeneous stochastic model previously developed (Pimblott and LaVerne, 1997). Figure 12.24 shows that the short-time

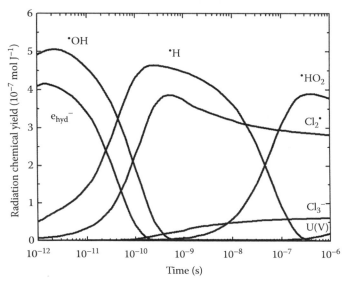

FIGURE 12.24 Simulations of the time dependences of radical yields obtained in the radiolysis of O$_2$-saturated aqueous solutions at pH = 0 with 2 mol L^{-1} Cl$^-$. (Reprinted from Atinault, E. et al., *J. Phys. Chem. A*, 114, 2080, 2009a; Atinault, E. et al., *J. Phys. Chem. A*, 113, 949, 2009b. With permission.)

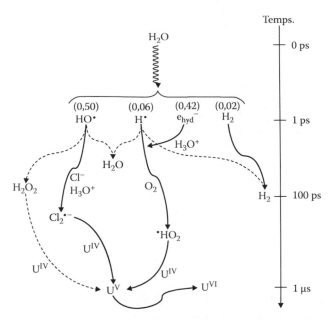

FIGURE 12.25 Scheme showing the oxidation mechanism of U^{IV} in O_2-saturated aqueous solutions at pH 0 containing 5×10^{-3} mol L^{-1} of U^{IV} and 2 mol L^{-1} of Cl^{-}. (Reprinted from Atinault, E. et al., *J. Phys. Chem. A*, 114, 2080, 2009a; Atinault, E. et al., *J. Phys. Chem. A*, 113, 949, 2009b. With permission.)

kinetics of the spur is essentially complete within a microsecond with the formation of the $Cl_2^{\cdot-}$, Cl_3^{-}, and $^{\cdot}HO_2$ oxidizing species. Oxidation of U^{IV} to U^{V} is just beginning on this timescale (Figure 12.25). The total oxidizing equivalent predicted by the simulations at 1 microsecond is 8.46×10^{-7} mol J^{-1}, which is slightly lower than the experimental value of 8.7×10^{-7} mol J^{-1}. With the same conditions, a value of 9.7×10^{-7} mol J^{-1} for the sum of the yield of the three radicals (e_{hyd}^{-}, $H^{\cdot}$, $OH^{\cdot}$) at one picosecond is predicted by the model simulations. That means that in the oxidation mechanism of U^{IV}, only 10% of the oxidizing species do not participate in the reactions, suggesting that oxidation of U^{IV} should provide a very good measure of the total initial yield of radicals produced in radiolysis of water. Uncertainty in the short-time measurement of the $OH^{\cdot}$ radical yield seems to be the limiting factor in the determination of the radiolytic decomposition of water at 1 ps.

12.6 CONCLUDING REMARKS

Great strides have been made in the area of radiation chemistry during the last decades due to technological advances. But more work remains to be done to get a better insight into the nonhomogeneous chemistry induced by ionizing irradiation.

Femtosecond laser development allowed the study of formation and solvation dynamics of electron in various solvents at room temperature. Numerous data have been obtained and analyzed with various solvation models, leading to many kinetics schemes. Those analyses correspond to phenomenological and macroscopic views of the processes. In order to unify those approaches, it is necessary to get a microscopic and structural view of the phenomena, and for that to happen, molecular dynamics simulations are required. Such calculations should provide a detailed picture of the electron solvation mechanism and unravel the role of the molecular structure of the solvent in the process. In addition, the detection of the fast transient cation, H_2O^+, which has not yet been observed, should be possible with the last technological breakthroughs improving the time resolution.

The advent of picosecond pulse-radiolysis facilities opened new possibilities to study the radiation effects at increasingly shorter times from an energy deposition point of view. However, some results are still controversial and many experiments are in progress to improve our knowledge. In particular, for water radiolysis, there are some discrepancies between the values reported for the hydrated electron yield at times shorter than 10 ps; these values need to be confirmed by pump-probe measurements. Up to now, the time-dependent yield of OH$^{\bullet}$ radical has only been determined by scavenging methods because OH$^{\bullet}$ absorbs in the UV spectral domain with a low absorption coefficient. Nevertheless, progress in the UV detection system should allow direct observations of OH$^{\bullet}$ by transient absorption setup using picosecond pulse accelerators. Such measurements are crucial to improve the description of spur reactions.

As matters stand at present, calculation codes concerning the time-dependent yield of the different species generated by radiolysis could also be improved. New experimental results from picosecond pulse-radiolysis measurements should make such an improvement possible by providing the necessary data to adjust empirical parameters.

REFERENCES

Abramczyk, H. 1991. Absorption spectrum of the solvated electron. 1. Theory. *J. Phys. Chem.* 95: 6149–6155.
Abramczyk, H. and Kroh, J. 1992. Near-IR absorption spectrum of the solvated electron in alcohols and deuterated water. *Radiat. Phys. Chem.* 39: 99–104.
Abramczyk, H. and Kroh, J. 1994. Spectroscopic properties of the solvated electron in water, alcohols, amines, ethers and alkanes. *Radiat. Phys. Chem.* 43: 291–297.
Allen, O. A. 1948. Radiation chemistry of aqueous solutions. *J. Phys. Colloid. Chem.* 52: 479–490.
Allen, O. A. 1989. The story of radiation chemistry of water. In *Early Developments in Radiation Chemistry*, J. Kroh (ed.), Cambridge, U.K.: Royal Society of Chemistry.
Aoki, Y., Yang, J. F., Hirose, M., Sakai, F., Tsunemi, A., Yorozu, M., Okada, Y., Endo, A., Wang, X. J., and Ben-Zvi, I. 2000. A new chemical analysis system using a photocathode RF gun. *Nucl. Instrum. Methods Phys. Res. Sect. A: Accel. Spectrom. Detect. Assoc. Equip.* 455 (1): 99–103.
Assel, M., Laenen, R., and Laubereau, A. 1998. Dynamics of excited solvated electrons in aqueous solution monitored with femtosecond-time and polarization resolution. *J. Phys. Chem. A* 102: 2256–2262.
Assel, M., Laenen, R., and Laubereau, A. 1999. Retrapping and solvation dynamics after femtosecond UV excitation of the solvated electron in water. *J. Chem. Phys.* 111 (15): 6869–6874.
Atinault, E., De Waele, V., Schmidhammer, U., Fattahi, M., and Mostafavi, M. 2008. Scavenging of and OH$^{\bullet}$ radicals in concentrated HCl and NaCl aqueous solutions. *Chem. Phys. Lett.* 460 (4–6): 461–465.
Atinault, E., De Waele, D., Belloni, J., Le Naour, C., Fattahi, M., and Mostafavi, M. 2009a. Radiolytic yield of U^{IV} oxidation into U^{VI}: A new mechanism for U^{V} reactivity in acidic solution. *J. Phys. Chem. A* 114 (5): 2080–2085.
Atinault, E., De Waele, V., Fattahi, M., Laverne, J. A., Pimblott, S. M., and Mostafavi, M. 2009b. Aqueous solution of UCl$_6^{2-}$ in O$_2$-saturated acidic medium: An efficient system to scavenge all primary radicals in spurs produced by irradiation. *J. Phys. Chem. A* 113: 949–951.
Barr, N. F. and Schuler, R. H. 1959. The dependence of radical and molecular yields on linear energy transfer in the radiation decomposition of 0.8 N sulfuric acid solutions. *J. Phys. Chem.* 63: 808–812.
Bartels, D. M. and Crowell, R. A. 2000. Photoionization yield vs. energy in H$_2$O and D$_2$O. *J. Phys. Chem. A* 104: 3349–3355.
Bartels, D. M., Cook, A. R., Mudaliar, M., and Jonah, C. D. 2000. Spur decay of the solvated electron in picosecond radiolysis measured with time-correlated absorption spectroscopy. *J. Phys. Chem. A* 104 (8): 1686–1691.
Bartels, D. M., David G., and Jonah, C. D. 2001. Spur decay kinetics of the solvated electron in heavy water radiolysis. *J. Phys. Chem. A* 105 (34): 8069–8072.
Baxendale, J. H. and Wardman, P. 1971. Direct observation of solvation of the electron in liquid alcohols by pulse radiolysis. *Nature* 230: 449–450.
Belloni, J., Monard, H., Gobert, F., Larbre, J.-P., Demarque, A., De Waele, V., Lampre, I., Marignier, J.-L., Mostafavi, M., Bourdon, J.-C., Bernard, M., Borie, H., Garvey, T., Jacquemard, B., Leblond, B., Lepercq, P., Omeich, M., Roch, M., Rodier, J., and Roux, R. 2004. ELYSE—A picosecond electron accelerator for pulse radiolysis research. *Nucl. Instrum. Methods Phys. Res. Sect. A: Accel. Spectrom. Detect. Assoc. Equip.* 539 (3): 527–539.

Bonin, J., Lampre, I., Pernot, P., and Mostafavi, M. 2007. Solvation dynamics of electron produced by two-photon ionization of liquid polyols. II. Propanediols. *J. Phys. Chem. A* 111: 4902–4913.

Bonin, J., Lampre, I., Pernot, P., and Mostafavi, M. 2008. Solvation dynamics of electron produced by two-photon ionization of liquid polyols. III. Glycerol. *J. Phys. Chem. A* 112: 1880–1886.

Bragg, A. E., Verlet, J. R. R., Kammrath, A., Cheshnovsky, O., and Neumark, D. M. 2004. Hydrated electron dynamics: From clusters to bulk. *Science* 306: 669–671.

Bronskil, M. J., Taylor, W. B., Wolff, R. K., and Hunt, J. W. 1970. Design and performance of a pulse radiolysis system capable of picosecond time resolution. *Rev. Sci. Instrum.* 41 (3): 333.

Brozek-Pluska, B., Gliger, D., Hallou, A., Malka, V., and Gauduel, Y. A. 2005. Direct observation of elementary radical events: Low- and high-energy radiation femtochemistry in solutions. *Radiat. Phys. Chem.* 72 (2–3): 149–157.

Brun, R., Reichert, P., and Kunsch, H. R. 2001. Practical identifiability analysis of large environmental simulation models. *Water Resour. Res.* 37 (4): 1015–1030.

Buxton, G. W. 2004. The radiation chemistry of liquid water: Principles and applications. In *Charged Particle and Proton Interaction with Matter*, A. Mozunder, and Y. Hatano (eds.). New York: Marcel Dekker.

Chase, W. J. and Hunt, J. W. 1975. Solvation time of the electron in polar liquids. Water and alcohols. *J. Phys. Chem.* 79 (26): 2835–2845.

Cobut, V., Frongillo, Y., Patau, J. P., Goulet, T., Fraser, M. J., and Jay-Gerin, J. P. 1998. Monte Carlo simulation of fast electron and proton tracks in liquid water: I. Physical and physicochemical aspects. *Radiat. Phys. Chem.* 51 (3): 229–243.

Curie, P. and Debierne, A. 1901. Sur la radio-activité induite et les gaz activés par le radium. *C. R. Acad. Sci.* 132: 770–772.

D'Agostini, G. 2003. *Bayesian Reasoning in Data Analysis*. River Edge, NJ: World Scientific Publishing.

D'Avignon, D. A., Bretthorst, G. L., Holtzer, M. E., and Holtzer, A. 1998. Site-specific thermodynamics and kinetics of a coiled-coil transition by spin inversion transfer NMR. *Biophys. J.* 74: 3190–3197.

D'Avignon, D. A., Bretthorst, G. L., Holtzer, M. E., and Holtzer, A. 1999. Thermodynamics and kinetics of a folded-folded′ transition at valine-9 of a GCN4-like leucine zipper. *Biophys. J.* 76: 2752–2759.

De Waele, V., Sorgues, S., Pernot, P., Marignier, J. L., Monard, H., Larbre, J. P., and Mostafavi, M. 2006. Geminate recombination measurements of solvated electron in THF using laser-synchronized picosecond electron pulse. *Chem. Phys. Lett.* 423 (1–3): 30–34.

De Waele, V., Sorgues, S., Pernot, P., Marignier, J.-L., and Mostafavi, M. 2007. Kinetics study of the solvated electron decay in THF using laser-synchronised picosecond electron pulse. *Nucl. Sci. Tech.* 18 (1): 10–15.

De Waele, V., Schmidhammer, U., Marquès, J.-R., Monard, H., Larbre, J.-P., Bourgeois, N., and Mostafavi, M. 2009. Non-invasive single bunch monitoring for ps pulse radiolysis. *Radiat. Phys. Chem.* 78: 1099–1101.

du Penhoat, M. A. H., Goulet, T., Frongillo, Y., Fraser, M. J., Bernat, P., and Jay-Gerin, J. P. 2000. Radiolysis of liquid water at temperatures up to 300 degrees C: A Monte Carlo simulation study. *J. Phys. Chem. A* 104 (50): 11757–11770.

Duane, W. and Scheuer, O. 1913. Recherches sur la décomposition de l'eau par les rayons α. *Le Radium* 10: 33–46.

Ehrler, O. T. and Neumark, D. M. 2009. Dynamics of electron solvation in molecular clusters. *Acc. Chem. Res.* 42 (6): 769–777.

Emde, M. F., Baltuska, A., Kummrow, A., Pshenichnikov, M. S., and Wiersma, D. A. 1998. Ultrafast librational dynamics of the hydrated electron. *Phys. Rev. Lett.* 80 (21): 4645–4648.

Freeman, G. R. 1987. Introduction. In *Kinetics of Nonhomogeneous Processes. A Practical Introduction for Chemists, Biologists, Physicists, and Materials Scientists*, G. R. Freeman (ed.). New York: Wiley Interscience.

Fricke, H. 1934. The reduction of oxygen to hydrogen peroxide by the irradiation of its aqueous solution with x-rays. *J. Chem. Phys.* 2 (9): 556–559.

Frongillo, Y., Goulet, T., Fraser, M. J., Cobut, V., Patau, J. P., and Jay-Gerin, J. P. 1998. Monte Carlo simulation of fast electron and proton tracks in liquid water: II. Nonhomogeneous chemistry. *Radiat. Phys. Chem.* 51 (3): 245–254.

Gauduel, Y. 1995. Femtosecond optical spectroscopy in liquids: Applications to the study of reaction dynamics. *J. Mol. Liq.* 63 (1–2): 1–54.

Gauduel, Y., Pommeret, S., Migus, A., and Antonetti, A. 1990. Some evidence of ultrafast H_2O^+-water molecule reaction in femtosecond photoionization of pure liquid water: Influence on geminate pair recombination dynamics. *Chem. Phys.* 149 (1–2): 1–10.

Giesel, F. 1900. Einigesüber radium-barium-salze und derenstrahlen. *Verhandl. Deutsch Phys. Ges.* 2: 9–10.

Goulet, T. and Jay-Gerin, J.-P. 1989. Thermalization of subexcitation electrons in solid water. *Radiat. Res.* 118: 46–62.

Goulet, T., Pépin, C., Houde, D., and Jay-Gerin, J.-P. 1999. On the relaxation kinetics following the absorption of light by solvated electrons in polar liquids: Roles of the continuous spectral shifts and of the stepwise transition. *Radiat. Phys. Chem.* 54 (5): 441–448.

Gross, B. 1957. Irradiation effects in borosilicate glass. *Phys. Rev.* 107 (2): 368–373.

Gross, B. 1958. Irradiation effects in plexiglas. *J. Polym. Sci.* 27 (115): 135–143.

Hart, E. J. and Boag, J. W. 1962. Absorption spectrum of the hydrated electron in water and in aqueous solutions. *J. Am. Chem. Soc.* 84: 4090–4095.

Hertwig, A., Hippler, H., and Unterreiner, A.-N. 1999. Transient spectra, formation, and geminate recombination of solvated electrons in pure water UV-photolysis: An alternative view. *Phys. Chem. Chem. Phys.* 1 (24): 5633–5642.

Hirata, Y. and Mataga, N. 1990. Solvation dynamics of electrons ejected by picosecond dye laser pulse excitation of p-phenylenediamine in several alcoholic solutions. *J. Phys. Chem.* 94: 8503–8505.

Holpar, P., Megyes, T., and Keszei, E. 1999. Electron solvation in methanol revisited. *Radiat. Phys. Chem.* 55: 573–577.

Holtzer, M. E., Bretthorst, G. L., D'Avignon, D. A., Angeletti, R. H., Mints, L., and Holtzer, A. 2001. Temperature dependence of the folding and unfolding kinetics of the GCN4 Leucine Zipper via $^{13}C\alpha$-NMR. *Biophys. J.* 80: 939–951.

Huerta Parajon, M., Rajesh, P., Mua, T., Pimblott, S. M., and LaVerne, J. A. 2008. H atom yield in the radiolysis of water. *Radiat. Phys. Chem.* 77: 1203–1207.

Hunt, J. W., Wolff, R. K., Bronskill, M. J., Jonah, C. D., Hart, E. J., and Matheson, M. S. 1973. Radiolytic yields of hydrated electrons at 30 to 1000 picoseconds after energy absorption. *J. Phys. Chem.* 77 (3): 425–426.

Janik, I., Bartels, D. M., and Jonah, C. D. 2007. Hydroxyl radical self-recombination reaction and absorption spectrum in water up to 350°C. *J. Phys. Chem. A* 111 (10): 1835–1843.

Jay-Gerin, J. P. and Ferradini, C. 2000. A new estimate of the (OH)-O-center dot radical yield at early times in the radiolysis of liquid water. *Chem. Phys. Lett.* 317 (3–5): 388–391.

Jonah, C. D. 1975. Wide-time range pulse-radiolysis system of picosecond time resolution. *Rev. Sci. Instrum.* 46: 42–46.

Jonah, C. D. and Miller, J. R. 1977. Yield and decay of the hydroxyl radical from 200 ps to 2 ns. *J. Phys. Chem.* 81 (21): 1974–1976.

Jonah, C. D., Hart, E. J., and Matheson, M. S. 1973. Yields and decay of the hydrated electron at times greater than 200 picoseconds. *J. Phys. Chem.* 77 (15): 1838–1843.

Jonah, C. D., Matheson, M. S., Miller, J. R., and Hart, E. J. 1976. Yield and decay of hydrated electron from 100 ps to 3 ns. *J. Phys. Chem.* 80: 1267–1270.

Kadhum, A. A. H. and Salmon, G. A. 1984. Electron scavenging in the [gamma]-radiolysis of tetrahydrofuran. *Radiat. Phys. Chem.* (1977) 23 (1–2): 67–71.

Kambhampati, P., Son, D. H., Kee, T. W., and Barbara, P. F. 2002. Solvation dynamics of the hydrated electron depends on its initial degree of electron delocalization. *J. Phys. Chem. A* 106: 2374–2378.

Kashiwagi, S., Kuroda, R., Oshima, T., Nagasawa, F., Kobuki, T., Ueyama, D., Hama, Y., Washio, M., Ushida, K., Hayano, H., and Urakawa, J. 2005. Compact soft x-ray source using Thomson scattering. *J. Appl. Phys.* 98 (12): 123302–123306.

Kawaguchi, M., Ushida, K., Kashiwagi, S., Kuroda, R., Kuribayashi, T., Kobayashi, M., Hama, Y., and Washio, M. 2005. Development of compact picosecond pulse radiolysis system. *Nucl. Instrum. Methods Phys. Res. Sect. B: Beam Interact. Mater. Atoms* 236: 425–431.

Kee, T. W., Son, D. H., Kambhampati, P., and Barbara, P. F. 2001. A unified electron transfer model for the different precursors and excited states of the hydrated electron. *J. Phys. Chem. A* 105: 8434–8439.

Kernbaum, M. 1909. Action chimique sur l'eau des rayons pénétrants de radium. *C. R. Acad. Sci.* 148: 705–706.

Keszei, E. and Jay-Gerin, J.-P. 1992. On the role of the parent cation in the dynamics of formation of laser-induced hydrated electrons. *Can. J. Chem.* 70: 21–23.

Kloepfer, J. A., Vilchiz, V. H., Lenchenkov, V. A., Germaine, A. C., and Bradforth, S. E. 2000. The ejection distribution of solvated electrons generated by the one-photon photodetachment of aqueous I⁻ and two-photon ionization of the solvent. *J. Chem. Phys.* 113 (15): 6288–6307.

Kobayashi, H. and Tabata, Y. 1985. A 20 ps time resolved pulse-radiolysis using 2 linacs. *Nucl. Instrum. Methods Phys. Res. Sect. B: Beam Interact. Mater. Atoms* 10–11 (May): 1004–1006.

Kozawa, T., Mizutani, Y., Yokoyama, K., Okuda, S., Yoshida, Y., and Tagawa, S. 1999. Measurement of far-infrared subpicosecond coherent radiation for pulse radiolysis. *Nucl. Instrum. Methods Phys. Res. Sect. A: Accel. Spectrom. Detect. Assoc. Equip.* 429 (1–3): 471–475.

Kozawa, T., Mizutani, Y., Miki, M., Yamamoto, T., Suemine, S., Yoshida, Y., and Tagawa, S. 2000. Development of subpicosecond pulse radiolysis system. *Nucl. Instrum. Methods Phys. Res. Sect. A: Accel. Spectrom. Detect. Assoc. Equip.* 440 (1): 251–253.

Kraus, C. A. 1908. Solutions of metals in non-metallic solvents; IV. Material effects accompanying the passage of an electrical current through solutions of metals in liquid ammonia. Migration experiments. *J. Am. Chem. Soc.* 30: 1323–1344.

Lampre, I., Bonin, J., Soroushian, B., Pernot, P., and Mostafavi, M. 2008a. Formation and solvation dynamics of electrons in polyols. *J. Mol. Liq.* 141: 124–129.

Lampre, I., Pernot, P., Bonin, J., and Mostafavi, M. 2008b. Comparison of solvation dynamics of electrons in four polyols. *Radiat. Phys. Chem.* 77: 1183–1189.

LaVerne, J. A. and Pimblott, S. M. 1991. Scavenger and time dependences of radicals and molecular products in the electron radiolysis of water: Examination of experiments and models. *J. Phys. Chem.* 95 (8): 3196–3206.

LaVerne, J. A. and Pimblott, S. M. 2000. New mechanism for H_2 formation in water. *J. Phys. Chem. A* 104: 9820–9822.

Lee, I.-R., Lee, W., and Zewail, A. H. 2008. Dynamics of electrons in ammonia cages: The discovery system of solvation. *ChemPhysChem* 9: 83–88.

Lian, R., Crowell, R. A., and Shkrob, I. A. 2005. Solvation and thermalization of electrons generated by above-the-gap (12.4 eV) two-photon ionization of liquid H_2O and D_2O. *J. Phys. Chem. A* 109: 1510–1520.

Lin, M. Z., Mostafavi, M., Muroya, Y., Han, Z. H., Lampre, I., and Katsumura, Y. 2006. Time-dependent radiolytic yields of the solvated electrons in 1,2-ethanediol, 1,2-propanediol, and 1,3-propanediol from picosecond to microsecond. *J. Phys. Chem. A* 110 (40): 11404–11410.

Lindner, J., Unterreiner, A.-N., and Vöhringer, P. 2006. Femtosecond relaxation dynamics of solvated electrons in liquid ammonia. *ChemPhysChem* 7: 363–369.

Long, F. H., Lu, H., and Eisenthal, K. B. 1990. Femtosecond studies of the presolvated electron: An excited state of the solvated electron? *Phys. Rev. Lett.* 64 (12): 1469–1472.

Madsen, D., Thomsen, C. L., Thogersen, J., and Keiding, S. R. 2000. Temperature dependent relaxation and recombination dynamics of the hydrated electron. *J. Chem. Phys.* 113 (3): 1126–1134.

Magee, J. L. and Chatterjee, A. 1987. Track reactions of radiation chemistry. In *Kinetics of Nonhomogeneous Processes. A Practical Introduction for Chemists, Biologists, Physicists, and Materials Scientists*, G. R. Freeman (ed.). New York: Wiley Interscience.

Marignier, J. L., de Waele, V., Monard, H., Gobert, F., Larbre, J. P, Demarque, A., Mostafavi, M., and Belloni, J. 2006. Time-resolved spectroscopy at the picosecond laser-triggered electron accelerator ELYSE. *Radiat. Phys. Chem.* 75 (9): 1024–1033.

Martini, I. B., Barthel, E. R., and Schwartz, B. J. 2000. Mechanisms of the ultrafast production and recombination of solvated electrons in weakly polar fluids: Comparison of multiphoton ionization and detachment via the charge-transfer-to-solvent transition of Na- in THF. *J. Chem. Phys.* 113 (24): 11245–11257.

Migus, A., Gauduel, Y., Martin, J. L., and Antonetti, A. 1987. Excess electrons in liquid water: First evidence of a prehydrated state with femtosecond lifetime. *Phys. Rev. Lett.* 58 (15): 1559–1562.

Mirdamadi-Esfahani, M., Lampre, I., Marignier, J.-L., de Waele, V., and Mostafavi, M. 2009. Radiolytic formation of tribromine ion Br3- in aqueous solutions, a system for steady-state dosimetry. *Radiat. Phys. Chem.* 78 (2): 106–111.

Mizuno, M. and Tahara, T. 2003. Picosecond time-resolved resonance Raman study of the solvated electron in water. *J. Phys. Chem. A* 107: 2411–2421.

Mizuno, M., Yamaguchi, S., and Tahara, T. 2005. Relaxation dynamics of the hydrated electron: Femtosecond time-resolved resonance Raman and luminescence study. *J. Phys. Chem. A* 109: 5257–5265.

Mozumder, A. 2002. Ionization and excitation yields in liquid water due to the primary irradiation: Relationship of radiolysis with far UV-photolysis. *Phys. Chem. Chem. Phys.* 4: 1451–1456.

Mozumder, A. 2005. Electrons in condensed media. *Radiat. Phys. Chem.* 72: 73–78.

Muroya, Y., Lin, M., Watanabe, T., Wu, G., Kobayashi, T., Yoshii K., Ueda, K., Uesaka, M., and Katsumura, Y. 2001a. Ultra-fast pulse radiolysis system combined with a laser photocathode RF gun and a femtosecond laser *Nucl. Instrum. Methods Phys. Res. Sect. A: Accel. Spectrom. Detect. Assoc. Equip.* 489 (1–3): 554–562.

Muroya, Y., Watanabe, T., Wu, G., Li, X., Kobayashi, T., Sugahara, J., Ueda, T., Yoshii, K., Uesaka, M., and Katsumura, Y. 2001b. Design and development of a sub-picosecond pulse radiolysis system. *Radiat. Phys. Chem.* 60 (4–5): 307–312.

Muroya, Y., Meesungnoen, J., Jay-Gerin, J. P., Filali-Mouhim, A., Goulet, T., Katsumura, Y., and Mankhetkorn, S. 2002. Radiolysis of liquid water: An attempt to reconcile Monte Carlo calculations with new experimental hydrated electron yield data at early times. *Can. J. Chem.* 80: 1367–1374.

Muroya, Y., Lin, M., Iijima, H., Ueda, T., and Katsumura, Y. 2005a. Current status of the ultra-fast pulse radiolysis system at NERL, the University of Tokyo. *Res. Chem. Intermed.* 31 (1–3): 261–272.

Muroya, Y., Lin, M. Z., Wu, G. Z., Iijima, H., Yoshi, K., Ueda, T., Kudo, H., and Katsumura, Y. 2005b. A re-evaluation of the initial yield of the hydrated electron in the picosecond time range. *Radiat. Phys. Chem.* 72 (2–3): 169–172.

Muroya, Y., Lin, M. Z., Han, Z. H., Kumagai, Y., Sakumi, A., Ueda, T., and Katsumura, Y. 2008. Ultra-fast pulse radiolysis: A review of the recent system progress and its application to study on initial yields and solvation processes of solvated electrons in various kinds of alcohols. *Radiat. Phys. Chem.* 77 (10–12): 1176–1182.

Nagai, H., Kawaguchi, M., Sakaue, K., Komiya, K., Nomoto, T., Kamiya, Y., Hama, Y., Washio, M., Ushida, K., Kashiwagi, S., and Kuroda, R. 2007. Improvements in time resolution and signal-to-noise ratio in a compact pico-second pulse radiolysis system. *Nucl. Instrum. Methods Phys. Res. Sect. B: Beam Interact. Mater. Atoms* 265 (1): 82–86.

Nikogosyan, D. N., Oraevsky, A. A., and Rupasov, V. I. 1983. Two-photon ionization and dissociation of liquid water by powerful laser UV radiation. *Chem. Phys.* 77 (1): 131–143.

Nordlund, D., Ogasawara, H., Bluhm, H., Takahashi, O., Odelius, M., Nagasono, M., Pettersson, L. G. M., and Nilsson, A. 2007. Probing the electron delocalization in liquid water and ice at attosecond time scales. *Phys. Rev. Lett.* 99: 217406.

Okamoto, K., Saeki, A., Kozawa, T., Yoshida, Y., and Tagawa, S. 2003. Subpicosecond pulse radiolysis study of geminate ion recombination in liquid benzene. *Chem. Lett.* 32 (9): 834–835.

Oulianov, D. A., Crowell, R. A., Gosztola, D. J., Shkrob, I. A., Korovyanko, O. J., and Rey-de-Castro, R. C. 2007. Ultrafast pulse radiolysis using a terawatt laser wakefield accelerator. *J. Appl. Phys.* 101 (5): 053102–053109.

Paik, D. H., Lee, I.-R., Yang, D.-S., Baskin, J. S., and Zewail, A. H. 2004. Electrons in finite-sized water cavities: Hydration dynamics observed in real time. *Science* 306: 672–675.

Pastina, B., LaVerne, J. A., and Pimblott, S. M. 1999. Dependence of molecular hydrogen formation in water on scavengers of the precursor to the hydrated electron. *J. Phys. Chem. A* 103: 5841–5846.

Pépin, C., Goulet, T., Houde, D., and Jay-Gerin, J.-P. 1994. Femtosecond kinetic measurements of excess electrons in methanol: Substantiation for a hybrid solvation mechanism. *J. Phys. Chem.* 98: 7009–7013.

Pépin, C., Goulet, T., Houde, D., and Jay-Gerin, J.-P. 1997. Observation of a continuous spectral shift in the solvation kinetics of electrons in neat liquid deuterated water. *J. Phys. Chem. A* 101 (24): 4351–4360.

Pimblott, S. M. and LaVerne, J. A. 1997. Stochastic simulation of the electron radiolysis of water and aqueous solutions. *J. Phys. Chem. A* 101 (33): 5828–5838.

Pommeret, S., Gobert, F., Mostafavi, M., Lampre, I., and Mialocq, J.-C. 2001. Femtochemistry of the hydrated electron at decimolar concentration. *J. Phys. Chem. A* 105: 11400–11406.

Pshenichnikov, M. S., Baltuska, A., and Wiersma, D. A. 2004. Hydrated electron population dynamics. *Chem. Phys. Lett.* 389: 171–175.

Reuther, A., Laubereau, A., and Nikogosyan, D. N. 1996. Primary photochemical processes in water. *J. Phys. Chem.* 100 (42): 16794–16800.

Saeki, A., Kozawa, T., Kashiwagi, S., Okamoto, K., Isoyama, G., Yoshida, Y., and Tagawa, S. 2005. Synchronization of femtosecond UV-IR laser with electron beam for pulse radiolysis studies. *Nucl. Instrum. Methods Phys. Res. Sect. A: Accel. Spectrom. Detect. Assoc. Equip.* 546 (3): 627–633.

Saeki, A., Kozawa, T., and Tagawa, S. 2006. Picosecond pulse radiolysis using femtosecond white light with a high S/N spectrum acquisition system in one beam shot. *Nucl. Instrum. Methods Phys. Res. Sect. A: Accel., Spectrom., Detect. Assoc. Equip.* 556 (1): 391–396.

Saeki, A., Kozawa, T., Ohnishi, Y., and Tagawa, S. 2007. Reactivity between biphenyl and precursor of solvated electrons in tetrahydrofuran measured by picosecond pulse radiolysis in near-ultraviolet, visible, and infrared. *J. Phys. Chem. A* 111 (7): 1229–1235.

Salmon, G. A., Seddon, W. A., and Fletcher, J. W. 1974. Pulse radiolytic formation of solvated electrons, ion-pairs, and alkali metal anions in tetrahydrofuran. *Can. J. Chem.* 52: 3259–3268.

Sander, M., Brummund, U., Luther, K., and Troe, J. 1992. Fast processes in UV-two-photon excitation of pure liquids. *Ber. Bunsenges. Phys. Chem.* 96 (10): 1486–1490.

Sander, M. U., Luther, K., and Troe, J. 1993. On the photoionization mechanism of liquid water. *Ber. Bunsenges. Phys. Chem.* 97: 953–961.

Sauer, M. C., Shkrob, I. A., Lian, R., Crowell, R. A., Bartels, D., Chen, X., Suffern, D., and Bradforth, S. E. 2004. Electron photodetachment from aqueous anions. 2. Ionic strength effect on geminate recombination dynamics and quantum yield for hydrated electron. *J. Phys. Chem. A* 108 (47): 10414–10425.

Scheidt, T. and Laenen, R. 2003. Ionization of methanol: Monitoring the trapping of electrons on the fs time scale. *Chem. Phys. Lett.* 371 (3–4): 445–450.

Shi, X., Long, F. H., Lu, H., and Eisenthal, K. B. 1995. Electron solvation in neat alcohols. *J. Phys. Chem.* 99 (18): 6917–6922.

Shkrob, I. A. 2006. Ammoniated electron as a solvent stabilized multimer radical anion. *J. Phys. Chem. A* 110: 3967–3976.

Silva, C., Walhout, P. K., Reid, P. J., and Barbara, P. F. 1998a. Detailed investigations of the pump-probe spectroscopy of the equilibrated solvated electron in alcohols. *J. Phys. Chem. A* 102: 5701–5707.

Silva, C., Walhout, P. K., Yokoyama, K., and Barbara, P. F. 1998b. Femtosecond solvation dynamics of the hydrated electron. *Phys. Rev. Lett.* 80 (5): 1086–1089.

Sivia, D. S. 1996. *Data Analysis: A Bayesian Tutorial*. Oxford, U.K.: Clarendon Press.

Soroushian, B., Lampre, I., Bonin, J., Pernot, P., Pommeret, S., and Mostafavi, M. 2006. Solvation dynamics of the electron produced by two-photon ionization of liquids polyols. I. Ethylene glycol. *J. Phys. Chem. A* 110 (5): 1705–1717.

Sumiyoshi, T. and Katayama, M. 1982. The yield of hydrated electron at 30 ps. *Chem. Lett.* 12: 1887–1890.

Sumiyoshi, T., Tsugaru, K., Yamada, T., and Katayama, M. 1985. Yields of solvated electrons at 30 picoseconds in water and alcohols. *Bull. Chem. Soc. Jpn.* 58 (11): 3073–3075.

Swiatla-Wojcik, D. and Buxton, G. V. 1995. Modeling of radiation spur processes in water at temperatures up to 300°C. *J. Phys. Chem.* 99 (29): 11464–11471.

Swiatla-Wojcik, D. and Buxton, G. V. 2000. Diffusion-kinetic modelling of the effect of temperature on the radiation chemistry of heavy water. *Phys. Chem. Chem. Phys.* 2 (22): 5113–5119.

Symons, M. C. R. 1976. Solutions of metals: Solvated electrons. *Chem. Soc. Rev.* 5: 337–358.

Tabata, Y., Kobayashi, H., Washio, M., Tagawa, S., and Yoshida, Y. 1985. Pulse-radiolysis with picosecond time resolution. *Radiat. Phys. Chem.* 26 (5): 473–479.

Tauber, M. J. and Mathies, R. A. 2002. Resonance Raman spectra and vibronic analysis of the aqueous solvated electron. *Chem. Phys. Lett.* 354: 518–526.

Tauber, M. J. and Mathies, R. A. 2003. Structure of the aqueous solvated electron from resonance Raman spectroscopy: Lessons from isotopic mixtures. *J. Am. Chem. Soc.* 125: 1394–1402.

Tauber, M. J., Stuart, C. M., and Mathies, R. A. 2004. Resonance Raman spectra of electrons solvated in liquid alcohols. *J. Am. Chem. Soc.* 126: 3414–3415.

Thaller, A., Laenen, R., and Laubereau, A. 2006. The precursors of the solvated electron in methanol studied by femtosecond pump-repump-probe spectroscopy. *J. Chem. Phys.* 124: 024515.

Thomas, J. M., Edwards, P. P., and Kuznetsov, V. L. 2008. Sir Humphry Davy: Boundless chemist, physicist, poet and man of action. *ChemPhysChem* 9:59–66.

Turi, L., Holpar, P., and Keszei, E. 1997. Alternative mechanisms for solvation dynamics of laser-induced electrons in methanol. *J. Phys. Chem. A* 101: 5469–5476.

Uesaka, M., Tauchi, K., Kozawa, T., Kobayashi, T., Ueda, T., and Miya, K. 1994. Generation of a subpicosecond relativistic electron single bunch at the S-band linear accelerator. *Phys. Rev. E* 50 (4): 3068.

Uesaka, M., Sakumi, A., Hosokai, T., Kinoshita, K., Yamaoka, N., Zhidkov, A., Ohkubo, T., Ueda, T., Muroya, Y., Katsumura, Y., Iijima, H., Tomizawa, H., and Kumagai, N. 2005. New accelerators for femtosecond beam pump-and-probe analysis. *Nucl. Instrum. Methods Phys. Res. Sect. A: Accel. Spectrom. Detect. Assoc. Equip.* 241 (1–4): 880–884.

Walhout, P. K., Alfano, J. C., Kimura, Y., Silva, C., Reid, P. J., and Barbara, P. F. 1995. Direct pump/probe spectroscopy of the near-IR band of the solvated electron. *Chem. Phys. Lett.* 232: 135–140.

Weiss, J. 1944. Radiochemistry in aqueous solutions. *Nature* 153: 748–750.

Wishart, J. F., Cook, A. R., and Miller, J. R. 2004. The LEAF picosecond pulse radiolysis facility at Brookhaven National Laboratory. *Rev. Sci. Instrum.* 75 (11): 4359–4366.

Wojcik, M., Tachiya, M., Tagawa, S., and Hatano, Y. 2004. Electron-ion recombination in condensed matter: Geminate and bulk recombination processes. In *Charged Particles and Photon interactions with Matter*, A. Mozumder and Y. Hatano (eds.). New York: Marcel Dekker.

Wolff, R. K., Bronskill, M. J., Aldrich, J. E., and Hunt, J. W. 1973. Picosecond pulse radiolysis. IV. Yield of the solvated electron at 30 picoseconds. *J. Phys. Chem.* 77 (11): 1350–1355.

Wolff, R. K., Aldrich, J. E., Penner, T. L., and Hunt, J. W. 1975. Picosecond pulse radiolysis. V. Yield of electrons in irradiated aqueous solution with high concentrations of scavenger. *J. Phys. Chem.* 79 (3): 210–219.

Yang, J. F., Kondoh, T., Kozawa, T., Yoshida, Y., and Tagawa, S. 2006. Pulse radiolysis based on a femtosecond electron beam and a femtosecond laser light with double-pulse injection technique. *Radiat. Phys. Chem.* 75 (9): 1034–1040.

Yokoyama, K., Silva, C., Son, D. H., Walhout, P. K., and Barbara, P. F. 1998. Detailed investigation of the femtosecond pump-probe spectroscopy of the hydrated electron. *J. Phys. Chem. A* 102: 6957–6966.

13 Radiation Chemistry of Liquid Water with Heavy Ions: Steady-State and Pulse Radiolysis Studies

Shinichi Yamashita
Japan Atomic Energy Agency
Tokai, Japan

Mitsumasa Taguchi
Japan Atomic Energy Agency
Takasaki, Japan

Gérard Baldacchino
Commissariat à l' Énergie Atomique et aux Énergies Alternatives
Saclay, France
and
Centre National de la Recherche Scientifique
Gif-sur-Yvette, France

Yosuke Katsumura
The University of Tokyo
Tokyo, Japan
and
Japan Atomic Energy Agency
Tokai, Japan

CONTENTS

13.1 INTRODUCTION

13.1.1 Importance of Heavy-Ion Beams

The term *ion beam* involves, in a broad sense, beams of not only positively charged atomic ions but also of positively/negatively charged molecular and cluster ions. In this chapter, this term is used to collectively designate beams of positively charged atomic ions. It is phenomenologically well known that irradiation of fast heavy ions leads to much different results compared to γ-rays, x-rays, and fast electrons, which are categorized as low-LET radiations. Heavy ions deposit their energies much more densely than low-LET radiations, and are categorized as high-LET radiations. Here, LET is the acronym for linear energy transfer, and is practically used as an indicator of 1D energy deposition density along the radiation trajectory, similarly as stopping power ($-dE/dx$), although their strict definitions are different. It is worth noting that one of the significant criteria between high- and low-LET radiations can be estimated by considering the initial size of spur at 1 ps and the diffusion distance during intra-spur reactions terminating by about 100 ns. The former is 3 nm or less in deviation, and the latter is briefly estimated to be 16–34 nm by using the diffusion coefficient of the water molecule, 2 nm^2/ns, and that of H$_3$O$^+$, 9 nm^2/ns. Then, spurs located within 40 nm would overlap each other during intra-spur processes. In addition, the average energy necessary to produce a single spur is about 60 eV. Thus, one of the significant criteria is 60 eV/40 nm = 1.5 eV/nm, at which or higher LET the intra-spur processes are affected by neighboring spurs. Due to their high-LET feature, heavy ions induce much denser ionizations and excitations in matter than low-LET radiations, leading to distinctive irradiation effects, such as high efficiency in induction of lethal damage to cells and so on. In this chapter, radiation chemical effects of heavy ions onto water are focused on not only because water is one of the simplest and ubiquitous molecules on Earth, but also because it is one of the main components of our bodies.

Such distinctiveness is now utilized in many applications (see Chapters 20, 21, and 24), for example, proton and C ions of several hundreds of MeV per nucleon (MeV/u) are used in the therapy of cancer typically located at deep positions inside human bodies. After intensive clinical studies and confirmation of the powerfulness and safety or risk of the application, the actual utilization of heavy-ion therapy has begun; however, a detailed mechanism that would lead to distinctive biological effects suitable for therapy is not yet known well. Human bodies would be exposed to heavy-ion beams during space activities, which are increasing day by day. Examples of such heavy-ion beams are space radiations such as solar particle events (SPEs) and galactic cosmic rays (GCRs), which mainly consist of a wide variety of atomic ions. It is inevitably important to assess the exposed dose of astronauts during their space missions and to maintain risk to the lowest possible level. Due to the wide variety of ion types and energy involved in space radiations, it is desired to establish and sophisticate a universal theory that can be applied to a wide range of ion beams. The distinctiveness of ion beams is also utilized in analytical techniques such as particle-induced x-ray emission (PIXE), in surface processing such as focused ion beam (FIB) (see Chapter 26), and in inducing mutation in plant breeding to give a wide variety of plants (see Chapter 34). In addition, understanding irradiation effects induced by ion beams is important in nuclear engineering (see Chapter 35). For example, very heavy ions of a few mega electron volts per nucleon are produced as fission fragments (FFs), and ejected alpha particles as He ions of a few mega electron volts. Ejected fast neutrons are also associated with ion beams because they lose their energies mainly through elastic collision with hydrogen nuclei of water molecules, resulting in the generation of recoil protons possessing half of the initial neutron energy at an average. It is recommended that the chapters of this book indicated in parentheses above be referred appropriately. Note that understanding irradiation effects of atomic ions also helps understand those of molecular and cluster ions, because these basically convert into atomic ions after initial collision processes.

The energy unit of kinetic energy divided by the mass number of ions, that is, MeV/u, is very useful and widely employed in ion beam research, because the same value gives the same velocity even for a different kind of ion.

13.1.2 IRRADIATION EFFECTS OF HEAVY-ION BEAMS

13.1.2.1 Physics: Energy Deposition and Appearance of Track

The above-mentioned applications utilize, more or less, distinctive features of heavy-ion beams in physics, chemistry, and biology. When a fast charged particle has a large amount of energy, it deposits its energy through Coulombic interactions with matter. These consist of electronic and nuclear interactions, which are defined as interactions with bond electrons and atomic nuclei in matter, respectively. For a high-energy heavy ion, the former interaction is predominant and results in ejection of high-energy secondary electrons. These secondary electrons also deposit their energy through Coulombic interactions, and *tertiary* electrons are produced; however, electrons of all generations produced in these processes are collectively referred to as secondary electrons. Energy depositions from both the primary projectile ion and secondary electrons lead to the production of electrons and holes as well as excited molecules, which participate in consequent chemical reactions and would be responsible for subsequent biological responses. Thus, the patterns of energy deposition and distributions of chemical species are required to comprehend consequent chemistry, biology, and so on, and are called *track* or *track structure*. In order to compass the whole picture of energy deposition processes initiated by the fast heavy ion, cross sections for energy transfer between secondary electrons and bond electrons as well as between the heavy ion and electrons are necessary. These cross sections have intensively been investigated, and several reviews on them and theories to correlate them to the stopping power have been proposed (see Chapters 2 through 5 and 13 (Mozumder, 2004; Toburen, 2004) and references therein).

The stopping power or LET is the most widely and conventionally used parameter to indicate how densely the ionizing radiation deposits its energy. The most important and basic physical parameters that represent particle radiations are kinetic energy (E), velocity (β), and effective charge (Z_{eff}), which are used in the estimation of stopping power. These parameters are related to each other by the following equations:

$$\beta = \sqrt{1 - \left(\frac{Mc^2}{Mc^2 + E} \right)^2} \tag{13.1}$$

$$Z_{eff} = Z\left[1 - \exp(-125\beta Z^{-(2/3)}) \right] \tag{13.2}$$

where

   M and Z are the mass and the atomic number of the projectile ion, respectively
   c is the light velocity in vacuum

A few theories have been put forward to determine the effective charge. Note that Equation 13.2 is given by the Barkas theory (Barkas, 1963), and that high-energy ions of 0.4 MeV/u in β or of 85 MeV/u in E are fully stripped (Z_{eff}/Z is higher than 0.9988) based on this equation if Z is 20 or smaller. Neglecting several assumptions, stopping powers for a proton and ions can be represented in a compact form by the Bethe equations (Berger et al., 1993):

$$\left(-\frac{dE}{dx} \right)_{ion} = \frac{4\pi e^4}{mc^2} (N_{tar} Z_{tar}) \left(\frac{Z_{eff}^2}{\beta^2} \right) \ln \frac{2mc^2\beta^2}{I} \tag{13.3}$$

$$\left(-\frac{dE}{dx} \right)_{proton} = \frac{4\pi e^4}{mc^2} (N_{tar} Z_{tar}) \left(\frac{1}{\beta^2} \right) \ln \frac{2mc^2\beta^2}{I} \tag{13.4}$$

where

   m and e are the mass and the charge of the electron, respectively
   N_{tar} and Z_{tar} are the number density and the atomic number of atoms in the target material, respectively
   I is the mean excitation potential of the target material

Thus, the ratio of the stopping power for ions to that for the proton at the same velocity is given by the square of effective charge:

$$\left(-\frac{dE}{dx} \right)_{ion} \Bigg/ \left(-\frac{dE}{dx} \right)_{proton} = Z_{eff}^2 \tag{13.5}$$

In fact, effective charge is originally defined by this equation to extrapolate the stopping power for a proton to that for heavier ions, and can be different from the average charge, which is simply defined as the average of the charge. Going back to Equation 13.3, the stopping power of ions is approximately proportional to the inverse of the square of β as long as β is not extremely small. Thus, the stopping power is low when the ion energy is high and vice versa. In other words, just after penetrating into matter, the stopping power becomes relatively low and the projectile ion loses its energy slowly. With decreasing velocity, the stopping power increases, which accelerates the increase of the stopping power itself as well as the decrease of the velocity. As a result, the heavy ion shows a drastic decline in the velocity as well as a drastic rise in the stopping power just before the position where it stops, which is referred to as the Bragg peak. For example, stopping powers for C ions of 100, 10, and 1 MeV/u in water are calculated as 25.9, 175, and 745 eV/nm, respectively, by the SRIM code (Ziegler, 1998). In cancer therapy with ion beams, irradiation is conducted with putting the Bragg peak on cancer, leading to the desired dose distribution in human body. In surface processing, relatively low-energy ions are used and their penetration depths are controlled by controlling the ion energy.

13.1.2.2 Chemistry: Intra-Track Reactions and Product Yields

After energy deposition events, physicochemical processes follow, in which, for example, a secondary electron is thermalized, trapped, and hydrated, and thus it converts into a hydrated electron (e_{aq}^-). Hole of water ($H_2O^{\bullet+}$) gives protons (H^+) to surrounding water molecules and becomes a hydroxyl radical ($^\bullet OH$), and thus H_3O^+ is produced. Excited water molecules (H_2O^*) dissociate into hydrogen atoms ($H^\bullet$) and $^\bullet OH$ or into smaller amounts of hydrogen molecules (H_2) and oxygen atoms (O).

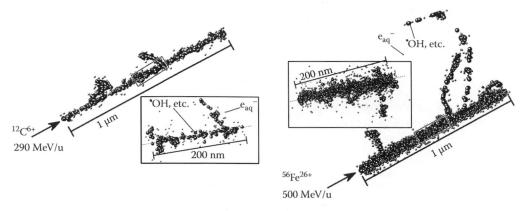

FIGURE 13.1 Track structures simulated with the code named IONLYS-IRT (Meesungnoen and Jay-Gerin, 2005a). Left and right panels show track structures at 1 ps after passages of C and Fe ions of 290 and 500 MeV/u, respectively, and insets in the panels are blowups of corresponding regions surrounded by squares.

These processes, having completed within 1 ps, are categorized as physicochemical stages and thought to be common for high-energy heavy ions and electrons, because most of the energy is commonly deposited via secondary electrons. However, the spatial distribution of these initial products must strongly depend on the type and the velocity of the projectile ion, which determine the frequency of energy deposition as well as ejection angles and energies of secondary electrons. Note that the positions of the initial products can be slightly different from the positions where energy depositions occur, because their flights can occur during the physicochemical stage. Figure 13.1 shows two examples of track structures for C and Fe ions of 290 and 500 MeV/u, respectively, at 1 ps after their passages were obtained by the Monte Carlo simulation with the IONLYS-IRT code (see Chapter 15 and Meesungnoen and Jay-Gerin (2005a)). As seen in this figure, the track of the Fe ion is much wider and denser than that of the C ion, and it is clear that the track structure is strongly related to radiation quality. After the physicochemical stage, water decomposition products begin to react with each other parallel to their diffusions from the positions they are produced. Such reactions are called *intra-track reactions*. The denser the initial distribution, the more frequently the intra-track reactions occur. Thus, chemical yields after or during intra-track reactions depend strongly on the type and the energy of heavy ions as well. In general, yields of molecular products, such as H_2 and hydrogen peroxide (H_2O_2), increase with increasing LET, while radical yields except for those of $HO_2^\bullet$ and $O_2^{\bullet-}$ decrease.

13.1.2.3 Biology: RBE, OER, etc.

Although it is phenomenologically well known that the distinctiveness of heavy ions in energy deposition and chemical yields leads to different biological effects, these effects are not explained well in terms of such distinctiveness. For example, some special cancers are radio resistant against low-LET radiations but not against heavy ions. Relative biological effectiveness (RBE) is a commonly used barometer to represent biological features of radiation, and is defined as the dose for 250 kV x-rays divided by the dose for the radiation of interest to attain the same biological effects such as cell death, mutation, and strand break. A higher value of RBE indicates that a smaller dose is required as compared with the dose for the 250 kV x-rays. Oxygen enhancement ratio (OER) is also a commonly used barometer, which indicates how much a certain biological effect is increased by the presence of oxygen. C ion irradiation gives the maximum value of RBE at an LET of about 200 eV/nm, which is almost at the end of the range of the ion, and OER decreases with increasing LET and reaches the minimum value of 1 at almost the same LET. These biological characteristics have significant advantages in the treatment of cancer because the concentration of O_2 in tumors is very low, as given in Chapter 24, and are well established, but only phenomenologically. In other words, a detailed mechanism through which physical and chemical features of ion beams lead to these characteristics is not yet clarified, and it is essential to deeply understand their basic phenomena and their correlations.

13.1.3 History: Earlier Studies and Accelerators

The earliest study on water radiolysis with heavy ions was just after the discovery of radiation itself. However, only alpha particles were available during the former half of the previous century. In the latter half of the century, the development of ion accelerators and experimental techniques helped researchers use other ions and conduct more precise investigations. However, most of these investigations were conducted with relatively lighter ions (H, D, and He ions) of lower energies (typically 10 MeV/u or less) under highly acidic conditions. Since the mechanisms for both the Fricke and cerium dosimeters were well established with low-LET radiations, the combination of these two dosimeters could be used to evaluate the yields of water decomposition products in ion beam radiolysis. Only 20–30 years ago, eventually, ion beams of heavier ions of higher energies have become available and investigations with these beams under neutral conditions are still on their way. Figure 13.2 shows a plot of research results on water radiolysis with heavy ions under neutral conditions. Although not many studies have been conducted under neutral conditions, much understanding has been derived over the past decade.

Only four groups conducted investigations on water radiolysis with steady-state ion beams: those from the Brookehaven National Laboratory (BNL) (Appleby et al., 1985, 1986); the Notre Dame Radiation Laboratory (NDRL) (LaVerne, 1989; LaVerne and Yoshida, 1993; Pastina and LaVerne, 1999; LaVerne et al., 2005); Quantum Beam Sciences (QuBS), Japan Atomic Energy Agency (JAEA) (Taguchi and Kojima, 2005, 2007; Taguchi et al., 2009); and the University of Tokyo and Advanced Science Research Center (ASRC), JAEA (Yamashita et al. 2008a,b,c; Baldacchino et al., 2009). Concerning pulse radiolysis studies, only some groups have succeeded in heavy-ion pulse radiolysis until now, because of the constraints of accelerator specifications and experimental difficulties, which are different from those of an electron beam, for example, shorter range

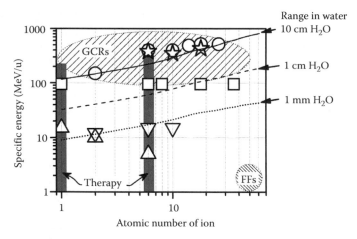

FIGURE 13.2 An overview of studies on radiolysis of neutral water with heavy ions. Circles, squares, triangles, inverse triangles, and stars show beam conditions used in studies of Yamashita et al. and Baldacchino et al. at the HIMAC facility (Yamashita et al., 2008a,b,c; Baldacchino et al., 2009), Baldacchino et al. and Wasselin-Trupin et al. at the GANIL facility (Baldacchino et al. 1997, 1998a,b, 1999, 2001, 2003, 2004, 2006b; Wasselin-Trupin et al., 2000, 2002), LaVerne et al. at NDRL (LaVerne 1989; LaVerne et al. 2005), Taguchi et al. at the TIARA facility (Taguchi and Kojima 2005, 2007; Taguchi et al., 2009), and Appleby et al. at the BEVALAC facility (Appleby et al., 1985, 1986), respectively. Shaded areas correspond to typical energy and atomic number of heavy ions contained in GCRs and FFs, and gray areas correspond to proton and carbon ion beams used in actual cancer therapy. Solid, dashed, and dotted lines show energies for each ion to obtain ranges of 10, 1, and 0.1 cm in water, respectively, which are calculated with the SRIM code (Ziegler, 1998).

TABLE 13.1
List of High-Energy Heavy-Ion Facilities and Their Specifications

Institute (City/Country)	Facility	Ions	Energy	Pulse Width	Intensity
CEA/CNRS (Caen/France)	GANIL	C–Kr	75–95 MeV/u	1 ns–1 ms	3–4 μA
CNRS (Nante/France)	ARRONAX	H, He	70 MeV	1 ns	100–750 μA
CNRS (Orléans/France)	CEMTHI	H, D, He	2.5–40 MeV	1 μs–CW	15–40 μA
JAEA (Takasaki/Japan)	TIARA	H–Au	2.5–27 MeV/u	1 μs–1 ms CW	3 μA
NIRS (Chiba/Japan)	HIMAC	He–Xe	6 MeV/u 100–500 MeV/u	5 μs CW	0.1 μA ≤1–4 nA
NDRL (Notre Dame/United States)	FN Tandem Van de Graaff	H–C	15 MeV/u	CW	~1 nA
GSI (Darmstadt/Germany)	FAIR	H–U	2–10 GeV/u	Pulse	10^{12}–10^{13} ions/pulse
BNL (Brookehaven/ United States)	NSRL	Ne–Au	0.3–1.5 GeV/u	CW	0.7–27 nA

than 1 mm in water and lower yield of reactive species. Burns (Burns et al., 1981) and Sauer (Sauer et al., 1977, 1983) had succeeded in the measurement with H^+ and He^{2+} in 1977. Since the 3 MeV H^+ ion that Burns et al. used has the shorter range in water, they performed the measurement using a jet without a cell (without a glass window). Since then, no group has succeeded in such system development. Research of heavy-ion pulse radiolysis, however, prospered from around 1997 due to the development of the latest accelerator technology. The types of incident ions, energy, time resolution, etc., of the newly developed system are summarized in Table 13.1. Katsumura et al. succeeded in the measurement of $(SCN)_2^{\cdot-}$ with He ions with the low-energy linear accelerator at the HIMAC facility (Chitose et al., 1999a,b, 2001). They took a Ti foil extraction window for the beam as a wall of sample cell, and the horizontal beam was injected directly into the water sample through the foil. On the other hand, Baldacchino et al. and Wasselin-Trupin et al. succeeded in the measurement of e_{aq}^-, H_2O_2, and $(SCN)_2^{\cdot-}$ using the high-energy heavy ion of GeV class at the GANIL facility (Baldacchino et al., 1997, 1998a,b, 1999, 2001, 2003, 2004, 2006b; Wasselin-Trupin et al., 2000, 2002; Baldacchino and Katsumura, 2010). They took advantage of the high-energy heavy ion having a longer range in water and produced almost constant LET along the probed sample, which produced results comparable to the Monte Carlo simulations. Moreover, several years later, Taguchi et al. developed a spectroscopic system with superhigh sensitivity under a pulsed heavy ion of several 100 MeV injected into the water surface at the TIARA facility (Taguchi et al.). Yoshida et al. have developed a ns-time-resolved optical absorption measurement system using the luminescence of a scintillator (Kondoh et al., 2008). Katsumura et al. used a 6 MeV/u He ion with pulse widths of 5 and 10 μs from a linear accelerator equipped in front of the main synchrotron accelerators at the HIMAC facility. Moreover, Taguchi et al. used ions in the range of 20 MeV/u H ions to 17.5 MeV/u Ne ions in pulse widths of μs–ms from the AVF cyclotron at the TIARA facility. On the other hand, Baldacchino et al. irradiated a GeV-class heavy ion with a pulse width from 1 ns to ms using the tandem-connected cyclotrons at the GANIL facility. Pulses consisting of very short fine pulses in a continuous row were used for irradiation, except for the 1 ns pulse used by Baldacchino et al. Although a short fine pulse irradiation can also be used as a specification of the accelerator, the dose per pulse is too small due to such a short fine pulse, and it is not easy to perform the transient absorption measurement. As mentioned above, if the lower yields of the reactive species are considered, such as $\cdot OH$ and e_{aq}^-, in heavy-ion irradiation, as compared with electron irradiation, 10–100 Gy is required for the dose per pulse. The kHz repetition rate and million data averaging (Baldacchino et al., 2004) make the experiment possible even with lower doses (a few cGy). On the other hand, if the behaviors of reactive species are followed by the luminescence, measurement by an ns pulse with a low dose is easier than that by absorption (Taguchi et al., 1997; Wasselin-Trupin et al., 2000, 2002).

Furthermore, whether irradiation can be done with a vertical or a horizontal beam can also be an experimental constraint. For example, even low-energy ions of very short range in water can be irradiated if an overhead irradiation system of a vertical beam is available. On the other hand, if only a horizontal irradiation system is available, such low-energy ions cannot reach the sample solution because there must be a wall of an irradiation vessel before the sample. A similar topic is being discussed in cancer therapy. For example, there are three irradiation rooms equipped with a vertical irradiation system, a horizontal one, or both of them for actual treatment at HIMAC. This is because it is inevitably important to reduce exposure of normal cells, especially of radiosensitive organs, such as blood-forming organs and gonad tissues, to ionizing radiations as well as to give sufficient dose to cancer. Some of the other therapeutic facilities have a gantry to adjust the irradiation angle, most appropriately, patient by patient, or irradiation by irradiation.

In the following text, emphasis will be placed on advances in 5 years in experimental studies on radiation chemistry with heavy ions. Especially, Section 13.2 for steady-state radiolysis studies and Section 13.3 for pulse radiolysis studies should be read complementarily because they are strongly related to each other, although they are separately summarized. Studies on radiation chemical effects and track structures of heavy ions until 2004 are well summarized in several reviews, and it is recommended that readers refer to them as necessary (LaVerne, 1996, 2000, 2004; Pimblott and Mozumder, 2004). It is also recommended to refer to Chapter 14, which is closely related to these studies.

13.2 STEADY-STATE RADIOLYSIS STUDIES

13.2.1 STRATEGY AND TECHNIQUES

In steady-state radiolysis studies, scavengers have conventionally been used not only to convert transient water radicals such as e_{aq}^- and $\cdot OH$ to easily detectable and stable species, but also to avoid reactions between water radicals and relatively stable accumulated products such as H_2O_2 and H_2. The scavenging reaction with the bimolecular rate constant, k_S ($dm^3/mol/s$), is given as follows:

$$R + S \rightarrow P \quad k_S\, dm^3/mol/s \qquad (13.6)$$

where R, S, and P are radical, scavenger, and product, respectively. As far as the dose and dose rate are low enough, the concentration of the radical, $[R]$ mol/dm^3, is much lower than that of the scavenger, $[S]$ mol/dm^3. Then, the scavenger concentration can be regarded as constant compared to the radical concentration, which can drastically vary as a function of time. In most of the studies using scavengers, it is assumed that the scavenging reaction can be regarded as a pseudo-first-order reaction with a reaction rate equal to a product of k_S and $[S]$, as follows:

$$R \rightarrow P \quad k_S[S]\, s^{-1} \qquad (13.7)$$

This reaction rate is called the scavenging capacity, and its inverse gives an approximate timescale of the scavenging reaction. Then, by selecting a scavenger and its concentration, one can control the scavenging timescale when transient water radicals are converted into products. It is noted that, strictly speaking, the timescale should be carefully interpreted based on the Laplace transformation; however, the above-mentioned approximation is valid for a timescale normally later than 10^{-8} s, because intra-spur and intra-track reactions are not so significant at and after 10^{-8} s (Pimblott and Laverne, 1994; Yamashita et al., 2008b). In addition, this method is also applicable to stop chemical reactions that may perturb the yield measurement. For example, H_2O_2 can be destroyed through reactions with e_{aq}^- and $\cdot OH$ as follows:

$$H_2O_2 + e_{aq}^- \rightarrow {}^\bullet OH + OH^- \quad 1.3 \times 10^{10} \ dm^3/mol/s \tag{13.8}$$

$$H_2O_2 + H^\bullet \rightarrow HO_2^\bullet + H_2 \quad 1.0 \times 10^{10} \ dm^3/mol/s \tag{13.9}$$

$$H_2O_2 + {}^\bullet OH \rightarrow HO_2^\bullet + H_2O \quad 2.9 \times 10^7 \ dm^3/mol/s \tag{13.10}$$

By scavenging both e_{aq}^- and ${}^\bullet OH$, such destructive reactions can be inhibited.

The production yield is briefly defined as the number of products produced divided by the absorbed energy. More strictly, the production yield of species X is represented by the following relationship:

$$G(X) = 100 \cdot e \cdot N_A \cdot \frac{1}{\rho} \cdot \lim_{\Delta D \to 0} \frac{\Delta C(X)}{\Delta D} \tag{13.11}$$

where

$G(X)$ and $C(X)$ are the production yield and the concentration of species X in units of $(100 \, eV)^{-1}$ and mol/dm^3, respectively

ρ (kg/dm^3), D (Gy), and N_A (mol^{-1}) are the density of irradiated matter, the absorbed dose, and the Avogadro number, respectively

Thus, dosimetry as well as the quantification of product concentration is essential in determining the product yield. Throughout this chapter, the radiation chemical yield is expressed in the form of a G-value, which is defined as the number of produced or decomposed species per energy absorption of $100 \, eV$ and can be converted into SI units by using the relationship $G(100 \ eV)^{-1} = G/(100 e N_A) \ mol/J = 1.036 \times 10^{-7} \ mol/J$.

There are several methods used in dosimetry for heavy-ion irradiation. A Faraday cup and a secondary electron monitor are normally used to measure the absolute and relative values of beam current, respectively. If the energy deposition of the ion is known, the absorbed energy is obtainable from the number of irradiated ions, which is obtained from the beam current. The absolute dose is also measurable by an ionization chamber, which can count the number of ion pairs produced in the chamber. With the W-value, which is defined as the averaged energy to create a single ion pair, the dose is determined as a product of the W-value and the number of ionizations. Note that the so-called recombination correction is normally necessary for a heavy ion. Track detectors such as CR39 are also useful in determining the number of ions and their LET values based on the number of holes and their sizes, respectively. In addition, chemical dosimeters are also useful if G-values corresponding to the chemical change are known. The Fricke dosimeter focuses on the G-value of a ferric ion (Fe^{3+}), which is produced through the oxidation of ferrous ions (Fe^{2+}) by oxidative water decomposition products, such as ${}^\bullet OH$, $HO_2^\bullet$, and H_2O_2. It is remarkable that $G(Fe^{3+})$ in the radiolysis of the Fricke dosimeter with a different ion beam has been intensively investigated by a group from BNL (Appleby and Christman, 1974; Christman et al., 1981) and by a group from NDRL (LaVerne and Schuler, 1987, 1994; Pimblott and LaVerne, 2002).

13.2.2 Track-Segment, Track-Differential, and Track-Averaged Yields

As described in Section 13.1.2.1, the energy deposition profile of a heavy ion varies with the energy and the type of ions. As can be seen in Figure 13.2, a C ion of about 20 MeV/u has a range of 1 mm in water. With much higher energy, the C ion can pass completely through a water sample of 1 mm thickness. In this case, the physical properties of the projectile ion are almost constant and the measured yield is called the *track-segment* yield for an almost constant LET value. On the other hand, with lower energy, the C ion cannot pass through and completely stops within the sample solution. In other words, the Bragg peak appears within the sample and the physical properties of the ion drastically change in the sample solution. The simplest way to express the overall physical properties of such low-energy ions is to show values averaged over their ranges. For example, the track-averaged LET is defined as follows:

$$\overline{\text{LET}} = \frac{\int_0^{x_0} \text{LET}\, dx}{x_0} = \frac{E_0}{x_0} \tag{13.12}$$

where

x and x_0 are the penetration depth and the range of the projectile ion, respectively
E_0 is the kinetic energy of the ion at the entrance of the sample

Because there are not many ion accelerators capable of accelerating ions up to several hundred MeV/u, radiation chemical yields for heavy-ion irradiation are commonly determined as *track-averaged* yields for track-averaged LET or for other parameters represented as track-averaged values. It is worth noting that LaVerne and Schuler developed a mathematical analysis to differentiate the track-averaged yields for a variety of ion energies into *track-differential* yields, which are basically equivalent to track-segment yields (LaVerne and Schuler, 1985). In addition, track-segment yields including track-differential yields are important in developing a theoretical model that is reliable enough to predict the track structure appropriately. Recently, radiation chemical studies with heavy ions have been conducted to measure track-segment or track-differential yields, as opposed to earlier studies until the 1980s, in which track-averaged yields were measured.

13.2.3 Primary Yields of Major Products Related to Track Structures

13.2.3.1 Primary Yield and Its Significance

The most intensively investigated water decomposition products are e_{aq}^-, $^{\bullet}$OH, H_2O_2, and H_2. The former two radical products are initially produced during physical and physicochemical stages occurring within 1 ps, and their initial yields at 1 ps are highest among all products. Based on the facts mentioned above, the following three reactions are most significant among all intra-spur and intra-track reactions:

$$e_{aq}^- + e_{aq}^- (+2H_2O) \rightarrow H_2 + 2OH^- \quad 5.5 \times 10^9 \ \text{dm}^3/\text{mol/s} \tag{13.13}$$

$$e_{aq}^- + {}^{\bullet}\text{OH} \rightarrow OH^- \quad 2.5 \times 10^{10} \ \text{dm}^3/\text{mol/s} \tag{13.14}$$

$${}^{\bullet}\text{OH} + {}^{\bullet}\text{OH} \rightarrow H_2O_2 \quad 6.0 \times 10^9 \ \text{dm}^3/\text{mol/s} \tag{13.15}$$

In these reactions, the initially produced e_{aq}^- and $^{\bullet}$OH react with themselves, leading to the production of H_2 and H_2O_2. As a result, the latter two molecular products are mainly produced from intra-spur and intra-track reactions during the chemical stage occurring from 1 ps to 100 ns. It is noted that the term *spur* represents a relatively narrow region where a few ionizations and excitations occur in a compact manner, while the term *track* represents a relatively wide region along the projectile trajectory where a large number of spurs are conjugated. It is well known that intra-spur reactions almost terminate by 100 ns after the irradiation of low-LET radiations. Then, the product yield at 100 ns, referred to as the primary yield, is valuable because it bridges the absorbed dose and the amount of water decomposition products having escaped from the spur to the bulk. Although it is reported that intra-track reactions for high-LET radiations do not terminate by 100 ns (LaVerne et al., 2005), the primary yield is still useful because it is an indicator inherently involving the information of the track structure or the overlapping of spurs in the track.

13.2.3.2 Hydrated Electron (e_{aq}^-)

A hydrated electron is the strongest reducing species among all water decomposition products and shows strong optical absorption in the visible and near-IR region (λ_{max} = 720 nm, ε_{max} = 18,500 dm³/mol/cm). This initiated intensive studies on the e_{aq}^- yield and its time profile as an anchor for understanding radiation chemical phenomena. The e_{aq}^- is protonated under acidic conditions to convert into H•, leading to lower and higher yields of e_{aq}^- and H•, respectively, than under neutral conditions. While many works on e_{aq}^- produced in low-LET radiolysis have been accumulated, those with high-LET radiations are still scarce. The first reason is that the accessibility to ion accelerators is much more limited than to electron accelerators. The second is low-LET radiation sources as well as that most of the ion accelerators cannot provide an ultra-short pulse with a duration of few tens of ns. At present, the heavy-ion beam pulse radiolysis with very high time resolution shorter than μs is available only at the GANIL facility, and thus, the direct observation of e_{aq}^- is limited in the timescale of >μs. Thus, the behavior of e_{aq}^- produced in steady-state heavy-ion radiolysis has been evaluated with indirect measurements, in which e_{aq}^- is converted into stable, easily detectable, and accurately measurable products through the scavenging reaction. Track-averaged yields of e_{aq}^- have been measured by Appleby and Schwarz for the first time (Appleby and Schwarz, 1969), and there is a review by LaVerne (2004). On the other hand, many measurements of track-segment or track-differential yields of e_{aq}^- have been performed, and only few have been conducted under neutral conditions (Appleby et al., 1986; LaVerne et al., 2005; Yamashita et al., 2008a, 2008c). Reported primary yields of e_{aq}^- in these three studies are summarized in Figure 13.3. It should be noted that the results of other studies mentioned above are not shown here because experimental conditions were highly acidic or because yields were measured as track-averaged yields and cannot be directly compared. It is also noted that H• yields measured by Schwarz et al. (1959) and by Anderson and Hart (1961) would be summations of yields of H• and e_{aq}^- because these studies were performed in advance of the discovery of e_{aq}^- in 1962. There are three important findings seen in this figure. First, it is clear that the primary yield of e_{aq}^- decreases with increasing LET. This decrease is due to the increased consumption of e_{aq}^- during intra-track reactions, such as reactions (13.13) and (13.14). Second, focusing on data from each report separately, it is commonly seen that e_{aq}^- yields for radiolysis with different ions of comparable LET are different and lighter-ion irradiation tends to give smaller yields. Although such a tendency was already observed in the radiolysis of the Fricke dosimeter, slower reactions significant at 100 ns or longer can interfere with the correlation between the track structure and observed yields in this chemical system. Anyway, this tendency indicates that a smaller ion generates a narrower and denser track than a heavier ion with the same LET value. Third, the yields obtained by Yamashita et al. with He ions shown as open squares are slightly larger than those obtained by LaVerne et al. with the same ions, shown as a dashed line. This disagreement can be explained by the difference in scavenging timescales, or might be explained by the difference in chemical and irradiation systems, or by the difference between track-segment and track-differential yields. The scavenging timescale in the former study is 100 ns while that in the latter is 230 ns. The consumption of e_{aq}^- would keep on going slowly but continuously during 100–230 ns, as mentioned by LaVerne et al. LaVerne and coworkers performed measurements with different scavenger concentrations, and e_{aq}^- yields at 23 ns and 2.3 μs are also determined, the former of which are shown in Figure 13.4. As is clear from the figure, there is a consistent tendency that lighter-ion irradiation results in a lower e_{aq}^- yield than heavier-ion irradiation of comparable LET. In addition, there is a considerable difference between the yield for fast electrons and that for heavy ions of LET higher than 100 eV/nm even at only a few ns after irradiation. It is often assumed that the initial yields for highly energetic charged particles are almost constant, because, in general, yields of ionizations and excitations as a first step of radiolysis do not depend on the radiation type, as well as because most of the energy depositions occur through highly energetic secondary electrons. LaVerne et al. insisted that the geminate recombination, $H_2O^{•+} + e_{pre}^- \rightarrow H_2O^*$, before the thermalization and solvation of e_{pre}^- would be significant for heavy ions of a few MeV/u or less, leading to smaller initial yields

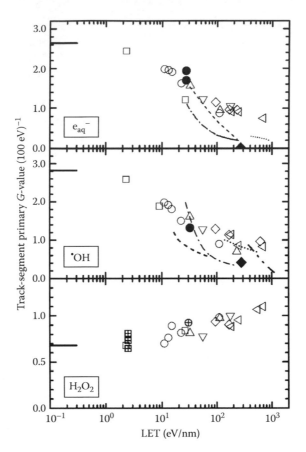

FIGURE 13.3 Primary yields of e_{aq}^-, $^\bullet OH$, and H_2O_2 as a function of LET. Top, middle, and bottom panels shows primary yields of e_{aq}^-, $^\bullet OH$, and H_2O_2, respectively. Solid lines on the left side of all the panels indicate primary yields for low-LET radiations (Elliot, 1994). Open squares, circles, triangles, inverse triangles, diamonds, and left triangles show data reported by Yamashita et al. with He, C, Ne, Si, Ar, and Fe ions from HIMAC, respectively (Yamashita et al., 2008a,c). Solid circles and diamonds show data reported by Baldacchino et al. with C and Ar ions from GANIL, respectively (Baldacchino et al., 1997, 1998a, 1999, 2003, 2004, 2006a,b). Squares and circles with a cross inside are data reported by Wasselin-Trupin et al. with He and C ions from GANIL, respectively (Wasselin-Trupin et al., 2002). Dashed, dotted, and dash-dotted lines in the top panel are track-differential e_{aq}^- yields for H, He, and C ions reported by a group from NDRL, respectively (LaVerne et al., 2005). The dash-dotted line in the middle panel is the track-differential $^\bullet OH$ yield with He ions at NDRL (LaVerne, 1989). Dashed, dotted, and dash-dotted lines are track-differential $^\bullet OH$ yields with He, C, and Ne ions from the TIARA facility reported by Taguchi et al. (Taguchi and Kojima, 2005, 2007; Taguchi et al., 2009), respectively. (From Yamashita, S. et al., *Radiat. Phys. Chem.*, 77, 1224, 2008a. With permission.)

of e_{aq}^- and $^\bullet OH$ and their consequent yields at later timescales (LaVerne et al., 2005). Note that geminate recombination in water is much faster than in nonpolar liquids, and whether or not there is geminate recombination in water and whether or not it can lead to a change in initial yields are important but still open questions. This will be a future subject.

13.2.3.3 Hydroxyl Radical ($^\bullet OH$)

A hydroxyl radical ($^\bullet OH$) is the strongest oxidizing species among all products in water radiolysis, which is assumed to be responsible for lethal damage to DNA through indirect actions of ionizing radiations. In spite of its importance, $^\bullet OH$ is difficult to be observed directly by pulse radiolysis because of weak absorption in the UV region ($\lambda_{max} \sim 240\,nm$, $\varepsilon_{max} \sim 600\,dm^3/mol/cm$). Hence, conversion with scavengers has been employed in $^\bullet OH$ yield measurement both in pulse radiolysis and

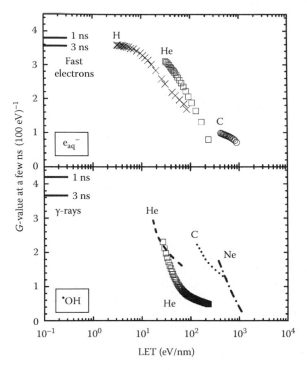

FIGURE 13.4 Yields of e_{aq}^{-} and ˙OH at a few ns after irradiation obtained with steady-state beams. Top and bottom panels show track-differential yields of e_{aq}^{-} and ˙OH, respectively. Solid lines on the left side of both panels indicate yields at 1 and 3 ns for low-LET radiations evaluated by LaVerne et al. (LaVerne, 1989; LaVerne et al., 2005). Crosses, open squares, and open circles show data at 2.3 ns obtained with H, He, and C ions by LaVerne et al. (2005). Open squares in the bottom panel show data at 3 ns obtained with He ions by LaVerne (1989), and dashed, dotted, and dash-dotted lines show data at 1.5 ns obtained with He, C, and Ne ions from the TIARA facility by Taguchi et al. (Taguchi and Kojima, 2005, 2007; Taguchi et al., 2009), respectively.

in steady-state radiolysis. Among these studies, only five studies have focused on track-segment or track-differential yields of ˙OH. Appleby et al. and Yamashita et al. determined the track-segment ˙OH yields for C, Ne, and Ar ions at the BEVALAC facility and those for He, C, Ne, Si, Ar, and Fe ions at the HIMAC facility, respectively (Appleby et al., 1985; Yamashita et al., 2008a,c). In these studies, aerated aqueous solutions of formate and bromide anions were commonly used, and the difference in the production of H_2O_2 in these solutions was taken as the ˙OH yield. Baldacchino et al. used a fluorescent probe to detect the final product after the scavenging reaction for ˙OH produced by the irradiation of C and Ar ions at the HIMAC facility (Baldacchino et al., 2009). Although this method needs high performance liquid chromatography (HPLC) to separate the parent scavenger and the final fluorescent product, its sensitivity is quite high and γ-irradiation of 10 cGy or even lower dose is detectable. Taking the high sensitivity as an advantage, they varied the scavenging timescale from a few ns to μs. The track-differential ˙OH yield for irradiation of He ions was determined by LaVerne at NDRL (LaVerne, 1989). Because formic acid is used as a scavenger for ˙OH, this is not directly comparable to the ˙OH yield measure under neutral conditions. Since e_{aq}^{-} is protonated into H˙ in acidic solutions, the consumption of ˙OH in intra-track reactions becomes minor because the reaction between H˙ and ˙OH is slower than that between e_{aq}^{-} and ˙OH. Recently, Taguchi et al. evaluated the track-differential ˙OH yields for the irradiation of He, C, and Ne ions under neutral conditions at the TIARA facility (Taguchi and Kojima, 2005, 2007; Taguchi et al. 2009). In their studies, phenol was taken as the scavenger for ˙OH and the production yields of the main products, catechol, hydroquinone, and resorcinol, were measured. The primary yields of ˙OH in these studies are summarized in Figure 13.3. Similarly as e_{aq}^{-}, there are three important findings: in short, the

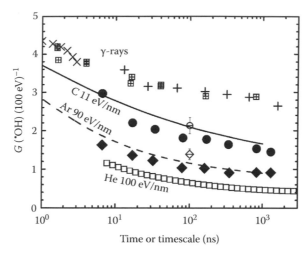

FIGURE 13.5 Time profile of ˙OH yield. Open squares with a plus inside and crosses show reported ˙OH yield for γ-rays (Draganić and Draganić, 1973; Jonah and Miller, 1977). Pluses, solid circles, and diamonds show experimentally determined ˙OH yields with γ-rays, 400 MeV/u C ions, and 500 MeV/u Ar ions from the HIMAC facility, respectively (Baldacchino et al., 2009). Solid and dashed lines show Monte Carlo simulation results for 400 MeV/u C and 500 MeV/u Ar ions, respectively (Meesungnoen and Jay-Gerin, 2005a; Yamashita et al., 2008b). Open squares show experimentally reported yield for 1.1 MeV/u He ions, the LET of which is comparable to 500 MeV/u Ar ions (LaVerne, 1989). (From Baldacchino, G. et al., *Chem. Phys. Lett.*, 468, 275, 2009. With permission.)

decrease of the ˙OH yield with increasing LET, smaller ˙OH yields for lighter ions than those for heavier ions of comparable LET, and slight differences between ˙OH yields in different studies.

The yields of ˙OH at a few ns have also been estimated by LaVerne (1989), Taguchi et al. (Taguchi and Kojima, 2005, 2007; Taguchi et al., 2009), and Baldacchino et al. (2009), which are shown in Figure 13.4. There is a wide range of ˙OH yields at this timescale, that is, approximately from 3 to 0.2 $(100 \text{ eV})^{-1}$. Repeatedly, it is also clearly shown that the irradiation of a lighter ion tends to lead to a smaller ˙OH yield than that of a heavier ion of comparable LET even at this timescale. These findings indicate that intra-track reactions take place significantly within a few ns and that they are more significant for lower energy ions within a few ns, resulting in a wide range of ˙OH yields even at a few ns after irradiation. In addition, although there is an overlapped LET region (20–70 eV/nm) in the data series for He ions reported by LaVerne and by Taguchi et al., the agreement between these two data series is not always good. This would imply that the mathematical treatment employed in both studies might lead to less reliable track-differential yields in the highest LET region or near the Bragg peaks.

In Figure 13.5, temporal information of the ˙OH yield for γ-rays and C and Ar ions from a few ns to μs evaluated using a fluorescent probe by Baldacchino et al. (2009) is summarized in comparison to similar data for He ions reported by LaVerne (1989) and primary g-values reported by Yamashita et al. (2008a,c) as well as the results of the Monte Carlo simulation (Meesungnoen and Jay-Gerin, 2005a; Yamashita et al., 2008b). There is a perfect agreement between the ˙OH yields estimated by Baldacchino et al., those by Yamashita et al., and the Monte Carlo simulation although the evaluation with fluorescence spectroscopy seems to result in relatively smaller ˙OH yields than the absorption analysis employed in the studies of Yamashita et al. This might be because of the error in the conversion ratio from ˙OH to the final fluorescent product. Baldacchino et al. assumed that 5.6% of the scavenged amount of ˙OH leads to the production of the final fluorescent product.

13.2.3.4 Hydrogen Peroxide (H_2O_2)

Hydrogen peroxide is a key species for understanding the entire concept of water radiolysis. It itself is a reactive oxygen species (ROS) and can be a source of other ROS, such as ˙OH and $O_2^{\cdot-}$. Because

H_2O_2 is not so reactive compared to e_{aq}^- and $\cdot OH$, and is much more soluble in water than H_2, it can be determined rather easily. Only two groups have reported on track-segment yields of H_2O_2. Yamashita et al. measured H_2O_2 yields for He, C, Ne, Si, Ar, and Fe ions of several $100\,MeV/u$ at the HIMAC facility by applying the Ghormley tri-iodide method (Yamashita et al., 2008a,c). They employed the conventional off-line absorption spectroscopy after the conversion of survived H_2O_2 into I_3^- for the quantification of produced H_2O_2. Wasselin-Trupin et al. applied emission spectroscopy for irradiations of C and Ar ions of $95\,MeV/u$ at the GANIL facility, although a pulsed beam, and not a steady-state beam, was used for irradiation. They detected chemiluminescence emitted after luminol reactions for the quantification of H_2O_2 (Wasselin-Trupin et al., 2002). Track-segment primary yields of H_2O_2 reported in these studies are also shown in Figure 13.3.

Compared to radical products, H_2O_2 yields are less labile to variation, and differences in radiolysis with different ions are less clear. However, focusing on the two data series obtained with C and Fe ions, the H_2O_2 yields in Fe ion radiolysis seem to be shifted to a higher-LET region from those in C ion radiolysis. For example, H_2O_2 yields in C ion radiolysis at an LET of about $20\,eV/nm$ are about $0.8\ (100\,eV)^{-1}$, almost the same as that in Fe ion radiolysis at an LET of about $200\,eV/nm$. In short, heavier-ion radiolysis of much higher LET can lead to almost the same H_2O_2 yield as obtained in lighter-ion radiolysis of much lower LET. This indicates that an apparent radical density in the track can be regarded as equivalent for these two ions in terms of H_2O_2 production, which is consistent with the tendency mentioned in Sections 13.2.3.2 and 13.2.3.3 that a lighter ion tends to create a narrower and denser track structure. One might think that the so-called higher-order reactions to consume already produced H_2O_2 might be effective in a heavier ion of higher LET; however, this speculation is inconsistent with the above-mentioned finding seen in radical yields. Anyway, there are only few studies reporting on track-segment or track-differential H_2O_2 yields. Pastina and LaVerne measured H_2O_2 yields for H, He, and C ions of $2-15\,MeV/u$ as track-averaged yields at neutral pH (Pastina and LaVerne, 1999). Of course, their results cannot be directly compared to track-segment yields in other reports; however, it is worth noting that they found that H_2O_2 yields for C ions monotonically decrease with increasing LET from 629 to $787\,eV/nm$, while those for H and He ions monotonically increase with increasing LET from 10 to 34.8 and from 78 to $156\,eV/nm$, respectively. This tendency is explained as follows: the importance of short-time reactions, which would be higher-order reactions such as $H_2O_2 + e_{aq}^-$ and $H_2O_2 + \cdot OH$, increases with increasing LET for C ion radiolysis. A detailed mechanism for this tendency of the H_2O_2 yield is not clarified yet, and another possible explanation is reported theoretically by incorporating multiple ionizations (Meesungnoen and Jay-Gerin, 2005b).

13.2.3.5 Hydrogen Molecule (H_2)

Similar to H_2O_2, the hydrogen molecule is also a key species for understanding the entire concept of water radiolysis. It is also important to investigate the initial processes induced in water radiolysis within $1\,ps$. LaVerne and Pimblott proposed a new mechanism, so-called dissociative electron attachment, for H_2 formation in water radiolysis (LaVerne and Pimblott, 2000). They insisted that some precursors of e_{aq}^- (e_{pre}^-) attach to the surrounding water molecule to produce a water radical anion ($H_2O^{\cdot-}$), leading to the production of H_2 in the following reactions:

$$e_{pre}^- + H_2O \rightarrow H_2O^{\cdot-} \tag{13.16}$$

$$H_2O^{\cdot-} \rightarrow H^- + \cdot OH \tag{13.17}$$

$$H^- + H_2O \rightarrow H_2 + OH^- \tag{13.18}$$

Geminate recombination might lead to H_2 production in the dissociation of the excited state of the water molecule as follows, although the existence of such a pathway has not been clearly found experimentally:

$$e_{pre}^- + H_2O^{\bullet+} \rightarrow H_2O^*$$ (13.19)

$$H_2O^* \rightarrow H_2 + O$$ (13.20)

The significance of these processes on the initial yields of e_{aq}^- and H_2 is still not well established and is one of the open questions to be most intensively scrutinized.

13.2.4 A Parameter Alternative to LET: $(Z_{eff}/\beta)^2$

As is seen in Figures 13.3 and 13.4, irradiations with different ions of comparable LET result in different radiation chemical yields due to the difference in 3D track structures, and it is indicated that lighter ions tend to create narrower and denser tracks than heavier ions of comparable LET. Thus, LET is not always the best parameter to represent and reflect track structures. Of course, this insufficiency would lead to uncertainties in predicting subsequent radiobiological effects based on LET. LaVerne proposed to use MZ_{eff}^2/E_0 as a universal parameter and succeeded in unifying, with this parameter, a huge amount of track-averaged yields of e_{aq}^-, H_2, $H^{\bullet}$, OH, H_2O_2, and the summation of $HO_2^{\bullet} + O_2$ obtained with a wide variety of ion types and energies (LaVerne 2004). Although this parameter works quite well when unifying track-averaged yields, it does not do so when trying to unify track-segment and track-differential yields. Instead of MZ_{eff}^2/E_0, $(Z_{eff}/\beta)^2$ was proposed by Christman et al. (1981) and recently by Ohno et al. (2001). In principle, these two parameters are used based on the same strategy as follows. Z_{eff}^2 and β^2 are responsible for the mean free path for the physical interactions of the projectile ion with matter and for the highest possible energy of the secondary electron, respectively. Thus, Z_{eff}^2 and $1/\beta^2$ can be regarded as indicators of the density of track along and perpendicular to the projectile trajectory, respectively. Combining them, the 3D track density is briefly estimated, leading to a better parameter than LET in unifying product yields, which are strongly affected by the 3D track density. Figure 13.6 shows the primary track-segment yields of e_{aq}^-, $^{\bullet}OH$, and H_2O_2 as a function of $(Z_{eff}/\beta)^2$. For simplicity, only the data reported by Yamashita et al. (2008a,c) are shown in the figure. The same equation used by LaVerne (2004), given below, is applied to fitting in the figure:

$$G(x) = G_0 + \{G_\infty - G_0\} \frac{(ax)^n}{1 + (ax)^n}$$ (13.21)

where

x is $(Z_{eff}/\beta)^2$
G_0 is the G-value for low-LET radiations
G_∞, a, and n are fitting parameters

Coincidently, $(Z_{eff}/\beta)^2$ was a better parameter than MZ_{eff}^2/E_0 when unifying the track-segment and track-differential yields. Although not shown here, if the same data in Figure 13.6 are plotted as a function of MZ_{eff}^2/E_0, the data are more scattered and a universal curve can hardly be drawn. Thus, a universal plot as a function of MZ_{eff}^2/E_0 and $(Z_{eff}/\beta)^2$ would be useful to predict track-averaged yields in radiolysis with low-energy ions and track-segment yields in radiolysis with high-energy ions, respectively, and a criterion between high and low energies in this regard is about 10 MeV/u.

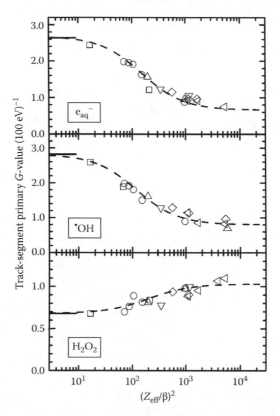

FIGURE 13.6 Primary yields of e_{aq}^-, $^{\bullet}OH$, and H_2O_2 as a function of $(Z_{eff}/\beta)^2$. Note that only track-segment yields reported by Yamashita et al. (2008a,c) are shown in the figure for better viewability and the same symbols as used in Figure 13.3 are used. Dashed lines are universal curves drawn by fitting with a function explained in the text. (From Yamashita, S. et al., *Radiat. Phys. Chem.*, 77, 1224, 2008a. With permission.)

13.2.5 INTRA-TRACK DYNAMICS

As long as direct observation focusing on chemical species of interest is difficult, the yield measurement is conducted with a scavenger, which is used normally as much as possible so as not to affect chemical conditions except for expected scavenging reactions. This is because emphasis is often placed on what happens in pure water. However, it is also important to investigate what would happen if chemical conditions are willingly disturbed. With increasing scavenger concentration, the scavenging reaction becomes to occur earlier and earlier, and it competes with intra-track reactions and diffusions more and more significantly. Then, the information of such competition is extracted by increasing scavenger concentration and intra-track dynamics can be investigated. Yamashita et al. selected a deaerated aqueous solution of 2.5×10^{-4} mol/dm^3 methyl viologen cation (MV^{2+}) containing various concentrations of formate anion (HCOO$^-$) as the chemical system for this kind of investigation (Yamashita et al., 2008b). In the radiolysis of this solution, the following reactions are expected to occur:

$$e_{aq}^- + MV^{2+} \rightarrow MV^{\bullet+} \quad 8.3 \times 10^{10} \text{ dm}^3/\text{mol/s} \tag{13.22}$$

$$^{\bullet}OH + HCOO^- \rightarrow {}^{\bullet}COO^- + H_2O \quad 3.2 \times 10^9 \text{ dm}^3/\text{mol/s} \tag{13.23}$$

$$H^{\bullet} + HCOO^- \rightarrow {}^{\bullet}COO^- + H_2 \quad 2.4 \times 10^8 \text{ dm}^3/\text{mol/s} \tag{13.24}$$

$$^{\bullet}COO^- + MV^{2+} \rightarrow MV^{\bullet+} + CO_2 \quad 1.0 \times 10^{10} \text{ dm}^3/\text{mol/s} \tag{13.25}$$

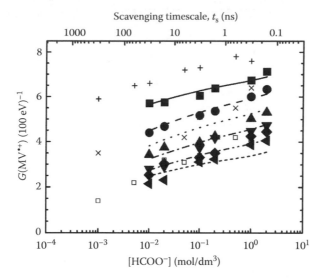

FIGURE 13.7 $G(MV^{\bullet+})$ as a function of scavenging capacity for $^{\bullet}OH$. Solid squares, circles, triangles, inverse triangles, diamonds, and left triangles show data obtained with steady-state beams of 150 MeV/u He, 290 MeV/u C, 400 MeV/u Ne, 490 MeV/u Si, 500 MeV/u Ar, and 500 MeV/u Fe ions, respectively, reported by Yamashita et al. (2008b). Solid, dashed, dotted, dash-dotted, dash-dot-dotted, and short-dashed lines show corresponding results of the Monte Carlo simulation (Yamashita et al., 2008b). Pluses, crosses, and open squares show data for γ-rays, pulsed 6 MeV/u H, and pulsed 6 MeV/u He ions, respectively, reported by Chitose et al. (1998, 1999b, 2001). Note that all ion beams are provided at the HIMAC facility and the G-value of MV$^{\bullet+}$ was determined after stabilization of the radiation chemical change. (From Yamashita, S. et al., *Radiat. Res.*, 170, 521, 2008b. With permission. Courtesy of Radiation Research Society.)

Thus, all the major water radicals, e_{aq}^{-}, $^{\bullet}OH$, and H$^{\bullet}$, contribute to the reduction of MV^{2+} to MV$^{\bullet+}$. The reduced form of methyl viologen, MV$^{\bullet+}$, is stable under the deoxygenated condition and has strong absorption in the visible region ($\lambda_{max} = 605$ nm, $\varepsilon_{max} = 13,100$ dm^3/mol/cm); thus, its concentration after the stabilization of chemical reactions is easily measured by absorption. G-values of MV$^{\bullet+}$ measured for He, C, Ne, Si, Ar, and Fe ions of 150, 290, 400, 490, 500, and 500 MeV/u, respectively, are shown as a function of formate anion concentration in Figure 13.7 in comparison to those reported with ^{60}Co-rays and 6 MeV/u H and He ions by Chitose et al. (1998, 2001). There are three findings depicted in the figure. First, $G(MV^{\bullet+})$ for each radiation increases with increasing HCOO$^-$ concentration or with decreasing scavenging timescale. The faster the scavenging reactions occur, the more radicals are scavenged. This is because they are consumed with time during intra-track reactions if there is no or less scavenging. Second, with increasing LET, the $G(MV^{\bullet+})$ value decreases almost monotonically due to increased radical density in the track and due to the significance of intra-track reactions. Third, irradiation with different ions of much different LET can give comparable $G(MV^{\bullet+})$ values. For example, the results for 500 MeV/u Ar ions and those for 6 MeV/u He ions are almost the same. This is consistent with the experimentally determined primary yields of e_{aq}^{-} and $^{\bullet}OH$ shown in Figure 13.3.

The authors successfully reproduced the experimentally obtained $G(MV^{\bullet+})$ values by the Monte Carlo simulation using the IONLYS-IRT code, which was developed by a group from Université de Sherbrooke (the authors of Chapter 14). In addition to the four expected reactions (13.22) through (13.25), they added 13 other reactions with known rate constants. One of such reactions is as follows:

$$^{\bullet}COO^- + {}^{\bullet}COO^- \rightarrow products \quad 7.5 \times 10^8 \text{ dm}^3/\text{mol/s} \quad (13.26)$$

Furthermore, they incorporated the following reactions in order to reproduce experimental results more precisely:

$$^{\bullet}OH + MV^{\bullet+} \rightarrow products \quad (3.0 \pm 0.5) \times 10^{10} \ dm^3/mol/s \tag{13.27}$$

$$^{\bullet}OH + {^{\bullet}COO^-} \rightarrow products \quad (5.0 \pm 0.5) \times 10^{10} \ dm^3/mol/s \tag{13.28}$$

Although the above two reactions are not yet reported, their rate constants could be adjusted to match the simulation results to the experimental ones as much as possible. As can be seen in Figure 13.7, the simulation results shown as lines are in good agreement with the experimental ones. The reliability of the simulation would be supported by an excellent reproduction of the experimental data obtained under various conditions, such as ion types, LET values, and scavenger concentrations.

In the simulations, the contributions of all incorporated reactions are calculated and relative importance is compared as a function of time after ion passage. For example, the amount of $^{\bullet}COO^-$ produced only through a scavenging reaction for $^{\bullet}OH$ (reaction 13.23) can be extracted separately from its production and consumption through the other reactions. For convenience, such an amount of product X in a certain reaction N is expressed as $G(X)_N$ from here on. In an ideal case, $G(^{\bullet}COO^-)_{13.23}$ in formate solutions at the end of the simulation can be interpreted as $g(^{\bullet}OH)$ in pure water at a timescale of the inverse of scavenging capacity, $(k_S[S])^{-1}$. Such an estimation of the temporal behavior of $^{\bullet}OH$ is shown in Figure 13.8 in comparison to the time profile of $^{\bullet}OH$ reported in simulations of pure water radiolysis. The $^{\bullet}OH$ yields estimated by the two simulations agree well with each other down to a few ns; however, at earlier time, the $^{\bullet}OH$ yields estimated from the simulations of $MV^{2+}/HCOO^-$ solutions are 20%–30% smaller than true values, which are obtained in pure water simulations. This implies that there is a limitation of the scavenging method in estimating product yields at an early stage within a few ns or earlier because of the competition between scavenging reactions and intra-track reactions.

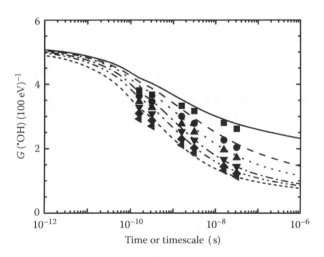

FIGURE 13.8 Comparison of the time profile of $^{\bullet}OH$ yields in Monte Carlo simulations. Note that all data shown in this figure are results of Monte Carlo simulations with the IONLYS-IRT code (Meesungnoen and Jay-Gerin, 2005b; Yamashita et al., 2008b). Solid, dashed, dotted, dash-dotted, dash-dot-dotted, and short-dashed lines show $^{\bullet}OH$ yields obtained from the simulation of pure water radiolysis with 150 MeV/u He, 290 MeV/u C, 400 MeV/u Ne, 490 MeV/u Si, 500 MeV/u Ar, and 500 MeV/u Fe ions, respectively. Solid squares, circles, triangles, inverse triangles, diamonds, and right triangles show corresponding $^{\bullet}OH$ yields estimated from the simulation of radiolysis of an aqueous solution of methyl viologen and formate, and details of the estimation are explained in Section 13.2.5. (From Yamashita, S. et al., *Radiat. Res.*, 170, 521, 2008b. With permission. Courtesy of Radiation Research Society.)

Although it was expected that the summation of scavenged amounts of e_{aq}^-, $^\bullet OH$, and $H^\bullet$ is equivalent to the final production of $MV^{\bullet+}$, the expected amount becomes higher than the experimental results with increasing LET and/or formate concentration. It was found that reaction (13.26) occurs significantly when the $HCOO^-$ concentration or LET increases, due to considerable loss of $^\bullet COO^-$. In the work of Yamashita et al., further discussion related to biological features of heavy ions is also attempted (Yamashita et al., 2008b).

13.3 PULSE RADIOLYSIS STUDIES

13.3.1 Strategy

The reactions of the reactive species in the track have been investigated by steady-state radiolysis, as described in Section 13.2. The yields of the reactive species were estimated for specified ions, energies, and elapsed times just after irradiation using the scavenging reactions. The chemical reactions were analyzed theoretically based on the track structure theory. This has helped in obtaining a clear overview of track reactions. However, the difficulties in evaluating the effects of nonhomogeneous and high-density radical distributions and potentially important side reactions on track reactions remain, because the specified scavenging reactions were used for steady-state radiolysis. Furthermore, there is a limit for the concentration of scavenger reagents not to be made high in order to keep a sample system, considering the principle of a scavenging reaction.

Pulse radiolysis is a very suitable technique for pursuing reactions directly with the time of the sample irradiated with the pulsed ionizing radiation. Especially, electron pulse radiolysis has been dynamically utilized all over the world, and is a very powerful tool to elucidate the mechanism of radiation-induced reactions. The reaction rate constants of various elementary processes have been also determined in this way. Furthermore, many reactions/phenomena can be predicted based on the above obtained knowledge. The reaction system, where the water radicals generate and react, is relatively homogeneous under pulsed electron irradiation. In the case of heavy-ion irradiation, reactive species are produced high-density and nonhomogeneous distribution in the track, which is usually expected from the result of a microdosimetry or track model calculation. So, these chemical reactions are more complex than those by pulsed electron irradiation. Therefore, the results obtained by electron pulse radiolysis are not directly applied in understanding the heavy-ion irradiation effects using LET or other parameters.

The reactions occurring near the track center cannot be clearly understood only from the result of a steady-state radiolysis or an extension of electron pulse radiolysis. Moreover, other methods cannot measure the reactive species qualitatively and quantitatively with a high time resolution. Heavy-ion pulse radiolysis systems are entreated in order to investigate the effect of high-density and nonhomogeneous distribution of the reactive species. Heavy-ion pulse radiolysis, however, possesses some important experimental difficulties: a limited facility providing the high intensity and short pulse of heavy ions, shorter ranges in aqueous solutions, smaller G-values of the reactive species evaluated by the scavenger method than those of the electron beam, and so on. For these reasons, only some groups have succeeded in pulse radiolysis measurement (Burns et al., 1981; Baldacchino et al., 1997, 1998a,b, 1999, 2001, 2003, 2004, 2006b, 2009; Chitose et al., 1999a,b, 2001; Sauer et al., 1977, 1983; Baldacchino and Katsumura, 2010; Taguchi et al., 2009), and their time resolutions were mainly in the order of μs. The originally devised heavy-ion pulse radiolysis system, which surmounts many difficulties, and important data are described in the following text.

13.3.2 Detection Method and Measurement System

The transient absorbance measurements with a high time resolution such as ns by using a high-energy electron accelerator are developed all over the world (Wojcik et al., 2004). In the case of the electrons, the measurement stage has somewhat shifted from the ns to the ps. The basic design of the irradiation and optical measurement system of pulsed heavy ions is similar to that of pulsed electrons. A sample solution is irradiated with pulsed electrons/heavy ions, and the absorbance of transient species is evaluated by decreasing the probe light intensity. The heavy ions generated in the ion source are accelerated and transported into a vacuum beam line. Then, the heavy ion is taken out in the atmosphere through a thin metal foil, and a sample solution is irradiated. One way to generate heavy-ion bunches is to use a chopper synchronizing with the accelerator at a high precision. It can be installed in the upper and/or lower stream of the accelerator and form a fine pulse of the heavy ion. The control signal also turns into a start signal of an oscilloscope or transient digitizer of the measurement system. The irradiated sample solution is analyzed with a direct current or pulsed probe light, and the decrease in the light intensity due to the optical absorption of the reactive species is detected by a photodiode and recorded on the fast digitizer. The reactions of the reactive species, which are generated by the pulsed heavy-ion irradiation, are mainly observed by the optical absorption spectroscopy in the UV-visible region. The reactive species can be identified from the structure of the absorption spectrum and analyzed qualitatively from the absorbance. More quantitative analyses can be carried out by combining the obtained result with a Monte Carlo simulation. Furthermore, time-resolved chemiluminescence was used as a more sensitive method (Wasselin-Trupin et al., 2000).

In the case of the GeV-class energy heavy ion, since the range is in about mm, an optical measurement system similar to that of the electron pulse radiolysis is applicable (Baldacchino et al., 1997, 1998a,b, 1999, 2001, 2003, 2004, 2006b; Baldacchino and Katsumura, 2010). The energy loss in a sample solution is small, so the yields of the reactive species under a constant LET value are measurable. However, in the case of MeV-class heavy ions, since they have several hundred μm ranges in water, a specially devised optical system is needed. The circuit diagram of the system developed by Taguchi et al. is shown in Figure 13.9 (Taguchi et al., 2009). They took out the pulsed heavy ion in the atmosphere and reduced the energy loss of the heavy ion by using a very thin glass cell window. They conquered the difficulty of the optical system alignment by letting the probe light path align with the ion beam on the same axis (collinear probing). However, since the effective light path length becomes the range of the heavy ion in this case, the light path length is short (several hundred μm), and absorbance becomes very small as compared with the 1 or 2 cm optical path length in electron pulse radiolysis. Therefore, they conquered this difficulty by constructing a highly sensitive detector system.

On the other hand, Katsumura et al. connected the sample cell directly to the extract window of the ion beam from vacuum to the atmosphere, without taking a beam out in the atmosphere

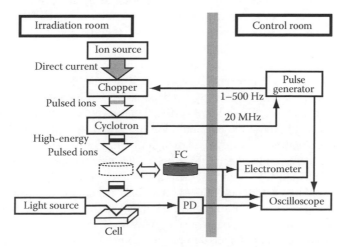

FIGURE 13.9 Electric circuit for pulsed heavy-ion irradiation and transient absorption measurement system constructed at the TIARA facility.

(Chitose et al., 1999a,b, 2001). They aligned the laser light as the monitor light just on the downstream side of the extract window. The optical path length can be elongated to the size of the ion beam, and thus a sufficient optical path length can be obtained. However, the difficulty that an alignment of an optical system is very delicate is not considered. In order to conquer this difficulty, Yoshida et al. are using the luminescence of a scintillator as probe light to develop the ns-time-resolved single-ion optical absorption measurement system in recent years (Kondoh et al., 2008).

In order to analyze reactive species quantitatively, dosimetry is very important in pulse radiolysis as well as in steady-state radiolysis. Except for the GeV heavy ion, an ion beam stops in the aqueous sample solution. Therefore, the dose correction for each irradiation cannot be performed by a current value measured with the Faraday cup set in the lower stream of a beam line to an irradiation cell, as used by electron pulse radiolysis. Moreover, although it is possible to put a metal foil to measure the beam current on the upstream side of the sample cell, the problem of energy loss by the foil remains. The beam current is also measured by the Faraday cup on the upstream side of a sample cell just before and after irradiation, and the stability of the beam must be checked. Measurement data is also selected.

13.3.3 Water Radical Kinetics

13.3.3.1 Applicable Scope of Direct Observation

In this section, we discuss in detail a few results concerning the water radicals directly observed by pulse radiolysis. Some of them are easier to detect than others because they have a great absorption coefficient (e_{aq}^-) or a poor reactivity, which gives the possibility of integrating the transmitted signal and, therefore, obtaining a good signal-to-noise ratio ($O_2^{\cdot-}$). Most of the other studies under heavy-ion irradiation were performed by steady-state radiolysis by using scavenging reagents, for instance, the glycylglycine (LaVerne et al., 2005) or $S_2O_8^{2-}$ (Chitose et al., 1999a,b). These results generally reproduce the tendency of the time dependence of G-values, but probing the earliest time stage needs a very high concentration of the scavenging reagents. This could imply a direct ionization effect or at least a strong reactivity, and always the ionic strength change.

13.3.3.2 Hydrated Electron (e_{aq}^-)

The hydrated electron (e_{aq}^-) has been extensively studied for a long time (since its discovery in the 1960s) with low-LET particles (high-energy electron and γ-rays). Its wide spectrum makes it easy to detect in the visible and near-IR domain (Spinks and Woods, 1990). Its high absorptivity at 720 nm of about 2×10^4 dm^3/mol/cm allows its detection at very low concentration levels. This is an essential species also to provide information on the ionization track structure. It is a good candidate to be a probe of the track at an early time after its formation, since its formation after thermalization and solvation was extensively studied both experimentally and theoretically.

A hydrated electron is commonly detected by pulse radiolysis with electron beams. In the case of a highly structured track, this very reducing species reacts easily with oxidant species like $\cdot OH$ in its vicinity at the earliest time after the ionization track is formed. This is the reason why it is a real challenge to detect this species with heavy-ion irradiation. To overcome this challenge, the time resolution must be sufficient in order to increase the signal level corresponding to e_{aq}^-. Providing a G-value remains difficult because the concentrations are lower than 10^{-7} mol/dm^3, and the dose must be measured with a high accuracy. As is shown in Figure 13.10, with a pulsed e-beam, e_{aq}^- follows a slow time dependence and only reactions occurring after the microsecond stage convert it into H$^\cdot$, OH$^-$, or H$_2$. With high-LET particles, the kinetics of e_{aq}^- is typical of a dense concentration time dependence or similar to a high dose rate of a low-LET irradiation: the spatial structure at the earliest time after ionization strongly influences the kinetics in the first 100 ns since the concentration of initial e_{aq}^- is at least decayed by a factor of 2. Afterward, the concentration seems to begin a slower decay. This evolution is well depicted in recent papers (Baldacchino et al., 2003, 2004) and a book chapter (Baldacchino and Katsumura, 2010). The Monte Carlo simulation reproduces the time dependence (Frongillo et al., 1996, 1998; Cobut et al., 1998). The comparison of the G-value is

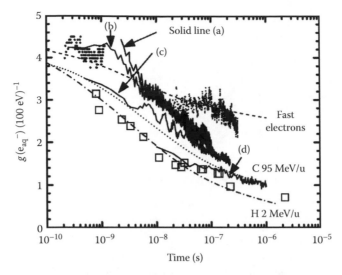

FIGURE 13.10 Measured and calculated time-dependence of e_{aq}^- decay. Small dots and dashed line show data experimentally and theoretically reported with fast electrons by Muroya et al. (2005), respectively. Solid lines (a) through (d) and dotted line show data experimentally and theoretically reported with 95 MeV/u C ions of 1 ns pulse width by Baldacchino et al. (1998a, 2004), respectively. Solid squares and dash-dotted line show data reported with 2 MeV/u H ions experimentally by Burns et al. (1977) and corresponding simulation results, respectively. Note that solid lines (a) through (d) were obtained with different time resolutions.

not reliable and an experimental G-value higher than 4.3 $(100\,eV)^{-1}$ at the earliest time is of course suspected (LaVerne et al., 2005). Actually a limiting value of 4 $(100\,eV)^{-1}$ is a commonly accepted value, deduced from the ionization potential of water. A better accuracy in the dosimetry is needed. In this experiment, the improvement of the detection system has contributed to the detection of almost 10^{-9} mol/dm^3 of e_{aq}^-. The acquisition system allowed an average of about 1 million of kinetics with 1 kHz of repetition rate during the flow of the solution. As shown in Figure 13.10, the three experiments that detected e_{aq}^- by using this method have obtained similar kinetics with various ion beams: ns protons (Burns et al., 1977, 1981; Rice et al., 1982); µs pulses of $^4He^{2+}$ and $^2H^+$ (Sauer et al., 1977, 1983); and ns pulses of C^{6+}, Ar^{18+}, and O^{8+} of 95 MeV/u (Baldacchino et al., 2003, 2004). The results from the studies of Sauer et al. have also investigated the LET dependence by translating the analysis along the penetration depth of the ion beam. They could plot for two ions the dependence of the yield with $(Z_{eff}/\beta)^2$, which is a common parameter used instead of LET.

13.3.3.3 Hydroxyl Radical (•OH)

A hydroxyl radical is detectable only in deep UV regions (between 200 and 290 nm with a maximum of around 220 nm) with a low absorption coefficient (about 600 dm^3/mol/cm at 210 nm). It is mainly detected by using scavengers. There still does not exist any direct detection of •OH. By using pulse radiolysis and the thiocyanate anion (Milosavljevic and LaVerne, 2005), •OH radicals can be specifically scavenged to give $(SCN)_2^{•-}$, which is more conveniently detected at the µs stage in the visible range with a relatively high extinction coefficient (around 6000 dm^3/mol/cm at 500 nm). In this way, one can observe an absorption signal like that shown in Figure 13.11, obtained with the setup at the TIARA facility at Takasaki, Japan. This indirect determination of the •OH yield seems to give satisfactory results if the concentration of thiocyanate is not too high (<10^{-2} mol/dm^3). It was shown that the scavenger molecule can be subject to multiple possible reactions in the track of heavy ions even by using other scavenger species, like the bromide anion (Baldacchino et al., 2006b). Other results tended to show that the increase of the •OH yield with the increase of the scavenging capacity can saturate or even decrease after a maximum value of the yield for pulsed He ion beams (Chitose et al., 1999b) and proton beams (Chitose et al., 2001). After the conversion of concentrations into time, the

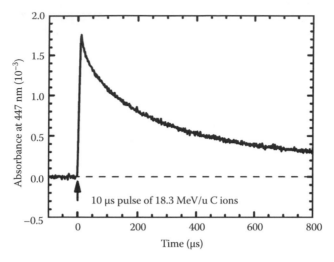

FIGURE 13.11 Time-dependent absorbance recorded at 447 nm for a 10^{-2} mol/dm^3 aqueous KSCN solution saturated with air by 18.3 MeV/u C ion irradiation with 10 μs pulse widths.

kinetics of the ·OH radicals shows a saturation and decrease at the earliest times (see Baldacchino et al. (2006a)). One can notice that this effect was not observed by Sauer et al. since they investigated only one concentration of thiocyanate, 10^{-2} mol/dm^3 (Sauer et al., 1983), which gave a similar value of the ·OH yield, around 1.5 (100 eV)$^{-1}$ at 8 μs for an LET value of 30 eV/nm. At the earliest time, the scavenging of ·OH is probably disturbed by nonlinear effects due to interfering reactions in the mechanism of the disproportionation of $(SCN)_2$·$^-$. The size of the track can actually play an important role in the shape of the scavenging plot. Therefore, it remains difficult to give the influence of heavy ions on the ·OH yield as long as there is no direct determination of its yield without using a scavenger. Another chemical system has been tested to detect ·OH (for high LET values only) by using a light-emitting chemical process. This is also a scavenging method associated with the pulsed heavy-ion beam (Wasselin-Trupin et al., 2000, 2001). In the chemical mechanism of luminol, the luminol molecule must react with ·OH and O_2·$^-$ to form a 3-aminophthalate molecule in its excited state, which relaxes by emitting fluorescence at around 420 nm. At a high LET value, the yield of ·OH becomes lower than the yield of O_2·$^-$. Therefore, in this case, the ·OH yield is the limiting value. In this way, at an LET value of 280 eV/nm with Ar^{18+} ions, the ·OH yield is found equal to 0.21 (100 eV)$^{-1}$.

13.3.3.4 Superoxide Radicals (HO_2·/O_2·$^-$)

Superoxide radicals (HO_2·/O_2·$^-$) are a peculiar case in water radiolysis since the formation yield of a superoxide radical increases with the LET value in a deaerated condition. It could also be formed by the reaction of e_{aq}^- with the molecular oxygen in an aerated condition. The general trend is that the radical recombination increases the production of molecular species (H_2 and H_2O_2). Currently, there are three studies considering the direct detection of a superoxide radical by using pulse radiolysis with heavy-ion beams. The first one (Sauer et al., 1978) used 200 μs pulses of 10 MeV/u deuterons and 10 MeV/u alpha with LET values below 25 eV/nm. Later, Baldacchino et al. used 2 ms pulses of 2–3 GeV Ar^{18+} and S^{16+} (LET values are 290 and 250 eV/nm, respectively) and performed detection in the UV region in order to determine the HO_2·/O_2·$^-$ yield (Baldacchino et al., 1998b, 2001). This is currently the only method of dissociating the yield of the superoxide and the eventual molecular oxygen primary yield. The difficulty arises also from the very low level of the yields even with high LET values: at 250 eV/nm, the yield reaches a value of about 0.06 (100 eV)$^{-1}$. Nevertheless, the low reactivity of this species in pure water is essentially due to its slow dismutation (reaction 13.29), which makes it easy to escape from the track recombination and to persist in the diffusion step (Bielski and Allen, 1977):

$$HO_2^{\cdot} + O_2^{\cdot-} (+H_2O) \rightarrow H_2O_2 + O_2 + OH^- \quad 9.7 \times 10^7 \ dm^3/mol/s \quad (13.29)$$

If this radical species is formed during the ionization stage, it is the only one that is still living at the ms stage. Since the $^{\cdot}$OH radicals that could react efficiently with the superoxide in high-LET-particles' irradiation are depleted very early after the ionizations, the superoxide has no other partner to recombine. In this way, for very high-LET irradiation, it is a very good probe of the earliest track history.

Recently, pulse radiolysis with Kr ions having a few thousands of eV/nm of the LET value along a few mm (this track-segment length allows suitable detection) was performed and the acquisition of the UV time-resolved spectra was made. Figure 13.12 gives this 3D plot in the case of 100 μs pulses of Kr ions. This figure reveals some difficulties to get the spectral and kinetics information in the UV range. Actually, with this very high-LET particle, the production of H_2O_2 is very efficient and the absorption is increased and superimposed with $HO_2^{\cdot}/O_2^{\cdot-}$. For these reasons, shorter pulses and better time resolution is also necessary. There is a similar problem in the end track segments (the Bragg peak region). This has several potential consequences in biology and medicine. Some simulation studies have tried to explain how a superoxide radical is formed in the tracks of heavy ions, and many mechanisms have been suggested in the past. The multiple-ionizations model is one of the most probable models because the huge energy deposited in the medium is considerably greater than the total energy needed to ionize the total number of water molecules along the ion track. This has been recently exploited in the Monte Carlo simulations (Champion et al., 2005; Gervais et al., 2005, 2006; Gaigeot et al., 2007). This model seems to be in good agreement with the few experimental results.

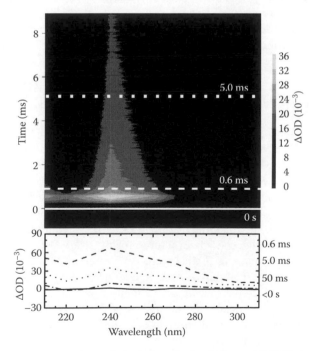

FIGURE 13.12 Time-resolved UV absorption spectra obtained after a 6 GeV Kr ion of about 3000 eV/nm of LET. The pulse duration is 30 μs and the solution is aerated. The signal corresponds to superimposition of the superoxide radical spectrum and the hydrogen peroxide spectrum. The latter is formed in a non-neglected quantity. The dose was around 1 kGy/pulse.

13.4   SUMMARY: CURRENT SUBJECTS AND FUTURE PERSPECTIVES

In this chapter, the studies that have been experimentally clarified recently for water radiolysis with positively charged atomic heavy ions are described. Due to the recent studies, track-segment and track-differential yields of water decomposition products are being accumulated day by day, while earlier studies developed the information of track-averaged yields for low-energy ions. In terms of timescale, the focus is now shifting to the ns region or earlier from the μs region or later. Theoretical studies are also being improved, as described in Chapter 14, owing to the improvement in the numeric capacity of computers. These improvements commonly attempt to clarify what occurs in the track at a very early timescale, what is the track structure like, and what is the origin of distinctive irradiation effects of ion beams in the subsequent chemical and biological stages.

For both steady-state and pulse radiolysis studies, the limitation of the ion accelerator is the highest barrier for such improvements. The heavy-ion pulse radiolysis systems presently under operation are at the GANIL and TIARA facilities, and these facilities are planning the advancement of the pulse properties. Simultaneously, a synchrotron is reconstructed at the HIMAC facility to succeed in the drawing of a nanosecond pulse (Noda et al., 2005). Constructions of the new pulse radiolysis system at a worldwide scale are foreseen. With the increase of high-energy ion sources, it is desired to clarify the contributions of fragmentation reactions in more detail. However, there is a lack of cross sections for nuclear interaction even for a high-energy proton impact, and the situation is much more severe for heavier-ion irradiations.

Although the behaviors of major products have been clarified, those of minor products are still not clarified. For example, yields of $O_2^{\cdot-}$ and O_2 are strongly related to multiple ionizations in physical interactions. However, their yields at neutral pH are not known well and only discussed based on experimental evidence in highly acidic conditions or based on theoretical calculations. These minor products are key species in radiation biology as well. The oxygen-in-track hypothesis is one possible explanation of the low-OER property of heavy ions.

Demands on interdisciplinary studies with physics, biology, engineering, and medicine are increasing and pervasive effects to the fields are expected and desired by understanding in more detail radiation chemical effects of heavy ions.

ACKNOWLEDGMENTS

The authors thank a number of collaborators, particularly, T. Maeyama, D. Hiroishi, C. Matsuura (University of Tokyo), Dr. Murakami, and operators at HIMAC (NIRS) for their support for heavy-ion irradiation at HIMAC, NIRS; Dr. S. Kurashima for his excellent contribution for the pulse radiolysis system at the TIARA facility (JAEA); and Dr. B. Hickel for his valuable advices and discussions, and for the management of the first studies in France with pulsed heavy ions. They also thank J. Meesungnoen and J.-P. Jay-Gerin (Université de Sherbrooke), who are authors of Chapter 14, for kindly allowing one of the authors (Shinichi Yamashita) to use the (unpublished) simulated track structures for 400 MeV/u C and 500 MeV/u Fe ions shown in Figure 13.1. The studies in France were performed at GANIL in collaboration with Dr. S. Bouffard (Commissariat à l'Énergie Atomique et aux Énergies Alternatives, Caen). The authors would like to thank him for sparing his valuable time and for sharing his great knowledge of the beam control. They also acknowledge the GANIL staff and the colleagues in the Laboratoire de Radiolyse at Saclay for their great help during the runs: Dr. S. Pin, G. Vigneron, Dr. J. P. Renault, Dr. S. Le Caër, Dr. S. Pommeret, and Dr. J.-C. Mialocq. The first experiments at the GANIL facility were also performed in collaboration with Prof. M. Gardès-Albert Laboratory (Paris 5 University), and Prof. S. Deycard (Caen University). The authors are thankful for their help and discussion.

REFERENCES

Anderson, A. R. and Hart, E. J. 1961. Molecular product and free radical yields in decomposition of water by protons, deuterons, and helium ions. *Radiat. Res.* 13: 689–704.

Appleby, A. and Christman, E. A. 1974. Radiation chemical studies with 3.9 GeV N^{7+} ions. *Radiat. Res.* 60: 34–41.

Appleby, A. and Schwarz, H. A. 1969. Radical and molecular yields in water irradiated by gamma rays and heavy ions. *J. Phys. Chem.* 73: 1937–1941.

Appleby, A., Christman, E. A., and Jayko, M. 1985. Radiation-chemistry of high-energy carbon, neon, and argon ions — Hydroxyl radical yields. *Radiat. Res.* 104: 263–271.

Appleby, A., Christman, E. A., and Jayko, M. 1986. Radiation-chemistry of high-energy carbon, neon, and argon ions — Hydrated electron yields. *Radiat. Res.* 106: 300–306.

Baldacchino, G. and Katsumura, Y. 2010. Recent process in heavy ion tracks. In *Recent Trends in Radiation Chemistry*, J. F. Wishart and B. S. M. Rao (eds.), pp. 231–254. Singapore: World Scientific Publishing Company.

Baldacchino, G., Bouffard, S., Balanzat, E. et al. 1997. LET effect on the radiolytic yield and the kinetics of the hydrated electron generated by 75 MeV/A C^{6+} pulses. *J. Chim. Phys.* 94: 200–204.

Baldacchino, G., Bouffard, S., Balanzat, E. et al. 1998a. Direct time-resolved measurement of radical species formed in water by heavy ions irradiation. *Nucl. Instrum. Methods Phys. Res. Sect. B-Beam Interact. Mater. Atoms* 136: 528–532.

Baldacchino, G., Le Parc, D., Hickel, B. et al. 1998b. Direct observation of HO_2/O_2^- free radicals generated in water by a high-linear energy transfer pulsed heavy-ion beam. *Radiat. Res.* 139: 128–133.

Baldacchino, G., Trupin, V., Bouffard, S. et al. 1999. LET effects in water radiolysis. Pulse radiolysis experiments with heavy ions. *J. Chim. Phys.* 96: 50–60.

Baldacchino, G., Trupin-Wasselin, V., Bouffard, S. et al. 2001. Production of superoxide radicals by linear-energy-transfer pulse radiolysis of water. *Can. J. Physiol. Pharm.* 79: 180–183.

Baldacchino, G., Vigneron, G., Renault, J. P. et al. 2003. A nanosecond pulse radiolysis study of the hydrated electron with high energy carbon ions. *Nucl. Instrum. Methods Phys. Res. Sect. B-Beam Interact. Mater. Atoms* 209: 219–223.

Baldacchino, G., Vigneron, G., Renault, J. P. et al. 2004. A nanosecond pulse radiolysis study of the hydrated electron with high energy ions with a narrow velocity distribution. *Chem. Phys. Lett.* 385: 66–71.

Baldacchino, G., De Waele, V., Monard, H. et al. 2006a. Hydrated electron decay measurements with picosecond pulse radiolysis at elevated temperatures up to 350 degrees C. *Chem. Phys. Lett.* 424: 77–81.

Baldacchino, G., Vigneron, G., Renault, J. P. et al. 2006b. Hydroxyl radical yields in the tracks of high energy $^{13}C^{6+}$ and $^{36}Ar^{18+}$ ions in liquid water. *Nucl. Instrum. Methods Phys. Res. Sect. B-Beam Interact. Mater. Atoms* 245: 288–291.

Baldacchino, G., Maeyama, T., Yamashita, S. et al. 2009. Determination of the time-dependent OH-yield by using fluorescent probe. Application to heavy ion irradiation. *Chem. Phys. Lett.* 468: 275–279.

Barkas, W. H. 1963. *Nuclear Research Emulsions—I. Techniques and Theory*. New York: Academic Press.

Berger, M. J., Inokuti, M., Anderson, H. H. et al. 1993. Stopping powers and ranges for protons and alpha particles. ICRU Report 49.

Bielski, B. H. J. and Allen, A. O. 1977. Mechanism of disproportionation of superoxide radicals. *J. Phys. Chem.* 81: 1048–1050.

Burns, W. G., May, R., Buxton, G. V., and Tough, G. S. 1977. Yield and decay of hydrated electron in proton tracks—Pulse-radiolysis study. *Faraday Discuss.* 63: 47–54.

Burns, W. G., May, R., Buxton, G. V., and Wilkinsontough, G. S. 1981. Nanosecond proton pulse-radiolysis of aqueous-solutions. 2. Improved measurements and isotope effects. *J. Chem. Soc. Faraday Trans.* 77: 1543–1551.

Champion, C., L'Hoir, A., Politis, M. F., Fainstein, P. D., Rivarola, R. D., and Chetioui, A. 2005. A Monte Carlo code for the simulation of heavy-ion tracks in water. *Radiat. Res.* 163: 222–231.

Chitose, N., LaVerne, J. A., and Katsumura, Y. 1998. Effect of formate concentration on radical formation in the radiolysis of aqueous methyl viologen solutions. *J. Phys. Chem. A* 102: 2087–2090.

Chitose, N., Katsumura, Y., Domae, M., Zuo, Z., and Murakami, T. 1999a. Radiolysis of aqueous solutions with pulsed helium ion beams. 2. Yield of SO_4^- formed by scavenging hydrated electron as a function of $S_2O_8^{2-}$ concentration. *Radiat. Phys. Chem.* 54: 385–391.

Chitose, N., Katsumura, Y., Domae, M., Zuo, Z. H., Murakami, T., and LaVerne, J. A. 1999b. Radiolysis of aqueous solutions with pulsed helium ion beams. 3. Yields of OH radicals and the sum of e_{aq}^- and H atom yields determined in methyl viologen solutions containing formate. *J. Phys. Chem. A* 103: 4769–4774.

Chitose, N., Katsumura, Y., Domae, M. et al. 2001. Radiolysis of aqueous solutions with pulsed ion beams. 4. Product yields for proton beams in solutions of thiocyanate and methyl viologen/formate. *J. Phys. Chem. A* 105: 4902–4907.

Christman, E. A., Appleby, A., and Jayko, M. 1981. Radiation-chemistry of high-energy carbon, neon, and argon ions—Integral yields from ferrous sulfate-solutions. *Radiat. Res.* 85: 443–457.

Cobut, V., Frongillo, Y., Patau, J. P., Goulet, T., Fraser, M. J., and Jay-Gerin, J. P. 1998. Monte Carlo simulation of fast electron and proton tracks in liquid water. I. Physical and physicochemical aspects. *Radiat. Phys. Chem.* 51: 229–243.

Draganić, Z. D. and Draganić, I. G. 1973. Studies on the formation of primary yields of hydroxyl radical and hydrate in the γ-radiolysis of water. *J. Phys. Chem.* 77: 765–772.

Elliot, A. J. 1994. Rate constants and G-values for the simulation of the radiolysis of light water over the range 0–300 degree Celsius. AECL-11073, COG-94-167.

Frongillo, Y., Fraser, M. J., Cobut, V., Goulet, T., JayGerin, J. P., and Patau, J. P. 1996. Evolution of the species produced by the slowing down of fast protons in liquid water: Simulation based on the independent reaction times approximation. *J. Chim. Phys.* 93: 93–102.

Frongillo, Y., Goulet, T., Fraser, M. J., Cobut, V., Patau, J. P., and Jay-Gerin, J. P. 1998. Monte Carlo simulation of fast electron and proton tracks in liquid water. II. Nonhomogeneous chemistry. *Radiat. Phys. Chem.* 51: 245–254.

Gaigeot, M. P., Vuilleumier, R., Stia, C. et al. 2007. A multi-scale ab initio theoretical study of the production of free radicals in swift ion tracks in liquid water. *J. Phys. B: At. Mol. Opt. Phys.* 40: 1–12.

Gervais, B., Beuve, M., Olivera, G. H., Galassi, M. E., and Rivarola, R. D. 2005. Production of HO_2 and O_2^- by multiple ionization in water radiolysis by swift carbon ions. *Chem. Phys. Lett.* 410: 330–334.

Gervais, B., Beuve, M., Olivera, G. H., and Galassi, M. E. 2006. Numerical simulation of multiple ionization and high LET effects in liquid water radiolysis. *Radiat. Phys. Chem.* 75: 493–513.

Jonah, C. D. and Miller, R. 1977. Yield and decay of the OH radical from 200 ps to 3 ns. *J. Phys. Chem.* 81: 1974–1976.

Kondoh, T., Yang, J., Kan, K. et al. 2008. Extension of the heavy ion beam pulse radiolysis using scintillators. *JAERI Rev.* 2008-055.

LaVerne, J. A. 1989. The production of OH radicals in the radiolysis of water with ^{4}He ions. *Radiat. Res.* 118: 201–210.

LaVerne, J. A. 1996. Development of radiation chemistry studies of aqueous solutions with heavy ions. *Nucl. Instrum. Methods Phys. Res. Sect. B-Beam Interact. Mater. Atoms* 107: 302–307.

LaVerne, J. A. 2000. Track effects of heavy ions in liquid water. *Radiat. Res.* 153: 487–496.

LaVerne, J. A. 2004. Radiation chemical effects of heavy ions. In *Charged Particle and Photon Interactions with Matter*, A. Mozumder and Y. Hatano (eds.), pp. 403–429. New York: Marcel Dekker.

LaVerne, J. A. and Pimblott, S. M. 2000. New mechanism for H_2 formation in water. *J. Phys. Chem. A* 104: 9820–9822.

LaVerne, J. A. and Schuler, R. H. 1985. Production of HO_2. In the track of high-energy carbon-ions. *J. Phys. Chem.* 89: 41713173.

LaVerne, J. A. and Schuler, R. H. 1987. Radiation chemical studies with heavy-ions—Oxidation of ferrous ion in the Fricke dosimeter. *J. Phys. Chem.* 91: 5770–5776.

LaVerne, J. A. and Schuler, R. H. 1994. Track effects in water radiolysis—Yields of the Fricke dosimeter for carbon-ions with energies up to 1700 MeV. *J. Phys. Chem.* 98: 4043–4049.

LaVerne, J. A. and Yoshida, H. 1993. Production of the hydrated electron in the radiolysis of water with helium-ions. *J. Phys. Chem.* 97: 10720–10724.

LaVerne, J. A., Stefanic, I., and Pimblott, S. M. 2005. Hydrated electron yields in the heavy ion radiolysis of water. *J. Phys. Chem. A* 109: 9393–9401.

Meesungnoen, J. and Jay-Gerin, J. P. 2005a. High-LET radiolysis of liquid water with ^{1}H$^+$, ^{4}He^{2+}, ^{12}C^{6+}, and ^{20}Ne^{9+} ions: Effects of multiple ionization. *J. Phys. Chem. A* 109: 6406–6419.

Meesungnoen, J. and Jay-Gerin, J. P. 2005b. Effect of multiple ionization on the yield of H_2O_2 produced in the radiolysis of aqueous 0.4 M H_2SO_4 solutions by high-LET ^{12}C^{6+} and ^{20}Ne^{9+} ions. *Radiat. Res.* 164: 688–694.

Milosavljevic, B. H. and LaVerne, J. A. 2005. Pulse radiolysis of aqueous thiocyanate solution. *J. Phys. Chem. A* 109: 165–168.

Mozumder, A. 2004. Interaction of fast charged particles with matter. In *Charged Particle and Photon Interactions with Matter*, A. Mozumder and Y. Hatano (eds.), pp. 9–29. New York: Marcel Dekker.

Muroya, Y., Lin, M. Z., Wu, G. Z. et al. 2005. A re-evaluation of the initial yield of the hydrated electron in the picosecond time range. *Radiat. Phys. Chem.* 72: 169–172.

Noda, K., Tann, D., Uesugi, T., Shibuya, S., Honma, T., and Hashimoto, Y. 2005. Production of short-pulsed beam for ion-beam pulse radiolysis. *Nucl. Instrum. Methods Phys. Res. Sect. B-Beam Interact. Mater. Atoms* 240: 18–21.

Ohno, S., Furukawa, K., Taguchi, M., Kojima, T., and Watanabe, H. 2001. An ion-track structure model based on experimental measurements and its application to calculate radiolysis yields. *Radiat. Phys. Chem.* 60: 259–262.

Pastina, B. and LaVerne, J. A. 1999. Hydrogen peroxide production in the radiolysis of water with heavy ions. *J. Phys. Chem. A* 103: 1592–1597.

Pimblott, S. M. and Laverne, J. A. 1994. Diffusion-kinetic theories for LET effects on the radiolysis of water. *J. Phys. Chem.* 98: 6136–6133.

Pimblott, S. M. and LaVerne, J. A. 2002. Effects of track structure on the ion radiolysis of the Fricke dosimeter. *J. Phys. Chem. A* 106: 9420–9427.

Pimblott, S. M. and Mozumder, A. 2004. Modeling of physicochemical and chemical processes in the interactions of fast charged particles with matter. In *Charged Particle and Photon Interactions with Matter*, A. Mozumder and Y. Hatano (eds.), pp. 75–103. New York: Marcel Dekker.

Rice, S. A., Playford, V. J., Burns, W. G., and Buxton, G. V. 1982. Nanosecond proton pulse-radiolysis. *J. Phys. E Sci. Instrum.* 15: 1240–1243.

Sauer, M. C., Schmidt, K. H., Hart, E. J., Naleway, C. A., and Jonah, C. D. 1977. LET dependence of transient yields in pulse-radiolysis of aqueous systems with deuterons and alpha-particles. *Radiat. Res.* 70: 91–106.

Sauer, M. C., Schmidt, K. H., Jonah, C. D., Naleway, C. A., and Hart, E. J. 1978. High-LET pulse-radiolysis— O_2^- and oxygen production in tracks. *Radiat. Res.* 75: 519–528.

Sauer, M. C., Jonah, C. D., Schmidt, K. H., and Naleway, C. A. 1983. LET dependences of yields in the pulse-radiolysis of aqueous systems with $^2H^+$ and $^4He^{2+}$. *Radiat. Res.* 93: 40–50.

Schwarz, H. A., Caffrey, J. M., and Scholes, G. 1959. Radiolysis of neutral water by cyclotron produced deuterons and helium ions. *J. Am. Chem. Soc.* 81: 1801–1809.

Spinks, J. W. T., and Woods, R. J. 1990. *An Introduction to Radiation Chemistry*, 3rd edn. New York/Chichester/Brisbane/Toronto/Singapore: John Wiley & Sons, Inc.

Taguchi, M. and Kojima, T. 2005. Yield of OH radicals in water under high-density energy deposition by heavy-ion irradiation. *Radiat. Res.* 163: 455–461.

Taguchi, M. and Kojima, T. 2007. Yield of OH radicals in water under heavy ion irradiation. dependence on mass, specific energy, and elapsed time. *Nucl. Sci. Tech.* 18: 35–38.

Taguchi, M., Aoki, Y., Namba, H., Watanabe, R., Matsumoto, Y., and Hiratsuka, H. 1997. Fast fluorescence decay of naphthalene induced by Ar ion irradiation. *Nucl. Instrum. Methods Phys. Res. Sect. B-Beam Interact. Mater. Atoms* 132: 135–131.

Taguchi, M., Kimura, A., Watanabe, R., and Hirota, K. 2009. Estimation of yields of hydroxyl radicals in water under various energy heavy ions. *Radiat. Res.* 171: 254–263.

Taguchi, M., Baldacchino, G., Kurashima, S. et al. 2009. Transient absorption measurement system using pulsed energetic ion. *Radiat. Phys. Chem.* 78: 1169–1174.

Toburen, L. H. 2004. Ionization and secondary electron production by fast charged particles. In *Charged Particle and Photon Interactions with Matter*, A. Mozumder and Y. Hatano (eds.), pp. 31–74. New York: Marcel Dekker.

Wasselin-Trupin, V., Baldacchino, G., Bouffard, S. et al. 2000. A new method for the measurement of low concentrations of OH/O_2^- radical species in water by high-LET pulse radiolysis. A time-resolved chemiluminescence study. *J. Phys. Chem. A* 104: 8709–8713.

Wasselin-Trupin, V., Baldacchino, G., and Hickel, B. 2001. Detection of OH and O_2^- radicals generated by water radiolysis of water. *Can. J. Physiol. Pharm.* 79: 171–175.

Wasselin-Trupin, V., Baldacchino, G., Bouffard, S., and Hickel, B. 2002. Hydrogen peroxide yields in water radiolysis by high-energy ion beams at constant LET. *Radiat. Phys. Chem.* 65: 53–61.

Wojcik, M., Tachiya, M., Tagawa, S., and Hatano, Y. 2004. Electron-ion recombination in condensed matter: Geminate and bulk recombination processes. In *Charged Particle and Photon Interactions with Matter*, A. Mozumder and Y. Hatano (eds.), pp. 259–300. New York: Marcel Dekker.

Yamashita, S., Katsumura, Y., Lin, M., Muroya, Y., Maeyama, T., and Murakami, T. 2008a. Water radiolysis with heavy ions of energies up to 28 GeV-2: Extension of primary yield measurements to very high LET values. *Radiat. Phys. Chem.* 77: 1224–1229.

Yamashita, S., Katsumura, Y., Lin, M. et al. 2008b. Water radiolysis with heavy ions of energies up to 28 GeV. 3. Measurement of $G(MV^{•+})$ in deaerated methyl viologen solutions containing various concentrations of sodium formate and Monte Carlo simulation. *Radiat. Res.* 170: 521–533.

Yamashita, S., Katsumura, Y., Lin, M., Muroya, Y., Miyazakia, T., and Murakami, T. 2008c. Water radiolysis with heavy ions of energies up to 28 GeV. 1. Measurements of primary g values as track segment yields. *Radiat. Phys. Chem.* 77: 439–446.

Ziegler, J. F. 1998. RBS/ERD simulation problems: Stopping powers, nuclear reactions and detector resolution. *Nucl. Instrum. Methods Phys. Res. Sect. B-Beam Interact. Mater. Atoms* 137: 131–136.

14 Radiation Chemistry of Liquid Water with Heavy Ions: Monte Carlo Simulation Studies

Jintana Meesungnoen
Université de Sherbrooke
Sherbrooke, Québec, Canada

Jean-Paul Jay-Gerin
Université de Sherbrooke
Sherbrooke, Québec, Canada

CONTENTS

14.1 ON THE IMPORTANCE OF AQUEOUS RADIATION CHEMISTRY

A large fraction of radiation chemistry studies have been concerned with the studies of water and aqueous solutions because of the unique importance of water in biological systems and because of its relevance to a number of practical applications, including (1) various areas of nuclear science and technology such as water-cooled nuclear power reactors (where the radiolytic processes need to be carefully controlled to avoid deleterious effects), (2) radiation effects in space, (3) radiotherapy and diagnostic radiology, and (4) the environmental management of radioactive waste materials (LaVerne, 2004; Garrett et al., 2005). Water also serves as the standard reference material for clinical radiation therapy because its absorption properties for ionizing radiation are similar to those of biological tissue (Medin et al., 2006). The study of the radiolysis* of water has been actively examined for more than a century. In fact, Curie and Debierne (1901), Giesel (1902, 1903), and Ramsay and Soddy (1903) were the first to observe that dissolved radium salts decompose aqueous solutions by liberating hydrogen and oxygen gases continuously (due principally to the release of α-particles from the radium). Excellent accounts of the history of aqueous radiation chemistry have been given by Kroh (1989), Jonah (1995), Ferradini and Jay-Gerin (1999), LaVerne (2000), Zimbrick (2002), and Buxton (2004).

All biological systems are damaged by ionizing radiation. Since living cells and tissue consist mainly of water (~70%–85% by weight), the knowledge of the radiation chemistry of aqueous solutions is critical to our understanding of early stages in the complicated chain of radiobiological events that follows the passage of radiation. In this context, reactive species produced by water radiolysis in the cellular environment are likely to be major contributors to the induction of chemical modifications and changes in cells that may subsequently act as triggers of signaling or damaging effects (Muroya et al., 2006), and ultimately lead to the observation of a biological response (Hall and Giaccia, 2006; von Sonntag, 2006; Tubiana, 2008). Moreover, cellular radiobiological phenomena depend not only on the absorbed dose but also on the quality of radiation employed, a measure of which is given by the "linear energy transfer" (LET) that represents, to a first approximation, the nonhomogeneity of the energy deposition on a submicroscopic scale. For example, for similar absorbed doses of different types of radiation, a less uniform geometric distribution of absorbed energy produces biological changes more efficiently (e.g., Barendsen, 1968; Tsuruoka et al., 2005). The finding of a relationship between the "relative biological effectiveness" (RBE), defined in radiobiology to describe these differences,[†] and radiation quality (LET) emphasizes the importance of the initial spatial distribution of reactants—commonly referred to as the "track structure"—in the response of irradiated cells.

From the viewpoint of pure aqueous radiation chemistry, the successful prediction of the effects of radiation type and energy in radiolysis requires not only a realistic description of the early physical aspects of the radiation track structure, but also an accurate modeling of the temporal, three-dimensional development of the track, in which the various water decomposition products are specified and allowed to diffuse and react with one another or with the milieu (Muroya et al., 2006).

---

* The term *radiolysis* refers to any chemical changes induced by ionizing radiation, and includes synthesis as well as degradation (Hart and Platzman, 1961).
† The RBE of some test radiation Y is defined as the ratio of the absorbed dose of a "standard radiation" (customarily 250 kV x-rays) to that of the radiation Y required to produce the same specific biological effect (Hall and Giaccia, 2006). The RBE applies to all ionizing radiation, whether photon or particle, and depends on several factors such as the biological system or end point studied, the absorbed dose, the irradiation parameters (e.g., type of radiation, its energy and dose rate), and the environmental conditions (e.g., the level of oxygenation).

Therefore, it is critical to understand how the beam quality and the irradiation conditions affect the subsequent water decomposition products and their space distribution. Finally, it is also important to know how the initial, spatially nonhomogeneous distribution of reactive species relaxes in time toward a homogeneous distribution.

14.2 ENERGY DEPOSITION EVENTS AND CREATION OF TRACKS BY CHARGED PARTICLES

14.2.1 INTERACTION OF IONIZING RADIATION WITH MATTER

Ionizing radiations are defined as those types of energetic particle and electromagnetic radiations that, either directly or indirectly, cause ionization of a medium, i.e., the removal of a bound orbital electron from an atom or a molecule and, thereby, the production of a residual positive ion. Some molecules, instead of being ionized, may also be excited to upper electronic states (e.g., Evans, 1955; Anderson, 1984; Mozumder, 1999; Toburen, 2004). Directly ionizing radiations are fast-moving charged particles (e.g., electrons, protons, α-particles, stripped nuclei, or fission fragments) that produce ionizations through direct Coulomb interactions. In this case, note that particle–particle contact is not necessary since the Coulomb force acts at a distance. Indirectly ionizing radiations are energetic electromagnetic radiations (like x- or γ-ray photons) or neutrons that can also liberate bound orbital electrons, but secondarily to a preliminary interaction. For photons, this interaction is predominantly via the production of Compton electrons and photoelectrons (and, if the incident photon energy is greater than 1.02 MeV, the production of electron–positron pairs). Neutrons interact with matter through elastic nuclear scattering, resulting in the production of energetic recoil positively charged nuclei (ions), protons and oxygen ions in water, which can go on to generate ionized and excited molecules along their paths. Regardless of the type of ionizing radiation, the final common result in all the modes of the absorption of ionizing radiation is thus the formation of tracks of energy-loss events in the form of ionization and excitation processes and in a geometrical pattern that depends on the type of radiation involved.

Generally, the electrons ejected in the ionization events may themselves have sufficient energy to ionize one or more other molecules of the medium. In this way, the primary high-energy electron can produce a large number ($\sim 4 \times 10^4$ by a 1 MeV particle) of secondary or higher-order generation electrons* along its track as it gradually slows down (ICRU Report 31, 1979). From atomic physics, it is known that most energy-loss events by fast electrons involve small transfers of energy. In fact, the probability of a given energy transfer Q varies inversely with the square of that energy loss (Evans, 1955). "Distant" or "soft" collisions, in which the energy loss is small, are therefore strongly favored over "close" or "hard" collisions, in which the energy loss is large (Mozumder, 1999). The vast majority of these secondary electrons have low initial kinetic energies, with a distribution that lies essentially below 100 eV, and a most probable energy around 9–10 eV (LaVerne and Pimblott, 1995; Cobut et al., 1998; Pimblott and LaVerne, 2007). In most cases, they lose all their excess energy by multiple quasi-elastic (i.e., elastic plus phonon excitations) and inelastic interactions with their environment, including ionizations and/or excitations of electronic, intramolecular vibrational or rotational modes of the target molecules (Michaud et al., 2003), and quickly reach thermal equilibrium (i.e., they are "thermalized"). Determining exactly which of these competing interaction types will take place is a complex function of the target medium and the energy range of the incident electron. By definition, a measure of the probability that any particular one of these interactions will occur is called the "cross section" (expressed in units of area) for that particular interaction type. The total interaction cross section σ, summed over all considered individual processes i, is used to determine the distance to the next interaction, and the relative contributions σ_i to σ are used to

---

* It is customary to refer to all electrons that are not primary as "secondary."

determine the type of interaction. Actually, the mean distance between two consecutive interactions or "mean free path" λ is defined by

$$\lambda = \frac{1}{N\sigma}, \tag{14.1}$$

where N is the number of atoms or molecules per unit volume, and

$$\sigma = \sum_i \sigma_i. \tag{14.2}$$

In a dilute aqueous environment, thermalized electrons undergo trapping and hydration in quick succession (within ~10^{-12} s) as a result of the water electric dipoles rotating under the influence of the negative charge (Bernas et al., 1996). Some electrons that have kinetic energies lower than the first electronic excitation threshold of the medium, the so-called subexcitation electrons (Platzman, 1955), may also undergo, prior to thermalization, prompt geminate ion recombination (Freeman, 1987) or induce the production of energetic (~1–5 eV) anion fragments via formation of dissociative negative ion states (resonances) (i.e., dissociative electron attachment, DEA) (Christophorou et al., 1984; Bass and Sanche, 2003). As a consequence of the energy gained by the medium, a sequence of very fast reactions and molecular rearrangements leads to the formation of new, highly nonhomogeneously distributed chemical species in the system, such as charged and/or neutral molecular fragments, reactive free radicals, and other excited chemical intermediates. The trail of the initial physical events, along with the chemical species, is generally referred to as the "track" of a charged particle, and its overall detailed spatial distribution, including contributions from secondary electrons, is commonly known as "track structure" (e.g., Magee and Chatterjee, 1987; Paretzke, 1987; Kraft and Krämer, 1993; Paretzke et al., 1995; Mozumder, 1999; LaVerne, 2000a, 2004; Dingfelder, 2006; Muroya et al., 2006).

A charged particle is called "heavy" if its rest mass is large compared with that of an electron. Thus, protons ($^1H^+$), α-particles ($^4He^{2+}$) and their near relatives ($^2H^+$, $^3H^+$, $^3He^{2+}$), higher-charged particles (i.e., $^{12}C^{6+}$, $^7Li^{3+}$, $^{20}Ne^{10+}$, $^{56}Fe^{26+}$, etc.), and fission fragments are all heavy-charged particles. As most radiation chemical studies involve ions with initial energy of a few MeV per nucleon or more depending on the source, the main interaction responsible for energy loss at these energies is the Coulomb force between the charge of the incident heavy ion and that of the bound electrons of the medium. The tracks of heavy ions can thus be explained in basically the same manner as that for fast electrons. A heavy-charged particle traversing matter loses energy primarily through the ionization and excitation of target atoms or molecules. This proceeds until the ions are so much slowed down that they may be neutralized by electron capture, whereby their travel is stopped. For sufficiently fast ions and electrons, there are, however, some obvious differences due almost entirely to the small mass of the electron. For example, a (nonrelativistic) heavy ion (which has mass M, velocity v_i, and kinetic energy $E_i = Mv_i^2/2$) is restricted to a maximum allowed energy loss (Q_{max}) to an initially free stationary atomic electron (which has mass m_0) given by (Evans, 1955; Anderson, 1984)

$$Q_{max} \approx E_i \left[\frac{4Mm_0}{(M+m_0)^2} \right] \approx 4E_i \left(\frac{m_0}{M} \right) = 2m_0 v_i^2 \quad (M \gg m_0), \tag{14.3}$$

whereas a fast incident electron can transfer all its initial kinetic energy in a single collision with a bound electron. As seen in Equation 14.3, the energy that a heavy-charged particle loses in each event and the corresponding change in its momentum amount to only a small fraction (proportional to m_0/M) of its total energy and momentum. Because the heavier mass of fast ions prevents them from being scattered as much as electrons, heavy ions move in a relatively straight line (their deflections in collisions are negligible) until very near the end of their range. By contrast, electrons can undergo large-angle scattering in collisions with medium electrons, leading to extremely torturous paths.

14.2.2 LINEAR ENERGY TRANSFER AND INTERACTION CROSS SECTIONS FOR HEAVY CHARGED PARTICLES

The earliest experiments on the radiation–chemical effects in liquid water showed that there were differences in the yields of radiolytic products for various types of radiation (Kernbaum, 1910; Usher, 1911; Duane and Scheuer, 1913; Debierne, 1914). The amounts of changes observed in the quantities and proportions of the chemical species produced depend both on the total energy deposition and the spatial distribution of the energy. The LET, first introduced by Zirkle et al. (1952), is a measure of the rate of energy deposition along and within the track of a penetrating charged particle. This quantity is especially important in evaluating the overall chemical effect in the applications of radiation to biological systems because the concentration of the deposited energy is significant in those cases. LET is defined as

$$\text{LET} = -\frac{dE}{dx}, \tag{14.4}$$

where dE is the average energy locally (i.e., in the vicinity of the particle track) imparted to the medium by the particle in traversing a distance dx (ICRU Report 16, 1970). In this way, the (sometimes termed "energy-unrestricted," i.e., when all permissible energy transfers are included) LET is equivalent to the "stopping power" (which is commonly used in the domain of radiation physics) of the medium traversed (Anderson, 1984; Mozumder, 1999). Usually, LET values are in units of keV/µm (the conversion to SI units is: 1 keV/µm ≈ 1.602×10^{-19} J/nm).

The Bethe theory of stopping power describes the average energy loss due to the electromagnetic interactions between fast charged particles and the electrons in absorber atoms. For ions whose kinetic energies are small compared to their rest mass energy, the nonrelativistic stopping power formula of Bethe (Bethe, 1930; Bethe and Ashkin, 1953) is given by (in SI units)

$$-\frac{dE}{dx} = \left(\frac{1}{4\pi\varepsilon_0}\right)^2 \frac{4\pi Z^2 e^4}{m_0 V^2} N \ln\left(\frac{2m_0 V^2}{I}\right), \tag{14.5}$$

where

Ze is the charge on the incident ion
V is the ion velocity
m_0 is the mass of an electron
N is the number of electrons per cubic meter of the absorbing medium
I is the mean of all the ionization and excitation potentials of the bound electrons in the absorber

In most cases, I is determined by fitting the theory to experimental results related to the stopping power. For liquid water, $I = 79.7 \pm 0.5$ eV (Bichsel and Hiraoka, 1992). Based on the lowest-order (or first "Born") approximation in the electromagnetic interaction between the incident particle and the atomic electrons, the Bethe equation has a wide range of validity except for slow, highly charged heavy particles such as fission fragments.

As we can see from Equation 14.5, for any incident particle, the LET depends on the particle's kinetic energy only through its velocity. It is approximately inversely proportional to V^2, but the mass of the fast-moving particle does not appear. Hence, there are really no isotope effects, provided the ions have the same velocity. For example, protons ($^1H^+$) and deuterons ($^2H^+$) of equal velocity

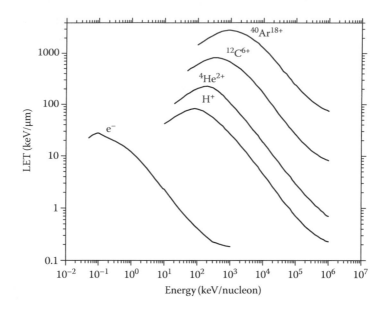

FIGURE 14.1 LET of some heavy ions and electrons in liquid water as a function of energy. The LET plotted here is the average energy dissipated per unit track length by the charged particles having the energy shown (and not the energy loss averaged over the whole track length). Note that the value of LET for ^{60}Co γ-rays (photon energies of 1.17 and 1.33 MeV) is ~0.3 keV/μm; this value applies also to x-rays and fast electrons of the same energies. (Data from Watt, D.E., *Quantities for Dosimetry of Ionizing Radiations in Liquid Water*, Taylor & Francis, London, U.K., 1996. With permission.)

give similar radiation chemical yields (Anderson and Hart, 1961). Moreover, the velocity term in the numerator of the logarithm term and in the denominator of the pre-logarithmic term gives rise to the familiar Bragg peak, i.e., with decreasing velocity of the incident particle, the LET increases to a maximum and then decreases at lower velocities (LaVerne, 2000a, 2004). Specifically, the Bragg peak refers to the increase in the density of ionizations as the incident heavy ion approaches the end of its track (range) and slows down (the slower the speed, the more time the particle spends in the vicinity of an atom and, therefore, the greater the probability of interaction and resulting ionization and excitation). In Figure 14.1, we show the LET as a function of incident energy for electrons and four different ions in liquid water. We see the large difference between the LET of electrons and heavy particles of the same energy (due to the difference in the velocity).

As Equation 14.5 also shows, the LET is proportional to the square of the projectile charge number Z^2. An important implication of this result is the following. For example, a 1 MeV proton and a 26 MeV He^{2+} ion have approximately the same (instantaneous) LET in water (~26.4 and 26.7 keV/μm, respectively). In order for them to have the same LET, the He^{2+} ion has to have a velocity twice that of the proton. Hence, the heavier particle (helium in this case) will produce higher energy δ-rays (secondary electrons that have sufficient energy to form tracks of their own, branching from the primary track). This results in the formation of a broader track structure for the He^{2+} ion.

An important feature of the first Born approximation is that it also provides a convenient framework to obtain the interaction cross sections of different, sufficiently fast projectile ions from one another by a simple Z^2 scaling operation. This follows from the similarity between the theoretical formulation of the stopping power and of the total cross section for inelastic scattering. In fact, within this theory, the cross section for the collision of a charged projectile with a single electron (assumed to be initially at rest) in an atom (known as the "Rutherford" cross section) behaves essentially as $(Z/V)^2$, where Z is the projectile charge number and V is the impact speed, and does not depend upon the mass of the particle (Inokuti, 1971; McDaniel et al., 1993; ICRU Report 55, 1996). This Z^2 scaling is particularly useful for providing cross sections for inelastic scattering by "bare" (i.e., fully ionized or stripped) ion projectiles, especially as there are only limited experimental data available involving ions heavier than proton or helium in collision with molecular targets of biological interest (e.g., H_2O). The "reference" cross-section data generally used in this context are those for proton impact, owing to the fact that protons represent, by far, the most comprehensive database of collision cross sections for bare ions (Rudd, 1990; Cobut et al., 1998; Dingfelder et al., 2000; Toburen, 2004). In other words, the cross sections for ionization or excitation by heavy-charged particles of charge number Z are approximately Z^2 times the cross sections for proton impact *at the same velocity.* This simple scaling procedure holds only at sufficiently high energies. Deviations from that rule occur at lower energies (below approximately 1 MeV/nucleon), where the first Born approximation—which rests on the assumption that the projectile velocity is large compared to the orbital speeds of the valence electrons in the target—is no longer satisfied. Such slow ions are usually incompletely stripped and can undergo successive electron capture and loss events contributing to a changing equilibrium charge state that depends on the velocity (e.g., ICRU Report 55, 1996; LaVerne, 2004). On the average, the net positive (or "effective") charge on an incident ion decreases when the speed decreases. This charge exchange also complicates the derivation of relevant cross sections for heavy charged, partially "dressed" particles in the low-velocity regime.

14.3 STRUCTURE OF CHARGED-PARTICLE TRACKS IN LIQUID WATER

14.3.1 TRACK STRUCTURE IN RADIATION CHEMISTRY AND RADIOBIOLOGY

A great many experimental and theoretical studies have shown that the quantities and proportions of the chemical products formed in the radiolysis of water are highly dependent on the distances separating the primary radiolytic species from each other along the track of the ionization radiation. The distribution of separations (i.e., the "track structure") is determined to a large extent by the distribution of the physical energy deposition events and their geometrical dispositions, or, in other words, by the quality of the radiation. In fact, track structure effects are also usually called "LET effects" as most of the early studies used this parameter to characterize the different radiation chemical yields (or "G-values")* resulting from various irradiating ions in liquid water. The radiation track structure is of crucial importance in specifying the precise spatial location and identity of all the radiolytic species and free-radical intermediates generated in the tracks, and their subsequent radiobiological action at the molecular and cellular levels. Track structure, combined with a reaction scheme and yields of primary species, forms the basis of radiation–chemical theory (Mozumder, 1999). It is now well accepted by the scientific community that differences in the biochemical and biological effects (e.g., damage to DNA, changes in cell signaling, etc.) of different qualities of radiation must be analyzed in terms of track structure (Chatterjee and Holley, 1993; Muroya et al., 2006).

14.3.2 SPATIAL ASPECTS OF TRACK STRUCTURES

14.3.2.1 Low-LET Radiation and Track Entities

The average LET of a 1 MeV electron in water is ~0.3 keV/μm. The track-averaged mean energy loss per collision event by such a fast electron is in the region ~48–65 eV (LaVerne and Pimblott, 1995; Cobut et al., 1998; Mozumder, 1999). This means that the energy-loss events are, on the average, separated by distances of about 2000 Å. This nonhomogeneous distribution of energy deposition events in space gives rise to the "spur" theory for low-LET track structure (Kara-Michailova and Lea, 1940; Allen, 1948; Samuel and Magee, 1953; Ganguly and Magee, 1956),[†] according to which

---

* In radiation chemistry, G-values are defined as the number of molecules, ions, or excited species formed or destroyed per 100 eV of energy absorbed.

† Sometimes called the "string-of-beads" model of a track.

the entire track is to be viewed as a random succession of (more or less spherical) spurs, or *spatially localized* energy-loss events (it is assumed that irradiating particles are isolated from each other, an assumption not necessarily correct at very high dose rates or with very short pulses of intense beams). The few tens of electronvolts deposited in a spur cause a secondary electron to be ejected from a molecule. As the ejected electron moves away, it undergoes collisions with surrounding water molecules, loses its excess energy, and becomes thermalized (~0.025 eV) within about 80–120 Å of its geminate positive ion (Goulet and Jay-Gerin, 1988; Muroya et al., 2002; Meesungnoen et al., 2002a; Pimblott and Mozumder, 2004; Uehara and Nikjoo, 2006). This electron thermalization distance or "penetration range" can be viewed as an estimate of the average radius of the spurs in the first stages of their development. Thus, the individual spurs produced by low-LET radiation (so-called sparsely ionizing radiation) are so far apart along the track that they are not initially overlapping (but they will overlap somewhat later as they develop in time).

In their pioneering work to model the radiation–chemical consequences of different energy-loss processes, Mozumder and Magee (1966a,b) considered, somewhat arbitrarily, a low-LET track as composed of a random sequence of three types of essentially nonoverlapping entities: "spurs, blobs, and short tracks" (Figure 14.2). The spur category contains all track entities created by the energy losses between the lowest excitation energy of water and 100 eV; in most cases, there are one to three ion pairs in such isolated spatial areas and about the same number of excited molecules (Pimblott and Mozumder, 1991). Blobs were defined as track entities with energy transfers between 100 and 500 eV, and short tracks as those with energy transfers between 500 eV and 5 keV. Secondary electrons produced in energy transfers above 5 keV were considered as "branch tracks."* Short and branch tracks are, collectively, described as δ-rays. This old concept of track entities proved to be very helpful in greatly facilitating the visualization of track processes and in modeling radiation–chemical kinetics. It is still a useful approach for the classification of track structures, since it takes into account the spatial arrangements of initial species, which affect their subsequent reactions.

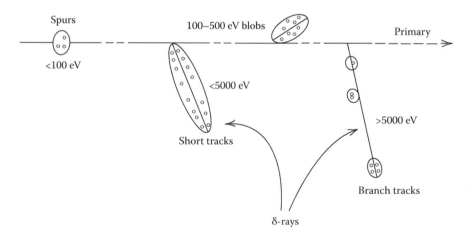

FIGURE 14.2 Classification of energy deposition events in water by track structure entities so-called spurs (spherical entities, up to 100 eV), blobs (spherical or ellipsoidal, 100–500 eV), and short tracks (cylindrical, 500 eV–5 keV) for a primary high-energy electron (not to scale). Short and branch tracks are, collectively, described as δ-rays. (From Burton, M. *Chem. Eng. News*, 47, 86, 1969. With permission.)

---

* The energy partition between these three track entities strongly depends on the incident particle energy, dividing approximately as the ratio of 0.75:0.12:0.13 between the spur, blob, and short track fractions for a 1 MeV electron in liquid water (Pimblott et al., 1990).

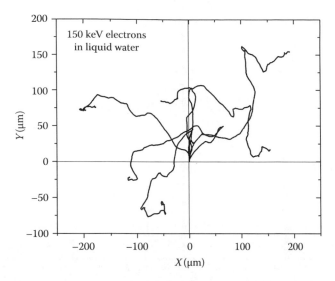

FIGURE 14.3 Simulated track histories (projected into the *XY* plane of figure) of six electrons incident on liquid water, showing the stochastic nature of paths. Initial energy of the electrons is 150 keV. Each electron is generated at the origin and starts moving along the *Y* axis.

For the sake of illustration of low-LET tracks, Figure 14.3 shows an example of the complete tracks of six 150 keV electrons and the secondary electrons they produce in water, calculated from our Monte Carlo track structure simulations.

14.3.2.2 Structure of High-LET Radiation Tracks

With increasing LET, the distantly spaced, nearly spherical spurs are formed increasingly closer together and eventually overlap (for LET greater than ~10–20 keV/μm) to form dense continuous columns of species consisting initially of a cylindrical "core" of high LET produced by the heavy-particle track itself and a surrounding region traversed by the emergent, comparatively low-LET secondary electrons (δ-rays) called the "penumbra" (Mozumder et al., 1968; Chatterjee and Schaefer, 1976; Ferradini, 1979; Magee and Chatterjee, 1987; Mozumder, 1999; LaVerne, 2000a, 2004). Figure 14.4 illustrates typical two-dimensional representations of short (1–5 μm) track segments of $^1H^+$, $^4He^{2+}$, $^{12}C^{6+}$, and $^{20}Ne^{10+}$ ions, calculated with our own Monte Carlo simulation code called IONLYS (see below) under the same LET conditions (~70 keV/μm). As one can see, these tracks can be considered as straight lines (see above). It is also seen that the ejected high-energy secondary electrons travel to a greater average distance away from the track core as the velocity of the incident ion increases, from protons to neon ions. In other words, even though all those particles are depositing the same amount of energy per unit path length, that energy is lost in a volume that increases in the order $^1H^+ < ^4He^{2+} < ^{12}C^{6+} < ^{20}Ne^{10+}$, indicating that the higher-Z particle has the lower mean density of reactive species. This irradiating-ion dependence of the track structure at a given LET (i.e., tracks of different ions with the same LET have different radial profiles) is in accord with Bethe's theory of stopping power* and indicates, as it has been frequently noted (e.g., Schuler and Allen, 1957; Miller, 1958; Sauer et al., 1977; Kaplan and Miterev, 1987; LaVerne and Schuler, 1987a; Ferradini and Jay-Gerin, 1999; Pimblott and LaVerne, 2002; LaVerne, 2000a, 2004), that LET is not a unique descriptor of the radiation chemical effects within heavy-charged particle tracks. Attempts have been made to introduce other comparative characteristics of radiation in place of LET like, for instance, the factor Z^2/β^2 (where Z is the ion charge number and β is the ratio of its velocity to that of light) (Katz, 1970) or, equivalently, the parameter MZ^2/E (where M is the heavy ion mass and

---

* It follows from Equation 14.5 that for two different ions of equal LET, the one with the higher charge will have the higher velocity.

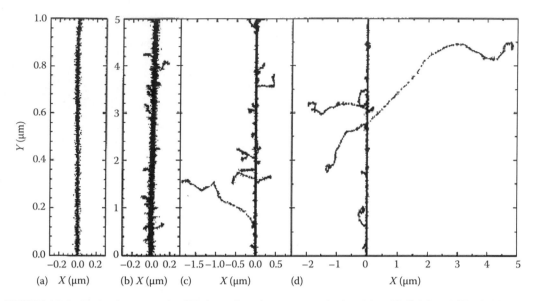

FIGURE 14.4 Projections over the XY plane of track segments calculated (at $\sim 10^{-13}$ s) for (a) $^1H^+$ (0.15 MeV), (b) $^4He^{2+}$ (1.75 MeV/nucleon), (c) $^{12}C^{6+}$ (25.5 MeV/nucleon), and (d) $^{20}Ne^{10+}$ (97.5 MeV/nucleon) impacting ions. Ions are generated at the origin and along the Y axis in liquid water under identical LET conditions (~ 70 keV/μm). Dots represent the energy deposited at points where an interaction occurred. (From Muroya, Y. et al., *Radiat. Res.*, 165, 485, 2006. With permission.)

$E = \frac{1}{2}MV^2$ its kinetic energy) (Pimblott and LaVerne, 2002; LaVerne, 2004). Several sets of radiation chemical data appear to be better unified using these parameters instead of LET, others do not. Following Pimblott and LaVerne (2002), it should be recognized, however, that no deterministic parametrization can realistically represent a phenomenon (namely, the effect of radiation quality on early radiation chemistry) that is stochastic in nature. Katz (1978) also indicated that no fewer than two parameters are needed in order to speak of a track structure and that single parameter reductions do not fully describe observed effects. Nevertheless, despite its limitations, LET still continues to be a dominant parameter in the radiation chemistry of heavy ions.

14.4 RADIOLYSIS OF LIQUID WATER AND AQUEOUS SOLUTIONS

14.4.1 RADIOLYSIS OF WATER AT LOW LET

The complex succession of events that follow the irradiation of water is usually divided into three, more or less distinct, temporal stages (Platzman, 1958): (1) deposition of radiant energy and formation of initial products in a specific, highly nonhomogeneous track structure geometry ("physical" stage), (2) establishment of thermal equilibrium in the bulk medium with reactions and reorganization of initial products to give stable molecules and chemically reactive species such as free atoms and radicals ("physicochemical" stage), and (3) thermal chemistry during which the various reactive species diffuse and react with one another (or with the environment) ("chemical" stage). The radiolysis of water by low-LET, sparsely ionizing radiation (e.g., energetic photons, such as γ-rays from ^{60}Co or ^{137}Cs or hard x-rays, or high-energy charged particles, such as fast electrons or protons generated by a particle accelerator) is generally well understood. It leads to the formation of the free radicals and molecular products e_{aq}^-, H^+, OH^-, $H^\bullet$, H_2, $^\bullet OH$, H_2O_2, $O_2^{\bullet-}$ (or $HO_2^\bullet$), etc. (e.g., Allen, 1961; Draganić and Draganić, 1971; Spinks and Woods, 1990; Ferradini and Jay-Gerin, 1999; Buxton, 2004). Under ordinary irradiation conditions (i.e., at modest dose rates), these products are generated nonhomogeneously on subpicosecond time scales in spurs along the track of the incident radiation (e.g., Plante et al., 2005; Muroya et al., 2006). Owing to diffusion from their initial positions, the radiolytic products then either react within the spurs as they develop in time or escape into the bulk solution. At ambient temperature and pressure, the so-called spur expansion is essentially complete by ~10^{-6} s after the initial energy deposition. At this time, the species that have escaped from spur reactions become homogeneously distributed throughout the bulk of the solution (also referred to as the "background") and the track of the radiation no longer exists. The radical and molecular products, considered as additions to the background, are then available to react with solutes (treated as spatially homogeneous) present in moderate concentrations.

For low-LET radiation, the radiolysis of pure, deaerated (air-free) liquid water can be adequately described by the following *global* equation, written for an absorbed energy of 100 eV (Ferradini and Jay-Gerin, 1999):

$$G_{-H_2O}H_2O \rightsquigarrow G_{e_{aq}^-}e_{aq}^- + G_{H\cdot}H^{\bullet} + G_{\cdot OH}{}^{\bullet}OH + G_{H^+}H^+ + G_{OH^-}OH^-$$

$$+ G_{H_2}H_2 + G_{H_2O_2}H_2O_2 + \cdots, \tag{14.6}$$

where the coefficients G_X (symbol G with the formula of the species as subscript) are the so-called primary free-radical and molecular yields ("long-time" or "escape" yields)* representing the numbers of the various radiolytic species X that remain throughout the bulk of the solution at the end of the nonhomogeneous chemical stage (i.e., after spur expansion). At this stage, G_{-H_2O} denotes the corresponding yield for net water decomposition. For ^{60}Co γ-radiation and neutral water at room temperature, the most recently reported values of the primary yields (expressed here in units of molecules/100 eV)† are (LaVerne, 2004)

$$G_{e_{aq}^-} = 2.50 \quad G_{H\cdot} = 0.56 \quad G_{H_2} = 0.45$$

$$G_{\cdot OH} = 2.50 \quad G_{H_2O_2} = 0.70 \quad G_{HO_2^{\bullet}/O_2^{\bullet-}} = 0.02^{\ddagger} \quad G_{-H_2O} = 3.94 \tag{14.7}$$

14.4.2 TIME SCALE OF EVENTS AND FORMATION OF PRIMARY FREE-RADICAL AND MOLECULAR PRODUCTS IN NEUTRAL WATER RADIOLYSIS

The overall process of producing chemical changes by ionizing radiation starts with the bombardment of water by the high-energy radiation and terminates with the reestablishment of chemical equilibrium. As mentioned above, this process can be divided into three temporal stages (the time scale of events that occur in the radiolysis of water is shown in Figure 14.5) that we briefly describe below.

14.4.2.1 The "Physical" Stage

The *physical stage* consists of the phenomena by which energy is transferred from the incident ionizing radiation to the system. It lasts not more than ~10^{-16} s. This time is too short for molecular motions of any kind, and only electronic processes are possible. The result of this energy absorption is the production, along the path of the radiation, of a large number of ionized and electronically

---

* Some authors prefer to use the lowercase symbol $g(X)$, rather than G_X, to represent the primary yield of the species X. Observed or final yields are always given in the form $G(X)$. Note that $G(X)$ can have many values because it refers to a specific time and set of conditions as specified by the author.

† The conversion to SI units is: 1 molecule/100 eV (abbreviated, in this work, molec./100 eV) $\approx 1.0364 \times 10^{-7}$ mol/J.

‡ Note that, for low-LET radiolysis, $HO_2^{\bullet}/O_2^{\bullet-}$ has an extremely small yield in comparison to the other radiolytic species and can be normally ignored (Bjergbakke and Hart, 1971).

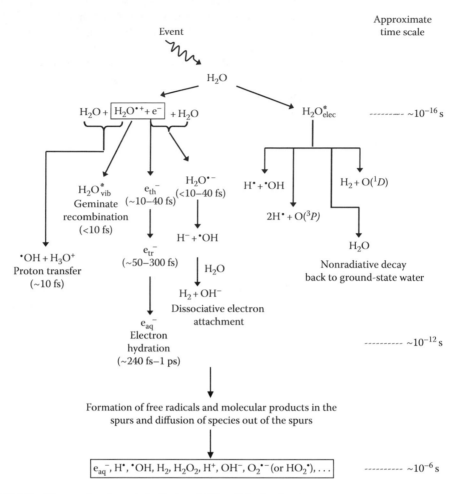

FIGURE 14.5   Time scale of events in the low-LET radiolysis of water.

excited water molecules (noted $H_2O^{\bullet+}$ and $H_2O^*_{elec}$, respectively). Note that $H_2O^*_{elec}$ represents here the many excited states, including the so-called superexcitation states (Platzman, 1962a) and the excitations of collective electronic oscillations of the "plasmon" type (Heller et al., 1974; Bednář, 1985; Kaplan and Miterev, 1987; LaVerne and Mozumder, 1993; Wilson et al., 2001). The earliest processes in the radiolysis of water are

$$H_2O \rightsquigarrow H_2O^{\bullet+} + e^- \tag{14.8}$$

$$H_2O \rightsquigarrow H_2O^*_{elec}. \tag{14.9}$$

Generally, the electron ejected in the ionization event has sufficient energy to ionize or excite one or more other water molecules in the vicinity, and this leads to the formation of track entities that contain the products of the events and that have been called "spurs" (see above).

14.4.2.2   The "Physicochemical" Stage

The *physicochemical stage* consists of the processes that lead to the establishment of thermal equilibrium in the system. Its duration is $\sim 10^{-12}$ s. During this stage, the secondary electron ejected from an ionized water molecule undergoes scattering as it moves away from its parent ion. It transfers energy to the molecules with which it collides and eventually reaches thermal equilibrium with the liquid. Once it has slowed down to thermal energy (e_{th}^-), it can be localized or trapped (e_{tr}^-) in a preformed potential energy well of appropriate depth in the liquid, before it reaches a fully relaxed, hydrated state (e_{aq}^-) as the dipoles of the surrounding molecules orient in response to the negative charge of the electron (Klassen, 1987). Thermalization, trapping, and hydration can then follow in quick succession (less than $\sim 10^{-12}$ s) (Mozumder, 1999)*:

$$e^- \to e_{th}^- \to e_{tr}^- \to e_{aq}^-. \tag{14.10}$$

The ejected electron that escapes process (14.10), after reaching the final stages of energy degradation, can also be temporarily captured resonantly by a water molecule to form a transient molecular anion. This anion then undergoes dissociation mainly into H^- and $^{\cdot}OH$ according to

$$e^- + H_2O \to H_2O^{\cdot-} \to H^- + {}^{\cdot}OH, \tag{14.11}$$

followed by the reaction of the hydride anion with another water molecule through a fast proton transfer:

$$H^- + H_2O \to H_2 + OH^-. \tag{14.12}$$

This "dissociative electron attachment" or DEA process has been observed in amorphous solid water at ~ 20 K for electron energies between ~ 5 and 12 eV (Rowntree et al., 1991). Reactions (14.11) and (14.12) are considered to be responsible, at least in part, for the production of the so-called nonscavengeable molecular hydrogen observed experimentally in the radiolysis of liquid water at very early times (Platzman, 1962b; Faraggi and Désalos, 1969; Goulet and Jay-Gerin, 1989; Kimmel et al., 1994; Cobut et al., 1996, 1998; Pastina et al., 1999; Meesungnoen and Jay-Gerin, 2005a).

At the end of the physical stage, a substantial yield of $H_2O^{\cdot+}$ radical ions is formed within Coulomb interaction distance of the slowing-down ("dry") electron (prior to its thermalization). This Coulomb attraction between the sibling ion and electron tends to draw them back together to undergo electron–cation "geminate" recombination:

$$e^- + H_2O^{\cdot+} \to H_2O_{vib}^*. \tag{14.13}$$

As the electron is recaptured, the parent ion is transformed into a (vibrationally) excited neutral molecule.

---

* In liquid water at 25°C, time-resolved femtosecond laser spectroscopic studies have revealed that electron "localization" and "hydration" occur on time scales of ~ 50–300 fs and ~ 240 fs–1 ps, respectively (e.g., Jay-Gerin et al., 2008 and references therein). The observation of the short-lived, weakly bound (i.e., "incompletely relaxed") electron (e_{tr}^-, the precursor to e_{aq}^-) is perhaps one of the most important discoveries made in the area of the radiolysis of water over the past two decades. It should be noted that, as soon as the thermalized electron becomes trapped in the liquid, its localized nature may allow it to undergo chemical reactions before settling into the "fully relaxed" e_{aq}^- state (e.g., Hunt, 1976; Jonah et al., 1977; Ferradini and Jay-Gerin, 1990; Pimblott and LaVerne, 1998; Lu et al., 2004). In particular, Jay-Gerin and Ferradini (1990) suggested that it could react in water according to

$$e_{tr}^- + H_2O \to H^{\cdot} + OH^-.$$

This reaction would occur on a time scale comparable to the duration of electron hydration, and would be in competition with the latter. It would offer a new mechanism by which an "initial" yield of $H^{\cdot}$ atoms would be produced in water (note that the primary source of atomic hydrogen in water radiolysis is generally considered at these early times as arising from the dissociation of excited water molecules; see below).

In the time scale of $\sim 10^{-14}$ s (consistent with the characteristic time of a vibrational period of a water molecule*) (Mozumder and Magee, 1975), the positive ions $H_2O^{\bullet+}$ undergo a proton transfer reaction with neighboring H_2O molecules:

$$H_2O^{\bullet+} + H_2O \rightarrow H_3O^+ + {}^\bullet OH, \tag{14.14}$$

where H_3O^+ (or equivalently, H_{aq}^+) represents the hydrated proton. However, Ogura and Hamill (1973) pointed out that $H_2O^{\bullet+}$ may migrate (randomly) during its very short lifetime (<10 fs) by means of a sequence of resonant electron transfers from neighboring water molecules. In fact, the results found by these authors suggest that the positive hole jumps on the average about 20 times before reaction (14.14) occurs. The estimates of the jump time are 10^{-15} s (Mozumder and Magee, 1975), and the ranges of a migrating hole are a few molecular diameters (Cobut et al., 1998).

Excited molecules may be produced directly in an initial act [reaction (14.9)] or by the neutralization of an ion [reaction (14.13)]. We have little knowledge, from both a theoretical and an experimental point of view, about the decay channels for excited water molecules in the liquid phase and the branching ratios associated with each of them. Fortunately, the contribution of the water excited states to the primary free-radical and molecular products in water radiolysis is of relatively minor importance in comparison with that of the ionization processes, so that the lack of information about their decomposition has only limited consequences. Consequently, the competing deexcitation mechanisms of H_2O^* are generally assumed to be essentially the same as those reported for an isolated water molecule (note that the same decay processes have been reported to occur for the electronically and vibrationally excited H_2O molecules in the gas phase), namely (e.g., Swiatla-Wojcik and Buxton, 1995; Cobut et al., 1998; Meesungnoen and Jay-Gerin, 2005a),

$$H_2O^* \rightarrow H^\bullet + {}^\bullet OH \tag{14.15}$$

$$H_2O^* \rightarrow H_2 + O(^1D) \tag{14.16}$$

$$H_2O^* \rightarrow 2H^\bullet + O(^3P) \tag{14.17}$$

$$H_2O^* \rightarrow H_2O + \text{release of thermal energy,} \tag{14.18}$$

where $O(^1D)$ and $O(^3P)$ represent oxygen atoms in their singlet ^1D excited state and triplet ^3P ground state, respectively (see Figure 14.5). Note that the dissociation of excited water molecules via reaction (14.15) is generally considered as the main source of the initial yield of $H^\bullet$ atoms. As for the different branching ratios (or decay probabilities) associated with reactions (14.15) through (14.18), they are chosen in order to consistently match the observed picosecond G-values of the various spur species (Muroya et al., 2002; Meesungnoen and Jay-Gerin, 2005a). It should be recalled here that the $O(^1D)$ atoms produced in reaction (14.16) react very efficiently with water to form H_2O_2 or possibly also $2^\bullet OH$ (Taube, 1957; Biedenkapp et al., 1970). In contrast, the ground-state $O(^3P)$ atoms in aqueous solution are rather inert to water but react with most additives (Amichai and Treinin, 1969).

14.4.2.3 The "Nonhomogeneous Chemical" Stage

The *nonhomogeneous chemical stage* consists of diffusion and reactions of the reactive species leading to the reestablishment of chemical equilibrium. As stated above, the various "initial" decomposition products present at $\sim 10^{-12}$ s following the passage of the radiation (namely, e_{aq}^-, $^\bullet OH$, $H^\bullet$, H_2, H_3O^+, OH^-, $^\bullet O^\bullet$,...) are distributed nonhomogeneously with high concentrations in the center of spurs or along the axis of tracks. They then proceed to diffuse away from the site

---

* Recall here that the O–H stretching frequency of H_2O in the liquid phase is ~ 3400 cm^{-1}.

where they were originally produced according to macroscopic diffusion laws and to react with themselves or with dissolved solutes (if any) present at the time of irradiation, until all spur or track reactions are complete. A number of like radicals will combine to form the molecular products H_2 and H_2O_2; a number will combine to re-form H_2O, while the remainder will diffuse out into the bulk of the solution. Table 14.1 gives the set of reactions that are likely to occur while the spurs expand. The time for completion of spur processes is generally taken to be ~10^{-6} s. By this time, the spatially nonhomogeneous distribution of reactive species has relaxed. Basically, it is the competition between reaction and escape, which determines the "primary" or "escape" yields of the radical and molecular products (see Section 14.4.1).

TABLE 14.1
Main Reaction Scheme and Rate Constants (k) Used in Our Simulations of the Radiolysis of Pure Liquid Water at 25°C

Reaction	k (M^{-1} s^{-1})	Reaction	k (M^{-1} s^{-1})
$H^{\bullet} + H^{\bullet} \to H_2$	5.03×10^9	$e_{aq}^- + e_{aq}^- \to H_2 + 2OH^-$	5.0×10^9
$H^{\bullet} + {}^{\bullet}OH \to H_2O$	1.55×10^{10}	$e_{aq}^- + H^+ \to H^{\bullet}$	2.11×10^{10}
$H^{\bullet} + H_2O_2 \to H_2O + {}^{\bullet}OH$	3.5×10^7	$e_{aq}^- + O_2^{\bullet-} \to H_2O_2 + 2OH^-$	1.3×10^{10}
$H^{\bullet} + e_{aq}^- \to H_2 + OH^-$	2.5×10^{10}	$e_{aq}^- + HO_2^- \to O^{\bullet-} + OH^-$	3.51×10^9
$H^{\bullet} + OH^- \to H_2O + e_{aq}^-$	2.51×10^7	$e_{aq}^- + O^{\bullet-} \to 2OH^-$	2.31×10^{10}
$H^{\bullet} + O_2 \to HO_2^{\bullet}$	2.1×10^{10}	$e_{aq}^- + H_2O \to H^{\bullet} + OH^-$	15.8
$H^{\bullet} + HO_2^{\bullet} \to H_2O_2$	1.0×10^{10}	$e_{aq}^- + O_2 \to O_2^{\bullet-}$	1.74×10^{10}
$H^{\bullet} + O_2^{\bullet-} \to HO_2^-$	1.0×10^{10}	$e_{aq}^- + HO_2^{\bullet} \to HO_2^-$	1.28×10^{10}
$H^{\bullet} + HO_2^- \to {}^{\bullet}OH + OH^-$	1.46×10^9	$e_{aq}^- + O(^3P) \to O^{\bullet-}$	2.0×10^{10}
$H^{\bullet} + O(^3P) \to {}^{\bullet}OH$	2.02×10^{10}	$e_{aq}^- + O_3 \to O_3^{\bullet-}$	3.6×10^{10}
$H^{\bullet} + O^{\bullet-} \to OH^-$	2.0×10^{10}	$H^+ + O^{\bullet-} \to {}^{\bullet}OH$	4.78×10^{10}
$H^{\bullet} + O_3 \to O_2 + {}^{\bullet}OH$	3.7×10^{10}	$H^+ + O_2^{\bullet-} \to HO_2^{\bullet}$	4.78×10^{10}
$H^{\bullet} + O_3^{\bullet-} \to OH^- + O_2$	1.0×10^{10}	$H^+ + OH^- \to H_2O$	1.12×10^{11}
${}^{\bullet}OH + {}^{\bullet}OH \to H_2O_2$	5.5×10^9	$H^+ + O_3^{\bullet-} \to {}^{\bullet}OH + O_2$	9.0×10^{10}
${}^{\bullet}OH + H_2O_2 \to HO_2^{\bullet} + H_2O$	2.87×10^7	$H^+ + HO_2^- \to H_2O_2$	5.0×10^{10}
${}^{\bullet}OH + H_2 \to H^{\bullet} + H_2O$	3.28×10^7	$OH^- + O(^3P) \to HO_2^-$	4.2×10^8
${}^{\bullet}OH + e_{aq}^- \to OH^-$	2.95×10^{10}	$OH^- + HO_2^{\bullet} \to O_2^{\bullet-} + H_2O$	6.3×10^9
${}^{\bullet}OH + OH^- \to O^{\bullet-} + H_2O$	6.3×10^9	$O_2 + O^{\bullet-} \to O_3^{\bullet-}$	3.7×10^9
${}^{\bullet}OH + HO_2^{\bullet} \to O_2 + H_2O$	7.9×10^9	$O_2 + O(^3P) \to O_3$	4.0×10^9
${}^{\bullet}OH + O_2^{\bullet-} \to O_2 + OH^-$	1.07×10^{10}	$HO_2^{\bullet} + O_2^{\bullet-} \to HO_2^- + O_2$	9.7×10^7
${}^{\bullet}OH + HO_2^- \to HO_2^{\bullet} + OH^-$	8.32×10^9	$HO_2^{\bullet} + HO_2^{\bullet} \to H_2O_2 + O_2$	8.3×10^5
${}^{\bullet}OH + O(^3P) \to HO_2^{\bullet}$	2.02×10^{10}	$HO_2^{\bullet} + O(^3P) \to O_2 + {}^{\bullet}OH$	2.02×10^{10}
${}^{\bullet}OH + O^{\bullet-} \to HO_2^-$	1.0×10^9	$HO_2^{\bullet} + H_2O \to H^+ + O_2^{\bullet-}$	1.29×10^4
${}^{\bullet}OH + O_3^{\bullet-} \to O_2^{\bullet-} + HO_2^{\bullet}$	8.5×10^9	$O_2^{\bullet-} + O^{\bullet-} \to O_2 + 2OH^-$	6.0×10^8
${}^{\bullet}OH + O_3 \to O_2 + HO_2^{\bullet}$	1.11×10^8	$O_2^{\bullet-} + H_2O \to HO_2^{\bullet} + OH^-$	0.075
$H_2O_2 + e_{aq}^- \to OH^- + {}^{\bullet}OH$	1.1×10^{10}	$O_2^{\bullet-} + O_3 \to O_3^{\bullet-} + O_2$	1.5×10^9
$H_2O_2 + OH^- \to HO_2^- + H_2O$	4.75×10^8	$HO_2^- + H_2O \to H_2O_2 + OH^-$	3.83×10^4
$H_2O_2 + O(^3P) \to HO_2^{\bullet} + {}^{\bullet}OH$	1.6×10^9	$HO_2^- + O^{\bullet-} \to O_2^{\bullet-} + OH^-$	3.5×10^8
$H_2O_2 + O^{\bullet-} \to HO_2^{\bullet} + OH^-$	5.55×10^8	$HO_2^- + O(^3P) \to O_2^{\bullet-} + {}^{\bullet}OH$	5.3×10^9
$H_2 + O(^3P) \to H^{\bullet} + {}^{\bullet}OH$	4.77×10^3	$O^{\bullet-} + O^{\bullet-} \to H_2O_2 + 2OH^-$	1.0×10^8
$H_2 + O^{\bullet-} \to H^{\bullet} + OH^-$	1.21×10^8	$O^{\bullet-} + O_3^{\bullet-} \to 2O_2^{\bullet-}$	7.0×10^8
$O(^3P) + O(^3P) \to O_2$	2.2×10^{10}	$O^{\bullet-} + H_2O \to {}^{\bullet}OH + OH^-$	1.02×10^6
$O(^3P) + H_2O \to 2\,{}^{\bullet}OH$	1.9×10^3	$O_3^{\bullet-} + H_2O \to O^{\bullet-} + O_2$	48.0

Finally, we should note that, in the time domain beyond a few microseconds, the reactions that occur in the bulk solution can usually be well described with conventional homogeneous chemistry methods (Pastina and LaVerne, 1999).

14.4.3 CONTROVERSIAL ISSUES: PRIMARY YIELDS OF $HO_2^{\bullet}/O_2^{\bullet-}$ AND H_2O_2 VERSUS LET AND THE PRODUCTION OF O_2 IN THE HEAVY-ION RADIOLYSIS OF WATER AT HIGH LET

Under heavy-ion irradiation conditions, the general trends on increasing LET of the radiation are to give lower free-radical and higher molecular primary yields (e.g., Allen, 1961; Anderson and Hart, 1961; Pucheault, 1961; Burns and Sims, 1981; Appleby, 1989; Elliot et al., 1996; McCracken et al., 1998; Ferradini and Jay-Gerin, 1999; LaVerne, 2000a, 2004). This behavior is explained by the increased intervention of radical–radical reactions as the local concentration of radicals along the track of the impacting ion is high and many radical interactions occur before the products can escape into the bulk solution. This should permit fewer radicals to escape combination and recombination reactions during the expansion of the tracks and in turn lead to the formation of more molecular products. However, there are two important exceptions to this rule:

1. Unlike the behavior of other radicals, the primary yield of the superoxide anion radical $O_2^{\bullet-}$ and its conjugate acid $HO_2^{\bullet}$ (hydroperoxyl radical) ($O_2^{\bullet-}$ is always in a pH-dependent equilibrium with $HO_2^{\bullet}$, $pK_a = 4.8$; see: Bielski et al., 1985) rises sharply with LET (Donaldson and Miller, 1956; Lefort and Tarrago, 1959; Appleby and Schwarz, 1969; Bibler, 1975; Baverstock and Burns, 1976; Sims, 1978; Burns and Sims, 1981; LaVerne et al., 1986; LaVerne and Schuler, 1987b; Baldacchino et al., 1998a,b; McCracken et al., 1998; Wasselin-Trupin et al., 2000). As mentioned earlier, with fast electrons and γ-rays, the production of $HO_2^{\bullet}/O_2^{\bullet-}$ is normally neglected because its very small yield of ~0.02 molecule/100 eV accounts for less than 1% of the other primary radiolytic species. However, with high-LET heavy ions, it is the major radical product that survives spur/track expansion, since the other primary radical G-values, i.e., those of $^{\bullet}OH$, $H^{\bullet}$, and e_{aq}^-, are virtually zero. The origin of these $HO_2^{\bullet}/O_2^{\bullet-}$ radicals is, as yet, not clearly established, even though there are a number of suggested (but not completely adequate) mechanisms to account for their formation and their increase with increasing LET (for reviews, see Sims, 1978; LaVerne, 1989; Ferradini and Jay-Gerin, 1998).
2. A second peculiar feature is that the primary yield of H_2O_2 rises with increasing LET to a maximum, after which it falls (Sims, 1978; Burns and Sims, 1981; Pastina and LaVerne, 1999; Wasselin-Trupin et al., 2002). It has been shown that the maximum in the value of $G_{H_2O_2}$ occurs precisely at the point where $G_{HO_2^{\bullet}/O_2^{\bullet-}}$ begins to rise sharply, suggesting that the yields of $HO_2^{\bullet}/O_2^{\bullet-}$ and H_2O_2 are closely linked (Sims, 1978). Apart from calculations using the Schwarz diffusion model (Schwarz, 1969; Burns and Sims, 1981), which predict a slight maximum in the H_2O_2 yield, but at lower LET than is found experimentally,* there is hitherto no suitable explanation for the presence of such a maximum in $G_{H_2O_2}$ as a function of LET.

Another controversial issue in the radiolysis of water at high LET concerns the hypothesis of the generation of O_2 in heavy-ion tracks. Oxygen is a powerful radiation sensitizer; the biological response to irradiation is greater under oxygenated conditions than under hypoxic conditions (e.g., Hall and Giaccia, 2006; von Sonntag, 2006). This ability of oxygen to sensitize cells to damage from ionizing radiation has been known for many years.[†] The degree of sensitization can be quantified by

---

* In addition, the Schwarz model cannot account for the $HO_2^{\bullet}/O_2^{\bullet-}$ yield at high LET, suggesting that, at high LET, additional mechanisms, which are not included in the model, become important.

[†] For an account of the historical development of the subject, see, e.g., Gray et al., 1953; Patt, 1953; Howard-Flanders, 1958.

the "oxygen enhancement ratio" (OER), the ratio of the dose without oxygen to the dose with oxygen needed to achieve the *same* biological effect. The OER is recognized as a dose-modifying factor of fundamental importance in radiobiology as well as of practical relevance in radiotherapy (since many tumors contain high levels of hypoxia). For low-LET radiations with conventional dose rates, the OER values are generally found between 2 and 3. For most cellular organisms, the oxygen effect shows a strong dependence on radiation quality; the OER decreases progressively with increasing LET of the radiation used, approaching its minimum value of 1 when the LET becomes greater than 100–200 keV/μm (e.g., Barendsen, 1968; Alper and Bryant, 1974). In other words, for high-LET radiation, the survival of tumor cells is practically the same in the presence or in the absence of O_2 (Curtis et al., 1982). To account for this experimental finding, Scholes and Weiss (1959), Swallow and Velandia (1962), Neary (1965), and Alper and Bryant (1974) invoked the "oxygen-in-the-track" hypothesis, suggesting that O_2 generated in the tracks of densely ionizing particles, as they traverse the biological material, is responsible for the decrease of OER with LET. In other words, cells exposed to high-LET radiation exhibit an "oxygenated" microenvironment around the particle track, even when they are irradiated under anoxic (i.e., no oxygen) conditions (Baverstock and Burns, 1976, 1981). This hypothesis of the generation of molecular oxygen in heavy-ion tracks, which presupposes that O_2 is a product of the radiolysis of water at high LET (for radiation of low LET, O_2 is *not* a radiolytic product), has often been invoked for a variety of aqueous biological systems, but other mechanisms have also been considered to explain the observed decrease in the OER with LET (see below), and a definite conclusion supporting this hypothesis has not yet been forthcoming.

14.4.4 HYPOTHESIS OF MULTIPLE IONIZATION OF WATER IN SINGLE COLLISIONS UNDER HEAVY-ION BOMBARDMENT

When an impacting heavy-charged particle (particularly a multiply charged heavy ion) collides violently with a multielectron target molecule, some of the translational kinetic energy of the ion is deposited in the electronic degrees of freedom of the target, with the result that the target emerges with an electronic excitation or ionization. Collisions at sufficiently large impact parameters will produce only singly ionized and excited molecules. As the impact parameter becomes smaller, multiple ionization (MI) of the target's outer (loosely bound) electron shells in a "single act" occurs with appreciable probabilities. For small impact parameters, inner-shell electrons can be ejected along with the multiple ionization of the outer shells (for reviews, see Cocke and Olson, 1991; McDaniel et al., 1993; ICRU Report 55, 1996; Toburen, 2004). Although inner-shell ionization (followed by Auger relaxation) effects are well known,* the consequences of direct multiple outer-shell ionization with two (or more) outgoing electrons in the final state have not often been considered in the models of radiation chemistry and biology. In fact, more than 50 years ago, Platzman (1952) concluded, in a discussion of the possible role of direct, multiple target ionization in radiation action, that these processes, although infrequent relative to single ionization events, should be "extremely effective chemically" owing to the high instability of the multiply ionized molecules produced.† In this same radiation–chemical context, the hypothesis of "multiple ionization of a

---

* Creation of an initial inner-shell vacancy (hole) by a swift charged particle is followed by the emission of one (or several) Auger electrons (commonly called an Auger cascade) in the ensuing "rearrangement" of the electron cloud as the excited atom/molecule relaxes to the ground state. Typically, the energy transferred in such a process amounts to some hundreds of electronvolts (i.e., the K-shell ionization energies of a C, N, and O atom are 283.8, 401.6, and 532 eV, respectively). Auger relaxation can produce doubly (or multiply) charged target ions (e.g., Chattorji, 1976).

† Platzman (1952) also pointed out the possibility of formation, by the incident particle, of multiply "excited" atoms or molecules in single collisions. By multiple excitation is meant the *simultaneous* excitation, by a passing charged particle, of several electrons in a single atom or in atoms closely coupled together in a molecule. According to this author, this process should be less likely to have a significant influence on chemical or biological effects than does multiple ionization. Multiple excitation is ignored in the present work.

single molecule or of near neighbors" was also proposed by Gäumann and Schuler (1961) as a possible explanation for the large increase in the yield of H_2 in the radiolysis of benzene by densely ionizing radiations (relative to that produced by fast electrons). Recently, Ferradini and Jay-Gerin (1998) reconsidered this earlier hypothesis and suggested that MI could play a significant role in the heavy-ion radiolysis of liquid water at high LET. In particular, they proposed that MI could intervene to explain the large yields of $HO_2^{\bullet}/O_2^{\bullet-}$ observed experimentally in high-LET water radiolysis (see above).

Multiple ionization has long been recognized as an important process in atomic physics, and experimental data have accumulated for many years. Most experiments have been performed with light ions in the gas phase. The importance of the direct outer-shell ionization in fast ion–atom collisions was first clearly demonstrated by the measurements of Manson et al. (1983), who studied 0.5–4 MeV protons incident on neon. They showed, at least for that target, that direct double ionization was the dominant Ne^{2+} production process, far more probable than inner-shell ionization (the Auger channel). In fact, inner-shell ionization cross sections are generally several orders of magnitude smaller than direct outer-shell ionization, that is, inner-shell ionization is a "rare" event (Toburen, 2004). The multiple ionization and the possible avenues of fragmentation (following MI) of polyatomic molecular targets induced by the interaction with heavy ions have been investigated intensively only recently. For water vapor, experimental studies have been reported so far by several groups (e.g., Werner et al., 1995a,b; Olivera et al., 1998; Siegmann et al., 2001; Pešić et al., 2004; Legendre et al., 2005; Montenegro and Luna, 2005; Alvarado et al., 2005; Alvarado, 2007; for a recent review, see Stolterfoht et al., 2007). Specifically, it is found that the cross section for double ionization is usually more than an order of magnitude less than for single ionization. Triple ionization is usually more than an order of magnitude less than double ionization. The similar ionization of higher multiplicity is much less probable (e.g., LaVerne, 2000a; Champion, 2003; Champion et al., 2005; Gervais et al., 2005, 2006). For such collisions, the double-, triple-, and quadruple ionization thresholds for H_2O are ~40, 65, and 88 eV, respectively (Meesungnoen and Jay-Gerin, 2005a). To our best knowledge, only limited data exist for studies of MI under highly charged ion impact for molecular systems of chemical or biological importance in the "condensed" phase. This is explained by the fact that such experiments are extremely difficult and further complicated by the occurrence of many possible molecular break-up channels following ionization. Theoretically, a detailed description of MI is also a difficult task due to the complex, quantum-mechanical many-body nature of the scattering mechanisms involved. Nevertheless, a few attempts have been made recently to simulate the role of multiple ionization in *liquid* water with the aim of evaluating its consequences in the heavy-ion radiation chemistry of water (Meesungnoen et al., 2003; Meesungnoen and Jay-Gerin, 2005a,b, 2009; Gervais et al., 2005, 2006; Gaigeot et al., 2007).

14.5 MONTE CARLO SIMULATIONS OF HIGH-LET WATER RADIOLYSIS

The complex sequence of events that follow absorption in liquid water of ionizing radiation can be modeled successfully by the use of Monte Carlo simulation techniques. Such a procedure is well adapted to account for the *stochastic* nature of the phenomena, provided that realistic probabilities and cross sections for all possible events are adequately known. The simulation then allows one to reconstruct the intricate action of the radiation. It also offers a powerful tool for appraising the validity of different assumptions, for making a critical examination of proposed reaction mechanisms, and for estimating some unknown parameters. The accuracy of these calculations is best determined by comparing their predictions with experimental data on well-characterized chemical systems that have been examined with a wide variety of incident radiation particles and energies.

Turner and his collaborators at the Oak Ridge National Laboratory (Oak Ridge, Tennessee) jointly with Magee and Chatterjee at Lawrence Berkeley Laboratory (Berkeley, California) were the first to use Monte Carlo simulations to derive computer-plot representations of the chemical evolution of a few keV electron tracks in liquid water at times between $\sim 10^{-12}$ and 10^{-7} s (Turner et al., 1981, 1983, 1988). Following this pioneering work, stochastic simulation codes employing Monte Carlo procedures were used with success by a number of investigators to study the relationship between the initial track structure and the ensuing chemical processes that occur in the radiolysis of both pure water and water-containing solutes (for reviews, see, e.g., Ballarini et al., 2000; Uehara and Nikjoo, 2006; Kreipl et al., 2009). Two main approaches have been widely used: (1) the "step-by-step" (or random flights Monte Carlo simulation) method, in which the trajectories of the diffusing species of the system are modeled by time-discretized random flights and in which reaction occurs when reactants undergo pairwise encounters, and (2) the "independent reaction times" (IRT) method (Clifford et al., 1986; Pimblott et al., 1991; Pimblott and Green, 1995), which allows the calculation of reaction times without having to follow the trajectories of the diffusing species. Among the stochastic approaches, the most reliable is certainly the full random flights simulation, which is generally considered as a measure of reality. However, this method can be exceedingly consuming in computer time when large systems (such as complete radiation tracks or track segments) are studied. The IRT method was devised to achieve much faster realizations than are possible with the full Monte Carlo model. In essence, it relies on the approximation that the distances between pairs of reactants evolve independently of each other, and, therefore, the reaction times of the various potentially reactive pairs are independent of the presence of other reactants in the system. For every pair, a reaction time is stochastically sampled according to the time-dependent survival function (Green et al., 1990; Goulet and Jay-Gerin, 1992; Frongillo et al., 1998) that is appropriate for the type of reaction considered. This function depends on the initial (or zero-time) distance separating the species, their diffusion coefficients, their Coulomb interaction, their reaction radius, and the probability of reaction during one of their encounters. The first reaction time is found by taking the minimum of the resulting ensemble of reaction times and allowing the corresponding pair of species to react at this time. This procedure for modeling reaction is continued either until all reactions are completed or until a predefined cut-off time is reached. The IRT simulation technique also allows one to incorporate, in a simple way, pseudo first-order reactions of the radiolytic products with various scavengers that are homogeneously distributed in the medium (such as H^+, OH^-, and H_2O itself, or more generally any solutes for which the relevant reaction rates are known). The ability of the IRT method to give accurate time-dependent chemical yields has been well validated by comparison with full random flight Monte Carlo simulations that do follow the reactant trajectories in detail (Pimblott et al., 1991; Goulet et al., 1998; Plante, 2009).

14.5.1 THE IONLYS-IRT SIMULATION CODE

In a program begun in the summer of 1988 in collaboration with Jean Paul Patau (Université Paul-Sabatier, Toulouse, France) and Christiane Ferradini (Université René-Descartes, Paris, France), the Sherbrooke group has developed and progressively refined, with very high levels of detail, several Monte Carlo codes that simulate the track structure of ionizing particles in water, the production of the various ionized and excited species, and the subsequent reactions of these species in time with one another or with available solutes (Cobut, 1993; Cobut et al., 1998; Frongillo et al., 1998; Hervé du Penhoat et al., 2000; Meesungnoen et al., 2001, 2003; Meesungnoen and Jay-Gerin, 2005a,b, 2009; Muroya et al., 2002, 2006; Plante et al., 2005; Autsavapromporn et al., 2007; Guzonas et al., 2009). A most recent version of the Sherbrooke codes, called IONLYS-IRT (Meesungnoen and Jay-Gerin, 2005a,b), is used in the present work.

Briefly, the IONLYS step-by-step simulation program models all the events of the physical and physicochemical stages in the track development. To take into account the effects of multiple ionizations under high-LET heavy-ion impact, the model incorporates double-, triple-, and quadruple ionization processes in single ion–water collisions (see above). The double ionization of water is assumed to lead to the production of $HO_2^{\bullet}/O_2^{\bullet-}$ through the intervention of oxygen atoms

(Kuppermann, 1967; LaVerne et al., 1986; Gardès-Albert et al., 1996; Olivera et al., 1998) formed in their ^3P ground state (Ferradini and Jay-Gerin, 1998), according to*

$$H_2O^{++} + 2H_2O \rightarrow 2H_3O^+ + O(^3P) \tag{14.19}$$

at very short times, followed by

$$^\bullet OH + O(^3P) \rightarrow HO_2^\bullet \tag{14.20}$$

As for the triple and quadruple ionizations of water molecules, they are assumed to lead directly to the formation of $HO_2^\bullet/O_2^{\bullet-}$ and O_2, respectively, by acid–base re-equilibration reactions, according to (Ferradini and Jay-Gerin, 1998; Meesungnoen and Jay-Gerin, 2005a)

$$H_2O^{\bullet3+} + 4H_2O \rightarrow 3H_3O^+ + HO_2^\bullet \tag{14.21}$$

and

$$H_2O^{4+} + 5H_2O \rightarrow 4H_3O^+ + O_2. \tag{14.22}$$

The complex spatial distribution of reactants at the end of the physicochemical stage, which is provided as an output of this program, is used directly as the starting point for the nonhomogeneous chemical stage. This third and final stage is covered by the program IRT, which employs the IRT method to model the chemical development that occurs during this stage and to simulate the formation of the measurable yields of chemical products. Its detailed implementation has been given previously (Frongillo et al., 1998). Only slight adjustments have been made in some reaction rate constants and the diffusion coefficients of reactive species to take account of the latest data available from the literature. The reaction scheme used in our IRT program for pure liquid water is given in Table 14.1.

To simulate the radiolysis of (deaerated) water under *acidic* conditions with IONLYS-IRT, we have used 0.4 M H_2SO_4 aqueous solutions[†] and supplemented the pure-water reaction scheme to include the seven reactions listed in Table 14.2, which account for the species present in irradiated sulfuric acid solutions. The details of our computational modeling are given elsewhere (Meesungnoen et al., 2001; Meesungnoen and Jay-Gerin, 2005b; Autsavapromporn et al., 2007). To summarize, the effects of the background concentrations of H^+ in solutions were added to the IRT program as pseudo first-order reactions. In addition, we have introduced the effects due to the ionic strength of the solutions for all reactions between ions, except for the peculiar bimolecular self-recombination of e_{aq}^- for which there is no evidence of any ionic strength effect (Schmidt and Bartels, 1995). The rate constants for those various reactions, corrected for these ionic strength effects, were used in the

---

* A fragmentation mechanism, based on the "Coulomb explosion" of multiply ionized water molecules observed in the gas phase, has also been proposed recently to explain the formation of $HO_2^\bullet/O_2^{\bullet-}$ and O_2 in the heavy-ion radiolysis of liquid water at high LET (Gervais et al., 2005, 2006; Gaigeot et al., 2007). This mechanism is discussed below in relation with reactions (14.19) through (14.22).

† At such a high concentration of H_2SO_4 (pH 0.46), the H^+ ions capture most, if not all, of the hydrated electrons in the expending spurs/tracks to form $H^\bullet$ atoms (e.g., Ferradini and Jay-Gerin, 2000):

$$e_{aq}^- + H^+ \rightarrow H^\bullet \tag{14.23}$$

with a bimolecular rate constant of ~1.12×10^{10} $M^{-1}s^{-1}$ (value corrected to take into account ionic strength effects at this acid concentration) (Meesungnoen et al., 2001).

TABLE 14.2
Reactions Added to the Pure Water Reaction Scheme to Simulate the Radiolysis of Deaerated Aqueous 0.4 M H$_2$SO$_4$ Solutions, at 25°C^a

Reaction	k (M^{-1} s^{-1})
$H^\bullet + SO_4^{\bullet-} \rightarrow HSO_4^-$	1.0×10^{10}
$H^\bullet + S_2O_8^{2-} \rightarrow SO_4^{\bullet-} + HSO_4^-$	2.5×10^7
$^\bullet OH + HSO_4^- \rightarrow H_2O + SO_4^{\bullet-}$	1.5×10^5
$e_{aq}^- + S_2O_8^{2-} \rightarrow SO_4^{\bullet-} + SO_4^{2-}$	1.2×10^{10}
$H_2O_2 + SO_4^{\bullet-} \rightarrow HO_2^\bullet + HSO_4^-$	1.2×10^7
$OH^- + SO_4^{\bullet-} \rightarrow {}^\bullet OH + SO_4^{2-}$	8.3×10^7
$SO_4^{\bullet-} + SO_4^{\bullet-} \rightarrow S_2O_8^{2-}$	4.4×10^8

Source: Autsavapromporn, N. et al., *Can. J. Chem.*, 85, 214, 2007.

^{a} The rate constants given here for the reactions between ions are at ionic strength equal zero.

calculations. Finally, in our simulations, the direct action of ionizing radiation on the sulfuric acid anions (mainly HSO$_4^-$) has been neglected, which is a reasonably good approximation.*

The influence of the LET on the yields of the various radiation-induced species produced in neutral water and in 0.4 M H$_2$SO$_4$ aqueous solutions at 25°C was investigated by varying the impacting ion energy from ~300 to 0.15 MeV (~0.3–70 keV/μm) for protons,† ~300 to 0.3 MeV/nucleon (~1.25–213 keV/μm) for ^{4}He^{2+}, ~300 to 1.25 MeV/nucleon (~12–604 keV/μm) for ^{12}C^{6+}, and ~300 to 3.2 MeV/nucleon (~25–938 keV/μm) for ^{20}Ne^{9+}. The calculations were performed by simulating short (~1.5–100 μm) ion track segments, over which the energy and LET of the ion are well defined and remain nearly constant. Typically, ~5,000–400,000 reactive chemical species were generated in those simulated track segments (depending on ion type and energy), thus ensuring only small statistical fluctuations in the determination of average yields. Finally, at the incident ion energies of interest here, interactions involving electron capture and loss by the moving ion (charge-changing collisions) have been neglected (Pimblott and LaVerne, 2002). The nuclear fragmentation of incident beam projectiles is also ignored.

14.5.2 MULTIPLE IONIZATION CROSS SECTIONS AND THE FATE OF MULTIPLY IONIZED WATER MOLECULES

14.5.2.1 Cross Sections for Multiple Ionization of Water Molecules

The successful Monte Carlo modeling of heavy-ion radiation effects in water relies on the knowledge of the cross sections of all interaction processes. The multiple ionization of a single water molecule induced by swift heavy-ion impact has been observed in many circumstances. However,

---

* In fact, in 0.4 M H$_2$SO$_4$, only ~3.5% of the total energy expended in the solution is initially absorbed by direct action on HSO$_4^-$ (rather than on H$_2$O) (assuming that the energy absorbed by each component is proportional to its electron fraction).

† To reproduce the effects of fast electron or ^{60}Co γ-radiolysis, we used ~300 MeV protons whose average LET value obtained in the simulations was ~0.3 keV/μm.

a survey of the literature shows that most of experiments have been performed with ions in water vapor but not in *liquid* water, as measurements with liquid are either impractical or very difficult. As well, all existing heavy-ion cross section calculations have been performed for water vapor. Due to this scarcity of reliable "condensed-phase" cross-section data, a quantitative description of MI effects in liquid water represents a challenging problem. In an effort to overcome these difficulties, we infer here the MI cross sections *indirectly* from measurements made in the liquid phase. As described previously (Meesungnoen and Jay-Gerin, 2005a), our strategy mainly consists in treating the ratio of the double-to-single ionization cross sections ($\alpha = \sigma_{di}/\sigma_{si}$, where σ_{si}, the "single" ionization cross section, is known) in our track simulations as an adjustable parameter chosen to best fit the available experimental values of $(G_{HO_2^{\bullet}} + G_{O_2})$ as a function of LET in the heavy-ion radiolysis of air-free aqueous $FeSO_4$ (1 mM)–$CuSO_4$ (10 mM) solutions under acidic (0.005 M H_2SO_4, pH ~ 2.1) conditions* (LaVerne et al., 1986; LaVerne and Schuler, 1987b; LaVerne, 1989). As for the triple and quadruple ionizations of water (which are shown, under the conditions of this study, to contribute only weakly to the production of $HO_2^{\bullet}/O_2^{\bullet-}$ and O_2), σ_{ti} and σ_{qi} are assumed to be equal to $\alpha^2\sigma_{si}$ and $\alpha^3\sigma_{si}$ ($\alpha < 0.5$), respectively, in accordance with the calculated ratios of σ_j/σ_1 (where σ_j is the cross section for the ejection of j electrons) for collisions of various heavy ions on gas-phase H_2O molecules (Champion, 2003; Champion et al., 2005; Gervais et al., 2005, 2006). The values of α so obtained are shown in Figure 14.6 for the four impacting ions considered in this

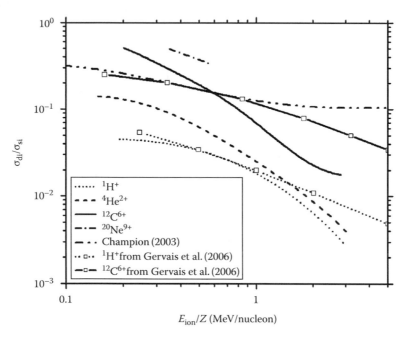

FIGURE 14.6 Plot of the ratio of double-to-single ionization cross sections ($\alpha = \sigma_{di}/\sigma_{si}$) against E_{ion}/Z, where E_{ion} is the ion energy per nucleon and Z is the projectile charge number. The short-dot, dash, solid, and dash-dot lines represent, respectively, the α values for $^1H^+$, $^4He^{2+}$, $^{12}C^{6+}$, and $^{20}Ne^{9+}$ impacting ions, obtained from our Monte Carlo simulations (see text for explanation). The dash-dot-dot line shows the α values reported by Champion (2003), and the symbols ($\square$) with dot and solid lines represent the α values calculated by Gervais et al. (2006) for protons and carbon ions, respectively. (From Meesungnoen, J. and Jay-Gerin, J.-P., *J. Phys. Chem. A*, 109, 6406, 2005a. With permission.)

---

* The principle of this (indirect) method to measure $HO_2^{\bullet}$ is to generate molecular oxygen by the reaction of $HO_2^{\bullet}$ with cupric ion (e.g., Hart, 1955; Donaldson and Miller, 1956). It should be noted here that any primary O_2, produced directly or by track reactions, will also be measured using this method. Therefore, the yields generally quoted for $HO_2^{\bullet}$ under these experimental conditions also include O_2 (Baverstock and Burns, 1976; LaVerne, 2004; Meesungnoen and Jay-Gerin, 2005a).

work (namely, $^1H^+$, $^4He^{2+}$, $^{12}C^{6+}$, and $^{20}Ne^{9+}$) as a function of (E_{ion}/Z), the ion energy per nucleon divided by the projectile charge number. The corresponding σ_{di}/σ_{si} values reported by Champion (2003) and Gervais et al. (2006) for various ions and gaseous water are also included in the figure for the sake of comparison. It is seen that our values of α depend on the type of the projectile ion (in contrast to Champion's values, which follow a unique law, independent of the impacting ion) and, for a given value of E_{ion}/Z, increase from protons ($Z = 1$) to neon ions ($Z = 9$). As can also be seen from Figure 14.6, our ratios of σ_{di}/σ_{si} compare reasonably well with those obtained from the MI cross section calculations of Gervais et al. (2006) for protons and carbon ions ($Z = 6$). Although such an accord is satisfactory here, it should be emphasized that the process of multiple ionization of polyatomic molecules (including H_2O) in heavy-ion collisions and in particular the effects due to the phase, are an area where fundamental information is incomplete or even completely lacking in some cases, and where more experimental and theoretical work is expected and clearly desirable.

14.5.2.2 Fate of Multicharged Water Cations

14.5.2.2.1 Acid–Base Re-Equilibration Reactions

Very little is known about the fate of multiply ionized water molecules *in solution*. In this work, the rearrangement of these thermodynamically unstable multiply charged water cations is treated following the general mechanism proposed by Ferradini and Jay-Gerin (1998), which assumes that, in *liquid* water, H_2O^{n+} dissociates through acid–base re-equilibration processes. For example, for the predominantly formed doubly ionized water molecules, the overall dissociation reaction

$$H_2O^{2+} + 2H_2O \rightarrow 2H_3O^+ + O, \tag{14.19}$$

can be regarded as resulting from a two-step process involving two deprotonation reactions (the two protons being taken away by water molecules), namely,

$$H_2O^{2+} + H_2O \rightarrow H_3O^+ + OH^+,$$

followed by

$$OH^+ + H_2O \rightarrow H_3O^+ + O.$$

Similar reactions can be written for the triply and quadruply charged water cations [see reactions (14.21) and (14.22)], and for higher degree of ionization ($n > 4$) (see Table 14.3).

14.5.2.2.2 Fragmentation Caused by "Coulomb Explosions"

In *gaseous* water, MI is followed by the fragmentation of the ionized target as a result of the so-called Coulomb explosion.* The mechanism of Coulomb explosion (e.g., Gemmell, 1980; Latimer, 1993) is based on the simplified scenario where two or more electrons from a molecule are suddenly removed by the incident projectile, so that the transiently formed highly ionized parent ion becomes unstable (due to the Coulomb repulsion between the positive atomic fragment ions (holes) that can repel one another apart) and rapidly dissociates into both charged and neutral fragments by bond breaking.

According to experimental results in the gas phase (e.g., Werner et al., 1995a,b; Olivera et al., 1998; Gobet et al., 2001; Siegmann et al., 2001; Pešić et al., 2004; Alvarado et al., 2005; Legendre et al., 2005; Luna and Montenegro, 2005; Sobocinski et al., 2006; Alvarado, 2007; Stolterfoht et al., 2007, and references therein), three main, competing water dication fragmentation channels have been identified (by the coincident detection of correlated fragments from a particular molecular breakup), namely,

---

* Fragmentation originating from violent binary collisions (involving small impact parameters) has also been observed (Stolterfoht et al., 2007).

TABLE 14.3
Radiolytic Products Inferred from the
Dissociation of Multiply Charged Water
Cations*

n	Products
1	$H_3O^+ + \cdot OH$
2	$2H_3O^+ + O$
3	$3H_3O^+ + HO_2\cdot$
4	$4H_3O^+ + O_2$
5	$5H_3O^+ + \cdot OH + O_2$
6	$6H_3O^+ + O + O_2$
7	$7H_3O^+ + HO_2\cdot + O_2$
8	$8H_3O^+ + 2O_2$
9	$9H_3O^+ + \cdot OH + 2O_2$
10	$10H_3O^+ + O + 2O_2$

Source: Ferradini, C., unpublished results, January 1994.
* (H_2O^{n+}, $n = 1-10$) (the molecule of water has 10 bound
electrons), as obtained by acid–base re-equilibration, in
the heavy-ion radiolysis of pure liquid water at high LET.

$$H_2O^{2+} \rightarrow H^+ + H^+ + O \quad (14.24)$$

$$H_2O^{2+} \rightarrow H^+ + OH^+ \quad (14.25)$$

$$H_2O^{2+} \rightarrow H^+ + O^+ + H\cdot, \quad (14.26)$$

while for ionization of higher multiplicity ($q \geq 3$), the dominating fragmentation pathway involves a three-body breakup of the multi-ionized water molecule as follows:

$$H_2O^{q+} \rightarrow H^+ + H^+ + O^{(q-2)+}. \quad (14.27)$$

Only very few data exist on the cross sections (or probabilities) for these various possible water polycation dissociation channels. For instance, Olivera et al. (1998) reported the values ~6.3×10^{-15}, 1.3×10^{-15}, and 1.5×10^{-15} cm^2 (with an estimated uncertainty of ~30%) for the fragment production channels (14.24) through (14.26) coming from the double ionization of water under 6.7 MeV/ nucleon Xe^{44+} ion impact, respectively. These values are found to be ~4–18 times less than the corresponding single ionization cross section measured by these authors (~2.4×10^{-14} cm^2). As for the triple ionization ($q = 3$), the cross section for the $H^+ + H^+ + O^+$ fragmentation channel (14.27) obtained by Olivera et al. (1998) amounts to 2.5×10^{-15} cm^2, an order of magnitude smaller than the cross section for single ionization fragment production. Werner et al. (1995b) also reported that, for 100–350 keV proton–H_2O collisions, the fragmentation channel (14.24) has a cross section about twice higher than the $H^+ + H^+ + O^+$ fragmentation (14.27), a result that is consistent with Olivera et al.'s (1998) data. In another, more recent study, Pešić et al. (2004) investigated the fragmentation of water molecules following interaction with 2–90 keV Ne^{p+} ions (p = 1, 3, 5, 7, and 9). These authors observed a strong dependence of the fragment ion production cross sections on the projectile charge state. They also found that their cross sections are almost two orders of magnitude larger than those reported by Werner et al. (1995b) for the fragmentation of H_2O following proton impact. Although a great deal of attention is currently being devoted by different laboratories to the studies of the influence of the type, energy, and charge state of the impinging ion on the ion-impact-induced fragmentation of water vapor, a detailed knowledge of the basic features of the explosive fragmentation pattern of multiply charged water cations remains fragmentary. Clearly, there is a need to explore these phenomena further.

In the *liquid* phase, it has been suggested that, following (single) ionization,* the $H_2O^{•+}$ hole can migrate rapidly by resonance electron transfer with neighboring water molecules until the ion–molecule reaction (14.14) localizes the hole (see above).† This proton transfer reaction (14.14) occurs on the time scale of $\sim 10^{-14}$ s (i.e., ~ 10 fs), time required for a single vibration of a water molecule. Unfortunately, very little is known about the different dissociation pathways of H_2O^{n+} ($n \geq 2$) in liquid water. To the best of our knowledge, the only previous studies devoted to this question are those of Olivera et al. (1998), Gaigeot et al. (2007), and more recently Tavernelli et al. (2008). Using arguments based on the ability of the medium to neutralize a charged species, Olivera et al. (1998) tentatively extrapolated their water vapor multi-ionization results to liquid water. Among other things, they concluded that, at least for the process of double ionization, the result of the multi-ionization in liquid water may not be a direct Coulomb explosion of the molecule, but rather a quick neutralization of the water polycation by the surrounding medium. In essence, this conclusion was based on a comparison of the time involved in a dissociative Coulomb explosion process‡ and some estimates of characteristic time scales of the competitive neutralization processes, such as hole migration, geminate electron–cation recombination, and proton transfer reactions (~ 1–10 fs; see, e.g., Mozumder and Magee, 1975; Goulet et al., 1990; Meesungnoen et al., 2002b), in combination with energy transfer to the various modes of excitation of the surrounding medium. One further argument against Coulomb explosion-induced fragmentation of the molecule (at least for the lowest degree of ionization) is the "cage effect" observed for molecules in the condensed phase (Franck and Rabinowitsch, 1934), which forces the ejected fragments to stay together (Olivera et al., 1998). This cage recombination effect may significantly alter product yields between the two phases (gas phase versus liquid).

In their recent molecular dynamics simulations of a doubly ionized water molecule in the condensed phase, Gaigeot et al. (2007) and Tavernelli et al. (2008) were able to confirm the formation of two H_3O^+ ions and an oxygen atom assuming a Coulomb explosion mechanism of fragmentation. Based on those calculations, these authors showed that double ionization leads to two main dissociation pathways (in proportion to 84% of all H_2O^{2+} fragmentations) for the production of oxygen atoms in liquid water, namely,

$$H_2O^{2+} \rightarrow H^+ + H^+ + O \xrightarrow{2H_2O} 2H_3O^+ + O \quad (55\%) \tag{14.28}$$

$$H_2O^{2+} \rightarrow H^+ + OH^+ \xrightarrow{2H_2O} 2H_3O^+ + O \quad (29\%) \tag{14.29}$$

---

* From the Heisenberg uncertainty principle $\Delta t \Delta E \sim \hbar$, the collision time required to know that an energy loss has taken place in ionization is, typically, $\sim 10^{-16}$ s.

† By itself, $H_2O^{•+}$ is stable, but in the presence of water, it dissociates, giving H_3O^+ and an $•OH$ radical.

‡ Most recent values of this time for the explosive dissociation of H_2O^{2+} in liquid water, estimated from molecular dynamics calculations and time-dependent density functional theory-based molecular dynamics simulations, are >1–15 fs (Gaigeot et al., 2007) and <4 fs (Tavernelli et al., 2008), respectively.

These authors also considered the three-body breakup of the doubly ionized water molecule (a minor process, contributing only 16% to the total fragmentation):

$$H_2O^{2+} \rightarrow H^+ + O^+ + H^{\cdot} \xrightarrow{3H_2O} 2H_3O^+ + H^{\cdot} + {\cdot}OH + O, \tag{14.30}$$

which, in addition to O atoms, leads to the formation of $H^{\cdot}$ and ${\cdot}OH$ radicals. Finally, for a higher degree of ionization ($q \geq 3$), they proposed the following fragmentation channel:

$$H_2O^{q+} \rightarrow H^+ + H^+ + O^{(q-2)+} \xrightarrow{2(q-1)H_2O} qH_3O^+ + (q-2){\cdot}OH + O. \tag{14.31}$$

Gervais et al. (2005, 2006) used this dissociation scheme (14.28) through (14.31) to simulate the formation of $HO_2^{\cdot}/O_2^{\cdot-}$ and O_2 in the heavy-ion radiolysis of liquid water at high LET.

    Whatever the mechanism that actually controls the dissociation of the multicharged water cations *in solution* (either acid–base re-equilibration or Coulomb explosion), it is worth noting that the fragments $2H_3O^+ + O$ originating from the two most dominant Coulomb-induced reaction channels (14.28) and (14.29) are identical to those formed via the corresponding dissociation pathway (14.19) based on acid–base re-equilibration reactions. However, these fragmentation mechanisms differ at the end in the spatial distribution of the H_3O^+ and atomic oxygen species taking account of the energy distributions that are involved in the Coulomb explosion processes. In fact, the Coulomb breakup of H_2O^{2+} leads to target fragments that should have some kinetic energy* and, therefore, move farther away from the interaction point than they would do in the case of acid–base re-equilibration (or proton transfer) processes. In the case of the two-body H^+–OH^+ dissociation channel (14.29), the total kinetic energy of the H^+ and OH^+ fragments ejected from collisions of H_2O with 5.9 MeV/nucleon Xe^{18+} and Xe^{43+} ions was measured to be about 6.5 eV (Siegmann et al., 2001). Similar results were obtained by Sobocinski et al. (2006) in He^{2+}–H_2O collisions at impact energies of 1 and 5 keV. The three-body dissociation channel $H^+ + H^+ + O$ (14.28) was also observed by these latter authors, who showed that it gives rise to H^+ fragments with kinetic energies also near 5 eV. For 4–23 keV H^+ and He^+ projectiles, more energetic fragment protons (14.5 eV) were identified by Alvarado et al. (2005) in the (minor) three-body H_2O^{2+} asymmetric breakup channel (14.30). These results tend to indicate that fragment ions emitted in the Coulomb explosion of H_2O^{2+} will most probably have energy lower than the lowest ionization and electronic excitation thresholds of (condensed) water, which are ~9 and 7.3 eV (Michaud et al., 1991; Bernas et al., 1997; Cobut et al., 1998), respectively. This in turn implies that these fragments should not be energetic enough to generate their own physical tracks in the radiolysis of water and, as a consequence, only their subsequent chemistry should be taken into account in the radiolytic processes.

14.5.3 Yields of $HO_2^{\cdot}/O_2^{\cdot-}$ Radicals

Although the $HO_2^{\cdot}/O_2^{\cdot-}$ radical is especially interesting in the context of high-LET, heavy-ion tracks (it is the major radical product), the reaction mechanism for its production at high LET is probably the most uncertain in water radiolysis. Several hypotheses have been proposed and discussed previously (see, e.g., Sims, 1978; Burns and Sims, 1981; Burns et al., 1981; LaVerne et al., 1986; LaVerne, 1989; Ferradini and Jay-Gerin, 1998; Baldacchino et al., 1998a; Olivera et al., 1998). Early studies assumed $HO_2^{\cdot}$ to be produced in the reaction as follows:

$$\cdot OH + H_2O_2 \rightarrow HO_2^{\cdot} + H_2O, \quad k = 2.87 \times 10^7 \text{ M}^{-1}\text{s}^{-1} \tag{14.32}$$

---

* The mutual Coulomb energy can be estimated from the simple model of a point-charge interaction between charges [i.e., $V = q_1q_2/(4\pi\varepsilon_0)\varepsilon_s r$, where ε_0 is the vacuum permittivity, ε_s is the dielectric constant of the medium, and r is the separation distance between the charges q_1 and q_2. Note that charges do not interact as strongly in a solvent of high dielectric constant (such as liquid water, with $\varepsilon_s \approx 79$ at room temperature)] under the assumption that charges are localized and can be treated as point charges. This potential energy is converted into kinetic energy of the separating fragments as the Coulomb explosion develops.

but diffusion–kinetic models of track theory (Appleby and Schwarz, 1969) and Monte Carlo simulations (Frongillo et al., 1996) have shown that this reaction is not fast enough to account quantitatively for the totality of the observed yields at high LET. Without excluding definitely other possible mechanisms, the alternative hypothesis of Ferradini and Jay-Gerin (1998), proposing that the increased production of $HO_2^{\bullet}/O_2^{\bullet-}$ by a multicharged heavy ion involves the multiple ionization of water, has recently received strong support from Monte Carlo track structure simulations (Frongillo et al., 1997; Meesungnoen et al., 2003; Meesungnoen and Jay-Gerin, 2005a,b; Gervais et al., 2005, 2006). This is clearly shown in Figure 14.7, which compares the experimental $HO_2^{\bullet}/O_2^{\bullet-}$ yields of Baldacchino et al. (1998a,b) in deaerated *neutral* water with our $G_{HO_2^{\bullet}/O_2^{\bullet-}}$ values calculated (at 10^{-6} s) with IONLYS-IRT, with and without multiple ionization of water, for the case of irradiating $^1H^+$, $^4He^{2+}$, $^{12}C^{6+}$, and $^{20}Ne^{9+}$ ions up to ~900 keV/μm, at 25°C. As can be seen, our curves of $G_{HO_2^{\bullet}/O_2^{\bullet-}}$ versus LET, obtained by ignoring the mechanism of MI of water, cannot account for more than a small fraction of the measured $HO_2^{\bullet}/O_2^{\bullet-}$ escape yields at high LET, even if reaction (14.32) and all other possible reactions that can produce $HO_2^{\bullet}/O_2^{\bullet-}$ such as, e.g.:

$$e_{aq}^- + O_2 \rightarrow O_2^{\bullet-} \quad k = 1.74 \times 10^{10} \ M^{-1} s^{-1} \tag{14.33}$$

$$H^{\bullet} + O_2 \rightarrow HO_2^{\bullet} \quad k = 2.1 \times 10^{10} \ M^{-1} s^{-1} \tag{14.34}$$

$$^{\bullet}OH + O(^3P) \rightarrow HO_2^{\bullet} \quad k = 2.02 \times 10^{10} \ M^{-1} s^{-1} \tag{14.35}$$

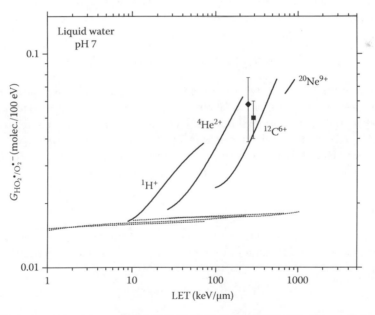

FIGURE 14.7 Variation of the primary $HO_2^{\bullet}/O_2^{\bullet-}$ yield (in molecule/100 eV) of the radiolysis of deaerated liquid water by $^1H^+$, $^4He^{2+}$, $^{12}C^{6+}$, and $^{20}Ne^{9+}$ ions as a function of LET up to ~900 keV/μm, at neutral pH and 25°C. The solid lines represent the results of our Monte Carlo simulations incorporating the double-, triple-, and quadruple ionizations of water molecules, obtained at 10^{-6} s (see text). The short-dot lines correspond to our $G_{HO_2^{\bullet}/O_2^{\bullet-}}$ values calculated as a function of LET without including the mechanism of MI of water. Experimental yields (pH ≈ 7): (♦) $^{36}S^{16+}$ ions (LET ~ 250 keV/μm) (Baldacchino et al., 1998a), and (■) $^{40}Ar^{18+}$ ions (LET ~ 290 keV/μm) (Baldacchino et al., 1998b). (From Meesungnoen, J. and Jay-Gerin, J.-P., *J. Phys. Chem. A*, 109, 6406, 2005a. With permission.)

(see Table 14.1) are included in the simulations. However, when the mechanism of MI is accounted for, we observe a marked increase of $G_{HO_2^\bullet/O_2^{\bullet-}}$ with increasing LET, which is in very good agreement with experiment.* In essence, $HO_2^\bullet/O_2^{\bullet-}$ is found to be produced *mainly* at an early stage of radiolysis by $O(^3P)$ atoms, formed predominantly from the double ionization of water,† reacting with $^\bullet OH$ in the track of heavy ions [see reactions (14.19) and (14.20)]. It is worth noting that such a mechanism involving, at high LET, the production of oxygen atoms followed by their intratrack reaction with $^\bullet OH$ was originally proposed by Kuppermann as early as 1967, and later by LaVerne et al. (1985).

$HO_2^\bullet/O_2^{\bullet-}$ appears as an exception to the general observation that an increase of the radiation LET increases the molecular yields of H_2 and H_2O_2, at the expense of the radical yields e_{aq}^-, $^\bullet OH$, and $H^\bullet$. In fact, $HO_2^\bullet/O_2^{\bullet-}$ is a radical, but it behaves like a molecular product, increasing in yield with increasing LET. When compared with such highly reactive species as $^\bullet OH$, the hydroperoxyl and superoxide anion radicals are far less reactive with DNA in aqueous solution. However, the radiobiological consequences of the production of $HO_2^\bullet/O_2^{\bullet-}$ could be important since both species can generate more reactive species. For example, the dismutation of $HO_2^\bullet/O_2^{\bullet-}$ generates H_2O_2 and O_2 in a pH-dependent fashion (e.g., Bielski et al., 1985). In living cells, the presence of Cu/Zn-superoxide dismutase (SOD) greatly accelerates this $HO_2^\bullet/O_2^{\bullet-}$ dismutation and then facilitates the cytotoxicity of H_2O_2 via the generation of $^\bullet OH$ radicals by the superoxide-assisted Fenton reaction:

$$Fe^{2+} + H_2O_2 \rightarrow Fe^{3+} + {}^\bullet OH + OH^- \tag{14.36}$$

or, equivalently, the so-called transition metal (iron/copper)-catalyzed Haber–Weiss reaction:

$$H_2O_2 + O_2^{\bullet-} \rightarrow O_2 + {}^\bullet OH + OH^- \tag{14.37}$$

(e.g., Halliwell and Gutteridge, 1999). Another cytotoxic species is *peroxynitrite* ($ONOOH/ONOO^-$; $pK_a = 6.8$), produced *in vivo* by the fast reaction of $HO_2^\bullet/O_2^{\bullet-}$ with nitric oxide ($^\bullet NO$) that is released in $^\bullet NO$-generating cells in response to oxidative stress:

$$HO_2^\bullet/O_2^{\bullet-} + {}^\bullet NO \rightarrow ONOOH/ONOO^-. \tag{14.38}$$

The rate constant of this reaction (~ 4–$6 \times 10^9 \, M^{-1} \, s^{-1}$) is comparable to that for the enzymatic dismutation (in the presence of SOD) of $O_2^{\bullet-}$ at physiological pH ($\sim 4 \times 10^9 \, M^{-1} \, s^{-1}$) (e.g., Jay-Gerin and Ferradini, 2000; Lymar and Poskrebyshev, 2003). $ONOO^-$ and its protonated form (peroxynitrous acid) are powerful oxidants, capable of attacking a wide range of biological targets (e.g., Beckman and Koppenol, 1996; Halliwell and Gutteridge, 1999; Radi et al., 2000; von Sonntag, 2006). A component of superoxide toxicity is almost certainly due to peroxynitrite.

14.5.4 Yields of H_2O_2

Hydrogen peroxide is the main oxidizing molecular product formed during the radiolysis of water. It is formed primarily by combination reactions of two $^\bullet OH$ radicals produced in the radiolytic decomposition of water:

$$^\bullet OH + {}^\bullet OH \rightarrow H_2O_2. \tag{14.39}$$

---

* In their recent Monte Carlo studies, Gervais et al. (2005, 2006) also found a very smooth increase in the primary yields of $HO_2^\bullet/O_2^{\bullet-}$ with increasing LET in the absence of multiple ionization of water. However, upon incorporation of the MI mechanism, their calculations show, in close similarity with our results, a sharp increase in $G_{HO_2^\bullet/O_2^{\bullet-}}$ above ~ 100–$200 \, keV/\mu m$ (depending on the impinging ion studied), in good agreement with experiment.

† Independent simulations incorporating the sole mechanism of double ionization of water molecules have shown that triply and quadruply charged water ions make, in fact, only a minor contribution to $G_{HO_2^\bullet/O_2^{\bullet-}}$ ($\sim 15\%$ for $\sim 555 \, keV/\mu m$ $^{12}C^{6+}$ ions) (Meesungnoen and Jay-Gerin, 2005a). Similar findings have also been obtained by Gervais et al. (2005, 2006) in their Monte Carlo simulations of water radiolysis by swift ions.

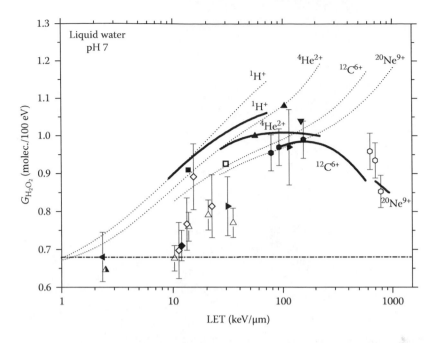

FIGURE 14.8 Variation of the primary H_2O_2 yield (in molecule/100 eV) of the radiolysis of deaerated liquid water by $^1H^+$, $^4He^{2+}$, $^{12}C^{6+}$, and $^{20}Ne^{9+}$ ions as a function of LET up to ~900 keV/μm, at *neutral* pH and 25°C. The solid lines represent the results of our Monte Carlo simulations obtained at 10^{-6} s (see text) incorporating the double, triple, and quadruple ionizations of water molecules. The short-dot lines correspond to our $G_{H_2O_2}$ values calculated as a function of LET without including the mechanism of MI of water. The symbols are experimental data from various laboratories [see Meesungnoen and Jay-Gerin (2005a) for references], including the recently measured $G_{H_2O_2}$ values of Yamashita et al. (2008) for $^4He^{2+}$ (◄), $^{12}C^{6+}$ (◇), and $^{20}Ne^{10+}$ (►) ions. The dash-dot line represents the limiting primary H_2O_2 yield obtained with ^{60}Co γ-rays or fast electrons (~0.68 molecule/100 eV) (Ferradini and Jay-Gerin, 1999).

From our Monte Carlo track structure simulations (Meesungnoen and Jay-Gerin, 2005a,b), performed at neutral pH (Figure 14.8) and also in aqueous 0.4 M H_2SO_4 solutions (pH ~ 0.46) (Figure 14.9), it is shown that the yields of H_2O_2 are sensitive to multiple ionization for irradiating high-LET heavy ions. As can be seen in Figures 14.8 and 14.9, significant differences exist between calculations including MI and those restricted to single ionizations. In the absence of MI, our calculated values of $G_{H_2O_2}$ continuously increase with increasing LET. However, when the mechanism of MI of water is incorporated in the simulations, the curves for $G_{H_2O_2}$ as a function of LET, first rise, then bend downward (in the case of incident protons) and, at higher LET, reach a maximum (in the case of impacting $^4He^{2+}$ and $^{12}C^{6+}$ ions), after which they fall (see Figure 14.8). While this maximum is relatively narrow for $^4He^{2+}$ ions, it is more pronounced for $^{12}C^{6+}$ ions. Its position also slightly shifts to higher LET as the ion charge increases. In fact, if we consider the curve of $G_{H_2O_2}$ versus LET that includes the ensemble of our four calculated $G_{H_2O_2}$(LET) curves for the different ions studied ($^1H^+$, $^4He^{2+}$, $^{12}C^{6+}$, and $^{20}Ne^{9+}$), the overall maximum can be estimated around 100–200 keV/μm, which is in good accord with experiment (Bibler, 1975; Sims, 1978; Burns and Sims, 1981; Pastina and LaVerne, 1999; Wasselin-Trupin et al., 2002). Moreover, for each ion investigated, this maximum of $G_{H_2O_2}$ occurs precisely at the point where $G_{HO_2^\bullet/O_2^{\bullet-}}$ begins to rise sharply, showing, in agreement with previous experimental data (Sims, 1978; Burns and Sims, 1981), that the yields of $HO_2^\bullet/O_2^{\bullet-}$ and H_2O_2 are closely linked. This is readily explained by the fact that H_2O_2 is formed within the tracks mainly by the combination reaction (14.39). As $^\bullet OH$ reacts with $O(^3P)$ by reaction (14.35), this latter reaction competes with reaction (14.39), thereby causing a drop in the observed hydrogen peroxide yields at high LET.

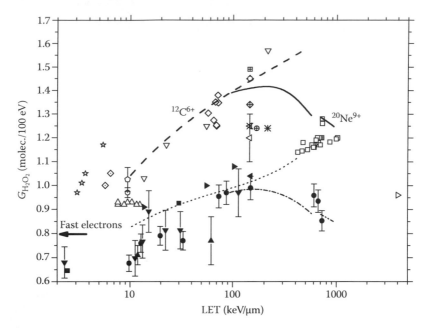

FIGURE 14.9 Variation of the primary H_2O_2 yield (in molecule/100 eV) of the radiolysis of deaerated 0.4 M H_2SO_4 aqueous solutions (pH 0.46) by $^{12}C^{6+}$ and $^{20}Ne^{9+}$ ions as a function of LET up to ~900 keV/μm at 25°C. The solid lines represent the results of our Monte Carlo simulations incorporating the double-, triple-, and quadruple ionizations of water molecules, obtained at 10^{-6} s. The dashed line corresponds to our $G_{H_2O_2}$ values for impacting $^{12}C^{6+}$ ions calculated as a function of LET without including the mechanism of MI of water (corresponding results for $^{20}Ne^{9+}$ are not shown in the figure for clarity). For the sake of comparison, our primary H_2O_2 yields of the radiolysis of deaerated neutral liquid water by $^{12}C^{6+}$ (dash-dot line) and $^{20}Ne^{9+}$ (dash-dot-dot line) ions, calculated when multiple ionization of water is incorporated in the simulations (see Figure 14.8), are also included in the figure. The dotted line corresponds to our results for $^{12}C^{6+}$ ions obtained in the absence of MI of water (see Figure 14.8). The open and closed symbols are experimental data obtained from various laboratories for 0.4 M H_2SO_4 solutions and for neutral liquid water, respectively [see Meesungnoen and Jay-Gerin (2005a,b) for references; note that the experimental $G_{H_2O_2}$ values of Yamashita et al. (2008) for neutral water (see Figure 14.8) are indicated here by the symbol (▼)]. The arrow represents the limiting primary H_2O_2 yield obtained with ^{60}Co γ-rays or fast electrons (~0.80 molecule/100 eV) (Ferradini and Jay-Gerin, 2000).

As is seen in Figures 14.8 and 14.9, there is a significant amount of scatter in the yields of H_2O_2 measured by different investigators as a function of LET (see, for reviews, McCracken et al., 1998; LaVerne, 2004). This uncertainty in H_2O_2 yields after exposure to high-LET radiation becomes even greater when pH effects are considered (Figure 14.9). However, a wide variety of studies indicate that, at a given LET, the production of H_2O_2 is higher in acidic solutions than in neutral water. In fact, the measurements of the H_2O_2 yield as a function of LET in deaerated 0.4 M H_2SO_4 aqueous solutions (pH 0.46) clearly show a well-defined maximum of ~1.4 molecule/100 eV (i.e., ~45% greater in magnitude than that found in neutral solutions) around ~180–200 keV/μm, with the yield decreasing thereafter to the value of 0.96 molecule/100 eV (Bibler, 1975) at an LET close to ~4000 keV/μm. As can be seen in Figure 14.9, this dependence of $G_{H_2O_2}$ on LET is *quantitatively* reproduced by our Monte Carlo simulations using irradiating $^{12}C^{6+}$ and $^{20}Ne^{9+}$ ions and incorporating the mechanism of multiple ionization of water. Furthermore, the extrapolation of our H_2O_2 escape yield values to higher LET reproduces very well Bibler's (1975) value of $G_{H_2O_2}$ for dissolved ^{252}Cf fission recoil fragments (LET ~ 4000 keV/μm). As can also be seen in the figure, our curve for $G_{H_2O_2}$ as a function of LET for $^{12}C^{6+}$ ions calculated without including MI increases continuously with increasing LET, in disagreement with experiment. Since our simulations show a clear decrease of $G_{H_2O_2}$ above ~200 keV/μm (for acidic solutions particularly), one could wonder whether the limit of H_2O_2 yields would not be zero for infinite LET. This would apparently differ from a recent assumption by LaVerne (2004), suggesting that the limiting value of $G_{H_2O_2}$ at infinite LET should be ~1 molecule/100 eV, somewhat similar for both neutral and acidic water. In the absence of more experimental work, we are led to conclude that the question of the limiting value of $G_{H_2O_2}$ at very high LET for both neutral and acidic liquid water is still open.

Similar findings have been obtained by Gervais et al. (2006) from their simulations of the radiolysis of neutral liquid water with carbon and argon ions. These authors found that, above ~100 keV/μm, $G_{H_2O_2}$ is sensitive to multiple ionization. In fact, they observed a marked decrease in the values of $G_{H_2O_2}$ (up to 30% for their largest LET investigated) upon incorporating MI in the calculations. However, in contrast to our results, they did not find any maximum in their $G_{H_2O_2}$ curves versus LET calculated in the presence of the MI mechanism. The source of such a difference in the dependence of $G_{H_2O_2}$ on LET between our work and that of Gervais et al. (2006) is examined below.

As seen above, the three (Coulomb-induced) fragmentation channels (14.28) through (14.30) that Gervais et al. (2006) considered for the (predominantly formed) doubly ionized water molecules all lead to the production of atomic oxygen, either directly or after neutralization of O^+ with water. This is essentially equivalent to the acid–base re-equilibration reaction (14.19) used in our study and which also produces O atoms. In other words, the difference in the formation mechanism of O atoms should not be the reason of the observed difference in $G_{H_2O_2}$ between our work and that of Gervais et al. (2006). We have confirmed this point by incorporating in our simulation program the three dissociation channels of H_2O^{2+} used by Gervais et al. (2006) [in place of reaction (14.19)]; indeed, no significant change in our $G_{H_2O_2}$ values was observed.*

In another facet of our work, we have assumed that all O atoms produced in reaction (14.19) are in their ^3P ground state. By contrast, Gervais et al. (2006) tentatively considered in their simulations that 85% of O atoms are generated in an excited state, while 15% are left in the ^3P state. These latter authors also showed that changing the proportion of excited states from 80% to 20% does not change the variation of the $HO_2^{\cdot}/O_2^{\cdot-}$ and O_2 yields with the LET, but only changes their absolute value by a small constant factor (Gervais et al., 2005, 2006). This is readily explained by the fact that when the concentration of radicals becomes larger and larger, the reaction of an oxygen atom, either in an excited state or in its ground state, with another radical becomes more and more likely. Also, at high LET, the relative number of available water molecules should decrease because of the high density of interactions of water molecules with impacting ions along their paths.

Actually, the difference between our Monte Carlo simulations and those of Gervais et al. (2006) regarding the LET dependence of $G_{H_2O_2}$ can be shown to originate mainly from the different ratios for the double-to-single ionization cross sections ($\alpha = \sigma_{di}/\sigma_{si}$) used by the two groups (see Figure 14.6). As can be seen from this figure for the case of $^{12}C^{6+}$ ions (shown for the sake of comparison), at low impacting ion energies (i.e., at high LET), our σ_{di}/σ_{si} values are higher than those used by Gervais et al. (2006). As a result, our simulations produce more doubly ionized water molecules and, therefore, more O atoms. This increased formation of oxygen atoms (as compared to Gervais et al.'s (2006) simulations) leads to an increased intervention of reaction (14.35) with an exacerbated competition with reaction (14.39), thereby causing the drop in the hydrogen peroxide yields, which is seen at high LET (Figures 14.8 and 14.9).

14.5.5 Production of O_2 and the "Oxygen-in-the-Track" Hypothesis

An especially important aspect of the action of radiations on living systems—in relation to the radiotherapy of tumors—is that cells that are hypoxic at the time of irradiation are generally much less damaged by a given dose of x- or γ-radiation than those that are well oxygenated. In other

---

* It is important to note that the respective proportions (or branching ratios) of the various fragmentation pathways are not critical for the simulations, mainly because they all generate atomic oxygen.

words, the absence of oxygen enhances the resistance of cells to low-LET ionizing radiations. By contrast, for high-LET radiation, the survival of tumor cells is practically the same in the presence or in the absence of O_2. As mentioned above, several different hypotheses (not all mutually exclusive) have been invoked in the literature to account for this experimental finding. At present, however, the mechanism of radiobiological action underlying the reduction of the oxygen enhancement ratio with increasing LET is not yet completely clear. Among the hypotheses advanced to explain the LET dependence of the OER either independently, or as parts of a single mechanism, the "oxygen-in-the-track" hypothesis proposes that O_2 is generated *in situ* by heavy-ion tracks passing through water, in quantities that depend on the radiation quality: the denser the track, the greater the effective concentration of oxygen. This hypothesis has often been invoked for a variety of biological systems. Other possible hypotheses have also been considered, including "interacting-radicals" (Alper, 1956; Alper and Howard-Flanders, 1956; Howard-Flanders, 1958), "oxygen depletion in the vicinity of heavy-ion tracks" (Kiefer, 1990; Stuglik, 1995), and, more recently, the "lesion complexity" (Ward, 1994) and "radical multiplicity" (Michael and Prise, 1996). In addition to these proposed radiation–chemical mechanisms and irrespective of which of them is true, there is evidence pointing to the involvement of biological factors, such as cellular (enzymatic) repair processes, to explain, at least in part, why the OER goes down as LET rises (e.g., Kiefer, 1990). Even though a definite conclusion has not yet been forthcoming, it has been suggested that there is probably more than one mechanism responsible for the lower OER found at high LET.

From the viewpoint of pure radiation chemistry, the "oxygen-in-the-track" hypothesis presupposes that O_2 is a product of the radiolysis of water at high LET (we recall that, for radiation of low LET, oxygen is *not* a radiolytic product). Oxygen generated in this way has been identified in several previous experiments (Lefort, 1955; Allen, 1961; Bibler, 1975; Baverstock and Burns, 1976; Burns et al., 1981; LaVerne and Schuler, 1987b, 1996). Most remarkably, Bibler (1975) estimated from his studies of the radiolysis of 0.4 M H_2SO_4 solutions with ^{252}Cf fission fragments that G_{O_2} could be as high as ~0.3–0.8 molecule/100 eV at an LET of ~4000 keV/μm (based on the material balance). For the biological point of view, such "track" oxygen would then be available immediately after the passage of the incident ion to react with adjacent potential cellular lesions (mainly measured as alterations in chromosomal DNA) formed by the same ionizing particle (Baverstock and Burns, 1981). These combination events of oxygen with DNA (bases or deoxyribosyl backbone) radicals (converting them into the corresponding peroxyl radicals)

$$DNA^{\bullet} + O_2 \rightarrow DNA\text{-}O_2^{\bullet} \tag{14.40}$$

result in "nonrestorable" lesions (oxygen is said to "fix" or make permanent the radiation lesion) (e.g., Ewing, 1998; Hall and Giaccia, 2006; von Sonntag, 2006).* This "nonrestorability" of DNA lesions formed with oxygen's chemical participation ultimately increases the amount of stable DNA damage, and thus the extent of cellular lethality, independently of external (or added) oxygen concentration. Hence, the radiolytic formation of O_2 in the tracks of heavy ions would likely be a determinant of increased radiation sensitivity.

In support to the "oxygen-in-the-track" hypothesis, our Monte Carlo simulations suggest that there is, indeed, an excess production *in situ* of molecular oxygen in high-LET, heavy-ion tracks at early time, which is not observed with lower LET radiations (Meesungnoen and Jay-Gerin, 2005a, 2009). They show, in particular, that the mechanism of multiple (mainly double) ionization of water, even though infrequent relative to single ionization events, is responsible for such an O_2 production

---

* The "oxygen fixation" hypothesis is widely regarded as the most satisfactory explanation of why O_2 is a radiation sensitizer. According to this hypothesis, oxygen sensitizes cells because DNA lesions that are produced by ionizing radiation with the participation of oxygen are difficult or impossible to restore chemically back to an original, undamaged state (repair enzymes cannot adequately work on such sites).

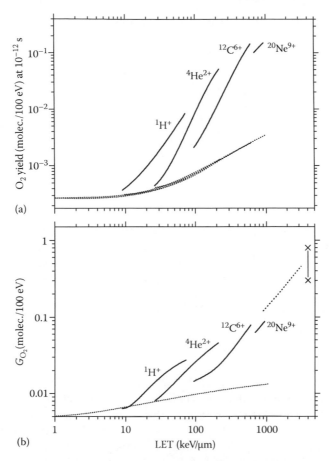

FIGURE 14.10 Variation of the initial (a) and primary (b) yields of O_2 (obtained at 10^{-12} and 10^{-6} s, respectively) (in molecule/100 eV) of the radiolysis of pure, deaerated liquid water by $^1H^+$, $^4He^{2+}$, $^{12}C^{6+}$, and $^{20}Ne^{9+}$ ions as a function of LET, at neutral pH and 25°C. The solid lines represent the results of our Monte Carlo simulations including the double-, triple-, and quadruple ionizations of water molecules. The dot lines correspond to our calculated values in the absence of MI of water. The yields $G_{O_2} \sim 0.3$ and 0.8 molecule/100 eV estimated by Bibler (1975) (×) in the ^{252}Cf fission fragment radiolysis of 0.4 M H_2SO_4 solutions at very high LET (~4000 keV/μm) are also shown in part (b) of the figure. The dot line (more widely spaced dots) is drawn here as a guide for the eye to show that extrapolation to higher LET of our calculated O_2 escape yields for the four ions studied reproduces well Bibler's estimates. (From Meesungnoen, J. and Jay-Gerin, J.-P., *J. Phys. Chem. A*, 109, 6406, 2005a. With permission.)

(through the formation of oxygen atoms).* This is clearly shown in Figure 14.10, which compares, for the four irradiating ions studied here ($^1H^+$, $^4He^{2+}$, $^{12}C^{6+}$, and $^{20}Ne^{9+}$), the initial (at 10^{-12} s) (Figure 14.10a) and the primary (at 10^{-6} s) (Figure 14.10b) yields of O_2 as a function of LET, obtained with and without the incorporation of multiple ionization of water. As we can see, in the absence of the multiple ionization mechanism, these yields remain very small throughout the range of LET studied and show only a slight gradual increase with increasing LET. In contrast, upon incorporation of the double-, triple-, and quadruple ionizations of water molecules in the simulations, we find a steep increase in both the initial and primary O_2 yield values at high LET. In fact, as shown in our

---

* Similar findings have also been obtained by Gervais et al. (2005, 2006) in their recent Monte Carlo simulations of water radiolysis by swift ions.

previous work (Meesungnoen and Jay-Gerin, 2005a, 2009), there is a chemical production of O_2 at early times arising from the reactions:

$$O(^3P) + O(^3P) \rightarrow O_2 \qquad (14.41)$$

or

$$^\bullet OH + O(^3P) \rightarrow HO_2^\bullet \qquad (14.42)$$

followed by

$$HO_2^\bullet + O(^3P) \rightarrow O_2 + {^\bullet OH}, \qquad (14.43)$$

whereas in the time interval ~10^{-12} to 10^{-6} s (i.e., during spur/track expansion), O_2 is formed mainly by

$$^\bullet OH + HO_2^\bullet \rightarrow O_2 + H_2O \qquad (14.44)$$

and, but to a much lower extent, by

$$^\bullet OH + O_2^{\bullet -} \rightarrow O_2 + OH^-. \qquad (14.45)$$

In addition, as observed with $HO_2^\bullet/O_2^{\bullet -}$ and H_2O_2, the O_2 yields at early times as well as at the microsecond time scale are both LET and irradiating-ion dependent. As is clearly seen in Figure 14.10, upon incorporating the MI mechanism, our calculations indicate that for different incident ions of equal LET but different velocities, G_{O_2} decreases as the ion velocity increases, a behavior that is akin to the other molecular yields (note that, according to the Bethe stopping power equation (see Equation 14.5), for two different ions of equal LET, the one with the higher charge will have the higher velocity).

To further test the veracity of the "oxygen-in-the-track" hypothesis, we need to know the extent to which O_2 is formed in heavy-ion tracks. This information is obviously of great importance for biologically based treatment planning in radiotherapy (recall here that, for low-LET radiation, high levels of tumor hypoxia have a positive correlation with treatment failure for many human cancers) (e.g., Hall and Giaccia, 2006). If it is correct to invoke that the greater efficiency of high-LET radiation for the inactivation of hypoxic tumor cells can be attributed to the formation of oxygen in tracks, then estimates of the track concentration of O_2 as a function of time could help to quantify and predict changes in the observed differences in radiosensitivity for hypoxic and better-oxygenated tumor cells. Using the G-values for O_2 obtained from our Monte Carlo simulations of the radiolysis of deaerated, neutral water for two representative irradiating particles, namely, 24 MeV $^{12}C^{6+}$ (LET ~ 490 keV/μm) and 4.8 MeV $^4He^{2+}$ (LET ~ 94 keV/μm) ions, including the mechanism of MI of water, we can calculate the concentrations of O_2 generated *in situ* in the tracks of those ions as a function of time (Meesungnoen and Jay-Gerin, 2009). In fact, assuming that the oxygen molecules are produced *evenly* in a cylinder whose initial radius r_0 is equal to the radius of the physical track "core" (which corresponds to the tiny radial region within the first few nanometers around the impacting ion path, at ~10^{-13} s) (Magee and Chatterjee, 1987; Mozumder, 1999), the track concentrations of O_2 are simply derived from (Baverstock and Burns, 1976; Pimblott and LaVerne, 2002; Meesungnoen and Jay-Gerin, 2005a, 2009)

$$[O_2] \approx G(O_2) \times \left(\frac{LET}{\pi r^2} \right) \qquad (14.46)$$

where

$$r^2 \approx r_0^2 + 4Dt \qquad (14.47)$$

represents the change with time of r_0 due to the diffusive expansion of the track. Here, r_0 obtained from our simulations is taken to be ~2.0 nm for the two irradiating ions considered (Muroya et al., 2006), t is the time, and D is the diffusion coefficient of O_2 [$D = 2.42 \times 10^{-9}$ m²/s at 25°C (Hervé du Penhoat et al., 2000)]. The values of [O_2] so obtained are shown in Figure 14.11b along with the corresponding values of $G(O_2)$ (Figure 14.11a) for the two different heavy ions studied, as well as for 300 MeV protons (corresponding to the low-LET limiting case of ^{60}Co γ-radiolysis or fast electrons, shown in the figure for the sake of comparison), as a function of time over the range 10^{-12} to 10^{-6} s. As can be seen from Figure 14.11a and b, $G(O_2)$ and [O_2] at early times (~10^{-12} s) are, respectively, ~0.1 molecule/100 eV and 63 mM for 24 MeV ^{12}C^{6+} ions, and ~0.0087 molecule/100 eV and 1.1 mM for 4.8 MeV ^{4}He^{2+} ions. These initial track concentrations of O_2 are much higher (in fact, more than three orders of magnitude for the carbon ions) than the oxygen levels present in normally oxygenated tumor regions (which are found to vary widely, from zero to above 20 μM) and in hypoxic tumor regions (where a large proportion of the vessels have near zero oxygenation) (Pogue et al., 2001), as well as in (most) normal human cells (typical values are around 30 μM) (Halliwell and Gutteridge, 1999; Hall and Giaccia, 2006; von Sonntag, 2006). The results in Figure 14.11b also show that, for the ^{12}C^{6+} and ^{4}He^{2+} ions, the concentration of O_2 remains almost constant as a function of time

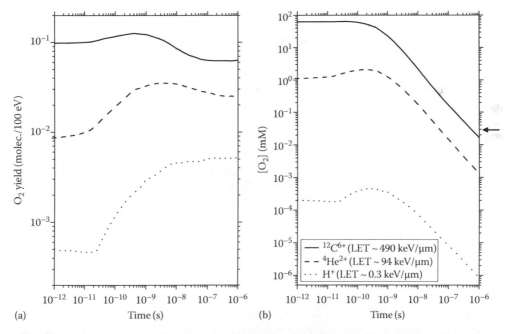

FIGURE 14.11 (a) Time dependences of the O_2 yields (in molecule/100 eV) calculated from our Monte Carlo simulations of the radiolysis of pure, air-free liquid water at pH 7, 25°C, in the interval 10^{-12} to 10^{-6} s, for impacting 24 MeV ^{12}C^{6+} (LET ~ 490 keV/μm, solid line) and 4.8 MeV ^{4}He^{2+} (LET ~ 94 keV/μm, dash line) ions, including multiple ionization of water molecules. The dot line corresponds to our calculated $G(O_2)$ values for the low-LET limiting case of 300 MeV protons (LET ~ 0.3 keV/μm) and is shown in the figure for the sake of comparison. (b) Time dependences of the corresponding track concentrations of O_2 (in mM) calculated from Equations 14.46 and 14.47 for the different irradiating ions considered, using r_0 ~ 2.0 nm, $D(O_2) = 2.42 \times 10^{-9}$ m²/s, and the $G(O_2)$ values shown in (a). Typical O_2 concentrations in normal human cells (~30 μM) are shown by the arrow on the right of the figure. (From Meesungnoen, J. and Jay-Gerin, J.-P., *Radiat. Res.*, 171, 379, 2009. With permission.)

between $\sim 10^{-12}$ and $(1-3) \times 10^{-10}$ s, and then decreases steeply to ~ 2.2 mM and 175 μM at $\sim 10^{-8}$ s, ~ 165 and 15 μM at $\sim 10^{-7}$ s, and ~ 17 and 1.3 μM at the microsecond time scale, respectively. These predictions of the track oxygen concentrations in fact compare very well with the experimental estimates of Baverstock and Burns (1976, 1981) and Baverstock et al. (1978) who found that, for ~ 100 keV/μm α-particles, $[O_2]$ would be 4.2 mM at early times and 51 μM after 10^{-8} s.* These O_2 concentrations provide clear evidence that supports the hypothesis that O_2 is engendered in substantial yields in the tracks of densely ionizing radiation. As a consequence, the present calculations strongly suggest that this excess production *in situ* of O_2 is a key intervening factor in the increased biological efficiency of radiations of high LET with, in turn, profound consequences in radiobiology, oxidative processes, and other applications. In particular, they largely plead in favor of the "oxygen in the heavy-ion track" hypothesis to account (at least in part) for the observed decrease of OER with LET.

14.6 CONCLUDING REMARKS

The Monte Carlo track structure simulations presented in this chapter strongly support the importance of the role of multiple ionization of water in the high-LET, heavy-ion radiolysis of pure, deaerated liquid water. In particular, the mechanism of double ionization is found to largely predominate at high LET, whereas it is insignificant at low LET. Upon incorporation of the multiple ionizations of water, a good overall agreement is obtained between the calculated yields and the available experimental data. Most remarkably, Monte Carlo calculations quantitatively reproduce the large increase observed in the $HO_2^{\bullet}/O_2^{\bullet-}$ yield with increasing LET, confirming a hypothesis put forward several years ago by Ferradini and Jay-Gerin (1998) that MI of water should intervene in the high-LET, heavy-ion radiolysis of water in the form of the production of $HO_2^{\bullet}/O_2^{\bullet-}$ radicals. With the exception of protons, our simulations also simultaneously predict a maximum in the curve for $G_{H_2O_2}$ as a function of LET around 180–200 keV/μm, in accord with experiment (although there are large uncertainties in measured H_2O_2 yields for different kinds of radiation). In addition, for each ion investigated, this maximum of $G_{H_2O_2}$ occurs precisely at the point where $G_{HO_2^{\bullet}/O_2^{\bullet-}}$ begins to rise sharply, suggesting, in agreement with previous experimental data, that the yields of $HO_2^{\bullet}/O_2^{\bullet-}$ radicals and H_2O_2 are closely linked. In accordance with experiment, we find that the maximum observed in $G_{H_2O_2}$ in the case of the radiolysis of 0.4 M H_2SO_4 solutions is more pronounced than that found in neutral solutions and about 45% greater in magnitude. Further experiments on the H_2O_2 yields for LET above 1000 keV/μm would be highly desirable to help clarify the question of the limiting value of $G_{H_2O_2}$ at very high LET for both neutral and acidic liquid water.

Another point needing special attention in the problem at hand is the determination of reliable "condensed-phase" multiple ionization cross-section data. In fact, existing heavy-ion cross sections have generally been obtained in water vapor but not in liquid water. To overcome these difficulties, we have adopted in our work a strategy consisting of inferring MI cross sections *indirectly* from measurements ($G_{HO_2^{\bullet}} + G_{O_2}$ in the heavy-ion radiolysis of air-free aqueous $FeSO_4$–$CuSO_4$ solutions at pH ≈ 2.1) made in the *liquid* phase. Using these cross sections in our Monte Carlo simulations allowed us to reproduce, among other things, the maximum in $G_{H_2O_2}$ versus LET that is observed experimentally. Other published reports have used cross-section data derived directly from gas-phase measurements or from gas phase-based calculations with various theoretical adjustments to take into account the effects due to the phase, but they could not reproduce any maximum in the variation of $G_{H_2O_2}$ with LET. More experimental and theoretical work is clearly needed in this area.

---

* Incidentally, another interesting remark that can be inferred from Figure 14.11b concerns the time dependence of $[O_2]$ for 300 MeV incident protons (LET ~ 0.3 keV/μm). In fact, our calculated O_2 concentrations in the tracks vary in the range ~ 0.2–0.45 μM between 10^{-12} and $\sim 2.5 \times 10^{-9}$ s and then fall to reach a value of $\sim 8.9 \times 10^{-4}$ μM at 10^{-6} s. As expected, these concentrations are too low to sensitize cells at low LET, based on the findings of Millar et al. (1979), who found that, for cobalt-60 γ radiation, sensitization of Chinese hamster cells (line V-79-753B) by oxygen occurs progressively as the concentration of oxygen (present at the time of irradiation) is increased from ~ 0.46 μM to 1.4 mM (see also von Sonntag, 2006).

For four representative irradiating ions (^{1}H$^+$, ^{4}He^{2+}, ^{12}C^{6+}, and ^{20}Ne^{9+}), our results show, upon incorporation of the mechanism of MI of water in the simulations, a substantial production *in situ* of "track" O$_2$ that is available immediately after the passage of the ion, pleading in favor of the radiobiological "oxygen in the heavy-ion track" hypothesis. Such an excess production of molecular oxygen in high-LET, heavy-ion tracks is not observed with lower LET radiations. We have shown, in particular, that the dissociation of doubly ionized water molecules is responsible, to a large extent, for this O$_2$ production *via* the formation of oxygen atoms. The radiolytic formation, at high LET, of substantial amounts of O$_2$ in heavy-ion tracks can have profound consequences in radiobiology and, in turn, be of great practical relevance in radiotherapy. In fact, being immediately available after the passage of the incident ion, this track oxygen can readily react, in *cells*, with adjacent potentially lethal lesions formed by the same ionizing particle (e.g., the so-called O$_2$ fixation of DNA damage) to produce the observed reduction in the OER with increasing LET. Even if the present calculations cannot exclude the possibility of other mechanisms that may also be involved to fully account for the decrease of OER with LET (or equivalently for the increased efficiency of high-LET radiations for the inactivation of tumor hypoxic cells), they appear to clearly indicate that the excess production *in situ* of O$_2$ in the high-LET tracks is a key intervening factor.

In contrast to $G_{HO_2^\bullet/O_2^{\bullet-}}$, $G_{H_2O_2}$, and G_{O_2}, the incorporation of the mechanism of MI in the simulations has almost no effect on the primary yields of e$_{aq}^-$, $^\bullet$OH, H$^\bullet$, and H$_2$ (data not shown here) (Meesungnoen and Jay-Gerin, 2005a; Gervais et al., 2006). As expected, $G_{e_{aq}^-}$ and $G_{\cdot OH}$ diminish steeply as the LET is increased, and for the highest LET studied, there is almost no e$_{aq}^-$ and $^\bullet$OH surviving at the microsecond time scale. Our calculated primary H$^\bullet$ atom yield values present a slight maximum near 6.5 keV/μm before decreasing steeply at higher LET (similarly to the behavior of other radicals). As for molecular hydrogen, G_{H_2} is found to rise monotonically with increasing LET for the different impacting ions studied. The available experimental data are generally well reproduced by our calculated escape yields over the whole LET range studied.

As a consequence of the fact that the LET does not exactly or uniquely define the densities of species in heavy-ion tracks and, therefore, the subsequent radiation–chemical effects, we observe, for all the species formed during the radiolysis, an irradiating-ion dependence of the yields at a given LET. In other words, two different incident ions of equal LET but different velocities produce different yields due to differences in the effects of track structure on the radiation chemistry.

Finally, *under high-LET irradiation conditions*, Equation 14.6 (traditionally given for low-LET radiolysis of water) should be rewritten as follows:

$$G_{-H_2O}H_2O \rightsquigarrow G_{e_{aq}^-}e_{aq}^- + G_{H^\bullet}H^\bullet + G_{\cdot OH}\,^\bullet OH + G_{H^+}H^+ + G_{OH^-}OH^-$$

$$+ G_{H_2}H_2 + G_{H_2O_2}H_2O_2 + G_{HO_2^\bullet/O_2^{\bullet-}}HO_2^\bullet/O_2^{\bullet-} + G_{O_2}O_2 + \cdots \tag{14.48}$$

(with the same conventions as before). The overall electroneutrality and stoichiometry equations connecting these product yields then are

$$G_{e_{aq}^-} + G_{OH^-} = G_{H^+} \tag{14.49}$$

$$G_{e_{aq}^-} + G_{H^\bullet} + 2G_{H_2} = G_{\cdot OH} + 2G_{H_2O_2} + 3G_{HO_2^\bullet/O_2^{\bullet-}} + 4G_{O_2} + \cdots, \tag{14.50}$$

while the net yield of water decomposition is given by

$$G_{-H_2O} = G_{e_{aq}^-} + G_{H^\bullet} + 2G_{H_2} - G_{HO_2^\bullet/O_2^{\bullet-}} - 2G_{O_2}$$

$$= G_{\cdot OH} + 2G_{H_2O_2} + 2G_{HO_2^\bullet/O_2^{\bullet-}} + 2G_{O_2}. \tag{14.51}$$

Of course, the material balance does not specify the source of $HO_2^\bullet/O_2^{\bullet-}$ and O_2. Our present Monte Carlo track structure simulations show that multiple ionization of water can successfully explain both the large yields of $HO_2^\bullet/O_2^{\bullet-}$ and the excess O_2 produced in the heavy-ion radiolysis of water at high LET.

ACKNOWLEDGMENTS

This work is dedicated to the memory of Professeur Christiane Ferradini, who together with her colleague Jacques Pucheault, was a pioneer in the study of the radiolysis of water with heavy ions.

The authors are grateful to Dr. Norman V. Klassen for reviewing the manuscript. This work is supported by the Natural Sciences and Engineering Research Council of Canada.

REFERENCES

Allen, A. O. 1948. Radiation chemistry of aqueous solutions. *J. Phys. Colloid Chem.* 52: 479–490.

Allen, A. O. 1961. *The Radiation Chemistry of Water and Aqueous Solutions.* Princeton, NJ: D. Van Nostrand Co.

Alper, T. 1956. The modification of damage caused by primary ionization of biological targets. *Radiat. Res.* 5: 573–586.

Alper, T. and Bryant, P. E. 1974. Reduction in oxygen enhancement ratio with increase in LET: Tests of two hypotheses. *Int. J. Radiat. Biol.* 26: 203–218.

Alper, T. and Howard-Flanders, P. 1956. Role of oxygen in modifying the radiosensitivity of *E. coli* B. *Nature (London)* 178: 978–979.

Alvarado, F. 2007. Ion induced radiation damage on the molecular level. PhD thesis, University of Groningen, Groningen, the Netherlands.

Alvarado, F., Hoekstra, R., and Schlathölter, T. 2005. Dissociation of water molecules upon keV H^+- and He^{q+}-induced ionization. *J. Phys. B: At. Mol. Opt. Phys.* 38: 4085–4094.

Amichai, O. and Treinin, A. 1969. Chemical reactivity of $O(^3P)$ atoms in aqueous solution. *Chem. Phys. Lett.* 3: 611–613.

Anderson, D. W. 1984. *Absorption of Ionizing Radiation.* Baltimore, MD: University Park Press.

Anderson, A. R. and Hart, E. J. 1961. Molecular product and free radical yields in the decomposition of water by protons, deuterons, and helium ions. *Radiat. Res.* 14: 689–704.

Appleby, A. 1989. Effects of early track structure on the radiation chemistry of water irradiated with heavy ions. *Radiat. Phys. Chem.* 34: 121–127.

Appleby, A. and Schwarz, H. A. 1969. Radical and molecular yields in water irradiated by γ rays and heavy ions. *J. Phys. Chem.* 73: 1937–1941.

Autsavapromporn, N., Meesungnoen, J., Plante, I., and Jay-Gerin, J.-P. 2007. Monte Carlo simulation study of the effects of acidity and LET on the primary free-radical and molecular yields of water radiolysis. Application to the Fricke dosimeter. *Can. J. Chem.* 85: 214–229.

Baldacchino, G., Le Parc, D., Hickel, B., Gardès-Albert, M., Abedinzadeh, Z., Jore, D., Deycard, S., Bouffard, S., Mouton, V., and Balanzat, E. 1998a. Direct observation of HO_2/O_2^- free radicals generated in water by a high-linear energy transfer pulsed heavy-ion beam. *Radiat. Res.* 149: 128–133.

Baldacchino, G., Bouffard, S., Balanzat, E., Gardès-Albert, M., Abedinzadeh, Z., Jore, D., Deycard, S., and Hickel, B. 1998b. Direct time-resolved measurement of radical species formed in water by heavy ions irradiation. *Nucl. Instrum. Methods Phys. Res. B* 146: 528–532.

Ballarini, F., Biaggi, M., Merzagora, M., Ottolenghi, A., Dingfelder, M., Friedland, W., Jacob, P., and Paretzke, H. G. 2000. Stochastic aspects and uncertainties in the prechemical and chemical stages of electron tracks in liquid water: A quantitative analysis based on Monte Carlo simulations. *Radiat. Environ. Biophys.* 39: 179–188.

Barendsen, G. W. 1968. Responses of cultured cells, tumours and normal tissues to radiations of different linear energy transfer. *Curr. Top. Radiat. Res. Q.* 4: 293–356.

Bass, A. D. and Sanche, L. 2003. Dissociative electron attachment and charge transfer in condensed matter. *Radiat. Phys. Chem.* 68: 3–13.

Baverstock, K. F. and Burns, W. G. 1976. Primary production of oxygen from irradiated water as an explanation for decreased radiobiological oxygen enhancement at high LET. *Nature (London)* 260: 316–318.

Baverstock, K. F. and Burns, W. G. 1981. Oxygen as a product of water radiolysis in high-LET tracks. II. Radiobiological implications. *Radiat. Res.* 86: 20–33.

Baverstock, K. F., Burns, W. G., and May, R. 1978. The oxygen in the track hypothesis: Microdosimetric implications. In *Sixth Symposium on Microdosimetry*, Vol. 2, Brussels, Belgium, May 22–26, 1978, J. Booz and H. G. Ebert (eds.), pp. 1151–1158. London, U.K.: Harwood Academic Publishers.

Beckman, J. S. and Koppenol, W. H. 1996. Nitric oxide, superoxide, and peroxynitrite: The good, the bad, and the ugly. *Am. J. Physiol.: Cell Physiol.* 271: C1424–C1437.

Bednář, J. 1985. Electronic excitations in condensed biological matter. *Int. J. Radiat. Biol.* 48: 147–166.

Bernas, A., Ferradini, C., and Jay-Gerin, J.-P. 1996. Électrons en excès dans les milieux polaires homogènes et hétérogènes. *Can. J. Chem.* 74: 1–23.

Bernas, A., Ferradini, C., and Jay-Gerin, J.-P. 1997. On the electronic structure of liquid water: Facts and reflections. *Chem. Phys.* 222: 151–160.

Bethe, H. 1930. Zur Theorie des Durchgangs schneller Korpuskularstrahlen durch Materie. *Ann. Physik (Leipzig)* 5: 325–400.

Bethe, H. A. and Ashkin, J. 1953. Passage of radiations through matter. In *Experimental Nuclear Physics*, Vol. 1, E. Segrè (ed.), pp. 166–357. New York: Wiley.

Bibler, N. E. 1975. Radiolysis of 0.4 M sulfuric acid solutions with fission fragments from dissolved californium-252. Estimated yields of radical and molecular products that escape reactions in fission fragment tracks. *J. Phys. Chem.* 79: 1991–1995.

Bichsel, H. and Hiraoka, T. 1992. Energy loss of 70 MeV protons in elements. *Nucl. Instrum. Methods Phys. Res. B* 66: 345–351.

Biedenkapp, D., Hartshorn, L. G., and Bair, E. J. 1970. The $O(^1D) + H_2O$ reaction. *Chem. Phys. Lett.* 5: 379–380.

Bielski, B. H., Cabelli, D. E., Arudi, R. L., and Ross, A. B. 1985. Reactivity of HO_2/O_2^- radicals in aqueous solution. *J. Phys. Chem. Ref. Data* 14: 1041–1100.

Bjergbakke, E. and Hart, E. J. 1971. Oxygen formation in the γ-ray irradiation of Fe^{2+}-Cu^{2+} solutions. *Radiat. Res.* 45: 261–273.

Burns, W. G. and Sims, H. E. 1981. Effect of radiation type in water radiolysis. *J. Chem. Soc. Faraday Trans. 1* 77: 2803–2813.

Burns, W. G., May, R., and Baverstock, K. F. 1981. Oxygen as a product of water radiolysis in high-LET tracks. I. The origin of the hydroperoxyl radical in water radiolysis. *Radiat. Res.* 86: 1–19.

Burton, M. 1969. Radiation chemistry: A godfatherly look at its history and its relation to liquids. *Chem. Eng. News* 47: 86–96.

Buxton, G. V. 2004. The radiation chemistry of liquid water: Principles and applications. In *Charged Particle and Photon Interactions with Matter: Chemical, Physicochemical, and Biological Consequences with Applications*, A. Mozumder and Y. Hatano (eds.), pp. 331–363. New York: Marcel Dekker.

Champion, C. 2003. Multiple ionization of water by heavy ions: A Monte Carlo approach. *Nucl. Instrum. Methods Phys. Res. B* 205: 671–676.

Champion, C., L'Hoir, A., Politis, M. F., Fainstein, P. D., Rivarola, R. D., and Chetioui, A. 2005. A Monte Carlo code for the simulation of heavy-ion tracks in water. *Radiat. Res.* 163: 222–231.

Chatterjee, A. and Holley, W. R. 1993. Computer simulation of initial events in the biochemical mechanisms of DNA damage. *Adv. Radiat. Biol.* 17: 181–226.

Chatterjee, A. and Schaefer, H. J. 1976. Microdosimetric structure of heavy ion tracks in tissue. *Radiat. Environ. Biophys.* 13: 215–227.

Chattorji, D. 1976. *The Theory of Auger Transitions*. New York: Academic Press.

Christophorou, L. G., McCorkle, D. L., and Christodoulides, A. A. 1984. Electron attachment processes. In *Electron-Molecule Interactions and Their Applications*, L. G. Christophorou (ed.), Vol. 1, pp. 477–617. Orlando, FL: Academic Press.

Clifford, P., Green, N. J. B., Oldfield, M. J., Pilling, M. J., and Pimblott, S. M. 1986. Stochastic models of multi-species kinetics in radiation-induced spurs. *J. Chem. Soc. Faraday Trans. 1* 82: 2673–2689.

Cobut, V. 1993. Simulation Monte Carlo du transport d'électrons non relativistes dans l'eau liquide pure et de l'évolution du milieu irradié: rendements des espèces créées de 10^{-15} à 10^{-7} s. PhD thesis, Université de Sherbrooke, Sherbrooke, Canada.

Cobut, V., Jay-Gerin, J.-P., Frongillo, Y., and Patau, J. P. 1996. On the dissociative electron attachment as a potential source of molecular hydrogen in irradiated liquid water. *Radiat. Phys. Chem.* 47: 247–250.

Cobut, V., Frongillo, Y., Patau, J. P., Goulet, T., Fraser, M.-J., and Jay-Gerin, J.-P. 1998. Monte Carlo simulation of fast electron and proton tracks in liquid water. I. Physical and physicochemical aspects. *Radiat. Phys. Chem.* 51: 229–243.

Cocke, C. L. and Olson, R. E. 1991. Recoil ions. *Phys. Rep.* 205: 153–219.

Curie, P. and Debierne, A. 1901. Sur la radio-activité induite et les gaz activés par la radium. *Comptes Rendus Acad. Sci. Paris* 132: 768–770.

Curtis, S. B., Schilling, W. A., Tenforde, T. S., Crabtree, K. E., Tenforde, S. D., Howard, J., and Lyman, J. T. 1982. Survival of oxygenated and hypoxic tumor cells in the extended-peak regions of heavy charged-particle beams. *Radiat. Res.* 90: 292–309.

Debierne, A. 1914. Recherches sur les gaz produits par les substances radioactives. Décomposition de l'eau. *Ann. Phys. (Paris)* 2: 97–127.

Dingfelder, M. 2006. Track structure: Time evolution from physics to chemistry. *Radiat. Prot. Dosim.* 122: 16–21.

Dingfelder, M., Inokuti, M., and Paretzke, H. G. 2000. Inelastic-collision cross sections of liquid water for interactions of energetic protons. *Radiat. Phys. Chem.* 59: 255–275.

Donaldson, D. M. and Miller, N. 1956. The action of α-particles on solutions containing ferrous and cupric ions. *Trans. Faraday Soc.* 52: 652–659.

Draganić, I. G. and Draganić, Z. D. 1971. *The Radiation Chemistry of Water*. New York: Academic Press.

Duane, W. and Scheuer, O. 1913. Recherches sur la décomposition de l'eau par les rayons α. *Le Radium (Paris)* 10: 33–46.

Elliot, A. J., Chenier, M. P., Ouellette, D. C., and Koslowsky, V. T. 1996. Temperature dependence of g values for aqueous solutions irradiated with 23 MeV $^2H^+$ and 157 MeV $^7Li^{3+}$ ion beams. *J. Phys. Chem.* 100: 9014–9020.

Evans, R. D. 1955. *The Atomic Nucleus*. Malabar, FL: Krieger Publishing Company.

Ewing, D. 1998. The oxygen fixation hypothesis: A reevaluation. *Am. J. Clin. Oncol.* 21: 355–361.

Faraggi, M. and Désalos, J. 1969. Effect of positively charged ions on the "molecular" hydrogen yield in the radiolysis of aqueous solutions. *Int. J. Radiat. Phys. Chem.* 1: 335–344.

Ferradini, C. 1979. Actions chimiques des radiations ionisantes. *J. Phys. Chim.* 76: 636–644.

Ferradini, C. and Jay-Gerin, J.-P. 1990. Hypothesis of a possible chemical fate for the incompletely relaxed electron in water and alcohols. *Chem. Phys. Lett.* 167: 371–373.

Ferradini, C. and Jay-Gerin, J.-P. 1998. Does multiple ionization intervene for the production of $HO_2^•$ radicals in high-LET liquid water radiolysis? *Radiat. Phys. Chem.* 51: 263–267.

Ferradini, C. and Jay-Gerin, J.-P. 1999. La radiolyse de l'eau et des solutions aqueuses: historique et actualité. *Can. J. Chem.* 77: 1542–1575.

Ferradini, C. and Jay-Gerin, J.-P. 2000. The effect of pH on water radiolysis: A still open question. A minireview. *Res. Chem. Intermed.* 26: 549–565.

Franck, J. and Rabinowitsch, E. 1934. Some remarks about free radicals and the photochemistry of solutions. *Trans. Faraday Soc.* 30: 120–131.

Freeman, G. R. 1987. Ionization and charge separation in irradiated materials. In *Kinetics of Nonhomogeneous Processes*, G. R. Freeman (ed.), pp. 19–87. New York: Wiley.

Frongillo, Y., Fraser, M.-J., Cobut, V., Goulet, T., Jay-Gerin, J.-P., and Patau, J. P. 1996. Évolution des espèces produites par le ralentissement de protons rapides dans l'eau liquide: Simulation fondée sur l'approximation des temps de réaction indépendants. *J. Chim. Phys.* 93: 93–102.

Frongillo, Y., Goulet, T., Fraser, M.-J., Jay-Gerin, J.-P., Cobut, V., and Patau, J. P. 1997. The effect of the radiation LET on the production of HO_2 radicals in irradiated liquid water: A Monte Carlo simulation study. In Paper Presented at the *45th Annual Meeting of the Radiation Research Society*, Providence, RI (Book of Abstracts, p. 81).

Frongillo, Y., Goulet, T., Fraser, M.-J., Cobut, V., Patau, J. P., and Jay-Gerin, J.-P. 1998. Monte Carlo simulation of fast electron and proton tracks in liquid water. II. Nonhomogeneous chemistry. *Radiat. Phys. Chem.* 51: 245–254.

Gaigeot, M. P., Vuilleumier, R., Stia, C., Galassi, M. E., Rivarola, R., Gervais, B., and Politis, M. F. 2007. A multi-scale ab initio theoretical study of the production of free radicals in swift ion tracks in liquid water. *J. Phys. B: At. Mol. Opt. Phys.* 40: 1–12.

Ganguly, A. K. and Magee, J. L. 1956. Theory of radiation chemistry. III. Radical reaction mechanism in the tracks of ionizing radiations. *J. Chem. Phys.* 25: 129–134.

Gardès-Albert, M., Jore, D., Abedinzadeh, Z., Rouscilles, A., Deycard, S., and Bouffard, S. 1996. Réduction du tétranitrométhane par les espèces primaires formées lors de la radiolyse de l'eau par des ions lourds Ar^{18+}. *J. Chim. Phys.* 93: 103–110.

Garrett, B. C., Dixon, D. A., Camaioni, D. M., Chipman, D. M., Johnson, M. A., Jonah, C. D., Kimmel, G. A., Miller, J. H., Rescigno, T. N., Rossky, P. J., Xantheas, S. S., Colson, S. D., Laufer, A. H., Ray, D., Barbara, P. F., Bartels, D. M., Becker, K. H., Bowen, K. H., Jr., Bradforth, S. E., Carmichael, I., Coe, J. V., Corrales, L. R.,

Cowin, J. P., Dupuis, M., Eisenthal, K. B., Franz, J. A., Gutowski, M. S., Jordan, K. D., Kay, B. D., LaVerne, J. A., Lymar, S. V., Madey, T. E., McCurdy, C. W., Meisel, D., Mukamel, S., Nilsson, A. R., Orlando, T. M., Petrik, N. G., Pimblott, S. M., Rustad, J. R., Schenter, G. K., Singer, S. J., Tokmakoff, A., Wang, L.-S., Wittig, C., and Zwier, T. S. 2005. Role of water in electron-initiated processes and radical chemistry: Issues and scientific advances. *Chem. Rev.* 105: 355–389.

Gäumann, T. and Schuler, R. H. 1961. The radiolysis of benzene by densely ionizing radiations. *J. Phys. Chem.* 65: 703–704.

Gemmell, D. S. 1980. Determining the stereochemical structures of molecular ions by "Coulomb-explosion" techniques with fast (MeV) molecular ion beams. *Chem. Rev.* 80: 301–311.

Gervais, B., Beuve, M., Olivera, G. H., Galassi, M. E., and Rivarola, R. D. 2005. Production of HO_2 and O_2 by multiple ionization in water radiolysis by swift carbon ions. *Chem. Phys. Lett.* 410: 330–334.

Gervais, B., Beuve, M., Olivera, G. H., and Galassi, M. E. 2006. Numerical simulation of multiple ionization and high LET effects in liquid water radiolysis. *Radiat. Phys. Chem.* 75: 493–513.

Giesel, F. 1902. Ueber Radium und radioactive Stoffe. *Ber. Deutsch. Chem. Ges. (Berlin)* 35: 3608–3611.

Giesel, F. 1903. Ueber den Emanationskörper aus Pechblende und über Radium. *Ber. Deutsch. Chem. Ges. (Berlin)* 36: 342–347.

Gobet, F., Farizon, B., Farizon, M., Gaillard, M. J., Carré, M., Lezius, M., Scheier, P., and Märk, T. D. 2001. Total, partial, and electron-capture cross sections for ionization of water vapor by 20–150 keV protons. *Phys. Rev. Lett.* 86: 3751–3754.

Goulet, T. and Jay-Gerin, J.-P. 1988. Thermalization distances and times for subexcitation electrons in solid water. *J. Phys. Chem.* 92: 6871–6874.

Goulet, T. and Jay-Gerin, J.-P. 1989. Thermalization of subexcitation electrons in solid water. *Radiat. Res.* 118: 46–62.

Goulet, T. and Jay-Gerin, J.-P. 1992. On the reactions of hydrated electrons with $^{\cdot}OH$ and H_3O^+. Analysis of photoionization experiments. *J. Chem. Phys.* 96: 5076–5087.

Goulet, T., Patau, J. P., and Jay-Gerin, J.-P. 1990. Influence of the parent cation on the thermalization of subexcitation electrons in solid water. *J. Phys. Chem.* 94: 7312–7316.

Goulet, T., Fraser, M.-J., Frongillo, Y., and Jay-Gerin, J.-P. 1998. On the validity of the independent reaction times approximation for the description of the nonhomogeneous kinetics of liquid water radiolysis. *Radiat. Phys. Chem.* 51: 85–91.

Gray, L. H., Conger, A. D., Ebert, M., Hornsey, S., and Scott, O. C. A. 1953. The concentration of oxygen dissolved in tissues at the time of irradiation as a factor in radiotherapy. *Br. J. Radiol.* 26: 638–648.

Green, N. J. B., Pilling, M. J., Pimblott, S. M., and Clifford, P. 1990. Stochastic modeling of fast kinetics in a radiation track. *J. Phys. Chem.* 94: 251–258.

Guzonas, D., Tremaine, P., and Jay-Gerin, J.-P. 2009. Chemistry control challenges in a supercritical water-cooled reactor. *PowerPlant Chem.* 11: 284–291.

Hall, E. J. and Giaccia, A. J. 2006. *Radiobiology for the Radiologist*, 6th edn. Philadelphia, PA: Lippincott, Williams & Wilkins.

Halliwell, B. and Gutteridge, J. M. C. 1999. *Free Radicals in Biology and Medicine*, 3rd edn. Oxford, NY: Oxford University Press.

Hart, E. J. 1955. Radiation chemistry of aqueous ferrous sulfate-cupric sulfate solutions. Effect of γ-rays. *Radiat. Res.* 2: 33–46.

Hart, E. J. and Platzman, R. L. 1961. Radiation chemistry. In *Mechanisms in Radiobiology*, Vol. 1, M. Errera and A. Forssberg (eds.), pp. 93–257. New York: Academic Press.

Heller, J. M., Jr., Hamm, R. N., Birkhoff, R. D., and Painter, L. R. 1974. Collective oscillation in liquid water. *J. Chem. Phys.* 60: 3483–3486.

Hervé du Penhoat, M.-A., Goulet, T., Frongillo, Y., Fraser, M.-J., Bernat, Ph., and Jay-Gerin, J.-P. 2000. Radiolysis of liquid water at temperatures up to 300°C: A Monte Carlo simulation study. *J. Phys. Chem. A* 104: 11757–11770.

Howard-Flanders, P. 1958. Physical and chemical mechanisms in the injury of cells by ionizing radiations. *Adv. Biol. Med. Phys.* 6: 553–603.

Hunt, J. W. 1976. Early events in radiation chemistry. In *Advances in Radiation Chemistry*, Vol. 5, M. Burton and J. L. Magee (eds.), pp. 185–315. New York: Wiley.

ICRU Report 16. 1970. *Linear Energy Transfer*. Washington, DC: International Commission on Radiation Units and Measurements.

ICRU Report 31. 1979. *Average Energy Required to Produce an Ion Pair*. Washington, DC: International Commission on Radiation Units and Measurements.

ICRU Report 55. 1996. *Secondary Electron Spectra from Charged Particle Interactions*. Bethesda, MD: International Commission on Radiation Units and Measurements.

Inokuti, M. 1971. Inelastic collisions of fast charged particles with atoms and molecules: The Bethe theory revisited. *Rev. Mod. Phys.* 43: 297–347.

Jay-Gerin, J.-P. and Ferradini, C. 2000. Are there protective enzymatic pathways to regulate high local nitric oxide (NO) concentrations in cells under stress conditions. *Biochimie* 82: 161–166.

Jay-Gerin, J.-P., Lin, M., Katsumura, Y., He, H., Muroya, Y., and Meesungnoen, J. 2008. Effect of water density on the absorption maximum of hydrated electrons in sub- and supercritical water up to 400°C. *J. Chem. Phys.* 129: 114511.

Jonah, C. D. 1995. A short history of the radiation chemistry of water. *Radiat. Res.* 144: 141–147.

Jonah, C. D., Miller, J. R., and Matheson, M. S. 1977. The reaction of the precursor of the hydrated electron with electron scavengers. *J. Phys. Chem.* 81: 1618–1622.

Kaplan, I. G. and Miterev, A. M. 1987. Interaction of charged particles with molecular medium and track effects in radiation chemistry. *Adv. Chem. Phys.* 68: 255–386.

Kara-Michailova, E. and Lea, D. E. 1940. The interpretation of ionization measurements in gases at high pressures. *Proc. Cambridge Phil. Soc.* 36: 101–126.

Katz, R. 1970. RBE, LET and z/β^α. *Health Phys.* 18: 175.

Katz, R. 1978. Track structure theory in radiobiology and in radiation detection. *Nucl. Track Detect.* 2: 1–28.

Kernbaum, M. 1910. Sur la décomposition de l'eau par divers rayonnements. *Le Radium (Paris)* 7: 242.

Kiefer, J. 1990. *Biological Radiation Effects*. Berlin, Germany: Springer-Verlag.

Kimmel, G. A., Orlando, T. M., Vézina, C., and Sanche, L. 1994. Low-energy electron-stimulated production of molecular hydrogen from amorphous water ice. *J. Chem. Phys.* 101: 3282–3286.

Klassen, N. V. 1987. Primary products in radiation chemistry. In *Radiation Chemistry: Principles and Applications*, Farhataziz and M. A. J. Rodgers (eds.), pp. 29–64. New York: VCH Publishers.

Kraft, G. and Krämer, M. 1993. Linear energy transfer and track structure. *Adv. Radiat. Biol.* 17: 1–52.

Kreipl, M. S., Friedland, W., and Paretzke, H. G. 2009. Time- and space-resolved Monte Carlo study of water radiolysis for photon, electron and ion irradiation. *Radiat. Environ. Biophys.* 48: 11–20.

Kroh, J. (ed.) 1989. *Early Developments in Radiation Chemistry*. Cambridge, U.K.: The Royal Society of Chemistry.

Kuppermann, A. 1967. Diffusion model of the radiation chemistry of aqueous solutions. In *Radiation Research: Proceedings of the Third International Congress of Radiation Research*, Cortina d'Ampezzo, Italy, June 26–July 2, 1966, G. Silini (ed.), pp. 212–234. Amsterdam, The Netherlands: North-Holland.

Latimer, C. J. 1993. The dissociative ionization of simple molecules by fast ions. *Adv. At. Molec. Opt. Phys.* 30: 105–138.

LaVerne, J. A. 1989. Radical and molecular yields in the radiolysis of water with carbon ions. *Radiat. Phys. Chem.* 34: 135–143.

LaVerne, J. A. 2000. Track effects of heavy ions in liquid water. *Radiat. Res.* 153: 487–496.

LaVerne, J. A. 2004. Radiation chemical effects of heavy ions. In *Charged Particle and Photon Interactions with Matter: Chemical, Physicochemical, and Biological Consequences with Applications*, A. Mozumder and Y. Hatano (eds.), pp. 403–429. New York: Marcel Dekker.

LaVerne, J. A. and Mozumder, A. 1993. Concerning plasmon excitation in liquid water. *Radiat. Res.* 133: 282–288.

LaVerne, J. A. and Pimblott, S. M. 1995. Electron energy-loss distributions in solid, dry DNA. *Radiat. Res.* 141: 208–215.

LaVerne, J. A. and Schuler, R. H. 1987a. Radiation chemical studies with heavy ions: Oxidation of ferrous ion in the Fricke dosimeter. *J. Phys. Chem.* 91: 5770–5776.

LaVerne, J. A. and Schuler, R. H. 1987b. Track effects in radiation chemistry: Production of $HO_2^{\cdot}$ in the radiolysis of water by high-LET ^{58}Ni ions. *J. Phys. Chem.* 91: 6560–6563.

LaVerne, J. A. and Schuler, R. H. 1996. Radiolysis of the Fricke dosimeter with ^{58}Ni and ^{238}U ions: Response for particles of high linear energy transfer. *J. Phys. Chem.* 100: 16034–16040.

LaVerne, J. A., Burns, W. G., and Schuler, R. H. 1985. Production of HO_2 within the track core in the heavy particle radiolysis of water. *J. Phys. Chem.* 89: 242–243.

LaVerne, J. A., Schuler, R. H., and Burns, W. G. 1986. Track effects in radiation chemistry: Production of $HO_2^{\cdot}$ within the track core in the heavy-particle radiolysis of water. *J. Phys. Chem.* 90: 3238–3242.

Lefort, M. 1955. Chimie des radiations des solutions aqueuses. Aspect actuel des résultats expérimentaux. In *Actions Chimiques et Biologiques des Radiations*, Vol. 1, M. Haïssinsky (ed.), pp. 93–204. Paris, France: Masson.

Lefort, M. and Tarrago, X. 1959. Radiolysis of water by particles of high linear energy transfer. The primary chemical yields in aqueous acid solutions of ferrous sulfate, and in mixtures of thallous and ceric ions. *J. Phys. Chem.* 63: 833–836.

Legendre, S., Giglio, E., Tarisien, M., Cassimi, A., Gervais, B., and Adoui, L. 2005. Isotopic effects in water dication fragmentation. *J. Phys. B: At. Mol. Opt. Phys.* 38: L233–L241.

Lu, Q.-B., Baskin, J. S., and Zewail, A. H. 2004. The presolvated electron in water: Can it be scavenged at long range? *J. Phys. Chem. B* 108: 10509–10514.

Luna, H. and Montenegro, E. C. 2005. Fragmentation of water by heavy ions. *Phys. Rev. Lett.* 94: 043201.

Lymar, S. V. and Poskrebyshev, G. A. 2003. Rate of ON–OO⁻ bond homolysis and the Gibbs energy of formation of peroxynitrite. *J. Phys. Chem. A* 107: 7991–7996.

Magee, J. L. and Chatterjee, A. 1987. In *Kinetics of Nonhomogeneous Processes*, G. R. Freeman (ed.), pp. 171–214. New York: Wiley.

Manson, S. T., Dubois, R. D., and Toburen, L. H. 1983. Multiple-ionization mechanisms in fast proton-neon collisions. *Phys. Rev. Lett.* 51: 1542–1545.

McCracken, D. R., Tsang, K. T., and Laughton, P. J. 1998. Aspects of the physics and chemistry of water radiolysis by fast neutrons and fast electrons in nuclear reactors. Report AECL-11895. Chalk River, Canada: Atomic Energy of Canada Ltd.

McDaniel, E. W., Mitchell, J. B. A., and Rudd, M. E. 1993. *Atomic Collisions: Heavy Particle Projectiles.* New York: Wiley.

Medin, J., Ross, C. K., Klassen, N. V., Palmans, H., Grusell, E., and Grindborg, J.-E. 2006. Experimental determination of beam quality factors, k_Q, for two types of Farmer chamber in a 10 MV photon and a 175 MeV proton beam. *Phys. Med. Biol.* 51: 1503–1521.

Meesungnoen, J. and Jay-Gerin, J.-P. 2005a. High-LET radiolysis of liquid water with $^{1}H^{+}$, $^{4}He^{2+}$, $^{12}C^{6+}$, and $^{20}Ne^{9+}$ ions: Effects of multiple ionization. *J. Phys. Chem. A* 109: 6406–6419.

Meesungnoen, J. and Jay-Gerin, J.-P. 2005b. Effect of multiple ionization on the yield of H_2O_2 produced in the radiolysis of aqueous 0.4 M H_2SO_4 solutions by high-LET $^{12}C^{6+}$ and $^{20}Ne^{9+}$ ions. *Radiat. Res.* 164: 688–694.

Meesungnoen, J. and Jay-Gerin, J.-P. 2009. High-LET ion radiolysis of water: Oxygen production in tracks. *Radiat. Res.* 171: 379–386.

Meesungnoen, J., Benrahmoune, M., Filali-Mouhim, A., Mankhetkorn, S., and Jay-Gerin, J.-P. 2001. Monte Carlo calculation of the primary radical and molecular yields of liquid water radiolysis in the linear energy transfer range 0.3–6.5 keV/μm: Application to ^{137}Cs gamma rays. *Radiat. Res.* 155: 269–278.

Meesungnoen, J., Jay-Gerin, J.-P., Filali-Mouhim, A., and Mankhetkorn, S. 2002a. Low-energy electron penetration range in liquid water. *Radiat. Res.* 158: 657–660.

Meesungnoen, J., Jay-Gerin, J.-P., Filali-Mouhim, A., and Mankhetkorn, S. 2002b. On the temperature dependence of the primary yield and the product Ge_{max} of hydrated electrons in the low-LET radiolysis of liquid water. *Can. J. Chem.* 80: 767–773.

Meesungnoen, J., Filali-Mouhim, A., Snitwongse Na Ayudhya, N., Mankhetkorn, S., and Jay-Gerin, J.-P. 2003. Multiple ionization effects on the yields of $HO_2^{\bullet}/O_2^{\bullet-}$ and H_2O_2 produced in the radiolysis of liquid water with high-LET $^{12}C^{6+}$ ions: A Monte Carlo simulation study. *Chem. Phys. Lett.* 377: 419–425.

Michael, B. D. and Prise, K. M. 1996. A multiple-radical model for radiation action on DNA and the dependence of OER on LET. *Int. J. Radiat. Biol.* 69: 351–358.

Michaud, M., Cloutier, P., and Sanche, L. 1991. Low-energy electron-energy-loss spectroscopy of amorphous ice: Electronic excitations. *Phys. Rev. A* 44: 5624–5627.

Michaud, M., Wen, A., and Sanche, L. 2003. Cross sections for low-energy (1–100 eV) electron elastic and inelastic scattering in amorphous ice. *Radiat. Res.* 159: 3–22.

Millar, B. C., Fielden, E. M., and Steele, J. J. 1979. A biphasic radiation survival response of mammalian cells to molecular oxygen. *Int. J. Radiat. Biol.* 36: 177–180.

Miller, N. 1958. Radical yield measurements in irradiated aqueous solutions. II. Radical yields with 10.9-MeV deuterons, 21.3- and 3.4-MeV alpha particles, and B(n,α)Li recoil radiations. *Radiat. Res.* 9: 633–646.

Montenegro, E. C. and Luna, H. 2005. Primary radicals production from water fragmentation by heavy ions. *Brazilian J. Phys.* 35: 927–932.

Mozumder, A. 1999. *Fundamentals of Radiation Chemistry*. San Diego, CA: Academic Press.

Mozumder, A. and Magee, J. L. 1966a. Model of tracks of ionizing radiations for radical reaction mechanisms. *Radiat. Res.* 28: 203–214.

Mozumder, A. and Magee, J. L. 1966b. Theory of radiation chemistry. VII. Structure and reactions in low LET tracks. *J. Chem. Phys.* 45: 3332–3341.

Mozumder, A. and Magee, J. L. 1975. The early events of radiation chemistry. *Int. J. Radiat. Phys. Chem.* 7: 83–93.

Mozumder, A., Chatterjee, A., and Magee, J. L. 1968. Theory of radiation chemistry. IX. Model and structure of heavy particle tracks in water. *Adv. Chem. Series* 81: 27–48.

Muroya, Y., Meesungnoen, J., Jay-Gerin, J.-P., Filali-Mouhim, A., Goulet, T., Katsumura, Y., and Mankhetkorn, S. 2002. Radiolysis of liquid water: An attempt to reconcile Monte Carlo calculations with new experimental hydrated electron yield data at early times. *Can. J. Chem.* 80: 1367–1374.

Muroya, Y., Plante, I., Azzam, E. I., Meesungnoen, J., Katsumura, Y., and Jay-Gerin, J.-P. 2006. High-LET ion radiolysis of water: Visualization of the formation and evolution of ion tracks and relevance to the radiation-induced bystander effect. *Radiat. Res.* 165: 485–491.

Neary, G. J. 1965. Chromosome aberrations and the theory of RBE. 1. General considerations. *Int. J. Radiat. Biol.* 9: 477–502.

Ogura, H. and Hamill, W. H. 1973. Positive hole migration in pulse-irradiated water and heavy water. *J. Phys. Chem.* 77: 2952–2954.

Olivera, G. H., Caraby, C., Jardin, P., Cassimi, A., Adoui, L., and Gervais, B. 1998. Multiple ionization in the earlier stages of water radiolysis. *Phys. Med. Biol.* 43: 2347–2360.

Paretzke, H. G. 1987. Radiation track structure theory. In *Kinetics of Nonhomogeneous Processes*, G. R. Freeman (ed.), pp. 89–170. New York: Wiley.

Paretzke, H. G., Goodhead, D. T., Kaplan, I. G., and Terrissol, M. 1995. Track structure quantities. In *Atomic and Molecular Data for Radiotherapy and Radiation Research*, IAEA-TECDOC-799, pp. 633–721. Vienna, Austria: International Atomic Energy Agency.

Pastina, B. and LaVerne, J. A. 1999. Hydrogen peroxide production in the radiolysis of water with heavy ions. *J. Phys. Chem. A* 103: 1592–1597.

Pastina, B., LaVerne, J. A., and Pimblott, S. M. 1999. Dependence of molecular hydrogen formation in water on scavengers of the precursor to the hydrated electron. *J. Phys. Chem. A* 103: 5841–5846.

Patt, H. M. 1953. Protective mechanisms in ionizing radiation injury. *Physiol. Rev.* 33: 35–76.

Pešić, Z. D., Chesnel, J.-Y., Hellhammer, R., Sulik, B., and Stolterfoht, N. 2004. Fragmentation of H_2O molecules following the interaction with slow, highly charged Ne ions. *J. Phys. B: At. Mol. Opt. Phys.* 37: 1405–1417.

Pimblott, S. M. and Green, N. J. B. 1995. Recent advances in the kinetics of radiolytic processes. *Res. Chem. Kinet.* 3: 117–174.

Pimblott, S. M. and LaVerne, J. A. 1998. On the radiation chemical kinetics of the precursor to the hydrated electron. *J. Phys. Chem. A* 102: 2967–2975.

Pimblott, S. M. and LaVerne, J. A. 2002. Effects of track structure on the ion radiolysis of the Fricke dosimeter. *J. Phys. Chem. A* 106: 9420–9427.

Pimblott, S. M. and LaVerne, J. A. 2007. Production of low-energy electrons by ionizing radiation. *Radiat. Phys. Chem.* 76: 1244–1247.

Pimblott, S. M. and Mozumder, A. 1991. Structure of electron tracks in water. 2. Distribution of primary ionizations and excitations in water radiolysis. *J. Phys. Chem.* 95: 7291–7300.

Pimblott, S. M. and Mozumder, A. 2004. Modeling of physicochemical and chemical processes in the interactions of fast charged particles with matter. In *Charged Particle and Photon Interactions with Matter: Chemical, Physicochemical, and Biological Consequences with Applications*, A. Mozumder and Y. Hatano (eds.), pp. 75–103. New York: Marcel Dekker.

Pimblott, S. M., LaVerne, J. A., Mozumder, A., and Green, N. J. B. 1990. Structure of electron tracks in water. 1. Distribution of energy deposition events. *J. Phys. Chem.* 94: 488–495.

Pimblott, S. M., Pilling, M. J., and Green, N. J. B. 1991. Stochastic models of spur kinetics in water. *Radiat. Phys. Chem.* 37: 377–388.

Plante, I. 2009. Développement de codes de simulation Monte Carlo de la radiolyse de l'eau par des électrons, ions lourds, photons et neutrons. Applications à divers sujets d'intérêt expérimental. PhD thesis, Université de Sherbrooke, Sherbrooke, Canada.

Plante, I., Filali-Mouhim, A., and Jay-Gerin, J.-P. 2005. SimulRad: A Java interface for a Monte Carlo simulation code to visualize in 3D the early stages of water radiolysis. *Radiat. Phys. Chem.* 72: 173–180.

Platzman, R. L. 1952. On the primary processes in radiation chemistry and biology. In *Symposium on Radiobiology. The Basic Aspects of Radiation Effects on Living Systems*, J. J. Nickson (ed.), pp. 97–116. New York: Wiley.

Platzman, R. L. 1955. Subexcitation electrons. *Radiat. Res.* 2: 1–7.

Platzman, R. L. 1958. The physical and chemical basis of mechanisms in radiation biology. In *Radiation Biology and Medicine. Selected Reviews in the Life Sciences*, W. D. Claus (ed.), pp. 15–72. Reading, MA: Addison-Wesley.

Platzman, R. L. 1962a. Superexcited states of molecules. *Radiat. Res.* 17: 419–425.

Platzman, R. L. 1962b. Dissociative attachment of subexcitation electrons in liquid water, and the origin of radiolytic "molecular" hydrogen. In *Abstracts of Papers, 2nd International Congress of Radiation Research*, Harrogate, U.K., August 5–11, 1962, p. 128.

Pogue, B. W., O'Hara, J. A., Wilmot, C. M., Paulsen, K. D., and Swartz, H. M. 2001. Estimation of oxygen distribution in RIF-1 tumors by diffusion model-based interpretation of pimonidazole hypoxia and Eppendorf measurements. *Radiat. Res.* 155: 15–25.

Pucheault, J. 1961. Action des rayons alpha sur les solutions aqueuses. In *Actions Chimiques et Biologiques des Radiations*, Vol. 5, M. Haïssinsky (ed.), pp. 31–84. Paris, France: Masson.

Radi, R., Denicola, A., Alvarez, B., Ferrer-Sueta, G., and Rubbo, H. 2000. The biological chemistry of peroxynitrite. In *Nitric Oxide: Biology and Pathobiology*, L. J. Ignarro (ed.), pp. 57–82. San Diego, CA: Academic Press.

Ramsay, W. and Soddy, F. 1903. Experiments in radioactivity, and the production of helium from radium. *Proc. R. Soc. London* 72: 204–207.

Rowntree, P., Parenteau, L., and Sanche, L. 1991. Electron stimulated desorption via dissociative attachment in amorphous H_2O. *J. Chem. Phys.* 94: 8570–8576.

Rudd, M. E. 1990. Cross sections for production of secondary electrons by charged particles. *Radiat. Prot. Dosim.* 31: 17–22.

Samuel, A. H. and Magee, J. L. 1953. Theory of radiation chemistry. II. Track effects in radiolysis of water. *J. Chem. Phys.* 21: 1080–1087.

Sauer, M. C., Jr., Schmidt, K. H., Hart, E. J., Naleway, C. A., and Jonah, C. D. 1977. LET dependence of transient yields in the pulse radiolysis of aqueous systems with deuterons and α particles. *Radiat. Res.* 70: 91–106.

Scholes, G. and Weiss, J. 1959. Oxygen effects and formation of peroxides in aqueous solutions. *Radiat. Res.* Suppl. 1: 177–189.

Schmidt, K. H. and Bartels, D. M. 1995. Lack of ionic strength effect in the recombination of hydrated electrons: $(e^-)_{aq} + (e^-)_{aq} \rightarrow 2(OH^-) + H_2$. *Chem. Phys.* 190: 145–152.

Schuler, R. H. and Allen, A. O. 1957. Radiation chemistry studies with cyclotron beams of variable energy: Yields in aerated ferrous sulfate solution. *J. Am. Chem. Soc.* 79: 1565–1572.

Schwarz, H. A. 1969. Applications of the spur diffusion model to the radiation chemistry of aqueous solutions. *J. Phys. Chem.* 73: 1928–1937.

Siegmann, B., Werner, U., Lutz, H. O., and Mann, R. 2001. Multiple ionization and fragmentation of H_2O in collisions with fast highly charged Xe ions. *J. Phys. B: At. Mol. Opt. Phys.* 34: L587–L593.

Sims, H. E. 1978. Some effects of linear energy transfer on the radiolysis of aqueous solutions. PhD thesis, University of Salford, Salford, U.K.

Sobocinski, P., Pešić, Z. D., Hellhammer, R., Klein, D., Sulik, B., Chesnel, J.-Y., and Stolterfoht, N. 2006. Anisotropic proton emission after fragmentation of H_2O by multiply charged ions. *J. Phys. B: At. Mol. Opt. Phys.* 39: 927–937.

Spinks, J. W. T. and Woods, R. J. 1990. *An Introduction to Radiation Chemistry*, 3rd edn. New York: Wiley.

Stolterfoht, N., Cabrera-Trujillo, R., Hellhammer, R., Pešić, Z., Deumens, E., Öhrn, Y., and Sabin, J. R. 2007. Charge exchange and fragmentation in slow collisions of He^{2+} with water molecules. *Adv. Quantum Chem.* 52: 149–170.

Stuglik, Z. 1995. On the "oxygen in heavy-ion tracks" hypothesis. *Radiat. Res.* 143: 343–348.

Swallow, A. J. and Velandia, J. A. 1962. Oxygen effect as an explanation of differences between the action of α-particles and x- or γ-rays on aqueous solutions of amino-acids and proteins. *Nature (London)* 195: 798–800.

Swiatla-Wojcik, D. and Buxton, G. V. 1995. Modeling of radiation spur processes in water at temperatures up to 300°C. *J. Phys. Chem.* 99: 11464–11471.

Taube, H. 1957. Photochemical reactions of ozone in solution. *Trans. Faraday Soc.* 53: 656–665.

Tavernelli, I., Gaigeot, M.-P., Vuilleumier, R., Stia, C., Hervé du Penhoat, M.-A., and Politis, M.-F. 2008. Time-dependent density functional theory molecular dynamics simulations of liquid water radiolysis. *ChemPhysChem* 9: 2099–2103.

Toburen, L. H. 2004. Ionization and secondary electron production by fast charged particles. In *Charged Particle and Photon Interactions with Matter: Chemical, Physicochemical, and Biological Consequences with Applications*, A. Mozumder and Y. Hatano (eds.), pp. 31–74. New York: Marcel Dekker.

Tsuruoka, C., Suzuki, M., Kanai, T., and Fujitaka, K. 2005. LET and ion species dependence for cell killing in normal human skin fibroblasts. *Radiat. Res.* 163: 494–500.

Tubiana, M. (ed.) 2008. *Radiobiologie*. Paris, France: Hermann.

Turner, J. E., Magee, J. L., Hamm, R. N., Chatterjee, A., Wright, H. A., and Ritchie, R. H. 1981. Early events in irradiated water. In *Seventh Symposium on Microdosimetry*, Oxford, U.K., September 8–12, 1980, J. Booz, H. G. Ebert, and H. D. Hartfiel (eds.), pp. 507–520. London, U.K.: Harwood Academic.

Turner, J. E., Magee, J. L., Wright, H. A., Chatterjee, A., Hamm, R. N., and Ritchie, R. H. 1983. Physical and chemical development of electron tracks in liquid water. *Radiat. Res.* 96: 437–449.

Turner, J. E., Hamm, R. N., Wright, H. A., Ritchie, R. H., Magee, J. L., Chatterjee, A., and Bolch, W. E. 1988. Studies to link the basic radiation physics and chemistry of liquid water. *Radiat. Phys. Chem.* 32: 503–510.

Uehara, S. and Nikjoo, H. 2006. Monte Carlo simulation of water radiolysis for low-energy charged particles. *J. Radiat. Res.* 47: 69–81.

Usher, F. L. 1911. Die chemische Einzelwirkung und die chemische Gesamtwirkung der α- und der β-Strahlen. *Jahrb. Radioakt. Elektron.* 8: 323–334.

von Sonntag, C. 2006. *Free-Radical-Induced DNA Damage and Its Repair*. Berlin, Germany: Springer-Verlag.

Ward, J. F. 1994. The complexity of DNA damage: Relevance to biological consequences. *Int. J. Radiat. Biol.* 66: 427–432.

Wasselin-Trupin, V., Baldacchino, G., Bouffard, S., Balanzat, E., Gardès-Albert, M., Abedinzadeh, Z., Jore, D., Deycard, S., and Hickel, B. 2000. A new method for the measurement of low concentrations of OH/O_2^- radical species in water by high-LET pulse radiolysis. A time-resolved chemiluminescence study. *J. Phys. Chem. A* 104: 8709–8714.

Wasselin-Trupin, V., Baldacchino, G., Bouffard, S., and Hickel, B. 2002. Hydrogen peroxide yields in water radiolysis by high-energy ion beams at constant LET. *Radiat. Phys. Chem.* 65: 53–61.

Watt, D. E. 1996. *Quantities for Dosimetry of Ionizing Radiations in Liquid Water*. London, U.K.: Taylor & Francis.

Werner, U., Beckord, K., Becker, J., Folkerts, H. O., and Lutz, H. O. 1995a. Ion-impact-induced fragmentation of water molecules. *Nucl. Instrum. Methods Phys. Res. B* 98: 385–388.

Werner, U., Beckord, K., Becker, J., and Lutz, H. O. 1995b. 3D imaging of the collision-induced Coulomb fragmentation of water molecules. *Phys. Rev. Lett.* 74: 1962–1965.

Wilson, C. D., Dukes, C. A., and Baragiola, R. A. 2001. Search for the plasmon in condensed water. *Phys. Rev. B* 63: 121101.

Yamashita, S., Katsumura, Y., Lin, M., Muroya, Y., Maeyama, T., and Murakami, T. 2008. Water radiolysis with heavy ions of energies up to 28 GeV – 2: Extension of primary yield measurements to very high LET values. *Radiat. Phys. Chem.* 77: 1224–1229.

Zimbrick, J. D. 2002. Radiation chemistry and the Radiation Research Society: A history from the beginning. *Radiat. Res.* 158: 127–140.

Zirkle, R. E., Marchbank, D. F., and Kuck, K. D. 1952. Exponential and sigmoid survival curves resulting from alpha and X irradiation of aspergillus spores. *J. Cellular Comp. Physiol.* 39 (Suppl. 1): 75–85.

15 Radiation Chemistry of High Temperature and Supercritical Water and Alcohols

Mingzhang Lin
Japan Atomic Energy Agency
Tokai, Japan

Yosuke Katsumura
The University of Tokyo
Tokyo, Japan
and
Japan Atomic Energy Agency
Tokai, Japan

CONTENTS

15.1 INTRODUCTION

A supercritical fluid is a highly compressed material that combines the properties of gases and liquids in an intriguing manner, at a temperature and pressure above its critical point. Additionally, close to the critical point, small changes in pressure or temperature result in large changes in density, allowing many properties to be "tuned." Due to their peculiar solvent properties, supercritical fluids offer a range of unusual chemical possibilities including environmentally benign separations and destruction of hazardous waste, as well as for new materials synthesis (Jessop et al., 1994; Savage et al., 1995; Eckert et al., 1996; Darr and Poliakoff, 1999; Akiya and Savage, 2002). Figure 15.1 shows the schematic representation of phase diagrams for water and carbon dioxide. The thermophysical properties of water and of many other fluid systems have been formulated and/or compiled by the National Institute of Standards and Technology (NIST) and the International Association for the Properties of Water and Steam (IAPWS). In Table 15.1, the critical properties are shown for some components that are commonly used as supercritical fluids.

Studies of the radiolysis of supercritical water ($t_c = 374°C$, $P_c = 22.1\,MPa$) were motivated by the development of one of the next-generation (GenIV) nuclear reactors—the supercritical water-cooled reactor (SCWR) (Oka and Kataoka, 1992; Squarer et al., 2003). This new-concept reactor has the advantages of higher thermal conversion efficiency, simplicity in structure, safety, etc. In these reactors, as in boiling water (BWR) and pressurized water reactors (PWR), light water or heavy water is used as both a coolant and a moderator. The water is exposed to a strong radiation field (>10 kGy/s) composed of γ-rays and 2 MeV fast neutrons, etc. As is well known in the BWR and PWR, two radiolysis products of water, O_2 and H_2O_2, strongly affect the corrosion of structural materials in the reactors. Proper water chemistry control, in particular the injection of H_2 into the coolant to convert O_2 and H_2O_2 into H_2O by radiolytic processes, effectively reduces the electrochemical corrosion

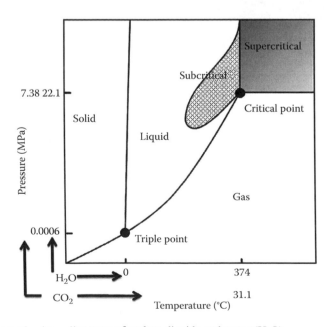

FIGURE 15.1 Schematic phase diagrams of carbon dioxide and water (H_2O).

TABLE 15.1
Critical Properties of Various Solvents

Solvent	Molecular Weight g/mol	Critical Temperature (t_c) K	Critical Pressure (P_c) MPa	Critical Density (ρ_c) g/cm³
Carbon dioxide (CO_2)	44.01	304.1	7.38	0.469
Water (H_2O)	18.015	647.096	22.064	0.322
Methanol (CH_3OH)	32.04	512.6	8.09	0.272
Ethanol (C_2H_5OH)	46.07	513.9	6.14	0.276
1-Propanol (C_3H_7OH)	60.10	263.6	5.175	0.274
2-Propanol (C_3H_7OH)	60.10	235.2	4.762	0.273
1-Butanol (C_4H_9OH)	74.12	289.9	4.423	0.270
Methane (CH_4)	16.04	190.4	4.60	0.162
Ethane (C_2H_6)	30.07	305.3	4.87	0.203
Propane (C_3H_8)	44.09	369.8	4.25	0.217
Ethylene (C_2H_4)	28.05	282.4	5.04	0.215

potential (ECP), and represents the key to maintaining the integrity of the reactors. Computer simulations are usually required to help predict the concentrations of water decomposition products. These simulations require a knowledge of the temperature-dependent G-values (denoting the experimentally measured radiolytic yields, with units of molecules/100 eV in this context) and rate constants for about 50 reactions. The rate constants and G-values of the radiolysis of light and heavy water over the range 0°C–300°C have been compiled by Elliot (1994) (Elliot et al., 1996) and reviewed by Buxton (2001), although most reactions were investigated only up to 200°C. It is assumed that a similar simulation of water radiolysis in the SCWR will be required. As is known, BWRs or PWRs are operated at constant temperature (280°C–325°C) and pressure (15–20 MPa). However, according to the current conceptual design of SCWR, the inlet temperature is 280°C and the outlet temperature is >500°C with a fixed pressure of 25 MPa (Oka and Kataoka, 1992; Squarer et al., 2003). Thus, it is necessary to accumulate the basic data on water radiolysis above 325°C, especially for supercritical water that possesses many unusual properties.

In supercritical water, the dielectric constant dramatically decreases from 78 at room temperature to 2.6 at 400°C/25 MPa, which is similar to that of benzene or toluene. Thus many organic compounds can be easily dissolved in SCW although these compounds have rather low solubility at room temperature. Contrarily, inorganic salts are difficult to dissolve in SCW due to their extremely small ion products. It has been reported that the water structure exhibits a remarkable change in SCW (Tucker and Maddox, 1998). For liquid water, structural studies by both neutron and x-ray scattering indicate that when water is placed under pressure, the number of hydrogen bonds per water molecule does not change by any appreciable amount relative to the ambient state; however, they are now bent out of their ideal orientation and are correspondingly weaker energetically. Liquid structure at high pressure is nearly independent of temperature variation. But for supercritical water, investigation extends over isochores between 0.05 and 1 g/mL and isotherms between 473 and 1273 K. As water is heated into subcritical and supercritical regimes, the number of hydrogen bonds per water molecule has been shown to decrease (Hoffmann and Conradi, 1997). The temperature at which the total breakdown of hydrogen bonding could be observed has been debated for many years, with most recent experimental studies, including neutron scattering (Soper et al., 1997), Raman spectroscopy (Ikushima et al., 1998), solution x-ray scattering (Ohtaki et al., 1997), NMR (Hoffmann and Conradi, 1997; Matubayasi et al., 1997), and microwave spectroscopy (Okada et al., 1997), suggesting that tetrahedral bonding persists to at least 650 K and possibly up to 770 K at 100 MPa. The normally compact tetrahedral-like water structure is decomposed and long-distance water–water interactions increase. Thus, supercritical water consists of small clusters, and much smaller aggregates such as oligomers, and even monomeric gas-like water molecules (Ohtaki et al., 1997). It is expected that the changes of these properties would change G-values, rate constants, and the spectral properties of the transient species of water radiolysis.

In this chapter, we summarize the most recent results obtained in studies of the radiation chemistry of SCW, especially the estimation of G-values, the reaction rate constants, and the absorption spectral properties of some transient species. For comparison or better understanding of SCW, some results concerning the radiolysis of alcohols are also presented.

15.2 INITIAL PROCESSES OF WATER RADIOLYSIS

Irradiation with electron beam, γ-rays, or high-energy charged particles leads to the decomposition of water molecules through excitation and ionization, and the initial processes of water radiolysis can be summarized as (Swiatla-Wojcik and Buxton, 1995)

$$H_2O \xrightarrow{\gamma, e^-} H_2O^+ + e_{se}^- \rightarrow H_2O^*_{vib}$$

$$H_2O^*_{vib} \rightarrow \begin{array}{l} OH + H \\ H_2 + O(^1D) \quad H_2 + H_2O_2 \ (\text{or } 2OH) \\ 2H + O(^3P) \end{array}$$

$$H_2O^+ \rightarrow H_{aq}^+ + OH$$

$$e_{se}^- + H_2O \rightarrow OH + H^- \rightarrow OH + H_2 + OH^-$$

$$e_{se}^- + H_2O \rightarrow e_{aq}^-$$

$$H_2O^*_{elec} \rightarrow \begin{array}{l} OH + H \\ H_2 + O(^1D) \rightarrow H_2 + H_2O_2 \ (\text{or } 2OH) \\ 2H + O(^3P) \end{array}$$

Electrons are stabilized by the surrounding water molecules to form hydrated electrons in less than 1 ps. The yields at this stage are defined as initial yields. The resulting transient species such as e_{aq}^-, $\cdot$H, and $\cdot$OH are distributed locally along the track where the energy deposits, called a spur. These products then diffuse randomly and either react together or escape into the bulk solution. After the completion of spur processes, which take place within 10^{-7}–10^{-6} s, the products homogeneously distribute in the solution, and the yields at this time are known as primary yields. Since the increase of temperature and changes in water properties and structure strongly affect the spur reactions, some interesting changes in the radiolytic yields under supercritical conditions are expected.

15.3 EXPERIMENTAL METHODS

Currently, most transient species kinetic studies are carried out using nanosecond pulse radiolysis techniques associated with spectroscopic detection methods. Only a few reports are of muonium reactions (Percival et al., 1999, 2000; Ghandi and Percival, 2003), steady state radiolysis (Miyazaki

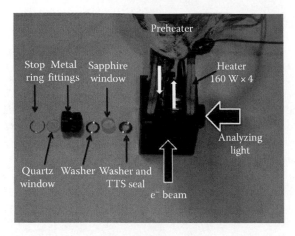

FIGURE 15.2 Picture of the HTHP cell used at the University of Tokyo and the sealing mechanism of the optical windows.

et al., 2006a,b; Janik et al., 2007a), laser photolysis (Lin et al., 2006a; Han et al., 2008), and pico-second pulse radiolysis (Baldacchino et al. 2006). Since conventional pulse radiolysis techniques are well known, we just briefly introduce here the high-temperature high pressure (HTHP) system for pulse radiolysis (Ferry and Fox, 1998; Takahashi et al., 2000; Wu et al., 2000). The structure of the HTHP optical cell is shown in Figure 15.2.

The size, structure, and sealing mechanism of the optical windows of the cell may vary, but the HTHP system usually consists of a preheater and an optical cell made of high-strength and corrosion-resistant alloys such as Hastelloy or SUS316 with sapphire windows. At least one thermocouple should be put into the sample solution inside the cell to monitor the temperature. Some special consideration and sometimes compromise are necessary for obtaining a system with good signal-to-noise (S/N) ratio, less dead volume, quick temperature equilibration, faster flushing of the cell, etc.

The main difficulties for the pulse radiolysis experiments of SCW are listed below:

- *Corrosion and damage of the sapphire windows*: The sapphire window is easily corroded in SCW, especially under acidic or alkaline conditions, or in the presence of some additives, such as O_2.
- *Thermal stability of chemical reagents*: Most organic compounds are thermally unstable at elevated temperatures.
- *Solubility of inorganic compounds*: The dissociation constant is rather small for salts under supercritical conditions.
- *Limitation of detection techniques*: Dramatically decreasing signal intensity for low-density SCW; many reactions become much faster than at room temperature, requiring a higher-time-resolution pulse radiolysis system.

15.4 OPTICAL ABSORPTION SPECTRAL PROPERTIES OF FREE RADICALS IN HIGH TEMPERATURE AND SUPERCRITICAL WATER AND ALCOHOLS

A pulse radiolysis system combined with a time-resolved optical absorption measurement allows us to study radiation chemical processes such as the radiolytic yields of transient species and reaction rate constants. The absorption spectrum that is useful for the identification of the intermediate radical is usually investigated as a first step. In cases of supercritical water and alcohols, studies on spectral properties are particularly important because the solvent properties may change with temperature and pressure. In turn, the identification of spectral changes might be helpful for a better understanding of solvent properties and solvent molecular structure at elevated temperatures.

15.4.1 SOLVATED ELECTRONS

The solvated electron is an ubiquitous species in radiation chemistry, resulting from secondary electrons generated in polar solvents by ionizing radiations such as fast electrons and ^{60}Co γ-rays (Hart and Anbar, 1970; Bernas et al., 1996). In fact, secondary (or "dry") electrons slow down to subexcitation energies, and following thermalization become localized and eventually solvated. In liquid water at 25°C, spectroscopic studies have revealed that electron "localization" and "hydration" occur on timescales of ~110–300 and ~240–620 fs, respectively (Bernas et al., 1996; Jay-Gerin, 1997; Muroya et al., 2002; Wang and Lu, 2007). The optical absorption of e_{aq}^- is characterized by a broad, intense, and featureless spectrum that covers most of the visible, tails into the near-infrared, and exhibits a maximum at $\lambda_{max} = 718$ nm (or $E_{max} = 1.73$ eV) in H_2O and $\lambda_{max} = 699$ nm (or $E_{max} = 1.77$ eV) in D_2O at 25°C (Bartels et al., 2005). The hydrated electron was discovered by transient absorption measurements in the pulse radiolysis of water about 50 years ago (Hart and Boag, 1962; Boag and Hart, 1963; Keene, 1963). Then, it was soon detected in various solvents through its intense optical absorption band in the visible or near-infrared spectral domain (Jou and Freeman, 1979a,b). The properties of the solvated electron depend on several factors such as the solvent, the temperature, or the pressure. Much attention has been paid to the temperature effects on the absorption spectra of the solvated electron since the discovery of the hydrated electron. However, there was no systematic study of the solvated electron in supercritical water and alcohols until 10 years ago.

15.4.1.1 In Pure Water

Temperature effects on e_{aq}^- have been extensively studied using the pulse radiolysis technique and the absorption maximum of e_{aq}^- spectrum is found to shift to longer wavelengths with increasing temperature. Typical temperature-dependent absorption spectra of e_{aq}^- in D_2O at 25 MPa from room temperature to 390°C is given in Figure 15.3. The absorbance largely decreases with temperature because the water density decreases and the decay of e_{aq}^- is accelerated with increasing temperature, as shown in the inset. In the 1970s and 1980s, the investigated temperatures were usually

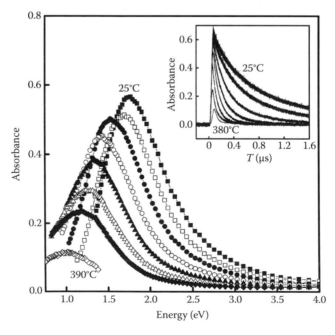

FIGURE 15.3 Absorption spectra of hydrated electron in D_2O under a pressure of 25 MPa at temperatures: 25°C, 50°C, 100°C, 150°C, 200°C, 250°C, 300°C, and 390°C. Inset: time profiles at different temperatures of hydrated electron at λ_{max} of each temperature.

within 100°C–300°C, not exceeding the critical temperature of water (Gottschall and Hart, 1967; Dixon and Lopata, 1978; Jou and Freeman, 1979b; Christensen and Sehested, 1986; Shiraishi et al., 1994). There was one report on the e_{aq}^- spectrum at temperatures up to 390°C below the critical pressure, but the information provided was very limited (Michael et al., 1971). In 2000, Wu et al. reported temperature dependences of the e_{aq}^- spectrum and $G\varepsilon_{max}$ by the pulse radiolysis method over a temperature range of 25°C–400°C including the supercritical condition (Wu et al., 2000). With increasing temperature, the absorption peak λ_{max} of e_{aq}^- shifts significantly to longer wavelengths. The value of $G\varepsilon_{max}$ in supercritical water is considerably smaller than in liquid water at room temperature. A later study reported by Bartels et al. showed that the e_{aq}^- spectrum at supercritical temperatures shifts slightly to the red as density decreases (Bartels et al., 2005). With the application of spectral moment theory, Bartels et al. estimated the average size of the electron wave function and of its kinetic energy. It appears that for water densities below about $0.6\,g/cm^3$, and down to below $0.1\,g/cm^3$, the average radius of gyration for the electron remains constant at around $3.4\,\text{Å}$, and its absorption maximum is near $0.9\,eV$. For higher densities, the electron is squeezed into a smaller cavity and the spectrum is shifted to the blue.

The fact that e_{aq}^- does exist in SCW and even persists at densities as low as $\sim 8 \times 10^{-3}\,g/cm^3$ (limit of the study) (Jortner and Gaathon, 1977) indicates that the electron experiences a strong interaction with the neighboring water molecules, implying a dominant role of the short-range molecular structure in the microscopic description of the electron localization and hydration mechanisms.

15.4.1.2 Density Effects on the Absorption Spectra of e_{aq}^-

From a microscopic perspective, many of the unique features of SCW are due, in large part, to the changes that take place in the intermolecular structure and hydrogen bonding of water at elevated temperatures. In fact, a wide variety of experimental investigations as well as molecular-based computer simulations in the last decade have shown that, at supercritical conditions, the infinite H-bond network of the molecules present in ambient water crosses a percolation transition, that is, breaks down to form small clusters of bonded water molecules in various tetrahedral configurations surrounded by nonbonded gas-phase-like molecules (Ohtaki et al., 1997; Tucker and Maddox, 1998; Partay and Jedlovszky, 2005; Wernet et al., 2005). As a result, the instantaneous picture of SCW can be viewed as that of an inhomogeneous medium with coexisting high- and low-density regions.

The absorption spectra of the hydrated electron have been measured by the electron pulse radiolysis techniques in supercritical water (D_2O) at different temperatures and densities (or pressures) (Jay-Gerin et al., 2008). Over the density range studied (~ 0.2–$0.65\,g/cm^3$), the e_{aq}^- absorption maximum is found to shift only slightly to the red with decreasing density, in agreement with previous work. Assembling the present data together with those already reported in the literature in subcritical and supercritical water shows that $E_{A_{max}}$ varies linearly (in a double logarithmic plot) with density for the various temperatures investigated (namely, 350°C, 375°C, 380°C, 390°C, and 400°C) and that the resulting lines are all parallel (within the experimental uncertainties). The temperature dependence of $E_{A_{max}}$ in subcritical and supercritical D_2O further reveals that, at a fixed pressure (25 MPa), $E_{A_{max}}$ decreases monotonically with increasing temperature in passing through the liquid-SCW phase transition at t_c, but exhibits a minimum at a fixed density (0.2 and $0.65\,g/cm^3$) as the water passes above t_c into SCW, as shown in Figure 15.4. These behaviors can be understood by means of simple microscopic arguments based on the changes that occur in the water properties and water structure in the subcritical and supercritical water regimes. Most importantly, the role of local density and molecular configurational fluctuations (associated with criticality) in providing preexisting polymeric clusters that act as trapping sites for the excess electron is a pivotal point in the interpretation of the data. By comparison with the $(H_2O)_n^-$ cluster data of Ayotte and Johnson (1997) and Coe (2001), in SCW at 400°C, the average cluster size is estimated to be ~ 32 water molecules for $\rho = 0.65\,g/cm^3$ and ~ 26 for $\rho = 0.2\,g/cm^3$, respectively. These cluster size values are consistent with reported experimental clustering data on $(H_2O)_n^-$ ions, indicating that interior-bound excess electron states are energetically favored in these subcritical and supercritical regions. Electrons

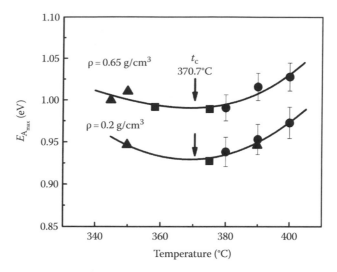

FIGURE 15.4 Energy of maximum absorption ($E_{A_{max}}$) for e_{aq}^- in subcritical and supercritical water (D_2O) as a function of temperature at two different water densities: 0.2 and 0.65 g/cm³. (●) Jay-Gerin et al. (2008), (■) Bartels et al. (2005), and (▲) Jortner and coworkers (Gaathon et al., 1973; Jortner and Gaathon, 1977).

residing in such clusters can, alternatively, be viewed as microscopic probes of the local structure of their host environment, and as such, electron solvation experiments present a powerful tool for future studies of the localization of excess electrons in subcritical and supercritical water where the dominant role of short-range electron–water interactions is clearly affirmed.

15.4.1.3 In Concentrated Aqueous Inert Salts

The absorption spectra of the hydrated electron in deuterated water solutions containing different concentrations of LiCl, LiClO$_4$, MgCl$_2$, or Mg(ClO$_4$)$_2$ were measured by pulse radiolysis techniques from room temperature to 300°C at a pressure of 25 MPa (Lin et al., 2007a; Kumagai et al., 2008). The results show that when the temperature is increased and the density is decreased, the absorption spectrum of the electron in the presence of a metallic cation is shifted to lower energies. Quantum classical molecular dynamics (QCMD) simulations (Nicolas et al., 2003) of an excess electron in bulk water and in the presence of a lithium cation have been performed to compare with the experimental results. The excess electron was treated only quantum mechanically, using the Born–Oppenheimer approximation. The other interactions were described by various empirical models. According to the simulations, the change in the shape of the spectrum is due to one of the three p-like excited states of e_{aq}^- destabilized by core repulsion. The study of $s \rightarrow p$ transition energies for the three p excited states reveals that for temperatures higher than room temperature, there is a broadening of each individual $s \rightarrow p$ absorption band due to a less structured water solvation shell. The increase in the temperature and the increase in the concentration of the nonreactive metal cation have opposite effects. With increase in temperature, the absorption spectra of the solvated electron shifts to the red whereas in the presence of a salt it shifts to the blue. When both effects are present, the shift intensity and the shape of the absorption spectra are modified. However, the effect of the temperature on the absorption spectrum of the solvated electron is stronger than that of the presence of nonreactive metal cation even up to 4 M Li$^+$ ion.

15.4.1.4 In Simple Alcohols

Although the absorption spectra of the solvated electron in simple alcohols were extensively studied in the 1970s, especially by the group of G.R. Freeman, no attention was paid to their behaviors under supercritical conditions. Recently, the absorption spectra in methanol, ethanol, 1-propanol, 2-propanol, and 1-butanol were measured from room temperature to supercritical conditions by Han et al. (Han, 2005; Han et al., 2005, 2008) by using both electron beam pulse radiolysis and laser flash photolysis techniques. In laser photolysis, the solvated electron is produced by photodetachment of electrons from the I⁻ anion dissolved in the alcohols. Because of a redshift of the charge-transfer-to-solvent absorption band of the I⁻ anion, the absorbance of the solvated electron increases with temperature. This is a great advantage for the studies of the spectral shift of the solvated electron in alcohols because its yield and absorption coefficient are much lower than that of the hydrated electron and the decay is much more accelerated due to reactions with the solvent molecules. Nevertheless, the two approaches give the same results, on both the spectral shape and absorption maximum position (EA_{max}). The EA_{max} of the solvated electron in these alcohols decreases linearly with increasing temperature up to supercritical conditions. The temperature coefficients, that is, dEA_{max}/dT, of primary alcohols are higher than that of secondary alcohols.

15.4.1.5 In Polyalcohols

The temperature-dependent absorption spectra of the solvated electron in ethane-1,2-diol (12ED), propane-1,2-diol (12PD), and propane-1,3-diol (13PD) have been investigated up to 300°C (Mostafavi et al., 2004; Lampre et al. 2005; Lin et al., 2006b, 2007a,b). At a given temperature, the shape of the spectrum in a given solvent is independent of time. The absorption band maximum at room temperature is around 570, 565, and 575 nm, for 12ED, 12PD, and 13PD, respectively. These values are in agreement with previously reported ones. By increasing the temperature, the decays become faster and most of the solvated electrons disappear within the electron pulse due to geminate recombination, resulting in a drop in the measured absorbance. However, it can be clearly observed that the maximum of the solvated electron absorption band shifts to the longer wavelength with the temperature rise. The transition energy at the absorption maximum plotted as a function of temperature is correctly fitted by a straight line for each solvent (Figure 15.5). From the slope of the curves, the temperature coefficients (dE_{max}/dT) of -2.5×10^{-3} eV/K in 12ED, -3.1×10^{-3} eV/K in

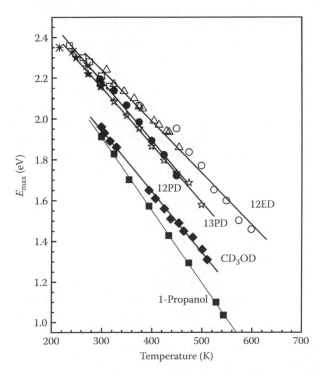

FIGURE 15.5 Energy of the maximum of the solvated electron absorption band (E_{max}) as a function of temperature in different solvents: 12PD (●, Lampre et al., 2005), (✳, Okazaki et al., 1984); 13PD (☆, Lampre et al., 2005), (★, Okazaki et al., 1984); 12ED (□, Okazaki et al., 1984), (△, Chandrasekhar and Krebs, 2000), (○, Lampre et al., 2005); CD₃OD (◆, Herrmann and Krebs, 1995); 1-propanol (■, Han et al., 2008).

12PD, and -2.8×10^{-3} eV/K in 13PD are determined. At lower temperature, Okazaki et al. reported the values of -2.7×10^{-3} and -3.0×10^{-3} eV/K, for 12PD and 13PD, respectively (Okazaki et al., 1984), but if we add their data to Figure 15.5, we note a good agreement with our results and deduce a common temperature coefficient of -2.9×10^{-3} eV/K in 12PD and -2.8×10^{-3} eV/K in 13PD. Those latter values are very close to each other, but are lower than the value in methanol (-3×10^{-3} eV/K, Herrmann and Krebs, 1995) and in 1-propanol (-3.5×10^{-3} eV/K) (Han et al., 2008).

The results show that even if the behavior of the solvated electron in these three polyols as a function of temperature is similar, a few differences exist. As the length of the aliphatic chain and the number of hydroxyl groups are the same for both solvents, our observations emphasize the significant influence of the distance between the two OH groups on the behavior of the trapped electron. At room temperature, the energy of the absorption maximum is higher in 12PD than in 13PD and the absorption band of the solvated electron is narrower in 12PD than in 13PD. These observations indicate that the two neighboring OH groups create deeper electron traps and a narrower distribution of traps in 12PD as compared to 13PD. These results are in agreement with those reported for 12ED at high temperature and for 12ED and 13PD glasses. However, the traps in 12PD appear less deep than in 12ED since the energy of the absorption maximum of the solvated electron measured at a given temperature is lower in 12PD than in 12ED. This shows an influence of the additional methyl group on the solvent structure, in particular on the created three-dimensional networks of hydrogen-bonded molecules. Moreover, an increase in temperature affects electron trapping in 12PD more than in 13PD since the temperature coefficient is more negative. The shape of the spectrum changes only in 12PD. Taking into account that the decrease in viscosity versus temperature is also faster in 12PD, the different behavior of the two solvents results from larger modifications of the solvent structure and molecular interactions for 12PD.

15.4.2 •OH Radical

At room temperature the •OH radical has a broad absorption band with a maximum around 230 nm and a long-wavelength tail that extends beyond 320 nm. The absorption band at deep UV (low intensity of analyzing light) and the low absorption coefficient become an obstacle for precise measurements. The use of a double monochromator is highly recommended due to the scattering of the longer wavelength light. In their very first paper on high-temperature pulse radiolysis, Christensen and Sehested reported •OH spectrum changes from 20°C to 200°C (Christensen and Sehested, 1980). A decade later, Buxton et al. reported the temperature-dependent absorption spectra of •OH and •OD and claimed no significant changes with temperature (Elliot and Buxton, 1992; Buxton et al., 1998). Recently, Janik et al. reported their pulse radiolysis study of N_2O-saturated water up to 350°C (Janik et al., 2007a,b). The UV absorption spectra of •OH showed a decrease in the primary band at 230 nm and growth of a weak band at 310 nm at elevated temperatures, with an isosbestic point near 305 nm. They interpreted the 230 nm band as due to hydrogen-bonded •OH, and the 310 nm band corresponding to "free" •OH. A decrease in the absorption coefficient of •OH at elevated temperatures was also reported.

15.4.3 Other Transient Species

Table 15.2 lists the spectral shifts of some of the intermediate radicals at room temperature and elevated temperatures (in most cases, above t_c), respectively. Generally, the anion or cation radicals show a redshift while the neutral radical of aromatic compounds show a blueshift, except that $CO_3^{•-}$ (Wu et al., 2002b) and $MV^{•+}$ (Lin et al., 2004) exhibit no change with increasing temperature. This is probably due to their fairly good symmetric molecular structure and the delocalization of the electric charge. In general, the optical absorption spectral shift is in fact a problem of solute–solvent interaction. The direction of shift should be largely related to the nature of the intermediate radical. The spectrum shift reflects a change in the energy difference between the ground state and the excited state with the changes of solvent environment. These energies reflect the differences in solvation where the solvent responds to a solute by means of lowering the energy of the system. Recently, Wu et al. have

TABLE 15.2
Spectral Shifts of Various Intermediate Radicals

Chemical	Radical Form	λ_{max} at RT	λ_{max} in SCW[a]	Reference
Benzophenone	$\phi_2\text{·COH}$	545 nm	525 nm (400°C)	Wu et al. (2002a,b)
	$\phi_2\text{·CO}^-$	610 nm	650 nm (300°C)	Wu et al. (2002a,b)
Thiocynate	$(\text{SCN})_2^{·-}$	470 nm	510 nm (400°C)	Wu et al. (2001a,b)
Carbonate	$\text{CO}_3^{·-}$	600 nm	600 nm (400°C)	Wu et al. (2002a,b)
Methyl viologen	$\text{MV}^{·+}$	605 nm	605 nm (400°C)	Lin et al. (2004)
Silver	$\text{Ag}°$	355 nm	370 nm (200°C)	Mostafavi et al. (2002)
	Ag_2^+	310 nm	380 nm (380°C)	Mostafavi et al. (2002)
4,4'-bipyridyl	$44\text{BpyH}^·$	540 nm	525 nm (400°C)	Lin et al. (2005)
Formate	$\text{CO}_2^{·-}$	235 nm	275 nm (380°C)	Lin et al. (2008)

[a] λ_{max} is density dependent in SCW; here we choose a typical value.

qualitatively interpreted the redshift of the benzophenone anion and the blueshift of the neutral ketyl radical in terms of electrostatic forces and the hydrogen bonding of water (Wu et al., 2002a,b). Boutin et al. have quantitatively explained the redshifts of $\text{Ag}°$ and Ag_2^+ using QCMD simulations (Boutin et al., 2005). Nevertheless, a more general and precise model remains to be developed, taking into account the hydrogen bonding network, the dielectric constant, the electrostatic forces and polarity of solute, the clustering effects of water molecules under supercritical conditions, etc.

15.5 RADIOLYTIC YIELDS OF WATER DECOMPOSITION PRODUCTS

The estimation of G-values of water decomposition products can be done by pulse radiolysis techniques or steady state radiolysis with product analysis methods. For pulse radiolysis, a direct measurement of the transient species is desirable. This is difficult for nanosecond pulse radiolysis because of the rapidity of spur reactions and the limitation of detection techniques (e.g., the absorption of the OH radical is in the far UV region with a rather small absorption coefficient). One is forced to adopt the scavenging method, that is, to use a chemical additive to react with the transient species that forms another easy-to-detect, relatively stable product. In this section, we introduce the estimation of the G-values of water decomposition products by pulse radiolysis, with the support of steady state γ-radiolysis of some aromatic compounds.

15.5.1 $G(e_{aq}^-)$

One suitable scavenging system involves the use of 0.5 mM methyl viologen (MV^{2+}) in the presence of *tert*-butanol as an OH radical scavenger, under neutral pH conditions (Lin et al., 2004). Another involves the use of 0.5 mM 4,4'-bipyridyl (4,4'-bpy) together with *tert*-butanol in alkaline solution (pH > 11) (Lin et al., 2005). In both cases, ideally only e_{aq}^- reacts with MV^{2+} or 4,4'-bpy to form a radical cation $\text{MV}^{·+}$ or radical anion 4,4'-bpy$^{·-}$, with fairly strong absorption at 605 and ~540 nm, respectively. The temperature-dependent $G(e_{aq}^-)$ from 25°C to 500°C at a fixed pressure of 25 MPa is shown in Figure 15.6a. $G(e_{aq}^-)$ increases with temperature up to 300°C, which agrees well with the previous reports. It decreases to a minimum near t_c before jumping to a rather high value at 400°C, and then it decreases again with increasing temperature up to 500°C. An independent measurement of $G(e_{sol}^-)$ in methanol shows a similar trend (Figure 15.6b).

The big change in $G(e_{sol}^-)$ at constant pressure is due to density effects. As displayed in the 3D plots of Figure 15.7, under supercritical conditions at a fixed density, $G(e_{sol}^-)$ decreases with increasing temperature while at a fixed temperature, $G(e_{sol}^-)$ decreases with increasing water density. Around t_c, the density effect is the most significant, but it becomes less important as temperature increases.

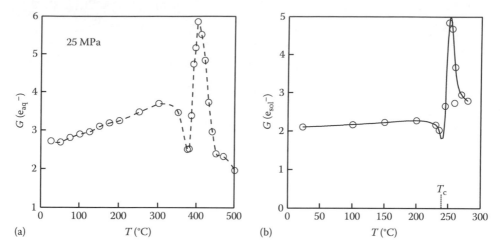

FIGURE 15.6 (a) $G(e_{aq}^{-})$ as a function of temperature at 25 MPa. (b) $G(e_{sol}^{-})$ of methanol as a function of temperature at 9.0 MPa.

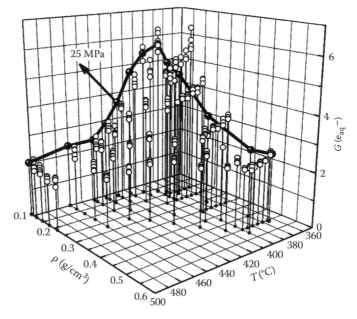

FIGURE 15.7 3D plots of $G(e_{aq}^{-})$ as a function of temperature and water density. The black curve shows the values at a fixed pressure of 25 MPa.

From the viewpoint of the W-value and the initial yield of the hydrated electron (recent values are 4.0–4.2 molecules/100 eV (Bartels et al., 2000; Muroya et al., 2002), $G(e_{aq}^{-})$ in low-density SCW (e.g., at 400°C/25 MPa, $G(e_{aq}^{-})$ is about 5.8 molecules/100 eV) seems to be overestimated. This could be due to the use of an extrapolated absorption coefficient for MV$^{•+}$ or incomplete scavenging of the H atom by *tert*-butanol (Lin et al., 2004). However, an increase in $G(e_{aq}^{-})$ due to water structure changes near t_c, such as the formation of monomers, dimers, and small clusters should not be excluded.

In a recent report, Sims (2006) reanalyzed the data reported by W.G. Burns and W.R. Marsh (Burns and Marsh, 1981) and gave a value of $G(e_{aq}^{-})$ for 400°C at a water density of 0.45 g/cm^3 as 0.8 molecules/100 eV. This is only about one third of the value reported by Lin et al. In their β/γ radiolysis of water combined with gas product analysis by mass spectrometry (Janik et al., 2007a,b), Janik et al. reported that their measured $G(e_{aq}^{-})$ at 380°C and 400°C agreed with that of Lin et al. (2004), but there was a big discrepancy in low-density water and the density effect on $G(e_{aq}^-)$ was not very significant. Although they claimed that the ratio of $G(\cdot H)/G(e_{aq}^-)$ agreed with their previous work (Cline et al., 2002), there was an apparent discrepancy with the kinetic measurements of density effect on the yield (relative intensity of absorption signals at 1200 nm by 4 ns pulse radiolysis) of e_{aq}^- at 380°C in the same report. In fact, the trend of their "direct" measurements of the decay of e_{aq}^- is in better agreement with Lin et al. (2004). A direct measurement with higher time resolution, that is, picosecond or sub-picosecond pulse radiolysis, might eventually resolve the discrepancy.

15.5.2 $\{G(e_{aq}^-) + G(OH) + G(H)\}$

Two scavenging systems have been used to evaluate $\{G(e_{aq}^-) + G(H) + G(OH)\}$ (denoted as G_{sum} hereafter). One is 0.5 mM MV^{2+} with 0.2 M ethanol, another is 0.5 mM 4,4'-bpy in the presence of 10 mM HCOONa (Lin et al., 2004). The solutions are deaerated with Ar gas. Ethanol MV^{2+} and $HCOO^-$ are used to convert $\cdot OH$ radical and $\cdot H$ atom to ethanol radical and $COO^{\cdot-}$, which will subsequently reduce MV^{2+} and 4,4'-bpy to form $MV^{\cdot+}$ and 4,4'-bpyH, the same products produced by e_{aq}^-. Therefore, the total yields of $MV^{\cdot+}$ or 4,4'-bpyH should be equal to the total yield G_{sum}. Figure 15.8a

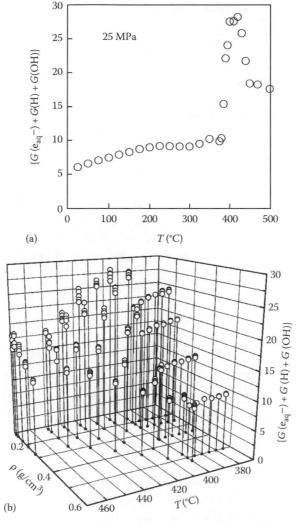

FIGURE 15.8 (a) $\{G(e_{aq}^-) + G(H) + G(OH)\}$ as a function of temperature at 25 MPa. (b) 3D plots of $\{G(e_{aq}^-) + G(H) + G(OH)\}$ as a function of temperature and water density.

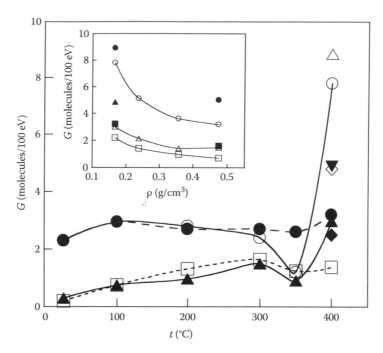

FIGURE 15.9 *G*-Values of benzophenone (BP) consumption and phenol formation at various temperatures and pressures by γ-radiolysis of BP in aqueous solutions. (○) Ar, 25 MPa, BP consumption; (▲) Ar, 25 MPa, phenol formation; (●) Ar, 35 MPa, BP consumption; (□) 35 MPa, phenol formation; (△) N_2O, 25 MPa, BP consumption; (◇) N_2O, 25 MPa, phenol formation; (◆) N_2O, 35 MPa, BP consumption; and (▼) N_2O, 35 MPa, phenol formation. Inset: Density dependence at 400°C of total *G*-values of BP consumption and products formation. (○) Ar, BP consumption; (△) Ar, phenol formation; (□) Ar, 9-fluorenone formation; (●) N_2O, BP consumption; (▲) N_2O, phenol formation; and (■) N_2O, 9-fluorenone formation.

shows the temperature dependence of G_{sum} at 25 MPa from room temperature to 500°C and Figure 15.8b displays the 3D plots of G_{sum} as a function of temperature and water density. Their trends are very similar to that of $G(e_{aq}^-)$. An estimation of the yield of COO$^{\bullet-}$ up to 400°C by pulse radiolysis of 10 mM sodium formate in N_2O-saturated solution supported these results (Lin et al., 2008).

Figure 15.9 shows the experimental results of γ-radiolysis of benzophenone from room temperature to 400°C, in deaerated solutions or N_2O-saturated solutions (Miyazaki et al., 2006a). Apparently, the tendency of temperature dependence and density effects (inset figure) on the yields of benzophenone decomposition and product formation is similar to that of $G(e_{aq}^-)$ and G_{sum}. The studies of other aromatic compounds such as phenol and benzene gave similar results. All these strongly imply that the decomposition of the aromatic compounds is related to the yields of water decomposition under irradiation. However, it should be pointed out that the reactions during steady state radiolysis are much more complicated. Some reverse reactions even give back the reactant. Thus the yields of solute decomposition are generally much lower than those of water decomposition products (Miyazaki et al., 2006a,b).

15.5.3 *G*(OH)

The estimation of *G*(OH) has been carried out using an aerated solution of 100 mM $NaHCO_3$ or a deaerated solution of 100 mM $NaHCO_3$ in the presence of 1 mM $NaNO_3$. In these systems, hydrated electrons are scavenged by O_2 or NO_3^- while the reaction between H atoms and HCO_3^- is rather slow. Consequently, the yield of $CO_3^{\bullet}$ would correspond to *G*(OH). Figure 15.10 shows *G*(OH) as a function of temperature. From room temperature to 380°C, the pressure is 25 MPa

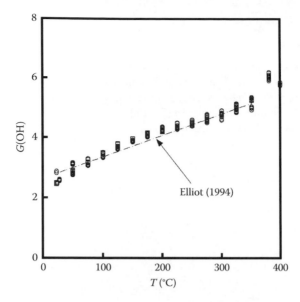

FIGURE 15.10 $G(OH)$ as a function of temperature at 25 MPa (35 MPa at 400°C). The line is based on the equation given by Elliot (1994).

while at 400°C it is 35 MPa because the solubility of NaHCO$_3$ at 400°C/25 MPa is too small to perform the measurements.

15.5.4 $G(H)$

$G(H)$ has not been measured directly by pulse radiolysis, but it can be calculated by a subtraction of $G(e_{aq}^-)$ and $G(OH)$ from G_{sum}. The trend is that it slightly increases as temperature increases up to 200°C and drops to a minimum around t_c, and then increases sharply. At 400°C/25 MPa ($\rho = 0.167$ g/cm^3), it reaches a value of 7.5 molecules/100 eV. If this value is divided by $G(e_{aq}^-)$, then we have a ratio of $G(H)/G(e_{aq}^-) \approx 1.3$. Considering that $G(e_{aq}^-)$ might be overestimated, the ratio $G(H)/G(e_{aq}^-)$ could be higher than 1.3. As is known, at room temperature, $G(H)/G(e_{aq}^-)$ is about 0.16. This qualitatively agrees with the result reported by Cline et al., who derived the ratio $G(H)/G(e_{aq}^-)$ from the fitting of the decay kinetics of hydrated electrons scavenged by SF$_6$ under alkaline conditions (Cline et al., 2002). They found that at 380°C, the ratio is greater than unity and becomes larger as density decreases. In low-density supercritical water, the initial yield of H atoms seems to be roughly five to six times the yield of hydrated electrons. Considering the uncertainty of the fittings, these two results seem to be consistent.

The yield of molecular products, especially $G(H_2)$ and $G(H_2O_2)$, is difficult to evaluate by pulse radiolysis techniques. This must be clarified by experiments, although H$_2$O$_2$ might be thermally unstable under supercritical conditions. At least one of these two molecular products should be measured by product analysis, the other component could be calculated by the material balance equation:

$$G(-H_2O) = G(e_{aq}^-) + G(H) + 2G(H_2) = G(OH) + 2G(H_2O_2)$$

15.6 RATE CONSTANTS

In this section, we do not attempt to describe in detail all the reactions that have been studied. Instead we focus on the common features of the radiation induced reactions with the two most important species of water radiolysis—the hydrated electron and hydroxyl radical.

15.6.1   REACTIONS WITH e_{aq}^-

15.6.1.1   Ionic Reactants

The temperature-dependent reaction rates of e_{aq}^- with H^+, NO_3^-, and NO_2^- have been reported (Cline et al., 2002; Takahashi et al., 2002; Wu et al., 2002a,b). As mentioned above, the dielectric constant of water is similar to nonpolar organic solvents, and the dissociation constants of inorganic salts are extremely small around the supercritical point. It is thought that these properties affect those ionic reactions that are coulombic-force-influenced. As shown in Figure 15.11, for the e_{aq}^- reaction with H^+, the rate strongly increases between 250°C and 350°C (Cline et al., 2002; Wu et al., 2002a,b). For NO_3^-, the rate constant increases as the temperature increases, reaches a maximum around 125°C–200°C, and then decreases (Takahashi et al., 2002).

The rate constants for diffusion-controlled reactions can be described by the Debye–Smoluchowski equation as follows:

$$k_{\text{diff}} = 4\pi RDF_{\text{D}},$$

$$F_{\text{D}} = \frac{r_{\text{c}}/R}{\exp(r_{\text{c}}/R) - 1}$$

where

    F_{D} is the Debye factor
    R is the reaction distance
    $r_{\text{c}} \equiv e^2/\varepsilon k_{\text{B}}T$ (e, electronic charge; ε, dielectric constant; k_{B}, Boltzmann constant) represents the critical distance at which the electron-positive-ion potential energy is numerically equal to $k_{\text{B}}T$

Because the dielectric constant of water decreases as temperature increases, the coulomb potential between reactants will change significantly. Thus the rate of the proton–electron reaction is

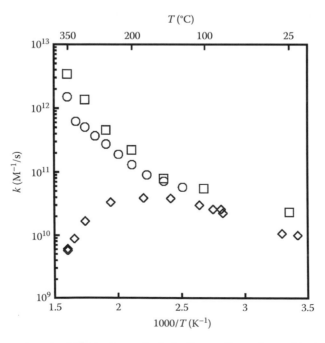

FIGURE 15.11 Temperature-dependent rate constants for the reactions of e_{aq}^- with nitrate ion ($\diamond$, 25.7 MPa, Takahashi et al., 2002) and proton ($\circ$, 25.7 MPa, Takahashi et al., 2002; $\square$, 25 MPa, Wu et al., 2002a,b).

accelerated further by attraction, while the nitrate–electron reaction is retarded due to repulsion between the two anions.

For a non-diffusion-controlled reaction, the Noyes equation is applied as follows:

$$\frac{1}{k_{obs}} = \frac{1}{k_{diff}} + \frac{\exp(r_c/R)}{k_{act}}$$

where k_{act} is the activated rate constant. There are two limiting cases, when $k_{act} \gg k_{diff}$, the reaction will occur on every encounter, so that $k_{obs} \approx k_{diff}$; when $k_{act} \ll k_{diff}$, the reaction rate will be given by $k_{obs} \approx k_{act}$.

15.6.1.2 Hydrophobic or Neutral Species

Reaction rates of e_{aq}^- with O_2, SF_6, N_2O, nitrobenzene, and 4,4'-Bpy have been investigated (Cline et al., 2002; Marin et al., 2002; Takahashi et al., 2004; Lin et al., 2005). For temperatures <300°C, the rate constants for scavenging by O_2 or SF_6 follow Arrhenius behavior but become increasingly dependent on water density (pressure) at higher temperatures. At fixed temperatures of 360°C, 370°C, 380°C, and 400°C, the rate constants for these reactions reach a distinct minimum near 0.45 g/cm³ (Figure 15.12). This behavior has been interpreted in terms of the potential of mean force separating an ion (OH⁻ or e_{aq}^-) from a hydrophobic species (H, O_2, or SF_6) in the compressible fluid (Cline et al., 2002; Marin et al., 2002). As is well known, around the supercritical point, the inhomogeneity of water density or the clustering effect of water molecules becomes a very important feature, which would affect the distribution of the hydrophobic reactant and hydrated electron. For example, the hydrophobic molecules such as O_2 would exist in the lower density region with higher probability. This kind of inhomogeneous distribution will hinder the encounter of hydrophobic molecules with hydrated electrons, thus the reaction rates decrease dramatically.

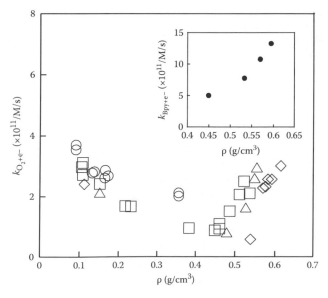

FIGURE 15.12 Density effect on the rate constant of the reaction of e_{aq}^- with O_2 at ($\triangle$) 360°C, ($\lozenge$) 370°C, ($\square$) 380°C, and ($\bigcirc$) 400°C (Cline et al., 2002). Inset: density effect on the rate constant of the reactions of e_{aq}^- with 4,4'-Bpy at 380°C (Lin et al., 2005).

15.6.2 Reactions with ·OH Radical

The ·OH radical is probably the most important oxidizing species related to the corrosion of structural materials in nuclear reactors. The studies of the reactions of ·OH radical at elevated temperatures or in supercritical water are thus essential. Most such studies have been carried out with aromatic compounds, or simple molecules such as CO_3^{2-} or HCO_3^-, partly due to their excellent thermal stability (Ferry and Fox, 1998, 1999; Feng et al., 2002; Marin et al., 2002; Wu et al., 2002a,b; Ghandi and Percival, 2003). Figure 15.13 shows a typical example of the temperature-dependent behavior for the reactions of ·OH radical with aromatic compounds. The measured bimolecular rate constant of ·OH radical with nitrobenzene shows distinctly non-Arrhenius behavior below 300°C and an increase in the subcritical and supercritical region. Feng et al. (2002) succeeded in modeling these data with a three-step reaction mechanism originally proposed by Ashton et al. (1995), while Ghandi and Percival (2003) claimed to have developed a so-called multiple collisions model to predict the rates for the reactions of the ·OH radical in subcritical and supercritical water.

It is worth mentioning that the rate constant for the reaction of ·OH radical with H_2, which is one of the most important reactions in water chemistry, has also been studied up to 350°C by competition kinetics using nitrobenzene as an ·OH scavenger (Marin et al., 2003). At higher temperatures, the value of the rate constant is lower than the extrapolation of the Arrhenius plot and actually decreases in value above 275°C. At 350°C, the measured rate constant is more than a factor of 5 below the Arrhenius extrapolation. This implies that the amount of hydrogen injection calculated by the current model of water radiolysis in a nuclear reactor might need reconsideration.

In a recent study by Janik et al., the rate constant for the self-recombination of ·OH in aqueous solution giving H_2O_2 was measured from 150°C to 350°C by direct measurement of the OH radical transient optical absorption at 250 nm (Janik et al., 2007b). The values of the rate constant are smaller than predicted previously in the 200°C–350°C range and show no apparent change in this regime. In combining their measurements with previous results, the non-Arrhenius behavior can be well described in terms of the Noyes equation $k_{obs}^{-1} = k_{act}^{-1} + k_{diff}^{-1}$, using the diffusion-limited rate constant k_{diff} estimated from the Smoluchowski equation and an activated barrier rate k_{act} nearly equal to the gas-phase high-pressure limiting rate constant for this reaction.

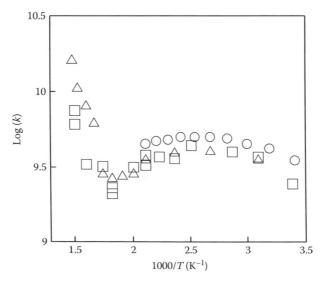

FIGURE 15.13 The bimolecular rate constant of ·OH with nitrobenzene shows non-Arrhenius behavior below 300°C and an increase in the subcritical and supercritical region. (○) Ashton et al. (1995); (□) Feng et al. (2002); and (△) Marin et al. (2002).

15.7   MONTE CARLO AND MOLECULAR DYNAMICS SIMULATIONS

15.7.1   SOLVATION DYNAMICS OF e_{aq}^- IN SCW

Boero et al. performed a first principles study of the hydrated electron in water at ordinary and supercritical conditions (Boero et al., 2003). According to their study, for normal water, the hydrogen bond (H-bond) network needs about 1.6 ps to accommodate the additional e^-. Subsequently, the delocalized state becomes localized, the H bonds of the system rearrange, and water molecules reorient in such a way that a cavity forms in the system. Six water molecules reorient in such a way that all of them point at least one H toward the electronic cloud, thus forming a solvation shell and the electron distribution is assumed to be ellipsoidal. Those cavities survive for a period ranging from ~50 fs to a maximum of about 90 fs. In contrast, under supercritical conditions the H-bond network is not continuous and the electron localizes in preexisting cavities in a more isotropic way. Four water molecules form the solvation shell and the localization time shortens significantly.

15.7.2   SPECTRAL SHIFT OF SOLVATED ELECTRON

Boutin et al. have performed quantum-classical molecular-dynamics (QCMD) simulations to study the temperature and density dependence of the absorption spectra of the hydrated electron at elevated temperatures and supercritical conditions (Nicolas et al., 2003; Boutin et al., 2005). Their simulation quantitatively explains the experimental data to some extent. When extending these simulations to very low densities corresponding to supercritical conditions, the results display a progressive and continuous "desolvation" of the hydrated electron. It was also suggested that the observed shift could be a density rather than a temperature effect. However, these expectations are in marked contrast to experimental results that exhibit only weak density dependence in a wide density interval of SCW, from ~0.1 to 0.6 g/cm^3 (Jortner and Gaathon, 1977; Bartels et al., 2005; Jay-Gerin et al., 2008).

15.7.3   TEMPORAL BEHAVIORS OF THE HYDRATED ELECTRON

The radiolytic yield is time dependent and at any given time $G(e_{aq}^-)$ is the result of the competition between diffusion into the bulk medium and intra-spur reactions. Both diffusion and reaction are known to be temperature dependent, but which process dominates at high temperatures is unknown. Since these processes occur at the beginning, a few tens of seconds after the formation of e_{aq}^-, such a short temporal domain is not easily accessible experimentally. Thus computer simulations are very helpful.

Two approaches have been reported. One is the deterministic diffusion-kinetic modeling approach and the other is the stochastic Monte Carlo simulations approach. For the deterministic approach, the spur diffusion model is based on the concept of an average spur with a deposited energy of 62.5 eV. This produces an initial yield of electrons and ionized and excited water molecules that undergo thermalization through random collisions. The Monte Carlo simulations are more complex, and start from the "physical stage"—consisting of the period prior to 10^{-14} s, during which energy is deposited by the primary radiation particles and by all of the secondary (tertiary, and so forth) electrons that result from the ionization of the water molecules. The subsequent non-homogeneous chemical evolution of the reactive species formed in the spurs (or tracks) was simulated by using the independent reaction times approximation. Both methods have been reported to give satisfactory results when compared to existing experimental data for the radiolytic yields of water decomposition products up to 300°C. However, there is some difference in the temperature-dependent temporal behaviors of e_{aq}^-, especially in the first few nanoseconds. There is also a contradictory opinion about the temperature dependence on the thermalization distances of electrons. By Monte Carlo simulations, du Penhoat et al. (2000) demonstrated that the best agreement with experimental results occurs when the distances of electron thermalization decrease with increasing temperature while the spur model suggests an increase with temperature (Laverne and Pimblott, 1993; Swiatla-Wojcik and Buxton, 1995). All of these should be examined by experiments with better time resolution.

Recently a kinetic analysis of the hydrated electron was performed up to 350°C in the picosecond time range (Baldacchino et al., 2006). The results showed increasing competition between the initial size of the spurs, the accelerated diffusion of the species and the temperature-activated recombination. Thus, the radiolytic yield of the hydrated electron remains stable between 100 ps and 3 ns at 350°C (Baldacchino et al., 2006).

15.8 CONCLUDING REMARKS

Under subcritical or supercritical conditions, the radiolytic yields of water decomposition products, the reaction rate constants, and the spectral properties of transient species show significant differences from those of water at ambient condition or even at elevated temperatures below 300°C. These properties are not only dependent on temperature but also greatly affected by water density in SCW, or in other words, they exhibit nonlinear behaviors or even non-monotonic functions with temperature at a fixed pressure. This would strongly suggest a reconsideration of the current water radiolysis model in nuclear reactors, especially for the future SCW reactors. On the other hand, these unusual behaviors are certainly related to the peculiar water properties and water structure of SCW such as dielectric constant, hydrogen bonding network, clustering, and density inhomogeneity. A thorough understanding of this field will require the involvement of gas-phase chemistry, cluster chemistry, computational chemistry, etc. For radiation chemistry, the fundamental studies of temperature and density effects of electron solvation, spur reaction processes and other more general chemical interests in these intriguing reaction media are fairly meaningful. To achieve these goals, higher performance equipment such as picosecond or sub-picosecond pulse radiolysis systems and more sophisticated theoretical calculations are essential.

ACKNOWLEDGMENTS

We would like to thank Professor Y. Hatano for his continuous support of our studies on the radiation chemistry of high temperature and supercritical fluids as well as his encouragement to write this chapter. We are also grateful to T. Ueda and Professor M. Uesaka for their technical assistance in pulse radiolysis experiments and encouragement. We are also thankful to Professor M. Mostafavi, Professor J.P. Jay-Gerin, Dr. Y. Muroya, Dr. G. Wu, Dr. T. Miyazaki, Dr. H. He, Dr. Z. Han, and many other students who have contributed to these studies.

REFERENCES

Akiya, N. and Savage, P. E. 2002. Roles of water for chemical reactions in high-temperature water. *Chem. Rev.* 102 (8): 2725–2750.

Ashton, L., Buxton, G. V., and Stuart, C. R. 1995. Temperature-dependence of the rate of reaction of OH with some aromatic-compounds in aqueous-solution — Evidence for the formation of a π-complex intermediate. *J. Chem. Soc. Faraday Trans.* 91 (11): 1631–1633.

Ayotte, P. and Johnson, M. A. 1997. Electronic absorption spectra of size-selected hydrated electron clusters: $(H_2O)_n^-$, $n = 6$–50. *J. Chem. Phys.* 106 (2): 811–814.

Baldacchino, G., De Waele, V., Monard, H., Sorgues, S., Gobert, F., Larbre, J. P., Vigneron, G., Marignier, J. L., Pommeret, S., and Mostafavi, M. 2006. Hydrated electron decay measurements with picosecond pulse radiolysis at elevated temperatures up to 350°C. *Chem. Phys. Lett.* 424 (1–3): 77–81.

Bartels, D. M., Cook, A. R., Mudaliar, M., and Jonah, C. D. 2000. Spur decay of the solvated electron in picosecond radiolysis measured with time-correlated absorption spectroscopy. *J. Phys. Chem. A* 104 (8): 1686–1691.

Bartels, D. M., Takahashi, K., Cline, J. A., Marin, T. W., and Jonah, C. D. 2005. Pulse radiolysis of supercritical water. 3. Spectrum and thermodynamics of the hydrated electron. *J. Phys. Chem. A* 109 (7): 1299–1307.

Bernas, A., Ferradini, C., and Jay-Gerin, J. P. 1996. Excess electrons in homogeneous and heterogeneous porous media. *Can. J. Chem.* 74 (1): 1–23.

Boag, J. W. and Hart, E. J. 1963. Absorption spectra in irradiated water and some solutions. *Nature* 197: 45–47.

Boero, M., Parrinello, M., Terakura, K., Ikeshoji, T., and Liew, C. C. 2003. First-principles molecular-dynamics simulations of a hydrated electron in normal and supercritical water. *Phys. Rev. Lett.* 90 (22): 226403.

Boutin, A., Spezia, R., Coudert, F. X., and Mostafavi, M. 2005. Molecular dynamics simulations of the temperature and density dependence of the absorption spectra of hydrated electron and solvated silver atom in water. *Chem. Phys. Lett.* 409 (4–6): 219–223.

Burns, W. G. and Marsh, W. R. 1981. Radiation-chemistry of high-temperature (300–410°C) water.1. Reducing products from gamma-radiolysis. *J. Chem. Soc. Faraday Trans. I* 77: 197–215.

Buxton, G. V. 2001. High temperature water radiolysis, in *Radiation Chemistry–Present Status and Future Trends*. Eds. Jonah, C. D. and Rao, B.S.M. Elsevier: Tokyo. pp. 145–162.

Buxton, G. V., Lynch, D. A., and Stuart, C. R. 1998. Radiation chemistry of D_2O — Time dependence of the yields of the radicals at ambient temperature and rate constants for the reactions ˙OD + ˙OD and ˙D + ˙OD up to 200°C. *J. Chem. Soc. Faraday Trans.* 94 (16): 2379–2383.

Chandrasekhar, N. and Krebs, P. 2000. The spectra and the relative yield of solvated electrons produced by resonant photodetachment of iodide anion in ethylene glycol in the temperature range 296 < T < 453 K. *J. Chem. Phys.* 112 (13): 5910–5914.

Christensen, H. and Sehested, K. 1980. Pulse-radiolysis at high-temperatures and high-pressures. *Radiat. Phys. Chem.* 16 (2): 183–186.

Christensen, H. C. and Sehested, K. 1986. The hydrated electron and its reactions at high temperatures. *J. Phys. Chem.* 90 (1): 186–190.

Cline, J., Takahashi, K., Marin, T. W., Jonah, C. D., and Bartels, D. M. 2002. Pulse radiolysis of supercritical water. 1. Reactions between hydrophobic and anionic species. *J. Phys. Chem. A* 106 (51): 12260–12269.

Coe, J. V. 2001. Fundamental properties of bulk water from cluster ion data. *Int. Rev. Phys. Chem.* 20 (1): 33–58.

Darr, J. A. and Poliakoff, M. 1999. New directions in inorganic and metal-organic coordination chemistry in supercritical fluids. *Chem. Rev.* 99 (2): 495–541.

Dixon, R. S. and Lopata, V. J. 1978. Spectrum of the solvated electron in heavy water up to 445 K. *Radiat. Phys. Chem.* 11 (3): 135–137.

du Penhoat, M. A. H., Goulet, T., Frongillo, Y., Fraser, M. J., Bernat, P., and Jay-Gerin, J. P. 2000. Radiolysis of liquid water at temperatures up to 300°C: A Monte Carlo simulation study. *J. Phys. Chem. A* 104 (50): 11757–11770.

Eckert, C. A., Knutson, B. L., and Debenedetti, P. G. 1996. Supercritical fluids as solvents for chemical and materials processing. *Nature* 383 (6598): 313–318.

Elliot, A. J. 1994. Rate constants and G-values for the simulation of the radiolysis of light water over the range 0–300°C. Report No. AECL-11073, COG-94-167.

Elliot, A. J. and Buxton, G. V. 1992. Temperature-dependence of the reactions $OH + O_2$ and $OH + HO_2$ in water up to 200°C. *J. Chem. Soc. Faraday Trans.* 88 (17): 2465–2470.

Elliot, A. J., Ouellette, D. C., and Stuart, C. R. 1996. The temperature dependence of the rate constants and yields for the simulation of the radiolysis of heavy water. Report No. AECL-11658, COG-96-390-1.

Feng, J., Aki, S. N. V. K., Chateauneuf, J. E., and Brennecke, J. F. 2002. Hydroxyl radical reactivity with nitrobenzene in subcritical and supercritical water. *J. Am. Chem. Soc.* 124 (22): 6304–6311.

Ferry, J. L. and Fox, M. A. 1998. Effect of temperature on the reaction of HO• with benzene and pentahalogenated phenolate anions in subcritical and supercritical water. *J. Phys. Chem. A* 102 (21): 3705–3710.

Ferry, J. L. and Fox, M. A. 1999. Temperature effects on the kinetics of carbonate radical reactions in near-critical and supercritical water. *J. Phys. Chem. A* 103 (18): 3438–3441.

Ghandi, K. and Percival, P. W. 2003. Prediction of rate constants for reactions of the hydroxyl radical in water at high temperatures and pressures. *J. Phys. Chem. A* 107 (17): 3005–3008.

Gottschall, W. C. and Hart, E. J. 1967. The effect of temperature on the absorption spectrum of the hydrated electron and on its bimolecular recombination reaction. *J. Phys. Chem.* 71 (7): 5.

Han, Z. 2005. Temperature and pressure dependent absorption spectra, kinetics and yields of solvated in alcohols from room temperature to supercritical condition studied by pulse radiolysis and laser photolysis. PhD thesis, The University of Tokyo, Tokyo, Japan.

Han, Z., Katsumura, Y., Lin, M., He, H., Muroya, Y., and Kudo, H. 2005. Temperature and pressure dependence of the absorption spectra and decay kinetics of solvated electrons in ethanol from 22 to 250°C studied by pulse radiolysis. *Chem. Phys. Lett.* 404 (4–6): 267–271.

Han, Z., Katsumura, Y., Lin, M., He, H., Muroya, Y., and Kudo, H. 2008. Effect of temperature on the absorption spectra of the solvated electron in 1-propanol and 2-propanol: Pulse radiolysis and laser photolysis studies at temperatures up to supercritical condition. *Radiat. Phys. Chem.* 77 (4): 409–415.

Hart, E. J. and Anbar, M. 1970. *The Hydrated Electron*. New York: Wiley-Interscience.

Hart, E. J. and Boag, J. W. 1962. Absorption spectrum of the hydrated electron in water and in aqueous solutions. *J. Am. Chem. Soc.* 84: 4090–4095.

Herrmann, V. and Krebs, P. 1995. Temperature-dependence of optical absorption spectra of solvated electrons in CD_3OD for $T \leq T_c$. *J. Phys. Chem.* 99 (18): 6794–6800.

Hoffmann, M. M. and Conradi, M. S. 1997. Are there hydrogen bonds in supercritical water? *J. Am. Chem. Soc.* 119 (16): 3811–3817.

Ikushima, Y., Hatakeda, K., Saito, N., and Arai, M. 1998. An in situ Raman spectroscopy study of subcritical and supercritical water: The peculiarity of hydrogen bonding near the critical point. *J. Chem. Phys.* 108 (14): 5855–5860.

Janik, D., Janik, I., and Bartels, D. M. 2007a. Neutron and beta/gamma radiolysis of water up to supercritical conditions. 1. Beta/gamma yields for H_2, $H^\bullet$ atom, and hydrated electron. *J. Phys. Chem. A* 111 (32): 7777–7786.

Janik, I., Bartels, D. M., and Jonah, C. D. 2007b. Hydroxyl radical self-recombination reaction and absorption spectrum in water up to 350°C. *J. Phys. Chem. A* 111 (10): 1835–1843.

Jay-Gerin, J. P. 1997. Correlation between the electron solvation time and the solvent dielectric relaxation times τ_2 and τ_{L1}, in liquid alcohols and water: Towards a universal concept of electron solvation? *Can. J. Chem.* 75 (10): 1310–1314.

Jay-Gerin, J. P., Lin, M., Katsumura, Y., He, H., Muroya, Y., and Meesungnoen, J. 2008. Effect of water density on the absorption maximum of hydrated electrons in sub- and supercritical water up to 400°C. *J. Chem. Phys.* 129 (11): 114511.

Jessop, P. G., Ikariya, T., and Noyori, R. 1994. Homogeneous catalytic-hydrogenation of supercritical carbondioxide. *Nature* 368 (6468): 231–233.

Jortner, J. and Gaathon, A. 1977. Effects of phase density on ionization processes and electron localization in fluids. *Can. J. Chem.* 55 (11): 1801–1819.

Jou, F. Y. and Freeman, G. R. 1979a. Band resolution of optical spectra of solvated electrons in water, alcohols, and tetrahydrofuran. *Can. J. Chem.* 57: 591–597.

Jou, F. Y. and Freeman, G. R. 1979b. Temperature and isotope effects on the shape of the optical absorption spectrum of solvated electron in water. *J. Phys. Chem.* 83 (18): 2383–2387.

Keene, J. P. 1963. Optical absorptions in irradiated water. *Nature* 197: 47–48.

Kumagai, Y., Lin, M., Lampre, I., Mostafavi, M., Muroya, Y., and Katsumura, Y. 2008. Temperature effect on the absorption spectrum of the hydrated electron paired with a metallic cation in deuterated water. *Radiat. Phys. Chem.* 77: 1198–1202.

Lampre, I., Lin, M., He, H., Han, Z., Mostafavi, M., and Katsumura, Y. 2005. Temperature dependence of the solvated electron absorption spectra in propanediols. *Chem. Phys. Lett.* 402: 192–196.

Laverne, J. A. and Pimblott, S. M. 1993. Diffusion-kinetic modeling of the electron radiolysis of water at elevated-temperatures. *J. Phys. Chem.* 97 (13): 3291–3297.

Lin, M., Katsumura, Y., Muroya, Y., He, H., Wu, G., Han, Z., Miyazaki, T., and Kudo, H. 2004. Pulse radiolysis study on the estimation of radiolytic yields of water decomposition products in high-temperature and supercritical water: Use of methyl viologen as a scavenger. *J. Phys. Chem. A* 108 (40): 8287–8295.

Lin, M., Katsumura, Y., He, H., Muroya, Y., Han, Z., Miyazaki, T., and Kudo, H. 2005. Pulse radiolysis of 4,4′-bipyridyl aqueous solutions at elevated temperatures: Spectral changes and reaction kinetics up to 400°C. *J. Phys. Chem. A* 109 (12): 2847–2854.

Lin, M., Katsumura, Y., Muroya, Y., and He, H. 2006a. A high temperature/pressure optical cell for absorption studies of supercritical fluids by laser photolysis. *Ind. J. Radiat. Res.* 3 (2–3): 69–77.

Lin, M., Mostafavi, M., Muroya, Y., Han, Z. H., Lampre, I., and Katsumura, Y. 2006b. Time-dependent radiolytic yields of the solvated electrons in 1,2-ethanediol, 1,2-propanediol, and 1,3-propanediol from picosecond to microsecond. *J. Phys. Chem. A* 110 (40): 11404–11410.

Lin, M., Kumagai, Y., Lampre, I., Coudert, F. X., Muroya, Y., Boutin, A., Mostafavi, M., and Katsumura, Y. 2007a. Temperature effect on the absorption spectrum of the hydrated electron paired with a lithium cation in deuterated water. *J. Phys. Chem. A* 111: 3548–3553.

Lin, M., Mostafavi, M., Muroya, Y., Lampre, I., and Katsumura, Y. 2007b. Time-dependent radiolytic yields at room temperature and temperature-dependent absorption spectra of the solvated electrons in polyols. *Nucl. Sci. Tech.* 18 (1): 2–9.

Lin, M., Katsumura, Y., Muroya, Y., He, H., Miyazaki, T., and Hiroishi, D. 2008. Pulse radiolysis of sodium formate aqueous solution up to 400°C: Absorption spectra, kinetics and yield of carboxyl radical $CO_2^{\cdot-}$. *Radiat. Phys. Chem.* 77: 1208–1212.

Marin, T. W., Cline, J. A., Takahashi, K., Bartels, D. M., and Jonah, C. D. 2002. Pulse radiolysis of supercritical water. 2. Reaction of nitrobenzene with hydrated electrons and hydroxyl radicals. *J. Phys. Chem. A* 106 (51): 12270–12279.

Marin, T. W., Jonah, C. D., and Bartels, D. M. 2003. Reaction of •OH radicals with H_2 in sub-critical water. *Chem. Phys. Lett.* 371 (1–2): 144–149.

Matubayasi, N., Wakai, C., and Nakahara, M. 1997. NMR study of water structure in super- and subcritical conditions. *Phys. Rev. Lett.* 78 (13): 2573.

Michael, B. D., Hart, E. J., and Schmidt, K. H. 1971. The absorption spectrum of e_{aq}^- in the temperature range −4 to 390°C. *J. Phys. Chem.* 75 (18): 2798–2805.

Miyazaki, T., Katsumura, Y., Lin, M., Muroya, Y., Kudo, H., Asano, M., and Yoshida, M. 2006a. γ-Radiolysis of benzophenone aqueous solution at elevated temperatures up to supercritical condition. *Radiat. Phys. Chem.* 75 (2): 218–228.

Miyazaki, T., Katsumura, Y., Lin, M., Muroya, Y., Kudo, H., Taguchi, M., Asano, M., and Yoshida, M. 2006b. Radiolysis of phenol in aqueous solution at elevated temperatures. *Radiat. Phys. Chem.* 75 (3): 408–415.

Mostafavi, M., Lin, M., Wu, G., Katsumura, Y., and Muroya, Y. 2002. Pulse radiolysis study of absorption spectra of Ag^0 and Ag_2^+ in water from room temperature up to 380°C. *J. Phys. Chem. A* 106 (13): 3123–3127.

Mostafavi, M., Lin, M., He, H., Muroya, Y., and Katsumura, Y. 2004. Temperature-dependent absorption spectra of the solvated electron in ethylene glycol at 100 atm studied by pulse radiolysis from 296 to 598 K. *Chem. Phys. Lett.* 384 (1–3): 52–55.

Muroya, Y., Meesungnoen, J., Jay-Gerin, J. P., Filali-Mouhim, A., Goulet, T., Katsumura, Y., and Mankhetkorn, S. 2002. Radiolysis of liquid water: An attempt to reconcile Monte Carlo calculations with new experimental hydrated electron yield data at early times. *Can. J. Chem.* 80 (10): 1367–1374.

Nicolas, C., Boutin, A., Levy, B., and Borgis, D. 2003. Molecular simulation of a hydrated electron at different thermodynamic state points. *J. Chem. Phys.* 118 (21): 9689–9696.

Ohtaki, H., Radnai, T., and Yamaguchi, T. 1997. Structure of water under subcritical and supercritical conditions studied by solution X-ray diffraction. *Chem. Soc. Rev.* 26 (1): 41–51.

Oka, Y. and Kataoka, K. 1992. Conceptual design of a fast breeder reactor cooled by supercritical steam. *Ann. Nucl. Energy* 19 (4): 243–247.

Okada, K., Imashuku, Y., and Yao, M. 1997. Microwave spectroscopy of supercritical water. *J. Chem. Phys.* 107 (22): 9302–9311.

Okazaki, K., Idriss-Ali, K. M., and Freeman, G. R. 1984. Temperature and molecular structure dependences of optical spectra of electrons in liquid diols. *Can. J. Chem.* 62: 2223–2229.

Partay, L. and Jedlovszky, P. 2005. Line of percolation in supercritical water. *J. Chem. Phys.* 123 (2): 024502.

Percival, P. W., Brodovitch, J. C., Ghandi, K., Addison-Jones, B., Schuth, J., and Bartels, D. M. 1999. Muonium in sub- and supercritical water. *Phys. Chem. Chem. Phys.* 1 (21): 4999–5004.

Percival, P. W., Ghandi, K., Brodovitch, J. C., Addison-Jones, B., and McKenzie, I. 2000. Detection of muoniated organic free radicals in supercritical water. *Phys. Chem. Chem. Phys.* 2 (20): 4717–4720.

Savage, P. E., Gopalan, S., Mizan, T. I., Martino, C. J., and Brock, E. E. 1995. Reactions at supercritical conditions—Applications and fundamentals. *AIChE J.* 41 (7): 1723–1778.

Shiraishi, H., Sunaryo, G. R., and Ishigure, K. 1994. Temperature dependence of equilibrium and rate constants of reactions inducing conversion between hydrated electron and atomic hydrogen. *J. Phys. Chem.* 98 (19): 5164–5173.

Sims, H. E. 2006. Yields of radiolysis products from gamma-irradiated supercritical water-A re-analysis data by W. G. Burns and W. R. Marsh. *Radiat. Phys. Chem.* 75(9): 1047–1050.

Soper, A. K., Bruni, F., and Ricci, M. A. 1997. Site-site pair correlation functions of water from 25 to 400°C: Revised analysis of new and old diffraction data. *J. Chem. Phys.* 106 (1): 247–254.

Squarer, D., Schulenberg, T., Struwe, D., Oka, Y., Bittermann, D., Aksan, N., Maraczy, C., Kyrki-Rajamaki, R., Souyri, A., and Dumaz, P. 2003. High performance light water reactor. *Nucl. Eng. Des.* 221 (1–3): 167–180.

Swiatla-Wojcik, D. and Buxton, G. V. 1995. Modeling of radiation spur processes in water at temperatures up to 300°C. *J. Phys. Chem.* 99 (29): 11464–11471.

Takahashi, K., Cline, J. A., Bartels, D. M., and Jonah, C. D. 2000. Design of an optical cell for pulse radiolysis of supercritical water. *Rev. Sci. Instrum.* 71 (9): 3345–3350.

Takahashi, K., Bartels, D. M., Cline, J. A., and Jonah, C. D. 2002. Reaction rates of the hydrated electron with NO_2^-, NO_3^-, and hydronium ions as a function of temperature from 125 to 380°C. *Chem. Phys. Lett.* 357 (5–6): 358–364.

Takahashi, K. J., Ohgami, S., Koyama, Y., Sawamura, S., Marin, T. W., Bartels, D. M., and Jonah, C. D. 2004. Reaction rates of the hydrated electron with N_2O in high temperature water and potential surface of the N_2O^- anion. *Chem. Phys. Lett.* 383 (5–6): 445–450.

Tucker, S. C. and Maddox, M. W. 1998. The effect of solvent density inhomogeneities on solute dynamics in supercritical fluids: A theoretical perspective. *J. Phys. Chem. B* 102 (14): 2437–2453.

Wang, C.-R. and Lu, Q.-B. 2007. Real-time observation of a molecular reaction mechanism of aqueous 5-halo-2′-deoxyuridines under UV/ionizing radiation. *Angew. Chem. Int. Ed.* 46 (33): 6316–6320.

Wernet, P., Testemale, D., Hazemann, J.-L., Argoud, R., Glatzel, P., Pettersson, L. G. M., Nilsson, A., and Bergmann, U. 2005. Spectroscopic characterization of microscopic hydrogen-bonding disparities in supercritical water. *J. Chem. Phys.* 123 (15): 154503.

Wu, G., Katsumura, Y., Muroya, Y., Li, X., and Terada, Y. 2000. Hydrated electron in subcritical and supercritical water: A pulse radiolysis study. *Chem. Phys. Lett.* 325 (5–6): 531–536.

Wu, G., Katsumura, Y., Muroya, Y., Li, X., and Terada, Y. 2001a. Pulse radiolysis of high temperature and supercritical water: Experimental setup and e_{aq}^- observation. *Radiat. Phys. Chem.* 60 (4–5): 395–398.

Wu, G., Katsumura, Y., Muroya, Y., Lin, M., and Morioka, T. 2001b. Temperature dependence of $(SCN)_2^{\cdot-}$ in water at 25–400°C: Absorption spectrum, equilibrium constant, and decay. *J. Phys. Chem. A* 105 (20): 4933–4939.

Wu, G., Katsumura, Y., Lin, M., Morioka, T., and Muroya, Y. 2002a. Temperature dependence of ketyl radical in aqueous benzophenone solutions up to 400°C: A pulse radiolysis study. *Phys. Chem. Chem. Phys.* 4 (16): 3980–3988.

Wu, G., Katsumura, Y., Muroya, Y., Lin, M., and Morioka, T. 2002b. Temperature dependence of carbonate radical in $NaHCO_3$ and Na_2CO_3 solutions: Is the radical a single anion? *J. Phys. Chem. A* 106 (11): 2430–2437.

16 Radiation Chemistry of Water with Ceramic Oxides

Jay A. LaVerne
University of Notre Dame
Notre Dame, Indiana

CONTENTS

16.1 INTRODUCTION

One of the new challenges in radiation chemistry is to understand the radiolysis of water under the various and often extreme environmental conditions associated with nuclear technology. The radiation chemistry of liquid water and aqueous solutions has been thoroughly examined and some of the basic aspects have been summarized in recent reviews (Buxton, 2004; Garrett et al., 2005). Energy deposition by ionizing radiation is sufficient to populate a variety of ionic and excited states leading to the formation of radicals that decay by a range of chemical reactions over a large timescale. Initial water decomposition occurs on the ultrafast timescale making water radiolysis studies very difficult and many details have yet to be resolved (Garrett et al., 2005; De Waele et al., 2009). The radiolytic decomposition of water adsorbed on solid surfaces can be even more demanding than bulk water because of the heterogeneity of the system. One of the fundamental questions in heterogeneous water radiolysis is whether the radiation-induced decomposition of water is different for molecules adsorbed on surfaces or in a near-surface layer as compared to the bulk liquid. The presence of a solid interface can lead to catalytic, steric, or other effects that alter the water decomposition. There is also the heterogeneous nature of the energy deposition since energy can transfer between the solid phase and the water to enhance or hinder the water decomposition, modify product yields, and even alter the solid surface. Migration of energy can lead to problems in determining the "effective" dosimetry, which is equivalent to the quantity of energy available for radiation effects in the adsorbed water. The transport of charge, excited states, even atoms or molecules to or through the water–solid interface can lead to significant chemical consequences beyond that observed in bulk water alone.

    Differences between the radiolysis of bulk liquid water and that of water in the vicinity of solid interfaces have long been observed, but not vigorously examined until recently. This chapter will present some of the newer results on the mechanisms that explain how the presence of a ceramic oxide interface can affect the outcome of water radiolysis. Most of the discussion will concern the radiolytic effects on water absorbed on or very near to the surface of a ceramic oxide interface. A thorough review of the radiolytic effects of organic molecules absorbed in or on solid oxides such as silica, alumina, and the zeolites has been given elsewhere (Thomas, 1993). Low-weight-percent aqueous suspensions of the solid essentially behave like bulk water or dilute solutions and will not be covered here because of the minimal contribution to the overall radiolysis by the heterogeneous nature of the system.

Most of the systems discussed in this chapter consist of small particles in the nanometer size range. These so-called nanoparticles are used because of their large surface area and rarely because of the quantum effects observed at small dimensions. A recent survey discusses the radiation effects of nanoparticle suspensions, so the observations on those systems will not be discussed here as thoroughly as in that work (Meisel, 2003). Only ceramic oxides as the solid phase will be covered here because these are commonly encountered in the nuclear technology field. Numerous summaries have been published on the radiolysis of water associated with metals, especially the noble metals, and their applications range from solar cells to photography and nanoparticle fabrication. Ceramic oxides can be good insulators and they often do not follow the same redox chemistry as observed with semiconductors. In addition, the surface of ceramic oxides is rarely inert and it can contribute significantly to the radiation chemistry of surrounding water.

16.2   APPLICATIONS

There are a number of very important practical reasons for performing radiolysis studies on water with ceramic oxides. Many parts of some nuclear reactors, including fuel rods (ZrO_2), control elements (BeO), and cooling coils (Al_2O_3, Fe_2O_3, etc.), contain ceramic oxides that are in constant contact with water. Water decomposition products such as the OH radical or H_2O_2 are responsible for initiating and propagating much of the corrosion of reactor components and yet there is little information on the radiation-driven processes occurring at the critical water–solid interfaces (Roberto and de la Rubia, 2006). Heterogeneous radiation chemistry studies are required to predict the aging of the existing reactors and evaluate how long they may safely operate. The next generation of reactors must have radiation chemistry data to be correctly engineered against radiation damage in the harsh environments expected to be encountered.

Decommissioned weapons and recycled fuel elements will require long-term storage facilities that are often enclosed and constructed of stainless steel and a variety of other oxides (National Research Council, 2001, 2002). Shipping containers are almost always sealed and held to higher standards than storage facilities. These containers may pass through populated areas and they are required to be able to withstand extreme temperature ranges and mechanical stress. Water from the humidity in the air will settle on the surface of the waste materials during processing and packaging. Self-radiolysis will then lead to water decomposition over the lifetime of the container. Variations in the yields of the potentially explosive product H_2 (or the ratio of H_2 to O_2) can lead to long-term management problems.

Nuclear power is being promoted more regularly as a source of the energy that will be required to meet global energy needs. No toxic gases are produced by using this technology in its intended manner and the contribution to global warming is minimal. However, radiation effects still present serious challenges to an expansion in the use of nuclear power reactors. A variety of workshops sponsored by the U.S. Department of Energy have contained significant discussions related to the problems due to the radiolysis of water in the presence of a solid phase (U.S. Department of Energy, 2003, 2006, 2008). Very little information can be found in the open literature on the interaction of the solid interface with water in a radiation environment. However, this research topic has recently become more popular. Practical solutions have been made in some instances within the nuclear power industry, but the underlying mechanisms describing the role of water radiolysis in the corrosion processes are still to be resolved. Concurrent with corrosion is the dissolution of the solid phase into the liquid (Nechaev, 1986; Christensen et al., 1994; Christensen and Sunder, 1996; Corbel et al., 2001; Sattonnay et al., 2001). This process degrades fuel elements and disperses radioactive materials throughout the complex. Much time and money has been spent to slow dissolution processes. Dissolution may be a critical obstacle in the containment of waste materials in wet storage environments. Water in these scenarios has an important role as a solvent and also in the redox chemistry.

Targeting radiation specifically to tumor sites has long been a goal in radiation therapy in order to minimize damage to normal cells. Nanoparticles are now being examined with this goal in mind (Juzenas et al., 2008). The scheme is to selectively deposit nanometer-sized materials at the site of a tumor and irradiate. The high local density of the nanoparticle will absorb more energy than the surrounding tissue. Low-energy electrons can then escape the nanoparticle to locally irradiate the nearby tumor, thereby minimizing damage to normal surrounding tissue. Optimization of these techniques will require much more information on the mechanisms of energy transfer between phases in radiolysis.

16.3 FUNDAMENTAL PROCESSES

16.3.1 ENERGY DEPOSITION

Rutherford was the first to determine the fundamental interactions of ions as they pass through matter (Rutherford, 1911). The simple cross section developed by Rutherford is not sufficient to calculate the rate of energy loss or stopping power ($-dE/dx$) because it does not take into account the properties of the medium. A classical formula for the stopping power of matter was developed by Bohr, which was followed by a quantum-mechanical description by Bethe (Bohr, 1913; Bethe, 1930). Thorough discussions of the passage of ions in matter and the tracks they form are given in several reviews (Bohr, 1948; Bethe and Ashkin, 1953; Fano, 1963; Mozumder, 1969; LaVerne, 2000). All charged particles interact mainly with the electrons of the medium. The exceptions are very low-velocity, highly charged ions, which undergo nuclear ballistic collisions (Bohr, 1948).

Most of the influence of the medium on the stopping power is contained in a parameter called the mean excitation energy, but this dependence is logarithmic. Even though mean excitation energies can vary widely for different media, the effect on the stopping power is relatively small (*ICRU Report* 37 1984). For example, the energy corresponding to the median density of energy loss events for 1 MeV electrons in SiO_2 is calculated to be 27 eV as compared to 22 eV for liquid water even though the mean excitation energies are 139.2 and 71.6 eV, respectively (*ICRU Report* 37 1984; Milosavljevic et al., 2004). The distribution of secondary electrons is shifted to slightly higher energies in silica than in water, but the effect on the chemistry is expected to be negligible. The individual energy loss events are nearly the same in either phase. However, the stopping power formalism dictates that the number of events depends directly on the electron density of the medium. Electron densities are very nearly the same for most media, for instance relative to water = 1.00, $SiO_2 = 0.90$, and $ZrO_2 = 0.82$. Therefore, the energy loss of ions in each component of a heterogeneous medium is roughly proportional to their weight fraction. At a relative humidity of less than about 75%, the weight fraction of adsorbed water on many ceramic oxides is one percent or less, so energy loss is by far more predominant in the solid phase (LaVerne and Tandon, 2002).

The average gamma ray energy from the cobalt-60 sources typically used in radiation chemistry studies is about 1.25 MeV. Compton scattering is the dominant energy loss process for photons of this energy and leads to the production of two secondary electrons. These electrons lose energy in the different components of the medium by Coulombic interactions as discussed above. However, because many ceramic oxides are composed of very high Z materials, the gamma ray cross sections will be slightly different than in the aqueous media often used for dosimetry, the Fricke dosimeter for instance. Studies with cobalt-60 gamma rays used ionization chambers to show that the energy absorbed by ZrO_2 is about 10% greater than that with the Fricke dosimeter (Sagert and Robinson, 1968; Wong and Willard, 1968). Monte Carlo calculations were used to estimate that a correction factor of 15% should be added to the dose measured using the Fricke dosimeter to arrive at the dose in pure SiO_2 (Rotureau et al., 2005). Accordingly, a suitable correction to the dose determined with aqueous dosimeters must be applied to systems of water and ceramic oxides. No thorough study seems to have been made using mass attenuation coefficients for the different compounds. A cursory examination of the mass attenuation coefficients for a variety of ceramic oxides suggests that a correction in dose of 10%–15% to the Fricke dosimeter is appropriate (Hubbell and Seltzer, 1996).

Energy deposition in the different phases is relatively straightforward to predict, but an estimation of the ionization produced in each medium is more problematic. Monte Carlo track calculations predict an ionization yield of 4.11 per 100 eV of energy deposited in SiO_2, which is very similar to that of liquid water (Milosavljevic et al., 2004). The energy required to produce an electron-hole pair in SiO_2 is estimated to be between 17 and 30 eV (Petr, 1985). These values give a range of ionization yields of 3–6 per 100 eV, which is comparable to the Monte Carlo predictions. A previous work proposed that the average energy to produce an electron-hole pair is approximately equal to $2.73E_p + 0.55$ eV, where E_p is the band gap (Alig and Bloom, 1975). The band gap for SiO_2 is 11 eV, resulting in a predicted ionization yield of 3.3 per 100 eV. ZrO_2 has a band gap of 5 eV and the above formula would estimate an ionization yield of 7 per 100 eV. Experiments have suggested the ionization yield in ZrO_2 to be 10 per 100 eV, but the chemistry of those systems may be misleading (Meisel, 2003). The ionization yield is probably not solely dependent on the band gap energy due to a variety of other processes such as trapping and recombination. Clearly, better estimates of the ionization yields in various ceramic oxides are required to gain a complete understanding of charge production and transport in the radiolysis of heterogeneous media.

16.3.2 WATER AT INTERFACES

Two very detailed compilations have been published on the adsorption of water on a variety of surfaces (Thiel and Madey, 1987; Henderson, 2002). Physisorbed water may be expected to behave similarly to that of bulk water especially if there is a large amount of water present. However, water fragments such as OH, H, and O will have quite different chemistry if the amount of water available is not sufficient to support complete hydration. Water dissociatively adsorbs on many ceramic oxide surfaces (Boehm, 1971). The two OH groups formed by the water molecule are not equivalent on the surface. The OH anion will be bound to a surface cation site to form a terminal OH group while the proton will attach to a surface oxygen anion to form a bridged OH site, as shown in Figure 16.1. Even more types of OH groups are possible by binding to multiple sites or defects at the surface.

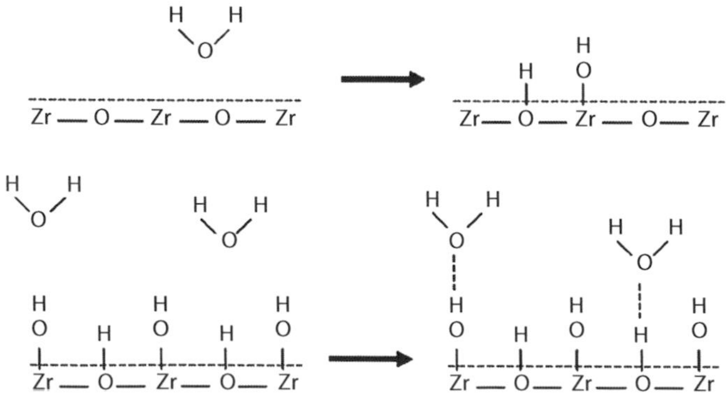

FIGURE 16.1 Schematic for the dissociative water addition to ZrO_2 and formation of ice-like layer.

A variety of complementary techniques can be used to make and probe the OH sites on the surface of ceramic oxides. Clean surfaces can be made in ultra high vacuum (UHV) chambers and water vapor slowly deposited and examined. Surfaces can be completely reduced with H_2 or D_2 and the addition of D_2O can be used for isotopic studies. UHV chambers allow the use of electron spectroscopy and electron scattering for the examination of OH groups on particle surfaces (Freund et al., 1996; Hendrich and Cox, 1996; Vickerman, 1997; Henderson, 2002). Some of the oxides examined for surface water dissociation are listed in the compilation of Henderson and include: Al_2O_3, CaO, CeO_2, MgO, SnO_2, $SrTiO_3$, TiO_2, UO_2, and ZrO_2 (Henderson, 2002). Several of these studies examine the kinetics of the reversible and irreversible water dissociation and give valuable information on the species identity and their expected stability on the ceramic oxide surface. Many of the early studies were interested in water on single crystal metal oxides and much of the discussion concerned whether the most reactive sites are actually defect sites and if water dissociation is associated with these sites only (Hendrich and Cox, 1996). Studies have shown that water can dissociate at non-defect and defect sites, although the latter seem to be more active. Most water dissociation on ceramic oxides is reversible, which means that heating leads to the reformation and desorption of the water molecules.

A variety of infrared spectroscopic techniques can be used to examine oxide surfaces including transmission/absorption, total reflection, and diffuse reflection (Busca, 1996). The oxides may be in high vacuum or under controlled atmosphere conditions. Diffuse reflection infrared Fourier transform (DRIFT) techniques have proven to be very useful for the surface examination of ceramic oxide powders. Powder surfaces are different than single crystals in that they have a variety of facets and defect sites. They are probably much more representative of the surfaces encountered in practical applications. In the DRIFT technique, the physisorbed water is driven off by heating to several hundred degrees to reveal the underlying surface OH groups. In many respects, this technique is complementary to thermal desorption studies that examine the release of gases as a function of temperature programming. DRIFT spectra have been measured for Al_2O_3, SiO_2, TiO_2, and ZrO_2 powders (McDonald, 1958; Yates, 1961; Peri, 1965; Lavalley et al., 1988; Merle-Mejean et al., 1998; Szczepankiewicz et al., 2000; Al-Abadleh and Grassian, 2003; Roscoe and Abbatt, 2005; Takeuchi et al., 2005; Murakami et al., 2006; Thomas and Richardson, 2006; Carrasco-Flores and La Verne, 2007). Typical DRIFT spectra for ZrO_2 are shown in Figure 16.2. The broad OH stretch band

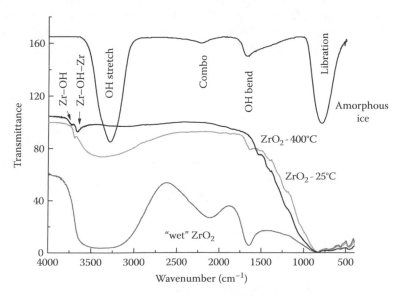

FIGURE 16.2 DRIFT spectra for amorphous ice (shifted up) (Hudgins et al., 1993) for ZrO_2 at 25°C and 400°C (Carrasco-Flores and La Verne, 2007), and for "wet" ZrO_2.

at 3500 cm^{-1} due to the physisorbed water decreases on heating to reveal two OH surface bands at 3780 and 3680 cm^{-1}. These bands are due to the terminal and bridged OH groups, respectively (Merle-Mejean et al., 1998). Absorption at the low energies is usually due to solid lattice modes.

Proton NMR has been used to identify the OH groups on the surface of SiO_2. Silica can have several different OH groups depending on whether silane or siloxane groups are present. Thomas has used this technique to observe three different OH groups (Thomas, 1993). Ceramic oxide powder surfaces have also been probed with neutron and Raman scattering to characterize the OH surface sites (Ozawa et al., 1997). However, highly penetrating probes cannot distinguish whether the observed sites are on the surface or in the bulk. Water trapped in interstitial cavities or co-crystallized during powder formation will be observed with NMR, EPR, x-ray, or neutron techniques.

The dissociation of water at a particular site on the ceramic oxide is due to the specific activity at that site. For instance, ZrO_2 possesses acidic, basic, oxidizing, and reducing properties (Tanabe, 1985). Both Lewis and Bronsted acid sites have been identified on ZrO_2 (Hertl, 1989). A wide variety of ceramic oxides exhibit both acid and basic character (Boehm, 1971). Sustained heating will drive the water off the surface site leaving a Lewis acid due to the unsaturated cation site. Attachment of an available anion will quickly follow. Irradiation may also lead to dissociation of a surface group leaving a reactive cation site.

Surface-bound OH groups essentially form a layer of H atoms covering the ceramic oxide. The next layer of water molecules is highly influenced by this interface and forms a structured water–ice layer through hydrogen bonding, Figure 16.1. Infrared observation of Al_2O_3 surfaces with increasing amounts of relative humidity show that the next 2–3 water layers adsorbed on top of the monolayer of OH groups are very ice-like. The infrared spectra show a very structured ice-like layer due to the hydrogen bonding network. A further increase in relative humidity results in a more disordered transition layer in which the water is more liquid-like (Al-Abadleh and Grassian, 2003; Thomas and Richardson, 2006). A somewhat similar result was observed in PuO_2 (Stakebake, 1967, 1981; Stakebake and Dringman, 1968a,b).

Water has three degrees of freedom so a typical IR spectrum should reveal two stretching modes and a bending mode. The two stretching modes are close together and often merge into one broad peak. Figure 16.2 shows that the absorbances due to the stretching and bending modes are readily observable in amorphous ice at 3280 and 1680 cm^{-1}, respectively (Hudgins et al., 1993). A combination peak is also observed at 2220 cm^{-1} and a libration due to hindered rotation is seen at 790 cm^{-1}. Many of the features observed in amorphous ice are also observed from water adsorbed on ZrO_2, Figure 16.2, and other ceramic oxides. These results suggest that the first few water layers on these oxides are very ice-like; that is, they have hindered mobility. The amount of physisorbed water increases with increasing water content and the spectra reveal more bulk water features. Figure 16.2 shows the DRIFT spectra for ZrO_2 on the addition of a drop of water to the surface to give "wet" ZrO_2. The distinctive water adsorption peaks for physisorbed water are broadened compared to that for ice due to the disorder of the liquid-like phase. The change from the OH surface layer through the transition ice layer to the bulk water layer is not abrupt. Different ceramic oxides are expected to have different thicknesses of the transition ice-like layer. However, almost all ceramic oxides can be expected to have a layered structure of water near its surface. These layers are expected to affect the radiation chemistry of the associated water molecules and to influence the reactions of other molecular species migrating toward the interface from the liquid phase.

16.4 RADIOLYSIS OF AQUEOUS SUSPENSIONS

The main stable reducing species in the radiolysis of water is H_2, while the main oxidizing species is H_2O_2. Both products are due to intra-track reactions of species formed by the water decomposition. At about 1 μs following the passage of a gamma ray, the yield of H_2 is about 0.45 molecule/100 eV of energy absorbed in liquid water (Pastina et al., 1999). The yield of H_2O_2 under similar radiolytic conditions is about 0.7 molecule/100 eV (Pastina and LaVerne, 1999). These yield values are maxima.

At longer times, the reactions of H_2 with OH radicals and H_2O_2 with hydrated electrons or H atoms lead to a series of reactions to reform water (Pastina and LaVerne, 2001). These back reactions can be suppressed if the H_2 escapes from the liquid water phase, for instance by vaporization into a headspace or purged away by a gas flow. Selected scavengers are routinely added to aqueous solutions in radiation chemistry studies to stop the back reactions and allow the measurement of stable product yields. Essentially no H_2 is observed at long times in the radiolysis of closed systems of pure bulk liquid water with gamma rays (Allen, 1961).

A large number of the early studies on aqueous suspensions focused on the photolysis of water with metal colloids because of the importance in solar energy conversion and in environmental remediation by photooxidation (Rabani et al., 1988). Many of the first papers on the radiolysis of aqueous solutions of metals and metal oxides were logical extrapolations of these early systems, especially the TiO_2 and SiO_2 suspensions. An obvious advantage to examining aqueous suspensions is that pulsed radiolysis techniques can be utilized to examine transient species. The radiolysis of acidic solutions of alcohols containing TiO_2 finds that excess electrons can be transferred from the aqueous phase to the solid. Both the hydrated electron and the alcohol radical promote electrons to the TiO_2 conduction band. Conduction band electrons can then react with added solutes. These results are important because they prove that the transfer of energy and charge through the water–oxide interface is possible (Safrany et al., 2000; Gao et al., 2002, 2003). Interestingly, these studies show no effect due to particle size. The excess electrons in the conduction band can even reduce aqueous Pt to form Pt clusters on the TiO_2 surface (Kasarevic-Popovic et al., 2004). The Pt clusters can then reduce aqueous H^+ to H_2 at the surface of the TiO_2 (Behar and Rabani, 2006). The scavenging of hydrated electrons or radicals by TiO_2 particles is not that surprising. With the right redox potentials, the particle is just behaving like a typical aqueous solute. However, the potential for initiating new chemistry at the surface of the particles makes these systems very interesting.

As one increases the concentration of particles in aqueous suspensions, more energy will be deposited initially into the solid phase. A fundamental inquiry concerns whether that energy can migrate to the surface and become available for initiating chemistry in the aqueous phase. Meisel and coworkers used pulse radiolysis techniques to examine the hydrated electron yields in the aqueous phase with variation of SiO_2 concentration and particle size (Schatz et al., 1998). Some of the data of that work for 7 nm SiO_2 is shown in Figure 16.3. The system was probed using pulsed electron radiolysis techniques with a fixed sample size. With increasing concentration of SiO_2, the relative electron density of the sample increases and the amount of energy deposited by the electron beam pulse increases proportionately. In converse, the volume of water in the sample must decrease with increasing SiO_2 concentration. The upper and lower solid lines of Figure 16.3 show these values, respectively. Hydrated electrons, which are obviously in the water phase, have yields that track with the electron density of the total medium. If energy deposited initially within the oxide remained there, then the expected hydrated electron yields should vary with water content as predicted by the lower curve in Figure 16.3. The results show that not only do the electrons escape the SiO_2, but the ionization yield within the oxide must be very similar to that within water. Monte Carlo track calculations indeed predict that the average ionization yield of SiO_2 is 4.11 ions/100 eV, which is in excellent agreement with the measured yield of hydrated electrons in water (Milosavljevic et al., 2004). The pulse radiolysis experiments found no significant variation in hydrated electron yields with particle sizes of 7–22 nm, suggesting that the electrons readily escape into the aqueous phase. No fast time resolved experiments on the observation of the hydrated electron have been published, but preliminary experiments have shown that the hydrated electron is produced within 8 ps in aqueous suspensions of up to 40% SiO_2, suggesting that electrons escape the SiO_2 with more than just thermal energy (LaVerne et al., 2003). In other words, the carrier of the charge or energy within the particle must migrate very quickly to and through the interface with water.

Concurrent with the production of an electron, a hole must also be formed in the solid phase. Holes within the solid phase are difficult to examine in aqueous suspensions, but should the holes migrate to the particle surface, they may react with solutes adsorbed on that surface. The oxidation

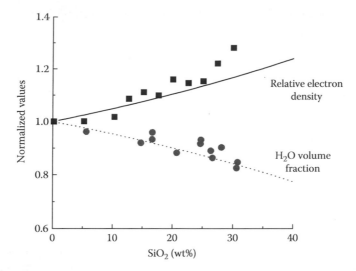

FIGURE 16.3 The radiolysis of aqueous SiO_2 suspensions relative to that for pure water as a function of SiO_2 weight percent: (■) hydrated electron concentration (Schatz et al., 1998); (●) hole production (Dimitrijevic et al., 1999); (solid line) electron density; (dotted line) water volume fraction. (Reprinted from Schatz, T. et al., *J. Phys. Chem. B*, 102, 7225, 1998; Dimitrijevic, N.M. et al., *J. Phys. Chem. B*, 103, 7073, 1999. With permission.)

of $(SCN)_2^-$ and $Fe(CN)_6^{3-}$ on the surface of SiO_2 by radiolytically generated holes has been examined using electron pulse radiolysis techniques (Dimitrijevic et al., 1999). The oxidized products are shown in Figure 16.3 and they can be seen to follow the volume fraction of the water in these systems. These results suggest that the holes produced in the SiO_2 make it to the surface, but do not cross into the aqueous phase. At the surface, they can initiate the oxidation of surface-bound species (Dimitrijevic et al., 1999). Most commercial suspensions have substantial concentrations of stabilizers that may react at the surface and interfere with the desired chemistry. Long time studies of the hydrated electron will exhibit a pseudo first-order decay due to the reaction with additives.

Molecular hydrogen is the most common stable product examined in aqueous suspensions. The driving force for many of these experiments is the same as in photochemical studies—the conversion of radiant energy into chemical energy. A series of experiments have observed the change in the production of H_2 in aqueous suspensions as a function of added TiO_2 and Al_2O_3 (Yamamoto et al., 1999; Seino et al., 2000, 2001a,b; Chitose et al., 2003). In all cases, an increase in the amount of H_2 generation over that of water alone is observed. Particle size ranged from 7 to 200 nm with slurries of 0.01% to 2% weight of ceramic oxide. The production of H_2 is independent of particle size for particles less than a few tens of nm, but a substantially reduced yield of H_2 is observed at 200 nm. Somewhat similar results are observed with SiO_2 suspensions, where above-normal yields of H_2 are observed for 7 and 12 nm particles and a decreased yield is observed for 22 nm particles (Meisel, 2003). Yields are found to be linearly dependent on surface area suggesting that H_2 is formed at the water–oxide interface (Seino et al., 2001a,b). However, large surface areas can be attributed to small particle sizes and the observed increase in H_2 with decreasing particle size may be due to the increased probability for H_2 precursor escape to or through the interface. High dose rates decrease H_2 yields, presumably by increased second order reactions. The slurries of TiO_2 are very efficient oxidizing agents in phenol degradation (Chitose et al., 2003). Studies with aqueous suspensions containing a variety of ceramic oxides show that the efficiency of H_2 production increases with particle surface area and that the yields are greater than that of normal water (Cecal et al., 2008). In all of the studies on suspensions, an increase in H_2 yield is found with increasing particle concentration. Aqueous suspensions of TiO_2 with molar concentrations of methanol are found to give more H_2

than without methanol (Jung et al., 2003). However, in these systems, methanol is probably being sacrificed at the surface of the particles and it is the source of H_2 rather than a modification in the water chemistry.

16.5 RADIOLYSIS OF ADSORBED WATER

The yields of H_2 in the radiolysis of suspensions and slurries of ceramic oxides are enhanced compared to that found in bulk liquid water. They also show a direct dependence on the particle surface area suggesting that the water–oxide interface has an important effect on the radiation chemistry of the surrounding water. However, even the most concentrated slurries have far too much water to observe the chemistry occurring at the water–oxide interface without using special techniques such as probe molecules on the particle surface. Pulsed radiolysis studies on aqueous suspensions suggest that many of the electrons formed in SiO_2 migrate into the bulk water while the holes tend to remain on the particle surface (Schatz et al., 1998; Dimitrijevic et al., 1999). However, the actual yield of electron-hole pairs formed in ceramic oxides is not known. The recombination of an electron-hole pair on a water molecule at the surface of a particle can lead to H_2 production by a dissociative process (LaVerne and Pimblott, 2000). Even a single low-energy electron can lead to H_2 formation from water by a dissociative attachment reaction (Rowntree et al., 1991; Kimmel et al., 1994; Corbut et al., 1996). Such a process has been attributed to the observation of H_2 from water on silicon surfaces (Klyachko et al., 1997). Exciton (electron-hole pair) formation and migration to the solid surface is another potential source of H_2 production. The results are conclusive in that chemistry can occur at the water–oxide interface, but the details remain unresolved. An excellent method to examine radiation effects occurring on or very near the interface of water and ceramic oxides is to remove most of the excess water and observe the chemistry at surfaces with a few layers of adsorbed water.

The first studies on the radiolysis of adsorbed molecules examined the decomposition of organics on a variety of mineral solids, especially silica gel, in part because of interest in application to petroleum refining. Allen and coworkers found the decomposition of the organics to be enhanced in the adsorbed state as compared to the liquid or gas state and attributed the increase in radiolytic yields to an energy-transfer process involving an exciton (Caffrey and Allen, 1958; Sutherland and Allen, 1961; Rabe et al., 1964, 1966). They noted that the enhancement effects were greatest with wide band gap solids and almost nonexistent with semiconductors. Sagert and coworkers noticed that replacement of silanol by siloxane groups on silica gel had a large effect on radiation yields. This work and several others attributed the increase in the decomposition of the organics to electron-transfer processes from the solid to the adsorbants (Rojo and Hentz, 1966; Wong and Willard, 1968). Khare and Johnson used an inorganic matrix system to definitively show that the transfer of energy can depend on the band gap and illustrated the role of excitons in the process (Khare and Johnson, 1970).

Most of the early studies on the radiolysis of adsorbed water were performed in the former Soviet Union. The first published studies on the radiolysis of adsorbed water were performed on semiconductors and silica gel (Bubyreva et al., 1966; Krylova and Dolin, 1966). This work has been followed by others examining the radiolysis of water adsorbed on a variety of surfaces and the production of H_2 has been observed (Garibov et al., 1982a,b, 1984, 1987a,b, 1990, 1991a,b, 1992; Rustamov et al., 1982; Garibov, 1983; Nakashima and Tachikawa, 1983, 1987; Nakashima and Aratono, 1993; Nakashima and Masaki, 1996). All of these studies found that the type of oxide has a huge effect on H_2 yields. The production of H_2 is by far the most common process examined in the radiolysis of adsorbed water. This product is stable and relatively easy to measure. The main stable oxidizing species produced in the radiolysis of water, H_2O_2, is known to react thermally at ceramic oxide surfaces (Hiroki and LaVerne, 2005). Few studies have been reported on the radiolysis of H_2O_2 in the presence of ceramic oxides even though the process may be extremely important in nuclear power reactors.

A very thorough examination of the effects of various types of ceramic oxides on the production of H_2 from adsorbed water has been performed (Aleksandrov et al., 1991; Petrik et al., 1999, 2001). These studies find that most ceramic oxides fall into three categories when the yield of H_2 is determined with respect to the energy deposited in the water phase alone. These categories are as follows: oxides in which the yield of H_2 is lower than that of water (MnO_2, Cr_2O_3, Co_2O_3, and ZnO) with yields of 0.001–0.01 molecule of H_2 per 100 eV; oxides in which the yield of H_2 is comparable to that for water (SiO_2, V_2O_5, NiO, CuO, CdO, TiO_2, and the alkaline earth oxides) with yields of 0.1–1 molecule of H_2 per 100 eV; and oxides in which the yield of H_2 is greater than that of water (La_2O_3, MgO, ZrO_2, and Er_2O_3) with yields of 50–100 eV per 100 eV. The researchers conclude that the physisorbed water has little effect on H_2 production and that surface OH groups also have little effect. They claim that most of the H_2 is due to chemisorbed water, the highly structured ice-like layer between OH group surface layer and the physisorbed liquid-like layer.

A few observations should be made about the determination of H_2 yields from adsorbed water. The previous section on the radiolysis of aqueous suspensions noted that the density of the system, and thereby the energy absorbed, increases with increasing oxide concentration. The dose in aqueous suspensions is usually reported with respect to the energy deposited to the total system, oxide and water. Product yields are expected to change with variation in oxide concentration simply because of density and volume effects. The yields from the radiolysis of adsorbed water often report the dose with respect to the energy absorbed by the water alone. This method is often favored in systems of adsorbed water because a direct representation of energy transfer between the water–oxide phases is obtained. A yield of H_2 lower than that observed in liquid water suggests energy or charge transport into the oxide from the liquid or some sort of quenching process by the oxide surface. A yield of H_2 greater than that observed in liquid water indicates energy transfer from the solid to the liquid or other enhancement in water decomposition.

The amount of water on an oxide surface is usually only a few weight percent at most and difficult to measure accurately (LaVerne and Tandon, 2002). A major problem in dosimetry with respect to the energy deposited directly into the water is quantitatively determining the amount of adsorbed water for each type of ceramic oxide and in each experimental condition. Figure 16.4 shows a typical result for the yield of H_2 from water adsorbed on ZrO_2 as determined by the energy deposited in the

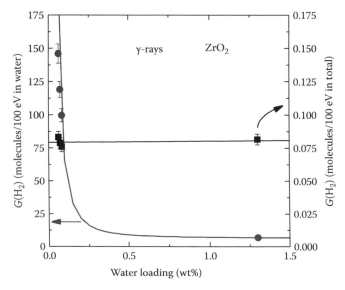

FIGURE 16.4 Yield of H_2 in the gamma ray radiolysis of the water adsorbed on ZrO_2 as a function of the water weight percent: (●) yield relative to the energy deposited in the water alone; (■) yield relative to the energy deposited in both the water and oxide.

whole system (ceramic oxide and water) or deposited in the water alone (LaVerne and Tandon, 2002). The amount of water in this example is equal to the physisorbed water as determined by the weight gain following adsorption from constant humidity chambers. The yield of H_2 with respect to energy deposited directly into the water is on the order of a few hundred molecules per 100 eV of energy. This yield is huge and it is too large to be due only to water decomposition by the energy directly deposited in that phase. A yield of 100 molecules/100 eV corresponds to 1 eV for the formation of one H_2 molecule. This amount of energy is much too small to account for bond breakage and atomic rearrangement and indicates that energy is being transferred from the solid phase to the water. A yield of 100 molecules/100 eV would seem to imply the prolific production of H_2, remember that the yield for bulk water is 0.45 molecules/100 eV. However, the amount of water in these systems is extremely small. On an absolute scale, the yield of H_2 is only about 0.08 molecules/100 eV when the dose is determined with respect to the energy deposited in the whole system (ceramic oxide and water), Figure 16.4. Dosimetry in the radiolysis of adsorbed water is commonly reported using both methods for energy deposition, in the water alone or in the water and ceramic oxide, depending on the point being made.

The yields of H_2 exhibit interesting trends as a function of the amount of water loading that can be used to show the effects due to the type of ceramic oxide. Figure 16.5 gives the yields of H_2 from water adsorbed on ZrO_2 and CeO_2 as a function of the water weight percent (LaVerne and Tandon, 2002). Dosimetry in this example is with respect to the total energy deposited in both the water and the ceramic oxide. The results show a slight increase in H_2 yield with increasing water fraction, and the yields with ZrO_2 are distinctly higher than that observed with CeO_2. Another interesting feature in this presentation of the data is the extrapolated yield of H_2 at zero water fraction. Obviously, no H_2 can be produced without some form of water on the particle surface. The water in this example, Figure 16.5, is determined from the weight gain of physisorbed water. A dissociated water layer and an ice-like layer are strongly bound to the oxide surface beneath the physisorbed water layer. These layers of water near to the particle surface appear to be responsible for a large portion of the observed

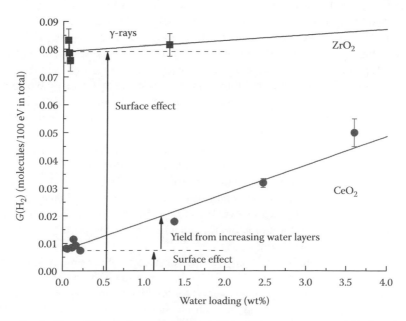

FIGURE 16.5 Production of H_2 relative to the amount of total energy (water + oxide) deposited by gamma rays as a function of the weight percent of physisorbed water on the oxides: (●) CeO_2, (■) ZrO_2 (LaVerne and Tandon, 2002). The extrapolated H_2 yields at zero water loading represent the production due to dissociated water and ice-like water at or near the oxide surface.

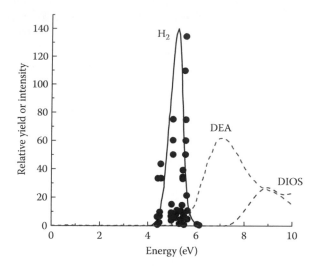

FIGURE 16.6 The relative yield of H_2 (●) (Petrik et al., 2001) determined from the energy deposited in the water alone as a function of the band gap energy for several oxides. The dashed lines are the relative cross sections for dissociative electron attachment (Klyachko et al., 1997) and the differential dipole oscillator strength distribution for amorphous ice at specific energies (LaVerne and Mozumder, 1986). (Reprinted from Petrik, N.G. et al., *J. Phys. Chem. B*, 105, 5935, 2001. With permission.)

difference in the yields of H_2. Water in the physisorbed layer is probably behaving like normal bulk water when irradiated. Clearly, the water–oxide interface is playing an important role in water decomposition and in the production of H_2.

Much of the literature on the radiolysis of adsorbed water has focused on the effects of the type of oxide on the production of H_2 at the water–oxide interface. Following an extensive investigation of many types of oxides, the observation was made that oxides within a narrow band gap energy at about 5 eV can lead to a significant increase in H_2 (Petrik et al., 1999, 2001). The dose in these studies was determined with respect to energy deposited in water alone and some of those results are reproduced in Figure 16.6. The narrow band gap energy response for the large increase in H_2 yield suggests that a resonance process is responsible. An exciton may possibly migrate to the particle surface and resonantly couple to a specific state of a water molecule that is favorable toward the production of H_2. However, not all oxides within this band gap lead to an increase in H_2 yields, so the band gap cannot be the sole characteristic for enhanced H_2 production. Furthermore, a resonance process implies a coupling between the oxide and the water. Figure 16.6 shows the dissociative electron absorption (DEA), cross section, and the differential dipole oscillator distribution (DIOS) for water ice as functions of the absorbed energy (LaVerne and Mozumder, 1986; Klyachko et al., 1997). The energy dependence of neither of these processes matches well with the proposed band gap resonance energy. Special states of water may be formed when they are in the vicinity of certain ceramic oxides. Of course, resonance and a variety of other processes may be contributing in various magnitudes to the radiolytic formation of H_2 from adsorbed water. Further work is required to understand this important phenomenon.

The mechanism for the transport of energy to the surface of the ceramic oxide is still in question. The transport medium in the radiolysis of ZrO_2 has been suggested to be excitons, which are electron-hole pairs (Petrik et al., 2001). One method for identifying the nature of the energy carrier is by the addition of dopants that can trap the transient species within the particle. An increase in the number of trapping sites should decrease the probability of excitons reaching the surface and thereby decrease the observed H_2 production. The substitution of Nb^{5+} for Zr^{4+} sites in ZrO_2 at an amount of only 0.1% by mass gives rise to a dramatic decrease in H_2 yields to essentially zero. Substitution of a small amount of Li^+ or Rb^+ ions in ZrO_2 is found to increase the H_2 production by a factor of 2 (Petrik et al., 2001). These results suggest that the precursor to H_2 production is due to an electronic excitation such as an exciton that diffuses through the solid to the interface. The attenuation of secondary electrons would require much larger concentrations of dopants than observed. However, these results have to be resolved with the radiolysis of suspensions of SiO_2, where there is strong evidence that nonthermal electrons directly escape the particle. A competition must exist between exciton transport to the surface and trapping by defect sites within the particle. Excitons are self-trapped in SiO_2 within 250 fs so they obviously cannot travel a large distance through the particle (Saeta and Greene, 1993; Zhang et al., 1997). The different energy carriers may exist in most of the ceramic oxides, but they may have dissimilar degrees of importance and future studies will have to resolve the issue.

Studies on the production of H_2 from SiO_2 of different surface areas suggest that the transport is limited to the outer layer of 100 nm thickness (Aleksandrov et al., 1991). Other studies with SiO_2 suggest that particles smaller than 22 nm have complete transfer of energy to or through the water–oxide interface (Milosavljevic et al., 2004). Thomas and coworkers found a dramatic decrease in H_2 yields on variation of SiO_2 particles size from 6 to 100 nm (Zhang et al., 1997). Exciton migration distance is estimated to be only 5 nm in ZrO_2 (Petrik et al., 2001). Nonthermal electrons may travel relatively large distances in solid media, whereas excitons must be formed very near to the surface in order to react with adsorbed molecules. The transport layer thickness may also depend on the morphology of a given ceramic oxide. Differences in electronic transport are observed in the monoclinic and tetragonal forms of ZrO_2 (Lemaignan, 2002). These differences may explain the increased production of H_2 observed in tetragonal ZrO_2 as compared to the monoclinic form (LaVerne, 2005). This latter dependence of yields on morphology is especially important in nuclear reactor technology as radiation is known to change the microscopic crystalline structure of certain ceramic oxides (Lemaignan, 2002). Radiation aging of reactor components could modify the chemistry at the water–oxide interface and thereby greatly increase the difficulty in making long-term predictions.

16.6 RADIOLYSIS OF CONFINED WATER

Water can be in intimate contact with surfaces while confined to pores or cavities within the ceramic oxide. This environment is different than that of adsorbed water in that the chemistry of the normally reactive species such as the hydrated electron, H atom, and OH radical can be modified due to the confined space. Diffusion-controlled reactions of radicals can be restricted within a cavity because of the small number of molecules. The passage of species in and out of cavities may be limited by the diameter of the cavity or the size of the openings to the cavity. Reaction of species between cavities can be prohibited or predetermined by some characteristic such as size or local acidity. Many natural materials such as the zeolites have distinctive cavity sizes and shapes and were among the first to be studied using radiolysis (Zhang et al., 1998). The interest in these materials followed naturally from their use as catalysts in a wide variety of circumstances especially in the petroleum industry. Recent fabrication techniques have led to an array of porous materials specifically designed for research and industrial applications.

The first instances of the radiolysis of confined water made use of wet silica gel, alumina, and quartz (Emmett et al., 1962; Weeks and Abraham, 1965). Gamma radiolysis of these materials at 80 K is relatively easy to perform and results in the trapping of the H atoms produced by the water decomposition. Electron paramagnetic resonance (EPR) techniques give a readily identifiable H atom signal because of their huge coupling constant. Interstitial sites have been identified as Bronsted or Lewis acid sites.

The radiolytic production of the H atoms observed in the EPR experiments is derived from the water incorporated in the cavities. Specifically, the adsorbed water is in the form of siloxyl groups ($\equiv Si-OH$). EPR studies on silica have provided extensive knowledge on the formation, migration, and reaction of H atoms (Shkrob and Trifunac, 1996, 1997; Chemerisov and Trifunac, 2001;

Chemerisov et al., 2001). The solid phase makes up most of the mass of these systems so the initial radiolysis process in silica is the production of the electron and hole, $SiO_2 \rightarrow e^- + h^+$, followed by $e^- + h^+ \rightarrow {}^3exciton$. Combination reactions of the electron and hole lead to the triplet exciton formation with a mean lifetime of about 150 fs (Audebert et al., 1994). The excitons, electrons, and holes are the main mobile species on the very short timescale. They will be trapped by interstitial sites or diffuse to the surface where they may initiate chemical reactions with the confined water. As mentioned before, self-trapping of the exciton in SiO_2 occurs within 250 fs, so excitons will not move far (Saeta and Greene, 1993). On the other hand, many porous materials have very large surface-to-volume ratios so excitons can readily reach the water–oxide interface.

At low temperatures, excitons produced by the radiation migrate to the silanol group giving the H atom and an oxygen hole center, $\equiv SiOH + {}^3exciton \rightarrow \equiv SiO^\bullet + H^\bullet$. The H atoms are trapped at temperatures below about 150 K and are observable using EPR. As the temperature is raised, H atoms become very mobile in silica and they decay mainly by recombining with oxygen hole centers or polarons in silica without added water (Shkrob and Trifunac, 1996). H atoms may be produced by reactions of thermal electrons with interstitial protons at higher temperatures, $\equiv SiOH^+ Si\equiv + e^- \rightarrow \equiv SiOSi\equiv + H^\bullet$, or $\equiv SiOH_2^+ + e^- \rightarrow \equiv SiOH + H^\bullet$. Even "dry" silicates produce a significant amount of H_2 due to reactions of H atoms produced from the hydroxyl surface groups (Rotureau et al., 2005). H atoms and excitons can migrate to cavity surfaces and initiate chemical reactions with confined water (Shkrob et al., 1999). Electrons and holes are more likely to recombine in the bulk solid unless they are formed very near the surface. Obviously, energy directly deposited in the confined water will also lead to product formation. Competition reactions leading to product formation seem to be controlled by migration of species between the phases as well as pore size, shape, and water fraction.

Further evidence that free electrons are involved in the radiolysis of interstitial water is shown by the production of N_2 from zeolites with N_2O (Nakazato and Masuda, 1986; Aoki et al., 1988). More importantly for water radiolysis, the yield of H_2 is found to decrease with the addition of N_2O and increase with the addition of ammonia. The N_2O scavenges the electron precursors to H_2, while the ammonia scavenges the positive hole centers to decrease recombination reactions. Zeolite slurries are observed to give H_2 yields of about 1–2 molecules of energy adsorbed in the system (zeolite + water), which is a considerable increase from the 0.45 value for pure water (Cecal et al., 2008). Smaller-sized cavities were found to give higher H_2 yields than larger ones, suggesting that cavity size and not just surface area is important. Kinetic studies to determine the reactivity of the electrons in the cavities suggest that the electrons are not hydrated (Nakazato and Masuda, 1986). However, spectroscopic studies have identified hydrated electrons in the radiolysis of water in certain types of zeolites (Liu et al., 1995).

A recent series of papers have thoroughly examined the decomposition of water confined in a variety of porous silica and mesoporous molecular sieves (Foley et al., 2005; Le Caer et al., 2005a,b; Rotureau et al., 2005). The formation of H_2 in large cavities filled with water mainly occurs by the same mechanisms as in bulk water radiolysis, that is, by reactions of H atoms and hydrated electrons. The systems were examined as a function of water content within the cavities and also as a function of cavity size. As water is depleted from the cavity, a void is formed in the center of the cavity. The remaining water adsorbed on the walls is more stable due to the ice-like structure discussed in the previous section. H_2 yields were found to increase dramatically with decreasing water content. Yields of H_2 are as high as 30 molecules/100 eV energy initially deposited in the water phase (Rotureau et al., 2005). The production of H_2 from the adsorbed water appears to be much more efficient than in bulk water, even bulk water confined to a cavity. Hot electrons or H atoms may pass through the interface and produce H_2 by dissociative attachment reactions (Rotureau et al., 2005). Excitons at the surface may also transfer energy to the confined water leading to H_2 production by excited-state dissociation. A variety of mechanisms may occur, but no experimental evidence exists on the relative importance of these reactions leading to H_2 production. Interestingly, EPR and IR techniques show more defects in irradiated dry silica than when water is present (Le Caer et al., 2005a,b). Transfer of energy from the solid phase to the water is preferred, even to the point of depletion of the available water molecules.

In normal bulk water radiolysis, OH radical scavengers such as bromide anion or alcohol are added to prevent depletion of the H_2 yield. The addition of bromine anion, formate, or isopropanol has no effect on the production of H_2 from water in silica cavities (Rotureau et al., 2005). Radicals may be separated by the pores thereby decreasing combination reactions or the OH radicals may be more reactive with something else in the cavity than with H_2. OH radicals are produced in cavities as evidence by their scavenging with coumarin (Foley et al., 2005). As with H_2 yields, OH radical yields increase with decreasing water content of the cavities suggesting that adsorbed water is very efficient in their production. In fact, the scavenger studies suggest that a significant amount of OH radicals are produced at the silica–water interface. Higher yields of OH radicals are observed with increasing areas of the interface, inferring the transfer of energy from the silica to the interface is responsible for at least part of the OH production. Some OH radicals may be scavenged by wall sites, but a significant fraction combines to form H_2O_2. The yield of H_2O_2 has been determined using scavenger techniques and found to agree well with H_2 yields (Le Caer et al., 2007). The results suggest that the complementary reducing and oxidizing stable species of H_2 and H_2O_2 are formed in equal yields in the radiolysis of confined water. If so, then both oxidizing and reducing species must be transported through the silica–water interface, leading to the decomposition of water.

16.7 CONCLUSIONS

The radiolysis of water in association with interfaces is becoming one of the new challenges in association with nuclear technology. Understanding the radiolytic decomposition of water adsorbed on or very near to solid surfaces can be even more demanding than bulk water because of the transport of charge and energy to and through the water–solid interface. Many studies have shown that the radiolysis of water is different at a solid interface than in the bulk liquid, although the mechanisms are still not resolved. The nature of the energy carrier and the mode of transport in the solid are still in question. Of particular importance is to identify the specific properties of the solid responsible for the transport of energy and charge. The nature of the water at or near solids and the effect of the interface on its radiolytic decomposition still require investigation.

ACKNOWLEDGMENTS

This contribution is NDRL-4810 from the Notre Dame Radiation Laboratory, which is supported by the Office of Basic Energy Sciences of the U.S. Department of Energy.

REFERENCES

Al-Abadleh, H. A. and Grassian, V. H. 2003. FT-IR study of water adsorption on aluminum oxide surfaces. *Langmuir* 19: 341–347.

Aleksandrov, A. B., Byakov, A. Y., Vall, A. I., Petrik, N. G., and Sedov, V. M. 1991. Radiolysis of adsorbed substances on oxide surfaces. *Russ. J. Phys. Chem.* 65: 847–849.

Alig, R. C. and Bloom, S. 1975. Electron-hole-pair creation energies in semiconductors. *Phys. Rev. Lett.* 35: 1522–1525.

Allen, A. O. 1961. *The Radiation Chemistry of Water and Aqueous Solutions.* New York: Van Nostrand.

Aoki, M., Nakazato, C., and Masuda, T. 1988. Hydrogen formation from water adsorbed on zeolite during gamma-irradiation. *Bull. Chem. Soc. Jpn.* 61: 1899–1902.

Audebert, P., Daguzan, P., Dos Santos, A., Gauthier, J. C., Geindre, J. P., Guizard, S., Hamoniaux, G., Krastev, K., Martin, P., Petite, G., and Antonetti, A. 1994. Space-time observation of an electron gas in SiO_2. *Phys. Rev. Lett.* 73: 1990–1993.

Behar, D. and Rabani, J. 2006. Kinetics of hydrogen production upon reduction of aqueous TiO_2 nanoparticles catalyzed by Pd^0, Pt^0, or Au^0 coating and an unusual hydrogen abstraction; steady state and pulse radiolysis study. *J. Phys. Chem. B* 110: 8750–8755.

Bethe, H. 1930. Zur Theorie des Durchgangs Schneller Korpuskularstrahlen Durch Materie. *Ann. Phys.* 5: 325–400.

Bethe, H. and Ashkin, J. 1953. Passage of radiations through matter. In *Experimental Nuclear Physics*, E. Segre (ed.). New York: Wiley.

Boehm, H. P. 1971. Acidic and basic properties of hydroxylated metal oxide surfaces. *Disc. Faraday Soc.* 53: 264–289.

Bohr, N. 1913. On the theory of the decrease of velocity of moving electrified particles on passing through matter. *Phil. Mag.* 25: 10–31.

Bohr, N. 1948. The penetration of atomic particles through matter. *Det. Klg. Danske Videnskabernes Selskab* 18: 1–144.

Bubyreva, N. S., Dolin, P. I., Kononovich, A. A., and Rozenblyum, N. D. 1966. The radiolysis of water vapor in the presence of the semiconductor oxides ZnO and V_2O_5. *Kinet. Catal.* 7: 846–848.

Busca, G. 1996. The use of vibrational spectroscopies in studies of heterogeneous catalysis by metal oxides: An introduction. *Catal. Today* 27: 323–352.

Buxton, G. V. 2004. The radiation chemistry of liquid water: Principles and applications. In *Charged Particle Interactions with Matter: Chemical, Physicochemical, and Biological Consequences with Applications.* A. Mozumder and Y. Hatano (eds.), pp. 331–363. New York: Marcel Dekker.

Caffrey, J. M. J. and Allen, A. O. 1958. Radiolysis of pentane adsorbed on mineral surfaces. *J. Phys. Chem.* 62: 33–37.

Carrasco-Flores, E. A. and La Verne, J. A. 2007. Surface species produced in the radiolysis of zirconia nanoparticles. *J. Chem. Phys.* 127: 234703.

Cecal, A., Macovei, A., Tamba, G., Hauta, O., Popa, K., Ganju, D., and Rusu, I. 2008. On the hydrogen production by catalyzed radiolysis of water. *Rev. Roumaine Chim.* 53: 203–206.

Chemerisov, S. D. and Trifunac, A. D. 2001. Probing nanoconfined water in zeolite cages: H atom dynamics and spectroscopy. *Chem. Phys. Lett.* 347: 65–72.

Chemerisov, S. D., Werst, D. W., and Trifunac, A. D. 2001. Formation, trapping and kinetics of H atoms in wet zeolites and mesoporous silica. *Radiat. Phys. Chem.* 60: 405–410.

Chitose, N., Ueta, S., Seino, S., and Yamamoto, T. A. 2003. Radiolysis of aqueous phenol solutions with nanoparticles. 1. Phenol degradation and TOC removal in solutions containing TiO_2 induced by UV, γ-ray and electron beams. *Chemosphere* 50: 1007–1013.

Christensen, H. and Sunder, S. 1996. An evaluation of water layer thickness effective in the oxidation of UO_2 fuel due to radiolysis of water. *J. Nucl. Mater.* 238: 70–77.

Christensen, H., Sunder, S., and Showsmith, D. W. 1994. Oxidation of nuclear fuel (UO_2) by the products of water radiolysis: Development of a kinetic model. *J. Alloys Compd.* 213/214: 93–99.

Corbel, C., Sattonnay, G., Lucchini, J. F., Ardois, C., Barthe, M. F., Huet, F., Dehaudt, P., Hickel, B., and Jegou, C. 2001. Increase of the uranium release at an UO_2/H_2O interface under He^{2+} ion beam irradiation. *Nucl. Instrum. Methods Phys. Res. Sect. B-Beam Interact. Mater. Atoms* 179: 225–229.

Corbut, V., Jay-Gerin, J.-P., Frongillo, Y., and Patau, J. P. 1996. On the dissociative electron attachment as a potential source of molecular hydrogen in irradiated liquid water. *Radiat. Phys. Chem.* 47: 247–250.

De Waele, V., Lampre, I., and Mostafavi, M. 2010. Time-resolved study on nonhomogeneous chemistry induced by ionizing radiation with low linear energy transfer in water and polar solvents at room temperature. In *Charged Particle Interactions with Matter: Recent Advances, Applications, and Interfaces.* Y. Hatano, Y. Katsumura, and A. Mozumder (eds.), pp. 289–324, Boca Raton, Florida: CRC press.

Dimitrijevic, N. M., Henglein, A., and Meisel, D. 1999. Charge separation across the silica nanoparticle/water interface. *J. Phys. Chem. B* 103: 7073–7076.

Emmett, P. H., Livingston, R., Zeldes, H., and Kokes, R. J. 1962. Formation of hydrogen atoms in irradiated catalysts. *J. Phys. Chem.* 66: 921–923.

Fano, U. 1963. Penetration of protons, alpha particles, and mesons. *Ann. Rev. Nucl. Sci.* 13: 1–66.

Foley, S., Rotureau, P., Pin, S., Baldacchino, G., Renault, J.-P., and Mialocq, J.-C. 2005. Radiolysis of confined water: Production and reactivity of hydroxyl radicals. *Angew. Chem. Int. Ed.* 44: 110–112.

Freund, H.-J., Kuhlenbeck, H., and Staemmler, V. 1996. Oxide surfaces. *Rep. Prog. Phys.* 59: 283–347.

Gao, R., Safrany, A., and Rabani, J. 2002. Fundamental reactions in TiO_2 nanocrystallite aqueous solutions studied by pulse radiolysis. *Radiat. Phys. Chem.* 65: 599–609.

Gao, R., Safrany, A., and Rabani, J. 2003. Reactions of TiO_2 excess electron in nanocrystalline aqueous solutions studied in pulse and gamma-radiolysis systems. *Radiat. Phys. Chem.* 67: 25–39.

Garibov, A. A. 1983. Water radiolysis in the presence of oxides. In *Proceedings of the Fifth Tihany Symposium on Radiation Chemistry*, Budapest, Hungary, Akademiai Kiado.

Garibov, A. A., Melikzade, M. M., Bakirov, M. Y., and Ramazanova, M. K. 1982a. Effect of cations on the catalytic properties of silica gel in the radiolysis of adsorbed water. *High Energy Chem.* 16: 101–105.

Garibov, A. A., Melikzade, M. M., Barikov, M. Y., and Ramazanova, M. K. 1982b. Radiolysis of adsorbed water molecules on the oxides. *High Energy Chem.* 16: 177–179.

Garibov, A. A., Bakirov, M. Y., Velibekova, G. Z., and Elchiev, Y. M. 1984. Radiation catalytic properties of natural zeolite of the mordenite type in the radiolysis of water. *High Energy Chem.* 18: 398–401.

Garibov, A. A., Gezalov, K. B., and Kasumov, R. D. 1987a. Radiation-induced heterogeneous reactions in $BeO + H_2O$ using EPR. *Radiat. Phys. Chem.* 30: 197–199.

Garibov, A. A., Gezalov, K. B., Velibekova, G. Z., Khudiev, A. T., Ramazanova, M. K., Kasumov, R. D., Agaev, T. N., and Gasanov, A. M. 1987b. Heterogeneous radiolysis of water: Effect of the concentration of water in the adsorbed phase on the hydrogen yield. *High Energy Chem.* 21: 416–420.

Garibov, A. A., Velibekova, G. Z., Kasumov, R. D., Gezalov, K. B., and Agaev, T. N. 1990. Characteristics of energy transfer in heterogeneous radiolysis of water in the presence of amorphous aluminosilicate. *High Energy Chem.* 24: 174–179.

Garibov, A. A., Agaev, T. N., and Kasumov, R. D. 1991a. Effect of concentration of aluminum on the radiolytic-catalytic activity of aluminosilicate in obtaining hydrogen from water. *High Energy Chem.* 25: 337–341.

Garibov, A. A., Parmon, V. N., Agaev, T. N., and Kasumov, R. D. 1991b. Influence of the polymorphous forms of the oxide and the temperature on the transfer of energy during radiation-induced heterogeneous processes in the $Al_2O_3 + H_2O$ system. *High Energy Chem.* 25: 86–90.

Garibov, A. A., Velibekova, G. Z., Agaev, T. N., Dzhafarov, Y. D., and Gadzhieva, N. N. 1992. Radiation heterogeneous processes in contact of aluminum with water. *High Energy Chem.* 26: 184–187.

Garrett, B. C., Dixon, D. A., Camaioni, D. M., Chipman, D. M., Johnson, M. A., Jonah, C. D., Kimmel, G. A., Miller, J. H., Rescigno, T. N., Rossky, P. J., Xantheas, S. S., Colson, S. D., Laufer, A. H., Ray, D., Barbara, P. F., Bartels, D. M., Becker, K. H., Bowen, K. H., Bradforth, S. E., Carmichael, I., Coe, J. V., Corrales, L. R., Cowin, J. P., Dupuis, M., Eisenthal, K. B., Franz, J. A., Gutowski, M. S., Jordan, K. D., Kay, B. D., LaVerne, J. A., Lymar, S. V., Madey, T. E., McCurdy, C. W., Meisel, D., Mukamel, S., Nilsson, A. R., Orlando, T. M., Petrik, N. G., Pimblott, S. M., Rustad, J. R., Schenter, G. K., Singer, S. J., Tokmakoff, A., Wang, L.-S., Wittig, C., and Zwier, T. S. 2005. The role of water on electron-initiated processes and radical chemistry: Issues and scientific advances. *Chem. Rev.* 105: 355–389.

Henderson, M. A. 2002. The interaction of water with solid surfaces: Fundamental aspects revisited. *Surf. Sci. Rep.* 46: 1–308.

Hendrich, V. E. and Cox, P. A. 1996. *The Surface Science of Metal Oxides*. Cambridge, U.K.: Cambridge University Press.

Hertl, W. 1989. Surface chemistry of zirconia polymorphs. *Langmuir* 5: 96–100.

Hiroki, A. and LaVerne, J. A. 2005. Decomposition of hydrogen peroxide at water-ceramic oxide interfaces. *J. Phys. Chem. B* 109: 3364–3370.

Hubbell, J. H. and Seltzer, S. M. 1996. *Tables of X-Ray Mass Attenuation Coefficients and Mass Energy-Absorption Coefficients Report NISTIR* 5632. Gaithersburg, MD: National Institute of Standards and Technology.

Hudgins, D. M., Sandford, S. A., Allamandola, L. J., and Tielens, A. G. G. M. 1993. Mid- and far-infrared spectroscopy of ices: Optical constants and integrated absorbances. *Astrophys. J. Suppl. Ser.* 86: 713–870.

ICRU. 1984. *Stopping Powers for Electron and Positrons*. Bethesda, MD: International Commission on Radiation Units and Measurements.

Jung, J., Jeong, H. S., Chung, H. H., Lee, M. J., Jin, J. H., and Park, K. B. 2003. Radiocatalytic H_2 production with gamma-irradiation and TiO_2 catalysts. *J. Radioanal. Nucl. Chem.* 258: 543–546.

Juzenas, P., Chen, W., Sun, Y. P., Coelho, M. A. N., Generalov, R., Generalova, N., and Christensen, I. L. 2008. Quantum dots and nanoparticles for photodynamic and radiation therapies of cancer. *Adv. Drug Del. Rev.* 60: 1600–1614.

Kasarevic-Popovic, Z., Behar, D., and Rabani, J. 2004. Role of excess electrons in TiO_2 nanoparticles coated with Pt in reduction reactions studied in radiolysis of aqueous solutions. *J. Phys. Chem. B* 108: 20291–20295.

Khare, M. and Johnson, E. R. 1970. Energy transfer in alkali halide matrices. *J. Phys. Chem.* 74: 4085–4091.

Kimmel, G. A., Orlando, T. M., Vezina, C., and Sanche, L. 1994. Low-energy electron-stimulated production of molecular hydrogen from amorphous water ice. *J. Chem. Phys.* 101: 3282–3286.

Klyachko, D. V., Rowntree, P., and Sanche, L. 1997. Dynamics of surface reactions induced by low-energy electrons oxidation of hydrogen-passivated Si by H_2O. *Sur. Sci.* 389: 29–47.

Krylova, Z. L. and Dolin, P. I. 1966. Radiolysis of water adsorbed on silica gel. *Kinet. Catal.* 7: 840–844.

Lavalley, J.-C., Bensitel, M., Gallas, J. P., Lamotte, J., Busca, G., and Lorenzelli, V. 1988. FT-IR study of the $\delta(OH)$ mode of surface hydroxyl groups on metal oxides. *J. Mol. Struct.* 175: 453–458.

LaVerne, J. A. 2000. Track effects of heavy ions in liquid water. *Radiat. Res.* 153: 487–496.

LaVerne, J. A. 2005. H_2 formation from the radiolysis of liquid water with zirconia. *J. Phys. Chem. B* 109: 5395–5397.

LaVerne, J. A. and Mozumder, A. 1986. Effect of phase on the stopping and range distribution of low-energy electrons in water. *J. Phys. Chem.* 90: 3242–3247.

LaVerne, J. A. and Pimblott, S. M. 2000. New mechanism for H_2 formation in water. *J. Phys. Chem. A* 104: 9820–9822.

LaVerne, J. A. and Tandon, L. 2002. H_2 production in the radiolysis of water on CeO_2 and ZrO_2. *J. Phys. Chem. B* 106: 380–386.

LaVerne, J. A., Meisel, D., and Jonah, C. 2003. unpublished results.

Le Caer, S., Rotureau, P., Brunet, F., Charpentier, T., Blain, G., Renault, J. P., and Mialocq, J. C. 2005a. Radiolysis of confined water: Hydrogen production at low dose rate. *Chem. Phys. Chem.* 6: 2585–2596.

Le Caer, S., Rotureau, P., Vigneron, G., Blain, G., Renault, J. P., and Mialocq, J. C. 2005b. Irradiation of controlled pore glasses with 10 MeV electrons. *Rev. Adv. Mater. Sci.* 10: 161–165.

Le Caer, S., Renault, J. P., and Mialocq, J. C. 2007. Hydrogen peroxide formation in the radiolysis of hydrated nanoporous glasses: A low and high dose rate study. *Chem. Phys. Lett.* 450: 91–95.

Lemaignan, C. 2002. Physical phenomena concerning corrosion under irradiation of Zr alloys. In *Zirconia in the Nuclear Industry: Thirteenth International Symposium, ASTM STP* 1423, G. D. Moan and P. Rudling (eds.), pp. 20–29. West Conshohocken, PA: ASTM International.

Liu, X., Zhang, G., and Thomas, J. K. 1995. Spectroscopic studies of electron trapping in zeolites: Cation cluster trapped electrons and hydrated electrons. *J. Phys. Chem.* 99: 10024–10034.

McDonald, R. S. 1958. Surface functionality of amorphous silica by infrared spectroscopy. *J. Phys. Chem.* 62: 1168–1178.

Meisel, D. 2003. Radiation effects in nanoparticle suspensions. In *Nanoscale Materials*, L. M. Liz-Maran and P. V. Kamat (eds.), pp. 119–134. Norwell, MA: Academic Publishers.

Merle-Mejean, T., Barberis, P., Othmane, S. B., Nardou, F., and Quintard, P. E. 1998. Chemical forms of hydroxyls on/in zirconia: An FT-IR study. *J. Eur. Ceram. Soc.* 18: 1579–1586.

Milosavljevic, B. H., Pimblott, S. M., and Meisel, D. 2004. Yields and migration distances of reducing equivalents in the radiolysis of silica nanoparticles. *J. Phys. Chem. B* 108: 6696–7001.

Mozumder, A. 1969. Charged particle tracks and their structure. In *Advances in Radiation Chemistry*, M. Burton and J. L. Magee (eds.), pp. 1–102. New York: Wiley-Interscience.

Murakami, Y., Kenji, E., Nosaka, A. Y., and Nosaka, Y. 2006. Direct detection of OH radicals diffused to the gas phase from the UV-irradiated photocatalytic TiO_2 surfaces by means of laser-induced fluorescence spectroscopy. *J. Phys. Chem. B* 110: 16808–16811.

Nakashima, M. and Tachikawa, E. 1983. Hydrogen evolution from tritiated water on silica gel by gamma-irradiation. *Radiochim. Acta* 33: 217–222.

Nakazato, C. and Masuda, T. 1986. Reactivity of electrons produced in gamma-irradiated zeolite toward several electron scavengers. *Bull. Chem. Soc. Jpn.* 59: 2237–2239.

Nakashima, M. and Aratono, Y. 1993. Radiolytic hydrogen gas formation from water adsorbed on type A zeolites. *Radiat. Phys. Chem.* 41: 461–465.

Nakashima, M. and Masaki, N. M. 1996. Radiolytic hydrogen gas formation from water adsorbed on type Y zeolites. *Radiat. Phys. Chem.* 47: 241–245.

Nakashima, M. and Tachikawa, E. 1987. Radiolytic gas production from tritiated water adsorbed on molecular sieve 5A. *J. Nucl. Sci. Tech.* 24: 41–46.

National Research Council. 2001. *Improving Operations and Long-Term Safety of the Waste Isolation Pilot Plant.* Washington, DC: National Research Council.

National Research Council. 2002. *Research Opportunities for Managing the Department of Energy's Transuranic and Mixed Wastes.* Washington, DC: National Research Council.

Nechaev, A. 1986. Radiation induced processes on solid surfaces: General approach and outlook. *Radiat. Phys. Chem.* 28: 433–436.

Ozawa, M., Suzuki, S., Loong, C.-K., and Nipko, J. C. 1997. Neutron and Raman scattering studies of surface adsorbed molecular vibrations and bulk phonons in ZrO_2 nanoparticles. *Appl. Surf. Sci.* 121/122: 133–137.

Pastina, B. and LaVerne, J. A. 1999. Hydrogen peroxide production in the radiolysis of water with heavy ions. *J. Phys. Chem. A* 103: 1592–1597.

Pastina, B. and LaVerne, J. A. 2001. Effect of molecular hydrogen on hydrogen peroxide in water radiolysis. *J. Phys. Chem. A* 105: 9316–9322.

Pastina, B., LaVerne, J. A., and Pimblott, S. M. 1999. Dependence of molecular hydrogen formation in water on scavengers of the precursor to the hydrated electron. *J. Phys. Chem. A* 103: 5841–5846.

Peri, J. B. 1965. A model for the surface of γ-alumina. *J. Phys. Chem.* 69: 220–230.

Petr, I. 1985. Production of electron-hole pairs in SiO_2 films. *J. Radioanal. Nucl. Chem. Lett.* 95: 195–200.

Petrik, N. G., Alexandrov, A. B., Orlando, T. M., and Vall, A. I. 1999. Radiation-induced processes at oxide surfaces and interfaces relevant to spent nuclear fuel. *Trans. Am. Nucl. Soc.* 81: 101–102.

Petrik, N. G., Alexandrov, A. B., and Vall, A. I. 2001. Interfacial energy transfer during gamma radiolysis of water on the surface of ZrO_2 and some other oxides. *J. Phys. Chem. B* 105: 5935–5944.

Rabani, J., Fessenden, R. W., and Sassoon, R. E. 1988. A pulse radiolysis study to probe reduced platinum colloids in aqueous solution using optical and conductivity techniques. *J. Phys. Chem.* 92: 2379–2385.

Rabe, J. G., Rabe, B., and Allen, A. O. 1964. Radiolysis in the adsorbed state. *J. Am. Chem. Soc.* 86: 3887–3888.

Rabe, J. G., Rabe, B., and Allen, A. O. 1966. Radiolysis and energy transfer in the adsorbed state. *J. Phys. Chem.* 70: 1098–1107.

Roberto, J. and de la Rubia, T. D. 2006. *Basic Research Needs for Advance Nuclear Energy Systems*. Washington, DC: Office of Science U.S. Department of Energy

Rojo, E. A. and Hentz, R. R. 1966. The reaction of isopropylbenzene on γ-irradiated silica gels. *J. Phys. Chem.* 70: 2919–2925.

Roscoe, J. M. and Abbatt, J. P. D. 2005. Diffuse reflectance FTIR study of the interaction of alumina surfaces with ozone and water vapor. *J. Phys. Chem. A* 109: 9028–9034.

Rotureau, P., Renault, J. P., Lebeau, B., Patarin, J., and Mialocq, J. C. 2005. Radiolysis of confined water: Molecular hydrogen formation. *Chem. Phys. Chem.* 6: 1316–1323.

Rowntree, P., Parenteau, L., and Sanche, L. 1991. Electron stimulated desorption via dissociative attachment in amorphous H_2O. *J. Chem. Phys.* 94: 8570–8576.

Rustamov, V. R., Bugaenko, L. T., Kurbanov, M. A., and Kerimov, V. K. 1982. Energy-transfer performance in the heterogeneous radiolysis of water. *High Energy Chem.* 16: 148–150.

Rutherford, E. 1911. The scattering of a and b particles by matter and the structure of the atom. *Phil. Mag.* 21: 669–688.

Saeta, P. N. and Greene, B. I. 1993. Primary relaxation processes at the band edge of SiO_2. *Phys. Rev. Lett.* 70: 3588–3591.

Safrany, A., Gao, R., and Rabani, J. 2000. Optical properties and reactions of radiation induced TiO_2 electrons in aqueous colloid solutions. *J. Phys. Chem. B* 104: 5848–5853.

Sagert, N. H. and Robinson, R. W. 1968. Radiolysis in the adsorbed state. II. N_2O adsorbed on silica gel and zirconia. *Can. J. Chem.* 46: 2075–2080.

Sattonnay, G., Ardois, C., Corbel, C., Lucchini, J. F., Barthe, M. F., Garrido, F., and Gosset, D. 2001. Alpha-radiolysis effects on UO_2 alteration in water. *J. Nucl. Mater.* 288: 11–19.

Schatz, T., Cook, A. R., and Meisel, D. 1998. Charge carrier transfer across the silica nanoparticle/water interface. *J. Phys. Chem. B* 102: 7225–7230.

Seino, S., Fujimoto, R., Yamamoto, T., Katsura, M., and Okuda, S. 2000. Hydrogen gas evolution from water-dispersed titania and alumina nanoparticles by γ-ray irradiation. *Radioisotopes* 49: 354–358.

Seino, S., Yamamoto, T. A., Fugimoto, R., Hashimoto, K., Katsura, M., Okuda, S., Okitsu, K., and Oshima, R. 2001a. Hydrogen evolution from water dispersing nanoparticles irradiated with gamma-ray/size effect and dose rate effect. *Scr. Mater.* 44: 1709–1712.

Seino, S., Yamamoto, T. A., Fujimoto, R., Hashimoto, K., Katsura, M., Okuda, S., and Okitsu, K. 2001b. Enhancement of hydrogen evolution yield from water dispersing nanoparticles irradiated with gamma ray. *J. Nucl. Sci. Tech.* 38: 633–636.

Shkrob, I. A. and Trifunac, A. D. 1996. Time-resolved EPR of spin-polarized mobile H atoms in amorphous silica: The involvement of small polarons. *Phys. Rev. B* 54: 15073–15078.

Shkrob, I. A. and Trifunac, A. D. 1997. Spin-polarized H/D atoms and radiation chemistry in amorphous silica. *J. Chem. Phys.* 107: 2374–2385.

Shkrob, I. A., Tadjikov, B. M., Chemerisov, S. D., and Trifunac, A. D. 1999. Electron trapping and hydrogen atoms in oxide glasses. *J. Chem. Phys.* 111: 5124–5140.

Stakebake, J. L. 1981. Kinetic-studies of the reaction of plutonium hydride with oxygen. *Nucl. Sci. Eng.* 78: 386–392.

Stakebake, J. L. 1967. Low temperature adsorption of oxygen on plutonium dioxide. *J. Nucl. Mater.* 23: 154–162.

Stakebake, J. L. and Dringman, M. R. 1968a. Desorption from plutonium dioxide. US Atomic Energy Commission Report RFP-1248.

Stakebake, J. L. and Dringman, M. R. 1968b. Hydroscopy of plutonium dioxide. US Atomic Energy Commission Report RFP-1056.

Sutherland, J. W. and Allen, A. O. 1961. Radiolysis of pentane in the adsorbed state. *J. Am. Chem. Soc.* 83: 1040–1047.

Szczepankiewicz, S. H., Colussi, A. J., and Hoffman, M. R. 2000. Infrared spectra of photoinduced species on hydroxylated titania surfaces. *J. Phys. Chem. B* 104: 9842–9850.

Takeuchi, M., Martra, G., Coluccia, S., and Anpo, M. 2005. Investigations of the structure of H_2O clusters adsorbed on TiO_2 surfaces by near-infrared absorption spectroscopy. *J. Phys. Chem. B* 109: 7387–7391.

Tanabe, K. 1985. Surface and catalytic properties of ZrO_2. *Mater. Chem. Phys.* 13: 347–364.

Thiel, P. A. and Madey, T. E. 1987. The interaction of water with solid surfaces—Fundamental aspects. *Surf. Sci. Rep.* 7: 211–385.

Thomas, J. K. 1993. Physical aspects of photochemistry and radiation chemistry of molecules adsorbed on SiO_2, γ-Al_2O_3, zeolites, and clays. *Chem. Rev.* 93: 301–320.

Thomas, A. C. and Richardson, H. H. 2006. 2D-IR correlation analysis of thin film water adsorbed on α-Al_2O_3(0001). *J. Mol. Struct.* 799: 158–162.

U.S. Department of Energy. 2003. *Basic Research Needs to Assure a Secure Energy Future*. Washington, DC: Office of Basic Energy Sciences, U.S. Department of Energy.

U.S. Department of Energy. 2006. *Basic Research Needs for Advanced Nuclear Energy Systems*. Washington, DC: Office of Basic Energy Sciences, U.S. Department of Energy.

U.S. Department of Energy. 2008. *Basic Research Needs for Materials under Extreme Environments*. Washington, DC: Office of Basic Energy Sciences, U.S. Department of Energy.

Vickerman, J. C. Ed. 1997. *Surface Analysis*. New York: John Wiley & Sons.

Weeks, R. A. and Abraham, M. 1965. Electron spin resonance of irradiated quartz: Atomic hydrogen. *J. Chem. Phys.* 42: 68–71.

Wong, P. K. and Willard, J. E. 1968. Evidence for electron migration during γ irradiation of silica gels. Reactions of adsorbed electron scavengers. *J. Phys. Chem.* 72: 2623–2627.

Yamamoto, T. A., Seino, S., Okitsu, K., Oshima, R., and Nagata, Y. 1999. Hydrogen gas evolution from alumina nanoparticles dispersed in water irradiated with γ-ray. *Nanostruct. Mater.* 12: 1045–1048.

Yates, D. J. C. 1961. Infrared studies of the surface hydroxyl groups on titanium dioxide, and of the chemisorption of carbon monoxide and carbon dioxide. *J. Phys. Chem.* 65: 746–753.

Zhang, G., Mao, Y., and Thomas, J. K. 1997. Surface chemistry induced by high energy radiation in silica of small particle structures. *J. Phys. Chem. B* 101: 7100–7113.

Zhang, G., Liu, X., and Thomas, J. K. 1998. Radiation induced physical and chemical processes in zeolite materials. *Radiat. Phys. Chem.* 51: 135–152.

17 Ionization of Solute Molecules at the Liquid Water Surface, Interfaces, and Self-Assembled Systems

Akira Harata
Kyushu University
Kasuga, Japan

Miki Sato
Kyushu University
Kasuga, Japan

Toshio Ishioka
Kyushu University
Kasuga, Japan

CONTENTS

17.1   GENERAL COMMENTS

17.1.1   Introduction

Liquid interfaces are of fundamental and practical importance in science and technology (Birdi, 1997, 1999; Jungwirth et al., 2006). Understanding the molecular behavior at the interface involves the fundamental sciences of molecular interaction and assists with the resolution of a variety of problems related to chemistry, physics, and life and environmental sciences. Among liquid interfaces, the air–water interface (water surface) is the most universal and may be the most important.

Many studies have targeted the molecular behavior at liquid interfaces experimentally (Benjamin, 2006; Davidovits et al., 2006; Donaldson and Vaida, 2006; Gopalakrishnan et al., 2006; Shen and Ostroverkhov, 2006; Winter and Faubel, 2006) and theoretically (Benjamin, 2006; Chang and Dang, 2006; Garret et al., 2006; Jungwirth and Tobias, 2006; Mundy and Kuo, 2006; Perry et al., 2006). These studies have clarified that molecules at a solution interface behave differently from those of the same chemical species in a bulk solution. It is natural to consider that, primarily, the inhomogeneous and thermally fluctuating structure in the vicinity of the top surface of the half-space filled with solvent molecules dominates the interfacial properties of adsorbed solute molecules. Additionally, solute–solute interaction in the interface region influences the molecular behavior.

Fundamental interests in solute molecules at the liquid water surface include depth distribution, orientation, molecular interaction, adsorption–desorption dynamics, surface diffusion, and the influence of the bulk properties of solvent liquid on these from both the static (time-averaged) and dynamic points of view. For applications to chemical processing and material fabrication using a liquid surface, such as the Langmuir–Blodgett technique, the precise control of the molecular behavior at the interface might provide a novel way to control the properties of the materials produced.

To precisely understand and control the molecular behavior at the interface, methodologies for observing molecules at the interface play important roles. Interface-selective observation methods are required because it is typical for a large amount of the same species of the target solute molecules to be in a bulk solution. Furthermore, a highly sensitive performance is preferable for this method because the absolute amount of the interface-adsorbed species is smaller than 1 nmol/cm^2.

17.1.2   Thermodynamic View of the Water Surface

The water surface is thermally fluctuating. Thermodynamic theory predicts a time-averaged view of the liquid surface (Benjamin, 1996, 2006). As shown in Figure 17.1, the water surface has a transition region of mass density with a finite thickness. The density distribution along the surface normal is described as a co-error function with a parameter representing the half-width of the transition region: the thickness of the water surface can be defined as 2σ, which is given as

$$2\sigma = \sqrt{\frac{kT}{\pi\gamma} \ln \frac{1+(2\pi l_c/\xi)^2}{1+(2\pi l_c)^2/S}}, \tag{17.1}$$

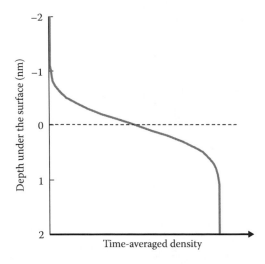

FIGURE 17.1 Time-average mass density distribution at the pure water surface at 300 K.

where

γ is the surface tension

S is the surface area of liquid

k is the Boltzmann constant

T is the temperature

ξ is the bulk correlation length (0.6–0.9 nm)

Here l_c is the capillary length equal to $(\gamma/\rho_0 g)^{1/2}$, with ρ_0 the mass density, and g the acceleration due to gravity

The thickness of the pure water surface is estimated to be 0.8 nm at 300 K, which is comparable to the size of an organic molecule. Surface tension has a strong influence on the thickness. A smaller surface tension, typically induced by the addition of organic molecules in the water, makes the transition region thicker. It is notable that these molecules are dynamically moving in, adsorbing to, and desorbing from the transition region.

Traditionally, surface tension measurements have been made to thermodynamically understand a state of solute molecules adsorbed at the water surface. Surface excess Γ is a function of the surface tension and the solute concentration C in a bulk solution as (Motomura, 1978; Motomura et al., 1981)

$$\Gamma = -\frac{C}{\nu RT}\left(\frac{\partial \gamma}{\partial C}\right)_{T,P},$$ (17.2)

where

R is the universal gas constant

P is the pressure

ν represents the number of ions contributing to the surface tension decrease per single solute molecule

The surface density of solute molecules N is equal to Γ for most organic molecules because most of the excess molecules are expected to be in a nanometer region of the surface although, strictly speaking, the definitions of N and Γ differ. Generally, for water-soluble molecules, the Langmuir isotherm given as

$$\frac{1}{\theta} = 1 + \frac{1}{K_{ad}C}$$ (17.3)

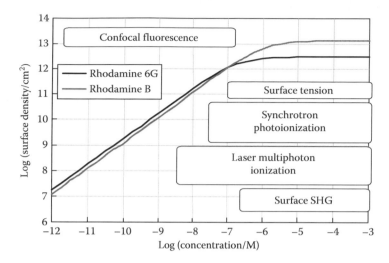

FIGURE 17.2 Relationship between concentration and surface density for rhodamine dyes. The concentration ranges for an observation with a variety of methods are shown. The surface density ranges should be calculated from the concentration range with the relationship between the concentration and surface density (solid curves).

provides a good approximation for the relation between $N = N_{max}\theta$ and C, where θ is the surface coverage, K_{ad} is the adsorption constant, and N_{max} is the maximum value of surface density. Assuming the Langmuir adsorption isotherm, we get

$$\gamma = \gamma_w - \nu RTN_{max}\ln(1 + K_{ad}C), \tag{17.4}$$

where γ_w is the surface tension of pure water.

The typical relation between the concentration and surface density is shown in Figure 17.2 for two types of rhodamine dyes estimated from the surface tension-dependence on the bulk concentration using Equation 17.4. It is evident that the surface density of an adsorbed solute molecule saturates at a specific concentration. The applicable concentration ranges of a variety of methods are also indicated (*vide infra*).

Surface tension measurements are valuable because the surface density of adsorbed solute molecules can be estimated through them. However, a molecular-scale view of the interface is, unfortunately, not easy to obtain; this is especially true for a condition with low adsorption coverage.

17.1.3 INTERFACE-SELECTIVE SPECTROSCOPIC METHODS

Interface-selective spectroscopic methods have been applied to provide a variety of information on the solute molecules adsorbed at the water surface. Every interaction between light and surface molecules can be utilized to observe molecules at the interface, as shown in Figure 17.3.

Linear spectroscopic methods, such as reflection spectroscopy, including Brewster angle microscopy (Hoenig and Moebius, 1991; Lheveder et al., 2000), and fluorescence spectroscopy, including confocal fluorescence microscopy (Li et al., 1998a,b; Zheng et al., 2000, 2001, 2004; Zheng and Harata, 2001), are general and powerful. Nonlinear spectroscopic methods, such as resonance optical second harmonic generation (SHG) (Shen, 1989; Zhao et al., 1990; Eisenthal, 1992; Slyadneva et al., 1999, 2003; Tsukanova et al., 2000; Benjamin, 2006) and vibrational sum-frequency generation (Gopalakrishnan et al., 2006; Perry et al., 2006; Shen and Ostroverkhov, 2006), are powerful tools for the surface selective observation of molecular electronic and vibrational states, respectively. However, it is often the case that their sensitivity

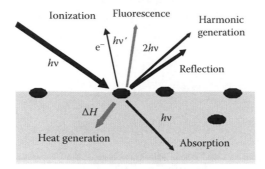

FIGURE 17.3   Interactions of laser light with molecules at an interface.

to detect solute molecules adsorbed at the water surface is not sufficient, especially when the surface coverage is less than 5% of the monolayer coverage.

The applicable concentration ranges of a variety of methods are summarized in Figure 17.2; the findings in this figure are based solely on our experience and are one of the guidelines. Of course, the range depends on the target molecules and experimental conditions. However, among a variety of spectroscopic methods for observing molecules at the air–water interface, the photoionization method surely has good potential to detect adsorbed solute molecules with surface selectivity and sensitivity. Only the fluorescence detection of the surface-adsorbed molecules is highly sensitive. When confocal fluorescence microscopy is used, the surface-selective detection of adsorbed dye molecules is possible up to a surface coverage of 10^{-6} level even if the dye molecule is water-soluble. However, under a high-concentration condition, surface selectivity is lost because the ratio of surface-adsorbed to bulk molecules decreases.

17.1.4   PHOTOIONIZATION OF MOLECULES AT THE WATER SURFACE

When photo-emitted electrons from the molecules at the water surface are collected in the upper gas phase, as in the total current yield detection method used in x-ray spectroscopy, the photoionization current flowing from the solution to the ground reflects the photoionization of molecules only at the surface (Figure 17.4).

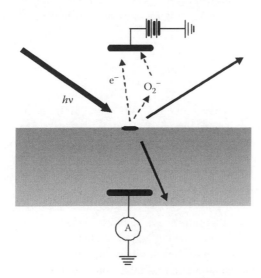

FIGURE 17.4   Schematic illustration for the detection of photoionization current from molecules adsorbed at the water surface.

The signal intensity Q of the photoionization current for the m-photon process is described as

$$Q = AI^m \left[\eta_{surf}\sigma_{surf}N + \eta_{bulk}\sigma_{bulk}\int_0^\infty dz \int_0^{\pi/2} d\varphi\{C(z)\sin\varphi\exp(-z/\lambda\cos\varphi)\} \right], \qquad (17.5)$$

where

A is a proportionality constant

I is the pump pulse energy

η_{surf} and η_{bulk} are the effective detection efficiencies of photoemitted electrons from the dye molecules at the surface and in the bulk solution, respectively

σ_{surf} and σ_{bulk} are the photoionization cross sections of the dye molecules at the surface and in the bulk solution, respectively

N is the surface density of the dye molecules

$C(z)$ is the dye concentration depending on depth z from the surface

λ is the escape length of the photoelectrons from the bulk solution, which has been considered to be a few nanometers for electrons (<1 eV) in water (Jo and White, 1991)

φ is the angle between the surface normal and the direction of initially emitted electrons

If the bulk concentration is depth independent, as $C(z) = C$, the double integral in Equation 17.5 equals $C\lambda/2$. The small escape depth ensures surface selectivity.

When the Langmuir adsorption isotherm (Equation 17.3) is applicable to describe the adsorption of dye molecules at the water surface, a relation between N and C is given by

$$N = \frac{N_{max}K_{ad}C}{1 + K_{ad}C}, \qquad (17.6)$$

where

N_{max} is the maximum surface density of dye molecules so that surface coverage is $\theta = N/N_{max}$

K_{ad} is the adsorption equilibrium constant

For dye molecules with surface activity, N is much larger than $C\lambda$ because λ is as small as a few nanometers. Then, we can ignore the second term in the right-hand side of Equation 17.5, and the expression of the concentration-dependent photoionization signal intensity for the solution surface observation is given as

$$Q = \frac{Q_{max}K_{ad}C}{1 + K_{ad}C}, \qquad (17.7)$$

where $Q_{max} = AI^m\eta_{surf}\sigma_{surf}N_{max}$ is the signal intensity at the high-concentration limit.

The photoionization phenomena in the gas and condensed phases as well as at the surface have been investigated from a variety of viewpoints, such as chemical physics, physical chemistry, biological chemistry, environmental science, and solar energy conversion (Hatano, 1999, 2001a,b; Ogawa, 1999; Kouchi and Hatano, 2004). The light sources used are conventional vacuum ultraviolet light, synchrotron radiation, and laser.

Vacuum ultraviolet light with photon energy over the ionization potential of a molecule makes the molecule ionized via a one-photon process. A pulsed laser, which generates a light beam of a large photon flux, easily makes a molecule ionized through a multiphoton ionization (MPI) process. Resonant MPI measurements have good sensitivity, wide applicability and, species selectivity; for these reasons, they have been selected as a profitable analytical method.

The photoionization of the water-soluble molecules adsorbed at the aqueous solution surface has been studied (Watanabe et al., 1980, 1998; Watanabe and Ikeda, 1983; Watanabe, 1994, 2006;

Winter and Faubel, 2006) because there is a wide variety of surface-active molecules and their adsorption behavior has great importance in physical chemistry, biochemistry, environmental chemistry, and technology. However, it is not always apparent how much the photoionization process is affected by the interaction of the solute with the solvent molecules, how far the solute molecule are from the surface, or how long a molecules stays at the surface.

17.2 EXPERIMENTAL SETUPS

17.2.1 PHOTOIONIZATION-CURRENT SPECTRA OF MOLECULES AT THE WATER SURFACE

A typical experimental setup for the measurements of photoionization-current spectra is illustrated in Figure 17.5 (Inoue et al., 1998; Ogawa et al., 1999; Seno et al., 2002). The monochromated synchrotron light (3.5–9.2 eV) was obtained from the beamline 1B at the Ultraviolet Synchrotron Orbital Radiation (UVSOR) facility of the Institute for Molecular Sciences (Okazaki, Japan) with a Seya–Namioka monochromator. The light was emitted from the end chamber of the beamline to the He-purged cell through an MgF_2 window. A quartz plate was inserted, if necessary, to cut high-energy photons caused by higher-order diffraction from the grating.

The emitted light was reflected on an Al mirror and vertically irradiated on the aqueous solution surface through a Cu-mesh electrode. The electrode was set at 5 mm above the liquid surface and high voltage (400–500 V) was applied to the electrode so that emitted electrons were collected, for which bias dependence was checked to collect electrons as much as possible with no abrupt discharge. The photocurrent (~100 fA) was measured with a picoammeter (Model 617, Keithley). The incident photon intensity was monitored by measuring the fluorescence intensity from a sodium salicylate plate with a photomultiplier tube (1P28, Hamamatsu Photonics).

A sample of aqueous solution was poured in a vessel made of platinum (30 mm diameter, 6 mm depth, and 7.07 cm² surface area) or stainless steel (34 mm diameter, 6 mm depth, and 9.08 cm² surface area) and set on an electrode in the cell box. The typical volume of the sample solution was 4 mL.

17.2.2 LASER MULTIPHOTON IONIZATION CURRENT OF MOLECULES AT THE WATER SURFACE

A typical experimental apparatus for the multiphoton ionization experiments for observing molecules at the water surface is shown below (Masuda et al., 1993; Chen et al., 1994; Inoue et al., 1994; Ogawa et al., 1994a,b). The third harmonics of Nd: yttrium aluminum garnet laser (Minilite Mini II, Continuum; wavelength, 355 nm; pulse duration, 4 ns; pulse energy, 10 mJ/pulse) was operated at a

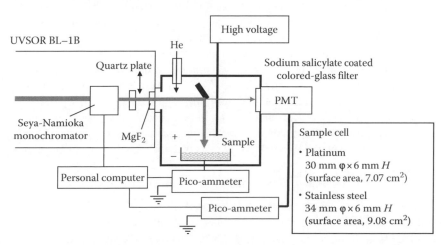

FIGURE 17.5 Experimental setup for measuring the photoionization-current spectra of molecules adsorbed at the water surface.

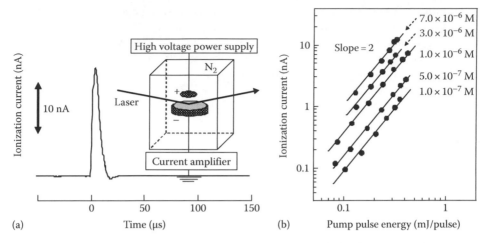

FIGURE 17.6 (a) Typical time profile of a measured laser two-photon ionization current of an aqueous solution of rhodamine B chloride (10 μM). The laser pulse energy was 0.25 mJ/pulse. The peak current was 22 nA, and the ionization charge determined as time-integrated current was 160 fC. Inset: Schematic illustration of the experimental setup around the sample. (b) Pump pulse-energy dependence of the ionization current amplitude for an aqueous solution of rhodamine B chloride. The excitation wavelength was 355 nm.

repetition rate of 2 Hz. After the pulse energy was adjusted to a certain value less than 0.5 mJ/pulse with a set of a half-wave plate and a glass plate, the laser beam was focused softly with a quartz lens (focal length, 300 mm) on the liquid solution surface at an incident angle of 84°. The polarization plane of the laser beam was controlled with a half-wave plate.

The 4.0 mL sample solution was poured into a stainless steel vessel (34 mm diameter, 6 mm depth, 9.08 cm^2 surface area) and set on an electrode in a cell box purged with nitrogen gas, as shown in the inset of Figure 17.6a. Another disk electrode (12 mm diameter) was placed 8 mm above the solution surface and positively biased at 2.0 kV with a high-voltage power supply (C3350, Hamamatsu). The photocurrent fed from the sample vessel to a current amplifier (428, Keithley; gain, 10^7 V/A) connected to the ground was measured with a digital oscilloscope (DCS-8200, Kenwood). The signal was accumulated for 64 pulses of the laser. Figure 17.6a shows a typical time profile of laser two-photon ionization current measured for an aqueous solution of rhodamine B chloride (RhB). The peak current or charge, determined as the time-integrated current, was used as the ionization signal intensity. Generally, no significant difference between the peak current and the charge was found when analyzing the photoionization signal.

Figure 17.6b shows the pump pulse-energy dependence of the ionization current amplitude of an aqueous solution of RhB. The photoionization signal intensity was quadratically proportional to the laser pulse energy for every concentration investigated regardless of the polarization of the excitation laser. Similar results were obtained for rhodamine 6G chloride (R6G). The photoionization of the rhodamine molecules adsorbed on the water surface in a two-photon process under the excitation wavelength of 355 nm was confirmed. This is reasonable because the photon energy at 355 nm is 3.5 eV and the photoionization threshold energy of RhB and R6G on the water surface has been reported to be 5.6 eV (Seno et al., 2002). The quadratic dependence shows that the space charge effect is not significant under the experimental conditions and the signal intensity reflects the number of photo-ionized molecules.

17.2.3 Laser Second Harmonic Generation and Surface Tension Measurements

Second harmonic generation measurements were carried out with an experimental setup described elsewhere (Slyadneva et al., 1999, 2003; Tsukanova et al., 2000). Briefly, a frequency-doubled beam of a pulsed Nd^{3+}: yttrium aluminum garnet laser (PY61, Continuum; wavelength, 532 nm; pulse duration, 40 ps; repetition rate, 10 Hz; maximum pulse energy, 30 mJ/pulse), whose polarization plane was rotated by a half-wave plate to obtain an s- or a p-polarized beam, was softly focused by a quartz lens (focus length 90 cm) at the sample surface, resulting in a beam waist area of $0.25 \, \text{cm}^2$. The incident angle was 45° from the surface normal. The energy density per pulse at the surface was $2 \, \text{mJ/cm}^2$. The SHG light reflected from the solution surface was separated from the incident laser beam by optical cut-off and band-pass filters and detected with a photomultiplier tube (R585, Hamamatsu). The signal intensity was recorded with a digital oscilloscope (TDS 380, Tektronix). The SHG signal for each measurement under the condition of a fixed concentration was averaged over 2 min.

A similar expression to Equation 17.7 is obtained for the case of SHG measurements. If it is assumed that the SHG signal is quadratically proportional to the number density of dye molecules at the surface, the SHG signal intensity S is given as

$$S = S_{\text{max}} \left[\frac{K_{\text{ad}} C}{1 + K_{\text{ad}} C} \right]^2 + S_{\text{w}}, \tag{17.8}$$

where

S_{max} is the signal intensity at a high-concentration limit
S_{w} is the SHG signal intensity of the water surface

Here, we simply add S_{w} because SHG from the water surface is non-resonant while SHG from the dye molecules at the water surface is resonant at 532 nm; thus, a $\pi/2$-phase difference is roughly expected.

The surface tension of aqueous solutions of rhodamine dye was measured with a surface-pressure meter (Model HMB, Kyowakaimen) of the Wilhelmy type.

17.3 PHOTOIONIZATION SPECTRA AND PHOTOIONIZATION THRESHOLD OF MOLECULES AT THE WATER SURFACE

17.3.1 Photoionization-Current Spectra of Molecules at the Water Surface

The photocurrent spectra at the surface of both pure water and the aqueous solutions of 1-aminopyrene are shown in Figure 17.7, in which the photocurrents are normalized with the ring current of the synchrotron orbital. No significant photocurrents for the surface of pure water and buffer solutions are detected. Thus, the photocurrents for the surfaces of the 1-aminopyrene aqueous solutions are successfully measured for two pH values of the solution. The current shows increases with the photon energy in the region

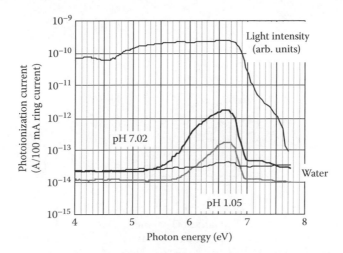

FIGURE 17.7 Photocurrent spectra of 1-aminopyrene at the aqueous solution surfaces with different pH values. The photocurrents are normalized with the ring current but not with light intensity measured. High-energy photons are cut with a quartz plate.

above the photoionization thresholds: the photocurrent shows a similar threshold behavior to that of the bulk solution. Similar spectra are observed for a variety of molecules adsorbed on the water surface.

It is known that the photocurrent, I, near the photoionization threshold, E_{th}, can be represented with an empirical power law (Holroyd et al., 1983; Hoffman and Albrecht, 1991),

$$S = \sum S_i \left(h\nu - E_{th}^{(i)} \right)^n \quad \text{for } h\nu > E_{th}^{(i)}, \tag{17.9}$$

where

$E_{th}^{(i)}$ is the threshold energy of photoionization for the ith component

$h\nu$ is the photon energy

n is an exponent representing the power law with respect to the excess energy $h\nu - E_{th}^{(i)}$

The summation of the number of ionization transitions should be conducted.

The photocurrent spectra of 1-aminopyrene at the water surface in Figure 17.7 show positive, although not strong, evidence of pH dependence. Both changes in intensity and a small shift in photoionization thresholds are observed (*vide infra*). The spectral shape also changes slightly: a shoulder at 6.1 eV is observed at pH 7.02. It is reasonable because 1-aminopyrene at pH 7.02 has a different chemical form (neutral) from that at pH 1.05 (cationic). However, pyrenebutyric acid and pyrenehexadecanoic acid show pH-independent spectral shapes although the chemical forms are neutral at pH 3.11 and anionic at pH 11.19 while small shifts in photoionization thresholds are observed (*vide infra*). The signal intensity of pyrenehexadecanoic acid is pH-independent. This suggests that solubility might be important for changes in spectral shapes to take place.

A large pH-dependent change in spectral shape is observed for R6G, as shown in Figure 17.8. Apparently, the spectral shape at pH 1 is different from those under the other pH conditions although

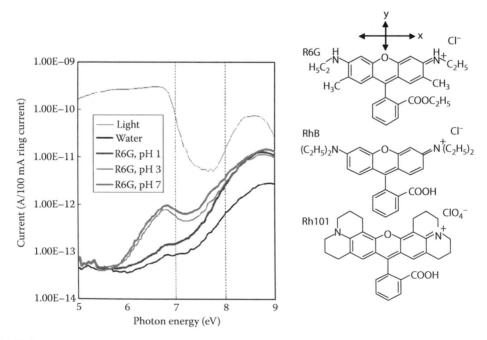

FIGURE 17.8 Photocurrent spectra of rhodamine 6G chloride (R6G) at the aqueous solution surfaces with different pH values. The photocurrents are normalized with the ring current but not with light intensity. The decrease in the light intensity around 6.8–8.4 eV is due to light absorption by water vapor. The signal around 5.0–5.4 eV is due to the insufficient cut of second-order light at 10.0–10.8 eV because a quartz plate was not inserted in the light path. The chemical structures and direction of the molecular axis are shown for rhodamine 6G as well as rhodamine B chloride (RhB) and rhodamine 101 perchlorate (Rh101) for later discussion.

it is known that R6G in a bulk aqueous solution always has a cationic form for these pH ranges (Zheng et al., 2004). The cause of the pH-dependent changes in spectral shape has not been specified.

Unfortunately, it is not easy to discuss the spectral shapes; therefore, at the present stage, the photoionization threshold is the only good indicator of photoionization spectra for understanding the state of adsorbed solute molecules. In Section 17.3.3, the photoionization-threshold dependence on subphase pH is discussed.

17.3.2 DETERMINATION OF THE PHOTOIONIZATION THRESHOLD

The empirical relation in Equation 17.9 has been generally used to determine photoionization thresholds. Some problems in determining the photoionization thresholds using a fitting procedure are represented in Figure 17.9.

First of all, the validity of Equation 17.9 is not self-evident and careful consideration is required to adapt the value of the exponent n (Hoffman and Albrecht, 1991). It is not simple to determine threshold energy E_{th} and exponent n at the same time with an experimentally obtained photoionization spectrum even if a single component with one ionization threshold is assumed. Furthermore, the number of

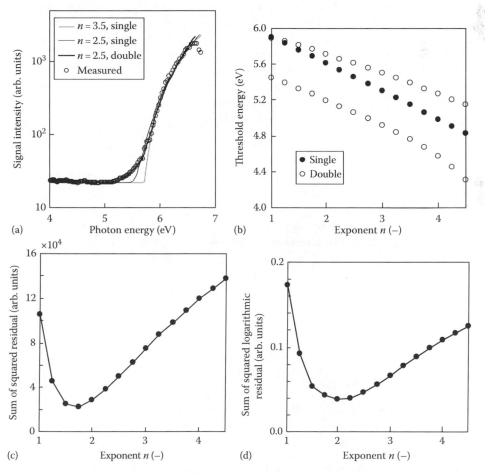

FIGURE 17.9 Fitting of a photocurrent spectrum of 1-aminopyrene to the empirical equation: (a) photocurrent spectrum and fitting results with single ($n = 2.5, 3.5$) and double ($n = 2.5$) components; (b) dependence of the estimated photoionization threshold energy on exponent n; (c) dependence of the summation of the squared residual errors on exponent n in the single component analysis; and (d) dependence of the summation of the squared logarithmic residual errors on exponent n in the single-component analysis.

ionization transitions is generally unknown for the medium-sized molecules that we are interested in, and a single species of molecule may have more than one chemical form at the water surface.

Figure 17.9 shows the results of fitting experimental data to the empirical equation. As shown in Figure 17.9a, for the single component analysis, there are small fitting errors near the threshold energy, independently of the exponent value; a double component fitting gives better results although there is no reliable explanation for the components being double. What is important is that the threshold energy systematically depends on the exponent values, as shown in Figure 17.9b, for both single and double component fittings. It is concluded that the accuracy of the determined ionization threshold is influenced by the exponent value used.

It is true that the summation of the residual errors in the fitting has a minimum, as shown in Figure 17.9c and d, but the position of the minima is not on $n = 2.5$ or 3.5, which is the value typically used as the exponent value. In summary, for the determination of ionization thresholds, it is realistic and effective to assume the number of components and the exponent to fixed values, respectively, but accuracy limitations should be taken into account.

17.3.3 Photoionization Thresholds of Molecules at the Water Surface

The determined photoionization thresholds (eV) of chemicals on the water surface are summarized in Table 17.1 for a few pH values of the water subphase. The reference values of the ionization potential (Farrell and Newton, 1965; Timoshenko et al., 1981) are given in Table 17.1. In the analysis, the exponent of 2.5 is used, and a single component for the ionization threshold is assumed so that some systematic errors could be included in the threshold energy.

The photoionization threshold, E_{th}, of a solute in a bulk solution is related to the ionization potential at the gas phase, IP, as follows (Holroyd et al., 1983):

$$E_{th} = IP + P^+ + V_0, \qquad (17.10)$$

TABLE 17.1
Photoionization Thresholds (eV) of Chemicals on the Water Surface Determined with the Exponent of 2.5

1-Aminopyrene		Rhodamine 6G	
pH 1.05 (cation)	5.7_2	pH 1 (cation)	5.7_9
pH 7.02 (neutral)	5.4_5	pH 3 (cation)	5.7_4
1-Pyrenebutyric acid		pH 7 (cation)	5.6_7
pH 3.11 (neutral)	6.0_9	Rhodamine B	
pH 11.19 (anion)	5.8_8	pH 7 (zwitterion)	5.6
1-Pyrenehexadecanoic acid		Rhodamine 101	
pH 3.11 (neutral)	6.0_7	pH 7 (zwitterion)	5.3
pH 11.19 (anion)	5.8_7		

Note: The pH value of the water subphase is adjusted. The major chemical form is indicated in parentheses. The molecular density is controlled: coverage, <0.74%. 1-Aminopyrene and 1-pyrenebutyric acid are slightly soluble in water, 1-pyrenehexadecanoic acid is insoluble, and rhodamine 6G is soluble.

Refs.: Ionization potential (eV)

Aminopyrene	6.8 ± 0.1	(CTS, Farrell and Newton, 1965)
Pyrene	7.426	(NIST Chemistry WebBook, Online, 2002)
Rhodamine B	6.7	(Timoshenko et al., 1981)

where

   P^+ is the polarization energy of the resultant positive ion
   V_0 is the electron affinity of the solvent

The polarization energy P^+ can be calculated with Born's equation (Born, 1920),

$$P^+ = \frac{-e^2}{8\pi\varepsilon_0 r}\left(1 - \frac{1}{\varepsilon_r}\right), \tag{17.11}$$

where

   e is the elemental charge of electron
   ε_0 is the dielectric constant of vacuum
   r is an effective radius of the ion
   ε_r is the relative dielectric constant of the medium

For the ionization of molecules at the liquid surface, V_0 in Equation 17.10 seems negligible because the photoelectron is directly emitted to the gas phase.

The experimental value of P^+ for RhB at the aqueous solution surface is -1.1 eV. It is suggested that the probe molecule for photoionization is positioned in a space where the photogenerated positive ion can sufficiently interact with the solvent water molecules. On the other hand, the value of P^+ in the aqueous solution estimated with Equation 17.11 is -0.5 eV, where the effective radius r of the RhB positive ion is roughly estimated to be 0.6 nm with the Corey–Pauling model (Koltun, 1965) of the RhB molecule. The large P^+ value experimentally observed suggests that V_0 cannot be ignored even at the surface or that the effective radius of the resultant positive RhB ion is smaller than 0.6 nm.

As for the pH-dependence, shown in Table 17.1, the tendencies observed for pyrene derivatives are summarized as follows: more positively charged forms tend to have higher thresholds, the threshold downshifts 0.20–0.27 eV per electron charge, and the shifts of insoluble species (pyrene-hexadecanoic acid) and less soluble ones (pyrenebutyric acid) are smaller than those of soluble ones (aminopyrene). These results are not surprising when the charge density of the pyrene unit is considered. The small shift for insoluble species could come from immovability in the depth position, indicating a constant degree of solvation between charged and neutral forms.

Rhodamine 6G (soluble species) has the same chemical forms in the bulk aqueous solution throughout the pH range investigated (Zheng et al., 2004), but the threshold shift is observed: a lower pH results in a higher threshold. The degree of protonation seems to cause the results. At pH 1, a different chemical form from that at a higher pH is expected for R6G only at the water surface. This is already suggested in Figure 17.8. Although the expected surface-specific chemical form is not identified, a similar proposal has been reported in a confocal fluorescence study of R6G molecules at the water surface (Zheng et al., 2004).

17.3.4 Photoionization of Rhodamine B at the Water Surface under Self-Assembled Layer

Photoelectron emission from a liquid solution surface covered with a self-assembled layer of aliphatic acid has been investigated (Ishioka et al., 2003; Ishioka and Harata, 2004). The sample solutions investigated were composed of a surface-active dye (RhB, 10 µM), a buffer electrolyte (HCl, pH 1.0), and water. The aqueous solution surface was modified with arachidic acid ($C_{19}H_{39}COOH$) by spreading as a benzene solution. The added amount was approximately within two monolayers at the maximum surface density, which is calculated by the assumption that a close-packed layer was formed on the aqueous solution surface.

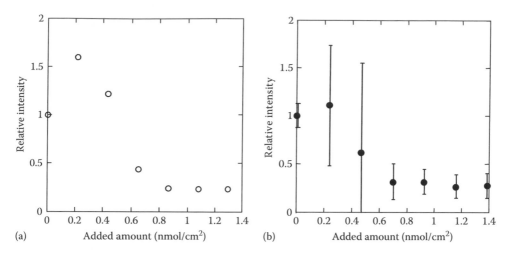

FIGURE 17.10 (a) Effect of arachidic acid addition on the photocurrent from the surface of rhodamine B aqueous solution measured by single-photon ionization and (b) measured by two-photon ionization. Error bars indicate deviations of current intensities for each laser pulse.

Similar shapes of photoionization spectra were observed for all the surface density of arachidic acid: no significant change in photoionization thresholds was observed while the signal intensities were increased or decreased. Figure 17.10a shows the plot of the photocurrent intensity at 6.67 eV measured by single-photon ionization against the added amount of arachidic acid. The current intensity increases remarkably when a small amount of arachidic acid is added (~0.2 monolayer) and then decreases to a constant value above the monolayer formation level. These experimental results cannot be explained by a simple model based on uniform monolayer formation and simple electron scattering through the aliphatic monolayer because this model needs a monotonous decrease of the photoionization current upon film formation.

Figure 17.10b shows the same experiment as described above except that the measurement was conducted by laser two-photon ionization under a beam-focusing condition. The pulse-to-pulse deviation of the signal intensity is remarkable at the concentration range of the photocurrent increase. The increased fluctuation suggests that a small domain formed by the arachidic acid layers has a comparable size to that of the focused area (10–100 μm). Aggregate formation and the accumulation of dyes around the aggregates are the most probable explanations for the enhancement of the photoionization current by arachidic acid addition.

A similar experiment was performed with a series of aliphatic acids having different CH-chain lengths. The details of the analyses will be published elsewhere.

17.4 PHOTOIONIZATION-CURRENT DEPENDENCE ON THE SURFACE DENSITY OF SOLUTE MOLECULES AND EFFECT OF INCIDENT LIGHT POLARIZATION

17.4.1 CONCENTRATION DEPENDENCE OF SURFACE-SELECTIVE SIGNALS OF RHODAMINE DYES AT THE WATER SURFACE: LASER MULTIPHOTON IONIZATION AND SECOND HARMONIC GENERATION STUDIES

Resonant laser multiphoton ionization is superior to one-photon ionization because it has species selectivity as well as surface selectivity. Resonant second harmonic generation has both species selectivity and surface selectivity as well. Data given in this section are based on the measurements of laser multiphoton ionization, second harmonic generation, and surface tension (Seno et al., 2001; Harata et al., 2009).

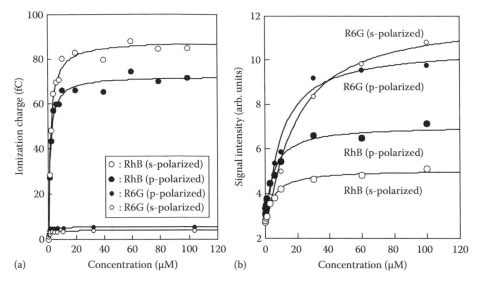

FIGURE 17.11 (a) Concentration dependence of the ionization charge of an aqueous solution of rhodamine dyes when s- or p-polarized laser light excited the surfaces of aqueous solutions of rhodamine B chloride (RhB) and rhodamine 6G chloride (R6G). The curves are the fitting results of the experimental data to the Langmuir isotherm. (b) Concentration dependence of the second harmonic generation signal intensity from the surface of aqueous solutions of rhodamine dyes when s- or p-polarized laser light at 532 nm excited the surfaces of aqueous solutions of RhB and R6G. The curves are the fitting results of the experimental data to the Langmuir isotherm. Polarization is that of the excitation laser. The polarization of the second harmonic light at 266 nm is not selected.

The concentration dependences of the photoionization charge of aqueous solutions of RhB and R6G are shown in Figure 17.11a. The intensity of the signal generated with s- and p-polarized laser light excitation at 355 nm is plotted. All the data show the saturation behavior of the photoionization signal at high-concentration regions, as expected from Equation 17.7. The curves are fitting results of the experimental data to the Langmuir isotherm, for which the fitting parameters of K_{ad} (M^{-1}) and Q_{max} (fC) are summarized in Table 17.2. Equation 17.7 represents the experimental data well. The results suggest that rhodamine molecules adsorbed at the water surface are selectively detected and that the photoionization signal from rhodamine molecules in the bulk solution is negligible, as is that from water molecules. As the magnitudes of Q_{max} in Table 17.2 demonstrate, the photoionization signal of RhB is larger for s-polarized than for p-polarized excitation, while R6G shows the reverse order. It is suggested that the transition moment of the two-photon photoionization (TPI) is relatively parallel and perpendicular to the water surface for RhB and R6G, respectively. The polarization dependence of K_{ad} observed for both RhB and R6G suggests that σ_{surf} and/or η_{surf} varies with the bulk concentration, meaning that the molecular orientation at the surface changes with the bulk concentration. Details of the values K_{ad} and Q_{max} are discussed in Section 17.4.4.

Figure 17.11b shows the concentration dependence of the SHG signal intensity from the surface of aqueous solutions of rhodamine dyes when s- or p-polarized laser light at 532 nm excites the surfaces of aqueous solutions of RhB and R6G. All the data show the saturation behavior of the SHG signal at high-concentration regions, as expected from Equation 17.8. The curves are fitting the results of the experimental data to Equation 17.8, for which the fitting parameters of K_{ad} (M^{-1}) and S_{max} (arb. units) are listed in Table 17.2. Equation 17.8 represents the experimental data well, as shown in Figure 17.11a; however, K_{ad} of R6G is one order of magnitude smaller than that obtained in photoionization experiments. Moreover, the polarization dependence of S_{max} is quite different from that of the photoionization measurements. Details of the values K_{ad} and S_{max} are discussed in Section 17.4.4.

TABLE 17.2

Adsorption Equilibrium Coefficients K_{ad} and Signal Intensity Factors (Q_{max} and S_{max}) Evaluated from Each Experiment

Dye	Polarization	Q_{max} (fC), S_{max} (arb. units)	K_{ad} (10^5 M^{-1})
Two photon ionization measurements			
Rhodamine B	s	87	6.1
	p	72	7.4
Rhodamine 6G	s	4.5	21
	p	6.0	27
Second harmonic generation measurements			
Rhodamine B	s	2.3	4.1
	p	3.9	4.0
Rhodamine 6G	s	9.3	1.2
	p	7.7	2.2

Note: Q_{max}, signal intensity factors of two-photon ionization measurements with one unit of fC. The polarization is of the excitation laser. S_{max}, signal intensity factors of second harmonic generation measurements with an arbitrary unit.

17.4.2 MEASUREMENTS OF SURFACE TENSION AND ULTRAVIOLET-VISIBLE ABSORPTION SPECTRA

Surface tension and ultraviolet absorption measurements provide important reference data for following discussion on solute dye molecules adsorbed at the water surface (Seno et al., 2002; Harata et al., 2009).

The dependence of the surface tension of rhodamine solutions on the concentration is shown in Figure 17.12a. The surface tension of each rhodamine aqueous solution decreases with the bulk concentration. This clearly indicates that each rhodamine molecule is positively adsorbed to the

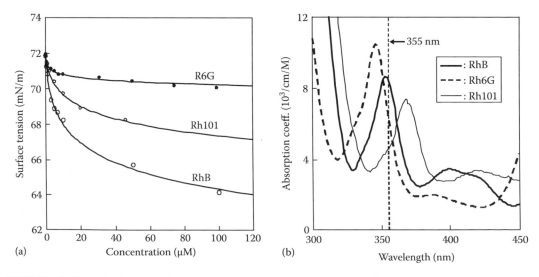

FIGURE 17.12 (a) Concentration dependence of the surface tension for aqueous solutions of rhodamine B chloride (RhB), rhodamine 6G chloride (R6G), and rhodamine 101 perchlorate (Rh101) at 25°C. (b) Ultraviolet–visible absorption spectra of aqueous solutions of rhodamine dyes.

TABLE 17.3

Some Properties of Rhodamine Dyes for Adsorption at the Water Surface at Room Temperature (25°C) Determined with Surface Tension Measurements by Assuming the Langmuir Adsorption Isotherm

Dye	K_{ad} (10^5 M^{-1})	N_{max} (10^{-7} mol/m^2)	A_{min} (nm^2/molecule)
Rhodamine B	8.8	6.7	2.5
Rhodamine 6G	58	1.0	16.5
Rhodamine 110	6.3	4.3	3.8

Note: K_{ad}, equilibrium constant; N_{max}, molecular density at the adsorption saturation; and A_{min}, specific molecular area at the adsorption saturation.

water surface. The curves are fitting results of the experimental data to Equation 17.4. In Table 17.3, the fitting parameters K_{ad} (M^{-1}) and N_{max} (mol/m^2) are summarized for each rhodamine dye with the specific molecular area A_{min} (nm^2/molecule) corresponding to N_{max}. In calculating N_{max}, $v = 1$ is assumed. At a neutral pH, the principal chemical forms of RhB, R6G, and Rh101 are zwitterion, cation, and zwitterion, respectively. The assumption $v = 1$ means that the surface activity of protons, chloride ions, and perchloride ions is neglected.

It has been reported that the surface pressure–area isotherm of dioctadecyl-rhodamine B (RhB-based water-insoluble rhodamine dye having two long hydrocarbon chains on two monoethylamino groups) shows a collapse pressure at 0.26 nm^2/molecule (Slyadneva et al., 2001). This value is as small as 10%, 1.6%, and 0.68% of A_{min} of RhB, R6G, and Rh101, respectively. The fact indicates that each rhodamine dye, especially R6G, has enough space at the water surface even when adsorption is saturated. Strong interaction between dye molecules at the water surface could be expected.

The ultraviolet absorption spectra of rhodamine dyes in Figure 17.12b shows that multiphoton ionization at 355 nm of each dye investigated is under a resonant condition. Although there is a possibility that the absorption peak of the rhodamine dyes at the water surface shifts from those of bulk aqueous solution, we disregard the possibility because no large spectral shift is expected.

In discussing molecular orientation, it is important as well as useful to consider the direction of the effective transition moment for TPI of rhodamine dyes because of the complexity of the two-photon process. All the rhodamine dyes used have an absorption peak around 540 nm, attributed to a transition from the ground state to the first singlet excitation state S_1, whose transition dipole moment is nearly parallel to the xanthene ring (x-direction in Figure 17.8) (Jakobi and Kuhn, 1962; Peterson and Harris, 1989; Drexhage, 1997). The second singlet excitation state S_2 lies at around 350 nm, and the transition dipole moment is nearly perpendicular to the xanthene ring (y-direction in Figure 17.7) (Jakobi and Kuhn, 1962; Peterson and Harris, 1989; Drexhage, 1997). The excitation wavelength of 355 nm in the photoionization experiment is resonant with the transition to the second singlet excitation state S_2.

The resonant two-photon photoionization process may be caused by both a simultaneous two-photon process and a stepwise two-photon process, and the contributing ratio of both processes depends on the intensity and pulse width of the excitation laser. In the simultaneous process, two-photon absorption of the ground state molecules directly into a pre-ionization state via the intermediate S_2 state should be considered. In the stepwise process, three processes, that is, photoexcitation to S_2, relaxation from S_2 to S_1, and photoionization from S_1 should be taken into consideration in combination. Unfortunately, little is known of the directions of the transition moments for the simultaneous two-photon absorption and stepwise S_2 absorption as well as of their contributing ratio. Strictly speaking, we cannot identify the direction of the transition moment for the TPI of the rhodamine dyes. However, a simple assumption can be considered: the direction is the same as that of resonance photo-absorption, namely, the y-direction. This might be reasonable if the resonance process dominates the two-photon absorption in the simultaneous process and if the transition from S_2 to the pre-ionization state does not have large anisotropy.

A similar discussion can be made for the direction of the transition moment for SHG. The direction is assumed to be parallel to that of the resonance photo-absorption, namely, the x-direction. Under these assumptions, we could determine the averaged direction of adsorbed rhodamine molecules at the water surface. Such a discussion has been reported for 1-pyrenehexadecanoic acid on the water surface (Sato et al., 2001).

17.4.3 CONCENTRATION-DEPENDENT ORIENTATION OF DYE MOLECULE AT THE WATER SURFACE

Figure 17.13a shows the dependence of the photoionization signal intensity $I(\theta)$ on the polarization angle θ of the excitation laser with $84°$ of the incident angle (Seno et al., 2001; Harata et al., 2009). Four different-concentration data sets are shown of aqueous solutions of each rhodamine dye. $\theta = 0°$ and $\theta = 90°$ correspond to p- and s-polarized light, respectively: $I_p = I(\theta = 0°)$ and $I_s = I(\theta = 90°)$. The curves are the fitting results of the experimental data to sinusoidal functions. As seen, this function describes each of the polarization dependences well. The arrows in Figure 17.13 indicate the order of the sample concentrations noted on the right-hand side. It is notable that the polarization dependence does not monotonically change with the concentration.

In Figure 17.13b, the ratio I_p/I_s of the photoionization signal intensity, representing the polarization dependence, is plotted as a function of the surface excess for aqueous solutions of RhB and R6G. The surface excess is calculated with the concentration of the solutions using the adsorption properties listed in Table 17.3. Notable features are that for each rhodamine dye, the I_p/I_s values show a maximum or minimum around a point where the surface excess is just saturated and the I_p/I_s values decrease or increase gradually even under a much lower surface-excess condition than the extreme point. The former indicates that the direction of the effective transition moment changes with the surface excess, and the latter means that molecules on the water surface significantly interact with each other even when they have enough free area for moving on the surface.

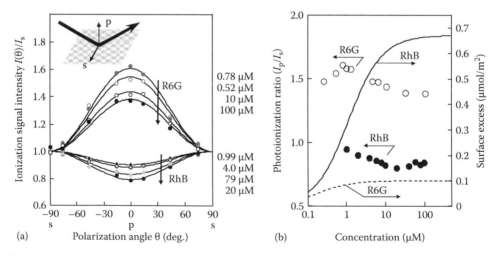

FIGURE 17.13 (a) Dependence of the photoionization signal intensity $I(\theta)$ on the polarization angle θ of the excitation laser at 355 nm and $84°$ of incident angle for 4 sets of concentrations of an aqueous solution of rhodamine dyes: rhodamine B chloride (RhB) and rhodamine 6G chloride (R6G). $\theta = 0°$ and $\theta = 90°$ for p- and s-polarized light, respectively, and $I_s = I(\theta = 90°)$. The curves are the fitting results of the experimental data to sinusoidal functions. The arrows indicate the order of the sample concentrations listed at the their right-hand side. (b) Concentration dependence of the photoionization ratio I_p/I_s and surface excess for an aqueous solution of rhodamine dyes, RhB and R6G. The surface excess is calculated with the adsorption properties listed in Table 17.3.

17.4.4 EFFECTIVE ADSORPTION EQUILIBRIUM CONSTANTS

It is not necessary for the adsorption equilibrium constants (effective adsorption equilibrium constants), experimentally determined with a variety of methods, to agree with each other because different types of interactions are used for these measurements. Surface tension measurements providing surface excess are based on the thermodynamic equilibrium, for which the spatial distributions of the concentration and orientation of molecules in the depth direction are not considered. On the other hand, photoionization and SHG signals depend on the spatial distribution of molecules. As for photoemission, when the photoemitting groups of the dye molecules are positioned as higher on the water surface, a larger photoionization signal is detected because of a smaller disturbance from water molecules. As for SHG, when photoabsorbing chromospheres of the dye molecules are positioned closer to the air–water boundary and are better oriented at the boundary, a larger resonant-SHG signal is detected because of a larger second-order nonlinear susceptibility. Consequently, the effective adsorption equilibrium constants determined with a variety of methods and experimental conditions reflect the state and behavior of molecules at the water surface.

As summarized in Tables 17.2 and 17.3, the adsorption equilibrium constants K_{ad} of RhB determined by measurements of the surface tension, TPI and SHG, have the same order of magnitude but K_{ad} by TPI and SHG is 69% and 47% for s-polarized excitation (84% and 45% for p-polarized excitation) of that by the surface tension, respectively (Seno et al., 2001; Harata et al., 2009). The results have meaningful implications: the surface density of RhB molecules increases as the bulk concentration increases even after surface-excess saturation, suggesting the existence of a concentration gradient under the surface whose magnitude depends on the concentration, and the orientation of dye molecules at the water surface depends on the bulk concentration, as discussed before. Figure 17.14a shows the dependence of the photoionization charge on the surface excess of RhB on the water surface. Photoionization charges are larger than expected from a linear relation between the surface excess and the charges both for s- and p-polarized excitation. It is qualitatively explainable that, as bulk concentration increases, the RhB molecules rise up and adhere closer to the gas phase.

The difference between the determined K_{ad} values of R6G is more significant than RhB: K_{ad} by TPI and SHG is 36% and 2.1% for s-polarized excitation (47% and 3.8% for p-polarized excitation) of that by surface tension, respectively. Figure 17.14b shows the dependence of photoionization charge on the surface excess of R6G on the water surface. Photoionization charges seem quadratically dependent on

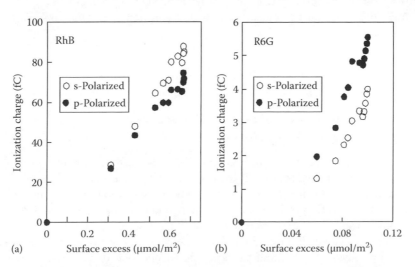

(a) Surface excess (μmol/m²)   (b) Surface excess (μmol/m²)

FIGURE 17.14 (a) Dependence of the photoionization charge on the surface excess of rhodamine B on water. The surface excess is calculated with the adsorption properties listed in Table 17.2. (b) Dependence of the photoionization charge on the surface excess of rhodamine 6G on the water.

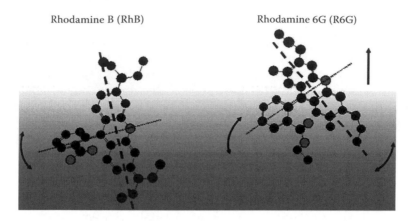

FIGURE 17.15 Rough sketch of rhodamine dyes at the water surface. The arrows indicate the direction of molecular motion in increasing concentration in bulk solution. The short axis of the rhodamine B (RhB) molecule is relatively parallel to the water surface, while the long and short axes of the rhodamine 6G (R6G) molecule have large angles with respect to the water surface. With a concentration increase, RhB changes the direction of the short axis, and R6G changes the directions of the long and short axes as well as the depth position.

the surface excess and are much larger than expected from a linear relation both for s- and p-polarized excitation. The results can be explained by the same mechanism as RhB, although the concentration-dependent changes of the concentration gradient and molecular orientation are large for R6G.

17.4.5 Rhodamine Dyes at the Water Surface

A rough sketch, based on the discussion above concerning the rhodamine dyes adsorbed at the water surface, is shown in Figure 17.15 (Harata, 2003). As the time and ensemble average, RhB and R6G at the water surface have different orientations. The concentration-dependent behavior, indicated by arrows in the figure, is also different. Even if two dye molecules are far apart from each other at the water surface, some effective interaction is expected. It is noteworthy, of course, that further discussion along with precise experiments and analyses is required even for the time- and ensemble-averaged view of surface-adsorbed molecules. A dynamic view with fluctuation of the orientation and three-dimensional position of molecules and with adsorption–desorption behavior will be presented in further studies.

17.5 DETECTION AND MONITORING OF MOLECULES AT THE WATER SURFACE

17.5.1 Detection Limit of Molecules at the Water Surface by Photoionization Measurements

Laser-based spectroscopic techniques generally provide high sensitivity in molecule detection even for molecules adsorbed at the water surface. However, when surface-selective detection is required, as in the case of detecting adsorbed liquid-soluble molecules on the liquid solution surface, three main techniques are currently available: confocal laser fluorescence, SHG, and TPI. For nonfluorescent or less fluorescent molecules, there are two techniques. Second harmonic generation spectroscopy, the most popular technique for studying molecules on a liquid surface, is particularly sensitive to the first layer of the liquid surface. The second one is laser two-photon ionization spectroscopy, which is sensitive for detection of probe molecules both in a bulk solution (Voigtman et al., 1981; Voigtman and Winefordner, 1982; Yamada et al., 1982, 1983, 1986; Ogawa et al., 1992a, 1995) and on the surface of a solution (Masuda et al., 1993; Chen et al., 1994; Inoue et al., 1994;

TABLE 17.4
Detection Limit (3σ) of Some Water-Soluble Molecules at the Water Surface Shown in Bulk Concentration

Compound	Excitation Wavelength (nm)	Molar Absorptivity (M^{-1} cm^{-1})	E_{excess}(eV)	Log P(o/w)	Detection Limit (nM)
Rhodamine B	337	2,000	0.20	2.13	41
Pyrene	337	15,000	0.90	5.18	0.044
1-Aminopyrene	337	13,200	0.62	3.72	0.094
Anthracene	337	5,900	1.85	4.49	0.39
	355	5,100	1.07		0.40
1-Chloroanthracene	355	3,600	0.62	5.20	0.22
9,10-Dichloroanthracene	355	5,000	0.90	5.92	0.20

Note: Excess energy is calculated as $E_{excess} = (E_{S_1} + h\nu) - E_{th}$, where E_{S_1} is the energy of the first excited singlet state determined from absorption spectra, $h\nu$ is the photon energy, and E_{th} is the energy of the photoionization threshold determined from the photoionization spectra. The partition coefficients P(o/w) between n-octanol and water were estimated with Chem3D Ultra6.0 (Cambridge Soft Corp.). The values of the detection limits, molar absorptivity, and excess energy for the excitation wavelength of 337 and 355 nm are referred from Inoue et al. (1994) and Chen et al. (1994), respectively.

Ogawa et al., 1994a,b; Gridin et al., 1998) or metal (Ogawa et al., 1992b). The sensitivities of these two techniques are, however, different, as shown in Figure 17.2. The typical detection limit of a solute molecule with the TPI technique was 0.25 pmol/cm^2 (Ogawa et al., 1992b), while that with the SHG technique was 5.0 pmol/cm^2 (Inoue et al., 1995).

It is important to estimate the detection limit of a target molecule theoretically or empirically. The detection limit of molecules at the water surface by TPI measurements depends on the properties of target molecules as well as the experimental conditions, which include the absorption coefficient ε at the excitation wavelength, the photoionization threshold E_{th}, the energy of the first excited singlet state E_{S_1}, and the adsorption coefficient K_{ad}. It is natural to assume that the detection limit expressed as the bulk concentration is inversely proportional to $\varepsilon E_{excess}{}^n K_{ad}$, where n is the exponent given in Equation 17.9 and E_{excess} is excess energy in photoionization relating to photon energy $h\nu$. For a stepwise TPI process, $E_{excess} = (E_{S_1} + h\nu) - E_{th}$, and $E_{excess} = 2h\nu - E_{th}$ for simultaneous TPI. Even if well-defined experimental conditions are adjusted, it is often the case that some of the properties of the target molecules are unknown. Especially, few reference values are available for K_{ad} of targets. It is not easy to obtain solubility data either. To roughly evaluate K_{ad}, n-octanol and the water partition coefficient P(o/w) may be useful because P(o/w) is easily estimated as an indicator of water solubility.

Table 17.4 is a summary of the detection limits of some chemicals on the aqueous solution surface (Inoue et al., 1994; Maeda et al., 2005; Maeda, 2006), along with the molecular absorptivity, excess energy, and n-octanol and water partition coefficient. As shown, there is no good correlation between the detection limit and ε, E_{th}, or log P(o/w), but some tendencies to a higher detection limit are seen for chemicals with higher ε and higher log P(o/w). This could be a measure of estimating the detection limit, which may be applicable even for molecules with unknown properties.

17.5.2 Acid–Base Equilibrium Constant of Molecules at the Water Surface

Two-photon photoionization measurements have been used to determine the acid–base equilibrium constants of the molecules on the water surface and the distribution coefficients of neutral species and ionized species between the surface and bulk water (Sato et al., 2000, 2004). The photoionization

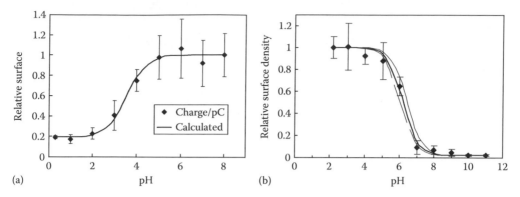

FIGURE 17.16 pH-dependence of the 2-photon ionization signal for (a) 1-aminopyrene and (b) 1-pyrenebutyric acid on the water surface. The subphase pH and ionic strength are adjusted with NaH_2PO_4, HCl, and KCl.

signal of 1-aminopyrene (4.0 nmol in the 0.600 mL vessel) on the water surface was measured for aqueous solutions with various pH values. Figure 17.16a shows the result, in which the ordinate is normalized to the value of the signal for aminopyrene at pH 8 for simplicity. The photoionization signal of aminopyrene on the water surface tends to increase with increasing subphase pH, meaning that the surface concentration of aminopyrene increases with the pH. Although both neutral and ionic forms of aminopyrene may stay in the bulk and on the surface of water in principle, neutral species are expected to prefer to stay on the surface, and the ionic species, in bulk water. This is reasonable because ionic species are more hydrophilic than neutral species. It is natural to assume that signals observed around pH 0–1 are mainly due to cationic species on the water surface and those around pH 6–8 are mainly due to neutral species.

The pH dependence of the photoionization signal can be described with the following equilibrium constants: the acid–base dissociation constants for solution surface (K_{a_s}) and for bulk solution (K_{a_b}), and the distribution coefficients between the surface and bulk of neutral species (K_{DA}) and cationic species (K_{DHA}) of solute molecules. These constants are defined by

$$K_{a_s} = \frac{[A]_s[H^+]_s}{[HA^+]_s},$$ (17.12a)

$$K_{a_b} = \frac{[A]_b[H^+]_b}{[HA^+]_b},$$ (17.12b)

$$K_{DHA} = \frac{[HA^+]_s}{[HA^+]_b},$$ (17.12c)

$$K_{DA} = \frac{[A]_s}{[A]_b},$$ (17.12d)

where the subscript "s" denotes the quantity on the surface and where the surface concentration means an effective concentration within the region of the escape depth of a photoelectron. It is possible that $[H^+]_s$ is different from $[H^+]_b$ due to the existence of a surface charge (Zhao et al., 1990), but, because the absolute number density of the ionized species on the surface is small in this case, the difference in pH between the surface and the bulk is negligible: $[H^+]_s = [H^+]_b$. Among four equilibrium constants, K_{a_b} is estimated to be 3.6 ± 0.1 for aminopyrene, using fluorometry of bulk solution (Sato et al., 2004); the distribution coefficients are estimated to be $K_{DHA} = 2.5 \times 10^{-7}$ m and $K_{DA} = 9.6 \times 10^{-7}$ m by comparison of the photoionization signal intensity of aminopyrene with that of insoluble pyrene derivatives of pyrenehexadecanoic acid.

The observed photoionization signal is proportional to $\eta_A[A]_s + \eta_{HA}[HA^+]_s$, where η_A and η_{HA} are the photoionization efficiencies of neutral and ionized forms of aminopyrene on the water surface, respectively. The ionization efficiencies, η_A and η_{HA}, of neutral forms and ionized forms, respectively, are of critical importance in solving Equations 17.12 (Sato et al., 2000). The ratio, η_{HA}/η_A, is estimated to be 9.4 on the basis of assumption that the ionization efficiency of the two forms is dominated only by a resonant condition, namely, the absorption coefficient at the excitation wavelength (Sato et al., 2004).

With the above consideration, pK_{a_s} is determined to be 2.0. The solid curve in Figure 17.16a is theoretically calculated with these equilibrium constants, which agrees well with the experimental data. A decrease in the surface concentration with decreasing pH is attributed to the decrease in neutral species; the neutral form of aminopyrene is more likely to stay on the surface than the ionized form.

A similar analysis was performed for 1-pyrenebutyric acid, as shown in Figure 17.16b (Sato et al., 2000). The distribution coefficients for neutral and ionic species are reported to be $(5.9 \pm 2.8) \times 10^{-2}$ m and $(4.82 \pm 0.1) \times 10^{-5}$ m, respectively, and pK_{a_s} is 7.85 ± 0.13. The acid–base equilibrium constants of adsorbed molecules are quite different from those in bulk solutions, even under the low surface-density conditions.

17.5.3 Monitoring of Volatile Molecules Spread on the Water Surface

The adsorption–desorption dynamics of molecules between liquid and gas phases are of fundamental and practical importance. Desorption of volatile molecules from the liquid to gas phase was investigated with ultraviolet single-photon photoionization measurements (Harata et al., 2002). Figure 17.17 shows the time-dependent photoionization spectra of the water surface after spreading 10 μL *p*-xylene on a 9.08 cm² area. The photoionization signal intensity decreased with time.

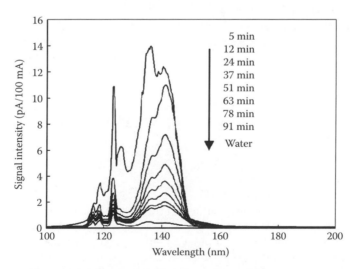

FIGURE 17.17 Photoionization spectra of *p*-xylene at the water surface after 10 μL *p*-xylene is spread on the 9.08 cm² area of the water surface. The photocurrents are normalized with the ring current but not with light intensity. The spectra shape reflects the photoabsorption of water vapor in the gas phase above the water as well as the spectrum of the light source and the photoionization threshold of *p*-xylene.

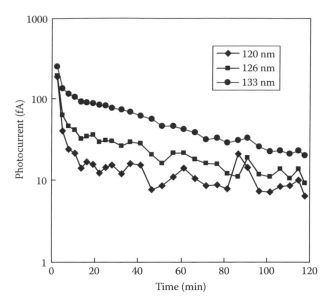

FIGURE 17.18   Photoionization current at the water surface after $3\,\mu L$ benzene is spread on the $9.08\,cm^2$ area of the water surface. The photocurrents are normalized with the ring current but not with light intensity.

It changed rapidly in the initial 12 min and continued to decrease gradually over 90 min. This reflects the desorption of p-xylene from the water surface. Although p-xylene in the gas phase could contribute to the ionization signal in this case, the amount of molecules in the gas phase is expected to be much smaller than that at the water surface. Unfortunately, the spectral shapes in Figure 17.17 are strongly influenced by water vapor, and the complementation for the real photoionization spectra of p-xylene at the water surface is not an easy task because an estimation of light intensity at the water surface is required. However, monitoring could be performed.

In Figure 17.18, time traces are shown of the photoionization current at the water surface after $3\,\mu L$ benzene was spread over a $9.08\,cm^2$ area. The responses of the initial 5 min are dependent on the excitation wavelength, but it is clearly evident that the signal caused by the spread benzene remains on the water surface for 2 h after spreading.

Benzene is the most preferred solvent for spreading organic molecules on the water surface because it has an appropriate balance of surface tension and interface tension (vs. water) to expand a benzene solution rapidly over the water surface. Thus, benzene has been used as a spreading solvent in a monolayer film formation, such as the Langmuir–Blodgett technique, even though it is a toxic solvent. On the basis of surface tension measurements, benzene is assumed to evaporate within 5 min after a base is spread (Pagano and Gershfeld, 1972; Birdi, 1999, p. 37). The result in Figure 17.18 disproves common sense.

17.6   SUMMARY

In this chapter, we describe the photoionization measurements of solute molecules adsorbed at the water surface. The molecular-level understanding of molecules at the water surface is still developing, mainly due to lack in methodology for observations with good surface- and species selectivity and sensitivity. Powerful tools for interface selective observation, such as second harmonic generation and sum-frequency generation, sometimes lack sensitivity. The fluorescence measurement of a powerful tool for sensitive observation often suffers from surface selectivity and adaptability to less fluorescent molecules. Thus, a photoionization measurement with moderate sensitivity, fine surface selectivity, and good adaptability would be an attractive technique.

Some discussion on photoionization thresholds, polarization- and concentration-dependence, and surface molecular monitoring of surface-adsorbed molecules has been presented; however, the details of discussion are still insufficient. Especially, we cannot still experimentally determine the average depth of adsorbed solute molecules, which relates to the degree of hydration of the adsorbed molecules.

In this chapter, we have introduced only a few studies on self-assembled layers on the water surface and no studies on immiscible liquid–liquid interfaces (Tashiro et al., 2003; Inoue et al., 2005). Theoretical studies also provide valuable information (Benjamin, 2006; Chang and Dang, 2006; Garret et al., 2006; Jungwirth and Tobias, 2006; Mundy and Kuo, 2006; Perry et al., 2006); thus, a comparison between theory and experiments is essential. The molecular-level understanding of interfaces has attracted increasing attention, and many types of spectroscopic methodologies should be used to enhance such an understanding. Among them, photoionization measurements will provide valuable information.

ACKNOWLEDGMENTS

A part of this work was financially supported by a grant from the Global-Center of Excellence in Novel Carbon Resource Sciences, Kyushu University. The authors would like to express their gratitude to Yoshihiko Hatano, professor emeritus of Tokyo Institute of Technology and former professor at Kyushu University, for his participation in helpful discussions.

REFERENCES

Benjamin, I. 1996. Chemical reactions and solvation at liquid interfaces: A microscopic perspective. *Chem. Rev.* 96: 1449–1475.
Benjamin, I. 2006. Static and dynamic electronic spectroscopy at liquid interfaces. *Chem. Rev.* 106: 1212–1233.
Birdi, K. S. (ed.), 1997. *Handbook of Surface and Colloid Chemistry*. Boca Raton, FL: CRC Press.
Birdi, K. S. 1999. *Self-Assembly Monolayer Structures of Lipids and Macromolecules at Interfaces*. New York: Kluwer Academic/Plenum Publishers.
Born, M. 1920. Volumes and hydration warmth of ions. *Z. für Phys.* 1: 45–48.
Chang, T.-M. and Dang, L. X. 2006. Recent advances in molecular simulations of ion solvation at liquid interfaces. *Chem. Rev.* 106: 1305–1322.
Chen, H., Inoue, T., and Ogawa, T. 1994. Solvents effects in determination of anthracene on solution surfaces by laser two-photon ionization technique. *Anal. Chem.* 66: 4150–4153.
Davidovits, P., Kolb, C. E., Williams, L. R., Jayne, J. T., and Worsnop, D. R. 2006. Mass accommodation and chemical reactions at gas-liquid interfaces. *Chem. Rev.* 106: 1323–1354.
Donaldson, D. J. and Vaida, V. 2006. The influence of organic films at the air-aqueous boundary on atmospheric processes. *Chem. Rev.* 106: 1445–1461.
Drexhage, K. H. 1997. Structure and properties of laser dyes. In *Dye Lasers*, 2nd edn., F. P. Schäfer (ed.), p. 167, Berlin, Germany: Springer.
Eisenthal, K. B. 1992. Equilibrium and dynamic processes at interfaces by second harmonic and sum frequency generation. *Annu. Rev. Phys. Chem.* 43: 627–661.
Farrell, P. G. and Newton, J. 1965. Ionization potentials of aromatic amines. *J. Phys. Chem.* 69: 3506–3509.
Garret, B. C., Schenter, G. K., and Morita, A. 2006. Molecular simulation of the transport of molecules across the liquid/vapor interface of water. *Chem. Rev.* 106: 1355–1374.
Gopalakrishnan, S., Liu, D., and Allen, H. C. 2006. Vibrational spectroscopic studies of aqueous interfaces: Salt, acids, bases, and nanodrops. *Chem. Rev.* 106: 1155–1175.
Gridin, V. V., Litani-Barzilai, I., Kadosh, M., and Schechter, I. 1998. Determination of aqueous solubility and surface adsorption of polycyclic aromatic hydrocarbons by laser multiphoton ionization. *Anal. Chem.* 70: 2685–2692.
Harata, A. 2003. Ultrasensitive and ultrafast spectroscopies for investigating dynamic behavior of molecules in a nanospace at interfaces. *Housyasen Kagaku* 75: 12–19 (in Japanese).
Harata, A., Ishioka, T., and Hatano, Y. 2002. Photoionization of molecules adsorbed at the water surface: Acidity dependence of photoionization thresholds studied with a synchrotron light source. In *Abstract of The 1st Saga Synchrotron Light Symposium SSLS2002*, Tosu, Saga, Japan, pp. 21–22.

Harata, A., Seno, K., Slyadneva, O. N., and Hatano, Y. 2009. Unpublished results.

Hatano, Y. 1999. Interaction of vacuum ultraviolet photons with molecules. Formation and dissociation dynamics of molecular superexcited states. *Phys. Rep.* 313: 109–169.

Hatano, Y. 2001a. Interaction of VUV photons with molecules: Spectroscopy and dynamics of molecular superexcited states. *J. Electron Spectrosc. Relat. Phenom.* 119: 107–125.

Hatano, Y. 2001b. Spectroscopy and dynamics of molecular superexcited molecules. In *Chemical Applications of Synchrotron Radiation*, T. K. Sham (ed.), pp. 55–111. Singapore: World Scientific Publishing Co.

Hoenig, D. and Moebius, D. 1991. Direct visualization of monolayers at the air-water interface by Brewster angle microscopy. *J. Phys. Chem.* 95: 4590–4592.

Hoffman, G. J. and Albrecht, A. C. 1991. Near-threshold photoionization spectra in nonpolar liquids: A geminate pair based model. *J. Phys. Chem.* 95: 2231–2241.

Holroyd, R. A., Preses, J. M., and Zevos, N. 1983. Single-photon induced conductivity of solutes in nonpolar solvents. *J. Chem. Phys.* 79: 483–487.

Inoue, T., Masuda, K., Nakashima, K., and Ogawa, T. 1994. Highly sensitive detection of aromatic molecules by laser two-photon ionization on the surface of water in ambient air. *Anal. Chem.* 66: 1012–1014.

Inoue, T., Moriguchi, M., and Ogawa, T. 1995. Highly sensitive determination of dye molecules on a surface by the second harmonic generation technique. *Anal. Sci.* 11: 671–672.

Inoue, T., Sasaki, S., Tokeshi, M., and Ogawa, T., 1998. Photoionization threshold of perylene on water surface as measured by synchrotron radiation. *Chem. Lett.* 7: 609–610.

Inoue, T., Nishi, H., Kitauira, T., Kurauchi, Y., and Ohga, K. 2005. Highly sensitive determination of aromatic molecules in the interface region of an oil/water system using laser two-photon ionization. *Bunseki Kagaku* 54: 467–471.

Ishioka, T. and Harata, A. 2004. Nonuniform distribution of adsorbed dye at the water surface probed by photoionization methods. *UVSOR Act. Rep.* 2003: 54.

Ishioka, T., Harata, A., and Hatano, Y. 2003. Photoionization of adsorbed dye below aliphatic acid monolayer at the aqueous solution surface. *UVSOR Act. Rep.* 2002: 108–109.

Jakobi, H. and Kuhn, H. 1962. The orientation of the transition moments of absorption bands of acridine-phenazine, phenoxazine, and xanthine dyes from dichroism and fluorescence-polarization studies. *Z. Elektrochem. Angew. Phys. Chem.* 66: 46–53.

Jo, S. K. and White, J. M. 1991. Low energy (<1 eV) electron transmission through condensed layers of water. *J. Chem. Phys.* 94: 5761–5765.

Jungwirth, P. and Tobias, D. J. 2006. Specific ion effects at air/water interface. *Chem. Rev.* 106: 1259–1281.

Jungwirth, P., Finlayson-Pitts, B. J., and Tobias, D. J. 2006. Introduction: Structure and chemistry at aqueous interfaces. *Chem. Rev.* 106: 1137–1139.

Koltun, W. L. 1965. Precision space-filling atomic models. *Biopolymers* 3: 665–679.

Kouchi, N. and Hatano, Y. 2004. Interaction of photons with molecules: Photoabsorption, photoionization and photodissociation. In *Charged Particle and Photon Interactions with Matter: Chemical, Physicochemical, and Biological Consequences with Applications*, A. Mozumder and Y. Hatano (eds.), pp. 105–120, New York: Marcel Dekker.

Lheveder, C., Meunier, J., and Henon, S. 2000. Brewster angle microscopy. In *Physical Chemistry of Biological Interfaces*, A. Baszbin and W. Norde (eds.), pp. 559–575. New York: Marcel Dekker.

Li, Y.-Q., Sasaki, S., Inoue, T., Harata, A., and Ogawa, T. 1998a. Spatial imaging identification in a fiber-bundle confocal fluorescence microspectrometer. *Appl. Spectrosc.* 52: 1111–1114.

Li, Y.-Q., Slyadnev, M. N., Inoue, T., Harata, A., and Ogawa, T. 1998b. Spectral fluctuation and heterogeneous distribution of porphine on a water surface. *Langmuir* 15: 3035–3037.

Maeda, Y. 2006. Chemical state analysis of environmental molecules adsorbed at the water surface by using photoionization methods. Master thesis, Kyushu University, Fukuoka, Japan, pp. 1–40 (in Japanese).

Maeda, Y., Isoda (Sato), M., Ishioka, T., and Harata, A. 2005. Analysis of environmental chloride and nitride pollutants at the water surface by photoionization methods using laser and synchrotron radiation. In *Books of Abstracts for 7th Cross Straits Symposium on Material, Energy, and Environmental Sciences*, Fukuoka, Japan, pp. 251–252.

Masuda, K., Inoue, T., Yasuda, T., Nakashima, K., and Ogawa, T. 1993. Highly sensitive detection of pyrene by laser multi-photon ionization on the surface of water. *Anal. Sci.* 9: 297–298.

Motomura, K. 1978. Thermodynamic studies on adsorption at interfaces. I. General formulation. *J. Colloid Interface Sci.* 64: 348–355.

Motomura, K., Iwanaga, S., Hayami, Y., Uryu, S., and Matuura, R. 1981. Thermodynamic studies on adsorption at interfaces. IV. Dodecylammonium chloride at water/air interface. *J. Colloid Interface Sci.* 80: 32–38.

Mundy, C. J. and Kuo, I.-F. W. 2006. First principles approaches to the structure and reactivity of atmospherically relevant aqueous interfaces. *Chem. Rev.* 106: 1282–1304.

Ogawa, T. 1999. Laser 2-photon ionization in solution and on surface in ambient air: Investigations through conductivity measurement. In *Photoionization and Photodetachment*, C.-Y. Ng (ed.), pp. 601–633. Singapore: World Scientific Publishing Co.

Ogawa, T., Kise, K., Yasuda, T., Kawazumi, H., and Yamada, S. 1992a. Trace determination of benzene and aromatic molecules in hexane by laser two-photon ionization. *Anal. Chem.* 64: 1217–1220.

Ogawa, T., Yasuda, T., and Kawazumi, H. 1992b. Laser two-photon ionization detection of aromatic molecules on a metal surface in ambient air. *Anal. Chem.* 64: 2615–2617.

Ogawa, T., Chen, H., Masuda, K., and Inoue, T. 1994a. Highly sensitive determination of aromatic molecules on the water surface by laser two-photon ionization at 355 nm. *Anal. Sci.* 10: 219–221.

Ogawa, T., Chen, H., Inoue, T., and Nakashima, K. 1994b. Laser two-photon ionization spectrometry and photoionization threshold of perylene on the surface of water. *Chem. Phys. Lett.* 229: 328–332.

Ogawa, T., Sato, M., Tachibana, M., Ideta, K., Inoue, T., and Nakashima, K. 1995. Dependence of the laser two-photon ionization signal of anthracene on the electron mobility and the excess energy in non-polar solvents. *Anal. Chim. Acta* 299: 355–359.

Ogawa, T., Sasaki, S., Tokeshi, M., Inoue, T., and Harata, A. 1999. One-photon photoionization thresholds of aromatic molecules at water/air surface. *UVSOR Act. Rep.* 1998: 68–69.

Pagano, R. E. and Gershfeld, N. L. 1972. Millidyne film balance for measuring intermolecular energies in lipid films. *J. Colloid Interface Sci.* 41: 311–317.

Perry, A. P., Neipert, C., Space, B., and Moor, P. B. 2006. Theoretical modelling of interface specific vibrational spectroscopy: Methods and applications to aqueous interfaces. *Chem. Rev.* 106: 1234–1258.

Peterson, E. S. and Harris, C. B. 1989. A new technique for the determination of surface adsorbate geometries utilizing second harmonic generation and absorption band shifts. *J. Chem. Phys.* 91: 2683–2688.

Sato, M., Kaieda, T., Ohmukai, K., Kawazumi, H., Harata, A., and Ogawa, T. 2000. Acid-base equilibrium constants at the water surface and distribution coefficients between the surface and the bulk as studied by the laser two-photon ionization technique. *J. Phys. Chem.* 104: 9873–9877.

Sato, M., Akagishi, H., Harata, A., and Ogawa, T. 2001. Polarization and surface density dependence of pyrene-hexadecanoic acid at the air-water interface under compression studied by a laser two-photon ionization. *Langmuir* 17: 8167–8171.

Sato, M., Harata, A., Hatano, Y., Ogawa, T., Kaieda, T., Ohmukai, K., and Kawazumi, H. 2004. Acid-base equilibrium constants and distribution coefficients of aminopyrene between the surface and bulk of liquid water as studied by a laser two-photon ionization technique. *J. Phys. Chem.* 108: 12111–12115.

Seno, K., Ishioka, T., Harata, A., and Hatano, Y. 2001. Multi-photon ionization of rhodamine dyes adsorbed at the aqueous solution surfaces. In *Abstracts for ASIANALYSIS VI*, Yoyogi, Japan, p. 28.

Seno, K., Ishioka, T., Harata, A., and Hatano, Y. 2002. Photoionization of rhodamine dyes adsorbed at the aqueous solution surfaces investigated by synchrotron radiation. *Anal. Sci.* 17: i1177–i1179.

Shen, Y. R. 1989. Optical second harmonic generation at interfaces. *Annu. Rev. Phys. Chem* 40: 327–350.

Shen, Y. R. and Ostroverkhov, V. 2006. Sum-frequency vibrational spectroscopy on water interfaces: Polar orientation of water molecules at interfaces. *Chem. Rev.* 106: 1140–1154.

Slyadneva, O. N., Slyadnev, M. N., Tsukanova, V. M., Inoue, T., Harata, A., and Ogawa, T. 1999. Orientation and aggregation behavior of rhodamine dye in insoluble film at the air–water interface under compression: Second harmonic generation and spectroscopic studies. *Langmuir* 15: 8651–8658.

Slyadneva, O. N., Slyadnev, M. N., Harata, A., and Ogawa, T. 2001. Second harmonic generation from the mixed films of rhodamine dye and arachidic acid at the air-water interface: Correlation with spectroscopic properties, role of spacer molecules in the monolayer. *Langmuir* 17: 5329–5336.

Slyadneva, O. N., Harata, A., and Hatano, Y. 2003. Second harmonic generation from insoluble films of a rhodamine dye at the air/water interface: Effect of sodium dodecylsulfate. *J. Colloid Interface Sci.* 260: 142–148.

Tashiro, K., Ishioka, T., and Harata, A. 2003. Analysis of the oil/water interface probed by laser-multiphoton ionization of rhodamine dyes. In *Proceedings of the 5th Cross Straits Symposium on Materials, Energy and Environmental Sciences*, Pusan, Korea, pp. 61–62.

Timoshenko, M. M., Korkoshko, I. V., Kleimenov, V. I., Petrachenko, N. E., Chizhov, Y. V., Ryl'kov, V. V., and Akopyan, M. E. 1981. Ionization potentials of rhodamine dyes. *Doklady Akademii Nauk SSSR* 260: 138–140.

Tsukanova, V., Harata, A., and Ogawa, T. 2000. Second-harmonic probe of pressure-induced structure ordering within long-chain fluorescence monolayer at the air/water interface. *J. Phys. Chem. B* 104: 7707–7712.

Voigtman, E. and Winefordner, J. D. 1982. Two-photon photoionization detection of polycyclic aromatic hydrocarbons and drugs in a windowless flow cell. *Anal. Chem.* 54: 1834–1989.

Voigtman, E., Jurgensen, A., and Winefordner, J. D. 1981. Comparison of laser excited fluorescence and photoacoustic limits of detection for static and flow cells. *Anal. Chem.* 53: 1921–1923.

Watanabe, I. 1994. Optical oxidation potential: A new measure for reducing power. *Anal. Sci.* 10: 229–239.

Watanabe, I. 2006. Analytical methods for surfaces, interfaces and boundary regions: Photoelectron as a probe for the solution surface study. *Bunseki* 3: 102–108 (in Japanese).

Watanabe, I. and Ikeda, S. 1983. Study of solutions by photoemission spectroscopy. *Kagaku (Kyoto, Japan)* 38: 14–20 (in Japanese).

Watanabe, I., Flanagan, J. B., and Delahay, P. 1980. Vacuum ultraviolet photoelectron emission spectroscopy of water and aqueous solutions. *J. Chem. Phys.* 73: 2057–2062.

Watanabe, I., Takahashi, N., and Tanida, H. 1998. Dehydration of iodide segregated by tetraalkylammonium at the air/solution interface studied by photoelectron emission spectroscopy. *Chem. Phys. Lett.* 287: 714–718.

Winter, B. and Faubel, M. 2006. Photoemission from liquid aqueous solutions. *Chem. Rev.* 106: 1176–1211.

Yamada, S., Kano, K., and Ogawa, T. 1982. Laser two-photon ionization technique for a highly sensitive detection of pyrene. *Bunseki Kagaku* 31: E247–E250.

Yamada, S., Hino, A., Kano, K., and Ogawa, T. 1983. Laser two-photon ionization for determination of aromatic molecules in solution. *Anal. Chem.* 55: 1914–1917.

Yamada, S., Ogawa, T., and Zhang, P. 1986. Trace determination of some aromatic molecules by laser two-photon ionization. *Anal. Chim. Acta* 183: 251–256.

Zhao, X., Subrahmanyan, S., and Eisenthal, K. B. 1990. Determination of pK_a at the air/water interface by second harmonic generation. *Chem. Phys. Lett.* 171: 558–562.

Zheng, X.-Y. and Harata, A. 2001. Confocal fluorescence microscope studies of the adsorptive behavior of dioctadecyl-rhodamine B molecules at a cyclohexane-water interface. *Anal. Sci.* 17: 131–135.

Zheng, X.-Y., Harata, A., and Ogawa, T. 2000. Fluorescence photon bursts from low surface-density dioctadecyl-rhodamine B molecules at the air-water surface observed with a confocal fluorescence photon-counting microscope. *Chem. Phys. Lett.* 316: 6–12.

Zheng, X.-Y., Harata, A., and Ogawa, T. 2001. Study of the adsorptive behavior of water-soluble dye molecules (rhodamine 6G) at the air-water interface using confocal fluorescence microscope. *Spectrochim. Acta Part A* 57: 315–322.

Zheng, X.-Y., Wachi, M., Harata, A., and Hatano, Y. 2004. Acidity effects on the adsorptive behavior and fluorescence properties of rhodamine 6G molecules at the air-water interface studied with confocal fluorescence microscopy. *Spectrochim. Acta Part A* 60: 1085–1090.

18 Low-Energy Electron-Stimulated Reactions in Nanoscale Water Films and Water–DNA Interfaces

Gregory A. Grieves
Georgia Institute of Technology
Atlanta, Georgia

Jason L. McLain
Georgia Institute of Technology
Atlanta, Georgia

Thomas M. Orlando
Georgia Institute of Technology
Atlanta, Georgia

CONTENTS

18.1 INTRODUCTION

When high-energy ions, electrons, x-rays, and gamma rays strike material surfaces, physical and chemical transformations are often induced. Although linear energy and momentum transfer plays a significant role in producing track structures and displacement-driven damage, we concentrate on low-energy secondary electrons that are very important in stimulating nonthermal chemical changes in materials. For heterogeneous surfaces, mixed interfaces, and soft targets (i.e., molecular solids), conventional track structures may not describe the primary inelastic events and radiation-driven chemistry occurring under highly nonequilibrium conditions. In general, it is important to realize that radiation chemistry is essentially governed by the physics associated with the initial ionization channels and the subsequent inelastic scattering of low-energy electrons (LEEs).

Low-energy (1–100 eV) electron–molecule and electron–atom collisions are very important as these (1) initiate and drive many of the processes leading to radiation damage of biological media (Zamenhof et al., 1958; Ahnstrom, 1974; Van der Schans et al., 1989; Sognier et al., 1991; Ahmed, 1993; Beach et al., 1994; Banath et al., 1999; Chen and Chao, 2004), (2) are largely responsible for the nonthermal chemical synthesis/destruction in naturally occurring plasmas, and (3) are key to the formation of planetary atmospheres within and beyond our solar system (Johnson and Quickenden, 1997; Madey et al., 2002; Johnson et al., 2005, 2006). Electron collisions are also of obvious technological and environmental concern, since they are involved in modifying materials and surfaces present in radioactive containment facilities, plasma processing of substrates and interfaces used in the electronics device industry, and in electron-beam-based lithography. Despite the general importance of electron scattering, a complete understanding and accurate theoretical description only exists for electron–atom collisions. Good approximations exist to describe electron scattering with diatomic targets, and progress is being made in understanding and describing electron interactions with polyatomic collision partners. The latter is complicated due to additional degrees of freedom (vibration and rotation) and the subsequent dissociation events that occur as a result of the energy exchanged during the collision. The reactive scattering of atomic and molecular fragments and the production of electronically excited states lead to important chemical transformations.

The reactions of LEEs in dense matter often differ greatly from those in a low-pressure gas. In the latter case, the electrons are normally free and the collision mean free path, l, is much longer than the de Broglie wavelength, while in the former case, l is smaller than the de Broglie wavelength and as the medium density increases, the electrons become constrained to move within bands of a finite width. These differences introduce changes in the symmetries of electron–molecule scattering resonances (Azria et al., 1987), polarization effects that can enhance resonance lifetimes (Sanche, 1995a,b; Nagesha and Sanche, 1998; Lu and Sanche, 2004), suppression of autoionizing states (Kimmel et al., 1997), removal of high n Rydberg states (Rowntree et al., 1993), and new channels that involve reactive scattering of the ion and neutral (radical) products (Sieger et al., 1998). Methods that utilize state-of-the-art laser detection of the quantum-state distributions of neutral fragments and measurements of threshold/energy-resolved desorption/damage events are new techniques that can help unravel the overall interactions involved in the inelastic scattering of LEEs at interfaces and in the condensed phase (Kimmel and Orlando, 1995, 1996; Alexandrov et al., 2001). These new approaches are discussed and illustrated in this chapter with regard to the LEE beam–induced damage of nanoscale films of water adsorbed on metal or metal-oxide surfaces and LEE interactions at hydrated biological interfaces such as DNA. The chapter also involves a discussion of a theoretical formalism to describe multiple elastic scattering of LEEs within condensed-phase targets or at interfaces. The theoretical approach is a modification of the scattering theory used to describe photoelectron diffraction (Chen et al., 1998) and X-ray absorption fine structure (XAFS)

(Rehr and Albers, 1990). It is most applicable to structurally ordered targets and should be generally useful for describing electron interactions with many biomolecules.

18.2 LOW-ENERGY ELECTRONS AND ENERGY-LOSS CHANNELS

The emission of LEEs produced during high-energy electron radiation of surfaces and interfaces is very well known. The secondary-electron yields have been measured from many solid-phase targets, including surfaces containing multilayers of ice. The LEE yields depend primarily upon stopping power, ionization probability, and the material's work function. The LEEs produced in condensed water and released from nanoscale water interfaces or surfaces are shown approximately in Figure 18.1 as $N(E)$. As these electrons move through the media, they undergo elastic and inelastic energy-loss scattering with cross sections that depend upon the electron kinetic energy. This is shown in Figure 18.1 as $\sigma(E)$ (circles) measured by Sanche and coworkers (Michaud et al., 2003). Superimposed for comparison is the electronic scattering cross section of gas-phase water (triangles) (Ness and Robson, 1988).

As shown in Figure 18.1, the electron energy distribution is weighted by LEEs, and, thus, the convolution of $N(E) \cdot \sigma(E)$ yields a damage probability function that is weighted by the LEE distribution and the excited states and resonances that occur at excitation energies <20 eV. The inelastic

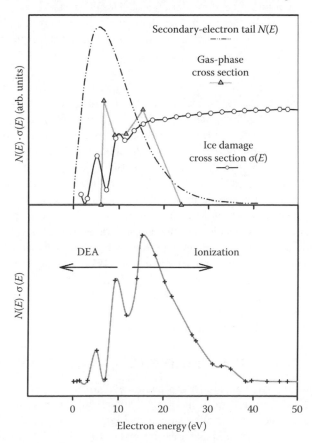

FIGURE 18.1 Upper frame: Schematic representation of the secondary-electron yields, $N(E)$, and the cross section for excitation/energy loss as a function of incident electron energy, (E). Also shown is the total cross section, $\sigma(E)$, for elastic and inelastic scattering of LEEs with amorphous water ice target (Michaud et al., 2003) and for gas-phase water (Ness and Robson, 1988). Lower frame: The convolution of $N(E) \cdot \sigma(E)$ yields a damage probability function that is weighted by the LEE distribution and the excited states and resonances that occur at excitation energies <20 eV.

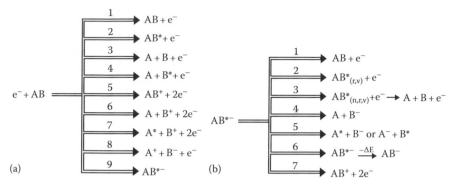

FIGURE 18.2 The decay channels that can result from LEE interactions with an arbitrary molecule AB. (a) Shows direct decay channels, while (b) illustrates those that proceed via TNI. (From Lane, C.D. and Orlando, T.M., *Applied Surface Science*, Elsevier, 2007. With permission.)

channels primarily involve excitations of the outer and inner valence levels and the formation of complicated, configurationally mixed, excited electronic states. When these complicated excitations localize at a surface or interface, dissociation and desorption can occur. As described in more detail in Section 18.5, the excited states can be parent states to core-excited Feshbach resonances. These resonances can decay to form neutral and anionic molecular fragments. In general, the damage probability function illustrates the general importance of LEEs and the need to unravel the details of all the energy-loss channels within this energy range.

A general schematic of some of the interactions that occur during electron scattering in this energy range is presented in Figure 18.2a. As an incident electron collides with a molecule (AB), pathways include (1) elastic scattering, (2) inelastic scattering, (3) dissociation into neutrals, (4) dissociation into a neutral and an excited-state neutral, (5) ionization of the parent molecule, (6) dissociation into a neutral and cation, (7) dissociation into an excited-state neutral and cation, (8) dipolar dissociation, and (9) electron attachment forming a transient negative ion (TNI). A summary of several important decay pathways of TNIs is shown in Figure 18.2b. These include (1) detachment without transfer of energy; (2) detachment leaving the neutral molecule in a rotationally and vibrationally excited state; (3) detachment leaving the neutral molecule in an electronically, rotationally, and vibrationally excited state that can predissociate; (4) dissociation producing a neutral and an anion; (5) dissociation producing an excited-state neutral and an anion; (6) energy transfer that stabilizes the TNI; and (7) ionization through electron emission (Sanche, 1995a,b; Chutjian et al., 1996; Christophorou and Olthoff, 2002). When a TNI decays to form a stable negative-ion fragment, the process is normally referred to as dissociative electron attachment (DEA). DEA channels have been examined extensively for many molecules in the gas phase (Christophorou and Olthoff, 2004) and, more recently, at interfaces (thin films) and in condensed molecular solids (Sanche, 2002).

18.3   OVERVIEW OF EXPERIMENTAL DESIGNS, SAMPLES, AND STRATEGIES

The experimental systems used to study the chemical transformations brought about by the inelastic scattering of LEEs with thin films of molecular solids such as water require ultrahigh vacuum (UHV) systems, such as the ones shown in Figures 18.3 and 18.4. These are custom-designed UHV systems with a typical base pressure of 1×10^{-10} Torr. The chamber in Figure 18.3 is equipped with a compressed He or liquid-nitrogen-cooled rotatable sample holder, a pulsed LEE gun, a quadrupole mass spectrometer (QMS), and a time-of-flight (ToF) mass spectrometer. Coverages ranging from submonolayer to several hundred monolayers are usually studied. The surfaces are usually prepared under UHV conditions and are bombarded with LEEs. Though continuous-beam measurements are common, pulsed beams coupled to ToF methods tend to be more sensitive and reduce the probability of charge buildup on or within the film. When using

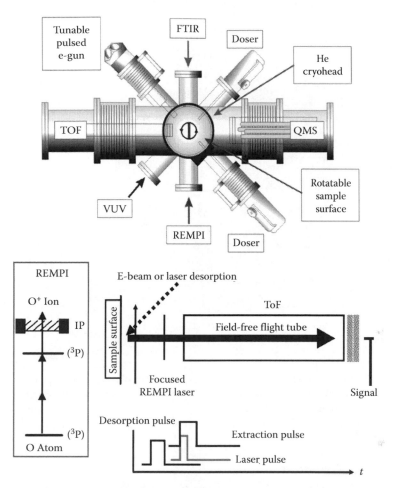

FIGURE 18.3 Upper frame: Typical UHV system developed to understand the role of LEE interactions with nanoscale water films. Lower frame: Illustration of the ToF mass spectrometer setup used for resonance enhanced multi photon ionization (REMPI) and vacuum ultra violet single photon ionization (VUV SPI) detection.

continuous-beam methods, the desorbing cations or anions are typically detected with a QMS. The pulsed methods also allow the use of the ToF, and depending upon the signal intensity, data can be taken under applied field-free conditions so that the kinetic energy distributions of the desorbing ions can be directly measured. Using a well-controlled electron beam, the yields of the ions can also be measured as a function of the incident electron energy. This allows the determination of threshold energies and is very helpful with regard to assessing the states most important in the LEE-induced dissociation and desorption events.

18.3.1 SAMPLE PREPARATION

Nanoscale films of pure water or solutions containing biological molecules such as DNA are typically deposited on very clean and ordered single crystals of metals, metal oxides, or graphite sample substrates. These substrates are usually sputtered and annealed, and then cleaned by flash heating to above 1000 K prior to film deposition. In order to study the phase dependence of the damage probability of ice, a substrate that allows the nucleation and growth of crystalline ice (CI) should be used (Lofgren et al., 2003; Ahlstrom et al., 2004). In this case, we use a Pt(111) crystal. In general, single crystals are not necessary since many layers of amorphous ice are often studied. Amorphous ice is chosen since the geometric structure factors and electronic properties are reminiscent of liquid water.

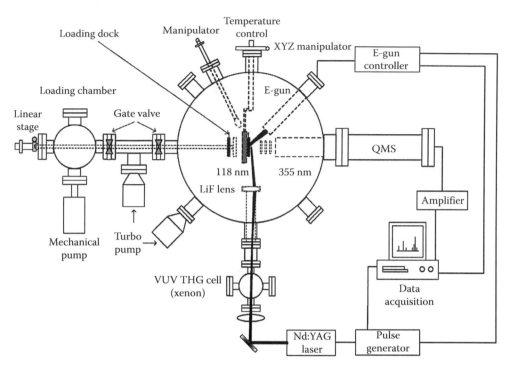

FIGURE 18.4 Custom UHV system developed to understand the role of LEEs in the damage of biological targets such as DNA. The system is unique since it contains an antechamber region for the deposition of biomolecules and a VUV photon source for neutral product detection using SPI. The system can also use laser detection schemes such as REMPI. (From Chen, Y.F. et al., *Int. J. Mass. Spectrom.*, 277, 314, 2008. With permission.)

18.3.1.1 Water Ice Preparation

Water ice samples are deposited in the UHV chamber using background and directional dosing. The water samples are normally purified by several freeze-pump-thaw cycles prior to dosing. Water condensation at temperatures below 100 K forms microporous amorphous ice (Mayer and Pletzer, 1986). We refer to this as porous amorphous solid water (PASW), which is characterized by a very high surface area ($\sim$640 m^2/g) and low density. Sintering the microporous ice between 100 and 120 K or deposition at these temperatures forms normal amorphous ice or simply amorphous solid water (ASW). Films grown above 140 K form cubic, Ic(001), which we refer to as simply crystalline ice (CI). Amorphous ice annealed at 150 K undergoes a phase transition to polycrystalline cubic ice (Smith et al., 1996) and sublimates rapidly in vacuum above 165 K. The samples discussed in this chapter are PASW, ASW, and CI grown on either Pt(111), Cu(111), ZrO_2, or graphite surfaces. Water coverage is usually calibrated relative to the temperature-programmed desorption (TPD) of water on Pt(111). A minimum of 40 ML was found to be sufficient to decouple the surface of the ice from the substrate, and all samples were >50 ML to ensure the measurements were not affected by the underlying substrate.

The porosity of the ice can also be controlled using directional dosing techniques (Stevenson et al., 1999). Ices produced in this way are highly porous and can adsorb large amounts of materials due to the large effective surface area. As described below, the pore surfaces can also provide "buried" interfaces, surfaces, and free-space volumes for reactions within the condensed films. Since amorphous ice has an intrinsically higher defect density resulting from increased disorder, the ionization efficiency is expected to be higher and the resultant chemical transformations may have higher yields. This is described in more detail in Section 18.6.2.1.

18.3.1.2 DNA Sample Preparation

The DNA films were deposited in a differentially pumped antechamber that was initially backfilled with gas-phase nitrogen (Figure 18.4). These films were deposited on a 5×5 mm tantalum substrate using a nebulizer spray or a calibrated syringe. The Ta substrate was cleaned using acetone and methanol, and sonicated in nanopure water for 10 min prior to film deposition. Aliquots of only a few microliters of solution were deposited per sample. The solutions contained P14 double-stranded circular DNA derived from pBluescript SK II (-). These P14 strands contained 6360 base pairs with a molecular weight of $\sim 4.2 \times 10^6$ Da. Because some biomolecules chemically decompose upon adsorption onto metal surfaces (Sanche, 2005), relatively thick films were typically studied. The thick DNA layers (>100 nm) also prevent effects of back-scattered electrons from the metal substrate. Since the penetration depth/mean free path of 5–100 eV electrons is in the range of 15–35 nm in liquid water or amorphous ice (Meesungnoen et al., 2002), the thick films ensured that the measured signals were produced from electron interaction with the DNA molecules within the condensed overlayer. The DNA sample was either held in a desiccator for 10 min and then moved into the loading chamber to be vacuum-dried or deposited directly onto the sample held in a nitrogen background. The solid DNA films were then transferred into the analysis chamber for LEE irradiation experiments.

The purified P14 plasmids were present as fully solvated species in a dilute (200 mg of DNA/mL) aqueous buffer solution. DNA films with good alignment and unitary orientation can be formed by the simple deposition and drying techniques described above, provided the concentration allows initial formation of a nematic-like crystalline phase (Morii et al., 2004). The deposited and vacuum-dried samples were characterized by a scanning electron microscope and were found to be polycrystalline with some aggregation. Under the sample deposition conditions described above, it is impossible to remove all of the water from DNA at room temperature. Thus, the DNA samples studied contain trapped water mainly in the form of primary and secondary waters of hydration.

18.3.2 Neutral Detection Using Time-of-Flight Techniques

In order to detect the neutral fragments from ice or other complicated targets, we routinely utilize a very sensitive laser detection technique known as resonance-enhanced multiphoton ionization (REMPI)/time-of-flight (ToF) mass spectrometry (Kimmel and Orlando, 1995; Kimmel et al., 1995; Orlando et al., 1999) or single-photon ionization (SPI) (Chen et al., 2008). For REMPI, the laser wavelengths that we use vary from 198 to 250 nm and are generated by frequency doubling or tripling the outputs of Nd:YAG-pumped dye lasers or Nd:YAG-pumped master oscillator power oscillator (MOPO) solid-state lasers. For SPI, we typically triple the 355 nm (or other wavelengths) in Xe (or other rare gas mixtures) using nonlinear third harmonic generation. The approach using 355 nm beams produces coherent 118 nm light. The ToF we use is shown in more detail in Figure 18.3b. Since we use a pulsed electron beam, varying the delay time between the electron-beam pulse and the focused laser beam allows us to map out a quantum-resolved velocity distribution of the desorbing neutrals. Typically, all desorbing ions are removed by an initial pulsed extraction field followed by an ionization of the neutrals under constant field conditions. The detection sensitivities of this arrangement when using REMPI are $\sim 10^6$ atoms/quantum state/cm^3. The detection sensitivities using SPI are similar, but, unlike REMPI, the detection is not state specific. Also, our SPI scheme can only detect molecules having ionization potentials (IPs) below 10.5 eV. Since all the atomic and molecular products of electron-induced dissociation have IPs above 10.5 eV, in this chapter, we illustrate the utility of neutral detection using REMPI through a discussion on the neutral H(^{1}S) and O(^{3}P$_J$), and O(^{1}D) and H$_2$ $\left(^1\Sigma_g^+ \right)$ yields from ice. The two-photon-resonant, one-photon ionization (2 + 1) REMPI schemes utilized to detect D(^{2}S), O(^{3}P$_{2,1,0}$), and O(^{1}D$_2$) atomic fragments were the $3s\ ^2S \leftarrow 1s\ ^2S$ at 205.1 nm, $3p\ ^3P_{2,1,0} \leftarrow 2p\ ^3P_{2,1,0}$ at 225.7–226.2 nm, and $3p\ ^1F_3 \leftarrow ^1D_2$ at 203.8 nm, respectively. The $(2 + 1)$ REMPI scheme via the $E,F\,^1\Sigma_g^+$ intermediate state was used to detect the $D_2(H_2)$ products in the ground $(X\,^1\Sigma_g^+)$ state.

The LEEs are typically supplied using pulsed electron beams generated by means of commercially available low-energy (5–1000 eV) electron guns. The typical full width at half maximum (FWHM) of the electron energy distribution is <0.5 eV. The focused beam size is normally ~1 mm, and varies little with the electron energy. Most experiments use a variable pulsed (20–200 Hz, 100 ns–25 μs) electron beam that gives electron doses of 10^{-5}–10^{-3} electrons/surface molecule/s. This dose is quite small so that all desorption and dissociation events observed are due to direct electronic transitions and have cross sections greater than ~10^{-23} cm². The electron flux is measured by picoammeters and is always kept below 10^{14} electrons/cm²/s to eliminate or greatly reduce any potential charging effects. The variation of the electron flux is performed by changing the gun emission current; all desorption and dissociation events discussed in this chapter have a linear dependence on the electron-beam flux. This indicates that the observed dissociation and desorption channels arise solely from single scattering events. When necessary, the pulsed electron beam can be focused on the target and controlled by x–y deflection to scan the entire substrate surface. Since the electron-beam spot size and flux are measured accurately, absolute cross sections can be obtained.

18.3.3 Postirradiation Temperature-Programmed Desorption

In addition to the measurement of accurate stimulated desorption and dissociation cross sections, the subsequent chemistry initiated by the LEE inelastic energy-loss channels can be examined by a technique known as postirradiation temperature-programmed desorption. This is a straightforward approach that involves irradiating the sample with a fixed dose or flux at a given electron impact energy. After the irradiation, the film is heated slowly (typically 8 K/min) and the desorbed products are detected with a QMS as a function of substrate temperature. These experiments are then carried out for several different electron impingement energies and total doses. The results, when compared to TPD studies of similar targets that have not been irradiated, yield information regarding the electron-induced chemistry within the ice films. A disadvantage of this technique is that it is difficult to examine purely the radiation-driven processes, since the results are a convolution of radiation and thermally induced reactions. Although this topic will not be discussed in detail in this chapter, we note that an Fourier transform infra red (FTIR) spectroscopic study of the films before, during, and after irradiation helps to overcome this problem.

18.3.4 Postirradiation Gel Electrophoresis

Another novel technique for the analysis of DNA damage is to remove the irradiated samples from the UHV chamber and then apply conventional wet chemical methods to determine the extent of damage. The technique was originally developed by Sanche and coworkers (Boudaiffa et al., 2000). After electron irradiation, the biological materials (e.g., DNA) are removed from the chamber, redissolved into water, and examined ex situ using a technique known as agarose gel electrophoresis. The supercoiled DNA single-strand breaks (SSBs) and double-strand breaks (DSBs) can easily be separated spatially based on their different migrating abilities in agarose. Quantitative analysis of SSBs and DSBs can be determined by integrating the light intensities of corresponding ethidium-stained DNA bands, thus allowing the measurement of the SSB and DSB probabilities as a function of electron-beam energy and dose. For the water–DNA damage studies, total electron doses of only ~10^{13} electrons/DNA sample are used. This dose is two orders of magnitude lower than that known to cause charging in DNA films (Zheng et al., 2006) and assured a dose response in the linear regime. The amount of DNA is quantified by measuring absorbance at 260 nm before and after irradiation. Usually, more than 95% of the deposited mass of DNA is recovered. Three controlled DNA samples are usually analyzed together with postirradiation samples: (1) an original plasmid DNA solution; (2) a DNA sample that was dried, transferred in vacuum, and recovered in water without irradiation; and (3) DSB DNA made by the restriction enzyme Sca I.

18.4   A USEFUL MULTIPLE SCATTERING FORMALISM

The theory we utilize to investigate the LEE-induced damage of water and water–DNA interfaces was developed to examine diffraction effects in stimulated desorption (Sieger et al., 1999; Orlando et al., 2004; Oh et al., 2006) and is a modification of the scattering theory used to describe photo-electron diffraction (Chen et al., 1998) and XAFS (Rehr and Albers, 1990). A brief description of the most relevant aspects of the calculation is given below. Specifically, the electron wavefunction is given by

$$\phi(\mathbf{r}) = \phi^0(\mathbf{r}) + \phi^1(\mathbf{r}) + \phi^2(\mathbf{r}) + \cdots + \phi^N(\mathbf{r}) \tag{18.1}$$

where
  $\phi^0(\mathbf{r})$ and $\phi^N(\mathbf{r})$ represent the non-scattered
  Nth order scattered components, respectively

The first-order scattering component can be represented as

$$\phi^1(\mathbf{r}) = \sqrt{4\pi} \sum_i \sum_L G_{0L}\left(\vec{\rho}_{ri}\right) t_\ell^i(\mathbf{R}_i) Y_L^*(\hat{\mathbf{k}} \, \mathrm{B}) e^{i\mathbf{k} \cdot \mathbf{R}_i} \tag{18.2}$$

In order to solve the free-space electron propagator, $G_{LL'}$, the separable representation has been introduced with the expression

$$G_{LL'}(\vec{\rho}) = \frac{e^{i\rho}}{\rho} \sum_\lambda \tilde{\Gamma}_\lambda^L(\vec{\rho}) \Gamma_\lambda^{L'}(\vec{\rho}) \tag{18.3}$$

where λ is the matrix index for quasi-angular momentum. After some mathematical treatment, the first-order scattering component can be expressed as

$$\phi^1(\mathbf{r}) = \sum_i \sum_\lambda \mathbf{M}_\lambda(\vec{\rho}_{ri}) \mathbf{P}_\lambda(\vec{\rho}_{ri}, \hat{\mathbf{k}}) \frac{e^{i\rho_{ri}}}{\rho_{ri}} e^{i\mathbf{k} \cdot \mathbf{R}_i} \tag{18.4}$$

where

$$\mathbf{P}_\lambda(\vec{\rho}_{ji}, \hat{\mathbf{k}}) = \sqrt{4\pi} \sum_L \Gamma_\lambda^L(\vec{\rho}_{ji}) t_\ell^i Y_L^*(\hat{\mathbf{k}}) \tag{18.5}$$

is defined as the introduction matrix and $\mathbf{M}_\lambda(\vec{\rho}_{ri}) = \tilde{\Gamma}_\lambda^0(\vec{\rho}_{ri})$ is the termination matrix (Rehr and Albers, 1990; Rehr et al., 1995).

18.5   LOW-ENERGY ELECTRON INTERACTIONS WITH WATER INTERFACES

Before we discuss the primary desorption and dissociation channels involved in LEE interactions with nanoscale films of water and water–DNA interfaces, a brief review of the electronic structure and excited states of condensed-phase water is necessary. A more detailed description of the electronic structure of water can be found elsewhere (Ramaker, 1983; Sieger et al., 1997; Herring et al., 2004).

18.5.1 ELECTRONIC STRUCTURE OF WATER THIN FILMS

The real-space orbitals of water can be represented as a linear combination of the valence molecular orbitals: the $1b_2$ and $2a_1$ orbitals are the primary constituents of the O–H bonds, while the $1b_1$ and a_1 make up the oxygen lone-pair orbitals (Jorgensen and Salem, 1973). The four lowest unoccupied molecular orbitals are $4a_1$, $2b_2$, $2b_1$, and $5a_1$. The $4a_1$ and $5a_1$ orbitals are strongly antibonding, and occupying these states leads to breaking of the O–H bond. The $4a_1$ orbital mixes with the 3s Rydberg state in the gas phase, and is sometimes referred to as $3s4a_1$. Calculations and photoemission data show that the electronic structure of condensed ice retains much of the gas-phase character with some broadening and shifting of the energy levels. Since the molecular orbitals retain much of their gas-phase character, the peaks in the ice valence-band density of states are usually labeled by the same spectroscopic notation as free water. The molecular orbitals, the gas-phase photoelectron spectrum, and condensed-phase photoemission data are shown schematically in Figure 18.5. The unoccupied $4a_1$ orbital is in the bandgap, spatially extended, and is mixed with the $2b_1$ level. The conduction band of ice is very narrow, with the band minimum less than 1 eV below the vacuum level. The Fermi level is estimated to be 5 eV above the $1b_1$ band maximum. The optical absorption spectra of ice show a pronounced peak at ~8.6 eV, corresponding primarily to a $1b_1 \rightarrow 4a_1$ transition, well below the photoelectric threshold (Kobayashi, 1983). In the solid state, the unoccupied molecular orbitals (i.e., the $4a_1$) may be best described as Frenkel excitons, in which the Coulomb attraction localizes the electron–hole pair on the water molecule. This may be particularly important at surfaces and terminal sites.

The largest change in the condensed phase occurs at the levels of a_1 symmetry, which are significantly broadened (compared to the $1b_1$ and $1b_2$ levels), and are most affected by hydrogen bonding. Calculations of the ice band structure indicate that the a_1 bands have the most dispersion, while the $1b_1$ and $1b_2$ bands are virtually dispersionless. One would then expect that electronic excitations involving a_1 bands would be the most sensitive to changes in the hydrogen-bonding environment. Specifically, a reduction of the bandwidth should occur if the

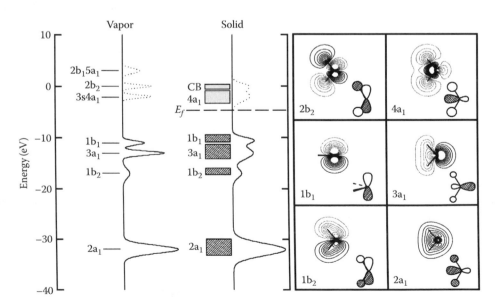

FIGURE 18.5 Left frame: Comparative relationship between the gas-phase and condensed-phase photoelectron spectra of water. Right frame: The similarity allows for extension of the gas-phase state labels to the solid phase. Molecular orbital density plots of the valence levels for the water molecule.

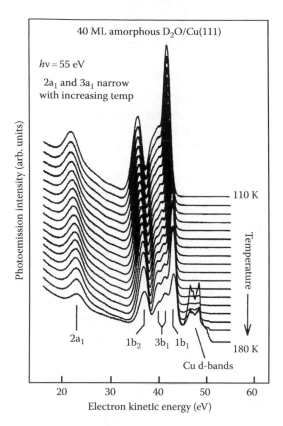

FIGURE 18.6 Temperature-dependent photoemission data of 40 ML amorphous ice deposited on Cu(111) from 110 to 180 K.

hydrogen bonding is weakened, and the lifetimes of electrons and holes in these bands should increase as a consequence.

The effect of temperature on the photoemission spectra of ice has been measured and is shown in Figure 18.6. It is clear that the a_1 bands are the broadest; however, the broadening associated with the $3a_1$ level emerges into a reasonably well-defined splitting as the temperature is increased. The $3a_1$ molecular orbital is composed primarily of the oxygen lone-pair orbitals, is strongly involved in hydrogen bonding, and is very sensitive to the local geometry of nearby water molecules. The splitting and narrowing of the $3a_1$ level is likely due to the reduced perturbation on the a_1 levels and the general return of the atomic P_z character as the number of hydrogen bonds drops with increasing temperature. For a symmetric pair of water molecules in a unit cell of ice, the $3a_1$ orbitals directly interact with each other giving rise to a bonding and antibonding combination (Nordlund et al., 2008). Alternatively, this splitting can also be described in terms of Davydov splitting due to the fact that ice has a nonzero entropy even at 0 K (Petrenko and Whiteworth, 2002). Thus, adjacent unit cells in the lattice are symmetry-inequivalent and can have nonzero overlap integrals allowing for mixing of the a_1 bands between cells (Winter et al., 2004).

Though it is difficult to discern this in Figure 18.6, the bands, particularly the $2a_1$ level, shift ~1 eV to lower energy (relative to the Cu substrate) as the temperature increases. This can be attributed to the change in the work function. Note that the orientation of water molecules in the surface dipole layer collectively contributes to the work function. As the temperature increases, the net orientation of these molecules changes, giving rise to many more water molecules with dangling OH bonds pointing into the vacuum.

18.6 STIMULATED DISSOCIATION AND DESORPTION OF WATER THIN FILMS AND INTERFACES

18.6.1 LEE-STIMULATED DESORPTION OF ANIONS

TNI resonances result when the ground or excited electronic states of molecular collision partners temporarily capture an incident electron during the electron–molecule collision. The simplest TNI is produced when the incident electron is trapped by the centrifugal potential arising from the interaction between the incident electron and the neutral molecule in its ground electronic state. These are referred to as single-particle shape resonances. Shape resonances usually lie at low energies (~E_i < 5 eV) and are energetically above the potential energy surfaces of the neutral molecule ground states. These resonances typically decay via autodetachment leaving the neutral molecule with vibrational and rotational excitation. Another important decay pathway is the DEA, which involves the formation of anionic and neutral fragments. Shape resonances can also involve electronic excitations, and these core-excited shape resonances involve an attractive interaction between the incident electron and the excited state of the target molecule. Since the potential barrier is strongly dependent on the angular momentum value of the excited-state orbitals, p-, d-, and f-waves are expected. These resonances also lie above the neutral excited-state energies and decay by autodetachment or DEA.

Another type of a core-excited resonance is referred to as the Feshbach (Type-I) resonance. This involves coupling of the kinetic energy of the captured electron to nuclear motion. If the incident electron excites a valence or shallow core level, a dynamic polarization is produced and the electron is temporarily trapped in the dipole field of the excited target molecule. Electron correlation and reduced screening allow the incoming electron to couple to the excited electron and a slightly positive core, producing a one-hole, two-electron core-excited Feshbach resonance. These resonances lie below the corresponding excited states (i.e., they have a positive electron affinity). The energy gained in this correlation is referred to as the Feshbach decrement, and is typically ~0.5 eV. As depicted in Figure 18.7, the transition is typically mediated by Franck–Condon factors and the cross section is mediated by lifetimes against autodetachment and dissociative attachment (Ingolfsson et al., 1996). The negative-ion fragments are often produced with an excess kinetic energy, which can be partitioned into reactive scattering events with coadsorbed molecules or surface terminal

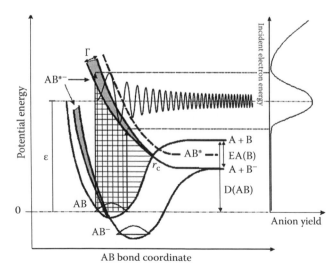

FIGURE 18.7 Schematic of the potential energy surfaces accessed during DEA processes. Inelastic electron scattering results in a transition to an excited (dissociative) potential surface that is asymptotic to the anionic decay channels.

groups. As illustrated in Figure 18.7, the conservation of energy and momentum yields the most probable kinetic energy of the departing anion fragment, E_K^-, as

$$E_K^- = (1-\beta) [\varepsilon + EA(B) - D(AB) - D(AX) - E_n] \tag{18.6}$$

where
   β is the ratio of the mass of B^- to that of AB
   ε is the captured electron's energy
   EA(B) is the electron affinity of B
   D(AB) is the bond energy of AB
   E_n is the excitation energy of the fragments (Christophorou et al., 1984)

It is apparent that DEA favors the loss of the light fragment.

The effects of the condensed phase on the DEA and TNI resonances have been studied and several aspects are well understood. Specifically, the DEA and TNI resonances can be drastically altered due to third-body interactions that open pathways for energy dissipation (Bass and Sanche, 1998, 2003a,b; Sanche, 2000). These interactions can have an effect during the lifetime of the TNI as well as alter products after the TNI dissociates. Most notably, dissociation from the low-energy shape resonances is suppressed, while dissociation from the Feshbach resonances can be enhanced by the surrounding medium. Although there is no detection of anions from the low-energy shape resonance, dissociation could still occur where the products simply do not gain sufficient kinetic energy to escape the surface potential. Mechanisms for enhancement in dissociation can be explained by the effects of the medium on the electron autodetachment probability. One mechanism for dissociation enhancement through the Feshbach resonance is the lowering in energy of the anionic potential energy surface via the substrate and surrounding medium polarization interaction (Balog et al., 2004). Due to the lowering of the ionic-state potential, the wavepacket on the TNI potential spends less time traversing it, thus lowering the autodetachment probability. Another explanation for enhanced dissociation in the condensed phase is the more efficient conversion of an open-channel resonance into a Feshbach resonance due to interactions with the medium (Balog et al., 2004). An open-channel resonance lies higher in energy than the Feshbach resonance and only requires a one-electron transition to return to the associated neutral molecule compared to the two-electron transitions needed for the Feshbach resonance. The electron autodetachment probability is higher for the one-electron transition compared to the two-electron transitions; thus, the Feshbach resonance is more likely to dissociate than autodetach.

The formation of $H^-(D^-)$ from the DEA of water H_2O (D_2O) was studied extensively in the gas phase (Compton and Christophorou, 1967; Melton, 1972; Jungen et al., 1979; Curtis and Walker, 1992; Fedor et al., 2006), and some of these recent results from Fedor et al. are shown in Figure 18.8a (Fedor et al., 2006). There is a strong resonance with a threshold energy near ~5 eV, a peak at 6.7 eV, and a weak resonance with a peak near 9.7 eV. The yield of $H^-(D^-)$ from ASW (i.e., nanoscale multilayers of water) condensed on Pt(111) has also been studied over the past several years. Our results from these experiments are shown in Figure 18.8b (Simpson et al., 1997). The main feature peaking near 6.7 eV remains although the peak shifts to ~7.3 eV in the condensed phase. The width of this feature is also broader, and there is a resolvable weak structure near 10 eV.

The first feature observed in Figure 18.8a involves excitation of the $1b_1$ electron into the mixed $3s4a_1$ level, giving rise to a $^2B_1$, one-hole, two-electron ($...1b_1^{-1}3s4a_1^2$) core-excited Feshbach resonance. Since the $4a_1$ level is dissociative, this leads to facile bond breakage. The detailed dynamics of the $^2B_1$ DEA channel has been calculated using accurate potential surfaces (Haxton et al., 2007a,b). The second, much weaker feature, may be due to the excitation of the $3a_1$ electron into the mixed $3s4a_1$ level, giving rise to a $^2A_1$, one-hole, two-electron ($...3a_1^{-1}1b_1^23s4a_1^2$) core-excited Feshbach resonance. This state mixes with the lower energy $^2B_1$ state via nonadiabatic curve crossings, and also dissociates due to the $4a_1$ antibonding level and probably some admixture containing the $2b_1$ character.

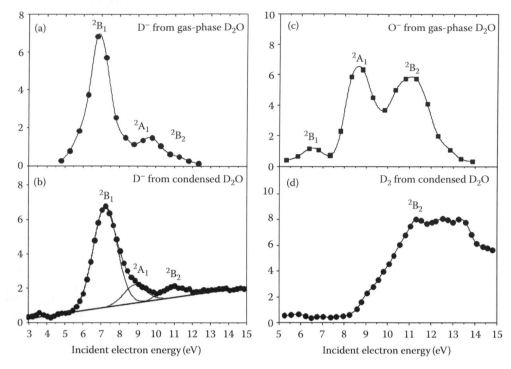

FIGURE 18.8 Comparison of DEA resonances for gas-phase water (Fedor et al., 2006) and ice films (Simpson et al., 1997). (a) and (b) show a highly correlated resonance structure for D⁻ desorption. (c) and (d) show that D_2 desorption from a condensed phase correlates with excitation to the $^2B_2$ state (Kimmel and Orlando, 1996).

As can be seen in Figure 18.8, the yields of H⁻(D⁻) in both the gas phase and the condensed phase are very low for excitation energies above 10 eV (Simpson et al., 1997). This is largely due to the fact that the higher-energy resonances associated with excitations from the $3a_1$ and $1b_2$ levels decay to primarily form O⁻ and H_2. The O⁻ yield has been measured in the gas phase (Compton and Christophorou, 1967; Melton, 1972; Jungen et al., 1979; Curtis and Walker, 1992; Fedor et al., 2006), whereas the neutral molecular hydrogen channel was measured from multilayer water samples condensed at 90 K. These are also shown in Figure 18.8c. The two resonances dominating the O⁻ yields in Figure 18.8 are one-hole, two-electron core-excited Feshbach resonances, and are formally assigned to the $^2A_1$ and $^2B_2$ states. We have demonstrated that the molecular hydrogen produced from water multilayers via the $^2A_1$ and/or $^2B_2$ resonances contains only even quanta of vibrational energy (Kimmel and Orlando, 1996). Since the vibrational and rotational quantum-state distributions of the departing H_2 are identical from $E_i = 8–14$ eV, it is very likely that the $^2A_1$ and/or $^2B_2$ resonances are strongly mixed, particularly in the condensed phase.

Based upon the β parameter discussed above, the O⁻ is not expected to leave the surface of multilayer water ice. However, it is very probable that this product, as well as the OH product, reacts within the ice or water directly at an interface or surface. In the case of water at a DNA interface, this can lead to facile DEA-induced damage of DNA.

18.6.2 LEE-STIMULATED DESORPTION OF CATIONS

Figure 18.9 shows the cations produced and desorbed during electron impact of pristine ice. A detailed description of the cation yields from pristine ice can be found elsewhere (Herring-Captain et al., 2005). Briefly, the cation yield is dominated by H⁺ with a much smaller yield of H_2^+. The H⁺ and H_2^+ yields increase with increasing growth and substrate temperature with ratios of 1:2:3 (PASW:ASW:CI) for H⁺ and 1:1:1.5 for H_2^+. There is also a very reproducible protonated

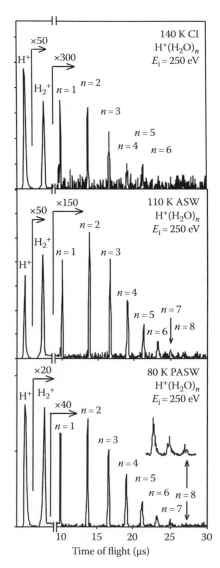

FIGURE 18.9 Cation ion yields from crystalline, amorphous, and porous amorphous ice films. Pulsed irradiation stimulated desorption using a 250 eV incident electron energy produce protonated water clusters $H^+(H_2O)_n$. The yields dramatically increase with the disorder in the film structure.

cluster ion signal from PASW, ASW, and CI. The cluster yield from PASW is typically five to six times larger than CI, and the ASW cluster yield is about two to three times larger than CI (Herring-Captain et al., 2005).

The threshold energies for electron-stimulated production and desorption of H^+ and H_2^+ from CI, and H^+ and $H^+(H_2O)$ from PASW and ASW are 22 ± 3 eV. There is also a H_2^+ yield increase near 40 ± 3 eV and an ~70 eV threshold for $H^+(H_2O)_{n=2-8}$ for all the phases of ice studied. These threshold measurements are shown in Figure 18.10 and are very useful in determining the dominant excited states and excitations involved in cation desorption.

18.6.2.1 Mechanisms for LEE-Stimulated Cation Desorption

Of the two-hole and two-hole, one-electron configurations, at least four are known to produce protons from ice with kinetic energies ranging from 0 to >7 eV (Sieger et al., 1997; Sieger and Orlando, 2000;

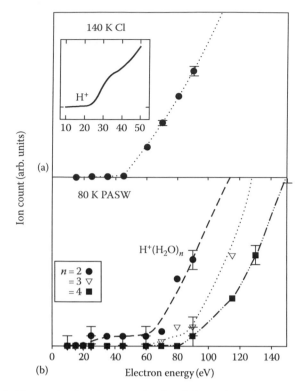

FIGURE 18.10 Threshold energies for electron-stimulated desorption of protonated water cation clusters from crystalline and amorphous ice.

Herring et al., 2004; Herring-Captain et al., 2005). These configurationally mixed two-hole excited states and two-hole, one-electron excited states have excitation energies from 21 to 70 eV. The linear dependence of the proton yield on dose as well as the threshold energy of 21–25 eV indicated that low-kinetic-energy (~4 eV) proton desorption is the result of two-hole states occurring at the surface of the ice. These have been assigned to $(3a_1)^{-1}(1b_1)^{-1}(4a_1)^1$ final states of terminal water molecules with dangling bonds. There is a secondary threshold energy of ~40 eV, which may correspond to the $(3a_1)^{-2}(4a_1)^1$ and $(1b_1)^{-2}(4a_1)^1$ states of water, the latter producing protons with kinetic energy up to 7 eV. These energetic protons (>7 eV) can also be produced from the $(2a_1)^{-2}$ configuration, which can be produced at incident electron energies of >70 eV.

All of these two-hole excited states and two-hole, one-electron excited states involve bands of a_1 symmetry. As previously discussed, levels with a_1 character should be sensitive to the local bonding environment of the water molecule (Herring-Captain et al., 2005). Therefore, the temperature dependence of the proton yield (not shown) gives insight into the local configuration of the surface of the ice and the dangling bonds responsible for proton production. At low dosing temperatures, the random orientation of the surface-water molecules leads to a low proton yield for PASW compared to CI, most likely because the proton undergoes reactive scattering across the ice surface or into the bulk (Sieger and Orlando, 1997). As the temperature increases, the degree of coupling between water molecules decreases and the mobility of surface-water molecules increases (i.e., they have sufficient energy to rotate and start to diffuse) (Akbulut et al., 1997). The reduced coupling and emergence of a split $3a_1$ level discussed above lead to band narrowing, increased excited-state lifetimes, and increased proton emission (Sieger et al., 1997; Herring et al., 2004).

H_2^+ desorption has similar threshold values compared to H^+, but the yields are much less. Although the branching ratio is small, some of the above-mentioned two-hole states and two-hole, one-electron states responsible for proton desorption can decay to form H_2^+. This has been observed in the time-correlated dissociation of doubly ionized states of water, most likely through a dissociative triplet state (Tan et al., 1978). Photoionization near the O 1s → 2b$_2$ resonance (Piancastelli et al., 1999) was reported to produce H$_2^+$. The absorption of four photons to produce H$_2^+$ $\left(X\,^1\Sigma_g^+\right)$ was also reported in a recent multiphoton ionization study of gas-phase water (Rottke et al., 1998). The H$_2^+$ yield observed from these ices is large relative to the gas phase studies. This indicates that many body interactions and orientational defects may play a role in its production. Reactive scattering of the energetic protons may also be a source of H$_2^+$.

The mechanism for cluster ion desorption involves the production of two holes and a Coulomb explosion resulting from these holes in neighboring water molecules. A detailed description of this model can be found elsewhere (Herring-Captain et al., 2005). Briefly, the weak ~25 eV threshold for H$^+$(H$_2$O) formation is most likely due to the reactive scattering of a proton (Bernholc and Phillips, 1986). The primary 70 eV threshold for clusters with 1–7 water molecules may initially correspond to the 2a$_1^{-2}$ state. Coupling of this doubly ionized water molecule to a neighboring water molecule leads to a final state containing one positive charge on each of the two water molecules, most likely with a OH$^+$...H$^+$(H$_2$O) configuration. The surrounding water molecules respond to the charges by reorienting in an attempt to solvate the charges, and finally the unstable configuration undergoes a Coulomb explosion, resulting in desorption of a protonated cluster.

The temperature/phase dependence of the cluster ion yield provides insight into the local hydrogen bonding in the terminal water layers. While the proton yield reflects the behavior of the surface molecules in a direct line of sight to the vacuum, the clusters reflect the behavior of the terminal 1–3 water layers. At low dosing temperatures, the water molecules are randomly oriented and there is no long-range order. The localization of two holes can cause dissociation and proton transfer (or hole hopping) along the hydrogen bond, forming a hydronium ion. The holes are then localized on neighboring water molecules. Due to the lack of an ordered hydrogen-bonding network in PASW, the holes are less likely to be able to hop to a screening distance (1–2 nm) before a Coulomb explosion occurs. This "localized hole hopping," where a hole hops to a neighboring water molecule but not far enough to be screened, occurs in the terminal 1–3 water layers of PASW and results in a large cluster yield. As the temperature increases, a more crystalline structure is formed (Smith et al., 1996; Sieger et al., 1997) and the holes are more likely to hop further away from each other through the hydrogen-bonding network. CI has the most extensive hydrogen-bonding network of the samples studied here and the lowest cluster yield, due to the efficient hole hopping through this network.

18.6.3 LEE-Stimulated Desorption of Neutral Fragments and Products

It is well known that ion desorption accounts for only a small fraction of the total mass loss during electron bombardment of materials and surfaces. In ice and more strongly coupled materials, this is due to hole hopping, efficient autoionization, and electron–hole recombination. In the latter case, the lowest-energy "excitons" formed as a result of hole (ion)–electron recombination can localize at the surface or in the near surface zone, and dissociate to produce atomic fragments such as H (^{1}S) and O (^{3}P, ^{1}D, etc.) (Kimmel and Orlando, 1995; Orlando and Kimmel, 1997) and molecular products such as hydrogen (Orlando et al., 1999) and OH. The threshold energies for producing the neutral atomic hydrogen and oxygen fragments are shown in Figure 18.11. The experimentally observed value of ~6.5 eV (relative to the vacuum level) is lower than the 9.5 and 11.5 eV thermodynamic energies required to produce O(^{3}P$_J$) + 2D(^{2}S) and O(^{1}D) + 2D(^{2}S), respectively. The low threshold value indicates that the formation of atomic oxygen must occur via a pathway that involves a molecular elimination step. Since the molecular hydrogen is in the $\left(X\,^1\Sigma_g^+\right)$ ground state, the observation of both singlet and triplet atomic oxygen demonstrates that singlet and triplet excitons are involved. This is consistent with exciton formation via hole–electron recombination and the fact that the triplet states can be directly excited during electron scattering since the transitions are not governed by optical selection rules.

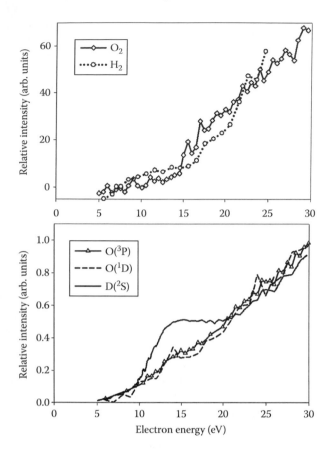

FIGURE 18.11   Electron stimulated desorption (ESD) thresholds for neutral desorbed atoms and molecules. Molecular hydrogen and oxygen share the same threshold as the ground and electronically excited atomic fragments, suggesting that hot-atom reactive scattering is the source term for their production.

The atomic oxygen and OH radicals can react further to produce stable products and, eventually, molecular oxygen (Sieger et al., 1998; Orlando and Sieger, 2003; Johnson et al., 2005). As shown in Figure 18.11, the thresholds for producing O_2 and H_2 are also close to 7 eV, indicating that reactive scattering of atomic oxygen is possible (Kimmel et al., 1994). Oxygen molecules are not produced immediately upon electron bombardment; an "incubation dose" is required before the yield becomes appreciable. A similar incubation dose has been observed in ion (Reimann et al., 1990; Johnson et al., 2005) and vacuum ultraviolet (VUV) photon-stimulated "sputtering" of ice (Westley et al., 1995), and implies a two-step, precursor-mediated mechanism. A dose-, energy-, and temperature-dependent study on the electron-beam-induced production and release of O_2 from ice lead to the formulation of a simple mechanism that initially involves reactive scattering of atomic oxygen or OH radicals to produce a stable precursor (on the timescale of at least an hour at 120 K), such as H_2O-O, HO_2, or H_2O_2. Careful molecular-beam dosing studies using isotope ($H_2^{16}O$ versus $H_2^{18}O$)-labeled water and kinetic modeling indicate that most of the oxygen is produced at the vacuum–surface interface and may likely involve H_2O_2. We note that H_2O-O, HO_2, and H_2O_2 are all known to be produced in irradiated ice, but the yields are highly dependent on the dose and substrate temperature. At low temperature (<120 K), an H_2O-O species can be trapped, and a kinetic model involving direct excitation of this precursor to form O_2 has also been proposed. We note that the trapped O atom can be formed from intrinsic ice defects and it can by easily converted to H_2O_2. Thus, the peroxide route may dominate at temperatures above 120 K and at high electron doses.

We have also measured the quantum-state distributions of the molecular hydrogen produced and desorbed from ice using sensitive REMPI techniques (Kimmel et al., 1995; Orlando et al., 1999). The formation of extremely vibrationally and rotationally excited molecular hydrogen during 100 eV electron bombardment was attributed primarily to the formation and decay of surface excitons with triplet character. In the lower-energy region, vibrationally excited hydrogen was also observed with a distinct propensity for even vibrational quantum numbers. This propensity, along with the energy dependence of the yields, was interpreted in terms of the decay of core-excited Feshbach resonances, particularly the $^2B_2$ state, which is thought to asymptotically produce $O^- + H_2$ $\left(X\,^1\Sigma_g^+\right)$. This state can also undergo autodetachment, leaving a highly excited electronic state that can undergo molecular elimination.

Unfortunately, little is known regarding the nature of the excited states that can lead to the direct formation of molecular hydrogen. The production of atomic oxygen and molecular hydrogen following UV/VUV photoexcitation of gas-phase water has received limited attention. Initial work (Slanger and Black, 1982) demonstrated that approximately 10% of the dissociation cross section using 121.6 nm excitation leads to $O + H_2$. This was an indirect measurement obtained by analyzing the OH A-X emission. The formation of $O(^3P_J\,^1D)$ is likely correlated with a conical intersection on the B-state surface. A very recent quantum mechanical calculation that includes nonadiabatic effects has reported cross sections for the formation of $O(^1D)$ and rovibrational distributions of the H_2 following excitation of water vapor with 121.6 nm photons (van Harrevelt and van Hemert, 2008). We assume that similar decay pathways are available for water molecules at surfaces and buried interfaces, and that the decay of these "exciton-like" states leads to the formation of $O(^3P_J,\,^1D)$ and vibrationally and rotationally excited H_2. This important channel is currently under further investigation.

18.6.4 Effect of Porosity on LEE-Stimulated Production and Release of O_2 and H_2

Molecular ices, particularly polar ices and those capable of hydrogen bonding, facilitate chemistry in a number of ways, both physical and electronic. The diffusion of sparsely distributed molecules trapped within ice matrices is unlikely to occur at low temperatures. This would suggest that the morphology of ice, in particular, its porosity, is of critical importance to ice chemistry. The physical morphology of porous ice also affects the energetic processes governing the excitation and transport of energy through the matrix. We have demonstrated the production and release of molecular hydrogen and oxygen from electron-beam-irradiated low-temperature ice. The formation and release of molecular hydrogen was found to be strongly correlated with the presence of pores. The yields of H_2 and O_2 increase with the degree of disorder in the ice, which implies that their production occurs primarily at interfaces and grain boundaries. After irradiation, the release of H_2 occurs via diffusion through micropores and when macropores collapse ($\sim$80 K). O_2 is retained until ice undergoes an amorphous-to-crystalline phase transition ($\sim$160 K), a point where grain boundaries sinter and clathrate hydrates decompose.

18.7 LOW-ENERGY-ELECTRON-INDUCED DAMAGE OF DNA

Biologically important molecules such as DNA, RNA, amino acids, and nucleobases are constantly bombarded with ionizing radiation from a variety of sources both anthropogenic and non-anthropogenic. A complete description of the effects of ionizing radiation on DNA in living cells is fundamental to radiobiology, public health, and clinical applications such as radiation protection, chemotherapy, and radiosensitizer development. It is well known that high-energy ionization radiation can cause sugar–phosphate cleavage (strand breaks) in DNA, which might be critical DNA lesions responsible for toxic, mutagenic, and cell malfunction, or death. Depending on the site of chemical bond dissociation on double-helix backbones, the DNA damage could be either in the form of an SSB or a DSB.

Radiation-induced DNA damage is usually divided into two groups: (1) "direct" damage, resulting from direct ionization of DNA and/or the closely bound water molecule, and (2) "indirect" damage, resulting from radicals generated by energy deposition in water molecules or other biological molecules surrounding the DNA (Purkayastha and Bernhard, 2004). It is estimated that the direct/indirect DNA damage ratio is about 1:2 in all cellular damage (Michael and O'Neill, 2000). Most of the energy deposited in living cells by ionizing radiation produces secondary species, such as ballistic electrons (1–20 eV), neutrals, or ionic radicals (Sonntag, 1987). Electrons are the most abundant secondary species with an estimated yield of about 10^4 electrons per MeV, and most have initial kinetic energies below 20 eV (Cobut et al., 1998). Pioneering work by Sanche and coworkers has demonstrated that LEEs could localize on various DNA components (base, deoxyribose, phosphate, or waters of hydration) to form TNIs. The decay of the local transient anions into dissociating pathways, such as DEA or autoionization, directly leads to the DNA strand breakage (Boudaiffa et al., 2000).

It has been reported that strand breaks in supercoiled DNA were induced by free ballistic electrons (3–20 eV) (Boudaiffa et al., 2000). The electron energy dependence of SSB and DSB presented strong resonance features in the energy range of 5–15 eV, with a maximum around 8–10 eV. These results provided a fundamental challenge to the traditional view that genotoxic damage by secondary electrons can only occur at energies above the onset of ionization, or upon solvation when they become slowly reacting chemical species. Recently, the DEA of related molecules (water, bases, deoxyribose analog, and phosphate) in the condensed phase has been intensively investigated by measuring electron-energy-dependent desorption yields of negative ions (H^-, OH^-, or R^-) (Sieger et al., 1998; Antic et al., 1999; Boudaiffa et al., 2000; Abdoul-Carime et al., 2001; Zheng et al., 2005). The yields of desorbing anions below 15 eV supported the hypothesis that electron resonance can contribute to DNA damage (Huels et al., 2003; Pan et al., 2003).

One factor that could contribute significantly to the investigation of DNA damage is the effects of water on the formation and reaction of radiation-induced species. It has been reported that water molecules in DNA hydration shells have unique properties (Tao et al., 1989; Swarts et al., 1992). They are less mobile than free water, more mobile than ice water molecules, and impermeable to cations. The unique properties of waters of hydration in DNA could play a major role in the interaction between slow electrons and DNA. In other words, DNA damage resulting from LEE irradiation could be greatly affected by intrinsic water molecules close to DNA. In the studies of LEE-induced DNA damage, plasmid DNA molecules were usually transferred into a UHV chamber for electron bombardment and mass spectrometric analysis. Under these conditions, most of the free water molecules around DNA have been removed by evaporation. However, water molecules tightly bound to or trapped in supercoiled plasmid DNA likely remained. Limited studies have been performed to assess the role of water content and water-plasmid DNA interactions under UHV condition. This is critical to understand the mechanisms of LEE-induced strand breaks.

This problem has now been approached from both a theoretical and an experimental perspective. We have carried out elastic scattering calculations to describe potential diffraction effects involving intrinsic waters within the major and minor grooves. We have also carried out experiments on LEE-induced DNA SSBs and DSBs (Orlando et al., 2008). These approaches are explicitly described and compared in the remaining part of the chapter.

18.7.1 ELASTIC SCATTERING CALCULATION

The elastic scattering calculation uses the multiple scattering formalism described previously and does not require extensive computational resources. The A-tract DNA 5′-D(CGCGAATTCGCG)-3′ and B-DNA 5′-D(CCGGCGCCGG)-3′ targets and collision geometries used in our calculations are shown in Figure 18.12. Note that C, G, A, and T represent cytosine, guanine, adenine, and thymine, respectively. Because the inelastic mean free path is short for LEEs (we use 10.6 Å for biological molecules), the dimensions of these 10 or 12 base-pair DNA strands are large enough to sample

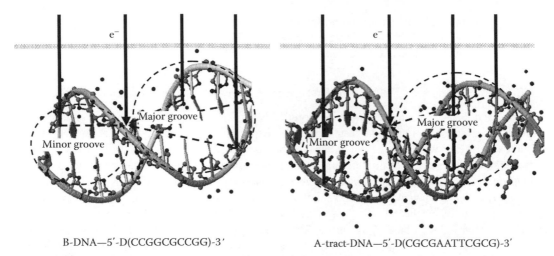

B-DNA—5′-D(CCGGCGCCGG)-3′   A-tract-DNA—5′-D(CGCGAATTCGCG)-3′

FIGURE 18.12 Geometric configurations for the DNA targets used in the scattering calculations. (From Orlando, T.M. et al., *J. Chem. Phys.*, 128, 195102/1, 2008. With permission.)

all atomic scattering centers and guarantee convergence. We introduce a spherically symmetric muffin-tin potential to represent each scatterer of the condensed DNA target. We also use an inner potential of ~6 eV, a value derived from electron holography measurements on biological targets (Stoyanova et al., 1968) and the maximum value of the optical potential used in previous R-matrix calculations on a model DNA target (Caron and Sanche, 2003, 2004, 2005).

In order to compare our calculations with experimental results, the DNA strand is directed so that its symmetry axis is normal to the electron *k*-vector. Phosphate-, sugar-, or water-absorber sites are chosen, and charge densities at each atom that constitute functional groups are summed together to give the total localized charge density around the respective groups. For water, we sum the charge on the oxygen. We examined theoretically only molecules in the primary shell, specifically those in the major and minor grooves and those coupled to the bases and sugar–phosphate groups. For example, 8 waters within the minor groove and 14 waters within the major groove were used in the calculation for the B-DNA 5′-D(CCGGCGCCGG)-3′ target. For the A-tract DNA 5′-D(CGCGAATTCGCG)-3′ target, 14 waters within the minor groove and 24 waters within the major groove were used.

The directly irradiated component is obtained by fixing the phase and decaying the amplitude at the absorber. First-order scattered amplitudes are then calculated and phase summed with the non-scattered amplitude. Complex phase shifts for each atomic scatterer are calculated with the modified FEFF 8.2 code developed by Ankudinov et al. (1996). The elastically scattered electron intensities (not shown) are calculated to be highest when examining the sugar groups, and the overall scattered intensity is greatest for electrons with incident energy less than 15 eV. There is only a very weak structure superimposed upon the sugar components between 22 and 30 eV. The scattered electron intensity is similar for the phosphates; however, there is a very weak structure observable between 8–12 and 22–30 eV for the B-DNA 5′-D(CCGGCGCCGG)-3′ target and 12–20 eV for the A-tract DNA 5′-D(CGCGAATTCGCG)-3′ target. Although the overall scattered intensity is less for structural waters, the constructive buildup of electron intensity due to elastic scattering is most discernable for the water components. We therefore determined the contributions of the minor and major groove waters separately, and these are shown in Figure 18.13. There is essentially no structure observed for what has been assigned as the minor groove waters, whereas a considerable structure in the charge density buildup is associated with the assigned major groove waters (solid line).

The theoretical results can be summarized as follows: (1) There is considerable LEE intensity present on the phosphates, sugars, and structural waters, especially for incident electron energies below 15 eV. (2) The resonance structure is primarily associated with the structural water within the major grooves. (3) The amplitude buildup occurs between 5–10, 12–18, and 22–28 eV for the B-DNA 5′-D (CCG GCGCCGG)-3′ target and between 7–11, 12–18, and 18–25 eV for the A-tract DNA 5′-D(CGCGAATT CGCG)-3′ target.

18.7.2 Comparison of Theory and Experimental Measurements of SSBs and DSBs

Although the DNA sequences examined theoretically are only surrogates for more complex DNA targets, it is useful to compare the results to recent experiments on LEE-induced SSBs and DSBs. These experiments were carried out using the apparatus already described above. A comparison of the calculated electron density buildup on the major groove waters with our measured SSB and DSB probabilities as a function of incident electron energy is shown in Figure 18.13. Although the samples studied are much more complicated than the targets used in the calculation, the comparison serves to qualitatively illustrate the potential importance of diffraction and structural water in the overall electron-induced break probability. The calculated charge density buildups on the major groove waters of hydration for both DNA targets are shown on the topmost part of Figure 18.13. The measured SSB (open diamonds) and DSB (open triangles) yields as a function of incident electron energy are shown in the center part of Figure 18.13. Each point was taken on a separate DNA film

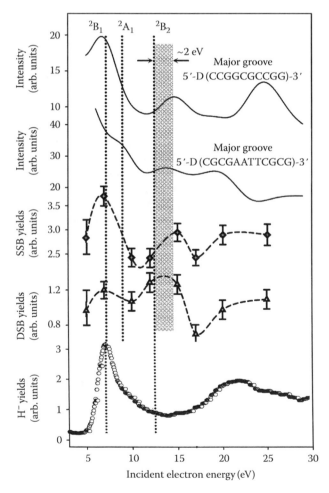

FIGURE 18.13 Upper plots show the resultant scattering intensities for major groove absorber sites for the two DNA oligomers studied. SSB (diamonds) and DSB (triangles) yields (Chen et al., 2008). For comparison, the H⁻ DEA spectrum is also shown (Simpson et al., 1997).

and each point is the average of three or four separate measurements. Although the density of points is low, there is a general correspondence of the SSB and DSB probabilities with the calculated interference structure in the scattered electron intensity associated with the structural waters found in the major groove. These results are also generally consistent with the first experiments by Sanche and colleagues on DNA plasmid thin films (Boudaiffa et al., 2000) and more recent experiments demonstrating maxima in the SSB and DSB yields at ~9.5 eV in linear and supercoiled DNA (Huels et al., 2003; Pan et al., 2003).

18.7.3   DNA Damage

Figure 18.13 shows a resonant structure in the SSB and DSB probabilities, which is higher in energy than the single-particle shape resonances discussed above. SSBs and DSBs in this energy range can be associated with core-excited Feshbach resonances and dissociative excitation of phosphates, sugars, and bases. As already discussed, the core-excited Feshbach resonances can decay via DEA. In fact, recent studies on the DEA of DNA bases in the gas phase have revealed the formation of many fragment anions, especially at energies between 5 and 15 eV (Huels et al., 1998; Denifl et al., 2004; Ptasińska et al., 2005a,b). Similar experiments on condensed films of cytosine and thymine also revealed DEA resonances; however, the overall fragmentation appeared to be less due to interactions with the surroundings (Huels et al., 1995). There is recent work on the DEA of gas-phase thymidine (thymine bound to 2-deoxyribose) that addresses coupling of the base resonances with the attached sugar (Ptasińska et al., 2006), and work on condensed short GCAT oligomers (Zheng et al., 2006) that examines base coupling as well as energy transfer to the sugar and phosphate groups. There are not many studies on the DEA of ribose sugars comprising DNA (Ptasińska et al., 2004), nor has there been much work on isolated phosphates or phosphoric acid (Konig et al., 2006; Pan and Sanche, 2006). In the case of phosphoric acid and aligned self-assembled monolayers of DNA films (Pan and Sanche, 2005), the resonance between 4 and 12 eV leading to O^- and OH^- desorption was attributed to a core-excited resonance localized on the phosphate group. This was investigated further in an attempt to experimentally address the role of water in DNA damage (Ptasińska and Sanche, 2007). The data suggested that the water coupling was with the phosphate groups or bases and that compound states may be important in DNA damage.

18.7.4   The Role of Water and Diffraction in DNA Damage

Since water is likely involved in electron-induced damage of DNA, we recall the primary electronic excitations that lead to the DEA of water. The positions of the DEA states in nanoscale water films are shown in Figure 18.13, while the $H^-(D^-)$ yield is shown in the lower part. The vertical lines indicate potential correspondences with features in the SSB and DSB yields. Specifically, the SSB and DSB features between 5–10 and 10–17 eV may potentially correlate with the $^2B_1$ and $^2B_2$ resonances, respectively (Kimmel and Orlando, 1996). As noted in Figure 18.13, the damage features near 10–17 eV are ~2 eV higher in energy and broader than the condensed water $^2B_2$ resonance. They are also higher in energy than the resonance known to produce O^- directly via DEA of the phosphate group (Ptasińska and Sanche, 2006), but seem to be in reasonable agreement with the position of the compound resonance reported from hydrated GCAT (Ptasińska and Sanche, 2007). The observed energy shifts may be a consequence of stronger coupling relative to hydrogen bonding in water and modified Franck–Condon factors associated with compound states involving water and DNA constituents, such as sugar–phosphate groups and bases/counter ions. The reproducible shifts to higher energy (relative to DEA resonances of water and phosphate) are not due to charging.

Calculations indicate the presence of resonances in phosphoric acid at 7.7 and 12.5 eV (Ptasińska and Sanche, 2006), and experiments indicate prominent DEA resonances peaking at 5 and 10 eV for most bases (Ptasińska and Sanche, 2006). Since these states are nearly degenerate with the DEA resonances of water, they can easily mix, and, depending upon the lifetimes, the coupled states could lead to enhanced DEA or increased autodetachment (i.e., electron emission). Therefore, SSBs can be partially attributed to the formation of H⁻, O⁻, OH⁻, H, O, or OH via the decay of the compound (H_2O:DNA) DEA resonances and dissociative excited states. The presence of these reactive anions and radicals in the vicinity of the phosphate–sugar or within the major groove leads to facile bond breakage and is a well-known damage mechanism for RNA (Soukup and Breaker, 1999).

Calculations indicate that facile C–O σ bond cleavage between the sugar and phosphate groups can occur as a result of excitation of the low-lying phosphate σ* or π* orbitals. This excitation could initially involve the π* level of the base, but could be transferred through the sugar to the phosphate. If LEEs (i.e., those emitted during decay of resonances at and above 5 eV) undergo inelastic scattering within their mean free paths, they can excite a phosphate in the near vicinity or on the opposite strand. Since both the initial excitation and the secondary scattering can lead to strand breaks, this could be an efficient path for DSBs. This process is consistent with a DSB threshold energy of ~5 eV and indicates that DSBs may actually require major groove waters and a compound resonance that simultaneously forms very reactive fragments and an LEE. Overall, our results indicate that the diffraction amplitude associated with the major groove waters can enhance both the SSB and DSB probabilities by providing a spatially varying charge density at the energy necessary for excitation of water–DNA compound resonances.

18.8 SUMMARY

In this chapter, we have discussed and summarized the role of LEEs in the nonthermal dissociation and desorption of nanoscale water films and water–DNA interfaces. Initially, we discussed the convolution integral of $N(E) \cdot \sigma(E)$ and the nature of the accessible electronic states in condensed-phase water between 5 and 25 eV. The importance of the temperature dependence of the a_1-type bands in ice was pointed out and the effect on hole lifetimes was shown. The role of core-excited Feshbach resonances in water (i.e., $^2B_1$, $^2A_1$, and $^2B_2$) and their decay via DEA have been shown to lead to the formation and release of reactive negative ions (H⁻, D⁻) and neutral fragments. Complex multihole states can result in the formation and release of protons at excitation energies above 22–25 eV. Inelastic channels that primarily involve excitations of the outer and inner valence levels, and the formation of complicated, configurationally mixed excited electronic states were discussed. When these complicated excitations localize at a surface or interface, dissociation and desorption can occur. The excited states can be parent states for core-excited Feshbach resonances. These resonances can decay to form neutral and anionic molecular fragments. In general, the damage probability function illustrates the general importance of LEEs and the need to unravel the details of all the energy-loss channels within this energy range.

Many of the recent experimental techniques used to examine the energy-loss features, which involved the production and release of anions, cations, and neutrals, were discussed. UHV systems are typically used to study the chemical transformations initiated by the inelastic scattering of LEEs with thin films of molecular solids. Samples are usually prepared under UHV conditions and are bombarded with LEEs. Typically, these systems are custom-designed and equipped with a cryo-cooled rotatable sample holder to utilize multiple analysis techniques. The detection of charged species is often accomplished with either a quadrupole or a ToF mass spectrometer. Ion yields can be measured as a function of the incident electron energy, allowing threshold energies to be determined, and are very helpful with regard to assessing the states most important in the LEE-induced dissociation and desorption events. The detection of neutral fragments from ice or other complex targets can be accomplished by utilizing very sensitive laser detection techniques, for example, laser REMPI/ToF spectrometry and SPI.

The chemistry initiated by the LEE inelastic energy-loss channels was examined by postirradiation TPD, and postirradiation gel electrophoresis has been shown to be a useful technique for determining quantitatively SSB and DSB probabilities, thus allowing measurements of the SSB and DSB probabilities as a function of electron-beam energy and dose. In addition, multiple scattering formalisms can be used to investigate diffraction effects in LEE-induced damage of water and water–DNA interfaces. Specifically, these calculations use the separable representation of a free-space electron propagator and a curved-wave multiple scattering formalism. They reveal constructive interference features arising from diffraction when examining the structural waters within the major groove that are localized within an 8–10 Å sphere around each base and the sugar–phosphate backbone units. Since the diffraction intensity is localized, compound (H_2O:DNA) states involving core-excited Feshbach-type resonances may contribute to the LEE-induced SSB and DSB probabilities. These results have illustrated the potential importance of diffraction and structural water in the overall electron-induced break probability, and suggested that the water coupling was with the phosphate groups or bases. SSB mechanisms have been shown to be partially attributed to the formation of H^-, O^-, OH^-, H, O, or OH via decay of the compound (H_2O:DNA) resonance. The presence of these reactive anions and radicals in the vicinity of the phosphate–sugar or within the major groove facilitates bond breakage. Electron autodetachment has been shown as an important decay channel that can lead to secondary scattering and excitation of lower-lying σ^* or π^* orbitals of the phosphate. Radiation damage involves the phased sum of multiple scattering events; however, internal diffraction may enhance the DNA damage probability, especially for electron energies below 15 eV.

ACKNOWLEDGMENTS

Much of the work discussed in this chapter was supported by the U.S. Department of Energy, Office of Science, Contract DE-FG02-02ER15337. We thank Matt Sieger and T. C. Chiang for supplying the photoemission data presented in Figure 18.6.

REFERENCES

Abdoul-Carime, H., Cloutier, P., and Sanche, L. 2001. Low-energy (5–40 eV) electron-stimulated desorption of anions from physisorbed DNA bases. *Radiat. Res.* 155 (4): 625–633.

Ahlstrom, P., Lofgren, P., Lausma, J., Kasemo, B., and Chakarov, D. 2004. Crystallization kinetics of thin amorphous water films on surfaces: Theory and computer modeling. *Phys. Chem. Chem. Phys.* 6 (8): 1890–1898.

Ahmed, F. E. 1993. Urea facilitates quantitative estimation of DNA damage in situ by preventing its degradation. *Anal. Biochem.* 210 (2): 253–257.

Ahnstrom, G. 1974. Repair processes in germinating seeds. Caffeine enhancement of damage induced by gamma-radiation and alkylating chemicals. *Mutat. Res.* 26 (2): 99–103.

Akbulut, M., Madey, T. E., and Nordlander, P. 1997. Low energy (<5 eV) F^+ and F^- ion transmission through condensed layers of water. *J. Chem. Phys.* 106 (7): 2801–2810.

Alexandrov, A., Piacentini, M., Zema, N., Felici, A. C., and Orlando, T. M. 2001. Role of excitons in electron- and photon-stimulated desorption of neutrals from alkali halides. *Phys. Rev. Lett.* 86 (3): 536–539.

Ankudinov, A. L., Zabinsky, S. I., and Rehr, J. J. 1996. Single configuration Dirac-Fock atom code. *Comput. Phys. Commun.* 98 (3): 359–364.

Antic, D., Parenteau, L., Lepage, M., and Sanche, L. 1999. Low-energy electron damage to condensed-phase deoxyribose analogues investigated by electron stimulated desorption of H- and electron energy loss spectroscopy. *J. Phys. Chem. B* 103 (31): 6611–6619.

Azria, R., Parenteau, L., and Sanche, L. 1987. Dissociative attachment for condensed molecular oxygen: Violation of the selection rule $\Sigma^- \leftrightarrow \Sigma^+$. *Phys. Rev. Lett.* 59 (6): 638–640.

Balog, R., Langer, J., Gohlke, S., Stano, M., Abdoul-Carime, H., and Illenberger, E. 2004. Low energy electron driven reactions in free and bound molecules: From unimolecular processes in the gas phase to complex reactions in a condensed environment. *Int. J. Mass Spectrom.* 233 (1–3): 267–291.

Banath, J. P., Wallace, S. S., Thompson, J., and Olive, P. L. 1999. Radiation-induced DNA base damage detected in individual aerobic and hypoxic cells with endonuclease III and formamidopyrimidine-glycosylase. *Radiat. Res.* 151 (5): 550–558.

Bass, A. D. and Sanche, L. 1998. Absolute and effective cross-sections for low-energy electron-scattering processes within condensed matter. *Radiat. Environ. Biophys.* 37 (4): 243–257.

Bass, A. D. and Sanche, L. 2003a. Dissociative electron attachment and charge transfer in condensed matter. *Radiat. Phys. Chem.* 68 (1–2): 3–13.

Bass, A. D. and Sanche, L. 2003b. Reactions induced by low energy electrons in cryogenic films (review). *Low Temp. Phys.* 29 (3): 202–214.

Beach, C., Fuciarelli, A. F., and Zimbrick, J. D. 1994. Electron migration along 5-bromouracil-substituted DNA irradiated in solution and in cells. *Radiat. Res.* 137 (3): 385–393.

Bernholc, J. and Phillips, J. C. 1986. Kinetics of cluster formation in the laser vaporization source: Carbon clusters. *J. Chem. Phys.* 85 (6): 3258–3267.

Boudaiffa, B., Cloutier, P., Hunting, D., Huels, M. A., and Sanche, L. 2000. Resonant formation of DNA strand breaks by low-energy (3 to 20 eV) electrons. *Science* 287 (5458): 1658–1660.

Caron, L. G. and Sanche, L. 2003. Low-energy electron diffraction and resonances in DNA and other helical macromolecules. *Phys. Rev. Lett.* 91 (11): 113201/1–113201/4.

Caron, L. and Sanche, L. 2004. Diffraction in resonant electron scattering from helical macromolecules: A- and B-type DNA. *Phys. Rev. A* 70 (3): 032719/1–032719/10.

Caron, L. and Sanche, L. 2005. Diffraction in resonant electron scattering from helical macromolecules: Effects of the DNA backbone. *Phys. Rev. A* 72 (3, Pt. A): 032726/1–032726/6.

Chen, H.-Y. and Chao, I. 2004. Ionization-induced proton transfer in model DNA base pairs: A theoretical study of the radical ions of the 7-azaindole dimer. *ChemPhysChem* 5 (12): 1855–1863.

Chen, Y., Garcia de Abajo, F. J., Chasse, A., Ynzunza, R. X., Kaduwela, A. P., Van Hove, M. A., and Fadley, C. S. 1998. Convergence and reliability of the Rehr-Albers formalism in multiple-scattering calculations of photoelectron diffraction. *Phys. Rev. B* 58 (19): 13121–13131.

Chen, Y. F., Aleksandrov, A., and Orlando, T. M. 2008. Probing low-energy electron induced DNA damage using single photon ionization mass spectrometry. *Int. J. Mass Spectrom.* 277 (1–3): 314–320.

Christophorou, L. G. and Olthoff, J. K. 2002. Electron interactions with plasma processing gases: Present status and future needs. *Appl. Surf. Sci.* 192 (1–4): 309–326.

Christophorou, L. G. and Olthoff, J. K. 2004. *Fundamental Electron Interactions with Plasma Processing Gases, Physics of Atoms and Molecules.* New York: Kluwer Academic/Plenum Publishers.

Christophorou, L. G., McCorkle, D. L., and Christodoulides, A. A. 1984. *Electron–Molecule Interactions and Their Applications*, Vol. 1. New York: Academic Press Inc.

Chutjian, A., Garscadden, A., and Wadehra, J. M. 1996. Electron attachment to molecules at low electron energies. *Phys. Rep.* 264 (6): 393–470.

Cobut, V., Frongillo, Y., Patau, J. P., Goulet, T., Fraser, M. J., and Jay-Gerin, J. P. 1998. Monte Carlo simulation of fast electron and proton tracks in liquid water: I. Physical and physicochemical aspects. *Radiat. Phys. Chem.* 51 (3): 229–243.

Compton, R. N. and Christophorou, L. G. 1967. Negative-ion formation in H_2O and D_2O. *Phys. Rev.* 154 (1): 110–116.

Curtis, M. G. and Walker, I. C. 1992. Dissociative electron attachment in water and methanol (5–14 eV). *J. Chem. Soc. Faraday Trans.* 88 (19): 2805–2810.

Denifl, S., Ptasińska, S., Probst, M., Hrušák, J., Scheier, P., and Märk, T. D. 2004. Electron attachment to the gas-phase DNA bases cytosine and thymine. *J. Phys. Chem. A* 108 (31): 6562–6569.

Fedor, J., Cicman, P., Coupier, B., Feil, S., Winkler, M., Gluch, K., Husarik, J., Jaksch, D., Farizon, B., Mason, N. J., Scheier, P., and Märk, T. D. 2006. Fragmentation of transient water anions following low-energy electron capture by H_2O/D_2O. *J. Phys. B: At., Mol. Opt. Phys.* 39 (18): 3935–3944.

Haxton, D. J., McCurdy, C. W., and Rescigno, T. N. 2007a. Dissociative electron attachment to the H_2O molecule. I. Complex-valued potential-energy surfaces for the $^2B_1$, $^2A_1$, and $^2B_2$ metastable states of the water anion. *Phys. Rev. A: At., Mol. Opt. Phys.* 75 (1, Pt. A): 012710/1–012710/25.

Haxton, D. J., Rescigno, T. N., and McCurdy, C. W. 2007b. Dissociative electron attachment to the H_2O molecule. II. Nuclear dynamics on coupled electronic surfaces within the local complex potential model. *Phys. Rev. A: At., Mol. Opt. Phys.* 75 (1, Pt. A): 012711/1–012711/24.

Herring, J., Aleksandrov, A., and Orlando, T. M. 2004. Stimulated desorption of cations from pristine and acidic low-temperature water ice surfaces. *Phys. Rev. Lett.* 92 (18):187602.

Herring-Captain, J., Grieves, G., Alexandrov, A., Sieger, M. T., Chen, H., and Orlando, T. M. 2005. Low-energy (5–250 eV) electron stimulated desorption of H^+, H_2^+ and $H^+(H_2O)_n$ from low-temperature water ice surfaces. *Phys. Rev. B* 72: 035431-1.

Huels, M. A., Parenteau, L., Michaud, M., and Sanche, L. 1995. Kinetic-energy distributions of O^- produced by dissociative electron attachment to physisorbed O_2. *Phys. Rev. A* 51 (1): 337–349.

Huels, M. A., Hahndorf, I., Illenberger, E., and Sanche, L. 1998. Resonant dissociation of DNA bases by subionization electrons. *J. Chem. Phys.* 108 (4): 1309–1312.

Huels, M. A., Boudaiffa, B., Cloutier, P., Hunting, D., and Sanche, L. 2003. Single, double, and multiple double strand breaks induced in DNA by 3–100 eV electrons. *J. Am. Chem. Soc.* 125 (15): 4467–4477.

Ingolfsson, O., Weik, F., and Illenberger, E. 1996. The reactivity of slow electrons with molecules at different degrees of aggregation: Gas phase, clusters and condensed phase. *Int. J. Mass Spectrom. Ion Processes* 155 (1/2): 1–68.

Johnson, R. E. and Quickenden, T. I. 1997. Photolysis and radiolysis of water ice on the outer solar system bodies. *J. Geophys. Res., [Planets]* 102 (E5): 10985–10996.

Johnson, R. E., Cooper, P. D., Quickenden, T. I., Grieves, G. A., and Orlando, T. M. 2005. Production of oxygen by electronically induced dissociations in ice. *J. Chem. Phys.* 123 (18): 184715/1–184715/8.

Johnson, R. E., Luhmann, J. G., Tokar, R. L., Bouhram, M., Berthelier, J. J., Sittler, E. C., Cooper, J. F., Hill, T. W., Smith, H. T., Michael, M., Liu, M., Crary, F. J., and Young, D. T. 2006. Production, ionization and redistribution of O_2 in Saturn's ring atmosphere. *Icarus* 180 (2): 393–402.

Jorgensen, W. L. and Salem, L. 1973. *The Organic Chemist's Book of Orbitals.* New York: Academic Press.

Jungen, M., Vogt, J., and Staemmler, V. 1979. Feshbach resonances and dissociative electron attachment of water. *Chem. Phys.* 37 (1): 49–55.

Kimmel, G. A. and Orlando, T. M. 1995. Low-energy (5–120 eV) electron-stimulated dissociation of amorphous D_2O ice: $D(^2S)$, $O(^3P_{2,1,0})$, and $O(^1D_2)$ yields and velocity distributions. *Phys. Rev. Lett.* 75 (13): 2606–2609.

Kimmel, G. A. and Orlando, T. M. 1996. Observation of negative ion resonances in amorphous ice via low-energy (5–40 eV) electron-stimulated production of molecular hydrogen. *Phys. Rev. Lett.* 77 (19): 3983–3986.

Kimmel, G. A., Orlando, T. M., Vezina, C., and Sanche, L. 1994. Low-energy electron-stimulated production of molecular-hydrogen from amorphous water ice. *J. Chem. Phys.* 101 (4): 3282–3286.

Kimmel, G. A., Tonkyn, R. G., and Orlando, T. M. 1995. Kinetic and internal energy distributions of molecular hydrogen produced from amorphous ice by impact of 100 eV electrons. *Nucl. Instrum. Methods Phys. Res. Sect. B* 101 (1,2): 179–183.

Kimmel, G. A., Orlando, T. M., Cloutier, P., and Sanche, L. 1997. Low-energy (5–50 eV) electron-stimulated desorption of atomic hydrogen and metastable emission from amorphous ice. *J. Phys. Chem. B* 101 (32): 6301–6303.

Kobayashi, K. 1983. Optical spectra and electronic structure of ice. *J. Phys. Chem.* 87 (21): 4317–4321.

Konig, C., Kopyra, J., Bald, I., and Illenberger, E. 2006. Dissociative electron attachment to phosphoric acid esters: The direct mechanism for single strand breaks in DNA. *Phys. Rev. Lett.* 97 (1): 018105/1–018105/4.

Lofgren, P., Ahlstrom, P., Lausma, J., Kasemo, B., and Chakarov, D. 2003. Crystallization kinetics of thin amorphous water films on surfaces. *Langmuir* 19 (2): 265–274.

Lu, Q. B. and Sanche, L. 2004. Enhancements in dissociative electron attachment to CF_4, chlorofluorocarbons and hydrochlorofluorocarbons adsorbed on H_2O ice. *J. Chem. Phys.* 120 (5): 2434–2438.

Madey, T. E., Johnson, R. E., and Orlando, T. M. 2002. Far-out surface science: Radiation-induced surface processes in the solar system. *Surf. Sci.* 500 (1–3): 838–858.

Mayer, E. and Pletzer, R. 1986. Astrophysical implications of amorphous ice: A microporous solid. *Nature* 319 (6051): 298–301.

Meesungnoen, J., Jay-Gerin, J.-P., Filali-Mouhim, A., and Mankhetkorn, S. 2002. Low-energy electron penetration range in liquid water. *Radiat. Res.* 158 (5): 657–660.

Melton, C. E. 1972. Cross sections and interpretation of dissociative attachment reactions producing OH^-, O^-, and H^- in water. *J. Chem. Phys.* 57 (10): 4218–4225.

Michael, B. D. and O'Neill, P. 2000. Molecular biology: A sting in the tail of electron tracks. *Science* 287 (5458): 1603–1604.

Michaud, M., Wen, A., and Sanche, L. 2003. Cross sections for low-energy (1–100 eV) electron elastic and inelastic scattering in amorphous ice. *Radiat. Res.* 159 (1): 3–22.

Morii, N., Kido, G., Suzuki, H., Nimori, S., and Morii, H. 2004. Molecular chain orientation of DNA films induced by both the magnetic field and the interfacial effect. *Biomacromolecules* 5 (6): 2297–2307.

Nagesha, K. and Sanche, L. 1998. Effects of band structure on electron attachment to adsorbed molecules: Cross section enhancements via coupling to image states. *Phys. Rev. Lett.* 81 (26, Pt. 1): 5892–5895.

Ness, K. F. and Robson, R. E. 1988. Transport-properties of electrons in water-vapor. *Phys. Rev. A* 38 (3): 1446–1456.

Nordlund, D., Odelius, M., Bluhm, H., Ogasawara, H., Pettersson, L. G. M., and Nilsson, A. 2008. Electronic structure effects in liquid water studied by photoelectron spectroscopy and density functional theory. *Chem. Phys. Lett.* 460 (1–3): 86–92.

Oh, D., Sieger, M. T., and Orlando, T. M. 2006. Zone specificity in low energy electron stimulated desorption of Cl^+ from reconstructed $Si(111)$ 7×7:Cl surfaces. *Surf. Sci.* 600 (19): L245–L249.

Orlando, T. M. and Kimmel, G. A. 1997. The role of excitons and substrate temperature in low-energy (5–50 eV) electron-stimulated dissociation of amorphous D_2O ice. *Surf. Sci.* 390 (1–3): 79–85.

Orlando, T. M. and Sieger, M. T. 2003. The role of electron-stimulated production of O_2 from water ice in the radiation processing of outer solar system surfaces. *Surf. Sci.* 528 (1–3): 1–7.

Orlando, T. M., Kimmel, G. A., and Simpson, W. C. 1999. Quantum-resolved electron stimulated interface reactions: D_2 formation from D_2O films. *Nucl. Instrum. Methods Phys. Res., Sect. B* 157 (1–4): 183–190.

Orlando, T. M., Oh, D., Sieger, M. T., and Lane, C. D. 2004. Electron collisions with complex targets: Diffraction effects in stimulated desorption. *Phys. Scr.* T110: 256–261. (*XXIII International Conference on Photonic, Electronic and Atomic Collisions*, Stockholm, Sweden, 2003)

Orlando, T. M., Oh, D., Chen, Y., and Aleksandrov, A. B. 2008. Low-energy electron diffraction and induced damage in hydrated DNA. *J. Chem. Phys.* 128 (19): 195102/1–195102/7.

Pan, X. and Sanche, L. 2005. Mechanism and site of attack for direct damage to DNA by low-energy electrons. *Phys. Rev. Lett.* 94 (19): 198104/1–198104/4.

Pan, X. and Sanche, L. 2006. Dissociative electron attachment to DNA basic constituents: The phosphate group. *Chem. Phys. Lett.* 421 (4–6): 404–408.

Pan, X., Cloutier, P., Hunting, D., and Sanche, L. 2003. Dissociative electron attachment to DNA. *Phys. Rev. Lett.* 90 (20): 208102.

Petrenko, V. and Whiteworth, R. W. 2002. *Physics of Ice.* New York: Oxford University Press.

Piancastelli, M. N., Hempelmann, A., Heiser, F., Gessner, O., Rudel, A., and Becker, U. 1999. Resonant photofragmentation of water at the oxygen K edge by high-resolution ion-yield spectroscopy. *Phys. Rev. A* 59 (1): 300–306.

Ptasińska, S. and Sanche, L. 2006. On the mechanism of anion desorption from DNA induced by low energy electrons. *J. Chem. Phys.* 125 (14): 144713/1–144713/9.

Ptasińska, S. and Sanche, L. 2007. Dissociative electron attachment to hydrated single DNA strands. *Phys. Rev. E* 75 (3): 031915/1–031915/5.

Ptasińska, S., Denifl, S., Scheier, P., and Märk, T. D. 2004. Inelastic electron interaction (attachment/ionization) with deoxyribose. *J. Chem. Phys.* 120 (18): 8505–8511.

Ptasińska, S., Denifl, S., Scheier, P., Illenberger, E., and Märk, T. D. 2005a. Bindungs- und ortsselektive abspaltung von H-atomen aus nucleobasen, induziert durch elektronen sehr niedriger energie (<3 eV). *Angew. Chem.* 117 (42): 6949–7125.

Ptasińska, S., Denifl, S., Scheier, P., Illenberger, E., and Märk, T. D. 2005b. Bond- and site-selective loss of H atoms from nucleobases by very-low-energy electrons (<3 eV). *Angew. Chem., Int. Ed.* 44 (42): 6941–6943.

Ptasińska, S., Denifl, S., Gohlke, S., Scheier, P., Illenberger, E., and Märk, T. D. 2006. Decomposition of thymidine by low-energy electrons: implications for the molecular mechanisms of single-strand breaks in DNA. *Angew. Chem., Int. Ed.* 45 (12): 1893–1896.

Purkayastha, S. and Bernhard, W. A. 2004. What is the initial chemical precursor of DNA strand breaks generated by direct-type effects? *J. Phys. Chem. B* 108 (47): 18377–18382.

Ramaker, D. E. 1983. Comparison of photon-stimulated dissociation of gas-phase, solid and chemisorbed water. *Chem. Phys.* 80 (1–2): 183–202.

Rehr, J. J. and Albers, R. C. 1990. Scattering-matrix formulation of curved-wave multiple-scattering theory: Application to x-ray-absorption fine structure. *Phys. Rev. B* 41 (12): 8139.

Rehr, J. J., Zabinsky, S. I., Ankudinov, A., and Albers, R. C. 1995. Atomic-XAFS and XANES. *Physica B* 208 & 209 (1–4): 23–26.

Reimann, C. T., Brown, W. L., Nowakowski, M. J., Cui, S. T., and Johnson, R. E. 1990. Ejection of argon dimers from solid argon films electronically excited by MeV helium ions. *Springer Ser. Surf. Sci.* 19 (Desorption Induced Electron. Transitions DIET 4): 226–234.

Rottke, H., Trump, C., and Sandner, W. 1998. Multiphoton ionization and dissociation of H_2O. *J. Phys. B* 31 (5): 1083–1096.

Rowntree, P., Sambe, H., Parenteau, L., and Sanche, L. 1993. Formation of anionic excitations in the rare-gas solids and their coupling to dissociative states of adsorbed molecules. *Phys. Rev. B* 47 (8): 4537–4554.

Sanche, L. 1995a. Interactions of low-energy electrons with atomic and molecular solids. *Scanning Microsc.* 9 (3): 619–656.

Sanche, L. 1995b. Transmission through organic thin-films. *Phys. Rev. Lett.* 75 (15): 2904–2904.

Sanche, L. 2000. Electron resonances in DIET. *Surf. Sci.* 451 (1–3): 82–90.

Sanche, L. 2002. Nanoscopic aspects of radiobiological damage: Fragmentation induced by secondary low-energy electrons. *Mass Spectrom. Rev.* 21 (5): 349–369.

Sanche, L. 2005. Low energy electron-driven damage in biomolecules. *Eur. Phys. J. D* 35 (2): 367–390.

Sieger, M. T. and Orlando, T. M. 1997. Effect of surface roughness on the electron-stimulated desorption of D^+ from microporous D_2O ice. *Surf. Sci.* 390 (1–3): 92–96.

Sieger, M. T. and Orlando, T. M. 2000. Probing low-temperature water ice phases using electron-stimulated desorption. *Surf. Sci.* 451 (1–3): 97–101.

Sieger, M. T., Simpson, W. C., and Orlando, T. M. 1997. Electron-stimulated desorption of D^+ from D_2O ice: Surface structure and electronic excitations. *Phys. Rev. B: Condens. Matter* 56 (8): 4925–4937.

Sieger, M. T., Simpson, W. C., and Orlando, T. M. 1998. Production of O_2 on icy satellites by electronic excitation of low-temperature water ice. *Nature* 394 (6693): 554–556.

Sieger, M. T., Schenter, G. K., and Orlando, T. M. 1999. Stimulated desorption by surface electron standing waves. *Phys. Rev. Lett.* 82 (16): 3348–3351.

Simpson, W. C., Parenteau, L., Smith, R. S., Sanche, L., and Orlando, T. M. 1997. Electron-stimulated desorption of D^- (H^-) from condensed D_2O (H_2O) films. *Surf. Sci.* 390 (1–3): 86–91.

Slanger, T. G. and Black, G. 1982. Photodissociative channels at 1216 Å. for water, ammonia, and methane. *J. Chem. Phys.* 77 (5): 2432–2437.

Smith, R. S., Huang, C., Wong, E. K. L., and Kay, B. D. 1996. Desorption and crystallization kinetics in nanoscale thin films of amorphous water ice. *Surf. Sci.* 367 (1): L13–L18.

Sognier, M. A., Eberle, R. L., Zhang, Y., and Belli, J. A. 1991. Interaction between radiation and drug damage in mammalian cells. V. DNA damage and repair induced in LZ cells by adriamycin and/or radiation. *Radiat. Res.* 126 (1): 80–87.

Sonntag, C. V. 1987. *The Chemical Basis for Radiation Biology.* London, U.K.: Taylor & Francis.

Soukup, G. A. and Breaker, R. R. 1999. Relationship between internucleotide linkage geometry and the stability of RNA. *RNA* 5 (10): 1308–1325.

Stevenson, K. P., Kimmel, G. A., Dohnalek, Z., Smith, R. S., and Kay, B. D. 1999. Controlling the morphology of amorphous solid water. *Science* 283 (5407): 1505–1507.

Stoyanova, I. G., Anaskin, I. F., and Shpagina, M. D. 1968. Mean inner potential of biological and organic microobjects measured in an electron interference microscope. *Biofizika* 13 (3): 521–524.

Swarts, S. G., Sevilla, M. D., Becker, D., Tokar, C. J., and Wheeler, K. T. 1992. Radiation-induced DNA damage as a function of hydration. I. Release of unaltered bases. *Radiat. Res.* 129 (3): 333–344.

Tan, K. H., Brion, C. E., Van der Leeuw, P. E., and Van der Wiel, M. J. 1978. Absolute oscillator strengths (10–60 eV) for the photoabsorption, photoionization and fragmentation of water. *Chem. Phys.* 29 (3): 299–309.

Tao, N. J., Lindsay, S. M., and Rupprecht, A. 1989. Structure of DNA hydration shells studied by Raman spectroscopy. *Biopolymers* 28 (5): 1019–1030.

Van der Schans, G. P., Van Loon, A. A. W. M., Groenendijk, R. H., and Baan, R. A. 1989. Detection of DNA damage in cells exposed to ionizing radiation by use of anti-single-stranded DNA monoclonal antibody. *Int. J. Radiat. Biol.* 55 (5): 747–760.

van Harrevelt, R. and van Hemert, M. C. 2008. Quantum mechanical calculations for the $H_2O + hv \rightarrow O(^1D) + H_2$ photodissociation process. *J. Phys. Chem. A* 112 (14): 3002–3009.

Westley, M. S., Baragiola, R. A., Johnson, R. E., and Baratta, G. A. 1995. Ultraviolet photodesorption from water ice. *Planet. Space Sci.* 43 (10/11): 1311–1315.

Winter, B., Weber, R., Widdra, W., Dittmar, M., Faubel, M., and Hertel, I. V. 2004. Full valence band photoemission from liquid water using EUV synchrotron radiation. *J. Phys. Chem. A* 108 (14): 2625–2632.

Zamenhof, S., De Giovanni, R., and Greer, S. 1958. Induced gene unstabilization. *Nature* 181 (4612): 827–829.

Zheng, Y., Cloutier, P., Hunting, D. J., Sanche, L., and Wagner, J. R. 2005. Chemical basis of DNA sugar-phosphate cleavage by low-energy electrons. *J. Am. Chem. Soc.* 127 (47): 16592–16598.

Zheng, Y., Cloutier, P., Hunting, D. J., Wagner, J. R., and Sanche, L. 2006. Phosphodiester and N-glycosidic bond cleavage in DNA induced by 4–15 eV electrons. *J. Chem. Phys.* 124 (6): 064710/1–064710/9.

19 Physicochemical Mechanisms of Radiation-Induced DNA Damage

David Becker
Oakland University
Rochester, Michigan

Amitava Adhikary
Oakland University
Rochester, Michigan

Michael D. Sevilla
Oakland University
Rochester, Michigan

CONTENTS

19.1 INTRODUCTION: OVERVIEW OF RADIATION INTERACTION WITH BIOMOLECULES

In this chapter, recent findings are presented that elaborate the free radical and molecular mechanisms of radiation damage to DNA. This chapter gives an overview of this area emphasizing recent findings. A series of previous reviews and books should be consulted for a more comprehensive understanding of the field (Bernhard, 1981; Becker and Sevilla, 1993, 1998, 2008; Ward, 2000; Bernhard and Close, 2004; Sevilla and Becker, 2004; von Sonntag, 2006; Becker et al., 2007, 2010; Close, 2008; Kumar and Sevilla, 2008a). The findings covered in this chapter come from a variety of laboratories and lead to an improved understanding of the processes that lead to radiation-induced damage to DNA, beginning with initial energy deposition events, the subsequent formation of radical intermediates, and, ultimately, the formation of diamagnetic products and strand breaks. Although the physics of energy deposition of ionizing radiation and the eventual biologically significant endpoints have received substantial attention, it is the intervening radiation-induced free radical mechanisms and subsequent molecular product formation that have been less investigated and are the focus of this chapter. Direct-type effects of radiation on DNA are emphasized as these events account for about half the amount of the cellular DNA damage for low linear energy transfer (LET) radiation and become the dominant effect, as the LET of the radiation increases (Bernhard and Close, 2004; Sevilla and Becker, 2004; von Sonntag, 2006; Becker et al., 2007, 2010; Becker and Sevilla, 2008; Close, 2008). Direct-type effects include both the direct effect, which is the direct ionization of DNA, and the quasi-direct effects, which include those ionizations of the first hydration shell and its near environment that transfer holes and non solvated electrons to DNA (Becker and Sevilla, 1993, 1998, 2008; Becker et al., 2007, 2010; Close, 2008). ESR studies suggest that the waters of hydration that contribute to the quasi-direct effect extend out to 9 or 10 waters per nucleotide (La Vere et al., 1996; Becker and Sevilla, 1998; Ward, 2000). The indirect effect of radiation chiefly results from hydroxyl radicals produced from radiation damage to water molecules beyond 10 waters per nucleotide but still within a few nanometers of the DNA strand (O'Neill, 2001; von Sonntag, 2006). Electrons and a few hydrogen atoms formed on water ionization also contribute to the indirect effect (O'Neill, 2001; von Sonntag, 2006). The radiation-produced hydroxyl radicals are the chief damaging agent from the indirect effect (O'Neill, 2001; von Sonntag, 2006). However, each of the water radical species is largely scavenged by the high concentration of histone and other proteins in the DNA environment (Becker and Sevilla, 1993; Jones et al., 1993; O'Neill, 2001; von Sonntag, 2006).

19.1.1 RADIATION DEPOSITION

As high-energy radiation (photon, electron, or heavy-ion particle) passes through matter, energy loss occurs in the form of ionizations and excitations (Bethe equation) (Magee and Chatterjee, 1987; Chatterjee and Holley, 1993; Kraft and Kramer, 1993; Goodhead et al., 1994; Schuhmacher and Dangendorf, 2002; Friedland et al., 2003; Jakob et al., 2003; Nikjoo and Uehara, 2004; Swiderek, 2006), in approximate proportion to electron density at the specific site. For low LET radiations, about 90% of the energy loss is in the form of ionizations, with 10% in the form of excitations. Low LET electromagnetic radiations such as γ-radiation and x-rays ionize and excite molecules largely via the Compton effect, with the photoelectric effect contributing at lower energies (Magee and Chatterjee, 1987; Chatterjee and Holley, 1993; Kraft and Kramer, 1993; Goodhead et al., 1994; Schuhmacher and Dangendorf, 2002; Friedland et al., 2003; Jakob et al., 2003; Nikjoo and Uehara, 2004; Swiderek, 2006). Low LET radiations are relatively sparsely ionizing, creating local clusters of ions called spurs, blobs, and short tracks. Secondary-electron track ends create short tracks of higher LET damage (Magee and Chatterjee, 1987; Chatterjee and Holley, 1993). High LET charged particles, such as α-particles, protons, and ion-beams, produce a dense trail of ionizations that have a core of ionized and excited molecules and a surrounding penumbra of damage by secondary electrons (Magee and Chatterjee, 1987; Chatterjee and Holley, 1993; Bernhard and Close, 2004; von Sonntag, 2006; Becker et al., 2007, 2010; Becker and Sevilla, 2008; Close, 2008). In the penumbra, damage similar to that produced by γ-rays is seen, especially as the distance from the core increases (Magee and Chatterjee, 1987; Chatterjee and Holley, 1993; Bernhard and Close, 2004; von Sonntag, 2006; Becker et al., 2007, 2010; Becker and Sevilla, 2008; Close, 2008). The distinct spatial characteristics of the energy deposition, that is, track structure, from these various radiations (high LET vs. low LET) have significant effects on the resulting damage (Magee and Chatterjee, 1987; Chatterjee and Holley, 1993; Kraft and Kramer, 1993; Goodhead et al., 1994; O'Neill, 2001; Schuhmacher and Dangendorf, 2002; Friedland et al., 2003; Jakob et al., 2003; Nikjoo and Uehara, 2004; Swiderek, 2006; von Sonntag, 2006) (also see Chapters 13 and 14 for more details).

Although the initial ionizations by γ-radiation and heavy ions are directly related to the electron density, the subsequent ionizations by secondary electrons, as they lose energy, eventually come from only from the outer valence electrons (Becker et al., 2007). This is because the initial ionizations induce a cascade of secondary electrons that have lower and lower energies, so that eventually all ionizations are from the outer valence electrons (Becker et al., 2007).

19.1.2 ION-RADICAL FORMATION

Ionization creates sites of electron loss (a hole or cation radical) and an ejected electron. Most ejected electrons, after thermalization, will contain electron-gain sites (radical anion) (Bernhard, 1981; Becker and Sevilla, 1993, 1998, 2008; La Vere et al., 1996; Ward, 2000; Bernhard and Close, 2004; Sevilla and Becker, 2004; von Sonntag, 2006; Becker et al., 2007, 2010; Li and Sevilla, 2007; Close, 2008; Kumar and Sevilla, 2008a). The radical anions and cations formed account for over 80% of the radical species formed in DNA by low LET radiation (La Vere et al., 1996; von Sonntag, 2006; Becker et al., 2007, 2010; Li and Sevilla, 2007; Becker and Sevilla, 2008; Close, 2008; Kumar and Sevilla, 2008a). Excitations alone do not lead to a measurable radical formation for low LET radiation; however, recent evidence shows that excitations of ion-radicals induce biologically significant damaging events including strand breaks (Becker et al., 2007, 2010; Becker and Sevilla, 2008; Kumar and Sevilla, 2008a). Low energy electrons (LEE) of less than 20 eV have the potential to directly get added to DNA and undergo dissociative electron attachment (DEA) (Boudaïffa et al., 2000; Bernhard and Close, 2004; Sevilla and Becker, 2004; Bald et al., 2006, 2008; Simons, 2006;

Swiderek, 2006; von Sonntag, 2006; Becker et al., 2007, 2010; Becker and Sevilla, 2008; Close, 2008; Kumar and Sevilla, 2008a; Sanche, 2008). Although the initial radical formation from DEA for low LET radiation may be only a few percent of the total production, LEE can target sites such as the sugar–phosphate backbone and create biologically significant damage sites (Boudaïffa et al., 2000; Bernhard and Close, 2004; Bald et al., 2006, 2008; Simons, 2006; Swiderek, 2006; von Sonntag, 2006; Becker et al., 2007, 2010; Li and Sevilla, 2007; Becker and Sevilla, 2008; Close, 2008; Kumar and Sevilla, 2008a; Sanche, 2008).

19.1.2.1 Oxidative Pathway, Hole Transfer, and Reaction

Radiation induction of one-electron oxidations results in a number of major reaction pathways for the DNA hole that are now well established (Bernhard and Close, 2004; Cai and Sevilla, 2004; Sevilla and Becker, 2004; von Sonntag, 2006; Becker et al., 2007; Li and Sevilla, 2007; Becker and Sevilla, 2008; Close, 2008; Kumar and Sevilla, 2008a). Instantaneously on ionization, holes are formed on the bases and the sugar–phosphate backbone, in proportion to the electron density at that site (see Scheme 19.1). Much work has focused on the hole in DNA because it is thought to be the major damaging entity and this will be discussed in more detail in this chapter. In this section, we present a brief overview. For the holes on the DNA bases, a fast hole-transfer process localizes the hole preferentially to a nearby guanine base. Much work has established that stacks of Gs, that is, GG, GGG, etc, are ultimately the most stable hole sites after long range hole transfer and these G stacks form loci for DNA base damage (Arkin et al., 1997; Gasper and Schuster, 1997; Saito et al., 1998; Núñez et al., 1999). The guanine hole itself is in a prototropic equilibrium between the parent that is, the guanine cation radical ($G^{\bullet+}$), and its corresponding deprotonated species, $G(-H)^{\bullet}$; however, $G(-H)^{\bullet}$ is favored by several fold over $G^{\bullet+}$ in double-stranded DNA owing to intra-base pair proton transfer (Steenken, 1989, 1992, 1997; Adhikary et al., 2009; Kumar and Sevilla, 2009b). However, nucleophilic addition of water takes place only with $G^{\bullet+}$, forming the hydroxyl adduct at C-8, that is, $GOH^{\bullet}$ (Cullis et al., 1996; Shukla et al., 2004a, 2007). One-electron oxidation and reduction of $GOH^{\bullet}$ results in the molecular damage products 8-oxo-G and fapyG, respectively (Ravanat et al., 2003; von Sonntag, 2006). These molecular products, 8-oxo-G and fapyG, are the chief products formed from the hole in DNA.

For the sugar–phosphate backbone, hole transfer to a nearby DNA base competes with rapid deprotonation from sugar to water. It is estimated that about 2/3 of the holes on the sugar group transfer to the base (Swarts et al., 1992, 1996; Purkayastha et al., 2006a); the remaining holes undergo deprotonation resulting in various neutral sugar radicals (C1'', C3'', C5'') (Becker et al., 2007; Becker and Sevilla, 2008; Kumar and Sevilla, 2008a). C3'' and C5'' further react to form single strand breaks whereas C1'' leads to form unaltered base release sites (von Sonntag, 2006). Although the C4'' is produced by the hydroxyl radical via the indirect effect, there is no evidence at this time that direct type-effects of radiation produces this species in significant quantities (Razskazovskiy et al., 2003a; Adhikary et al., 2006a).

19.1.2.2 Reductive Pathway, One-Electron Adduct Formation

Direct ionization of DNA results in excess electrons that lose their kinetic energy, by energy transfer to the surroundings, into molecular electronic and vibrational energy levels and, at lower energies, into the phonon modes of the medium (Bernhard, 1981; Becker and Sevilla, 1993, 1998, 2008; Ward, 2000; Bernhard and Close, 2004; Sevilla and Becker, 2004; von Sonntag, 2006; Becker et al., 2007; Li and Sevilla, 2007; Close, 2008; Kumar and Sevilla, 2008a). After thermalization, these electrons add to DNA bases, preferentially at C and T (Bernhard, 1981; Becker and Sevilla, 1993, 1998, 2008; Ward, 2000; Bernhard and Close, 2004; Sevilla and Becker, 2004; von Sonntag, 2006; Becker et al., 2007; Close, 2008). Solvated electrons produced via the indirect effect also attach to the DNA bases and result in the C and T anion radicals (Scheme 19.2) (O'Neill, 2001; von Sonntag, 2006). Solvated and fully thermalized electron addition to DNA does not result in DNA strand breaks (O'Neill,

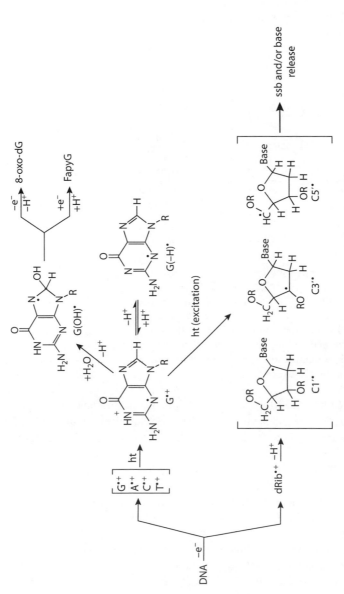

SCHEME 19.1 The one-electron loss pathway in DNA is depicted in this scheme. Sugar radicals are produced from direct ionization of the deoxyribose phosphate (dRib) backbone or excitation of the base cation radical. The symbols: ht, hole transfer; ssb, single strand break.

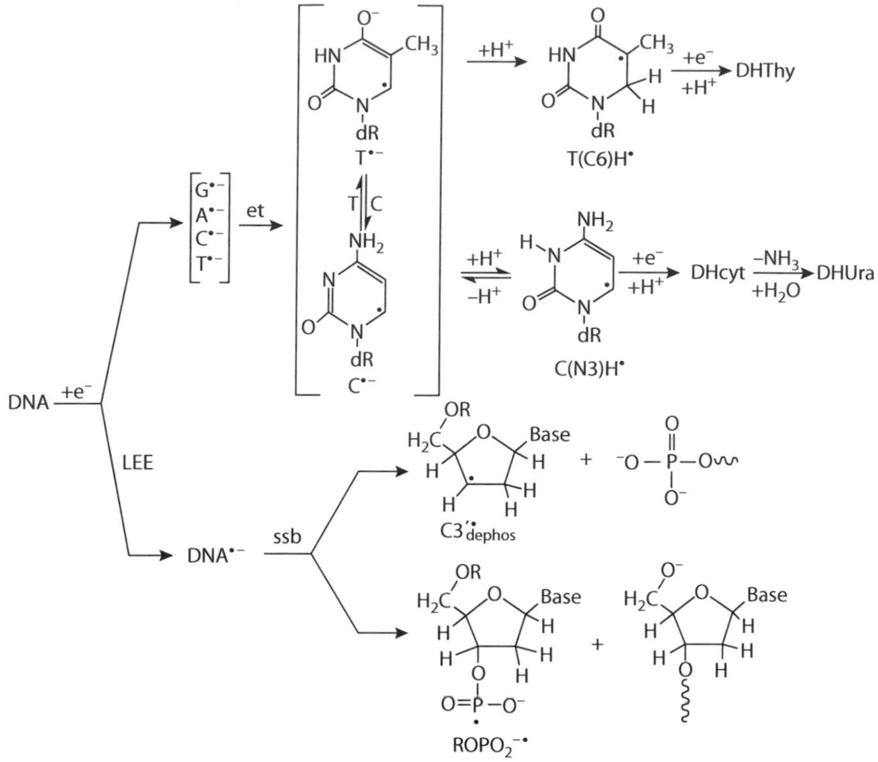

SCHEME 19.2 The reductive (one-electron addition) pathway in DNA is depicted in this scheme. Thermalized electrons add to the DNA bases and transfer to thymine and cytosine bases, where they undergo reversible and irreversible protonation reactions. Strand cleavage and sugar phosphate radicals are not produced by thermalized electrons, however, low energy electrons (LEE) with energies from 1 to 20 eV produced from direct ionization do cleave the DNA backbone, yielding radical species. The symbols: et, electron transfer; ssb, single strand break.

2001; von Sonntag, 2006). After formation, the C and T anion radicals undergo protonation reactions (Bernhard, 1981; Becker and Sevilla, 1993, 1998, 2008; Bernhard and Close, 2004; Sevilla and Becker, 2004; von Sonntag, 2006; Becker et al., 2007; Close, 2008). For C, this is a fast reversible protonation at the N3 position ($pK_a = 13$) and for T, a fast reversible protonation at O6 ($pK_a = 7$) (von Sonntag, 2006). Thus, the higher pK_a for the C anion radical initially favors the stabilization of the excess electron on C. Both C and T anion radicals or their reversibly protonated species also undergo irreversible protonations at the C6 position (and also C5 for C) to form neutral carbon-centered radical species (T(C6)H$^{\bullet}$ and C(C6)H$^{\bullet}$), which on one-electron reduction form dihydrothymine (DHThy) and dihydrocytosine (DHCyt) as molecular products (Scheme 19.2) (Bernhard, 1981; Becker and Sevilla, 1993, 1998; von Sonntag, 2006). DHCyt is not stable as a molecular product and slowly converts to dihydrouracil (DHUra) (von Sonntag, 2006). T(C6)H$^{\bullet}$ and C(C6)H$^{\bullet}$ have redox potentials that allow them to oxidize other radical species and thus prone to the formation of the dihydroproducts (von Sonntag, 2006). However, C(C5)H$^{\bullet}$ is reducing to most species and this results in its oxidation to form 5,6-dihydro 6-hydroxyl cytosine (von Sonntag, 2006).

Energetic electrons produced on or near DNA via direct-type effects can also add to DNA before thermalization and solvation. Such species in the energy range 0–20 eV are called low energy electrons (LEE) (Boudaïffa et al., 2000; Bald et al., 2006, 2008; Simons, 2006; Swiderek, 2006; Li and Sevilla, 2007; Kumar and Sevilla, 2008a; Sanche, 2008). LEE, possessing only 1–5 eV have been found to directly induce DNA single strand breaks whereas those with over 5 eV have been found to directly induce both single and double strand breaks in DNA (Boudaïffa et al., 2000; Bald et al.,

2006, 2008; Simons, 2006; Sanche, 2008). The mechanism of formation has been a focus of active interest (Boudaïffa et al., 2000; Bald et al., 2006, 2008; Simons, 2006; Swiderek, 2006; Li and Sevilla, 2007; Kumar and Sevilla, 2008a; Sanche, 2008). The two radicals likely to be formed by electron fragmentation of the DNA sugar–phosphate backbone, $C3'_{dephos}$ and $ROPO_2^{-•}$, have both been observed in ion-beam irradiated DNA samples and were attributed to LEE effects (Scheme 19.2) (Becker et al., 1996, 2003; Becker and Sevilla, 1998; Sevilla and Becker, 2004; Becker et al., 2007; Becker and Sevilla, 2008). ESR studies estimated that the production of the carbon-centered radical $C3'_{dephos}$ dominates by 20-fold over the phosphorous radical ($ROPO_2^{-•}$) (Becker et al., 1996, 2003). This result is in agreement with the cleavage products found in product analyses studies by Sanche and coworkers (Sanche, 2008). However, the fact that the formation of these strand-break radicals is found to be 10-fold lesser by low LET radiation than high LET radiation suggests that energy density is a critical factor in the formation of these radicals (Becker et al., 1996, 2003). It is possible that the electronic and vibrational excitations coupled with thermalized electrons or even electron adducts to DNA bases may induce strand cleavage (Li and Sevilla, 2007; Kumar and Sevilla, 2008a). The two sugar–phosphate backbone radicals, $C3'_{dephos}$ and $ROPO_2^{-•}$, clearly and directly result from strand fragmentations without further chemistry. The $C3'_{dephos}$ radical is likely to react further by beta phosphate cleavage to remove one additional base from its strand (Becker et al., 1996, 2003).

19.2 PATHWAYS IN DIRECT-TYPE EFFECTS

19.2.1 DNA Holes (Electron Loss) Trapped Radicals after Low LET Irradiation at Low Temperatures

19.2.1.1 Radicals Found after Irradiation at 77 K

This portion of the chapter focuses on those radiation-produced entities that can be detected by ESR spectroscopy, that is, free radicals. After irradiation of hydrated salmon sperm DNA at 77 K, the ESR spectrum obtained consists of a composite spectrum of mainly eight free radicals on the DNA bases: $G^{•+}$, $T^{•-}$, and $C(N3)H^•$; and on the sugar–phosphate backbone, $C1'^•$, $C3'^•$, $C5'^•$, $C3'_{dephos}$, and $RO_2PO^{•-}$ (Schemes 19.1 and 19.2) (Wang et al., 1994; Shukla et al., 2005; Becker et al., 2007; Becker and Sevilla, 2008). Of these, four are from the electron loss (oxidative) path, $G^{•+}$, $C1'^•$, $C3'^•$, and $C5'^•$ (Scheme 19.1), and the remaining four from the electron-gain (reductive) path, $T^{•-}$, $C(N3)H^•$, $C3'_{dephos}$, and $RO_2PO^{•-}$ (Scheme 19.2). It is possible, using proper benchmark spectra, to analyze the composite spectra found for the amount of each radical present (Table 19.1). The relative G-values are estimated to have an error limit of ±10%.

TABLE 19.1
Radicals Found in DNA γ-Irradiated at 77 K

Radical (Electron Loss)	$G^{a,b}$	Percent	Radical (Electron Gain)	$G^{a,b}$	Percent
$G^{•+}$	0.11	43	$dC(N3)H^•$	0.055[c]	22
$\Sigma C1'^•, C3'^•, C5'^•$	0.034	13	$T^{•-}$	0.055[c]	22
			$C3'_{dephos}$	—	ca. 0.5[d]
			$ROPO_2^{•-}$	—	0.02–0.04
Total percentage		56			45

[a] μmol/J, for DNA hydrated to $\Gamma = 14 \pm 2$ D_2O/nucleotide, based on total sample mass.

[b] Shukla et al. (2005).

[c] Assumes equal amounts of $C(N3)H^•$ and $T^{•-}$, based on Wang et al. (1994).

[d] Estimated from Becker et al., (2003).

The fact that only a few radicals from each path are found has profound implications regarding hole and electron transfer processes in irradiated DNA, as does the distribution of radicals between the bases and the deoxyribose sugars. In this section, we focus on the oxidative path and consider the location of ionization events in photon irradiated DNA.

For typical γ-radiation, most of the molecular damage from low LET radiation is effected through ionizations caused by the cascade of electrons of various energies that follow the initial ionization by γ-photons (Section 19.1.1). The total yield of ionizations in DNA, in the first 10^{-16}–10^{-15} s after irradiation, has been estimated to be ca. 1.2 μmol/J (Bernhard and Close, 2004). The total yield of free radicals trapped at 77 K for hydrated salmon sperm DNA is ca. 0.25 μmol/J (Shukla et al., 2005). This value is actually dependent on the level of hydration (Wang et al., 1993). For dry salmon sperm DNA (Γ = 2.5 D$_2$O/nucleotide), G = 0.13 μmol/J (Wang et al., 1993); for desiccated herring sperm DNA with residual H$_2$O, rather than D$_2$O, G = 0.41 μmol/J has been reported; and for desiccated calf thymus DNA, 0.29 μmol/J (Spalletta and Bernhard, 1992). Although there is some difference in these results, it is quite clear that most of the initially formed radicals recombine in early spurs, blobs, and short tracks.

Recombinations have the potential to result in excited states, but, unfortunately, little is known about the consequences of recombinations for DNA radiation damage.

Relatively high free radical yields at cryogenic temperatures have been reported for x-ray irradiated hydrated plasmid DNA. For pUC18 DNA, the total yield of trapped free radicals ranged from G = 0.302 μmol/J at Γ = 2.5 H$_2$O/nucleotide, to G = 0.718 μmol/J at Γ = 22.5 H$_2$O/nucleotide and showed a monotonic increase as hydration levels increased (Purkayastha et al., 2007). For pEC plasmids, G = 0.71 μmol/J was also reported (Purkayastha et al., 2005). These high yields of trapped free radicals are ascribed to the dense packing of supercoiled plasmid DNA.

19.2.1.2 The Role of Hole Transfer

If it is assumed that the aforementioned recombinations are random, roughly half of the surviving ionizations will be located on the DNA bases and roughly half on the sugar–phosphate backbone. However, as can be seen from Table 19.1, ca. 77% of the trapped electron-loss radicals at 77 K reside on the bases, and the remaining 23% originate with backbone radicals. It is quite evident that there is a substantial hole transfer from the backbone to the bases before the electron loss radicals are trapped. From the values in Table 19.1, and the assumption that about half the ionizations occur on the bases and half on the sugar–phosphate backbone (Section 19.1.2), at 77 K, there is a significant hole transfer from A$^{\bullet+}$, C$^{\bullet+}$, and T$^{\bullet+}$ to G to form G$^{\bullet+}$, and, in addition, significant hole transfer from the sugar–phosphate backbone to G to form G$^{\bullet+}$ (Scheme 19.3).

A very similar conclusion was reached by Swarts et al., who performed a more detailed analysis of oxidative path reactions at room temperature, in which the specific fates of holes on the different bases and sugars were determined by quantification of base release and base damage (Swarts et al., 1992, 1996). Here, it was found that ca. 13% of the holes on the deoxyribose sugars transferred to the bases and that 39% of the holes on adenine, cytosine, and thymine transferred to guanine before

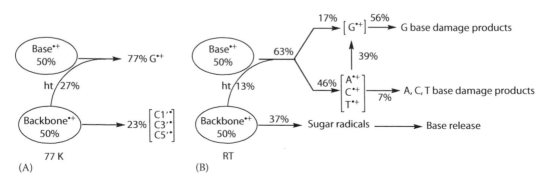

SCHEME 19.3 Scheme showing the hole transfer and reaction pathways at 77 K (A) and room temperature (B).

guanine base damage products (8-oxo-guanine, fapyguanine) were formed. A scant 7% of the cation radicals $A^{\cdot+}$, $C^{\cdot+}$, and $T^{\cdot+}$ were able to survive long enough to form their own base damage products (Swarts et al. (1996) should be consulted for details). In summary, the conclusions drawn from two very disparate experiments at two very different temperatures are remarkably similar. After irradiation, most of the holes formed (63%–75%) end up on the bases, predominantly on guanine, with the remainder consigned to the sugars.

19.2.1.3   Multiple Oxidations Centered on a Single dG$^{\cdot+}$

In a chemically oxidizing environment, $dG^{\cdot+}$ may undergo a series of reactions that might serve as a radiation protection mechanism (Shukla et al., 2004a, 2007). It is currently thought that the principal (but not sole) source of direct-type effect radiation-induced strand-break formation is the electron loss path (Becker and Sevilla, 1998, 2008; Ward, 2000; Bernhard and Close, 2004; Sevilla and Becker, 2004; von Sonntag, 2006; Li and Sevilla, 2007; Becker et al., 2007; Close, 2008; Kumar and Sevilla, 2008a). Any factors that might impede this path might be viewed as being protective, and, indeed, such factors do exist. In a study of the reactions of $dG^{\cdot+}$ using double stranded DNA, samples were annealed step wise from 77 to 258 K and the ESR spectra obtained were analyzed for possible intermediates (Shukla et al., 2004a, 2007). It was found that at least three one-electron oxidation steps occurred in transforming the original $G^{\cdot+}$ to possible diamagnetic molecular products (Shukla et al., 2004a, 2007). This is shown in Scheme 19.4. In Scheme 19.4A, the structures of the actual radical intermediates observed using ESR spectroscopy ($G^{\cdot+}$, $GOH^{\cdot}$, 8-oxo-$G^{\cdot+}$) are shown along with the reaction sequence that occurs. Scheme 19.4B summarizes the free radical reaction process found. Formation of $GOH^{\cdot}$ results in a damage site that is itself easily oxidized. The oxidizing species is additional $G^{\cdot+}$ (E_{red} = 1.29 V (Steenken, 1989, 1992, 1997)), which by hole transport can be collected at such reactive sites in irradiated DNA. Each step thereafter is also downhill and should be rapid, for example, 8-oxo-G is also easily oxidized (E_{red} of one-electron oxidized 8-oxo-dG at pH 7 at ambient temperature in aqueous solution is 0.74 V (Steenken et al., 2000)).

Recalling that the electron loss path can be thought of as a major damage pathway, Scheme 19.4A shows that four potential damages can be intercepted and channeled via three one-electron oxidation events into a single likely repairable base damage. This mechanism might be very effective in "simplifying" the complex damage in clusters of proximate radicals. Such clusters are thought to be particularly harmful since clustering of damage may impede normal cellular repair processes. Proof that this mechanism occurs in cells has not yet been found.

19.2.1.4   Influence of the Protonation State of the One-Electron Oxidized Guanine in DNA on the Mechanism of the Formation of 8-oxo-G Cation Radical

It has already been mentioned in Section 19.2.1.3 that ESR studies employing systematic thermal annealing of γ-irradiated hydrated DNA samples to higher temperatures (195–258 K) (Shukla et al.,

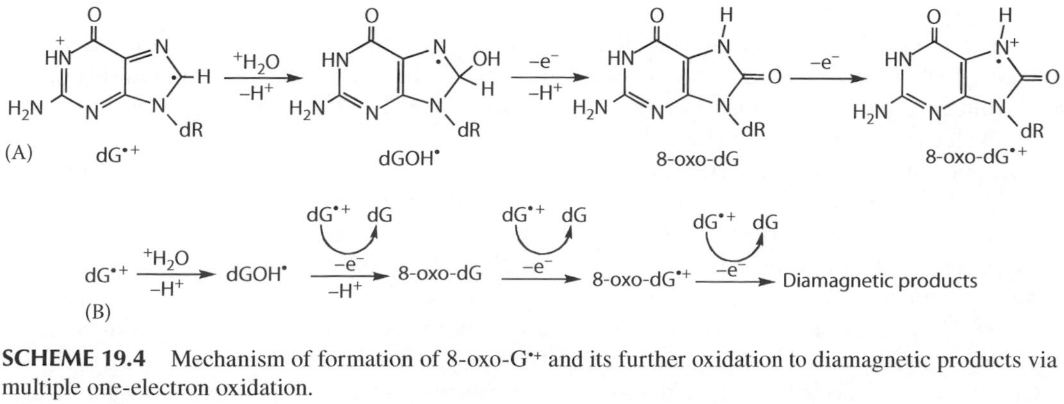

SCHEME 19.4   Mechanism of formation of 8-oxo-$G^{\cdot+}$ and its further oxidation to diamagnetic products via multiple one-electron oxidation.

2004a, 2007) have shown that 8-oxo-G cation radical is formed at $\geq 200\,K$. Cullis and coworkers have established that, for the formation of 8-oxo-G, one-electron oxidized guanine must remain in its cation radical state ($G^{\bullet+}$:C) (Cullis et al., 1996). Only in the protonated state, does the nucleophilic addition of water take place at the C-8 atom in the guanine ring—which is the first step in the formation of 8-oxo-G (see Schemes 19.1 and 19.4A).

Recent work employing double stranded DNA-oligomers incorporated with C-8-deuterated guanine at specific locations have clearly established that, at 77 K, the protonation state of the one-electron oxidized guanine in double stranded DNA is entirely the proton transferred radical (C($+H^+$):G($-H$)$^\bullet$) (discussed in Section 19.5.5). This means, the proton transferred state (C($+H^+$):G($-H$)$^\bullet$) is the thermodynamically stable state (Adhikary et al., 2009). From this work (Adhikary et al., 2009), and also from published work on electron and hole transfer by thermal annealing in DNA (Spalletta and Bernhard, 1992; Becker and Sevilla, 1998, 2008; Bernhard and Close, 2004; Cai and Sevilla, 2004; Sevilla and Becker, 2004; Shukla et al., 2004a, 2007; von Sonntag, 2006; Becker et al., 2007), it is concluded that barrier to the reverse intra-base pair proton transfer from C($+H^+$):G($-H$)$^\bullet$ to $G^{\bullet+}$:C is small. In agreement, a recent theoretical report suggests the barriers to forward and reverse proton transfer are 1.4 and 2.6 kcal/mol, respectively (Kumar and Sevilla, 2009a). Thus, at elevated temperatures ($\geq 200\,K$), the prototropic equilibrium between C($+H^+$):G($-H$)$^\bullet$ and $G^{\bullet+}$:C is established (see Section 19.5.5). At these temperatures, the presence of $G^{\bullet+}$:C allows for the nucleophilic addition of H_2O, thereby providing the first step as shown in Scheme 19.4 toward the formation of 8-oxo-G in DNA.

19.2.2 DNA Electron-Gain Trapped Radicals after Low LET Irradiation at 77 K

19.2.2.1 Base Radicals Found after Irradiation at 77 K

Table 19.1 and Scheme 19.2 show the principal electron-gain radicals found at 77 K after low LET radiation. As can be seen from Table 19.1, the preponderance of the trapped radicals is on the pyrimidine bases, in the form of $T^{\bullet-}$ and C(N3)H$^\bullet$. There is no evidence as yet that these two radicals are responsible for strand-break formation, and much evidence to the contrary. However, they are not biologically meaningless. On warming to room temperature, $T^{\bullet-}$ forms T(C6)H$^\bullet$, and ultimately, dihydrothymine (DHThy), and C(N3)H$^\bullet$ forms dihydrocytosine (DHCyt). Both these lesions may occur in significant amounts in anaerobic cells subjected to radiation (Dawidzik et al., 2004). Although both lesions are typically repairable through a base excision process (Cadet et al., 2000), when in a damage cluster, they are not easily repaired and can prevent repair of nearby strand breaks (Hada and Georgakilas, 2008).

19.2.2.2 Low Energy Electron (LEE): Dissociative Electron Attachment (DEA)

The two reductive path backbone radicals found, $RO_2PO^{\bullet-}$ and $C3'^\bullet_{dephos}$, are present in only small amounts with low LET radiation, and in larger relative concentrations with ion-beam irradiations (Section 19.4). However, they are significant because they are immediate strand-break radicals, that is, they are the direct result of a strand break. Early ESR investigations of small phosphate esters and diphosphate esters concluded that radicals of this type originate with DEA (Nelson and Symons, 1977; Nelson et al., 1993; Sanderud and Sagstuen, 1996). In DNA, the two radicals originate from divergent DEA paths as illustrated in Scheme 19.5. The LEE (low energy electron) is a thermal electron ($\leq$ca. 20 eV) (Sanche, 2008).

The percentages given for each path are derived from the approximate relative concentrations of $C3'^\bullet_{dephos}$ (95%) and $ROPO_2^{\bullet-}$ (5%) found in ion-beam irradiated samples (Becker et al., 2003).

A direct and striking proof that LEE cause induced DNA strand breaks was first reported in a classic paper by Sanche and coworkers (Boudaïffa et al., 2000). In this work, pGEM 3Zf(–) DNA isolated from *E scherichia coli* and hydrated to $\Gamma = 2.5\ H_2O$/nucleotide was subjected to a beam of LEE in the 3–20 eV range. Subionization (<ca. 7.5 eV) electrons were able to cause substantial strand breaks, including both single and double strand breaks. The yield information in this work (ca. 10×10^{-4} sb total/e^- for a 10 eV electron) suggested a small but significant probability for strand-break formation.

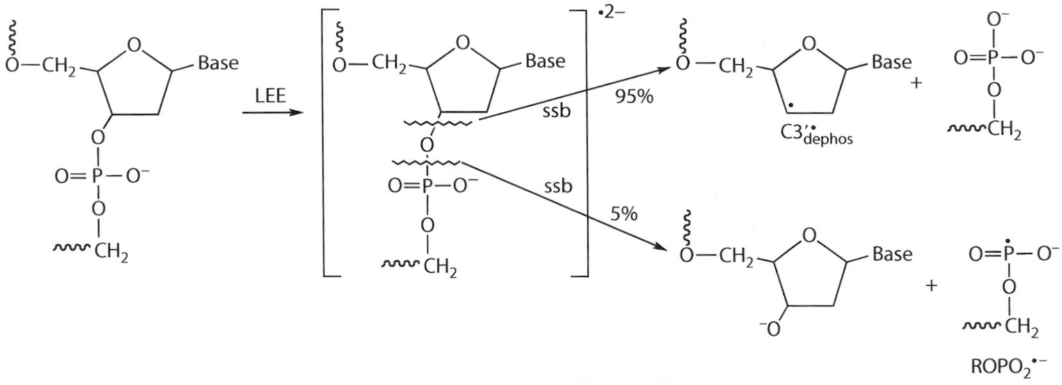

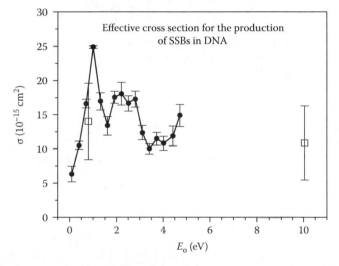

SCHEME 19.5 Low Energy Electrons (LEE) induce cleavage of DNA backbone resulting in specific strand-break radicals. (Adapted from Becker, D. et al., *Radiat. Res.*, 160, 174, 2003. With permission.)

FIGURE 19.1 Effective cross sections for formation of single strand breaks in pGEM 3Zf(–) plasmid DNA, determined using two different analytical methods. See reference Panajotovic for details. (From Panajotovic, R. et al., *Radiat. Res.*, 165, 452, 2006. With permission.)

Sanche and coworkers later refined their studies by showing that very low energy electrons, down to near 0 eV, could also cause single strand breaks with yields similar to those at far higher energies (Figure 19.1) (Burrow et al., 2006; Pan and Sanche, 2006; Sanche, 2008). For example, the yield of ssb at 1 eV was even greater than that at 10 eV (Burrow et al., 2006). Furthermore, in more recent efforts, a base sequence dependence on sites of the LEE induced cleavage in DNA oligomers have also been demonstrated (Sanche, 2008). Furthermore, other groups have recently elucidated mechanisms of LEE damage (Bald et al., 2006, 2008; Swiderek, 2006; Sanche, 2008). Overall, these results suggest that LEE addition to both the base and the phosphate group are pathways to DNA damage.

19.2.2.3 Theoretical Treatment of LEE-Induced Strand Breaks

A great deal of theoretical work has focused on the mechanisms for LEE-induced strand-break formation, and at least three mechanisms have been proposed in the literature (Simons, 2006; Kumar and Sevilla, 2007, 2008a,b, 2009b; Li and Sevilla, 2007). In the first mechanism, Simons suggested that a shape resonance results from an LEE initially attaching to a LUMO (a π^*-MO localized on the DNA base). Subsequently, during C3′–O3 bond elongation (upper path Scheme 19.5), the electron transfers to the phosphate group and a strand break results (Simons, 2006). Simons suggested that this

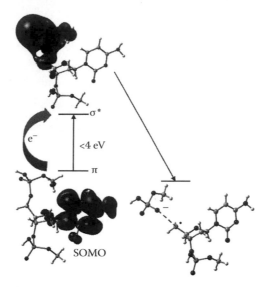

FIGURE 19.2 Low energy electrons interact with DNA to create various excited anion radical states. Here the excited anion radical of the DNA model system thymidine results in a σ* state that leads to strand-break formation. For details of the calculation, see Kumar and Sevilla (2009b). (From Kumar, A. and Sevilla, M.D., *Chem. Phys. Chem.*, 10, 1426, 2009b. With permission.)

process would likely be too slow by thermal activation alone to explain the effect of LEE on DNA, and recent theoretical work (Pan and Sanche, 2006; Sanche, 2008) has confirmed that a strand break from this mechanism is unlikely because a significant activation barrier is found, ca. 19 kcal/mol for the transient anion radical and nearly 30 kcal/mol for the adiabatic anion radical (Kumar and Sevilla, 2007, 2008a). Although LEE attachment to the LUMO confines the electron to the base, LEE addition to higher energy unoccupied molecular orbitals (UMOs) (<4 eV) form electronically excited anion radical states, which have energy in excess of the activation energy toward DEA. These higher energy UMOs can be localized on the phosphate group, the base, or both (Kumar and Sevilla, 2008b, 2009b). Thus, the second mechanism proposed occurs through a shape resonance formed by an LEE addition to an available UMO of higher energy than the LUMO. The excited anion radical state (likely a π* state) thus formed can couple with the dissociative σ* state on phosphate (see Figure 19.2) and lead directly to C–O bond cleavage (Burrow et al., 2008; Kumar and Sevilla, 2008b, 2009b). A third mechanism was briefly presented by Li and Sevilla, who suggested that a vibrational excitation of the C–O bond could itself capture an electron and lead to a strand break (see Li and Sevilla, 2007, p. 80). Along these lines, vibrational Feshbach resonances that couple dipole-bound and valence-anion states in LEE interactions with DNA bases have been implicated in bond fragmentation via σ* pathways (Burrow et al., 2006). In each of the mechanisms proposed above, the captured electron must couple to the σ* dissociative state involved in the C–O bond cleavage of the sugar–phosphate DNA backbone (shown in Scheme 19.5 upper pathway). There are mechanisms available at higher energies (>4 eV) that involve other pathways. For example, LEE induced radical anion excited states known as core excited states (e.g., LEE interaction with a molecule to form an UMO filled with two electrons and an inner shell hole (core hole)) are available at energies above 4 eV (Panajotovic et al., 2006; Sanche, 2008; Kumar and Sevilla, 2009b). Because of their higher energies and reactive nature, core excited states lead to molecular fragmentation reactions resulting in H⁻, O⁻, and OH⁻ in model systems and single and double strand breaks in DNA (Panajotovic et al., 2006; Sanche, 2008).

19.2.3 STRAND BREAKS FROM DIRECT-TYPE EFFECTS

Strand breaks from direct-type effects at low LET have been measured in a variety of DNA and DNA model compounds by a variety of methods (Table 19.2). Table 19.2 is not meant to be

TABLE 19.2
Strand Breaks in Hydrated DNA

DNA	Hydration	G(ssb, frank)	G(dsb, frank) fr	G(ssb)/G(dsb)	References
pEC plasmid	22	0.08	0.01	8	Purkayastha et al. (2005)
pUC18 plasmid	22	0.12	0.006	20	Purkayastha et al. (2005)
pUC18 plasmid	24 ± 7	0.071	0.006	12	Yokoya et al. (2002)
B-DNA	22	0.09[a]	—	—	Swarts et al. (1992, 1996)

[a] Base release.

comprehensive, but rather to give a general feel for these phenomena; the references cited should be consulted for details and further elaboration regarding strand-break phenomena under different hydration levels, and for the effect of enzyme sensitive and heat labile sites of strand breaks.

Strand breaks have also been measured for the following oligomers at a hydration of $\Gamma = 8\text{--}9$ H_2O/nucleotide: d(CGCG)$_2$, $G = 0.11$ μmol/J (Razskazovskiy et al., 2003b); d(CGCGCG)$_2$, $G = 0.085$ μmol/J (Sharma et al., 2007); d(CGCACG):d(CGTGCG), $G = 0.13$ μmol/J (Razskazovskiy et al., 2003b); and d(CACGCG):d(CGCGTG), $G = 0.055$ μmol/J (Razskazovskiy et al., 2003b).

19.2.3.1 Direct-Type Effect Strand Breaks and Free Radical Yields

It has been a long standing assumption of DNA radiation chemistry that prompt strand breaks from ionizing radiation originate with free radicals (von Sonntag, 2006). This has been one of the principal motivations underlying the need to measure and detect the free radicals that form after irradiation. A recent series of papers reports that, with direct-type effects, the yield of trapped radicals at 4.2 K is lower than the yield of strand breaks found when the same samples are warmed and strand breaks measured (Purkayastha et al., 2006b). This, of course, is not necessarily inconsistent with the paradigm that strand breaks largely originate with free radicals when low LET radiation is employed. It does suggest that some significant chemistry occurs, even at 4.2 K. From earlier investigations, it was determined that the yield of frank strand breaks, *as measured by loss of supercoiled pUC18 plasmid DNA*, exceeded the yield of free radicals at low hydration levels (Table 19.3); later work by the same group expanded this to all hydration levels studied (vide infra) (Purkayastha et al., 2005).

As can be seen, the yields of prompt strand breaks at room temperature (RT) are higher than the yield of trapped free radicals (4.2 K) for hydration levels of $\Gamma \leq 11.5$ H_2O/nucleotide. To explain these results, a double oxidation mechanism is proposed, in which for the initial oxidation, ionization at

TABLE 19.3
Free Radical vs. Strand-Break Yields, pUC18 Plasmid DNA

Γ[a]	G(Sugar Free Radicals)[b]	G(ssb)[c]	G(dsb)[c]
2.5	0.033	0.069	0.0052
7.5	0.040	0.058	0.0046
11.5	0.056	0.060	0.0044
15.0	0.069	0.063	0.0047
22.5	0.079	0.054	0.0039

[a] (H_2O/nucleotide).

[b] μmol/J, 4.2 K.

[c] μmol/J, room temperature frank strand breaks, from loss of supercoiled DNA.

a deoxyribose sugar followed by rapid deprotonation occurs to form a neutral sugar radical (S˙). Instead of being trapped and observed at 4.2 K, it is presumed that a nearby base cation radical oxidizes S˙, producing a carbocation, which then forms a strand break on warming. Thus, a double oxidation mechanism produces a strand break without producing any trapped radicals. The authors correctly note that at hydration levels above $\Gamma \geq 11.5$, some ˙OH will be formed. At 4.2 K, however, the ˙OH will not mobilize to from DNA radicals. On warming to measure strand breaks, a fraction of these will form strand breaks due to the indirect effect.

The same researchers, in a later report, noted a puzzling feature regarding base damage yields in pUC18 plasmids as a function of hydration (Purkayastha et al., 2007). The damage was detected using the base excision repair enzymes endonuclease III (Nth) and formamidopyrimidine-DNA glycolase (Fpg), which convert most base damage sites on purines (Nth) and pyrimidines (Fpg) to single strand breaks. In the hydration range $\Gamma = 2.5$ H$_2$O/nucleotide to $\Gamma = 22.5$ H$_2$O/nucleotide, the base damage yields (RT) decreased by a factor of 3.2 as the yield of trapped DNA radicals (4.2 K) increased by a factor of 2.4, contrary to the expected relationship. The hypothesis that this discrepancy is caused by an undercounting of single strand breaks using standard methods of analysis (loss of supercoiled DNA and Poisson statistics) was then tested by using base release, G(fbr) in µmol/J, to measure single strand-break formation and assuming that G(fbr) $\approx G$(sb) (Sharma et al., 2008). For this determination, the multiplicity of single strand breaks m(ssb), defined as the number of ssb formed per super coiled DNA lost, was determined. The formation of strand breaks from ˙OH at higher hydrations was subtracted so that direct-type effects only were considered, and measuring base release scores a single dsb as two ssb. At $\Gamma = 2.5$ H$_2$O/nucleotide, m(ssb) = 1.4 ± 0.2, and at $\Gamma = 22.5$ H$_2$O/nucleotide, m(ssb) = 2.8–3.0 ± 0.5, indicating that, for pUC18 under the conditions used, there is more than one ssb per supercoiled DNA lost. In addition, this work rectified the aforementioned conundrum regarding strand breaks and trapped radicals, since both now increased in going from $\Gamma = 2.5$ H$_2$O/nucleotide to $\Gamma = 22.5$ H$_2$O/nucleotide. Most importantly, these results now indicate that the excess of strand breaks over trapped free radicals, G(diff) = G(ssb) – G(fr), persists at high hydration levels, contrary to the results obtained if sb are scored by loss of supercoiled DNA. G(diff) was relatively constant in the range of 0.090–0.110 µmol/J for the two hydration levels used.

19.3 CHARGE TRANSFER PROCESSES IN DNA: SUMMARY OF THE MECHANISMS

In irradiated DNA, both DNA-cation and -anion radicals are produced and the transfer of these species through DNA is of significance regarding the ultimate location of DNA damage (Bernhard, 1981; Becker and Sevilla, 1993, 1998, 2008; Ward, 2000; Bernhard and Close, 2004; Sevilla and Becker, 2004; von Sonntag, 2006; Becker et al., 2007; Li and Sevilla, 2007; Close, 2008; Kumar and Sevilla, 2008a). A number of experimental studies based on UV–vis photoexcitation of dsDNA-oligomers with specific sequences having donors and acceptors at fixed distances, followed by damage analysis by gel electrophoresis at ambient temperatures have suggested that charge, that is, holes (or cation radicals) and electrons, can migrate through long distances (Arkin et al., 1997; Henderson et al., 1999; Giese, 2000, 2002; Lewis, 2005; Lewis et al., 2000, 2001; Schuster, 2000; Drummond et al., 2003; Wagenknecht, 2003). In addition, ultrafast (from pico- to femtoseconds) laser flash photolysis of DNA with donors and acceptors has further elucidated the time dependence of these charge transfer processes in DNA (Wan et al., 1999, 2000; Lewis, 2005; Lewis et al., 2000, 2001; Wagenknecht, 2003).

While the transfer of charge in DNA is a complex process, it can be simplified to two principal mechanisms: tunneling and hopping.

The tunneling mechanism involves the direct (i.e., single step) charge transfer between the donor and the acceptor due to the electronic coupling (π-way) between the donor and the acceptor via the intervening base sequence. This step is relatively temperature independent and is effective at even 4 K. The tunneling rate constant can be described by the relation, $k = k_0 e^{-\beta D}$, which gives the fall off in tunneling rate constant with distance decay constant (β) and distance (D) between the donor and the acceptor. β is dependent upon the nature of the bridge (i.e., the intervening sequence) and the electronic coupling between the donor and acceptor. For tunneling of hole and excess electron transfer within DNA, β values ranging from 0.6–1.1 Å^{-1} have been observed in low temperature work, which isolate this process and in fast kinetic studies for short distances (Lewis et al., 2000, 2001; Lewis, 2005) where this process dominates. Note that, higher β values (ca. 1–1.5 Å^{-1}) are observed for charge transfer processes in proteins (Nocek et al., 1996).

While tunneling is always in effect, it is rapid only for short distances. For long distance action, the tunneling-hopping mechanism is needed (Jortner et al., 1998; Grozema and Siebbeles, 2007). In the "hopping" mechanism, charge transfer along DNA has been proposed to occur in discrete thermally activated steps from the donor through intervening way stops to the acceptor (Bernhard and Close, 2004; Cai and Sevilla, 2004; Sevilla and Becker, 2004; von Sonntag, 2006; Becker et al., 2007; Becker and Sevilla, 2008; Close, 2008; Kumar and Sevilla, 2008a). This process is not effective in DNA systems until approximately at 200 K. This thermal activation is necessary to overcome the stabilization by polarization of the medium around the hole and excess electron adduct as well as proton transfer processes within DNA (Cai and Sevilla, 2004). These create a potential barrier to transfer via hopping. Indeed, polarization of the media alone by the trapped charges can be a limiting factor in the hopping process (Cai and Sevilla, 2004). Another related mechanism for transfer involving polarization is "polaron transport" in DNA (Henderson et al., 1999). If the polaron is only a single relaxed ion radical with media polarization around it, then charge transfer via this polaron is the usual hopping case. However, if the hole or electron is delocalized within a stack of bases (such as A-stacks), then this larger entity can migrate. Usually, the size of the polaron in DNA is limited to 3–4 base pairs because the polarization stabilization is strongest for smaller ionic species and drops off as the ion radius increases (Conwell and Rakhmanova, 2000; Conwell, 2005). Therefore, delocalization decreases the stabilization of the polarization. Theoretical studies have shown that solvent polarization and nuclear reorganization prevent extensive base-to-base hole delocalization for bases other than adenine (Adhikary et al., 2008a), and as a result, such polaron formation, in the case of holes, occurs only in A-stacks (Adhikary et al., 2008a). For hopping or polaron-assisted hopping, the tunneling equation mentioned above is not strictly applicable, as a very weak distance dependence is expected and the rate of charge transfer process should not necessarily decay exponentially with distance. For such cases, an apparent low value of β (e.g., 0–0.2 Å^{-1}) has been reported from various laboratories (Arkin et al., 1997; Fink and Schönenberger, 1999; Henderson et al., 1999; Ly et al., 1999; Porath et al., 2001; Drummond et al., 2003). Much effort has shown that base sequence is especially significant for such long range tunneling (Giese, 2002). As an example, a G placed in between every several base pairs allows for long distance hole "hopping" from G to G; tracks of A show little or no distance dependence (Giese et al., 2001; Giese, 2002; Shao et al., 2004; Kawai and Majima, 2005; Joy et al., 2006; Augustyn et al., 2007; Lewis et al., 2008). Intra-base pair proton transfer processes have been suggested to play a role of "gating switch" to both hole and electron transfer through DNA (Steenken, 1989, 1992, 1997; Adhikary et al., 2009; Kumar and Sevilla, 2009a). Ultimately, the transfer distances of both holes and electrons are limited by irreversible protonation of thymine and cytosine anion radicals and reaction of guanine cation radicals with water described in Schemes 19.1 and 19.2.

In cellular systems, DNA is found in higher levels of organization in chromatin, in the form of nucleosomes, solenoids, and fibers, which place DNA double strands in close proximity and excludes much of the bulk water from the vicinity of DNA. The close packaging increases the possibility of inter-duplex charge transfer in addition to charge transfer within the duplex. Low temperature ESR work using crystalline DNA (Debije and Bernhard, 2000), or hydrated DNA pellets and DNA-ice samples (Pezeshk et al., 1996; Cai and Sevilla, 2000, 2004) have provided evidence for such inter-duplex transfer. Cai and Sevilla have proposed a three-dimensional model that accounts for both the intra- and inter-duplex charge transfer processes within nucleohistone (Cai et al., 2001; Cai and Sevilla, 2004). The works of Barton and coworkers employing isolated nuclei (Núñez et al., 2001) and nucleosome core particles (Núñez et al., 2002) indicate formation of oxidative damage products via charge transfer.

In the last 10 years, a number of reviews as well as books have described both the hole and the excess electron transfer processes in DNA (Schuster, 2004; Wagenknecht, 2005).

19.4 LINEAR ENERGY TRANSFER EFFECTS

19.4.1 INTRODUCTION

It is well established that higher LET radiations usually produce more severe and less repairable biological damage than do lower LET radiations (Terato et al., 2008). For α particles, it has been suggested that up to 70% of the lesions in V79-4 Chinese Hamster cells are caused by direct-type effects (de Lara et al., 1995). It has also been noted that the track structure of radiation is important in determining the resulting radiation damage to DNA (Hill, 1999). With the development of the International Space Station, the contemplation of manned space missions to Mars, and the increasing use of ion beams in radiation therapy, a full understanding of the physicochemical track structure, that is, the physical and chemical events that occur in the track of radiations of various quality, and the spatial organization of these events, is quite relevant.

19.4.2 FREE RADICAL YIELDS AT 77 K

Ion-beam irradiation of hydrated DNA at low temperature results in substantial differences in the free radical yields and dose–response behavior for DNA free radicals relative to that found using low LET radiation. These effects depend not only on the radiation quality, but also on the hydration level and the identity of the irradiating ion. Typically, however, the overall yield of radicals is lower for high LET radiation than for low LET radiation. Table 19.4 shows this for dry DNA using various quality radiations, but a similar phenomenon occurs for hydrated DNA.

Significantly, the relative and/or absolute amount of neutral sugar radicals (C1$'^{\bullet}$, C2$'^{\bullet}$, C3$'^{\bullet}$, C3$'^{\bullet}_{dephos}$) and phosphorus-centered radicals are also much higher in ion-beam irradiated DNA relative to γ-irradiated samples (Becker et al., 1996, 2003). As an example, Table 19.5 shows the yields of radicals for DNA hydrated to $\Gamma = 12 \pm 2$ D_2O/nucleotide after argon ion irradiation with an ion energy of 75 MeV/u and initial LET of 600 keV/μm.

For this particular experiment, the total yield of base radicals [G$^{\bullet+}$, C(N3)H$^{\bullet}$, T$^-$] equals about 0.10 μmol/J. For a γ-irradiated sample under identical conditions, the total yield of radicals is

---

TABLE 19.4
Free Radical Yields in Dry DNA

Irradiation	LET	G (μmol/J)	References
Gamma	—	0.133	Wang et al. (1993)
X-ray	—	0.122	Weiland and Hüttermann (1999)
	—	0.29	Spalletta and Bernhard (1992)
	—	0.41	Spalletta and Bernhard (1992)
Ti	1,521	0.064	Weiland and Hüttermann (1999)
Zn	3,861	0.035	Weiland and Hüttermann (1999)
Au	11,650	0.101	Weiland and Hüttermann (1999)
Bi	12,440	0.042	Weiland and Hüttermann (1999)

TABLE 19.5

Approximate Radical Yields Found in DNA ($\Gamma = 12 \pm 2$ D_2O/Nucleotide) Ar Ion-Irradiated at 77 K

Radical (Electron Loss)	G^a	Percent	Radical (Electron Gain)	G^a	Percent
$G^{\bullet+}$	0.043	32	$C(N3)H^{\bullet}$	0.032	24
$A^{\bullet+}$	—	<5	$T^{\bullet-}$	0.029	21
Σ C1'$^{\bullet}$, C3'$^{\bullet}$, C5'$^{\bullet}$	0.032	20–22	$C3'^{\bullet}{}_{dephos}$	—	2–4
			$RO_2PO^{\bullet-}$	—	0.1–0.2
Total percentage		ca. 50			ca. 50

Source: Becker, D. et al., *Radiat. Res.*, 160, 174, 2003.

[a] μmol/J, based on total sample mass, $\pm10\%$.

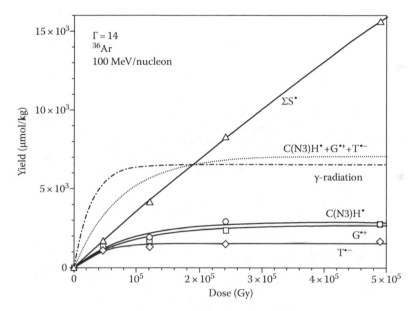

FIGURE 19.3 Dose-response of individual base radicals and $\Sigma S^{\bullet} = C1'^{\bullet}+C3'^{\bullet}+C5'^{\bullet}$. The high dose yield of the sum of the base radicals is very close to the high dose yield for all radicals for a similar γ-irradiated sample. This is for a hydrated ($\Gamma = 14$ D_2O/nucleotide) salmon sperm DNA sample irradiated with an Ar-36 100 MeV/u beam. The LET is 400 keV/μ. (From Becker, D. et al., *Radiat. Res.*, 160, 174, 2003. With permission.)

ca. 0.21 μmol/J and that of base radicals is ca. 90% of this, or 0.19 μmol/J. It was also noted for ion-beam irradiated samples that the shape of the base radical dose response curve was similar to that found for γ-irradiated samples, but with lower initial yields (G in μmol/J) (Figure 19.3).

In addition, the plateau yield of base radicals is, within experimental limits of error, the same for argon ion-beam irradiated samples as for γ-irradiated samples. These factors are all consistent with a model that invokes the partition of energy between a core and a penumbra (Section 19.1.1) of the ion track (Figure 19.4) as the principal underlying reason for these experimental observations. The penumbra is formed from secondary electrons emitted from the core, very much like the secondary electrons formed through γ-irradiation, and, as in γ-irradiated samples, the energy of the penumbra is laid down in spurs, blobs, and short tracks. Thus, the processes in the penumbra lead to radical formation, very similar to those in γ-irradiated samples, with formation, predominantly, of base radicals.

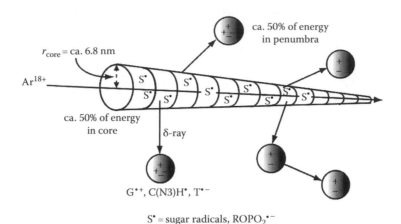

FIGURE 19.4 Schematic depiction of chemical track structure from Ar ion beam irradiation of hydrated DNA.

In the core, energy is deposited by direct interactions with the ion (Section 19.1.1), and the density of energy deposition is much higher than in the penumbra. Anionic and cationic base radicals form very close to each other and columbic forces drive substantial recombination of these radicals before they are trapped at 77 K. Neutral sugar radicals, however, are able to survive core recombinations more easily and as a consequence, most of the trapped neutral sugar radicals are produced in the core.

With the presumption that most of the base radicals are produced in the penumbra, it is possible to use their yields to determine the partition of energy between the core and the penumbra. For example, in the sample illustrated in Figure 19.3 and Table 19.5, one can conclude that 53% of the ion energy (0.10/0.19) is deposited in the penumbra, and the remainder in the core. For a series of experiments using argon ion irradiation, it was concluded that for five samples with hydration $9 \geq \Gamma \geq 16$ D_2O/nucleotide and LET ≤ 650 keV/μm, the percentage energy deposited in the penumbra ranged from 49% to 56%, or roughly 50%.

The large absolute Gs for neutral sugar radical production with ion-beams are not simply explained by the partition of energy. A further hypothesis regarding the effects of track structure on yields is that excited state reactions in the core are partially responsible for the relatively high yield of sugar radicals (Becker et al., 2003). The high density of energy deposition in the core could clearly lead to excited states. For those excited states that are created on ion radicals, excited state chemical reaction channels likely exist. A striking confirmation of this hypothesis was obtained in a series of papers in which the direct excitation of $G^{\bullet+}$ in DNA and in model compounds and of $A^{\bullet+}$ in model compounds leads to sugar radicals (Becker et al., 2007; Becker and Sevilla, 2008; Kumar and Sevilla, 2008a). This work is described in detail in Section 19.5.

Further LET effects with regard to the irradiation of hydrated DNA are observed in the yields of the prompt strand-break radicals $ROPO_2^{\bullet-}$ and $C3'_{dephos}$. The yield of these radicals with argon and oxygen ion-beams is ca. 10 times higher than that found with γ-irradiation (Tables 19.1 and 19.5). The pathway for formation of these radicals originates with LEE, so a question arises regarding what factor is causing a approximate 10-fold increase in the number of such radicals. The likely answer is that an electronic and/or vibrational excited state is involved in strand-break formation. Thus, LEE capture induces excited anionic states that are one source of $ROPO_2^{\bullet-}$ and $C3'_{dephos}$. A second source can arise from excited states of existing anionic radicals. For example, the higher density of energy deposition in the core facilitates formation of excited states in proximity to existing radicals. DFT calculations also indicate that LEE capture by DNA can be modeled by calculations of excited states of the anion radicals. Thus excited states are very much involved in LEE capture by DNA and clearly lead to subsequent strand breaks (Kumar and Sevilla, 2007, 2008b, 2009b).

Because a strand break that is part of a multiply damaged site (damage cluster) is likely to be less repairable than an isolated strand break (vide infra), the spatial clustering of radicals is highly

TABLE 19.6
Multiply Damaged Sites in Ar-Irradiated DNA

Track Parameter	Value for Doses 1.7–50 kGy
Radicals in cluster	17.7 ± 0.7
Cluster radius (spherical cluster assumed)	6.8 nm
Radical concentration in cluster	$18.8 \, \mu mol/cm^3$
Radical concentration for whole sample (dose dependent)	$0.20 \, \mu mol/cm^3$

relevant to understanding damage pathways. A pulsed electron double resonance (PELDOR) study of argon ion irradiated hydrated DNA found that clusters of radicals with a local radical concentration higher than the overall radical concentration do exist (Bowman et al., 2005). In this study, the observing magnetic field was set outside the range of the base radical ESR spectra, but on the neutral sugar radical spectrum, and a strong pulse was applied. The pulse causes transitions in most of the radicals present. The effect of the pulse on the neutral sugar radicals is then observed. It was concluded that clusters of radicals do exist, as shown in Table 19.6.

What is most remarkable is that in the 1.7 kGy dose sample, although the average radical concentration is $0.20 \, \mu mol/cm^3$, the cluster radical concentration is $18.8 \, \mu mol/cm^3$. This is confirmation of the existence of significant multiply damaged sites in Ar-irradiated DNA. Since the observing magnetic field detects effects on the neutral sugar radicals, and the experimental technique used detects mainly a local concentration of radicals, the spherical cluster radius observed is actually a measure of the core radius (Figure 19.4). In addition, because the core is largely populated with neutral sugar radicals, it is likely that more than one of the radicals in a cluster will lead to a strand break. Since two strand breaks within 10–20 base pairs (3.4–6.8 nm) on opposite strands is generally regarded to constitute a double strand break, if two breaks on opposite strands occur within a cluster, a double strand break will almost certainly result.

The track parameters reported, that is, the core radius, the radical concentration in the core, and the energy partition between the core and penumbra all agreed well with track-structure calculations (Magee and Chatterjee, 1987; Chatterjee and Holley, 1993; Chatterjee, 2006).

19.4.3 STRAND BREAKS AND DAMAGE CLUSTERS FROM HEAVY-ION DIRECT EFFECTS

Very little work has been published (also see Chapter 21 for further information) regarding strand breaks from ion-beam irradiation using experimental protocols that assure direct-type effects are being measured. Using hydrated ($\Gamma = 35$ H_2O/nucleotide) pUC18 prepared in a tris buffer, Urushibara et al. (2008) determined the yield of ssb, dsb, and damage clusters in pUC18 DNA using α-radiations of varying LET at 5.6°C. Using α-radiation, it was found that the yield of frank ssb decreased as LET increased (Table 19.7) and that heat treatment did not change the yield of ssb. Contrary to the behavior of ssb, the yield of dsb increased to a maximum at an LET of 141 keV/μm, and then dropped at 148 keV/μm.

Determination of the yield of newly formed ssb after enzymatic treatment with Nth and Fpg to reveal isolated base damage indicated the presence of such significant damage at the lower LETs used, but the yield of these sites decreased markedly as the LET increased. In addition, measurement of newly formed dsb after treatment with Nth and Fpg, to reveal damage clusters, indicated the presence of significant yields of such clusters, which also appeared to decrease as LET increased.

This latter decrease was interpreted as possibly being an artifact caused by the failure of the BER glycolases to function properly, rather than as an actual decrease in base lesions. However, as indicated earlier (vide infra), there actually is a substantial decrease in initial trapped base radical yields as LET increases, caused by the partition of energy in a core and a penumbra, and the effects

TABLE 19.7
Yields of Strand Breaks in α-Irradiated pUC18 DNA at Varying LET[a]

LET (keV/μm)	SSB[a]	DSB[a]	G(SSB)[b]
ca. 0.4 (γ)	7.2 ± 0.5	0.72 ± 0.07	0.72 ± 0.05
19 (α)	6.9 ± 0.2	0.37 ± 0.16	0.69 ± 0.10
63 (α)	7.2 ± 0.2	0.75 ± 0.12	0.72 ± 0.02
95 (α)	6.7 ± 0.3	1.09 ± 0.21	0.67 ± 0.03
121[c]	4.8 ± 0.5	0.44 ± 0.16	0.48 ± 0.05
141 (Pu-α)	6.0 ± 0.9	1.7 ± 1.0	0.60 ± 0.09
148	3.8 ± 0.3	0.97 ± 0.07	0.38 ± 0.03

Source: Urushibara, A. et al., *Int. J. Radiat. Biol.*, 84, 23, 2008.

[a] Break/Gy/Da × 10^{11}.

[b] μmol/J.

[c] The datum at 121 keV/μm was thought to have a potential systematic error.

thereof. It may be the case that the loss of enzyme sensitive sites is real and due to track-structure effects. Earlier experimental evidence from the same group using γ-irradiation showed that the yields of ssb and dsb do not increase for $\Gamma \geq 15$, suggesting that the indirect effect is suppressed in the samples (Yokoya et al., 2002).

In earlier experiments by the same team, α-particle irradiated DNA was investigated at three hydration levels, from dry DNA to $\Gamma = 35$ H_2O/nucleotide (Yokoya et al., 2003). The dsb yield increased from the first to the second hydration level, and then appeared to plateau for fully hydrated DNA. At all three hydration levels used, the yield of dsb was two to three times higher than that found for γ-irradiated samples, indicating clusters of damage originated in the ion track core. In a manner consistent with the work cited above, fewer Nth and Fpg enzyme sensitive sites were detected than one might expect. As discussed earlier, it may be the case that the lack of enzyme sensitive sites is actually due to fewer base lesions being formed rather than the inability of the BER enzymes to properly catalyze strand cleavage.

19.4.4 DAMAGE BY SOFT AND ULTRASOFT X-RAYS

Extensive literature exists regarding the damage caused by soft and ultrasoft x-rays. For these radiations, the photoelectric effect induces inner shell ionizations of light atoms, and damage is caused both by the original ionization of the target atom as well as by the resulting auger and secondary electrons as well. These beams have LETs higher than that of γ-rays or hard x-rays. A discussion of these effects is outside the scope of this chapter, but recent reviews (Yokoya et al., 2006) as well Chapter 21 in this book can be consulted.

19.5 EXCITED STATES OF ION-RADICALS AS DNA DAMAGE PRECURSORS

19.5.1 SUGAR–PHOSPHATE DAMAGE VIA EXCITED STATES OF DNA BASE ION-RADICALS

Early ESR studies using γ-irradiated hydrated salmon sperm DNA (Wang et al., 1994) showed that the yield of neutral sugar radicals increased with radiation dose, in apparent direct correlation with the loss of existing one-electron oxidized guanine radicals (Wang et al., 1994). This observation suggested that, in these γ-irradiated hydrated DNA samples, a portion of the

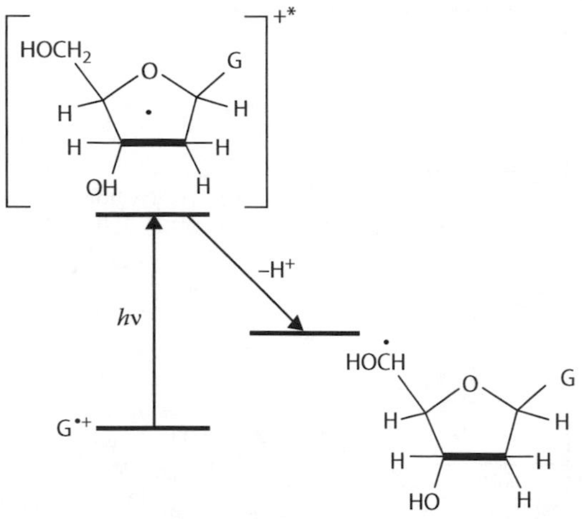

SCHEME 19.6 C5′⁺ formation via photoexcitation of G⁺⁺. (Reprinted from Khanduri, D. et al., *J. Phys. Chem. B*, 112, 2168, 2008. With permission.)

sugar radicals was formed from secondary radiation effects as well as by the reactions of the primary radical species. This observation led to the hypothesis that radiation-induced excitation of one-electron oxidized guanine in DNA might induce the formation of sugar radicals (Wang et al., 1994).

This hypothesis was tested and verified in a number of subsequent experiments. Shukla et al., first reported the formation of a neutral sugar radical, C1′⁺, via photoexcitation of pre-existing guanine cation radicals (G⁺⁺) in hydrated highly polymerized double stranded (ds) DNA at 77 K (Shukla et al., 2004b). The proposed mechanism for sugar radical formation suggested that UV light (<300 nm) was involved. Later work showed that light >300 nm was most effective in the photoconversion to sugar radicals (Adhikary et al., 2005). These authors also showed that photoexcitation of G⁺⁺ in 3′-dGMP, and 5′-dGMP at 77 K produced carbon-centered neutral sugar radicals albeit to small extents (ca. 15%–30%) (Adhikary et al., 2005). The mechanism for sugar radical formation in this system is illustrated in Scheme 19.6 (Khanduri et al., 2008).

As shown in Scheme 19.6, the proposed mechanism of sugar radical formation via photoexcitation of G⁺⁺ involves formation of a transient excited state, which delocalizes a substantial extent of spin and charge from the base moiety of G⁺⁺ on to its sugar moiety, S⁺⁺. Subsequent deprotonation of the transient S⁺⁺ at specific sites of hole localization in the sugar moiety (e.g., at C5′-atom as shown in Scheme 19.6) leads to the formation of the neutral carbon-centered sugar radical at that site.

To test the generality of this mechanism of sugar radical formation, the following systems have been investigated at low temperatures: (1) DNA-nucleosides and -tides (Adhikary et al., 2005, 2006b); (2) DNA-oligomers (Adhikary et al., 2006a, 2007); (3) highly polymerized double stranded salmon sperm DNA (Adhikary et al., 2005, 2007); (4) RNA-nucleosides and -tides (Adhikary et al., 2008b; Khanduri et al., 2008); and (5) RNA-oligomers (Khanduri et al., 2008). In each case, sugar radical formation was found on excitation of the purine cation radicals. These results are described in more detail below.

19.5.2 Formation of One-Electron Oxidized Purines

In Figure 19.5, ESR spectra are presented showing formation of guanine cation radicals in dGuo (Adhikary et al., 2005) via one-electron oxidation by $Cl_2^{\cdot-}$ after γ-irradiation of 7.5 M LiCl glassy solutions containing the nucleoside and persulfate as an electron scavenger.

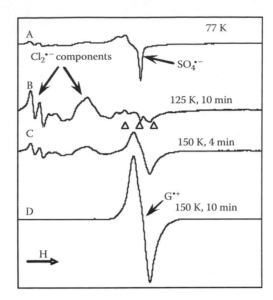

FIGURE 19.5 (A) ESR spectrum showing formation of $SO_4^{\bullet-}$ and $Cl_2^{\bullet-}$ in a γ-irradiated (2.5 kGy) sample of 5′,5′-D,D-dGuo in a 7 M $LiCl/D_2O$ glass in the presence of K2S2O8. (B) After annealing to 125 K for ca. 10 min. (C) After further annealing at 150 K for 4 min. (D) After annealing for another 6 min (i.e., total 10 min) at 150 K. Each spectrum was recorded at 77 K. (Reprinted from Adhikary, A. et al., *Nucleic Acids Res.*, 33, 5553, 2005. With permission.)

The formation of $G^{\bullet+}$ shown in Figure 19.5 results due to thermal annealing of the glasses from 77 to 155 K, where one-electron oxidation of dGuo (Adhikary et al., 2005) or dAdo (Adhikary et al., 2008a) by $Cl_2^{\bullet-}$ occurs on softening of the glass. This chemistry is shown in Scheme 19.7.

This technique has been employed in a series of investigations on DNA and RNA nucleosides, nucleotides, and oligomers with success (Shukla et al., 2004a,b; Adhikary et al., 2005, 2006a,b,c, 2007, 2008a,b 2009; Khanduri et al., 2008).

Using this technique involving one-electron oxidation by $Cl_2^{\bullet-}$, the isolation of $G^{\bullet+}$ and $A^{\bullet+}$ is unexpected as the pK_a values of these cation radicals at ambient temperature in aqueous solutions are ca. 3.9 and ≤1, respectively (Steenken, 1989, 1992, 1997). However, the pK_a value of $G^{\bullet+}$ (for dGuo) in this system (glassy homogeneous solution at low temperature) is found to be between 5 and 6 (Adhikary et al., 2006c), and the pH of the matrix (i.e., 7.5 M LiCl solution in D_2O or in H_2O) is found to be ca. 5 at ambient temperature. Thus, roughly equal amounts of $G^{\bullet+}$ and its N–1 deprotonated species $G(-H)^{\bullet}$ are found (Section 19.5.4).

The pK_a value of $A^{\bullet+}$ (for dAdo or Ado) has been determined in the glassy homogeneous D_2O solution at low temperature as ca. 8 (Adhikary et al., 2008a) and not ≤1 as found at room temperature (Steenken, 1989, 1992, 1997). Owing to the low temperature in glassy solutions, aggregation of dAdo

$$h^+ + 2Cl^- \rightarrow Cl_2^{\bullet-}$$

$$e^- + S_2O_8^{2-} \rightarrow SO_4^{2-} + SO_4^{\bullet-}$$

$$SO_4^{\bullet-} + 2Cl^- \rightarrow SO_4^{2-} + Cl_2^{\bullet-}$$

$$Cl_2^{\bullet-} + G(\text{or A}) \rightarrow G^{\bullet+}(\text{or A}^{\bullet+}) + 2Cl^-$$

SCHEME 19.7 Formation of $G^{\bullet+}$ or $A^{\bullet+}$ via one-electron oxidation by $Cl_2^{\bullet-}$. h+ represents a matrix hole formed by gamma irradiation. (Adapted from Adhikary, A. et al., *Nucleic Acids Res.*, 33, 5553, 2005.)

takes place, and as a result, the adenine cation radical is thought to be stabilized by charge–resonance interaction within the aggregated adenine stacks (Adhikary et al., 2008a). In agreement, theoretical calculations predict that the adenine dimer cation radical ($A_2^{\bullet+}$) is stabilized by ca. 12–16 kcal/mol relative to its monomers $A^{\bullet+}$ and A (Adhikary et al., 2008a). This stabilization occurs as a result of the delocalization of the charge (hole). This delocalization also accounts for the experimental increase in pK_a for monomeric $A^{\bullet+}$ from ≤ 1 (Steenken, 1989, 1992, 1997) to ca. 8 for the stacked system. In reasonable agreement with the experimentally obtained pK_a values, theoretical calculations predict the pK_a of $A_2^{\bullet+}$ as ca. 7 and that of monomeric $A^{\bullet+}$ as -0.3 (Adhikary et al., 2008a).

19.5.3 FORMATION OF SUGAR RADICALS

19.5.3.1 Monomeric Model Compounds

Employing the technique described above, sugar radical formation by photoexcitation of DNA and RNA base cation radicals has been reported in a number of recent investigations (Adhikary et al., 2005, 2006a,b, 2007, 2008b; Khanduri et al., 2008). Table 19.8 shows selected results for visible photoexcitation of $G^{\bullet+}$ in dGuo, and its nucleosides and nucleotides.

It is well established that in DNA or in RNA, sugar rings can adopt various conformations as defined by its pseudo rotation cycle (Adhikary et al., 2005, 2008b; Khanduri et al., 2008 and references therein). Therefore, simply based on the values of the hyperfine splittings alone, identification of a particular sugar radical formed at a specific site becomes problematic in an amorphous glass and is prone to error. In order to identify the specific sugar radical, photoexcitation studies were carried out by employing a number of dGuo derivatives with deuteration at specific sites (C3'- and C5'-) in the sugar moiety (Adhikary et al., 2005). Similar studies were carried out in Guo with selective deuteration at C1', C2', C3', and at C5' (Khanduri et al., 2008); in Ado (at C1', C2', C5') (Adhikary et al., 2008b); and also ^{13}C labeling at C5' (Adhikary et al., 2006b). ^{13}C isotopic substitution at C5' in dAdo (Adhikary et al., 2006b) was employed to ascertain whether the prominent anisotropic doublet (ca. 19–21 G)

TABLE 19.8

Sugar Radicals Formed on Photoexcitation of $G^{\bullet+a,b}$

Compound	Temperature (K)[c]	Percent Converted[d]	C1'[•e]	C3'[•e]	C5'[•e]
dGuo	143	90	10%	35%	55%
	77	30	10	40	50
5'-dGMP	143	95	95	5	—
	77	30	15	30	55
3'-dGMP	143	85	40	—	60
	77	15	40	—	60
dsDNA (ice)	143	ca. 50	100	—	—
	77	ca. 50	100	—	—

Source: Reprinted from Adhikary, A. et al., *Nucleic Acids Res.*, 33, 5553, 2005. With permission.

[a] Percentage expressed to ±5% relative error.

[b] All glassy samples are at the native pH of 7 M LiCl (ca. 5). For DNA, the pH of the aqueous solution before freezing was 7 and $G^{\bullet+}$ was produced directly by γ-irradiation.

[c] Temperature at which $G^{\bullet+}$ was illuminated.

[d] Percentage of $G^{\bullet+}$ that converts to sugar radicals. The total spectral intensities before and after illumination were the same, within experimental uncertainties.

[e] Each calculated as percentage of total sugar radical concentration; these sum to 100%.

observed at the center in many sugar radical cohorts was from the C5'• or C4'•. DFT calculations (B3LYP/6-31G(d), fully optimized) have shown that the ^{13}C hyperfine couplings at C5' in C5'• differ substantially from the corresponding ^{13}C hyperfine couplings at C5' in C4'• (see Scheme 19.8). The comparison of experimental and calculated values clearly pointed to the radical site in the sugar moiety of dAdo as the C5'• site.

From these experiments, the assignment of spectra to specific sugar radicals was made and a benchmark spectrum for each sugar radical found; these benchmarks were then employed to determine the contribution of each sugar radical in the sugar radical cohort (Table 19.8).

For dAdo and its deoxynucleotides (Adhikary et al., 2006b), near complete (ca. 80%–100%) conversion to sugar radicals (primarily C5'•, and C3'•) after photoexcitation of A•+ (Adhikary et al., 2008a) has been observed. This was also found for Ado and in its nucleotides (Adhikary et al., 2008b).

Interestingly, the C5'•-spectrum in 3'-dAMP and in 3'-AMP were found to differ greatly (Adhikary et al., 2008b). For 3'-dAMP, the ESR spectrum shows only a 19 G doublet from the C5' alpha hydrogen, which was not pH dependent. However, for 3'-AMP, a large increase in the β-coupling of the C4'-H atom from ca. 6 G (unresolved) to 34.5 G was observed upon increasing the pH from 6 to 9 (Adhikary et al., 2008b). For C5'• in 3'-AMP, DFT calculations (B3LYP/6-31G*) show the existence of two different conformations of C5'• (see Figure 19.6). At higher pHs, a new and strong hydrogen bond formation between the O5'-H and the 3'-phosphate dianion of C5'• in 3'-AMP results, which leads to the new conformation of C5'• where the large β-coupling of the C4'-H atom is observed (see Figure 19.6).

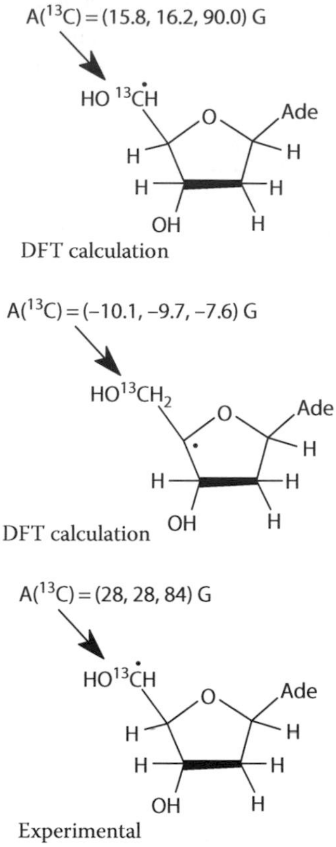

Absence of C4'• in DNA model compounds and C4'• and C2'• in RNA model compounds. It is interesting to note that the formation of C4'• has not been observed from photoexcitation of G•+ in dGuo or Guo or in its nucleotides (Adhikary et al., 2005; Khanduri et al., 2008), nor in dAdo or Ado and their nucleotides via photoexcitation of A•+

$A(^{13}C) = (15.8, 16.2, 90.0)$ G

DFT calculation

$A(^{13}C) = (-10.1, -9.7, -7.6)$ G

DFT calculation

$A(^{13}C) = (28, 28, 84)$ G

Experimental

SCHEME 19.8 Comparison of DFT calculated and experimental hyperfine coupling constants of $^{13}C5'$ for C5'• and C4'• in dAdo. (Reprinted from Adhikary, A. et al., *Nucleic Acids Res.*, 34, 1501, 2006b. With permisssion.)

(Adhikary et al., 2006b, 2008b). Theoretical calculations point out that in DNA, C4'• is energetically as stable as C1'• and C3'• (Li et al., 2005) and for RNA, C4'• and C2'• are energetically as stable as C1'• and C5'• (Li et al., 2006). Thus, based on bond energies alone, C4'• is expected to be formed in DNA (Li et al., 2005) and on similar arguments, formation of C4'• and C2'• are expected for RNA systems (Li et al., 2006). Their lack of observation stems from the mechanism of formation of the sugar radicals.

Studies with deuterated derivatives of dGuo (Adhikary et al., 2005), of Guo (Khanduri et al., 2008), and of Ado (Adhikary et al., 2008b) have shown very small deuterium isotope effects for sugar radical formation at C3' in 3'-D-dGuo and 3'-D-Guo; as well as at C5' in 5',5'-D,D-dGuo, 5',5'-D,D-Guo (Adhikary et al., 2005; Khanduri et al., 2008); and 5',5'-D,D-Ado (Adhikary et al., 2008b). These observations suggest that the geometry of the transition state for deprotonation from the sugar moiety must be close to the starting geometry of the cation radical in the excited state (Scheme 19.6). The charge and spin density at each site in the sugar ring in the excited cation radical likely control the formation of neutral sugar radical since deprotonation occurs most readily from sites of high positive charge (spin) density. Effectively, the pK_a of the H-atom attached to each carbon atom site in the sugar moiety in the excited cation radical will be affected by the charge distribution, and the charge distribution crucially influences the formation of the neutral sugar radical via Scheme 19.6.

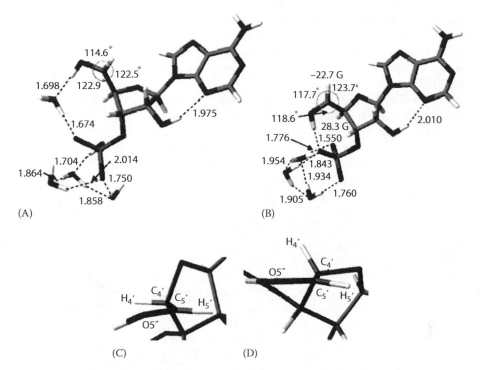

FIGURE 19.6 B3LYP/6-31G* optimized geometries of C5′• in (A) monoprotonated phosphate (PO4H-1) in the presence of four waters and (B) deprotonated form (PO_4^{-2}) in the presence of three waters. Figures C and D are the exposed views of the atoms O5′, C5′, H5′, and C4′ shown in parts A and B. Part C shows that the C4′-H atom in part A is in the nodal plane of the p-orbital of the C5′•, whereas part D points out that it is lifted significantly out of the nodal plane and results in a large beta proton hyperfine coupling from the C4′-H atom in the C5′•. (Reprinted from Adhikary, A. et al., *J. Phys. Chem. B*, 112, 15844, 2008b. With permission.)

Theoretical studies. Theoretical studies employing time-dependent density functional theory (TD-DFT) for one-electron oxidized DNA base radicals in nucleosides such as G•+ in dGuo or Guo (Adhikary et al., 2005; Khanduri et al., 2008), as well as A•+ and A(-H)• in dAdo (Adhikary et al., 2006b), or Ado (Adhikary et al., 2008b) suggest that core excitations (i.e., excitations from the inner shell molecular orbitals to the singly occupied molecular orbital (SOMO)) are responsible for the delocalization of charge and spin into the sugar moiety (see Figure 19.7 as an example).

The molecular orbital plots in Figure 19.7 show that, in the ground state, the SOMO is localized on the adenine base in the Ado cation radical, in agreement with experimental observations. During electronic excitations, the hole effectively transfers from the adenine base to the sugar moiety. Thus, theory supports the proposed mechanism of neutral sugar radical production via an excited cation radical originally proposed from experimental observations and shown in the Scheme 19.6.

19.5.3.2 DNA- and RNA-Oligomers and DNA

19.5.3.2.1 DNA-Dinucleoside Phosphates

Experimental work regarding sugar radical formation via photoexcitation of G•+ and A•+ in monomeric DNA- and RNA-model systems was extended to a dinucleoside phosphate, TpdG (Adhikary et al., 2006a). In TpdG samples, photoexcitation of G•+ at 143 K leads to ca. 85% sugar radicals (75% C1′• and 10% C3′•) (Adhikary et al., 2006a). The ESR spectrum of the sugar radical cohort produced from TpdG via photoexcitation of G•+ matches that produced from 5′-dGMP via photoexcitation of $G^{•+}$ (see Figure 19.8). These observations point out that the thymine moiety at the 5′-site in TpdG has little influence in dictating the extent of and the formation of either $C1'^{•}$ or $C3'^{•}$. In fact, no line component(s) that would be expected for thymine radicals were detected in TpdG samples.

TD-DFT/6-31G(d) calculations carried out for $G^{•+}$ in TpdG (Adhikary et al., 2006a) indicate that at longer wavelengths (≥500 nm) base-to-base $\pi–\pi^*$ hole transfer should dominate. So, as shown in Figure 19.9, the hole would be transferred from $G^{•+}$ to T (e.g., see transition S_1) and not to the sugar–phosphate moiety. As a result of this base-to-base $\pi–\pi^*$ hole transfer, base radical formation on thymine is expected, but no thymine radicals are found experimentally. These calculations were extended to dinucleoside phosphates (dGpdG, dApdA, dApdT, TpdA, and dGpT) with similar findings (Kumar and Sevilla, 2006).

19.5.3.2.2 Longer DNA-Oligomers

The ESR spectra found for single stranded (ss) DNA-oligomers (e.g., TpdG, TGT, TGGT, 5′-p-TGGT, TGGGT, TTGTT, TTGGTT, TTGGTTGGTT) on one-electron oxidation with $Cl_2^{•-}$ (Adhikary et al., 2007) have been observed to match that of one-electron oxidized G shown in Figure 19.5D. Taking TGT as a model for ssDNA-oligomer, it was recently found that the one-electron oxidized guanine moiety exists as an equal mixture of $G^{•+}$ and $G(-H)^•$ (Adhikary et al., 2009), just as found for the monomer dGuo. On the other hand, in a dsDNA oligomer d[GCGCGCGC]$_2$, the one-electron oxidized guanine exists only in the $G(-H)^•$ form at low temperatures (Adhikary et al., 2009) (see Section 19.5.5 for details).

In longer ssDNA-oligomers (e.g., TGT, TGGT, TGGGT, TTGTT, TTGGTT, TTGGTTGGTT), a considerable extent (55%–95%) of sugar radical formation via photoexcitation of $G^{•+}$ at ca. 143 K has been reported (Adhikary et al., 2007). The sugar radical cohort in these DNA-oligomers has been found to consist primarily of $C5'^{•}$ and $C1'^{•}$ along with an observable (albeit small) extent of $C3'^{•}$ (Adhikary et al., 2007).

In ssDNA-oligomers, the yields of sugar radicals formed via photoexcitation of $G^{•+}$ were found to decrease with increasing chain length in a wavelength-dependent manner. In shorter DNA-oligomers, formation of the sugar radicals was observed at wavelengths ≥540 nm. However, with increasing chain length of the oligomers, a decline in sugar radical yields was found at ≥540 nm (Adhikary et al., 2007). This wavelength dependence of sugar radical formation in longer DNA-oligomers was attributed to base-to-base hole transfer, because TD-DFT calculations (see Figure 19.9) suggest base-to-base hole transfer should compete effectively with base-to-sugar hole-transfer processes (Adhikary et al., 2007).

It has also been observed that the initial rate of sugar radical formation and the final sugar radical cohort in TGGT showed no observable difference with that found for 5′-p-TGGT. This indicates that the presence of the phosphate moiety at the 5′-site did not have a significant influence on the rate of production or on the type of sugar radical formed via photoexcitation (Adhikary et al., 2007). We note that studies by Osakada et al. (2008) and Liu and Barton (2005) on hole transfer in dsDNA-oligomers suggest that the "sticky end" (i.e., 5′-phosphate) also did not have any significant influence on the extent and rate of photoinduced hole transfer.

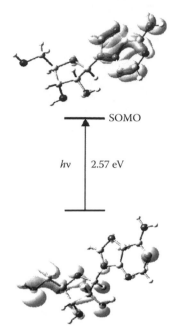

hv | 2.57 eV

SOMO

FIGURE 19.7 Representative TD-B3LYP/6-31G(d) calculated electronic transition (10th) occurring from inner core doubly occupied molecular orbital to the SOMO (singly occupied molecular orbital). This transition, as most others, shows significant hole localization at C5′ in the excited state. After the electronic transition, the lower figure becomes the SOMO. Thus, the transition moves the hole from base to sugar. (Reprinted from Adhikary, A. et al., *J. Phys. Chem. B*, 112, 15844, 2008b. With permission.)

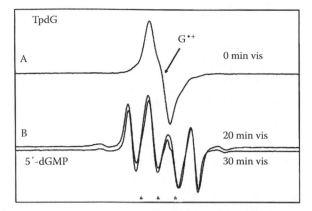

FIGURE 19.8 (A) ESR spectrum of $G^{\cdot+}$ obtained from one-electron oxidation of TpdG. (B) ESR spectra showing sugar radical formation on photoexcitation of $G^{\cdot+}$ in TpdG at ca. 145K. C1$'^{\cdot}$ (prominent quartet in the middle) and C3$'^{\cdot}$ (line components at the wings) are formed. The sugar radical spectrum obtained from 5$'$-dGMP (lower spectrum in B) (Adhikary et al. 2005) is also presented here for comparison. Each spectrum was recorded at 77 K. (Adapted from Adhikary, A. et al., *Radiat. Res.*, 165, 479, 2006a. With permission.)

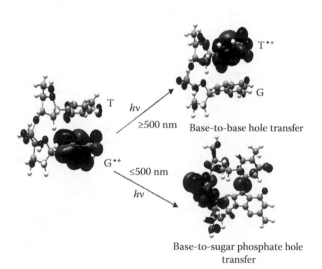

FIGURE 19.9 Schematic representation of base-to-base hole transfer, which is favored at longer wavelengths, and base-to-sugar phosphate hole transfer, which is favored at shorter wavelengths, on photoexcitation of TpdG$^{\cdot+}$. Two of the 20 transitions calculated are shown for the 0.76 eV (upper) and 2.79 eV (lower) transitions. The MOs were obtained by employing TD-DFT/B3LYP/6-31G(d) calculations. (Adapted from Adhikary, A. et al., *Radiat. Res.*, 165, 479, 2006a. With permission.)

19.5.3.2.3 Highly Polymerized DNA

Photoexcitation of one-electron oxidized guanine in γ-irradiated, hydrated ($\Gamma = 12$ D$_2$O/nucleotide) highly polymerized salmon sperm DNA has been observed to form C1$'^{\cdot}$ in the wavelength range 310–480 nm (Adhikary et al., 2005). In this wavelength range, substantial conversion (ca. 50%) to C1$'^{\cdot}$ was observed without a significant temperature dependence in the temperature range 77–180 K (Adhikary et al., 2007). However, formation of sugar radicals are not observed at wavelengths $\geq$500 nm (Adhikary et al., 2005). These results for highly polymerized DNA are expected from the wavelength dependence already observed for sugar radical yields in DNA-oligomers (Adhikary et al., 2007).

At ambient temperatures, photoinduced hole transfer in dsDNA oligomers in aqueous solution has been reported to occur through long uninterrupted A tracts with little distance dependence (Giese et al., 2001; Giese, 2002; Shao et al., 2004; Kawai and Majima, 2005; Joy et al., 2006; Augustyn et al., 2007; Lewis et al., 2008).

19.5.4 The Protonation State of One-Electron Oxidized 2′-Deoxyguanosine

The DNA cation radicals $G^{•+}$ and $A^{•+}$ are well known to undergo transformation via two types of competitive reactions: (1) reversible deprotonation reactions forming neutral radicals ($G(-H)^•$ and $A(-H)^•$), and (2) irreversible nucleophilic addition reaction of water (Steenken, 1989, 1992, 1997; Becker and Sevilla, 1993, 1998, 2008; Bernhard and Close, 2004; Sevilla and Becker, 2004; von Sonntag, 2006; Becker et al., 2007; Li and Sevilla, 2007; Close, 2008; Kumar and Sevilla, 2008a). The pK_a of $G^{•+}$ in dGuo in aqueous solution at ambient temperature has been found to be ca. 3.9 (Steenken, 1989, 1992, 1997); hence, at neutral pH, $G^{•+}$ in dGuo undergoes fast deprotonation to the surrounding water molecules ($1.8 \times 10^7 \, s^{-1}$) (Kobayashi and Tagawa, 2003). The slow hydration reaction at C-8 in $G^{•+}$ occurs with the $G^{•+}$ in prototropic equilibrium with the dominant species $G(-H)^•$.

Pulse radiolysis studies using optical detection and various analogs of dGuo had suggested that deprotonation of $G^{•+}$ in dGuo could occur via loss of a proton from N1 to give $G(N1-H)^•$ (Scheme 19.9) (Steenken, 1989, 1992, 1997) and, at high pH > ca. 11, $G(N1-H)^•$ further deprotonates to produce $G(-2H)^{•-}$ (Scheme 19.9) with $pK_a = 10.8$ (Steenken, 1989). These studies have shown that in nucleotides of dGuo (e.g., in 5′-dGMP), the presence of the phosphate group did not significantly affect the prototropic equilibria shown in Scheme 19.9 (Steenken, 1989). Furthermore, the pK_a of the exocyclic amine group in 1-methylguanosine cation radical (1-Me-Guo$^{•+}$) is found to be 4.7 at room temperature (Steenken, 1989). This implies that, in dGuo, deprotonation from the exocyclic nitrogen (N2) should be competitive with deprotonation from N1 (Scheme 19.9). Theoretical calculations have shown that the energies involved in producing the N1 deprotonated $G^{•+}$, that is, $G(N1-H)^•$, or the N2 deprotonated $G^{•+}$, that is, $G(N2-H)^•$, are very similar (Mundy et al., 2002; Luo et al., 2005; Naumov and von Sonntag, 2008). Theoretical calculations had also suggested that $G(N1-H)^•$ would be favored over $G(N2-H)^•$ in environments with high dielectric constants, such as water (von Sonntag, 2006; Naumov and von Sonntag, 2008).

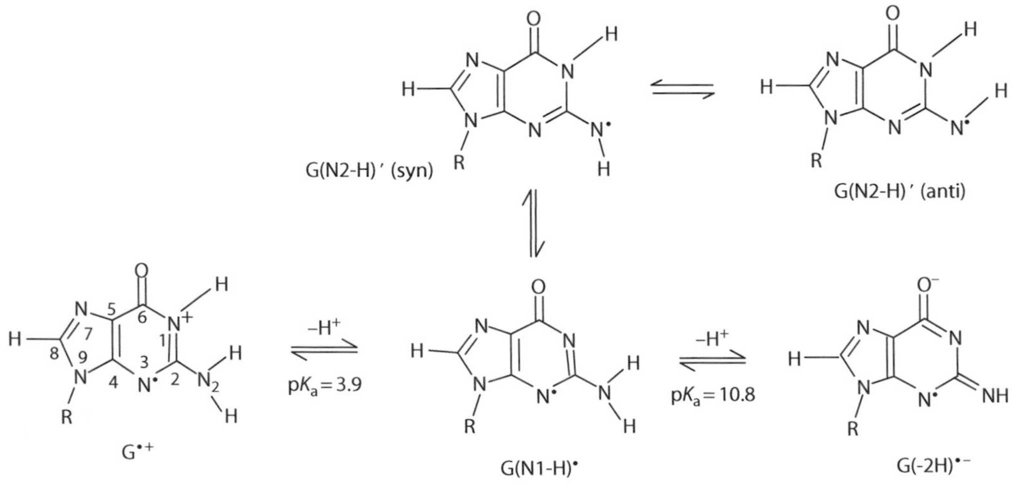

SCHEME 19.9 The numbering scheme and prototropic equilibria of one-electron oxidized guanine cation radical ($G^{•+}$), the mono-deprotonated species ($G(N1-H)^•$ and $G(N2-H)^•$ in syn and anti-conformers with respect to the N3 atom), and the di-deprotonated species ($G(-2H)^{•-}$). (Adapted from Adhikary, A. et al., *J. Phys. Chem. B*, 110, 24171, 2006c.)

ENDOR studies of x-ray-irradiated single crystals of Guo, dGuo, 5'-dGMP and 3',5'-cyclic guanosine monophosphate clearly show that in these crystalline environments, deprotonation of $G^{•+}$ results in $G(N2-H)^{•}$ rather than in $G(N1-H)^{•}$ (Bernhard and Close, 2004; Close, 2008) and references therein.

Thus, even though a considerable number of experimental and theoretical studies had been carried out on the deoxyguanosine cation radical ($G^{•+}$) and its deprotonated species ($G(-H)^{•}$), the specific site of deprotonation (at N1 or at the exocyclic N-atom (N2)) was still not clear (Adhikary et al., 2006c). However, in a recent ESR study of the prototropic equilibrium between $G^{•+}$ and $G(-H)^{•}$ in dGuo, specific isotopic substitution of deuterium at C-8-H in the guanine ring and of ^{15}N at N1, N2, and N3 atoms in the guanine ring, as well as, theoretical calculations (Adhikary et al., 2006c) provided some insights to this issue.

Glassy homogenous aqueous (D_2O) solutions of dGuo (2'-dG) in 7.5 M LiCl were investigated using ESR and UV–Vis spectroscopy in the pH range 3–12 (Adhikary et al., 2006c). In D_2O, exchangeable hydrogen atoms (N1-H and two N2-H atoms) do not show observable hyperfine coupling (Adhikary et al., 2006c, 2008a, 2009). The three prototropic forms of one-electron-oxidized guanine ($G^{•+}$, $G(-H)^{•}$, and $G(-2H)^{•-}$) in dGuo (Scheme 19.9) were distinguished from each other in the pH ranges (3–5), (7–9), and (>11), respectively, and characterized.

Deuteration of the C8 position in the guanine ring allowed the observation of the N^{14} nitrogen atom hyperfine couplings at different states of the prototropic equilibria of $G^{•+}$. ^{15}N isotopic substitutions at N1, N2, and N3 in the guanine ring were used to verify these couplings and their assignments to specific nitrogen atom sites. With knowledge of these couplings, ESR spectra simulations allowed the determination of the C8(H) hyperfine couplings in different protonation states of one-electron oxidized guanine (Adhikary et al., 2006c) and gave strong evidence that the N1 site is the site of deprotonation in $G^{•+}$.

Ab initio DFT calculations for various one-electron oxidized guanine radicals with 7–10 waters of hydration, confirmed assignments of experimentally observed hyperfine couplings to nitrogen atoms. A comparison of the calculated hyperfine couplings of $G(N1-H)^{•}$ + 7 H_2O and $G(N2-H)^{•}$ + 7 H_2O with experimental values clearly confirm that the site of deprotonation in $G^{•+}$ is at N1 in aqueous solution (Adhikary et al., 2006c).

Previous studies using pulse radiolysis with optical detection of 8-bromo-dGuo along with theoretical modeling (Chatgilialoglu et al., 2005, 2006) suggested that at first, $G^{•+}$ in dGuo decays to $G(N2-H)^{•}$, which via subsequent water-assisted tautomerization is converted to $G(N1-H)^{•}$. The work described above (Adhikary et al., 2006c) suggests that with the inclusion of additional waters, $G^{•+}$ in dGuo decays directly to $G(N1-H)^{•}$.

19.5.5 The Protonation State of One-Electron Oxidized Guanine in dsDNA-Oligomers: The Role of Base Pairing

In dsDNA-oligomers or in DNA, the prototropic equilibrium of one-electron oxidized guanine, Scheme 19.10, is controlled by two important proton transfer processes: (1) the intra-base proton transfer process within the $G^{•+}$:C base pair (Steenken, 1989, 1992, 1997; Becker and Sevilla, 1993,

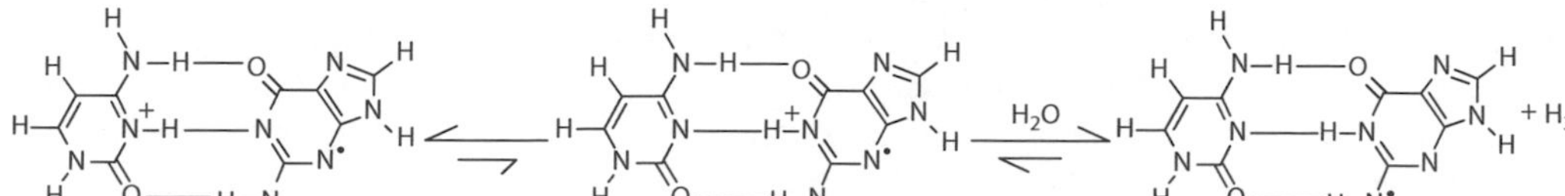

SCHEME 19.10 Schematic representation of prototropic equilibria in the $G^{•+}$:C pair. Both intra-base pair proton transfer within the $G^{•+}$:C base pair and proton transfer between the $G^{•+}$:C base pair and surrounding water are shown.

1998, 2008; Bernhard and Close, 2004; Sevilla and Becker, 2004; von Sonntag, 2006; Becker et al., 2007; Li and Sevilla, 2007; Close, 2008; Kumar and Sevilla, 2008a), and (2) the proton transfer processes between the $G^{\bullet+}$:C base pair and the surrounding water (Shafirovich et al., 2001; Kobayashi and Tagawa, 2003; Kobayashi et al., 2008; Lee et al., 2008).

The extent of these proton transfer processes depends on the pK_a of the ion-radicals involved. It is also evident from Scheme 19.10 that the intra-base pair proton transfer leads to separation of charge from spin. Based upon the pK_a values of $G^{\bullet+}$ in dGuo (ca. 3.9) and $C(H^+)$ in 2′-deoxy-cytidine (ca. 4.5), Steenken had estimated the equilibrium constant for this intra-base pair proton transfer process, K_{eq}, to be ca. 2.5 (Steenken, 1989, 1992, 1997; Adhikary et al., 2009). From this value of K_{eq}, at ambient temperature, both cation radical ($G^{\bullet+}$:C) (30%) and deprotonated neutral radical ($G(-H)^{\bullet}$:C($+H^+$)) (70%) forms should exist (Adhikary et al., 2009). This equilibrium constant suggests that $\Delta G^o = -0.5$ kcal/mol at 298 K (Steenken, 1989, 1992, 1997). For $G^{\bullet+}$:C, theoretical studies carried out in the gas phase have shown that for the intra-base pair proton transfer along N1–H bond shown in Scheme 19.10, ΔS (proton transfer) = ca. 0 cal/mol/K, and ΔG (proton transfer) = 1.4 kcal/mol at 298 K (Hütter and Clark, 1996; Bertran et al., 1998; Li et al., 2001). Thus, theory predicts that $G^{\bullet+}$:C is thermodynamically stable whereas experimental results mentioned above suggest that $G(-H)^{\bullet}$:C($+H^+$) is favored.

Two recent works have provided some additional understanding of the intra-base proton transfer process (Adhikary et al., 2009; Kumar and Sevilla, 2009a). ESR investigations of C8-deuterated dGuo (Adhikary et al., 2006c) (see Section 19.5.4) clearly showed that selective deuteration at C-8 on the guanine moiety in dGuo (G*) resulted in an ESR signal from the guanine cation radical ($G^{*\bullet+}$), which is easily distinguishable from that of the N1-deprotonated radical, $G*(N1-H)^{\bullet}$ (Adhikary et al., 2006c). Based upon these results, ESR studies employing G* incorporated ssDNA-oligomer TG*T and the dsDNA-oligomer d[G*CG*CG*CG*C]$_2$ have established that in ssDNA, the one-electron oxidized guanine exists as an equal mixture of $G^{\bullet+}$ and $G(N1-H)^{\bullet}$, whereas in dsDNA, only $G(N1-H)^{\bullet}$ is found, or more accurately, $G(-H)^{\bullet}$:C($+H^+$) as shown in Scheme 19.10 (Adhikary et al., 2009).

We note here that, in ssDNA-oligomer, owing to the lack of base pairing, the formation of $G(N1-H)^{\bullet}$ occurs because of the deprotonation of $G^{\bullet+}$ to the solvent; whereas, in the dsDNA-oligomer d[GCGCGCGC]$_2$, deprotonation takes place from $G^{\bullet+}$ to the base paired C by intra-base pair proton transfer (Adhikary et al., 2009). While the deprotonation in the dsDNA oligomer was complete, for the ssDNA oligomer it was only 50%. This shows that the base paired cytosine is a better proton acceptor than the solvent.

The second work that aids our understanding of this system involves a theoretical treatment of the GC cation radical (Kumar and Sevilla, 2009a). As mentioned above, previous theoretical calculations for the gas phase $G^{\bullet+}$:C pair had predicted an unfavorable intra-base pair proton transfer energy (Hütter and Clark, 1996; Bertran et al., 1998; Li et al., 2001). However, recent theoretical calculations that included explicit waters of hydration show that, on inclusion of the hydration shell (ca. 11 waters), the free energy change (ΔG) for intra-base pair proton transfer in the $G^{\bullet+}$:C pair becomes favorable ($\Delta G = -0.65$ kcal/mol) (Kumar and Sevilla, 2009a). Thus both experimental (Adhikary et al., 2009) and theoretical results (Kumar and Sevilla, 2009a) are in accord and agree with Steenken (1989) who suggested a favorable ΔG for the intra-base pair proton transfer from guanine N1 to cytosine N3 in the $G^{\bullet+}$:C pair.

While it is now clear that the site of interbase proton transfer $G^{\bullet+}$:C is from N1 in guanine, the site of deprotonation of the DNA cation radical ($G^{\bullet+}$:C) to water is still uncertain. A few recent pulse radiolysis studies in aqueous solution at ambient temperature (Anderson et al., 2006; Chatgilialoglu et al., 2006) have proposed that the likely site for deprotonation of the one-electron oxidized guanine in DNA to water is at the exocyclic amine N2 (see Scheme 19.10). DFT calculations (B3LYP/DZP++) in the gas phase also show that $G(N2-H)^{\bullet}$:C is indeed slightly favored over $G(N1-H)^{\bullet}$:C by about 1 kcal/mol (Bera and Schaefer, 2005). Thus, the site of deprotonation of one-electron oxidized guanine in dsDNA to water is likely to be in an equilibrium between N1 and N2 deprotonation, slightly favoring N2 (Scheme 19.10) but this needs further confirmation.

19.5.6 FACTORS AFFECTING THE SUGAR RADICAL FORMATION FROM EXCITATION OF CATION RADICALS

Studies employing DNA- and RNA-nucleosides, nucleotides, and DNA- and RNA-oligomers, as well as highly polymerized double stranded salmon sperm DNA (see Sections 19.5.3.1 and 19.5.3.2), have established the following features: (1) the identity of the sugar radical produced via deprotonation at specific sites in the sugar moiety and (2) the overall yield of the sugar radical cohort is critically influenced by a number of factors. These are discussed below.

19.5.6.1 Site of Phosphate Substitution

Theoretical calculations (Colson and Sevilla, 1995a,b) predict that phosphate substitution at 3′- or at 5′-site deactivates the formation of a sugar radical at that site by a small increase in the C–H bond energy. It is evident from Table 19.8 that in dGuo, C5′•, C3′•, and C1′• are produced via photoexcitation of G•+; whereas, in 5′-dGMP, preferential formation of C1′• occurs along with a small extent of C3′• production. In 3′-dGMP, predominant formation of C5′• and observable amount of C1′• production have been observed. Similarly, in 5′-GMP, primarily C1′• formation is observed (Khanduri et al., 2008). In accordance with theory, both 3′-dAMP and 3′-AMP, show primarily C5′• production (Adhikary et al., 2006b, 2008b). For 5′-dAMP, both C5′• and C3′• formation in almost equal amounts have been observed (Adhikary et al., 2006b). Thus, the theoretical prediction regarding deactivation of formation of the sugar radical site owing to phosphate substitution at that site seems to hold for DNA- and RNA-monomers (Adhikary et al., 2005, 2006b, 2008b; Khanduri et al., 2008) and for DNA-dinucleoside phosphate TpdG (Adhikary et al., 2006a).

Contrary to the theoretical predictions (Colson and Sevilla, 1995a,b), significant C5′• formation has been observed in ssDNA-oligomers of various chain lengths (Adhikary et al., 2007). Although, in highly polymerized double stranded salmon sperm DNA, only C1′• formation has been found (Adhikary et al., 2005). Moreover, in ssDNA-oligomers, the presence of a 5′-phosphate in 5′-p-TGGT did not alter the rate and extent of C5′• formation (Adhikary et al., 2007). Also, in the RNA-oligomers, for example, UGGGU, photoexcitation of the guanine cation radical yields primarily C1′• along with an unidentified radical of ca. 28 G doublet (Khanduri et al., 2008). These findings clearly suggest the contribution of other factors apart from the site of phosphate substitution in determining the extent and type of the sugar radicals produced in the cohort as discussed below.

19.5.6.2 Protonation State of One-Electron Oxidized Purine and pH

Production of sugar radicals via photoexcitation of one-electron oxidized G in dGuo and in Guo was found to be pH dependent. It has been observed that both in dGuo (Adhikary et al., 2005) and in Guo (Khanduri et al., 2008), sugar radical formation occurs via photoexcitation of the cation radical (G•+). However, at pH ≥9 where only G(-H)• is present, no sugar radical formation occurs. For dAdo and its derivatives, selective formation of C3′• is observed via A(−H)• even at pH ca. 12 (Adhikary et al., 2006b).

In dsDNA-oligomers containing G, it has been shown that the one-electron oxidized G is the N1-deprotonated neutral radical, that is, G(-H)• (Section 19.5.5) (Adhikary et al., 2009). Thus, no photoinduced sugar radical formation is expected. Surprisingly, photoexcitation of G(-H)• in double stranded DNA oligomers was indeed found to result in the formation of sugar radicals (Adhikary et al., 2005). Arguing from the results for dGuo, the one-electron oxidized guanine base should be protonated for sugar radical formation to take place (Adhikary et al., 2005). It is therefore likely that reprotonation of the guanine base in the dG(-H)•:dC(+H+) in the excited state would precede the deprotonation from the sugar. In other words, in the excited state, after hole transfer to the sugar, the equilibrium shown in Scheme 19.10 should rapidly shift to reprotonate guanine, that is, to produce a ribose cation radical, (dR•+)G:dC, thereby allowing deprotonation from the sugar moiety. Proton transfer in excited states has been extensively studied in aqueous solutions and the rate of the proton transfer in excited states is often very fast (ca. <ps range) (Agmon, 2005) and thus, the reprotonation of the guanine base in the dG(-H)$^•$:dC(+H$^+$) in the excited state should be rapid ($\leq 10^{-12}$ s).

19.5.6.3 Photoexcitation: Effect of Temperature

Table 19.8 clearly demonstrates that the extent of sugar radical production via photoexcitation of the cation radical in glassy systems increases substantially (from 15%–30% at 77 K to ≥80% at 143 K) with temperature. This quantitative increase in the extent of the sugar radical formation has been found to be a general observation for all the DNA- and RNA-monomers (nucleosides and -tides) (Adhikary et al., 2005, 2006b, 2008b; Khanduri et al., 2008).

However, Table 19.8 shows that, photoexcitation of G$^{•+}$ in 5′-dGMP or 5′-GMP samples at 77 K leads to predominant formation of C5′$^•$ along with an observable amount of C3′$^•$ and small amounts of C1′$^•$ (Adhikary et al., 2005; Khanduri et al., 2008), whereas at 143 K, almost exclusively C1′$^•$ is formed. We note that this type of change in the sugar radical cohort with temperature during photoexcitation of G$^{•+}$ has not been observed for other nucleosides and nucleotides such as dGuo, 3′-dGMP, Guo, dAdo, 3′-dAMP, Ado, and 3′-AMP (Adhikary et al., 2005, 2006b, 2008b; Khanduri et al., 2008). These results suggest that the molecular relaxation of the excited cation radical at higher temperature may change the site of deprotonation in the sugar moiety. This has not yet been tested theoretically.

A recent report that shows an increase in the rate and extent of intramolecular proton transfer with decreasing viscosity in protic media provides an explanation for the increasing yield found with elevated temperatures in glassy systems (Yushchenko et al., 2007). However, this was not found to be the case for frozen aqueous solutions of dsDNA. As discussed in Section 19.5.3.2, the initial rate of C1$^•$ formation from the photoexcitation of one electron oxidized dsDNA (salmon sperm) has been found to be independent of temperature in the range 77–180 K (Adhikary et al., 2007). This difference is not yet explained but the rigidity of the glass at 77 K is a likely factor.

19.5.6.4 Wavelength of Photoexcitation

Production of sugar radicals via photoexcitation of one-electron oxidized DNA- and RNA-monomers (nucleosides and nucleotides) and also in smaller DNA oligomers, for example TGGT, was found to be relatively independent of wavelength in the range 310–650 nm (Adhikary et al., 2005, 2006b, 2008b; Khanduri et al., 2008). However, C1′$^•$ formation via photoexcitation of one-electron oxidized guanine in dsDNA is wavelength dependent (Adhikary et al., 2005). C1′$^•$ formation has been observed in the wavelength range 310–480 nm but not for longer wavelengths (Adhikary et al., 2005). TD-DFT calculations (Adhikary et al., 2005, 2006a; Kumar and Sevilla, 2006), suggest that this wavelength dependence is likely a result of base-to-base hole transfer during photoexcitation (see Figure 19.9).

19.6 CONCLUSION

As described in this chapter, over the last several years, studies regarding our understanding of the physicochemical mechanisms of radiation-induced DNA damage have been extended well beyond the initial ionization event to include the effects of LEE and hole-excited states in the production of DNA damages including strand breaks. Experimental studies at low and at room temperatures along with ab initio quantum chemical studies have provided new understandings of these processes. The site of deprotonation on the sugar moiety in the transient hole-excited states found by experiment is predicted theoretically to be at those sites with high localization of the hole. Similarly, addition of a low energy electron to a DNA component resulting in DNA strand breaks is predicted to be equivalent to the formation of an excited anion radical state, which localizes the electron to a σ* dissociative state. Excess DNA sugar radicals produced by heavy ions are suggested to be driven by track-structure processes that produce excitations of ion-radicals. Both electronic and vibrational excitations have been suggested to drive ion-radicals to produce sugar radicals and subsequent strand breaks. Of course, with respect to the in vivo systems (e.g., see Chapter 22), these mechanistic explanations obtained via employing in vitro model systems are tentative and will need further studies to ascertain the validity and the extent of their application.

ACKNOWLEDGMENTS

The authors thank the National Cancer Institute of the National Institutes of Health (Grant RO1CA45424) for support and Dr. Anil Kumar for his help in preparing this manuscript.

REFERENCES

Adhikary, A., Malkhasian, A. Y. S., Collins, S., Koppen, J., Becker, D., and Sevilla, M. D. 2005. UVA-visible photo-excitation of guanine radical cations produces sugar radicals in DNA and model structures. *Nucleic Acids Res.* 33: 5553–5564.

Adhikary, A., Kumar, A., and Sevilla, M. D. 2006a. Photo-induced hole transfer from base to sugar in DNA: Relationship to primary radiation damage. *Radiat. Res.* 165: 479–484.

Adhikary, A., Becker, D., Collins, S., Koppen, J., and Sevilla, M. D. 2006b. C5′- and C3′-sugar radicals produced via photo-excitation of one-electron oxidized adenine in 2′-deoxyadenosine and its derivatives. *Nucleic Acids Res.* 34: 1501–1511.

Adhikary, A., Kumar, A., Becker, D., and Sevilla, M. D. 2006c. The guanine cation radical: Investigation of deprotonation states by ESR and DFT. *J. Phys. Chem. B* 110: 24171–24180.

Adhikary, A., Collins, S., Khanduri, D., and Sevilla, M. D. 2007. Sugar radicals formed by photo-excitation of guanine cation radical in oligonucleotides. *J. Phys. Chem. B* 111: 7415–7421.

Adhikary, A., Kumar, A., Khanduri, D., and Sevilla, M. D. 2008a. The effect of base stacking on the acid-base properties of the adenine cation radical [A⁺] in solution: ESR and DFT studies. *J. Am. Chem. Soc.* 130: 10282–10292.

Adhikary, A., Khanduri, D., Kumar, A., and Sevilla, M. D. 2008b. Photo-excitation of adenine cation radical [A⁺] in the near UV-vis region produces sugar radicals in adenosine and in its nucleotides. *J. Phys. Chem. B* 112: 15844–15855.

Adhikary, A., Khanduri, D., and Sevilla, M. D. 2009. Direct observation of the hole protonation state and hole localization site in DNA-oligomers. *J. Am. Chem. Soc.* 131: 8614–8619.

Agmon, N. 2005. Elementary steps in excited-state proton transfer. *J. Phys. Chem. A* 109: 13–35.

Anderson, R. F., Shinde, S. S., and Maroz, A. 2006. Cytosine-gated hole creation and transfer in DNA in aqueous solution. *J. Am. Chem. Soc.* 128: 15966–15967.

Arkin, M. R., Stemp, E. D. A., Pulver, S. C., and Barton, J. K. 1997. Long-range oxidation of guanine by Ru(III) in duplex DNA. *Chem. Biol.* 4: 389–400.

Augustyn, K. E., Genereux, J. C., and Barton, J. K. 2007. Distance-independent DNA charge transport across an adenine tract. *Angew. Chem. Int. Ed.* 46: 5731–5733.

Bald, I., Kopyra, J., and Illenberger, E. 2006. Selective excision of C5 from D-ribose in the gas phase by low-energy electrons (0–1 eV): Implications for the mechanism of DNA damage. *Angew. Chem. Int. Ed.* 45: 4851–4855.

Bald, I., Dabkowska, I., and Illenberger, E. 2008. Probing biomolecules by laser-induced acoustic desorption: Electrons at near zero electron volts trigger sugar-phosphate cleavage. *Angew. Chem. Int. Ed.* 47: 8518–8520.

Becker, D. and Sevilla, M. D. 1993. The chemical consequences of radiation-damage to DNA. *Adv. Radiat. Biol.* 17: 121–180.

Becker, D. and Sevilla, M. D. 1998. Radiation damage to DNA and related biomolecules. In *Royal Society of Chemistry Specialist Periodical Report: Electron Spin Resonance*, B. C. Gilbert, M. J. Davies, and D. M. Murphy (eds.), Vol. 16, pp. 79–115. Cambridge, U.K.: The Royal Society of Chemistry.

Becker, D. and Sevilla, M. D. 2008. EPR studies of radiation damage to DNA and related molecules. In *Royal Society of Chemistry Specialist Periodical Report: Electron Spin Resonance*, B. C. Gilbert, M. J. Davies, and D. M. Murphy (eds.), Vol. 21, pp. 33–58. Cambridge, U.K.: The Royal Society of Chemistry.

Becker, D., Razskazovskii, Y., Callaghan, M. U., and Sevilla, M. D. 1996. Electron spin resonance of DNA irradiated with a heavy-ion beam ($^{16}O^{8+}$): Evidence for damage to the deoxyribose phosphate backbone. *Radiat. Res.* 146: 361–368.

Becker, D., Bryant-Friedrich, A., Trzasko, C., and Sevilla, M. D. 2003. Electron spin resonance study of DNA irradiated with an argon-ion beam: Evidence for formation of sugar phosphate backbone radicals. *Radiat. Res.* 160: 174–185.

Becker, D., Adhikary, A., and Sevilla, M. D. 2007. The role of charge and spin migration in DNA radiation damage. In *Charge Migration in DNA Physics, Chemistry and Biology Perspectives*, T. Chakraborty (ed.), p. 139. Berlin/Heidelberg, Germany: Springer-Verlag.

Becker, D., Adhikary, A., and Sevilla, M. D. 2010. Mechanism of radiation induced DNA damage: Direct effects. In *Recent Trends in Radiation Chemistry*, B. S. M. Rao and J. Wishart (eds.), pp. 509–542. Singapore/New Jersey/London, U.K.: World Scientific Publishing Co.

Bera, P. P. and Schaefer III, H. F. 2005. (G-H)·-C and G-(C-H)· radicals derived from the guanine-cytosine base pair cause DNA subunit lesions. *Proc. Natl. Acad. Sci. USA* 102: 6698–6703.

Bernhard, W. A. 1981. Solid-state radiation of DNA: The bases. *Adv. Radiat. Biol.* 9: 199–280.

Bernhard, W. A. and Close, D. M. 2004. DNA damage dictates the biological consequences of ionizing irradiation: The chemical pathways. In *Charged Particle and Photon Interactions with Matter. Chemical, Physicochemical and Biological Consequences with Applications*, A. Mozumdar and Y. Hatano (eds.), pp. 431–470. New York/Basel, Switzerland: Marcel Dekker, Inc.

Bertran, J., Oliva, A., Rodriguez-Santiego, L., and Sodupe, M. 1998. Single versus double proton-transfer reactions in Watson-Crick base pair radical cations. A theoretical study. *J. Am. Chem. Soc.* 120: 8159–8167.

Boudaïffa, B., Cloutier, P., Hunting, D., Huels, M. A., and Sanche, L. 2000. Resonant formation of DNA strand breaks by low-energy (3–20 eV) electrons. *Science* 287: 1658–1660.

Bowman, M. K., Becker, D., Sevilla, M. D., and Zimbrick, J. D. 2005. Track structure in DNA irradiated with heavy ions. *Radiat. Res.* 163: 447–454.

Burrow, P. D., Gallup, G. A., Scheer, A. M., Denifl, S., Ptasinska, S., Mark, T., and Scheier, P. 2006. Vibrational Feshbach resonances in uracil and thymine. *J. Chem. Phys.* 124,124310.

Burrow, P. D., Gallup, G. A., and Modelli, A. 2008. Are there pi* shape resonances in electron scattering from phosphate groups? *J. Phys. Chem. A* 112: 4106–4113.

Cadet, J., Bourdat, A.-G., D'Ham, C., Duarte, V., Gasparutto, D., Romieu, A., and Ravanat, J.-L. 2000. Oxidative base damage to DNA: Specificity of base excision repair enzymes. *Mutat. Res.* 462: 121–128.

Cai, Z. and Sevilla, M. D. 2000. Electron spin resonance study of electron transfer in DNA: Inter-double-strand tunneling processes. *J. Phys. Chem. B* 104: 6942–6949.

Cai, Z. and Sevilla, M. D. 2004. Studies of excess electron and hole transfer in DNA at low temperature. In *Long Range Transfer in DNA II: Topics in Current Chemistry*, G. B. Shuster (ed.), Vol. 237, pp. 103–127. Berlin/Heidelberg, Germany/New York: Springer-Verlag.

Cai, Z., Gu, Z., and Sevilla, M. D. 2001. Electron spin resonance study of electron and hole transfer in DNA: Effects of hydration, aliphatic amine cations, and histone proteins. *J. Phys. Chem. B* 105: 6031–6041.

Chatgilialoglu, C., Caminal, C., Guerra, M., and Mulazzani, Q. G. 2005. Tautomers of one-electron-oxidized guanosine. *Angew. Chem. Int. Ed.* 44: 6030–6032.

Chatgilialoglu, C., Caminal, C., Altieri, A., Vougioukalakis, G. C., Mulazzani, Q. G., Gimisis, T., and Guerra, M. 2006. Tautomerism in the guanyl radical. *J. Am. Chem. Soc.* 128: 13796–13805.

Chatterjee, A. 2006. Importance of collaborative research between theoretical modelers and experimentalists, In *Abstracts, Fifty-Third Annual Meeting of the Radiation Research Society*, Philadelphia, PA, p. 3.

Chatterjee, A. and Holley, W. R. 1993. Computer-simulation of initial events in the biochemical-mechanisms of DNA-damage. *Adv. Radiat. Biol.* 17: 181–226.

Close, D. M. 2008. From the primary radiation induced radicals in DNA constituents to strand breaks: Low temperature EPR/ENDOR studies. In *Radiation Induced Molecular Phenomena in Nucleic Acid: A Comprehensive Theoretical and Experimental Analysis*, M. K. Shukla and J. Leszczynski (eds.), pp. 493–529. Berlin/Heidelberg, Germany/New York: Springer-Verlag.

Colson, A.-O. and Sevilla, M. D. 1995a. Elucidation of primary radiation damage in DNA through application of *ab initio* molecular orbital theory. *Int. J. Radiat. Biol.*, 67: 627–645.

Colson, A.-O. and Sevilla, M. D. 1995b. Structure and relative stability of deoxyribose radicals in a model DNA backbone: *ab initio* molecular orbital calculations. *J. Phys. Chem.* 99: 3867–3874.

Conwell, E. M. 2005. Charge transport in DNA in solution: The role of polarons. *Proc. Natl. Acad. Sci. USA* 102: 8795–8799.

Conwell, E. M. and Rakhmanova, S. V. 2000. Polarons in DNA. *Proc. Natl. Acad. Sci. USA* 97: 4556–4560.

Cullis, P. M., Malone, M. E., and Merson-Davies, L. A. 1996. Guanine radical cations are precursor of 7,8-dihydro-8-oxo-2′-deoxyguanosine but are not precursor of immediate strand breaks in DNA. *J. Am. Chem. Soc.* 118: 2775–2781.

Dawidzik, J. B., Budzunzki, E. E., Patrzyc, H. B., Cheng, H.-C., Iijima, H., Alderfer, J. L., Tabaczynski, W. A., Wallace, J. C., and Box, H. C. 2004. Dihydrothymine lesion in x-irradiated DNA: Characterization at the molecular level and detection in cells. *Int. J. Radiat. Biol.* 80: 355–361.

de Lara, C. M., Jenner, T. J., Townsend, K. M., Marsden, S. J., and O'Neill, P. 1995. The effect of dimethyl sulfoxide on the induction of DNA double-strand breaks in V79-4 mammalian cells by alpha particles. *Radiat. Res.* 144: 43–49.

Debije, M. G. and Bernhard, W. A. 2000. Electron and hole transfer induced by thermal annealing of crystalline DNA x-irradiated at 4 K. *J. Phys. Chem. B* 104: 7845–7851.

Drummond, T. G., Hill, M. G., and Barton, J. K. 2003. Electrochemical DNA sensors. *Nat. Biotechnol.* 21: 1192–1199.

Fink, H.-W. and Schönenberger, C. 1999. Electrical conduction through DNA molecules. *Nature* 398: 407–410.

Friedland, W., Jacob, P., Bernhardt, P., Paretzke, H. G., and Dingfelder, M. 2003. Simulation of DNA damage after proton irradiation. *Radiat. Res.* 159: 401–410.

Gasper, S. M. and Schuster, G. B. 1997. Intramolecular photoinduced electron transfer to anthroquinones linked to duplex DNA: Effect of gaps and traps on long range of radical cation migration. *J. Am. Chem. Soc.* 119: 12762–12771.

Giese, B. 2000. Long distance charge transport in DNA: The hopping mechanism. *Acc. Chem. Res.* 33: 631–636.

Giese, B. 2002. Long-distance electron transfer through DNA. *Annu. Rev. Biochem.* 71: 51–70.

Giese, B., Amaudrut, J., Köhler, A.-K., Spormann, M., and Wessely, S. 2001. Direct observation of the hole transfer through DNA by hopping between adenine bases and by tunneling. *Nature* 412: 318–320.

Goodhead, D. T., Leenhouts, H. P., Paretzke, H. G., Terrisol, M., Nikjoo, H., and Blaauboer, R. 1994. Track structure approaches to the interpretation of radiation effects on DNA. *Radiat. Prot. Dosimetry* 52: 217–223.

Grozema, F. C. and Siebbeles, L. D. 2007. Mechanism and absolute rates of charge transfer through DNA. In *Charge Migration in DNA Physics, Chemistry and Biology Perspectives*, T. Chakraborty (ed.), pp. 21–43. Berlin/Heidelberg, Germany: Springer-Verlag.

Hada, M. and Georgakilas, A. G. 2008. Formation of clustered DNA damage after high-LET irradiation: A review. *J. Radiat. Res.* 49: 203–210.

Henderson, P. T., Jones, D., Hampikian, G., Kan, Y. Z., and Schuster, G. B. 1999. Long-distance charge transport in duplex DNA: The phonon-assisted polaron-like hopping mechanism. *Proc. Natl. Acad. Sci. USA* 96: 8353–8358.

Hill, M. A. 1999. Radiation damage to DNA: The importance of track structure. *Radiat. Meas.* 31: 15–23.

Hütter, M. and Clark, T. 1996. On the enhanced stability of the guanine-cytosine base-pair radical cation. *J. Am. Chem. Soc.* 118: 7574–7577.

Jakob, B., Scholz, M., and Taucher-Scholz, G. 2003. Biological imaging of heavy charged-particle tracks. *Radiat. Res.* 159: 676–684.

Jones, G. D., Milligan, J. R., Ward, J. F., Calabro-Jones, P. M., and Aguilera, J. A. 1993. Yield of strand breaks as a function of scavenger concentration and LET for SV40 irradiated with ^{4}He ions. *Radiat. Res.* 136: 190–196.

Jortner, J., Bixon, M., Langenbacher, T., and Michel-Beyerle, M. E. 1998. Charge transfer and transport in DNA. *Proc. Natl. Acad. Sci. USA* 95: 12759–12765.

Joy, A., Ghosh, A. K., and Schuster, G. B. 2006. One-electron oxidation of DNA oligomers that lack guanine: Reaction and strand cleavage at remote thymines by long-distance radical cation hopping. *J. Am. Chem. Soc.* 128: 5346–5347.

Kawai, K. and Majima, T. 2005. Spectroscopic investigation of oxidative hole transfer via adenine hopping in DNA. In *Charge Transfer in DNA: From Mechanism to Application*, H.-A. Wagenknecht (ed.), p. 117. Weinheim, Germany: Wiley-VCH Verlag GmbH & Co. KGaA.

Khanduri, D., Collins, S., Kumar, A., Adhikary, A., and Sevilla, M. D. 2008. Formation of sugar radicals in RNA model systems and oligomers via excitation of guanine cation radical. *J. Phys. Chem. B* 112: 2168–2178.

Kobayashi, K. and Tagawa, S. 2003. Direct observation of guanine radical cation deprotonation in duplex DNA using pulse radiolysis. *J. Am. Chem. Soc.* 125: 10213–10218.

Kobayashi, K., Yamagami, R., and Tagawa, S. 2008. Effect of base sequence and deprotonation of guanine cation radical in DNA. *J. Phys. Chem. B* 112: 10752–10757.

Kraft, G. and Kramer, M. 1993. Linear-energy-transfer and track structure. *Adv. Radiat. Biol.* 17: 1–52.

Kumar, A. and Sevilla, M. D. 2006. Photoexcitation of dinucleoside radical cations: A time-dependent density functional study. *J. Phys. Chem. B* 110: 24181–24188.

Kumar, A. and Sevilla, M. D. 2007. Low-energy electron attachment to 5″-thymidine monophosphate: Modeling single strand breaks through dissociative electron attachment. *J. Phys. Chem. B* 111: 5464–5474.

Kumar, A. and Sevilla, M. D. 2008a. Radiation effects on DNA: Theoretical investigations of electron, hole and excitation pathways to DNA damage. In *Radiation Induced Molecular Phenomena in Nucleic Acid: A Comprehensive Theoretical and Experimental Analysis*, M. K. Shukla and J. Leszczynski (eds.), pp. 577–617. Berlin/Heidelberg, Germany/New York: Springer-Verlag.

Kumar, A. and Sevilla, M. D. 2008b. The role of pi sigma* excited states in electron-induced DNA strand break formation: A time-dependent density functional theory study. *J. Am. Chem. Soc.* 130: 2130–2131.

Kumar, A. and Sevilla, M. D. 2009a. Influence of hydration on proton transfer in the guanine-cytosine radical cation (G($\cdot^+$)-C) base pair: A density functional theory study. *J. Phys. Chem. B* 113(33): 11359–11361. Article ASAP DOI: 10.1021/jp903403d, Publication Date (Web): June 1, 2009.

Kumar, A. and Sevilla, M. D. 2009b. Role of excited states in low-energy electron (LEE) induced strand breaks in DNA model systems: Influence of aqueous environment. *ChemPhysChem.* 10: 1426–1430.

La Vere, T., Becker, D., and Sevilla, M. D. 1996. Yields of ·OH in gamma-irradiated DNA as a function of DNA hydration: Hole transfer in competition with ·OH formation. *Radiat. Res.* 145: 673–680.

Lee, Y. A., Durandin, A., Dedon, P. C., Geacintov, N. E., and Shafirovich, V. 2008. Oxidation of guanine in G, GG, and GGG sequence contexts by aromatic pyrenyl radical cations and carbonate radical anions: Relationship between kinetics and distribution of alkali-labile lesions. *J. Phys. Chem. B* 112: 1834–1844.

Lewis, F. D. 2005. DNA molecular photonics. *Photochem. Photobiol.* 81: 65–72.

Lewis, F. D., Llu, X., Llu, J., Miller, S. E., Hayes, R. T., and Wasielewski, M. R. 2000. Direct measurement of hole transport dynamics in DNA. *Nature* 406: 51–53.

Lewis, F. D., Letsinger, R. L., and Wasielewski, M. R. 2001. Dynamics of photoinduced charge transfer and hole transport in synthetic DNA hairpins. *Acc. Chem. Res.* 34: 159–170.

Lewis, F. D., Daublain, P., Zhang, L., Cohen, B., Vura-Weis, J., Wasielewski, M. R., Shafirovich, V., Wang, Q., Raytchev, M., and Fiebig, T. 2008. Reversible bridge-mediated excited-state symmetry breaking in stilbene-linked DNA dumbbells. *J. Phys. Chem. B* 112: 3838–3843.

Li, X. F. and Sevilla, M. D. 2007. DFT treatment of radiation produced radicals in DNA model systems. *Adv. Quantum Chem.* 2007, 52: 59–87.

Li, X., Cai, Z., and Sevilla, M. D. 2001. Investigation of proton transfer within DNA base pair anion and cation radicals by density functional theory (DFT). *J. Phys. Chem. B* 105: 10115–10123.

Li, M.-J., Liu, L., Fu, Y., and Guo, Q.-X. 2005. Development of an ONIOM-G3B3 method to accurately predict C-H and N-H bond dissociation enthalpies of ribonucleosides and deoxyribonucleosides. *J. Phys. Chem. B* 109: 13818–13826.

Li, M.-J., Liu, L., Fu, Y., and Guo, Q.-X. 2006. Significant effects of phosphorylation on relative stabilities of DNA and RNA sugar radicals: Remarkably high susceptibility of H-2″ abstraction in RNA. *J. Phys. Chem. B* 110: 13582–13589.

Liu, T. and Barton, J. K. 2005. DNA electrochemistry through the base pairs not the sugar-phosphate backbone. *J. Am. Chem. Soc.* 127: 10160–10601.

Luo, Q., Li, Q. S., Xie, Y., and Schaefer III, H. F. 2005. Radicals derived from guanine: Structures and energetics. *Collect. Czech. Chem. Commun.* 70: 826–836.

Ly, D., Sanii, L., and Schuster, G. B. 1999. Mechanism of charge transport in DNA: Internally-linked anthraquinone conjugates support phonon-assisted polaron hopping. *J. Am. Chem. Soc.* 121: 9400–9410.

Magee, J. L. and Chatterjee, A. 1987. Track reactions of radiation chemistry. In *Kinetics of Nonhomogenous Processes*, G. R. Freeman (ed.), pp. 171–214. New York: Wiley.

Mundy, C. J., Colvin, M. E., and Quong, A. A. 2002. Irradiated guanine: A Car-Parrinello molecular dynamics study of dehydrogenation in the presence of an OH radical. *J. Phys. Chem. A* 106: 10063–10071.

Naumov, S. and von Sonntag, C. 2008. Guanine-derived radicals: Dielectric constant-dependent stability and UV/Vis spectral properties: A DFT study. *Radiat. Res.* 169: 364–372.

Nelson, D. and Symons, M. C. R. 1977. Unstable intermediates. Part 169. Electron capture processes in organic phosphates: An electron spin resonance study. *J. Chem. Soc. Perkin II* 286–293.

Nelson, D. J., Symons, M. C. R., and Wyatt, J. L. 1993. Electron-paramagnetic-resonance studies of irradiated D-glucose-6-phosphate ions: Relevance to DNA. *J. Chem. Soc. Faraday Trans.* 89: 1955–1958.

Nikjoo, H. and Uehara, S. 2004. Track structure studies of biological systems. In *Charged Particle and Photon Interactions with Matter. Chemical, Physicochemical and Biological Consequences with Applications*, A. Mozumdar and Y. Hatano (eds.), pp. 491–531. New York/Basel, Switzerland: Marcel Dekker, Inc.

Nocek, J. M., Zhou, J. S., de Forest, S., Priyadarshy, S., Beratan, D. N., Onuchic, J. N., and Hoffman, B. M. 1996. Theory and practice of electron transfer within protein-protein complexes: Application to the multi-domain binding of cytochrome *c* by cytochrome *c* peroxidase. *Chem. Rev.* 96: 2459–2489.

Núñez, M. E., Hall, D. B., and Barton, J. K. 1999. Long-range oxidative damage to DNA: Effects of distance and sequence. *Chem. Biol.* 6: 85–97.

Núñez, M. E., Holmquist, G. P., and Barton, J. K. 2001. Evidence for DNA charge transport in the nucleus. *Biochemistry* 40: 12465–12471.

Núñez, M. E., Noyes, K. T., and Barton, J. K. 2002. Oxidative charge transport through DNA in nucleosome core particles. *Chem. Biol.* 9: 403–415.

O'Neill, P. 2001. Radiation-induced damage in DNA. In *Radiation Chemistry: Present Status and Future Trends*, C. D. Jonah and B. S. M. Rao (eds.), pp. 585–622. Amsterdam, the Netherlands: Elsevier.

Osakada, Y., Kawai, K., Fujitsuka, M., and Majima, T. 2008. Charge transfer in DNA assemblies: Effects of sticky ends. *Chem. Commun.* 23: 2656–2658.

Pan, X. and Sanche, L. 2006. Dissociative electron attachment to DNA basic constituents: The phosphate group. *Chem. Phys. Lett.* 421: 404–408.

Panajotovic, R., Martin, F., Cloutier, P. C., Hunting, D., and Sanche, L. 2006. Effective cross sections for production of single-strand breaks in plasmid DNA by 0.1 to 4.7 eV electrons. *Radiat. Res.* 165: 452–459.

Pezeshk, A., Symons, M. C. R., and McClymont, J. D. 1996. Electron movement along DNA strands: Use of intercalators and electron paramagnetic resonance spectroscopy. *J. Phys. Chem.* 100: 18562–18566.

Porath, D., Bezryadin, A., de Vries, S., and Dekker, C. 2001. Direct measurements of electrical transport through DNA molecules. *Nature* 403: 635–637.

Purkayastha, S., Milligan, J. R., and Bernhard, W. A. 2005. Correlation of free radical yields with strand break yields produced in plasmid DNA by the direct effect of ionizing radiation. *J. Phys. Chem. B* 109: 16967–16973.

Purkayastha, S., Milligan, J. R., and Bernhard, W. A. 2006a. The role of hydration in the distribution of free radical trapping in directly ionized DNA. *Radiat. Res.* 166: 1–8.

Purkayastha, S., Milligan, J. R., and Bernhard, W. A. 2006b. An investigation into the mechanisms of DNA strand breakage by direct ionization of variably hydrated plasmid DNA. *J. Phys. Chem. B* 110: 26286–26291.

Purkayastha, S., Milligan, J. R., and Bernhard, W. A. 2007. On the chemical yield of base lesions, strand breaks, and clustered damage generated in plasmid DNA by the direct effect of x rays. *Radiat. Res.* 168: 357–366.

Ravanat, J.-L., Saint-Pierre, C., and Cadet, J. 2003. One electron oxidation of the guanine moiety of 2'-deoxyguanosine: Influence of 8-oxo-7,8-dihydro-2'-deoxyguanosine. *J. Am. Chem. Soc.* 125: 2030–2031.

Razskazovskiy, Y., Debije, M. G., and Bernhard, W. A. 2003a. Strand breaks produced in X-irradiated crystalline DNA: Influence of base sequence. *Radiat. Res.* 159: 663–669.

Razskazovskiy, Y., Debije, M. G., Howerton, S. B., Williams, L. D., and Bernhard, W. A. 2003b. Strand breaks in x-irradiated crystalline DNA: Alternating CG oligomers. *Radiat. Res.* 160: 334–339.

Saito, I., Nakamura, T., Nakatani, K., Yoshioka, Y., Yamaguchi, K., and Sugiyama, H. 1998. Mapping of the hot spots for DNA damage by one-electron oxidation: Efficacy of GG doublets and GGG triplets as a trap in long-range hole migration. *J. Am. Chem. Soc.* 120: 12686–12687.

Sanche, L. 2008. Low energy electron damage to DNA. In *Radiation Induced Molecular Phenomena in Nucleic Acid: A Comprehensive Theoretical and Experimental Analysis*, M. K. Shukla and J. Leszczynski (eds.), pp. 531–575. Berlin/Heidelberg, Germany/New York: Springer-Verlag.

Sanderud, A. and Sagstuen, E. 1996. EPR study of X-irradiated hydroxyalkyl phosphate esters: Phosphate radical formation in polycrystalline glucose phosphate, ribose phosphate and glycerol phosphate salts at 77 and 295 K. *J. Chem. Soc. Faraday Trans.* 92: 995–999.

Schuhmacher, H. and Dangendorf, V. 2002. Experimental tools for track structure investigations: New approaches for dosimetry and microdosimetry. *Radiat. Prot. Dosimetry.* 2002, 99: 317–323.

Schuster, G. B. 2000. Long-range charge transfer in DNA: Transient structural distortions control the distance dependence. *Acc. Chem. Res.* 33: 253–260.

Schuster, G. B. (ed.) 2004. *Long Range Charge Transfer in DNA. I and II, Topics in Current Chemistry*. Berlin/Heidelberg, Germany: Springer-Verlag.

Sevilla, M. D. and Becker, D. 2004. ESR studies of radiation damage to DNA and related biomolecules. In *Royal Society of Chemistry Specialist Periodical Report: Electron Spin Resonance*, B. C. Gilbert, M. J. Davies, and D. M. Murphy (eds.), Vol. 19, pp. 243–278. Cambridge, U.K.: The Royal Society of Chemistry.

Shafirovich, V., Dourandin, A., and Geacintov, N. E. 2001. Proton-coupled electron-transfer reactions at a distance in DNA duplexes: Kinetic deuterium isotope effect. *J. Phys. Chem. B* 105: 8431–8435.

Shao, F., O'Neill, M. A., and Barton, J. K. 2004. Long-range oxidative damage to cytosines in duplex DNA. *Proc. Natl. Acad. Sci. USA* 101: 17914–17919.

Sharma, K. K., Purakayastha, S., and Bernhard, W. A. 2007. Unaltered free base release from d(CGCGCG)$_2$ produced by the direct effect of ionizing radiation at 4 K and room temperature. *Radiat. Res.* 167: 501–507.

Sharma, K. K., Milligan, J. R., and Bernhard, W. A. 2008. Multiplicity of DNA single-strand breaks produced in pUC18 exposed to the direct effects of ionizing radiation. *Radiat. Res.* 170: 156–162.

Shukla, L. I., Adhikary, A., Pazdro, R., Becker, D., and Sevilla, M. D. 2004a. Formation of 8-oxo-7,8-dihydro-guanine-radicals in gamma-irradiated DNA by multiple one-electron oxidations. *Nucleic Acids Res.* 32: 6565–6574.

Shukla, L. I., Pazdro, R., Huang, J., DeVreugd, C., Becker, D., and Sevilla, M. D. 2004b. The formation of DNA sugar radicals from photoexcitation of guanine cation radicals. *Radiat. Res.* 161: 582–590.

Shukla, L. I., Pazdro, R., Becker, D., and Sevilla, M. D. 2005. Sugar radicals in DNA: Isolation of neutral radicals in gamma-irradiated DNA by hole and electron scavenging. *Radiat. Res.* 163: 591–602.

Shukla, L. I., Adhikary, A., Pazdro, R., Becker, D., and Sevilla, M. D. 2007. Formation of 8-oxo-7,8-dihydro-guanine-radicals in gamma-irradiated DNA by multiple one-electron oxidations. *Nucleic Acids Res.* 35: 2460–2461.

Simons, J. 2006. How do low-energy (0.1–2 eV) electrons cause DNA-strand breaks? *Acc. Chem. Res.* 39: 772–779.

Spalletta, R. A. and Bernhard, W. A. 1992. Free radical yields in A:T polydeoxynucleotides, oligodeoxynucleotides, and monodeoxynucleotides at 4 K. *Radiat. Res.* 130: 7–14.

Steenken, S. 1989. Purine bases, nucleosides and nucleotides: Aqueous solution redox chemistry and transformation reactions of their radical cations and e$^-$ and $\cdot$OH adducts. *Chem. Rev.* 89: 503–520.

Steenken, S. 1992. Electron-transfer-induced acidity/basicity and reactivity changes of purine and pyrimidine bases. Consequences of redox processes for DNA base pairs. *Free. Radical Res. Commun.* 16: 349–379.

Steenken, S. 1997. Electron transfer in DNA? Competition by ultra-fast proton transfer? *Biol. Chem.* 378: 1293–1297.

Steenken, S., Jovanovic, S. V., Bietti, M., and Bernhard, K. 2000. The trap depth (in DNA) of 8-oxo-7,8-dihydro-2'-deoxyguanosine as derived from electron transfer equilibria in aqueous solution. *J. Am. Chem. Soc.* 122: 2373–2374.

Swarts, S. G., Sevilla, M. D., Becker, D., Tokar, C. J., and Wheeler, K. T. 1992. Radiation-induced DNA damage as a function of hydration. I. Release of unaltered bases. *Radiat. Res.* 129: 333–344.

Swarts, S. G., Becker, D., Sevilla, M. D., and Wheeler, K. T. 1996. Radiation-induced DNA damage as a function of hydration. II. Base damage from electron-loss centers. *Radiat. Res.* 145: 304–314.

Swiderek, P. 2006. Fundamental processes in radiation damage of DNA. *Angew. Chem. Int. Ed.* 45: 4056–4059.

Terato, H., Tanaka, R., Nakaarai, Y., Nohara, T., Doi, Y., Iwai, S., Hirayama, R., Furusawa, Y., and Ide, H. 2008. Quantitative analysis of isolated and clustered DNA damage induced by gamma-rays, carbon ion beams, and iron ion beams. *J. Radiat. Res.* 49: 133–146.

Urushibara, A., Shikazono, N., O'Neill, P., Fujii, K., Wada, S., and Yokoya, A. 2008. LET dependence of the yield of single-, double-strand breaks and base lesions in fully hydrated plasmid DNA films by (^{4}He^{2+}) ion irradiation. *Int. J. Radiat. Biol.* 84: 23–33.

von Sonntag, C. 2006. *Free-Radical-Induced DNA Damage and Its Repair*, pp. 335–447. Berlin/Heidelberg, Germany: Springer-Verlag.

Wagenknecht, H.-A. 2003. Reductive electron transfer and transport of excess electrons in DNA. *Angew. Chem. Int. Ed.* 42: 2454–2460.

Wagenknecht, H.-A. (ed.) 2005. *Charge Transfer in DNA: From Mechanism to Application*. Weinheim, Germany: Wiley-VCH Verlag GmbH & Co. KGaA.

Wan, C., Fiebig, T., Kelly, S. O., Treadway, C. R., Barton, J. K., and Zewail, A. H. 1999. Femtosecond dynamics of DNA-mediated electron transfer. *Proc. Natl. Acad. Sci. USA* 96, 6014–6019.

Wan, C., Fiebig, T., Schiemann, O., Barton, J. K., and Zewail, A. H. 2000. Femtosecond direct observation of charge transfer between bases in DNA. *Proc. Natl. Acad. Sci. USA* 97: 14052–14055.

Wang, W., Becker, D., and Sevilla, M. D. 1993. The influence of hydration on the absolute yields of primary ionic free radicals in gamma-irradiated DNA at 77 K. I. Total radical yields. *Radiat. Res.* 135: 146–154.

Wang, W., Yan, M., Becker, D., and Sevilla, M. D. 1994. The influence of hydration on the absolute yields of primary free radicals in gamma-irradiated DNA at 77 K. II. Individual radical yields. *Radiat. Res.* 137: 2–10.

Ward, J. F. 2000. Complexity of damage produced by ionizing radiation. *Cold. Spring. Harb. Symp. Quant. Biol.* 65: 377–382 and references therein.

Weiland, B. and Hüttermann, J. 1999. Free radicals from lyophilized "dry" DNA bombarded with heavy-ions as studied by electron spin resonance spectroscopy. *Int. J. Radiat. Biol.* 75: 1169–1175.

Yokoya, A., Cunniffe, S. M. T., and O'Neill, P. 2002. Effect of hydration on the induction of strand breaks and base lesions in plasmid DNA films by gamma-radiation. *J. Am. Chem. Soc.* 124: 8859–8866.

Yokoya, A., Cuniffe, S. M. T., Stevens, D. L., and O'Neill, P. 2003. Effects of hydration on the induction of strand breaks, base lesions, and clustered damage in DNA films by alpha-radiation. *J. Phys. Chem. B* 107: 832–837.

Yokoya, A., Fujii, K., Usigome, T., Shikazono, N., Urushibara, A., and Watanabe, R. 2006. Yields of strand breaks and base lesions induced by soft x-rays in plasmid DNA. *Radiat. Prot. Dosimetry* 122: 86–88.

Yushchenko, D. A., Shvadchak, V. V., Klymchenko, A. S., Duportail, G., Pivovarenko, V. G., and Mély, Y. 2007. Modulation of excited-state intramolecular proton transfer by viscosity in protic media. *J. Phys. Chem. A* 111: 10435–10438.

20 Spectroscopic Study of Radiation-Induced DNA Lesions and Their Susceptibility to Enzymatic Repair

Akinari Yokoya
Japan Atomic Energy Agency
Tokai, Japan

Kentaro Fujii
Japan Atomic Energy Agency
Tokai, Japan

Naoya Shikazono
Japan Atomic Energy Agency
Tokai, Japan

Masatoshi Ukai
Tokyo University of Agriculture and Technology
Tokyo, Japan

CONTENTS

20.1　INTRODUCTION

Ionizing radiation induces chemically stable molecular changes in cellular DNA (DNA damage). The biological effects of ionizing radiation, such as cell killing, induction of mutations, chromosome aberrations, transformation, and the adaptive responses of irradiated cells, are thought to arise from the formation of such DNA damage. However, it has been gradually recognized that living cells can effectively repair DNA damage through several enzymatic pathways. DNA double-strand breaks (DSBs), for instance, are thought to be one of the main types of DNA damage that induce the cell-killing effect. Most DSBs (80%–90%) initially produced in a cell can be rejoined even in cell extracts (de Lara et al., 1995), but the boundary line that divides repairable and unrepairable DNA damage has not yet been clarified. A long-standing and important question remains as to which kinds of DNA damage slip through the defenses of the enzymatic repair system in a living cell, ultimately causing deleterious radiation effects. It has been suggested that the susceptibility of DNA damage to repair strongly depends on the track structure of the radiation that induced the damage, represented by a linear energy transfer (LET) value. LET is defined as the ratio of dE to dl (dE/dl), where dE is the mean energy and dl is the distance traversed by the charged particle (ICRU, 1970). Monte Carlo track simulation studies have predicted that radiation with higher LET has a greater tendency to produce clusters of isolated DNA lesions by dense ionization/excitation within a distance of a few nanometers than radiation with lower LET (Nikjoo et al., 1994, 1999). These clustered sites of damage can be categorized into two groups: DSB and non-DSB types of damage. DSB damage consists of two or more single-strand breaks (SSBs) produced within about six base pairs' separation in DNA (Hanai et al., 1998). Non-DSB damage, on the other hand, comprising two or more lesions, including nucleobase lesions, abasic sites (AP sites), and SSBs formed within about 10 base pairs' separation by a single radiation track, has been proposed to be more biologically relevant (Ward, 1988; Goodhead, 1994a). It is becoming clear that clustered sites of DNA damage are less readily repaired than isolated nucleobase lesions, SSBs, or AP sites and, therefore, may induce serious genetic changes in cells. In order to understand the initial processes of induction of DNA damage, including the site of clustering of isolated lesions, and the susceptibility of this damage to repair, spectroscopic techniques combined with biochemical assays have been used. In Chapter 19, the molecular mechanisms of radiation damage to DNA are extensively presented in terms of their physicochemical aspects. In this chapter, recent studies aiming at clarifying the nature of DNA damage in terms of its biological reparability, particularly by base-excision repair (BER) proteins, are highlighted. New approaches using spectroscopic techniques combined with a brilliant synchrotron radiation source to reveal the role of the photoelectric effect on DNA induction are also introduced.

20.2　DIRECT AND INDIRECT EFFECTS ON INDUCTION OF DNA STRAND BREAKS AND NUCLEOBASE LESIONS

DNA damage is thought to be induced by both direct energy deposition on DNA (direct effect) (see review by Bernhard and Close, 2003) and reactions with diffusible water radicals (indirect effect) (see review by O'Neill and Fielden, 1993). Most mechanistic studies to date (von Sonntag, 1987; O'Neill, 2001) have focused on the indirect effects using dilute, aqueous solutions containing DNA. The results indicated that the hydroxyl radical (OH·) is the main water radiolysis species that induces SSBs and DSBs in DNA, whereas hydrated electrons, H atoms, and hydroxyl radicals induce DNA nucleobase lesions. Experimental (de Lara et al., 1995) and theoretical (Nikjoo et al., 2002) studies have indicated that in living cells or under highly scavenging conditions similar to those involved in OH· scavenging in the cell, ~40% of the lesions induced in DNA by low-LET radiation could be ascribed to direct effects, increasing to ~70% for high-LET α-particles. In the pioneering work by Krisch et al. (1991), the yields of SSBs and DSBs by both direct and indirect effects were determined using SV40 DNA-irradiated γ-rays under conditions of greatly varying radical scavenger concentration. They also concluded that in high scavenging conditions, DSBs from indirect effects are produced predominantly by local clusters of OH· from single energy deposition events. The experimentally obtained ratios between direct and indirect effects examined by various biological end points have been summarized by Becker and Sevilla (1993).

When comparing the yield of DNA damage among various irradiation experiments, one should focus on both the track structure of radiation and the scavenging capacity of the sample. In addition to the SSB and DSB yields, heat-labile site and base lesions also contribute to total damage yields. In order to figure out the nature of various radiations that induce DNA damage, a few studies have been performed using the same DNA sample and the same scavenger condition as those used in previous studies. The yield of DNA damage induced by x-rays (150 kVp) has been measured using closed circular plasmid (pUC18) DNA (Yokoya et al., 2006) to compare the yield induced by lower-LET γ-rays or higher-LET Al_K soft x-rays (1.8 keV) (Fulford et al., 2000, 2001). The common experimental method for detecting DNA damage using plasmid DNA is shown in Figure 20.1. Several DNA solutions with three kinds of radical scavenger capacities and fully hydrated DNA samples were irradiated to determine the indirect contribution by the reaction of diffusible water radicals, such as OH·, with the direct action of secondary electrons. The yield of prompt SSBs summarized in Table 20.1 decreased with increasing levels of scavenging capacity (Figure 20.2), indicating that the fraction of damage attributable to direct effects under these cell-mimetic conditions was about 30%. Heat-labile sites were hardly detected under the two conditions of higher scavenging capacity. The yields of oxidative nucleobase lesions revealed by the treatments of Nth or Fpg proteins (experimental details are described below) are summarized in Table 20.2. The yields of nucleobase lesions decreased with increasing levels of scavenging capacity (Figure 20.3), although a large fraction of direct effects was obtained under the cell-mimetic conditions: 60% for treatment with Nth and 40% for treatment with Fpg.

In order to know how the damage yields depend on the radiation track structure, the data is compared with those obtained by irradiation with lower-LET γ-rays and higher-LET soft x-rays (1.8 keV) and shown in Figure 20.2. The three lines have different slopes and intersect around the cell-mimetic conditions. The extrapolated values of the 150 kVp soft x-ray data under the condition of highest scavenging capacity (hydration condition) are larger than those for the other two

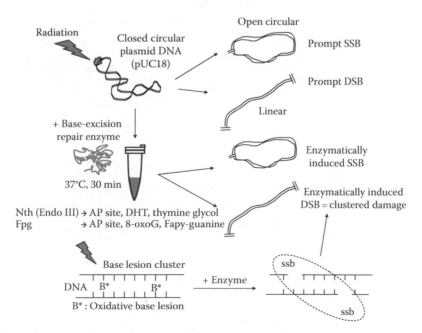

FIGURE 20.1 Schematic illustration of the experimental method for detection of DNA damage using BER proteins and closed circular plasmid DNA as the substrate of the enzymes.

TABLE 20.1

Yields of Strand Breaks Induced in pUC18 Plasmid DNA Irradiated with Soft X-Rays under Various Scavenging Conditions

Tris Concentration (mM)[a]	Scavenging Capacity (s⁻¹)	SSB (Gy/Da)	DSB (Gy/Da)	SSB/DSB
0.2	3×10^5	2.28×10^{-8}	ND	—
2.5	4×10^6	5.02×10^{-9}	3.04×10^{-10}	17
50	7×10^7	9.61×10^{-10}	6.61×10^{-11}	19
Hydrated[b]	—	1.48×10^{-10}	1.82×10^{-11}	8

SSB, single-strand breaks; DSB, double-strand breaks.

[a] Concentrations of other solutes in the sample solution: 50 μg/mL pUC18 DNA; 1 mM EDTA (Yokoya et al., 2006).

[b] Fully hydrated samples were prepared by exposing dry pUC18 DNA film to a condition of 97% relative humidity over 16 h. The hydration level was estimated to be 35 water molecules per nucleotide (Yokoya et al., 2002).

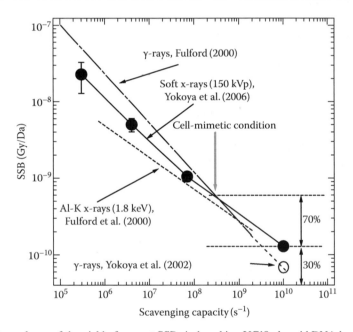

FIGURE 20.2 Dependence of the yield of prompt SSBs induced in pUC18 plasmid DNA by 150 kVp soft x-ray irradiation on the scavenging capacity at 277 K under an aerobic condition. The data points represented as closed circles are cited from our previous report (Yokoya et al., 2006). Previous data for γ-irradiation (Fulford, 2000) and Al-K soft x-ray (1.8 keV) irradiation (Fulford et al., 2001) are shown by dotted lines and are compared with our data for 150 kVp soft x-ray irradiation (closed circles). The yield of a hydrated pUC18 sample irradiated with γ-rays at 277 K under an aerobic condition is shown in open circles for comparison, assuming that the scavenging capacity of the hydrated sample was 1×10^{10} s⁻¹.

irradiations, and the data points for the hydrated sample show the expected SSB yields. The SSB yields demonstrate a rather complicated relationship:

$SSB_{\gamma\text{-rays}} > SSB_{150\,kVp} > SSB_{1.8\,keV}$ under dilute conditions

$SSB_{\gamma\text{-rays}} \sim SSB_{150\,kVp} > SSB_{1.8\,keV}$ under cell-mimetic conditions

$SSB_{\gamma\text{-rays}} > SSB_{150\,kVp} \sim SSB_{1.8\,keV}$ under conditions in which the direct effect is dominant

TABLE 20.2
Yields of Heat-Labile Sites and Nucleobase Lesions Visualized
by Enzymatic Treatment under Various Scavenging Conditions

Tris Concentration (mM)[a]	Heat-Labile Sites	Nth-Sensitive Sites	Fpg-Sensitive Sites
0.2	1.03×10^{-8}	2.31×10^{-8}	2.99×10^{-8}
2.5	4.80×10^{-10}	4.11×10^{-9}	3.69×10^{-9}
50	ND	8.33×10^{-10}	8.13×10^{-10}
Hydrated[b]	ND	5.19×10^{-10}	3.16×10^{-10}

[a] Concentrations of other solutes in the sample solution: $50\,\mu g/mL$ pUC18 DNA; 1 mM EDTA (Yokoya et al., 2006).

[b] Fully hydrated samples were prepared by exposing dry pUC18 DNA film to a condition of 97% relative humidity over 16 h. The hydration level was estimated to be 35 water molecules per nucleotide (Yokoya et al., 2002).

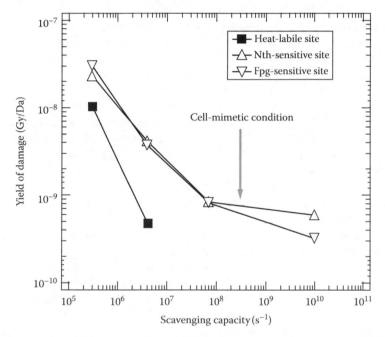

FIGURE 20.3 Dependence of the yields of heat-labile and enzyme-sensitive sites in pUC18 plasmid DNA induced by soft x-ray irradiation on the scavenging capacity at 277 K under an aerobic condition. (■) Heat-labile sites; (△) Nth-sensitive sites; (▽) Fpg-sensitive sites. (Data from Yokoya, A. et al., *Radiat. Prot. Dosim.*, 122, 86, 2006.)

This should be kept in mind when the yield of DNA damage is compared with that induced by other radiations, because the yield strongly depends on the radical scavenging conditions during irradiation.

20.3 CLUSTERED DNA DAMAGE AND ITS SUSCEPTIBILITY TO ENZYMATIC REPAIR

The fact that additional DSBs are generated after exposure of cells to ionizing radiation has been known for over 30 years (Dugle et al., 1976; Ahnstrom and Bryant, 1982), and this is now considered a strong piece of evidence supporting the *in vivo* processing of clustered damage. Blaisdell and Wallace

(2001) were the first to demonstrate that in *Escherichia coli* cells, the level of additional DSBs generated after irradiation was dependent on the amount of glycosylase (Fpg) and, thus, was related to the amount of clustered damage. Yang et al. further demonstrated that the induction of DSBs, mutations, and lethality all strongly correlated with the expression levels of glycosylases (hOGG1 and hNTH1) after exposure of human B lymphoblastoid cells to 3 Gy of γ-rays (Yang et al., 2004, 2006). It is important to note that in these studies, substantial amounts of additional DSBs were observed at high doses (~500 Gy) or at high levels of glycosylases. On the other hand, at sublethal doses, only a fraction (~10%) of the clustered DNA damage sites was converted to DSBs in nonhomologous end-joining (NHEJ)-deficient CHO xrs-5 cells after irradiation with γ-rays (Gulston et al., 2004), and DSB-repair-proficient Chinese hamster V79 cells and human hematopoietic cells did not generate additional DSBs after postirradiation by γ-rays (Georgakilas et al., 2004; Gulston et al., 2004). Collectively, these results suggest that (1) an abortive enzymatic repair of clusters predominates under high doses or high levels of glycosylases in vivo and (2) at sublethal doses, cells have mechanisms to avoid the formation of DSBs or to quickly rejoin de novo DSBs through NHEJ repair and alternative pathways. It is important to note that enzymatic processing of clustered lesions has not been observed after irradiation with high-LET radiation. In the case of wild-type Chinese hamster V79 cells, the amount of DSBs remained essentially constant after α-irradiation (Gulston et al., 2004).

Recently, BER proteins have been widely used as enzymatic probes to detect the nucleobase lesions induced by various radiation sources (Melvin et al., 1998; Prise et al., 1999; Milligan et al., 2000; Sutherland et al., 2000). Endonuclease III (Nth) and formamidopyrimidine-DNA glycosylase (Fpg) are typical enzymes used for these studies. The Nth protein excises mainly pyrimidine base lesions including ring-saturated pyrimidines, such as 5,6-dihydrothymine (DHT), thymine glycol, and AP sites (Demple and Linn, 1980; Breimer and Lindahl, 1984; Dizdaroglu et al., 1993; Hatahet et al., 1994). The Fpg protein, on the other hand, excises mainly purine base lesions, such as 2,6-diamino-4-hydroxyl-5-*N*-methylformamidopyrimidine, 7,8-dihydro-8-oxo-2′deoxyguanine (8-oxoG), and AP sites (Chetsanga and Lindahl, 1979; Tchou et al., 1991; Boiteux et al., 1992). These enzymatic probes convert nucleobase lesions into readily detectable SSBs. Using this method, it was revealed that nucleobase lesions were produced considerably more frequently by irradiation than strand breaks, that is, the yield of total nucleobase lesions induced in dry DNA films was about threefold larger than that of SSBs (Yokoya et al., 2002).

The yields of the nucleobase lesions induced in a diluted solution showed similar values for Nth- and Fpg-treatment (Yokoya et al., 2008a, Figure 20.3). On the other hand, these nucleobase lesions were induced with different efficiencies in the hydrated sample (see Table 20.3). These results indicate that the direct energy deposition of soft x-ray irradiation induced damage in the plasmid DNA through different mechanisms from that induced by reaction with water radicals.

The yields of clustered damage, revealed as the yields of the additional DSBs induced by Nth- and Fpg-treatment, were similar to those for spontaneous DSBs under conditions of lower scavenging capacity, although they were significantly larger than those under conditions of higher scavenging capacity (Figure 20.4). At the highest scavenging capacity (fully hydrated sample), these yields were nearly twice of those for DSBs (see Table 20.3). These results show that direct effects might efficiently induce oxidative nucleobase lesions and clustered damage sites (including nucleobase lesions) rather than DSBs. A theoretical study has also predicted that most SSBs are induced concomitantly with nucleobase lesions (Nikjoo et al., 2001). The hydrating water shell surrounding a DNA molecule helps in inducing oxidative nucleobase lesions, but does not play a significant role in inducing SSBs (Yokoya et al., 2002). Thus, direct energy deposited onto the hydration shell might be transferred to DNA, inducing a large amount of nucleobase lesions as well as base lesion clusters.

In order to elucidate the precursors of DNA damage, a low-temperature electron paramagnetic resonance (EPR) method has been applied to crystalline or dry samples of DNA and its components to investigate free radicals induced by irradiation. These studies have shown that base and sugar radicals were induced by electron trapping (radical anions) or hole transfer (radical cations) (see review by Bernhard and Close, 2004); however, much less is known about the mechanisms of the direct effect,

TABLE 20.3
Yields of DSBs and Enzymatically Visualized Clustered Damage under Various Scavenging Conditions

Tris Concentration (mM)[a]	Prompt DSBs	Prompt DSBs + Heat-Labile DSBs	Nth-Induced DSBs	Fpg-Induced DSBs
0.2	ND	3.33×10^{-9}	1.77×10^{-9}	2.99×10^{-9}
2.5	3.04×10^{-10}	2.32×10^{-10}	2.22×10^{-10}	2.41×10^{-10}
50	6.61×10^{-11}	3.15×10^{-11}	6.47×10^{-11}	1.09×10^{-10}
Hydrated[b]	1.82×10^{-11}	1.60×10^{-11}	3.42×10^{-11}	3.29×10^{-11}

DSB, double-strand breaks.

[a] Concentrations of other solutes in the sample solution: 50 μg/mL pUC18 DNA; 1 mM EDTA (Yokoya et al., 2006).

[b] Fully hydrated samples were prepared by exposing dry pUC18 DNA film to a condition of 97% relative humidity over 16 h. The hydration level was estimated to be 35 water molecules per nucleotide (Yokoya et al., 2002).

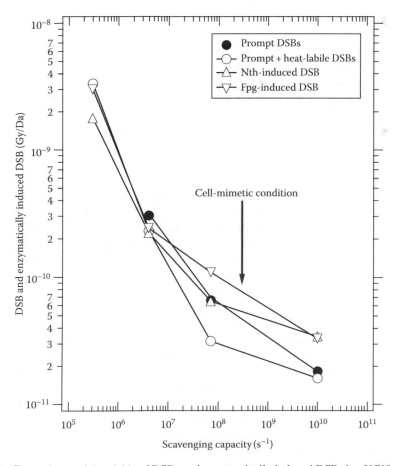

FIGURE 20.4 Dependence of the yields of DSBs and enzymatically induced DSBs in pUC18 plasmid DNA irradiated by soft x-ray on the scavenging capacity at 277 K under an aerobic condition. (●) Prompt DSBs; (○) Prompt + heat-labile DSBs; (△) Nth-induced DSBs; (▽) Fpg-induced DSBs. (Reproduced from Yokoya, A. et al., *Radiat. Phys. Chem.*, 77, 1280, 2008b. With permission.)

because these radical yields were not fully consistent with the final induced biological effects. It has been widely recognized that densely ionizing radiation, such as α-particles, induces cell killing more efficiently than sparsely ionizing radiation, such as γ- or x-rays. The relative biological effectiveness (RBE) of the lethal effect on cells irradiated with α-particles when compared with γ-irradiation is about two. Becker et al. (2003) have reported, however, that when hydrated DNA samples at 77 K are irradiated with high-LET argon ions (60 and 100 MeV/nucleon), nucleobase ion radicals were induced at a lower yield than that obtained with γ-irradiation, although the neutral deoxyribose radicals, which are thought to accompany an SSB, were induced with a higher yield than with γ-irradiation.

Few studies have investigated the nucleobase lesions induced by direct energy deposition from an ion track to a DNA molecule at ambient temperatures. Recent studies using hydrated DNA have shown that the yield of DSBs induced by α-irradiation (Yokoya et al., 2003; LET = 140 keV/μm) was twice that induced by γ-irradiation (Yokoya et al., 2002), indicating that dense ionization or excitation events along the α-particle tracks are more effective at inducing clustered types of strand breaks. Furthermore, irrespective of the hydration level, α-particle irradiation of DNA induced more complex types of clusters of DNA damage, including nucleobase lesions, than photon irradiations such as γ- or hard x-rays. Recently, the track size of high-LET argon ions has been experimentally determined using pulsed electron double resonance spectroscopy (PELDOR) techniques by Bowman et al. (2005). These authors reported that the number of clustered radicals within a 1 nm length of DNA was 2.6, which is consistent with the distribution of lesions in a cluster damage site, that is, 4–10 lesions over several base pairs (Nikjoo et al., 2001).

In order to focus on the clustered damage sites induced by irradiation with high-LET ion particles, the yields of nucleobases have been determined by enzymatic probes that convert nucleobase lesions into readily detectable SSBs (as described previously). In these studies, plasmids or phage DNA were used as models. The yields of damage were measured as a function of the ionization density of the radiation using ion particles from accelerator facilities with LET values. The yields of nucleobase lesions detected by the enzymatic treatments, as well as strand breaks, varied depending on the experimental conditions. In dilute DNA solutions, the yields of SSBs and DSBs decreased with increasing LET, although the ratio of DSBs to SSBs increased with increasing LET (Taucher-Scholz and Kraft, 1999). Similarly, both the nucleobase cluster lesions revealed by Fpg or Nfo proteins and the prompt DSBs induced in T7 DNA decreased with increasing LETs of the ion particles (Hada and Sutherland, 2006). In the dilute DNA solutions, diffusible OH• produced by radiolysis of water contributed primarily to the induction of DNA damage. It has been well understood that the yield of OH• decreases with increasing LET because the higher density of the radicals in the track leads to a higher probability of an intratrack recombination process between the radicals than that for a low-LET radiation track. Recently, Yamashita et al. (2008) discussed the intratrack reactions along the ion-particle tracks using a model experimental system for the formation of methyl viologen cation radicals. On the other hand, when DNA was irradiated with helium ions under hydrated conditions, the yield of prompt SSBs did not depend significantly on the LET of the helium ions, whereas the yield of DSBs increased with increasing LET. The yields of isolated nucleobase lesions revealed by Nth and Fpg as additional SSBs decreased drastically with increasing LET (Figure 20.5A, Urushibara et al., 2008), and very few enzyme-sensitive sites were induced at 120 keV/μm. These results indicate that a cluster of nucleobase lesions induced at 120 keV/μm is less readily repaired by the BER proteins than that induced at a lower-LET region and, therefore, may show a high RBE value for cell killing. The sum of the yields of DSBs and additional DSBs revealed by Nth and Fpg increased with increasing LET. These studies concluded that the yields of clustered damage, revealed as DSB and non-DSB clustered damage sites but not isolated lesions (i.e., SSBs), increased with increasing ionization density of the He ions under 140 keV/μm (Figure 20.5B). Chang et al. (2005) observed no enhancement of cell killing after exposing an Fpg-overexpressing *E. coli* strain to α-rays. These results may reflect the greater complexity of clustered damage sites generated by high-LET radiation, in which the lesions are processed sequentially to avoid the DSB formation as described in Section 20.4.

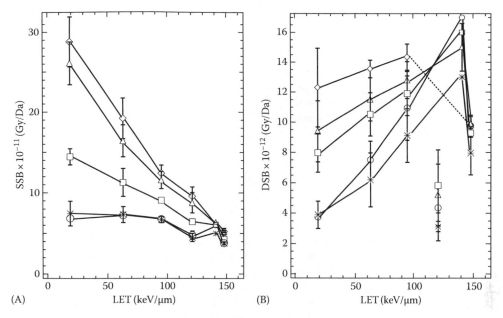

FIGURE 20.5 Dependence of the yield of SSBs (A) and DSBs (B) on LET for fully hydrated DNA irradiated with $^4He^{2+}$ ions at 279 K (○) or after a postirradiation incubation for 30 min at 310 K in the absence (✳) or presence of either Nth (△), Fpg (□), or both Nth and Fpg (◇). The vertical error bars are ±SD for the SSB values determined from the slope of the dose-response curves from three independent experiments. The data at 141 keV/μm taken from the previous report (Yokoya et al., 2003) for α-irradiation are shown for comparison, assuming that the LET of Pu-α particles in the sample film comprising hydrated plasmid DNA and TE buffer solutes (ρ = 1.2) is estimated at 141 keV/μm. (Data from Urushibara, A. et al., *Int. J. Radiat. Biol.*, 84, 23, 2008.)

20.4 BIOLOGICAL RELEVANCE OF CLUSTERED DAMAGE

Individual lesions, whether they are clustered or isolated, are thought to be recognized and processed mainly by base-excision repair (BER). In the initial phase of BER, DNA glycosylases recognize the damaged base and excise the N-glycosidic bond to generate an AP site. A variety of nucleobase lesions are recognized by several DNA glycosylases, each of which shows broad substrate specificities. The DNA backbone at the AP site is excised by the accompanying AP lyase activity of the glycosylase or by the AP endonucleases (step 1) to generate an SSB. Blocks at the 3′-end of the SSB generated by the AP lyase activity are also removed by AP endonucleases. One or more nucleotides are inserted at the SSB (step 2), followed by displacement of the strand (step 3), cleavage of the displaced nucleotide(s), and ligation of the break (step 4). Depending on the number of nucleotides inserted, BER proceeds through either of the two subpathways: short-patch repair or long-patch repair. The major enzymes from different species involved in the process are listed in Table 20.4.

Although a considerable amount of clustered DNA damage appears to be induced in cells, how and to what extent clustered DNA damage is processed in vivo and, perhaps more importantly, is related to biological consequences have long remained unknown. Apart from the difficulties in interpreting the results of clustered damage revealed as DSBs by glycosylases or AP endonucleases, this is probably because (1) radiation-induced clustered DNA lesions are difficult to fully detect experimentally and (2) they have random configurations in terms of the types, numbers, and relative positions of the lesions. To overcome the difficulty posed by the random nature of radiation-induced lesions, many research groups have used synthetic clusters, in which the type, the number, and the relative position of the lesions are specified, to examine how the clusters are processed. Whether a clustered DNA damage site is indeed relevant to biological end points has mostly been investigated using plasmid-based assays with model synthetic lesions. A plasmid-based assay, in principle,

TABLE 20.4
Enzymes Associated with BER and SSBR

	Enzymes	Species	Functions
Glycosylase	Fpg	*E. coli*	Removal of damaged purines
	yOgg1	*S. cerevisiae*	
	hOGG1	human	
Glycosylase	Nth	*E. coli*	Removal of damaged
	Ntg1	*S. cerevisiae*	pyrimidines
	Ntg2	*S. cerevisiae*	
	yNTH1	Human	
Glycosylase	Nei	*E. coli*	Removal of damaged
	hNEIL1	Human	pyrimidines
	hNEIL2	Human	
AP endonuclease	Xth	*E. coli*	Incision at AP sites
	Nfo	*E. coli*	(including oxidized AP)
	APE1	*S. cerevisiae*	
		Human	
Polymerase	PolI	*E. coli*	Insertion of nucleotides
	Polβ	Human	
	Polδ	Human	
dRPase/FLAP endonuclease	PolI	*E. coli*	Removal of
	Polβ	Human	deoxyribophosphate/FLAP
	Polδ	Human	
	FEN1	Human	
Ligase	Ligase	*E. coli*	Sealing of breaks
	LigaseI	Human	
	LigaseIIIα	Human	

BER, base-excision repair; SSBR, single-strand break repair; AP, abasic sites.

transforms cells with model clusters that are ligated into the plasmids. Transformation efficiencies as well as mutation frequencies are monitored to evaluate the ability of the plasmids with clusters to replicate and generate mutations in cells. Various types of synthetic nucleobase lesions, such as 8-oxoG, thymine glycol (Tg), DHT, dihydrouracil (DHU), 5-hydroxyuracil (5-OHU), as well as AP sites, have been subjected to this analysis.

Several studies indicated that the mutagenic potential of an 8-oxoG is enhanced in *E. coli* cells by the presence of another nucleobase lesion or an AP site within close proximity on the opposite strand (Malyarchuk et al., 2003, 2004; Pearson et al., 2004; Shikazono et al., 2006; Bellon et al., 2009). An 8-oxoG is a mutagenic lesion, and *E. coli* is known to avoid mutagenesis by this lesion with at least two DNA glycosylases: Fpg, which excises 8-oxoG residues from DNA, and MutY, which excises adenine residues incorporated opposite 8-oxoG after DNA synthesis or replication. With clusters containing 8-oxoG, a lack of MutY led to a marked increase in mutation frequency compared with that in wild type. The importance of MutY in mutagenesis suggests that 8-oxoG:adenine base pairs can arise from the presence of 8-oxoG at replication because of retardation or absence of excision in the cluster; however, they are readily repairable, because, after replication, they are no longer part of a cluster. The frequencies of mutation are very similar for the DHT/8-oxoG cluster and the 8-oxoG/8-oxoG cluster in the same bacterial strains, suggesting that the same key intermediate (AP/8-oxoG or SSB/8-oxoG) is involved. In the studies with clusters containing 8-oxoG, the majority of the mutations found were G-to-T transversions at the 8-oxoG sites, with few deletions. In addition, the transformation efficiency of bacteria by plasmids carrying these clusters was comparable to that by undamaged DNA, which indicated that most of these clusters were not processed into DSBs during repair. The in vivo results from these model bistranded damage sites indicate that (1) the repair of an 8-oxoG is retarded so that some of the clusters remain partly unrepaired by the time of replication, (2) lesions are repaired sequentially and the formation of DSBs is minimized, and (3) there appears to be a preferential order for excising different lesions (Figure 20.6). These suggestions have recently been reinforced by the results of a study by Kozmin et al., in which they used up to five different lesions in yeast (Kozmin et al., 2009). The existence of a hierarchy points toward the biological importance of the complex interplay between various repair proteins to avoid the generation of DSBs that could be deleterious to the cells. In contrast, the mutagenic potential of an 8-oxoG placed in tandem with an AP site remained similar to that of a single 8-oxoG in wild-type cells of *E. coli*, and even decreased in *fpgmutY* cells (Cunniffe et al., 2007). This implies that the position of the base damage greatly influenced the mutagenic consequences of a cluster. A model that explains this difference in mutagenesis is shown as follows. In a bistranded cluster, the strand opposite 8-oxoG could undergo a long-patch repair or its rate of amplification could be reduced, so that the 8-oxoG-carrying strand serves as the major template for DNA synthesis (Figure 20.6A). This might lead to an enhanced mutagenic potential of 8-oxoG. It is important to note that with a long-patch repair, as long as the repair patch size is larger than the distance between lesions, a nucleotide will always be inserted opposite 8-oxoG by repair synthesis. In the case of tandem clusters with an adjacent lesion of 8-oxoG positioned on the same strand, a similar scheme results in the induction of fewer or no mutations because an undamaged strand is the more preferred template for DNA synthesis.

With bistranded uracils, which are thought to be quickly converted into bistranded AP sites in vivo, the induction of DSBs through the processing of clustered DNA damage in *E. coli* has been inferred from the formation of deletions (Dianov et al., 1991). In addition, a reduction of transformation efficiency, implying the formation of DSBs, was also observed when the two uracils (Us) or AP sites were positioned in close proximity on both strands in *E. coli* (separated by less than 7–8 bp) (D'Souza and Harrison, 2003; Shikazono and O'Neil, 2009), as well as in

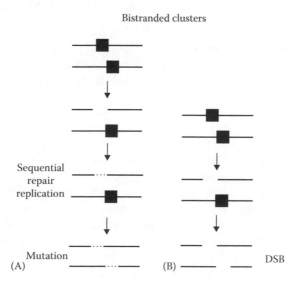

FIGURE 20.6 Pathways of the processing of bistranded clusters in vivo. Black squares indicate DNA lesions. (A) Base lesions are processed and repaired sequentially by BER to avoid the formation of a DSB. The unrepaired base damage may serve as a template for repair synthesis or replication, resulting in mutations. (B) AP sites are processed simultaneously before completion of repair, resulting in the formation of a DSB.

yeast (Kozmin et al., 2009). Interestingly, *in E. coli*, a reduced transformation efficiency was still found in the absence of the enzymatic activities, such as those of AP endonucleases (Xth, Nfo, and endoV), AP lyases (Nth, Fpg, and Nei), and nucleotide excision repair (NER) (UvrA), needed for the incision of an AP site (Harrison et al., 2006). A similar situation with low transformation efficiency of bistranded Us or AP sites in mutants deficient for AP endonuclease (apn1apn2 double mutant) was also found in yeast (Kozmin et al., 2009). These results suggest that bistranded clusters comprising lesions that block replication strongly diminish the transformation efficiency. It has recently been shown in mouse cells that two opposed tetrahydrofurans could be cleaved into a DSB by AP endonuclease(s), and a fraction of the lesions could be inaccurately repaired by NHEJ repair resulting in deletions (Malyarchuk et al., 2008). Tetrahydrofuran is a stable AP-site analog. This formation of DSBs in mammalian cells is consistent with the results from in vitro processing of bistranded AP sites in cell extracts. The bistranded clusters comprising an AP site and a one-nucleotide gap (GAP) have also been shown to have very low transformation efficiencies, similar to those of bistranded Us or AP sites (Shikazono and O'Neil, 2009). Taking these results together, it was inferred that bistranded nucleobase lesions, that is, single-base lesions on each strand but not clusters containing only AP sites and strand breaks, are repaired in a coordinated manner; thus, the formation of DSBs is avoided. In other words, when either nucleobase lesion is initially excised from a bistranded damage site, the remaining nucleobase lesion will only rarely be converted into an AP site or an SSB in vivo.

All of the above information suggests that the presence of one or more vicinal lesions affects the rate, fidelity, and pathways of DNA repair and determines the outcome of processing. It is worth noting that in most, if not all, cases, the repair of clustered lesions was retarded compared with the repair of isolated lesions. The less effective repair of lesions in a cluster should not simply be considered deleterious or harmful to cells, as it may often protect against the formation of lethal DSBs. A major consequence of retarded processing of clustered DNA damage is the extended lifetime of lesions within the cluster, which allows the cluster to be present at replication.

20.5 INTERMEDIATE SPECIES AND DEGRADATION PROCESSES OF DNA REVEALED BY SYNCHROTRON SOFT X-RAYS AS PROBES

In order to overcome the difficulty posed by the random nature of the radiation damage to DNA, atom-selective ionization is another powerful tools that induces spatially regulated energy deposition in DNA. In general, as discussed above, it is difficult to induce specific lesions in DNA using conventional radiation sources such as γ- or hard x-rays, which nonspecifically ionize the DNA's constituent atoms and cause damage with random configurations. In order to achieve atom-specific ionization, many studies have used soft x-ray photons, particularly those with energies below 10 keV. Soft x-ray photons mainly interact with matter in a living cell through a photoelectric process: as a consequence, a photoelectron and Auger electrons are ejected from the atom that absorbs the soft x-ray photon. The biological effects induced by soft x-irradiation are thought to arise from the formation of DNA damage through both the ionization of DNA and the impact of the secondary electrons, which ionize or excite nearby molecules through inelastic scattering. The main products resulting from such electron loss are oxidized nucleobases, such as 8-oxoG.

Synchrotron radiation has been used as an intense soft x-ray source. Monochromatic photons from a high-resolution monochromator can be used to induce inner-shell photoionization at a particular atom in order to explore the details of the photoelectric effect in biological samples (Hieda and Ito, 1991). The atom-selective irradiation of a biological system provides potentially new insights into the role of photoabsorption events for each constituent element of the DNA in the induction of the final biological effects (Figure 10.7). *K*-shell ionizations of carbon, nitrogen, and oxygen atoms in DNA have been explored using monochromatic soft x-ray photons around the *K*-edge energies: carbon, 284 eV; nitrogen, 410 eV; and oxygen, 543 eV. Early studies, however, reported that the yields of DNA strand breaks were almost constant or slightly enhanced at

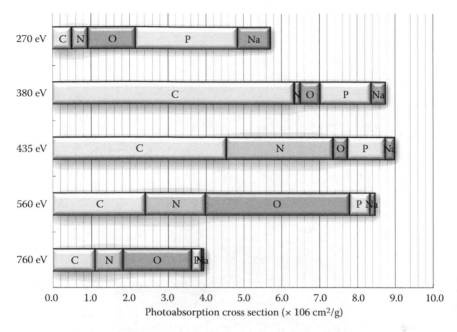

FIGURE 20.7 Photoabsorption cross section of pUC18 plasmid DNA at the energies above phosphorus *L*-edge (below carbon *K*-edge, 270 eV), above carbon *K*-edge (380 eV), above nitrogen *K*-edge (435 eV), above oxygen *K*-edge (560 eV) and below phosphorus *K*-edge (760 eV). The partial cross sections corresponding to each atom are shown by hatched areas.

the oxygen *K*-edge energy (Yokoya et al., 1999; Fayard et al., 2002). Recent studies also showed a similar enhancement of the yield of DSBs above the oxygen *K*-edge region by a factor of 1.4–2 (Eschenbrenner et al., 2007; Agrawala et al., 2008). However, these energy dependencies of the yields of DNA strand breaks are not larger than those initially expected. As indicated by Goodhead, how inner-shell ionization can cause specific biological and biochemical effects has been a long-standing question (Goodhead, 1994b and Goodhead et al., 1981). In order to answer this question and understand the mechanism by which soft x-irradiation produces such a high efficiency of biological effects, not only DNA strand breaks but also oxidized nucleobases as discussed above should be examined.

Knowledge about the yield of nucleobase lesions has been, however, very scarce in the studies of DNA damage by soft x-irradiation. Only the yield of nucleobase lesions detected by Fpg treatment has been reported in a DNA film irradiated with monochromatic soft x-rays (250, 380, and 760 eV) (Agrawala et al., 2008). We still lack clear evidence to determine whether the photoelectric effect on the DNA's constituent atoms induces particular types of nucleobase lesions as well as DNA strand breaks.

Recently, Fujii and coworkers determined the yields of pyrimidine and purine nucleobase lesions as well as SSBs and DSBs induced in dry DNA films by soft x-irradiation using monochromatic synchrotron radiation in the carbon, nitrogen, and oxygen *K*-edge regions (280–760 eV) (Fujii et al., 2009) using a soft x-ray beamline at SPring-8 (Figure 20.8). The role of the photoelectric effect on the DNA's constituent atoms was discussed in terms of the yield of nucleobase lesions relative to strand breaks. They also showed the potential utility of the selective induction of a specific lesion achieved by irradiation with highly monochromatic synchrotron radiation in the research field of DNA damage and its enzymatic repair processes. The dependence of the damage yield on photon energy is shown in Figure 20.9. The main findings from their study on the direct ionization of DNA's constituent atoms are as follows: (1) The yields of both oxidative pyrimidine and purine nucleobase lesions, revealed as Nth- and Fpg-sensitive sites, respectively, were strikingly enhanced by oxygen

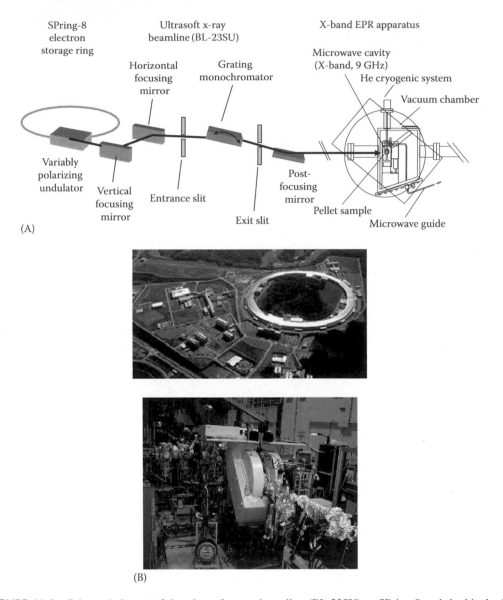

(A)

(B)

FIGURE 20.8 Schematic layout of the ultrasoft x-ray beamline (BL-23SU) at SPring-8 and the biophysics end-station (A) and the EPR apparatus installed in the end-station (B). A variably polarizing undulator (APPLE II type) was used as an intense photon source of ultrasoft x-rays. High-energy resolution ($E/\Delta E \sim 10,000$) of the ultrasoft x-ray energy selected by rotating the valid-line-spacing grating (Saitoh et al., 2001) was achieved.

K-ionization at 560 eV and were much lower at the photon energy just below the nitrogen K-edge (380 eV) than those observed at the other photon energies tested. (2) The yield of prompt SSBs was also enhanced by oxygen K-ionization, but not as distinctly when compared with the yield of nucleo-base lesions. (3) The yield of the purine lesions (Fpg-sensitive sites) induced by 760 eV photons was significantly lower than that of pyrimidine lesions (Nth-sensitive sites). (4) Finally, clustered lesions detected as DSBs after treatment with Nth or Fpg were similarly enhanced by oxygen K-ionization (560 eV), and also showed large yields from 270 eV photons, although the yield of prompt DSBs was only enhanced by oxygen K-ionization. Recently, Agrawala et al. (2008) also reported similar values for Fpg-sensitive sites from three soft x-ray energies (1.05, 0.49, and 0.92×10^{-11} Gy^{-1}/Da for 250, 380, and 760 eV photons, respectively).

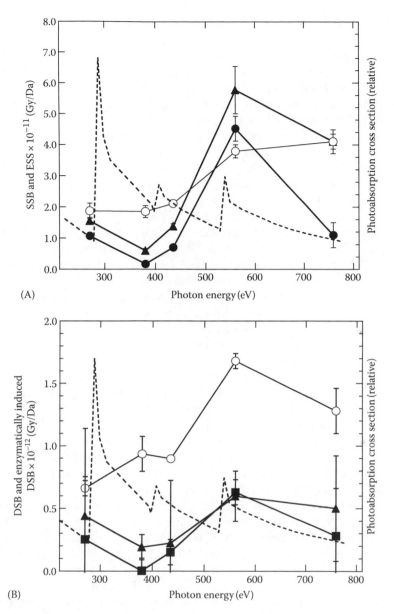

FIGURE 20.9 Dependence of the yield of Nth- and Fpg-sensitive sites and prompt SSBs (A) and prompt and enzymatically induced DSBs (B) induced by irradiation of pUC18 plasmid DNA on the photon energy at 277 K under a vacuum. The dotted lines are photoabsorption cross sections of DNA. (Reproduced from Fujii, K. et al., *J. Phys. Chem.*, 113, 16007, 2009. With permission.)

The irradiation of nucleobases with ionizing radiation is known to produce a variety of unpaired electron species observed as free radicals by EPR or electron and nuclear double resonance (ENDOR) methods (see review by Bernhard and Close, 2003). Purkayastha et al. (2006) have studied the *G*-values of free radicals trapped in hydrated DNA that were directly ionized by x-irradiation (70 kV, tungsten target) using a low-temperature (4 K) EPR technique. They suggested that the observed free radicals were mainly localized at the nucleobases (80%–90%). These trapped radicals were divided into two groups: purine radical cations resulting from one-electron oxidations or holes, and pyrimidine radical anions resulting from one-electron attachments (Bernhard and Close, 2003). The pathways that produce the chemically stable nucleobase lesions initiated by the production of oxidative or reductive radicals are reviewed in the previous chapter (see Schemes 19.1, 19.2, and 19.4).

One of the major nucleobase radicals is the cytosine anion radical formed by trapping one electron at low temperature (4–10 K). At higher temperature (>180 K), it is primarily thymine that traps a single electron to produce the thymine anion radical. This radical easily converts to 5,6-dihydro-thymine-5-yl- radical (5-thymyl radical) and shows a specific eight-line EPR spectrum (Ormerod, 1965). The 5-thymyl radical is likely to be a precursor of DHT, which is one of the primary substrates of the Nth protein. When a K-shell electron in the constituent atoms of thymine is ionized by soft x-irradiation, one or two holes are left in the thymine as a consequence of the Auger process, and an electron adduct is not likely to be produced at the thymine. Nevertheless, the 5-thymil radical was one of the main products induced by oxygen K-shell photoabsorption in a thymine pellet sample that was irradiated with 538 eV soft x-rays in a vacuum at 77 K (Figure 20.10A; Akamatsu et al., 2004a). The dose response of the total spin number obtained from the EPR spectrum was similar to that found using a 407 eV photoirradiation (nitrogen K-shell photoabsorption) in the low dose range (Figure 20.10B). An high performance liquid chromatography (HPLC) analysis revealed that the yield of DHT produced in the thymine-pellet sample irradiated in a vacuum with 538 eV soft x-rays was slightly lower than for irradiation with soft x-rays (395 or 407 eV) below the oxygen K-edge (Figure 20.11; Akamatsu et al., 2004b). These experimental results suggest that the production of DHT is not enhanced by oxygen K-ionization when thymine exists alone, and not as a part of DNA. In other words, either oxygen K-ionization induced in not only thymine but also in the sugar-phosphate backbone, or hydrating water molecules, which bind tightly to DNA even in vacuum, was responsible for the strong enhancement of the yield of Nth-sensitive sites. Furthermore, we must examine the source of hydrogen atoms that are bound to the C6 or both the C5 and C6 positions in thymine to produce a 5-thymyl radical or DHT. Fujii et al. (2004a) reported that H$^+$ is the major ion desorbed by a 538 eV soft x-irradiation of a thin film of thymine as well as 2-deoxy-D-ribose, thymidine (dThd), and thymidine 5′-monophosphate (dTMP) (Figure 20.12). These results suggest that H$^+$ was overwhelmingly produced in the sample by oxygen K-ionization. In addition, the 2.5 hydrating water molecules per nucleotide, which inevitably exist in DNA samples even under high vacuum (Tao et al., 1989), could be the hydrogen source. Thus, we propose a model of production of Nth-sensitive sites enhanced by oxygen K-ionization, as shown in Scheme 20.1.

Similarly, oxygen K-ionization of the functional groups surrounding guanine or adenine in DNA could be responsible for the production of Fpg-sensitive sites. Guanine is known to be a major hole-trapping site in DNA because it has the lowest ionization potential (in the gas phase) among the nucleobases (see the review by Bernhard and Close, 2003). Not only oxygen K-ionization of the carbonyl group at the C6 position but also hole transfer from the other ionizing sites to guanine may produce a guanine cation radical. The production of the oxidative guanine lesions, 8-oxoG or 2,6-diamino-4-hydroxy-5-formamindopyrimidine (Fapy-Gua), which are substrates of the Fpg protein, requires the addition of an oxygen atom to an additional carbonyl group at the C8 position of guanine (or the guanine cation radical). This oxygen atom should come from outside of the parent guanine molecule. The hydrating water molecule proximately located to guanine is probably a source of the oxygen atom. K-ionization of the oxygen atom in the hydrating water leaves a chemically reactive oxygen anion (or radical), which could easily react with guanine to produce the oxidative guanine lesions (Scheme 20.2). The yields of enzymatically induced DSBs were enhanced by not only oxygen K-ionization at 530 eV but also irradiation with 270 eV (below the carbon K-edge) photons. On the other hand, the prompt DSBs induced by 270 eV photons showed the lowest yield of the energies tested and were significantly enhanced by oxygen K-ionization. As presented above, oxygen K-ionization increases the yield of nucleobase lesions. Consequently, the yields of the clustered nucleobase lesions are thought to increase above the oxygen K-edge energy. The clustering of nucleobase lesions might be induced through valence electron ionization or phosphorus L-ionization by 270 eV photons (below the carbon K-edge). The ionization of phosphorus L-shell electrons produces phosphorus LMM Auger electrons (~120 eV) (Watanabe et al., 2004). Yokoya et al., have

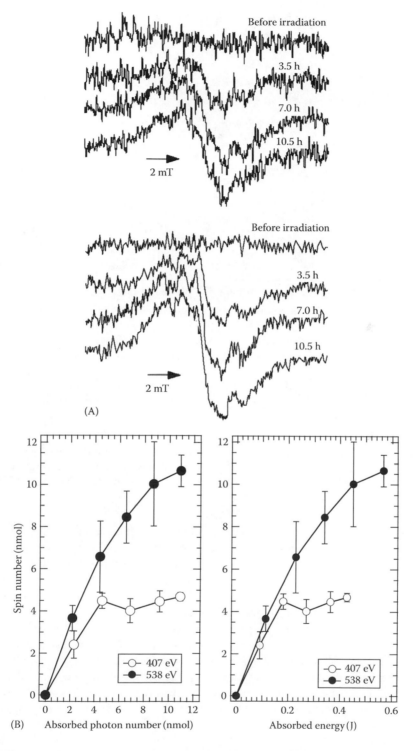

FIGURE 20.10 (A) EPR signal changes of thymine irradiated with 407 eV (upper) and 538 eV (lower) soft x-rays. The irradiation periods (h) are shown in the figure. (B) Dose-response relationship between the spin number obtained from EPR spectra of irradiated thymine with 407 (O) and 538 eV (●) photons at 77 K in a vacuum against absorbed photon number (left) and absorbed energy (right). (Reproduced from Akamatsu, K. et al. *Int. J. Radiat. Biol.*, 80, 849, 2004a. With permission.)

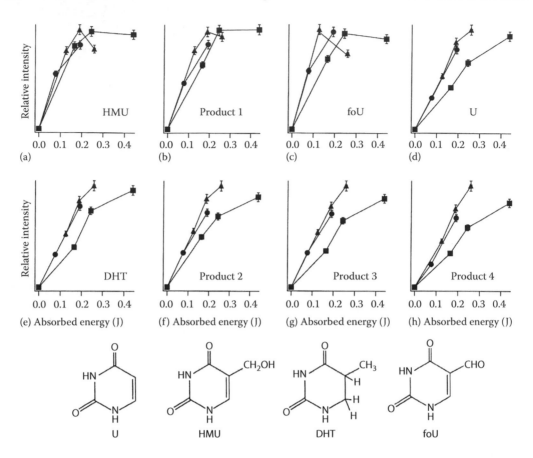

FIGURE 20.11 Dose-response profiles and molecular structures of the thymine decomposition products analyzed by the HPLC method. The thymine pellet was irradiated with soft x-rays at 395 eV (●), 407 eV (▲), and 538 eV (■) at 298 K under vacuum. (Reproduced from Akamatsu, K. et al., *Radiat. Res.*, 162, 469, 2004b. With permission.)

recently reported that these LMM Auger electrons were responsible for the clustering of the nucleobase lesions induced by phosphorus K-ionization (Yokoya et al., 2009a). However, irradiation with 380 eV photons also produced photoelectrons of similar energy (~100 eV) from the carbon K-shell. Although the detailed mechanism has not yet been clarified, not only the core ionizations but also the valence electron ionization and the resulting hole in the DNA may be involved in causing serious DNA damage, such as the clustering of nucleobase lesions.

Recently, Yokoya et al. (2008a) have studied short-lived unpaired electron species produced in DNA films by soft x-ray irradiation using an EPR spectrometer installed in a synchrotron soft x-ray beamline (Figure 20.8). They reported that a significant EPR signal from the short-lived transient species was observed only during soft x-irradiation (Figure 20.13; Yokoya et al., 2008a). Although the short-lived EPR spectrum has not yet been assigned to any specific molecular species, one of the possible origins of the induction of the sharp lines (see Figure 20.13A) is that the low-energy electrons generated as a consequence of inelastic scattering of photo- or Auger electrons in the sample might be trapped into a specific molecular site that attracts an electron (Yokoya et al., 2009b). The EPR signal intensity was significantly enhanced at the oxygen K-edge but not so significantly at the nitrogen K-edge (Figure 20.14). They also found the short-lived species arising in evaporated films of guanine and adenine (Yokoya et al., 2009b). In this case, the photon energy dependence of the EPR intensities of the short-lived species coincided with the photoabsorption spectra of the nucleobases, showing significant enhancement of the EPR intensity at the nitrogen K-edge region (Figure 20.15A).

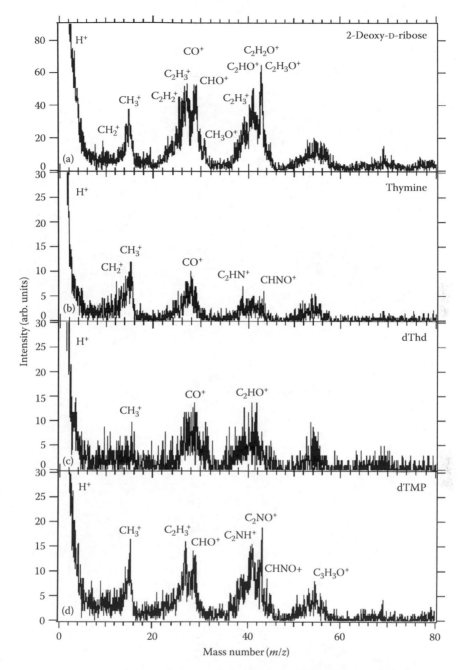

FIGURE 20.12   Mass spectra of positive ions desorbed from a thin film of DNA constituent molecules irradiated with soft x-rays (538 eV; oxygen K-edge) at 298 K under vacuum. (Reproduced from Fujii, K. et al., *Radiat Res.*, 161, 435, 2004a. With permission.)

Interestingly, the enhancement of the EPR intensity of the evaporated adenine film was reduced almost completely at the nitrogen K-edge region after a slight exposure to water vapor in the vacuum. Instead, a significant oxygen K-edge structure appeared in the spectrum of the photon energy dependence, although adenine does not have any oxygen atoms (Figure 20.15B). These results obtained from the EPR experiments also support the idea discussed above that the water molecules hydrating the DNA play an important role in the induction of purine nucleobase lesions. We have reported

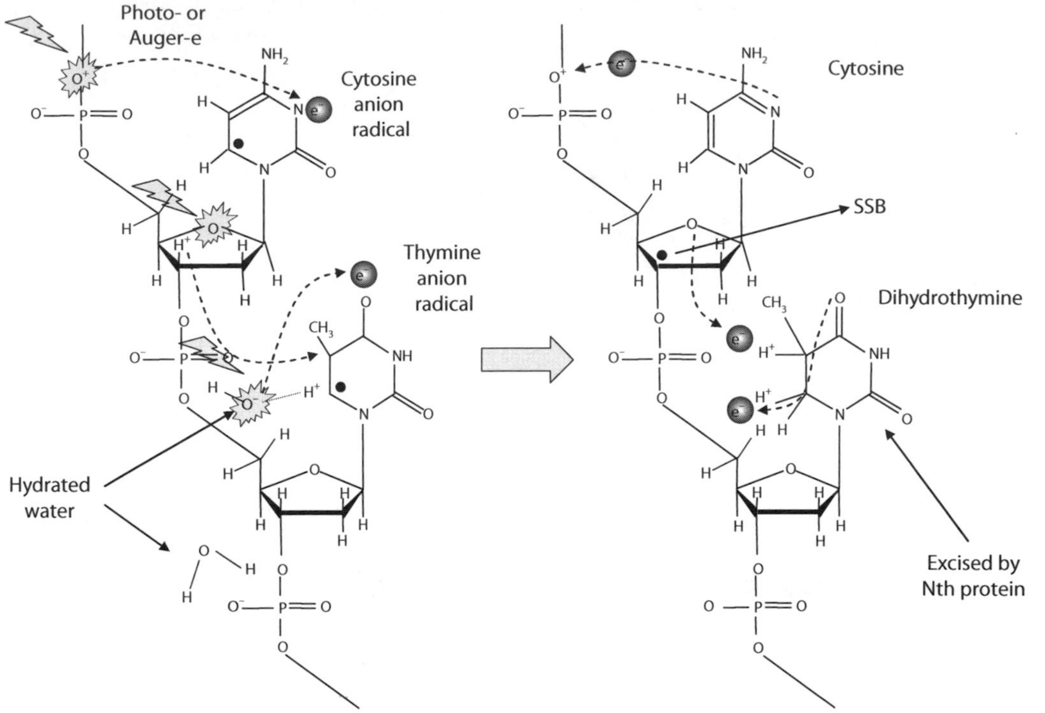

SCHEME 20.1 Induction of pyrimidine base damage by oxygen K-ionization.

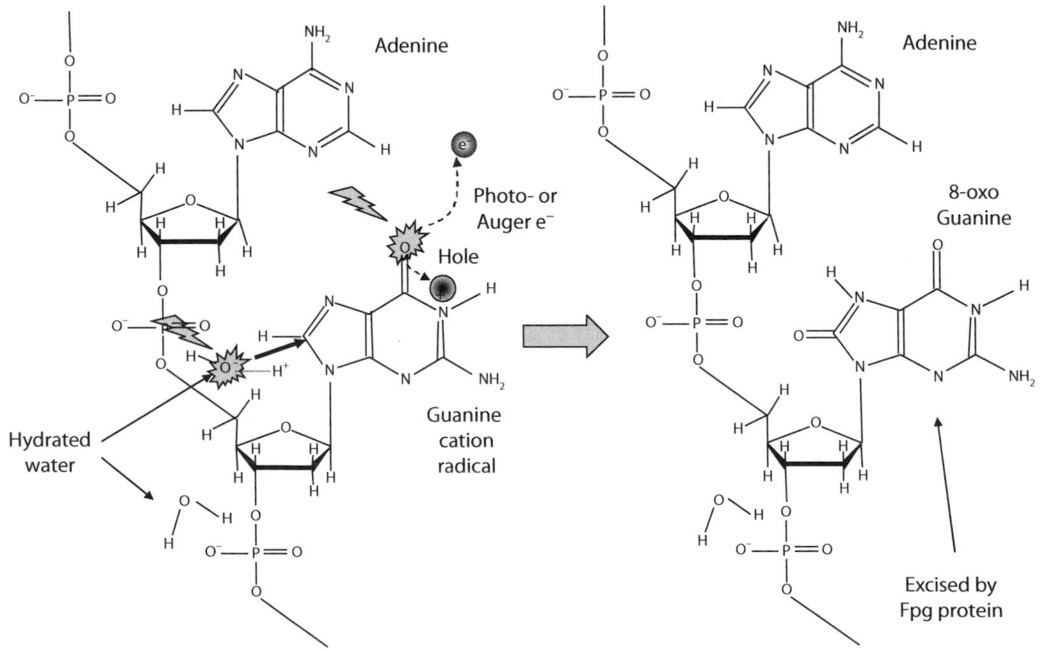

SCHEME 20.2 Induction of purine base damage by oxygen K-ionization.

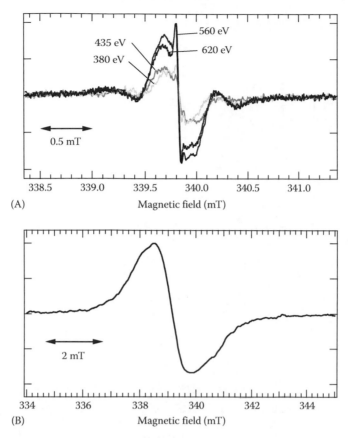

FIGURE 20.13 (A) EPR spectra of short-lived unpaired electron species obtained by in situ EPR measurement. Thin films of calf thymus DNA were irradiated with various energies of soft x-rays at 298 K under vacuum. (B) EPR spectrum of stable radical induced in a DNA pellet sample obtained after 150 kVp soft x-ray irradiation. (Data from Yokoya, A. et al., *Int. J. Radiat. Biol.*, 84, 1069, 2008a.)

that hydrating water molecules surrounding the DNA significantly increased the yield of nucleobase lesions, but not that of SSBs when irradiated with γ-radiation (Yokoya et al., 2002).

The yield of the prompt SSBs is also enhanced by oxygen K-ionization, but not as remarkably when compared with the yield of nucleobase lesions. Ejected electrons in DNA would also cause molecular damage, including not only the normal ionization of DNA but also the resonantly dissociative attachment process. The latter process preferentially induces DNA strand breaks (Boudaïffa et al., 2000). The effect of electron impacts would be nonspecifically induced regardless of the atom from which the secondary electrons were ejected. In our previous studies, the prompt SSB yield for dry pUC18 plasmid DNA was almost constant for a variety of radiations ($0.5–1.2 \times 10^{-10}$ SSB/Gy/Da), indicating that the DNA strand breaks were mainly induced by random hits of the secondary electrons. Ito and Saito (1988) reported that the main target for creating SSBs in solid-state DNA was the pentose ring (deoxyribose). We also concluded that the deoxyribose was a more fragile site than the nucleobases, as revealed by an ion desorption mass spectroscopy study using soft x-rays at the oxygen K-edge (Fujii et al., 2004a). The induction of SSBs would not be very sensitive to the soft x-ray energy.

Thus, irradiation with monochromatic soft x-rays allows us to induce quasi-selective damage in DNA following the characteristic branching ratios, as shown in Figure 20.16, by tuning the energy to specific K-edge regions. SSBs are predominantly induced at 380 eV, and nucleobase lesions, in particular Fpg-sensitive sites, are poorly induced. On the other hand, both SSBs and nucleobase

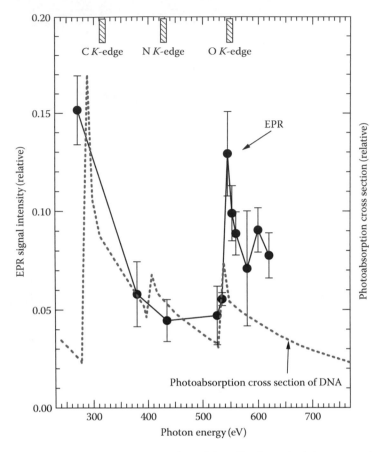

FIGURE 20.14 Dependence of the spin concentration of short-lived unpaired electron species induced in thin films of calf thymus DNA on soft x-ray photon energy at 298 K under vacuum. The dashed line shows the photoabsorption cross section of DNA. (Reproduced from Yokoya, A. et al., *Int. J. Radiat. Biol.*, 84, 1069, 2008a. With permission.)

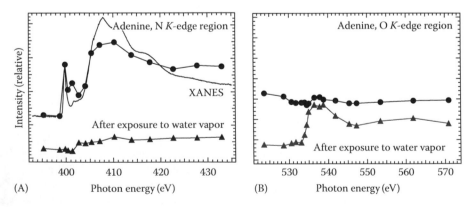

FIGURE 20.15 Dependence of the spin concentration of short-lived unpaired electron species induced in an evaporated adenine thin film on the soft x-ray photon energy around the nitrogen (A) and oxygen (B) *K*-edges (●) at 298 K under vacuum (Yokoya et al., 2009b). The solid red line in (A) shows the XANES spectrum of an adenine thin film cited from Fujii et al. (2004b). The XANES spectrum around the oxygen *K*-edge is not shown in (B) because adenine has no oxygen atoms. The spin concentration obtained by irradiation of the sample with soft x-rays after exposing the sample to water vapor at low pressure of water vapor (~10^{-5} Pa for over 6 h) is plotted for comparison (▲). (Reproduced from Yokoya, A. et al., *Radiat. Phys. Chem.*, 78, 1211, 2009b.)

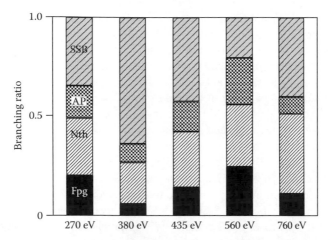

FIGURE 20.16 Dependence of the relative yields of DNA lesions on soft x-ray photon energy at 298 K under vacuum. The total yield of damage is normalized to one. SSB, single-strand breaks; Nth, base lesion excised by Nth protein; Fpg, base lesion excised by Fpg protein. The yield of the common substrates AP sites was estimated using the data obtained by simultaneous treatment with Nth and Fpg (Fujii et al., 2009), according to the following equation: yield of common substrates = Nth-sensitive sites + Fpg-sensitive sites − (yield of additionally induced SSB by Nth + Fpg treatment).

lesions are satisfactorily induced above the oxygen K-edge region. In particular, Nth-sensitive sites are the most probable damage induced by irradiation with 760 eV photons. The characteristic distribution of damage depending on the photon energy might offer a new technique for studying enzymatic DNA repair processes in living cells.

20.6 ROLE OF WATER MOLECULES SURROUNDING DNA AND PHOTOELECTRON SPECTROSCOPY FOR DNA SOLUTIONS

As described above, water molecules, in particular, the hydration layer that is located very close to the DNA, are heavily involved in the induction of nucleobase lesions as well as strand breaks and non-DSB types of clustered damage. The geometric structures of DNA with hydrated water molecules in solution and their electronic states are of great importance in exploring the interaction of radiation with DNA and in understanding the mechanisms of damage induction and of spatial- and time-propagation of unstable species in solution. The primary processes induced by the interaction of radiation with biochemical molecules in water solution can provide a proper prototype of the above processes in cell-mimetic conditions. It can be safely assumed that the conformational structure of hydrated DNA is determined by the hydrogen-bonded "hydration network" of the water molecules surrounding it. These conformational structures further determine the electronic energies of the DNA–water complex and, presumably, the chemical reaction pathways from bulk water to DNA. Becker et al. (1994) and La Vere et al. (1996) proposed for the first time that the hydration network surrounding DNA plays an important role in terms of the hole transfer from the first glassy hydration layer to DNA and hydroxyl radical formation in the surrounding water layers. This assumption also allows us to have the following thermodynamic perspective on the damage caused by radiation to DNA and its electronic or chemical restoration.

The intact genomic DNA in a cell is in its thermochemically and biologically stable state defined by the hydration-network-aided natural conformation, namely, the B-form or the so-called Watson–Crick conformation. The primary irradiation products, that is, the ionic and excited sites of DNA induced by the primary processes during interaction with radiation, are regarded as both the cores of the remaining radiation damage and localized excess energies. Some of the excess energies diffuse into the surroundings with time, which either gives rise finally to a metastable thermal state involving stable damage or restores the original state of the natural molecular form. The hydration network plays an important role both as the intermediary of heat transport to the environment and as the structural memory of the original hydrogen-bonded conformation in the restoration processes. It is important to note the necessity of experimentally observing the stable and/or dynamic hydrated structures of DNA in solution from the perspective of its electronic properties. The hydration of DNA should also influence the eigenstate energy of the excited molecular orbital more than the valence or inner-shell orbitals, so that the photoabsorption spectrum, presenting the excited orbitals, should show an environmental "chemical shift" for molecules in water solution. Recently, a new method of spectroscopy for molecules in liquid solution using free liquid jet samples in vacuum has been proposed combined with soft x-ray synchrotron radiation. The technique of liquid beams in vacuum (Fuchs and Legge, 1979; Faubel et al., 1988; Faubel and Kisters, 1989; Faubel and Steiner, 1992) has been employed in several experiments using physicochemical spectroscopy (Siegbahn, 1974; Faubel et al., 1997, 1999; Morgner, 1998). However, biological applications are few.

The function of the liquid jet technique is to delay the actual phase transition and achieve a practical size for liquid samples used during experiments in vacuum vessels. Let us briefly consider the phase state of the liquid water jet sample in the vacuum, which will be further examined below along with the result of the temperature-dependent experiment. Figure 20.17 illustrates the phase diagram of water and the change in the thermal state of the liquid water beam. The liquid water introduced through the micrometer-sized orifice into the vacuum with a significant amount of stagnation pressure finds itself in the gas phase because of the abrupt decrease in atmospheric pressure. However, because the amount of internal energy required to release each molecule in the liquid phase into the gas phase is not contained in the thin liquid beam, substantial evaporation occurs

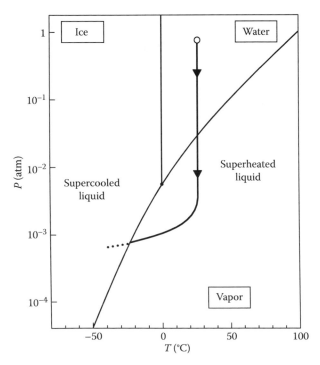

FIGURE 20.17 Schematic phase diagram for a liquid water beam. Thin solid curves indicating the surface between the phases are calculated theoretically on the basis of the Clausius–Clapeyron equation. The thick solid curve approximates the thermal state of a liquid water beam starting at 1×10^5 Pa and 300 K in the liquid phase and ending in the solid phase in vacuum.

only from the surface of the beam. The release of a certain amount of heat of evaporation gradually decreases the temperature of the liquid as a function of the flight distance of the liquid beam in vacuum (see Fuchs and Legge, 1979; Faubel et al., 1988; Faubel and Kisters, 1989; Faubel and Steiner, 1992), so that most of the water molecules remain in the liquid or in the superheated liquid phase. Further decreases in the temperature moderate the emission of heat of evaporation, and then the molecules remain in a supercooled state even when they find themselves in the solid phase. Thus, from an experimental perspective, a certain retention time or retention distance is available until the liquid is finally caught in the solid phase.

For reference, on the basis of the above consideration of the thermodynamic behavior of a liquid in vacuum, Ukai et al. (2008) calculated the approximate temperature of the liquid beam as a function of the flight distance in vacuum. For example, the temperatures of the liquid water admitted through a 10 µm orifice into vacuum with stagnation pressures of 1, 3, or 8 MPa at 300 K, thus calculated, are shown in Figure 20.18 together with the x-ray absorption near edge structure (XANES) spectra for the total photoelectron yields of water beams in the same conditions.

The total photoelectron yields for a liquid water beam are measured as a function of both the incident photon energy of 531–547 eV and the flight distance of 1–17 mm, which is the distance from the nozzle of the liquid beam to the point of irradiation on the beam in the vacuum. The results obtained with a stagnation pressure of 3 MPa at 300 K using the 10 µm nozzle are shown in Figure 20.18. As described above, they represent the temperature dependence of the XANES spectrum in the vicinity of the oxygen K-edge. The main features of the XANES spectrum of water are a broad

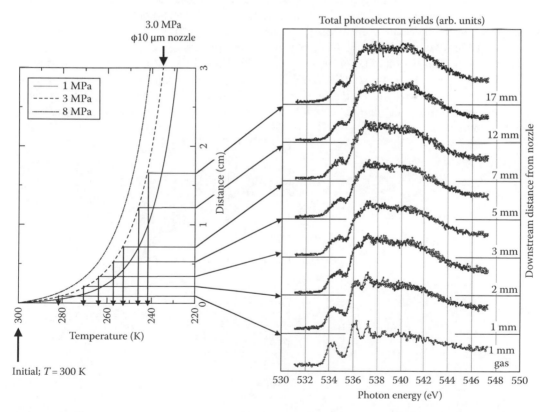

FIGURE 20.18 Left: Approximate temperature of a liquid water beam emitted through a 10 µm nozzle with stagnation pressures of 1, 3, and 8 MPa at 300 K. Right: The XANES spectra for a liquid water beam as a function of flight distance of it at 1, 2, 3, 5, 7, 12, and 17 mm downstream from the nozzle in vacuum. The XANES spectrum for water vapor is also shown at the bottom, which was obtained at a distance at 1 mm downstream from the nozzle and at 1 mm off the liquid beam.

enhancement of the yields at photon energies of 537–540 eV and a small enhancement at the photon energy around 535 eV, the latter of which is called the "pre-edge peak," although the K-shell ionization potential of water in liquid has not yet been reported. It is readily seen that the appearance and diminished size of the post-edge peaks and the shift of the pre-edge peak are the result of "contamination" from the coexisting water vapor when the temperature of the liquid beam is relatively high (close to 300 K). At a flight distance of more than 10 mm, a new peak appears to grow at the photon energy of 541 eV. The XANES spectrum for ice presents a strong peak at this energy (Wernet et al., 2004), so the appearance of the new peak signifies the growth of micro ices in the liquid beam. It is, therefore, concluded that at a flight distance of a few millimeters, the water in the present beam is in the liquid phase, where the present XANES spectrum is in good accord with the previous XANES spectra obtained by a total photoelectron measurement (Wilson et al., 2004) and by a total x-ray fluorescence measurement (Kashtanov et al., 2004).

In this liquid water beam condition, Ukai et al. (2008) reported the first results of the total photoelectron yields for a DNA constituent molecule in water. They measured the yield of guanosine 5′-monophosphate (GMP)/water as a function of the incident photon energies of 396–416 eV (shown in Figure 20.19), from which a much greater fraction of continuous electron yields is subtracted. Small enhancements are seen in the electron yields, the intensities of which are about 15% of the total electrons. The photoionization cross sections of a water molecule monotonously decrease with increasing photon energy in the region of 396–416 eV (Berkowitz, 2002). Furthermore, the absorption edge for the carbon K-shell is at a much lower energy by about 100 eV and that for oxygen is much greater by 130 eV. It is therefore natural to conclude that the features observed in the present spectra of photoelectron yields are due to the photoabsorption and photoionization of the nitrogen K-shell electrons in the guanine site of GMP, so that the features are regarded as the XANES spectrum of GMP in water solution in the vicinity of the nitrogen K-edge.

Let us briefly consider the ratio of the electron yields from the guanine site to the continuous yields for other elements obtained in the XANES spectrum in the vicinity of the nitrogen K-edge. The continuous electron yields are the result of the ionization of the valence orbital electrons of water and the ionizations due to the 2s and 2p orbitals of the oxygen atoms, the 1s orbital of the

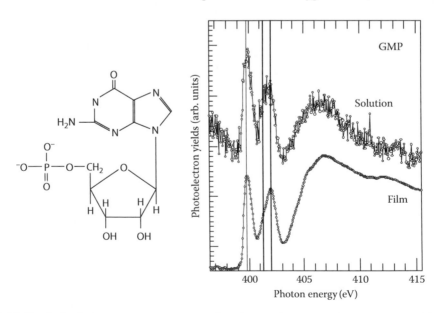

FIGURE 20.19 Left: Structure of GMP (guanosine-5′-monophospate). Right: The XANES spectrum for GMP in the vicinity of the nitrogen K-edge. "Solution" shows the XANES spectrum for the beam of a 10 g/100 mL GMP solution in water. The large amplitudes of continuous electron yield from water molecules and other elements are subtracted. "Film" shows the XANES spectrum for a solid film for reference.

carbon atoms, and the 2s and 2p orbitals of the phosphorus atom in GMP. Although the molecular photoabsorption cross sections for water or the nucleotide in liquid water have not yet been obtained experimentally or theoretically, the sum of the atomic photoabsorption cross sections can provide reasonable estimates. Because the cross section for light atoms in the region of K-shell absorption is about 1 Mb ($10-18\,cm^2$) at maximum, the photoabsorption cross sections for other atoms at a photon energy around 400 eV are about 0.5 Mb for carbon, 0.06 Mb for oxygen, and 0.7 Mb for phosphorus (Yeh and Lindau, 1985). If we consider the case of a mass concentration of GMP/water of 10 g/100 mL (corresponding to a molecular ratio of 1/213), the net absorption cross section for continuous electron yields relative to one guanine unit containing five nitrogen atoms amounts to 18 Mb. Thus, a value of about 15% for the ratio of the electron yields due to nitrogen atoms in GMP to the continuous electron yields due to water and other elements is reasonably explained by the fact that GMP molecules are not concentrated in the middle or the surface of the liquid beam, but are equally dispersed in water.

The main features of the XANES spectrum for GMP in water solution consist of relatively sharp peaks at photon energies of 400 and 402 eV and a much broader peak in the region of photon energy around 407 eV. The peaks at 400 and 402 eV are normally ascribed to the excitation of nitrogen K-shell electrons into the vacant antibonding orbital(s). The XANES spectrum for GMP in the form of a thin solid film is also shown in Figure 20.19. The GMP in the solid film is thought to be neat GMP or GMP containing the fewest number of water molecules under vacuum. The XANES spectrum for a GMP/water solution is, in general, very similar to that for the solid film. A similar reference is also available showing the XANES spectrum for a guanine thin solid film (Fujii et al., 2004b), which is also similar to the present XANES spectra for the GMP/water solution and GMP in the solid film.

Because excitation in the region near the nitrogen K-edge takes place mainly in the frame of the guanine site, the XANES spectrum is thought to be sensitive to the influence of water molecules surrounding the guanine site. Both the K-shell and antibonding orbitals of GMP in solution are subject to the influence of the external Coulombic field from the surrounding water molecules. Because the antibonding orbital is thought to be widely spread around the nucleobase site, its energy is more sensitive to the external Coulombic field than that of the K-shell orbital; thus, strong hydration should give rise to an environmental chemical shift. However, a significant chemical shift is not observed in the peak energies in the XANES spectrum for GMP in a water solution when compared with those for a solid film. Fujii et al. (2004b), therefore, tentatively conclude that strong attractive interactions at the guanine site from the water molecules are absent, which is consistent with the common understanding of the hydrophobic properties of nucleobases, although the guanine site is thought to be embedded in the bulk water molecules (as discussed above) because of the ratio of the yield of electrons ejected from nitrogen atoms in GMP to those from other elements. In other words, the hydrophobic interaction of GMP at the guanine site should increase the interaction energy with an approaching water molecule and repel the water molecule at a relatively large intermolecular distance, which is accessible with thermal energy. However, this tentative conclusion as regards the interaction of the guanine site in GMP with water molecules should be more clearly examined with higher-resolution chemical-shift measurements. Photoelectron emission spectra from DNA-related molecules in solution should be analyzed in future using an electron energy analyzer combined with the liquid water beam technique. Ukai et al. (2009) have developed a photoelectron energy analyzer for liquid beam samples. They reported that partial electron yields for the K^{-1} $1b_1$ $1b_1$ Auger transition were obtained for the first time by measuring the electrostatically dispersed electron kinetic energy spectra as a function of photon energy.

20.7 SUMMARY

Experimental evidence introduced in this chapter suggests the following: (1) The yield of DNA damage demonstrates a rather complicated relationship when compared with that induced by other radiations, because the yield strongly depends on the radical scavenging conditions during irradiation.

(2) The yields of base lesions and clustered damage sites including base lesions have been revealed using enzymatic probes of BER proteins for low-LET irradiation. (3) Higher-LET irradiation, however, induces more complicated damage, which cannot be recognized by BER enzymes. (4) Thus, some of the complex damage sites, which consist of two or more base lesions, are more poorly repaired by BER, and, ultimately, induce mutagenesis in living cells. (5) Oxygen K-ionization in DNA (and, presumably, water hydrated to DNA) plays an important role in inducing base lesions. (6) Finally, a new water beam technique combined with synchrotron radiation has been developed to explore the electric properties of DNA in water.

REFERENCES

Agrawala, P. K., Eschenbrenner, A., Hervè Du Penhoat, M. A., Boissière, A., Eot-Houllier, G., Abel, F., Politis, M. F., Touati, A., Sage, E., and Chetioui, A. 2008. Induction and repairability of DNA damage caused by ultrasoft X-rays: Role of core events. *Int. J. Radiat. Biol.* 84: 1093–1103.

Ahnstrom, G. and Bryant, P. E. 1982. DNA double-strand breaks generated by the repair of X-ray damage in Chinese hamster cells. *Int. J. Radiat. Biol. Relat. Stud. Phys. Chem. Med.* 41: 671–676.

Akamatsu, K., Fujii, K., and Yokoya, A. 2004a. Low-energy Auger- and photo-electron effects on the degradation of thymine by ultrasoft X-irradiation. *Int. J. Radiat. Biol.* 80: 849–853.

Akamatsu, K., Fujii, K., and Yokoya, A. 2004b. Qualitative and quantitative analyses of the decomposition products that arise from the exposure of thymine to monochromatic ultrasoft X rays and 60Co gamma rays in the solid state. *Radiat. Res.* 162: 469–473.

Becker, D. and Sevilla, M. D. 1993. The chemical consequences of radiation damage to DNA. *Adv. Radiat. Biol.* 17: 121–180.

Becker, D., La Vere, T., and Sevilla, M. D. 1994. ESR detection at 77 K of the hydroxyl radical in the hydration layer of gamma-irradiated DNA. *Radiat. Res.* 140: 123–129.

Becker, D., Bryant-Friedrich, A., Trzasko, C., and Sevilla, M. D. 2003, Electron spin resonance study of DNA irradiated with an argon-ion beam: evidence for formation of sugar phosphate backbone radicals. *Radiat. Res.* 160: 174–185.

Bellon, S., Shikazono, N., Cunniffe, S., Lomax, M., and O'Neill, P. 2009. Processing of thymine glycol in a clustered DNA damage site: Mutagenic or cytotoxic. *Nucl. Acids Res.* 37: 4430–4440.

Berkowitz, J. 2002. *Atomic Molecular Photoabsorption: Absolute Total Cross Sections*, J. Berkowitz (ed.) (Argonne National Laboratory). San Diego, CA: Academic Press.

Bernhard, W. A. and Close, D. M. 2004. DNA damage dictates the biological consequence of ionizing radiation: The chemical pathways. In *Charged Particle Interactions with Matter: Chemical, Physicochemical, and Biological Consequences with Applications*. A. Mozumder and Y. Hatano (eds.), pp. 471–489. New York: Marcel Dekker.

Blaisdell, J. O. and Wallace, S. S. 2001. Abortive base-excision repair of radiation-induced clustered DNA lesions in *Escherichia coli*. *Proc. Natl. Acad. Sci. USA* 98: 7426–7430.

Boiteux, S., Gajewski, E., Laval, J., and Dizdaroglu, M. 1992. Substrate specificity of the *Escherichia coli* Fpg protein (formamidopyrimidine-DNA glycosylase): Excision of purine lesions in DNA produced by ionizing radiation or photosensitization. *Biochemistry* 31: 106–110.

Boudaïffa, B., Cloutier, P., Hunting, D., Huels, M. A., and Sanche, L. 2000. Resonant formation of DNA strand breaks by low-energy (3 to 20 eV) electrons. *Science* 287: 1658–1660.

Bowman, M. K., Becker, D., Sevilla, M. D., and Zimbrick, J. D. 2005. Track structure in DNA irradiated with heavy ions. *Radiat. Res.* 163: 447–454.

Breimer, L. H. and Lindahl, T. 1984. DNA glycosylase activities for thymine residues damaged by ring saturation, fragmentation, or ring contraction are functions of endonuclease III in *Escherichia coli*. *J. Biol. Chem.* 259: 5543–5548.

Chang, P. W., Zhang, Q. M., Takatori, K., Tachibana, A., and Yonei, S. 2005. Increased sensitivity to sparsely ionizing radiation due to excessive base excision in clustered DNA damage sites in *Escherichia coli*. *Int. J. Radiat. Biol.* 81: 115–123.

Chetsanga, C. J. and Lindahl, T. 1979. Release of 7-methylguanine residues whose imidazole rings have been opened from damaged DNA by a DNA glycosylase from *Escherichia coli*. *Nucl. Acids Res.* 6: 3673–3684.

Cunniffe, S. M., Lomax, M. E., and O'Neill, P. 2007. An AP site can protect against the mutagenic potential of 8-oxoG' when present within a tandem clustered site in *E. coli*. *DNA Repair (Amst.)* 6: 1839–1849.

de Lara, C. M., Jenner, T. J., Townsend, K. M., Marsden, S. J., and O'Neill, P. 1995. The effect of dimethyl sulfoxide on the induction of DNA double-strand breaks in V79-4 mammalian cells by alpha particles. *Radiat. Res.* 144: 43–49.

de Lara, C. M., Hill, M. A., Jenner, T. J., Papworth, D., and O'Neill, P., 2001. Dependence of the yield of DNA double-strand breaks in Chinese hamster V79-4 cells on the photon energy of ultrasoft X-rays. *Radiat. Res.* 155: 440–448.

Demple, B. and Linn, S. 1980. DNA N-glycosylases and UV repair. *Nature* 287: 203–208.

Dianov, G. L., Timchenko, T. V., Sinitsina, O. I., Kuzminov, A. V., Medvedev, O. A., and Salganik, R. I. 1991. Repair of uracil residues closely spaced on the opposite strands of plasmid DNA results in double-strand break and deletion formation. *Mol. Gen. Genet.* 225: 448–452.

Dizdaroglu, M., Laval, J., and Boiteux, S. 1993. Substrate specificity of the *Escherichia coli* endonuclease III: Excision of thymine- and cytosine-derived lesions in DNA produced by radiation-generated free radicals. *Biochemistry* 32: 12105–12111.

D'Souza, D. I. and Harrison, L. 2003. Repair of clustered uracil DNA damages in *Escherichia coli*. *Nucl. Acids Res.* 31: 4573–4581.

Dugle, D. L., Gillespie, C. J., and Chapman, J. D. 1976. DNA strand breaks, repair, and survival in x-irradiated mammalian cells. *Proc. Natl. Acad. Sci. USA* 73: 809–812.

Eschenbrenner, A., Hervè Du Penhoat, M. A., Boissière, A., Eot-Houllier, G., Abel, F., Politis, M. F., Touati, A., Sage, E., and Chetioui, A. 2007. Strand breaks induced in plasmid DNA by ultrasoft X-rays: Influence of hydration and packing. *Int. J. Radiat. Biol.* 83: 687–697.

Faubel, M. and Kisters, T. 1989. Non-equilibrium molecular evaporation of carboxylic acid dimmers. *Nature* 339: 527–529.

Faubel, M. and Steiner B. 1992. Strong bipolar electrokinetic charging of thin liquid jets emerging from 10 μm PtIr nozzles. *Ber. Bunsenges. Phys. Chem.* 96: 1167–1172.

Faubel, M., Shlemmer, S., and Toennies, J. P. 1988. A molecular beam study of the evaporation of water from a liquid jet. *Z. Phys. D* 10: 269–277.

Faubel, M., Steiner, B., and Toeneies, J. P. 1997. Photoelectron spectroscopy of liquid water, some alcohols, and pure nonane in free micro jet. *J. Chem. Phys.* 106: 9013–9031.

Faubel, M., Steiner, B., and Toeneies J. P. 1999. Measurements of He I photoelectron spectra of liquid water, formamide and ethylene glycol in fast-flowing microjets. *J. Electron Spectrosc. Rel. Phenom.* 95: 159–169.

Fayard, B., Touati, A., Abel, F., Herve du Penhoat, M. A., Despiney-Bailly, I., Gobert, F., Ricoul, M., L'hoir, A., Polities, M. F., Hill, M. A., Stevens, D. L., Sabatier, L., Sage, E., Goodhead, D. T., and Chetioui, A. 2002. Cell inactivation and double-strand breaks: The role of core ionization, as probed by ultrasoft X rays. *Radiat. Res.* 157: 128–140.

Fuchs, H. and Legge, H. 1979. Flow of a water jet into vacuum. *Acta Astronaut.* 6: 1213–1226.

Fujii, K., Akamatsu, K., and Yokoya, A. 2004a. Ion desorption from DNA components irradiated with 0.5 keV ultrasoft X-ray photons. *Radiat Res.* 161: 435–441.

Fujii, K., Akamatsu, K., and Yokoya, A. 2004b. Near-edge x-ray absorption fine structure of DNA nucleobases thin film in the nitrogen and oxygen *K*-edge region. *J. Phys. Chem. B* 108: 8031–8035.

Fujii, K., Shikazono, N., and Yokoya, A. 2009. Nucleobase lesions and strand-breaks in dry DNA thin film selectively induced by monochromatic soft X-rays. *J. Phys. Chem.* 113: 16007–16015.

Fulford, J. 2000. Quantification of complex DNA damage by ionising radiation: An experimental and theoretical approach. PhD thesis, University of Brunel, London, U.K.

Fulford, J., Nikjoo, H., Goodhead, D. T., and O'Neill, P. 2001. Yields of SSB and DSB induced in DNA by AlK X-rays and α-particles: Comparison of experimental and simulated yields. *Int. J. Radiat. Biol.* 77: 1053–1066.

Georgakilas, A. G., Bennett, P. V., Wilson III D. M., and Sutherland, B. M. 2004. Processing of bistranded abasic DNA clusters in gamma-irradiated human hematopoietic cells. *Nucl. Acids Res.* 32: 5609–5620.

Goodhead, D. T. 1994a. Initial events in the cellular effects of ionizing radiations: Clustered damage in DNA. *Int. J. Radiat. Biol.* 65: 7–17.

Goodhead, D. T. 1994b. In *Synchrotron Radiation in the Biosciences*, B. Chance, J. Deisenhofer, S. Ebashi, D. T. Goodhead, J. R. Helliwell, H. E. Huxley, T. Iizuka et al. (eds.), pp. 683–705. Oxford, U.K.: Clarendon Press.

Goodhead, D. T., Thacker, J., and Cox, R. 1981. Is selective absorption of ultrasoft x-rays biologically important in mammalian cells? *Phys. Med. Biol.* 26: 1115–1127.

Gulston, M., de Lara, C., Jenner, T., Davis, E., and O'Neill, P. 2004. Processing of clustered DNA damage generates additional double-strand breaks in mammalian cells post-irradiation. *Nucl. Acids Res.* 32: 1602–1609.

Hada, M. and Sutherland, B. M. 2006. Spectrum of complex DNA damages depends on the incident radiation. *Radiat Res*. 165: 223–230.

Hanai, R., Yazu, M., and Hieda, K. 1998. On the experimental distinction between SSB and DSB in circular DNA. *Int. J. Radiat. Biol*. 73: 457–479.

Harrison, L., Brame, K. L., Geltz, L. E., and Landry, A. M. 2006. Closely opposed apurinic/apyrimidinic sites are converted to double strand breaks in *Escherichia coli* even in the absence of exonuclease III, endonuclease IV, nucleotide excision repair and AP lyase cleavage. *DNA Repair (Amst.)* 5: 324–335.

Hatahet, Z., Kow, Y. W., Purmal, A. A., Cunningham, R. P., and Wallace, S. S. 1994. New substrates for old enzymes. 5-Hydroxy-2'-deoxycytidine and 5-hydroxy-2'-deoxyuridine are substrates for *Escherichia coli* endonuclease III and formamidopyrimidine DNA N-glycosylase, while 5-hydroxy-2'-deoxyuridine is a substrate for uracil DNA N-glycosylase. *J. Biol. Chem*. 269: 18814–18820.

Hieda, K. and Ito, T. 1991. Radiobiological experiments in the X-ray region with synchrotron radiation. In: *Handbook on Synchrotron Radiation*, Vol. 4, S. Ebashi, M. Koch, and E. Rubenstein (eds.), p. 431. Amsterdam, the Netherlands: North-Holland.

International Commission on Radiation Units and Measurements. 1970. Linear Energy Transfer. ICRU Report 16. Washington, DC.

Ito, T. and Saito, M. 1988. Degradation of oligonucleotides by vacuum-UV radiation in solid: role of the phosphate group and bases. *Photochem. Photobiol*. 48: 567–572.

Kashtanov, S., Augustsson, A., Luo, Y., Guo, J. H., Såthe, C., Rubensson, J. E., Siegbahn, H., Nordgren, J., and Ågren, H. 2004. Local structure of liquid water studied by x-ray emission spectroscopy. *Phys. Rev. B* 69: 024201-1–024201-8.

Kozmin, S. G., Sedletska, Y., Reynaud-Angelin, A., Gasparutto, D., and Sage, E. 2009. The formation of double-strand breaks at multiply damaged sites is driven by the kinetics of excision/incision at base damage in eukaryotic cells. *Nucl. Acids Res*. 37: 1767–1777.

Krisch, R. E., Flick, M. B., and Trumbore, C. N. 1991. Radiation chemical mechanism of single- and double-strand break formation in irradiated SV40 DNA. *Radiat. Res*. 126: 251–259.

La Vere, T., Becker, D., and Sevilla, M. D. 1996. Yields of OH in gamma-irradiated DNA as a function of DNA hydration: Hole transfer in competition with OH formation. *Radiat. Res*. 145: 673–680.

Malyarchuk, S., Youngblood, R., Landry, A. M., Quillin, E., and Harrison, L. 2003. The mutation frequency of 8-oxo-7,8-dihydroguanine (8-oxodG) situated in a multiply damaged site: comparison of a single and two closely opposed 8-oxodG in *Escherichia coli*. *DNA Repair (Amst.)* 2: 695–705.

Malyarchuk, S., Brame, K. L., Youngblood, R., Shi, R., and Harrison, L. 2004. Two clustered 8-oxo-7,8-dihydroguanine (8-oxodG) lesions increase the point mutation frequency of 8-oxodG, but do not result in double strand breaks or deletions in *Escherichia coli*. *Nucl. Acids Res*. 32: 5721–5731.

Malyarchuk, S., Castore, R., and Harrison, L. 2008. DNA repair of clustered lesions in mammalian cells: Involvement of non-homologous end-joining. *Nucl. Acids Res*. 36: 4872–4882.

Melvin, T., Cunniffe, S. M., O'Neill, P., Parker, A. W., and Roldan-Arjona, T. 1998. Guanine is the target for direct ionisation damage in DNA, as detected using excision enzymes. *Nucl. Acid Res*. 26: 4935–4942.

Milligan, J. R., Aguilera, J. A., Nguyen, T. T., Paglinawan, R. A., and Ward, J. F. 2000. DNA strand-break yields after post-irradiation incubation with base excision repair endonucleases implicate hydroxyl radical pairs in double-strand break formation. *Int. J. Radiat. Biol*. 76: 1475–1483.

Morgner, H. 1998. Electron spectroscopy for the determination of concentration depth profoles and for the investigation of local electric fields. *Surf. Investig*. 13: 463–474.

Nikjoo, H., Charlton, D. E., and Goodhead, D. T. 1994. Monte Carlo track structure studies of energy deposition and calculation of initial DSB and RBE. *Adv. Space Res*. 14: (10)161–(10)180.

Nikjoo, H., O'Neill, P., Goodhead, D. T., and Terrisol, M. 1997. Computational modelling of low-energy electron-induced DNA damage by early physical and chemical events. *Int. J. Radiat. Biol*. 71: 467–483.

Nikjoo, H., Uehara, S., Wilson, W. E., Hoshi, M., and Goodhead, D. T. 1998. Track structure in radiation biology: Theory and applications. *Int. J. Radiat. Biol*. 73: 355–364.

Nikjoo, H., O'Neill, P., Terrisol, M., and Goodhead, D. T. 1999. Quantitative modelling of DNA damage using Monte Carlo track structure method. *Radiat. Environ. Biophys*. 38: 31–38.

Nikjoo, H., O'Neill, P., Wilson, W. E., and Goodhead, D. T. 2001. Computational approach for determining the spectrum of DNA damage induced by ionizing radiation. *Radiat. Res*. 156: 577–583.

Nikjoo, H., Bolton, C. E., Watanabe, R., Terrisol, M., O'Neill, P., and Goodhead, D. T. 2002. Modelling of DNA damage induced by energetic electrons (100 eV to 100 keV). *Radiat. Prot. Dosim*. 99: 77–80.

O'Neill, P. 2001. Radiation-induced damage in DNA. In *Radiation Chemistry*, pp. 585–622. Dordrecht, the Netherlands: Elsevier Science.

O'Neill, P. and Fielden, E. M. 1993. Primary free radical processes in DNA. *Adv. Radiat. Biol.* 17: 53–120.

Ormerod, M. G. 1965. Free-radical formation in irradiated deoxyribonucleic acid. *Int. J. Radiat. Biol. Relat. Stud. Phys. Chem. Med.* 9: 291–300.

Pearson, C. G., Shikazono, N., Thacker, J., and O'Neill, P. 2004. Enhanced mutagenic potential of 8-oxo-7,8-dihydroguanine when present within a clustered DNA damage site. *Nucl. Acids Res.* 32: 263–270.

Prise, K. M., Pullar, C. H., and Michael, B. D. 1999. A study of endonuclease III-sensitive sites in irradiated DNA: Detection of alpha-particle-induced oxidative damage. *Carcinogenesis* 20: 905–909.

Purkayastha, S., Milligan, J. R., and Bernhard, W. A. 2006. The role of hydration in the distribution of free radical trapping in directly ionized DNA. *Radiat. Res.* 166: 1–8.

Saitoh, Y., Nakatani, T., Matsushita, T., Agui, A., Teraoka, Y., and Yokoya, A. 2001. First results from the actinide science beamline BL23SU at SPring-8. *Nucl. Instrum. Methods A* 474: 253–258.

Shikazono, N. and O'Neil, P. 2009. Biological consequences of potential repair intermediates of clustered base damage site in *Escherichia coli*. *Mutat. Res.* 669: 162–168.

Shikazono, N., Pearson, C., O'Neill, P., and Thacker, J. 2006. The roles of specific glycosylases in determining the mutagenic consequences of clustered DNA base damage. *Nucl. Acids Res.* 34: 3722–3730.

Siegbahn, K. 1974. Electron spectroscopy — An outlook. *J. Electron Spectrosc. Relat. Phenom.* 5: 3–97.

Sutherland, B. M., Bennett, P. V., Sidorkina, O., and Laval, J. 2000. Clustered DNA damages induced in isolated DNA and in human cells by low doses of ionizing radiation. *Proc. Natl. Acad. Sci. USA* 97: 103–108.

Tao, N. J., Lindsay, S. M., and Rupprecht, A. 1989. Structure of DNA hydration shells studied by Raman spectroscopy. *Biopolymers* 28: 1019–1030.

Taucher-Scholz, G. and Kraft, G. 1999. Influence of radiation quality on the yield of DNA strand breaks in SV40 DNA irradiated in solution. *Radiat. Res.* 151: 595–604.

Tchou, J., Kasai, H., Shibutani, S., Chung, M. H., Laval, J., Grollman, A. P., and Nishimura, S. 1991. 8-Oxoguanine (8-hydroxyguanine) DNA glycosylase and its substrate specificity. *Proc. Natl. Acad. Sci. USA* 88: 4690–4694.

Ukai, M., Yokoya, A., Fujii, K., and Saitoh, Y. 2008. X-ray absorption spectrum for guanosine-5'-monophosphate in water solution in the vicinity of the nitrogen K-edge observed in free liquid jet in vacuum. *Radiat. Phys. Chem.* 77: 1265–1269.

Ukai, M., Yokoya, A., Nonaka, Y., Fujii, K., and Saitoh, Y. 2009. Synchrotron radiation photoelectron studies for primary radiation effects using a liquid water jet in vacuum: total and partial photoelectron yields for liquid water near the oxygen K-edge. *Radiat. Phys. Chem.* 78: 1202–1206.

Urushibara, A., Shikazono, N., O'Neill, P., Fujii, K., Wada, S., and Yokoya, A. 2008. LET dependence of the yield of single-, double-strand breaks and base lesions in fully hydrated plasmid DNA films by ^{4}He^{2+} ion irradiation. *Int. J. Radiat. Biol.* 84: 23–33.

von Sonntag, C. 1987. *The Chemical Basis of Radiation Biology*. London, U.K.: Taylor & Francis.

Ward, J. F. 1988. DNA damage produced by ionizing radiation in mammalian cells: Identities, mechanisms of formation, and reparability. *Prog. Nucl. Acid Res. Mol. Biol.* 35: 95–125.

Watanabe, R., Yokoya, A., Fujii, K., Saito, K. 2004. DNA strand breaks by direct energy deposition by Auger and photo-electrons ejected from DNA constituent atoms following K-shell photoabsorption. *Int. J. Radiat. Biol.* 80: 823–832.

Wernet, Ph., Nordlund, D., Bergmann, U., Cavalleri, M., Odelius, M., Ogasawara, H., and Näslund, L. Å. 2004. The structure of the first coordination shell in liquid water. *Science* 304: 995–999.

Wilson, K. R., Rude, B. S., Smith, J., Cappa, C., Co, D. T., Schaller, R. D., Larsson, M., Catalano, T., and Saykally, R. J. 2004. Investigation of volatile liquid surfaces by synchrotron x-ray spectroscopy of liquid microjet. *Rev. Sci. Instrum.* 75: 725–736.

Yamashita, S., Katsumura, Y., Lin, M., Muroya, Y., Miyazaki, T., Murakami, T., Meesungnoen, J., and Jay-Gerin, J. P. 2008. Water radiolysis with heavy ions of energies up to 28 GeV. 3. Measurement of G(MV^{*+}) in deaerated methyl viologen solutions containing various concentrations of sodium formate and Monte Carlo simulation. *Radiat. Res.* 170: 521–533.

Yang, N., Galick, H., and Wallace, S. S. 2004. Attempted base excision repair of ionizing radiation damage in human lymphoblastoid cells produces lethal and mutagenic double strand breaks. *DNA Repair (Amst.)* 3: 1323–1334.

Yang, N., Chaudhry, M. A., and Wallace, S. S. 2006. Base excision repair by hNTH1 and hOGG1: A two edged sword in the processing of DNA damage in gamma-irradiated human cells. *DNA Repair (Amst.)* 5: 43–51.

Yeh, J. J. and Lindau, I. 1985. Atomic subshell photoionization cross sections and asymmetry parameters: $1 \leq Z \leq 103$. *At. Data Nucl. Data Tables* 32: 1–155.

Yokoya, A., Watanabe, R., and Hara, T. 1999. Single- and double-strand breaks in solid pBR322 DNA induced by ultrasoft x-rays at photon energies at 388, 435 and 573 eV. *J. Radiat. Res.* 40: 145–158.

Yokoya, A., Cunniffe, C. M. T., and O'Neill, P. 2002. Effect of hydration on the induction of strand breaks and nucleobase lesions in plasmid DNA films by gamma-radiation. *J. Am. Chem. Soc.* 124: 8859–8866.

Yokoya, A., Cunniffe, C. M. T., Stevens, D. L., and O'Neill, P. 2003. Effect of hydration on the induction of strand breaks, nucleobase lesions and clustered damage in DNA films by α-radiation. *J. Phys. Chem. B* 124: 832–837.

Yokoya, A., Fujii, K., Ushigome, T., Shikazono, N., Urushibara, A., and Watanabe, R. 2006. Yields of soft X-ray induced strand breaks and nucleobase lesions in plasmid DNA. *Radiat. Prot. Dosim.* 122: 86–88.

Yokoya, A., Fuji, K., Shikazono, N., Akamatsu, K., Urushibara, A., and Watanabe, R. 2008a. Studies of soft X-ray-induced Auger effect on the induction of DNA damage. *Int. J. Radiat. Biol.* 84: 1069–1081.

Yokoya, A., Shikazono, N., Fujii, K., Urushibara, A., Akamatsu, K., and Watanabe, R. 2008b. DNA damage induced by the direct effect of radiation, *J. Radiat. Phys. Chem.* 77: 1280–1285.

Yokoya, A., Cunniffe, S. M. T., Watanabe, R., Kobayashi, K., and O'Neill, P. 2009a. Induction of DNA strand breaks, nucleobase lesions, and clustered damage sites in hydrated plasmid DNA films by ultrasoft x-rays around the phosphorus K-edge. *Radiat. Res.* 172: 296–305.

Yokoya, A., Fujii, K., Fukuda, Y., and Ukai, M. 2009b. EPR study of radiation damage to DNA irradiated with synchrotron Soft X-rays around nitrogen and oxygen K-edge. *Radiat. Phys. Chem.* 78: 1211–1215.

21 Application of Microbeams to the Study of the Biological Effects of Low Dose Irradiation

Kevin M. Prise
Queen's University Belfast
Belfast, United Kingdom

Giuseppe Schettino
Queen's University Belfast
Belfast, United Kingdom

CONTENTS

21.1 INTRODUCTION

Humans are continually exposed to ionizing radiation from a range of sources including environmental and occupational sources. The doses from these various sources are very low and generally of low dose rates. At the level of individual cells within the human body, this can equate to only single tracks of radiation crossing a cell over periods of weeks or years. Our understanding of the effects of these low doses requires technological approaches that can deliver highly localized radiation beams into biological models. This chapter reviews the use of novel microbeam technologies that allow relevant doses to be tested in biological systems. We outline some of the key technological approaches used to produce microbeams using different types of ionizing radiations and give examples of some of the biological results that have been obtained with them.

21.2 LOW DOSE RADIATION EFFECTS AND RADIATION RISK

Humans are exposed to multiple physical, chemical, and biological agents during their lifetime. Of these, ionizing radiation(s) has long been known to be deleterious after high dose exposure (>100 mSv), predominantly due to cancer induction; although very high dose exposures cause tissue damage and ultimately death (see Hall, (2000) for a general textbook). Ionizing radiations are widely used in society, play a key role in the treatment of cancer, and are an important diagnostic tool. Despite a century of study, risk estimates for cancer induction in humans to be used for radiation protection purposes are extrapolated from the Japanese atomic bomb survivors, who were exposed to both relatively high doses and dose rates. Around 120,000 survivors have been followed up to the present date; 20,000 of these had received doses of greater than 5 mSv. By 1990, around 6000 deaths resulted from cancer, of which ~400 were considered to be deaths due to excess radiation. Other exposed populations have been studied, including those treated with radiation for various medical conditions such as ankylosing spondylitis, conditions related to leukemia risk, and other tumors. Several studies of radiation workers have been undertaken as these populations were exposed to protracted low-dose exposures (Cardis et al., 1995; Muirhead et al., 1999). From these epidemiological data, a simple extrapolation of risk has been made to low doses generally found in environmental and most occupational exposures. This has been called the linear non-threshold model (LNT), which assumes a linear dose–response relationship between dose and risk. Currently, with the exception of radiotherapy, the doses that members of the population can be typically exposed to are lower than the doses received by the bomb survivors and are therefore areas where epidemiological data is scarce. Against a typical background dose of around 3 mSv/year, examples of routine medical exposures include 3 mSv for a breast mammogram and 0.7 mSv for a dental x-ray (Brenner et al., 2003). From this background, a range of experimental approaches have been taken to study the effects of low dose and low dose rates of radiation exposure and to test the assumptions of the LNT model. Ionizing radiations are typically compared on the basis of their linear energy transfer (LET). Sparsely ionizing radiations such as γ-rays and x-rays are classified as low LET and densely ionizing radiations such as alpha-particles and neutrons are classified as high LET. For low-LET radiations, a single radiation track will deposit around 1 mGy in a human cell. A novel approach has been the use of microbeams that allow radiation to be precisely targeted to a particular biological target of interest. They have contributed enormously to our understanding of the effects of low dose radiation exposure and this will be the subject of the rest of this chapter.

21.3 INTRODUCTION TO MICROBEAMS

Charged-particle microbeams have been used since the 1960s for quantitative elemental analysis of geological, historical, and biological samples (Watt and Grime, 1987) where two-dimensional elemental maps can be obtained by scanning a small ion beam across a sample and monitoring the x-rays produced by the sample elements (Watt et al., 1982). However, it was only toward the end of the 1990s that microbeams have attracted the interest of the radiobiology community and been developed into specific tools to investigate the effect of ionizing radiation on living samples. Modern "radiobiological microbeams" are instruments capable of delivering accurate predetermined doses of ionizing radiation to individual cells (or parts of cells) and which can assess the damage induced on a cell-by-cell basis. The advantages of a deterministic irradiation achieved by targeting and analyzing cells individually have been recognized since the beginning of radiobiological studies. Using basic setups, Zirkle and colleagues (Zirkle and Bloom, 1953) tried to correlate radiation-induced cell damage to the type and energy of radiation, the number of ions per cell, and even the subcellular compartment irradiated. Despite the limited control and precision offered by their polonium-tipped syringe needle, important observations were made regarding the nuclear and cytoplasmic sensitivity, strengthening the hypothesis that considerable benefit could be achieved with a deterministic irradiation approach. However, it was only in the last decade of the twentieth century that improvements in radiation production, detection and delivery, image processing, and micropositioning have the required precision and speed been achieved to successfully develop radiobiological microbeam facilities.

21.4   HISTORY, RATIONALE, AND KEY ASPECTS

The first microbeam experiment has to be attributed to Zirkle and Bloom (Zirkle and Bloom, 1953) who in 1953 used a 2 MV Van de Graaff accelerator and microcollimators to study the process of cell division following proton exposure. Their collimator consisted of two metal plates with a groove etched on one, clamped together to achieve a 2.5 μm beam or adjustable cross slits to form a nominal proton beam of any size from a few microns to a few millimeters. Later a 25 μm microbeam was developed using the cyclotron facility at the Brookhaven National Laboratories using a similar collimator approach and 11 MeV/amu proton or 22 MeV/amu deuterons (Baker et al., 1961), to investigate radiation damage to mouse-brain cells (Zeman et al., 1959). These first microbeams were limited to relatively high doses and impossibility to control and/or determine the number of particles reaching the samples. As a solution, the GSI Darmstadt microbeam (Kraske et al., 1990) based on a UNILAC accelerator to collimate ions from carbon to uranium (1.4 MeV/amu) used plastic track detectors to monitor the particle traversals and, a posteriori, determine the traversals experienced by each sample. Such an approach, however, was extremely time consuming and only 20 cells could be irradiated with a single ion during 10 h of beam time.

The incentive to develop modern microbeams came mainly from the necessity to evaluate the biological effects of exactly one particle traversal in order to evaluate environmental and occupational radiation risks, where only a few cells in the human body experience isolated traversals by charged particles (Brenner et al., 1995) separated by intervals of months or years (see Figure 21.1). Due to the random Poisson distribution of tracks, such scenarios cannot be simulated in vitro using conventional broad field techniques, and current excess cancer risks associated with exposure to very low doses of ionizing radiation are estimated by extrapolating high dose data. These data obtained from in vitro experiments or from epidemiological data from the atomic bomb survivors, however, suffer from limited statistical power and are unable to resolve uncertainties from confounding factors, forcing the adoption of the precautionary LNT model. Moreover, there is experimental evidence that non-linear effects may play a considerable role at low doses. Genomic instability (Kadhim et al., 1995), hypersensitivity (Marples and Joiner, 1993), and bystander effect (Prise et al., 2005; Morgan and Sowa, 2007), for example, may increase the initial radiation risk, while the adaptive response (Boreham et al., 2000) may act as a protective mechanism reducing the overall risks at low doses. Charged-particle microbeams therefore represent a unique tool. They allow targeting of single cells and analysis of the induced damage on a cell-by-cell basis, which is critical to assess the shape of the dose–response curve in the low-dose region. A second driving question was related to the radiation-sensitivity of the whole cell and the cell nucleus in particular. Traditional radiobiological theories were based on the assumption that the DNA was the only critical target and that any biological effect observed was entirely the result of DNA alteration. Following some observation of nonuniformity of cellular response (Datta et al., 1976), the investigation of radiosensitivity across the cell nucleus was considered to be of great interest in improving radiobiology models. Modern microbeam facilities were developed almost in parallel at the Pacific Northwest Laboratories (United States) (Braby and Reece, 1990), Columbia University (United States) (Michael et al., 1995), and the Gray Laboratory (UK)

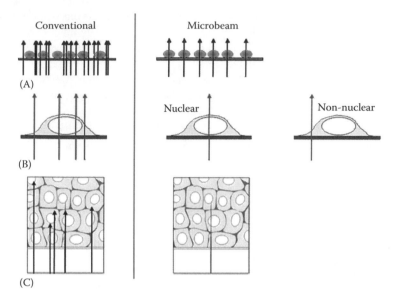

FIGURE 21.1 Schematic showing differences between conventional radiation exposure experiments and those using a microbeam. (A) depicts the ability to deliver uniform numbers of charged-particle tracks to individual cells, including that of a single track. (B) represents the ability to target radiation to different subcellular locations. (C) represents the ability to deliver localized radiation to defined depths within tissue.

(Folkard et al., 1997a,b). They all adopted, at least initially, a collimating approach using either adjustable knife-edges, sets of microapertures, etched grooves, or microcapillaries to produce micron size beams of proton and helium ions. The main difference between these microbeams and the previous facilities were the ability to a priori determine the number of particles to be delivered to each sample, and the individual cell or part of it to be irradiated (see Figure 21.1). This has been achieved using custom-designed radiation detectors (see below for details) coupled to fast electrostatic beam defectors able to shut down the beam when the desired number of particles have been delivered. The availability of accurate micropositioning stages and computer resources for automatically driving the experiments were also essential elements.

The initial success of these first microbeam facilities was related to the ability to measure radiation effects at very low doses with great accuracy. The Columbia charged-particle microbeam was used to measure the oncogenic transforming efficiency of the nuclear traversal of exactly one α particle (Miller et al., 1999). It was found to be significantly lower than that predicted by a Poisson-distribution delivery by an average of one α particle. Such findings suggested that multiple traversals dominate the biological response, and extrapolation from multiple particle traversals may overestimate the single traversal risk. However, similar experiments (Hei et al., 1997) highlighted how even a single α-particle traversal has considerable toxic (~20%) and mutagenic probability (average 110 mutants per 10^5 survivors). Similarly, using 3.2 MeV protons, Prise and coworkers at the Gray Laboratory (Prise et al., 2000) used the micronucleus assay as a measure of predominantly lethal chromosome damage showing a linear dose response in the range 1–30 protons per nucleus with a single proton responsible for micronuclei formation in less than 2% of exposed cells. A single-particle traversal was also shown to induce a significant increase in the proportion of aberrant human T-lymphocyte cells, 12–13 population doublings after exposure (Kadhim et al., 2001). The unstable phenotype indicated by the high level of chromatid-type aberrations suggested that a single α particle through the cell nucleus can also induce genomic instability. Moreover, by targeting the cytoplasm, microbeams have shown that intracellular signaling between the cytoplasm and the nucleus can also cause DNA damage, undermining therefore the fundamental paradigm of radiobiology that considers direct DNA exposure a prerequisite for the manifestation of the radiation effects. This is well summarized in the work of Wu and colleagues (Wu et al., 1999), where cytoplasmic irradiation by α particles was found to induce mutations in human hamster hybrid (A_L) cells, indicating that the cytoplasm is also a critical target. By comparing the mutant fraction induced by nuclear and cytoplasmic α-particle traversals for an equitoxic dose, cytoplasmic irradiation was found to be as much as seven times more mutagenic (due to the low cell killing) than nuclear exposure and therefore potentially more harmful. Moreover, it was noticed that the spectrum of aberrations induced by the cytoplasm irradiation was quite different from that produced by nuclear irradiation, suggesting that different mutagenic mechanisms may be involved. The importance of cytoplasmic irradiation has also been highlighted by other microbeam experiments directed at investigating the bystander phenomenon. Using both α particles (Shao et al., 2004) and soft x-rays, increased micronuclei formation and decreased clonogenic survival have been measured following the irradiation of one or a few cells through their cytoplasm. This finding showed that direct DNA damage is not required for switching on of important intra cell-signaling mechanisms. Another advantage offered by the microbeam approach is the possibility of assessing radiation damage on a cell-by-cell basis, thus avoiding the statistical uncertainty that affects some conventional assays. For conventional assays, accurate measurements of the radiation effect may be limited by both the uncertainty in the number of samples exposed and the dose delivered. This is a particular problem in the low-dose region, where only small effects are expected. However, using the Gray Laboratory charged-particle microbeam, it has been possible to precisely measure the survival of V79 cells exposed to 3.2 MeV protons at doses below 1 Gy (Schettino et al., 2001). As a result of the precise particle delivery system and knowing the number of cells exposed, it was possible to detect small variations in the initial slope of the survival curve indicating the presence of a hypersensitivity region that had been shown previously using x-rays (Marples and Joiner, 1993) and proving that microbeams are ideally suited to investigate cell survival at very low doses.

Since then, microbeams have established themselves as a powerful, often unique, radiobiological tool facilitating our understanding of a variety of biological responses. For example, the microbeam approach is considered particularly appropriate to investigate the so-called bystander effect (Morgan and Sowa, 2007), where radiation damage is being detected in unexposed samples following the irradiation of neighboring cells. Even though this type of study could be undertaken using basic shielding arrangements, comprehensive investigations could only be done using the precise pre-selected dose delivery and the individual cell system assay offered by the microbeam. Using the Gray Laboratory microbeam (Prise et al., 1998), it was possible to demonstrate that the level of bystander damage is independent of the number of cells targeted (up to 25%) and of the number of particle traversals leading to a binary all-or-nothing interpretation of the bystander response (Schettino et al., 2005). Bystander studies have now been extended to tissues and 3D cell models in order to understand the role of cell-to-cell communication and the implication of tissue architecture on radiation response. It is anticipated that the new generation of microbeams, employing longer-range radiations and deep-tissue imaging techniques, may offer a unique contribution. Moreover, increasing scientific interest toward subcellular targets and signaling pathways that regulate the various radiobiological responses greatly benefits from the use of microbeam facilities (see Figure 21.1). Only by using facilities that are able to localize the radiation dose to specific areas of interest, is it possible to address where the triggering effects for specific biological responses (Lyng et al., 2006) (such as apoptosis or the bystander effect) are initiated and whether the nonuniformity of the damage induction may be responsible for the differential damage expression.

Following the success of the first microbeams at the end of the twentieth century, considerable interest has been generated in the scientific communities regarding microbeam technologies, resulting in a sharp increase in the number of facilities worldwide (Gerardi et al., 2005). Many different experimental setups have been exploited in order to assess the damage induced following irradiation by a micrometer (or submicrometer in some cases) beam that targets individual cells or part of cells. Some of the most popular approaches are reviewed and discussed below. In general, however, there are a few basic requirements that a microbeam facility has to fulfill in order to perform accurate radiobiological experiments.

These are as follows:

1. Production of a stable radiation beam of micron or submicron size.
2. Suitable dose rate to assure a fast and accurate dose delivery.
3. Radiation detectors able to monitor, with high efficiency, the dose delivered to the samples and trigger the beam-stop mechanism when the desired dose has been reached.
4. Specially designed cell holder to maintain the sample in a suitable sterile environment while allowing cell visualization and irradiation.
5. Image system for sample localization. This will have to be supported by appropriate software for image analysis and coordinate recording.
6. Micropositioning stage to align the samples with the radiation probe with high spatial resolution and reproducibility. The stage speed is also critical in assuring the high throughput necessary to investigate low frequency events (i.e., oncogenic transformation).

21.5 CHARGED-PARTICLE MICROBEAMS

21.5.1 MICROBEAM PRODUCTION

The key aspect for any microbeam facility is the ability to spatially limit the irradiation to a specific sample or part of it. There are two methods to reduce an accelerated beam of charged particles to a micron or submicron sized spot: collimation and focusing.

Collimation is by far the more straightforward way offering a number of advantages such as beam location, easy extraction into air, and reduction of the particle flux to radiobiologically relevant dose-rate ranges. As the samples need to be placed as close as possible to the collimator end, this is normally achieved by the optical devices used to image the sample allowing therefore a very precise beam location. Moreover, the size of the collimators represents a natural way to cut off the high flux of particles generated by the accelerators making it easy to reduce the fluency to ~100s of particles per second. Such a rate is ideal for irradiating a large number of samples in a short period of time while achieving a very high accuracy in the number of particles delivered (see particle detection & delivery). The main limitation of the collimation approach is particle scattering, which allows low energy particles to emerge with a wider lateral angle than the primary beam (this effect is inversely proportional to the radius–length ratio of the collimator). Scattering can be minimized by optimizing the collimator material, geometry, and beam alignment, all of which require significant design and skills. At the Gray Laboratory, several collimator options have been evaluated with the best being a 1 mm length of thick walled (outside diameter of 225 μm + 10 μm polyimide sheath) fused silica tubing with apertures as small as 1 μm in diameter (Folkard et al., 1997a,b). The collimator was mounted into a gimballing assembly (which allowed a precise beam–collimator alignment) and capped by a 3 μm Mylar film that acts as a vacuum window. The whole assembly was motorized in the vertical direction to move the end of the collimator closer to the sample during the irradiation and back down to allow the next sample positioning. The collimator performance was assessed in terms of quality of the energy and spatial spread. The first was done by measuring the energy spectrum of the emerging particles, resulting in a full width at half-maximum of 47 keV, while the latter was calculated by measuring the distance between particles tracks produced by irradiating CR-39 track-etch detectors. Using protons, it was possible to confine 90% of the particles within a 2 μm spot and 96% within 5 μm, while using $^3He^{2+}$, an accuracy of 99% <2 μm was achieved (Folkard et al., 2005). An alternative collimation system is represented by the use of a pair of aligned microapertures, where the first one defines the beam size and the second (usually larger) acts as an antiscatter device stopping the particles scattered by the first aperture. The distance between the two apertures and the relative alignment with the beam are critical parameters. Such a system has been adopted by many laboratories such as the JAERI (Takasaki, Japan) (Kobayashi et al., 2004), the INFN-LNL (Italy) (Gerardi et al., 2005), and the Columbia University where a pair of laser-drilled microapertures (5 and 6 μm) separated by 300 μm was assembled at the end of the beam line using two orthogonal micrometers (Randers-Pehrson et al., 2000). Monte Carlo modeling indicates that 91% of the emerging particles are within the unscattered area (5 μm diameter) with 7% of the beam diffused throughout an 8 μm halo.

If collimation was the preferred method of the first modern microbeams, the focusing approach has rapidly increased in popularity, mainly driven by already existing facilities (previously used for nuclear microscopy) and the need to investigate the sensitivity of subcellular targets such as the mitochondria (Tartier et al., 2007). There is little doubt that the narrowest charged-particle beams (down to a few 10s of nanometers) and high targeting accuracy are achieved in a vacuum by magnetic or electrostatic focusing devices as they don't suffer from collimator scattering problems. However, radiobiological microbeams require the radiation beam to be extracted in air in order to irradiate living samples in their natural surroundings. As the focused beam passes through the vacuum window, significant scattering is introduced, which compromises and limits the achievable beam spot size. Focused spots of 1 μm or less on the samples are, however, achievable with a very low fraction of scattered particles of reduced energy reaching the targets. Other relevant factors to consider for microbeam facilities based on focusing devices are a usually high beam current (which may limit their application for low-dose studies), the nonnegligible costs of the devices, and the length required to achieve significant focusing. The latter point is usually a strong constraint that limits the adoption of focusing lenses for radiobiological microbeams on a wide scale. Finally, a further complication of focusing-based microbeams is the localization of the radiation probe. While this is a straightforward operation when using collimators (whose end can be visualized using the same optics used for the biological sample), localizing the focused radiation spot in 3D requires extra and more complex steps. Once the ion beam focus has been achieved (this by itself may require iterative sequences measuring the ion beam size for combinations of the electrostatic lens voltages), the X–Y beam location is generally determined by scanning a knife edge (or a known shape object) across the beam profile while monitoring the particle energy-loss. When the knife edge is placed at the center of the ion beam, it is visualized registering the actual beam position to which to align the samples. Successful charged-particle microbeams based on focus systems have been developed and are routinely used in Germany at the PTB (Greif et al., 2006) and at the GSI (Heiss et al., 2004), in Japan at the NIRS (Imaseki et al., 2007), and at the Columbia University (United States). Brenner and colleagues (Bigelow et al., 2009) have adopted a compound lens consisting of two electrostatic quadrupole triplets with "Russian symmetry" to ensure a circular beam spot of 0.8 μm for 6 MeV He ions. The lenses have been realized using four 1 cm diameter ceramic rods, gold plated in bands to create positive and negative electrodes whose strengths are arranged following the (+A, −B, +B, −A) pattern for each quadruplet and (+A, −B, +C), (−C, +B, −A) for the triplets with voltage up to 15 kV. Each triplet is 0.3 m long and they are placed 1.9 m apart.

21.5.2 BEAM DETECTION

Another key feature of the modern microbeam facilities is the ability to detect and therefore control the number of particles delivered to the samples. In order to achieve this, it is necessary to develop a high-efficiency detection system (which will trigger the signal when the preset number of events has been reached) coupled to a very fast beam deflection system to terminate the irradiation. Particle detection is probably the feature that differs most between the microbeam facilities developed so far. They basically can be divided into two categories according to whether the detection occurs before or after the particle reaches the biological sample.

By placing the detector between the vacuum window and the samples, no further constraints are imposed on the sample holder or the cell environment. Moreover, it is possible to use particles of energy low enough to stop within the cell. The main issue with this type of approach is the limited space available and the inevitable detector–beam interaction, which reduces the quality and accuracy of the exposure. In order to minimize energy loss in the detector, only thin, transmission-type detectors are appropriate. These abrogate the effects of increased beam size and reduced targeting accuracy due to scattering. These detectors are generally thin films of plastic scintillators whose light flashes generated by the particle traversals are collected by a photomultiplier and then processed. It is also critical to choose a scintillator with a short decay time (<100 ns) to avoid light flashes generated by different particles that may overlap resulting in false signals. Available scintillators are not very efficient, and their reduced thickness leads to very weak light signals. A 5 MeV α particle loses ~300 keV in a 5 μm plastic scintillator, of which only ~4% are converted in 2000–3000 isotropically emitted photons of 3–5 eV. It is therefore necessary to collect most of the emitted photons using a very efficient photomultiplier and placing them close to the scintillator without interfering with the sample irradiation. This approach was first successfully adopted by the Gray Laboratory (Folkard et al., 1997a,b) using a 18 μm BC400 scintillator and an Hamamatsu PM-tube switched into position of the microscope objective (i.e., immediately above the scintillator and the samples) during the irradiation. By enclosing the irradiation area in a light-proof cage, a detection efficiency of >99% was achieved for both protons and $^3He^{2+}$ ions. Slight variation of the above method has been followed at the CENBG. Here, a custom-designed low pressure transmission gas chamber (3.5 mm thick, filled with 10 mbar isobutene) with 100 nm Si_3N_4 windows is placed in the beam path upstream of the samples. The system has been reported to monitor the particle traversals with greater efficiency and minimum interference (Barberet et al., 2005).

The alternative configuration consists of placing the detector behind the sample holder and therefore monitoring the particles after they have been delivered to the cells. Using this approach, there is no extra scattering introduced by the detector, and in theory better targeting accuracy can be reached. Such configurations benefit also from less space constraints for the detector, and more refined devices such as silicon detectors and gas chambers can be easily used. The detector of the Columbia University microbeam consists of a pulsed ion counter filled with P10 gas and built into the microscope objective used for the cell localization (Randers-Pehrson et al., 2001), allowing simultaneous viewing of the targeted samples and monitoring of the particles delivered. Such configurations, however, require that the delivered particles have enough energy to pass through the samples and escape with sufficient energy to be accurately detected, setting a limit on the low energy usable. In many cases, it is also necessary to remove the culture medium, requiring additional procedures (such as humidity control devices) to keep the cells viable during the irradiation process.

Finally, an interesting approach has been adopted at the GSI, where the use of heavy ions makes placing the detector either above or below the sample less than optimal. In order to preserve the quality of the focused beam without imposing strong constraints on the sample holder, Heiss and colleagues (Heiss et al. 2006) used an in-vacuum channeltron to collect secondary electrons produced by the heavy ions crossing the Si_3N_4 vacuum window that was coated with a thin layer of 50 μg/cm² CsI to increase the secondary electron emission. As all the detection elements are in vacuum, this configuration has the advantage of minimal interference with both the radiation beam and the sample holder/imaging apparatus. A detection efficiency of 99.5% for carbon ions has been claimed.

21.5.3 BEAM ORIENTATION AND SAMPLE HOLDER

One of the major difficulties in developing a microbeam facility for radiobiological studies is to create and maintain suitable conditions for the biological sample for the duration of the irradiation experiment. Despite sounding trivial, this is very complex considering all the other requirements and constraints imposed by the radiation delivery, detection, and cell imaging. Considerable efforts have been devoted to developing appropriate sample holders that, while supporting the cells without causing extra damage/or stress, would not interfere significantly with the radiation beam. The main requirements for a microbeam sample holder are the following: (1) the ability of samples to firmly attach to the holder to achieve accurate individual targeting of multiple targets can only be achieved for firmly attached samples, (2) the property of being optically transparent and non-UV fluorescent (see Section 21.5.4), (3) the property of minimum interference with the radiation beam and easily accessible on both sides by the radiation probe and the optical devices. This has led to the popular choice of thin (<5 μm) polyester films (Mylar or prolypropylene) glued or clamped into conventional or custom-made Petri dishes on which the cells are seeded and through which the radiation is delivered. Other materials used as a cell substrate are 100s nm thin Si_3N_4 membranes and track plastic detectors (CR39 and LR115) etched to submillimeter thickness. The latter also offer the option of visualizing the track traversals and relating them to the sample positions. Although this option can be used to check the number and site of traversals, it cannot be used to predetermine the number of particles delivered. The sample holder design is however strongly influenced by the orientation of the radiation probe. While microbeams expressly designed for radiobiological studies tend to have a vertical beam, most of the existing facilities are horizontally oriented, being adapted from existing accelerator beamlines. Taking advantage of gravity, vertical microbeams can generally adopt simpler sample holders to keep cells in an optimal environment for a long period of time. By contrast, horizontal beams require more complex designs to either enclose cells in a leak proof chamber that may limit or preclude access to the sample for the objective/detector, or control the humidity level with medium flow or humidified gas. This can limit these to short irradiation times.

21.5.4 Optics and Sample Preparation

In order to irradiate individual biological targets with high accuracy, the samples need to be identified with high resolution. Microbeam facilities, therefore, make effective use of state-of-the-art imaging devices and techniques. The most common approach is to use commercially available microscopes (often custom modified to be integrated into the radiation setup) and fluorescence microscopy. By taking advantage of the variety of molecular fluorescent probes available, subcellular targets of interest are labeled using specific fluorescent dyes and subsequently viewed in situ following snapshot UV illumination. Images are usually collected using CCD devices and analyzed by appropriate software for target identification and coordinates recording. Using this method, very fast and accurate target localization can be achieved due to the excellent quality of the images collected and the ability to perform such an operation immediately before or during the irradiation procedure. Despite using very low concentrations of dye (down to the μM range (Randers-Pehrson et al., 2001)) for minimum incubation periods (<1 h) and short UV exposures (<100 ms (Folkard et al., 1997a,b)), the influence of these factors on the cellular response cannot be completely eliminated, and accurate control (sham irradiation) experiments are necessary to check their negligible contribution and correct for it. Typical dyes used for microbeam experiments are the DNA binding compound Hoechst 33258 (is excited at 350 nm and emits at 461 nm) for the nucleus and Nile Red (excites at 485 nm and emits at 525 nm) for the cytoplasm. However, the advent of green fluorescent proteins (GFP) approaches has redefined the need for fluorescent microscopy. GFP can be expressed in different biological structures enabling morphological distinctions to be made, and as they are usually much less harmful when illuminated in living cells, they offer great potential for automated live cell fluorescence microscopy and microbeam studies.

An alternative approach using phase-contrast microscopy technique has also been exploited successfully by the INFN-LNL laboratories (Gerardi et al., 2005). The great benefit of this method is the lack of external insults for biological samples as only visible light is used. However, in order to obtain high quality images easily processable by software, phase-contrast microscopy requires the illumination source and microscope objective to be placed at opposite sides of the samples. This is usually a challenging issue for microbeam facilities as physical constraints on one side of the sample are imposed by the radiation delivery setup. To overcome such limitations, the INFN-LNL adopted an off-line sample recognition procedure that relies on a sophisticated translation of the sample holder to match the target coordinates with the radiation probe.

The ability to quickly and accurately locate biological targets is also critical for performing good microbeam experiments. The speed of image acquisition and processing is usually the determining factor influencing the cell throughput. Using relatively low microscope objectives (×20 to ×40) and relying on good CCD devices, large field images of samples are generally acquired with a resolution of the order of 1 μm. This allows irradiation throughput in the range 5,000–15,000 cells/h when supported by automated cell recognition algorithms. Cell density and dose rate also influence the final throughput. Experiments with yields as low as 10^{-4} (i.e., mutagenic and oncogenic end points) are therefore possible. It has to be noted that the accuracy with which biological targets are identified is just as important as other aspects of the microbeam experiment. Depending on the end point investigated and the experimental design, considerable uncertainties can be introduced if not all viable targets are correctly identified. These include missing viable cells, generally caused by poor image quality due to nonuniform UV illumination or dye uptake and/or inability to resolve two or more samples in close proximity, and mistaking background noise for biological targets.

21.5.5 POSITIONING STAGE

The accuracy to which samples can be aligned with the radiation probe has to at least match the spatial resolution of the beam. This is generally achieved by using automated translation stages with submicrometer precision and high reproducibility. Stage speed also plays an important role in the scanning and location of the target samples, contributing to the final irradiation throughput achievable.

21.6 X-RAY MICROBEAMS

Following the success of charged-particle microbeams, x-ray and electron microbeams have also started to be developed providing quantitative and mechanistic radiobiological information that complement charged-particle studies. Unlike charged particles, soft and ultrasoft x-rays (i.e., ~100s eV–10s keV photons) interact almost exclusively by photoelectric effects and therefore don't suffer from scattering issues. Higher energy x-rays (>10 keV) are not suitable for microbeam applications as the range of secondary electrons produced by the photon absorption is greater than the x-ray spot resolution, invalidating the efforts made to reduce the radiation beam to micron or submicron size. Even if 10 keV x-rays are focused into a micron spot inside a cell, the range of the secondary electrons produced will spread the dose delivered over a volume considerably larger (the range of 10 keV electron in tissue is ~2.5 μm). While extreme care has to be taken for charged-particles microbeams to ensure that the radiation beam accuracy is not compromised by scattering through the collimator, vacuum window, sample holder, and/or detector devices, it is in theory possible, with x-rays, to achieve radiation spots in air of an order of magnitude or more smaller than those so far achieved with ion beams. Moreover, such high spatial resolution is maintained as the x-ray beam penetrates through cells, making it possible to irradiate targets that are several tenths or hundreds of microns deep inside the samples with micron and submicron precision. The lack of scattering and the superior spatial resolution offered by x-ray microbeams make them the perfect tool to investigate questions regarding the role of subcellular targets in 2D cell cultures as well as in more complex 3D cell structures. Finally, another critical role covered by x-ray (and electron) microbeams concern the biological effectiveness of different LET radiations. It is commonly accepted that radiation effects such as apoptosis, mutation, and chromosomal instability are highly LET dependent (Lorimore et al., 1998; Kadhim et al., 1992). Electrons produced by the absorption of soft x-rays will produce ionization patterns that closely resemble the track structure of a low-LET irradiation in a way that is not possible to achieve using charged particles. In this respect, x-ray microbeams complement the work done with charged-particle facilities to investigate the LET dependence.

21.6.1   X-Ray Source

Generally, x-ray microbeams are based on similar concepts elucidated earlier for the charged-particle facilities; however, they face different technical issues due to the radiation production and the different nature of its interaction with matter. Some of the approaches described above for sample preparation, holding, imaging, and positioning are also applicable for x-ray microbeams. However while charged-particle microbeams require the use of particle accelerators, x-ray micro-beams can be developed using table-top sources (Folkard et al., 2001). Clear advantages of such an approach are the reduced costs, the simpler operation (experiments can be performed by a single scientist), and the possibility of building a compact facility. Compact table-top x-ray sources that have been used for microbeam facilities have been of the electron bombardment type, where the electron beam generated by a conventional electron gun is accelerated up to 20–30 kV and focused onto a solid target producing a "point-like" x-ray course (see Figure 21.2). The electron beam is generally produced by heated tungsten filaments (or more recently using single crystal filaments with brightness above 1×10^6 A/cm^2 sr) and focused either by permanent magnets or by electro-static lenses. Despite electron bombardment x-ray sources being an established technology, the requirement of focusing a high electron current into a small spot (in order to achieve a bright x-ray point source) is not easily achievable and still represents a major limitation for x-ray microbeams. Currently the best performances have been achieved by the Queen's University x-ray microbeam where up to 1 mA of 15 keV electrons have been squeezed into <20 μm diameter spot. The radiation spectrum generated by electron bombardment sources is composed of Bremsstrahlung as well as characteristic x-ray lines of the target element upon which the electrons impinge. In order to obtain a monochromatic x-ray beam, which is required both for optimum working of the x-ray focusing devices as discussed below and for a correct interpretation of the biological effects caused, the Bremsstrahlung component is removed either by filtration and/or by reflection. Depending on the x-ray energy of interest, low energy Bremsstrahlung can be removed by using appropriate filters while the high-energy component is usually eliminated by using a reflection approach. Taking advantage of the total reflection phenomenon that occurs for low energy photons below a critical angle, highly polished flat surface mirrors can be used to attenuate the high-energy component

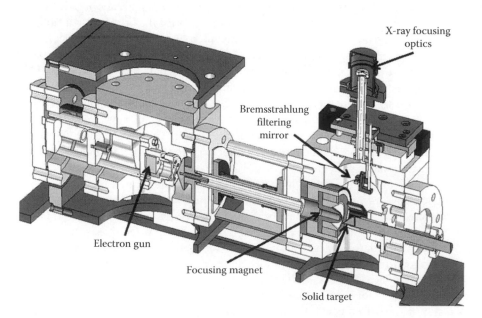

FIGURE 21.2   The x-ray source of the Queen's University x-ray microbeam, showing the key elements of an electron gun, focusing magnet, the target chamber, and the x-ray focusing optics.

while reflecting a significant fraction of low energy x-rays. The final result is the production of a nearly monochromatic x-ray beam corresponding to the brightest characteristic x-ray line of the electron bombarded target. As mentioned, the major limitation of the electron bombardment x-ray sources is represented by their limited brightness and size. The spot size into which it is possible to focus electrons is ultimately limited by the electron beam current as space charge effects will produce a repulsive force that will compromise the focusing action of permanent or electromagnetic lenses. Moreover, the x-ray production by electron bombardment is not a very efficient process with only a small percentage of the electron beam energy actually transformed into x-ray production. Most of the energy deposited by the electrons into solid targets will be dissipated as heat. Special care, therefore, is to be taken when designing solid targets to avoid high temperature damage. Using fixed thick targets, it is generally necessary to compromise between high electron currents and the size of the electron focus as the maximum power that can be possibly dissipated on a thick target depends on the power density of the impinging beam (i.e., electron current × accelerating electron voltage/area electron focus). Cooled or rotating targets are other approaches that have been investigated to increase the electron current and therefore the x-ray production rate. However, both approaches present technical issues for their implementation in microbeam facilities. Cooling the target is an efficient method but only if the heat extraction device (generally water or gas) can be placed very close to the electron spot. As the electrons are focused into a micron sized spot and have limited penetration in the target, it is basically impossible to efficiently cool the area of the target where the electrons impinge. Such an approach will only bring minor improvements in terms of x-ray production. Rotating targets offer more promising benefits and have indeed been used to produce x-ray sources for microbeam applications (Lekki et al., 2009). Higher electron currents with higher voltages and small spot sizes (~1 µm) have been reported. However, due to the rotation of the target, this type of x-ray source generally suffers from mechanical vibrations that may ultimately affect the resolution and accuracy of the final x-ray focused beam. The smallest x-ray spot size achieved using a rotating target x-ray source is 7 µm. Another limitation of the electron bombardment sources is the range of x-ray energies that can be possibly generated. X-ray energy is determined by the energy of the electrons and the target material. In order to produce high-energy x-rays, high Z materials and high-energy electrons have to be used. While high-energy electrons are harder to focus and dissipate higher power on the target, high Z materials offer lower cross section values for the x-ray production. Both factors conspire against the production of suitable monochromatic high-energy x-ray beams and, to date, x-ray microbeam facilities have successfully been developed only for energies up to 4.5 keV (characteristic K_α titanium line). Some of the electron bombardment source limitations may be overcome using proton bombardment sources. The advantage of using protons instead of electrons lies in the higher x-ray production cross section (i.e., more energy is transformed into x-rays) and the reduced (in some cases negligible) Bremsstrahlung component that does not require beam filtration. The use of protons, however, requires linear accelerators and although higher proton beam currents can be achieved and focused into smaller spot sizes, the maximum power that can be possibly dissipated into the target will ultimately limit their applicability for microbeam systems.

Synchrotron radiation generated by high-energy electrons accelerated in storage-type circular accelerators also provides a very interesting source of x-rays for microbeams applications. Despite the difficulties in working with a horizontal beam line (see Section 21.5.3), the high brightness, wide energy range, and nearly parallel directional beam of synchrotron facilities offer great potential from both the technical development aspect as well as for radiobiological applications. The nearly parallel and monochromatic beam assures that the best focusing resolution can be achieved by the employed x-ray optical devices, while the tunable x-ray energy is an incredibly powerful tool to address critical radiobiological phenomena. However, the cost of running a synchrotron-based microbeam and the restricted beamlines available are considerable limiting factors, and so far the only microbeam facility taking advantage of synchrotron light is based at the Photon Factory in Japan (Kobayashi et al., 2009), used to irradiate biological samples with a 5 µm x-ray beam of

4–20 keV. Finally, it has to be mentioned that laser plasma or electrostatic pinch plasma sources (Blackborow et al., 2007) are also being considered as possible x-ray sources for microbeams. Such sources are characterized by small size and high brightness and may provide valuable input in microbeam development.

21.6.2   X-Ray Optics

The rapid increase in the development of x-ray microbeams is due to the high spatial resolution possible, with which to focus x-rays. Sophisticated x-ray optics have been under development world-wide for many years for lithography and astrophysics applications. Microbeam facilities can there-fore rely on a considerable pool of expertise and technical devices.

X-ray beams can be focused using one of two different principles: reflection or diffraction. Focusing methods based on reflection use curved surfaces coated with a single layer (mirror) or several layers (multilayers) to bend the x-ray beam and force it to converge in a single spot. They can be used for a wide energy range, have large acceptance (i.e., can collect a large fraction of the x-ray beam), and benefit from high efficiencies even for high-energy x-rays (>10 keV). The most com-monly used configuration is the so-called Kirkpatrick–Baez design where two orthogonal parabolic mirrors are used to successively reflect the x-ray beam, achieving a two-dimensional focused spot. This approach provides high accuracy and flexibility as it offers the possibility to vary the focal distance and focus different x-ray energies by "simply" changing the mirror's orientation to alter the x-ray incident angle. However, the mirror alignment/tuning is a very critical step usually performed by automated precision motors to optimize the eight degrees of freedom of the system. The coalign-ment of many reflecting surfaces to form an optimum image is a difficult process as aberrations are easily introduced and the final focusing accuracy compromised. Another factor limiting their use for microbeam applications is the typically long focal distance required by the reflecting mirrors to produce optimum results.

The majority of the x-ray microbeam facilities, however, employ diffraction techniques to bring the x-ray beam to a fine submicron spot. Fine x-ray probes (<50 nm diameter) are achieved in x-ray microscopy using Fresnel zone plate lenses that are circular diffraction gratings with increasing line density (see Figure 21.3). Being a diffraction device, zone plates will focus the incident x-ray radiation into a series of orders progressively closer to the zone plate itself. In order to stop all but the first order focusing from reaching the sample, an arrangement of masks have to be accurately aligned. The mask is usually an axial disk mounted on the zone plate itself and an equal size pin-hole. The first order diffraction focus is generally chosen as it is the most intense, while the zero (i.e., undiffracted) and secondary orders have to be removed as they would reach the sample unfo-cused, therefore spoiling the beam resolution. Due to diffraction principle constraints (i.e., gap between transmitting and opaque zones needs to be of the order of the wavelength of the incident x-ray radiation), zone plates are very small devices (a few hundreds of microns in diameter) with fine structures down to 50 nm. Combined with their low focusing efficiency (theoretical max for low energy x-ray ~20% and rapidly decreasing at higher energies), this means that zone plates are only able to focus a very small fraction of the available x-ray beam. Despite these limitations, the superior spatial resolution offered by the zone plates and the relatively straightforward applica-tion make them particularly suitable for microbeam applications (Folkard et al., 2001; Thompson et al., 2004).

Both focusing methods described above require a parallel incident x-ray beam in order to pro-duce the finest possible spot. In case of non parallel beams, zone plate and mirror work can be compared to that of a normal optical lens; they basically demagnify the x-ray source. In practice, the final x-ray spot is therefore determined by the size of the x-ray source and its distance from the focusing device rather than the focusing device itself. Another requirement unrelated to the focusing approach is that microbeam x-ray sources be point-like and bright enough to illuminate the optical device with suitable photon fluxes.

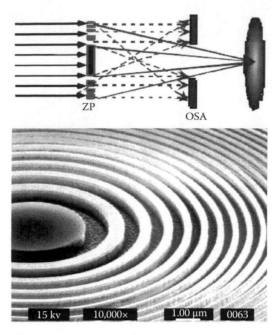

FIGURE 21.3 Schematic illustration of the focusing principles of a zone plate—An order sorting aperture (OSA) prevents alignment with all but the first order diffraction prevented from reaching the sample. A detailed example of a zone plate lens is also shown.

21.6.3 X-Ray Detection and Dosimetry

Due to the nature of the x-ray interaction with matter, it is not possible to detect the same photons that have or will interact with the biological samples. For x-ray microbeams, the irradiation is basically a time exposure determined on pre-irradiation characterization of the emerging beam. The focused x-ray beam is usually characterized using an x-ray detector (proportional chamber or photodiode) operating in a photon counting mode and scanning a knife edge mask across the beam to determine the profile and its location. Considering the low dose rate generally available for x-ray microbeams (fraction of Gy/s) and the large number of photons necessary to deliver a suitable dose (for carbon K_α x-rays ~10,000 photons are necessary to deliver 1 Gy to a typical mammalian cell), a fast mechanical shutter is appropriate to assure an accurate dose delivery and uncertainties due to the variation in the flux of photons generated (which follows a Poisson distribution) is negligible. The methods used to report the dose absorbed by the samples exposed to localized and/or partially penetrating radiations are often the basis for discussion and it may present problematic issues such as the critical mass, which should be considered for dosimetry calculations. In the case of charged-particle microbeams, the number of particle traversals is usually reported as well as the average nuclear dose, calculated considering the energy deposited and the entire cell nucleus mass. For x-ray microbeams, the number of photons that deposit energy inside the samples cannot be precisely estimated, and despite the high non homogeneous energy deposition (especially for ultrasoft x-rays that are highly attenuated), the average nuclear dose is usually reported. It must also be noted that cell morphology is a critical parameter, and precise information on the cytoplasm and cell nucleus thickness must be taken into consideration when assessing the effects induced by x-ray microbeam irradiation.

21.7 ELECTRON MICROBEAMS

Charged-particle microbeams cover predominantly the high LET spectrum of ionizing radiation. The energy of x-ray facilities is still too high to be classified as low LET, and therefore, the ideal way to produce a low-LET microbeam is to use a fine beam of energetic electrons (<100 keV).

As for x-ray, electron microbeams are generally self-contained units with relatively low development and maintenance costs. They rely on standard electron guns and electrostatic devices to produce and accelerate energetic electron beams that are subsequently reduced to micrometer size by the use of collimators (generally laser-drilled apertures) or electromagnetic focusing. The use of collimating apertures is often employed (Sowa et al., 2005) due to their straightforward application. As the electron penetration is very limited, only thin apertures are required, making the alignment process much easier than what was discussed earlier for charged particles, and the low beam fluency is suitable for low-dose experiments. However, since electrons are easily scattered as they exit the vacuum system, the distance and the amount of material between the collimator and the samples have to be minimized. As electrons also scatter as they interact with the biological samples, it is impossible for electron microbeams to achieve targeting resolutions of the micron or submicron level despite the actual size of the focused beam. Microdosimetry studies (Wilson et al., 2001) indicate that the energy deposited by a 25 keV electron track is barely contained within a typical mammalian cell and the ionizations are nonuniformly distributed across its nucleus. As the energy increases, the electron range also increases, but the later component of the scattering actually reduces in favor of the forward component. The maximum later scattering is calculated to occur at ~50 keV when about 16% of the energy deposited in the targeted cell leaks into the neighboring samples (Miller et al., 2001). A great advantage of the electron microbeam is also the ability to easily vary the electron energy (and therefore the LET) in order to investigate the relative biological importance of various parts of the electron track. Using electron energies ranging from 25 to 80 keV, Sowa Resat and colleagues (Sowa Resat and Morgan, 2004) showed that the peak in micronuclei formation is shifted to lower doses as the electron energy decreases. This has been successfully used to strengthen the hypothesis that the electron track end is mainly responsible for the induction of biological damage. Finally, for more energetic electrons, considerable penetration can be achieved (path length of 90 keV electron exceeds 100 μm) allowing irradiation of thick tissue samples of 5–6 cell layers.

21.8   FUTURE DEVELOPMENTS

The interest in the microbeam technique is such that the number of facilities developed worldwide has increased from the initial three of the mid 1990s to over thirty acknowledged in 2006. The great contribution to radiobiology offered by precise and accurate irradiation of the samples justifies the efforts devoted by several laboratories in developing new methodology and indeed improving the existing microbeam facilities. Some of the future developments that are already under investigations are reported below.

21.8.1   BEAM SPOT SCANNING

One of the great advantages offered by the electrostatic focusing of charged particles over the collimator approach is the possibility to precisely deflect the radiation beam. Accurate beam spot scanning can be achieved by simply varying the voltage of the electrostatic lenses and moving therefore the radiation probe to the sample position in contrast to what is currently done. As electrostatic beam movement can be done much faster than mechanical stage movements, this approach has the potential to speed-up the irradiation process and significantly increases the cell throughput. However, radiation beams can only be scanned over very small distances without losing focusing accuracy, limiting the exploitable area of the sample holder. A practical solution would be to couple the electrostatic beam scanning to a coarse stage movement in order to be able to use a larger sample area. Finally, the beam scanning irradiation approach necessitates accurate calibration procedures to determine the correct parameters of the electrostatic lens necessary to move the beam to the desired sample position. While beam scanning is an established technique, the submicron precision required by a microbeam experiment represents a technical issue that has prevented such techniques to be successfully employed so far.

21.8.2 Non-Staining Imaging

Of great interest for the new generation of microbeam facilities is the use of more sophisticated optical devices to both locate the biological targets of interest and, following the radiation exposure, determine the damage induced at a single-cell level. Current biological targets used for the microbeam experiments are considerably larger than the radiation probe. For example, the cell nucleus and cytoplasm are several microns in diameter against the submicron resolution achieved by some facilities. In order to take full advantage of the microbeam targeting properties, it is crucial to be able to visualize subcellular compartments other than the nucleus with high precision. For the microbeam technique to be capable of detecting very small cellular responses, the potential confounding insult of external agents such as UV illumination and dye staining has to be eliminated/minimized. A few laboratories have already started to employ non fluorescent techniques (see Section 21.5.4), and a very promising approach is represented by the use of quantitative phase microscopy (QPM). Using QPM, high resolution images can be obtained by analyzing three images acquired with conventional light microscopy. Commercial software that performs such analysis is already available and only requires one in focus image and two slightly defocused (positively and negatively) images. The considerable benefit of being able to use conventional light microscopy on line for microbeam facilities has however been judged against the resolution of the images obtained and the processing speed.

Alongside the biological demands of performing microbeam experiments in tissues and 3D cell systems, there is a need to equip the microbeam facilities with deep imaging techniques such as multiphoton microscopy. Multiphoton fluorescence microscopy is an imaging technique that allows imaging of living tissue samples up to a few millimeters thick. It is based on the principle that two long-wavelength photons (700–1000 nm) simultaneously excite the same fluorophore into a higher energy state than they would individually. The fluorophores then decay, emitting a fluorescent signal relative to its energetic state (400–500 nm). The main differences between multiphoton and conventional fluorescent microscopy are the longer excitation wavelength, which requires use of lasers, required to produce a fluorescent signal and the localized nature of the emission signal that is only emitted from the fluorophores where two exiting photons are focused. As a consequence, the insult induced by multiphoton imaging is considerably lower, and high resolution images at depth of several hundred microns can be obtained. Moreover, the use of tunable lasers allows multiphoton microscopes to observe post-irradiation cellular dynamics by acquiring time-lapse images (Bigelow et al., 2009).

21.8.3 Hard X-Ray Optics

As illustrated in Section 21.6.2, the finest x-ray focusing is currently achieved with diffraction elements (i.e., zone plates) that are wavelength dependent and characterized by low efficiencies that severely limit their application with conventional lab bench x-ray sources. The use of reflecting optics overcomes such problems, although available devices described earlier suffer from complex alignment routines, considerable length, and limited focusing ability. However, new grazing incident reflecting optic devices for x-rays (i.e., polycapillary systems based on commercial microchannel plates originally designed as electron multipliers) are being designed and manufactured offering promising alternatives for x-ray microbeams. In particular, the MOXI device (microfabricated optical array for x-ray imaging) consists of two symmetrical arrays of microchannels etched into silicon wafers and arranged in a circular pattern (see Figure 21.4). By precisely aligning the two micro-channel arrays, a focusing action can be achieved with considerable efficiency and minimum aberrations when one or both arrays are appropriately bent (Prewett and Michette, 2001). The benefits offered by the MOXI devices lie in having a much larger numerical aperture compared to zone plates (i.e., they can collect a larger fraction of the produced x-ray beam), a greater focusing efficiency, and the possibility to use higher energy x-rays. Moreover, the focal length of

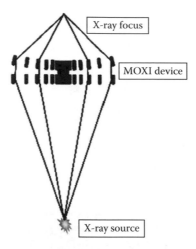

FIGURE 21.4 Schematic illustration of the working principle of a MOXI lens.

such micro-optical devices can be varied providing zoom lens capability. Realization of MOXIs, however, presents some technical challenges due to the required smoothness of the micro-channels arrays and the ability to control their individual bending.

21.8.4 BIOLOGICAL ADVANCES

Alongside the advances in imaging mentioned earlier, a major development is the move from cell targeting studies to tissue and ultimately in vivo approaches. A range of studies have now been reported in 3-D model systems using microbeams. These have been relatively simple approaches where a single region of a tissue reconstruct is irradiated and then measurements made at later times using conventional approaches of fixing these tissues, sectioning these into multiple slices, and looking at distance dependence of various biological responses (Belyakov et al., 2005; Sedelnikova et al., 2007). The coupling of multiphoton imaging approaches to microbeam exposure will allow sophisticated real time imaging of responses at depth to delineate multicellular signaling pathways. With higher energy microbeams, the opportunity to do in vivo exposures, especially in mouse models, will be possible. This is particularly timely as recent data has suggested long range bystander responses can be observed using in vivo models, using relatively crude shielding opportunities and there is an urgent need to test the responses using targeted microbeam approaches (Mancuso et al., 2008). Live cell imaging for cellular studies will also develop rapidly alongside multiple new fluorescent markers of response. A key requirement will be to utilize time-lapse approaches for following changes after microbeam irradiation. For high throughput, these will have to be done offline from the microbeam irradiation station. This requires good coordinate transfer approaches so that irradiated cells can be revisited and followed offline.

21.9 SUMMARY

A range of approaches have been developed for producing microbeams for radiation biology research. These include the following: charged particles, x-rays, and electron microbeams. Significant technological development in radiation delivery, beam detection, and positioning, alongside advances in imaging has made microbeams powerful tools for assessing the effects of radiation exposure in a range of biological models. Much of this has concentrated on the effects of low- dose exposures of relevance to environmental and occupational exposures received by human populations. With advances in both technology and biology, microbeam approaches will continue to make a significant contribution to our knowledge of radiation effects in the future.

ACKNOWLEDGMENTS

The authors are grateful to Cancer Research UK [CUK] grant number C1513/A7047, the European Union NOTE project (FI6R 036465), and the U.S. National Institutes of Health (5P01CA095227-02) for funding their work.

REFERENCES

Baker, C. P., Curtis, H. J., Zeman, W., and Woodley, R. G. 1961. The design and calibration of a deuteron microbeam for biological studies. *Radiat. Res.* 15: 489–495.

Barberet, P., Balana, A., Incerti, S., Michelet-Habchi, C., Moretto, P., and Pouthier, T. 2005. Development of a focused charged particle microbeam for the irradiation of individual cells. *Rev. Sci. Instrum.* 76(1): 015101–015106.

Belyakov, O. V., Mitchell, S. A., Parikh, D., Randers-Pehrson, G., Marino, S. A., Amundson, S. A., Geard, C. R., and Brenner, D. J. 2005. Biological effects in unirradiated human tissue induced by radiation damage up to 1 mm away. *Proc. Natl. Acad. Sci. USA* 102(40): 14203–14208.

Bigelow, A., Garty, G., Funayama, T., Randers-Pehrson, G., Brenner, D., and Geard, C. 2009. Expanding the question-answering potential of single-cell microbeams at RARAF, USA. *J. Radiat. Res. (Tokyo)* 50 Suppl A: A21–28.

Blackborow, P. A., Gustafson, D. S., Smith, D. K., Besen, M. M., Horne, S. F., D'Agostino, R. J., Minami, Y., and Denbeaux, G. 2007. Application of the Energetiq EQ-10 electrodeless Z-Pinch EUV light source in outgassing and exposure of EUV photoresist. In *Emerging Lithographic Technologies XI (Proceedings of the SPIE)*, M. Lercel (ed.). Bellingham, WA: SPIE.

Boreham, D. R., Dolling, J.-A, Broome, J., Maves, S. R., Smith, B. P., and Mitchel, R. E. J. 2000. Cellular adaptive response to single tracks of low-LET radiation and the effect on nonirradiated neighboring cells. *Radiat. Res.* 153: 230–231.

Braby, L. A. and Reece, W. D. 1990. Studying low dose effects using single particle microbeam irradiation. *Radiat. Prot. Dosimetry* 31(1–4): 311–314.

Brenner, D. J., Doll, R., Goodhead, D. T., Hall, E. J., Land, C. E., Little, J. B., Lubin, J. H., Preston, D. L., Preston, R. J., Puskin, J. S., Ron, E., Sachs, R. K., Samet, J. M., Setlow, R. B., and Zaider, M. 2003. Cancer risks attributable to low doses of ionizing radiation: assessing what we really know. *Proc. Natl. Acad. Sci. USA* 100(24): 13761–13766.

Brenner, D. J., Miller, R. C., Huang, Y., and Hall, E. J. 1995. The biological effectiveness of radon-progeny alpha particles. III. Quality factors. *Radiat. Res.* 142(1): 61–69.

Cardis, E., Gilbert, E. S., Carpenter, L., Howe, G., Kato, I., Armstrong, B. K., Beral, V., Cowper, G., Douglas, A., Fix, J., Fry, S. A., Kaldor, J., Lave, C., Salmon, L., Smith, P. G., Voelz, G. L., and Wiggs, L. D. 1995. Effects of low doses and low dose rates of external ionizing radiation: Cancer mortality among nuclear industry workers in three countries. *Radiat. Res.* 142(2): 117–132.

Datta, R., Cole, A., and Robinson, S. 1976. Use of track-end alpha particles from 241Am to study radiosensitive sites in CHO cells. *Radiat. Res.* 65: 139–151.

Folkard, M., Vojnovic, B., Hollis, K. J., Bowey, A. G., Watts, S. J., Schettino, G., Prise, K. M., and Michael, B. D. 1997a. A charged-particle microbeam: II. A single-particle micro-collimation and detection system. *Int. J. Radiat. Biol.* 72(4): 387–395.

Folkard, M., Vojnovic, B., Prise, K. M., Bowey, A. G., Locke, R. J., Schettino, G., and Michael, B. D. 1997b. A charged-particle microbeam: I. Development of an experimental system for targeting cells individually with counted particles. *Int. J. Radiat. Biol.* 72(4): 375–385.

Folkard, M., Schettino, G., Vojnovic, B., Gilchrist, S., Michette, A. G., Pfauntsch, S. J., Prise, K. M., and Michael, B. D. 2001. A focused ultrasoft x-ray microbeam for targeting cells individually with submicrometer accuracy. *Radiat. Res.* 156: 796–804.

Folkard, M., Prise, K. M., Schettino, G., Shao, C., Gilchrist, S., and Vojnovic, B. 2005. New insights into the cellular response to radiation using microbeams. *Nucl. Instrum. Methods B* 231: 189–194.

Gerardi, S., Galeazzi, G., and Cherubini, R. 2005. A microcollimated ion beam facility for investigations of the effects of low-dose radiation. *Radiat. Res.* 164(4 Pt 2): 586–590.

Greif, K., Beverung, W., Langner, F., Frankenberg, D., Gellhaus, A., and Banaz-Yasar, F. 2006. The PTB microbeam: A versatile instrument for radiobiological research. *Radiat. Prot. Dosimetry* 122(1–4): 313–315.

Hall, E. J. 2000. *Radiobiology for the Radiologist*. Philadelphia, PA: Lippincott Williams & Wilkins.

Hei, T. K., Wu, L. J., Liu, S. X., Vannais, D., Waldren, C. A., and Randers-Pehrson, G. 1997. Mutagenic effects of a single and an exact number of alpha particles in mammalian cells. *Proc. Natl. Acad. Sci. USA* 94(8): 3765–3770.

Heiss, M., Fischer, B. E., and Cholewa, M. 2004. Status of the GSI microbeam facility for cell irradiation with single ions. *Radiat. Res.* 161: 98–99.

Heiss, M., Fisher, B. E., Jakob, B., Fournier, C., Becker, G., and Taucher-Scholz, G. 2006. Targeted irradiation of mammalian cells using a heavy-ion microprobe. *Radiat. Res.* 165(2): 231–239.

Imaseki, H., Ishikawaa, T., Isoa, H., Konishia, T., Suyaa, N., Hamanoa, T., Wanga, X., Yasudaa, N. and Yukawaa, M. 2007. Progress report of the single particle irradiation system to cell (SPICE). *Nucl. Instrum. Methods Phys. Res. Sect. B: Beam Interact. Mater. Atoms* 260(1): 81–84.

Kadhim, M. A., MacDonald, D. A., Goodhead, D. T., Lorimore, S. A., Marsen, S. J., and Wright, E. G. 1992. Transmission of chromosomal instability after plutonium alpha-particle irradiation. *Nature* 355: 738–740.

Kadhim, M. A., Lorimore, S. A., Townsend, K. M., Goodhead, D. T., Buckle, V. J., and Wright, E. G. 1995. Radiation-induced genomic instability: Delayed cytogenetic aberrations and apoptosis in primary human bon marrow cells. *Int. J. Radiat. Biol.* 67(3): 287–293.

Kadhim, M. A., Marsden, S. J., Malcomson, A. M., Goodhead, D. T., Prise, K. M., and Michael, B. D. 2001. Long-term genomic instability in human lymphocytes induced by single-particle irradiation. *Radiat. Res.* 155: 122–126.

Kobayashi, Y., Funayama, T., Wada, S., Furusawa, Y., Aoki, M., Shao, C., Yokota, Y., Sakashita, T., Matsumoto, Y., Kakizaki, T., and Hamada, N. 2004. Microbeams of heavy charged particles. *Biol. Sci. Space* 18(4): 235–240.

Kobayashi, Y., Funayama, T., Hamada, N., Sakashita, T., Konishi, T., Imaseki, H., Yasuda, K., Hatashita, M., Takagi, K., Hatori, S., Suzuki, K., Yamauchi, M., Yamashita, S., Tomita, M., Maeda, M., Kobayashi, K., Usami, N., and Wu, L. 2009. Microbeam irradiation facilities for radiobiology in Japan and China. *J. Radiat. Res. (Tokyo)* 50 Suppl A: A29–47.

Kraske, F., Ritter, S., Scholz, M., Schneider, M., Kraft, G., Weisbrod, U., and Kankeleit, E. 1990. Directed irradiation of mammalian cells by single charged particles with a given impact parameter. *Radiat. Prot. Dosimetry* 31: 315–318.

Lekki, J., Bielecki, J., Bozek, S., Stachura, Z., and Kwiatek, W. M. 2009. Design of the Krakow x-ray microprobe facility for targeted X-ray irradiations of biological objects. *J. Radiat. Res.* 50(Suppl.): A98.

Lorimore, S. A., Kadhim, M. A., Pocock, D. A., Papworth, D., Stevens, D. L., Goodhead, D. T., and Wright, E. G. 1998. Chromosomal instability in the descendants of unirradiated surviving cells after alpha-particle irradiation. *Proc. Natl. Acad. Sci. USA* 95: 5730–5733.

Lyng, F. M., Maguire, P., Kilmurray, N., Mothersill, C., Shao, C., Folkard, M., and Prise, K. M. 2006. Apoptosis is initiated in human keratinocytes exposed to signalling factors from microbeam irradiated cells. *Int. J. of Radiat. Biol.* 82(6): 393–399.

Mancuso, M., Pasquali, E., Leonardi, S., Tanori, M., Rebessi, S., Di Majo, V., Pazzaglia, S., Toni, M. P., Pimpinella, M., Covelli, V., and Saran, A. 2008. Oncogenic bystander radiation effects in patched heterozygous mouse cerebellum. *Proc. Natl. Acad. Sci. USA* 105(34): 12445–12450.

Marples, B. and Joiner, M. C. 1993. The response of Chinese hamster V79 cells to low radiation doses: evidence of enhanced sensitivity of the whole cell population. *Radiat. Res.* 133: 41–51.

Michael, B. D., Prise, K. M., Folkard, M., Brocklehurst, B., Munro, I. H., and Hopkirk, A. 1995. Critical energies for ssb and dsb induction in plasmid DNA: Studies using synchrotron radiation. In *Radiation Damage in DNA: Structure/Function Relationships at Early Times*, ed. A. Fuciarelli, and J. Zimbrick, pp. 251–258. Columbus, Ohio: Battelle Press.

Miller, R. C., Randers-Pehrson G., Geard, C. R., Hall, E. J., and Brenner, D. J. 1999. The oncogenic transforming potential of the passage of single alpha-particles through mammalian cell nuclei. *Proc. Natl. Acad. Sci. USA* 96: 19–22.

Miller, J. H., Wilson, W. E., Lynch, D. E., Resat, M., and Trease, H. E. 2001. Computational dosimetry for electron microbeams: Monte Carlo track simulation combined with confocal microscopy. *Radiat. Res.* 156: 438–439.

Morgan, W. F. and Sowa, M. B. 2007. Non-targeted bystander effects induced by ionizing radiation. *Mutat. Res.* 616(1–2): 159–164.

Muirhead, C. R., Goodill, A. A., Haylock, R. G., Vokes, J., Little, M. P., Jackson, D. A., O'Hagan, J. A., Thomas, J. M., Kendall, G. M., Silk, T. J., Bingham, D., and Berridge, G. L. 1999. Occupational radiation exposure and mortality: Second analysis of the National Registry for Radiation Workers. *J. Radiol. Prot.* 19(1): 3–26.

Prewett, P. D. and Michette, A. G. 2001. MOXI: A novel microfabricated zoom lens for x-ray imaging. *Proc. SPIE* 4145: 180–187.

Prise, K. M., Belyakov, O. V., Folkard, M., and Michael, B. D. 1998. Studies of bystander effects in human fibroblasts using a charged particle microbeam. *Int. J. Radiat. Biol.* 74(6): 793–798.

Prise, K. M., Folkard, M., Malcolmson, A. M., Pullar, C. H., Schettino, G., Bowey, A. G., and Michael, B. D. 2000. Single ion actions: The induction of micronuclei in V79 cells exposed to individual protons. *Adv. Space Res.* 25(10): 2095–2101.

Prise K. M., Schettino, G., Folkard, M., and Held, K. D. 2005. New insights on cell death from radiation exposure. *Lancet Oncol.* 6: 520–528.

Randers-Pehrson, G., Geard, C., Johnson, G., and Brenner, D. 2000. Technical characteristics of the Columbia University single-ion microbeam. *Radiat. Res.* 153: 221–223.

Randers-Pehrson, G., Geard, C. R., Johnson, G., Elliston, C. D., and Brenner, D. J. 2001. The Columbia University single-ion microbeam. *Radiat. Res.* 156: 210–214.

Schettino, G., Folkard, M., Prise, K. M., Vojnovic, B., Bowey, A. G., and Michael, B. D. 2001. Low-dose hypersensitivity in Chinese hamster V79 cells targeted with counted protons using a charged-particle microbeam. *Radiat. Res.* 156: 526–534.

Schettino, G., Folkard, M., Michael, B. D., and Prise, K. M. 2005. Low-dose binary behaviour of bystander cell killing after microbeam irradiation of a single cell with focused Ck x-rays. *Radiat. Res.* 163: 332–336.

Sedelnikova, O. A., Nakamura, A., Kovalchuk, O., Koturbash, I., Mitchell, S. A., Marino, S. A., Brenner, D. J., and Bonner, W. M. 2007. DNA double-strand breaks form in bystander cells after microbeam irradiation of three-dimensional human tissue models. *Cancer Res.* 67(9): 4295–4302.

Shao, C., Folkard, M., Michael, B. D., and Prise, K. M. 2004. Targeted cytoplasmic irradiation induces bystander responses. *Proc. Natl. Acad. Sci. USA* 101(37): 13495–13500.

Sowa Resat, M. and Morgan, W. F. 2004. Microbeam developments and applications: A low linear energy transfer perspective. *Cancer Metastasis Rev.* 23: 323–331.

Sowa, M. B., Murply, M. K., Miller, J. H., McDonald, J. C., Strom, D. J., and Kimmel, G. A. 2005. A variable-energy electron microbeam: A unique modality for targeted low-LET radiation. *Radiat. Res.* 164: 695–700.

Tartier, L., Gilchrist, S., Folkard, M., and Prise, K. M. 2007. Cytoplasmic irradiation induces 53BP1 protein relocalization in irradiated and bystander cells. *Cancer Res.* 67(12): 5872–5879.

Thompson, A. C., Blakely, E. A., Bjornstad, K. A., Chang, P. Y., Rosen, C. J., Sudar, D., and Schwarz, R. I. 2004. Microbeam studies of low-dose x-ray bystander effect on epithelial cells in fibroblasts using synchrotron radiation. *Radiat. Res.* 161: 101–102.

Watt, F. and Grime, G. W. 1987. *Principles and Applications of High-Energy Ion Microbeams*. Bristol, U.K.: Hilger.

Watt, F., Grime, G. W., Blower, G. D., and Takacs, J. 1982. The Oxford 1 micron proton microprobe. *Nucl. Instrum. Methods* 197: 65–77.

Wilson, W. E., Lynch, D. J., Wei, K., and Braby, L. A. 2001. Microdosimetry of a 25 keV electron microbeam. *Radiat. Res.* 155: 89–94.

Wu, L. J., Randers-Pehrson, G., Xu, A., Waldren, C. A., Geard, C. R., Yu, Z., and Hei, T. K. 1999. Targeted cytoplasmic irradiation with alpha particles induces mutations in mammalian cells. *Proc. Natl. Acad. Sci. USA* 96: 4959–4964.

Zeman, W., Curtis, H. J., Gebhard, E. L., and Haymaker, W. 1959. Tolerance of mouse-brain tissue to high-energy deuterons. *Science* 130: 1760–1761.

Zirkle, R. E. and Bloom, W. 1953. Irradiation of parts of individual cells. *Science* 117: 487–493.

22 Redox Reactions of Antioxidants: Contributions from Radiation Chemistry of Aqueous Solutions

K. Indira Priyadarsini
Bhabha Atomic Research Centre
Mumbai, India

CONTENTS

22.1 FREE RADICALS: INTRACELLULAR GENERATION AND REACTIONS

The general term *free radical* represents a molecule or atom having an unpaired electron, and reactive free radicals are those in which the electron is readily available to react with other molecules. Oxygen-derived free radicals that have drawn the initial attention in biology are superoxide ($O_2^{\cdot-}$), hydroxyl ($^{\cdot}OH$), peroxyl ($ROO^{\cdot}$), oxyl ($RO^{\cdot}$), singlet oxygen ($^1O_2$), and hydroperoxyl ($HOO^{\cdot}$) radicals. The term *reactive oxygen species* (ROS) was meant to denote these radicals collectively (Halliwell, 1990; Halliwell and Gutteridge, 1993). With the discovery of nitric oxide (NO) and its role in essential physiology, radicals centered on nitrogen, like nitric oxide (NO), nitrogen dioxide (NO_2), and peroxynitrite ($ONOO^-$), are also included in the term ROS (Fang et al., 2002; Finkel and Holbrook, 2000; Kohen and Nyska, 2002; Winterbourn, 2008). Some authors prefer to use the term *reactive nitrogen species* to denote radicals derived from nitric oxide, etc. Other important radicals that are also crucial in biology are those derived from sulfur, halogens, and carbon. In this chapter, all these radicals are collectively termed as ROS.

The intracellular production of free radicals is mediated through oxygen-centered $O_2^{\bullet-}$ radicals and the $NO^\bullet$ radical. The $O_2^{\bullet-}$ radical is produced through multiple biological processes like mitochondrial respiration, phagocytosis, etc. During mitochondrial respiration, $O_2^{\bullet-}$ is generated by single-electron transfer to oxygen during the process of ATP synthesis (Sies, 1997; Cadenas and Davies, 2000; Finkel and Holbrook, 2000; Fang et al., 2002). $O_2^{\bullet-}$ radicals are also generated by enzymatic reactions like NADPH oxidation by NADPH oxidase, oxidation of hypoxanthine, and xanthine by xanthine oxidase, at times by chemical reactions like autoxidation of monoamines, etc. (Kohen and Nyska. 2002). Under normal physiological conditions, nearly 2% of the oxygen consumed by the body is converted into $O_2^{\bullet-}$ and other radicals. Its percentage increases during conditions such as in competitive sports, activation of immune system, and exposure to environmental factors like pollutants, radiation, UV light, etc.

Nitric oxide ($NO^\bullet$) is produced from L-arginine by the three isoforms of nitric oxide synthase (NOS) enzymes (Fukuto and Ignarro, 1997; Koppenol, 1998; Priyadarsini, 2000). Both $NO^\bullet$ and $O_2^{\bullet-}$ radicals are not very reactive, but they are converted to powerful oxidizing free radicals like $^\bullet OH$, $ROO^\bullet$, $RO^\bullet$, $^1O_2$, etc., by complex transformation reactions. Some of the radical species produce high amounts of molecular oxidants like H_2O_2, peroxynitrite ($ONOO^-$), hypochlorous acid (HOCl), etc. (Finkel and Holbrook, 2000; Winterbourn, 2008). H_2O_2 is produced from many different processes, including dismutation of $O_2^{\bullet-}$ radicals, the two-electron reduction of oxygen by cytochrome P-450, during glycine metabolism pathway, etc. HOCl is produced by the reaction of H_2O_2 and Cl^- by myeloperoxidase (MPO) in immunologically activated macrophages or phagocytes. $ONOO^-$ is generated by the diffusion-controlled reaction of $NO^\bullet$ with $O_2^{\bullet-}$. Under favorable conditions, these molecular species act as sources of free radicals. H_2O_2 is converted to $^\bullet OH$ radicals by the presence of trace metals and HOCl produces singlet oxygen. $ONOO^-$ at physiological pH is in equilibrium with the protonated form peroxynitrous acid (ONOOH), which reacts with substrates through the $^\bullet OH$ radical and the nitrogen dioxide ($NO_2^\bullet$) radical (Goldstein and Czapski, 1995; Goldstein et al., 1996; Koppenol, 1998), and at physiological concentrations of carbon dioxide, $ONOO^-$ is converted to another powerful oxidant, the carbonate radical anion ($CO_3^{\bullet-}$) (Augusto et al., 2002; Bartesaghi et al., 2004; Goldstein and Merenyi, 2008). Free radicals like $^\bullet OH$ radicals, on reacting with cellular organelles, produce different types of secondary radicals, and sometimes these radicals react with oxygen to produce peroxyl radicals. All these secondary and primary reactions lead to the production of several different types of peroxyl radicals inside cells, such as those obtained from lipids, DNA bases, amino acid radicals, thiyl radicals, etc. Peroxyl radicals often induce chain reactions (Sonntag, 1987; Cadenas and Davies, 2000; O'Neill and Wardman, 2009). In Scheme 22.1, different primary and secondary pathways leading to intracellular generation of ROS are given.

Free-radical induced chemical reactions inside cells are quite complex and the outcome of such free-radical initiated cellular changes depend on several factors like site of generation, nature, and type of cell, expression of redox proteins, activation of repair mechanisms, and many other factors (Halliwell and Gutteridge, 1993; Finkel and Holbrook, 2000; Valko et al., 2006; Winterbourn, 2008). For example, free-radical reactions with membrane lipids cause lipid peroxidation, which leads to changes in membrane fluidity and structure. Free-radical reactions with proteins can lead to their denaturation and loss of vital enzyme activity. Free-radical reactions with DNA cause base modification, single and double strand breaks, and mutations. Under normal conditions, most of the free-radical mediated changes in biomolecules are reversed by the repair mechanisms, but under pathological conditions, the damage may become irreversible and lead to a disease.

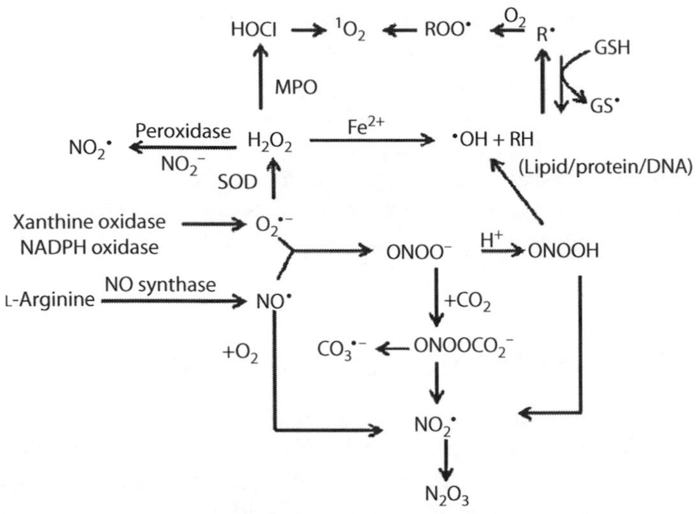

SCHEME 22.1 Reactions leading to the intracellular formation of ROS.

22.2 OXIDATIVE STRESS AND ANTIOXIDANTS

In living cells, the redox environment is strictly controlled. The word "redox state" is used to describe the cellular redox potential, which fluctuates significantly during the cellular functions (Schafer and Buettner, 2001; Kohen and Nyska. 2002; Mats et al., 2008). When less reactive free radicals (less powerful oxidants) are converted to more reactive species inside cells, there is a shift from reducing status to oxidative status. This shift toward an oxidative environment leads to *oxidative stress*. Change in the cellular redox environment also affects signal transduction pathways, DNA, RNA and protein synthesis, and regulation of cell cycle. Oxidative stress has now been implicated in many diseases such as atherosclerosis, Parkinson's disease, Alzheimer's disease, cancer, etc., in humans (Halliwell and Gutteridge, 1993; Finkel and Holbrook, 2000; Fang et al., 2002; Valko et al., 2006; Winterbourn, 2008). Cells can overcome small perturbations in oxidative status and regain their original status by endogenous antioxidant systems involving superoxide dismutase (SOD), catalase, glutathione peroxidase, and low molecular weight antioxidants like ascorbate, tocopherols, reduced coenzyme, urate, glutathione, etc. These systems either prevent the production of free radicals or inhibit free-radical mediated damage to biomolecules, or they repair the damaged biomolecule. When the endogenous defense fails, cells need to be supplemented with exogenous agents known as antioxidants. An *antioxidant* is therefore a reducing agent that is capable of preventing the pro-oxidation process, or oxidative damage in cells. The recent definition of an antioxidant is "a compound that when present in low concentrations significantly prevents or delays oxidation of the bulk oxidizable substrate."

22.3 CLASSIFICATION OF ANTIOXIDANTS

Low molecular weight antioxidants are classified into many different types (Sies, 1997; Shi and Noguchi, 2000; Fang et al., 2002). Depending on their solubility and polarity, they are either lipid-soluble or water-soluble antioxidants. Vitamin E is an example of a lipid-soluble antioxidant, and vitamin C or ascorbic acid is an example of a water-soluble antioxidant. They are also classified as synthetic and natural antioxidants. Synthetic antioxidants are much less in number as compared to the natural ones. Vitamins, flavonoids, carotenoids, plant polyphenols, etc., are natural antioxidants. Butylated hydroxy toluene (BHT) and tertiary butylhydroquinone (TBHQ) are examples of synthetic antioxidants. Depending on their mechanism of action, they are classified as preventive or chain-breaking antioxidants. Preventive antioxidants like desferrioxamine are compounds that form chelates with transition metals, thereby helping in the prevention of free-radical production. Radical-scavenging antioxidants are the most important class of antioxidants. Ascorbic acid, being a water-soluble compound, can scavenge ˙OH radicals that are produced in the cytosol, but lipid-soluble compounds like vitamin-E, quercetin, etc., scavenge the lipid peroxyl radicals; therefore these compounds are also termed as chain-breaking antioxidants. Vitamin E is one of the first reported lipid-soluble chain-breaking antioxidants, which is often considered as a standard to compare the antioxidant capacity of new molecules (Pryor, 1989). Antioxidants like ascorbic acid help in recycling chain-breaking antioxidants.

22.4 REDOX REACTIONS INVOLVING FREE RADICALS

It is obvious that the function of an antioxidant is to minimize the oxidative stress, and an important requirement for an antioxidant is to convert reactive free radicals and molecular oxidants to less reactive species (Halliwell and Gutteridge, 1993). Due to the presence of an unpaired electron, a reactive free radical in general participates in electron and hydrogen transfer reactions, causing both oxidation and reduction. The oxidizing and reducing ability of a free radical not only depends on its reduction potential, but also on the energy and electronic state of the molecule with which it is reacting. Another parameter that decides the reactivity of a free radical is the rate constant for a particular reaction. Therefore, knowledge of both thermodynamic and kinetic parameters is necessary for diverting the course of a free-radical reaction (Wardman, 1989; Buettner, 1993; Kohen and Nyska. 2002). The most powerful oxidizing radical is the ˙OH radical and the most powerful reducing radical is the hydrated electron (e_{aq}^-), and the reactions with these radicals take place with rate constants close to diffusion-controlled limits (Sonntag, 1987; Buxton et al., 1988). Peroxyl radicals are of many different types and their reduction potentials vary from 0.4 to 1.3 V vs. normal hydrogen electrode (NHE) (Jonsson, 1996), and their rate constants with antioxidants can vary from 10^3 to $10^8 M^{-1} s^{-1}$ (Neta et al., 1990). $O_2^{˙-}$ radicals can either act as an oxidant or a reductant, and their reaction rate constants are much less than those with ˙OH radicals (Sonntag, 1987). The molecular oxidants can induce both one- and two-electron oxidation reactions (Koppenol, 1994; Furtmuller et al., 2005). H_2O_2 is a powerful two-electron oxidant, but reacts slowly with many molecules and does not directly react with several biomolecules. But HOCl, with a lower reduction potential, reacts much faster with many biomolecules. NO is a poor oxidant, but reacts with $O_2^{˙-}$ to produce peroxynitrite, a powerful oxidant. Peroxynitrite reacts with biomolecules both by one-electron and two-electron transfer. In addition to the thermodynamic and kinetic properties, another important factor that is to be considered while extrapolating these results to real biological systems is the diffusion coefficient of the radical. $O_2^{˙-}$ radicals have very low membrane permeability and react in the same reaction region where it is generated. H_2O_2 is cell permeable and therefore can travel through different reaction zones. Other oxygen free radicals with short lifetimes like ˙OH radicals have much restricted reaction zones (Finkel and Holbrook, 2000; Winterbourn, 2008). Table 22.1 lists different types of ROS important in biology along with their one-electron reduction potentials.

22.5 RADIATION CHEMISTRY IN THE STUDY OF FREE RADICALS AND ANTIOXIDANTS

Since the last half century, radiation chemists have contributed significantly to the understanding of fundamental biological processes, and thousands of research papers were published in the literature in the interfacial research areas of chemistry and biology. Radiolysis of water was the common interlink between these two branches of science, as living cells contain nearly 70% of water and radiation induced chemical changes being non-specific, cause radiolysis of water (Bensasson et al., 1983; Sonntag, 1987; Buxton et al., 1988; Richter, 1998). Decades of research by the radiation chemists

TABLE 22.1
Redox Potentials of Some Important Reactive Oxygen Species

Name of the Radical	Symbol	E^7 (V vs. NHE)	Couple
Hydroxyl	$^{\bullet}OH$	1.9	$^{\bullet}OH/OH^-$
Superoxide	$O_2^{\bullet-}$	−0.33	$(O_2/O_2^{\bullet-})$
		1.80	$(O_2^{\bullet-}, 2H^+/H_2O + O_2)$
Singlet oxygen	$^1O_2$	0.65	$^1O_2/O_2^{\bullet-}$
Alkoxyl	$RO^{\bullet}$	1.0–1.6	$(RO^{\bullet}, H^+/ROH)$
Alkyl peroxyl	$ROO^{\bullet}$	~0.43–1.26	$ROO^{\bullet}, H^+/ROOH)$
Nitric oxide	$NO^{\bullet}$	1.21	NO^+/NO
		0.39	NO/NO^-
Nitrogen dioxide	$NO_2^{\bullet}$	0.99	NO_2/NO_2^-
Carbonate	$CO_3^{\bullet-}$	1.78	$CO_3^{\bullet-}/CO_3^{2-}$
Lipid peroxyl	$LOO^{\bullet}$	1.0	$LOO^{\bullet}/LOOH$
Hydrogen peroxide	H_2O_2	1.32	$H_2O_2/2H_2O$
		0.30	$H_2O_2/H_2O, ^{\bullet}OH$
Hypochlorous acid	$HOCl$	1.08	$HOCl/H_2O, Cl^-$
		0.17	$HOCl, H^+/H_2O, Cl^{\bullet}$
Peroxynitrite	$ONOO^-$	0.2	$ONOO^{\bullet}/ONOO^-$
		1.6	$ONOO^-/NO_2^{\bullet}$

have led to the complete understanding of the primary radiolytic processes in water. This also provided a unique method to generate almost all the ROS exclusively and study their reactions with antioxidants. Since several chapters in this book discuss the radiation chemistry of water in detail, it is mentioned only briefly in this chapter.

Interaction of ionizing radiation with dilute aqueous solutions causes initial excitation and ionization of water molecules, followed by ion–molecule reactions, dissociation reactions, solvation reactions, spur expansion reactions, etc., and at ~10^{-7} s after initial interaction of ionizing radiation, the species present in a homogeneous distribution during the radiolysis of water can be represented by Equation 22.1 (Buxton et al., 1988; Richter, 1998):

$$H_2O \rightsquigarrow e_{aq}^-, \,^{\bullet}OH, H^{\bullet}, H_2O_2, H_2, H_3O^+ \quad (22.1)$$

Out of these radical species, $H^{\bullet}$ and e_{aq}^- are reducing in nature, while $^{\bullet}OH$ radicals are oxidizing in nature. Employing suitable additives, such as inorganic salts and dissolved gases, it is possible to convert these primary radicals of water radiolysis to new radicals (Buxton et al., 1988; Neta and Huie, 1988; Neta et al., 1990). The chemical reactions leading to the conversion of these primary species into different oxidizing and reducing radicals are given in Scheme 22.2 and the relevant reactions and their corresponding rate constants are listed in Table 22.2. With the development of the pulse radiolysis technique, it was possible to monitor the free-radical reactions with antioxidants in real time scales. The pulse radiolysis technique utilizes short pulses of charged particles and high-energy photons from accelerators. These short pulses of electrons/charged particles or photons induce a non-equilibrium situation in a small reaction zone in a very short time scale, such that a sufficiently high concentration of transient free-radical species is formed. These free-radical species are detected within their short lifetimes, by following changes in spectroscopic properties, electrochemical properties, or changes in spin density (Baxendale and Busi, 1981; Bensasson et al., 1983). Although modern pulse radiolysis techniques are capable of producing much shorter pulses (<10^{-12} s), most of the relevant studies with antioxidants have been performed in the time scale of nanoseconds to a few seconds.

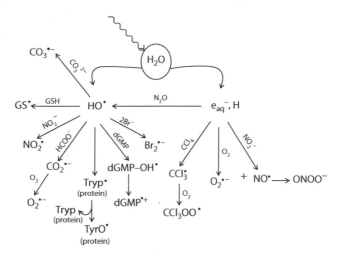

SCHEME 22.2   Generation of ROS by radiolysis of aqueous solutions.

TABLE 22.2
Important Radiation Chemical Reactions along with Rate Constants for the Generation of Reactive Oxygen Species

S.No.	Reaction	Rate Constant $(M^{-1}\,s^{-1})$
1	$N_2O + e_{aq}^- \longrightarrow {}^\bullet OH + OH^- + N_2$	9.1×10^9
2	${}^\bullet OH + (CH_3)_3 C - OH \longrightarrow {}^\bullet CH_2 C(CH_3)_2 - OH + H_2O$	6.0×10^8
3	${}^\bullet OH + HCOO^- \longrightarrow CO_2^{\bullet-} + H_2O$	3.2×10^9
4	${}^\bullet H + HCOO^- \longrightarrow CO_2^{\bullet-} + H_2$	2.1×10^8
5	$CO_2^{\bullet-} + O_2 \longrightarrow O_2^{\bullet-} + CO_2$	3.5×10^9
6	$e_{aq}^- + O_2 \longrightarrow O_2^{\bullet-}$	2×10^{10}
7	${}^\bullet OH + (CH_3)_2 CH - OH \longrightarrow (CH_3)_2 C^\bullet - OH + H_2O$	1.9×10^9
9	$NO_2^- + {}^\bullet OH \longrightarrow NO_2 + OH^-$	6.0×10^9
10	$NO_3^- + e_{aq}^- \longrightarrow NO_3^{2-}$	9.2×10^9
11	$e_{aq}^- + NO_2^- \longrightarrow NO_2^{2-}$	4.1×10^9
12	${}^\bullet H + NO_2^- \longrightarrow {}^\bullet NO + OH^-$	7.1×10^8
13	${}^\bullet OH + CO_3^{2-} \longrightarrow CO_3^{\bullet-} + OH^-$	3.0×10^8
14	${}^\bullet OH + HCO_3^- \longrightarrow CO_3^{\bullet-} + H_2O$	8.5×10^6
15	${}^\bullet OH + N_3^- \longrightarrow N_3 + OH^-$	1.4×10^{10}
16	${}^\bullet OH + 2Br^- \longrightarrow Br_2^{\bullet-} + OH^-$	1.1×10^{10}
17	$e_{aq}^- + Thymine(T) \longrightarrow T^{\bullet-}$	1.7×10^{10}
18	${}^\bullet OH + dGMP \to dGMP - OH^\bullet$	6.8×10^9
19	${}^\bullet OH + TRPH \to TRP^\bullet + H_2O$	1.3×10^{10}
20	${}^\bullet OH + GSH \to GS^\bullet + H_2O$	2.3×10^{10}

Recently, many laboratories involved in radiation chemistry programs have taken up antioxidant research and published a number of papers in the literature. The rate constants for the scavenging reactions of oxidizing and reducing free radicals by antioxidants were determined at near physiological pH conditions. The antioxidant radicals have been characterized by transient spectra, decay kinetics, prototropic equilibrium constants (pK_a), conductivity changes, and reactions with other molecules like oxygen. Such studies have been found to be useful to quantify reaction kinetics, estimate the reactivity of antioxidant substances, and also in identifying the site of free-radical attack on the antioxidant molecule. One of the greatest contributions of pulse radiolysis is the estimation of one-electron reduction potentials of transient free radicals. Establishing reversible electron transfer between two couples (Wardman, 1989), one-electron reduction potentials of several antioxidants have been reported.

In addition to the radicals given in Scheme 22.2 and Table 22.2, radiation chemists have also designed methods to study reactions of secondary radicals from amino acids of proteins and DNA radicals with antioxidants (Butler et al., 1984; Solar et al., 1984; O'Neill and Chapman, 1985; Sonntag, 1987; Asmus and Bonifacic, 1999; Li et al., 2000; Santus et al., 2001; Filipe et al., 2002, 2004; Zhao et al., 2002, 2003). The most commonly employed aromatic amino acid radicals are the indolyl radicals of tryptophan (TRP$^\bullet$) and the tyrosine phenoxyl radicals (TYRO$^\bullet$) and sulfur-centered radicals from amino acids like methionine, glutathione, etc. Reactions of antioxidants with these radicals have been used to evaluate their ability to protect proteins from oxidative damage, especially for hydrophobic antioxidants, which show preferential affinity toward proteins.

To evaluate the ability of an antioxidant to protect DNA from oxidative damage, studies on the repair and electron transfer reactions of several antioxidants have been carried out with secondary DNA radicals (O'Neill and Chapman, 1985; Zhao et al., 2001, 2002, 2003). The following are important DNA radicals that could be generated by pulse radiolysis: $^\bullet$OH radical adducts of deoxyguanosine monophosphate (dGMP-OH$^\bullet$), deoxyadenosine monophosphate (dAMP-OH$^\bullet$), polyadenylic acid (poly A-OH$^\bullet$), polyguanylic acid (poly G-OH$^\bullet$) and single- or double-stranded DNA (DNA-OH$^\bullet$), radical anions of thymine (T$^{\bullet-}$) and thymidine monophosphate (TMP$^{\bullet-}$), and radical cations of dGMP and dAMP (dGMP$^{\bullet+}$ and dAMP$^{\bullet+}$). Another major contribution of the radiation chemists to antioxidant research has been in studying $O_2^{\bullet-}$ radical reactions and development of SOD mimics.

In the present chapter, an attempt has been made to summarize the important aspects of the research carried out in the last three decades on pulse radiolysis studies of antioxidants. Complete coverage of the reactions of antioxidants is beyond the scope of this chapter. However, studies with the most promising antioxidant systems have been discussed in detail. It is expected that the information provided in the chapter would be useful for a new radiation chemist to initiate work in this multidisciplinary research area.

22.6 REDOX STUDIES OF ANTIOXIDANTS BY PULSE RADIOLYSIS

The redox reactions of antioxidants studied by employing pulse radiolysis belong to two broad and general types: natural antioxidants and synthetic antioxidants. Of the two, natural products have attracted the attention of many more scientists as they are present in several food products, and such antioxidants can be developed as neutraceuticals because they are consumed through the diet (Shi and Noguchi, 2000). Natural antioxidants can be further classified as phenolic and non-phenolic compounds; the phenolic compounds outnumber the non-phenolic compounds and are therefore discussed separately in detail.

22.6.1 Phenolic Antioxidants

Important antioxidants belonging to the class of phenolic compounds that have been studied by pulse radiolysis are simple phenols, benzoic acid derivatives, cinnamic acid derivatives (Lin et al., 1998; Bors et al., 2003), methoxy-phenols (Priyadarsini et al., 1998; Bors et al., 2002; Mercero et al., 2002), salicylic acid derivatives (Joshi et al., 2005b), gallic acid derivatives (Bors and Michel, 1999), tocopherols (Packer et al., 1979), flavonoids (Bors et al., 1994, 1995; Jovanovic et al., 1994, 1996; Li and Fang, 1998; Miao et al., 2001a,b), curcuminoids (Priyadarsini, 1997), resveratrol (Stojanovic and Brede, 2002), ellagic acid (Priyadarsini et al., 2002), sesamol (Joshi et al., 2005a), xanthones (Mishra et al., 2006a), anthocyanins, and tannins (Bors et al., 2001a,b), folate (Joshi et al., 2001), and many other plant-derived polyphenols (Bors et al., 2001a,b; Shi and Noguchi, 2000; Singh et al., 2009).

The antioxidant action of phenols proceeds by donating an electron or hydrogen atom to ˙OH radicals, specific one-electron oxidants, inorganic radicals, and chain propagating peroxyl radicals. Most of these reactions produce a phenoxyl radical as the phenolic OH group in these compounds has maximum electron density and the O–H bond is the weakest to break. Depending on the interaction of the unpaired electron on the phenoxyl radical with other substituents, the phenoxyl radical gets resonance stabilized and the stability of this radical decides the efficacy of the phenol as an antioxidant. In addition to this, a free radical may add to the aromatic ring of phenols producing radical adducts, which may react with oxygen producing peroxyl radicals. For example, ˙OH radicals add to the aromatic ring of simple phenol, where the majority of it goes to ortho, para positions, while a minor fraction goes to meta and ipso positions (Mvula et al., 2001). These radical adducts can participate in different reactions like acid- and base-catalyzed elimination of water to yield phenoxyl radicals, absorbing at ~400 nm or reacting with oxygen to form peroxyl radicals (Alfassi et al., 1994; Mvula et al., 2001). In general, the phenoxyl radicals do not react with oxygen and therefore are not converted to peroxyl radicals. The formation of phenoxyl radicals accounts for the antioxidant action, as this reaction converts the more reactive ˙OH radicals to the less reactive phenoxyl radicals, but formation of peroxyl radicals may implicate a pro-oxidant effect. Therefore, one has to look into all these aspects while exploring phenolic compounds as antioxidants. The reaction of ˙OH radical with phenol is summarized in Scheme 22.3.

There are many natural phenolic compounds, which show excellent antioxidant activity, and their ability to neutralize free-radical oxidants is dependent on the substitutions. Compounds like cinnamic acid, caffeic acid, and chlorogenic acid have two hydroxyl groups, and pulse radiolysis studies have been employed to confirm their excellent free-radical-scavenging activity (Kono et al., 1997; Lin et al., 1998; Lu and Liu, 2002; Bors et al., 2003). Several natural antioxidant compounds have methoxy phenolic acids, compounds like ferulic acid, eugenol, isoeugenol, and vanillin are ortho-methoxy phenolic acids found in food spices, exhibiting potent antioxidant activity. The compounds react with most of the oxidizing free radicals including chain-breaking peroxyl radicals (Priyadarsini et al., 1998; Guha and Priyadarsini, 2000; Bors, 2001; Mercero, 2002; Barik et al., 2004). In most of these reactions, phenoxyl radicals are produced that get stabilized by resonance through the aromatic ring and also by interacting with the lone pair on the methoxy oxygen.

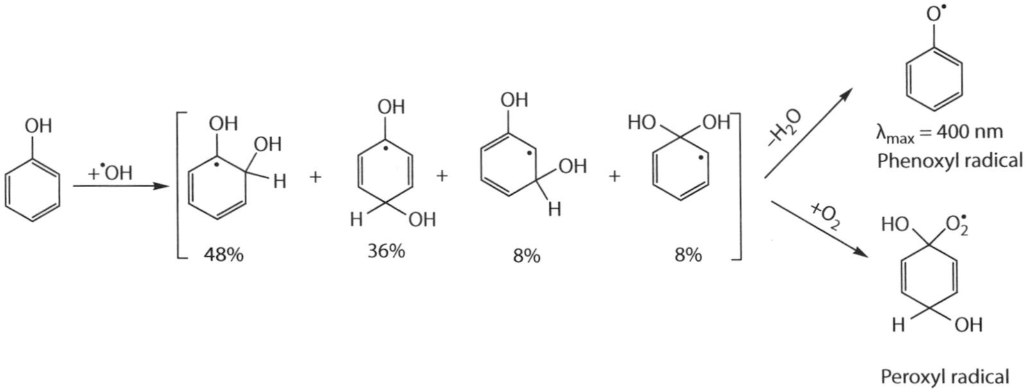

SCHEME 22.3   Hydroxyl radical reactions with phenol formation of phenoxyl and peroxyl radicals.

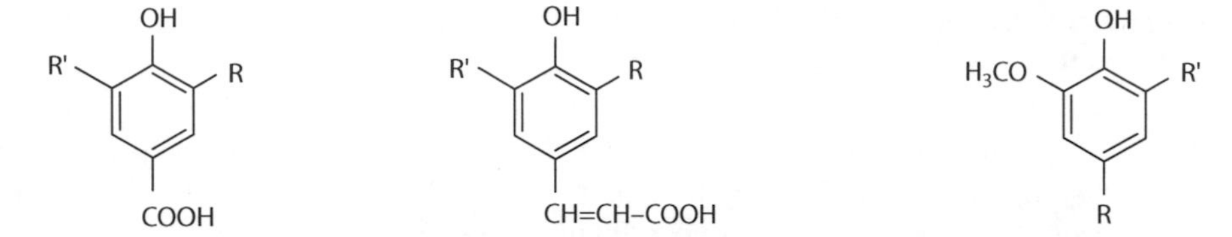

Benzoic acid derivatives | Cinnamic acid derivatives | Methoxyl phenols

R = –H; R' = $–OCH_3$; Vanillic acid
R = $–OCH_3$; R' = $–OCH_3$; Syringic acid
R = –OH; R' = –OH; Gallic acid

R = –H; R' = $–OCH_3$; Ferulic acid
R = $–OCH_3$; R' = $–OCH_3$; Sinapic acid
R = –OH; R' = –H; Caffeic acid

R = $–CH_2–CH=CH_2$, Eugenol; R' = H
R = $–CH_2=CH–CH_3$, Isoeugenol; R' = H
R' = –CHO, o-Vanillin

SCHEME 22.4 Some important dietary phenolic antioxidants.

Further, compounds like syringic acid and sinapic acid having two methoxy groups ortho to the phenolic OH groups have been found to exhibit very good antioxidant activity. In such compounds, substitution of additional methoxy groups not only helps in the stabilization of the phenoxyl radicals but also increases their lipid solubility. Chemical structures of important phenolic antioxidants are included in Scheme 22.4.

Among the natural vitamins, vitamin C (ascorbic acid) and vitamin E (α-tocopherol) are the most promising antioxidants and are some of the first compounds recognized. Pulse radiolysis contributions on these two compounds are discussed in detail.

22.6.1.1 Vitamin C and Vitamin E

Vitamin C occurs as L-ascorbic acid and dehydroascorbic acid in fruits and vegetables. Ascorbic acid is the most effective water-soluble antioxidant in the plasma and it is an excellent free-radical scavenger. The chemical structure of ascorbic acid is such that it can be considered as both a phenolic and a non-phenolic antioxidant. In biological systems, oxidation of ascorbic acid to dehydroascorbic acid proceeds through the formation of ascorbyl radicals. Radiation chemical experiments performed several decades ago (Bielski et al., 1971, 1981; Schuler, 1977) using pulse radiolysis confirmed that ascorbic acid radicals produced during the oxidation of ascorbic acid by $^{\bullet}OH$ radicals and many other oxidizing radicals showed two pK_a values of 1.10 and 4.25 with the protonated form ($AH_2^{\bullet+}$), neutral radical ($AH^{\bullet}$), and ascorbate radical anions ($A^{\bullet-}$) in prototropic equilibrium (Scheme 22.5). At neutral pH, $A^{\bullet-}$ radicals absorb at 360 nm with a molar extinction coefficient of 3300 M^{-1} cm^{-1}. The decay of ascorbate radicals is very well documented, and from the studies on effect of pH, ionic strength, and buffers, it is proposed that $A^{\bullet-}$ radical is in equilibrium with a dimer. The dimer reacts with other proton donors like water and buffers, and produces products like ascorbate ion and dehydroascorbic acid by disproportionation. At pH 7, the reduction potentials of the couple $A^{\bullet-}$, H^+/AH^- is ~0.32 V vs. NHE and that for dehydroascorbic acid/$A^{\bullet-}$ is ~−0.17 V vs. NHE (Buettner, 1993). Due to this low redox potential, ascorbic acid can act both as a reducing agent and an oxidizing agent.

The reactions of $A^{\bullet-}$ with various biological molecules have also been investigated using pulse radiolysis (Kobayashi et al., 1991). The $A^{\bullet-}$ radical reacted with fully reduced and semiquinone

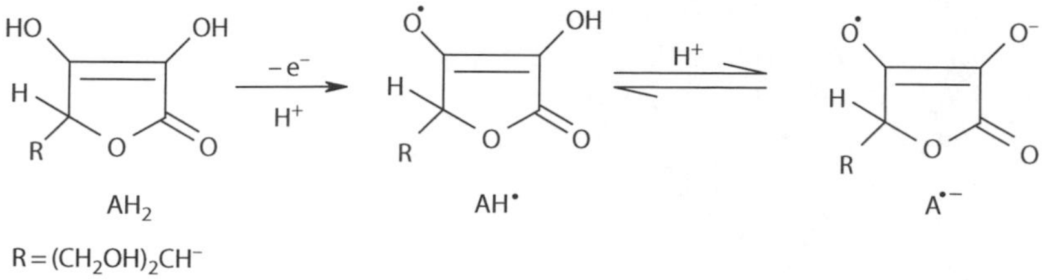

SCHEME 22.5 Ascorbic radicals generated from the one-electron oxidation of ascorbic acid.

forms of hepatic NADH-cytochrome $b5$ reductase with second-order rate constants of 4.3×10^6 and $3.7 \times 10^5 \, M^{-1} \, s^{-1}$, respectively. It however did not react with the ferrous form of cytochrome $b5$, whereas it induced oxidation of cytochrome $b5$ in the presence of ascorbate oxidase, suggesting that the rate constant of $A^{\cdot-}$ with the ferrous cytochrome $b5$ must be several orders of magnitude smaller than that of the disproportionation of $A^{\cdot-}$. On the other hand, $A^{\cdot-}$ could reduce Fe^{3+}-EDTA with a second-order rate constant of $4.0 \times 10^6 \, M^{-1} \, s^{-1}$ but did not reduce ferric hemoproteins.

Due to the low redox potential, the ascorbate ion acts as a source of ·OH radicals and participates in Fenton reactions. This led to the speculation whether ascorbate is a pro-oxidant or an antioxidant. It is now confirmed that ascorbate definitely acts as an antioxidant at low concentrations but may become pro-oxidant at high concentrations especially in presence of free metal ions (Yen et al., 2002).

Vitamin E, obtained from nuts, seeds, and cereals, is a collective term for eight compounds: α-, β-, γ-, and δ-tocopherol and α-, β-, γ-, and δ-tocotrienols, but α-tocopherol accounts for 90% of vitamin E. All the tocopherols contain a phenolic-OH group that enables them to donate hydrogen to a free radical. Vitamin E is readily incorporated into cell membranes and it is the first known chain-breaking antioxidant. Pulse radiolysis studies of vitamin E have been reported several years ago (Packer et al., 1979; Jore et al., 1986; Bisby, 1990). Vitamin E reacts with almost all the oxidizing free radicals and the phenoxyl radicals produced during oxidation reactions absorb at ~425 nm. Since vitamin E is insoluble in water, most of the radiation chemical studies could be carried out either in alkaline water or in aqueous–organic solutions or in micellar solutions. Vitamin E reacts with many different types of peroxyl radicals and the rate constants range from 10^4 to $10^8 \, M^{-1} \, s^{-1}$, depending on the solvent and hydrophobicity of the peroxyl radical. The phenoxyl radicals of vitamin E have long lifetimes and decay by second-order reactions with 2k values of ~$10^3 \, M^{-1} \, s^{-1}$, and the lifetime varied significantly with the polarity of the medium. The one-electron reduction potential of vitamin E is ~0.50 V vs. NHE (Buettner, 1993). The chemical structure of the vitamin E phenoxyl radical is given in Scheme 22.6.

The phenoxyl radicals of vitamin E, sometimes called as α-tocopheroxyl radicals, are highly stabilized and several research groups have investigated quantitative structure–activity studies (Buettner, 1993; VanAcker et al., 1993). Based on these studies, the structural features responsible for the stability of the phenoxyl radical have been understood as follows: In the lipid bilayer, vitamin E is orientated with the chroman head group toward the surface and the hydrophobic phytyl side chain buried within the hydrocarbon region. The phenoxyl radical produced after donating hydrogen atom to the lipid peroxyl radical acquires stability by sharing its electron with nearby atoms. The lone pair containing the p-orbital of the heterocyclic oxygen is almost perpendicular to the aromatic plane. This lone pair overlaps with the singly occupied molecular orbital of the phenoxyl radical and provides extra stability. The polarity of the vitamin E phenoxyl radical is such that it moves to the surface of the lipid, where it is regenerated by vitamin C.

The regeneration reaction of vitamin E phenoxyl radicals back to vitamin E by the water-soluble antioxidant vitamin C was first reported by pulse radiolysis, in which direct decay of the phenoxyl radicals of vitamin E absorbing at 425 nm was followed by simultaneous formation of ascorbyl radicals absorbing at 360 nm, and the rate constant for the regeneration reaction was found to be $1.55 \times 10^6 \, M^{-1} \, s^{-1}$ (Packer et al., 1979). The long lifetime of phenoxyl radicals allows the

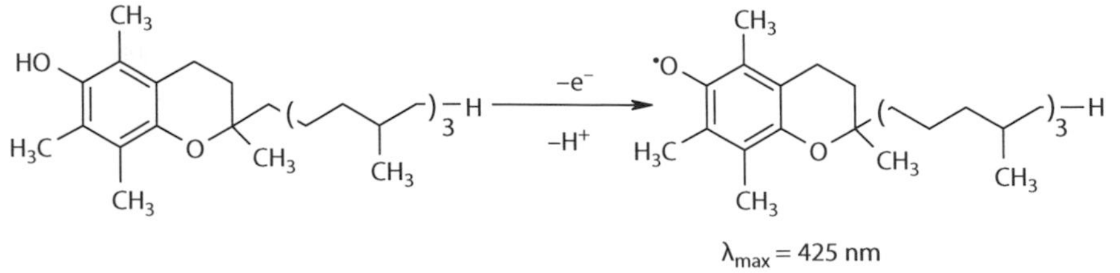

$\lambda_{max} = 425 \, nm$

SCHEME 22.6 Phenoxyl radical generated from the one-electron oxidation vitamin E.

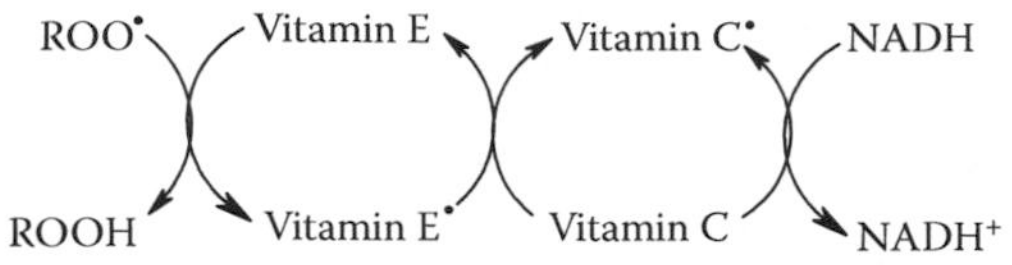

SCHEME 22.7 Redox recycling of antioxidants: Vitamin E and vitamin C.

regeneration reaction to compete with other radical reactions. The resultant ascorbic acid radical produced during the regeneration process is either recycled by NADH or converted to non-reactive products (Kobayashi et al., 1991). The overall chain-breaking antioxidant mechanism involving lipid-soluble vitamin E and water-soluble vitamin C in cells is represented by the sequential electron transfer process as represented in Scheme 22.7.

Based on the kinetic and thermodynamic parameters for the reactions of polyunsaturated fatty acid (PUFA-H) with its peroxyl radical, PUFA-OO$^{\bullet}$, and with vitamin E, it was possible to provide a convincing explanation on how vitamin E when present in very small concentrations (even at a ratio of PUFA and vitamin E as 1000:1) prevents the oxidation of bulk fatty acids and protects the cell membrane very effectively (Buettner, 1993).

22.6.1.2 Water-Soluble Analogues of Vitamin E

Trolox C is a water-soluble analogue of vitamin E, which also exhibits similar scavenging reactions with several types of peroxyl radicals and oxidizing free radicals (Davies et al., 1988). The studies employing pulse radiolysis showed that trolox C rapidly undergoes electron transfer to produce phenoxyl radicals absorbing at 435 nm. The one-electron reduction potential for the conversion of trolox C to its phenoxyl radicals is 0.48 V vs. NHE. Trolox C phenoxyl radicals are readily repaired by ascorbate with a rate constant of $8.3 \times 10^6 M^{-1} s^{-1}$ and also by some thiols with rate constants of $<10^5 M^{-1} s^{-1}$. Kinetic measurements also confirmed that trolox C repairs oxidized proteins. The phenoxyl radicals of trolox C react with $O_2^{\bullet-}$ radicals (Cadenas et al., 1989) with a rate constant of $4.5 \times 10^8 M^{-1} s^{-1}$, by electron transfer, converting back to trolox C. The chemical structure of trolox C and its phenoxyl radicals are given in Scheme 22.8.

Both vitamin E and trolox C are used as standards for evaluating the antioxidants. A term called trolox equivalent antioxidant capacity (TEAC) is used to compare the antioxidant ability of new compounds (Kohen and Nyska, 2002;). Recently, a glycosylated derivative of α-tocopherol has been synthesized and the pulse radiolysis studies have been reported in aqueous solutions. The studies confirmed that the free-radical chemistry of the glycosylated derivative is similar to that of α-tocopherol (Kapoor et al., 2002).

22.6.1.3 Flavonoids

Flavonoids are another large and diverse group of naturally occurring phenolic compounds ubiquitous in plants and common in a great variety of fruits, vegetables, and beverages. The basic structure of flavonoids consists of two aromatic rings linked through a furan ring, denoted by the three rings A, B, and C, and their chemical structures as given in Scheme 22.9 differ mainly in the presence of double bond at 2,3-position and the 3-OH substitution. Examples of flavonoids that have been

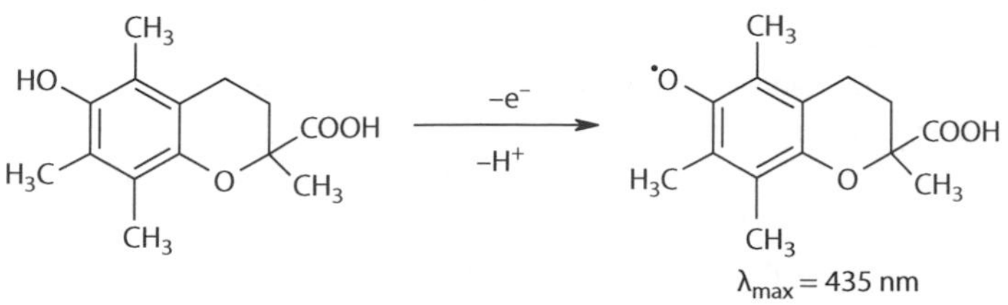

SCHEME 22.8 Phenoxyl radicals generated from the one-electron oxidation of trolox C.

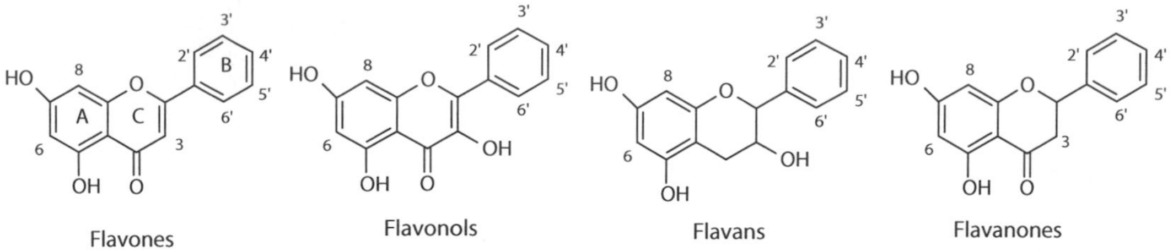

SCHEME 22.9 Basic structure of different classes of flavonoids.

extensively studied are green tea polyphenols, catechins, epicatechins, hesperidine, taxifolin, quercetin, rutin, silybins, etc. This list does not include other similar class of compounds like isoflavones, neoflavones, anthocyanins, xanthones, or biflavonoids.

Flavonoids participate in redox reactions with both reducing and oxidizing free radicals and also act as metal chelators. With suitable hydroxyl substitutions, flavonoids act as excellent hydrogen donors to reactive oxygen free radicals (Bors and Saran, 1987; Bors et al., 1994, 2001a,b; Jovanovic et al., 1994, 1995, 1996; Li and Fang, 1998; Shi and Noguchi, 2000). Using the pulse radiolysis technique, reactions of oxidizing radicals with several structurally related hydroxy-flavonoids have been studied and the absorption spectra of the flavonoid phenoxyl radicals have been reported (Bors and Saran, 1987; Bors et al., 1994, 2001a,b; Jovanovic et al., 1994, 1996. Mishra et al., 2003, 2006a; Zielonka et al., 2003; Tamba and Torreggiani, 2004; Fu et al., 2008). Depending on the hydroxyl group substitution, the absorption spectrum of the phenoxyl radical showed distinct features. Although the available literature on the absorption spectral details cannot be used with certainty as a characteristic tool to identify different flavonoid phenoxyl radicals, a generalized and empirical approach could be adopted to know whether the phenoxyl radical is derived from the A ring or the B ring. For example, if the phenoxyl radical is from the catechol of the B ring, it absorbs at 340–390 nm and if the phenoxyl radical is derived from the resorcinol moiety of the A ring, it absorbs at 420–480 nm (Jovanovic et al., 1994, 1996; Cren-Olive et al., 2002). Depending on the structure, the phenoxyl radicals have a pK_a between ~4 and 6, and the absorption maximum of the anionic form of the radical is redshifted. The absorption spectrum in the 350 nm region is quite sharp but that in the visible region is broad and extends up to 600 nm depending on the substitution. In flavonoids like catechins, where the conjugation through the C ring is not present, two distinct absorption bands due to two different phenoxyl radicals were observed (Cren-Olive et al., 2002; Tamba and Torreggiani, 2004).

Although hydroxyl substitution on the three rings makes flavonoids reactive toward peroxyl radicals, the catechol hydroxyl group of the B-ring makes it more feasible to undergo oxidation. The one-electron reduction potentials for the formation of phenoxyl radicals for a number of hydroxy flavonoids have been reported at neutral pH conditions and the values ranged between 0.3 and 0.7 V vs. NHE (Jovanovic et al., 1994, 1996; Bors et al., 1995). The reduction potential is significantly reduced when the 3-hydroxy group in the C ring is in conjugation with the catechol moiety. DFT calculations of different flavonoids also supported the results from pulse radiolysis results (Zielonka et al., 2003).

Flavonoid phenoxyl radicals in aqueous solutions showed lifetimes of several hundred milliseconds. The long lifetime and the reduction potential values suggest that flavonoids can be regenerated back from their phenoxyl radicals by vitamin C and vitamin E. Indeed, both the reactions could be observed by pulse radiolysis. Although, thermodynamically, reactions of flavonoid phenoxyl radical with vitamin E are feasible, such reactions would not be expected at the physiologically relevant concentrations of vitamin E in the cell membrane. However, regeneration by ascorbate inside the cells provides synergism like that observed with vitamin E. Because of this ascorbate-protective role, flavonoids are sometimes called as vitamin P (Bors et al., 1995). Although the long lifetime of the flavonoid phenoxyl radical favors the regeneration reaction, it is also proposed that this may also allow some of the chain propagating reactions like lipid peroxidation, indicating its contribution to the pro-oxidant effects.

Based on all these pulse radiolysis studies and many relevant redox reactions, three structural moieties have been identified to be important for antioxidant and radical-scavenging activity of flavonoids. These are: an *o*-hydroxy group in the B-ring, a 2,3-double bond combined with 4-oxo group in the C-ring, and hydroxyl group at positions 3 and 5 of ring A (Bors et al., 2001a,b). Accordingly, quercetin, with all these structural features showed promising antioxidant effects, both in animals and humans. The pulse radiolysis studies on quercetin have been briefly summarized here.

Reactions of radicals like $^\bullet$OH, $N_3^\bullet$, NO_2, $O_2^{\bullet-}$, peroxyl radicals, 1-hydroxylethyl radicals, etc., with quercetin have been studied and the transients characterized by absorption spectroscopy, conductivity measurements, and Raman spectroscopy (Jovanovic et al., 1994, 1996; Miao et al., 2001a,b; Marfak et al., 2004; Torreggiani et al., 2005). Two types of phenoxyl radicals were observed (as shown in Scheme 22.10) during one-electron oxidation of quercetin and from pH-dependent absorption changes, a pK_a of 5, for the deprotonation of the catechol hydroxyl group has been proposed. The reduction potential of quercetin at pH 7 is 0.33 V vs. NHE. Therefore, quercetin is not a good electron donor to ascorbate, but it can reduce vitamin E radicals thereby helping in the regeneration of vitamin E. Quercetin could repair TRP$^\bullet$ radicals produced during the oxidation of low density lipoprotein (LDL) and human serum albumin (HSA) (Filipe et al., 2002; Santus et al., 2001). HSA-bound quercetin could repair urate radicals by electron transfer, both in presence and absence of copper (II) (Filipe et al., 2004). Quercetin could repair the $T^{\bullet-}$ radical and dGMP-OH radical adduct (Zhao et al., 2001, 2002, 2003), indicative of its ability to repair damaged DNA.

While the antioxidant effects of flavonoids are related to the phenolic moieties, their pro-oxidant effects are related to the reactions leading to the production of semiquinone and quinonoid products, $O_2^{\bullet-}$, and H_2O_2 generation. Unlike the oxidation studies, not many reports are available on one-electron reduction of flavonoids, mainly because the anion radicals show weak absorption in the UV–VIS region. The reaction of e_{aq}^- and reducing radicals with a number of flavonoids indicated that the anion radicals of all the flavonoids showed absorption at wavelength <400 nm (Cai et al., 1999; Fu et al., 2008). Flavonoids like naringin and quercetin containing C4 keto group showed highest reactivity toward e_{aq}^- radicals, while hydroxyl substitution in the B ring, and 2,3-double bond, which are crucial for antioxidant activity, showed no influence on the e_{aq}^- scavenging activity.

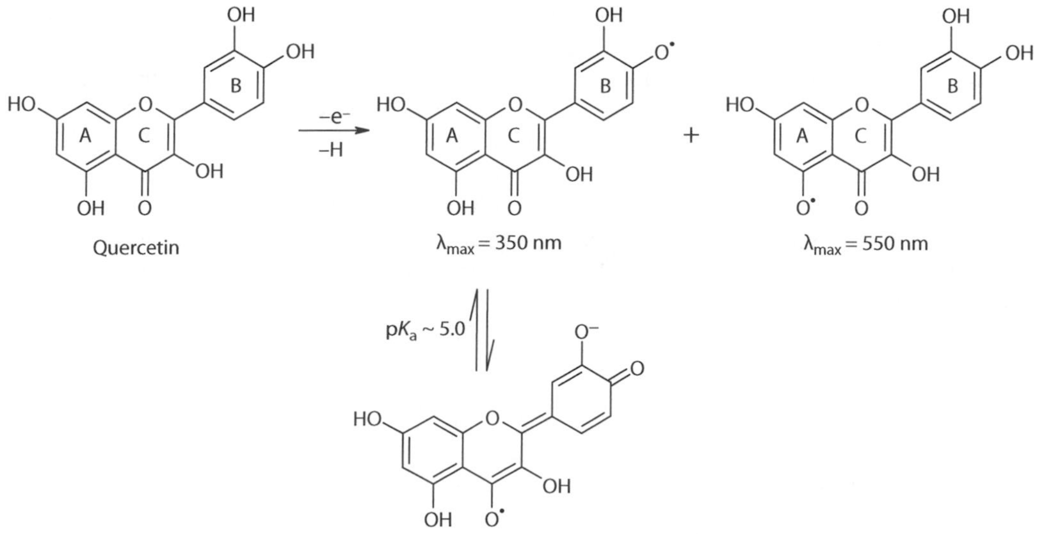

SCHEME 22.10 Different types of quercetin phenoxyl radicals formed by one-electron oxidation.

22.6.1.4 Resveratrol

Trans-resveratrol (trans-3,5,4′-trihydroxystilbene), a non-flavonoid polyphenol responsible for the antioxidant activity of red wine, is found in grapes, mulberries, and other food products (Stojanovic and Brede, 2002; Stojanovic et al., 2001; Mahal and Mukherjee, 2006). In addition to antioxidant activity, resveratrol could inhibit platelet aggregation and showed anticancer activity. The phenoxyl radicals of resveratrol produced during oxidation by ·OH radicals, one-electron oxidants, and peroxyl radicals showed absorption maximum at 410 nm. Comparing the spectral and kinetic properties of the transients derived from trans-resveratrol and its analogues, it has been concluded that in the neutral and acidic solution, the para hydroxy group of trans-resveratrol is more reactive than the meta-hydroxy groups (Scheme 22.11). Quantum chemical studies on resveratrol derivatives confirmed that the 4′-hydroxyl group of resveratrol is more reactive than the ones at the 3- and 5- positions because of the resonance effects (Cao et al., 2003). Resveratrol could repair the nucleic acid radicals and amino acid radicals with rate constants ranging from 10^9 to $10^8 M^{-1} s^{-1}$ (Mahal and Mukherjee, 2006). Some of these reports also indicate that trans-resveratrol is a better radical scavenger than vitamins E and C and is as efficient as some of the flavonoids.

22.6.1.5 Curcumin

Curcumin is a major phenolic pigment derived from turmeric, which is commonly used as a spice and as a household medicine in India. Recent scientific research has shown that curcumin is a ten times more effective antioxidant than vitamin E. It is also a potent antitumor agent and at present several phase I and phase II clinical trials are being carried out on curcumin for the treatment of different types of cancers. The greatest advantage of curcumin is the pharmacological safety to humans even at a dose of 8 g day^{-1} (Shishodia et al., 2005). Pulse radiolysis studies on reactions of $O_2^{·-}$ radicals, $CCl_3OO^·$ radicals, lipidperoxyl, methyl, and methyl peroxyl radicals, glutathione radicals, tryptophan radicals, etc., with curcumin have been reported (Gorman et al., 1994; Priyadarsini, 1997; Khopde et al., 1999; Jovanovic et al., 2001; Kapoor and Priyadarsini, 2001; Priyadarsini et al., 2003). The rate constants for the reactions of these radicals and several other oxidants have been found to be in the range of 10^5 to $10^9 M^{-1} s^{-1}$ in aqueous/aqueous–organic solutions. In all these reactions, a transient showing a strong absorption band with a maximum at 500 nm and another weak band in the UV region was observed. The 500 nm absorbing transient was characterized as the phenoxyl radical, which acquires resonance stabilization through the α,β-unsaturated β-diketone structure. The lifetime of the phenoxyl radical is a few hundred milliseconds in membrane models (Priyadarsini, 1997). The phenoxyl radicals of curcumin could be converted back to the parent curcumin by ascorbic acid and the rate constant is comparable to that of vitamin E. Curcumin has two possible sites for free-radical attack, these are: the central methylenic (CH_2) group and the phenolic OH group (Scheme 22.12). Coupling pulse radiolysis studies on free-radical reactions, with in vitro and in vivo antioxidant activities and quantum chemical calculations with several curcumin derivatives, it has been confirmed that the phenolic-OH is mainly involved in the free-radical-scavenging activity and the antioxidant activity of curcumin.

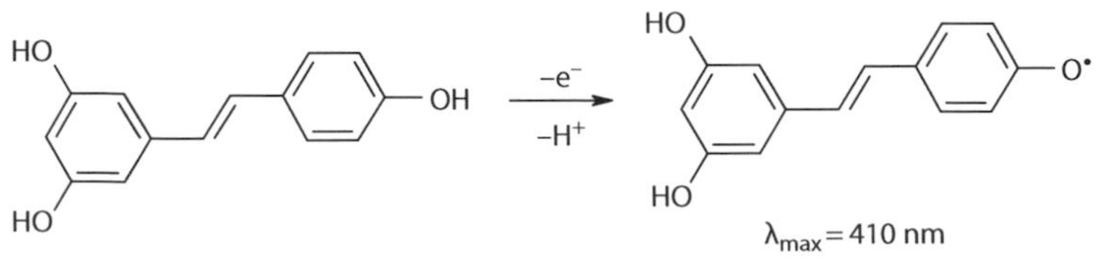

SCHEME 22.11 Phenoxyl radicals, generated from the one-electron oxidation of resveratrol.

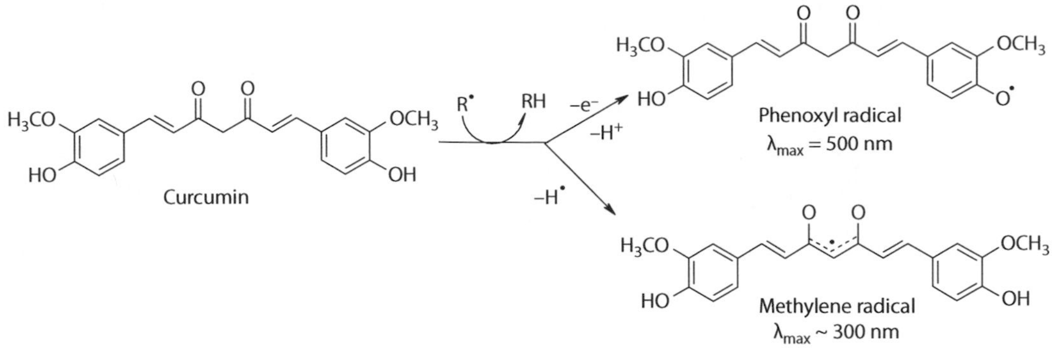

SCHEME 22.12 Possible free radical induced oxidation reaction pathways on curcumin.

22.6.1.6 Folic Acid

Folic acid, also known as vitamin B9, is essential for normal body functions and is involved in many biochemical processes including nucleotide synthesis. Folate, the mono anion of folic acid is present in leafy vegetables and fruits. Although folic acid was recognized as an essential vitamin, its role as an antioxidant was recognized much later. Folic acid exists in keto–enol tautomerism in solution. Using the pulse radiolysis technique, the reactions of oxidizing free radicals $CCl_3O_2^{\cdot}$, $N_3^{\cdot}$, $SO_4^{\cdot-}$, $Br_2^{\cdot-}$, $^{\cdot}OH$, and $O^{\cdot-}$ with folic acid were studied at different pH conditions (Joshi et al., 2001; Patro et al., 2005). All these radicals react with folic acid to produce a phenoxyl radical having absorption maximum around 430 nm, which undergoes cleavage to form smaller products. Additionally it repairs thiyl radicals at physiological pH, a reaction important in contributing to the antioxidant mechanism of folic acid. The one-electron reduction of folic acid by hydrated electron and isopropylketyl radicals was studied (Moorthy and Hayon, 1976), in the pH range 0–14, and reported formation of transient exhibiting characteristic spectra in the UV–Vis region with prominent maximum ~465 nm. The one-electron reduced radicals have four pK_a values at 1, 6.6, 8.0, and 10.3, and the radicals act as powerful reducing agents. The oxidation reactions of folic acid at pH 7 are summarized in Scheme 22.13.

22.6.2 Non-Phenolic Antioxidants

Unlike phenolic antioxidants, non-phenolic natural products with antioxidant capacity are much less in number. Compounds like melatonin, carotenoids, retinal, and thiols, and synthetic compounds like cyclic nitroxides, thiols, and selenium compounds are some of the well-studied systems reported in literature. The radiation chemical contributions to some of these systems are summarized below.

22.6.2.1 Melatonin

Melatonin (N-acetyl-5-methoxytryptamine) is an endogenous compound whose antioxidant potential has been recognized recently. It is a hormone secreted by the pineal gland of vertebrates. It plays an important role in regulating circadian rhythms and sleep. Melatonin exhibits immunomodulatory properties and is a potent antioxidant and protects organisms from free-radical damage

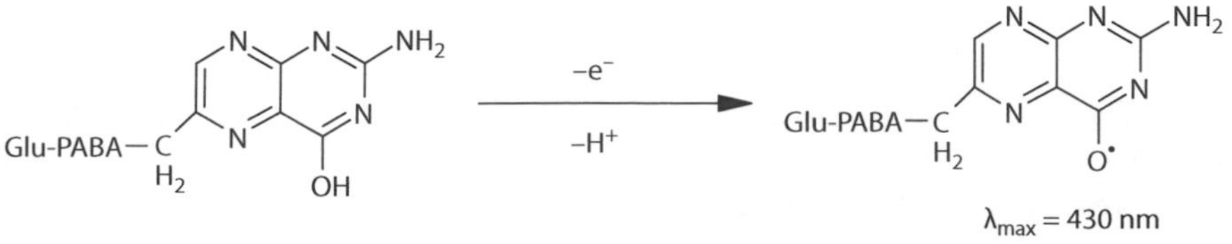

SCHEME 22.13 Phenoxyl radicals generated from the one-electron oxidation of folic acid at neutral pH.

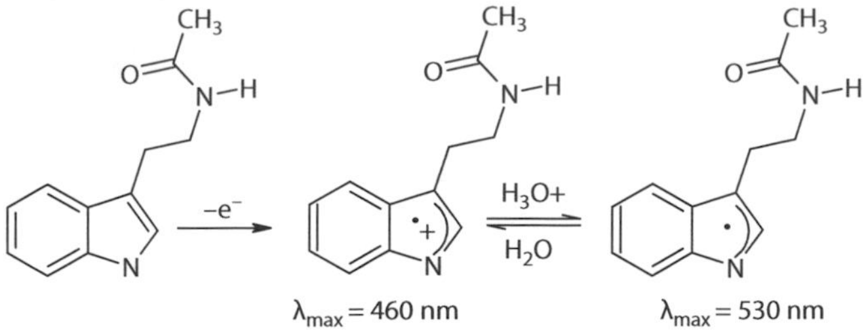

SCHEME 22.14 Radicals produced during one-electron oxidation of melatonin.

(Goldman, 1995). Melatonin reacts with $^{\bullet}$OH radicals, peroxyl radicals, and other oxidizing radicals with rate constants comparable to many important antioxidants like ascorbate. The $^{\bullet}$OH radical reacts with melatonin through addition followed by elimination to produce indolyl radicals absorbing at 330 and 520 nm with radical pK_a at ~4.5 (Roberts et al., 1988; Stasica et al., 1998; Mahal et al., 1999). No direct reaction of melatonin with $O_2^{\bullet-}$ radicals was observed by pulse radiolysis. Some reports indicate that melatonin induces formation of $O_2^{\bullet-}$ radicals, a reaction considered to be responsible for its tumorigenic response. Melatonin reacts with reducing radicals like e_{aq}^-, H atom with rate constants less than diffusion-controlled limits. The rate constants for the regeneration of melatonin from the melatonin radical by ascorbate and urate were determined to be 5 7 × 10⁷ and $4 \times 10^7\,M^{-1}\,s^{-1}$, respectively. Melatonin reacts with guanosine radicals with a rate constant of 3×10^9 $M^{-1}\,s^{-1}$, indicating its ability to repair damaged bases in DNA. Scheme 22.14 shows the structures of one-electron oxidation of melatonin.

22.6.2.2 β-Carotene

Carotenoids represent one large class of natural non-phenolic antioxidants present in vegetables, fruits, and flowers. Around 600 different types of carotenoids are identified; important among these are all-trans-β-carotene, zeaxanthin, lycopene, canthaxanthin, lutein, etc. (Pryor et al., 2000). Carotenoids are practically insoluble in water, but are soluble in surfactant solutions. β-Carotene is one of the most important among the carotenoids and an unusual compound that is extensively studied both as an antioxidant and as a pro-oxidant. Initial reports were aimed at understanding its potent antioxidant behavior. β-Carotene has been considered to be an excellent chain-breaking antioxidant and scavenges peroxyl radicals with rate constants varying from 10⁶ to $10^9\,M^{-1}\,s^{-1}$ (Hill et al., 1995). It has also been reported to neutralize thiyl, thiyl peroxyl radicals, NO_2 radicals, $ONOO^-$, and $^1O_2$ (Hill et al., 1995; Everett et al., 1996; Mortensen et al., 1997; Bohm et al., 1998; Edge et al., 2000; Agamey et al., 2004). The β-carotene radical cation, produced by electron transfer to these oxidizing radicals, exhibits strong absorption at 950 nm (Chauvet et al., 1983). The one-electron reduction potential of β-carotene in triton X-100 surfactant solutions was reported to be 1.06 V vs. NHE (Edge et al., 2000; Getoff, 2000). β-Carotene radical cations are repaired by water-soluble vitamin C by a rate constant of $10^7\,M^{-1}\,s^{-1}$. All these studies initially led to the assumption that electron transfer reactions contribute to the antioxidant activity of β-carotene. However, a randomized double blind study on the effect of β-carotene on the incidence of lung cancer suggested that male smokers who received β-carotene showed higher incidence of cancer; this led to doubts on the antioxidant ability of β-carotene (Heinonen and Albanes, 1994). The possibility of a pro-oxidant effect of β-carotene as a contributor to this effect has been proposed. Based on pulse radiolysis studies with a variety of peroxyl radicals in polar and non-polar solvents and also by identification of the other products, it has been concluded that the reactivity and mode of reaction with the peroxyl radical depends on the reduction potential of the radical and the polarity of the medium (Everett et al., 1996; Khopde et al., 1998; Agamey et al., 2004). Three different types of radicals have been identified during these reactions, radical cations absorbing at 950–1000 nm, a radical adduct absorbing at 750–850 nm, and neutral radicals formed from hydrogen abstraction absorbing in the UV–Vis

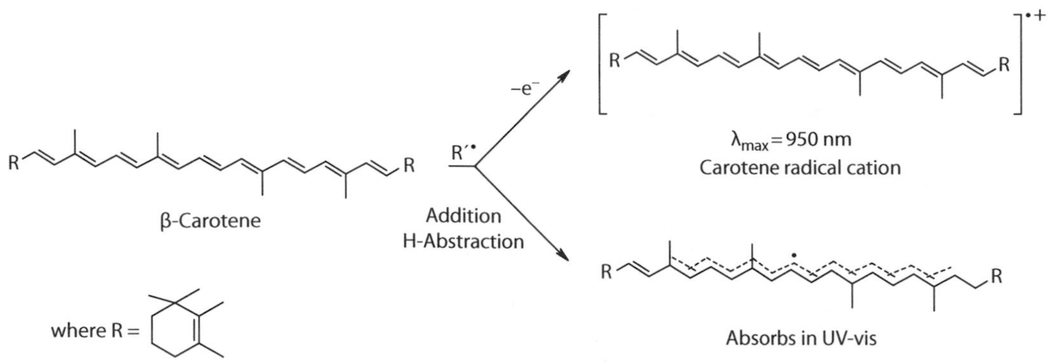

SCHEME 22.15 Possible free radical induced oxidation reaction pathways on β-carotene.

region (Scheme 22.15). Formation of radical cations may most likely lead to antioxidant effects as they are regenerated back to β-carotene by ascorbate and vitamin E. Some of the radical cations could oxidize tyrosine and cysteine, which may affect the function of proteins, thereby inducing pro-oxidant effects. Neutral carotenoid radicals and the fragmentation products may subsequently react with molecular oxygen to produce peroxyl radicals that may be responsible for the pro-oxidant activity of β-carotene. Although some indirect evidence for the differential antioxidant and pro-oxidant behavior of carotenoids has been provided by pulse radiolysis studies, much needs to be verified through further research.

22.6.2.3 Cyclic Nitroxides

Cyclic nitroxides ($>NO^{\bullet}$) are stable free radicals, which are often used as biophysical markers for monitoring membrane fluidity, as contrast agents for in vivo imaging. They have been found to reduce the oxidative stress in different cellular models and provide protection through oxidation of reduced transition metals, detoxification of semiquinones, and scavenging of ROS (Zhang et al., 1999; Samuni et al., 2002; Goldstein et al., 2003a,b, 2004, 2006). Unlike most low molecular weight antioxidants, which are depleted while attenuating oxidative damage, nitroxides can be recycled. The antioxidant activity of nitroxides is associated with electron transfer induced switching between the oxidized and reduced forms. Pulse radiolysis studies on reactions of five and six member cyclic nitroxides with peroxyl radicals, $^{\bullet}OH$, $O_2^{\bullet-}$, NO_2, and $CO_3^{\bullet-}$ radicals have been reported. Depending on the substitution and type of radical, the electron transfer rate constants with a number of oxidizing free radicals varied between 10^7 and $10^8 M^{-1} s^{-1}$. The one-electron oxidation reactions with nitroxides produced oxoammonium cations ($>N=O+$), and the reduction potentials for the couple

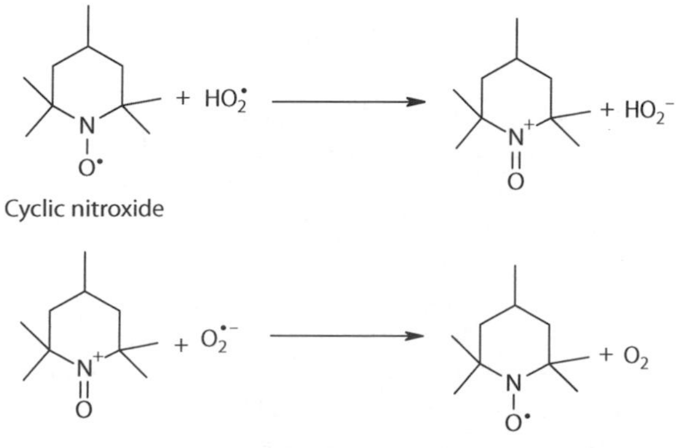

SCHEME 22.16 Reactions between nitroxide and superoxide radicals leading to SOD activity.

>N=O+/>NO, ranged from 0.7 to 0.9 V vs. NHE depending on the ring size and substitution. Nitroxides with a less positive reduction potential provide greater protection against cellular damage. The easy switching between the oxidized and reduced forms of these nitroxides made them potential candidates for antioxidant enzyme mimics (Zhang et al., 1999; Goldstein et al., 2003b) such as the SOD enzyme. The reactions leading to SOD activity are given in Scheme 22.16.

22.6.3 SULFUR AND SELENIUM ANTIOXIDANTS

It is well known that low molecular weight thiols (RSH) protect cells from oxidative processes and free-radical attack. Glutathione (GSH) is the major component of intracellular thiols, which is present in millimolar concentrations in mammalian cells (Sonntag, 1987; Halliwell and Gutteridge, 1993; O'Neill and Wardman, 2009), and should provide a first line of defense for oxidative stress. The protection rendered by thiols (RSH) is conceptualized through a repair mechanism, where they donate a hydrogen atom to the damaged target as the S–H bond energy is less than the C–H bond energy, and in biology many of the damaged targets are C-centered. Radiation chemists with the help of pulse radiolysis had provided evidence for this repair process (Sonntag, 1987). Thiols, either in the free or protein-bound form, on reaction with free-radical oxidants produce thiyl radicals (Equation 22.2) (Tamba et al., 1995; Asmus and Bonifacic, 1999; O'Neill and Wardman, 2009). The fate of thiyl radicals inside the cells is a much-discussed topic. The rich chemistry of S-centered radicals and the contributions of pulse radiolysis along with detailed quantum chemical calculations have been compiled in a book edited by Alfassi (1999). Therefore, in this chapter, only a few essential reactions related to antioxidant chemistry are briefly mentioned.

Thiyl radicals are oxidizing in nature and the reduction potential of glutathione couple (GS$^\bullet$, H$^+$/GSH) is ~0.8–0.9 V (Buettner, 1993; Tamba et al., 1995). The most probable reaction for thiyl radicals is reaction with another thiolate anion (RS$^-$) to form disulfide anion (RSSR$^{\bullet-}$) (Equation 22.3). RSSR$^{\bullet-}$ is reducing in nature and reacts with oxygen to form O$_2^{\bullet-}$ radicals (Equation 22.4), a reaction considered as an important radical sink in biology (Wardman, 1989). Other probable reactions competing for thiyl radicals are reactions with molecular oxygen and ascorbate anion. By reaction with oxygen, a thiyl radical is converted to thiyl peroxyl radicals (RSOO$^\bullet$) (Equation 22.5). Thiyl peroxyl radicals are oxidizing in nature. The reduction potential of peroxyl radicals of GSH are estimated to be ≤0.9 V vs. NHE for the couple GSOO$^\bullet$/GSOO$^-$ (Tamba et al., 1995):

$$RSH + {}^\bullet OH \rightarrow RS^\bullet + H_2O \tag{22.2}$$

$$RS^\bullet + RS^- \rightarrow RSSR^{\bullet-} \tag{22.3}$$

$$RSSR^{\bullet-} + O_2 \rightarrow RSSR + O_2^{\bullet-} \tag{22.4}$$

$$RS^\bullet + O_2 \rightarrow RSOO^\bullet \tag{22.5}$$

Thiyl peroxyl radicals can induce peroxidation of lipids, leading to the speculation by some researchers that thiols may not actually act as antioxidants in the sense of the term. However, some others have argued that at physiological levels of oxygen, formation of O$_2^{\bullet-}$ radicals and peroxyl radicals from thiyl radicals might not be important. Accordingly, the most crucial redox reaction of GSH in cells is its oxidation and conversion to GSSG.

Since several proteins contain thiols, thiyl radicals are important intermediates in the oxidative denaturation of peptides and proteins. Of many important amino acids, the oxidation of methionine plays a significant role during in vivo oxidative denaturation as it can influence the stability, conformation, and activity of the proteins (Schoneich, 2003). Methionine undergoes both two electron and one-electron oxidation: the former produces methionine sulfoxide, while the latter gives a radical cation. Methionine sulfoxide is reduced back to methionine by methionine sulfoxide reductase, by which the deactivation of the protein is inhibited. The radical cations of methionine produced by one-electron oxidation in peptides are stabilized with electron rich heteroatoms such as N, O, S, P, and Se. The radical cation undergoes a number of irreversible reactions like loss of carbon dioxide, formation of carbon-centered radicals and peroxyl radicals, which may induce chain reactions of protein oxidation, leading to protein damage.

Thiols are also important as radioprotectors, which are agents employed to minimize the damage caused by ionizing radiation to normal cells (Weiss and Landauer, 2003). Simple thiols like cysteine and many synthetic thiols have also been examined as radioprotectors. Aminothiols and their phosphothiolate derivatives have been evaluated as radioprotectors. The clinically acceptable radioprotector is an aminothiol, amifostine also known as WR-2721. Mechanism of action of radioprotection by thiols includes scavenging of radicals like $^{\bullet}OH$ and chemical repair of damaged target like DNA by hydrogen donation.

Selenium is a trace mineral, an essential micronutrient, and also an active component of the mammalian selenoproteins (Jacob et al., 2003; Papp et al., 2007). The best characterized and widely studied selenoprotein is glutathione peroxidase (GPx), which is a redox enzyme. GPx is also an antioxidant enzyme and catalyzes the reduction of hydroperoxides at the expense of GSH. Because of its catalytic role and ability to undergo easy oxidation, selenium compounds are being explored as a new class of antioxidants. Unlike the radiation chemistry of sulfur compounds, the radiation chemistry of selenium compounds is not so much studied and the reactions of selenium-centered radicals are not fully understood. In a recent pulse radiolysis study of selenomethionine, the (Se $\therefore$ N) bonded radical cations have been characterized and their subsequent secondary reactions have been studied (Mishra et al., 2006b, 2009). A significant difference between the radical cations of methionine and selenomethionine is the longer lifetime of selenium centered radicals. Although both the radicals undergo loss of carbon dioxide from the radical cation, the yield is nearly half for the selenium analogue as compared to methionine. Because of this, formation of selenoxide is more favorable than the sulfoxide from the radical cation. This reaction in particular makes selenomethionine a better antioxidant than methionine as the selenoxide can be easily reduced back to selenomethionine by GSH and this reaction does not require the reductase enzyme. Based on this reaction, it is proposed that selenomethionine could provide an important line of defense against oxidative damage in proteins. This distinctive feature of selenoxide has been used to develop synthetic selenium compounds with GPx activity as antioxidants. Compounds like ebselen and selenocystine derivatives showed good antioxidant and radioprotecting ability (Kunwar et al., 2007; Tak and Park, 2009).

22.7 $O_2^{\bullet-}$ RADICAL REACTIONS AND DEVELOPMENT OF SOD MIMICS

$O_2^{\bullet-}$ radicals act both as mild oxidants and reductants (Bielski et al., 1985). Pulse radiolysis has been found to be a very convenient method for studying the reactions of $O_2^{\bullet-}$ radicals as they can be generated exclusively in high yields and detected by transient absorption at 250 nm. The prototropic equilibrium constant (pK_a) for the equilibrium between $O_2^{\bullet-}$ and $HO_2^{\bullet}$ radicals is 4.7 (Bielski et al., 1985; Sonntag, 1987). The $HO_2^{\bullet}$ radicals are more reactive than $O_2^{\bullet-}$ radicals and react with substrates by hydrogen abstraction or by addition to the double bonds. $O_2^{\bullet-}$ radicals do not cause hydrogen abstraction reactions but participate in a number of redox reactions with metal ions and substrates like quinone, ascorbate, etc. (Buettner, 1993). The rate constants for most of these reactions vary from 10^3 to $10^7 M^{-1} s^{-1}$ (Bielski et al., 1985) and for SOD-like molecules, rate constants almost close to diffusion-controlled limits have also been reported. In the absence of other reactions, $O_2^{\bullet-}$ radicals decay either by a radical–radical dismutation reaction or by the reaction of $O_2^{\bullet-}$ with $HO_2^{\bullet}$, producing H_2O_2; the respective rate constants for these reactions are 0.39 and $10^8 M^{-1} s^{-1}$ (Sonntag, 1987). Important antioxidants like vitamin E, curcumin, quercetin, and ascorbic acid react with $O_2^{\bullet-}$ radicals with rate constants of 5.8×10^3, 4.6×10^4, 4.7×10^4, and $5.0 \times 10^4 M^{-1} s^{-1}$, respectively (Bielski et al., 1985; Jovanovic et al., 1994; Mishra et al., 2004).

Pulse radiolysis has also been found to be useful to follow the mechanisms involved in the SOD enzyme activities of native and synthetic SOD mimcs. SOD is an important antioxidant enzyme that catalyzes the disproportionation of $O_2^{\bullet-}$ radicals to H_2O_2 and O_2. Different types of SODs are present in eukaryotic cells such as Mn-SOD in mitochondria and cytoplasm and Cu, Zn-SOD in the cytosol and in extracellular surfaces (Halliwell and Gutteridge, 1993). In prokaryotic cells, in addition to Mn-SOD, Fe-SOD is also observed. In the SOD enzyme, the dismutation reaction of $O_2^{\bullet-}$ radicals takes place in two steps (Equations 22.6 and 22.7), involving diffusion-controlled cycling between the oxidized and reduced forms of the metal centers:

$$M^{n+}SOD + O_2^{\bullet-} \rightarrow M^{(n-1)+}SOD + O_2 \tag{22.6}$$

$$M^{(n-1)+}SOD + O_2^{\bullet-} + 2H^+ \rightarrow M^{n+}SOD + H_2O_2 \tag{22.7}$$

In addition to this, another alternate mechanism is proposed, involving formation of inhibition complex followed by the release of the peroxide as in the following steps (Equations 22.8 and 22.9):

$$M^{n+}SOD + O_2^{\bullet-} \rightarrow M^{(n+)}SOD - O_2^{-2} \tag{22.8}$$

$$M^{(n+)}SOD - O_2^{-2} + 2H^+ \rightarrow M^{n+}SOD + H_2O_2 \tag{22.9}$$

Pulse radiolysis studies with many different types of SODs have been performed and the data was useful to understand the role of these different steps.

The catalytic dismutation reaction of $O_2^{\bullet-}$ radicals by SOD is one of the fastest reactions in biology. The first accurate estimation of the rate constant for this reaction with Cu,Zn-SOD was reported by pulse radiolysis to be $2.3 \times 10^9 M^{-1} s^{-1}$ (Klug et al., 1973). Later, this value was further confirmed by many other methods. In one recent study involving superoxide reductase (SOR) from *Desulfoarculus baarsii*, the precise step responsible for the catalytic action was examined (Niviere et al., 2004). Its active site contains an unusual mononuclear ferrous center. The studies confirmed that the reaction of SOR with $O_2^{\bullet-}$ radicals involves two reaction intermediates, an iron(III)-peroxo species and iron(III)-hydroperoxo species. *Deinococcus radiodurans* (*Drad*) is a radio-resistant bacterium with an extraordinary capacity to tolerate high levels of ionizing radiation, and it possesses only Mn-SOD. Pulse radiolysis studies coupled with steady state measurements have been used to demonstrate the effectiveness of *Drad* Mn-SOD at high $O_2^{\bullet-}$ fluxes, which could also be responsible for its resistance to oxidative stress induced by high levels of radiation (Abreu et al., 2008).

Under oxidative stress conditions, such as ischemic reperfusion injury, the endogenous SODs are inadequate and they need to be supplemented. However, the native enzyme suffers from several disadvantages such as low shelf life and nonavailability, therefore new SOD mimics have been developed as alternatives. Although most of the antioxidants could scavenge $O_2^{\bullet-}$ radicals, not many showed SOD like catalytic activity. Antioxidants having different functional groups that could undergo both oxidation and reduction showed SOD mimicking activity (Mishra et al., 2004).

Since the catalytic effects in the native enzymes are due to reversible redox reactions in metal centers, simple salts of copper and manganese have been examined for SOD activity. Such salts although showing excellent SOD-like behavior can be highly toxic as they induce oxidative stress by participating in metal catalyzed reactions. Recently, efficient SOD like activity was observed in osmium tetroxide (OsO_4) (Goldstein et al., 2005) and manganous phosphate (Barnese et al., 2008). The catalytic rate constant for OsO_4 was determined by pulse radiolysis to be $1.43 \times 10^9 M^{-1} s^{-1}$ (Goldstein et al., 2005). Similarly, a suitable mechanism for the SOD activity in manganous phosphate and its absence in manganous chloride, sulfate, or pyrophosphate has been proposed (Barnese et al., 2008). Mn^{+2} in manganous phosphate reacts with $O_2^{\bullet-}$ with a rate constant of $2.8 \times 10^7 M^{-1} s^{-1}$ to form MnO_2^+, which undergoes rapid disproportionation to give manganous phosphate, oxygen, and hydrogen peroxide. The presence of the phosphate ion provides extra stabilization to the MnO_2^+ ion.

In addition to simple salts, several transition metal complexes of natural antioxidants have also been examined as new SOD mimics (Hirano et al., 2000; Vajragupta et al., 2003; Barik et al., 2005; Etcheverry et al., 2008; Maroz et al., 2008). Reactions of $O_2{}^{\bullet-}$ radicals with such complexes and the determination of their one-electron redox potentials helped in evaluation of such complexes as SOD mimics. For example, using pulse radiolysis, the mechanism of SOD mimicking activity of Mn(II) pentaazanmacrocycle compounds was reported (Maroz et al., 2008) and three important steps have been identified as the oxidation of the Mn(II)-complex to intermediate Mn(III)-complex by $O_2{}^{\bullet-}$, followed by the rate determining reduction of the intermediate complex by the reaction of another $O_2{}^{\bullet-}$ radical, and finally by the proton-assisted release of H_2O_2 with the concomitant release of the starting compound. In another recent study, unique SOR activity has been reported from an adduct of a penta-coordinated ferrous iron complex $[Fe^{+2}(N\text{-}His)_4(S\text{-}Cys)]$ and ferrocyanide. The adduct while reducing $O_2{}^{\bullet-}$ radicals did not generate H_2O_2; since H_2O_2 formation could also contribute to oxidative stress, this particular adduct may be superior to the other conventional SOD mimics (Molina-Heredia et al., 2006).

22.8 CONCLUSIONS

Oxidative stress, a pathological state associated with excessive free-radical production, actually changes the overall potential of the cells, from a more negative to a less negative side, and physiological processes are very sensitive to such changes. For example, a change in the half cell reduction potential of the cell even by 30–40 mV can make proliferating cells to undergo differentiation and a further change in potential may induce apoptosis. Intracellular thiols try to maintain the redox balance in the cells, and the ratio of concentration of reduced glutathione (GSH) and oxidized glutathione (GSSG) is an indicator of the redox status of the cells. Most healthy cells have a GSH and GSSG ratio of 100:1. Increase in the levels of oxidative stress causes imbalance between the production of oxidants and the system's ability to detoxify them, resulting in irreversible changes in crucial biological machinery, including damage to proteins, lipids, and DNA; necrosis; ATP depletion; and prevention of controlled apoptotic death.

Under pathological situations, when the cells are subjected to oxidative stress and the endogenous antioxidant system fails to maintain the redox balance, cells need to be supplemented with external antioxidant agents. The primary role of an antioxidant is therefore to maintain this redox balance either by preventing free-radical formation or by direct scavenging of free-radical oxidants. Radiation chemistry with the help of pulse radiolysis has been found to be extremely powerful in understanding the latter process in real-time scales. Kinetic and thermodynamic properties of antioxidants determined from such studies provided useful information to predict the feasibility of the radical-scavenging ability, site of free-radical attack on the antioxidant molecules, and the subsequent reactions of antioxidants. Estimation of one-electron reduction potentials of antioxidants using the pulse radiolysis technique has been crucial in understanding the fate of the primary and secondary free-radical electron transfer reactions.

When a molecule reacts with an oxidizing free radical, a new radical is generated. The fate of the resultant radical decides whether a compound can actually act as a potent antioxidant or a prooxidant, in spite of being an efficient scavenger of primary free radicals like $^{\bullet}OH$ radicals. If the reduction potential of the antioxidant radical is high (more positive), it can cause further oxidation of some biomolecules. If the reduction potential is too low, it can undergo auto-oxidation and occasionally induce the production of $O_2{}^{\bullet-}$ radicals. If the lifetime of the antioxidant radical is too short, the regeneration reaction cannot compete with the radical decay and the antioxidant is not recycled. If the antioxidant radical is too long lived, the long lifetime can allow other unwanted secondary reactions like induction of lipid peroxidation, etc. Another key factor to be considered in this regard is the nature of the products formed during the antioxidant process. If the products are less reactive and non-toxic, such compounds are ideal, but if the products are reactive like for example, semiquinones, or quinines, they can participate in other reactions like depletion of thiol, thereby elevating

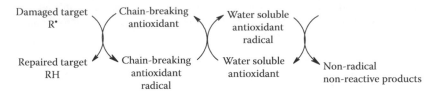

SCHEME 22.17   Redox processes indicating the repair of damaged biomolecules by antioxidants.

oxidative stress. The site of radical attack in the antioxidant molecule also decides the efficacy of these compounds. If the resultant antioxidant radical is a carbon-centered radical, it can add to molecular oxygen, thereby converting the antioxidant radical to oxidizing peroxyl radicals. Not only carbon-centered, but some of the sulfur-centered radicals may also form peroxyl radicals. Oxygen-centered radicals are generally not reactive to oxygen and therefore do not generate peroxyl radicals. This is one of the reasons why phenolic compounds are some of the most powerful antioxidants. All these factors put together decide the overall efficacy of an antioxidant.

The crucial sequential steps involved in the antioxidant action are summarized in Scheme 22.17. An antioxidant protects the biomolecules by donating a hydrogen atom and inhibiting the chain reaction, producing a less reactive antioxidant radical. This antioxidant radical is regenerated by a water-soluble antioxidant, and the water-soluble antioxidant is then converted to non-reactive products.

Decades of experimental research by radiation chemists with the help of pulse radiolysis were central to all these studies and provided a wealth of information. Such studies have not only helped in the evaluation of antioxidants at the molecular level, but they were also useful in the design of new leads. However, most of these studies were limited to aqueous solutions or a few bio-mimetic systems. Although such studies formed the basis for in vivo examinations, they could not be extrapolated directly to real biological models. Therefore, future experiments should be oriented to estimating the transient redox properties of antioxidants in models that are close to cellular systems, so that full advantage of this powerful tool can be made in the understanding of this multidisciplinary research area.

ACKNOWLEDGMENTS

The author would like to thank Drs. T. Mukherjee and S. K. Sarkar for the encouragement and support and Dr. B. Mishra for many useful suggestions in the preparation of the manuscript.

REFERENCES

Abreu, I. A., Hearn, A., Haiqain, A., Harry, S., Silverman, D. N., and Cabelli, D. E. 2008. The kinetic mechanism of manganese-containing superoxide dismutase from *Deinococcus radiodurans*: A specialized enzyme for the elimination of high superoxide concentrations. *Biochemistry* 47: 2350–2356.

Agamey, A. El., Lowe, G. M., McGarvey, D. J., Mortensen, A., Phillip, D. M., Truscott, T. G., and Young, A. J. 2004. Carotenoid radical chemistry and oxidant/pro-oxidant properties. *Arch. Biochem. Biophys.* 430: 37–48.

Alfassi, Z. B. 1999. *S-Centered Radicals*. New York: Wiley.

Alfassi, Z. B., Marguet, S., and Neta, P. 1994. Formation and reactivity of phenylperoxyl radicals in aqueous solutions. *J. Phys. Chem.* 98: 8019–8023.

Asmus, K. D. and Bonifacic, M. 1999. In *Sulfur-Centered Reactive Intermediates as Studied by Radiation Chemical and Complimentary Techniques: S-Centered Radicals*, Z. B. Alfassi (ed.), p. 142. New York: Wiley.

Augusto, O., Bonini, M. G., Amanso, A. M., Linares, E., Santos, C. C., and Menezes, S. D. 2002. Nitrogen dioxide and carbonate radical anion: Two emerging radicals in biology. *Free Radical Biol. Med.* 32: 841–859.

Barik, A., Priyadarsini, K. I., and Mohan, H. 2004. Redox reactions of 2-hydroxy-3-methoxybenzaldehyde (o-vanillin) in aqueous solution. *Rad. Phys. Chem.* 70: 687–696.

Barik, A., Mishra, B., Shen, L., Mohan, H., Kadam, R. M., Dutta, S., Zhang, H. Y., and Priyadarsini, K. I. 2005. Evaluation of a new copper (II)-curcumin complexes as superoxide dismutase mimic and its free radical reactions. *Free Radic. Biol. Med.* 39: 811–822.

Barnese, K., Gralla, E. B., Cabelli, D. E., and Valentine, J. S. 2008. Manganous phosphate acts as a superoxide dismutase. *J. Am. Chem. Soc.* 130: 4604–4606.

Bartesaghi, S., Trujillo, M., Denicola, A., Folkes, L., Wardman, P., and Radi, R. 2004. Reactions of desferrioxamine with peroxynitrite-derived carbonate and nitrogen dioxide radicals. *Free Radical Biol. Med.* 36: 471–481.

Baxendale, J. H. and Busi, F. 1981. The study of fast processes and transient species by electron pulse radiolysis. In *Proceedings of the NATO Advanced Study Institute, Italy*. Boston, MA: D. Reidel Publishing Co.

Bensasson, R. V., Land, E. J., Truscott, T. G. 1983. *Flash Photolysis and Pulse Radiolysis, Contributions to the Chemistry of Biology and Medicine*. London, U.K.: Pergamon Press.

Bielski, B. H. J., Comstock, D. A., and Bowen, R. A. 1971. Ascorbic acid free radicals I. Pulse radiolysis study of optical absorption and kinetic properties. *J. Am. Chem. Soc.* 93: 5624–5629.

Bielski, B. H. J., Allen, A. O., and Schwarz, H. A. 1981. Mechanism of disproportionation of ascorbic acid radicals. *J. Am. Chem. Soc.* 103: 3516–3518.

Bielski, B. H. J., Cabelli, D. E., Arudi, R., and Ross, A. 1985. Reactivity of HO_2/O_2^- radicals in aqueous solutions. *J. Phys. Chem. Ref. Data* 14: 1041–1100.

Bisby, R. H. 1990. Interaction of vitamin E with free radicals and membrane. *Free Radic. Res. Com.* 8: 299–306.

Bohm, F., Edge, R., Mc Garvey, D. J., and Truscott, T. G. 1998. β-carotene with vitamin E and C offers synergestic cell protection against Nox. *FEBS Lett.* 436: 387–389.

Bors, W. and Michel, C. 1999. Antioxidant capacity of flavanols and gallate esters: Pulse radiolysis studies. *Free Radic. Biol. Med.* 27: 1413–1426.

Bors, W. and Saran, M. 1987. Radical scavenging by flavonoid antioxidants. *Free Radic. Res. Commun.* 2: 289–294.

Bors, W., Michel, C., and Saran, M. 1994. Flavonoid antioxidants: Rate constants for reactions with oxygen radicals. *Methods Enzymol.* 234: 420–429.

Bors, W., Michel, C., and Schikora, S. 1995. Interaction of flavonoids with ascorbate and determination of their univalent redox potentials: A pulse radiolysis study. *Free Radic. Biol. Med.* 19: 45–52.

Bors, W., Michel, C., and Stettmaier, K. 2001a. Structure-activity relationships governing antioxidant capacities of plant polyphenols. *Methods Enzymol.* 335: 166–180.

Bors, W., Foo, L. Y., Hertkorn, N., Michel, C., and Stettmaier, K. 2001b. Chemical studies of proanthocyanidins and hydrolyzable tannins. *Antioxid. Redox Signal.* 3: 995–1008.

Bors, W., Kazazic, S. P., Michel, C., Kortenska, V. D., Stettmaier, K., and Klasinc, L. 2002. Methoxyphenols-antioxidant principles in food plants and spices: Pulse radiolysis, EPR spectroscopy and DFT calculations. *Int. J. Quant. Chem.* 90: 969–979.

Bors, W., Michel, C., Stettmaier, K., Lu, Y., and Yeap Foo, L. 2003. Pulse radiolysis, electron paramagnetic resonance spectroscopy and theoretical calculations of caffeic acid oligomer radicals. *Biochim. Biophys. Acta* 1620: 97–107.

Buettner, G. R. 1993. The pecking order of free radicals and antioxidants: Lipid peroxidation, α-tocopherol and ascorbate. *Arch. Biochem. Biophys.* 300: 535–543.

Butler, J., Land, E. J., Swallow, A. J., and Prutz, W. 1984. The azide radicals and its reaction with tryptophan and tyrosine, *Rad. Phys. Chem.* 23: 265–270.

Buxton, G. V., Greenstock, C. L., Helman, W. P., and Ross, A. B. 1988. Critical review of rate constants for reactions of hydrated electron, hydrogen atom and hydroxyl radicals in aqueous solutions. *J. Phys. Chem. Ref. Data* 17: 513–886.

Cadenas, E. and Davies, K. J. A. 2000. Mitochondrial free radical generation, oxidative stress and ageing. *Free Radic. Biol. Med.* 29: 222–230.

Cadenas, E., Merenyi, G., and Lind, J. 1989. Pulse radiolysis study of the reactivity of Trolox C phenoxyl radical with superoxide anion. *FEBS Lett.* 253: 235–238.

Cai, Z., Li, X., and Katsumura, Y. 1999. Interaction of hydrated electron with dietary flavonoids and phenolic acids: Rate constants and transient spectra studied by pulse radiolysis. *Free Radic. Biol. Med.* 27: 822–829.

Cao, H., Pan, X., Li, C., Zhou, C., Deng, F., and Li, T. 2003. Density functional theory calculations for resveratrol. *Bioorg. Med. Chem. Lett.* 13: 1869–1871.

Chauvet, J. P., Vlovy, R., Land, E. J., Santus, R., and Truscott, T. G. 1983. One-elelctron oxidation of carotene and electron transfers involving carotene cations in micelles. *J. Phys. Chem.* 87: 592–601.

Cren-Olive, C., Hapiot, P., Pinson, J., and Rolando, C. 2002. Free radical chemistry of Flavan-3-ols: Determination of thermodynamic parameters and of kinetic reactivity from short (ns) to long (ms) time scale. *J. Am. Chem. Soc.* 124: 14027–14038.

Davies, M. J., Forni, L. G., and Wilson, R. L. 1988. Vitamin E analogue Trolox c. ESR and pulse radiolysis studies of free radical reactions. *Biochem. J.* 255: 513–522.

Edge, R., Land, E. J., McGarvey, J., Burke, M., and Truscott, T. G. 2000. The reduction potentials of β-carotene+/ β-carotene couple in an aqueous micro-heterogeneous environment. *FEBS Lett.* 471: 125–127.

Etcheverry, S. B., Ferrer, E. G., Naso, L., Rivadeneira, J., Salinas, V., and Williams, P. A. M. 2008. Antioxidant effects of the VO(IV) hesperidin complex and its role in cancer chemoprevention. *J. Biol. Inorg. Chem.* 13: 435–447.

Everett, S. A., Dennis, M. F., and Patel, K. B. 1996. Scavenging of nitrogen dioxide, thiyl and sulfonyl radicals by nutritional antioxidant β-carotene. *J. Biol. Chem.* 271: 3983–3994.

Fang, Y., Yang, S., and Wu, G. 2002. Free radicals, antioxidants and nutrition. *Nutrition* 18: 872–879.

Filipe, P., Morliere, P., Patterson, L. K., Hug, G. L., Maziere, J. C., Maziere, C., Freitas, J. P., Fernandes, A., and Santus, R. 2002. Repair of amino acid radicals of apolipoprotein B100 of low-density lipoproteins by flavonoids. A pulse radiolysis study with quercetin and rutin. *Biochemistry* 41: 11057–11064.

Filipe, P., Morliere, P., Patterson, L. K., Hug, G. L., Maziere, J. C., Freitas, J. P., Fernandes, A., and Santus, R. 2004. Oxygen-copper (II) interplay in the repair of semi-oxidized urate by quercetin bound to human serum albumin. *Free Radic. Res.* 38: 295–301.

Finkel, T. and Holbrook, N. J. 2000. Oxidants, oxidative stress and biology of ageing. *Nature* 408: 239–247.

Fu, H., Katsumura, Y., Lin, M., and Muroya, Y. 2008. Laser photolysis and pulse radiolysis studies on silybin. *Radiat. Phys. Chem.* 77: 1300–1305.

Fukuto, J. M. and Ignarro, L. J. 1997. In vivo aspects of nitric oxide chemistry: Does peroxynitrite play a major role in cytotoxicity. *Acc. Chem. Res.* 30: 149–152.

Furtmuller, P. G., Arnhold, J., Jantschko, W., Zederbauer, M., Jakopitsch, C., and Obinger, C. 2005. Standard reduction potentials of all couples of the peroxidase cycle. *J. Inorg. Chem.* 99: 1220–1229.

Getoff, N. O. 2000. Pulse radiolyisis studies of β-carotene in oxygenated DMSO solution. *Radiat. Res.* 154: 692–696.

Goldman, A. 1995. Melatonin, a review. *Brit. J. Clin. Pharmacol.* 19: 258–260.

Goldstein, S. and Czapski, G. 1995. The reaction of NO with O_2^- and HO_2: A pulse radiolysis study. *Free Radic. Biol. Med.* 19: 505–510.

Goldstein, S. and Merenyi, G. 2008. The chemistry of peroxynitrite: Implications for biological activity. *Methods Enzymol.* 436: 49–61.

Goldstein, S., Squadrito, G. L., Pryor, W. A., and Czapski, G. 1996. Direct and indirect oxidation by peroxynitrite neither involving the hydroxyl radicals. *Free Radic. Biol. Med.* 21: 965–974.

Goldstein, S., Samuni, A., and Russo, A. 2003a. Reaction of cyclic nitroxides with nitrogen dioxide: The intermediacy of oxoammonium cations. *J. Am. Chem. Soc.* 125: 8364–8370.

Goldstein, S., Merenyi, G., Russo, A., and Samuni, A. 2003b. The role of oxoammonium cation in the Sod-mimic activity of cyclic nitroxides. *J. Am. Chem. Soc.* 125: 789–795.

Goldstein, S., Samuni, A., and Merenyi, G. 2004. Reaction of NO, peroxynitrite and carbonate radicals with nitroxide and their oxonium cations. *Chem. Res. Toxicol.* 17: 250–257.

Goldstein, S., Czapski, G., and Heller, A. 2005. Osmium tetroxide, used in the treatment of arthritic joints, is a fast mimic of superoxide dismutase. *Free Radic. Biol. Med.* 38: 839–845.

Goldstein, S., Samuni, A., Hideg, K., and Merenyi, G. 2006. Structure-activity relationship of cyclic nitroxides as SOD mimics and scavengers of nitrogendioxide and carbonate radicals. *J. Phys. Chem. A* 110: 3679–3685.

Gorman, A. A., Hamblett, V. S., Srinivasan, V. S., and Wood, P. D. 1994. Curcumin derived transients: A pulsed laser and pulse radiolysis study. *Photochem. Photobiol.* 59: 389–398.

Guha, S. N. and Priyadarsini, K. I. 2000. Kinetic and redox characteristics of phenoxyl radicals of eugenol and isoeugenol: A pulse radiolysis study. *Int. J. Chem. Kinetics* 32: 17–23.

Halliwell, B. 1990. How to characterize a biological antioxidant? *Free Radic. Res. Commun.* 9: 1–32.

Halliwell, B. and Gutteridge, J. M. C. 1993. *Free Radicals in Biology and Medicine.* Oxford, U.K.: Clarendon Press.

Heinonen, O. P. and Albanes, D. 1994. The effect of vitamin E and beta carotene on the incidence of lung cancer and other cancers in male smokers. *New Eng. J. Med.* 330: 1029–1035.

Hill, T. J., Land, E. J., McGarvey, D. J., Schalch, W., Tinkler, J. H., and Truscott, T. G. 1995. Interactions between carotenoids and the CCl_3O_2·radical. *J. Am. Chem. Soc.* 117: 8322–8326.

Hirano, T., Hirobe, M., Kobayashi, K., Odani, A., Yamauchi, O., Ohsawa, M., Satow, Y., and Nagano, T. 2000. Mechanism of superoxide dismutase-like activity of Fe(II) and Fe(III) complexes of tetrakis-N,N,N',N'(2-pyridylmethyl)ethylenediamine. *Chem. Pharm. Bull.* 48: 223–230.

Jacob, C., Giles, G. I., Giles, N. M., and Sies, H. 2003. Sulfur and selenium: The role of oxidation state in protein structure and function. *Angew. Chem. Int. Ed.* 42: 4742–4758.

Jonsson, M. 1996. Thermochemical properties of peroxides and peroxyl radicals. *J. Phys. Chem.* 100: 6814–6818.

Jore, D., Ferradini, C., and Patterson, L. K. 1986. Gamma and pulse radiolytic study of the antioxidant activity of vitamin E. *Radiat. Phys. Chem.* 28: 557–558.

Joshi, R., Adhikari, S., Patro, B. S., Chattopadhyay, S., and Mukherjee, T. 2001. Free radical scavenging behaviour of folic acid: Evidence for possible antioxidant activity. *Free Radic. Biol. Med.* 30: 1390–1399.

Joshi, R., Kumar, M. S., Satyamoorthy, K., Unnikrisnan, M. K., and Mukherjee, T. 2005a. Free radical reactions and antioxidant activities of sesamol: Pulse radiolytic and biochemical studies. *J. Agric. Food Chem.* 53: 2696–2703.

Joshi, R., Kumar, S., Unnikrishnan, M. K., and Mukherjee, T. 2005b. Free radical scavenging reactions of sulfasalazine, 5-aminosalicylic acid and sulfapyridine: Mechanistic aspects and antioxidant activity. *Free Radic. Res.* 39: 1163–1172.

Jovanovic, S. V., Steenken, S., Tosic, M., Morjanovic, B., and Simic, M. G. 1994. Flavonoids as antioxidants. *J. Am. Chem. Soc.* 116: 4846–4851.

Jovanovic, S. V., Hara, Y., Steenken, S., and Simic, M. G. 1995. Antioxidant potential of gallocatechins. A pulse radiolysis and laser photolysis study. *J. Am. Chem. Soc.* 117: 9881–9888.

Jovanovic, S. V., Steenken, S., Hara, Y., and Simic, M. G. 1996. Reduction potentials of flavonoid and model phenoxyl radicals. Which ring in flavonoids is responsible for antioxidant activity? *J. Chem Soc. Perkin Trans.* 2 (11): 2497–2504.

Jovanovic, S. V., Boone, C. W., Steenken, S., Trinoga, M., and Kaskey, R. B. 2001. How curcumin works preferentially with water soluble antioxidants. *J. Am. Chem. Soc.* 123: 3064–3068.

Kapoor, S. and Priyadarsini, K. I. 2001. Protection of radiation-induced protein damage by curcumin. *Biophys. Chem.* 92: 119–126.

Kapoor, S., Mukherjee, T., Kagiya, T. V., and Nair, C. K. K. 2002. Redox reactions of tocopherol monoglucoside in aqueous solutions: A pulse radiolysis study. *J. Radiat. Res.* 43: 99–106.

Khopde, S. M., Priyadarsini, K. I., Mukherjee, T., Kulkarni, P. B., Satav, J. G., and Bhattacharya, R. K. 1998. Does β-carotene protect membrane lipids from peroxidation? *Free Radic. Biol. Med.* 25: 66–71.

Khopde, S. M., Priyadarsini, K. I., Venkatesan, P., and Rao, M. N. A. 1999. Free radical scavenging ability and antioxidant efficiency of curcumin and its substituted analogue. *Biophys. Chem.* 80: 85–91.

Klug, D., Rabani, J., and Fridovich, I. 1973. A direct demonstration of the catalytic action of superoxide dismutase through the use of pulse radiolysis. *J. Biol. Chem.* 247: 2645–2649.

Kobayashi, K., Harada, Y., and Hayashi, K. 1991. Kinetic behavior of the monodehydroascorbate radical studied by pulse radiolysis. *Biochemistry* 30: 8310–8315.

Kohen, R. and Nyska, A. 2002. Oxidation of biological systems: Oxidative stress phenomena, antioxidants, redox reactions and methods of quantification. *Toxicol. Pathol.* 30: 620–650.

Kono, Y., Kobayashi, K., Tagawa, S., Adachi, K., Ueda, A., Sawa, Y., and Shibata,, H. 1997. Antioxidant activity of polyphenolics in diets. Rate constants of reactions of chlorogenic acid and caffeic acid with reactive species of oxygen and nitrogen. *Biochim. Biophys. Acta* 1335: 335–342.

Koppenol, W. H. 1994. Thermodynamic consideration on the formation of reactive species from hypochloride. *FEBS Lett.* 347: 5–8.

Koppenol, W. H. 1998. The basic chemistry of nitrogen monoxide and peroxynitrite. *Free Radic. Biol. Med.* 25: 385–391.

Kunwar, A., Mishra, B., Barik, A., Kumbhare, L. B., Jain, V. K., and Priyadarsini, K. I. 2007. 3,3'-Diselenodipropionic acid, an efficient peroxyl radical scavenger and a GPx mimic, protects erythrocytes (RBCs) from AAPH induced hemolysis. *Chem. Res. Toxicol.* 20: 1482–1487.

Li, F. and Fang, X. 1998. Pulse radiolysis of epicatechin in aqueous solution. *Radiat. Phys. Chem.* 52: 405–408.

Li, Y.-M., Han, Z.-H., Jiang, S.-H., Jiang, Y., Yao, S.-D., and Zhu, D.-Y. 2000. Fast repairing of oxidized OH radical adducts of dAMP and dGMP by phenylpropanoid glycosides from *Scrophalaria ningpoensis* Hemsl. *Acta Pharmacologica Sinica* 21: 1125–1128.

Lin, W., Navaratnam, S., Yao, S., and Lin, N. 1998. Antioxidant principles of hydroxycinnamic acid derivatives and a phenyl propanoid glucoside. A pulse radiolysis study. *Rad. Phys. Chem.* 53: 425–430.

Lu, C. and Liu, Y. 2002. Interactions of lipoic acid radical cations with vitamins C and E analogue and hydroxycinnamic acid derivatives. *Arch. Biochem. Biophys.* 406: 78–84.

Mahal, H. S. and Mukherjee, T. 2006. Scavenging of reactive oxygen radicals by resveratrol: Antioxidant effect. *Res. Chem. Intermed.* 32: 59–71.

Mahal, H. S., Sharma, H. S., and Mukherjee, T. 1999. Antioxidant properties of melatonin. *Free Radic. Biol. Med.* 26: 557–565.

Marfak, A., Trouillas, P., Allais, D. P., Calliste, C. A., Cook-Moreau, J., and Duroux, J. L. 2004. Reactivity of flavonoids with 1-hydroxyethyl radical: A γ-radiolysis study *Biochim. Biophys. Acta* 1670: 28–39.

Maroz, A., Kelso, G. F., Smith, R. A. J., Ware, D. C., and Anderson, R. F. 2008. Pulse radiolysis investigation on the mechanism of the catalytic action of Mn(II)-pentaazamacrocycle compounds as superoxide dismutase mimetics. *J. Phys. Chem. A.*, 112: 4929–4935.

Mats, J. M., Segura, J. A., Alonso, F. J., and Marquez, J. 2008. Intracellular redox status and oxidative stress: Implications for cell proliferation, apoptosis, and carcinogenesis. *Arch. Toxicol.* 82: 273–299.

Mercero, J. M., Matxain, J. M., Lopez, X., Fowler, J. E., and Ugalde, J. M. 2002. Methoxyphenols-antioxidant principles in food plants and spices: Pulse radiolysis, EPR spectroscopy, and density functional theory calculations. *Int. J. Quantum Chem.* 90: 969–979.

Miao, J. L., Wang, W. F., Pan, J. X., Li, C. Y., Lu, R. Q., and Yao, S. D. 2001a. The scavenging reactions of nitrogen dioxide radical and carbonate radical by tea polyphenol derivatives: A pulse radiolysis study. *Rad. Phys. Chem.* 60: 163–168.

Miao, J. L., Wang, W., Pan, J., Han, Z., and Yao, S. 2001b. Pulse radiolysis study on the mechanisms of reactions of CCl3OO2 radical with quercetin, rutin and epigallocatechin gallate. *Sci. China Ser. B: Chem.* 44: 353–359.

Mishra, B., Priyadarsini, K. I., Sudheer Kumar, M., Unnikrishnan, M. K., and Mohan, H. 2003. Effect of o-glycosilation on the antioxidant activity and free radical reactions of a plant flavonoid chrysoeriol. *Bioorg. Med. Chem.* 11: 2677–2685.

Mishra, B., Priyadarsini, K. I., Bhide, M. K., Kadam, R., and Mohan, H. 2004. Reactions of superoxide radicals with curcumin: Probable mechanisms by optical spectroscopy and EPR. *Free Radic. Res.* 38: 355–362.

Mishra, B., Priyadarsini, K. I., Sudheerkumar, M., Unnikrishhnan, M. K., and Mohan, H. 2006a. Pulse radiolysis studies of mangiferin: A C-glycosyl xanthone isolated from *Mangifera indica. Rad. Phys. Chem.* 75: 70–77.

Mishra, B., Priyadarsini, K. I., and Harimohan. 2006b. Effect of pH on one-electron oxidation chemistry of organoselenium compounds in aqueous solutions. *J. Phys. Chem.* 110: 1894–1900.

Mishra, B., Sharma, A., Naumov, S., and Priyadarsini, K. I. 2009. Novel reactions of one-electron oxidized radicals of selenomethionine in comparison with methionine. *J. Phys. Chem. B* 113: 7709–7715.

Molina-Heredia, F. P., Houee-Levin, C., Berthomieu, C., Touati, D., Tremey, E., Favaudon, V. Adam, V., and Niviere, V. 2006. Detoxification of superoxide without production of H2O2: Antioxidant activity of superoxide reductase complexed with ferrocyanide. *Proc. Natl. Acad. Sci. USA* 103: 14750–14755.

Moorthy, P. N. and Hayon, E. 1976. One-electron redox reactions of water-soluble vitamins. II. Pterin and folic acid. *J. Org. Chem.* 41: 1607–1613.

Mortensen, A., Skibsted, L. H., Sampson, J., Rce-Evans, C., and Everett, S. A. 1997. Comparative mechanism and rates of free radical scavenging by carotenoid antioixdants. *FEBS Lett.* 418: 91–97.

Mvula, E., Scuchmann, M. N., and von Sonntag, C. 2001. Reactions of phenol-OH adduct radicals. Phenoxyl radical formation by water elimination vs. oxidation by dioxygen. *J. Chem. Soc. Perkin Trans.* 2: 264–268.

Neta, P. and Huie, R. E. 1988. Rate constants for reactions of inorganic radicals in aqueous solutions. *J. Phys. Chem. Ref. Data* 17: 1027–1284.

Neta, P., Huie, R. E., and Ross, A. 1990. Rate constants for reactions of peroxyl radicals in fluid solutions. *J. Phys. Chem. Ref. Data* 19: 413–513.

Niviere, V., Asso, M., Weill, C. O., Lombard, M., Guigliarelli, B., Favaudon, V., and Houee-Levin, C. 2004. Superoxide reductase from *Desulfoarculus baarsii*: Identification of protonation steps in the enzymatic mechanism. *Biochemistry* 43: 808–818.

O'Neill, P. and Chapman, P. W. 1985. Potential repair of free radical adducts of dGMP and dG by a series of reductants pulse radiolytic study. *Int. J. Radiat. Biol.* 47: 71–80.

O'Neill, P. and Wardman, P. 2009. Radiation chemistry comes before radiation biology. *Int. J. Radiat. Biol.* 85: 9–25.

Packer, J. E., Slater, T. F., and Wilson, R. L. 1979. Direct observation of a free radical interaction between vitamin E and vitamin C. *Nature* 278: 737–738.

Papp, L. V., Lu, J., Holmgren, A., and Khanna, K. K. 2007. From selenium to selenoproteins: Synthesis, identity, and their role in human health. *Antioxid. Redox Signal.* 9: 775–793.

Patro, B. S., Adhikari, S., Mukherjee, T., and Chattopadhyay, S. 2005. Possible role of hydroxyl radicals in the oxidative degradation of folic acid. *Bioorg. Med. Chem. Lett.* 15: 67–71.

Priyadarsini, K. I. 1997. Free radical reactions of curcumin in membrane models. *Free Radic. Biol. Med.* 23: 838–843.

Priyadarsini, K. I. 2000. Characteristic chemical reactions of nitric oxide. *Proc. Natl. Acad. Sci. India* 70: 339–352.

Priyadarsini, K. I., Guha, S. N., and Rao, M. N. A. 1998. Physico-chemical properties and anti-oxidant activity of methoxy phenols. *Free Radic. Biol. Med.* 24: 933.

Priyadarsini K. I., Khopde S. M., Kumar S. S., and Mohan H. 2002. Free radical studies of ellagic acid, a natural phenolic antioxidant. *J. Agric. Food Chem.* 50: 2200–2206.

Priyadarsini, K. I., Maity, D. K., Naik, G. H., Sudheer Kumar, M., Unnikrishnan, M. K., Satav, J. G., and Mohan, H. 2003. Role of phenolic O-H and methylene hydrogen on the free radical reactions and antioxidant activity of curcumin. *Free Radic. Biol. Med.* 35: 475–484.

Pryor, W. A. 1989. Vitamin E: The status of current research suggestions for future studies. *Ann. N. Y. Acad. Sci.* 570: 400–405.

Pryor, W. A., Stahl, W., and Rock, C. L. 2000. Beta-carotene from biochemistry to clinical trials. *Nutr. Rev.* 58: 53–59.

Richter, H. W. 1998. Radiation chemistry: Principles and applications. In *Photochemistry and Radiation Chemistry*, J. F. Wishart and D. G. Nocera (eds.), pp. 5–33. Washington, DC: American Chemical Society.

Roberts, J. E., Hu, D., and Wishart, J. F. 1988. Pulse radiolysis studies of melatonin and chloromelatonin. *J. Photochem. Photobiol. B: Biol.* 42: P125–P132.

Samuni, A., Goldstein, S., Russo, A., Mitchell, J. B., Krishna, M. C., and Neta, P. 2002. Kinetics and Mechanism of OH-radical and OH-adduct radical reactions of nitroxides with their hydroxyl amines. *J. Am. Chem. Soc.* 124: 8719–8724.

Santus, R., Patterson, L. K., Filipe, P., Morlire, P., Hug, G. L., Fernandes, A., and Mazire, J. C. 2001. Redox reactions of the urate radical/urate couple with the superoxide radical anion, the tryptophan neutral radical and selected flavonoids in neutral aqueous solutions *Free Radic. Res.* 35: 129–136.

Schafer, F. Q. and Buettner, G. R. 2001. Redox environment of the cell as viewed through the redox state of the glutathione disulfide/glutathione couple. *Free Radic. Biol. Med.* 30: 1191–1212.

Schoneich, C. 2003. Methionine oxidation by reactive oxygen species: Reaction mechanisms and relevance to Alzheimer's disease. *Biochim. Biophys. Acta* 1703: 111–119.

Schuler, R. H. 1977. Oxidation of ascorbate anion by electron transfer to phenoxyl radicals. *Radiat. Res.* 69: 417–433.

Shi, H. and Noguchi, N. 2000. Introducing natural antioxidants. In *Antioxidants in Food*, J. Pokorny, N. Yanishlieva, and M. Gordon (eds.), pp. 147–158. Washington, DC: CRC Press.

Shishodia, S., Sethi, G., and Aggarwal, B. B. 2005. Curcumin: Getting back to the roots. *Ann. N. Y. Acad. Sci.* 1056: 206–217.

Sies, H. 1997. Oxidative stress: Oxidants and antioxidants. *Exp. Physiol.* 82: 291–295.

Singh, U., Barik, A., and Priyadarsini, K. I. 2009. Reactions of hydroxyl radical with bergenin, a natural poly phenol studied by pulse radiolysis. *Bioorg. Med. Chem.* 17: 6008–6014.

Solar, S., Solar, W., and Getoff, N. 1984. Reactivity of OH with tyrosine in aqueous solution studied by pulse radiolysis. *J. Phys. Chem.* 88: 2091–2095.

Sonntag, C. V. 1987. *The Chemical Basis of Radiation Biology*. London, U.K.: Taylor & Francis.

Stasica, P., Ulanski, P., and Rosiak, J. M. 1998. Reactions of melatonin with radicals in deoxygenated aqueous solutions. *J. Radioanal. Nucl. Chem.* 232: 107–113.

Stojanovic, S. and Brede, O. 2002. Elementary reactions of the antioxidant action of trans-stilbene derivatives: Resveratrol, pinosylvin and 4-hydroxystilbene. *Phys. Chem. Chem. Phys.* 4: 757–764.

Stojanovic, S., Sprinz, H., and Brede, O. 2001. Efficiency and mechanism of the antioxidant action of trans-resveratrol and its analogues in the radical liposome oxidation. *Arch. Biochem. Biophys.* 391: 79–89.

Tak, J. K. and Park, J. W. 2009. The use of ebselen for radioprotection in cultured cells and mice. *Free Radic. Biol. Med.* 46: 1177–1185.

Tamba, M. and Torreggiani, A. 2004. Radiation induced effects in the electron beam irradiation of dietary flavonoids. *Radiat. Phys. Chem.* 71: 21–25.

Tamba, M., Torregiani, A., and Tubertini, O. 1995. Thiyl and thiyl-peroxyl radicals produced from the irradiation of antioxidant thiol compounds. *Radiat. Phys. Chem.* 46: 569–574.

Torreggiani, A., Trinchero, A., Tamba, M., and Taddei, P. 2005. Raman and pulse radiolysis studies of the antioxidant properties of quercetin: Cu(II) chelation and oxidizing radical scavenging. *J. Raman Spectrosc.* 36: 380–388.

Vajragupta, O., Boonchoong, P., Sumanont, Y., Watanabe, H., Wongkrajang, Y., and Kammasud, N. 2003. Manganese-based complexes of radical scavengers as neuroprotective agents. *Bioorg. Med. Chem.* 11: 2329–2337.

Valko, M., Rhodes, C. J., Moncol, J., Izakovic, M., and Mazur, M. 2006. Free radicals, metals and antioxidants in oxidative stress induced cancer. *Chem.-Biol. Interact.* 160: 1–40.

VanAcker, S. A. B. E., Koymans, L. M. H., and Bast, A. 1993. Molecular pharmacology of vitamin E, structural aspects of antioxidant activity. *Free Radic. Biol. Med.*, 15: 311–328.

Wardman, P. 1989. Reduction potentials of one-electron couples involving free radicals in aqueous solutions. *J. Phys. Chem. Ref. Data* 18: 1637–1755.

Weiss, J. F. and Landauer, M. R. 2003. Protection against radiation by antioxidant nutrients and phytochemicals. *Toxicology* 189: 1–20.

Winterbourn, C. C. 2008. Reconciling the chemistry and biology of reactive oxygen species, *Nat. Chem. Biol.* 4: 278–286.

Yen, G. C., Duh, P. D., and Tsai, H. L. 2002. Antioxidant and pro-oxidant properties of ascorbic acid and gallic acid. *Food Chem.* 79: 307–313.

Zhang, R., Goldstein, S., and Samuni, A. 1999. Kinetics of superoxide induced exchange and nitroxide antioxidant and their oxidized and reduced forms. *Free Radic. Biol. Med.* 26: 1245–1252.

Zhao, C., Shi, Y., Wang, W., Lin, W., Fan, B., Jia, Z., Yao, S., and Zheng, R. 2001. Fast repair of the radical cations of dCMP and poly C by quercetin and rutin. *Mutagenesis* 16: 271–275.

Zhao, C., Shi, Y., Wang, W., Lin, W., Fan, B., Jia, Z., Yao, S., and Zheng, R. 2002. Fast repair activities of quercetin and rutin toward dGMP hydroxyl radical adducts. *Radiat. Phys. Chem.* 63: 137–142.

Zhao, C., Shi, Y., Wang, W., Jia, Z., Yao, S., Fan, B., and Zheng, R. 2003. Fast repair of deoxythymidine radical anions by two polyphenols: Rutin and quercetin. *Biochem. Pharmacol.* 65: 1967–1971.

Zielonka, J., Gebicki, J., and Grynkiewicz, G. 2003. Radical scavenging properties of genistein. *Free Radic. Biol. Med.* 35: 958–965.

23 Computational Human Phantoms and Their Applications to Radiation Dosimetry

Kimiaki Saito
Japan Atomic Energy Agency
Tokai, Japan

CONTENTS

23.1 INTRODUCTION

Various computational human phantoms modeling human bodies have been developed for radiation dosimetry. The most important field of application of computational human phantoms is in radiation protection. The purpose of radiation protection is to protect people and the environment against the detrimental effects of radiation exposure without unduly limiting the desirable human actions that may be associated with such exposure (ICRP, 2007). In radiation protection, exposures to ionizing radiation are managed and controlled so that deterministic effects are prevented, and the risks of stochastic effects are reduced to the extent that can be reasonably achieved. For these purposes, radiation dose limits are set for workers and the public, and the so-called optimization of protection is performed to reduce exposure as low as can be reasonably achieved. To control and optimize doses, the effective dose defined as the sum of risk-weighted organ doses has been used. Various procedures to estimate radiation doses using experimental, theoretical, and simulation approaches are designated as radiation dosimetry.

It is difficult to measure the organ dose that is the total energy deposited in an organ or tissue divided by its mass; therefore, organ doses have been evaluated on computational human models called phantoms with simulation. Exposure conditions are classified into two dominant categories of external exposure and internal exposure. External exposure is the case where radiation sources exist outside of a human body, while internal exposure is the case where a human body is exposed to radiation sources inside the body due to the intake or inhalation of radionuclides. In both cases, dose coefficients that convert measurable or easily estimable quantities to organ doses and effective doses are prepared from simulations assuming typical exposure conditions.

A computational human phantom needs to have a body structure properly modeling a human body: the shapes and weights of dominant organs and tissues need to be modeled properly, and realistic elemental compositions and densities need to be assigned to the organs and tissues to ensure that radiation interactions inside the body and consequent energy depositions be simulated with demanded accuracy. In radiation protection, stylized phantoms in which the shapes of the body, organs, and tissues are expressed by combinations of mathematical equations have been widely used for a long time and provided many essential data. On the other hand, in recent years, voxel phantoms in which the organs and tissues are expressed as assemblies of small rectangular units called voxels have come to be developed and used in many ways. Voxel phantoms are usually constructed on the bases of images from computed tomography (CT) or magnetic resonance imaging (MRI); thus, realistic modeling can be attained.

In radiation protection, basic dosimetric data have been prepared for the reference Caucasian persons defined on the statistical analysis of comprehensive anatomical and physiological data for Caucasians (ICRP, 1975, 2003). Consequently, phantoms have been developed mostly for Caucasians. This motivated some researchers to investigate difference in doses due to race by developing Asian phantoms. It would be anticipated that external doses for Japanese are greater than those for Caucasians generally, since the Asian body size is smaller than that of Caucasians and the self-shielding effect is smaller for Asians. The Japan Atomic Energy Agency (JAEA) has developed several Japanese voxel phantoms (Saito et al., 2001, 2008; Sato et al., 2007a,b, 2009) and has investigated dose variation according to races and individuals. Korean (Lee et al., 2006) and Chinese voxel phantoms (Zhang et al., 2007) have also been developed in recent years. Phantoms developed in the world so far are reviewed in several publications (ICRU, 1992; Petoussi-Henss, 2002; Lemosquet et al., 2003; Zaidi and Xu, 2007).

The main target of phantoms has been to obtain reference data used in radiation protection; further, the utility of phantoms has been extended to a variety of research fields. These include medical applications both in diagnostics and therapy, dosimetry in radiation accidents, dosimetry for animals aimed at development of radiotherapeutics or environmental protection, evaluation of exposure to electromagnetic fields, etc.

This chapter will describe various phantoms utilized for radiation protection and other purposes, present examples of basic data obtained from simulation using phantoms, and discuss the present features and prospects.

23.2 DOSIMETRY IN RADIATION PROTECTION

In radiation protection, a specific dosimetric system has been developed on the basis of the absorbed dose averaged over an organ or tissue (organ dose). In this dosimetry system, the organ dose is modified considering biological effectiveness of radiations and organ-specific sensitivity so as to relate the radiation dose to the radiation risk (detriment) concerning stochastic radiation effects such as cancer and hereditary effects. The effective dose E, which was introduced in ICRP Publication 60 (ICRP, 1991) and in the succeeding Publication 103 (ICRP, 2007) is defined as:

$$E = \sum_{T} w_T H_T = \sum_{T} w_T \sum_{R} w_R D_{T,R} \tag{23.1}$$

where
   $D_{T,R}$ is the average absorbed dose in the volume of a specific organ or tissue T due to radiation of type R
   w_R is the radiation weighting factor for radiation R
   w_T is the tissue weighting factor for tissue T
   $\Sigma w_T = 1$
   H_T is the equivalent dose

The sum is performed over all organs and tissues of the human body considered in the effective dose. The unit of effective dose is $J\ kg^{-1}$ with the special name sievert (Sv). The basic concept on this kind of risk-weighted dose was already accomplished in Publication 26 (ICRP, 1977) as the effective dose equivalent H_E and was revised as the effective dose E.

The values of w_R shown in Table 23.1, taken from Publication 103 (ICRP, 2007), are based on the relative biological effectiveness (RBE) of the different radiations. However, basic data concerning RBE are not absolutely sufficient to determine reliable w_R especially for charged particles; further studies would be desirable.

Ideally, the radiation weighting to consider differences in biological effects due to radiation quality should be made at interaction points. In fact, in Publication 26 (ICRP, 1977) radiation weighting was recommended to be performed at the interaction point using the $Q(L)$ function, indicating relative biological effectiveness as a function of linear energy transfer (LET). However, in order to avoid the misunderstanding that the $Q(L)$ function does exactly reflect the biological radiation effects, a macroscopic manner was instead developed in Publication 60 (ICRP, 1991) and inherited.

Presently, in dose calculations, values of radiation weighting factors are determined at positions where the radiations enter a human body in external exposures, or at positions of emissions from radionuclides in internal exposures. This method is somewhat controversial, a radiation is considered to exert a different effect on a human body according to the initial position and direction, because the ways to deliver its energy to important organs are different according to the initial conditions.

The radiation weighting factor for photons and electrons is assumed to be 1 regardless of energy. It is recognized that biological effectiveness for low-energy electrons is higher than that

TABLE 23.1
Recommended Radiation Weighting Factors

Radiation Type	Radiation Weighting Factor, w_R
Photons	1
Electrons and muons	1
Protons and charged pions	2
Alpha particles, fission fragments, heavy ions	20
Neutrons	A continuous function of neutron energy

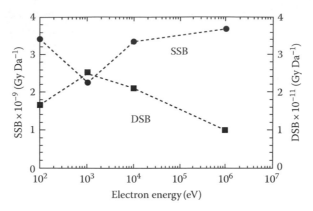

FIGURE 23.1 Yield of double strand breaks (DSBs) as a function of electron energy.

for high-energy electrons because of substantially higher LET. Figure 23.1 gives the yield of DNA double strand break (DSB) for electrons obtained from microscopic simulation on track structures (Watanabe and Saito, 2002) in a manner similar to those described in the references (Nikjoo and Uehara, 2004; Pimblott and Mozumder, 2004). Complex DNA damage represented by DSB is considered to be important in terms of biological effects, and the yield obviously changes with electron energy. The DSB yield for photons also changes according to energy because of the alteration in secondary electron energy spectrum. Further, it is reported that significant DNA damage tends to arise in case that Auger electrons are emitted after photoelectric absorption by elements constituting DNA (Watanabe et al., 2000; Kobayashi, 2004). Nevertheless, in radiation protection, the radiation weighting factor for photons and electrons is taken to be constant as 1 for simplicity.

The tissue weighing factors shown in Table 23.2 (ICRP, 2007) were determined to represent the contributions of individual organs and tissues to overall radiation detriment from stochastic effects. On the basis of epidemiological studies on cancer induction in exposed populations and risk assessments for heritable effects, a set of w_T values was chosen considering radiation detriment. They represent mean values for humans averaged over both sexes and a wide range of ages. To obtain effective doses, organ doses are calculated using computational human phantoms and weighted according to Equation 23.1.

Here, the basic dose quantities used in general prospective radiation protection are introduced. However, required dose quantities are different according to purpose. For example, in retrospective dosimetry for evaluating consequences of accidents, absorbed doses for specific persons should be

TABLE 23.2
Recommended Tissue Weighting Factors

Tissue	w_T	Σw_T
Bone marrow (red), colon, lung, stomach, breast, remainder tissues[a]	0.12	0.72
Gonads	0.08	0.08
Bladder, esophagus, liver, thyroid	0.04	0.16
Bone surface, brain, salivary glands, skin	0.01	0.04
	Total	1.00

[a] Remainder tissues: Adrenals, extrathoracic (ET) region, gall bladder, heart, kidneys, lymphatic nodes, muscle, oral mucosa, pancreas, prostate (♂), small intestine, spleen, thymus, uterus/cervix (♀).

evaluated but not effective doses for representative humans. In radiation therapy, dose distribution in and around a tumor is critical and organ doses are not always necessary.

23.3 COMPUTATIONAL HUMAN PHANTOMS

Computational human phantoms are required to fill certain conditions. First, the body structures need to properly model human anatomies, considering required dose specificity and accuracy. Generally, in radiation protection, effective doses and organ doses are calculated for the reference Caucasian male and female that have representative body sizes and organ masses defined by ICRP Publication 89 (ICRP, 2003). In this publication, a comprehensive amount of anatomical data for the Caucasian population was accumulated and analyzed, and the reference values of body height, weight, organ, and tissue masses were defined on the basis of these analyses. Phantoms used for radiation protection purposes basically need to have the reference anatomy.

Another important requirement is that the elemental compositions and densities of organs and tissues be appropriately considered. It depends on the type and energy of radiation considered in simulation and how precise the data must be given on elemental compositions. Generally speaking, skeletal parts need to be given their specific elemental compositions because they are obviously different from other soft tissues; they contain a relatively large number of heavy atoms that cause radiation interactions that are different from those caused by light atoms. The precise elemental compositions of organs and tissues are tabulated in ICRP Publication 89 (ICRP, 2003). Compositions of important tissues are tabulated in Table 23.3.

Computational phantoms are classified into two categories: one is stylized phantoms and the other is voxel phantoms. In this chapter, each type of phantom will be introduced with several typical examples.

23.3.1 Stylized Phantoms

In a stylized phantom, the outer shapes of the body, organs, and tissues are expressed by a combination of mathematical equations showing a plane, sphere, ellipsoid, cylinder, elliptical cylinder, cone, etc. For example, in Adam and Eva developed at GSF (Kramer et al., 1986) (Figure 23.2), the trunks and arms are represented by elliptical cylinders, and the legs by truncated circular cones. The special features of these kinds of phantoms are that the sizes of the body and organs can be easily changed by adjusting parameters used in the mathematical equations, and that organs can be located as desired provided they do not overlap. These features enable one to easily construct phantoms that have reference Caucasian anatomical properties defined by ICRP. Several stylized phantoms of the reference Caucasian have been constructed. This kind of phantom was developed as the MIRD-5 phantom at ORNL for internal dosimetry (Snyder et al., 1969), and since then many stylized phantoms have been constructed and employed in different ways.

TABLE 23.3
Composition of Human Tissues

Organ/Tissue	Elemental Composition (% by Mass)											
	H	C	N	O	Na	Mg	P	S	Cl	K	Ca	Fe
Fat	11.4	59.8	0.7	27.8	0.1		0.1	0.1				
Skin	10.0	20.4	4.2	64.5	0.2		0.1	0.2	0.3	0.1		
Muscle	10.2	14.3	3.4	71.0	0.1		0.2	0.3	0.1	0.4		
Lung	10.3	10.5	3.1	74.9	0.2		0.2	0.3	0.3	0.2		
Active marrow	10.5	41.4	3.4	43.9	0.1	0.2	0.2	0.2				0.1
Bone mineral	3.5	16.0	4.2	44.5	0.3	0.2	9.5	0.3			21.5	

Relations among some typical stylized phantoms are shown in Figure 23.3. The MIRD-5 phantom is a hermaphrodite (Snyder et al., 1969): it has the body size of the reference male and testis, ovaries, uterus, but not female breast. Cristy (1976) developed several phantoms with different sizes representing different ages for internal dosimetry; they were modified by Yamaguchi (1992) for external dosimetry. Japanese phantoms were developed by Kerr et al. (1976) to evaluate external organ doses in atomic bombardments at Hiroshima and Nagasaki. Sex-specific phantoms were constructed by Kramer et al. (1986) for external photon dosimetry and by Stewart et al. (1993) for external neutron dosimetry. A pregnant woman was modeled by Kai (1985) intended for emergency dosimetry in the environment. These stylized phantoms have provided precious data from the viewpoint of radiation protection.

23.3.2 Voxel Phantoms

According to the method employed, stylized phantoms have limits in modeling human anatomy: it is difficult to model realistic shapes and positional relations of organs and tissues in the body. While three-dimensionally precise CT or MRI medical images have become easily available, voxel phantoms based on the medical images for specific persons have come to be developed. Voxel phantoms consist of small rectangular units called voxels (volume pixels). The identification (ID) number for an organ or tissue is assigned to each voxel, and the assembly of voxels having the same ID number represent the organ or tissue. An assembly of voxels

FIGURE 23.2 Stylized phantoms Adam developed at GSF. (Courtesy of Maria Zankl, Helmholtz Zentrum Munich, Germany.)

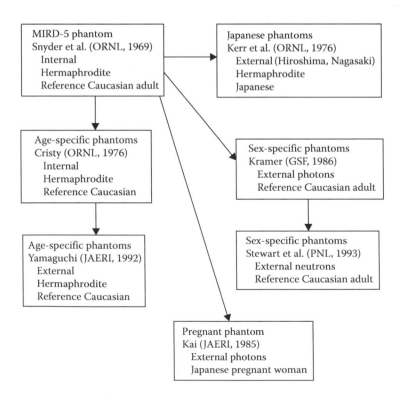

FIGURE 23.3 Typical stylized phantoms with different characteristics.

can model any structure within the resolution of the voxel size, and the voxel phantom can express human anatomy realistically because it is based on medical images of a real human.

The first voxel phantoms intended for radiation protection uses were Baby and Child developed at GSF (Zankl et al., 1988; Veit et al., 1989). Baby was constructed from the CT data of a dead 8-week-old baby; Child was constructed from the CT data of a 7-year-old child who suffered from leukemia. They have been used for external exposure evaluation in fundamental conditions (Zankl et al., 1988; Veit et al., 1989) and also in environmental conditions (Saito et al., 1990). After 2000, the number of voxel phantoms has increased gradually, especially in recent years, many voxel phantoms have been constructed. The main voxel phantoms developed so far are tabulated on the Web site of the Consortium of Computational Human Phantoms (Xu, 2008).

Most of the phantoms for radiation protection purposes, both stylized phantoms and voxel phantoms, have been constructed for Caucasians. Researchers at JAEA desired to clarify the difference in doses due to the anatomical differences between Caucasians and Asians. From this viewpoint, JAEA constructed the first Asian voxel model, Otoko (Saito et al., 2001). Since then, five Japanese voxel phantoms in total have been completed from CT data: three male adult phantoms and two female adult phantoms (Saito et al., 2001, 2008; Sato et al., 2007a,b, 2009). All the CT data were taken for healthy volunteers at the Fijita Health University Hospital after receiving the Ethics Committee's approval.

Pictures of voxel phantoms that have been developed are shown in Figure 23.4, and the physical characteristics are listed in Table 23.4. In these phantoms, compact bone and bone marrow are separately modeled in each skeletal voxel according to the CT value after the GSF method (Zankl et al., 1988; Veit et al., 1989), enabling users to calculate doses taking the bone marrow distribution in the body into account.

The first generation phantoms Otoko (Saito et al., 2001) and Onago (Saito et al., 2008) have a voxel size of $0.98 \times 0.98 \times 10\,\text{mm}^3$, while the second generation phantoms JM (Sato et al., 2007a)

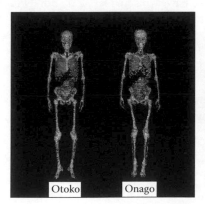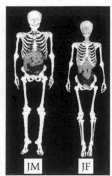

FIGURE 23.4 Japanese voxel phantoms developed at JAEA.

TABLE 23.4
Physical Characteristics of Voxel Phantoms Developed at JAEA

	Otoko	Onago	JM	JF	Asian Reference Man	
Gender	Male	Female	Male	Female	Male	Female
Weight (kg)	65	57	66	44	64	46
Height (cm)	170	162	171	152	170	155
Slice thickness (mm)	10	10	1	1	—	—
Pixel side length (mm)	0.98	0.98	0.98	0.98	—	—

and JF (Sato et al., 2009) have a finer voxel size of $0.98 \times 0.98 \times 1\,mm^3$. The performance of CT scanners greatly improved in a short period and made it possible to take CT pictures at a high resolution with less exposure, resulting in the development of the high-resolution phantoms.

Both the male phantoms Otoko and JM have body sizes close to the Asian Reference Man (ARM) defined by Tanaka (Tanaka and Kawamura, 1996). The body size of JF is also close to the Asian Reference Man, Female (ARMF), but Onago has larger body size than ARMF. We compared the body thicknesses and widths of the developed voxel phantoms with the reference values for the thorax, the abdomen, and the buttocks; they are all within 2σ deviations around the reference values, and it was confirmed that the developed voxel phantoms do not much deviate from the average body shapes of the Japanese.

In radiation protection, dose calculations are usually performed using an assumed upright position of human bodies, while voxel phantoms have been constructed from CT data taken in lying position. The structures of a human body and its organs are considered to change slightly according to the posture; therefore, to investigate the effect of posture on organ doses, the voxel phantom JM2 was constructed from CT data taken in an upright position for the same person as JM (Sato et al., 2007b). According to comparison of doses between JM and JM2, it was concluded that the effect of body posture is not significant from a viewpoint of radiation protection, though apparent dose difference is observed in some cases (Sato and Endo, 2008; Sato et al., 2008).

23.3.3 ICRP Reference Phantoms

In Publication 103, ICRP (2007) concretely defined the reference phantoms that should be used in dosimetry. Currently ICRP has not defined any specific reference phantoms even though the committee has published data on the reference anatomy (ICRP, 1975, 2003). Therefore, various reference phantoms had been used to obtain fundamental dosimetric quantities; some are hermaphrodite and some are sex specific; some have arms and some do not. Now ICRP specified the reference voxel phantoms of the adult reference male and female developed by Zankl et al. (ICRP, 2010). Therefore these voxel phantoms are required usage to obtain basic dosimetric data from now on. Since the phantoms were originally constructed from CT data for individuals, they had anatomies different from the reference male and female. Thus, the ICRP reference phantoms were adjusted by image processing to have body sizes and organ masses approximately the same as the reference Caucasian male and female. Namely, the reference body heights are 176 and 163 cm and the reference body weights are 73 and 60 kg for male and female, respectively (ICRP, 2003).

23.4 DOSE CALCULATION FOR RADIATION PROTECTION PURPOSES

23.4.1 Monte Carlo Radiation Transport Calculation

In order to obtain organ doses and related quantities using computational phantoms, radiation interactions inside phantoms are simulated using Monte Carlo transport calculation. For example, when a photon enters the phantom, it produces secondary electrons and deposits energy mainly through secondary electrons. In Monte Carlo simulation, all the trajectories of primary and secondary radiations are followed. Then an organ dose is calculated as the sum of energy deposited by all these radiations inside the organ divided by the mass. Radiation transport calculations can be carried out for different types of radiations by selecting an appropriate simulation code. Here, some explanations will be given about photon–electron transport calculation with reference to EGS4 (Nelson et al., 1985).

Interaction processes considered in EGS4 are photoelectric absorption, Compton scattering, pair production, coherent scattering for photons, energy loss by collisions and Bremsstrahlung, Moliere multiple scattering, Moller and Bhabha scatterings for electrons and positrons. When a pair production takes place, a positron–electron pair is produced and the positron emits two annihilation gamma gays of 0.511 MeV after it stops moving. Basically radiation interactions are probabilistic events submitted to specific distributions; therefore, in simulation, properties of radiation interactions and emitted secondary particles are determined randomly using random numbers and specific distributions of physical quantities. Then, energy deposited, to calculate the organ dose in an organ, is the sum of a large number of radiation incidence.

Monte Carlo code systems combined with the voxel phantoms for calculating doses and related quantities have been constructed (Funabiki et al., 2000, 2001; Saito et al., 2001; Kinase et al., 2003, 2007) as user codes of EGS4. In the whole code systems, we added the function to calculate radiation transport in voxel geometry that EGS4 originally did not contain.

23.4.2 External Dose Calculation

In this section, some typical results for external exposures are shown and discussed. In radiation protection for external exposures, usually monoenergetic irradiations of broad-parallel beams in idealized geometries are assumed. The geometries generally assumed are anterior posterior (AP), posterior anterior (PA), lateral (LAT, which represents both left lateral [LLAT] and right lateral [RLAT]), rotational (ROT), isotropic (ISO), which are schematically shown in Figure 23.5.

23.4.2.1 Photon

In Figure 23.6, examples of effective dose curves for external photon exposures are indicated (ICRP, 1996). Effective doses and organ doses are often normalized to air kerma that is a kind of absorbed dose in air assuming that secondary electrons entirely lose their energies at the origin points. From the figure, some typical features of effective doses are observed. Effective doses decrease rapidly below several tens of keV because of the self-shielding effect of the human body. Low-energy photons are absorbed by tissues at the body surface with a high probability; therefore, it is difficult for these to reach important organs existing in the middle of the body. On the other hand, in a high-energy region, photons can enter the body deeply and deposit energy rather homogeneously.

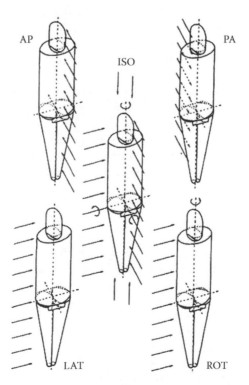

FIGURE 23.5 Idealized irradiation geometries of external exposure assumed in radiation protection.

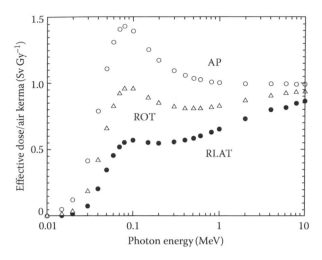

FIGURE 23.6   Variation in effective doses for external photons.

Thus, the effective dose curves are rather flat in the MeV region. The curve has a peak just below 100 keV, resulting from a build-up effect; contributions from scattered photons and secondary electrons are maximum in this energy region.

Further, it is clear that effective doses are different depending on geometry. Among the six geometries considered here, the effective dose is the highest in AP irradiation. This is because dominant organs having high radiation sensitivities exist mostly at the front side of the body and the AP beam directly irradiates these organs, while the doses are small in lateral geometries where photons need to travel long distances in the body to reach the dominant organs. The effective dose curves for PA, ROT, and ISO exist between those for AP and LAT.

Variations in organ doses for external photons were investigated using the developed voxel phantoms at JAEA (Sato et al., 2008). Organ doses were calculated for six typical external irradiation conditions of AP, PA, LLAT, RLAT, ISO, and ROT with 25 kinds of monoenergetic photons ranging from 0.01 to 10 MeV. In order to clarify the difference in organ doses between the Japanese and the Caucasian phantoms, analyses were made using the calculated data (Sato et al., 2008; Saito et al., 2009). The ratio of an organ dose in a Japanese phantom to that in the Caucasian phantom of the corresponding gender was calculated for the lungs, stomach, bladder wall, liver, esophagus, thyroids, skin, brain, colon, testes, breast, and ovaries. Organ doses for Otoko and JM were normalized to that for Rex (Schlattl et al., 2007), and organ doses for Onago and JF were normalized to that for Regina having body characteristies close to the references (Schlattl et al., 2007).

Figure 23.7 shows the maximum and minimum values of the ratios for AP, ROT, and LLAT geometries in the energy range of 50 keV to 3 MeV. This energy range was selected because the typical exposure in the environment is due to photons within this range. The data indicate that organ dose differences between the Japanese and the Caucasian phantoms vary within a factor of 2 except for lateral geometries. On an average, organ doses were found to be slightly larger for Japanese, which is considered to be due to differences in body size; and relatively large differences are observed in lateral geometries. However, the difference in organ doses averaged over diverse organs and conditions was found to be very small.

Figure 23.8 shows energy dependencies of thyroid doses for Onago, JF, and Regina in LLAT geometry. The positions of the thyroids tend to change greatly according to individuals, and this could vary the dose. The positions are clearly different even among Japanese phantoms. The thyroids of Regina are expected to locate at deeper positions where the shielding effect by the shoulder is significant in lateral irradiation geometries, though the detailed anatomical data have not been released on Regina. It is not yet clear if the large difference is due to racial difference or due to individual difference, and further analysis is necessary for understanding the systematic difference.

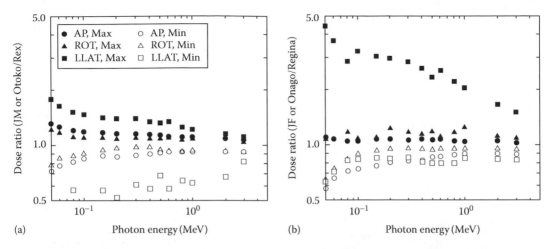

FIGURE 23.7 Range of the organ dose ratios of the Japanese phantoms to the Caucasian phantoms. (a) Male and (b) female.

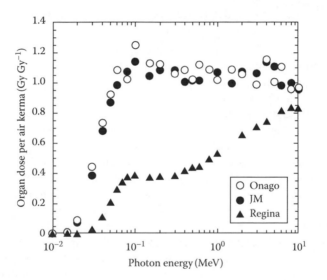

FIGURE 23.8 Comparison of thyroid doses among Onago, JF, and Regina for external photon exposure in LLAT geometry.

In total, concerning individual organs, the dose difference among different phantoms comes up to a several factor. The difference in effective doses between the Japanese and the Caucasian phantoms was within 10% for almost all the cases. It must be noted that, in case of partial irradiation by narrow low-energy beams, organ doses change drastically by the incident position and angle (Ohnishi et al., 2004). In such cases, an appropriate individual phantom is considered necessary to obtain realistic organ doses and effective doses.

23.4.2.2 Electron

Features of energy deposition by incidences of electrons or charged particles are greatly different from those of photons. Electrons have finite penetration ranges and it makes a large difference whether they reach target organs or not. Saito et al. (2008) calculated external electron doses for Otoko and Onago in AP, PA, and ISO geometries and compared the data with a MIRD-type phantom by Ferrari et al. (1997). Here, organ doses were normalized to incident electron fluence.

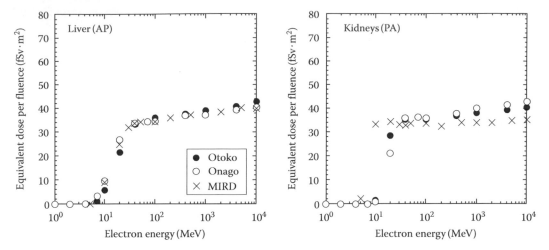

FIGURE 23.9 Examples of organ doses for external electron exposure calculated using Otoko, Onago, and MIRD phantoms. (From Saito, K. et al., *Jpn. J. Health Phys.*, 43, 122, 2008. With permission.)

Two examples of the dose coefficients for the three phantoms are shown in Figure 23.9. The liver doses in AP geometry show a good agreement among the three phantoms; while the kidney doses in PA geometry show apparent difference between the MIRD phantom and the voxel phantoms. In external electron exposure, organ depth is a dominant factor to determine the energy dependency of the dose coefficients. The figure indicates that the effective depth is quite similar for three phantoms in the case of liver, but obviously different in the case of kidney.

Generally, the energy dependency curve of organ dose for electrons consists of three regions with different tendencies: (1) At low energy, the organ dose is very small; (2) at middle energy, the curve steeply increases with energy; and (3) at high energy, the dose is nearly constant or slightly increases with energy.

In region (1), the organ dose is close to zero, since electrons do not reach the target organ because their penetration ranges are too short. However, even in this energy region an extremely small amount of energy is delivered to the target organ. This is considered mainly due to Bremsstrahlung radiation emitted by electron transport. In region (2), the dose increases rapidly with energy, because electron penetration becomes great enough that some of the electrons can reach the organ and deposit energy directly. The electron energy at which this rapid increase starts varies according to the depth of the organ from the body surface. In region (3), the dose saturates and exhibits a slower increase. In this region the electron penetration range exceeds the order of body thickness, and electrons deposit their energy rather homogeneously in the body. The slow increase of dose is considered mainly due to buildup of Bremsstrahlung radiation.

Figure 23.10 shows organ doses for Otoko and Onago normalized to those for the MIRD phantom. Large differences between different phantoms are observed around 10 MeV where the electron range is several cm; dominant organs exist mostly at depths of several cm. The maximum dose difference observed in the study was a factor of 50 in kidney dose in PA geometry. A factor of 50 does not have any concrete meaning, but the electron dose for an individual organ sometimes shows a large discrepancy between different phantoms. The maximum difference observed in effective doses was a factor of two, which coincides with data presented by Kramer (Kramer, 2005).

23.4.3 INTERNAL DOSE CALCULATION

23.4.3.1 Absorption Fraction and Specific Absorption Fraction

Figure 23.11 gives a scheme concerning internal exposure. Radionuclides accumulated in a source organ emit gamma rays, beta rays, and in some cases alpha rays. The emitted radiations deposit their energy inside the source organ itself and also in other target organs. Then, the fraction of

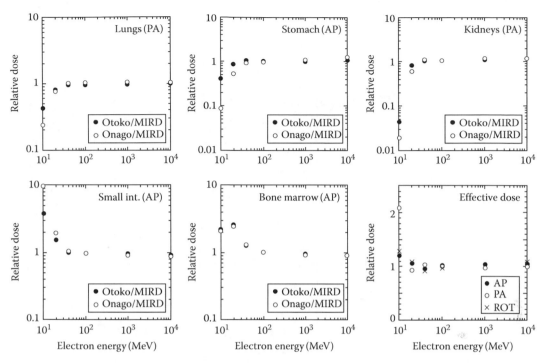

FIGURE 23.10   Organ doses and effective doses for Otoko and Onago normalized to those for a MIRD phantom in external electron exposure. (From Saito, K. et al., *Jpn. J. Health Phys.*, 43, 122, 2008. With permission.)

energy per emitted energy deposited in the target organ is designated as absorption fraction (AF), and the AF divided by the mass of the target organ is the specific absorption fraction (SAF). The AF and SAF are basic quantities for the evaluation of internal doses for monoenergetic radiations prepared using computational phantoms. Further, by integrating AFs or SAFs considering types and energy distributions of emitted radiations, S-values are obtained to convert radionuclide concentration to organ doses or effective doses.

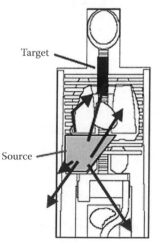

AF:
Fraction of energy absorbed in the target organ per energy emitted from the source organ

The SAFs obtained from calculations using MIRD phantoms have been currently used. However, positional relation among organs, which is an important factor to determine SAFs, could change significantly according to the phantom; MIRD-type phantoms may not reflect positional relations among organs in human bodies accurately. Thus, SAF calculations using sophisticated phantoms are desired to accurately evaluate internal doses.

Sato et al. (2005) calculated self-absorption fractions (self-AFs) using Otoko, JM, and a MIRD phantom. In self-AFs, the target organ is taken as the same as the source organ. Examples of self-AFs are shown for six organs in Figure 23.12. It was found that self-AFs are similar among different phantoms for large organs having simple identical shapes, such as brain and kidneys. In case of small organs, however, the sizes and shapes affect the self-AFs significantly. When the shapes of small organs are identical, the portion of energy absorbed by the organ itself is largely affected by the size. The difference in spleen doses could be explained by the sizes, while in pancreas, thyroid, and urinary bladder wall, the self-AFs are not explained by the size, and this indicated that shape is an important factor for these cases.

FIGURE 23.11   Schematic representation of the absorption fraction (AF) in internal exposure.

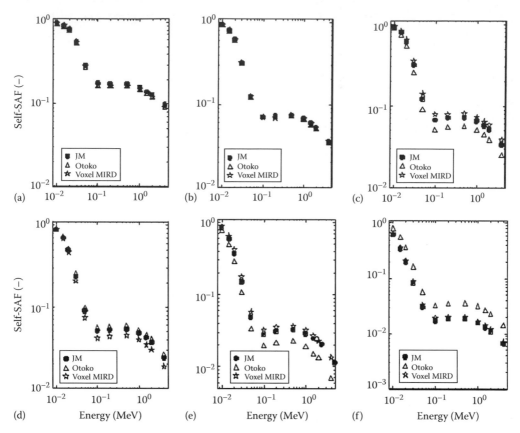

FIGURE 23.12   Self-AFs for six different organs calculated using Otoko, JM, and MIRD phantoms. (a) Brain, (b) kidneys, (c) spleen, (d) pancreas, (e) thyroid, and (f) urinary bladder wall.

Sato et al. (2005) examined the relation of SAFs to organ distance using Otoko, Onago, JF, and JM. In Figure 23.13, SAFs for several different combinations of source and target organs in the four Japanese voxel phantoms are plotted together as functions of distance between the centers of gravity. The distance between the centers of gravity may not be the most suitable parameter to analyze SAFs in every case; nevertheless, it was confirmed that SAFs would be roughly expressed as a function of this distance.

23.4.3.2   S-Value

Kinase and Saito (2007) and Kinase et al. (2004) evaluated S-values, which are the mean absorbed doses per unit cumulated activity to the target organ, from uniformly distributed radioactivity within the source organ, for several beta-ray emitters. S-values have been used for dose estimates in radiological protection and also in medical diagnostic and treatment procedures. In particular, S-values for positron emitters in the brain, heart, and urinary bladder play an important role for accurate dose evaluation of patients with administered radiopharmaceutical for clinical PET imaging. S-values for positron emitters within the urinary bladder are currently derived on the simple assumption that the dose at the surface of the content is approximately half the dose within their volume.

Figure 23.14 shows the ratios of the S-values derived from the currently used simple assumption by ICRP to those by Monte Carlo calculations using Otoko and Onago. The S-values for the urinary bladder wall for several beta-ray emitters in the bladder contents have been evaluated. The ratios increase as beta-ray energy decreases. The discrepancies between Otoko and Onago may be explained by the different ratios of the bladder wall mass to the bladder contents mass. It was

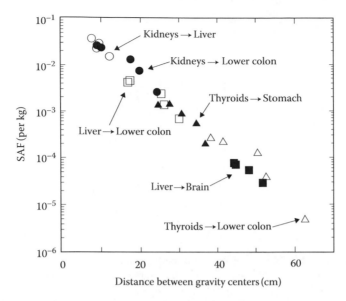

FIGURE 23.13 SAFs as a function of distance between the gravity centers. Data calculated using Otoko, Onago, JM, and JF are shown together.

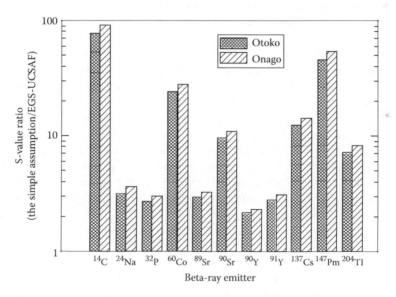

FIGURE 23.14 Ratios of the S-values for positron emitters derived from the currently used method to those by Monte Carlo calculations with voxel models. (From Kinase, S. et al., *J. Nucl. Sci. Technol. Suppl.*, 4, 136, 2004. With permission.)

confirmed that the S-values from the simple assumption currently used by ICRP are conservative. However, the difference is quite large from the realistic value, and realistic dose evaluation would be desirable in the future.

23.4.4 Environmental Exposure

Usually in radiation protection dosimetry for external exposure, it is assumed that reference adults are irradiated by monoenergetic radiations in simplified geometries. When we consider exposure in the environment, the situations are different. Radionuclides are distributed in the environment

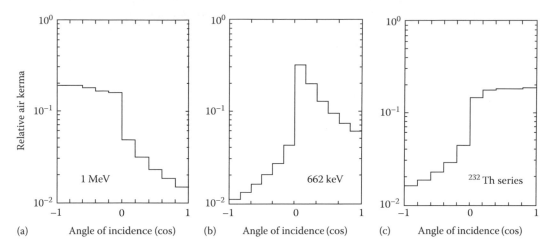

FIGURE 23.15 Angular distributions of environmental gamma rays at 1 m above the ground for three typical environmental source distributions. (a) Volume source in air, (b) plane source in ground, and (c) volume source in ground.

with their specific distributions depending on the origin; thus, the photons incident to a human body have angular and energy distributions even for the case where sources emit monoenergetic photons. Further, people covering a wide range of age exist in the environment, and this is different from exposure in nuclear facilities.

Saito et al. simulated exposure of adults and children in the environment assuming three typical source distributions: (a) uniformly distributed volume source in the air; (b) uniformly distributed plane source in the ground; and (c) uniformly distributed volume source in the ground (Saito et al., 1990, 1991, 1998; Petoussi et al., 1991; Zankl et al., 1997). Sources (a) and (b) are modeled on anthropogenic radionuclides released from nuclear facilities. Soon after the release of radionuclides into the environment, they stay in the air; source (a) simulates this situation. If some successive release of radionuclides arises, they deposit on the ground; source (b) simulates this situation. Source (c) models natural radionuclides of ^{238}U series, ^{232}Th series, and ^{40}K in the soil. A human body was assumed to stand vertically on the ground, and transport of photons in the environment and in the body were both accurately simulated.

Figure 23.15 gives angular distributions of air kerma at 1 m above the ground as cosines to a vector normal to the ground. Each source has its specific angular distribution. In case of volume sources, that is sources (a) and (c), gamma rays enter the body in isotropic directions from the half space where source exists, and small amounts of scattered gamma rays enter the body from the other half space. In plane source on the ground, the dominant part of gamma rays enters the body from the horizontal directions. This is because the amount of source per unit solid angle increases greatly at horizontal directions.

In Figure 23.16 are indicated effective doses normalized to air kerma in sources (a) and (b) for adults, a 7-year-old child, and an 8-week-old baby. It was confirmed that effective doses are higher for infants, and the maximum dose difference between the adult and the baby is a factor of two above 50 keV where exposure in the environment is generally significant. Further, the maximum dose difference for a single organ was found to be a factor of 3 (Saito et al., 1991). In source (c), effective doses for the baby were found, some 70% higher that those for adult. Further, the variation in effective doses due to several reasons such as bias of source distribution, human body posture, and relational change to the source have been also investigated (Saito et al., 1998).

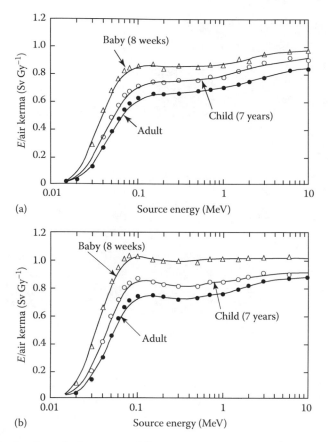

FIGURE 23.16 Effective doses for humans at different ages due to environmental gamma rays. (a) Volume source in air and (b) plane source in ground.

23.5 APPLICATION TO OTHER FIELDS

23.5.1 APPLICATION TO RADIATION THERAPY

Radiation therapy is an important application of dose calculation technique using voxel phantoms with which accurate individual doses can be obtained. It is recommended that the dose accuracy be within 5% in radiation therapy (ICRU, 1976), and this means that the net error in dose calculation should be within a few percent. This dose accuracy would be difficult to attain by conventional analytical methods in some cases where the elemental composition and density change drastically according to position in the patient body, since the disequilibrium of electrons could lead to significant errors in the dose calculation.

Some commercial dose planning systems have been released that utilize Monte Carlo calculations potentially able to perform accurate simulations. However, Monte Carlo calculations need enormous computational time, and also users need some basic knowledge to properly perform the calculations. These requirements have prevented the Monte Carlo systems from wide spread use in Japan.

Saito et al. (2004, 2006) developed a system for providing accurate dose distributions through networks to multiple medical facilities based on voxel phantoms and Monte Carlo calculations. The basic concepts of the system design are to provide reliable accurate doses for any conditions so that the calculated doses can be used as standard values and to construct an environment where users

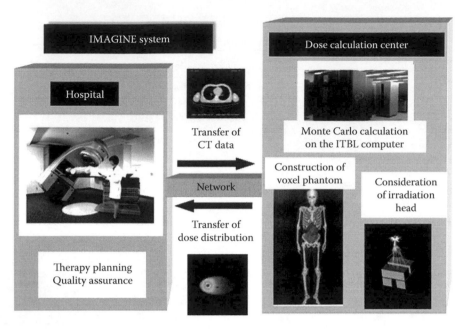

FIGURE 23.17 Schematic diagram of the dose calculation system IMAGINE for supporting medical facilities through the network.

without basic knowledge can easily employ the system. The schematic diagram of the system is shown in Figure 23.17. It is designed so that all dose calculations are performed at the dose calculation center using a high-performance computer. First, a user sends a data set of CT images and a treatment plan to the center through the network; then, a simplified voxel phantom of the patient is automatically constructed in a short time; and a Monte Carlo calculation is performed utilizing super-parallel computing; then the calculated dose distribution is sent back to the clinic and used for therapy planning or QA/QC.

Concerning voxel phantoms on patients, we investigated how many tissues having different elemental compositions are necessary to obtain sufficient dose accuracy. Consequently, it was confirmed that a phantom consisting of six tissues of skin, muscle, adipose, lung, cortical bone, and bone marrow can give doses with very high accuracy. Obviously inaccurate doses are obtained when other tissues are substituted for skeletal tissue that has a high proportion of the relatively high atomic number element Ca. Further, it was found that substitution of adipose tissue by other tissues could cause significant error in some cases.

23.5.2 Dosimetry in Accidents

Electron spin magnetic resonance (ESR) dosimetry using tooth enamel has been utilized in retrospective dose assessments in accidents where no other dosimetry method is available. Accumulated dose in tooth enamel can be estimated from the ESR signals. However, the information needed in accidents is not the enamel dose but doses for other important organs; then, conversion coefficients from the tooth enamel dose to organ doses must be prepared. Further, the variations of the enamel dose due to tooth position, irradiation conditions such as photon energy and incident direction should be investigated for reliable measurements. For this purpose, a series of studies using phantoms have been carried out.

Takahashi et al. (2002, 2003) investigated the characteristics of tooth doses using both a physical phantom and computational phantoms. A voxel phantom was constructed from CT images of a realistic physical phantom having real human teeth. An example of CT images on this phantom is

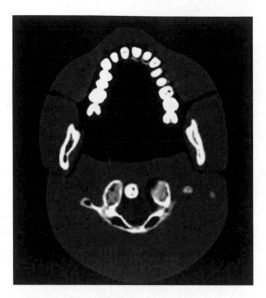

FIGURE 23.18   CT image of the physical phantom used for analyses of tooth enamel doses for external photon exposure.

shown in Figure 23.18. Thermoluminescence dosimeters (TLDs) were placed at the different tooth positions in the physical phantom to evaluate enamel doses. Then irradiation experiments were carried out and compared with simulation using the constructed voxel phantom. They showed good agreement, further, various characteristics of tooth doses were clarified by simulations under various conditions. Also, conversion coefficients from the tooth enamel dose to important organ doses have been prepared. This study was extended by Ulanovsky et al. (2005).

23.5.3   ANIMAL PHANTOMS

In recent years, animal phantoms have been developed and used for diverse purposes. A mouse phantom was developed by Dogdas et al. (2007) at the University of South California and modified by Kinase et al. (2008) at JAEA for dosimetry use. The phantom called Digimouse was segmented from CT and cryosection data. The voxel size is $100 \times 100 \times 100 \,\mu m^3$ and the body weight is 28 g. Further, Kinase (2008) constructed a frog phantom based on segmented images from cryosection data provided on a website of Lawrence Berkley National Laboratory (Johnson and Robertson, 2008). The frog phantom shown in Figure 23.19 has a voxel size of $250 \times 250 \times 250 \,mm^3$, body weight of 35 g with 15 organs and tissues.

The mouse phantom is intended for use mainly for radiopharmaceutical development (Kinase et al., 2009). In developing a pharmaceutical, the effect examined using animal experiments needs to be extrapolated to humans, considering the doses delivered to the tumor and organs. Then, accurate doses should be estimated both for the animal and humans.

The objective of the frog phantom is to provide fundamental data on environmental protection for animals and plants, the necessity of which was clearly stated in the new ICRP Publication (ICRP, 2007). The frog is among the reference animals and plants for the purpose of environmental protection (Pentreath, 2005). Presently, dose evaluations for the reference animals and plants are performed using extremely simplified phantoms, which, except for some exceptions, do not have explicit organs. Thus, internal dosimetry is largely impossible with these simplified phantoms.

Figure 23.20 compares AFs for photons and electrons among the mouse, the frog, and two human phantoms (Kinase, 2008; Kinase et al., 2008, 2009). Here, kidneys were taken as the source and

FIGURE 23.19 Frog phantom developed by Kinase at JAEA. (From Kinase, S., *J. Nucl. Sci. Technol.*, 45, 1049, 2008. With permission.)

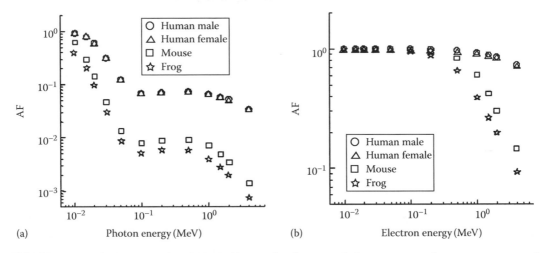

FIGURE 23.20 Comparison of self-AFs in kidneys for photons and electrons among humans, a mouse and a frog. (a) Photon and (b) electron.

target organs. The self-AFs for photons show a large difference between the animal phantoms and the human phantoms reflecting their organ sizes. While the self-AFs for electrons are identical at low energy below a few hundred keV because the electron penetration range is very short, they deviate greatly as the electron energy increases. When comparing animal doses to human doses, these features must be taken into account.

23.6 SUMMARY

In radiation protection, computational phantoms have played an important role in helping to obtain fundamental data necessary for dosimetry. The use of phantoms started with stylized phantoms that express human bodies with mathematical equations, and developed into more realistic voxel phantoms based on medical images. These phantoms have been employed to calculate organ doses and effective doses with Monte Carlo simulations both in external and internal exposure. In radiation protection, dose quantities have been prepared mostly for the reference Caucasian adults, while Japanese voxel phantoms have been developed to investigate difference between the Caucasian and the Japanese phantoms. It has been confirmed that the dose difference due to race is not significant from a viewpoint of radiation protection. On the other hand, individual organ doses can vary greatly according to conditions, and this suggests that sophisticated phantoms must be used if accurate doses for specific individuals are needed.

Voxel phantom techniques have proved to be quite useful and applied to various fields. These cover quite a wide range of fields such as medical applications both in diagnostics and therapy, retrospective dosimetry in accidents, exposure assessment of electromagnetic fields, environmental studies, and the development of radiopharmaceuticals. Thus, many voxel phantoms have been developed including animal phantoms. Obviously one important application is in the medical field. An example of the dose calculation system for x-ray therapy was shown in this chapter; meanwhile, a therapy-planning system for boron neutron capture therapy (BNCT) has been developed at JAEA and played an important role in BNCT in Japan, though details are not described here. Further, dose evaluation for radiation diagnoses would also be important, since advanced diagnoses leading to relatively high doses such as CT, PET, and SPECT have begun to be used.

Furthermore, various advanced phantoms are being developed, though all of these could not be discussed in this chapter. These include 4D phantoms that move with time, hybrid phantoms that have merits of both stylized and voxel phantoms, multi-scale phantoms that consider finer structures of organ and tissues, etc. Among these, the development of multi-scale phantoms that perform microdosimetric simulation at cellular levels besides current macroscopic simulation is an important subject to relate dosimetry more directly to radiation effects; and in this simulation the knowledge compiled in this book would be widely utilized.

On the basis of these versatile technical improvements and expanding demands, applications of computational human phantoms would be expected to increase more and more in the future.

ACKNOWLEDGMENTS

The author would like to thank Dr. K. Sato, Dr. S. Kinase, Dr. F. Takahashi, and Dr. R. Watanabe for their kind support in preparing materials. The author is grateful to Dr. A. Endo for his valuable advice on this chapter.

REFERENCES

Cristy, M. 1976. Mathematical phantoms representing children of various ages. ORNL/NUREG/TM-5336, Oak Ridge National Laboratory, Oak Ridge, TN.

Dogdas, B., David, S., Chatziioannou, A. F., and Leahy, R. M. 2007. Digimouse: A 3D whole body mouse atlas from CT and cryosection data. *Phys. Med. Biol.* 52: 577–587.

Ferrari, A., Pelliccioni, M., and Pillon, M. 1997. Fluence to effective dose and effective dose equivalent conversion coefficients for electrons from 5 MeV to 10 GeV. *Radiat. Prot. Dosimetry* 69: 97–104.

Funabiki, J., Terabe, M., Zankl, M., Koga, S., and Saito, K. 2000. An user code with voxel geometry and a voxel phantom generation system. In *KEK Proceedings 2000-20*, Tsukuba, Japan, pp. 56–63.

Funabiki, J., Saitoh, H., Sato, O., Takagi, S., and Saito, K. 2001. An EGS4 Monte Carlo user code for radiation therapy planning. In *KEK Proceedings 2001-22*, Tsukuba, Japan, pp. 80–86.

ICRP (International Commission on Radiological Protection). 1975. Report of the Task Group on Reference Man. Publication 23.

ICRP (International Commission on Radiological Protection). 1977. The 1977 Recommendation of the International Commission on Radiological Protection, Publication 26.

ICRP (International Commission on Radiological Protection). 1991. The 1990 Recommendation of the International Commission on Radiological Protection, Publication 60.

ICRP (International Commission on Radiological Protection). 1996. Conversion coefficients for use in radiological protection against external radiation. Publication 74.

ICRP (International Commission on Radiological Protection). 2003. Basic anatomical and physiological data for use in radiological protection: Reference values, Publication 89.

ICRP (International Commission on Radiological Protection). 2007. The 2007 Recommendation of the International Commission on Radiological Protection, Publication 103.

ICRP (International Commission of Radiological Protection). 2009. Adult reference computational phantoms. Publication 110.

ICRU (International Commission on Radiation Units and Measurements). 1976. Determination of absorbed dose in a patient irradiated by beams of X or gamma rays in radiotherapy procedures. ICRU Report 24, ICRU, Bethesda, MD.

ICRU (International Commission on Radiation Units and Measurements). 1992. Phantoms and computational models in therapy, diagnosis and protection, ICRU Report 48. ICRU, Bethesda, MD.

Johnson, W. and Robertson, D. 2008. Whole Frog Project. http://froggy.lbl.gov/

Kai, M. 1985. Estimation of embryonic and fetal doses from accidentally released radioactive plumes. *Radiat. Prot. Dosimetry* 2: 91–94.

Kerr, G. D., Hwang, J. M. L., and Jones, R. M. 1976. A mathematical model of a phantom developed for use in calculation of radiation dose to the body and major internal organs of a Japanese adult. ORNL/TM-5536, Oak Ridge National Laboratory, Oak Ridge, TN.

Kinase, S. 2008. Voxel-based frog phantom for internal dose evaluation. *J. Nucl. Sci. Technol.* 45: 1049–1052.

Kinase, S. and Saito, K. 2007. Evaluation of self-dose S values for positron emitters in voxel phantoms. *Radiat. Prot. Dosimetry* 127: 197–200.

Kinase, S., Zankl, M., Kuwabara, J., Sato, K., Noguchi, H., Funabiki, J., and Saito, K. 2003. Evaluation of specific absorbed fraction in voxel phantoms using Monte Carlo simulation. *Radiat. Prot. Dosimetry* 105: 557–563.

Kinase, S., Zankl, M., Funabiki, J., Noguchi, H., and Saito, K. 2004. Evaluation of S values for beta-ray emitters within the urinary bladder. *J. Nucl. Sci. Technol.* Suppl. 4: 136–139.

Kinase, S., Takagi, S., Noguchi, H., and Saito, K. 2007. Application of voxel phantoms and Monte Carlo method to whole-body counter calibration. *Radiat. Prot. Dosimetry* 125: 189–193.

Kinase, S., Takahashi, M., and Saito, K. 2008. Evaluation of self-absorbed doses for the kidneys of a voxel mouse. *J. Nucl. Sci. Technol.* (Suppl. 5): 268–270.

Kinase, S., Matsuhashi, S., and Saito, K. 2009. Interspecies scaling of self-organ doses from a voxel mouse to voxel humans. *Nucl. Technol.* 168: 1–4.

Kobayashi, K. 2004. Photon-induced biological consequences. In *Charged Particles Interactions with Matter: Chemical, Physicochemical, and Biological Consequences with Applications.* A. Mozumder and Y. Hatano, (eds.), pp. 471–489. New York: Marcel Dekker.

Kramer, R. 2005. Effective dose ratios for tomographic and stylized models from external exposure to electrons. In *Proceedings of the Monte Carlo Method*: *Versatility Unbounded in A Dynamic Computing World Chattanooga*, Chattanooga, TN, April 17–21, 2005, on CD-ROM, LaGrange Park, IL: American Nuclear Society.

Kramer, R., Zankl, M., Williams, G., and Drexler, G. 1986. The calculation of dose from external photon exposures using reference human phantoms and Monte Carlo Methods. Part I: The male (Adam) and female (Eva) adult mathematical phantoms. GSF-Bericht S-885, GSF—National Research Center for Environment and Health, Neuherberg, Germany.

Lee, C., Lee, C., Park, S. H., and Lee, J. K. 2006. Development of the two Korean tomographic computational phantoms for organ dosimetry. *Med. Phys.* 33: 380–390.

Lemosquet, A., De Carlton, L., and Clairand, I. 2003. Voxel anthropomorphic phantoms: Review of models used for ionizing radiation dosimetry. *Radioprotection* 38: 509–528.

Nelson, W. R., Hirayama, H., and Rogers, D. W. O. 1985. The EGS4 code system. SLAC Report SLAC-265, Stanford Linear Accelerator Center, Stanford University, Stanford, CA.

Nikjoo, H. and Uehara, S. 2004. Track structure studies of biological systems. In *Charged Particles Interactions with Matter: Chemical, Physicochemical, and Biological Consequences with Applications.* A. Mozumder and Y. Hatano (eds.), pp. 491–531. New York: Marcel Dekker.

Ohnishi, S., Odano, N., Nariyama, N., and Saito, K. 2004. Analysis of localised dose distribution in human body by Monte Carlo code system for photon irradiation. *Radiat. Prot. Dosimetry* 111: 65–71.

Pentreath, R. J. 2005. Concept and use of reference animals and plants. In *Proceedings of the Protection of the Environment from the Effects of Ionizing Radiation*, Stockholm, Sweden, IAEA-CN-109. Vienna, Austria: IAEA, pp. 411–420.

Petoussi, N., Zankl, M., Jacob, P., and Saito, K. 1991. Organ doses for foetuses, babies, children and adults from environmental gamma rays. *Radiat. Prot. Dosimetry* 37: 31–41.

Petoussi-Henss, N., Zankle, M., Fill, U., and Regulla, D. 2002. The GSF family of voxel phantoms. *Phys. Med. Biol.* 47: 89–106.

Pimblott, S. M. and Mozumder, A. 2004. Modeling of physicochemical and chemical processes in the interactions of fast charged particles with matter. In *Charged Particles Interactions with Matter: Chemical, Physicochemical, and Biological Consequences with Applications*. A. Mozumder and Y. Hatano (eds.), pp. 75–120. New York: Marcel Dekker.

Saito, K., Petoussi-Henss, N., Zankl, M., Veit, R., Jacob, P., and Drexler, G. 1990. Calculation of organ doses from environmental gamma rays using human phantoms and Monte Carlo methods. Part I: Monoenergetic sources and natural radionuclides in the ground. GSF-Bericht 2/90, GSF—National Research Center for Environment and Health, Neuherberg, Germany.

Saito, K., Petoussi-Henss, N., Zankl, M., Veit, R., Jacob, P., and Drexler, G. 1991. Organ doses as a function of body weight for environmental gamma rays. *J. Nucl. Sci. Technol.* 28: 627–641.

Saito, K., Petoussi-Henss, N., and Zankl, M. 1998. Calculation of the effective dose and its variation from environmental gamma ray sources. *Health Phys.* 74: 698–706.

Saito, K., Wittmann, A., Koga, S., Ida, Y., Kamei, T., Funabiki, J., and Zankl, M. 2001. Construction of a computed tomographic phantom for a Japanese male adult and dose calculation system. *Radiat. Environ. Biophys.* 40: 69–76.

Saito, K., Kunieda, E., Narita, Y., Deloar, H. M., Kimura, H., Hirai, M., Fujisaki, T., Myojoyama, A., and Saitoh, H. 2004. Development of the accurate dose calculation system IMAGINE for remotely aiding radiotherapy. In *KEK Proceedings 2003-15*, Tsukuba, Japan, pp. 81–87.

Saito, K., Kunieda, E., Narita, Y., Kimura, H., Hirai, M., Deloar, H. M., Kaneko, K.,Ozaki, M., Fujisaki, T., Myojoyama, A., and Saitoh, H. 2006. Dose calculation system for remotely supporting radiotherapy. *Radiat. Prot. Dosimetry* 116: 190–195.

Saito, K., Koga, S., Ida, Y., Kamei, T., and Funabiki, J. 2008. Construction of a voxel phantom based on CT data for a Japanese female adult and its use for calculation of organ doses from external electrons. *Jpn. J. Health Phys.* 43: 122–130.

Saito, K., Sato, K., Endo, A., and Kinase, S. 2009. Recent progress on Japanese voxel phantoms and related techniques at JAEA. *Nucl. Technol.* 168: 213–219.

Sato, K. and Endo, A. 2008, Analysis of effects of posture on organ doses by internal photon emitters using voxel phantoms. *Phys. Med. Biol.* 53: 4555–4572.

Sato, K. Noguchi, H., Emoto, Y., Koga, S., and Saito, K. 2005. Development of a Japanese adult female voxel phantom. In *2005 Annual meeting of Atomic Energy Society of Japan*, Hiratsuka, Japan, March 29–31, 2005.

Sato, K., Noguchi, H., Emoto, Y., Koga, S., and Saito, K. 2007a. Japanese adult male voxel phantom constructed on the basis of CT images. *Radiat. Prot. Dosimetry* 123: 337–344.

Sato, K., Noguchi, H., Endo, A., Emoto, Y., Koga, S., and Saito, K. 2007b. Development of a voxel phantom of Japanese adult male in upright posture. *Radiat. Prot. Dosimetry* 127: 205–208.

Sato, K., Endo, A., and Saito, K. 2008. Dose conversion coefficients calculated using a series of adult Japanese voxel phantoms against external photon exposure. JAEA-Data/Code 2008-016, Japan Atomic Energy Agency, Japan.

Sato, K., Noguchi, H., Emoto, Y., Koga, S., and Saito, K. 2009. Development of a Japanese adult female voxel phantom. *J. Nucl. Sci. Technol.* 46: 907–913.

Schlattl, H., Zankl, M., and Petoussi-Henss, N. 2007. Organ dose conversion coefficients for voxel models of the reference male and female from idealized photon exposures. *Phys. Med. Biol.* 52: 2123–2145.

Snyder, W. S., Fisher, H. L., Ford, M. R., and Warner, G. G. 1969. Estimate of absorbed fractions for monoenergetic photon sources uniformly distributed in various organs of a heterogeneous phantom, MIRD Pamphlet No. 5. *J Nucl. Med.* 10 (Suppl. 3): 7.

Stewart, R. D., Tanner, J. E., and Leonowich, J. A. 1993. An extended tabulation of effective dose equivalent from neutrons incident on a male and anthropomorphic phantom. *Health Phys.* 65: 405–413.

Takahashi, F., Yamaguchi, Y., Iwasaki, M., Miyazawa, C., Hamada, T., and Saito, K. 2002. Conversion from tooth enamel dose to organ doses for electron spin resonance dosimetry. *J. Nucl. Sci. Technol.* 39: 964–971.

Takahashi, F., Yamaguchi, Y., Iwasaki, M., Miyazawa, C., Hamada, T., Funabiki, J., and Saito, K. 2003. Analyses of absorbed dose to tooth enamel against external photon exposure, *Radiat. Prot. Dosimetry* 103: 125–130.

Tanaka, G. and Kawamura, H. 1996. Anatomical and physiological characteristics for Asian Reference Man. NIRS-M-115, National Institute of Radiological Sciences, Hitachinaka, Japan.

Ulanovsky, A., Wieser, A., Zankl, M., and Jacob, P. 2005. Photon dose conversion coefficients for human teeth in standard irradiation geometries. *Health Phys.* 89: 645–659.

Veit, R., Zankl, M., Petoussi, N., Mannweiler, E., Williams, G., and Drexler, G. 1989. Tomographic anthro-pomorphic models. Part I: Construction technique and description of models of an 8 week old baby and a 7 year old child. GSF-Bericht 3/89, GSF—National Research Center for Environment and Health, Neuherberg, Germany.

Xu, X. G. 2008. Consortium of Computational Human Phantoms. http://www.virtualphantoms.org/

Watanabe, R. and Saito, K. 2002. Monte Carlo simulation of strand-break induction on plasmid DNA in aque-ous solution by monoenergetic electrons. *Radiat. Environ. Biophys.* 41: 207–215.

Watanabe, R., Yokoya, A., and Saito, K. 2000. Modeling of production process of DNA damage by irra-diation with monochromatic X-rays around the K-edge of phosphorus. In *Proceedings of 10th International Congress of the International Radiation Protection Association*, Hiroshima, Japan, May 14–19, 2000.

Yamaguchi, Y. 1992. A computer code to calculate photon external doses using age-specific phantoms. *Hoken Butsuri* 27: 305–312.

Zaidi, H. and Xu, X. G. 2007. Computational anthropomorphic models of the human anatomy: The path to realistic Monte Carlo modeling in radiological sciences. *Annu. Rev. Biomed. Eng.* 9: 471–500.

Zankl, M., Veit, R., Williams, G., Schneider, K., Fendel, H., Petoussi, N., and Drexler, G. 1988. The con-struction of computer tomographic phantoms and their application in radiology and radiation protection. *Radiat. Environ. Biophys.* 27: 153–164.

Zankl, M., Drexler, G., Petoussi-Henss, N., and Saito, K. 1997. The calculation of dose from external photon exposures using reference human phantoms and Monte Carlo methods. Part VII: Organ doses due to parallel and environmental exposure geometries. GSF-Bericht 8/97, GSF—National Research Center for Environment and Health, Neuherberg, Germany.

Zhang, B. Q., Ma, J. Z., Liu, L. Y., and Cheng, J. P. 2007. CNMAN: A Chinese adult male voxel phantom constructed from color photographs of a visible anatomical data set. *Radiat. Prot. Dosimetry* 124: 130–136.

24 Cancer Therapy with Heavy-Ion Beams

Koji Noda
National Institute of Radiological Sciences
Chiba, Japan

Tadashi Kamada
National Institute of Radiological Sciences
Chiba, Japan

CONTENTS

24.1 INTRODUCTION

Heavy-ion beams have drawn growing interest for their use in cancer treatment not only due to their high dose localization at the Bragg peak, but also due to the high biological effect in this region. In 1946, R.R. Wilson proposed the clinical applications of protons and heavier ions in the treatment of human cancer (Wilson, 1946). Pioneering work on clinical applications of proton and helium beams was carried out (Tobias et al., 1952) using the 184 in. synchrocyclotron at the Lawrence Berkeley National Laboratory (LBNL). More than 1000 patients had been treated since 1957. Between 1977 and 1992, 433 patients were treated with heavy-ion beams accelerated by the Bevalac (Alonso, 1993). Encouraged by the prospective results of heavy-ion radiotherapy at LBNL and the recent progresses in the accelerator and beam-delivery technologies (Chu et al., 1993), the National Institute of Radiological Sciences (NIRS) decided to carry out heavy-ion radiotherapy. The construction project of the Heavy-Ion Medical Accelerator in Chiba (HIMAC) (Hirao et al., 1992) was promoted by NIRS as one of the projects of "Comprehensive 10-year strategy for cancer control" that was initiated by the Japanese government in 1984. The construction of the HIMAC facility was completed in 1993, and it was the first heavy-ion medical accelerator facility dedicated to cancer radiotherapy in the world. On June 21, 1994, NIRS began heavy-ion radiotherapy using carbon-ion beams generated by the HIMAC accelerator complex. Since then, clinical studies to develop safe and secure irradiation technologies and optimized dose fractionation for various diseases have been conducted. More than 4500 patients have been treated with HIMAC at NIRS, and the clinical efficacy of carbon-ion radiotherapy has been demonstrated for various diseases.

Owing to high dose localization and high biological effect of the carbon-ion beam used in cancer radiotherapy, the carbon-ion beam is expected to be effective against intractable, photon-resistant cancers, and to greatly reduce the treatment period compared with photon or proton radiotherapy. The results of our previous and ongoing clinical studies, and a number of irradiation technologies developed, suggest that we are indeed reaching these goals. Nonetheless, the data collected to date may not be sufficient to fully prove the efficacy of carbon-ion radiotherapy, and more clinical data are still needed.

At present, carbon-ion radiotherapy is carried out at three facilities in the world (two in Japan and one in China), five new facilities (three in Germany, one in Italy, and one in Japan) are under construction, and at least four more facilities in European Union and in Japan are very close to starting construction. These facilities, either performing or planning carbon-ion radiotherapy, have been collaborating in their research and exchanging results with each other.

In this chapter, we introduce the past, present, and future of carbon-ion radiotherapy in conjunction with the development of the accelerator and beam-delivery systems, especially at NIRS.

24.2 CLINICAL TRIAL OF CANCER THERAPY WITH CARBON BEAM

A carbon-ion beam produces maximum ionization at the Bragg peak. It allows a high localized dose to be given safely even with critical organs in close proximity to the target lesion. In addition to this physical selectivity, a carbon beam possesses high radiobiological effectiveness at the Bragg peak, such as a small difference in radiation sensitivity through the cell cycle. Carbon beams are therefore expected to be less toxic to the surrounding normal tissues and have higher effects on intractable malignant tumors. These features of carbon beams are most advantageous in cancer therapy. The purpose of the clinical research at NIRS on carbon beams is to prove the efficacy and safety of carbon therapy in cancer treatment. All the carbon therapies were performed as prospective phase I/II dose escalation studies and phase II fixed-dose clinical trials in an attempt to identify tumor sites suitable for this treatment, including radio-resistant tumors, and to determine optimal dose fractionation, especially for hypofractionation in common cancers.

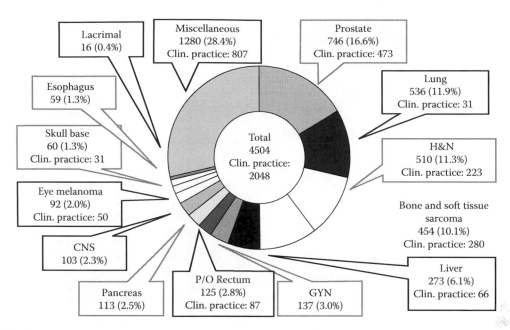

FIGURE 24.1 Distribution of tumors treated for patients enrolled for carbon-ion radiotherapy (June 1994 to February 2009).

24.2.1 CLINICAL TRIALS WITH CARBON BEAM

Carbon-ion radiotherapy has focused mainly on malignant diseases that are difficult to cure by conventional radiotherapy, among which are head and neck cancer, skull base tumor, lung cancer, liver cancer, prostate cancer, bone and soft tissue tumor, postoperative local recurrence of rectal cancer, and so on. The total number of patients enrolled by February 2009 was 4504, with various types of tumors being treated (Figure 24.1).

In the phase I/II studies, to confirm the safety of carbon-ion radiotherapy and obtain information concerning the antitumor effect, the number of fractions and treatment period were fixed for each disease and the total dose was gradually increased by 5%–10%. When the recommended dose was determined during the phase I/II studies, it was incorporated into the phase II studies. As shown in Figure 24.2, the number of patients enrolled has increased year by year. This increase occurred in conjunction with the treatment regimen becoming established and smoothly executed, and, as a result, the number of fractions and treatment period per patient could be greatly reduced.

Since a carbon-ion beam has the therapeutically advantageous feature of a higher biological effect at deeper sites from the surface of the body, treatment can be completed in a shorter period. For stage-I non-small-cell lung cancer (NSCLC) and liver cancer, for example, treatment can be completed in one or two irradiation fractions. For prostate cancer and bone/soft tissue tumors, it is completed in 16 fractions, which is about half the number of fractions employed in x-ray and proton beam radiotherapy. The average number of treatment fractions per patient is now 12 and the treatment period is about 3 weeks at NIRS.

24.2.2 RESULTS OF CLINICAL TRIALS WITH CARBON BEAM

More than 50 protocols for various tumors were designed and used in clinical studies. Details of treatment results of head and neck cancer, skull base tumor, lung cancer, liver cancer, prostate cancer, and bone and soft tissue tumor are summarized below.

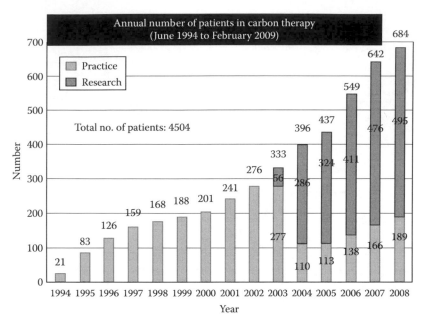

FIGURE 24.2   Number of patients enrolled for carbon-ion radiotherapy (June 1994 to February 2009).

24.2.2.1 Head and Neck Cancers

Carbon-ion radiotherapy was first used for head and neck cancers in 1994. Most patients had locally advanced lesions or postoperative recurrence, with little expectation of cure from other therapies. In the first phase I/II study, a fractionation regimen of 18 fractions over 6 weeks was employed, after which the second phase I/II study using a fractionation regimen of 16 fractions over 4 weeks was introduced. The comparison of the results of these two studies showed no difference in terms of side effects and local control (Mizoe et al., 2004). Since April 1997, phase II studies using 57.6–64.0 GyE in 16 fractions over 4 weeks, the optimal dose determined in the second study, have been conducted, and almost no severe side effects have been observed to date.

With respect to effectiveness, local control has been favorable, being 80%–90% in adenocarcinomas, adenoid cystic carcinomas, and malignant melanomas. For the treatment of bone and soft tissue sarcomas in the head and neck region, it was found necessary to increase the total dose to improve local control. Thus, a total dose of 70.4 GyE in 16 fractions was used, and local control has so far been quite satisfactory.

24.2.2.2 Skull Base Tumors

As the term implies, these tumors develop in the base of the skull, and include primarily chordomas and chondrosarcomas. Because the tumors frequently develop in the vicinity of the brain stem, optic nerve, and large vessels, complete surgical resection is usually quite difficult. While the advent of proton therapy has improved the treatment results, it has become clear that the recurrence of chordomas 5 years or more after treatment is not rare (Munzenrider and Liebsch, 1999). In this respect, carbon-ion radiotherapy has the potential to improve long-term survival (Schulz-Ertner and Tsujii, 2007).

The most common types of tumors treated with carbon-ions at NIRS are chordomas and meningiomas. As the dose was increased in dose escalation studies, an improvement in tumor control was observed, and there were no severe adverse reactions. A fractionation regimen of 60.8 GyE in 16 fractions over 4 weeks yielded the best local control with acceptable morbidity.

24.2.2.3   Lung Cancer (Non-Small-Cell Lung Cancer)

NSCLCs were divided according to tumor location into peripheral type and central type. It was thought necessary to change the fractionation regimen depending on the location, since the tolerance of the surrounding normal tissue would be different between the two. Clinical studies were conducted to establish a short-course radiotherapy regimen for stage-I (T1-2/N0/M0) peripheral-type cancer. The treatment was started with the use of 18 fractions over 6 weeks, and then the number of fractions and treatment time were carefully reduced.

We conducted a clinical study to develop a short-course, hypofractionated radiotherapy. Treatment regimens of nine fractions over 3 weeks and four fractions over 1 week were evaluated, with both showing little severe toxicity and a local control rate of >90% (Miyamoto et al., 2007).

A phase I/II dose escalation study using a single fraction of carbon beam has been under investigation for peripheral-type stage-I NSCLC. Single-fraction irradiation is the ultimate regimen utilizing the features of a carbon beam. We intend to adopt this regimen as a clinical practice as early as possible. The respiratory-gated irradiation method, a technique that uses synchronization with the respiratory movement of target organs, was developed and effectively used to reduce unnecessary dose to the surrounding normal lung in these clinical trials.

24.2.2.4   Liver Cancer (Hepatocellular Carcinoma)

All the clinical studies of liver cancer dealt with patients for whom other therapies were not expected to provide sufficient effects or were not indicated. In the first phase I/II study, patients were treated with irradiation in 15 fractions over 5 weeks, and safety and efficacy were confirmed. The 3- and 5-year local control rates were both 81% (Kato et al., 2004). In the second phase I/II study, the dose was escalated using increasingly shorter irradiation times of 12 fractions over 3 weeks, 8 fractions over 2 weeks, and 4 fractions over 1 week, for the purpose of developing a short-course irradiation method. All of the fractionation regimens were confirmed to be safe in this study (Kato, 2004).

On the basis of these results, a phase II study (third protocol) to examine the efficacy of treatment with 52.8 GyE in four fractions over 1 week was conducted, and there were no or only minimal changes in the liver function after the treatment. The 3- and 5-year local control rates were both 96%, and the 3- and 5-year cumulative crude survival rates were 58% and 35%, respectively, indicating an excellent fractionation regimen in terms of safety and efficacy (Kato et al., 2005a).

Finally, a phase I/II study with the fourth protocol (very-short-course irradiation of two fractions over 2 days) was conducted, and no severe side effects were observed (Kato et al., 2005b). This fractionation regimen has now been adopted as a clinical practice.

When the side effects after performing carbon-ion radiotherapy were analyzed, there were few limitations due to liver dysfunction or tumor diameter, and more than 90% of the patients were almost free of symptoms during and after the treatment. Patients having cancers with no contact between the tumor and the gastrointestinal tract, moderate or higher liver function, and a tumor diameter of 10 cm or less are considered to be good candidates for undergoing carbon-ion radiotherapy.

24.2.2.5   Prostate Cancer

After the early dose escalation study was terminated, 66 GyE in 20 fractions over 5 weeks was used as the standard irradiation regimen in prostate cancer. For a further decrease in the incidence of adverse reactions, the dose was reduced to 63 GyE. However, we have also been conducting treatment with 57.6 GyE in 16 fractions over 4 weeks in parallel with the standard regimen in order to reduce the treatment period, and we recently employed this fractionation regimen as the standard method.

Patients with prostate cancer were classified into two groups, high-risk and low-risk groups, based on factors before treatment, including PSA, Gleason score, and TNM classification, to determine the need for the combination with the endocrine treatment for the initial 10 years. The high-risk group was treated with carbon-ion radiotherapy in combination with the endocrine therapy, and the low-risk group was treated with carbon-ion radiotherapy alone (Akakura et al., 2004; Tsuji et al., 2005; Ishikawa et al., 2006). Patients considered to not require long-term hormone therapy were separated from the high-risk group as a medium-risk group, and the period of their hormone therapy was reduced to 6 months for the last couple of years.

The incidence of adverse reactions of Grade 3 or more in the rectum or lower urinary tract (bladder/urethra) in all patients was 1.5%, but no reactions of Grade 3 or more were observed in any patients after the optimal irradiation dose was established. The incidence of adverse reactions of Grade 2 with the current technique was as low as 1.5% in the rectum and 5.7% in the lower urinary tract, and it has recently tended to decrease after the total dose was reduced to 63 GyE. Treatment with 57.6 GyE in 16 fractions has produced no reactions of Grade 2 during observation for 6 months or more, and this regimen is considered to be safer than that of 63 GyE in 20 fractions. According to the recent survival analysis in these patients, the 5-year survival rate was 91.6%, the cause-specific survival rate was 98.5%, the 5-year local control rate was 99.1%, and the biochemical nonrecurrence rate was 88.5%. A comparison of the biochemical nonrecurrence rate in patients with a PSA level of 20 ng/mL or higher before treatment and in those receiving other radiotherapies showed a remarkably higher nonrecurrence rate in those receiving carbon-ion radiotherapy. The high nonrecurrence rate should also be associated with the effects of our sound use of combination hormone therapy, but a comparison with clinical studies in Europe and North America combining hormone therapy and x-ray radiotherapy showed that the survival rate in our results was 10%–15% higher, confirming that the high local effect of carbon-ion radiotherapy led to good treatment results.

Treatment with 57.6 GyE in 16 fractions, the present standard regimen for prostate cancer at NIRS, is expected to produce results better than those produced with an irradiation of 20 fractions over 5 weeks in terms of the antitumor effect with a lower risk of adverse reactions, and shows great promise for future long-term results.

24.2.2.6 Bone and Soft Tissue Tumors

In bone and soft tissue tumors, a phase I/II dose escalation study was followed by a phase II fixed-dose study, and it is now being performed as a clinical practice.

The total dose started at 52.8 GyE in 16 fractions over 4 weeks, and was increased to 73.6 GyE in the initial study. The local control rate improved as the dose increased, but some patients in the group who were given the largest dose developed severe skin and soft tissue reactions (Kamada et al., 2002).

In the phase II fixed-dose study, the 3- and 5-year local control rates were both 84% and the 3- and 5-year survival rates were 68% and 49%, respectively. Skin and soft tissue toxicities, as severe side effects, developed at an incidence of about 3%, but recently, as a result of decreased skin doses, almost no side effects have been observed. In patients with osteosarcomas in the pelvis or spine, whose resection was difficult, the 5-year survival rate was 25%. In patients with chordomas other than those developing in the skull base, the 5-year local control rate was 96% and the 5-year survival rate was 81%. Chordomas of the sacral bone were reported in *Clinical Cancer Research*, 2004, and appeared in the *Year Book of Oncology*, 2006 (Imai et al., 2006; Leoehrer et al., 2006).

Bone and soft tissue tumors are considered one of the best indications for carbon-ion radiotherapy. Although long-term observation should be continued, carbon-ion radiotherapy may replace surgical resection in elderly patients and in patients whose function would be greatly reduced with resection, as well as provide a treatment for patients for whom resection is not indicated.

More than 4500 patients have already received carbon-ion radiotherapy at NIRS. The results have been revolutionary for cancer treatment. It has proven effective for cancers in the head and neck, lung, liver, prostate, sarcomas, etc. Many of the diseases targeted in these clinical trials were considered unlikely to be effectively treatable with other therapies. Although it is difficult to carry out direct comparative studies with other therapeutics, it has become increasingly evident that carbon-ion radiotherapy is capable of curing cancers that are incurable by other treatments, as well as in a shorter time and more safely than when treated by other modalities.

24.2.3 ANALYSIS OF TUMOR CONTROL PROBABILITY

On the basis of dose escalation studies, the clinical results at HIMAC were analyzed in order to reduce the number of fractionated irradiations. The clinical dose distributions of therapeutic carbon-ion beam, currently used at NIRS HIMAC, are based on in vitro human salivary gland (HSG) cell survival response and clinical experience from fast-neutron radiotherapy. The moderate radiosensitivity of HSG cells is expected to be a typical response of tumors to carbon beams. At first, the biological-dose distribution is designed so as to cause a flat biological effect on HSG cells in the spread-out Bragg peak (SOBP) region. Then, the entire biological-dose distribution is evenly raised in order to attain an RBE (relative biological effectiveness) = 3.0 at a depth where dose-averaged LET (linear energy transfer) is 80 keV/μm. At this point, biological experiments have shown that carbon-ions can be expected to have a biological effect identical to fast neutrons, which showed a clinical RBE of 3.0 for fast-neutron radiotherapy at NIRS.

The resulting clinical dose distribution in this approximation is not dependent on the dose level, the tumor type, or the fractionation scheme, and thus reduces the unknown parameters in the analysis of the clinical results. The width of SOBP and the clinical/physical dose at the center of SOBP specify the dose distribution.

The clinical results of NSCLC treated by HIMAC beams were analyzed. They depicted a very conspicuous dose dependency of the local control rate. A dose escalation study was performed with a treatment schedule of 18 fractions in 6 weeks (Miyamoto et al., 2003). The dose dependency of the tumor control probability (TCP) with the photon beam was given by the following formula proposed by Webb:

$$\text{TCP} = \sum_i \frac{1}{\sqrt{2\pi}\sigma} \left\{ -\frac{(\alpha_i - \alpha)^2}{2\sigma^2} \right\} \cdot \exp\left[-N \exp\left\{ -n\alpha d \left(1 + d/(\alpha + \beta)\right) + \frac{0.693(T - T_k)}{T_p} \right\} \right] \quad (24.1)$$

α and β are coefficients of the linear quadratic (LQ) model of the cell survival curve. The cell survival rate, $S(D_p)$, is defined as a function of the absorbed dose (physical dose), D_p:

$$S(D_p) = \exp(-\alpha \cdot D_p - \beta \cdot D_p^2) \quad (24.2)$$

In the analysis, α and β values of HSG cells were used. σ is the standard deviation of the coefficient α, which reflects the patient-to-patient variation of radiosensitivity. N is the number of clonogens in tumor (a fixed value of 10^9 was used). n and d are the total fraction number and the fractionated dose, respectively. T (42 days), T_k (0 days), and T_p (7 days) are the overall time for treatment, the kick-off time, and the average doubling time of tumor cells, respectively. Values used in the analysis are shown in brackets. The analysis was carried out for 18 fractionations in order to determine the tumor-specific radiosensitivity parameter, α, and its variation, σ. Once the parameters are fixed, it is possible to estimate the dose response curve in various fraction schedules. Figure 24.3 shows an example of the estimation of the dose response curve of NSCLC. TCP curves of 1, 4, and 9 fractions are estimated from the parameters derived from the data of 18 fractions. This approach would be useful in estimating appropriate prescribed doses when initiating hypofractionated radiotherapy.

24.3 RADIATION QUALITY OF CARBON-ION

When incident heavy ions travel in a patient's body, some of them suffer fragmentation reactions with target nuclei, and various species of fragments are generated and become widely distributed. The biological effectiveness of heavy ions is affected not only by the deposited energy, but also by the particle species (Blakely et al., 1984). In heavy-ion therapy, therefore, precise calculations of the spatial distribution of radiation quality such as dose, dose-averaged LET, fluence, and energy distributions for each species of particles play a highly important role. In passive beam

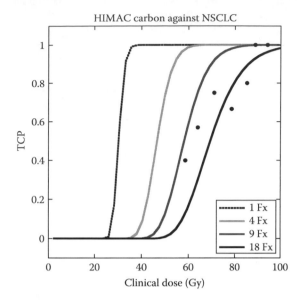

FIGURE 24.3 TCP curve of NSCLC by carbon-ion radiotherapy. TCP curves of 1, 4, and 9 fractionations are calculated by clinical data derived from 18 fractionations.

delivery, lateral distribution of the radiation quality is, at the first approximation, regarded as constant due to the equilibrium between incoming and outgoing particles in the region of interest in the irradiation field. Figures 24.4 and 24.5 show an axial fluence and the LET distribution of the therapeutic carbon beam measured at HIMAC.

In order to estimate the distribution of radiation quality in the irradiation field in detail, or in order to shape the irradiation field by a scanning method with pencil beams, it is strongly required

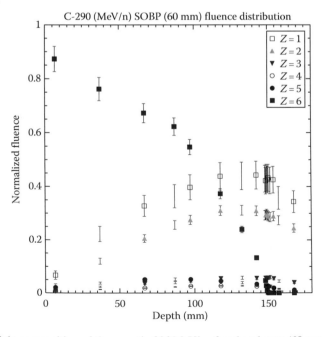

FIGURE 24.4 Particle composition of therapeutic 290 MeV/n of carbon beam (60 mm SOBP) as a function of thickness in water.

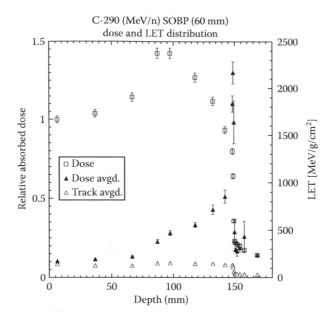

FIGURE 24.5 Axial LET distribution of therapeutic 290 MeV/n of carbon beam (60 mm SOBP) in water.

to understand the spatial distribution of the radiation quality. The deflection of primary particles in a thick medium has been well described by Moliere's multiple scattering theory (Molière, 1948; Wong et al., 1990). On the other hand, multiple scattering alone is not sufficient to account for the distribution of fragment particles. Our recent study revealed that the large deflection of fragment particles in a substance could be accounted for in the multiple scattering formula by considering one additional term representing a lateral "kick" at the production point of the fragment (Matsufuji et al., 2005). This additional term can be explained as a transfer of the intra-nucleus Fermi momentum of a projectile to the fragment, and its extent obeys the expectation derived from the Goldhaber model (Goldhaber, 1974). The angular distribution of fragment particles was measured with a monoenergetic 290 MeV/n ^{12}C beam through a nuclear reaction in a thick water target (Matsufuji et al., 2005). Matsufuji et al. determined a parameter describing the extent of the transferred momentum in the Goldhaber model so as to reproduce the observed angular distributions. On the basis of these studies, a semi-analytical beam transportation code was developed for energetic heavy-ion beams, in which the 3D distribution of radiation quality can be calculated for each species of particles (Inaniwa et al., 2007b). The productions of secondary and tertiary fragments are considered, and the effects of Fermi momentum transfer are taken into account at their production point. Despite its simplicity, the developed code could reproduce the experimental result well. We expect that it can be used in treatment planning to evaluate the biological effectiveness of heavy-ion beams precisely.

Especially for 3D scanning with a pencil beam, it is essential to estimate the biological-dose distribution given by one pencil beam. The biological-dose distribution, $D_b(\mathbf{r})$, is expressed as the product of the physical-dose distribution, $D_p(\mathbf{r})$, and RBE($\mathbf{r}$):

$$D_b(\mathbf{r}) = D_p(\mathbf{r}) \times \text{RBE}(\mathbf{r}) \tag{24.3}$$

The RBE is obtained using measured α and β values in the LQ model. On the other hand, the $D_p(\mathbf{r})$ is directly measured. Figure 24.6 shows the integrated depth-dose distribution measured by a large-area parallel-plate ionization chamber. It is noted that the Bragg peak of the pristine beam (monoenergetic carbon-ions with 350 MeV/n) is slightly broadened by the ridge filter for

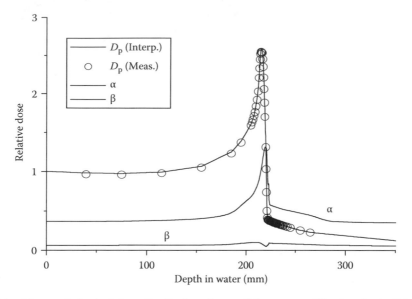

FIGURE 24.6 Measured physical-dose distribution of a pencil beam, α and β values in the LQ model, as a function of the depth in water.

3D scanning. The α and β values are also shown in Figure 24.6, as a function of the depth in water. Figure 24.7 shows the lateral size of the dose profile in one standard deviation under various range-shifter thicknesses as a function of the depth in water. With increasing the depth, the lateral size of the dose profile is gradually increased due to multiple scattering. On the other hand, the size is rapidly increased beyond the primary beam range, which is caused by fragment nuclei.

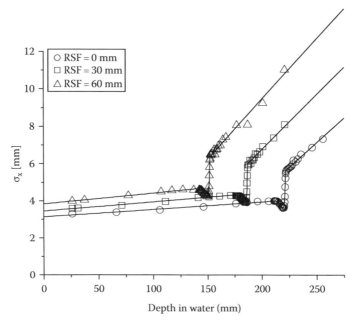

FIGURE 24.7 Lateral size of physical-dose profile as a function of the depth in water for various ranges. Circles, squares, and triangles represent the measured size with range shifter plates of 0, 30, and 60 mm thickness, respectively.

Another approach to derive the spatial distribution of radiation quality in matter is the use of Monte Carlo codes. Although Monte Carlo codes are time consuming and, currently, not practical for implementation into a treatment-planning system of heavy-ion therapy, some articles have reported that they can provide precise estimates of radiation quality in matter (Kameoka et al., 2008; Nose et al., 2009).

24.4 DEVELOPMENT OF ACCELERATOR AND BEAM-DELIVERY SYSTEMS

24.4.1 HEAVY-ION THERAPY FACILITY

The existing carbon-ion radiotherapy facilities are HIMAC, the Hyogo Ion Beam Medical Center (HIBMC), the Institute of Modern Physics in China (IMP), and the Gesellschaft fuer Schwerionenforschung (GSI). At HIBMC, treatments were carried out using both protons and carbon-ions, and 454 patients were treated with carbon-ion beams. At IMP, using a carbon-ion beam with 100 MeV/n, 103 patients with superficially placed tumors have been treated since 2006. Further, six patients with deeply seated tumors have been successfully treated with a 400 MeV/n carbon-ion beam in March 2009. GSI has successfully carried out cancer treatments for around 400 patients with carbon-ion beams, applying the intensity controlled raster-scanning method, since 1997.

As a typical carbon-ion radiotherapy facility, the HIMAC facility is introduced here. The design parameters of HIMAC are based on the radiological requirements. Although the ion species ranged from He to Ar in the design stage, those from p to Xe can be delivered at present. The beam energy was designed to vary from 100 to 800 MeV/n for efficient treatment. HIMAC consists of an injector linac cascade, dual synchrotron rings with independent vertical and horizontal beam lines, and three treatment rooms equipped with the beam-delivery system. For carbon-ion radiotherapy, a C^{2+} beam as produced by the 10 GHz electron cyclotron resonance (ECR) ion source is injected into the linac cascade consisting of RFQ and Alvarez linacs, and is accelerated up to 6 MeV/n. After the carbon beam is fully stripped with a carbon-foil stripper, the beam is injected into the synchrotron rings by the multiturn injection scheme, and is slowly extracted after acceleration to a desired energy. Finally, the carbon beam extracted from the synchrotron is delivered to the beam-delivery system. Figure 24.8 shows a bird's eye view of the HIMAC facility.

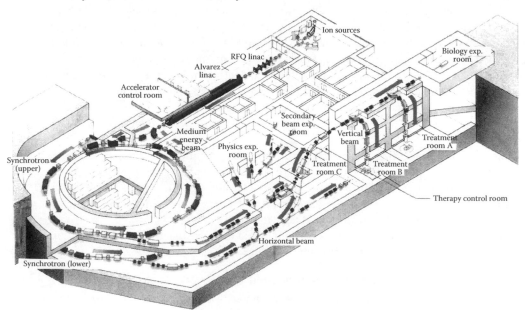

FIGURE 24.8 Bird's eye view of the HIMAC facility for heavy-ion cancer therapy. (Adapted from Hirao, Y.E. et al., *Nucl. Phys. A.*, 538, 541C, 1992.)

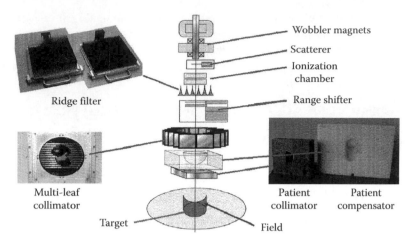

FIGURE 24.9 Schematic diagram of the HIMAC beam-delivery system.

The HIMAC facility employs a passive beam-delivery method (Torikoshi et al., 2007), as shown in Figure 24.9: single beam-wobbling for forming the lateral dose distributions and ridge-filter methods for forming the SOBP as a distal dose distribution, respectively. The pencil beam delivered from the accelerator is wobbled by a pair of dipole magnets placed in tandem with their field directions orthogonal to one another. The magnets are sinusoidally excited with an alternating electric current of the same frequency of 56.4 Hz, but with a 90° phase shift between them. Controlling the amplitudes of the magnetic fields, the pencil beam moves on a circular orbit around the original beam axis. At the same time, the pencil beam is broadened with a scatterer, placed just downstream from the wobbler magnets. With an appropriate combination of the amplitudes of wobbling and thickness of the scatterer, a uniform dose distribution could be obtained at the isocenter.

A bar-ridge filter has been used at HIMAC in order to spread out the Bragg peak, so as to match its size with the target thickness. Passing each monoenergetic ion through a different thickness of the bar-ridges, the ion looses its energy in proportion to the thickness under the high incident energy. The cross-sectional shape of the bar-ridge is designed to deliver a biological dose to the target region homogeneously. In the case of a heavy-ion beam, the SOBP is composed of various LET components with different weighting factors at each depth. The survival rate under a mixing LET radiation field can be described by a formalism proposed in the theory of dual radiation action on the basis of the LQ model (Kanai et al., 1997), and was experimentally proven. We finally designed the SOBP of mixed ions with different LET according to the procedure proposed (Chu et al., 1993) and the formalism (Kanai et al., 1997).

Using the beam-wobbling and ridge-filter methods, the maximum lateral field and the SOBP size are designed to be 22 cm in diameter with ±2.5% uniformity at the isocenter and 15 cm, respectively. A multi-leaf collimator (MLC) and a patient collimator are used in order to match the lateral dose distribution precisely with a lateral target shape. On the other hand, a bolus collimator is used for precisely matching the SOBP with a distal target shape.

At HIMAC, beam-delivery methods were improved in order to increase the irradiation accuracy. These are (1) the respiratory-gated irradiation method and (2) the layer-stacking irradiation method.

24.4.1.1 Respiratory-Gated Irradiation Method

Damage to normal tissues around a tumor was inevitable in the treatment of the tumor moving along with the patient's respiration. A respiratory-gated irradiation method, therefore, which can respond quickly to irregular respiration, was strongly required. In this method, the irradiation-gate signal is generated only when the target is located at the design position and the synchrotron can extract a beam. The beam is delivered according to the gate signal. Thus, one of the key technologies for the respiratory-gated irradiation is the beam extraction method from the synchrotron according to the gate signal. For this purpose, the radio frequency knockout (RF-KO) slow extraction method, which utilizes a transverse beam heating by an RF field tuned with a wave number of a horizontal betatron oscillation, was developed (Noda et al., 1996). Advantages of the RF-KO extraction are a quick response within sub-milliseconds to a signal of an extraction start/stop and a low horizontal emittance as compared with that in the ordinary slow extraction method, due to a constant separatrix. In accordance with the RF-KO method, the respiratory-gated irradiation system was developed (Minohara et al., 2000). In this system, the respiration signal is generated by the observation of the movement of an LED set on the surface of the patient's body through a position-sensitive detector, and a beam can be delivered only with the gate-signal. The respiratory-gated irradiation system has been used for treatments of liver and lung tumors since 1996.

24.4.1.2 Layer-Stacking Irradiation Method

By the conventional method, the fixed SOBP, which is produced by a ridge filter, results in undesirable dosage to the normal tissue in front of the target, because the width of an actual target varies within the irradiation field. In order to suppress this undesirable dosage, the layer-stacking irradiation method was proposed (Kanai et al., 1983), and the HIMAC irradiation system has been upgraded to put the technique including the treatment planning (Kanematsu et al., 2002) into practice. In this method, as shown in Figure 24.10, a small SOBP, several mm in water equivalent length (WEL), which is produced by a single ridge filter, is longitudinally scanned over the target volume in a stepwise manner by dynamically controlling the conventional beam-modifying devices. Concerning the lateral dose distribution, the target volume is longitudinally divided into slices, and each slice is irradiated by conforming the beam to cover each cross-sectional shape with the MLC. The performance of this method was experimentally verified (Futami et al., 1999; Kanai et al., 2006), and it has been used mainly for the treatment of the head and neck cancer.

24.4.2 Development of Compact Carbon-Ion Radiotherapy Facility

24.4.2.1 Standard-Type Carbon-Ion Radiotherapy Facility in Japan

For the purpose of the widespread use of carbon-ion radiotherapy in Japan, the NIRS decided to design a standard-type carbon-ion radiotherapy system in order to reduce construction costs by downsizing the accelerator complex and beam-delivery systems. Based on the design study by NIRS (Noda et al., 2007), a standard-type facility has been under construction at the Gunma University

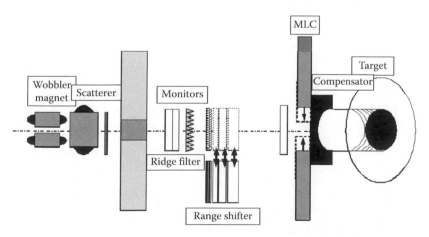

FIGURE 24.10 Schematic diagram of the layer-stacking irradiation method. (Adapted from Kanai, T. et al., *Med. Phys.*, 33, 2989, 2006.)

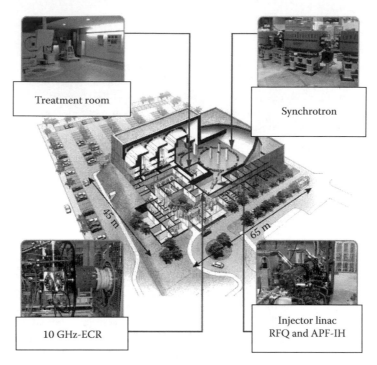

FIGURE 24.11   Image view of the standard-type carbon-ion radiotherapy facility at the Gunma University. (Adapted from Noda, K. et al., *J. Radiat Res.*, 48, A43, 2007.)

since 2006. This facility consists of an ECR ion source, an RFQ and an APF-IH linac, a synchrotron ring, three treatment rooms, and one experimental room for basic research. In this facility, a C^{4+} beam, which is generated by a compact 10 GHz ECR ion source, is accelerated to 4 MeV/n through the injector linac cascade. After fully stripped, the C^{6+} beam is injected into the synchrotron by a multiturn injection scheme and is accelerated to 400 MeV/n at maximum. All of the magnets in the beam transport lines are made of laminated steel in order to change the beam line quickly within 1 min. The beam-delivery system employs a spiral beam-wobbling method (Komori et al., 2004; Yonai et al., 2008) for forming a uniform lateral dose distribution with a relatively thin scatterer. The facility size was downsized to be one-third of the HIMAC facility. An image of the Gunma University facility with installed devices is shown in Figure 24.11.

24.4.2.2   Compact Heavy-Ion Radiotherapy Facility in Europe

In Europe, the Heidelberg ion therapy (HIT) facility was constructed in Heidelberg, Germany, based on the developments and experiences at GSI. Further, the Proton and Ion Medical Machine Study (PIMMS) project was also started in 1996. Based on the PIMMS design modified by the PMMS/TERA project, the Centro Nazionale di Adroterapia Oncologica (CNAO) facility has been under construction in Pavia, Italy, since 2003.

24.4.2.2.1   HIT Facility

The HIT facility, which is a hospital-based light-ion accelerator facility for the clinic in Heidelberg, was proposed (Eickhoff et al., 2004). At present, almost all beam commissionings have been successfully completed. The HIT facility consists of two ECR ion sources, an RFQ and IH linac cascade, a synchrotron ring, and two treatment rooms with a horizontal beam-delivery system and one rotating gantry. Light ions (p, He, C, and O) generated by the ECR ion sources are accelerated to 7 MeV/n by the RFQ and IH linac cascade and are injected into the synchrotron. The ions extracted slowly by the RF-KO method are delivered to each treatment room. The residual range of C^{6+} ions is designed to be 30 cm at maximum, which corresponds to 430 MeV/n. The main characteristic of this facility is the application of the raster-scanning method with an active variation in intensity, energy, and beam size both in the two horizontal irradiation ports and in the rotating gantry. Using the developed technologies in the HIT project, two other facilities (Marburg and Kiel) are under construction in Germany.

24.4.2.2.2 CNAO Facility

The Italian hadron therapy center, CNAO, is presently under construction in Pavia, Italy. The CNAO project will be devoted to the treatment of deeply seated tumors with proton and carbon-ion beams and to clinical and radiobiological research (Rossi, 2006). The CNAO accelerator is designed based on the modified PIMMS. The facility consists of two ECR ion sources, a 7 MeV/n injector linac cascade designed by GSI, and a 400 MeV/n synchrotron ring. The CNAO synchrotron consists of two symmetric achromatic arcs connected by two dispersion-free straights and has a circumference of approximately 78 m. The extraction scheme employs both the acceleration-driven and RF-KO methods under the third-order slow extraction condition. In the final phase, the CNAO project will have five treatment rooms (three rooms with fixed beams and two rooms with gantries) and one experimental room. In the first phase, three treatment rooms will be equipped with four fixed beams, three horizontal and one vertical. As a beam-delivery method, CNAO will employ the active scanning method that was developed at GSI.

24.4.3 DEVELOPMENT OF NEXT-GENERATION IRRADIATION SYSTEMS

In the HIMAC treatments, we have sometimes observed shrinkage in the size of the target as well as a change in its shape during the course of the treatment. In order to keep the sophisticated conformations of the dose distributions even in such cases, a strict requirement has been that treatment planning be carried out just before each fractional irradiation, which is called adaptive therapy. For this purpose, 3D scanning with a pencil beam should be employed, as it does not use a bolus and patient collimators requiring long manufacturing time. It is also well known that 3D scanning has brought about high treatment accuracy in the case of a fixed target (Kanai et al., 1980; Haberer et al., 1993; Pedroni et al., 1995). However, this method has not yet been put into practical use for the treatment of a target moving with respiration. Since the HIMAC facility should carry out treatments not only for fixed targets but also for moving targets, we have developed a 3D scanning method that can treat both moving and fixed targets. Then, in cooperation with a rotating gantry, the 3D scanning method can achieve a higher accuracy of treatment even for a target close to critical organs by the multi-field optimization method (Lomax, 1999), as compared with conventional carbon-ion radiotherapy. On the basis of these new developments, we have also designed and developed a rotating gantry incorporating the 3D scanning method (Furukawa et al., 2008a).

24.4.3.1 Development of Phase-Controlled Rescanning Method

In 3D scanning, an interplay effect between the scanning motion and the target motion brings about hot and/or cold spots in the target volume, even when using the respiration-gated irradiation method, because the sizes of the distal and lateral dose profiles of the pencil beam are comparable to the residual motion range. Therefore, the phase-controlled rescanning (PCR) method (Furukawa et al., 2007), which is combined with the rescanning technique and gated irradiation, is employed in order to avoid producing hot/cold spots. In the PCR method, rescanning on a slice is completed during one gate generated in the expiration period of the patient's breathing. Since the moving-target position is averaged in both the lateral and distal directions, the hot and/or cold spots are not produced. A simulation study revealed that the PCR method provided a feasible solution with a dynamic beam-intensity control technique (Sato et al., 2007) based on the RF-KO slow extraction method. In the PCR method, it is noted that the irradiation time for each depth slice should be adjusted to be within 1–2 s of the respiration-gate duration. Consequently, we obtained a practicable solution for moving-target irradiation by the fast raster-scanning method with rescanning and gating functions.

24.4.3.2 Development of Fast 3D Scanning Method

In the PCR method, fast scanning is an essential technology in order to complete one fractional irradiation within a tolerable period even under several-times rescanning. For this purpose, we developed (1) new treatment planning, (2) modified synchrotron operation, and (3) fast-scanning magnet.

24.4.3.2.1 New Treatment Planning

A new treatment-planning system (Inaniwa et al., 2007a, 2008) has been developed for 3D scanning with a pencil beam. A biological-dose distribution of a pencil beam is obtained by combining the measured physical-dose distribution and the RBE obtained through measured α and β values in the LQ model, as mentioned in Section 24.3. Using the biological-dose distribution, the new treatment planning optimizes the assignment of spot positions and their weights so as to give the prescribed dose on a target and to significantly reduce the dose on surrounding normal tissues.

For fast 3D scanning, raster scanning is employed instead of spot scanning, in order to save the beam-off time during beam movement between spot positions. However, raster scanning causes an extra dose due to beam movement between spot positions, and the extra dose is proportional to the beam intensity delivered. On the other hand, the fast scanning requires high beam intensity. As a result, it has been difficult for fast raster scanning to yield uniform dose distribution due to the high extra dose. However, since the HIMAC synchrotron has delivered a highly reproducible beam with a uniform time structure (Sato et al., 2007), the extra dose can be predicted and taken into account in the treatment planning. Consequently, the new treatment planning can increase the scanning speed by around five times. Applying a modified "travelling-salesman problem," further, the path length of raster scanning can be shortened by 20%–30%. The new treatment planning was developed according to the above considerations.

24.4.3.2.2 Modified Operation of Synchrotron

On the basis of the high beam-utilization efficiency of around 100% in the scanning method and an intensity upgrade to 2×10^{10} carbon-ions by suppressing both the transverse and longitudinal space–charge effects, we can complete the single-fractional irradiation of almost all treatment procedures in a single-operation cycle of the synchrotron. This single-cycle operation can significantly reduce dead time, such as beam injection, acceleration, and deceleration. As a result, we have proposed extended flattop operation in the single-cycle operation of the synchrotron. In this operation mode, stability of the beam was tested, and it was verified that the position and profile stabilities were less than ±0.5 mm at the isocenter during 100 s of extended flattop operation. This extended flattop operation can shorten the irradiation time by a factor of 2 and has been routinely utilized for 3D scanning experiments.

24.4.3.2.3 High-Speed Scanning Magnet

The scanning speed is designed to be 100 and 50 mm/ms in horizontal and vertical directions, respectively, faster by around one order than that of conventional spot scanning. In order to increase the scanning speed, we designed the scanning magnet with slits at both ends of the magnetic pole, based on a thermal analysis including an eddy current loss and a hysteresis loss. The result of a preliminary test showed a temperature rise of around 30 degree at maximum, which was consistent with the thermal analysis, under the designed scanning speed.

24.4.3.3 Fast 3D Raster-Scanning Experiment

At the first stage, we carried out a fast raster-scanning experiment using the HIMAC spot-scanning test line (Urakabe et al., 2001). The irradiation control system was modified so as to

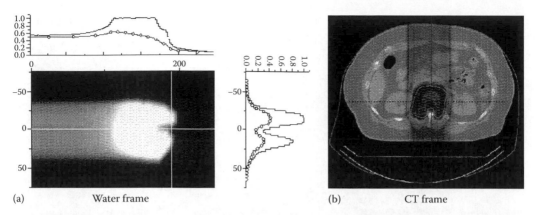

(a)   Water frame   (b)   CT frame

FIGURE 24.12   Dose distribution obtained by new treatment planning. (a) Dose distribution in water frame. Solid lines with circles (measured physical doses) represent the physical-dose distribution designed to obtain the biological distribution (black line). (b) Dose distribution reprocessed on a CT image. (From Inaniwa, T. et al., *Nucl. Instrum. Meth. B.*, 266, 2194, 2008. With permission.)

be capable of raster-scanning irradiation instead of spot-scanning irradiation. In the experiment, we adapted the measured dose response of the pencil beam with an energy of 350 MeV/n, corresponding to a 22 cm range in water. The beam size at the entrance and the width of the Gaussian-shaped mini-SOBP were 3.5 and 4 mm at one standard deviation, respectively. As a result of the experiment, it was verified that the new treatment planning could accurately provide a 3D dose distribution, as shown in Figure 24.12. In the PCR experiment, the dose distributions with/without the PCR method were measured two dimensionally. As shown in Figure 24.13, it was confirmed that the PCR method provided a uniform dose distribution even for a moving target (Furukawa et al., 2008b). Owing to both the new treatment planning and extended flattop operations, the irradiation time was shortened by around ten times compared with that of the conventional spot scanning.

We constructed a test irradiation port, which was installed with the fast-scanning magnets, in order to verify the design goal of fast 3D scanning. Figure 24.14 shows the test port for the fast 3D scanning experiment. In a preliminary test, we irradiated a spherically shaped target with a 6 cm diameter and obtained uniform 3D dose distribution within 10 s even under 10 times rescanning.

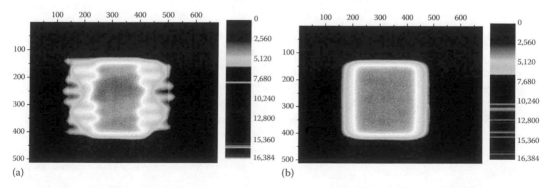

(a)   (b)

FIGURE 24.13   Dose distributions by the PCR method. (a) Without PCR. (b) With PCR (eight times rescanning). (From Furukawa, T. et al., Development of scanning system at HIMAC, in *Proceedings of the 11th EPAC*, Genoa, Italy, pp. 1794–1796, 2008b.)

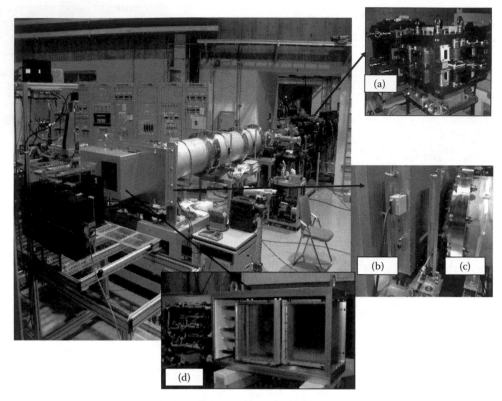

FIGURE 24.14 Test irradiation port of fast 3D raster-scanning experiment: (a) scanning magnets, (b) position monitor, (c) dose monitor, (d) range shifter.

24.5 THE FUTURE OF CARBON-ION RADIOTHERAPY

While the clinical studies conducted at NIRS over the past 15 years established the basic usefulness of carbon-ion radiotherapy, clinical studies and basic research aimed at developing the next generation of therapies have already started. Our specific approaches to future research on carbon-ion radiotherapy are as follows.

24.5.1 CONCOMITANT CHEMO-CARBON-ION RADIOTHERAPY

Local control is mandatory for curing cancer, and a good local control has been achieved with carbon beams. Depending on the type of cancer, however, improving local control does not always prolong the survival period. In locally advanced malignant melanoma, in particular, long-term survival was not always obtained despite the fact that local control was significantly improved through carbon-ion radiotherapy. There are patients who, although good local control was obtained, do not attain long-term survival because distant metastasis develops or second and third foci (colonies) sometimes appear. To improve long-term survival, we have used carbon-ion radiotherapy in combination with chemotherapy and have obtained favorable results. We have established safe methods for carbon-ion radiotherapy for many patients with inoperable cancers. As the next step, it will be necessary to develop therapies for improving survival rates, bearing in mind that cancer is a systemic disease.

24.5.2 DEVELOPMENT OF NOVEL IRRADIATION TECHNIQUES

A great advantage of carbon-ion radiotherapy is that it permits a conformed irradiation of various target sites compared to other radiotherapies. Layer-stacking irradiation, a new scanning irradiation method adaptive to respiratory movement, and an on-demand irradiation system sensitive to day-to-day or time-to-time changes in the tumor or body have been under development to obtain further improvements in carbon-ion dose distribution.

24.5.3 PROMOTION OF SHORT-COURSE HYPOFRACTIONATION IN CARBON-ION RADIOTHERAPY

Carbon-ion radiotherapy is performed in only two institutions in Japan, and as the number of patients treated continues to rise at a remarkable rate, there is an increasing demand for greater treatment efficiency. In this respect, short-course hypofractionation is effective and greatly contributes to the efficient utilization of restricted resources. Very-short-course irradiation consisting of only one or two sessions developed for lung and liver cancer has been acknowledged internationally, and the extension of this technique to other cancers is the next step. Studies have already been initiated to develop short-term irradiation modes for other regions by analyzing data obtained from previous application instances of a clinical practice. The specific practice is to set up groups with different irradiation periods and different numbers of irradiation fractions for the same disease, and then conduct randomized comparative intergroup studies to determine the most efficient irradiation period.

24.5.4 DIAGNOSTIC IMAGING FOR TREATMENT AND BIOLOGICAL STUDIES

Advanced diagnostic techniques have played a major role in the improvement of therapeutic results, and their further advancement is highly desirable. The introduction of new diagnostic equipment, such as PET, multi-detector computed tomography (MDCT), and high-T MRI, enables early small cancers to be accurately diagnosed. To utilize the information for actual carbon-ion radiotherapy, it is necessary to consider not only the relationship between the focus (tumor) and surrounding organs, but also the integration of different images and the deformation and movement of organs associated with respiration. Specific studies may include early assessment of therapeutic effects and analysis of prognostic factors by combining various diagnostic imaging techniques and creating fusion images. The studies may also include advancement of treatment planning by utilizing imaging equipment that temporally traces the dynamics of fusion images and 4D CT. Recently, imaging of the state of oxygen as well as sugar and amino acid metabolism in the tumor has become possible, and an important research issue in the near future will be how these techniques can be applied to treatment.

In various biological studies, to further enhance the effectiveness of carbon-ion radiotherapy and verify the safety, samples and data obtained from clinical studies on carbon-ion radiotherapy are indispensable, and close cooperation is important. Clinical application of the knowledge obtained both from these studies and from drugs (e.g., anticancer agents, sensitizers, and protective agents) to improve treatment outcomes is also a target of clinical research.

24.5.5 COMPARATIVE STUDIES

Comparative studies, already feasible at present, to further clarify the usefulness (indications) of carbon-ion radiotherapy include those in which the same protocol is applied to other therapies, the backgrounds of subjects are matched, and the treatment desired by study participants is offered. In such cases, consent from study participants is easily obtained, study costs are low, and agreement among participating facilities is relatively easily obtained, since the treatment desired by each patient is provided on the basis of cooperation with other institutions. Another method is to determine the inclusion criteria for patients already treated, and then to comparatively analyze the therapeutic results in a matched-pair study. This is feasible if consent is obtained from the institutions performing the therapies being compared.

When comparing various therapies, the results of randomized controlled studies provide the strongest evidence at present. Theodore S. Lawrence, editor of the *Journal of Clinical Oncology*, has made the following statement about the need for randomized controlled studies when comparing different therapies, or different types of therapeutic equipment, in the field of radiation oncology (Lawrence et al., 2007):

> In medical oncology, there is a need to compare drug treatment A to drug treatment B. It is not possible to determine which is better without carrying out a randomized trial. In radiation oncology, one can know, based on physics, that protons deliver a better dose distribution than photons or that IMRT is superior to three-dimensional conformal therapy. If treatment planning and delivery are carried out properly, which can be defined by a set of rules, there is no debate; this is a matter of physics. The big question is: is the clinical improvement worth the added expense? What kind of trial needs to be designed to answer this question? It will be difficult to run a randomized trial in the United States asking whether a treatment that is superior based on physics is worth it. Would patients permit themselves to be randomly assigned to the standard but less expensive therapy?

Previous progress in radiation oncology is inextricably linked to the development of irradiation equipment, but no randomized controlled study has been conducted for the purpose of the introduction of new therapeutic equipment. No randomized controlled study was conducted in the shift from cobalt irradiation equipment to linac x-ray systems, but if that had been done, there would have been little difference between cobalt and linac in the treatment results in patients with early laryngeal glottic cancer. Cobalt irradiation has advantages in terms of both cost and equipment maintenance, and the results of linac x-ray irradiation may be poor unless the correct energy is selected. Cobalt irradiation is sufficient to treat laryngeal cancer, and it is unnecessary to introduce linac irradiation to achieve successful treatment for this cancer. However, almost no treatment with cobalt is performed in developed countries. The distribution of the very-high-energy x-rays of linac allows safer treatment of foci (tumors) existing at a deep site, and the design of a study to determine the difference between cobalt and linac irradiation in patients with laryngeal cancer is itself a problem.

Randomized controlled study protocols should be considered for studies aimed at making the usefulness of carbon-ion radiotherapy clearer and which may lead to more efficient operation, such as extended indications of short-term hypofractionation, or to more effective treatment and improved local effect. However, at present, most of the patients receiving carbon-ion radiotherapy visit the clinic looking for this treatment, and it is difficult to obtain consent for a randomized controlled study from patients. It is difficult to conduct a randomized study between different types of equipment even in the United States, where the use of randomized controlled studies is advanced; it is equally, if not more, difficult in Japan.

TABLE 24.1
Carbon-Ion Radiotherapy Facilities in the World

Under Operation	Start	No. of Patients	Until
NIRS (Japan)	1994	4504	February 2009
GSI (Germany)	1997	384	October 2007
HIBMC (Japan)	2002	454	December 2008
IMP (China)	2006	109	March 2009
		5451	

Sources: Amaldi, U. and Kraft, G., *J. Radiat. Res.* 48, A27, 2007; Noda, K., *J. Radiat. Res.*, 48, A43, 2007.

Notes: Under construction: Gunma (Japan), HIT (Germany), CNAO (Italy), Marburg (Germany), and Kiel (Germany). Plan in progress: ETOILE (France), Med-Austron (Austria), Saga (Japan), and Kanagawa (Japan).

Accordingly, future comparative studies on heavy-ion radiotherapy will follow these guidelines:

1. Clinical studies will use the same protocol that is applied to other therapies
2. Matched-pair controlled studies in subjects will be matched to those receiving other therapies
3. Studies will compare different charged particle therapies

24.6 CONCLUSION

More than 5000 patients were treated with carbon-ion radiotherapy at four facilities worldwide by the end of 2008. The therapeutic results are internationally acknowledged. A considerable number of carbon-ion radiotherapy systems are under construction in European countries and in Japan, as summarized in Table 24.1, offering the prospect of achieving more reliable and safer cancer therapy in the near future (Amaldi and Kraft, 2007).

In industrialized countries, the number of patients with cancer is increasing steadily with the aging of the population. The development of therapies for curing cancer with safety and assurance, and with less suffering, even among the elderly, is strongly desired by such countries. Owing to the results of previous studies, carbon-ion radiotherapy has been recognized as a treatment that can achieve this goal. Studies conducted over the past 10 years have made it possible to reduce the cost of a carbon-ion radiotherapy system to about one-third of that of HIMAC. Next-generation carbon-ion radiotherapy equipment will provide more advanced treatment at a drastically reduced cost. To achieve this, cooperation between equipment manufacturers and researchers in this field is indispensable, and government support is critical.

Future carbon-ion radiotherapy will provide cancer treatment with less suffering for many intractable cancers using more advanced systems and, hopefully, recent molecular biological techniques.

ACKNOWLEDGMENTS

The authors would like to express their gratitude to Dr. N. Matsufuji for useful discussion on radiation quality and TCP in carbon-ion radiotherapy, and to the other members of the Research Center of Charged Particle Therapy at NIRS for their useful discussion and warm support.

REFERENCES

Akakura, K., Tsujii, H., Morita, S., Tsuji, H., Yagishita, T., Isaka, S., Ito, H., Akaza, H., Hata, M., Fujime, M., Harada, M., and Shimazaki, J. 2004. Phase I/II clinical trials of carbon ion therapy for prostate cancer. *Prostate* 58: 252–258.

Alonso, J. R. 1993. Synchrotron: The American experience. In *Hadrontherapy in Oncology*, U. Amaldi and B. Larsson (eds.), pp. 266–281. Amsterdam, the Netherlands: Elsevier.

Amaldi, U. and Kraft, G. 2007. European developments in radiotherapy with beams of large radiobiological effectiveness. *J. Radiat. Res.* 48: A27–A41.

Blakely, E. A., Ngo, F. Q., Curtis, S. B., and Tobias, C. A. 1984. Heavy ion radiotherapy: Cellular studies. *Adv. Radiat. Biol.* 11: 295–390.

Chu, W. T., Ludewigt, B. A., and Renner, T. R. 1993. Instrumentation for treatment of cancer using proton and light-ion beams. *Rev. Sci. Instrum.* 64: 2055–2122.

Eickhoff, H., Haberer, Th., Schlitt, B., and Weinrich, U. 2004. HICAT—The German hospital-based light ion cancer therapy project. In *Proceedings of the Ninth EPAC*, Lucerne, Switzerland, pp. 290–294.

Furukawa, T., Inaniwa, T., Sato, S., Tomitani, T., Minohara, S., Noda, K., and Kanai, T. 2007. Design study of a raster scanning system for moving target irradiation in heavy-ion radiotherapy. *Med. Phys.* 34: 1085–1097.

Furukawa, T., Inaniwa, T., Sato, S., Iwata, Y., Fujimoto, T., Minohara, S., Noda, K., and Kanai, T. 2008a. Design study of a rotating gantry for the HIMAC new treatment facility. *Nucl. Instrum. Meth. B* 266: 2186–2189.

Furukawa, T., Inaniwa, T., Sato, S., Saotome, N., Takei, Y., Iwata, Y., Nagano, A., Mori, S., Minohara, S., Shirai, T., Murakami, T., Takada, E., Noda K., and Kanai, T. 2008b. Development of scanning system at HIMAC. In *Proceedings of the 11th EPAC*, Genoa, Italy, pp. 1794–1796.

Futami, Y., Kanai, T., Fujita, M., Tomura, H., Higashi, A., Matsufuji, N., Miyahara, N., Endo, M., and Kawachi, K. 1999. Broad-beam three-dimensional irradiation system for heavy-ion radiotherapy at HIMAC. *Nucl. Instrum. Meth. A* 430: 143–153.

Goldhaber, A. S. 1974. Statistical models of fragmentation process. *Phys. Lett. B* 53: 306–308.

Haberer, Th., Becher, W., Shardt, D., and Kraft, G. 1993. Magnetic scanning system for heavy ion therapy. *Nucl. Instrum. Meth. A* 330: 296–305.

Hirao, Y., Ogawa, H., Yamada, S., Sato, Y., Yamada, T., Sato, K., Itano, A., Kanazawa, M., Noda, K., Kawachi, K., Endo, M., Kanai, T., Kohno, T., Sudou, M., Minohara, S., Kitagawa, A., Soga, F., Takada, E., Watanabe, S., Endo, K., Kumada, M., and Matsumoto, S. 1992. Heavy ion synchrotron for medical use. *Nucl. Phys. A* 538: 541c–550c.

Imai, R., Kamada, T., Tsuji, H., Tsujii, H., Tsuburai, Y., and Tatezaki, S. 2006. Cervical spine osteosarcoma treated with carbon ion radiotherapy. *Lancet Oncol.* 7: 1034–1035.

Inaniwa, T., Furukawa, T., Tomitani, T., Sato, S., Noda, K., and Kanai, T. 2007a. Optimization for fast-scanning irradiation in particle therapy. *Med. Phys.* 34: 3302–3311.

Inaniwa, T., Furukawa, T., Matsufuji, N., Kohno, T., Sato, S., Noda, K., and Kanai, T. 2007b. Clinical ion beams: Semi-analytical calculation of their quality. *Phys. Med. Biol.* 52: 7261–7279.

Inaniwa, T., Furukawa, T., Sato, S., Tomitani, T., Kobayashi, M., Minohara, S., Noda, K., and Kanai, T. 2008. Development of treatment planning for scanning irradiation at HIMAC. *Nucl. Instrum. Meth. B* 266: 2194–2198.

Ishikawa, H., Tsuji, H., Kamada, T., Yanagi, T., Mizoe, J., Kanai, T., Morita, S., Wakatsuki, M., Shimazaki, J., and Tsujii, H. 2006. Carbon ion radiation therapy for prostate cancer. Results of a prospective phase II study. *Radiother. Oncol.* 81: 57–64.

Kamada, T., Tsujii, H., Tsuji, H., Yanagi, T., Mizoe, J., Miyamoto, T., Kato, H., Yamada, S., Morita, S., Yoshikawa, K., Kandatsu, S., and Tateishi, A. 2002. Efficacy and safety of carbon ion radiotherapy in bone and soft tissue sarcomas. *J. Clin. Oncol.* 22: 4472–4477.

Kameoka, S., Amako, K., Iwai, G., Murakami, K., Sasaki, T., Toshito, T., Yamashita, T., Aso, T., Kimura, A., Kanai, T., Komori, M., Takei, Y., Yonai, S., Tashiro, M., Koikegami, H., Tomita, H., and Koi, T. 2008. Dosimetric evaluation of nuclear interaction models in the Geant4 Monte Carlo simulation toolkit for carbon-ion radiotherapy. *Radiol. Phys. Tech.* 1: 183–187.

Kanai, T., Kawachi, K., Kumamoto, Y., Ogawa, H., Yamada, T., Matsuzawa, H., and Inada, T. 1980. Spot scanning system for proton radiotherapy. *Med. Phys.* 7: 365–369.

Kanai, T., Kawachi, K., Matsuzawa, H., and Inada, T. 1983. Broad beam three dimensional irradiation for proton radiotherapy. *Med. Phys.* 10: 344–346.

Kanai, T., Furusawa, Y., Fukutsu, K., Itsukaichi, H., Eguchi-Kasai, K., and Ohara, H. 1997. Irradiation of mixed beam and designing of spread-out Bragg peak for heavy-ion radiotherapy. *Radiat. Res.* 147: 78–85.

Kanai, T., Kanematsu, N., Minohara, S., Komori, M., Torikoshi, M., Asakura, H., Ikeda, N., Uno, T., and Takei, T. 2006. Commissioning of a conformal irradiation system for heavy-ion radiotherapy using a layer-stacking method. *Med. Phys.* 33: 2989–2997.

Kanematsu, N., Endo, M., Futami, Y., and Kanai, T. 2002. Treatment planning for the layer-stacking irradiation system for three-dimensional conformal heavy-ion radiotherapy. *Med. Phys.* 29: 2823–2829.

Kato, H. 2004. Clinical study of carbon ion radiotherapy for hepatocellular carcinoma. In *Hepatocellular Carcinoma Screening, Diagnosis, and Management, NIH Conference*, Bethesda, MD, pp. 195–196.

Kato, H., Tsujii, H., Miyamoto, T., Mizoe, J., Kamada, T., Tsuji, H., Yamada, S., Kandatsu, S., Yoshikawa, K., Obata, T., Ezawa, H., Morita, S., Tomizawa, M., Morimoto, N., Fujita, J., and Ohto, M. 2004. Results of the first prospective study of carbon ion radiotherapy for hepatocellular carcinoma with liver cirrhosis. *Int. J. Radiat. Oncol. Biol. Phys.* 59: 1468–1476.

Kato, H., Yamada, S., Yasuda, S., Maeda, Y., Kamada, T., Mizoe, J., Ohto, M., and Tsujii, H. 2005a. Phase II study of short-course carbon ion radiotherapy (52.8GyE/4-fraction/1-week) for hepatocellular carcinoma. *Hepatology* 42 (Suppl. 1): 381A.

Kato, H., Yamada, S., Yasuda, S., Yamaguchi, K., Kitabayashi, H., Kamada, T., Mizoe, J., Ohto M., and Tsujii, H. 2005b. Two-fraction carbon ion radiotherapy for hepatocellular carcinoma. Preliminary results of a phase I/II clinical trial. *J. Clin. Oncol.* 23 (Suppl. S): 338S.

Komori, M., Furukawa, T., Kanai, T., and Noda, K. 2004. Optimization of spiral wobbler system for heavy-ion radiotherapy. *Jpn. J. Appl. Phys.* 43: 6463–6467.

Lawrence, T. S., Petrelli, N. J., Li, B. D., and Galvin, J. M. 2007. Think globally, act locally. *J. Clin. Oncol.* 25: 924–930.

Leoehrer, P., Arececi, R., Glatstein, E., Gordon, M., Hanna, N., Morrow, M., and Thigpen, T. (eds.). 2006. *The Year Book of Oncology 2006*, pp. 368–370. Philadelphia, PA: Elsevier/Mosby.

Lomax, A. 1999. Intensity modulation methods for proton radiotherapy. *Phys. Med. Biol.* 44: 185–205.

Matsufuji, N., Komori, M., Sasaki, H., Akiu, K., Ogawa, M., Fukumura, A., Urakabe, E., Inaniwa, T., Nishio, T., Kohno, T., and Kanai, T. 2005. Spatial fragment distribution from a therapeutic pencil-like carbon beam in water. *Phys. Med. Biol.* 50: 3393–3403.

Minohara, S., Kanai, T., Endo, M., Noda, K., and Kanazawa, M. 2000. Respiration gated irradiation system for heavy-ion radiotherapy. *Int. J. Radiat. Oncol. Biol. Phys.* 47: 1097–1103.

Miyamoto, T., Yamamoto, N., Nishimura, H., Koto, M., Tsujii, H., Mizoe, J., Kamada, T., Kato, H., Yamada, S., Morita, S., Yoshikawa, K., Kandatsu, S., and Fujisawa, T. 2003. Carbon ion radiotherapy for stage I non-small cell lung cancer. *Radiother. Oncol.* 66: 127–140.

Miyamoto, T., Baba, M., Yamamoto, N., Koto, M., Sugawara, T., Yashiro, T., Kadono, K., Ezawa, H., Tsujii, H., Mizoe, J., Yoshikawa, K., Kandatsu, S., and Fujisawa, T. 2007. Curative treatment of stage I non-small cell lung cancer with carbon beams using a hypo-fractionated regimen. *Int. J. Radiat. Oncol. Biol. Phys.* 67: 750–758.

Mizoe, J., Tsujii, H., Kamada, T., Matsuoka, Y., Tsuji, H., Osaka, Y., Hasegawa, A., Yamamoto, N., Ebihara, S., and Konno, A. 2004. Dose escalation study of carbon ion radiotherapy for locally advanced head and neck cancer. *Int. J. Radiat. Oncol. Biol. Phys.* 60: 358–364.

Molière, G. 1948. Theorie der Streuung Schneller Geladener Teilchen. Mehrfach-und Vielfachstreuung Z Naturforsch. 3a: 78–97.

Munzenrider, J. E. and Liebsch, N. J. 1999. Proton therapy for tumors of the skull base. *Strahlenther Onkol.* 175(Suppl. 2): 57–63.

Noda, K., Kanazawa, M., Itano, A., Takada, E., Torikoshi, M., Araki, N., Yoshizawa, J., Sato, K., Yamada, S., Ogawa, H., Itoh, H., Noda, A., Tomizawa, M., and Yoshizawa, M. 1996. Slow beam extraction by a transverse RF field with AM and FM. *Nucl. Instrum. Meth. A* 374: 269–277.

Noda, K., Furukawa, T., Fujisawa, T., Iwata, Y., Kanai, T., Kanazawa, M., Kitagawa, A., Komori, M., Minohara, S., Murakami, T., Muramatsu, M., Sato, S., Takei, Y., Tashiro, M., Torikoshi, M., Yamada, S., and Yusa, K. 2007. New accelerator facility for carbon-ion cancer-therapy. *J. Radiat. Res.* 48: A43–A54.

Nose, H., Kase, Y., Matsufuji, N., and Kanai, T. 2009. Field size effect of radiation quality in carbon therapy using passive method. *Med. Phys.* 36: 870–875.

Pedroni, E., Bacher, R., Blattmann, H., Boehringer, T., Coray, A., Lomax, A., Lin, S., Munkel, G., Scheib, S., Schneider, U., and Tourovsky, A. 1995. The 200-MeV proton therapy project at the Paul Scherrer Institute: Conceptual design and practical realization. *Med. Phys.* 22: 37–53.

Rossi, S. 2006. Developments in proton and light-ion therapy. In *Proceedings of the 10th EPAC*, Edinburgh, U.K., pp. 3631–3635.

Sato, S., Furukawa, T., and Noda, K. 2007. Dynamic intensity control system with RF-knockout slow-extraction at HIMAC synchrotron. *Nucl. Instrum. Meth. A* 574: 226–231.

Schulz-Ertner, D. and Tsujii, H. 2007. Particle radiation therapy using proton and heavier ion beams. *J. Clin. Oncol.* 10: 953–964.

Tobias, C. A., Anger, H. O., and Lawrence, J. H. 1952. Radiological use of high energy deuterons and alpha particles. *Am. J. Roentgenol. Radium. Ther. Nucl. Med.* 67: 1–27.

Torikoshi, M., Monohara, S., Kanematsu, N., Komori, M., Kanazawa, M., Noda, K., Miyahara, N., Itoh, H., Endo, M., and Kanai, T. 2007. Irradiation system for HIMAC. *J. Radiat. Res.* 48: A15–A25.

Tsuji, H., Yanagi, T., Ishikawa, H., Kamada, T., Mizoe, J., Kana, T., Morita, S., and Tsujii, H. 2005. Hypofractionated radiotherapy with carbon beams for prostate cancer. *J. Radiat. Oncol. Biol. Phys.* 32: 1153–1160.

Urakabe, E., Kanai, T., Kanazawa, M., Kitagawa, A., Noda, K., Tomitani, T., Suda, M., Iseki, Y., Hanawa, K., Sato, K., Shinbo, M., Mizuno, H., Hirata, Y., Futami, Y., Iwashita, Y., and Noda, A. 2001. Spot scanning using radioactive 11C beams for heavy-ion therapy. *Jpn. J. Appl. Phys.* 40: 2540–2548.

Wilson, R. R. 1946. Radiological use of fast protons. *Radiology* 47: 487–491.

Wong, M., Schimmerling, W., Phillips, M. H., Ludewigt, B. A., Landis, D. A., Walton, J. T., and Curtis, S. B. 1990. The multiple Coulomb scattering of very heavy charged particles. *Med. Phys.* 17: 163–171.

Yonai, S., Kanematsu, N., Komori, M., Kanai, T., Takei, Y., Takahashi, O., Isobe, Y., Tashiro, M., Koikegami, H., and Tomita, H. 2008. Evaluation of beam wobbling methods for heavy-ion radiotherapy. *Med. Phys.* 35: 927–938.

25 Nanoscale Charge Dynamics and Nanostructure Formation in Polymers

Akinori Saeki
Osaka University
Osaka, Japan

Shu Seki
Osaka University
Osaka, Japan

Kazuo Kobayashi
Osaka University
Osaka, Japan

Seiichi Tagawa
Osaka University
Osaka, Japan

CONTENTS

25.1 INTRODUCTION

The first systematic study of irradiation of polymers was carried out in order to determine the radiation damage to polymers for use in radiation fields of the Manhattan project during World War II. The Manhattan project had a very great influence on all fields of nuclear research, including radiation chemistry. The peaceful uses of the nuclear reactor in the field of radiation effects on polymers started before the "Atoms for Peace" address delivered by President Eisenhower in 1953. The first systematic study of radiation effects on polyethylene thin films by Dole and Rose, just after World War II, was carried out using nuclear reactors. Their research was published as a master's thesis in 1948 and, later, as a published paper (Dole et al., 1954). However, non-soluble and non-melting cross-linked polyethylene irradiated by a nuclear reactor was demonstrated by Charlesby in 1952, and it had a huge impact on industry (Charlesby, 1952). Today, the economic scale of utilization of radiation in industry is very large and comparable to the economic scale of utilization of nuclear energy in Japan and the United States (Tagawa et al., 2002; Yanagisawa et al., 2002). The economic scale of radiation use in polymer fields, such as radiation cross-linking for the production of improved cables, heat-shrinkable packaging films and tubes, cross-linked automobile tire components, resist materials for semiconductor device fabrication, polymers for biomedical applications, radiation sterilization of medical plastic supplies, radiation curing etc., is now well established (Tagawa et al., 2002; Yanagisawa et al., 2002).

This chapter is presented in four sections. Section 25.2 reviews the nanoscale dynamics of charged species studied by pulse radiolysis. Pulse radiolysis studies on relaxation processes of polymer radical ions, highly sensitive nanospace reactions through the combination of dissociative electron attachment with geminate recombination, recent progress in σ-conjugated polymer research based on the direct observation of radical anions and cations of σ-conjugated polymers, and the first detection and dynamics of radical cations and anions of DNA in an aqueous solution at room temperature in connection with the initial step of the radiation damages to DNA are reviewed. In Section 25.3, nanoscale dynamics of charge carriers by time-resolved microwave conductivity (TRMC) is discussed. TRMC is an electrodeless method that detects the intrinsic nature of charge carrier transport in the nanospace of organic semiconductor materials, comparable with field-effect transistor (FET) and time-of-flight (TOF) methods, which are dependent on grain boundary, impurities, and structural defects. Section 25.4 deals with the "unique" ion beam (IB)-induced nanostructure formation in polymers, that is, nanostructure formation by high-energy IBs. A limited spatial distribution of the transient species induced by IBs gives "unique" nanomaterials as the final products via nonhomogeneous chemical processes, which could not be realized by other chemical reactions. We can produce nanostructures based on any cross-linkable polymer materials.

25.2 NANOSCALE DYNAMICS OF CHARGED SPECIES STUDIED BY PULSE RADIOLYSIS

25.2.1 PULSE RADIOLYSIS STUDIES ON CHARGED SPECIES OF POLYMERS

After the start of systematic studies on industrial applications of radiation of polymers (Charlesby, 1952; Dole et al., 1954), the mechanisms of radiation-induced polymer reactions have been extensively studied. A direct detection of reactive intermediates is essential to elucidate the detailed mechanisms of radiation-induced polymer reactions. Pulse radiolysis is the most powerful method for detecting radiation-induced reactive intermediates of polymers directly and for investigating the dynamics of reactive charged species (Tagawa, 1986, 1991a,b, 1993). A large number of unsuccessful pioneering researches had been carried out to detect the transient absorption of both polymer and monomer radical ions as initiators of both polymer and polymerization reactions by pulse radiolysis techniques after the introduction of pulse radiolysis in radiation chemistry in the late 1950s. The first detection of the radical ions of polymers by pulse radiolysis was carried out for polyvinylcarbazole in 1974

(Tagawa et al., 1974). Also, the first detection of the monomer radical cation as the initiator of polymerization was carried out in 1972 (Tagawa et al., 1972). After the first detection of polymer radical ions (Tagawa et al., 1974), many kinds of reactive radical ions of polymers (Tagawa, 1986, 1991a,b, 1993), as initiators of radiation-induced polymer reactions, have been observed by pulse radiolysis at room temperature, including IB pulse radiolysis of polymers (Tagawa, 1993) as well as electron-beam (EB) pulse radiolysis of polymers (Tagawa, 1991a,b). The spectroscopic studies on radical ions of typical polymers, such as polystyrene- and polyethylene-related polymers, polysilanes, and polymethylmethacrylate (PMMA), have been reviewed (Tagawa et al., 2004). Here, we review pulse radiolysis studies on nanospace dynamics of charged species in polymers: at first, nanospace relaxation processes of polymer ion radicals, and then, highly sensitive decomposition of polymers and highly sensitive reactions of EB, x-ray, and extreme ultraviolet (EUV) resists through nanospace reactions. Polystyrene-related polymers have been extensively studied in connection with industrial applications such as highly sensitive radiation-induced cross-linking and degradation initially, and highly sensitive resist materials later. Pulse radiolysis and laser flash photolysis have been extensively employed to clarify the detailed mechanisms of highly sensitive radiation-induced cross-linking and degradation and highly sensitive resist processes. In all cases, the combination of highly effective dissociative electron attachment with geminate ion recombination plays an important role in highly sensitive nanospace reactions. This combination is only one large-scale industrial application of radiation-induced nanoscale charge dynamics, as far as we know.

25.2.1.1 Relaxation Processes of Polymer Radical Ions

After the first detection of polymer radical cation (Tagawa et al., 1974), the detailed dynamics of polyvinylcarbazole radical cation was studied by nanosecond pulse radiolysis, and the difference between the reactivity of ethylcarbazole (the monomer unit of polymer) and polyvinylcarbazole radical cations was clearly shown (Washio et al., 1981). The stabilization of the polymer radical cation decreases with its reactivity (Washio et al., 1981). Relaxation kinetics of the polymer excited states of polyvinylcarbazole studied by picosecond pulse radiolysis (Tagawa et al., 1979) was studied in tandem with the intramolecular excimer formation of polyvinylcarbazole in the late 1970s. A large number of picosecond and femtosecond laser flash photolysis studies on the dynamics of polymer excited states have been carried out in 1980s, 1990s, and 2000s. A comprehensive survey on this subject is beyond the scope of this chapter. Recently, systematic picosecond pulse radiolysis studies on the relaxation of cation radicals of polystyrene and related polymers have been performed (Okamoto et al., 2001, 2006). Deprotonation reactions have also been studied with respect to the development of EUV resists (Okamoto et al., 2008, 2009). Picosecond pulse radiolysis studies on two "unique" relaxation processes for both polymer excited states and radical ions have been carried out (Matsui et al., 2002b; Ohnishi et al., 2009). The intermolecular energy migration between conjugated segments in a polymer chain of σ-conjugated polymer excited states (Matsui et al., 2002b) and the conformational relaxation of σ-conjugated polymer radical anions, which was observed for the first time by nanosecond pulse radiolysis in 1987 (Ban et al., 1987), have also been clearly observed by picosecond pulse radiolysis in 2009 (Ohnishi et al., 2009).

25.2.1.2 Highly Sensitive Decomposition of Polymers through Combination of Dissociative Electron Attachment with Geminate Ion Recombination

Polystyrene-related polymers in a solid state or in a solution of dioxane and cyclohexane are generally not sensitive to radiation. But these become very sensitive to radiation and are easily decomposed by irradiation in some chlorinated solvents. The reason why radiation-resistant polystyrene-related polymers change to radiation-sensitive polymers in some chlorinated solvents was made clear by the researches in laser flash photolysis (Tagawa and Schnabel, 1980a,b, 1983; Tagawa et al., 1984; Tagawa, 1986) and pulse radiolysis (Tagawa et al., 1981; Itagaki et al., 1983, 1987, 1992; Washio et al., 1983; Tagawa, 1986; Zezin et al., 1995; Zhang and Thomas, 1996; Okamoto et al., 2001, 2006).

In the case of photolysis of polystyrene and polystyrene-related polymers in cyclohexane, the photodecomposition of polymers does not occur effectively, and singlet monomer and excimer states (Tagawa and Schnabel, 1980a) and also triplet states (Tagawa et al., 1984) of polystyrene are observed by laser flash photolysis. But, in the case of photolysis of polystyrene and poly-α-methylstyrene in chlorinated solvents, the photodecomposition of polymers occurs very effectively and polymer excited states are converted to charge-transfer (CT) complexes. The CT complex was observed by laser flash photolysis (Tagawa and Schnabel, 1980b, 1983). In the case of radiolysis of polystyrene-related polymers in cyclohexane, the photodecomposition of polymers does not occur effectively, and singlet monomer and excimer states of polystyrene and poly-α-methylstyrene are observed by picosecond pulse radiolysis (Tagawa et al., 1981; Itagaki et al., 1983, 1987, 1992). The CT complex was also observed in the case of pulse radiolysis of polystyrene and poly-α-methylstyrene in chlorinated solvents (Tagawa et al., 1981; Washio et al., 1983). Based on these studies, the combination of very effective dissociative electron attachment with geminate ion recombination produces a CT complex and then polymer radicals, and plays an important role in the initial reactions of highly sensitive decomposition of polystyrene in chlorinated solvents with oxygen.

25.2.1.3 Highly Sensitive Electron Beam, X-Ray, and EUV Resists through the Combination of Dissociative Electron Attachment with Geminate Ion Recombination

Photo and radiation induced decomposition of polymers occurs very effectively in polystyrene related polymer solutions in chlorinated solvents, but photo and radiation induced polymer cross-linking occurs very effectively in chlorinated polystyrene-related polymers in the solid state. Chlorinated polystyrene-related polymer films were used for EB and excimer negative resists. In the 1980s, the increasing density of VLSI required the development of submicron lithography, such as excimer, x-ray, EB, and IB lithography. At that time, much attention was devoted to negative EB resists (Imamura, 1979; Imamura et al., 1982; Kamoshida et al., 1983) containing chlorinated or chloromethylated phenyl side chains. This type of enhancement is observed in other resists containing chlorinated or chloromethylated silicon-containing resists (Morita et al., 1983, 1984). EB (Tabata et al., 1984, 1985; Tagawa, 1986, 1987) and IB (Tagawa, 1993) pulse radiolysis and excimer laser flash photolysis (Tagawa, 1986, 1987) studies on resist materials were carried out in order to elucidate the detailed cross-linking reaction of chlorinated or chloromethylated excimer, x-ray, EB, and IB resists. The same CT complex is confirmed in both photo- and radiation-induced chlorinated or chloromethylated resist reactions. Similar reactions of iodination (Shiraishi et al., 1980) of phenyl rings are used for EB resists. In both photolysis and radiolysis, the generation of two polymer radicals and acid through dissociative electron attachment and decomposition of the CT complex is observed. But the initial steps of EB and x-ray resist reactions are mainly induced by ionization and are different from the initial steps of photoresists induced by excited states of molecules. The combination of geminate ion recombination with dissociative electron attachment induces pair production of polymer radicals, which in turn induces highly effective cross-linking even for chloromethylpoly-α-methylstyrene, although poly-α-methylstyrene is well known to be decomposed on irradiation. The fundamental radiation chemistry of nanospace reactions due to the geminate recombination in organic materials, such as liquid alkanes (Thomas et al., 1968; Sauer and Jonah, 1980; van den Ende et al., 1980; Tagawa et al., 1983, 1989; Yoshida et al., 1987; Le Motais and Jonah, 1989; Saeki et al., 2002, 2004, 2005a) and liquid aromatics (Okamoto et al., 2003, 2007), and also the formation processes of the CT complex of polystyrene-related polymers through dissociative electron attachment and geminate recombination (Okamoto et al., 2001, 2006) have been studied extensively by nanosecond, picosecond, and subpicosecond pulse radiolysis.

Recently, a single-component chemically amplified resist based on dehalogenation of polymers has been reported (Yamamoto et al., 2007a). The reaction mechanism of this resist is acid formation through dissociative electron attachment and geminate recombination (Yamamoto et al., 2005a,b), similar to the acid formation of halogenated polystyrene-related resists (HPRR) (Tabata et al., 1984, 1985; Tagawa, 1986,1987). Both acid formation processes produce two polymer radicals. Chemically amplified resists use acid for acid-catalyzed reactions, and HPRRs use a polymer radical for cross-linking. The difference between the acid formation processes for HPRR and brominated single-component chemically amplified resist (BSCAR) is explained as follows. In both cases, the initial steps are ionization, emission of secondary electrons, and ionization and excitation of molecules, and then thermalization of electrons. The processes of polymer radical cation formation by ionization and halogen anion formation by dissociative electron attachment are the same. The stability of polymer radical cations is different. Polymer radical cations are stable in HPRR but unstable in BSCAR, where the protons and polymer radicals are generated through the deprotonation of polymer radical cations. In both processes, geminate recombination occurs. In BSCAR, geminate pairs are proton and halogen anion and geminate recombination produces acid. In HPRR, geminate pairs are polymer radical cation and halogen anion and geminate recombination produces a CT complex, which in turn produces acid and a polymer radical. Thus, the final products are the same, but the formation processes and distances between two polymer radicals are different. Distance of two polymer radicals is very short in HPRR, but is fairly long in BSCAR. The dissociative electron attachment is competitive between halogenated resist polymers and the acid generators in BSCAR (Yamamoto et al., 2009).

25.2.2 Pulse Radiolysis Studies on Dynamics of Charged Species in σ-Conjugated Systems

σ-Conjugated polymers are of current interest because of their characteristic electron delocalization along their conjugated backbones (Rice, 1979) and associated optoelectronic properties (Trefonas et al., 1985; Kajzar et al., 1986; West, 1986). The dynamics of excess electrons and holes on the conjugated skeletons have been investigated vigorously in view of their potential application in electroluminescent diodes and as photoconductors (Kepler et al., 1987; Suzuki et al., 1996), on the basis of their utility as positive charge conductors. The dominant process of charge carrier transport in σ-conjugated polymers is the hopping of holes between localized states originating from domain-like subunits along the chain (Abkowitz et al., 1987; Fujino, 1987). The mean length of the segments is controlled by steric hindrance of the side chains and/or thermal molecular motion. Several groups have reported the synthesis of a series of polysilanes or polygermanes with backbone conformations varying from random coil to stiff and rod-like conformations by changing the polymer substitution pattern (Fujiki, 1994, 1996; Koe et al., 2000). The clear correlation between the conformation of the backbones and the characteristic electronic transition observed at near-UV region has been reported. This indicates that one can easily estimate the conformation of the σ-conjugated polymers in a solvent at any temperature by a simple spectroscopy of σ-conjugated polymer solutions. Despite positive charge being the dominant carrier in the transport process, there have been few dynamics or quantitative analyses on the electronic state of positive charges on the Si or Ge chain other than by transient spectroscopy (Ban et al., 1987; Seki et al., 2004a) or electron spin resonance (Kumagai et al., 1995; Seki et al., 1998). The localization of the charge carriers was revealed to be suppressed in the σ-conjugated polymers bearing bulky pendant groups, suggesting not only that the localization in typical dialkyl-substituted polymers arises from the flexibility of Si(Ge) catenation, but also that delocalization occurs in σ-conjugated polymers with stiff or rod-like skeletons. However, the quantitative correlation between the molecular stiffness and the degree of positive charge delocalization has not been elucidated to date.

This chapter summarizes the recent results on the direct observation of anion and cation radicals of σ-conjugated polymers by pulse radiolysis. Pulse-radiolysis transient absorption spectroscopy (PR-TAS) is a very powerful and useful technique for achieving selective formation of ion radicals in matrices (solvents) and for tracing reaction kinetics. To date, quantitative tracing of the anion and cation radicals has been very difficult because of their very low ionization potentials (<6 eV) (Seki et al., 1988; Ishii et al., 1995). The PR-TAS is a unique technique that enables the quantitative formation of charged radicals of the polymers without associated counter ions. The intrinsic electronic structures of the charged radicals of σ-conjugated polymers, including the molar extinction coefficient and oscillator strength, were determined fully experimentally by both spectroscopic techniques and an efficient CT reaction between the radical ions and some electron acceptors (donors) (Seki et al., 1999a, 2001a,b, 2002, 2004a; Kawaguchi et al., 2003). The degree of charge delocalization on σ-conjugated chains is discussed in terms of the molecular stiffness of σ-conjugated polymers, and the role of pendant rings in the charge carrier transport processes in polymer materials is addressed.

The details of the synthesis of monomers and polymers have been described elsewhere (Seki et al., 1999a, 2004a). The chemical structures of the synthesized polymers are shown in Figure 25.1.

The UV-vis absorption spectra of three series of polysilanes (n-alkylphenyl [PHn], di-n-alkyl [PDn], and methyl-n-alkyl [MEn] substituted: the value of n represents the number of carbon atoms in the n-alkyl substituents) are shown in Figure 25.2. The intense UV absorption band observed for steady-state polysilane solutions is ascribed to the transition between the valence band (VB) and the lowest excitonic states (ES) of the Si backbones (Tachibana et al., 1990). The molar extinction coefficient (ε^{abs}) and oscillator strength (f_{VB-ES}) of the absorption are listed in Table 25.1. The values of both ε^{abs} and f_{VB-ES} depend strongly on the substitution patterns, with a considerable increase accompanying the change from asymmetric alkyl to symmetric alkyl substitution. The values also increase dramatically with the elongation of n-alkyl substituents from methyl to n-octyl in the case of poly(n-alkylphenylsilane)s, with only a slight drop with further elongation beyond n-octyl. Recently, Fujiki reported an empirical relationship between ε_{abs} and the viscosity index, α, reflecting the geometric structure of the polymer main chain. The following empirical formula was obtained for the relationship between α and ε_{abs} (Matsui et al., 2002a):

$$\varepsilon_{abs} = 1130 \times e^{2.9\alpha} \tag{25.1}$$

The relationship between gyration length (R_g) and α is as follows:

$$R_g = \kappa M^\nu \tag{25.2}$$

where
κ is a constant
M is the molecular weight of the polymer
$\nu = (\alpha + 1)/3$

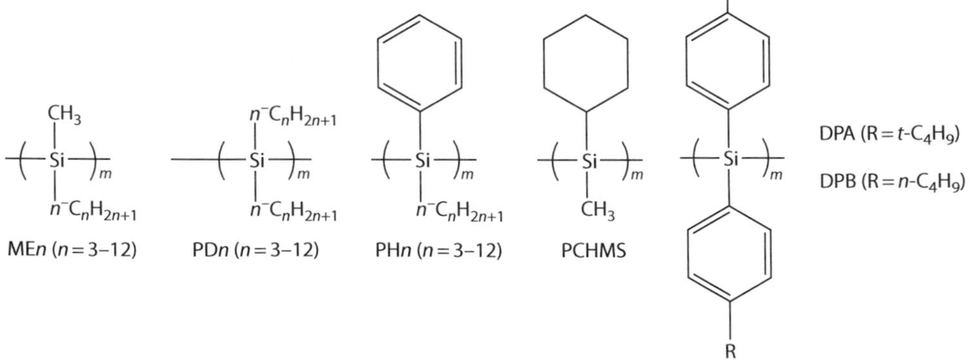

FIGURE 25.1 Chemical structure of the series of polysilanes employed.

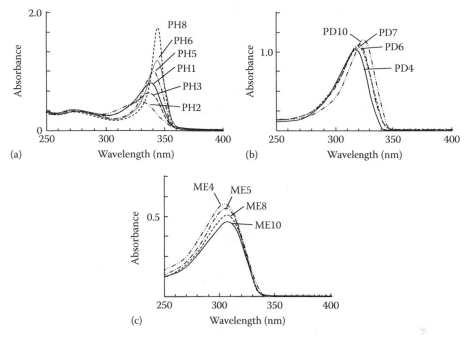

FIGURE 25.2 UV-vis absorption spectra of (a) PHn, (b) PDn, and (c) MEn. Spectra were recorded at 25°C for solutions of polysilanes in THF at 1.0×10^{-4} mol dm^{-3} (base mol unit). (Reprinted from Seki, S. et al., *J. Am. Chem. Soc.*, 126, 3521, 2004a. With permission.)

The estimated values of α are shown in Figure 25.3, varying from 0.46 for PH2 to 0.94 for PH8. Assuming a constant degree of polymerization for the polysilanes, R_g is expected to increase with ε_{abs}. The actual Kuhn segment lengths (q) of the polymers were found to be $q = 1.1$ nm for PH1 (Seki et al., 2002), 3.0 nm for PD6 (Cotts et al., 1991; Shukla et al., 1991), and 4.5 nm for DPB, determined by light-scattering experiments (Koe et al., 2001; Seki et al., 2002). All the polysilanes prepared in the present study have high molecular weights ($>10^4$), low polydispersities (<4), and monomodal distributions similar to the Schulz–Zimm distribution. The polysilanes with high values of α, such as DPB or PH5–PH12, expected to exhibit longer segment lengths, have molecular weights higher than 10^5. A flexible wormlike chain model with a persistence length of ~10 nm successfully reproduced the global dimension of the Si chain in DPB with a molecular weight of ~10^5 (Cotts et al., 1990). Thus, for almost all of the polysilanes in the present study, flexible Kuhn chains are acceptable as the first approximation model of their chain configurations on the basis of α and molecular weight. The gyration length of the flexible Kuhn chain is given by

$$R_g = \left(\frac{\langle r^2 \rangle_0}{6} \right)^{1/2} \tag{25.3}$$

where $\langle r^2 \rangle_0$ is the mean-square end-to-end distance under θ conditions. The value of q at the chosen molecular weight is derived from the mean-square end-to-end distance ($\langle r^2 \rangle$) as follows:

$$q = \langle r^2 \rangle \left(\frac{M_L}{2M} \right) \tag{25.4}$$

where M_L is the mass per unit length. The excluded volume parameter a, defined as $a^2 \equiv \langle r^2 \rangle / \langle r^2 \rangle_0$, has been reported to be less than 1.2 for DPB at 25°C in toluene, which is a better solvent than THF (Shukla et al., 1991). The low second virial coefficient of PD6 in THF (1.14×10^{-4} mL mol g^{-2}) also

TABLE 25.1

UV Absorption, Fluorescence, and Optical Properties of Cation Radicals of Poly(n-Alkylphenylsilane)s

Entry	λ_{max}^{abs} (nm)	$\varepsilon^{abs\ a}$ (mol^{-1} dm^3 cm^{-1})	$f_{VB-ES}^{\ b}$	λ_{max}^{fl} (nm)	$\lambda_{max}^{\bullet-}$ (nm)	$\varepsilon^{\bullet-}$ (10^5 mol^{-1} dm^3 cm^{-1})	$f^{\bullet-}$	$\lambda_{max}^{\bullet+}$ (nm)	$\varepsilon^{\bullet+\ c}$ (10^4 mol^{-1} dm^3 cm^{-1})	$f^{\bullet+}$
PH1	339	7,900	0.091	365	369	1.6	0.68	365	9.4	0.49
PH2	331	4,300	0.0547	366	365	1.5	—	360	6.9	0.38
PH3	336	5,700	0.0642	366	—	—	—	363	7.1	0.42
PH4	341	8,000	0.0774	366	370	2.1	—	365	9.8	0.51
PH5	342	9,900	0.0914	367	—	—	—	369	12	0.59
PH6	347	11,700	0.11	369	370	3.0	—	372	15	0.65
PH7	348	14,300	0.11	369	—	—	—	372	16	0.59
PH8	348	17,500	0.13	370	375	3.6	—	372	18	0.63
PH10	348	16,600	0.13	369	—	—	—	370	17	0.66
PH12	348	12,800	0.110	369	—	—	—	370	16	0.60
PD4	317	10,400	0.053	346	314	1.2	—	346	4.5	0.31
PD5	318	10,400	0.056	347	—	—	—	346	4.7	0.32
PD6	319	10,600	0.056	347	318	1.3	—	344	5.1	0.37
PD7	321	10,900	0.057	348	—	—	—	340	5.2	0.38
PD8	321	11,600	0.057	348	—	—	—	342	5.2	0.39
PD10	324	11,700	0.057	349	—	—	—	342	5.5	0.41
PD12	325	10,600	0.050	350	—	—	—	342	5.5	0.46
ME3	307	5,200	0.043	338	358	1.1	—	348	3.8	0.24
ME4	306	6,400	0.045	336	—	—	—	342	4.1	0.25
ME5	307	6,100	0.045	336	—	—	—	346	4.5	0.28
ME6	307	5,600	0.043	336	360	1.1	—	344	5.3	0.33
ME8	309	5,500	0.041	336	—	—	—	338	4.5	0.32
ME10	310	5,300	0.040	337	—	—	—	338	3.3	0.26
ME12	311	4,700	0.034	336	358	0.85	—	342	3.5	0.24
PCHMS	326	7,320	0.054	347	366	1.3	—	355	4.9	0.33
DPA	377	7,600	0.0750	—	410	1.8	0.73	392	17	0.65
DPB	393	13,300	0.11	—	405	2.4	1.9	<405	>20	>0.80

Sources: Data from Seki, S. et al., *Macromolecules*, 32, 1080, 1999a; Seki, S. et al., *Radiat. Phys. Chem.*, 60, 411, 2001a; Seki, S. et al., *J. Am. Chem. Soc.*, 126, 3521, 2004a; Kawaguchi, T. et al., *Chem. Phys. Lett.*, 374, 353, 2003.

[a] Molar extinction coefficient per Si unit.

[b] Oscillator strength obtained by numerical integration.

[c] Molar extinction coefficient per radical cation at the transient absorption maximum.

supports the small contributions of excluded volume effects on the conformations of polysilanes with longer segment lengths than PD6 (Cotts et al., 1991; Shukla et al., 1991). Thus, the assumption of $a = 1$ gives an appropriate estimate of the segment length for DPB, PH5–PH12 and PD6–PD12, although the segment length of polysilanes with highly flexible backbones may be underestimated. On the basis of the Si–Si bond length (0.2414 nm) and Si–Si–Si bond angle (114.4°) (Koe et al., 1998), the estimated Kuhn segment length varies from 0.55 to 8.6 nm (PH2–PH8) for poly(n-alkylphenylsilane)s, from 0.48 to 1.1 nm (ME3–ME10) for poly(methyl-n-alkylsilane)s, and from 2.4 to 6.3 nm (PD4–PD12) for poly(di-n-alkylsilane)s (Table 25.2).

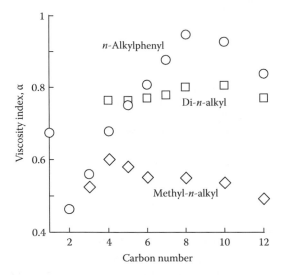

FIGURE 25.3   Viscosity index (α) of PHn, PDn, and MEn as a function of length of n-alkyl substituents. Values of α are estimated from Equation 25.1 based on ε^{abs}. (Reprinted from Seki, S. et al., *J. Am. Chem. Soc.*, 126, 3521, 2004a. With permission.)

TABLE 25.2
UV Absorption, Fluorescence, Molecular Weights, Degree of Polymerization, and Isotropic Mobility of Poly(n-Alkylphenylsilane)

	λ_{max}^{abs} (nm)	ε^{abs} (mol^{-1} dm^{-3} cm^{-1})	λ_{max}^{fl} (nm)	Mw ($\times 10^4$)	Mn ($\times 10^4$)	N	μ_{1D}^{+} (cm^2 V^{-1} s^{-1})
PH1-1	339	7,900	365	49	22	1700	~0.03
PH1-2	339	7,900	365	17	7.5	590	~0.02
PH1-3	339	7,900	365	4.9	3.2	250	~0.02
PH1-4	339	7,800	365	1.7	1.1	87	~0.02
PH1-5	338	7,800	365	0.96	0.80	63	~0.02
PH6-1	347	12,800	369	98	58	930	0.30
PH6-2	347	12,800	369	57	32	510	0.25
PH6-3	347	12,000	369	18	8.3	130	0.26
PH6-4	347	11,700	370	12	6.1	98	0.15
PH6-5	347	10,500	372	9.4	4.2	67	0.092

The segment length also reflects the degree of delocalization of the lowest ES along the Si chains, and it is described by the following empirical relationship (Schreiber and Abe, 1993; Fujiki, 1996; Seki et al., 2002):

$$\varepsilon_{abs} = 330 \times L \tag{25.5}$$

where L is the segment length in Si repeating numbers. Equation 25.5 gives the shortest value of $L \cong 13$ for PH2 and the longest value of $L \cong 53$ for PH8 in the present study, consistent with the changes in $f_{VB\text{-}ES}$.

In the pulse radiolysis experiment, the incident electron pulse in the solution produces cation radicals (Bz$^{\bullet+}$), excited states (Bz*), and electrons, as follows:

$$Bz \rightarrow Bz^{\bullet+}, Bz^{\bullet}, e^{-} \tag{25.6}$$

In deaerated solutions containing triethylamine (TEA), $Bz^{\bullet+}$ is scavenged by TEA and the excess electrons react with polysilane molecules (PS) according to

$$Bz^{\bullet+} + TEA \longrightarrow [Bz...TEA]^{\bullet+}$$

$$e^- + PS \xrightarrow{k_1} PS^{\bullet-} \tag{25.7}$$

giving polysilane anion radicals ($PS^{\bullet+}$) without contribution from counter cations. The excited states and excess electrons are rapidly scavenged in oxygen-saturated solutions ($[O_2] = 11.9$ mM at 1 atm, 25°C), giving singlet oxygen molecules ($^1O_2$) and oxygen anions (O_2^-) (Candeias et al., 2000; Kawaguchi et al., 2003). PS in the solution react with $Bz^{\bullet+}$ via $Bz_2^{\bullet+}$, yielding polymer cation radicals ($PS^{\bullet-}$):

$$Bz_2^{\bullet+} + PS \xrightarrow{k_2} 2Bz + PS^{\bullet+} \tag{25.8}$$

Tagawa and Seki have already discussed the transient spectra of radical ions of polysilanes, including PH1, PD6, etc. (Ban et al., 1987; Seki et al., 1998, 1999a, 2000, 2001a, b, 2002, 2004a; Kawaguchi et al., 2003). $PS^{\bullet-}$ and $PS^{\bullet+}$ were found to exhibit two absorption bands in the near-UV (350–400 nm) and infrared (IR) (>1600 nm) regions. The transient spectra of $PS^{\bullet-}$ and $PS^{\bullet+}$ suggest the presence of an interband level occupied by an excess electron (IBL$^-$) or by a hole (IBL$^+$). The UV and IR bands in $PS^{\bullet-}$ are due to the transition from VB to IBL$^-$, and from IBL$^-$ to the conduction band (CB), respectively. This is also the case of $PS^{\bullet+}$, attributing the UV and IR bands to transition from IBL$^+$ to CB, and from VB to IBL$^+$, respectively. The presence of IBL$^-$ or IBL$^+$ is interpreted well as a delocalized negative (positive) polaron state (Rice and Phillpot, 1987; Abkowitz et al., 1990; Tachibana et al., 1990; Seki et al., 1999a, 2004a) and/or the Anderson localization state on a Si segment (Ichikawa et al., 1999a,b). Figure 25.4 shows a series of transient absorption spectra of anion radicals observed in PH1-12 as examples of transient absorption spectra. The spectra observed for anion radicals of MEn and PDn are almost identical, with only a small dependence on the chain length of n-alkyl substituents. In contrast, PHn, in Figure 25.4, exhibits dramatic changes in the IR band with a considerable red shift upon elongation of the n-alkyl chains. The IR band of cation radicals of PHn also tends to shift dramatically with an elongation of n-alkyl substituents, despite a negligibly small change observed for the IR bands of MEn and PDn anion radicals [55]. The peak energy of the IR band in relation to the binding energy of negative or positive polarons on the Si chains (Pitt, 1977; Abkowitz et al., 1987) is given by

$$\delta\varepsilon \approx \left(\frac{\Delta V}{2A}\right)\left(\frac{a}{\xi_p}\right)^2 \tag{25.9}$$

where

$\delta\varepsilon$ denotes the binding energy of a polaron
a denotes a lattice unit of a trans-chain segment
ξ_p is the polaron width
V is the matrix element describing the interaction between two atomic orbitals consisting of a covalent bond
Δ denotes the matrix element between two atomic orbitals of a Si atom

The parameter A is given by

$$A \equiv [V(l) - \Delta]$$

$$l = \frac{a}{\sin\theta} \tag{25.10}$$

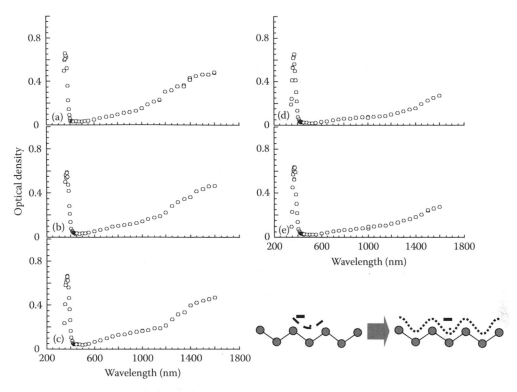

FIGURE 25.4 Transient absorption spectra of anion radicals of (a) PH1, (b) PH2, (c) PH4, (d) PH6, and (e) PH8, in Bz at 5.0×10^{-3} mol dm^{-3} (base mol unit) with TEA at $RT = 2.0 \times 10^{-2}$ mol dm^{-3} conc. Spectra were recorded 10 ns after electron pulse irradiation.

where 2θ is the tetrahedral bond angle. Thus, Δ is a parameter specifying the degree of delocalization of σ electrons on a σ-conjugated segment, while V describes the localization of a pair of electrons in a local bond. The relative polaron width on the polymers can be estimated based on Equation 25.8. With the previously reported values of Δ in poly(dimethylsilane) (Kumada and Tamao, 1968; Pollard and Lucovsky, 1982) and V in poly(dichlorosilane) (Koe et al., 1998), the value of ($\Delta V/2A$) can be estimated to be ca. 1.6 eV. The transition energy of the IR band observed for PH2 cation radicals is 0.77 eV, the highest example value for all the polysilanes, yielding $\xi_p = {\sim}2$ Si repeating units as the minimum value. The considerable red shift of the IR band observed for PHn induces a dramatic increase in the polaron width, and hence a highly delocalized negative (anion radicals) or positive (cation radicals) charge state.

Figure 25.5 shows the transient UV band observed for the ME3 solution. The spectra appear to suggest the simultaneous formation of the radical anion and cation in the Ar-saturated solution, leading to overlap of the two transient absorption bands at 358 and 348 nm, which have been determined independently (Seki et al., 2001b, 2004a; Kawaguchi et al., 2003). In contrast to the Ar-saturated solution, only the radical anion or the radical cation is selectively formed in the Ar-saturated solution with 20 mM TEA or in the O$_2$-saturated solution. The kinetic trace is recorded over a wide range of observation time from a few ns to μs. The extinction coefficients of anion radicals ($\varepsilon^{\cdot-}$) were $\sim$ determined by using the electron-transfer reaction between PS and pyrene (Py) as follows:

$$PS^{\cdot-} + Py \longrightarrow PS + Py^{\cdot-} \tag{25.11}$$

The kinetic traces of PS$^{\cdot-}$ and Py$^{\cdot-}$ ($\varepsilon^{\cdot-}$ of Py = 5.0×10^4 M^{-1} cm^{-1} at 492 nm (Gill et al., 1964)) gave the $\varepsilon^{\cdot-}$ values of polysilanes, as summarized in Table 25.1. The extinction coefficient of the cation

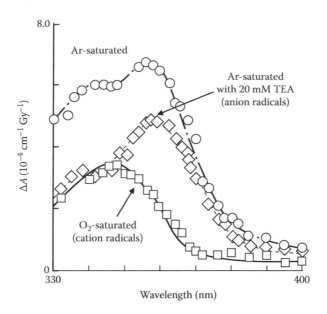

FIGURE 25.5 UV band in the transient absorption spectra of ME3 in Bz at 5.0×10^{-3} mol dm^{-3} (base mol unit). Spectra were recorded 100 ns after electron pulse irradiation. Solid (squares), dashed (diamonds), and dot-dashed (circles) lines are spectra observed in O$_2$-saturated, Ar-saturated with TEA at 2.0×10^{-3} mol dm^{-3}, and Ar-saturated solutions, respectively. (Reprinted from Seki, S. et al., *J. Am. Chem. Soc.*, 126, 3521, 2004a. With permission.)

radicals can also be determined by the CT reaction between PS$^{\bullet+}$ and N,N,N',N'-tetramethyl-p-phenylene-diamine (TMPD) as follows:

$$PS^{\bullet+} + TMPD \longrightarrow PS + TMPD^{\bullet+} \tag{25.12}$$

The formation process of TMPD$^{\bullet+}$ can be observed as an increasing process in the transient absorption at 570 nm, at which TMPD$^{\bullet+}$ in Bz peaks with an extinction coefficient of 1.2×10^4 M^{-1} cm^{-1} (Stegman and Cronkright, 1970). The fitting of the formation of the kinetic trace at 570 nm and the decay trace at the absorption maximum of PS$^{\bullet+}$ based on the extinction coefficient of TMPD$^{\bullet+}$ gives PS$^{\bullet+}$ extinction coefficients ($\varepsilon^{\bullet+}$) of 3.3×10^4–2.0×10^5 M^{-1} cm^{-1}, as listed in Table 25.1.

In the quantitative elucidation of the degree of charge delocalization on the Si skeleton, the authors have already reported the degree of charge delocalization (n_{del}) determined through simultaneous observation of transient bleaching of the lowest excitonic backbone peak (Δ_{OD}^{Bl}) and the formation of the UV bands ($\Delta_{OD}^{\bullet\pm}$), as given by (Seki et al., 2001a, 2004a; Kawaguchi et al., 2003)

$$n_{del} = \frac{\Delta_{OD}^{Bl} \cdot \varepsilon^{\bullet\pm}}{\Delta_{OD}^{\bullet\pm} \cdot \varepsilon_{ES}} \tag{25.13}$$

The empirical relationship between the apparent oscillator strength of the UV band ($f^{\bullet\pm}$) and the value of n_{del} has been obtained, expressed as (Seki et al., 1999a, 2001a,b, 2004a)

$$n_{del} \propto f^{\bullet\pm} \tag{25.14}$$

$$f^{\bullet\pm} = 4.32 \times 10^{-9} \int \varepsilon^{\bullet\pm} d\nu \tag{25.15}$$

The values of $f^{\bullet-}$ and $f^{\bullet+}$ can be obtained by a numerical integration of the Gaussian fit to the IBL$^+$–CB transitions. The obtained values of $f^{\bullet-}$ and $f^{\bullet+}$ are summarized in Table 25.1. The half-Gaussian (high-energy side), half-Lorentzian (low-energy side) fit gives lower values of $f^{\bullet-}$ and $f^{\bullet+}$, but the difference between the two fittings is less than 20% for all spectra. The values of $f^{\bullet+}$ for a series of

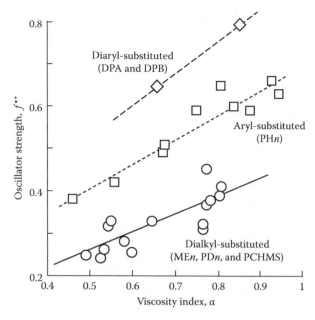

FIGURE 25.6 Oscillator strength ($f^{\cdot+}$) of UV band versus viscosity index (α). (Reprinted from Seki, S. et al., *J. Am. Chem. Soc.*, 126, 3521, 2004a. With permission.)

MEn, PDn, PHn, and poly(arylsilane)s are plotted as a function of α in Figure 25.6. The values of $\varepsilon^{\cdot+}$ principally reflect the values of α, and exhibit a similar dependence on the degree of delocalization of ES on the Si backbone. However, $\varepsilon^{\cdot+}$ and $f^{\cdot+}$ values observed for the phenyl-substituted polysilanes are clearly higher than those for the dialkyl-substituted polysilanes. In contrast to cation radicals, the anion radicals of all the polysilanes represent a clear correlation between the values of $f^{\cdot-}$ and α, without contribution from the presence of phenyl pendant groups in their side chains.

The values of both q and L derived from Equations 25.4 and 25.5 clearly indicate that the physical segmentation length in Si backbones of the ground-state or excited-state polysilanes is well interpreted in terms of the molecular stiffness, α. In the case of cation radicals, the dependence of $f^{\cdot+}$, or n_{del}, on the molecular stiffness can be apparently categorized into three series: polysilanes bearing no, one, or two phenyl rings, despite the no categories presenting in anion radicals. It should be noted that the transient spectra in the present study were recorded after intra-chain CT processes, which were expected to occur within a few ns (Matsui et al., 2002b). Thus, the spectra occur mainly due to energetically favorable segments in polysilane backbones. Despite the fact that PH2 exhibits almost identical Stokes shifts to those of any poly(dialkylsilanes), a relatively high value of $f^{\cdot+}$ is observed for PH2. This is strongly suggestive of a crucial contribution from the phenyl rings only in the delocalization of positive charges. Takeda et al. suggested, based on theoretical calculations, that the valence state and the ES of steady-state polysilanes bearing phenyl rings are considerably affected by σ-π mixing. The present results also indicate that this is the case, giving rise to IBL$^+$. A delocalized IBL$^+$ over the phenyl ring is a good explanation for the increase in $f^{\cdot+}$ for poly(n-alkylphenyl-silane), though an IBL$^-$ is considered to be localized onto Si chains without spreading over phenyl substituents. This explanation is supported by models of the spread of the singly occupied molecular orbital (SOMO) state into substituents, as calculated theoretically only for polysilane cation radicals (Kumagai et al., 1995; Seki et al., 1999a, 2002). Therefore, the UV band includes both the transition from IBL$^+$ to CB, and the transition from IBL$^+$ to pseudo-π at an energy of ~0.2 eV below the CB.

In summary, results of the spectroscopic measurement on ion radicals of σ-conjugated polymers were quantitatively discussed in terms of oscillator strength, hence the degree of delocalization of excess charges along their σ-conjugated skeletons. The principal role of side-chain pendant groups was suggested for the delocalization of positive charge, despite the insulating nature of the side chains for electrons delocalized over Si catenations. The quantitative analysis of the transient absorption spectra gave precise number (concentration) of charge carriers generated in the media.

25.2.3 NANOMETER-SCALE DYNAMICS OF THE HOLE AND THE ELECTRON IN DNA

The direct ionization of DNA produces electrons and holes randomly throughout the DNA structure and its intimate surroundings nearly in proportion to the number of valence electrons at each site. However, electron spin resonance (ESR) experiments have shown that the final damage to DNA is not a random distribution (Bernhard and Close, 2004; Schuster and Landman, 2004; Becker et al., 2007). The sugar and phosphate radicals are produced in smaller abundance than expected from the 50% fraction of ionization that occurs on the sugar–phosphate backbone. Rather, the majority of the radicals are on the DNA bases. Furthermore, the radical is not randomly distributed among the bases. Among the bases, the localization is not statistical, but shows a high degree of specificity. The electrons are trapped initially on cytosine (C) and later on thymine (T) moieties. The electron "hole" is localized almost exclusively on guanine (G), which has the lowest oxidation potential, among the four bases. The identification of the DNA sites that trap these holes and electrons is essential to understand radiation-induced damage. This section describes the dynamics of cation and anion radicals of DNA in an aqueous solution at room temperature using the technique of pulse radiolysis (Kobayashi and Tagawa, 2003; Kobayashi et al., 2008; Yamagami et al., 2008).

25.2.3.1 Radical Cation

For the observation of the radical cation generated in DNA, aqueous solutions of oligonucleotides (ODNs) are irradiated in the presence of persulfate ($S_2O_8^{2-}$). The hydrated electron (e_{aq}^-) reacts with $S_2O_8^{2-}$ to give $SO_4^{-\bullet}$. Since the $SO_4^{-\bullet}$ (2.5–3.1 V) is powerful enough to oxidize all four bases (Candeias and Steenken, 1993; Steenken and Javanovic, 1997), the radical cations of bases are formed unselectively within double-stranded ODN. However, the guanine radical cation ($G^{+\bullet}$) is exclusively formed, whereas the spectra in the single-stranded ODN exhibit a composite spectra of four nucleic acid base species. The essentially quantitative formation of $G^{+\bullet}$, based on the initial yield of $SO_4^{-\bullet}$, suggests that hole transfer occurs efficiently through 13–15 mer ODN. This indicates that hole transfer from the initially formed loss centers to G sites occurs along double-stranded ODN very rapidly ($<2 \times 10^7 s^{-1}$). By introducing mismatch, the formation of $G^{+\bullet}$ was diminished, suggesting that local perturbation of the duplex structure results in incomplete hole migration.

The $G^{+\bullet}$ of deoxyguanosine (dG) has a pK_a of 3.9, and it rapidly loses the N1 proton with a rate constant of $1.8 \times 10^7 s^{-1}$ at a pH of 7.0 (Kobayashi and Tagawa, 2003). In the guanine radical cation/cytosine base pair ($G^{+\bullet}$-C), it is not clear whether or not the proton of N1 of G shifts to C. Based on the pK_a values of the N1 proton of $G^{+\bullet}$ ($pK_a = 3.9$) and N3-protonated C ($pK_a = 4.3$), the equilibrium for the proton transfer lies slightly to the right. The value of K_{eq} for proton transfer suggests that the base pair ($G^{+\bullet}$-C) in DNA has only partial proton transfer and both cation radical (30%) and deprotonated neutral radical (70%) forms exist. There exists extensive discussion regarding the protonation state of the base pair ($G^{+\bullet}$-C) (Steenken, 1989, 1997; Kobayashi et al., 2003, 2008; Lee et al., 2008). Recently, ESR studies at 77 K clearly show that one-electron-oxidized G in double-stranded ODN exists as the deprotonated neutral radical $G(-H)^\bullet$ (Adhikary et al., 2009).

$$(25.16)$$

The $G^{+\bullet}$ in ODN, produced by oxidation with $SO_4^{-\bullet}$, deprotonates to form $G(-H)^{\bullet}$ with a half time of 35 ns (Kobayashi et al., 2003, 2008). The spectrum of $G^{+\bullet}$ exhibits absorption around 450 nm, whereas $G(-H)^{\bullet}$ has absorption around 380–500 nm, together with a characteristic shoulder in the spectral range of 600–700 nm (Figure 25.7). The deprotonation of $G^{+\bullet}$ is monitored at 625 nm (Figure 25.8).

The deprotonation of $G^{+\bullet}$ in double-stranded DNA is composed of several steps, including proton transfer from G to C in the base-pair radical cation ($G^{+\bullet}$-C) (step i) and the release of the proton from $C(+H^+)$ into the solution (step ii) (Scheme 25.1): K_{eq} ($=k_i/k_{-i}$) is the equilibrium constant and k_{ii} is the rate constant of the release of proton. Since proton transfer from the N1 proton of G to the N3 proton of C could, in principle, occur very rapidly on the order of $10^{14}\,s^{-1}$ (Douhal et al., 1995), the rate-limiting step in these processes is the deprotonation of $C(+H^+)$ by a water molecule ($k_{ii} \ll k_{-i}$).

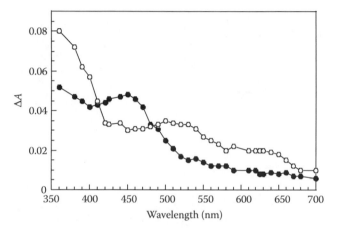

FIGURE 25.7 Kinetic difference spectra of pulse radiolysis of ODN G_{3AA} monitored at 50 ns (•) and 500 ns (○) after pulse radiolysis.

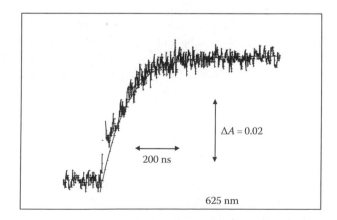

FIGURE 25.8 Kinetics of absorbance change after pulse radiolysis of ODN G_{3AA}.

$$[G^{+\bullet} ----- C \xrightleftharpoons[(i)]{K_{eq}} G(-H)^{\bullet} ----- C(+H^+)] \xrightarrow[(ii)]{k_{ii}} G(-H)^{\bullet} ----- C$$

$$H_2O \qquad H_3O^+$$

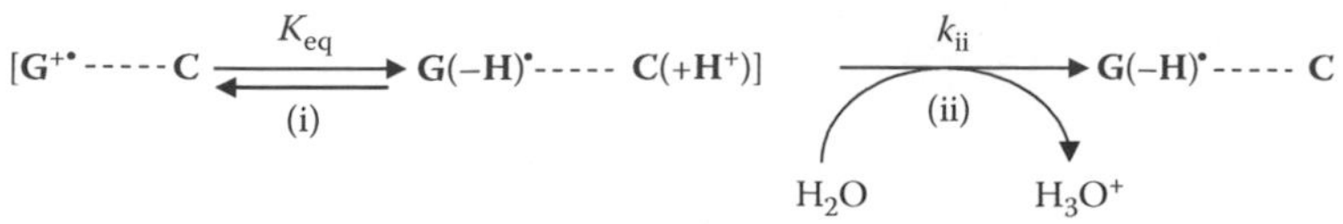

SCHEME 25.1

TABLE 25.3
Sequences of the ODNs

Name	Sequence (5′ ⟶ 3′)
G_{3AA}	5′ AAAAAGGGAAAAA 3′
	3′ TTTTTCCCTTTTT 5′
^{Br}C	5′ AAAAAAGAAAAAA 3′
	3′ TTTTTT^{Br}CTTTTTT 5′
^{CH_3}C	5′ AAAAAAGAAAAAA 3′
	3′ TTTTTT^{CH_3}CTTTTTT 5′

^{Br}C ^{CH_3}C

The rate constant of the deprotonation can be expressed as $k = k_{ii}K_{eq}$. This suggests that the absorbance changes observed would correspond to the proton loss from the oxidized proton-shifted resonance structure ($G^{+\bullet}:C \leftrightarrow G(-H)^{\bullet}:C(+H^+)$) by a water molecule. This is further supported by the effect of the substitution by the methyl or bromine group for the cytosine C5 hydrogen in selected positions complementary to G (Table 25.3) and by the solvent deuterium isotope effect (Kobayashi et al., 2008). A bromo substituent on C, as an electron-accepting group, suppressed the process ca. 20-fold compared with a methyl substituent on C, as an electron-donating group. The proton transfer from $G^{+\bullet}$ ($pK_a = 3.9$) to ^{Br}C is thermodynamically unfavorable, since the pK_a of ^{Br}dC is 2.8. The equilibrium of ^{Br}C in Equation 25.1 is far to the left, and the equilibrium constant (K_{eq}) of the proton transfer in $G^{+\bullet}$-^{Br}C is estimated to be 0.08, suggesting that a large portion of the oxidized G in ^{Br}C remains protonated after oxidation. Thus, the difference in the rate constants can be explained primarily by the pK_a values of C. For the solvent deuterium isotope effect, it should be noted that a kinetic isotope effect ($k_H/k_D = 3.8$) of the deprotonation in ODN was about twice that of the kinetic isotope effect on the deprotonation of free dG ($k_H/k_D = 1.7$). This suggests that the formation of $G(-H)^{\bullet}$ in ODN may be associated with the loss of tightly bound protons in DNA. Thus, this process is consistent with step (ii) in Scheme 25.1.

It is of particular interest to study ODN sequences showing the spectroscopic changes and the rate constants of the deprotonation of $G^{+\bullet}$ (Kobayashi et al., 2008). The characteristic absorption maxima of $G^{+\bullet}$ intermediate around 450 nm were found to differ. These spectra shifted to longer wavelengths in the order of G < GG < GGG. The spectral shift can be understood as the stacking interaction of two or three consecutive guanine bases. These results demonstrate that the transiently formed $G^{+\bullet}$ is stabilized by base pairing with C and the stacking interaction of neighborhood nucleobases. These findings provide direct spectroscopic evidence of the delocalization of the positive charge along the extended π orbitals of DNA bases (Saito et al., 1998; Yoshioka et al., 1999; Shao et al., 2004). In contrast, the spectra of $G(-H)^{\bullet}$ in ODNs are essentially identical to that of dG and are not affected by the ODN sequence. This strongly suggests that the radical orbital of $G(-H)^{\bullet}$ is essentially localized on a preferential specific guanine base site.

The rate constants of deprotonation are affected by ODN sequences. The differences among G, GG, and GGG were not distinct, but rather depended on the neighboring bases. The rate constant increases in the order CGC > TGT > AGA > AGG > GGG, an order that may correlate with the calculated ionization potential of the nucleobases (Yoshioka et al., 1999; Voityuk et al., 2000). The finding that the rate decreases as the oxidation potential decreases may reflect on the stability of the radical cation.

25.2.3.2 Radical Anion

The radical anions of nucleotides in DNA are produced by the reaction of hydrated electrons (e_{aq}^-) with ODN. Hydrated electrons (e_{aq}^-) can react with all four nucleobases, and the resulting products of the nucleic acids have been identified spectroscopically. The differences in the dynamics of ODNs and corresponding isolated nucleotides are compared. However, unlike positive holes, electrons are not trapped at specific sites of DNA. Here, we present the dynamics of electron adducts of ODNs containing A·T and G·C base pairs with those of single-stranded ODNs and the isolated nucleotides.

Radical anions of double-stranded ODNs containing A·T bases were produced by pulse radiolysis of a deaerated aqueous solution. Figure 25.9 shows transient absorption spectra after pulse radiolysis of ODN **AT**. For comparison, the spectra of single-stranded ODN (5′-TAATTTAATAT-3′), dT, and dA are shown in Figure 25.10. The spectrum of single-stranded ODN, which has no absorption around 480 nm, is essentially similar to that of dT. In single-stranded ODN, electron transfer from the initially formed transients to T sites is completed within 10 ns. In double-stranded ODN, on the other hand, the initial transient obtained after the complete decay of e_{aq}^-, and characterized by an absorption at around 480 nm and a broad absorption from 600 to 900 nm, is apparently complex because of the presence of A and T bases. However, this spectrum was not reproduced as a set of benchmark spectra for dA and dT. These findings, therefore, suggest that the transiently formed (A·T)$^{·-}$ is stabilized by hydrogen-bond base-pairing, and the unpaired electron is delocalized to the A site.

Following the spectral change, other spectral changes took place in the time range of microseconds. The absorption around 480 nm, associated with the charge delocalization on the A site, disappeared with the decay of absorption at 350 nm (Figure 25.9). The spectrum of the resulting product at 20 μs after the pulse, which has an absorption maximum at 350 nm, is similar to that of T$^{·-}$. This process follows a decrease of absorbance at 480 nm (inset of Figure 25.9) and obeys

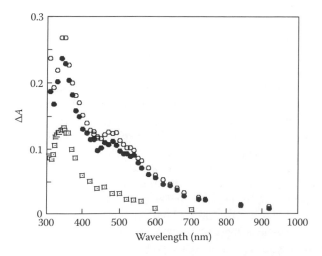

FIGURE 25.9   Kinetic difference spectra of pulse radiolysis of ODN **AT** at 50 ns (○), 1 μs (•), and 20 μs (□).

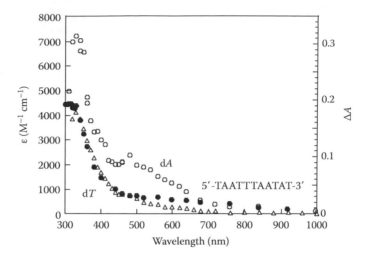

FIGURE 25.10 Absorption spectra at 200 ns after pulse radiolysis of dA (○), dT (●), and single-stranded AT (△). Note that the right-hand ordinate is in units of dA corresponding to the kinetic difference spectrum of ODN, whereas the left-hand ordinate is in units of extinction coefficients of dA and dT.

first-order kinetics with a rate constant for ODN **AT** of $1.6 \times 10^6 \, s^{-1}$, and it is independent of ODN concentration. Therefore, the spectroscopic changes in Figure 25.8 may reflect progress from the transient $(A·T)^{·-}$ to $T^{·-}$ in the A·T anion complex. The nature of the species generating the 20 μs spectrum suggests that the electron adduct center became localized mainly at the T site (Gu et al., 2005).

Proton-transfer reactions are important for the stabilization of initial DNA ion radicals (Barnes et al., 1991; Steenken et al., 1992; Steenken, 1997; Kobayashi et al., 2008). In regard to excess electrons of DNA, attention has been focused on the protonation state and protonation sites of C and T. In a G-C pair, $C^{·-}$ rapidly acquires a proton at N3 through hydrogen bonding with the HN1 of G. From the pK_a values of the N3-protonated $C(H)^·$ ($pK_a > 13$) and the N1 proton of G ($pK_a = 9.5$) (Steenken et al., 1992), the equilibrium of $GC^{·-}$ in ODN **GC** is far to the right, and the spectrum observed is regarded as corresponding to $G(–H)^-$-$C(H)^·$. Moreover, the protonation of free dC occurs before 10 ns. Such a rapid reaction may be due to the protonation by water-bound N3 of C. Similar results were observed using conductance techniques (Hissung and von Sonntag, 1979) and in ESR/electron nuclear double resonance (ENDOR) studies of cytosine monohydrate single crystals irradiated at 10 K (Sagstuen et al., 1992). In contrast, since $T^{·-}$ is only a weak base, it was present transiently in a neutral aqueous solution and protonated to form the neutral radical $T(H)^·$. Thus, the protonation of dT occurs reversibly at O4 by the direct combination of $T^{·-}$ with protons in solution. In double-stranded ODNs, since the pK_a values of the O4 of protonated $T(H)^·$ ($pK_a = 6.9$) and the N6 of A ($pK_a > 13.75$), $T^{·-}$ is not protonated by its complementary base, A. Thus, the spectrum of the electron adduct of ODN **AT** corresponds to A-$T^{·-}$. The process of protonation of T in ODN **AT** is thus consistent with a slower process of spectroscopic change observed on the scale of hundreds of microseconds. Remarkably, however, this process revealed a substantial difference between the ODN **AT** and free dT. At a pH of 7.0, the rate constant of the former ($4 \times 10^4 \, s^{-1}$) was 10-fold slower than that of the latter ($4.3 \times 10^5 \, s^{-1}$). In addition, the rates of ODN **AT** are independent of a change in the pH from 5.8 to 8.3. This difference may be due to the differences in the protonation sites of T in DNA and free dT. ESR studies using low-temperature glasses yield strong evidence that when DNA is exposed to ionizing radiation at 77 K, an ESR doublet associated with $T^{·-}$ is converted into a readily identifiable eight-line spectrum of the 5,6-dihydrothymine-5-yl radical, a conversion due to the irreversible protonation of $T^{·-}$ at C6. A water molecule bound to DNA would participate in this protonation. Thus, irreversible protonation by water at C6 is both a pH independent and a relatively slow process.

25.3 NANOSCALE DYNAMICS OF CHARGE CARRIERS BY TIME-RESOLVED MICROWAVE CONDUCTIVITY

The π- and σ-conjugated materials are well-known organic semiconductors due to their robust nature, multiplicity of chemical constitution, and capability of charge transport that can be tailored via judicious functionalization. They have demonstrated feasibility for industrial applications, such as organic thin-film transistors [TFT] (Sirringhaus et al., 1998, 1999), light-emitting diodes [LED] (Kido et al., 1995; Sheats et al., 1996), photovoltaic cell [PV] (Sariciftci et al., 1992; Yu et al., 1995), and radio-frequency identification [RFID] tag (Baude et al., 2003; Myny et al., 2008). The possibility of solution process opens a versatile route toward the fabrication of large-area devices on flexible substrates. Alongside these benefits, considerable efforts have been devoted not only to the development of novel conjugated materials, but also to the understanding of their optical, electric, and optoelectronic properties. The mobility of charge carriers as well as their polarity (positive or negative) play a key role in the performance of organic electronic devices, as these relate to the response speed, the efficiency of charge carrier transport, and the realization of complementary circuits. Generally, the mobility is measured by direct-current (DC) techniques, for example, FET and TOF; however, these are considerably dependent on the film morphology and sample purity like grain boundary, impurities, and structural defects, which are undesirable features hiding the intrinsic nature of charge carrier transport. The interface between the electrodes and the semiconductors causes problems of barrier against charge injection and strong electric field, which disturbs the thermal motion of charge carriers. The long-range transport property estimated by the DC technique is important for the actual electrical devices; however, to reveal the principal character, without being affected by complicated factors, will offer chances to mitigate barriers for the development of novel organic electronics and for their performance optimization.

In order to perform direct observation of the drift mobility of charge carriers, free from the disturbance at interfaces, such as electrode contacts and domain boundaries, the TRMC technique based on microwave absorption has been developed (de Haas et al., 1975; Warman et al., 2002). TRMC, where the probe is an alternating-current (AC) electromagnetic wave, can overcome the drawbacks of DC measurements, as follows: (1) As it does not necessitate the fabrication of electrodes in contact with polymers, contact and surface interactions can be avoided. (2) The high frequency (GHz) of microwaves extends the observable region of charge carrier mobility to a very short regime (nanometer scale). Thus, the mobility obtained at the end of a pulse is not significantly affected by the grain and/or domain boundary conditions. (3) The low magnitude of the electric field makes it possible to observe intrinsic charge-transport phenomena in well-organized and/or highly extended conjugated segments of polymers without perturbing the thermal drift of the charge carriers. Because of these unique properties, the mobilities of polymer matrices obtained by TRMC are typically orders of magnitude higher than those obtained by the DC technique, in which charge migration is dominated by interchain transport and other undesirable factors.

The microwave absorption technique was initially proposed by Biondi et al. in 1949, who measured electron concentrations in the gas phase (Biondi and Brown, 1949; Biondi, 1951). Time-resolved experiments with an enhanced S/N ratio and wide dynamic ranges were realized by the improvement of microwave circuits and the use of pulsed high-energy radiations (mainly, EB or converted x-ray). Fessenden et al. at the University of Notre Dame have conducted investigations on dissociative electron attachment in the gas phase (Warman and Fessenden, 1968; Warman et al., 1972). Shimamori and Hatano reported an electron thermalization process in the gas phase and the reaction between electrons and molecules (Shimamori and Hatano, 1977). The researches were expanded to transient change in the inter- or intramolecular dipole moment (Fessenden et al., 1979; Fessenden et al., 1982). The separation of real and imaginary parts of the signals has been discussed in a gas phase (Suzuki and Hatano, 1986a,b). Investigations on energy loss processes (Shimamori and Sunagawa, 1997) and excited states in the liquid phase (Fessenden and Hitachi, 1987) have been performed. Using pulse-radiolysis time-resolved microwave conductivity (PR-TRMC), Warman et al.

at the Delft University of Technology have vigorously investigated charge dynamics in organic liquids (van den Ende et al., 1982), in TiO_2 (Warman et al., 1984), and in conjugated materials (van de Craats et al., 1996; Hoofman et al., 1998a; Warman et al., 2005). Siebbeles and Grozema et al. at Delft extended the researches further to charge transport along conjugated molecular wires (Prins et al., 2006; Pieter et al., 2007; Feng et al., 2009; Kocherzhenko et al., 2009).

By combining flash-photolysis time-resolved microwave conductivity (FP-TRMC) and flash-photolysis transient absorption spectroscopy (FP-TAS), nanosecond EB PR-TAS, and DC current integration upon photoirradiation (DC-CI), we have investigated charge carrier dynamics in various organic/inorganic semiconductors, such as liquid crystals (Li et al., 2008; Sakurai et al., 2008; Motoyanagi et al., 2009), self-assembled nanotubes (Yamamoto et al., 2006a,b, 2007b), conjugated polymers (oligomers) (Acharya et al., 2005a,b; Saeki et al., 2005b, 2008a; Umemoto et al., 2008), DNA (Yamagami et al., 2006), organic crystals (Imahori et al., 2007; Hisaki et al., 2008; Saeki et al., 2008b; Amaya et al., 2009), inorganic nanorods (Nagashima et al., 2008), and dendrimers (Seki et al., 2005a, 2008). PR-TRMC can directly assess the charge carrier mobility from the obtained conductivity ($\Delta\sigma = e\Sigma\mu N$, where $\Delta\sigma$, change of conductivity; e, an elementary of charge; $\Sigma\mu$, sum of mobilities; N, generated charge carrier concentration). N, generated by a high-energy EB, can be, thanks to its nonselective ionization process, estimated by an empirical formula considering the bandgap energy of the semiconductor and its density (Alig et al., 1980; Warman et al., 2004). However, the N of FP-TRMC, namely, ϕ, the quantum yield of charge carrier generation for one photon absorption at a given time resolution, varies considerably among materials. The photoresponse property of organic semiconductors is a matter of great importance, especially in organic photovoltaics (OPV). The problem in the estimation of ϕ is addressed quantitatively using FP-TAS, DC-CI, or other techniques. There is a case where the thin film of FP-TAS shows transient absorption maxima that are attributed to a radical cation (hole) or a radical anion (electron). Otherwise, the incorporation of an electron acceptor (donor) molecule into the donor (acceptor) host material is an alternative way, which leads not only to an increase of photoconductivity utilizing enhanced electron-transfer yield between donor and acceptor, but also to the possibility of spectroscopic detection of the acceptor radical anion (Saeki et al., 2008a). These situations allow us to obtain ϕ on the basis of the extinction coefficient of the radical ion, which is known or estimated by PR-TAS. On the other hand, the DC-CI technique, analogous to TOF, characterizes the photocurrent upon exposure to pulsed laser, giving the total charge number collected by the electrodes by integrating the transient current over the time.

In this fashion, AC mobilities of charge carriers have been measured and found to be on the orders of 10^{-3}–10^0 cm^2 V^{-1} s^{-1}, which is, in most of the cases, a few orders of magnitude higher than those estimated by the DC technique, demonstrating the potential feasibility for use as efficient organic semiconducting materials. The kinetics of transient conductivity is sensitive to film morphology, grain size of thin films, bicontinuous ordering of liquid crystal, and π-stacking distance (e.g., side-chain length of a conjugated polymer). The photoconductivity intensities are dependent significantly on nanometer-scale ordering, such as amorphous versus crystal; random stacking versus ordered stacking; pristine polymer film versus film mixed with an additive, which disturbs the stacking; donor-type materials versus acceptor-incorporated donor materials; and random chain versus rod-like backbone of conjugated polymers. These differences are sometimes not observed in DC measurements because of unexpected factors like impurities and film/device conditions.

Another advantage over DC techniques is the implementation of nanometer-scale anisotropic conductivity measurements with high angle resolution, performed just by rotating a sample in a resonant cavity, where the direction of the microwave electric field is fixed (Hoofman et al., 1998b; Grozema et al., 2001). A self-assembled hexabenzocoronene nanotube shows a 10 times higher conductivity along the parallel direction of a macroscopically aligned nanotube relative to the perpendicular direction (Yamamoto et al., 2006b). Anisotropies of organic semiconducting crystals, such as rubrene (2.3 times) (Saeki et al., 2008b), dehydrobenzoannulene (12 times) (Hisaki et al., 2008), and sumanene (9.2 times) (Amaya et al., 2009), were revealed, which were rationalized by tighter intermolecular packing along the crystalline axis affordable for more efficient charge carrier transport.

FIGURE 25.11   Chemical structure of rubrene. The picture shows single crystals of rubrene grown by physical vapor transport in a stream of Ar gas. The length of each crystal is about 1 cm at maximum.

This chapter introduces the charge carrier transport in single-crystal rubrene through a combination of FP-TRMC and FP-TAS. We focus on intermolecular charge transport in single-crystal rubrene, which is discussed in terms of anisotropy, ambipolarity, second-order charge recombination, and exciton–exciton annihilation. The latter provides an insight into the intramolecular charge transport along the molecular wires in conjunction with their backbone conformations and lengths.

Single-crystal rubrene, a tetraphenyl derivative of tetracene, shown in Figure 25.11, is one of the most appropriate benchmarks for the investigation of the intrinsic nature of organic electronics, due to the elimination of grain boundary and minimization of charge-trapping sites (Podzorov et al., 2004; Sundar et al., 2004; Takenobu et al., 2007). Single-crystal rubrene was grown by physical vapor transport in a stream of Ar gas (Takeya et al., 2004). The crystal, the size of which is approximately 1 (*a*-axis) × 9 (*b*-axis) mm × 8 μm (thickness), was placed on a quartz plate without any adhesives and used for both TRMC and TAS experiments. The third harmonic generation of the nanosecond Nd:YAG laser (355 nm) was used as an excitation pulse for the experiments, where the incident photon density was varied from 0.59 to 17.8 × 10^{15} photon^{-1} cm^{-2} pulse^{-1}. The TAS experiments were performed using a wide-dynamic-range streak camera. All experiments were performed at room temperature.

Figure 25.12 shows the transient photoabsorption (PA) spectrum together with steady-state PA and emission spectra of single-crystal rubrene. The transient PA spectrum indicates a broad feature

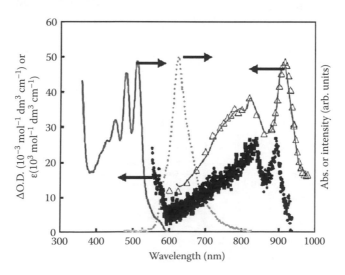

FIGURE 25.12   End-of-pulse transient PA spectrum of single-crystal rubrene upon exposure to a 355 nm nanosecond laser. It is depicted as closed black circles, and its unit is 10^{-3} ΔO.D. The gray solid and dotted lines are normalized steady-state PA and emission spectra, respectively. The open triangles interpolated by a gray line represent the sum of PA spectra of the radical cation and anion (1:1) of rubrene in a unit of 10^3 mol^{-1} dm^3 cm^{-1}, which was assessed by nanosecond EB pulse radiolysis.

from the visible to the near-IR region with two distinct peaks (835 and 895 nm). In order to assess PA spectra of the radical cation (Ru$^{•+}$, hole) and the radical anion (Ru$^{•-}$, one excess electron) of the rubrene molecule and their extinction coefficients, we performed a nanosecond EB pulse radiolysis experiment, using pyrene and biphenyl as mediators of electrons and holes, respectively. The sum of the spectra composed of an equivalent ratio of Ru$^{•-}$ and Ru$^{•+}$ is superimposed in Figure 25.12, since the hole and the electron are equally produced via exciton dissociation upon exposure to a laser pulse. The spectrum looks exactly like the transient PA spectrum in single crystals, and therefore it is ascribable to the overlap of Ru$^{•-}$ and Ru$^{•+}$. The contribution of triplet excited state is ruled out as its PA peak locates at ca. 500 nm (Saeki et al., 2008b).

The kinetic traces of transient PA are shown in Figure 25.13a. The inset provides the quantum efficiency, ϕ, of charge carrier generation at the pulse end, estimated by using the transient optical density, the extinction coefficient, and the reported procedure (Saeki et al., 2006a, 2007). With an increase in excitation photon density (I_0), the ϕ decreases, which is caused by exciton–exciton annihilation (mainly, singlet–singlet annihilation), leading to a decrease in the efficiency of charge separation from excitons. This is supported by the dependence of emission intensity on photon density, where the normalized emission intensity divided by I_0 is identical to the dependence of ϕ on I_0 (Saeki et al., 2008b). The validity of exciton–exciton annihilation is also strongly corroborated by the relatively long radiative lifetime of the singlet ($\tau_0 = 25.9$ ns), estimated from the steady-state PA corresponding to a highest occupied molecular orbital (HOMO) $\rightarrow$ lowest unoccupied molecular orbital (LUMO) transition in solution and the relationship $1/\tau_0 = \upsilon_0^2 \cdot f = \upsilon_0^2 \cdot 4.32 \times 10^9 \int \varepsilon \, d\upsilon$, where f and υ_0 are the

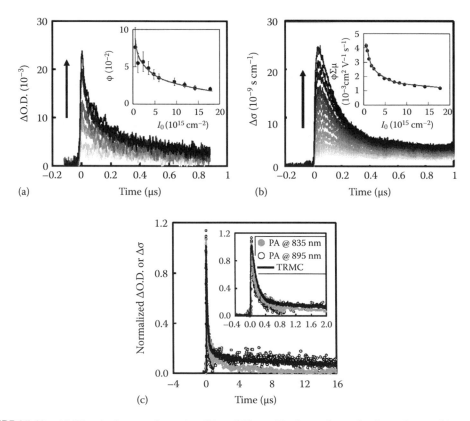

FIGURE 25.13 (a) Kinetic decays of transient PA at 835 nm. The inset shows the dependence of ϕ on excitation photon density (I_0). The arrow indicates the increase of I_0. (b) Kinetic decays of transient conductivity ($\Delta\sigma$). $\phi\Sigma\mu$, the product of quantum efficiency and charge carrier mobility, is plotted in the inset. The arrow indicates the increase of I_0. (c) Comparison of TRMC (black solid line) and transient PA kinetics at 835 (gray closed circles) and 895 (open circles) nm on the long timescale. The inset is the magnification.

oscillator strength and the center of emission (cm^{-1}), respectively. The singlet lifetime might be shortened, particularly on the surface, by oxygen under the atmosphere; however, exciton migration would be extremely enhanced by the well-defined lattice environment, resulting in the enhancement of the exciton–exciton annihilation efficiency.

An interesting fact is that the decay of transient PA is accelerated by increased I_0. This tendency is linked to the second-order reaction between holes and electrons. By a linear function fitting to an inverse of decay, we obtained a second-order bimolecular rate constant of $(2.0 \pm 0.3) \times 10^{10}\,mol^{-1}\,dm^3\,s^{-1}$. It should be noted that such a clear acceleration of decay with photon density is not always observed for other organic semiconductors, for example, pentacene thin films deposited on a quartz substrate by thermal vapor deposition (Saeki et al., 2006b), and amorphous conjugated polymer films (Pieter et al., 2007). The single-crystal nature is anticipated to account for the clear dependence on photon density, while many grain boundaries and chemical/physical defects in usual organic semiconductors would hide this dependence.

Figure 25.13b demonstrates the change of conductivity ($\Delta\sigma$) measured by TRMC, where $\Delta\sigma$ is derived from charge carriers generated by irradiation of the laser pulse. The b-axis of the single crystal is placed in the resonant cavity parallel to the direction of the microwave electric field. The kinetic decays are in good agreement with the PA kinetics, suggesting that TRMC and PA signals originate from charge carriers. The triplet is assumed not to contribute to TRMC as much, because there is no significant dipole change between S_0 and T_1 states. The $\Delta\sigma$ is converted into $\phi\Sigma\mu$: the product of quantum efficiency of charge carrier generation, ϕ, and the sum of charge carrier mobilities, $\Sigma\mu$ ($=\mu_+ + \mu_-$). In other words, $\phi\Sigma\mu$ is in proportion to $\Delta\sigma$ divided by I_0. The dependence of $\phi\Sigma\mu$ on photon density (shown in the inset of Figure 25.13b) corresponds to that of ϕ (shown in the inset of Figure 25.13a), implying almost constant $\Sigma\mu$ over the whole photon densities examined.

Nanoscale charge carrier mobility, $\Sigma\mu$, obtained by a fully experimental protocol, is calculated by dividing $\phi\Sigma\mu$ with ϕ at the pulse end. We found the TRMC mobility to be $(5.2 \pm 0.7) \times 10^{-2}\,cm^2\,V^{-1}\,s^{-1}$ by averaging over the examined photon densities. Normalized TRMC and PA kinetics are shown in Figure 25.13c to facilitate the comparison in the long time range. Although the kinetic traces at 835 (mainly attributed to $Ru^{\bullet-}$) and 895 (mainly attributed to $Ru^{\bullet+}$) are identical at less than 1 μs of delay time, the former disappeared at the later than 10 μs of delay. This suggests that electrons are more likely to be trapped by oxygen and other chemical impurities than holes even in a single crystal. The TRMC mobility of holes obtained at 10 μs of delay was found to be $(3.6 \pm 0.8) \times 10^{-2}\,cm^2\,V^{-1}\,s^{-1}$, which is about 70% of that at the pulse end. This result leads to the separation of the sum of the charge carrier mobility at the pulse end, $\Sigma\mu$, into $\mu_+ = 3.6 \times 10^{-2}$ and $\mu_- = 1.6 \times 10^{-2}\,cm^2\,V^{-1}\,s^{-1}$, demonstrating the ambipolar nature of single-crystal rubrene. These values are, however, approximately three orders of magnitude smaller than the highest reported FET mobility in single-crystal rubrene. The reasons for the low TRMC mobility are speculated as follows: (1) underestimation of the extinction coefficient of the hole in a crystal, (2) increase of the trapping site for the larger single crystal used in the TRMC measurement, (3) AC characteristics for highly mobile charge carriers (Prins et al., 2006), (4) insufficiency of time resolution and microwave frequency, (5) small carrier density enough not to fill most of the carrier traps, and (6) the difference of charge transport efficiency between the surface and the inside of the single crystal. Presently, we cannot settle this issue, and thus, further investigations are required, such as those on changes of microwave frequency, sample thickness, and excitation wavelength.

The mean concentration of the pair of positive and negative charge carriers was estimated to be on the order of $10^{16-17}\,cm^{-3}$, which implies one pair of charge carriers per 10^{3-4} rubrene molecules. This carrier density is, however, a few orders of magnitude smaller than those usually accumulated in FET devices. It has been reported that the trap density in rubrene crystals ranges from $10^{19}\,cm^{-3}\,eV^{-1}$ on the surface to $10^{15-17}\,cm^{-3}\,eV^{-1}$ in the bulk (Goldmann et al., 2006). Therefore, the low carrier density in the present study did not reach the situation of fully occupied traps, and thus, the estimated TRMC mobility might be lower even though the trapping effect is reduced by the nanosecond time resolution of TRMC (reasons (4) and (5)). The bimolecular rate constant leads to the sum of charge carrier mobility as ca. $10^{-3}\,cm^2\,V^{-1}\,s^{-1}$. This mobility is about one order of magnitude lower than

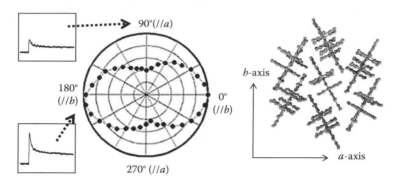

FIGURE 25.14 Anisotropic conductivity of single-crystal rubrene. The 0° corresponds to the case when the electric field is parallel to the *b*-axis. The figure to the right illustrates the crystal structure, where rubrenes in the *b*-axis are more preferable π–π stackings than in the *a*-axis.

TRMC mobility, because the former is reflected from the diffusive motion of charge carrier during the recombination process, which is more affected by traps than the latter. Due to the exponential profile of photo charge carrier generation in the direction of the depth, the transport property on the surface would contribute more to the TRMC results than that inside the bulk; therefore, TRMC mobility is thought to be lowered (reason (6)), which may be inferred from the report that transport efficiency inside a crystal is superior to that on the surface (Takeya et al., 2007).

The anisotropic conductivity of single-crystal rubrene was investigated by rotating the sample stage relative to the direction of the microwave electric field in the resonant cavity, as demonstrated in Figure 25.14. The highest Δσ was observed when the *b*-axis of a crystal was parallel to the electric field. The Δσ in the *b*-axis direction was found to be about 2.3 times larger than that in the *a*-axis. This anisotropy is almost equal to the reported values assessed by FET (Podzorov et al., 2004), suggesting that the anisotropy of charge carrier motion on the nanometer scale is achieved also on the micrometer scale. Since the microwave power is small enough in order not to deviate the charge carrier generation efficiency, the observed anisotropy of Δσ is readily translated into the anisotropy of charge carrier mobility. The appearance of anisotropic conductivity predicts the band-like motion of charge carrier transport in single crystals.

In conclusion, we have elucidated the dynamics of charge carriers in rubrene single crystals through the combination of FP-TRMC and FP-TAS without using an attached electrode. By comparing the kinetics and estimating independently the extinction coefficients of charged species, the 9.1 GHz AC mobility of the charge carrier was found to be $(5.2 \pm 0.7) \times 10^{-2}\,cm^2\,V^{-1}\,s^{-1}$, which consists of approximately 70% holes and 30% electrons. This mobility is, however, expected to be the minimum value because of several reasons, such as much smaller carrier density than those in FET devices, effect of surface traps, and underestimation of extinction coefficients of charged species in the crystal phase. We also demonstrated an anisotropy ratio of 2.3 of charge carrier mobility along the *b*-axis relative to the *a*-axis. The optoelectronic property of single-crystal rubrene clarified here is of great importance for further investigation on the intrinsic nature of charge carriers.

25.4 NANOSTRUCTURE FORMATION

In this section, the radiation sensitivity of σ-conjugated polymers is discussed, especially upon irradiation to high-energy charged particles. The radiation-induced reactions in σ-conjugated polymers have been revealed to depend strongly on the nature of radiation sources exhibited, for example, by the linear energy transfer (LET) that represents the concentration of energy deposited by radiations along their paths (trajectories) in the media (Chatterjee and Magee, 1980; Magee and Chatterjee, 1980; Seki et al., 2004b). Polysilanes were cross-linked by high-LET radiations including high-energy charged particles, despite predominant main-chain scission reactions observed for low-LET radiations or photons (Seki et al., 1996a, 1999b). It should be noted that the cross-linking reactions depended strongly on the density of neutral reactive intermediates, that is, silyl radicals (Maeda et al., 2001), and the reactions seemed to occur within a nanometer-scaled cylindrical space along an incident particle trajectory where the intermediates are distributed densely and nonhomogeneously (Chatterjee and Magee, 1980; Magee and Chatterjee, 1980; Seki et al., 2004b). This process will potentially give an insoluble "nanogel" along each corresponding particle, and produce wire-like 1D nanostructures via the isolation of the "nanogel" on a substrate by removing soluble non-cross-linked parts (Seki et al., 2001c, 2003, 2005b, 2006; Tsukuda et al., 2004a,b, 2005a,b, 2006).

Simple procedures were employed to cause cross-linking reactions: coating of the polymers onto Si substrates at 0.2–1.0 μm thickness, irradiation to a variety of MeV-order high-energy charged particles from several accelerators in vacuum chambers, and washing the film in solvents to remove non-cross-linked parts of the film (Seki et al., 2006). After the washing and drying procedures, the surface structure of the substrate was observed directly by an atomic force microscope (AFM).

High-energy charged particle irradiation of PH1 films was found to cause gelation of the polymers for all particles, all energies, and all molecular weights of PH1. Figure 25.15 shows the evolution curves of the gel volume for irradiation of PH1 films with 2 MeV He⁺ particles. In the figure, the gel fraction corresponds to the normalized thickness, and the evolution curve was calibrated by the absorbed dose (D). According to the statistical theory of cross-linking and scission of polymers induced by radiation, the G-values of cross-linking (efficiency of the cross-linking reaction, $G(x)$) and main-chain scission ($G(s)$) are expressed by the Charlesby–Pinner relationship as follows (Charlesby, 1954; Charlesby and Pinner, 1959):

$$s + s^{1/2} = \frac{p}{q} + \frac{m}{q} M_n D \qquad (25.17)$$

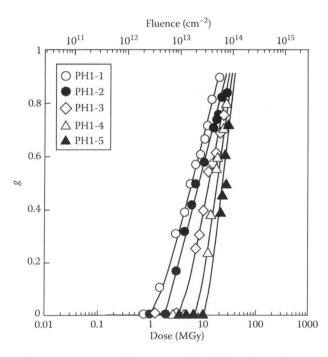

FIGURE 25.15 Sensitivity curves (gel evolution curves) for PH1s with various molecular weights under irradiation with a 2 MeV ⁴He⁺ beam. (Reprinted from Seki, S. et al., *Chem. Lett.*, 34, 1690, 2005a; Seki, S. et al., *Macromolecules*, 38, 10164, 2005b. With permission.)

$$s = 1 - g \tag{25.18}$$

$$G(x) = 4.8 \times 10^3 q \tag{25.19}$$

$$G(s) = 9.6 \times 10^3 p \tag{25.20}$$

where
  p is the probability of scission
  q is the probability of cross-linking
  s is the sol fraction
  g is the gel fraction
  m is the molecular weight of a unit monomer
  M_n is the number average molecular weight before irradiation

The cross-linking G-values calculated using these equations for irradiation of PH1 with 2 MeV He$^+$, H$^+$, C$^+$, and N$^+$ particles are compared in Table 25.4. The values of $G(x)$ for high-molecular-weight PH1 are much lower than those for low-molecular-weight PH1. Besides chain length, there are no differences in the chemical structures of the polymers, indicating that the efficiency of cross-linking should be identical for all series of polymers to a first approximation. The effects of the molecular weight distribution on the radiation-induced gelation of a real polymer system were considered by Saito (1958) and Inokuti (1960), who traced the changes in distribution due to simultaneous reactions of main-chain scission and cross-linking. However, in the present case, the molecular weight distributions of the target polymers are reasonably well controlled to be less than 1.2, and the initial distributions are predicted not to play a major role in gelation. The simultaneous change in the molecular weight distribution due to radiation-induced reactions also results in a nonlinearity of Equation 25.17. Therefore, the following equations are proposed to extend the validity of the relationship by introducing a deductive distribution function of molecular weight on the basis of an arbitrary distribution (Olejniczak et al., 1991):

$$s + s^{1/2} = \frac{p}{q} + \frac{(2 - p/q)(D_V - D_g)}{D_V - D} \tag{25.21}$$

TABLE 25.4
Values of $G(x)$ Determined for Various High-Energy Particles and Polymer-Chain Lengths

Entry	2 MeV H 15 eV nm^{-1}	2 MeV He 220 eV nm^{-1}	0.5 MeV C 410 eV nm^{-1}	2 MeV C 720 eV nm^{-1}	2 MeV N 790 eV nm^{-1}
PH1-1	0.0018	0.0049	0.021	0.072	0.082
	(0.0021)[a]	(0.0052)[a]	(0.022)[a]	(0.0079)[a]	(0.0095)[a]
PH1-2	0.0021	0.0095	0.052	0.081	0.15
PH1-3	0.0030	0.019	0.07	0.18	0.21
PH1-4	0.0075	0.021	0.075	0.20	0.26
PH1-5	0.019	0.061	0.18	0.27	0.34
	(0.021)[a]	(0.089)[a]	(0.19)[a]	(0.33)[a]	(0.42)[a]

Sources: Data from Seki, S. et al., *Radiat. Phys. Chem.*, 48, 539, 1996a; Seki, S. et al., *Adv. Mater.*, 13, 1663, 2001c; Seki, S. et al., *Chem. Lett.*, 34, 1690, 2005a; Seki, S. et al., *Macromolecules*, 38, 10164, 2005b.

[a] Values in parentheses estimated by Equations 25.21 and 25.22.

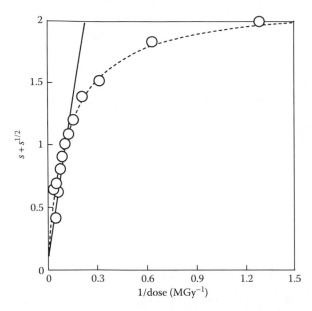

FIGURE 25.16 Charlesby–Pinner plot of $s + s^{1/2}$ versus $1/D$ for 2 MeV ^{4}He$^+$ irradiation of PH1-1. The solid line denotes the fit by Equation 25.17 in the high-dose region for the estimation of $G(x)$ in Table 25.4. The dashed line is given by Equation 25.23 for the entire dose region. (Reprinted from Seki, S. et al., *Chem. Lett.*, 34, 1690, 2005a; Seki, S. et al., *Macromolecules*, 38, 10164, 2005b. With permission.)

$$D_V = \frac{4\left((1/uu_n) - (1/u_w)\right)}{3q} \tag{25.22}$$

where
  D_g is the gelation dose
  u is the degree of polymerization

Equation 25.21 provides a better fit to the observed values of s at high doses than Equation 25.17, as shown in Figure 25.16. However, the $G(x)$ derived from this fit are almost identical to those in Table 25.4, depending on the molecular weight, because the values are estimated in the low-dose region where Equation 25.17 is sufficiently linear.

The effects of the polymer structure (Seki et al., 1998) and the cross-linking precursor species (Seki et al., 1996b) have been considered in previous studies, where it was revealed that there was less of a dependence of $G(x)$ on the molecular weight in the case of γ-rays or EB irradiation. The cross-linking points in polymer materials are distributed nonhomogeneously on an ion track along an ion trajectory. A schematic of this type of nonhomogeneous distribution is shown in Figure 25.17. Intramolecular cross-linking, which constitutes a greater contribution to $G(x)$ in the inner region of ion tracks associated with higher densities of deposited energy, is not taken into account by the statistical treatment in Equations 25.17 through 25.22, and it is this case that gives rise to an underestimate of the number of cross-links. As the polymer film used in the present study is sufficiently thin to allow the change in the kinetic energy of the incident particle to be neglected, a model of cylindrical energy deposition is sufficient, without needing to refer to the dependence of the radial energy distribution on the direction of the ion trajectory. The total yield of gels generated by charged particle irradiation can thus be treated using a simplified expression (Seki et al., 1997, 1999b, 2004b, 2005b), as follows:

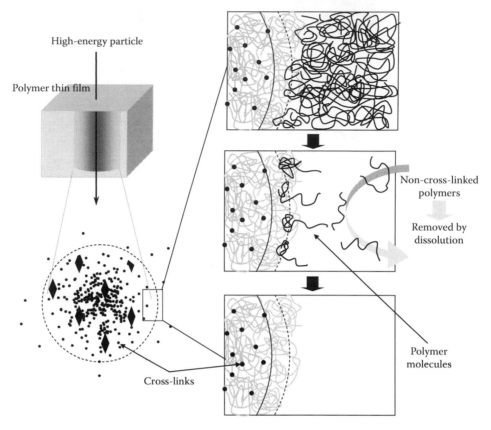

FIGURE 25.17 Schematic of the nonhomogeneous distribution of cross-links in an ion track. Dissolution of non-cross-linked molecules results in the formation of isolated nanowires.

$$g = 1 - \exp\left[-n\pi r_{cc}^{2}\right] \qquad (25.23)$$

where

r_{cc} is the radius of the cross section of the cylindrical region (chemical core)

n represents the fluence of incident ions

The effect of diffusion of reactive intermediates, determined using a low-energy charged particle, has been formulated as the term δr (Seki et al., 2004b):

$$r_{cc} = r' + \delta r'. \qquad (25.24)$$

The value of $\delta r'$ in PH1 was determined to be 0.5–0.7 nm (Seki et al., 2004b). Based on traces of the gel fraction by Equation 25.23, the estimated values of r_{cc} are summarized for a variety of high-energy charged particles in Table 25.5. The value of r_{cc} increases substantially with the molecular weight in all cases except for irradiation with 2 MeV H^+, suggesting that the cylindrical scheme may not be applicable for the 2 MeV H^+ because the deposited energy density is too low to promote the cylindrical distribution of cross-links in an ion track.

AFM images of the cylindrical gel area are shown in Figure 25.18 for comparison with traces of the gel fraction. Equation 25.23 provides a good fit for the trace of the gel fraction, and the estimated values of r_{cc} for both particles correspond to the values observed from the AFM micrographs. Figure 25.19 shows a series of AFM micrographs observed for the irradiation of PH1-1 and

TABLE 25.5

Values of r_{cc} for Various High-Energy Particles and Polymer-Chain Lengths

Ions	Energy (MeV)	LET (eV nm⁻¹)	r_{cc} for PH1-1 (nm)	r_{cc} for PH1-2 (nm)	r_{cc} for PH1-3 (nm)	r_{cc} for PH1-4 (nm)	r_{cc} for PH1-5 (nm)
¹H[a]	2.0	15	0.15	0.15	0.14	0.14	0.15
⁴He[a]	2.0	220	1.2	0.95	0.76	0.60	0.52
¹²C[a]	0.50	410	2.2	2.1	1.6	1.1	0.94
¹²C[a]	2.0	720	5.0	4.3	4.0	3.4	2.7
¹⁴N[a]	2.0	790	5.1	4.8	4.2	3.8	3.0
¹⁴N[b]	2.0	790	4.8	4.8	4.3	3.6	2.8
⁵⁶Fe[b]	5.1	1,550				5.5	
²⁸Si[b]	5.1	1,620				5.9	
²⁸Si[b]	10.2	2,150				6.1	
⁴⁰Ar[b]	175	2,200	8.2	7.3		6.1	4.0
⁵⁶Fe[b]	8.5	2,250				5.8	
⁵⁶Fe[b]	10.2	2,600				7.7	
⁸⁴Kr[a]	520	4,100	10.7	9.6	8.6	7.9	6.4
⁸⁴Kr[b]	520	4,100	10.2	9.2	8.3		6.1
¹²⁹Xe[b]	450	8,500	12.1	10.5	9.3		6.9
¹⁹²Au[b]	500	11,600	19.4		16.2	12.5	10.7

Sources: Data from Seki, S. et al., *Adv. Mater.*, 13, 1663, 2001c; Seki, S. et al., *Chem. Lett.*, 34, 1690, 2005a; Seki, S. et al., *Macromolecules*, 38, 10164, 2005b.

[a] Values estimated from gel traces by Equation 25.23.

[b] Values estimated by direct AFM observation.

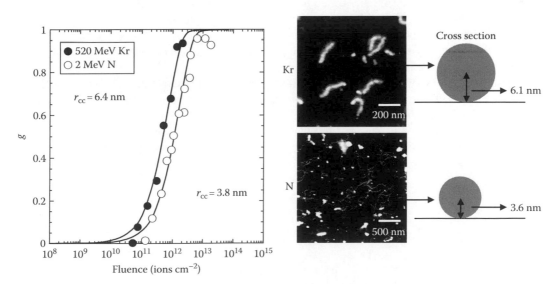

FIGURE 25.18 Gel evolution curves recorded for 520 MeV ⁸⁴Kr irradiation of PH1-5 and 2 MeV ¹⁴N irradiation of PH1-4. Solid lines denote the fit for the respective gel fraction based on Equation 25.23. The estimated values of r_{cc} are 6.4 and 3.8 nm for the Kr and N particles. AFM micrographs were observed for the same set of polymers and particles. The fluence of Kr and N ions was set at 1.4×10^{10} and 6.4×10^{9} cm⁻², respectively. (Reprinted from Seki, S. et al., *Chem. Lett.*, 34, 1690, 2005a; Seki, S. et al., *Macromolecules*, 38, 10164, 2005b. With permission.)

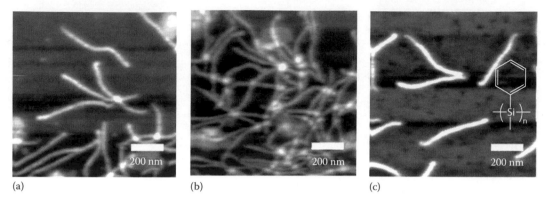

FIGURE 25.19 AFM micrographs of nanowires based on PH1s showing the variation in size with molecular weight and number density on the substrate. The nanowires were formed by a 500 MeV ^{192}Au-beam irradiation to (a,b) PH1-3 and (c) PH1-1 thin films at (a) 3.0×10^9, (b) 5.0×10^9, and (c) 1.0×10^9 ions cm^{-2}, respectively. The thicknesses of the target films were (a,b) 350 nm and (c) 250 nm.

PH1-3 thin films with varying number of incident particles. These images clearly reveal 1D rod-like structures (nanowires) on the substrate, which is ascribed to the cylindrical formation of nanogel along ion trajectories. It should be noted that the density of the nanowires on the substrate increased clearly with an increase in the number of the incident particles, and the observed number density of the nanowires coincided with the number density of the incident particle. This is also suggestive that "one nanowire" is produced corresponding to one incident particle along its trajectory, which is a clear evidence of the model described above. As shown in Figure 25.19, the length of the nanowires is uniform in each image, and is consistent precisely with the initial thickness of the film. This is due to the geometrical limitation of the distribution of the cross-liking reaction: the gelation occurs from the top surface to the bottom of the polymer film. Thus, the length of the nanowires can be perfectly controlled by the present technique. Based on the measurement of the cross-sectional trace of the nanowires by AFM, the radial distribution of cross-links in the nanowires is discussed in terms of r_{cc} defined as the radius of the cross section. It is clear that the value of r_{cc} depends on the molecular weight of the target polymer, as shown in Figure 25.20, and that r_{cc} also changes with the LET of the incident particle, as summarized in Table 25.5.

The authors previously reported that cross-linking reactions are mainly promoted by side-chain-dissociated silyl radicals, and that the predominant reaction is determined by the radical concentration in the ion tracks (Cotts et al., 1991; Seki et al., 1996a,b, 1999b). Thus, the distribution of cross-links in an ion track is expected to reflect the radial dose (deposited energy density) distribution, where ρ_{cr} is the critical energy density for the predominance of cross-linking in PH1. The radial dose distribution in an ion track is thus given by Chatterjee and Magee (1980) and Magee and Chatterjee (1980)

$$\rho_c = \frac{LET}{2} \left[\pi r_c^2 \right]^{-1} + \frac{LET}{2} \left[2\pi r_c^2 \ln\left(\frac{e^{1/2} r_p}{r_c} \right) \right]^{-1} \quad r \leq r_c \tag{25.25}$$

$$\rho_p(r) = \frac{LET}{2} \left[2\pi r^2 \ln\left(\frac{e^{1/2} r_p}{r_c} \right) \right]^{-1} \quad r_c < r \leq r_p \tag{25.26}$$

where
ρ_c is the deposited energy density in the core area
r_c and r_p are the radii of the core and penumbra area

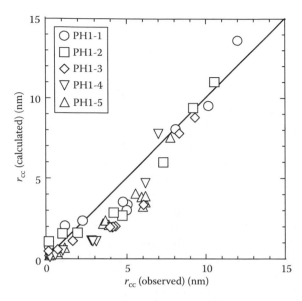

FIGURE 25.20 Correlation between r' values estimated experimentally and those calculated using Equation 25.29. (Reprinted from Seki, S. et al., *Chem. Lett.*, 34, 1690, 2005a; Seki, S. et al., *Macromolecules*, 38, 10164, 2005b. With permission.)

For gel formation in a polymer system, it is necessary to introduce one cross-link per polymer molecule. Assuming a sole contribution from the cross-linking reactions in the chemical core, ρ_{cr} is given by

$$\rho_{cr} = \frac{100\rho A}{G(x)mN} \qquad (25.27)$$

where
  A is Avogadro's number
  N is the degree of polymerization

The value of $mN/\rho A$ reflects the volume of a polymer molecule. Substituting $\rho_p(r)$ in Equation 25.26 with ρ_{cr} gives the following requirement for r_{cc} (Seki et al., 2004b; Tsukuda et al., 2005a):

$$r'^2 = \frac{LET \cdot G(x)mN}{400\pi\rho A}\left[\ln\left(\frac{e^{1/2}r_p}{r_c}\right)\right]^{-1} \qquad (25.28)$$

Using the reported value of $G(x) = 0.12$, derived from radiation-induced changes in the molecular weight (Seki et al., 1996a), the values of r_{cc} calculated by the nonempirical formulation of Equation 25.28 are compared with the experimental values showing a good consistency with the experimental values for polymers with sufficient chain lengths (PH1-1 and PH1-2). However, a considerable discrepancy occurs between the calculated and experimental results for polymers with shorter chains. The global configuration of the polymer molecules depends greatly on the length of the polymer chains, leading to a transformation from random coil (long-chain) to rod-like (short-chain) conformations. The gyration radius of a polymer molecule, which determines the size of a molecule spreading in the media, is correlated with this transformation. The correlation between R_g and N is discussed previously in this chapter, giving Equation 25.2. Based on the persistence length of PH1 (1.1 nm, see Section 25.4), the scaling law for a helical worm-like chain model (Yamakawa et al., 1994) results in an index, α, of 0.410, 0.419, 0.451, 0.485, and 0.492 for PH1-1–5, respectively. Thus, the effective volume of a polymer chain can be simply calculated as $4/3\pi R_g^3$, and the substitution of $mN/\rho A$ in Equation 25.28 with the effective volume leads to the expression (Seki et al., 2005b, 2006)

$$r'^2 = \frac{\text{LET} \cdot G(x) N^{3\alpha}}{400\pi\beta} \left[\ln\left(\frac{e^{1/2} r_p}{r_c} \right) \right]^{-1} \tag{25.29}$$

where β is the effective density parameter of the monomer unit (kg m^{-3}). Based on Equation 25.29, the calculated value of r' is plotted against the experimental values in Figure 25.20. All polymers, with a variety of molecular weights, follow a single trend, and the calculated values display good correspondence when $r' > 7$ nm. The underestimate of r' by Equation 25.29 for $r' < 7$ nm suggests that the initial deposition of energy and the radial dose distribution estimated by Equations 25.25 and 25.26 do not account for the radial distribution of chemical intermediates, and thus, cannot model the concentration of cross-linking in the core of the ion track. The value of $G(x)$ increases dramatically with an increase in the density of reactive intermediates. Based on the assumption that $G(x)$ is a function of the density of deposited energy, the present results indicate that the yield of the chemical reaction is dependent on the energy density. Cross-linking reactions in ion tracks therefore have the potential for not only single-particle fabrication with sub-nanometer-scale spatial resolution for any kind of cross-linking polymer, but also the study of nanoscale distributions of radial dose and chemical yield in an ion track.

REFERENCES

Abkowitz, M. A., Knier, F. E., Yuh, H. J., Weagley, R. J., and Stolka, M. 1987. Electronic transport in amorphous silicon backbone polymers. *Sol. Stat. Commun.* 62: 547–550.

Abkowitz, M. A., Rice, M. J., and Stolka, M. 1990. Electronic transport in silicon backbone polymers. *Philos. Mag. B* 61: 25–64.

Acharya, A., Seki, S., Saeki, A., Koizumi, Y., and Tagawa, S. 2005a. Study of transport properties in fullerene-doped polysilane films using flash photolysis time-resolved microwave technique. *Chem. Phys. Lett.* 404: 356–360.

Acharya, A., Seki, S., Koizumi, Y., Saeki, A., and Tagawa, S. 2005b. Photogeneration of charge carriers and their transport properties in poly[bis(p-n-butylphenyl)silane]. *J. Phys. Chem. B* 109: 20174–20179.

Adhikary, A., Khanduri, D., and Sevilla, M. D. 2009. Direct observation of the hole protonation state and hole localization site in DNA-oligomers. *J. Am. Chem. Soc.* 131: 8614–8619.

Alig, R. C., Bloom, S., and Struck, C. W. 1980. Scattering by ionization and phonon emission in semiconductors. *Phys. Rev. B* 22: 5565–5582.

Amaya, T., Seki, S., Moriuchi, T., Nakamoto, K., Nakata, T., Sakane, H., Saeki, A., Tagawa, S., and Hirao, T. 2009. Anisotropic electron transport properties in sumanene crystal. *J. Am. Chem. Soc.* 131: 408–409.

Ban, H., Sukegawa, K., and Tagawa, S. 1987. Pulse radiolysis study on organopolysilane radical anions. *Macromolecules* 20: 1775–1778.

Barnes, J., Bernhard, W. A., and Mercer, K. R. 1991. Distribution of electron trapping in DNA-protonation of one-electron reduced cytosine. *Radiat Res* 126: 104–107.

Baude, P. F., Ender, D. A., Haase, M. A., Kelley, T. W., Muyres, D. V., and Theiss, S. D. 2003. Pentacene-based radio-frequency identification circuitry. *Appl. Phys. Lett.* 82: 3964–3966.

Becker, D., Adhikary, A., and Sevilla, M. D. 2007. The role of charge and spin migration in DNA radiation damage. In *Charge Migration in DNA: Physics, Chemistry and Biology Perspectives*, T. Chakraborty (ed.), pp. 139–175. Berlin/Heidelberg/New York: Springer-Verlag.

Bernhard, W. A. and Close, D. M. 2004. *Charged Particle Interactions with Matter: Chemical, Physicochemical, and Biological Consequences with Applications*, A. Mozunder and Y. Hatano (eds.), pp. 431–470. New York/Basel: Marcel Dekker, Inc.

Biondi, M. A. 1951. Measurement of the electron density in ionized gases by microwave techniques. *Rev. Sci. Instrum.* 22: 500–502.

Biondi, M. A. and Brown, S. C. 1949. Measurements of ambipolar diffusion in helium. *Phys. Rev.* 75: 1700–1705.

Candeias, L. P. and Steenken, S. 1993. Electron-transfer in di(deoxy)nucleoside phosphates in aqueous solution-rapid migration of oxidative damage (via adenine) to guanine. *J. Am. Chem. Soc*. 115: 2437–2440.

Candeias, L. P., Wildeman, J., Hadziioannou, G., and Warman, J. M. 2000. Pulse radiolysis–optical absorption studies on the triplet states of *p*-phenylenevinylene oligomers in solution. *J. Phys. Chem. B* 104: 8366–8371.

Charlesby, A. 1952. Cross-linking of polyethylene by pile radiation. *Proc. R. Soc.* (*Lond.*) A215: 187–214.

Charlesby, A. 1954. The cross-linking and degradation of paraffin chains by high-energy radiation. *Proc. R. Soc. Lond. Ser. A* 222: 60–74.

Charlesby, A. and Pinner, S. H. 1959. Analysis of the solubility behaviour of irradiated polyethylene and other polymers. *Proc. R. Soc. Lond. Ser. A* 249: 367–386.

Chatterjee, A. and Magee, J. L. 1980. Radiation chemistry of heavy-particle tracks. 2. Fricke dosimeter system. *J. Phys. Chem.* 84: 3537–3543.

Cotts, P. M., Miller, R. D., and Sooriyakumaran, R. 1990. Configurational properties of a polysilane. In *Silicon Based Polymer Science*, p. 397. Washington, DC: American Chemical Society.

Cotts, P. M., Ferline, S., Dagli, G., and Pearson, D. S. 1991. Solution properties of poly(di-n-hexylsilane): An estimate of its unperturbed dimensions. *Macromolecules* 24: 6730–6735.

de Haas, M. P., Warman, J. M., Infelta, P. P., and Hummel, A. 1975. The direct observation of a highly mobile positive ion in nanosecond pulse irradiated liquid cyclohexane. *Chem. Phys. Lett.* 31: 382–386.

Dole, M., Keeling, C. D., and Rose, D. G. 1954. The pile irradiation of polyethylene. *J. Am. Chem. Soc.* 76: 4304–4311.

Douhal, A., Kim, S. K., and Zewail, A. H. 1995. Femtosecond molecular dynamics of tautomerization in model base pairs. *Nature* 378: 260–263.

Feng, X., Marcon, V., Pisula, W., Hansen, M. R., Kirkpatrick, J., Grozema, F., Andrienko, D., Kremer, K., and Müllen, K. 2009. Towards high charge-carrier mobilities by rational design of the shape and periphery of discotics. *Nat. Mater.* 8: 421–426.

Fessenden, R. W. and Hitachi, A. 1987. A study of the dielectric relaxation behavior of photoinduced transient species. *J. Phys. Chem.* 91: 3456–3462.

Fessenden, R. W., Carton, P. M., Paul, H., and Shimamori, H. 1979. Detection of transient species by measurement of dielectric loss. *J. Phys. Chem.* 83: 1676–677.

Fessenden, R. W., Carton, P. M., Shimamori, H., and Scaiano, J. C. 1982. Measurement of the dipole moments of excited states and photochemical transients by microwave dielectric absorption. *J. Phys. Chem.* 86: 3803–3811.

Fujiki, M. 1994. Ideal exciton spectra in single- and double-screw-sense helical phlysilanes. *J. Am. Chem. Soc.* 116: 6017.

Fujiki, M. 1996. A correlation between global conformation of polysilane and UV absorption characteristics. *J. Am. Chem. Soc.* 118: 7424.

Fujino, M. 1987. Photoconductivity in organopolysilanes. *Chem. Phys. Lett.* 136: 451–453.

Gill, D., Jagur-Grodzinsky, J., and Swarc, M. 1964. Chemistry of radical-ions. Electron-transfer reactions: –DD→dimethyl anthracene and –DD→pyrene. *Trans. Faraday Soc.* 60: 1424–1431.

Goldmann, C., Krellner, C., Pernstich, K. P., Haas, S., Gundlach, D. J., and Batlogg, B. 2006. Determination of the interface trap density of rubrene single-crystal field-effect transistors and comparison to the bulk trap density. *J. Appl. Phys.* 99: 034507-1–034507-8.

Grozema, F. C., Savenije, T. J., Vermeulen, J. W., Siebbeles, L. D. A., Warman, J. M., Meisel, A., Neher, D., Nothofer, H. G., and Scherf, U. 2001. Electrodeless measurement of the in-plane anisotropy in the photoconductivity of an aligned polyfluorene film. *Adv. Mater.* 13: 1627–1630.

Gu, J. D., Xie, Y. M., and Schaefer, H. F. 2005. Structural and energetic characterization of a DNA nucleoside pair and its anion: Deoxyriboadenosine (da)–deoxyribothymidine (dt). *J. Phys. Chem. B* 109: 13067–13075.

Hisaki, I., Sakamoto, Y., Shigemitsu, H., Tohnai, N., Miyata, M., Seki, S., Saeki, A., and Tagawa, S. 2008. Superstructure-dependent optical and electrical properties of an unusual face-to-face, π-stacked, one-dimensional assembly of dehydrobenzo[12]annulene in the crystalline state. *Chem. Eur. J.* 14: 4178–4187.

Hissung, A. and von Sonntag, C. 1979. The reaction of solvated electrons with cytosine, 5-methyl cytosine and 2'-deoxycytidine in aqueous solution the reaction of the electron adduct intermediates with water, p-nitroacetophenone and oxygen a pulse spectroscopic and pulse conductometric study. *Int. J. Radiat. Biol.* 35: 449–458.

Hoofman, R. J. O. M., de Hass, M. P., Siebbeles, L. D. A., and Warman, J. M. 1998a. Highly mobile electrons and holes on isolated chains of the semiconducting polymer poly(phenylene vinylene). *Nature* 392: 54–56.

Hoofman, R. J. O. M., Siebbeles, L. D. A., de Haas, M. P., and Hummel, A. 1998b. Anisotropy of the charge-carrier mobility in polydiacetylene crystals. *J. Chem. Phys.* 109: 1885.

Ichikawa, T., Yamada, Y., Kumagai, J., and Fujiki, M. 1999a. Suppression of the Anderson localization of charge carriers on polysilane quantum wire. *Chem. Phys. Lett.* 306: 275–279.

Ichikawa, T., Sumita, M., and Kumagai, J. 1999b. Relation between the helicity and the ESR *g*-anisotropy of σ-conjugated quantum wires. *Chem. Phys. Lett.* 307: 81.

Imahori, H., Ueda, M., Kang, S., Hayashi, H., Hayashi, S., Kaji, H., Seki, S., Saeki, A., Tagawa, S., Umeyama, T., Matano, Y., Yoshida, K., Isoda, S., Shiro, M., Tkachenko, N. V., and Lemmetyinen, H. 2007. Effects of porphyrin substituents on film structure and photoelectrochemical properties of porphyrin/fullerene composite clusters electrophoretically deposited on nanostructured SnO_2 electrodes. *Chem. Eur. J.* 13: 10182–10193.

Imamura, S. 1979. Chloromethylated polystyrene as a dry etching-resistant negative resist for sub-micron technology. *J. Electrochem. Soc.* 126: 1628–1630.

Imamura, S., Tamamura, T., Harada, K., and Sugawara, S. 1982. High-performance electron negative resist, chlorinated polystyrene – A study on molecular-parameters. *J. Appl. Polym. Sci.* 27: 937–949.

Inokuti, M. 1960. Molecular size distribution and gelation of irradiated copolymers. *J. Chem. Phys.* 33: 1607–1615.

Ishii, H., Yuyama, A., Norioka, S., Seki, K., Hasegawa, S., Fujino, M., Isaka, H., Fujiki, M., and Matsumoto, N. 1995. Photoelectron spectroscopy of polysilanes, polygermanes and related compounds. *Synth. Met.* 69: 595–596.

Itagaki, H., Horie, K., Mita, I., Washio, M., Tagawa, S., and Tabata, Y. 1983. Intramolecular excimer formation of oligostyrenes from dimer to tridecamer—The measurements of rate constants for excimer formation, singlet energy migration and relaxation of internal-rotation. *J. Chem. Phys.* 79: 3996–4005.

Itagaki, H., Horie, K., Mita, I., Washio, M., Tagawa, S., Tabata, Y., Sato, H., and Tanaka, Y. 1987. Local molecular-motion of polystyrene model compounds measured by using picosecond pulse-radiolysis 1. Diastereoisomeric styrene dimmers—Multicomponent fluorescence decay curves, concentration-dependence, and alkyl end-group effect on excimer formation. *Macromolecules* 20: 2774–2782.

Itagaki, H., Washio, M., Washio, M., and Tagawa, S. 1992. Rates for intramolecular excimer formation of poly(alpha-methylstyrene) measured by using picosecond pulse-radiolysis. *Polym. Bull.* 28: 197–202.

Kajzar, F., Messier, J., and Rosilio, C. 1986. Nonlinear optical properties of thin films of polysilane. *J. Appl. Phys.* 60: 3040.

Kamoshida, Y., Koshida, M., Yoshimoto, H., Harita, K., and Harada, K. 1983. Application of chlorinated poly-methylstyrene, CPMS, to electron-beam lithography. *J. Vac. Sci. Technol.* B1: 1156–1159

Kawaguchi, T., Seki, S., Okamoto, K., Saeki, A., Yoshida, Y., and Tagawa, S. 2003. Pulse radiolysis study of radical cations of polysilanes. *Chem. Phys. Lett.* 374: 353–357.

Kepler, R. G., Zeigler, J. M., Harrah, L. A., and Kurtz, S. R. 1987. Photocarrier generation and transport in σ-bonded polysilanes. *Phys. Rev. B* 35: 2818.

Kido, J., Kimura, M., and Nagai, K. 1995. Multilayer white light-emitting organic electroluminescent device. *Science* 267: 1332–1334.

Kobayashi, K. and Tagawa, S. 2003. Direct observation of guanine radical cation deprotonation in duplex DNA using pulse radiolysis. *J. Am. Chem. Soc.* 125: 10213–10218.

Kobayashi, K., Yamagami, R., and Tagawa, S. 2008. Effect of base sequence and deprotonation of guanine cation radical in DNA. *J. Phys. Chem. B* 112: 10752–10757.

Kocherzhenko, A. A., Patwardhan, S., Grozema, F. C., Anderson, H. L., and Siebbeles, L. D. A. 2009. Mechanism of charge transport along zinc porphyrin-based molecular wires. *J. Am. Chem. Soc.* 131: 5522–5529.

Koe, J. R., Powell, D. R., Buffy, J. J., Hayase, S., and West R., 1998. Perchloropolysilane: X-ray structure, solid-state ^{29}Si NMR spectroscopy, and reactions of $[SiCl_2]_n$. *Angew. Chem. Int. Ed.* 37: 1441.

Koe, J. R., Fujiki, M., Nakashima, H., and Motonaga, M. 2000. Temperature-dependent helix–helix transition of an optically active poly(diarylsilylene). *Chem. Commun.* 389–390.

Koe, J. R., Fujiki, M., Motonaga, M., and Nakashima, H. 2001. Cooperative helical order in optically active poly(diarylsilylenes). *Macromolecules* 34: 1082–1089.

Kumada, M. and Tamao, K. 1968. Aliphatic organopolysilanes. *Adv. Organomet. Chem.* 6: 80.

Kumagai, J., Yoshida, H., and Ichikawa, T. 1995. Electronic structure of oliosilane and polysilane radical cations as studied by electron spin resonance and electronic absorption spectroscopy. *J. Phys. Chem.* 99: 7965.

Le Motais, B. C. and Jonah, C. D. 1989. Picosecond pulse-radiolysis studies of the primary processes in hydro-carbon radiation-chemistry. *Radiat. Phys. Chem.* 33: 505–517.

Lee, Y. A., Durandin, A., Dedon, P. C., Geacintov, N. E., and Shafirovich, V. 2008. Oxidation of guanine in G, GG, and GGG sequence contexts by aromatic pyrenyl radical cations and carbonate radical anions: Relationship between kinetics and distribution of alkali-labile lesions. *J. Phys. Chem. B* 112: 1834–1844.

Li, W. S., Yamamoto, Y., Fukushima, T., Saeki, A., Seki, S., Tagawa, S., Masunaga, H., Sasaki, S., Takata, M., and Aida, T. 2008. Amphiphilic molecular design as a rational strategy for tailoring bicontinuous electron donor and acceptor arrays: Photoconductive liquid crystalline oligothiophene-C_{60} dyads. *J. Am. Chem. Soc.* 130: 8886–8887.

Maeda, K., Seki, S., Tagawa, S., and Shibata, H. 2001. Radiation effects on branching polysilanes. *Radiat. Phys. Chem.* 60: 461–466.

Magee, J. L. and Chatterjee, A. 1980. Radiation chemistry of heavy-particle tracks. 1. General considerations. *J. Phys. Chem.* 84: 3529–3536.

Matsui, Y., Nishida, K., Seki, S., Yoshida, Y., Tagawa, S., Yamada, K., Imahori, H., and Sakata, A. 2002a. Direct observation of intra-molecular electron transfer from excess electrons in σ-conjugated main chain to porphyrin side chain in polysilanes having tetraphenylporphyrin side chain by pulse radiolysis technique. *Organometallics* 21: 5144.

Matsui, Y., Seki, S., and Tagawa, S. 2002b. Direct observation of intra-molecular energy migration in σ-conjugated polymer by femto-second laser flash photolysis. *Chem. Phys. Lett.* 357: 346–350.

Morita, M., Tanaka, A., Imamura, S., Tamamura, T., and Kogure, O. 1983. High-resolution double-layer resist system using new silicone based negative resist (SNR). *Jpn. J. Appl. Phys. Part* 2(22): L659–L660.

Morita, M., Imamura, S., Tanaka, A., and Tamamura, T. 1984. A new silicone-based negative resist (SNR) for 2-layer resist system. *J. Electrochem. Soc.* 131: 2402–2406.

Motoyanagi, J., Yamamoto, Y., Saeki, A., Alam, M. A., Kimoto, A., Kosaka, A., Fukushima, T., Seki, S., Tagawa, S., and Aida, T. 2009. Unusual side-chain effects on charge-carrier lifetime in discotic liquid crystals. *Chem. Asian J.* 4: 876–880.

Myny, K., Steudel, S., Vicca, P., Genoe, J., and Heremans, P. 2008. An integrated double half-wave organic Schottky diode rectifier on foil operating at 13.56 mHz. *Appl. Phys. Lett.* 93: 093305-1–093305-3.

Nagashima, K., Yanagida, T., Tanaka, H., Seki, S., Saeki, A., Tagawa, S., and Kawai, T. 2008. Effect of the heterointerface on transport properties of in situ formed mgo/titanate heterostructured nanowires. *J. Am. Chem. Soc.* 130: 5378–5382.

Ohnishi, Y., Saeki, A., Seki, S., and Tagawa, S. 2009. Conformational relaxation of sigma-conjugated polymer radical anion on picosecond scale. *J. Chem. Phys.* 130: 204907.

Okamoto, K., Kozawa, T., Yoshida, Y., and Tagawa, S. 2001. Study on intermediate species of polystyrene by using pulse radiolysis. *Radiat. Phys. Chem.* 60: 417–422.

Okamoto, K., Saeki, A., Kozawa, T., Yoshida, Y., and Tagawa, S. 2003. Subpicosecond pulse radiolysis study of geminate ion recombination in liquid benzene. *Chem. Lett.* 32: 834–835.

Okamoto, K., Kozawa, T., Miki, M., Yoshida, Y., and Tagawa, S. 2006. Pulse radiolysis of polystyrene in cyclohexane – Effect of carbon tetrachloride on kinetic dynamics of dimmer radical cation. *Chem. Phys. Lett.* 426: 306–310.

Okamoto, K., Kozawa, T., Saeki, A., Yoshida, Y., and Tagawa, S. 2007. Subpicosecond pulse radiolysis in liquid methyl-substituted benzene derivatives. *Radiat. Phys. Chem.* 76: 818–826.

Okamoto, K., Kozawa, T., Natsuda, K., Seki, S., and Tagawa, S. 2008. Formation of intramolecular poly(4-hydroxystyrene) dimer radical cation. *J. Phys. Chem.* 112: 9275–9280.

Okamoto, K., Tanaka, M., Kozawa, T., and Tagawa, S. 2009. Dynamics of radical cation of poly(4-hydroxystyrene) and its copolymer for extreme ultraviolet and electron beam resists. *Jpn. J. Appl. Phys. Part 2* 48: 06FC06.

Olejniczak, K., Rosiak, J., and Charlesby, A. 1991. Gel/dose curves for polymers undergoing simultaneous crosslinking and scission. *Radiat. Phys. Chem.* 37: 499–504.

Pieter, A. C., Sweelssen, J., Koetse, M. M., Savenije, T. J., and Siebbeles, L. D. A. 2007. Formation and decay of charge carriers in bulk heterojunctions of MDMO-PPV or P3HT with new n-type conjugated polymers. *J. Phys. Chem. C* 111: 4452–4457.

Pitt, C. G. 1977. Conjugative properties of polysilanes and catenated σ-systems. In *Homoatomic Rings, Chains, and Macromolecules of Main Group Elements*, A. L. Rheingold (ed.). Amsterdam, the Netherlands: Elsevier.

Podzorov, V., Menard, E., Borissov, A., Kiryukhin, V., Rogers, J. A., and Gershenson, M. E. 2004. Intrinsic charge transport on the surface of organic semiconductors. *Phys. Rev. Lett.* 93: 086602-1–086602-4.

Pollard, W. B. and Lucovsky, G. 1982. Phonons in polysilane alloys. *Phys. Rev. B* 26: 3172–3180.

Prins, P., Grozema, F. C., Schins, J. M., Patil, S., Scherf, U., and Siebbeles, L. D. A. 2006. High intrachain hole mobility on molecular wires of ladder-type poly(p-phenylenes). *Phys. Rev. Lett.* 96: 146601-1–146601-4.

Rice, M. 1979. Charged pi-phase kinks in lightly doped polyacetylene. *Phys. Lett.* 71A: 152.

Rice, M. J. and Phillpot, S. R. 1987. Polarons and bipolarons in a model tetrahedrally bonded homopolymer. *Phys. Rev. Lett.* 58: 937–940.

Saeki, A., Kozawa, T., Yoshida, Y., and Tagawa, S. 2002. Study on radiation-induced reaction in microscopic region for basic understanding of electron beam patterning in lithographic process (II)—Relation between resist space resolution and space distribution of ionic species. *Jpn. J. Appl. Phys.* 41: 4213–4216.

Saeki, A., Kozawa, T., Yoshida, Y., and Tagawa, S. 2004. Adjacent effect on positive charge transfer from radical cation of n-dodecane to scavenger studied by picosecond pulse radiolysis, statistical model, and Monte Carlo simulation. *J. Phys .Chem. A* 108: 1475–1481.

Saeki, A., Kozawa, T., Yoshida, Y., and Tagawa, S. 2005a. Multi spur effect on decay kinetics of geminate ion recombination using Monte Carlo technique. *Nucl. Instrum. Meth. B* 234: 285–290.

Saeki, A., Seki, S., Koizumi, Y., Sunagawa, T., Ushida, K., and Tagawa, S. 2005b. Increase in the mobility of photogenerated positive charge carriers in polythiophene. *J. Phys .Chem. B* 109: 10015–10019.

Saeki, A., Seki, S., Sunagawa, T., Ushida, K., and Tagawa, S. 2006a. Charge-carrier dynamics in polythiophene films studied by in-situ measurement of flash-photolysis time-resolved microwave conductivity (FP-TRMC) and transient optical spectroscopy (TOS). *Philos. Mag.* 86: 1261–1276.

Saeki, A., Seki, S., and Tagawa, S. 2006b. Electrodeless measurement of charge carrier mobility in pentacene by microwave and optical spectroscopy techniques. *J. Appl. Phys.* 100: 023703–023708.

Saeki, A., Seki, S., Koizumi, Y., and Tagawa, S. 2007. Dynamics of photogenerated charge carrier and morphology dependence in polythiophene films studied by in situ time-resolved microwave conductivity and transient absorption spectroscopy. *J. Photochem. Photobiol. A* 186: 158–165.

Saeki, A., Ohsaki, S., Seki, S., and Tagawa, S. 2008a. Electrodeless determination of charge carrier mobility in poly(3-hexylthiophene) films incorporating perylenediimide as photoconductivity sensitizer and spectroscopic probe. *J. Phys. Chem. C* 112: 16643–16650.

Saeki, A., Seki, S., Takenobu, T., Iwasa, Y., and Tagawa, S. 2008b. Mobility and dynamics of charge carriers in rubrene single crystals studied by flash-photolysis microwave conductivity and optical spectroscopy. *Adv. Mater.* 20: 920–923.

Sagstuen, E., Hole, E. O., Nelson, W. H., and Close, D. M. 1992. Protonation state of radiation-produced cytosine anions and cations in the solid state: EPR/ENDOR of cytosine monohydrate single crystals x-irradiated at 10 K. *J. Phys. Chem.* 96: 8269–8276.

Saito, O. 1958. Effects of high energy radiation on polymers II. End-linking and gel fraction. *J. Phys. Soc. Jpn.* 13: 1451–1464.

Saito, I., Nakamura, T., Nakatani, K., Yoshioka, Y., Yamaguchi, K., and Sugiyama, H. 1998. Mapping of the hot spots for DNA damage by one-electron oxidation: Efficacy of GG doublets and GGG triplets as a trap in long-range hole migration. *J. Am. Chem. Soc* 120: 12686–12687.

Sakurai, T., Shi, K., Sato, H., Tashiro, K., Osuka, A., Saeki, A., Seki, S., Tagawa, S., Sasaki, S., Masunaga, H., Osaka, K., Takata, M., and Aida, T. 2008. Prominent electron transport property observed for triply fused metalloporphyrin dimer: Directed columnar liquid crystalline assembly by amphiphilic molecular design. *J. Am. Chem. Soc.* 130: 13812–13813.

Sariciftci, N. S., Smilowitz, L., Heeger, A. J., and Wudl, F. 1992. Photoinduced electron transfer from a conducting polymer to buckminsterfullerene. *Science* 258: 1474–1476.

Sauer, M. C. and Jonah, C. D. 1980. Kinetics of solute excited-state formation in the pulse-radiolysis of liquid alkanes—Comparison between theory and experiment. *J. Phys. Chem.* 84: 2539–2544.

Schreiber, M. and Abe, S. 1993. The effect of disorder on optical spectra of conjugated polymers. *Synth. Met.* 55–57: 50–55.

Schuster, G. B. and Landman, U. 2004. The mechanism of long-distance radical cation transport in duplex DNA: Ion-gated hopping of polaron like distorsion. In *Long Range Charge Transfer in DNA. I and II, Topics in Current Chemistry*, G. B. Shuster (ed.), pp. 139–161. Berlin/Heidelberg, Germany: Springer-Verlag.

Seki, K., Mori, T., Inokuchi, H., and Murano, K. 1988. Electronic structure of poly(dimethylsilane) and polysilane studied by XPS, UPS, and band calculation, Bull. *Chem. Soc. Jpn.* 61: 351–358.

Seki, S., Shibata, H., Ban, H., Ishigure, K., and Tagawa, S. 1996a. Radiation effects of ion beams on poly(methylphenylsilane). *Radiat. Phys. Chem.* 48: 539–544.

Seki, S., Tagawa, S., Ishigure, K., Cromack, K. R., and Trifunac, A. D. 1996b. Observation of silyl radical in γ-radiolysis of solid poly(dimethylsilane). *Radiat. Phys. Chem.* 47: 217–219.

Seki, S., Kanzaki, K., Yoshida, Y., Shibata, H., Asai, K., Tagawa, S., and Ishigure, K. 1997. Positive-negative inversion of silicon based resist materials: Poly(di-n-hexylsilane) for ion beam irradiation. *Jpn. J. Appl. Phys.* 36: 5361–5364.

Seki, S., Cromack, K. R., Trifunac, A. D., Yoshida, Y., Tagawa, S., Asai, K., and Ishigure, K. 1998. Radical stability of aryl substituted polysilanes with linear and planar silicon skeleton structures. *J. Phys. Chem. B* 102: 8367–8371.

Seki, S., Yoshida, Y., Tagawa, S., and Asai, K. 1999a. Electronic structure of radical anions and cations of polysilanes with structural defects. *Macromolecules* 32: 1080–1086.

Seki, S., Maeda, K., Kunimi, Y., Tagawa, S., Yoshida, Y., Kudoh, H., Sugimoto, M., Morita, Y., Seguchi, T., Iwai, T., Shibata, H., Asai, K., and Ishigure, K. 1999b. Ion beam induced crosslinking reactions in poly(di-n-hexylsilane). *J. Phys. Chem. B* 103: 3043–3408.

Seki, S., Kunimi, Y., Nishida, K., Aramaki, K., and Tagawa, S. 2000. Optical properties of pyrrolyl-substituted polysilanes. *J. Organomet. Chem.* 611: 62–68.

Seki, S., Yoshida, Y., and Tagawa, S. 2001a. Charged radicals of polysilane derivatives studied by pulse radiolysis. *Radiat. Phys. Chem.* 60: 411–415.

Seki, S., Kunimi, Y., Nishida, K., Yoshida, Y., and Tagawa, S. 2001b. Electronic state of radical anions on poly(methyl-*n*-propylsilane) studied by low temperature pulse radiolysis. *J. Phys. Chem. B* 105: 900.

Seki, S., Maeda, K., Tagawa, S., Kudoh, H., Sugimoto, M., Morita, Y., and Shibata, H. 2001c. Formation of quantum wires along ion projectiles in Si backbone polymers. *Adv. Mater.* 13: 1663–1665.

Seki, S., Matsui, Y., Yoshida, Y., Tagawa, S., Koe, J. R., and Fujiki, M. 2002. Dynamics of charge carriers on poly[bis(p-alkylphenyl)silane]s by electron beam pulse radiolysis. *J. Phys. Chem. B* 106: 6849–6852.

Seki, S., Tsukuda, S., Yoshida, Y., Kozawa, T., Tagawa, S., Sugimoto, M., and Tanaka, M. 2003. *Jpn. J. Appl. Phys.* 43: 4159.

Seki, S., Koizumi, Y., Kawaguchi, T., Habara, H., and Tagawa, S. 2004a. Dynamics of positive charge carriers on Si chains of polysilanes. *J. Am. Chem. Soc.* 126: 3521–3528.

Seki, S., Tsukuda, S., Maeda, K., Matsui, Y., Saeki, A., and Tagawa, S. 2004b. Inhomogeneous distribution of crosslinks in ion tracks in polystyrene and polysilanes. *Phys. Rev. B* 70: 144203.

Seki, S., Acharya, A., Koizumi, Y., Saeki, A., Tagawa, S., and Mochida, K. 2005a. Mobilities of charge carriers in dendrite and linear oligogermanes by flash photolysis time-resolved microwave conductivity technique. *Chem. Lett.* 34: 1690–1691.

Seki, S., Tsukuda, S., Maeda, K., Tagawa, S., Shibata, H., Sugimoto, M., Jimbo, K., Hashitomi, I., and Koyama, A. 2005b. Effects of backbone configuration of polysilanes on nanoscale structures formed by single-particle nanofabrication technique. *Macromolecules* 38: 10164–10170.

Seki, S., Tsukuda, S., Tagawa, S., and Sugimoto, M. 2006. Correlation between width roughness of nanowires and backbone conformation of polymer materials. *Macromolecules* 39: 7446–7450.

Seki, S., Saeki, A., Acharya, A., Koizumi, Y., Tagawa, S., and Mochida, K. 2008. Intra-molecular mobility of charge carriers along oligogermane backbones studied by flash photolysis time-resolved microwave conductivity and transient optical spectroscopy techniques. *Radiat. Phys. Chem.* 77: 1323–1327.

Shao, F., O'Neill, M. A., and Barton, J. K. 2004. Long damage to cytosines in duplex DNA. *Proc. Natl. Acad. Sci. USA* 101: 17914–17919.

Sheats, J. R., Antoniadis, H., Hueschen, M., Leonard, W., Miller, J., Moon, R., Roitman, D., and Stocking, A. 1996. Organic electroluminescent devices. *Science* 273: 884–888.

Shimamori, H. and Hatano, Y. 1977. Thermal electron attachment to O_2 in the presence of various compounds as studied by a microwave cavity technique combined with pulse radiolysis. *Chem. Phys.* 21: 187–201.

Shimamori, H. and Sunagawa, T. 1997. Electron energy loss rates in gaseous argon determined from transient microwave conductivity. *J. Chem. Phys.* 106: 4481–4490.

Shiraishi, H., Taniguchi, Y., Horigome, S., and Nonogaki, S. 1980. Iodinated polystyrene—An ion-millable negative resist. *Polym. Eng. Sci.* 20: 1054–1057.

Shukla, P., Cotts, P. M., Miller, R. D., Russel, T. P., Smith, B. A., Wallraff, G. M., Baier, M., and Thiyagarajan, P. 1991. Conformational transition studies of organosilane polymers by light and neutron scattering. *Macromolecules* 24: 5606–5613.

Sirringhaus, H., Tessler, N., and Friend, R. H. 1998. Integrated optoelectronic devices based on conjugated polymers. *Science* 280: 1741–1744.

Sirringhaus, H., Brown, P. J., Friend, R. H., Nielsen, M. M., Bechgaard, K., Langeveld-Voss, B. M. W., Spiering, A. J. H., Janssen, R. A. J., Meijer, E. W., Herwing, P., and de Leeuw, D. M. 1999. Two-dimensional charge transport in self-organized, high-mobility conjugated polymers. *Nature* 401: 685–688.

Steenken, S. 1989. Purine-bases, nucleosides, and nucleotides-aqueous solution redox chemistry and transformation reactions of their radical cations and e-and OH adducts. *Chem. Rev.* 89: 503–520.

Steenken, S. 1997. Electron transfer in DNA? Competition by ultra-fast proton transfer. *Biol. Chem.* 378: 1293–1297.

Steenken, S. and Javanovic, S. V. 1997. How easily oxidizable is DNA? One-electron reduction potentials adenosine and guanosine radicals in aqueous solution. *J. Am. Chem. Soc.* 119: 617–618.

Steenken, S., Telo, J. P., Novais, H. M., and Candeias, L. P. 1992. One-electron reduction potentials of pyrimidine-bases, nucleosides, and nucleotides in aqueous-solution-consequences for DNA redox chemistry. *J. Am. Chem. Soc.* 114: 4701–4709.

Stegman, J. and Cronkright, W. 1970. Formation of Wurster's blue in benzene at 25 deg. *J. Am. Chem. Soc.* 92: 6736–6743.

Sundar, V. C., Zaumseil, J., Podzorov, V., Menard, E., Willett, R. L., Someya, T., Gershenson, M. E., and Rogers, J. A. 2004. Elastomeric transistor stamps: Reversible probing of charge transport in organic crystals. *Science* 303: 1644–1646.

Suzuki, E. and Hatano, Y. 1986a. Electron thermalization processes in rare gases with the Ramsauer minimum. *J. Chem. Phys.* 84: 4915–4918.

Suzuki, E. and Hatano, Y. 1986b. Electron thermalization processes in a He–Kr bicomponent system and a Ne pure system. *J. Chem. Phys.* 85: 5341–5344.

Suzuki, H., Meyer, H., Hoshino, S., and Haarer, D. 1996. Charge carrier and exciton dynamics in polysilane-based multilayer light-emitting diodes as monitored with electroluminescence. *J. Lumin.* 66–67: 423–428.

Tabata, Y., Tagawa, S., and Washio, M. 1984. Pulse-radiolysis studies on the mechanism of the high-sensitivity of chloromethylated polystyrene as an electron negative resist. *ACS Symp. Ser.* 255: 151–163.

Tabata, Y., Tagawa, S., Washio, M., and Hayashi, N. 1985. Study on sensitized degradation and crosslinking of polymers by means of picosecond pulse-radiolysis. *Radiat. Phys. Chem.* 25: 305–316.

Tachibana, H., Kawabata, Y., Koshihara, S., and Tokura, Y. 1990. Exciton states of polysilanes as investigated by electro-absorption spectra. *Sol. State Commun.* 75: 5–9.

Tagawa, S. 1986. Pulse-radiolysis and laser photolysis studies on radiation-resistance and sensitivity of polystyrene and related polymers. *Radiat. Phys. Chem.* 27: 455–459.

Tagawa, S. 1987. Main reactions of chlorine-containing and silicon-containing electron and deep-UV (excimer laser) negative resists. *ACS Symp. Ser.* 346: 37–45.

Tagawa, S. 1991a. Pulse radiolysis studies of polymers. *ACS Symp. Ser.* 475: 2–30.

Tagawa, S., 1991b. Pulse radiolysis studies on polymers. In: *CRC Handbook of Radiation Chemistry*, Y. Tabata, Y. Ito, and S. Tagawa (eds.). Boca Raton, FL: CRC Press.

Tagawa, S. 1993. Radiation effects of ion beams on polymers. *Adv. Polym. Sci.* 105: 99–116.

Tagawa, S. and Schnabel, W. 1980a. Laser flash-photolysis studies on excited single-states of benzene, toluene, para-xylene, polystyrene, and poly-alpha-methylstyrene. *Chem. Phys. Lett.* 75: 120–122.

Tagawa, S. and Schnabel, W. 1980b. On the mechanism of the photolysis of polystyrene in chloroform solution—Laser flash-photolysis studies. *Macromol. Chem., Rapid Commun.* 1: 345–350.

Tagawa, S. and Schnabel, W. 1983. On the kinetics of polymer degradation. 10. Laser flash-photolysis of poly-alpha-methylstyrene in chloroform solution. *Polym. Photochem.* 3: 203–209.

Tagawa, S., Arai, S., Kira, A., Arai, S., Imamura, M., Tabata, Y., and Oshima, K. 1972. Pulse radiolysis study of polymerization of vinylcarbazole in benzonitrile solutions. *Polym. Sci.* C10: 295–299.

Tagawa, S., Tabata, Y., Arai, S., and Imamura, M. 1974. Solvent effects on both transient optical-spectra and products in radiation-induced polymerization and dimerization of vinylcarbazole. *J. Polym. Sci.* C12: 545–548.

Tagawa, S., Washio, M., and Tabata, Y. 1979. Picosecond time-resolved fluorescence studies of poly(n-vinylcarbazole) using a pulse-radiolysis technique. *Chem. Phys. Lett.* 68: 276–281.

Tagawa, S., Schnabel, W., Washio, M., and Tabata, Y. 1981. Picosecond pulse-radiolysis and laser flash-photolysis studies on polymer degradation of polystyrene and poly-alpha-methylstyrene. *Radiat. Phys. Chem.* 18: 1087–1095.

Tagawa, S., Washio, M., Kobayashi, H., Katsumura, Y., and Tabata, Y. 1983. Picosecond pulse-radiolysis studies on geminate ion recombination in saturated-hydrocarbon. *Radiat. Phys. Chem.* 21: 45–52.

Tagawa, S., Nakashima, N., and Yoshihara, K. 1984. Absorption-spectrum of the triplet-state and the dynamics of intramolecular motion of polystyrene. *Macromolecules* 17: 1167–1169.

Tagawa, S., Hayashi, N., Yoshida, Y., Washio, M., and Tabata, Y. 1989. Pulse-radiolysis studies on liquid alkanes and related polymers. *Radiat. Phys. Chem.* 34: 503–511.

Tagawa, S., Kashiwagi, M., Kamada, T., Sekiguchi, M., Hosobuchi, K., Tominaga, H., Ooka, N., and Makuuchi, K. 2002. Economic scale of utilization of radiation (I) Industry comparison between Japan and the USA. *J. Nucl. Sci. Technol.* 39: 1002–1007.

Tagawa, S., Seki, S., and Kozawa, T. 2004. Charged particle and photon-induced reactions in polymers. In *Charged Particle Interactions with Matter: Chemical, Physicochemical, and Biological Consequences with Applications*, A. Mozumder and Y. Hatano (eds.), pp. 551–578. New York: Marcel Dekker.

Takenobu, T., Takahashi, T., Takeya, J., and Iwasa, Y. 2007. Effect of metal electrodes on rubrene single-crystal transistors. *Appl. Phys. Lett.* 90: 013507-1–013507-3.

Takeya, J., Nishikawa, T., Takenobu, T., Kobayashi, S., Iwasa, Y., Mitani, T., Goldmann, C., Krellner, C., and Batlogg, B. 2004. Effects of polarized organosilane self-assembled monolayers on organic single-crystal field-effect transistors. *Appl. Phys. Lett.* 85: 5078–5080.

Takeya, J., Kato, J., Hara, K., Yamagishi, M., Hirahara, R., Yamada, K., Nakazawa, Y., Ikehata, S., Tsukagoshi, K., Aoyagi, Y., Takenobu, T., and Iwasa, Y. 2007. In-crystal and surface charge transport of electric-field-induced carriers in organic single-crystal semiconductors. *Phys. Rev. Lett.* 98: 196804-1–196804-4.

Thomas, J. K., Johnson, K., Klippert, T., and Lowers, R. 1968. Nanosecond pulse radiolysis studies of reaction of ions in cyclohexane solutions. *J. Chem. Phys.* 48: 1608.

Trefonas, P., West, R., and Miller, R. D. 1985. Polysilane high polymers: Mechanism of photodegradation. *J. Am. Chem. Soc.* 107: 2737.

Tsukuda, S., Seki, S., Saeki, A., Kozawa, T., Tagawa, S., Sugimoto, M., Idesaki, A., and Tanaka, S. 2004a. Precise control of nanowire formation based on polysilane for photoelectronic device application. *Jpn. J. Appl. Phys.* 43: 3810.

Tsukuda, S., Seki, S., Tagawa, S., Sugimoto, M., Idesaki, A., Tanaka, S., and Ohshima, A. 2004b. Fabrication of nano-wires using high-energy ion beams. *J. Phys. Chem. B* 108: 3407.

Tsukuda, S., Seki, S., Sugimoto, M., and Tagawa, S. 2005a. Nanowires with controlled sizes formed by single ion track reactions in polymers. *Appl. Phys. Lett.* 87: 233119.

Tsukuda, S., Seki, S., Sugimoto, M., and Tagawa, S. 2005b. Formation of nanowires based on π-conjugated polymers by high-energy ion beam irradiation. *Jpn. J. Appl. Phys.* 44: 5839–5842.

Tsukuda, S., Seki, S., Sugimoto, M., and Tagawa, S. 2006. Customized morphologies of self-condensed multi-segment polymer nanowires. *J. Phys. Chem. B* 110: 19319–19322.

Umemoto, Y., Ie, Y., Saeki, A., Seki, S., Tagawa, S., and Aso, Y. 2008. Electronegative oligothiophenes fully annelated with hexafluorocyclopentene: Synthesis, properties, and intrinsic electron mobility. *Org. Lett.* 10: 1095–1098.

van de Craats, A. M., Warman, J. M., de Haas, M. P., Adam, D., Simmerer, J., Haarer, D., and Schuhmacher, P. 1996. The mobility of charge carriers in all four phases of the columnar discotic material hexakis(heylthio) triphenylene: Combined TOF and PR-TRMC results. *Adv. Mater.* 8: 823–826.

van den Ende, C. A. M., Nyikos, L., Warman, J. M., and Hummel, A. 1980. Geminate ion decay kinetics in nanosecond pulse irradiated cyclohexane solutions studied by optical and microwave-absorption. *Radiat. Phys. Chem.* 15: 273–281.

van den Ende, C. A. M., Luthjens, L. H., Warman, J. M., and Hummel, A. 1982. Geminate ion recombination in irradiated liquid CCl_4. *Radiat. Phys. Chem.* 19: 455–466.

Voityuk, A. A., Jortner, J., Bixon, M., and Rösch, N. 2000. Energetics of hole transfer in DNA. *Chem. Phys. Lett.* 324: 430–434.

Warman, J. M. and Fessenden, R. W. 1968. Three body electron by nitrous oxide. *J. Chem. Phys.* 49: 4718–4719.

Warman, J. M., Fessenden, R. W., and Bakale, G. 1972. Dissociative attachment of thermal electrons to N_2O and subsequent electron detachment. *J. Chem. Phys.* 57: 2702–2711.

Warman, J. M., de Haas, M. P., Grätzel, M., and Infelta, P. P. 1984. Microwave probing of electronic processes in small particle suspensions. *Nature* 310: 306–308.

Warman, J. M., Gelinck, G. H., and de Haas, M. P. 2002. The mobility and relaxation kinetics of charge carriers in molecular materials studied by means of pulse-radiolysis time-resolved microwave conductivity: Dialkoxy-substituted phenylene-vinylene polymers. *J. Phys.: Condens. Matter* 14: 9935–9954.

Warman, J. M., de Haas, M. P., Dicker, G., Grozema, F. C., Piris, J., and Debije, M. G. 2004. Charge mobilities in organic semiconducting materials determined by pulse-radiolysis time-resolved microwave conductivity: π-Bond-conjugated polymers versus π–π-stacked discotics. *Chem. Mater.* 16: 4600–4609.

Warman, J. M., Piris, L., Pisula, W., Kastler, M., Wasserfallen, D., and Müllen, K. 2005. Charge recombination via intercolumnar electron tunneling through the lipid-like mantle of discotic hexa-alkyl-hexa-peri-hexabenzocoronenes. *J. Am. Chem. Soc.* 127: 14257–14262.

Washio, M., Tagawa, S., and Tabata, Y. 1981. Pulse-radiolysis studies on poly(n-vinylcarbazole) cation. *Polym. J.* 13: 935–938.

Washio, M., Tagawa, S., and Tabata, Y. 1983. Pulse-radiolysis of polystyrene and benzene in cyclohexane, chloroform and carbon-tetrachloride. *Radiat. Phys. Chem.* 21: 239–243.

West, R. 1986. The polysilane high polymers. *J. Organomet. Chem.* 300: 327–346.

Yamagami, R., Kobayashi, K., Saeki, A., Seki, S., and Tagawa, S. 2006. Photogenerated hole mobility in DNA measured by time-resolved microwave conductivity. *J. Am. Chem. Soc.* 128: 2212–2213.

Yamagami, R., Kobayashi, K., and Tagawa, S. 2008. Formation of spectral intermediate G-C and A-T anion complex in duplex DNA studied by pulse radiolysis. *J. Am. Chem. Soc.* 130: 14772–14777.

Yamakawa, H., Abe, F., and Einaga, Y. 1994. Second virial coefficient of oligo- and polystyrenes near the θ temperature. More on the coil-to-globule transition. *Macromolecules* 27: 5704–5712.

Yamamoto, H., Kozawa, T., Nakano, A., Okamoto, T., Tagawa, S., Ando, T., Sato, M., and Komano, H. 2005a. Reaction mechanisms of brominated chemically amplified resists. *Jpn. J. Appl. Phys.* 44: L842–L844.

Yamamoto, H., Kozawa, T., Nakano, A., Okamoto, T., Tagawa, S., Ando, T., Sato, M., and Komano, H. 2005b. Study on acid generation from polymer. *J. Vac. Sci. Technol. B* 23: 2728–2732.

Yamamoto, Y., Fukushima, T., Suna, Y., Ishii, N., Saeki, A., Seki, S., Tagawa, S., Taniguchi, M., Kawai, T., and Aida, T. 2006a. Photoconductive coaxial nanotubes of molecularly connected electron donor and acceptor layers. *Science* 314: 1761–1764.

Yamamoto, Y., Fukushima, T., Jin, W., Kosaka, A., Hara, T., Nakamura, T., Saeki, A., Seki, S., Tagawa, S., and Aida, T. 2006b. A glass hook allows fishing of hexa-peri-hexabenzocoronene graphitic nanotubes: fabrication of a macroscopic fiber with anisotropic electrical conduction. *Adv. Mater.* 18: 1297–1300.

Yamamoto, H., Kozawa, T., Tagawa, S., Ohmori, K., Sato, M., and Komano, H. 2007a. Single-component chemically amplified resist based on dehalogenation of polymer. *Jpn. J. Appl. Phys.* 46: L648–L650.

Yamamoto, Y., Fukushima, T., Saeki, A., Seki, S., Tagawa, S., Ishii, N., and Aida, T. 2007b. Molecular engineering of coaxial donor-acceptor heterojunction by coassembly of two different hexabenzocoronenes: Graphitic nanotubes with enhanced photoconducting properties. *J. Am. Chem. Soc.* 129: 9276–9277.

Yamamoto, H., Kozawa, T., Saeki A., Tagawa, S., Mimura, T., Yukawa, H., and Onodera, J. 2009. Reactivity of halogenated resist polymer with low-energy electrons. *Jpn. J. Appl. Phys.* 48: 06FC09.

Yanagisawa, K., Kume, T., Makuuchi, K., Tagawa, S., Chino, M., Inoue, T., Takehisa, M., Hagiwara, M., and Shimizu, M. 2002. An economic index regarding market creation of products obtained from utilization of radiation and nuclear energy (IV) – Comparison between Japan and the USA. *J. Nuclear Sci. Technol.* 39: 1120–1124.

Yoshida, Y., Tagawa, S., Kobayashi, H., and Tabata, Y. 1987. Study of geminate ion recombination in a solute solvent system by using picosecond pulse-radiolysis. *Radiat. Phys. Chem.* 30: 83-87.

Yoshioka, Y., Kitagawa, Y., Takano, Y., Yamaguchi, K., Nakamura, T., and Saito, I. 1999. Experimental and theoretical studies on the selectivity of GGG triplets toward one-electron oxidation in B-form DNA. *J. Am. Chem. Soc.* 121: 8712–8719.

Yu, G., Gao, J., Hummelen, J. C., Wudl, F., and Heeger, A. J. 1995. Polymer photovoltaic cells: Enhanced efficiencies via a network of internal donor-acceptor heterojunctions. *Science* 270: 1789–1791.

Zezin, A. A., Feldman, V. I., and Sukhov, F. F. 1995. Effect of electron scavengers on the formation of paramagnetic species upon radiolysis of polystyrene and its low-molecular-weight analogs. *High Energy Chem.* 29: 154–158.

Zhang, G. H. and Thomas, J. K. 1996. Energy transfer via ionic processes in polymer films irradiated by 0.4 meV electrons. *J. Phys. Chem.* 100: 11438–1145.

26 Radiation Chemistry of Resist Materials and Processes in Lithography

Takahiro Kozawa
Osaka University
Osaka, Japan

Seiichi Tagawa
Osaka University
Osaka, Japan

CONTENTS

26.1 INTRODUCTION

The initial steps in electron beam (EB) and extreme ultraviolet (EUV) resist reactions are mainly induced by ionization and are different from the initial steps of photoresists, induced by excited states of molecules. In the1980s, the increasing density of VLSI necessitated the development of submicron lithography, such as excimer, x-ray, EB, and ion-beam (IB) lithography. At that time, much attention was devoted to negative EB resists containing chlorinated or chloromethylated phenyl side chains, which exhibit high sensitivity without losing resolution. Polymer cross-linking occurs very effectively in these resists. This type of enhancement is also observed in other resists containing chlorinated or chloromethylated silicon-containing resists. Similar reactions of iodination and bromination of phenyl rings are used. The first systematic study on radiation chemistry of EB and x-ray resists (Tabata et al., 1984; Tagawa, 1986, 1987) was carried out by pulse radiolysis in order to make clear the mechanisms of high-sensitivity and high-resolution chlorinated resists, such as effective dissociative electron attachment, formation of charge transfer complexes through geminate ion recombination, and polymer radical and acid generation processes. Excimer laser flash photolysis studies on resist materials were also carried out for chlorinated resists (Tagawa, 1986, 1987). In both photolysis and radiolysis, polymer radical and acid generation takes place through the same charge transfer complex as the initiator of resist reactions. But the initial steps of EB and x-ray resist reactions are mainly

711

induced by ionization and are different from the initial steps of photoresists induced by excited states of molecules. The combination of geminate ion recombination and highly effective dissociative electron attachment is important in highly sensitive EB resists and also produces acid (Tabata et al., 1984; Tagawa, 1986, 1987). A systematic research on radiation chemistry of chemically amplified resists started with the proposal of the mechanisms of chemically amplified EB and x-ray resists (Kozawa et al., 1992), and a systematic research on the mechanisms of EB and deep UV chemically amplified resists was carried out by product analysis, pulse radiolysis, and laser flash photolysis (Tagawa et al., 2000). The radiation chemistry of chemically amplified EB and x-ray resists was reviewed in 2004 (Tagawa et al., 2004). At that time, the limit of the resolution of chemically amplified resists was believed to be around 60 nm. But then EUV lithography began to develop, for which the resists showing high performance, especially high resolution and high sensitivity, have been the key technology. The best data of the resolution of chemically amplified EUV resists has now reached sub-20-nm with high dose, the best data of the resolution of chemically amplified EUV resists has now reached sub-20-nm. Many serious problems, especially a trade-off relationship among resolution, sensitivity, and line-edge roughness (LER), described in the following text, have been encountered in the development of high-performance chemically amplified EUV resists. A systematic fundamental research, especially on radiation chemistry of chemically amplified EUV resists, has been performed. The nanospace reactions through the combination of geminated ion recombination and highly effective dissociative electron attachment play a very important role in chemically amplified EUV and EB resists, similar to halogenated nonchemically amplified resists. The geminate pairs are anion and proton in chemically amplified resists and the geminate pairs are anion and radical cations in halogenated nonchemically amplified resists. Products of geminate ion recombination are acid and two isolated radicals (distance between two radicals is long) in halogenated nonchemically amplified resists, although products of geminate ion recombination are acid and pair of polymer radicals (distance between pair radicals is very short). This chapter summarizes the recent progress in radiation chemistry of chemically amplified EUV and EB resist materials and processes.

The concept of "chemical amplification" (Ito and Willson, 1983; Ito, 2005) has made a great impact on resist designs. Chemically amplified resists utilize acid-catalytic reactions to form latent images before development. Acids produced by exposure diffuse in the resist matrix and catalyze chemical reactions, such as deprotection. In the early stage of the development of these resists, there occurred many problems associated with the acid-catalytic reaction coupled with acid diffusion. The instability of resist performance and the formation of surface insoluble layers are among these. Owing to their worldwide intensive development, the performance of chemically amplified resists had been improved for practical use. Chemically amplified resists were first deployed in the mass production lines of semiconductor devices during the transition of the exposure tool from the i line of a Hg lamp to a KrF excimer laser. Since then, chemically amplified resists have been used as a mainstream resist technology in KrF, ArF, and ArF immersion lithography. Despite acid diffusion, sub-60-nm features have been resolved in the mass production lines of semiconductor devices.

Thus far, the miniaturization of feature sizes has been achieved mainly by shortening the wavelength of exposure tools. The shortening of the wavelength implies the increase in the photon energy of exposure tools. This trend will continue to enhance the ionization potential of resist materials with the deployment of EUV exposure tools. The concept of chemical amplification will remain a key to the realization of EUV lithography. Owing to the intensive development of chemically amplified resist materials and exposure tools, the resolution of EUV lithography has reached the sub-30-nm region with 10–20 mJ cm^{-2} sensitivity (Oizumi et al., 2007; Shimizu et al., 2007; Thackeray et al., 2008; Yamashita et al., 2008; Itani, 2009). For the realization of EUV lithography, which targets a 22 nm resolution mass production, the performance of resist materials is however still inadequate. Another important ionizing radiation for lithography is EB. EB lithography had been a preferred method for the next-generation lithography until the development of EB projection lithography. However, EB lithography is an indispensable technology for the fabrication of photomasks. 4× or 5× photomasks have been used for the mass production lines of semiconductor devices. The requirement for the fabrication accuracy using EB lithography also becomes increasingly strict with the reduction of feature size and the increasing complexity of resolution enhancement technology, such as the optical phase shift.

In the development of resists for EUV lithography, a trade-off relationship was found among resolution, sensitivity, and LER (Gallatin, 2005). This has become the most serious problem in resist development. A premise for this trade-off relationship is that pattern formation efficiency is constant. The pattern formation efficiency is determined by absorption efficiency of incident energy, quantum yield of acids, and efficiency of catalytic chain reaction. However, the energy absorption efficiency is limited by sidewall degradation (Kozawa and Tagawa, 2008; Kozawa et al., 2008a). The quantum yield is limited by secondary electron emission efficiency (Hirose et al., 2008a). The efficiency of the catalytic chain reaction is limited by the diffusion-controlled rate of the chemical reaction (Kozawa et al., 2008b,c). Therefore, the orthodox strategy to circumvent this problem is to increase the pattern formation efficiency to its physical limits.

The understanding of the energy consumption of incident radiation in resist materials is essential to the efficient use of the energy of incident radiation. In resist pattern formation of chemically amplified resists, first, an accumulated energy profile is formed by the energy deposition from an exposure tool. Using the accumulated energy, acids are then generated. The accumulated energy profile is converted into a latent acid image. The acids catalyze the pattern formation reaction during post-exposure baking (PEB) and create a so-called latent image. By developing this image, a resist pattern is formed. Although the photochemistry of acid generators has been intensively investigated (Houlihan et al., 1988; Aoai et al., 1989; Buhr et al., 1989; Dektar and Hacker, 1990a,b; Sakamizu et al., 1993; Ortica et al., 2000), the sensitization mechanism of acid generators in chemically amplified resists markedly changes above the ionization potential (Kozawa et al., 1992; Tagawa et al., 2000). First of all, the energy deposition becomes a nonselective process with the increase in the energy. In photoresists, acid generators are mainly decomposed from their excited states via direct excitation or sensitization by excited polymer molecules (Dektar and Hacker, 1990a). It has been, however, well known that chemically amplified photoresists are generally sensitive to ionizing radiation, such as an EB. In the energy deposition of ionizing radiation into solute-matrix systems, such as chemically amplified resists, the direct electronic excitation and ionization of solute molecules, such as acid generators, are generally ineffective because the energy deposition is nonselective, unlike in the case of photons. This is also true for EUV lithography. It has been pointed out that the direct absorption of EUV photons by acid generators is negligible in typical EUV resists (Dentinger et al., 2000; Hirose et al., 2007). The reason why acids are efficiently generated in chemically amplified EB and EUV resists, even if the direct energy absorption of acid generators is small, is that acid generators can react with low-energy electrons (~0 eV) generated by ionization events. However, it has been pointed out that the reaction with low-energy electrons causes significant blurring of latent acid images (Kozawa et al., 2005, 2008d). Thus, low-energy electrons play an important role in the pattern formation of chemically amplified EB and EUV resists. In this chapter, radiation chemistry in resist materials and processes in lithography have been reviewed.

26.2 ENERGY DEPOSITION

The energy deposition on resist materials has been intensively investigated using conventional non-chemically amplified resists, such as poly(methyl methacrylate) (PMMA). When a high-energy electron enters the material, the electron loses its energy sporadically through inelastic scattering in the material and excites or ionizes molecules. The energy deposition is expressed along the electron trajectories using Bethe's continuous slowing down model for inelastic scattering (Bethe, 1930):

$$\frac{dE}{ds} = -\frac{2\pi e^4}{E} nZ \left(\frac{1.166E}{J} \right) \tag{26.1}$$

Here, E, s, e, Z, n, and J are the electron energy, the distance of the electron trajectory, the elementary electric charge, the atomic number, the volume concentration of atoms, and the mean excitation potential, respectively. In the relativistic energy region, a modified form of the Bethe equation is used (Shimizu et al., 1975):

$$\frac{dE}{ds} = -\frac{2\pi e^4}{m_0 v^2} nZ \left[\ln \frac{m_0 v^2 E}{2J^2 \left(1-\beta^2\right)} - \left\{ 2\sqrt{1-\beta^2} - \left(1-\beta^2\right) \right\} \ln 2 + \left(1-\beta^2\right) + \frac{1}{8} \left(1-\sqrt{1-\beta^2}\right)^2 \right]$$

(26.2)

Here, m_0, v, and β are the electron rest mass, the electron velocity, and the ratio of v to the velocity of light in vacuum, respectively. The relativistic effect on the resist sensitivity has been investigated. Because the electrons are decelerated in the resist and the underlying substrate, the results were actually controversial (Ryan et al., 1995). The relativistic effect has recently been observed (Kim et al., 2005).

In EB lithography, the electrons backscattered from the substrate irradiate resist materials and deform the designed patterns. Process simulators have been utilized for the correction of proximity effects in the photomask design. The calculation of accumulated energy profiles is a key element of the process simulation for EB lithography. The accumulated energy is generally accurately estimated by tracking the electron trajectories by the calculation of elastic and inelastic scattering in resist materials (Hawryluk et al., 1974; Shimizu et al., 1975; Murata et al., 1981; Williams et al., 1989; Aktary et al., 2006). The electron trajectories are calculated, for example, using Rutherford's differential cross-section model for elastic scattering, Moller's cross-section model for fast secondary electron generation, and Bethe's continuous slowing down model for inelastic scattering. This method requires a full-scale Monte Carlo simulation and is time consuming. A point spread function instead is used for the calculation of accumulated energy profiles so that the processing time can be reduced down to an acceptable value (Chang, 1975; Parikh and Kyser, 1979; Rishton and Kern, 1987; McMillan et al., 1989; Koba et al., 2005).

Several forms of point spread functions for conventional resists, PSF_{conv}, have been reported. One of the well-known forms is the so-called double Gaussian (Chang, 1975; Parikh and Kyser, 1979):

$$PSF_{conv}(r) = \frac{1}{2\pi\beta_f^2} \exp\left(-\frac{r^2}{2\beta_f^2}\right) + \frac{\eta}{2\pi\beta_b^2} \exp\left(-\frac{r^2}{2\beta_b^2}\right)$$

(26.3)

where
  β_f is the forward-scattering range
  β_b is the backscattering range
  η is the ratio of the energy intensity of backscattering to forward scattering

When a Gaussian beam,

$$B(r) = \frac{1}{2\pi\beta_e^2} \exp\left(-\frac{r^2}{2\beta_e^2}\right)$$

(26.4)

is assumed, the point spread function including beam blur is expressed as

$$PSF'_{conv}(r) = \int\limits_{x',y'} B\left(\sqrt{\left(x'-x\right)^2 + \left(y'-y\right)^2}\right) PSF_{conv}\left(\sqrt{x'^2 + y'^2}\right) dx'dy'$$

$$= \frac{1}{2\pi\left(\beta_f^2 + \beta_e^2\right)} \exp\left(-\frac{r^2}{2\left(\beta_f^2 + \beta_e^2\right)}\right) + \frac{\eta}{2\pi\left(\beta_b^2 + \beta_e^2\right)} \exp\left(-\frac{r^2}{2\left(\beta_b^2 + \beta_e^2\right)}\right)$$

(26.5)

Here, β_e is the beam size. The parameters β_f and β_b are either calculated by Monte Carlo simulation (Parikh and Kyser, 1979; McMillan et al., 1989) or are experimentally obtained, typically by analyzing resist patterns delineated by a point beam (Chang, 1975; Rishton and Kern, 1987). In Monte Carlo simulation, the energy is mainly accumulated through ionization and electronic excitation events when the energy threshold for the termination of the calculation is set at 10–20 eV. Therefore, Equation 26.1 or Equation 26.5 represents the probability density distribution of ionization and electronic excitation events. This is also true when a point spread function is evaluated using a point beam and a PMMA resist. In PMMA resists, the molecular weight of ionized molecules is reduced through side-chain detachment and subsequent main-chain scission near the ionization and electronic excitation points (Kozawa et al., 2007a).

In a Monte Carlo simulation for the calculation of the point spread function, the electron trajectories are generally tracked until the electron energy is reduced down to 20 eV. The inelastic mean free path of 20 eV electrons is less than 1 nm in organic materials (Tanuma et al., 1994). When pattern formation reactions, such as main-chain scission and cross-linking, take place at ionized molecules, the threshold of 20 eV gives sufficient accuracy even for nanoscale simulation. The relation between experimental and simulated resist profiles has been intensively investigated using conventional non-chemically amplified resists (Hawryluk et al., 1974; Shimizu et al., 1975). For chemically amplified EB resists, acid generators are sensitized by secondary electrons similarly as PMMA. The difference between sensitization mechanisms of chemically amplified resists and PMMA is that acid generators are decomposed through the reaction with low-energy electrons as low as thermal energy (~25 meV) (Kozawa et al., 1992, 2003). Although the inelastic mean free path of 20 eV electrons is less than 1 nm, the thermalization distance reaches 3–7 nm on an average in organic materials, as discussed in the following text. This fact indicates that the accumulated energy profile does not correspond to the initial acid distribution (Kozawa et al., 2002; Saeki et al., 2002).

Although the sensitization mechanism of acid generators in EUV resists is analogous to that of EB resists (Kozawa et al., 2006a), the energy deposition mechanisms on resist materials in EUV resists are different from those in EB resists. The absorption of EUV photons is expressed by Lambert's law similar to that of the KrF and ArF excimer laser:

$$\frac{dI(z)}{dz} = -\alpha I(z) \tag{26.6}$$

Here, I, z, and α are the light intensity, the distance from the surface in the direction of depth, and the absorption coefficient of the resist film. This absorption coefficient is determined by the absorption coefficients of polymers and acid generators. In the resists for KrF lithography, acid generators are also sensitized through energy or electron transfer from the excited states of polymers (Hacker and Welsh, 1991; Cameron et al., 2001). However, the direct excitation of acid generators is a primary process for acid generation. Therefore, a tremendous effort has been devoted to the reduction of the absorption coefficient of polymers during the development of KrF, ArF, and F_2 resists without exception. On the basis of the Lambert–Beer law, the photosensitization of conventional resist materials has been formulated by Dill et al. (1975). Acid generation in chemically amplified photoresists is also expressed by the formulation of Dill et al. However, the reaction mechanisms after photoabsorption significantly differ between deep ultraviolet (DUV) and EUV exposures.

When EUV photons are absorbed by resist molecules, photoelectrons are emitted. The photoelectrons with excess energy induce further ionization and electronic excitation. The number of ionization events induced by a single EUV photon can be estimated using the W value. The W value is defined as the average energy required to produce an ion pair by ionizing radiations. The W value of atoms and molecules has been intensively investigated by a variety of measurements for different radiations above several keV (Inokuti, 1975; Kimura et al., 1991). It is well known that the W value is insensitive to quality and energy for the radiations above a few keV. W values of some gases, such as rare gases, propane, and ethylene, have been reported for the energy region below 1 keV (Suzuki and Saito, 1985, 1987; Saito and Suzuki, 2001). The W value for electrons increases with decreasing electron energy, because the ratio of the probability for excitation to that for ionization in a single collision of an electron with a gas molecule increases below 1 keV. The increase in the W value for photons is small compared with that for electrons except for the region just above the absorption edge of molecules. The K edges of carbon and oxygen atoms are 284 and 547 eV, respectively. Therefore, the W value of a typical organic resist polymer for 92.5 eV photons is only slightly higher than that for 1 keV photons or electrons. The W value of poly(4-hydroxystyrene) (PHS) is 22.2 eV for a 75 keV EB (Kozawa et al., 2006b). Thus, approximately four electrons are generated in PHS by a single EUV photon.

26.3 THERMALIZATION AND INITIAL CONFIGURATION OF SPURS

The mean free path of secondary electrons is one of the factors that determine the resolution of lithography using ionizing radiation. The effects of high-energy secondary electrons on the resolution of EB lithography have been well investigated, but are not discussed here. In EUV lithography, the energy of photoelectrons is approximately 80 eV. The elastic and inelastic mean free paths of 80 eV electrons are less than 1 nm. The mean free path in EUV resists does not seem to matter for resolution. However, the acid generators in chemically amplified resists are decomposed through dissociative electron attachment, and the targeted resolution of EUV lithography is below 22 nm. Therefore, the thermalization distance is not negligible in the next-generation lithography. Thermalization mechanisms have been intensively investigated (Goulet et al., 1990). Although it is still controversial, the thermalization mechanism of ejected electrons is considered to be as follows. The electrons ejected by ionization rapidly lose their energy through the ionization and electronic excitation. When the energy of electrons becomes subexcitational, the energy loss is mainly due to the excitation of intramolecular vibration. Below subvibrational energy, electrons are thermalized by, for example, the processes of energy exchange and the excitation of intermolecular vibrations and phonons. Rassolov and Mozumder suggested that most of the thermalization range is accrued in the subvibrational stage because of the ineffectiveness of the energy loss process, and thereby, the distribution would be relatively insensitive to the initial electron energy (Rassolov and Mozumder, 2001). Because the thermalization is generally completed within 1 ps in condensed matter, the thermalization process cannot be directly observed. However, the thermalization distance can be estimated by assuming the initial distribution and applying Onsager's theory (Onsager, 1938). The thermalization distances in many aromatic compounds range from 3 to 5 nm, and those in many alkanes range from 5 to 7 nm (Jay-Gerin et al., 1993; Mozumder, 2002). The inelastic mean free path of 20–100 eV is less than 1 nm in organic materials. The thermalization distance is significantly longer than the inelastic mean free path because of the ineffectiveness of energy loss below the electronic excitation energy of molecules. The typically assumed forms of initial distributions are exponential and Gaussian distributions. To explain the details of experimental results, the modified forms of these distributions have also been proposed. The thermalization distance depends on the distribution profile.

The initial distribution of thermalized electrons can also be estimated by observing the early processes in radiation chemistry (Yoshida et al., 1987). The initial distribution in model organic compounds associated with resist materials has been investigated using high-time-resolution pulse radiolysis of benzene derivatives (Okamoto et al., 2003, 2007) and dodecane (Saeki et al., 2002). Pulse radiolysis is a method for observing directly the dynamics of short-lived intermediates generated by a pulsed beam. The dynamics is observed by, for example, photoabsorption spectroscopy. A high-energy (>MeV) EB is generally used as an excitation source for pulse radiolysis (Kozawa et al., 2000; Yoshida et al., 2001). In the MeV energy region, the average distance between spurs is 100–500 nm. The ejected electron is thermalized near its parent radical cation (ionized molecule) compared with the other radical cations. The thermalized electron migrates under the strong electric field produced by its parent radical cation. The Onsager length, at which the potential energy of electric field corresponds to the thermal energy, is expressed as

$$r_c = \frac{e^2}{\varepsilon k_B T} \tag{26.7}$$

Here, ε, k_B, and T are the dielectric constant, the Boltzmann constant, and the absolute temperature, respectively. Because the dielectric constant of benzene derivatives and dodecane is 2, the Onsager length is approximately 28 nm. The Onsager length is defined as the length at which the thermal energy of the electron corresponds to the energy of the electron under the Coulomb potential. Most electrons are thermalized within the Onsager length in these model compounds, since the thermalization distance is 3–7 nm. Therefore, the recombination reaction between thermalized electrons and radical cations is geminate recombination between a thermalized electron and its parent radical cation in the early stages in radiation chemistry of these compounds. It has been demonstrated that the exponential distribution well reproduces the dynamics of intermediates in the model compounds (Saeki et al., 2002; Okamoto et al., 2003, 2007). Assuming an exponential distribution, the thermalization distances in PHS (Kozawa et al., 2006b) and PMMA films (Kozawa et al., 2007a) have been estimated to be approximately 4 and 6 nm, respectively, by analyzing the dependence of acid yields on acid generator concentration.

An ionized molecule with a positive charge (radical cation) and an electron with a negative charge interact through the Coulomb interaction. When spurs overlap, the electron dynamics changes and affects the chemical yield and distribution. The phenomenon that describes the interaction between packed ion pairs is called the "multispur effect." Although the influence of the multispur effect on the recombination reaction and radiation-chemical yields has been investigated from the viewpoint of the fundamental science of electron–material interaction (Clifford et al., 1982; Bartczak and Hummel, 1987, 1993; Green et al., 1989; Bartczak et al., 1990; Wojcik et al., 1992; Siebbeles et al., 1997; Saeki et al., 2005), the details are still unclear. To discuss the effect of spur overlapping in resist materials, two different cases have to be considered. One is the case when the spurs generated by the same incident electron (or photon) are overlapped. The other is the case when the spurs generated by different incident electrons (or photons) are overlapped. In the latter case, the effect depends on the exposure dose rate. Although the exposure-dose-rate dependence of chemically amplified resists has not been reported, it may become a problem when a high-power EUV source is developed. However, we discussed the former case here. Because the former case is induced by one incident electron (or photon), the effect is essential to the sensitization mechanisms of chemically amplified resists used for ionizing radiation. A typical configuration of ionized molecules (parent radical cations) and thermalized electrons is shown in Figure 26.1 (Kozawa et al., 2009). Within the Onsager length, charged species interact through the Coulomb interaction. The Onsager length is approximately 14 nm in PHS, which is a typical backbone polymer for chemically amplified resists used in EB and EUV lithography.

The distribution of overlapped spurs in EB and EUV resists has been investigated by a Monte Carlo simulation. In the simulation, the inelastic scattering of 100,000 electrons and the subsequent trajectories of secondary electrons were calculated. A histogram showing the numbers of ion pairs generated is shown in Figure 26.2 (Kozawa et al., 2009). The distributions are significantly different between EB and EUV exposures. The difference is mainly caused by the fact that the "first" secondary electron is generated through inelastic scattering in EB resists, although it is generated through photoelectron emission in EUV resists. The probability of generating a single spur in EB resists was 0.54. The probability of generating a multispur ($N = 2$) was 0.41. A significant number of spurs also overlapped in EB resists as well as in EUV resists. The average number of ion pairs in an isolated space is 1.5 in EB resists and 4.2 in EUV resists.

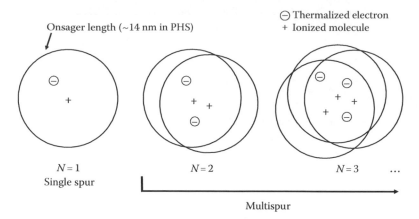

FIGURE 26.1 Typical configuration of ionized molecules (parent radical cations) and thermalized electrons. The Onsager length is defined as the length at which the thermal energy of an electron corresponds to the Coulomb potential. This length is approximately 14 nm in PHS. (From Kozawa, T. et al., *Jpn. J. Appl. Phys.*, 48, 056508, 2009. With permission.)

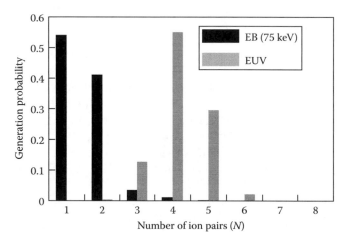

FIGURE 26.2 Probability of generating each configuration shown in Figure 26.1 in EB and EUV resists. The average number of ion pairs in an isolated space is 1.5 for EB and 4.2 for EUV. The energy of incident electrons is 75 keV. (From Kozawa, T. et al., *Jpn. J. Appl. Phys.*, 48, 056508, 2009. With permission.)

26.4 ELECTRON MIGRATION AND MULTISPUR EFFECT

The thermalized electrons migrate in the resist matrix until they find the localization sites such as acid generators. In the deceleration process and the following migration, the secondary electrons excite the acid generators or attach to the acid generators (Kozawa et al., 1992). The ratio of acid yields generated through direct excitation and dissociative electron attachment depends on the acid generator concentration (Kozawa et al., 2006b, 2007b). The acid yield generated through the electronic excitation of acid generators has been reported to account for approximately 10% of the total acid yield at 10 wt% acid generator concentration (Kozawa et al., 2006b). Because the acid generator is used in the concentration range of 5–10 wt%, the decomposition of acid generators is mostly induced by the electron attachment to the acid generators.

The relationship between the acid yield and physical properties of acid generators has been investigated from the viewpoint of the reactivity of acid generators with thermalized electrons in resist materials. The correlation between reductive potential and relative acid yield has been reported for

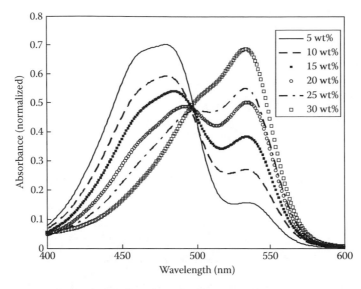

FIGURE 26.3 Absorption spectra of PHS films with TPS-tf and 5 wt% C6 upon exposure to 75 keV EB at a dose of 10 μC cm^{-2}. The TPS-tf concentrations were 5, 10, 15, 20, 25, and 30 wt%. Absorbance was normalized with the thickness of resist film (1 μm). (From Natsuda, K. et al., *Jpn. J. Appl. Phys.*, 48, 06FC05, 2009. With permission.)

sulfonium and iodonium salts as well as nonionic acid generators (Nagahara, et al., 2000; Masuda et al., 2006). Also, the reactivity of acid generators with solvated electrons formed in solutions has been measured (Tsuji et al., 2000; Nakano et al., 2006a). The reactivity estimated from the rate constant for the reaction with methanol-solvated electrons (Tsuji et al., 2000) well explains the acid generation efficiency for triphenylsulfonium triflate (TPS-tf) upon exposure to EB (Kozawa et al., 2006b) and EUV (Kozawa et al., 2007c; Hirose et al., 2008b). However, the same accuracy does not always apply to a wide range of acid generators. It has neither been proved that reductive potential applies to a wide range of acid generators.

Recently, the C_{37} parameter has been proposed as a possible indicator for the reactivity of acid generators, which may correlate with acid generation efficiency in chemically amplified EB resists (Natsuda et al., 2008). Figure 26.3 shows the dependence of the absorption spectra of exposed PHS films with TPS-tf and 5 wt% Coumarin 6 (C6) on TPS-tf concentration (Natsuda et al., 2009). The absorption bands at approximately 460 and 533 nm are due to neutral forms and proton adducts of C6, respectively (Pohlers et al., 1997; Yamamoto et al., 2004a). The absorption intensity at approximately 460 nm decreased and that at approximately 533 nm increased with increasing TPS-tf concentration. The concentration of C6 proton adducts was estimated using the Lambert–Beer law from the difference in absorption intensity of spectra before and after exposure. In this estimation, an absorption coefficient of 7.8×10^4 M^{-1} cm^{-1} at 533 nm was used (Yamamoto et al., 2004b). The dependence of acid (C6 proton adduct) yield on acid generator concentration in PHS films containing an acid generator and C6 is shown in Figure 26.4 for 14 acid generators. The acid yield tended to saturate at a high acid-generator concentration for all samples. The dependence of the acid yield on the acid generator concentration has been investigated in detail using TPS-tf (Kozawa et al., 2006b). An acid yield of 0.01 M corresponds to a G-value of acids of approximately 1. The G-value is the number of molecules generated by an absorbed energy of 100 eV. For all acid generators except for diphenyliodonium triflate (DPI-tf) and bis(*p-tert*-butylphenyl)iodonium triflate (BBI-tf), the G-value saturated at approximately 5. The G-value of ionization has been reported to be 4.5 for PHS (Kozawa et al., 2006b). The G-value of acids generated through the direct electronic excitation of TPS-tf has been estimated to be approximately 1 at a TPS-tf concentration of 0.6 M (Kozawa et al., 2006b). The sum of two G-values is 5.5. Therefore, almost all secondary electrons are considered

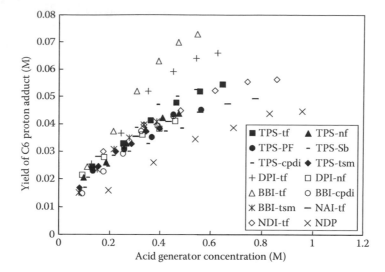

FIGURE 26.4 Relationship between the concentration of acid generators and the yield of C6 proton adducts (acids). (From Natsuda, K. et al., *Jpn. J. Appl. Phys.*, 48, 06FC05, 2009. With permission.)

to be scavenged at acid yields of approximately 5. The observation of abnormality of some diphenyliodonium salts agrees with that in previous studies using γ-ray (Nagahara et al., 2000; Nakano et al., 2005), EUV (Brainard et al., 2008; Hirose et al., 2008a), and deep UV (Ablaza et al., 2000) as exposure tools. Side reactions induced by reactive decomposition products have been suggested as the reason for the high efficiency of DPI-tf.

The C_{37} parameter has been estimated by pulse radiolysis for acid generators. Figure 26.5 shows representative kinetic traces of the decay of tetrahydrofuran-(THF) solvated electrons with and without triphenylsulfonium tris(trifluoromethansulfonyl)methane (TPS-tsm) (Natsuda et al., 2009). The decrease in the initial yield and the increase in the decay rate were observed with increasing TPS-tsm concentration. The increase in the decay rate indicates the reaction of acid generators with solvated electrons. The rate constants for the reactions of acid generators with solvated electrons have been used to estimate the effective reaction radii of acid generators (Kozawa et al., 2004).

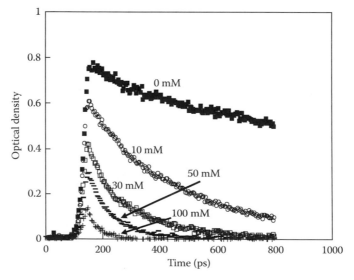

FIGURE 26.5 Kinetic traces at 1300 nm in a TPS-tsm solution in THF. The concentrations of TPS-tsm are 0, 10, 30, 50, and 100 mM. (From Natsuda, K. et al., *Jpn. J. Appl. Phys.*, 48, 06FC05, 2009. With permission.)

The radii estimated from the reaction with solvated electrons did not have sufficient accuracy to explain the difference between acid generators as described above. This is because the acid generators react not with solvated electrons but with (unsolvated) thermalized electrons in chemically amplified resists upon exposure to ionizing radiation.

The decrease in the initial yield is expressed by the C_{37} parameter. Hunt and coworkers (Wolff et al., 1970, 1975; Aldrich et al., 1971; Lam and Hunt, 1975; Hunt and Chase, 1977) focused on the decrease in the initial yield of hydrated electrons in water upon addition of various solutes and demonstrated that the fraction of decrease, f, is expressed by an exponential as a function of solute concentration:

$$f = \exp\left(\frac{-[S]}{C_{37}}\right) \tag{26.8}$$

where
   $[S]$ is the concentration of the scavenger
   C_{37} is a constant that depends on the solvent and the solute

The C_{37} parameter can be understood as the solute concentration at which the initial yield of solvated electrons is decreased to $1/e$ (37%). C_{37} is useful when the concentration is rather high. When an electron pulse enters the THF solution, THF molecules are ionized and secondary electrons are emitted. The emitted electrons lose their kinetic energy through their interaction with surrounding molecules and are finally thermalized. The thermalized electrons are solvated with THF molecules and become detectable, as shown in Figure 26.5. Considering the time resolution of the pulse radiolysis system used, the initial decrease can be assumed to occur before solvation. The C_{37} parameter is believed to reflect the reactions with electrons before solvation (Hamill, 1969; Lam and Hunt, 1975; Hunt and Chase, 1977; Jonah et al., 1977, 1989; Lewis and Jonah, 1986; Chernovitz and Jonah, 1988; Saeki et al., 2007). In THF, presolvated states, such as an excited state of solvated electrons, have not been reported. Therefore, C_{37} in THF is likely related to the reaction with thermalized and possibly epithermal electrons. Figure 26.6 shows the dependence of the initial yield of THF-solvated electrons on the TPS-tsm concentration

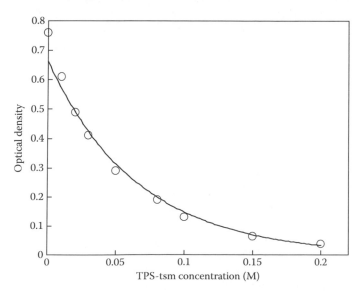

FIGURE 26.6 Relationship between the initial yield of THF-solvated electrons and TPS-tsm concentration. (From Natsuda, K. et al., *Jpn. J. Appl. Phys.*, 48, 06FC05, 2009. With permission.)

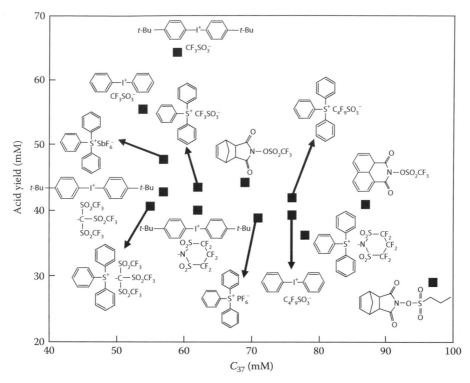

FIGURE 26.7 Relationship between the C_{37} parameter and the acid yield generated in PHS films with 5 wt% C6 and 0.4 M acid generators. (From Natsuda, K. et al., *Jpn. J. Appl. Phys.*, 48, 06FC05, 2009. With permission.)

(Natsuda et al., 2009). The C_{37} parameter can be calculated from the slopes in Figure 26.6. It has been reported that the C_{37} parameter does not correlate to the rate constant for the reaction with solvated electrons (Natsuda et al., 2008). Figure 26.7 shows the relationships between the C_{37} parameter and the acid yield. The acid yield was evaluated at an acid generator concentration of 0.4 M. A good correlation was observed between the C_{37} parameter and the acid yield, although chemical reactions in the solid phase are generally different from those in the liquid phase. Because the thermalization of secondary electrons is completed within 0.1–1 ps (Rassolov and Mozumder, 2001), this thermalization and the immediate reactions of the secondary electrons before solvation are unlikely to be significantly affected by the translation and rotation of molecules even in the liquid phase. The C_{37} parameter is a good indicator for estimating the reactivity of acid generators with thermalized electrons generated in resist films. Also, the observed correlation demonstrated the validity of previously reported reaction mechanisms of chemically amplified resists (Kozawa et al., 1992).

In EB lithography, a 2–100 keV EB is generally used. A single-spur model (Kozawa et al., 2004), which assumes that each pair does not interact, has been reported to closely reproduce the acid yields generated in chemically amplified resists upon exposure to a 75 keV EB (Kozawa et al., 2006b, 2007a). The acid generation efficiency per ionization is calculated to be 0.74 in a PHS film with 10 wt% TPS-tf (Kozawa et al., 2005). The single-spur model for chemically amplified EB resists was constructed on the basis of the fact that it closely reproduces the dynamics of intermediate species generated in the model solutions of resist materials upon exposure to a 28 MeV EB (Saeki et al., 2002; Okamoto et al., 2003, 2007).

The thermalized electrons migrate in the resist matrix under electric fields produced by charged species within the Onsager length. Acid generators react with these electrons and release anions. The mobile electrons are replaced with immobile anions, which act as counteranions of the acid.

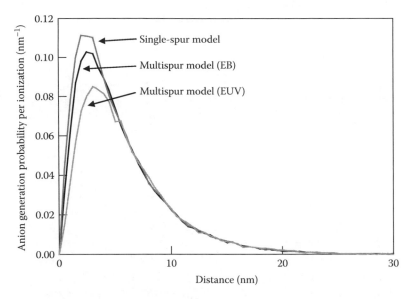

FIGURE 26.8 Distribution of probability of anion generation per spherical-shell thickness (per ionization). The horizontal axis represents the distance from the first ionization point (the origin). (From Kozawa, T. et al., *Jpn. J. Appl. Phys.*, 48, 056508, 2009. With permission.)

The calculated distribution of counteranions is shown in Figure 26.8 (Kozawa et al., 2009). The distribution calculated using the single-spur model is also shown in Figure 26.8. The integral of the graph indicates the acid generation efficiency, which decreased when multispurs were taken into account. The relationship between the average number of overlapped ion pairs and acid generation efficiency (per ionization) was calculated as shown in Figure 26.9. The relationship with the resolution blur is also shown in Figure 26.9. The acid generation efficiency decreases with the increase in the average number of overlapped ion pairs because of the strong electric fields generated by multiple cations and the cross recombination. The strong electric fields promote the recombination between cations and electrons thermalized near cations and decrease the probability of the reaction of acid generators with electrons. Similarly, the resolution blur increases with the average number of ion pairs. The acid generation efficiency was calculated to be 0.70 per ionization in EB resists at 10 wt% TPS-tf loading (0.74 for the single-spur model). The resolution blur was 5.8 nm (5.6 nm for the single-spur model).

In nanoscale patterning using a highly sensitive resist, such as a chemically amplified resist, the statistical effect is not negligible. The fact that ion pairs overlap indicates an increase in the inhomogeneity of the distribution of intermediate species. However, the long electron migration in the subexcitational energy region moderates this inhomogeneity, although it induces resolution blur (Kozawa and Tagawa, 2006). The subsequent proton migration also contributes to reducing the inhomogeneity of the acid distribution (Saeki et al., 2006).

26.5 DEPROTONATION OF RADICAL CATIONS

26.5.1 Dynamics of PHS Radical Cation (PHS•+)

When onium salts react with low-energy electrons generated by ionization, they decompose into a neutral radical and an anion. This decomposition path of onium salts is similar to that induced by the electron transfer from the excited state of polymers observed in photoresists in KrF excimer laser lithography (Hacker et al., 1992; Cameron et al., 2001). The decomposed products of neutral triphenylsulfonium radicals were suggested by Dektar and Hacker (1990a). Considering that the

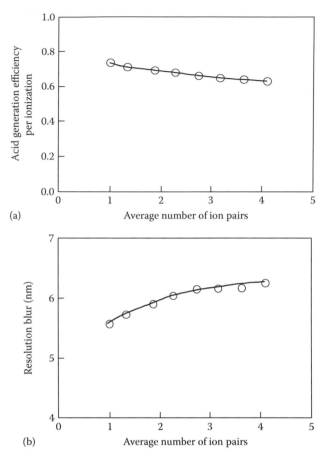

FIGURE 26.9 Relationship of average number of ion pairs in an isolated space with (a) acid generation efficiency (per ionization) and (b) resolution blur. (From Kozawa, T. et al., *Jpn. J. Appl. Phys.*, 48, 056508, 2009. With permission.)

decomposed products are diphenylsulfides and benzenes (phenyl radicals), it is clear that protons are not generated from onium salts in this reaction path. Instead, protons are generated from the radical cations of polymers (Kozawa et al., 1997). Figure 26.10 shows the transient absorption spectra obtained in the pulse radiolysis of 100 mM of PHS-benzonitrile solution, monitored at the pulse end, and 200 and 400 ns after the electron pulse (Nakano et al., 2006b). Note that the spectrum at the pulse end largely overlapped with the transient absorption spectra of solvent intermediates (Tagawa et al., 1972). An absorption peak was observed at 390 nm and 200 and 400 ns after the pulse. The absorption at 390 nm does not decay within the observed time range. The absorption from 400 to 550 nm shows a slight decay. Comparing the absorptions at 200 and 400 ns, this decaying component is considered to have an absorption maximum at 430–450 nm. By adding triethylamine (TEA) to the sample, changes in transient spectra were observed, as shown in Figure 26.11 (Nakano et al., 2006b). Although the absorption intensity decreased, the kinetics at 390 nm did not change from 200 to 400 nm after the pulse. The decaying component observed between 400 and 550 nm almost disappeared with TEA addition.

Although the transient absorption spectra of PHS intermediates have not yet been reported, the radiation chemistry of phenol derivatives has been well studied. Phenoxy radicals, which are stable decomposition intermediates of phenols, were first observed in the 1950s (Porter and Wright, 1955). Land et al. reported that phenoxy radicals are very weak bases (Land et al., 1961) and show a characteristic absorption band at 400 nm (Land and Ebert, 1967). The absorption maxima of phenoxy

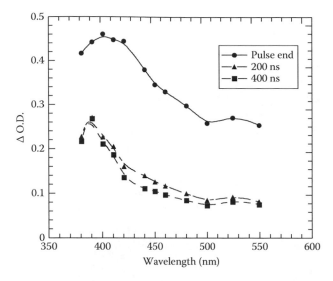

FIGURE 26.10 Transient absorption spectra obtained in pulse radiolysis of 100 mM of PHS solution in benzonitrile. The sample was deaerated by SF_6 bubbling before irradiation. (From Nakano, A. et al., *Jpn. J. Appl. Phys.*, 45, 6866, 2006b. With permission.)

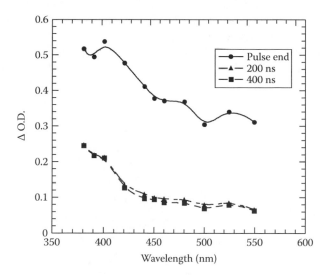

FIGURE 26.11 Transient absorption spectra obtained in pulse radiolysis of 100 mM of PHS solution in benzonitrile with 2.5 mM TEA. The sample was deaerated by SF_6 bubbling before irradiation. (From Nakano, A. et al., *Jpn. J. Appl. Phys.*, 45, 6866, 2006b. With permission.)

radicals of phenol and *p*-cresol in water are 400 and 405 nm, respectively (Land and Porter, 1963; Land and Ebert, 1967). Radical cations of phenol derivatives, which are precursors of phenoxy radicals in many cases, have been more elusive than phenoxy radicals (Land et al., 1961; Dixon and Murphy, 1976, 1978; Holton and Murphy, 1979; Maruyama et al., 1986; Tripathi and Schuler, 1987; Shoute and Neta, 1990; Kesper et al., 1991; Brede et al., 1996). The absorption maxima of radical cations of phenol and *p*-cresol are 440 and 430 nm, respectively (Ganapathi et al., 2000). These reports are consistent with the results for PHS. The absorption at 390 nm and the decaying component were assigned to the phenoxy radical and the radical cation of PHS, respectively.

The deprotonation of radical cations of phenol derivatives has been investigated. Dixon and Murphy reported that the pK_a values of radical cations of phenol and *p*-cresol were −2.0 and −1.6,

respectively (Dixon and Murphy, 1978). Also, Bordwell and Cheng reported that those of phenol and
p-cresol were −8.1 and −7.1, respectively (Bordwell and Cheng, 1991). The radical cation of 2,4-dime-
thoxyphenol was reported to be deprotonated by 2,6-lutidine at a rate constant of 6.5×10^8 M^{-1} s^{-1}
in an acetonitrile solution (Gadosy et al., 1999). The radical cation of phenol is deprotonated by
ethanol at a rate constant of 9.16×10^{10} M^{-1} s^{-1} (Ganapathi et al., 2001). This is in contrast to the
reaction of the poly(4-methoxystyrene) (PMS) radical cation with ethanol, as discussed in the following text. For p-cresol, the absorption coefficients of the phenoxy radical and the radical cation
are 2.30×10^3 and 2.88×10^4 M cm^{-1} at each peak wavelength (Ganapathi et al., 2000). For the
other phenol derivatives, the absorption coefficients of radical cations observed at 400–600 nm are
generally larger than those of phenoxy radicals observed at 400–500 nm. Therefore, the radical cation of PHS is considered to be mostly deprotonated 200 ns after the pulse. Considering that
ethanol and p-cresol have similar proton affinities (PAs), the efficient deprotonation of PHS is
probably due to the ion-molecular reaction between base units.

26.5.2 Dynamics of PMS Radical Cation (PMS$^{\bullet+}$)

PMS is a model compound of a typical resist polymer (partially protected PHS). The pulse radiolysis of a PMS solution in dichloromethane has been reported. Figure 26.12 shows the transient
absorption spectra obtained in the pulse radiolysis of 100 mM of PMS solution in dichloromethane deaerated by Ar bubbling (Nakano et al., 2006b). The transient absorption spectra shown in
Figure 26.12 have a peak at approximately 460 nm. The peak did not decay within the observed
time range. The transient absorption spectra of PMS have not been reported. However, the dynamics of radical cations of methoxylated benzenes, such as methoxybenzene (anisole), methoxytoluene, and dimethoxybenzene, have been intensively investigated using pulse radiolysis (O'Neill
et al., 1975; Holcman and Sehested, 1976; Takamuku et al., 1989; Jonsson et al., 1993), photolysis (Ito et al., 1989; Baciocchi et al., 1993), chemical oxidations (Forbes and Sullivan, 1966;
Sullivan and Brette, 1975; Schlesener and Kochi, 1984), and electrochemical methods (Ronlan
et al., 1974; Buck and Wagoner, 1980). Radical cations of p-methoxytoluene are observed at 440
(Baciocchi et al., 1993) and 450 nm (Ito et al., 1989) in acetonitrile and at 460 nm in 1,2-dichlo-
roethane (Takamuku et al., 1989), and those of methoxybenzene are observed at 430 nm in water
(O'Neill et al., 1975; Holcman and Sehested, 1976; Jonsson et al., 1993). Also, in an aromatic

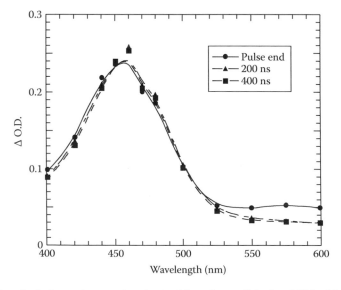

FIGURE 26.12 Transient absorption spectra obtained in pulse radiolysis of 100 mM of PMS solution in
dichloromethane. (From Nakano, A. et al., *Jpn. J. Appl. Phys.*, 45, 6866, 2006b. With permission.)

compound solution in a halocarbon, a π-molecular complex forms through a reaction with a radical cation of aromatic compounds with a chlorine anion. The absorption maxima of the π-molecular complex band of methoxybenzene are 509 nm in dichloromethane (Zweig, 1963) and 524 nm in carbon tetrachloride (Raner et al., 1989). As for methoxytoluene, Takamuku et al. reported that the π-molecular complex of a p-isomer is not observed in the visible wavelength region, although those of o- and m-isomers are observed at 550 and 490 nm, respectively (Takamuku et al., 1989). To clarify the assignment of the absorption at 460 nm (Figure 26.12), TEA was added to the solution. TEA is a well-known cation scavenger. Figure 26.13 shows the transient absorption spectra of the solution with 10 mM of TEA and the kinetic curves of absorption at 460 nm at several TEA concentrations. The decay of the absorption curve became faster with an increase in the TEA concentration. The observed rate constants are plotted in the inset of Figure 26.13b. These rate constants show a linear dependence on the TEA concentration. The bimolecular rate constant calculated from the slope is 1.3×10^8 M^{-1} s^{-1}. By analogy with previous research results mentioned

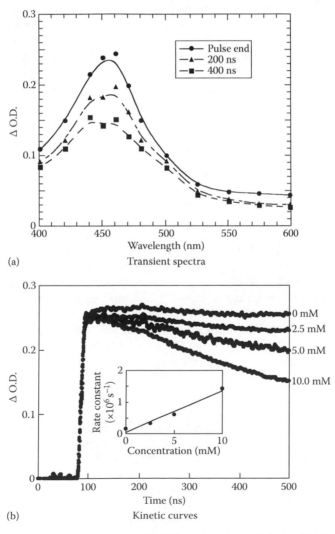

(a) Transient spectra

(b) Kinetic curves

FIGURE 26.13 (a) Transient absorption spectra obtained in pulse radiolysis of 100 mM of PMS solution in dichloromethane with 10 mM TEA and (b) changes in kinetic curves at 460 nm induced by TEA addition (0, 2.5, 5.0, and 10.0 mM). The inset shows the relationship between the observed rate constant and the concentration of TEA. (From Nakano, A. et al., *Jpn. J. Appl. Phys.*, 45, 6866, 2006b. With permission.)

above and from our experimental results, we conclude that the absorption band at 460 nm is ascribed not to the π-molecular complex but to the radical cation of PMS.

The deprotonation of radical cations of methoxylated benzenes has been investigated (Schlesener and Kochi, 1984; Baciocchi et al., 1993). The pK_a of the dimethoxytoluene radical cation is 0.45 in acetonitrile (Baciocchi et al., 1993), which is significantly higher than those of phenol and p-cresol. Dimethoxytoluene radical cations are deprotonated at a rate constant of 3.0×10^6 M^{-1} s^{-1} by 2,6-lutidine (Baciocchi et al., 1993), whose PA is 955 kJ mol^{-1} (Lias et al., 1984). This is two orders lower than that for the reaction of 2,4-dimethoxyphenol radical cations with 2,6-lutidine. It is important to determine whether the PMS radical cation is deprotonated in the absence of strong proton acceptors, such as amine. Using ethanol as a proton acceptor, the deprotonation of the PMS radical cation has been investigated (Nakano et al., 2006b). Ethanol has a PA (788 kJ mol^{-1}) similar to that of the oxygen atom of p-cresol (PA = 756 kJ mol^{-1}) (van Beelen et al., 2004), which is a model compound of PHS. By adding ethanol to 100 mM of PMS-dichloromethane solution, it was examined whether PMS$^{\bullet+}$ is deprotonated. However, no evident decay was observed within the observed time range. This means that PMS$^{\bullet+}$ does not release a proton, at least spontaneously. Also, because the PA of p-methoxytoluene is 801 kJ mol^{-1} (van Beelen et al., 2004), the deprotonation through ion-molecular reaction with a neighboring unit is considered to be ineffective. This is consistent with the experimental result that PMS$^{\bullet+}$ hardly decays within the observed time range, as shown in Figure 26.12. It has been confirmed that the acid yield in the PHS film with TPS-tf was about two times higher than that in the PMS film (Nakano et al., 2006b). The decrease of the acid yield agrees with the difference in deprotonation dynamics between PMS and PHS radical cations observed in the pulse radiolysis experiments.

On the basis of the experiments in solutions, the deprotonation mechanism of PHS is suggested as follows:

$$MOH \rightarrow MOH^{\bullet+} + e^- \tag{26.9}$$

$$MOH^{\bullet+} + MOH \rightarrow MO^{\bullet} + MOH_2^{+} \tag{26.10}$$

Here, MOH, MOH$^{\bullet+}$, MO$^{\bullet}$, and MOH$_2^{+}$ represent a PHS molecule, its radical cation, its phenoxy radical, and its proton adduct, respectively. This deprotonation mechanism is similar to that of novolak (Kozawa et al., 1997). The PHS radical cation generated by ionization is deprotonated by inter- and intramolecular ion-molecular reactions. The reaction mechanism in the solution may be different from that in the solid film because of the different conditions of polymers. However, most PHS hydroxyl groups form inter- and intramolecular hydrogen bonds with a neighboring hydroxyl group in solid resist films (Li and Brisson, 1998; Singh et al., 2005). Therefore, reaction (26.10) is considered to efficiently take place even in solid films.

26.5.3 PMMA

Upon exposure to ionizing radiation, PMMA molecules are either ionized or excited. For keV electrons, the cross section of ionization is larger than that for excitation. Through ionization, a PMMA radical cation and an electron with excess energy are generated. The ejected electron reacts with PMMA molecules or recombines with its parent radical cation after losing a sufficient amount of energy. A PMMA anion radical is formed through the reaction of PMMA with an electron (Ogasawara et al., 1987; Nakano et al., 2004a). An excited state of PMMA is also formed through the recombination of an electron with its parent radical cation. The radical cations (Ichikawa and Yoshida, 1990), radical anions (Tabata et al., 1983; Sakai et al., 1995), and excited PMMA molecules (Fox et al., 1963; Gupta et al., 1980; Torikai et al., 1990) are reported to decompose. In either case, the side chain is first detached and a macroradical is formed. This macroradical results in β-scission. During the decomposition, protons are considered to be generated. It is known that PMMA becomes acidic upon irradiation. Proton generation has been observed in PMMA films without an acid generator. The G-value of acid generation is 1.4 without any acid generators (Nakano et al., 2005). Although the counteranion has not been identified, it is possibly a formic (Tatro et al., 2003) or acetic (Gupta et al., 1980) anion. The main-chain scission of PMMA is also induced by UV light (Fox et al., 1963; Gupta et al., 1980; Torikai et al., 1990). PMMA decomposes through the n-π^* singlet state. Upon exposure to ionizing radiation, the contribution of the excited state of PMMA is, however, low because the recombination reaction (the main generation path for the excited state) is suppressed due to the reaction of electrons with PMMA. In the presence of acid generators, the generation of an excited state through a recombination reaction is suppressed even more because acid generators are strong electron scavengers. The decomposition of PMMA radical anions also leads to main-chain scission (Sakai et al., 1995). However, the efficiency has not been reported. If the decomposition of PMMA anion radicals is effective, this path causes significant resolution blur, because the ejected electrons migrate in the PMMA matrix before they are trapped by PMMA. However, PMMA is known as the highest-resolution resist. The ultimate resolution of the PMMA resist has been intensively investigated and is reported to be 3–5 nm (Chen and Ahmed, 1993; Cumming et al., 1996; Yasin et al., 2001). Therefore, PMMA radical anions are unlikely to decompose effectively. This is also consistent with the fact that PMMA radical anions are observed to be relatively stable (Nakano et al., 2004a).

PMMA radical cations, which are obviously a primary product brought about by ionization, have not been observed even at 4.2 K (Tanaka et al., 1990). The identification of PMMA radical cations was previously reported (Tabata et al., 1983). They later turned out to be PMMA radical anions (Ogasawara et al., 1987). This elusiveness of PMMA radical cations is widely believed to be due to their short lifetime. Although the decomposition path has not been completely clarified, some intermediates and products have been identified. Besides PMMA radical anions, neutral radical species such as main-chain radical $-C^\bullet H-$, side-chain radical $-COOC^\bullet H_2-$, and β-scission products $-CH_2-C^\bullet(CH_3)COOCH_3$ have been identified (Geuskens and David, 1973). It has been widely accepted that the main-chain radical leads to main-chain scission. The G-values of main-chain scission reported by many researchers were scattered. The average of the reported values is 1.9 ± 0.3 (Chapiro, 1962). The main-chain radical is generated through the detachment of the side chain. Among these reported neutral radicals, the side-chain radical is a direct product generated through the deprotonation of PMMA radical cations. The G-value of $-COOC^\bullet H_2-$ has been reported to be approximately 2 (Tanaka et al., 1990). The G-value of trifluoromethanesulfonic acids in PMMA with 10 wt% TPS-tf is shown in Figure 26.14 (Kozawa et al., 2007a). The G-value of

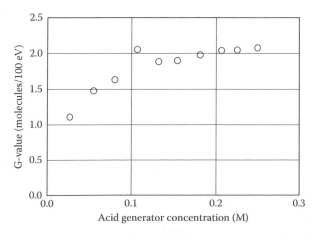

FIGURE 26.14 Dependence of the G-value of acid generation in PMMA on acid generator concentration. (From Kozawa, T. et al., *J. Photopolym. Sci. Technol.*, 20, 577, 2007a. With permission.)

acid generation is saturated at the G-value of main-chain scission or –COOC·H$_2$–. Considering the above-mentioned decomposition mechanism and acid yields with and without acid generators, the acid generation in PMMA consists of two parts, namely, polymer-limited and acid-generator-limited regions. Below 0.1 mol dm^{-3}, the yield of trifluoromethanesulfonic acids is limited by the reaction of acid generators with electrons. Generally, the yield of cationic species is increased by the addition of an electron scavenger (acid generators are typical electron scavengers), because the recombination reaction of cationic species with electrons is suppressed. In PMMA, the ejected electrons are trapped by PMMA itself. Therefore, the addition of acid generators does not noticeably affect the proton yield. Therefore, the observed acid yield corresponded to the decomposition yield of acid generators (Kozawa et al., 2007a). Above 0.1 mol dm^{-3}, the proton yield was less than the counteranion yield produced by acid generator decomposition (Kozawa et al., 2007a). In this region, the acid generation efficiency is restricted by the deprotonation efficiency of PMMA radical cations. The deprotonation efficiency of PMMA radical cations is lower than that of PHS radical cations owing to the difference in the deprotonation mechanisms. It has also been reported that the proton yield strongly depends on the ester group (Nakano et al., 2004b). All these facts suggest that protons are mainly generated by the deprotonation of side-chain (radical) cations.

26.6 SUMMARY

The next-generation lithography requires extremely strict specifications for resist materials. In particular, the trade-off relationship among resolution, sensitivity, and LER is the most serious problem in the development of next-generation materials. Because the enhancement of pattern formation efficiency is an essential solution for this problem, the understanding of the energy flow from the exposure tool to chemical reactions induced in the materials is important for the development of resist materials. Knowledge on radiation physics and chemistry will be a prerequisite in the fields of the mass production of semiconductor devices.

REFERENCES

Ablaza, S. L., Cameron, J. F., Xu, G., and Yueh, W. 2000. The effect of photoacid generator structure on deep ultraviolet resist performance. *J. Vac. Sci. Technol. B* 18: 2543–2550.

Aktary, M., Stepanova, M., and Dew, S. K. 2006. Simulation of the spatial distribution and molecular weight of polymethylmethacrylate fragments in electron beam lithography exposures. *J. Vac. Sci. Technol. B* 24: 768–779.

Aldrich, J. E., Bronskill, M. J., Wolff, R. K., and Hunt, J. W. 1971. Picosecond pulse radiolysis. 3. Reaction rates and reduction in yields of hydrated electrons. *J. Chem. Phys.* 55: 530–536.

Aoai, T., Umehara, A., Kamiya, A., Matsuda, N., and Aotani, Y. 1989. Application of silicon polymer as positive photosensitive material. *Polym. Eng. Sci.* 29: 887–890.

Baciocchi, E., Delgiacco, T., and Elisei, F. 1993. Proton-transfer reactions of alkylaromatic cation radicals. The effect of alpha-substituents on the kinetic acidity of *p*-methoxytoluene cation radicals. *J. Am. Chem. Soc.* 115: 12290–12295.

Bartczak, W. M. and Hummel, A. 1987. Computer-simulation of ion recombination in irradiated nonpolar liquids. *J. Chem. Phys.* 87: 5222–5228.

Bartczak, W. M. and Hummel, A. 1993. Computer-simulation study of spatial-distribution of the ions and electrons in tracks of high-energy electrons and the effect on the charge recombination. *J. Phys. Chem.* 97: 1253–1255.

Bartczak, W. M., Tachiya, M., and Hummel, A. 1990. Triplet formation in the ion recombination in irradiated liquids. *Int. J. Radiat. Appl. Instrum., Part C* 36: 195–198.

Bethe, H. A. 1930. The theory of the passage of rapid neutron radiation through matter. *Ann. Phys.* 5: 325–400.

Bordwell, F. G. and Cheng, J.-P. 1991. Substituent effects on the stabilities of phenoxyl radicals and the acidities of phenoxyl radical cations. *J. Am. Chem. Soc.* 113: 1736–1743.

Brainard, R., Higgins, C., Hassanein, E., Matyi, R., and Wuest, A. 2008. Film quantum yields of ultrahigh PAG EUV photoresists. *J. Photopolym. Sci. Technol.* 21: 457–464.

Brede, O., Orthner, H., Zubarev, V., and Hermann, R. 1996. Radical cations of sterically hindered phenols as intermediates in radiation-induced electron transfer processes. *J. Phys. Chem.* 100: 7097–7105.

Buck, R. P. and Wagoner, D. E. 1980. Selective anodic oxidation of *p*-alkylaryl ethers—pathways and products. *J. Electroanal. Chem.* 115: 89–113.

Buhr, G., Dammel, R., and Lindley, C. 1989. Nonionic photoacid generating compounds. *Polym. Mater. Sci. Eng.* 61: 269.

Cameron, J. F., Chan, N., Moore, K., and Pohlers, G. 2001. Comparison of acid generating efficiencies in 248 and 193 nm photoresists. *Proc. SPIE* 4345: 106–118.

Chang, T. H. P. 1975. Proximity effect in electron-beam lithography. *J. Vac. Sci. Technol.* 12: 1271–1275.

Chapiro, A. 1962. *Radiation Chemistry of Polymeric Systems.* New York: Wiley, p. 515.

Chen, W. and Ahmed, H. 1993. Fabrication of 5–7 nm wide etched lines in silicon using 100 keV electron-beam lithography and polymethylmethacrylate resist. *Appl. Phys. Lett.* 62: 1499–1501.

Chernovitz, A. C. and Jonah, C. D. 1988. Isotopic dependence of recombination kinetics in water. *J. Phys. Chem.* 92: 5946–5950.

Clifford, P., Green, N. J. B., and Pilling, M. J. 1982. Stochastic-model based on pair distribution-functions for reaction in a radiation-induced spur containing one-type of radical. *J. Phys. Chem.* 86: 1318–1321.

Cumming, D. R. S., Thoms, S., Beaumont, S. P., and Weaver, J. M. R. 1996. Fabrication of 3 nm wires using 100 keV electron beam lithography and poly(methyl methacrylate) resist. *Appl. Phys. Lett.* 68: 322–324.

Dektar, J. L. and Hacker N. P. 1990a. Photochemistry of triarylsulfonium salts. *J. Am. Chem. Soc.* 112: 6004–6015.

Dektar, J. L. and Hacker, N. P. 1990b. Photochemistry of diaryliodonium salts. *J. Org. Chem.* 55: 639–647.

Dentinger, P., Cardinale, G., Henderson, C., Fisher, A., and Ray-Chaudhuri, A. 2000. Photoresist film thickness for extreme ultraviolet lithography. *Proc. SPIE* 3997: 588–599.

Dill, F. H., Hornberger, W. P., Hauge, P. S., and Shaw, J. M. 1975. Characterization of positive photoresist. *IEEE Trans. Electron Devices* 22: 445–452.

Dixon, W. T. and Murphy, D. 1976. Determination of the acidity constants of some phenol radical cations by means of electron spin resonance. *J. Chem. Soc., Faraday Trans. 2* 72: 1221–1230.

Dixon, W. T. and Murphy, D. 1978. Determination of acid dissociation constants of some phenol radical cations. *J. Chem. Soc., Faraday Trans. 2* 74: 432–439.

Forbes, W. F. and Sullivan, P. D. 1966. The use of aluminum chloride-nitromethane for the production of cation radicals. *J. Am. Chem. Soc.* 88: 2862–2863.

Fox, R. B., Isaacs, L. G., and Stokes, S. 1963. Photolytic degradation of poly(methyl methacrylate). *J. Polym. Sci. A* 1: 1079–1086.

Gadosy, T. A., Shukla, D., and Johnston, L. J. 1999. Generation, characterization, and deprotonation of phenol radical cations. *J. Phys. Chem. A* 103: 8834–8839.

Gallatin, G. M. 2005. Resist blur and line edge roughness. *Proc. SPIE* 5754: 38–52.

Ganapathi, M. R., Hermann, R., Naumov, S., and Brede, O. 2000. Free electron transfer from several phenols to radical cations of non-polar solvents. *Phys. Chem. Chem. Phys.* 2: 4947–4955.

Ganapathi, M. R., Naumov, S., Hermann, R., and Brede, O. 2001. Nucleophilic effects on the deprotonation of phenol radical cations. *Chem. Phys. Lett.* 337: 335–340.

Geuskens, G. and David, C. 1973. Identification of free radicals produced by the radiolysis of poly(methyl methacrylate) and poly(methyl acrylate) at 77°K. *Makromol. Chem.* 165: 273–280.

Goulet, T., Mattei, I., and Jay-Gerin, J. P. 1990. A description of random-walks with collision anisotropy and with a nonconstant mean free-path. *Can. J. Phys.* 68: 912–917.

Green, N. J. B., Pilling, M. J., Pimblott, S. M., and Clifford, P. 1989. Stochastic-models of diffusion-controlled ionic reactions in radiation-induced spurs. 2. Low-permittivity solvents. *J. Phys. Chem.* 93: 8025–8031.

Gupta, A., Liang, R., Tsay, F. D., and Moacanin, J. 1980. Characterization of a dissociative excited state in the solid state: Photochemistry of poly(methyl methacrylate). Photochemical processes in polymeric systems. 5. *Macromolecules* 13: 1696–1700.

Hacker, N. P. and Welsh, K. M. 1991. Photochemistry of triphenylsulfonium salts in poly[4-[(*tert*-butoxycarbonyl)oxy]styrene]: Evidence for a dual photoinitiation process. *Macromolecules* 24: 2137–2139.

Hacker, N. P., Hofer, D. C., and Welsh, K. M. 1992. Photochemical and photophysical studies on chemically amplified resists. *J. Photopolym. Sci. Technol.* 5: 35–46.

Hamill, W. H. 1969. A model for radiolysis of water. *J. Phys. Chem.* 73: 1341–1349.

Hawryluk, R. J., Hawryluk, A. M., and Smith, H. I. 1974. Energy dissipation in a thin polymer film by electron-beam scattering. *J. Appl. Phys.* 45: 2551–2566.

Hirose, R., Kozawa, T., Tagawa, S., Kai, T., and Shimokawa, T. 2007. Dependence of absorption coefficient and acid generation efficiency on acid generator concentration in chemically amplified resist for extreme ultraviolet lithography. *Jpn. J. Appl. Phys.* 46: L979–L981.

Hirose, R., Kozawa, T., Tagawa, S., Kai, T., and Shimokawa, T. 2008a. Dependence of acid generation efficiency on molecular structures of acid generators upon exposure to extreme ultraviolet radiation. *Appl. Phys. Express* 1: 027004.

Hirose, R., Kozawa, T., Tagawa, S., Shimizu, D., Kai, T., and Shimokawa, T. 2008b. Difference between acid generation mechanisms in poly(hydroxystyrene)- and polyacrylate-based chemically amplified resists upon exposure to extreme ultraviolet radiation. *Jpn. J. Appl. Phys.* 47: 7125–7127.

Holcman, J. and Sehested, K. 1976. Anisole radical cation reactions in aqueous solution. *J. Phys. Chem.* 80: 1642–1644.

Holton, D. M. and Murphy, D. 1979. Determination of acid dissociation constants of some phenol radical cations. Part 2. *J. Chem. Soc., Faraday Trans.* 2 75: 1637–1642.

Houlihan, F. M., Shugard, A., Gooden, R., and Reichmanis, E. 1988. Nitrobenzyl ester chemistry for polymer processes involving chemical amplification. *Macromolecules* 21: 2001–2006.

Hunt, J. W. and Chase, W. J. 1977. Temperature and solvent dependence of electron scavenging efficiency in polar liquids: Water and alcohols. *Can. J. Chem.* 55: 2080–2087.

Ichikawa, T. and Yoshida, H. 1990. Mechanism of radiation-induced degradation of poly(methyl methacrylate) as studied by ESR and electron spin echo methods. *J. Polym. Sci. A* 28: 1185–1196.

Inokuti, M. 1975. Ionization yields in gases under electron-irradiation. *Radiat. Res.* 64: 6–22.

Itani, T. 2009. Recent status and future direction of EUV resist technology. *Microelectron. Eng.* 86: 207–212.

Ito, H. 2005. Chemical amplification resists for microlithography. In *Microlithography/Molecular Imprinting.* Advances in Polymer Science Series. Vol. 172, pp. 37–245. Berlin/Heidelberg, Germany: Springer.

Ito, H. and Willson, C. G. 1983. Chemical amplification in the design of dry developing resist materials. *Polym. Eng. Sci.* 23: 1012–1018.

Ito, O., Akiho, S., and Iino, M. 1989. Kinetic study for reactions of nitrate radical (NO$_3$) with substituted toluenes in acetonitrile solution. *J. Org. Chem.* 54: 2436–2440.

Jay-Gerin, J. P., Goulet, T., and Billard, I. 1993. On the correlation between electron-mobility, free-ion yield, and electron thermalization distance in nonpolar dielectric liquids. *Can. J. Chem.* 71: 287–293.

Jonah, C. D., Miller, J. R., and Matheson, M. S. 1977. Reaction of precursor of hydrated electron with electron scavengers. *J. Phys. Chem.* 81: 1618–1622.

Jonah, C. D., Bartels, D. M., and Chernovitz, A. C. 1989. Primary processes in the radiation-chemistry of water. *Int. J. Radiat. Appl. Instrum., Part C* 34: 145–156.

Jonsson, M., Lind, J., Reitberger, T., Eriksen, T. E., and Merenyi, G. 1993. Redox chemistry of substituted benzenes: The one-electron reduction potentials of methoxy-substituted benzene radical cations. *J. Phys. Chem.* 97: 11278–11282.

Kesper, K., Diehl, F., Simon, J. G. G., Specht, H., and Schweig, A. 1991. Resonant two-photon ionization of phenol in methylene chloride doped solid argon using 248 nm KrF laser and 254 nm Hg lamp radiation, a comparative study. The UV/VIS absorption spectrum of phenol radical cation. *Chem. Phys.* 153: 511–517.

Kim, B.-S., Lee, H.-S., Wi, J.-S., Jin, K.-B., and Kim, K.-B. 2005. Sensitivity characteristics of positive and negative resists at 200 kV electron-beam lithography. *Jpn. J. Appl. Phys.* 44: L95–L97.

Kimura, M., Kowari, K., Inokuti, M., Bronic, I. K., Srdoc, D., and Obelic, B. 1991. Theoretical-study of W values in hydrocarbon gases. *Radiat. Res.* 125: 237–242.

Koba, F., Yamashita, H., and Arimoto, H. 2005. Highly accurate proximity effect correction for 100 kV electron projection lithography. *Jpn. J. Appl. Phys.* 44: 5590–5594.

Kozawa, T. and Tagawa, S. 2006. Resolution blur of latent acid image and acid generation efficiency of chemically amplified resists for electron beam lithography. *J. Appl. Phys.* 99: 054509.

Kozawa, T. and Tagawa, S. 2008. Side wall degradation of chemically amplified resists based on poly (4-hydroxystyrene) for extreme ultraviolet lithography. *Jpn. J. Appl. Phys.* 47: 7822–7826.

Kozawa, T., Yoshida, Y., Uesaka, M., and Tagawa, S. 1992. Radiation-induced acid generation reactions in chemically amplified resists for electron beam and x-ray lithography. *Jpn. J. Appl. Phys.* 31: 4301–4306.

Kozawa, T., Nagahara, S., Yoshida, Y., Tagawa, S., Watanabe, T., and Yamashita, Y. 1997. Radiation-induced reactions of chemically amplified x-ray and electron beam resists based on deprotection of t-butoxycarbonyl groups. *J. Vac. Sci. Technol. B* 15: 2582–2586.

Kozawa, T., Mizutani, Y., Miki, M., Yamamoto, T., Suemine, S., Yoshida, Y., and Tagawa, S. 2000. Development of subpicosecond pulse radiolysis system. *Nucl. Instrum. Methods Phys. Res. Sect. A* 440: 251–254.

Kozawa, T., Saeki, A., Yoshida, Y., and Tagawa, S. 2002. Study on radiation-induced reaction in microscopic region for basic understanding of electron beam patterning in lithographic process (I). *Jpn. J. Appl. Phys.* 41: 4208–4212.

Kozawa, T., Saeki, A., Nakano, A., Yoshida, Y., and Tagawa, S. 2003. Relation between spatial resolution and reaction mechanism of chemically amplified resists for electron beam lithography. *J. Vac. Sci. Technol. B* 21: 3149–3152.

Kozawa, T., Saeki, A., and Tagawa, S. 2004. Modeling and simulation of chemically amplified electron beam, x-ray, and EUV resist processes. *J. Vac. Sci. Technol. B* 22: 3489–3492.

Kozawa, T., Yamamoto, H., Saeki, A., and Tagawa, S. 2005. Proton and anion distribution and line edge roughness of chemically amplified electron beam resist. *J. Vac. Sci. Technol. B* 23: 2716–2720.

Kozawa, T., Tagawa, S., Oizumi, H., and Nishiyama, I. 2006a. Acid generation efficiency in a model system of chemically amplified extreme ultraviolet resist. *J. Vac. Sci. Technol. B* 24: L27–L30.

Kozawa, T., Shigaki, T., Okamoto, K., Saeki, A., Tagawa, S., Kai, T., and Shimokawa, T. 2006b. Analysis of acid yield generated in chemically amplified electron beam resist. *J. Vac. Sci. Technol. B* 24: 3055–3060.

Kozawa, T., Tagawa, S., Kai, T., and Shimokawa, T. 2007a. Sensitization distance and acid generation efficiency in a model system of chemically amplified electron beam resist with methacrylate backbone polymer. *J. Photopolym. Sci. Technol.* 20: 577–580.

Kozawa, T., Tagawa, S., and Shell, M. 2007b. Theoretical study on relationship between acid generation efficiency and acid generator concentration in chemically amplified extreme ultraviolet resists. *Jpn. J. Appl. Phys.* 46: L1143–L1145.

Kozawa, T., Tagawa, S., Cao, H. B., Deng, H., and Leeson, M. J. 2007c. Acid distribution in chemically amplified extreme ultraviolet resist. *J. Vac. Sci. Technol. B* 25: 2481–2485.

Kozawa, T., Okamoto, K., Nakamura, J., and Tagawa, S. 2008a. Feasibility study on high-sensitivity chemically amplified resist by polymer absorption enhancement in extreme ultraviolet lithography. *Appl. Phys. Express* 1: 067012.

Kozawa, T., Tagawa, S., Santillan, J. J., Toriumi, M., and Itani, T. 2008b. Effects of rate constant for deprotection on latent image formation in chemically amplified extreme ultraviolet resists. *Jpn. J. Appl. Phys.* 47: 4926–4931.

Kozawa, T., Tagawa, S., Santillan, J. J., Toriumi, M., and Itani, T. 2008c. Feasibility study of chemically amplified extreme ultraviolet resists for 22 nm fabrication. *Jpn. J. Appl. Phys.* 47: 4465–4468.

Kozawa, T., Tagawa, S., and Shell, M. 2008d. Resolution degradation caused by multispur effect in chemically amplified extreme ultraviolet resists. *J. Appl. Phys.* 103: 084306.

Kozawa, T., Okamoto, K., Saeki, A., and Tagawa, S. 2009. Difference of spur distribution in chemically amplified resists upon exposure to electron beam and extreme ultraviolet radiation. *Jpn. J. Appl. Phys.* 48: 056508.

Lam, K. Y. and Hunt, J. W. 1975. Picosecond pulse-radiolysis. 6. Fast electron reactions in concentrated solutions of scavengers in water and alcohols. *Int. J. Radiat. Phys. Chem.* 7: 317–338.

Land, E. J. and Ebert, M. 1967. Pulse radiolysis studies of aqueous phenol – water elimination from dihydroxy-cyclohexadienyl radicals to form phenoxyl. *Trans. Faraday Soc.* 63: 1181–1190.

Land, E. J. and Porter, G. 1963. Primary photochemical processes in aromatic molecules. Part 7.—Spectra and kinetics of some phenoxyl derivatives. *Trans. Faraday Soc.* 59: 2016–2026.

Land, E. J., Porter, G., and Strachan, E. 1961. Primary photochemical processes in aromatic molecules. 6. Absorption spectra and acidity constants of phenoxyl radicals. *Trans. Faraday Soc.* 57: 1885–1893.

Lewis, M. A. and Jonah, C. D. 1986. Evidence for 2 electron-states in salvation and scavenging processes in alcohols. *J. Phys. Chem.* 90: 5367–5372.

Li, D. and Brisson, J. 1998. Hydrogen bonds in poly(methyl methacrylate) poly(4-vinyl phenol) blends: 1. Quantitative analysis using FTIR spectroscopy. *Polymer* 39: 793–800.

Lias, S. G., Liebman, J. F., and Levin, R. D. 1984. Evaluated gas-phase basicities and proton affinities of molecules: Heats of formation of protonated molecules. *J. Phys. Chem. Ref. Data* 13: 695–808.

Maruyama, K., Furuta, H., and Osuka, A. 1986. Cidnp study on porphyrin-photosensitized reactions with phenol and quinone: Dimerization of 4-methoxyphenol and cross coupling of benzoquinone to porphyrins covalently linked with phenol group. *Tetrahedron* 42: 6149–6155.

Masuda, S., Kawanishi, Y., Hirano, S., Kamimura, S., Mizutani, K., Yasunami, S., and Kawabe, Y. 2006. The material design to reduce outgassing in acetal based chemically amplified resist for EUV lithography. *Proc. SPIE* 6153: 615342.

McMillan, J. A., Johnson, S., and MacDonald, N. C. 1989. Simulation of electron-beam exposure of sub-micron patterns. *J. Vac. Sci. Technol. B* 7: 1540–1545.

Mozumder, A. 2002. Free-ion yield and electron mobility in liquid hydrocarbons: A consistent correlation. *J. Phys. Chem. A* 106: 7062–7067.

Murata, K., Kyser, D. F., and Ting, C. H. 1981. Monte Carlo simulation of fast secondary-electron production in electron-beam resists. *J. Appl. Phys.* 52: 4396–4405.

Nagahara, S., Sakurai, Y., Wakita, M., Yamamoto, Y., Tagawa, S., Komura, M., Yano, E., and Okazaki, S. 2000. Methods to improve radiation sensitivity of chemically amplified resists by using chain reactions of acid generation. *Proc. SPIE* 3999: 386–394.

Nakano, A., Okamoto, K., Kozawa, T., and Tawaga, S. 2004a. Pulse radiolysis study on proton and charge transfer reactions in solid poly(methyl methacrylate). *Jpn. J. Appl. Phys.* 43: 4363–4367.

Nakano, A., Okamoto, K., Kozawa, T., and Tagawa, S. 2004b. Effects of ester groups on proton generation and diffusion in polymethacrylate matrices. *Jpn. J. Appl. Phys.* 43: 3981–3983.

Nakano, A., Okamoto, K., Yamamoto, Y., Kozawa, T., Tagawa, S., Kai, T., and Nemoto, H. 2005. Dependence of acid yield on acid generator in chemically amplified resist for post-optical lithography. *Jpn. J. Appl. Phys.* 44: 5832–5835.

Nakano, A., Kozawa, T., Tagawa, S., Szreder, T., Wishart, J. F., Kai, T., and Shimokawa, T., 2006a. Reactivity of acid generators for chemically amplified resists with low-energy electrons. *Jpn. J. Appl. Phys.* 45: L197–L200.

Nakano, A., Kozawa, T., Okamoto, K., Tagawa, S., Kai, T., and Shimokawa, T. 2006b. Acid generation mechanism of poly(4-hydroxystyrene)-based chemically amplified resists for post-optical lithography: Acid yield and deprotonation behavior of poly(4-hydroxystyrene) and poly(4-methoxystyrene). *Jpn. J. Appl. Phys.* 45: 6866–6871.

Natsuda, K., Kozawa, T., Saeki, A., Tagawa, S., Kai, T., and Shimokawa, T. 2008. Study of the reaction of acid generators with epithermal and thermalized electrons. *Jpn. J. Appl. Phys.* 47: 4932–4935.

Natsuda, K., Kozawa, T., Okamoto, K., Saeki, A., and Tagawa, S. 2009. Correlation between C_{37} parameters and acid yields in chemically amplified resists upon exposure to 75 keV electron beam. *Jpn. J. Appl. Phys.* 48: 06FC05.

Ogasawara, M., Tanaka, M., and Yoshida, H. 1987. Reaction of solvated electron with poly(methyl methacrylate) and substituted poly(methyl methacrylate) in hexamethylphosphoramide studied by pulse radiolysis. *J. Phys. Chem.* 91: 937–941.

Oizumi, H., Tanaka, Y., Kumise, T., Shiono, D., Hirayama, T., Hada, H., Onodera, J., Yamaguchi, A., and Nishiyama, I. 2007. Evaluation of new molecular resist for EUV lithography. *J. Photopolym. Sci. Technol.* 20: 403–410.

Okamoto, K., Saeki, A., Kozawa, T., Yoshida, Y., and Tagawa, S. 2003. Subpicosecond pulse radiolysis study of geminate ion recombination in liquid benzene. *Chem. Lett.* 32: 834–835.

Okamoto, K., Kozawa, T., Saeki, A., Yoshida, Y., and Tagawa, S. 2007. Subpicosecond pulse radiolysis in liquid methyl-substituted benzene derivatives. *Radiat. Phys. Chem.* 76: 818–826.

O'Neill, P., Steenken, S., and Schulte-Frohlinde, D. 1975. Formation of radical cations of methoxylated benzenes by reaction with hydroxyl radicals, thallium(2+), silver(2+), and peroxysulfate (SO4⁻) in aqueous solution. Optical and conductometric pulse radiolysis and in situ radiolysis electron spin resonance study. *J. Phys. Chem.* 79: 2773–2779.

Onsager, L. 1938. Initial recombination of ions. *Phys. Rev.* 54: 554–557.

Ortica, F., Scaiano, J. C., Pohlers, G., Cameron, J. F., and Zampini, A. 2000. Laser flash photolysis study of two aromatic N-oxyimidosulfonate photoacid generators. *Chem. Mater.* 12: 414–420.

Parikh, M. and Kyser, D. F. 1979. Energy deposition functions in electron resist films on substrates. *J. Appl. Phys.* 50: 1104–1111.

Pohlers, G., Scaiano, J. C., and Sinta, R. 1997. A novel photometric method for the determination of photo-acid generation efficiencies using benzothiazole and xanthene dyes as acid sensors. *Chem. Mater.* 9: 3222–3230.

Porter, G. and Wright, F. J. 1955. Primary photochemical processes in aromatic molecules. 3. Absorption spectra of benzyl, anilino, phenoxy and related free radicals. *Trans. Faraday Soc.* 51: 1469–1475.

Raner, K. D., Lusztyk, J., and Ingold, K. U. 1989. Ultraviolet visible spectra of halogen molecule/arene and halogen atom/arene π-molecular complexes. *J. Phys. Chem.* 83: 564–570.

Rassolov, V. A. and Mozumder, A. 2001. Monte Carlo simulation of electron thermalization distribution in liquid hydrocarbons: Effects of inverse collisions and of an external electric field. *J. Phys. Chem. B* 105: 1430–1437.

Rishton, S. A. and Kern, D. P. 1987. Point exposure distribution measurements for proximity correction in electron-beam lithography on a sub-100 nm scale. *J. Vac. Sci. Technol. B* 5: 135–141.

Ronlan, A., Coleman, J., Hammerich, O., and Parker, V. D. 1974. Anodic oxidation of methoxybiphenyls: Effect of the biphenyl linkage on aromatic cation radical and dication stability. *J. Am. Chem. Soc.* 96: 845–849.

Ryan, J. M., Hoole, A. C. F., and Broers, A. N. 1995. A study of the effect of ultrasonic agitation during development of poly(methylmethacrylate) for ultrahigh resolution electron-beam lithography. *J. Vac. Sci. Technol. B* 13: 3035–3039.

Saeki, A., Kozawa, T., Yoshida, Y., and Tagawa, S. 2002. Study on radiation-induced reaction in microscopic region for basic understanding of electron beam patterning in lithographic process (II). *Jpn. J. Appl. Phys.* 41: 4213–4216.

Saeki, A., Kozawa, T., Yoshida, Y., and Tagawa, S. 2005. Multi spur effect on decay kinetics of geminate ion recombination using Monte Carlo technique. *Nucl. Instrum. Methods Phys. Res. Sect. B* 234: 285–290.

Saeki, A., Kozawa, T., Tagawa, S., and Cao, H. B. 2006. Line edge roughness of a latent image in post-optical lithography. *Nanotechnology* 17: 1543–1546.

Saeki, A., Kozawa, T., Ohnishi, Y., and Tagawa, S. 2007. Reactivity between biphenyl and precursor of solvated electrons in tetrahydrofuran measured by picosecond pulse radiolysis in near-ultraviolet, visible, and infrared. *J. Phys. Chem. A* 111: 1229–1235.

Saito, N. and Suzuki, I. H. 2001. Photon W-value for Ar in the sub-keV x-ray region. *Radiat. Phys. Chem.* 60: 291–296.

Sakai, W., Tsuchida, A., Yamamoto, M., and Yamauchi, J. 1995. Main chain scission reaction of poly(methyl methacrylate) caused by two-photon ionization of dopant. *J. Polym. Sci. A* 33: 1969–1978.

Sakamizu, T., Yamaguchi, H., Shiraishi, H., Murai, F., and Ueno, T. 1993. Development of positive electron-beam resist for 50 kV electron-beam direct-writing lithography. *J. Vac. Sci. Technol. B* 11: 2812–2817.

Schlesener, C. J. and Kochi, J. K. 1984. Stoichiometry and kinetics of *p*-methoxytoluene oxidation by electron transfer: Mechanistic dichotomy between side chain and nuclear substitution. *J. Org. Chem.* 49: 3142–3150.

Shimizu, R., Ikuta, T., Everhart, T. E., and Devore, W. J. 1975. Experimental and theoretical-study of energy dissipation profiles of keV electrons in polymethylmethacrylate. *J. Appl. Phys.* 46: 1581–1584.

Shimizu, D., Maruyama, K., Saitou, A., Kai, T., Shimokawa, T., Fujiwara, K., Kikuchi, Y., and Nishiyama, I. 2007. Progress in EUV resist development. *J. Photopolym. Sci. Technol.* 20: 423–428.

Shoute, L. C. T. and Neta, P. 1990. Bromine atom complexes with bromoalkanes: Their formation in the pulse radiolysis of di-, tri-, and tetrabromomethane and their reactivity with organic reductants. *J. Phys. Chem.* 94: 2447–2453.

Siebbeles, L. D. A., Bartczak, W. M., Terrissol, M., and Hummel, A. 1997. Computer simulation of the ion escape from high-energy electron tracks in nonpolar liquids. *J. Phys. Chem. A* 101: 1619–1627.

Singh, L., Ludovice, P. J., and Henderson, C. L. 2005. The effect of film thickness on the dissolution rate and hydrogen bonding behavior of photoresist polymer thin films. *Proc. SPIE* 5753: 319–328.

Sullivan, P. D. and Brette, N. A. 1975. Temperature-dependent splitting constants in the electron spin resonance spectra of cation radicals. V. Methylidyne protons in some tetrasubstituted benzenes. *J. Phys. Chem.* 79: 474–479.

Suzuki, I. H. and Saito, N. 1985. Effect of inner-shell excitations on the W-value of propane. *Bull. Chem. Soc. Jpn.* 58: 3210–3214.

Suzuki, I. H. and Saito, N. 1987. Oscillatory variation near the C-K edge in the photon W-value of ethylene. *Bull. Chem. Soc. Jpn.* 60: 2989–2992.

Tabata, M., Nilsson, G., Lund, A., and Sohma, J., 1983. Ionic species in irradiated poly(methyl methacrylate): A pulse radiolysis and ESR study. *J. Polym. Sci. Polym. Chem. Ed.* 21: 3257–3268.

Tabata, Y., Tagawa S., and Washio M. 1984. Pulse-radiolysis studies on the mechanism of the high-sensitivity of chloromethylated polystyrene as an electron negative resist. *ACS Symposium Series* 255: 151–163.

Tagawa, S. 1986. Pulse-radiolysis and laser photolysis studies on radiation-resistance and sensitivity of polystyrene and related polymers. *Radiat. Phys. Chem.* 27: 455–459.

Tagawa S. 1987. Main reactions of chlorine-containing and silicon-containing electron and deep-UV (excimer laser) negative resists. *ACS Symposium Series* 346: 37–45.

Tagawa, S., Arai, S., Kira, A., Imamura, M., Tabata, Y., and Oshima, K. 1972. Pulse radiolysis study of polymerization of vinylcarbazole in benzonitrile solutions. *J. Polym. Sci., Part B* 10: 295–297.

Tagawa, S., Nagahara, S., Iwamoto, T., Wakita, M., Kozawa, T., Yamamoto, Y., Werst, D., and Trifunac, A. D. 2000. *Proc. SPIE* 3999: 204.

Tagawa, S., Seki, S., and Kozawa, T. 2004. Charged particle and photon-induced reactions in polymers. In *Charged Particle Interactions with Matter: Chemical, Physicochemical, and Biological Consequences with Applications*. A. Mozumder and Y. Hatano (eds.), pp. 551–578. New York: Marcel Dekker.

Takamuku, S., Komitsu, S., and Toki, S. 1989. Radical cations of anisole derivatives: Novel complex-formation. *Radiat. Phys. Chem.* 34: 553–559.

Tanaka, M., Yoshida, H., and Ichikawa, T. 1990. Thermal and photo-induced reactions of polymer radicals in γ-irradiated poly(alkyl methacrylate). *Polym. J.* 22: 835–841.

Tanuma, S., Powell, C. J., and Penn, D. R. 1994. Calculations of electron inelastic mean free paths. 5. Data for 14 organic-compounds over the 50–2000 eV range. *Surf. Interface Anal.* 21: 165–176.

Tatro, S., Baker, G., Bisht, K., and Harmon, J. 2003. A MALDI, TGA, TG/MS, and DEA study of the irradiation effects on PMMA. *Polymer* 44: 167–176.

Thackeray, J. W., Aqad, E., Cronin, M. F., and Spear-Alfonso, K. 2008. Design of faster high resolution resists: Getting more acid yield from EUV photons. *J. Photopolym. Sci. Technol.* 21: 415–420.

Torikai, A., Ohno, M., and Fueki, K. 1990. Photodegradation of poly(methyl methacrylate) by monochromatic light: Quantum yield, effect of wavelengths, and light intensity. *J. Appl. Polym. Sci.* 41: 1023–1032.

Tripathi, G. N. R. and Schuler, R. H. 1987. Resonance Raman spectra of *p*-benzosemiquinone radical and hydroquinone radical cation. *J. Phys. Chem.* 91: 5881–5885.

Tsuji, S., Kozawa, T., Yamamoto, Y., and Tagawa, S. 2000. Studies on reaction mechanisms of EB resist by pulse radiolysis. *J. Photopolym. Sci. Technol.* 13: 733–738.

van Beelen, E. S. E., Koblenz, T. A., Ingemann, S., and Hammerum, S. 2004. Experimental and theoretical evaluation of proton affinities of furan, the methylphenols, and the related anisoles. *J. Phys. Chem. A* 108: 2787–2793.

Williams, L. E., Callcott, T. A., Ashley, J. C., and Anderson, V. E. 1989. Interactions of electrons with poly(methyl-methacrylate). *J. Electron Spectrosc. Relat. Phenom.* 49: 323–334.

Wojcik, M., Bartczak, W. M., and Hummel, A. 1992. Computer-simulation of electron scavenging in multipair spurs in dielectric liquids. *J. Chem. Phys.* 97: 3688–3695.

Wolff, R. K., Bronskill, M. J., and Hunt, J. W. 1970. Picosecond pulse radiolysis studies. 2. Reactions of electrons with concentrated scavengers. *J. Chem. Phys.* 53: 4211–4217.

Wolff, R. K., Aldrich, J. E., Penner, T. L., and Hunt, J. W. 1975. Picosecond pulse-radiolysis. 5. Yield of electrons in irradiated aqueous-solution with high-concentrations of scavenger. *J. Phys. Chem.* 79: 210–219.

Yamamoto, H., Nakano, A., Okamoto, K., Kozawa, T., and Tagawa, S. 2004a. Polymer screening method for chemically amplified electron beam and x-ray resists. *Jpn. J. Appl. Phys.* 43: 3971–3973.

Yamamoto, H., Kozawa, T., Nakano, A., Okamoto, K., Yamamoto, Y., Ando, T., Sato, M., Komano, H., and Tagawa, S. 2004b. Proton dynamics in chemically amplified electron beam resists. *Jpn. J. Appl. Phys.* 43: L848–L850.

Yamashita, K., Kamimura, S., Takahashi, H., and Nishikawa, N. 2008. A resist material study for LWR and resolution improvement in EUV lithography. *J. Photopolym. Sci. Technol.* 21: 439–442.

Yasin, S., Hasko, D. G., and Ahmed, H. 2001. Fabrication of <5 nm width lines in poly(methylmethacrylate) resist using a water: Isopropyl alcohol developer and ultrasonically-assisted development. *Appl. Phys. Lett.* 78: 2760–2762.

Yoshida, Y., Tagawa, S., Kobayashi, H., and Tabata, Y. 1987. Study of geminate ion recombination in a solute solvent system by using picosecond pulse-radiolysis. *Radiat. Phys. Chem.* 30: 83–87.

Yoshida, Y., Mizutani, Y., Kozawa, T., Saeki, A., Seki, S., Tagawa, S., and Ushida, K. 2001. Development of laser-synchronized picosecond pulse radiolysis system. *Radiat. Phys. Chem.* 60: 313–318.

Zweig, A. 1963. Molecular complexes of methoxybenzenes. *J. Phys. Chem.* 67: 506–508.

27 Radiation Processing of Polymers and Its Applications

Masao Tamada
Japan Atomic Energy Agency
Takasaki, Japan

Yasunari Maekawa
Japan Atomic Energy Agency
Takasaki, Japan

CONTENTS

27.1 BRIEF INTRODUCTION TO BASIC CHEMISTRY OF CROSS-LINKING, GRAFTING, AND DEGRADATION

27.1.1 RADIATION-INDUCED CROSS-LINKING

Cross-linking is the most well-known radiation effect on polymers in industrial applications such as manufacture of rubber tire and covered electric wire. In the case of rubber tire, the cross-linking increases the mechanical strength and reduces the tack strength as pre-cross-linking. Electric wire covered with polyethylene and polyvinylchloride can be improved in terms of thermal resistance. There is a linear relationship between such radiation effects and dose. After cross-linking, viscosity, molecular weight, and branching increase with the degree of resulting cross-linking and these phenomena are induced by a strong interaction of polymer chains.

Cross-linking occurs in the radical reaction of the neighboring molecules. The G-value of cross-linking is estimated by a famous Charlesby–Pinner plot in which gel percentage, molecular weight, and dose are measured (Charlesby and Pinner, 1959). The degree of cross-linking is affected by molecular weight, chemical structure, and irradiation conditions such as temperature and atmosphere. Though the radical number in polymers is determined by dose, the cross-linking effect appears as gelation in proportion of a radical number in weight average molecular weight (Mahmudi et al., 2007). Polymers having a polyethylene-like chemical structure of $-(CH_2-CH_2)-$ are classified as cross-linkable polymers. Vinylidene types, $-(CH_2-CR_2)-$, belong to the group of degradable polymers. Both cross-linking and degradation occurs in the vinyl type, $-(CH_2-CRH)-$. However, polyvinyl chloride is a typical cross-linking polymer that is utilized in the coating of electric wires. In the presence of oxygen gas, however, degradation becomes the main reaction in polyvinylchloride. Polytetrafluoroethylene (PTFE) reduces the mechanical strength by irradiation, but it can be cross-linked at a small range of temperature 13°C higher than melting point (327°C) (Sun et al., 1994; Oshima et al., 1995). Water-soluble polysaccharides such as carboxymethyl cellulose and starch can be cross-linked at a high concentration of aqua solution. The radiation-induced cross-linking of PTFE enhances several polymer properties such as mechanical strength, radiation resistance, heat deflection temperature, and wear resistance.

27.1.2 Radiation-Induced Graft Polymerization

Radiation-induced graft polymerization is a convenient method to impart a desired function into the polymers. The barrier membrane in the button-type battery is synthesized by grafting an ionic monomer into the polyethylene membrane. The chemical filter used in semiconductor factories uses a fibrous adsorbent prepared by the grafting of acidic monomers to remove the trace ammonia released from the human body.

The important factor for graft polymerization is to select a functional monomer and trunk polymer and to optimize the conditions such as dose, solvent, reaction temperature, reaction time, etc. Generally, the trunk polymer is irradiated and grafted in inert nitrogen gas. Various shapes of trunk polymers such as fabric (Lee et al., 2008), film (Choi and Nho, 2000), hollow fiber (Saito et al., 1999), particle, and fiber are available. Grafting needs the chemical propagation reaction of monomer on trunk polymer. There are two methods distinguished by irradiation step in radiation-induced graft polymerization. In the first method, mutual grafting, a trunk polymer is irradiated in a monomer solution. Thus, the process of grafting is quite simple. However, a large amount of homopolymer is produced since the monomer solution is irradiated at the same time. Accordingly, the dose rate affects the monomer utilization to grafting. The grafting rate is limited by monomer diffusion into the trunk polymer. In the second method, preirradiation grafting, the trunk polymer is irradiated in advance and is then put into a monomer solution. The radiation and grafting steps are discrete, so that lesser homopolymer is created. The total grafting yields in both grafting methods are determined by irradiation dose. For an industrial application of grafting, therefore, the preirradiation grafting is generally selected since the cleaning of homopolymer from grafted trunk polymer is an extremely cumbersome process. The industrial gas adsorbent manufacturing process has adopted a continuous process of preirradiation grafting (Fujiwara, 2007).

27.1.3 Radiation-Induced Degradation

Degradation occurs simultaneously during cross-linking. The general classification of cross-linkable and degradable polymers was described in Section 27.1.1. PTFE is classified into typical degradable polymer. The changes of the thermal properties are reviewed in the different dose rate and irradiation atmospheres (Schierholz et al., 1999; Briskman and Tlebaev, 2007). After degradation, polytetrafluoroethylene has been applied as a fine powder used as a chemically and thermally stable lubricant agent for gaskets.

As new applications of degraded polymers are developed, polysaccharides gain importance as an attractive material. Decomposed marine polysaccharides such as alginate and chitosan play the role of plant growth promoters. The radiation degradation of such polysaccharides has been investigated from the viewpoint of solvent (Hien et al., 2000) and mechanism (Wasikiewicz et al., 2005b). The industrialization of degraded polysaccharides has been supported by IAEA as several coordinated research projects (CRP) (Haji-Saeid et al., 2007).

27.2 NEW MATERIAL PRODUCTION USING RADIATION-INDUCED CROSS-LINKING

27.2.1 POLYMER GELS

Polysaccharides and their derivatives were well-known degradation polymers in radiation processing. These materials transformed to hydrogel by adding sodium salt (Mitsumata et al., 2003) and polyamines (Miani et al., 2004). It was found that water-soluble polysaccharide derivatives like carboxymethyl cellulose (CMC) could be cross-linked when irradiated in a highly concentrated solution (paste-like condition), without any additives. CMC is a cellulose derivative that is synthesized by reacting cellulose with chloroacetic acid in alkaline condition. The chemical structure of CMC is a polymer of β-(1-4)-D-glucopyranose, as shown in Figure 27.1. CMC is used as a food additive, for its properties as a viscosity modifier and thickener. It is also a component of many nonfood products such as toothpaste, laxatives, diet pills, water-based paints, and detergents. This is because it has a high viscosity, is nontoxic, and is generally nonallergenic.

The kneading of CMC powders with water at concentrations higher than 10% gives a homogeneous paste-like state. When CMC in the paste-like state is irradiated by electron beams and γ-rays, cross-linking occurs and biodegradable hydrogels are formed (Wach et al., 2003). If the CMC concentration is lesser than 10% and higher than 70%, degradation is preceded and no gel fraction is obtained. This result indicates that the distances between CMC molecules are not enough for the cross-linking reaction at concentrations less than 10%, and CMC is not homogenously distributed in water and some insoluble regions of polymer exists at concentrations higher than 70%.

There are many CMC grades having different degrees of substitution (DS), since the CMC molecule has three hydroxyl parts in one unit of glucopyranose. The gel fractions of cross-linked CMC having different DS are shown in Figure 27.2. A high degree of substitution leads to higher cross-linking at 20% CMC aqua paste. This phenomenon implies that carboxymethyl groups are related to radiation-induced cross-linking. When the obtained hydrogel was soaked in water, swelling reached an equilibrium after soaking for 4h. The swelling, in grams of absorbed solvent per gram of dried gel, was calculated as follows:

$$\text{Degree of swelling} = \frac{(G_{\text{s}} - G_{\text{d}})}{G_{\text{d}}}$$

where G_{s} and G_{d} are the weight of the hydrogel in a swollen and dry state, respectively. The degree of swelling was 400 and 120 in water and NaCl solution, respectively, since ionic strength retarded the swelling of the gel.

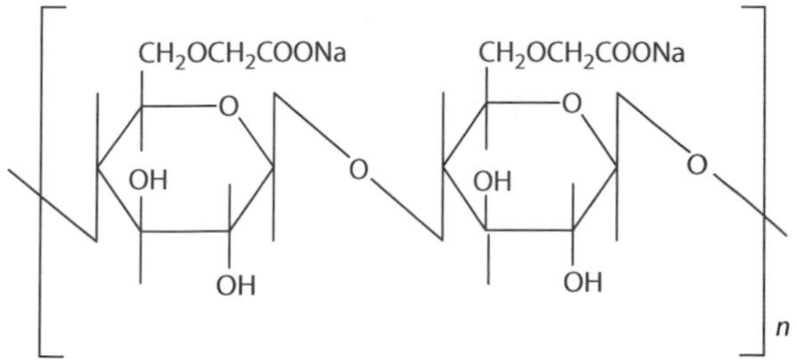

FIGURE 27.1 Chemical structure of CMC.

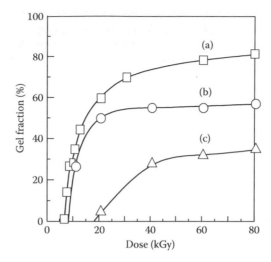

FIGURE 27.2 Gel fraction of cross-linked CMC having different degrees of substitution (DS) at concentration of 20%; (a) DS = 2.2, (b) DS = 1.3, and (c) DS = 0.86. (From Fei, B. et al., *J. App. Polym. Science*, 278, 2000. With permission.)

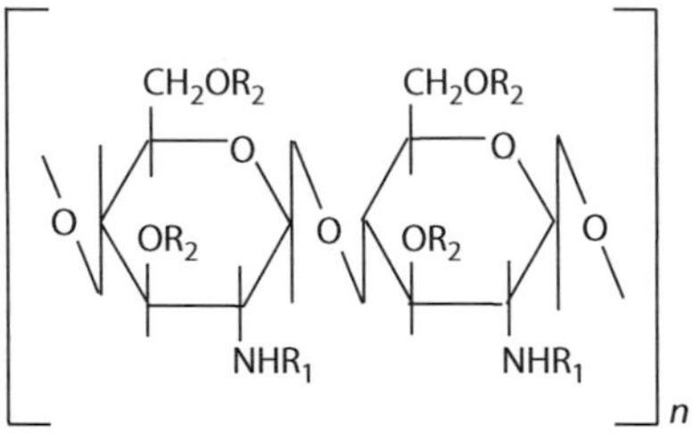

FIGURE 27.3 Chemical structures of carboxymethyl chitin and chitosan; R_1 = H or $COCH_3$ and R_2 = CH_2COONa or H.

Chitin is poly-β-(1-4)-*N*-acetyl-D-glucosamine, which is the main component of the external skeleton in Arthropoda and the periostracum in Mollusca. Its deacetylated derivative is chitosan. Chitin and chitosan are not soluble in water. After carboxymethylation, the resulting carboxymethyl chitin, CMCht, and carboxymethyl chitosan, CMChts, become soluble in water. Their chemical structures are shown in Figure 27.3. The chemically modified CMCht and CMChts can be cross-linked in these paste-like states (Wasikiewicz et al., 2006), though intrinsic chitin and chitosan are decomposed by high energy irradiation. Figure 27.4 shows the effect of concentration on gel fraction. EB irradiation of 50 kGy gave gel fractions of 65% and 40% for CMCht and CMChts, respectively, in the 30%–40% concentration range. At lower and higher concentrations, the gel fraction went down.

In the swelling, the correlations were divided into two categories. In both the cases of CMCht and CMChts, the hydrogels show similar tendencies. There are linear relationships between the gel fraction and swelling of CMCht and CMChts. Though the swelling decreased with an increase of the gel fraction, CMChts hydrogels show lower swelling than CMCht hydrogels at the same gel fraction.

27.2.1.1 Applications of Polymer Gels

Polysaccharides are well-known biodegradable materials. This feature is maintained even after cross-linking with high energy radiation. The obtained gel can be degraded by microorganisms after disposal. This characteristic is useful for environment-friendly products such as bedsore prevention mats (Chmielewski, 2006), coolants, and water absorbents for livestock excrement treatment.

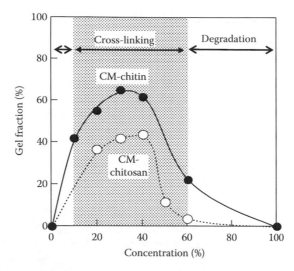

FIGURE 27.4 Effect of concentration on gel fraction in cross-linked CM-chitin and CM-chitosan; CM-chitin and CM-chitosan were irradiated with 50 kGy at 30% aqua paste-like state.

Bedsore prevention requires the bedsore prevention mat to be suitably elastic for promoting blood circulation. CMC hydrogel was applied as bedsore prevention mat, which is used as bed mats during the surgical operation procedures (Figure 27.5). Before the operation is conducted, the mat is pre-heated to body temperature (37°C) using an oven heater. Temperature could be maintained during operation. Clinical tests proved that CMC mat was very effective in the prevention of bedsores. The comparison of blood flow with CMC mat, with those of other mats such as board and general mattress revealed that CMC hydrogel mat reached the highest value. The hydrogel can disperse the body pressure and improve circulation of blood during operation. Thus, it could prevent bedsores in patients.

CMC hydrogel has also been applied as a coolant. This hydrogel coolant has a high capacity to retain low temperatures longer than general liquid coolants. Fish and vegetables could be transported from the production district to distant markets using a hydrogel coolant. This product uses swollen CMC dry gel.

CMC dry gel can absorb water, several hundreds times its own weight. The addition of CMC dry gel controls the water content in livestock excreta and accelerates its composting process. Field heaping and open storage of livestock excreta have been legally prohibited in Japan, since the release of excrement and urine contaminates groundwater and river water. The high water content of around 90% in excrement retards its fermentation to fertilizer. The addition of 0.2% CMC dry gel to the excrement as a water absorbent can reduce the water content by up to 70% and can enhance

FIGURE 27.5 Bedsore prevention mat composed of hydrogel obtained by cross-linking of CMC.

the fermentation. Previously, saw dust was used for this purpose. Usage of CMC dry gel minimized the consumption of saw dust to one-sixth. The advantages of this novel technology are to reduce storage area for saw dust, heavy work, and odor diffusion to environment.

CMCht and CMChts are considered to be the most abundant natural amino polysaccharide and have versatile applications in cosmetics, food, and medical materials (Kumar, 2000). They were used as metal-ion adsorbents as they carry an amine group that has affinity toward metal ions such as copper, zinc, and mercury (Benavente, 2008). Such commercialized, adsorbent resins were generally synthesized by copolymerizing the precursor monomer, chloromethylstyrene, and cross-linker, divinylbenzene. The resulting cross-linked resin has the convertible site of $-CH_2Cl$ in the chloromethylstyrene moiety to the functional group having affinity toward metal ions. This precursor resin can be chemically modified to a metal-ion adsorbent having $-SO_3$, $-COOH$, and $-NH_2$ groups. However, these resins are derived from petroleum products. To secure petroleum resources and decrease environmental burdens, the naturally occurring polymers such as CMCht and CMChts should be used as raw materials for synthesizing metal-ion adsorbents.

The metal adsorption of CMCht and CMChts hydrogels shows the adsorption of various metal ions as shown in Figure 27.6 (Wasikiewicz et al., 2005a). The CMCht hydrogel adsorbed Pd, Au, Cd, V, and Pt in 60 min. In the case of CMChts, Au and Pt were preferentially adsorbed in 120 min. Adsorbed metals on CMCht and CMChts hydrogels can be eluted by dilute hydrochloric acid solution. After disposal of CMCht and CMChts, the size and thickness of the films in soil become smaller. Figure 27.7 shows CMCht and CMChts hydrogel films, 1 mm thick, before and after keeping

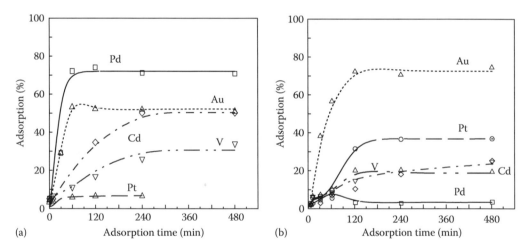

FIGURE 27.6 Metal adsorption of CM-chitin and chitosan gels at pH 3.9; (a) CM-chitin hydrogel and (b) CM-chitosan hydrogel. (Modified from Wasikiewicz, J.M. et al., *Nucl. Instrum. Methods Phys. Res. Sect. B*, 236, 617, 2005a.)

FIGURE 27.7 Degradation of CM-chitin and chitosan gels kept in soil for 10 weeks; (a) CM-chitin hydrogel and (b) CM-chitosan hydrogel. (From Wasikiewick, J.M. et al., *J. Appl. Polym. Sci.*, 102, 758, 2006. With permission.)

in soil for 10 weeks from October to December in Japan. In the case of CMChts, a number of small holes are noticeable. It can be concluded that prolonged storing of these films in the ground would cause their complete disintegration as a result of bacterial activity. Natural polymers including chitin and chitosan are renewable resources and are spontaneously degraded by naturally occurring microorganisms. Such polymers do not increase the burden to the natural environment since they do not produce any toxic waste products during the degradation process.

27.2.2 BIODEGRADABLE PLASTICS

Poly(L-lactic acid), PLA, is a transparent and hard plastic and is produced by condensation polymerization of lactic acid obtained by fermentation of starch (Reddy et al., 2008). In this regard, PLA is a typical renewable plastic. In the near future, nonbiodegradable engineering plastics will be replaced by PLA on the viewpoint of environmental preservation. One of the promising applications of PLA is in thermally molded products such as those used in food packaging, bottles, medical, and pharmaceutical products, which require high thermal stability. However, PLA is thermally deformed at temperatures higher than its glass transition temperature of 60°C, though it has a high melting point of 175°C.

PLA is not cross-linked by irradiation without a certain cross-linker, since PLA is degraded by ionizing radiation. It was found that the addition of polyfunctional monomers (PFM) as cross-linkers to PLA could induce cross-linking (Charlesby, 1981). PFM such as triallyl isocyanurate (TAIC), trimethallyl isocyanurate (TMAIC), trimethylolpropane triacrylate (TMPTA), trimethylolpropane trimethacrylate (TMPTMA), 1,6-hexanediol diacrylate (HDDA), and ethylene glycol bis[pentakis(glycidyl allyl ether)] ether (EG) have been widely used as cross-linkers for polyolefins, owing to their high reactivity against polymer chains.

After screening of various PFM, it was found that TAIC is a suitable cross-linker for PLA. Figure 27.8 shows the effect of dose on the gel fraction of irradiated PLA with 1% and 3% TAIC. The gel fraction of the cross-linked PLA was estimated by the weight of its insoluble part after immersion in chloroform for 48 h. The gel fraction reached 83.3% for 3% of TAIC after irradiation of 50 kGy. TMAIC gave a slightly lower gel fraction. In the cases of TMAPT, TMPTMA, HDDA, and EG, the gel fractions did not reach 30%. TAIC was the most effective cross-linker for PLA since it has three functional groups of C=C and an isocyanuric ring that achieves a greater three-dimensional network by irradiation than that achieved by acrylate-type PFM.

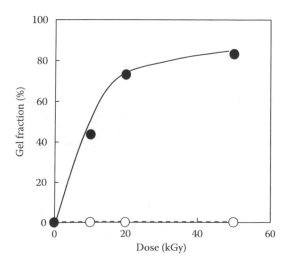

FIGURE 27.8 Effect of dose on cross-linking of PLA with TAIC; (○) 1% TAIC and (●) 3% TAIC. (Modified from Nagasawa, N. et al., *Nucl. Instrum. Methods Phys. Res. Sect. B*, 236, 611, 2005.)

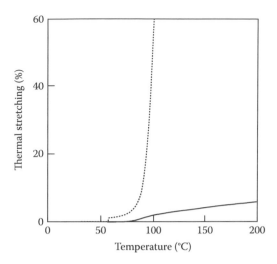

FIGURE 27.9 Thermal deformation of PLA and cross-linked PLA; PLA was kneaded with 3% of TAIC and then irradiated with 50 kGy; broken line is for PLA and dotted line cross-linked PLA. (Modified from Nagasawa, N. et al., *Nucl. Instrum. Methods Phys. Res. Sect. B*, 236, 611, 2005.)

Figure 27.9 shows the thermal deformation of cross-linked PLA with 3% TAIC and PLA without cross-linking (Nagasawa et al., 2005). The thermal deformation was evaluated on PLA films heated from room temperature up to 200°C in a nitrogen atmosphere under a constant load of 0.5 g at a heating rate of 10°C/min. Under thermal stretching, the PLA started to elongate at the glass transition temperature of 60°C, and then thermal stretching reached 60% at 100°C. The cross-linked PLA showed no elongation at glass transition temperature and has low elongation of thermal stretching less than 10% even at 200°C. This result revealed that radiation-induced cross-linking induced the thermal resistance of PLA, high enough for versatile applications. Enzymatic degradation of PLA was evaluated by degradation using proteinase K enzyme from *Tritirachium album*. The weight loss of PLA was 60% after 140 h incubation, whereas the cross-linked PLA with TAIC were degraded 30% in weight losses. After cross-linking, the diffusion of the enzyme is considered to be disturbed by the network structure of PLA. However, enzyme degradability was still maintained at half of intrinsic PLA. This result implies that PLA is an environmentally acceptable material, even after cross-linking.

Cross-linked PLA can be applied to tableware such as cups and plates. PLA containing 3% TAIC was molded to cups and plates by an extruder and then irradiated with 50 kGy to induce cross-linking. The obtained tableware were transparent since PLA is an intrinsically colorless polymer. When boiling water was poured into cups with and without cross-linked PLA, the cup without PLA cross-linking deformed and changed to white as shown in Figure 27.10. But the cross-linked PLA

(a)   (b)

FIGURE 27.10 Effect of cross-linking on thermal property of PLA; (a) PLA and (b) cross-linked PLA. (Modified from Nagasawa, N. et al., *Nucl. Instrum. Methods Phys. Res. Sect. B*, 236, 611, 2005.)

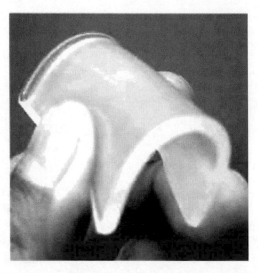

FIGURE 27.11 Elasticity appeared by holding plasticizer in cross-linked PLA. (From Nagasawa, N. et al., High Functionalization and Recycling Technology for Bioplastic NTS Inc. pp. 162–169, 2008. With permission.)

cup maintained its original transparency and shape. The color change from transparent to milky was caused by the crystallization of PLA. After cross-linking, the movement of PLA molecules was restricted by the cross-linking point. Due to this reason, the cross-linked PLA cup could maintain transparency when it was heated close to 100°C.

Cross-linked PLA can be used in heat-shrinkable tubes. The resulting cross-linked PLA tube is expanded diametrically to double its size at 200°C and then holds the expanded shape when it cools down to room temperature. This is a heat-shrinkable tube that thermally shrinks to unexpanded tube size. PLA without cross-linking cannot be expanded at the temperature over its melting point of 160°C. Such heat-shrinkable polyethylene tube have already been commercialized. However, PLA heat-shrinkable tubes have several advantages such as biodegradability, high heat resistance, and transparency.

If the elastic property can be imparted into PLA, new applications such as renewable lapping films and impact absorbing liners can be realized. Plasticizers are used to induce the elastic property into hard plastics like PLA. When PLA is simply mixed with plasticizer, PLA temporarily becomes elastic. However, the plasticizer is gradually expelled from the PLA. In the case of PLA without cross-linking, the plasticizer is not maintained in the PLA for 30 min at 100°C. This is because the PLA crystallizes at this temperate and simultaneously the plasticizer is expelled from PLA. After cross-linking, PLA can contain the plasticizer up to the concentration of 35% and obtains stable elasticity as shown in Figure 27.11.

27.3 NEW MATERIAL PRODUCTION USING RADIATION-INDUCED GRAFTING

27.3.1 GENERAL APPLICATION

The membranes prepared by radiation-induced graft polymerization (radiation-grafted membranes) have been widely applied to separation processes (pervaporation, reverse osmosis, and electrodialysis), electrochemical processes (bottom battery and fuel cells), and adsorbents for metal ions (Figure 27.12) (Saito and Sugo, 2001; Gupta et al., 2004; Nasef and Hegazy, 2004). These membranes, in general, consist of similar components, hydrophilic grafted polymers (grafts) attached into hydrophobic trunk polymers (substrates), namely, the so-called anion-exchange membrane structures. The membranes containing the grafts with acids such as carboxylic, phosphoric, and sulfonic acids and those with bases such as amino groups and ammonium salts are called "cation- and anion-exchange membranes," respectively. Hydrophobic polymer substrates can be classified as

Separation process
- Pervaporation (alcohol, organic chemicals from water)
- Reverse osmosis
- Metal adsorbents (hazardous metal separation or precious metal correction)

Electrochemical processes
- Electrodialysis (production of water, NaCl)
- Chloroalkali (production of NaOH, Cl_2)

Battery/energy system
- Bottom battery
- Polymer electrolytes for fuel cell

FIGURE 27.12 Application fields of radiation-grafted membranes.

conventional polyolefin films (polyethylene (PE) and polypropylene (PP)), perfluoro- and partially fluorinated-polymer films (poly(tetrafluoroethylene)) (PTFE), poly(tetrafluoroethylene-*co*-hexafluoropropylene) (FEP), poly(tetrafluoroethylene-*co*-perfluorovinyl ether) (PFA), poly(vinylidene fluoride) (PVDF) and poly(ethylene-*co*-tetrafluoroethylene) (ETFE), and natural polymer resins (wool, cotton, and paper), as shown in Figure 27.13. To the above substrates, acrylic acid (AAc), methyl methacrylate (MMA), styrene derivatives, vinyl pyridine, and the corresponding salts are introduced as hydrophilic graft polymers (grafts), and subsequent chemical transformation results in radiation-grafted membranes.

The grafting of poly(copper acrylate), poly(vinylpyridinium salts), and phosphonium salts gave antibacterial natural polymers. The grafted membranes, consisting of poly(AAc) incorporated into

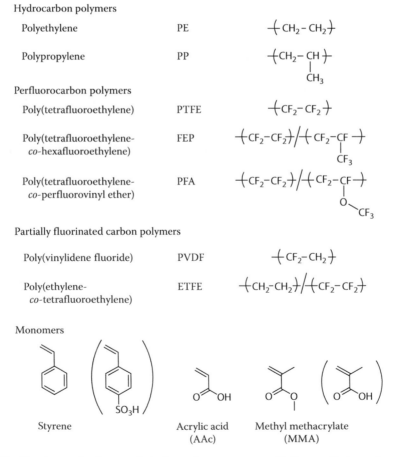

FIGURE 27.13 Structures of polymer substrates and monomers constituting grafted membranes.

PE, can be applied to pervaporation membranes to remove only water from aqueous alcohol solutions owing to higher affinity of the poly(AAc) grafts to water than alcohols (Gupta et al., 2004). The membranes having carboxylic, sulfonic, and phosphoric acids in the grafts into PE can act as metal-ion adsorbents owing to the high affinity of the grafts to the metal ions in aqueous solutions (Saito and Sugo, 2001). Furthermore, the radiation-grafted membranes, possessing poly(AAc) grafts into PE can be commercialized as a battery separator of bottom cells because only hydroxide ions, but not metal ions $(Ag(OH)_2^-)$ can pass through the membranes (Ishigaki et al., 1982a,b; Hsiue and Huang, 1985).

After the successful utilization of radiation-grafted membranes in battery cells, the grafted membranes consisting of fluorinated-polymer substrates, having higher mechanical and thermal strengths, and the grafts, having more acidic sulfonic acid groups, have been developed for polymer electrolyte membranes (PEM) for fuel cells, which require higher conductivity and durability at higher temperatures. Sections 27.3.2 and 27.3.3 focus on recent advancements of radiation-grafted membranes as metal adsorbents to recover precious metal ions and in battery applications such as a bottom cells and fuel cells.

27.3.2 Fibrous Metal-Ion Adsorbents

To synthesize the metal adsorbent, the functional groups having strong affinity toward metal ions should be imparted into these trunk polymers by grafting, as shown in Figure 27.14. Other functional groups were reviewed by Smith and Alexangratos (2000). When a monomer has a chelating group in its side chain, the metal adsorbent is directly synthesized only by grafting. In the case of a monomer having a precursor for coordination, chemical modification is necessary after its grafting (Seko et al., 2005). Table 27.1 lists the representative functional group for metal adsorption, corresponding grafting monomer, and chemical reagent (Basuki et al., 2003).

GMA is a useful monomer for the precursor of metal adsorbents. The grafting of GMA generally is carried out using organic solvents such as methanol (Kavalla et al., 2004) and dimethyl sulfoxide (Aoki et al., 2001). It was found that grafting yield was dramatically enhanced when GMA was emulsified by surfactants in water instead of organic solvents (Seko et al., 2007). This aqueous GMA emulsion grafting on PE fibers gave a degree of grafting of 130% in preirradiation conditions of 10 kGy, grafting temperature of 40°C, and grafting time of 2 h.

Nonwoven fabric made of polyethylene is used as a trunk polymer. This is because the resulting fabric adsorbent causes swift adsorption of metal ions and ensures easy handling in the adsorption process. For example, the metal adsorbed in the fabric adsorbent can be picked up from the solution by forceps after it is dipped into the metal solution for a few minutes. Additionally, a conventional adsorbent resin is used in a volume-normalized flow rate, namely space velocity, around $10 h^{-1}$. At flow rates far more than $10 h^{-1}$, the adsorption capacity dramatically decreases in a column mode adsorption. However, the adsorbent fabric can be used in the space velocity more than $1000 h^{-1}$ (Jyo et al., 2003).

As applications of metal adsorbents, toxic and rare metals were attempted to be collected for environmental preservation and metal resource security, respectively. During scallop processing, the midgut gland is discarded since this part contains 20–40 ppm of cadmium. The discarded mid-gut gland

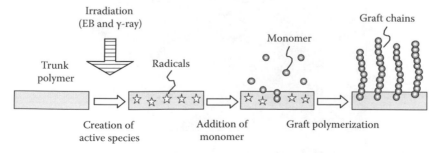

FIGURE 27.14 Schematic diagram of radiation-induced graft polymerization.

TABLE 27.1

Functional Groups Imparted by Graft Polymerization, Grafting Monomer and Chemical Reagents for Synthesis of Metal Adsorbents

Functional Group	Grafting Monomer	Chemical Reagent	References
Amidoxime	Acrylonitrile	Hydoxylamine	Seko et al. (2005)
Amines	Glycidyl methacrylate	Diethylamine	Kabay et al. (1993)
	N-Vinyl formamide	Sodium hydroxide	Abrol et al. (2007)
Iminodiethanol	Glycidyl methacrylate	2,2′-Iminodiethanol	Awual et al. (2008)
Iminodiacetic acid	Glycidyl methacrylate	Sodium iminodiacetate	Ozawa et al. (2000)
Glucamine	Glycidyl methacrylate	N-Methyl glucamine	Choi and Nho (1999)
Sulfonic acid	Styrene	Sulfuric acid	Hoshina et al. (2007)
	Glycidyl methacrylate	Sodium sulfite	Vahdat et al. (2007)
Phosphoric acid	Glycidyl methacrylate	Phosphoric acid	Kim and Saito (2000)
	2-Hydroxyethyl methacrylate phosphoric acid		Lee et al. (2002)

Source: Tamada, M. K., *Kobunshi (High Polymers)*, 58, 397, 2009.

is incinerated even though it contains a large amount of nutrients such as straight-chain fatty acids and fat. The mid-gut gland is treated with amidoxime adsorbent (Shiraishi et al., 2003). The bench scale plants for removal of cadmium from mid-gut glands was operated using iminodiacetic acid adsorbent and cadmium-free mid-gut gland could be used as fertilizer and animal feed. Iminodiacetic acid adsorbent was used in the bench scale plant because the pH of the eluent was 3. In this pH range, the iminodiacetic acid adsorbent has a higher selectivity to cadmium than the amidoxime adsorbent.

Uranium collection from seawater has been researched from the viewpoint of resource security of atomic power generation. Amidoxime-type adsorbents synthesized by graft polymerization can adsorb uranium occurring in an extremely low concentration, 3 ppb, in seawater with the coexistence of sodium and magnesium ions.

The performance of uranium adsorbent fabrics was evaluated using 350 kg adsorbent stacks by soaking in the ocean in the offing of Mutsu-Sekine, Aomori, Japan, as shown in Figure 27.15.

FIGURE 27.15   Collection process of uranium adsorbent stacks at marine experiment. (From Seko, N. et al., *Nucl. Technol.*, 144, 274, 2003. With permission.)

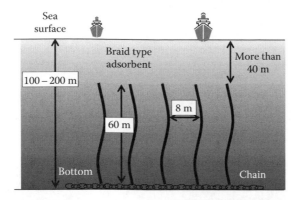

FIGURE 27.16   Mooring state of braid adsorbent for practical system of uranium collection. (From Tamada, M. et al., *Trans. At. Energy Soc. Jpn.*, 5, 358, 2006. With permission.)

One kilogram of uranium was successfully collected in 3 years' marine experiment in the form of a yellow cake (Seko et al., 2003). To make uranium collection from seawater cost effective, a braid-type adsorbent was developed. The mooring system for the braid-type adsorbent does not need the float on the sea surface and a heavy adsorbent bed. The braid-type adsorbent is fabricated by braiding the amidoxime adsorbent fibers. The recovery cost of uranium from seawater was evaluated on the basis of the recovery system of the braid-type adsorbent. Figure 27.16 shows the assembly of braid-type adsorbent for a supposed annual collection of 1200 t uranium. When a performance of 4 g of uranium/kg of adsorbent, which can be reused 18 or more times is achieved, the uranium cost reduces to 25,000 ¥/kg of uranium (96 $/lb-$U_3O_8$) (Tamada et al., 2006). The 4 g uranium/kg of adsorbent is the most promising performance of uranium adsorbent in marine experiments.

27.3.3   POLYMER ELECTROLYTE MEMBRANES FOR BATTERY AND FUEL CELLS

27.3.3.1   AAc-Grafted PE: Bottom Battery

The separator membranes in compact-bottom-type silver oxide cells were previously made of regenerated cellulose and porous PP films. However, the conventional silver oxide cells had drawbacks, so that silver hydroxide ion ($Ag(OH)_2^-$) generated as a by-product on anode passes through the separator and was reduced to precipitated metal Ag, resulting in self-discharge and power deterioration. To solve the above problems, radiation grafting was applied to the separation membranes; here, poly(acrylic acid) (poly(AAc)) was grafted onto PE, which is very cheap, good film forming, and chemically stable (Ishigaki et al., 1982a,b; Hsiue and Huang, 1985).

Hsiue et al. reported that the specific resistivity of the films rapidly decreased with increasing grafting degrees up to 20% and reached 50 Ω cm ($\sigma = 0.02$ S/cm) (Hsiue and Huang, 1985). In real production processes, PE roll films are irradiated with an EB accelerator to generate radicals and subsequently immersed in an aqueous AAc monomer solution to introduce poly(AAc) grafts homogeneously through the PE films (Figure 27.17). The obtained membranes act as high performance battery separators because only hydroxide ions (OH$^-$) but not by-product $Ag(OH)_2^-$ can pass through the grafted membranes, which suppress self-discharge, resulting in the higher shelf life. To develop the radiation-grafting technique, the bottom-type battery cells prepared using radiation grafting can be used for a long time and in a stable manner, and the above problem can be solved. Thus, the AAc-grafted PE achieved 100% of market share in Japan.

27.3.3.2   Fluoropolymers for Fuel Cell PEM

As a result of the successful application of the poly(AAc)-grafted PE to bottom cells, radiation grafting has been applied to the preparation of PEM for hydrogen type (polymer electrolyte fuel cell (PEFC)) and direct methanol type fuel cell (DMFC) in the last 10–15 years (Table 27.2).

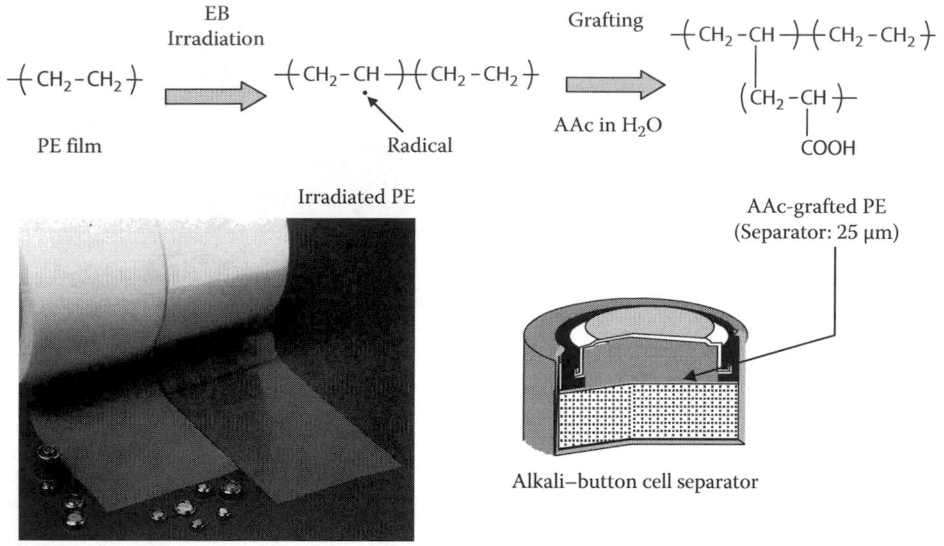

FIGURE 27.17 Preparation of AAc-grafted PE for bottom cell separator (roll film).

TABLE 27.2
Type and Characteristics of Fuel Cells

FC Type	PEFC	DMFC
Fuel	Hydrogen	Methanol
Application	Fuel cell vehicle	Mobile phone
	Residential FC	Note PC
Temperature	70°C–130°C	r.t. –50°C

PEFC has been gradually commercialized for residential cogeneration systems and is expected to be utilized in fuel cell hybrid vehicles (FCHV), while DMFC has been announced to be commercialized for mobile phone batteries and notebook type PCs from several companies. In order to apply radiation-grafted membranes to fuel cell PEM, the membranes must have higher acidic groups than carboxylic acid, to show higher proton (ion) conductivity, which is directly related to higher fuel cell power density and possess size stability in fuels (water and aqueous methanol) (Figure 27.18). Thus, recently, there have been many attempts for fuel cell PEM prepared by grafting styrene derivatives into fluorinated-polymer substrates such as cross-linked PTFE (cPTFE), FEP, ETFE, and PVDF (Figure 27.19) (Gubler et al., 2005b; Gursel et al., 2008). These membranes have been examined as electrolyte membranes for both PEFC and DMFC, for which the membranes must be durable up to at least 60°C for DMFC and 80°C for PEFC.

27.3.3.2.1　Radiation Grafting into Fluoropolymers

The first radiation-induced grafting of fluorinated-polymer substrates had been reported in 1959 and summarized in a review in 1962 by Chapiro (Chapiro, 1959, 1962). He reported that styrene and MMA could be grafted into the deep inside of PTFE (Teflon) films by simultaneous irradiation with a low dose rate. Since Teflon substrates are not swelled in the monomer solutions at all, the author proposed the so-called grafting front mechanism. The graft polymerization of styrene commences at the surface region of Teflon films to give a grafting layer consisting of polystyrene. The newly generated grafting layer, in which propagation of styrene polymerization continues, expands and induces deformation of polymer chains of Teflon. As a result, the stain between grafts

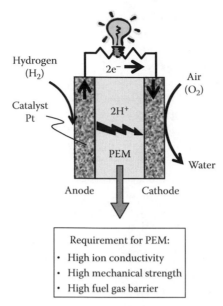

FIGURE 27.18 Schematics of fuel cells.

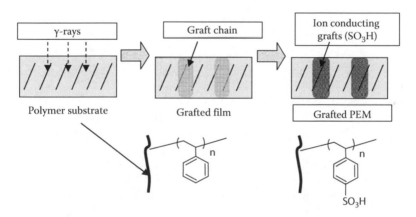

FIGURE 27.19 Preparation scheme of radiation-grafted PEM for fuel cells.

and substrate interfaces due to the deformation of Teflon polymer chains leads to further advance of monomers into the inner part of Teflon.

Later, Hegazy et al. reported that acrylic acid could be grafted into Teflon (Hegazy et al., 1981a,b). The AAc-grafted PTFE films with grafting degrees above 30% showed high ion conductivity ($>10^{-2}$ S/cm). They confirmed that the films with high conductivity have a homogeneous distribution of poly(AAc) grafts in Teflon substrates. Bozzi et al. had also reported the radiation grafting of AAc into FEP (Bozzi and Chapiro, 1988). However, there had been no reports from the viewpoint of fuel cell applications at that time.

27.3.3.2.2 PEM Consisting of Perfluorinated-Polymer Substrates

The preparation of polymer electrolyte membranes using radiation-induced grafting and their evaluation as a fuel cell membrane was reported by the group of PSI in 1993 (Gupta et al., 1993b; Rouilly et al., 1993). Later, there have been many reports of the evaluation of PEFC (hydrogen fuel type) and DMFC (methanol fuel type) performance of the single fuel cells using the radiation-grafted membranes. They prepared the PEM by radiation-grafting of styrene into FEP substrates with thickness of mainly 25 and 50 μm and successive sulfonation of the polystyrene grafts to obtain the PEM consisting of poly(styrenesulfonic acid) (PSSA) grafted into FEP. They precisely investigated the preparation conditions such as absorbed doses, grafting solvents, and the electrolyte, chemical, and thermal properties (Gupta et al., 1993a, 1994; Gupta and Scherer, 1994). The advantages of this method, they claimed, are the low cost and the simple procedures of production processes compared with fully fluorinated Nafion® membranes. Even with grafting degrees less than 20%, the FEP-based PEM showed a higher ion conductivity than that of Nafion.

The fuel cell performances of the MEA consisting of the FEP-based PEM were also investigated by the same group (Büchi et al., 1995a,b; Felix et al., 1995). The radiation-grafted membrane consisting of PSSA with a grafting degree of 18% and divinylbenzene as a cross-linker (10 wt% of monomers) exhibited a fuel cell performance similar to the one while using same operation conditions using Nafion. The structure and amount effects of cross-linkers on the PEM properties were carefully examined; they concluded that the usage of TAC with DVB improved the durability to be stable for 1400 h in the PEFC operation at 80°C. They recently reported that the FEP-based PEM with 25 μm thickness with a DVB cross-linker showed a stable operation more than 2500 h at 60°C–80°C.

Even though PTFE (Teflon) based PEM had been prepared by the radiation-grafting of styrene and the subsequent sulfonation of the polystyrene grafts, the PTFE-based PEM showed quite high swelling in water and aqueous methanol, mainly because of the radiation sensitive (degradable) property of a PTFE substrate. Consequently, it was difficult to give high conductivity by increasing IEC (namely, with high grafting degrees) of PTFE-based PEM. On the other hand, it was reported that PTFE films were cross-linked by γ-rays or EB at temperatures higher than its melting point (320°C), while below it, PTFE is well known as a typical radiation degradable polymer. After the new findings, two groups independently reported the PEM consisting of cross-linked PTFE (cPTFE) substrate and PSSA grafts with cross-linkers (Figure 27.20) (Sato et al., 2003; Yamaki et al., 2003).

The cPTFE membranes showed several excellent polymer properties such as mechanical strength, radiation resistance, heat deflection temperature, and wear resistance. Thus, cPTFE films kept their mechanical strength after irradiation for grafting even with quite high doses, resulting in cPTFE-based PEM with higher grafting degrees, namely, higher conductivity ($\sigma = 2.5$ S/cm) and IEC (3.0 mmol/g). The cPTFE-based PEM is dimensionally quite stable in bulk methanol and aqueous methanol solutions owing to the cross-linking structure. Furthermore, recently, the PEM was revealed to have quite low permeability to methanol and water with superior proton conductivity compared with Nafion. Thus, the cPTFE-based PEM should be promising, especially in DMFC applications (Sawada et al., 2008). The PEFC performance of the cPTFE-based PEM have been examined. The MEA of the cPTFE-based PEM with higher grafting degrees (higher IEC) showed a similar FC performance compared to that of Nafion-based MEA. After optimization of structures and amounts of cross-linkers, the MEA maintained stable operation for several hundred hours at 80°C (Li et al., 2006).

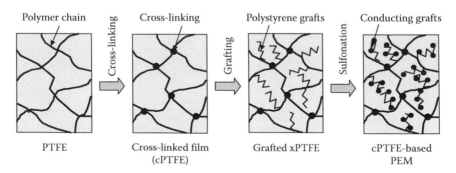

FIGURE 27.20 Preparation and grafting of cPTFE by radiation technique.

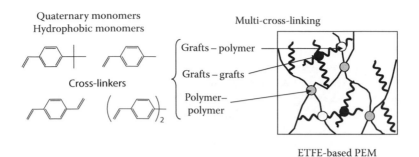

ETFE-based PEM

FIGURE 27.21 Preparation of ETFE-based PEM for DMFC.

27.3.3.2.3 PEM Consisting of Partially Fluorinated-Polymer Substrates

The group of PSI had reported the preparation of PEM with ETFE as a substrate using the same technique and compared them with the FEP-based PEM (Brack and Scherer, 1997; Gubler et al., 2005a). The ETFE-based PEM showed similar conductivity with grafting degrees of 20%–30% and was superior to the FEP-based PEM in terms of mechanical strength. Below 20%, the grafting in the middle of the grafted film was not sufficient; above 30%, the grafted film became brittle and had lesser mechanical strength. The ETFE-based PEM consisting of 10% DVB showed good chemical and physical durability; the MEA consisting of the above PEM showed stable operation for 770 h at 80°C with fully humidified hydrogen fuel in a single fuel cell device.

Recently, Yoshida et al. had precisely evaluated the effects of graft monomers and cross-linking procedures on PEM properties (Figure 27.21) (Chen et al., 2006a,b). They clearly showed that graft monomer structures, cross-linker structures, and radiation-induced cross-linking procedures were important parameters for oxidative stability of the PEM (3% H_2O_2 at 60°C), which was one of the most distinct characteristics to evaluate PEM stability in fuel cell operation. Compared with the PSSA grafts, the ETFE-based PEM with the co-graft polymers, poly(methyl styrene–co-t-butylstyrene sulfonic acid), showed four times longer durability periods. They introduced 1,2-distyrylethane (DVPE) as a flexible cross-linker and the small amount of DVPE with DVB enhanced the durability of the PEM in the oxidative condition. Furthermore, the post radiation-cross-linking after grafting of styrene monomers, which induced multi-cross-links of substrate–substrate, graft–graft, and substrate–graft, possessed three times longer durability than the PEM without the post cross-linking. Accordingly, they concluded that the ETFE-based PEM with hydrophobic styrene monomers and a small amount of cross-linkers with post-cross-linking process exhibit quite high DMFC performance with high oxidative stability.

27.3.3.3 Aromatic Hydrocarbon Polymers for Fuel Cell PEM

27.3.3.3.1 Poly(Ether Ether Ketone)-Based PEM

Fluorinated PEM such as Nafion-type and Even graft-type PEM have drawbacks such as low mechanical strength at higher operating temperatures and unsatisfactory gas barrier properties. In this respect, the sulfonated form of aromatic hydrocarbon polymers, so-called "super engineering plastics" such as poly(ether ether ketone) (PEEK), polyimide (PI), and poly(sulfone) (PSU), have been given attention owing to their excellent mechanical properties at elevated temperature as well as excellent barrier properties against fuels (methanol, H_2) and oxygen. Thus, the aromatic polymer-based PEM are actively investigated for hydrogen fuel type fuel cells (PEFC), which are utilized in residential cogeneration system fuel cells and fuel cell hybrid vehicles (FCHV).

The super engineering plastic films also have high chemical resistance; thus it is difficult to introduce grafting monomers into the films. Recently, we had reported that the grafting of styrene into the PEEK film pre-irradiated with 30 kGy was accelerated in 1-propanol as a grafting solvent at 80°C to obtain the styrene-grafted PEEK with grafting degrees more than 60% (Hasegawa et al., 2008).

However, the sulfonation of the polystyrene grafts with chlorosulfonic acid in a dichloroethane solution proceeded not only at the grafts layers but also at the crystalline and amorphous phases of PEEK. In consequence, the PEEK-based PEM with conductivity of more than 0.01 S/cm are subject to severe deformation due to higher water uptake.

To prevent the damage of a PEEK substrate during sulfonation process, the graft polymerization of sulfo-containing styrene, ethyl 4-styrenesulfonate (E4S), into two PEEK substrates with high and low crystallinity, h-PEEK and l-PEEK (degree of crystallinity: 32% and 11%), was examined via preirradiation and post grafting method because the obtained grafted PEEK films should be converted to PEEK-based PEM only by hydrolysis of the sulfonic acid ester containing precursor grafts. Graft polymerization of E4S hardly proceeded into h-PEEK, whereas, it gradually proceeded into l-PEEK with grafting degrees of more than 50% at 80°C for 72 h. The PEEK-based PEM with 0.08 S/cm and water uptake of <50% can be obtained by aqueous hydrolysis of grafted films and exhibited mechanical strength (95 MPa) of 62% of original PEEK substrates.

We found that a small amount (<10 wt%) of divinylbenzene (DVB) could be introduced into l-PEEK via thermal grafting. Surprisingly, the introduction of DVB via thermal pre-grafting enhanced the grafting rate of E4S, which is a precursor of poly(styrenesulfonic acid), in radiation-induced grafting; thus, the grafting degrees of DVB-PEEK reached 100% and above at 80°C for 10 h. As a consequence of the enhancement of grafting rates, the DBV-PEEK-based PEM, after hydrolysis, showed quite high conductivity (0.12 S/cm) with low water uptake (<50%) (Figure 27.22) (Chen et al., 2008). In comparison with the commercial Nafion, the obtained PEEK-based PEM possesses superior proton conductivity and mechanical property at 80°C under high relative humidity of 100 RH%. Here, the grafted layers introduced by thermal treatment enhanced the radiation-induced grafting to give the electrolyte membrane exhibiting 1.5 times higher conductivity and 2.3 times higher mechanical strength, compared with conventional fluorinated electrolyte membranes.

The single fuel cell device consisting of the developed electrolyte membrane maintained stable operation over 1000 h at 95°C under relative humidity of 80% RH with pure hydrogen and oxygen gases (Chen et al., 2009). Furthermore, under severe conditions (95°C, 40% RH), in which Nafion obviously degraded, these membranes can maintain their output voltages for more than 250 h. Thus, these membranes can be applied to fuel cell vehicles, which should contribute to solve the current environmental problem, judging from the fact that no appreciable degradation of these membranes occurs even under the operating conditions of lower humidity (Figure 27.23).

27.3.3.3.2 Other Aromatic Hydrocarbon PEM

Recently, we reported the preparation of polyamide-based PEM by radiation grafting of sodium styrenesulfonate (3S) in dimethylsulfoxide (Li et al., 2009). There have been many attempts to use a 3S monomer as a grafting monomer because of no extra treatment to obtain sulfonic acid groups after grafting is required and also due to the relatively low cost. However, a 3S monomer is soluble

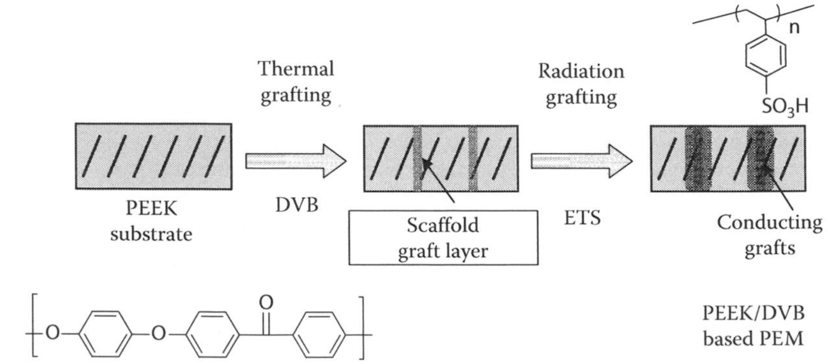

FIGURE 27.22 Thermal/radiation-induced two-step grafting using PEEK.

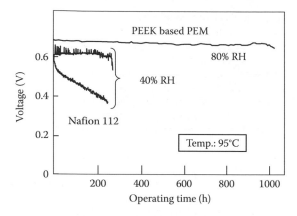

FIGURE 27.23 Fuel cell performance of PEEK-based PEM.

in aqueous solvents. Thus, radiation grafting of 3S into hydrophobic polymer substrates hardly proceeds, probably due to less swelling of the substrates in aqueous media. However, when dimethylsulfoxide is utilized as a grafting solvent, the monomer is completely dissolved in the solvent and the polyamide films are slightly swelled in the 50% monomer solution. An aromatic polyamide, poly(m-xylene adipamide) (Nylon-MXD6), films were pre-irradiated with 60 kGy under argon atmosphere and then immersed in DMSO at 60°C for 10 h to give the 3S-grafted Nylon-MXD6 film with a grafting degree of 49%. The obtained PEM showed conductivity of 0.083 S/cm with water uptake of 70%.

The alicyclic polyimide (A-PI) film, consisting of aromatic dianhydride and alicyclic diamines, was utilized as a substrate of radiation-induced polymerization to prepare a PI-based PEM for fuel cells. We successfully prepared the graft-type PEM by radiation-induced grafting of styrene into A-PI films and subsequent sulfonation (A-PI-PEM). The obtained A-PI-PEM has higher ion conductivity and mechanical and thermal properties compared with conventional PEM (Nafion) (Park et al., 2009). However, these PEM consisting of polyamide and polyimide substrates are not stable in aqueous media at temperatures higher than 80°C. In order to utilize the polycondensation type aromatic polymer substrates, we must introduce less hydrolyzed structures in the main chains.

27.4 NEW MATERIAL PRODUCTION USING RADIATION-INDUCED DEGRADATION

27.4.1 PLANT GROWTH PROMOTER

Oligosaccharides obtained by radiation-induced degradation stimulate plant growth in shoot elongation and productivity. Oligosaccharides can be degraded by acid hydrolysis (Holtan et al., 2006) and enzymatic reaction (Zhang et al., 2004). However, the plant growth promoter prepared by radiation degradation shows higher activity than that obtained by acid hydrolysis (Luan et al., 2006). This is because the oligosaccharide molecular weight from 1000 to 3000 is produced at a higher yield by radiation degradation than by acid hydrolysis. This component has the highest activity against the plant growth promoter (Matsuhashi and Kume, 1997). Oligosaccharides such as radiation-degraded alginate, chitosan, and carrageenan have been tested on the growth of crops and vegetables.

REFERENCES

Abrol, K., Qazi, G. N., and Ghosh, A. K. 2007. Characterization of an anion-exchange porous polypropylene hollow fiber membrane for immobilization of ABL lipase. *J. Biotechnol.* 128: 838–848.

Aoki, S., Saito, K., Jyo, A., Katakai, A., and Sugo, T. 2001. Phosphoric acid fiber for extremely rapid elimination of heavy metal ions from water. *Anal. Sci.* 17(Suppl.): i205–i208.

Awual, Md. R., Urata, S., Jyo, A., Tamada, M., and Katakai, A. 2008. Arsenate removal from water by a weak-base anion exchange fibrous adsorbent. *Water Res.* 42: 689–696.

Basuki, F., Seko, N., Tamada, M., Sugo, T., and Kume, T. 2003. Direct synthesis of adsorbent having phosphoric acid with radiation induced graft polymerization. *J. Ion Exchange* 14 (Suppl.): 209–212.

Benavente, M. 2008. Adsorption of metallic ions onto chitosan: Equilibrium and kinetic studies, Licentiate thesis, Royal Institute of Technology, Stockholm, Sweden.

Bozzi, A. and Chapiro, A. 1988. Synthesis of perm-selective membranes by grafting acrylic acid into air-irradiated Teflon-FEP films. *Radiat. Phys. Chem.* 32: 193–196.

Brack, H.-P. and Scherer, G. G. 1997. Modification and characterization of thin polymer films for electrochemical applications. *Macromol. Symp.* 126: 25–49.

Briskman, B. A. and Tlebaev, K. B. 2007. Radiation effects on thermal properties of polymers. II. Polytetrafluoroethylene. *High Perform. Polym.* 20: 86–114.

Büchi, F. N., Gupta, B., Haas, O., and Shcerer, G. G. 1995a. Study of radiation-grafted FEP-g-polystyrene membranes as polymer electrolytes in fuel cells. *Electrochim. Acta* 40: 345–353.

Büchi, F. N., Gupta, B., Haas, O., and Shcerer, G. G. 1995b. Performance of differently cross-linked, partially fluorinated proton exchange membranes in polymer electrolyte fuel cells. *J. Electrochem. Soc.* 142: 3044–3048.

Chapiro, A. 1959. Preparation des copolymers greffes du polytetrafluoroethylene (Teflon) par vie radiochimique. *J. Polym. Sci.* 34: 481–501.

Chapiro, A. 1962. *Radiation Chemistry of Polymeric Systems*, Chapter XII. New York: Interscience Publishers, John Wiley & Sons.

Charlesby, A. 1981. Crosslinking and degradation of polymers. *Radiat. Phys. Chem.* 18: 59–66.

Charlesby, A. and Pinner, S. H. 1959. Analysis of the solubility behaviour of irradiated polyethylene and other polymers. *Proc. Roy. Soc.* A249: 367–386.

Chen, J. H., Asano, M., Yamaki, T., and Yoshida, M. 2006a. Chemical and radiation crosslinked polymer electrolyte membranes prepared from radiation-grafted ETFE films for DMFC applications. *J. Power Sources* 158 (1): 69–77.

Chen, J. H., Asano, M., Yamaki, T., and Yoshida, M. 2006b. Improvement of chemical stability of polymer electrolyte fuel cell membranes by grafting of new substituted styrene monomers into ETFE films. *J. Mater. Sci.* 41 (4): 1289–1292.

Chen, J., Asano, M., Maekawa, Y., and Yoshida, M. 2008. Fuel cell performance of polyetheretherketone-based polymer electrolyte membranes prepared by a two-step grafting method. *J. Membr. Sci.* 319 (1–2): 1–4.

Chen, J., Zhai, M., Asano, M., Huang, L., and Maekawa, Y. 2009. Long-term performance of polyetheretherketone-based polymer electrolyte membrane in fuel cells at 95°C. *J. Mater. Sci.* 44: 3674–3681.

Chmielewski, A. G. 2006. Worldwide developments in the field of radiation processing of materials in the down of 21st century, *Nukleonika* 51 (1): S3–S9.

Choi, S. and Nho, Y. 1999. Adsorption of Co^{2+} and Cs^{1+} on polyethylene membrane with iminodiacetic acid and sulfonic acid modified by radiation-induced graft copolymerization. *J. Appl. Polym. Sci.* 71: 999–1006.

Choi, S. H. and Nho, Y. C. 2000. Radiation-induced graft polymerization of binary monomer mixture containing acrylonitrile onto polyethylene films. *Radiat. Phys. Chem.* 58: 157–168.

Felix, N., Büchi, F. N., Marek, A., and Scherer, G. G. 1995. In situ membrane resistance measurements in polymer electrolyte fuel cells by fast auxiliary current pulses. *J. Electrochem. Soc.* 142: 1895–1901.

Fujiwara, K. 2007. Separation functional fibers by radiation induced graft polymerization and application. *Nucl. Instrum. Methods Phys. Res., Sect. B* 265: 517–525.

Gubler, L., Prost, N., Gursel, S. A., and Scherer, G. G. 2005a. Proton exchange membranes prepared by radiation grafting of styrene/divinylbenzene onto poly(ethylene-alt-tetrafluoroethylene) for low temperature fuel cells. *Solid State Ionics* 176: 2849–2860.

Gubler, L., Gursel, S. A., and Scherer, G. G. 2005b. Radiation grafted membranes for polymer electrolyte fuel cells. *Fuel Cells* 5 (3): 317–335.

Gupta, B. and Scherer, G. G. 1994. Cation exchange membranes by pre-radiation grafting of styrene onto FEP films. *J. Polym. Sci., Polym. Chem.* 32: 1931–1938.

Gupta, B., Büchi, F. N., and Scherer, G. G. 1993a. Proton exchange membranes prepared by radiation grafting of styrene onto FEP films. I. Thermal characteristics of copolymer membranes. *J. Appl. Polym. Sci.* 50: 2129–2134.

Gupta, B., Büchi, F. N., Scherer, G. G., and Chapiro, A. 1993b. Materials research aspects of organic solid proton conductors. *Solid State Ionics* 61 (1–3): 213–218.

Gupta, B., Highfield, J. G., and Scherer, G. G. 1994. Proton exchange membranes prepared by radiation grafting of styrene onto FEP films. II. Mechanical of thermal degradation in copolymer membranes. *J. Appl. Polym. Sci.* 51: 1659–1666.

Gupta, B., Anjum, N., Jain, R., Revagade, N., and Singh, H. 2004. Development of membranes by radiation-induced graft polymerization of monomers onto polyethylene films. *J. Macromol. Sci., Polym. Rev.* C44 (3): 275–309.

Gursel, S. A., Gubler, L., Gupta, B., and Shcerer, G. G. 2008. Radiation grafted membranes. *Adv. Polym. Sci.* 215: 157–217.

Haji-Saeid, M., Sampa, M. H., Ramamoorthy, N., Güven, O., and Chmielewski, A. G. 2007. The role of IAEA in coordinating research and transferring technology in radiation chemistry and processing of polymers. *Nucl. Instrum. Methods Phys. Res., Sect. B* 265: 51–57.

Hasegawa, S., Suzuki, Y., and Maekawa, Y. 2008. Preparation of poly(ether ether ketone)-based polymer electrolytes for fuel cell membranes using grafting technique. *Radiat. Phys. Chem.* 77 (5): 617–621.

Hegazy, E. A., Ishigaki, I., and Okamoto, J. 1981a. Radiation grafting of acrylic acid onto fluorine-containing polymers. I. Kinetic study of preirradiation grafting onto poly(tetrafluoroethylene). *J. Appl. Polym. Sci.* 26 (9): 3117–3124.

Hegazy, E. A., Ishigaki, I., Rabie, A., Dessouki, A. M., and Okamoto, J. 1981b. Study on radiation grafting of acrylic acid onto fluorine-containing polymers. II. Properties of membrane obtained by preirradiation grafting onto poly(tetrafluoroethylene). *J. Appl. Polym. Sci.* 26 (11): 3871–3883.

Hien, N. Q., Nagasawa, N., Tham, L. X., Yoshii, F., Dang, V. H., Mitomo, H., Makuuchi, K., and Kume, T. 2000. Growth-promotion of plants with depolymerized alginates by irradiation. *Radiat. Phys. Chem.* 59: 97–101.

Holtan, S., Zhang, Q., Strand, W. I., and Skjak-Braek, G. 2006. Characterization of the hydrolysis mechanism of polyalternating alginate in weak acid and assignment of the resulting MG-oligosaccharides by NMR spectroscopy and ESI-mass spectrometry. *Biomacromolecules* 7: 2108–2121.

Hoshina, H., Seko, N., Ueki, Y., and Tamada, M. 2007. Synthesis of graft adsorbent with N-methyl-D-glucamine for boron adsorption. *J. Ion Exchange* 18: 236–239.

Hsiue, G. and Huang, W. 1985. Preirradiation grafting of acrylic and methacrylic acid onto polyethylene films: Preparation and properties. *J. Appl. Polym. Sci.* 30: 1023–1033.

Ishigaki, I., Sugo, T., Senoo, K., Okada, T., Okamoto, J., and Machi, S. 1982a. Graft polymerization of acrylic acid onto polyethylene film by preirradiation method. I. Effects of preirradiation dose, monomer concentration, reaction temperature, and film thickness. *J. Appl. Polym. Sci.* 27: 1033–1041.

Ishigaki, I., Sugo, T., Takayama, T., Okada, T., Okamoto, J., and Machi, S. 1982b. Graft polymerization of acrylic acid onto polyethylene film by preirradiation method. II. Effects of oxygen at irradiation, storage time after irradiation, Mohr's salt, and ethylene dichloride. *J. Appl. Polym. Sci.* 27: 1043–1051.

Jyo, A., Okada, K., Nakao, M., Sugo, T., Tamada, M., and Katakai, A. 2003. Bifunctional phosphonate fiber derived from vinylbiphenyl-grafted polyethylene coated polypropylene fiber for extremely rapid removal of iron (III). *J. Ion Exchange* 14 (Suppl.): 69–72.

Kabay, N., Katakai, A., Sugo T., and Egawa, T. 1993. Preparation of fibrous adsorbents containing amidoxime groups by radiation-induced grafting and application to uranium recovery from sea water. *J. Appl. Polym. Sci.* 49: 599–607.

Kavakli, P. A., Seko, N., Tamada, M., and Güven, O. 2004. A highly efficient chelating polymer for the adsorption of uranyl and vanadyl ions at low concentrations. *Sep. Sci. Technol.* 39: 1631–1644.

Kim, M. and Saito, K. 2000. Radiation-induced graft polymerization and sulfonation of glycidyl methacrylate on to porous hollow-fiber membranes with different pore sizes. *Radiat. Phys. Chem.* 57: 167–172.

Kumar, M. N. R. 2000. A review of chitin and chitosan applications. *React. Funct. Polym.* 46: 1–27.

Lee, K., Choi, S., and Kang, H. 2002. Preparation and characterization of polyvalence membranes modified with four different ion-exchange groups by radiation-induced graft polymerization. *J. Chromatogr. A* 948: 129–138.

Lee, S. W., Bondar, Y., and Han, D. H. 2008. Synthesis of a cation-exchange fabric with sulfonate groups by radiation-induced graft copolymerization from binary monomer mixtures. *React. Funct. Polym.* 68: 474–482.

Li, J., Matsuura, A., Kakigi, T., Miura, T., Oshima, A., and Washio, M. 2006. Performance of membrane electrode assemblies based on proton exchange membranes prepared by pre-irradiation induced grafting. *J. Power Sources* 161 (1): 99–105.

Li, D., Chen, J., Zhai, M., Asano, M., Maekawa, Y., Oku, H., and Yoshida, M. 2009. Hydrocarbon proton-conductive membranes prepared by radiation-grafting of styrenesulfonate onto aromatic polyamide films. *Nucl. Instrum. Methods Phys. Res., Sect. B* 267: 103–107.

Luan, L. Q., Nagasawa, N., Tamada, M., and Nakanishi, T. M. 2006. Enhancement of plant growth activity of irradiated chitosan by molecular weight fractionation. *Radioisotopes* 55: 21–27.

Mahmudi, N., Sen, M., Rendevski, S., and Güven, O. 2007. Radiation synthesis of low swelling acrylamide based hydrogels and determination of average molecular weight between cross-links. *Nucl. Instrum. Methods Phys. Res., Sect. B* 265: 375–378.

Matsuhashi, S. and Kume, T. 1997. Enhancement of antimicrobial activity of chitosan by irradiation. *J. Sci. Food Agric.* 73: 237–241.

Miani, M., Gianni, R., Liut, G., Rizzo, R., Toffanin, R., and Deleben, F. 2004. Gel beads from novel ionic polysaccharides, *Carbohydr. Polym.* 55: 163–169.

Mitsumata, T., Suemitsu, Y., Fujii, K., Fujii, T., Taniguchi, T., and Koyama, K. 2003. pH-Response of chitosan, k-carrageenan, carboxymethyl cellulose sodium salt complex hydrogels. *Polymer* 44: 7103–7111.

Nagasawa, N., Kaneda, A., Kanazawa, S., Yagi, T., Mitomo, H., Yoshii, F., and Tamada, M. 2005. Application of poly(lactic acid) modified by radiation crosslinking. *Nucl. Instrum. Methods Phys. Res., Sect. B* 236: 611–616.

Nasef, M. M. and Hegazy, E. S. A. 2004. Preparation and applications of ion exchange membranes by radiation-induced graft copolymerization of polar monomers onto non-polar films. *Prog. Polym. Sci.* 29 (6): 499–561.

Oshima, A., Tabata, Y., Kudo, H., and Seguchi, T. 1995. Radiation induced crosslinking of polytetrafluoroethylene. *Radiat. Phys. Chem.* 45: 269–273.

Ozawa, I., Saito, K., Sugita, K., Sato, K., Akiba, M., and Sugo, T. 2000. High-speed recovery of germanium in a convection-aided mode using functional porous hollow-fiber membranes. *J. Chromatogr. A* 888: 43–49.

Park, J., Enomoto, K., Yamashita, T., Takagi, Y., Todaka, K., and Maekawa, Y. 2009. Long-lived intermediates in radiation-induced reactions of alicyclic polyimides films. *J. Photopolym. Sci. Technol.* 22: 285–287.

Reddy, G., Altaf, Md., Naveena, B. J., Venkateshwar, M., and Kumar, E. V. 2008. Amylolytic bacterial lactic acid fermentation. *Biotechnol. Adv.* 26: 22–34.

Rouilly, M. V., Kotz, E. R., Haas, O., Scherer, G. G., and Chapiro, A. 1993. Proton-exchange membranes prepared by simultaneous radiation grafting of styrene onto teflon-FEP films: Synthesis and characterization. *J. Membr. Sci.* 81 (1–2): 89–95.

Saito, K. and Sugo, T. 2001. High-performance polymeric materials for separation and reaction, prepared by radiation-induced graft polymerization. In *Radiation Chemistry: Present Status and Future Trends*, C. D. Jonah and M. Rao (eds.), pp. 671–704. Amsterdam, the Netherlands: Elsevier.

Saito, K., Tsuneda, S., Kim, M., Kubota, N., Sugita, K., and Sugo, T. 1999. Radiation-induced graft polymerization is the key to develop high-performance functional materials for protein purification. *Radiat. Phys. Chem.* 54: 167–172.

Sato, K., Ikeda, S., Iida, M., Oshima, A., Tabata, Y., and Washio, M. 2003. Study on poly-electrolyte membrane of crosslinked PTFE by radiation-grafting. *Nucl. Instrum. Methods Phys. Res., Sect. B* 208: 424–428.

Sawada, S., Yamaki, T., Nishimura, H., Asano, M., Suzuki, A., Terai, T., and Maekawa, Y. 2008. Water transport properties of crosslinked-PTFE based electrolyte membranes. *Solid State Ionics* 179: 1611–1614.

Schierholz, K., Lappan, U., and Lunkwitz, K. 1999. Electron beam irradiation of polytetrafluoroethylene in air: Investigations on the thermal behaviour. *Nucl. Instrum. Methods Phys. Res., Sect. B* 151: 232–237.

Seko, N., Katakai, A., Hasegawa, S., Tamada, M., Kasai, N., Takada, H., Sugo, T., and Saito, K. 2003. Aquaculture of uranium in seawater by fabric-adsorbent submerged system, *Nucl. Technol.* 144: 274–278.

Seko, N., Tamada, M., and Yoshii, F. 2005. Current status of adsorbent for metal ions with radiation grafting and crosslinking techniques, *Nucl. Instrum. Methods Phys. Res., Sect. B* 236: 21–29.

Seko, N., Bang, L. T., and Tamada, M. 2007. Syntheses of amine-type adsorbents with emulsion graft polymerization of glycidyl methacrylate. *Nucl. Instrum. Methods Phys. Res., Sect. B* 265: 146–149.

Shiraishi, T., Tamada, M., Saito, K., and Sugo, T. 2003. Recovery of cadmium from waste of scallop processing with amidoxime adsorbent synthesized by graft-polymerization. *Radiat. Phys. Chem.* 66: 43–47.

Smith S. D. and Alexangratos, S. 2000. Ion-selective polymer-supported reagents. *Solvent Extr. Ion Exch.* 18: 779–807.

Sun, J., Zhang, Y., Zhong, X., and Zhu, X. 1994. Modification of polytetrafluoroethylene by radiation, 1. Improvement in high temperature properties and radiation stability. *Radiat. Phys. Chem.* 44: 655–659.

Tamada, M., Seko, N., Kasai, N., and Shimizu, T. 2006. Cost estimation of uranium recovery from seawater with system of braid type adsorbent, *Trans. At. Energy Soc. Jpn.* 5: 358–363.

Tamada, M. K., 2009. Fibrous metal ion adsorbent synthesized by radiation processing. *Kobunshi* (*High Polymers*) 58: 397–400.

Vahdat, A., Bahrami, H., Ansari, N., and Ziaie, F. 2007. Radiation grafting of styrene onto polypropylene fibers by a 10 MeV electron beam. *Radiat. Phys. Chem.* 76: 787–793.

Wach, R. A., Mitomo, H., Nagasawa, N., and Yoshii, F. 2003. Radiation crosslinking of carboxymethylcellulose of various degree of substitution at high concentration in aqueous solutions of natural pH. *Radiat. Phys. Chem.* 68: 771–779.

Wasikiewicz, J. M., Nagasawa, N., Tamada, M., Mitomo, H., and Yoshii, F. 2005a. Adsorption of metal ions by carboxymethylchitin and carboxymethylchitosan hydrogels. *Nucl. Instrum. Methods Phys. Res., Sect. B* 236: 617–623.

Wasikiewicz, J. M., Yoshii, F., Nagasawa, N., Wach, R. A., and Mitomo, H. 2005b. Degradation of chitosan and sodium alginate by gamma radiation, sonochemical and ultraviolet methods. *Radiat. Phys. Chem.* 73: 287–295.

Wasikiewicz, J. M., Mitomo, H., Nagasawa, N., Yagi, T., Tamada, M., and Yoshii, F. 2006. Radiation crosslinking of biodegradable carboxymethylchitin and carboxy-methylchitosan. *J. Appl. Polym. Sci.* 102: 758–767.

Yamaki, T., Asano, M., Maekawa, Y., Morita, Y., Suwa, T., Chen, J. H., Tsubokawa, N., Kobayashi, K., Kubota, H., and Yoshida, M. 2003. Radiation grafting of styrene into crosslinked PTEE films and subsequent sulfonation for fuel cell applications. *Radiat. Phys. Chem.* 67: 403–407.

Zhang, Z., Yu, G., Guan, H., Zhao, X., Du, Y., and Jiang, X. 2004. Preparation and structure elucidation of alginate oligosaccharides degraded by alginate lyase from Vibro sp. 510. *Carbohydr. Res.* 339: 1475–1481.

28 UV Molecular Spectroscopy from Electron Impact for Applications to Planetary Atmospheres and Astrophysics

Joseph M. Ajello
California Institute of Technology
Pasadena, California

Rao S. Mangina
California Institute of Technology
Pasadena, California

Robert R. Meier
George Mason University
Fairfax, Virginia

CONTENTS

28.1 INTRODUCTION

In the upper atmospheres and torus regions of the terrestrial and Jovian planets and in the interstellar medium (ISM), an important mechanism for energy transfer and diagnostic spectroscopy is electron collision processes with both neutral and ionic species leading to the emission of electromagnetic radiation. Six of the planets (Earth, Mars, Jupiter, Saturn, Uranus, and Neptune) are known to have internal magnetic fields that lead to particle acceleration and energy deposition into a planetary atmosphere (Bagenal et al., 2007). Mercury also has an intrinsic magnetic field with a tenuous atmosphere that is a planetary exosphere. The ubiquitous presence of energetic electron-excited ultraviolet (UV) dayglow and aurora in the solar system (Broadfoot et al., 1979, 1981a,b, 1989; Sandel et al., 1979; Yung et al., 1982; Meier, 1991; Ajello et al., 1998b, 2001, 2005a; Gustin et al., 2002, 2004) has been studied spectroscopically over the last 30 years using observations from interplanetary spacecraft beginning with the Voyager Grand Tour mission and Mars and Venus Mariner missions, and the earth-orbiting satellites beginning with the Orbiting Geophysical Observatories (OGOs) and International Ultraviolet Explorer (IUE). Simultaneously, astronomical observations of H_2 Rydberg band emissions from Herbig-Haro and T-Tauri stellar objects (Raymond et al., 1997; Herczeg et al., 2002, 2004; Bergin et al., 2004) and models of H_2 Rydberg band emissions generated in the interior of molecular clouds within the ISM (Gredel et al., 1987, 1989; Liu and Dalgarno, 1996) have been achieved, first with the Copernicus, which was equipped with a UV telescope (Grewing et al., 1978, Snow, 1979), and followed at higher spectral resolution by the Hubble Space Telescope (HST) with its corrected optics (Petersen and Brandt, 1995).

The importance of electron impact excitation of molecules was dramatized during the Voyager mission due to findings of rich dayglow and auroral spectra at each of the outer planets with strong magnetospheres and thick H_2 atmospheres. For example, the Voyager 1 spacecraft equipped with the ultraviolet spectrometer (UVS), having a detection range of 50–170 nm, arrived at Jupiter in January 1979 and at Saturn in November 1980, and made spectacular discoveries related to the physical processes involving electron acceleration and corotating plasma that control the atmosphere and magnetosphere (Broadfoot et al., 1979, 1981a,b, 1989; Sandel et al., 1979). Voyager discovered that Jupiter's UV aurora is the second brightest source of UV in the solar system (second to the Sun), with an emission intensity of ~10^{13-14} W in the H_2 Rydberg bands. Its ultimate power source is the rotational energy of the planet as well as plasma processes in the near corotating middle magnetosphere. The breadth of objects studied with these telescopes is shown in Figure 28.1.

In the early 1980s, the initial attempts in modeling the extreme ultraviolet (EUV) auroral spectra obtained by Voyager and other missions were not very successful due to lack of spectroscopic signatures and their reliable cross section data. At this point in time, reliable electron excitation cross sections can only be provided through laboratory measurements, but seldom by theory. Useful earlier reviews can be found for terrestrial auroral spectroscopy in *Auroral Physics* (Meng et al., 1991), and for emission cross section measurements prior to 1998 for some of the planetary species in Avakyan et al., (1998).

The analysis of observations of planetary atmospheres made by HST, Cassini and Far Ultraviolet Spectroscopic Explorer (FUSE), and Earth-orbiting spacecraft Thermosphere Ionosphere Mesosphere Energetics and Dynamics/Global Ultraviolet Imager (TIMED/GUVI), Midcourse Space Experiment (MSX), Defense Meteorological Satellite Program (DMSP), POLAR spacecraft, and Imager for Magnetopause-to-Aurora Global Exploration (IMAGE) require accurate collision

FIGURE 28.1 Typical types of objects that are studied by spacecraft equipped with UV observatories are the Earth (Terrestrial Objects), Jupiter (Jovian planets) and the Horse Head nebula (ISM).

cross sections. With the advent of the newest generation of high-resolution UV imaging space instruments (e.g., Space Telescope Imaging Spectrograph (STIS) and Faint Object Spectrograph (FOS) with UV spectral resolving power $\approx 10^5$ onboard HST, the emissions of the outer planets have been examined in much greater detail at better resolution and in different spectral regions than Voyager.

28.2 UV SPECTROSCOPY OF MOLECULES IN PLANETARY ATMOSPHERES

To illustrate the key relationships among spectroscopic observations, molecular parameters, and planetary constituents, we present the formalism for the Jovian aurora. Airglow processes are similar except that precipitating particle fluxes are replaced by photoelectron fluxes. A schematic of the Jovian magnetic field interacting with the ionosphere is shown in the upper panel of Figure 28.2 along with a model of the primary + secondary differential electron flux, (electrons/s/cm²/eV) at various altitudes in the Jovian aurora shown in the lower panel (Ajello et al., 2001, 2005a; Grodent et al., 2001). The auroral UV spectrum produced by the precipitating electrons can be used to estimate Q, the precipitating electron energy flux of the primary particles, and E_o, their characteristic energy (Strickland et al., 1995). The modeling of the dynamical magnetosphere–ionosphere coupling producing the Jovian aurora has been recently described (Bunce and Cowley, 2001; Cowley and Bunce, 2001; Hill, 2001). These authors suggest that the auroral oval indicates the presence of a global-scale Birkeland current system that maps to $\sim 30 R_j$. This current system passes planetary angular momentum to the outward moving plasma sheet maintaining the middle magnetosphere in near corotation (Hill, 2001). In the region of upward currents (downward electrons), field-aligned potentials accelerate electrons to

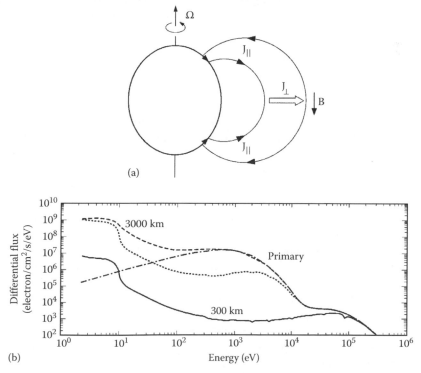

FIGURE 28.2 (a) The model of the Jupiter ionosphere–magnetosphere coupling showing the currents for the aurora (Cowley and Bunce, 2001; Hill, 2001), and (b) the electron differential energy distribution of primary + secondary electrons at four altitudes in the Jupiter aurora are distinguished by dashed or solid curves (Ajello et al., 2001, 2005a). The unmarked dotted curve is at 800 km.

auroral energies of 10–100 keV (Mauk et al., 2002), which in turn generate UV and visible aurora through electron–molecule collisions that are observed by spacecraft at the Jovian planets.

The volume emission rate, $V_\lambda(z)$ (photons/s/cm³), for a simple molecule H_2 spectral line at wavelength, λ, emitted in the aurora of the Jovian planets can be written (Ajello et al., 2001, 2005a) as

$$V_\lambda(z) = g_j(z)\omega_{jm}(\lambda)(1-\eta_j)\,N(z,T) \qquad (28.1)$$

where $g_j(z)$ is the excitation rate (excitations per molecule per second) at altitude z into an upper electronic state j characterized by α, v, and J which are principal quantum number, vibrational level quantum number, and rotational level quantum number, respectively. The index m refers to the quantum numbers β (principle), v (vibrational), and J (rotational) for the lower electronic state. $N(z,T)$ is the atmospheric H_2 density at altitude z with a local kinetic temperature T. The quantity η_j is the nonradiative yield of predissociation and pre-ionization for the upper state. $\omega_{jm}(\lambda)$ is the emission branching ratio for the transition $j \rightarrow m$ at wavelength λ and is given by

$$\omega_{jm}(\lambda) = \frac{A_{jm}}{A_j} \qquad (28.2)$$

where
A_{jm} is the Einstein spontaneous transition probability from the upper state j to a lower state m
A_j is the total emission transition probability to all lower states

The excitation rate, $g_j(z)$, is proportional to the sum of the individual excitation rates from a ground state rotational-vibrational level, $X(v_i, J_i)$. The individual excitation rates equal to the product of the fraction of molecules in the initial ground state rotational-vibrational level, $X(v_i, J_i)$, times the fine structure excitation cross section σ_{ij} from level i to level j for an electron impact energy, ε, times the electron flux. It is written as

$$g_j(z) = \int F(\varepsilon, z) \left[\sum_i f_{X_i}(T)\sigma_{ij}(\varepsilon) \right] d\varepsilon, \qquad (28.3)$$

where f_{X_i} is the fraction of the H_2 molecules in the initial ground vibrational level v_i and rotational level J_i (see Ajello et al., 2001, 2005a for notation). The sum extends over the fine structure rotational branches for Σ and Π transitions of H_2 in an upper atmosphere. $F(\varepsilon, z)$ is the precipitating electron flux distribution function in electrons/sec/cm²/eV at an energy ε and altitude z.

The first step in modeling the Jovian aurora requires careful laboratory studies to determine $\sigma_{ij}(\varepsilon)$ for each rotational line following electron impact excitation of H_2 (Liu et al., 1998; Glass-Maujean et al., 2009). UV emissions from the outer planets observed by Voyager and IUE were explained with electron transport models of high-energy electron impact (1–100 keV primary electron flux at the top of the atmosphere) of H_2 that included all Rydberg states from principal quantum number, $n = 2$–5 (Yung et al., 1982; Shemanksy and Ajello, 1983; Ajello et al., 1984). Modeling to date (Jonin et al., 2000; Liu et al., 2000, 2002, 2003; Glass-Maujean et al., 2009) now takes into account the rotational-electronic coupling among the $n = 2$ through 5 states, *ungerade* B, B′, B″, C, D, D′, D″ ($^1\Sigma_u^+$, $^1\Pi_u$) $n = 2, 3, 4, 5 \rightarrow$ X $^1\Sigma_g^+$ and can be used for accurate modeling of the aurora spectra.

We show in Figure 28.3a the first model analysis (Yung et al., 1982) (Equations 28.1 through 28.3) of the Jupiter auroral far ultraviolet (FUV) spectrum from IUE. A comparison of this spectrum obtained at 0.1 nm resolution with a model provided convincing proof that most of the emission features come from H_2 (Yung et al., 1982). Comparison of a Saturnian auroral spectrum obtained by Voyager in the EUV at 3.3 nm resolution with an auroral model is shown in Figure 28.3b that all of the features could be identified with molecular hydrogen as well (Shemansky and Ajello, 1983). Gustin et al. (2009) has recently analyzed the Cassin ultraviolet imaging spectrograph (UVIS)

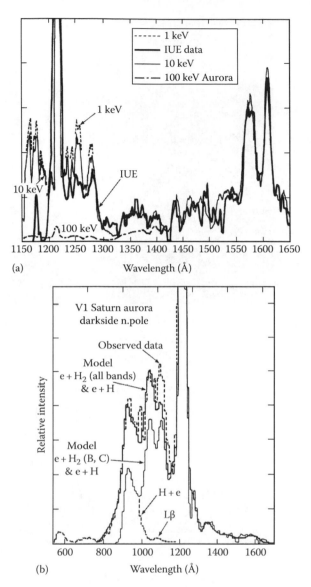

FIGURE 28.3 (a) IUE observation of Jupiter with three auroral models (1, 10, 100 keV primary electrons) in the FUV (Yung et al., 1982). (b) Shown in dash are the Voyager 1 and 2 observation of prominent UV radiation from the atmosphere of Saturn corresponding to auroral emission, which is concentrated in the polar regions, and is excited by high-energy particle precipitation along the magnetic field lines (Shemansky and Ajello, 1983). The strong disk and auroral feature at 1216 Å, which runs off scale, is the hydrogen Ly α line. The H_2 band emission between 900 and 1130 Å is the characteristic of the auroral region. We indicate two models of the observations: B and C-states ($n = 2$ only) and all Rydberg states ($n = 2–4$).

auroral spectrum of Saturn at much higher resolution of 0.4 nm FWHM. The IUE analysis provided the magnitude of both E_0 and Q. The characteristic energy for the primary electron flux was estimated to be ~10 keV. A typical characteristic energy ranging from 5 to 100 keV was found by the UVS in the Galileo orbiter mission to Jupiter (Ajello et al., 1998b, 2001, 2005a) and subsequently by Cassini and HST analyses (Dols et al., 2000; Gustin et al., 2002, 2004; Grodent et al., 2003a,b; Ajello et al., 2005a). The globally averaged electron energy flux precipitating into the Jupiter atmosphere was 0.5–2 erg/cm²/s at the time of the IUE observation (Yung et al., 1982).

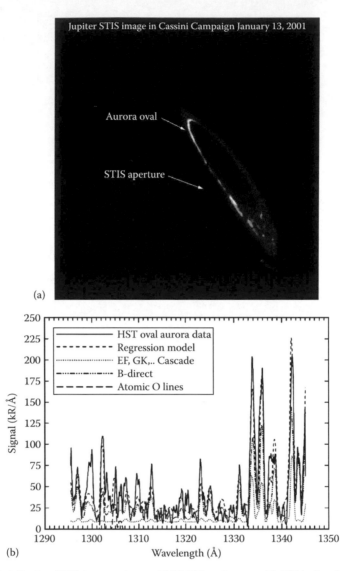

FIGURE 28.4 (a) A Jupiter STIS image taken at 16:50 UT on January 13, 2001, showing the three auroral zones of polar cap, auroral oval and limb with the 52×0.5 arcs^2 slit projected on the image (Ajello et al., 2005a). (b) The HST STIS G140M medium-resolution grating relative photon intensity short wavelength spectrum (1295–1345 Å) for the north aurora of January 13, 2001 fitted in linear regression. The linear regression analysis with independent vectors of (1) a direct excitation optically thin spectrum of the Rydberg bands of H_2 by a monoenergetic electron distribution at 100 eV, (2) a cascade excitation optically thin spectrum of the Rydberg bands of H_2 by a monoenergetic electron distribution at 100 eV, and (3) an atomic oxygen multiplet at 1304 Å. The regression model to the observational data is based upon a linear regression analysis with these three independent vectors and transmission through a slab of CH_4. We were able to estimate the total contributions from direct excitation and cascading (Ajello et al., 2005a). The atomic O 130.4 nm emission is an artifact of the Earth's dayglow superimposed on the Jupiter observation.

The STIS performed medium- and high-resolution spectral image observations (FWHM = 10^{-2} to 10^{-3} nm) of Jupiter in the FUV between 115 and 170 nm, revealing the rotational structure of the principal thermospheric gas, H_2, in the aurora oval, in the polar cap, and in the Io flux tube. A STIS medium-resolution spectrum and H_2 model are shown in Figure 28.4 (Ajello et al., 2005a). The medium-resolution (0.09 nm FWHM, G140 M grating) FUV observations 129.5–134.5 nm by STIS on January 13, 2001 were analyzed using a recently developed high-spectral-resolution model for the electron-excited H_2 rotational lines that considered the Lyman Band spectrum (B $^1\Sigma_u^+ \rightarrow$ X $^1\Sigma_g^+$) to be composed of an allowed direct excitation component (X $^1\Sigma_g^+ \rightarrow$ B $^1\Sigma_u^+$) and an optically forbidden component (X $^1\Sigma_g^+ \rightarrow$ EF, GK, H$\bar{\text{H}}$,... $^1\Sigma_g^+$ followed by the cascade transition EF, GK, H$\bar{\text{H}}$,... $^1\Sigma_g^+ \rightarrow$ B $^1\Sigma_u^+$). The medium-resolution spectral regions for the Jupiter aurora were carefully chosen to emphasize the cascade component (Ajello et al., 2005a).

28.3 APPARATUS AND EXPERIMENTAL METHODS

In response to the need for accurate collision cross sections to model spectroscopic observations of the terrestrial and Jovian planetary systems, the Emission Spectroscopy Laboratory (ESL) at Jet Propulsion Laboratory (JPL) has established five unique instruments for routinely measuring the absolute emission cross sections of stable and radical gases from the broad spectral region of the UV to the Visible-Optical-Infrared (VOIR) (40–1100 nm). Three of the five instruments are shown in Figure 28.5 and are identified as follows: (1) atomic O (Johnson et al., 2003a,b, 2005), (2) 3 m high resolution (Liu et al., 1995), and (3) atomic H (James et al., 1998a) apparatuses. Not shown in the figure are the VOIR (Aguilar et al., 2008; Ajello et al., 2008; Mangina et al., 2010) and the large chamber for studying the long-lived metastable emissions (Kanik et al., 2003). The 3 m optical spectrometer system is capable of high spectral resolution with a resolving power of $\lambda/\Delta\lambda = 50,000$ and is equipped with a Codacon 1340×400 array detector for studying the rotational structure and kinetic energy (line profiles) of excited fragments (Ajello and Ciocca, 1996a). Each UV spectrometer has dual exit ports and dual grating holders to allow scanning over two wavelength ranges, e.g., the extreme ultraviolet (EUV from 40 to 120 nm), far ultraviolet (FUV from 110 to 310 nm), or visible-optical-near IR (VOIR from 300 to 1100 nm), without breaking the vacuum.

Using this instrumentation, ESL has carried out measurements consisting of a calibrated primary data set of optically thin UV and VOIR fluorescence spectra (50–1100 nm) at spectral resolutions between 0.002 and 1 nm and absolute excitation cross sections at electron impact energies 0–2 keV for several stable gases such as H_2 (Ajello et al., 1995b, 1996b; James et al., 1998b; Liu et al., 1998, 2000, 2002, 2003; Dziczek et al., 2000; Jonin et al., 2000; Aguilar et al., 2008; Glass-Maujean et al., 2009); HD (Ajello et al., 2005b); D_2 (Ciocca et al., 1997a; Abgrall et al., 1999); He (Shemansky et al., 1985); Ar (Ajello et al., 1990); Ne (Kanik et al., 1996); CO (Ciocca et al., 1997b; Zetner et al., 1998; Beegle et al., 1999); CO_2 (Kanik et al., 1993); H_2O (Makarov et al., 2004), O_2 (Noren et al., 2001b; Kanik et al., 2003; Terrell et al., 2004); N_2 (Ajello et al., 1998a; Liu et al., 2008, Mangina et al., 2010; Young et al., 2010), SO_2 (Ajello et al., 2002a, 2008; Vatti Palle et al., 2004), NO (Ajello et al., 1989a), NO_2 (Young et al., 2009), N_2O (Malone et al., 2008), and for the radical atomic gases H (James et al., 1997, 1998a) and O (Noren et al., 2001a; Johnson et al., 2003a,b, 2005a). A large number of analyses of data from the wide variety of satellite missions listed above have used the ESL measured cross sections and line profiles (Hord et al., 1992; Ciocca et al., 1997a,b; Prange et al., 1997; Feldman et al., 2001; Gustin et al., 2002; 2004; Esposito et al., 2004). Likewise, the mission planning and instrument calibration phases of the UV instruments on board Cassini and the Pluto New Horizons (Stern et al., 2008) interplanetary spacecraft depended on cross sections, spectra, and spectral line profiles established by the ESL (Ajello et al., 1995a,b).

The emission cross sections of the majority of neutral and single-ionized planetary gases have been reviewed by Avakyan et al. (1998) and Majeed and Strickland (1997), and for molecular hydrogen and its isotopes by Tawara et al. (1990). For atomic oxygen, one of the most important planetary gases, and oxygen-bearing molecule's reviews of cross sections have been given recently by Johnson et al. (2005) and McConkey et al. (2008).

The experimental technique developed at JPL for the measurements of electron-impact-induced UV emission cross sections and spectra of stable atoms and molecules has been described in Ajello et al. (1989b, 2002a,b) and references therein, and is shown schematically in Figure 28.6. In brief, UV emission spectra are generated from collision of a collimated beam of energetic electrons with

The atomic O experiment

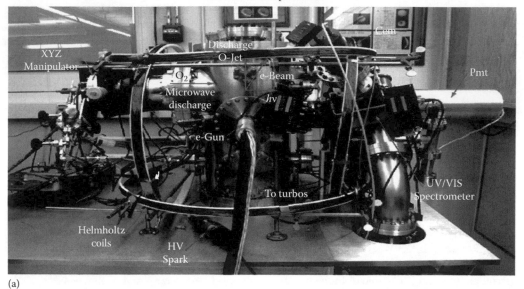

(a)

High resolution 3m UV spectrometer   The atomic H experiment

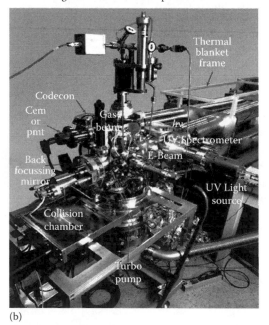   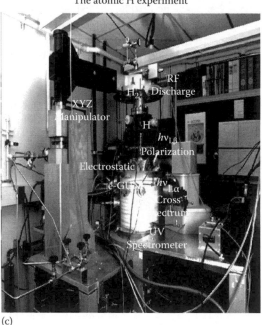

(b)   (c)

FIGURE 28.5   The JPL instrumentation consisting of: (a) an atomic O radical source with a low-resolution spectrometer, $\lambda/\Delta\lambda = 1000$ (Johnson et al., 2005); (b) a high-resolution 3 m imaging spectrograph with Codacon MCP detectors similar to the detector flown on Cassini UVIS, $\lambda/\Delta\lambda = 50,000$ (Liu et al., 1995); and (c) the atomic H radical source with low-resolution spectrometer, $\lambda/\Delta\lambda = 1000$ (James et al., 1997).

a beam of target gas (produced by a capillary array) in a crossed-beams geometry at a background pressure that ensures optically thin, single-scattering conditions. The interaction volume of the crossed beams is approximately $2\,mm^3$. Emitted photons corresponding to the radiative decay of collisionally excited states of the target species are detected by the UV spectrometer with its optic axis at 54.74° (magic angle) or 90° to the plane containing the crossed electron- and target molecular beams.

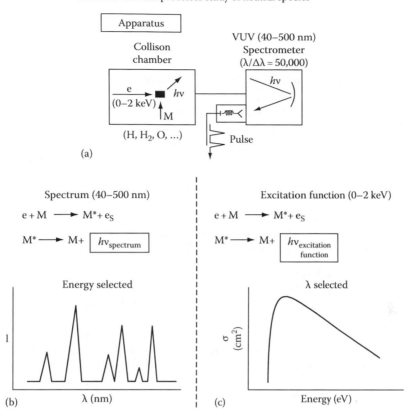

Electron collision processes study of neutral species

FIGURE 28.6 (a) A schematic of the crossed electron beam-molecular-beam apparatus for (b) measuring electron-impact-induced fluorescence spectra and (c) emission cross sections as a function of excitation energy measurement by continually scanning electron energy, e.g., from 0 to 2 keV.

In an electron-impact-induced emission experiment, a molecule in a ground electronic X-state is excited to an electronic state α. A model of the irradiance from the interaction region for the transition $|\alpha, v, J\rangle \rightarrow |X, v_f, J_f\rangle$ from a target molecule, such as H_2 or CO, is based on the calculated transition probabilities (e.g., for H_2 Lyman and Werner bands use Abgrall et al. (1993a,b,c, 1997, 1999, 2000) and the rotational line positions of Roncin and Launay (1994)). For the CO Fourth Positive system, we use the transition probabilities and molecular constants given in Morton and Noreau (1994). The model for electron-excited H_2 molecules has been presented in previous papers (Liu et al., 1995, 1998, 2000, 2003; Jonin et al., 2000). We will briefly describe it here. The populations of the ground-state rotational levels are controlled by the gas temperature and nuclear spin of the molecule. The ground-state molecules in vibrational and rotational thermal equilibrium are excited into the various rovibronic states according to the excitation rate $g'(\alpha, v, J, E_e)$. The photoemission intensity into the various branches from rovibronic state $|\alpha, v, J\rangle$ is partitioned according to the emission branching ratio, predissociation yield, and pre-ionization yield. The photoemission volume intensity, V' (photons/s/cm^3), at any given point (x, y, z) in the interaction region defined by the intersection of electron beam of energy E_e and molecular beam, in the laboratory, including self-absorption and predissociation, is given by

$$V'(X,\alpha,v_f,v,J_f,J) = \frac{A(X,\alpha,v_f,v,J_f,J)}{A(X,\alpha,v,J)} \cdot g'(\alpha,v,J,E_e) \cdot (1-\eta_P(\alpha,v,J)) \cdot \mathrm{Tr}(X,\alpha,v,J,T) \quad (28.4)$$

$$A(X,\alpha,v,J) = \sum_{v_f,J_f} A(X,\alpha,v_f,v,J_f,J) \tag{28.5}$$

where

$A(X, \alpha, v_f, v, J_f, J)$ is the Einstein A-coefficient for spontaneous transition from the excited state $|\alpha, v, J\rangle$ back to $|X, v_f, J_f\rangle$ of the X $^1\Sigma_g^+$ ground state

$A(X, \alpha, v, J)$ is the total emission probability (including transitions to lower singlet-*gerade* states and to the continuum levels of the X $^1\Sigma_g^+$ state)

$\eta_P(\alpha, v, J)$ is the predissociation + pre-ionization yield

$\mathrm{Tr}(X, \alpha, v, J, T)$ is the transmission function for self-absorption by resonance lines

The volumetric excitation rate (excitations/s/cm^3) at that point (x, y, z) is given by

$$g'(\alpha,v,J,E_e) = F_e \sum_i N_i \sigma_{ij}(E_e) \tag{28.6}$$

We use the index i to represent the initial electronic state $|X, v_i, J_i\rangle$, the index j to represent the excited electronic state $|\alpha, v, J\rangle$, $N_i = N_o \times f_{x_i}$ and the excitation cross section, $\sigma_{ij} = \sigma(v_i, v, J_i, J)$. The volumetric excitation rate $g'(\alpha, v, J, E_e)$ is proportional to the population density N_i of the molecule in the initial vibrational–rotational level v_i, J_i, and the monoenergetic electron-impact flux F_e due to the electron beam current (Glass-Maujean et al., 2009). Equations 28.4 and 28.6 can be used to define a P, Q, or R branch rotational line emission cross section (without cascade) between two electronic states X and α:

$$\sigma_{em}(\alpha,X,v,v_f,J,J_f) = \frac{A(X,\alpha,v_f,v,J_f,J)}{A(X,\alpha,v,J)}(1-\eta_J)\sum_{v_i,J_i}\sigma_{ex}(X,\alpha,v_i,v,J_i,J,E_e) \tag{28.7}$$

The total thermally averaged excitation cross section for an electronic band system is defined by Glass-Maujean et al. (2009) as

$$\sigma_{ex} = \frac{\left(\sum_{i,j} N_i \sigma_{ij}(E_e)\right)}{N_0} \tag{28.8}$$

and the corresponding thermally averaged emission cross section is given by

$$\sigma_{em} = \frac{\sum_{i,j} N_i \sigma_{ij}(E_e)(1-\eta_j)}{N_0} \tag{28.9}$$

The individual rovibronic excitation cross sections $\sigma_{ij}(E_e)$ can be calculated from a known transition probability and a measured excitation function for H$_2$ (Liu et al., 1998). For negligible self-absorption, the total band system volumetric photoemission intensity in Equation 28.4 is proportional to the total emission cross section, σ_{em}, and the neutral gas density, N_0. The σ_{em} is related to σ_{ex} as indicated by comparing Equations 28.8 and 28.9. The ratio σ_{em} divided by σ_{ex} gives the thermally averaged emission yields for each of the *ungerade* Rydberg states in Table 28.1 (Glass-Maujean et al., 2009).

TABLE 28.1
Electronic-Band Cross Sections and Emission Yields of H_2 Singlet-*Ungerade* States[a]

State	Present σ_{ex}	Previous σ_{ex}	Present σ_{em}	Previous σ_{em}	Present Emission Yield (%)	Previous Emission Yield (%)
B $^1\Sigma_u^+$	264[b]	262[c]	263	262[c]	99[b]	100
C $^1\Pi_u$	244[b]	241[c]	249[b]	241[c]	98[b]	100
B' $^1\Sigma_u^+$	36[b]	38[d,e]	21	21[d]	53	56
D $^1\Pi_u^+$	25	24[d]	11	11[d]	43	46
D $^1\Pi_u^-$	21	18[d]	21	18[d]	100	100
B'' B $^1\Sigma_u^+$	11	>4[d]	2.2	1.6[d]	20	<40
D' $^1\Pi_u^+$	9.3	7.1[d]	1.6	1.0[d]	18	14
D' $^1\Pi_u^-$	7.3	≥5.3[d]	5.7	5.3[d]	78	≤100
D'' $^1\Pi_u$	3.2	>0.6	0.9	0.6	28	—
5pσ $^1\Sigma_u^+$	—	—	1.1	—	—	—
6pσ $^1\Sigma_u^+$	—	—	0.6	—	—	—
6pπ $^1\Pi_u$	—	—	0.9	—	—	—
7pπ $^1\Sigma_u^+$	—	—	0.6	—	—	—

Source: Glass-Maujean, M. et al., *Astrophys. J. Suppl.*, 180, 38, 2009. With permission.

[a] $E = 100\,eV$ and $T = 300\,K$. Unit is $10^{-19}\,cm^2$. σ_{ex} and σ_{em} denote excitation and emission cross sections, respectively. Certain numbers may not add up due to roundings. See Section 28.5.3 (Glass-Maujean et al., 2009) for estimated errors in cross sections.

[b] Excitation cross sections include the excitation into the H(1s) + H(2l) continuum, which is estimated from the calculation of Glass-Maujean (1986). Emission cross sections exclude emission from the H(1s)+H(2l) continuum levels, but include continuum emission from the excited discrete levels into the continuum levels of the X $^1\Sigma_g^+$ state. Transitions to the X $^1\Sigma_g^+$ continuum contribute 27.5% and 1.5%, respectively, to total emission cross sections of B $^1\Sigma_u^+ - X\ ^1\Sigma_g^+$ and C $^1\Pi_u - X\ ^1\Sigma_g^+$ (Abgrall et al. 1997). Note correction to B' $^1\Sigma_u^+$ present σ_{ex}, as suggested by Liu (private communication, 2010).

[c] From Liu et al. (1998).

[d] From Jonin et al. (2000).

[e] Include excitations into the continuum levels of the B $^1\Sigma_u^+$ state.

The UV wavelength calibration methodology developed at JPL has been used by several spacecraft missions (Hord et al., 1992; Esposito et al., 2004). This technique coupled with a relative flow or swarm gas calibration developed at JPL (Ajello et al., 1989b) allows the determination of absolute emission cross sections σ_{em} from any of the ESL optical instrument systems. Recent work by the ESL has established benchmark standards for absolute emission cross sections by electron-impact fluorescence measurements at 100 eV for reference gas H_2 (121.6 nm) $\sigma_{em} = 7.03 \pm 0.47 \times 10^{-18}\,cm^2$ (McConkey et al., 2008) and reference gas N_2 (120 nm) $\sigma_{em} = 3.7 \pm 0.5 \times 10^{-18}\,cm^2$ (Malone et al., 2008).

The absolute cross section for the emission of a particular spectral line λ induced by electron impact on a target species M in a swarm gas experiment at 90° to the electron beam can be measured at low pressure as (Johnson et al., 2003a,b)

$$\sigma(M)_\lambda = \frac{KS_\lambda(1-(p_\lambda/3))}{\xi b_\lambda PI_e} \tag{28.10}$$

where

S_λ is the photon signal

K is a constant related to the geometry of the detector

b_λ is the sensitivity of the detector

ξ is the instrumental polarization sensitivity of the system

p_λ is the polarization of the emitted radiation

P is the gas pressure

I_e is the electron beam current

The trapping of resonance radiation can reduce the emission rate significantly at high gas pressure. To avoid this complication, the gas pressure range is maintained at a sufficiently low level. K is determined from an intensity measurement of the known N_2 120 nm or H_2 121.6 nm emissions, since the instrumental factors are common to both target species (standard and target gas M). At 100 eV, the atomic multiplets from dissociative excitation are unpolarized since many repulsive states contribute to the emissions.

28.4 PRESENT STATUS OF H_2, N_2, CO, AND SO_2

28.4.1 H_2–UV

Hydrogen is by far the most abundant element in the universe, playing a pivotal role in many physical and chemical processes. For example, in diffuse molecular clouds of the ISM and stellar atmospheres, hydrogen chemistry permeates astronomical changes and provides the markers of stellar evolution (Dalgarno, 1993, 1995). Over the past two decades, the observations of the ISM have shown that H_2 is an active component of star formation. The changes that a star undergoes during the formation and dying process are truly dramatic. These result in the most important interactions between a star and its environment. Indeed, it is in this area of research that some of the most challenging astrophysical problems remain unanswered. UV and near-IR emissions from H_2 are among the principal ways the interstellar gas cools following gravitational collapse during the star formation (Lepp and Dalgarno, 1996; Lepp et al., 2002). During the last 10 years, the observations of the distribution of H_2 gas throughout the galaxy by FUSE have contributed to our understanding of stellar evolution (Moos et al., 2000). Hence, H_2 has a unique and extraordinary position in astronomy by virtue of its UV spectroscopic signature of diverse energetic environments.

Molecular hydrogen is the simplest molecule from a structural point of view, but its band spectra are quite complex and extend from UV to near-IR wavelengths due to the relatively large values of rotational constants for all the electronic states. An accurate model of the H_2 spectrum has been a fundamental building block for understanding the chemistry of the solar system and ISM. Until recently, a 50%–200% uncertainty existed for some of the excitation cross sections and transition probabilities of the singlet-*gerade* (even) states of H_2 and HD, i.e., the states that provide VOIR cascade excitation to the Lyman and Werner bands (Ajello et al., 2005b; Aguilar et al., 2008). The recent study of H_2 emission cross sections (Aguilar et al., 2008) is the first of the VOIR wavelength range (300–1100 nm) in 50 years, since the pioneering work of Dieke and coworkers (Dieke, 1958; Dieke and Cunningham, 1965; Crosswhite, 1972) who demonstrated the existence of over 100,000 rotational lines and transitions in this wavelength region involving 15 electronic states of H_2 (Crosswhite, 1972). The complete single-scattering VOIR spectra of the H_2 and HD *gerade–ungerade* band systems had never been studied in the laboratory, nor have the oscillator strengths been accurately calculated until recently. The theoretical oscillator strength study by the Meudon Observatory (Aguilar et al., 2008 and references there in) involves detailed calculations of emission transition probabilities and line positions of individual rotational lines of the nine coupled EF, GK, HH, K, P $^1\Sigma_g^+$ states and I, R $^1\Pi_g$ and J, S $^1\Delta_g^+$ states. All of these coupled states contribute heavily to the UV spectrum through cascading. Comparing the laboratory spectra to model calculations based on the theoretical oscillator strengths, many irregularities (intensity and wavelength positions) in the VOIR were explained (Aguilar et al., 2008), although many remain unaccounted for and are being reevaluated.

The ESL has provided the same molecular parameters for the singlet-*ungerade* (odd) states of H_2 and HD (and even of D_2) to an accuracy of 10% (i.e., states that lead to the direct excitation of the Lyman and Werner bands) (Liu et al., 1995; 1998; 2000; 2002; 2003; Abgrall et al., 1999; Ajello et al., 2005b; Glass-Maujean et al., 2009). These will aid in the studies of the ISM and planetary atmospheres, where both types of electron-excited transitions take place.

Very accurate synthetic spectral models of H_2 for UV astronomy have been developed recently (Dols et al., 2000; Jonin et al., 2000; Liu et al., 2003; Gustin et al., 2004; Glass-Maujean et al., 2009). These models properly account for cascade, predissociation, and resonance effects by utilizing high-resolution measurements of spectra and cross sections (Glass-Maujean et al. 2009). We now understand that the complexity of the H_2 band system arises from intense configuration interaction, predissociation, and autoionization that are present in the 11–16 eV electronic energy region of the Rydberg and valence (RV) states. A simplified adiabatic energy level diagram of H_2 exhibiting the strongest allowed excitation process producing the B → X (Lyman bands) and the strongest optically forbidden process producing the EF → B (Lyman band cascade) is shown in Figure 28.7.

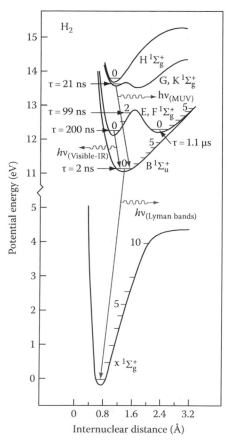

FIGURE 28.7 A partial energy level diagram for H_2 showing the energy regions for the VOIR (*gerade-ungerade*) singlet transitions and the UV (*ungerade-gerade*) transitions (Aguilar et al., 2008). The lowest-lying *ungerade–gerade* transition is B $^1\Sigma_u^+ \rightarrow$ X $^1\Sigma_g^+$ (direct excitation of the Lyman bands) and the lowest-lying *gerade–ungerade* transition is E, F $^1\Sigma_g^+ \rightarrow$ B $^1\Pi_u \rightarrow$ X $^1\Sigma_g^+$ (cascade excitation of the Lyman bands). The approximate lifetimes for some of the direct (dipole-allowed) and cascade (optically forbidden) vibronic states are listed.

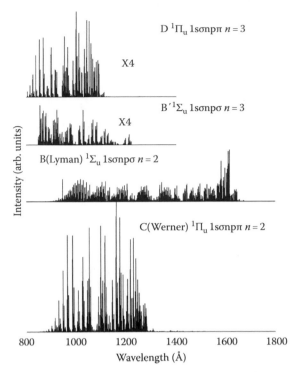

FIGURE 28.8 Model line spectrum of the B $^1\Sigma_g^+ \to$ X $^1\Sigma_g^+$, C $^1\Pi_u \to$ X $^1\Sigma_g^+$, B′ $^1\Sigma_u^+ \to$ X $^1\Sigma_g^+$, and D $^1\Pi_u \to$ X $^1\Sigma_g^+$ band systems of H_2 at 300 K and 100 eV without self-absorption. The model is based on the line positions and transition probabilities of Abgrall et al. (1993a,b,c, 1994, 1997) and the electron emission cross sections of Liu et al. (1998).

Shemansky and Ajello (1983) identified, for the first time, the presence of the two most important UV emissions in the observations by the Voyager I, II spacecraft for application to the outer planets. These are the Rydberg series of H_2, namely, $^1\Sigma_u^+$ 1sσnpσ (B, B′, B″, $n = 2, 3, 4$) $\to$ X $^1\Sigma_g^+$ and $^1\Pi_u$1sσnpπ (C, D, D′, D″, $n = 2, 3, 4, 5$) $\to$ X $^1\Sigma_g^+$ through $n = 5$, along with optically forbidden excitation of X $^1\Sigma_g^+ \to$ EF $^1\Sigma_g^+$ followed by dipole-allowed cascade of EF $^1\Sigma_g^+ \to$ B $^1\Sigma_u^+$, which were found to be the indicators of electron energy and were the source of the principal (overlapping) spectral contributions to the Voyager UV spectrum (Shemansky and Ajello, 1983; Ajello et al., 1984; Jonin et al., 2000). Prior to this work, the Lyman and Werner band systems ($n = 2$) were thought to be the only bands involved in the Voyager analysis (Broadfoot et al., 1979, 1981a,b), as shown in Figure 28.3a. In Figure 28.8, we show (to scale) a composite of the first four ($n = 2, 3$) of the 15 singlet-state band systems contributing directly to the H_2 UV spectrum (Liu et al., 1995; Ajello et al., 2001).

Over the last 20 years, a study has been carried out of the remaining singlet-*ungerade* states in the high-resolution EUV emission spectra of molecular hydrogen excited by electron impact at 100 eV under optically thin, single-scattering experimental conditions (Jonin et al., 2000; Liu et al., 2000; Glass-Maujean et al., 2009). A portion (94.5–96 nm) of the high resolution spectrum (FWHM = 0.0085 nm) spanning the wavelength range of 80–115 nm is shown in Figure 28.9. The total emission cross sections for D $^1\Pi_u$, D′ $^1\Pi_u$, D″ $^1\Pi_u$, B′ $^1\Sigma_u^+$, B″ $^1\Sigma_u^+$ states for the $n = 3$ and four transitions to the ground state were obtained at 100 eV by measuring the emission cross section of each rotational line. The Lyman and Werner bands have the largest emission cross sections at 100 eV with values of 2.64×10^{-17} cm^2 and 2.44×10^{-17} cm^2, respectively (Liu et al., 1998; Jonin et al., 2000; Glass-Maujean et al., 2009). Glass-Maujean et al. (2009) now give the 100 eV electronic Rydberg band system emission and excitation cross sections through $n = 7$. We list these cross sections in Table 28.1.

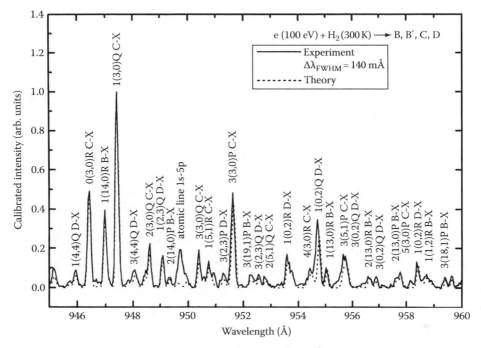

FIGURE 28.9 Over-plot of the observed spectra (FWHM = 140 m Å) and model spectra of H_2 for the low pressure regime of background pressure of 1.2×10^{-5} Torr. The high-pressure spectrum spans the wavelength range from 945 to 960 Å. The model uses the transition probabilities of Abgrall et al. (1993a,b,c, 1994, 1997) with a transmission function for self-absorption at 100 eV electron-impact energy and a gas temperature of 300 K.

For J values of 1–4 at laboratory temperatures of ~300 K, the emission yields of each rovibronic level of the $np\sigma\ ^1\Sigma_u^+$ and $np\pi\ ^1\Pi_u$ states are determined by comparing observed and calculated rotational spectra (Glass-Maujean et al., 2009). Since Jovian aurora take place at elevated rotational temperatures of 500–1200 K (Gustin et al., 2004), models of the EUV require predissociation yields to high J-value (~J = 5–10) (Glass-Maujean et al., 2009). In summary, the mean emission yields of the B $^1\Sigma_u^+$, C $^1\Pi_u$, B′ $^1\Sigma_u^+$, D $^1\Pi_u^+$, D $^1\Pi_u^-$, B″ $^1\Sigma_u^+$, D′ $^1\Pi_u^+$, D′ $^1\Pi_u^-$, D″ $^1\Pi_u$ states, defined previously, are 99%, 98%, 53%, 43%, 100%, 20%, 18%, 78%, 28%, respectively, at 100 eV and 300 K (Glass-Maujean et al., 2009) (see Table 28.1).

Using the Q1(1,4) Werner rovibronic line and the P1,2,3(8,14) Lyman rovibronic line, we developed an accurate modified Born model for the excitation cross section (without cascade) of the Lyman and Werner bands for use in the electron transport codes of planetary atmospheres and astrophysics. The model for 0–1.2 keV electron impact energy is shown in Figure 28.10 with the updated Lyman and Werner 100 eV cross sections (Liu et al., 1998; Glass-Maujean et al., 2009). Glass-Maujean et al. (2009) give estimates for excitation functions for the B″ and D′-states.

An accurate model of a 100 eV high-resolution laboratory electron-impact-induced fluorescence spectrum, based on the calculated transition probabilities and predissociation yields of Abgrall et al., (1993a,b,c) and Glass-Maujean et al., (2007a,b,c,d, 2008), verifies adiabatic transition probabilities for RV rovibronic states of n = 4–8 and nonadiabatic transition probabilities for n = 2–4. The synthetic spectrum is capable of modeling over 98% of the laboratory e + H_2 emission spectrum at room temperature 300 K and 100 eV electron energy. Furthermore, the Lyman and Werner emission cross section energy dependence from Liu et al. (1998) for the B and C states, and the rotational line positions from Roncin and Launay (1994) have allowed us to generate an accurate (~15% accuracy from 79 to 90 nm and ~5% from 90 to 175 nm) synthetic high-resolution rotational line spectrum of the singlet-ungerade states in the UV with electron energies of 10–1000 eV. Using the recent work

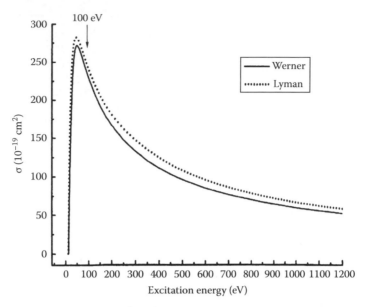

FIGURE 28.10 Total cross section (in 10^{-19} cm^2) of the Lyman and Werner band systems at 300 K. The cross section of the Lyman band system is represented by dots, while that of the Werner band system is shown by the solid line (Liu et al., 1998; Glass-Maujean et al., 2009).

of Glass-Maujean et al. (2007a,b,c,d, 2009) and comparing the synthetic spectra through $n = 8$ with the ESL high-resolution spectra, it has been possible to accurately model direct excitation to the high-lying, *ungerade-singlet* Rydberg states. The high-lying states contribute significantly to the EUV spectrum below 90 nm.

28.4.2 H$_2$–VOIR

The electron-impact-induced fluorescence spectrum of H$_2$ from 330 to 1000 nm at 20 and 100 eV was reported for the first time (Aguilar et al., 2008) at high resolution ($\lambda/\Delta\lambda = 10,000$). The spectrum contains the *gerade* Rydberg series of singlet states EF $^1\Sigma_g^+$, GK $^1\Sigma_g^+$, H$\bar{\text{H}}$ $^1\Sigma_g^+$, I $^1\Pi_g$, J $^1\Delta_g \ldots \rightarrow$ B $^1\Sigma_u^+$, C $^1\Pi$ and the Rydberg series of triplet states dominated by the d $^3\Pi_u$, k $^3\Pi_u$, j $^3\Delta_g \rightarrow$ a $^3\Sigma_g^+$, C $^3\Pi_u$. These VOIR bands were recently observed by the Cassini Imaging Subsystem (ISS) in the visible/near IR filters viewing the Saturn aurora (Dyudina et al., 2007). STIS or Cosmic Origins Spectrograph (COS) might also be able to observe these directly on the outer planets. A model VOIR spectrum of H$_2$ from 750 to 1000 nm, based on newly calculated transition probabilities and line positions, including rovibrational coupling for the singlet-*gerade* states, is in excellent agreement with the observed intensities. Figure 28.11 shows the experimental data of high-resolution (0.07 nm FWHM), electron-impact-induced UV-VOIR emission spectra from 330 to 1200 nm at 100 eV.

The absolute emission cross section values for excitation to the singlet-*gerade* states at 100 eV for optically thin, single-scattering condition was measured to be $4.58 \pm 1.37 \times 10^{-18}$ cm^2; for excitation to the triplet states at 20 eV, the cross section was $1.38 \pm 0.41 \times 10^{-19}$ cm^2 (Aguilar et al., 2008). The singlet-*gerade* emission cross sections are due to cascading into the UV (including the Lyman and Werner band systems), and the triplet-state emission cross sections are due to cascade dissociation of the H$_2$(a $^3\Sigma_g^+$ – b $^3\Sigma_u^+$) continuum that produces fast hydrogen atoms (Ajello and Shemansky, 1993).

In a complementary type of experiment to the VOIR, a newly devised pulsed spectroscopic technique demonstrated (Dziczek et al., 2000; Liu et al., 2002) that the *gerade* series (EF, GK, H$\bar{\text{H}}$, I, J...) make a significant contribution (~50% at 20 eV) to the UV spectrum of H$_2$, by virtue of their cascade to states that correlate to H(1s) and H(2s,2p) (i.e., the upper states of the Lyman and Werner

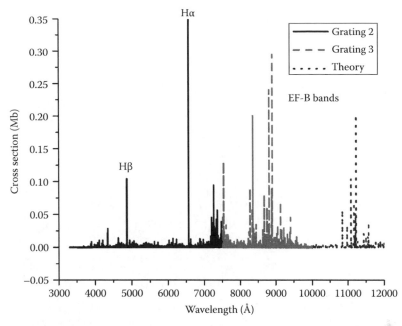

FIGURE 28.11 High-resolution spectrum with intensity in units of cross section (cm^2) for the electron-impact-induced fluorescence spectrum from 3,300 to12,000 Å of H$_2$ at 100 eV in three wavelength regions: (1) grating-2 (3300–7500 Å) in black, grating-3 (7,500–10,000 Å) in long-dash light gray, and theory (10,000–12,000 Å) in short-dash dark gray. The Hα multiplet is off scale (Aguilar et al., 2008).

bands, respectively). This large cascade cross section is important because the mean secondary electron energy in planetary thermosphere and in cosmic-ray-induced ionization in molecular cloud lies between 20 and 100 eV (Gredel et al., 1989; Ajello et al., 2002a, 2005b). The intensity of UV resonance transitions excited by electron impact is determined by both the direct and cascade processes. The lifetime (τ) for decay by spontaneous emission from a dipole-allowed transition is typically short (<10 ns, see Figure 28.7). To measure the electron-impact cascade spectrum of the $n = 2$ H$_2$ Lyman band system (B $^1\Sigma_u^+$ – X $^1\Sigma_g^+$), we use the longer lifetimes (>30 ns, see Figure 28.7) for cascade from higher lying states (EF, GK, HH̄, … $^1\Sigma_g^+$). The pulsed gun technique takes advantage of the drastic difference in lifetimes (~1 ns vs. ~100 ns) between the *ungerade* (direct) states and singlet-*gerade* (cascade) states, respectively. The first laboratory studies and modeling of the UV spectrum of H$_2$ attributed to cascade used a pulsed gun technique to separate cascade and direct excitation effects (Dziczek et al., 2000; Liu et al., 2002). Pulsing the electron gun and gating the photon detector to measure the cascade spectrum after the directly excited population decays also allows a determination of the cascade cross section because the spectral pattern from direct excitation and cascade are different. Direct excitation produces a large population in the B-state centered at $v' = 7$, whereas cascade populates the lower vibrational levels most strongly, beginning at $v' = 0$. By studying Figure 28.12, we clearly see that there are regions in the FUV spectrum dominated by cascading. The most important two wavelength regions that are exclusively (more than 90%) due to cascade lie near the 133–135 and 139–142 nm. These regions correspond to the rotational lines of the (0,1) and (0,2) vibrational bands of the B–X Lyman bands, the two strongest bands of the $v' = (0, v'')$ progression. The strongest feature in the cascade spectrum occurs at 161 nm and involves the superposition of rovibronic transitions from $v' = 4$, 5, and 6.

The medium-resolution spectrum at 14 eV is shown in Figure 28.13b in both the FUV- and middle ultraviolet (MUV) portion of the VOIR extending from 100 to 530 nm, including the Lyman bands and the H$_2$(a $^3\Sigma_g^+$ → b $^3\Sigma_u^+$) continuum (James et al., 1998b). The first detection of the H$_2$ a-b continuum MUV emission in astronomy (Pryor et al., 1998) was made through comparison to our

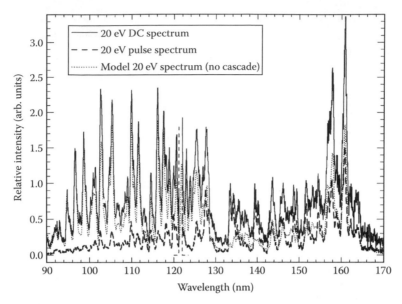

FIGURE 28.12 The 20 eV steady state (cascade + direct) spectrum and linear regression fit using the 20 eV pulsed-gun cascade spectrum and 20 eV model direct excitation spectrum. H Lα is included in the model as a monochromatic line. The photon gate delay for the pulsed-gun spectrum is 135 ns (Dziczek et al., 2000).

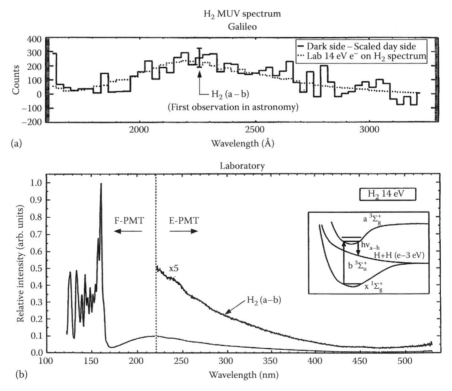

FIGURE 28.13 (a) The first observation in astronomy of the H_2 a-b continuum. Jupiter darkside aurora spectrum overplotted with a 14 eV laboratory spectrum (Proyor et al., 1998). (b) The combined FUV spectrum, measured with an EMR F-photomultiplier (CeTe photocathode), and the MUV spectrum, measured with an EMR –E-photomultiplier, of H_2 corrected for instrument sensitivity produced by electron impact at 14 eV measured at 1.7 nm FWHM at 300 K and 2×10^{-4} Torr background gas pressure in the crossed-beams mode (James et al., 1998a,b).

laboratory spectrum (James et al., 1998b). We show the comparison of the Galileo observation of the Jupiter dark side aurora with the laboratory spectrum in Figure 28.13a.

The set of triplet states above the b $^1\Sigma_u^+$ repulsive state leads to continuous emission with high-velocity H-atoms (~3 eV per atom) formed in the spontaneous dissociation. This process of electron excitation is shown in the inset to Figure 28.13b. The b $^3\Sigma_u^+$ state, which is the lowest lying repulsive state (1sσ) (2pσ), can be excited to the continuum by direct excitation from the ground X $^1\Sigma_g^+$ state or via cascade from the a $^3\Sigma_g^+$ – b $^3\Sigma_u^+$ continuum. The a $^3\Sigma_g^+$ state produces the strongest triplet-state emission from the process H$_2$ (a $^3\Sigma_u^+ \to$ b $^3\Sigma_g^+$), leading to the famous a-b continuum, for electron energies below 30 eV (Ajello and Shemansky, 1993). The a $^3\Sigma_u^+$ state is strongly populated by cascade from the c-, d-, and e-states. The excitation function for the a-b continuum is shown in Figure 28.14. The figure (see inset) also shows a 20 eV spectrum indicating relative importance and wavelength region for strong cascade and a-b continuum in the FUV. The triplet states are the major source of dissociation of the hydrogen molecule by electron impact.

Optical excitation functions of the triplet states have also been measured by few experimenters in the VOIR. Dieke (Crosswhite, 1972) has shown the presence of many triplet band systems in his discharge emission experiments with the strongest and most extensive to be the Fulcher-α band system (3pπ d $^3\Pi_u \to$ 2sσ a $^3\Sigma_g^+$). The emission cross sections of the Fulcher-α diagonal bands ($\Delta v = 0$) have been studied by Möhlmann and de Heer (1976). Those processes leading to triplet emissions arise first from singlet–triplet excitation. The excitation occurs by electron exchange, which is characterized by a fast rise and decrease in the emission cross section within a few eV of threshold (see Figure 28.14). Tawara et al. (1990) have reviewed the excitation cross sections of the X $\to$ 2sσ a $^3\Sigma_g^+$, 2pσ b $^3\Sigma_u^+$, 2pπ c $^3\Pi_u$, and 3pσ e $^3\Sigma_u^+$ states by electron-energy-loss experimental techniques. Emission from the e $^3\Sigma_u^+$ triplet state e $\to$ a transition was observed by Dieke (1958) and Dieke and Cunnigham (1965) as well as by Dieke in a couple of early publications as referenced by Huber and Herzberg (1979) and Crosswhite (1972). The c-state, whose $v' = 0$ level lies below the a-state, is forbidden for transitions to the b- and X-state. Thus, there are two Rydberg series terminating in the two bound triplet states,

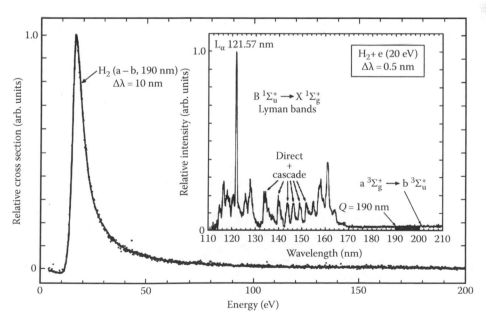

FIGURE 28.14 The H$_2$(a $^3\Sigma_g^+ \to$ b $^3\Sigma_u^+$) excitation function from 0 to 200 eV (Ajello and Shemansky, 1993). The inset is a combined FUV and MUV spectrum of H$_2$(a $^3\Sigma_g^+ \to$ b $^3\Sigma_u^+$) continuum at 20 eV. The cross-hatched area shows the band pass of the spectrometer for the excitation function measurement.

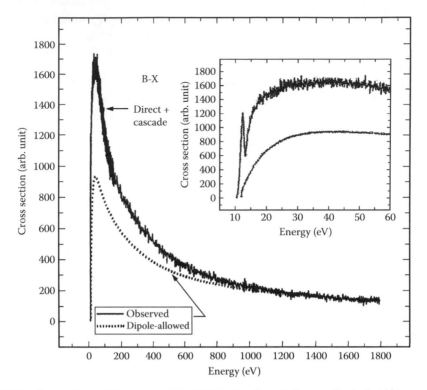

FIGURE 28.15 Comparison of the observed H_2 B-X dipole-allowed direct + dipole-forbidden cascade excitation function for $v = 0$ (solid), compared to the $v > 7$ dipole-allowed excitation function (dotted). The inset shows the threshold behavior of the direct excitation and total cross section. The total cross section demonstrates the existence of resonance excitation and cascading for the low vibrational levels (0–4) near 15 eV (Liu et al., 2003).

a and c. Excited triplet states undergoing transitions to either the a- or c-state are thereby forbidden by the g ↔ u rule for transitions to both final states.

An additional important topic related to low-energy electron excitation of the RV states of H_2 involves resonance excitation, especially of the Lyman bands. We have recently published an analytic model for the $n = 2$ Lyman band system (B $^1\Sigma_u^+ \rightarrow$ X $^1\Sigma_g^+$) of H_2 (Liu et al., 2003) that is accurate at threshold. We have shown in Figure 28.15 (inset) that for the B $^1\Sigma_u^+$ state, the measurement of the UV excitation function of a single rotational line of a low vibrational level (0–4) contains near threshold structure arising from a combination of resonance, dipole-forbidden, and dipole-allowed components. Figure 28.15 shows a measurement of the B $^1\Sigma_u^+$ (0,4) P3 high-resolution excitation function. The $v' = 0$ excitation function is composed of three processes: (1) direct excitation (see dotted component in Figure 28.15); (2) resonance excitation of H_2^- autoionizing states (see first peak in inset at ~13 eV); and (3) a dipole-forbidden excitation from ($n = 2$) EF, ($n = 3$) GK, ($n = 4$) H$\overline{\text{H}}$ $^1\Sigma_g^+ \rightarrow$ B $^1\Sigma_u^+$ *gerade* state cascading. The competition between UV emission production by the band systems of the singlet states and dissociative production of fast H(1s) atoms is a very sensitive function of electron energy in the threshold energy region of 10–50 eV. This allows the mean electron impact energy to be unfolded from astronomical regimes, as was done with Voyager spectral observations of the outer planets.

28.4.3 N_2–EUV

The strongest dipole-allowed transitions of N_2 occur in the EUV (Ajello et al., 1989b). The excitation of the N_2 RV states present in the EUV and FUV (80–140 nm) plays a role in establishing the physical composition of an N_2-bearing atmosphere (Earth, Titan, Triton, and Pluto). The electronic transitions proceed from the X $^1\Sigma_g^+$ ground state to nine closely spaced (12–15 eV) RV states, which are the source of the molecular emissions in the EUV observed by spacecraft (Ajello et al., 2007). Three of these RV states, b $^1\Pi_u$, b′ $^1\Sigma_u^+$, and c′₄ $^1\Sigma_u^+$, are highly perturbed, weakly-to-strongly predissociated, and have significant emission cross sections (e.g., James et al., 1990). The other two RV states, c₃ $^1\Pi_u$ and o₃ $^1\Pi_u$, are nearly 100% predissociated by the triplet C $^3\Sigma_u^+$ and C′ $^3\Pi_u$ states (Lewis et al., 2005). When these five singlet-*ungerade* states predissociate, they eject fast N-atoms (>1 eV) through the N(^{4}S°) + N(^{2}D°) and N(^{4}S°) + N(^{2}P°) dissociation limits located at 12.1 and 13.3 eV, respectively. The energy level diagram of the RV states and the names of the emission band systems are shown in Figure 28.16. The emission spectrum of the singlet-*ungerade* RV states

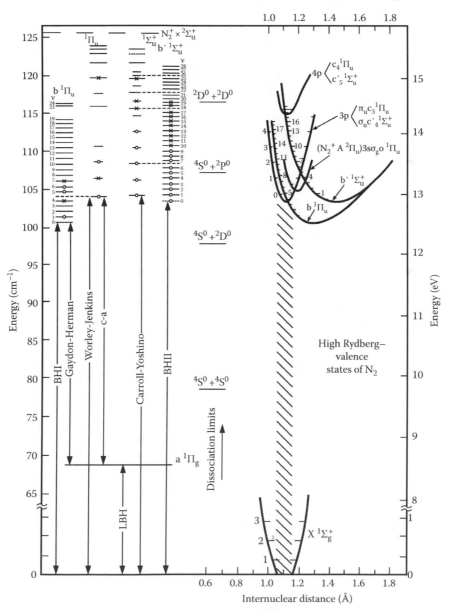

FIGURE 28.16 Partial energy-level diagram for N_2 emphasizing the 12–15 eV energy region of the RV states. The right side of the figure shows the diabatic potential curve, and the left-hand side the observed vibrational levels. The circles and x's in the figure are explained in Ajello et al. (1989b).

TABLE 28.2
The Cross Sections for Low Lying Five RV States of N_2

State	Q_{ex} (10^{-19} cm^2)	Q_{em} (10^{-19} cm^2)	Q_{pre} (10^{-19} cm^2)	Q^f_{em} (10^{-19} cm^2)
c'_4 $^1\Sigma^+_u$	158[a,b,d]	125[a,c,d]	34[a,b,c]	122[f]
b' $^1\Sigma^+_u$	128[a,e]	15.5[a]	112.5[a,e]	19.3[f]
b $^1\Pi_u$	121[a]	6.1[a]	115[a]	8.4[f]
c_3 $^1\Pi_u$	161[a]	0[a]	161[a]	7.4[f]
o_3 $^1\Pi_u$	75[a]	0[a]	75[a]	5.7[f]
Totals	643	147	497	163[f]

Note: Predissociation (fast N I atom) yield (Q_{pre}/Q_{ex}) = 497/643 = 77% and EUV photon yield = 23% (Ajello et al., 2007, columns 2–4).

[a] Where $Q_{ex} = Q_{em} + Q_{pre}$.
[b] $v' = 0$ at 150 K, $\eta_{pre}(c'_4) = 23\%$, at 300 K, $\eta_{pre}(c'_4) = 26\%$, where η is predissociation yield.
[c] Includes additional c'_4 $(0, v'' = 6–12)$ bands (Bishop et al., 2003; Ajello et al., 2007).
[d] Includes revised $Q_{ex}(v' = 0) = 125$ (units of 10^{-19} cm^2).
[e] b' $^1\Sigma^+_u$ value based on correction to $v' = 9$ and 11 emission cross sections (Walters et al., 1994; Ajello et al., 2007).
[f] Preliminary high resolution emission cross sections (Ajello, 2010) in column 5.

displays many irregularities due to homogeneous RV interactions within the $^1\Sigma^+_u$ and $^1\Pi_u$ manifolds and $^1\Sigma^+_u \sim {}^1\Pi_u$ p-complex heterogeneous interactions (Liu et al., 2008). The recently revised emission cross sections of the highly perturbed b $^1\Pi_u$, b' $^1\Sigma^+_u$, and c'_4 $^1\Sigma^+_u$ states that are weakly-to-strongly predissociated are large; the excitation, emission and predissociation cross sections for the low-lying five RV states are summarized in Table 28.2. Table 28.2 adapted from the work of Ajello et al., 2007, including minor changes, in columns 2 to 4, respectively; and the emission cross section from the recent higher resolution work of Ajello (2010) is given in column 5.

Ever since the Voyager 1 (V1) encounter with Saturn in 1980, the EUV airglow of Titan has challenged attempts to explain both its spectral content and its excitation source. Because of the similarity to optically thin laboratory spectra from electron impact on N_2 (Fischer et al., 1980; Ajello et al., 1989b), most early analyses of the V1 UVS data argued that Titan's EUV airglow was dominated by the N_2 c'_4 $^1\Sigma^+_u(0) \rightarrow X$ $^1\Sigma^+_g(0)$, i.e., $c'_4(0,0)$ band near 95.8 nm and the $c'_4(0,1)$ band near 98 nm (Broadfoot et al., 1981a,b, Strobel and Shemansky, 1982). Though readily excited by photoelectron impact, the earliest work on the Titan airglow noted, however, that the resonant $c'_4(0,0)$ band was optically thick near peak photoelectron excitation. An excitation source driven by the Sun was therefore ruled out, since the $c'_4(0,0)$ emission band would be radiatively trapped, so a magnetospheric source near Titan's exobase was proposed instead (Strobel and Shemansky, 1982).

The issue was studied by Stevens et al. (1994), who developed a $c'_4(0, v'')$ multiple scattering model for the terrestrial atmosphere and showed that $c'_4(0,0)$ should be weak or undetectable near peak photoelectron excitation and that $c'_4(0,1)$ should dominate over $c'_4(0,0)$. A similar analysis was done for Titan's airglow by Stevens (2001, 2002) and Stevens et al. (2003), who argued that $c'_4(0,0)$ was misidentified at Titan and two prominent N I multiplets (95.2 and 96.4 nm) produced primarily by photodissociative ionization (PDI) of N_2 were present instead. This meant that the Titan EUV dayglow could be excited exclusively by the Sun. The key N I emissions that could not be conclusively identified by UVS because of its low spectral resolution (3 nm) have now been identified with the higher spectral resolution (0.56 nm) of the UVIS instrument on the Cassini spacecraft.

The UVIS disk-averaged dayglow spectrum in the spectral range 90–114 nm is shown in Figure 28.17 (top panel) at 0.56 nm FWHM. The 16 indicated dayglow features are identified in Table 28.3. The identifications are based on the work of Ajello et al. (1989b), James et al. (1990), and

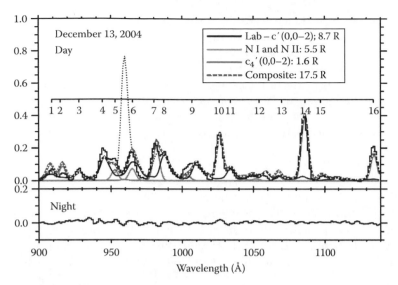

FIGURE 28.17 (Top panel) Regression analysis to UVIS dayglow spectrum from December 13, 2004. The regression model fit consists of three independent vectors: (1) an optically thin 20 eV electron-impact lab fluorescence spectrum, with c_4' (0,0) dotted; (2) the calculated multiple scattered emergent spectrum of the c_4' $(0,v'' = 0-2)$ progression transmitted through an optically thick medium; and (3) a spectrum of N I and N II emissions (Ajello et al., 2007). (Lower panel) Cassini UVIS nightglow spectrum.

TABLE 28.3
Identification of Strongest Titan Dayglow Emission Features from Cassini UVIS on December 13, 2004

Feature	λ (Å)	$4\pi I^{b,c}$ (R)	Identification[a]
1	909	0.50	$c'(4,1)$, $N(^4S\text{-}^4P)$, $b'(16,2)$, $b'(12,1)$
2	915	0.32	$N(^3P\text{-}^3P^o)$
3	928	0.41	$b'(9,1)$, $c'(6,4)$
4	944	1.56	$c'(4,3)$, $c'(3,2)$, $b'(9,2)$, $c'(6,5)$, $b'(16,4)$
5	953	0.84	$N(^4S^o\text{-}^4D)$, $N(^4S^o\text{-}^4P)$
6	965	1.39	$c'((1,1)$, $N(^4S^o\text{-}^4P)$, $c'(3,3)$, $c'(4,4)$
7	980	1.47	$c'(0,1)$
8	987	1.45	$c'(3,4)$, $c'(4,5)$, $c'(6,7)$
9	1007	1.46	$c\ ('0,2)$, $c'(3,5)$, $c'(4,6)$, $b(1,1)$, $c'(6,8)$, $b'(9,5)$
10	1026	1.71	H Ly-β
11	1034	0.46	$b(1,2)$, $b'(9,6)$
12	1054	0.56	$N(^2D^o\text{-}^2P)$, $b'(11,7)$, $b(1,3)$, $b'(3,5)$
13	1070	0.33	$N(^2D^o\text{-}^{2,4}D)$, $N(^2D^o\text{-}^{2,4}P)$
14	1086	2.31	$N^+(^3P\text{-}^3D^o)$
15	1118	0.51	$N(^2D^o\text{-}^2D)$, $N(^2D^o\text{-}^2P)$, $b'(9,9)$
16	1135	1.33	$N(^4S^o\text{-}^4P)$

Source: Ajello, J.M. et al., *Geophys. Res. Lett.*, 34, L24204, 2007.

[a] Identifications from Ajello et al. (1989b), James et al. (1990), Bishop and Feldman (2003), and unpublished high-resolution (0.2 Å FWHM) laboratory spectra.

[b] Total observed EUV integrated disk intensity (900–1140 Å) in Rayleighs (R) is 16.6 R.

[c] VI UVS: UVIS comparison; total observed UVIS intensity (920–1015 Å) is 8.6 R vs. total modeled VI intensity 920–1015 Å is 20.9 R.

unpublished, laboratory high-resolution spectra Ajello (2010), as well as the terrestrial airglow spectra of Bishop and Feldman (2003). Laboratory spectra (0.02 nm FWHM) have shown the presence of about 200 spectral features from electron-excited N_2 over the same EUV spectral range (Ajello et al., 2009). On average, there are some 10 emissions for each feature number in Figure 28.17 (top panel); we list only the strong ones in Table 28.3. The strongest is feature number 14, the N II multiplet (^3P-$^3D^o$) near 108.5 nm with an intensity of 2.3 Rayleigh (R). Whereas Voyager 1 only observed a few blended EUV features, the UVIS dayglow spectrum indicates 16 features, with the c_4' (0,0) band conspicuous by its absence. For completeness, the nightglow spectrum is included in the lower panel of Figure 28.17.

There have been no high-resolution (<0.01 nm FWHM) laboratory studies of the optically thin, electron-impact N_2 fluorescence EUV spectrum from 50 to 120 nm since the medium-resolution (0.03 nm FWHM) study 20 years ago (Ajello et al., 1989b). ESL's effort for higher resolution (FWHM $\approx$ 0.002–0.010 nm) studies is now underway. Our more recent high-resolution laboratory measurements have found a total of nine RV states contributing to the N_2 EUV emission spectra from 80 to 140 nm; see Table 28.4 (Ajello, 2010). These states have principal quantum numbers through $n = 6$ and contribute to the emission and predissociation cross sections of N_2 in the EUV.

We show examples of high-resolution laboratory spectrum measured at ESL in Figures 28.18 and 28.19. Figure 28.18 shows the c_4' $^1\Sigma_u^+$ ($v' = 0$) $\rightarrow$ X $^1\Sigma_g^+$ ($v'' = 1$) band model at 98 nm (Liu et al., 2008), the strongest band in the Titan EUV airglow (i.e., feature 7 in Figure 28.17a). The c_4' (0,1) band analysis was used to determine predissociation yields for each rotational level and a band transition moment. There are now full spectral models for the rotational structure of the c_4' $^1\Sigma_u^+$ ($v' = 0$) $\rightarrow$ X $^1\Sigma_g^+$ ($v'' = 0,1,2$) progression at 95.9, 98.0, and 100.3 nm (Stevens, 2001; Liu et al., 2005, 2008) with which we can synthesize optically thick photoelectron-excited spectral lines using a multiple scattering model. Figure 28.19 shows a medium-resolution laboratory spectrum (0.02 nm FWHM) at both 20 and 100 eV compared to a Cassini EUV (90–115 nm) spectrum (dashed-line) indicating that there are many N_2 bands (approximately 200) contributing to each of the observed 16 Cassini features in Table 28.3. At this point, it is necessary to study the rotational cross sections and predissociation yields of the strongest remaining vibrational band features identified in Table 28.3.

TABLE 28.4
List of N_2 Rydberg and Valence Electronic Band Systems Observed in EUV Spectra of Electron-Impact-Induced Fluorescence

Electronic Transition	T_e (cm^{-1})
b' $^1\Sigma_u^+ \rightarrow$ X $^1\Sigma_g^+$-valence	104,498
b $^1\Pi_u \rightarrow$ X $^1\Sigma_g^+$-valence	101,675
c_4' 3pσ $^1\Sigma_u^+ \rightarrow$ X $^1\Sigma_g^+$ Rydberg	104,519
o_3 3sσ $^1\Pi_u \rightarrow$ X $^1\Sigma_g^+$-core excited	105,869
c_5' 4pσ $^1\Sigma_u^+ \rightarrow$ X $^1\Sigma_g^+$	115,876
c_6' 5pσ $^1\Sigma_u^+ \rightarrow$ X $^1\Sigma_g^+$	—
c_3 3pπ $^1\Pi_u \rightarrow$ X $^1\Sigma_g^+$	104,476
c_4 4pπ $^1\Pi_u \rightarrow$ X $^1\Sigma_g^+$	115,636
c_5 5pπ $^1\Pi_u \rightarrow$ X $^1\Sigma_g^+$	—

Source: Ajello, J.M. High resolution spectra of N_2, 2010, in preparation. See Huber and Herzberg, 1979 for electronic energies of each state.

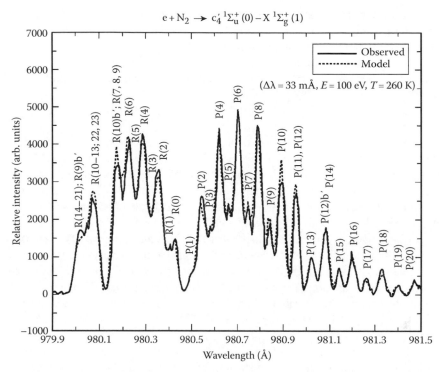

FIGURE 28.18 Comparison of the high-resolution laboratory and close-coupled-Schrödinger spectral model for the c'_4 (0)-X(1) transition rotational structure to obtain predissociation yields and diabatic transition moments (Liu et al., 2008).

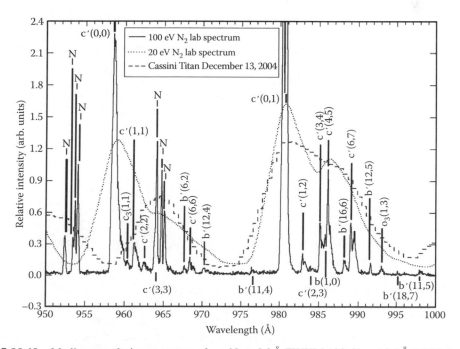

FIGURE 28.19 Medium-resolution spectrum of e + N_2 at 0.2 Å FWHM 100 eV and 0.6 Å FWHM at 20 eV compared to the Cassini UVIS spectrum of December 13, 2004 at 4 Å FWHM. The identification of the molecular and atomic features is from Ajello (2010).

Additionally, the low temperature of Titan's atmosphere prompts the need for laboratory measurements of cross sections at low temperatures from 120 K found near the mesopause (500 km) to 186 K at the exobase at 1400 km (Wilson and Atreya, 2005). We have perfected expansion cooling of molecules through effusive nozzle for molecular-beam–electron-beam interaction to match Titan's atmospheric temperatures (Ajello et al., 1998a) and intend to carry out such studies.

28.4.4 N_2–FUV

A variety of space missions have been flown to observe terrestrial N_2 emissions in the FUV spectral regime. These include the MSX, TIMED, POLAR, IMAGE, and DMSP satellites. Moreover, solar UV spectral irradiance measurements, important for establishing the radiative energy input to the Earth's upper atmosphere, are currently being obtained by instruments on board the TIMED and SORCE satellites. This suite of instruments allows the interaction between the Sun and the Earth to be studied in unprecedented detail over a solar cycle.

The measurement goals of these missions are to achieve an accuracy of better than 10% in defining the atmospheric parameters (temperature, composition, density, etc.) on a global scale and to determine radiative, chemical, and dynamical energy sources. For remote sensing of the upper atmosphere, there are four principal wavelength intervals in UV imaging satellites that have been used to observe the distribution of N_2 and O. Typical of these bands are those of the Global UltraViolet Imager (GUVI) instrument on TIMED: O I (130.4 nm), O I (135.6 nm), N_2 (141–153 nm), and N_2 (167–181 nm). The latter two wavelength intervals are referred to as Lyman–Birge–Hopfield (LBH) short (LBH_S) and LBH long (LBH_L), respectively. They arise from the transition of N_2 (a $^1\Pi_g \to$ X $^1\Sigma_g^+$). As an example, we show in Figure 28.20 an MSX FUV auroral spectrum (Paxton and Meng, 1999) with the LBH bands and atomic oxygen emissions indicated. Because of their importance in atmospheric remote sensing, we review the current understanding of LBH excitation and emission processes in some detail in this chapter.

The analysis of the remote sensing spectra aimed at unraveling the behavior of the major constituents of the upper thermosphere, N_2, O_2, and O, and the auroral energy input depends on the details of the N_2 LBH, and O emission cross sections, as well as the absorption cross section of O_2. The fundamental excitation and emission processes involved for N_2 and O and their cross section definitions are

1. $e^- (E_e) + N_2 \to e^- (E_e - \Delta E_e) + N_2{}^* \to e^- + N_2 + h\nu$ (a $^1\Pi_g$ – X $^1\Sigma_g^+$)
   - σ_{ex} (LBH) is the cross section for direct excitation of the optically forbidden ($\tau \sim 55\,\mu s$) a $^1\Pi_g$ – X $^1\Sigma_g^+$ LBH band system from 120 to 210 nm, including LBH_S (141–153 nm) and LBH_L (167–181 nm).
2. $e^- (E_e) + N_2 \to e^- (E_e - \Delta E_e) + N_2{}^* \to e^- + N_2 + h\nu$ (a′ $^1\Sigma_u$ and w $^1\Delta_u \to$ a $^1\Pi_g$)
   - σ_{casc} (LBH) for cascade emission from optically forbidden ($\tau > 1$ ms) cascade transitions (a′ $^1\Sigma_u$ and w $^1\Delta_u \to$ a $^1\Pi_g$) to the LBH band system.
3. $e^- (E_e) + O \to e^- (E_e - \Delta E_e) + O^* \to e^- + O + h\nu$ ($^3P_2 \to {}^5S_2{}^o$ at 135.6 nm)
   - σ_{ex} for optically forbidden emission ($\tau \sim 180\,\mu s$) of OI (135.6 nm).
4. $e^- (E_e) + O \to e^- (E_e - \Delta E_e) + O^* \to e^- + O + h\nu$ (3s $^5S_2{}^o \to$ 3p $^5P_{1,2,3}$ and 3s $^3S_1{}^o \to$ 3p $^3P_{0,1,2}$)
   - σ_{casc} (O I) for dipole-allowed $h\nu$ ($\tau \sim 1$–10 ns) of O I (777.4 and 844.6 nm and higher order states).

The history of LBH cross sections by electron-impact measurements was reviewed by van der Burgt et al. (1989) and (Meier, 1991). Shown in Table 28.5 are the N_2 LBH cross section data reported in the literature. There are considerable differences among the cross sections, with values for direct excitation to the a-state differing by almost a factor of two in some cases. The discrepancies could be due to experimental limitations in fully capturing the long-lived emitting states, improper accounting of cascade contributions, or both. The shape of the excitation function peaking at about 18 eV was measured by Ajello and Shemansky (1985). This has been the standard cross section used in the

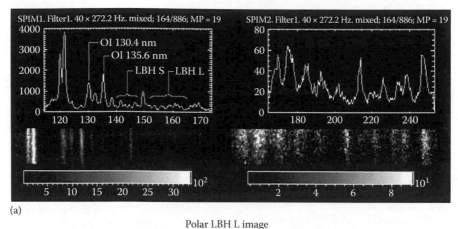

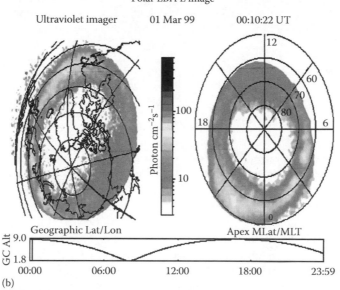

FIGURE 28.20 (a) MSX FUV spectrum showing N_2 LBH and OI multiplets (Fischer et al., 1980); (b) POLAR spacecraft image of LBH image. Note that for GUVI, the LBHL band covers 167–181 nm (Paxton and Meng, 1999).

Aeronomy community since then. But recent remeasurements by Johnson et al. (2005b) and Young et al. (2010) found the peak nearer 20 eV, with a different energy function. The implications of the new measurement are discussed below.

Strickland et al. (1995) have shown that satellite observations of the thermospheric ratio of O I 135.6 to N_2 LBH are closely related to the O/N_2 column density ratio, which itself is a good indicator of thermospheric dynamical conditions, especially during geomagnetic storms (Meier et al., 2005; Crowley and Meier, 2008). The N_2 LBH cross sections of Ajello and Shemansky (1985 or AS85) have been extensively used in establishing this relationship. Various modelers have scaled the AS85 cross section by as much as a factor of 2 both to adjust for revisions in the absolute standard for the measurements of direct excitation to the a-state and to account for the estimated effects of cascade to the a-state from the a' $^1\Sigma_u$ and w $^1\Delta_u$ states (Meier, 2008). Because of the importance of the LBH cross section to remote sensing, we review the basis of Meier (2008) for estimating the best cross section to use until new measurements become available.

AS85 obtained a value of 3.02×10^{-17} cm^2 at 18 eV for the peak excitation cross section from a model fit to the laboratory data (their Table 5a, column 2). They found that predissociation accounts

TABLE 28.5
Comparison of N_2 LBH Cross Sections (in Units of 10^{-17} cm²)

Reference	Peak Excitation Cross Section	Cascade Contribution Estimates (% of Excitation)	Peak Emission Cross Section, Including Cascades Estimate
Ajello (1970) PE	3.85		
Brinkmann and Trajmar (1970) ES	4.50		
Cartwright (1977;1978) ES	2.72, 3.02[a]	73	4.63
Ajello and Shemansky (1985) PE	2.65[b]	Not observed	2.35
Mason and Newell (1987) MP	3.50		
Brunger and Teubner (1990) ES	4.24		
Eastes and Dentamaro (1996), Eastes (2000) model	3.02	55	4.15
Budzien et al. (1994)	2.70	45	3.50
Strickland et al. (2004) estimate	2.71	40	3.79
Campbell et al. (2001)			4.69
Johnson et al. (2005) EI			2.10
Meier (2008) model analysis	2.64	74	4.08
Young et al. (2010)	1.98	29–48	2.22–2.57
Present estimate	1.98	31	2.28

Note: PE: photoemission measurement, MP: metastable particle measurement, ES: electron scattering, EI: electron impact.

[a] This value is the result of a scaling by a factor of 0.90 following the work of Trajmar et al. (1983).

[b] This value is the result of a scaling by a factor of 0.875 due to the reevaluation of the H Lyman-α standard (van der Burgt et al., 1989).

for a loss of 12.29% of all excitations to the a-state; the emission branching ratio for total emission is therefore 0.8771 (ignoring vibrational/rotational dependence). To calibrate their emission cross section measurements, AS85 used as their standard, the cross section for electron impact on H_2 yielding H Lyman alpha. AS85 adopted a cross section of 0.818×10^{-17} cm² at 100 eV (or 0.578×10^{-17} cm² at 200 eV). More recently, Liu et al. (1998) measured an H_2 Lyman alpha cross section of 0.716×10^{-17} cm² (near the value of 0.703×10^{-17} cm² recommended by McConkey et al. (2008)). Consequently, we reduce the AS85 peak LBH excitation cross section to 2.64×10^{-17} cm² (=$3.02 \times 0.716/0.818$). For emission without cascade, this becomes $2.64 \times 10^{-17} \times B = 2.32 \times 10^{-17}$ cm².

Figure 28.21 shows the electron impact cross sections for excitation of a, a′, and w states. The scaled AS85 cross section is shown as a solid line, and the data of Young et al. are given by the individual points. The dashed line is a fit to the Young et al. data obtained by scaling and stretching the AS85 energy function. The Cartwright et al. (1977) a′ and w cross sections, often used for dayglow modeling (Strickland et al. 1999), are shown as solid lines. More recently, Johnson et al. (2005b) have remeasured these cross sections, finding them to be lower than Cartwright et al. Their data are plotted as asterisks. The dashed line is an attempt to fit the Young et al. data by adjusting the Cartwright et al. energy function. The dashed lines in Figure 28.21 do not fit the data very well for energies greater than about 60 eV, but they should be sufficiently accurate for dayglow modeling because the photoelectron fluxes peak at much lower energies.

It has been known (at least) since the work of Freund (1972) that cascade takes place between the a $^1\Pi_g$ state and both the a′ $^1\Sigma_u$ and w $^1\Delta_u$ states. Later evaluations of cascade were published by Cartwright (1977, 1978) and Eastes (2000). Both Cartwright and Eastes carried out detailed calculations of the interactions among the various states, including laddering back and forth, enhanced ground state vibrational populations, threshold effects, etc. Because a correct calculation of the a $^1\Pi_g$

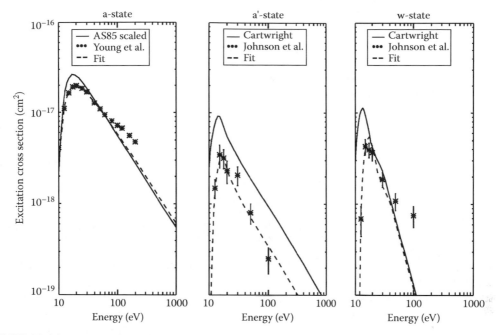

FIGURE 28.21 Electron impact excitation cross sections relevant to the N_2 Lyman-Birge-Hopfield band system (a $^1\Pi_g \rightarrow X\ ^1\Sigma_g^+$) in nitrogen atmospheres. Left panel: direct excitation to the a-state. Middle panel: excitation to the cascading a′-state. Right panel: excitation to the cascading w-state. See text for details.

population rate, including interactions with the a′ $^1\Sigma_u$ and w $^1\Delta_u$ states is very involved, it is useful to derive an effective emission cross section to correct the direct a $^1\Pi_g$ population rate for the effects of cascade. This has been a common approach where a fully developed cascade model is computationally prohibitive for routine processing of satellite databases.

Because transitions of a′ and w to the X-state of N_2 are forbidden, we assume that all excitations of these states end up as a-state emissions. The total volume emission rate from the a-state (ignoring the details of populating the individual vibrational states) can be written as the product of the excitation (population) rates and the branching ratio, B, for emission:

$$j_{em}^a(z) = B j_{ex}^a(z) + B j_{ex}^{a'}(z) + B j_{ex}^w(z) \tag{28.11}$$

where j is the number of emissions (excitations) cm³/s and B is the branching ratio for emission (1 −the probability of predissociation). Equation 28.11 can be rewritten as

$$j_{em}^a(z) = B j_{ex}^a(z) \left[1 + \frac{j_{ex}^{a'}(z)}{j_{ex}^a(z)} + \frac{j_{ex}^w(z)}{j_{ex}^a(z)} \right] \tag{28.12}$$

If the a, a′, and w volume excitation rates have similar altitude dependences, the quantity in brackets is constant. (See below for more details on this assumption.)

The direct volume excitation rate is the product of the excitation rate (g-factor) per second per molecule and the N_2 number density, n:

$$j_{ex}(z) = g_{ex}(z) n_{N_2}(z) \tag{28.13}$$

with the g-factor defined as

$$g_{ex}(z) = \int \sigma_{ex}(E) F(E)\ dE \tag{28.14}$$

where

σ_{ex} is the excitation cross section
F is the photoelectron flux, both evaluated at energy, E

The cross section can be written as the product of the peak cross section and ϕ, the normalized energy function: $\sigma = \sigma^{peak}\phi(E)$. Then

$$g_{ex}(z) = \int \sigma_{ex}^{peak}\phi(E)F(z,E)\ dE = \sigma_{ex}^{peak}\gamma(z) \tag{28.15}$$

where $\gamma(z)$ describes the altitude dependence of the g-factor. If γ is the same for excitation and emission (i.e., the energy functions are the same or the energy dependence of F does not change with altitude), then the a-state volume emission rate can be defined in a similar manner:

$$j_{em}^{a}(z) \equiv n_{N_2}(z)g_{em}(z) = n_{N_2}(z)\int \sigma_{eff}^{a}\phi(E)F(z,E)\ dE = n_{N_2}(z)\sigma_{eff}^{a}\gamma(z) \tag{28.16}$$

where σ_{eff}^{a} is defined as the effective peak cross section for emission from the a state, including cascade components.

Substitution of Equations 28.13 through 28.16 into Equation 28.12 gives the following relationship for the effective emission cross section:

$$\sigma_{eff}^{a}\gamma(z)n_{N_2}(z) = B\sigma_{ex}^{a}\gamma(z)n_{N_2}(z)\left[1 + \frac{j_{ex}^{a'}(z)}{j_{ex}^{a}(z)} + \frac{j_{ex}^{w}(z)}{j_{ex}^{a}(z)}\right] \tag{28.17}$$

or

$$\sigma_{eff}^{a} = B\sigma_{ex}^{a}\left[1 + \frac{j_{ex}^{a'}(z)}{j_{ex}^{a}(z)} + \frac{j_{ex}^{w}(z)}{j_{ex}^{a}(z)}\right] \tag{28.18}$$

Next the Atmospheric Ultraviolet Radiance Integrated Code (AURIC; Strickland et al., 1999) model was used to calculate thermospheric volume excitation rates (without cascade) using the latest Young et al. (2010) and Johnson et al. (2005b) cross sections for a, a' and w states at several solar activity and illumination conditions (Evans, 2010). We found that the ratios for a'/a and w/a (using shorthand notation for the ratios of volume excitation rates) are nearly independent of altitude between 150 and 250 km (to within ±5%) where the bulk of the dayglow originates, and equal 0.134 and 0.172, respectively. Substituting these into Equation 18, the effective a-state emission cross section becomes $\sigma_{eff} = 0.8771 \times 1.98 \times 10^{-17}$ cm$^2 \times (1 + 0.134 + 0.174) = 1.74 \times 10^{-17}$ cm$^2 \times [1 + 0.31]$, or $\sigma_{eff} = 2.28 \times 10^{-17}$ cm^2. Thus, our estimate indicates an a-state emission cross section enhancement of 31% by cascade.

Our simplified estimate can be compared with the much more detailed calculations of Cartwright (1978) and Eastes (2000). Cartwright (1978) modeled the vibrational populations of a variety of N_2 states in an IBC II Aurora. His direct a-state excitation cross section was 3.02×10^{-17} cm^2 (surprisingly the same as AS85). His Figure 28.18, left panel, displays the vibrational populations of the a-state for direct excitation and cascade from the a'- and w-states, for altitudes of 110 km and >130 km. Cartwright found little change in the percentage cascade contributions and fractional populations above 130 km, thereby supporting the assumption used in our derivation of the effective cross section. Using Cartwright's >130 km case, we can estimate an effective emission cross section from his work. Summing the population rates for each vibrational level of the three states, we find total population rate ratios of a'/a = 0.60 and w/a = 0.13. The total is 0.73, more than twice our value, as expected from his larger cross sections. We obtain a similar value for the total cascade enhancement using the Trajmar et al. (1983) renormalization of the Cartwright cross sections, although the individual contributions from a' and w are different.

Eastes (2000) carried out a more detailed calculation of the LBH emission rate that included cascade. According to his Table 28.1 a peak excitation cross section of 2.69 was used. The conclusions of his paper state that the calculated emission rate with cascade is 55% larger than with excitation alone (i.e., without cascade). The effective cross section becomes $2.69 \times 1.55 \times 0.8771 = 3.66$, somewhat larger than ours. But using the Johnson et al. a' and w cross sections, his cascade would become 22% and the effective cross section becomes 2.9×10^{-17} cm², closer to our derived value.

Cascade rates from a'- and w-states to the a-state shown in Figure 28.4 of Eastes (2000) vary strongly with altitude below about 200 km. This is discrepant with Cartwright (i.e., small variations above 130 km) and with the ratios of excitation rates from AURIC (although there may be some altitude-dependent effect in the cascade rates that could cause the emission rate ratios to differ from the excitation rate ratios). This discrepancy has not been resolved. A possible explanation is that Eastes used the Continuous-Slowing-Down photoelectron model of Jasperse (1976), whereas the AURIC photoelectron model is much more physically realistic. If the Jasperse model produces much greater altitude variations in the photoelectron flux energy dependence, it could account for the different behavior. Eastes also calculated the effect of collision-induced electronic transitions, but these should not be of much significance in the dayglow that is produced at higher altitudes where collisions are infrequent.

Budzien et al. (1994) analyzed dayglow observations of the LBH band system. They used the AS85 excitation cross section corrected for the revised H_2 standard (2.7×10^{-17} cm²). They found that the total excitation rate had to be increased by a factor of 1.45, which when adjusted for predissociation, yields an effective emission cross section of $1.45 \times 2.7 \times 10^{-17} \times 0.8771 = 3.5 \times 10^{-17}$ cm², larger than ours by 54%.

Young et al. (2010) estimate the effect of cascade based on the ratios of cross sections: (a' + w)/a = 0.48 at 15 eV and 0.29 at 20 eV. Our value using volume emission rate ratios of 0.31 is quite close to their estimate. Their effective emission cross section is given in Table 28.5.

In summary, we have made a simple estimate of the effective emission cross section that can be used to adjust the a-state excitation cross section for the effects of cascade from the a'- and w-states. Other determinations of the emission cross section based on either AS85 or Cartwright range are mostly higher than our derivation, because the more recent cross section measurements are lower than they used. Consequently, we recommend the use of $\sigma_{eff} = 2.28 \times 10^{-17}$ cm² at the peak at 20 eV as the effective LBH emission cross section to account for the effect of cascade into the a-state.

Clearly, new measurements are needed to quantify the effect of cascade in the LBH system. As well, more modeling efforts are called for. Our estimate was based on a variety of simplifying assumptions and a single dayglow case. The work of Cartwright and Eastes needs to be repeated with more accurate photoelectron and auroral fluxes under a variety of geophysical and solar conditions.

Another potential source of error in applying LBH cross sections to conditions where the energetic electron flux is changing rapidly with energy is the threshold effect. Thresholds for different vibrational levels are spread from 9 to 15 eV for the $v'' = 0 \to v'$ manifold. With increasing electron-impact energy, the cross section for each vibrational level rises very steeply in the low-energy region where measurement uncertainties are generally large (Note that the Young et al. cross section data in Figure 24.21 are for the (3.0) vibrational transition.). Accurate laboratory measurements are needed near threshold energies using a high-resolution electron gun with electron beam energy spread of $\Delta E_e \approx 100$ meV.

Although this chapter focuses on molecular process, we digress briefly to consider the status of electron-impact excitation of atomic oxygen to the $^5S^o$ state. The subsequent emission at 135.6 nm is used by aeronomers along with the LBH bands to measure the O/N_2 column density in the thermosphere. Its absolute signal is a measure of the solar energy input to the atmosphere in the Earth's dayglow. A literature review shows that there exists only one measurement (Stone and Zipf, 1974) and one theoretical calculation (Julienne and Davis, 1976) of OI 135.6 nm cross section of O reported more than 30 years ago and they differ by a factor of two. Meier (1991) scaled the Stone and Zipf measurement downward by a factor 0.36 to account for the improved determinations of the OI

130.4 nm emission cross section that was measured simultaneously with that for 135.6 nm by Stone and Zipf. His recommended value for the peak cross section is 9×10^{-18} cm^2 at 16 eV. A validation of the old laboratory measurement of the emission cross section of O I (135.6 nm) is strongly needed as is an evaluation of the large cascade contributions. Since electron impact is the dominant excitation process in the dayglow and aurora, its cross section remains one of the outstanding missing parameters needed for reliable aeronomical remote sensing. Given the revision of the effective LBH emission cross section from that of AS85, the airglow algorithms need to be reconsidered.

28.4.5 CO

As the most abundant interstellar molecule after H_2, CO plays a very important role in the photo-chemistry of the ISM. The abundance ratio of CO to H_2 is difficult to obtain from observations of the ISM, but can be determined from theoretical models. The models involve chemical reactions in which photodissociation by vacuum ultraviolet (VUV) radiation is the main destruction mechanism for CO, particularly in the range between 91.1 nm, the edge of atomic H absorption continuum, and 111.8 nm, which is the dissociation limit of CO into ground state atoms. The rate of photodissocia-tion of CO by EUV radiation is one of the major uncertainties in these models. In view of these uncertainties and the importance of CO as a tracer molecule, a large number of experimental stud-ies aimed at finding coincidences between CO molecular absorption lines and molecular hydrogen emission lines have been performed (Ciocca et al., 1997a,b).

A large disparity in the values of the oscillator strengths of the A–X, B–X, C–X, and E–X exists in the literature. To resolve the discrepancies, we have carried out high-resolution EUV measurements of these states to determine predissociation yields and oscillator strengths (Ciocca et al, 1997a,b; Beegle et al., 1999). In Figure 28.22, we show for states we have studied, the potential energy diagram of Rydberg states lying above the A $^1\Pi$ valence state. The internuclear distance of the minima in the poten-tial curves of the Rydberg states overlies exactly the minima of the ground state, resulting in intense (0,0) bands. With a full-width-at-half-maximum (FWHM) of 0.0036 nm, we can resolve the rotational structure of the B(0,0), C(0,0), and E(0,0) bands. We show the rotational structure for these (0,0) diagonal bands in Figure 28.23. We also show a model of these bands. A simple model based on Honl-London factors (Ciocca et al., 1997a,b) and rotational constants matches the observed spectra. The predissocia-tion yield of 88% is found for the E $^1\Pi$ state, an average of 85% for the Π^+ state, and 91% for the Π^- state.

The CO A–X band system emission spectrum has been observed in the airglow spectrum of Mars and Venus between 120 and 180 nm (Feldman et al., 2000). Longward of 125 nm, the UV spectra of both planets are dominated by the emission of CO fourth positive band system (A $\rightarrow$ X) and strong O I and C I multiplets. In addition, CO bands, B $^1\Sigma^+$–X $^1\Sigma^+$(0, 0) at 115.1 nm and C $^1\Sigma^+$–X $^1\Sigma^+$(0, 0) at 108.8 nm, are detected, and in the spectrum of Venus, there is a weak indication of the E $^1\Pi$–X $^1\Sigma^+$(0, 0) band at 107.6 nm. The production mechanism of excited CO(A) molecules in the thermo-sphere is attributed to the photodissociation of CO_2 by solar EUV shortward of 92 nm. The CO A–X band system is excited by the impact of photoelectrons and absorption of solar UV photons by both CO and CO_2 (Barth et al., 1992). We have reported electron-impact-induced, medium-resolution fluorescence spectra (0.0031 and 0.00366 nm FWHM) of CO at 100 eV over the spectral region of 130–205 nm (Beegle et al., 1999). The features in the FUV emission spectra correspond to the fourth positive band system (A $^1\Pi$–X $^1\Sigma^+$), atomic multiplets from C and O, and their ions. The absolute electronic transition moment was determined as a function of internuclear distance from relative band intensities. The excitation function of the (0,1) band (159.7 nm) was measured from electron impact in the energy range from threshold to 750 eV and placed on an absolute scale at 100 eV, as discussed in Section 28.3. The CO A–X band system emission cross section was established from a measurement of the relative band intensities at 100 eV. We have obtained high-resolution (~0.0034 nm FWHM), second-order spectra of the A(0,1) band at 159.7 nm from CO by direct exci-tation and from CO_2 by dissociative excitation. The fourth positive bands produced by dissociative excitation are significantly broader and hotter (300 K vs. 1400 K) than direct excitation at laboratory

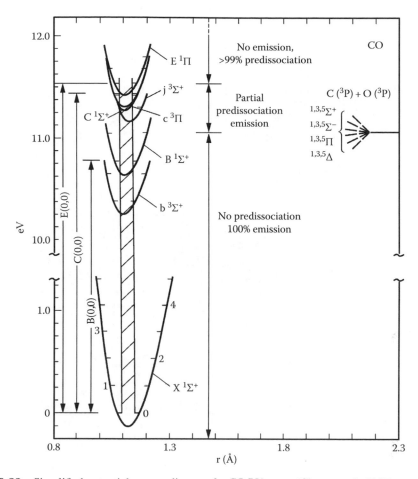

FIGURE 28.22 Simplified potential energy diagram for CO RV states (Ciocca et al., 1997).

temperatures. A resolution of ~0.2 nm of CO band structure is sufficient to distinguish the two types of excitation mechanisms in a planetary atmosphere of known thermospheric temperature.

28.4.6 SO$_2$

Jupiter's Io tenuous atmosphere is dominated by molecular SO$_2$ and its dissociation products SO, S, O, and probable emissions from sulfur allotropes, S$_x$(S, S$_2$,...). A detailed model of the auroral processes at Io requires, as a first step, a medium-resolution laboratory study of both S$_x$ and SO$_2$ as a function of energy to match the spectral dependencies of ground-based spectral observations (Bouchez et al., 2000), as well as the Galileo SSI observations of the blue Io auroral glow on E15 (Geissler et al., 1999). To date, S$_x$ has not been successfully studied in the UV.

Atomic and molecular data have been used most recently in the analysis of the Galileo Solid State Imaging observations of Io (Geissler et al., 1999), Cassini spacecraft observations of Io by the ISS (Porco et al., 2004; Geissler et al., 2004) and the UVIS (Esposito et al., 2005). The ISS is equipped with 15 filter combinations that span the wavelength range of 235–1100 nm. From the data acquired during the Jupiter Millennium Cassini encounter, Geissler et al. (2004) performed a detailed comparison of laboratory SO$_2$ spectra with the Cassini ISS observations and inferred that a mixture of gases contribute to the equatorial glow. The equatorial glows were particularly bright in the near- UV wavelengths (230–500 nm) filters UV1, UV2, UV3, and Blue1. Based on the laboratory work of Ajello et al. (1992a,b), the relative emission intensities within each band pass confirm

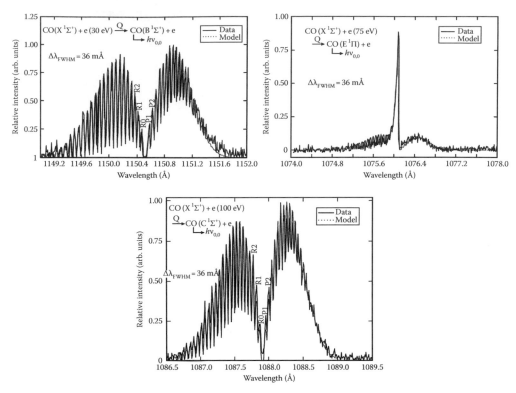

FIGURE 28.23 High-resolution EUV spectra of CO (B,C,E $\rightarrow$ X) (0,0) bands: data(solid) and model (dash). The FWHM is 36 mÅ (Ciocca et al., 1997).

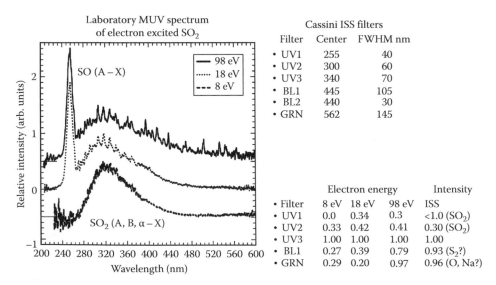

Cassini ISS filters		
Filter	Center	FWHM nm
• UV1	255	40
• UV2	300	60
• UV3	340	70
• BL1	445	105
• BL2	440	30
• GRN	562	145

	Electron energy			Intensity
• Filter	8 eV	18 eV	98 eV	ISS
• UV1	0.0	0.34	0.3	<1.0 (SO_2)
• UV2	0.33	0.42	0.41	0.30 (SO_2)
• UV3	1.00	1.00	1.00	1.00
• BL1	0.27	0.39	0.79	0.93 (S_2?)
• GRN	0.29	0.20	0.97	0.96 (O, Na?)

FIGURE 28.24 The calibrated SO_2 electron-impact-induced fluorescence spectra were obtained at 8, 18, and 98 eV electron-impact energy and a spectral resolution of 1.8 nm over the wavelength range 200–600 nm. The two tables present the Cassini ISS filter band passes and the fraction of electron-impact-induced fluorescence of SO_2 emission signal expected in each filter band pass normalized to unity at filter UV3 (Geissler et al., 2004; Ajello et al., 2002b).

the presence of molecular SO_2 in the Io atmosphere. The laboratory SO_2 electron-impact-induced fluorescence spectrum is shown in Figure 28.24 along with a table that lists the UV, middle UV, and visible filters, and a comparison of the laboratory relative intensities at 8, 18, and 98 eV normalized to UV3 intensities. There is an agreement between the measured output of 100 kR from filter UV3 and the 8 and 18 eV relative intensities within the same band-pass, e.g., the laboratory peak intensity at 320 nm agrees closely with the UV3 filter centered at nearly the same wavelength. There is a Blue1 discrepancy and Green elevated due to the possible atomic emissions indicated in the table.

The laboratory work of Ajello et al. (2002b) used in the Cassini ISS comparison only covered the wavelength range of 200–600 nm. Most of the ISS filters from the blue to the near-IR (500–1200 nm) range show substantial emission intensities as well. We have recently expanded our wavelength coverage to include the entire range of 200–1100 nm using newly acquired laboratory spectroscopic instrumentation that closely matches the Cassini ISS wavelength capability (Ajello et al., 2008). This added coverage is the first study of SO_2 from the UV, visible, optical to near-IR range, i.e., the entire VOIR, which is needed to confirm the model analysis of Geissler et al. (2004).

The Cassini spacecraft at Io and Jupiter during the Cassini Campaign has repeated a spectacular series of visible/near-IR auroral observations similar to those obtained by Galileo SSI but at improved spectral resolution. We show in Figure 28.25 the initial multi-spectral image of Io

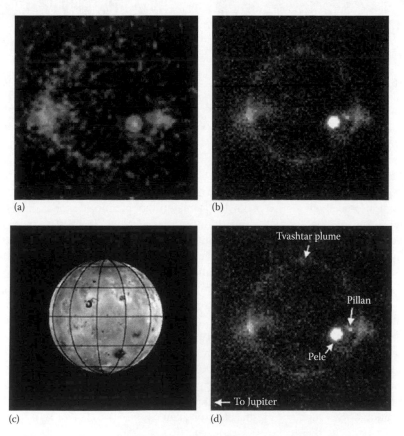

(a)　(b)

(c)　(d)

FIGURE 28.25 Multispectral Image of Io during eclipse of January 1, 2001. (a) False color composite made up of IR4, CB1, and UV3 filter-images portrayed in color in the original work of Geissler et al. 2004 as red, green and blue, respectively. We portray the multi-spectral image in a grey scale with the small circular volcanic feature in right-center from Pele as 'red' IR4, outer ansa and circular limb glow is mostly 'green' UV3, inner ansa glow is 'blue' CB1. (b) CB1 (red) and UV3 (blue) images superposed on clear filter image. (c) Location reference map with grid lines at 30 degree intervals. (d) Annotated clear-filter image showing locations of volcanoes and plume glows discussed (Geissler et al., 2004).

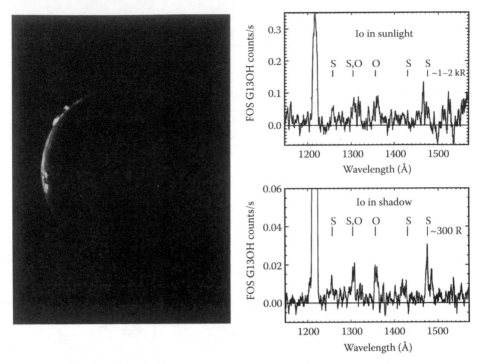

FIGURE 28.26 Voyager visible image of Io by HST showing an active volcano and HST GHRS FUV spectrum of Io with tick marks at strong atomic emission lines of sulfur and oxygen (Clarke et al., 1998).

obtained during the eclipse of January 1, 2002 showing a false color composite of the equatorial and limb emissions (Geissler et al., 2004).

Besides being a scientific destination in its own right (Galileo and JUNO), Jupiter is used for gravitational assist trajectories to the outer solar system (e.g., the Cassini and New Horizons missions). Consequently, UV observations of the Jovian system are also taking place periodically as secondary mission objectives. Active magnetospheres are coupled to the ionospheres of the giant planets' systems and produce particle-excited aurora and airglow. Trapped particle impact on a planetary satellite atmosphere can result in global excitation, both day and night. For Io, having a volcanically generated SO_2 atmosphere, dissociative excitation by the magnetospheric Jovian plasma torus results in excited SO, S, and O, in addition to excited SO_2 and ions, all of which emit radiation. It is believed that the bulk of Io's atomic emission is powered by electron excitation of neutral S and O directly, rather than by the electron dissociative excitation of SO_2 (Ballester et al., 1996). A visible image of Io is shown in Figure 28.26 along with an HST/GHRS spectrum in the FUV with spectral identifications (Ajello et al., 1992a,b; Clarke et al., 1994). There is an evidence, however, that both direct excitation of S and O and dissociative processes upon SO_2 contribute to the Io emission spectrum: (1) Oliversen et al. (2001) indicate that short-term fluctuations in the O I 630 nm intensity is the evidence for a high-energy, nonthermal plasma tail (~30 eV) for a one-step process, and (2) Ballester (1998) indicates that neither electron excitation of O I nor electron dissociative excitation of SO_2 alone by plasma electrons can explain the HST or IUE observations.

28.5 CONCLUSIONS

UV emission spectroscopy by electron-impact-induced fluorescence has been reviewed for a few molecules of major importance to astronomy and planetary atmospheres. The laboratory goals are to (1) measure electron-impact emission cross sections (0–2 keV) and fluorescence spectra (50–300 nm) for important atoms and molecules key to remote sensing observations of FUSE

(Far Ultraviolet Explorer), HST (Hubble Space Telescope), HUT (Hopkins Ultraviolet Telescope), TIMED (Thermosphere Ionosphere Mesosphere Energetics and Dynamics), Galileo, Cassini, and many other spacecrafts; (2) emphasize recently observed and relevant UV transitions of cosmically abundant species of Jovian and terrestrial planetary systems, ISM, and comets (e.g., H, H_2, HD, N_2, SO_2, O, CO, and O_2); (3) provide collision strengths in analytical form or tables of cross sections for UV radiative processes for electron energy loss transport codes and for comparison to *ab initio* calculations; and (4) study corresponding VOIR cascade transitions (300–1100 nm or 1–5 eV excitation energy).

A summary of salient planetary observation results from recent studies of the important atoms and molecules employing the ESL database are listed below:

- The first analysis of the Cassini UVIS observations of the Saturn H_2 aurora (Esposito et al., 2005) and the first analysis of the Jupiter millennium H_2 aurora observations (Ajello et al., 2005a).
- The first analysis of the Cassini Io visible aurora observations of predominantly SO_2 by the ISS (Geissler et al., 2004).
- The development of atomic and molecular hydrogen models used in the interpretation of low-resolution Galileo spectra, medium-resolution HUT spectra, and high-resolution FUSE and HST UV spectra of Jupiter (Dols et al., 2000; Ajello et al., 2002a, 2005a; Gustin et al., 2002, 2004).
- The analysis of Galileo Solid State Imaging (SSI) H_2 VOIR observations of Jupiter (Vasavada et al., 1999).
- The modeling of Cassini and STIS/GHRS (Goddard High Resolution Spectrometer) UV observations of Io, Europa, and Ganymede (Trafton et al., 2007; Noren et al., 2001a,b; Vatti Palle et al., 2004; Hansen et al., 2005).
- Resolved longstanding V1 UVS ambiguity on Titan EUV spectral content showing that the N_2 Carroll-Yoshino c_4' (0,0) band (Broadfoot et al., 1981b) was undetectable and N I photodissociative ionization (PDI) multiplets were present instead (Ajello et al., 2007; Stevens et al., 1994).

ACKNOWLEDGMENTS

The research described in this chapter was carried out at the Jet Propulsion Laboratory, California Institute of Technology, and was sponsored by the NASA Planetary Atmospheres, Astronomy and Physics Research and Analysis, Cassini Data Analysis and Heliophysics Research Geospace Science Program Offices, and NSF Aeronomy Program Office Under Grant Number 0850396.

REFERENCES

Abgrall, H., Roueff, E., Launay, F., Roncin, J. Y., and Subtil, J. L. 1993a. Table of the Lyman Band system of molecular hydrogen. *Astron. Astrophys. Suppl.* 101: 273–321.

Abgrall, H., Roueff, E., Launay, F., Roncin, J. Y., and Subtil, J. L. 1993b. Table of the Werner Band system of molecular hydrogen. *Astron. Astrophys. Suppl.* 101: 323–362.

Abgrall, H., Roueff, E., Launay, F., Roncin, J. Y., and Subtil, J. L. 1993c. The Lyman and Werner Band systems of molecular hydrogen. *J. Mol. Spectrosc.* 157: 512–523.

Abgrall, H., Roueff, E., Launay, F., and Roncin, J. Y. 1994. The B' $^1\Sigma_u^+ \rightarrow$ X $^1\Sigma_g^+$ and D $^1\Pi_u \rightarrow$ X $^1\Sigma_g^+$ band systems of molecular hydrogen, *Can. J. Phys.* 72: 856–8.

Abgrall, H., Roueff, E., Liu, X., and Shemansky, D. E. 1997. The emission continuum of electron-excited H_2. *Astrophys. J.* 481: 557–566.

Abgrall, H., Roueff, E., Liu, X., Shemansky, D., and James, G. 1999. High-resolution far ultraviolet emission spectra of electron-excited molecular deuterium. *J. Phys. B: At. Mol. Opt. Phys.* 32: 3813–3838.

Abgrall, H., Roueff, E., and Drira, I. 2000. Total transition probability of B, C, B' and D states of molecular hydrogen. *Astron. Astrophys. Suppl. Ser.* 141: 297–300.

Aguilar, A., Ajello, J. M., Mangina, R. S., James, G. K., Abgrall, H., and Roueff, E. 2008. The electron excited middle UV to near-IR spectrum of H_2: Cross sections and transition probabilities. *Astrophys. J. Supp. Ser.* 177: 388–407.

Ajello, J. M. 1970. Emission cross sections of N_2 in the vacuum ultraviolet by electron impact Source. *J. Chem. Phys.* 53: 1156–1165.

Ajello, J. M. 2010. High resolution spectra of N_2. (in preparation).

Ajello, J. M. and Shemansky, D. E. 1985. A re-examination of important N_2 cross sections by electron impact with application to the dayglow: The Lyman-Birge Hopfield band system and NI (119.99 nm). *J. Geophys. Res.* 90: 9845–9861.

Ajello, J. M. and Shemansky, D. E. 1993. Electron excitation of the H_2 (a $^3\Sigma_g^+ - ^3\Sigma_g^+$) continuum in the vacuum ultraviolet. *Astrophys. J.* 407: 820–825.

Ajello, J. M. and Ciocca, M. 1996a. Fast nitrogen atoms from dissociative excitation of N_2 by electron impact. *J. Geophys. Res.* 101: 18953–18960.

Ajello, J. M., Shemansky, D., Kwok, T. L., and Yung, Y. L. 1984. Studies of extreme ultraviolet emission from Rydberg series of H_2 by electron impact. *Phys. Rev.* 29: 636–653.

Ajello, J. M., Pang, K. D., Franklin, B. O., Howell, S. K., and Bowring, N. J. 1989a. A study of electron impact excitation of NO: The vacuum ultraviolet from 40 to 170 nm. *J. Geophys. Res.* 94: 9093–9103.

Ajello, J. M., James, G. K., Franklin, B. O., and Shemansky, D. E. 1989b. Medium resolution studies of EUV emission from N_2 by electron impact: Vibrational perturbations and cross sections of the c_4' $^1\Sigma_g^+$ and b' $^1\Sigma_u^+$ states. *Phys. Rev. A.* 40: 3524–3556.

Ajello, J. M., James, G. K., Franklin, B., and Howell, S. 1990. Study of electron impact excitation of Argon in the EUV: Emission cross sections of resonance lines of Ar I, Ar II. *J. Phys. B: At. Mol. Opt. Phys.* 23: 4355–4376.

Ajello, J. M., James, G. K., Kanik, I., and Franklin, B. O. 1992a. The complete UV spectrum of SO_2 by electron impact: Part 1: The vacuum ultraviolet spectrum. *J. Geophys. Res.* 97: 10473–10500.

Ajello, J. M., James, G. K., and Kanik, I. 1992b. The complete UV spectrum of SO_2 by electron impact: Part 2: The middle ultraviolet spectrum. *J. Geophys. Res.* 97: 10501–10512.

Ajello, J. M., Ahmed, S., Kanik, I., and Multari, R. 1995a. Kinetic energy distribution of H(2p) atoms from dissociative excitation of H_2. *Phys. Rev. Lett.* 75: 3261–3264.

Ajello, J. M., Kanik, I., Ahmed, S. M., and Clarke, J. T. 1995b. The line profile of H Lyman-α from dissociative excitation of H_2 with application to Jupiter. *J. Geophys. Res.* 100: 26411–26420.

Ajello, J. M., Ahmed, S. M., and Liu, X. 1996b. Line profile of H-Ly-β from dissociative excitation H_2. *Phys. Rev.* 53: 2303–2308.

Ajello, J. M., James, G. K., and Ciocca, M. 1998a. High resolution EUV spectroscopy of N_2 c' $^1\Sigma_u^+$ v' = 3 and 4 levels by electron impact. *J. Phys. B.* 31: 2437–2448.

Ajello, J. M., Shemansky, D., Pryor, W., Tobiska, K., Hord, C., Stephens, S., Stewart, A. I., Clarke, J., Simmons, K., Gebben, J., Miller, D., McClintock, W., Barth, C., and Sandel, B. 1998b. Galileo Orbiter ultraviolet observations of Jupiter aurora. *J. Geophys. Res.* 103: 20125–20148.

Ajello, J., Shemansky, D. E., Pryor, W., Stewart, A. I., Simmons, K., Majeed, T., Waite, H., and Gladstone, G. 2001. Spectroscopic evidence for high altitude aurora from Galileo extreme ultraviolet and Hopkins ultraviolet telescope observations. *Icarus* 152: 151–171.

Ajello, J., Vatti Palle, V. P., and Osinski, G. 2002a. UV spectroscopy by electron impact for planetary astronomy and astrophysics. In *Current Developments in Atomic, Molecular Physics*, M. Mohan (ed.), pp. 143–150. New York: Kluwer Academic Press.

Ajello, J. M., Hansen, D., Beegle, L. W., Terrell, C., Kanik, I., James, G., and Makarov, O. 2002b. The middle ultraviolet and visible spectrum of SO_2 by electron impact. *J. Geophys. Res.* 107: SIA2.1–SIA2.7, doi:10.1029/2001JA000122.

Ajello, J., Pryor, W., Esposito, L., Stewart, I., McClintock, W., Gustin, J., Grodent, D., Gérard, J.-C., and Clarke, J. 2005a. The Cassini Campaign observations of the Jupiter aurora by the ultraviolet imaging spectrograph and the space telescope imaging spectrograph. *Icarus* 178: 327–345.

Ajello, J., Vatti Palle, P., Abgrall, H., Roueff, E., Bhardwaj, A., and Gustin, G. 2005b. The electron excited UV spectrum of HD: Cross sections and transition probabilities. *Astrophys. J. Suppl.* 159: 314–330.

Ajello, J. M., Stevens, M. H., Stewart, I., Larsen, K., Esposito, L., Colwell, J., Mcclintock, W., Holsclaw, G., Gustin, J., and Pryor, W. 2007. Titan airglow spectra from Cassini ultraviolet imaging spectrograph (UVIS): EUV analysis. *Geophys. Res. Lett.* 34: L24204, doi:10.1029/2007GL031555.

Ajello, J. M., Aguilar, A., Mangina, R. S., James, G. K., Geissler, P., and Trafton, L. 2008. The middle UV to near IR spectrum of electron excited SO_2. *J. Geophys. Res.-Planets* 113: E03002, doi:10.1029/2007je002921.

Avakyan, S. V., Ii'In, R. N., Lavrov, V. M., and Ogurstov, G. N. 1998. *Collision Processes and Excitation of UV Emission from Planetary Atmospheric Gases: A Handbook of Cross Sections*, pp. 1–342. Amsterdam, the Netherlands: Gordon and Breach Science Publishers.

Bagenal, F., Dowling, T., and McKinnon, W. 2007. *Jupiter, The Planet, Satellites and Magnetosphere* (*Cambridge Planetary Science*), pp. 1–684. Cambridge, U.K.: Cambridge University Press.

Ballester, G. E., Clarke, J. T., Rego, D., Combi, M., Larsen, N., Ajello, J., Strobel, D. F., Schneider, N. M., and McGrath, M. 1996. Characteristics of Io's far-UV neutral oxygen and sulfur emissions derived from recent HST observations. *Bull. Am. Astron. Soc.* 28: p1156.

Ballester, G. E. 1998. *Ultraviolet Astrophysics Beyond the IUE Final Archive*, pp. 21–28. Noordwijk, the Netherlands: ESA Publications.

Barth, C. A., Stewart, A. I. F., and Brougher, S. W. 1992. *Mars*, pp. 1054–1065. Tucson, AZ: University of Arizona Press.

Beegle, L., Ajello, J. M., James, G. K., Dziczek, D., and Alvarez, M. 1999. The emission spectrum of the CO (A $^1\Pi$-X $^1\Sigma^+$) fourth positive system by electron impact. *Astron. Astrophys.* 347: 375–390.

Bergin, E., Calvet, N., Sitko, M. L., Abgrall, H., D'Alessio, P., Herczeg, G. J., Roueff, E., Qi, C., Lynch, D. K., Russell, R. W., Brafford, S. M., and Perry, R. B. 2004. New probe of planet forming region in T Tauri Disks. *Astrophys. J.* 614: L133–136.

Bishop, J. and Feldman, P. 2003. Analysis of the Astro-1/Hopkins Ultraviolet Telescope EUV-FUV dayside nadir spectral radiance measurements. *J. Geophys. Res.* 108: 1243, doi:10.1029/2001JA000330.

Bouchez, A. H., Brown, M. E., and Schneider, N. 2000. Eclipse spectroscopy of Io's atmosphere. *Icarus* 148: 316–319.

Brinkmann, R. T. and Trajmar, S. 1970. Electron impact excitation of N_2. *Annales de Geophysique* 26: 201–207.

Broadfoot, A. L., Belton, M. J. S., Takacs, P. J., Sandel, B. R., Shemansky, E. E., Holberg, J. B., Ajello, J. M., Atreya, S. K., Donahue, T. M., Moos, H. W., Bertaux, J. L., Blamont, J. E., Strobel, D. F., McConnell, J. C., Dalgarno, A., Goody, R., and McElroy, M. B. 1979. Extreme ultraviolet observations from Voyager 1 encounter with Jupiter. *Science* 204: 979–982.

Broadfoot, A. L., Sandel, B. R., Shemansky, D. E., McConnell, J. C., Smith, G. R., Holberg, J. B., Atreya, S. K., Donahue, T. M., Strobel, D. F., and Bertaux, J. L. 1981a. Overview of the Voyager ultraviolet spectrometry results through Jupiter encounter. *J. Geophys. Res.* 86: 8259–8284.

Broadfoot, A. L., Sandel, B. R., Shemansky, D. E., Holberg, J. B., Smith, G. R., Strobel, D. F., McConnell, J. C., Kumar, S., Hunten, D. M., Atreya, S. K., Donahue, T. M., Moos, H. W., Bertaux, J. L., Blamont, J. E., Pomphrey, R. B., and Linick, S. 1981b. Extreme UV observations from Voyager 1 encounter of Saturn. *Science* 212: 206–211.

Broadfoot, A. L., Atreya, S. K., Bertaux, J. L., Blamont, J. E., Dessler, A. J., Donahue, T. M., Forrester, W. T., Hall, D. T., Herbert, F., Holberg, J. B., Hunten, D. M., Krasnopolsky, V. A., Linick, S., Lunine, J. I., Mcconnell, J. C., Moos, H. W., Sandel, B. R., Schneider, N. M., Shemansky, D. E., Smith, G. R., Strobel, D. F., and Yelle, R. V. 1989. Ultraviolet spectrometer observations of Neptune and Triton. *Science* 246: 1459–1466.

Brunger, M. J. and Teubner, P. J. O. 1990. Differential cross sections for electron-impact excitation of the electronic states of N_2 source. *Phys. Rev. A* 41: 1413–1426.

Bunce, E. J. and Cowley, S. W. H. 2001. Divergence of the equatorial current in the dawn sector of Jupiter's magnetosphere: Analysis of Pioneer and Voyager magnetic field data. *Planet. Space Sci.* 49: 1089–1113.

Budzien, S. A., Feldman, P. D., and Conway, R. R. 1994. Observations of the far ultraviolet airglow by the ultraviolet limb imaging experiment on STS-39. *J. Geophys. Res.* 99: 23275–23287.

Campbell, L., Brunger, M. J., Nolan, A. M., Kelly, L. J., Wedding, A. B., Harrison, J., Teubner, P. J. O., Cartwright, D. C., and McLaughlin, B. 2001. Integral cross sections for electron impact excitation of electronic states of N_2. *J. Phys. B* 34: 1185–1199.

Cartwright, D. C. 1978. Vibrational populations of the excited states of N_2 under auroral conditions. *J. Geophys. Res.* 83: 517–531.

Cartwright, D. C., Chutjian, A., Trajmar, S., and Williams, W. 1977. Electron-impact excitation of electronic states of N_2: 1. Differential cross-sections at incident energies from 10 to 50 eV. *Phys. Rev. A* 16: 1013–1040.

Ciocca, M., Ajello, J. M., Liu, X., and Maki, J. 1997a. Kinetic energy distribution of D(2p) atoms from analysis of the D Lyman- α line profile. *Phys. Rev. A* 56: 1929–1937.

Ciocca, M., Kanik, I., and Ajello, J. M. 1997b. High resolution studies extreme ultraviolet emission from CO. *Phys. Rev. A* 55: 3547–3556.

Clarke, J. T., Ajello, J., Luhmann, J., Scheider, N., and Kanik, I. 1994. HST UV spectral observations of io passing into eclipse. *J. Geophys. Res.* 99: 8387–8402.

Clarke J. T., Ballester, G., Trauger, J., Ajello, J., Pryor, W., Tobiska, K., Connerney, J. E. P., Gladstone, G. R., Waite, J. H., Ben Jaffe, L., and Gerard, J.-C. 1998. Hubble space telescope imaging of Jupiter's UV aurora during the Galileo orbiter mission. *J. Geophys. Res.* 103: 20217–20236.

Cowley, S. W. H. and Bunce, E. 2001. Origin of the main oval in Jupiter's coupled magnetosphere-ionospheric system. *Planet. Space Sci.* 49: 1067–1088.

Crosswhite, H. M. 1972. *The Hydrogen Molecule Wavelength Tables of Gerhard Heinrich Dieke*, pp. 1–616. New York: Wiley-Interscience.

Dalgarno, A. 1993. The chemistry of astronomical environments. *J. Chem. Soc. Faraday Trans.* 89: 2111–2117.

Dalgarno, A. 1995. Infrared emission from molecular hydrogen. In *Physics of the Interstellar Medium and Intergalactic Medium.* ASP Conference Series: 80, A. Ferrara, C. F. McKee, C. Heiles, P. R. Shapiro (eds.), pp. 37–44. San Francisco, CA: Astronomical Society of the Pacific.

Dieke, G. H. 1958. The molecular spectrum of hydrogen and its isotopes. *J. Mol. Spectrosc.* 2: 494–517.

Dieke, G. H. and Cunningham, S. P. 1965. Bands of D_2 and T_2 Originating from the Lowest Excited $^1\Sigma_g$ States $(1s\sigma)$ $(2s\sigma)$ $^1\Sigma_g$ and $(2p\sigma)$ $2^1\Sigma_g*$. *J. Mol. Spectrosc.* 18: 288–320.

Dols, V., Gérard, J.-C., Clarke, J. T., Gustin, J., and Grodent, D. 2000. Diagnostics of the Jovian aurora deduced from ultraviolet spectroscopy: Model and HST/GHRS observations. *Icarus* 147: 251–266.

Dyudina, U. A., Ingersoll, A. P., and Ewald, S. P. 2007. Aurora at the North Pole of Saturn as seen by Cassini ISS. *Eos Trans. AGU* 88 (Fall Meet. Suppl. 52): P31A–0188.

Dziczek, D., Ajello, J., James, G., and Hansen, D. 2000. Cascade contribution to the H_2 Lyman Band system from electron impact. *Phys. Rev. A* 61: 64702–64706.

Eastes, R. W. 2000. Modeling the N_2 Lyman-Birge-Hopfield bands in the dayglow: Including radiative and collisional cascading between the singlet states. *J. Geophys. Res.* 105: 18557–18573.

Eastes, R. W. and Dentamaro, A. V. 1996. Collision-induced transitions between the a $^1\Pi_g$, a' $^1\Sigma_u^-$, w $^1\Delta_g$ states of N_2: Can they effect auroral N_2 Lyman-Birge-Hopfiled ban emissions? *J. Geophys. Res.* 101: 26931–26940.

Esposito, L. W., Barth, C. A., Colwell, J. E., Lawrence, G. M., McClintock, W. E., Stewart, A. I. F., Keller, H. U., Korth, A., Lauche, H., Festou, M. C., Lane, A. L., Hansen, C. J., Maki, J. N., West, R. A., Jahn, H., Reulke, R., Warlich, K., Shemansky, D. E., and Yung, Y. L. 2004. The Cassini ultraviolet imaging spectrograph investigation. *Space Sci. Rev.* 115: 299–361.

Esposito, L. W., Colwell, J. E., Hallett, J. T., Hansen, C. J., Hendrix, A. R., Keller, H. U., Korth, A., Larsen, K., McClintock, W. E., Pryor, W. R., Reulke, R., Shemansky, D. E., Stewart, A. I. F., West, R. A., Ajello, J. M., and Yung, Y. L. 2005. Ultraviolet imaging spectroscopy shows and active Saturnian system. *Science* 307: 1251–1255.

Evans, Scott. 2010. Private Communication.

Feldman, P. D., Burgh, E. B., Durrance, S. T., and Davidsen, A. F. 2000. Far ultraviolet spectroscopy of Venus and Mars at 4 Å resolution with the Hopkins ultraviolet telescope on Astro 2. *Astrophys. J.* 538: 395–400.

Feldman, P., Sahnow, D., Kruk, J., Murphy, E., and Moos, W. 2001. High-resolution spectroscopy of the terrestrial day airglow with the far ultraviolet spectroscopic explorer. *J. Geophys. Res.* 106: 8119–8129.

Fischer, F., Stasek, G., and Schmidtke, G. 1980. Identification of auroral EUV emissions. *Geophys. Res. Lett.* 7: 1003–1006.

Freund, R. S. 1972. Radiative lifetime of N_2 (a $^1\Pi_g$) and formation of metastable N_2. *J. Chem. Phys.* 56: 4344–4351.

Geissler, P. E., McEwen, A. S., Ip, W., Belton, M. J. S., Johnson, T. V., Smyth W. H., and Ingersoll, A. P. 1999. Galileo imaging of atmospheric emissions from Io. *Science* 285: 870–874.

Geissler, P., McEwen, A., Porco, C., Strobel, D., Soar, J., Ajello, J., and West, R. 2004. Cassini observations of Io's Visible Aurora. *Icarus* 172: 127–140.

Glass-Maujean, M. 1986. Photodissociation of vibrationally excited H_2 by absorption into the continua of B, C, and B' systems. *Phys. Rev. A* 33: 342–345.

Glass-Maujean, M., Klumpp, S., Werner, L., Ehresmann, A., and Schmoranzer, H. 2007a. Observation of the oscillating absorption spectrum of a double-well state: The B"$\overline{B}^1\Sigma_u^+$ state of H_2. *J. Phys. B: At. Mol. Opt. Phys.* 40: F19.

Glass-Maujean, M., Klumpp, S., Werner, L., Ehresmann, A., and Schmoranzer, H. 2007b. Study of the B"$\overline{B}^1\Sigma_u^+$ state of H_2: Transition probabilities from the ground state, dissociative widths, and Fano parameters. *J. Chem. Phys.* 126: 144303–144308.

Glass-Maujean, M., Klumpp, S., Werner, L., Ehresmann, A., and Schmoranzer, H. 2007c. Transition probabilities from the ground state of the [image omitted] states of H_2. *Mol. Phys.* 105: 1535–1542.

Glass-Maujean, M., Klumpp, S., Werner, L., Ehresmann, A., and Schmoranzer, H. 2007d. Cross sections for the ionization continuum of H_2 in the 15.3–17.2 eV energy range. *J. Chem. Phys.* 126: 094306, doi:10.1063/1.2435345.

Glass-Maujean, M., Klumpp, S., Werner, L., Ehresmann, A., and Schmoranzer, H. 2008. The study of the fifth $^1\Sigma_u^+$ state (5pσ) of H_2: Transition probabilities from the ground state, natural line widths and predissociation yields. *J. Mol. Spectrosc.* 249: 51–59.

Glass-Maujean, M., X. Liu, and Shemansky, D. E. 2009. Analysis of electro-impact excitation and emission of the npσ $^1\Sigma_u^+$ and npπ $^1\Pi_u$ Rydberg Series of H_2. *Astrophys. J. Suppl.* 180: 38–53.

Gredel, R., Lepp, S., and Dalgarno, A. 1987. The C/CO ratio in interstellar clouds. *Astrophys. J.* 323: L137–L139.

Gredel, R., Lepp, S., Dalgarno, A., and Herbst, E. 1989. Cosmic ray induced photodissociation and photoionization rates of interstellar molecules. *Astrophys. J.* 347: 289–293.

Grewing, M., Boksenberg, A., Seaton, M. J., Snijders, M. A. J., Wilson, R., Boggess, A., Bohlin, R. C., Perry, P. M., Schiffer III, I. H., Gondhalekar, P. M., Macchetto, F., Savage, B. D., Jenkins, E. B., Johnson, H. M., Perinotto, M., and Whittet, D. C. B. 1978. IUE observations of the interstellar medium. *Nature* 275: 394–400.

Grodent, D., Waite, J. H., and Gérard, J. C. 2001. A self-consistent model of the jovian auroral thermal structure. *J. Geophys. Res.* 106: 12933–12952.

Grodent, D., Clarke, J. T., Waite, J. H., Cowley, S. W., Gérard, J.-C., and Kim, J. 2003a. Jupiter's polar auroral emissions. *J. Geophys. Res.* 108: 1366–1388.

Grodent, D., Clarke, J. T., Waite, J. H., Cowley, S. W., Gérard, J.-C., and Kim, J. 2003b. Jupiter's main aurora oval. *J. Geophys. Res.* 108: 1389–1396.

Gustin, J., Grodent, D., Gerard, J., and Clarke, C. 2002. Spatially resolved far ultraviolet spectroscopy of the Jovian Aurora. *Icarus* 157, 91–103.

Gustin, J., Feldman, P. D., Gérard, J.-C., Vidal-Madjar, A., Ben Jaffel, L., Grodent, D., Moos, H. W., Sahnow, D. J., Weaver, H. A., Wolven, B. C., Ajello, J. M., Waite, J. H., Roueff, E., and Abgrall, H. 2004. Jovian auroral spectroscopy with FUSE: Analysis of self-absorption and implications on electron precipitation. *Icarus* 171: 336–355.

Gustin, J., Gerard, J.-C., Pryor, W., Feldmanc, P. D., Grodent, D., and Holsclaw, G. 2009. Characteristics of Saturn's polar atmosphere and auroral electrons derived from HST/STIS, FUSE and Cassini/UVIS spectra. *Icarus* (in press).

Hansen, C., Shemansky, D. E., and Hendrix, A. R. 2005. Cassini UVIS observations of Europa's oxygen atmosphere and torus. *Icarus* 176: 305–315.

Herczeg, G., Linsky, J. L., Valenti, J. A., Johns-Krull, C. M., and Wood, B. E. 2002. A high resolution UV spectrum of the pre-main sequence star TW Hydrae. I. Observations of H_2 fluorescence. *Astrophys. J.* 572: 310–325.

Herczeg, G., Wood, B. E., Linsky, J. L., Valenti, J. A., and Johns-Krull, C. M. 2004. Far UV spectrum of TW Hydrae : II Models of H_2 fluorescence in a disc. *Astrophys. J.* 607: 369–383.

Hill, T. W. 2001. The Jovian aurora oval. *J. Geophys. Res.* 106: 8101–8107.

Hord, C. W., McClintock, W. E., Stewart, A. I. F., Barth, C. A., Esposito, L. W., Thomas, G. E., Sandel, B. R., Hunten, D. M., Broadfoot, A. L., Shemansky, D. E., Ajello, J. M., Lane, A. L., and West, R. A. 1992. The Galileo ultraviolet spectrometer experiment. *Space Sci. Rev.* 60: 503–530.

Huber, K. P. and Herzberg, G. 1979. *Constants of Diatomic Molecules*, pp. 1–716. New York: Van Nostrand Co.

James, G. K., Ajello, J. M., Franklin, B. O., and Shemansky, D. E. 1990. Medium resolution studies of EUV emission from N_2 by electron impact: The effect of predissociation on emission cross sections of the b$^1\Pi_u$ state. *J. Phys. B: At. Mol. Phys.* 23: 2055–2082.

James, G. K., Slevin, J. A., Shemansky, D. E., McConkey, J. W., Bray, I., Dziczek, D., Kanik, I., and Ajello, J. M. 1997. Optical excitation function of H(1s-2p) by electron impact from threshold to 1.8 keV. *Phys. Rev. A* 55: 1069–1087.

James, G. K., Slevin, J. A., Dziczek, D., McConkey, J. W., and Bray, I. 1998a. Polarization of H-Lα from atomic H by e-Impact. *Phys. Rev. A* 57: 1787–1797.

James, G. K., Ajello, J. M., and Pryor, W. R. 1998b. The MUV-visible spectrum of H_2 excited by electron impact. *J. Geophys. Res.* 103: 20113–20123.

Jasperse, J. R. 1976. Boltzmann-Fokker-Planck model for the electron distribution function in the Earth's ionosphere. *Planet. Space Sci.* 24: 33–40.

Johnson, P. V., Kanik, I., Shemansky, D. E., and Liu, X. 2003a. Electron-impact cross sections of atomic oxygen. *J. Phys. B: At. Mol. Opt. Phys.* 36: 3203–3218.

Johnson, P. V., Kanik, I., Khakoo, M. A., McConkey, J. W., and Tayal, S. S. 2003b. Low energy differential and integral electron-impact cross sections for the $2s^22p^4$ ^{3}P $\rightarrow$ $2p^33s$ ^{3}S^o excitation in atomic oxygen. *J. Phys. B: At. Mol. Opt. Phys.* 36: 4289–4300.

Johnson, P. V., McConkey, J. W., Tayal, S. S., and Kanik, I. 2005a. Collisions of electrons with atomic O-current status. *Can. J. Phys.* 83: 589–616.

Johnson, P. V., C. P. Malone, I. Kanik, K. Tran, and M. A. Khakoo. 2005b, Integral cross sections for the direct excitation of the A $^3\Sigma_u^+$, B $^3\Pi_g$, W $^3\Delta_u$, B' $^3\Sigma_u^-$, a' $^1\Sigma_u^-$, a $^1\Pi_g$, w $^1\Delta_u$, and C $^3\Pi_u$ electronic states in N_2 by electron impact *J. Geophys. Res.*, 110, A11311.

Jonin, C., Liu, X., Ajello, J. M., James, G. K., and Abgrall, H. 2000. The high resolution EUV spectrum of H_2 by electron impact: Cross sections and predissociation yields. *Astrophys. J. Suppl.* 129: 247–256.

Julienne, P. S. and Davis, J. 1976. Cascade and radiation trapping effects on atmospheric atomic oxygen emission excited by electron impact. *J. Geophys. Res.* 81: 1397–1403.

Kanik, I., Ajello, J. M., and James, G. K. 1993. Extreme ultraviolet emission spectrum of CO_2 induced by electron impact at 200eV. *Chem. Phys. Lett.* 211: 523–528.

Kanik, I., Ajello, J. M., and James, G. K. 1996. Electron-impact-induced emission cross sections of neon in the extreme ultraviolet. *J. Phys. B: At. Mol. Opt. Phys.* 29: 2355–2366.

Kanik, I., Noren, C., Makarov, O., Vatti Palle, P., Ajello, J., and Shemansky, D. 2003. Electron impact dissociative excitation of O_2. II. Absolute emission cross sections of OI(130.4nm) and OI(135.6nm). *J. Geophys. Res.* 108: E1151256, doi:10.1029/2000JE001423.

Lepp, S. and Dalgarno, A. 1996. X-ray induced chemistry of interstellar clouds. *Astron. Astrophys.* 306: L21–L24.

Lepp, S., Stancil, P. C., and Dalgarno, A. 2002. Atomic and molecular processes in the early universe. *J. Phys. B: At. Mol. Opt. Phys.* 35: R57–R80.

Lewis, B. R., Gibson, S. T., Zhang, W., Lefebvre-Brion, H., and Robbe, J.-M. 2005. Predissociation mechanisms for the lowest $^1\Pi_u$ states of N_2. *J. Chem. Phys.* 122 (14): 4302.

Liu, W. and Dalgarno, A. 1996. The ultraviolet spectrum of the Jovian aurora. *Astrophys. J.* 467: 446–453.

Liu, X., Ahmed, S., Multari, R., James, G., and Ajello, J. M. 1995. High resolution electron impact study of the FUV emission spectrum of molecular hydrogen. *Astrophys. J. Suppl.* 101: 375–399.

Liu, X., Shemansky, D., Ahmed, S., James, G., and Ajello, J. 1998. Excitation Lyman and Werner systems of molecular hydrogen. *J. Geophys. Res.* 103: 26739–26758.

Liu, X., Shemansky, D. E., Ajello, J. M., Hansen, D. L., Jonin, C., and James, G. K. 2000. High resolution electron-impact emission spectrum of H_2 II. 760–900Å. *Astrophys. J. Suppl.* 129: 267–280.

Liu, X., Shemansky, D., Abgrall, H., Roueff, E., Dziczek, D., Hansen, D., and Ajello, J. 2002. Time-resolved electron impact study of excitation of H_2 singlet-gerade states from cascade emission in the vacuum ultraviolet region. *Astrophys. J. Suppl.* 138: 229–245.

Liu, X., Shemansky, D. E., Abgrall, H., Roueff, E., Ahmed, S. M., and Ajello, J. M. 2003. Electron impact excitation of H_2: Resonance excitation of B $^1\Sigma_u^+(J_j = 2, v_j = 0)$ and effective excitation function of EF $^1\Sigma_g^+$. *J. Phys. B: At. Mol. Phys.* 36: 173–196.

Liu, X., Shemanksy, D. E., Ciocca, M., Kanik, I., and Ajello, J. M. 2005. Analysis of the physical properties of the N_2 c' $^1\Sigma_u^+(0) - X \Sigma_g^+(0)$ transition. *Astrophys. J.* 623; 579–584.

Liu, X., Shemansky, D. E., Malone, C., Johnson, P., Ajello, J. M., Kanik, I., Lewis, B., Gibson, S., and Stark, G. 2008. Experimental and coupled channels investigation of the radiative properties of the N_2 c'$_4$ $^1\Sigma_u^+ - X \Sigma_g^+$ band system. *J. Geophys. Res.* 113: A02304, doi:10.1029/2007JA012787.

Majeed, T. and Strickland, D. 1997. New survey of electron impact cross sections for photoelectron and auroral electron energy loss calculations. *J. Chem. Phys. Chem. Ref. Data* 26: 335–349.

Makarov, O., Kanik, I., Ajello, J. M., and Prahlad, V. 2004. Kinetic energy distributions and line profile measurements of dissociation products of water upon electron impact. *J. Geophys. Res.* 109: A09303, doi:10.1029/2002JA009353.

Malone, C. P., Johnson, P. V., McConkey, J. W., Ajello, J. M., and Kanik, I. 2008. Dissociative excitation of N_{2O} by electron impact. *J. Phys. B: At. Mol. Opt. Phys.* 41: 095201, doi:10.1088/0953-4075/41/9/095201.

Mangina, R. S., Ajello, J. M., West, R. A., and Dziczek, D. 2010. High resolution electron impact emission spectra and cross sections for N_2 from 330–1100 nm. *Astrophys. J. Supp.* (in submission).

Mason, N. J. and Newell, W. R. 1987. Electron impact total excitation cross section of the a $^1\Pi_g$ state of N_2. *J. Phys. B: At. Mol. Phys.* 20: 3913–3921.

Mauk, B. H., B. J. Anderson, and R. M. Thorne. 2002. Magnetosphere-ionosphere coupling at Earth, Jupiter and beyond. In *Atmospheres in the Solar System: Comparative Aeronomy,* Geophysical Monograph 130, M. Mendillo, A. Nagy, and J. H. Waite (eds.), pp. 97–114. Washington, DC: AGU.

McConkey, J. W., Malone, C. P., Johnson, P. V., Winstead, C., McKoy, V., and Kanik, I. 2008. Electron impact dissociation of oxygen-containing molecules–A critical review. *Phys. Rep.* 466: 1–103.

Meier, R. R. 1991. Ultraviolet spectroscopy and remote sensing of the upper atmosphere. *Space Sci. Rev.* 58: 1–185.

Meier, R. R., G. Crowley, D. J. Strickland, A. B. Christensen, L. J. Paxton, D. Morrison, and C. L. Hackert, 2005. First look at the November 20, 2003 super storm with TIMED/GUVI: Comparisons with a thermospheric global circulation model, *J. Geophys. Res.*, 110, A09S41.

Meng, C. I., Rycroft, M. J., and Frank, L. A. 1991. *Auroral Physics*. Cambridge, U.K.: Cambridge University Press.

Möhlmann, G. R. and de Heer, F. 1976. Emission cross sections of the $H_2(3p^3\Pi_u \rightarrow 2s^3\Sigma^+_g)$ transition for electron impact on H_2. *Chem. Phys. Lett.* 43: 240–244.

Moos, H. W., Cash, W. C., Cowie, L., Davidsen, A. F., Dupree, A. K., Feldman, P. D., Friedman, S. D., Green, J. C., Green, R. F., Gry, C., Hutchings, J. B., Jenkins, E. B., Linsky, J. L., Malina, R. F., Michalitsianos, A. G., Savage, B. D., Shull, J. M., Siegmund, O. H. W., Snow, T. P., Sonneborn, G., Vidal-Madjar, A., J. Willis, A., Woodgate, B. E., York, D. G., Ake, T. B., Andersson, B.-G., Andrews, J. P., Barkhouser, R. H., Bianchi, L., Blair, W. P., Brownsberger, K. R., Cha, A. N., Chayer, P., Conard, S. J., Fullerton, A. W., Gaines, G. A., Grange, R., Gummin, M. A., Hebrard, G., Kriss, G. A., Kruk, J. W., Mark, D., McCarthy, D. K., L. Morbey, C., Murowinski, R., Murphy, E. M., Oegerle, W. R., Ohl, R. G., Oliveira, C., Osterman, S. N., Sahnow, D. J., Saisse, M., Sembach, K. R., Weaver, H. A., Welsh, B. Y., Wilkinson, E., and Zheng, W. 2000. Overview of far ultraviolet spectroscopic explorer mission. *Astrophys. J.* 538: L1–L6.

Morton, D. C. and Noreau, L. 1994. A compilation of electronic transitions in the CO molecule and the interpretation of some puzzling interstellar absorption features, *Astrophys. J. Supp.* 95: 301–312.

Noren, C., Kanik, I., Johnson, P. V., McCartney, P., James, G. K., and Ajello, J. M. 2001a. Electron-impact studies of atomic oxygen: II. Emission cross section measurements of the O I $^3S^o \rightarrow {}^3P$ transition (130.4 nm). *J. Phys B: At. Mol. Opt. Phys.* 34: 2667–2677.

Noren, C., Kanik, I., Ajello, J. M., McCartney, P., and Makarov, O. P. 2001b. Emission cross section OI (135.6 nm) dissociative excitation of O_2. *Geophys. Res. Lett.* 28: 1379–1382.

Oliversen R. J., Scherb, F., Smyth, W. H., Freed, M. E., Woodward, R. C., Marconi M. L., Retherford, K. D., Lupie, O. L., and Morgan, J. P. 2001. Sunlit Io atmospheric O I 6300 Å emission and the plasma thorus. *J. Geophys. Res.* 106: 26183–26193.

Paxton, L. J. and Meng, C. I. 1999. Auroral imaging and space-based optical remote sensing. *APL Technol. Dig.* 20: 556–569.

Petersen, C. and Brandt, J. C. 1995. *Hubble Vision, Astronomy with the Hubble Space Telescope*, pp. 1–272. Cambridge, NY: Cambridge University Press.

Prange, R., Rego, D., Pallier, L., Jaffel, L. B., Emerich, C., Ajello, J., Clarke, J. T., and Ballester, G. E. 1997. Detection of self-reversed Lyα lines from the Jovian Aurorae with the Hubble Space Telescope. *Astrophys. J. Lett.* 484: L169–L173.

Pryor, W. R., Ajello, J. M., Tobiska, W. K., Shemansky, D. E., James, G. K., Hord, C. W., Stephans, S. K., West, R. A., Stewart, A. I. F., McClintock, W. E., Simmons, K. E., Hendrix, A. R., and Miller, D. A. 1998. Galileo ultraviolet spectrometer observations of Jupiter's auroral spectrum from 1600–3200 Å. *J. Geophys. Res.* 103: 20149–20158.

Raymond, J., Blair, W. P., and Long, K. S. 1997. Hopkins telescope observations of H_2 emission from HH2. *Astrophys. J.* 489: 314–318

Roncin, J.-Y. and Launay, F. 1994. *Atlas of the Vacuum Ultraviolet Emission Spectrum of Molecular Hydrogen*. Monographs 4. New York: American Institute of Physics.

Sandel, B. R., Shemansky, D. E., Broadfoot, A. L., Bertaux, J. L., Blamont, J. E., Belton, M. J., Ajello, J. M., Holberg, J. B., Atreya, S. K., Donahue, T. M., Moos, H. W., Strobel, D. F., McConnell, J. C., Dalgarno, A., Goody, R., McElroy, M. B., and Takacs, P. Z. 1979. Extreme ultraviolet observations from Voyager 2 encounter with Jupiter. *Science* 206: 962–966.

Shemansky, D. E. and Ajello, J. M. 1983. The Saturn spectrum in the EUV-electron excited hydrogen. *J. Geophys. Res.* 88: 459–464.

Shemansky, D. E., Ajello, J. M., Hall, D. T., and Franklin, B. 1985. Vacuum ultraviolet studies of electron impact on helium: Excitation of He n $^1P^o$ Rydberg series and ionization-excitation of He^+ nl Rydberg series. *Astrophys. J.* 296: 774–783.

Snow, T. P. 1979. Ultraviolet observations of interstellar molecules and grains from spacelab. *Astrophys. Space Sci.* 66: 453–466.

Stern, S. A., Slater, D. C., Scherrer, J., Stone, J., Dirks, G., Versteeg, M., Davis, M., Randall Gladstone, G., Parker, J. W., Young, L. A., and Siegmund, O. H. W. 2008. ALICE: The ultraviolet imaging spectrograph aboard the New Horizons Spacecraft Pluto–Kuiper Belt Mission. *Space Sci. Rev.* 140: 155–187.

Stevens, M. H. 2001. The EUV airglow of Titan: Production and loss of NB_{2B} $c'B_{4B}(0)$-X. *J. Geophys. Res.* 106: 3685–3689.

Stevens, M. 2002. The extreme ultraviolet airglow of N_2 atmospheres. In *Atmospheres in the Solar System: Comparative Aeronomy*. Geophysical Monograph, M. Mendillo, A. Nagy, and J. H. Waite (eds.), p. 319. Washington, DC: American Geophysical Union.

Stevens, M. H., Meier, R. R., Conway, R., and Strobel, D. 1994. A resolution of the N_2 c$'B_{4B}$(0)-X and problem in the Earth's atmosphere. *J. Geophys. Res.* 99: 417–433.

Stevens, M. H., Bishop, J., and Feldman, P. D. 2003. A new view of Titan's EUV airglow. In *DPS Meeting Abstract*, 931 pp. Monterey, CA, 35.

Stone, E. J. and Zipf, E. C. 1974. Electron-impact excitation of the $^3S^0$ and $^5S^0$ states of atomic oxygen. *J. Chem. Phys.* 60: 4237–4243.

Strickland, D. J., Evans, J. S., and Paxton, L. J. 1995. Satellite remote sensing of thermospheric O/N_2 and solar EUV. 1. Theory. *J. Geophys. Res.* 100: 12217–12226.

Strickland, D. J., J. Bishop, J. S. Evans, T. Majeed, P. M. Shen, R. J. Cox, R. Link, and Huffman, R. E. 1999, Atmospheric Ultraviolet Radiance Integrated Code (AURIC): Theory, software architecture, inputs, and selected results, *J. Quant. Spectrosc. Radiat. Transfer*, 62, 689–742.

Strickland, D. J., Meier, R. R., Walterscheid, R. L., Craven, J. D., Christensen, A. B., Paxton, L. J., Morrison, D., and Crowley, G. 2004. Quiet time seasonal behavior of the thermosphere seen in the far ultraviolet dayglow. *J. Geophys. Res.* 109: A01302, doi:10.1029/2003JA010220.

Strobel, D. F. and Shemansky, D. E. 1982. EUV emission from Titan's upper atmosphere: Voyager 1 encounter. *J. Geophys. Res.* 87: 1361–1368.

Tawara, H., Itakawa, Y., Nishimura, H., and Yoshino, M. 1990. *J. Phys. Chem. Ref. Data* 19: 617–636.

Terrell, C. A., Hansen, D. L., and Ajello, J. M. 2004. The middle ultraviolet and visible spectrum of O_2 by electron impact. *J. Phys. B* 37: 1931–1950.

Trafton, L. M., Moore, C. H., Goldstein, D. B., Varghese, P. L., and Walker, A. C. 2007. Modeling Io's UV-V Eclipse Aurorae from the Joint HST-Galileo Io Campaign. In *Magnetospheres of the Outer Planets*, June 25–29 2007, San Antonio, TX, Abstract booklet, p. 115.

Trajmar, S., Register, D. F., and Chutjian, A. 1983. Electron-scattering by molecules: 2. Experimental methods and data. *Phys. Rep.* 97: 221–356.

van der Burgt, P. J. M., Westerveld, W. B., and Risley, J. S. 1989. Photoemission cross sections for atomic transitions in the extreme ultraviolet due to electron collisions with atoms and molecules. *J. Phys. Chem. Ref. Data* 18: 1757–1805.

Vasavada, A. R., Bouchez, A. H., Ingersoll, A. P., Little, B., and Anger, C. D. 1999. Jupiter's visible aurora and Io footprint. *J. Geophys. Res.* 104: 27133–27142.

Vatti Palle, P. V., Ajello, J. M., and Bhardwaj, A. 2004. The high resolution spectrum of electron-excited SO_2. *J. Geophys. Res.* 109: A02310, doi: 10.1029/2003JA009828.

Walter, C. W., Cosby, P. C., and Helm, H. 1994. Predissociation quantum yields of singlet nitrogen. *Phys. Rev. A* 50: 2930–2936.

Wilson, E. H. and Atreya, S. K. 2005. Current states of modeling the photochemisty of Titan's mutually dependent atmosphere and ionosphere. *J. Geophys. Res.* 109: E06002, doi: 10.1029/2003JE002181.

Young, J. A., Malone, C. P., Johnson, P. V., Liu, X., Ajello, J. M., and Kanik, I. 2009. Dissociative excitation of NO_2 by electron impact. *J. Phys. B: At. Mol. Opt. Phys.* 42:185201-1–185201-12.

Young, J. A., Malone, C. P., Johnson, P. V., Ajello, J. M., Liu, X., and Kanik, I. 2010. Lyman-Birge-Hopfield emissions from electron impact excited N_2. *J. Phys. B: At. Mol. Opt. Phys.* 43:135201-1–135201-16.

Yung, Y. L., Gladstone, G. R., Chang, K. M., Ajello, J. M., and Srivastava, S. K. 1982. H_2 fluorescence spectrum from 1200 to 1700 Å by electron impact: Laboratory study and application to Jovian Aurora. *Astrophys. J.* 254: L65–L70.

Zetner, P. W., Kanik, I., and Trajmar, S. 1998. Electron impact excitation of the a $^3\Pi$, a$'$ $^3\Sigma^+$, d $^3\Delta$, and A $^1\Pi$ states of CO at 10.0, 12.5 and 15.0 eV impact energies. *J. Phys. B: At. Mol. Opt. Phys.* 31: 2395–2414.

29 Chemical Evolution on Interstellar Grains at Low Temperatures

Kenzo Hiraoka
University of Yamanashi
Kofu, Japan

CONTENTS

29.1 INTRODUCTION

Dense interstellar clouds are the birthplaces of stars of all masses and their planetary systems. Interstellar molecules and dust become the building blocks for protostellar disks, from which planets, comets, asteroids, and other macroscopic bodies eventually form. It has been recognized that dust is ubiquitous in the Galaxy. Light essentially cannot penetrate dense interstellar clouds, and temperature inside dark clouds remains at ~10 K. About 120 interstellar molecules have been observed by millimeter and submillimeter spectroscopy in such dark clouds. It is becoming increasingly clear that gas-phase ion–molecule reactions—the dominant processes in diffuse inter-stellar clouds—are less important in the dense clouds. Instead, chemical reactions on cosmic dust grains play an important role (Millar and Williams, 1993). Interstellar ices, which grow on solid dust particles in cold, dense clouds, become the substrates for various chemical reactions. The Infrared Space Observatory (ISO) has provided much information about interstellar and cometary ices (Ehrenfreund and Schutte, 2000).

Unfortunately, direct observations of the solid-phase chemistry of dense clouds are difficult to come by. In the absence of direct evidence, we have to resort to laboratory studies. A wealth of data has been gathered on gas-phase ion–molecule reactions, but investigations on the solid-phase reactions of various molecules remain scarce (Greenberg and Pirronello, 1991), although there have been experimental studies of molecular hydrogen formation on the surface of cosmic dust analogues (Pirronello et al., 2000).

Hydrogen atoms are believed to play a particularly important role in the chemical evolution of dense clouds because of their high mobility in solids. Because of their wave nature and low atomic weight, hydrogen atoms can "tunnel" through seemingly insurmountable barriers (Yang and Rabitz, 1994). In a solid, a hydrogen atom can be visualized as a quantum liquid that creeps quickly from one site to another and participates in tunneling reactions whenever it encounters suitable reactions.

In the following sections, the role of dust grains for the formation of hydrocarbons, formaldehyde, methanol, HCN/HNC, and ammonia will be described. The importance of the H-atom tunneling reactions and radiation chemistry induced by the cosmic rays in the dark clouds will be underlined.

29.2 SIMULATION OF THE CHEMICAL EVOLUTION TAKING PLACE ON THE DUST GRAINS IN THE DARK CLOUDS

The conceptual idea of the apparatus is shown in Figure 29.1 (Hiraoka et al., 2002a). The cryocooler (Iwatani Plantech, type D310) and a quadrupole mass spectrometer (Leda Mass, Microvision 300D) are housed in a vacuum manifold. The vacuum chamber was evacuated by two turbomolecular pumps (ULVAC, UTM-500, $500\,L\,s^{-1}$ and Seiko Seiki, STP-H200, $200\,L\,s^{-1}$) connected in tandem.

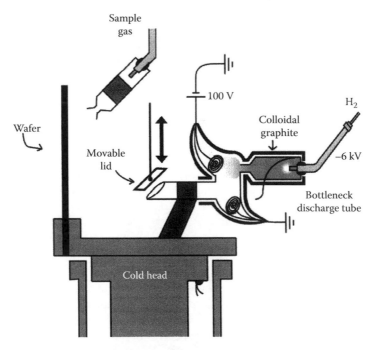

FIGURE 29.1 Schematic diagram of the experimental system (not to scale). (From Hiraoka, K. et al., *Astrophys. J.*, 577, 265, 2002a. Reproduced with permission of the AAS.)

The base pressure of the vacuum system under the current experimental conditions was ~5 × 10^{-10} Torr, after baking the vacuum manifold at about 130°C for 72 h. The major residual gas components are H_2O, N_2, O_2, and H_2. The pressure rise over the lifetime of the experiment was ~1 × 10^{-9} Torr. The original base pressure was resumed after overnight evacuation at room temperature.

The sample gas was deposited on the silicon substrate ([100] surface with the size of 30 × 50 × 0.5 mm³), which was firmly pressed to the cold head of the cryocooler using indium foil between the mating surfaces. The sample gas was introduced through a calibrated stainless steel capillary (internal diameter of 0.1 mm and 1 m long) onto the cooled silicon substrate in the vacuum chamber. After the deposition of the sample, H (or D) atoms produced by the dc discharge of H_2 (or D_2) were sprayed over the sample film. The base pressure rose to ~1 × 10^{-6} Torr by the H_2 gas introduction.

The dc discharge was generated by applying −6 kV to the stainless steel capillary in a bottleneck discharge tube. The sample film was prevented from being bombarded by the charged particles and UV photons produced by the plasma. The electron current effusing out of the dc discharge tube over the silicon substrate was measured by an electrometer. By adjusting the voltage applied to the coil electrodes inside the optical traps of the bottleneck discharge tube, the electron current was suppressed below the minimum current that could be measured by the electrometer (<10^{-12} A). That is, the flux of electrons sprayed over the film was lower than ~10^6 electrons cm^{-2} s^{-1}. Thus, the chemical reactions induced by the electron bombardment of the sample film should be negligible. The discharge tube that was held tight by the copper sleeve connected to the cold head was kept at ~27 K when the cold head was cooled to 10 K. Thus, the temperature of H atoms sprayed over the sample film may have been about 27 K.

In order to prevent the contamination of the gas line of the H-atom spray system, the exit of the bottleneck discharge tube for the H-atom spray was plugged by using a movable lid every time for gas introduction and for the thermal desorption experiment. In order to sputter the residual molecules adsorbed on the inner wall of the discharge tube, H_2-plasma was always generated for 30 min before and after the experiment. These procedures are necessary in order to perform the experiments on the low-temperature solid-phase reactions in which the amounts of reaction products are only limited to less than an equivalent monolayer.

The introduction rate of H_2 into the discharge tube was about 10^{16} molecules s^{-1}. The measurement of the flux of H atoms sprayed over the solid film was not made in this experiment. Cherigier et al. (1999) reported that a few percent of reagent gas is decomposed into atoms in the ordinary glow discharge plasma. The H-atom flux was estimated to be ≤10^{13} atoms cm^{-2} s^{-1} under the present experimental conditions (Hiraoka et al., 2002a). To a crude approximation, the irradiation of H atoms with the flux of ~10^{13} cm^{-2} s^{-1} for 300 min spray may correspond to ~5 × 10^4 years of H dose in the dark cloud, assuming that the H-atom number density is ~1 cm^{-3}, the H-atom temperature is 70 K, and the dust grain size is 0.1 μm in the dark cloud.

The electrical power necessary to decompose H_2 molecules completely to H atoms, with the flow rate of ~1 × 10^{16} H_2 s^{-1}, is less than 0.1 W. In a separate experiment, we used a microwave plasma with the power of ~10 W as the H-atom source. We found that the wall of the discharge tube was seriously corroded by plasma etching (sputtering) and various radiation products such as solid particulates (amorphous silicon), SiH_4, H_2O, CO, etc., were deposited on the cold silicon substrate. The contamination from the discharge products will totally destroy the very delicate experiments because the present experiments dealt with a few to tens of monolayers (ML) sample films. The present dc discharge with the electric power of 0.2 W (Figure 29.1) is almost free from contamination. Besides, the dc discharge tube was maintained at about 27 K. This must suppress the etching of the Pyrex discharge tube.

In our work, the quantitative analysis of reaction products was performed by thermal desorption spectrometry (TDS) with the programming rate of 6 K min^{-1}, taking the mass cracking patterns and the relative ionization cross sections into account for all gaseous products. The yields of the products were calculated from the ion intensities of the gaseous products and those of the reactant recovered.

To perform in situ and real-time product analysis in the low-temperature solid-phase reactions, the infrared absorption spectra of the deposited film reacted with H (or D) were measured using a FT-IR spectrometer (Nicolet, Magna–IR 760), with a resolution of 4 cm^{-1} in combination with a KBr beam splitter and a liquid N$_2$-cooled MCT (HgCdTe) detector. The quantitative analyses of reaction products were made by FT-IR spectra, but the reliability is less than those obtained by TDS due to weak absorption intensities.

No gaseous products were detected by the mass spectrometer during the H-atom spray over the solid sample in the temperature range investigated. This indicates that the desorption of reactants and products during the H-atom spray was negligible, suggesting that the reactions proceed via the Langmuir–Hinshellwood mechanism. This is reasonable because the temperature of H atoms sprayed over the film was as low as 27 K.

29.3   REACTIONS OF H AND D ATOMS WITH SOLID C$_2$H$_2$, C$_2$H$_4$, AND C$_2$H$_6$ AT CRYOGENIC TEMPERATURES

29.3.1   In Situ and Real-Time Observation of Products Formed from Reactions of H Atoms with Solid C$_2$H$_4$ and C$_2$H$_2$

It is well known that the formation of saturated hydrocarbons in the interstellar medium is difficult to explain by the gas-phase ion–molecule reactions, because the carbenium ions cease to react with H$_2$ in the unsaturated forms (Ikezoe et al., 1987). Consequently, the observation of saturated hydrocarbons, either frozen on grains or as gases, would provide strong evidence that reactions on the dust grains play important roles in the interstellar chemical evolution.

Cometary nuclei are the least modified solar system bodies extant from the time of planetary formation. Their current compositions are thought to be representative of the materials from which they formed (Mumma et al., 1996). The native ices are thought to have originated either in the interstellar dense cloud core that preceded the solar system or at distances beyond Jupiter in the nebula surrounding the protosun. In comet C/1996 B2 Hyakutake, hydrocarbons C$_2$H$_2$, C$_2$H$_6$, and CH$_4$ along with CO and H$_2$O were detected with the use of a high-resolution infrared telescope on Mauna Kea (Mumma et al., 1996). The abundances of C$_2$H$_6$, CH$_4$, CO, and H$_2$O were in the proportions C$_2$H$_6$:CH$_4$:CO:H$_2$O = 0.4:0.7:5.8:100. Mason et al. reported the detection of C$_2$H$_2$ in Hyakutake at an abundance of 0.1%–0.4% relative to H$_2$O (Mason et al., 1996).

Comet C/1996 01 Hale–Bopp was observed at wavelengths from 2.4 to 195 μm with ISO when the comet was about 2.9 AU from the sun (Crovisier et al., 1997; Crovisier, 1998). The gas-phase abundance of C$_2$H$_2$ and C$_2$H$_6$ were found to be about the same and their ratios with H$_2$O were about 0.5%.

In the observation of comets Hyakutake and Hale–Bopp, it should be noted that C$_2$H$_4$ had not been detected by infrared spectroscopy despite the reasonably strong intensities of C$_2$ hydrocarbons, C$_2$H$_2$, and C$_2$H$_6$.

The abundances of C$_2$H$_4$ are generally observed to be low in the solar system; for example, C$_2$H$_4$ is missing in Saturn and its moon Titan, where other hydrocarbons with two carbon atoms are common. To find out why, we must consider the channels through which these molecules form.

In order to perform in situ and real-time product analysis in the low-temperature solid-phase reactions, we measured the infrared absorption spectra of C$_2$H$_4$ and C$_2$H$_2$ films being reacted with H atoms at cryogenic temperatures (Hiraoka et al., 2000). Figure 29.2a shows the FT-IR spectra for a 20 monolayers (ML) thick C$_2$H$_4$ sample film sprayed by H atoms for 3 h at 10 K. The bottom spectrum is that for the 20 ML thick C$_2$H$_4$ before the reaction. The spectra from the bottom to the top were measured with a time interval of 15 min. Steady growth in the absorption of the ν$_{10}$ band (2974 cm^{-1}) and ν$_5$ band (2881 cm^{-1}) of C$_2$H$_6$ can be seen with the reaction time. This clearly indicates the occurrence of the reaction of H with solid C$_2$H$_4$ to form C$_2$H$_6$ at 10 K. However, no C$_2$H$_5$ radicals could be detected by FT-IR. The H atom being sprayed over the solid C$_2$H$_4$ film is itself

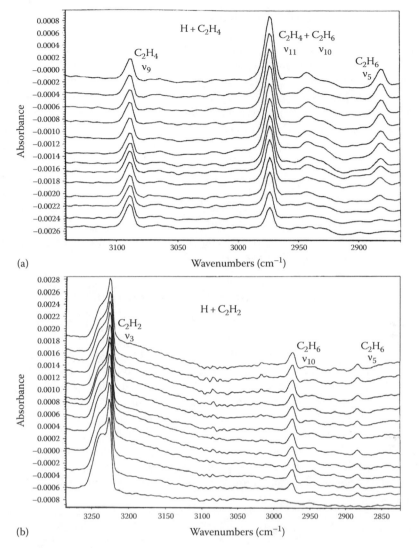

FIGURE 29.2 (a) Change of the FT-IR spectra when the 20 ML thick C_2H_4 film was sprayed by H atoms for 3 h at 10 K. The bottom spectrum is that for the 20 ML thick C_2H_4 before reaction. The spectra from the bottom to the top were measured with the time interval of 15 min. (b) Change of the FT-IR spectra when the 10 ML thick C_2H_2 film was reacted with H atoms for 5 h at 10 K. The bottom spectrum is that for the 10 ML thick C_2H_2 film before reaction. The spectra from the bottom to the top were measured with the time interval of about 30 min. (From Hiraoka, K. et al., *Astrophys. J.*, 532, 1029, 2000. Reproduced with permission of the AAS.)

a radical scavenger. Thus, the steady-state concentration of C_2H_5 in the C_2H_4 matrix must be kept low during the H-atom spray over the C_2H_4 solid film because of the occurrence of the efficient recombination or abstraction reaction between H and C_2H_5. No radical species could be detected by FT-IR in the reactions of H atoms with either C_2H_2 or C_2H_6.

Figure 29.2b displays the FT-IR spectra for a 10 ML thick C_2H_2 film reacted with H atoms for 5 h at 10 K. The bottom spectrum is that for the 10 ML thick C_2H_2 before reaction. The spectra from the bottom to the top were measured with the time interval of about 25 min. Steady growths in the absorption of the v_{10} band (2974 cm^{-1}) and v_5 band (2881 cm^{-1}) of C_2H_6 can be seen with the H-atom spray time. In the reaction of H with C_2H_2 to form C_2H_6, C_2H_4 must be formed as an intermediate product. However, no absorption due to C_2H_4 is observed in Figure 29.2b, nor was C_2H_4 detected by the much more sensitive TDS. The fact that no intermediate products could be detected except for the final C_2H_6 product indicates that the initial reaction of H with C_2H_2 to form C_2H_3 is the rate-controlling process and the following reactions to form the final C_2H_6 proceed much faster. This finding is in accord with the observation of comets Hyakutake and Hale–Bopp; that is, C_2H_2 and C_2H_6 but no C_2H_4 were detected in the comae of these comets. Because C_2H_2 is known to be formed by the gas-phase reactions, the C_2H_2 observed in comets is likely to originate from the gas-phase reaction. In contrast, the saturated hydrocarbons cannot be formed by the gas-phase reactions. The presence of C_2H_6 is consistent with the production of C_2H_6 in icy grain mantles in the natal cloud, either by the photolysis of CH_4-rich ice or by hydrogen-addition reactions to C_2H_2 condensed from the gas phase (Gerakines et al., 1996). The model of gas-grain chemistry in dense interstellar clouds predicts that the grain-surface concentrations of C_2H_2 and C_2H_6 are considerably greater than that of C_2H_4 (Hasegawa et al., 1992). This result is derived from the assumed larger activation energy for H + C_2H_2 (10.0 kJ mol^{-1}) than that for H + C_2H_4 (6.2 kJ mol^{-1}). Bennett and Mile investigated the reaction of H with solid C_2H_2 and C_2H_4 at 77 K by electron spin resonance using a rotating cryostat (Bennett and Mile, 1973). They found that the ratio of the rate constants k(H + C_2H_4)/k(H + C_2H_2) is about 6×10^3 in solid-phase reactions at 77 K. The result is in line with findings that C_2H_4 was not detected as an intermediate product for the reaction of H with C_2H_4.

29.3.2 Temperature Dependence on the Yield of Reaction Products

Figure 29.3 displays the relationship between the yields of C_2H_6 (%) and the reaction temperature for reactions H + C_2H_4 and H + C_2H_2. The yield (%) corresponds to the amount of product with respect to the sample deposited on the substrate. In this experiment, the sample C_2H_4 or C_2H_2 was deposited first at 10 K and then H atoms were sprayed over the solid film after the substrate temperature was raised from 10 K to the reaction temperature. The experiment was not performed above 50 K because the reactant and C_2 hydrocarbon products start to sublime above ~55 K.

For reaction H + C_2H_4 in Figure 29.3, the yield (%) is temperature independent in the range of 50–40 K but it shows a sharp increase below 40 K. The yield reaches the maximum (38%) at 12 K and decreases below 12 K. For reaction H + C_2H_2, no reaction product could be detected at 50 K. At 40 and 35 K, the yields of C_2H_6 were nearly three orders of magnitude smaller than that at 10 K (24%). With further decrease of temperature, the yield of C_2H_6 becomes about 0.1% at 30 K and

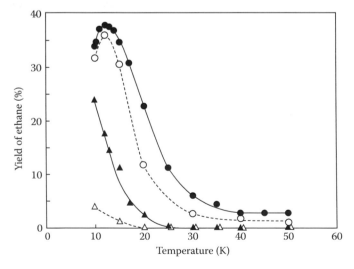

FIGURE 29.3 Temperature dependence on the yields of ethane (%) from reactions of H + C_2H_4 (filled circles), D + C_2H_4 (open circles), H + C_2H_2 (filled triangles), and D + C_2H_2 (open triangles). H- and D-atom spray time, 60 min; thickness of the sample film, 10 ML. (From Hiraoka, K. et al., *Astrophys. J.*, 532, 1029, 2000. Reproduced with permission of the AAS.)

shows a steep increase below 25 K. The fact that at 50 K about 3% of C_2H_4 was converted to C_2H_6 for reaction $H + C_2H_4$, whereas no C_2H_6 could be detected for reaction $H + C_2H_2$, suggests that rate constant $k(H + C_2H_4)$ may be orders of magnitude larger than $k(H + C_2H_2)$ at this temperature. This is in agreement with the finding by Bennett and Mile (1973).

Villa et al. studied theoretically the tunneling reaction $H + C_2H_4 \rightarrow C_2H_5$ (Villa et al., 1998a,b). They predicted the classical barrier height from 5.0 to 11.3 kJ mol^{-1}. They pointed out that tunneling becomes important below ~500 K. Knyazev et al. performed experimental and theoretical studies of the tunneling reaction $H + C_2H_2$ (Knyazev and Slagle, 1996; Knyazev et al., 1996). They found that the tunneling effect for the reaction is evident and recommended the value of the classical barrier for the reaction to be 16.7 kJ mol^{-1}. The larger value of the classical barrier for reaction $H + C_2H_2$ than that for $H + C_2H_4$ is in line with the much greater rate for reaction $H + C_2H_4$ than for $H + C_2H_2$ observed by Bennett and Mile (1973) and by us (Hiraoka et al., 2000).

The rate for the formation of C_2H_6 is a function of the concentration of H on the solid film (i.e., [H]). The results in Figure 29.3 may be interpretable by the increase of the sticking probability of H atoms at lower temperature. Becker et al. investigated the absorption of H_2 by condensed solids of Ar, Kr, Xe, CO_2, and C_3H_8 as a function of temperature (Becker et al., 1972). They found that the adsorption capacity shows a steep increase starting at about 30–40 K and after reaching the plateau at about 15 K, it decreases below ~15 K. Since the polarizability of the H atom (0.67×10^{-30} m^3) is smaller than that of the H_2 molecule (0.8×10^{-30} m^3), H may have a smaller sticking probability than H_2. It is highly probable that the temperature dependence of the sticking probability of H atoms reflects the C_2H_6 yields in Figure 29.3.

The decrease in the yield of C_2H_6 below 12 K may be mainly due to the decrease of diffusion rates of H atoms in the solid C_2H_4 film. The lack of thermal energy at lower temperatures may diminish the penetration of H atoms inside the film. Actually, Bhattacharya et al. (1981) found that the trapped H atoms in CH_4 and CD_4 matrices become largely immobile below 12 K.

29.3.3 Reactions of D Atoms with Solid C_2H_4, C_2H_2, and C_2H_6: Isotope Effect

In the low-temperature tunneling reactions, remarkable isotope effects have been found for many systems (Miyazaki, 1991). In order to obtain information on the isotope effect, the reactions of D atoms with C_2H_4, C_2H_2, and C_2H_6 at 10 K were studied (Hiraoka et al., 2000). The D atoms were produced by introducing D_2 (99.5%) in the bottleneck discharge tube in Figure 29.1. The quantitative analyses of the D-atom-substituted products ($C_2H_{6-n}D_n$) were performed by TDS.

When a 10 ML thick C_2H_4 was reacted with D atoms at 10 K, $C_2H_4D_2$ was formed as a major product and $C_2H_{6-n}D_n$ as minor products with $n = 3$–6. The yields of $C_2H_{6-n}D_n$ are summarized in Table 29.1 as a function of the D-atom spray time.

The formation of $C_2H_{6-n}D_n$ with $n \geq 3$ by the reactions of D atoms with C_2H_4 (Table 29.1) indicates that the H atoms of the reactant C_2H_4 were substituted by the D atoms. If the H-atom abstraction, $H + C_2H_4 \rightarrow C_2H_3 + H_2$, and its analogue take place to some extent, the formation of $C_2H_{6-n}D_n$ with $n \geq 3$ can be explained. However, this is less likely because Bennett and Mile (1973) found that the H-atom addition reaction to the double bonds takes place exclusively and the H-atom abstraction from C_2H_4 was not observed in the reactions of H atoms with olefins. Thus, the formation of $C_2H_{6-n}D_n$ with $n \geq 3$ can be attributed only to the occurrence of the abstraction reaction of D with ethyl radical, that is, $D + C_2H_4D \rightarrow C_2H_3D + HD$ and their analogues $D + C_2H_{5-n}D_n \rightarrow C_2H_{4-n}D_n + HD$ with $n = 2$–4.

In Table 29.1, there is a trend that the relative yields of $C_2H_{6-n}D_n$ with larger n increase with D-atom spray time. The total yield of $C_2H_{6-n}D_n$ increases with D-atom spray time and more than 40% of the reactant C_2H_4 is converted to $C_2H_{6-n}D_n$ for a 300 min reaction time. At the early stages of reaction, the sample surface is mainly composed of C_2H_4, and thus D atoms may have no difficulty encountering the reactant C_2H_4 on the surface to form an ethyl radical. The D atoms on the surface may be more mobile than inside the film; consequently, they may have less difficulty approaching the ethyl radical to form ethane by recombination. As the reaction goes on, the surface of the C_2H_4

TABLE 29.1

D-Atom Spray Time Dependence on the Yields of $C_2H_{6-n}D_n$ Formed from Reaction of D with C_2H_4

Time (min)	Yield (%)	Relative Yield (%)				
		$C_2H_4D_2$	$C_2H_3D_3$	$C_2H_2D_4$	C_2HD_5	C_2D_6
1	2	91	9	—	—	—
5	9	79	15	3	2	1
30	25	70	18	7	3	2
60	33	62	22	9	4	3
180	43	61	22	10	4	3
300	44	58	21	9	6	5

Source: From Hiraoka, K. et al., *Astrophys. J.*, 532, 1029, 2000. Reproduced with permission of the AAS.

Note: Film thickness, 10 ML; reaction temperature, 10 K.

film is being contaminated by the ethane molecules, and D atoms must diffuse inside the film to react with ethylene. For the recombination of a D atom with an ethyl radical to occur inside the film, the atom and the ethyl radical must approach each other, whereas for abstraction only the H-atom transfer is required, which may be preferred by tunneling in the solid film. This explains why the D-atom enrichment becomes more favorable with longer reaction time.

In Table 29.2, the temperature dependence of the yields of $C_2H_{6-n}D_n$ with $n \geq 2$, formed by the reaction of D with 10 ML C_2H_4 film for 1 h, is summarized. In Figure 29.3, the total yields of $C_2H_{6-n}D_n$ with $n \geq 2$ by the reaction $D + C_2H_4$ are shown as a function of reaction temperature. They show a similar trend as C_2H_6 from the reaction $H + C_2H_4$, although they are somewhat smaller than those of C_2H_6, that is, an isotope effect. In Table 29.2, $C_2H_4D_2$ is the major product above 30 K, and the relative amounts of $C_2H_{6-n}D_n$ with larger n increase rapidly with decrease of temperature. With decrease of temperature, the sticking probability of D atoms increases, which results in the accumulation of $C_2H_{6-n}D_n$ ($n \geq 2$) products near the top surface of the film. The slower diffusion of D atoms inside the film results in the more preferable occurrence of the abstraction reaction than that of the recombination. This explains the enrichment of D atoms in $C_2H_{6-n}D_n$ at a lower temperature in Table 29.2.

TABLE 29.2

Temperature Dependence of the Yields of $C_2H_{6-n}D_n$ Formed by Reaction of D with C_2H_4

Temperature (K)	Yield (%)	Relative Yield (%)				
		$C_2H_4D_2$	$C_2H_3D_3$	$C_2H_2D_4$	C_2HD_5	C_2D_6
10	32	61	21	10	5	3
12	36	63	22	8	4	3
15	31	70	19	7	2	2
20	12	80	15	3	1	1
30	3	~100	—	—	—	—
40	2	~100	—	—	—	—
48	1	~100	—	—	—	—

Source: From Hiraoka, K. et al., *Astrophys. J.*, 532, 1029, 2000. Reproduced with permission of the AAS.

Note: Film thickness, 10 ML; D-atom spray time, 60 min.

TABLE 29.3
D-Atom Spray Time Dependence on the Yields of $C_2H_{6-n}D_n$ Formed from Reaction of D with C_2H_2

Time (min)	Yield (%)	Relative Yield (%)		
		$C_2H_2D_4$	C_2HD_5	C_2D_6
1	0.4	~100	—	—
15	0.7	84	16	—
60	4	72	25	3
180	12	58	30	12
300	18	55	33	12

Source: From Hiraoka, K. et al., *Astrophys. J.*, 532, 1029, 2000. Reproduced with permission of the AAS.
Note: Film thickness, 10 ML; reaction temperature, 10 K.

When the 10 ML thick C_2H_2 was reacted with D atoms, $C_2H_2D_4$ was formed as a major product and C_2HD_5 and C_2D_6 as minor products. The total yields of $C_2H_{6-n}D_n$ with $n = 4$–6 are summarized in Table 29.3 as functions of D-atom spray time. As seen in Table 29.3, the yields of $C_2H_{6-n}D_n$ from reaction $D + C_2H_2$ are considerably smaller than that of C_2H_6 from the reaction $H + C_2H_2$. This indicates that the constant of the rate-controlling reaction, $D + C_2H_2 \rightarrow C_2H_2D$, is smaller than that of the reaction, $H + C_2H_2 \rightarrow C_2H_3$. Figure 29.3 shows that the yields of $C_2H_{6-n}D_n$ from the reaction $D + C_2H_2$ are much smaller than those of C_2H_6 from $H + C_2H_2$. $C_2H_{6-n}D_n$ could not be detected at reaction temperatures above 20 K for the reaction $D + C_2H_2$, despite the fact that C_2H_6 could be detected up to about 40 K for the reaction $H + C_2H_2$. This suggests that the isotope effect in reactions $H/D + C_2H_2$ is much more prominent than in the reactions $H/D + C_2H_4$.

In Table 29.3, the relative yields of $C_2H_{6-n}D_n$ ($4 \geq n$), with larger n, increase with D-atom spray time. The enrichment of D content in $C_2H_{6-n}D_n$ may be reasonably accounted for by the occurrence of the sequential abstraction reactions of D with vinyl and ethyl radicals.

Table 29.4 summarizes the relative yields of $C_2H_{6-n}D_n$ from reaction $D + C_2H_2$ as a function of reaction temperature. The number of D atoms in $C_2H_{6-n}D_n$ increases with decrease of reaction temperature. This trend is similar to the case of reaction $D + C_2H_4$ as shown in Table 29.2.

TABLE 29.4
Temperature Dependence on the Yields of $C_2H_{6-n}D_n$ Formed from Reaction of D with C_2H_2

Temperature (K)	Yield (%)	Relative Yield (%)		
		$C_2H_2D_4$	C_2HD_5	C_2D_6
10	4	72	25	3
15	1	76	24	3
20	0.1	~100	—	—
30	~0	—	—	—
40	~0	—	—	—
50	~0	—	—	—

Source: From Hiraoka, K. et al., *Astrophys. J.*, 532, 1029, 2000. Reproduced with permission of the AAS.
Note: Film thickness, 10 ML; D-atom spray time, 60 min.

The association reaction, $C_2H_3 + H_2 \rightarrow C_2H_5$, may be one of the candidates for the formation of ethane. However, it has been widely accepted that this reaction is slow (Knyazev and Slagle, 1996; Knyazev et al., 1996), for example, the rate constant of the reaction is of the order of $10^{-16}\,cm^3$ molecule^{-1} s^{-1} even at ~600 K. This is mainly due to the large energy barrier for the reaction (>33 kJ mol^{-1}) (Knyazev and Slagle, 1996; Knyazev et al., 1996). The contribution of this reaction may be minor for the formation of ethane.

In order to examine the reactivity of the H atom with C_2H_6, the reaction of D atoms with a C_2H_6 film was investigated. If reaction $D + C_2H_6 \rightarrow C_2H_5 + HD$ takes place, C_2H_5D will be formed as a primary product. When a 10 ML thick C_2H_6 film was sprayed by D atoms for 1 h at 10 K, 0.2% of the reactant C_2H_6 was converted to C_2H_5D. This indicates the occurrence of abstraction reaction, $D + C_2H_6 \rightarrow C_2H_5 + HD$, but the yield of C_2H_5D (0.2%) is much lower than the ethane yields from the reactions of D with C_2H_4 (33%) or C_2H_2 (4%) at 10 K for 1 h D-atom spray. At temperatures above 20 K, no C_2H_5D could be detected. The extremely low yield of C_2H_5D indicates the much smaller rate constant for $D + C_2H_6$ than those for $D + C_2H_4/C_2H_2$ at 10 K.

The negative temperature dependences for the rates of reactions $H/D + C_2H_4/C_2H_2$ in Figure 29.3 may be reasonably explained either by the increase of the sticking probability of H/D atoms on solid films or by the slower rate of the H/D atom diffusion on and in the solid film at lower temperatures. It is highly probable that the temperature dependence of the sticking probability and the diffusion rate of H atoms reflect the yields in Figure 29.3. During the H-atom spray over the solid film, the number of H atoms sprayed over the film surface per unit time (d[H]/dt) is equal to the sum of the rates of the H-atom annihilation due to recombination reaction, $H + H \rightarrow H_2$, the reaction of H with the reactant, and the H-atom desorption from the surface. That is, the following relationship should hold:

$$d[H]/dt = \text{constant} = k_1[H]^2 + k_2[H][\text{reactant}] + k_d[H] \qquad (29.1)$$

where k_1, k_2, and k_d are the rate constant for reaction $H + H \rightarrow H_2$, that for the reaction of H with the reactant, and that for the desorption of H from the surface, respectively. The experimental fact that the rates of the tunneling reactions increase with decrease of temperature (Figure 29.3) means that the value of $k_2[H][\text{reactant}]$ in Equation 29.1 increases with decrease of temperature. That is, k_2 and/or [H] increase with decrease of temperature. In Equation 29.1, k_d must be highly temperature dependent, that is, $k_d = A \exp(-E_d/RT)$ (A, pre-exponential factor; E_d, activation energy for the desorption of the H atoms from the surface). At high temperature, the larger k_d keeps the steady-state concentration of H ([H]) on the solid surface extremely low. With decrease of temperature, k_d decreases exponentially, and concomitantly [H] on the sample surface increases. Besides, the slower diffusion rate of H atoms on and in the film at lower temperatures results in the smaller value of k_1 in Equation 29.1. This also leads to the increase of the H-atom concentration at lower temperatures.

In Figure 29.3, the yield of C_2H_6 from reaction $H + C_2H_2$ is of the same order as that from $H + C_2H_4$ at 10 K. In case k_2 for $H + C_2H_4$ is much larger than k_2 for $H + C_2H_2$, as predicted by Bennett and Mile (1973), [H] on the C_2H_2 film must be much larger than [H] on the C_2H_4 film at ~10 K. In order to examine this possibility, the reactions of H atoms with a few ML thick C_2H_2 or C_2H_4 deposited on the C_3H_6 were studied (Hiraoka and Sato, 2001).

If the sticking probability of H atoms on the C_2H_2 is much higher than that on the C_2H_4 film, the conversion of C_3H_6 to C_3H_8 for the film C_2H_2/C_3H_6 would become much larger than that for C_2H_4/C_3H_6. The yields of C_3H_8 were 22% and 28% for the C_2H_2 (3 ML)/C_3H_6 (10 ML) and C_2H_4 (3 ML)/C_3H_6 (10 ML) films, respectively. This suggests that the steady-state concentrations [H] on the C_2H_2 and C_2H_4 films are of the same order. Thus, the steep increase of the yield of C_2H_6 from reaction $H + C_2H_2$ could possibly be due to the negative temperature-dependent rate constant k_2 for $H + C_2H_2$. A negative temperature dependence of the rate constants for the low-temperature tunneling reactions has been predicted theoretically by Takayanagi and Sato (1990). They performed the bending-corrected rotating-linear-model calculations of the rate constants for the $H + H_2$ reaction and its isotopic variants at low temperatures and examined the effect of the van der Waals well. They found that van der

Waals wells included in both potential surfaces significantly affected the calculated rate constants at temperatures lower than 10 K.

29.4 FORMATION OF FORMALDEHYDE BY THE TUNNELING REACTIONS OF H WITH SOLID CO AT 10 K

Formaldehyde (H_2CO) is one of the most complex molecules for which specific gas-phase reaction $CH_3 + O \rightarrow H_2CO + H$ has been proposed (Watson, 1977). The production of H_2CO is then crucially dependent on the adequate production of CH_3. On the other hand, the significant abundances of H_2CO in the diffuse envelopes of dark clouds cannot be explained by purely gas-phase processes. The production of CH_3 is insufficient in these regions to drive the reaction to form enough H_2CO (Sen et al., 1992).

After H_2, CO is the most abundant molecule in dense clouds and thus has special importance (Tielens et al., 1991). Tielens et al. (1991) suggested that most of the CO accreted in H_2O-rich mantles has reacted with other species on the grain surface. When the H-atom accretion rate is high, this leads to the formation of HCO, H_2CO, and possibly CH_3OH, accounting for the observed large abundance of CH_3OH in grain mantles (Tielens, 1989) and comets (Crovisier, 1998).

$$CO \underset{}{\overset{H}{\rightleftharpoons}} HCO \underset{}{\overset{H}{\rightleftharpoons}} H_2CO \underset{}{\overset{H}{\rightleftharpoons}} H_3CO \underset{}{\overset{H}{\rightleftharpoons}} CH_3OH \tag{29.2}$$

Van Ijzendoorn et al. (1983) measured the absorption spectrum of HCO in Ar, Kr, Xe, CH_4, CO, and N_2 matrixes. They found that the absorbance of HCO in Ar, Kr, and CH_4 matrices grew several-fold while that for trapped H atoms decreased on warm up from 10 to 15 K.

In Sections 29.4.1 through 29.4.3, it will be shown that the low-temperature tunneling reaction of H with CO may not be the major source for the ubiquitous interstellar methanol (Hiraoka et al., 2002a).

29.4.1 Formation of H_2CO from the Reaction of H with Solid CO

It should be noted that the CO molecule is highly adsorptive (Hiraoka et al., 2002a). The inner wall of the H-atom spray bottleneck discharge was found to be rather quickly contaminated by the CO molecules. With the contamination of the discharge tube, several hydrogenated products such as CH_4 and CH_3OH were formed when the H_2 plasma was generated in the bottleneck discharge tube. The gaseous products formed in the H_2 plasma were sprayed over the silicon substrate and condensed there as contaminants. Thus, great care must be taken for the measurement of the reaction of H with CO. Whenever the sample gas was deposited on the silicon substrate, the exit of the H-atom spray bottleneck discharge tube was plugged by the flat stainless steel lid that prevented the sample gas molecules from entering into the discharge tube (see Figure 29.1). By using this experimental system, no discharge products were detected for a long period of the repetitive experiments.

Figure 29.4 shows the TDS spectra of the 10 ML thick solid CO sample reacted with H atoms for 1 h at 10 K. The peak at m/z 30 appearing at ~130 K is due to the formation of H_2CO. If methanol is formed as the reaction product, the base peak at m/z 31 for methanol must appear in the desorption mass spectrum. However, no ion signals with m/z 31 could be detected by TDS as shown in Figure 29.4. This indicates that the reaction of H atom with H_2CO to form CH_3OH is negligible under the present experimental conditions.

We investigated the reaction of H atoms with C atoms seeded in the CO solid (Hiraoka et al., 1998). It was found that the H atoms diffuse deep inside the CO solid film and hydrogenate the C atoms embedded in the CO matrix efficiently. This clearly indicates that the H atoms can migrate in the CO matrix and have ample chance to react with seeded C atoms to form CH_4. In other words, the reactivity of H atom toward the CO molecule is low enough to allow the diffusion of H atom in the CO matrix.

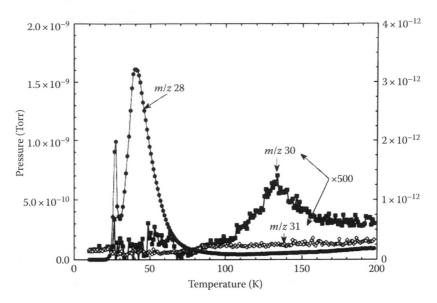

FIGURE 29.4 TDS spectra for the 10 M thick CO solid sample reacted with H atoms for 1 h at 10 K. (From Hiraoka, K. et al., *Astrophys. J.* 577, 265, 2002a. Reproduced with permission of the AAS.)

29.4.2 Temperature Dependence of the Yield of H_2CO from the Reaction $H + CO$

Figure 29.5 displays the relationship between the yields of H_2CO (%) and the reaction temperature for reaction $H + CO$ together with the yields of C_2H_6, C_2H_6, and solid product (mainly polysilane) for reactions of $H + C_2H_2$, $H + C_2H_4$ (Hiraoka et al., 2000), and $H + SiH_4$ (Hiraoka et al., 2001a), respectively, obtained under similar experimental conditions. The experiment above 25 K for $H + CO$ was not made because the reactant CO starts to sublime above ~28 K. The yield of H_2CO in Figure 29.5 shows a steep increase with decrease of temperature from 0.01% at 25 K to 0.08% at 10 K. In Figure 29.5, one sees a general trend that the rates of all tunneling reactions dealt with increase with a decrease of temperature. In general, a steeper increase of the rates with a decrease of temperature was observed for molecules whose rates of reactions with H are small at higher temperature regions, for example, C_2H_2 and CO in Figure 29.5. The observed increase of the rates of reactions $H + C_2H_2$ and $H + CO$ in Figure 29.5 is much steeper than the increase of the sticking probability of H atoms on amorphous carbon with a decrease of temperature (Pirronello et al., 2000), that is, the observed increases of the yields of C_2H_6 and H_2CO in Figure 29.5 may not be explainable only by the increase of the sticking probability of the H atoms on the solid surface.

29.4.3 Does the van der Waals Solid Film Erode during the H-Atom Spray?

In the low-temperature reactions, only exothermic reactions can take place. A greater part of the heat of reaction evolved will eventually degrade to the phonon energy. There is ample evidence that multiphonon processes for the dissipation of the vibrational energy are slow because of the large differences in the vibrational and cohesive energies (Dressler et al., 1975; Oehler et al., 1977). Calaway and Ewing (1975) found that the lifetime of the decay of $v = 1-0$ vibrational levels of N_2 is as long as 1.5 ± 0.5 s in liquid nitrogen. In the present experiment, the H-atom recombination reaction $H + H \rightarrow H_2$ with heat of reaction of 435 kJ mol^{-1}, as well as other exothermic reactions such as reactions (29.2), take place on the solid surface. The occurrence of exothermic reactions on the solid surface may result in the local heating of the van der Waals solid. This could lead to the erosion of the sample film. Actually, this process is considered to be one possible mechanism for the desorption of adsorbed molecules into the gas phase in the dark clouds at 10 K.

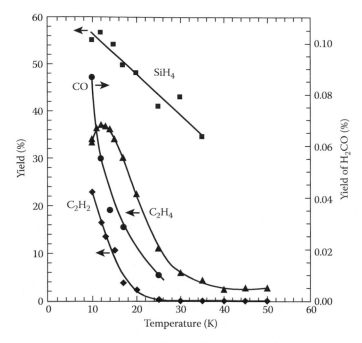

FIGURE 29.5 Relationship between the yield of H_2CO (%) and the reaction temperature for reaction H + CO. The yields of C_2H_6, C_2H_6, and solid product (mainly polysilane) for reactions $H + C_2H_2$, $H + C_2H_4$ (Hiraoka et al., 2000) and $H + SiH_4$ (Hiraoka et al., 2001a), respectively, are also shown. These experiments were made under similar experimental conditions. Film thickness, 10 ML; H-atom spray time, 1 h. (From Hiraoka, K. et al., *Astrophys. J.*, 577, 265, 2002a. Reproduced with permission of the AAS.)

In order to check whether the desorption of CO molecules takes place during the H-atom spray over the solid CO film, the recovery of deposited CO was measured after the solid CO film was sprayed by the H atoms for 1 h by means of TDS. Figure 29.6 shows the temperature dependence of the ratio of the CO recovered after the H-atom spray for 1 h at a certain temperature to that for the deposited at 10 K without spraying the H atoms. As seen in Figure 29.6, the ratio is almost independent of the temperature up to 25 K. This suggests that the relaxation of the vibrational energy of H_2^* formed by the recombination reaction $H + H \rightarrow H_2^*$ to the lattice phonon is inefficient. This finding is in accord with the theoretical calculation by Takahashi et al. (1999). They investigated the time and space dependence of the local temperature increase of icy mantles caused by the release of H_2 formation energy in the vicinity of H_2-forming sites using classical molecular dynamics computational simulations. They predicted that the H_2 formation energy was partitioned to the vibrational energy (78.5%), the rotational energy (9.9%), and the translational energy (7.4%) for the desorbing H_2 molecule and only 4.4% was absorbed to the amorphous ice. The contribution of the highly exothermic reaction $H + H \rightarrow H_2$ to the desorption of the grain mantle may not be as large as thought before.

29.5 REACTION OF H AND D ATOMS WITH SOLID FORMALDEHYDE AND METHANOL AT CRYOGENIC TEMPERATURES

In Section 29.4, studies on the reaction of H with solid CO at 10 K (Hiraoka et al., 2002a) have been described. The CO molecules were found to be converted to H_2CO with a very low efficiency. Besides, no CH_3OH was detected within the experimental error.

In this section, the study on the reactions of H and D atoms with solid H_2CO and CH_3OH films will be discussed (Hiraoka et al., 2005). It would give us more information on the reaction mechanisms for the formation of H_2CO and CH_3OH from reaction H + CO in the dark clouds because H_2CO is the intermediate product for the reaction of H with CO to form CH_3OH.

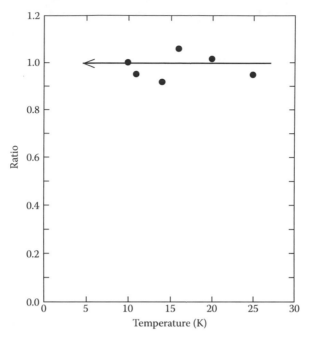

FIGURE 29.6　Temperature dependence of the ratio of the CO recovered after the reaction of CO film with H atoms for 1 h at a certain reaction temperature to that for the CO deposited at 10 K without spraying the H atoms. (From Hiraoka, K. et al., *Astrophys. J.*, 577, 265, 2002a. Reproduced with permission of the AAS.)

The sample gas H_2CO was prepared by heating the paraformaldehyde powder (Aldrich, 95%) in a glass bulb at about 70°C–100°C. The moisture in the formaldehyde vapor was removed by putting some unhydrous magnesium sulfate together with the paraformaldehyde powder in the glass bulb. The formaldehyde or methanol sample was degassed by repeating the freeze-pump-thaw cycles.

The product analysis was performed by TDS. For the H_2CO sample, less-volatile paraformaldehyde was found to be the major product by FT-IR. The yield of paraformaldehyde was calculated by comparing the TDS spectra of the neat sample (not reacted) with that being irradiated by H atoms.

29.5.1　Reaction of H with Solid H_2CO

Figure 29.7a shows the FT-IR spectra of 20 ML thick H_2CO sample sprayed by H atoms for 300 min at 10 K. The bottom spectrum corresponds to that for the 20 ML thick H_2CO before the H-atom spray. With increase in reaction time, the intensities of H_2CO absorption (1182 and 1246 cm^{-1}) decrease gradually and the growth of new absorption for CH_3OH (1035 cm^{-1}), paraformaldehyde (1126 cm^{-1}), and CO (2140 cm^{-1}) is observed.

The chemical reactions envisaged to occur are summarized in Table 29.5. The formation of methanol suggests the occurrence of reactions (29.3) through (29.5) in Table 29.5. Reaction (29.3b) is more exothermic than reaction (29.3a) by about 41 kJ mol^{-1} because the bond energy of $H–CH_2OH$ (393 kJ mol^{-1}) is smaller than that of $CH_3O–H$ (435 kJ mol^{-1}). However, Sosa and Schlegel (1986) theoretically predicted that the barrier height for $H + H_2CO \rightarrow CH_3O$ (19.2 kJ mol^{-1}) is considerably lower than that for $H + H_2CO \rightarrow CH_2OH$ (43.9 kJ mol^{-1}) at the MP4SDTQ/6-31(d) after spin annihilation. Woon (2002) performed an ab initio calculation and predicted that reaction (29.3b) has a barrier about 42 kJ mol^{-1} higher than reaction (29.3a) that excludes reaction (29.3b). At present, no information is available on the branching ratio of reaction (29.3a) and (29.3b).

The appearance of CO in Figure 29.7a suggests the occurrence of H-atom abstraction reactions (29.6) followed by reactions (29.7) and/or (29.8a). Here, reaction (29.7) is the H-atom abstraction reaction while reaction (29.8a) is the recombination reaction of two radicals, H and HCO, to form the intermediate activated complex [H₂CO]* followed by its unimolecular dissociation. In reaction (29.9), the intermediate complex [H₂CO]* is stabilized through the excess energy dissipation to the surrounding matrix molecules. It should be noted that the reaction $H_2CO \rightarrow H_2 + CO$ is exothermic by $2 \, kJ \, mol^{-1}$, that is, the decomposition of H_2CO into H_2 and CO is thermochemically favorable. However, the molecular fragmentation channel (29.8a) is found to be minor relative to the atomic fragmentation channel (29.8b) in the photoexcitation fragmentation of H_2CO (Fleck et al., 1991). The branching ratio of reactions (29.7) through (29.8a) could not be determined in the present experiment.

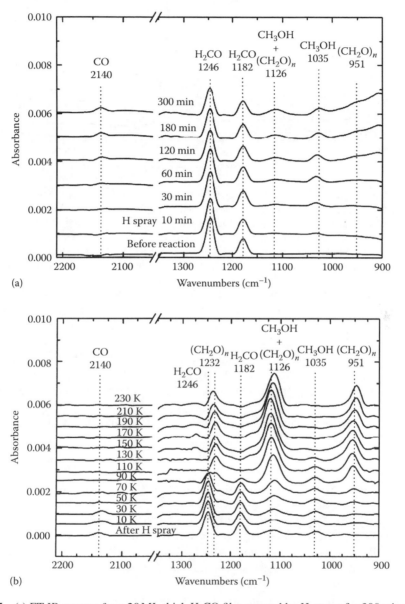

FIGURE 29.7 (a) FT-IR spectra for a 20 ML thick H₂CO film sprayed by H atoms for 300 min at 10 K. The bottom spectrum corresponds to that for the 20 ML thick H₂CO film before the reaction. (b) Temperature dependence of the FT-IR spectra for the 20 ML thick H₂CO film after the film was sprayed with H atoms for 300 min at 10 K. The substrate temperature was increased continuously from 10 to 230 K with a heating rate of $6 \, K \, min^{-1}$. The rapid growth of peaks at ~951 and ~1126 cm⁻¹ observed at ~70 K is due to the radical-induced polymerization reaction of H₂CO to form paraformaldehyde.

(continued)

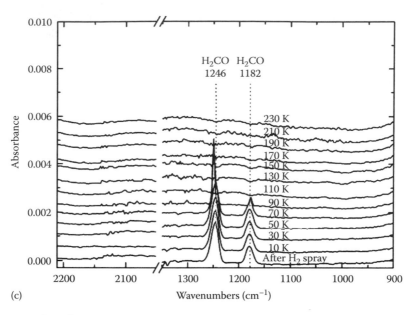

FIGURE 29.7 (continued) (c) Temperature dependence of FT-IR spectra for the 20 ML thick H_2CO film after the film was sprayed with H_2 molecules without generating the plasma for 300 min at 10 K. Absence of peaks for paraformaldehyde at ~1126 and ~951 cm^{-1} above ~70 K indicates that spontaneous polymerization reaction does not take place in the neat H_2CO film. The sharpening of the peaks at 1246 and 1182 cm^{-1} for H_2CO at ~70 K may be due to the change of morphology of the H_2CO film (probably from amorphous to crystalline). (From Hiraoka, K. et al., *Astrophys. J.*, 620, 542, 2005. Reproduced with permission of the AAS.)

Figure 29.8a shows the TDS spectra for a 20 ML H_2CO film being sprayed by H atoms for 300 min at 10 K. The formation of CO and CH_3OH are clearly discernible from peaks with m/z 12 appearing at ~35 K and with m/z 31 at ~140 K, respectively. The CO was monitored by tracing C$^+$ (m/z 12) but not CO$^+$ (m/z 28). This is because in the thermal desorption mass spectrum, CO$^+$ overlaps with N_2^+ originating from the background impurity N_2 gas. The C$^+$ peaks appearing at ~75 K and ~97 K are due to the desorption of CO_2 (residual gas) and H_2CO (reagent), respectively. Without spraying the H atoms, no trace amounts of CO and CH_3OH were detected in the TDS spectra.

Figure 29.9 represents the temperature dependence on the yields (%) of CO and CH_3OH formed from the reaction of H with 20 ML thick solid H_2CO. As seen in the figure, the yields of CH_3OH and CO increase steeply with decrease in temperature. It should be noted that the yields at 10 K (~1%) are much lower than those of C_2H_6 from reactions H + C_2H_2 (23%) and H + C_2H_4 (33%) (Hiraoka et al., 2000) and that of solid product (amorphous silicon) from H + SiH_4 (56%) (Hiraoka et al., 2001a). It is likely that the rate constant for reaction H + H_2CO is much smaller than those for H + C_2H_4, H + C_2H_2, and H + SiH_4 at ~10 K.

The reaction of H with the solid sample is affected by many factors, for example, the morphology of the film, substrate temperature, sample deposition rate, the flux of H atoms, H-atom adsorption and desorption, H-atom surface and internal diffusion, the local heating of the film by the dissipation of the excess energy of the excited molecules, the buildup of the reaction products near the surface of the film, etc. For further investigation, it would be profitable to study the low-temperature solid-phase reactions by deconvoluting these factors independently. Recently, Roser et al. (2002, 2003) measured the kinetic energy of hydrogen molecules formed by the recombination of hydrogen atoms desorbing from amorphous ice at 10 K. They found that the effective kinetic temperature is ~16 K. They suggested that the internal energy of the newly formed H_2 molecule is thermally accommodated to the ice at 10 K. It is likely that the nascent excited H_2 molecule suffers from multiple collisions on the surface of the amorphous ice and dissipates its energy to the ice

TABLE 29.5
Possible Reactions Taking Place in the Reactions of H and D
with H$_2$CO and CH$_3$OH

Reaction	($\Delta H°$)	Type of Reaction
H + H$_2$CO → CH$_3$O (29.3a)	(−94)	Addition
→ CH$_2$OH (29.3b)	(−135)	Addition
H + CH$_3$O → CH$_3$OH (29.4)	(−435)	Recombination
H + CH$_2$OH → CH$_3$OH (29.5)	(−393)	Recombination
H + H$_2$CO → HCO + H$_2$ (29.6)	(−64)	Abstraction
H + HCO → CO + H$_2$ (29.7)	(−373)	Abstraction
H + HCO → [H$_2$CO]* → CO + H$_2$ (29.8a)	(−373)	Unimolecular dissociation
H + HCO → [H$_2$CO]* → HCO + H (29.8b)		Back dissociation
H + HCO → [H$_2$CO]* → H$_2$CO (29.9)	(−371)	Recombination
D + H$_2$CO → CH$_2$DO (29.10a)		Addition
→ CH$_2$OD (29.10b)		Addition
D + CH$_2$DO → CH$_2$DOD (29.11)		Recombination
D + CH$_2$OD → CH$_2$DOD (29.12)		Recombination
D + H$_2$CO → HCO + HD (29.13)		Abstraction
D + HCO → CO + HD (29.14)		Abstraction
D + HCO → [HDCO]* → CO + HD (29.15a)		Unimolecular dissociation
→HCO (or DCO) + D (or H) (29.15b)		Back dissociation
D + HCO → [HDCO]* → HDCO (29.16)		Recombination
H + CH$_3$OH → CH$_2$OH + H$_2$ (29.17)	(−42)	Abstraction
H + CH$_3$OH → CH$_3$O + H$_2$ (29.18)	(+2.5)	Abstraction
H + CH$_2$OH → H$_2$CO + H$_2$ (29.19)	(−301)	Abstraction
H + CH$_2$OH → CH$_3$OH (29.20)	(−393)	Recombination
D + CH$_3$OH → CH$_2$OH + HD (29.21)		Abstraction
D + CH$_2$OH → CH$_2$DOH (29.22)		Recombination

Source: From Hiraoka, K. et al., *Astrophys. J.*, 620, 542, 2005. Reproduced with permission of the AAS.

Note: $\Delta H°$ denotes the enthalpy change of reaction in kJ mol^{-1}.

matrix. Their finding clearly indicates that the morphology of the solid film is a crucial factor for the dissipation of the nascent excited product molecules. Hornekaer et al. (2003) performed detailed laboratory experiments on the formation of HD from atom recombination on amorphous solid water films. They found that the recombination process is extremely efficient in the temperature range of 8–20 K, implying fast mobility of H and D atoms at this temperature. These findings implicitly suggest that the H-atom flux should be suppressed as low as possible to investigate the low-temperature solid-phase reactions in order to suppress the local heating of the film. Otherwise, the reaction temperature cannot be defined precisely.

As shown in Figure 29.7a, the absorption of paraformaldehyde appears during the H-atom spray at 10 K. This suggests that the polymerization reaction involving the matrix H$_2$CO molecules proceeds even at 10 K. When the 20 ML H$_2$CO film was warmed from 10 to 250 K without being sprayed by H atoms, no absorption of paraformaldehyde appeared (Figure 29.7c). This indicates that polymerization is initiated by residual radicals such as HCO, CH$_3$O, CH$_2$OH, etc., formed in the H$_2$CO matrix. The steady-state concentration of radicals must be kept low because these radicals should be annihilated efficiently by the reactions with H atoms. In fact, we could not detect any intermediate radicals such as HCO, CH$_2$OH, CH$_3$O, etc., in the infrared spectra.

Figure 29.7b shows the temperature dependence of the FT-IR spectra when the temperature of the film, having been sprayed by H atoms for 300 min at 10 K, was increased up to 230 K. The absorption of paraformaldehyde at 1232, 1126, and 951 cm^{-1} starts to grow steeply with increasing temperature above ~70 K. This temperature just corresponds to that for the start of rapid decrease of absorption of H$_2$CO at ~1182 and ~1246 cm^{-1}. This coincidence clearly indicates that the self-diffusion of H$_2$CO molecules in solid promotes the polymerization of H$_2$CO.

The yield of paraformaldehyde was estimated from the TDS of H$_2$CO recovered, compared to that of the deposited H$_2$CO not being sprayed by H atoms. About 10%–20% of deposited H$_2$CO

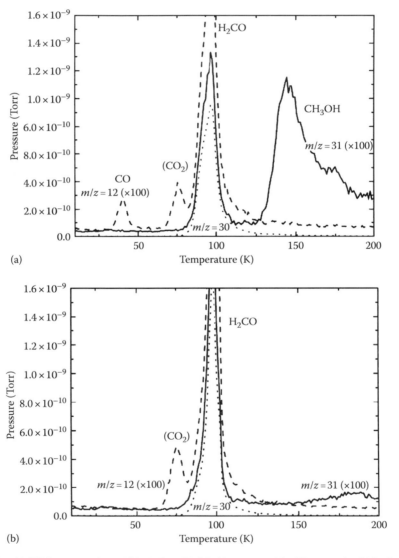

FIGURE 29.8 (a) TDS spectra for a 20 ML dose H$_2$CO film sprayed by H atoms for 300 min at 10 K. The peak at m/z 12 at ~35 K is due to desorption of the reaction product CO. The peak at m/z 12 at ~75 K is probably due to desorption of CO$_2$ (residual impurity gas in the vacuum chamber). The peaks with m/z 12, 30, and 31 at ~97 K are due to desorption of the reagent H$_2$CO. The peak with m/z 31 at ~97 K originates from the isotope peaks of H$_2^{13}$CO$^+$ and HD^{12}CO$^+$. The peak with m/z 31 at ~140 K is due to desorption of the reaction product CH$_3$OH (m/z 31 is the base peak for CH$_3$OH). (b) TDS spectra for the 20 ML dose H$_2$CO film sprayed by H$_2$ molecules for 300 min at 10 K without generating the plasma. The absence of peaks at m/z 12 at ~35 K and m/z 31 at ~140 K indicates that the CO and CH$_3$OH detected in (a) are products from the reaction H + H$_2$CO.

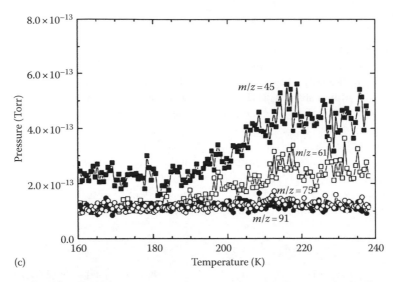

(c)

FIGURE 29.8 (continued) (c) TDS spectra for the 10 ML thick H_2CO film reacted with H atoms for 300 min at 10 K for higher masses. Peaks with m/z 45, 61, 75, and 91 start to grow at about ~190 K. This is likely due to desorption of paraformaldehyde, $CH_3OCH_2-O-CH_2-O-CH_2-O-\cdots$, formed by radical-induced polymerization reactions in the solid H_2CO film. (From Hiraoka, K. et al., *Astrophys. J.*, 620, 542, 2005. Reproduced with permission of the AAS.)

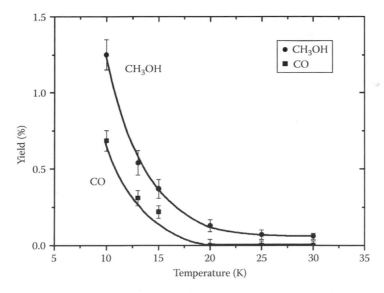

FIGURE 29.9 Relationship between the percentage yields of CH_3OH and CO and the reaction temperature for the reaction $H + H_2CO$. Film thickness of H_2CO: 20 ML. The data points in the figure were obtained by an H-atom spray for 300 min over the H_2CO film at fixed temperatures in the range of 10–30 K. After the H-atom spray for 300 min at the fixed sample temperature shown in the figure, the sample temperature was decreased to 10 K, and a TDS measurement was made for a quantitative analysis of the reaction products, CO and CH_3OH. (From Hiraoka, K. et al., *Astrophys. J.*, 620, 542, 2005. Reproduced with permission of the AAS.)

(20 ML) was found to be converted to paraformaldehyde at elevated temperatures. The observed high yield suggests that the radical-initiated polymerization of H_2CO is a very efficient process as already reported by Goldanskii et al. (1973).

Preliminary results from the positive ion cluster composition analyzer (PICC) on the Giotto spacecraft during the encounter with the comet Halley have been presented by Huebner et al. (1987).

They pointed out that the sequence of profiles with peaks at m/z 45, 61, 75, 91, and 105 has differences of 16, 14, 16, and 14 u, and the intensities decrease smoothly with increasing mass. These peaks just correspond to the fragment ions from paraformaldehyde, $CH_3OCH_2-O-CH_2-O-CH_2-$ O– ···. The formation of this type of paraformaldehyde may be initiated by the CH_3O radical that attacks the C atom of the H_2CO molecule. Figure 29.8c shows the thermal desorption mass spectra for the 10 ML H_2CO film, reacted with H atoms for 300 min. The peaks with m/z 45, 61, 75, and 105 start to grow at about ~190 K. This is likely due to the desorption of paraformaldehyde formed by the CH_3O-initiated polymerization.

Schutte et al. (1993) found that a trace amount of NH_3 in H_2CO ($NH_3/H_2CO = 0.05$) is enough to trigger the polymerization of H_2CO to form paraformaldehyde spontaneously (without the presence of radicals) at temperatures as low as 40 K. As described, gaseous formaldehyde is readily formed by heating paraformaldehyde above 70°C. Thus, the formaldehyde observed in the comae of comets may partly be due to the thermal decomposition of paraformaldehyde formed in the comet's ice.

29.5.2   Reaction of D with Solid H_2CO

As described in Section 29.3, the reactivity of D with C_2H_2 was considerably lower than that of H. Such a remarkable isotope effect was not found for C_2H_4. It would be interesting to examine the reactivities of H and D atoms toward H_2CO. In the reactions of D with H_2CO, reactions (29.10) through (29.16) may take place.

In reaction H + H_2CO, the recombination reaction (29.9) regenerates H_2CO that cannot be distinguished from the reagent H_2CO. In contrast, reaction (29.16) forms HDCO that can be distinguished from H_2CO through mass spectrometry. Thus, the branching ratio of [reactions (29.14) and (29.15a) for the formation of CO] to [reaction (29.16) for the formation of HDCO] can be determined in the reaction D + H_2CO. The quantitative analysis for HDCO was performed by measuring the m/z 31 ion signal for $DHCO^+$. The contribution from the isotope peak of $H_2^{13}CO^+$ was corrected.

Figure 29.10 shows the temperature dependence on the yields (%) of CO, HDCO, and CH_2DOD formed from the reaction of D with solid H_2CO (20 ML) for 300 min of D-atom irradiation. An increase

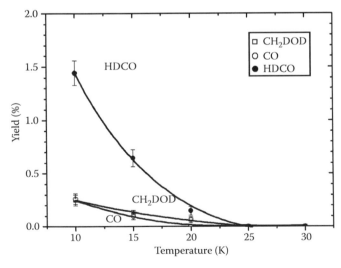

FIGURE 29.10 Relationship between the percentage yields of CH_2DOD, HDCO, and CO and the reaction temperature for the reaction D + H_2CO. Film thickness of H_2CO, 20 ML. The data points in the figure were obtained by D-atom spray for 300 min over an H_2CO film at fixed temperatures in the range of 10–30 K. After the D-atom spray for 300 min at the fixed sample temperature shown in the figure, the sample temperature was decreased to 10 K, and a TDS measurement was made for a quantitative analysis of the reaction products. (From Hiraoka, K. et al., *Astrophys. J.*, 620, 542, 2005. Reproduced with permission of the AAS.)

of the product yields with decrease of temperature is reproduced as in the case of H + H_2CO. While the D-atom addition reactions (29.10a) and (29.10b) followed by reactions (29.11) and (29.12) form CH_2DOD, the H-atom abstraction reaction (29.13) leads to the formation of CO and HDCO. Thus, the ratio of [the yield of CH_2DOD (~0.25%)] to [those of CO and HDCO (~0.25 + ~1.4 ≈ ~1.7%)], that is, ~0.25:~1.7 ≈ 1:7, will give an approximate branching ratio of [the D-atom addition reactions (29.10a) and (29.10b)] to [the H-atom abstraction reaction (29.13)] (reaction temperature, 10 K). It is apparent that the abstraction reaction (29.14) is much more favorable than the addition reaction (29.10).

The intermediate radical HCO formed by reaction (29.13) has two chemical channels, that is, the formation of CO by reactions (29.14) and (29.15a) and the formation of HDCO by reaction (29.16). The much higher yield of HDCO (~1.4%) than that of CO (~0.25%) indicates that the recombination reaction (29.16) is a main channel for reaction D + HCO. This is reasonable because reaction (29.16) is a radical–radical recombination reaction whose energy barrier should be much smaller than that of an abstraction reaction (29.14).

It should be noted that the yield of CO from D + H_2CO in Figure 29.10 is much lower than that from H + H_2CO in Figure 29.9. This suggests that reaction (29.13) followed by reactions (29.14) and (29.15a) is less efficient than reaction (29.6) followed by reaction (29.7) and (29.8a). The yield of deuterated methanol CH_2DOD formed by addition reactions (29.10) through (29.12) is considerably smaller than that of CH_3OH formed from reactions (29.3) through (29.5). It is evident that both D-atom addition and abstraction reactions are less efficient than the H-atom reactions.

29.5.3 Reactions of H and D with Solid CH_3OH

In the previous section, it was shown that the reaction H + H_2CO leads to the formation of both CO and CH_3OH. We further studied whether reaction H + CH_3OH produces H_2CO by the consecutive H-atom abstraction reactions, $CH_3OH \rightarrow CH_2OH \rightarrow H_2CO$. In the reaction H + CH_3OH, reactions (29.17) through (29.20) may take place. In these reactions, abstraction reaction (29.18) may be negligible at cryogenic temperatures because it is slightly endothermic and has a high energy barrier of 58.9 kJ mol^{-1} (theoretical value due to Jodkowski et al. (1999)). Despite the scrutiny of the FT-IR spectra for the CH_3OH film that reacted with H atoms for 300 min, no absorption of H_2CO and CO appeared. These compounds were not detected by the much more sensitive TDS technique either. Therefore, the generation of H_2CO from reaction H + CH_3OH is concluded to be negligible.

As mentioned above, the reaction of H with CH_3OH does not produce any detectable reaction products. However, this does not necessarily mean that abstraction reaction (29.17) does not take place because the intermediate product CH_2OH may regenerate the original CH_3OH by reaction (29.20). In order to examine whether or not reaction (29.17) followed by (29.20) takes place, the reaction of D with CH_3OH was examined. If CH_2DOH is formed, this will be evidence for the occurrence of reaction (29.21) followed by reaction (29.22).

Figure 29.11 shows the TDS spectra for 20 ML CH_3OH film, reacted with D atoms for 300 min at 10 K. The appearance of the peak with m/z 33 due to CH_2DOH was clearly observed at 140 K. This indicates the occurrence of reactions (29.21) and (29.22). The yield of CH_2DOH was 1.6% for a 20 ML thick CH_3OH film sprayed by D atoms for 300 min. No reaction products other than CH_2DOH were detected by TDS. The low yield of CH_2DOH is reasonable because the large energy barrier of ~38 kJ mol^{-1} for reaction (29.17) has been predicted theoretically (Blowers et al., 1998; Jodkowski et al., 1999).

The yields of reaction products for the reactions of H or D with 20 ML thick H_2CO and CH_3OH for the H- and D-atom spray of 300 min at 10 K are summarized in Figure 29.12. In Figure 29.12a and b, the reactions of H and D with H_2CO lead to the formation of CO and CH_3OH, that is, both addition and abstraction reactions take place. From Figure 29.12b for D + H_2CO, the branching ratio of the abstraction reaction to the addition one can be estimated to be about 7:1 (1.7:0.25), that is, the abstraction reaction is the prevalent process over addition. For D + HCO, the branching ratio of the reaction channel to form CO, to that to form HDCO is about 1:6 (0.25:1.4), indicating that the

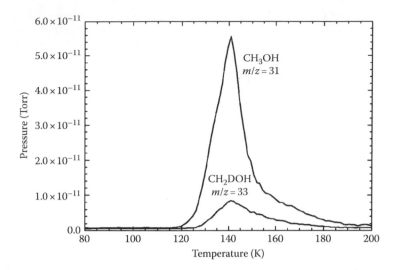

FIGURE 29.11 TDS spectra for a 20 ML thick CH_3OH film reacted with D atoms for 300 min at 10 K. The peak at m/z 33 appearing at ~140 K is due to the formation of monodeuterated methanol, CH_2DOH. (From Hiraoka, K. et al., *Astrophys. J.*, 620, 542, 2005. Reproduced with permission of the AAS.)

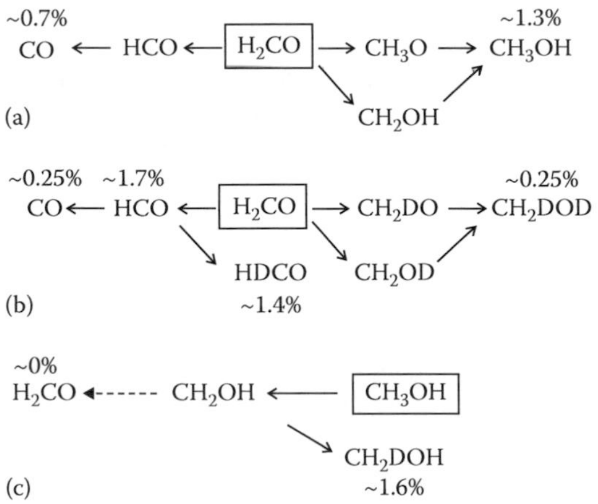

FIGURE 29.12 Reaction channels for reactions of (a) H-atom spray on H_2CO (20 ML), (b) D-atom spray on H_2CO (20 ML), and (c) D-atom spray on CH_3OH (20 ML). The H/D-atom spray time was 300 min, the film thickness of H_2CO was 20 ML, and the reaction temperature was 10 K. The yields in the figure were determined by TDS. (From Hiraoka, K. et al., *Astrophys. J.*, 620, 542, 2005. Reproduced with permission of the AAS.)

radical–radical recombination reaction (29.16) takes place more preferably than the H-atom abstraction reaction (29.14). In Figure 29.12a and b, the reactivity of H toward H_2CO is higher than that of D. In the reactions of H and D with CH_3OH, no H_2CO was detected as a reaction product. Thus CH_3OH may be regarded as the terminal product in the consecutive reactions of H with CO because no H_2CO was detected for H + CH_3OH.

From the experimental results described above, we conjecture that the consecutive hydrogenation reactions CO → HCO → H_2CO → $CH_3CO(CH_2COH)$ → CH_3OH may not be sufficient to explain the abundance of methanol found in the interstellar ices because the reactivities of H toward CO and H_2CO are found to be rather low. From the number density of H atoms (~1 cm^{-3}), UV photon flux (~10^3 photons cm^{-2} s^{-1}) (D'Hendecourt and Allamandola, 1986; Lanzerotti and Johnson, 1986; Allamandola et al., 1988; Grim et al., 1989; Johnson, 1990; Shalabiea and Greenberg, 1994; Kaiser and Roessler, 1998), and cosmic-ray flux (~10 particles cm^{-2} s^{-1}) (Lanzerotti and Johnson,

1986; Johnson, 1990; Kaiser and Roessler, 1998) in the dense molecular cloud, the time interval for irradiation by a hydrogen atom, an UV photon, and a cosmic particle on the dust grain (0.1 μm in diameter, cross section of ~10^{-10} cm^2) can be crudely estimated to be once every day, once every 100 days, and once every 10,000 days, respectively. The contribution of a cosmic ray should be as important as that of UV photons because a cosmic ray can generate about 100 suprathermal species in solid phase (Kaiser and Roessler, 1998). Formaldehyde would be formed on the dust grains in the dark clouds by the cooperative action of H-atom adsorption, UV photons, and cosmic rays: none of these would be negligible because the reaction H + CO is quite inefficient.

Since the rate of reaction of H with CO (Hiraoka et al., 2002a) is much lower than those with paraffins (Hiraoka et al., 1992), olefins (Hiraoka et al., 2000, 2002b; Hiraoka and Sato, 2001), and SiH$_4$ (Hiraoka et al., 2001a), the chance for H atoms to react with CO will decrease drastically when the mantle of the dust grain is being contaminated by other molecules, that is, the H atoms adsorbed on the dust grains may well be annihilated by reactions with more reactive contaminants and the less reactive CO will be left intact in the mantle. This may be the reason why the natal CO is well preserved in the mantles of the dust grains and comets. We think that the role of CO to form CH$_3$OH becomes increasingly less important with the proceeding of the chemical evolution in the dark clouds. Another source for the formation of CH$_3$OH such as UV-photon and/or cosmic-ray induced formation of CH$_3$OH in the dirty H$_2$O ice, containing some carbon source (e.g., CH$_4$), must be invoked (Allamandola et al., 1988; Kaiser and Roessler, 1998; Moore and Hudson, 1998; Hudson and Moore, 1999). In Section 29.6, the interactions of low-energy electrons with CH$_4$ seeded in H$_2$O will be described in order to investigate the role of the cosmic rays on the formation of methanol on the dust grains.

29.6 METHANOL FORMATION FROM ELECTRON-IRRADIATED MIXED H$_2$O/CH$_4$ ICE AT 10 K

While the formation of many interstellar molecules are reasonably explained by the gas-phase reactions, solid-phase reactions must be invoked for some molecules, such as saturated hydrocarbons, H$_2$CO, CH$_3$OH, NH$_3$, etc., which are of paramount importance for the evolution of life. As described above, the origin of ubiquitous formaldehyde and methanol in interstellar objects is controversial. Interstellar dust grains, comets, and icy satellites are subject to cosmic-ray irradiation. By far the most important process caused by the interaction between the cosmic rays and matter is ionization. The ejected electron may have enough energy to further ionize and excite the ambient molecules resulting in second-generation ions, radicals, electrons, photons, and rovibronically excited species. The primary ions may participate in various ion–molecule reactions and eventually be neutralized by secondary electrons to form reactive neutral species or stable molecules. The formed radicals may react with other radicals or molecules to form the terminal products.

The laboratory studies of energetic-particle irradiation on ices relevant to astrochemical interest have been extensively carried out over the last decade (Kaiser and Roessler, 1998; Moore and Hudson, 1998; Hudson and Moore, 1999; Roser et al., 2002, 2003). However, studies on the electron irradiation of solid films is only very limited. In order to investigate the role of secondary electrons formed by the cosmic rays in dust grains that may play major roles in the chemical evolution in cold interstellar medium (ISM), we studied the low-energy (10–300 eV) electron irradiation on the water ice containing 10% methane at 10 K (Wada et al., 2006).

Figure 29.13 shows the conceptual idea of the apparatus (ARIOS, Inc. Akishima, Tokyo). The H$_2$O and CH$_4$ gases were pre-mixed in the stainless steel gas reservoir (300 cm^3) in a predetermined mixing ratio (H$_2$O/CH$_4$ = 10/1). The sample gases were deposited on the cold substrate. The hot-filament electron gun was installed on the vacuum manifold. The electron beam was raster-scanned in order to irradiate the sample surface homogeneously. Electron flux was calibrated by

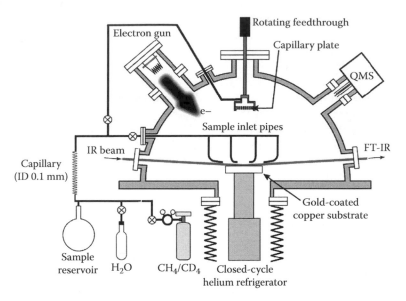

FIGURE 29.13 Schematic diagram of the experimental apparatus. QMS: quadrupole mass spectrometer. (From Wada, A. et al., *Astrophys. J.*, 644, 300, 2006. Reproduced with permission of the AAS.)

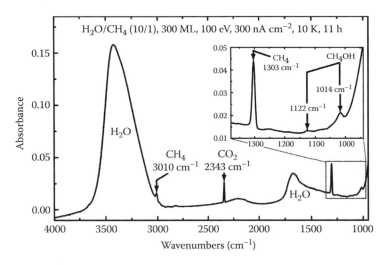

FIGURE 29.14 Infrared absorption spectrum of H_2O/CH_4 (10/1) ice irradiated continuously by 100 eV electrons with flux of 300 nA cm^{-2} during the sample deposition for 11 h at 10 K. Total thickness of the sample is 300 ML. The fluence is 8 eV $molecule^{-1}$. (From Wada, A. et al., *Astrophys. J.*, 644, 300, 2006. Reproduced with permission of the AAS.)

using a Faraday cup. The angle of the electron beam to the substrate was 45° to the surface normal. The quantitative analysis of reaction products was performed by TDS.

The real-time and in situ observation of electron-induced reactions was made by FT-IR (Nicolet Nexus 670) in the region of interest, 900–4000 cm^{-1}. Figure 29.14 shows the IR spectrum, obtained by simultaneous sample deposition/e^--irradiation experiments for 11 h. As seen in the figure, the dominant peak at ~3400 cm^{-1}, and small peaks at 3030 cm^{-1} and 1303 cm^{-1} are due to the reactants, water and methane, respectively. The formation of methanol as a major product is easily recognized by the absorption peaks appearing at 1014 and 1122 cm^{-1} (see inset in Figure 29.14).

Figure 29.15 shows the dependence of yields of reaction products on the electron flux for the sample H_2O/CH_4 (10/1), using electron energy of 100 eV (electron irradiation time of 22 min). At the

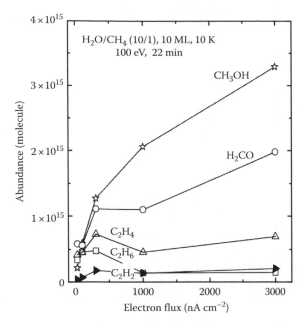

FIGURE 29.15 Yields of CH_3OH, H_2CO, C_2H_4, C_2H_6, and C_2H_2 as a function of electron flux. Electron energy, 100 eV. Simultaneous electron irradiation during deposition of the sample at 10 K for 22 min. The total sample thickness: 10 ML. (From Wada, A. et al., *Astrophys. J.*, 644, 300, 2006. Reproduced with permission of the AAS.)

lowest electron flux, that is, $30\,nA\,cm^{-2}$ ($1.5\,eV$ molecule^{-1}), H_2CO is the most abundant product, followed by C_2H_4, C_2H_6, CH_3OH, and C_2H_2. The predominance of the formation of C2 hydrocarbons with low-electron flux may be due to the enriched CH_4 on the solid surface due to its segregation from the water ice. Methanol shows a steady increase with electron flux. The delayed appearance of C_2H_2 suggests the occurrence of successive dehydrogenation reactions, $C_2H_6 \rightarrow C_2H_4 \rightarrow C_2H_2$.

Figure 29.16 shows the yield of methanol as a function of the incident electron energy obtained by the simultaneous sample deposition/e$^-$-irradiation experiments. Methanol, which could not be detected with electron energy of 10 eV, started to be detected at 30 eV and its yield increases monotonically up

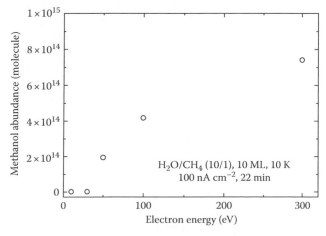

FIGURE 29.16 Yield of CH_3OH as a function of electron energy. Simultaneous electron irradiation with flux of $100\,nA\,cm^{-2}$ during deposition of H_2O/CH_4 (10/1) at 10 K for 22 min. The total sample deposited, 10 ML. (From Wada, A. et al., *Astrophys. J.*, 644, 300, 2006. Reproduced with permission of the AAS.)

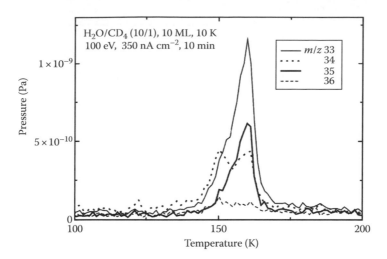

FIGURE 29.17 TPD spectra for e⁻-irradiated H_2O/CD_4 (10/1). Simultaneous 100 eV electron irradiation with the flux of 350 nA cm⁻² (8 eV molecule⁻¹) at 10 K. The total sample thickness, 10 ML. (From Wada, A. et al., *Astrophys. J.*, 644, 300, 2006. Reproduced with permission of the AAS.)

to 300 eV. The steady increase of the yield is likely due to the occurrence of more frequent cascade scattering of primary electrons, resulting in the denser track formation in the solid sample.

In order to obtain more detailed information on the reaction mechanisms for the formation of methanol, CD_4 was used instead of CH_4 as the mixing partner with H_2O. The TDS spectra obtained for 100 eV e⁻-irradiated H_2O/CD_4 (10/1), with a flux of 100 nA cm⁻², are shown in Figure 29.17. The appearance of the peak with m/z 35 at 162 K clearly indicates the formation of CD_3OH. In Figure 29.17, the peak with m/z 34 is approximately 70% of that with m/z 35, which is much larger than the expected value for the fragment ion from CD_3OH. This strongly suggests the formation of the less deuterated methanol, that is, CHD_2OH. Though the cracking pattern of CHD_2OH is not available, it is reasonable to assume that the ion with m/z 34 is a fragment originating from CHD_2OH. This is because the molecular ion with m/z 35 is strongly observed for CD_3OH, and it should hold for CHD_2OH also. The relative abundances of the products, CD_3OH and CHD_2OH, were about the same.

In this work, the electron flux was measured by using a Faraday cup and thus the quantitative determination of the abundance of methanol could be made. As shown in Figure 29.15, the yield of methanol increases with the electron flux. The initial slope may make it possible to estimate the yield of methanol molecules with respect to the electron numbers irradiated. By the method of least squares applied to Figure 29.15, the yield of one CH_3OH molecule per 60 incident electrons (100 eV) could be derived. This corresponds to a G-value of 0.017.

The methanol formation may proceed along the following pathways: ionization of H_2O and CH_4 followed by rapid ion–molecule reactions, and neutralization with excess electrons to form reactive radicals (Roser et al., 2003). The net reactions may be simplified as follows.

$$H_2O \rightarrow H + OH \tag{29.23}$$

$$CH_4 \rightarrow CH_3 + H \tag{29.24}$$

$$CH_3 + OH \rightarrow CH_3OH \tag{29.25}$$

The above mechanism is supported by the experiment using mixed H_2O/CD_4 ice in which CD_3OH has been detected as one of the major deuterated methanol products.

$$CD_3 + OH \rightarrow CD_3OH \tag{29.26}$$

The mechanism (29.26) is consistent with the previous work of the proton irradiation experiment (Moore and Hudson, 1998). In addition to CD_3OH, CHD_2OH was formed in about equal amounts. This suggests the occurrence of the methylene radical insertion reaction (29.27),

$$CD_2 + H_2O \rightarrow CHD_2OH \qquad (29.27)$$

Kaiser and Roesller (1998) suggested that the CH_2 radicals are formed very efficiently in pure solid CH_4 irradiated by protons and alpha particles. They found that the long-chain hydrocarbons dominate the reaction products. In contrast, limited formation of hydrocarbons larger than C2 could be detected in this work. This is reasonable because the polymerization reactions induced by the hydrocarbon radicals are largely suppressed by the primary radicals of OH and H formed from the major component, H_2O, which act as radical scavengers. About equal amounts of CD_3OH and CHD_2OH formed from the e^--irradiated H_2O/CD_4 (10/1) suggest that the relative contribution of reactions (29.26) and (29.27) for the formation of methanol is about equal.

The rather facile formation of H_2CO found in this work is noticeable. As shown in Figure 29.15, the yield of H_2CO shows a rather steep increase in the range of low-electron dose, 30–300 nA cm^{-2}. This trend is similar to that observed for methanol. This suggests that H_2CO is not a secondary but a primary product. The C-atom addition to H_2O, followed by the intramolecular isomerization, may explain the direct formation of H_2CO.

$$C + HOH \rightarrow HCOH \quad \text{(formation of carbene)} \qquad (29.28)$$

$$HCOH \rightarrow H_2CO \quad \text{(intramolecular isomerization)} \qquad (29.29)$$

In fact, the mechanism proposed above is confirmed by the experiment using mixed CD_4/H_2O ice. That is, only H_2CO but no HDCO or D_2CO could be detected from this system.

It would be meaningful to consider the relative importance of the possible sources for methanol in ISM, for example, cosmic-ray induced radiation chemistry taking place in the mantles of dust grains. The methanol was found to be formed from the mixed H_2O/CO (10/1) ice irradiated by the 0.8 MeV protons at a dose of 22 eV molecule^{-1} (Hudson and Moore, 1999). The column density of CH_3OH molecules formed was 1.23×10^{16} molecule cm^{-2}. That is, 6.4 methanol molecules were formed per one proton bombardment. The components of cosmic rays consist of 97%–98% protons and 2%–3% alpha particles. In the energy range of 1–10 MeV, the cosmic-ray flux is 10 particles cm^{-2} s^{-1}. For the average size of the dust grain with 0.1 μm in diameter (cross section of $\sim 10^{-10}$ cm^2), the frequency of cosmic irradiation is 1 particle per 10,000 days. Thus, the rates of methanol generation in ISM can be estimated as 6.4 for every 10,000 days. In the interstellar environment, the sticking probability of a H atom on the dust grain becomes increasingly smaller above 20 K (Hiraoka et al., 2006). Therefore, the contribution to methanol formation by H-atom addition reactions to CO on the dust grain may be much smaller than that induced by cosmic rays. Discussing the routes of formation of CH_3OH in quantitative terms is not an easy task, because the parameters appearing in rate equations, such as the fluxes of cosmic rays, H atoms, and UV photons and rate constants for various possible reactions, have large uncertainties. Moreover, the interstellar media are spatially not uniform even in molecular clouds and diffuse clouds. Although reliable estimation on the relative contribution of various sources for the formation of methanol is difficult at present, we believe that the cosmic-ray induced radiation chemistry plays an important role for the formation of methanol because dirty H_2O-base ice mantles on the dust grains can survive up to ~ 100 K in the diffuse clouds where they must suffer from intense irradiation by cosmic rays and UV photons. In the diffuse clouds, H-atom grain-surface reactions may not play important roles because the sticking probability of H atoms on the dust grains is much less than that in the dark clouds at ~ 10 K.

29.7 SOLID-PHASE REACTIONS OF D WITH CN TO FORM DNC AND DCN AT CRYOGENIC TEMPERATURES

HNC was first observed in the gas phase by radio astronomers (Snyder and Hollis, 1976; Irvine et al., 1998). The HNC/HCN ratio, which would be almost zero at thermal equilibrium, can be greater than unity in cold molecular clouds, and it exhibits a striking inverse temperature dependence. Hirota et al. (1998) determined the abundance ratio of HNC/HCN to be 0.54–4.5 in dark cloud cores. The large value of the HNC/HCN abundance ratio in cold molecular clouds has been considered to be due to gas-phase reactions (29.30) through (29.32), (Watson, 1974, 1980; Irvine et al., 1996, 1998),

$$H_3^+ + HCN \rightarrow HCNH^+ + H_2 \tag{29.30}$$

$$HCNH^+ + e^- \rightarrow HNC + H \tag{29.31}$$

$$HCNH^+ + e^- \rightarrow HCN + H \tag{29.32}$$

and due to reactions (29.33) and (29.34) (Green and Herbst, 1979)

$$C^+ + NH_3 \rightarrow H_2NC^+ + H \tag{29.33}$$

$$H_2NC^+ + e^- \rightarrow HNC + H \tag{29.34}$$

It has been shown that the HNC and HCN are produced with nearly equal abundances from reaction, $HCNH^+ + e^-$ (Watson, 1974; Herbst, 1978; Taketsugu et al., 2004).

In the diffuse interstellar clouds, the first interstellar molecules, CN, CH, and CH$^+$, were detected via their characteristic absorption spectra in the visible region of the spectrum by using ground-based telescopes. Prasad et al. (1987) discussed the chemical evolution in the interstellar clouds. They pointed out that CN is the terminal product for gas-phase nitrogen chemistry and its coupling with carbon chemistry. In cold molecular clouds, CN radicals formed in the gas phase will be adsorbed on the dust grains and they might be involved in the grain-surface reactions, for example, the association of CN with H to form HNC and HCN.

We studied the association reactions (29.35) and (29.36) in solid phase (Hiraoka et al., 2006).

$$D + CN \rightarrow DNC \quad (\Delta H^\circ = \sim -456 \text{ kJ mol}^{-1}) \tag{29.35}$$

$$D + CN \rightarrow DCN \quad (\Delta H^\circ = \sim -518 \text{ kJ mol}^{-1}) \tag{29.36}$$

In this work, CN radicals were generated by decomposing HCN in a dc discharge plasma. The HCN reagent gas was synthesized by mixing NaCN with sulfuric acid (100%). The HCN gas was mixed with N_2 major gas with the ratio of $HCN/N_2 = 1/50$.

The N_2 gas containing 2% of HCN was excited by dc discharge plasma generated in a straight-type bottleneck discharge tube shown in Figure 29.1.

During the deposition, D atoms produced by the dc discharge of D_2 gas using another discharge tube shown in Figure 29.1 were simultaneously sprayed over the substrate for 60 min. The dose of N_2 sprayed on the metal target is estimated to be about 10^4 ML.

Figure 29.18a shows the FT-IR spectrum for N_2 containing 2% of HCN deposited on the metal substrate (not discharge ignited). Only the absorption peaks at 2096 and 3286 cm^{-1} due to the reactant HCN are observed. Figure 29.18b through d show the FT-IR spectra for the dc-plasma activated N_2/HCN (50/1) deposited on the gold-plated copper substrate being reacted with D simultaneously at 10, 15, and 20 K. The absorption intensities of the reactant HCN at 3286 cm^{-1} in Figure 29.18b

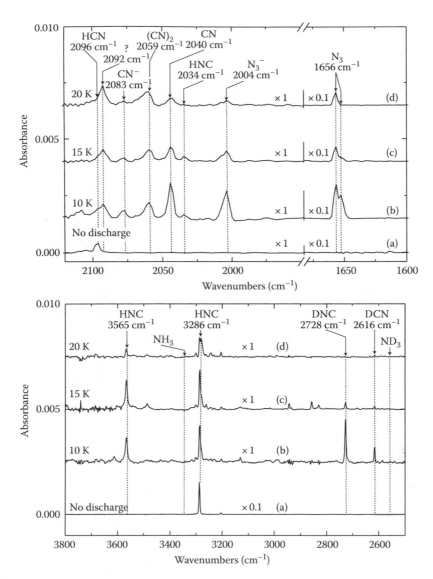

FIGURE 29.18 (a) FT-IR spectrum for the N_2 containing 2% of HCN deposited on the metal substrate with no discharge ignited. Only a strong peak of HCN is observed. Figures (b), (c), and (d) show the FT-IR spectra for the dc-plasma activated mixed gas N_2/HCN (50/1) deposited on the gold-plated copper substrate at 10, 15, and 20 K, respectively. During the deposition, D atoms were sprayed simultaneously over the deposited film for 60 min. The dose of N_2 molecules sprayed over the metal substrate is about 10^4 monolayers (ML). (From Hiraoka, K. et al., *Astrophys. J*, 643, 917, 2006. Reproduced with permission of the AAS.)

through d become about 10% of that in Figure 29.18a. This indicates that a greater amount of HCN supplied into the discharge tube has been decomposed and converted to the discharge products. In fact, the inside of the discharge tube was found to be coated by yellow film products (tholin) after hours of operation. In order to obtain reproducible data, laborious effort was necessary to clean the discharge tube from time to time. In Figure 29.18b, absorptions due to N_3, N_3^-, HNC, CN, $(CN)_2$, CN^-, DCN, and DNC have been observed as major product peaks. The appearance of negative ions, N_3^- and CN^-, suggests some leakage of electrons from the straight-type bottleneck discharge tube shown in Figure 29.1. The high electron affinity (*EA*) of CN (3.7 eV, (Lias et al., 1988)) explains the formation of CN^-. The *EA* of N_3 has not been measured. The appearance of N_3^- in Figure 29.18 suggests that N_3 has a positive *EA*.

The appearance of strong DNC and DCN in Figure 29.18b at 10 K is likely due to the occurrence of the association reactions of CN with D to form DNC and DCN on the surface of the film. The ratio DNC/DCN is about 3 at 10 K in Figure 29.18b. With increase of temperature $10 \rightarrow 15 \rightarrow 20$ K, the peak intensities of DNC and DCN decreased rapidly and became negligible at 20 K. This is mainly due to the decrease of the sticking probability of D atoms on the solid surface. The total absence of DNC and DCN at 20 K suggests that the residence time of D atoms on the solid surface is too short to encounter with the trapped CN radicals. The grain-surface reactions in which H atoms participate would become increasingly less important with increase of temperature above 20 K in ISM.

The bond energies of D–NC ($\sim$456 kJ mol^{-1}) and D–CN ($\sim$518 kJ mol^{-1}) are larger than that of D_2 (443 kJ mol^{-1}). Thus, D-atom abstraction reactions are both endothermic, that is, these reactions must be negligible at cryogenic temperatures.

$$D + DNC \rightarrow D_2 + CN \quad (\Delta H^\circ \approx +13 \text{ kJ mol}^{-1}) \tag{29.37}$$

$$D + DCN \rightarrow D_2 + CN \quad (\Delta H^\circ \approx +75 \text{ kJ mol}^{-1}) \tag{29.38}$$

In Figure 29.18, a small but a distinct peak of CN$^-$ at 2083 cm^{-1} was observed. In the gas phase, an associative electron detachment reaction is known to take place with a collision rate.

$$CN^- + H \text{ (or D)} \rightarrow HCN \text{ (or DCN)} + e^- \quad (\Delta H^\circ = -159 \text{ kJ mol}^{-1}) \tag{29.39}$$

$$\rightarrow HNC \text{ (or DNC)} + e^- \quad (\Delta H^\circ = -92 \text{ kJ mol}^{-1}) \tag{29.40}$$

The branching ratio of these reactions has not been determined. DNC and DCN observed in Figure 29.18 may partly be formed by these reactions.

The isomerization via reaction (29.41) could take place during the D-atom spray over the DNC and HNC molecules trapped in the N_2 matrix.

$$D + CND \rightarrow [D-CND]_{TS} \rightarrow DCN + D \tag{29.41a}$$

$$D + CNH \rightarrow [D-CNH]_{TS} \rightarrow DCN + H \tag{29.41b}$$

The associated complex, HCNH, is predicted to locate 110 kJ mol^{-1} lower in energy than the isolated H and CNH (Talbi et al., 1996). Since the energy barrier for $H + CNH \rightarrow [H-CNH]_{TS} \rightarrow HCNH$ is calculated to be only 17.6 kJ mol^{-1}, DCND and/or DCNH may form under the present experimental conditions through tunneling processes. However, we could not identify any probable candidates for DCND and DCNH in the infrared absorption spectra. This suggests either that the formation of the species DCND (or DCNH) is largely suppressed by the presence of the energy barrier (17.6 kJ mol^{-1}) or that the species DCND and DCNH are quickly converted to CND (or CNH) and DCN by the D-atom abstraction reactions below:

$$D + DCND \text{ (or DCNH)} \rightarrow CND \text{ (or CNH)} + D_2 \quad (\Delta H^\circ \approx -159 \text{ kJ mol}^{-1}) \tag{29.42}$$

$$\rightarrow DCN + D_2 \text{ (or HD)} \quad (\Delta H^\circ \approx -322 \text{ kJ mol}^{-1}) \tag{29.43}$$

Since the latter reaction is much more exothermic than the former, the formation of DCN may be more favorable than DNC.

The intensities of the absorption peaks of products in Figure 29.18 were found to decrease with increase of temperature but they persisted to appear at 20 K except for DNC and DCN. This is reasonable because these products, other than DNC and DCN, are mainly formed in the gas-phase plasma of mixed N_2/HCN gas and are deposit–frozen in the N_2 matrix. Gradual decrease in the absorption intensities for these products is likely to decrease in the sticking probability of N_2 molecules (i.e., matrix gas) on the solid surface with increase of temperature.

Although the present result doesn't give any quantitative information whether HCN and HNC production favorably competes with other routes of HCN/HNC formation, the grain-surface reactions of H atoms with CN radicals trapped in the dust grains may explain some part of the abundance of HNC and HCN in the cold molecular clouds since CN radicals, as the precursors for the formation of these cyanide compounds, are present with rather high abundance in ISM (Prasad et al., 1987).

29.8 IS NH_3 FORMED IN SOLID-PHASE REACTIONS?

In Figure 29.18b through d, N_3 is observed as the strongest absorption peak. Interestingly, however, neither DN_3 nor ND_3 was observed in Figure 29.18. This suggests that the N atom and also the N_3 radical do not play important roles for the formation of ND_3 and DN_3. This is rather puzzling because both N and N_3 have unpaired electrons and association reactions (29.44) and (29.45) are highly exothermic.

$$H \text{ (or D)} + N \rightarrow NH \text{ (or ND)} \quad (\Delta H^\circ = -314 \text{ kJ mol}^{-1}) \qquad (29.44)$$

$$H \text{ (or D)} + N_3 \rightarrow NH_3 \text{ (or } DN_3) \quad (\Delta H^\circ = -397 \text{ kJ mol}^{-1}) \qquad (29.45)$$

In low-temperature solid-phase reactions, the rate of reaction is strongly dependent on the energy dissipation processes. The energy relaxation of vibrational energy to phonon energy is quite inefficient because vibrational quanta are more than one order of magnitude larger than phonon quanta as already mentioned. In this regard, high exothermic reactions do not necessarily take place with high efficiencies in the low-temperature solid-phase reactions.

The N-atom abstraction reaction (29.46) may be another channel for reaction of H (or D) with N_3.

$$H \text{ (or D)} + N_3 \rightarrow NH \text{ (or ND)} + N_2 \quad (\Delta H^\circ = -259 \text{ kJ mol}^{-1}) \qquad (29.46)$$

Reaction (29.46) might be more favorable than reaction (29.45) in solid-phase reactions because reaction (29.46) is less exothermic than reaction (29.45) and the heat of reaction (29.46) can be dissipated as the kinetic energies of the products, NH and N_2.

The N_3 radical may be formed by the reaction of electronically excited N atoms, for example, $N(^2D)$, with a N_2 molecule (ground state $X^1\Sigma_g^+$ or excited states, e.g., $A^3\Sigma_u^+$).

$$N(^2D) + N_2 \rightarrow N_3(X^2\Pi_g) \quad (\Delta H^\circ \approx -284 \text{ kJ mol}^{-1} \text{ for } N_2 : X^1\Sigma_g^+) \qquad (29.47)$$

The standard heat of formation of N_3 was calculated to be $\Delta H_f^\circ \approx 418 \text{ kJ mol}^{-1}$ by ab initio calculations with CCSD(T)/cc-pVTZ basis set (Hiraoka et al., 2006). This value leads to the enthalpy changes of reactions (29.46) and (29.47) to be −259 and −284 kJ mol^{-1}, respectively. If NH is formed by reactions (29.44) and/or (29.46), NH_3 may be formed as a final product. Strangely, however, ND_3 could not be detected by FT-IR as already pointed out. One possible reason for the absence of ND_3 is the existence of the energy barrier that prohibits the reaction (29.46) to proceed at cryogenic temperatures. Another reason is that even if NH radical is formed in reactions (29.44) and (29.46), it does not lead to the formation of the final product ammonia due the occurrence of H-atom abstraction reactions (29.48) and (29.49).

$$H \text{ (or D)} + HN \text{ (or DN)} \rightarrow H_2 \text{ (or } D_2) + N \quad (\Delta H^\circ = -121 \text{ kJ mol}^{-1}) \qquad (29.48)$$

$$H \text{ (or } D) + H_2N \text{ (or } D_2N) \rightarrow H_2 \text{ (or } D_2) + NH \text{ (or } ND) \quad (\Delta H^\circ = -29 \text{ kJ mol}^{-1}) \quad (29.49)$$

$$H \text{ (or } D) + H_3N \text{ (or } D_3N) \rightarrow H_2 \text{ (or } D_2) + NH_2 \text{ (or } ND_2) \quad (\Delta H^\circ = +17 \text{ kJ mol}^{-1}) \quad (29.50)$$

In our previous work (Hiraoka et al., 1995), the reaction of H atoms with N atoms trapped in an N_2 matrix was studied. Ammonia was formed but its yield was found to be very low. Besides, only N and H atoms were observed but no radicals such as NH and NH_2 could be detected by the electron spin resonance spectroscopy when the solid N_2/H_2 (100/5) was irradiated by x-ray at 4.2 K, and the sample temperature was increased up to 30 K. The absence of NH and NH_2 in the electron spin resonance spectra suggests that sequential association reactions (29.44) and (29.51) are less likely.

$$H + NH \rightarrow NH_2 \quad (\Delta H^\circ = -405 \text{ kJ mol}^{-1}) \quad (29.51)$$

The absence of NH_3 may be due to the inefficient association reactions (29.44) and (29.51) and/or the favorable H-atom abstraction reactions (29.48) and (29.49). These predictions are not very unreasonable because the highly exothermic reactions (29.44) and (29.51) must experience appreciable centrifugal barriers when two reactants come close to each other. In contrast, abstraction reactions (29.48) and (29.49) are less exothermic and may proceed more efficiently because the heats of reactions can be carried away as kinetic energies of two separating products. Actually, the energy dissipation plays a crucial role in the solid-phase low-temperature reactions. Tielens (1989) pointed out that heat transfer in the amorphous materials changes abruptly in character at about 25 K. Below that temperature, the energy is carried by low-frequency phonons ($<10^{-12}$ s), which have a long mean free path ($\sim 10^5$ Å). With increasing temperature, the heat is carried by high-frequency phonons with a mean free path of about 10 Å. Thus, the energy relaxation becomes increasingly less favorable at lower temperatures. This may result in the less exothermic H-atom abstraction reactions being more favorable than the more exothermic association reactions at lower temperatures. Besides, the conservation of angular momentum for reactions (29.48) and (29.49) is much more easily realized than that for reactions (29.44) and (29.51).

In the separate experiment, using the apparatus shown in Figure 29.13, OCN^-, CO, CO_2, and C_2H_6 were formed as major reaction products, but no NH_3 could be detected by FT-IR when the mixed $CH_4/H_2O/N_2$ (1/1/9) film was irradiated by 100 eV electrons at 10 K. In order to obtain more direct information on the formation of NH_3 in the solid-phase reactions, the H atoms and N atoms were sprayed simultaneously over the metal substrate at 10 K. The H and N atoms were generated by flowing the H_2 and N_2 gases through two discharge tubes, respectively, as shown in Figure 29.1. Surprisingly, no NH_3 could be detected by FT-IR though a strong absorption due to N_3 radicals could be detected in the deposited sample. These results again indicate that the consecutive hydrogenation reactions, $N \rightarrow NH \rightarrow NH_2 \rightarrow NH_3$, are quite inefficient in the solid phase.

In low-temperature solid-phase reactions, the reactants basically interact with each other by the remote-encounter fashion because the close encounter is largely prohibited by the presence of matrix molecules. In such a case, the H-atom tunneling reactions such as reactions (29.48) and (29.49) may become more favorable than reactions (29.44) and (29.51). As was described, the deuterated ethane $C_2H_{6-n}D_n$ with n up to 6 were formed. This clearly indicates the occurrence of the H-atom abstraction reaction, ethyl radical $+ D \rightarrow$ ethylene $+ HD$ (or D_2). This result is only an example demonstrating that H-atom abstraction reactions play important roles in solid–phase reactions. Such H-atom abstraction reactions taking place on the dust grains may be one possible source for the formation of hydrogen molecules in the universe as was already pointed out by Tielens (1989).

The less efficient NH_3 formation in solid-phase reactions underlines the importance of the gas-phase synthesis of NH_3 in ISM. Herbst and Klemperer (1973) proposed the consecutive ion–molecule reactions of NH_n^+ ($n = 0 \rightarrow 3$) with H_2 to form the final product ion NH_4^+, followed by the recombination of NH_4^+ with an electron to form NH_3. The bottleneck for these consecutive reactions is that with $n = 3$, that is, $NH_3^+ + H_2 \rightarrow NH_4^+ + H$, the rate constant was measured to be 1.8×10^{-13} cm^3 molecule^{-1}s^{-1}

at 85 K (Adams and Smith, 1984).This value is about four orders of magnitude smaller than the collision rate. The gas-phase ion–molecule routes for the formation of NH_4^+ may be reconciled by the alternative gas-phase reactions of NH_3^+ with ubiquitous H_2O and CH_4,

$$NH_3^+ + H_2O \rightarrow NH_4^+ + OH \tag{29.52}$$

$$NH_3^+ + CH_4 \rightarrow NH_4^+ + CH_3 \tag{29.53}$$

The rate constants for reactions (29.52) and (29.53) have been measured to be 4×10^{-10} and 5×10^{-10} cm^3 molecule^{-1} s^{-1}, respectively (Ikezoe et al., 1987), which are close to the collision rates. We conjecture that H_2O and CH_4 play an important role as reactants not only in the solid phase but also in the gas phase.

29.9 CONCLUSION

Because of their high mobility in solid, H atoms play an important role in the chemical evolution of dense clouds at 10 K via tunneling processes. In all the reactions investigated, the product yields increased drastically with decrease of temperature from 50 to 10 K. This finding led us to conclude that the chemical evolution, taking place on the dust grains via H-atom tunneling reactions, becomes efficient only at cryogenic temperatures. If the temperature of the molecular clouds is higher than 10 K, the rate of the chemical evolution would become slower due to the short residence time of H atoms on the dust grains (Hiraoka et al., 2001b).

In the reactions of D with solid C_2H_2 and C_2H_4, the fully deuterated product, C_2D_6, is formed. This indicates the occurrence of the solid-phase abstraction reactions of D atoms with C_2H_3 and C_2H_5 to form C_2HD and C_2H_3D, respectively. In the reaction of D with H_2CO, H-atom abstraction reaction $D + H_2CO \rightarrow HD + HCO$ is much more favored than the addition reaction, $D + H_2CO \rightarrow H_2DCO$ (or CH_2OD). In the reaction of H with N atoms seeded in the solid matrix, no NH_3 was detected as a reaction product. This indicates that the major processes for the reaction of H with NH_n with $n = 1$ and 2 are the H-atom abstraction reactions, $H + NH_n \rightarrow NH_{n-1} + H_2$. As such, the abstraction reactions are found to be one of the major processes in the solid-phase reactions of H with relevant molecules. The favorable abstraction reactions may be due to the efficient dissipation of internal energies of the intermediate products as kinetic energies of the separating products and the presence of large channels for the conservation of angular momentum of the reaction system.

While the consecutive hydrogenation reactions $CO \rightarrow HCO \rightarrow H_2CO \rightarrow CH_3CO(CH_2COH) \rightarrow CH_3OH$ are less likely to explain the ubiquitous methanol in the interstellar medium, it was found that methanol is the major product from the electron-irradiated water ice containing carbon source contaminants. From this result, the paramount importance of ubiquitous water ice can be envisaged for the chemical evolution in cold ISM. In addition to its role as a matrix to trap various volatile molecules (i.e., nature's solvent), it generates highly reactive species such as OH and H in the interaction with cosmic rays and UV photons. Moreover, the water molecule acts as a reactant itself to form astrochemically relevant molecules such as methanol and formaldehyde as discussed in Section 29.6. Water is indispensable not only for the evolution of life but also for the chemical evolution in the ISM.

REFERENCES

Adams, N. G. and Smith, D. 1984. A study of the reactions of NH_3^+ and ND_3^+ with H_2 and D_2 at several temperatures. *Int. J. Mass Spectrom. Ion Processes* 61:133–139.

Allamandola, L. J., Sandford, S. A., and Valero, G. J. 1988. Photochemical and thermal evolution of interstellar/precometary ice analogs. *Icarus* 76: 225–252.

Becker, K., Klipping, G., Schonherr, W. D., Schulze, W., and Tolle, V. 1972. Adsorption characteristics of condensed Ar, C_2H_6, NH_3 and CO_2 layers with respect to cryopumping. In *Proceedings of the Fourth International Cryogenic Engineering Conference*, Eindhoven, the Netherlands, pp. 323–326.

Bennett, J. E. and Mile, B. 1973. Studies of radical-molecule reactions using a rotating cryostat. Reactions of hydrogen atoms with organic substrates at 77 K. *J. Chem. Soc. Faraday Trans.* 1 69: 1398–1414.

Bhattacharya, D., Wang, H.-Y., and Willard, J. E. 1981. Trapped hydrogen atoms, deuterium atoms, and methyl radicals in methane and methane-d4 at 5–50 K. Yields from photolysis of HX and from radiolysis; decay mechanisms; reactions with oxygen and carbon monoxide. *J. Phys. Chem.* 85: 1310–1323.

Blowers, P., Ford, L., and Masel, R. 1998. Ab Initio calculations of the reactions of hydrogen with methanol: A comparison of the role of bond distortions and Pauli repulsions on the intrinsic barriers for chemical reactions. *J. Phys. Chem. A* 102: 9267–9277.

Calaway, W. F. and Ewing, G. E. 1975. Vibrational relaxation in liquid nitrogen. *Chem. Phys. Lett.* 30: 485–489.

Cherigier, L., Czarnetzki, U., Luggenholscher, D., der Gathen, V. S., and Dobele, H. F. 1999. Absolute atomic hydrogen densities in a radio frequency discharge measured by two-photon laser induced fluorescence imaging. *J. Appl. Phys.* 85: 696–702.

Crovisier, J. 1998. Physics and chemistry of comets: Recent results from comets Hyakutake and Hale-Bopp answers to old questions and new enigmas. *Faraday Dis.* 109: 437–452.

Crovisier, J., Leech, K., Bockelee-Morvan, D., Brooke, T. Y., Hanner, M. S., Altieri, B., Uwe Keller, H., and Lellouch, E. 1997. The spectrum of comet Hale-Bopp (C/1995 O1) observed with the infrared space observatory at 2.9 astronomical units from the sun. *Science* 275: 1904–1907.

D'Hendecourt, L. B. and Allamandola, L. J. 1986. Time dependent chemistry in dense molecular clouds. III: Infrared band cross sections of molecules in the solid state at 10 K. *Astron. Astrophys. Suppl. Ser.* 64: 453–467.

Dressler, K., Oehler, O., and Smith, D. A. 1975. Measurement of slow vibrational relaxation and fast vibrational energy transfer in solid N_2. *Phys. Rev. Lett.* 34: 1364–1367.

Ehrenfreund, P. and Schutte, W. A. 2000. Infrared observations of interstellar ices. In *Proceedings of IAU Symposium 197*, Seogwipo, South Korea, p. 135.

Fleck, L. E., Feehery, W. F., Plummer, E. W., Ying, Z. C., and Dai, H. L. 1991. Laser-induced polymerization of submonolayer formaldehyde on silver (111). *J. Phys. Chem.* 95: 8428–8430.

Gerakines, P. A., Schutte, W. A., and Ehrenfreund, P. 1996. Ultraviolet processing of interstellar ice analogs. I. Pure ices. *Astron. Astrophys.* 312: 289–305.

Goldanskii, V. I., Frank-Kamenetskii, M. D., and Barkalov, I. M. 1973. Quantum low-temperature limit of a chemical reaction rate. *Science* 182: 1344–1345.

Green, S. and Herbst, E. 1979. Metastable isomers—A new class of interstellar molecules. *Astrophys. J.* 229: 121–131.

Greenberg, J. M. and Pirronello, V. 1991. *Chemistry in Space*. Berlin, Germany: Springer.

Grim, R. J. A., Greenberg, J. M., de Groot, M. S., Baas, F., Schutte, W. A., and Schmitt, B. 1989. Infrared spectroscopy of astrophysical ices—New insights in the photochemistry. *Astron. Astrophys. Suppl. Ser.* 79: 161–186.

Hasegawa, T. I., Herbst, E., and Leung, C. M. 1992. Models of gas-grain chemistry in dense interstellar clouds with complex organic molecules. *Astrophys. J. Suppl. Ser.* 82: 167–195.

Herbst, E. 1978. What are the products of polyatomic ion-electron dissociative recombination reactions. *Astrophys. J.* 222: 508–516.

Herbst, E. and Klemperer, W. 1973. The formation and depletion of molecules in dense interstellar clouds. *Astrophys. J.* 185: 505–534.

Hiraoka, K. and Sato, T. 2001. Laboratory simulation of tunneling reactions in interstellar ices. *Radiat. Phys. Chem.* 60: 389–393.

Hiraoka, K., Matsunaga, K., Shoda, T., and Takimoto, H. 1992. Reaction of hydrogen atoms with solid unsaturated hydrocarbons at 77 K. *Chem. Phys. Lett.* 197: 292–296.

Hiraoka, K., Yamashita, A., Yachi, Y., Aruga, K., Sato, T., and Muto, H. 1995. Ammonia formation from the reactions of H atoms with N atoms trapped in a solid N_2 matrix at 10–30 K. *Astrophys. J.* 443: 363–370.

Hiraoka, K., Miyagoshi, T., Takayama, T., Yamamoto, K., and Kihara, Y. 1998. Gas-grain processes for the formation of CH_4 and H_2O: Reactions of H atoms with C, O, and CO in the solid phase at 12 K. *Astrophys. J.* 498: 710–715.

Hiraoka, K., Takayama, T., Euchi, A., Handa, H., and Sato, T. 2000. Study of the reactions of H and D atoms with solid C_2H_2, C_2H_4, and C_2H_6 at cryogenic temperatures. *Astrophys. J.* 532: 1029–1037.

Hiraoka, K., Sato, T., Sato, S., Hishiki, S., Suzuki, K., Takahashi, Y., Yokoyama, T., and Kitagawa, S. 2001a. Formation of amorphous silicon by the low-temperature tunneling reaction of H atoms with solid thin film of SiH_4 at 10 K. *J. Phys. Chem. B* 105: 6950–6955.

Hiraoka, K., Sato, T., and Takayama, T. 2001b. Tunneling reactions in interstellar ices. *Science* 292: 869–870.

Hiraoka, K., Sato, T., Sato, S., Sogoshi, N., Yokoyama, T., Takashima, H., and Kitagawa, S. 2002a. Formation of formaldehyde by the tunneling reaction of H with solid CO at 10 K revisited. *Astrophys. J.* 577: 265–270.

Hiraoka, K., Sato, T., Sato, S., Takayama, T., Yokoyama, T., Sogoshi, N., and Kitagawa, S. 2002b. Study on the tunneling reaction of H atoms with a solid thin film of C_3H_6 at 10 K. *J. Phys. Chem. B* 106: 4974–4978.

Hiraoka, K., Wada, A., Kitagawa, H., Kamo, M., Unagiike, H., Ueno, T., Sugimoto, T., Enoura, T., Sogoshi, N., and Okazaki, S. 2005. The reactions of H and D atoms with thin films of formaldehyde and methanol at cryogenic temperatures. *Astrophys. J.* 620: 542–551.

Hiraoka, K., Ushiama, S., Enoura, T., Unagiike, H., Mochizuki, N., and Wada, A. 2006. Solid-phase reactions of D with CN to form DNC and DCN at cryogenic temperatures. *Astrophys. J.* 643: 917–922.

Hirota, T., Yamamoto, S., Mikami, H., and Ohishi, M. 1998. Abundances of HCN and HNC in dark cloud cores. *Astrophys. J.* 503: 717–728.

Hornekaer, L., Baurichter, A., Petrunin, V. V., Field, D., and Luntz, A. C. 2003. Importance of surface morphology in interstellar H_2 formation. *Science* 302: 1943–1946.

Hudson, R. L. and Moore, M. H. 1999. Laboratory studies of the formation of methanol and other organic molecules by water + carbon monoxide radiolysis: relevance to comets, icy satellites, and interstellar ices. *Icarus* 140: 451–461.

Huebner, W. F., Boice, D. C., and Sharp, C. M. 1987. Polyoxymethylene in comet halley. *Astrophys. J.* 320: L149–L152.

Ikezoe, Y., Matsuoka, S., Takebe, M., and Viggiano, A. 1987. *Gas Phase Ion-Molecule Reaction Rate Constants through 1986*. Tokyo, Japan: Maruzen.

Irvine, W. M., Bockelee-Morvan, D., Lis, D. C., Matthews, H. E., Biver, N., Crovisier, J., Davies, J. K., Dent, W. R. F., Gautier, D., Godfrey, P. D., Keene, J., Lovell, A. J., Owen, T. C., Phillips, T. G., Rauer, H., Schloerb, F. P., Senay, M., and Young, K. 1996. Spectroscopic evidence for interstellar ices in comet Hyakutake. *Nature* 383: 418–420.

Irvine, W. M., Dickens, J. E., Lovell, A. J., Schloerb, F. P., Senay, M., Bergin, E. A., Jewitt, D., and Matthews, H. E. 1998. Chemistry in cometary comae. *Faraday Disc.* 109: 475–492.

Jodkowski, J. T., Rayez, M.-T., Rayez, J.-C., Berces, T., and Dobe, S. 1999. Theoretical study of the kinetics of the hydrogen abstraction from methanol. 3. Reaction of methanol with hydrogen atom, methyl, and hydroxyl radicals. *J. Phys. Chem. A* 103: 3750–3765.

Johnson, R. E. 1990. *Energetic Charged-Particle Interactions with Atmospheres and Surfaces*. Berlin, Germany: Springer-Verlag.

Kaiser, R. I. and Roessler, K. 1998. Theoretical and laboratory studies on the interaction of cosmic-ray particles with interstellar ices. III. Suprathermal chemistry-induced formation of hydrocarbon molecules in solid methane (CH_4), ethylene (C_2H_4), and acetylene (C_2H_2). *Astrophys. J.* 503: 959–975.

Knyazev, V. D. and Slagle, I. R. 1996. Experimental and theoretical study of the $C_2H_3 \rightleftarrows H + C_2H_2$ reaction. Tunneling and the shape of falloff curves. *J. Phys. Chem.* 100: 16899–16911.

Knyazev, V. D., Bencsura, A., Stoliarov, S. I., and Slagle, I. R. 1996. Kinetics of the $C_2H_3 + H_2 \rightleftarrows H + C_2H_4$ and $CH_3 + H_2 \rightleftarrows H + CH_4$ reactions. *J. Phys. Chem.* 100: 11346–11354.

Lanzerotti, L. J. and Johnson, R. E. 1986. Astrophysical implications of ions incident on insulators. In *Ion Beam Modification of Insulators*, P. Mazzoldi and G. W. Arnold (eds.), pp. 631–644. Amsterdam, the Netherlands: Elsevier.

Lias, S. G., Bartmess, J. E., Liebman, J. F., Holmes, J. L., Levin, R. D., and Mallard, W. G. 1988. Gas-phase ion and neutral thermochemistry. *J. Phys. Chem. Ref. Data* 17(Suppl. 1): 1–861.

Mason, C. G., Gehrz, R. D., Jones, T. J., Mergen, J., Williams, D., Tokunaga, A. T., Brooke, T. Y., Weaver, H. A., Crovisier, J., and Bockelee-Morvan, D. 1996. Comet C/1996 B2 (Hyakutake). *IAU Circ.* 6378: 1

Millar, T. J. and Williams, D. A. 1993. *Dust and Chemistry in Astronomy*. Bristol, U.K.: CRC Press.

Miyazaki, T. 1991. Reaction of hydrogen atoms produced by radiolysis and photolysis in solid phase at 4 and 77 K. *Radiat. Phys. Chem.* 37: 635–642.

Moore, M. H. and Hudson, R. L. 1998. Infrared study of ion-irradiated water-ice mixtures with hydrocarbons relevant to comets. *Icarus* 135: 518–527.

Mumma, M. J., DiSanti, M. A., Russo, N. D., Fomenkova, M., Magee-Sauer, K., Kaminski, C. D., and Xie, D. X. 1996. Detection of abundant ethane and methane, along with carbon monoxide and water, in comet C/1996 B2 Hyakutake: Evidence for interstellar origin. *Science* 272: 1310–1314.

Oehler, O., Smith, D. A., and Dressler, K. 1977. Luminescence spectra of solid nitrogen excited by electron impact. *J. Chem. Phys.* 66: 2097–2107.

Pirronello, V., Biham, O., Manico, J., Roser, J. E., and Vidali, G. 2000. Laboratory studies of molecular hydrogen formation on surfaces of astrophysical interest. In *Molecular Hydrogen in Space*, F. Combes and G.P.D. Forêts (eds.), pp. 71–84. Cambridge, U.K.: Cambridge University Press.

Prasad, S. S., Tarafdar, S. P., Villere, K. R., and Huntress, W. T., Jr. 1987. Chemical evolution of molecular clouds. In *Interstellar Processes*, D. J. Hollenbach and H. A. Thronson (eds.), pp. 631–666. Dordrecht, the Netherlands: Reidel.

Roser, J. E., Manico, G., Pirronello, V., and Vidali, G. 2002. Formation of molecular hydrogen on amorphous water ice: Influence of morphology and ultraviolet exposure. *Astrophys. J.* 581: 276–284.

Roser, J. E., Swords, S., Vidali, G., Manico, G., and Pirronello, V. 2003. Measurement of the kinetic energy of hydrogen molecules desorbing from amorphous water ice. *Astrophys. J. Lett.* 596: L55-L58.

Schutte, W. A., Allamandola, L. J., and Sandford, S. A. 1993. An experimental study of the organic molecules produced in cometary and interstellar ice analogs by thermal formaldehyde reactions. *Icarus* 104: 118–137.

Sen, A. D., Anicich, V. G., and Federman, S. R. 1992. Formaldehyde reactions in dark clouds. *Astrophys. J.* 391: 141–143.

Shalabiea, O. M. and Greenberg, J. M. 1994. Two key processes in dust/gas chemical modelling: Photoprocessing of grain mantles and explosive desorption. *Astron. Astrophys.* 290: 266–278.

Snyder, R. J. and Hollis, J. M. 1976. HCN, X-ogen /HCO+/, and U90.66 emission spectra from L134 (interstellar molecules). *Astrophys. J. Lett.* 204: L139-L142.

Sosa, C. and Schlegel, H. B. 1986. Ab initio calculations on the barrier height for the hydrogen addition to ethylene and formaldehyde. The importance of spin projection. *Int. J. Quantum Chem.* 29: 1001–1015.

Takahashi, J., Masuda, K., and Nagaoka, M. 1999. Product energy distribution of molecular hydrogen formed on icy mantles of interstellar dust. *Astrophys. J.* 520: 724–731.

Takayanagi, T. and Sato, S. 1990. The bending-corrected-rotating-linear-model calculations of the rate constants for the $H+H_2$ reaction and its isotopic variants at low temperatures: The effect of van der Waals well. *J. Chem. Phys.* 92: 2862–2868.

Taketsugu, T., Tajima, A., Ishii, K., and Hirano, T. 2004. Ab initio direct trajectory simulation with nonadiabatic transitions of the dissociative recombination reaction $HCNH^+ + e^- \rightarrow HNC/HCN + H$. *Astrophys. J.* 608: 323–329.

Talbi, D., Ellinger, Y., and Herbst, E. 1996. On the HCN/HNC abundance ratio: A theoretical study of the H + CNH ↔ HCN + H exchange reactions. *Astron. Astrophys.* 314: 688–692.

Tielens, A. G. G. M. 1989. Dust in dense clouds. In *IAU Symposium 135, Interstellar Dust*, L. A. Allamandola and A. G. G. M. Tielens (eds.), Santa Clara, CA, pp. 239–262. Dordrecht, the Netherlands: Kluwer.

Tielens, A. G. G. M., Tokunaga, A. T., Geballe, T. R., and Baas, F. 1991. Interstellar solid CO: Polar and nonpolar interstellar ices. *Astrophys. J.* 381: 181–199.

Van Ijzendoorn, L. J., Allamandola, L. J., Baas, F., and Greenberg, J. M. 1983. Visible spectroscopy of matrix isolated HCO: The $^2A''(\Pi) \leftarrow X\ ^2A'$ transition. *J. Chem. Phys.* 78: 7019–7028.

Villa, J., Corchado, J. C., Gonzalez-Lafont, A., Lluch, J. M., and Truhlar, D. G. 1998a. Explanation of deuterium and muonium kinetic isotope effects for hydrogen atom addition to an olefin. *J. Am. Chem. Soc.* 120: 12141–12142.

Villa, J., Gonzalez-Lafont, A., Lluch, J. M., and Truhlar, D. G. 1998b. Entropic effects on the dynamical bottleneck location and tunneling contributions for $C_2H_4 + H \rightarrow C_2H_5$: Variable scaling of external correlation energy for association reactions. *J. Am. Chem. Soc.* 120: 5559–5567.

Wada, A., Mochizuki, N., and Hiraoka, K. 2006. Methanol formation from electron-irradiated mixed H_2O/CH_4 ice at 10 K. *Astrophys. J.* 644: 300–306.

Watson, W. D. 1974. Ion-molecule reactions, molecule formation, and hydrogen-isotope exchange in dense interstellar clouds. *Astrophys. J.* 188: 35–42.

Watson, W. D. 1977. Interstellar chemistry. *Acc. Chem. Res.* 10: 221–226.

Watson, W. D. 1980. Molecule formation in cool, dense interstellar clouds. In *Interstellar Molecules, IAU Symposium 87*, B. H. Andrew(ed.), Quebec, Canada, pp. 341–350. Dordrecht, the Netherlands: Reidel.

Woon, D. E. 2002. Modeling gas-grain chemistry with quantum chemical cluster calculations. I. Heterogeneous hydrogenation of CO and H_2CO on icy grain mantles. *Astrophys. J.* 569: 541–548.

Yang, K. and Rabitz, H. 1994. Quantum effects in the surface penetration of energetic hydrogen atoms. *J. Chem. Phys.* 101: 8205–8213.

30 Radiation Effects on Semiconductors and Polymers for Space Applications

Takeshi Ohshima
Japan Atomic Energy Agency
Takasaki, Japan

Shinobu Onoda
Japan Atomic Energy Agency
Takasaki, Japan

Yugo Kimoto
Japan Aerospace Exploration Agency
Tsukuba, Japan

CONTENTS

30.1 INTRODUCTION

Technology for space exploration and development constantly progresses. Many space missions have been completed and even more are planned. For example, the space station project has been realized; as a result, one can stay beyond the bounds of the earth for long periods. The uses and benefits from man-made satellites are expansive, including weather forecast, broadcast, resource

FIGURE 30.1   Photograph of a vacuum chamber for space solar cell evaluation installed at Japan Atomic Energy Agency (JAEA), Takasaki. The chamber is connected with a beam line of the AVF Cyclotron.

radiation degradation can be completed. In an effort to increase the turnaround of samples, evaluation techniques that can measure the electrical performance of solar cells during irradiation experiments have been developed (Ohshima et al., 1996). Figure 30.1 shows a vacuum chamber installed at Japan Atomic Energy Agency (JAEA), Takasaki, for space solar cell experiments. Since the irradiation chamber has an AM0 solar simulator, the degradation behavior of the electrical performance of solar cells can be evaluated during proton irradiation experiments (simultaneous technique). In addition, the degradation behavior of solar cells under low temperature irradiation has become a topic of interest, specifically for missions to Saturn and Jupiter. To tackle such varied missions, a sample holder with a temperature control system was installed in the chamber (Ohshima et al., 2005; Harris et al., 2008). Figure 30.2 shows the comparison of remaining factors of the electrical performance (short current circuit, I_{SC}, open circuit voltage, V_{OC}, and maximum power, P_{MAX}) of space triple-junction solar cells between RT and 175 K irradiations (Ohshima et al., 2005).

In addition to ground experiments, on-orbit flight demonstrations have also been conducted (Imaizumi et al., 2005). The satellite named MDS-1 (Mission Demonstration-test Satellite No. 1) was launched in 2002. The MDS-1 was placed in a geosynchronous transfer orbit (GTO) with apogee of 36,000 km and perigee of 500 km. Typical solar cells designed for terrestrial applications were mounted on the MDS-1, and their on-orbit electrical performance was measured. A GTO was selected for this flight experiment because it crosses the Van Allen radiation belts which consist of high-energy charged particles, mainly electrons and protons. Thus, solar cells were subjected to high concentrations of electrons and protons for a short time. Figure 30.3 shows the remaining factor of I_{SC} for solar cells on the MDS-1 after launch (Imaizumi et al., 2005). The I_{SC} for all cell types, except Cu(In,Ga)Se$_2$ (CIGS), decreased within days after launch as shown in Figure 30.3. Since MDS-1 cuts across the Van Allen belts, the degradation behavior for solar cells without CIGS can be explained in terms of heavy damage created by proton and electron irradiations. For CIGS solar cells, it has been reported that electrical performance recovers due to the annealing of defects even at room temperature (Kawakita et al., 2002). Therefore, the apparent lack of degradation for CIGS solar cells on MDS-1 is attributed to on-orbit defect annealing.

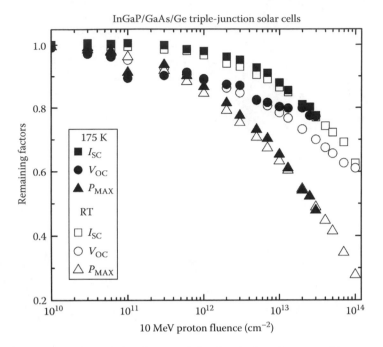

FIGURE 30.2 Comparison of remaining factors of the electrical performance (short current circuit, I_{SC}, open circuit voltage, V_{OC}, and maximum power, P_{MAX}) of space InGaP/GaAs/Ge triple-junction solar cells between RT and 175 K irradiations. (Reprinted from Ohshima, T. et al., Evaluation of the electrical characteristics of III-V compounds solar cells irradiated with proton at low temperature, *Proceedings of 31st IEEE Photovoltaic Specialists Conference*, Lake Buena Vista, FL, 2005, pp. 806–809. With permission.)

30.2.2 Basic Mechanism of Radiation Degradation of Solar Cells

Solar cells generate electric power when photo-induced minority carriers in a base layer migrate and reach a top layer before recombination. Thus, the diffusion length (lifetime) of minority carriers is one of the most important parameters for the solar cell performance. As mentioned above, charged particles create defects in semiconductors. Since some of the radiation-induced defects act as recombination centers, a decrease in diffusion length (lifetime) of minority carriers due to irradiation is the most dominant degradation effect in solar cells. In this region, experimental results can be described by the following semi-empirical equation (Tada et al., 1982),

$$\left(\frac{1}{L_{\Phi}}\right)^2 = \left(\frac{1}{L_0}\right)^2 - K_L \Phi, \tag{30.1}$$

where

 L_{Φ} and L_0 are the minority carrier diffusion length after and before irradiation, respectively
 K_L and Φ are the damage coefficient of minority carrier and fluence

Figure 30.4 shows the remaining factors of triple-junction solar cells irradiated with protons of various energies (Sumita et al., 2003). As shown in the figure, the fitting lines calculated using the semi-empirical equation are in agreement with experimental results. From the fundamental point of view, radiation-induced defects in new materials are intensively investigated because many candidates for high efficiency solar cells have been proposed and properties of defects in such new materials are not yet fully understood (Walters et al., 2001; Dharmarasu et al., 2002; Soga et al., 2003).

As the second irradiation effect, majority carrier removal occurs in high-fluence regions, and the reduction of majority carriers can be empirically expressed as

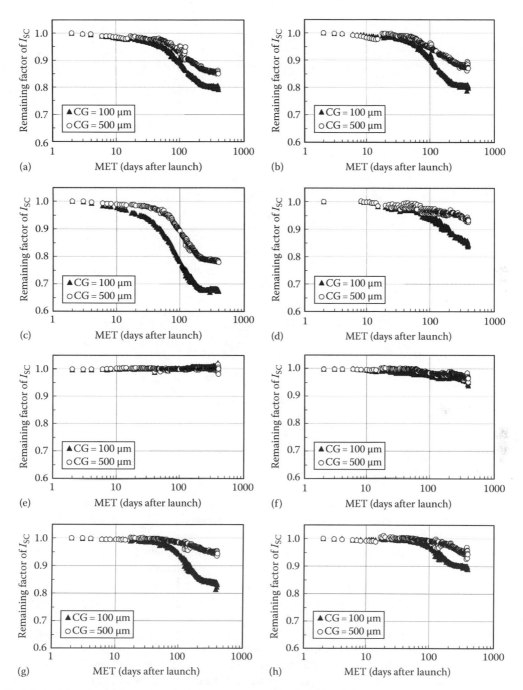

FIGURE 30.3 Remaining factor of I_{SC} for various solar cells on the MDS-1 after launch. (a) U poly-crystalline silicon cell. (b) Poly-crystalline silicon cell. (c) N-type base single crystal silicon cell. (d) InGaP/GaAs dual-junction tandem cell. (e) Large-area CIGS cell. (f) High-efficiency CIGS cell. (g) Single crystal silicon space cell (2 Ω cm). (h) Single crystal silicon space cell (10 Ω cm). (Reprinted from Imaizumi, M. et al., *Prog. Photovolt: Res. Appl.*, 13, 93, 2005. With permission. John Wiley & Sons, Ltd.)

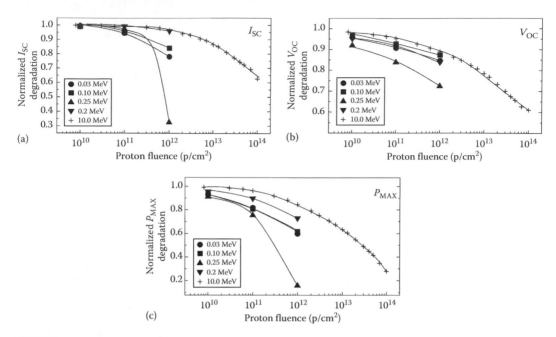

FIGURE 30.4 Remaining factors of I_{SC}, V_{OC}, and P_{MAX} for InGaP/GaAs/Ge triple-junction solar cells irradiated with protons with various energies as a function of proton fluence. Symbols and lines represent experimental and fitting results, respectively. (Reprinted from Sumita, T. et al., *Nucl. Instrum. Methods B*, 206, 448, 2003. With permission.)

$$p_\Phi = p_0 - R_C \Phi \quad \text{or} \quad p_\Phi = p_0 \exp\left(\frac{-R_C \Phi}{p_0}\right), \tag{30.2}$$

where

p_Φ and p_0 are majority carriers after and before irradiations, respectively
R_C is a "carrier removal rate"

The carrier concentration in the base layer of solar cells decreases, when the carrier removal effect appears. Since the depletion layer length in a solar cell depends on the difference of carrier concentration between the top and base layers, the decrease in carrier concentration in the base layer causes the depletion layer to extend. As a result, the number of minority carriers that can reach the top layer from the base layer increases. Thus, this indicates that the I_{SC} increases with increasing fluence. Of course, this very unique increase in I_{SC} is a temporary behavior, and with further increase of fluence, the value of I_{SC} suddenly drops to almost zero because the base layer becomes intrinsic or type conversion occurs in the base layer; as a result, no pn junction exists (Matsuura et al., 2006). Figure 30.5 shows the remaining factor of I_{SC} for Si solar cells irradiated with either 10 MeV protons or 1 MeV electrons as a function of fluence. In a low fluence region, the value of I_{SC} gradually decreases with increasing fluence. This decrease suggests that the dominant effect is the decrease in the diffusion length of minority carriers. On the other hand, in the higher fluence region, the increase in I_{SC} before sudden dropping down is observed. This behavior can be explained in terms of majority carrier removal. The excellent agreement of the above-mentioned mechanism with experimental results has been reported in the literature (Ohshima et al., 1996; Yamaguchi et al., 1996).

The decreases in minority carrier diffusion length and majority carrier concentration are basically the main degradation effects of solar cells due to irradiation. However, in some cases, the recovery of the electrical performance is also observed. For amorphous Si, the value of photoconductivity becomes higher than the initial value by irradiation (Amekura et al., 2000). The electrical

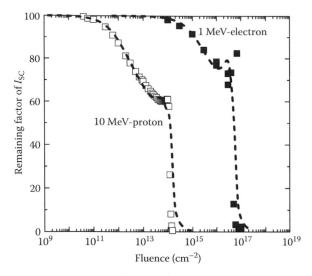

FIGURE 30.5 Remaining factor of I_{SC} for Si solar cells irradiated with either 10 MeV protons or 1 MeV electrons as a function of fluence. Symbols and lines represent experimental and fitting results, respectively.

properties of CIGS irradiated with charged particles recovered even close to room temperature annealing (Kawakita et al., 2002). For InP and related materials, the recovery of the electrical properties by minority carrier injection has been reported (Yamaguchi et al., 1995, 1997). These are very unique behaviors. However, the details of defect annealing/creation have not yet been fully clarified in these materials. Further investigation by many researchers is currently underway.

30.2.3 SPACE SOLAR CELLS

A brief history of space photovoltaics will be described before discussing the current status. Early on, simple single pn junction solar cells based on crystalline Si were used. The efficiency of Si solar cells installed on Vanguard I, launched in 1954, was 7%–8% under AM0 illumination. With improving device fabrication processes, the design of Si space solar cells was modified, resulting in a conversion efficiency increase. A back surface field (BSF), which is a p⁺ layer, was fabricated behind a p-type base layer to reduce the surface recombination of photo-induced carriers. To avoid heating solar cells under illumination at long wavelengths, a back surface reflector (BSR) layer which consists of a thin metal layer was applied to space solar cells. In addition to these modifications, inverse pyramid shaped textures creating a non reflective surface (NRS) were fabricated at the surface of the Si solar cells (Katsu et al., 1994). Then, the conversion efficiency of Si solar cells increased up to 18% under AM0. Single-junction-GaAs-based solar cells were also applied to space applications because of their relatively high conversion efficiency (18%–20%) (Matsuda et al., 1994).

From the point of view of high radiation resistance, InP and related materials have been proposed for space solar cells (Yamaguchi et al., 1984, 1995; Yamaguchi, 2001). Figure 30.6 shows the remaining factor of P_{MAX} for InP, GaAs, and Si solar cells as a function of 1 MeV electron fluence (Yamaguchi, 2001). InP solar cells obviously have higher radiation resistance than GaAs and Si solar cells. In addition, depending on irradiation conditions, InP solar cells show different radiation degradation. From DLTS studies, the variety of the degradation behaviors for InP solar cells is interpreted in terms of the reduction of a major radiation-induced center, which acts as a hole trap by minority carrier injection (Yamaguchi et al., 1986). Although the reason of this reduction has not yet been fully clarified, a reformation of defects in InP by energy transfer from injected carriers has been proposed, most likely because the migration energy of vacancies at both In and P lattice sites are lower than another III-V compounds such as GaAs (Yamaguchi, 2001). Further investigation is necessary to more fully understand the detailed mechanism of this minority carrier injection

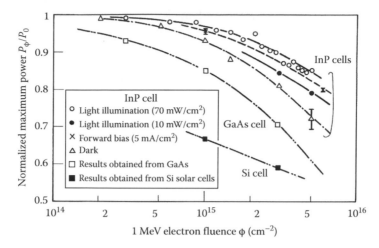

FIGURE 30.6 Remaining factor of P_{MAX} for InP, GaAs, Si solar cells as a function of 1 MeV electron fluence. For InP solar cells, results obtained from irradiation under various conditions are depicted. (Reprinted from Yamaguchi, M., *Sol. Energy Mater. Sol. Cells*, 68, 31, 2001. With permission.)

effect. The radiation resistance of InP in space was observed with no significant degradation of I_{SC} for about 1000 days after launch (in fact, a small increase by the injection effect was reported.) (Weinberg, 1991).

Multi-junction solar cells based on III-V compounds are the main streams for space applications nowadays because they combine high conversion efficiency and high radiation resistance. Figure 30.7 shows the schematic cross section of a typical InGaP/GaAs/Ge triple-junction solar cell (3J solar cell). The 3J solar cells are grown on Ge substrates by an organometallic vapor phase epitaxy (OMVPE). As shown in Figure 30.7, the 3J solar cells consist of three solar cells (InGaP, GaAs, and Ge sub cells) with tunnel junction diodes located between each sub cell. Since each sub cell can absorb light at specific wavelengths as shown in Figure 30.8, very high efficiency ~28% can be achieved with AM0 illumination. In addition, new compounds such as AlInGaP and AlGaAs are utilized in the top and middle sub cells of state-of-the-art 3J solar cells, to further increase in the conversion efficiency (Takamoto et al., 2005). Nitride-based top cells have also been investigated in an attempt to improve conversion efficiency (Dimroth et al., 2005). Furthermore, using an inverted metamorphic (IMM) approach to overcome lattice mismatch issues, very high efficiency 3J solar cells have been achieved (Geisz et al., 2008).

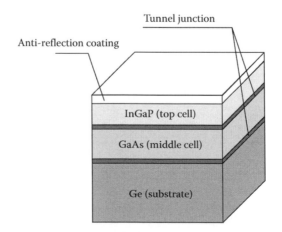

FIGURE 30.7 Schematic cross section of InGaP/GaAs/Ge 3 junction solar cells.

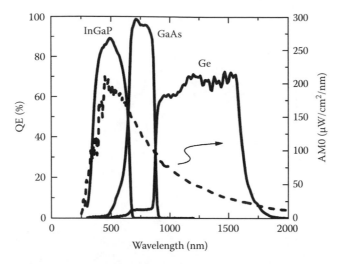

FIGURE 30.8 Quantum efficiency (QE) of InGaP/GaAs/Ge 3J solar cells as a function of wave length. The spectrum of AM0 is also shown in the figure.

The thin-film solar cells based on new materials such as CIGS and amorphous Si (s-Si) have been studied for space applications (Kawakita et al., 2002; Granata et al., 2005). These solar cells are thought to be very promising candidates because light weight and flexible paddles can be designed to take advantage of their unique characteristics. The technologies for these solar cells are improved day-by-day, and mass production for terrestrial application has already started. However, as mentioned above, since these solar cells show unique recovery behaviors, the mechanism of radiation effects should be revealed to predict accurate lifetime in a space environment before commercial use for space applications.

30.2.4 RADIATION DEGRADATION OF TRIPLE-JUNCTION SOLAR CELLS

As mentioned above, the degradation of solar cells in space is caused by the decreases in minority carrier diffusion length and majority carrier concentration due to charged particle irradiation. The decrease in these parameters can be scaled using K_L and R_C, respectively, and these values depend on the material. Thus, in the case of multi-junction solar cells, each sub cell has its own K_L as well as R_C values, and as a result, the degradation behavior of a sub cell is different from that of another. This suggests that the degradation behavior of multi-junction solar cells is not simple like that of a single junction solar cell because the radiation resistance is different between component sub cells. For V_{OC}, the degradation behavior of multi-junction solar cells is the sum of the degradation of each sub cell connected in series. On the other hand, the degradation behavior of I_{SC} for multi-junction solar cells is equal to the degradation behavior of a sub cell with the lowest I_{SC} since the I_{SC} for multi-junction solar cells is limited by a sub cell with lowest I_{SC} (current-limiting cell). Figure 30.9 shows the relative damage coefficient of V_{OC} for InGaP/GaAs/Ge 3J solar cells as a function of the energy of irradiated protons. The relative damage coefficient is defined as the degradation value normalized by the degradation value due to 10 MeV proton (or 1 MeV electron) irradiation. For example, the relative damage coefficient of 10 means that the decrease in the electrical performance is 10 times larger than that in the case of 10 MeV at the same fluence. In Figure 30.9, the peaks are observed around 0.03, 0.25, and 2 MeV. This indicates that the degradation becomes worse by proton irradiation at these energies. Calculations using a Monte Carlo code, SRIM (Ziegler et al., 2008), show that the projection range (penetration depth) of protons with these energies corresponds to the position of pn junction of sub cells. Since protons create heavy damage at the tail end of their projection range, the three peaks can be interpreted in terms of the heavy damage creation in each sub cell by protons with these energies.

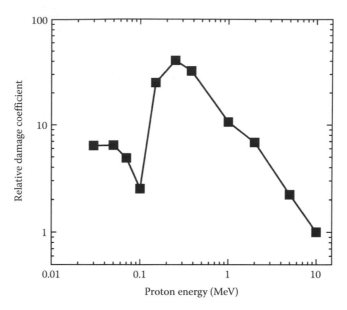

FIGURE 30.9 Relative damage coefficient of V_{OC} for InGaP/GaAs/Ge 3J solar cells as a function of irradiated proton energy.

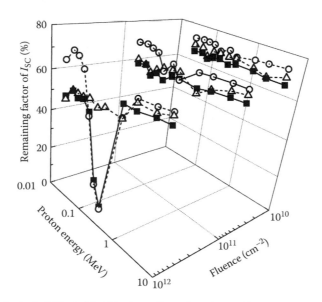

FIGURE 30.10 I_{SC} for InGaP/GaAs/Ge 3J solar cells as functions of irradiated proton energy and fluence (closed squares). For comparison, the values of I_{SC} for InGaP (open triangles) and GaAs (open circles) sub cells estimated from spectral sensitivity measurements are also plotted in the figure.

The unique degradation behavior is also observed in I_{SC} for 3J solar cells by proton irradiation with various energies. Figure 30.10 shows the value of I_{SC} for InGaP/GaAs/Ge 3J solar cells as functions of irradiated proton energy and fluence (closed squares). The values of I_{SC} for InGaP (open triangles) and GaAs (open circles) sub cells, estimated from spectral sensitivity measurements, are plotted in Figure 30.10. At a fluence of 10^{10} cm^{-2}, the value of I_{SC} for the 3J solar cells does not depend on the value of proton energy, and is almost the same as the value of I_{SC} for InGaP solar cells. This indicates that the InGaP cell is the current-limiting cell in this fluence region. The degradation of I_{SC} for GaAs solar cells irradiated with protons at hundreds of keV is larger than that of GaAs solar cells irradiated at another energy. The value for GaAs solar cells irradiated with protons at hundreds of keV becomes almost the same value for InGaP solar cells irradiated at a fluence of 10^{11} cm^{-2}. The I_{SC} for GaAs solar cells dramatically decreases and becomes the lowest value at 10^{12} cm^{-2}. Thus, since the current-limiting cell changes to the GaAs cell from the InGaP cell, the value of I_{SC} for 3J solar cell becomes the same as that of the GaAs cell. This faster decrease in I_{SC} for the GaAs sub cell by irradiation of protons with hundreds of keV is explained as follows. The protons with given energy stop around the pn junction in the GaAs sub cell. This indicates that heavy damage is created in this region by the irradiation. Since the radiation resistance of GaAs is not higher than that of InGaP, the degradation of the GaAs sub cell is much faster than the InGaP sub cell by hundreds keV range proton irradiation. From this result, it can be concluded that the improvement of the radiation hardness of the GaAs sub cell is important to improve the overall radiation hardness of 3J solar cells.

30.2.5 Prediction Methodology of Triple-Junction Solar Cell Degradation

Protons and electrons with a very wide range of energies exist in space. However, available energies for ground-based irradiation experiments are limited by the capability of accelerators. Therefore, it is important to develop accurate prediction methods on the radiation degradation of solar cells in real space, using results obtained from ground tests. Presently, the damage-equivalent fluence method, which was proposed by the U.S. Jet Propulsion Laboratory (JPL) more than 20 years ago, is the most popular prediction method (Tada et al., 1982). According to the JPL method, a result obtained by 1 MeV electron irradiation testing is standard, and the relative damage coefficient is estimated from the comparison of results using electrons/protons with other energies to the result obtained by 1 MeV electron/proton irradiation. In the case of single junction solar cells, the JPL method is thought to be a reasonable way for the lifetime prediction of solar cells, since a single peak is usually observed in the relative damage coefficient as a function of proton/electron energy. But, for triple-junction solar cells, as shown in Figure 30.9, three peaks appear in the proton/electron energy dependence of the relative damage coefficient. This indicates that proton/electron irradiation experiments using various energies are necessary to clarify the relationship between proton/electron energy and the relative damage coefficient for 3J junction solar cells. From the standpoint of saving time and money, increasing the number of irradiation experiments is not a good idea. Instead of fluence, displacement damage dose (D_d), based on the concept of non ionizing energy loss (NIEL), has been proposed by the U.S. Naval Research Laboratory (NRL) (Summers et al., 1987; Messenger et al., 2001). The NIEL is a concept for the energy reduction of incident particles in materials, which is for displacement of atoms at lattice sites and is estimated as D_d and is defined as

$$D_d = \mathrm{NIEL}(E) \times \text{fluence} \times \left(\frac{\mathrm{NIEL}(E)}{\mathrm{NIEL}(E_{ref})} \right)^{n-1} \tag{30.3}$$

where
NIEL(E_{ref}) is NIEL for a standard energy such as 1 MeV electrons
n is a correction number usually between 1 and 2 (Messenger et al., 2001)

If the D_d is applied as the value of damage creation, since the D_d does not depend on the energy of incident protons/electrons, the degradation curves for solar cells irradiated with protons/electrons at various energies should be single line. Figure 30.11 shows the degradation curve for V_{OC} of InGaP/GaAs/Ge 3J solar cells as a function of D_d. Two degradation curves are shown in the figure. Since the projection range for protons with lower than 100 keV is within the InGaP sub cell, the obtained result can be explained in terms of the curve obtained by low energy proton irradiation for the InGaP sub cell, and the curve obtained by high-energy proton irradiation for GaAs and Ge sub cells. From this result, it is apparent that the degradation of V_{OC} can be described by D_d on the basis of the NIEL concept. However, the degradation of I_{SC} for 3J solar cells is not simple, even if D_d is applied, as shown in

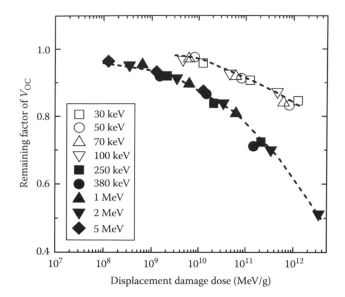

FIGURE 30.11 Degradation curve of V_{OC} for InGaP/GaAs/Ge 3J solar cells irradiated with protons with various energies as a function of D_d.

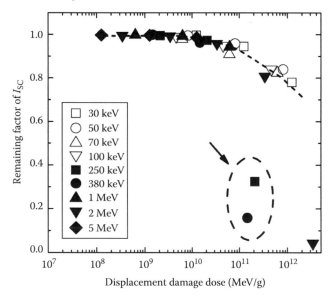

FIGURE 30.12 Degradation curve of I_{SC} for InGaP/GaAs/Ge 3J solar cells irradiated with protons with various energies as a function of D_d.

Figure 30.12. The results at D_d around 10^{11} MeV/g obtained by 250 and 380 keV proton irradiations are not plotted on the line. Since proton irradiation at these energies creates damage into GaAs sub cells, the obtained result is attributed to the remarkable degradation of I_{SC} by the current-limiting cell changing from the GaInP subcell to the GaAs subcell. This result suggests that it is difficult to describe the degradation of I_{SC} using D_d, if the change occurs in the current-limiting cell. Thus, although the NRL method on the basis of NIEL concept is very convenient to reduce irradiation tests, we need to know the D_d or fluence value for the change in current-limiting cell by some other method.

Modeling of degradation behavior of InGaP/GaAs/Ge 3J solar cells using spectra response (quantum efficiency) was proposed by JAEA and Japan Aero eXploration Agency (JAXA) (Sato et al., 2009a,b). In this modeling, the change in quantum efficiency for 3J solar cells due to proton/electron

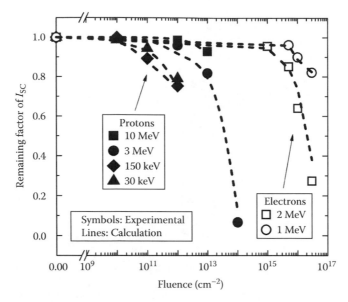

FIGURE 30.13 Degradation curves of I_{SC} for 3J solar cells irradiated with either protons or electrons. Symbols and lines represent experimental and calculated results, respectively.

irradiation is measured, and the simulation of the measurement result is carried out to obtain L and p of each sub cell. Then, the K_L and R_C for each sub cell are estimated from the fluence dependence of L and p. Since I_{SC} and V_{OC} for 3J solar cells can be estimated from the simulation using K_L and R_C, the degradation curves for I_{SC} and V_{OC} are obtained. Figure 30.13 shows the degradation curves of I_{SC} for 3J solar cells irradiated with either protons or electrons. The simulation results mentioned above are also depicted as solid lines. Excellent agreement between experimental and calculated results is observed as shown in Figure 30.13. This indicates that the degradation curve for 3J solar cells can be described by this model, even if the current-limiting cell changes from the InGaP cell to the GaAs cell. Furthermore, it was found that the value of K_L and R_C can be described as functions of NIEL (Sato et al., 2009b). This indicates that the degradation curve for 3J solar cells can be predicted using the K_L and R_C as a function of NIEL, which can be obtained from minimal irradiation testing. Since a wide variety of materials are used in solar cells and the structure becomes more complicated, the degradation modeling is more important from the viewpoint of lifetime prediction and development of new space solar cells. To obtain highly accurate modeling of radiation degradation of solar cells, further investigation is necessary, especially for unique effects such as room temperature annealing effect and minority carrier injection effect.

30.3 ELECTRONIC DEVICES

30.3.1 OVERVIEW OF SINGLE-EVENT EFFECTS IN SPACE

The electronic devices in the artificial satellites are exposed to high levels of space radiations. The high-energy heavy ions in cosmic rays especially cause the malfunction of electronic devices known as SEEs. It is well known that the energy of heavy ions in space exceeds GeV (Xapsos, 2006). As an example, the energy spectrum of iron ions at a geostationary orbit is shown in Figure 30.14 (Dodd et al., 1998). The energy at the peak flux for iron ion is around 300 MeV/amu. The atomic mass unit (amu) for iron is 56; therefore, the energy of iron at the peak flux is 16.8 GeV. The experiments, using the high-energy heavy ions provided by accelerators, are carried out on the ground to understand the mechanism of SEEs and to predict the probability of SEEs during space missions. Figure 30.14 shows typical ion irradiation facilities for lower and higher energies. In the case of the test facility

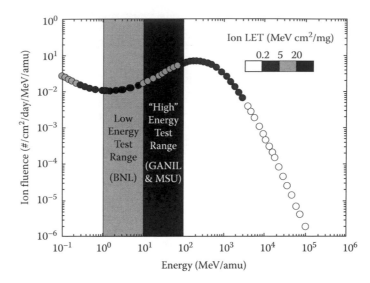

FIGURE 30.14 Energy spectrum of iron ion in a geostationary orbit at solar minimum conditions behind 100 mil Al shielding. Ion LET is indicated by the gray of the symbols. Typical heavy-ion testing is performed at energies of 1–10 MeV/amu. (Reprinted from Dodd, P.E. et al., *IEEE Trans. Nucl. Sci.*, 45, 2483, 1998. With permission.)

called Takasaki Ion Accelerators for advanced Radiation Application (TIARA) of JAEA, ion energy, accelerated by the AVF cyclotron, ranges from 2.5 to 27 MeV/amu (Onoda et al., 2006; Kurashima et al., 2007). In this figure, the LET of iron ion along the curve is indicated with gray coding.

A SEE is categorized as having either nondestructive or destructive effects. The former is often referred to as "soft error," while the latter is referred to as "hard error" (Petersen, 2008). A soft error occurs when the amount of collected charge is large enough to reverse or flip the data state of a memory cell, register, latch, or flip-flop. The error is "soft" because the circuit/device itself is not permanently damaged by the radiation. If new data is written to the bit, the device will store it correctly. The single-event upset (SEU) and the multiple-bit upset (MBU) in a static random access memory (SRAM) and a dynamic random access memory (DRAM) are notable examples. Other good example is the single-event functional interrupt (SEFI) in field programmable gate array (FPGA) or DRAM control circuitry. The single-event transient (SET) is also a serious problem in analog electronics and digital logic cells. In general, SETs in analog electronics are referred to as ASETs while those in digital combinatorial logic are referred to as DSETs. In contrast, a "hard" error is manifested when the device is physically damaged such that improper operation occurs, data is lost, and the damaged state is permanent. The single-event latch-up (SEL), the single-event burnout (SEB), and the single-event gate rupture (SEGR) in power electronic devices provide examples of the hard errors.

The SEEs in devices, circuits, and applications used in radiation environments are too complicated to be discussed here in detail. In this section, therefore, the DSET will be discussed with a focus on the kind of mechanism in operation. With this aim, three factors seem to be helpful in attempting to sketch its mechanism: the first is the generation of charge and ion track when an ion passes through a semiconductor, the second is the transient currents resulting from the charge transportation in semiconductor, and the third is the propagation of transient voltage pulse in logic circuits.

30.3.2 Generation of Charge and Ion Tracks in Semiconductors

When the incident ion passes through a semiconductor, it loses kinetic energy predominantly through interactions with the orbital electrons of the semiconductor crystal. As a result, an ion leaves dense electron-hole (e-h) pairs along its trajectory. Of course, the ion can also interact directly with crystal atoms but the probability of this reaction is orders of magnitude less than the electronic interaction. If the generated charges (dense electrons and holes) are collected at an electrode and exceed a threshold, then SEE is initiated. The charge collection threshold for the SEE is called the critical charge.

The probability of SEE depends on how much energy is deposited (thus, how many charges are created) in the sensitive region of electronic devices. The deposited energy can be expressed as a function of LET. The LET is defined as the average energy loss per unit path length, normalized by the density of the target material, ρ, and its unit is typically MeV cm²/mg. The LET can be calculated by using the advanced computer codes that employ Monte Carlo techniques and detailed physical models. One of the most popular simulators in the SEE community is SRIM code (Ziegler et al., 2008). The deposited energy, E, over a path length, l, is expressed as follows:

$$E = \rho \int_l l \cdot \mathrm{LET}(x)\,dx. \tag{30.4}$$

The LET of an ion can be assumed to be a constant value when the ion has a long range compared to sensitive volume geometry and when multiple scattering is negligible. The above equation reduces to

$$E = \rho \cdot l \cdot \mathrm{LET}(x). \tag{30.5}$$

It has long been considered that the LET is the most important key factor to dominate the mechanism of SEE. At the same time, ion track distributed along the trajectory of the ion in three dimensions has been thought to be a second-order effect. In principle, different track structures lead to a different charge collection and SEE response in spite of the same LET. At a given LET, the higher energy ion produces a longer, wider, and less dense track than the lower energy ion. Various ion track models have been suggested and one of these is by Kobetich and Katz (1968). Since this model overestimates the e-h density at the core of the ion track, an improved model using empirical equations was presented by Fageeha et al. (1994). These models use the delta-electron production under MeV ion penetration and the keV delta-electron transmission theory in the matter to calculate the initial radial track distribution. Figure 30.15 shows the predictions using a theoretical model for two ions with the same LET (11 MeV cm²/mg) (Dodd et al., 1998). As shown, the ion track radius of Kr at 5.04 GeV is wider than that of Cr at 210 MeV. The tracks reach radii of over μm.

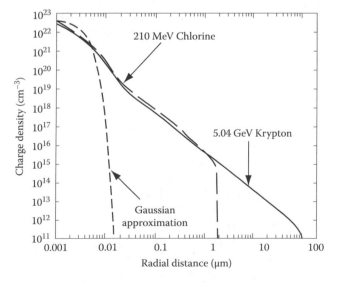

FIGURE 30.15 Radial track structure of low and high-energy ions with the same incident LET. Also shown is a Gaussian approximation to the track center region. All three curves have the same integral charge. (Reprinted from Dodd, P.E. et al., *IEEE Trans. Nucl. Sci.*, 45, 2483, 1998. With permission.)

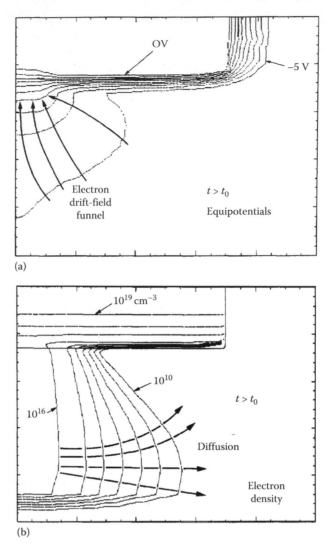

FIGURE 30.16 Counter maps of potential (a) and electron density (b) being calculated by using two-dimensional numerical codes in cylindrical coordinate after ion track was formed. The direction of carrier motion by drift and diffusion is indicated by means of arrows. (Reprinted from Zoutendyk, J.A. et al., *IEEE Trans. Nucl. Sci.*, 35, 1644, 1988. With permission.)

If the radial extent of the ion track is larger than the sensitive volume geometry, then one could imagine that less charge would be collected. On the other hand, the charge outside the sensitive volume is transported laterally by diffusion toward other electrodes. Figure 30.16a shows the calculated results of potential represented in counter by using two-dimensional numerical codes in cylindrical coordinate (Zoutendyk et al., 1988). The direction of carrier motion by drift is indicated by means of arrows. In contrast, the counter map of electron density is represented in Figure 30.16b. As shown, the electrons move laterally by diffusion. As a result of lateral charge transport, the MBU occurs near the position where the ion hits. Figure 30.17 shows the typical MBU cluster induced in 256 k DRAM caused by a normal incidence Kr at 295 MeV (Zoutendyk et al., 1989). It follows from what has been said, that the ion track plays an important role in SEE. However, in some cases, no significant difference was observed between high and low energy ions exhibiting the same LET (Dodd et al., 1998). Dodd et al. (1998) concluded that the track structure is not important. They concluded that the central cores of the track structures were nearly identical for low and high-energy ions.

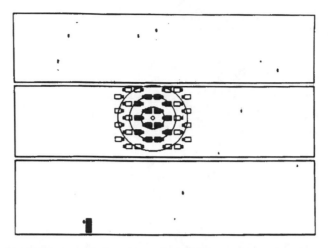

FIGURE 30.17 Typical MBU cluster induced in 256k DRAM caused by a normal incidence Kr at 295 MeV. (Reprinted from Zoutendyk, J.A. et al., *IEEE Trans. Nucl. Sci.*, 36, 2267, 1989. With permission.)

Three-dimensional simulations confirmed that charge collection was similar in the two cases. This is partially because the sensitive volumes do not have an abrupt boundary, and the initial ion track expands very rapidly by radial diffusion. In any case, the influence of ion track on the charge collection and SEEs is still one of the most interesting issues because of these contrasting observations.

30.3.3 Transient Currents Occurring Subsequent to Ion Track Generation

The dense carriers (electrons and holes) created by an ion are transported to an electrode by drift and diffusion processes in semiconductors. As a result, transient current flows in electronic devices. Since the behavior of the transient current strongly depends on the impact point of ions in semiconductor devices, the size of the ion beam as well as the information on the impact point of ions are necessary to clarify the detailed generation mechanism of SEEs.

With this aim, Wagner et al. (1987) have developed the measurement system using collimated beam and wide bandwidth electronics. The beam provided by the accelerator was radially scattered by a Pt thin film, and the scattered beam penetrated a Pt micro-collimator at a farther, angled location. Using this method, though the parallel pencil beam can be obtained easily, the beams contain the spattered atoms with low energies as mixtures, and its energy spectrum becomes slightly broad and shifts toward the left. Since the LET generally decreases with increasing ion energy after reaching the maximum at several MeV, the low energy ions including the spattered ions induce a larger transient current than scattered ions, which leads to the lower signal-to-noise ratio. In the 1990s a MeV range-focused microbeam at TIARA facility was used for transient current measurement instead of a collimated beam (Nashiyama et al., 1993; Laird et al., 2001). In this system, ion beams with a spot size of 1 μm at energies from 6 to 18 MeV (less than about 1 MeV/amu) can be irradiated into samples. Combining the focused microbeam with wide bandwidth electronics, the transient current with a spatial imaging can be measured. This system is referred to as transient ion beam induced current (TIBIC) system. A similar system called time-resolved ion beam induced current (TRIBIC) has been also developed at Sandia National Laboratories (SNL) (Schone et al., 1998). Both methods allow one to map the spatiotemporal response of a semiconductor device and examine ultrafast charge collection dynamics (Breese, 1993). These systems have a key advantage in being able to probe a specific region in a device at high resolution; however, its energy is much lower than that of the cosmic ray in which we are interested, as shown in Figure 30.14. To overcome this problem, the high-energy focused microbeam connected to the AVF Cyclotron at JAEA TIARA facility has been successfully

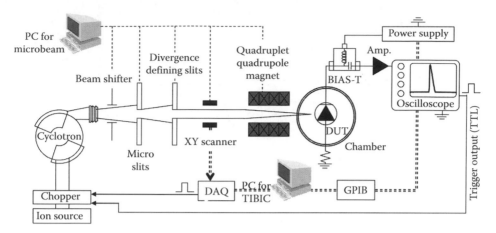

FIGURE 30.18 Schematic diagram of the high-energy TIBIC system connected with the AVF Cyclotron. (Reprinted from Hirao, T. et al., *Nucl. Instrum. Methods. B*, 267, 2216, 2009. With permission.)

developed in the 2000s (Oikawa et al., 2003, 2007). The 260 MeV ^{20}Ne^{7+} and the 520 MeV ^{40}Ar^{14+} ions are available for microbeam. The details of the high-energy TIBIC system are given below.

Figure 30.18 shows the schematic diagram of the high-energy TIBIC system (Hirao et al., 2009; Vizkelethy et al., 2009). The TIBIC system contains a beam chopping system; a beam focusing system including a beam shifter, micro-slits, a scanner, and a quadruplet quadrupole magnets; an irradiation chamber; and electronics for TIBIC measurements. In this system, the beam chopping system is used for controlling the beam current. In addition, the beam current can be reduced by the beam attenuators located next to the chopper. The beams accelerated by the AVF Cyclotron are optimized by the focusing system. The focused microbeam is transported into the irradiation chamber. The scintillator film (ZnS) and the #1000 Cu mesh lie in the same plane of the sample holder in the irradiation chamber. These are used for detecting the microbeam position and for evaluating the full width at half maximum (FWHM) of the microbeam spot size, respectively. Software control sequence for TIBIC imaging proceeds as follows. First, a constant reverse bias is applied to the sample. During this period, the beam is removed using a chopping system controlled by a transistor–transistor logic (TTL) signal provided by a device acquisition (DAQ) board. Next, the microbeam is electrostatically scanned by the raster method with a speed of about 330 s/frame. The beam flux is controlled to remain at several tens of ions per second. When a transient current is generated by an ion strike, the digital storage oscilloscope (DSO) is triggered and the beam is again stopped by the TTL signal from DSO. The DAQ board also receives the TTL signal from DSO and simultaneously records the x–y positions. After the transient current and positions are stored in PC, the process is repeated over the complete scan area. Finally, we perform the pulse shape analysis (PSA) on each transient current and generate images based on extracted PSA parameters, such as the charge, the peak current, the fall time, and the rise time.

Figure 30.19 shows the typical TIBIC peak image observed from a Si pin photodiode by using high-energy TIBIC at JAEA (Vizkelethy et al., 2009). This image indicates that that the diode region covered with the circular electrode is SEE sensitive. The peak current at the edge of electrode is lower than that at the center region. Although not shown here, it is known that no remarkable difference between TIBIC charge images at the edge and the center is observed. In contrast with the edge, both the peak current and the charge at the bonding pad are much higher than those at the center. The reason why the charges at the bonding pad is larger than that at the center is simply interpreted in terms that a penetrated ion loses its energy in the bonding wire and the ion having lower energy generates the charge in the depletion layer. In other words, the obtained result indicates that the charge generated in Si under the bonding pad at the thickness of 20 μm can be evaluated by using ions with several hundreds of MeV. Thus, this result suggests that integrated

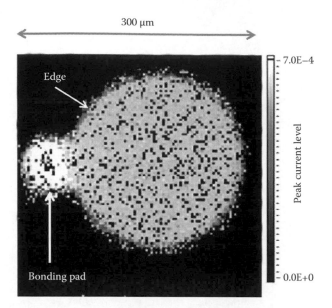

FIGURE 30.19 Typical TIBIC peak image observed from Si pin photodiodes at a reverse bias of 20 V by using 520 MeV Ar microbeam. This image indicates that that the diode region covered with the circular electrode is SEE sensitive. (Reprinted from Hirao, T. et al., *Nucl. Instrum. Methods. B*, 267, 2216, 2009. With permission.)

circuits with multiple wiring can be evaluated by the developed TIBIC system. Figure 30.20a and b show the typical transient current at a reverse bias of 20 V when Ne at 260 MeV and Ar at 520 MeV strike at the edge, the center, and the bonding pad of electrode, respectively (Vizkelethy et al., 2009). The oscillation around 3 ns is probably caused by the impedance miss-matching at the feedthrough terminal on the irradiation chamber. Figure 30.20c represents the transient currents at the center on a log scale. The carrier dynamics in Si pin photodiodes after an ion strike has been studied by a numerical device simulator, the so-called technology computer aided design (TCAD) (Weatherford, 2002; Laird et al., 2005; Onoda et al., 2006). As a result, the carrier dynamics can be simplified into the ambipolar, the bipolar, and the diffusion phases. These phases are distinguished in Figure 30.20c. As shown, the plasma induced by Ar at 520 MeV makes the ambipolar phase longer due to dense plasma. Because of the oscillation, it is difficult to evaluate the boundary between the bipolar phase and the diffusion phase. However it is considered that the overall current is less affected by the diffusion current. Figure 30.21a shows another example of transient currents measured at the source and drain electrodes of a 0.25 μm bulk n-MOSFETs by using TRIBIC system at SNL (Ferlet-Cavrois et al., 2004). The Cl ion at 35 MeV strikes the drain region with the transistor biased either in the OFF- or TG-state. In the OFF-state, the drain is biased and the other electrodes are grounded. The OFF-state corresponds to a common logic sensitive configuration. In the TG-state, both the source and drain are biased, while the other electrodes are grounded. In both cases, the drain current has the typical shape of charge collection in a reverse biased junction as shown in Figure 30.20. The charge collection on the source and drain electrodes can be calculated by integrating the transient currents and these are plotted in Figure 30.21b. For both the OFF- and TG-states, a positive charge is collected on the drain. The drain charge increases monotonically and saturates after 5 ns.

30.3.4 MODELING OF TRANSIENT VOLTAGE PULSE IN LOGIC CIRCUITS

The transient voltage pulse, which is referred to as DSET, is likely to become the dominant source of soft errors for advanced CMOS logic very-large-scale-integrations (VLSIs) (Buchner

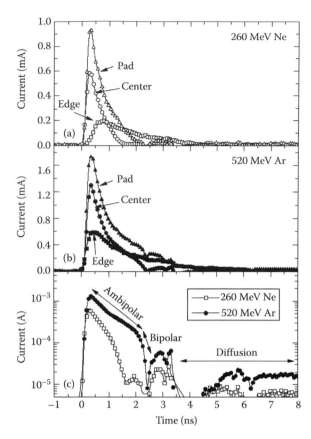

FIGURE 30.20 Typical transient currents at a reverse bias of 20 V induced by 260 MeV Ne (a) and 520 MeV Ar (b), when the ion strikes at the edge, the center, and the bonding pad. The transient currents obtained from the center of electrode on a log scale (c). The ambipolar, the bipolar, and the diffusion phases are denoted in this figure. (Reprinted from Hirao, T. et al., *Nucl. Instrum. Methods. B*, 267, 2216, 2009. With permission.)

and Baze, 2001; Benedetto et al., 2004; Gadlage et al., 2004; Uemura et al., 2006). For example, Figure 30.22 shows the individual contributions from an inverter and a latch to the overall error rate (Buchner et al., 1997). At low frequencies, the SEU rate is dominated by those originating in the latch and at high frequency by those in the inverter.

In this section, I will shift the emphasis away from transient current in single MOSFET (See Figure 30.21) to transient voltage pulse in an inverter chain. The inverter chain consists of n-MOSFETs and p-MOSFETs as shown in Figure 30.23a (Kobayashi et al., 2007a). In principle, the transient current in single n-MOSFET differs from that in n-MOSFET of inverter. Figure 30.24 shows an example of numerically calculated transient currents in single and integrated SOI MOSFETs (Ferlet-Cavrois et al., 2006). The difference between these is due to the bias condition. During transient current measurements on single n-MOSFET, the drain voltage is artificially forced. On the contrary, the drain of the n-MOSFET in inverter chain is floating, and therefore the drain voltage is not forced to a fixed value. The look up table (LUT) technique, proposed by D. Kobayashi, acts as a bridge between them (Kobayashi et al., 2007a,b). Figure 30.23b and c show how to estimate the transient voltage pulse by using the transient currents on the single MOSFET. The transient currents, I_{HI}, are measured under various constant drain voltage conditions. The measured data are replotted as a function of drain voltage, V_D, for any time. The I_{HI} at 4 ps is represented in the bottom left of Figure 30.23b. On the other hand, the recovery current, I_R, versus drain voltage are calculated for each time. The recovery current is the sum of p-MOSFET drain current, I_p, and

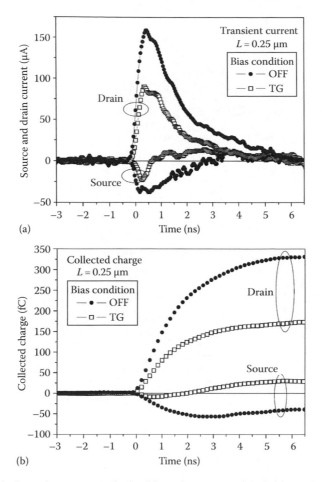

FIGURE 30.21 Typical transient currents obtained from the source and drain electrodes of bulk n-MOSFETs biased in the OFF-state or TG-state during irradiation (a). The charge collection on the source and drain, which is calculated by integrating the transient currents is represented in (b). (Reprinted from Ferlet-Cavrois, V. et al., *IEEE Trans. Nucl. Sci.*, 51, 3255, 2004. With permission.)

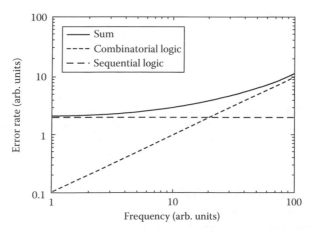

FIGURE 30.22 Error rate as a function of frequency for combinational and sequential logic elements as well as their sum. (Reprinted from Buchner, S. et al., *IEEE Trans. Nucl. Sci.*, 44, 2209, 1997. With permission.)

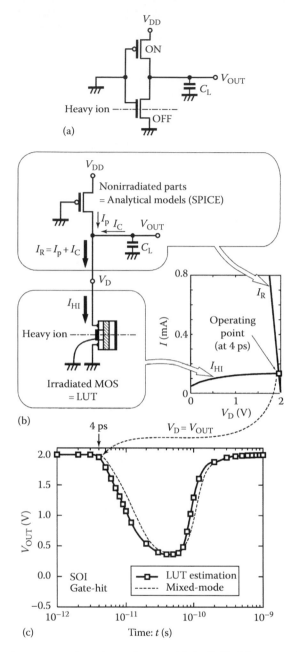

FIGURE 30.23 SET-pulse construction. A test inverter (a) is divided into two circuit blocks: irradiated n-MOSFET and nonirradiated parts (b). The inverter's operating point is determined by the intersection of I_R-V_D curve and I_{HI}-V_D curve. SET pulse is constructed from the operating points (c). A mixed-mode simulation result with the same device model and circuit configuration is superimposed for reference. (Reprinted from Kobayashi, D. et al., *IEEE Trans. Nucl. Sci.*, 54, 2347, 2007a. With permission.)

capacitance discharge current, I_C. The inverter's operating voltage is determined by the intersection of I_{HI}-V_D curve and I_R-V_D curve. Figure 30.23c shows the transient voltage pulse in the inverter chain by using the LUT technique. A simulation result by TCAD is also shown as a broken line. The LUT estimation agrees rather well with the simulated result. Thus, this indicates that the transient voltage pulse in inverter chain can be described using this modeling method.

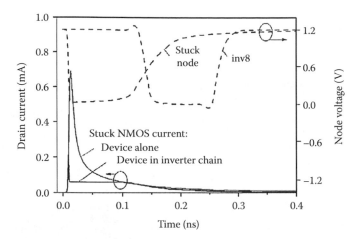

FIGURE 30.24 Simulation of a 130 nm SOI n-MOSFET struck by a 5 MeV cm²/mg ion in the center of the gate. The struck transistor is either simulated as a device alone in the OFF-state or integrated in an inverter chain with mixed-mode simulation. The current driven by the struck device is shown on the left axis. The output inverter voltage are indicated on the right axis for the struck node and the eighth inverter. (Reprinted from Ferlet-Cavrois, V. et al., *IEEE Trans. Nucl. Sci.*, 53, 3242, 2006. With permission.)

30.4 POLYMERS

Polymer materials have been widely used for structure and thermal control in spacecraft because of their light weight, workability, flexibility, and electrical insulation properties. Recently, polymer materials have been used for their enhanced and sophisticated performance due to their high specific strength, high specific stiffness, high thermal stability, and high dimensional stability. Meanwhile, these materials are exposed to severe and complex space-environment conditions. As a result, some material properties—mechanical, thermal, and optical—are degraded. One of the polymer material degrading factors is space radiation particles. Another degrading factor is intense solar ultraviolet (UV) rays, atomic oxygen (AO) in low earth orbit (LEO), and thermal cycle. There have been cases where volatile compounds from polymers in vacuum contaminated a high-profile sensor. A synergistic effect may be caused by a combination of these factors.

One approach to solving these problems is the evaluation of the material properties on the ground in the required mission period. For such evaluations, we utilize our combined space effects test facility, which accommodates the irradiation of independent or coincidental electron beams, ultraviolet (UV) rays, and atomic oxygen.

Furthermore, the material property data related to the effects of the natural space environment is necessary for performing the evaluations on the ground. Space materials exposure experiments have been done not only for this purpose but also for demonstration for space use.

The radiation effects on polymers, especially on aromatic polyimides which are typically used in spacecraft material, are summarized briefly, and then recent ground evaluation facilities are described. Finally, space materials exposure experiment projects and results concerning radiation effects are summarized.

30.4.1 GENERAL RADIATION EFFECTS ON POLYMERS

Several authors have described degradation in polymers by means of bar charts showing the stages of degradation versus dose (Holmes-Siedle and Adams, 2000). This type of chart is a plot of the growth of damage versus dose, or a "radiation index" (RI) such as the logarithm of the dose in Gy at which elongation at break, E, a well-defined mechanical property, is halved. Figure 30.25a and b are classifications of materials in decreasing order of their radiation resistance (Tavlet et al., 1998). Figure 30.25a and b list rigid thermoplastics, thermoset resins and composites with respect

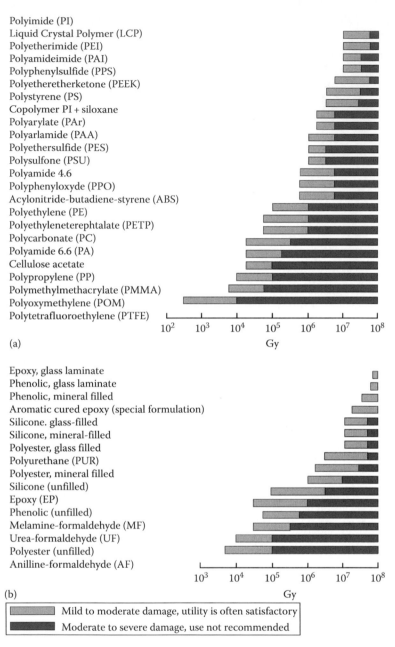

FIGURE 30.25 General classification of radiation resistance. (a) Rigid thermoplastics. (b) Thermoset resins and composites. (Reprinted from Tavlet, M. et al., Compilation of radiation damage test data, Part II, 2nd edn, Thermoset and thermoplastic regins; composite materials, Report No. CERN-98-01, CERN, Geneva Switzerland. With permission.)

to their radiation resistance. This classification gives the order of magnitude of the maximum dose of usability of the materials; it corresponds to long-term irradiations. In particular, thermoplastics (Figure 30.25a) may be very sensitive to oxidation and hence to dose-rate effects; during long-term irradiation in the presence of oxygen, their degradation starts at a much lower dose (Tavlet et al., 1998). This datum is acquired in a nuclear reactor with a high dose-rate irradiation in air, corresponding to the popular threshold dose limit. It is necessary to consider the "real radiation filed" where polymers are used and screened.

In the aromatic polyimides, the imide rings and the linking groups between the aromatic groups within the polymer chains, principally determine the chemical, thermal, and radiation stability (Inoue et al., 1987). In addition, the concentration of the aromatic groups influences the sensitivity because these groups offer protection against structural damage due to exposure to electron or gamma irradiations (Inoue et al., 1987). This aromatic protective effect is believed to occur via the ability of aromatic groups to dissipate absorbed energy to heat through their manifold vibrational states (O'Donnell and Sangster, 1970; Megusar, 1997). So, polyimide is widely used for thermal control films and lubricants for space use because it has a high radiation tolerance.

The mechanical properties have been examined by means of tensile tests after high dose (5 kGy/s) electron irradiation, which is similar to a vacuum environment, and a gamma ray irradiation test under pressure with 0.7 MPa oxygen has been performed for 12 kinds of polyimides such as Kapton, polyetheretherketones (PEEK); the results are reported in Sasuga and Hagiwara (1985, 1987). Figure 30.26 shows stress–strain diagrams after electron irradiation of Kapton H, which is a typical example of polymer that expands without constriction, and amorphous PEEK which is a typical example of polymer that expands with constriction. The property value of breakdown strength and ultimate elongation is greatly influenced by the irradiation compared with the property value of yield strength and Young's modulus. Figure 30.27

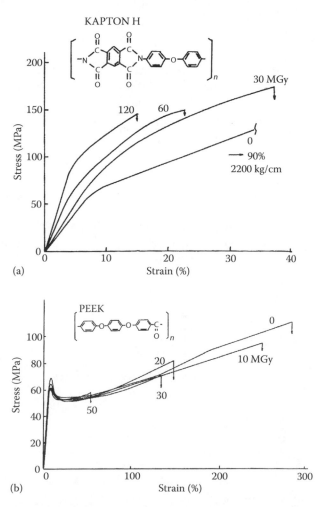

FIGURE 30.26 The stress–strain curves of (a) polyimide "Kapton" and (b) non-crystalline PEEK irradiated by various doses. (Reprinted from Sasuga, T. and Hagiwara, M., *Polymer*, 26, 1039, 1985. With permission.)

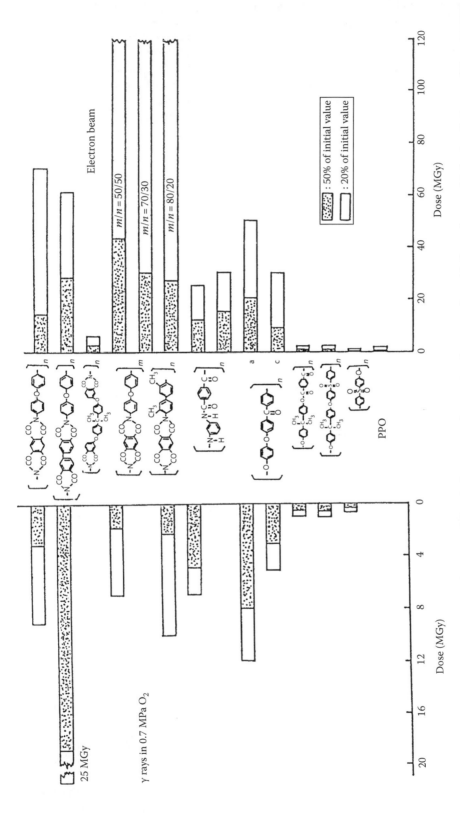

FIGURE 30.27 Relation between radiation resistance and chemical structure of aromatic polymers under oxidative and non-oxidative irradiation conditions. ▓ 50% of the initial value and ☐ 20% of the initial value. (Reprinted from Hayakawa, J., *Useful Life and Its Prediction for Polymer Material* [in Japanese], IPC, Japan, 1989. With permission.)

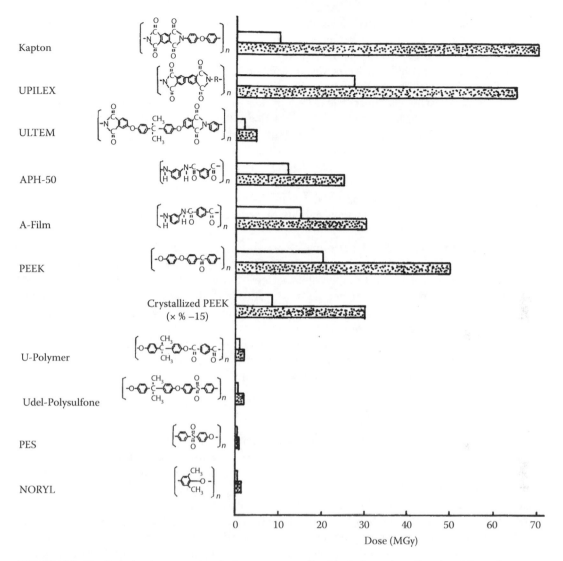

FIGURE 30.28 Relation between chemical structures and residual elongation. (Reprinted from Sasuga, T. and Hagiwara, M., *Polymer*, 28, 1915, 1987. With permission.)

shows the dose at 50% and 20% elongation compared to the initial values for polymers under oxidative and non-oxidative irradiation conditions. When even the aromatic polymers are irradiated under oxygen environment, they are degraded at the dose from one-tenth to one-fifth compared to a non-oxygen environment (Sasuga, 1988). The radiation tolerance for the aromatic polymers is very different based on the chemical structure. Figure 30.28 shows the residual elongation of 50% and 20% of the initial value as a function of dose for each polymer. The radiation stability of the units that make up the main chain is estimated by comparison of the chemical structure and the residual elongation as follows (Sasuga and Hagiwara, 1985):

1. Polyimides such as Kapton and UPILEX show extensive radiation resistivity, indicating that the aromatic imide ring has excellent radiation stability.
2. The two aromatic polyimides, A-Film and APH-50, show relatively high stability for radiation, so the aromatic amide is a highly resistive structure for radiation.

3. Non-crystalline PEEK also shows very high radiation resistivity, indicating that the aromatic ether and aromatic ketone are stable under irradiation.
4. The polyetherimide ULTEM shows radiation resistivity of only several MGy, although it consists of a radiation resistive imide ring and aromatic ether. Similarly, the polyarylate U-Polymer, is spite of containing a stable aromatic ether and ester, shows low radiation resistivity. These two polymers contain bis-phenol A group in the main chain. It is considered that the bis-phenol A group is a low radiation stable structure.
5. Since the two polysulfones and the aromatic sulfones show very low resistivity, the aromatic sulfone group does not have high radiation stability.
6. The modified poly (phenylene oxide) NORYL shows low radiation resistivity. The phenylene oxide unit cannot be considered a weak structure for radiation on the analogy with the case of aromatic ketone and ether. The modification, by blending with polystyrene for example, may weaken the radiation stability of this polymer.

The mechanical properties of high performance polyimide, which has optically transparent characteristics, are studied (Devasahayam et al., 2005). The mechanical properties of four optically transparent polyimides prepared from the dianhydrides ODPA and 6FDA and the diamines ODA and DAB were assessed. The four polyimides were synthesized with the objective of obtaining optimum transparency for space applications where high-energy radiation doses of 15–20 MGy may be expected in geosynchronous orbits over a life span of 20 years. The polyimides were shown to maintain good optical and tensile properties at temperatures up to about 450 K for a dose of 18.5 MGy, but above this temperature the module of the polymers began to deteriorate and there was a small decrease in the transmittance of the exposed polymer films.

30.4.2 GROUND EVALUATION FACILITY

In ground simulation tests, single, sequential, and simultaneous irradiation tests using AO, UV, and radiation sources have been carried out to simulate the space-environmental degradation of polymer materials (Kanazawa et al., 1992; Marco and Remaury, 2004; Miyazaki and Shimamura, 2007; Shimamura and Miyazaki, 2009).

Equipped with AO, vacuum ultraviolet (VUV), and electron beam (EB) sources, the Combined Space Effects Test Facility in JAXA can irradiate these beams simultaneously into an irradiation chamber with a high vacuum of about 10^{-5} Pa (Miyazaki and Shimamura, 2007; Shimamura and Miyazaki, 2009). The standard specifications of the facility are listed in Table 30.1; its schematic view and the visual appearance of the facility are shown in Figure 30.29. The linear motion feedthrough is installed in the irradiation chamber, which is located over the sample holder and which can move parallel to the sample holder's plane. A photodiode or a quartz crystal microbalance (QCM) can be attached on the linear motion feedthrough. The AO generation in the facility is based on the laser detonation phenomenon invented by Physical Science, Inc. The AO beam source is composed of a pulsed valve and a pulsed CO_2 laser at a wavelength of 10.6 μm with a pulse energy of about 10 J. The translational energy of the hyperthermal AO produced by the facility is controlled at approximately 5 eV to replicate the LEO environment. The VUV sources, a total of 48 deuterium lamps (L2581, Hamamatsu Photonics K.K.), are installed in the lamp chamber, which has been purged using high-purity Ar gas. The VUV beam from the lamps is concentrated by the CaF_2 lens and a collection mirror coated with Al and MgF_2. It is then injected into the irradiation chamber through an MgF_2 plate separating the lamp chamber and the irradiation chamber.

The facility named SEMIRAMIS, which was designed at ONERA in France for the evaluation of thermal control coatings in a simulated space environment, is often used to simulate LEO or GEO orbits (Marco and Remaury, 2004). Its main characteristics are cleanliness (very low organic residual partial pressures in vacuum) and reliability (samples in vacuum for several months). Cleanliness is achieved by the use of stainless steel, the use of a majority of metallic seals, and the exclusive

TABLE 30.1
Standard Specifications of the Combined Space Effects Test Facility

Items	Specification
Vacuum	10^{-5} Pa (10^{-3}–10^{-2} Pa during AO irradiation)
Sample	Size: 25 mm diam. × max 3 mm thick
	Exposure area: 20 mm diam.
	Quantity: 16 samples per holder
AO	Method: laser detonation
	Laser: pulsed CO_2 laser
	Laser wavelength: 10.6 µm
	Pulse rate: 12 Hz
	Laser output power: ~10 J/pulse
	Beam velocity: 8 km/s
	Translational energy: 5 eV
	Flux: ~5 × 10^{15} atoms/cm²/s
VUV	Source: 30 W deuterium lamps
	Number of lamps: 48
	Lamp current: 250–350 mA
	Tube voltage: 70–90 V
	Flux: 0.3–0.5 mW/cm²
	(Integral intensity at a wavelength of 120–200 nm)
EB	Accelerating voltage: 100–500 kV
	Source current: 0.1–2.0 mA
Sample temperature	−150°C–80°C

Source: Shimamura, H. and Miyazaki, E., *J. Spacecr. and Rockets*, 46, 241–247, 2009.

use of cryogenic pumping units, each equipped with a gate valve. The reliability is increased by the implementation of each function in a different vacuum chamber separated by its own gate valve and redundant pumping units whose associated gate valve is automatically closed upon failure. The main facility is composed of four vacuum chambers. The lower part is the irradiation chamber containing three irradiation ports, one for protons normal to the samples, and the others for electrons and UV in two symmetrical directions at an incidence angle of 30°. At the upper stage, is a chamber used for measurements and the ultimate safeguard of the vacuum in which the sample holder can be introduced by a vertical translation from the irradiation chamber. On one side of the upper chamber, an optical measurement device storage chamber is mounted, and on one side of the lower chamber there is a contamination device storage chamber. The proton and electron beams are supplied by 2.5 and 2.7 MeV Van de Graaff accelerators, respectively. The protons are obtained from plasma of pure hydrogen and separated from other charged species by a magnetic mass analysis after acceleration. In order to irradiate the samples, the protons are swept across the sample holder surface. In the case of electrons, the beam is diffused through a thin aluminum window. The solar UV generator is based on short arc xenon sources (4000 or 6500 W) whose spectral distribution in UV is close to that of the sun. The source emission is filtered in order to reject all visible and infrared light above 400 nm and reproduce only the UV outer atmosphere emission of the Sun. Multiple solar constants calculated in the range of 200–380 nm can be applied over the sample holder surface (1 Thekaekara solar constant is 9.46 mW/cm²). The number of laboratory irradiation hours multiplied by the number of used solar constants gives the equivalent amount of sun hours (ESH) of space. Figure 30.30 shows the visual appearance of the SEMIRAMIS (ONERA, http://www.onera.fr/desp-en/technical-resources/semiramis.php).

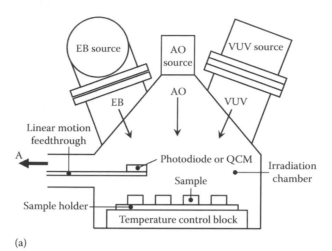

(a)

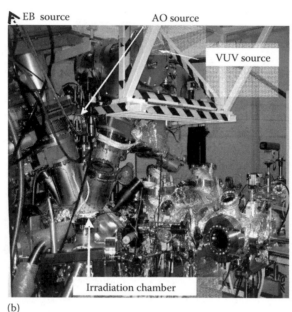

(b)

FIGURE 30.29 Schematic view (a) and visual appearance (b) of the combined space effects test facility. (Reprinted from Shimamura, H. and Miyazaki, E., *J. Spacecr. Rockets*, 46, 241, 2009; Miyazaki, E. and Shimamura, H., Current performance and issues of the combined space effects test facility, JAXA-RM-07-004 [in Japanese], 2007. With permission.)

30.4.3 SPACE MATERIALS EXPOSURE EXPERIMENT

Understanding the degradation of spacecraft materials can be achieved through actual space exposure and ground laboratory studies. Each method has advantages and disadvantages. Opportunities for examining space flown materials, through retrieved flight hardware or dedicated experiments, are rare. Dedicated space experiments are expensive and require long lead times from planning to flight. Differences between the experimental environment, intended mission environment, and synergistic environmental effects require cautious interpretation of results. In cases where it is not possible to retrieve spacecraft hardware, data available from satellite operations such as power output and spacecraft surface temperatures can also be used to assess material performance to some extent. When severe degradation is evident, however, it is often not possible to conclusively identify exploration, broadband communication, global positioning, and so on. Missions for deep space as well as near the sun are also planned by the United States, Europe, and Japan.

Many materials and devices that support missions with high reliability are used for space applications. Since the reliability and durability of such materials and devices directly relate to the lifetime of space applications, it is important to understand the characteristics of such materials and devices before their launch. Ground testing is critical in the analysis and understanding of these materials. In space, many kinds of high-energy radiations exist. For example, protons and heavy ions with high energies come from the sun and from outside of our galaxy. High concentrations of protons and electrons are trapped by the magnetic field of the earth (Van Allen belts). Secondary radiations such as x-rays and neutrons are also produced by the interaction between highly energetic particles and space applications. In addition, highly reactive species such as atomic oxygen exist and attack the surface of space-borne materials. Since the properties of materials and devices are affected by their incidence, irradiation experiments on the ground must be carried out to predict their lifetime in a space environment. In this chapter, radiation effects on semiconductor devices and polymers are described from the point of view of space applications. First, the degradation of the electrical performance of space solar cells is considered, and then, radiation effects on electronic devices are discussed. Finally, polymers for space applications are introduced.

30.2 SPACE SOLAR CELLS

30.2.1 Radiation Degradation of Solar Cells in Space

Three major effects, single-event effects (SEEs), total ionizing dose (TID) effects, and displacement damage effects are observed in semiconductor devices by the incidence of charged particles. SEEs are malfunctions of electronic devices such as large scale integration (LSIs) and power devices. There are both nondestructive and destructive SEE failures. More comprehensive details of SEEs will be described in Section 30.3. The TID effect only affects metal-insulator-semiconductor (MIS) structures, such as metal-oxide-semiconductor (MOS) devices, because of electron-hole pair generated in insulator layers due to irradiation. If the TID effect occurs, the electrical characteristics of MIS devices gradually degrade with increasing dose of radiations, until a destructive malfunction occurs.

When charged particles are introduced into semiconductors, atoms at lattice sites are scattered into non-lattice sites (knock-on effects); as a result, vacancies are created in semiconductors. In fact, many kinds of residual defects such as divacancies, vacancy clusters, and vacancy-impurity complexes are created by irradiation because generated vacancies thermally migrate and transform into more stable defects. Since such defects act as scattering/recombination centers to free carriers, electrical properties of semiconductors are degraded by damage generation due to charged particle irradiation. This degradation is called the displacement damage effect. Similar to the TID effect, the degradation behavior of semiconductors by displacement damage effect worsens with increasing charged particle fluence until the fatal malfunction finally occurs.

Displacement damage is the most dominant effect for space solar cells since the electrical performance is degraded by crystal damage due to proton and electron irradiations. Therefore, electron/proton irradiation experiments using accelerators are conducted on the ground to understand the decrease of power generation of solar cells during missions. For the evaluation on the ground, the electrical performance of solar cells is generally measured at a separate facility before/after proton irradiation experiments (sequential technique). Additional photovoltaic characterization such as spectral response and photo luminescence are also performed because these measurements can be done independently from irradiation experiments. However, it is important to note that solar cells often become radioactive after high-energy (such as 10 MeV) proton irradiation. In this case, it is difficult to handle irradiated solar cells for the measurement of the post-irradiation electrical performance. Thus, we must wait for radioactivity of the solar cells to subside before the evaluation of

FIGURE 30.30 SEMIRAMIS. (Reprinted from ONERA/DESP, Home page http://www.onera.fr/desp-en/technical-resources/semiramis.php. With permission.)

the cause or mechanism of degradation, since comprehensive analysis of materials is not possible. Ground laboratory studies can be used to examine individual environmental effects or a combination of environmental effects. Laboratory tests can be conducted in a timely manner by using accelerated levels for some environmental effects; but due to the difficulties in simulating the space effects with accuracy, complex calibrations and cautious interpretation of the results are required. A combination of space exposures, ground laboratory studies, and computational modeling is most useful for assuring durability of spacecraft materials (Kutz, 2005).

Evaluation of oxygen integration with materials (EOIM)-1, -2, and -3 experiments were conducted in the United States in the 1980s. The materials were exposed for approximately one week for the mission period of the space shuttle (Brinza et al., 1994). The Long Duration Exposure Facility (LDEF) was launched by the space shuttle on April 7, 1984. The retrieval was delayed by 69 months due to the Challenger disaster (January 1986). A lot of valuable data about the influence of space environment was obtained (LDEF Report, 1991, 1992, 1993). The European Retrievable Carrier (Eureca) was launched with the space shuttle in the exposure satellite experiment that Europe carried out in July 1992, and it was retrieved to the ground in July 1993 after about one year (ESA, 1994). An experiment that used the Mir space station of the former Soviet Union was also conducted. The Passive Optical Sample Assembly (POSA) I and II experiments were part of an International Space Station risk-mitigation study placed on the exterior of the MIR Space Station (Zwiener, 1998; Gary Pippin et al., 2000). The National Space Development Agency of Japan (NASDA), the forerunner of JAXA, tested space-material exposure in the STS-85/Evaluation of Space Environment and Effects on Materials (ESEM) mission in 1997 (Imagawa, 2002), and in the Exposed Facility Flyer Unit (EFFU) of the Space Flyer Unit (SFU) in 1996 (Fukatsu, 1997).

The International Space Station (ISS) is a huge manned construction located about 400 km above the Earth. The ISS will be assembled in space step by step, combining components carried by more than 40 launch missions, and will be completed in 2010. The United States has implemented a series of Material International Space Station Experiments (MISSE) on the ISS. MISSE-1 and 2 were attached to the exterior of the ISS during the STS 105 mission (August 10, 2001). Then, MISSE-1 and -2 were retrieved during the STS 114 mission (July 26, 2005) and the candidate materials returned to the experiment investigators for analysis. MISSE-6A and 6B were mounted on the Station's exterior on a truss segment (March 2008). The samples for MISSE-6A and 6B include over 400 new and affordable materials that may be used in advanced reusable launch systems and advanced spacecraft systems including optics, sensors, electronics, power, coatings, structural materials, and protection for the next generation of spacecraft. The samples are installed in holders and placed in experiment trays, called passive experiment containers (PECs). ISS crewmembers will retrieve MISSE-6A and 6B during an extra vehicular activity (EVA) after approximately one year of exposure (deGroh et al., 2001; Banks et al., 2006; Finckenor, 2006; JAXA, 2009; Kleiman et al., 2009). The Micro-Particles Capturer and Space Environment Exposure Device (MPAC&SEED) experiment that used the Service Module (SM) of Russia on the ISS was executed from 2001 to 2005. This was the first experiment that prepared the same samples in three sets and evaluated the relation between material deterioration and exposure period (10 months (0.9 years), 28 months (2.4 years), and 46 months (3.8 years)) (JAXA, 2009). In the SEED experiment, we were able to get data on space-environmental effects on the material, as well as orbital induced environment effect data such as contamination effect data (Baba and Kimoto, 2009; Kimoto et al., 2009; Miyazaki and Yamagata, 2009; Shimamura and Yamagata, 2009; Steagall et al., 2009). Figure 30.31 shows examples of space materials exposure experiments. The Hubble Space Telescope (HST) is one of NASA's most successful and long-lasting science missions. The HST was launched on April 25, 1990 into low earth orbit. The HST was designed to be serviced on orbit to upgrade its scientific capabilities. During the second servicing mission (SM2) in February 1997, astronauts noticed that the multilayer insulation (MLI) covering the telescope was damaged. Large pieces of the outer layer of MLI (Aluminized Teflon fluorinated ethylene propylene (Al-FEP)) were cracked in several

(a)

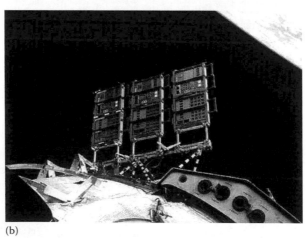

(b)

FIGURE 30.31 Space materials exposure experiment. (a) MISSE 1 being examined during a spacewalk in January 2003. (b) MPAC&SEED on SM. (Reprinted from JAXA, *Proceedings of International Symposium on SM/MPAC&SEED Experiment*, JAXA-SP-08-015E, Tsukuba, Japan, 2009. With permission.)

FIGURE 30.32   Large cracks in HST MLI after 6.8 years in space. (Reprinted from de Groh, K.K. et al., The effect of heating on the degradation of ground laboratory and space irradiated Teflon® FEP, NASA/TM-2002-211704, 2002, 2 (Fig. 1). With permission.)

locations around the telescope. A piece of curled-up Al-FEP was retrieved by the astronauts and was found to be severely embrittled, as witnessed by ground testing. Figure 30.32 shows large cracks on the outer layer of the solar facing MLI on the HST as observed during SM2, after 6.8 years in space (de Groh et al., 2002).

Extensive testing was conducted on the retrieved HST MLI, combined with ground-based testing, which revealed that embrittlement of FEP on HST is caused by radiation exposure (electron and proton radiation, with contributions from solar flare x-rays and ultraviolet (UV) radiation) combined with thermal effects (deGroh, 2001, 2002). But it has been shown that accelerated laboratory testing often does not simulate accurately the degradation caused by combined on-orbit environmental exposure (Townsend et al., 1998).

REFERENCES

Amekura, H., Kishimoto, N., Kono, K., and Kondo, A. 2000. Persistent excited conductivity and the threshold fluence in a-Si:H under 17 MeV proton irradiation. *J. Non-Cryst. Solids* 266–269: 444–449.

Baba, N. and Kimoto, Y. 2009. Contamination growth observed on the micro-particles capturer and space environment exposure device. *J. Spacecr. Rockets* 46: 33–38.

Banks, B. A., deGroh, K. K., and Miller, S. K. 2006. MISSE scattered atomic oxygen characterization experiment. NASA/TM-2006-214355, NASA, Cleveland, OH.

Benedetto, J., Eaton, P., Avery, K., Mavis, D., Gadlage, M., Turflinger, T., Dodd, P. E., and Vizkelethy, G. 2004. Heavy ion-induced digital single-event transients in deep submicron processes, *IEEE Trans. Nucl. Sci.* 51: 3365–3368.

Breese, M. B. H. 1993. A theory of ion beam induced charge collection. *J. Appl. Phys.* 74: 3789–3799.

Brinza, D. E., Chung, S. Y., Minton, T. K., and Liang, R. H. 1994. The NASA/JPL Evaluation of Oxygen Interactions with Materials-3 (EOIM-3). NASA, CR-198865, NASA, Pasadena, CA.

Buchner, S. P. and Baze, M. P. 2001. *IEEE Nuclear Space Radiation Effects Conference Short Course Notebook*, Section V, Vancouver, Canada.

Buchner, S., Baze, M., Brown, D., McMorrow, D., and Melinger, J. 1997. Comparison of error rates in combinational and sequential logic. *IEEE Trans. Nucl. Sci.* 44: 2209–2216.

de Groh, K. K., Banks, B. A., Hammerstrom, A. M., Youngstrom, E. E., Kaminski, C., Marx, L. M., Fine, E. S., Gummow, J. D., and Wright, D., 2001. MISSE PEACE polymers: An international space station environmental exposure experiment. NASA/TM-2001-211311, NASA, Cleveland, OH.

de Groh, K. K. and Martin, M. 2002. The effect of heating on the degradation of ground laboratory and space irradiated Teflon® FEP. NASA/TM-2002-211704, NASA, Cleveland, OH.

Devasahayam, D., Hill, D. J. T., and Connell, J. W. 2005. Effect of electron beam radiolysis on mechanical properties of high performance polyimides. A comparative study of transparent polymer films. *High Perform. Polym.* 17: 547–559.

Dharmarasu, N., Yamaguchi, M., Bourgoin, J., Takamoto, T., Ohshima, T., Itoh, H., Imaizumi, M., and Matsuda, S. 2002. Majority and minority carrier deep level traps in proton-irradiated n^+/p-InGaP space solar cells. *Appl. Phys. Lett.* 81: 64–66.

Dimroth, F., Baur, C., Bett, A. W., Meusel, M., and Strobl, G. 2005. 3–6 Junction photovoltaic cells for space and terrestrial concentrator applications. *Proceedings of 31st Photovoltaic Specialists Conference*, Lake Buena Vista, FL, pp. 525–529.

Dodd, P. E., Musseau, O., Shaneyfelt, M. R., Sexton, F. W., D'hose, C., Hash, G. L., Martinez, M., Loemker, R. A., Leray, J.-L., and Winokur, P. S. 1998. Impact of ion energy on single-event upset. *IEEE Trans. Nucl. Sci.* 45: 2483–2491.

ESA, 1994. Eureca the European Retrievable Carrier Technical Report. ESA WPP-069, ESA, Noordwijk, the Netherlands.

Fageeha, O., Howard, J., and Block, R. C. 1994. Distribution of radial energy deposition around the track of energetic charged particles in silicon. *J. Appl. Phys.* 75: 2317–2321.

Ferlet-Cavrois, V., Vizkelethy, G., Paillet, P., Torres, A., Schwank, J. R., Shaneyfelt, M. R., Baggio, J., du Port de Pontcharra, J., and Tosti, L. 2004. Charge enhancement effect in NMOS bulk transistors induced by heavy ion irradiation: Comparison with SOI. *IEEE Trans. Nucl. Sci.* 51: 3255–3262.

Ferlet-Cavrois, V., Paillet, P., Gaillardin, M., Lambert, D., Baggio, J., Schwank, J. R., Vizkelethy, G., Shaneyfelt, M. R., Hirose, K., Blackmore, E. W., Faynot, O., Jahan, C., and Tosti, L. 2006. Statistical analysis of the charge collected in SOI and bulk devices under heavy ion and proton irradiation—Implications for digital SETs. *IEEE Trans. Nucl. Sci.* 53: 3242–3252.

Finckenor, M. M. 2006. The Materials on International Space Station Experiment (MISSE): First results from MSFC investigations. AIAA 2006-472, NASA, Huntsville, AL.

Fukatsu, T. 1997. Postflight analysis of the exposed materials on EFFU. *Proceedings of the 7th ISMSE* 1997, Toulouse, France, pp. 287–292.

Gadlage, M. J., Schrimpf, R. D., Benedetto, J. M. et al. 2004. Single event transient pulse widths in digital microcircuits. *IEEE Trans. Nucl. Sci.* 51: 3285–3290.

Gary Pippin, H., Woll, S. L. B., Loebs, V. A., and Bohnhoff-Hlavacek, G., 2000. Contamination effects on the passive optical sample assembly experiments. *J. Spacecr. Rockets* 37: 567–572.

Geisz, J. F., Friedman, D. J., Ward, J. S., Duda, A., Olavarria, W. J., Moriarty, T. E., Kiehl, J. T., Romero, M. J., Norman, A. G., and Jones, K. M. 2008. 40.8% Efficient inverted triple-junction solar cell with two independently metamorphic junctions. *Appl. Phys. Lett.* 93: 123505-1–123505-3.

Granata, J. E., Sahlstrom, T. D., Hausgen, P., Messenger, S. R., Walters, R. J., and Lorentzen, J. R. 2005. Thin-film photovoltaic radiation testing and modeling for a MEO orbit. *Proceedings of the 31st Photovoltaic Specialists Conference*, Lake Buena Vista, FL, pp. 607–610.

Harris, R. D., Imaizumi, M., Walters, R. J., Lorentzen, J. R., Messenger, S. R., Tischler, J. G., Ohshima, T., Sato, S., Sharps, P. R., and Fatemi, N. S. 2008. In situ irradiation and measurement of triple junction solar cells at low intensity, low temperature (LILT) conditions. *IEEE Trans. Nucl. Sci.* 55: 3502–3507.

Hayakawa, J. 1989 *Useful Life and Its Prediction for Polymer Material*. Japan: IPC [in Japanese].

Hirao, T., Onoda, S., Oikawa, M., Satoh, T., Kamiya, T., and Ohshima, T. 2009. Transient current mapping obtained from silicon photodiodes using focused ion microbeams with several hundreds of MeV. *Nucl. Instrum. Methods B* 267: 2216–2218.

Holmes-Siedle, A. and Adams, L. 2000. *Handbook of Radiation Effects*. New York: Oxford University Press, p. 368.

Imagawa, K. 2002. Evaluation and analysis of parts and materials installed on MFD-ESEM. NASDA-TMR-000011, NASDA, Japan.

Imaizumi, M., Sumita, T., Kawakita, S., Aoyama, K., Anzawa, O., Aburaya, T., Hisamatsu, T., and Matsuda, S. 2005. Results of flight demonstration of terrestrial solar cells in space. *Prog. Photovolt: Res. Appl.* 13: 93–102.

Inoue, H., Okamoto, H., and Hiraoka, Y. 1987. Effect of the chemical structure of acid dianhydride in the skeleton on the thermal property and radiation resistance of polyimide. *Int. J. Radiat. Appl. Instrum. Part C: Radiat. Phys. Chem.* 29: 283–288.

JAXA, 2009. *Proceedings of International Symposium on "SM/MPAC&SEED Experiment"*. JAXA-SP-08-015E, Tsukuba, Japan.

Kanazawa, T., Haruyama, Y., and Yotsumoto, K. 1992. JAERI-M 1992-062. Takasaki, Japan: Japan Atomic Energy Research Institute Publication.

Katsu, T., Shimada, K., Washio, H., Tonomura, Y., Hisamatsu, T., Kamimura, K., Saga, T., Matsutani, T., Suzuki, A., Kawasaki, O., Yamamoto, Y., and Matsuda, S. 1994. Development of high efficiency silicon space solar cells. *Proceedings of 1994 IEEE 1st World Conference on Photovoltaic Energy Conversion (WCPEC), Hawai, US, Conference Record of the 24th IEEE Photovoltaic Specialist Conference*, Waikoloa, HI, pp. 2133–2136.

Kawakita, S., Imaizumi, M., Yamaguchi, M., Kushiya, K., Ohshima, T., Itoh, H., and Matsuda, S. 2002. Annealing enhancement effect by light illumination on proton irradiated Cu(In,Ga)Se$_2$ thin-film solar cells. *Jpn. J. Appl. Phys.* 41: L797–L799.

Kimoto, Y., Yano, K., Ishizawa, J., Miyazaki, E., and Yamagata, I. 2009. Passive space-environment-effect measurement on the international space station. *J. Spacecr. Rockets* 46: 22–27.

Kleiman, J. I., Iskanderova, Z., Issoupov, V., Grigorevskiy, A. V., Kiseleva, L. V., Finckenor, M., Naumov, S. F., Sokolova, S. P., and Kurilenok, A. O. 2009. The Results of Ground-based and In-flight Testing of Charge-dissipative and Conducting EKOM Thermal Control Paints, *Proceedings of the 9th International Conference on "Protection of Materials and Structures from Space Environment" AIP Conference Proceedings*, Vol. 1087, pp. 610-620, Toronto, Canada.

Kobayashi, D., Hirose, K., Makino, T., Ikeda, H., and Saito, H. 2007a. Feasibility study of a table-based SET-pulse estimation in logic cells from heavy-ion-induced transient currents measured in a single MOSFET. *IEEE Trans. Nucl. Sci.* 54: 2347–2354.

Kobayashi, D., Saito, H., and Hirose, K. 2007b. Estimation of single event transient voltage pulses in VLSI circuits from heavy-ion-induced transient currents measured in a single MOSFET. *IEEE Trans. Nucl. Sci.* 54: 1037–1041.

Kobetich, E. J. and Katz, R. 1968. Energy deposition by electron beams and δ rays. *Phys. Rev.* 170: 391–396.

Kurashima, S., Miyawaki, N., Okumura, S., Oikawa, M., Yoshida, K., Kamiya, T., Fukuda, M., Satoh, T., Nara, T., Agematsu, T., Ishibori, I., Yokota, W., and Nakamura, Y. 2007. Improvement in beam quality of the JAEA AVF cyclotron for focusing heavy-ion beams with energies of hundreds of MeV. *Nucl. Instrum. Methods B* 260: 65–70.

Kutz, M. 2005. *Handbook of Environmental Degradation of Material*. Norwich, NY: William Andrew Publishing, p. 465.

Laird, J. S., Hirao, T., Mori, H., Onoda, S., Kamiya, T., and Itoh, H. 2001. Development of a new data collection system and chamber for microbeam and laser investigations of single event phenomena. *Nucl. Instrum. Methods B* 181: 87–94.

Laird, J. S., Toshio, H., Shinobu, O., and Hisayoshi, I., 2005. High-injection carrier dynamics generated by MeV heavy ions impacting high-speed photodetectors. *J. Appl. Phys.* 98: 013530-1–013530-14.

LDEF 69 Months in Space. First Post-Retrieval Symposium, 1991, NASA-CP-3134, Kissimmee, FL.

LDEF 69 Months in Space. Second Post-Retrieval Symposium, 1992, NASA-CP-3194, San Diego, CA.

LDEF 69 Months in Space. Third Post-Retrieval Symposium, 1993, NASA-CP-3275, Williamsburg, VA.

Marco, J. and Remaury, S. 2004. Evaluation of thermal control coatings degradation in simulated geo-space environment. *High Perform. Polym.* 16: 177–196.

Matsuda, S., Yamamoto, Y., and Kawasaki, O. 1994. GaAs space solar cells. *Optoelectron.: Devices Technol.* 9: 561–576.

Matsuura, H., Iwata, H., Kagamihara, S., Ishihara, R., Komeda, M., Imai, H., Kikuta, M., Inoue, Y., Hisamatsu, T., Kawakita, S., Ohshima, T., and Itoh, H., 2006. Si substrate suitable for radiation resistant space solar cells. *Jpn. J. Appl. Phys.* 45: 2648–2655.

Megusar, J. 1997. Low temperature fast-neutron and gamma irradiation of Kapton® polyimide films. *J. Nucl. Mater.* 245: 185–190.

Messenger, S. R., Summers, G. P., Burke, E. A., Walters, R. J., and Xapsos, M. A. 2001. Modeling solar cell degradation in space: A comparison of the NRL displacement damage dose and the JPL equivalent fluence approaches. *Prog. Photovolt: Res. Appl.* 9: 103–121.

Miyazaki, E. and Shimamura, H. 2007. Current performance and issues of the combined space effects test facility. JAXA-RM-07-004 [in Japanese].

Miyazaki, E. and Yamagata, I. 2009. Results of space-environment exposure of the flexible optical solar reflector. *J. Spacecr. Rockets* 46: 28–32.

Nashiyama, I., Hirao, T., Kamiya, T., Yutoh, H., Nishijima, T., and Sekiguti, H. 1993. Single-event current transients induced by high energy ion microbeams. *IEEE Trans. Nucl. Sci.* 40: 1935–1940.

O'Donnell, J. H. and Sangster, D. S. 1970. *Principals of Radiation Chemistry.* London, U.K.: Edword Armold.

Ohshima, T., Morita, Y., Nashiyama, I., Kawasaki, O., Hisamatsu, T., Nakao, T., Wakow, Y., and Matsuda, S. 1996. Mechanism of anomalous degradation of silicon solar cells subjected to high-fluence irradiation. *IEEE Trans. Nucl. Sci.* 43: 2990–2997.

Ohshima, T., Sumita, T., Imaizumi, M., Kawakita, S., Shimazaki, K., Kuwajima, S., Ohi, A., and Itoh, H., 2005. Evaluation of the electrical characteristics of III-V compounds solar cells irradiated with proton at low temperature. *Proceedings of 31st IEEE Photovoltaic Specialists Conference*, Lake Buena Vista, FL, pp. 806–809.

Oikawa, M., Kamiya, T., Fukuda, M., Okumura, S., Inoue, H., Masuno, S., Umemiya, S., Oshiyama, Y., and Taira, Y. 2003. Design of a focusing high-energy heavy ion microbeam system at the JAERI AVF cyclotron. *Nucl. Instrum. Methods B* 210: 54–58.

Oikawa, M., Satoh, T., Sakai, T., Miyawaki, N., Kashiwagi, H., Kurashima, S., Okumura, S., Fukuda, M., Yokota, W., and Kamiya, T. 2007. Focusing high-energy heavy ion microbeam system at the JAEA AVF cyclotron. *Nucl. Instrum. Methods B* 260: 85–90.

ONERA home page http://www.onera.fr/desp-en/technical-resources/semiramis.php

Onoda, S., Hirao, T., Laird, J. S., Mishima, K., Kawano, K., and Itoh, H. 2006. Transient currents generated by heavy ions with hundreds of MeV. *IEEE Trans. Nucl. Sci.* 53: 3731–3737.

Petersen, E. 2008. *IEEE Nuclear Space Radiation Effects Conference Short Course Notebook,* Section III, Tucson, AZ.

Sasuga, T. 1988. Oxidative irradiation effects on several aromatic polyimides. *Polymer* 29: 1562–1568.

Sasuga, T. and Hagiwara, M. 1985. Degradation in tensile properties of aromatic polymers by electron beam irradiation. *Polymer* 26: 1039–1045.

Sasuga, T. and Hagiwara, M. 1987. Radiation deterioration of several aromatic polymers under oxidative conditions. *Polymer* 28: 1915–1921.

Sato, S., Miyamoto, H., Imaizumi, M., Shimazaki, K., Morioka, C., Kawano, K., and Ohshima, T. 2009a. Degradation modeling of InGaP/GaAs/Ge triple-junction solar cells irradiated with various-energy protons. *Sol. Energy Mater. Sol. Cells* 93: 768–773.

Sato, S., Ohshima, T., and Imaizumi, M. 2009b. Modeling of degradation behavior of InGaP/GaAs/Ge triple-junction space solar cell exposed to charged particles. *J. Appl. Phys.* 105: 044504-1–044504-6.

Schone, H., Walsh, D. S., Sexton, F. W. et al. 1998. Time-resolved ion beam induced charge collection (TRIBICC) in micro-electronics. *IEEE Trans. Nucl. Sci.* 45: 2544–2549.

Shimamura, H. and Yamagata, I. 2009. Degradation of mechanical properties of polyimide film exposed to space environment. *J. Spacecr. Rockets* 46: 15–21.

Shimamura, H. and Miyazaki, E. 2009. Investigations into synergistic effects of atomic oxygen and vacuum ultraviolet. *J. Spacecr. Rockets* 46: 241–247.

Soga, T., Chandrasekaran, N., Imaizumi, M., Inuzuka, Y., Taguchi, H., Jimbo, T., and Matsuda, S. 2003. High-radiation resistance of GaAs solar cell on Si substrate following 1 MeV electron irradiation. *Jpn. J. Appl. Phys.* 42: L1054–L1056.

Steagall, C. A., Smith, K., Soares, C., Mikatarian, R., and Baba, N. 2009. Induced-contamination predictions for the micro-particle capturer and space environment exposure device. *J. Spacecr. Rockets* 46: 39–44.

Sumita, T., Imaizumi, M., Matsuda, S., Ohshima, T., Ohi, A., and Itoh, H. 2003. Proton radiation analysis of multi-junction space solar cells. *Nucl. Instrum. Methods B* 206: 448–451.

Summers, G. P., Burke, E. A., Dale, C. J., Wolicki, E. A., Marshall, P. W., and Gehlhausen, M. A. 1987. Collection of particle-induced displacement damage in silicon. *IEEE Trans. Nucl. Sci.* NS34: 1134–1139.

Tada, H. Y., Carter, Jr., J. R., Anspaugh, B. E., and Dowing, R. G. 1982. *Solar Cell Radiation Handbook,* 3rd edn. pp. 82–96, Pasadena, CA: JPL Publications.

Takamoto, T., Agui, T., Washio, H., Takahashi, N., Nakamura, K., Anzawa, O., Kaneiwa, M., Kamimura, K., Okamoto, K., and Yamaguchi, M. 2005. Future development of InGaP/(In)GaAs based multijunction solar cells. *Proceedings of 31st Photovoltaic Specialists Conference*, Lake Buena Vista, FL, pp. 519–524.

Tavlet, M., Fontaine, A., and Schonbacher, H. 1998. Compilation of radiation damage test data, Part II, 2nd edn: Thermoset and thermoplastic regins; composite materials. Report No. CERN-98-01. Geneva Switzerland: CERN.

Townsend, J. A., Powers, C., Viens, M., Ayres-Treusdell, M., and Munoz, B. 1998. Degradation of teflon FEP following charged particle radiation and rapid thermal cycling. *Proceedings of Space Simulation Conference*, NASA CR-1998-208598, Annapolis, MD, pp. 201–209.

Uemura, T., Tosaka, Y., and Satoh, S. 2006. Neutron-induced soft-error simulation technology for logic circuits. *Jpn. J. Appl. Phys.* 45: 3256–3259.

Vizkelethy, G., Onoda, S., Hirao, T., Ohshima, T., and Kamiya, T. 2009. Time resolved ion beam induced current measurements on MOS capacitors using a cyclotron microbeam. *Nucl. Instrum. Methods B* 267: 2185–2188.

Wagner, R. S., Bradley, J. M., Bordes, N., Maggiore, C. J., Sinha, D. N., and Hammond, R. B. 1987. Transient measurements of ultrafast charge collection in semiconductor diodes. *IEEE Trans. Nucl. Sci.* 34: 1240–1245.

Walters, R. J., Messenger, S. R., Summers, G., Romero, M., Al-Jassim, M., Araujo, D., and Garcia, R. 2001. Radiation response of n-type base InP solar cells. *J. Appl. Phys.* 90: 3558–3565.

Weatherford, T. 2002. *IEEE Nuclear Space Radiation Effects Conference Short Course Notebook, Section IV*, Phoenix, AZ.

Weinberg, I. 1991. Radiation damage in InP solar cells. *Sol. Cells* 31: 331–348.

Xapsos, M. 2006. *IEEE Nuclear Space Radiation Effects Conference Short Course Notebook*, Section II, Ponte Vedra Beach, FL.

Yamaguchi, M. 2001. Radiation-resistant solar cells for space use. *Sol. Energy Mater. Sol. Cells* 68: 31–53.

Yamaguchi, M., Uemura, C., and Yamamoto, A. 1984. Radiation damage in InP single crystals and solar cells. *J. Appl. Phys.* 55: 1429–1436.

Yamaguchi, M., Itoh, Y., Ando, K., and Yamamoto, A. 1986. Room-temperature annealing effects on radiation-induced defects in InP crystals and solar cells. *Jpn. J. Appl. Phys.* 25: 1650–1656.

Yamaguchi, M., Takamoto, T., Ikeda, E., Kurita, H., Ohmori, M., Ando, K., and Vargas-Aburto, C. 1995. Radiation resistance of InP-related materials. *Jpn. J. Appl. Phys.* 34: 6222–6225.

Yamaguchi, M., Tayler, S., Matsuda, S., and Kawasaki, O. 1996. Mechanism for the anomalous degradation of Si solar cells induced by high fluence 1 MeV electron irradiation. *Appl. Phys. Lett.* 68: 3141–3143.

Yamaguchi, M., T. Okuda, T,, and Tayler, S. 1997. Minority-carrier injection-enhanced annealing of radiation damage to InGaP solar cells. *Appl. Phys. Lett.* 170: 2180–2182.

Ziegler, J. F., Biersack, J. P., and Ziegler, M. D. 2008. *SRIM–The Stopping and Range of Ions in Matter*. Chester, MD: SRIM Co.

Zoutendyk, J. A., Schwartz, H. R., and Nevill, L. R. 1988. Lateral charge transport from heavy-ion tracks in integrated circuit chips. *IEEE Trans. Nucl. Sci.* 35: 1644–1647.

Zoutendyk, J. A., Edmonds, L. D., and Smith, L. S. 1989. Characterization of multiple-bit errors from single-ion tracks in integrated circuits. *IEEE Trans. Nucl. Sci.* 36: 2267–2274.

Zwiener, J. M., 1998. Contamination observed on the passive optical sample assembly (POSA)-I experiment. *SPIE Proc.* 3427: 186–195.

31 Applications of Rare Gas Liquids to Radiation Detectors

Satoshi Suzuki
Waseda University
Tokyo, Japan

Akira Hitachi
Kochi Medical School
Nankoku, Japan

CONTENTS

31.1 INTRODUCTION

Rare gas liquids (RGLs) have excellent properties such as those required for radiation detectors, for instance, high electron mobility, low *W*-value, high photon production rate, high density, high Z number with no permanent radiation damage, and practically no limit for the size. In addition, ultrahigh purification is attainable in RGLs. Furthermore, electron and photon multiplications are possible in liquid xenon (LXe).

Section 31.2 discusses basic radiation properties of RGLs for detector media, which are considerably different from those in water and in molecular liquids (Mozumder and Hatano, 2004). Evidence observed in simple RGLs may help study the effects of ionizing radiation in matter. Next, new detector research and development for ionization and scintillation detectors and current topics are introduced. The following text discusses RGL detectors designed for applications in accelerator physics, non-accelerator physics, and medical imaging. These include various phenomena, such as neutrino physics, "$\mu \rightarrow e\gamma$" decay, dark matter search and, for medical imaging applications, positron emission tomography (PET). Some calorimeters, the time projection chamber (TPC), and scintillation detectors are also introduced.

We have introduced only a few topics, which we found to be unique or pioneering, because of space limitations. With regard to numerous experimental proposals using RGL detectors, especially for rare event searches, such as dark matter or neutrinoless double-beta decay, the recent fast evolution of RGL detectors and related technologies deserve special mention. These are the particle detection technique by simultaneous observation of ionization and scintillation signals or a waveform analysis for scintillation signals, VUV photon detection at low temperature, purification techniques, etc. Furthermore, large detection technologies have been developed, the ICARUS* liquid argon (LAr) detector and the 20t XMASS† LXe detector, their estimated volume approaching the world annual production volume of Ar and Xe.

The development of the detectors also produced interesting results in the fields of radiation physics and chemistry as a by-product. Dark matter searches, for instance, encouraged new, elaborated measurements in the interaction of very-low-energy ions with condensed media. The atomic collisions in the estimated energy range have been studied almost half a century ago. However, radiation effects have not been discussed in detail. Slow energy collision is also theoretically quite difficult to deal with. For example, the Thomas–Fermi model, a major method of dealing with slow ion collisions, becomes uncertain in Xe–Xe collisions below 10 keV (Lindhard et al., 1963). New results that would be obtained with dark matter searches will give materials to develop a new collision theory in the extreme low-energy region.

Here, we refer to RGL as Ar, Kr, and Xe, unless otherwise stated. Ionization and scintillation properties of He and Ne are different from those of Ar, Kr, and Xe, and are not discussed here.

31.2 BASIC PROPERTIES OF RARE GAS LIQUIDS FOR DETECTOR MEDIA

The energy levels in solid argon are shown schematically in Figure 31.1, and those for RGLs are basically the same. One of the remarkable features in the condensed phase is the existence of the conduction band. The bandgap energies for LAr and LXe are 14.3 and 9.28 eV, respectively, and are considerably lower than the ionization potentials of 15.75 and 12.13 eV, respectively, in the gas phase. Further, the excitonic levels appear instead of the excited states of atoms (Baldini, 1962; Beaglehole, 1965; Steinberger and Schnepp, 1967; Asaf and Steinberger, 1971; Laporte et al., 1980), which show the exciton mass, m_{ex}, to be 1–5 m_e, where m_e is the electron mass.

The drift velocities, v_d, for electrons in LAr and LXe, shown in Figure 31.2, in the condensed phase, are much higher than the corresponding values in the gas phase, normalized by E/N, because of the formation of the conduction band. Here, E is the electric field and N is the number density of atoms. The electron mobilities in LAr, liquid krypton (LKr), and LXe are discussed elsewhere (Holroyd, 2004; Wojcik et al., 2004), and therefore, only briefly mentioned here. Review articles for RGLs are also found in Schwentner et al. (1985), Christophorou (1988), Holroyd and Schmidt (1989), and Lopes and Chepel (2003). The properties of LAr and LXe as detector media are listed in Table 31.1.

The ion drift velocity is not so high as that for electrons, but ~2–3 times larger than the value estimated from the self-diffusion coefficient, D_{self} (Hilt et al., 1994). The drift velocity is also observed to be high in the solid (Le Comber et al., 1975). The positive charge carrier is not a free hole. The ion R^+ produced by the ionizing particles self-traps immediately to form R_2^+. The binding energy, E_b, for R_2^+ is only several tens of meV (Le Comber et al., 1975), which is quite small compared with corresponding values of about 1 eV in the gas phase (Mulliken, 1964; Kuo and Keto, 1983). The reduction in E_b

---

* Imaging Cosmic and Rare Underground Signals, http://icarus.lngs.infn.it/

† Xenon MASSive detector for solar neutrino (pp/7Be), Xenon neutrino MASS detector (ββ decay), Xenon detector for Weakly Interacting MASSive Particles (DM search), http://www-sk.icrr.u-tokyo.ac.jp/xmass/index-e.html

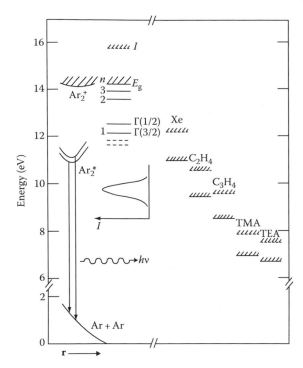

FIGURE 31.1 Schematic diagram for condensed Ar. The inset is the emission spectrum of Ar_2^*. Ionization potentials in the gas phase (upper) and in LAr (lower) for various molecules are also shown.

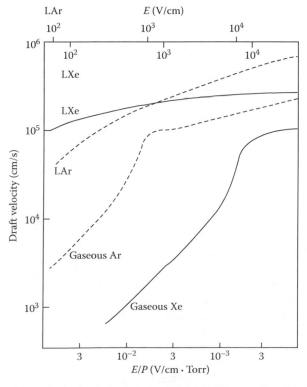

FIGURE 31.2 Electron drift velocity in liquid and gaseous Ar and Xe as a function of reduced electric field (Doke, 1981; Miller et al., 1968; Pack et al., 1962; Yoshino et al., 1976).

TABLE 31.1
Properties of LAr and LXe as Detector Media

		LAr	LXe	Comment
E_g	eV	14.3	9.28	
W	eV	23.6	15.6	
W_{ph}	eV	19.5	14	rel. H.I.
N_{ex}/N_i		0.21	0.06, 0.13	
μ_e	cm^2/V/s	475	1900	At T.P.
μ_h	10^{-3} cm^2/V/s		3.5	At T.P.
$D_\perp$	cm^2/s	~20	~80	At 1 kV/cm
λ	nm	127	178	
τ_S	ns	7	4.3	
τ_T	ns	1600	22	
τ_{rec}	ns		45	Apparent
τ_{th}	ns	0.9	6.5	
l_{ph}	cm	66	29–50	
T.P.	K	84	161	
D_{self}	cm^2/s	2.43×10^{-5}		
ρ	g/cm^3	1.40	2.94	At B.P.
n			1.54–1.70	
ε_0		1.52	1.94	
Radiation length[a]	cm	14	2.8	At 1 GeV
Moliere radius[b]	cm	10	5.7	

Note: Doke (1981), Hitachi et al., (2005), Jortner (1965), Lopes and Chepel (2003), Rabinovich et al. (1988), Rahman (1964), Seidel (2002), Sinnock (1980).

[a] Bremsstrahlung and electron pair production are the dominant processes for high-energy electrons and photons. The radiation length, X_0, is defined as the length that the high-energy electron travels in matter losing its energy by bremsstrahlung to 1/e. A value for 1 GeV is usually taken since the radiation length, X_0, becomes almost independent of energy above 1 GeV.

[b] Moliere radius is given by the radiation length, X_0, and the atomic number, Z, as

$$R_M = 0.0265 X_0 (Z + 1.2).$$

R_M is a good scaling constant of a material in describing the transverse dimension of electromagnetic showers.

may be attributed to the formation of the conduction band. The small polaron (R_2^+, the carrier self-trapped in its polarization field) moves from one site to another by a tunneling or a hopping process.

The thermalization time, τ_{th}, for the excess electron is quite large in RGLs, because of the lack of an effective energy loss mechanism. τ_{th} in LAr and LXe have been measured using the microwave technique, and the reported values are 0.9 and 6.5 ns, respectively (Sowada et al., 1982). Consequently, the thermalization length, l_{th}, is also quite long compared with the Onsager radius (the distance from the positive ion at which the Coulomb energy is equal to the thermal energy) of about 40–110 nm. The ions produced are no longer regarded as isolated even for the minimum ionizing particles. The distributions of the electrons and ions are quite different following electron thermalization. Also, the mobilities of the electron and of the positive charge are orders of magnitude different, as shown in Table 31.1. The basic recombination theories of Jaffe (1913), Onsager (Onsager, 1938; Hung et al., 1991), and Kramers (1952) do not apply to the condensed rare gases. The Mozumder model (Mozumder, 1995) deals with this problem; however, the direction of the field is restricted to be parallel to the electron track. The Thomas model (Thomas et al., 1988) conforms the charge yield fairly well with several parameters; however, the physical picture is not clear.

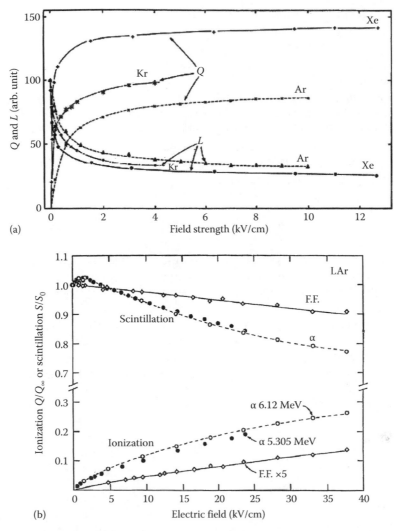

FIGURE 31.3 (a) Ionization, Q, and scintillation, L, yields for electrons as a function of the applied electric field in LAr, LKr, and LXe. (From Kubota, S. et al., *Phys. Rev. B.*, 20, 3486, 1979. With permission.) (b) Ionization, Q, and scintillation, S, yields for α-particles and fission fragments in LAr as a function of the applied electric field (Hitachi et al., 1987).

Both ionization and scintillation can be observed in RGLs. The ionization, Q, and scintillation, S, yields observed for electrons, α-particles, and fission fragments are shown in Figure 31.3, as a function of the applied electric field. The W-value, that is, the average energy required to produce an ion pair, is smaller in the condensed phase than in the gas phase. The energy balance equation (Platzman, 1961) expresses the W-value as

$$W = \frac{E_0}{N_i} = E_i + \left(\frac{N_{ex}}{N_i}\right)E_{ex} + \varepsilon \tag{31.1}$$

where
E_0 is the energy of the ionizing particle
E_i and E_{ex} are the average energies expended for ionization and excitation, respectively, producing N_i ionizations and N_{ex} excitations
ε is the average energy spent as the kinetic energy of sub-excitation electrons

The N_{ex}/N_i ratio in LAr is estimated to be 0.21 using the optical approximation and agrees well with a measured value of 0.19 observed in the Penning ionization (Kubota et al., 1976). However, a value of 0.06 estimated in LXe is much smaller than a value of 0.13 obtained from the variation of the charge and scintillation relation with the applied field. W-values measured are 23.6 eV (Miyajima et al., 1974) and 15.6 eV (Takahashi et al., 1975) for LAr and LXe, respectively; these are smaller than the corresponding values of 26.4 and 22.0 eV, respectively, in the gas phase.

The molecular ion R_2^+ recombines with a thermalized electron, producing R_2^{**}. R_2^{**} relaxes non-radiatively to the lowest R_2^* (self-trapped exciton or dimer) levels via exciton levels. The ionizing particles also produce excitons, R^* and R^{**}, and these again relax non-radiatively to R_2^*. The origin of the vuv scintillation is almost the same as gas at high pressure, the transition from self-trapped excitons ($^1\Sigma_u^+$ and $^3\Sigma_u^+$) to the repulsive ground state, $^1\Sigma_g^+$, and gives a broad, structureless vuv band (Jortner et al., 1965), as shown in Figure 31.1.

$$R_2^* \rightarrow R + R + h\nu \tag{31.2}$$

The self-trapping time of excitons is quite short, of the order of 1 ps (Martin, 1971). However, this is long enough for the free excitons to play an important role in reactions such as the Penning ionization (Kubota et al., 1976; Hitachi, 1984) and in high-excitation-density quenching (Hitachi et al., 1992b). Thermal excitons can be very fast, since the exciton mass is only a few times the electron mass. On the other hand, the energy transfers from R_2^* and R_2^+ follow usual diffusion-controlled mechanisms (Hirschfelder et al., 1954; Yokota and Tanimoto, 1967; Hitachi, 1984).

The scintillation decay shape in RGLs depends on the kind of ionizing radiation, as shown for LXe in Figure 31.4. The electron excitation gives an apparent decay of 45 ns (Hitachi et al., 1983).

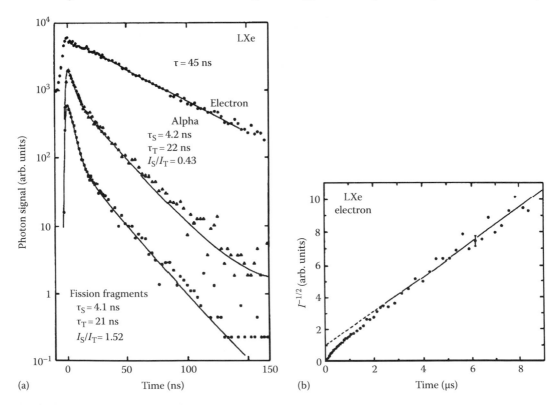

FIGURE 31.4 (a) Typical decay curves for the scintillation from LXe excited by electrons, α-particles, and fission fragments. (b) Variation of $I^{-1/2}$, where I is the scintillation intensity, as a function of time due to electron excitation for LXe (Hitachi et al., 1983).

It has a slow, non-exponential tail representing homogeneous recombination, as shown in Figure 31.4b. These components disappear under the electric field and the decay shows two exponentials, as those for α-particle and fission fragment excitation; therefore, the slow decay has been attributed to recombination (Kubota et al., 1979, 1980). High-LET (linear energy transfer) particles, α-particles and fission fragments, show two decay components of 4 and 22 ns, corresponding to the singlet and triplet lifetimes, respectively. LAr shows basically only two exponential decay components of 7 ns and 1.6 μs, representing the singlet and the triplet, respectively, due to electron, α-particle, and fission fragment excitation. In addition, the time profile also shows a rise time of ~1 ns and a few ns, respectively, for electron excitation in LAr and LXe, corresponding to the thermalization times (Hitachi et al., 1983). The decay shapes also show that the recombination in a heavy ion track occurs faster than the thermalization in LAr and LXe. It should be mentioned that the purification of liquids is quite important for the decay time measurements.

The singlet-to-triplet intensity ratio, S/T, depends on the LET. S/T plays an important role in particle discrimination, as discussed in Section 31.4.3. In RGLs, it shows an opposite trend in LET dependence as that for organic liquids. S/T increases with LET in RGLs. Strictly speaking, the LS coupling (Russell–Saunders notation) is not adequate for heavy atoms such as Ar and Xe. Therefore, the triplet states are not real metastables. The mechanisms proposed for S/T in organic liquids, such as spur recombination (Magee and Huang, 1972), do not apply to RGLs.

The energy resolution, ΔE, for ionization particles of energy, E_0, is given by

$$\Delta E = 2.36\sqrt{FWE_0} \tag{31.3}$$

where F is the Fano factor, which can be smaller than 1 because the energy of the incident particle is fixed. The estimated F values become smaller in the condensed phase, because the existence of the conduction band reduces the N_{ex}/N_i values considerably. The values of F estimated for LAr and LXe are 0.12 and 0.06, respectively (Doke et al., 1976). However, such a good resolution has not been reported so far. In fact, the measured resolutions are even worse than the values predicted by the Poisson statistics. The best values obtained so far are 2–8 times those predicted by the Poisson statistics. The reason for this is not understood as yet. The degradation of energy resolution in the gas, as a function of density up to $1.7\,g/cm^3$, has been reported (Bolotnikov and Ramsey, 1997).

The ionization, Q, and scintillation, S, yields are complementary to each other, as shown in Figure 31.3. We have (Masuda et al., 1989b)

$$Q + aS = \text{const.} \tag{31.4}$$

This is because there is no non-radiative relaxation path from the lowest molecular states, R_2^*, to the ground state. An ionization or an excitation gives one free electron or one scintillation photon in an RGL. The energy resolution obtained by the sum signal is much better than either of the two. However, the number of photons produced is not easy to estimate because of the geometrical factor and the quantum efficiency of the photomultiplier (PMT). In addition, the electron escape probability, χ, for electron and γ excitations (Doke et al., 1985) and the quenching factor, q, for heavy ions (Hitachi et al., 1987) are needed to determine a. The charge scintillation anti-correlation brings about improvement in the resolution as a value reported of $\sigma = 1.7\%$ for 662 keV γ rays at an electric field as low as 1 kV/cm in LXe, as shown in Figure 31.5 (Aprile et al., 2007).

A good energy resolution can also be obtained by doping a trace of molecules in an RGL. The method, that is, the photo-mediated ionization or the photoionization detection (Anderson, 1986; Suzuki et al., 1986a,b; LaVerne et al., 1996), is particularly useful for heavy ions in which charge collection is difficult. The idea is, instead of taking electrons from a strong cylindrical electric field of positive ions in the heavy ion track (Mozumder, 1999), to spread out the electron–ion pairs into the entire detector volume and to isolate them from each other. The scintillation vuv photons produced in the ion track core carry the energy and transfer it to a doped molecule far from the track.

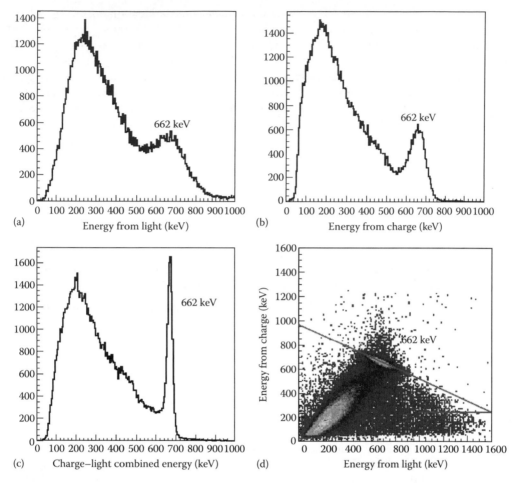

FIGURE 31.5 The energy resolution obtained by ionization (a), scintillation (b), the sum signal (c), and a scatter plot (d) in LXe at an applied field of 1 kV/cm (Aprile et al., 2007). (Courtesy of XENON.)

The charge collection increases drastically by the addition of a few ppm of photoionizing molecules. The charge yield, $Q_d(E)$, for α-particles in doped LAr or LXe is given by that for pure liquid, $Q_\alpha(E)$, as (Hitachi et al., 1997)

$$Q_d(E) = Q_\alpha(E) + \eta'[qN_i - Q_\alpha(E)]Y_{iso}(E) + \eta' qN_{ex}Y_{iso}(E) \tag{31.5}$$

with the apparent photoionization yield, η', expressed as

$$\eta' = g\phi_{vuv}\phi \tag{31.6}$$

where
   g is the fraction of vuv photons absorbed in the detector area
   ϕ_{vuv} is the quantum yield for vuv emission (~1)

$Y_{iso}(E)$ is the fraction of collected charge expected for isolated ion pairs in pure liquid. Since no value for isolated ion pairs is available, that for 1 MeV electrons is substituted. Equation 31.5 conforms well with the experimental values except at a low electric field, as shown in Figure 31.6.

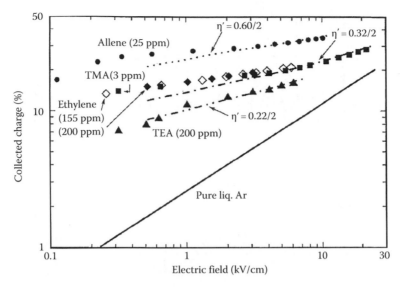

FIGURE 31.6 Collected charge, Q/N_i, where N_i is the total ionization (in unit of electrons) for α-particles as a function of the applied electric field in LAr and doped LAr (Hitachi et al., 1997). Curves give the calculated results from Equation 31.5.

A large increase in charge collection can be obtained even at an electric field as low as 1–2 kV/cm. A good energy resolution of 1.4% FWHM (full width at half maximum) is reported for α-particles in 4 ppm of allene-doped LAr (Masuda et al., 1989a). The best resolution reported is 0.37% FWHM for 33.5 MeV/n O ions in 80 ppm of allene-doped LAr (Hitachi et al., 1994). However, these resolutions are still 3–8 times worse than the values predicted by the Poisson statistics.

The photoionization quantum yield, ϕ, for doped molecules is quite large in LAr and LXe (Hitachi et al., 1997) than that in the gas phase at the same excess energy, E_Δ:

$$E_\Delta = h\nu - I_g^l \tag{31.7}$$

where, I_g^l is the ionization energy in the RGL. In fact, the values obtained for trimethylamine (TMA) and triethylamine (TEA) are 0.8 in LXe in spite of the excess energy being only ~1 eV. One reason may be attributed to the cage effect (Hynes, 1985; Schriever et al. 1989). The ϕ values of TMAE (tetrakis-dimethylaminoethylene) in supercritical LXe are also reported to be large compared with the values in gas (Nakagawa et al., 1993). Few measurements are reported for ϕ in organic liquids because of experimental difficulties (Koizumi, 1994). More studies are needed for the photoionization of molecules in liquids.

The scintillation yield, that is, the scintillation per unit energy deposit, in LAr is shown as a function of LET in Figure 31.7. The scintillation yields obtained for relativistic heavy ions (~GeV/n), Ne, Fe, Kr, and La are the same and do not show quenching (i.e., quenching factor $q = 1$). RGLs are the strongest scintillators against high-excitation-density quenching among scintillators in solid and liquid phases. The scintillation yields for α-particles, fission fragments, and relativistic Au are smaller because of quenching. Similar results have been obtained for LXe (Tanaka et al., 2001). The scintillation yield for electrons and H ions are also smaller not because of quenching but because of insufficient recombination (escaping electrons). The thermalization length in RGLs is large, and some electrons do not recombine within the usual observation time or go to the wall. The sum signals for these low-LET particles become the same as those for the relativistic heavy ions with no quenching.

The Birks model of scintillation quenching (Birks, 1964) does not apply to RGLs. A biexcitonic collision model has been proposed. The lifetimes of the singlet and triplet states do not depend on LET; therefore, excimer states, R_2^*, are not responsible for quenching. The "free" exciton plays an important

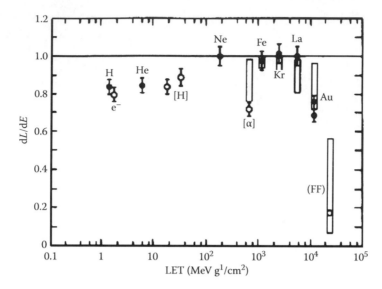

FIGURE 31.7   The scintillation yields for various particles in LAr as a function of LET. Circles show experimental results (Doke et al., 1988). Closed circles are for relativistic ions. Vertical boxes show a diffusion reaction model based on biexcitonic quenching (upper end: a dipole–dipole collision cross section, lower end: the hard-sphere collision cross section divided by 4 (Hitachi et al., 1992b).

role. The quenching mechanism that consists of exciton–exciton collisions and diffusion is discussed elsewhere (Hitachi et al., 1992b). The proposed quenching mechanism is biexcitonic collision:

$$R^* + R^* \rightarrow R + R^+ + e^- \ (K.E.) \tag{31.8}$$

where R^* is the free exciton. The ejected electron, e^-, will lose the kinetic energy close to the ionization energy before it recombines with an ion, again to produce excitation. The overall effect is that two excitons are required for one photon.

The total energy, T, given to the liquid by a heavy ion is divided into the core, T_c, and the penumbra, T_p (Mozumder et al., 1968; Mozumder, 1999). The quenching model assumes that quenching takes place exclusively in the high-excitation-density track core. The energy, T_s, available for scintillation is

$$T_s = qT = q_c T_c + T_p \tag{31.9}$$

where q and q_c are the overall quenching factor and the quenching factor in the core, respectively. The quenching factors are defined as 1 for no quenching and 0 for total quenching. Typically, T_c/T values are 0.5 and 0.7 for relativistic heavy ions and α-particles, respectively. The hard-sphere collision and a dipole–dipole mechanism (Watanabe and Katuura, 1967) were considered for the cross sections, σ, for process in Equation 31.8. The rate constant, k, is given by $k = \sigma v$, where v is the thermal velocity of collision partners ($v \sim 1.2 \times 10^7$ cm/s). Diffusion reaction equations were solved by the method of prescribed diffusion. The results calculated for are compared with experimental results in Figure 31.7.

An increase in scintillation has been reported in LAr with the application of an electric field of a few kV/cm (Hitachi et al., 1987, 1992a, 2002; LaVerne et al., 1996). The increase occurs at a field much lower than that required for proportional scintillation (10^6 V/cm), and it is attributed to the recovery of quenching in the high-excitation-density region of the heavy ion track. The applied electric field may shift the distributions of negative and positive charges, thereby consequently reducing the excitation density, as a result, quenching.

The attenuation lengths, the length at which the number of photons or electrons reduces to $1/e$, for photons (l_{ph}) and electrons (l_e) are important factors for large-scale detectors. An l_e value of more than 1 m (Masuda et al., 1981), which corresponds to a free-electron lifetime of ~ ms, has been obtained. However, l_{ph} reported for LAr and LXe are only 66 and 30–50 cm, respectively (Ishida et al., 1997). Rayleigh scattering seems to limit l_{ph} (Seidel et al., 2002).

When an RGL is used as an ionization detector, it is important to remove the electron-attaching compounds, such as oxygen and fluorinated or chlorinated compounds. The electron attachment to an impurity, M, can be described by the following reaction:

$$e^- + M \underset{k_M}{\Rightarrow} M^- \tag{31.10}$$

where k_M is the rate constant. The number of electrons, N_{e0}, produced by the passage of an ionizing particle at time $t = 0$ is reduced to N_e after time t:

$$N_e = N_{e0} e^{-t/\tau_e} \tag{31.11}$$

where τ_e is the electron lifetime and is related to the rate constant and the impurity concentration, $[M]$, by

$$\tau_e = \frac{1}{k_M[M]} \tag{31.12}$$

The rate constant is a function of the applied electric field, as is shown in Figure 31.8 for O_2, N_2O, and SF_6 (Bakale et al., 1976). The attenuation length, l_{att}, is related to the lifetime by the electron drift velocity, v_d:

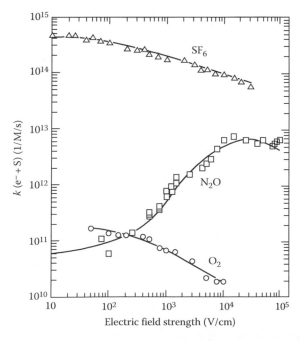

FIGURE 31.8 The electron attachment rate constant, k_M, as a function of the electric field for O_2, N_2O, and SF_6 (Bakale et al., 1976).

$$l_{att} = v_d \tau_e \qquad (31.13)$$

An alternative way of expressing the above equation is

$$l_{att} = K_M \frac{E}{[M]} \qquad (31.14)$$

where

E is the electric field

$[M]$ is the impurity concentration in ppm

K_M is the trapping constant, which has been measured to be (0.15 ± 0.03) ppm cm^2/kV for LAr
(Buckley et al., 1989)

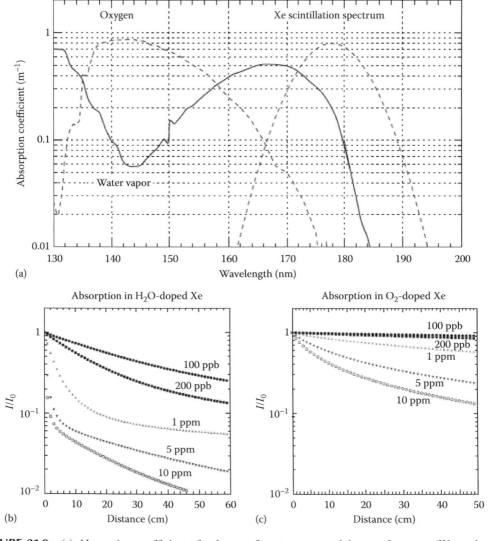

FIGURE 31.9 (a) Absorption coefficients for 1 ppm of water vapor and 1 ppm of oxygen (Watanabe and Zelikoff, 1953; Watanabe et al., 1953) The xenon scintillation spectrum is superimposed. (b) Scintillation light intensity as a function of the distance from the light source for various concentrations of water in LXe. (c) Same as (b), but for oxygen. (From Baldini, A. et al., *Nucl. Instrum. Methods Phys. Res. Sect. A.*, 545, 753, 2005a. With permission.)

An RGL should in principle be transparent to its own scintillation light due to the scintillation mechanism through the excimer state, R_2^*, as shown in Figure 31.1. However, possible contaminants in an RGL, such as water and oxygen at ppm level, considerably absorb scintillation photons (Watanabe and Zelikoff, 1953; Watanabe et al., 1953). In Figure 31.9a, the absorption coefficients for vuv light are shown for 1 ppm contamination of water vapor and oxygen. The absorption spectra of water and oxygen largely overlap with the xenon scintillation spectrum. Given these absorption coefficients and neglecting the scattering ($l_{abs} < l_{Ray}$), the light intensity as a function of the distance from the light source for various concentrations of the contaminant is shown in Figure 31.9b and c. Apparently, water is the worst contaminant. Since water tends to absorb light with shorter wavelengths, only a component with longer wavelengths survives for a long distance.

Furthermore, removal of radioactive impurities is very important for rare event searches, such as neutrinoless double-beta decay and dark matter search. A detailed example is described in Section 31.4.2.

It should be remembered that the energy loss distribution by a relativistic particle (~GeV/n) with $Z \leq 2$ does not follow a Maxwell distribution but a Landou–Vavilov distribution (Badwar, 1973; Ahlen, 1980; Shibamura et al., 1987).

31.3 APPLICATIONS IN ACCELERATOR PHYSICS

31.3.1 IONIZATION CALORIMETER

The first use of RGLs in particle physics experiments was in the form of a calorimeter,[*] which is a device that measures the energy of high-energy particles or γ rays. A calorimeter must fully absorb incoming particles and all of secondary particles within its sensitive volume. A RGL ionization calorimeter was developed as a sampling calorimeter consisting of a multi-parallel-plate electrode system (passive medium) immersed in the RGL (active medium), as shown in Figure 31.10. LAr, among RGLs, is the one most often used for this type of detector because it is available in large volumes due to its cheaper cost. The advantage of the RGL ionization calorimeter is that any configuration of the detector can be formed depending on the experiment at will. Furthermore, it is easy to determine the positional distribution of energy deposition given by incident particles in a detector by using strip electrodes of different orientations coated on the insulated plate, such as a printed

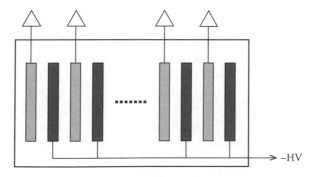

FIGURE 31.10 Multilayer configuration for the RGL sampling calorimeter.

---

[*] Calorimeters are usually classified as electromagnetic or hadron calorimeters. An electromagnetic calorimeter is one specifically designed to measure the energy of particles that interact primarily via electromagnetic interaction, while a hadronic calorimeter is one designed to measure particles that interact via a strong nuclear force. The same detector media and similar readout techniques can be used, but the responses are very different in both cases.

circuit board. Then, a very accurate determination of the incident direction, the conversion point of the particle, and particle identification can be provided.

The energy resolution in a sampling calorimeter is determined mainly by sampling fluctuation. A formula for sampling fluctuation was derived from an analogy with Fano's formalism for the straggling of the total ionization produced by ionization radiation with a fixed energy, and it is expressed as follows (Hoshi et al., 1982):

$$\frac{\sigma_s}{E_d} = \sqrt{2\left(1 - \frac{E_d}{E_0}\right)^2 \frac{\Delta E_{cell}}{E_0}} \qquad (31.15)$$

where

E_d is the total energy deposited in the active medium
σ_s is its standard deviation
ΔE_{cell} is the average energy loss per unit cell in the calorimeter
E_0 is the energy of the incident particles

It is obvious from Equation 31.15 that the energy resolution improves by increasing E_d or decreasing ΔE_{cell}. There is a technical limit up to which ΔE_{cell} can be decreased. On the other hand, it is possible to increase E_d, namely, by thinner electrodes and thicker gaps of the RGL. This point of view ultimately leads to the idea of a (quasi-) homogeneous calorimeter. Homogeneous calorimeters are much larger devices compared with sampling calorimeters. For reducing the volume of the device, LKr and LXe are more attractive media because of their shorter radiation lengths*; in addition, a better energy resolution is achievable owing to their lower W-values. The LKr calorimeter is beginning to be used in practical experiments (Fanti et al., 2007; Peleganchuk, 2009), and the study of the LXe calorimeter has also started as test devices (Baranov et al., 1990; Okada et al., 2000).

The disadvantage of RGL ionization calorimeters is their slow time response, that is, the integration time to collect full charge, which originates from the electrons' drifting time (t_d) in the liquid gap, is quite long (Figure 31.2). The liquid gap of ordinary RGL ionization calorimeters is a few mm; therefore, the electron drifting time is usually of the order of μsec. The so-called initial current measurement, which uses much a shorter integration time than t_d, was introduced to refine this weak point (see Figure 31.11 (Aubert et al., 1991)). There is not a great difference in the energy resolution between the methods of full integration and the initial current measurement. Hence, the (quasi-) homogeneous RGL calorimeter was established.

31.3.2 NA48 LKr Calorimeter

A good example of the working of a quasi-homogeneous LKr calorimeter can be seen from the NA48 experiment performed at CERN[†] using its super proton synchrotron (SPS) accelerator. The aim of the NA48 experiment was the precise measurement of direct CP violation in the system of neutral kaons as well as the measurement of the very rare K_S and charged kaons decays (Fanti et al., 1999; Lai et al., 2003; Batley et al., 2009). Neutral kaons are produced at two targets located at different distances from the detector, thus creating simultaneously a K_L and a K_S beam for the CP violation experiment. The experiment is performed to measure the decays of kaons from these two beams into two neutral or two charged pions. The detection of photons from a neutral pion decay is achieved by means of the LKr calorimeter. In order to suppress the huge background from $K_L^0 \rightarrow 3\pi^0$

---

[*] Bremsstrahlung and electron pair production are the dominant processes for high-energy electrons and photons. Radiation length, X_0, is defined as the length that the high-energy electron travels in matter losing its energy by bremsstrahlung to 1/e. A value for 1 GeV is usually taken, since the radiation length, X_0, becomes almost independent of energy above 1 GeV.

[†] European Organization for Nuclear Research.

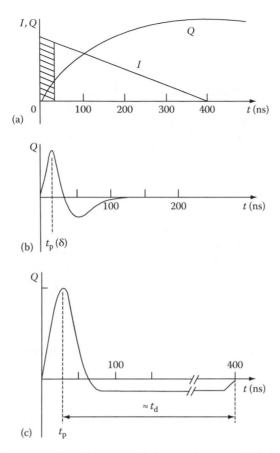

FIGURE 31.11 (a) Drift current, I, and integrated charge, Q, versus drift time for ionized electrons. The maximum drift time is 400 ns. (b) Response of a shaping amplifier to a very short current pulse (δ). (c) Response of a shaping amplifier to the current form shown in (a). (From Aubert, B. et al., *Nucl. Instrum. Methods Phys. Res. A.*, 309, 438, 1991. With permission.)

decays, very good spatial, energy, and time resolutions are required. The LKr volume is 125 cm deep (~27 radiation lengths) along the direction of the beam and forms an octagon of about a 5.5 m² cross section. The anodes and cathodes are 40 μm thick copper–beryllium ribbons running almost parallel to the beam (Figure 31.12). They form 13,500 readout cells having a cross section of $2 \times 2\,\text{cm}^2$. Cold preamplifiers are directly connected to one end of the anode ribbons (downstream side). In order to stand the high rate of particles (~1 MHz) from the K_L beam and to still provide the good time resolution required for K_S/K_L tagging, an initial current readout with fast shaping by means of a linear amplifier is used. The shaped pulse has a width of only 70 ns, while the drift time in the calorimeter cell is over 3 μs. The performance of the LKr calorimeter was tested and the results were as follows (in units of GeV, cm, and ns) (Barr et al., 1996; Costantini, 1998; Jeitler, 2002):

$$\frac{\sigma_E}{E} = \frac{0.032}{\sqrt{E}} \oplus \frac{0.09}{E} \oplus 0.0042$$

$$\sigma_{X,Y} = \frac{0.42}{\sqrt{E}} \oplus 0.06 \tag{31.16}$$

$$\sigma_t = \frac{2.5}{\sqrt{E}}$$

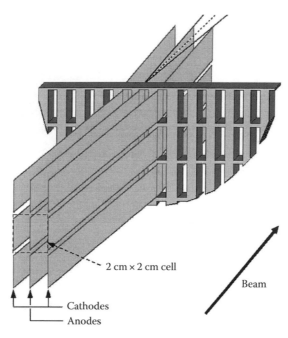

FIGURE 31.12 Cell geometry of NA48 LKr calorimeter. (From Costantini, F., *Nucl. Instrum. Methods Phys. Res. A.*, 409, 570, 1998. With permission.)

31.3.3 ATLAS LAr Calorimeter

The LHC (Large Hadron Collider) experiments at CERN began in 2008. The optimization of the detector is driven by the requirement to detect new particles, in particular, the Higgs boson* decaying to two photons or to four electrons, in the mass range from 90 to 180 GeV. The ATLAS experiment should be capable of measuring the Higgs mass with 1% precision using the calorimeter system alone. This translates into the requirements of a sampling term of less than $10\%/\sqrt{E}$, associated with a global constant term of 0.7% and an angular resolution better than $50 \, \text{mrad}/\sqrt{E}$. The technology chosen for the ATLAS electromagnetic calorimeter is Pb/LAr sampling with accordion-shaped electrodes and absorbers. LAr has been chosen because of its intrinsic linear behavior, stability of the response over time, and radiation tolerance. In addition, the accordion geometry minimizes inductances in the signal path, allowing the use of the fast shaping (40 ns) needed for operation with 25 ns bunch intervals between collisions at the LHC.

The ATLAS LAr calorimeter system consists of an electromagnetic barrel calorimeter and two end-cap cryostats with electromagnetic (EMEC), hadronic (HEC), and forward (FCal) calorimeters (Figure 31.13). The barrel, covering a pseudo-rapidity† range of $|\eta| < 1.475$, shares its cryostat with the superconducting solenoid, the calorimeter being behind the solenoid. Each wheel consists of 16 modules of 1024 lead absorbers interleaved with readout electrodes. Each end-cap, covering the pseudo-rapidity range of $1.4 < |\eta| < 3.2$, consists of eight modules (see Figure 31.14b). To correct for the energy lost in the dead material in front of the calorimeter, both end-caps and barrel are completed with pre-sampler detectors, covering the $|\eta| < 1.8$ range. These pre-samplers are thin layers of argon equipped with readout electrodes, but without an absorber. Figure 31.14a shows the segmentation in the (η, ϕ) plane and along the longitudinal electromagnetic shower development in the barrel. The HEC is structured into two wheels: the front, HEC1, and the rear,

---

* The Higgs boson is the only Standard Model particle that has not yet been observed. Experimental detection of the Higgs boson would help explain the origin of mass in the universe.

† $\eta = -\ln \tan(\theta/2)$, where θ is the polar angle of the produced particles against beam axis. The azimuthal angle ϕ is measured around the beam axis.

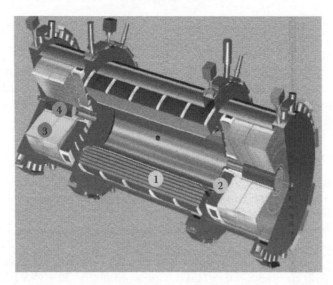

FIGURE 31.13 The ATLAS LAr calorimeter: (1) the electromagnetic barrel; (2) the EMEC end-cap; (3) the HEC end-cap; (4) the FCal end-cap. (From Aharrouche, M., *Nucl. Instrum. Methods Phys. Res. Sect. A.*, 581, 373, 2007. With permission.)

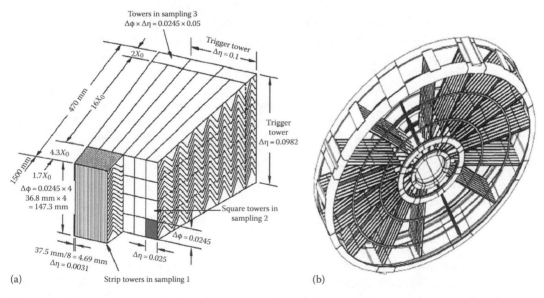

FIGURE 31.14 (a) Projection of the electromagnetic module showing the longitudinal and lateral segmentation. (b) Schematic view of the electromagnetic end-cap wheel (Aubert et al., 2005; Aharrouche, 2007). (From Aharrouche, M., *Nucl. Instrum. Methods Phys. Res. Sect. A.*, 581, 373, 2007. With permission.)

HEC2, with 32 modules each, covering the pseudo-rapidity range of $1.5 < |\eta| < 3.2$. It consists of copper-plate absorbers of 25 mm (50 mm) thickness for HEC1 (HEC2) interleaved with electrodes. The readout technique (electrostatic transformer) of the HEC splits the gap between two absorbers into four sub-gaps, each with a width of 1.85 mm. The readout granularity is $\Delta\eta \times \Delta\phi = 0.1 \times 0.1$ for $1.5 < |\eta| < 2.5$ and 0.2×0.2 for $2.5 < |\eta| < 3.2$. The HEC is designed to measure the jets in the forward direction. The FCal shares the same cryostat with the EMEC and the HEC and covers the pseudo-rapidity range of $3.2 < |\eta| < 4.9$. It consists of three consecutive modules along the beam line. The modules consist of a tungsten matrix in FCal2 and FCal3 (copper matrix in FCal1) housing cylindrical electrodes, consisting of copper rods. The LAr gaps are very small: 0.250, 0.375, and 0.500 mm, for FCal1, FCal2, and FCal3, respectively. The readout granularity is $\Delta\eta \times \Delta\phi = 0.2 \times 0.2$. The design of the FCal was constrained by the high radiation level in the most forward region. The detailed studies of the construction and performance are described in Aubert et al. (2005).

31.3.4 LXe SCINTILLATION CALORIMETER (MEG EXPERIMENT)

The MEG experiment searches for a rare muon decay, $\mu^+ \to e^+\gamma$, which is forbidden in the standard model. Several new theories beyond the standard model predict the branching ratio of the decay just below the current experimental limit, that is, 1.2×10^{-12} (Brooks et al., 1999). The MEG experiment aims a sensitivity of 10^{-14} (Mori, 1999), where most predictions are covered. The discovery of the muon decay is a probe for new physics beyond the standard model. A $\mu^+ \to e^+\gamma$ decay event is characterized by the clear two-body final state where the decay positron and the γ ray are emitted in opposite directions with energies equal to half the muon mass ($E = 52.8$ MeV). While positrons of this energy are abundant from the standard Michel decay of muons, γ rays with such high energies are very rare. Therefore, the key requirement for the MEG detector is a high energy resolution for γ rays, since the accidental background rate decreases, at least, with the square of the energy resolution (Mori, 1999).

The excellent properties of LXe as a scintillator and the developments of new PMTs with high quantum efficiency, fast timing, and their possible operation in LXe motivated the choice of the detector for the MEG experiment. With the Xe scintillation light detected by a large number of PMTs immersed in the liquid volume, the detector's energy resolution is proportional to $1/\sqrt{N_{pe}}$, where N_{pe} is the number of photoelectrons, and was estimated to be better than 1% (1σ) for 52.8 MeV γ rays (Doke and Masuda, 1999). In addition, the γ ray interaction points can also be determined from the light distribution on each PMT (Mori, 1999). Following the R&D with a small prototype (Mihara et al., 2002) and with further improvement of the metal channel-type PMT, a large LXe prototype detector with 228 PMTs of 2 in. was constructed and tested with 10–83 MeV γ rays (Mihara et al., 2004). The detector and associated cooling and purification apparatus, shown schematically in Figure 31.15, were tested at the Paul Scherrer Institute in Switzerland where the final MEG experiment is located. A pulse tube refrigerator (PTR) was used for the first time to condense Xe gas and maintain the liquid temperature stable over a long period. The PTR, optimized at the LXe temperature, had a cooling power of 189 W (at 165 K) (Haruyama, 2002). The prototype was large enough to test both the reliability and stability of the cryogenic system based on the PTR, and the efficiency of the purification system required for the calorimeter's performance.

While the transmission of the scintillation light is affected by Rayleigh scattering, this process does not cause any loss of light, and hence does not affect the energy resolution. Two types of circulation systems for Xe purification were developed to improve the light attenuation length. One is gaseous purification using a heated metal getter, in which the evaporated Xe is brought to the hot getter system, and then the Xe is recondensed in LXe in the detector (Baldini et al., 2005a). This method is effective in removing all types of impurities, but is not efficient for water, which was identified as the major contributor to light absorption, because its vapor pressure at the LXe temperature is too low. Furthermore, the purification speed is limited by the cooling power. The other method was developed specifically to remove water at a much higher rate (100 L/h) by circulating LXe with a cryogenic centrifugal pump. Water is efficiently removed by a filter containing molecular sieves. With this method, the total impurity concentration was reduced in 5 h from 250 to 40 ppb in the total 100 L LXe volume of the MEG prototype detector (Mihara et al., 2006).

A novel calibration technique was developed and applied to the prototype detector. A lattice of α point sources deposited on thin (100 μm diameter) gold-plated tungsten wires was permanently

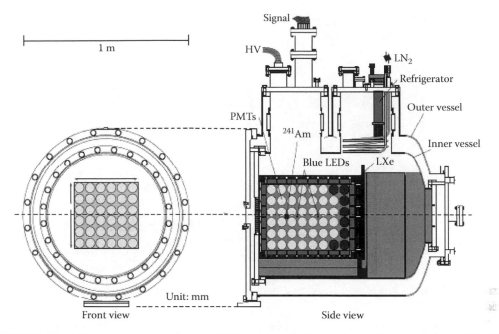

FIGURE 31.15   LXe γ ray prototype detector. PMTs are installed on six faces of a rectangular solid holder. (From Mihara, S. et al., *Cryogenics*, 44, 223, 2004. With permission.)

suspended in the volume and fixed at the surfaces of the large vessel containing the LXe (Baldini et al., 2006). This method was successfully implemented and used to determine the relative quantum efficiencies of all the PMTs and monitor the stability of the detector during the experiment. In the final MEG detector, the PMT configuration is similar to that in the prototype; hence, the methods of reconstructing the γ ray energy and interaction points were proven with the prototype detector. The reconstructed energy distribution for 55 MeV γ rays from π⁰ decay is shown in Figure 31.16 (Baldini et al., 2005b). The spectrum is asymmetric with a low-energy tail, caused mainly by γ ray interactions with materials in front of the sensitive region and by the leakage of shower

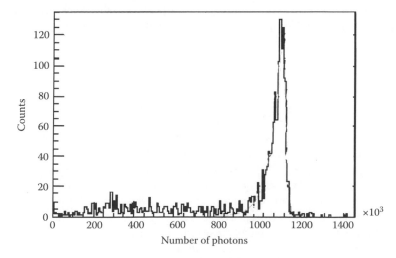

FIGURE 31.16   Energy spectrum for 54.9 MeV γ rays measured by the large prototype detector. (From Baldini, A. et al., Transparency of a 100 liter liquid xenon scintillation calorimeter prototype and measurement of its energy resolution for 55 MeV photons. In 2005 IEEE International Conference on Dielectric liquids, June 26–July 1, 2005 (ICDL 2005), Coimbra, Portugal, pp. 337–340. With permission.)

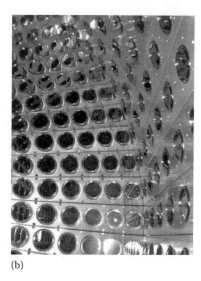

(a) (b)

FIGURE 31.17 (a) Picture of the LXe scintillation calorimeter. (b) Inside the detector.

components. The estimated energy resolution at the upper edge is $1.2 \pm 0.1\%$ (1σ) (Baldini et al., 2005b), as shown in Figure 31.16, which is roughly equal to the value obtained by extrapolation of the small prototype data. The events that remain above 55 MeV are known to be due to the mistagging of π^0 events. The position and time resolutions are evaluated to be 5 mm and ~120 ps (FWHM), respectively (Baldini et al., 2006). The MEG cryostat, consisting of cold (inner) and warm (outer) vessels, is designed to reduce passive materials at the γ ray entrance. The same refrigerator tested on the prototype will be used for cooling, with an emergency cooling system based on liquid nitrogen flowing through pipes installed both inside the cold vessel and on its outer warm vessel.

Gaseous and liquid purification systems are located on site for removing impurities possibly retained after liquefaction. The construction of the calorimeter was completed in the autumn of 2007. Its calibration started in the same year and physical data taking continues to date. The MEG γ ray scintillation calorimeter, with a total LXe mass of almost 2.7 t and 846 PMTs, is currently the largest LXe detector in operation worldwide (Figure 31.17). The many new technologies that were brought by the MEG LXe detector have a great influence on the detector, such as those developed for dark matter searches discussed in the following text.

31.4 APPLICATIONS IN NON-ACCELERATOR PHYSICS

31.4.1 Liquid Argon Time Projection Chamber (ICARUS)

The technology of the LAr TPC was first proposed by C. Rubbia in 1977 (Rubbia, 1977). It combines the characteristics of a bubble chamber with the advantages of the electronic readout: it is fully and continuously sensitive, self-triggering, and able to provide three-dimensional views of ionizing events with particle identification from dE/dx and range measurement, and it can be operated over large active volumes. This detector is also a superb calorimeter of very fine granularity and high accuracy. The ICARUS project envisages the usage of LAr TPC detectors for studies of neutrinos from different sources as well as for searches for proton decay. This detection technique offers 3D imaging with a granularity for tracking and calorimetric measurements, with a fully electronic detector. Owing to this, it gives good measurements in a wide range of neutrino energies or background-free and high-efficiency measurements, for example, for proton decay.

A configuration of the electrodes suitable for the 3D readout of the events was proposed by Gatti et al. (1979). A voltage is applied to each wire plane in order to ensure the transparency of the grids

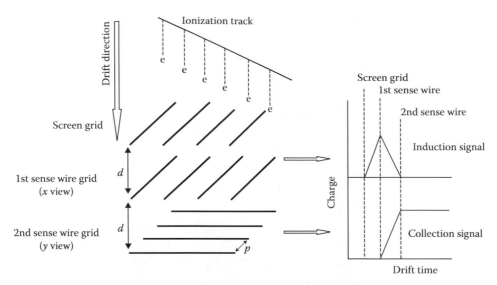

FIGURE 31.18 Configuration of the anode wire plane and expected signals.

to the drifting electrons (Bunemann et al., 1949). The position and charge of the track elements are measured by means of the current induced in each wire. The drifting electrons reach and cross in sequence following the wire planes (Figure 31.18): (1) A plane of wires running in the y-direction (x and y are the two orthogonal directions along which the wires are placed), functioning as a screening grid. (2) A plane of wires running again in the y-direction, located below the screening grid; they measure, by induction, the x-coordinate. (3) A plane of wires running in the x-direction, located below the previous plane; they measure the y-coordinate. Since this is the last sensitive plane, the electric field is such that this plane collects the drifting electrons. An electron drifting in the sensitive volume above the screening grid does not produce any current in the coordinate planes (in the case of perfect shielding). A positive induced current starts in the coordinate plane when the electron crosses the screening grid, becomes negative when the electron crosses the coordinate plane, and ends when the electron crosses the y-coordinate plane. The charge signal from the y-coordinate plane has similar behavior, with the exception that only the positive current is present. Thus, the segmentation of the readout system provides x–y coordinates. The coordinate along the drift field (z-direction) can be inferred from the drift time, measured as the time delay of the arrival of electrons to the readout system with respect to a trigger signal (e.g., by direct scintillation). From this information, the three-dimensional reconstruction of the ionization particle track can be obtained. Figure 31.19 shows a cosmic ray–induced shower as observed in the ICARUS LAr TPC.

A fundamental requirement of the LAr TPC is that electrons produced by the ionizing particles can travel unperturbed from the point of production to the collecting wires. To this effect, the system has to be able to purify and maintain pure LAr with a concentration of electronegative impurities lower than 1 ppb of O_2 equivalent for the longest period of time possible. By using standard procedures for high vacuum (suitable materials, cleaning, design of internal components) and vacuum conditioning of the internal surfaces, a high LAr purity after filling can be achieved. The pollution of LAr is mainly due to the outgassing of inner surfaces in contact with gaseous Ar, while the contamination from objects immersed in the liquid is negligible. In the absence of large convective motions, recirculation of the gas is sufficient to maintain the LAr purity. Commercial filters provide a purity level that is well above the experimental requirements.

A purity monitor chamber for the measurement and control of the electron lifetime is developed (Bettini et al., 1991). This chamber is a doubly girded ionization chamber, as shown schematically in Figure 31.20. The anode (A) and the cathode (K) are almost completely screened by two grids (GA and GK) located a short distance from each electrode. Field intensities ($E1$, $E2$, and $E3$)

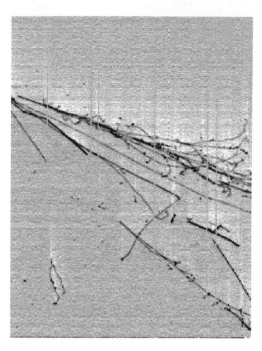

FIGURE 31.19 Electronic image of a cosmic-ray shower as observed in the ICARUS 3 t prototype. The drift direction is along the horizontal axis, spanning a distance of about 40 cm. The vertical axis corresponds to 192 sense wires with a 2 mm pitch. (From Benetti, P. et al., *Nucl. Instrum. Methods Phys. Res. Sect. A.*, 332, 395, 1993b. With permission.)

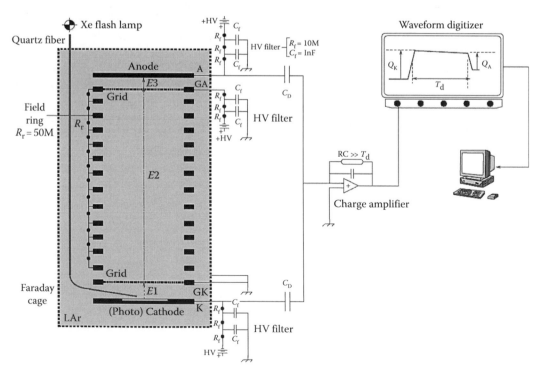

FIGURE 31.20 Layout of the purity monitor and readout scheme for the ICARUS T600 detector (Bettini et al., 1991; Amerio et al., 2004). (From Amerio, S. et al., *Nucl. Instrum. Methods Phys. Res. Sect. A*, 527, 329, 2004. With permission.)

are chosen to ensure complete transparency of the grids. The two grids are separated by a long drift space. The attenuation due to electronegative impurities of an electron cloud drifting over this distance is measured by comparing the charge leaving the cathode with that reaching the anode. The currents from the cathode and the anode feed the same charge amplifier. Free electrons are produced by a short pulse Xe flash lamp (or UV laser) brought to a gold deposit on the cathode through an optical fiber. The electron cloud, photoproduced by the light pulse, drifts initially to the first grid producing a positive current in the amplifier. This current vanishes at the instant the electrons cross the cathode grid. The charge integrated by the amplifier is then equal to the charge Q_K produced at the cathode. While the charge drifts in the uniform field between the two grids, no current is experienced by the amplifier. If electronegative impurities are present, the amount of charge diminishes. When the charge finally crosses the anode grid, it produces a negative current in the input of the amplifier and a negative step in the output (charge) signal; the height of the step is equal to the surviving charge (Q_A) reaching the anode. The free-electron lifetime τ can simply be expressed as follows:

$$\frac{Q_A}{Q_K} = \exp\left(-\frac{T_d}{\tau}\right) \tag{31.17}$$

where T_d is the electron drift time. Obviously, the sensitivity is higher for longer drift times, which can be obtained by working with low electric field intensities.

The performance of the LAr TPC has been studied through the analysis of a wide variety of events occurring in the detector (Cennini et al., 1994).

31.4.1.1 Space Resolution

Cosmic muons crossing the detector have been used to evaluate the single-point space resolution along the drift coordinate (σ_z). Values of about 150 μm are the norm and appear to be independent on the field. Instead, σ_z strongly depends on the signal-to-noise ratio (≈ 10 in the case of ICARUS). The space resolutions on the other two coordinates are strictly related to the wire pitch (p): $\sigma_{x,y} \approx p/\sqrt{12}$.

31.4.1.2 Energy Resolution

By means of cosmic muons, the distribution of the charge deposited over 2 mm (wire pitch) was measured. Its width is the convolution of a Landau function with the Gaussian electronic noise and is directly related to the resolution of the energy deposited by ionization. A comparison with a test pulse distribution shows that the noise is the main contribution to the width. Hence, an energy resolution of $\approx 10\%$ over a 2 mm track is the typical value in the 3 t prototype detector. An energy resolution in the MeV range has also been evaluated by studying the Compton spectrum and the pair production peak produced by a 4.43 MeV monochromatic γ ray source. The best fit gives a resolution of 7% at 4 MeV in agreement with the calorimetric measurement found in the literature.

31.4.1.3 Particle Identification

Particle identification, a vital feature, is directly related to the ability of measuring the dE/dx value along the range of a stopping particle. A study of cosmic muons and protons stopping in the 3 t prototype has shown that the charge deposited along the track is not proportional to the energy deposited because of the electron–ion recombination effect that strongly depends on the ionization density. This nonlinear response could degrade the particle identification capability. Based on the fact that the electron–ion recombination has the result of producing UV scintillation photons, a solution to recover linearity has been found and consists in dissolving in LAr a photosensitive dopant able to convert back the scintillation light due to recombination into free electrons (see Section 31.2) (Cennini et al., 1995). Tetra-methyl-germanium (TMG) is one of the suitable dopants, because it has a large photoabsorption cross section (62 Mb) and a large quantum yield (close to 100%); this

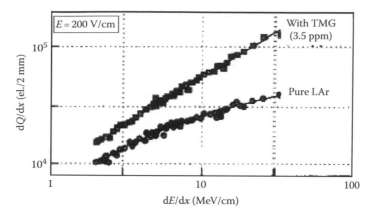

FIGURE 31.21 dQ/dx versus dE/dx from stopping muons and protons. Charge saturation is evident in pure LAr. Linearity is recovered with TMG. (From Cennini, P. et al., *Nucl. Instrum. Methods Phys. Res. Sect. A.*, 355, 660, 1995. With permission.)

implies that small quantities of TMG are enough to convert all the scintillation photons into electrons in the vicinity of the ionizing track, that is, without spoiling the space resolution. Figure 31.21 illustrates the dQ/dx versus dE/dx relation before and after the doping with TMG.

The ICARUS project has demonstrated the feasibility of the technology by an extensive R&D program, which included 10 years of study on small LAr volumes (proof of principle, LAr purification methods, readout schemes, and electronics) and 5 years of study with several prototypes of increasing mass (purification technology, collection of physics events, pattern recognition, long duration tests, and readout technology). The largest of these devices had a mass of 3 t of LAr (Benetti et al., 1993a,b; Cennini et al., 1994) and had operated continuously for more than 4 years, collecting a large sample of cosmic-ray and γ source events. Furthermore, a smaller device (50 L) of LAr (Arneodo et al., 2006) was exposed to the CERN neutrino beam, demonstrating the high recognition capability of the technique for neutrino interaction events.

Thereafter, the project has entered an industrial phase, a necessary path to follow toward the realization of large volume detectors for astroparticle and neutrino physics. The first step was the realization of a large cryogenic prototype (14 t of LAr) (Arneodo et al., 2003) to test the final industrial solutions that were adopted. The second step was represented by the construction of the T600 module: a detector employing about 600 t of LAr to be operated at the National Laboratory of Gran Sasso (LNGS), Italy (see Figure 31.22). This stepwise strategy allowed for the progressive development of the necessary know-how to build a multi-kton LAr detector (Amerio et al., 2004). The T600 was transported to the LNGS, where it will start taking data from atmospheric and solar neutrino interactions.

The LXe TPC is also developed to study γ ray astronomy in the energy range of MeV. The energy and direction of an incident γ ray are reconstructed from the 3D locations and energy deposits of individual interactions taking place in the homogeneous detector volume. While the charge signals provide energy information and x–y positions, the fast xenon scintillation signal is used to trigger the detector. The drift time measurement, referred to as the time of the trigger signal, gives the z position with the known drift velocity (Aprile et al., 2008).

31.4.2 XMASS Project

An advantage of LXe is that Xe does not have an isotope with a lifetime longer than a couple of months. The longest isotope is ^{127}Xe with a half-life of 36.4 days, which decays through electron capture with a Q-value of 0.664 MeV. The second longest isotope is ^{131}Xe with the lifetime of 11.8 days. Thus, experiments on rare phenomena (such as dark matter searches and double-beta decay) may be carried out shortly after moving the xenon underground from the surface. Therefore, the cosmogenic production of Xe isotopes is not a serious problem. One of the main sources of the background

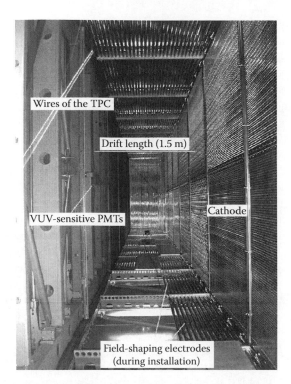

FIGURE 31.22 Picture of the internal detector layout inside the half-module: the cathode divides the volume in two symmetric sectors. The picture refers to the left sector where wires and the mechanical structure of the TPC and some PMTs are visible. (From Amerio, S. et al., *Nucl. Instrum. Methods Phys. Res. Sect. A.*, 527, 329, 2004. With permission.)

is the presence of trace amounts of radioactive impurities, such as ^{85}Kr. The commercially available xenon contains ~1 ppb ~ppm of Kr, and such purity is sufficient for most of the applications of Xe. ^{85}Kr is a radioactive nucleus that decays into ^{85}Rb with a half-life of 10.76 years, and emits γ rays with a maximum energy of 687 keV and a 99.57% branching ratio. Large quantities of ^{85}Kr are produced artificially in nuclear fission. Xe is usually abstracted from air. The concentration of Xe in air is ~0.1 ppm, while the concentration of Kr is ~1 ppm. This constitutes the main source of ^{85}Kr in air. The concentration of ^{85}Kr in air is measured to be ~1 Bq/m^3 (Igarashi et al., 2000; Cauwels et al., 2001), which corresponds to ^{85}Kr/Kr = ~10^{-11}. The possible methods for removing Kr from Xe are distillation and adsorption. These are commonly used for industrial processes, but systems that could reduce Kr sufficiently to meet the requirements of experiments never existed. Further studies for both adsorption-based chromatography and distillation followed the McCabe–Thiele method that succeeded in reaching ~ppt levels (McCabe and Smith, 1976; Bolozdynya et al., 2007).

The XMASS project aims at a multipurpose detection, which observes not only dark matter, but also pp and ^{7}Be solar neutrinos, which the Super-Kamiokande cannot observe, as well as ^{136}Xe neutrinoless double-beta decay to measure the neutrino mass, with a 20 t LXe scintillation detector in the Kamioka observatory. The possibility of using ultrapure LXe as a low-energy solar neutrino detector by means of ν + e scatterings has been evaluated. A detector possibly with 10 t of fiducial volume* will give ~14 events for pp neutrinos and ~6 events for ^{7}Be neutrinos with the energy threshold at 50 keV. The high density and high Z of LXe would provide very good self-shields against the incoming backgrounds originating from the container and outer environments (PMTs, cables, etc.). A 30 cm thick self-shield is equivalent to 4 m of water shield for the electromagnetic component.

---

* Fiducial volume (or mass) is the effective volume (or mass) for particle detection in a 3D detector. The events that happen in the volume are considered to be reliable.

In the first phase of the XMASS program, a 100 kg prototype for dark matter detection was built and deployed at the Kamioka underground laboratory. The 30 cm cubic detector was contained in a low-activity copper vessel further shielded for the reduction of γ and neutron backgrounds. Nine PMTs are attached on each face of the vessel, and scintillation photons are detected by a total of 54 low-background PMTs of 2 in. (Hamamatsu R8778) through MgF_2 windows. The PMT photocathode coverage for light detection was 17%. The background spectra were measured before and after the distillation, as shown in Figure 31.23. The background rate around 100–300 keV before distillation is higher than that after distillation by ~5 × 10^{-2} kg/keV/day. This background level corresponds to a Kr concentration of 2–3 ppb, assuming a ^{85}Kr/Kr ratio of 1.2 × 10^{-11}. The dashed curve shows the expected ^{85}Kr β spectrum, assuming a Kr concentration of 3 ppb. The Kr concentration of purified Xe was measured to be 3.3 ± 1.1 ppt using a gas chromatography apparatus and an atmospheric pressure ionization (API) mass spectrometer. The results confirmed that the cryogenic distillation system can reduce the amount of Kr in Xe gas by up to three orders of magnitude (Abe et al., 2009).

The second phase, currently under construction (Suzuki, 2008), consists of an 856 kg LXe detector with the main goal of physically detecting dark matter (Figure 31.24). The large liquid volume is viewed by 642 hexagonal PMTs, arranged in an approximately spherical shape (radius ~43 cm) with 67% photocathode coverage. A light yield of 4.4 photoelectrons/keV was predicted by simulations. The fiducial volume is a region inside the radius of 20 cm from the center, and the mass of LXe is 100 kg. A new low-background PMT, Hamamatsu R10789, with a hexagonal photocathode shape was developed, and the total radioactive contamination was reduced by a factor of 10 from R8778 (used for the 100 kg prototype). The measured activities per PMT are 0.37 × 10^{-3} Bq for the U-chain, 1.0 × 10^{-3} Bq for the Th-chain, 5.9 × 10^{-3} Bq for ^{40}K, and 3.1 × 10^{-3} Bq for ^{60}Co. In order to reduce the environmental radioactive background, such as those due to γ rays and fast neutrons, the detector will have to be installed in the center of a water tank (10 m in diameter and 10 m high). Twenty-inch PMTs will have to be arranged on the inner wall of the tank to observe water Cherenkov signals for anti-coincidence counting. The sensitivity of the

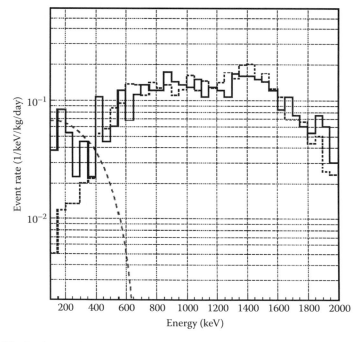

FIGURE 31.23 The background spectra measured before distillation (solid histogram) and after distillation (dotted histogram). The dashed curve shows the expected ^{85}Kr beta spectrum assuming a Kr concentration of 3 ppt. (From Abe, K. et al., *Astropart. Phys.*, 31, 290, 2009. With permission.)

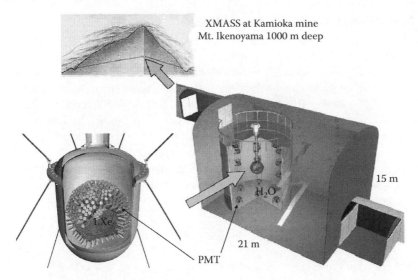

FIGURE 31.24 The XMASS detector in Kamioka. An LXe detector (bottom left) is set in the center of the water tank. PMTs of 20 in. are installed on the wall of the water tank. LXe is surrounded by 642 PMTs. The whole detector is installed 1000 m below the mountain. (©Kamioka Observatory, Institute for Cosmic Ray Research (ICRR). The University of Tokyo, Tokyo, Japan.)

detector for 5 years of running is expected to reach $\sim 10^{-45}$ cm^2 for spin-independent interactions and $\sim 10^{-39}$ cm^2 for spin-dependent interactions.

A similar experiment using liquid neon (LNe) is also proposed by the cryogenic low energy astrophysics with noble gases (CLEAN) project (Coakley and McKinsey, 2004).

31.4.3 DARK MATTER SEARCH

Most of the constituent matter in the universe is dark with no emission, no absorption, and even no scattering of light. It is, in fact, transparent. In 1933, Zwicky observed that the rotational velocity of galaxies in the Coma cluster is greater than that expected from the virial theorem, and announced that the universe contains more matter than is inferred from optical observation (Zwicky, 1933). Further scientific evidence confirmed his view: the rotational velocity in a galaxy, gravitational lensing, the microwave radiation background, and even the shape of thin galaxies require unseen mass to prevent a galaxy from collapsing. The unseen dark matter accounts for a quarter of the universe. Ordinary matter makes up only 4%. The rest, ~73%, is supposed to be dark energy.

The leading candidate is WIMPs (weakly interacting massive particles), cold massive elementary particles with a typical mass of several tens of GeV. The supersymmetric theorem (SUSY) predicts a neutralino as a favorable candidate (Ellis and Flores, 1991). Review articles for dark matter candidates and other dark matter detectors are found in Lewin and Smith (1996), Sumner (2002), Freedman and Turner (2003), Gaitskell (2004), and Spooner (2007).

Our solar system travels around the galactic center at about 230 km/s (Spergel, 1988). WIMPs are surrounding the Galaxy. Scientists are in search of recoil nuclei of a few tens of keV energy, produced by elastic scattering with WIMPs. However, the event is quite rare ($<10^{-2}$ to 10^{-4}/kg/day). Almost all the signals produced even in an underground detector are due to the background, mostly γ rays. The resulting kinetic energy spectrum for recoil ions will not be monochromatic, but may be described as similar to the exponential. Both the exposure (mass × time) and the ability to distinguish the WIMP signal from the background is essential. RGLs have an excellent capability of discriminating nuclear recoil signals from the γ background using both the ionization and the scintillation, and can provide a strong possibility of the direct observation of dark matter in the Galaxy.

31.4.3.1 Interaction of Slow Ions with Matter

The interaction mechanism for recoil ions of a few tens of keV energy with matter is considerably different from that of fast ions described in Section 31.2. The nuclear stopping, S_{nc}, which can be ignored in fast ion collisions, becomes comparable to or larger than the electronic stopping, S_{el}. The total stopping, S_T, can be expressed as the sum of the two. Apart from the difficulty in obtaining values for S_{nc} and S_{el} for slow collisions, a simple relation, $S_T = S_{nc} + S_{el}$, produces another complexity. The interaction of slow ions with matter has been extensively studied by Lindhard et al. (1963). The nuclear process follows the usual procedure of a screened Rutherford scattering. The electronic interaction is based on the Thomas–Fermi model. S_e is expressed as $(d\varepsilon/d\rho)_e = k\varepsilon^{1/2}$, where ε and ρ are the dimensionless reduced energy and reduced range. When the projectile and the target are the same element ($Z_1 = Z_2$; indices 1 and 2 stand for the projectile and the target atom, respectively), k is given by the atomic number Z and mass A, as follows:

$$k = 0.133 Z_2^{2/3} A_2^{-1/2} \tag{31.18}$$

The secondary ions will enter into the collision processes again, and this will go on. After the cascade processes for stopping collisions take place, most of the ion energy is spent in the atomic motion, ν, and wasted as heat in ordinary detectors. Only the energy η is utilized in electronic excitation and contributes to the ionization or scintillation. The ratio η/ε is called the nuclear quenching factor, q_{nc}, or the Lindhard factor. The asymptotic equation for ν ($=\varepsilon - \eta$) for recoil ions in a homonuclear medium ($Z_1 = Z_2$) is given as follows (Lindhard et al., 1963):

$$\nu = \frac{\varepsilon}{1 + k \cdot g(\varepsilon)} \tag{31.19}$$

An approximate expression for $g(\varepsilon)$ is given by (Lewin and Smith, 1996)

$$g(\varepsilon) = 3\varepsilon^{0.15} + 0.7\varepsilon^{0.6} + \varepsilon \tag{31.20}$$

Once a value for q_{nc} is obtained, it can be used for the study of the track structure and quenching. The electronic LET ($LET_{el} = -d\eta/dR$), that is, the electronic energy deposited per unit length along the ion track, can be an important parameter to consider the radiation effects due to slow recoil ions (Hitachi, 2005, 2008). The value of LET_{el} obtained for recoil ions is close to the LET value for α-particles. This makes α-particles a good measure for decay times, the S/T ratio, etc., for recoil ions, as discussed in Section 31.2.

The excitation density in the recoil ion track can be as high as that in an α-track; therefore, scintillation quenching, q_{el}, is expected, as discussed in Section 31.2. The total quenching factor, q_T, is expressed as $q_{nc} \times q_{el}$. The track structure can be considered by using LET_{el} and ion velocity. Then, the value of q_{el} is estimated as discussed in Section 31.2. We have $T = q_{nc}E$ in Equation 31.9. Most of the δ rays produced by recoil ions do not have sufficient energy to escape from the core and form an undifferentiated core, that is, $T_c/T = 1$ for recoil ions. The quenching factors calculated for recoil ions in LAr and LXe (Hitachi, 2005) are shown in Figure 31.25 together with reported experimental values (Arneodo et al., 2000; Bernabei et al. 2002; Akimov et al., 2002; Aprile et al., 2005; Chepel et al., 2006).

Practically, the recoil ion-to-γ ratio, RN/γ, is a good parameter for the design and development of a detector for dark matter searches, and it can be expressed as (Hitachi, 2005)

$$\frac{RN}{\gamma} = \frac{q_{nc} \cdot q_{el}}{L_\gamma} \tag{31.21}$$

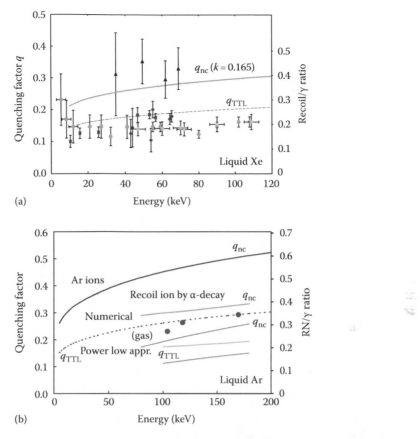

FIGURE 31.25 The nuclear, q_{nc}, and the total, $q_{TTL} = q_{nc} \times q_{el}$, quenching factors as a function of recoil ion energy in (a) LXe (Hitachi, 2005) and in (b) LAr. The values of RN/γ ratios are shown on the right axis. (a) The points on the graph denote experimental values (Arneodo et al., 2000; Akimov et al., 2002; Bernabei et al., 2002; Aprile et al., 2005; Chepel et al., 2006). (b) q values estimated for very heavy recoil ions (Pb) in α-decay are also shown for 80–180 keV, ● denotes experimental q_{nc} values in gas.

where L_γ is the scintillation efficiency for γ rays. It is less than 1 because of a reduced efficiency for the electron–ion recombination (see Figure 31.7). The value of q_{nc} is larger for light atoms as shown in Table 31.2. However, the value of W_{ph}, the (electronic) energy required to produce one photon, is larger for light atoms. These factors compensate each other. As a result, the number of photons, N_{ph}, per keV is typically 10–13 at about 50 keV, and it does not change much for LAr and LXe, as shown in Table 31.2. q_{el} for LNe was estimated simply by taking into consideration LET_{el} and the scintillation efficiency. The N_{ph} value estimated for LNe is small mainly because of a large W_{ph} value due to a low luminescence yield and electronic quenching. The main contribution to the error in RN/γ comes from an uncertainty in L_γ. Experimental values are shown in Figure 31.25a for LXe. The values reported are 0.26 for 387 keV recoil ions in LNe (Nikkel et al., 2008) and 0.28 for 65 keV recoil ions in LAr (Brunetti et al., 2005), and compare well with the estimated values.

Natural Ar and Xe are contaminated with ^{222}Rn. Daughter nuclei of Rn are produced in the ionized state, and drift to the electrodes or attach onto the wall. Their decay produces αs, βs, and very heavy recoil ions ($A \sim 200$). The quenching factor estimated by using the numerical calculation and the low-power approximation (Lindhard et al., 1963) for very heavy recoil ions in LAr is shown in Figure 31.25b. Very heavy ions of 100–200 keV recoiled in α-decay give a charge distribution quite similar to that for ions recoiled by WIMPs in detector media (Hitachi, 2008).

TABLE 31.2
Estimated Scintillation for Recoil Ions in RGLs

Liquid	Units	LNe	LAr	LXe[a]
Energy	keV	20	40	60
Range	μm	0.1	0.13	0.07
$q_{nc} = \eta/E$		0.4	0.37	0.28
$q_T = q_{nc} \times q_{el}$		(0.3)	0.22	0.19
RN/γ		(0.3)	0.26	0.24
N_{ph}	Photon/keV	(2)	11	13

Note: Values in parenthesis obtained by assuming $L_\gamma = 1$.
[a] Hitachi (2005).

31.4.3.2 Detectors for Dark Matter Searches

Dark matter detectors are installed deep underground to avoid cosmic rays, as shown, for example, in Figure 31.24 for the XMASS dark matter detector being constructed at Kamioka mine. High-energy neutrons from muon spallation in the rock are shielded by a water tank. The main detector is an 80 cm diameter 800 kg LXe. XMASS relies on shielding for background rejection. The outer layer of 20 cm LXe is used as an active shield. XMASS is a multipurpose detector aiming also at solar neutrinos and double-beta decay, as discussed in detail in Section 31.4.2. The final backgrounds are low-energy neutrinos or double-beta decay. Theoretically, γ's and recoil ions can be separated by observing the difference in the decay shape, as discussed in Section 31.2. A waveform analysis has been tested in LXe (ZEPLIN I, Alner, 2005). However, it is difficult to separate these effectively by the scintillation decay in LXe. The reason may be the fact that the number of photoelectrons observed is not enough. Also, the main difference in decay is in the recombination process, which can change with particle energy in LXe. Other LXe dark matter detectors rely on the scintillation and the charge for particle identification.

The discrimination by the decay shape is possible in LAr and LNe. The decay does not show the recombination component but the singlet and triplet states. The decay times for these states differ by several orders of magnitude. This difference can be exploited for particle identification and background rejection. Both, the differences in scintillation decay, and scintillation and charge yields for γ's and recoil ions can be simultaneously used in LAr. The principle for a dual-phase detector, a TPC with photon detection capability, is described in the following text.

The WArP prototype detector (WARP, 2008) uses 2.3 L (3.2 kg) of LAr and is in operation in the INFN LNGS. WArP uses a large difference in decay times in LAr, namely, 7 ns and 1.6 μs (Section 31.2.2), for the singlet and triplet states, respectively. The wavelength for LAr scintillation is quite short, and so, a wavelength shifter, such as tetra-phenyl-butadiene (TPB), is needed. The presence of cosmic [39]Ar is considered to be a major source of background. [39]Ar is a 565 keV β emitter. [39]Ar-depleted Ar is available via centrifugation or thermal diffusion; however, the method is expensive, particularly for a ton-scale detector. It has been found that Ar contained in underground wells shows a highly suppressed [39]Ar contamination (WARP, 2008). The 140 kg active target detector is under construction at LNGS. The detector is surrounded by a neutron shield consisting of 70 cm thick polyethylene and a 4π active neutron veto (8 t LAr with 303 PMTs). The inner dual-phase detector provides the x–y position by 37 PMTs. The z-coordinate is given by the time delay between S1 and S2 signals, as discussed in the following text. The particle discrimination capability using a waveform is shown in Figure 31.26.

ArDM (Rubbia, 2006; Otyugova, 2008) based on the dual-phase method and a system of large electron multiplier (LEM) is implemented for electron multiplication in the gas phase. ArDM is in the research and development stage. A ton-scale detector has a LAr target volume of 850 kg with a vertical drift length of 120 cm. A voltage up to ~400 kV is supplied by a Greinacher (Cockroft–Walton) chain.

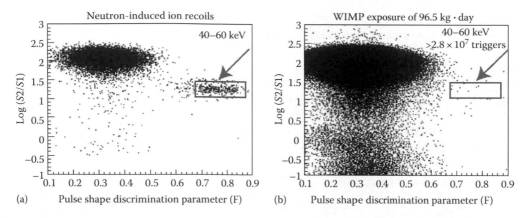

FIGURE 31.26 Particle discrimination capability in a LAr dark matter detector using the scintillation pulse shape (a) obtained with an Am–Be source and (b) obtained from WIMP exposure. The box shown is indicative (WARP, 2008). (From Benetti, P. et al., *Astroparticle Physics*, 28, 495, 2008. With permission.)

The DEAP/CLEAN (Boulay and Hime, 2006) detectors at SNOLAB, located in an active nickel mine in Sudbury, Canada, observe only the scintillation light from LAr or LNe. DEAP-1 uses 7 kg of LAr viewed by 2 PMTs of 5 in. LNe contains no long-lived radio isotopes and is easily purified using cold-traps. Mini-CLEAN is a planned WIMP detector containing 100 kg of LNe or LAr viewed by 32 PMTs (McKinsey, 2007). The ability to exchange the liquids having different sensitivities with WIMPs and fast neutrons will allow both of these event populations to be distinguished and characterized.

A typical liquid–gas dual-phase detector, XMASS II, for WIMP searches is shown in Figure 31.27. It is a position-sensitive LXe TPC having scintillation-observing capability. The difference in scintillation and charge yields due to γ's and recoil ions (Figure 31.3) is used for particle identification (Benetti et al., 1993a,b; Aprile et al., 2006). The liquid phase is basically a 3D detector with an allay of PMTs for scintillation detection. It uses the primary scintillation, $S1$, and the charge signal

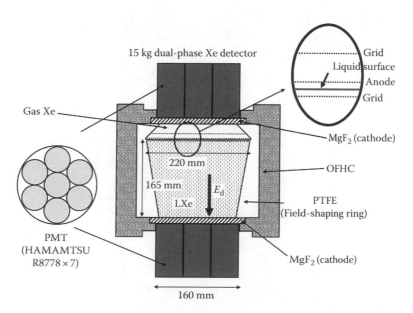

FIGURE 31.27 Schematic view of XMASS II showing a dual-phase LXe detector.

via the proportional scintillation, $S2$, in the gas phase for particle discrimination. The $S1$ signal also gives time zero for the detector. The side wall is made of polytetrafluoroethylene (PTFE), which has a high reflectivity for vuv light. The electrons produced by ionizing particles drift upward through the grid. A grid is placed just above the liquid surface to pick electrons from the liquid to the gas phase (Schmidt, 1997). The gas phase is a multiwire proportional scintillation counter where the electrons are accelerated by a high electric field to produce scintillation photons, $S2$. The photons are detected mainly by the PMT arrays above.

XENON10 (Angle et al., 2008a,b), the leading detector for the WIMP searches at LNGS, is shown in Figure 31.28. The particle discrimination capability of the dual-phase method has been demonstrated by XENON10, as shown in Figure 31.29.

The XENON100 uses 170 kg of LXe, which has a drift length of 30 cm viewed by 178 PMTs of 1 in. from the top and the bottom. Sixty-four veto PMTs view the outer layer of the several-cm-thick active LXe shield, which has a fiducial mass of 70 kg. The LXe detector is surrounded by polyethylene and lead, which act as a passive shield to reduce neutrons and γ's. The upgraded XENON100 version and XENON1T use the new QUPID, quartz photon-intensifying detectors (Fukasawa et al., 2010). The QUPID has a photocathode inside a semisphere quartz, where a high voltage of about $-10\,kV$ is applied. The photoelectrons are focused onto the APD at the center, which is set to the ground level. The QUPID is position sensitive and shows a fast response with clear single-photon separation.

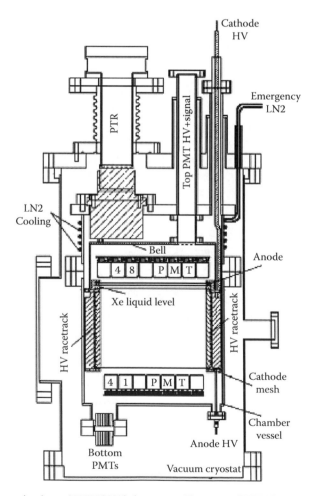

FIGURE 31.28   Schematic view of XENON10 detectors. (Courtesy of XENON.)

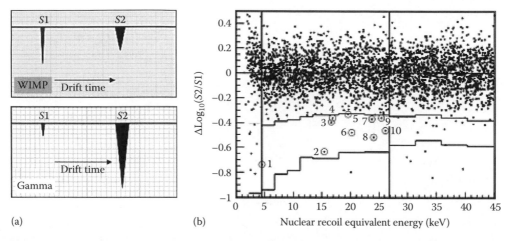

(a) (b) Nuclear recoil equivalent energy (keV)

FIGURE 31.29 A dual-phase LXe detector's ability to distinguish a recoil nucleus WIMP signal from the γ background. (a) The $S1/S2$ ratio is used to separate the WIMP signal from the γ background. (b) Results from WIMP search in XENON10 (Angle et al., 2008a,b). The WIMP signal region is defined between the two vertical lines (4.5–27 keV) and two zigzag lines. The 10 events are in the defined area; however, all are likely leakage events. (From Angle, J. et al., *Phys. Rev. Lett.*, 100, 021303, 2008a. With permission.)

ZEPLIN III (Akimov, 2009) and LUX use basically the same principle of operation as XMASS II or XENON. ZEPLIN III uses a high-purity Cu container to reduce the background from the container and is set at the U.K. Boulby mine. The target LXe has the shape of a pancake, 38 cm in diameter and 3.5 cm thick. A thin layer of LXe makes it possible to apply a high field to gain a large separation between γ's and nuclear recoils. LUX is to be constructed in the Homestake mine, United States, and is a 300 kg dual-phase LXe detector having a fiducial mass of 100 kg; a mass of 200 kg is used for self-shielding to reduce the large background contribution from PMTs.

31.5 APPLICATIONS IN MEDICAL IMAGING (PET)

PET is an imaging technique extensively applied in the field of medicine. γ rays emitted by the annihilation of positrons from emitters dispersed throughout the body are measured and, subsequently, the pattern of positron distribution is reconstructed. Usually, PET systems are constructed from many crystal scintillators surrounding the body to be imaged. Pairs of γ rays emitted at almost 180° from the positron annihilation point are detected by a pair of detectors that are placed on both sides of the annihilation point. If a photon pair is detected simultaneously in both detectors, the positron interaction point exists somewhere on the line between the two detectors (except scatter and random events).

31.5.1 LIQUID XENON COMPTON PET

High detection efficiency and high counting rate capability are essentially required for a PET detector to suppress the radiation dose for a patient. Additional requirements are good energy, position, and time resolutions to discriminate photons scattered in a patient and to reduce false coincidences. A LXe TPC with a scintillation trigger is suitable for this purpose. It is desirable to have 3D position information on the interaction event instead of the two-dimensional one, which is traditionally provided. This additional feature is fundamental for solving the parallax error, which occurs if the positron annihilation point is out of the detector (ring) center and causes a deterioration of the position resolution.

The first experiments used a prototype detector shown in Figure 31.30 (Chepel et al., 1995, 1997; Lopes and Chepel, 2003). The proposed design of the LXe PET ring is composed of about 200 multiwire cells disposed, as shown in Figure 31.30a. The detector is a LXe multiwire detector consisting of

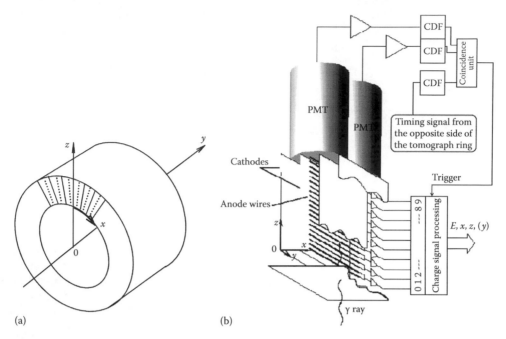

FIGURE 31.30 (a) Design of the positron tomograph ring based on multiwire cells. (b) Multiwire ionization cell, a basic element of the LXe PET detector (Lopes and Chepel, 2003). (Courtesy of LIP Coimbra.)

six ionization cells, each formed by two parallel cathode plates with a multiwire anode in the middle. Negative voltage is applied to the cathode, with the anode wires (50 μm diameter, 2.5 mm spacing) at ground. The 20 anode wires were connected in pairs to reduce the number of readout channels (see Figure 31.30b). The signal from each charge channel was fed to a low-noise charge-sensitive preamplifier followed by a linear amplifier, connected to a leading edge discriminator and a peak-sensing analog to digital converter (ADC). A typical noise level of 800 electrons FWHM was achieved for the charge readout. Two PMTs are fixed above the cells and partly (at least the photon entrance window) immersed in the LXe to improve light collection. The signals from the PMTs are used to select coincidences with a resolution of about 1 ns (FWHM) and provide a fast trigger for acquiring the signals from the anodes. The x-axis of the interaction point is determined with a precision better than 1 mm (FWHM) by measuring the electron drift time, triggered by the light signal. The identification of the pair of wires on which the charge is collected gives the depth of the interaction (z-direction), with a resolution estimated at 5 mm (twice the wire spacing). For the measurement of the y-direction, at first, the ratio of the amplitudes of the two PMTs was used but the position resolution was only ~10 mm (1σ). To improve this resolution, the use of a 2D mini-strip plate was tested with 122 keV γ rays and an improved position resolution of better than 2 mm (FWHM) was demonstrated (Solovov et al., 2002a,b, 2003).

While single-point position resolution is an important parameter of a LXe detector for PET, the double-point resolution, that is, the minimum distance required between two sources still to be separated in the image, is desirable. As discussed by Giboni et al. (2007), this double-point resolution depends not only on position resolution, but also on the statistics of events from the sources, as well as on the background from unrelated events. In an image, these factors determine the contrast besides the background from wrong Compton sequencing. There is a very strong background from genuine positron annihilation events with at least one γ ray Compton scattering even before leaving the patient (Figure 31.31). Most of these scatterings are forward-peaked, with a small angle to the original direction, and with only a small change in γ ray energy. To identify and reject this background as far as possible, energy resolution is by far the most important factor. The LXe TPC, proposed as Compton PET, relies on the improved energy resolution that results from the demonstrated sum of the anti-correlated ionization and scintillation signals (Aprile et al., 2007). The estimated FWHM energy

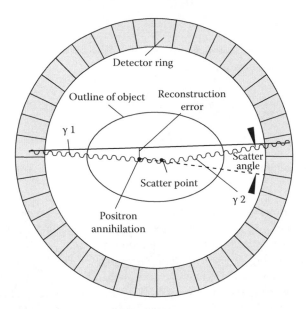

FIGURE 31.31 Measurement geometry in a PET detector ring. One γ ray (γ 2) scatters within the body, leading to the indicated reconstruction error. (From Giboni, K. et al., Compston positron emission tomography with a liquid xenon time projection chamber, JINST2:P10001. With permission.)

resolution of 4% is much better than crystal scintillators, for example, GSO (Gd_2SiO_5) is 15% and BGO ($Bi_4Ge_3O_{12}$) is 25%.

Compton kinematics not only determines the sequence of interaction points, but also constrains the direction of the incoming photon to the surface of a cone. The opening angle of the cone is the Compton scattering angle in the first interaction point. If the incoming directions are thus determined for both annihilation γ rays, even if one of the γ rays scatters inside the body of a patient, it is still possible to determine the original line on which the annihilation γ rays were emitted (Figure 31.32). The number of events with one of the γ rays scattering once in the body amounts to about twice the number of nonscattered events. Therefore, with Compton reconstruction, the number of acceptable events is nearly a factor 3 larger than the number of good events in a standard PET detector (Giboni et al., 2007). This increased detection efficiency results in a reduction of the radiation dose for the patient.

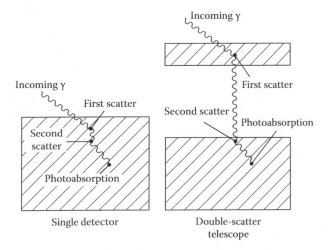

FIGURE 31.32 Principle of operation of a double-scatter telescope. (From Giboni, K. et al., Compston position emission tomography with a liquid xenon time projection chamber, JINST2:P10001. With permission.).

31.5.2 Liquid Xenon TOF-PET

The time-of-flight (TOF) system can measure the difference between the arrival times from two annihilation points using the following equation:

$$X = \frac{c \times (T_1 - T_2)}{2}$$

(31.22)

where

X is the position annihilation point
T_1 and T_2 are the arrival times of γ rays
c is the light velocity

The accuracy of the determination of the annihilation position is determined by the time resolution of the detectors. For example, to achieve a position resolution of 1.5 cm, a time resolution of 100 ps (FWHM) is required. However, the present time resolutions of actual detectors are over 350 ps, corresponding to a position resolution of more than 5 cm. Therefore, the TOF information does not improve the position resolution, but can be used to improve the signal-to-noise ratio in the reconstructed image.

LXe as a scintillator is one of the candidates for TOF-PET because of its fast response and high photon yield. A prototype, consisting of two LXe chambers, was constructed to study the feasibility. Each chamber has a LXe sensitive volume of $120 \times 60 \times 60 \, mm^3$ viewed by 32 square PMTs of 1 in. (Hamamatsu R5900-06AL12S-ASSY), with one side left uncovered as an entrance window for γ rays (Figure 31.33a, Nishikido et al., 2004). A schematic of the whole setup is shown in Figure 31.33b. The distance between the entrance planes is 70 cm. A test on the prototype was carried

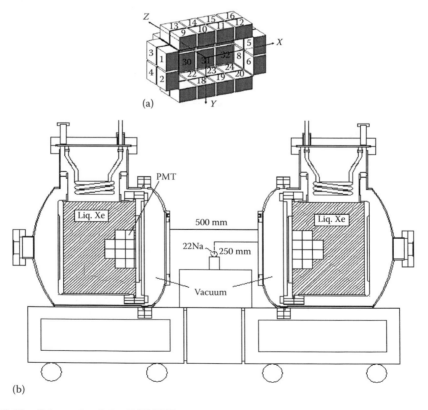

FIGURE 31.33 Schematic of the TOF-PET prototype. (a) Arrangement of 32 PMTs in LXe. (b) Cross-sectional view. (From Nishikido, F. et al., *Jpn. J. Appl. Phys.*, 43, 779, 2004. With permission.)

out with a ^{22}Na source placed in the center of the setup. The output signals from the PMTs of each chamber were recorded by charge-sensitive analog-to-digital converters, and the TOF information was recorded with time-to-digital converters. The conversion point of a γ ray was determined by calculating the center of gravity of the light output from the individual PMTs after applying a non-uniformity correction with the alpha source.

$$N = \sum N_j$$

$$X = \left(\frac{1}{N}\right)\sum X_j N_j$$

$$Y = \left(\frac{1}{N}\right)\sum Y_j N_j \tag{31.23}$$

$$Z = \left(\frac{1}{N}\right)\sum Z_j N_j$$

where
  N is the total number of photoelectrons
  N_j is the number of photoelectrons of the individual PMTs
  X_j, Y_j, and Z_j are the positions of the individual PMTs

The projection on the X-Y, X-Z, and Y-Z planes of distributions of conversion points from annihilation γ rays inside the effective volume are shown in Figure 31.34.

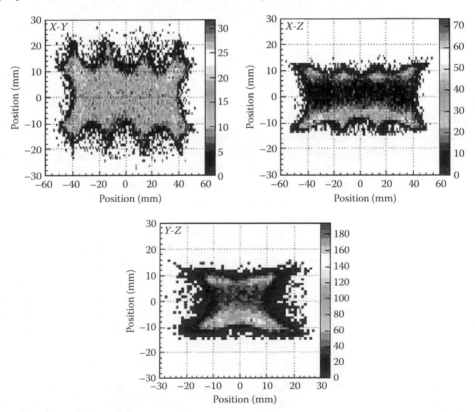

FIGURE 31.34 Position distributions of conversion points for annihilation γ rays in X-Y, X-Z, and Y-Z planes inside the sensitive area. (From Doke, T. et al., *Nucl. Instrum. Methods Phys. Res. A.*, 569, 863, 2006. With permission.)

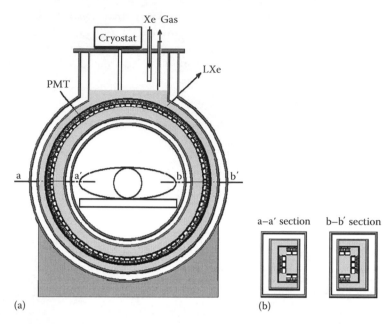

FIGURE 31.35 (a) Cross-sectional view of LXe scintillation TOF-PET seen from the front. (b) Two cross-sectional views seen from the top of TOF-PET. (From Doke, T. et al., *Nucl. Instrum. Methods Phys. Res. A.*, 569, 863, 2006. With permission.)

The energy resolution obtained from the sum of 32 PMTs was 26.0% (FWHM) (Doke et al., 2006). The resolution of the time-difference spectrum between the two detectors was 514 ps (FWHM) for annihilation γ rays. For a central volume of $5 \times 5 \times 5\,mm^3$, a better time resolution of 253 ps was obtained. It is expected that the time resolution improvement be inversely proportional to $\sqrt{N_{pe}}$, that is, the timing improves by an increase in the quantum efficiency of the PMT or the light collection efficiency of detectors. This is within reach, based on available PMTs with significantly better quantum efficiency, compared to the PMTs used in the prototype experiment. The number of background events effectively reduces with a time resolution of 300 ps. A full-scale TOF-PET, with an axial sensitive length of 24 cm, has been proposed by Nishikido et al. (2005) (see Figure 31.35).

Another LXe PET was proposed by Gallin-Martel et al. (2009), applying a light division method to measure the position in the axial direction. However, the reported position resolution was only about 10 mm (FWHM) in the central region, not competitive with what is achieved in commercially available crystal PET. Simulation results show much better performance, but the large discrepancy is not understood or explained yet.

Clearly, more studies are required to establish LXe Compton/TOF detectors as an alternative to classical crystal PET systems. LXe-based detectors require associated systems (from cryogenics to gas handling and purification), which are typically perceived as complex and costly, leading to some reluctance in their use, especially as diagnostic tools in medicine. In the last few years, however, many significant technological advances have resulted in large LXe detector systems operating for many months with excellent and stable performance. In particular, the LXe detectors deployed underground for dark matter searches have proven that long-term stability, reliability, and also safety issues can be confidently solved.

REFERENCES

Abe, K., Hosaka, J., Iida, T. et al. 2009. Distillation of liquid xenon to remove krypton. *Astropart. Phys.* 31: 290–296.
Aharrouche, M. 2007. The ATLAS liquid argon calorimeter: Construction, integration, commissioning and combined test beam results. *Nucl. Instrum. Methods Phys. Res. Sect. A* 581: 373–376.
Ahlen, S. P. 1980. Theoretical and experimental aspects of the energy loss of relativistic heavy ionizing particles. *Rev. Modern Phys.* 52: 121–173.

Akimov, D. 2009. Detectors for dark matter search (review). *Nucl. Instrum. Methods Phys. Res. Sect. A* 598: 275–281.

Akimov, D., Bewick, A., Davidge, D. et al. 2002. Measurement of scintillation efficiency and pulse shape for low energy recoils in liquid xenon. *Phys. Lett. B* 524: 245–251.

Alner, G. J., UK Dark Matter Collaboration. 2005. First limits on nuclear recoil events from ZEPLIN I galactic dark matter detector. *Astropart. Phys.* 23: 444–462.

Amerio, S., Amoruso, S., Antonello, M. et al. 2004. Design, construction and tests of the ICARUS T600 detector. *Nucl. Instrum. Methods Phys. Res. Sect. A* 527: 329–410.

Anderson, D. F. 1986. Photosensitive dopants for liquid argon. *Nucl. Instrum. Methods Phys. Res. Sect. A* 242: 254–258; New photosensitive dopants for liquid argon. *ibid.* 245: 361–365.

Angle, J. et al. (XENON Collaboration). 2008a. First results from the XENON10 dark matter experiment at the Gran Sasso National Laboratory. *Phys. Rev. Lett.* 100: 021303.

Angle, J. et al. 2008b. Limits on spin-dependent WIMP-nucleon cross sections from the xenon10 experiment. *Phys. Rev. Lett.* 101: 091301-1-5.

Aprile, E., Giboni, K. L., Majewski, P. et al. 2005. Scintillation response of liquid xenon to low energy nuclear recoils. *Phys. Rev. D* 72: 072006.

Aprile, E., Dahl, C. E., de Viveiros, L. et al. 2006. Simultaneous measurement of ionization and scintillation from nuclear recoils in liquid xenon for a dark matter experiment. *Phys. Rev. Lett.* 97: 081302-1-4.

Aprile, E., Giboni, K. L., Majewski, P. et al. 2007. Observation of anticorrelation between scintillation and ionization for MeV gamma rays in liquid xenon. *Phys. Rev. B* 76: 014115.

Aprile, E., Curioni, A., Giboni, K. L. et al. 2008. Compton imaging of MeV gamma-rays with the liquid xenon gamma-ray imaging telescope (LXeGRIT). *Nucl. Instrum. Methods Phys. Res. Sect. A* 593: 414–425.

Arneodo, F., Baiboussinov, B., Badertscher, A. et al. 2000. Scintillation efficiency of nuclear recoil in liquid xenon. *Nucl. Instrum. Methods Phys. Res. Sect. A* 449:147–157.

Arneodo, F., Benetti, P., Bonesini, M. et al. 2003. Performance of the 10 m^3 ICARUS liquid argon prototype. *Nucl. Instrum. Methods Phys. Res. Sect. A* 498: 292–311.

Arneodo, F., Benetti, P., Bonesini, M. et al. 2006. Performance of a liquid argon time projection chamber exposed to the CERN west area neutrino facility neutrino beam. *Phys. Rev. D* 74: 112001.

Asaf, U. and Steinberger, I. T. 1971. Wannier excitons in liquid xenon. *Phys. Lett.* 34A:207–208.

Aubert, B., Bazan, A., Cavanna, F. et al. 1991. Performance of a liquid argon electromagnetic calorimeter with an "accordion" geometry. *Nucl. Instrum. Methods Phys. Res. A* 309: 438–449.

Aubert, B., Ballansat, J., Colas, J. et al. 2005. Development and construction of large size signal electrodes for the ATLAS electromagnetic calorimeter. *Nucl. Instrum. Methods Phys. Res. A* 539: 558–594.

Badwar, G. D. 1973. Calculation of the Vavilov distribution allowing for electron escape from the absorber. *Nucl. Instrum. Methods* 109: 119–123.

Bakale, G., Sowada, U., and Schmidt, W. F. 1976. Effect of an electric field on electron attachment to SF_6, N_2O, and O_2 in liquid argon and xenon. *J. Phys. Chem.* 80: 2556–2559.

Baldini, G. 1962. Ultraviolet absorption of solid argon, krypton, and xenon. *Phys. Rev.* 128: 1562–1556.

Baldini, A., Bemporad, C., Cei, F. et al. 2005a. Absorption of scintillation light in a 100L LXe γ-ray detector and expected detector performance. *Nucl. Instrum. Methods Phys. Res. Sect. A* 545: 753–764.

Baldini, A., Bemporad, C., Cei, F. et al. 2005b. Transparency of a 100 liter liquid xenon scintillation calorimeter prototype and measurement of its energy resolution for 55 MeV photons. In *2005 IEEE International Conference on Dielectric Liquids,* June 26–July 1, 2005 (*ICDL 2005*), Coimbra, Portugal, pp. 337–340.

Baldini, A., Bemporad, C., and Cei, F. 2006. A radioactive point-source lattice for calibrating and monitoring the LXe calorimeter of the MEG experiment. *Nucl. Instrum. Methods Phys. Res. Sect. A* 565: 589–598.

Baranov, A., Baskakov, V., Bondarenko, G. et al. 1990. A liquid-xenon calorimeter for the detection of electromagnetic showers. *Nucl. Instrum. Methods Phys. Res. A* 294: 439–445.

Barr, G. D., Bruschini, C., Bocquet, C. et al. 1996. Performance of an electromagnetic liquid krypton calorimeter based on a ribbon electrode tower structure. *Nucl. Instrum. Methods Phys. Res. A* 370: 413–424.

Batley, J. R., Culling, A. J., Kalmus, G. et al. 2009. Precise measurement of the $K^\pm \to \pi^\pm e^+ e^-$ decay. *Phys. Lett. B* 677: 246–254.

Beaglehole, D. 1965. Reflection studies of excitons in liquid and solid xenon. *Phys. Rev. Lett.* 15: 551–553.

Benetti, P., Bettini, A., Calligarich, E. et al. 1993a. A simple and effective purifier for liquid xenon. *Nucl. Instrum. Methods Phys. Res. A* 329: 361–364.

Benetti, P., Bettini, A., Calligarich, E. et al. 1993b. A three-ton liquid argon time projection chamber. *Nucl. Instrum. Methods Phys. Res. Sect. A* 332: 395–412.

Bernabei, R. et al. 2002. Results with the DAMA/LXe experiment at LNGS. *Nucl. Phys. B* (*Proc. Suppl.*) 110: 88–90.

Bettini, A., Braggiotti, A., Casagrande, F. et al. 1991. A study of the factors affecting the electron life time in ultra-pure liquid argon. *Nucl. Instrum. Methods Phys. Res. Sect. A* 305: 177–186.

Birks, J. B. 1964. *The Theory and Practice of Scintillation Counting*. Oxford, U.K.: Pergamon.

Bolotnikov, A. and Ramsey, B. 1997. The spectroscopic properties of high-pressure xenon. *Nucl. Instrum. Methods Phys. Res. Sect. A* 396: 360–370.

Bolozdynya, A. I., Brusov, P. P., Shutt, T. et al. 2007. A chromatographic system for removal of radioactive [85]Kr from xenon. *Nucl. Instrum. Methods Phys. Res. Sect. A* 579: 50–53.

Boulay, M. G. and Hime, A. 2006. Technique for direct detection of weakly interacting massive particles using scintillation time discrimination in liquid argon. *Astropart. Phys.* 25: 179–182.

Brooks, M. L. et al. (MEGA Collaboration). 1999. New limit for the lepton-family number nonconserving decay μ^+ ($e^+\gamma$). *Phys. Rev. Lett.* 83: 1521–1524.

Brunetti, R. et al. 2005. WARP liquid argon detector for dark matter survey. *New Astron. Rev.* 49: 265–269.

Buckley, E., Campanella, M., Carugno, G. et al. 1989. A study of ionization electrons drifting over large distances in liquid Argon. *Nucl. Instrum. Methods Phys. Res. Sect. A* 275: 364–372.

Bunemann, O., Cranshaw, T. E., and Harvey, J. A. 1949. *Can. J. Res. Sect. A* 27: 191–206.

Cauwels, P. et al. 2001. Study of the atmospheric [85]Kr concentration growth in Gent between 1979 and 1999. *Radiat. Phys. Chem.* 61: 649–651.

Cennini, P., Cittolin, S., Revol, J. P. et al. 1994. Performance of a three-ton liquid argon time projection chamber. *Nucl. Instrum. Methods Phys. Res. Sect. A* 345: 230–243.

Cennini, P., Revol, J. P., Rubbia, C. et al. 1995. Improving the performance of the liquid argon TPC by doping with tetra-methyl-germanium. *Nucl. Instrum. Methods Phys. Res. Sect. A* 355: 660–662.

Chepel, V. Y. U., Lopes, M. I., Araújo, H. M. et al. 1995. Liquid xenon multiwire chamber for positron tomography. *Nucl. Instrum. Methods Phys. Res. A* 367: 58–61.

Chepel, V., Lopes, M. I., Kuchenkov, A. et al. 1997. Performance study of liquid xenon detector for PET. *Nucl. Instrum. Methods Phys. Res. A* 392: 427–432.

Chepel, V., Solovov, V., Neves, F. et al. 2006. Scintillation efficiency of liquid xenon for nuclear recoils with the energy down to 5 keV. *Astropart. Phys.* 26: 58–63.

Christophorou, L. G. 1988. Gas/liquid transition: Interphase physics. In *The Liquid State and Its Electrical Properties*, E. E. Kunhardt, L. G. Christophorou, and L. H. Lussen (eds.), pp. 283–316. New York: Plenum Pub. Corp.

Coakley, K. J. and McKinsey, D. N. 2004. Spatial methods for event reconstruction in CLEAN. *Nucl. Instrum. Methods Phys. Res. A* 522: 504–520.

Costantini, F. 1998. The liquid krypton calorimeter of NA48: First operation results. *Nucl. Instrum. Methods Phys. Res. A* 409: 570–574.

Doke, T. 1981. Fundamental properties of liquid argon, krypton and xenon as radiation detector media. *Port. Phys.* 12: 9–48.

Doke, T. and Masuda, K. 1999. Present status of liquid rare gas scintillation detectors and their new application to gamma-ray calorimeters. *Nucl. Instrum. Methods Phys. Res. A* 420: 62–80.

Doke, T., Hitachi, A., Kubota, S., Nakamoto, A., and Takahashi, T. 1976. Estimation of Fano factors in liquid argon, krypton, xenon and xenon-doped liquid argon. *Nucl. Instrum. Methods* 134: 353–357.

Doke, T., Hitachi, A., Kikuchi, J. et al. 1985. Estimation of the fraction of electrons escaping from recombination in the ionization of liquid argon with relativistic electrons and heavy ions. *Chem. Phys. Lett.* 115: 164–166.

Doke, T., Crawford, H., Hitachi, A. et al. 1988. LET dependence of scintillation yields in liquid argon. *Nucl. Instrum. Methods Phys. Res. A* 269: 291–296.

Doke, T., Kikuchi, J., and Nishikido, F. 2006. Time-of-flight positron emission tomography using liquid xenon scintillation. *Nucl. Instrum. Methods Phys. Res. A* 569: 863–871.

Ellis, J. and Flores, R. A. 1991. Elastic supersymmetric relic-nucleus scattering revisited. *Phys. Lett. B* 263: 259–266.

Fanti, V., Lai, A., Marras, D. et al. NA48 Collaboration. 1999. A new measurement of direct CP violation in two pion decays of the kaon. *Phys. Lett. B* 465: 335–348.

Fanti, V., Lai, A., Marras, D. et al. 2007. The beam and detector for the NA48 neutral kaon CP violation experiment at CERN. CERN-PH-EP/2007-006.

Freedman, W. and Turner, M. 2003. Colloquium: Measuring and understanding the universe. *Rev. Mod. Phys.* 75: 1433–1447.

Fukasawa, A., Arisaka, K., Wang, H., and Suyama, M. 2010. QUIPID, a single photon sensor for extremely low radioactivity. *Nucl. Instrum. Methods. Phys. Res. Sect. A*, in press.

Gaitskell, R. 2004. Direct detection of dark matter. *Annu. Rev. Nucl. Part. Sci.* 54: 315–359.

Gallin-Martel, M. L., Gallin-Martel, L., Grondin, Y. et al. 2009. A liquid xenon positron emission tomograph for small imaging: First experimental results of prototype cell. *Nucl. Instrum. Methods Phys. Res. A* 599: 275–283.

Gatti, E., Padovini, G., Quartapelle, L. et al. 1979. Considerations for the design of a time projection liquid argon ionization chamber. *IEEE Trans. Nucl. Sci.* 26: 2910–2932.

Giboni, K., Aprile, E., Doke, T. et al. 2007. Compton positron emission tomography with a liquid xenon time projection chamber. *JINST* 2: P10001.

Haruyama, T. 2002. Boiling heat transfer characteristics of liquid xenon. *Adv. Cryog. Eng.* 47: 1499–1506.

Hilt, O., Schmidt, W. F., and Khrapak, A. G. 1994. Ionic mobilities in liquid xenon. *IEEE Trans. Dielectr. Electr. Insul.* 1: 648–656.

Hirschfelder, J. O., Curtiss, C. F., and Bird, R. B. 1954. *Molecular Theory of Gases and Liquids.* New York: Wiley.

Hitachi, A. 1984. Exciton kinetics in condensed rare gases. *J. Chem. Phys.* 80: 745–748.

Hitachi, A. 2005. Properties of liquid xenon scintillation for dark matter searches, *Astropart. Phys.* 24: 247–256.

Hitachi, A. 2008. Bragg-like curve for dark matter searches: Binary gases. *Radiat. Phys. Chem.* 77: 1311–1317, A W-value for CF_4 used in this reference was too large. A proper value of 34.3 eV (G.F. Reinking et al., *J. Appl. Phys.* 60, 499, 1986) makes the ionization 1.57 times that shown in Figure 31.7 (the right axis) of this reference.

Hitachi, A., Takahashi, T., Funayama, N. et al. 1983. Effect of ionization density on the time dependence of luminescence from liquid argon and xenon. *Phys. Rev. B* 27: 5279–5285.

Hitachi, A., Yunoki, A., Doke, T., and Takahashi, T. 1987. Scintillation and ionization yield for α particles and fission fragments in liquid argon. *Phys. Rev. A* 35: 3959–3958.

Hitachi, A., LaVerne, J. A., and Doke, T. 1992a. Effect of an electric field on luminescence quenching in liquid argon. *Phys. Rev. B* 18: 540–543.

Hitachi, A., Doke, T., and Mozumder, A. 1992b. Luminescence quenching in liquid argon under charged-particle impact: Relative scintillation yield at different liner energy transfers. *Phys. Rev. B* 18: 11463–11470.

Hitachi, A., LaVerne, J. A., Kolata, J. J., and Doke, T. 1994. Energy resolution of allene doped liquid argon detectors for ions of energy 23–34 MeV/amu. *Nucl. Instrum. Methods Phys. Res. Sect. A*: 546–550.

Hitachi, A., Ichinose, H., Kikuchi, J. et al. 1997. Photoionization quantum yield of organic molecules in liquid argon and xenon. *Phys. Rev. B* 55, 5742–5748.

Hitachi, A., LaVerne, J., Kolata, J. J., and Doke, T. 2002. Field effects on ionic and excitonic quenching for heavy ions in liquid Ar. *IEEE Trans. Dielectr. Electr. Insul.* 9, 45–47.

Hitachi, A., Chepel, V., Lopes, M. I., and Solovov, V. N. 2005. New approach to the calculation of refractive index of liquid and solid xenon. *J. Chem. Phys.* 123: 234508-1–234508-6, and references therein.

Holroyd, R. A. 2004, Electrons in nonpolar liquids. In *Charged Particle and Photon Interactions with Matter*, A. Mozumder and Y. Hatano (eds.). New York: Marcel Dekker, Inc.

Holroyd, R. A. and Schmidt, W. F. 1989. Transport of electrons in nonpolar fluids. *Annu. Rev. Phys. Chem.* 40: 439–468.

Hoshi, Y., Masuda, K., and Doke, T. 1982. Theoretical estimation of energy resolution in liquid argon sampling calorimeter. *Jpn. J. Appl. Phys.* 21:1086–1094.

Hung, S. S. S., Gee, N., and Freeman, G. R. 1991. Ionization of liquid argon by x-rays: Effect of density on electron thermalization and free ion yields. *Radiat. Phys. Chem.* 37: 417–421.

Hynes, J. T. 1985. Chemical reaction dynamics in solution. *Annu. Rev. Phys. Chem.* 36: 573–597.

Igarashi, Y. et al. 2000. Radioactive noble gases in surface air monitored at MRI, Tsukuba, before and after the JCO accident. *J. Environ. Radioact.* 50:107–118.

Ishida, N., Chen, M., Doke, T. et al. 1997. Attenuation length measurements of scintillation light in liquid rare gases and their mixtures using an improved reflection suppresser. *Nucl. Instrum. Methods Phys. Res. A* 384: 380–386.

Jaffe, G. 1913. Zur Theorie der Ionisation in Kolonner. *Annalen der Physik* 347: 303–344.

Jeitler, M. 2002. The NA48 liquid-krypton calorimeter. *Nucl. Instrum. Methods Phys. Res. A* 494: 373–377.

Jortner, J., Meyer, L., Rice, S. A., and Wilson, E. G. 1965. Localized excitons in condensed Ne, Ar, Kr, and Xe. *J. Chem. Phys.* 42: 4250–4253.

Koizumi, H. 1994. Photoionization quantum yield for liquid squalane and squalene estimated from photoelectron emission yield. *Chem. Phys. Lett.*, 219: 137–142.

Kramers, H. A. 1952. On a modification of Jaffe's theory of column-ionization. *Physica* 18: 665–675.

Kubota, S., Nakamoto, A., Takahashi, T. et al. 1976. Evidence of the existence of exciton states in liquid argon and exciton-enhanced ionization from xenon doping. *Phys. Rev. B* 13: 1649–1653.

Kubota, S., Hishida, M., and Ruan (Gen), J. 1978. Evidence for a triplet state of the self-trapped exciton states in liquid argon, krypton and xenon. *J. Phys. C* 11: 2645–2651.

Kubota, S., Hishida, M., Suzuki, M., and RuanGen, J. Z. 1979. Dynamical behavior of free electrons in the recombination process in liquid argon, krypton and xenon. *Phys. Rev. B* 20: 3486–3496.

Kubota, S. et al. 1980. Specific-ionization-density effect on the time dependence of luminescence in liquid xenon. *Phys. Rev. B* 21: 2632–2634.

Kuo, C.-Y. and Keto, J.W. 1983. Dissociable recombination in electron-beam excited argon at high pressure. *J. Chem. Phys.* 78: 1851–1860.

Lai, A., Marras, D., Batley, J. R. et al. 2003. Precise measurements of the $K_S \rightarrow \gamma\gamma$ and $K_L \rightarrow \gamma\gamma$ decay rates. *Phys. Lett. B* 551: 7–15.

Laporte, P. et al. 1980. Intermediate and Wannier excitons in fluid xenon. *Phys. Rev. Lett.* 45: 2138–2140.

LaVerne, J., Hitachi, A., Kolata, J. J., and Doke, T. 1996. Scintillation and ionization in allene-doped liquid argon irradiated with [18]O and [36]Ar ions of 30 MeV/u. *Phys. Rev. B* 54: 15724–15729.

Le Comber, P. G., Loveland, R. J., and Spear, W. E. 1975. Hole transport in the rare-gas solids Ne, Ar, Kr, and Xe. *Phys. Rev. B* 11: 3124–3130.

Lewin, J. D. and Smith, P. F. 1996. Review of mathematics, numerical factors, and corrections for dark matter experiments based on elastic nuclear recoil. *Astropart. Phys.* 6: 87–112.

Lindhard, J., Nielsen, V., Sharff, M., and Thomsen, P. V. 1963. Integral equations governing radiation effects. *Matematisk-fysiske Meddelelser Danske Videnskabernes Selskab* 33(10): 1–42.

Lopes, M. I. and Chepel, V. 2003. Liquid rare gas detectors: Recent development and applications. *IEEE Trans. Dielectr. Electr. Insul.* 10: 994–1005.

Magee, J. L. and Huang, J. T. J. 1972. Triplet formation in ion recombination in spurs. *J. Chem. Phys.* 76: 3801–3805.

Martin, M. 1971. Exciton self-trapping in rare-gas crystals. *J. Chem. Phys.* 54: 3289–3299.

Masuda, K., Doke, T., and Takahashi, T. 1981. A liquid xenon position sensitive gamma-ray detector for positron annihilation experiments. *Nucl. Instrum. Methods* 188: 629–638.

Masuda, K., Doke, T., Hitachi, A. et al. 1989a. Energy resolution for alpha particles in liquid argon doped with allene. *Nucl. Instrum. Methods Phys. Res. A* 279: 560–566.

Masuda, K., Shibamura, E., Doke, T. et al. 1989b. Relation between scintillation and ionization produced by relativistic heavy ions in liquid argon. *Phys. Rev. A* 39: 4732–4734.

McCabe, W. L. and Smith, J. C. 1976. *Unit Operations of Chemical Engineering*, 3rd edn. New York: McGraw-Hill.

McKinsey, D. N. 2007. The Mini-CLEAN experiment. *Nucl. Phys. B* (Proc. Suppl.) 173: 152–155.

Mihara, S., Doke, T., Kamiya, Y. et al. 2002. Development of a liquid Xe photon detector for $\mu^+ \rightarrow e^+\gamma$ decay search experiment at PSI. *IEEE Trans. Nucl. Sci.* 49: 588–591.

Mihara, S., Doke, T., Haruyama, T. et al. 2004. Development of a liquid-xenon photon detector towards the search for a muon rare decay mode at Paul Scherrer Institute. *Cryogenics* 44: 223–228.

Mihara, S., Haruyamma, T., Iwamoto, T. et al. 2006. Development of a method for liquid xenon purification wring a cryogenic centrifugal pump. *Cryogenics* 46: 688–693.

Miller, L. S., Howe, S., and Spear, W. E. 1968. Charge transport in solid and liquid Ar, Kr and Xe. *Phys. Rev.* 166: 871–878.

Miyajima, M. et al. 1974. Average energy expended per ion pair in liquid argon. *Phys. Rev. A* 9: 1438–1443. *ibid.* 10: 1452–1452.

Mori, T. MEG collaboration. 1999. Search for $\mu^+ \rightarrow e^+\gamma$ down to 10^{-14} branching ratio. Research Proposal to Paul Scherrer Institut.

Mozumder, A. 1995. Free-ion yield and electron-ion recombination rate in liquid xenon. *Chem. Phys. Lett.* 245: 359–363.

Mozumder, A. 1999. *Fundamentals of Radiation Chemistry*. San Diego, CA: Academic Press.

Mozumder, A. and Hatano, Y. (eds.). 2004. *Charged Particle Interactions with Matter: Chemical, Physicochemical, and Biological Consequences with Applications*. New York: Marcel Dekker.

Mozumder, A., Chatterjee, A., and Magee, J. L. 1968. Theory of radiation chemistry IX. Model and structure of heavy particle tracks in water. In *Radiation Chemistry*, Advances in Chemistry Series 81, R. F. Gould (ed.), p. 27. Washington, DC: American Chemical Society.

Mulliken, R. S. 1964. Rare-gas and hydrogen molecule electronic states, noncrossing rule, and recombination of electrons with rare-gas and hydrogen ions. *Phys. Rev.* 136: 962–965.

Nakagawa, K., Kimura, K., and Ejiri, A. 1993. Photoionization quantum yield of TMAE (tetrakis-dimethyl-amino-ethylene) in supercritical xenon fluid. *Nucl. Instrum. Methods Phys. Res. Sect. A* 327: 60–62.

Nikkel, J. A. et al. 2008. Scintillation of liquid neon from electronic and nuclear recoils. *Astropart. Phys.* 29: 161–169.

Nishikido, F., Doke, T., Kikuchi, J. et al. 2004. Performance of a prototype of liquid xenon scintillation detector system for positron emission tomography. *Jpn. J. Appl. Phys.* 43:779–784.

Nishikido, F., Doke, T., Kikuchi, J. et al. 2005. Performance of a prototype of liquid xenon scintillation detector system for time of flight type positron emission tomography with improved photomultipliers. *Jpn. J. Appl. Phys.* 44: 5193–5198.

Okada, H., Doke, T., Kashiwagi, T. et al. 2000. Liquid Xe homogeneous electro-magnetic calorimeter. *Nucl. Instrum. Methods Phys. Res. A* 451: 427–438.

Onsager, L. 1938. Initial recombination of ions. *Phys. Rev.* 54: 554–557.

Otyugova, P. 2008. (ArDM collaboration), The ArDM, a ton-scale liquid argon experiment for direct dark matter detection. *J. Phys.: Conf. Ser.* 120: 042023 (1–3).

Platzman, R. L. 1961. Total ionization in gases by high-energy particles: An appraisal of our understanding. *J. Appl. Radiat. Isot.* 10: 116–127.

Pack, J. L., Voshall, R. E., and Phelps, A. V. 1962. Drift velocity of slow electrons in krypton, xenon, deuterium, carbon monoxide, carbon dioxide, water vapor, nitrous oxide and ammonia. *Phys. Rev.* 127: 2084–2089.

Peleganchuk, S. 2009. Liquid noble gas calorimeters at Budker INP. *Nucl. Instrum. Methods Phys. Res. A* 598: 248–252.

Rabinovich, V. A., Vasserman, A. A., Nedostup, V. I., and Veksler, L. S. 1988. *Thermophysical Properties of Neon, Argon, Krypton and Xenon*. Washington, DC: Hemisphere Pub. Corp.

Rahman, A. 1964. Correlations in the motion of atoms in liquid argon. *Phys. Rev. A* 136: 405–411.

Rubbia, C. 1977. The liquid-argon time projection chamber: A new concept for Neutrino Detector. CERN-EP/77-08.

Rubbia, C. 2006. Proc. TAUP 2005. *J. Phys.: Conf. Ser.* 36: 133–135.

Schmidt, W. F. 1997. *Liquid State Electronics of Insulating Liquids*. New York: CRC, Chapter 6.

Schriever, R., Chergui, M., Kunz, H., Stepanenko, V., and Schwentner, N. 1989. Cage effect for the abstraction of H from H_2O in Ar matrices. *J. Chem. Phys.* 91: 4128–4133.

Schwentner, N., Koch, E. E., and Jortner, J. 1985. *Electronic Excitation in Condensed Rare Gases*. Berlin, Germany: Springer-Verlag.

Seidel, G. M., Lanou, R. E., and Yao, W. 2002. Rayleigh scattering in rare-gas liquids. *Nucl. Instrum. Methods Phys. Res. Sect. A* 489: 189–194.

Shibamura, E., Crawford, H. J., Doke, T. et al. 1987. Ionization and scintillation produced by relativistic Au, He and H ions in liquid argon. *Nucl. Instrum. Methods Phys. Res. Sect. A* 260: 437–442.

Sinnock, A. C. 1980. Refractive indices of the condensed rare gases, argon, krypton and xenon. *J. Phys. C: Solid. Phys.* 13: 2375–2391.

Solovov, V., Chepel, V., Pereira, A. et al. 2002a. Two-dimensional readout in a liquid xenon ionisation chamber. *Nucl. Instrum. Methods Phys. Res. A* 477: 184–190.

Solovov, V., Chepel, V., Lopes, M. I. et al. 2002b. Liquid-xenon γ-camera with ionisation readout. *Nucl. Instrum. Methods Phys. Res. A* 478: 435–439.

Solovov, V., Chepel, V., Lopes, M. I. et al. 2003. Mini-strip ionization chamber for γ-ray imaging. *IEEE Trans. Nucl. Sci.* 50: 122–125.

Sowada, U., Warman, J. M., and de Haas, M. P. 1982. Hot-electron thermalization in solid and liquid argon, krypton, and xenon. *Phys. Rev. B* 25, 3434–3437.

Spergel, D. N. 1988. Motion of the Earth and the detection of weakly interacting massive particles. *Phys. Rev. D* 37, 1353–1355.

Spooner, N. J. C. 2007. Direct dark matter searches. *J. Phys. Soc. Jpn.* 76: 111016.

Steinberger, I. T. and Schnepp, O. 1967. Wannier excitons in solid xenon. *Solid State Commun.* 5: 417–418.

Sumner, T. J. 2002. Experimental searches for dark matter. http://www.livingreviews.org/Articles/Volume5/2002–4sumner/.

Suzuki, Y. 2008. XMASS experiment. In *IDM Conference*, Stockholm, Sweeden.

Suzuki, S., Doke, T., Hitachi, A. et al. 1986a. Photoionization effect in liquid xenon doped with triethylamine (TEA) or trimethylamine (TMA). *Nucl. Instrum. Methods Phys. Res. Sect. A* 245: 78–81.

Suzuki, S., Doke, T., Hitachi, A. et al. 1986b. Photoionization in liquid argon doped with trimethylamine or triethylamine. *Nucl. Instrum. Methods Phys. Res. Sect. A* 245: 366–372.

Takahashi, T., Konno, S., Hamada, T. et al. 1975. Average energy expended per ion pair in liquid xenon. *Phys. Rev. A* 12, 1771–1775.

Tanaka, M., Doke, T., Hitachi, A. et al. 2001. LET dependence of scintillation yields in liquid xenon. *Nucl. Instr. Methods A* 457: 454–463, The LET values for α particles in Figures 31.4 and 31.12 of this article were wrong. They should be ~400 MeV · g^{-1} · cm^2 in liquid Xe.

Thomas, D., Imel, A., and Biller, S. 1988. Statistics of charge collection in liquid argon and liquid xenon. *Phys. Rev. A* 38: 5793–5800.

WARP. 2008. First results from a dark matter search with liquid argon at 87 K in the Gran Sasso underground laboratory. *Nucl. Instrum. Methods Phys. Res. Sect. A* 587: 46–51.

Watanabe, K. and Zelikoff, M. 1953. Absorption coefficients of water vapor in vacuum ultraviolet. *J. Opt. Soc. Am.* 43: 753–754.

Watanabe, T. and Katuura, K. 1967. Ionization of atoms by collisions with excited atoms. II. A formula without the rotating atom approximation. *J. Chem. Phys.* 47: 800–810.

Watanabe, K., Edward, C., Inn, Y., and Zelikoff, M. 1953. Absorption coefficients of oxygen in the vacuum ultraviolet. *J. Chem. Phys.* 21: 1026–1030.

Wojcik, M., Tachiya, M., Tagawa, S., and Hatano, Y. 2004. Electron-ion recombination in condensed matter. Geminate and bulk recombination processes. In *Charged Particle Interactions with Matter: Chemical, Physicochemical, and Biological Consequences with Applications.* A. Mozumder and Y. Hatano (eds.). New York: Marcel Dekker, Inc.

Yokota, M. and Tanimoto, O. 1967. Effects of diffusion on energy transfer by resonance. *J. Phys. Soc. Jpn.* 22: 779–784.

Yoshino, K., Sowada, U., and Schmidt, W. F. 1976. Effect of molecular solutes on the electron drift velocity in liquid Ar, Kr and Xe. *Phys. Rev. A* 14: 438–444.

Zwicky, F. 1933. Die Rotverschiebung von extragalaktischen Nebeln. *Helv. Phys. Acta* 6: 110–127.

32 Applications of Ionizing Radiation to Environmental Conservation

Koichi Hirota
Japan Atomic Energy Agency
Takasaki, Japan

CONTENTS

32.1 INTRODUCTION

New technologies and processes have promoted our economic growth and have been remarkably changing our lifestyles. They have created a sophisticated society and we have been enjoying comfortable lives. However, technological innovation is not always welcome. It destroys the environment and puts us in a dangerous situation. The emission of inorganic pollutants such as nitrogen oxides (NO_X) and sulfur dioxide (SO_2) has been remarkably reduced by state-of-the-art technologies during the past 50 years. However, industrial countries have been using a lot of oil and coal to obtain energy for electricity, vehicles, chemical products, etc. The combustion of oil and coal produces large amounts of carbon dioxide, which causes the greenhouse effect. For this reason, energy sources have been diversified from conventional oil and coal to atomic, solar, and biomass energies, so as to reduce the emission of carbon dioxide and geopolitical risk against the energy supply. The EU-15 emission trend for NO_X and SO_2 shows emission reductions of 31% and 70%, respectively, from 1990 to 2004 (European Environmental Agency, 2006). Although the developing countries also try to increase the consumption of environment-friendly energies instead of fossil fuels, the steep demand for energy due to the recent economic growth makes it difficult to reduce the usage of fossil fuels. This causes a large amount of emission of NO_X, SO_2, as well as carbon dioxide to the atmosphere (Streets et al., 2000; Kato and Akimoto, 2007; Gaffney and Marley, 2009; Ramanathan and Feng, 2009). According to estimated long-term SO_2 emission trends for Asia (Carmichael et al., 2002), even if various chemical processes emitting SO_2 are operated under current legislation, the emission of SO_2 will increase. In particular, China is the major source of NO_X and SO_2 emissions. The emission of SO_2 in 1995 was about 25,200 kt in China. It will increase by about 60,700 kt, if emission control is not implemented. Even with emission control, the SO_2 emission would be expected to be 30,600 kt (Streets and Waldhoff, 2000). The emission of NO_X and SO_2 leads to acid rain that is detrimental to forests, crops, and building structures (Gaffney et al., 1987; Irwin and Williams, 1988; Xie et al., 2004). While the industrial countries have effected a sufficient reduction of NO_X and SO_2 emissions, it is still an urgent issue for the developing countries.

Volatile organic compounds (VOCs) are essential chemicals in various manufacturing processes such as surface painting, petroleum refining, and metal cleaning (Theloke and Friedrich, 2007; Celebi and Vardar, 2008; Jia et al., 2008). More than 300 VOCs were found at both indoor and outdoor locations in the ambient environment. Although the VOC concentration range in the indoor and outdoor environments is quite low in the order of parts per billion (ppb) (Shah and Singh, 1988; Mohamed et al., 2002), people spend a lot of time under indoor environments, where harmful VOCs such as formaldehyde, toluene, and chlorophenol exist. They cause heavy damage to the body and the environment. Long exposure even at a low concentration of VOC may cause cancer. Headache, nausea, and stimulation of the eyes and nose are acute toxic symptoms of VOCs. Volatile organic compounds also have an influence on the environment, where photochemical oxidant formation, stratospheric ozone depletion, and tropospheric ozone formation take place. Photochemical oxidant formation, caused by tropospheric ozone formation, is accelerated as hydrocarbon concentrations increase. Stratospheric ozone depletion leads to excess radiation of UV rays on the earth, which induces skin cancer. Many OECD countries have made legislations for the emission control of VOCs (Ministry of the Environment, 2009a), as shown in Table 32.1. They established their own standards for target facilities in large scale, since middle- and small-scale manufactures cannot afford to cope with the reduction of VOCs. In contrast, there is no national legislation in China, where about 20.1 Mt VOC were emitted in 2005. The major sources of the VOC emissions are road transportation, industrial solvent use, and fuel combustion. The VOC emission from gasoline and diesel vehicles used in road transportation accounts for about 25% of the total emission (Wei et al., 2008). Also, the VOC emission in China will increase by 20.7 Mt in 2030, which is about 11 times higher than that in Japan (Klimont et al., 2001). In response to the OECD recommendation, the Government of Japan issued a new legislation of Pollutant Release and Transfer Register (PRTR) in 1999 (Ministry of the Environment, 2009b). Under the PRTR legislation, industrial corporations have to periodically report on the nature and quantity of pollutants emitted and the

TABLE 32.1
OECD Legislation for VOC Emission Control

Nation	Legislation	Number of Target Chemicals	Target Facilities	Beginning Period
United States	TRI[a]	667	M.I.[f]	1987
Canada	NPRI[b]	273	M.I.[f]	1993
Australia	NPI[c]	90	M.I.[f]	1998
United Kingdom	PI[d]	170	M.I.[f]	1991
Holland	IEI[e]	67	Facilities[g]	1976
Japan	PRTR	354	M.I.[f]	2001

Source: Hirota, K. et al., *Radiat. Phys. Chem.*, 45, 649, 1995. With permission.

[a] Toxic release inventory.
[b] National pollutant release inventory.
[c] National pollutant inventory.
[d] Pollution inventory.
[e] Integrated emission inventory.
[f] Manufacturing industries and others determined by the number of employee and annual transaction volume.
[g] Facilities required for the allowance under environmental management low.

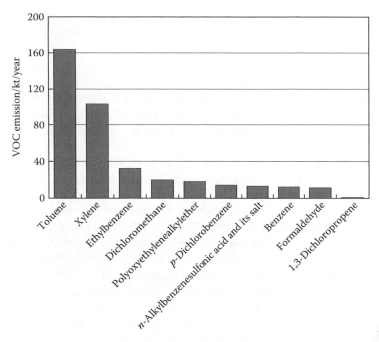

FIGURE 32.1   Emission amount of the 10 highest-ranked chemicals in 2006.

environmental media they are released into. The 2006 PRTR in Japan has reported that 0.46 Mt/year of registered chemicals were totally emitted from notified and un-notified facilities, mobile sources, and households to the environment (Ministry of the Environment, 2009c). Figure 32.1 shows the emission amounts of the 10 highest-ranked chemicals in Japan (Ministry of the Environment, 2009c). The total emission of these chemicals, of which nine are VOCs, accounts for 86% of the total emission. The emission control of VOCs is not sufficient to save our lives and the environment.

Dioxin is one of the most toxic man-made chemicals, involving polychlorinated dibenzo-p-dioxins (PCDD) and polychlorinated dibenzofurans (PCDF). There are 210 PCDD/F isomers, the toxicity of which depends on the number of substituent Cl and their positions on the rings. The toxic equivalent (TEQ) is defined to evaluate the toxicity of the isomers because not all the isomers show the same amount of risk. The total TEQ of a dioxin mixture is calculated using the toxic equivalency factors (TEFs). The TEF of 2,3,7,8-tetra-chlorinated dibenzo-p-dioxin (TeCDD), the most toxic dioxin of all, is defined as 1. Thus, the TEQ concentration of a dioxin mixture (ng I(international)-TEQ) can be calculated as follows:

$$I - TEQ = \sum c_i TEF \tag{32.1}$$

where c_i is the mass concentration of each isomer (ng). When co-planar polychlorinated biphenyls (co-PCB) is included in dioxin, the TEQ concentration is expressed in a unit of ng WHO (World Health Organization)-TEQ.

Dioxins have been anthropogenically emitted from many sources such as incineration facilities and industrial sites (Carroll et al., 2001; Dyke and Amendola, 2007; Mininni et al., 2007; Choi et al., 2008). The emission of dioxin from the major source, municipal solid waste incinerators and hospital waste incinerators, has been drastically reduced worldwide. For instance, about 4000 g I-TEQ of dioxin was released to the environment from municipal solid waste incinerators in Europe in 1995. It decreased by about 200 g I-TEQ in 2005. In the case of hospital waste incinerators, about 90% reduction of dioxin emission was achieved in 2005, compared with a emission of 2000 g I-TEQ in 1985 (Quaβ et al., 2004). Japan also had made many efforts to comply with a new law on the emission of PCDD/F established in 1999 by the Government of Japan. The municipal solid waste incinerator is the largest source for emitting dioxin, because about 80% of solid waste is incinerated in Japan (Kishimoto et al., 2001). A successful reduction from about 3000 g WHO-TEQ in 1999 to 290 g WHO-TEQ in 2007 was achieved by the replacement of electric precipitators with bag filters. The improvement of incineration conditions such as turbulence, temperature, and time also contributed to reduction of PCDD/F emissions (Ministry of the Environment, 2009d). Although dioxin emissions have progressively reduced worldwide, the risk of dioxin to the body and the environment still remains (Li et al., 2008; Zhang et al., 2008; Venier et al., 2009). Some of the local governments and the small islands in Japan cannot afford to upgrade existing municipal solid waste incinerators or install new ones. They need to request neighboring governments to incinerate waste, which is a big burden financially.

It is well known that endocrine disrupting chemicals (EDCs) such as bisphenol A, nonylphenol, and estradiol have been found in river water, which are detrimental to water creatures because of their hormone-mimicking and hormone-antagonizing effects in these organisms (Jackson and Sutton, 2008; Zhao et al., 2009). Even trace concentrations of EDCs have affected all kinds of animals on the earth, through the food chain. EDCs have been released from industrial process and residential waste. So far, bisphenol-A, a starting chemical for the synthesis of polycarbonate and epoxy resins, was widely used in the manufacture of plastic dishes, baby bottles, etc. However, many manufactures have replaced these with glass-made ones, because unreacted bisphenol-A is dissolved out with hot foods and milk contained in the dishes and bottles. Some documents sound the alarm about the accumulation of bisphenol-A in pregnant women consuming canned food and drink. The intake of EDCs causes female reproductive disorders in mammals, which is accelerated for infants (Crain et al., 2008; Mariscal-Arcas et al., 2009).

Analytical techniques advanced well in the twentieth century, and we can detect hazardous chemicals such as dioxin, EDCs, and VOCs at quite low concentrations. Although some of the existing technology can treat these chemicals, there still remain problems of cost, lower efficiency, and influence on the environment. In this section, the application of ionizing radiation, one of the advanced oxidation technology, is introduced to help the readers understand the importance of ionizing radiation for environmental conservation.

32.2 RADIATION TREATMENT

Among various sources of ionizing radiation, gamma-ray and electron beams can substantially treat environmental pollutants. Any ion beam generated from accelerators cannot be applied to the treatment from energetic and economic points of view. The shorter range of ions in polluted media such as exhaust gas and waste water is also a disadvantage for the treatment of pollutants. This section describes the characteristics of electron beams and gamma-rays for environmental conservation.

32.2.1 ELECTRON BEAMS

Electron beams are generated from an electron accelerator which can accelerate electrons with a voltage of 10–10,000 kV. The electron beam treatment is based on radical reactions. When electron beams are incident in air containing environmental pollutants, more than 99% of the energy is absorbed in the main components of air, such as nitrogen, oxygen, water, and carbon dioxides. The energy absorption to the pollutants is negligible because their concentrations are usually quite low compared to the air components. There are two ways to lose energy as described below:

(a) Excitation and/or ionization by coulombic interaction
(b) Moderation by coulomb field of atomic core

In case of (b), bremsstrahlung, equivalent to the energy loss is emitted. The energy losses (a) and (b) are equal when the value of EZ is about 800, where E and Z denote electron energy (in MeV) and atomic number, respectively. Thus, when electron beams are incident in air, electrons lose their energies by excitation and ionization.

Electron-beam treatment is based on reactions with charged molecules and atoms. When air containing gaseous pollutants is irradiated with electron beams, active species such as N, O, N_2^+, O_2^+, e, OH, and O_3 are produced by the ionization and excitation of the basic components of air, namely nitrogen, oxygen, and H_2O (Mätzing, 1992). Among these species, gaseous pollutants can be oxidized through reactions with hydroxyl oxides (OH). This is similar to those observed in the atmosphere under the sunlight, where the fates of gaseous pollutants have been determined by the rate constants for reactions with OH radicals (Finlayson-Pitss and Pitts, 1986; Atkinson, 1987). The important pathway for generating OH radicals under the irradiation is as follows:

$$N_2^+ + H_2O \rightarrow N_2 + H_2O^+ \tag{32.2}$$

$$N_2^+ + O_2 \rightarrow N_2 + O_2^+ \tag{32.3}$$

$$O_2^+ + H_2O \rightarrow O_2 + H_2O^+ \tag{32.4}$$

$$H_2O^+ + H_2O \rightarrow H_3O^+ + OH \tag{32.5}$$

Charge-transfer reactions among the two charged molecules and water form hydroxyl radicals (Willis et al., 1970; Willis and Boyd, 1976). Because the OH radicals are the predominant species for the decomposition of organic pollutants, their OH rate constants are related to decomposition G-values. Figure 32.2 shows correlation of rate constants for reactions with OH radicals and G values. Ninety chemicals in aromatics, aliphatics, and alicyclics were irradiated with electron beams (Hirota et al., 2004). Interestingly, the data points obtained from the aromatics of benzene, toluene, xylene, and ethylbenzene were fitted on a straight line. The data points of aliphatics and alicyclics also gave straight lines. We can estimate electron-beam decomposition G-values of chemicals from their chemical structures and OH rate constants.

Dose rate (kGy/mA) is the one of the important parameters for treating air pollutants using an electron beam. Although electron beams with a high dose rate can treat exhaust gas at a high flow rate (~10^5 m³/h), the high dose irradiation frequently leads to lower treatment performance than lower dose irradiation. Figure 32.3 shows the decomposition profile for xylene by an electron beam generated with 1.0 MV and 170 kV accelerators. Both the experiments

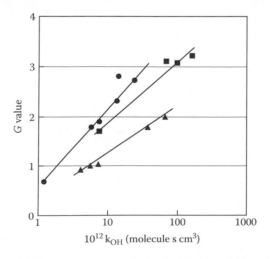

FIGURE 32.2 Correlation of OH rate constants and G values: (■) aromatics, (●) alicyclics, (▲) aliphatics. (From Hirota, K. et al., *Ind. Eng. Chem. Res.*, 43, 1185, 2004. With permission.)

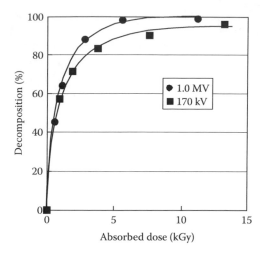

FIGURE 32.3   Decomposition efficiency of xylene by electron-beam irradiation.

were conducted using the same irradiation reactor. It is obvious that irradiation by an 1 MV accelerator caused a higher decomposition of xylene over the investigated doses. For instance, xylene was completely oxidized at a dose of 6 kGy when treated with irradiation from a 1 MV accelerator. On the other hand, electrons operated on 170 kV decomposed about 90% at the same dose. The dose rate at a dose of 6 kGy was 26.1 kGy/s for the 170 kV accelerator, which is about 10 times higher than that of the 1 MV accelerator. This indicates that electron energy absorbed in the gases per unit time was different. The irradiation at the high dose rate produced excess amount of active species for the xylene decomposition, leading to their recombination. Irradiation with low dose rate is a key for reducing the recombination, namely effective treatment of gaseous pollutants.

Although one of the better ways to obtain a low dose rate using the low energy electron accelerator is to enlarge the irradiation reactor in the direction of the irradiation; much attention should be paid to the range of the electron beam. Figure 32.4 shows the dose depth curve obtained for low energy electron accelerators operated at 170 and 300 kV. Supposing exhaust gases at an ambient temperature (density $\rho = 1.2 \times 10^{-3}$ g/cm^3) are treated by the two electron accelerators equipped with a Ti foil ($\rho = 4.5$ g/cm^3) window in a thickness of 13 μm, the range of electrons generated from 175 and 300 kV are 18.5 and 53.5 cm, respectively. The depth of the reactor in the irradiation direction must be less than these values to avoid non-irradiation space in the reactor.

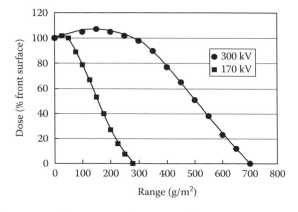

FIGURE 32.4   Depth dose curve of low energy electron accelerators.

32.2.2 GAMMA RAYS

In the case of wastewater treatment, gamma-rays generated from ^{60}Co are applied to produce active species for decomposing organic pollutants, because gamma-rays can penetrate deeply into water. Cobalt-60 is an artificial radioisotope that is produced in a nuclear reactor. It decays to nickel via beta-radioactivity with the emission of gamma-rays at 1.17 and 1.33 MeV. The thickness of water that reduces the intensity of the gamma-rays by one-tenth is about 34 cm. The range of electron beams at 2 MeV is about 0.9 cm. When treating wastewater with electron beams, its thickness needs to be controlled.

When the energy of gamma-rays is absorbed in water, various ion and active species are produced through ionization and excitation (Burton and Magee, 1976; Getoff, 1996):

$$H_2O \longrightarrow\!\!\!\sim\!\!\!\sim\!\!\!\sim\!\!\!\rightarrow e_{aq}^-, H_3O^+, H, OH, H_2, H_2O_2, HO_2 \qquad (32.6)$$

Among these products, the hydrated electron (e_{aq}^-), the hydrogen atom (H), and the hydroxyl radical (OH) contribute to the decomposition of organic pollutants in wastewater. For example, halogenated compounds (RX) are subject to be de-halogenated through reactions with hydrated electrons that are produced with a G-value of 2.7:

$$RX + e_{aq}^- \rightarrow R + X^- \qquad (32.7)$$

Hydroxyl radicals with a G-value of 2.8 can decompose organic pollutants through addition or H abstraction:

$$(32.8)$$

The reaction of hydrogen atom with alcohol proceeds through H abstraction from the alkoxy group:

$$H + C_2H_5OH \rightarrow \dot{C}_2H_4OH + H_2 \qquad (32.9)$$

The hydrogen atom is produced with a G-value of 0.6. In general, hydrated electrons and hydrogen atoms undergo reactions with oxygen dissolved in wastewater and produce HO_2 radicals that have lesser reactivity than OH radicals. Hydroxyl radicals play an important role in the treatment of organic pollutants in wastewater.

32.3 NO_X AND SO_2 IN FLUE GASES

To meet the strict regulations established by the Japanese government and the local governments, the wet lime scrubber method and the selective catalytic reduction method have been applied to treat SO_2 and NO_X. However, the wet lime scrubber method requires wastewater treatment, and the catalysts have to be replaced periodically. New technology is expected for simple and simultaneous treatment processing of the pollutants. The electron-beam irradiation process for flue gas purification has been proposed as an efficient method because it has the following advantages:

- Has simultaneous denitrification and desulfurization possibilities
- Requires no wastewater treatment due to dry process
- Requires no expensive catalysts
- Has simple process and operation
- Produces profitable products

The application of electron beams to treat flue gases, such as removing sulfur dioxide, was started at Ebara Corporation in Japan. They demonstrated in 1971 that sulfur dioxide at an initial concentration of 1000 ppm was removed at 80% with a dose of 28 kGy. The Japan Atomic Energy Research Institute and the Ebara Corporation have started collaborative research on electron-beam treatment of NO_X and SO_2 in flue gases from heavy oil combustion, where sufficient removal of NO_X and SO_2 was simultaneously achieved at a dose of 40 kGy (Machi et al., 1977). In parallel, various basic studies were conducted for the reduction of required dose, the effect of dose rate, and a dry process (Tokunaga et al., 1978a; Tokunaga and Suzuki, 1984; Gentry et al., 1988). It is demonstrated that the addition of ammonia enhanced the removal of NO_X and SO_2 and produced ammonium nitrate and ammonium sulfate as fertilizers (Tokunaga et al., 1978b). Various radicals produced by the electron-beam irradiation of the flue gases react with nitric and sulfuric oxides to produce nitric and sulfuric acids. These acids then react with the added ammonia and convert to the final products, ammonium nitrate and ammonium sulfate, as follows:

$$NO_X, \ SO_2 \xrightarrow{\ OH, \ O, \ HO_2\ } HNO_3, \ H_2SO_4 \qquad (32.10)$$

$$HNO_3, \ H_2SO_4 \xrightarrow{\ NH_3\ } H_4NO_3, \ \left(NH_4\right)_2 SO_4 \qquad (32.11)$$

The enhancement of the removal of NO_X and SO_2 in the presence of ammonia is important to reduce the operation cost. The irradiation at a lower temperature led to a higher removal of NO_X, because NH_4NO_3 is produced in an exothermic process (Machi et al., 1977). In addition to the effect of the gas temperature, multi-stage irradiation improved NO_X removal. Figure 32.5 shows the removal profile for NO_X at three different irradiation stages. The experiments were performed using a special irradiation reactor having a few chambers for the multi-irradiation experiments (Tokunaga et al., 1993). In the case of triple-stage irradiation, simulated flue gas was irradiated three times with electron beams, when flowing into each chamber in the irradiation reactor. There are non-irradiation zones between the chambers. It was found that higher NO_X removal was obtained from double- and triple-stage irradiation. This is probably due to the effective consumption of OH radicals for reactions with intermediates.

It is very difficult to examine nitrogen balance before and after irradiation, because NO, N_2, and NH_3 are the nitrogen sources in flue gases. Namba et al. (1990) have conducted the material balance of nitrogen and sulfur compounds using ^{15}N-labeled NO for the electron-beam treatment of simulated flue gas (Namba et al., 1990). Electron beams were irradiated at 80°C to a mixture of gases containing 250 ppm [15]NO, 220 ppm SO_2, 17.8% O_2, and 82.1% N_2 in the presence of 690 ppm NH_3.

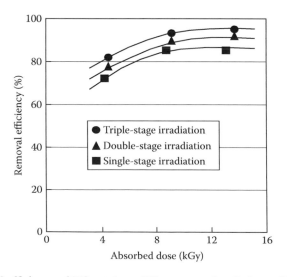

FIGURE 32.5   Removal efficiency of NO_X at three different stage irradiations. (From Tokunaga, O. et al., *Non-Thermal Plasma Techniques for Pollution Control*, NATO ASI Series, Penetrante, B. M. and Schultheis, S. E. (Eds.), Vol. 34, Part B, Springer-Verlag, Berlin Heidelberg, pp. 55–62, 1993. With permission.)

The results show that about 55% of ^{15}NO was converted to $^{15}NO_3^-$. Interestingly, other nitrogen compounds such as $^{15}N^{14}N$, $^{15}N^{14}NO$, and $^{15}N^{15}N$ were also formed, indicating that ^{15}NO reacted with $^{14}N_2$ and/or $^{14}NH_3$ as follows:

$$^{15}NO + {^{14}N} \rightarrow {^{15}N^{14}N} + O \tag{32.12}$$

$$^{15}NO + {^{14}NH_2} \rightarrow {^{15}N^{14}N} + H_2O \tag{32.13}$$

$$^{15}NO_2 + {^{14}NH_2} \rightarrow {^{15}N^{14}NO} + H_2O \tag{32.14}$$

The radicals of ^{14}N and $^{14}NH_2$ are produced by the radiolysis of nitrogen and ammonia.

As described earlier, SO_2 is removed through reaction with OH radicals produced by electron-beam irradiation. This is a similar mechanism for the removal of NO_X. However, in the electron-beam treatment process, SO_2 can be removed through direct reactions with NH_3 in the absence of electron-beam irradiation. The reactions, called thermal reactions, can also save the operation cost for the electron-beam process. Laboratory-scale studies have been carried out for the reactions of SO_2 with NH_3 (Hirota et al., 1996). The results show that the reactions occur in a water layer formed on an inner surface of reactor by water condensation out of the gas phase. The reactions were initiated by the dissolution of SO_2 and NH_3 into the water layer. The dissolution of SO_2 forms hydrogen sulfite and sulfite ions, which are heterogeneously oxidized to hydrogen sulfate and sulfate ions by oxygen and NO in the flue gas, respectively. The resulting hydrogen sulfate and sulfate ions in the aqueous layer react with ammonium ions to form $(NH_4)_2SO_4$. Oxygen and NO play an oxidative role in SO_2.

Many basic research studies on electron-beam treatment of NO_X and SO_2 have opened up the challenge of application to real flue gases. Pilot-scale tests had started at Shin-Nagoya Power Plant of Chubu Electric Power Company to treat an actual coal-fired flue gas of 150–230 ppm NO_X and 650–950 ppm SO_2 by an electron beam (Tokunaga, 1998a). Ammonia was added to a flue gas of 12,000 m^3/h, after the reduction of gas temperature by 65°C. Then the flue gases were irradiated with electron beams using an electron accelerator, operating at 800 kV with a current of 135 mA. Three irradiation heads (45 mA × 3) of the accelerator made it possible to do three stage irradiations for high performance of NO_X removal. The results showed that a dose of 11 kGy led to more than 90% removal of NO_X and SO_2. It is possible to obtain 90% removal of SO_2 using thermal reactions without irradiation.

On the basis of laboratory- and pilot-scale tests, industrial-scale processes have been constructed in China, Poland, and Bulgaria. Half of the 600,000 m^3/h flue gases from a coal-fired power generation boiler in China was treated using two electron accelerators (800 kV 400 mA), where 1800 ppm SO_2 was removed by 80% at about 4 kGy (Tokunaga, 1998b). A test in an electron-beam facility in Poland showed that removal performance of NO_X is deeply related with its initial concentration, temperature, ammonia stoichiometry, as well as dose and temperature (Chmielewski et al., 2004). The accelerator used for the treatment of flue gases at the Shin-Nagoya Power Plant of Chubu Electric Power Company was delivered to the Maritsa East coal-fired power plant for the treatment of flue gas containing about 150 ppm NO_X and 5000 ppm SO_2. A dose of 4–5 kGy achieved 90% removal of NO_X and SO_2 and the test was successfully finished in 2003.

32.4 VOCS IN EXHAUST GASES

The VOC-treatment method can be categorized into two technologies for recovering and destroying. Recovering technology comprises adsorption (Ads), absorption (Abs), and condensation (Cnd). Figure 32.6 shows the process capability of representative VOC-treatment methods. The adsorption method can treat 1–100,000 m^3/h VOC stream at a concentration of 10–10,000 ppm, where VOCs are trapped by activated carbon or zeolites, and then stripped by steam for reuse. The replacement of adsorbing materials is required in every 4–5 years to maintain the high adsorption performance. In the absorption method, VOCs absorbed into water or organic liquid are separated by decantation

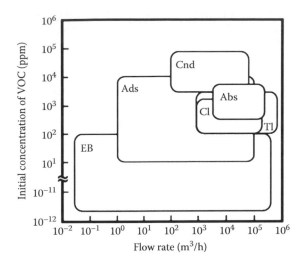

FIGURE 32.6 Characteristics of representative VOC-treatment technologies. (From Hirota, K. et al., *Ind. Eng. Chem. Res.*, 43, 1185, 2004. With permission.)

or distillation. A higher concentrated VOC stream can be treated by the condensation method. The destroying technologies comprise thermal incineration (TI), catalytic incineration (CI), and electron beam (EB) treatment. Incineration is costly and generates a great amount of CO_2 due to the supplementary fuel required for combustion of VOCs. Catalytic incineration cannot treat VOC stream containing sulfur, phosphorous, halogens, etc., because they deactivate the catalysts. The electron beam method is the preferable method to treat VOCs at less than a few hundred ppm.

Hydrocarbons are oxidized through reactions with hydroxyl radicals in electron processing. It indicates that a higher decomposition is obtained from target chemicals having higher OH rate constants. Figure 32.7 shows the decomposition profile for xylene mixture by an electron beam. Although the reagent xylene liquid contained ethylbenzene, the decomposition profiles for *o*-, *m*-, and *p*-xylenes, and ethylbenzene were roughly in proportion to the OH rate constants of *o*- (14.7×10^{-12} cm^3/mol/s), *m*- (24.5×10^{-12}), and *p*-xylenes (15.2×10^{-12}), and ethylbenzene (7.5×10^{-12}), respectively (Atkinson, 1986). In addition, mass balance in the xylene-irradiation demonstrated that the ring cleavage occurred through reactions with OH radicals. Table 32.2 lists identified products in xylene-irradiation. The mass

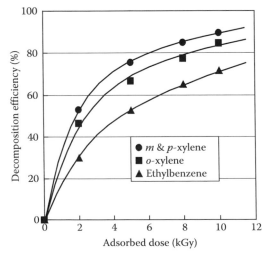

FIGURE 32.7 Decomposition efficiencies of xylene mixture by electron-beam irradiation. (From Hirota, K. et al., *Radiat. Phys. Chem.*, 45, 649, 1995. With permission.)

TABLE 32.2
Carbon Mass Balance in Xylene Irradiation

Dose (kGy)	HCOOH (mg C/m³)	CH₃COOH (mg C/m³)	C₂H₅COOH (mg C/m³)	C₃H₇COOH (mg C/m³)	CO (mg C/m³)	CO₂ (mg C/m³)	Aerosols (mg C/m³)	Unreacted Xylene (mg C/m³)	Input (mg C/m³)	Recovery (%)
2	1.51	3.12	19.21	n.d.	2.52	0.11	22.20	49.01	93.75	104.2
5	3.41	6.80	26.59	n.d.	4.27	2.70	34.48	29.20	98.27	109.9
8	2.59	5.79	6.05	1.58	6.43	2.37	38.69	18.86	95.65	86.10
10	2.48	6.26	2.58	1.76	5.25	6.31	40.56	13.98	97.02	81.60

balance in xylene-irradiation was examined on the basis of C-balance (mg C/m^3) by comparison of input xylene to irradiation products such as formic, acetic, propionic, and butyric acids, CO, CO_2, and aerosols (Hirota et al., 1995). Aerosols were formed by the coagulation of low volatile compounds produced by the irradiation of VOCs. Lower recovery at higher dose was due to the increment of fragmentation that could not be detected. The highest concentrations of the carboxylic acids were obtained at 5 kGy, except for butyric acids, indicating that irradiation products were decomposed as the dose was increased. Actually, the concentration of carbon dioxide increased with dose.

In the case of chlorobenzene, irradiation at an initial concentration of 20 ppm, mass balance of chlorine was examined on the basis of absolute amount of chlorine (mg Cl) for a certain period at doses of 4 and 8 kGy, and 4 kGy in the presence of NH_3. Excellent recoveries (100% ± 2%) were obtained for the three runs and the majority of chlorine in the reacted chlorine was found in the gaseous products (Hirota et al., 1999). Figure 32.8a through c show the weight ratio of organic and inorganic chlorine to reacted chlorine in the chlorobenzene molecule for gaseous products, aerosols, and residues in the three runs. Inorganic and organic chlorine found in the gaseous products accounted for more than 60% of the reacted chlorine. The addition of NH_3 reduced the organic-chlorine ratio of the gaseous products, suggesting that the added ammonia reacted with chlorine dissociated from carbon during irradiation. The reaction occurred before the recombination of dissociated chlorine with carbon and NH_4Cl particles formed, which increased the inorganic chlorine in aerosols, as shown in Figure 32.8b. Ammonia played a role in the dechlorination of the gaseous products. On the other hand, double energy input from 4 to 8 kGy caused a little change in the inorganic and organic chlorine ratios in the gaseous products, and not in the aerosols. The results indicate that the extra energy cannot effectively enhance dechlorination. Abundant radicals generated by the extra energy were consumed through reactions with chlorobenzene and intermediates. The extra energy input

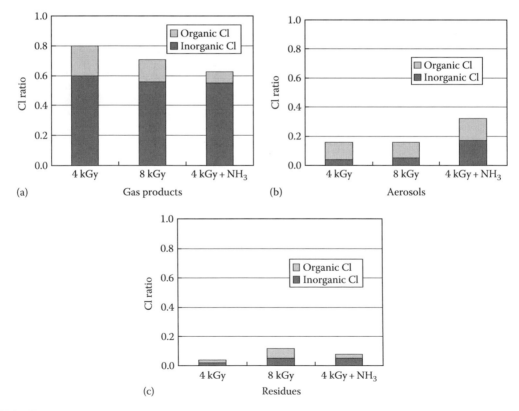

FIGURE 32.8 Organic and inorganic chlorine found in (a) gas products, (b) aerosols, and (c) residues. (From Hirota, K. et al., *Radiat. Phys. Chem.*, 57, 63, 1999. With permission.)

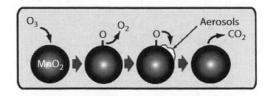

FIGURE 32.9 Oxidation of aerosols on the surface of MnO_2 catalyst.

increased decomposition ratio of chlorobenzene from 40% to 58%, with the aerosol concentration from 17.8 to 34.3 mg/m³. Dechlorination cannot take place during the first attack of OH radical to chlorobenzene. It was found that about 65% of reacted chlorobenzene was dechlorinated at doses of 4 and 8 kGy. Ammonia enhanced the dechlorination up to about 80%.

To commercialize the application of electron beam for the treatment of VOCs, the mineralization of aerosols is the most challenging task. Manganese dioxide is a useful catalyst for the oxidation of aerosols. It decomposes ozone produced by the irradiation of electron beams to target gases and produces active oxygen and oxygen molecules on its surface. The active oxygen oxidizes aerosols deposited on its surface (Hakoda et al., 2008), as shown in Figure 32.9. The Japan Atomic Energy Agency has been developing the hybrid system of electron accelerator and manganese dioxide to treat VOC exhaust gas at a flow rate of 500–2000 m³/h.

32.5 DIOXIN IN MUNICIPAL WASTE INCINERATOR GASES

As described previously, many municipal solid waste incinerators in Japan have replaced electric precipitators with bag filters to meet a new regulation for PCDD/F emission. This is because dioxins are synthesized at the temperature where electric precipitators can effectively trap fly ash and activated carbon adsorbing PCDD/Fs. In the case of bag filters, the adsorbents are trapped by filters at about 453 K. However, there still remains a concern of secondary treatment of PCDD/Fs adsorbed on the surface of added activated carbons.

One of the alternative ways is to destroy dioxins in incinerator gases by an ionizing radiation. A few studies have been experimentally and theoretically carried out for the electron-beam treatment of PCDD/Fs in Germany (Paur et al., 1991, 1998), where more than 90% decomposition was obtained at 6 kGy for PCDD and at 15 kGy for PCDF at a temperature of 358 K in the pilot plant. When considering the installation of the electron beam process to existing municipal solid waste incinerators, an electron accelerator should be placed downstream of the electric precipitator, where the temperature of incinerator gases is about 473 K in Japan. A pilot-scale facility for electron-beam treatment for PCDD/Fs was constructed at a site of Takahama Clean Center treating 450 t/day (150 t/day × 3 furnaces) solid waste in Japan (Hirota et al., 2003). The schematic flowchart of the facility is shown in Figure 32.10. The incineration of 150 t solid wastes emits about 40,000 m³$_N$/h incinerator gases.

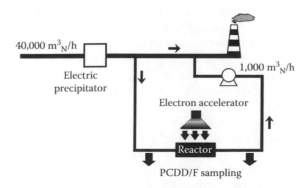

FIGURE 32.10 Schematic flow of PCDD/F treatment facility with electron beams at the Takahama Clean Center.

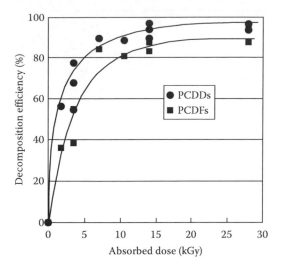

FIGURE 32.11   Decomposition efficiency of PCDD/Fs by electron-beam irradiation.

An incinerator gas of $1000\,m^3_N/h$ was diverted for the test facility from the downstream of electric precipitators and backed into the main stream incinerator after electron-beam treatment. Electron beams were generated by an electron accelerator supplied at 300 kV with a current of up to 40 mA. The incinerator gases containing PCDD/Fs were isokinetically and simultaneously sampled upstream and downstream of the reactor during irradiation to evaluate the decomposition ratios of PCDD/Fs. Although PCDD/F concentrations in the incinerator gases depended on the contents of garbage and the incineration conditions, etc., the concentrations of PCDD/F before the electron-beam irradiation were 0.22–0.88 and 0.35–1.24 ng-TEQ/m^3_N, respectively. Figure 32.11 shows the decomposition efficiencies of PCDD/Fs in the incinerator gases with electron beams. The decomposition efficiencies increased with the dose. A dose of 12 kGy led to 90% decomposition efficiency of PCDDs. In the case of PCDFs, decomposition efficiency was about 10% lower than that of PCDD and reached 90% at a dose of 16 kGy. Totally, the TEQ concentration of PCDD/Fs was reduced by 90% at a dose of 14 kGy. Additional energy of electron beams is required to treat PCDD/Fs in incinerator gases at higher temperatures than in the previous studies (Paur et al., 1998).

A kinetic approach of the decomposition of PCDD/F homologues allowed us to know if the OH radical is the main oxidant. From the comparison of PCDD G-values obtained from the initial slopes of their plots of absorbed dose vs. decomposition efficiency, the following relation was obtained (Hirota et al., 2003):

$$1.2k_{H8CDD} < k_{H4CDD} < 1.4k_{H5CDD} < 1.6k_{H7CDD} < 2.2k_{H6CDD} \qquad (32.15)$$

This relation is roughly satisfied with the OH rate constants for reactions with PCDD homologues, where the OH rate constant increases as the number of chlorine atoms on the aromatic rings decreases (Kwok et al., 1995). However, the lower chlorinated homologues required much more energy to be decomposed in the case of PCDF. This is because the dechlorination as well as oxidation of PCDF homologues occurred through reactions with activated species produced by the irradiation. The dechlorination caused an increase in the concentration of PCDF homologues such as 2,4,6,7-TeCDF, 2,3,7,8-TeCDF, and 1,2,3,4,9-PeCDF after irradiation at lower doses (Hirota et al., 2003; Hirota and Kojima, 2005).

Tsinghai University in China recently showed interest in electron-beam technology to treat dioxins emitted from medical waste incinerators. There are about 150 medical waste incinerators in China, where a bag filter system reduces dioxin emission to 1–5 ng-TEQ/m^3. However, the Chinese Government is planning to tighten the dioxin emission control to 0.1 ng-TEQ/m^3. More than 85% of the existing medical waste incinerators cannot find an avenue for coping with the new legislation. The electron-beam technology is expected to reduce the dioxin emission to a legislation-satisfying level.

32.6 ORGANIC POLLUTANTS IN WASTEWATER

Compared to the treatment of gaseous pollutants, there are many more research studies and developments analyzing ionizing radiation for aqueous pollutants. This is probably due to easier access for analyzing reaction products after ionizing radiation treatment. An on-line system is necessary to measure gaseous pollutants during irradiation. In contrast, irradiation and analysis can be separately done in the case of research on the treatment of wastewater.

In general, household effluent is treated by activated sludge. In the case of industrial wastewater, physicochemical methods such as precipitation, filtration, and adsorption are added to remove heavy metals and inorganic compounds. The combination of biological and physicochemical methods is necessary to treat industrial wastewaters. One of the methods for treating the contaminated water is by ionizing radiation (Getoff, 1986; Gehringer et al., 1988; Cooper et al., 1993). The gamma-ray irradiation of tri- and tetra-chloroethylene in the order of ppb showed that more than 99% of the target chemicals were decomposed in drinking water at a dose of 1.6 kGy (Gehringer et al., 1985). In case of purified water, a dose of 0.1 kGy was enough to obtain the sufficient performance. Ozone helped the decomposition efficiency for tri- and tetra-chloroethylene (Gehringer et al., 1987). Interestingly, the addition of metal ions also influenced the treatment of the organics, where higher decomposition efficiency of trichloroethylene was obtained in the order of $Fe^{2+} > Ni^{2+} > Cr^{3+} >$ only gamma-ray $> Co^{2+} > Zn^{2+} > Cu^{2+} > Mn^{2+}$ (Huang et al., 2009). A similar tendency was observed in the experiments on gamma-ray treatment of tetrachloroethylene.

In recent years, people are paying much attention to the contamination of environment with chemicals having biological effects on the body. Endocrine disrupting chemicals (EDCs) are very detrimental to aquatic creatures because of their hormone mimicking and antagonizing action in their organism. For example, 17 β-estradiol is a steroid hormone produced in female and male reproductive systems. However, once 17 β-estradiol is released to the environment and delivered to the body, it interferes with the normal physiological process. To identify the crucial chemicals for oxidizing EDCs, gamma-ray irradiation was carried under the condition of various atmospheres in a laboratory scale. The sample solutions were saturated with air, oxygen, helium, and nitrous oxide before irradiation. There was no difference in 17 β-estradiol decomposition except under N_2O condition (Kimura et al., 2004). The saturation of N_2O in a sample solution accelerated the oxidation of 17 β-estradiol because N_2O reacts with the hydrated electron in the solution to produce hydroxyl radicals (Janata and Schuler, 1982):

$$N_2O + e_{aq}^- + H_2O \rightarrow OH + N_2 + OH^- \tag{32.16}$$

As described previously, G-values for generation of hydrated electron and hydroxyl radical are 2.7 and 2.8, respectively. Thus, the concentration of OH was double in the N_2O-saturated solution, where the higher decomposition efficiency was obtained. It was found that the decomposition of 17 β-estradiol occurred through reaction with hydroxyl radicals. In general, hydrocarbons in N_2O-saturated water readily decomposed because the hydroxyl radical is a crucial chemical for their oxidation. When hydroxyl radicals attack target pollutants in water, many intermediates are produced. It should be noted that the toxicity of the intermediates may be higher than the target pollutants. To examine the toxicity of solution before and after irradiation, chemical and biological decomposition behavior of nonylphenol in water was investigated. Although a dose of 3 kGy decomposed nonylphenol completely, its estrogen activity still remained due to intermediates produced in the γ-irradiation of nonylphenol (Kimura et al., 2006). However, the estrogen activity was eliminated at irradiations over 5 kGy. Ionizing radiation can eliminate chemical and biological influence of pollutants in water.

Pilot-scale tests for the treatment of wastewater were conducted using ionizing radiation in a few countries, in which dyeing complex wastewater was treated with electron beams in combination with biological processes in Korea. An effluent of 1000 m^3/h was irradiated using an electron accelerator supplied with 1 MV with a beam power of 40 kW, followed by biological treatment. A dose of 1–2 kGy to the effluents before biological process enhanced the reduction of COD and BOD by 30%–40% compared to that of only biological processes (Han et al., 1998, 2002). Treatment period was also reduced from 17 to 8 h by an electron-beam irradiation. Brazil has treated wastewater using an electron accelerator of 1.5 MV with 25 mA at a flow rate of 3–4 m³/h (Sampa et al., 2001). The results showed that about 80% decomposition efficiency was achieved at a dose of 2 kGy with aeration for the major components of the wastewater such as chloroform, carbon tetrachloride, trichloroethylene, etc. In contrast, chemical oxygen demand (COD) was reduced by 30% at the same dose. With the increase of absorbed dose, COD increased by 5%–8% up to 15 kGy and then decreased again. Only 35% reduction of COD was obtained at a dose of 20 kGy. Fragmentation of parent pollutants temporarily increased COD in the wastewater.

REFERENCES

Atkinson, R. 1986. Kinetics and mechanism of the gas-phase reactions of the hydroxyl radicals with organic compounds under atmospheric condition. *Chem. Rev.* 86: 69–201.

Atkinson, R. 1987. Estimation of OH radical reaction rate constants and atmospheric lifetimes for polychlorobiphenyls, dibenzo-*p*-dioxins, and dibenzofurans. *Environ. Sci. Technol.* 21: 305–307.

Burton, M. and Magee, J. L. 1976. *Advances in Radiation Chemistry*, Vol. 1. New York: Wiley-Interscience.

Carmichael, G. R., Streets, D. G., Calori, G., Amann, M., Jacobson, M. Z., Hansen, J., and Ueda, H. 2002. Changing trends in sulfur emission in Asia: Implications for acid deposition, air pollution, and climate. *Environ. Sci. Technol.* 36: 4707–4713.

Carroll, W. F. Jr., Berger, T. C., Borrelli, F. E., Garrity, P. J., Jacobs, R. A., Ledvina, J., Lewis, J. W., McCreedy, R. L., Smith, T. P., Tuhovak, D. R., and Weston, A. F. 2001. Characterization of emissions of dioxins and furans from ethylene dichloride, vinyl chloride monomer and polyvinyl chloride facilities in the United States. Consolidated report. *Chemosphere* 43: 689–700.

Celebi, U. B. and Vardar, N. 2008. Investigation of VOC emissions from indoor and outdoor painting process in shipyards. *Atmos. Environ.* 42: 5685–5695.

Chmielewski, A. G., Licki, J., Pawelec, A., Tyminski, B., and Zimek, Z. 2004. Operational experience of the industrial plant for electron beam flue gas treatment. *Radiat. Phys. Chem.* 71: 439–442.

Choi, K. I., Lee, S. H., and Lee, D. H. 2008. Emissions of PCDD/Fs and dioxin-like PCBs from small waste incinerators in Korea. *Atmos. Environ.* 42: 940–948.

Cooper, W. J., Meacham, D. E., Nickelsen, M. G., Lin, K., Ford, D. B., Kurucz, C. N., and Waite, T. D. 1993. The removal of tir-(TCE) and tetrachloroethylene (PCE) from aqueous solution using high energy electrons. *J. Air Waste Manag. Assoc.* 43: 1358–1366.

Crain, D. A., Janssen, S. J., Edwards, T. M., Heindel, J., Ho, S., Hunt, P., Iguchi, T., Juul, A., McLachlan, J. A., Schwartz, J., Skakkebaek, N., Soto, A. M., Swan, S., Walker, C., Woodruff, T. K., Woodruff, T. J., Giudice, L. C., and Guillette, L. J. Jr., 2008. Female reproductive disorder: The roles of endocrine-disrupting compounds and developmental timing. *Fertil. Steril.* 90: 911–940.

Dyke, P. H. and Amendola, G. 2007. Dioxin releases from U.S. chemical industry sites manufacturing or using chlorine. *Chemosphere* 67: S125–S134.

European Environmental Agency. 2006. Annual European Community LRTAP Convention Emission Inventory 1990–2004. EEA Technical Report No 8/2006, <www.eea.europa.eu/publications/technical_report_2006_8>.

Finlayson-Pitss, B. J. and Pitts, J. N. Jr., 1986. *Atmospheric Chemistry*. New York: John Wiley & Sons, Inc.

Gaffney, J. S. and Marley, N. A. 2009. The impact of combustion on air quality and climate—From coal to biofuels and beyond. *Atmos. Environ.* 43: 23–36.

Gaffney, J. S., Streit, G. E., Spall, W. D., and Hall, J. H. 1987. Beyond acid rain. Do soluble oxidations and organic toxins interact with SO_2 and NO_X to increase ecosystem effects? *Environ. Sci. Technol.* 21: 519–524.

Gehringer, P., Proksch, E., and Szinovatz, W. 1985. Radiation–induced degradation of trichloroethylene and tetrachloroethylene in drinking water. *Int. J. Appl. Radiat. Isot.* 36: 313–314.

Gehringer, P., Proksch, E., Szinovatz, W., and Eschweiler, H. 1987. Decomposition of trichloroethylene and tetrachloroethylene in drinking water by a combined radiation/ozone treatment. *Water Res.* 22: 645–646.

Gehringer, P., Proksch, E., Szinovatz, W., and Eschweiler, H. 1988. Radiation-induced decomposition of aqueous trichloroethylene solutions. *Appl. Radiat. Isot.* 39: 1227–1231.

Gentry, J. W., Paur, H.-R., Mäting, H., and Bauman, W. 1988. A modelling study of the dose rate effect on the efficiency of the EBDS-process. *Radiat. Phys. Chem.* 31: 95–100.

Getoff, N. 1986. Radiation induced decomposition of biological resistant pollutants. *Appl. Radiat. Isot.* 37: 1103–1109.

Getoff, N. 1996. Radiation–induced degradation of water pollutants—State of the art. *Radiat. Phys. Chem.* 47: 581–593.

Hakoda, T., Matsumoto, K., Shimada, A., Narita, T., Kojima, T., and Hirota, K. 2008. Application of ozone decomposition catalysts to electron-beam irradiated xylene/air mixtures for enhancing carbon dioxide production. *Radiat. Phys. Chem.* 77: 585–590.

Han, B., Kim, D. K., and Pikaev, A. K. 1998. Research activities of Samsung heavy industries in the conservation of the environment. In *Radiation Technology for Conservation of the Environment, Proceedings of an International Symposium*, IAEA-TECDOC-1023, Zakopane, Poland, pp. 339–347.

Han, B., Ko, J., Kim, J., Kim, Y., Chung, W., Makarov, I. E., Ponomarev, A. V., and Pikaev, A. K., 2002. Combination electron-beam and biological treatment of dyeing complex wastewater. Pilot plant experiments. *Radiat. Phys. Chem.* 64: 53–59.

Hirota, K. and Kojima, T. 2005. Decomposition behavior of PCDD/F isomers in incinerator gases under electron-beam irradiation. *Bull. Chem. Soc. Jpn.* 78: 1685–1690.

Hirota, K., Mätzing, H., Paur, H.-R., and Woletz, K. 1995. Analyses of products formed by electron beam treatment of VOC/air mixtures. *Radiat. Phys. Chem.* 45: 649–655.

Hirota, K., Mäkelä, J., and Tokunaga, O. 1996. Reactions of sulfur dioxide with ammonia: Dependence on oxygen and nitric oxide. *Ind. Eng. Chem. Res.* 35: 3362–3368.

Hirota, K., Hakoda, T., Arai, H., and Hashimoto, S. 1999. Dechlorination of chlorobenzene in air with electron beam. *Radiat. Phys. Chem.* 57: 63–73.

Hirota, K., Hakoda, T., Taguchi, M., Takigami, M, Kim, H., and Kojima, T. 2003. Application of electron beam for the reduction of PCDD/F emission from municipal solid waste incinerators. *Environ. Sci. Technol.* 37: 3164–3170.

Hirota, K., Sakai, H., Washio, M., and Kojima, T. 2004. Application of electron beams for the treatment of VOC streams. *Ind. Eng. Chem. Res.* 43: 1185–1191.

Huang, S. K., Hsieh, L. L., Chen, C. C., Lee, P. H., and Hsieh, B. T. 2009. A study on radiation technology degradation of organic chloride wastewater—Exemplified TCE and PCE. *Appl. Radiat. Isot.* 67: 1493–1498.

Irwin, J. G. and Williams, M. L. 1988. Acid rain: Chemistry and transport. *Environ. Pollut.* 50: 29–59.

Jackson, J. and Sutton, R. 2008. Sources of endocrine-disrupting chemicals in urban wastewater, Oakland, CA. *Sci. Total Environ.* 405: 153–160.

Janata, E. and Schuler, R. H. 1982. Rate constant for scavenging e_{aq}^- in N_2O-saturated solution. *J. Phys. Chem.* 86: 2078–2084.

Jia, C., Battermann, S., and Godwin, C. 2008. VOCs in industrial, urban and suburban neighborhoods. Part 1: Indoor and outdoor concentrations, variation, and risk drivers. *Atmos. Environ.* 42: 2083–2100.

Kato, N. and Akimoto, H. 2007. Anthropogenic emissions of SO_2 and NO_X in Asia: Emission inventory. *Atmos. Environ.* 41: S171–S191.

Kimura, A., Taguchi, M., Arai, H., Hiratsuka, H., Namba, H., and Kojima, T. 2004. Radiation-induced decomposition of trace amounts of 17 β-estradiol in water. *Radiat. Phys. Chem.* 69: 295–301.

Kimura, A., Taguchi, M., Ohtani, Y., Takigami, M., Shimada, Y., Kojima, T., Hiratsuka, H., and Namba, H. 2006. Decomposition of *p*-nonylphenols in water and elimination of their estrogen activities by ^{60}Co γ-irradiation. *Radiat. Phys. Chem.* 75: 61–69.

Kishimoto, A., Oka, T., Yoshida, K., and Nakanishi, J. 2001. Cost effectiveness of reducing dioxin emissions from municipal solid waste incinerators in Japan. *Environ. Sci. Technol.* 35: 2861–2866.

Klimont, Z., Cofla, J., Schöpp, W., Amann, M., Streets, D. G., Ichikawa, Y., and Fujita, S. 2001. Projections of SO_2, NO_X, NO_3 and VOC emissions in East Asia up to 2030. *Water Air Soil Pollut.* 130: 193–198.

Kwok, E. S. C., Atkinson, R., and Arey, J. 1995. Rate constants for the gas-phase reactions of the OH radical with dichlorobiphenyls, 1-chlorodibenzo-p-dioxin, 1,2-dimethoxybenzene, and diphenyl ether: Estimation of OH radical reaction rate constants for PCBs, PCDDs, and PCDFs. *Environ. Sci. Technol.* 29: 1591–1598.

Li, Y., Jiang, G., Wang, Y., Cai, Z., and Zhang, Q. 2008. Concentrations, profiles and gas-particle partitioning of polychlorinated dibenzo-*p*-dioxins and dibenzofurans in the ambient air of Beijing, China. *Atmos. Environ.* 42: 2037–2047.

Machi, S., Tokunaga, O., Nishimura, K., Hashimoto, S., Kawakami, W., and Washio, M. 1977. Radiation treatment of combustion gases. *Radiat. Phys. Chem.* 9: 371–388.

Mariscal-Arcas, M., Rivas, A., Granada, A., Monteagudo, C., Murcia, M. A., and Olea-Serrano, F. 2009. Dietary exposure assessment of pregnant women to bisphenol-A from cans and microwave containers in Southern Spain. *Food. Chem. Toxicol.* 47: 506–510.

Mätzing, H. 1992. Model studies of flue gas treatment by electron beams. In *Proceedings of an International Symposium on Applications of Isotopes and Radiation in Conservation of the Environment*, IAEA-SM-325/186, Karlsruche, Germany, pp. 115–124.

Mininni, G., Sbrilli, A., Braguglia, C. M., Guerriero, E., Marani, D., and Rotatori, M. 2007. Dioxins, furans and polycyclic aromatic hydrocarbons emissions from a hospital and cemetery waste incinerator. *Atmos. Environ.* 41: 8527–8536.

Ministry of the Environment. 2009a. PRTR information hiroba (in Japanese). <www.env.go.jp/chemi/prtr/about/about-3.html>.

Ministry of the Environment. 2009b. Background to Japanese PRTR. <www.env.go.jp/en/chemi/prtr/about/index.html>.

Ministry of the Environment. 2009c. PRTR information hiroba (in Japanese). <www2.env.go.jp/chemi/prtr/prtrinfo/contents/prtr.jsp>.

Ministry of the Environment. 2009d. Dioxin emission inventory (in Japanese). <www.env.go.jp/press/press.php?serial = 10531>.

Mohamed, M. F., Kang, D., and Aneja, V. P. 2002. Volatile organic compounds in some urban locations in United States. *Chemosphere* 47: 863–882.

Namba, H., Tokunaga, O., Suzuki, R., and Aoki, S. 1990. Material balance of nitrogen and sulfur components in simulated flue gas treated by an electron beam. *Appl. Radiat. Isot.* 41: 569–573.

Paur, H.-R., Mätzing, H., and Schikarski, W. 1991. Removal of chlorinated dioxins and furans from simulated incinerator flue gas by electron beam. In *Proceedings of Chlorinated Dioxins and Furans from Simulated Engineering*, Karlsruhe, Germany, pp. 452–456.

Paur, H.-R., Baumann, W., Mätzing, H., and Jay, K. 1998. Electron beam induced decomposition of chlorinated aromatic compounds in waste incinerator off-gas. *Radiat. Phys. Chem.* 52: 355–359.

Quaβ, U., Fermann, M., and Bröker, G. 2004. The European dioxin air emission inventory project—Final results. *Chemosphere* 54: 1319–1327.

Ramanathan, V. and Feng, Y. 2009. Air pollution, greenhouse gases and climate change: Global and regional perspectives. *Atmos. Environ.* 43: 37–50.

Sampa, M. H. O., Rela, P. R., Duarte, C. L., Borrely, S. I., Oikawa, H., Somessari, E. S. R., Silveira, C. G., and Costa, F. E. 2001. Electron beam wastewater treatment in Brazil. In *Use of Irradiation for Chemical and Microbial Decontamination of Water, Wastewater and Sludge*, IAEA-TECDOC-1225, IAEA, Vienna, Austria, pp. 65–86.

Shah, J. J. and Singh, H. B. 1988. Distribution of volatile organic chemicals in outdoor and indoor air: A national VOCs data base. *Environ. Sci. Technol.* 22: 1381–1388.

Streets, D. G. and Waldhoff, S. T. 2000. Present and future emissions of air pollutions in China: SO_2, NO_X, and CO. *Atmos. Environ.* 34: 363–374.

Streets, D. G., Tsai, N. Y., Akimoto, H., and Oka, K. 2000. Sulfur dioxide emissions in Asia in the period 1985–1997. *Atmos. Environ.* 34: 4413–4424.

Theloke, J. and Friedrich, R. 2007. Compilation of database on the composition of anthropogenic VOC emissions for atmospheric modeling in Europe. *Atmos. Environ.* 41: 4148–4160.

Tokunaga, O. 1998a. Electron-beam technology for purification of flue gas. In *Environmental Application of Ionizing Radiation*, W. J. Cooper, R. D. Curry, and K. E. O'Shea (eds.), pp. 99–112. New York: John Wiley & Sons, Inc.

Tokunaga, O. 1998b. A technique for desulfurization and denitration of exhaust gases using an electron beam. *Sci. Technol. Jpn.* 16: 47–50.

Tokunaga, O. and Suzuki, N. 1984. Radiation chemical reactions in NO_X and SO_2 removals from flue gas. *Radiat. Phys. Chem.* 24: 145–165.

Tokunaga, O., Nishimura, K., Suzuki, N., and Washio, M. 1978a. Radiation treatment of exhaust gases. IV. Oxidation of NO in the moist mixture of O_2 and N_2. *Radiat. Phys. Chem.* 11: 117–122.

Tokunaga, O., Nishimura, K., Suzuki, N., Machi, S., and Washio, M. 1978b. Radiation treatment of exhaust gases. V. Effect of NH_3 on the removal of NO in the moist mixture of O_2 and N_2. *Radiat. Phys. Chem.* 11: 299–303.

Tokunaga, O., Namba, H., and Hirota, K., 1993. Experiments on chemical reactions in electron-beam-induced NO_x/SO_2 removal. In Penetrante, B. M. and Schultheis, S. E. (Eds.), *Non-Thermal Plasma Techniques for Pollution Control*, NATO ASI Series Vol. 34. Part B Springer-Verlag, Berlin Heidelberg, pp. 55–62.

Venier, M., Ferrario, J., and Hites, R. A. 2009. Polychlorinated dibenzo-*p*-dioxins and dibenzofurans in the atmosphere around the Great Lakes. *Environ. Sci. Technol.* 43: 1036–1041.

Wei, W., Wang, S., Chatani, S., Klimont, Z., Cofala, J., and Hao, J. 2008. Emission and speciation of non-methane volatile organic compounds from anthropogenic sources in China. *Atmos. Environ.* 42: 4976–4988.

Willis, C. and Boyd, A. W. 1976. Excitation in the radiation chemistry of inorganic gases. *Int. J. Radiat. Phys. Chem.* 8: 71–111.

Willis, C., Boyd, A. W., and Young, M. J. 1970. Radiolysis of air and nitrogen-oxygen mixtures with intense electron pulses: Determination of a mechanism by comparison of measured and computed yields. *Can. J. Chem.* 48: 1515–1525.

Xie, S., Qi, L., and Zhou, D. 2004. Investigation of the effects of acid rain on the deterioration of cement concrete using accelerated tests established in laboratory. *Atmos. Environ.* 38: 4457–4466.

Zhang, H., Ni, Y., Chen, J., Su, F., Lu, X., Zhao, L., Zhang, Q., and Zhang, X. 2008. Polychlorinated dibenzo-*p*-dioxins and dibenzofurans in soils and sediments from Daliano River Basin, China. *Chemosphere* 73: 1640–1648.

Zhao, J. L., Ying, G. G., Wang, L., Yang, J. F., Yang, X. B., Yang, L. H., and Li, X. 2009. Determination of phenolic endocrine disrupting chemicals and acidic pharmaceuticals in surface water of the Pearl Rivers in South China by gas chromatography-negative chemical ionization-mass spectrometry. *Sci. Total Environ.* 407: 962–974.

33 Applications to Biotechnology: Ion-Beam Breeding of Plants

Atsushi Tanaka
Japan Atomic Energy Agency
Takasaki, Japan

Yoshihiro Hase
Japan Atomic Energy Agency
Takasaki, Japan

CONTENTS

33.1 INTRODUCTION

Since the observation of mutation induction with x-rays in *Drosophila* by Muller (1927) and with x-rays and gamma rays in maize and barley by Stadler (1928), a number of plant mutations have been induced by ionizing radiation and used for plant science and breeding. The biological effects of ion beams have also been investigated and it has been found that ion beams show a high relative biological effectiveness (RBE) in lethality, mutation, and so on compared to low linear energy transfer (LET) radiation such as gamma rays, x-rays, and electrons (Blakely, 1992). As ion beams deposit high energy on a local target, it is suggested that ion beams predominantly induce single- or double-strand DNA breaks with damaged end groups whose reparability would be low (Goodhead, 1995)

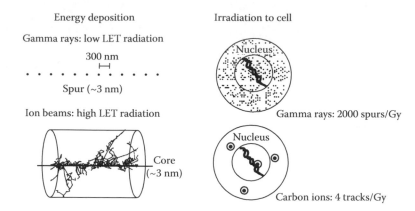

FIGURE 33.1 Energy transfer of gamma rays and ion beams on a cell.

(Figure 33.1). Therefore, it seems plausible that ion beams can frequently produce large DNA alterations such as inversions, translocations, and deletions rather than point mutations, producing characteristic mutants induced by ion beams. However, the characteristics of ion beams for mutation induction have not been elucidated yet. On the other hand, more than 2200 mutant varieties of cereals, beans, flowers, etc., have been officially registered to FAO/IAEA organization (FAO/IAEA, 2006), most of which are induced by ionizing radiations, chiefly gamma rays. As saturations for gamma- or x-ray mutanization have been realized in several kinds of mutant phenotypes because of its long history of about 80 years, new chemical or physical mutagens were expected to come into existence.

Based on suggestions of the consultative committee for the advanced radiation technology in Japan, the Takasaki Ion Accelerator Advanced Radiation Application (TIARA) was established and basic research on induction of plant mutations by ion beams was started in 1991 (Fukuda et al., 2003). We first investigated the characteristics of ion beams for mutation induction. For over 17 years, biological effects of ion beams have been greatly elucidated, and novel mutants and varieties of crops have been consistently and efficiently induced by ion beams. At present, more than 100 research activities have started to utilize the ion-beam irradiation method in several irradiation facilities and research organizations.

33.2 ION-BEAM IRRADIATION

Currently, there are four facilities available for plant ion-beam breeding in Japan. Across the world, there are few facilities for ion-beam breeding, but some exist in China, although they remain obscure. The four facilities are as follows: TIARA of the Japan Atomic Energy Agency, RIKEN RI Beam Factory (RIBF), the Wakasa Wan Energy Research Center Multi-purpose Accelerator with Synchrotron and Tandem (W-MAST), and the Heavy Ion Medical Accelerator in Chiba (HIMAC) of the National Institute of Radiological Sciences (NIRS). Table 33.1 shows the physical properties of the ion beams frequently used in these facilities. Here, we describe the protocol of the ion-beam irradiation in TIARA, which was originally described in the report by Tanaka et al. (1997a).

In the case of carbon ions, these are generated by the electron cyclotron resonance (ECR) ion source and accelerated by an azimuthally varying field (AVF) cyclotron to obtain 18.3 MeV/u $^{12}C^{+5}$ ions. At the target surface, the energy of the carbon ions slightly decreases to 17.4 MeV/u, resulting in the estimated 122 keV/μm mean LET in the target material (0.5 mm thickness) as water equivalent, and the range of ions is ca. 1.1 mm. These physical properties were calculated using an ELOSS code program, a type of modified OSCAR code program (Tanaka et al., 1997c). Particle fluences of the ions were determined using a diethyleneglycol-bis-allylcarbonate (CR-39) film track detector. In general, ion beams are scanned at around 70×70 mm; these exit the vacuum chamber through a beam window of 30 μm titanium foil (Figure 33.2). The sample is placed under the beam window

TABLE 33.1
Ion-Beam Irradiation Facilities and the Physical Parameters of the Radiations

Facility	Ion Beam	Energy (MeV/u)	LET (keV/μm)	Range (mm)
TIARA,	He	12.5	19	1.6
Japan Atomic Energy Agency	He	25.0	9	6.2
(http://www.taka.jaea.go.jp/index_e.html)	C	18.3	122	1.1
	C	26.7	86	2.2
	Ne	17.5	441	0.6
RIBF,	C	135	23	43
RIKEN Nishina Center	N	135	31	37
(http://www.rarf.riken.go.jp/Eng/index.html)	Ne	135	62	26
	Ar	95	280	9
	Fe	90	624	6
W-MAST,	H	200	0.5	256
The Wakasa Wan Energy Research Center	C	41.7	52	5.3
(http://www.werc.or.jp/english/index.htm)				
HIMAC,	C	290	13	163
National Institute of Radiological Sciences	Ne	400	30	165
(http://www.nirs.go.jp/ENG/index.html)	Si	490	54	163
	Ar	500	89	145
	Fe	500	185	97

Note: Representative ion radiations are shown in each facility.

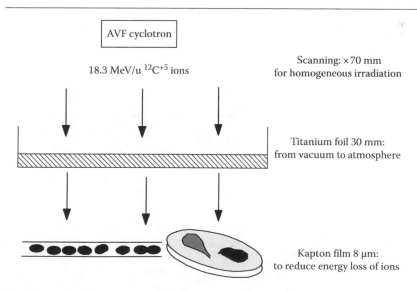

FIGURE 33.2 Schematic of ion-beam irradiation.

and irradiated in the atmosphere. In the case of *Arabidopsis* or tobacco seeds, 100–3000 seeds are sandwiched between Kapton films (8 μm thickness) to make a seed monolayer for homogeneous irradiation. In the case of rice or barley seeds, the embryo side faces the ion beams. Tissue cultures, such as those of ornamental explants, calluses, and shoot primordia, are contained in an aseptic petri dish that is irradiated directly, except that the lid of the dish is replaced by the Kapton film in order to decrease the loss of ion-beam energy. A sample is irradiated within 2 min for any dose.

33.3 BIOLOGICAL EFFECTS OF ION BEAMS

Several biological endpoints such as survival, growth, and chromosomal aberration have been investigated in order to elucidate the effects of ion beams on plants. Most efforts were focused on the LET dependence of survival of plants from treated seeds (Hirono et al., 1970; Mei et al., 1994). In general, all experiments showed a peak RBE for the survival of treated seeds above ca. 230 keV/μm. Using carbon ions with different LET, Hase et al. (2002) found that the RBE for plant survival and chromosome aberrations in tobacco increased with increasing LET and showed the highest value at 230 keV/μm. Shikazono et al. (2002) found that the LET of the peak RBE for lethality of *Arabidopsis* was over 221 keV/μm for carbon ions and over 350 keV/μm for neon and argon ions. These data suggest that the LET for maximum RBE for plant lethality is higher in seeds than in mammalian cells (Figure 33.3).

The effects of LET on chromosomal aberrations were also investigated. The frequencies of mitotic cells with chromosome aberrations, such as chromosome bridges, acentric fragments, and lagging chromosomes, were much higher for ion beams than for gamma rays (Hase et al., 1999). The highest RBE was 52.5 at 230 keV/μm of LET (Hase et al., 2002). The frequency of cells with chromosome aberrations did not decrease after fractionated irradiation with carbon ions, although a clear decrease was observed after exposure to electrons (Shimono et al., 2001).

For mutation induction, Kazama et al. (2008) showed that a LET of 30 keV/μm (N ion) was most effective for inducing albino plants of *Arabidopsis*, indicating that the relationships with LET may differ between mutation induction, plant lethality, chromosome aberrations, and double-strand breaks (DSBs).

Recently, DNA DSBs of tobacco protoplasts were quantified by pulsed-field gel electrophoresis (Yokota et al., 2007). Initial DSB yields were dependent on LET and the highest RBE was obtained at 124 and 241 keV/μm carbon ions. It is interesting that ion beams yielded short DNA fragments more frequently than gamma rays. The DSB rejoining efficiency, however, is comparable to mammalian cells (CHO-K1) (Yokota et al., 2009). These observations are well-grounded for the effective induction of biological endpoints caused by ion beams.

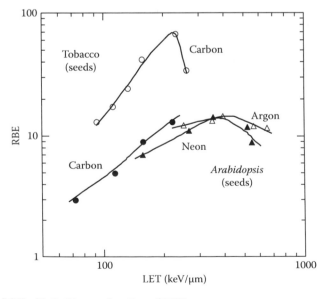

FIGURE 33.3 The RBE of lethality as a function of LET.

33.4 MUTATION INDUCTION

33.4.1 DETERMINATION OF IRRADIATION DOSE

Determination of the irradiation dose for irradiation processes is the most important step and involves hard work. Although the best irradiation dose could be determined as that which causes the highest mutation rate of phenotype interest, determining it is very difficult because it takes much time and labor. Alternatively, to determine appropriate doses for mutation induction, survival rate, growth rate, or chlorophyll mutation, etc. could be used as indicators.

Figure 33.4 shows the survival curves of *Arabidopsis* dry seeds for several ion beams and electrons as a control of low LET radiations. The effect of ion beams is higher than that of electrons but it varies by energy and kind of ions. Until now, 18.3 MeV/u carbon ions are well used and show high mutation rates and acquisition of novel mutants. However, it is not fully understood what kind of ion and ion energy is the best for mutation induction. It likely depends on plant species and structures, especially on genome size and ploidy, water content, and on what kind of mutation a researcher wants to induce. According to several results obtained so far, it is suggested that effects of ion beams on mutation induction are not due to the kind of ion but roughly due to its LET. Ion beams with LET of ~10 to ~500 keV/μm would be suitable.

As for the dose, 50% of lethal dose, namely LD_{50}, was previously thought to be the best choice for mutation induction for x-ray or gamma-ray irradiation. Recently, Yamaguchi et al. (2009a) clearly showed that the dose at the shoulder end of survival curves is sufficient to efficiently induce chlorophyll mutation not only by ion beams but also by low LET radiations in rice. In *Arabidopsis* seeds, 18.3 MeV/u carbon ions, 150 Gy, a dose that is three quarters of the shoulder end of the survival, was frequently used for mutation induction (Hase et al., 2000; Tanaka et al., 2002; Sakamoto et al., 2003). Irradiation doses that show 100%–80% growth rate (around the shoulder end of the growth curve) were well utilized for mutation induction from plantlet samples. Irradiation doses that show more than ~80% regeneration or growth rate of calli were used in the case of tissue culture (data not shown).

33.4.2 MUTATION FREQUENCY

Mutation frequency was investigated on a gene locus basis using known visible *Arabidopsis* mutant phenotypes, such as transparent testa (*tt*), in which the seed coat is transparent because of the lack of pigments, and glabrous (*gl*), in which no trichomes are produced on leaves and stems (Shikazono

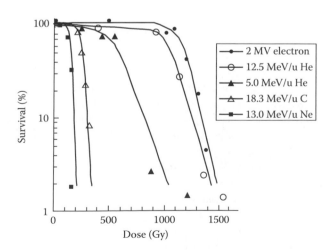

FIGURE 33.4 Survival curves of *Arabidopsis* seeds irradiated with electrons and ion beams.

TABLE 33.2
Mutation Frequency Induced by Carbon Ions and Electrons

Mutagen (Dose)	No. of M1 Plants	No. of M2 Plants	Mutant Group (Loci)	Mutation Frequency ($\times 10^{-6}$) (/Locus/Diploid Cell/Dose (Gy))
Carbon ions (150 Gy)	26,200	104,088	tt (tt3–tt7, tt18, tt19) gl (gl1–gl3, ttg1, ttg2)	1.9 (20 times)
Electrons (750 Gy)	ca. 17,600	80,827	tt (tt3–tt7, tt18, tt19) gl (gl1–gl3, ttg1, ttg2)	0.097

et al., 2003). Mutation rate of *tt* and *gl* loci was 1.9×10^{-6} per diploid cell per locus per dose (Shikazono et al., 2005) (Table 33.2). As the dose was used at 150 Gy, mutation rate is ca. 2.85×10^{-4}. The average mutation frequencies of *tt* and *gl* loci induced by C ions were 20-fold higher than those induced by electrons. Mutation frequency is generally calculated as a unit per dose for radiation-induced mutations. However, it is important to compare the mutation frequency as the number of mutants per irradiated population when considering the use of mutants for practical purposes, for example, in agriculture. Carbon ions can produce *Arabidopsis* mutants at a rate four times higher than by electrons, because carbon ions need one-fifth of the dose (i.e., RBE = 5) to induce the same biological effects as electrons (Tanaka et al., 1997a).

Mutation rates were also investigated in other plants with regard to phenotype. In chrysanthemum, the mutation rates of flower color induced by carbon ions, with floral petal and leaf as irradiation samples, were ca. 16% and 7%, respectively, which are approximately half the rates induced by gamma rays (Nagatomi et al., 1997); whereas mutation frequencies of flower color and shape in carnation were 2.8%, 2.3%, and 1.3% for ion beams, gamma rays, and x-rays, respectively (Okamura et al., 2003). For hairless mutation in sugarcane, the rate was 17.1% for 80 Gy of helium ions, which is almost double the maximum rate for 200 Gy of gamma rays (Degi et al., 2004). In rice, chlorophyll mutations were investigated in detail (Yamaguchi et al., 2009a). The mutation frequency increased linearly with dose by the irradiation of 18.3 and 26.7 MeV/u carbon ions, 25 MeV/u helium ions, and gamma rays. Mutation frequency per M_2 plant was equal to or higher in those three ion beams compared to gamma rays in several doses investigated.

33.4.3 Determination of Population Size for Mutation Induction

Since mutation basically occurs at random under the laws of probability, it is clear that the maximum number of samples possible should be used for irradiation. If the mutation frequency for a particular gene is known, the minimum size to induce at least one mutant for the desired gene can be calculated. Dr. Redei and Dr. Koncz (Redei and Koncz, 1992) suggested that 23,025 seeds of *Arabidopsis* should be enough only if all the treated seeds survive and produce large M_2 families, when the mutation rate is 1×10^{-4} and genetically effective cell number (GECN) is 2. As the mutation rate is ca. 2.85×10^{-4} for 18.3 MeV/u carbon ions at 150 Gy as above, about 8,000 seeds are necessary to induce at least one desired mutant. Whereas the mutation rate is 2.85×10^{-4} and the number of *Arabidopsis* genes is thought to be about 27,000, it may be calculated that about 3,500 seeds are necessary on an average to obtain one mutant for certain loci and that about eight genes will be mutated at the same time by this irradiation.

Based on our experience, the minimum population size to isolate one phenotypic mutation (not one gene) is likely to be around 2000–5000 M_1 seeds for *Arabidopsis*, rice, and other crops (unpublished results). However, it is not yet fully understood how many seeds will be needed for plants with different genome size, gene number, and ploidy. On the other hand, it seems that a smaller population size would be enough for mutation induction in explants or tissue cultures.

33.4.4 MUTATION SPECTRUM

The mutation spectrum has been investigated first on the pink flower color of chrysanthemum cv. Taihei by Nagatomi et al. (1997) (Table 33.3). The mutation induction of the regenerated plants from irradiated explants of floral petals was investigated. Most flower-color mutants induced by gamma rays were light pink, and a few were dark pink in color. In contrast, the color spectrum of the ion-beam-induced mutants shifted from pink to white, yellow, and orange (Table 33.3). Furthermore, flower mutants induced by C ions showed complex patterns and striped color types that have never been obtained by gamma-ray irradiation in this cultivar. Mutation spectra of flower color and shape of carnation were also investigated by Okamura et al. (2003) (Table 33.4). When carnation variety Vital, whose character is spray type and cherry pink flowers with frilly petals, was tested, flower-color mutants such as pink, white, and red were obtained by x-ray irradiation, while the color spectra of the mutants obtained by carbon ion irradiation were widely variant such as pink, light pink, salmon, red, yellow, complex, and striped types. It was suggested that the mutation spectrum of flower color induced by ion beams is broad and that novel mutation phenotypes can be inducible.

On the other hand, no distinct difference of spectrum for visible mutants has been found in *Arabidopsis* (Shikazono et al., 2003, 2005), rose (Yamaguchi et al., 2003), and rice (Yamaguchi et al., 2009a). It is not apparent whether this difference in spectrum by ion beams is due to the kind of plants investigated or to the phenotypes or genes investigated. Further examples and their analyses are needed to obtain the general feature of ion beams in mutation spectrum.

TABLE 33.3
Mutation Spectrum of Flower Color in Chrysanthemum

	Mutation Frequency (%)					
Mutagen	White	Light Pink	Dark Pink	Orange	Yellow	Complex/ Stripe
Unirradiated	0	0.3	0	0	0	0
Gamma rays	0	27.7	2.1	0	0	0
Carbon ions	0.3	4.6	0.3	0.3	0.2	10.2

Note: Original variety "Taihei" with pink-color petals was used.

TABLE 33.4
Mutation Spectrum of Flower Color in Carnation

	Mutation Frequency ($\times 10^{-1}$%)									
Mutagen	Light Pink	Pink	Dark Pink	Red	Salmon	Yellow	Cream	Stripe	Minute Striped	Complex
EMS	0	5.2	0	1.0	0	0	0	3.1	0	0
Soft x-rays	1.7	8.4	0	3.4	0	0	0	0	0	0
Gamma rays	1.7	2.6	0	1.7	0	0	0	0	11.3	0
Carbon ions	2.4	4.7	2.4	3.5	2.4	2.4	1.2	3.5	0	2.4

Note: Original variety "Vital" with cherry color and serrated petals was used.

TABLE 33.5
Characteristics of Mutation-Induced Carbon Ions and Electrons

	Carbon Ions		Electrons	
tt,gl loci	Point-Like Mutation	Large DNA Arrangement	Point-Like Mutation	Large DNA Arrangement
Mutation	48% (as 100%)	52%	75% (as 100%)	25%
Deletion	79%		44%	
Base substitution	14%		44%	
Insertion	7%		11%	
Breakpoint		(as 100%)		(as 100%)
Deletion		65%		13%
Duplication		24%		75%

Recently, the chimerical structure in axillary buds of chrysanthemum has been investigated (Yamaguchi et al., 2009b). Although no significant difference has been found in mutation frequency between gamma rays and ion beams, all the flower-color mutants induced by gamma rays were periclinal chimeras, whereas solid mutants were frequently induced by ion beams. This result suggests that not only mutation induction in a cell but also selection of mutation in a multicellular system should be considered for mutation frequency and spectrum.

33.4.5 MOLECULAR NATURE OF MUTATION INDUCED BY ION BEAMS

For analyzing mutations at a molecular level, several techniques such as chromosomal mapping, polymerase chain reaction (PCR), TAIL-PCR, cloning, and sequencing have been carried out (Shikazono et al., 2001, 2003, 2005) (Table 33.5). In the case of C ions, 14 loci out of 29 possessed intragenic point-like mutations, such as base substitutions, or deletions from several to hundreds of bases. Fifteen out of 29 loci, however, possessed intergenic DNA rearrangement ("large mutations") such as chromosomal inversions, translocations, and deletions. In the case of electrons, nine alleles out of twelve loci had point-like mutations and three out of twelve loci had DNA rearrangements. Sequence analysis revealed that C-ion-induced small mutations were mostly short deletions. Furthermore, analysis of chromosome breakpoints in large mutations revealed that C ions frequently deleted small regions around the breakpoints, whereas electron irradiation often duplicated these regions. Most of these break-and-rejoin sites for rearrangements have microhomologies, suggesting that nonhomologous end-joining (NHEJ) plays an important role for the rejoining of DSBs induced by these radiations. Thus, rearrangements induced by carbon ions have a preferably different molecular nature than those induced by electrons at a low LET (Figure 33.5).

33.5 NOVEL MUTANTS INDUCED BY ION BEAMS

Ion beams induce novel and new phenotypes, and mutations have been induced in most of the plants and crops investigated (Figure 33.6). Here, we will introduce their characteristics.

33.5.1 MODEL PLANTS

In an attempt to isolate novel mutants induced by ion beams, Arabidopsis (*Arabidopsis thaliana* (L.) Heynh.) was used as a model plant because thousands of *Arabidopsis* mutants were already induced by chemical mutagens, x-rays, T-DNA, etc. As one of such mutant phenotypes, ultraviolet

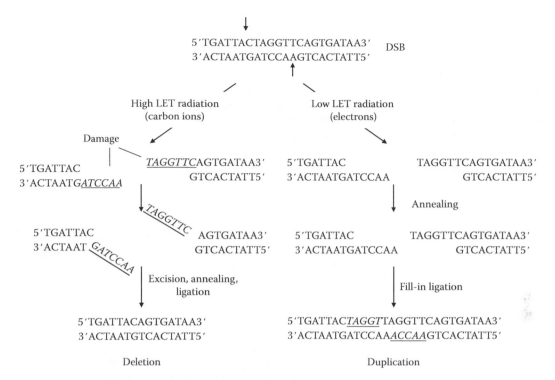

FIGURE 33.5 A model of DNA DSB rejoining after irradiation by high LET radiation such as carbon ions, and by irradiation of low LET radiation such as gamma rays, x-rays, and electrons. The ends of carbon-ion-induced DSBs are excised, followed by annealing of the ends by ligation. By contrast, the ends of electron-induced DSBs are annealed using complementary sequences and are subsequently duplicated by fill-in ligation.

FIGURE 33.6 Novel mutants and varieties induced by ion beams. *From top left to right:* 1 month old *Arabidopsis* plants, wild type (upper rank) and UV-B resistant mutants (lower rank) under high UV-B condition; flavonoid accumulating seed of the *Arabidopsis tt19* mutant; potato virus Y resistant tobacco; new barley mutant resistant to barley yellow mosaic virus. *From bottom left to right:* new chrysanthemum complex-color variety, "Ion-no-Seiko"; new rose-flower type carnation variety; new chrysanthemum variety "Aladdin," which has reduced axillary flower buds; new variety "KNOX" of *Ficus pumila*, which has a high capability for the uptake and assimilation of atmospheric nitrogen dioxide.

light-B (UV-B) resistant or sensitive mutants were obtained by irradiation with 18.3 MeV/u carbon ions, and the genes responsible have been identified (Tanaka et al., 2002; Sakamoto et al., 2003; Hase et al., 2006). New anthocyanin-accumulating or anthocyanin-defective mutants were also obtained (Tanaka et al., 1997b; Shikazono et al., 2003; Kitamura et al., 2004). A novel flower mutant, *frill*1, which has serrated petals and sepals, and the gene responsible for this phenotype have been found (Hase et al., 2000, 2005). A novel auxin mutant, the *aar*1-1, was also obtained by irradiation with carbon ions (Rahman et al., 2006). In *Lotus japonicus*, which is used as a model leguminous plant, a novel hypernodulation mutant named *klavier* (*klv*) was isolated following irradiation with helium ions (Oka-Kira et al., 2005). Thus, not only new mutants but also new genes can be discovered using ion-beam-induced mutagenesis.

33.5.2 CROPS

As the first challenge for mutation induction by ion beams, mutants resistant to bacterial leaf blight and blast disease were induced in rice. Higher mutation frequency was found in the ion-beam treatment compared to gamma rays or thermal neutrons (Nakai et al., 1995). Two mutant lines of yellow mosaic virus–resistant barley were found in a screen of ca. 50,000 M2 families (Kishinami et al., 1994). By exposure of tobacco anthers to ion beams, high frequency (2.9%–3.9%) of mutants resistant to potato virus Y was obtained for carbon and helium ions irradiation (Hamada et al., 1999). Recently, banana mutants tolerant to black Sigatoka in vitro were induced by carbon ions (Reyes-Borja et al., 2007).

In addition to disease resistance, chlorophyll mutant phenotypes were frequently observed in these crops. About 2.1% of the M2 generation derived from barley seed exposed to carbon ions were chlorophyll deficient mutants (Kishinami et al., 1996). An albino mutant of tobacco was obtained by irradiation with nitrogen ions at an early stage of embryonic development (Bae et al., 2001). A high frequency (11.6%) of chlorophyll deficient mutants, including an albino mutant, was obtained in rice by irradiation with neon ions (Abe et al., 2002). A variegated yellow leaf mutant of rice, which can be induced by the activation of endogenous transposable element, was also induced by carbon ions (Maekawa et al., 2003).

33.5.3 ORNAMENTAL FLOWERS

As described above, complex and stripe types of flower-color have been obtained in chrysanthemum (Nagatomi et al., 1997). Morphological mutant phenotypes have also been observed in chrysanthemum (Ikegami et al., 2005). One of these mutations, a reduced axillary flower bud mutant, was induced by carbon ions for the first time (Ueno et al., 2002). Recently, by means of re-irradiation of this mutant with carbon ions, the ideal characters of not only a few axillary flower buds but also low temperature flowering were obtained (Ueno et al., 2004). In addition to the carnation varieties described above (Okamura et al., 2003), mutants of petunia with altered flower color and form have been induced by ion beams (Okamura et al., 2006). In rose, mutants with more intense flower colors or mutants in the number of petals, flower size, and shape have been obtained (Yamaguchi et al., 2003). Mutants induced by nitrogen or neon ions in *Torenia* include two groups of flower-color mutants, one that lacks genes required for color or pigment production and another in which their expression is altered (Miyazaki et al., 2006). In cyclamen, ion-beam irradiation of the tuber was found to be more useful for changing flower characteristics than irradiation of other structures such as callus, somatic embryo, and so on (Sugiyama et al., 2008).

33.5.4 TREES

Ion beams have been used for generating mutants for tree breeding. Wax mutants and chlorophyll mutants such as Xanta and Albino were obtained in the forest tree, Hinoki cypress (Ishii et al., 2003). Shoot explants of *Ficus pumila* were irradiated with several kinds of ion beams to increase the capability of plants to assimilate atmospheric nitrogen dioxide (Takahashi et al., 2005). A mutant variety with 40%–80% greater capability to assimilate atmospheric nitrogen dioxide has been induced. Completely thornless Yuzu (*Citrus junos*) mutants together with several mutants with weak thorns or lesser number of thorns have been obtained by the irradiation of 26.7 MeV/u carbon ions (Matsuo et al., 2008).

33.6 OTHER BIOLOGICAL EFFECTS OF ION BEAMS

33.6.1 Cross Incompatibility

In plant breeding, obtaining an interspecific hybrid is important for introducing desirable genes from wild species to cultivated species. However, it is very difficult to obtain a viable hybrid plant between widely related species. These problems would predominantly come from sexual reproduction such as cross incompatibility, hybrid inviability, and so on.

The wild species of tobacco, *Nicotiana gossei*, has been reported to be resistant to more diseases and insects than other *Nicotiana* species; several attempts to hybridize it with *N. tabacum* (cultivar) have been made. In *N. gossei* crossed with *N. tabacum,* it was easy to get hybrid seeds with conventional cross, but the hybrid plants did not survive. In order to overcome the cross incompatibility, mature pollens from *N. tabacum* cv Bright Yellow 4 were irradiated with helium ions from tandem accelerator (6 MeV). Two viable hybrid plants were obtained at the rate of 1.1×10^{-3}. Both the hybrids were resistant to tobacco mosaic virus and aphid although the degree of resistance varied. With varying energy and kind of ions, hybrid production rate ranged from 10^{-3} to 10^{-2} (Inoue et al., 1994; Yamashita et al., 1995, 1997). The highest rate, 3.6×10^{-2}, was obtained with 10 Gy of 100 MeV helium ion beams. In the case of gamma irradiation, the rate of producing hybrids was 3.7×10^{-5}. Thus, ion beams are powerful tools to overcome hybrid incompatibility.

33.6.2 Sex Modification

Sex modification of a dioecious species, spinach, has been first observed when exposed to ion beams (Komai et al., 2003). Not only gynomonoecious but also andromonoecious plants have been induced from female plant seeds irradiated with helium ions. As the frequencies of sex modification are very high, it is plausible that ion irradiation induces some kind of stress, consistent with previous findings that photoperiod, temperature, and subirrigation conditions could affect sex expression. This is very useful for spinach breeding to maintain homogeneous female plants.

33.6.3 Activation of Endogenous Transposable Element

The possibility of activation of endogenous transposable element by ion beams was first suggested in variegated yellow leaf mutant in rice (Maekawa et al., 2003). When a rice variety with recessive stable yellow leaf (yl) was irradiated with carbon ions, variegated leaves were found in M2 with a good fit to 3:1. Albino plants were obtained in segregated M2 or M3 generations. From the genetic analysis, the gene conversion of recessive to dominant at germ line as well as somatic line would be induced by carbon ions, strongly suggesting the activation of an inactivated endogenous transposable element in rice.

33.7 CONCLUSION AND PERSPECTIVE

From a number of investigations discussed above, it is plausible that the characteristics required of ion beams for mutation induction are to induce mutants with a high frequency, to show a broad mutation spectrum, and, therefore, to produce novel mutants. We hypothesize that chemical mutagens, such as EMS, and low LET ionizing radiations, such as gamma rays and electrons will predominantly induce many minor modifications or DNA damages on DNA strands, thus producing

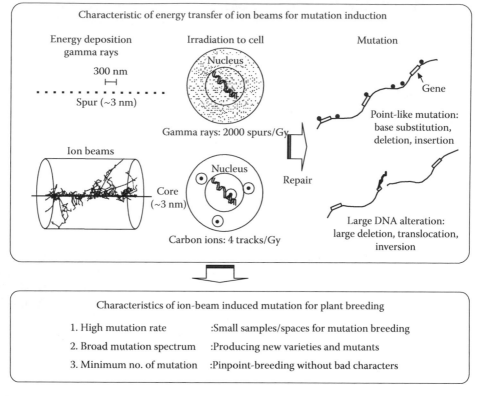

FIGURE 33.7   Characteristics of ion-beam-induced mutation and its application for plant breeding.

several point-like mutations on the genome. In contrast, ion beams as high LET ionizing radiations will efficiently cause fewer but larger, irreparable DNA damages locally, resulting in a limited number of null mutations (Figure 33.7).

For basic research, the usage of ion beams as a new mutagen casts new light on creating novel mutants that are necessary to step up to the next level of plant molecular biology. One of the most important objectives of the postgenome age is to elucidate the function of genes. Unfortunately, it is very difficult to apply the gene-knockout system that was used in mammals or bacteria in the plant kingdom. Therefore, ion beams can create thousands of novel mutants that provide information on the function of the mutated gene.

Another important point of view is the application of these new mutants for plant breeding, especially for practical use. Several flower varieties have already been commercially utilized. For example, new varieties of carnation with new flower colors or new petal shapes have been produced not only in Japan but also in Europe. A new chrysanthemum variety with a few axillary flower buds, named Aladdin, is currently being produced by more than 40 companies and associations. Japanese as well as other Asian groups have started to use ion beams for breeding purposes.

Although ion beams have become established as a new mutagen, the present characterization of ion-beam-induced mutations have been acquired almost exclusively using carbon ions with LET of ca. 100–200 keV/μm. Some information was obtained with other ions or at other energies. It is necessary to evaluate the effect of ion beams with different kinds of ions and at different energies. Another question is whether mutagens such as radiation could induce directed or adaptive mutation (Cairns et al., 1988). Nagatomi et al. showed that the mutation rates of the flower color of plants regenerated from floral petal was higher than those from leaf when both ion beams and gamma rays were used (Nagatomi et al., 1997). The reason for this is thought to be that the genes of flower color in floral petal cells are more active than those in leaf cells, and this may lead to higher mutation rate of plants regenerated from floral petals. A similar phenomenon was also found in inducing a few auxiliary flower bud mutants. If directed mutation exists in plants, a combination of the best ion with appropriate energy and doses, and a choice of the best organ or tissues of plant will cause a high induction of mutants that would be sufficient for obtaining the desired mutations.

ACKNOWLEDGMENTS

We would like to thank N. Shikazono for sharing his original model of ion-beam-induced mutation mechanism, and M. Okamura for sharing his data of mutation spectrum of flower color and shapes produced by ion beams. Ion beam breeding (IBB) described here could not have been achieved without great support from Dr. H. Watanabe, S. Nagatomi, and Dr. M. Inoue. We would like to dedicate this chapter to Dr. S. Tano, one of the founders of IBB, who passed away in July 2006.

REFERENCES

Abe, T., Matsuyama, T., Sekido, S., Yamaguchi, I., Yoshida, S., and Kameya, T. 2002. Chlorophyll-deficient mutants of rice demonstrated the deletion of a DNA fragment by heavy-ion irradiation. *J. Radiat. Res.* 43: 157–161.

Bae, C. H., Abe, T., Matsuyama, T., Fukunishi, N., Nagata, N., Nakano, T., Kaneko, Y., Miyoshi, K., Matsushima, H., and Yoshida, S. 2001. Regulation of chloroplast gene expression is affected in *ali*, a novel tobacco albino mutant. *Ann. Bot.* 88: 545–553.

Blakely, E. A. 1992. Cell inactivation by heavy charged particles. *Radiat. Environ. Biophys.* 31: 181–196.

Cairns, J., Overbaugh, J., and Miller, S. 1988. The origin of mutants. *Nature* 335: 142–145.

Degi, K., Morishita, T., Yamaguchi, H., Nagatomi, S., Miyagi, K., Tanaka, A., Shikazono, N., and Hase, Y. 2004. Development of the efficient mutation breeding method using ion bema irradiation. *JAERI-Rev.* 2004-025: 48–50.

FAO/IAEA. 2006. FAO/IAEA Mutant Varieties Database. Plant breeding and genetics section Vienna, Austria. http://www-mvd.iaea.org/MVD/default.htm.

Fukuda, M., Itoh, H., Ohshima, T., Saidoh, M., and Tanaka, A. 2004. New applications of ion beams to material, space, and biological science and engineering. In *Charged Particle Interactions with Matter: Chemical, Physicochemical, and Biological Consequences with Applications.* A. Mozumder and Y. Hatano (eds.), pp. 813–859. New York: Marcel Dekker.

Goodhead, D. T. 1995. Molecular and cell models of biological effects of heavy ion radiation. *Radiat. Environ. Biophys.* 34: 67–72.

Hamada, K., Inoue, M., Tanaka, A., and Watanabe, H. 1999. Potato virus Y-resistant mutation induced by the combination treatment of ion beam exposure and anther culture in *Nicotiana tabacum* L. *Plant Biotechnol.* 16: 285–289.

Hase, Y., Shimono, K., Inoue, M., Tanaka, A., and Watanabe, H. 1999. Biological effects of ion beams in *Nicotiana tabacum* L. *Radiat. Environ. Biophys.* 38: 111–115.

Hase, Y., Tanaka, A., Baba, T., and Watanabe, H. 2000. *FRL1* is required for petal and sepal development in *Arabidopsis. Plant J.* 24: 21–32.

Hase, Y., Yamaguchi, M., Inoue, M., and Tanaka, A. 2002. Reduction of survival and induction of chromosome aberrations in tobacco irradiated by carbon ions with different LETs. *Int. J. Radiat. Biol.* 78: 799–806.

Hase, Y., Fujioka, S., Yoshida, S., Sun, G., Umeda, M., and Tanaka, A. 2005. Ectopic endoreduplication caused by sterol alteration results in serrated petals in *Arabidopsis. J. Exp. Bot.* 56: 1263–1268.

Hase, Y., Trung, K. H., Matsunaga, T., and Tanaka, A. 2006. A mutation in the *uvi4* gene promotes progression of endo-reduplication and confers increased tolerance towards ultraviolet B light. *Plant J.* 46: 317–326.

Hirono, Y., Smith, H. H., Lyman, J. T., Thompson, K. H., and Baum, J. W. 1970. Relative biological effectiveness of heavy ions in producing mutations, tumors, and growth inhibition in the crucifer plant, *Arabidopsis. Radiat. Res.* 44: 204–223.

Ikegami, H., Kunitake, T., Hirashima, K., Sakai, Y., Nakahara, T., Hase, Y., Shikazono, N., and Tanaka, A. 2005. Mutation induction through ion beam irradiations in protoplasts of chrysanthemum. *Bull. Fukuoka Agric. Res. Cent.* 24: 5–9 (in Japanese).

Inoue, M., Watanabe, H., Tanaka, A., and Nakamura, A. 1994. Interspecific hybridization between *Nicotiana gossei* Domin and *N. tabacum* L., using 4He2+ irradiated pollen. *JAERI TIARA Annu. Rep.* 1993: 44–45.

Ishii, K., Yamada, Y., Hase, Y., Shikazono, N., and Tanaka, A. 2003. RAPD analysis of mutants obtained by ion beam irradiation to Hinoki cypress shoot primordial. *Nucl. Instrum. Methods Phys. Res. B* 206: 570–573.

Kazama, Y., Saito, H., Yoshiharu, Y., Yamamoto, Y., Hayashi, Y., Ichida, H., Ryuto, H., Fukunishi, N., and Abe, T. 2008. LET-dependent effects of heavy-ion beam irradiation in *Arabidopsis thaliana*. *Plant Biotechnol.* 25: 113–117.

Kishinami, I., Tanaka, A., and Watanabe, H. 1994. Mutations induced by ion beam (C5+) irradiation in barley. *JAERI TIARA Annu. Rep.* 3: 30–31.

Kishinami, I., Tanaka, A., and Watanabe, H. 1996. Studies on mutation induction of barley by ion beams. *Abstracts of the 5th TIARA Research Review Meeting* 165: 6 (in Japanese).

Kitamura, S., Shikazono, N., and Tanaka, A. 2004. TRANSPARENT TESTA 19 is involved in the accumulation of both anthocyanins and proanthocyanidins in Arabidopsis. *Plant J.* 37: 104–114.

Komai, F., Shikazono, N., and Tanaka, A. 2003. Sexual modification of female spinach seeds (*Spinacia oleracea* L.) by irradiation with ion particles. *Plant Cell Rep.* 21: 713–717.

Maekawa, M., Hase, Y., Shikazono, N., and Tanaka, A. 2003. Induction of somatic instability in stable yellow leaf mutant of rice by ion beam irradiation. *Nucl. Instrum. Methods Phys. Res. B* 206: 579–585.

Matsuo, Y., Hase, Y., Yokota, Y., Narumi, I., and Ohyabu, E. 2008. Induction of thornless Yuzu mutant by heavy ion beam irradiation. *JAERI-Rev.* 2007-060: 79.

Mei, M., Deng, H., Lu, Y., Zhuang, C., Liu, Z., Qiu, Q., Qiu, Y., and Yang, T. C. 1994. Mutagenic effects of heavy ion radiation in plants. *Adv. Space Res.* 14: 363–372.

Miyazaki, K., Suzuki, K., Iwaki, K., Kusumi, T., Abe, T., Yoshida, S., and Fukui, H. 2006. Flower pigment mutations induced by heavy ion beam irradiation in an interspecific hybrid of *Torenia*. *Plant Biotechnol.* 23: 163–167.

Muller, H. J. 1927. Artificial transmutation of the gene. *Science* 66: 84–87.

Nagatomi, S., Tanaka, A., Kato, A., Watanabe, H., and Tano, S. 1997. Mutation induction on chrysanthemum plants regenerated from in vitro cultured explants irradiated with C ion beam. *JAERI-Rev.* 96-017: 50–52.

Nakai, H., Watanabe, H., Kitayama, A., Tanaka, A., Kobayashi, Y., Takahashi, T., Asai, T., and Imada, T. 1995. Studies on induced mutations by ion beam in plants. *JAERI-Rev.* 19: 34–36.

Oka-Kira, E., Tateno, K., Miura, K., Haga, T., Hayashi, M., Harada, K., Sato, S., Tabata, S., Shikazono, N., Tanaka, A., Watanabe, Y., Fukuhara, I., Nagata, T., and Kawaguchi, M. 2005. *klavier* (*klv*), a novel hypernodulation mutant of Lotus japonicus affected in vascular tissue organization and floral induction. *Plant J.* 44: 505–515.

Okamura, M., Yasuno, N., Ohtsuka, M., Tanaka, A., Shikazono, N., and Hase, Y. 2003. Wide variety of flower-color and -shape mutants regenerated from leaf cultures irradiated with ion beams. *Nucl. Instrum. Methods Phys. Res. B* 206: 574–578.

Okamura, M., Tanaka, A., Momose, M., Umemoto, N., da Silva, J. T., and Toguri, T. 2006. Advances of mutagenesis in flowers and their industrialization. In *Floriculture, Ornamental and Plant Biotechnology*, Vol. I, pp. 619–628. London, U.K.: Global Science Books.

Rahman, A., Nakasone, A., Chhun, T., Ooura, C., Biswas, K. K., Uchimiya, H., Tsurumi, S., Baskin, T. I., Tanaka, A., and Oono, Y. 2006. A small acidic protein 1 (SMAP1) mediates responses of the Arabidopsis root to the synthetic auxin 2,4-dichlorophenoxyacetic acid. *Plant J.* 47: 788–801.

Redei, G. P. and Koncz, C. 1992. Classical mutagenesis. In *Methods in Arabidopsis Research*, C. Koncz, N. H. Chua and J. Schell (eds.). Singapore: World Scientific Publishing Co. Pte. Ltd.

Reyes-Borja, W. O., Sotomayor, I., Garzon, I., Vera, D., Cedeno, M., Castillo, B., Tanaka, A., Hase, Y., Sekozawa, Y., Sugaya, S., and Gemma, H. 2007. Alteration of resistance to black Sigatoka (*Mycosphaerella fijiensis* Morelet) in banana by *in vitro* irradiation using carbon ion-beam. *Plant Biotechnol.* 24: 349–363.

Sakamoto, A., Lan, V. T. T., Hase, Y., Shikazono, N., Matsunaga, T., and Tanaka, A. 2003. Disruption of the AtREV3 gene causes hypersensitivity to ultraviolet B light and gamma-rays in Arabidopsis: Implication of the presence of a translation synthesis mechanism in plants. *Plant Cell* 15: 2042–2057.

Shikazono, N., Tanaka, A., Watanabe, H., and Tano, S. 2001. Rearrangements of the DNA in carbon ion-induced mutants of *Arabidopsis thaliana*. *Genetics* 157: 379–387.

Shikazono, N., Tanaka, A., Kitayama, S., Watanabe, H., and Tano, S. 2002. LET dependence of lethality in *Arabidopsis thaliana* irradiated by heavy ions. *Radiat. Environ. Biophys.* 41: 159–162.

Shikazono, N., Yokota, Y., Kitamura, S., Suzuki, C., Watanabe, H., Tano, S., and Tanaka, A. 2003. Mutation rate and novel tt mutants of *Arabidopsis thaliana* induced by carbon ions. *Genetics* 163: 1449–1455.

Shikazono, N., Suzuki, C., Kitamura, S., Watanabe, H., Tano, S., and Tanaka, A. 2005. Analysis of mutations induced by carbon ions in *Arabidopsis thaliana*. *J. Exp. Bot.* 56: 587–596.

Shimono, K., Shikazono, N., Inoue, M., Tanaka, A., and Watanabe, H. 2001. Effect of fractionated exposure to carbon ions on the frequency of chromosome aberrations in tobacco root cells. *Radiat. Environ. Biophys.* 40: 221–225.

Stadler, L. J. 1928. Mutations in barley induced by x-rays and radium. *Science* 68: 186–187.

Sugiyama, M., Saito, H., Ichida, H., Hayashi, Y., Ryuto, H., Fukunishi, N., Terakawa, T., and Abe, T. 2008. Biological effects of heavy-ion beam irradiation on cyclamen. *Plant Biotechnol.* 25: 101–104.

Takahashi, M., Kohama, S., Kondo, K., Hakata, M., Hase, Y., Shikazono, N., Tanaka, A., and Morikawa, H. 2005. Effects of ion beam irradiation on the regeneration and morphology of *Ficus thunbergii* Maxim. *Plant Biotechnol.* 22: 63–67.

Tanaka, A., Shikazono, N., Yokota, Y., Watanabe, H., and Tano, S. 1997a. Effects of heavy ions on the germination and survival of Arabidopsis thaliana. *Int. J. Radiat. Biol.* 72: 121–127.

Tanaka, A., Tano, S., Chantes, T., Yokota, Y., Shikazono, N., and Watanabe, H. 1997b. A new Arabidopsis mutant induced by ion beams affects flavonoid synthesis with spotted pigmentation in testa. *Genes Genet. Syst.* 72: 141–148.

Tanaka, S., Fukuda, M., Nishimura, K., Watanabe, H., and Yamano, N. 1997c. IRACM: A code system to calculated induced radioactivity produced by ions and neutrons. *JAERI-Data/Code* 97-019.

Tanaka, A., Sakamoto, A., Ishigaki, Y., Nikaido, O., Sung, G., Hase, Y., Shikazono, N., Tano, S., and Watanabe, H. 2002. An ultraviolet-B-resistant mutant with enhanced DNA repair in Arabidopsis. *Plant Physiol.* 129: 64–71.

Ueno, K., Nagayoshi, S., Shimonishi, K., Hase, Y., Shikazono, N., and Tanaka, A. 2002. Effects of ion beam irradiation on chrysanthemum leaf discs and sweet potato callus. *JAERI-Rev.* 2002-035: 44–46.

Ueno, K., Nagayoshi, S., Hase, Y., Shikazono, N., and Tanaka, A. 2004. Additional improvement of chrysanthemum using ion beam re-irradiation. *JAERI-Rev.* 25: 53–55.

Yamaguchi, H., Nagatomi, S., Mrishita, T., Degi, K., Tanaka, A., Shikazono, N., and Hase, Y. 2003. Mutation induced with ion beam irradiation in rose. *Nucl. Instrum. Methods Phys. Res. B* 206: 561–564.

Yamaguchi, H., Hase, Y., Tanaka, A., Shikazono, N., Degi, K., Shimizu, A., and Morishita, T. 2009a. Mutagenic effects of ion beam irradiation on rice. *Breed. Sci.* 59: 169–177.

Yamaguchi, H., Shimizu, A., Hase, Y., Degi, K., Tanaka, A., and Morishita, T. 2009b. Mutation induction with ion beam irradiation of lateral buds of chrysanthemum and analysis of chimeric structure of induced mutants. *Euphytica* 165: 97–103.

Yamashita, T., Inoue, M., Watanabe, H., Tanaka, A., and Tano, S. 1995. Comparative analysis of interspecific hybrids between *Nicotiana gossei* Domin and *N. tabacum* L., obtained by the cross with $^4He^{2+}$-irradiated pollen. *JAERI-Rev.* 95-019: 37–39.

Yamashita, T., Inoue, M., Watanabe, H., Tanaka, A., and Tano, S. 1997. Effective production of interspecific hybrids between *Nicotiana gossei* Domin and *N. tabacum* L., by the cross with ion beam-irradiated pollen. *JAERI-Rev.* 96-017: 44–46.

Yokota, Y., Yamada, S., Hase, Y., Shikazono, N., Narumi, I., Tanaka, A., and Inoue, M. 2007. Initial yields of DNA double-strand breaks and DNA fragmentation patterns depend on linear energy transfer in tobacco BY-2 protoplasts irradiated with helium, carbon and neon ions. *Radiat. Res.* 167: 94–101.

Yokota, Y., Wada, S., Hase, Y., Funayama, T., Kobayashi, Y., Narumi, I., and Tanaka, A. 2009. Kinetic analysis of double-strand break rejoining reveals the DNA reparability of γ-irradiated tobacco cultured cells. *J. Radiat. Res.* 50: 171–175.

34 Radiation Chemistry in Nuclear Engineering

Junichi Takagi
Toshiba Corporation
Yokohama, Japan

Bruce J. Mincher
Idaho National Laboratory
Idaho Falls, Idaho

Makoto Yamaguchi
Japan Atomic Energy Agency
Tokai, Japan

Yosuke Katsumura
The University of Tokyo
Tokyo, Japan
and
Japan Atomic Energy Agency
Tokai, Japan

CONTENTS

34.1　INTRODUCTION

In recent years, nuclear power has been reevaluated as a sustainable and environmentally benign energy source to avoid global warming because of low emission of CO_2. In developed countries, there is renewed interest in further construction of nuclear power plants, and many developing countries are planning to introduce nuclear energy. Nuclear energy is always accompanied by radiation and radioactive materials, and there exist many radiation effects in nuclear reactors and spent fuel reprocessing. These radiation effects are normally detrimental to the performance and functionality of the facility and much care should be taken. Therefore, an understanding of radiation-induced phenomenon is essential to avoid the reduction of the integrity of the facilities.

This chapter is composed of three subjects in nuclear technology: water chemistry in nuclear power plants, reprocessing of spent nuclear fuel, and high-level radioactive waste. It is well known that coolant water in nuclear reactors receives high radiation doses under a strong mixed field of gamma rays and fast neutrons at high temperature and pressure. Radiation controls the chemical condition of the coolant and precise prediction of radiolysis effects is essential to avoid detrimental radiation-induced corrosion processes such as stress corrosion cracking (SCC).

The efficiency of nuclear energy production from uranium may be increased by the reprocessing of spent nuclear fuel. Reprocessing allows the recovery of fissionable Pu and U for use in new reactor fuel. In addition, further separation of the minor actinides, and possibly some fission products such as Sr and Cs is desired for future reprocessing. Future reprocessing will also be designed to ensure the nonproliferation of nuclear material.

After spent fuel reprocessing, high-level radioactive waste is separated and must be stored safely. This is an important issue in the field of nuclear technology. Again, radiation effects should be considered. The geological repository is assumed to be a most promising and acceptable method for disposition and current research relevant to the radiation-induced effects in the geological repository are also reviewed here.

34.2 APPLICATION OF RADIATION CHEMISTRY TO POWER PLANTS

34.2.1 INTRODUCTION

It has been widely recognized that intergranular stress corrosion cracking (IGSCC) is a key issue for boiling water reactors (BWRs) and occurs depending on the three factors of material susceptibility, residual stress, and environmental condition. Most of the SCC countermeasures have been taken from the viewpoint of material selection and stress improvement. On the other hand, SCC mitigation by changing chemistry has also been adopted by eliminating impurities and decreasing oxidizing species concentrations to achieve low corrosion potential. In order to evaluate the corrosion environment from the viewpoint of both modeling and experimental approaches, the application of radiation chemistry is of vital importance in understanding water radiolysis in the nuclear core. In this chapter, the radiation chemistry of water in BWRs and its implications for corrosion are described.

34.2.2 WATER CHEMISTRY CONTROL IN LIGHT WATER REACTORS

Most commercial nuclear reactors are light water reactors (LWRs) and are categorized into two types: BWRs and pressurized water reactors (PWRs). Proper coolant water chemistry control is required to minimize corrosion and to reduce radiation exposure to personnel.

34.2.2.1 BWR Water Chemistry

34.2.2.1.1 Objective of BWR Water Chemistry Control

The objectives of BWR coolant water chemistry control are to maintain material integrity of both fuel materials and structural materials, to reduce personnel radiation exposure, and to reduce radioactive waste generation. To achieve these goals, the principle of the BWR water chemistry control is to use pure neutral water to suppress corrosion of the materials. For example, the reactor water quality is defined as follows:

Reactor water conductivity (at 25°C): $1\,\mu S\ cm^{-1}$ or less
Reactor water chloride ion: 0.1 ppm or less
Reactor water pH: 5.6–8.6

34.2.2.1.2 Outline of BWR Primary Coolant Loop

In the BWR primary coolant loop, the steam generated at the nuclear core is transported to the turbine system and is condensed in the main condenser. The condensate is used as feedwater to return to the reactor core. Figure 34.1 shows the schematic of the BWR primary coolant loop (AESJ, 2000a). The total system is divided into three systems: the condensate and the feedwater system, the reactor coolant water system, and the main steam system. Chemistry specifications are determined for the condensate and the feedwater system and the reactor water system. The feedwater should be maintained as pure neutral water in order to avoid intrusion of any chemical species into the core. The reactor water is purified with the reactor water cleanup system, which has a capacity of treating 1%–7% of the feedwater flow. Examples of the water quality specifications for each system are shown in Tables 34.1 through 34.3 (AESJ, 2000a).

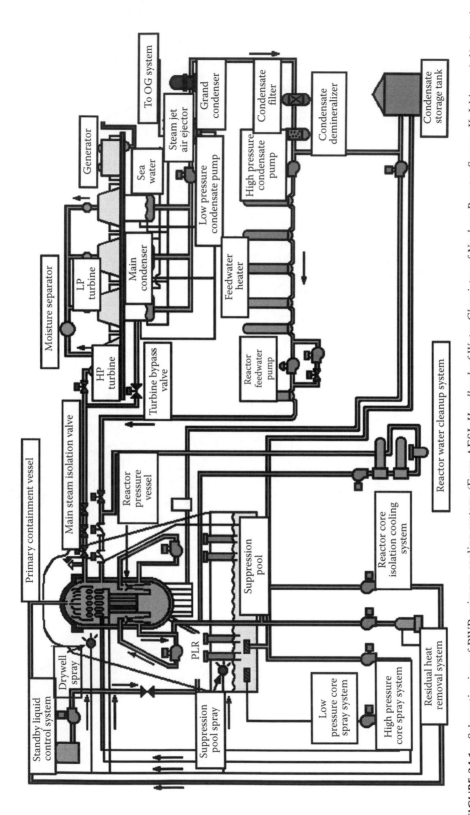

FIGURE 34.1 Schematic view of BWR primary cooling system. (From AESJ, *Handbook of Water Chemistry of Nuclear Reactor System*, K. Ishigure (ed.), Atomic Energy Society of Japan, Tokyo, Japan, 2000a. With permission.)

TABLE 34.1
An Example of BWR Reactor Water Quality Control Parameters

Items	Unit	Control Value	Standard Value
Conductivity (25°C)	$\mu S\ cm^{-1}$	<1	<10
pH (25°C)	—	5.6–8.6	4–10
Cl⁻ ion	ppb	<100	<500
Silica	ppm	<5	—

Source: AESJ, *Handbook of Water Chemistry of Nuclear Reactor System*, K. Ishigure (ed.), Atomic Energy Society of Japan, 2000c. With permission.

TABLE 34.2
An Example of BWR Feedwater Quality Control Parameters

Items	Unit	Control Value
Dissolved O_2	ppb	20–200
Metal impurities (as Fe, Cu, Ni, Cr)	ppb	<15
Total cupper (as Cu)	ppb	<2

Source: AESJ, *Handbook of Water Chemistry of Nuclear Reactor System*, K. Ishigure (ed.), Atomic Energy Society of Japan, 2000c. With permission.

TABLE 34.3
An Example of BWR Condensate Demineralizer Outlet Water Quality Control Parameters

Items	Unit	Control Value
(25°C)	$\mu S\ cm^{-1}$	<0.1
pH (25°C)	—	6.5–7.5
Cl⁻ ion	ppb	<100
Silica	ppb	<10
Metal impurity (as Fe, Cu, Ni, Cr)	ppb	<15
Total cupper (as Cu)	ppb	<2

Source: AESJ, *Handbook of Water Chemistry of Nuclear Reactor System*, K. Ishigure (ed.), Atomic Energy Society of Japan, 2000c. With permission.

34.2.2.2　PWR Water Chemistry

34.2.2.2.1　Objectives of PWR Primary Side Water Chemistry Control

The objectives of PWR primary side water chemistry control are to control the reactivity of the core, to maintain the material integrity of both fuels and structural materials, and to reduce personnel radiation exposure. Boron as boric acid is added to the reactor coolant water as a neutron absorber to control reactivity; lithium hydroxide is added as an alkaline agent to control the reactor water pH; and hydrogen gas is added to maintain a reducing environment to suppress SCC of the structural materials.

34.2.2.2.2 Outline of PWR Primary Coolant Loop

The PWR primary coolant loop is under non-boiling conditions and the generated heat is transferred to the secondary side through the steam generators. Figure 34.2 shows the schematic of the PWR primary coolant loop (AESJ, 2000b). The primary loop has the chemical volume control system (CVCS), which controls the volume change of the coolant by temperature change, regulates the concentrations of boric acid and lithium hydroxide, and controls the dissolved hydrogen concentration to suppress corrosion. An example of the water quality specification for the primary loop is shown in Table 34.4 (AESJ, 2000b).

34.2.3 EVALUATION OF CORROSION ENVIRONMENT IN BWRs

With the increase in operating years of commercial BWRs, it has been widely recognized that the role of preventive maintenance for plant materials is of great importance. In BWRs stainless steels and nickel-base alloys are used as structural materials. For plants with longer operating history, considerable attention should be paid to the maintenance of those components, especially the pressure boundary components and the reactor core internals.

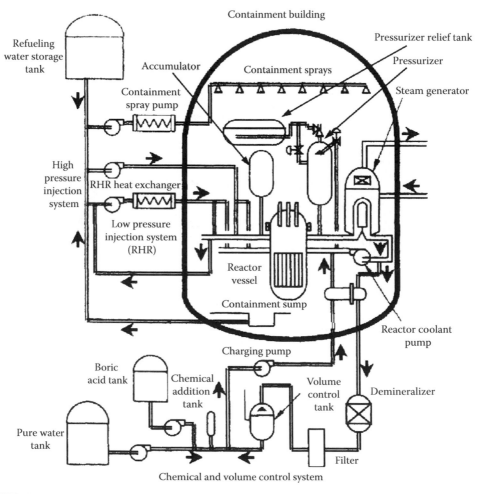

FIGURE 34.2 Schematic view of PWR primary cooling system. (From AESJ, *Handbook of Water Chemistry of Nuclear Reactor System*, K. Ishigure (ed.), Atomic Energy Society of Japan, Tokyo, Japan, 2000a. With permission.)

TABLE 34.4
An Example of PWR Primary Water Quality Control Parameters

Items	Unit	Control	Standard
Conductivity (25°C)	μS cm^{-1}	[a]	—
pH (25°C)	—	[a]	—
Boron	ppm	[b]	—
Cl$^-$ ion	ppm	≤0.05	≤0.15
F$^-$ ion	ppm	≤0.05	≤0.15
Dissolved O$_2$	ppm	≤0.005	≤0.1
Dissolved H$_2$	cc-STP kg$^{-1}\cdot$H$_2$O	25–35	≥15, ≤50
Li$^+$ ion	ppm	0.2–2.2	—

Source: AESJ, *Handbook of Water Chemistry of Nuclear Reactor System*, K. Ishigure (ed.), Atomic Energy Society of Japan, Tokyo, Japan, 2000b. With permission.

[a] Dependent on the combination of B and Li concentrations.
[b] Dependent on the operation.

IGSCC of austenitic stainless steels and nickel-base alloys has been a major issue in BWRs. This IGSCC occurs when three key factors exist simultaneously: material sensitization, residual stress, and an oxidizing environment. Countermeasures from the viewpoints of material and residual stress can be taken for the out-of-core components or the materials of good accessibility. The control of water chemistry, however, can reduce the possibility IGSCC occurrence in materials, which still have susceptibility. Figure 34.3 shows the basic concept of IGSCC prevention by removing one or more factors of IGSCC occurrence (AESJ, 2000c).

An approach to simulate the BWR core environment by water radiolysis modeling has attracted wide attention from the standpoint of IGSCC mitigation of the structural materials. As a result of the recent progress in modeling, water radiolysis in the BWR primary system is now well described by radiolysis calculation codes (AESJ, 2000c; Takagi et al., 1989, 1998, 2000; Ichikawa et al., 1992, 1994, 1996; Urata et al., 1998; Takagi and Ichikawa, 1999, 2001).

Corrosion potential has been regarded as an essential parameter to predict IGSCC behavior of the structural materials and much effort has been made to establish a prediction of corrosion

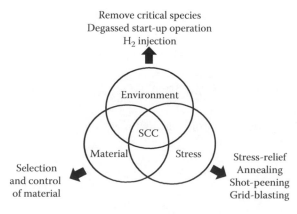

FIGURE 34.3 Three SCC factors and preventive protection of SCC (stress corrosion cracking). (From AESJ, *Handbook of Water Chemistry of Nuclear Reactor System*, K. Ishigure (ed.), Atomic Energy Society of Japan, Tokyo, Japan, 2000c. With permission.)

potential distribution in addition to radiolysis modeling. A mixed potential theory has therefore been applied to electrochemical corrosion potential (ECP) modeling.

An application of the radiolysis model and the ECP model to the prediction of the chemical species and ECP distributions in the primary system under normal water chemistry and hydrogen water chemistry conditions is briefly described in this chapter.

34.2.3.1   Preventive Water Chemistry

The importance of water chemistry control has been widely recognized from the viewpoint of mitigation of IGSCC especially for the structural materials in the BWR primary system. Hydrogen water chemistry (HWC) is one of the IGSCC countermeasures and has been applied to many BWR plants over the world.

According to the best current knowledge of material corrosion, corrosion potential control is critical in order to reduce the IGSCC susceptibility of stainless steels and nickel-base alloys. However corrosion potential is a function of material, water chemistry, and flow condition of the system. In this sense, corrosion potential is a key parameter, which connects the water chemistry and the material behavior as an IGSCC growth rate.

The local corrosion environment is different depending on its geometry, dose rate, and flow rate. As preventive maintenance is based on countermeasures to each of these, an understanding of the corrosion environment, especially under high flow condition or in a creviced region is of much interest. One of the possible approaches is to perform a three dimensional water flow and radiolysis calculation, which will be combined with an ECP model prediction. This development will utilize a more detailed and specific prediction of the corrosion potential of each component.

34.2.3.2   Basic Approaches to Control IGSCC

The behavior of material corrosion consists of two reactions. One is an anodic reaction of the metal dissolution or hydrogen oxidation which will donate electrons. The other is a cathodic reaction which is oxygen or hydrogen peroxide reduction which will receive electrons.

Generally a corrosion reaction is described electrochemically as follows:

Anodic reactions:

$$M = M^{n+} + ne^-$$

(34.1)

$$H_2 = 2H^+ + 2e^-$$

(34.2)

Cathodic reactions:

$$O_2 + 4H^+ + 4e^- = 2H_2O$$

(34.3)

$$H_2O_2 + 2H^+ + 2e^- = 2H_2O$$

(34.4)

Corrosion potential (E_{corr}) is defined as the potential at which the anodic current and the cathodic current are balanced.

$$i_a = i_c \quad \text{at } E = E_{corr},$$

(34.5)

where
   i_a is the total anodic current
   i_c is the total cathodic current
   E_{corr} is the corrosion potential

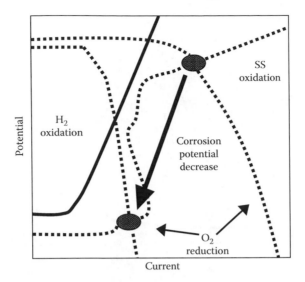

FIGURE 34.4 Comprehension of corrosion potential decrease by decrease in O_2 concentration on Evans diagram.

In order to reduce corrosion potential, there are basically two approaches. One is the cathodic reaction control such as hydrogen water chemistry, and the other is the anodic reaction control such as noble metal chemistry.

34.2.3.2.1 Cathodic Reaction Control

When hydrogen water chemistry is applied to a BWR plant, the concentration of the oxidizing species such as oxygen and hydrogen peroxide will be reduced depending on the injection rate of hydrogen. The principle of this approach is to reduce oxidizing species concentrations in the system. On an Evans diagram, it is shown as (1) a decrease in equilibrium potential, and thus (2) a decrease in limiting current.

As shown in Figure 34.4, the corrosion potential is expressed as the crossing point of the anodic curve and the cathodic curve, which becomes lower compared with the normal water chemistry condition. In this figure, the anodic curve that is metal dissolution is not changed and the corrosion current after hydrogen injection is reduced (Takagi and Ichikawa, 2001).

34.2.3.2.2 Anodic Reaction Control

When the material surface is treated by noble metal, the hydrogen oxidation reaction is much enhanced by a catalytic effect of the noble metal. Therefore, the total anodic current is largely increased. Figure 34.5 shows a schematic that illustrates the corrosion potential decrease on an Evans diagram (Takagi and Ichikawa, 2001).

The figure shows the corrosion potential decrease by enhancement of the anodic reaction of hydrogen. In this case, the exchange current density becomes large and controls the corrosion potential.

The corrosion current is increased when the potential reaches the corrosion potential. However, the corrosion potential in this system will be largely reduced. The reduction by anodic reaction control is sometimes larger than that by cathodic reaction control.

34.2.4 MODELING APPROACH TO BWR CORROSION ENVIRONMENT EVALUATION

34.2.4.1 General Modeling Method

In order to describe the IGSCC phenomena in BWRs, both crack initiation and crack propagation should be discussed. From a practical viewpoint, the quantitative analysis of crack propagation

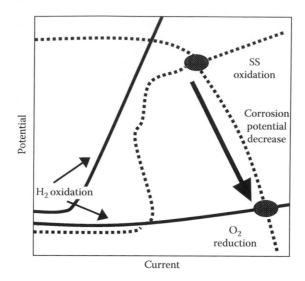

FIGURE 34.5 Comprehension of corrosion potential decrease by increase in H_2 anodic current on Evans diagram.

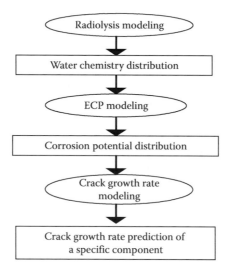

FIGURE 34.6 The basic concept of preventive water chemistry in the BWR primary system.

is recognized as the most important parameter when dealing with plant aging issues. Figure 34.6 shows the general modeling method used to describe the crack growth rate of a specific reactor component (Takagi and Ichikawa, 2001). This requires both radiolysis modeling and the ECP (electrochemical corrosion potential) modeling.

34.2.4.2 Radiolysis Modeling

Radiolysis modeling is done by solving a set of more than 30 differential equations representing chemical reactions of water radiolysis products. This model calculates the concentrations of oxygen, hydrogen, hydrogen peroxide, and other radicals along with the water flow in the primary loop of a BWR plant.

Water radiolysis by neutron and gamma rays produces the primary radiolysis species.

$$H_2O — (\text{neutrons, } \gamma\text{-rays}) \longrightarrow e_{aq}^-, H^+, {}^{\cdot}H, {}^{\cdot}OH, H_2, H_2O_2 \qquad (34.6)$$

Secondary reactions follow to produce molecular species like hydrogen and hydrogen peroxide. Those molecular species will react again with radicals.

$$^{\bullet}H + ^{\bullet}H \rightarrow H_2 \tag{34.7}$$

$$^{\bullet}H + ^{\bullet}OH \rightarrow H_2O \tag{34.8}$$

$$^{\bullet}OH + ^{\bullet}OH \rightarrow H_2O_2 \tag{34.9}$$

$$^{\bullet}H + H_2O_2 \rightarrow OH^{\bullet} + H_2O \tag{34.10}$$

$$^{\bullet}OH + H_2 \rightarrow ^{\bullet}H + H_2O \tag{34.11}$$

$$^{\bullet}OH + H_2O_2 \rightarrow HO_2^{\bullet} + H_2O \tag{34.12}$$

Generation of oxygen will occur in the later stage with HO_2 as a precursor.

$$HO_2^{\bullet} + ^{\bullet}H \rightarrow H_2O_2 \tag{34.13}$$

$$HO_2^{\bullet} + ^{\bullet}OH \rightarrow O_2 + H_2O \tag{34.14}$$

$$HO_2^{\bullet} + HO_2^{\bullet} \rightarrow O_2 + H_2O_2 \tag{34.15}$$

The effect of hydrogen injection is described by H atom reactions that recombine oxidizing species back into water molecules as follows:

$$^{\bullet}H + O_2 \rightarrow HO_2^{\bullet} \tag{34.16}$$

$$^{\bullet}H + HO_2^{\bullet} \rightarrow H_2O_2 \tag{34.13}$$

$$^{\bullet}H + H_2O_2 \rightarrow ^{\bullet}OH + H_2O \tag{34.10}$$

$$^{\bullet}H + ^{\bullet}OH \rightarrow H_2O \tag{34.8}$$

Finally, water radiolysis is shown as equilibrium reaction among water, hydrogen peroxide, and oxygen. The direction of the reactions determines whether the water chemistry condition is oxidizing or reducing.

$$H_2O \overset{neutron\ \gamma}{\longleftarrow} \overset{\bullet}{\longrightarrow} {}^{\bullet}OH \overset{H\ OH}{\longleftrightarrow} H_2O_2 \overset{H\ OH}{\longleftrightarrow} HO_2^{\bullet} \overset{H\ OH}{\longleftrightarrow} O_2 \tag{34.17}$$

The secondary reactions among the radiolysis products proceed according to initial yields (G-values) and the rate constants of the radiolytic product reactions.

The secondary reactions are described as

$$\frac{dC_i}{dt} = \frac{G_i P}{N} + \sum_{m,n} k_{mn} C_m C_b \tag{34.18}$$

where

   C_i is the concentration of species i (mol L^{-1})
   G_i is the G-value of species i (number 100 eV^{-1})
   P is the absorbed energy (100 eV s^{-1} L^{-1})
   N is the Avogadro's number
   k_{mn} is the reaction rate constant between species C_m and C_n

As input data, the G-values, reaction rate constants of the radiolysis products, neutron and gamma dose rates at each portion of the loop, flow velocity, and residence time are necessary. The BWR primary loop is divided into 11 sections. The concentration profiles of each radiolysis product are obtained at a certain hydrogen injection rate.

34.2.4.2.1 G-Values

G-Values are necessary as input data to the radiolysis model. Those at room temperature are known for both neutrons and gamma-rays, as shown in Table 34.5 (Burns and Moore, 1976). There is still a need for high-temperature G-value measurements for modeling work on water radiolysis in the BWR primary system. Because this is a complicated system under high temperature and pressure, very few reports are currently available, as shown in Table 34.6 (Elliot et al., 1993; Elliot, 1994; Sunaryo et al., 1995).

In general, water radiolysis is enhanced by 10%–30% with temperature increasing up to reactor temperature for both fast neutrons and gamma-rays (high-energy electrons). The G-values for the reducing species (e_{aq}^- and H) under neutron irradiation, however, are suppressed by about 30% at high temperature.

34.2.4.2.2 Reactions and Reaction Rate Constants

In order to perform a water radiolysis simulation, the secondary reactions of the radiolytic products must also be considered. In the current radiolysis model, 25 secondary reactions and 3 dissociation equilibrium equations are considered. Reverse reactions are also considered for three key reactions.

Table 34.7 shows the secondary reactions and their rate constants at high temperature (Takagi and Ichikawa, 2001). At high temperature, the rate constant is estimated by extrapolation from the room temperature value assuming Arrhenius behavior and an activation energy.

TABLE 34.5
G-Values of Water Decomposition Products at Room Temperature

		e_{aq}^-	H$^+$	$\cdot$H	$\cdot$OH	H$_2$	H$_2$O$_2$	HO$_2^{\cdot}$	References
n	25°C	0.96	0.96	0.52	1.13	0.91	1.03	0.04	Burns and Moore (1976)
Gamma	25°C	2.80	2.80	0.63	2.96	0.45	0.63	0.03	Burns and Moore (1976)

Note: Unit is (10^{-7} mol J^{-1}).

TABLE 34.6
G-Values of Water Decomposition Products at Elevated Temperatures

		e_{aq}^-	$\cdot$H$^+$	$\cdot$H	$\cdot$OH	H$_2$	H$_2$O$_2$	HO$_2^{\cdot}$	References
n	250°C	0.70	0.70	0.54	1.66	1.72	1.34		Sunaryo et al. (1995)
Gamma	250°C	3.67	3.67	0.97	3.61	0.58	1.10	0	Sunaryo et al. (1995)
n	300°C	0.63	0.63	0.35	2.09	1.31	0.67	0.05	Elliot et al. (1993)
Gamma	285°C	3.66	3.66	0.93	4.85	0.65	0.52	0	Elliot (1994), Elliot et al. (1993)

Note: Unit is (10^{-7} mol J^{-1})$^{-1}$.

TABLE 34.7
Reactions and Reaction Rate Constants at 285°C
Rate Constants Are Expressed in a Unit of M^{-1} s^{-1}

	Forward Reaction	Reverse Reaction	Forward	Reverse
H atom/hydrated electron equilibrium				
1	$e_{aq}^- + H_2O$	$^\bullet H + OH^-$	1.69E+02	6.88E+08
2	$e_{aq}^- + H^+$	$^\bullet H$	2.54E+11	1.00E+05
Radical–radical reaction				
3	$^\bullet H + ^\bullet H$	H_2	1.06E+11	—
4	$e_{aq}^- + ^\bullet H + H_2O$	$OH^- + H_2$	4.76E+09	—
5	$e_{aq}^- + e_{aq}^- + H_2O + H_2O$	$OH^- + H_2 + OH^-$	1.74E+07	—
6	$^\bullet H + ^\bullet OH$	H_2O	2.12E+11	—
7	$e_{aq}^- + ^\bullet OH$	OH^-	3.17E+11	—
8	$^\bullet OH + ^\bullet OH$	H_2O_2	4.76E+10	—
Radical–molecule reaction				
9	$^\bullet H + H_2O_2$	$^\bullet OH + H_2O$	3.07E+09	—
10	$e_{aq}^- + H_2O_2$	$^\bullet OH + OH^-$	1.38E+11	—
11	$e_{aq}^- + HO_2^- + H_2O$	$^\bullet OH + OH^- + OH^-$	6.67E+08	—
12	$^\bullet OH + H_2$	$^\bullet H + H_2O$	1.27E+09	1.10E+03
13	$^\bullet OH + H_2O_2$	$H_2O + HO_2^\bullet$	3.91E+08	—
HO_2 reaction				
14	$^\bullet H + HO_2^\bullet$	H_2O_2	2.11E+11	—
15	$^\bullet H + O_2^{-\bullet}$	HO_2^-	2.11E+11	—
16	$e_{aq}^- + HO_2^\bullet$	HO_2^-	2.12E+11	—
17	$e_{aq}^- + O_2^{-\bullet} + H_2O$	$HO_2^- + OH^-$	1.13E+08	—
18	$^\bullet OH + HO_2^\bullet$	$H_2O + O_2$	1.27E+11	—
19	$^\bullet OH + O_2^{-\bullet}$	$OH^- + O_2$	1.27E+11	—
20	$HO_2^\bullet + HO_2^\bullet$	$H_2O_2 + O_2$	9.29E+07	—
21	$HO_2^\bullet + O_2^-$	$O_2 + HO_2^-$	5.16E+08	—
22	$2\,O_2^{-\bullet} + 2H_2O$	$O_2 + H_2O_2 + 2OH^-$	1.93E+05	—
O_2 reaction				
23	$^\bullet H + O_2$	$HO_2^\bullet$	2.01E+11	—
24	$e_{aq}^- + O_2$	$O_2^{-\bullet}$	2.01E+11	—
Dissociation equilibrium				
25	$OH^- + H_2O_2$	$HO_2^- + H_2O$	1.72E+10	1.08E+05
26	$HO_2^\bullet$	$H^+ + O_2^{-\bullet}$	8.46E+06	5.29E+11
27	$H^+ + OH^-$	H_2O	1.48E+12	1.81E–01
H_2O_2 decomposition				
28	$2H_2O_2$	$2H_2O + O_2$	—	—

Source: Takagi, J. and Ichikawa, N., Water radiolysis modeling and the prediction of corrosion potentials at actual BWRs, in *Proceedings of the Seminar on Water Chemistry of Nuclear Reactor Systems 2001*, paper 23, Chung-Hwa Nuclear Society, Taipei, Taiwan, 2001, pp. 1–6. With permission.

$$k = A \exp\left(\frac{-E_a}{RT}\right), \tag{34.19}$$

where
    k is the reaction rate constant
    A is the frequency factor
    E_a is the activation energy
    R is the gas constant
    T is the absolute temperature

H_2O_2, HO_2, and H_2O each have dissociation constants and the dominant chemical form depends on pH. The backward reaction rate constants are shown in Table 34.7.

The following two reverse reactions are also added to the reaction scheme, as shown in Table 34.7:

$$^\bullet H \rightarrow e_{aq}^- + H^+ \tag{34.20}$$

$$^\bullet H + H_2O \rightarrow {}^\bullet OH + H_2 \tag{34.21}$$

A parametric study made under gamma-ray irradiation conditions demonstrated the importance of including these reverse reactions (Ichikawa and Takagi, 1998).

34.2.4.2.3 Radiolysis Simulation in BWR Primary Circuit

In order to perform the radiolysis simulation in the BWR primary circuit, not only the G-values and the rate constants but also the various plant parameters are needed. It is common to divide the coolant loop into several blocks (regions) depending on dose rate, flow rate, and residence time. In the core region, the void fraction distribution and the hydrogen and oxygen gas stripping rates from the liquid to the vapor phase are also necessary to simulate the two-phase flow.

An example of a block diagram for a jet-pump type BWR is shown in Figure 34.7 (AESJ, 2000c). Recently, efforts have been made to divide the loop into more detailed regions or to divide the

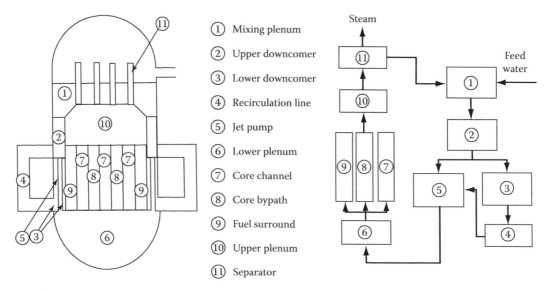

FIGURE 34.7 Block diagram of the radiolysis model for the BWR primary cooling system. (From AESJ, *Handbook of Water Chemistry of Nuclear Reactor System*, K. Ishigure (ed.), Atomic Energy Society of Japan, Tokyo, Japan, 2000c. With permission.)

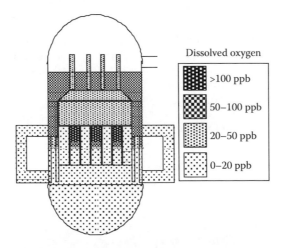

Dissolved oxygen

>100 ppb

50–100 ppb

20–50 ppb

0–20 ppb

FIGURE 34.8 An example of the radiolysis simulation result for the BWR primary circuit under the condition of 0.8 ppm hydrogen injection into the feedwater. (From AESJ, *Handbook of Water Chemistry of Nuclear Reactor System*, K. Ishigure (ed.), Atomic Energy Society of Japan, Tokyo, Japan, 2000c. With permission.)

downcomer region into several layers depending on the gamma-ray dose rate because of the large sensitivity of the model to recombination effects in the downcomer region.

The absorbed dose rate to the high-temperature water is important in the simulation as well as G-values. These dose rates are based on the neutron flux and gamma-ray distribution calculations, usually performed for the purpose of shielding calculations.

The flow rate and the residence time at each region are plant specific and are determined from the design parameters. Because the radiolytic reaction is fast compared to the residence time of the coolant at each region, an equilibrium always holds for each region depending on its dose rate.

Another parameter to be noted is the gas stripping rate from the liquid to the vapor phase at the boiling region, especially under hydrogen injection conditions. The distribution between the two phases is estimated to be largely different from the Henry's law constant because of the dynamics of core boiling. Therefore, those parameters are determined by benchmarking of the calculated hydrogen and oxygen concentrations in the main steam system to measured concentrations.

Figure 34.8 shows an example of the radiolysis simulation result for the BWR primary circuit (AESJ, 2000c). The stable species of hydrogen, oxygen, and hydrogen peroxide are shown. However, the concentrations of the short-lived species such as radicals are also numerically obtained.

34.2.4.3 ECP Modeling

Depending on the radiolysis modeling results, the ECP is calculated using mixed potential theory by combining the chemistry species distributions inside the BWR primary circuit and the hydrodynamic parameters such as flow rate and hydraulic diameter.

The corrosion potential (E_{corr}) is defined as the potential at which the anodic current (i_a) and the cathodic current (i_c) are balanced. The anodic reaction is described as metal dissolution and hydrogen oxidation, whereas the cathodic reaction is described as reducing reactions of oxygen and hydrogen peroxide, as shown in Section 34.2.3.2.

In order to reduce the corrosion potential to mitigate IGSCC, the cathodic current must be controlled by hydrogen injection or by controlling the anodic current, as shown in Figures 34.4 and 34.5.

The anodic polarization curve is derived from the measurement with stainless steel at high temperature. Under hydrogenated conditions, the anodic current of hydrogen oxidation will be added and in such cases it will become dominant. On the other hand, the cathodic polarization curve can be analytically obtained assuming appropriate parameters such as equilibrium potential, exchange current density, Tafel slope, and limiting current density.

34.2.4.3.1 Equilibrium Potential

The equilibrium potential of the corrosion equation is calculated using Nernst's equation. It depends on the chemical species concentration and the temperature. In the case of the oxygen reduction reaction, the equilibrium potential is described as shown below (AESJ, 2000c).

$$E = 1.02 + 0.028 \log P_{O_2} - 0.112 \text{pH}$$

$$= 0.289 \ (\text{V(SHE)}) \quad (O_2 = 100 \ \text{ppb, at } 290°C), \tag{34.22}$$

where

   E is the equilibrium potential
   P_{O_2} is the oxygen partial pressure
   pH is the pH of the solution

34.2.4.3.2 Exchange Current Density and Tafel Slope

Exchange current density is defined as a current density that occurs on the electrode surface under the equilibrium condition without over potential (MacDonald, 1992). Tafel slope is defined as a slope of potential against current where over potential exists by polarization. Charge transfer is the limiting process in this area. A slope of 0.23 V decade^{-1} at 290°C is reported from laboratory measurements.

34.2.4.3.3 Limiting Current Density

In the case of the ECP Model, a flow rate effect is taken into consideration in the formula of limiting current density together with the equivalent hydraulic diameter (Selman and Tobias, 1978).

$$I_{\text{lim}} = \frac{0.0165 n F D C_i Re^{0.86} Sc^{0.33}}{d}, \tag{34.23}$$

where

   F is the Faraday constant
   D is the diffusion constant
   C_i is the concentration of species i
   Re is the Reynolds number (Vd/v)
   Sc is the Schmidt number (v/D)
   d is the equivalent hydraulic diameter
   V is the flow velocity
   v is the kinematic viscosity

When considering the BWR primary loop condition, the corrosion potential becomes less under lower flow conditions due to a smaller limiting current density and higher under high flow conditions due to larger limiting current density. This means that flow velocity is one of the key parameters controlling the corrosion potential evaluation.

34.2.4.4 ECP Evaluation for BWR Primary Circuit

The same block diagram for the radiolysis modeling can be applied to the ECP simulation for the BWR primary loop. As input parameters, the concentrations of the chemical species like hydrogen, oxygen, and hydrogen peroxide are provided by the radiolysis modeling. The flow rate and the hydraulic diameter of each region of interest are necessary to carry out the ECP evaluation.

34.2.4.5 Advanced ECP Evaluation Using 3-D Flow Distribution

The corrosion potential of the structural materials at the jet pump outlet region has been evaluated by ECP modeling using a mixed potential theory (Ichikawa et al., 1992, 2002). One of the important parameters in this model is the limiting current density which is strongly affected by the water flow velocity. From the principle, it is expressed as

$$I_{\lim} = knFC_j,\qquad(34.24)$$

where
    k is the mass transfer coefficient
    n is the number of electrons transferred per mole
    F is the Faraday constant
    C_j is the bulk concentration of species j

The mass transfer coefficient is expressed as

$$k = \frac{Sh}{d},\qquad(34.25)$$

where
    Sh is the Sherwood number
    d is the hydraulic diameter

Two equations for evaluating Sh have been reported, as given below:

$$Sh = 0.023 Re^{0.8} Sc^{0.4} \quad (Sc < 100) \ (\text{Macadamas}, 1954)\qquad(34.26)$$

$$Sh = 0.0165 Re^{0.86} Sc^{0.33} \quad (1000 < Sc < 6000), \ (\text{Berger and Hau}, 1977)\qquad(34.27)$$

where
    Re is the Reynolds number (Ud/v)
    Sc is the Schmidt number (v/D)
    U is the flow velocity
    v is the kinematic viscosity
    D is the diffusivity

The Schmidt numbers for O_2, H_2, and H_2O_2 at 288°C are 0.34, 1.63, and 0.32, respectively. Equation 34.26 is suitable for BWR conditions.

Equations 34.24 and 34.25 make it necessary to have the hydraulic diameter (d) in order to calculate the limiting current density. A precise evaluation of hydraulic diameter, however, is impossible because of the very complex structure of the reactor pressure vessel (RPV) bottom region. One method of evaluating the mass transfer coefficient (k) without introducing the hydraulic diameter is to bring in the Stanton number (St). The Stanton number can be expressed as

$$St = \frac{Sh}{(ScRe)} = kU\qquad(34.28)$$

$$St = \frac{S}{(1 + PS1/2)\,Prt},\qquad(34.29)$$

where

    S is the coefficient of friction

    P is the P function $(P = 9.0(Sc/Prt-1)(Sc/Prt)-1/4)$

    Prt is the Prandtl number of turbulent flow

In this study, the Prandtl number was set at 0.9. This number simulates the Sh equation (34.26). On this basis, the limiting current density of species j can then be expressed as

$$I_{\lim} = StUnFC_j. \tag{34.30}$$

This approach has an advantage that, when the coefficient of friction (S) and the water flow velocity (U) are given and the bulk concentration of species j (C_j) is calculated using the radiolysis model, the Stanton number (St) and then the limiting current density at any location inside the primary loop can be estimated without the need for the hydraulic diameter.

34.2.5   Hydrogen Water Chemistry Application to Actual BWRs

34.2.5.1   Introduction

Hydrogen water chemistry (HWC) is a chemistry control technology used to reduce oxidizing species in reactor water. When hydrogen is injected from the feedwater and is introduced to the core, a recombination reaction of hydrogen with oxygen and hydrogen peroxide is enhanced under irradiation, resulting in the mitigation of corrosion by the scavenging of oxidizing species.

    In Japan, the first hydrogen water chemistry verification for commercial BWRs was carried out in 1992 (Ashida et al., 1992). It was a short-term program to identify the effectiveness of hydrogen water chemistry control and to measure the plant parameter responses including main steam system radiation level increases. That program was successfully done and the next verification program for a longer period was planned.

    In 1995, a long-term verification program on hydrogen water chemistry was started at another BWR (Takagi et al., 1998). In this program, it was planned to measure corrosion potentials at the bottom drain line of the RPV and to perform crack growth measurements in the out-of-core autoclaves for demonstration. The outline of the above verification program is described below.

34.2.5.2   Environmental Mitigation of IGSCC by Hydrogen Water Chemistry

In order to mitigate IGSCC of the primary system components in BWR plants, hydrogen water chemistry control has been implemented in several BWRs, especially in overseas plants. In those plants, it has been verified that reactor water oxygen and hydrogen peroxide are well suppressed by hydrogen addition. Efforts have also been made to measure corrosion potential directly in the primary system of some plants. The locations of the sensors are out-of-core piping, upper and lower core regions, and the bottom region of the reactor pressure vessel.

    When hydrogen is injected from the feedwater line, the recombination reaction of injected hydrogen and core-produced oxygen and hydrogen peroxide occurs in the so-called downcomer region, that is, the region between the core shroud and pressure vessel wall through which water flows downward with a rather moderate velocity. The gamma dose rate at that region enhances the radiolytic recombination reactions to produce water, which results in higher efficiency of oxygen and/or hydrogen peroxide suppression in the downcomer region and the out-of-core piping system connected to the downcomer. On the other hand, comparatively more hydrogen is required to protect core internal components.

    According to the in-plant autoclave type material testing results, it has been accepted that there is a threshold potential of $-230\,mV$ (SHE) (standard hydrogen electrode) for IGSCC occurrence. The target of hydrogen water chemistry is to reduce ECP of the key components below this threshold potential.

To evaluate the effectiveness of hydrogen injection, ECP is a useful indicator for the prediction of IGSCC prevention. The reduction of the concentration of oxidizing species such as oxygen and hydrogen peroxide will cause the reduction of ECP of the materials. In the case of austenitic stainless steels or nickel-based alloys, IGSCC initiation or propagation is mitigated under a certain threshold of corrosion potential. Therefore, estimation of the corrosion potential of the structural materials in the BWR primary system is needed to determine the required amount of hydrogen to be added to the system.

34.2.5.3　Program Description

34.2.5.3.1　Objectives

The main objective of this program is to verify the effectiveness of hydrogen water chemistry control as an IGSCC countermeasure and its applicability to BWRs. The goal is to evaluate IGSCC behavior of plant structural materials under hydrogen water chemistry. The program consists of three parts in order to accomplish this goal.

1. Water chemistry measurement
2. Corrosion potential measurement
3. Crack growth measurement

Figure 34.9 shows the evaluation flow for the IGSCC behavior of the materials under HWC conditions. Decreased dissolved oxygen and hydrogen peroxide concentrations due to hydrogen addition results in decreased corrosion potential. Therefore, the ECP measurement is regarded as a verification of mitigation. Corrosion potential is measured at the RPV bottom drain line to evaluate the vessel bottom corrosion environment.

Secondly, a decrease in corrosion potential results in a decrease in crack growth rate of the materials. Out-of-core in-plant material testing is performed as verification of IGSCC mitigation. Type 304 stainless steel and Inconel 182 are used as test specimens. The results are discussed elsewhere (Takagi et al., 1996).

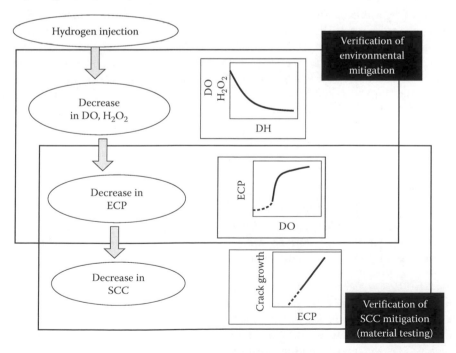

FIGURE 34.9　Evaluation flow of IGSCC behavior under hydrogen water chemistry in the verification program.

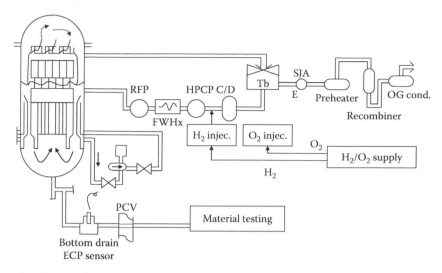

FIGURE 34.10   Description of the hydrogen water chemistry system in the verification program.

34.2.5.3.2   Plant Features

The features of the plant chosen for this program are summarized as follows:

- BWR Type-4
- Thermal power: 2381 MWth
- Rated power: 784 MWe
- Power density: 40 kW L^{-1}
- Commercial operation: since April 1978

34.2.5.3.3   Hydrogen and Oxygen Injection System

The schematic of the hydrogen water chemistry system used in this program is shown in Figure 34.10. Hydrogen gas is supplied from a gas loader system with an initial pressure of about 200 kg cm^{-2}, which is depressurized to about 8 kg cm^{-2} at the inlet of the hydrogen injection module. The injection point is at the outlet of the condensate demineralizer (at the suction header of the high-pressure condensate pumps) where the system pressure is the lowest at about 4 kg cm^{-2}. The maximum hydrogen injection rate was about 100 N m^3 h^{-1} (2 ppm in feedwater) and the injection rate for permanent injection is tentatively set at 25 N m^3 h^{-1} (0.5 ppm in feedwater).

Oxygen gas is supplied from a liquefied oxygen tank. As the injection point is at the inlet of off-gas preheater (upstream of the off-gas recombiner), high pressure is not required for the system. The oxygen injection rate is half of the hydrogen injection rate.

34.2.5.4   Experimental

34.2.5.4.1   Water Chemistry Measurement

Water chemistry measurements were conducted using existing plant sampling locations such as oxygen and hydrogen in reactor water, feedwater, condensate, and main steam; conductivity and pH in reactor water and feedwater; and other parameters taken by grab sampling.

34.2.5.4.2   Corrosion Potential Measurement

An in situ and high-flow corrosion potential measurement device was developed. A corrosion potential sensor was installed in a new sampling line branched from the existing RPV bottom drain line (see Figure 34.10), which contained the reactor water of the same water chemistry as that of the RPV bottom region. The decomposition of hydrogen peroxide between the vessel bottom and the sensor location is estimated to be about 10%.

The sensor unit consists of two kinds of electrodes: a silver/silver chloride reference electrode and a platinum electrode. The working electrode of this unit is the stainless steel piping (Type 316L) itself. The temperature and the pressure at the monitoring location are 275°C and 70 kg cm^{-2}, respectively. The flow velocity inside the piping is 3.1 m s^{-1}, which represents the highest flow condition inside the vessel along the bottom region.

34.2.5.5   Results and Discussion

34.2.5.5.1   Water Chemistry Results

The effect of hydrogen water chemistry is also measured by the decrease in reactor water dissolved oxygen, which is measured in a conventional sampling manner in the RPV bottom drain line, reactor water cleanup line and primary loop recirculation line. The results are shown in Figure 34.11. In general, the dissolved oxygen content in the recirculation line is lower than in the RPV bottom region because of effective recombination in the downcomer region. The dissolved oxygen in the RWCU line is the mixture of the recirculation line and the bottom drain line oxygen contents.

It is shown in Figure 34.11 that with 0.5 ppm hydrogen in feedwater, the reactor water dissolved oxygen is decreased to ca. 7 ppb. Note that at the end of the sampling line all the hydrogen peroxide thermally decomposes to oxygen and water, and the measured concentration of the dissolved oxygen is the sum of [O_2] and [H_2O_2]/2.

34.2.5.5.2   Water Chemistry Measurement Results

Figure 34.12 shows the relationship between the RPV bottom drain line corrosion potential and dissolved hydrogen in feedwater obtained in this program. Under normal water chemistry conditions, the corrosion potential at the bottom drain sensor unit shows almost +200 mV (SHE), which is slightly higher than conventional autoclave results. Then it decreases with an increase in feedwater hydrogen and reaches −230 mV (SHE) at a feedwater hydrogen concentration of 0.8 ppm.

At the hydrogen injection rate of 0.8 ppm in feedwater, the main steam system radiation level increase is by almost four times, as shown in Figure 34.12. To minimize the radiological impact due to N-16 carryover, low hydrogen water chemistry becomes an option. For example, 0.4–0.5 ppm hydrogen injection to feedwater will give a bottom drain line corrosion potential of −50 to −100 mV (SHE). It is expected that IGSCC crack growth behavior will be suppressed to some extent according to the degree of the potential reduction.

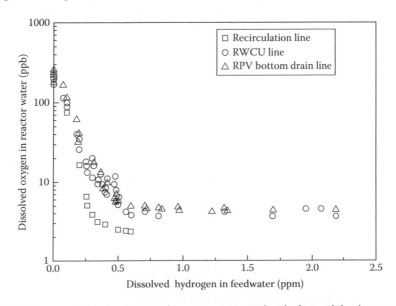

FIGURE 34.11   Behavior of dissolved oxygen in reactor water against hydrogen injection rate.

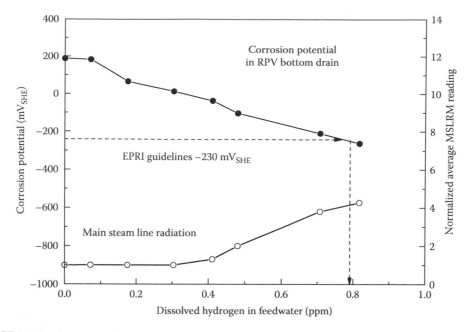

FIGURE 34.12 Behavior of bottom drain line corrosion potential against hydrogen injection rate.

The corrosion potential of the stainless steel at the ECP sensor unit decreases with the decrease in the dissolved oxygen concentration in the RPV bottom drain line. There is a unique relationship between the corrosion potential and the oxygen concentration, as shown in Figure 34.13. The corrosion potential decreased sharply in the range of the dissolved oxygen concentration below 10 ppb. However, it should be noted that the corrosion potential was measured at the ECP sensor unit location whereas the oxygen was measured at the end of the sampling line far apart from the sensor location.

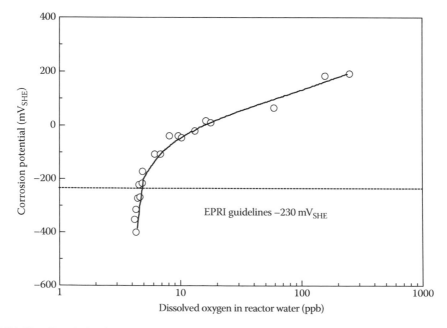

FIGURE 34.13 Correlation between corrosion potential and dissolved oxygen in bottom drain under hydrogen water chemistry.

The corrosion potential behavior has a strong dependency on the flow rate (Ichikawa et al., 1992, 1994). The relationship between the corrosion potential and the dissolved oxygen (in Figure 34.6) was obtained with a flow rate of $3.1\,m\,s^{-1}$. In the slower flow rate region, the corrosion potential decreases faster with the decrease of oxygen.

34.2.5.6 Hydrogen Water Chemistry Summary

Through the long-term verification program, hydrogen water chemistry control has been verified as an environmental countermeasure against potential IGSCC issues for BWRs. In order to evaluate the effectiveness of hydrogen water chemistry control on the protection of the vessel bottom region, a high flow and in situ corrosion potential measurement device was developed and installed in a branched sampling line from the RPV bottom drain line. In this program, 0.8 ppm hydrogen in feedwater was necessary to achieve a corrosion potential of $-230\,mV$ (SHE) at the vessel bottom region. The flow rate effect on the corrosion potential should be taken into account to predict corrosion potential distributions in the BWR primary system.

34.2.6 Development of an ECP Sensor for BWR Applications

An experimental approach to directly measure the corrosion potential at BWRs is important to verify the modeling approach. In order to measure the corrosion potential, ECP sensors are used as reference electrodes and working electrodes.

34.2.6.1 Reference Electrode

A reference electrode is absolutely necessary for the ECP measurement. It shows its own potential, which is not affected by water chemistry or flow changes. In BWR systems, a reference electrode, such as silver/silver chloride (Ag/AgCl) or iron/iron oxide (Fe/Fe_3O_4), has been used in high-temperature water.

Under hydrogen water chemistry control conditions, a platinum electrode is used as a reference electrode because it works as a theoretical hydrogen electrode. It also has a longer lifetime compared with the conventional reference electrodes. Therefore, it has long been accepted that both conventional reference and platinum reference electrodes can be used.

34.2.6.2 Working Electrode

Stainless steel or nickel-based alloys are used as working electrodes that behave similar to plant structural materials. Attention should be paid to the initial value before the electrode is prefilmed. The plant structural material itself can be regarded as a working electrode when the ground potential from the plant is measured against the reference electrode.

34.2.7 Radiation Effects on Corrosion in Nuclear Reactor Systems

The radiation effects on material corrosion in nuclear reactor systems is typically classified as a direct radiation effect on materials and an indirect effect through the change of the water chemical environment. The former is often recognized as radiation damage by neutrons or heavy ions, and the latter is comprehended as enhanced corrosion in liquid under irradiation. In this chapter, the relationship between the corrosion behavior and water radiolysis is discussed based on laboratory experimental findings.

34.2.7.1 Radiolysis Effects on the Water Environment

34.2.7.1.1 Production of Oxidant Species by Radiation

Oxidizing radiolytic products enhance material corrosion due to their contributions to the cathodic reactions.

$$\text{Anodic reactions:} \quad M = M^{n+} + ne^- \tag{34.1}$$

$$\text{Cathodic reactions:} \quad O_2 + 4H^+ + 4e^- = 2H_2O \tag{34.3}$$

$$H_2O_2 + 2H^+ + 2e^- = 2H_2O \tag{34.4}$$

$$OH + e^- = OH^- \tag{34.31}$$

$$HO_2 + H^+ + e^- = H_2O_2 \tag{34.32}$$

$$O_2^- + 2H^+ + e^- = H_2O_2 \tag{34.33}$$

Under irradiation conditions, various kinds of oxidizing radicals or species are produced, which contribute to the cathodic reactions and enhance the corrosion reaction. The distribution of the yield of each species is dependent on the quality of radiation known as linear energy transfer (LET).

When reducing species like hydrogen coexist in the system, the radiation effect does not always enhance the corrosion but sometimes reduces the oxidizing species concentrations due to the recombination effect. As a result, the cathodic reactions are suppressed and so is the corrosion reaction.

34.2.7.1.2 Corrosion Behavior by Radiolytic Oxidant Species

34.2.7.1.2.1 Stainless Steel Corrosion behavior of materials in reactor systems is due to the concentrations of the radiolytic oxidizing species such as oxygen, hydrogen peroxide, and radical species. Those concentrations are influenced by radiation. The oxidizing power of the system is described by an ECP on the material surface, which is a function of not only the chemical species concentrations but also the oxide film and hydrodynamic conditions.

In the case of stainless steel corrosion in high-temperature water, both the general corrosion rate and the SCC susceptibility are high at high corrosion potentials. Figure 34.14 shows the radiation

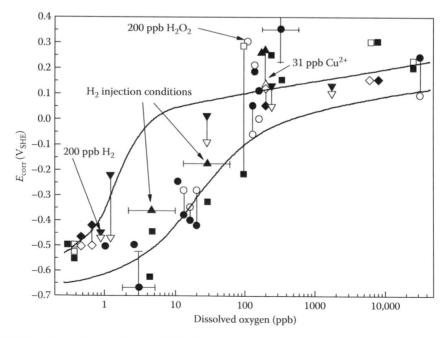

FIGURE 34.14 The radiation effect on ECP of stainless steel.

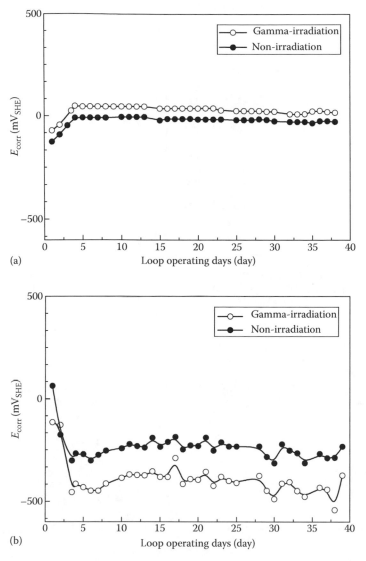

FIGURE 34.15 The gamma-rays effect on ECP of stainless steel. (a) Oxygen excess condition and (b) hydrogen excess condition.

effect on the ECP of stainless steel (Andresen, 1992). Figure 34.15 also shows the gamma-ray effect on the ECP of stainless steel, which increases under excess oxygen conditions due to the production of hydrogen peroxide, and decreases under excess hydrogen conditions due to the recombination reaction of the oxidizing species (Ichikawa et al., 1987).

The same tendency is shown in the case of SSRT (slow strain rate test) results to evaluate the IGSCC susceptibility of sensitized Type 304 stainless steel, as shown in Figure 34.16 (Saito et al., 1990). The IGSCC ratio on the fracture surface of the test specimen is reduced by gamma irradiation under hydrogenated conditions.

Figure 34.17 shows that the critical corrosion potential for IGSCC susceptibility is understood in relation to ECP and it is almost the same regardless of the gamma irradiation (Saito et al., 1997). However, for general corrosion, iron dissolution rates from stainless steel in high-temperature water are accelerated by gamma-ray irradiation, as shown in Figure 34.18 (Ishigure et al., 1980). Figure 34.19 shows that the metal dissolution rate from stainless steel is also accelerated by hydrogen peroxide addition and gamma-ray irradiation (Hemmi et al., 1994).

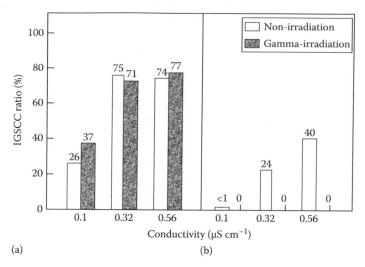

(a)   (b)

FIGURE 34.16 SSRT (slow strain rate test) experiment to evaluate IGSCC susceptibility of sensitized Type 304 stainless steel, sensitized at 650°C for 3 h. (a) BWR normal water chemistry condition at 250°C. (b) BWR hydrogen water chemistry condition at 250°C.

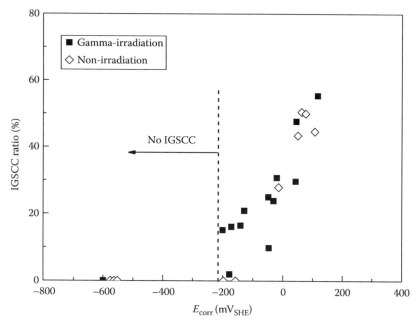

FIGURE 34.17 Critical corrosion potential of sensitized Type 304 stainless steel for IGSCC susceptibility at the SSRT speed of 4×10^{-7} s^{-1}. Sensitized condition is 24 h at 620°C.

34.2.7.1.2.2 Other Materials According to the above discussion, it seems acceptable to simulate the irradiation effect by adding suitable amounts of hydrogen peroxide to the system based on the finding that the oxidizing power generated by radiation is in most part due to the accumulated hydrogen peroxide concentration. The general corrosion rate of Alloy X750 (Ni-based alloy) and Stellite #6 (Co-based alloy) is reported to be enhanced by four times under the BWR core simulated condition with the addition of hydrogen peroxide (Hemmi et al., 1994). Zircaloy is used as fuel cladding material for both BWR and PWR reactors. It is reported that perhydroxy radical ($HO_2^{\bullet}$ or $O_2^{\bullet-}$) produced in the high radiation fields in the core plays a role in zircaloy corrosion (Hurst and Tyzack, 1974).

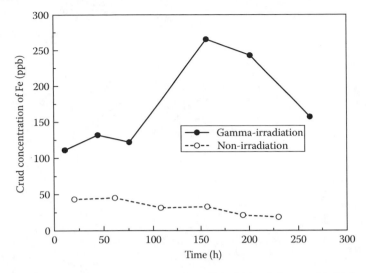

FIGURE 34.18 Acceleration of iron dissolution rate from Type 304 stainless steel in high-temperature water at 250°C under 200 ppb O_2.

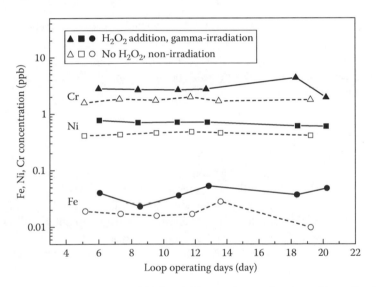

FIGURE 34.19 Effect of gamma-rays and H_2O_2 addition to the acceleration of iron dissolution rate from Type 316 stainless steel in high-temperature water containing 400 ppb of O_2 and 60 ppb of H_2 at 270°C.

34.2.8 RADIATION EFFECT ON MATERIALS

It is widely known that direct neutron irradiation effects to the structural materials should be considered in the core region. Austenitic stainless steel with solution heat treatment for an IGSCC countermeasure sometimes shows IGSCC under neutron irradiation. This is called irradiation-assisted intergranular stress corrosion cracking (IASCC). Although the mechanism is not yet fully understood, it is speculated that irradiation-assisted segregation in grain boundaries is a cause of corrosion property changes.

SCC susceptibility of austenitic stainless steel under neutron irradiation is shown in Figure 34.20 (Andresen et al., 1989). From the high-temperature water SSRT result with 32 ppm of dissolved oxygen, the IASCC susceptibility increases with more than 5×10^{24} nm^{-2} ($E > 1$ MeV) of neutron fluence and becomes significant with more than 2×10^{25} nm^{-2} ($E > 1$ MeV) of neutron fluence.

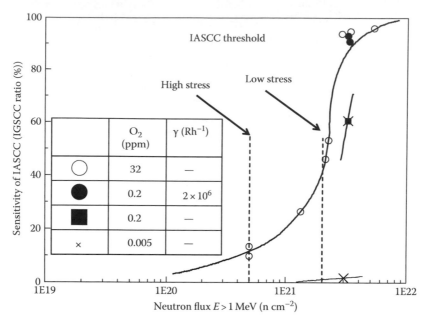

FIGURE 34.20   SCC susceptibility of Type 304 stainless steel under neutron irradiation at 288°C.

The essential countermeasure to this level of radiation damage must be through material improvement. However, research may eventually provide chemical control mitigations that will also work to increase corrosion resistance.

34.2.9 CONCLUSION

The evaluation of the corrosion environment in BWRs is discussed from the viewpoint of both modeling and experimental approaches. The comparison between measurement and calculation is of significance for establishing the countermeasures for plant aging issues.

Corrosion potential control is essential in reducing the IGSCC susceptibility of BWR primary loop components. Two basic approaches for reducing the corrosion potential are reviewed. One is cathodic reaction control and the other is anodic reaction control.

By applying the latest radiolysis and ECP models, normal water chemistry and hydrogen water chemistry control can be well evaluated.

34.3 RADIATION-INDUCED PROCESSES IN THE SPENT NUCLEAR FUEL REPROCESSING

Increased human populations and industrialization are causing increased energy demands at the same time that concerns about rising carbon dioxide concentrations in the atmosphere have made fossil fuels less attractive. Opposition to nuclear power has decreased with a renewed interest being expressed in many countries (Boullis, 2008). However, among the reasons cited by opponents against nuclear energy are difficulties with the disposition of nuclear waste. About 10,500 t of spent fuel are already discharged yearly from more than 400 nuclear reactors (IAEA, 2004). The rate of spent fuel generation could reach 15,000 t by 2050 (Laidler, 2008), and this spent fuel contains appreciable quantities of actinides and fission products, some with very long half-lives. For example, 1 t of UO_2 fuel, after 30 GW days of burnup, contains 950 kg U, 9 kg Pu, 75 g Np, 140 g Am, 47 g Cm, and 31 kg fission products (Benndict et al., 1981). With half-lives ranging from 433 years for [241]Am to 2.4 million years for [237]Np, the alpha-emitting actinide nuclides determine the radiotoxicity of the waste for periods of time on a geological scale. To address this, research programs in several countries are investigating the partitioning of these radionuclides and their transmutation to short-lived isotopes to significantly reduce the long-term hazard (OECD-NEA, 1999).

Aqueous solvent extraction is the most mature of these partitioning strategies, benefiting from more than 60 years of research and experience at the industrial scale (Mathur et al., 2001). As currently performed, an alkane solution of tributyl phosphate (TBP) is used to complex and extract U, Np, and/or Pu from nitric acid solutions of dissolved spent fuel. The process is called PUREX (Plutonium Uranium Refining by EXtraction) and remains successful because it provides high yields of the desired elements and high separation factors from the undesired elements, while producing only a small amount of secondary waste.

Looking to the future, many countries are currently investigating solvent extraction processes for the separation of the minor actinides Am and Cm from dissolved nuclear fuel. This separation would both decrease the radiotoxicity of waste prior to deep geological repository disposal and produce additional energy by burning these elements in a reactor. While none of the advanced processes have yet been implemented, their development has reached demonstration tests at the laboratory scale (Nash et al., 2006). The eventual application of such a partitioning strategy will result in natural uranium savings, uranium enrichment cost savings, simpler and safer waste disposal, and provide the potential for the recovery of other useful elements from spent fuel (Boullis, 2008).

The new extraction schemes vary in their details, but following the traditional PUREX extraction all would extract the minor actinides using hard donor (oxygen containing) ligands. Unfortunately, undesirable short-lived lanthanide fission products are also extracted and they are neutron poisons that must then be separated from Am and Cm, prior to incorporating the actinides in fuel. This second extraction employs soft donor ligands (usually nitrogen containing) capable of performing this difficult separation. Since these new complexing agents will be used in highly radioactive applications, the effects of radiation chemistry on their performance must be understood prior to their reliable application. Based on PUREX experience, the main effects may include decrease in extraction efficiency due to decomposition of the ligand, decrease in separation factors due to the accumulation of radiolysis products that are also complexing agents, and deteriorated phase separation performance due to ligand or diluent radiolysis products (Pikaev et al., 1988). In addition, the irradiation of aqueous nitric acid produces reactive species that can alter the oxidation states of the metals to be complexed, affecting their extraction efficiency (Pikaev et al., 1988). This chapter will examine the conditions and reactive species to be expected in the irradiated biphasic organic/aqueous environment of the solvent extraction process, and review what is known about radiolysis effects on selected ligands and metals during solvent extraction for nuclear reprocessing applications.

34.3.1 RADIOLYTICALLY PRODUCED REACTIVE SPECIES IN THE BIPHASIC SYSTEM

34.3.1.1 Produced Species in the Aqueous Phase

The direct radiolysis of ligands in a solvent extraction solution occurs in proportion to their abundance as constituents of that solution. Since ligands are normally present in millimolar concentrations, the diluent absorbs most of the radiation energy. Reactive species created by diluent radiolysis are largely responsible for the extent and nature of the degradation of ligands and changes the physical characteristics of the solvent. The direct radiolysis of alkanes and aqueous nitric acid results in the formation of electronically excited states, free radicals, and ions in spurs along the track of the incident particle. These reactive species undergo recombination to create molecular species, or they diffuse away from their point of origin to react with solutes in the bulk solution. The probability of recombination or diffusion depends on the LET (eV nm^{-1}) of the incident particle, which for x, γ, and β^- is much lower than for alpha particles. Reactive species that escape these spurs are the source of the secondary radiolysis reactions that degrade ligands or react with metal ions, and this phenomenon begins to occur at about 100 ns after the initial event.

For water irradiated by low LET particles such as x-rays, gamma-rays, or electrons (β^- particles) generated mainly by fission product decay, the species in Equation 34.34 are produced with the yields (G-values in μmol J^{-1}) shown in brackets (Buxton et al., 1988):

$$H_2O \longrightarrow [0.28]\ ^{\bullet}OH + [0.27]\ e_{aq}^- + [0.06]\ H^{\bullet} + [0.07]\ H_2O_2 + [0.27]\ H_3O^+ + [0.05]\ H_2 \qquad (34.34)$$

The most reactive species produced are the oxidizing hydroxyl radical ($^{\bullet}OH$) and hydrogen peroxide (H_2O_2), and the reducing aqueous electron (e_{aq}^-) and hydrogen atom ($H^{\bullet}$). Massive, highly charged particles such as alpha-particles (He^{2+}) generated mainly by actinide decay have short ranges and high LET (156 eV nm^{-1} for 5 MeV He^{2+}) (Pastina and LaVerne, 1999) and therefore deposit energy in closely spaced overlapping spurs. This results in high localized concentrations of reactive species, many of which undergo recombination before they can diffuse into the bulk solution. Thus, higher yields of molecular species and lower yields of radicals are found. Using the data of Lefort and Tarrago (1959) for 5.3 MeV ^{210}Po alpha-particles, the water radiolysis equation may be rewritten:

$$H_2O \longrightarrow [0.05]\ ^{\bullet}OH + [x]\ e_{aq}^- + [0.06]\ H^{\bullet} + [0.15]\ H_2O_2 + [x]\ H_3O^+ + [0.16]\ H_2 \qquad (34.35)$$

The yields for the aqueous electron and hydronium ion are not given because the measurements were made in acidic solution. It is seen that the yield of the molecular product hydrogen peroxide due to alpha radiolysis of water is double that for gamma-radiation. Comparatively, few studies have been done using alpha-radiolysis, and this chapter will cover mainly low LET effects.

Despite an initially equal production of oxidizing and reducing species, under the conditions of nuclear solvent extraction, the aerated, nitric acid system will be predominantly oxidizing due to fast scavenger reactions. The aqueous electrons and hydrogen atoms are scavenged according to

$$e_{aq}^- + H^+ \rightarrow H^{\bullet} \quad k = 2.3 \times 10^{10}\ M^{-1}\ s^{-1} \quad \text{(Buxton et al., 1988)} \qquad (34.36)$$

$$e_{aq}^- + O_2 \rightarrow\ ^{\bullet}O_2^- \quad k = 1.9 \times 10^9\ M^{-1}\ s^{-1} \quad \text{(Buxton et al., 1988)} \qquad (34.37)$$

$$e_{aq}^- + NO_3^- \rightarrow NO_3^{-2\bullet} \quad k = 9.7 \times 10^9\ M^{-1}\ s^{-1} \quad \text{(Elliot, 1989)} \qquad (34.38)$$

$$^{\bullet}H + O_2 \rightarrow\ ^{\bullet}HO_2 \quad k = 2.1 \times 10^{10} \quad \text{(Gordon et al., 1964)} \qquad (34.39)$$

$$^{\bullet}HO_2 \leftrightarrow H^+ +\ ^{\bullet}O_2^- \quad pK_a = 4.8 \quad \text{(Bielski et al., 1985)} \qquad (34.40)$$

This leaves H_2O_2 and the strongly oxidizing $^{\bullet}OH$ radical as the most important species in solution. The $^{\bullet}OH$ radical reacts with organic solutes by hydrogen abstraction, addition to unsaturated carbon bonds, or with organic and inorganic solutes by electron transfer reactions. Rate constants for $^{\bullet}OH$ radical reactions have been tabulated (Dorfman and Adams, 1973; Buxton et al., 1988). The $^{\bullet}HO_2$ radical produced in Equation 34.39 is also an oxidizing agent, although its reactions have not been well studied.

Additional reactive species are created by nitric acid radiolysis. A thorough review of the radiation chemistry of nitric acid is given by Katsumura (1998). Among the most important species, the oxidizing $^{\bullet}NO_3$ radical is produced indirectly by the reaction of the $^{\bullet}OH$ radical with undissociated nitric acid and directly by radiolysis of the nitrate anion dissociation product of nitric acid (Katsumura et al., 1991):

$$\cdot OH + HNO_3 \rightarrow \cdot NO_3 + H_2O \quad k = 5.3 \times 10^7 \, M^{-1} s^{-1} \tag{34.41}$$

$$NO_3^- \overset{\text{\tiny $\mathcal{M}$}}{\longrightarrow} e_{aq}^- + \cdot NO_3 \tag{34.42}$$

The reactions of the $\cdot NO_3$ radical are similar to those of the $\cdot OH$ radical, including hydrogen atom abstraction, addition, and electron transfer reactions, although with decreased electrophilic character and lower reaction rates (Neta and Huie, 1986; Shastri and Huie, 1990). Rate constants for its reaction with numerous species have been compiled (Neta et al., 1988).

In acidic solution, the $NO_3^{2-\cdot}$ radical product of Equation 34.38 will protonate (Grätzel et al., 1970):

$$NO_3^{2-\cdot} \leftrightarrow HNO_3^{-\cdot} \leftrightarrow H_2NO_3^{\cdot} \quad pKa_1 = 4.8, \quad pKa_2 = 7.5 \tag{34.43}$$

The product $H_2NO_3^{\cdot}$ decays to produce the $\cdot NO_2$ radical (Løgager and Sehested, 1993):

$$H_2NO_3^{\cdot} \rightarrow \cdot NO_2 + H_2O \quad k = 7 \times 10^5 \, s^{-1} \tag{34.44}$$

This species is less reactive than the $\cdot NO_3$ radical, but has been shown to add to unsaturated carbon bonds, or to carbon-centered radicals to produce nitrated derivatives of the original compound (Eberhardt, 1975; Olah et al., 1978; Dzengel et al., 1999; Ershov et al., 2007). Perhaps more importantly, the addition product of these two nitrogen-centered radicals decays to produce nitronium ion, a powerful nitrating species (Mincher et al., 2009b):

$$\cdot NO_2 + \cdot NO_3 \rightarrow N_2O_5 \quad k = 1.7 \times 10^9 \, M^{-1} s^{-1} \quad \text{(Katsumura et al., 1991)} \tag{34.45}$$

$$N_2O_5 \rightarrow NO_2^+ + NO_3^- \quad \text{(Sworski et al., 1968)} \tag{34.46}$$

Another important product of nitric acid radiolysis is nitrous acid. It is produced by direct and indirect effects:

$$HNO_3 \overset{\text{\tiny $\mathcal{M}$}}{\longrightarrow} \cdot O + HNO_2 \quad \text{(Nagaishi et al., 1994)} \tag{34.47}$$

$$NO_3^- \overset{\text{\tiny $\mathcal{M}$}}{\longrightarrow} \cdot O + NO_2^- \quad \text{(Daniels, 1968)} \tag{34.48}$$

$$NO_2^- + H^+ \leftrightarrow HNO_2 \quad pK_a = 3.2 \quad \text{(Lammel et al., 1990)} \tag{34.49}$$

$$\cdot NO_2 + \cdot NO_2 \rightarrow N_2O_4 \quad k_f = 4.5 \times 10^8 \, M^{-1} s^{-1} \quad \text{(Grätzel et al., 1969)} \tag{34.50f}$$

$$k_{reverse} = 6.0 \times 10^3 \, s^{-1} \quad \text{(Grätzel et al., 1969)} \tag{34.50r}$$

$$N_2O_4 + H_2O \rightarrow HNO_2 + HNO_3 \quad k = 18 \, M^{-1} s^{-1} \quad \text{(Olah et al., 1978)} \tag{34.51}$$

This species is important because of its affect on actinide oxidation states in irradiated aqueous nitric acid solution. This topic is discussed in more detail in Section 34.3.4.1. Its maximum concentration is limited by radiolytically produced H_2O_2, as shown (Bhattacharyya and Veeraraghavan, 1977; Vione et al., 2003):

$$H_2O_2 + HNO_2 \rightarrow H^+ + NO_3^- + H_2O \quad (34.52)$$

The loss of nitrous acid by reaction with radiolytically produced hydrogen peroxide is the main source of loss of nitrous acid in an irradiated system (Bhattacharyya and Veeraraghavan, 1977).

34.3.1.2 The Produced Species in the Organic Phase

Normal and branched-chain alkanes are the typical organic diluents for the ligands used in nuclear solvent extraction. The radiolysis of alkanes is represented by (Spinks and Woods, 1990):

$$CH_3(CH_2)_nCH_3^{\bullet} \xrightarrow{} e_{sol}^- + CH_3(CH_2)_nCH_3^{\bullet+} + CH_3(CH_2)_nCH_2^{\bullet} + {}^{\bullet}CH_3 + H^{\bullet} + H_2 \quad (34.53)$$

Specific yields for the products depend on the specific alkane irradiated. Branch-chain alkanes have higher product yields, in the range of 0.2–$0.6\,\mu mol\ J^{-1}$ for molecular hydrogen and 0.005–$0.1\,\mu mol\ J^{-1}$ for methane. Bishop and Firestone (1970) reported a yield of $H^{\bullet}$ atom of $0.07\,\mu mol\ J^{-1}$ for C_6–C_{10} hydrocarbons. The carbon-centered radical products may also be produced indirectly by hydrogen abstraction reactions with ${}^{\bullet}OH$ or ${}^{\bullet}NO_3$ radicals. Regardless of their origin, these carbon-centered radicals react with solutes by hydrogen atom abstraction or they may undergo radical–radical addition to create higher molecular weight products (Dewhurst, 1958):

$$CH_3(CH_2)_nCH_2^{\bullet} + CH_3(CH_2)_nCH_2^{\bullet} \rightarrow CH_3(CH_2)_{2n+2}CH_3 \quad (34.54)$$

As these higher molecular weight products accumulate, they change the physical characteristics of the solvent, including phase disengagement time, density, and viscosity. Alkane radicals may also undergo disproportionation to produce unsaturated products (Dewhurst, 1958):

$$CH_3(CH_2)_nCH_2^{\bullet} + CH_3(CH_2)_nCH_2^{\bullet} \rightarrow CH_3(CH_2)_nCH_3 + CH_3(CH_2)_{n-1}CH{=}CH_2 \quad (34.55)$$

Unsaturated products are susceptible to addition reactions by ${}^{\bullet}OH$ radical or N-centered radical species.

Among the most important of carbon-centered radical reactions is oxygen addition to produce peroxyl radicals (Alfassi, 1997):

$$R^{\bullet} + O_2 \rightarrow ROO^{\bullet} \quad (34.56)$$

Peroxyl radicals will undergo addition reactions to form tetroxides, which then decompose to produce aldehydes, ketones, and alcohols from the original compound (von Sonntag and Schuchmann, 1997). They are therefore important intermediates in the oxidative mineralization of organic compounds by radiolysis.

34.3.1.3 The Mixed Phase

Although the reactive species responsible for radiation chemical effects in solvent extraction systems can be readily identified, additional factors must be considered to provide a quantitative understanding of radiolysis in the biphasic system. Among these are that reactive species yields and reaction rates vary with solvents, although most are known only from the aqueous phase. Further, the transfer of produced reactive species across the aqueous/organic phase boundary must be considered.

Muroya et al. reported initial yields of solvated electrons of 0.4 and $0.2\,\mu mol\ J^{-1}$ for water and decanol, respectively (Muroya et al., 2008). Although the yield may be lower, reaction rates for electron attachment in nonpolar organic solutions are much higher than in aqueous solution (Borovkov, 2008). However, in benzene-, or dodecane-in-water microemulsions, Wu et al. (2001) found that the yield of aqueous electrons was identical to that of pure water. Electrons, being very mobile in nonpolar solution, apparently crossed the interface to the aqueous phase. This resulted in a constant yield of aqueous electrons despite changes in the proportion of water in the emulsion.

In contrast, it was determined in the same study (Wu et al., 2001) that water radiolysis was the only source of ˙OH radical in irradiated emulsion, indicating that radical cations produced in the organic phase did not cross the interface to oxidize water. The ˙OH yield in the emulsion was thus proportional only to the water content. The rate constants for several radical reactions, including the ˙OH radical reaction with benzene, were found to be similar in pure aqueous and microemulsion aqueous solution.

Some of the most important reactive species produced in the aqueous phase, such as ˙OH or ˙NO$_3$ radical, must cross the phase boundary prior to reacting with ligand molecules or their diluents. This should be relatively easy for neutral species and in fact the solvent extraction system is designed to move neutral species across the interface by providing intimate phase mixing using pulsed columns, mixer settlers, or centrifugal contactors. For these emulsions, it may be justifiable to neglect a phase transfer diffusion gradient and to assume uniform radiolysis. However, in reality, little information is available on the mass transfer rates of these species, and diffusion controlled regimes may dominate when dose rates are very high or with thick phase layers due to inadequate mixing (Macášek and Čech, 1984). In practice, the assumption is generally made that reactive species created in either phase are available for reaction during phase mixing in solvent extraction. The investigation of biphasic radiolysis reactions is an area in need of more detailed investigation.

34.3.2 PUREX PROCESS RADIATION CHEMISTRY

34.3.2.1 TBP Radiolysis

The PUREX process for the extraction of the major actinides consists of 30% TBP in alkane diluent. The process can either partition uranium separately or co-extract uranium, neptunium, and/or plutonium depending on how the valence states of the latter metals are set prior to extraction. The metal-loaded solvent is then stripped with a mildly acidic aqueous phase and recycled (Schultz and Navratil, 1984). However, its recycle potential is limited by the radiolytic degradation of TBP and its diluent. It has long been recognized that the major products of TBP radiolysis are hydrogen, methane, and dibutylphosphoric acid (HDBP), with monobutylphosphoric acid (H$_2$MBP) and phosphoric acid produced in lesser amounts. The radiation chemistry of TBP was recently reviewed and the following discussion is abbreviated from that source (Mincher et al., 2009a,b,c).

The accumulation of radiolytic degradation products in the PUREX solvent results in decreased extraction performance (Lane, 1963; Neace, 1983; Stieglitz and Becker, 1985). The acidic radiolysis products are complexing agents that interfere with uranium and plutonium stripping and fission product separation factors (Davis, 1984; Tripathi et al., 1999; Tripathi and Ramanujam, 2003). Interfacial crud formation and poor phase separation have been attributed to the formation of precipitable complexes of zirconium with H$_2$MBP and phosphoric acid (Rochoň, 1980; Stieglitz and Becker, 1985; Miyake et al., 1990; Sugai and Munakata, 1992; Egorov et al., 2002, 2005). The adverse affects of the buildup of these acidic phosphate products in the organic phase are mitigated during process extractions by solvent washing with aqueous Na$_2$CO$_3$ (Blake et al., 1963; Reif, 1988). However, with continued recycling, washing becomes less effective and the washed solvent shows increased retention of Pu, Zr, and Ru and increased solution viscosity (Tripathi et al., 2001a). This has been attributed to the accumulation of higher molecular weight radiolysis products with high organic phase solubility (Wagner and Towle, 1958; Stieglitz and Becker, 1985). The result is a permanently degraded and radioactively contaminated solvent, which is expensive to dispose.

Several mechanisms have been proposed to explain the production of HDBP in irradiated TBP solutions. Zaitsev and Khaikin (1994) reported that dissociative electron capture resulted in the production of the butyl radical and HDBP in irradiated neat TBP:

$$e_{sol}^- + (C_4H_9O)_3PO \rightarrow {}^{\bullet}C_4H_9 + (C_4H_9O)_2OPO^- \tag{34.57}$$

Jin et al. (1999) attributed the formation of HDBP under these conditions to a combination of dissociative electron capture and decay of excited TBP molecules, while Haase et al. (1973) reported that electron attachment could also result in free hydrogen atoms and a TBP carbon-centered radical. However, the electron-initiated reactions are unlikely to be of consequence in the acidic mixed phase due to the fast reactions shown in Equations 34.36 through 34.38, and the direct excitation of TBP becomes less important when TBP is dissolved in a diluent. These may not be important reactions in the solvent extraction process.

Burr (1958) proposed that HDBP was formed by the decay of the TBP carbon-centered radical:

$$(C_4H_9O)_2(^{\bullet}C_4H_8O)PO \rightarrow {}^{\bullet}C_4H_8{}^{+} + (C_4H_9O)_2OPO^{-} \tag{34.58}$$

This TBP radical could be produced by hydrogen atom abstraction (Burr, 1958) by reaction with either the radiolytically produced $^{\bullet}$H atom or $^{\bullet}$OH radical:

$$(^{\bullet}HO)\ {}^{\bullet}H + TBP \rightarrow (C_4H_9O)_2(^{\bullet}C_4H_8O)PO \rightarrow (H_2O)\ H_2$$

$$k_{OH} = 5.0 \times 10^9\ M^{-1}\,s^{-1} \quad \text{(Mincher et al., 2008)} \tag{34.59}$$

$$k_{H} = 1.8 \times 10^8\ M^{-1}\,s^{-1} \quad \text{(Mincher et al., 2008)}$$

Besides direct decay to HDBP, the TBP radical could also undergo hydrolysis to again produce HDBP (von Sonntag et al., 1972):

$$(C_4H_9O)_2(^{\bullet}C_4H_8O)PO + H_2O \rightarrow {}^{\bullet}C_4H_8OH + (C_4H_9O)_2POO^{-} + H^{+} \tag{34.60}$$

Khaikin (1998) suggested that HDBP was also the stable product of dissolved oxygen addition to the TBP radical, which gives the TBP peroxyl radical:

$$(C_4H_9O)_2(^{\bullet}C_4H_8O)PO + O_2 \rightarrow (C_4H_9O)_2P(O) - O - (C_4H_8)OO^{\bullet} \tag{34.61}$$

Superoxide elimination followed by hydrolysis produces HDBP and butyraldehyde, also a measured product (Clay and Witort, 1974).

$$(C_4H_9O)_2P(O) - O - (C_4H_8)OO^{\bullet} \rightarrow O_2^{-\bullet} + (C_4H_9O)_2P(O) - O - (C_4H_8)^{+} \tag{34.62}$$

$$(C_4H_9O)_2P(O) - O - (C_4H_8)^{+} + H_2O \rightarrow (C_4H_9O)_2P(O) - O^{-} + C_3H_7CHO + 2H^{+} \tag{34.63}$$

Wilkinson and Williams (1961) proposed yet another mechanism for HDBP formation based upon direct TBP radiolysis:

$$(C_4H_9O)_3PO \longrightarrow\!\!\!\!\bigwedge\!\!\!\!\longrightarrow e_{sol}^{-} + (C_4H_9O)_3PO^{\bullet+} \tag{34.64}$$

$$(C_4H_9O)_3PO^{\bullet+} \rightarrow (C_4H_9O)_2P(OH)OH^{+} + CH_2{=\!\!=}CHCH^{\bullet}CH_3 \tag{34.65}$$

$$(C_4H_9O)_2P(OH)OH^{+} \rightarrow (C_4H_9O)_2POOH + H^{+} \tag{34.66}$$

Intramolecular hydrogen bonding between the phosporyl oxygen and a butoxy hydrogen atom of the radical cation formed in Equation 34.64 forms a ring structure, which decays as shown in Equations 34.65 and 34.66. This mechanism, although dependent on direct TBP radiolysis cannot be entirely discounted since TBP is used at a rather high concentration of 30% in the PUREX process. The reactions described above all lead to the production of HDBP, and occur competitively. Continued irradiation produces H_2MBP and phosphoric acid from HDBP via analogous reactions.

Among the less abundant but still important radiolysis products are the higher molecular weight acid phosphates. These species with varying alkane chain lengths suggest that radical addition reactions occur between TBP, HDBP, and alkane solvent radicals, including the production of TBP dimers (Rochoń, 1980; Adamov et al., 1990). These higher molecular weight compounds are among those species that are not adequately removed from irradiated solvent by aqueous carbonate washing.

Additional products of TBP radiolysis in the presence of HNO_3 are nitrated phosphates, which also impede stripping efficiency. He et al. (2004) proposed that $^{\bullet}NO_3$ reacts with TBP by hydrogen atom abstraction, producing the TBP radical

$$^{\bullet}NO_3 + TBP \rightarrow (C_4H_9O)_2(^{\bullet}C_4H_8O)PO + HNO_3$$

$$k = 4.3 \times 10^6 \ M^{-1}s^{-1} \quad \text{(Mincher et al., 2008)} \tag{34.67}$$

The TBP radical was then postulated to undergo reaction with additional $^{\bullet}NO_3$ to produce nitrated TBP (He et al., 2004):

$$^{\bullet}NO_3 + (C_4H_9O)_2(^{\bullet}C_4H_8O)PO \rightarrow (C_4H_9O)_2(OC_4H_8NO_3)PO \tag{34.68}$$

The $^{\bullet}NO_2$ radical might be expected to add in the same way:

$$^{\bullet}NO_2 + (C_4H_9O)_2(^{\bullet}C_4H_8O)PO \rightarrow (C_4H_9O)_2(OC_4H_8NO_2)PO \tag{34.69}$$

Methylated, hydroxylated, and nitrated phosphates, resulting from radical addition reactions of methyl radical, hydroxyl radical, and the nitro-radicals shown above, have been identified in post-irradiation TBP solutions by numerous investigators (Nowak, 1977; Adamov et al., 1987; Lesage et al., 1997; Tripathi et al., 2001b). It can be seen that radical addition reactions can create a wide range of products, sometimes of high molecular weight.

34.3.2.2 PUREX Diluent Degradation

The TBP alkane diluent undergoes similar radiolytic degradation to generate metal complexing agents that do not wash out in solvent treatment with alkaline solutions (Lane, 1963). The decomposition of diluents in TBP solvent extraction was reviewed by Tahraoui and Morris (1995). Alkanes undergo radiolytic nitration of their carbon-centered radicals, in analogy with Equations 34.68 and 34.69 above. Stieglitz and Becker (1985) and Tripathi et al. (2001a) identified both nitro-, and nitrosoalkanes (RNO_2 and $RONO_2$) in alkanes irradiated in the presence of nitric acid. These nitroparaffins and their hydroxamic acid reaction products have been implicated in fission product complexation in the PUREX process. Nitroparaffins are thought to be converted to the complexing enol form by contact with the alkaline scrub intended to remove the acidic products of TBP decomposition (Blake et al., 1963):

$$RCH_2N{=}O(O) \leftrightarrow RCH{=}NO(O)^- \tag{34.70}$$

Hydroxamic acids (Lane, 1963; Egorov et al., 2002) are metal complexing and reducing agents formed from nitroparaffins by the Victor Meyer reaction:

$$RCH_2NO_2 \rightarrow RCONHOH \tag{34.71}$$

Although hydroxamic acids rapidly hydrolyze in acidic media, small but steady-state organic phase concentrations have been identified in irradiated solvent (Ohwada 1968; Zaitsev et al., 1987; Huang et al., 1989). Additional diluent radiolysis products identified in irradiated TBP–dodecane–nitric acid include alkane oligomers, aliphatic ketones, and acids (Becker et al., 1983). Thus, stripping difficulties and poor separation factors result from a combination of higher molecular weight acidic phosphates from TBP radiolysis and from compounds produced by diluent nitration.

Finally, it should be noted that the use of more stable TBP diluents results in higher yields of HDBP. This has been attributed to the diluent ionization potential (Tahraoui and Morris, 1995). Direct diluent radiolysis produces the diluent radical cation, as shown in Equation 34.53. A common radical cation stabilization route would be that of charge transfer by reaction with TBP:

$$[CH_3(CH_2)CH_3]^{\bullet +} + TBP \rightarrow CH_3(CH_2)_n CH_3 + TBP^{\bullet +} \tag{34.72}$$

Thus, the use of very stable diluents with high ionization potential may increase the rate of TBP degradation. It is not possible to completely eliminate adverse radiation chemical effects in an irradiated solvent extraction system.

34.3.3 Radiation Chemistry in the Future Fuel Cycle

In addition to uranium recovery by the PUREX process, future fuel cycle designs include solvent extraction steps for the recovery of americium and curium, the so-called minor actinides. The separation of the trivalent actinides from the trivalent lanthanides has long been one of the significant challenges in radiochemistry. The organophosphorous compound octylphenyldiisobutylcarbamoyl-methylphosphine oxide (CMPO) has been proposed for the group separation of the trivalent actinides and lanthanides in the United States (Kalina et al., 1981), whereas in European work tetraalkyl-diamides have been proposed for the co-extraction of the actinides and lanthanides. The DIAMEX process uses N,N'-dimethyl-N,N'-dioctyl-hexylethoxy-malonamide (DMDOHEMA) in an alkane diluent for this purpose (Sorel et al., 2008). Similarly, the tridentate amide N,N,N',N'-tetraoctyl-3-diglycolamide (TODGA) has been developed primarily in Japanese work. A 0.1 M dodecane solution of TODGA is capable of extracting U and Pu, as well as Am from nitric acid solution (Sasaki et al., 2008). Dialkylmonoamides, such as N,N-di-(2-ethylhexyl)isobutyramide (DEHiBA) have also been developed as suggested alternatives for TBP in U and Pu extraction (Miguirditchian et al., 2008). The potential advantages and uses of amides as actinide extractants have been reviewed by Gasparini and Grossi (1986). While the radiation chemistry of these newer solvent extraction processes has not been characterized as well as for TBP, a picture of their behavior is beginning to emerge.

34.3.3.1 Amide and Diamide Radiolysis

The general structures of dialkylmonoamides and tetralkyldiamides are shown in Figure 34.21. The effects of radiolysis on the solvent extraction behavior of these compounds have been investigated by numerous researchers. For example, Ruikar et al. (1993) reported increasing D_{Pu} and decreasing D_U over the absorbed dose range 0–1800 kGy for dialkylamides irradiated in dodecane solution after preequilibration with 3.6 M nitric acid. The decrease in amide concentration and ingrowth of the corresponding acids and amines were monitored by IR spectroscopy (Ruikar et al., 1991). The decrease in amide concentration satisfactorily explained the decrease in uranium extraction efficiency while increases in D_{Pu} were postulated to result from the participation of acidic degradation products in plutonium complexation. The distribution ratio for fission product zirconium, D_{Zr}, was also

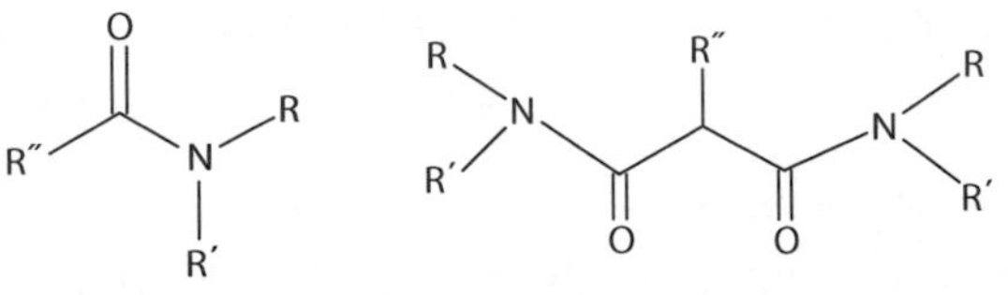

FIGURE 34.21 The general structure of dialkylmonamides (left), and tetraalkyldiamides (right).

elevated by irradiation. Increased distribution ratios for zirconium and plutonium may also be due to the nitrated products of diluent radiolysis, as has been reported for the PUREX solvent (Mincher et al., 2009a,b,c). The irradiation of 0.5 M of 11 different dialkylamides in benzene solution after pre-equilibration with 3.6 M nitric acid also resulted in degradation to the corresponding amines and carboxylic acids, with symmetrically substituted dialkylamides being more stable (Mowafy, 2004). In this work also, uranium extraction efficiency decreased with increasing absorbed dose, while D_{Zr} increased.

Mowafy (2007) concluded that within a class of compounds, stability decreased in the order: symmetrical amides > unsymmetrical amides > branched amides. The pentaalkylpropane diamides (malonamides) have the general structure $(RR'NCO)_2CHR''$, an example of which is shown in Figure 34.22. For the malonamide with $R = CH_3$, $R' = C_4H_9$ and $R'' = C_2H_4OC_2H_4OC_6H_{13}$ more than 30% was decomposed at an absorbed dose of 500 kGy for a 1 M t-butylbenzene solution irradiated in contact with 5 M nitric acid. There was a corresponding decrease in D_{Am}, although D_U and D_{Pu} were not adversely affected. Cuillerdier et al. (1991) concluded that stability as a function of R'' increased in the order: $H < C_2H_5 < C_2H_4OC_6H_{13} < C_2H_4OC_2H_4OC_6H_{13}$. Long oxyalky chains appear to protect extraction efficiency. Inclusion of a sacrificial ether linkage in the molecule may result in the generation of relatively harmless radiolysis products while maintaining diamide extraction capability. The radiolytic decomposition of a malonamide to the corresponding acid and amide is shown in Figure 34.23. A review of amide radiolysis is provided in reference (Mincher et al., 2009c).

34.3.3.2 DIAMEX and TODGA Process Radiolysis

The DIAMEX solvent extraction process for recovery of the minor actinides was developed using DMDOHEMA, shown in Figure 34.22. The gamma-radiolysis of this compound in alkane solution has been investigated by Berthon et al. (2001) using a suite of analytical techniques including gas chromatography, potentiometry, and mass spectrometry. Organic acids were a major diamide radiolysis product, and alcohols were also created by rupture of the ether linkages in the R″ group. No alcohols were detected in compounds without an oxygen atom in that chain. Products

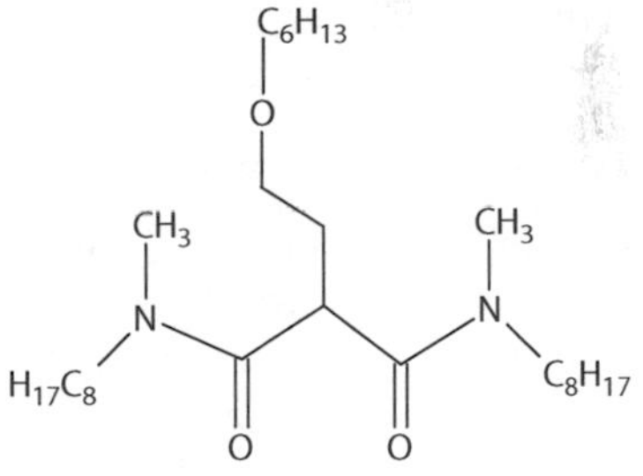

FIGURE 34.22 The pentaalkylpropane diamide (malonamide) dimethyl dioctyl hexylethoxymalonamide (DMDOHEMA).

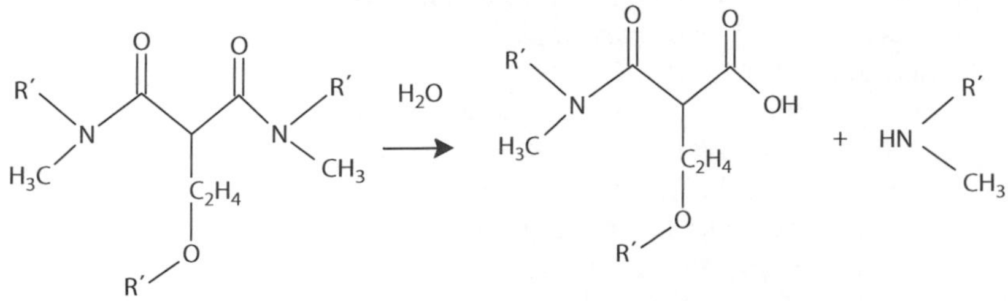

FIGURE 34.23 Radiolytic decomposition of a malonamide into the corresponding amine and carboxylic acid.

detected by Berthon et al. (2001, 2004) for DMDOHEMA radiolysis included the monoamide methyloctylhexyloxybutanamide, the acid amide methoxyoctylcarbamoyl 4-hexyloxybutanoic acid, the malonamides dimethyloctyl 2-hexyloxyethyl malonamide (DMOHEMA), and methyl-dioctyl 2-hexyloxyethyl malonamide (MDOHEMA) and the amine methyloctylamine. The loss of one amide function occurs by rupture of the carbamoyl C–N bond resulting in the amine and an acidic amide product (Berthon et al., 2001, 2004). This occurs only in the presence of acid, and is promoted by higher acid concentrations. Although significant products, monoamides are rapidly degraded in irradiated solution and the concentrations detected there are low, or sometimes undetectable.

For DMDOHEMA, with its eight carbon R′ group, the radiolysis products are soluble in the organic phase. A dose of 690 kGy to 1 M DMDOHEMA in dodecane in contact with 4 M nitric acid resulted in a final malonamide concentration of 0.59 M. The decrease in malonamide concentration was accompanied by a decrease in D_{Am}, D_{Eu}, and D_{Nd}. When solutions containing varying amounts of the major degradation products were prepared, all products were found to interfere with Am and Nd extraction, with the amine being the most harmful. An acidic solvent wash quantitatively removed the amine degradation product while an alkaline wash removed about 80% of the acidic amide (Nicol et al., 2000). The combination of an acid wash, followed by water scrubbing and then an alkaline wash restored the solvent's extraction capabilities to that expected for the decreased malonamide concentration (Berthon et al., 2004; Bisel et al., 2007).

Modolo et al. (2008) investigated the post-irradiation solvent extraction performance of 0.2 M TODGA in hydrocarbon diluent. The D_{Am} gradually decreased with absorbed dose for TODGA and TODGA/TBP mixtures irradiated in the presence and absence of nitric acid, while the D_{Eu} remained unchanged over the absorbed dose range 0–1000 kGy. The products of TODGA radiolysis include dioctylamine, dioctylacetamide, dioctylglycolamide, and dioctylformamide (Sugo et al., 2002, 2007). The presence of nitric acid did not enhance radiolytic degradation but did favor cleavage of the C–N bond to form the amine and acidic products over ether bond cleavage to form acetamide and glycolamide. These results are analogous to those described above for diamide radiolysis.

34.3.3.3 TRUEX Process Radiolysis

The compound octylphenyldiisobutylcarbamoylmethylphosphine oxide (CMPO), used in combination with TBP in an alkane diluent in the TRUEX (TRansUranic EXtraction) process, is a phosphorous-containing amide, and is shown in Figure 34.24. The irradiation of this solvent system in contact with an acidic aqueous phase results in decreased forward extraction distribution ratios for Am (D_{Am}), but not to the extent predicted by loss in CMPO concentration (Chiarizia and Horwitz, 1986). Stripping distribution ratios exceeded those of the forward extraction following irradiation to an absorbed gamma-dose of 195 kGy. Similar effects were encountered for only 36 kGy delivered from an alpha-source (Buchholz et al., 1996). Cleavage of the C–N bond was proposed to result in decomposition of CMPO to a carboxylic acid and an amine, followed by decarboxylation of the carboxylic acid to a phosphine oxide, in analogy with the reactions for DMDOHEMA discussed above. Oxidation of the phosphine oxide would produce a phosphinic acid. Products of this nature have been measured (Chiarizia and Horwitz, 1986; Nash et al., 1988, 1989; Buchholz et al., 1996). The occurrence of degradation product phosphine oxides may explain the reasonably high forward D_{Am} despite a loss in CMPO concentration, while the phosphinic acids would complex Am to prevent adequate stripping under low aqueous phase acidity conditions. The kinetic constants for the reactions of the major radicals produced by diluent radiolysis have not been measured for CMPO, TODGA, or DMDOHEMA.

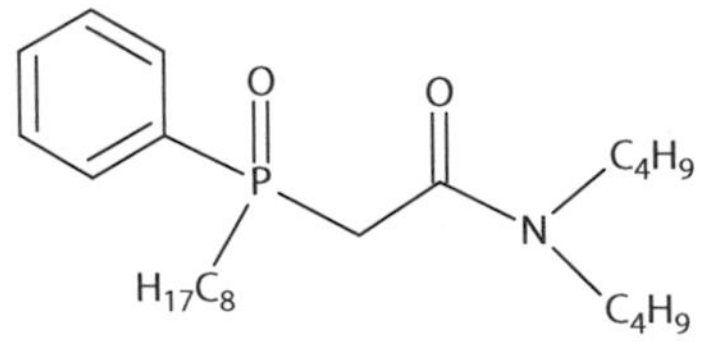

FIGURE 34.24 The structure of octylphenyldiisobutylcarbamoyl-methylphosphine oxide (CMPO).

34.3.4 Radiation Chemistry and Actinide Oxidation States

34.3.4.1 Radiolytic Production of Nitrous Acid

The radiolytic production of nitrous acid in irradiated nitric acid has consequences for nuclear solvent extraction because of its affect on metal valence. Probably, the most important actinide affected is Np. Highly extractable by the PUREX process in the tetra-, (Np^{4+}) and hexavalent (NpO_2^{2+}) states, Np is inextractable in the pentavalent (NpO_2^+) state. Attempts to co-extract Np with U and Pu, or to prevent Np extraction during the PUREX process are often confounded by changes in Np valency during the extraction. The oxidation state of neptunium in aqueous nitric acid depends on the concentration of nitrous acid according to Equation 34.73 (Siddall and Dukes, 1959):

$$NpO_2^+ + 3/2H^+ + 1/2NO_3^- \leftrightarrow NpO_2^{2+} + 1/2HNO_2 + H_2O \tag{34.73}$$

Neptunium(V) is stable in the absence of nitrous acid, however, it is oxidized to NpO_2^{2+} with small amounts of added nitrous acid, at increasing rates with increasing nitrous acid concentration. For example, 5×10^{-5} M HNO_2 was found to oxidize neptunium with a half-time of 22 min in 3 M HNO_3 (Siddall and Dukes, 1959). However, very high concentrations will shift the equilibrium of Equation 34.73 back to the left, favoring NpO_2^+.

The direct radiolytic production of nitrous acid was shown in Equations 34.47 and 34.48. Indirect sources were shown in Equations 34.49 through 34.51. The reaction of the hydroxyl radical with $^{\bullet}NO_3$ radical, shown in Equation 34.74 is an additional source (Matthews et al., 1972):

$$^{\bullet}OH + {^{\bullet}NO_3} \rightarrow HNO_4 \rightarrow O_2 + HNO_2 \tag{34.74}$$

Sources of the $^{\bullet}NO_2$ radical include the reaction of nitrate anion with the hydrogen atom in Equation 34.75 (Buxton et al., 1988), the hydrolysis of the $^{\bullet}NO_3$ radical, shown in Equation 34.76 (Vladimirova and Milovanova, 1972), and the $^{\bullet}NO_3$ radical addition reaction shown in Equation 34.77 (Matthews et al., 1972):

$$NO_3^- + {^{\bullet}H} \rightarrow {^{\bullet}NO_2} + OH^- \quad k = 1 \times 10^7 \, M^{-1} s^{-1} \tag{34.75}$$

$$^{\bullet}NO_3 + H_2O \rightarrow {^{\bullet}NO_2} + H_2O_2 \tag{34.76}$$

$$^{\bullet}NO_3 + {^{\bullet}NO_3} \rightarrow N_2O_6 \rightarrow O_2 + 2\,{^{\bullet}NO_2} \tag{34.77}$$

Production of the $^{\bullet}NO_2$ radical by any of the above routes is followed by the addition reaction to produce N_2O_4, ultimately resulting in nitrous acid (Sworski et al., 1968; Grätzel et al., 1969): This was shown in Equations 34.50 and 34.51.

Radiolytic reactions also deplete the nitrous acid concentration in irradiated nitric acid. The hydroxyl radical acts as a sink for nitrous acid produced during radiolysis (Bugaenko and Roshchektaev, 1971):

$$^{\bullet}OH + HNO_2 \rightarrow H_2O + {^{\bullet}NO_2} \quad k = 2.6 \times 10^9 \, M^{-1} s^{-1} \tag{34.78}$$

Hydroxyl radical addition reactions also oxidize $^{\bullet}NO_2$ radical to nitrate anion in Equation 34.79 (Vione et al., 2003), and add to generate hydrogen peroxide in Equation 34.80 (Buxton et al., 1988):

$$^{\bullet}OH + {^{\bullet}NO_2} \rightarrow HOONO \rightarrow H^+ + NO_3^- \quad k = 4.5 \times 10^9 \, M^{-1} s^{-1} \tag{34.79}$$

$$^{\bullet}OH + \, ^{\bullet}OH \rightarrow H_2O_2 \quad k = 5.5 \times 10^9 \, M^{-1} \, s^{-1} \tag{34.80}$$

The reaction of hydrogen peroxide with nitrous acid oxidizes nitrous acid back to nitric acid, as was shown in Equation 34.52 (Bhattacharyya and Veeraraghavan, 1977; Vione et al., 2003). Similarly, $^{\bullet}NO_3$ radical can oxidize nitrous acid (Bugaenko and Roshchektaev, 1971):

$$HNO_2 + \, ^{\bullet}NO_3 \rightarrow NO_3^- + H^+ + \, ^{\bullet}NO_2 \tag{34.81}$$

Maximum achievable concentrations of nitrous acid are limited by these oxidation reactions, and by the reversible decomposition to simple oxides of nitrogen, as shown in Equation 34.82 (Park and Lee, 1988):

$$2HNO_2 \leftrightarrow \, ^{\bullet}NO + \, ^{\bullet}NO_2 + H_2O \tag{34.82}$$

Park and Lee (1988) modeled that nitrous acid decomposition by the route in Equation 34.82 becomes significant at concentrations $>10^{-5}$ M. Additional loss of nitrous acid in open systems occurs through the volatilization of $^{\bullet}NO$ and $^{\bullet}NO_2$, which have limited nitric acid solubility (Andreichuk et al., 1984). Andreichuk et al. (1984) found that the loss of nitrous acid was reduced by limiting the volume of air in contact with the system. Thus, a complicated suite of reactions compete to produce and remove nitrous acid in irradiated aqueous nitric acid. The ability to model and predict its concentration under various solution conditions is limited by the lack of the fundamental bimolecular rate constants for many of these reactions.

However, empirical nitrous acid radiolysis yields, $G_{(HNO_2)}$, have been measured by several authors. The yield is affected by total absorbed dose, dose rate, radiation LET, and nitric acid concentration. Bhattacharyya and Saini (1973) reported that $G_{(HNO_2)}$ varied linearly with absorbed gamma-dose using a ^{60}Co source with a dose rate of 3 kGy h^{-1}. This resulted in a value of ~0.24 µmol J^{-1} in 3 M HNO$_3$. When $G_{(HNO2)}$ was evaluated versus nitric acid concentration, it was found to increase with increasing concentration, as shown in Figure 34.25. The yield was found to be the same in air-free and air-saturated solution. These data are considered maximum yields, as the produced nitrous acid was reacted in situ with the colorimetric reagent sulfanilamide for quantitation (Bhattacharyya and Saini, 1973). This eliminated competition from oxidizing agents that might limit the HNO$_2$ concentration and, therefore, the true yield will be lower due to the reactions shown in Equations 34.52, 34.78, and 34.81.

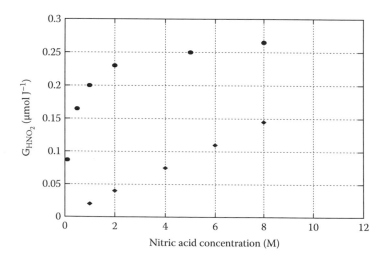

FIGURE 34.25 Nitrous acid yield in gamma-irradiated nitric acid. Curve a is data of Bhattacharyya and Saini (1973) (dose rate = 3 kGy h^{-1}) in the presence of sulfanilamide. Curve b is data of Vladimirova et al. (1969) (Dose rate = 6 kGy h^{-1}) without scavenger.

Also shown in Figure 34.25 are the data of Vladimirova et al. (1969), collected at a dose rate of 6 kGy h^{-1}. They irradiated nitric acid in the absence of sulfanilamide, and the actual $G_{(HNO_2)}$ was ~0.06 μmol J^{-1}, in 3 M HNO$_3$, about one-fourth of the value reported by Bhattacharyya and Saini (1973). The relationship between the yield of nitrous acid and the nitric acid concentration was linear. Bugaenko and Roshchektaev (1971) reported a linear relationship between total absorbed dose and nitrite ion concentration. They reported a $G_{NO_2^-}$ of 0.07 μmol J^{-1} for 3 M HNO$_3$ at a dose rate of 36 kGy h^{-1}. Similar results were reported by Jiang et al. (1994) also using gamma-irradiation. Production of nitrous acid was proportional to absorbed dose over the investigated range of 250–750 Gy, and to nitric acid concentration in the range 2–8 M. The maximum measured value for $G_{(HNO_2)}$ at a dose rate of 84 Gy h^{-1} was 0.1 μmol J^{-1} in 8 M nitric acid. These authors calculated a maximum possible yield of 0.32 μmol J^{-1}. As the concentration of nitric acid increases, the direct radiolysis of nitric acid becomes more important, resulting in the production of additional nitrous acid. The nitrous acid yield may be increased in the presence of $^{\bullet}$OH radical scavengers (Vladimirova and Milovanova, 1972), which would suppress the reactions shown in Equations 34.52 and 34.78 through 34.80. Such scavengers would include the organic diluents and complexing agents used in solvent extraction. Using the data of Vladimirova et al. (1969), it may be calculated that an absorbed dose of 10 kGy would generate about 7×10^{-4} M HNO$_2$ in 3 M HNO$_3$.

High LET alpha-radiolysis is of obvious concern in neptunium solvent extraction, and a number of studies have been performed using ^{244}Cm irradiation. Bibler (1974) concluded that the direct radiolysis of nitric acid proceeded in the same fashion for alpha-radiolysis as for gamma-radiolysis, resulting in the same oxygen (and therefore presumably the same nitrous acid according to Equation 34.47) yield. However, Vladimirova and Milovanova (1972) reported that the production of HNO$_2$ was greater for alpha-radiolysis. It depended positively on nitrate ion concentration and dose rate, but the yield decreased with increasing absorbed dose. The increase in hydrogen peroxide concentration with absorbed dose, for which the yield is greater for alpha-radiolysis (Equation 34.35), probably explains the decrease in nitrous acid generation. Andreichuk et al. (1984) reported similar results. Nitrous acid concentrations increased quickly with absorbed dose, but then leveled out to reach an equilibrium concentration that was positively dependent on the nitric acid concentration and the dose rate. Taking into account the higher hydrogen peroxide yield of high LET radiation, they calculated equilibrium concentrations of nitrous acid under various conditions. At a dose rate of 11.5 kGy h^{-1}, for example, this varied from 1×10^{-3} M HNO$_2$ in 1 M HNO$_3$ to 8.7×10^{-3} M HNO$_2$ in 7.5 M HNO$_3$. A value of 6.8×10^{-3} M HNO$_2$ was reported for 3.7 M HNO$_3$. According to Siddall and Dukes (1959), this concentration of nitrous acid is sufficient to oxidize Np(V) to Np(VI) with a half-time of only 6 min in 4.0 M HNO$_3$.

34.3.4.2 Actinide Reactions with Free Radicals

The reaction of neptunium with the primary oxidizing and reducing agents generated by water and nitric acid radiolysis may also alter neptunium oxidation states. The fundamental bimolecular rate constants for many neptunium (and the other actinide) reactions with radiolytically produced radicals have been measured (Table 34.8). Using Np as the example, those of interest in irradiated aqueous nitric acid are shown in Table 34.8 with a comprehensive collection provided by Mincher and Mezyk (2009). It can be seen that the aqueous electron is a powerful reducing agent with fast kinetics for reaction with neptunium cations. However, in the strongly acidic solutions used in solvent extraction, the reactions of the aqueous electron may be discounted due to its fast reactions shown in Equations 34.34. The produced hydrogen atom is less reactive, but is likely to be a reducing agent for neptunium. However, its ability to react with neptunium ions in solution will be limited by competition from oxygen to produce the less reactive hydroperoxyl radical, as shown in Equation 34.39. The hydrogen atom will also be scavenged by other dissolved metal ions such as uranium, which may be present at higher concentrations. Hydroxyl radical oxidizes Np^{4+} and NpO$_2^+$. However, its affects on the solvent extraction system will depend on the extent to which it is scavenged by organic compounds, for which

TABLE 34.8
Rate Constants for Reactions of Actinide Ions

Reaction	Rate Constant k (M^{-1} s^{-1})	References
$Np^{3+} + {}^{\bullet}OH \rightarrow Np(IV)$	4.00×10^9	Gogolev et al. (1988)
$Np(IV) + {}^{\bullet}OH \rightarrow NpO_2^+$	3.20×10^8 (pH 0)	Shilov et al. (1982)
$Np(V) + {}^{\bullet}OH \rightarrow ?$	4.60×10^7	Lierse et al. (1988)
$NpO_2^+ + {}^{\bullet}OH \rightarrow NpO_2^{2+}$	4.30×10^8 (pH 0)	Shilov et al. (1982)
$Np^{3+} + {}^{\bullet}H \rightarrow Np^{4+}$	6.00×10^7 (pH 0)	Gogolev et al. (1991)
$Np^{4+} + {}^{\bullet}H \rightarrow ?$	$<1 \times 10^6$ (pH 0)	Gogolev et al. (1991)
$NpO_2^+ + {}^{\bullet}H \rightarrow ?$	$<5 \times 10^6$ (pH 0)	Shilov and Pikaev (1982)
$NpO_2^{2+} + {}^{\bullet}H \rightarrow NpO_2^+$	$<1 \times 10^7$ (pH 1–3)	Schmidt et al. (1983)
$Np^{4+} + e_{aq}^- \rightarrow Np(III)$	6.00×10^{10} (pH 0)	Gogolev et al. (1991)
$NpO_2^+ + e_{aq}^- \rightarrow NpO2$	2.40×10^{10} (pH 3)	Schmidt et al. (1980)
$NpO_2^{2+} + e_{aq}^- \rightarrow NpO_2^+$	1.00×10^{11} (pH 2.5)	Schmidt et al. (1983)
$NpO_2^+ + {}^{\bullet}NO_3 \rightarrow NpO_2^{2+}$	8.10×10^8 (pH < 0)	Gogolev et al. (1986)

alkanes such as solvent extraction diluents typically react at rates of $\sim 10^9$ M^{-1} s^{-1}. For these reasons, radical redox reactions are probably less important than the nitrous acid concentration in establishing actinide oxidation states in irradiated aqueous nitric acid.

It can be seen from the foregoing discussions that a multitude of radiolytically induced reactions compete to set the neptunium oxidation state in irradiated acidic solution. Vladimirova (1995) provided the most comprehensive attempt to model the redox chemistry of neptunium in both alpha- and gamma-irradiated acidic solution. Taking into account reactions with nitrous acid and the major produced radicals, he calculated equilibrium concentrations of Np(IV), Np(V), and Np(VI) under various conditions of dose rate, absorbed dose, and nitric acid concentration. High dose rates favored the production of Np(IV), while high acidity resulted in rapid declines in Np(V) with increases in Np(IV) and Np(VI). The reduction of neptunium to the tetravalent oxidation state was attributed to disproportionation, and reaction with the $^{\bullet}$H atom. The rate constant of the reaction shown in Equation 34.83 was calculated to increase by a factor of 100 from $<5 \times 10^6$ M^{-1} s^{-1} to 2×10^8 M^{-1} s^{-1} with an increase in nitric acid concentration from 1 to 6M.

$$NpO_2^+ + {}^{\bullet}H + 2H^+ \rightarrow Np^{4+} + 2H_2O \qquad (34.83)$$

At a dose rate of 36 kGy h^{-1} and a nitric acid concentration of 3M, the equilibrium percentages of Np(IV), Np(V), and Np(VI) were 8, 72, and 20, respectively. The equilibrium yields of the three oxidation states were the same for both types of radiation. In additional work, this same author included refinements to account for the presence of organophosphorous extractants including TBP (Vladimirova et al., 1996). It was concluded that the presence of the organic compounds favored higher equilibrium concentrations of Np(IV) and Np(VI) at the expense of Np(V). Sensitivity analysis using a mathematical model suggested that the rate constants for the reduction of Np(VI) by nitrous acid, and the reaction of Np(IV) with Np(VI) to produce Np(V) were decreased, possibly due to complexation of Np(IV) and Np(VI) by the organic complexing agents.

34.4 GEOLOGICAL DISPOSAL OF HIGH-LEVEL RADIOACTIVE WASTE

34.4.1 Introduction

Management and geological disposal of high-level radioactive waste are among the most critical issues related to nuclear technology in recent years. High-level radioactive waste, which includes spent nuclear fuel discharged from commercial power reactors or vitrified waste arising from reprocessing of power reactor fuel, will be disposed several hundred meters (e.g., more than 300 m in Japan's case) below sea level in isolation from humans and their environment. This is referred to hereafter as geological disposal. The concept of geological disposal is being considered in many countries and is based on a system of multiple passive barriers, consisting of engineered barriers and the geological environment (natural barrier), as illustrated in Figure 34.26 (JNC, 2005; SKB, 2006a). In general, the engineered barriers consist of three different types of barriers, as described below.

The first barrier is the waste form itself since radioactive nuclides cannot be released until after being dissolved. The nature of the actual waste form will be dependent on whether the spent nuclear fuel is directly disposed (once-through) or reprocessed to recover fissile and fertile materials in order to provide fresh fuel for existing and future nuclear power plants (nuclear fuel cycle). In the former case, uranium dioxide pellets are encapsulated in a cladding material such as zircaloy.

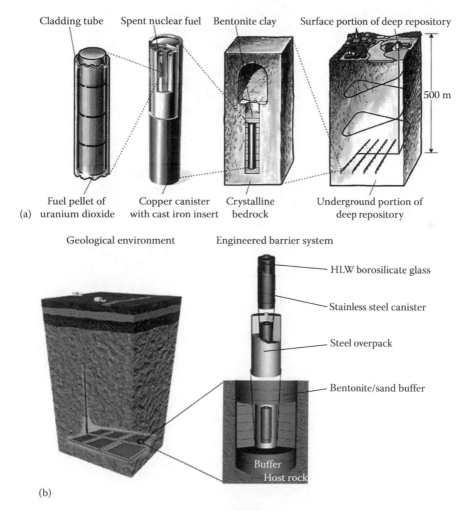

FIGURE 34.26 Conceptual design of geological disposal systems. (a) KBS-3 type spent nuclear fuel disposal. (Adapted from SKB, Long-term safety for KBS-3 reositories at Forsmark and Laxemar—a first evaluation, Main report of the SR-Can project, SKB Technical Report TR-06-09, Swedish Nuclear Fuel and Waste Management Co., Stockholm, Sweden, 2006a.) (b) HLW disposal in Japan. (Adapted from JNC, H17: Development and management of the technical knowledge base for the geological disposal of HLW, Knowledge Management Report, JNC-TN1400-2005-022, Japan Nuclear Cycle Development Institute, Tokai-mura, Japan, 2005.)

In the latter case, radioactive elements in the vitrified high-level radioactive waste are immobilized by matrices such as borosilicate glass, which provides a low solubility matrix in which the radionuclides are homogeneously distributed.

The second barrier is a waste package, which encapsulates the waste form to ensure its confinement for a certain period of time, especially so as to prevent groundwater from contact with the waste form. In some countries which are planning to implement direct disposal, spent nuclear fuel is packed in a copper canister with a carbon steel insert, while vitrified waste in a stainless steel canister is encapsulated in a carbon steel package. Although any metal inevitably corrodes even in the anoxic underground environment, a metallic waste package is expected to provide a period of complete containment during the radiogenic thermal transient, when conditions are rather dynamic and may be strongly coupled (e.g., thermal and water saturation profiles in the buffer).

The third barrier is a buffer material which surrounds the waste package. The buffer material is required to have favorable physical and chemical characteristics, such as self-sealing and mechanical buffering, good thermal conductivity, low hydraulic conductivity, colloid filtration function, and chemical buffering capacity to retard radionuclide migration. Materials like clay-based bentonites are typically used for the buffer due to their low hydraulic conductivity and large sorption capacity.

At present, bentonite is the most commonly available material for the buffer, whose mineral composition is shown in Table 34.9. Montmorillonite is the main component of bentonite and its properties are essential for the application of bentonite as the buffer material, including low permeabilities and high sorption capacities, which strongly retard migration of any radionuclides released from the waste package.

Figure 34.27a shows the structural arrangement of montmorillonite (Bradbury and Baeyens, 2002). The unit crystal lattice of montmorillonite is built with two layers of tetrahedral silicate units between which one layer of octahedral aluminate is sandwiched. A small part of Al^{3+} in the octahedral layer is substituted with Mg^{2+} and Fe^{2+} or Fe^{3+}, which results in negative charges in the layers. This charge deficiency is compensated by interlayer cations. As a result, the actual structure formula of montmorillonites from Wyoming, United States, and Tsukinuno, Japan, has been determined as follows:

TABLE 34.9
Mineralogical Composition of Bentonites

	Kunigel V1[a] (Tsukinuno, Japan)	MX-80[b] (Wyoming, the United States)
Montmorillonite	46–49	75
Kaolinite		<1
Mica		<1
Quartz/Calcedony	29–38	15.2
Feldspar	2.7–5.5	5–8
Calcite	2.1–2.6	0.7
Dolomite	2.0–2.8	
Analcime	3.0–3.5	
Siderite		0.7
Pyrite	0.5–0.7	0.3
Organic carbon	0.31–0.34	0.4

[a] Ito et al. (1993).
[b] Bradbury and Baeyens (2002).

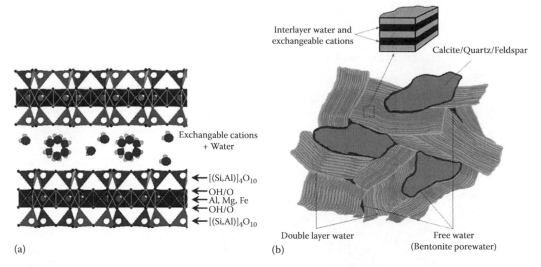

FIGURE 34.27 Structure of (a) montmorillonite and (b) compacted bentonite. (Adapted from Bradbury, M. H. and Baeyens, B., Pore water chemistry in compacted re-saturated MX-80 bentonite: Physicochemical characterization and geochemical modelling, PSI Bericht 02–10, Villigen PSI and NTB 01– 08, Nagra, Wettingen, Switzerland, 2002.)

Wyoming: $Na_{0.30}(Al_{1.55}Fe^{3+}{}_{0.20}Fe^{2+}{}_{0.01}Mg_{0.24})\ (Si_{3.96}Al_{0.04})O_{10}(OH)_2$

(Müller-Vonmoos and Kahr, 1983)

Tsukinuno: $(Ca_{0.05}Na_{0.38}K_{0.01})\ (Al_{1.55}Fe^{3+}{}_{0.09}Fe^{2+}{}_{0.01}Mg_{0.34}Ti_{0.01})\ (Si_{3.90}Al_{0.10})O_{10}(OH)_2$

(Suzuki et al., 1992)

Figure 34.27b shows a schematic illustration of compacted bentonite after water saturation (Bradbury and Baeyens, 2002). The typical size of a sheet of montmorillonite is about several hundred nanometers in diameter and 1 nm in thickness. Montmorillonite sheets form stacks and such secondary particles are randomly oriented in the compacted state together with mineral particles. Exchangeable cations in the stack are hydrated to some extent at ambient condition and their further hydration induces very strong volume swelling. As can be seen from the figure, any migration of ions and particles in this compacted bentonite would experience very tortuous pathways. Chemical reactions such as ion exchange with interlayer cations and surface complexation on edge sites of montmorillonite sheets are also important for retardation of radionuclide migration.

After the repository closure, the buffer material is gradually saturated with groundwater and chemical reactions in the buffer material and groundwater would occur. It is very difficult to experimentally determine chemical composition of porewater in bentonite, and therefore it has been estimated from geochemical modeling calculations. Table 34.10 shows the highest estimated concentration of the main chemical species in bentonite porewater (JNC, 2000). These results are summarized from various conditions of groundwater (e.g., freshwater/seawater origin, oxygen concentration, pH) and bentonite composition and reactions (Oda et al., 1999). Carbonates are commonly found in groundwater and the chloride concentration is high particularly in the case of groundwater with seawater origins. High concentrations of S(-II) are due to the dissolution of pyrite (FeS_2).

The bedrock surrounding the engineered barriers is expected to function as a natural barrier system since it further retards migration of radionuclides, which may be released through the buffer material.

The long-term performance of the multiple barrier system has been assessed in several countries in an attempt to demonstrate the technical feasibility of their disposal concepts. A number of

TABLE 34.10
Range of Chemical Species and pH
in Porewater of Bentonite

	Concentration (mol dm^{-3})
$HCO_3^-/CO_3^{2-}/H_2CO_3$	$<7.3 \times 10^{-2}$
SO_4^{2-}	$<6.1 \times 10^{-2}$
HS^-/H_2S	$<9.2 \times 10^{-2}$
Cl^-	$<5.9 \times 10^{-1}$
P (Total)	$<2.9 \times 10^{-6}$
NO_3^-	0.0
NH_3	$<1.6 \times 10^{-4}$
NH_4^+	$<5.1 \times 10^{-3}$
B (Total)	$<1.7 \times 10^{-3}$
pH	5.9–8.4

scenarios with specific features, events and coupled thermal, hydraulic, mechanical, and chemical processes have been considered. The source model includes the radioactivity of the waste form as it causes heat generation and ionizing radiation, which lead to material alteration and radiolysis of groundwater. Figure 34.28 shows changes in radioactivity with time for a typical spent nuclear fuel from an LWR (Poinssot et al., 2005). In the early post-closure stage, radioactivity of fission products is dominant. Since half-lives of most of fission products are shorter than those of actinides, radioactivity of actinides becomes dominant around several hundred years after repository closure.

From the viewpoint of system evolution, the impact of ionizing radiation from the waste form on the engineered barrier system can be categorized into the following two phases, as illustrated in Figure 34.29: In the earlier post-closure stage, the radioactivity of the waste is extremely high and the waste form itself is subjected to radiation damage including that of the damage by alpha recoil. The buffer material would be subjected to gamma radiation, which penetrates through the waste package. Pore fluids in the buffer material would be subject to gamma radiolysis and radiolytic products might result in oxidizing conditions in the porewater. This could lead to significant impacts

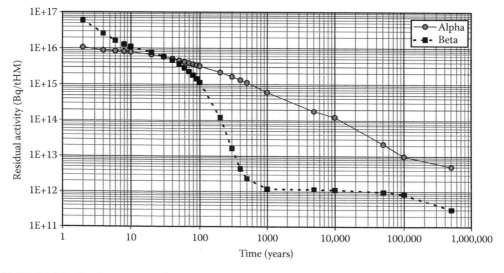

FIGURE 34.28 Evolution of α, β and γ radiations as a function of time for a 55 GWd t^{-1} UOX fuel. (Reprinted from Poinssot, C., et al., *J. Nucl. Mater*, 346, 66, 2005. With permission.)

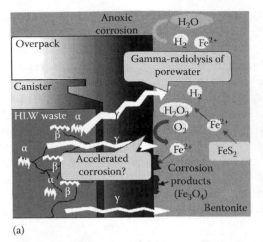
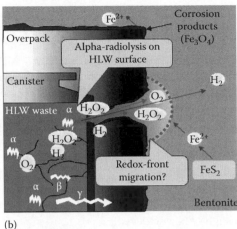

(a) (b)

FIGURE 34.29 Effects of ionizing radiation on a repository. (a) Before the waste package failure and (b) After the waste package failure. (Adapted from IRI, Radiation effects on barrier systems in high level radioactive waste disposal, Research Project Annual Report, Institute of Research and Innovation, Tokyo, Japan, 2005. (in Japanese).)

on the engineered barriers' function. For example, the corrosion of the waste package would be accelerated under oxidizing conditions, and if the extent is significant the waste package could be breached earlier than expected. The reactions of radiolytic products with the buffer materials might deteriorate their function of retarding radionuclide migration.

It is assumed in the scenario that the waste package would eventually be breached even under anoxic conditions due to metal corrosion and thus would fail more than several hundred years after repository closure. Groundwater would eventually contact the surface of the waste form and radionuclides would be released by dissolution of the waste form. Around this period, most of beta/gamma radionuclides would have decayed to a level several orders of magnitude lower than the initial level, and alpha radiation from the surface of the waste would induce radiolysis of water in a very limited region about 30 μm from that surface. In the case of spent nuclear fuel pellets, it is anticipated that the oxidative products of alpha-radiolysis of water would react with the uranium oxide matrix and thus increase its dissolution rate significantly. The radiolytic products might also react with the corrosion products of the waste package. Those having escaped from the reactions in the waste package might diffuse into the buffer material and also react with reductants such as Fe(II).

Those potential processes induced by the ionizing radiation in different phases of the repository system life are discussed in the following sections.

34.4.2 Effect of Gamma Radiation on Engineered Barrier Systems

In the near term after repository closure, gamma radiation is the dominant source of radiation impact on high-level waste forms and their immediate environment because alpha and beta radiation can be shielded completely by the metal waste package. The major source of gamma radiation is ^{137}Cs with a half-life is about 30 years. The impact of gamma radiation on the engineered barriers can be controlled to some extent by the engineering design. For this purpose, it has been proposed to design the waste package so as to reduce the dose rate at its outer surface to less than 1 Gy h^{-1} (Werme, 1998).

34.4.2.1 Effect on Clay Minerals

As noted above, the engineered barrier system would receive gamma radiation from the waste forms in the early phase after repository closure. Most of iron atoms exist as Fe(III) and they are reduced after gamma irradiation, as probed by Mössbauer and ESR spectra (Gournis et al., 2000; Plotze et al., 2003). Observed slight decreases in cation exchange capacities after long periods (22 months) may be attributed to dissipation of protons, lowering the layer charge (Plötze et al., 2003). No structural change was observed after gamma-irradiation up to 2 MGy (Negron et al., 2002).

Radiation-induced defects formed in the framework of smectite clays were assigned as trapped holes on oxygen atoms (Clozel et al., 1994, Allard and Calas, 2009). The relative amounts of the defects are dependent on the atomic composition and arrangements and they resulted in isotope exchange of hydroxyl ion and changes in specific surface area. However, those changes are small in the case of montmorillonite after irradiation up to 30 MGy (Pushkareva et al., 2002).

Most of the experiments summarized above were performed with powdered or dispersed smectites. Pusch et al. performed experiments to simulate the actual condition of compacted bentonite contacting steel under high-temperature and gamma-radiation conditions (Pusch et al., 1992). Commercial MX-80 was compacted to a dry density of 1.65 g cm^{-3} and saturated with simulated groundwater. Irradiation was continued for 1 year and the total dose was 33 MGy. Little change in the framework structure and physicochemical properties were observed, although Fe(III) ions may have slightly been reduced.

These results indicate physicochemical properties of montmorillonite as the buffer materials are little affected by gamma-ray exposure. However, significant damage leading to amorphization has been observed under electron-beam irradiation in electron micrographs when the accumulated dose is increased to the order of GGy (Gu et al., 2001, Sorieul et al., 2008). Threshold values for radiation-induced amorphization were strongly dependent on the temperature and structure of smectites: Threshold values are the largest at 300°C – 450°C, which coincides with the onset of dehydroxylation. At room temperature, thresholds for amorphization were over 10 GGy, which is comparable to cumulative doses estimated in HLW repositories after 10^6 years for radionuclides adsorbed to the bentonite buffer.

Several countries are planning to build repositories in clay formations and the effect of gamma radiation on natural clays has been studied. In particular, various tests have been performed in the underground research facility in the Boom clay in Belgium. Mössbauer spectra of the Boom clay indicated that while pyrite was completely oxidized after gamma-irradiation up to 30 MGy, Fe(II) remained, as in the case of the compacted bentonite (Ladriere et al., 2009). An in situ experiment in the underground research laboratory in the Boom clay formation was performed to develop a model to predict transient behavior of the disposal system. In the CERBERUS experiment, a ^{60}Co gamma-ray source and heaters were placed in the clay formation to simulate waste forms, and changes in E_h and pH were monitored (Zhang et al., 2008).

34.4.2.2 Effect on Corrosion of Waste Package

In the initial phase after the repository closure, the buffer material is unsaturated and therefore the gamma-radiolysis of air in pores of the buffers would generate nitrogen oxides, which eventually form nitric acid which might affect the corrosion of the canister. The amount of nitrate in the vessel is proportional to the gas volume against the liquid volume (Linacre and Marsh, 1981). However the estimated amount of generated nitric acid was negligible (King et al., 2001).

As the buffer materials are gradually saturated with groundwater, gamma-radiolysis products like H_2O_2 or O_2 are generated. These radiolytic products would induce oxidative corrosion of metal packages. Increased corrosion rates of carbon steel in saline water under gamma irradiation were observed in the 1980s in the United Kingdom. However, the effect was negligible at dose rate of about 3 Gy h^{-1} (Marsh et al., 1989). The effect was also observed in compacted bentonite saturated with the synthetic groundwater. They have performed 1D diffusion reaction model calculation to evaluate the effect of radiation by assuming cathodic reaction of radiolytically generated oxygen. Burns et al. have incorporated surface reactions of short-lived radiolytic products in models to calculate gas generation in gamma-irradiated steel vessels (Burns et al., 1983). A more elaborate mixed potential model was applied for the corrosion of HLW canisters in which homogenous and surface reactions of radiolytic products were fully taken into account (Macdonald and Urquidi-Macdonald, 1990). Recently, the corrosion of the carbon steel inserts of spent nuclear fuel canisters under gamma radiation was studied assuming cannsiter failure (Smart et al., 2008). The effect of gamma radiation was observed at the dose rate of 300 Gy h^{-1}. The predominant corrosion products were magnetite and oxyhydroxides (or maghemite), which are thought to be formed by anoxic corrosion of iron and subsequent dissolution of Fe^{2+} at pH < 7 (White et al., 1994).

$$3Fe + 4H_2O \rightarrow Fe_3O_4 + 4H_2 \tag{34.84}$$

$$Fe_3O_4 + 2H^+ \rightarrow \gamma\text{-}Fe_2O_3 + Fe^{2+} + H_2O \tag{34.85}$$

Copper is employed for the material for the canister in several countries and its corrosion under gamma radiation has also been studied (King et al., 2001). In the early stages of disposal, the oxygen concentration is still high and the corrosion in saline groundwater would proceed as follows:

$$\text{Anodic:} \quad Cu + 2Cl^- \rightarrow CuCl_2^- + e^- \tag{34.86}$$

$$\text{Cathodic:} \quad O_2 + 4e^- + 2H_2O \rightarrow 4OH^- \tag{34.87}$$

The effect of gamma radiation on the corrosion of copper embedded in a montmorillonite–sand mixture was studied (King et al., 1992; Shoesmith and King, 1999). The copper concentration in the mixture significantly decreased upon irradiation, presumably due to reduction of Cu(II) to Cu(I) which forms a weakly sorbed anionic complex. It was expected that the oxidizing species produced by groundwater radiolysis would accelerate the cathodic reaction. Although the observed reduction of copper by gamma-radiolysis was the opposite of expected, the effect was similar to the case of the reduction of iron in montmorillonite (Pusch et al., 1993). King et al. conjectured that electrons produced from montmorillonite ionization might have reduced cupric ion in the porewater. It has been proposed that the energy of ionizing radiation deposited in nanoparticles of metal oxides dispersed in the aqueous phase might be transferred to water to generate the hydrated electron, and eventually molecular hydrogen (Schatz et al., 1998; LaVerne and Tonnais 2003; LaVerne, 2005). The same reaction might have occurred in the porewater of montmorillonite under ionizing radiation as the thickness of a single layer of montmorillonite is about 1 nm and the secondary electrons produced might have easily escaped.

34.4.3 Effect on Dissolution of Waste Forms

After the assumed failure of a waste package, groundwater would penetrate to the surface of the waste form and eventually would receive alpha-radiolysis. In this section, the effects of water radiolysis on dissolution of the waste form are described. The dissolution rate of the waste form is a key parameter for the performance assessment of the disposal system as it determines the release rates of radionuclides which eventually diffuse into the engineered and natural barrier systems. While various factors such as pH, solutes, temperature, etc., affect the dissolution rates, the effect of ionizing radiation is particularly important in the case of geological disposal of spent nuclear fuel. Groundwater radiolysis would generate oxidizing species in an otherwise reductive environment and may increase the dissolution rate several orders of magnitude.

34.4.3.1 Dissolution of Spent Nuclear Fuel

In the case of direct disposal, most of the radioactive nuclides are retained in a uranium dioxide matrix and their release rates are determined by the dissolution of that matrix. Uranium dioxide is stable under the anoxic conditions of the repository and its solubility is extremely low. Figure 34.30 summarizes chemical processes on the surface of uranium dioxide (Poinssot et al., 2005; Martínez

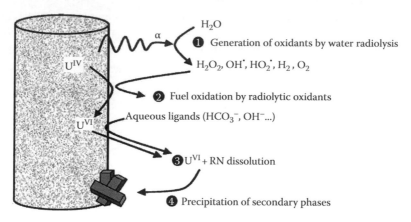

FIGURE 34.30   Radiation-induced processes on the surface of uranium dioxide. (Reprinted from Poinssot, C. et al., *J. Nucl. Mater.*, 346, 66, 2005. With permission.)

Esparza et al., 2005). Dissolution rates increase significantly when UO_2 is oxidized to UO_{2+x} ($0 < x < 1$) by dissolved oxygen or radiolytically generated oxidizing species. Dissolution of U(VI) is facilitated by complexation with carbonate ions. Dissolved U(VI) precipitates on the UO_2 pellet surface, although some amount would eventually migrate out of the canister to the engineered barrier systems.

Uranium dioxide is a p-type semiconducting material and oxidative dissolution of uranium dioxide can be regarded as corrosion with anodic and cathodic half reactions.

$$\text{Anodic:} \quad UO_2 \rightarrow UO_2^{2+} + 2e^- \tag{34.88}$$

$$\text{Cathodic:} \quad H_2O_2 + 2e^- \rightarrow 2OH^- \tag{34.89}$$

$$O_2 + 4e^- + 2H_2O \rightarrow 4OH^- \tag{34.87}$$

Under the condition in which fuel corrosion occurs, the redox potential of the groundwater (E_h) must be more positive than the equilibrium potential for fuel dissolution. When the groundwater and the spent nuclear fuel are contacted, corrosion potential (E_{corr}) is established at which the rates of the anodic and cathodic reactions will be equal (Shoesmith, 2000, 2007) and is measured to monitor the degree of oxidation of the sample (Johnson et al., 1983). Oxidation states of uranium on the surface can also be probed with x-ray photoelectron spectra (Sunder et al., 1990).

Gamma radiation has been commonly employed to study the effect of water radiolysis on oxidative dissolution of uranium oxide (Gromov, 1981; Sunder et al., 1992; Christensen and Sunder, 2000). However, alpha radiation is the main source of ionizing radiation after several hundred years and various methods have been developed to study its effect on radiation-induced oxidative dissolution of the fuel matrix.

The most straightforward method is to study dissolution of spent nuclear fuels, although currently available spent nuclear fuels still have high beta/gamma radioactivity (Bruno et al., 1999, 2003; Cera et al., 2006). This activity affects water radiolysis with different G-values than alpha-radiolysis. To eliminate the effect of beta/gamma radiation and complex chemical and morphological changes in spent nuclear fuels, uranium dioxide doped with alpha emitters (^{238}Pu, ^{233}U, etc.) has been utilized to study the effect of alpha radiation (Rondinella et al., 2000; Cobos et al., 2002; Cachoir and Lemmens, 2004; Jégou et al., 2004; Mennecart et al., 2004b; Muzeau et al., 2009). Dissolution rates were proportional to the dose rate of the samples, which were controlled by the addition of different amounts of alpha emitters. A simulated high-burnup spent nuclear fuel called SIMFUEL (Lucuta et al., 1991) was also employed for dissolution studies (Ollila, 1992).

Not only internal alpha sources but also external sources have been applied. Alpha emitter discs (^{241}Am and ^{210}Po) were employed in the first study of alpha-radiolysis affects on UO_2 dissolution by placing them close to the surface of UO_2 pellets (Bailey et al., 1985). In the presence of alpha-radiation, the corrosion potential of UO_2 increased to the same level as in the presence of dissolved O_2.

In recent years, helium ion beam irradiation has also been employed to simulate alpha-radiation. This method is advantageous in that high dose rates can be achieved and ion energy can be controlled to study the LET effect. Two different types of UO_2 are employed. The first is colloidal particles dispersed in aqueous solution (Mennecart et al., 2004a, Suzuki et al., 2006), while the other is UO_2 discs contacted with water on one face and being irradiated from the opposite face (Corbel et al., 2001; Sattonnay et al., 2001). The dissolved U(VI) concentration due to ion beam irradiation was higher than the value following the addition of the same amount of H_2O_2 generated by the ion beam, indicating the effect of radical species is not negligible (Corbel et al., 2006).

Radiation-induced oxidative dissolution of UO_2 was first modeled by taking into account the reactions of radical species with UO_2 (Christensen and Bjergbakke, 1987). Reaction rates at the UO_2 surface were estimated from the calculated number density of surface U(IV) atoms and homogeneous reaction rates for U(IV) in solution. The model was further refined (Christensen et al., 1994) and one-dimensional diffusion–reaction calculations of UO_2 dissolution were performed by setting different regions of water subjected to radiation (Christensen, 1998). The model was then extended to the case of UO_2 dissolution in salt brine (Kelm and Bohnert, 2000a,b). Dissolution rates of alpha-emitter-doped UO_2 were recently calculated (Christensen, 2006). The reaction rates of UO_2 and radical species were estimated to be about two orders of magnitude larger than for molecular species. Difficulties in setting appropriate values of surface-to-volume ratio were pointed out in the calculation of the dissolution rates of UO_2. One-dimensional diffusion–reaction model calculations of the UO_2 surface were performed to simulate helium ion beam irradiation experiments assuming alpha-dose rate profiles from the surface (Poulesquen and Jégou, 2007). The calculated values of U(VI) and H_2O_2 were lower than the measured values in both aerated and deaerated cases. Measured values remained relatively high even in the presence of hydrogen, in contrast to the case of dissolution experiments under dissolved hydrogen. The authors attributed this to a high dose rate used in the calculation, which may have overestimated radical–radical recombination to form H_2O_2.

Another model of radiolytic oxidative dissolution of UO_2 has been developed in which the reactions of UO_2 were limited to those with H_2O_2 and O_2 (Eriksen, 1996). The reaction of H_2O_2 is the most important and consumption rates of oxidants rapidly decreased possibly due to accumulated H_2, which scavenged OH$^•$ and suppressed H_2O_2 formation. They also measured dissolution rates of PWR spent nuclear fuel pellets (Eriksen et al., 1995). In the early stage of the experiments, measured concentrations of radiolytic products were lower than estimated by the model, and they were therefore assumed to have decomposed on the surface.

The model has been further developed by Jonsson in recent years (Jonsson et al., 2007). Oxidative dissolution was facilitated by U(VI) carbonate complex formation, and dissolution rates became proportional to surface oxidation rates when $[HCO_3^-] > 10^{-3}$ mol dm^{-3}, typical of groundwater. One-electron oxidation of UO_2 is considered to be the rate-determining step as the reaction rates of a series of oxidants with UO_2 were found to be logarithmically proportional to one-electron reduction potentials (Ekeroth and Jonsson, 2003). Based on UO_2 dissolution rates under gamma-irradiation and reaction model calculations, the relative contributions of various oxidants to UO_2 oxidation by water radiolysis were evaluated. It was concluded that contribution of H_2O_2 was more than 99.9% and other oxidant species were negligible, which supports the reaction model proposed by Eriksen (1996) and (Ekeroth et al., 2006).

One-dimensional reaction–diffusion model calculations for the surface were also performed. Dose rate distributions were provided from calculated inventories and geometrical considerations on the dissipation probability of alpha particles (Nielsen and Jonsson, 2006). Steady-state H_2O_2

concentrations were quickly achieved on the surface (Nielsen et al., 2008a), and this assumption was employed in the further model calculations of radiolytic oxidative dissolution (Nielsen et al., 2008b).

In the repository, it is predicted that anoxic corrosion of metal canisters by water will generate hydrogen and the concentration of dissolved hydrogen would be very high under confinement pressures up to 5 MPa. It has been shown that dissolved hydrogen significantly reduces the dissolution rates of spent nuclear fuels (Sunder et al., 1990; Spahiu et al., 2000, 2004a,b; Röllin et al., 2001; Ollila et al., 2003; Carbol et al., 2005, 2009; Loida et al., 2005, Muzeau et al., 2009). The corrosion potential of UO_2 in dissolved H_2 under gamma radiation decreased to a level lower than that of the unirradiated sample and reduction of the oxidized surface by hydrogen atom (King and Shoesmith, 2004; Shoesmith, 2008).

While the effect of dissolved hydrogen on dissolution rates has been experimentally established, its mechanism is still a matter of controversy. Since dissolved hydrogen showed almost no effect on the 5 MeV helium ion beam radiolysis of aqueous hydrogen peroxide solutions (Pastina and LaVerne, 2001), the effect probably involves surface reactions. The effect has been observed in the case of alpha-doped UO_2 indicating a direct effect of hydrogen (Spahiu et al., 2004a,b; Carbol et al., 2009). However, dissolved hydrogen showed no effect on the decomposition of H_2O_2 on neat UO_2 surfaces without ionizing radiation. Noble metals (Pd, Mo, Ru, Tc, Rh) as fission products in spent nuclear fuels are observed as nanosized metallic particles called ε-particles. Reduction of U(VI) in UO_2 was facilitated by ε-particles, which act as catalysts for the oxidation of dissolved hydrogen. This was observed as lower corrosion potentials of SIMFUEL in the presence of ε-particles (Broczkowski et al., 2005). The catalytic effect of noble metal particles was studied with UO_2 pellets doped with different amount of Pd particles (Trummer et al., 2008). The oxidative dissolution of UO_2 by H_2O_2 was reduced by increased concentrations of Pd particles under anoxic conditions. The oxidative dissolution of UO_2 by dissolved oxygen was also catalyzed by Pd particles (Nilsson and Jonsson, 2008). Radiolytic dissolution under gamma radiation was almost inhibited in 3 wt% Pd doped UO_2 in an N_2 atmosphere. The dissolution rate equation was expressed as the difference between the oxidation and reduction rates of UO_2, and the former was divided into direct and noble metal–catalyzed reactions. The reaction rate constants of oxidation and reduction were derived by comparison with the calculated and measured dissolution rates. They were close to diffusion-limited values (Trummer et al., 2009).

The effects of solutes in groundwater on the radiolytic dissolution of UO_2 have also been studied. Reduction by dissolved Fe^{2+} (Johnson and Shoesmith, 1988) acts as a radical recombiner as follows:

$$Fe^{2+} + {}^{\bullet}OH \rightarrow Fe^{3+} + OH^{-} \tag{34.90}$$

$$Fe^{3+} + {}^{\bullet}H \rightarrow Fe^{2+} + H^{+} \tag{34.91}$$

The same effect has recently been found to explain the effect of bromide ion which also suppressed the effect of dissolved hydrogen by recombining hydroxyl and hydrogen radicals in the same way (Metz et al., 2008). In aerated condition, 2-propanol increased the corrosion rate. The alcohol scavenges $OH^{\bullet}$ to form isopropyl radical, which then reduces dissolved oxygen to generate hydrogen peroxide. In contrast, 2 mol dm^{-3} Cl^{-} decreased the rate, which was attributed to lower $G_{(H_2O_2)}$ and lower dissolved oxygen concentration at higher ionic strength (Roth and Jonsson, 2009).

34.4.3.2 Dissolution of HLW Glass

In contrast to the case of spent nuclear fuel, the oxidation of the waste form is not critical with regard to dissolution of HLW glass. The effects of ionizing radiation on HLW glass were at first studied mainly under gamma radiation (McVay and Pederson, 1981; Pederson and McVay, 1983; Burns et al., 1982; Barkatt et al., 1983). Increased leach rates were attributed to increased acidity, possibly due to the formation of nitric acid in the gas phase of the vessel. A comprehensive review of radiation effects on HLW glasses are given by Weber (1988) and Weber et al. (1997). Changes in chemical bonding, migration of elements, gas generation, volume and specific surface area, and physical properties have been studied. Their effects on the leach rates were reviewed. It was concluded that the radiation effect on the leach rate would be less than a factor of 10, assuming uncertainties in the measured leach rates.

Effects of alpha radiation have also been studied and increased leach rates were observed in alpha-doped glasses (Weber et al., 1985). However, no enhancement of leach rates by alpha radiation was observed in ion-beam irradiation experiments (Abdelouas et al., 2004) or alpha-doped R7T7 glasses (Peuget et al., 2007).

Leach tests of HLW glasses doped with actinides were performed in situ, embedded in the backfill materials under Co-60 gamma radiation (European Commission, 2007). Certain differences were observed in the diffusion profiles of redox-sensitive nuclides in the buffer material resulting from gamma radiation. The leachate became slightly acidic under gamma-irradiation and the mechanism is not clearly understood.

34.4.4 REDOX FRONT MIGRATION INDUCED BY GROUNDWATER RADIOLYSIS

In a diffusion–reaction system, when oxidants/reductants migrate into the region in which reductants/oxidants are fixed, redox reactions occur almost instantaneously. A sharp front may be formed which separates the region in which redox reactive species are consumed and fixed. Formation and migration of such "redox fronts" in natural formations have been studied as natural analogues of the geological disposal systems that might experience changes in the redox condition during operation (e.g., Yoshida et al., 2008).

Redox fronts in the geological disposal system might be formed in the earliest phases of enclosure as the excavated walls of the repository are exposed to air and oxygen. Intrusion of oxygen containing groundwater to the repository might also form a redox front. Groundwater radiolysis was assumed to be another possible mechanism of redox front formation in the repository as follows (Neretnieks, 1983): After the waste package loses its integrity and porewater comes into contact with the waste form, alpha-radiolysis of water would occur. While radiolytically generated H_2 is relatively inert and would dissipate without chemical reactions, oxidizing species such as H_2O_2 and O_2 would gradually diffuse out of the waste package. They are then advectively transported by groundwater flow into the bedrock through fractures, and eventually diffuse into rock matrices through fracture surfaces. H_2O_2 and O_2 are reduced by Fe(II) contained in iron bearing minerals such as biotite in granite, and a redox front is formed in the rock matrix. The extent of the movement of the redox front in a repository was estimated to be about 60 m after 10^6 years, by assuming a H_2O_2 formation rate on the spent nuclear fuel surface by alpha-radiolysis, geometries of the fractures and groundwater flow rate, and the molar percentage of ferrous ion in the rock matrices. However, the generation of H_2O_2 might have been overestimated by using the G-value of pure water alpha-radiolysis, as radiolytic generation of H_2O_2 may be significantly suppressed with Fe^{2+} in groundwater (Christensen and Bjergbakke, 1982a,b). Decomposition of oxidants in backfill materials was also neglected, while they also contain iron-bearing minerals (e.g., pyrite) and smectites, which act as reductants for radiolytic oxidants. In an additional study, the migration distances of a redox front in the backfill materials surrounding a canister were estimated (Andersson et al., 1982).

The experimental verification of formation and migration of the radiolytic products of water in engineered barrier materials were performed. A beta-emitter (^{147}Pm) disk was contacted with a compacted bentonite disk saturated with synthetic groundwater and hydrogen generated by radiolysis of porewater in bentonite was measured after it diffused out of the opposite face of the bentonite disk (Eriksen and Jacobsson, 1983). The same setup was applied with an alpha-emitter (^{241}Am) disk (Eriksen et al., 1987, Eriksen and Ndalamba, 1988). The G-values of hydrogen peroxide generation and Fe^{2+} reduction in bentonite were estimated to be 0.069 and 0.40 μmol J^{-1}, respectively. These results are supposed to have provided a basis for performance assessments of geological disposal systems to consider the redox-front formation in the repository due to groundwater radiolysis (e.g., SKI, 1996).

The model of the reaction and migration of radiolytic products in the repository was further developed by taking into account various processes, e.g., the effect of dissolved oxygen, Fe^{2+}, the geometry of fuel cladding, precipitation of dissolved U(VI), and alpha-radiolysis of water in the buffer (Liu and Neretnieks 2001a,b,c, 2002; Liu et al., 2003). One-dimensional diffusion-reaction model calculations of the processes in the canister were also performed to estimate the dissolution rate by taking into account the corrosion of the iron canister material (Shoesmith et al., 2003). The effect of dissolved Fe^{2+} from corrosion of the waste package on the formation and decomposition of oxidative species by alpha-radiolysis was also considered, and the migration of redox fronts in the buffer material was predicted to be stopped (Johnson and Smith, 2000). However, redox front migration by groundwater radiolysis in the near field has not been considered in recent performance assessments. Based on recent results concerning spent nuclear fuel dissolution under reducing conditions, it is believed that radiation-induced oxidative dissolution is strongly suppressed (e.g., SKB, 2006b).

ACKNOWLEDGMENTS

Bruce J. Mincher's contribution to this work was supported by the U.S. Department of Energy, Office of Nuclear Energy, under DOE Idaho Operations Office Contract DE-AC07-05ID14517.

Makoto Yamaguchi's contribution was based in part on a project for assessment methodology development of chemical effects on geological disposal systems funded by the Ministry of Economy, Trade and Industry, Japan. He thanks Morimasa Naito for carefully reading a draft of the manuscript and giving him valuable comments and corrections.

REFERENCES

Abdelouas, A., Ferrand, K., Grambow, B., Mennecart, T., Fattahi, M., Blondiaux, G., and Houée-Lévin, C. 2004. Effect of gamma and alpha irradiation on the corrosion of the french borosilicate glass SON68. *Mater. Res. Soc. Symp. Proc.* 807: 175–180.

Adamov, V. M., Andreev, V. I., Belyaev, B. N., Lyubtsev, R. I., Markov, G. S, Polyakov, M. S., Ritari, A. É., Shil'nikov, A. Yu. Sov. Radiochem. 1987. Radiolysis of an extraction system based on solutions of tri-normal-butylphosphate in hydrocarbon diluents. *Sov. Radiochem.* 29: 775–781.

Adamov, V. M., Andreev, V. I., Belyaev, B. N., Markov, G. S., Polyakov, M. S., Ritari, A. É., Shil'nikov, and A. Yu. Kerntechnik. 1990. Identification of decomposition products of extraction systems based on tri-normal-butyl phosphate in aliphatic-hydrocarbons. *Kerntechnik* 55: 133–137.

AESJ. 2000a. Overview of nuclear power plants. In *Handbook of Water Chemistry of Nuclear Reactor System* (in Japanese), K. Ishigure (ed.). Tokyo, Japan: Atomic Energy Society of Japan.

AESJ. 2000b. Water chemistry control of PWR primary coolant, In *Handbook of Water Chemistry of Nuclear Reactor System* (in Japanese) K. Ishigure (ed.), Tokyo, Japan: Atomic Energy Society of Japan.

AESJ. 2000c. Water chemistry control of BWR primary coolant, In *Handbook of Water Chemistry of Nuclear Reactor System* (in Japanese) K. Ishigure (ed.). Tokyo, Japan: Atomic Energy Society of Japan.

Alfassi, Z. B. 1997. *Peroxyl Radicals*. New York: John Wiley & Sons.

Allard, Th. and Calas, G. 2009. Radiation effects on clay mineral properties. *Appl. Clay Sci.* 43:143–149.

Andersson, G., Rasmuson, A., and Neretnieks, I. 1982. Migration model for the near field. Final report. SKBF/KBS Teknisk rapport 82-24, Stockholm, Sweden.

Andreichuk, N. N., Rotmanov, K. V., Frolov, A. A., and Vasilev, V. Y. 1984. Effect of alpha-irradiation on the valence states of actinides. 4. Kinetics of HNO_2 formation in nitric-acid solutions. *Sov. Radiochem.* 26: 701–706.

Andresen, P. L. 1992. Irradiation-assisted stress–corrosion cracking. In *Stress–Corrosion Cracking*. R. H. Jones (ed.), pp. 181–210. Materials Park, OH: ASM International.

Andresen, P. L., Ford, F. P., Murphy, S. M., and Perks, J. M. 1989. State of knowledge of radiation effects on environmental cracking in light water reactor core materials. In *Proceedings of the Fourth International Symposium on Degradation of Materials in Nuclear Power Systems—Water Reactors*, D. Cubicciotti (ed.), Jekyll Island, GA, pp. 1-83–1-120. Houston, TX: NACE.

Ashida, S., Takagi, J., Ichikawa, N., and Morikawa, Y. 1992. First experience of hydrogen water chemistry at a Japanese BWR. In *Proceedings of the Water Chemistry of Nuclear Reactor Systems 6*, Bournemouth, U.K. vol. 2, pp. 103–110. London, U.K.: BNES.

Bailey, M. G., Johnson, L. H., and Shoesmith, D. W. 1985. The effects of the alpha-radiolysis of water on the corrosion of UO_2. *Corrosion Sci.* 25: 233–238.

Barkatt, A., Barkatt, A., and Sousanpour, W. 1983. Gamma radiolysis of aqueous media and its effects on the leaching processes of nuclear waste disposal materials. *Nucl. Technol.* 60: 218–227.

Becker, R., Stieglitz, L., and Bautz, H. 1983. Untersuchung der strahlenchemischen TBP-Zersetzung unter den Bedingungen des PURX-Prozesses. Report KfK3639.

Benndict, M., Pigford, T. H., and Levi, H. W. 1981. *Nuclear Chemical Engineering*, 2nd edn. New York: McGraw-Hill.

Berger, F. P. and Hau, K.-F. F.-L. 1977. Mass transfer in turbulent pipe flow measured by the electrochemical method. *Int. J. Heat Mass Transfer* 20:1185–1194.

Berthon, L., Morel, J. M., Zorz, N., Nicol, C., Virelizier, H., and Madic, C. 2001. Diamex process for minor actinide partitioning: Hydrolytic and radiolytic degradations of malonamide extractants. *Sep. Sci. Technol.* 36: 709–728.

Berthon, L., Journet, S., Lalia, V., Morel, J. M., Zorz, N., Berthon, C., and Amerkraz, B., 2004. Use of chromatographic techniques to study a degraded solvent for minor actinides partitioning: Qualitative and quantitative analysis. In *Atalante-2004*, P2-08, Nîmes, France, pp. 1–4.

Bhattacharyya, P. K. and Saini, R. D. 1973. Radiolytic yields $G(HNO_2)$ and $G(H_2O_2)$ in the aqueous nitric acid system. *Int. J. Radiat. Phys. Chem.* 5: 91–99.

Bhattacharyya, P. K. and Veeraraghavan, R. 1977. Reaction between nitrous acid and hydrogen peroxide in perchloric acid medium. *Int. J. Chem. Kinetics* 9: 629–640.

Bibler, N. E. 1974. Curium-244 alpha radiolysis of nitric acid. Oxygen production from direct radiolysis of nitrate ions. *J. Phys. Chem.* 78: 211–215.

Bielski, B. H. J., Cabelli, D. E., Arudi, R. L., and Ross, A. B. 1985. Reactivity of HO_2/O_2^- radicals in aqueous-solution. *J. Phys. Chem. Ref. Data* 14: 1041–1100.

Bisel, I., Camès, B., Faucon, M., Rudloff, D., and Saucerotte, B. 2007. DIAMEX-SANEX solvent behavior under continuous degradation and regeneration operation. In *Global-2007*, Boise, ID, 2007, pp. 1857–1860.

Bishop, W. P. and Firestone, R. F. 1970. Radiolysis of liquid n-pentane. *J. Phys. Chem.* 74: 2274–2284.

Blake, C. A., Davis, W. Jr., and Schmitt, J. M. 1963. Properties of degraded TBP-amsco solutions and alternative extractant-diluent systems. *Nucl. Sci. Eng.* 17: 626–637.

Borovkov, V. I. 2008. Excess electrons scavenging in *n*-dodecane solution: The role of tunneling of electron from its localized state to acceptor. *Radiat. Phys. Chem.* 77: 1190–1197.

Boullis, B. 2008. Future nuclear fuel cycles: Prospects and challenges. In *ISEC 2008, International Solvent Extraction Conference*, B. A. Moyer (ed.), Tucson, AZ, 2008, pp. 29–41. Tucson, AZ: Cognis Corporation.

Bradbury, M. H. and Baeyens, B., 2002. Pore water chemistry in compacted re-saturated MX-80 bentonite: Physicochemical characterization and geochemical modelling. PSI Bericht 02-10, Villigen PSI and NTB 01-08, Nagra, Wettingen, Switzerland.

Broczkowski, M. E., Noël, J. J., and Shoesmith, D. W. 2005. The inhibiting effects of hydrogen on the corrosion of uranium dioxide under nuclear waste disposal conditions. *J. Nucl. Mater.* 346: 16–23.

Bruno, J., Cera, E., Grivé E. M., Eklund, U.-B., and Eriksen, T. 1999. Experimental determination and chemical modeling of radiolytic processes at the spent fuel/water interface. SKB Technical Report TR-99–26, Swedish Nuclear Fuel and Waste Management Co., Stockholm, Sweden.

Bruno, J., Cera, E., Grivé, E. M., Duro, L., and Eriksen, T. 2003. Experimental determination and chemical modeling of radiolytic processes at the spent fuel/water interface. Experiments carried out in carbonate solutions in absence and presence of chloride. SKB Technical Report TR-03-03, Swedish Nuclear Fuel and Waste Management Co., Stockholm, Sweden.

Buchholz, B. A., Nunez, L., and Vandegrift, G. F. 1996. Effect of alpha-radiolysis on TRUEX-NPH solvent. *Sep. Sci. Technol.* 31: 2231–2243.

Bugaenko, L. T. and Roshchektaev, B. M. 1971. Radiolytic conversions of nitrate ion in nitric acid solutions. *Khimiya Vysokikh Energii* 5: 424–425.

Burns, W. G. and Moore, P. B. 1976. Water radiolysis and its effect upon in-reactor zircaloy corrosion. *Radiat. Effects* 30: 233–242.

Burns, W. G., Hughes, A. E., Marples, J. A. C., Nelson, R. S., and Stoneham, A. M. 1982. Effects of radiation on the leach rates of vitrified radioactive waste. *J. Nucl. Mater.* 107: 245–270.

Burns, W. G., Marsh, W. R., and Walters, W. S. 1983. The γ irradiation-enhanced corrosion of stainless and mild steels by water in the presence of air, argon and hydrogen. *Radiat. Phys. Chem.* 21: 259–279.

Burr, J. G. 1958. The radiolysis of tributyl phosphate. *Radiat. Res.* 8: 214–221.

Buxton, G. V., Greenstock, C. L., Helman, W. P., and Ross, A. B. 1988. Critical-review of rate constants for reactions of hydrated electrons, hydrogen-atoms and hydroxyl radicals ($^{\bullet}$OH/O$^-$) in aqueous-solution. *J. Phys. Chem. Ref. Data* 17: 513–886.

Cachoir, C. and Lemmens, K. 2004. Static dissolution of α-doped UO_2 in boom clay conditions: Preliminary results. *Mater. Res. Soc. Symp. Proc.* 807: 59–64.

Carbol, P., Cobos, J., Glatz, J. P., Ronchi, C., Rondinella, V., Wegen, D, Wiss, T., Loida, A., Metz, V., Kienzler, B., Grambow, B., Quinones, J., and Martinez, A. 2005. The effect of dissolved hydrogen on the dissolution of ^{233}U Doped UO_2(s). High burn-up spent fuel and MOX fuel. SKB Technical Report 05-09, Swedish Nuclear Fuel and Waste Management Co., Stockholm, Sweden.

Carbol, P., Fors, P., Gouder, T., and Spahiu, K. 2009. Hydrogen suppresses UO_2 corrosion. *Geochim. Cosmochim. Acta* 73: 4366–4375.

Cera, E., Bruno, J., Duro, L., and Eriksen, T. 2006. Experimental determination and chemical modeling of radiolytic processes at the spent fuel/water interface. Long contact time experiments. SKB Technical Report 06-07, Swedish Nuclear Fuel and Waste Management Co., Stockholm, Sweden.

Chiarizia, R. and Horwitz, E. P. 1986. Hydrolytic and radiolytic degradation of octyl(phenyl)-*N,N*-diisobutyl-carbamoylmethylphosphine oxide and related-compounds. *Solvent Extr. Ion Exc.* 4: 677–723.

Christensen, H. 1998. Calculations simulating spent-fuel leaching experiments. *Nucl. Technol.* 124: 165–174.

Christensen, H. 2006. Calculation of corrosion rates of alpha-doped UO_2. *Nucl. Technol.* 155: 358–364.

Christensen, H. and Bjergbakke, E. 1982a. Radiolysis of groundwater from HLW stored in copper canisters. SKBF/KBS Teknisk rapport 82-02, Swedish Nuclear Fuel Supply Co., Stockholm, Sweden.

Christensen, H. and Bjergbakke, E. 1982b. Radiolysis of ground water from spent fuel. SKBF/KBS Teknisk rapport 82–18, Swedish Nuclear Fuel Supply Co., Stochkolm, Sweden.

Christensen, H. and Bjergbakke, E. 1987. Radiation induced dissolution of UO_2. *Mater. Res. Symp. Proc.* 84: 115–122.

Christensen H. and Sunder, S. 2000. Current state of knowledge of water radiolysis effects on spent nuclear fuel corrosion. *Nucl. Technol.* 131: 102–123.

Christensen, H., Sunder, S., and Shoesmith, D. W. 1994. Oxidation of nuclear fuel (UO_2) by the products of water radiolysis: Development of a kinetic model. *J. Alloys Compd.* 213/214: 93–99.

Clay, P. G. and Witort, M. 1974. Radiolysis of tri-n-butyl phosphate in aqueous solution. *Radiochem. Radioanal. Lett.* 19: 101–107.

Clozel, B., Allard, T., and Muller, J.-P. 1994. Nature and stability of radiation-induced defects in natural kaolinites: New results and a reappraisal of published works. *Clays Clay Miner.* 42: 657–666.

Cobos, J., Havela, L., Rondinella, V. V., De Pablo, J., Gouder, T., Glatz, J. P., Carbol, P., and Matzke, Hj. 2002. Corrosion and dissolution studies of UO_2 containing α-emitters. *Radiochim. Acta* 90: 597–602.

Corbel, C., Sattonnay, G., Luccini, J. F., Ardois, C., Barthe, M. F., Huet, F., Dehaudet, P., Hicke, B., and Jégou, C. 2001. Increase of the uranium release at an UO_2/H_2O interface under He^{2+} ion beam irradiation. *Nucl. Instrum. methods Phys. Res. B* 179: 225–229.

Corbel, B., Sattonnay, G., Guilbert, S., Garrido, F., Barthe, M.-F., and Jégou, C. 2006. Addition versus radiolytic production effects of hydrogen peroxide on aqueous corrosion of UO_2. *J. Nucl. Mater.* 348: 1–17.

Cuillerdier, C., Musikas, C., Hoel, P., Nigond, L., and Vitart, X. 1991. Malonamides as new extractants for nuclear waste solutions. *Sep. Sci. Technol.* 26: 1229–1244.

Daniels, M. 1968. Radiolysis and photolysis of the aqueous nitrate system. *Adv. Chem. Ser.* 82: 153–163.

Davis, W. Jr. 1984. Radiolytic behavior. In *Science and Technology of Tributyl Phosphate*, W.W. Schulz and J.D. Navratil, (eds.), vol. 1, pp. 221–266. Boca Raton, FL: CRC Press.

Dewhurst, H. A. 1958. Radiation chemistry of organic compounds. III. Branched chain alkanes. *J. Chem. Soc.* 80: 5607–5610.

Dorfman, L. M. and Adams, G. E. 1973. Reactivity of the hydroxyl radical in aqueous solution. National Standard Reference Data Series, National Bureau of Standards (US), Washington, DC.

Dzengel, J., Theurich, J., and Bahnemann, D. W. 1999. Formation of nitroaromatic compounds in advanced oxidation processes: Photolysis versus photocatalysis. *Environ. Sci. Technol.* 33: 294–300.

Eberhardt, M. K. 1975. Radiation-induced homolytic aromatic substitution. III. Hydroxylation and nitration of benzene. *J. Phys. Chem.* 79: 1067–1069.

Egorov, G. F., Afanas'ev, O. P., Zilberman, B. Y., and Makarychev-Mikhailov, M. N. 2002. Radiation-chemical behavior of TBP in hydrocarbon and chlorohydrocarbon diluents under conditions of reprocessing of spent fuel from nuclear power plants. *Radiochemistry* 44: 151–156.

Egorov, G. F., Tkhorzhnitskii, G. P., Zilberman, B. Y., Schmidt, O. V., and Goletskii, N. D. 2005. Radiation chemical behavior of tributyl phosphate, dibutylphosphoric acid, and its zirconium salt in organic solutions and two-phase systems. *Radiochemistry* 47: 392–397.

Ekeroth, E., and Jonsson, M. 2003. Oxidation of UO_2 by radiolytic oxidants. *J. Nucl. Mater.* 322: 242–248.

Ekeroth, E., Roth, O., and Jonsson, M. 2006. The Relative impact of radiolysis products in radiation induced oxidative dissolution of UO_2. *J. Nucl. Mater.* 355: 38–46.

Elliot, A. J. 1989. A pulse-radiolysis study of the temperature-dependence of reactions involving H, OH and e_{aq}^- in aqueous-solutions. *Radiat. Phys. Chem.* 34: 753–758.

Elliot, A. J. 1994. Rate constants and G-values for the simulation of radiolysis of light water over the range 0–300°C. AECL-11073. Atomic Energy of Canada Ltd., Chalk River, ON.

Elliot, A. J., Chenier, M. P., and Quellette, D. C. 1993. Temperature dependence of g values for H_2O and D_2O irradiated with low linear energy transfer radiation. *J. Chem. Soc. Faraday Trans.* 89: 1193–1197.

Eriksen, T. 1996. Radiolysis of water within a ruptured fuel element. SKB Progress Report U-96-29, Swedish Nuclear Fuel and Waste Management Co., Stockholm, Sweden.

Eriksen, T. and Jacobsson, A. 1983. Radiation effects on the chemical environment in a radioactive waste repository. SKBF/KBS Teknisk rapport 83–27, Swedish Nuclear Fuel Supply Co., Stockholm, Sweden.

Eriksen, T. E. and Ndalamba, P. 1988. On the formation of a moving redox-front by α-radiolysis of compacted water saturated bentonite. SKB-Technical Report TR-88-27, Swedish Nuclear Fuel and Waste Management Co., Stockholm, Sweden.

Eriksen, T. E., Christensen, H., and Bjergbakke, E. 1987. Hydrogen production in α-irradiated bentonite. *J. Radioanal. Nucl. Chem.* 116: 13–25.

Eriksen, T., Eklund, U.-B., Werme, L., and Bruno, J. 1995. Dissolution of irradiated fuel: A radiolytic mass balance study. *J. Nucl. Mater.* 227: 76–82.

Ershov, B. G., Gordeev, A. V., Bykov, G. L., Zubkov, A. A., and Kosareva, I. M. 2007. Synthesis of nitromethane from acetic acid by radiation-induced nitration in aqueous solution. *Mendeleev Commun.* 17: 289–290.

European Commission. 2007. Integrated in situ corrosion test on α-active high-level waste (HLW) glass—Phase 2 (CORALUS-2). Contract No FIKW-CT-2000-00011 Final report.

Gasparini, G. M. and Grossi, G. 1986. Long-chain disubstituted aliphatic amides as extracting agents in industrial applications of solvent-extraction. *Solvent Extr. Ion Exc.* 4: 1233–1271.

Gogolev, A. V., Shilov, V. P., Fedoseev, A. M., and Pikaev, A. K. 1986. A pulse-radiolysis study of the reactivity of neptunoyl ions relative to inorganic free-radicals. *Bull. Acad Sci. USSR. Div. Chem. Sci.* 35, 422–424.

Gogolev, A. V., Shilov, V. P., Fedoseev, A. M., Makarov, I. E., and Pikaev, A. K. 1988. Reactivity of neptunium and plutonium with inorganic free radicals in aqueous solutions. *Sov. Radiochem.* 30: 721–725.

Gogolev, A. V., Shilov, V. P., Fedoseev, A. M., and Pikaev, A. K. 1991. The study of reactivity of actinide ions towards hydrated electrons and hydrogen atoms in acid aqueous solutions by a pulse radiolysis method. *Radiat. Phys. Chem.* 37: 531–535.

Gordon, S., Hart, E. J., and Thomas, J. K. 1964. The ultraviolet spectra of transients produced in the radiolysis of aqueous solutions. *J. Phys. Chem.* 68: 1262–1264.

Gournis, D., Mantaka-Marketou, A. E., Karakassides, M. A., and Petridis, D. 2000. Effect of γ-irradiation on clays and organoclays: A Mössbauer and XRD study. *Phys. Chem. Miner.* 27: 514–521.

Grätzel, M., Henglein, A., Lilie, J., and Beck, G. 1969. Pulsradiolysische Untersuchung einiger Elementarprozesse der Oxydation und Reduktion des Nitritions. *Ber. Bunsenges. Phys. Chem.* 73: 646–653.

Grätzel, M., Henglein, A., and Taniguchi, S. 1970. Pulsradiolsytische beobachtungen über die Reduktion des NO_3^- Ions und über Bildung und erfall der persalpetrigen Säure in wässriger Lösung. *Ber. Bunsenges. Phys. Chem.* 74: 292–298

Gromov, V. 1981. Dissolution of uranium oxides in the gamma-radiation field. *Radiat. Phys. Chem.* 18: 136–146.

Gu, B. X., Wang, L. M., Minc, L. D., and Ewing, R. C. 2001. Temperature effects on the radiation stability and ion exchange capacity of smectites. *J. Nucl. Mater.* 297: 345–354.

Haase, K. D., Schulte-Frohlinde, D., Kouřím, P., and Vacek, K. 1973. Low-temperature radiolysis of organic phosphates studied by electron spin resonance. *Radiat. Phys. Chem.* 5: 351–360.

He, H., Lin, M. Z., Muroya, Y., Kudo, H., and Katsumura, Y. 2004. Laser photolysis study on the reaction of nitrate radical with tributylphosphate and its analogues—Comparison with sulfate radical. *Phys. Chem. Chem. Phys.* 6: 1264–1268.

Hemmi, Y., Uruma, Y., and Ichikawa, N. 1994. *J. Nucl. Sci. Technology* 31:443–455.

Huang, H. X., Zhu, G. H., and Hou, S. B. 1989. Retention of ruthenium-nitrosyl complexes in 30-percent TBP-kerosene-laurohydroxamic acid. *Radiochim. Acta* 46: 159–162.

Hurst, P. and Tyzack, C. 1974. Water radiolysis its effect upon in-reactor zircaloy corrosion. In *Proceedings of BNES Conference on Effect of Environment on Material Properties in Nuclear Systems*, 37, London, U.K: BNES, London, U.K.

IAEA. 2004. Implications of partitioning and transmutation in radioactive waste management. Technical Report Series, Vienna, Austria.

Ichikawa, N. and Takagi, J. 1998. Effect of g-value and reactions on radiolysis simulation of high temperature water. In *Proceedings 1998 JAIF Water Chemistry in Nuclear Power Plants*, Kashiwazaki, Japan, pp. 854–857. Tokyo: Japan Atomic Industrial Forum.

Ichikawa, N., Hemmi, Y., Fukushima, T., Uruma, Y., Kamata, T., Nakayama, Y., and Nagao, H. 1987. Effect of dissolved oxygen on corrosion behavior of stainless steell in gamma ray irradiation environment. In *Proceedings of the Third International Symposium on Environmental Degradation of Materials in Nuclear Power Systems—Water Reactors*, G. J. Theus and J. R. Weeks (eds.), Traverse City, MI, pp. 323–331. AIME.

Ichikawa, N., Hemmi, Y., and Takagi, J. 1992. Estimation on corrosion potential of stainless steel in BWR primary circuit. In *Proceedings of the Water Chemistry of Nuclear Reactor Systems 6*, Bournemouth, U.K., vol. 2, pp. 127–132. London, U.K.: BNES.

Ichikawa, N., Hemmi Y., and Takagi, J. 1994. Effects of water chemistry and water flowrate on corrosion potential of stainless steel in high temperature water. In *Proceedings of the International Conference on Chemistry in Water Reactors: Operating Experience & New Developments*, vol. 1, Nice, France, pp. 307–314. Paris, France: SFEN.

Ichikawa, N., Hemmi, Y., and Takagi, J. 1996. Evaluation of hydrogen water chemistry effectiveness on materials in reactor pressure vessel bottom of a BWR. In *Proceedings of the Water Chemistry of Nuclear Reactor Systems 7*, vol. 2, Bournemouth, U.K., pp. 230–235. London, U.K.: BNES.

Ichikawa, N., Nakada, K., and Takagi, J. 2002. Precise evaluation of rater radiolysis and ECP under complex water flow condition. In *Proceedings of the Water Chemistry of Nuclear Reactor Systems 2002*, Avignon, France, paper 3-06. Paris, France: SFEN.

IRI. 2005. Radiation effects on barrier systems in high level radioactive waste disposal, Research Project Annual Report, Institute of Research and Innovation, Tokyo, Japan.

Ishigure, K., Fujita, N., Tamura, T., and Oshima, K. 1980. Effect of gamma radiation on the release rate of corrosion products from carbon steel and stainless steel in high temperature water. *Nucl. Technol.* 50: 169–177.

Ito, M., Okamoto, M., Shibata, M., Sasaki, Y., Danhara, T., Suzuki, K., and Watanabe, T. 1993. Mineral composition analysis of bentonite. PNC TN8430 93-003 (in Japanese). Power Reactor and Nuclear Fuel Development Co., Tokyo, Japan.

Jégou, C., Broudic, A., Poulesquen, A., and Bart, J. M. 2004. Effect of α and γ radiolysis of water on alteration of the spent UO_2 nuclear fuel matrix. *Mater. Res. Soc. Symp. Proc.* 807: 391–396.

Jiang, P. Y., Nagaishi, R., Yotsuyanagi, T., Katsumura, Y., and Ishigure, K. 1994. γ-radiolysis study of concentrated nitric-acid solutions. *J. Chem. Soc.-Faraday Trans.* 90: 93–95.

Jin, H. F., Wu, J. L., Zhang, X. J., Xingwang, F., Side, Y, Zhihua, Z., Lin, N., ianyun. 1999. The examination of TBP excited state by pulse radiolysis. *Radiat. Phys. Chem.* 54: 245–251.

JNC. 2000. H12: Project to establish the scientific and technical basis for HLW disposal in Japan. Supporting Report 2: Repository Design and Engineering Technology. JNC-TN1410-2000-003, Japan Nuclear Cycle Development Institute, Tokaimura, Japan.

JNC. 2005. H17: Development and management of the technical knowledge base for the geological disposal of HLW. Knowledge Management Report, JNC-TN1400-2005-022, Japan Nuclear Cycle Development Institute, Tokaimura, Japan.

Johnson L. H. and Shoesmith, D. W. 1988. Spent fuel. In *Radioactive Waste Forms for the Future*, W. B. Lutze and R. C. Ewing (eds.), pp. 635–698. Amsterdam, the Netherlands: Elsevier Scientific Publishers.

Johnson L. H. and Smith, P. A. 2000. The interaction of radiolysis products and canister corrosion products and the implications for spent fuel dissolution and radionuclide transport in a repository for spent fuel. Nagra NTB 00-04, Nagra, Wettingen, Switzerland.

Johnson, L. H., Stroes-Gascoyne, S., Shoesmith, D. W., Bailey, M. G., and Sellinger, D. M. 1983. Leaching and radiolysis studies on UO_2 fuel. AECL-8175, Atomic Energy of Canada Ltd., Ottawa, Canada.

Jonsson, M., Nielsen, F., Roth, O., Ekeroth, E., Nilsson, S., and Hossain, M. M. 2007. Radiation induced spent nuclear fuel dissolution under deep repository. *Environ. Sci. Technol.* 41: 7087–7093.

Kalina, D. G., Horwitz, E. P., Kaplan, L., and Muscatello, A. C. 1981. The extraction of Am(III) and Fe(III) by selected dihexyl *N,N*-dialkylcarbamoylmethyl-phosphonates, phosphinates and phosphine oxides from nitrate media. *Sep. Sci. Technol.* 16: 1127–1145.

Katsumura, Y. 1998. and radicals in radiolysis of nitric acid solutions. In *N-Centered Radicals*, Z. B. Alfassi (ed.), pp. 393–412. New York: John Wiley & Sons.

Katsumura, Y., Jiang, P. Y., Nagaishi, R., Oishi, T., Ishigure, K., and Yoshida, Y. 1991. Pulse-radiolysis study of aqueous nitric-acid solutions—Formation mechanism, yield, and reactivity of NO_3 radical. *J. Phys. Chem.* 95: 4435–4439.

Kelm, M. and Bohnert, E. 2000a. Radiation chemical effects in the near field of a final disposal site—I: Radiolytic products formed in concentrated NaCl solutions. *Nucl. Technol.* 129: 119–122.

Kelm, M. and Bohnert, E. 2000b. Radiation chemical effects in the near field of a final disposal site—II: Simulation of the radiolytic processes in concentrated NaCl solutions. *Nucl. Technol.* 129: 123–130.

Khaikin, G. I. 1998. Reactions of trialkyl phosphates with hydroxyl radicals and hydrated electrons. *High Energy Chem.* 32: 287–289.

King F. and Shoesmith, D. W. 2004. Electrochemical studies of the effect of H_2 on UO_2 dissolution. SKB Technical Report TR-04-20, Swedish Nuclear Fuel and Waste Management Co., Stockholm, Sweden.

King, F., Litke, C. D., and Ryan, S. R. 1992. A mechanistic study of the uniform corrosion of copper in compacted Na-montorillonite/sand mixtures. *Corrosion Sci.* 33: 1979–1995.

King, F. Ahonen, L., Taxén, C., Vuorinen, U., and Werme, L. 2001. Copper corrosion under expected conditions in a deep geologic repository. SKB Technical Report TR-01-23, Swedish Nuclear Fuel and Waste Management Co., Stockholm, Sweden.

Ladriere, J., Dussart, F., Dabi, J., Haulotte, O., Verhaeghe, S., and Regout, J. 2009. Mössbauer study of the boom clay, a geological formation for the storage of radioactive wastes in Belgium. *Hyperfine Interact.* 191: 1–9.

Laidler, J. J. 2008. An overview of spent-fuel processing in the global nuclear-energy partnership. In *ISEC 2008, International Solvent Extraction Conference*, B. A. Moyer (ed.), Tucson, AZ, pp. 695–702. Cognis Corporation.

Lammel, G., Perner, D., and Warneck, P. 1990. Decomposition of pernitric acid in aqueous-solution. *J. Phys. Chem.* 94: 6141–6144.

Lane, E. S. 1963. Performance and degradation of diluents for TBP and the cleanup of degraded solvents. *Nucl. Sci. Eng.* 17: 620–625.

LaVerne, J. A. 2005. H_2 formation from the radiolysis of liquid water with zirconia. *J. Phys. Chem. B* 109: 5395–5397.

LaVerne, J. A., and Tonnies, S. E. 2003. H_2 production in the radiolysis of aqueous SiO_2 suspensions and slurries. *J. Phys. Chem. B* 107: 7277–7280.

Lefort, M. and Tarrago, X. J. 1959. Radiolysis of water by particles of high linear energy transfer. The primary chemical yields in aqueous acid solutions of ferrous sulfate, and in mixtures of thallous and ceric ions. *J. Phys. Chem.* 63: 833–836.

Lesage, D., Virelizier, H., and Jankowski, C. K. 1997. Identification of minor products obtained during radiolysis of tributylphosphate (TBP). *Spectrose: Int. J.* 13: 275–290.

Lierse, C., Schmidt, K. H., and Sullivan, J. C. 1988. Reactions of radiolytically produced OH radicals with selected actinides in solution. *Radiochim. Acta* 44/45: 71–72.

Linacre, J. K. and Marsh, W. R. 1981. The radiation chemistry of heterogeneous and homogeneous nitrogen and water systems. AERE-R 10027, HMSO, London, U.K.

Liu, J. and Neretnieks, I. 2001a. Effect of water radiolysis caused by dispersed radionuclides on oxidative dissolution of spent fuel in a final repository. *Nucl. Technol.* 135: 154–161.

Liu, L. and Neretnieks, I. 2001b. A coupled model for oxidative dissolution of spent fuel and transport of radionuclides from an initially defective canister. *Nucl. Technol.* 135: 273–285.

Liu, L. and Neretnieks, I. 2001c. A reactive transport model for oxidative dissolution of spent fuel and release of nuclides within a defective canister. *Nucl. Technol.* 137: 228–240.

Liu, L. and Neretnieks, I. 2002. The effect of hydrogen on oxidative dissolution of spent fuel. *Nucl. Technol.* 138: 69–78.

Liu, J., Neretnieks, I., and Strömberg, B. H. E. 2003. Study on the consequences of secondary water radiolysis surrounding a defective canister. *Nucl. Technol.* 142: 294–305.

Løgager, T. and Sehested, K. 1993. Formation and decay of peroxynitrous acid: a pulse radiolysis study. *J. Phys. Chem.* 97: 6664–6669.

Loida, A., Metz, V., Kienzler, B., and Geckeis, H. 2005. Radionuclide release from high burnup spent fuel during corrosion in salt brine in the presence of hydrogen overpressure. *J. Nucl. Mater.* 346: 24–31.

Lucuta, P. G., Verrall, R. A., Matzke, Hj, and Palmer, B. J. 1991. Microstructural features of SIMFUEL—Simulated high-burnup UO_2-based nuclear fuel. *J. Nucl. Mater.* 178: 48–60.

Macadamas, W. H. 1954. *Heat Transmission*, 3rd edn. New York: McGraw Hill.

Macášek, F. and Čech, R. 1984. Macrokinetics of radiolysis in systems with liquid-liquid partition of substrates. 1. A general-approach to mathematical-models of simulated solvent-extraction systems. *Radiati Phys. Chem* 23: 473–479.

Macdonald, D. D. 1992. Viability of hydrogen water chemistry for protecting in-vessel components of boiling water reactors. *Corrosion* 48: 194–205.

Macdonald, D. D. and Urquidi-Macdonald, M. 1990. Thin-layer mixed-potential model for the corrosion of high-level nuclear waste canisters. *Corrosion* 46:380–390.

Marsh, G. P., Harker, A. H., and Taylor, K. J. 1989. Corrosion of carbon steel nuclear waste containers in marine sediment. *Corrosion* 45: 579–589.

Martínez Esparza, A., Cuñado, M. A., Gago, J. A., Quiñones, J., Iglesias, E., Cobos, J., González de la Huebra A., Cera, E., Merino, J., Bruno, J., de Pablo, J., Casas, I., Clarens, F., and Giménez, J. 2005. Development of a matrix alteration model(MAM). ENRESA PT 01/2005, Nacional Radioactive Waste Company, Madrid, Spain.

Mathur, J. N., Murali, M. S., and Nash, K. L. 2001. Actinide partitioning—A review, *Solvent Extr. Ion Exch.* 19: 357–390.

Matthews, R. W., Mahlman, H. A., and Sworski, T. J. 1972. Elementary processes in the radiolysis of aqueous nitric acid solutions. Determination of both G_{OH} and G_{NO_3}. *J. Phys. Chem.* 19: 2680–2684.

McVay, G. L. and Pederson, L. R. 1981. Effect of gamma radiation on glass leaching. *J. Am. Ceram. Soc.* 64: 154–158.

Mennecart, T., Grambow, B., Fattahi, M., Blondiaux, G., and Andriambololona, Z. 2004a. Effect of alpha radiolysis on UO_2 dissolution under reducing conditions. *Mat. Res. Soc. Symp. Proc.* 807: 403–408.

Mennecart, T., Grambow, B., Fattahi, M., and Andriambololona, Z. 2004b. Effect of alpha radiolysis on UO_2 dissolution under reducing conditions. *Radiochim. Acta* 92: 611–615.

Metz, V., Loida, A., Bohnert, E., Schild, D., and Dardenne, K. 2008. Effects of hydrogen and bromide on the corrosion of spent nuclear fuel and γ-irradiated $UO_2(s)$ in NaCl brine. *Radiochim. Acta* 96: 637–648.

Miguirditchian, M., Sorel, C., Camès, B., Bisel, I., and Baron, P., 2008. Extraction of uranium(VI) by N,N-di-(ethylhexyl)isobutyramide (DEHIBA): From the batch experimental data to the countercurrent process, In *ISEC 2008, International Solvent Extraction Conference*, B. A. Moyer (ed.), Tucson, AZ, pp. 721–726.

Mincher, B. J. and Mezyk, S. P. 2009. Radiation chemical effects on radiochemistry: A review of examples important to nuclear power. *Radiochim. Acta* 97: 519–534.

Mincher, B. J., Mezyk, S. P., and Martin, L. R. 2008. A pulse radiolysis investigation of the reactions of tributyl phosphate with the radical products of aqueous nitric acid irradiation. *J. Phys. Chem. A* 112: 6275–6280.

Mincher, B. J., Modolo, G., and Mezyk, S. P. 2009a. Review article: The effects of radiation chemistry on solvent extraction: 2. A review of fission-product extraction. *Solvent Extr. Ion Exch.* 27: 331–353.

Mincher, B. J., Elias, G., Martin, L. R., and Mezyk, S. P. 2009b. Radiation chemistry and the nuclear fuel cycle. *J. Radioanal. Nucl. Chem.* 282: 645–649.

Mincher, B. J., Modolo, G., and Mezyk, S. P. 2009c. The effects of radiation chemistry on solvent extraction: 3, A review of actinide and lanthanide extraction, *Solvent. Extr. Ion Exch.* 27: 579–606.

Miyake, C., Hirose, M., Yoshimura, T., Ikeda, M., Imoto, S., and Sano, M. 1990. The 3rd phase of extraction processes in fuel-reprocessing.1. Formation conditions, compositions and structures of precipitates in Zr-degradation products of TBP systems. *J. Nucl. Sci. Technol.* 27: 157–166.

Modolo, G., Vijgen, H., Malmbeck, R., Magnusson, D., and Sorel, C. 2008. Partitioning of trivalent actinides from a PUREX raffinate using a TODGA-based solvent-extraction process, In *ISEC 2008, International Solvent Extraction Conference*, B. A. Moyer (ed.), Tucson, AZ, pp. 521–526.

Mowafy, E. A. 2004. The effect of previous gamma-irradiation on the extraction of U(VI), Th(IV), Zr(IV), Eu(III) and Am(III) by various amides. *J. Radioanal. Nucl. Chem.* 260: 179–187.

Mowafy, E. A. 2007. Evaluation of selectivity and radiolysis behavior of some promising isonicotinamids and dipicolinamides as extractants. *Radiochim. Acta* 95: 539–545.

Müller-Vonmoos, M. and Kahr, G. 1983. Mineralogische untersuchungen von Wyoming bentonite MX-80 und Montigel. NAGRA Technischer Bericht 83–12, Wettingen, Switzerland.

Muroya, Y., Lin, M., Han, Z., Kumagai, Y., Sakumi, A., Ueda, T., and Katsumura, Y., 2008. Ultra-fast pulse radiolysis: A review of the recent system progress and its application to study on initial yields and solvation processes of solvated electrons in various kinds of alcohols. *Radiat. Phys. Chem.* 77: 1176–1182.

Muzeau, B., Jégou, C., Delaunay, F., Broudic, V., Brevet, A., Catalette, H., Simoni, E., and Corbel, C. 2009. Radiolytic oxidation of UO_2 pellets doped with alpha-emitters ($^{238/239}$Pu). *J. Alloy Comp.* 467: 578–589.

Nagaishi, R., Jiang, P. Y., Katsumura, Y., and Ishigure, K. 1994. Primary yields of water radiolysis in concentrated nitric-acid solutions. *J. Chem. Soc.-Faraday Trans.* 90: 591–595.

Nash, K. L., Gatrone, R. C., Clark, G. A., Rickert, P. G., and Horwitz, E. P. 1988. Hydrolytic and radiolytic degradation of O-Phi-D(Ib) Cmpo—continuing studies. *Sep. Sci. Technol.* 23: 1355–1372.

Nash, K. L., Rickert, P. G., and Horwitz, E. P. 1989. Degradation of truex-dodecane process solvent. *Solvent Extr. Ion Exch.* 7: 655–675.

Nash, K. L., Madic, C., Mathur, J., and Lacquement, J. 2006. Actinide separation science and technology. In *The Chemistry of the Actinide and Transactinide Elements*, 3rd edn., L. R. Morss, N. M. Edelstein, J. Fuger (eds.), Dordrecht, the Netherlands: Springer.

Neace, J. C. 1983. Diluent degradation products in the purex solvent. *Sep. Sci. Technol.* 18: 1581–1594.

Negron, A., Ramos, S., Blumenfeld, A. L., Pacheco, G., and Fripiat, J. J. 2002. On the structural stability of montmorillonite submitted to heavy γ-irradiation. *Clays Miner.* 50: 35–37.

Neretnieks, I. 1983. The movement of a redox front downstream from a repository for nuclear waste. *Nucl. Technol.* 62: 110–115.

Neta, P. and Huie, R. E. 1986. Rate constants for reactions of NO_3 radicals in aqueous-solutions. *J. Phys. Chem.* 90: 4644–4648.

Neta, P., Huie, R. E., and Ross, A. B. 1988. Rate constants for reactions of inorganic radicals in aqueous-solution. *J. Phys. Chem. Ref. Data* 17: 1027–1284.

Nicol, C., Cames, B., Margot, L., and Ramain, L. 2000. DIAMEX regeneration studies. In *Atalante-2000*, P3-22, Avignon, France, pp. 1–4.

Nielsen, F. and Jonsson, M. 2006. Geometrical α- and β-dose distributions and production rates of radiolysis products in water in contact with spent nuclear fuel. *J. Nucl. Mater.* 359: 1–7.

Nilsson, S. and Jonsson, M. 2008. On the catalytic effects of $UO_2(s)$ and Pd(s) on the reaction between H_2O_2 and H_2 in aqueous solution. *J. Nucl. Mater.* 372: 160–163.

Nielsen, F., Lundahl, K., and Jonsson, M. 2008a. Simulations of H_2O_2 concentration profiles in the water surrounding spent nuclear fuel. *J. Nucl. Mater.* 372: 32–35.

Nielsen, F., Ekeroth, E., Eriksen, T. E., and Jonsson, M. 2008b. Simulation of radiation induced dissolution of spent nuclear fuel using the steady-state approach. A comparison to experimental data. *J. Nucl. Mater.* 374: 286–289.

Nowak, Z. 1977. Radiolytic degradation of extractant-diluent systems used in the PUREX process. *Nukleonika* 22: 155–172.

Oda, C., Shibata, M., and Yui, M. 1999. Calculation of porewater chemistry in compacted bentonite for 2nd progress report. JNC TN8400 99-078 (in Japanese), Japan Nuclear Cycle Development Institute, Tokai-mura, Japan.

OECD-NEA. 1999. Actinide and Fission Product Partitioning and Transmutation—Status and Assessment Report, OECD Nuclear Energy Agency, Paris, France.

Ohwada, K. 1968. On the identification of hydroxamic acids formed by nitric acid degradation of kerosene and *i*-dodecane. *J. Nucl. Sci. Technol.* 5: 163–167.

Olah, G. A., Lin, H. C., Olah, J. A., and Narang, S. C. 1978. Electrophilic and free radical nitration of benzene and toluene with various nitrating agents. *Proc. Natl. Acad. Sci. U S A* 75: 1045–1049.

Ollila, K. 1992. SIMFUEL dissolution studies in granitic groundwater. *J. Nucl. Mater.* 190: 70–77.

Ollila, K., Albinsson, Y., Oversby, V., and Cowper, M. 2003. Dissolution rates of unirradiated UO_2, UO_2 doped with ^{233}U, and spent fuel under normal atmospheric conditions and under reducing conditions using an isotope dilution method. SKB Technical Report TR-03-13, Swedish Nuclear Fuel and Waste Management Co., Stockholm, Sweden.

Park, J. Y. and Lee, Y. N. 1988. Solubility and decomposition kinetics of nitrous-acid in aqueous-solution. *J. Phys. Chem.* 92: 6294–6302.

Pastina, B. and LaVerne, J. A. 1999. Hydrogen peroxide production in the radiolysis of water with heavy ions. *J. Phys. Chem. A* 103: 1592–1597.

Pastina, B. and LaVerne, J. A. 2001. Effect of molecular hydrogen on hydrogen peroxide in water radiolysis. *J. Phys. Chem. A* 105:9316–9322.

Pederson, L. R. and McVay, G. L. 1983. Influence of gamma irradiation on leaching of simulated nuclear waste glass: Temperature and dose rate dependence in deaerated water. *J. Am. Ceram. Soc.* 66: 863–867.

Peuget, S., Broudic, V., Jégou, C., Frudier, P., Roudil, D., Deschantels, X., Rabiller, H., and Noel, P. M. 2007. Effect of alpha radiation on the leaching behavior of nuclear glass. *J. Nucl. Mater.* 362: 474–479.

Pikaev, A. K., Kabakchi, S. A., and Egorov, G. F. 1988. Some radiation chemical aspects of nuclear-engineering. *Radiat. Phys. Chem.* 31: 789–803.

Plotze, M., Kahr, G., and Hermanns Stengele, R. 2003. Alteration of clay minerals-gamma-irradiation effects on physicochemical properties. *Appl. Clay Sci.* 23: 195–202.

Poinssot, C., Ferry, C., Lovera, P., Jegou, C., and Gras, J.-M. 2005. Spent fuel radionuclide source term model for assessing spent fuel performance in geological disposal. Part II: Matrix alteration model and global performance. *J. Nucl. Mater.* 346: 66–77.

Poulesquen, A. and Jégou, C. 2007. Influence of alpha radiolysis of water on UO_2 matrix alteration: Chemical/transport model. *Nucl. Technol.* 160: 337–345.

Pusch, R., Karnland, O., Lajudie, A., and Decarreau, A. 1992. MX80 clay exposed to high temperatures and gamma radiation. SKB Technical Report 93-03, Swedish Nuclear Fuel and Waste Management Co., Stockholm, Sweden.

Pushkareva, R., Kalinichenko, E., Lytovchenko, A., Pushkarev, A., Kadochnikov, V., and Plastynina, M. 2002. Irradiation effect on physic-chemical properties of clay minerals. *Appl. Clay Sci.* 21: 117–123.

Reif, D. J. 1988. Restoring solvent for nuclear separation processes. *Sep. Sci. Technol.* 23: 1285–1295.

Rochoñ, A. M. 1980. Study on composition of zirconium binding-compounds in irradiated purex-solvents. *Radiochem. Radioanal. Lett.* 44: 277–285.

Röllin, S., Spahiu, K., and Eklund, U.-B. 2001. Determination of dissolution rates of spent fuel in carbonate solutions under different redox conditions with a flow-through experiment. *J. Nucl. Mater.* 297: 231–243.

Rondinella, V. V., Matzke, Hj., Cobos, J., and Wiss, T. 2000. Leaching behavior of UO_2 containing α-emitting actinides. *Radiochim. Acta* 88: 527–531.

Roth, O. and Jonsson, M. 2009. On the impact of reactive solutes on radiation induced oxidative dissolution of UO_2. *J. Nucl. Mater.* 385: 595–600.

Ruikar, P. B., Nagar, M. S., and Subramanian, M. S. 1991. Quantitative-analysis of Some extractants by infrared spectrometry. *J. Radioanal. Nucl. Chem.-Lett.* 155: 371–376.

Ruikar, P. B., Nagar, M. S., and Subramanian, M. S. 1993. Extraction of uranium, plutonium and some fission-products with gamma-irradiated unsymmetrical and branched-chain dialkylamides. *J. Radioanal. Nucl. Chem.-Lett.* 176: 103–111.

Saito, N., Ichikawa, N., Hemmi, Y., Sudo, A., Ito, M., and Okada, T. 1990. Effects of gamma-ray irradiation and sodium sulfate on the intergranular stress corrosion cracking susceptibility of sensitized type 304 stainless steel in high-temperature water. *Corrosion* 46: 531–536.

Saito, N., Kikuchi, E., Sakamoto, H., Kuniya, J., and Suzuki, S. 1997. Susceptibility of sensitized type 304 stainless steel to intergranular stress corrosion cracking in simulated boiling-water reactor environments. *Corrosion* 53: 537–545.

Sasaki, Y., Kitatsuji, Y., Hirata, M., Kimura, T., and Yoshizuka, K. 2008. Extraction of actinides by multidentate diamides and their evaluation with computational molecular modeling, In *ISEC 2008, International Solvent Extraction Conference, 2008*, B. A. Moyer (ed.), Tucson, AZ, pp. 745–750.

Sattonnay, C., Ardois, C., Corbel, C., Lucchini, J. F., Barthe, M. F., Garrido, F., and Gosset, D. 2001. Alpha-radiolysis effects on UO_2 alteration in water. *J. Nucl. Mater.* 288: 11–19.

Schatz, T., Cook, A. R., and Meisel, D. 1998. Charge carrier transfer across the silica nanoparticle/water interface. *J. Phys. Chem. B* 102: 7225–7230.

Schmidt, K. H., Gordon, S., Thompson, R. C., and Sullivan, J. C. 1980. A pulse radiolysis study of the reduction of neptunium (V) by the hydrated electron. *J. Inorg. Nucl. Chem.* 42: 611–615.

Schmidt, K. H., Gordon, S., Thompson, M., Sullivan, J. C., and Mulac, W. A. 1983. The hydrolysis of neptunium(VI) and plutonium(VI) studied by the pulse radiolysis transient conductivity technique. *Radiat. Phys. Chem.* 21: 321–328.

Schulz, W. W. and Navratil, J. D. 1984. *Science and Technology of Tributyl Phosphate*, vol. 1, Boca Raton, FL: CRC Press.

Selman, J. R. and Tobias, C. W. 1978. Mass-transfer measurements by the limiting-current technique. In *Advances in Chemical Engineering*, T. B. Drew (ed.), vol. 10, pp. 212–216. New York: Academic Press.

Shastri, L. V. and Huie, R. E. 1990. Rate constants for hydrogen abstraction reactions of NO_3 in aqueous-solution. *Int. J. Chem. Kinet.* 22: 505–512.

Shilov, V. P. and Pikaev, A. K. 1982. Investigation of the reactivity of uranium and neptunium ions relative to hydrogen atoms in aqueous-solutions by the pulse radiolysis method. *High Energy Chem.* 16: 372–374.

Shilov, V. P., Fedoseev, A. M., and Pikaev, A. K. 1982. Study of reactivity of neptunium ions toward OH radicals in perchloric-acid solutions by pulse-radiolysis method. *Bull. Acad. Sci. USSR, Div. Chem. Sci.* 31: 832–834.

Shoesmith, D. W. 2000. Fuel corrosion processes under waste disposal conditions. *J. Nucl. Mater.* 282: 1–31.

Shoesmith, D. W. 2007. Used fuel on uranium dioxide dissolution studies—A review. NWMO TR-2007-03, Nuclear Waste Management Organization, Toronto, Canada.

Shoesmith, D.W. 2008. The role of dissolved hydrogen on the corrosion/dissolution of spent nuclear fuel. NWMO TR-2008-19, Nuclear Waste Management Organization, Toronto, Canada.

Shoesmith, D. W. and King, F. 1999. The effects of gamma radiation of the candidate materials for the fabrication of nuclear waste packages. AECL-11999, Atomic Energy of Canada Ltd., Pinawa, Canada.

Shoesmith, D. W., Kolar, M., and King, F. 2003. A mixed-potential model to predict fuel (uranium dioxide) corrosion within a failed nuclear waste container. *Corrosion* 59: 802–816.

Siddall, T. H. and Dukes, E. K. 1959. Kinetics of HNO_2 catalyzed oxidation of neptunium(V) by aqueous solutions of nitric acid. *J. Am. Chem. Soc.* 81: 790–794.

SKB. 2006a. Long-term safety for KBS-3 Reositories at Forsmark and Laxemar – a First Evaluation. Main Report of the SR-Can project. SKB Technical Report TR-06-09, Swedish Nuclear Fuel and Waste Management Co., Stockholm, Sweden.

SKB. 2006b. Fuel and canister process report for the safety assessment SR-Can. SKB Technical Report TR-06-22, Swedish Nuclear Fuel and Waste Management Co., Stockholm, Sweden.

SKI. 1996. Deep Repository Performance Assessment Project. SKI Report 96:36, Swedish Nuclear Power Inspectorate, Stockholm, Sweden.

Smart, N. R., Rance, A. P., and Werme, L. O. 2008. The effect of radiation on the anaerobic corrosion of steel. *J. Nucl. Mater.* 379: 97–104.

Sorel, C., Montuir, M., Espinoux, D., Lorrain, B., and Baron, P. 2008. Technical feasibility of the DIAMEX process. In *ISEC 2008, International Solvent Extraction Conference*, B. A., Moyer (ed.), Tucson, AZ, pp. 715–720.

Sorieul, S., Allard, Th., Wang, L. M., Grambin-Lapeyre, C., Lian, J., Calas, G., and Ewing, R. C. 2008. Radiation stability of smectite. *Environ. Sci. Technol.* 42: 8407–8411.

Spahiu, K., Werme, L., and Eklund, U.-B. 2000. The influence of near field hydrogen on actinide solubilities and spent fuel leaching. *Radiochim. Acta* 88: 507–511.

Spahiu K., Cui, D., and Lundström, M. 2004a. The fate of radiolytic oxidants during spent fuel leaching in the presence of dissolved near field hydrogen. *Radiochim. Acta* 92: 625–629.

Spahiu K., Devoy, J., Cui, D., and Lundström, M. 2004b. The reduction of U(VI) by near field hydrogen in the presence of UO_2(s). *Radiochim. Acta* 92: 597–601.

Spinks, J. W. T. and Woods, R. J. 1990. *An Introduction to Radiation Chemistry*, 3rd edn. New York: John Wiley & Sons.

Stieglitz, L. and Becker, R. 1985. Chemical and radiolytic solvent degradation in the purex process. *Atomkernenergie-Kerntechnik* 46: 76–80.

Sugai, H. and Munakata, K. 1992. Destruction of emulsions stabilized by precipitates of zirconium and tributylphosphate degradation products. *Nucl. Technol.* 99: 235–241.

Sugo, Y., Sasaki, Y., and Tachimori, S. 2002. Studies on hydrolysis and radiolysis of N,N,N',N'-tetraoctyl-3-oxapentane-1,5-diamide. *Radiochim. Acta* 90: 161–165.

Sugo, Y., Izumi, Y., Yoshida, Y., Nishijima, S., Sasaki, Y., Kimura, T., Sekine, T., and Kudo, H., 2007. Influence of diluent on radiolysis of amides in organic solution. *Radiat. Phys. Chem.* 76: 794–800.

Sunaryo, G. R., Katsumura, Y., and Ishigure, K. 1995. Radiolysis of water at elevated temperatures—III. Simulation of radiolytic products at 25 and 250°C under the irradiation with γ-rays and fast neutrons. *Radiat. Phys. Chem.* 45:703–14.

Sunder, S., Shoesmith, D. W., Christensen, H., and Miller, N. H. 1992. Oxidation of UO_2 fuel by the products of gamma radiolysis of water. *J. Nucl. Mater.* 190: 78–86.

Sunder, S., Boyer, G. D., and Miller, N. H. 1990. XPS studies of UO_2 oxidation by alpha radiolysis of water at 100°C. *J. Nucl. Mater.* 175: 163–169.

Suzuki, H., Shibata, M., Yamagata, J., Hirose, I., and Terakado K. 1992. Characteristic test of buffer material (I). PNC TN8410 92-057 (in Japanese), Power Reactor and Nuclear Development Fuel Corporation, Tokai-mura, Japan.

Suzuki, T., Abdelouas, A., Grambow, B., Mennecart, T., and Blondiaux, G. 2006. Oxidation and dissolution rates of UO_2(s) in carbonate-rich solutions under external alpha irradiation and initially reducing conditions. *Radiochim. Acta* 94: 567–573.

Sworski, T. J., Matthews, R. W., and Mahlman, H. A. 1968. Radiation chemistry of concentrated $NaNO_3$ solutions: Dependence of $G(HNO_2)$ on $NaNO_3$ concentration. *Adv. Chem. Ser.* 82: 164–181.

Tahraoui, A. and Morris, J. H. 1995. Decomposition of solvent-extraction media during nuclear reprocessing—Literature-review. *Sep. Sci. Technol.* 30: 2603–2630.

Takagi, J. and Ichikawa, N. 1999. Modeling technique for BWR primary system corrosion environment. In *Proceedings of the Seminar on Water Chemistry of Nuclear Reactor Systems'99*, Tokyo, Japan, pp. 57–61. Tokyo, Japan: Atomic Energy Society of Japan.

Takagi, J. and Ichikawa, N. 2001. Water radiolysis modeling and the prediction of corrosion potentials at actual BWRs. In *Proceedings of the Seminar on Water Chemistry of Nuclear Reactor Systems 2001*, paper 23, pp. 1–6. Taipei, Taiwan: Chung Hwa Nuclear Society.

Takagi, J., Ichikawa, N., Hemmim, Y., and Nagao, N. 1989. Study on radiolysis behavior and ECP measurement in BWR primary loop conditions. In *Proceedings of the Water Chemistry of Nuclear Reactor Systems 5*, vol. 2, Bournemouth, U.K., pp. 99–106. London, U.K.: NES.

Takagi, J., Morikawa, Y., Sakamoto, H., Ichikawa, N., Itow, M., Kawamura, S., and Takamori, K. 1996. Long term verification program on hydrogen water chemistry at a Japanese BWR. In *Proceedings of the Water Chemistry of Nuclear Reactor Systems 7*, pp. 489–495. London, U.K.: BNES.

Takagi, J., Ichikawa N., and Morikawa, Y. 1998. IGSCC mitigation concept for BWRs by preventive water chemistry. In *Proceedings of the 1998 JAIF Water Chemistry in Nuclear Power Plants*, Kashiwazaki, Japan, pp. 614–618. Tokyo, Japan: Japan Atomic Industrial Forum.

Takagi, J., Urata, H., and Ichikawa, N. 2000. Flow rate effect on corrosion potential of noble metal treated stainless steel. In *Proceedings of the Water Chemistry of Nuclear Reactor Systems 8*, vol. 2, Bournemouth, U.K., pp. 426–430. London, U.K.: BNES.

Tripathi, S. C. and Ramanujam, A. 2003. Effect of radiation-induced physicochemical transformations on density and viscosity of 30% TBP-*n*-dodecane-HNO_3 system. *Sep. Sci. Technol.* 38: 2307–2326.

Tripathi, S. C., Sumathi, S., and Ramanujam, A. 1999. Effects of solvent recycling on radiolytic degradation of 30% tributyl phosphate-*n*-dodecane-HNO_3 System. *Sep. Sci. Technol.* 34: 2887–2903.

Tripathi, S. C., Bindu, P., and Ramanujam, A. 2001a. Studies on the identification of harmful radiolytic products of 30% TNP-*n*-dodecane-HNO_3 by gas-liquid chromatography. I. Formation of diluent degradation products and their role in Pu retention behavior. *Sep. Sci. Technol.* 36: 1463–1478.

Tripathi, S. C., Ramanujam, A., Gupta, K. K., and Bindu, P. 2001b. Studies on the identification of harmful radiolytic products of 30% TBP-*n*-dodecane-HNO_3 by gas-liquid chromatography. II. Formation and characterization of high molecular weight organophosphates. *Sep. Sci. Technol.* 36: 2863–2883.

Trummer, M., Nilsson, S., and Jonsson, M. 2008. On the effects of fission product noble metal inclusions on the kinetics of radiation induced dissolution of spent nuclear fuel. *J. Nucl. Mater.* 378: 55–59.

Trummer, M., Roth, O., and Jonsson, M. 2009. H_2 inhibition of radiation induced dissolution of spent nuclear fuel. *J. Nucl. Mater.* 383: 226–230.

Urata, H., Ichikawa, N., Takagi, J., and Tanaka, N. 1998. Water flow velocity effect on corrosion potential of structural materials at bottom region of several BWR plants. In *Proceedings of the 1998 JAIF Water Chemistry in Nuclear Power Plants*, Kashiwazaki, Japan, pp. 850–853. Tokyo, Japan: Japan Atomic Industrial Forum.

Vione, D., Maurino, V., Minero, C., Borghesi, D., Lucchiari, M., and Pelizzetti, E. 2003. New processes in the environmental chemistry of nitrite. 2. The role of hydrogen peroxide. *Environ. Sci. Technol.* 37: 4635–4641.

Vladimirova, M. V. 1995. Mathematical-modeling of the radiation-chemical behavior of neptunium in HNO_3—Equilibrium states. *Radiochemistry* 37: 410–416.

Vladimirova, M. V. and Milovanova, A. S. 1972. α-radiolysis of HNO_3 solutions and acid $NaNO_3$ solutions. *Khimiya Vysokikh Energii* 6: 69–72.

Vladimirova, M. V., Kulikov, I. A., and Savel'ev, Y. I. 1969. Steady-state concentrations of nitrous acid in γ-irradiated nitric acid solutions. *High Energy Chem.* 3: 526–527.

Vladimirova, M. V., Fedoseev, D. A., and Artemova, L. A. 1996. Radiation-chemical behavior of neptunium in nitric acid solutions containing organic substances. *Radiochemistry* 38: 74–77.

von Sonntag, C. and Schuchmann, H.-P. 1997. Peroxyl radicals in aqueous solution. In *Peroxyl Radicals*, Z. B. Alfassi (ed.), pp. 173–234. New York: John Wiley & Sons.

von Sonntag, C., Ansorge, G., Sigimori, A., Omori, T., Koltzenburg, G., and Schulte-Frohlinde, D. Z. 1972. Alkyl phosphate cleavage of aliphatic phosphates induced by hydrated electrons and by OH radicals. *Naturforschung Teil B* 27: 71–472.

Wagner, R. M. and Towle, L. H., 1958. Radiation stability of organic liquids, Stanford Research Institute Report, Stanford Research Institute, Menlo Park, CA.

Weber, W. J. 1988. Radiation effects in nuclear waste glasses. *Nucl. Instrum Methods Phys. Res. B* 32: 471–479.

Weber, W. J., Wald, J. W., and McVay, G. L. 1985. Effect of α-radiolysis on leaching of a nuclear waste glass. *J. Am. Ceram. Soc.* 68: C253–255.

Weber, W. J., Ewing, R. C., Angell, C. A., Arnold, G. W., Cormack, A. N., Delaye, J. M., Griscom, D. L., Hobbs, L. W., Navrotsky, A., Price, D. L., Stoneham, A. M., and Weinberg, M. C. 1997. Radiation effects in glasses used for immobilization of high-level waste and plutonium disposition. *J. Mater. Res.* 12: 1946–1978.

Werme, L. 1998. Design premises for canister for spent nuclear fuel. SKB Technical Report 98-08, Swedish Nuclear Fuel and Waste Management Co., Stockholm, Sweden.

White, A. F., Peterson, M. L., and Hochella, M. F. Jr. 1994. Electrochemistry and dissolution kinetics of magnetite and ilmenite. *Geochim. Cosmochim. Acta* 58: 1859–1875.

Wilkinson, R. W. and Williams, T. F. 1961. The radiolysis of Tri-n-alkyl phosphates. *J. Chem. Soc.* 4098–4107.

Wu, G. Z., Katsumura, Y., Chitose, N., and Zuo, Z. H. 2001. A pulse radiolysis study of oil/water microemulsions. *Radiat. Phys. Chem.* 60: 643–650.

Yoshida, H., Metcalfe, R., Yamamoto, K., Murakami, Y., Hoshii, D., Kanekiyo, A., Naganuma, T., and Hayashi, T. 2008. Redox front formation in an uplifting sedimentary rock sequence: An analogue for redox-controlling processes in the geosphere around deep geological repositories for radioactive waste. *Appl. Geochem.* 23: 2364–2381.

Zaitsev, V. D. and Khaikin, G. I. 1994. Precursors of dibutylphosphoric acid upon radiolysis of tributyl-phosphate. *High Energy Chem.* 28: 269–272.

Zaitsev, V. D., Karasev, A. L., and Egorov, G. F. 1987. Formation of hydroxamic acids in the radiolysis of tri-normal-butyl phosphate containing nitric-acid. *Sov. Atom. Energy* 63: 524–527.

Zhang G., Samper, J., and Montenegro, L. 2008. Coupled hermo-hydro-bio-geochemical reactive transport model of CERBERUS heating and radiation experiment in boom clay. *Appl. Geochem.* 23: 932–949.

Index

Printed and bound by CPI Group (UK) Ltd, Croydon, CR0 4YY

24/10/2024

01778310-0003